LINDEN'S HANDBOOK OF BATTERIES

ABOUT THE EDITOR

THOMAS B. REDDY received a B.S. in Chemistry from Yale University and a Ph.D. in Physical Chemistry from the University of Minnesota where his thesis research was on the properties of nonaqueous electrolytes. He was a Post-Doctoral Research Associate at the University of Illinois and his work there included studies of thermal batteries. He served as a Member of the Technical Staff at Bell Laboratories, carrying out research on molten salt electrolytes. At the Central Research Laboratory of the American Cyanamid Company, he was a Senior Research Chemist and Group Leader and directed the group that developed the lithium/sulfur dioxide primary battery, widely used by the U.S. and foreign military. He subsequently served as the Director of Technology and a Vice-President of Power Conversion (PCI), later Hawker Eternacell, a leading manufacturer of lithium primary batteries for military and industrial applications. His responsibilities at PCI included R&D, quality control, and technology transfer. He completed his industrial career as the Vice-President of Engineering at Yardney Technical Products/Lithion where he directed R, D, & E programs on lithium-ion, silver, nickel and metal-air batteries. He served as the technical expert on batteries for NASA's New Millennium Program and directed development of lithium-ion and lithium primary batteries for JPL's Mars exploration programs, including the Mars Exploration Rovers.

Dr. Reddy is currently a Distinguished Visiting Scientist in the Department of Materials Science and Engineering at Rutgers University and an Adjunct Assistant Professor at the Robert Wood Johnson Medical School of the University of Medicine and Dentistry of New Jersey. He serves on U.S. Government and UL committees and consults for academia, government, and industry. Dr. Reddy was the coeditor with David Linden of the Third Edition of the *Handbook of Batteries*.

LINDEN'S HANDBOOK OF BATTERIES

Thomas B. Reddy Editor

David Linden Editor Emeritus

Fourth Edition

New York Chicago San Francisco Lisbon London Madrid
Mexico City Milan New Delhi San Juan Seoul
Singapore Sydney Toronto

The McGraw·Hill Companies

Cataloging-in-Publication Data is on file with the Library of Congress.

McGraw-Hill books are available at special quantity discounts to use as premiums and sales promotions, or for use in corporate training programs. To contact a representative please e-mail us at bulksales@mcgraw-hill.com.

Linden's Handbook of Batteries, Fourth Edition

Copyright ©2011, 2002, 1995, 1984 by The McGraw-Hill Companies, Inc. All rights reserved. Printed in the United States of America. Except as permitted under the United States Copyright Act of 1976, no part of this publication may be reproduced or distributed in any form or by any means, or stored in a data base or retrieval system, without the prior written permission of the publisher.

The Third Edition was published by McGraw-Hill in 2002 with the title *Handbook of Batteries*.

2 3 4 5 6 7 8 9 0 QVR/QVR 1 9 8 7 6 5 4 3 2

ISBN 978-0-07-162421-3
MHID 0-07-162421-X

The pages within this book were printed on acid-free paper.

Sponsoring Editor
Stephen S. Chapman

Acquisitions Coordinator
Alexis Richard

Editorial Supervisor
David E. Fogarty

Project Manager
Tania Andrabi

Copy Editor
Lunaea Weatherstone

Proofreaders
Constance Blazewicz and Andrew Lucas

Production Supervisor
Richard C. Ruzycka

Composition
Glyph International

Art Director, Cover
Jeff Weeks

Information contained in this work has been obtained by The McGraw-Hill Companies, Inc. ("McGraw-Hill") from sources believed to be reliable. However, neither McGraw-Hill nor its authors guarantee the accuracy or completeness of any information published herein, and neither McGraw-Hill nor its authors shall be responsible for any errors, omissions, or damages arising out of use of this information. This work is published with the understanding that McGraw-Hill and its authors are supplying information but are not attempting to render engineering or other professional services. If such services are required, the assistance of an appropriate professional should be sought.

*Dedicated
to the
Memory
of
David Linden
1923–2008*

CONTENTS

Contributors xix
Preface xxi

Part 1 Principles of Operation

Chapter 1. Basic Concepts *David Linden and Thomas B. Reddy* 1.3

1.1. Components of Cells and Batteries / *1.3*
1.2. Classification of Cells and Batteries / *1.4*
1.3. Operation of a Cell / *1.6*
1.4. Theoretical Cell Voltage, Capacity, and Energy / *1.9*
1.5. Specific Energy and Energy Density of Practical Batteries / *1.14*
1.6. Limits of Specific Energy and Energy Density / *1.16*
References / *1.17*

Chapter 2. Electrochemical Principles and Reactions *Mark Salomon* 2.1

2.1. Introduction / *2.1*
2.2. Thermodynamic Background / *2.3*
2.3. Electrode Processes / *2.4*
2.4. Electrical Double-Layer Capacity and Ionic Adsorption / *2.10*
2.5. Mass Transport to the Electrode Surface / *2.15*
2.6. Electrochemical Techniques / *2.19*
References / *2.35*
Bibliography / *2.36*

Chapter 3. Factors Affecting Battery Performance *David Linden* 3.1

3.1. General Characteristics / *3.1*
3.2. Factors Affecting Battery Performance / *3.1*
References / *3.22*

Chapter 4. Battery Standardization *Steven Wicelinski* 4.1

4.1. General / *4.1*
4.2. International Standards / *4.1*
4.3. Concepts of Standardization / *4.3*
4.4. IEC and ANSI Nomenclature Systems / *4.4*
4.5. Terminals / *4.7*
4.6. Electrical Performance / *4.8*
4.7. Markings / *4.9*
4.8. Cross-References of ANSI IEC Battery Standards / *4.9*

4.9. Listing of IEC Standard Round Primary Batteries / 4.9
4.10. Standard SLI and Other Lead-Acid Batteries / 4.12
4.11. Regulatory and Safety Standards / 4.19
Note / 4.20
References / 4.20

Chapter 5. Battery Design *Daniel D. Friel* 5.1

5.1. General / 5.1
5.2. Designing to Eliminate Potential Safety Problems / 5.1
5.3. Battery Safeguards When Using Discrete Batteries / 5.7
5.4. Battery Construction / 5.10
5.5. Design of Rechargeable Batteries / 5.14
5.6. Electronic Energy Management and Control Systems / 5.18
References / 5.23

Chapter 6. Mathematical Modeling of Batteries *Shriram Santhanagopalan and Ralph E. White* 6.1

6.1. Introduction / 6.1
6.2. Development of a Mathematical Model / 6.4
6.3. Building Empirical Models / 6.5
6.4. Mechanistic Models / 6.9
6.5. Kinetic Model of a Silver Vanadium Oxide Cell / 6.13
6.6. Modeling Porous Electrodes / 6.15
6.7. Lead-Acid Battery Model / 6.16
6.8. Intercalation in Porous Electrodes / 6.18
6.9. Energy Balance / 6.18
6.10. Degradation Models / 6.21
6.11. Determining the Right Model / 6.25
List of Symbols / 6.25
References / 6.27

Chapter 7. Battery Electrolytes *George E. Blomgren* 7.1

7.1. Introduction / 7.1
7.2. Aqueous Electrolytes / 7.1
7.3. Nonaqueous Electrolytes / 7.6
7.4. Ionic Liquids / 7.9
7.5. Solid Polymer Electrolytes / 7.10
7.6. Ceramic/Glassy Electrolytes / 7.11
References / 7.11

Part 2 Primary Batteries

Chapter 8. An Introduction to Primary Batteries *David Linden and Thomas B. Reddy* 8.3

8.1. General Characteristics and Applications of Primary Batteries / 8.3
8.2. Types and Characteristics of Primary Batteries / 8.4
8.3. Comparison of the Performance Characteristics of Primary Battery Systems / 8.8
8.4. Recharging Primary Batteries / 8.18

Chapter 9. Zinc-Carbon Batteries—Leclanché and Zinc Chloride Cell Systems   *Brooke Schumm, Jr.*   9.1

9.1. General Characteristics / 9.1
9.2. Chemistry / 9.4
9.3. Types of Cells and Batteries / 9.4
9.4. Construction / 9.6
9.5. Cell Components / 9.10
9.6. Performance Characteristics / 9.15
9.7. Special Designs / 9.33
9.8. Types and Sizes of Available Cells and Batteries / 9.36
References / 9.41

Chapter 10. Magnesium and Aluminum Batteries   *Patrick J. Spellman*   10.1

10.1. General Characteristics / 10.1
10.2. Chemistry / 10.2
10.3. Construction of Mg/MnO_2 Batteries / 10.4
10.4. Performance Characteristics of Mg/MnO_2 Batteries / 10.6
10.5. Sizes and Types of Mg/MnO_2 Batteries / 10.11
10.6. Other Types of Magnesium Batteries / 10.11
10.7. Aluminum Primary Batteries / 10.12
References / 10.12

Chapter 11. Alkaline-Manganese Dioxide Batteries   *John C. Nardi and Ralph J. Brodd*   11.1

11.1. General Characteristics / 11.1
11.2. Chemistry / 11.4
11.3. Cell Components and Materials / 11.7
11.4. Construction / 11.11
11.5. Evolta™ and Oxyride™ Batteries / 11.16
References / 11.17

Chapter 12. Mercuric Oxide Batteries   *Nathan D. (Ned) Isaacs*   12.1

12.1. General Characteristics / 12.1
12.2. Chemistry / 12.2
12.3. Cell Components / 12.3
12.4. Construction / 12.5
12.5. Performance Characteristics of Zinc/Mercuric Oxide Batteries / 12.7
12.6. Performance Characteristics of Cadmium/Mercuric Oxide Batteries / 12.12
References / 12.14

Chapter 13. Button Cell Batteries: Silver Oxide–Zinc and Zinc-Air Systems   *Joseph Passaniti, Denis Carpenter, and Rodney McKenzie*   13.1

Section A. Silver Oxide–Zinc Batteries / 13.1
13.1. General Characteristics of the Silver Oxide-Zinc Battery / 13.1
13.2. Battery Chemistry and Components / 13.2
13.3. Construction / 13.10
13.4. Performance Characteristics / 13.11
13.5. Cell Sizes and Types / 13.15

Section B. Zinc-Air Batteries / 13.16
13.6. General Characteristics of Button Cell Zinc/Air Batteries / *13.16*
13.7. Chemistry / *13.17*
13.8. Construction / *13.18*
13.9. Performance Characteristics / *13.21*
References / *13.36*
Bibliography / *13.37*

Chapter 14. Lithium Primary Batteries Thomas B. Reddy   14.1

14.1. General Characteristics / *14.1*
14.2. Chemistry / *14.4*
14.3. Characteristics of Lithium Primary Batteries / *14.8*
14.4. Safety and Handling of Lithium Batteries / *14.14*
14.5. Lithium/Sulfur Dioxide (Li/SO_2) Batteries / *14.16*
14.6. Lithium/Thionyl Chloride (Li/$SOCl_2$) Batteries / *14.26*
14.7. Lithium/Oxychloride Batteries / *14.42*
14.8. Lithium/Manganese Dioxide (Li/MnO_2) Batteries / *14.47*
14.9. Lithium/Carbon Monofluoride (Li/CFx) Batteries / *14.64*
14.10. Lithium/Iron Disulfide (Li/FeS_2) Batteries / *14.75*
14.11. Lithium/Copper Oxide (Li/CuO) Cells / *14.81*
14.12. Lithium/Silver Vanadium Oxide Batteries / *14.88*
14.13. Lithium/Water and Lithium/Air Batteries / *14.88*
References / *14.88*

Part 3 Secondary Batteries

Chapter 15. An Introduction to Secondary Batteries Thomas B. Reddy   15.3

15.1. General Characteristics and Applications of Secondary Batteries / *15.3*
15.2. Types and Characteristics of Secondary Batteries / *15.4*
15.3. Comparison of Performance Characteristics for Secondary Battery Systems / *15.9*
References / *15.19*

Chapter 16. Lead-Acid Batteries Alvin Salkind and George Zguris   16.1

16.1. General Characteristics / *16.1*
16.2. Chemistry / *16.7*
16.3. Constructional Features, Materials, and Manufacturing Methods / *16.17*
16.4. SLI (Automotive) Batteries: Construction and Performance / *16.37*
16.5. Deep-Cycle and Traction Batteries: Construction and Performance / *16.46*
16.6. Stationary Batteries: Construction and Performance / *16.56*
16.7. Charging and Charging Equipment / *16.67*
16.8. Maintenance, Safety, and Operational Features / *16.75*
16.9. Applications and Markets / *16.80*
References / *16.85*

Chapter 17. Valve Regulated Lead-Acid Batteries Kathryn R. Bullock and Alvin J. Salkind   17.1

17.1. General Characteristics / *17.1*
17.2. Chemistry / *17.3*
17.3. Cell Construction / *17.4*

17.4. Performance Characteristics / *17.8*
17.5. Charging Characteristics / *17.22*
17.6. Safety and Handling / *17.32*
17.7. Battery Types and Sizes / *17.33*
17.8. Applications of VRLA Batteries to Uninterruptible Power Supplies / *17.36*
17.9. Current Developments and Future Opportunities for VRLA Batteries / *17.38*
References / *17.39*

Chapter 18. Iron Electrode Batteries Gary A. Bayles   18.1

18.1. General Characteristics / *18.1*
18.2. Chemistry of Nickel-Iron Batteries / *18.1*
18.3. Conventional Nickel-Iron Batteries / *18.4*
18.4. Advanced Nickel-Iron Batteries / *18.12*
18.5. Iron/Air Batteries / *18.15*
18.6. Silver-Iron Battery / *18.17*
18.7. Recent Advances in Iron Anode Materials / *18.21*
18.8. Iron Materials as Cathodes / *18.21*
References / *18.24*

Chapter 19. Industrial and Aerospace Nickel-Cadmium Batteries
John K. Erbacher   19.1

19.1. Introduction / *19.1*
19.2. Chemistry / *19.3*
19.3. Construction / *19.3*
19.4. Performance Characteristics / *19.7*
19.5. Charging Characteristics / *19.12*
19.6. Sealed Nickel-Cadmium (SNC) Battery Technology / *19.13*
19.7. Fiber Nickel-Cadmium (FNC) Battery Technology / *19.13*
19.8. Manufacturers and Market Segments / *19.19*
19.9. Applications / *19.21*
Bibliography / *19.22*

Chapter 20. Vented Sintered-Plate Nickel-Cadmium Batteries R. David Lucero   20.1

20.1. General Characteristics / *20.1*
20.2. Chemistry / *20.1*
20.3. Construction / *20.3*
20.4. Performance Characteristics / *20.6*
20.5. Charging Characteristics / *20.15*
20.6. Maintenance Procedures / *20.19*
20.7. Reliability / *20.21*
20.8. Cell and Battery Designs / *20.24*
References / *20.28*

Chapter 21. Portable Sealed Nickel-Cadmium Batteries Joseph A. Carcone   21.1

21.1. General Characteristics / *21.1*
21.2. Chemistry / *21.1*
21.3. Construction / *21.3*
21.4. Performance Characteristics / *21.6*
21.5. Charging Characteristics / *21.19*
21.6. Special-Purpose Batteries / *21.25*

21.7. Battery Types and Sizes / *21.31*
21.8. Battery Sizes and Availability / *21.33*
Reference / *21.33*
Bibliography / *21.33*

Chapter 22. Nickel-Metal Hydride Batteries *Michael Fetcenko and John Koch* 22.1

22.1. Introduction / *22.1*
22.2. General Characteristics / *22.2*
22.3. NiMH Battery Chemistry / *22.2*
22.4. Cell Construction Types / *22.10*
22.5. Cell Design Issues / *22.15*
22.6. EV Battery Packs / *22.17*
22.7. Hybrid Battery Packs / *22.19*
22.8. Fuel Cell Startup and Power Assist / *22.21*
22.9. Consumer Batteries—Precharged NiMH / *22.21*
22.10. Discharge Performance / *22.22*
22.11. Charge Methods / *22.40*
22.12. Electrical Isolation / *22.47*
22.13. Next Generation NiMH / *22.48*
References / *22.50*

Chapter 23. Nickel-Zinc Batteries *Jeffrey Phillips and Samaresh Mohanta* 23.1

23.1. Introduction / *23.1*
23.2. Nickel-Zinc Chemistry / *23.2*
23.3. Cell Construction / *23.7*
23.4. Performance Characteristics / *23.11*
23.5. Applications / *23.20*
23.6. Environmental Aspects of the Nickel-Zinc Battery / *23.24*
References / *23.24*

Chapter 24. Nickel-Hydrogen Batteries *Jack N. Brill* 24.1

24.1. General Characteristics / *24.1*
24.2. Chemistry / *24.1*
24.3. Cell and Electrode-Stack Components / *24.3*
24.4. Ni-H_2 Cell Construction / *24.5*
24.5. Ni-H_2 Battery Design / *24.10*
24.6. Applications / *24.14*
24.7. Performance Characteristics / *24.17*
24.8. Advanced Designs / *24.23*
References / *24.27*
Bibliography / *24.28*

Chapter 25. Silver Oxide Batteries *Alexander P. Karpinski* 25.1

25.1. General Characteristics / *25.1*
25.2. Chemistry / *25.3*
25.3. Cell Construction and Components / *25.3*
25.4. Performance Characteristics / *25.8*
25.5. Charging Characteristics / *25.20*
25.6. Cell Types and Sizes / *25.21*
25.7. Special Features and Handling / *25.24*

25.8. Applications / *25.24*
25.9. Recent Developments / *25.27*
References / *25.29*

Chapter 26. Lithium-Ion Batteries   *Jeff Dahn and Grant M. Ehrlich*   26.1

26.1. General Characteristics / *26.1*
26.2. Chemistry / *26.3*
26.3. Construction / *26.41*
26.4. Characteristics and Performance / *26.47*
26.6. Safety Properties / *26.68*
26.7. Conclusions and Future Trends / *26.75*
Acknowledgments / *26.75*
References / *26.75*

Chapter 27. Rechargeable Lithium Metal Batteries (Ambient Temperature)
Daniel H. Doughty   27.1

27.1. General Characteristics / *27.1*
27.2. Chemistry / *27.3*
27.3. Characteristics of Lithium Rechargeable Batteries / *27.13*
27.4. Conclusions / *27.22*
Acknowledgments / *27.22*
References / *27.24*

Chapter 28. Rechargeable Zinc/Alkaline/Manganese Dioxide Batteries
Josef Daniel-Ivad and Karl Kordesch   28.1

28.1. General Characteristics / *28.1*
28.2. Chemistry / *28.1*
28.3. Construction / *28.3*
28.4. Performance / *28.4*
28.5. Charge Methods / *28.12*
28.6. Types of Cells and Batteries / *28.15*
References / *28.17*

Part 4   Specialized Battery Systems

Chapter 29. Batteries for Electric and Hybrid Vehicles   *Dennis A. Corrigan and Alvaro Masias*   29.3

29.1. Introduction / *29.3*
29.2. EV Battery Performance Targets / *29.12*
29.3. Batteries for Electric Vehicles / *29.15*
29.4. Other Energy Storage Technologies for Electric Vehicles / *29.21*
29.5. Hybrid Electric Vehicles / *29.22*
29.6. Types of Hybrid Electric Vehicles / *29.28*
29.7. Comparison of HEV Battery Performance Requirements / *29.35*
29.8. Vehicle Integration of HEV Batteries / *29.37*
27.9. Other Energy Storage Technologies for Hybrid Electric Vehicles / *29.46*
References / *29.46*

Chapter 30. Batteries for Electrical Energy Storage Applications
Abbas A. Akhil, John D. Boyes, Paul C. Butler, and Daniel H. Doughty 30.1

30.1. Introduction: Energy Storage on the Electric Grid / 30.1
30.2. Historical Perspective / 30.4
30.3. Battery Energy Storage for Electricity Applications: How Storage Systems Create Value / 30.5
30.4. Landmark Battery Energy Storage Systems / 30.18
30.5. Advanced Battery Technologies for Stationary Applications / 30.22
30.6. Flowing Electrolyte Batteries / 30.35
30.7. Conclusion / 30.45
Acknowledgments / 30.45
References / 30.45

Chapter 31. Batteries for Biomedical Applications *Randolph A. Leising, Nancy R. Gleason, Barry C. Muffoletto, and Curtis F. Holmes* 31.1

31.1. Applications and Requirements for Implantable Batteries / 31.1
31.2. Applications and Requirements for Batteries for Externally Powered Medical Devices / 31.7
31.3. Safety Considerations / 31.9
31.4. Reliability Considerations / 31.12
31.5. Characteristics of Batteries for Biomedical Applications / 31.15
References / 31.36

Chapter 32. Battery Selection for Consumer Electronics *John A. Wozniak* 32.1

32.1. Introduction / 32.1
32.2. Key Considerations in Selecting a Battery / 32.1
32.3. Typical Portable Applications / 32.2
32.4. Primary Battery Types and Applications / 32.3
32.5. Secondary Battery Types and Applications / 32.6
32.6. Specific Criteria for Battery Selection / 32.9
32.7. Decision Making and Trade-Offs / 32.17
32.8. Avoiding Common Pitfalls in Battery Selection / 32.20

Chapter 33. Metal/Air Batteries *Terrill B. Atwater and Arthur Dobley* 33.1

33.1. General Characteristics / 33.1
33.2. Chemistry / 33.4
33.3. Zinc/Air Batteries / 33.6
33.4. Aluminum/Air Batteries / 33.27
33.5. Magnesium/Air Batteries / 33.43
33.6. Lithium/Air Batteries / 33.44
References / 33.55

Chapter 34. Reserve Magnesium Anode and Zinc/Silver Oxide Batteries
R. David Lucero and Alexander P. Karpinski 34.1

Section A. Magnesium Water-Activated Batteries / 34.1
34.1. General Characteristics of Reserve Magnesium Batteries / 34.1
34.2. Chemistry / 34.2
34.3. Types of Water-Activated Batteries / 34.4
34.4. Construction / 34.6
34.5. Performance Characteristics / 34.10
34.6. Battery Applications / 34.22
34.7. Battery Types and Sizes / 34.24

Section B. Zinc/Silver Oxide Reserve Batteries / 34.26
34.8. General Characteristics of Zinc/Silver Oxide Reserve Batteries / 34.26
34.9. Chemistry / 34.26
34.10. Construction / 34.27
34.11. Performance Characteristics / 34.32
34.12. Cell and Battery Types and Sizes / 34.37
34.13. Special Features and Handling / 34.37
34.14. Cost / 34.40
References / 34.40
Bibliography / 34.41

Chapter 35. Reserve Military Batteries *David L. Chua, Benjamin M. Meyer, William J. Epply, Jeffrey A. Swank, and Michael Ding* 35.1

Section A. Ambient-Temperature Lithium Anode Reserve Batteries / 35.1
35.1. General Characteristics / 35.1
35.2. Chemistry / 35.2
35.3. Construction / 35.4
35.4. Performance Characteristics / 35.14
35.5. Applications / 35.19

Section B. Spin-Dependent Reserve Batteries / 35.20
35.6. General Characteristics / 35.20
35.7. Chemistry / 35.21
35.8. Design Considerations / 35.22
35.9. Performance Characteristics / 35.25
References / 35.30
Bibliography / 35.30

Chapter 36. Thermal Batteries *Charles M. Lamb* 36.1

36.1. General Characteristics / 36.1
36.2. Description of Electrochemical Systems / 36.3
36.3. Cell Chemistry / 36.7
36.4. Cell Construction / 36.10
36.5. Cell-Stack Designs / 36.13
36.6. Performance Characteristics / 36.15
36.7. Testing and Surveillance / 36.18
36.8. New Developments / 36.18
References / 36.18
Bibliography / 36.19

Part 5 Fuel Cells and Electrochemical Capacitors

Chapter 37. Introduction to Fuel Cells *David Linden and H. Frank Gibbard* 37.3

37.1. General Characteristics / 37.3
37.2. Operation of the Fuel Cell / 37.5
37.3. Subkilowatt Fuel Cells / 37.9
37.4. Innovative Subkilowatt Designs: Solid Oxide Fuel Cells / 37.13
References / 37.14

Chapter 38. Small Fuel Cells *Arthur Kaufman and H. Frank Gibbard* 38.1

38.1. General / 38.1
38.2. Applicable Fuel Cell Technologies / 38.2

38.3. Cell Electrochemical Operation / *38.3*
38.4. Cell Stacking Configurations / *38.5*
38.5. Fuel Selection / *38.6*
38.6. Fuel Processing and Storage Configurations / *38.6*
38.7. System Integration Requirements / *38.11*
38.8. Hardware and Performance / *38.13*
38.9. Prognosis / *38.20*
References / *38.20*

Chapter 39. Electrochemical Capacitors *Andrew F. Burke*   39.1

39.1. Introduction / *39.1*
39.2. Chemistry and Material Properties / *39.6*
39.3. Performance Characteristics of Devices / *39.10*
39.4. Electrochemical Capacitor Modeling / *39.13*
39.5. Testing Electrochemical Capacitors / *39.22*
39.6. Cost and System Considerations for Capacitors and Batteries / *39.33*
References / *39.41*

Appendices

Appendix A. Definitions   A.3

Appendix B. Standard Reduction Potentials   B.1

Appendix C. Electrochemical Equivalents of Battery Materials   C.1

Appendix D. Standard Symbols and Constants   D.1

Appendix E. Conversion Factors   E.1

Appendix F. Bibliography   F.1

Books / *F.1*
Periodicals / *F.2*
Proceedings of Annual/Biennial Conferences / *F.2*
Other Reference Sources: Handbooks and Bibliographies / *F.3*
Standards / *F.3*

Appendix G. Battery Manufacturers and R&D Organizations
Prepared by Vaidevutis Alminauskas   G.1

Appendix H. Methodologies for Battery Failure Analysis *Quinn Horn, Troy Hayes, Daren Slee, Kevin White, John Harmon, Ramesh Godithi, Ming Wu, Marcus Megerle, Surendra Singh, and Celina Mikolajczak*   H.1

Introduction / *H.1*
Collection of Background Information / *H.2*

Recovery of Physical Evidence or Samples / *H.3*
Host Device or System Examination / *H.3*
Battery and Cell Examination / *H.6*
Battery Management and Protection Circuitry Examination / *H.28*
Working with Data from OEMs and Battery Manufacturers / *H.30*
Conclusions / *H.31*
Acknowledgments / *H.32*
References / *H.32*

Index   I.1

CONTRIBUTORS

Abbas A. Akhil   *Sandia National Laboratories*

Vaidevutis Alminauskas   *Naval Sea Systems Command, Crane Division*

Terrill B. Atwater   *RDECOM, U.S. Army CERDEC*

Gary A. Bayles   *SAIC*

George E. Blomgren   *Blomgren Consulting Services*

John D. Boyes   *Sandia National Laboratories*

Jack N. Brill   *Eagle-Picher Technologies, LLC*

Ralph J. Brodd   *Broddarp of Nevada, Inc. (now with the Kentucky-Argonne Battery Manufacturing R and D Centre)*

Kathryn R. Bullock   *CoolOhm, Inc.*

Andrew F. Burke   *Institute of Transportation Studies, University of California-Davis*

Paul C. Butler   *Sandia National Laboratories*

Joseph A. Carcone   *PowerGenix, Inc.*

Denis Carpenter   *Rayovac Corp.*

David L. Chua   *Maxpower, Inc.*

Dennis A. Corrigan   *Wayne State University and DC Energy Consulting, LLC*

Jeffrey R. Dahn   *Departments of Physics and Chemistry, Dalhousie University, Nova Scotia, Canada*

Josef Daniel-Ivad   *Pure Energy Visions, Inc. Canada*

Michael Ding   *U.S. Army Research Laboratory*

Arthur Dobley   *Yardney Technical Products, Inc./Lithion Division*

Daniel H. Doughty   *Battery Safety Consulting, Inc.*

Grant M. Ehrlich   *Cantor Colburn, LLC*

William J. Epply   *Maxpower, Inc.*

John K. Erbacher   *U.S.A.F. Research Laboratory, Wright-Patterson AFB*

Michael Fetcenko   *Ovonic Battery Co.*

Daniel D. Friel   *Texas Instruments*

H. Frank Gibbard   *Gibbard R&D Corp.*

Nancy R. Gleason   *Greatbatch, Inc.*

Ramesh Godithi   *Exponent, Inc.*

John Harmon   *Exponent, Inc.*

Troy Hayes   *Exponent, Inc.*

Curtis F. Holmes   *Greatbatch, Inc. (retired)*

Quinn Horn *Exponent, Inc.*

Nathan D. (Ned) Isaacs *Ned Isaacs Battery Consulting*

Alexander P. Karpinski *Yardney Technical Products, Inc.*

Arthur Kaufman *Gibbard R&D Corp.*

John Koch *Ovonic Materials Div. Energy Conversion Devices*

Karl Kordesch *Technische Universitat, Graz, Austria*

Charles M. Lamb *Eagle-Picher Technologies, LLC*

Randolph A. Leising *Greatbatch, Inc.*

David Linden *Editor Emeritus, The Handbook of Batteries (now deceased)*

R. David Lucero *Eagle-Picher Technologies, LLC*

Alvaro Masias *Toyota Motors Research and Manufacturing-NA (now with Ford Motor Co.)*

Rodney McKenzie *Rayovac Corp.*

Marcus Megerle *Exponent, Inc.*

Benjamin M. Meyer *Maxpower, Inc.*

Celina Mikolajczak *Exponent, Inc.*

Samaresh (Sam) Mohanta *PowerGenix, Inc.*

Barry C. Muffoletto *Greatbatch, Inc.*

John C. Nardi *Energizer, Inc.*

Joseph L. Passaniti *Rayovac Corp.*

Jeffrey Phillips *PowerGenix, Inc.*

Thomas B. Reddy *Dept. of Materials Science and Engineering, Rutgers University and Robert W. Johnson Medical School, UMDNJ*

Alvin J. Salkind *Rutgers University and University of Medicine and Dentistry of New Jersey*

Mark Salomon *Maxpower, Inc.*

Shriram Santhanagopalan *Celgard, LLC (now with National Renewable Energy Laboratory)*

Brooke Schumm, Jr. *Eagle-Cliffs, Inc.*

Surendra Singh *Exponent, Inc.*

Daren Slee *Exponent, Inc.*

Patrick J. Spellman *Rayovac Corp. (retired)*

Jeffrey A. Swank *U.S. Army Research Laboratory*

Kevin White *Exponent, Inc.*

Ralph E. White *Dept. of Chemical Engineering, University of South Carolina*

Steven P. Wicelinski *Duracell, Inc.*

John A. Wozniak *Hewlett Packard Co.*

Ming Wu *Exponent, Inc.*

George Zguris *Hollingsworth and Vose Co.*

PREFACE

The first decade of the twenty-first century has seen a quantum leap in terms of the economic and technological significance of the battery industry to society worldwide. This has resulted in a complete revision of *Linden's Handbook of Batteries* for the Fourth Edition to include new information on emerging battery systems and their applications. Many new chapters have been added and others consolidated to improve access to information on related technologies and applications. Information on Fuel Cells has been updated, and a new chapter on Electrochemical Capacitors, which are of importance in hybrid power systems, is included. A new chapter has also been included on Battery Modeling, an area of increasing importance in battery technology. A chapter on Battery Electrolytes is also a new feature to summarize and augment the information presented in individual chapters on particular battery systems.

In the primary battery field, the emergence of applications needing high power, such as digital cameras, has resulted in the development of alkaline manganese designs meeting this requirement. The Oxyride battery has also been designed to meet these requirements by using a cathode combining manganese dioxide and nickel oxyhydroxide. The lithium/iron disulfide battery, with its capacity enhanced in the last decade, has emerged as the leading lithium primary battery for high-power applications, displacing lithium/manganese dioxide in some cases.

The explosion of the consumer electronics market for applications such as laptop computers, smart phones, and e-books has been fueled by improvements in the energy density and specific energy of lithium-ion batteries. Typical commercial 18650 cells now provide a capacity of 2.55 Ah, an energy density of 570 Wh/L, and a specific energy of 200 Wh/kg, using the lithium cobalt oxide/graphite chemistry. Such cells are being produced at the rate of 250 million/month and cost as little as $0.20/Wh. Advanced 18650 Li-ion cells provide 640 Wh/L and 240 Wh/kg, while the use of alloy anodes in place of graphitic carbons promises further improvements in performance. These advances are detailed in a completely updated and greatly expanded Chap. 26 on Lithium-ion Batteries. A new Chap. 32 on Battery Selection for Consumer Electronics has been added to detail the process by which electronics manufacturers select battery systems for use in their products.

The enhancements detailed above have also resulted in new propulsion systems for hybrid vehicles (HEVs), plug-in hybrid vehicles (PHEVs), and electric vehicles (EVs). Since the late 1990s, nickel-metal hydride batteries have been the system of choice for HEVs and have performed well in that application. With enhanced power capability, Li-ion batteries are providing a challenge to NiMH for use in HEVs. Lithium-ion has already been selected for use in the Chevy Volt PHEV and the Nissan Leaf EV, both of which are coming to market in the near future. It is also the battery system of choice for other vehicles in advanced development. Meanwhile, the lead-acid battery in new designs such as the Ultrabattery™ contain carbon in the Pb negative electrode, which provides a capacitive effect to reduce sulfation of the negative electrode. This allows the use of such batteries in microhybrid vehicles where the engine is turned off during periods of idling to reduce emissions and then restarted by the battery. All of these advancements are detailed in a new Chap. 29 on Batteries for Electric, Hybrid, and Plug-in Hybrid Vehicles.

Batteries for Electrical Energy Storage Systems are described in a new Chap. 30 which updates and consolidates information provided in several chapters from the Third Edition. Recent information from the U.S. Department of Energy indicates that lithium-ion may be the battery of choice for energy storage and power conditioning applications in the future.

In a similar vein, a new chapter on Batteries for Biomedical Applications has been added. This combines and updates information from the Third Edition and places greater emphasis on the end use and the selection of the battery required to meet such an application.

Another feature of the Fourth Edition is the consolidation of information from two or more chapters into one new chapter to provide easier access to similar battery systems or batteries for similar applications. Part 4 on Reserve Batteries has been eliminated, and reserve systems are covered in a new Part 4 on Specialized Battery Systems. Among the consolidated chapters is a new Chap. 13 on Button Cell Batteries. This combines information on silver/zinc and zinc/air button cells because of the similarity in sizes and applications for such batteries. Two chapters on Nickel-Metal Hydride Batteries have now been combined into a new Chap. 22 since the use of such batteries in vehicular applications is covered in a new Chap. 29. This combined chapter provides detailed information on improvements in this technology and the development of energy cells for consumer applications and power cells for HEVs. In the former category, the design of precharged cells for consumer applications is a major advancement.

Two new chapters on batteries for military and space applications have also been included in this edition. In the first case, two chapters on Magnesium Water-Activated Batteries and Zinc/Silver Oxide Reserve Batteries have been combined into a new Chap. 34 on Reserve Water-Activated Batteries. This has been carried out because these two systems are similar in their design and applications. Two earlier chapters on military application, one on Ambient-Temperature Lithium Anode Reserve Batteries and the other on Spin-Dependent Reserve Batteries, have also been combined into a new Chap. 35 on Reserve Military Batteries.

With the enormous increase in the use of high-energy battery systems in consumer electronics, field failures have become more common. A new Appendix H on Failure Analysis Methods in Battery Technology is another new feature of this edition.

The editor wishes to acknowledge the work of more than 60 contributors who have given of their time and energy to complete this volume. Without them, this new edition would not exist. The assistance of Mrs. Lois Kisch with typing is also acknowledged. The editor is grateful for the assistance provided by Drs. Dan Doughty, H. Frank Gibbard, Mark Salomon, and George Blomgren during the preparation of this edition. The editor also wishes to thank Stephen S. Chapman, executive editor at the McGraw-Hill Professional Book Group, for his advice and counsel during the planning and preparation phases of this work. The contributions of David Fogarty and his associates at McGraw-Hill and the Glyph International team during the production phase of this volume are also acknowledged.

This Handbook was initiated by David Linden, who served as editor of the First and Second Editions. At Dave's request, I served as coeditor of the Third Edition. Following his death, I have completed the Fourth Edition and sincerely hope that it meets the standards set by him in the earlier Editions.

THOMAS B. REDDY
Bronxville, New York

P · A · R · T · 1

PRINCIPLES OF OPERATION

CHAPTER 1
BASIC CONCEPTS

David Linden and Thomas B. Reddy

1.1 COMPONENTS OF CELLS AND BATTERIES

A battery is a device that converts the chemical energy contained in its active materials directly into electric energy by means of an electrochemical oxidation-reduction (redox) reaction. In the case of a rechargeable system, the battery is recharged by a reversal of the process. This type of reaction involves the transfer of electrons from one material to another through an electric circuit. In a non-electrochemical redox reaction, such as rusting or burning, the transfer of electrons occurs directly and only heat is involved. As the battery electrochemically converts chemical energy into electric energy, it is not subject, as are combustion or heat engines, to the limitations of the Carnot cycle dictated by the second law of thermodynamics. Batteries, therefore, are capable of having higher energy conversion efficiencies.

While the term "battery" is often used, the basic electrochemical unit being referred to is the "cell." A battery consists of one or more of these cells, connected in series or parallel, or both, depending on the desired output voltage and capacity.*

The cell consists of three major components:

1. The anode or negative electrode—the reducing or fuel electrode—which gives up electrons to the external circuit and is oxidized during the electrochemical reaction.

2. The cathode or positive electrode—the oxidizing electrode—which accepts electrons from the external circuit and is reduced during the electrochemical reaction.

3. The electrolyte—the ionic conductor—which provides the medium for transfer of charge, as ions, inside the cell between the anode and cathode. The electrolyte is typically a liquid, such as water or other solvents, with dissolved salts, acids, or alkalis to impart ionic conductivity. Some batteries use solid electrolytes or gel-type polymer electrolytes, which are ionic conductors at the operating temperature of the cell.

---

*Cell versus battery: A *cell* is the basic electrochemical unit providing a source of electrical energy by direct conversion of chemical energy. The cell consists of an assembly of electrodes, separators, electrolyte, container, and terminals. A *battery* consists of one or more electrochemical cells, electrically connected in an appropriate series/parallel arrangement to provide the required operating voltage and current levels, including, if any, monitors, controls, and other ancillary components (e.g., fuses, diodes), case, terminals, and markings. (In some publications, the term "battery" is considered to contain two or more cells.)

Popular usage considers the "battery" and not the "cell" to be the product that is sold or provided to the "user." In this 4th edition, the term "cell" will normally be used when describing the cell component of the battery and its chemistry. The term "battery" will be used when presenting performance characteristics, etc., of the product. Most often, the electrical data is presented on the basis of a single-cell battery. The performance of a multicell battery will usually be different than the performance of the individual cells or a single-cell battery (see Section 3.2.13).

The most advantageous combinations of anode and cathode materials are those that will be lightest and give a high cell voltage and capacity (see Sec. 1.4). Such combinations may not always be practical, however, due to reactivity with other cell components, polarization, difficulty in handling, high cost, and other deficiencies.

In a practical system, the anode is selected with the following properties in mind: efficiency as a reducing agent, high coulombic output (Ah/g), good conductivity, stability, ease of fabrication, and low cost. Hydrogen is attractive as an anode material, but obviously, must be contained by some means, which effectively reduces its electrochemical equivalence. Hydrogen is the active material in metal-hydride anodes (see Chap. 22). Practically, metals are mainly used as the anode material. Zinc has been a predominant anode because it has these favorable properties. Lithium, the lightest metal, with a high value of electrochemical equivalence, has become a very attractive anode as suitable and compatible electrolytes and cell designs have been developed to control its activity. With the development of intercalation electrodes, lithiated carbons are finding wide use in lithium-ion technology. Lithium alloys are also being explored for use as anodes in lithium-ion batteries.

The cathode must be an efficient oxidizing agent, be stable when in contact with the electrolyte, and have a useful working voltage. Oxygen can be used directly from ambient air being drawn into the cell, as in the zinc/air battery. However, most of the common cathode materials are metallic oxides. Other cathode materials, such as the halogens and the oxyhalides, sulfur and its oxides, are also used for special battery systems.

The electrolyte must have good ionic conductivity but not be electronically conductive, as this would cause internal short-circuiting. Other important characteristics are nonreactivity with the electrode materials, little change in properties with change in temperature, safety in handling, and low cost. Most electrolytes are aqueous solutions, but there are important exceptions as, for example, in thermal and lithium anode batteries, where molten salt and nonaqueous electrolytes are used to avoid the reaction of the anode with the electrolyte.

Physically, the anode and cathode electrodes are electronically isolated in the cell to prevent internal short-circuiting, but they are surrounded by the electrolyte. In practical cell designs, a separator material is used to separate the anode and cathode electrodes mechanically. The separator, however, is permeable to the electrolyte in order to maintain the desired ionic conductivity. In some cases, the electrolyte is immobilized for a nonspill design. Electrically conducting grid structures or materials may also be added to the electrodes to reduce internal resistance.

The cell itself can be built in many shapes and configurations—cylindrical, button, flat, and prismatic—and the cell components are designed to accommodate the particular cell shape. The cells are sealed in a variety of ways to prevent leakage and dry-out. Some cells are provided with venting devices or other means to allow accumulated gases to escape. Suitable cases or containers, means for terminal connection, and labeling are added to complete the fabrication of the cell and battery.

1.2 CLASSIFICATION OF CELLS AND BATTERIES

Electrochemical cells and batteries are identified as primary (nonrechargeable) or secondary (rechargeable), depending on their capability of being electrically recharged. Within this classification, other classifications are used to identify particular structures or designs. The classification used in this handbook for the different types of electrochemical cells and batteries is described in this section.

1.2.1 Primary Cells or Batteries

These batteries are not capable of being easily or effectively recharged electrically and, hence, are discharged once and discarded. Primary cells in which the electrolyte is contained by an absorbent or separator material (there is no free or liquid electrolyte) are termed "dry cells."

The primary battery is a convenient, usually inexpensive, lightweight source of packaged energy for portable electronic and electric devices, lighting, digital cameras, toys, memory backup, Global

Positioning System devices, and a myriad of other applications. The general advantages of primary batteries are good shelf life, high energy density at low to moderate discharge rates, little, if any, maintenance, and ease of use. Although large high-capacity primary batteries are used in military applications, signaling, and standby power, the vast majority of primary batteries are the familiar single-cell cylindrical and flat button batteries or multicell batteries using these component cells.

1.2.2 Secondary or Rechargeable Cells or Batteries

These batteries can be recharged electrically, after discharge, to their original condition by passing current through them in the opposite direction to that of the discharge current. They are storage devices for electric energy and are known also as "storage batteries" or "accumulators."

The applications of secondary batteries fall into two main categories:

1. Those applications in which the secondary battery is used as an energy-storage device, generally being electrically connected to and charged by a prime energy source and delivering its energy to the load on demand. Examples are automotive and aircraft systems, emergency no-fail and standby (UPS) power sources, hybrid electric vehicles, and battery energy storage systems (BESSs) for electric utility load leveling.

2. Those applications in which the secondary battery is used or discharged essentially as a primary battery, but recharged after use rather than being discarded. Secondary batteries are used in this manner as, for example, in portable consumer electronics, such as cell phones, laptop computers, power tools etc., for cost savings (as they can be recharged rather than replaced) and in applications requiring power drains beyond the capability of primary batteries. Electric vehicles (EVs) and plug-in hybrid PHEVs also fall into this category.

Secondary batteries are characterized (in addition to their ability to be recharged) by high power density, high discharge rate, flat discharge curves, and, in most cases, good low-temperature performance. Their energy densities are generally lower than those of primary batteries. Their charge retention also is poorer than that of most primary batteries, although the capacity of the secondary battery that is lost on standing can be restored by recharging.

Some batteries, known as "mechanically rechargeable types," are "recharged" by replacement of the discharged or depleted electrode, usually the metal anode, with a fresh one. Some of the metal/air batteries (Chap. 33) are representative of this type of battery.

1.2.3 Reserve Batteries

In these primary types, a key component is separated from the rest of the battery prior to activation. In this condition, chemical deterioration or self-discharge is essentially eliminated, and the battery is capable of long-term storage. Usually the electrolyte is the component that is isolated. In other systems, such as the thermal battery, the battery is inactive until it is heated, melting a solid electrolyte which then becomes conductive.

The reserve battery design is used to meet extremely long or environmentally severe storage requirements that cannot be met with an "active" battery designed for the same performance characteristics. These batteries are used, for example, to deliver high power for relatively short periods of time, in missiles, torpedoes, and other weapon systems. (See Chapters 34, 35, and 36.)

1.2.4 Fuel Cells

Fuel cells, like batteries, are electrochemical galvanic cells that convert chemical energy directly into electrical energy and are not subject to the Carnot cycle limitations of heat engines. Fuel cells are similar to batteries except that the active materials are not an integral part of the device (as in a battery) but are fed into the fuel cell from an external source when power is desired. The fuel cell

1.6 PRINCIPLES OF OPERATION

differs from a battery in that it has the capability of producing electrical energy as long as the active materials are fed to the electrodes (assuming the electrodes do not fail). The battery will cease to produce electrical energy when the limiting reactant stored within the battery is consumed.

The electrode materials of the fuel cell are inert in that they are not consumed during the cell reaction, but have catalytic properties which enhance the electroreduction or electro-oxidation of the reactants (the active materials).

The anode active materials used in fuel cells are generally gaseous or liquid (compared with the metal anodes generally used in most batteries) and are fed into the anode side of the fuel cell. As these materials are more like the conventional fuels used in heat engines, the term "fuel cell" has become popular to describe these devices. Oxygen or air is the predominant oxidant and is fed into the cathode side of the fuel cell.

Fuel cells have been of interest for over 160 years as a potentially more efficient and less polluting means for converting hydrogen and carbonaceous or fossil fuels to electricity compared to conventional engines. A well-known application of the fuel cell has been the use of the hydrogen/oxygen fuel cell, using cryogenic fuels, in space vehicles for over 50 years. Recent advances have revitalized interest in air-breathing systems for a variety of applications, including utility power, load leveling, dispersed or on-site electric generators, electric vehicles, and as a potential replacement for batteries in consumer electronics.

Fuel cell technology can be classified into two categories:

1. Direct systems where fuels, such as hydrogen, methanol and hydrazine, can react directly in the fuel cell
2. Indirect systems in which the fuel, such as natural gas or another fossil fuel, is first converted by reforming to a hydrogen-rich gas which is then fed into the fuel cell

Fuel cell systems can take a number of configurations depending on the combinations of fuel and oxidant, the type of electrolyte, the temperature of operation, and the application, etc.

Fuel cell technology is moving toward portable applications, historically the domain of batteries, with power levels from less than 1 to about 1000 W, blurring the distinction between batteries and fuel cells. Metal/air batteries (see Chap. 33), particularly those in which the metal is periodically replaced, can be considered a "fuel cell" with the metal being the fuel. Similarly, small fuel cells, now under development, which are "refueled" by replacing an ampule of fuel can be considered a "battery." One of these systems, the Direct Methanol Fuel Cell (DMFC), is a potential competitor for small batteries in consumer electronics.

Chapter 37 provides an introduction to fuel cells. Small to medium size fuel cells may become competitive with batteries for portable electronic and other applications. These portable devices are covered in Chap. 38. Information on the larger fuel cells for electric vehicles, utility power, etc., can be obtained from the references listed in Appendix F, "Bibliography."

1.3 *OPERATION OF A CELL*

1.3.1 Discharge

The operation of a cell during discharge is shown schematically in Fig. 1.1. When the cell is connected to an external load, electrons flow from the anode, which is oxidized, through the external load to the cathode, where the electrons are accepted and the cathode material is reduced. The electric circuit is completed in the electrolyte by the flow of anions (negative ions) and cations (positive ions) to the anode and cathode, respectively.

The discharge reaction can be written, assuming a metal as the anode material and a cathode material, such as chlorine (Cl_2), as follows:

Negative electrode: anodic reaction (oxidation, loss of electrons)

$$Zn \rightarrow Zn^{2+} + 2e$$

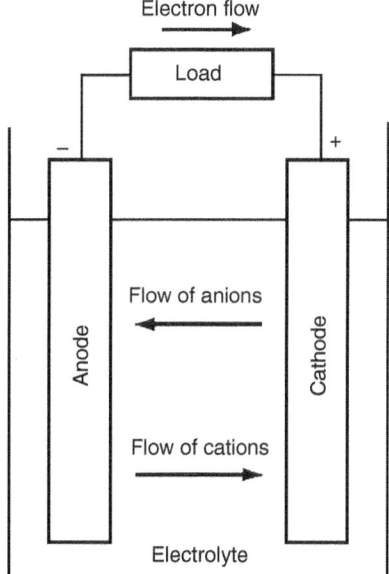

FIGURE 1.1 Electrochemical operation of a cell (discharge).

Positive electrode: cathodic reaction (reduction, gain of electrons)

$$Cl_2 + 2e \rightarrow 2Cl^-$$

Overall reaction (discharge):

$$Zn + Cl_2 \rightarrow Zn^{2+} + 2Cl^- (ZnCl_2)$$

1.3.2 Charge

During the recharge of a rechargeable or storage cell, the current flow is reversed and oxidation takes place at the positive electrode and reduction at the negative electrode, as shown in Fig. 1.2. As the anode is, by definition, the electrode at which oxidation occurs and the cathode the one where reduction takes place, the positive electrode is now the anode and the negative the cathode.

In the example of the Zn/Cl_2 cell, the reaction on charge can be written as follows:

Negative electrode: cathodic reaction (reduction, gain of electrons):

$$Zn^{2+} + 2e \rightarrow Zn$$

Positive electrode: anodic reaction (oxidation, loss of electrons):

$$2Cl^- \rightarrow Cl_2 + 2e$$

1.8 PRINCIPLES OF OPERATION

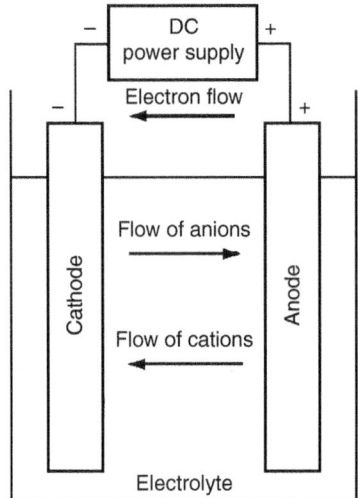

FIGURE 1.2 Electrochemical operation of a cell (charge).

Overall reaction (charge):

$$Zn^{2+} + 2Cl^- \rightarrow Zn + Cl_2$$

1.3.3 Specific Example: Nickel-Cadmium Cell

The processes that produce electricity in a cell are chemical reactions that either release or consume electrons as the electrode reaction proceeds to completion. This can be illustrated with the specific example of the reactions of the nickel-cadmium cell. At the anode (negative electrode), the discharge reaction is the oxidation of cadmium metal to cadmium hydroxide with the release of two electrons:

$$Cd + 2OH^- \rightarrow Cd(OH)_2 + 2e$$

At the cathode, nickel oxide (or more accurately, nickel oxyhydroxide) is reduced to nickel hydroxide with the acceptance of an electron:

$$NiOOH + H_2O + e \rightarrow OH^- + Ni(OH)_2$$

When these two "half-cell" reactions occur (by connection of the electrodes to an external discharge circuit), the overall cell reaction converts cadmium to cadmium hydroxide at the anode and nickel oxyhydroxide to nickel hydroxide at the cathode:

$$Cd + 2NiOOH + 2H_2O \rightarrow Cd(OH)_2 + 2Ni(OH)_2$$

This is the discharge process. If this were a primary nonrechargeable cell, at the end of discharge it would be exhausted and discarded. The nickel-cadmium battery system is, however, a secondary (rechargeable) system, and on recharge the reactions are reversed. At the negative electrode, the reaction is:

$$Cd(OH)_2 + 2e \rightarrow Cd + 2OH^-$$

At the positive electrode, the reaction is:

$$Ni(OH)_2 + OH^- \rightarrow NiOOH + H_2O + e$$

After recharge, the secondary battery reverts to its original chemical state and is ready for further discharge. These are the fundamental principles involved in the charge–discharge mechanisms of a typical secondary battery.

1.3.4 Fuel Cell

A typical fuel cell reaction is illustrated by the hydrogen/oxygen fuel cell. In this device, hydrogen is oxidized at the anode, electrocatalyzed by platinum or platinum alloys, while at the cathode, oxygen is reduced, again with platinum or platinum alloys as electrocatalysts. The simplified anodic reaction is:

$$2H_2 \rightarrow 4H^+ + 4e$$

while the cathodic reaction is:

$$O_2 + 4H^+ + 4e \rightarrow 2H_2O$$

The overall reaction is the oxidation of hydrogen by oxygen, with water as the reaction product:

$$2H_2 + O_2 \rightarrow 2H_2O$$

1.4 THEORETICAL CELL VOLTAGE, CAPACITY, AND ENERGY

The theoretical voltage and capacity of a cell are a function of the anode and cathode materials. (See Chap. 2 for detailed electrochemical theory.)

1.4.1 Free Energy

Whenever a reaction occurs, there is a decrease in the free energy of the system, which is expressed as:

$$\Delta G^0 = -nFE^0$$

where F = constant known as the Faraday ($\approx$96,500 C or 26.8 Ah)
n = number of electrons involved in stoichiometric reaction
E^0 = standard potential, V

1.4.2 Theoretical Voltage

The standard potential of the cell is determined by the type of active materials contained in the cell. It can be calculated from free-energy data or obtained experimentally. A listing of electrode potentials (reduction potentials) under standard conditions is given in Table 1.1. A more complete list is presented in Appendix B.

The standard potential of a cell can be calculated from the standard electrode potentials as follows (the oxidation potential is the negative value of the reduction potential):

Anode (oxidation potential) + cathode (reduction potential) = standard cell potential

For example, in the reaction Zn + Cl$_2$ → ZnCl$_2$, the standard cell potential is:

$$\begin{aligned} Zn &\rightarrow Zn^{2+} + 2e \quad -(-0.76 \text{ V}) \\ Cl_2 &\rightarrow 2Cl^- - 2e \quad \underline{1.36 \text{ V}} \\ E^\circ &= \quad 2.12 \text{ V} \end{aligned}$$

The cell voltage is also dependent on other factors, including concentration and temperature, as expressed by the Nernst equation (covered in detail in Chap. 2).

1.4.3 Theoretical Capacity (Coulombic)

The theoretical capacity of a cell is determined by the amount of active materials in the cell. It is expressed as the total quantity of electricity involved in the electrochemical reaction and is defined in terms of coulombs or ampere-hours. The "ampere-hour capacity" of a battery is directly associated with the quantity of electricity obtained from the active materials. Theoretically, 1 gram-equivalent weight of material will deliver 96,487 C or 26.8 Ah. (A gram-equivalent weight is the atomic or molecular weight of the active material in grams divided by the number of electrons involved in the reaction.)

The electrochemical equivalence of typical materials is listed in Table 1.1 and Appendix C.

TABLE 1.1 Characteristics of Typical Electrode Materials*

Material	Atomic or molecular weight, g	Standard reduction potential at 25°C, V	Valence change	Melting point, °C	Density, g/cm³	Electrochemical equivalents		
						Ah/g	g/Ah	Ah/cm³
Anode materials								
H$_2$	2.01	0	2	—	—	26.59	0.037	—
Li	6.94	−0.83†						
		−3.01	1	180	0.54	3.86	0.259	2.06
Na	23.0	−2.71	1	98	0.97	1.16	0.858	1.14
Mg	24.3	−2.38	2	650	1.74	2.20	0.454	3.8
		−2.69†						
Al	26.9	−1.66	3	659	2.69	2.98	0.335	8.1
Ca	40.1	−2.84	2	851	1.54	1.34	0.748	2.06
		−2.35†						
Fe	55.8	−0.44	2	1528	7.85	0.96	1.04	7.5
		−0.88†						
Zn	65.4	−0.76	2	419	7.14	0.82	1.22	5.8
		−1.25†						
Cd	112.4	−0.40	2	321	8.65	0.48	2.10	4.1
		−0.81†						
Pb	207.2	−0.13	2	327	11.34	0.26	3.87	2.9
(Li)C$_6$§	72.06	~−2.8	1	—	2.25	0.372	2.69	0.837
MH¶		−0.83†	2	—	—	0.305	3.28	—
CH$_3$OH	32.04	—	6	—	—	5.02	0.20	—

(Continued)

TABLE 1.1 Characteristics of Typical Electrode Materials* (*Continued*)

Material	Atomic or molecular weight, g	Standard reduction potential at 25°C, V	Valence change	Melting point, °C	Density, g/cm³	Electrochemical equivalents		
						Ah/g	g/Ah	Ah/cm³
Cathode materials								
CuF_2	101.5	3.55	2	—	—	0.528	1.89	
O_2	32.0	1.23	4	—	—	3.35	0.30	
		0.40†						
Cl_2	71.0	1.36	2	—	—	0.756	1.32	
SO_2	64.0	—	1	—	—	0.419	2.38	
MnO_2	86.9	1.28‡	1	—	5.0	0.308	3.24	1.54
NiOOH	91.7	0.49†	1	—	7.4	0.292	3.42	2.16
CuCl	99.0	0.14	1	—	3.5	0.270	3.69	0.95
FeS_2	119.9	—	4	—	—	0.89	1.12	4.35
AgO	123.8	0.57†	2	—	7.4	0.432	2.31	3.20
Br_2	159.8	1.07	2	—	—	0.335	2.98	
HgO	216.6	0.10†	2	—	11.1	0.247	4.05	2.74
Ag_2O	231.7	0.35†	2	—	7.1	0.231	4.33	1.64
PbO_2	239.2	1.69	2	—	9.4	0.224	4.45	2.11
$LiFePO_4$	163.8	~0.42	1	—	3.44	0.160	6.25	0.554
$LiMn_2O_4$ (spinel)	148.8	~1.2	1	—	4.1	0.120	8.33	0.492
Li_xCoO_2	98	~1.25	0.5	—	5.05	0.155	6.45	0.782
I_2	253.8	0.54	2	—	4.94	0.211	4.73	1.04

*See also Appendixes B and C.
†Basic electrolyte; all others, aqueous acid or non-aqueous electrolytes.
‡Based on density values shown.
§Calculations based only on weight of carbon.
¶Based on type AB_5 alloy.

The theoretical capacity of an electrochemical cell, based only on the active materials participating in the electrochemical reaction, is calculated from the equivalent weight of the reactants. Hence, the theoretical capacity of the Zn/Cl_2 cell is 0.394 Ah/g, that is,

$$Zn + Cl_2 \longrightarrow ZnCl_2$$

(0.82 Ah/g) (0.76 Ah/g)

1.22 g/Ah + 1.32 g/Ah = 2.54 g/Ah or 0.394 Ah/g

Similarly, the ampere-hour capacity on a volume basis can be calculated using the appropriate data for ampere-hours per cubic centimeter as listed in Table 1.1.

The theoretical voltages and capacities of a number of the major electrochemical systems are given in Table 1.2. These theoretical values are based on the active anode and cathode materials only. Water, electrolyte, or any other materials that may be involved in the cell reaction are not included in the calculation.

TABLE 1.2 Voltage, Capacity, and Specific Energy of Major Battery Systems—Theoretical and Practical Values

Battery type	Anode	Cathode	Reaction mechanism	Theoretical values[a] V	g/Ah	Ah/kg	Specific energy Wh/kg	Practical battery[b] Nominal voltage V	Specific energy Wh/kg	Energy density Wh/L
Primary batteries										
Leclanché	Zn	MnO_2	$Zn + 2MnO_2 \rightarrow ZnO \cdot Mn_2O_3$	1.6	4.46	224	358	1.5	85[f]	165[f]
Magnesium	Mg	MnO_2	$Mg + 2MnO_2 + H_2O \rightarrow Mn_2O_3 + Mg(OH)_2$	2.8	3.69	271	759	1.7	100[f]	195[f]
Alkaline MnO_2	Zn	MnO_2	$Zn + 2MnO_2 \rightarrow ZnO + Mn_2O_3$	1.5	4.46	224	358	1.5	154[f]	461[f]
Mercury	Zn	HgO	$Zn + HgO \rightarrow ZnO + Hg$	1.34	5.27	190	255	1.35	100[h]	470[h]
Mercad	Cd	HgO	$Cd + HgO + H_2O \rightarrow Cd(OH)_2 + Hg$	0.91	6.15	163	148	0.9	55[h]	230[h]
Silver oxide	Zn	Ag_2O	$Zn + Ag_2O + H_2O \rightarrow Zn(OH)_2 + 2Ag$	1.6	5.55	180	288	1.6	135[h]	525[h]
Zinc/O_2	Zn	O_2	$Zn + ½O_2 \rightarrow ZnO$	1.65	1.52	658	1085	—	—	—
Zinc/air	Zn	Ambient air	$Zn + (½ O_2) \rightarrow ZnO$	1.65	1.22	820	1353	1.5	415[h]	1350[h]
Li/$SOCl_2$	Li	$SOCl_2$	$4Li + 2SOCl_2 \rightarrow 4LiCl + S + SO_2$	3.65	3.25	403	1471	3.6	590[f]	1100[f]
Li/SO_2	Li	SO_2	$2Li + 2SO_2 \rightarrow Li_2S_2O_4$	3.1	2.64	379	1175	3.0	260[g]	415[g]
$LiMnO_2$	Li	MnO_2	$Li + Mn^{IV}O_2 \rightarrow Mn^{III}O_2(Li^+)$	3.5	3.50	286	1001	3.0	260[g]	546[g]
Li/FeS_2	Li	FeS_2	$4Li + FeS_2 \rightarrow 2Li_2S + Fe$	1.8	1.38	726	1307	1.5	310[g]	560[g]
Li/CF_x	Li	CF_x	$xLi + CF_x \rightarrow xLiF + xC$	3.1	1.42	706	2189	3.0	360[g]	540[g]
Li/I_2[e]	Li	$I_2(P2VP)$	$Li + ½I_2 \rightarrow LiI$	2.8	4.99	200	560	2.8	245	900
Reserve batteries										
Cuprous chloride	Mg	CuCl	$Mg + Cu_2Cl_2 \rightarrow MgCl_2 + 2Cu$	1.6	4.14	241	386	1.3	60[i]	80[i]
Zinc/silver oxide	Zn	AgO	$Zn + AgO + H_2O \rightarrow Zn(OH)_2 + Hg$	1.81	3.53	283	512	1.5	30[j]	75[j]
Thermal[d]	Li	FeS_2	See Section 36.3.1	2.1–1.6	1.38	726	1307	2.1–1.6	40[k]	100[k]

Secondary batteries

Battery	Anode	Cathode	Reaction	V		Ah/kg	Wh/kg	Wh/L	V	Wh/kg	Wh/L
Lead-acid	Pb	PbO_2	$Pb + PbO_2 + 2H_2SO_4 \rightarrow 2PbSO_4 + 2H_2O$	2.1	8.32	120	252	2.0	35	70[l]
Edison	Fe	Ni oxide	$Fe + 2NiOOH + 2H_2O \rightarrow 2Ni(OH)_2 + Fe(OH)_2$	1.4	4.46	224	314	1.2	30	55[l]
Nickel-cadmium	Cd	Ni oxide	$Cd + 2NiOOH + 2H_2O \rightarrow 2Ni(OH)_2 + Cd(OH)_2$	1.35	5.52	181	244	1.2	40	135[g]
Nickel-zinc	Zn	Ni oxide	$Zn + 2NiOOH + 2H_2O \rightarrow 2Ni(OH)_2 + Zn(OH)_2$	1.73	4.64	215	372	1.6	90	185
Nickel-hydrogen	H_2	Ni oxide	$H_2 + 2NiOOH \rightarrow 2Ni(OH)_2$	1.5	3.46	289	434	1.2	55	60
Nickel-metal hydride	MH[c]	Ni oxide	$MH + NiOOH \rightarrow M + Ni(OH)_2$	1.35	5.63	178	240	1.2	100	235[g]
Silver-zinc	Zn	AgO	$Zn + AgO + H_2O \rightarrow Zn(OH)_2 + Ag$	1.85	3.53	283	524	1.5	105	180[l]
Silver-cadmium	Cd	AgO	$Cd + AgO + H_2O \rightarrow Cd(OH)_2 + Ag$	1.4	4.41	227	318	1.1	70	120[l]
Zinc/chlorine	Zn	Cl_2	$Zn + Cl_2 \rightarrow ZnCl_2$	2.12	2.54	394	835	—	—	—
Zinc/bromine	Zn	Br_2	$Zn + Br_2 \rightarrow ZnBr_2$	1.85	4.17	309	572	1.6	70	60
Lithium-ion	Li_xC_6	$Li_{(1-x)}CoO_2$	$Li_xC_6 + Li_{(1-x)}CoO_2 \rightarrow LiCoO_2 + C_6$	4.1	9.14	109	448	3.8	200	570[g]
Lithium/manganese dioxide	Li	MnO_2	$Li + Mn^{IV}O_2 \rightarrow Mn^{IV}O_2(Li^+)$	3.5	3.50	286	1001	3.0	120	265
Lithium/iron disulfide[d]	Li(Al)	FeS_2	$2Li(Al) + FeS_2 \rightarrow Li_2FeS_2 + 2Al$	1.73	3.50	285	493	1.7	180[m]	350[m]
Sodium/sulfur[d]	Na	S	$2Na + 3S \rightarrow Na_2S_3$	2.1	2.65	377	792	2.0	170[m]	345[m]
Sodium/nickel chloride[d]	Na	$NiCl_2$	$2Na + NiCl_2 \rightarrow 2NaCl + Ni$	2.58	3.28	305	787	2.6	115[m]	190[m]

Fuel cells

H_2/O_2	H_2	O_2	$H_2 + \tfrac{1}{2}O_2 \rightarrow H_2O$	1.23	0.336	2975	3660	—	—	—	
H_2/air	H_2	Ambient air	$H_2 + (\tfrac{1}{2}O_2) \rightarrow H_2O$	1.23	0.037	26587	32702	—	—	—	
Methanol/O_2	CH_3OH	O_2	$CH_3OH + \tfrac{3}{2}O_2 \rightarrow CO_2 + 2H_2O$	1.24	0.50	2000	2480	—	—	—	
Methanol/air	CH_3OH	Ambient air	$CH_3OH + (\tfrac{3}{2}O_2) \rightarrow CO_2 + 2H_2O$	1.24	0.20	5020	6225	—	—	—	

[a]Based on active anode and cathode materials only, including O_2 but not air (electrolyte not included).
[b]These values are for single-cell batteries based on identified design and at discharge rates optimized for energy density, using midpoint voltage. More specific values are given in chapters on each battery system.
[c]MH = metal hydride, data based on type AB_5 alloy.
[d]High temperature batteries.
[e]Solid electrolyte battery (Li/I_2 (P2VP)).
[f]Cylindrical bobbin-type batteries.
[g]Cylindrical spiral-wound batteries.
[h]Button type batteries.
[i]Water-activated.
[j]Automatically activated 2- to 10-min rate.
[k]With lithium anodes.
[l]Prismatic batteries.
[m]Value based on cell performance, see appropriate chapter for details.

1.4.4 Theoretical Energy*

The capacity of a cell can also be considered on an energy (watthour) basis by taking both the voltage and the quantity of electricity into consideration. This theoretical energy value is the maximum value that can be delivered by a specific electrochemical system:

$$\text{Watthour (Wh)} = \text{voltage (V)} \times \text{ampere-hour (Ah)}$$

In the Zn/Cl_2 cell example, if the standard potential is taken as 2.12 V, the theoretical watthour capacity per gram of active material (theoretical gravimetric specific energy or theoretical gravimetric energy density) is

$$\text{Specific energy (watthours/gram)} = 2.12 \text{ V} \times 0.394 \text{ Ah/g} = 0.835 \text{ Wh/g or } 835 \text{ Wh/kg}$$

Table 1.2 also lists the theoretical specific energy of various electrochemical systems.

1.5 SPECIFIC ENERGY AND ENERGY DENSITY OF PRACTICAL BATTERIES

The theoretical electrical properties of cells and batteries are discussed in Sec. 1.4. In summary, the maximum energy that can be delivered by an electrochemical system is based on the types of active materials that are used (this determines the voltage) and on the amount of the active materials that are used (this determines ampere-hour capacity). In practice, only a fraction of the theoretical energy of the battery is realized. This is due to the need for electrolyte and nonreactive components (containers, separators, electrodes) that add to the weight and volume of the battery, as illustrated in Fig. 1.3. Another contributing factor is that the battery does not discharge at the theoretical voltage (thus lowering the average voltage), nor is it discharged completely to zero volts (thus reducing the delivered ampere-hours) (also see Sec. 3.2.1). Further, the active materials in a practical battery are usually not stoichiometrically balanced. This reduces the specific energy because an excess amount of one of the active materials is used.

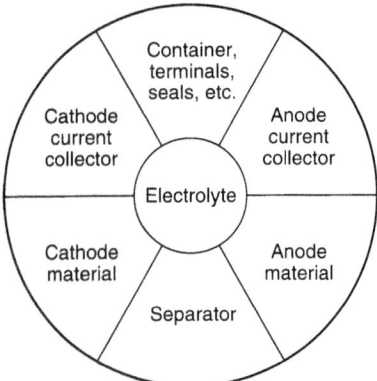

FIGURE 1.3 Components of a cell.

---

*The energy output of a cell or battery is often expressed as a ratio of its weight or size. The preferred terminology for this ratio on a weight basis, e.g., watthours/kilogram (Wh/kg), is "specific energy;" on a volume basis, e.g., watthours/liter (Wh/L), it is "energy density." Commonly, the term "energy density" may be used to refer to either ratio.

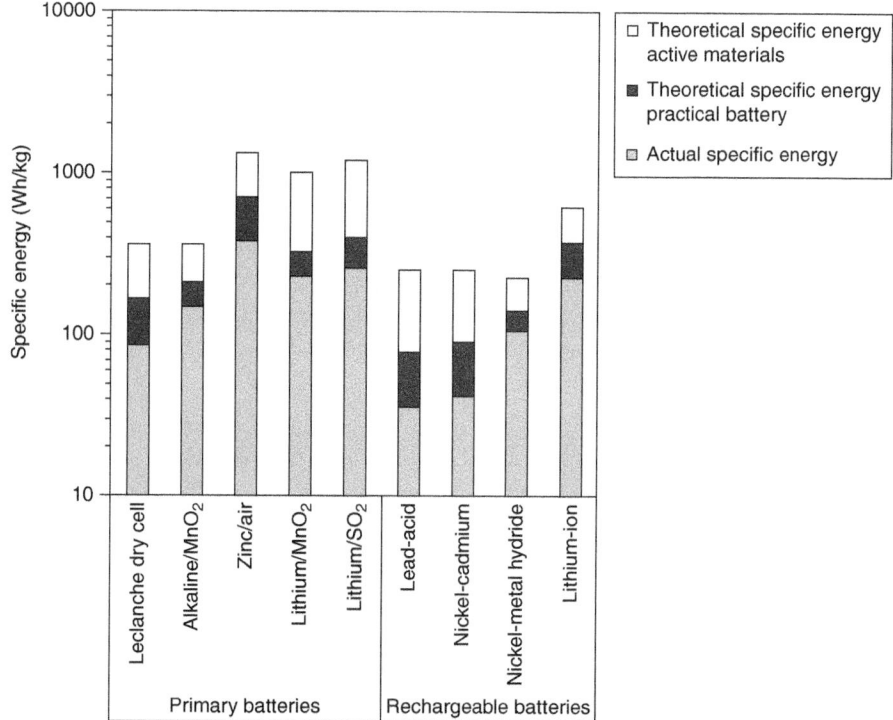

FIGURE 1.4 Theoretical and actual specific energy of battery systems.

In Fig. 1.4, the following values for some major batteries are plotted:

1. The theoretical specific energy (based on the active anode and cathode materials only)
2. The theoretical specific energy of a practical battery (accounting for the electrolyte and nonreactive components)
3. The actual specific energy of these batteries when discharged at 20°C under optimal discharge conditions

These data show:

- That the weight of the materials of construction reduces the theoretical energy density of the battery by almost 50%, and
- That the actual energy delivered by a practical battery, even when discharged under conditions close to optimum, may only be 50 to 75% of that lowered value

The development of "pouch cells" which use foil-laminate packaging has significantly reduced the weight and volume penalty associated with the container. (See Chapters 26 and 27.)

Thus, the actual energy that is available from a battery under practical, but close to optimum, discharge conditions is only about 25 to 35% of the theoretical energy of the active materials. Chapter 3 covers the performance of batteries when used under more stringent conditions.

These data are shown again in Table 1.2 which, in addition to the theoretical values, lists the characteristics of each of these batteries based on the actual performance of a practical battery. Again, these values are based on discharge conditions close to optimum for that battery.

1.16 PRINCIPLES OF OPERATION

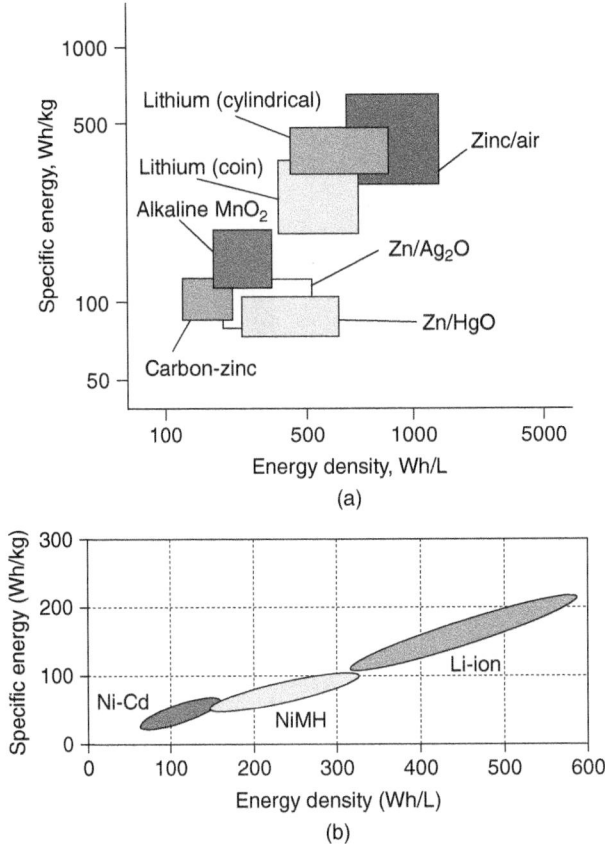

FIGURE 1.5 Comparison of the energy storage capability of various battery systems: (*a*) primary batteries, (*b*) rechargeable batteries.

The specific energy (Wh/kg) and energy density (Wh/L) delivered by the major battery systems are also plotted in Fig. 1.5(*a*) for primary batteries and 1.5(*b*) for rechargeable batteries. In these figures, the energy storage capability is shown as a field, rather than as a single optimum value, to illustrate the spread in performance of that battery system under different conditions of use.

In practice, as discussed in detail in Chap. 3, the electrical output of a battery may be reduced even further when it is used under more stringent conditions.

1.6 LIMITS OF SPECIFIC ENERGY AND ENERGY DENSITY

Many advances have been made in battery technology in recent years, as illustrated in Fig. 1.6, both through continued improvement of a specific electrochemical system and through the development and introduction of new battery chemistries. But batteries are not keeping pace with developments in electronics technology, where performance doubles every 18 months, a phenomenon known as Moore's Law. Batteries, unlike electronic devices, consume materials when delivering electrical energy and, as discussed in Secs. 1.4 and 1.5, there are theoretical limits to the amount of electrical energy that can be delivered electrochemically by the available materials.

As shown in Table 1.2 and other such tables in this book, except for some of the ambient air-breathing systems and the hydrogen/oxygen fuel cell, where the weight of the cathode active material

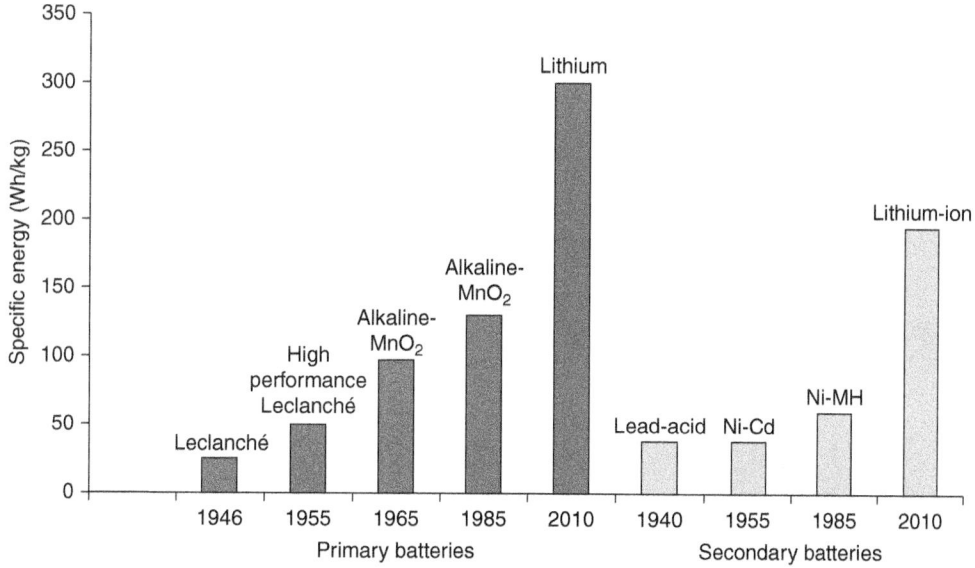

FIGURE 1.6 Advances in battery performance for portable applications.

is not included in the calculation, the values for the theoretical energy density do not exceed 1500 Wh/kg. Even the values for hydrogen/air and liquid fuel cells have to be lowered to include, at least, the weight and volume of suitable containers for these fuels.

The data in Table 1.2 also show that the specific energy delivered by these batteries, based on the actual performance when discharged under optimum conditions, does not exceed 600 Wh/kg, even including the air-breathing systems. Similarly, the energy density values do not exceed 1300 Wh/L. It is also noteworthy that the values for the rechargeable systems are lower than those of the primary batteries, due in part to a more limited selection of materials that can be recharged practically and the need for designs to facilitate recharging and cycle life.

Recently, there has been much interest in the lithium/air system, which has a theoretical specific energy of 11,000 Wh/kg, particularly as a rechargeable system for use in EVs. Primary cells have produced 800 Wh/kg (see Chap. 33).

Recognizing these limitations, while new battery systems will be explored, it will be more difficult to develop a new battery system that will have a significantly higher energy output and still meet the requirements of a successful commercial product, including availability of materials, acceptable cost, safety, and environmental acceptability.

Battery research and development will focus on reducing the ratio of inactive to active components to improve energy density, increasing conversion efficiency and rechargability, maximizing performance under the more stringent operating conditions and enhancing safety. The fuel cell is offering opportunities for powering electric vehicles, as a replacement for combustion engines, for use in utility power and possibly for the larger portable applications (see Chap. 38). The development of a fuel cell for small portable applications, such as cell phones, that will be competitive with batteries presents a formidable challenge. (see Chap. 37).

REFERENCES

1. D. Linden and T. B. Reddy, *Battery Power and Products Technology*, vol. 5, no. 2, pp. 10–12, March/April 2008.
2. M. Winter and R. Brodd, *Chemical Reviews*, vol. 104, 4245–4270, 2004.

CHAPTER 2
ELECTROCHEMICAL PRINCIPLES AND REACTIONS

Mark Salomon

2.1 INTRODUCTION

Batteries and fuel cells are electrochemical devices that convert chemical energy into electrical energy by electrochemical oxidation and reduction reactions, which occur at the electrodes. A cell consists of an anode where oxidation takes place during discharge, a cathode where reduction takes place, and an electrolyte which conducts the current (via ions) within the cell.

The maximum electric energy that can be delivered by the chemicals that are stored within or supplied to the electrodes in the cell depends on the change in Gibbs energy ΔG of the electrochemical couple, as shown in Eq. (2.5) and discussed in Sec. 2.2.

It would be desirable if during the discharge all of this energy could be converted to useful electric energy. However, losses due to polarization occur when a load current i passes through the electrodes, accompanying the electrochemical reactions. These losses include: (1) activation polarization, which drives the electrochemical reaction at the electrode surface, and (2) concentration polarization, which arises from the concentration differences of the reactants and products at the electrode surface and in the bulk as a result of mass transfer.

These polarization effects consume part of the energy, which is given off as waste heat, and thus not all of the theoretically available energy stored in electrodes is fully converted into useful electrical energy.

In principle, activation polarization and concentration polarization can be calculated from several theoretical equations, as described in later sections of this chapter, if some electrochemical parameters and the mass-transfer condition are available. However, in practice it is difficult to determine the values for both because of the complicated physical structure of the electrodes. As covered in Sec. 2.5, most battery and fuel cells electrodes are composite bodies made of active material, binder, performance enhancing additives, and conductive filler. They usually have a porous structure of finite thickness. It requires complex mathematical modeling with computer calculations to estimate the polarization components.

There is another important factor that strongly affects the performance or rate capability of a cell, the internal impedance of the cell. It causes a voltage drop during operation, which also consumes part of the useful energy as waste heat. The voltage drop due to internal impedance is usually referred to as "ohmic polarization" or *IR* drop and is proportional to the current drawn from the system. The total internal impedance of a cell is the sum of the ionic resistance of the electrolyte (within the separator and the porous electrodes), the electronic resistances of the active mass, the current collectors and electrical tabs of both electrodes, and the contact resistance between the active mass and the current collector. These resistances are ohmic in nature, and follow Ohm's law, with a linear relationship between current and voltage drop.

2.2 PRINCIPLES OF OPERATION

When connected to an external load R, the cell voltage E can be expressed as

$$E = E_0 - [(\eta_{ct})_a + (\eta_c)_a] - [(\eta_{ct})_c + (\eta_c)_c] - iR_i = iR \qquad (2.1)$$

where
- E_0 = electromotive force or open-circuit voltage of cell
- $(\eta_{ct})_a$, $(\eta_{ct})_c$ = activation polarization or charge-transfer overvoltage at anode and cathode
- $(\eta_c)_a$, $(\eta_c)_c$ = concentration polarization at anode and cathode
- i = operating current of cell on load
- R_i = internal resistance of cell

As shown in Eq. (2.1), the useful voltage delivered by the cell is reduced by polarization and the internal IR drop. It is only at very low operating currents, where polarization and the IR drop are small, that the cell may operate close to the open-circuit voltage and deliver most of the theoretically available energy. Figure 2.1 shows the relation between cell polarization and discharge current.

Although the available energy of a battery or fuel cell depends on the basic electrochemical reactions at both electrodes, there are many factors that affect the magnitude of the charge-transfer reaction, diffusion rates, and thus the magnitude of the energy loss. These factors include electrode formulation and design, electrolyte conductivity, and nature of the separators, among others. There exist some essential rules, based on the electrochemical principles, which are important in the design of batteries and fuel cells to achieve a high operating efficiency with minimal loss of energy.

1. The conductivity of the electrolyte should be high enough that the IR polarization is not excessively large for practical operation. Table 2.1 shows the typical ranges of specific conductivities for various electrolyte systems used in batteries. Batteries are usually designed for specific drain-rate applications ranging from microamperes to several hundred amperes. For a given electrolyte, a cell may be designed to have improved rate capability, with a higher electrode interfacial area and thin separator, to reduce the IR drop due to electrolyte resistance. Cells with a spirally wound electrode design are typical examples.

2. Electrolyte salt and solvents should have chemical stability to avoid direct chemical reaction with the anode or cathode materials.

3. The rate of electrode reaction at both the anode and the cathode should be sufficiently fast so that the activation or charge-transfer polarization is not too high to make the cell inoperable. A common method of minimizing the charge-transfer polarization is to use a porous electrode design. The porous electrode structure provides a high electrode surface area within a given geometric dimension of the electrode and reduces the local current density for a given total operating current.

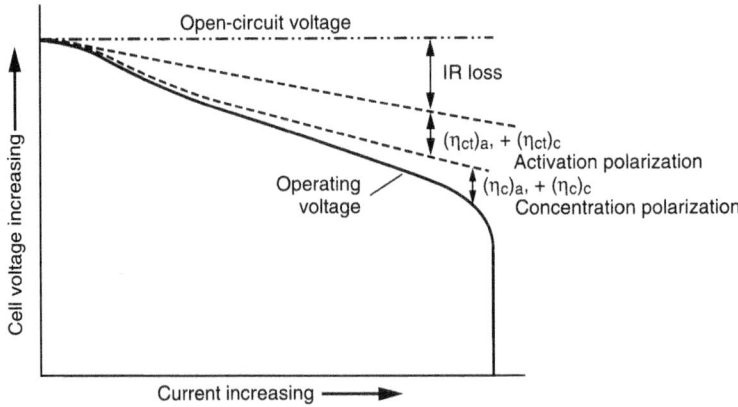

FIGURE 2.1 Cell polarization as a function of operating current.

TABLE 2.1 Conductivity Ranges of Various Electrolytes at Ambient Temperature

Electrolyte system	Conductivity/S cm^{-1}
Aqueous electrolytes	0.1–0.55
Molten salts	~10^{-1}
Inorganic electrolytes	10^{-2}–10^{-1}
Organic electrolytes	10^{-1}–10^{-2}
Ionic liquids	10^{-4}–10^{-2}
Polymer electrolytes	10^{-7}–10^{-3}
Inorganic solid electrolytes	10^{-8}–10^{-5}

4. In most battery and fuel cell systems, part or all of the reactants are supplied from the electrode phase, and part or all of the reaction products must diffuse or be transported away from the electrode surface. The cell should have adequate electrolyte transport to facilitate the mass transfer to avoid building up excessive concentration polarization. Proper porosity and pore size of the electrode, adequate thickness and structure of the separator, and sufficient concentration of the reactants in the electrolyte are very important to ensure functionality of the cell. Mass-transfer limitations should be avoided for normal operation of the cell.

5. The material of the current collector or substrate should be compatible with the electrode material and the electrolyte without causing corrosion problems. The design of the current collector should provide a uniform current distribution and low contact resistance to minimize electrode polarization during operation.

6. For rechargeable cells, it is preferable to have the reaction products remain at the electrode surface to facilitate the reversible reactions during charge and discharge. The reaction products should be stable mechanically as well as chemically with the electrolyte.

In general, the principles and various electrochemical techniques described in this chapter can be used to study all the important electrochemical aspects of a battery or fuel cell. These include the rate of electrode reaction, the existence of intermediate reaction steps, the stability of the electrolyte, the current collector, the electrode materials, the mass-transfer conditions, the value of the limiting current, the formation of resistive films on the electrode surface, the impedance characteristics of the electrode or cell, and the existence of the rate-limiting species.

2.2 THERMODYNAMIC BACKGROUND

In a cell, reactions essentially take place at two areas or sites in the device. These reaction sites are the electrode interfaces. In generalized terms, the reaction at one electrode (reduction in forward direction) can be represented by

$$a\text{A} + n\text{e} \rightleftharpoons c\text{C} \tag{2.2}$$

where a molecules of A take up n electrons e to form c molecules of C. At the other electrode, the reaction (oxidation in forward direction) can be represented by

$$b\text{B} \rightleftharpoons d\text{D} + n\text{e} \tag{2.3}$$

The overall reaction in the cell is given by addition of these two half-cell reactions

$$a\text{A} + b\text{B} \rightleftharpoons c\text{C} + d\text{D} \tag{2.4}$$

2.4 PRINCIPLES OF OPERATION

TABLE 2.2 Standard Reduction Potentials (Aqueous Solutions) of Electrode Reactions at 25°C

Electrode reaction	$E°/N$	Electrode reaction	$E°/N$
$Li^+ + e \rightleftharpoons Li$	−3.045	$CuCl + e \rightleftharpoons Cu + Cl^-$	0.121
$K^+ + e \rightleftharpoons K$	−2.925	$AgCl + e \rightleftharpoons Ag + Cl^-$	0.2223
$Na^+ + e \rightleftharpoons Na$	−2.714	$AgCl + e \rightleftharpoons Ag + Cl^-$ (seawater, pH 8.2)	0.2476
$Al^{3+} + 3e \rightleftharpoons Al$	−1.67	$Hg_2Cl_2 + 2e \rightleftharpoons 2Hg + 2Cl^-$	0.2682
$H_2O + e \rightleftharpoons \frac{1}{2}H_2 + OH^-$	0.8277	$Hg_2Cl_2 + 2e \rightleftharpoons 2Hg + 2Cl^-$ (satd KCl (SCE))	0.2412
$H_2O + e \rightleftharpoons \frac{1}{2}H_2 + OH^-$ (seawater, pH 8.2)	0.5325	$O_2 + 2H_2O + 4e \rightleftharpoons 4OH^-$	0.401
$Ni(OH)_2 + 2e \rightleftharpoons Ni + 2OH^-$	−0.72	$Cu^{2+} + Cl^- + e \rightleftharpoons CuCl$	0.559
$O_2 + H^+ + e \rightleftharpoons HO_2$	−0.046	$O_2 + 4H^+ + 4e \rightleftharpoons 2H_2O$ (pure water, pH 7)	0.815
$2H^+ + 2e \rightleftharpoons H_2$	0.000	$O_2 + 4H^+ + 4e \rightleftharpoons 2H_2O$	1.229
$HgO + H_2O + 2e \rightleftharpoons Hg + 2OH^-$	0.0977	$Cl_2 + 2e \rightleftharpoons 2Cl^-$	1.358

The change in the standard Gibbs energy $\Delta G°$ of this reaction is expressed as

$$\Delta G° = -nFE° \tag{2.5}$$

where F = constant known as the Faraday (96,487 coulombs equiv^{-1}) and $E°$ = standard electromotive force. Selected values of standard electrode potentials are given in Table 2.2, and additional values can be found in Appendix B.

When conditions are other than in the standard state, the voltage E of a cell is given by the Nernst equation,

$$E = E° - \frac{RT}{nF} \ln \frac{a_C^c a_D^d}{a_A^a a_B^b} \tag{2.6}$$

where a_i = activity of relevant species, R = gas constant (8.314 J K^{-1} mol^{-1}), and T is the absolute temperature in Kelvin.

The change in Gibbs energy $\Delta G°$ of a cell reaction is the driving force which enables a battery to deliver electrical energy to an external circuit. The measurement of the electromotive force, incidentally, also makes available data on changes in free energy, entropies, and enthalpies together with activity coefficients, equilibrium constants, and solubility products.

Direct measurement of single (absolute) electrode potentials is considered practically impossible.[1] To establish a scale of half-cell or standard potentials, a reference potential "zero" must be established against which single electrode potentials can be measured. By convention, the standard potential of the H_2/H^+ (aq) reaction is taken as zero and all standard potentials are referred to this potential. Table 2.2 and Appendix B list the standard potentials of a number of anode and cathode materials.

2.3 ELECTRODE PROCESSES

Reactions at an electrode are characterized by both chemical and electrical changes and are heterogeneous in type. Electrode reactions may be as simple as the reduction of a metal ion and incorporation of the resultant atom onto or into the electrode structure. Despite the apparent simplicity of the reaction, the mechanism of the overall process may be relatively complex and often involves several steps. Electroactive species must be transported to the electrode surface by migration or diffusion prior to the electron transfer step. Adsorption of electroactive material may be involved both prior to and after the electron transfer step. Chemical reactions may also be involved in the overall electrode reaction. As in any reaction, the overall rate of the electrochemical process is determined by the rate of the slowest step in the whole sequence of reactions.

The thermodynamic treatment of electrochemical processes presented in Sec. 2.2 describes the equilibrium condition of a system but does not present information on nonequilibrium conditions such as current flow resulting from electrode polarization (overvoltage) imposed to effect electrochemical reactions. Experimental determination of the current-voltage characteristics of many electrochemical systems has shown that there is an exponential relation between current and applied voltage. The generalized expression describing this relationship is called the Tafel equation,

$$\eta = a \pm b \log i \tag{2.7}$$

where η = overvoltage, i = current, and a and b are constants. Typically, the constant b is referred to as the Tafel slope. The Tafel relationship holds for a large number of electrochemical systems over a wide range of overpotentials. At low values of overvoltage, however, the relationship breaks down and results in curvature in plots of η versus $\log i$. Figure 2.2 is a schematic presentation of a Tafel plot, showing curvature at low values of overvoltage.

Success of the Tafel equation's fit to many experimental systems encouraged the quest for a kinetic theory of electrode processes. Since the range of validity of the Tafel relationship applies to high overvoltages, it is reasonable to assume that the expression does not apply to equilibrium situations but represents the current-voltage relationship of a unidirectional process. In an oxidation process, this means that there is a negligible contribution from reduction processes and vice versa. Rearranging Eq. (2.7) into exponential form, we have

$$i = \exp\left(\pm \frac{a}{b}\right) \exp \frac{\eta}{b} \tag{2.8}$$

To consider a general theory, one must consider both forward and backward reactions of the electroreduction process, shown in simplified form in Fig. 2.3. The reaction is represented by the equation

$$O + ne \rightleftharpoons R \tag{2.9}$$

where O = oxidized species
R = reduced species
n = number of electrons involved in electrode process

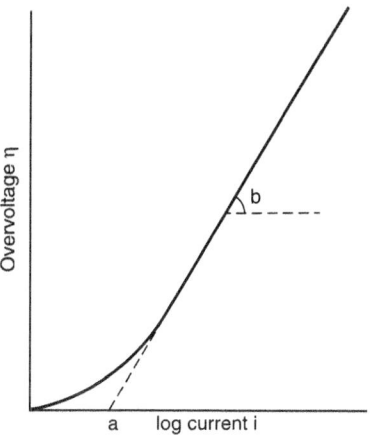

FIGURE 2.2 Schematic representation of a Tafel plot showing curvature at low overvoltage.

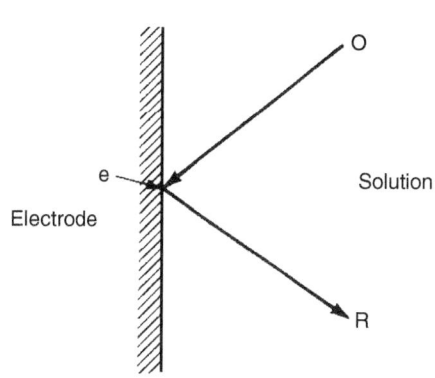

FIGURE 2.3 Simplified representation of electroreduction at an electrode.

The forward and backward reactions can be described by heterogeneous rate constants k_f and k_b, respectively. The rates of the forward and backward reactions are then given by the products of these rate constants and the relevant concentrations, which typically are those at the electrode surface. As will be shown later, the concentrations of electroactive species at the electrode surface often are dissimilar from the bulk concentration in solution. The rate of the forward reaction is $k_f C_O$ and that for the backward reaction is $k_b C_R$. For convenience, these rates are usually expressed in terms of currents i_f and i_b for the forward and backward reactions, respectively,

$$i_f = nFAk_f C_O \tag{2.10}$$

$$i_b = nFAk_b C_R \tag{2.11}$$

where A is the area of the electrode and F the Faraday.

Establishing these expressions is merely the result of applying the law of mass action to the forward and backward electrochemical processes. The role of electrons in the process is established by assuming that the magnitudes of the rate constants depend on the electrode potential. The dependence is usually described by assuming that a fraction αE of the electrode potential is involved in driving the reduction process, while the fraction $(1-\alpha)E$ is effective in making the reoxidation process more difficult. Mathematically, these potential-dependent rate constants are expressed as

$$k_f = k_f^0 \exp\left(\frac{-\alpha nFE}{RT}\right) \tag{2.12}$$

$$k_b = k_b^0 \exp\left(\frac{(1-\alpha)nFE}{RT}\right) \tag{2.13}$$

where α is the transfer coefficient and E the electrode potential relative to a suitable reference potential.

A little more explanation regarding what the transfer coefficient α (or the symmetry factor β as it is referred to in some texts) means in mechanistic terms is appropriate since this term is not implicit in the kinetic derivation.[2] The transfer coefficient determines what fraction of the electric energy resulting from the displacement of the potential from the equilibrium value affects the rate of electrochemical transformation. To understand the function of the transfer coefficient α, it is necessary to describe a potential energy diagram for the reduction-oxidation process. Figure 2.4 shows an approximate potential

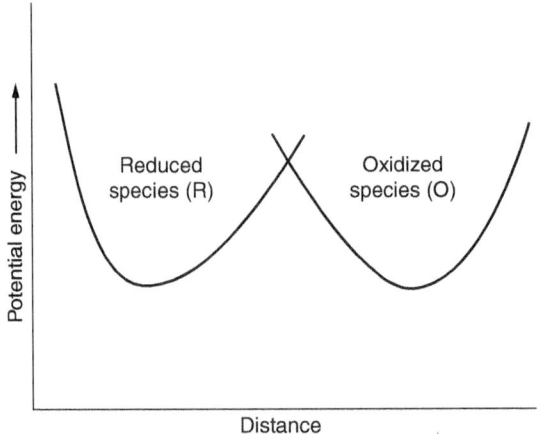

FIGURE 2.4 Potential energy diagram for a reduction-oxidation process taking place at an electrode.

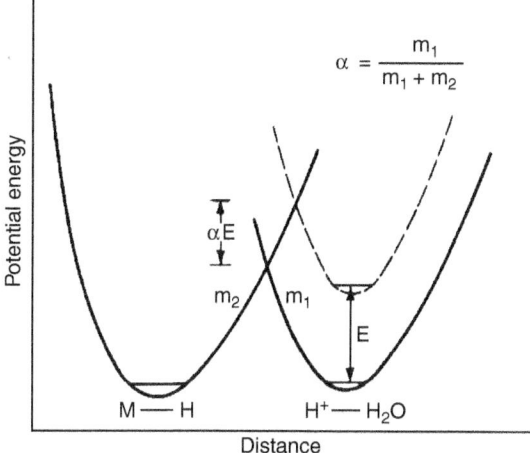

FIGURE 2.5 Potential energy diagram for reduction of H^+ at an electrode such as Pt. The dashed line represents the potential energy in the absence of an electrode field which is lowered when the ion is subjected to a potential $E°$. However, the activation energy is lowered by an amount by αE (usually half) as discussed in the text.

energy curve (Morse curve) for an oxidized species (e.g., solvated H^+) approaching an electrode surface together with the potential energy curve for the resultant reduced species (e.g., H adsorbed on a metal surface). For convenience, consider the hydrogen ion reduction at a solid electrode as the model for a typical electroreduction. According to Horiuti and Polanyi,[3] the potential energy diagram for reduction of the hydrogen ion (nominally H_3O^+) can be represented by Fig. 2.5 where the oxidized species O is the hydrated hydrogen ion and the reduced species R is a hydrogen atom bonded to the metal (electrode) surface. The effect of changing the electrode potential by a value of E is to raise the potential energy of the Morse curve of the hydrogen ion. The intersection of the two Morse curves forms an energy barrier, the height of which is αE. If the slope of the two Morse curves is approximately constant at the point of intersection, then α is defined by the ratio of the slope of the Morse curves at the point of intersection

$$\alpha = \frac{m_1}{m_1 + m_2} \tag{2.14}$$

where m_1 and m_2 are the slopes of the potential energy curves of the hydrated hydrogen ion and the hydrogen atom, respectively.

There are inadequacies in the theory of transfer coefficients. It assumes that α is constant and independent of $E°$. At present there are no data to prove or disprove this assumption. The other main weakness is that the concept is used to describe processes involving a variety of different species such as (1) redox changes at an inert electrode (Fe^{2+}/Fe^{3+} at Hg); (2) reactant and product soluble in different phases [$Cd^{2+}/Cd(Hg)$]; and (3) electrodeposition (Cu^{2+}/Cu). Despite these inadequacies, the concept and application of the theory are appropriate in many cases and represent the best understanding of electrode processes at the present time. Examples of selected electrode reactions at various metals are given in Table 2.3.[4]

From Eqs. (2.12) and (2.13) we can derive parameters useful for evaluating and describing an electrochemical system. Equations (2.12) and (2.13) are compatible both with the Nernst equation [Eq. (2.6)] for equilibrium conditions and with the Tafel relationships [Eq. (2.7)] for unidirectional processes. Under equilibrium conditions, both forward and backward currents

TABLE 2.3 Values of Transfer Coefficients α at 25°C for Selected Systems[4]

Electrode reaction	Metal	Electrode reaction	α
$H^+ + e \rightleftharpoons \tfrac{1}{2}H_2$	Pt (smooth)	1.0 mol dm^{-3} HCl	2.0
$H^+ + e \rightleftharpoons \tfrac{1}{2}H_2$	Ni	0.12 mol dm^{-3} NaOH	0.58
$H^+ + e \rightleftharpoons \tfrac{1}{2}H_2$	Hg	10.0 mol dm^{-3} HCl	0.61
$O_2 + 4H^+ + 4e \rightleftharpoons 2H_2O$	Pt	0.1 mol dm^{-3} H$_2$SO$_4$	0.49
$O_2 + 2H_2O + 4e \rightleftharpoons 4OH^-$	Pt	0.1 mol dm^{-3} NaOH	1.0
$Cd^{2+} + 2e \rightleftharpoons Cd$	Cd/Hg	10^{-3} mol dm^{-3} Cd(NO$_3$)$_2$ in 1 mol dm^{-3} KNO$_3$	5.0
$Cu^{2+} + 2e \rightleftharpoons Cu$	Cu	1 mol dm^{-3} CuSO$_4$	0.5

exist, but because the system is at equilibrium, the rates are equal and thus there is no net current flow; hence

$$i_f = i_b = i_o \tag{2.15}$$

where i_o is the exchange current. From Eqs. (2.10) to (2.13), together with Eq. (2.15), the following relationship is established:

$$C_O k_f^o \exp\left(\frac{-\alpha n F E_e}{RT}\right) = C_R k_b^o \exp\left(\frac{(1-\alpha)nFE_e}{RT}\right) \tag{2.16}$$

where E_e, is the equilibrium potential. Rearranging,

$$E_e = \frac{RT}{nF}\ln\left(\frac{k_f^o}{k_b^o}\right) + \frac{RT}{nF}\ln\left(\frac{C_O}{C_R}\right) \tag{2.17}$$

From this equation we can establish the definition of formal standard potential E_C^o, where concentrations are used rather than activities,

$$E_C^o = \frac{RT}{nF}\ln\left(\frac{k_f^o}{k_b^o}\right) \tag{2.18}$$

For convenience, the formal standard potential is often taken as the reference point of the potential scale in reversible systems. Combining Eqs. (2.17) and (2.18), we can show consistency with the Nernst equation,

$$E_e = E_C^o + \frac{RT}{nF}\ln\left(\frac{C_O}{C_R}\right) \tag{2.19}$$

except that this expression is written in terms of concentrations rather than activities. From Eqs. (2.10) and (2.12), at equilibrium conditions,

$$i_o = i_f = nFAC_O k_f^o \exp\left(\frac{-\alpha nFE_e}{RT}\right) \tag{2.20}$$

The exchange current as defined in Eq. (2.15) is a parameter of interest to researchers in the battery field. This parameter may be conveniently expressed in terms of the rate constant k by combining Eqs. (2.10), (2.12), (2.17), and (2.20),

$$i_o = nFAkC_O^{(1-\alpha)}C_R^\alpha \tag{2.21}$$

The exchange current i_o is a measure of the rate of exchange of charge between oxidized and reduced species at any equilibrium potential without net overall change. The rate constant k, however, has been defined for a particular potential, the formal standard potential of the system. It is not in itself sufficient to characterize the system unless the transfer coefficient is also known. However, Eq. (2.21) can be used in the elucidation of the electrode reaction mechanism. The value of the transfer coefficient can be determined by measuring the exchange current density as a function of the concentration of the reduction or oxidation species at a constant concentration of the oxidation of reduction species, respectively. A schematic representation of the forward and backward currents as a function of overvoltage, $\eta = E - E_e$, is shown in Fig. 2.6, where the net current is the sum of the two components.

For situations where the net current is not zero—that is, where the potential is sufficiently different from the equilibrium potential—the net current approaches the net forward current (or, for anodic overvoltages, the backward current). One can then write

$$i = nFAC_O k \exp\left(\frac{-\alpha nF\eta}{RT}\right) \tag{2.22}$$

Now when $\eta = 0$, $i = i_o$, then

$$i = i_o \exp\left(\frac{-\alpha nF\eta}{RT}\right) \tag{2.23}$$

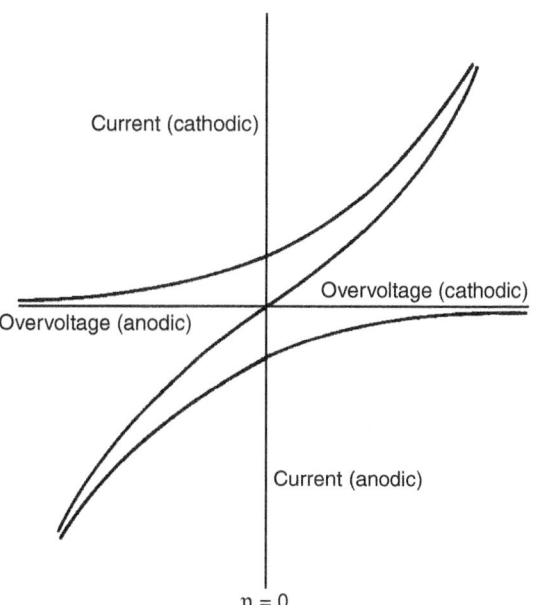

FIGURE 2.6 Schematic representation of relationship between overvoltage and current.

2.10 PRINCIPLES OF OPERATION

and

$$n = \frac{RT}{\alpha nF} \ln i_0 - \frac{RT}{\alpha nF} \ln i \tag{2.24}$$

which is the Tafel equation introduced earlier in a generalized form as Eq. (2.7). It can now be seen that the kinetic treatment here is self-consistent with both the Nernst equation [see Eq. (2.19)] (for equilibrium conditions) and the Tafel relationship [see Eq. (2.7)](for unidirectional processes). To present the kinetic treatment in its most useful form, a transformation into a net current flow form is appropriate. Using

$$i = i_f - i_b \tag{2.25}$$

substitute Eqs. (2.10), (2.13), and (2.18),

$$i = nFAk\left\{C_O \exp\left(\frac{-\alpha nFE_C^0}{RT}\right) - C_R \exp\left(\frac{(1-\alpha)nFE_C^0}{RT}\right)\right\} \tag{2.26}$$

When this equation is applied in practice, it is very important to remember that C_O and C_R are concentrations at the surface of the electrode, or are the effective concentrations. These are not necessarily the same as the bulk concentrations. Concentrations at the interface are often (almost always) modified by differences in electric potential between the surface and the bulk solution. The effects of potential differences that are manifest at the electrode-electrolyte interface are given in the following section.

2.4 ELECTRICAL DOUBLE-LAYER CAPACITY AND IONIC ADSORPTION

When an electrode (metal surface) is immersed in an electrolyte, the electronic charge on the metal attracts ions of opposite charge and orients the solvent dipoles. There exist a layer of charge in the metal and a layer of charge in the electrolyte. This charge separation establishes what is commonly known as the "electrical double layer."[5] Experimentally, the electrical double-layer effect is manifest in the phenomenon named "electrocapillarity." The phenomenon has been studied for many years, and there exist thermodynamic relationships that relate interfacial surface tension between electrode and electrolyte solution to the structure of the double layer. Typically the metal used for these measurements is mercury since it is the only conveniently available metal that is liquid at room temperature (although some work has been carried out with gallium, Wood's met al, and lead at elevated temperature).

Determinations of the interfacial surface tension between mercury and electrolyte solution can be made with a relatively simple apparatus. All that is needed are (1) a mercury-solution interface that is polarizable, (2) a nonpolarizable interface as reference potential, (3) an external source of variable potential, and (4) an arrangement to measure the surface tension of the mercury-electrolyte interface. An experimental system that will fulfill these requirements is shown in Fig. 2.7. The interfacial surface tension is measured by applying pressure to the mercury-electrolyte interface by raising the mercury "head." At the interface, the forces are balanced, as shown in Fig. 2.8. If the angle of contact at the capillary wall is zero (typically the case for clean surfaces and clean electrolyte), then it is a relatively simple arithmetic exercise to show that the interfacial surface tension is given by

$$\gamma = \frac{h\rho g r}{2} \tag{2.27}$$

where γ = interfacial surface tension, ρ = density of mercury, g = force of gravity, r = radius of capillary, and h = height of mercury column in capillary.

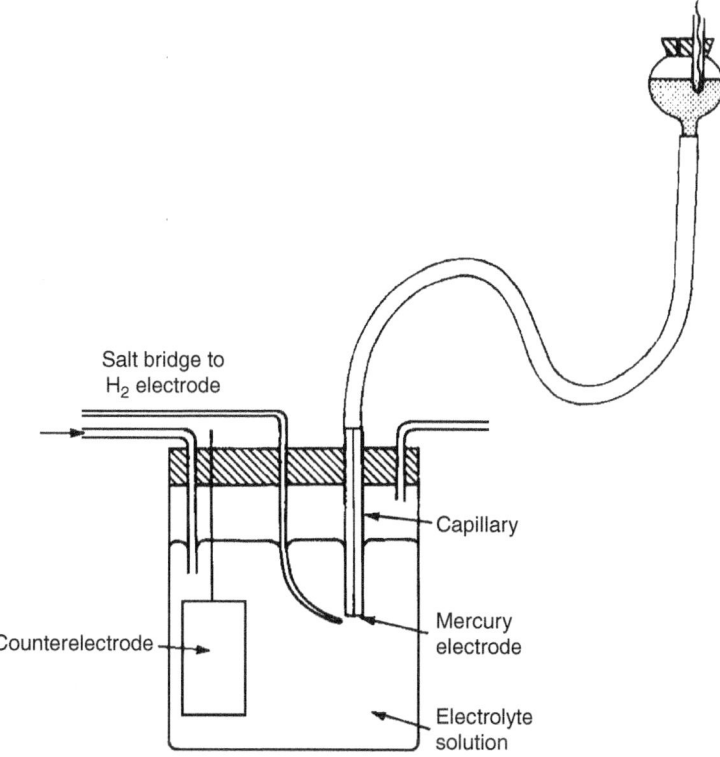

FIGURE 2.7 Experimental arrangement to measure interfacial surface tension at the mercury-electrolyte interface.

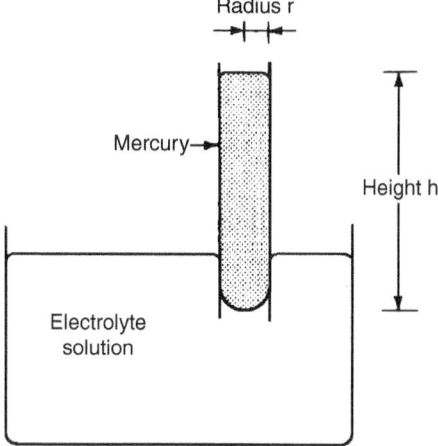

FIGURE 2.8 Close-up of the mercury-electrolyte interface in a capillary immersed in an electrolyte solution.

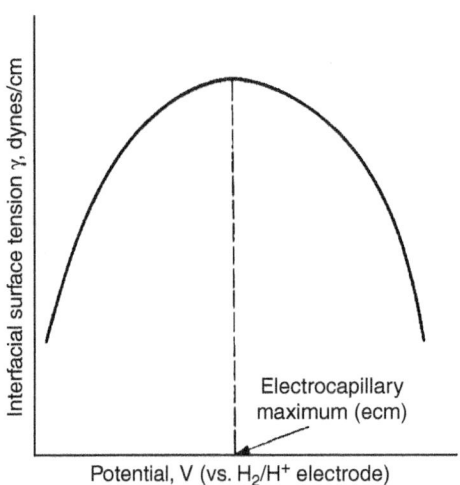

FIGURE 2.9 Generalized representation of an electrocapillary curve.

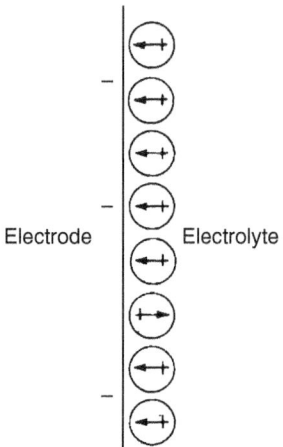

FIGURE 2.10 Orientation of water molecules in the electrical double layer at a negatively charged electrode.

The characteristic electrocapillary curve that one would obtain from a typical electrolyte solution is shown in Fig. 2.9. From such measurements and, more accurately, by AC impedance bridge measurements, the structure of the electrical double layer has been determined.[5]

Consider a negatively charged electrode in an aqueous solution of electrolyte. Assume that at this potential no electrochemical charge transfer takes place. For simplicity and clarity, the different features of the electrical double layer will be described individually. Orientation of solvent molecules, water for the sake of this discussion, is shown in Fig. 2.10. The water dipoles are oriented as shown in the figure, so that the majority of the dipoles are oriented with their positive ends (arrow heads) toward the surface of the electrode. This represents a "snapshot" of the structure of the layer of water molecules since the electrical double layer is a dynamic system that is in equilibrium with water in the bulk solution. Since the representation is statistical, not all dipoles are oriented the same way. Some dipoles are more influenced by dipole-dipole interactions than by dipole-electrode interactions.

Next, consider the approach of a cation to the vicinity of the electrical double layer. The majority of cations are strongly solvated by water dipoles and maintain a sheath of water dipoles around them despite the orienting effect of the double layer. With a few exceptions, cations do not approach right up to the electrode surface but remain outside the primary layer of solvent molecules and usually retain their solvation sheaths. Figure 2.11 shows a typical example of a cation in the electrical double layer. The establishment that this is the most likely approach of a typical cation comes partly from experimental AC impedance measurements of mixed electrolytes and mainly from calculations of the free energy of approach of an ion to the electrode surface. In considering water-electrode, ion-electrode, and ion-water interactions, the Gibbs energy of approach of a cation to an electrode surface is strongly influenced by the hydration of the cation. The general result is that cations of very large radius (and thus of low hydration) such as Cs^+ can contact/adsorb on the electrode surface, but for the majority of cations the change in free energy on contact absorption is positive and thus is against the mechanism of contact adsorption.[6] Figure 2.12 gives an example of the ion Cs^+ contact-adsorbed on the surface of an electrode.

It would be expected that because anions have a negative charge, contact adsorption of anions would not occur. In analyzing the free-energy balance of the anion system, it is found that anion-electrode contact is favored because the net Gibbs-energy balance is negative. Both from these calculations and from experimental measurements, anion contact adsorption is found to be relatively common. Figure 2.13 shows the generalized case of anion adsorption on an electrode. There are exceptions to this type of adsorption. Calculation of the Gibbs energy of contact adsorption of the

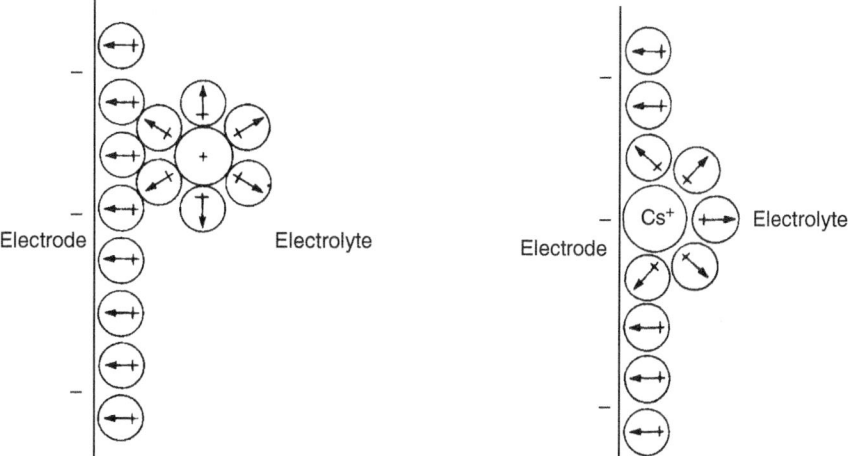

FIGURE 2.11 Typical cation situated in the electrical double layer.

FIGURE 2.12 Contact adsorption of Cs⁺ on an electrode surface.

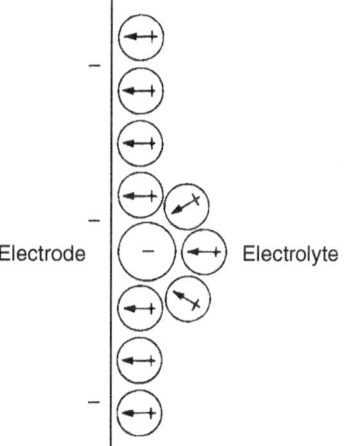

FIGURE 2.13 Contact adsorption of an anion on an electrode surface

fluoride ion is positive and unlikely to occur. This is supported by experimental measurement. This property is utilized, as NaF, as a supporting electrolyte to evaluate adsorption properties of surface-active species devoid of the influence of adsorbed supporting electrolyte.*

Extending out into solution from the electrical double layer (or the compact double layer, as it is sometimes known) is a continuous repetition of the layering effect, but with diminishing magnitude. This "extension" of the compact double layer toward the bulk solution is known as the Gouy-Chapman diffuse double layer.[5] Its effect on electrode kinetics and the concentration of electroactive species at the electrode surface is manifest when supporting electrolyte concentrations are low or zero. The end result of the establishment of the electrical double layer effect and the various types of ion contact adsorption

---

*A supporting electrolyte is a salt used in large excess to minimize internal resistance in electrochemical cells, but which does not enter into electrode reactions.

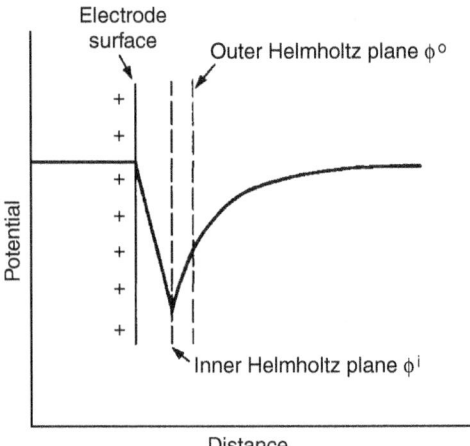

FIGURE 2.14 Potential distribution at a positively charged electrode.

is to influence directly the real (actual) concentration of electroactive species at an electrode surface and indirectly to modify the potential gradient at the site of electron transfer. In this respect, it is important to understand the influence of the electrical double layer and allow for it where and when appropriate.

The potential distribution near an electrode is shown schematically in Fig. 2.14. The inner Helmholtz plane corresponds to the plane that contains the contact-adsorbed ions and the innermost layer of water molecules. Its potential is defined as ϕ^i with the zero potential being taken as the potential of the bulk solution. The outer Helmholtz plane is the plane of closest approach of those ions which do not contact-adsorb but approach the electrode with a sheath of solvated water molecules surrounding them. The potential at the outer Helmholtz plane is defined as ϕ^o and is again referred to the potential of the bulk solution. In some texts ϕ^i is defined as ϕ^1 and ϕ^o as ϕ^2.

As mentioned previously, the bulk concentration of an electroactive species is often not the value to be used in kinetic equations. Species that are in the electrical double layer are in a different energy state from those in bulk solution. At equilibrium, the concentration C^e of an ion or species that is about to take part in the charge-transfer process at the electrode is related to the bulk concentration by

$$C^e = C^B \exp\left(\frac{-zF\phi^e}{RT}\right) \tag{2.28}$$

where z is the charge on the ion and ϕ^e the potential of *closest approach* of the species to the electrode. It will be remembered that the plane of closest approach of many species is the outer Helmholtz plane, and so the value of ϕ^e can often be equated to ϕ^o. However, as noted in a few special cases, the plane of closest approach can be the inner Helmholtz plane, and so the value of ϕ^e in these cases would be the same as ϕ^i. A judgment has to be made as to what value of ϕ^e should be used.

The potential that is effective in driving the electrode reaction is that between the species at its closest approach and the potential of the electrode. If E is the potential of the electrode, then the driving force is $E - \phi^e$. Using this relationship together with Eqs. (2.26) and (2.28), we have

$$\frac{i}{nFAk} = C_O \exp\left(\frac{-z_O E\phi^e}{RT}\right) \exp\left(\frac{-\alpha nF(E-\phi^e)}{RT}\right) \\ - C_R \exp\left(\frac{-z_R F\phi^e}{RT}\right) \exp\left(\frac{(1-\alpha)nF(E-\phi^e)}{RT}\right) \tag{2.29}$$

where z_O and z_R are the charges (with sign) of the oxidized and reduced species, respectively. Rearranging Eq. (2.28) and using

$$z_O - n = z_R \qquad (2.30)$$

yields

$$\frac{i}{nFAk} = \exp\left(\frac{(\alpha n - z_O)F\phi^e}{RT}\right)\left\{C_O\exp\left(\frac{-\alpha nFE}{RT}\right) - C_R\exp\left(\frac{(1-\alpha)nFE}{RT}\right)\right\} \qquad (2.31)$$

In experimental determination, the use of Eq. (2.26) will provide an apparent rate constant k_{app}, which does not take into account the effects of the electrical double layer. Taking into account the effects appropriate to the approach of a species to the plane of nearest approach,

$$k_{app} = k\exp\left(\frac{(\alpha n - z_O)F\phi^e}{RT}\right) \qquad (2.32)$$

For the exchange current the same applies,

$$(i_o)_{app} = i_o\exp\left(\frac{(\alpha n - z_O)F\phi^e}{RT}\right) \qquad (2.33)$$

Corrections to the rate constant and the exchange current are not insignificant. Several calculated examples are given in Bauer.[7] The differences between apparent and true rate constants can be as great as two orders of magnitude. The magnitude of the correction also is related to the magnitude of the difference in potential between the electrocapillary maximum for the species and the potential at which the electrode reaction occurs; the greater the potential difference, the greater the correction to the exchange current or rate constant.

2.5 MASS TRANSPORT TO THE ELECTRODE SURFACE

We have considered the thermodynamics of electrochemical processes, studied the kinetics of electrode processes, and investigated the effects of the electrical double layer on kinetic parameters. An understanding of these relationships is an important ingredient in the repertoire of the researcher of battery technology. Another very important area of study which has major impact on battery research is the evaluation of mass transport processes to and from electrode surfaces.

Mass transport to or from an electrode can occur by three processes: (1) convection and stirring, (2) electrical migration in an electric potential gradient, and (3) diffusion in a concentration gradient. The first of these processes can be handled relatively easily both mathematically and experimentally. If stirring is required, flow systems can be established, while if complete stagnation is an experimental necessity, this can also be imposed by careful design. In most cases, if stirring and convection are present or imposed, they can be handled mathematically.

The migration component of mass transport can also be handled experimentally (reduced to close to zero or occasionally increased in special cases) and described mathematically, provided certain parameters such as transport number or migration current are known. Migration of electroactive species in an electric potential gradient can be reduced to near zero by addition of an excess of inert "supporting electrolyte," which effectively reduces the potential gradient to zero and thus eliminates the electric field that produces migration. Enhancement of migration is more difficult. This requires that the electric field be increased so that movement of charged species is increased.

Electrode geometry design can increase migration slightly by altering electrode curvature. The fields at convex surfaces are greater than those at flat or concave surfaces, and thus migration is enhanced at convex curved surfaces.

The third process, diffusion in a concentration gradient, is the most important of the three processes and is the one which typically is dominant in mass transport in batteries. The analysis of diffusion uses the basic equation due to Fick[8] which defines the flux of material crossing a plane at distance x and time t. The flux is proportional to the concentration gradient and is represented by the expression:

$$q = D\frac{\delta C}{\delta x} \qquad (2.34)$$

where q = flux, D = diffusion coefficient, and C = concentration. The rate of change of concentration with time is defined by

$$\frac{\delta C}{\delta t} = D\frac{\delta^2 C}{\delta x^2} \qquad (2.35)$$

This expression is referred to as Fick's second law of diffusion. Solution of Eqs. (2.34) and (2.35) requires that boundary conditions be imposed. These are chosen according to the electrode's expected "discharge" regime dictated by battery performance or boundary conditions imposed by relevant electrochemical techniques.[9] Several of the electrochemical techniques are discussed in Sec. 2.6.

For application directly to battery technology, the three modes of mass transport have meaningful significance. Convective and stirring processes can be employed to provide a flow of electroactive species to reaction sites. Examples of the utilization of stirring and flow processes in batteries are the circulating zinc/air system, the vibrating zinc electrode, and the zinc-chlorine hydrate battery. In some types of advanced lead-acid batteries, circulation of acid is provided to improve utilization of the active materials in the battery plates. Migration effects are in some cases detrimental to battery performance, in particular those caused by enhanced electric fields (potential gradients) around sites of convex curvature. Increased migration at these sites tends to produce dendrite formations, which eventually lead to a short-circuit and battery failure.

2.5.1 Concentration Polarization

Diffusion processes are typically the mass-transfer processes operative in the majority of battery systems where the transport of species to and from reaction sites is required for maintenance of current flow. Enhancement and improvement of diffusion processes are an appropriate direction of research to follow to improve battery performance parameters. Equation (2.34) may be written in an approximate, yet more practical form, remembering that $i = nFq$, where q is the flux through a plane of unit area. Thus,

$$i = nF\frac{DA(C_B - C_E)}{\delta} \qquad (2.36)$$

where symbols are defined as before, and C_B = bulk concentration of electroactive species, C_E = concentration at electrode, A = electrode area, δ = boundary-layer thickness, that is, the layer at the electrode surface in which the majority of the concentration gradient is established (see Fig. 2.15). When $C_E = 0$, this expression defines the maximum diffusion current, i_L, that can be sustained in solution under a given set of conditions,

$$i_L = nF\frac{DAC_B}{\delta_L} \qquad (2.37)$$

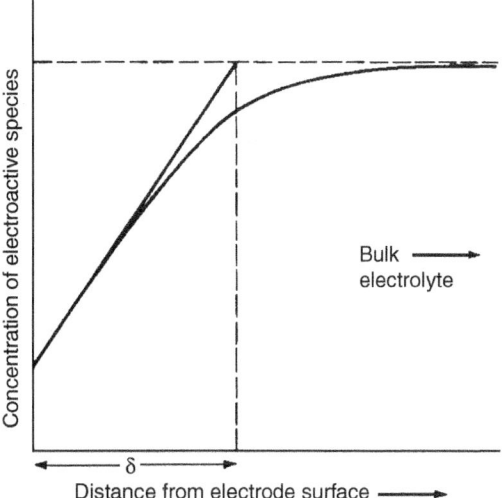

FIGURE 2.15 Boundary-layer thickness at an electrode surface.

where L is the boundary-layer thickness at the limiting condition. It tells us that to increase i_L, one needs to increase the bulk concentration, the electrode area, or the diffusion coefficient. In the design of a battery, an understanding of the implication of this expression is important. Specific cases can be analyzed quickly by applying Eq. (2.36), and parameters such as discharge rate and power densities of new systems can be estimated.

Assume that the thickness of the diffusion boundary layer does not change much with concentration. Then $\delta_L = \delta$ and Eq. (2.36) may be rewritten as

$$i = \left(1 - \frac{C_E}{C_B}\right) i_L \qquad (2.38)$$

The difference in concentration existing between the electrode surface and the bulk of the electrolyte results in a concentration polarization. According to the Nernst equation, the concentration polarization or overpotential produced from the change of concentration across the diffusion layer may be written as

$$\eta_c = \frac{RT}{nF} \ln \frac{C_B}{C_E} \qquad (2.39)$$

From Eq. (2.38) we have

$$\eta_c = \frac{RT}{nF} \ln \left(\frac{i_L}{i_L - i}\right) \qquad (2.40)$$

This gives the relation of concentration polarization and current for mass transfer by diffusion. Equation (2.40) indicates that as i approaches the limiting current i_L, theoretically the overpotential should increase to infinity. However, in a real process the potential will increase only to a point where another electrochemical reaction will occur, as illustrated in Fig. 2.16. Figure 2.17 shows the magnitude of the concentration overpotential as a function of i/i_L, with n = 2 at 25°C based on Eq. (2.40).

2.18 PRINCIPLES OF OPERATION

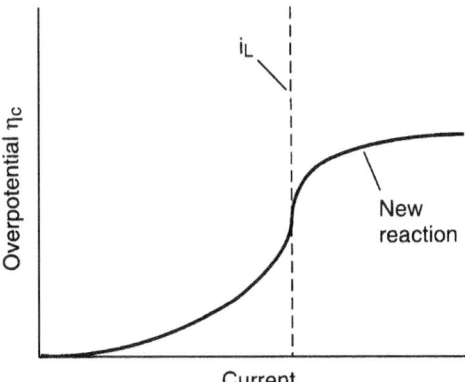

FIGURE 2.16 Plot of overpotential η_c vs. current i.

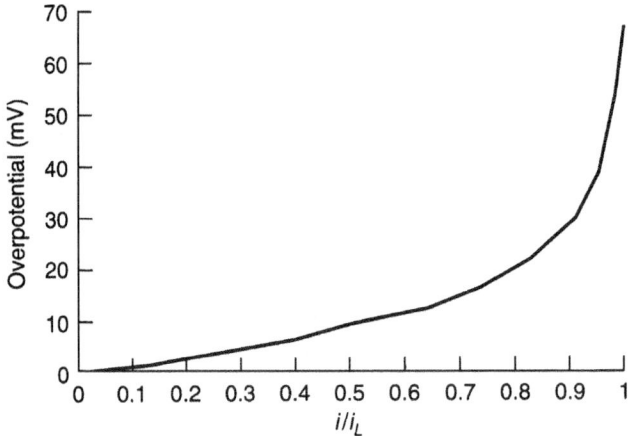

FIGURE 2.17 Magnitude of concentration overpotential as a function of i/i_L, with $n = 2$ at 25°C, based on Eq. (2.40).

2.5.2 Porous Electrodes

Electrochemical reactions are heterogeneous reactions that occur on the electrode-electrolyte interface. In fuel cell systems, the reactants are supplied from the electrolyte phase to the catalytic electrode surface. In battery systems, the electrodes are usually composites made of active reactants, binder, and conductive filler. In order to minimize the energy loss due to both activation and concentration polarizations at the electrode surface and to increase the electrode efficiency or utilization, it is preferred to have a large electrode surface area. This is accomplished with the use of a porous electrode design. A porous electrode can provide an interfacial area per unit volume several decades higher than that of a planar electrode (such as 10^4 cm^{-1}).

A porous electrode consists of porous matrices of solids and void spaces. The electrolyte penetrates the void spaces of the porous matrix. In such an active porous mass, the mass transfer condition in conjunction with the electrochemical reaction occurring at the interface is very complicated. In a given time during cell operation, the rate of reaction within the pores may vary significantly depending on the location. The distribution of current density within the porous electrode depends on the physical structure (such as tortuosity, pore sizes), the conductivity of the solid matrix and the electrolyte, and the electrochemical kinetic parameters of the electrochemical processes. A detailed treatment of such complex porous electrode systems can be found in Newman.[10]

2.6 ELECTROCHEMICAL TECHNIQUES

Many steady-state and transient electrochemical techniques are available to the experimentalist to determine electrochemical parameters and assist in both improving existing battery systems and evaluating couples as candidates for new batteries.[11] Some of these techniques are described in this section.

2.6.1 Cyclic Voltammetry

Of the electrochemical techniques, cyclic voltammetry (or linear sweep voltammetry) is one of the more versatile techniques available to the electrochemist. The derivation of the various forms of cyclic voltammetry can be traced to the initial studies of Matheson and Nicols[12] and Randles.[13] Essentially the technique applies a linearly changing voltage (ramp voltage) to an electrode. The scan of voltage might be ±3 V from an appropriate rest potential such that most electrode reactions would be encompassed. Commercially available instrumentation provides voltage scans as wide as ±5 V.

To describe the principles behind cyclic voltammetry, for convenience let us restate Eq. (2.9), which describes the reversible reduction of an oxidized species O,

$$O + ne \rightleftharpoons R \tag{2.9}$$

where O = oxidized species
R = reduced species
n = number of electrons involved in electrode process

In cyclic voltammetry, the initial potential sweep is represented by

$$E = E_i - \nu t \tag{2.41}$$

where E_i = initial potential
t = time
ν = rate of potential change or sweep rate (V/s)

The reverse sweep of the cycle is defined by

$$E = E_i + \nu' t \tag{2.42}$$

where ν' is often the same value as ν. By combining Eq. (2.42) with the appropriate form of the Nernst equation [Eq. (2.6)] and with Fick's laws of diffusion [Eqs. (2.34) and (2.35)], an expression can be derived that describes the flux of species to the electrode surface. This expression is a complex differential equation and can be solved by the summation of an integral in small successive increments.[14-16]

As the applied voltage approaches that of the reversible potential for the electrode process, a small current flows, the magnitude of which increases rapidly but later becomes limited at a potential slightly beyond the standard potential by the subsequent depletion of reactants. This depletion of reactants establishes concentration profiles which spread out into the solution, as shown in Fig. 2.18. As the concentration profiles extend into the solution, the rate of diffusive transport at the electrode surface decreases and with it the observed current. The current is thus seen to pass

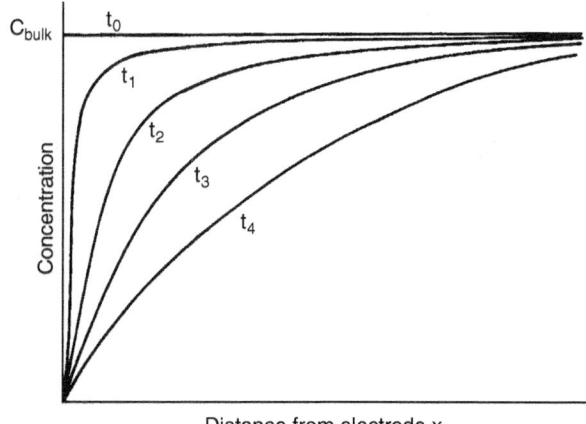

FIGURE 2.18 Concentration profiles for reduction of a species in cyclic voltammetry, $t_4 > t_0$.

through a well-defined maximum, as illustrated in Fig. 2.19. The peak current of the reversible reduction [Eq. (2.9)] is defined by

$$i_p = \frac{0.447 F^{3/2} A n^{3/2} D^{1/2} C_0 v^{1/2}}{R^{1/2} T^{1/2}} \tag{2.43}$$

The symbols have the same identity as before while i_p is the peak current and A the electrode area. It may be noted that the value of the constant varies slightly from one text or publication to another. This is because, as mentioned previously, the derivation of peak current height is performed numerically.

A word of caution is due regarding the interpretation of the value of the peak current. It will be remembered from the discussion of the effects of the electrical double layer on electrode kinetics that there is a capacitance effect at an electrode-electrolyte interface. Consequently the "true" electrode potential is modified by the capacitance effect as it is also by the ohmic resistance of the solution.

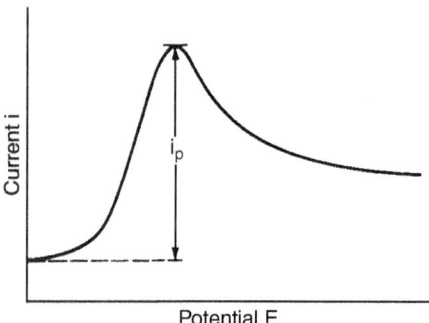

FIGURE 2.19 Cyclic voltammetry peak current for reversible reduction of an electroactive species.

Equation (2.41) should really be written in a form that describes these two components. Equation (2.44) shows such a modification,

$$E = E_i - vt + r(i_f + i_c) \qquad (2.44)$$

where r = cell resistance, i_f = faradaic current, and i_c = capacity current.

At small values of voltage sweep rate, typically below 1 mV/s, the capacity effects are small and in most cases can be ignored. At greater values of sweep rate, a connection needs to be applied to interpretations of i_p as described by Nicholson and Shain.[17] With regard to the correction for ohmic drop in solution, typically this can be handled adequately by careful cell design and positive feedback compensation circuitry in the electronic instrumentation.

Cyclic voltammetry provides both qualitative and quantitative information on electrode processes. A reversible, diffusion-controlled reaction such as presented by Eq. (2.9) exhibits an approximately symmetrical pair of current peaks, as shown in Fig. 2.20. The voltage separation ΔE of these peaks is

$$\Delta E = \frac{2.3RT}{nF} \qquad (2.45)$$

and the value is independent of the voltage sweep rate. In the case of the electrodeposition of an insoluble film, which can be subsequently reversibly reoxidized and which is not governed by diffusion to and from the electrode surface, the value of ΔE is considerably less than that given by Eq. (2.45), as shown in Fig. 2.21. In the ideal case, the value of ΔE for this system is close to zero. For quasi-reversible processes, the current peaks are more separated, and the shape of the peak is less sharp at its summit and is generally more rounded, as shown in Fig. 2.22. The voltage of the current peak is dependent on the voltage sweep rate, and the voltage separation is much greater than that given by Eq. (2.45).

A completely irreversible electrode process produces a single peak, as shown in Fig. 2.23. Again the voltage of the peak current is sweep-rate dependent, and, in the case of an irreversible charge-transfer process for which the back reaction is negligible, the rate constant and transfer coefficient can be determined. With negligible back reaction, the expression for peak current as a function of peak potential is[17]

$$i_p = 0.22nFC_0 k_{app} \exp\left[-\alpha \frac{nF}{RT}(E_m - E^o)\right] \qquad (2.46)$$

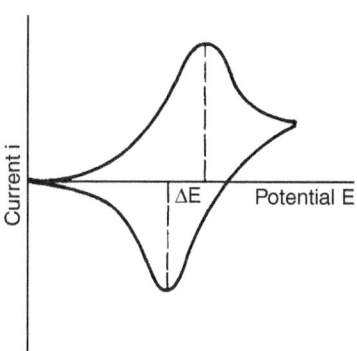

FIGURE 2.20 Cyclic voltammogram of a reversible diffusion-controlled process.

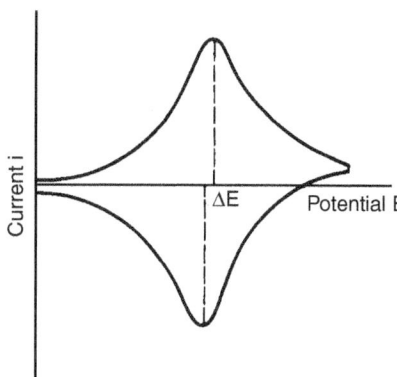

FIGURE 2.21 Cyclic voltammogram of electroreduction and reoxidation of a deposited, insoluble film.

2.22 PRINCIPLES OF OPERATION

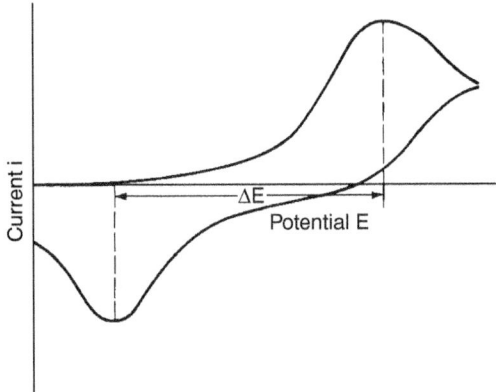

FIGURE 2.22 Cyclic voltammogram of a quasireversible process.

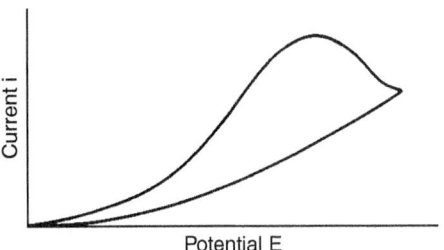

FIGURE 2.23 Cyclic voltammogram of an irreversible process.

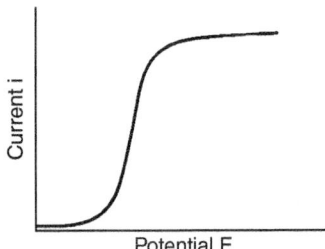

FIGURE 2.24 Cyclic voltammogram of electroreduction of a species controlled by a preceding chemical reaction.

where the symbols are as before and E_m is the potential of the current peak. A plot of E_m versus $\ln i_p$, for different values of concentration, gives a slope that yields the transfer coefficient α and an intercept that yields the apparent rate constant k_{app}. Though both α and k_{app} can be obtained by analyzing E_m as a function of voltage sweep rate v by a reiterative calculation, analysis by Eq. (2.46) (which is independent of v) is much more convenient.

For more complex electrode processes, cyclic voltammetric traces become more complicated to analyze. An example of one such case is the electroreduction of a species controlled by a preceding chemical reaction. The shape of the trace for this process is shown in Fig. 2.24. The species is formed at a constant rate at the electrode surface and, provided the diffusion of the inactive component is more rapid than its transformation to the active form, it cannot be depleted from the electrode surface. The "peak" current is thus independent of potential and resembles a plateau.

Cyclic voltammograms of electrochemical systems can often be much more complicated than the traces presented here. It often takes some ingenuity and persistence to determine which peaks belong to which species or processes. Despite these minor drawbacks, the cyclic voltammetric technique is a versatile, and relatively sensitive, electrochemical method appropriate to the analysis of systems of interest to battery development. The technique will identify reversible couples (desirable for secondary batteries), it provides a method for measuring the rate constant and transfer coefficient of an electrode process (a fast rate constant indicates a process of possible interest for battery development), and it can provide a tool to help unravel complex electrochemical systems.

2.6.2 Chronopotentiometry

Chronopotentiometry involves the study of voltage transients at an electrode upon which is imposed a constant current. It is sometimes alternately known as galvanostatic voltammetry. In this technique, a constant current is applied to an electrode, and its voltage response indicates the changes in electrode processes occurring at its interface. Consider, for example, the reduction of a species O as expressed by Eq. (2.9). As the constant current is passed through the system, the concentration of O in the vicinity of the electrode surface begins to decrease. As a result of this depletion, O diffuses from the bulk solution into the depleted layer, and a concentration gradient grows out from the electrode surface into the solution. As the electrode process continues, the concentration profile extends further into the bulk solution as shown in Fig. 2.25. When the surface concentration of O falls to zero (at time t_6 in Fig. 2.25), the electrode process can no longer be supported by electroreduction of O. An additional cathodic reaction must be brought into play and an abrupt change in potential occurs. The period of time between the commencement of electroreduction and the sudden change in potential is called the transition time τ. The transition time for electroreduction of a species in the presence of excess supporting electrolyte was first quantified by Sand,[18] who showed that the transition time τ was related into the diffusion coefficient of the electroactive species,

$$\tau^{1/2} = \frac{\pi^{1/2} n F C_o D^{1/2}}{2i} \tag{2.47}$$

where D is the diffusion coefficient of species O and the other symbols have their usual meanings.

Unlike cyclic voltammetry, the solution of Fick's diffusion equations [Eqs. (2.34) and (2.35)] for chronopotentiometry can be obtained as an exact expression by applying appropriate boundary conditions. For a reversible reduction of an electroactive species [Eq. (2.9)], the potential-time relationship has been derived by Delahay[19] for the case where O and R are free to diffuse to and from the electrode surface, including the case where R diffuses into a mercury electrode,

$$E = E_{\tau/4} + \frac{RT}{nf} \ln \frac{\tau^{1/2} - t^{1/2}}{t^{1/2}} \tag{2.48}$$

In this equation, $E_{\tau/4}$ is the potential at the one-quarter transition time and t is any time from zero to the transition time. The trace represented by this expression is shown in Fig. 2.26.

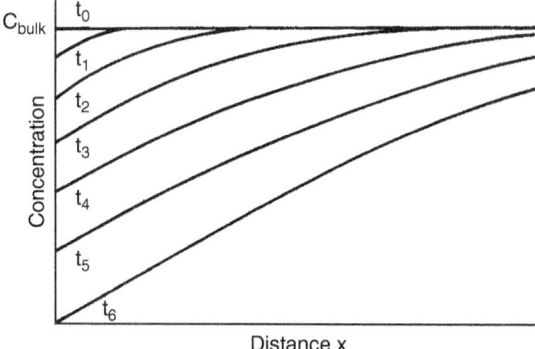

FIGURE 2.25 Concentration profiles extending into bulk solution during constant-current depletion of species at an electrode surface, $t_6 > t_0$.

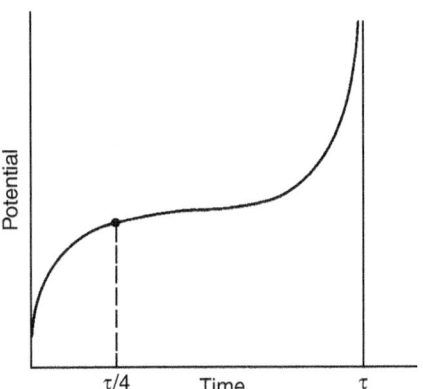

FIGURE 2.26 Potential curve at constant current for reversible reduction of an electroactive species.

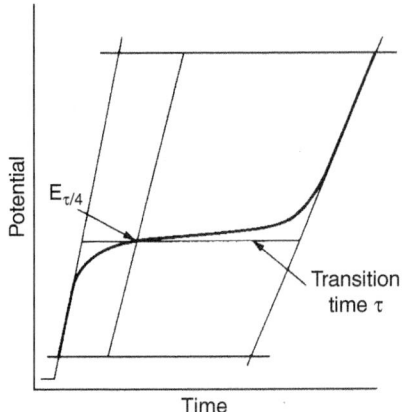

FIGURE 2.27 Construction of transition time τ for a chronopotentiogram.

The corresponding expression for an irreversible process[20] with one rate-determining step is

$$E = \frac{RT}{\alpha n_a F} \ln\left[\frac{nFC_0 k_{app}}{i}\right] + \frac{RT}{\alpha n_a F} \ln\left[1 - \left(\frac{t}{\tau}\right)^{1/2}\right] \quad (2.49)$$

where k_{app} is the apparent rate constant, n_a is the number of electrons involved in the rate-determining step (often the same as n, the overall number of electrons involved in the total reaction), and the other symbols have their usual meanings. A plot of the logarithmic term versus potential yields both the transfer coefficient and the apparent rate constant.

In a practical system, the chronopotentiogram is often less than ideal in the shape of the potential trace. To accommodate variations in chronopotentiometric traces, measurement of the transition time can be assisted by use of a construction technique, as shown in Fig. 2.27. The transition time is measured at the potential of $E_{\tau/4}$.

To analyze two or more independent reactions separated by a potential sufficient to define individual transition times, the situation is slightly more complicated than with cyclic voltammetry. Analysis of the transition time of the reduction of the nth species has been derived elsewhere[21,22] and is

$$(\tau_1 + \tau_2 + \cdots + \tau_n)^{1/2} - (\tau_1 + \tau_2 + \cdots + \tau_{n-1})^{1/2} = \frac{\pi^{1/2} nFD_n^{1/2} C_n}{2i} \quad (2.50)$$

As can be seen, this expression is somewhat cumbersome.

An advantage of the technique is that it can be used conveniently to evaluate systems with high resistance. The trace conveniently displays segments due to the *IR* component, the charging of the double layer, and the onset of the faradaic process. Figure 2.28 shows these different features of the chronopotentiogram of solutions with significant resistance. If the solution is also one that does not contain an excess of supporting electrolyte to suppress the migration current, it is possible to describe the transition time of an electroreduction process in terms of the transport number of the electroactive species[23,24]

$$\tau^{1/2} = \frac{\pi^{1/2} nFC_0 D_s^{1/2}}{2i(1-t_0)} \quad (2.51)$$

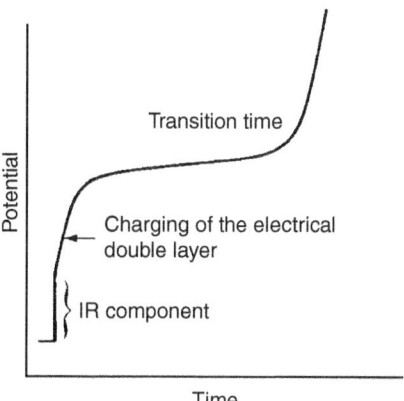

FIGURE 2.28 Chronopotentiogram of a system with significant resistance.

where D_s is the diffusion coefficient of the salt (not the ion) and t_o the transport number of the electroactive species. This expression can be useful in battery research since many battery systems do not have a supporting electrolyte.

2.6.3 Electrochemical Impedance Spectroscopy (EIS) Methods

The two preceding electrochemical techniques, one in which the measured value was the current during imposition of a potential scan and the other a potential response under an imposed constant current, owe their electrical response to the change in impedance at the electrode-electrolyte interface. A more direct technique for studying electrode processes is to measure the change in the electrical impedance of an electrode by electrochemical impedance spectroscopy (EIS). In this method, a small AC signal of ~5 to 10 mV is superimposed on an electrochemical cell at a finite DC bias potential or OCV, and the impedance Z (the equivalence of resistance R in DC measurements) is determined over a wide frequency range, normally between 0.01 Hz and 1 MHz. The resulting wave forms for current I and potential E are sinusoidal, as shown in Fig. 2.29. The two wave forms in Fig. 2.29 differ in magnitude as well as phase. If the system is purely resistive, i.e., without capacitive and other elements, the two wave forms will be in-phase. The potential sine wave and the current sine wave can be described, respectively, by

$$E_t = E_o \sin(\omega t) \tag{2.52}$$

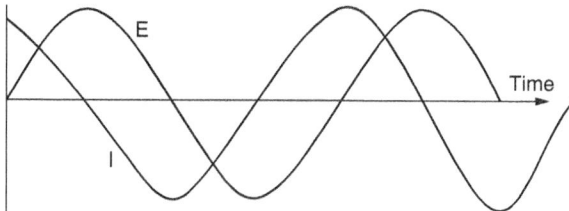

FIGURE 2.29 Sinusoidal current and potential wave forms at an electrode held at a specific DC (bias) potential or at OCV.

and

$$I_t = I_o \sin(\omega t + \phi) \tag{2.53}$$

where E_t and E_o are the potentials at time t and 0
I_t and I_o are the currents at time t and 0
ω is the frequency in radians/s (equal to $2\pi f$ where f units are Hz)

Using Euler's formula for trigonometric and complex functions, the impedance of the system can be represented by the following complex relation,[25,26]

$$Z(\omega) = \frac{E}{I} = Z_o \exp(j\phi) = Z_o(\cos\phi + \sin\phi) \tag{2.54}$$

where $j = \sqrt{-1}$. The components of $Z(\omega)$ thus consist of an imaginary part referred to as Z_i and a real part referred to as Z_r. When plotting Z_i versus Z_r, one obtains a semicircle called a "Nyquist" plot, as described below. Note that in the absence of capacitance and inductance, the Nyquist plot for a simple resistor would be a simple vertical straight line with the intercept on the Z_r axis representing the value of the resistor in ohms. For complex systems representing electrodes in cells and batteries, the Nyquist plots can be interpreted in terms of various electrode-electrolyte parameters such as solution resistance, kinetics (charge transfer), and capacitance; inductive effects are generally not observed in these electrochemical systems. To relate the complex impedance of the electrode-electrolyte interface to electrochemical parameters, it is necessary to model an equivalent circuit to represent the dynamic characteristics of the interface. The model consists of a number of impedance elements in networks based on series, parallel, or series/parallel combinations. For example, the total impedance for n elements in series is given by

$$Z_{total} = Z_1 + Z_2 + Z_3 + \cdots + Z_n \tag{2.55}$$

For n elements in parallel, the impedance will be given by

$$\frac{1}{Z_{total}} = \frac{1}{Z_1} + \frac{1}{Z_2} + \frac{1}{Z_3} + \cdots + \frac{1}{Z_n} \tag{2.56}$$

The important elements to be considered in modeling EIS data to electrodes are summarized in Table 2.4. A realistic model will then allow one to determine the electrochemical parameters for an electrode-electrolyte interface, which is discussed below.

In all of the equivalent circuits and Nyquist figures shown below, the double layer capacity is represented by C, which is the symbol used for a pure (ideal) capacitor. However, due to surface inhomogeneity and faradaic current (i.e., a "leaky" capacitor), for most electrochemical systems the

TABLE 2.4 Equivalent Circuit Elements

Circuit Element	Impedance
Resistance, R	R
Capacitance, C	$1/Cj\omega$
Constant phase element, Q (CPE)	$1/Q(j\omega)^\alpha$
Warburg impedance, W (infinite)	$1/Y(j\omega)^{1/2}$
Warburg impedance, W (finite)	$\tan[\delta D^{-1/2}(j\omega)^{-1/2}]/\gamma(j\omega)^{1/2}$
Inductance, L	$j\omega L$

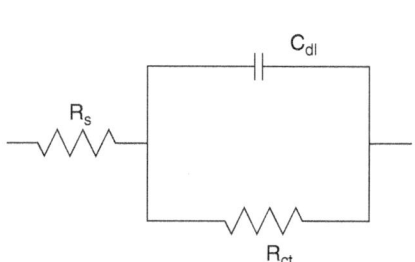

FIGURE 2.30 Randles' basic equivalent circuit for an electrode-electrolyte interface.

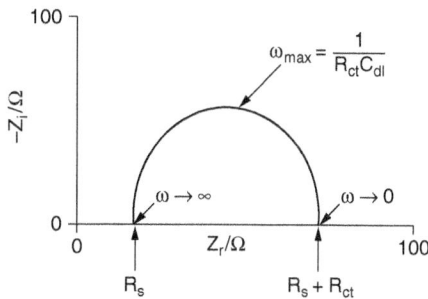

FIGURE 2.31 Schematic Nyquist plot for the Randles circuit in Fig. 2.30.

double layer rarely behaves as an ideal capacitor. In this case, the capacity C is replaced by a constant phase element (CPE) in which the impedance is given by

$$Z = \frac{1}{Q(j\omega)^\alpha} \qquad (2.57)$$

where α is an adjustable parameter (see Table 2.4). When $\alpha = 1$, the CPE acts as an ideal capacitor, i.e., $Q = C$, and when $\alpha = 0$, the CPE is equivalent to a pure resistor. As indicated in Table 2.4, there are two ways to represent the Warburg impedance based on an infinite or finite diffusion layer thickness. For the latter, the relation for the impedance contains the thickness of the diffusion layer (δ) and the diffusion coefficient (D) for the diffusing species.

In modeling the electrode-electrolyte interface for a single electrode (i.e., using a three-electrode cell comprised of a working electrode, a reference electrode, and a counter electrode), the adjustable (fitting) parameters include R, C, Q, Y, L, and α. It is therefore important to select a realistic model for the analyses of EIS data. The equivalent circuit for the basic model of the electrode-electrolyte originally proposed by Randles[27] is shown in Fig. 2.30, and the basic Nyquist plot for this equivalent circuit is shown in Figure 2.31, where R_s is the electrolyte solution resistance, C_{dl} is the double layer capacitance, and R_{ct} is the charge transfer resistance from which the exchange current density can be calculated.[25,26] If the system exhibits diffusion control, this can be accounted for by the circuit shown in Fig. 2.32 in which the Warburg impedance is added in series with R_{ct}. The corresponding Nyquist plot is shown in Fig. 2.33 where the Warburg impedance appears at low frequencies as a straight line with a slope of 45°.

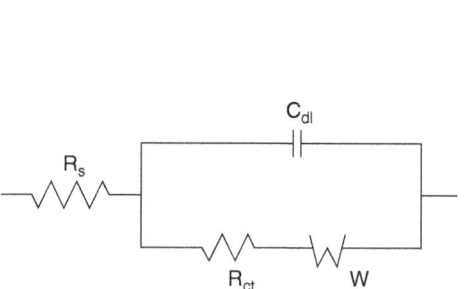

FIGURE 2.32 Randles' equivalent circuit for an electrode-electrolyte interface including the Warburg impedance.

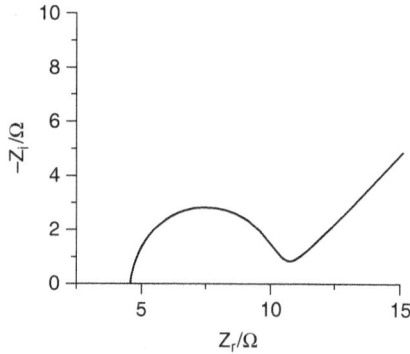

FIGURE 2.33 Schematic Nyquist plot for the equivalent circuit in Fig. 2.32.

2.28 PRINCIPLES OF OPERATION

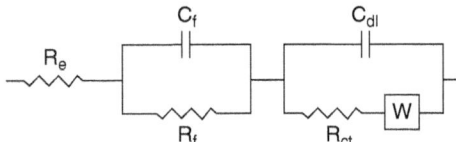

FIGURE 2.34 Equivalent circuit accounting for SEI formation

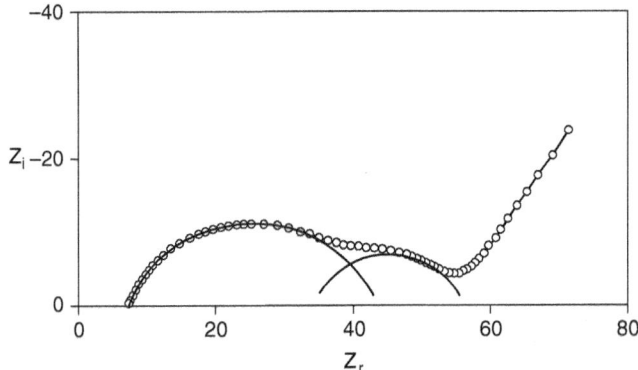

FIGURE 2.35 Schematic Nyquist plot for the equivalent circuit in Fig. 2.34.

Finally, for illustrative purposes, consider a system such as one based on a metallic Li anode or a LiC_6 anode in a Li-ion cell which reacts with the electrolyte solution to form a solid electrode interface (SEI). The equivalent circuit for this system is given in Fig. 2.34 where C_f and R_f represent, respectively, the capacity of the SEI film and the impedance of the SEI film. In this case, experimentally the Nyquest plot either shows two distinct time constants (two symmetric semicircles) in the spectra, or an unsymmetrical semicircle in which the two semicircles overlap as shown in Fig. 2.35.

Creating a reasonable model for the electrode-electrolyte interface is the first step in determining the parameters for a given system. The model can be any combination of impedances in series [Eq. (2.55)] or parallel [Eq. (2.56)] or any series and parallel combination of the elements given in Table 2.4. The next step is the deconvolution of the experimental data (frequency, Z_r and Z_i, separately for each bias potential or at OCV), which is most conveniently accomplished by a nonlinear least squares method (NLLSQ) such as that developed by Boukamp[28,29] or J. R. Macdonald's complex nonlinear least squares (CNLS) program LEVMW.[30] Conveniently, there are excellent commercial software programs available for fitting the experimental data to various combinations of the variables R, C, Q, Y, L, and α, such as Gamry's Echem Analytical EIS300 program,[31] Princeton Applied Research's ZSimWin program,[32] and Scribner Associates ZPlot and ZView programs.[33] These programs allow the user to select which elements in Table 2.4 are appropriate for a given model and then automatically proceed to determine all values of the selected variables with statistical error analyses.

2.6.4 Intermittent Titration Techniques

While the steady state electrochemical measurements reviewed above yield basic information that is a product of several variables such as concentration, diffusion, and kinetics, transient measurements such as the Galvanostatic Intermittent Titration Technique (GITT)[34] and Potentiostatic Intermittent Titration Technique (PITT)[35] offer more direct determination of kinetic and thermodynamic properties of single and multiphase electrodes (e.g., alloys and Li-insertion materials). Both methods enable one to determine the phase diagram of an electrode material as a function of capacity and voltage. For example, in a two-phase material at equilibrium at constant temperature and pressure, the potential of the electrode material will be independent of composition. However, when a phase transition occurs such as from a two-phase to single phase domain, the electrode potential will now vary as a function of composition. In addition, both GITT and PITT methods are convenient for determining the diffusion coefficient of the migrating ion in the various phases of the solid-state material. Solid-state diffusion coefficients can be determined from the Warburg impedance in EIS measurements as described above, and agreement is dependent upon the equivalent circuit model used in the deconvolution of the EIS data. Thus determination of the phases and diffusion coefficients by GITT and PITT are helpful in modeling equivalent circuits. Essentials of the GITT and PITT techniques are given below.

Galvanostatic Intermittent Titration Technique (GITT). The GITT method is a form of chronopotentiometry (discussed above), but it is simpler in determining diffusion coefficient for ions intercalating and deintercalating into and out of composite electrode material, i.e., materials based on at least two components. In this transient method, a constant current pulse is sequentially applied to the electrode for a given time t to remove or insert about 2 to 5% of the electrode's total capacity, e.g., x in Li_xCoO_2. The capacity inserted or deinserted is simply it (current vs. time), and the total number of constant current steps required to fully cover the stable range of x will depend upon the amount of capacity added to or removed from the electrode material from each constant current step. After each defined constant current pulse, the electrode potential is allowed to rest until it reaches a new equilibrium before the next current pulse is applied. Figure 2.36 is a schematic of a single step pulse for GITT

where τ = time over which the constant current pulse is applied starting from t_o
ΔE_t = transit voltage change during the pulse (without the IR drop)
ΔE_s = change in the steady state voltage (OCV)
E_1 = new OCV resulting from the insertion or deinsertion reaction (E_1 will of course not change for a two-phase material).

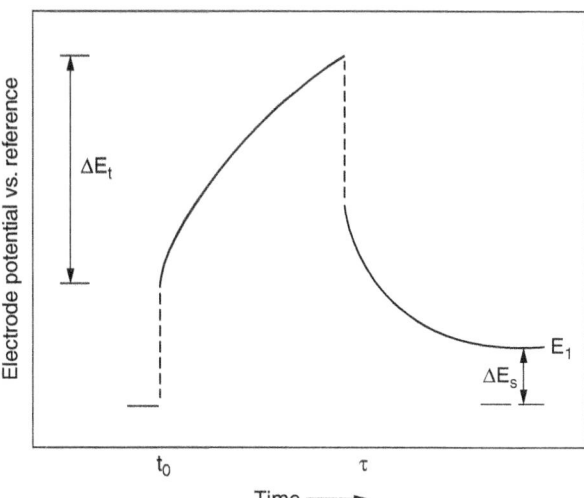

FIGURE 2.36 Typical single step constant current pulse for GITT.

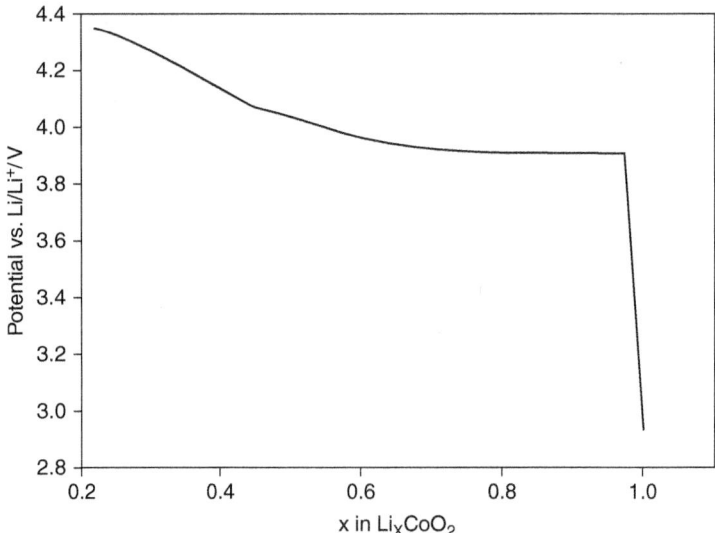

FIGURE 2.37 Potential vs. x in Li_xCoO_2 (based on data from Reference 36).

An example of determining phase transitions for Li_xCoO_2 from $x \approx 0.2$ to 1.0 is shown in Fig. 2.37 based on data published by Plichta et al.[36] There are clearly three major phases.

The figure shows that between ~0.2 and ~0.6 V vs. Li/Li+, there is a single-phase domain where the OCV varies with x in Li_xCoO_2 followed by a transition to a two-phase domain at around x ~0.6 to ~0.9 where the OCV is independent of x. Finally as x ~0.9 to 1.0, the OCV rapidly decreases, indicating a transition to a single-phase domain. The voltage versus capacity relationship varies somewhat between various authors, but the important phase regions identified using GITT are quite clear. However, the precise voltages at which these phase transitions occur is best determined by the PITT method as described below. For designing a battery based on the rate and capacity ability of various intercalating materials, the available capacity from each phase over the major stable region for x (Li+ in the above example) is of interest to battery developers, but so are the diffusion rates of the intercalating ion important for batteries specifically designed for high charge and discharge rates. A simple equation that can be used to calculate the chemical diffusion coefficient, D, is[34]

$$D = \frac{4}{\pi\tau}\left(\frac{m_b V_M}{M_b S}\right)^2 \left(\frac{\Delta E_s}{\Delta E_t}\right)^2 \quad (2.58)$$

where m_b = mass of the active material
V_M = molar volume of the active material
M_b = molecular mass of the active material
S = surface area of the electrode

The remaining terms in Eq. (2.58) are defined above. Note that the term in parentheses in Eq. (2.58) is the thickness, L, of the electrode, i.e.,

$$L = \frac{m_b V_M}{M_b S} \quad (2.59)$$

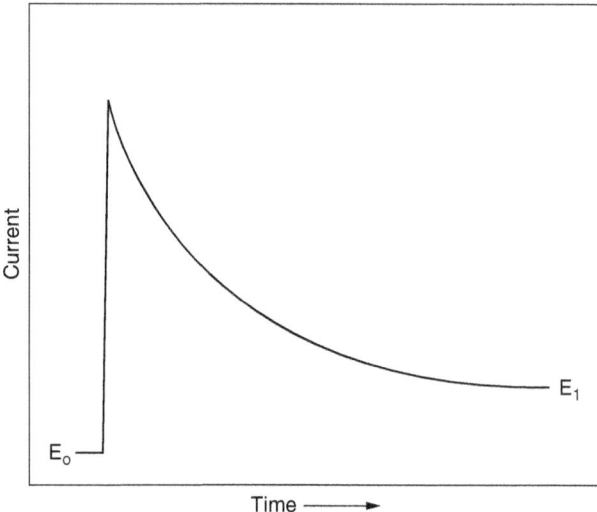

FIGURE 2.38 Typical single step constant potential pulse for PITT.

Potentiostatic Intermittent Titration Technique (PITT). The PITT method involves the application of a small amplitude voltage step to an electrode, typically around 10 mV, after which the current is recorded as a function of time, as shown in Fig. 2.38. Starting at an initial equilibrium voltage of E_0, the current decays to zero or very near (negligible) to zero, reaching a new equilibrium potential E_1. Subsequent potential pulses are imposed on the electrode to determine the incremental charge in coulombs, Q, associated with each pulse covering the whole range of capacity defining the phase diagram for the material under investigation. The total current I is recorded, and the incremental charge (or differential charge) for each potential pulse is given by

$$Q = \int_0^t I \, dt \tag{2.60}$$

Plotting the rest potential (the OCV) versus the composition of the electrode material results in a figure representing the phase diagram, i.e., similar to Fig. 2.37. Plots of the differential charge Q/E versus potential E results in sharp peaks precisely identifying the peak potentials for phase transitions. An example of this plot is shown in Fig. 2.39 for Li_xCoO_2.[36] Figure 2.39 is essentially identical to that obtained by cyclic voltammetry (Sec. 2.6.1 above), but there are important differences. The peak potentials for each phase are very sharp compared to what is observed in fast CV sweeps where the peak potentials are not precisely defined. By slowing down the sweep rate to ~0.001 mV/s, a cyclic voltammogram similar to Fig. 2.39 can be obtained,[37] but because of the very low currents for such a low sweep rate, determination of capacity is difficult and not nearly as accurate as those obtained by the PITT method.

As demonstrated for the GITT method[34], the PITT method is also useful for determining the chemical diffusion coefficient for the insertion or removal of an ion from its host material. Two approaches are based on the thickness of the electrode, L [see Eq. (2.59)] and the time, t, required to reach the equilibrium potential after each potential pulse.[36] The relations from Ref. 36 are given in Eqs. (2.61) and (2.62).

$$I(t) = \frac{QD^{1/2}}{L\pi^{1/2}}\left(\frac{1}{t^{1/2}}\right) \quad \text{if } t \ll L^2/D \tag{2.61}$$

2.32 PRINCIPLES OF OPERATION

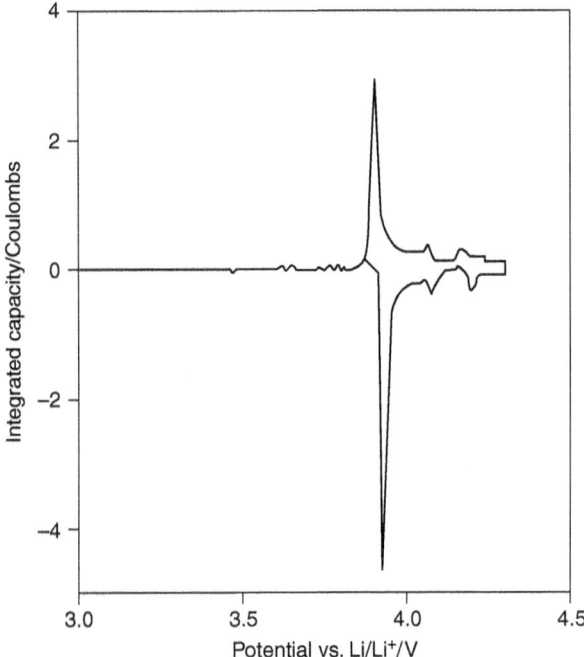

FIGURE 2.39 Plot of differential capacity Q versus potential for Li_xCoO_2 (based on data from Reference 36).

and

$$I(t) = \frac{2QD}{L^2} \exp\left(\frac{-\pi^2 Dt}{4L^2}\right) \quad \text{if } t \gg L^2/D \tag{2.62}$$

For very short times after a voltage step, the chemical diffusion coefficient can be determined from the slope of a plot of current versus $(1/t^{1/2})$ (Eq. 2.61), and for large times, the chemical diffusion coefficient can be determined from the slope of a plot of the logarithm of the current versus time t (Eq. 2.62).[34]

2.6.5 Thermodynamic Analyses of Phase Diagrams

Knowledge of Gibbs energies, enthalpies, and entropies are important for basic battery designs and battery modeling, which are, respectively, the subjects of Chapters 5 and 6 of this Handbook. These state functions have been important in developing alloy electrodes such as the "LiSb" systems[34] and "LiAl" systems[35,38] for use as anodes and reference electrodes over extremely wide temperature ranges. Enthalpies and entropies as a function of electrode composition supplement the determination of phase diagrams by providing additional details to those obtained from the GITT and PITT methods. Entropies are associated with the degree of disorder of the intercalating ion into the lattice of the host material, and enthalpies are associated with ion-host bond strength and repulsive interactions between intercalated ions.[39-42] As an example, consider a host material for Li such as Li_xM

where M represents carbon or graphite (LiC_6), or a metal oxide, such as Li_xCoO_2. In general, the electrode reaction can be written as

$$M + xLi^+ + xe^- \rightleftharpoons Li_xM \tag{2.63}$$

The Gibbs energies as a function of lithium content (x) is related to the OCV and state functions by

$$\Delta G(x) = -FE(x) = \Delta H(x) - T\Delta S(x) \tag{2.64}$$

where

$$\Delta S(x) = F\left(\frac{\delta E}{\delta T}\right)_x \tag{2.65}$$

and

$$\Delta H(x) = -FE(x) + TF\left(\frac{\delta E}{\delta T}\right)_x \tag{2.66}$$

The phase behavior in terms of entropies and enthalpies for $Li/Li_xMn_2O_4$ has been studied in detail by Thomas et al.[40] and Yamazi et al.[42] in half-cells using metallic Li as the negative electrode and the stoichiometric oxide as the positive electrode, and determining the OCV for various values of x as a function of temperature. The data for the OCV as a function of potential clearly shows two plateaus, indicating different two-phase domains. The first transition to a two-phase domain is at around $x = 0.2$, and the first plateau is observed for x ~0.3 to ~0.55, and the second plateau is observed for x ~0.6 to ~0.95, followed by a third region where the OCV drops rapidly as x approaches 1.0. The change in entropy calculated from Eq. (2.65) varies significantly with the state of charge as shown in Fig. 2.40. The entropy is positive for a nearly empty lattice for x ~0.25, and as lithium is intercalated into the host oxide, it rapidly decreases, eventually becoming negative when the available sites for lithium become nearly full. For the second transition to the second domain between x ~0.55 and ~0.95, the change in entropy starts to increase, reaching positive values as the filling of vacant sites proceeds before returning to negative values as the vacant sites become fully occupied. The changes in ΔS shown in Fig. 2.40 are not affected by the metallic Li anode since the Li atoms in the anode

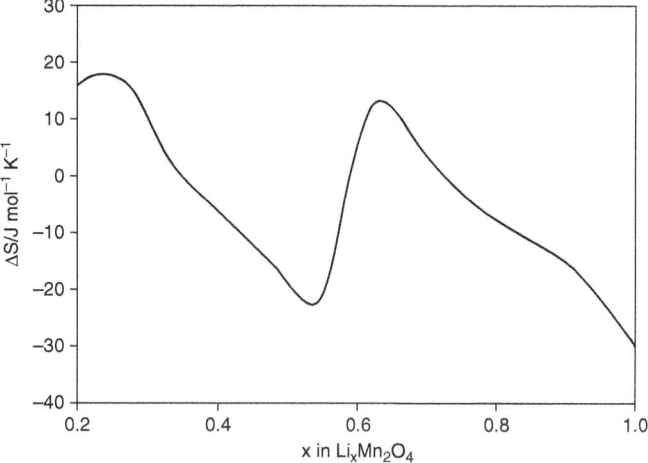

FIGURE 2.40 Entropy of lithium intercalation in $Li_xMn_2O_4$ (based on data from Reference 42).

are in the same environment at any state of charge.[41] The maximum for positive values of ΔS in phase diagrams as a function of x in Li_xM can also be correlated with higher diffusion coefficients and interstitial distances in the different phases as determined from crystal structures derived from X-ray diffraction data.[39]

2.6.6 Electrodes

Several electrochemical techniques have been discussed in the preceding sections, but little was mentioned about electrodes or electrode geometries used in the various measurements. This section deals with electrodes and electrode systems. A basic two-electrode cell contains an anode (negative electrode) and a cathode (positive electrode) and is typically used to study individual battery cell properties to determine Gibbs energies, energy densities, and specific energies. For studying the properties of a given electrode material by cyclic voltammetry and transient techniques, a three-electrode cell is preferred. Figure 2.41 is a schematic of the three-electrode cell and consists of the electrode material under study (the working electrode), a counter electrode, and the reference electrode.

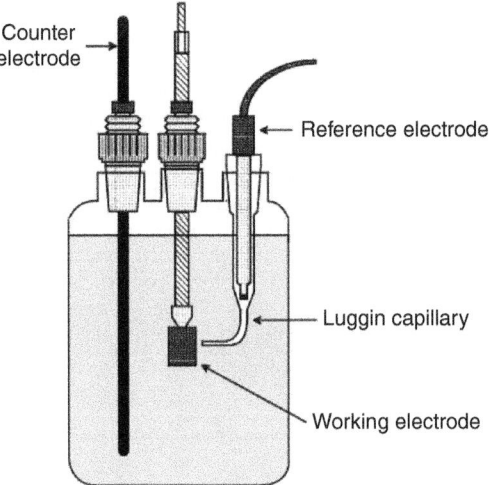

FIGURE 2.41 Three-electrode electrochemical cell.

The counter electrode is attached to the working electrode through a power supply or potentiostat enabling one to precisely vary the current or electrode potential. If the reaction products at the counter electrode could interfere with reaction occurring at the working electrode, the two electrodes could be separated by a sintered glass disk or some other porous medium that will allow ionic conduction but prevent gross mixing of solutions surrounding the counter and working electrodes. The potential of the working electrode is monitored at a nonpolarizable reference electrode. A Luggin capillary, shown in Fig. 2.41, can be used to reduce uncompensated IR drop when current is flowing between the counter and working electrodes. If a Luggin capillary is used, the reference electrode is placed in the Luggin as shown in Fig. 2.41, and the electrolyte solution inside the Luggin may be the same as the electrolyte solution in the cell or it may be a different electrolyte solution containing ions necessary to establish equilibrium at the reference electrode. For aqueous solutions, typical reference electrode systems such as Ag/AgCl, Hg/Hg_2Cl_2, and Hg/HgO are often used. For comprehensive treatises on the subject of reference electrodes, the reader is referred to two excellent texts by Ives and Janz[43] and Bard et al.[44]

For nonaqueous solutions, the reader is also referred to a chapter by Butler in volume 7 of the series Advances in Electrochemistry and Electrochemical Engineering.[45] For studies on Li and Li-ion batteries, the preferred reference electrode is metallic lithium, but use of metallic lithium is not always feasible, particularly when it is not stable in some electrolyte solutions such as several ionic liquids, and when temperatures exceed the melting point of pure lithium. In the latter case, a high melting point Li alloy such as LiAl[35,38] can be used, and in the former case a pseudo-reference can be used. Pseudo-reference electrodes typically used in solutions reactive with Li are metal wires (e.g., Al, Pt, or Ag) immersed in the electrolyte solution. Although such electrodes often provide a constant potential, they have no thermodynamic significance. Instead of a wire electrode, a good thermodynamically stable reference related to metallic lithium is a metal oxide such as $Li_4Ti_5O_{12}$ which has reversible potential of 1.55 V versus Li/Li$^+$.[46]

REFERENCES

1. J. O'M. Bockris and A. K. N. Reddy, *Modern Electrochemistry,* vol. 2, Plenum, New York, 1970, p. 644.
2. H. H. Bauer, *J Electroanal. Chem.* **16**:419 (1968).
3. J. Horiuti and M. Polanyi, *Acta Physicochim. U.S.S.R.* **2**:505 (1935).
4. J. O'M. Bockris and A. K. N. Reddy, op. cit., p. 918; see also J. O'M. Bockris, "Electrode Kinetics" in *Modern Aspects of Electrochemistry,* J. O'M. Bockris and B. E. Conway, eds., Butterworths, London, 1954, Chap. 2.
5. P. Delahay, *Double Layer and Electrode Kinetics,* Interscience, New York, 1965.
6. J. O'M. Bockris and A. K. N. Reddy, op. cit., p. 742.
7. H. H. Bauer, *Electrodics,* Wiley, New York, 1972, p. 54, table 3.2.
8. A. Fick, *Ann. Phys.* **94**:59 (1855).
9. P. Delahay, *New Instrumental Methods in Electrochemistry,* Interscience, New York, 1954.
10. J. S. Newman, *Electrochemical Systems,* 2d ed., Prentice Hall, Englewood Cliffs, NJ, 1991.
11. E. B. Yeager and J. Kuta, "Techniques for the Study of Electrode Processes," in *Physical Chemistry,* vol. IXA; *Electrochemistry,* Academic, New York, 1970, p. 346.
12. L. A. Matheson and N. Nichols, *J. Electrochem. Soc.* **73**:193 (1938).
13. J. E. B. Randles, *Trans. Faraday Soc.* **44**:327 (1948).
14. A. Sevcik, *Coll. Czech. Chem. Comm.* **13**:349 (1948).
15. T. Berzins and P. Delahay, *J. Am. Chem. Soc.* **75**:555 (1953).
16. P. Delahay, *J. Am. Chem. Soc.* **75**:1190 (1953).
17. R. S. Nicholson and I. Shain, *Anal. Chem.* **36**:706 (1964).
18. H. I. S. Sand, *Phil. Mag.* **1**:45 (1901).
19. P. Delahay, *New Instrumental Methods in Electrochemistry,* op. cit.
20. P. Delahay and T. Berzins, *J. Am. Chem. Soc.* **75**:2486 (1953).
21. C. N. Reilley, G. W. Everett, and R. H. Johns, *Anal. Chem.* **27**:483 (1955).
22. T. Kambara and L. Tachi, *J. Phys. Chem.* **61**:405 (1957).
23. M. D. Morris and J. J. Lingane, *J. Electroanal. Chem.* **6**:300 (1963).
24. J. Broadhead and G. J. Hills, *J. Electroanal. Chem.* **13**:354 (1967).
25. J. R. MacDonald, *Impedance Spectroscopy, Emphasizing Solid Materials and Systems,* Wiley, New York, 1987, pp. 154–155.
26. M. E. Orazem and B. Tribollet, *Electrochemical Impedance Spectroscopy,* ECS Series of Texts and Monographs, Wiley-Blackwell, (Oct. 2008).
27. J. E. B. Randles, *Disc. Faraday Soc.* **1**:11 (1947).
28. B. A. Boukamp, *Solid State Ionics,* **18**:136 (1986).

29. B. A. Boukamp, *Solid State Ionics*, **20**:31 (1986).
30. http://www.jrossmacdonald.com
31. http:/www.Gamry.com
32. http://www.princetonappliedresearch.com
33. http://www.scribner.com
34. W. Weppner and R. A. Huggins, *J. Electrochem. Soc.*, **124**:1569 (1977).
35. C. John Wen, B. A. Boukamp, and R. A. Huggins, *J. Electrochem. Soc.*, **126**:2558 (1979).
36. E. Plichta, S. Slane, M. Uchiyami, M. Salomon, D. Chua, W. B. Ebner, and H.-p. Lin, *J. Electrochem. Soc.*, **137**:1865 (1989).
37. K. West, B. Zachau-Christiansen, and T. Jacobsen, *Electrochim. Acta*, **28**:1829 (1983).
38. N. P. Yao, L. A. Heredy, and R. C. Saunders, *J. Electrochem. Soc.*, **118**:1039 (1971).
39. S. Bach, J. P. Pereira-Ramos, N. Baffier, and R. Messina, *J. Electrochem. Soc.*, **137**:1042 (1990).
40. K. E. Thomas, C. Bogatu, and J. Newman, *J. Electrochem. Soc.*, **148**:A570 (2001).
41. Y. F. Reynier, R. Yazami, and B. Fultz, *J. Electrochem. Soc.*, **151**:A422 (2004).
42. R. Yazami, Y. Reynier, and B. Fultz, *ECS Transactions*, **1(26)**:87 (2006).
43. D. J. G. Ives and G. J. Janz, *Reference Electrodes, Theory and Practice,* Academic, New York, 1961.
44. A. J. Bard, R. Parsons, and J. Jordan, *Standard Potentials in Aqueous Solutions*, Marcel Dekker, New York, 1985.
45. J. N. Butler, "Reference Electrodes in Aprotic Organic Solvents," in *Advances in Electrochemistry and Electrochemical Engineering*, P. Delahay, ed., vol. 7 (1970), pp. 77–175.
46. K. M. Colbow, J. R. Dahn, and R. R. Haering, *J. Power Sources*, **26**:397 (1989).

BIBLIOGRAPHY

General

J. O'M. Bockris and A. K. N. Reddy, *Modern Electrochemistry,* vols. 1 and 2, Plenum, New York, 1970.

B. E. Conway, *Theory and Principles of Electrode Processes,* Ronald Press, New York, 1965.

E. Gileadi, E. Kirowa-Eisner, and J. Penciner, *Interfacial Electrochemistry,* Addison-Wesley, Reading, MA, 1975.

A. J. Bard and L.R. Faulkner, *Electrochemical Methods: Fundamentals and Applications*, John Wiley, New York, 1980.

C. A. Vincent and B. Scrosati, *Modern Batteries*, 2nd ed. Butterworth-Heinemann, Oxford, 1997.

Transfer Coefficient

B. E. Conway, *Theory and Principles of Electrode Processes,* Ronald Press, New York, 1965.

H. H. Bauer, *J. Electroanal. Chem.* **16**:419 (1968).

J. O'M. Bockris and A. K. N. Reddy, *Modern Electrochemistry,* vol. 2, Plenum, New York, 1970.

Electrical Double Layer

V. S. Bogotsky, *Fundamentals of Electrochemistry*, 2nd ed. Wiley-Interscience, New York, 2005.

P. Delahay, *Double Layer and Electrode Kinetics,* Interscience, New York, 1965.

R. Parsons, "Equilibrium Properties of Electrified Interphases" in *Modern Aspects of Electrochemistry*, J. O'M. Bockris and B. E. Conway, eds., vol. 1, Butterworths, London, 1954, pp. 103–179.

Electrochemical Techniques

P. Delahay, *New Instrumental Methods in Electrochemistry,* Interscience, New York, 1954.

E. B. Yeager and J. Kuta, "Techniques for the Study of Electrode Processes," in *Physical Chemistry,* vol. IXA: *Electrochemistry,* Academic, New York, 1970.

D. T. Sawyer, A. Sobkowiak, and J. L. Roberts, *Experimental Electrochemistry for Chemists,* 2nd ed., Wiley, New York, 1995.

Reference Electrodes

D. J. G. Ives and G. J. Janz: *Reference Electrodes, Theory and Practice,* Academic, New York, 1961.

J. N. Butler, "Reference Electrodes in Aprotic Organic Solvents," in *Advances in Electrochemistry and Electrochemical Engineering*, P. Delahay, ed., vol. 7 (1970), pp. 77–175.

A. J. Bard, R. Parsons, and J. Jordan, *Standard Potentials in Aqueous Solutions*, Marcel Dekker, New York, 1985.

Electrochemistry of the Elements

A. J. Bard (ed.), *Encyclopedia of Electrochemistry of the Elements,* vols. I–XIII, Marcel Dekker, New York, 1979.

Organic Electrode Reactions

L. Meites and P. Zuman, *Electrochemical Data,* Wiley, New York, 1974.

AC Impedance Techniques

E. Barsoukov and J. R. Macdonald, *Impedance Spectroscopy: Theory, Experiment and Applications,* Wiley-Interscience, New York, 2005.

Gamry Application Note, *Basics of Electrochemical Impedance Spectroscopy,* http://www.Gamry.com

G. Gabrielli, *Identification of Electrochemical Processes by Frequency Response Analysis*, Tech. Rep. 004/83, Solartron Instruments, Billerica, MA, 1984.

CHAPTER 3
FACTORS AFFECTING BATTERY PERFORMANCE

David Linden

3.1 GENERAL CHARACTERISTICS

The specific energy of a number of battery systems is listed in Table 1.2. These values are based on optimal designs and discharge conditions. While these values can be helpful to characterize the energy output of each battery system, the performance of the battery may be significantly different under actual conditions of use, particularly if the battery is discharged under more stringent conditions than those under which it was characterized. The performance of the battery under the specific conditions of use should be obtained before any final comparisons or judgments are made.

3.2 FACTORS AFFECTING BATTERY PERFORMANCE

Many factors influence the operational characteristics, capacity, energy output, and performance of a battery. The effect of these factors on battery performance is discussed in this section. It should be noted that because of the many possible interactions, these effects can be presented only as generalizations and that the influence of each factor is usually greater under the more stringent operating conditions. For example, the effect of storage is more pronounced not only with high storage temperatures and long storage periods, but also under more severe conditions of discharge following storage. After a given storage period, the observed loss of capacity (compared with a fresh battery) will usually be greater under heavy discharge loads than under light discharge loads. Similarly, the observed loss of capacity at low temperatures (compared with normal temperature discharges) will be greater at heavy than at light or moderate discharge loads. Specifications and standards for batteries usually list the specific test or operational conditions on which the standards are based because of the influence of these conditions on battery performance.

Furthermore it should be noted that even within a given cell or battery design, there will be performance differences from manufacturer to manufacturer and between different versions of the same battery (such as standard, heavy-duty, or premium). There are also performance variables within a production lot, and from production lot to production lot, that are inherent in any manufacturing process. The extent of the variability depends on the process controls as well as on the application and use of the battery. Manufacturers' data should be consulted to obtain specific performance characteristics.

3.2 PRINCIPLES OF OPERATION

3.2.1 Voltage Level

Different references are made to the voltage of a cell or battery:

1. The *theoretical voltage* is a function of the anode and cathode materials, the composition of the electrolyte, and the temperature (usually stated at 25°C).
2. The *open-circuit voltage* is the voltage under a no-load condition and is usually a close approximation of the theoretical voltage.
3. The *closed-circuit voltage* is the voltage under a load condition.
4. The *nominal voltage* is one that is generally accepted as typical of the operating voltage of the battery as, for example, 1.5 V for a zinc-manganese dioxide battery.
5. The *working voltage* is more representative of the actual operating voltage of the battery under load and will be lower than the open-circuit voltage.
6. The *average voltage* is the voltage averaged during the discharge.
7. The *midpoint voltage* is the central voltage during the discharge of the cell or battery.
8. The *end* or *cutoff voltage* is designated as the end of the discharge. Usually it is the voltage above which most of the capacity of the cell or battery has been delivered. The end voltage may also be dependent on the application requirements.

Using the lead-acid battery as an example, the theoretical and open-circuit voltages are 2.1 V, the nominal voltage is 2.0 V, the working voltage is between 1.8 and 2.0 V, and the end voltage is typically 1.75 V on moderate and low-drain discharges and 1.5 V for engine-cranking loads. On charge, the voltage may range from 2.3 to 2.8 V.

When a cell or battery is discharged, its voltage is lower than the theoretical voltage. The difference is caused by *IR* (the product of the discharge current and the internal resistance) losses due to cell (and battery) resistance and polarization of the active materials during discharge. This is illustrated in Fig. 3.1. In the idealized case, the discharge of the battery proceeds at the theoretical voltage until the active materials are consumed and the capacity is fully utilized. The voltage then drops to zero. Under actual conditions, the discharge curve is similar to the other curves in Fig. 3.1. The initial voltage of the cell under a discharge load is lower than the theoretical value due to the internal cell resistance and the resultant *IR* drop as well as polarization effects at both electrodes. The voltage also drops during discharge as the cell resistance increases due to the accumulation of discharge products, activation and concentration, polarization, and related factors. Curve 2 is similar to curve 1, but represents a cell with a higher internal resistance or a higher discharge rate, or both, compared to the cell represented by curve 1. As the cell resistance or the discharge current is increased, the discharge voltage decreases and the discharge shows a more sloping profile.

The specific energy that is delivered by a battery in practice is, therefore, lower than the theoretical specific energy of its active materials, because:

1. The average voltage during the discharge is lower than the theoretical voltage.
2. The battery is not discharged to zero volts and all of the available ampere-hour capacity is not utilized.

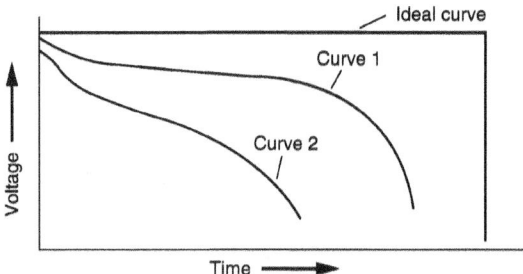

FIGURE 3.1 Characteristic discharge curves.

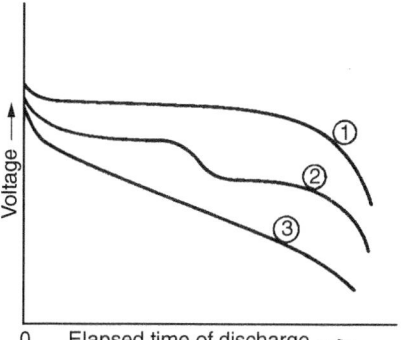

FIGURE 3.2 Battery discharge characteristics—voltage profiles.

As specific energy equals

$$\text{Watthours/gram} = \text{Voltage} \times \text{Ampere-hours/gram}$$

the delivered specific energy is lower than the theoretical energy as both of the components of the equation are lower.

The shape of the discharge curve can vary depending on the electrochemical system, constructional features, and other discharge conditions. Typical discharge curves are shown in Fig. 3.2. The flat discharge (curve 1) is representative of a discharge where the effect of change in reactants and reaction products is minimal until the active materials are nearly exhausted. The plateau profile (curve 2) is representative of two-step discharge indicating a change in the reaction mechanism and potential of the active material(s). The sloping discharge (curve 3) is typical when the composition of the active materials, reactants, internal resistance, and so on, change during the discharge, affecting the shape of the discharge curve similarly.

Specific examples of these curves and many others are presented in the individual chapters covering each battery system.

3.2.2 Current Drain of Discharge

As the current drain of the battery is increased, the *IR* losses and polarization effects increase, the discharge is at a lower voltage, and the service life of the battery is reduced. Figure 3.3a shows typical discharge curves as the current drain is changed. At extremely low current drains (curve 2) the discharge can approach the theoretical voltage and theoretical capacity. However, with very long discharge periods, chemical deterioration during the discharge can become a factor and cause a reduction in capacity (Sec. 3.2.12). With increasing current drain (curves 3–5), the discharge voltage decreases, the slope of the discharge curve becomes more pronounced, and the service life, as well as the delivered ampere-hour or coulombic capacity, are reduced.

If a battery that has reached a particular voltage (such as the cutoff voltage) under a given discharge current is used at a lower discharge rate, its voltage will rise and additional capacity or service life can normally be obtained until the cutoff voltage is reached at the lighter load. Thus, for example, a battery that has been used to its end-of-life in a flash camera (a high-drain application) can subsequently be used successfully in a quartz clock application, which operates at a much lower discharge rate. This procedure can also be used for determining the life of a battery under different discharge loads using a single test battery. As shown in Fig. 3.3b, the discharge is first run at the highest discharge rate to the specified end voltage. The discharge rate is then reduced to the next lower rate. The voltage increases and the discharge is continued again to the specified end voltage, and so on. The service life can be determined for each discharge rate, but the complete discharge

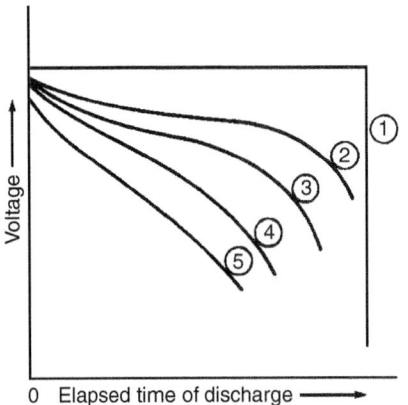 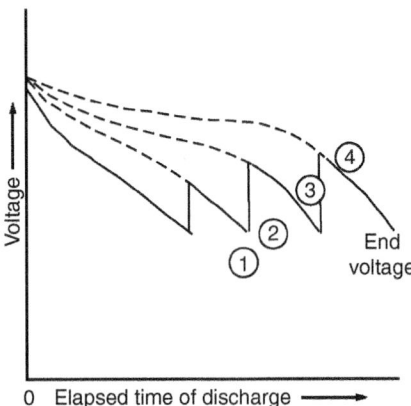

FIGURE 3.3 (*a*) Battery discharge characteristics—voltage levels. (*b*) Discharge characteristics of a battery discharged sequentially from high to lower discharge rates.

curve for the lower discharge rates, as shown by the dashed portion of the each curve, obviously is lost. In some instances, a time interval is allowed between each discharge for the battery to equilibrate prior to discharge at the progressively lower rates.

C-Rate.* A common method for indicating the discharge, as well as the charge current of a battery, is the C rate, expressed as

$$I = M \times C_n$$

where I = discharge current, A
C = numerical value of rated capacity of the battery, in ampere-hours (Ah)
n = time, in hours, for which rated capacity is declared
M = multiple or fraction of C

For example, the $0.1C$ or $C/10$ discharge rate for a battery rated at 5 Ah is 0.5 A. Conversely, a 250 mAh battery, discharged at 50 mA, is discharged at the $0.2C$ or $C/5$ rate, which is calculated as follows:

$$M = \frac{1}{C_n} = \frac{0.050}{0.250} = 0.2$$

Traditionally, the manufacturers and users of secondary alkaline cells and batteries have expressed the value of the current used to charge and discharge cells and batteries as a multiple of the capacity. For example, a current of 200 mA used to charge a cell with a rated capacity of 1000 mAh would be expressed as $C/5$ or $0.2\ C$ (or, in the European convention as $C/5$ A or 0, 2 CA). This method for designation of current has been criticized as being dimensionally incorrect in that a multiple of the capacity (e.g. ampere-hours) will be in ampere-hours and not, as required for current, in amperes. As a result of these comments, the International Electrotechnical Commission (IEC) Subcommittee SC-21A has published a "Guide to the Designation of Current in Alkaline Secondary Cell and Battery Standards (IEC 61434)," which describes a new method for so designating this current. In brief, the method states that the current (I) shall be expressed as

$$I_t (A) = C_n(Ah)/1(h)$$

where I = is expressed in amperes
C_n = is the rated capacity declared by the manufacturer in ampere-hours
n = is the time base in hours for which the rated capacity is declared

For example, a battery rated at 5 Ah at the 5 h discharge rate (C_5 (Ah)) and discharged at $0.1I_t$(A) will be discharged at 0.5 A or 500 mA.
For this Handbook, the method discussed in the text, and not in this footnote, will be used.

To further clarify this nomenclature system, the designation for a $C/10$ discharge rate for a battery rated at 5 Ah at the 5 h rate is

$$0.1\ C_5$$

In this example, the $C/10$ rate is equal to 0.5 A, or 500 mA.

It is to be noted that the capacity of a battery generally decreases with increasing discharge current. Thus the battery rated at 5 Ah at the $C/5$ rate (or 1 A) will operate for 5 h when discharged at 1 A. If the battery is discharged at a lower rate, for example the $C/10$ rate (or 0.5 A), it will run for more than 10 h and deliver more than 5 Ah of capacity. Conversely, when discharged at its C rate (or 5 A), the battery will run for less than 1 h and deliver less than 5 Ah of capacity.

Hourly Rate. Another method for specifying the current is the *hourly rate*. This is the current at which the battery will discharge for a specified number of hours.

E-Rate. The constant power discharge mode is becoming more popular for battery-powered applications. A method, analogous to the C rate, can be used to express the discharge or charge rate in terms of power

$$P = M \times E_n$$

where P = power (W)
 E = numerical value of the rated energy of the battery in watthours (Wh)
 n = time, in hours, at which the battery was rated
 M = multiple or fraction of E

For example, the power level at the $0.5E_5$ or $E_5/2$ rate for a battery rated at 1200 mWh, at the $0.2E$ or $E/5$ rate, is 600 mW.

3.2.3 Mode of Discharge (Constant Current, Constant Load, Constant Power)

The mode of discharge of a battery, among other factors, can have a significant effect on the performance of the battery. For this reason, it is advisable that the mode of discharge used in a test or evaluation program be the same as the one used in the application for which it is being tested.

A battery, when discharged to a specific point (same closed-circuit voltage, at the same discharge current, at the same temperature, etc.) will have delivered the same ampere-hours to a load regardless of the mode of discharge. However, as during the discharge, the discharge current will be different depending on the mode of discharge; the service time or "hours of discharge" delivered to that point (which is the usual measure of battery performance) will likewise be different.

Three of the basic modes under which the battery may be discharged are:

1. *Constant resistance.* The resistance of the load remains constant throughout the discharge (the current decreases during the discharge proportional to the decrease in the battery voltage).
2. *Constant current.* The current remains constant during the discharge.
3. *Constant power.* The current increases during the discharge as the battery voltage decreases, thus discharging the battery at constant power level (power = current × voltage).

The effect of the mode of discharge on the performance of the battery is illustrated under three different conditions in Figs. 3.4, 3.5, and 3.6.

Case 1: Discharge loads are the same for each mode of discharge at the start of discharge.

In Fig. 3.4, the discharge loads are selected so that at the start of the discharge, the discharge current and, hence, the power are the same for all three modes. Figure 3.4*b* is a plot of the voltage during discharge. As the cell voltage drops during the discharge, the current in the case of the constant resistance discharge, reflects the drop in the cell voltage according to Ohm's law:

$$I = V/R$$

This is shown in Fig. 3.4*a*.

3.6 PRINCIPLES OF OPERATION

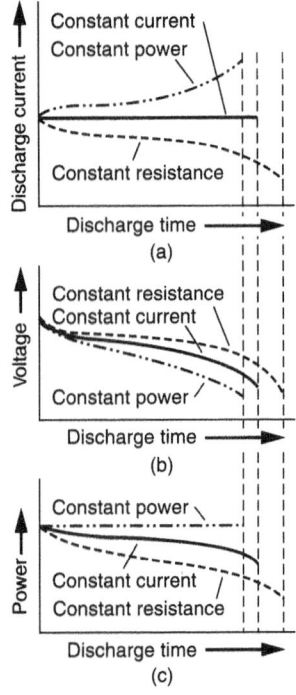

FIGURE 3.4 Discharge profiles under different discharge modes; same current and power at start of discharge. (*a*) Current profile during discharge. (*b*) Voltage profile during discharge. (*c*) Power profile during discharge.

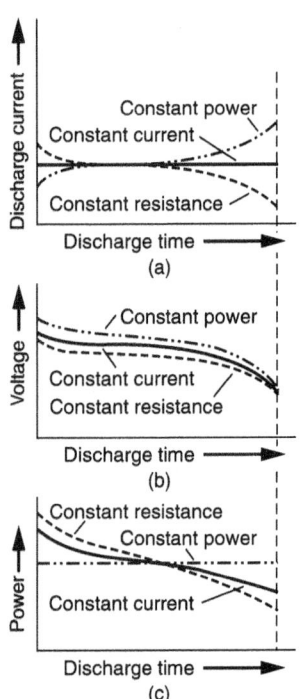

FIGURE 3.5 Discharge profiles under different discharge modes; same discharge time (*a*) Current profile during discharge. (*b*) Voltage profile during discharge. (*c*) Power profile during discharge.

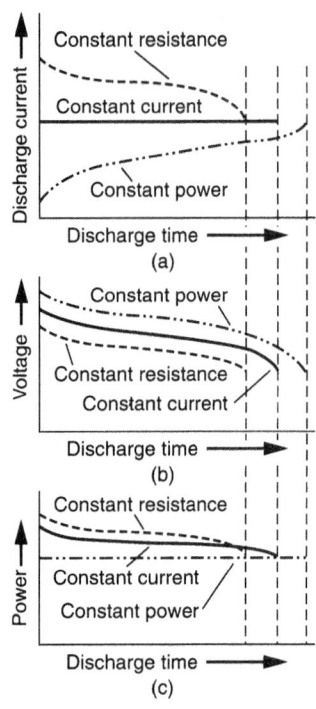

FIGURE 3.6 Discharge profiles under different discharge modes: same power at end of discharge. (*a*) current profile during discharge. (*b*) Voltage profile during discharge. (*c*) Power profile during discharge.

In the case of a constant-current discharge, the current remains the same throughout the discharge. However, the discharge time or service life is lower than for the constant-resistance case because the average current is higher. Finally, in the constant-power mode, the current increases with decreasing voltage according to the relationship

$$I = P/V$$

The average current is now even higher and the discharge time still lower.

Figure 3.4c is a plot of the power level for each mode of discharge.

Case 2: "Hours of discharge" is the same for each mode of discharge.

Figure 3.5 shows the same relationships but with the respective discharge loads selected so that the discharge time or "hours of service" (to a given end voltage) is the same for all three modes of discharge. As expected, the discharge curves vary depending on the mode of discharge.

Case 3: Power level is the same for each mode of discharge at the end of the discharge.

From an application point of view, the most realistic case is the assumption that the power under all three modes of operation is the same at the end of the discharge (Fig. 3.6). Electric and electronic devices require a minimum input power to operate at a specified performance level. In each case, the discharge loads are selected so that at the end of the discharge (when the cell reaches the cutoff voltage), the power output is the same for all of the discharge modes and at the level required for acceptable equipment performance. During the discharge, depending on the mode of discharge, the power output equals or exceeds the power required by the equipment until the battery reaches the cutoff voltage.

In the constant-resistance discharge mode, the current during the discharge (Fig. 3.6b) follows the drop in the battery voltage (Fig. 3.6a). The power, $I \times V$ or V^2/R, drops even more rapidly, following the square of the battery voltage (Fig. 3.6c). Under this mode of discharge, to ensure that the required power is available at the cutoff voltage, the levels of current and power during the earlier part of the discharge are in excess of the minimum required. The battery discharges at a higher current than needed, draining its capacity rapidly, which will result in a shorter service life.

In the constant-current mode, the current is maintained at a level such that the power output at the cutoff voltage is equal to the level required for acceptable equipment performance. Thus both current and power throughout the discharge are lower than for the constant-resistance mode. The average current drain on the battery is lower, and the discharge time or service life to the end of the battery life is longer.

In the constant-power mode, the current is lowest at the beginning of the discharge and increases as the battery voltage drops in order to maintain a constant-power output at the level required by the equipment. The average current is lowest under this mode of discharge, and hence, the longest service time is obtained.

It should be noted that the extent of the advantage of the constant-power discharge mode over the other modes of discharge is dependent on the discharge characteristics of the battery. The advantage is higher with battery systems that provide a wide voltage range to deliver their full capacity.

3.2.4 Example of Evaluation of Battery Performance Under Different Modes of Discharge

In evaluating or comparing the performance of batteries, because of the potential difference in performance (service hours) due to the mode of discharge, the mode of discharge used in the evaluation or test should be the same as that in the application. This is illustrated further in Fig. 3.7.

Figure 3.7a shows the discharge characteristics of a typical AA-size primary battery with the values for the discharge loads for the three modes of discharge selected so that the hours of discharge to a given end voltage (in this case, 1.0 V) are the same. This is the same condition shown in Fig. 3.5b. (This example illustrates the condition when a resistive load, equivalent to the average current, is used, albeit incorrectly, as a "simpler," less costly test to evaluate a constant-current or constant-power application.) Although the hours of service to the given end voltage are obviously the same because the loads were preselected, the discharge current versus discharge time and power versus discharge time curves (see Figs. 3.5a and c, respectively) show significantly different characteristics for the different modes of discharge.

Figure 3.7b shows the same three types of discharge as Fig. 3.7a, but on a battery that has about the same ampere-hour capacity (to a 1.0 V end voltage) as the battery illustrated in Fig. 3.7a. The battery illustrated in Fig. 3.7b, however, has a lower internal resistance and a higher operating voltage. Note, by comparing the voltage versus discharge time curves in Fig. 3.7b, that, although the voltage level is higher, the hours of discharge obtained on the constant-resistance discharge to the 1.0 V cutoff in Fig. 3.7b are about the same as obtained on Fig. 3.7a. However, the hours of service obtained on the constant-current discharge and, particularly, the constant-power discharge are significantly higher.

In Fig. 3.7a, the discharge loads were deliberately selected to give the same hours of service to 1.0 V at the three modes of discharge. Using these same discharge loads, but with a battery having different characteristics, Fig. 3.7b shows that different hours of service and performance are obtained with the different modes of discharge. To the specified 1.0 V end voltage, the longest hours of service are obtained with the constant-power discharge mode. The shortest service time is obtained with the constant-resistance discharge mode, and the constant-current mode is in the middle position. This clearly illustrates that, on application tests, where performance is measured in hours of service, erroneous results will be obtained if the mode of discharge used in the test is different from that used in the application.

3.8 PRINCIPLES OF OPERATION

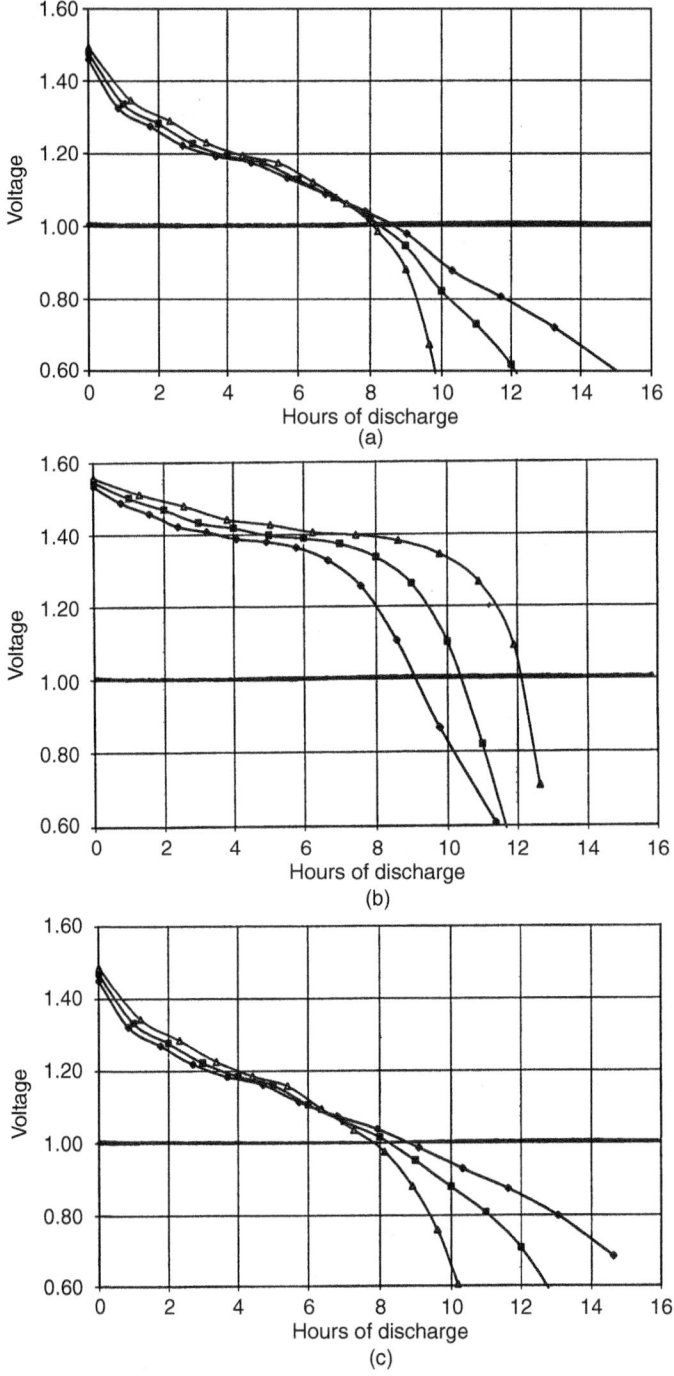

FIGURE 3.7 Characteristics of a AA-size primary battery discharged under constant-resistance, constant-current, and constant-power conditions at 5.9 ohms, ——●——; 200 milliamperes, ——■——; and 235 milliwatts, ——▲——. See Sec. 3.2.4 for detailed description.

It is recognized that the performance differences obtained when comparing batteries are directly dependent on the differences in battery design and performance characteristics. With batteries that are significantly different in design and characteristics, the performance obtained on test will be quite different as shown in the comparisons between Figs. 3.7a and 3.7b. When the batteries are similar, the performance differences obtained on any of the modes of discharge may not be large and may not appear to be significantly different. However, just because the differences in this case are small, it should not lead to the false assumption that testing under a discharge mode different from the application would give accurate results.

This is illustrated in Fig. 3.7c, which shows the discharge characteristics of another battery that has a slightly higher capacity and higher internal resistance than the one shown in Fig. 3.7a. Although the differences are small, a careful comparison of the Fig. 3.7a with Fig. 3.7c at the different modes of discharge does show a different behavior in the hours of discharge obtained to the specified 1.0 V end voltage. Under the constant-power mode, the hours of discharge show a slight decrease comparing Fig. 3.7c with Fig. 3.7a, while there is a slight increase under the constant-current and constant-resistance discharge modes.

(Note: The influence of end voltage should be noted. As 1.0 V was used as the end voltage in determining the load values for these examples, this end voltage should be used in making comparisons. If discharged to lower end voltages, the service life for the constant-resistance mode increases compared to the other modes because of the lower current and power levels. However, these lower values may be inadequate for the specified application).

3.2.5 Temperature of Battery During Discharge

The temperature at which the battery is discharged has a pronounced effect on its service life (capacity) and voltage characteristics. This is due to the reduction in chemical activity and the increase in the internal resistance of the battery at lower temperatures. This is illustrated in Fig. 3.8, which shows discharges at the same current drain but at progressively increasing temperatures of the battery (T_1 to T_4), with T_4 representing a discharge at normal room temperature. Lowering of the discharge temperature will result in a reduction of capacity as well as an increase in the slope of the discharge curve. Both the specific characteristics and the discharge profile vary for each battery system, design, and discharge rate, but generally best performance is obtained between 20 and 40°C. At higher temperatures, the internal resistance decreases, the discharge voltage increases and, as a result, the ampere-hour capacity and energy output usually increase as well. On the other hand, chemical activity also increases at the higher temperatures and may be rapid enough during the discharge (a phenomenon known as *self-discharge*) to cause a net loss of capacity. Again, the extent is dependent on the battery system, design, and temperature.

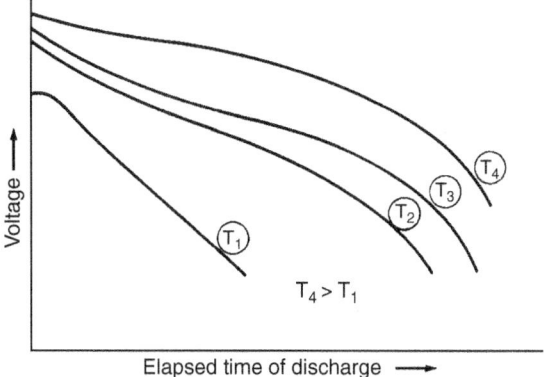

FIGURE 3.8 Effect of temperature on battery capacity. T_1 to T_4—increasing temperatures.

3.10 PRINCIPLES OF OPERATION

Figures 3.9 and 3.10 summarize the effects of temperature and discharge rate on the battery's discharge voltage and capacity. As the discharge rate is increased, the battery voltage (for example, the midpoint voltage) decreases; the rate of decrease is usually more rapid at the lower temperatures. Similarly, the battery's capacity falls off most rapidly with increasing discharge load and decreasing temperature. Again, as noted previously, the more stringent the discharge conditions, the greater the loss of capacity. However, discharging at high rates could cause apparent anomalous effects as the battery may heat up to temperatures much above ambient, showing the effects of the higher temperatures. Curve T_6 in Fig. 3.10 shows the loss of capacity at high temperatures at low discharge rates or long discharge times due to self-discharge or chemical deterioration. It also shows the higher capacity that may be obtained as a result of the battery heating at the high-rate discharge.

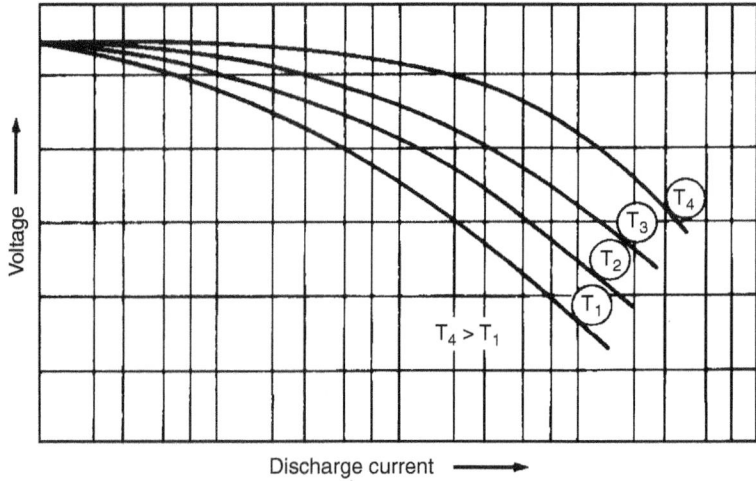

FIGURE 3.9 Effect of discharge load on battery midpoint voltage at various temperatures, T_1 to T_4—increasing temperatures; T_4—normal room temperature.

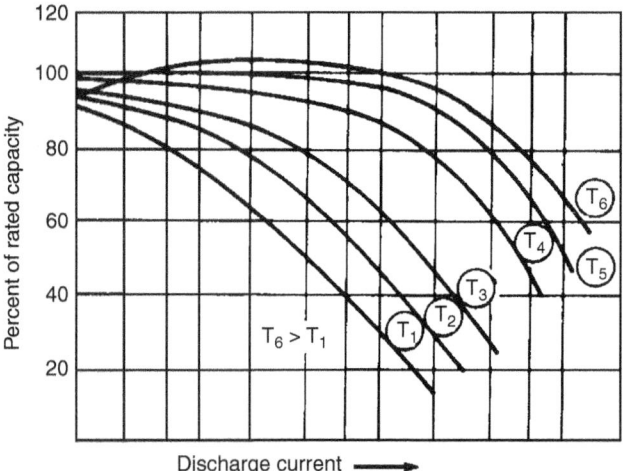

FIGURE 3.10 Effect of discharge load on battery capacity at various temperatures. T_1 to T_6—increasing temperatures; T_4—normal room temperature.

3.2.6 Service Life

A useful graph employed in this Handbook, summarizing the performance of each battery system, presents the service life at various discharge loads and temperatures, normalized for unit weight (amperes per kilogram) and unit volume (amperes per liter). Typical curves are shown in Fig. 3.11. In this type of presentation of data, curves with the sharpest slope represent a better response to increasing discharge loads than those which are flatter or flatten out at the high current drain discharges.

Data of this type can be used to approximate the service life of a given cell or battery under a particular discharge condition or to estimate the weight or size of a battery required to meet a given service requirement. In view of the linearity of these curves on a log-log plot, mathematical relationships have been developed to estimate the performance of batteries under conditions that are not specifically stated. Peukert's equation,

$$I^n t = C$$

or

$$n \log I + \log t = \log C$$

where I is the discharge rate, t the corresponding discharge time, and C is a constant, has been used in this manner to describe the performance of a battery. The value n is the slope of the straight line. The curves are linear on a log-log plot of discharge load versus discharge time, but taper off at both ends because of the battery's inability to handle very high rates and the effect of self-discharge at the lower discharge rates. A more detailed explanation of the use of these graphs in a specific example is presented in Fig. 14.13. Other mathematical relationships have been developed to describe battery performance and account for the nonlinearity of the curves.[1]

Other types of graphs are used to show similar data. A Ragone plot shows the specific energy or energy density of a battery system against the specific power or power density on a log-log scale. This type of graph effectively shows the influence of the discharge load (in this case, power) on the energy that can be delivered by a battery. Figure 32.2 shows a Ragone plot of specific power (W/kg) versus specific energy (Wh/kg) on a log-log scale for two primary battery systems. Figure 32.5 shows a Ragone plot for three secondary batteries. In some cases, The Ragone plot is on a log-linear scale.

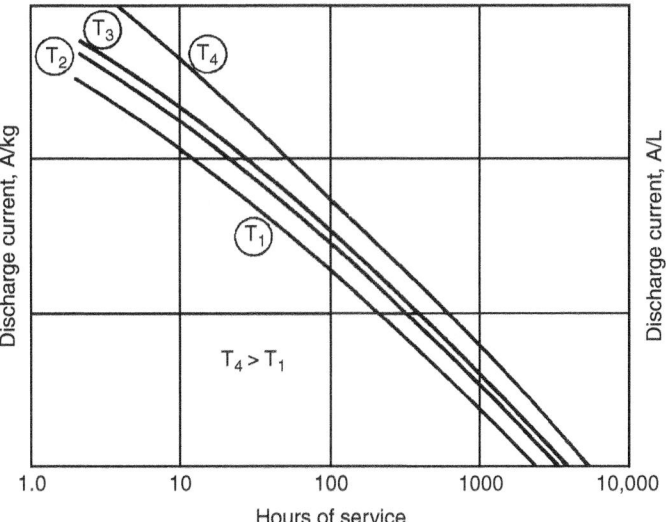

FIGURE 3.11 Battery service life at various discharge loads and temperatures (log-log scale). T_1 to T_4—increasing temperature.

3.12 PRINCIPLES OF OPERATION

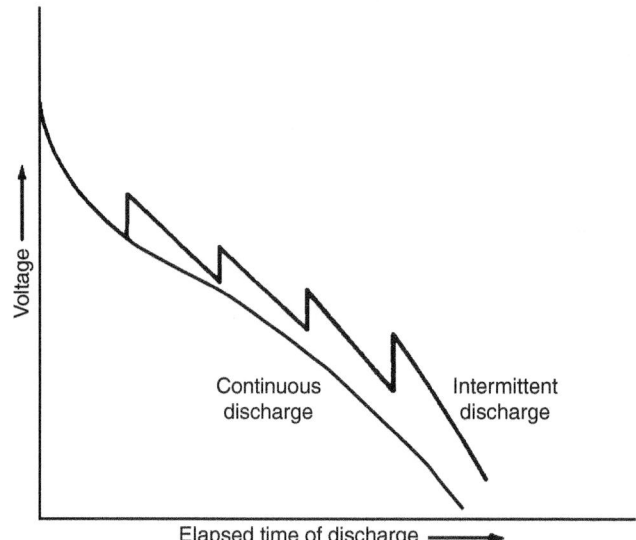

FIGURE 3.12 Effect of intermittent discharge on battery capacity.

3.2.7 Type of Discharge (Continuous, Intermittent, etc.)

When a battery stands idle after a discharge, certain chemical and physical changes take place which can result in a recovery of the battery voltage. Thus the voltage of a battery, which has dropped during a heavy discharge, will rise after a rest period, giving a sawtooth-shaped discharge, as illustrated in Fig. 3.12. This can result in an increase in service life. However, on lengthy discharges, capacity losses may occur due to self-discharge (see Sec. 3.2.12). This improvement, resulting from the intermittent discharge, is generally greater after the higher current drains (as the battery has the opportunity to recover from polarization effects that are more pronounced at the heavier loads). In addition to current drain, the extent of recovery is dependent on many other factors such as the particular battery system and constructional features, discharge temperature, end voltage, and length of recovery period.

The interactive effect on capacity due to the discharge load and the extent of intermittency is shown in Fig. 3.12. It can be seen that the performance of a battery as a function of duty cycle can be significantly different at low and high discharge rates. Similarly, the performance as a function of discharge rate can be different depending on the duty cycle.

3.2.8 Duty Cycles (Intermittent and Pulse Discharges)

Another consideration is the response of the battery voltage when the discharge current is changed during the discharge, such as changing loads from receive to transmit in the operation of a radio transceiver. Figure 3.13 illustrates a typical discharge of a radio transceiver, discharging at a lower current during the receive mode and at a higher current during the transmit mode. Note that the service life of the battery is determined when the cutoff or end voltage is reached under the higher discharge load. The average current cannot be used to determine the service life. Operating at two or more discharge loads is typical of certain electronic equipment because of the different functions they must perform during use.

Another example is a higher-rate periodic pulse requirement against a lower background current, such as backlighting for an LCD watch application, the audible trouble signal pulse in the operation

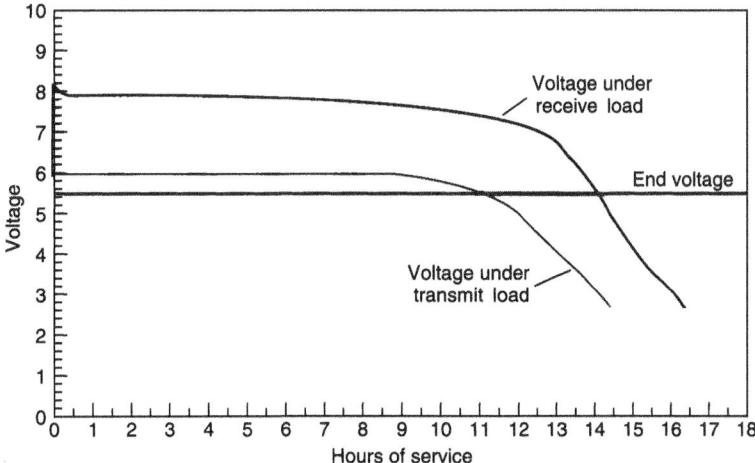

FIGURE 3.13 Typical discharge characteristics of a battery cycling between transmit and receive loads.

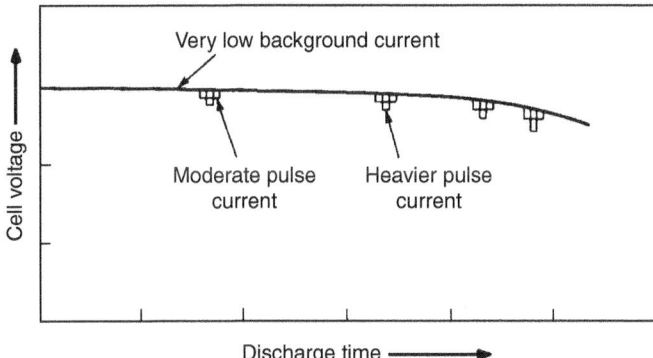

FIGURE 3.14 Typical discharge characteristics of a battery subjected to a periodic high-rate pulse.

of a smoke detector, or a high-rate pulse during the use of a cell phone or computer. A typical pulse discharge is plotted in Fig. 3.14. The extent of the drop in voltage depends on the battery design. The drop in voltage for a battery with lower internal resistance and better response to changes in load current will be less than one with higher internal resistance. In Fig. 3.14, note that the voltage spread widens as the battery is discharged due to the increase in internal resistance as the battery is discharged.

The shape of the pulse can vary significantly depending on the characteristics of the pulse and the battery. Figure 3.15 shows the characteristics of 9 V primary batteries subjected to the 100 ms audible trouble signal pulse in a smoke detector. Curve A shows the response of a zinc-carbon battery, the voltage dropping sharply initially and then recovering. Curves B and C are typical of the response of a zinc/alkaline/manganese dioxide battery, the voltage initially falling and either maintaining the lower voltage or dropping slowly as the pulse discharge continues.

The type of response shown in Fig. 3.15*a* is also typical of batteries that have developed a protective or passivating film on an electrode, the voltage recovering as the film is broken during the discharge (see Sec. 3.2.12 on voltage delay). The specific characteristics, however, are dependent

3.14 PRINCIPLES OF OPERATION

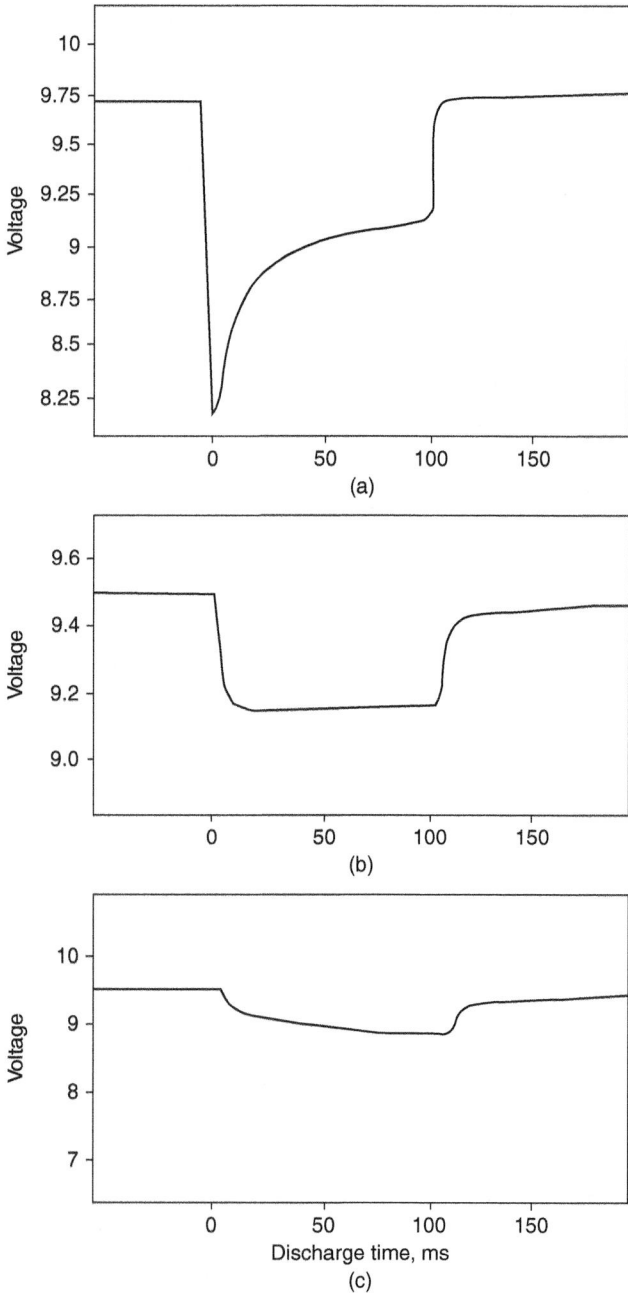

FIGURE 3.15 Discharge characteristics of a 9 V battery subjected to a 100 ms pulse (smoke detector pulse tests): (*a*) zinc-carbon battery; (*b*) and (*c*) zinc/alkaline/ manganese dioxide battery.

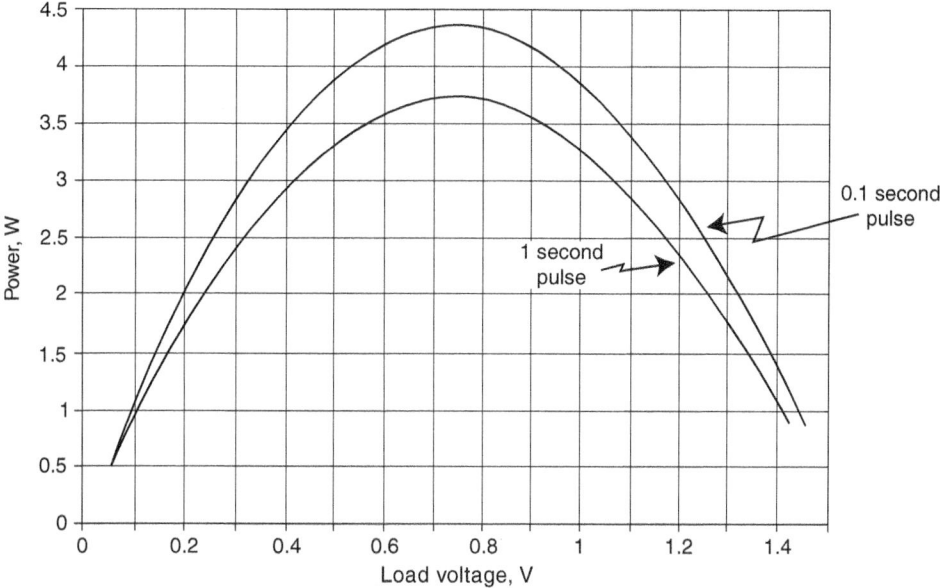

FIGURE 3.16 Power vs. load voltage at the end of a constant voltage pulses (undischarged zinc/alkaline/manganese dioxide AA-size battery. (*From Ref 2.*)

on the battery chemistry, design, state of discharge, and other factors related to the battery's internal resistance at the time of the pulse and during the pulse.

The performance of a battery under pulse conditions can be characterized by plotting the output power of the pulse against the load voltage, measuring the power delivered to the load by the short-term pulse over the range of open circuit to short circuit.[2] Peak power is delivered to the load when the resistance of the external circuit is equal to the internal resistance of the battery. Figure 3.16 is a power versus load voltage plot of the pulse characteristics of an undischarged zinc/alkaline/manganese dioxide battery (AA-size) at the end of constant voltage pulses of 0.1 and 1 s. The lower values of power for the longer pulse are indicative of the drop in voltage as the pulse length increases.

3.2.9 Voltage Regulation

The voltage regulation required by the equipment is most important in influencing the capacity or service life obtainable from a battery. As is apparent from the various discharge curves, design of the equipment to operate to the lowest possible end voltage and widest voltage range results in the highest capacity and longest service life. Similarly, the upper voltage limit of the equipment should be established to take full advantage of the battery characteristics.

Figure 3.17 compares two typical battery discharge curves: curve 1 depicts a battery having a flat discharge curve; curve 2 depicts a battery having a sloping discharge curve. In applications where the equipment cannot tolerate the wide voltage spread and is restricted, for example, to the −15% level, the battery with the flat discharge profile gives the longer service. On the other hand, if the batteries can be discharged to lower cutoff voltages, the service life of the battery with the sloping discharge is extended and could exceed that of the battery with the flat discharge profile.

Discharging multicell series-connected batteries to too low an end voltage, however, may result in safety problems. It is possible in this situation for the poorest cell to be driven into voltage reversal. With some batteries, such as the lithium-sulfur dioxide primary battery, this could result in venting or rupture.

3.16 PRINCIPLES OF OPERATION

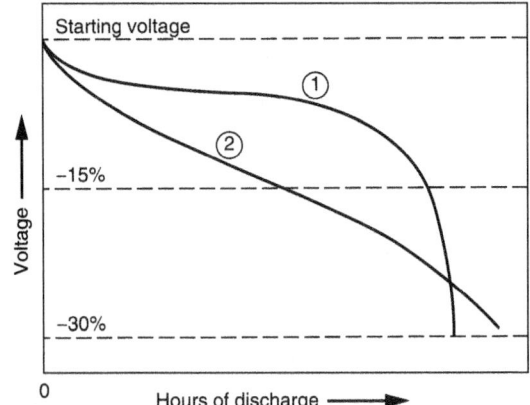

FIGURE 3.17 Comparison of flat ① and sloping ② discharge curves.

In applications where only a narrow voltage range can be tolerated, the selection of the battery may be limited to those having a flat discharge profile. An alternative is to use a voltage regulator to convert the varying output voltage of the battery into a constant output voltage consistent with the equipment requirements. In this way, the full capacity of the battery can be used, with inefficiency of the voltage regulator the only energy penalty. Figure 3.18 illustrates the voltage and current profiles of the battery and regulator outputs. The input from the battery to the regulator is at a constant power of 1 W, with the current increasing as the battery voltage drops. With an 84% conversion efficiency, the output from the regulator is constant at a predetermined 6 V and 140 mA (constant power = 840 mW).

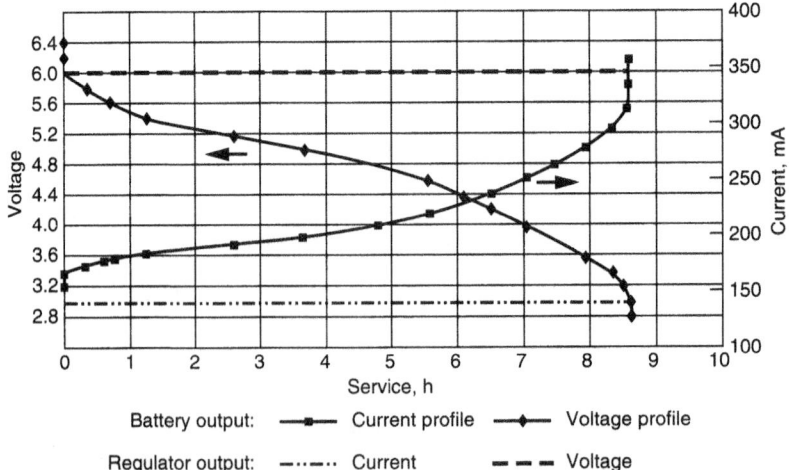

FIGURE 3.18 Characteristics of a voltage regulator. Battery output—1 W; regular output—840 mW.

3.2.10 Charging Voltage

If a rechargeable battery is used (for example, as a standby power source) in conjunction with another energy source that is permanently connected in the operating circuit, allowance must be made for the battery and equipment to tolerate the voltage of the battery on charge. Figure 3.19 shows the charge and discharge characteristics of such a battery. The specific voltage and the voltage profile on charge depend on such factors as the battery system, charge rate, temperature, and so on.

If a primary battery is used in a similar circuit (for example, as memory backup battery), it is usually advisable to protect the primary battery from being charged by including an isolating or protective diode in the circuit, as shown in Fig. 3.20. Two diodes provide redundancy in case one fails. The resistor in Fig. 3.20*b* serves to limit the charging current in case the diode fails.

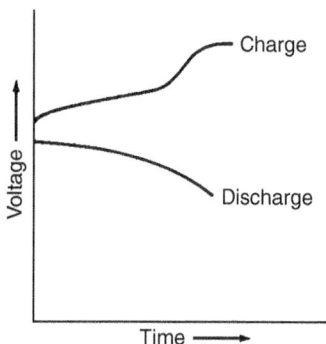

FIGURE 3.19 Typical voltage profile on charge and discharge.

The charging source must also be designed so that its output current is regulated during the charge to provide the needed charge control for the battery.

3.2.11 Effect of Cell and Battery Design

The constructional features of the cell and battery strongly influence their performance characteristics.

Electrode Design. Cells that are designed, for example, for optimum service life or capacity at relatively low or moderate discharge loads contain maximum quantities of active material. On the other extreme, cells capable of high-rate performance are designed with large electrode or reaction surfaces and features to minimize internal resistance and enhance current density (amperes per area of electrode surface), often at the expense of capacity or service life.

For example, two designs are used in cylindrical cells. One design, known as the bobbin construction, is typical for zinc-carbon and some alkaline-manganese dioxide cells. Here the electrodes are shaped into two concentric cylinders (Fig. 3.21*a*). This design maximizes the amount of active material that can be placed into the cylindrical can, but at the expense of surface area for the electrochemical reaction.

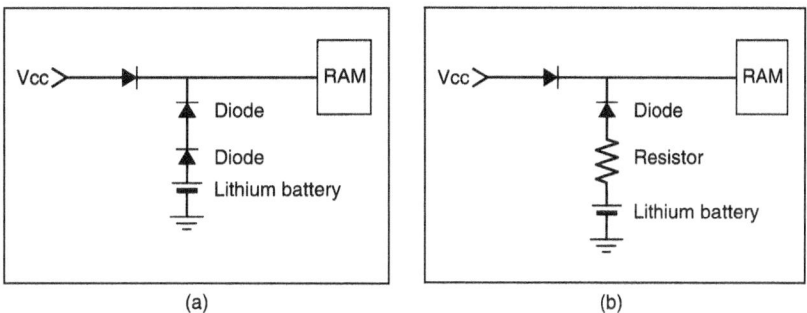

FIGURE 3.20 Protective circuits for memory backup applications. (*a*) Using two diodes. (*b*) Using diode and resistor.

3.18 PRINCIPLES OF OPERATION

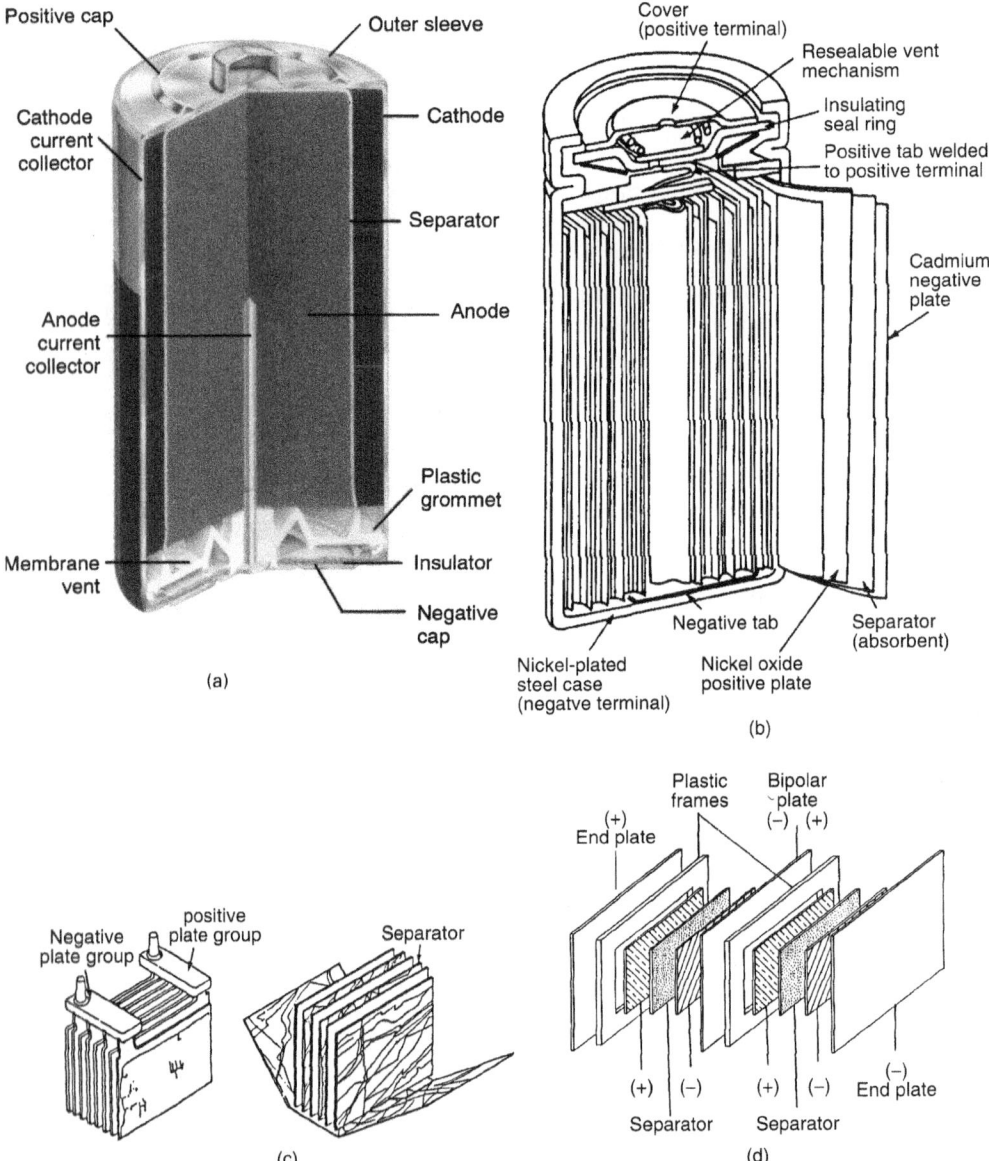

FIGURE 3.21 Cell design; typical internal configurations. (*a*) Bobbin construction. (*b*) Spiral wound construction. (*c*) Flat-plate construction. (*d*) Bipolar-plate construction.

The second design is the "spiral wound" electrode construction, typically used in sealed portable rechargeable batteries and high-rate primary and rechargeable lithium batteries (Fig. 3.21*b*). In this design, the electrodes are prepared as thin strips and then rolled, with a separator in between, into a "jelly roll" and placed into the cylindrical can. This design emphasizes surface area to enhance high-rate performance, but at the expense of active material and capacity. Lithium/sulfur dioxide primary cells employ this design.

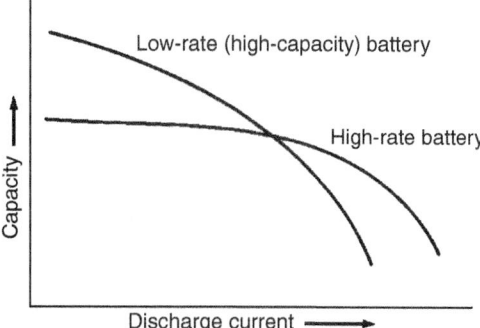

FIGURE 3.22 Comparison of performance of batteries designed for high- and low-rate service.

Another popular electrode design in the flat-plate construction, typically used in the lead-acid SLI (starting-lighting-ignition) and most larger storage batteries (Fig. 3.21c). This construction also provides a large surface area for the electrochemical reaction. As with the other designs, the manufacturer can control the relationship between surface area and active material (for example, by controlling the plate thickness) to obtain the desired performance characteristics.

A modification of this design is the bipolar plate illustrated in Fig. 3.21d. Here the anode and cathode are fabricated as layers on opposite sides of an electronically conductive but ion-impermeable material which serves as the intercell connector.

Most battery chemistries can be adapted to the different electrode designs, and some, in fact, are manufactured in different configurations. Manufacturers choose chemistries and designs to optimize the performance for the particular applications and markets in which they are interested.

In Fig. 3.22, the performance of a battery designed for high-rate performance is compared with one using the same electrochemical system, but optimized for capacity. The high-rate batteries have a lower capacity but deliver a more constant performance as the discharge rate increases.

Hybrid Designs. Hybrid designs, which combine a high energy power source with a high-rate power source, are becoming popular. These hybrid systems fulfill applications more effectively (e.g., higher total specific energy or energy density) than using a single power source. The high energy power source is the basic source of energy, but also charges a high-rate battery which handles any peak power requirement that cannot be handled efficiently by the main power source. Hybrid designs are being considered in many applications, ranging from combining a high-energy, low-rate metal/air battery or fuel cell with a high-rate rechargeable battery, such as a lithium-ion system. Hybrid electric vehicles use an efficient combustion engine with a rechargeable battery to handle starting, acceleration, and other peak power demands, and use regenerative braking to recharge the battery.

Shape and Configuration. The shape or configuration of the cell will also influence the battery capacity as it affects such factors as internal resistance and heat dissipation. For example, a tall, narrow-shaped cylindrical cell in a bobbin design will generally have a lower internal resistance than a wide, squat-shaped one of the same design and may outperform it, in proportion to its volume, particularly at the higher discharge rates. For example, a thin AA-size bobbin type cell will have proportionally better high-rate performance than a wider diameter D-size cell. Heat dissipation also will be better from cells with a high surface-to-volume ratio or with internal components that can conduct heat to the outside of the battery.

Volumetric Efficiency versus Energy Density. The size and shape of the cell or battery and the ability to effectively use its internal volume influence the energy output of the cell. The volumetric energy density (watthours per liter) decreases with decreasing battery volume as the percentage of

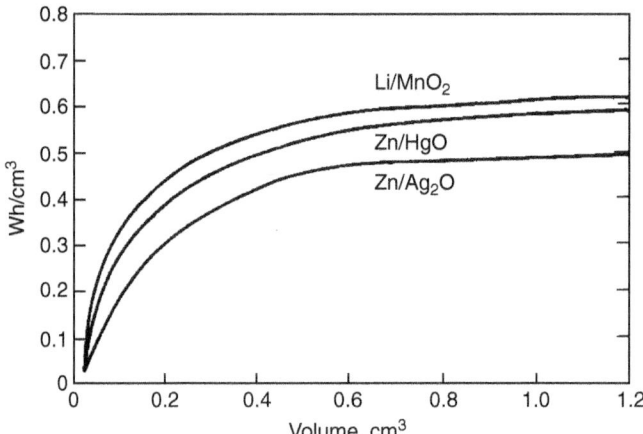

FIGURE 3.23 Energy density, in watthours per cubic centimeter, of button batteries as a function of cell volume. (See Ref. 3.)

"dead volume" for containers, seals, and so on increases for the smaller batteries. This relationship is illustrated for several button-type cells in Fig. 3.23. The shape of the cell (such as wide or narrow diameter) may also influence the volumetric efficiency as it relates to the amount of space lost for the seal and other cell construction materials.

Effect of Size on Capacity. The size of the battery influences the voltage characteristics by its effect on current density. A given current drain may be a severe load on a small battery, giving a discharge profile similar to curve 4 or 5 in Fig. 3.3, but it may be a mild load on a larger battery with a discharge curve similar to curve 2 or 3. Often it is possible to obtain more than a proportional increase in the service life by increasing the size of the battery (or paralleling cells) as the current density is lowered. The absolute value of the discharge current, therefore, is not the key influence, although its relation to the size of the battery—that is, the current density—is significant.

In this connection, the alternative of using a series-connected multicell battery versus a lower voltage battery, with fewer but larger cells and a voltage converter to obtain the required high voltage, should be considered. An important factor is the relative advantage of the potentially more efficient larger battery versus the energy losses of the voltage converter. In addition, the reliability of the system is enhanced by the use of a smaller number of cells. However, all pertinent factors must be considered in this decision because of the influences of cell and battery design, configuration, and so on, as well as the equipment power requirements.

3.2.12 Battery Age and Storage Condition

Batteries are a perishable product and deteriorate as a result of the chemical action that proceeds during storage. The design, electrochemical system, temperature, and length of storage period are factors that affect the shelf life or charge retention of the battery. The type of discharge following the storage period will also influence the shelf life of the battery. Usually the percentage charge retention following storage (comparing performance after and before storage) will be lower the more stringent the discharge condition. The self-discharge characteristics of several battery systems at various temperatures are shown in Fig. 32.11 as well as in the chapters on specific battery chemistries. As self-discharge proceeds at a lower rate at reduced temperatures, low-temperature storage extends the shelf life and is recommended for some battery systems. Batteries should be warmed before discharge to obtain maximum performance.

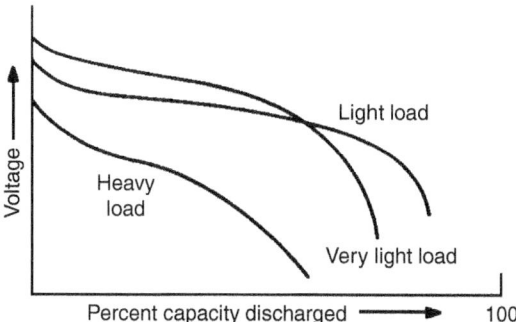

FIGURE 3.24 Effect of self-discharge on battery capacity.

Self-discharge can also become a factor during discharge, particularly on long-term discharges, and can cause a reduction in capacity. This effect is illustrated in Figs. 3.10 and 3.24. More capacity will be delivered on a discharge at a light load than on a heavy load. However, on an extremely light load over a long discharge period, capacity may be reduced due to self-discharge.

Some battery systems will develop protective or passivating films on one or both electrode surfaces during storage. These films can improve the shelf life of the battery substantially. However, when the battery is placed on discharge after storage, the initial voltage may be low due to the impedance characteristics of the film until it is broken down or depassivated by the electrochemical reaction. This phenomenon is known as "voltage delay" and is illustrated in Fig. 3.25. The extent of the voltage delay is dependent on and increases with increasing storage time and storage temperature. The delay also increases with increasing discharge current and decreasing discharge temperature.

The self-discharge characteristics of a battery that has been or is being discharged can be different from one that has been stored without having been discharged. This is due to a number of factors, such as the discharge rate and temperature, the accumulation of discharge products, the depth of discharge, or the partial destruction or reformation of the protective film. Some batteries, such as the magnesium primary battery (Chapter 10), may lose their good shelf-life qualities after being discharged because of the destruction of the protective film during discharge. Knowledge of the battery's storage and discharge history is needed to predict the battery's performance under these conditions.

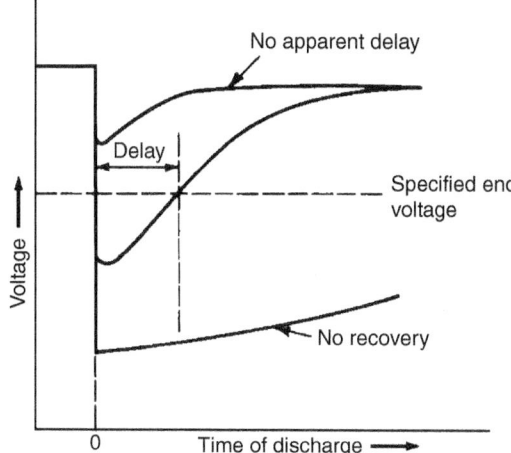

FIGURE 3.25 Voltage delay.

3.2.13 Effect of Battery Design

The performance of the cells in a multicell battery will usually be different than the performance of the individual cells. The cells cannot be manufactured identically and, although cells are selected to be "balanced," they each encounter a somewhat different environment in the battery pack.

The specific design of the multicell battery and the hardware that is used (such as packaging techniques, spacing between the cells, container material, insulation, potting compound, fuses and other electronic controls, etc.) will influence the performance as they effect the environment and temperature of the individual cells. Obviously, the battery materials add to its size and weight, and the specific energy or energy density of the battery will be lower than that of the component cells. Accordingly, when comparing values such as specific energy, in addition to being aware of the conditions (discharge rate, temperature, etc.) under which these values were determined, it should be ascertained whether the values given are for cells, single-cell batteries, or multicell batteries. Usually, as is the case in this Handbook, they are on the basis of a single-cell battery unless specified otherwise.

Battery designs that retain the heat dissipated by the cells can improve the performance at low temperatures. On the other hand, excessive buildup of heat can be injurious to the battery's performance, life, and safety. As much as possible, batteries should be thermally designed to maintain a uniform internal temperature and avoid "hot spots."

In the case of rechargeable batteries, cycling could cause the individual cells in a battery peak to become unbalanced and their voltage, capacity, or other characteristics could become significantly different. This could result in poor performance or safety problems, and end-of-charge or discharge control may be necessary to prevent this. Cell balancing techniques are employed with some systems, such as lithium-ion batteries.

The influence of battery design and recommendations for effective battery design are covered in Chap. 5.

Several recent papers review the current status and future prospects for improvements in battery performance.[4,5]

REFERENCES

1. R. Selim and P. Bro, "Performance Domain Analysis of Primary Batteries," *Electrochemical Technology, J. Electrochem. Soc.* **118**(5):829 (1971).
2. D. I. Pomerantz, "The Characterization of High Rate Batteries," *IEEE Transactions Electronics,* **36**(4):954 (1990).
3. P. Ruetschi, "Alkaline Electrolyte-Lithium Miniature Primary Batteries," *J. Power Sources,* **7**(1982).
4. M. Winter and B. Brodd, *Chem Revs.* **104**:4245–4270 (2004).
5. D. Linden and T. B. Reddy, *Battery Power and Products Technology,* **5**(2) (March/April 2008).

CHAPTER 4
BATTERY STANDARDIZATION

Steven Wicelinski

4.1 GENERAL

The standardization of batteries started in 1912, when a committee of the American Electrochemical Society recommended standard methods of testing dry cells. This eventually led to the first national publication in 1919, issued as an appendix to a circular from the National Bureau of Standards. It further evolved into the present American National Standards Institute (ANSI) Accredited Standards Committee C18 on Portable Cells and Batteries. Since then, other professional societies have developed battery-related standards. Many battery standards were also issued by international, national, military, and federal organizations. Manufacturers' associations, trade associations, and individual manufacturers have published standards as well. Related application standards published by the Underwriters Laboratories, the International Electrotechnical Commission, and other organizations that cover battery-operated equipment may also be of interest.

Tables 4.1a to d list some of the widely known standards for batteries. Standards covering the safety and regulation of batteries are listed in Table 4.11.

4.2 INTERNATIONAL STANDARDS

International standards are rapidly gaining in importance. This has been further accelerated by the creation of the European Common Market and the 1979 Agreement on Technical Barriers to Trade. The latter requires the use of international standards for world trade purposes.

The International Electrotechnical Commission (IEC) is the designated organization responsible for standardization in the fields of electricity, electronics, and related technologies. Promoting international cooperation on all questions of electrotechnical standardization and related matters is its basic mission. This organization was founded in 1906 and consists of over 70 national committees that represent more than 80% of the world's population and 95% of the world's production and consumption of electricity. The International Standards Organization (ISO) is responsible for international standards in fields other than electrical. IEC and ISO are gradually adopting equivalent development and documentation procedures while ever closer ties are being established between these two international organizations.

The American National Standards Institute (ANSI) is the sole U.S. representative of the IEC through the United States National Committee (USNC). This committee coordinates all IEC activities in the United States. It also serves as the U.S. interface with emerging regional standards-developing bodies such as CENELEC, PASC, CANENA, COPANT, ARSO, and other foreign and national groups. ANSI does not itself develop standards; rather it facilitates development by establishing consensus among accredited, qualified groups. These standards are published as U.S. National Standards (see Table 4.1b).

4.2 PRINCIPLES OF OPERATION

To further its overall mission, the objectives of the IEC are to:

1. Efficiently meet the requirements of the global marketplace
2. Ensure maximum use of its standards and conformity assessment schemes
3. Assess and improve the quality of products and services covered by its standards
4. Establish conditions for interchangeability
5. Increase the efficiency of electrotechnical industrial processes
6. Contribute to the improvement of human health and safety
7. Work towards protection of the environment

TABLE 4.1a International Standards (IEC—International Electrotechnical Commission)

Publication	Title	Electrochemical systems
IEC 60086-1, IEC 60086-2	Primary Batteries; Part 1, General, and Part 2, Specification Sheets	Zinc-carbon Zinc/air Alkaline-manganese dioxide Nickel oxyhydroxide Silver oxide Lithium/carbon monofluoride Lithium/manganese dioxide Lithium/chromium oxide Lithium/thionyl chloride
IEC 60086-3	Watch Batteries	
IEC 60095	Lead-Acid Starter Batteries	Lead-acid
IEC 60254	Lead-Acid Traction Batteries	Lead-acid
IEC 61951-1	Portable sealed rechargeable single cells; Part I: Nickel-cadmium	Nickel-cadmium
IEC 61960	Secondary lithium cells and batteries for portable applications	Lithium-ion
IEC 60622	Sealed Nickel-Cadmium Prismatic Rechargeable Single Cells	Nickel-cadmium
IEC 60623	Vented Nickel-Cadmium Prismatic Rechargeable Cells	Nickel-cadmium
IEC 60952	Aircraft Batteries	Nickel-cadmium Lead-acid
IEC 60896	Stationary Lead-Acid Batteries	Lead-acid
IEC 61056	General purpose lead-acid batteries	Lead-acid
IEC 61427	Secondary cells and batteries for photovoltaic energy systems	
IEC 61951-2	Portable sealed rechargeable single cells; Part 2: Nickel-metal hydroxide	Nickel-metal hydride
IEC 61959	Mechanical tests for sealed portable secondary cells and batteries	

Note: See Table 4.11a for IEC Safety Standards.

TABLE 4.1b National Standards (ANSI—American National Standards Institute)

Publication	Title	Electrochemical systems
ANSI C18.1M, Part 1	Standard for Portable Primary Cells and Batteries with Aqueous Electrolyte	Zinc-carbon Alkaline-manganese dioxide Silver oxide Zinc/air
ANSI C18.2M, Part 1	Standard for Portable Rechargeable Cells and Batteries	Nickel-cadmium Nickel-metal hydride Lithium-ion
ANSI C18.3M, Part 1	Standard for Portable Lithium Primary Cells and Batteries	Lithium/carbon monofluoride Lithium/manganese dioxide

Note: See Table 4.11a for ANSI Safety Standards.

TABLE 4.1c U.S. Military Standards (MIL)

Publication	Title	Electrochemical systems
MIL-B-18	Batteries Non-Rechargeable	Zinc-carbon
MIL-B-8565	Aircraft Batteries	Various
MIL-B-11188	Vehicle Batteries	Lead-acid
MIL-B-49030	Batteries, Dry, Alkaline (Non-Rechargeable)	Alkaline-manganese dioxide
MIL-B-55252	Batteries, Magnesium	Magnesium
MIL-B-49436	Batteries, Rechargeable, Sealed Nickel-Cadium	Nickel-cadmium
MIL-B-49450	Vented Aircraft Batteries	Nickel-cadmium
MIL-B-49458	Batteries, Non-Rechargeable	Lithium/manganese dioxide
MIL-B-49461	Batteries, Non-Rechargeable	Lithium/thionyl chloride
MIL-B-55130	Batteries, Rechargeable, Sealed Nickel-Cadmium	Nickel-cadmium
MIL-B-81757	Aircraft Batteries	Nickel-cadmium
MIL-PRF-49471	Batteries, Non-Rechargeable, High Performance	Various

TABLE 4.1d Manufacturers' and Professional Associations

Publication	Title	Battery type covered
Society of Automotive Engineers		
SAE AS 8033	Aircraft Batteries	Nickel-cadmium
SAE J 537	Storage Batteries	Lead-acid
Battery Council International	Battery Replacement Data Book	Lead-acid

The objectives of the international battery standards are to:

1. Define a standard of quality and provide guidance for its assessment
2. Ensure the electrical and physical interchangeability of products from different manufacturers
3. Limit the number of battery types
4. Provide guidance on matters of safety

The IEC sponsors the development and publication of standard documents. This development is carried out by working groups of experts from participating countries. These experts represent consumer, user, producer, academia, government, and trade and professional interests in the consensus development of these standards. The Groups of Experts in IEC working on battery standards are:

TC 21: Rechargeable Batteries

TC 35: Primary Batteries

The designation for the ANSI Committee on Portable Cells and Batteries is C18.

Table 4.1a lists the IEC standards that pertain to primary and secondary batteries. Many countries utilize these standards either by simply adopting them in toto as their national standards or by harmonizing their national standards to the IEC standards. Table 4.1b lists the ANSI battery standards. When feasible, the two groups harmonize the requirements in their standards.

4.3 CONCEPTS OF STANDARDIZATION

The objective of battery interchangeability is achieved by specifying the preferred values for the physical aspects of the battery, such as dimensions, polarity, terminals, nomenclature, and markings. In addition, performance characteristics, such as service life or capacity, may be described and specified with test conditions for verification.

4.4 PRINCIPLES OF OPERATION

It is the inherent nature of batteries, in particular primary batteries, that replacements will at some time be required. A third-party end-user of the equipment typically replaces the battery. It is therefore essential that certain characteristics of the battery be specified by standard values—size, shape, voltage, and terminals. Without a reasonable match of at least these parameters there can be no interchangeability. These characteristics are absolute requirements in order to fit the appliance receptacle, to make proper contact, and to provide the proper voltage. In addition to the end-user's need for replacement information, the original equipment manufacturer (OEM) appliance designer must have a reliable source of information about these parameters in order to design a battery compartment and circuits that will accommodate the tolerances on battery products available for purchase by the end-user.

4.4 IEC AND ANSI NOMENCLATURE SYSTEMS

It is unfortunate that the various standards identified in Table 4.1 do not share the same nomenclature system. The independent nomenclature systems of the various battery manufacturers even worsen this situation. Cross-references, however, are generally available from battery manufacturers.

4.4.1 Primary Batteries

The IEC nomenclature system for primary batteries, which became effective in 1992, is based on the electrochemical system and the shape and size of the battery. The letter designations for the electrochemical system and the type of cell remain the same as in the previous IEC system for primary batteries. The new numerical designations are based on a diameter/ height number instead of the arbitrary size classification used previously. The first digits specify the diameter of the cell in millimeters and the second the height of the cell (millimeters times 10). An example is shown in Table 4.2a. The codes for the shape and electrochemical system are given in Tables 4.2b and 4.2c,

TABLE 4.2a IEC Nomenclature System for Primary Batteries, Example

Nomenclature	Number of cells	System letter (Table 4.2c)	Shape (Table 4.2b)	Diameter, mm	Height, mm	Example
CR2025	1	C	R	20	2.5	A unit round battery having dimensions shown and electrochemical system letter C of Table 4.2c (Li /MnO$_2$)

TABLE 4.2b IEC Nomenclature for Shape, Primary Batteries

Letter designation	Shape
R	Round-Cylindrical
P	Non-Round
F	Flat (layer built)
S	Square (or Rectangular)

TABLE 4.2c Letter Codes Denoting Electrochemical System of Primary Batteries

ANSI	IEC	Negative electrode	Electrolyte	Positive electrode	Nominal voltage (V)
*	—	Zinc	Ammonium chloride, Zinc chloride	Manganese dioxide	1.5
	A	Zinc	Ammonium chloride, Zinc chloride	Oxygen (air)	1.4
LB	B	Lithium	Organic	Carbon monofluoride	3
LC	C	Lithium	Organic	Manganese dioxide	3
	E	Lithium	Non-aqueous inorganic	Thionyl chloride	3.6
LF	F	Lithium	Organic	Iron sulfide	1.5
	G	Lithium	Organic	Copper dioxide	1.5
A[†]	L	Zinc	Alkali metal hydroxide	Manganese dioxide	1.5
Z[‡]	P	Zinc	Alkali metal hydroxide	Oxygen (air)	1.4
SO[§]	S	Zinc	Alkali metal hydroxide	Silver oxide	1.55

Notes:
* No suffix — Carbon-zinc
  C — Carbon-zinc industrial
  CD — Carbon-zinc industrial, heavy duty
  D — Carbon-zinc, heavy duty
  F — Carbon-zinc, general purpose
[†] A — Alkaline
  AC — Alkaline industrial
[‡] Z — Zinc/air
  ZD — Zinc/air, heavy duty
[§] SO — Silver oxide
  N — Nickel oxyhydroxide

respectively. For reference, the ANSI letter codes for the electrochemical systems are also listed in Table 4.2c. The ANSI nomenclature system does not use a code to designate shape.

Nomenclature for existing batteries was grandfathered. Examples of the nomenclature for some of these primary cells and batteries are shown in Table 4.3a. Examples of the IEC nomenclature system for primary batteries are shown in Table 4.3b.

TABLE 4.3a IEC Nomenclature for Typical Primary Round, Flat, and Square Cells or Batteries*

IEC designation	Nominal battery dimensions, mm					ANSI designation	Common designation
	Diameter	Height	Length	Width	Thickness		
Round batteries							
R03	10.5	44.5				24	AAA
R1	12.0	30.2				—	N
R6	14.5	50.5				15	AA
R14	26.2	50.0				14	C
R20	34.2	61.5				13	D
R25	32.0	91.0				—	F
Flat cells							
F22			24	13.5	6.0		
Square batteries							
S4			125.0	57.0	57.0		

*Chart is not complete—only a sampling of sizes is shown. Dimensions are used for identification only: Complete dimensions can be found in the relevant specification sheets listed in IEC 60086-2.

TABLE 4.3b Examples of IEC Nomenclature for Primary Batteries

IEC Nomenclature	Number of cells	System letter (Table 4.2c)	Shape (Table 4.2b)	Cell (Table 4.3a)	C, P, S, X, Y	Parallel	Groups in parallel	Example
R20	1	None	R	20	*			A unit round battery using basic R20 type cell and electrochemical system letter (none) of Table 4.2c
LR20	1	L	R	20	*			Same as above, except using electrochemical system letter L of Table 4.2c
6F22	6	None	F	22	*			A 6-series multicell battery using flat F22 cells and electrochemical system letter (none) of Table 4.2c
4LR25-2	4	L	R	25	*		2	A multicell battery consisting of two parallel groups, each group having four cells in series of the R25 type and electrochemical system letter L of Table 4.2c
CR17345	1	C	R	See Section 4.4.1				A unit round battery, with a diam. of 17 mm and height of 34.5 mm, and electrochemical system letter C of Table 4.2c

*If required, letters C, P, or S will indicate different performance characteristics and letters X and Y different terminal arrangements.

4.4.2 Rechargeable Batteries

The documentation for standardization of rechargeable batteries is not as complete as the documentation for primary batteries. Most of the primary batteries are used in a variety of portable applications, using user-replaceable batteries. Hence, the need for primary battery standards to ensure interchangeability. Developing such standards have been active projects by both IEC and ANSI for many years.

The early use of rechargeable batteries was mainly with larger batteries, however, usually application specific and multicell. The large majority of rechargeable batteries were lead-acid manufactured for automotive SLI (starting, lighting, ignition) use. Standards for these batteries were developed by the Society for Automotive Engineers (SAE), the Battery Council International (BCI), and the Storage Battery Association of Japan. More recently, rechargeable batteries have been developed for portable applications, in many cases in the same cell and battery sizes as the primary batteries. Starting with the portable-sized nickel-cadmium batteries, IEC and ANSI are developing standards for the nickel-metal hydride and lithium-ion batteries. The currently available standards are listed in Tables 4.1a and 4.1b.

Table 4.4a lists the letter codes that are being used by IEC and those adopted by ANSI for secondary or rechargeable batteries. The IEC nomenclature system for nickel-metal hydride batteries is shown in Table. 4.4b. In this system, the first letter designates the electrochemical system, a second

TABLE 4.4a Letter(s) Denoting Electrochemical System of Secondary Batteries

ANSI	IEC*	Negative electrode	Electrolyte	Positive electrode	Nominal voltage (V)
H	H	Hydrogen absorbing alloy	Alkali metal hydroxide	Nickel oxide	1.2
K	K	Cadmium	Alkali metal hydroxide	Nickel oxide	1.2
P	PB	Lead	Sulfuric acid	Lead dioxide	2
I	IC	Carbon	Organic	Lithium cobalt oxide	3.6
I	IN	Carbon	Organic	Lithium nickel oxide	3.6
I	IM	Carbon	Organic	Lithium manganese oxide	3.6

*Proposed for portable batteries.

TABLE 4.4b IEC Nomenclature System for Rechargeable Nickel-Metal Hydride Cells and Batteries

Nomenclature*	System letter (Table 4.4a)	Shape (Table 4.2b)	Diameter, mm	Height, mm	Terminals	Example
HR 15/51 (R6)	H	R	14.5	50.5	CF	A unit round battery of the H system having dimensions shown, with no connecting tabs

*Nomenclature dimensions are shown rounded off. ( ) indicates interchangeable with a primary battery.
Source: IEC 61951-2.

letter the shape, the first number the diameter, and a second number the height. In addition, the letters L, M, and H may be used to classify arbitrarily the rate capability as low, medium, or high. The last part of the designation is reserved for two letters that indicate various tab terminal arrangements, such as CF—none, HH—terminal at positive end and positive sidewall, or HB—terminals at positive and negative ends, as shown in Table 4.4b.

4.5 TERMINALS

Terminals are another aspect of the shape characteristics for cells and batteries. It is obvious that without standardization of terminals and the other shape variables, a battery may not be available to match the receptacle facilities provided in the appliance. Some of the variety of terminal arrangements for batteries are listed in Table 4.5.

TABLE 4.5 Terminal Arrangements for Batteries

Cap and base	Terminals that have the cylindrical side of the battery insulated from the terminal ends
Cap and case	Terminals in which the cylindrical side forms part of the positive end terminal
Screw types	Terminals that have a threaded rod and accept either an insulated or a metal nut
Flat contacts	Flat metal surfaces used for electrical contact
Springs	Terminals that are flat metal strips or spirally wound wire
Plug-in sockets	Terminals consisting of a stud (nonresilient) and a socket (resilient)
Wire	Single or multistranded wire leads
Spring clips	Metal clips that will accept a wire lead
Tabs	Metal flat tabs attached to battery terminals

4.8 PRINCIPLES OF OPERATION

When applicable, the terminal arrangement is specified in the standard within the same nomenclature designators used for shape and size. The designators thus determine all interchangeable physical aspects of the cells and batteries in addition to the voltage.

4.6 ELECTRICAL PERFORMANCE

In terms of the requirement to provide fit and function in the end product, the actual appliance does not require specific values of electrical performance. The correct battery voltage, needed to protect the appliance from overvoltage, is assured by the battery designation. Batteries of the same voltage but having differences in capacity can be used interchangeably, but will operate for different service times. The minimum electrical performance of the battery is therefore cited and specified in the standards either by application or by capacity testing.

1. *Application tests.* This is the preferred method of testing the performance specified for primary batteries. Application tests are intended to simulate the actual use of a battery in a specific application. Table 4.6*a* illustrates typical application tests.
2. *Capacity (service output) tests.* A capacity test is generally used to determine the quantity of electric charge a battery can deliver under specified discharge conditions. This method is the one that has been generally used for rechargeable batteries. It is also used for primary batteries when an application test would be too complex to simulate realistically or too lengthy to be practical for routine testing. Table 4.6*b* lists some examples of capacity tests.

TABLE 4.6a Example of Application Tests for R20 Type Batteries

Nomenclature				R20P	R20S	LR20
Electrochemical system				Zinc-carbon (high power)	Zinc-carbon (standard)	Zinc/manganese dioxide
Nominal voltage				1.5	1.5	1.5
Application	Load, Ω	Daily period	End point	Minimum average duration†		
Portable lighting (1)	2.2	*	0.9	320 min	100 min	810 min
Tape recorders	3.9	1 h	0.9	11 h	4 h	11 h
Radios	10	4 h	0.9	32 h	18 h	81 h
Toys	2.2	1 h	0.8	5 h	2 h	15 h
Portable lighting (2)	1.5	**	0.9	135 min	32 min	450 min

*4 min beginning at hourly intervals for 8 h/day; **4 min/15 min, 8 h/day.
†For LR20: portable stereo test.

TABLE 4.6b Example of Capacity Tests

Nomenclature					SR54
Electrochemical system		(Refer to Table 4.3*b*)			S
Nominal voltage		(Refer to Table 4.2*c*)			1.55
		(Refer to Table 4.2*c*)			
Application*	Load, kΩ	Daily period	End point	Minimum average duration	
Capacity (rating) test	15	24 h	1.2	580 h	

*Application for this battery is watches. As an application test could take up to 2 years to test, a capacity test is specified.

TABLE 4.7 Marking Information for Batteries

Marking information	Primary batteries	Primary small batteries	Rechargeable round batteries
Nomenclature	×	×	×
Date of manufacture or code	×	××	×
Polarity	×	×	×
Nominal voltage	×	××	×
Name of manufacturer/supplier	×	××	×
Charge rate/time			×
Rated capacity			×

×—on battery.
××—on battery or package.

Test conditions in the standard must consider and therefore specify the following:

Cell (battery) temperature

Discharge rate (or load resistance)

Discharge termination criteria (typically loaded voltage)

Discharge duty cycle

If rechargeable, charge rate, termination criteria (either time or feedback of cell response) and other conditions of charge

Humidity and other conditions of storage may also be required.

4.7 MARKINGS

Markings on both primary and secondary (rechargeable) batteries may consist of some or all of the printed information given in Table 4.7 in addition to the form and dimension nomenclature discussed.

4.8 CROSS-REFERENCES OF ANSI IEC BATTERY STANDARDS

Table 4.8 lists some of the more popular ANSI battery standards and cross-references to the international standard publications for primary and secondary batteries.

4.9 LISTING OF IEC STANDARD ROUND PRIMARY BATTERIES

The eleventh edition of IEC 60086-2 for primary batteries lists over one hundred types with dimensional, polarity, voltage, and electrochemical requirements. The second edition of IEC 61951-1 for rechargeable nickel-cadmium cells (batteries) lists 25 sizes with diameter and height specified in chart form. Several rechargeable nickel-cadmium and nickel-metal hydride batteries are also packaged to be interchangeable with the popular sizes in the primary replacement market. These have

TABLE 4.8a ANSI/IEC Cross-Reference for Primary Batteries

ANSI	IEC	ANSI	IEC
13A	LR20	1137SO	SR48
13AC	LR20	1138SO	SR54
—	—	1139SO	SR42
13CD	R20C	1158SO	SR58
13D	R20C	1160SO	SR55
—	—	1162SO	SR57
14A	LR14	1163SO	SR59
14AC	LR14	1164SO	SR59
—	—	1165SO	SR57
14CD	R14C	1166A	LR44
14D	R14C	1170SO	SR55
—	—	1175SO	SR60
15A	LR6	1179SO	SR41
15AC	LR6	—	—
15CD	R6C	1406SO	4SR44
15D	R6C	1412A	4LR61
15N	ZR6	1414A	4LR44
24A	LR03	1604	6F22
24AC	LR03	1604A	6LR61
24D	R03	1604AC	6LR61
24N	ZR03	1604C	6F22
908A	4LR25X	1604CD	6F22
910A	LR1	1604D	6F22
918A	4LR25-2	5018LC	CR17345
918D	4R25-2	5024LC	CR-P2
1107SO	SR44	5032LC	2CR5
1131SO	SR44	7000ZD	PR48
1133SO	SR43	7002ZD	PR41
1134SO	SR41	7003ZD	PR44
1135SO	SR41	7005ZD	PR70
1136SO	SR48	—	—

TABLE 4.8b ANSI/IEC Select Cross-References for Rechargeable Batteries

ANSI	IEC
1.2H1	HR03
1.2H2	HR6
1.2H3	HR14
1.2H4	HR20

physical shapes and sizes that are identical to primary batteries and have equivalent voltage outputs under load. These batteries carry, in addition to the rechargeable nomenclature, the equivalent primary cell or battery size designations and therefore must comply with the dimensional requirements set forth for primary batteries. Table 4.9a lists the dimensions of select round primary batteries, and Table 4.9b lists some nickel-metal hydride rechargeable batteries that are interchangeable with the primary batteries.

TABLE 4.10 Standard SLI and Other Lead-Acid Batteries (*Continued*)

	BCI group numbers, dimensional specifications, and ratings								
	Maximum overall dimensions							Performance ranges	
	Millimeters			Inches			Assembly figure no.	Cold cranking performance amps. @ 0°F (−18°C)	Reserve capacity min @ 80°F (27°C)
BCI group number	L	W	H	L	W	H			
	Passenger car and light commercial batteries 12 V (6 cells) (*Continued*)								
40R	278	175	175	$10^{15}/_{16}$	$6^{7}/_{8}$	$6^{7}/_{8}$	15	590–600	110–120
41	293	175	175	$11^{9}/_{16}$	$6^{7}/_{8}$	$6^{7}/_{8}$	15	235–650	65–95
42	242	175	175	$9^{1}/_{2}$	$6^{13}/_{16}$	$6^{13}/_{16}$	15	260–495	65–95
43	334	175	205	$13^{1}/_{8}$	$6^{7}/_{8}$	$8^{1}/_{16}$	15	375	115
45	240	140	227	$9^{7}/_{16}$	$5^{1}/_{2}$	$8^{15}/_{16}$	10F	250–470	60–80
46	273	173	229	$10^{3}/_{4}$	$6^{13}/_{16}$	9	10F	350–450	75–95
47	242	175	190	$9^{1}/_{2}$	$6^{7}/_{8}$	$7^{1}/_{2}$	24(A, F)[a]	370–550	75–85
48	278	175	190	$12^{1}/_{16}$	$6^{7}/_{8}$	$7^{9}/_{16}$	24	450–695	85–95
49	353	175	190	$13^{7}/_{8}$	$6^{7}/_{8}$	$7^{9}/_{16}$	24	600–900	140–150
50	343	127	254	$13^{1}/_{2}$	5	10	10	400–600	85–100
51	238	129	223	$9^{3}/_{8}$	$5^{1}/_{16}$	$8^{13}/_{16}$	10	405–435	70
51R	238	129	223	$9^{3}/_{8}$	$5^{1}/_{16}$	$8^{13}/_{16}$	11	405–435	70
52	186	147	210	$7^{5}/_{16}$	$5^{13}/_{16}$	$8^{1}/_{4}$	10	405	70
53	330	119	210	13	$4^{11}/_{16}$	$8^{1}/_{4}$	14	280	40
54	186	154	212	$7^{5}/_{16}$	$6^{1}/_{16}$	$8^{3}/_{8}$	19	305–330	60
55	218	154	212	$8^{5}/_{8}$	$6^{1}/_{16}$	$8^{3}/_{8}$	19	370–450	75
56	254	154	212	10	$6^{1}/_{16}$	$8^{3}/_{8}$	19	450–505	90
57	205	183	177	$8^{1}/_{16}$	$7^{3}/_{16}$	$6^{15}/_{16}$	22	310	60
58	255	183	177	$10^{1}/_{16}$	$7^{3}/_{16}$	$6^{15}/_{16}$	26	380–540	75
58R	255	183	177	$10^{1}/_{16}$	$7^{3}/_{16}$	$6^{15}/_{16}$	19	540–580	75
59	255	193	196	$10^{1}/_{16}$	$7^{5}/_{8}$	$7^{3}/_{4}$	21	540–590	100
60	332	160	225	$13^{1}/_{16}$	$6^{5}/_{16}$	$8^{7}/_{8}$	12	305–385	65–115
61	192	162	225	$7^{9}/_{16}$	$6^{3}/_{8}$	$8^{7}/_{8}$	20	310	60
62	225	162	225	$8^{7}/_{8}$	$6^{3}/_{8}$	$8^{7}/_{8}$	20	380	75
63	258	162	225	$10^{3}/_{16}$	$6^{3}/_{8}$	$8^{7}/_{8}$	20	450	90
64	296	162	225	$11^{11}/_{16}$	$6^{3}/_{8}$	$8^{7}/_{8}$	20	475–535	105–120
65	306	192	192	$12^{1}/_{16}$	$7^{1}/_{2}$	$7^{9}/_{16}$	21	650–850	130–165
66	306	192	194	$12^{1}/_{16}$	$7^{9}/_{16}$	$7^{5}/_{8}$	13	650–750	130–140
70	208	180	186	$8^{3}/_{16}$	$7^{1}/_{16}$	$7^{5}/_{16}$	17	260–525	60–80
71	208	179	216	$8^{3}/_{16}$	$7^{1}/_{16}$	$8^{1}/_{2}$	17	275–430	75–90
72	230	179	210	$9^{1}/_{16}$	$7^{1}/_{16}$	$8^{1}/_{4}$	17	275–350	60–90
73	230	179	216	$9^{1}/_{16}$	$7^{1}/_{16}$	$8^{1}/_{2}$	17	430–475	80–115
74	260	184	222	$10^{1}/_{4}$	$7^{1}/_{4}$	$8^{3}/_{4}$	17	350–550	75–140
75	230	180	196[g]	$9^{1}/_{16}$	$7^{1}/_{16}$	$7^{11}/_{16}$[g]	17	430–690	90
76	334	179	216	$13^{1}/_{8}$	$7^{1}/_{16}$	$8^{1}/_{2}$	17	750–1075	150–175
78	260	180	186	$10^{1}/_{4}$	$7^{1}/_{16}$	$7^{5}/_{16}$	17	515–770	105–115
79	307	179	188	$12^{1}/_{16}$	$7^{1}/_{16}$	$7^{3}/_{8}$	35	770–840	140
85	230	173	203	$9^{1}/_{16}$	$6^{13}/_{16}$	8	11	430–630	90
86	230	173	203	$9^{1}/_{16}$	$6^{13}/_{16}$	8	10	430–640	90
90	242	175	175	$9^{1}/_{2}$	$6^{7}/_{8}$	$6^{7}/_{8}$	24	520–600	80

4.14 PRINCIPLES OF OPERATION

TABLE 4.10 Standard SLI and Other Lead-Acid Batteries (*Continued*)

	BCI group numbers, dimensional specifications, and ratings								
	Maximum overall dimensions							Performance ranges	
	Millimeters			Inches				Cold cranking	Reserve
BCI group number	L	W	H	L	W	H	Assembly figure no.	performance amps. @ 0°F (−18°C)	capacity min @ 80°F (27°C)
	Passenger car and light commercial batteries 12 V (6 cells) (*Continued*)								
91	278	175	175	11	6⁷/₈	6⁷/₈	24	600	100
92	315	175	175	12¹/₂	6⁷/₈	6⁷/₈	24	650	130
93	353	175	175	13⁷/₈	6⁷/₈	6⁷/₈	24	800	150
94R	315	175	190	12³/₈	6⁷/₈	7¹/₂	24	640–765	135
95R	394	175	190	15⁹/₁₆	6⁷/₈	7¹/₂	24	850–950	190
96R	242	175	175	9⁹/₁₆	6¹³/₁₆	6⁷/₈	15	590	95
97R	252	175	190	9¹⁵/₁₆	6⁷/₈	7¹/₂	15	557	90
98R	283	175	190	11³/₁₆	6⁷/₈	7¹/₂	15	620	120
99	207	175	175	8³/₁₆	6⁷/₈	6⁷/₈	34	360	50
100	260	179	188	10¹/₄	7	7⁵/₁₆	35	770	115
101	260	179	170	10¹/₄	7	6¹¹/₁₆	17	540	115
	Passenger car and light commercial batteries 6 V (3 cells)								
1	232	181	238	9¹/₈	7¹/₈	9³/₈	2	400–545	105–165
2	264	181	238	10³/₈	7¹/₈	9³/₈	2	475–650	136–230
2E	492	105	232	19⁷/₁₆	4¹/₈	9¹/₈	5	485	140
2N	254	141	227	10	5⁹/₁₆	8¹⁵/₁₆	1	450	135
17HF[b,d]	187	175	229	7³/₈	6⁷/₈	9	2B	—	—
	Heavy-duty commercial batteries 12 V (6 cells)								
4D[c]	527	222	250	20³/₄	8³/₄	9⁷/₁₆	8	490–1125	225–325
6D	527	254	260	20³/₄	10	10¹/₄	8	750	310
8D[c]	527	283	250	20³/₄	11¹/₈	9⁷/₁₆	8	850–1250	235–465
28	261	173	240	10⁵/₁₆	6¹³/₁₆	9⁷/₁₆	18	400–535	80–135
29H	334	171	232	13¹/₈	6³/₄	9¹/₈	10	525–840	145
30H	343	173	235	13¹/₂	6¹³/₁₆	9¹/₄	10	380–685	120–150
31A	330	173	240	13	6¹³/₁₆	9⁷/₁₆	18 (A,T)[a]	455–950	100–200
	Heavy-duty commercial batteries 6 V (3 cells)								
3	298	181	328	11³/₄	7¹/₈	9³/₈	2	525–660	210–230
4	334	181	328	13¹/₈	7¹/₈	9³/₈	2	550–975	240–420
5D	349	181	238	13³/₄	7¹/₈	9³/₈	2	720–820	310–380
7D	413	181	238	16¹/₄	7¹/₈	9³/₈	2	680–875	370–426
	Special tractor batteries 6 V (3 cells)								
3EH	491	111	249	19⁵/₁₆	4³/₈	9¹³/₁₆	5	740–850	220–340
4EH	491	127	249	19⁵/₁₆	5	9¹³/₁₆	5	850	340–420

TABLE 4.10 Standard SLI and Other Lead-Acid Batteries (*Continued*)

	BCI group numbers, dimensional specifications, and ratings								
	Maximum overall dimensions							Performance ranges	
	Millimeters			Inches			Assembly figure no.	Cold cranking performance amps. @ 0°F (−18°C)	Reserve capacity min @ 80°F (27°C)
BCI group number	L	W	H	L	W	H			
Special tractor batteries 12 V (6 cells)									
3EE	491	111	225	19$^{5}/_{16}$	4$^{3}/_{8}$	8$^{7}/_{8}$	9	260–360	85–105
3ET	491	111	249	19$^{5}/_{16}$	4$^{3}/_{8}$	9$^{3}/_{16}$	9	355–425	130–135
4DLT	508	208	202	20	8$^{3}/_{16}$	7$^{15}/_{16}$	16L	650–820	200–290
12T	177	177	202	7$^{1}/_{16}$	6$^{15}/_{16}$	7$^{15}/_{16}$	10	460	160
16TF	421	181	283	16$^{9}/_{16}$	7$^{1}/_{8}$	11$^{1}/_{8}$	10F	600	240
17TF	433	177	202	17$^{1}/_{16}$	6$^{15}/_{16}$	7$^{15}/_{16}$	11L	510	145
General-utility batteries 12 V (6 cells)									
U1	197	132	186	7$^{3}/_{4}$	5$^{3}/_{16}$	7$^{5}/_{16}$	10(X)[a]	120–375	23–40
U1R	197	132	186	7$^{3}/_{4}$	5$^{3}/_{16}$	7$^{5}/_{16}$	11(X)[a]	200–280	25–37
U2	160	132	181	6$^{5}/_{16}$	5$^{3}/_{16}$	7$^{1}/_{8}$	10(X)[a]	120	17
Electric golf car/utility batteries 6 V (3 cells)									
GC2	264	183	290	10$^{3}/_{8}$	7$^{3}/_{16}$	11$^{7}/_{16}$	2	*f*	*f*
GC2H[e]	264	183	295	10$^{3}/_{8}$	7$^{3}/_{16}$	11$^{5}/_{8}$	2	*f*	*f*
Electric golf car/utility batteries 8 V (4 cells)									
GC8	264	183	290	10$^{3}/_{8}$	7$^{3}/_{16}$	11$^{7}/_{16}$	31	—	—
Commercial batteries (deep cycle) 12 V (6 cells)									
920	356	171	311	14	6$^{3}/_{4}$	12$^{1}/_{2}$	37	—	—
921	397	181	378	15$^{3}/_{4}$	7$^{1}/_{8}$	14$^{7}/_{8}$	37	—	—
Marine/commercial batteries 8 V (4 cells)									
981	527	191	273	20$^{3}/_{4}$	7$^{1}/_{2}$	10$^{3}/_{4}$	8	—	—
982	546	191	267	21$^{1}/_{2}$	7$^{1}/_{2}$	10$^{1}/_{2}$	8	—	—
Ordnance batteries 12 V (6 cells)									
2H	260	135	227	10$^{1}/_{4}$	5$^{9}/_{16}$	8$^{15}/_{16}$	28	—	75
6T	286	267	230	11$^{1}/_{4}$	10$^{1}/_{2}$	9$^{1}/_{16}$	27	600–750	180–230

[a] Letter in parentheses indicates terminal type.
[b] Rod end types—Extend top ledge with holes for holddown bolts.
[c] Ratings for batteries recommended for motor coach and bus service are for double insulation. When double insulation is used in other types, deduct 15% from the rating values for cold cranking performance.
[d] Not in application section but still manufactured.
[e] Special-use battery not shown in application section.
[f] Capacity test in minutes at 75 A to 5.25 V at 80°F (27°C); cold cranking performance test not normally required for this battery.
[g] Maximum height dimension shown includes batteries with raised-quarter cover design. Flat-top design model height (minus quarter covers) reduced by approximately 3/8 in (10 mm).

TABLE 4.10 Standard SLI and Other Lead-Acid Batteries (*Continued*)

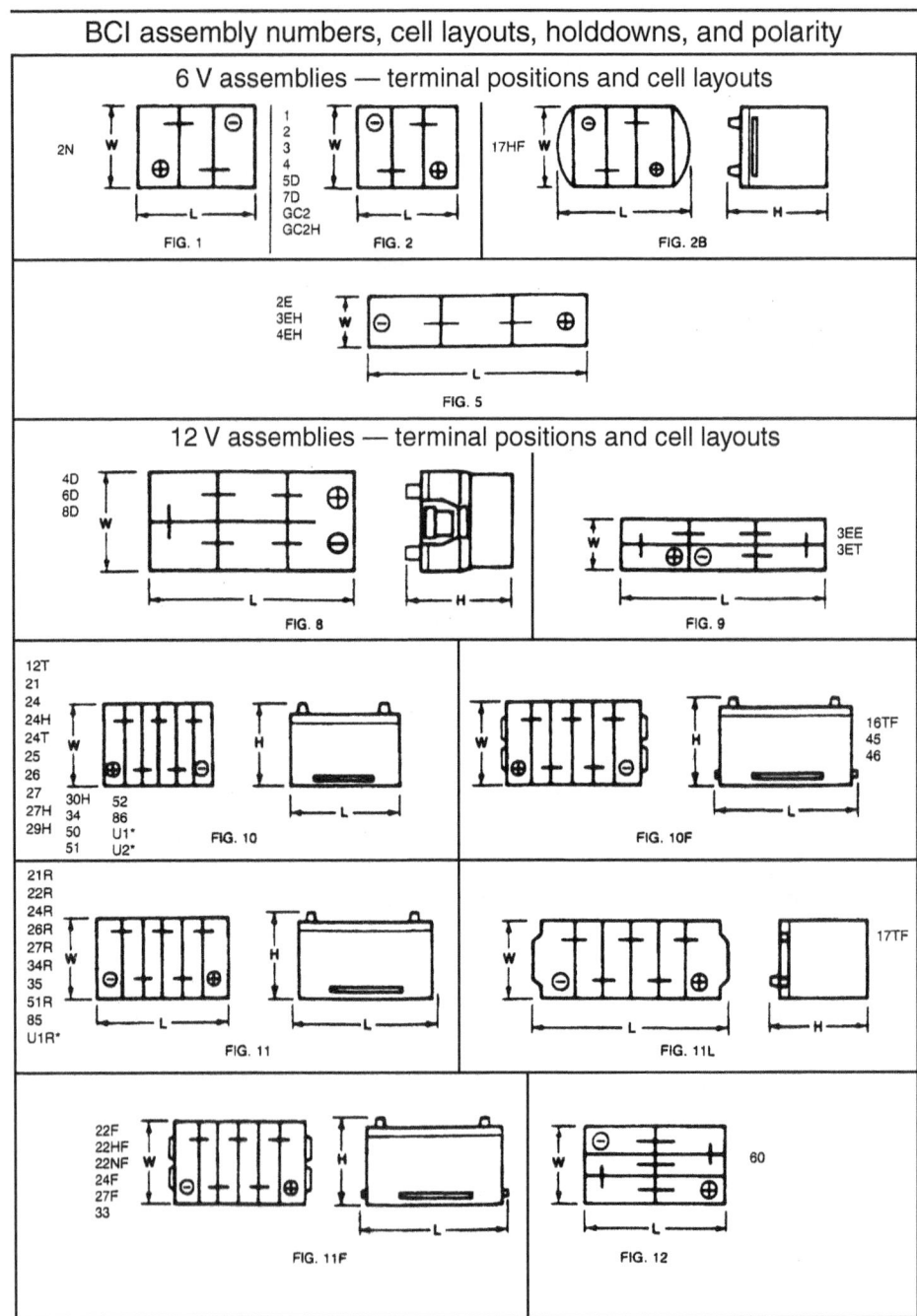

BATTERY STANDARDIZATION 4.17

TABLE 4.10 Standard SLI and Other Lead-Acid Batteries (*Continued*)

BCI assembly numbers, cell layouts, holddowns, and polarity

12 V assemblies — terminal positions and cell layouts

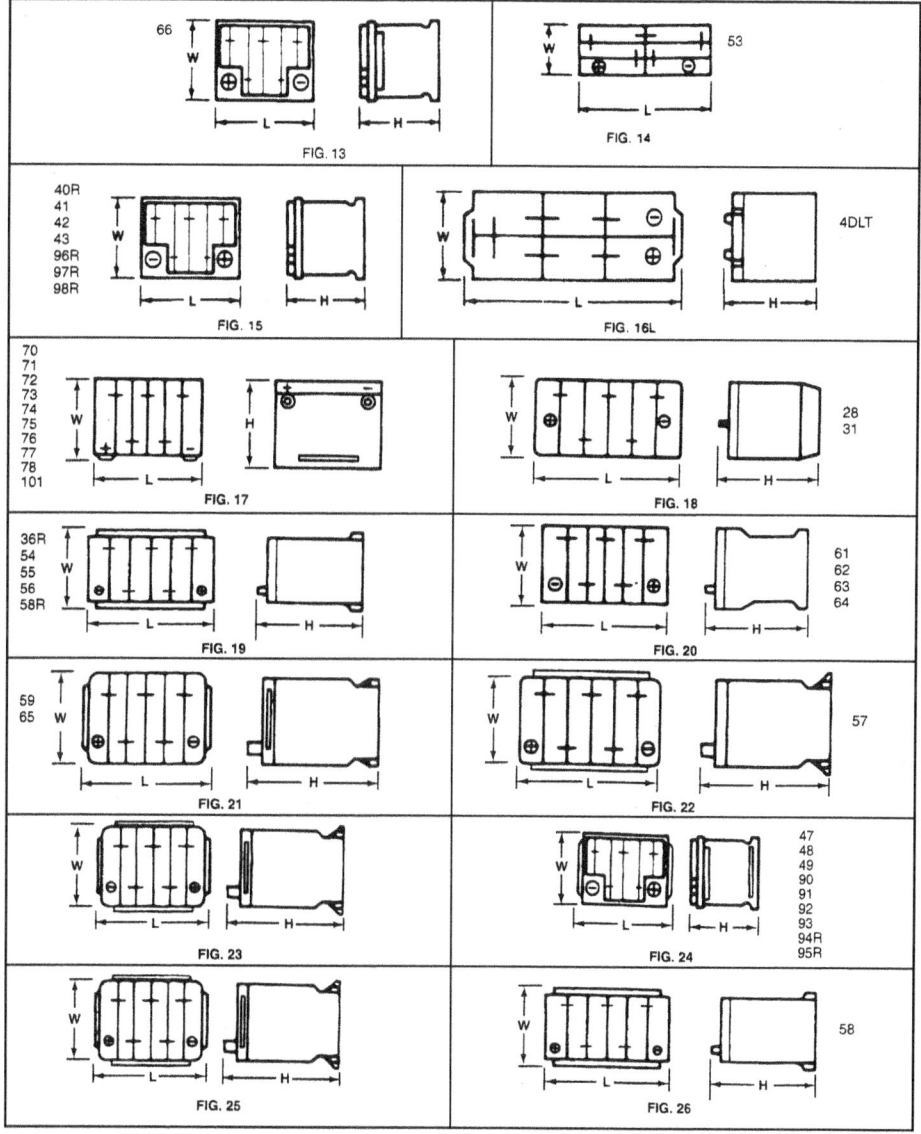

TABLE 4.10 Standard SLI and Other Lead-Acid Batteries (*Continued*)

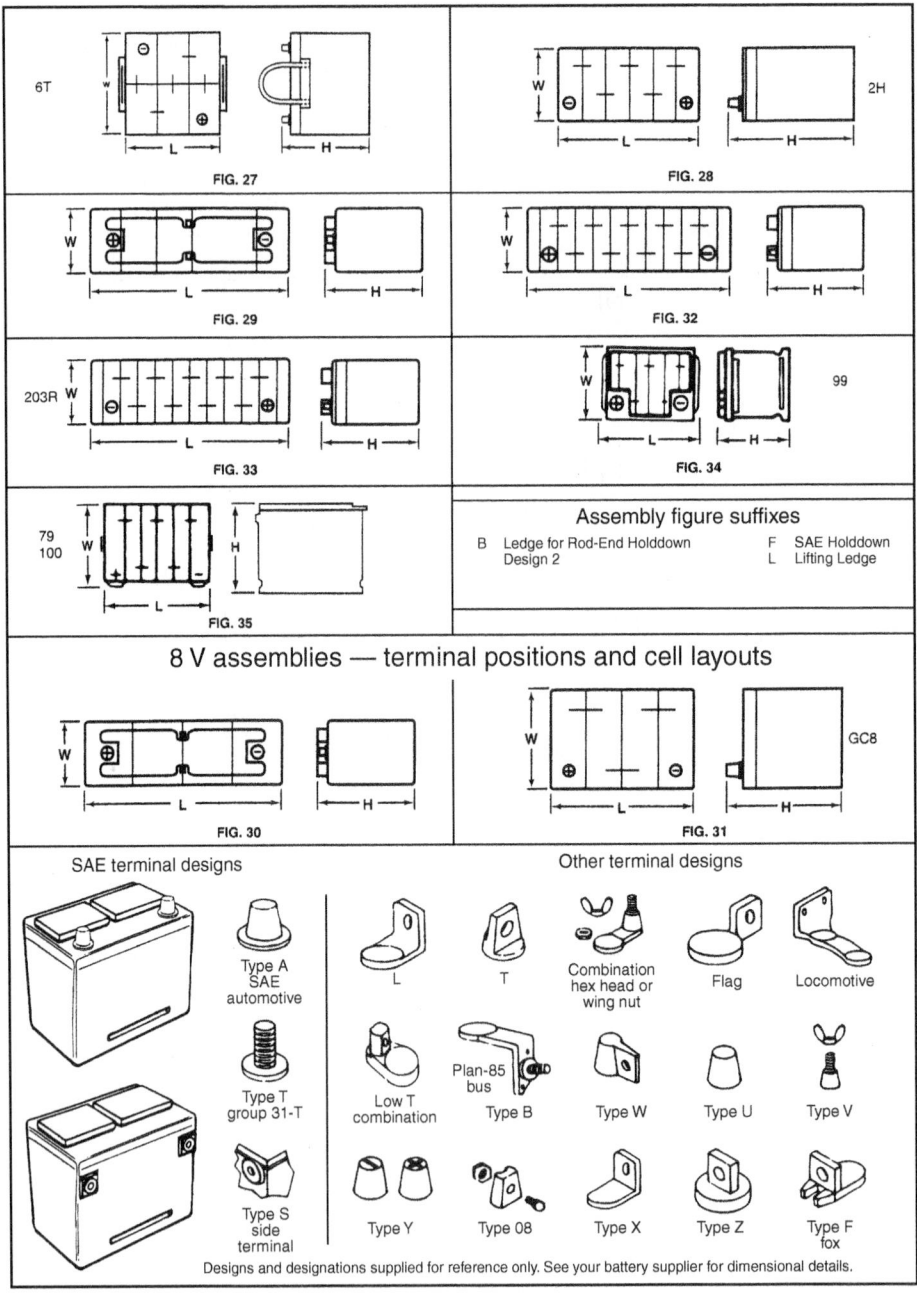

4.11 REGULATORY AND SAFETY STANDARDS

With the increasing complexity and energy of batteries and the concern about safety, greater attention is being given to developing regulations and standards with the goal to promote safe operation in use and transport. Stand-alone safety documents on primary and rechargeable batteries have been published by IEC and ANSI. In addition, the Underwriters Laboratories (UL) has published several battery safety standards aimed at the safe operation of UL-approved equipment.[4]

Table 4.11a is a list of organizations working on safety standards and the safety standards they prepared that cover various primary and secondary battery systems.

While the various groups involved in developing safety standards are dedicated to the principle of harmonization, there are still differences in the procedures, tests, and criteria between the various standards. It is recommended that users of these standards follow them on a judicious basis, and place their battery or application in the proper context.

Table 4.11b is a list of organizations that have focused on the safe transport of various goods and the regulations they have published. These regulations include procedures for the transport of batteries, including lithium batteries.

TABLE 4.11a Safety Standards

Publication	Title
American Standards Institute	
ANSI C18.1M, Part 2	American National Standard for Portable Primary Cells and Batteries with Aqueous Electrolyte—Safety Standard
ANSI C18.2M, Part 2	American National Standard for Portable Rechargeable Cells and Batteries—Safety Standard
ANSI C18.3M, Part 2	American National Standard for Portable Lithium Primary Cells and Batteries—Safety Standard
International Electrotechnical Commission	
IEC 60086-4	Primary Batteries—Part 4: Safety for Lithium Batteries
IEC 60086-5	Primary Batteries—Part 5: Safety of Batteries with Aqueous Electrolyte
IEC 62281	Safety of Primary and Secondary Lithium Batteries During Transport
IEC 62133	Safety for Portable Sealed Alkaline Secondary Cells and Batteries
Underwriters Laboratories	
UL1642	Standard for Lithium Batteries
UL2054	Standard for Household and Commercial Batteries

TABLE 4.11b Transportation Recommendations and Regulations

Organization	Title
Department of Transportation (DOT)	Code of Federal Regulations—Title 49 Transportation
Federal Aviation Administration (FAA)	TSO C042, Lithium Batteries (referencing RTCA Document DO-227 "Minimum Operational Performance Standards for Lithium Batteries")
International Air Transport Association (IATA)	Dangerous Goods Regulations
International Civil Aviation Association (ICAO)	Technical Instructions for the Safe Transport of Dangerous Goods
United Nations (UN)	Recommendations on the Transportation of Dangerous Goods Manual of Tests and Criteria

In the United States, this responsibility for regulating the transport of goods rests with the Department of Transportation (DOT) through its Research and Special Programs Administration (RSPA).[5] These regulations are published in the Code of Federal Regulations (CFR49), which include the requirements for the shipment and transport of batteries under all modes of transportation. Under the DOT, the Federal Aviation Administration (FAA) is responsible for the safe operation of aircraft and has also issued regulations covering the use of batteries in aircraft.[6,7] Similar organizations are part of the governments of most countries throughout the world.

Internationally, transport is regulated by such organizations as the International Civil Aviation Organization (ICAO),[8] the International Air Transport Association (IATA),[9] and the International Maritime Organization. Their regulations are guided by the United Nations (UN) through their Committee of Experts on the Transport of Dangerous Goods, which has developed recommendations for the transportation of dangerous goods. These recommendations, which also include tests and criteria,[10,11] are addressed to governments and international organizations concerned with regulating the transport of various products. Currently, the UN Committee of Experts has developed guidelines covering the transport of lithium primary and secondary batteries. The quantity of lithium or lithium equivalent content in each cell and battery determines which specific rules and regulations are applied concerning the packaging, mode of shipment, marking, and other special provisions.

As these standards, regulations, and guidelines can be changed on an annual or periodic basis, the current edition of each document should be used.

NOTE

It is imperative that only the latest version of each standard be used. Due to the periodic revision of these standards, only the latest version can be relied upon to provide reliable enforceable specifications of battery dimensions, terminals, marking, general design features, conditions of electrical testing for performance verification, mechanical tests, test sequences, safety, shipment, storage, use, and disposal.

REFERENCES

1. Society of Automotive Engineers, 400 Commonwealth Drive, Warrendale, PA 15096, www.sae.org.
2. Battery Council International, 401 North Michigan Ave., Chicago, IL 60611, www.batterycouncil.org.
3. Battery Council International, *Battery Replacement Data Book.*
4. Underwriters Laboratories, Inc., 333 Pfingsten Road, Northbrook, IL 60062.
5. Department of Transportation, Office of Hazardous Materials Safety, Research and Special Programs Administration, 400 Seventh St., SW, Washington, DC 20590.
6. Department of Transportation, Federal Aviation Administration, 800 Independence Ave., SW, Washington, DC 20591.
7. RTCA, 1828 L St., NW, Suite 805, Washington, DC 20036, info@rtca.org.
8. International Civil Aviation Organization, 1000 Sherbrooke St., W., Montreal, Quebec, Canada.
9. International Air Transport Association, 2000 Peel St., Montreal, Quebec, Canada.
10. United Nations, *Recommendation on the Transport of Dangerous Goods,* New York, NY, and Geneva, Switzerland.
11. United Nations, *Manual of Tests and Criteria,* New York, NY, and Geneva, Switzerland.

CHAPTER 5
BATTERY DESIGN

Daniel D. Friel

5.1 GENERAL

Proper design of the battery pack or the battery compartment is important to assure optimum, reliable, and safe operation. Many problems attributed to the battery may have been prevented had proper precautions been taken with both the design of the battery itself, any battery monitoring or protection devices or electronics, and how the battery pack is designed into the battery-operated equipment.

It is important to note that the performance of a cell in a battery can be significantly different from that of an individual cell depending on the particular environment of that cell in the battery. Specifications and data sheets provided by the manufacturers should only be used as a guide as it is not always possible to extrapolate the data to determine the performance of multicell batteries in a series/parallel configuration in a battery pack. Such factors as the cell uniformity, number of cells, series or parallel connections, battery case material and design, conditions of discharge and charge, and temperature, to name a few, influence the performance of the battery. The problem is usually exacerbated under the more stringent conditions of use, such as high-rate charging and discharging, operation, and extreme temperatures and other conditions that tend to increase the variability of the cells within the battery. Higher series/parallel configurations also present unique challenges which must be considered.

Further, specific energy and energy density data based on cell or single-cell battery performance have to be derated when the weight and volume of the battery case, battery assembly materials, and any ancillary equipment, such as monitoring or protection electronics, have to be considered in the calculation.

Another factor that must be considered, particularly with newly developing battery technologies, is the difficulty of scaling up laboratory data based on smaller individual batteries to multicell batteries using larger cells manufactured on a production line.

This chapter will address the issues that the product designer should consider. Cell and battery manufacturers should also be consulted to obtain specific details on their recommendations for the batteries they market.

5.2 DESIGNING TO ELIMINATE POTENTIAL SAFETY PROBLEMS

Batteries are sources of energy and when used properly will deliver their energy in a safe manner. There are instances, however, when a battery may vent, rupture, or even explode if it is abused. The design of the battery should include protective devices and other features that can prevent, or at least minimize, the problem.

Some of the most common causes for battery failure are:

1. Short-circuiting of battery terminals
2. Excessive high rate discharge or charge

3. Overdischarge below the minimum recommended operating voltage of the cell (also includes voltage reversal, or the discharging of the cell below 0 V)
4. Charging of primary batteries
5. Improper charge control when charging secondary batteries
6. Imbalance between series cells

These conditions may cause an internal pressure increase within the cells, resulting in an activation of the vent device or a rupture or explosion of the battery. Internal cell shorts can also cause failures, although these are rare. These may occur due to impurities being accidentally introduced into the cells during manufacture. There are a number of means to minimize the possibilities of these occurrences. Additional failure mechanisms can occur due to the assembly of individual cells into the battery pack improperly. For example, poor cell connector tab welds, lack of proper insulation between tabs, and improper case assembly can all lead to latent battery failures.

The use of high quality individual cells does not guarantee a safe battery pack assembly. All factors must be considered carefully, including the mechanical assembly of the pack, any internal protection devices or electronics, contacts, monitoring components, and the pack casing.

5.2.1 Charging Primary Batteries

All major manufacturers of primary batteries warn that the batteries may leak or explode if they are charged. As discussed in Sec. 8.4, some primary batteries can be recharged if done under controlled conditions. Nevertheless, charging primary batteries is not usually recommended because of the potential hazards.

Protection from External Charge. The simplest means of preventing a battery from being charged from an external power source is to incorporate a blocking diode in the battery pack, as shown in Fig. 5.1. The diode chosen must have a current rating in excess of the operating current of the device. It should be rated, at a minimum, at twice the operating current. The forward voltage drop of the diode should be as low as possible. Schottky diodes are commonly used because of their typical 0.5 V drop in the forward direction. Another consideration in selecting the diode is the reverse voltage rating. The peak inverse voltage (PIV) rating should be at least twice the voltage of the battery.

Protection from Charging within Battery. When multiple series stacks are paralleled within a battery pack, charging may occur when a defective or a low-capacity cell is present in one of the stacks (Fig. 5.2a). The remaining stacks of cells will charge the stack with the defective cell. At best this situation will discharge the good stack, but it could result in rupture of the cells in the weak stack. To avoid this, diodes should be placed in each series string to block charging currents from stack to stack (Fig. 5.2b).

When diode protection is used in each series stack, the diode will prevent the stack containing the defective cell from being charged. The diode should have the following characteristics:

1. Forward voltage drop should be as low as possible, Schottky type preferable.
2. Peak inverse voltage should be rated at twice the voltage of the individual series stack.
3. Forward current rating of the diodes should be a minimum of

$$I_{min} = \frac{I_{op}}{N} \times 2$$

where I_{op} = device operating current
N = number of parallel stacks

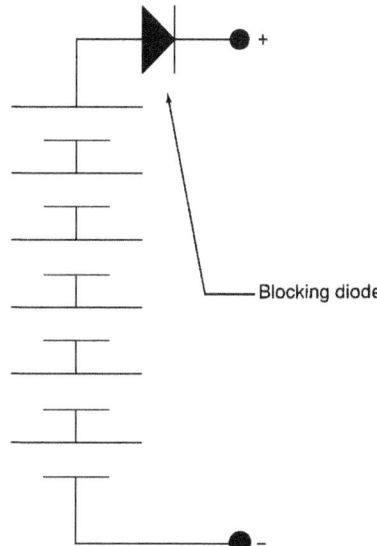
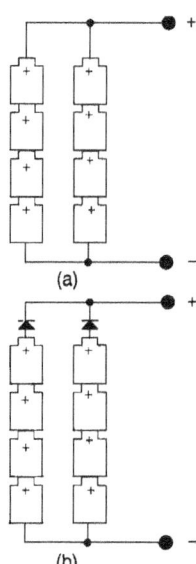

FIGURE 5.1 Battery circuit incorporating a blocking diode to prevent charge.

FIGURE 5.2 Series/parallel battery. (*a*) Without diode protection. (*b*) With diode protection.

5.2.2 Preventing Battery Short-Circuit Conditions

When a battery is short-circuited through the external terminals, the chemical energy is converted into heat within the battery. In order to prevent short-circuiting, the positive and negative terminals of the battery should be physically isolated. Effective battery design will incorporate the following:

1. The battery terminals should be recessed within the external case (Fig. 5.3*a*).
2. If connectors are used, the battery should incorporate the female connection. The connector should also be polarized to only permit correct insertion (Fig. 5.3*b*). This figure illustrates a multi-pin female connector that receives bladed connections for battery power and other signals (if used). Such a connector is often molded into the battery case.[1]

Short-Circuit Protection. In addition, it may be also necessary to include some means of circuit interruption. There are a number of devices that can perform this function, including:

1. Fuses or circuit breakers.
2. Thermostats designed to open the battery circuit when the temperature or current reaches a predetermined upper limit.
3. Positive-temperature-coefficient (PTC) devices that, at normal currents and temperatures, have a very low value of resistance. When excessive currents pass through these devices or the battery temperature increases, the resistance increases by orders of magnitude, limiting the current. These devices are incorporated internally in some cells by the cell manufacturer. When using cells with internal protection, it is advisable to use an external PTC selected to accommodate both the current and the voltage levels of the battery application (see Sec. 5.5.1). Note that PTC devices will not prevent a short-circuit condition from discharging the cell or pack fully. A continually shorted cell or pack will still be discharged through the PTC, although at a slow rate.

5.4 PRINCIPLES OF OPERATION

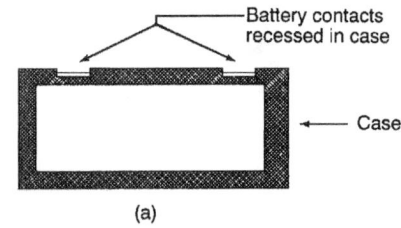

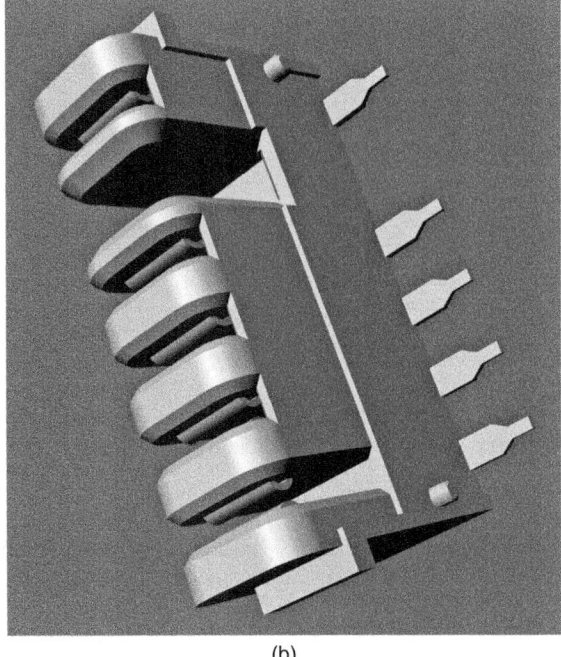

FIGURE 5.3 Battery pack terminal design options.

4. *Additional protection methods.* Protection methods go beyond external mechanisms. Proper battery pack assembly is also critical to preventing internal short circuits within the battery pack (see Sec. 5.4).

5.2.3 Voltage Reversal

Due to variability in manufacturing, capacities will vary from battery to battery. When discharged in a series configuration, the capacity of the weakest cell in the series string of a multicell battery will be depleted before the others. If the discharge is continued, the voltage of the low-capacity cell will reach 0 V and then reverse. The heat generated may eventually cause pressure buildup in the cell and subsequent venting or rupture. This process is sometimes referred to as "forced discharge."

A common test to determine the ability of cells to withstand voltage reversal is the forced-discharge test. The cells are deliberately discharged, at specified currents, to below 0 V by other cells in a series string or by an external power supply to determine whether a venting, rupture, or other undesired safety problem arises.

Discharge beyond the cell manufacturer's recommended operating range is not suggested, so precautions should be taken at the system level if such discharge is possible. If external conditions are likely to cause discharge below recommended limits, then protection within the battery pack may be required. This is often utilized in lithium rechargeable chemistries.

Some cells are designed to withstand a forced discharge to specified discharge currents. The cells may also be designed with internal protection, such as fuses or thermal cutoff devices, to interrupt the discharge if an unsafe condition develops.

This condition of cell unbalance could be exacerbated with rechargeable cells, as the individual cell capacities could change during cycling. To minimize this effect, rechargeable batteries should at least be constructed with "matched" cells, that is, cells having nearly identical capacities. Cells are sorted, within grades, by at least one cycle of charge and discharge. Typically cells are considered matched when the capacity range is within 3%. Recent advances in manufacturing control have reduced the number of cell grades. Some manufacturers have reached the optimal goal of one grade, which negates the need of matching. This information is readily available from the battery companies.

Cell imbalance, however, can occur after the cells are assembled into a battery pack and utilized in the end application. Such cell imbalance can result from uneven thermal gradients across the battery that cause some cells to reach higher temperatures than others. This temperature gradient will cause a difference in cell self-discharge, potentially leading to cell imbalance. If imbalance occurs within the battery pack, then corrective action must be taken to prevent an accumulation of imbalance. Some chemistries permit a low-rate overcharge to correct imbalance, while other chemistries, such as lithium-ion/polymer, require rebalancing by electronic methods.

Battery Design to Prevent Voltage Reversal. Even though matched cells are used, other battery designs or applications can cause an imbalance in cell capacity. One example is the use of voltage taps on cells of a multicell battery in a series string. In this design, the cells are not discharged equally.

Many early battery designs using Leclanché-type cells incorporated the use of voltage taps. Batteries with as many as 30 cells in series (45 V) were common, with taps typically at 3, 9, 13.5 V, and so on. When the cells with the lower voltage taps were discharged, they could leak. This leakage could cause corrosion, but usually these cells would not be prone to rupture. With the advent of the high-energy, tightly sealed cells, this is no longer the case. Cells driven into voltage reversal may rupture or explode. In order to avoid problems, the battery should be designed with electrically independent sections for each voltage output. If possible, the device should be designed to be powered by a single input voltage source. DC to DC converters can be used to safely provide for multiple voltage outputs. Converters are now available with efficiencies greater than 90%.

Modern electronic circuits that convert the primary battery cell's output to a usable system voltage may also include battery charge protection features. Fig. 5.4 shows a typical DC-to-DC converter designed for operation from single cell batteries and which includes cell-charging protection to prevent damage when used with primary cells.[3]

Parallel Diodes to Prevent Voltage Reversal. Some battery designers, particularly for multicell lithium primary batteries, add diodes in parallel to each cell to limit voltage reversal. As the cell voltage drops below 0 V and into reversal, the diode becomes conducting and diverts most of the

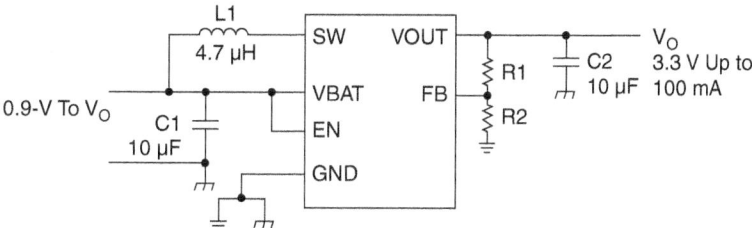

FIGURE 5.4 Typical DC:DC converter for use with primary battery cells.[2]

current from flowing through the cell. This limits the extent of the voltage reversal to that of the characteristic of the diode.

5.2.4 Protection of Primary Cells and Batteries from External Charge

Many battery-powered devices are also operated from rectified alternating-current (AC) sources. These could include devices that offer both AC and battery operation or devices that use the battery for backup when the AC power supply fails or is not available. Both primary lithium as well as other chemistries such as alkaline are commonly used in devices that operate from both battery power and AC power sources.

In the case where the battery is a backup for the main power supply as, for example, in memory backup, the primary battery must be protected from being charged by the main power supply. Typical circuits are shown in Fig. 5.5. In Fig. 5.5a, two blocking diodes are used for redundancy to provide protection in case of the failure of one. A resistor is used in Fig. 5.5b to limit the charge current if the diode fails in a closed position. This blocking diode should have the features of a low voltage drop in the forward direction to minimize the loss of battery backup voltage, and a low leakage current in the reverse direction to minimize the charging current.

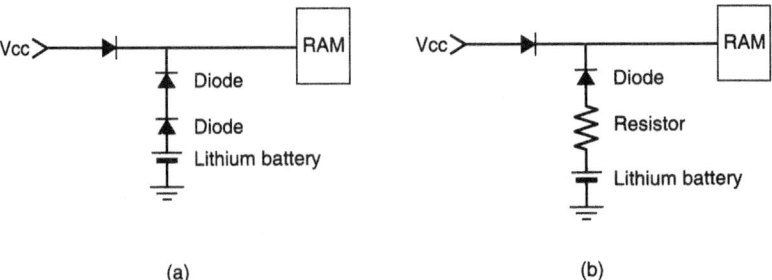

FIGURE 5.5 Protective circuitry for memory backup batteries. (*a*) Using two diodes. (*b*) Using diode and resistor, V_{cc} = power supply voltage.

5.2.5 Special Considerations When Designing Lithium Primary Batteries

Lithium primary batteries contain an anode of elemental lithium (see Chap. 14) and, because of the activity of this metal, special precautions may be required in the design and use of the batteries, particularly when multiple cells are used in the battery pack. Some of the special precautions that should be taken in the design of these batteries include the following:

1. When multiple cells are required, due to voltage and/or the capacity requirement of the application, they should be welded into battery packs, thus preventing the user mixing cells of different chemistries or capacities if replaceable cells were used.
2. A thermal disconnect device should be included to prevent the buildup of excessive heat. Many of the batteries now manufactured include a PTC or a mechanical disconnect, or both, within the cell. Additional protective thermal devices should be included, external to the cells, in the design of a multicell battery pack.
3. The following protective devices should be included:
    a. Series diode protection to prevent charging must be included.
    b. Cell bypass diode protection to prevent excessive voltage reversal of individual cells in a multicell series and/or series parallel configuration may be used.
    c. Short-circuit protection by means of a PTC, permanent fuse, or electronic means, or a combination of all three.

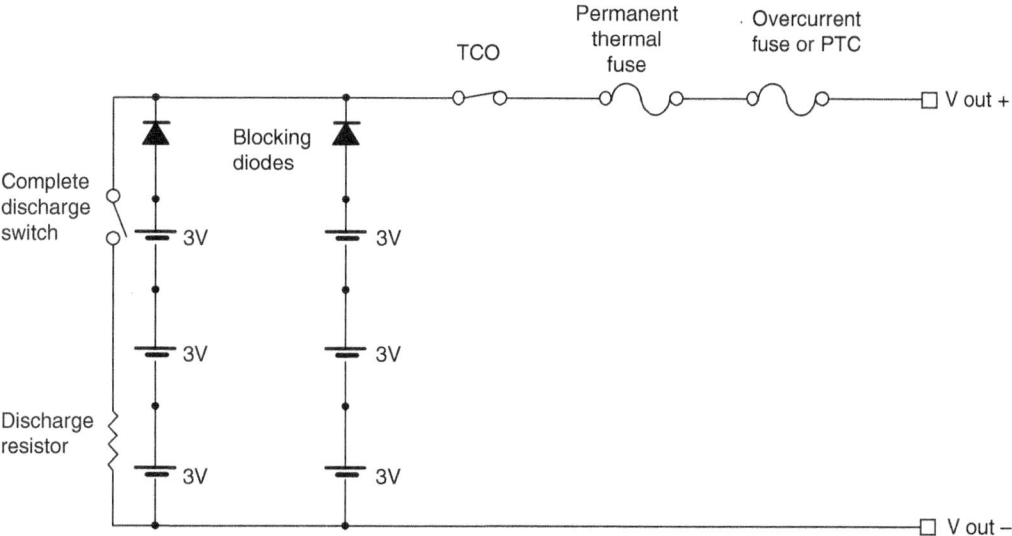

FIGURE 5.6 Typical lithium primary military battery.

4. In order to make the used battery safe for disposal, for some lithium batteries the remaining lithium within the battery must be depleted. This is accomplished by placing a resistive load across the cell pack to completely discharge the battery after use. The resistive load should be chosen to ensure a low current discharge, typically at a five (5) day rate of the original capacity of the battery. This feature has been used mainly in lithium primary military batteries.

Fig. 5.6 illustrates a typical schematic for a lithium primary military battery.

5.3 BATTERY SAFEGUARDS WHEN USING DISCRETE BATTERIES[3]

5.3.1 Design to Prevent Improper Insertion of Batteries

When designing products using individual single-cell batteries, special care must be taken in the layout of the battery compartment. If provisions are not made to ensure the proper placement of the batteries, a situation may result in which some of the batteries that are improperly inserted could be exposed to being charged. This could lead to leakage, venting, rupture, or even explosion. Figure 5.7 illustrates simple battery-holder concepts for cylindrical and button batteries, which will prevent the batteries from being inserted incorrectly. Figure 5.8 shows several other design options for preventing improper installation.

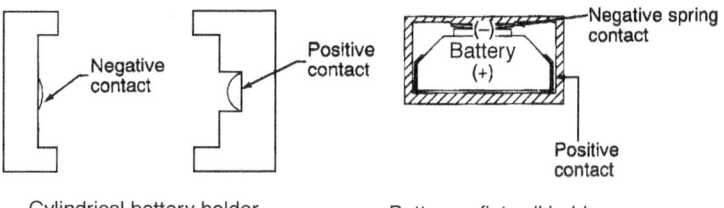

FIGURE 5.7 Battery holders. (Left) Cylindrical. (Right) Button or flat.

5.8 PRINCIPLES OF OPERATION

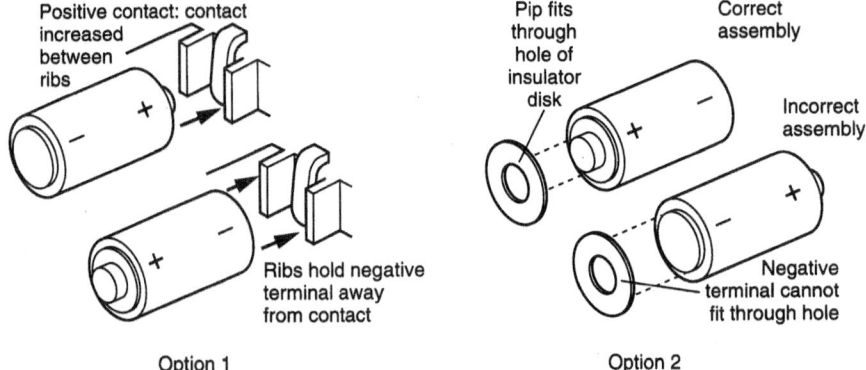

FIGURE 5.8 Battery contact designs that prevent reverse installation of cells.

Two commonly used battery circuits that are potentially dangerous without proper battery orientation are:

1. Series/parallel with one battery reversed (Fig. 5.9). In this circuit, battery 3 has been reversed. As a result, batteries 1–3 are now in series and are charging battery 4. This condition can be avoided, if possible, by using a single series string of larger batteries. Further, as discussed in Sec. 5.2.1, the use of diodes in each series section will at least prevent one parallel stack from charging the other.

2. Multicell series stack with one battery reversed in position (Fig. 5.10). The fourth battery is reversed and will be charged when the circuit is closed to operate the device. Depending on the magnitude of the current, the battery may vent or rupture. The magnitude of the current is dependent on the device load, the battery voltage, the condition of the reversed battery, and other conditions of the discharge.

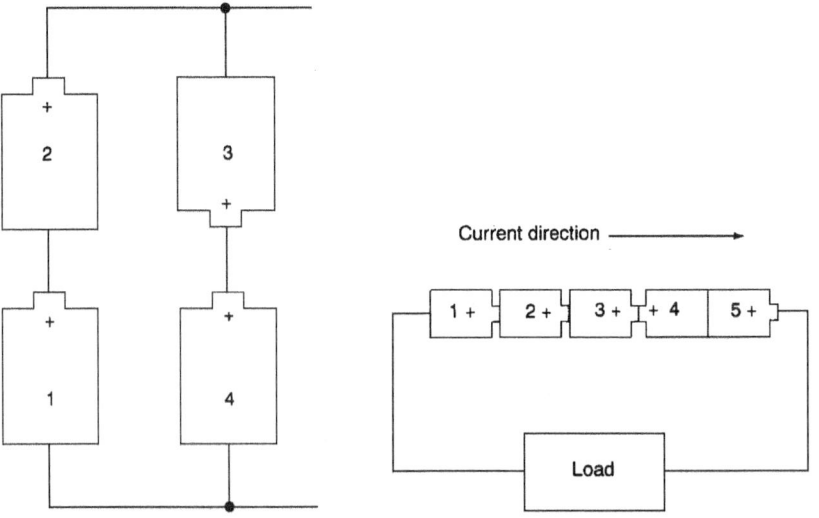

FIGURE 5.9 Series/parallel circuit; cell 4 being charged

FIGURE 5.10 One cell reversed in a series stack; cell 4 being charged.

To minimize the possibility of physically reversing a battery, the proper battery orientation should be clearly marked on the device, with simple and clear instructions. Blind battery compartments, where the individual batteries are not visible, should be avoided. The best practice is to use oriented or polarized battery holders, as discussed previously.

A suggested approach is to design the cell cavities for single cells so there are no strings of cells that could allow an incorrect reversal insertion of one cell. This does add cost to the device by requiring additional contacts, but it ensures that the circuit is correctly completed (by virtue of the physical connection of the cells by the device's circuit). Such a design is strongly suggested when a device can accept primary and rechargeable cells of a particular size, such as the AA or AAA sizes, which are commonly available in primary alkaline, rechargeable nickel, or primary lithium.

5.3.2 Battery Dimensions

At times, equipment manufacturers may design the battery cavity of their device around the battery of a single manufacturer. Unfortunately, the batteries made by the various manufacturers are not exactly the same size. While the differences may not be great, this could result in a cavity design that will not accept batteries of all manufacturers.

Along with variations in size, the battery cavity design must also be able to accommodate unusual battery configurations that fall within IEC standards. For example, several battery manufacturers offer batteries with negative recessed terminals that are designed to prevent contact when they are installed backward. Unfortunately, negative recessed terminals will mate only with contacts whose width is less than the diameter of the battery's terminal. Figure 5.11a illustrates the dimensional differences between cells with standard and recessed terminals.

The battery cavity should not be designed around the battery of a single manufacturer whose battery may be a unique size or configuration. Instead, cavity designs should be based on International Electrotechnical Commission (IEC) standards and built to accommodate maximum and minimum sizes. IEC and ANSI standards (see Chap. 4) provide key battery dimensions, including overall

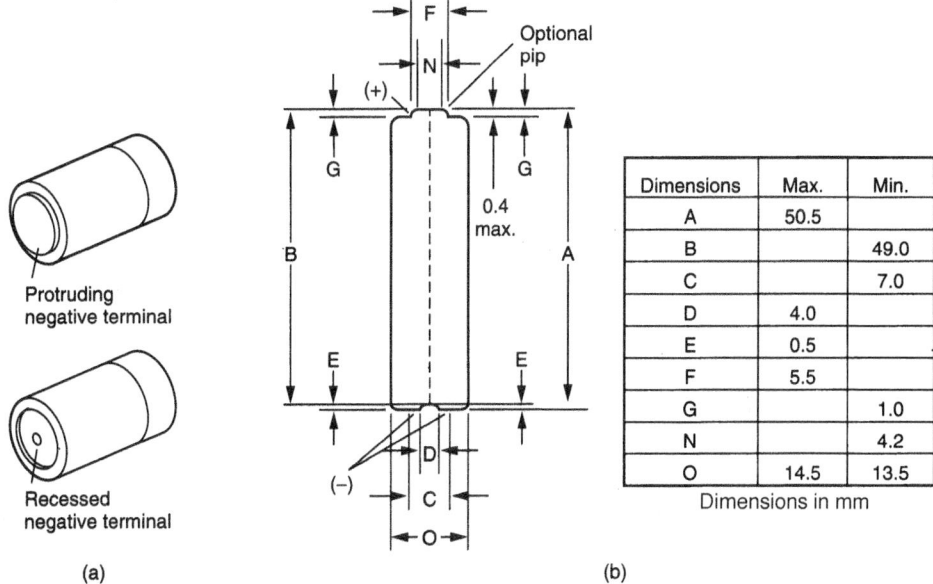

FIGURE 5.11 (a) Types of battery terminals falling within IEC standards. (b) Illustration of typical standard IEC dimensions.

height, overall diameter, pip diameter, pip height, and diameter of negative cap. Maximum and minimum values are usually specified, as shown in Fig. 5.11b. As these standards are revised periodically, the latest edition should be used.

5.4 BATTERY CONSTRUCTION

The following constructional features should also be considered in the design and fabrication of battery packs:

1. Intercell connections
2. Encapsulation of cells (should not inhibit vent activation)
3. Case configuration and materials
4. Terminals and contact materials

5.4.1 Intercell Connections

Soldering is the method of connection for batteries using Leclanché-type cells. Wires are soldered between the negative case and the adjoining positive cap. This effective method of construction for these cells is still widely used.

Welding of conductive tabs between cells is the preferred method of intercell connection for most of the other battery systems. The tab materials for most applications are either pure nickel or nickel-plated steel. The corrosion resistance of the nickel and its ease of welding result in reliable permanent connections. The resistance of the tab material must be matched to the application to minimize voltage loss. The resistance can be calculated from the resistivity of the material, which is normally expressed in ohm-centimeters,

$$\text{Resistance} = \frac{\text{resistivity} \times \text{length (cm)}}{\text{cross-sectional area (cm}^2)}$$

The resistivity values of nickel and nickel-plated steel are

Nickel $\qquad$ $6.84 \times 10^{-6}\ \Omega \cdot \text{cm}$

Nickel-plated steel $\qquad$ $10 \times 10^{-6}\ \Omega \cdot \text{cm}$

For example, the resistance of a tab with dimensions of 0.635-cm width, 0.0127-cm thickness, and 2.54-cm length is

for nickel,

$$\frac{6.84 \times 10^{-6} \times 2.54}{0.635 \times 0.0127} = 2.15 \times 10^{-3}\ \Omega$$

for nickel-plated steel,

$$\frac{10 \times 10^{-6} \times 2.54}{0.635 \times 0.0127} = 3.15 \times 10^{-3}\ \Omega$$

As is evident, the resistance of the nickel-plated steel material is 50% higher than that of nickel for an equivalent-size tab. Normally this difference is of no significance in the circuit, and nickel-plated steel is chosen due to its lower cost.

The resistance may be significant, however, if the tab structure is used as a fuse for high-rate discharge cells in larger arrays. Due to the need for high discharge currents, such as in power tools or electric vehicles, normal cell protection devices such as PTCs may not be utilized. However, to protect against a cell short or other potentially damaging cell failure, particularly in large parallel combinations of cells, the tab interconnect can be constructed to be a fuse element. Fused links such as this can be used to prevent adjacent parallel connected cells from delivering excessive current into a shorted cell.

Resistance spot welding is the welding method of choice. Care must be taken to ensure a proper weld without burning through the cell container. Excessive welding temperatures could also result in damage to the internal cell components and venting may occur. Typically AC or capacitance discharge welders are used.

In all instances the weld should have a clean appearance, with discoloration of the base materials kept to a minimum. At least two weld spots should be made at each connection joint. When the weld is tested by pulling the two pieces apart, the weld must hold while the base metal tears. For tabs, the weld diameter, as a rule of thumb, should be three to four times the tab thickness. For example, a 0.125-mm-thick tab should have a tear diameter of 0.375–0.5 mm. Statistical techniques of weld pull strength for process control are helpful, but a visual inspection of the weld diameter must accompany the inspection process. Figure 5.12 shows examples of poor welds that can cause latent failures in battery packs.

In addition to proper welds, the connection tabs must be kept clean and straight to avoid mechanical stress on the welds. Tab edges should also be prevented from cutting into the cells. The least preferred method of battery connection is the use of pressure contacts. Although this technique is used with some inexpensive consumer batteries, it can be the cause of battery failure where high reliability is desired. This type of connection is prone to corrosion at the contact points. In addition, under shock and vibration, intermittent loss of contact may result.

FIGURE 5.12 Examples of poor welds.[4]

5.4.2 Cell Encapsulation

Most applications require that the cells within the battery be rigidly fixed in position. In many instances this involves the encapsulation of the cells with epoxy, foams, tar, or other suitable potting materials. Electronic circuit boards contained within the battery pack may be similarly fixed. Conformal coating is recommended to protect the circuitry from cell venting events, which could cause shorting of the circuit components.

Care must be taken to prevent the potting material from blocking the vent mechanisms of the cells. A common technique is to orient the cell vents in the same direction and encapsulate the battery to a level below the vent, as shown in Fig. 5.13. If possible, the preferred method to keep the cells immobile, within the battery, is through careful case design without the use of potting materials. Although this method may increase initial tooling costs, future labor savings could be realized. Figure 5.13c shows an example of dangerous use of encapsulation since the vent ends of the cells are completed obstructed.[4]

5.12   PRINCIPLES OF OPERATION

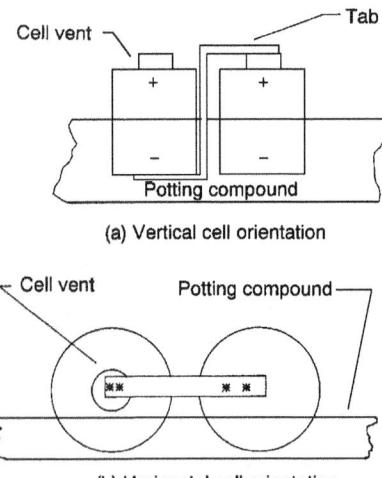

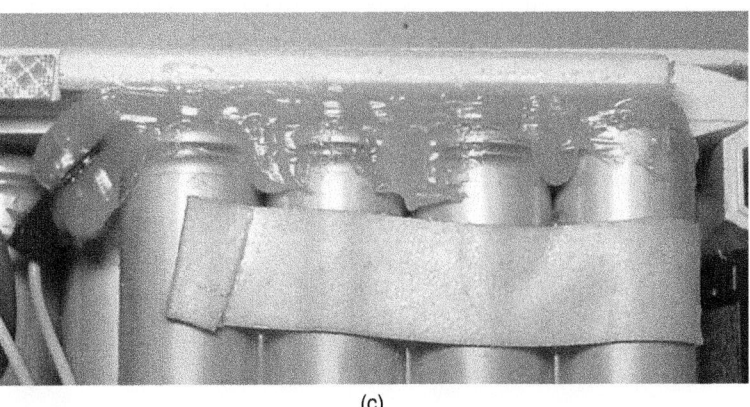

FIGURE 5.13 Battery encapsulation techniques. (*a*) Vertical cell orientation. (*b*) Horizontal cell orientation. (*c*) An example of poor use of encapsulation.[4]

5.4.3 Case Design

Careful design of the case should include the following:

1. Materials must be compatible with the cell chemistry chosen. For example, aluminum reacts with alkaline electrolytes and must be protected where cell venting may occur.

2. Flame-retardant materials may be required to comply with end-use requirements. Underwriters Laboratories, the Canadian Standards Association, and other agencies may require testing to ensure safety compliance.

3. Adequate battery venting must allow for the release of vented cell gases. In sealed batteries, this requires the use of a pressure relief valve or breather mechanisms. Case enclosures should also consider the expansion of cells during use. Lithium-ion polymer "bag" or "pouch" cells can expand during use and with age, thus increasing the size of the enclosure needed.

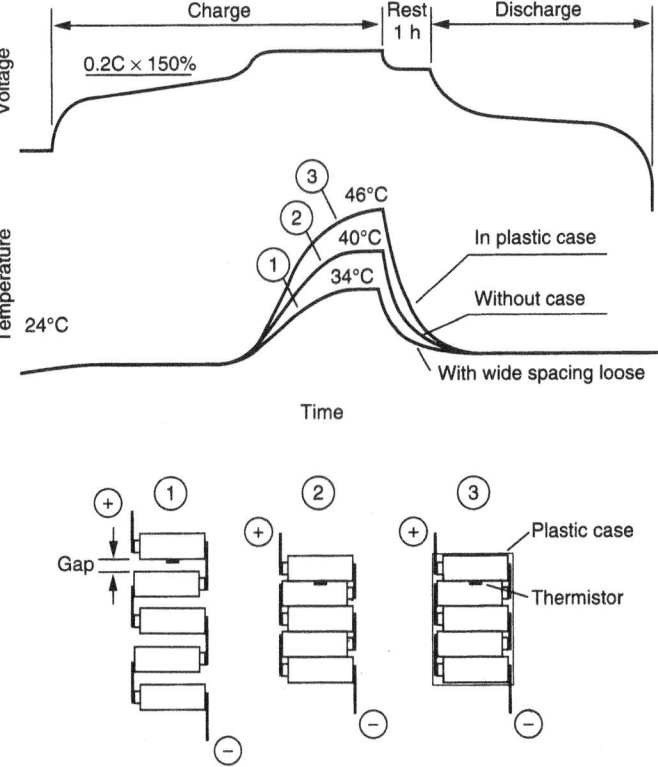

FIGURE 5.14 Temperature increase characteristics during charge of battery pack.

4. The design must provide for effective dissipation of heat to limit the temperature rise during use and especially during charge. High temperatures should be avoided as they reduce charge efficiency, increase self-discharge, could cause cell venting, and generally are detrimental to battery life. As previously mentioned, temperature gradients in the battery pack can also lead to cell imbalance, which can also degrade performance and safety. The temperature increase is greater for a battery pack than for an individual or separated cells as the pack tends to limit the dissipation of heat. The problem is exacerbated when the pack is enclosed in a plastic case. This is illustrated in Fig. 5.14, which compares the temperature rise of groups of cells with and without a battery case. Note that the internal cell temperatures can be higher than the measured skin temperatures of the cell, further increasing the detrimental effects.

5.4.4 Terminal and Contact Materials

Terminal material selection must be compatible with the environments of the battery contents as well as the surroundings. Noncorrosive materials should be selected. Nickel-cadmium and nickel-metal hydride batteries typically use solid nickel contacts to minimize corrosion at the terminal contacts.

A number of factors must be considered when specifying contact materials. Several principles apply to the substrate. The normal force provided by the contact must be great enough to hold the battery in place (even when the device is dropped) and to prevent electrical degradation and any resulting instability. Contacts must be able to resist permanent set. This refers to the ability of the

TABLE 5.1 Contact Materials

Gold plating	Provides the most reliable metal-to-metal contact under all environmental conditions
Nickel (solid)	Provides excellent resistance to environmental corrosion and is second only to gold plating as a contact material; material can be drawn or formed
Nickel-clad stainless steel	Performs almost as well as solid nickel, with excellent resistance to environmental corrosion
Nickel-plated stainless steel	Very good material; unplated stainless steel is not recommended due to the adverse impact of passive films that develop on stainless steel, resulting in poor electrical contact
Inconel alloy	Provides good electrical conductivity and good corrosion resistance; if manufacturers prefer to solder the contact piece in the circuit, soldering may be difficult unless an active flux is used
Nickel-plated cold-rolled steel	The most economical contact material; continuous, nonporous nickel plating of 200 μm is preferred

contact to resist permanent deformation with a set number of battery insertions. Temperature rise at high current drains due to the resistance of the contact material must be limited. Excessive temperature increase could lead to stress relaxation and loss in contact pressure, as well as to the growth of oxide films which raise contact resistance.

A common way to minimize contact resistance is to provide a wiping action of the device contact to the battery contact when the battery is inserted in place. Most notebook computer batteries incorporate this feature. Figure 5.3*b* illustrates a typical battery connector with wiping action receptacles.

Coatings should be selected to satisfy requirements not met by the substrate material, such as conductivity, wear, and corrosion resistance. Gold is an optimal coating due to its ability to meet most of the requirements. However, other materials may be used. Table 5.1 lists the characteristics of various materials used as contacts.

5.5 DESIGN OF RECHARGEABLE BATTERIES

All of the criteria addressed for the design of primary batteries should be considered for the design of rechargeable batteries.

In addition, multicell rechargeable batteries should be built using cells having matched capacities. In a series-connected multicell battery, the cell with the lowest capacity will determine the duration of the discharge, while the one with the highest capacity will control the capacity returned during the charge. If the cells are not balanced, the battery will not be charged to its designed capacity. To minimize the mismatch, the cells within a multicell battery should be selected from one production lot, and the cells selected for a given battery should have as close to identical capacities as possible. This is especially important with lithium-ion batteries because, due to the need for limiting current during charge, it is not possible to balance the capacity of the individual cells with a top-off or trickle charge.

Alternatively, techniques for minimizing imbalance should be employed as previously mentioned: reduce thermal gradients, limit differences in charge or discharge rates, etc.

Modern electronic circuits available today for lithium-ion and -polymer rechargeable batteries also include options to rebalance individual cells using resistive dissipative or active charge transfer methods.

Discharging rechargeable lithium cells beyond manufacturers' recommended limits should be minimized, particularly at low discharge rates. Brief excursions below minimum voltages at high discharge rates may be permitted. Furthermore, safeguards must be included to control charging to prevent damage to the battery due to abusive charging. Proper control of the charge and discharge process is critical to the ultimate life and safety of the battery. The two (2) major considerations to be addressed include:

1. Voltage and current control to prevent overcharge (overvoltage) and overdischarge (undervoltage). These controls can be located in the battery pack for redundancy or be part of the device's system design, which includes the charger.
2. Temperature sensing and response to maintain the battery temperature within the range specified by the battery manufacturers.

5.5.1 Charge Control

The controls for voltage and current during charge for most non-lithium rechargeable chemistry batteries are contained in the charger. Nickel-cadmium and nickel-metal hydride batteries may be charged over a fairly broad range of input current, ranging from less than a 0.05C rate to greater than 1.0C. As the charge rate increases, the degree of charger control increases. While a simple, constant current control circuit may be adequate for a battery being charged at a 0.05C rate, it would not suffice at a rate of 0.5C or greater. Protective devices may be installed within the battery pack to stop the charge in the event of an unacceptable temperature rise. The thermal devices that can be used include the following:

1. *Thermistor.* This device is a calibrated resistor whose value varies inversely with temperature. The nominal resistance is its value at 25°C. The nominal value is in the Kohm range with 10K being the most common. By proper placement within the battery pack, a measurement of the temperature of the battery is available and T_{max}, T_{min}, and $\Delta T/\Delta t$ or other such parameters can be established for charge control. In addition, the battery temperature can be sensed during discharge to control the discharge, e.g., turn off loads to lower the battery temperature, in the event that excessively high temperatures are reached during the discharge. Temperature measurement can also be used to determine when a nickel battery is fully charged: the rate of change of temperature when a sufficiently high charge current is used to charge the nickel pack will indicate completion.[5]

2. *Thermostat (temperature cutoff, TCO).* This device operates at a fixed temperature and is used to cut off the charge (or discharge) when a preestablished internal battery temperature is reached. TCOs are usually resettable. They are connected in series within the cell stack.

3. *Thermal fuse.* This device is wired in series with the cell stack and will open the circuit when a predetermined temperature is reached. Thermal fuses are included as a protection against thermal runaway and are normally set to open at approximately 30–50°C above the maximum battery operating temperature. They do not reset.

4. *Positive temperature coefficient (PTC) device.* This is a resettable device, connected in series with the cells, whose resistance increases rapidly when a preestablished temperature is reached, thereby reducing the current in the battery to a low and acceptable current level. The characteristics of the PTC device are shown in Fig. 5.15. It will respond to high circuit current beyond design limits (such as a short circuit) and acts like a resettable fuse. It will also respond to high

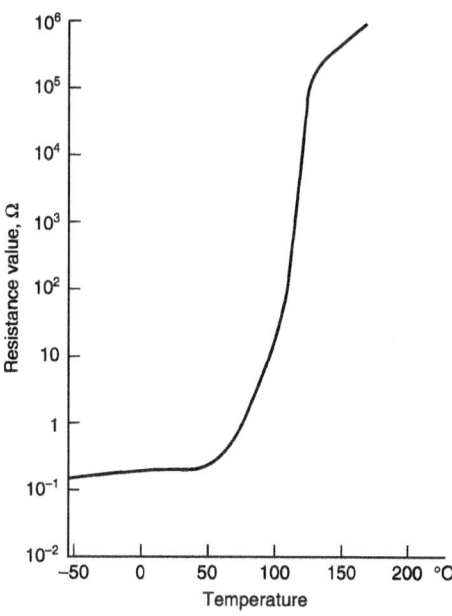

FIGURE 5.15 Characteristics of a typical positive temperature coefficient (PTC) device.

5.16 PRINCIPLES OF OPERATION

temperatures surrounding the PTC device, in which case it operates like a temperature cutoff (TCO) device.

Figure 5.16 shows a schematic of a typical nickel-chemistry battery circuit, indicating the electrical location of these protective devices. The location of the thermal devices in the battery assembly is critical to ensure that they will respond properly, as the temperature may not be uniform throughout the battery pack. Examples of recommended locations in a battery pack are shown in Fig. 5.17. Other arrangements are possible, depending on the particular battery design and application.

Details of the specific procedures for charging and charge control are covered in the various chapters on rechargeable batteries. For lithium-rechargeable chemistries, temperature is less useful as a charge control mechanism. Additionally, these chemistries have lower thermal run-away thresholds so thermal fuses, TCOs, and PTCs are not as useful. Cell voltage is the best control mechanism for the lithium rechargeable chemistries.

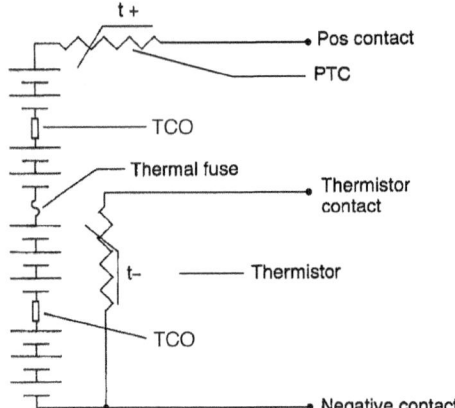

FIGURE 5.16 Protective devices for charge control in nickel-chemistry packs.

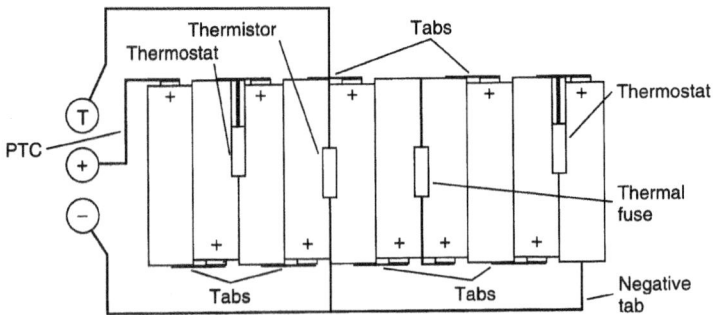

FIGURE 5.17 Location of protective devices in a nickel-chemistry battery pack.

5.5.2 Examples of Discharge and Charge Control

Electronic circuitry can be used to maximize battery service life by cutting off the discharge as close to the specified end or cut-off voltage as possible. Ending the discharge at too high a voltage will result in a loss of a significant amount of battery capacity; ending it at too low an end voltage and, thus, discharging the battery beyond its safe cut-off could cause permanent damage to the battery.[6] Similarly, on charge, accurate control, as discussed above, will enable a maximum charge under safe conditions without damage to the battery.

Modern battery packs and devices will utilize electronics in the device or battery to provide monitoring that ensures the operational limits of the battery pack are not exceeded. These electronic circuits can also provide enhanced safety and reliability, fuel-gauging, warranty data recording, and long-term battery health information. Some devices such as cellular phones, PDAs, MP3 players and digital cameras may use system-side monitoring and protection circuits, while larger devices such as notebook computers and power tools use monitoring circuits located inside the battery pack.

The ability to interrupt charge and discharge current to protect the battery may also reside in a separate location from where the monitoring occurs. Power tools, for example, may only interrupt charge current inside the charger, while a laptop computer may contain a battery pack that can interrupt both charge and discharge current inside the battery pack itself. A laptop computer, however, will often still contain circuitry to allow it to interrupt both charge and discharge currents independently.

Discharge and charge control are especially important for a lithium-ion battery for which the preferred charge protocol for a high rate charge is to start the charge at a relatively high, usually constant current to a given voltage and then taper charging at a constant voltage to a given current cutoff. Exceeding the maximum voltage is a potential safety hazard and could cause irreversible damage to the battery. Charging to a lower voltage will reduce the capacity of the battery, although in some applications such as uninterruptible power supply (UPS) systems, charging to a lower voltage is preferred. As the charge is continuous, the lower capacity is an accepted side effect in this usage since the life of the battery is more critical.[7]

Another interesting example is the control of charge for hybrid electric vehicles (HEVs). In this application, it is advantageous to obtain close to 100% charge efficiency or charge acceptance rather than maximum battery capacity since the battery is essentially used as a capacitor or power reservoir instead of an energy source. The charge acceptance for a nickel-metal hydride battery at the low state-of-charge (SOC) is close to 100%. As the nickel-metal hydride battery is charged, charge acceptance becomes progressively poorer, particularly above 80% SOC.[8] (At full charge, the charge acceptance is zero). In the HEV application using nickel-metal hydride batteries, the charge control keeps the state-of-charge, under normal driving and regenerative braking conditions, as close to 50% SOC as possible, and preferably within 30 to 70%. At these states-of-charge, the coulombic charge efficiency is very high.[9]

Other useful information can also be monitored and recorded by any charge or discharge control electronic circuitry either in the battery pack or on the device (system) side. This information can be used for warranty return analysis, life-time predictions, fuel-gauging, and similar advanced features.

5.5.3 Lithium-Ion Batteries

Special controls should be used with lithium-ion batteries for management of charge and discharge. Typically, the control circuit will address the following items that affect battery life and safety:

Cell Voltage. The voltage of each individual cell in the battery pack is monitored on a continuous basis. Due to safety concerns, secondary cell voltage monitors are often employed in the event that the primary monitor fails. These secondary monitors typically only look for an overvoltage condition and once detected, activate a permanent fuse in the charge current path. Depending on the specific lithium-ion battery chemistry that is used, the upper voltage limit on charge, as specified by the manufacturer, is usually limited between 4.1 to 4.3 V. On discharge, the cell voltage should not fall below 2.5 to 2.7 V. Newer lithium-ion and polymer formulations have significantly different overvoltage and undervoltage limits. Phosphate-based lithium-ion voltages have maximum limits near 3.8 V and minimums below 2.0 V. Titanate-based chemistries are even lower, with maximum limits near 2.5 V and minimums near 1.8 V. In all cases, the tolerance for primary detection is typically +/− 25 to 50mV for overvoltage conditions and +/− 50 to 100mV for undervoltage conditions.

Temperature Control. As with any battery system, high temperature will cause irreversible damage. With lithium-ion and polymer cells, temperature can alter how the cells should be charged or discharged. Guidelines from industry organizations limit charge currents when temperatures exceed suggested thresholds, while cell manufacturers often have similar limits for discharge currents at low temperatures.[10] Internal battery temperature, for most applications, should be kept below 75°C. Temperature cutoff, with a trip of 70°C and reset temperature in the range of 45 to 55°C, is routinely used. Temperatures in excess of 100°C could result in permanent cell damage. For this, permanent type fuses are used, typically set for 104°C with a tolerance of +/− 5°C. Temperature in lithium rechargeable chemistries is more difficult to detect compared to nickel-based chemistries', which exhibit a more linear trend. Internal cell temperatures are difficult to detect, and once a large rise occurs, it may be too late to take effective action. Thermal runaway can occur at temperatures as low as 130°C.

Short-Circuit Protection. Normally, current limits are incorporated into the protection circuits located in the battery pack or device. These circuits monitor the current in or out of the battery cells via a very low value series sense resistance placed in series with the power path. These circuits must be operating continuously and respond quickly to open a power MOSFET or similar device to interrupt the current. Short-circuit protection on discharge as well as overcurrent protection on charge (from a faulty charger) are often employed in lithium battery packs. As a backup, a PTC device or fuse is placed in series with the battery pack. It is advisable to place the PTC between the pack assembly and the output of the battery. By placing it at this point, the PTC will not interfere with the operation of the upper or lower voltage detection of the electronic control circuit. However, for some high-rate devices, such as power tools and electric vehicles, a PTC is not utilized since short duration peak currents must be tolerated. In these devices, the electronic overcurrent monitors may have multiple detection thresholds that are able to respond to not only the magnitude of the current, but also the time duration of the current.

5.6 ELECTRONIC ENERGY MANAGEMENT AND CONTROL SYSTEMS

In the mid-1990s, an important development in rechargeable battery technology was the introduction of the use of electronic microprocessors to optimize the performance of the battery pack, control charge and discharge, enhance safety, and provide the user with information on the condition of the battery. This microprocessor function can be incorporated into the battery (the "smart battery"), into the battery charger, or into the host battery-using equipment. Since their introduction, such smart battery electronics have become more sophisticated, more integrated, more precise, and less costly.[10]

Although generically referred to as a "smart battery," there are varying degrees of smart or intelligent battery systems. Specific examples detailed later include Smart Battery System (SBS SMBus) products that conform to a set of specifications for interoperability.[11,12]

But batteries of all sizes can incorporate electronic controls for enhanced performance, safety, and reliability. Guidelines for the proper design and use of rechargeable battery packs have expanded since some high-profile accidents in recent years.[13] These guidelines suggest the use of electronics for maintaining the operation of the battery cells within safe limits for particular applications such as cellular phones[14] and laptop computers.[15]

Electronics in a battery pack can range from basic protection functions that prevent or mitigate abusive conditions to sophisticated measurement, calculation, and communication engines that provide protection, monitoring, and communications to a host or end-user device. Such equipment is commonly found in battery packs using lithium rechargeable chemistries, such as notebook computers and smartphones, but it is also being utilized with other chemistries in handheld two-way radios, power tools, and hybrid-electric vehicles.

Some of the features and benefits of embedding advanced electronics into the battery pack, system or charger include:

- **Charge control:** Battery electronics can monitor the battery during charge and assist in controlling the charge rate and charge termination, such as time, Vmax, -deltaV, deltaTemp, and deltaTemp/deltatime, to cut off the charge or to switch to a lower charge rate or another charge method. Constant current to constant voltage charge can be controlled and options can be incorporated into the electronics for pulse charging, "reflex" charging (a brief periodic discharge pulse during charge), or other appropriate control features, such as pre-charging cells at a lower charge rate when at low capacity levels or low temperatures. Finally, charge protection from overcurrent and overvoltage conditions can also be included and tailored for the particular end products requirements. Charge control can occur within the battery pack or the charger.

- **Discharge control:** Discharge control is also provided to regulate such items as discharge rate, end-of-life cutoff voltage (to prevent overdischarge), cell equalization (balance) and assist with thermal management. Individual cells, as well as the entire battery pack, can be monitored for voltage or temperature during discharge and direct action taken to alter the discharge current or inform the host device to terminate or slow the discharge. Pack current can also be monitored to detect overcurrent and short-circuit conditions to prevent damaging the cells. Discharge control can be placed inside the battery pack or the host end-user equipment using current control devices such as power MOSFET switches.

- **Cell balancing:** Cell balancing can be used to improve the performance of multicell battery packs by maintaining all the series cell elements in voltage balance with one another. Maintaining cell balance therefore increases the usable capacity of the battery pack and improves cycle life. Cell imbalance is often the cause of poor battery pack life in multicell series packs, typically those greater than four series cells. Even if the individual cells are well matched when the battery pack is assembled, cells will diverge with time due to slight differences in capacity, self-discharge rate, etc. Temperature gradients across a large multicell battery pack will also alter the rate of divergence such that cell balancing or equalization is often required in applications such as electric and hybrid-electric vehicles. Some chemistries are easier to rebalance using overcharge techniques, while most lithium rechargeable chemistries require alternate approaches such as bypass balancing or charge transfer balancing.[10]

- **Communications:** Communicating the battery information to the end-user or host device can be accomplished by both simple and sophisticated methods, depending on the requirements. Basic measurement data, such as voltage, temperature, and current, can be relayed to the host device for use in charge or discharge control or to calculate battery state-of-charge (SOC). If the battery pack contains on-board calculation capability, then more information can be calculated locally and communicated to the end-user or host device. Information such as battery SOC information estimates the remaining battery capacity by factoring in such variables as the discharge rate and time, temperature, self-discharge, cell impedance, past history, charge rate, and charge duration. The SOC, remaining capacity, or run-time can be displayed locally on the battery pack by a sequence of illuminated LEDs or a LCD display. Detailed data can also be directly communicated to the host device via a standard communications link such as the Inter-Integrated Circuit (I2C) bus or the derivative System Management Bus (SMBus). This communications option provides significantly more information than can be displayed via a local method and is often utilized by the end-device for use in a more detailed graphic form, such as on a laptop computer's main screen. Alternatively, single-wire data communications (DQ-bus, HDQ, or others) or simple level-based analog threshold signals can also be used to communicate that the battery is operating outside of normal limits and that external action should be taken by the charger or the host end-use device.

- **Historical information:** Battery data is collected during the life of the battery and can be used to make changes to the operating algorithm to maintain optimum performance. Other information can also be collected beyond initial manufacturing information (date of manufacture, chemistry, configuration), including detailed battery history, cycle count, and other such data, which can provide a complete accounting of the battery's usage over time. Data such as maximum temperature, time at temperature, time at voltage, and similar parameters can also be utilized to determine battery aging. This data can be retained even if a battery failure occurs so that warranty returns can be properly evaluated and analyzed.

- **Customization:** The continued advancement of electronics technology allows the monitoring electronics to be custom tailored to the chemistry of the battery as well as the specific requirements of the cell manufacturer and host end-device. This improves the accuracy, safety, and reliability of the battery pack as well as the performance of the device using the battery. "Smart" battery electronics today range from simple protection circuits with limited communications to sophisticated all-in-one circuits that provide protection, fuel-gauging, balancing, charge control, history, and adaptive algorithms that compensate for the battery as it ages.

There are several main elements to consider in the design of batteries containing electronics. These are:

1. *Monitoring and measurement.* There are multiple parameters that can be measured directly by battery pack electronics to provide the basic information about the battery pack. These include cell and pack voltage, pack current, cell and/or pack temperature, and time.

   The accuracy of the measurements should fit the requirements of the chemistry and the intended application: Precise fuel-gauging and charge control often require accurate measurements while protection from abusive conditions may require less stringent monitoring. For high-reliability gas-gauging of remaining capacity and runtime, it is important that these measurements be made as accurately as possible to provide the best data for the control algorithms and predictive functions.

   In some cases, measurements are not required, only simple monitoring is needed. Signals are only generated when the monitored values (cell voltages, pack currents) exceed preset thresholds—otherwise, the exact measured values are not required. Simple comparator-based monitoring is often sufficient when protecting the battery from operation outside desired limits.

   Combinations of certain measurements can provide additional information, such as a cell's DC resistance or impedance for example, when voltage and current measurements are synchronous.

   Voltage measurements can be critical as the charge control and termination depend on the battery voltage, which, for some chemistries, should be accurate to 50 mV or better at the cell level. An inaccurate measurement could result in under- or overcharge, which could lead to short service life or damage to the battery. In the case of the lithium rechargeable cells, overcharge could be a safety hazard. Similarly, on discharge, terminating the discharge prematurely results in a shortened service life, while overdischarge could result in damage to the cells.

   Errors in the measurement of current affect not only the calculation of capacity and the state-of-charge "gas gauge," but influence the termination of charge and discharge, as the termination voltages may vary depending on the current, particularly in high-current devices such as power tools. Complicating this measurement is the fact that current is not a constant value during discharge; there are often multiple device power modes and high current pulses as short as milliseconds.

   Likewise, temperature is an important parameter, as the performance of batteries is highly temperature dependent, and exposure to high temperatures can cause irreversible damage to the cells. Temperature gradients across the battery pack that cause the cells to reach different temperatures will create cell imbalance, which limits battery life.

   The key points are to select monitoring electronics that match the application requirements. A nickel battery pack for a power tool may only require pack-level temperature and voltage monitoring, while the same application using rechargeable lithium may require individual cell voltage monitoring and limited temperature monitoring.

2. *Calculation.* Having calculation ability in the battery pack allows the system to adjust to the environment or usage conditions and permits safer operation and better performance. The level of calculation required can vary from a predefined logic state machine to a more versatile processor that can factor more variables when making adjustments.

   For example, consider the precharge function typically used with lithium rechargeable chemistries. When the cell voltage or temperature is below preset thresholds, a separate charge path that reduces the charge current is enabled until the temperature or cell voltage increases above the preset thresholds. Similar level-based conditions are often used for charging.

More sophisticated calculation can be required for accurate gas-gauging to represent battery state-of-charge, state-of-power (the battery's ability to provide high discharge currents), or state-of-health (battery cycle life information). Although some chemistries can provide relatively useful gas-gauging information by monitoring the open-circuit voltage of the battery, most lithium rechargeable chemistries require more sophisticated approaches.

Calculations transform any measured data through the use of simple or complex algorithms, depending on the host device application's requirements and the chemistry used. Prior knowledge of battery characteristics, such as capacity at various discharge loads and temperatures, charge acceptance, self-discharge, etc., are required to determine future battery performance. Early battery electronics used simple linear models for these parameters, which severely limited accuracy in predicting the battery's performance. As noted in the descriptions in various chapters in this Handbook, battery performance is often very nonlinear. Self-discharge, for example, is a complex relationship influenced at least by temperature, time, state-of-charge, and other factors. Further, the performance of even those batteries using the same chemistry varies with design, size, manufacturer, age, etc. A good calculation engine and algorithm will account for these relationships and help assure safe, reliable operation.

Calculations can also be used to maximize the performance from the battery pack during actual use by considering calculated values, such as cell impedance along with voltage, current, and temperature measurements. Techniques that operate the cells to the edge of their performance envelope require precise measurements but also well-known models of the cells' characteristics and performance under various usage conditions. Processors to perform these measurements and calculations in real time, while under heavy loads, can maximize the performance obtained from the cells in the battery pack. High-end power tool products and hybrid and electric vehicles often use such sophisticated calculations.

As with monitoring and measuring, properly matching the calculation requirements with the battery chemistry's needs and the end-application requirements is critical to a high performance, low cost, reliable design.

3. *Communication.* Just as measurements may vary between exact values and threshold monitoring, communications can range from detailed measurement data over a communications bus to a single line "go/no-go" signal that indicates that the battery pack is operating outside of preset limits.

Battery packs have for years used a single interface line to represent the temperature of the battery via the voltage on the line. The voltage is a representation of the pack temperature with a negative-temperature-coefficient (NTC) thermistor. The resistance of an NTC temperature sensing device located in the battery pack is monitored externally, often by the charger. Low resistances represent high temperatures and vice versa. Nickel-based chemistries often use this signaling approach to detect end-of-charge via a change in the rate of rise of the temperature. This same approach can still be utilized by chemistries that do not exhibit any temperature changes with full charge. For example, a common technique with lithium rechargeable chemistries is to simply mimic the temperature of a "hot" pack, which signals the charger to terminate charging.

When more information is to be conveyed between the battery, charger, and host device, a digital interface, such as the Inter-Integrated Circuit (I2C), Serial Peripheral Interface (SPI), DQ/HDQ, 1-Wire, or Systems Management Bus (SMBus) protocol, is often used. These are standardized data communications interfaces with low power characteristics well suited for battery applications. The electrical and data protocols are defined and available in many prepackaged parts for use in battery packs, chargers, and end-equipment devices. Automotive battery systems may utilize Local Interconnect Network (LIN) or Controller Area Network (CAN) bus interfaces for additional robustness.

Information that is often communicated between the battery and charge includes the required charge conditions, such as maximum charge current, maximum charge voltage, and perhaps maximum temperature to initiate charge separately from a maximum charge continuation temperature.

As previously mentioned in the precharge example, other charge-gating information may also be communicated relating to the specific conditions of the battery pack prior to initiation of charge.

During discharge, information from the battery to the end-equipment host device can be utilized to maximize the run-time of the device while also preventing abusive discharge conditions. In laptop computers, for example, various power management techniques are employed by the notebook systems as the battery's SOC decreases. High current loads, such as the spin-up current from a hard disk or DVD player can be delayed briefly while the notebook reduces loads elsewhere, perhaps momentarily dimming the screen backlight or powering down other subsections. Advanced smart batteries can provide information to the notebook system so that such decisions can be easily determined.

Gas-gauging that represents the run-time of the device in meaningful terms instead of SOC percentages can also be communicated. Similar application-specific information that brings more meaning to the end equipment and provides a more user-friendly experience is also possible. Smart batteries can provide information ranging from the time remaining during charge as well as discharge, the number of usage cycles, and the approximate remaining useful life. In the case of HEVs, the state-of-power is more valuable to the vehicle controller: can the battery support a high power load for enough time until the internal combustion engine can be restarted? If not, then the engine may not be stopped until the battery has reached a higher state-of-power capability.

In some devices, the reliability of the communication link is critical both to accurately convey the data and to prevent unauthorized access. (Some battery systems include encryption or challenge-response authentication to limit unauthorized battery packs to operate in the end-equipment or charger.)

5.6.1 The Smart Battery System (SBS)[11]

A formalized electronic battery management system was created by leading cell suppliers, laptop computer makers, and semiconductor manufacturers in 1995 to standardize the electrical interface between battery packs, chargers, and notebook computers. This Smart Battery System (also called SMBus System) has been widely adopted by notebook computer makers and other portable device manufacturers for many industrial and general purpose battery systems. The physical form factor of the battery packs is not standardized (although some standard sizes exist, such as the DR202). The standardization is only for the communications interface.

The System Management Bus (SMBus) defines additional protocols and electrical requirements on top of the I2C specification developed by Philips Corp. These protocols include error detecting mechanisms, minimum voltage levels, and similar timing and power requirements. Typical portable battery systems utilize SMBus V1.1, while fixed non-battery systems may also use SMBus V2.0 for other devices, such as backlight controllers found in a typical notebook computer.

The Smart Battery System also includes specifications for the data content and transfer between a host device such as a notebook computer, the smart battery, and a smart charger. The Smart Battery Data Specification and the Smart Battery Charger Specification detail the interaction and data requirements for each device. SBS smart batteries provide up to 34 data values, both measured and calculated, that can be utilized by the host device or charger to enhance battery performance and system power management. Similarly, there are three levels of smart chargers that can be utilized in SBS platforms.

The goal of the smart battery interface is to provide adequate information for power management and charge control regardless of the particular battery's chemistry. The smart battery consists of a collection of cells or single-cell batteries and is equipped with specialized hardware that provides present state, calculated, and predicted information to the Host. The electronics need not be inside the smart battery if the battery is not removable from the device.

Many semiconductor companies supply battery monitor, charger, or host controller products that comply to the various SBS standards to provide easy operability. However, testing should still be done with components since interpretation of the specifications has left some minor incompatibilities.

REFERENCES

1. Tyco Electronics, Battery Interconnection System Products Receptacle Assemblies, e.g., part numbers 1-1123688-7 or 1-1437118-0.
2. TPS6107x Boost Converter datasheet, Texas Instruments, March 2009, www.ti.com.
3. Duracell Alkaline Technical Bulletin, www.Duracell.com/OEM.
4. From "The Dangers of Counterfeit Battery Packs" by Micro Power Electronics Inc., www.Micro-Power.com.
5. Duracell NiMH Technical Bulletin, www.Duracell.com/OEM.
6. Sec. 21.4.2 of this Handbook.
7. Sec. 26.4 of this Handbook.
8. Sec. 21.5 and 25.5 of this Handbook.
9. Sec. 22.11.5 of this Handbook.
10. Datasheets for smart battery monitors, bq20z95, bq6400, bq78PL114, and bq77PL900. Texas Instruments, www.ti.com.
11. Smart Battery Data Specification, Rev. 1.1, System Management Bus Specification, www.sbs-forum.org; Smart Battery System Implementers Forum, part of the System Management Interface Forum, Inc.
12. D. Friel, "How Smart Should a Battery Be," *Battery Power Products and Technology*, March 1999.
13. "A Guide to the Safe Use of Secondary Lithium Ion Batteries in Notebook-type Personal Computers" and "Safe Use Manual for Lithium Ion Rechargeable Batteries in Notebook Computers," Japan Electronics and Information Technology Industries Association (JEITA), www.jeita.or.jp, and Battery Association of Japan (BAJ), www.baj.or.jp; April 2007.
14. IEEE ANSI STD. 1725(TM)-2006, "IEEE Standard for Rechargeable Batteries for Cellular Telephones," IEEE, 3 Park Avenue, New York, www.ieee.org.
15. IEEE Std. 1626(TM)-2008, "IEEE Standard for Rechargeable Batteries for Multi-Cell Mobile Computing Devices," IEEE, 3 Park Avenue, New York, www.ieee.org.

CHAPTER 6
MATHEMATICAL MODELING OF BATTERIES

Shriram Santhanagopalan and Ralph E. White

6.1 INTRODUCTION

Mathematical modeling of batteries can be described as a process of developing an equation or a set of equations to describe the performance of a battery. For example, a simple, single equation model can be used to predict the capacity of a battery as a function of the discharge current obtained from that battery. More complicated models can be developed based on equations used to describe the phenomena that occur between the current collectors of a single pair of electrodes. For example, a model for a lithium-ion cell with one spatial coordinate (from the anode to the cathode, say) could consist of the anode current collector (e.g., copper metal foil), an anode electrode coating made of carbon, a separator, a cathode electrode coating (e.g., made of $LiCoO_2$), and finally the cathode current collector (e.g., aluminum foil). Such a model would be based on the spatial coordinate between the current collectors and a unit of projected electrode coating area for the other two spatial coordinates. This model can be used, for example, to predict the performance of a lithium-ion jelly roll by appropriately accounting for the current collectors that are coated on both sides, the actual projected electrode area to form the cell, etc. A mathematical model of a battery with multiple electrode pairs could then be formed by internal or external connections between the cells in a series or parallel arrangement as needed for the voltage and capacity requirements.

The level of detail of a mathematical model depends on its intended use. For example, complex three-dimensional models (three spatial coordinates and time) with multiple electrode pairs have been developed to study the thermal characteristics of the electrode pairs and cells made from them. This chapter begins with a description of the evolution of battery models.

6.1.1 Evolution of Battery Models

The earliest mathematical models for batteries were simply empirical relationships between measured parameters, such as the battery voltage, overall resistance, density of the electrolyte, pressure within the can or temperature of the cell, versus the remaining capacity under different operating conditions. These models are still used today, and perhaps the best known example is Peukert's relationship.[1] This equation has been used to represent the discharge capacity as a function of discharge current for a lead-acid cell, for example, as shown in Fig. 6.1, which shows a comparison between the capacity predicted by Peukert's equation and the experimentally measured value for the cell capacity. This simple relationship has been used under a variety of different scenarios for several decades in the battery industry.

6.2 PRINCIPLES OF OPERATION

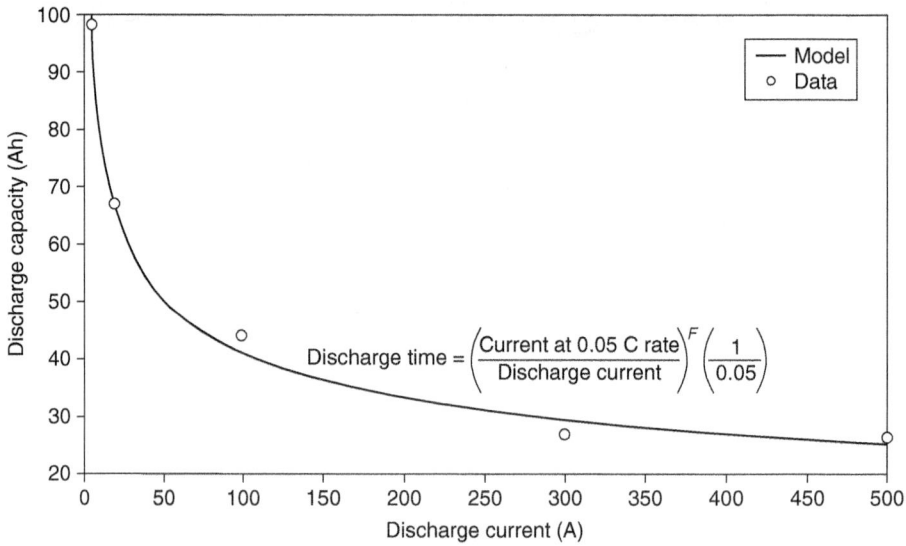

FIGURE 6.1 Cell capacity vs. discharge rate for a 100 Ah lead-acid battery following Peukert's equation with current at 0.05-C rate equal to 4.98 A and the Peukert coefficient (F) set to 1.3.

A second example of an empirical relationship that has been used to monitor the state-of-charge (SOC) of a battery is shown in Fig. 6.2. This figure presents the available capacity (SoC, %) from a lead-acid battery as a function of the specific gravity of the electrolyte in that cell during discharge. In this case, a simple correlation between the available capacity of the cell and the measured density can be seen. Later in this chapter, we will reinforce the basis for the validity of this largely experimental observation of the relationship between the electrolyte density and battery performance.

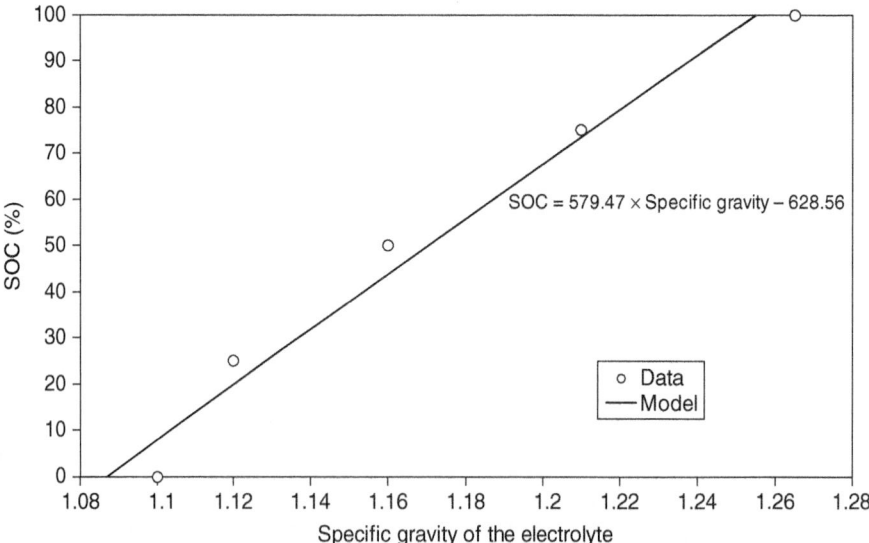

FIGURE 6.2 Linear relationship between the specific gravity of the electrolyte and the state of charge (SOC) of a lead-acid battery.

Due to the success of Peukert's equation, investigators started to determine what causes the capacity of the cell to drop and why the relationship between the discharge rate and the available capacity of the battery follows the relationship shown in Fig. 6.1.

One early step toward understanding limitations in battery performance was to quantify the loss in efficiency of the battery compared to theoretical expectations and to attribute such losses to different parts of the cells. The limitations within the cell were categorized into those arising from the electrodes, those arising from the electrolyte, and so on.

The use of AC impedance as a diagnostic tool for electrochemical systems emerged in the 1970s[2, 3] and with that evolved the concept of representing the battery as a circuit consisting of traditional electrical components, such as resistors and capacitors, as shown in Fig. 6.3.

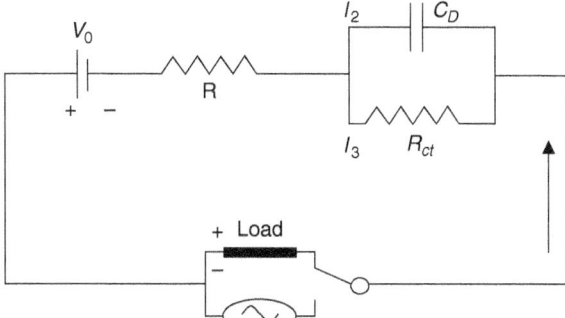

FIGURE 6.3 Equivalent circuit representation of a cell.

Figure 6.3 presents a commonly used circuit diagram to represent a single cell.[4] The behavior of a battery can be explained in terms of the values for the circuit elements shown in Fig. 6.3. The voltage source V_0 represents the open circuit voltage (OCV) of the battery, which can be interpreted as representing the thermodynamic limitations on the performance of the cell. The resistance term R refers to ohmic drop inside the battery that arises from current passing through the electrolyte, contact resistances, and the like. The two parameters R_{ct} and C_D together represent charge transport across the interface: R_{ct} refers to the Faradaic part of the charge transfer resistance, and C_D is a pseudocapacitance term often used to characterize mass transfer limitations. With these parameters in place, one can characterize the ideal behavior of the battery as the change in V_0 during a discharge. The other parameters shown in Fig.6.3 can be assigned different values to represent experimental data obtained, for example, at various discharge rates of the battery. Any deviation from the open circuit voltage is because of limitations in the transport of ions, ohmic drop, kinetics of the reaction, etc. Any such non-ideality can be captured as a change in one of the parameters. For example, difficulties in the electrochemical reaction are represented by changes in the value of R_{ct}.

This circuit analog approach to modeling of a battery was a big step toward mechanistic investigations, albeit the only progress from the theoretical front being made was the classification of observed changes in the battery from one experimental parameter to several empirical circuit parameters. Circuit analog models are still used in the battery industry because the relationship between the battery voltage and current drawn can often be expressed as an analytical expression and is hence easy to use, and the computational requirements for such models are so minimal that this can be used for rapid estimates of the state-of-charge and can be implemented in hardware. However, the utility of a circuit-based model in the design of a battery is limited because the circuit elements do not readily translate into physically meaningful parameters that can be used in battery design.

The need to link the battery design parameters such as electrode coating thickness to observed experimental behavior has led to the formulation of physics-based models that employ universal laws (e.g., charge, mass, momentum, and energy balances) to characterize the behavior of the individual components within a cell as functions of their material properties. These component models can then be integrated to form rigorous mechanistic models to describe the behavior of the cell. Approximation of properties of composite electrodes over macroscopic volume elements has been used to develop mathematical models that use practically measurable physical properties such as effective conductivities and diffusivities. Today, people in industry are using sophisticated physics-based models to determine the precise design measurements, such as actual geometry of the electrodes and the electronic circuitry needed to integrate batteries with other components in a device.[5]

The difficulties presented by the mathematical tedium of the models have by and large been removed by the development of excellent user-friendly interfaces such as Matlab™ and COMSOL Multiphysics™. As a result, the demand for development of universal standards for batteries and the use of models to determine such standards have reached an all-time high. Thus battery models have evolved significantly over the empirical relationships that served as rules of thumb in the battery industry. However, the principal objective behind the models has remained the same: to predict whether the battery can deliver the required output (in terms of energy or power) through the required amount of time. The following sections describe the process of developing a mathematical model for a cell, the choice of model equations and parameters, and some examples of implementing such models for optimal design of batteries.

6.2 DEVELOPMENT OF A MATHEMATICAL MODEL

Developing a mathematical model for a battery involves identifying the physical processes that take place during the operation of the battery and how each component responds to those processes. A systematic approach to represent the response of the components to the various processes is by using generalized laws that describe the behavior of the materials under different scenarios. The simplest example is the representation of current flow through a copper wire: When current is drawn or supplied to a battery, it invariably passes through the bus-bars or tabs that connect the electrodes to the external load (or power source). The first step is to identify the physical processes that happen across the copper cable once the battery is connected to the load: There is a flow of current across the cable; there is some heat-up of the cable—especially at the welded joints—that increases with increase in the flow of current, and so on. The second step is to identify the general rules that help us to quantify each of these phenomena. For example, the flow of current across the copper cable follows Ohm's law, which states that the potential drop (V) depends on the amount of current (I) that flows across a metallic cable and the resistance the metal offers to the flow of current:

$$V = IR \tag{6.1}$$

where R represents the resistance of the metal. Equation (6.1) makes it clear that if the value of the resistance R is higher, the potential drop required to pass a given amount of current out of the battery increases. This is in line with our observation that the voltage of a battery during discharge with a rusty plug (having a lower electronic conductivity) drops faster compared to that with proper contacts. The second phenomenon of interest in this example is the heat-up of the welds. The heat-up of a material due to passage of current was first quantified by Joule using the following relationship:

$$\Delta H = I^2 R t \tag{6.2}$$

Here, ΔH represents the amount of heat generated, R is the resistance at the weld (the heat is a function of the conductivity of the material that the weld is made up of), and t is the duration for which the current flows across the weld. Thus we see that Eqs. (6.1) and (6.2) can be used to adequately describe some physical observations during the passage of current through a cable; these equations constitute the mathematical model in this example. These equations apply to any material that the cable is made of, as long as one can measure the conductivity of the material experimentally and prove independently that each of the above laws holds true under the operating conditions of interest. The utility of the model is versatile. For example, one can choose the material of which the weld is made, by knowing the amount of heat that will be generated for a given discharge rate and duration. Alternatively, one can determine the maximum amount of current that can be passed safely without damaging the welds. Note that the values for I and t are specified depending on the situation of interest, in this case, the discharge rate and the duration of discharge, respectively. These variables specify the operating conditions. The variables V and ΔH represent the measured physical quantities. A parameter like the conductivity of the metal, on the other hand, remains constant for different durations or rates of discharge—it is an inherent property of the material. Every mathematical model thus comprises input variables, measured variables, and physical parameters.

The next task in the development of a mathematical model for a battery is to build such component models for other physical phenomena that take place within the battery during its operation. The complexity of a mathematical model depends on the number of physical processes we wish to describe and the accuracy with which these must be described. For example, if we had chosen to ignore the heat-up at the weld-joints, one may adequately describe the flow of current across the bus-bar using Eq. (6.1) only. On the other hand, for some materials, the resistance of the cable changes with temperature. If we wish to capture these effects very accurately, we need to use additional equations that describe the change in the parameter R shown in Eq. (6.2). An efficient model should thus strike a balance between the complexity associated with describing the data set and the need for insight gained from the results. The following sections describe complementary approaches used to develop such models. Finally, the various components are integrated to study the interaction among the different parts of the battery in response to the mutually interactive physical phenomena of interest. Based on the degree of comprehension the models provide, they are usually classified into empirical and mechanistic models.

6.3 BUILDING EMPIRICAL MODELS

An empirical model often assumes the form of an expression that relates the operating conditions such as the rate of discharge or the load across the battery with the measured quantities such as the temperature or voltage of the battery. Such an expression is often obtained by prior knowledge of the battery's behavior or by trial and error. However, a basic understanding of the various limitations within the cell serves as a guideline. For example, the equivalent circuit shown in Fig.6.3 was designed to account for the deviation of the cell voltage from the open circuit voltage due to several factors as described in Sec. 6.1.1. A popular objective of a battery model is to predict the cell voltage as a function of the state-of-charge (SOC) of the battery. This section demonstrates the development of such a relationship for the circuit shown in Fig. 6.3.

The voltage drop across each of the resistors R and R_{ct} follows Ohm's law [see Eq. (6.1)]. The rate of charge buildup in the capacitor C_D equals the current that flows through the capacitor. This is expressed mathematically as follows:

$$I_2 = \frac{dq}{dt} \tag{6.3}$$

Kirchoff's node-and-loop rules relate the currents that flow across the different branches of the circuit and the voltage across each branch. These rules state that the voltage across any branch is the sum of all voltage drops along the branch and that the sum of all currents that enter or exit a node on the circuit is zero. For example, the total current I that branches out into I_2 and I_3 as shown in Fig. 6.3. Hence according to Kirchoff's laws we have

$$I = I_2 + I_3 \tag{6.4}$$

6.6 PRINCIPLES OF OPERATION

The constraint on the voltage across each branch yields the following equations:

$$V = V_0 + IR + I_3 R_{ct} \tag{6.5}$$

$$V = V_0 + IR + \frac{q}{C_D} \tag{6.6}$$

Equations (6.3) through (6.6) above can be rearranged to obtain the relationship between the change in the applied current, I, with time, dI/dt, and the resultant voltage drop, $(V - V_0)$

$$R\frac{dI}{dt} + \frac{1}{C_D}\left(1 + \frac{R}{R_{ct}}\right)I = \frac{dV}{dt} + \frac{1}{R_{ct}C_D}(V - V_0) \tag{6.7}$$

Equation (6.7) now contains the component models [Eqs. (6.3) through (6.6)] for each element of the battery model. The mathematical solution for Eq. (6.7) can be found in standard references. For the case of constant current, the solution takes the following form:[4]

$$V = \frac{Q_0}{C_D} e^{-t/R_{ct}C} + V_0 + IR + IR_{ct}\left(1 - e^{-t/R_{ct}C_D}\right) \tag{6.8}$$

This model equation relates the change in cell voltage, V, to the input current, I. The parameter Q_0 refers to the total capacity of the battery. The change in cell capacity during a charge or discharge is calculated by integrating the current passed as follows:

$$Q = Q_0 - \int_0^t I\,dt \tag{6.9}$$

The values for the circuit elements such as V_0, C_D, R, and R_{ct} are adjusted to represent the experimental data of interest adequately.

A typical set of such values for the NiMH battery is shown in Table 6.1. These parameters were obtained by comparing results from the charging and discharging of a 10-cell battery pack with 85 Ah capacity, at 64 A current. Figure 6.4 shows the cell voltage versus the capacity of the cell during charge and discharge at various rates. Shown also are some experimental data for comparison. The results for a wide range of C-rates are obtained by utilizing the circuit parameters extracted using data at one discharge (or charge) rate.

TABLE 6.1 Parameters for the Equivalent Circuit Representation of a NiMH Battery[4]

Parameter	Discharge	Charge
E_a (kcal/mol)	6	6
r_0, mΩ	1.45	2.70
r_1, mΩ	—	−5.15
r_2, mΩ	—	6.23
nj (See expression for R_{ct})	0	2
τ	24	
C_D(F)	R_{ct}/τ	
R, mΩ	0.786	
R_{ct}, mΩ	$\left(\sum_{j=0}^{nj} r_j (SOC)^j\right) \exp\left(-E_a/RT\right) - R$	

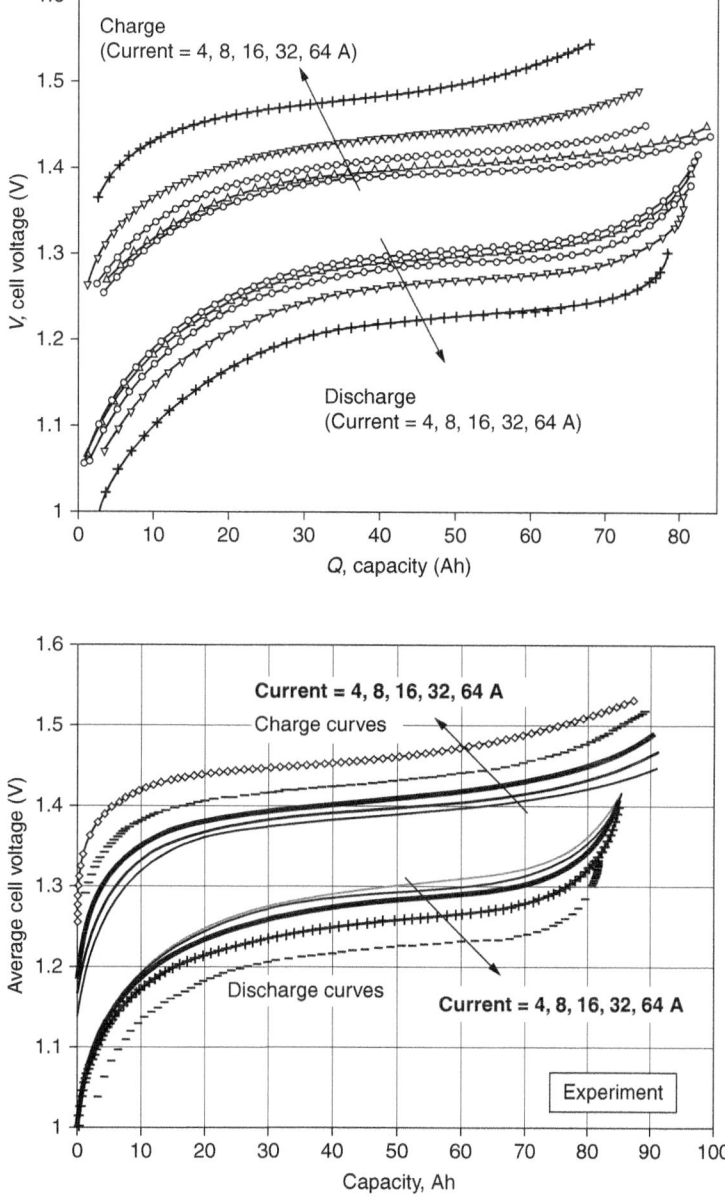

FIGURE 6.4 Model predictions for a NiMH cell. Results from a model using the empirical equivalent circuit shown in Fig. 6.3 are shown in the upper plot and the corresponding experimental data on the lower plot.[4]

Similar results can be obtained for constant power loads by replacing the current (I) in Eq. (6.7) with P/V where P is the power set by the load. Figure 6.5 shows comparison of model versus experimental data for the parameter set shown in Table 6.2 for a lithium-ion cell, using the same set of equations described above.

6.8 PRINCIPLES OF OPERATION

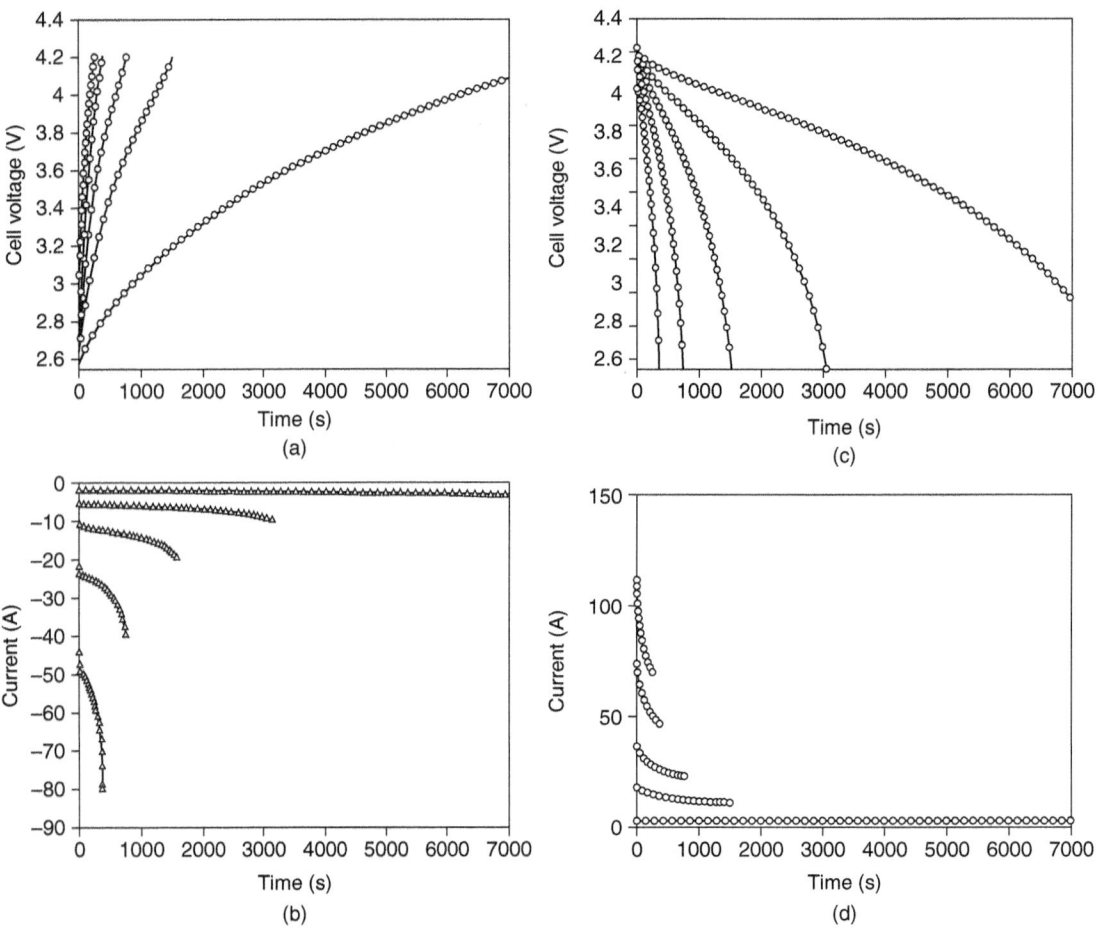

FIGURE 6.5 Model predictions vs. experimental data[6] from the equivalent circuit model under constant power discharge for a lithium-ion cell at 0°C. (a) Cell voltage during charge at constant power; the power supplied was set to 10, 25, 50, 100, 200 W, respectively, for each curve shown in the figure. (b) The charge current corresponding to the cell voltages shown in (a); since the voltage rises rapidly at higher rates of charge, the current tapers at the end of charge in order to maintain constant power during the entire process. (c) Cell voltage during discharge at constant power; the power drawn from the cell was set to the same values as described in (a). (d) Cell current during discharge, corresponding to (c). The symbols represent the experimental data and the solid lines the model predictions.

TABLE 6.2 Parameters[5] Used in the Equivalent Circuit Model to Predict the Response of a Lithium-Ion Cell as Shown in Fig. 6.5.

Parameter	Discharge	Charge
τ (s)	5	5
C_D (F)	12500	16667
R, mΩ	1.637	1.637
R_{ct}, mΩ	0.4	0.3

6.4 MECHANISTIC MODELS

Mechanistic models relate the battery characteristics to physical properties of the constituent materials. Such properties are usually measurable in independent experiments.

For example, a mechanistic model for the Ohm's law equation shown in Eq. (6.1) can be built by describing the resistance parameter R in terms of physically measurable properties of the copper bus-bar. In this case, the properties of interest are the electronic conductivity of the metal (σ_c), the cross-sectional area (A_c), and the length of the bus-bar (L). Each of these properties is characteristic of the bus-bar. The resistance R is related to these parameters as follows:

$$R = \frac{L}{\sigma_c A_c} \tag{6.10}$$

and hence Eq. (6.1) can be rewritten as

$$V = I \frac{L}{\sigma_c A_c} \tag{6.11}$$

Note that Eq. (6.11) can be used for a cable of any given dimension, made up of any material whose electronic conductivity is known. The use of Eq. (6.1), however, requires that we measure the resistance parameter R every time the bus-bar is replaced. Also, the use of a mechanistic model enables one to identify better materials (e.g., a bus-bar with higher electronic conductivity) suited for an application. We now proceed to developing mechanistic models to describe the other physical processes that take place within the battery. A few processes commonly encountered in batteries are movement of ions in the electrolyte, movement of electrons within the electrodes, and chemical and electrochemical reactions. See Chap. 1 for background information on the basic equations governing each of these processes. In this section, we employ these concepts to build a mechanistic model for a battery.

6.4.1 Charge Transport by Electrons

The total voltage across a cell (V) can be approximated as the sum of the potential drop across the electrodes, across the electrolyte, and other losses arising from contact resistances. In the following sections, subscript 1 will be used to denote the properties/variables in the electrodes and subscript 2 to represent the corresponding variables in the electrolyte. We already considered voltage drop due to the flow of electrons across metal cables in Eq. (6.11) above. An equivalent of this equation can be used to represent contact resistances. The voltage drop across the electrodes ($\nabla \phi_{1,j}$) is also governed by Ohm's law

$$\nabla \phi_{1,j} = -\frac{i_1}{\sigma_j^{\it eff}}, \quad j = n \text{ or } p \tag{6.12}$$

where i_1 is the current per unit area (called the current density) and $\sigma_j^{\it eff}$ is the effective electronic conductivity of the electrode material within electrode j ($j = n$ refers to the negative electrode or the anode and $j = p$ refers to the positive electrode or the cathode). Usually a battery electrode is comprised of several components such as solid solutions of different metals, or composites of active material, binders, and other components. The effective conductivity corrects for the additional components within the electrode. Usually the effective conductivity is calculated as the sum of the conductivities of the individual components, scaled in proportion to the composition of the electrode

$$\sigma_j^{\it eff} = \sum_k x_k \sigma_k \tag{6.13}$$

Here x_k is the proportion of the individual components k that constitute the electrode, and σ_k refers to the electronic conductivities of the pure components. Alternatively, $\sigma_j^{\it eff}$ can be measured directly after the electrode is assembled.

6.4.2 Charge Transport by Ions

A unique feature of electrochemical devices is the transport of the current by ions. Once the current moves past the electrodes and undergoes the electrochemical reaction, the charge transport from one electrode to another is facilitated by ions. This transport of charge by the movement of ions is more complicated than the current carrying mechanism by movement of electrons. Usually, there are several species of ions present in the electrolyte. The total current density carried by the electrolyte (i_2) across a unit normal area is the sum of the current density carried by each species k

$$i_2 = \sum_k i_k \tag{6.14}$$

The current density carried by species k, i_k is proportional to its flux[7]

$$i_k = F \sum_k N_k \tag{6.15}$$

The proportionality factor in Eq. (6.15) is the Faraday's constant, which is the amount of charge carried by each mole of the ion. The flux of ion k may be defined as the product of the number of k ions per unit volume of the electrolyte (i.e., the concentration of species k) and the velocity of each ion

$$N_k = c_k v_k \tag{6.16}$$

The concentration of the electrolyte is a readily measurable quantity; the velocity of an ion is proportional to the charge it carries (z_k) and the potential gradient in the solution phase ($\nabla \phi_2$) which is the electrical driving force for that ion to move

$$v_k = -u_k F z_k \nabla \phi_2 \tag{6.17}$$

The proportionality constant in Eq. (6.17) is called the mobility (u_k) of the ion and is obtained from equivalent conductance measurements. The negative sign indicates that the ions move from a region of higher potential to lower. Equations (6.14) to (6.17) can be rearranged to obtain.[7]

$$i_2 = \left(F^2 \sum_k c_k u_k z_k \right) \nabla \phi_2 \tag{6.18}$$

Equation (6.18) resembles Ohm's law closely, and the electrical conductivity for the electrolyte (κ) is now given by

$$\kappa = \left(F^2 \sum_k c_k u_k z_k \right) \tag{6.19}$$

As stated in the previous section (see Eq. [6.13]), Eq. (6.19) relates the properties of the component ions to the conductivity of the electrolyte. Hence knowing the composition of the electrolyte, one can model the movement of ions in the electrolyte. Alternatively, one can experimentally measure the electrical conductivity (κ) in Eq. (6.19).

In deriving Eq. (6.18), an implicit assumption that the concentration of the electrolyte was uniform precluded the effects of concentration gradients present within the cell. However, this assumption can be easily relaxed by incorporating flux terms arising from concentration differences using Fick's laws of diffusion. Equation (6.16) then becomes

$$N_k = c_k v_k - D_k \nabla c_k \tag{6.20}$$

where D_k is the diffusion coefficient of ions k. In the case of flow batteries, there is a convective velocity in addition to v_k in Eq. (6.20). Thus the modified flux is now given by

$$N_k = c_k(v_k + \mathbf{v}) - D_k \nabla c_k \tag{6.21}$$

Here $\mathbf{v}$ is the velocity of electrolyte flow. Combining Eqs. (6.17) and (6.21) gives[7]

$$N_k = -z_k u_k F c_k \nabla \phi - D_k \nabla c_k + c_k \mathbf{v} \tag{6.22}$$

The equation presented above represents the case of dilute electrolytic solutions. More sophisticated models that consider the mutual interaction of ions within the electrolyte and the effects of temperature on the conductivity of the electrolyte are available.[8]

6.4.3 Driving Forces for Charge Transfer Across the Interface

The unique component of storing charge in batteries is the conversion of chemical energy into electrical energy or vice versa. Faraday's law dictates the maximum amount of charge generated for a given amount of active material. At equilibrium (when no current flows across the plates of the battery), the driving force is referred to as the open circuit potential and is related to the free energy of the system by Faraday's law[9]

$$E^0 = -\frac{\Delta G}{nF} \tag{6.23}$$

The negative sign implies that the free energy is reduced when the battery is discharged. In practice, the generation of electrical energy from chemicals depends on the temperature and the concentration of the chemical species taking part in the reaction generating the electrical energy. The open circuit voltage (E) under practical conditions where the battery operates is related to the equilibrium value (E_0) for changes to temperature and concentrations is given by the Nernst equation

$$E = E^0 + \frac{RT}{nF} \ln\left(\frac{c_{\text{Oxd}}}{c_{\text{Red}}}\right) \tag{6.24}$$

where c_{Oxd} refers to the concentration at the electrode surface of the species that release the electrons to the external circuit of the battery for the current to flow, and c_{Red} refers to the surface concentration of the ions that complete the electric path by moving across the electrolyte from one electrode plate to another. More complicated models exist that relate the surface concentration of the reacting species to the open circuit voltage of the battery[10]

An alternative to rigorous relationships between the open circuit voltage and the surface concentrations is the use of empirical expressions. This is particularly true of intercalation electrodes used in the lithium-insertion batteries, which are claimed to exhibit a non-Nernstian behavior. The most popular approach to model the open circuit potential in such cases is to measure the voltage of the individual electrodes with respect to a standard reference, at a very slow charge or discharge rate. The concentration of the reacting species in this case is assumed to be uniform throughout the reference cell and is calculated by counting the coulombs and using Faraday's law. Figure 6.6 shows examples of such measurements.

6.4.4 Rate of Charge Transfer

Like any chemical reaction, the efficiency of charge transfer also depends on how fast the reaction can take place. The rate of reaction is related to the local overpotential at the reacting interface j, by the Butler-Volmer expression[7]

$$i_j = i_{0,j}\left[\exp\left(\frac{\alpha_{a,j} n_j F \eta_{s,j}}{RT}\right) - \exp\left(\frac{-\alpha_{c,j} n_j F \eta_{s,j}}{RT}\right)\right] \tag{6.25}$$

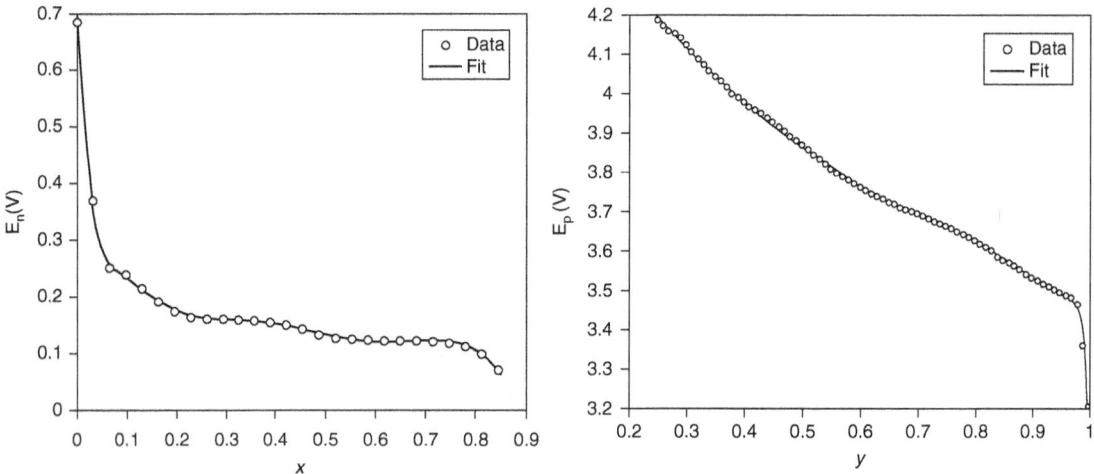

FIGURE 6.6 Open circuit voltage vs. stoichiometry of lithium in an intercalation electrode: the curve on the left shows the experimental data from an anode comprised of mesocarbon microbeads (MCMB) fit to an empirical expression, and the curve on the right shows the similar results from a nickel-cobalt-oxide (LiNiCoO$_2$) cathode.[23]

Here $i_{0,j}$ is the exchange current density; often this term includes concentration dependence of the reacting species at the interface and is written as follows:

$$i_{0,j} = i_{0,j}^{ref} f(c, c_s) = i_{0,j}^{ref} \left(\frac{c}{c_{ref}} \right)^\gamma \left(\frac{c_s}{c_{s,ref}} \right)^\delta \qquad (6.26)$$

The parameter $i_{0,j}^{ref}$ is the electrochemical analog of the rate constant for a chemical reaction. The function f relates the concentrations of the reacting species in the electrolyte and at the surface of the electrode (c and c_s, respectively) to the exchange current density. The superscript *ref* refers to these concentrations at the reference condition, and the parameters γ and δ correspond to the order of the reaction with respect to the participating species. In addition to a typical rate equation for a chemical reaction, the Butler-Volmer type reaction includes an exponential dependence of the current density to the local overpotential $\eta_{s,j}$ which is the difference between the potential at the electrode surface ($\phi_{1,s}$) and that in the electrolyte at the interface ($\phi_{2,s}$)

$$\eta_{s,j} = \phi_{1,s} - \phi_{2,s} \qquad (6.27)$$

Alternatively, the overpotential term may include a reference potential that accounts for the potential difference across the interface at the open-circuit conditions, wherein the term E [see Eq. (6.24)], corresponding to the electrode j, is subtracted from $\eta_{s,j}$

$$\eta_j = \eta_{s,j} - E_j \qquad (6.28)$$

If η_j [i.e., Eq. (6.28)] is used instead of $\eta_{s,j}$ [Eq. (6.27)] in the Butler-Volmer expression [Eq. (6.25)], the concentration dependence term f in Eq. (6.26) is modified accordingly[7] to accommodate concentration terms from Eq. (6.24). Similarly, if the reaction involves an intermediate step such as adsorption, the kinetic expression for each step of the mechanism is developed, and the final expression for the charge transfer reaction is usually expressed in the form of Eq. (6.25).

6.4.5 Distribution of Ions

Equation (6.24) relates the driving force for the generation of electricity to the concentration of the participating chemical species at the electrode-electrolyte interface. All the concentration terms (c and c_s) are defined at the reacting interface. It is difficult to monitor the concentration of chemicals at the electrode surface. A material balance for the ions relates the concentration in the bulk of the solution to that at the electrode surface. The material balance equation states that the concentration of the ion changes with time in accordance with the flux of the ions[11]

$$\frac{\partial c_k}{\partial t} = -\nabla \cdot (N_k) + R_k \quad (6.29)$$

The flux used in the material balance is consistent with the one used to determine the conductivity of the electrolyte [Eq. (6.22)]. The term R_k refers to change in concentration of the ion if it were consumed or generated in a reaction k. At the electrode-electrolyte interface, the change in concentration of the ions is because of the electrochemical reaction, and hence Eq. (6.15) is used to relate the amount of ions that are produced by the reaction to the amount of ions present at the electrode-electrolyte boundary. In the case of electrolytes at higher concentrations, interactions among the ions must be considered. For example, the diffusion of one species of ions depends on its interaction with ions of every other kind present within the electrolyte. Complexities such as this are usually handled by defining an *effective* property that considers such interactions. In this case, the following expression is used for the flux:

$$\hat{N} = c(\hat{v} + \hat{v}) + \hat{D}\nabla c \quad (6.30)$$

In Eq. (6.30), properties such as the diffusion coefficient are interpreted as effective properties. Note that the effective flux ($\hat{N}$) is now a function of the electrolyte concentration (c) and not the concentration of the individual ions (c_k). The velocity term $\hat{v}$ now relates to an effective field within the electrolyte and is usually expressed in terms of the transport number (t_+^0)

$$c\hat{v} = \left(1 - t_+^0\right)\frac{i_2}{F} \quad (6.31)$$

Expressions such as Eq. (6.30) enable one to obtain values for diffusivity or conductivity of the electrolyte experimentally measured using the actual mixture. The equations outlined in Sec. 6.4 constitute the mathematical framework for the mechanistic model of a battery. The following sections illustrate a few examples using these equations for some commonly encountered battery chemistries.

6.5 KINETIC MODEL OF A SILVER VANADIUM OXIDE CELL

The Silver Vanadium Oxide (SVO) cell (see Ch. 31) is commonly used as a primary cell in medical devices, for example. The cathode reaction can be written as follows[12]

$$Ag_2^+V_4^{5+}O_{11} + (x+y)Li^+ + (x+y)e^- \rightarrow Li_{x+y}^+Ag_{2-x}^+V_{4-y}^{5+}O_{11} + xAg^0 \quad (6.32)$$

It is assumed that two electrochemical reactions take place at the cathode

$$Ag_2^+V_4^{5+}O_{11} + xLi^+ + xe^- \rightarrow Li_x^+Ag_{2-x}^+V_4^{5+}O_{11} + xAg^0 \quad (6.33)$$

$$Ag_2^+V_4^{5+}O_{11} + yLi^+ + ye^- \rightarrow Li_y^+Ag_2^+V_{4-y}^{5+}O_{11} \quad (6.34)$$

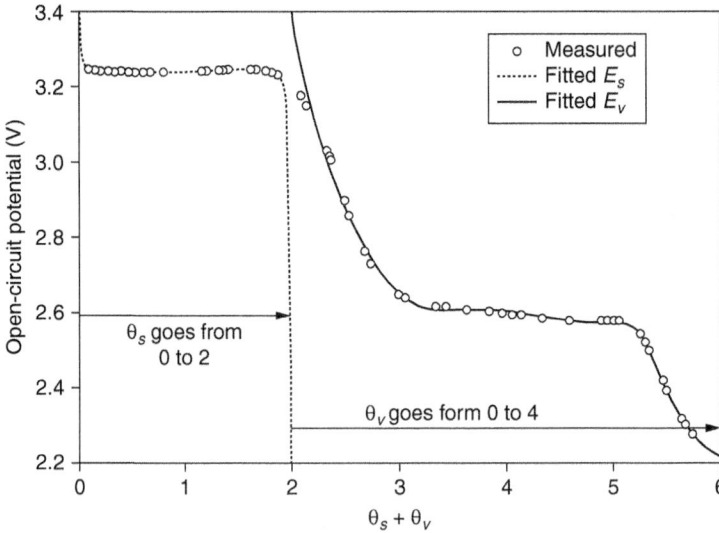

FIGURE 6.7 Open circuit voltage (OCV) for a silver vanadium oxide cathode. Empirical expressions are used to relate the OCV of the electrode to the stoichiometric coefficients.[12, 13]

The first reaction (6.33) corresponds to the reduction of silver, and the second (6.34) corresponds to the reduction of the vanadium ions. The stoichiometric parameter x varies from 0 to 2, and y varies from 0 to 4. The open circuit voltage as a function of composition is shown in Fig. 6.7.

A simple mechanistic model for these cells can exclude all transport limitations such as diffusion in the electrolyte. Since the operating current for these cells is low, these restrictions do not affect the cell performance. The cell is then said to be in the kinetics-controlled regime. The material balance Eq. (6.29) can then be written by setting the flux equal to zero and the reaction rate equal to the charge transfer reaction

$$\frac{\partial \theta_j}{\partial t} = \frac{\partial (c_j/c_{max})}{\partial t} = -\frac{aV}{n_j F c_{max}} i_j \tag{6.35}$$

where the subscript j is set to S for reaction (6.33) and to V for (6.34). The term aV refers to the area available for the reactions across the entire volume of the electrode V and is set equal to 2.0×10^4 cm^2/cm^3. The theoretical maximum for the concentration (c_{max}) is equal to 124.35 mol/cm^3. The Butler-Volmer equation for each of the reactions (6.33) and (6.34) is written using the parameters in Table 6.3. The total current density across the electrode is given as the sum of the individual reaction current densities (see Eq. [6.14]).

$$i_2 = i_s + i_v \tag{6.36}$$

TABLE 6.3 Parameters for the Silver Vanadium Oxide Battery Model[12]

Parameter	Silver reduction reaction	Vanadium reduction reaction
$i_{0,j}$ (A/cm^2)	$10^{-10}(2-\theta_s)^2$	10^{-8}
$\alpha_{a,j}$	0.5	0.5
$\alpha_{c,j}$	0.5	0.5
n_j	2	4
η_j (V)	$E - E_S$	$E - E_V$

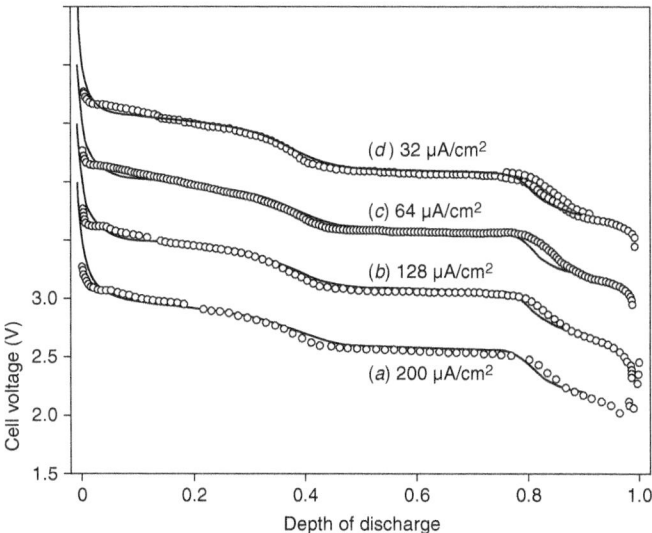

FIGURE 6.8 Predicted cell voltage (E) for different current densities (i_2) using model Eq. (6.36). The curves for different current densities are offset at 0.5 V intervals for clarity.

Equations (6.25), (6.35), and (6.36) are used to relate the cell voltage (E) to the applied current density i_2. Figure 6.8 shows good agreement between model predictions and the experimental data for different current densities.

6.6 MODELING POROUS ELECTRODES

Battery electrodes are often designed to be porous in order to improve the efficiency of the electrodes by providing access for the electrolyte to the active material in the electrode. In essence, the objective is to enhance the area accessible directly by the ions in the electrolyte, for the charge transfer reaction. At the same time, the potential drop across the solution phase in the electrode must be small. The material balance for transport of ions across a porous electrode closely follows Eq. (6.29). The concentration terms are now based on the fraction of the electrode volume occupied by the electrolyte; hence, a porosity term (ε) is used. To simulate the transport limitations along the tortuous path through the electrodes, effective properties similar to those discussed in Sec. 6.4.5 are used. For example, the conductivity of the electrolyte within a porous electrode is corrected for the geometric effects as follows:[14]

$$\kappa_{eff} = \varepsilon^b \hat{\kappa} \tag{6.37}$$

The superscript b, called the tortuosity factor, is often an empirical term. In a porous electrode, the reaction is distributed throughout the volume of the electrode. Hence, the flux of the ions and reaction rates are now measured as quantities averaged across the volume of the electrode (V). As a result, the material balance for the porous electrode becomes[15]

$$\frac{\partial(\varepsilon c)}{\partial t} = -\nabla \cdot (\bar{N}) + \bar{R} \tag{6.38}$$

where $\bar{N}$ is the volume averaged flux given by

$$\bar{N} = \frac{1}{V}\int_v \hat{N} dV \tag{6.39}$$

6.16 PRINCIPLES OF OPERATION

[See Eq. (6.30) for definition of $\hat{N}$]. The volume averaged reaction rate $\bar{R}$ is calculated using a similar expression. The current densities for the charge transfer reactions in the Butler-Volmer expression are all expressed per unit volume

$$j = \frac{di_2}{dx} = i_{0,j} a \left[\exp\left(\frac{\alpha_{a,j} n_j F \eta_{s,j}}{RT}\right) - \exp\left(\frac{-\alpha_{c,j} n_j F \eta_{s,j}}{RT}\right) \right] \quad (6.40)$$

where a is the area available for reaction, per unit volume of the electrode. The subscript 2 for the current density refers to the solution phase (i.e., the electrolyte).

6.7 LEAD-ACID BATTERY MODEL

A lead-acid battery model developed by Nguyen[16] has the following reactions taking place:

$$PbO_2 + HSO_4^- + 3H^+ + 2e^- \rightarrow PbSO_4 + 2H_2O \quad \text{(positive electrode)} \quad (6.41)$$

$$Pb + HSO_4^- \rightarrow PbSO_4 + 2e^- + H^+ \quad \text{(negative electrode)} \quad (6.42)$$

At each electrode, the material balance is given by Eq. (6.38). The reaction term $\bar{R}_k$ is given by the Butler-Volmer equation for reaction [Eq. (6.41)] at the positive electrode and [Eq. (6.42)] at the negative electrode. In addition, the volume of the product formed in these reactions (i.e., $PbSO_4$) is much higher than the reactants (PbO_2 or Pb). This induces a change in the porosity of the electrode during charge/discharge. This change in porosity is accounted for by using Faraday's law as follows:[17]

$$\frac{\partial \varepsilon}{\partial t} = \frac{1}{n_j F} \left[\left(\frac{M}{\rho}\right)_{Product} - \left(\frac{M}{\rho}\right)_{Reactant} \right] \left(\frac{di_j}{dx}\right) \quad (6.43)$$

The total current is carried across the electrode, from the separator/electrode interface to the current-collector/electrode interface, by both the electrons within the electrode matrix and by the ions in the electrolyte that fill the pores across the thickness of the electrode (see Sections 6.4.1 and 6.4.2)

$$i_{tot} = i_1 + i_2 \quad (6.44)$$

The current across the electrode matrix (i_1) is given by Eq. (6.12) and the current across the electrolyte (i_2) is given by Eq. (6.18), after modifying the conductivity as shown in Eq. (6.37). In addition, the influence of concentration gradients on the transport of ions is modeled using transport numbers. The resultant expression for the current in the electrolyte phase is given by[16]

$$i_2 = -\kappa_{eff} \left[\nabla \phi_2 + \frac{2RT}{F}(1 - 2t_+^0) \frac{\nabla c}{c} \right] \quad (6.45)$$

The resultant profiles for the distribution of the porosity across the thickness of the electrode are shown in Fig. 6.9. The porosity at the electrode/separator interface decreases for both the anode and the cathode, owing to the precipitation of $PbSO_4$ that occupies a higher volume than the active materials. The reduction in porosity is higher at the negative electrode since the difference in density between the reactant (metallic lead) and the product ($PbSO_4$) is higher. A direct effect of blocking the electrode surface is the restriction of access of the entire volume of the electrode to the electrolyte. As a result, the reaction distribution is highly nonuniform, as shown in Fig.6.10. The reaction rates at the current-collector end of the electrodes are close to zero, indicating poor utilization of the electrodes.

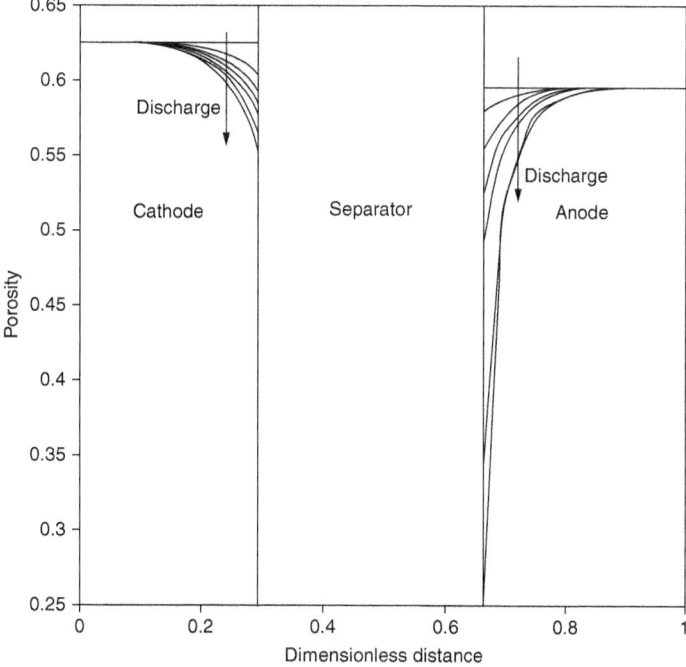

FIGURE 6.9 Porosity distribution across the thickness of the electrodes as a function of time during discharge in a lead-acid cell. The porosity at the electrode-separator interface decreases in both the anode and the cathode/separator interfaces due to the formation of $PbSO_4$.[17]

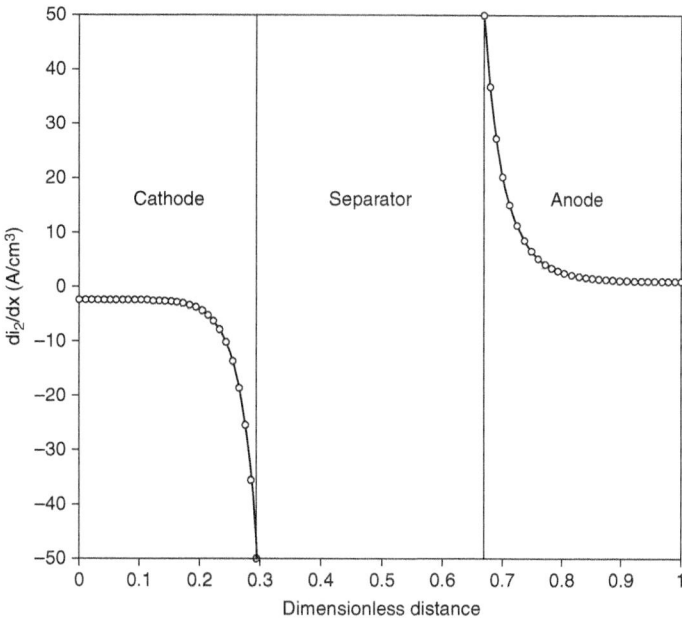

FIGURE 6.10 Reaction rate distribution across the thickness of a lead-acid cell [see Eq. (6.40)]. The blockage of the electrode surface by the formation of $PbSO_4$ creates a nonuniform distribution of the reaction across the thickness.[17]

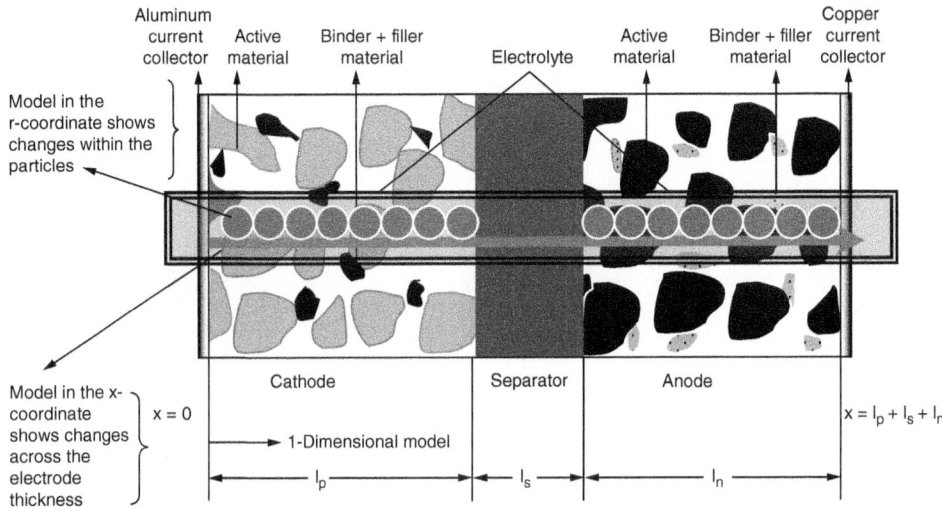

FIGURE 6.11 Schematic of a lithium-ion cell used to develop a one-dimensional model along the thickness of the electrodes.

6.8 INTERCALATION IN POROUS ELECTRODES

The intercalation mechanism in porous electrodes has been represented in mathematical models in several ways. The simplest treatment considers the phenomenon as diffusion of ions into a solid solution. Fick's law is used to represent this process. The electrode particles are usually assumed to be of a regular geometry (see Fig. 6.11). For example, the diffusion of ions within spherical particles is governed by the following equation:

$$\frac{\partial c_s}{\partial t} = D_s \left(\frac{\partial^2 c_s}{\partial r^2} + \frac{2}{r} \frac{\partial c_s}{\partial r} \right) \tag{6.46}$$

The subscript s in Eq. (6.46) is used to refer to the solid particles. The concentration of the ions at the surface of the particles is mathematically connected to the electrolyte concentration at the interface through the Butler-Volmer equation [(see Eq. (6.40)].

Figure 6.12 summarizes the utility of a mechanistic model in cell design. Several thought experiments can be simulated by altering the different design parameters, such as the particle size, as well as material properties, such as the conductivity. The model is used to identify the critical factor that limits performance of the cell at high rates.

6.9 ENERGY BALANCE

Temperature control has been a concern with a lot of battery chemistries. In some cases, the effect of an abnormal temperature is a reduction in performance, whereas in others it may lead to concerns over safe operation of the battery. Heat generation within a battery is usually modeled using an energy balance equation that relates the heat generated due to Joule heating, chemical/electrochemical reactions etc., to the heat exchange with the environment in which the battery operates. A general form of the material balance equation is shown below[11]

$$\frac{\partial(\rho c_P T)}{\partial t} = \nabla \cdot (\lambda \nabla T) + q \tag{6.47}$$

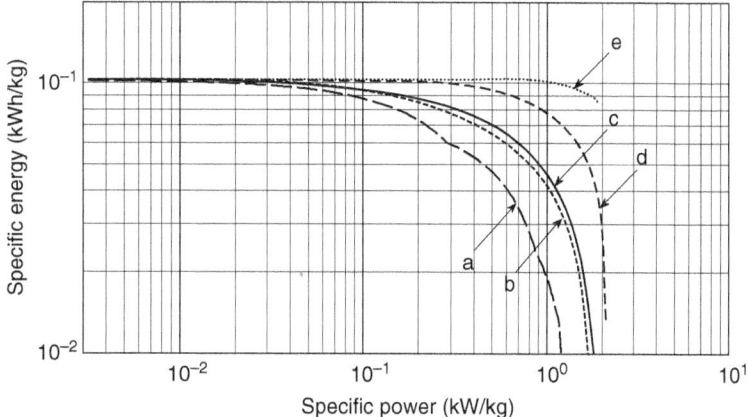

FIGURE 6.12 Simulated Ragone plots constructed using a mechanistic model. (a) Model predictions show that the original design cell delivers very low specific energies at high power applications. (b) Increasing the electronic conductivity of the cathode matrix (e.g., by addition of conductive carbon) shows some improvement. (c) Further enhancement in electronic conductivity shows little change in cell performance. (d) Diffusion limitations are relaxed by increasing the solid-phase diffusion coefficient at the cathode (e.g., by doping). (e) Further limitations within the solid phase are eliminated by reducing the particle size to a few nanometers.

The left side of Eq. (6.47) denotes the rate at which energy is consumed or generated per unit volume. The first term on the right side represents heat transfer by conduction following Fourier's law. The term q refers to heat generated or consumed due to the reactions that take place during the operation of the battery. Typically, for an electrochemical reaction, this term is expressed as follows:[19]

$$q = \frac{\partial i_2}{\partial x}\left[\phi_1 - \phi_2 - \left(E_j - T\frac{\partial E_j}{\partial T}\right)\right] - i_1\frac{\partial \phi_1}{\partial x} - i_2\frac{\partial \phi_2}{\partial x} \quad (6.48)$$

The first component of q represents the heat generated from the charge-transfer reaction, the second term is the Joule heat generated due to the current flow across the solid matrix, and the last term is the corresponding value for current flow in the electrolyte. Additional complexities, such as heat transfer due to differential phase changes, radiation effects, etc., can be treated by incorporating the amount of heat generated from such phenomena in Eq. (6.48). The term $T\frac{\partial E_j}{\partial T}$ corrects the open circuit voltage for changes in entropy with temperature. Open circuit voltages measured at different temperatures may be used to evaluate this term empirically. Changes in other properties such as the diffusivity or conductivity with temperature are often approximated by the Arrhenius equation

$$\Phi = \Phi_{ref} \exp\left[-\frac{E_a}{R}\left(\frac{1}{T} - \frac{1}{T_{ref}}\right)\right] \quad (6.49)$$

where Φ may represent D_{eff}, κ_{eff}, D_s, etc., Φ_{ref} represents the corresponding property measured at the reference temperature (T_{ref}), and E_a is the activation energy.

Figure 6.13 illustrates the change in battery performance with different degrees of convective cooling during a 3C-rate discharge of a lithium-ion cell. For the adiabatic case, increase in the cell temperature favors higher reaction rates and enhances transport within the electrolyte following [Eq. (6.49)]. The model predicts that without a cooling system in place, the difference in cell

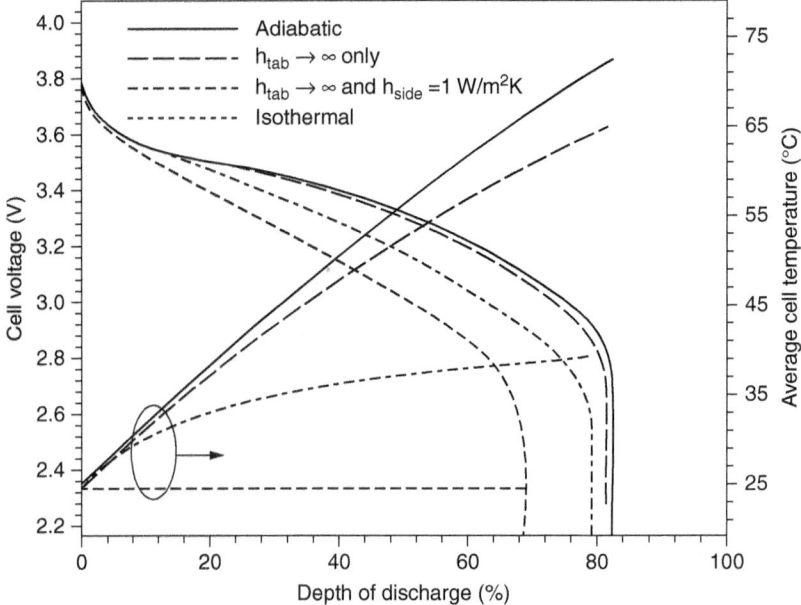

FIGURE 6.13 Comparison of the cell performance with various degrees of heat exchange with the environment.[19]

temperature from the ideal isothermal case can be as high as 45°C. The other two cases shown indicate the effect of using suitable packing material to implement rapid cooling of the substrate and the influence of additional convection along the side walls of the cell. Including a simple heat-transfer model for the walls thus provides significant insight into the design of an efficient cooling system, particularly for large format cells.

A second example is the case of a NiMH battery, in which reactions involving oxygen contribute to heat generation within the cell, in addition to the primary electrochemical reactions. The primary reactions are as follows:[20]

Positive electrode:

$$NiOOH + H_2O \underset{Charge}{\overset{Discharge}{\rightleftarrows}} Ni(OH)_2 + OH^- \tag{6.50}$$

Negative electrode:

$$MH + OH^- \underset{Charge}{\overset{Discharge}{\rightleftarrows}} H_2O + M + e^-$$

In addition, the inadequate utilization of one electrode during the end of charge leads to evolution of oxygen, which undergoes subsequent reactions

$$2OH^- \rightarrow \frac{1}{2}O_2 + H_2O + 2e^- \quad \text{(positive)} \tag{6.51}$$

$$4MH + O_2 \rightarrow 4M + 2H_2O \quad \text{(negative)}$$

Reactions of the following type involving phase changes within the metal hydride electrode leading to the formation of β-MH have also been proposed:

$$(x-y)H + MH_y \rightarrow MH_x \tag{6.52}$$

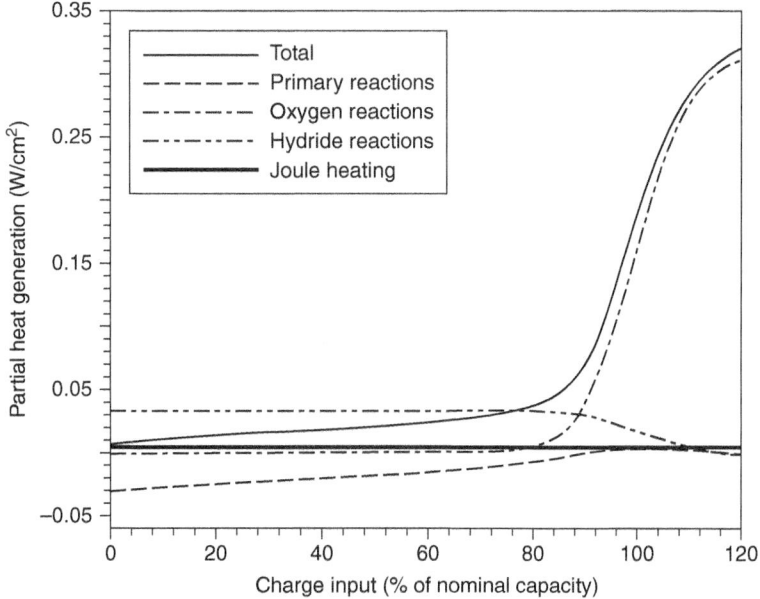

FIGURE 6.14 Contribution of different chemical reactions to heat generated within a NiMH cell.[20]

Reaction (6.52) is a chemical reaction, and hence the heat generated by this reaction is calculated as the product of the rate of the reaction and the enthalpy of reaction.

The heat generated from the above reactions is compared in Fig. 6.14 to the Joule heating terms [(see Eq. (6.48)]. The model simulates charging of a NiMH cell at the 1-C rate. At the beginning of charge, the heat generated from the MH reactions (6.52) is balanced by the endothermic primary reactions. Also, the oxygen evolution reaction does not take place to a significant extent; as a result, heat contributions from reaction (6.51) are negligible. Toward the end of charge, the enthalpy changes in favor of heat absorption in reactions [Eq. (6.50)]; in addition, overcharge leads to significant evolution of O_2. As a result, there is a dramatic increase in the heat generating within the cell toward the end of charge.

6.10 DEGRADATION MODELS

One important requirement of a mathematical model for a battery is the capability to provide some insight into the future performance of the battery. If a rechargeable lithium-ion cell is designed based on the results shown in Fig. 6.13, a higher performance due to elevated temperatures may be desirable for an application targeting a few cycles; however, it is experimentally observed that prolonged cycling under these conditions leads to faster deterioration of the cell performance. One must account for such phenomena in developing a life-model for a battery. Understanding the mechanism of degradation is a critical step in developing a model to predict life of a battery. For example, in the nickel-based electrodes used in a lithium-ion cell, surface oxidation of the particles results in an additional impedance created at the cathode, whereas such an increase in the impedance of the cathode for a cobalt-based system is attributed to phase changes at higher voltages. Similarly, dissolution of manganese ions is a major reason for capacity loss with cycling in a $LiMn_2O_4$-based electrode. Another major challenge in using physics-based models for life prediction arises from determining the values for the parameters used in the model. The number of parameters used in a physics-based model is considerably higher than an empirical fit. While most of these parameters are available from the operating conditions and the design of the cell components at the initial cycle, monitoring changes in several parameters over several cycles is not straightforward. Also, changes in many parameters depend on the operating conditions or the history of the battery.

The simplest life prediction models use linear extrapolation: plotting the capacity of the cell versus the cycle number and obtaining the slope and the intercept of a straight line by regression. When a cell is subjected to repeated cycling under mild operating conditions (e.g., shallow depths of discharge) that do not cause severe wear out of the cell until the end of life criterion is reached, linear extrapolation has been found to provide a good degree of confidence in predicting the end of life of the cell. The primary advantage of using this technique is the ease of extracting the coefficients. Depending on the range of operating conditions, more than one set of coefficients may be required for successful prediction of the cell performance. For example, if several cells are subject to different depths of discharge (DOD) at the end of each cycle, the degradation rates are different—and consequently, the coefficients in the empirical fits for each case are different. Some predictions made using empirical models are shown in Fig. 6.15. The accuracy of the method relies on the functional terms in the expression used. A complicated polynomial expression may provide a better prediction compared to a linear equation. Tools for nonlinear regression are also readily available in the form of commercial packages. The success of the technique in making life predictions depends entirely on the prior knowledge of the system at hand. In other words, the curve-fitting technique is used more often to interpolate to an unknown operating scenario rather than to make predictions beyond the limiting cases at which experimental data is available. This shortcoming is typical of any empirical prediction technique and yet does not prevent curve-fitting from being the most popular choice in the industry.

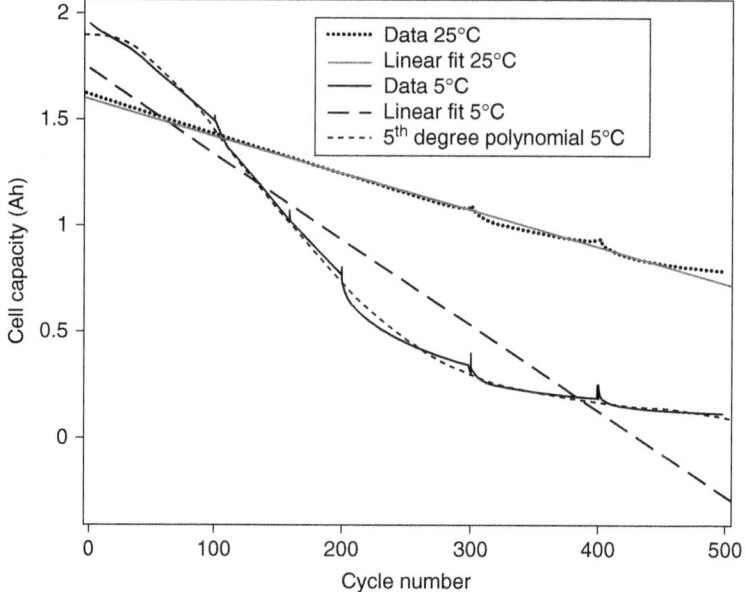

FIGURE 6.15 Cell capacity vs. cycle number data for a lithium-ion cell fit to empirical expressions. The linear equation fits the data at milder conditions (cycling at 25°C) better than the data at a more rigorous condition (cycling at 5°C). Predictions made using the linear expression are closer to the experimental observation for the data at 25°C. For the data collected at 5°C, a more complicated expression (in this case, a fifth-degree polynomial) is required to represent the data more accurately.[24]

TABLE 6.4 Summary of equations used to represent a porous intercalation electrode. The corresponding equation numbers from the text are shown in the last column.[18]

Variable	Governing equation	Equation number in the text
Solid phase potential ($\phi_{1,j}$)	$i_1 = -\sigma_j^{eff} \nabla \phi_{1,j}$	(6.12)
Solution phase potential ($\phi_{2,j}$)	$i_2 = -\kappa_{eff}\left[\nabla\phi_2 + \frac{2RT}{F}(1-2t_+^0)\frac{\nabla c}{c}\right]$	(6.45)
Solid phase current density (i_1)	$i_{tot} = i_1 + i_2$	(6.44)
Solution phase current density (i_2)	$\frac{di_2}{dx} = i_{0,j} a\left[\exp\left(\frac{\alpha_{a,j} n_j F \eta_{s,j}}{RT}\right) - \exp\left(\frac{-\alpha_{c,j} n_j F \eta_{s,j}}{RT}\right)\right]$	(6.40)
Solid phase concentration (c^s)	$\frac{\partial c_s}{\partial t} = D_s\left(\frac{\partial^2 c_s}{\partial r^2} + \frac{2}{r}\frac{\partial c_s}{\partial r}\right)$	(6.46)
Solution phase concentration (c_2)	$\varepsilon\frac{\partial c}{\partial t} = D_{eff}\frac{\partial^2 c}{\partial x^2} + \frac{(1-t_+)}{F}\frac{\partial i_2}{\partial x}$	(6.38)

An alternate approach is to use a mechanistic model similar to the one shown in Table 6.4 and adjust parameters, such as the diffusivity and the exchange current density, periodically as to obtain a good fit between the model predictions and experimentally observed performance.[21] Such an approach is usually referred to as a semi-empirical model. Figure 6.16 illustrates the change in a few parameters observed for a lithium-ion cell that loses about 30% of its initial capacity during the first 800 cycles.

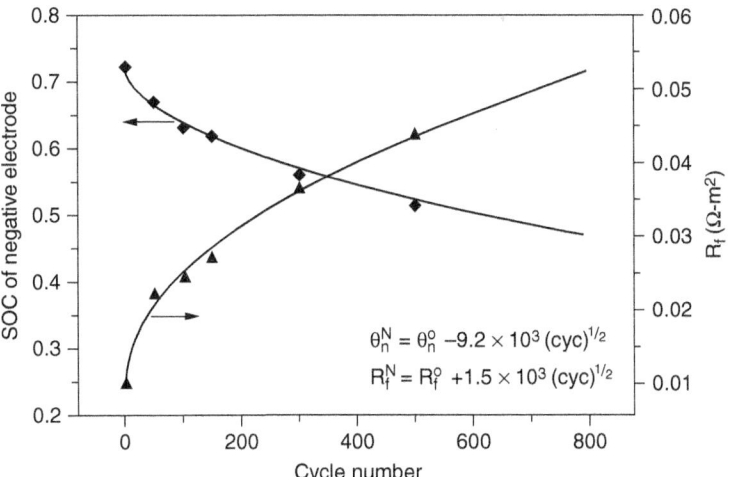

FIGURE 6.16 Change in parameters with cycling of a lithium-ion cell. Adjustable parameters in the semi-empirical model are the SOC of the negative electrode at the beginning of discharge (θ_n^N) and the resistance of the film formed on the anode (R_f^N). The change in these parameter values with cycle number were obtained by adjusting the experimental curves at various cycle numbers.[21]

A third approach is the development of a mechanistic model for the degradation process. An electrochemical reaction that consumes part of the cell's deliverable capacity over several cycles may be proposed. For example, the increase in resistance R_f^N is modeled as a film formed on the surface of the anode particles due to reduction of the solvent during charge.[22] Since this is a charge transfer reaction involving reduction of Li$^+$ into salts, a Butler-Volmer type equation assumes the following form:

$$\frac{di_{side}}{dx} = -i_{0,side} a \exp\left(-\frac{\alpha_{a,side} n_{side} F \eta_{side}}{RT}\right) \quad (6.53)$$

The formation of the film introduces an additional resistance at the surface of the anode particles and consequently a drop in the overpotential at the anode

$$\begin{aligned}\eta_n &= \phi_{1,s} - \phi_{2,s} - E_n - \frac{1}{a_n}\left(\frac{\partial i_2}{\partial x}\right)\left(\frac{\delta_{film}}{\kappa_{film}}\right) \\ \eta_{side} &= \phi_{1,s} - \phi_{2,s} - E_{side} - \frac{1}{a_n}\left(\frac{\partial i_{side}}{\partial x}\right)\left(\frac{\delta_{film}}{\kappa_{film}}\right)\end{aligned} \quad (6.54)$$

The thickness of the film δ_{film} is calculated knowing the amount of capacity lost, using Faraday's law

$$\frac{\partial \delta_{film}}{\partial t} = -\frac{1}{Fa_n}\left(\frac{\partial i_{side}}{\partial x}\right)\left(\frac{M_{side}}{\rho_{side}}\right) \quad (6.55)$$

where M_{side} and ρ_{side} represent the molecular weight of the side reaction product and the density of the film, respectively.

Figure 6.17 shows the thickness of film formed on the anode particle surface as predicted by the mechanistic model under different operating conditions. The model predicts a rapid growth of the film

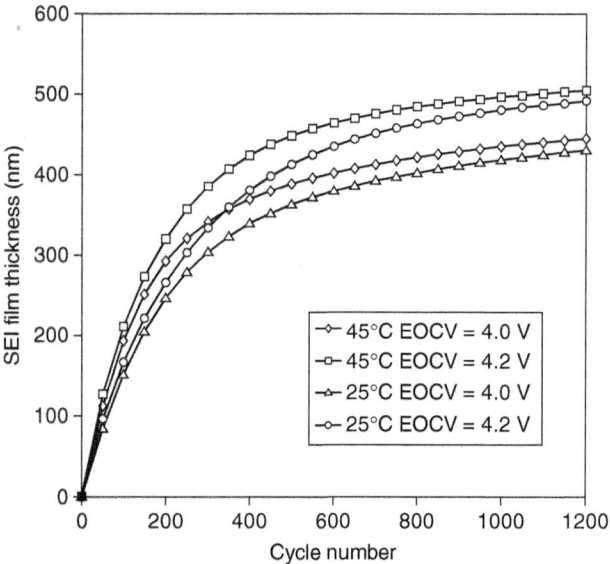

FIGURE 6.17 SEI film thickness as a function of the cycle number as predicted by a mechanistic model for a lithium-ion cell.

during the first few cycles that eventually levels off. For the higher end of charge voltage (EOCV), the anode is maintained at highly reducing potentials for longer periods of time, and as a result, the growth rate of the film is higher in a cell cycled with a higher EOCV. Under identical conditions, a cell cycled at a higher temperature has lower transport limitations, and hence the duration of a given cycle is longer than it would be at a lower temperature. This results in additional charging time; since the model for the film growth depends on the amount of time taken to charge the cell [(see (6.55)]; the cell cycled at 40°C shows higher thickness of the film during the first few cycles and hence more capacity fade. However, after about 300 cycles, the enhancements in charging time due to the temperature are offset by the additional resistance created by the growth of the film. As a result, the cell cycling at 25°C, but with a higher EOCV, loses more capacity. These results indicate that if operating at a relatively higher temperature for about 300 cycles, the cell can be programmed to have a higher EOCV (and hence deliver higher capacity), whereas if the application demands a longer cycle life, a more conservative EOCV should be used. Similar conclusions can be drawn from empirical models once data under different operating conditions is available, whereas such an insight into the physical phenomena as described above is unique to mechanistic models. Once the rate constant for the side-reaction and the conductivity of the film are determined by independent experiments, the model allows for cell design under a variety of different scenarios as long as the degradation mechanism proposed is valid.

6.11 DETERMINING THE RIGHT MODEL

A good mathematical model should strike a balance between the limited details associated with the input parameters available to the end-user and the amount of insight the model can provide to improve battery design. Limitations to the mechanistic models are attributed to the tedium involved in developing and solving the model equations as well as the large number of parameters required by such models. Often many of these parameters cannot be obtained from direct experimental measurements. On the other hand, circuit analog models provide limited insight into the physical phenomena that take place within the battery. For example, a drop in capacity at higher rates of discharge can be modeled as an increase in the pseudocapacitance parameter C_D shown in Fig. 6.3. One cannot determine whether this change is caused by a limitation in the diffusion of ions within the electrode or if the conductivity of the electrolyte has been reduced over time. Consequently, making improvements to the cell design based on circuit analog models is difficult; in this example, it is not obvious whether an increase in the porosity of the electrode plate would resolve the issue or if the electrolyte formulation would have to be modified.

Typically, fine tuning of the parameters such as the conductivity of the electrolyte or the porosity of the electrodes is carried out at the cell design phase. Hence, a mechanistic model is invaluable at this design phase. In all cases, the assumptions behind a mathematical model must be carefully explored before employing the conclusions made from simulations.

LIST OF SYMBOLS

a	Specific area (m^2/m^3)
A	Area of the electrode (m^2)
b	Tortuosity factor
c_k	Concentration of the ion k (mol/m^3)
c	Volume averaged concentration of the electrolyte (mol/m^3)
c^s	Volume averaged concentration of the ion within the electrode (mol/m^3)
c_p	Specific heat capacity (J/kg/K)
C_D	Double layer capacitance (F)
D	Diffusion coefficient of the ion in the electrolyte (m^2/s)

D_s	Diffusion coefficient of the ion in the electrode (m²/s)
E_a	Activation energy (J/mol)
E^0	Equilibrium potential (V) of electrode
F	Faraday's constant (96487 C/mol)
G	Gibb's free energy (J/mol)
ΔH	Amount of heat generated (J)
i	Current density (A/m²)
$i_{0,j}$	Exchange current density (A/m²)
I	Current (A)
j	Volumetric current density, A/m³
L	Length (m)
M	Molecular weight (kg/mol)
n	Number of electrons transferred
N	Flux (mol/m²/s)
$\bar{N}$	Volume averaged flux (mol/m³/s)
$\hat{N}$	Effective flux of an ion (mol/m²/s)
q	Charge (C)
Q	Capacity of the cell (Ah)
Q_0	Initial capacity of the cell (Ah)
r	Empirical parameter used in Table 6.1 (mOhm)
R	Ohmic resistance within a cell (Ohm)
R_{ct}	Charge transfer resistance (Ohm)
R_k	Rate of reaction involving ions k (mol/m³/s)
$\bar{R}$	Volume averaged reaction rate (mol/m³/s)
R	Universal gas constant (8.314 J/mol/K)
t	Time (s)
t_+^0	Transport number
T	Temperature (K)
u	Mobility of an ion (cm² mol/J/s)
v	Convective velocity (m/s)
V	Volume of the electrode (m³)
V	Cell voltage (V)
V_0	Open circuit voltage (OCV) of the cell (V)
x	Spatial variable (m)
z	Charge carried by an ion

Greek

α	Transfer coefficient
δ_{film}	Thickness of the SEI film (m)
ε	Porosity
κ	Ionic conductivity of the electrolyte (S/cm)

λ	Thermal conductivity (W/K)
η	Overpotential (V)
φ	Potential (V)
ρ	Density (kg/m^3)
σ	Electronic conductivity of the electrodes (S/cm)
τ	Time constant (s^{-1})
θ	Dimensionless concentration

Subscripts and Superscripts

c	Current collector
eff	Effective
n	Negative electrode
p	Positive electrode
ref	Reference condition
$side$	Side reaction
0	Initial or standard condition
1	Electrode matrix
2	Electrolyte
s	Solid phase
$\hat{\Lambda}$	Effective value of the parameter Λ

Abbreviations

CAD	Computer-aided design
DOD	Depth of discharge
MCMB	Meso-carbon micro beads
SEI	Solid electrolyte interface
SoC	State-of-charge

REFERENCES

1. Peukert, W., Über die Abhängigkeit der Kapazität von der Entladestromstärcke bei Bleiakkumulatoren. *Elektrotechnische Zeitschrift*, 1897. **20**.
2. Macdonald, D. D., Reflections on the History of Electrochemical Impedance Spectroscopy, *Electrochimica Acta*, 2006. **51**(8–9): p. 1376–1388.
3. De Levie, R., Response of Porous and Rough Electrodes. *Advances in Electrochemistry and Electrochemical Engineering*, eds. P. Delahay and C. W. Tobias. Vol. 6. 1971, New York: John Wiley & Sons.
4. Verbrugge, M. W., and R. S. Conell, Electrochemical and Thermal Characterization of Battery Modules Commensurate with Electric Vehicle Integration. *Journal of the Electrochemical Society*, 2002. **149**(1): p. A45–A53.
5. Bergeveld, H. J., W. S. Krujit, and P. H. L. Notten, *Battery Management Systems: Design by Modeling*. Phillips Research Book Series, ed. M. A. Norwell. 1999, The Netherlands: Kluwer Academic Publications.

6. Verbrugge, M., Adaptive Characterization and Modeling of Electrochemical Energy Storage Devices for Hybrid Electric Vehicle Applications. *Modern Aspects of Electrochemistry*, ed. M. Schlesinger. Vol. 43. 2008, New York: Springer-Verlag.

7. Newman, J., *Electrochemical Systems,* 2nd ed. 1991, New York: Prentice Hall.

8. Lin, C., R. E. White, and H. J. Ploehn, Modeling the Effects of Ion Association on Alternating Current Impedance of Solid Polymer Electrolytes. *Journal of the Electrochemical Society*, 2002. **149**(7): p. E242–E251.

9. McQuarrie, D., and J. D. Simon, *Molecular Thermodynamics.* 1999, Sausalito, CA: University Science Books.

10. Ohzuku, T., and A. Ueda, Phenomenological Expression of Solid-State Redox Potentials of $LiCoO_2$, $LiCo_{1/2}Ni_{1/2}O_2$ and $LiNiO_2$ Insertion Electrodes. *Journal of the Electrochemical Society*, 1997. **144**(8): p. 2780–2785.

11. Slattery, J. C., *Advanced Transport Phenomena.* 1999, New York: Cambridge University Press.

12. Gomadam, P. M., et al., Modeling Li/CFx-SVO Hybrid-Cathode Batteries. *Journal of the Electrochemical Society*, 2007. **154**(11): p. A1058–A1064.

13. Crespi, A. M., P. M. Skarstad, and H. W. Zandbergen, Characterization of Silver Vanadium Oxide Cathode Material by High-Resolution Electron Microscopy. *Journal of Power Sources*, 1995. **54**(1): p. 68–71.

14. Whitaker, S., Diffusion and Dispersion in Porous Media. *AIChE Journal*, 1967. **13**(3): p. 420–427.

15. Newman, J., and W. Tiedemann, Porous-Electrode Theory with Battery Applications. *AIChE Journal*, 1975. **21**(1): p. 25–41.

16. Nguyen, T. V., *A Mathematical Model for a Parallel Plate Electrochemical Reactor, CSTR, and Associated Recirculation System,* Ph.D. Dissertation, 1985, College Park, Texas A & M University.

17. Nguyen, T. V., R. E. White, and H. Gu, The Effects of Separator Design on the Discharge Performance of a Starved Lead-Acid Cell. *Journal of the Electrochemical Society*, 1990. **137**(10): p. 2998–3004.

18. Fuller, T. F., M. Doyle, and J. Newman, Simulation and Optimization of the Dual Lithium-Ion Insertion Cell. *Journal of the Electrochemical Society*, 1994. **141**(1): p. 1–10.

19. Gu, W., and C. Y. Wang. Thermal and Electrochemical Coupled Modeling of a Lithium-Ion Cell in Lithium Batteries, in *Proceedings of the Electrochemical Society, Vol. 99-25(1).* 2000, Pennington, NJ, Plenum.

20. Wang, C. Y., W. B. Gu, and B. Y. Liaw, Thermal-Electrochemical Modeling of Battery Systems, *Journal of the Electrochemical Society*, 2000. **147**(8): p. 2910–2922.

21. Ramadass, P., B. Haran, R. White, and B. N. Popov, Mathematical Modeling of the Capacity Fade of Li-Ion Cells. *Journal of Power Sources*, 2003. **123**(2): p. 230–240.

22. Ramadass, P., B. Haran, P. M. Gomadam, R. White, and B. N. Popov, Development of First Principles Capacity Fade Model for Li-Ion Cells. *Journal of the Electrochemical Society*, 2004. **151**(2): p. A196–A203.

23. Qi Zhang and R. E. White, Capacity Fade Analysis of a Lithium-Ion Cell. *Journal of Power Sources*, 2008. **179**: p. 793–298.

24. Santhanagopalan, S., J. Stockel, and R. E. White, Life Prediction for Lithium-Ion Batteries, in *Encyclopedia of Electrochemical Power Sources*, eds. J. Garche, C. Dyer, P. Moseley, Z. Ogumi, D. Rand, and B. Scrosati. Vol. 5, 2009, p. 418–437, Amsterdam: Elsevier Publications.

CHAPTER 7
BATTERY ELECTROLYTES

George E. Blomgren

7.1 INTRODUCTION

Previous editions of this Handbook did not include a chapter on electrolytes. The individual chapters on each battery type included information needed to understand the electrochemical operation of the systems under discussion. This edition also includes some information about the electrolytes for each battery type within the chapter describing the system, but this chapter attempts to broaden the horizon on the electrolyte component because of the key role of electrolytes as the bridging phase between the electrodes. Each electrode has a contribution to the battery impedance that is due to the electrode-electrolyte interface, as well as a separate contribution due to the impedance of the electrolyte itself beyond the region of the electrical double layer. This impedance term becomes especially important under high-current conditions when mass transport of ions through the electrolyte frequently becomes the limiting process through the Nernst equation relating the concentration of ions at the electrode interface to the electrode polarization.

Since most batteries of the 19th and much of the 20th century utilized aqueous electrolytes, the discussion begins with the various aqueous electrolytes of importance in batteries. Some specialized electrolytes are omitted in the discussion in order to focus on those of major importance.

Beginning in the late 1950s, electrolytes were developed that had good stability with lithium metal. This opened the door to lithium primary batteries, which became available in the 1970s. After major safety difficulties with rechargeable lithium metal batteries became evident, the lithium-ion battery, using lithiated carbon as the negative electrode, made its appearance in the early 1990s, and the rechargeable lithium-ion battery industry was born. An overview of the electrolytes used in both battery types is given. In addition, newly developed ionic liquid electrolytes, which offer the possibility of low flammability to enhance safety, are also discussed.

While lithium metal batteries did not achieve success as rechargeable wet cells, the use of ceramic or glassy solids as lithium-ion conductive electrolytes has permitted the use of lithium metal to produce high cycle-life rechargeable batteries, mostly as thin film cells (see Chap. 27). New work on these solid inorganic electrolytes offers the hope of making high-capacity lithium metal cells and will also be discussed. Similarly, new approaches to dry polymer electrolytes have renewed interest in this type of cell. Therefore, an introduction to these types of electrolytes is also presented.

7.2 AQUEOUS ELECTROLYTES

Aqueous electrolytes can be broken up on the pH scale into alkaline, neutral (or mildly acidic), and strong-acid electrolytes. The alkaline electrolytes are usually very strong with pH values close to 13. Neutral electrolytes are generally composed of salts of strong acids and bases. Additives of weaker bases cause lower pH values. For example, in the Leclanché electrolyte, zinc chloride is an important electrolyte component, and its complex equilibria in aqueous media shift the pH to mildly acidic conditions. Likewise, the ubiquitous presence of carbon dioxide dissolved in aqueous solutions shifts the pH to mild-acid conditions. It must always be borne in mind that the equilibrium voltage window for aqueous solutions is approximately 1.2 V (the actual value depends on electrolyte concentration, temperature and other factors) so that gassing of hydrogen with anodes more active than hydrogen and of oxygen for cathodes more active than oxygen will occur to some extent even in the presence of passive films on the electrodes. Thus, much of the electrolyte work in aqueous media deals with the twin problems of anode corrosion and cathode gassing and the means to control these deleterious reactions. The situation with rechargeable batteries is even more stringent because of the increased potential of both negative and positive electrodes during charge.

7.2.1 Alkaline Electrolytes

Alkaline electrolytes are utilized in a large variety of primary and rechargeable batteries. The most commonly used primary battery type is the so-called alkaline manganese-dioxide battery, described in Chap. 11 of this volume. Other primary batteries with alkaline electrolytes are the various button cells such as zinc-silver oxide and zinc-air (Chap. 13), and zinc-mercuric oxide, which are made in both button and cylindrical format (Chap. 12). Rechargeable cells using alkaline electrolytes include nickel-metal hydride (Chap. 22), nickel-cadmium (Chaps. 19–21), nickel-zinc (Chap. 23), and nickel-hydrogen (Chap. 24).

Alkaline electrolytes generally have considerably higher conductivity than neutral electrolytes because of the enhanced proton conductance of high pH electrolytes. For example, 20 to 40% solutions of NaOH or KOH are frequently used in batteries, giving pH values near 14. Proton conductance in alkaline electrolytes has been widely studied for many years. KOH is generally preferred to NaOH because of its higher conductivity at a given concentration and lower freezing points in the eutectic region.[1] Figure 7.1 shows the conductivity relationships of KOH and NaOH as a function of the weight percent of the hydroxide at 15 and 25°C, and clearly the KOH solutions are better by at least 40% in most concentration ranges.[2] Figure 7.2 shows the effect of the dissolution of ZnO on the conductivity of both KOH and NaOH.[3] The effect of ZnO dissolution is to diminish

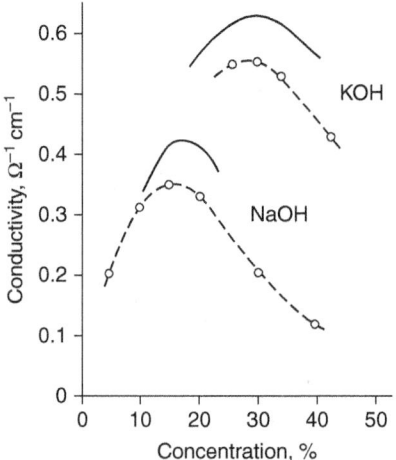

FIGURE 7.1 Specific conductance of NaOH and KOH aqueous solutions. Solid line is 25°C, dotted line is 15°C, Ref. 2.

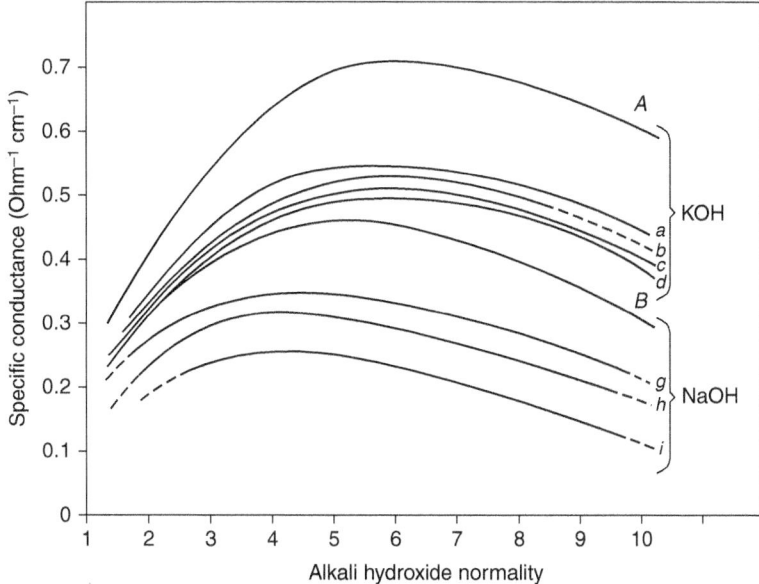

FIGURE 7.2 Specific conductance of NaOH and KOH containing different ratios of dissolved ZnO to alkali hydroxide at 30°C, Ref. 1.
   A. KOH only
      a. 1 mole ZnO:4.33 mole KOH
      b. 1 mole ZnO:3.71 mole KOH
      c. 1 mole ZnO:3.37 mole KOH
      d. 1 mole ZnO:3.00 mole KOH
   B. NaOH only
      g. 1 mole ZnO:4.05 mole NaOH
      h. 1 mole ZnO:3.03 mole NaOH
      i. 1 mole ZnO:1.76 mole NaOH

the conductivity due to the removal of hydroxide ion from the solution as the reaction, where M is either sodium or potassium

$$ZnO + 2MOH + H_2O \rightarrow M_2Zn(OH)_4 \tag{7.1}$$

Reaction (7.1) is very important for batteries with zinc anodes because zinc ions formed as the anodic reaction occurs are converted to zincate ions, $Zn(OH)_4^{-2}$, until saturation of the solution occurs.[4] Thereafter, the product is ZnO or $Zn(OH)_2$ in the solid phase, although supersaturation of the solution with ZnO is a frequent occurrence along with many complications in the species and solution structure.[5] Thus, the electrolyte changes throughout the early part of the discharge (unless it is presaturated with ZnO) until the zinc compound is precipitated, mainly in the anode compartment. The relatively high concentration of zincate ions in the electrolyte has important ramifications for the rechargeable zinc electrode, such as used in the nickel-zinc, silver oxide-zinc or the MnO_2-Zn battery, for example. Zinc deposition occurs mainly from the soluble species during charge and, as is common with deposition from soluble species of high concentration, this deposition can take a number of morphologies depending on the current density, zinc concentration, and other factors. It would be most desirable to deposit specular (layered) zinc to give even deposits and continuous contact to the substrate. Unfortunately, the overpotential relationships are very complicated and at least five different types of deposits have been identified under different conditions[6] with a great sensitivity to impurities and substrate as well as electrochemical and solution parameters. In the regimes of dendritic and mossy growth, it is inevitable that separator bridging and shorting will occur to limit the cycle life of the cell. Most commercial rechargeable zinc batteries have focused on additives to the KOH electrolyte to try to improve the morphology of the deposit, and reference is made to the individual chapters of this book for information.

The electrolyte for primary alkaline batteries usually has a gel-forming polymer added to the solution, although the earlier wet cells such as Zn-air and Zn-CuO cells used the liquid without gelation. The gel-forming polymer must of course be alkali stable. For many years, the favorite polymer was sodium carboxymethylcellulose, first used in the zinc-mercuric oxide primary cell to immobilize the electrolyte and convert it from a wet cell to a dry cell. This material is still used in some alkaline batteries, but is subject to oxidation by high potential cathodes. Polymers such as other cellulosic or starch derivatives, polyacrylates, or ethylene maleic anhydride copolymers are used in some alkaline batteries. The manufacturers generally keep their anode gel formulation as trade secrets, so not much information can be included in this section. An effect of the use of gelling agents is to lower the conductivity of the electrolyte phase and this must be taken into account in the cell design so that electrolyte path lengths are not too long. It should be mentioned that zero-added mercury cells have become predominant, especially in alkaline-MnO_2 cells, thus placing greater importance on electrolyte purity, particularly with regard to iron and chloride ion since these materials are corrosion accelerants. Of course, all of the battery materials are required to have good purity with regard to these and other elements that can accelerate corrosion. Group 3 and Group 5 elements such as indium and bismuth have taken the place of mercury in many alkaline types in order to reduce the landfill burden of this toxic element. In addition, certain organic additives, which have corrosion inhibition properties, are included in the electrolyte formulation. In fact, the use of mercuric oxide cells has declined substantially due to this concern, although the high-voltage stability of the Zn-HgO and Cd-HgO batteries has preserved some high-end electronic applications for these systems. Notably, the Cd-HgO battery can be regarded as one of the few electrolyte stable aqueous batteries with its operating voltage of about 0.9 V. This factor has permitted its use even in high-temperature applications where a zinc anode would corrode substantially.

Alkaline electrolytes are very important for rechargeable batteries and are frequently used in specially designed wet cells. The overpotential during charging can cause oxygen gassing with metal oxide positive electrodes, which is very deleterious to polymer gelling agents. For example with nickel oxyhydroxide positive electrodes, the desire to get high cycle-life mitigates against using most organic compounds in the electrolyte. The exception is for Zn-MnO_2 rechargeable cells (see Chap. 28) which have limited cycle life for other reasons and thus the oxidation of gelling agents, which is relatively slow, is not the limiting problem. In fact, the anode gel in this rechargeable cell is similar to that of the primary alkaline battery. Additives are also an important aspect of alkaline rechargeable cells. Lithium hydroxide is a common additive for nickel cadmium batteries, and sometimes cobalt salts are incorporated in the electrolyte for cells with nickel oxyhydroxide positive electrodes.[7] The history of nickel-zinc secondary cells is essentially that of various electrolyte additives, the main purpose of which is to suppress the solubility of zinc in the alkaline electrolyte. In the absence of such additives, the formation of zinc dendrites on charge severely limits the cycle life of the cell. These topics are dealt with in Chap. 23.

7.2.2 Neutral Electrolytes

The main battery type that uses neutral or slightly acidic electrolytes is the Leclanché or carbon-zinc battery. There are two main versions of this cell, namely the Leclanché electrolyte, which is a combination of zinc chloride and ammonium chloride salts dissolved in water, and the zinc chloride cell, which is mainly zinc chloride dissolved in water with a small amount of ammonium chloride sometimes added. A typical Leclanché electrolyte contains about 26% NH_4Cl, 9% $ZnCl_2$, and 65% water, while a typical zinc chloride electrolyte contains about 30–40% $ZnCl_2$ and 60–70% water. A small amount of corrosion inhibitor is also usually contained in the electrolyte for ultimate deposition on the zinc surface. The conductivity of zinc chloride solutions reaches a maximum of 0.107 S/cm at 3.7 M $ZnCl_2$ with a relatively slow decrease with increasing or decreasing concentrations.[8]

The addition of ammonium chloride increases the conductance substantially. The actual conductance in the cell is modified, however, by the inclusion of a gelling agent, such as starch, which immobilizes the electrolyte and minimizes leakage and orientation effects. This is part of the art of manufacturing carbon-zinc cells, and the reader is referred to Chap. 9 and references therein for details.

The zinc-bromine rechargeable cell, which is under development for large-scale energy storage applications, employs a near-neutral zinc bromide solution in the discharged state. During charge, one electrode provides a surface for zinc plating, while the second generates bromine which is dissolved in the electrolyte. An ammonium salt, which may have up to four organic substituents on the nitrogen (either cyclic or linear), is usually added to form a tribromide ion, which has a much lower vapor pressure than elemental bromine. The chemistry is rather complex, especially if chlorides are added, but much of the work on additives is proprietary so it is difficult to interpret the results. An additional system using neutral electrolytes for energy storage is the so-called polysulfide-bromine battery. This system uses a sodium bromide electrolyte on the positive electrode side and a sodium polysulfide electrolyte on the negative electrode side. Again, proprietary additives are common and analysis is difficult. The reader is referred to Chap. 30 for further information.

7.2.3 Acid Electrolytes

The main electrolyte in this category is sulfuric acid, which, although it has a long history in batteries, is now mainly employed in the lead-acid or more recently in the carbon-lead-acid cell. It can be argued that this electrolyte is the most important one because of the widespread use of this battery type and its worldwide economic importance. The lead-acid cell was invented in 1859 by Gaston Planté and utilized dilute sulfuric acid which today is usually fixed on construction of the cell at a concentration of 37% by weight in the fully charged condition. As is the case for alkaline batteries, the electrolyte is a reactant and therefore varies in concentration during charge and discharge; the electrode reactions given by

$$\text{Positive: } PbSO_4 + 5\,H_2O \rightleftharpoons PbO_2 + 3H_3O^+ + HSO_4^- + 2e^- \qquad E° = 1.685\text{ V} \qquad (7.2)$$

$$\text{Negative: } PbSO_4 + H_3O^+ + 2e^- \rightleftharpoons Pb + HSO_4^- + H_2O \qquad E° = -0.356\text{ V} \qquad (7.3)$$

$$\text{Cell reaction: } 2\,PbSO_4 + 4\,H_2O \rightleftharpoons Pb + PbO_2 + 2\,H_3O^+ + 2\,HSO_4^- \qquad E° = -2.041\text{ V} \qquad (7.4)$$

where the cell reaction shows that 2 moles of sulfuric acid are used during discharge to produce 2 moles of water. This utilization of sulfuric acid contributes substantially to the weight and volume of the battery and also results in electrolyte properties changing during the course of the discharge as the electrolyte concentration changes. The reader is referred to the chapters on lead-acid batteries (Chaps. 16 and 17) for further details, although we note that the conductivity of sulfuric acid in the 35% concentration range is of the order of 800 mS/cm, one of the most conductive electrolytes used in batteries at room temperature. It is noted from the individual electrode potentials given in Eqs. (7.2) and (7.3) that the electrodes are not stable with respect to oxygen evolution [Eq. (7.2)] and hydrogen evolution [Eq. (7.3)] in the fully charged condition. These reactions cause a steady decline in capacity of the cell as gassing occurs, which is one of the principle problem areas of the lead-acid battery. The degree to which hydrogen is evolved causes an irreversible increase in pressure since the oxygen evolution is balanced to some degree by recombination with the negative electrode material, while hydrogen combination with the positive or with oxygen is very slow. In some versions of VRLA cells, a hydrogen-oxygen recombination catalyst is included in the head space of the cell to reform water. In the absence of such a catalyst, the cell simply vents the gasses, thus causing a decrease in the amount of water in the cell (increasing the sulfuric acid concentration and lowering the electrolyte level) and severe stress on the electrodes. Venting of hydrogen-containing gasses can also cause hydrogen explosions if a spark is present and the gas mixture is within the explosion limit of hydrogen content (above 4%).

A variation on the lead-acid battery includes the addition of activated carbon to the lead negative electrode leading to a combined double-layer, Faradaic redox process. The double-layer capacitance can be quite large when very high surface area activated carbons are used and can have a much more rapid response than the lead electrode. The electrolyte phase is similar to that of the standard lead-acid battery so the reader is referred to Chaps. 16 and 17. Further information on the use of this type of battery in microhybrid vehicles is found in Chap. 29.

Sulfuric acid has also been used in the vanadium and other redox types of batteries. Here the solution also contains vanadium ions of different valence states (uranium ions and other ions have also been tested) with a vanadium (II–III) couple used on the negative electrode side and a vanadium (IV–V) couple used on the positive electrode side. A simple microporous separator can be used to separate the two solutions when they are in the interelectrode space, although sometimes an ion-specific separator is used (proton conductive). No particular harm occurs when the solutions interdiffuse except in lowering the round-trip energy efficiency of the process since both solutions contain vanadium ions. When the system is not undergoing charge or discharge, the solutions are pumped into storage containers, and when the system is operating, the solutions are pumped into the interelectrode space in a flow-through cell. Even in these systems, one must be concerned about gas evolution since the overpotential of the charging regime can carry the cell into the gassing range (see Chap. 30 for details).

Occasionally other acids have been employed for acid battery types. For example, a HBF_4 electrolyte has been used with lead/lead dioxide electrodes. These types are not in current production, however, so they will not be discussed.

7.3 NONAQUEOUS ELECTROLYTES

This section is arbitrarily separated into organic solvent-based electrolytes, inorganic solvent-based electrolytes, ionic liquids, solid polymer electrolytes, and ceramic/glassy electrolytes. The use of polymeric materials to cause gelation of the electrolyte (used mainly with organic solvents) has little effect on the basic electrolyte properties such as conductivity and diffusivity, except for a major effect on viscosity. However, the electrolyte in most modern batteries has very little convective flow, so viscosity has little effect on battery operation. The greatest development has been with organic solvent-based electrolytes as they have been used in numerous primary lithium batteries and many variations of the lithium-ion battery type. Inorganic solvent-based electrolytes have been used mainly in liquid cathode batteries. The others are still in the development stage, but will be discussed briefly.

7.3.1 Organic-Solvent Electrolytes

The greatest electrochemical use for organic-solvent electrolytes has been in the fields of primary lithium batteries and rechargeable lithium-ion batteries. The successful application in primary lithium batteries predates that of the rechargeable batteries by several decades, even though work was instituted on both primary and secondary batteries at about the same time in the early 1960s. Techniques for handling and purifying the organic liquids had to be developed first, with special attention to the contamination of water and other impurities. More than a decade of work was required to understand the importance of reducing impurities to the ppm level from salts and solvents. Cathode materials and other cell components often have much more adsorbed water than was commonly realized so that methods to remove water from all components had to be developed as well. A good compilation of the techniques required to study nonaqueous electrochemistry is given in Ref. 9. In addition, it was necessary to develop an understanding of the stability of purified solvents and salts with active materials of both electrodes, especially the negative electrode. Studies of the electrochemical window on materials such as conductive diamond, platinum, or glassy carbon during cyclic voltammetric sweeps in the negative direction (cathodic scans) required much interpretation because the film-forming properties of these materials was very different from that of lithium metal (or in later work on carbons or graphites). Likewise, anodic scans on inert substrates were found to depend on both the salt and solvent, and most interpretations of oxidation reaction mechanisms are still unresolved.[10]

Chapter 14 of this volume sets out some rules for electrolytes that are to be used in lithium anode batteries. The rules are:

1. The electrolyte must be aprotic, that is, have no reactive protons or hydrogen atoms, although hydrogen atoms may be in the molecule.
2. It must have low reactivity with lithium (or form a protective coating on the lithium surface to prevent further reaction) and the cathode.
3. It must have good ionic conductivity.
4. It should be liquid over a broad temperature range.
5. It should have favorable physical characteristics, such as low vapor pressure, stability, nontoxicity, and nonflammability.

To expand on these rules, the first is a criterion that has already been discussed in terms of water and other contaminants in the electrolyte, however, it is broader because any Brønsted acid can contribute protonic species which can penetrate passive films and corrode the lithium (or lithiated graphite or other alloys), thus making the system unstable. The formation of gaseous hydrogen at the anode surface further disrupts the passive film during bubble formation. The second rule is critical to any active metal anode, since the polar solvents useful for electrolytes are at best only metastable to the anode metal. The author has shown by thermodynamic calculations that even propylene and ethylene carbonates are capable of highly exothermic reactions with lithium metal to produce lithium carbonate and the corresponding alkene.[11] A recognition of the importance of the passivating film (or the Solid Electrolyte Interphase [SEI] layer) is demonstrated in the book *Lithium Batteries: Solid Electrolyte Interphase*[12] and many other writings referred to therein. Additives to the electrolyte have frequently been employed to improve the SEI.[13] Reactions with the cathode are also of great concern to the electrochemist studying lithium and lithium-ion batteries. Again, the solvents are only metastable as demonstrated by calculations of simple reactions[11] and the mechanisms are poorly understood. However, the exothermicity of solvent reaction with strong oxidants can be very great as can be seen by DSC measurements of electrolyte-charged positive electrode samples.[14] The third rule relates to the practicality of making cells with usable current output in high energy density (and power density) configurations for ambient operation. The conductivity of the electrolyte should be at least 3 mS/cm to make practical electrolytes, although higher values are very desirable and allow thicker, shorter electrodes to be used. Many liquid electrolytes exceed this value at room temperature and will be discussed later. However, the attempts to produce solid polymer electrolytes with adequate conductivity at room temperature have yet to succeed. The best of these types have achieved conductivities of the order of 0.1 mS/cm, but adequate designs have not yet been fully developed to utilize these interesting materials. Rule 4 is also important and has made necessary the use of solvent mixtures to satisfy both high and low temperature requirements. Even so, extreme high or low temperature operation has necessitated special purpose electrolytes, which do not perform well at the opposite extremes. Finally, rule 5 has many ramifications. The lack of toxicity has been a relative goal as some electrolytes evince moderate toxicity and may even have reaction products of high toxicity on exposure to ambient environments. An example is the frequently used $LiAsF_6$ salt, which can be converted to toxic arsenic oxide on exposure to room atmosphere. Low vapor pressure is moderated through the use of solvent mixtures. If a component has a high vapor pressure, it is usually found in modest concentration in the solvent mixture to suppress the vapor pressure, but may contribute to the level of conductivity of the electrolyte or improve low-temperature performance. Stability is really a reference to stability within the cell environment, as one of the most common salts used in lithium-ion cells, $LiPF_6$, is thermally, as well as photochemically, unstable in many solutions, but in the cell environment, it can maintain stability over many years of operation. Nonflammability or flame retardancy is generally not observed with organic solvents. However, phosphorus and halogen substituents (such as fluorine) can confer these properties, and a number of workers are active in this field at present. No agreed-upon materials have yet been found. As in other property studies, an admixture of a flame-retardant chemical is the most likely to be successful.[15]

Solvent properties of the most common solvents for lithium primary batteries are given in Chap. 14 of this volume, and the most common solvents for lithium-ion rechargeable batteries are given in Chap. 26. Additional information is also found in Chap. 27. A more complete listing of solvents can be found in Ref. 16. As alluded to earlier, mixtures of two or more of these solvents are usually employed to obtain the best combination of properties for the intended application. A high dielectric constant, high-viscosity solvent such as propylene carbonate (permittivity = 64.4, viscosity = 2.5 cP) can be mixed with a low dielectric constant, low-viscosity solvent such as dimethoxyethane (permittivity = 7.2, viscosity = 0.455 cP) to give a mixed solvent of intermediate dielectric constant and intermediate viscosity with very good solvation properties toward lithium salts. Ideal mixing rules adapted from traditional physical chemistry generally predict the mixed solvent properties to within a few percent for aprotic solvents.[17] Certain empirical parameters, such as donor and acceptor numbers, have been usefully employed to help choose cosolvents to improve properties, while concepts such as ion association from physical chemical studies may also be useful to understand the conductivity and viscosity of electrolyte solutions.[16]

The choice of electrolyte salt is also very important for thermal and electrochemical stability of the electrolyte solution as well as affecting the conductivity of the electrolyte solution. It is not widely recognized, but many times the anodic window of an electrolyte is determined by the reactivity of the anion, which can play a catalytic role in the oxidation of the solution. Thus, a one-electron oxidation of the anion at the positive electrode to form a neutral free radical leads to attack of a solvent molecule in a chain reaction. Chain termination usually results from radical combination, but often great damage is done to the electrolyte, which becomes evident on further cycling. These reactions are especially important for rechargeable systems.[10] The salt may also be unstable at the negative electrode. For example, the tetrachloroaluminate anion may deposit aluminum in an exchange reaction with lithium. Furthermore, the salt may effect the SEI of the graphite anode in a beneficial way as exemplified by the LiBOB (lithium bis-oxalatoborate) salt in some lithium-ion battery electrolytes (see Chap. 26). One of the most important effects of salts in lithium-ion batteries is in the effect on the collector substrate for the positive electrode. Aluminum is the most commonly used collector in lithium-ion batteries and the choice of salt is very important, particularly at longer times. The pitting corrosion potential is similar for a group of salts at short times, but the effect at longer times shows that only a few salts such as $LiPF_6$ and $LiBF_4$ are stable with aluminum.[18] Thermal stability is affected by the salt as well as the solvent choice. One of the critical thermal aspects is related to the onset temperature of the dissolution of the SEI layer, which is very sensitive to the choice of electrolyte salt. These matters are discussed in Ref. 15. In summary, rechargeable cells are the most sensitive to the choice of salt, with $LiPF_6$ the preferred salt in most cases. A wider choice is available for primary batteries, in part because the cathode is never charged and thus the collector is not exposed to high overpotentials. $LiCF_3SO_3$, $LiPF_6$, $LiBF_4$, LiBr, LiI, $LiN(CF_3SO_2)_2$, and $LiClO_4$ have all been used in primary lithium batteries. Dahn and Ehrlich discuss the conductivity of various electrolyte salts in different solvents in Chap. 26. It should be emphasized, however, that in all cases, the conductivity of electrolytes with organic solvents (as well as the inorganic solvent systems discussed below) are at least an order of magnitude poorer than aqueous electrolyte solutions, especially those with enhanced proton conductance such as acids and bases. This has a profound effect on cell design for the nonaqueous systems. For high-rate cells, electrodes are much thinner, the corresponding electrode areas are much greater, and separators are much thinner. This makes the cells more expensive to manufacture than aqueous cells of similar rate capability. However, because of the high voltage or high capacity of the chosen nonaqueous systems, the cost of a given cell size per watt hour of energy is mitigated.

Many additives have been developed for lithium-ion batteries for the purpose of improving safety, extending the calendar life, and extending cycle life of cells. The subject is complicated because the level of additive is generally small (1% or less of the electrolyte) unless the additive is for the purpose of adding flame retardancy to the electrolyte, when the level is generally 5% or higher. Furthermore, extensive chemical analysis before and after cycling has not often been performed, so the effects are generally left as empirical findings. Also, the long-term effects of the additive are generally not described. Their use in batteries is discussed in Chap. 26 of this volume as well as in Ref. 15.

7.3.2 Inorganic-Solvent Electrolytes

A class of liquid cathode primary batteries has been developed that uses purely inorganic solvents. These cells have very high energy density, in part because the electrolyte carries out the dual role of electrolyte solution and cathode active material. The main representatives are thionyl chloride, $SOCl_2$, and sulfuryl chloride, SO_2Cl_2, although a number of other solvents have appeared in the patent literature. Both types have been modified by the inclusion of additives in the electrolyte, BrCl in the thionyl chloride electrolyte and Cl_2 in the sulfuryl chloride electrolyte. The effect on the energy output of the cells due to the additives is small, but there are some advantages in resistance of the cells to abuse conditions. Generally, the shelf life of the cells is adversely affected by the additives. It is somewhat surprising that the conductivities of the electrolytes are relatively high (1M $LiAlCl_4$ solution in thionyl chloride = 14.6 mS/cm; in sulfuryl chloride = 7.4 mS/cm), since the dielectric constants of these solvents are low (permittivity of thionyl chloride = 9.25; of sulfuryl chloride = 9.15).[19] The preferred salt in all of these solutions is $LiAlCl_4$ although some work has been carried out with $LiGaCl_4$ to show improved conductivity and reduced passivation. These systems are surprisingly stable even though the oxidizing liquids are in direct contact with lithium metal and one would expect at least a strong reaction, if not an explosive one under these circumstances. Extensive study of the lithium surface has shown, however, that a tight, compact layer of lithium chloride is formed which impedes further reaction. The shelf life is accordingly very long (better than many organic solvent systems), although a delayed response in current is often shown after a long period of storage as the layer must be at least somewhat disrupted in order for current to pass. Chapter 14 of this work discusses these and other aspects of the systems in detail. Further insight into these systems can be derived from Refs. 20 and 21.

The lithium-sulfur dioxide liquid cathode system is very important in military and industrial applications. The electrolyte phase is a mixture of an organic solvent with condensed-phase sulfur dioxide. Acetonitrile is usually used as the organic solvent because of its high solubility for and stability with sulfur dioxide. Acetonitrile has a moderately high dielectric constant (35.95) and very low viscosity (0.341 cP). This combination (30/70 by volume) with a1M LiBr salt gives a conductivity of about 52 mS/cm at ambient temperature, approaching the conductivity of aqueous solutions (see Chap. 14 for the temperature dependence of conductivity). As with oxyhalide liquid cathode cells, a compact protective film on the lithium metal is the enabling feature of the electrolyte, only in this case the material formed is lithium dithionite ($Li_2S_2O_4$), which is also the reaction product of the cell. This material has very low electronic conductivity and forms a very compact layer, which on cell discharge must be somewhat disrupted to allow the lithium ions to enter the electrolyte. The cell has lower energy content than oxyhalide cells because of a lower voltage and the dilution of the electrolyte with acetonitrile. It also has to contain pressurized sulfur dioxide, which has a boiling point of −10°C. This electrolyte allows $LiSO_2$ cell to provide excellent low temperature performance to −40°C.

7.4 IONIC LIQUIDS

Ionic liquids are defined as liquids that are primarily dissociated into ions even though they have complex polyatomic structures of each of the ions. Many of these are actually liquid at room temperature and below and also support the dissolution of lithium salts at reasonable concentrations, such as 1 M. Because they have very low vapor pressures in general, they offer flame-retardant properties that few other electrolytes accomplish. Also, because of their high concentration of ions, the conductivities are comparable to many organic solvent systems, even though the viscosities tend to be much higher. The high viscosity can cause problems in filling cells in short time periods as well as creating wetting problems of electrodes and separators. Typical structures are shown in Fig. 7.3 from a recent paper on this topic.[22] An electrochemical difficulty with most ionic liquids occurs due to the fact that the onium cations are reduced at more positive potentials than lithium deposition or intercalation in graphite, and the SEI formed with these materials is frequently unstable due to dissolution in these

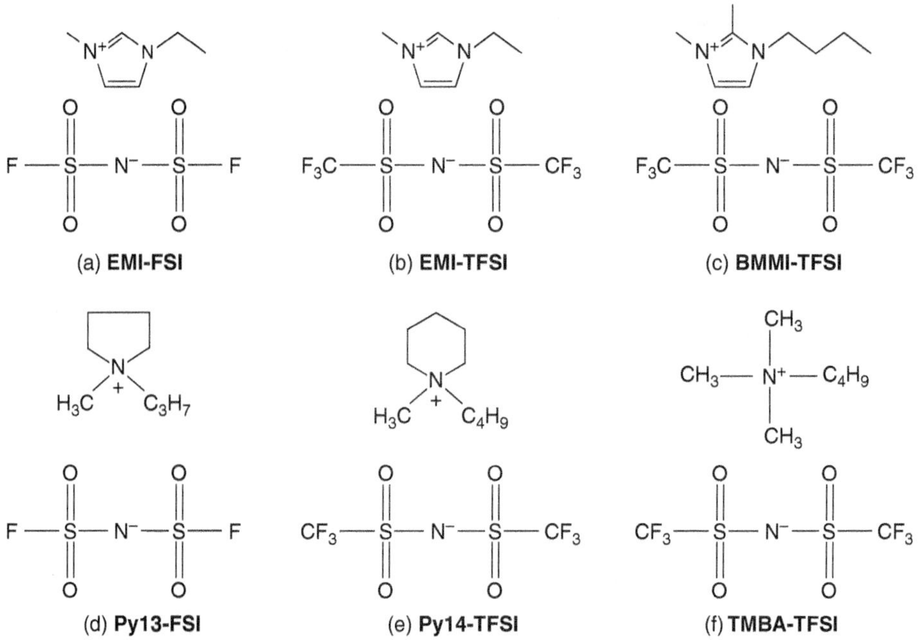

FIGURE 7.3 Chemical formulas and structures of some typical room temperature ionic liquids, Ref. 22.

excellent solvents. Therefore, the efficiency of lithium deposition or intercalation is reduced and the cycle life of the cell is correspondingly reduced. Additives that improve the SEI, such as vinylene chloride (VC), also help the cycle life, but not enough as yet to allow commercial production.[22] The reader is referred to Ref. 23 for further information.

7.5 SOLID POLYMER ELECTROLYTES

Batteries using solid polymer electrolytes (SPE) are discussed in Chap. 27. The electrolytes used initially were simple polymers of polyethylene oxide with salts such as $LiClO_4$ dissolved within. There were no solvents employed, but surprisingly large amounts of salt could be dissolved due to the large interaction energy of the ethereal oxygens in the polymer with the lithium ions of the salt. The anion was believed to be somewhat isolated from the lithium ion. Unfortunately, the conductivity of the SPE was too low at room temperature to be of use in a practical battery (of the order of 10^{-7} S/cm). Many efforts were made to increase the conductivity, including dissolution of a "plasticizer" solvent such as ethylene or propylene carbonate and backbone modification, including isomerization. Unfortunately, the methods that increased the conductivity also weakened the mechanical properties. Practical solutions to these problems have not yet been found, but several workers continue in the field. The incentive for the work is the possibility of using lithium metal as the anode material, which would confer high energy density and specific energy to the cell. One newer direction is graft blocks of different types of polymers such that one type imparts conductivity while the other contributes strength, each creating a continuous microphase to the structure. Various polymer combinations have been explored in the attempt to find a satisfactory combination of properties by different groups.[24–26]

7.6 CERAMIC/GLASSY ELECTROLYTES

Electrolytes of this type have, until recently, been employed primarily in thin-film batteries (about 10 μm stack thickness) (see Chap. 27). Glassy phosphorus oxysulfides were employed for lower-voltage couples such as Li/TiS_2, which gave many thousands of cycles even though no liquid electrolyte was present.[27] The short path lengths and good diffusion characteristics of the cathode material allowed cycling with very little loss in this thin-film configuration. The electrolyte phase was made by RF magnetron sputtering and had conductivities in the range of 10^{-3} S/cm. In later developments, a similar thin-film battery was made by utilizing an amorphous material called LIPON, which is a nonstoichiometric material of lithium, phosphorus, nitrogen, and oxygen. This electrolyte has higher voltage stability than the oxysulfide glasses, so it could accommodate higher voltage positive materials such as lithiated cobalt oxide or manganese oxide. The LIPON was also made by magnetron sputtering, in this case of lithium phosphate in a nitrogen atmosphere, but has much poorer conductivity in the range of 10^{-5} S/cm,[28] although still high enough for the thin layers employed. Other methods have subsequently been developed for making the LIPON, such as pulsed laser deposition.[29] Most methods for making the electrolytes and the electrodes have been relatively slow and use expensive machinery in the process. It is clearly the goal of the several companies trying to implement product development to find ways to lower the cost of the processes. To this end, recent work at Poly Plus has utilized a well-known ceramic lithium-ion conductor known as LISICON in combination with lithium anodes. LISICON has high conductivity (of the order of 1mS/cm), but is known to be unstable in direct contact with lithium metal. To prevent degradation of the anode, another material is interposed (in one case, a very thin crystalline electronic insulator such as Li_3N, which also conducts lithium ions[30] and in another, a thin separator containing solution[31]). Provision is made to completely encapsulate the lithium metal. Many positive electrodes are under investigation with this electrolyte, including aqueous air electrodes since the encapsulated lithium package is impervious to and stable in the presence of water. Another group from Mie University has published several studies utilizing a similar approach.[32] Because of the good conductivity of the LISICON, much thicker electrodes can be used, and practical high-capacity, high-energy density cells are a distinct possibility. Many details remain to be worked out, including cell designs, optimal cathode materials, as well as costs.

REFERENCES

1. E. A. Schumacher, *The Primary Battery, Vol. 1*, G. W. Heise and N. C. Cahoon Eds., John Wiley, New York, 1971, p. 179.
2. S. A. Megahed, J. Passaniti, and J. C. Springstead, *Handbook of Batteries*, 3rd Ed., D. Linden and T. B. Reddy, Eds., McGraw-Hill, New York, 2002, p. 12.9.
3. Schumacher, *The Primary Battery*, **1**, p. 180.
4. K. J. Cain, C. A. Mendres, and V. A. Maroni, *J. Electrochem. Soc.* **134**, 519 (1987) and references therein.
5. C. Debiemme-Chouvy, J. Vedel, M. Bellissent-Funel, and R. Cortes, *J. Electrochem. Soc.* **142**, 1359 (1995) and references therein.
6. R. Y. Wang, D. W. Kirk, and G. X. Zhang, *J. Electrochem. Soc.* **153**, C357 (2006).
7. F. Beck and P. Ruetschi, *Electrochim. Acta* **145**, 2467 (2000).
8. B. K. Thomas and D. J. Fray, *J. Applied Electrochem.* **12**, 1 (1982).
9. D. Aurbach and A. Zaban, Chap. 3 in *Nonaqueous Electrochemistry*, D. Aurbach, Ed., Marcel Dekker, Inc., New York, 1999, pp. 81–136.
10. D. Aurbach and Y. Gofer, Chap. 4, ibid., pp. 137–212.
11. G. E. Blomgren, Chap. 2 in *Lithium Batteries*, J. P. Gabano, Ed., Academic Press, New York, 1983, pp. 13–42.
12. P. B. Balbuena and Y. Wang, Eds., *Lithium Batteries: Solid Electrolyte Interphase*, Imperial College Press, London, 2004.

13. M. Winter, K.-C. Moeller, and J. O. Besenhard, Chap. 5 in *Lithium Batteries: Science and Technology*, G.-A. Nazri and G. Pistoia, Eds., Springer Science – Business Media, New York, 2009, pp. 144–194.
14. P. G. Balakrishnan, R. Ramesh, and T. P. Kumar, *J. Power Sources* **155**, 401 (2006).
15. J-i. Yamaki, Chap. 5 in *Advances in Lithium-Ion Batteries*, W. A. van Schalkwijk and B. Scrosati, Eds., Kluwer Academic/Plenum Publishers, New York, 2002, pp. 155–184.
16. G. E. Blomgren, Chap. 2 in *Nonaqueous Electrochemistry*, D. Aurbach, Ed., Marcel Dekker, Inc., New York, 1999, pp. 53–80.
17. G. E. Blomgren, *J. Power Sources* **14**, 39 (1985).
18. S.S. Zhang and T. R. Jow, *J. Power Sources* **109**, 458 (2002).
19. M. L. Kronenberg and G. E. Blomgren, Chap. 8 in *Comprehensive Treatise of Electrochemistry, Vol. 3*, J. O'M. Bockris, B. E. Conway, E. Yeager, and R. E. White, Eds., Plenum Press, New York, 1981, pp. 247–278.
20. E. Peled, Chap. 3 in *Lithium Batteries*, J-P. Gabano, Ed., Academic Press, New York, 1983, pp. 43–72.
21. C. R. Schlaikjer, Chap. 13, ibid., pp. 304–370.
22. A. Guerfi, M. Dontigny, P. Charest, M. Petitclerc, M. Lagacé, A. Vijh, and K. Zaghib, *J. Power Sources* **195**, 845 (2010).
23. A. Webber and G. E. Blomgren, Chap. 6 in *Advances in Lithium-Ion Batteries*, W. A. van Schalkwijk and B. Scrosati, Eds., Kluwer Academic/Plenum Publishers, New York, 2002, pp. 185–232.
24. P. E. Trapa, Y-Y. Won, S. C. Mui, E. A. Olivetti, B. Huang, D. R. Sadoway, A. M. Mayes, and S. Dallek, *J. Electrochem. Soc.* **152**, A1 (2005).
25. M. Singh, O. Odusanya, G. M. Wilmes, H. B. Etouni, E. D. Gomez, A. J. Patel, V. L. Chen, M. J. Park, P. Fragouli, H. Iatrou, N. Hadjichristidis, D. Cookson, and N. P. Balsara, *Macromolecules* **40**, 4578 (2007).
26. M. A. Meador, V. A. Cubon, D. A. Schelman, and W. R. Bennett, *Chem. Materials* **15**, 3018 (2003).
27. S. D. Jones and J. R. Akridge, *J. Power Sources* **44**, 505 (1993).
28. X. Yu, J. B. Bates, G. E. Jellison, Jr., and F. X. Hart, *J. Electrochem. Soc.* **144**, 524 (1997).
29. S. Zhao, Z. Fu, and Q. Qin, *Thin Solid Films* **415**, 108 (2002).
30. S. J. Visco, Y. S. Nimon, B. D. Katz, and L. C. De Jonghe, U. S. Patent 7,491,458, Feb. 17, 2009.
31. Ibid., U. S. Patent 7,282,295, Oct. 16, 2007.
32. T. Zhang, N. Imanishi, Y. Shimonishi, A. Hirano, J. Xie, Y. Takeda, O. Yamamoto, and N. Sammes, *J. Electrochem. Soc.* **157**, A214 (2010) and references therein.

PART 2

PRIMARY BATTERIES

CHAPTER 8
AN INTRODUCTION TO PRIMARY BATTERIES

David Linden and Thomas B. Reddy

8.1 GENERAL CHARACTERISTICS AND APPLICATIONS OF PRIMARY BATTERIES

The primary battery is a convenient source of power for portable electric and electronic devices, lighting, photographic equipment, PDAs (personal digital assistants), communication equipment, hearing aids, watches, toys, memory backup, and a wide variety of other applications, providing freedom from utility power. Major advantages of the primary battery are that it is convenient, simple, and easy to use, requires little, if any, maintenance, and can be sized and shaped to fit the application. Other general advantages are good shelf life, reasonable energy and power density, reliability, and acceptable cost.

Primary batteries have existed for over 100 years, but up to 1940, the zinc-carbon battery was the only one in wide use. During World War II and the postwar period, significant advances were made, not only with the zinc-carbon system, but with new and superior types of batteries. Capacity was improved from less than 50 Wh/kg with the early zinc-carbon batteries to more than 500 Wh/kg now obtained with lithium and zinc/air batteries. The shelf life of batteries at the time of World War II was limited to about 1 year when stored at moderate temperatures; the shelf life of present-day conventional batteries is from 2 to 5 years. The shelf life of the newer lithium batteries is as high as 10 years, with a capability of storage at temperatures as high as 70°C. Low-temperature operation has been extended from 0 to −40°C, and the power density has been improved manyfold.

Some of the advances in primary battery performance are shown graphically in Figs. 1.6 and 8.1.

Many of the significant advances were made during the 1970–90 period and were stimulated by the concurrent development of electronic technology, new demands for portable power sources, and support for space, military, and environmental improvement programs.

During this period, the zinc/alkaline manganese dioxide battery began to replace the zinc-carbon or Leclanché battery as the leading primary battery, capturing the major share of the U.S. market. Environmental concerns led to the elimination of mercury in most batteries without any impairment of performance, but also led to the phasing out of those batteries, zinc/mercuric oxide and cadmium/mercuric oxide, that used mercury as the cathodic active material. Fortunately, zinc/air and lithium batteries were developed that could successfully replace these "mercury" batteries in many applications. A major accomplishment during this period was the development and marketing of a number of lithium batteries, using metallic lithium as the anode active material. The high specific energy of these lithium batteries, at least twice that of most conventional aqueous primary batteries, and their superior shelf life opened up a wide range of applications—from small coin and cylindrical batteries for memory backup and cameras to very large batteries which were used for backup power for missile silos.

8.4 PRIMARY BATTERIES

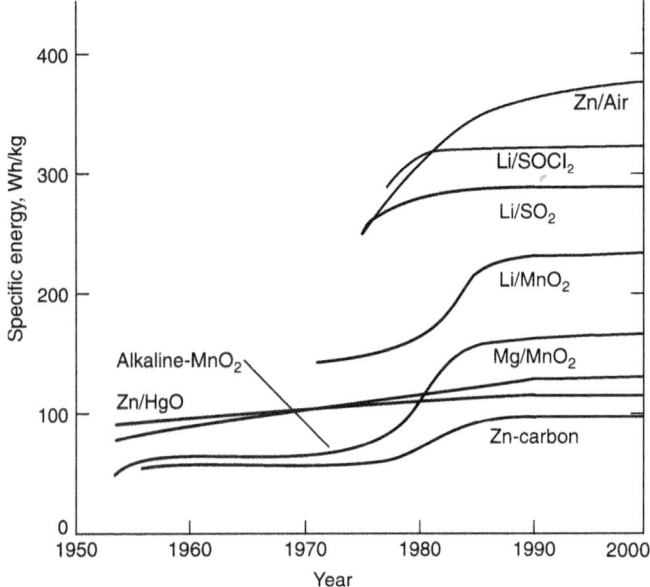

FIGURE 8.1 Advances in development of primary batteries in the 20th century. Continuous discharge at 20°C; 40–60 h rate; AA or similar size battery.

Increases in the energy density of primary batteries have tapered off as the existing battery systems have matured, and the development of new higher energy batteries is limited by the lack of new and/or untried battery materials and chemistries. Nevertheless, advances have been made in other important performance characteristics, such as power density, shelf life, and safety. Examples of these developments are the high power zinc/alkaline/manganese dioxide batteries for portable consumer electronics, the improvement of the zinc/air battery, and the introduction of new lithium batteries, such as Li/FeS_2, a 1.5 V, system.

These improved characteristics have opened up many new opportunities for the use of primary batteries. The higher energy density and specific energy have resulted in a substantial reduction in battery size and weight. This reduction, taken with the advances in electronics technology, has made many portable radio, communication, and electronic devices practical. The higher power density has made it possible to use these batteries in PDAs, transceivers, communication devices, digital cameras, and other high-power applications, which heretofore had to be powered by secondary batteries or utility power, which do not have the convenience and freedom from maintenance and recharging as do primary batteries. The long shelf life that is now characteristic of many primary batteries has similarly resulted in new uses in medical electronics, memory backup, and other long-term applications, as well as in an improvement in the lifetime and reliability of battery-operated equipment.

Figure 8.1 shows the development of primary battery systems through the year 2000. The current state-of-the art is given later in the chapter.

8.2 TYPES AND CHARACTERISTICS OF PRIMARY BATTERIES

Although a number of anode-cathode combinations can be used as primary battery systems (see Part 1), only a relatively few have achieved practical success. Zinc has been by far the most popular anode material for primary batteries because of its good electrochemical behavior, high electrochemical equivalence, compatibility with aqueous electrolytes, reasonably good shelf life, low cost, and availability. Aluminum is attractive because of its high electrochemical potential and electrochemical equivalence and availability, but due to passivation and generally limited electrochemical performance, it has not been developed successfully into a practical active primary battery system. Magnesium also has attractive electrical properties and low cost and has been used successfully in an active primary battery, particularly for military applications, because of its high energy density and good shelf life. Commercial interest has been limited and use by the U.S. military has terminated. Magnesium also is used as the anode in reserve batteries. There is an increasing focus on lithium, which has the highest gravimetric energy density and standard potential of all the metals. The lithium anode battery systems, using a number of different nonaqueous electrolytes in which lithium is stable and different cathode materials, offer the opportunity for higher energy density and power density in the performance characteristics of primary systems.

8.2.1 Characteristics of Primary Batteries

Typical characteristics and applications of the different types of primary batteries are summarized in Table 8.1.

TABLE 8.1 Major Characteristics and Applications of Primary Batteries

System	Characteristics	Applications
Zinc-carbon (Leclanché), Zinc/MnO_2	Common, low-cost primary battery; available in a variety of sizes	Flashlight, portable radios, toys, novelties, instruments
Magnesium (Mg/MnO_2)	High-capacity primary battery; long shelf life	Formerly used for military receiver-transmitters, and aircraft emergency transmitters (EPIRBs)
Mercury (Zn/HgO)	Highest capacity (by volume) of conventional types; flat discharge; good shelf life	Hearing aids, medical devices (pacemakers), photography, detectors, military equipment, but in limited use at present due to environmental hazard of mercury
Mercad (Cd/HgO)	Long shelf life; good low- and high-temperature performance; low energy density	Special applications requiring operation under extreme temperature conditions and long life; in limited use
Alkaline (Zn/alkaline/MnO_2)	Most popular general-purpose battery; good low-temperature and high-rate performance; low cost	Most popular primary battery; used in a variety of portable battery operated equipment
Silver/zinc (Zn/Ag_2O)	Highest capacity (by weight) of conventional types; flat discharge; good shelf life; costly	Hearing aids, photography, electric watches, missiles, underwater and space application (larger sizes)
Zinc/air (Zn/O_2)	Highest energy density; low cost; not independent of environmental conditions	Special applications, hearing aids, pagers, medical devices, military electronics
Lithium/soluble cathode	High energy density; long shelf life; good performance over wide temperature range	Wide range of applications requiring high energy density, long shelf life, e.g., from utility meters to military electronics applications
Lithium/solid cathode	High energy density; good rate capability and low-temperature performance; long shelf life; competitive cost	Replacement for conventional button and cylindrical cell applications, such as digital cameras
Lithium/solid electrolyte	Extremely long shelf life; low-power battery	Medical electronics

Zinc-Carbon Battery. The Leclanché or zinc-carbon dry cell battery has existed for over 100 years and had been the most widely used of all the primary batteries because of its low cost, relatively good performance, and ready availability. Cells and batteries of many sizes and characteristics have been manufactured to meet the requirements of a wide variety of applications. Significant improvements in capacity and shelf life were made with this battery system in the period between 1945 and 1965 through the use of new materials (such as chemical and electrolytic manganese dioxide and zinc chloride electrolyte) and cell designs (such as the paper-lined cell). The low cost of the Leclanché battery is a major attraction, but it has lost most of its market share, except in the developing countries, because of the newer primary batteries with superior performance characteristics, particularly alkaline manganese dioxide.

Zinc/Alkaline/Manganese Dioxide Battery. In the past two decades, an increasing portion of the primary battery market has shifted to the Zn/alkaline/MnO_2 battery. This system has become the battery of choice because of its superior performance at higher current drains and low temperatures and its better shelf life. While more expensive than the Leclanché battery on a unit basis, it is more cost-effective for those applications requiring the high-rate or low-temperature capability, where the alkaline battery can outperform the Leclanché battery by a factor of 2 to 10. In addition, because of the advantageous shelf life of the alkaline cell, it is often selected for applications in which the battery is used intermittently and exposed to uncontrolled storage conditions (such as consumer flashlights and smoke alarms), but must perform dependably when required. Most recent advances have been in the design of batteries, providing improved high-rate performance for use in digital cameras and other consumer electronics requiring this high power capability. Competition in the marketplace has also driven down the cost significantly.

Zinc/Mercuric Oxide Battery. The zinc/mercuric oxide battery was another important zinc anode primary system. This battery was developed during World War II for military communication applications because of its good shelf life and high volumetric energy density. In the postwar period, it was used in small button, flat, or cylindrical configurations as the power source in electronic watches, calculators, hearing aids, photographic equipment, and similar applications requiring a reliable long-life miniature power source. The use of the mercuric oxide battery in consumer application has ended due mainly to environmental problems associated with mercury and with its replacement by other battery systems, such as the zinc/air and lithium batteries, which have superior performance for many applications.

Cadmium/Mercuric Oxide Battery. The substitution of cadmium for the zinc anode (the cadmium/mercuric oxide cell) results in a lower-voltage but very stable system, with a shelf life of up to 10 years as well as performance at high and low temperatures. Because of the lower voltage, the watthour capacity of this battery is about 60% of the zinc/mercuric oxide battery capacity. Again, because of the hazardous characteristics of mercury and cadmium, the use of this battery is limited.

Zinc/Silver Oxide Battery. The primary zinc/silver oxide battery is similar in design to the small zinc/mercuric oxide button cell, but it has a higher specific energy and performs better at low temperatures. These characteristics make this battery system desirable for use in hearing aids, calculators, and electronic watches. However, because of its high cost and the development of other battery systems, the use of this battery system as a primary battery has been limited mainly to small button battery applications where the higher cost is justified. Larger cells continue to be employed for military applications.

Zinc/Air Battery. The zinc/air battery system is noted for its high energy density, but it had been formerly used only in large, low-power batteries for signaling and navigational-aid applications. With the development of enhanced air electrodes, the high-rate capability of the system was improved and small button-type batteries are now used widely in hearing aids, electronics, and similar applications. These batteries have a very high energy density as no active cathode material is needed. Wider use of this system and the development of larger batteries have been slow because of some of their performance limitations (sensitivity to extreme temperatures, humidity, and other environmental factors, such as carbonation, as well as poor activated shelf life and low power density). Nevertheless, because of their attractive energy density, zinc/air batteries are now being used for military applications (see Chap. 33).

Magnesium Batteries. While magnesium has attractive electrochemical properties, there has been relatively little commercial interest in magnesium primary batteries because of the generation of hydrogen gas during discharge and the relatively poor storageability of a partially discharged cell. Magnesium dry cell batteries have been used successfully in military communications equipment in the past, taking advantage of the long shelf life of a battery in an undischarged condition, even at high temperatures, and its higher energy density. Magnesium is still employed as an anode material for reserve type and metal/air batteries (see Chaps. 33 and 34).

Aluminum Batteries. Aluminum is another attractive anode material with a high theoretical energy density, but problems such as polarization and parasitic corrosion have inhibited the development of a commercial product. The best promise for its use is as a reserve or mechanically rechargeable battery (see Chaps. 10 and 34).

Lithium Batteries. The lithium anode batteries are a relatively recent development (since 1970). They have the advantage of high energy density and specific energy, as well as operation over a very wide temperature range with long shelf life, and are gradually replacing some conventional battery systems. However, except for camera, medical, watch, memory backup, military, and other niche applications, they have not yet captured the general purpose markets as was anticipated because of their higher cost.

As with the zinc systems, there are a large number of lithium batteries which have been used ranging in capacity from less than 5 mAh to 10,000 Ah, using various designs and chemistries, but having in common the use of lithium metal as the anode.

The lithium primary batteries can be classified into three categories (see Chap. 14). The smallest are the low-power solid-state batteries (see Chap. 31) with excellent shelf life, and are used in applications such as cardiac pacemakers. In the second category are the solid-cathode batteries, which are designed in coin or small cylindrical configurations. These batteries have replaced the conventional primary batteries in watches, calculators, memory circuits, photographic equipment, communication devices, and other such applications where its high energy density and long shelf life are critical. The soluble-cathode batteries (using liquid cathode materials) constitute the third category. These batteries are typically constructed in a cylindrical configuration, as flat disks, or in prismatic containers using flat plates. These batteries, up to about 35 Ah in size, are used in military and industrial applications, lighting products, and other devices where small size, low weight, and operation over a wide temperature range are important. The larger batteries have been used for special military applications or as standby emergency power sources.

Solid Electrolyte Batteries. The solid-electrolyte batteries are different from other battery systems in that they depend on the ionic conductivity, in the solid state, of an electronically nonconductive compound rather than the ionic conductivity of a liquid electrolyte. Batteries using these solid electrolytes are low-power (microwatt) devices, but have extremely long shelf lives and the capability of operating over a wide temperature range, particularly at high temperatures. These batteries have been used in medical electronics, for memory circuits, and for other such applications requiring a long-life, low-power battery. The first solid-electrolyte batteries used a silver anode and silver iodide for the electrolyte, but lithium is now the anode of choice for most of these batteries, offering a higher voltage and energy density. Current use is limited except in pacemakers where the LiI electrolyte forms "in situ".

8.3 COMPARISON OF THE PERFORMANCE CHARACTERISTICS OF PRIMARY BATTERY SYSTEMS

8.3.1 General

A qualitative comparison of the various primary battery systems is given in Table 8.2. This listing illustrates the performance advantages of the lithium anode batteries. Nevertheless, the conventional primary batteries, because of their low cost, availability, and generally acceptable performance in many consumer applications, still maintain a major share of the market.

The characteristics of the major primary batteries are summarized in Table 8.3. This table is supplemented by Table 1.2 in Chap. 1, which lists the theoretical and practical electrical characteristics of these primary battery systems. A graphic comparison of the theoretical and practical performance of various battery systems given in Fig. 1.4 shows that only about 25 to 35% of the theoretical capacity is attained under practical conditions as a result of design and discharge requirements.

It should be noted, as discussed in detail in Chaps. 1, 3 and 32, that most of these types of data and comparisons are based on the performance characteristics of single-cell batteries and are necessarily approximations, with each system presented under favorable discharge conditions. The specific performance of a battery system is very dependent on the cell and battery design and all of the specific conditions of use and discharge of the battery.

TABLE 8.2 Comparison of Primary Batteries*

System	Voltage	Specific energy (gravimetric)	Power density	Flat discharge profile	Low-temperature operation	High-temperature operation	Shelf life	Cost
Zinc/carbon	5	4	4	4	5	6	8	1
Zinc/alkaline/manganese dioxide	5	3	2	3	4	4	7	2
Magnesium/manganese dioxide	3	3	2	2	4	3	4	3
Zinc/mercuric oxide	5	3	2	2	5	3	4	5
Cadmium/mercuric oxide	6	5	2	2	3	2	3	6
Zinc/silver oxide	4	3	2	2	4	3	5	6
Zinc/air	5	2	3	2	5	5	—	3
Lithium/soluble cathode	1	1	1	1	1	2	1	5
Lithium/solid cathode	1	1	1	2	2	3	2	3

*1 to 8—best to poorest.

8.3.2 Voltage and Discharge Profile

A comparison of the discharge curves of the major primary batteries is presented in Fig. 8.2. The zinc anode batteries generally have a discharge voltage of between about 1.5 and 0.9 V. The lithium anode batteries, depending on the cathode, usually have higher voltages, many on the order of 3 V, with an end or cutoff voltage of about 2.0 V. The cadmium/mercuric oxide battery operates at the lower voltage level of 0.9 to 0.6 V. The discharge profiles of these batteries also show different characteristics. The conventional zinc-carbon and zinc/alkaline/manganese dioxide batteries have sloping profiles; the magnesium/manganese dioxide and lithium/manganese dioxide batteries have less of a slope (although at lower discharge rates, the lithium/manganese dioxide battery shows a flatter profile). Most of the other battery types have a relatively flat discharge profile.

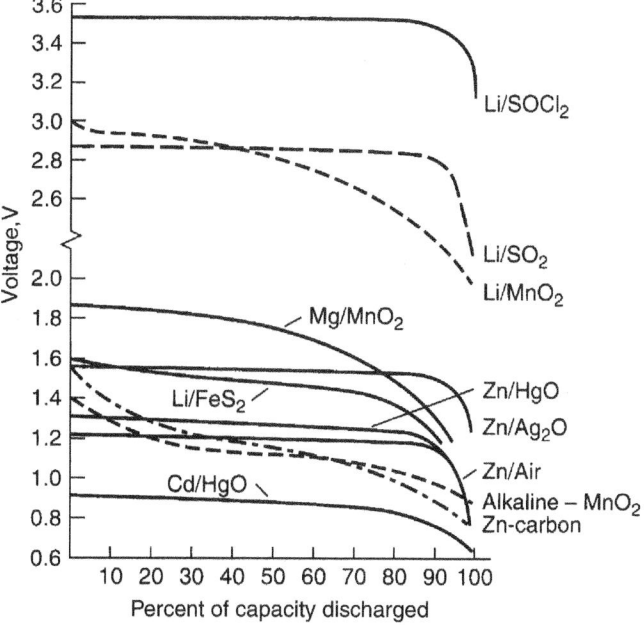

FIGURE 8.2 Discharge profiles of primary battery systems 30–100 h rate.

8.3.3 Specific Energy and Specific Power

Figure 8.3 presents a comparison of the specific energy (also called gravimetric energy density) of the different primary battery systems at various discharge rates at 20°C. This figure shows the hours of service each battery type (normalized to 1 kg battery weight) will deliver at various power (discharge current × midpoint voltage) levels to an end voltage usually specified for that battery type. The energy density can then be determined by

$$\text{Specific energy} = \text{specific power} \times \text{hours of service}$$

or

$$\text{Wh/kg} = \text{W/kg} \times h = \frac{A \times V \times h}{\text{kg}}$$

The conventional zinc-carbon battery has the lowest energy density of the primary batteries shown, with the exception, at low discharge rates, of the cadmium/mercuric oxide battery due to the low voltage of the latter electrochemical couple. The zinc-carbon battery performs best at light discharge loads. Intermittent discharges, providing a rest or recovery period at intervals during the discharge, improve the service life significantly compared with a continuous discharge, particularly at high discharge rates.

The ability of each battery system to perform at high current or power levels is shown graphically in Fig. 8.3 by the reduced slope at the higher discharge rates. The 1000 Wh/kg line indicates the slope at which the capacity or energy density of the battery remains constant at all discharge rates. The capacity of most battery systems decreases with increasing discharge rate, and the slope of the linear portion of each of the other lines is less than that of the theoretical 1000 Wh/kg line. Furthermore, as

8.10 PRIMARY BATTERIES

TABLE 8.3 Characteristics of Primary Batteries

System	Zinc-carbon (Leclanché)	Zinc-carbon (zinc chloride)	Mg/MnO$_2$	Zn/Alk./MnO$_2$	Zn/HgO	Cd/HgO
Chemistry:						
Anode	Zn	Zn	Mg	Zn	Zn	Cd
Cathode	MnO$_2$	MnO$_2$	MnO$_2$	MnO$_2$	HgO	HgO
Electrolyte	NH$_4$Cl and ZnCl$_2$ (aqueous solution)	ZnCl$_2$ (aqueous solution)	MgBr$_2$ or Mg(ClO$_4$) (aqueous solution)	KOH (aqueous solution)	KOH or NaOH (aqueous solution)	KOH (aqueous solution)
Cell voltage, V§:						
Nominal	1.5	1.5	1.6	1.5	1.35	0.9
Open-circuit	1.5–1.75	1.6	1.9–2.0	1.5–1.6	1.35	0.9
Midpoint	1.25–1.1	1.25–1.1	1.8–1.6	1.23	1.3–1.2	0.85–0.75
End	0.9	0.9	1.2	0.8	0.9	0.6
Operating temperature, °C	−5 to 45	−10 to 50	−40 to 60	−40 to 50	0 to 55	−55 to 80
Energy density at 20°C§:						
Button size:						
Wh/kg				81	100	55
Wh/L				361	470	230
Cylindrical size:						
Wh/kg	65	85	100	154	105	
Wh/L	100	165	195	461	325	
Discharge profile (relative)	Sloping	Sloping	Moderate slope	Moderate slope	Flat	Flat
Power density	Low	Low to moderate	Moderate	Moderate	Moderate	Moderate
Self-discharge rate at 20°C, % loss per year‡	10	7	3	3	4	3
Advantages	Lowest cost; good for noncritical use under moderate conditions; variety of shapes and sizes; availability	Low cost; better performance than regular zinc-carbon	High capacity compared with zinc-carbon; good shelf life (undischarged)	High capacity; good low-temperature and high-rate performance; low cost	High volumetric energy density; flat discharge; stable voltage	Good performance at high and low temperatures; long shelf life
Limitations	Low energy density; poor low-temperature, high-rate performance	High gassing rate; performance lower than premium alkaline batteries	High gassing (H$_2$) on discharge; delayed voltage	Electrolyte leakage may occur	Expensive; moderate gravimetric energy density, poor-low-temperature performance	Expensive; low-energy density
Status	High production, but losing market share	High production, but losing market share	NLA*	High production; most popular primary battery	Phased out because of toxic mercury	In limited production being phased out because of toxic components except for some special applications
Major types available	Cylindrical single-cell bobbin and multicell batteries (see Table 9.9)	Cylindrical single-cell bobbin and multicell batteries (see Table 9.9)	Cylindrical single-cell bobbin and multicell batteries (see Table 10.3) previously available	Button, cylindrical, and multicell batteries (see Tables 11.8 and 11.9)	NLA*	NLA*

*No longer readily available commercially.
†See Chap. 14 for other lithium primary batteries.
‡Rate of self-discharge usually decreases with time of storage.
§Data presented are for 20°C, under favorable discharge condition. See details in appropriate chapter.

Zn/Ag$_2$O	Zinc/air	Li/SO$_2$†	Li/SOCl$_2$†	Li/MnO$_2$†	Li/FeS$_2$†	Solid state
Ag$_2$O or AgO KOH or NaOH (aqueous solution)	Zn O$_2$ (air) KOH (aqueous solution)	Li SO$_2$ Organic solvent, salt solution	Li SOCl$_2$ SOCl$_2$ w/AlCl$_4$	Li MnO$_2$ Organic solvent, salt solution	Li FeS$_2$ Organic solvent, salt solution	Li I$_2$(P2VP) Solid
1.5 1.6 1.6–1.5 1.0 0 to 55	1.5 1.45 1.3–1.1 0.9 0 to 50	3.0 3.1 2.9–2.75 2.0 −55 to 70	3.6 3.65 3.6–3.3 3.0 −60 to 85	3.0 3.3 3.0–2.7 2.0 −20 to 55	1.5 1.8 1.6–1.4 1.0 −20 to 60	2.8 2.8 2.8–2.6 2.0 0 to 200
135 530	415 1350			230 545		
	Prismatic 500 Prismatic 1250	260 415	Bobbin 590 / Spiral-wound 495 1100 / 970	Bobbin 270 / Spiral-wound 261 620 / 546	310 560	220–280 820–1030
Flat	Flat	Very flat	Flat	Flat	Medium to high	Moderately flat (at low discharge rates)
Moderate	Low	High	Medium (but dependent on specific design)	Moderate	Medium to high	Very low
6	3 (if sealed)	2	1–2	1–2	1–2	<1
High energy density; good high-rate performance	High volumetric energy density; long shelf life (sealed)	High energy density; best low-temperature, high-rate performance; long shelf life	High energy density; long shelf life because of protective film	High energy density; good low-temperature, high-rate performance; cost-effective replacement for small conventional type cells	Replacement for Zu/alkaline/MnO$_2$ batteries for high-rate performance	Excellent shelf life (10–20 y); wide operating temperature range (to 200°C)
Expensive, but cost-effective on button battery applications	Not independent of environment—flooding, drying out; limited power output	Pressurized system; potential safety problems; toxic components; shipment regulated	Voltage delay after storage	Available in small sizes; larger sizes being considered; shipment regulated	Higher cost than alkaline batteries	For very low discharge rates; poor low temperature performance
In production	Moderate production; key use in hearing aids	Moderate production, mainly military	Produced in wide range of sizes and designs; mainly for special applications	Increasing consumer production	Produced in AAA and AA sizes; 9 V batteries also available	In production for special applications
Button batteries (see Table 13.3)	(See Tables 13.5 and 13.6, also Chap. 33)	Cylindrical batteries (see Tables 14.9 and 14.10)	(See Sec. 14.6 and Tables 14.11 to 14.13)	Button and small cylindrical batteries (see Tables 14.10 and 14.18–14.20)	Cylindrical cells and 9 V (see Table 14.23)	(See Sec 31.5.)

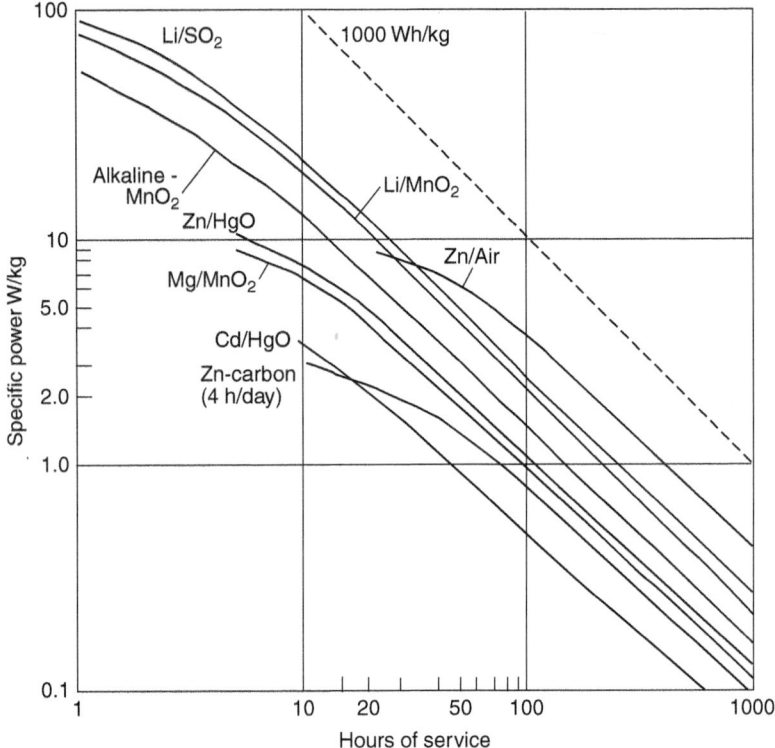

FIGURE 8.3 Comparison of typical performance of primary battery systems—specific power vs. hours of service.

the discharge rate increases, the slope drops off more sharply. This occurs at higher discharge rates for the battery types that have high-power capability.

The performance of the zinc-carbon battery falls off sharply with increasing discharge rate, although the heavy-duty zinc chloride version of the zinc-carbon battery (see Chap. 9) gives better performance under the more stringent discharge conditions. The zinc/alkaline/manganese dioxide battery, the zinc/mercuric oxide battery, the zinc/silver oxide battery, and the magnesium/manganese dioxide battery all have about the same specific energy and performance at 20°C. The zinc/air system has a higher specific energy at the low discharge rates, but falls off sharply at moderately high loads, indicating its low specific power. The lithium batteries are characterized by their high specific energy, due in part to the higher cell voltage. The lithium/sulfur dioxide battery and some of the other lithium batteries are distinguished by their ability to deliver this higher capacity at the higher discharge rates.

Volumetric energy density is, at times, a more useful parameter than gravimetric specific energy, particularly for button and small batteries, where the weight is insignificant. The denser batteries, such as the zinc/mercuric oxide battery, improve their relative position when compared on a volumetric basis, as shown in Fig. 8.9. Many chapters on the individual battery systems include a family of curves giving the hours of service each battery system will deliver at various discharge rates and temperatures.

8.3.4 Comparison of Performance of Representative Primary Batteries

Figure 8.4 compares the performance of a number of primary battery systems in a typical button cell configuration, size 44 IEC standard. The data are based on the rated capacity at 20°C at about the $C/500$ rate. The performance of the different systems can be compared, but one should recognize that

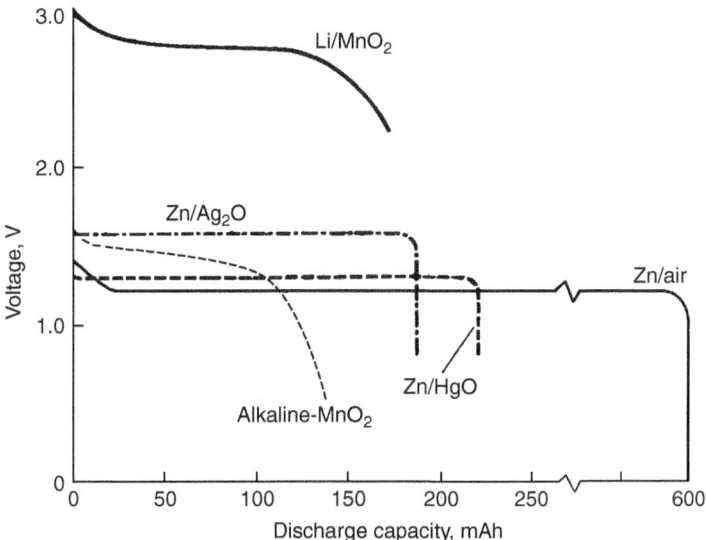

FIGURE 8.4 Typical discharge curves for primary battery systems, 11.6 mm diameter, 5.4 mm high, 20°C. (Li/MnO$_2$ battery is $\frac{1}{3}$ N size).

battery manufacturers may design and fabricate batteries, in the same size and with the same electrochemical system, with differing capacities and other characteristics, depending on the application requirements and the particular market segment the manufacturer is addressing.

Table 8.4 summarizes the typical performance obtained with the different primary battery systems for several cylindrical type batteries. The discharge curves for the AA-size batteries are shown in Fig. 8.5, those for the ANSI 1604 9-V batteries in Fig. 8.6.

8.3.5 Effect of Discharge Load and Duty Cycle

The effect of the discharge load on the battery's capacity was shown in Fig. 8.3 and is again illustrated for several primary battery systems in Fig. 8.7. The Leclanché zinc-carbon battery performs best under light discharge loads, but its performance falls off sharply with increasing discharge rates. The zinc/alkaline/manganese dioxide system has a higher energy density at light loads which does not drop off as rapidly with increasing discharge loads. The lithium battery has the highest energy density with reasonable retention of this performance at the higher discharge rates. For low-power applications, the service ratio of lithium:zinc (alkaline):zinc-carbon is on the order of 4:3:1. At the heavier loads, however, such as those required for toys, motor-driven applications, and pulse discharges such as digital cameras, the ratio can widen to 24:8:1 or greater. At these heavy loads, selection of premium batteries is desirable on both a performance and a cost basis.

8.3.6 Effect of Temperature

The performance of the various primary batteries over a wide temperature range is illustrated in Fig. 8.8 on a gravimetric basis and in Fig. 8.9 on a volumetric basis. The lithium/soluble-cathode systems (Li/SOCl$_2$ and Li/SO$_2$) show the best performance throughout the entire temperature range, with the higher-rate Li/SO$_2$ system having the best capacity retention at the very low temperatures. The zinc/air system has a high energy density at normal temperatures, but only at light discharge loads. The lithium/solid-cathode systems, represented by the Li/MnO$_2$ system, show high performance over a wide temperature range, superior to the conventional zinc anode systems.

TABLE 8.4 Comparison of Cylindrical-Type Primary Batteries[‡]

	Zinc-carbon (standard)	Zinc-carbon (heavy-duty $ZnCl_2$)	Zn/MnO_2* (alkaline)	Zn/HgO[‡]	Mg/MnO_2[‡]	Li/SO_2	$Li/SOCl_2$ (bobbin type)	Li/MnO_2 (spiral wound)	Li/FeS_2
Working voltage, V	1.2	1.2	1.2	1.25	1.75	2.8	3.3	2.8	1.5
				D-size cells					
Ah	4.5	7.0	18.5	14	7	7.75	19	11.1	
Wh	5.4	8.4	22.8	17.5	12.2	21.7	64.6	30.0	
Weight, g	85	93	148	165	105	85	93	115	
Wh/kg	65	90	154	105	115	255	695	261	
Wh/L	100	160	407	325	225	397	1235	546	
				N-size cells					
Ah	0.40		1.00	0.8	0.5				
Wh	0.48		1.20	1.0	0.87				
Weight, g	6.3		9.0	12	5.0				
Wh/kg	75		133	85	170				
Wh/L	145		364	330	290				
				AA-size cells					
Ah	0.8	1.05	2.80	2.5		0.95	2.4	1.4[†]	3.1
Wh	0.96	1.25	3.39	3.1		2.66	8.41	3.9	4.495
Weight, g	14.7	15	2.30	30		15	18	17	14.5
Wh/kg	65	84	1.47	103		177	467	235	310
Wh/L	125	162	4.18	400		334	1007	525	562

*Zn/MnO_2 (Alkaline) data to a 0.8 V cutoff.
[†]$\frac{2}{3}$ A size.
[‡]These batteries may no longer be available.

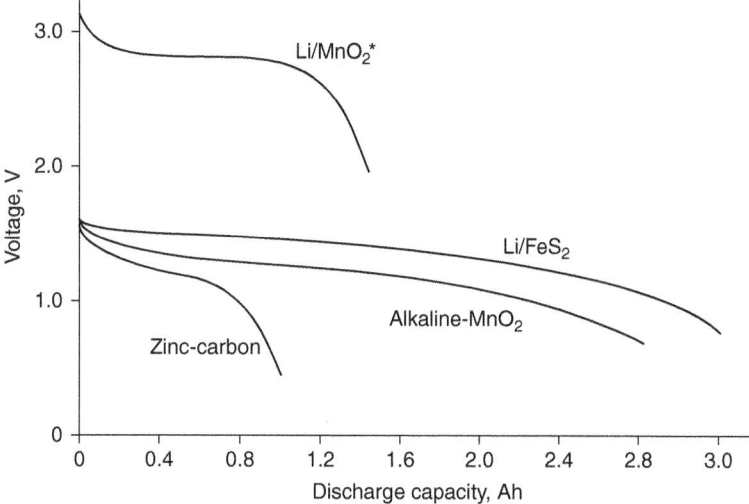

FIGURE 8.5 Typical discharge curves for primary battery systems. AA-size cells, approx. 20 mA discharge rate.
*$\frac{2}{3}$ A-size battery.

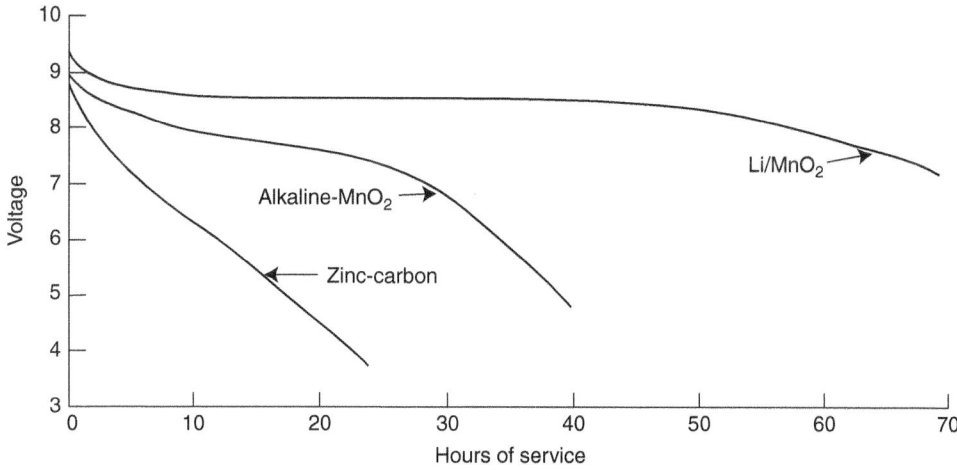

FIGURE 8.6 Typical discharge curves ANSI 1604 battery, 9V, 500 ohm discharge load 20°C.

Figure 8.9 shows an improvement in relative position of the denser, heavier battery systems when compared on a volumetric basis.

(Note: As stated earlier, these data are necessarily generalized and present each battery system under favorable discharge conditions. With the variability in performance due to manufacturer, design, size, discharge conditions, end voltage, and other factors, they may not apply under specific conditions of use. For these details, refer to the appropriate chapter for each battery system.)

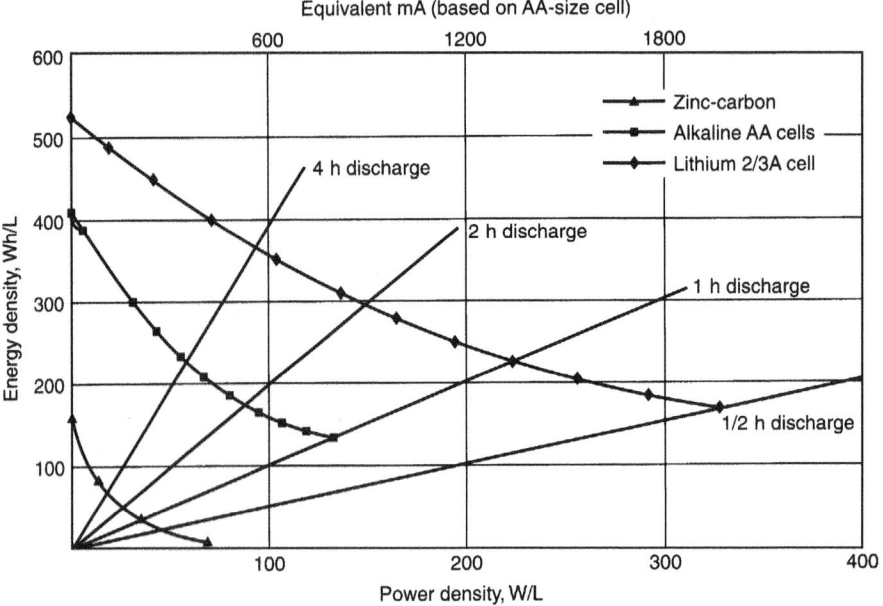

FIGURE 8.7 Comparison of primary battery systems under various continuous discharge loads at 20°C.

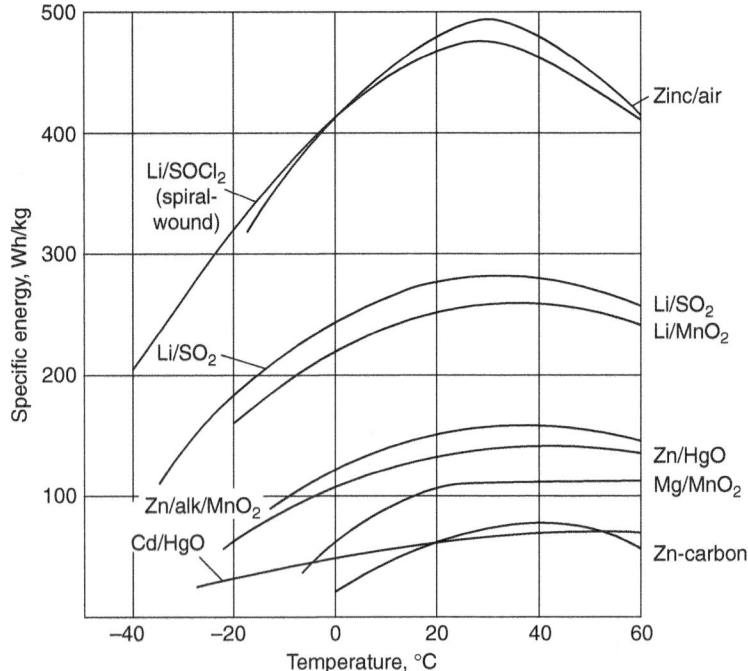

FIGURE 8.8 Specific energy of primary battery systems.

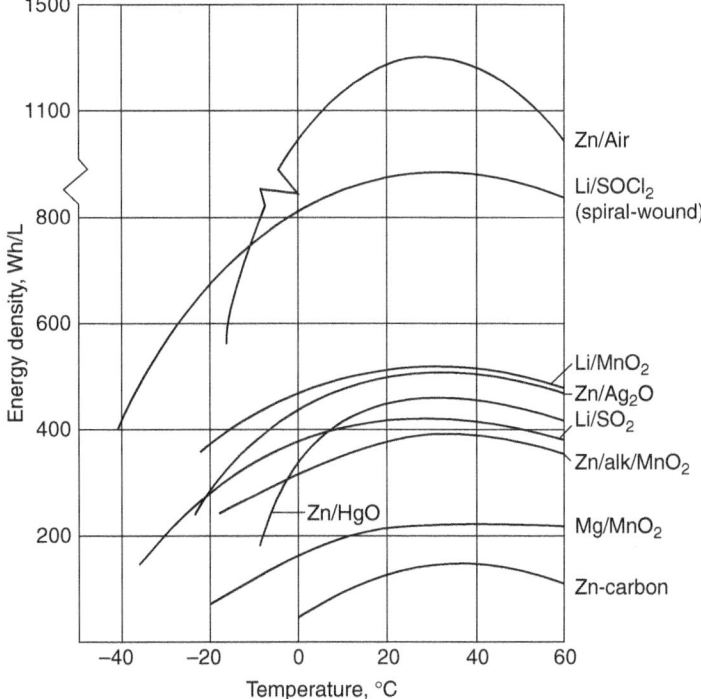

FIGURE 8.9 Volumetric energy density of primary battery systems.

8.3.7 Shelf Life of Primary Batteries

The shelf-life characteristics of the major primary battery systems are plotted in Fig. 8.10 and show the rate of loss (in terms of percentage capacity loss per year) from 20 to 70°C. The relationship is approximately linear when log capacity loss is plotted against log 1/T (temperature, °Kelvin). The data assume that the rate of capacity loss remains constant throughout the storage period, which is not necessarily the case with most battery systems. For example, as shown in Chap. 14 for several lithium batteries, the rate of loss tapers off as the storage period is extended. The data are also a generalization of the capability of each battery system under manufacturer-rated conditions because of the many variations in battery design and formulation. The discharge conditions and size also have an influence on charge retention. The capacity loss is usually highest under the more stringent discharge conditions.

The ability to store batteries improves as the storage temperature is lowered. Cold storage of batteries is used to extend their shelf life. Moderately cold temperature, such as 0°C, was usually used as freezing could be harmful for some battery systems or designs. As the shelf life of most batteries has been improved, manufacturers are no longer recommending cold storage but suggest room temperature storage is adequate provided that excursions to high temperature are avoided.

8.3.8 Cost

Consumer Reports conducts tests on different types of commercial primary and secondary batteries for applications such as digital cameras. The results are compared on a basis of the cost effectiveness of a particular type in the given application. The most recent study for digital camera is given in the Dec. 2009 issue, but results are updated regularly. See also Chap. 32.

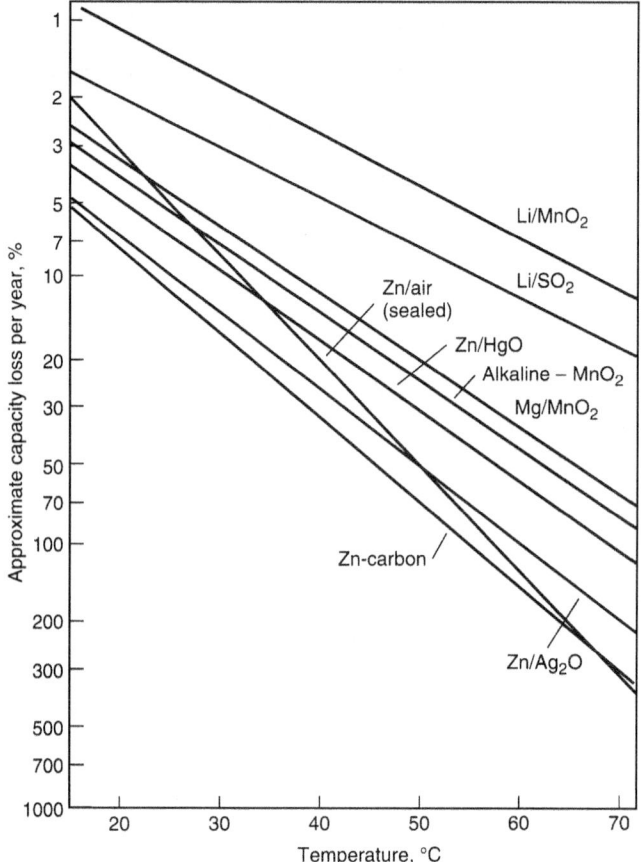

FIGURE 8.10 Shelf-life characteristics of primary battery systems.

8.4 RECHARGING PRIMARY BATTERIES

Recharging primary batteries is a practice that should be avoided because the cells are not designed for that type of use. In most instances, it is impractical, and it could be dangerous with cells that are tightly sealed and not provided with an adequate mechanism to permit the release of gases that form during charging. Such gassing could cause a cell to leak, rupture, or explode, resulting in personal injury, damage to equipment, and other hazards. Most primary batteries are labeled with a cautionary notice advising that they should not be recharged.

Technically, some primary cells can be recharged for several cycles under carefully controlled charging conditions and usually at low charge rates. However, even if successful, they may not deliver full capacity and may have poor charge retention after recharge. Primary batteries are not designed to be recharged, and charging should not be attempted with any primary battery, unless one is fully aware of the charging conditions, equipment, and risks.

The zinc/alkaline/manganese dioxide system has been designed in a rechargeable configuration, as described in Chap. 28.

CHAPTER 9
ZINC-CARBON BATTERIES—LECLANCHÉ AND ZINC CHLORIDE CELL SYSTEMS

Brooke Schumm, Jr.

9.1 GENERAL CHARACTERISTICS

Zinc-carbon batteries have been well known for over a hundred years. The two types of zinc-carbon batteries that are popular now are the Leclanché and zinc chloride systems. Both systems remain among the most widely used of all the primary battery systems worldwide although their use in the United States and Europe is declining. The use of flashlights, portable radios, and other moderate- and light-drain applications, as well as the absence of a high-drain device base, is stimulating the use of zinc-carbon batteries in emerging countries. The battery is characterized as having low cost, ready availability, and acceptable performance for a large number of applications.

The zinc-carbon battery industry continues to grow worldwide. The global battery market is expected to reach $75 billion or more in sales by the year 2015.[1] Some details of the zinc-carbon battery market and the global primary battery market as of 2002 are given in Table 9.1, estimating that Asia-Pacific would change the most from 2002 to 2012.

The current estimate of annual growth for the zinc-carbon global market through the year 2012 continues to be +5% per year. The expected decline in the zinc-carbon battery market was only realized in the United States, with a relatively constant −2% to −5% decline in sales volume per year. This is expected to continue. Asia, emerging countries, and Eastern European markets drove the global demand for the inexpensive zinc-carbon battery system. As an example, 80% of all primary batteries presently sold in Eastern and Central Europe are zinc-carbon types. Even in the United States, this system still shows substantial usage, with total U.S. sales in 1998 of $370 million dollars.[2,3] (See Fig. 9.1.)

New, heavier-drain toys, lighting, and communications devices entering the consumer market continue to stimulate an increased preference for zinc-alkaline cells. This has spawned a segmentation of the zinc-alkaline system, resulting in the design of increased power, heavy-duty, zinc-alkaline batteries for those applications. These new applications and continued impact from the use of rechargeable cells will be additional factors impacting zinc-carbon sales in the United States.

Historically, the first prototype of the modern dry cell was the Leclanché wet cell developed by a telegraph engineer, Georges-Lionel Leclanché, in 1866. The design resulted from the need to provide a more reliable and easily maintained power source for telegraph offices and railroad signaling. The cell was unique in that it was the first practical cell using a single low-corrosive fluid, ammonium chloride, as an electrolyte instead of the strong mineral acids in use at the time.

9.2 PRIMARY BATTERIES

TABLE 9.1 Zinc-Carbon Battery Market

Regional market location	Total primary battery market value by 2002 (billions $US)	Zinc-carbon battery market value by 2002 (billions $US)	Zinc-carbon battery as a percent of global market (%)
U.S. & Canada	4.4	0.3	6.8
Latin America	1.4	1.0	64.3
Western Europe	3.9	0.9	20.5
Eastern Europe	2.8	1.0	32.1
Asia-Pacific	8.6	4.0	45.3
Global Total	21.1	7.2	34.5

Source: Freedonia Group, 1999 battery market study.[2]

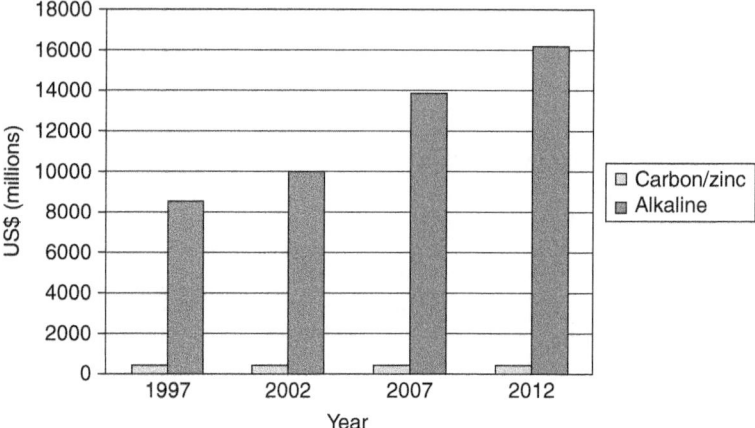

FIGURE 9.1 U.S. primary battery sales.[2]

This rendered the cells relatively inactive until the external circuit was connected. The cell was inexpensive, safe, easily maintained, and provided excellent shelf (storage) life with adequate performance characteristics. The cell consisted of an amalgamated zinc bar serving as the negative electrode anode, a solution of ammonium chloride as the electrolyte, and a one-to-one mixture of manganese dioxide and powdered carbon packed around a carbon rod as the positive electrode or cathode. The positive electrode was placed in a porous pot, which was, in turn, placed in a square glass jar along with the electrolyte and zinc bar. By 1876, Leclanché had evolved the design, removing the need for the porous pot by adding a resin (gum) binder to the manganese dioxide-carbon mix. In addition, he formed this composition into a compressed block by use of hydraulic pressure at a temperature of 100°C. Leclanché's inventiveness brought together the major components of today's zinc-carbon battery and set the stage for conversion from the "wet" cell to the "dry" cell concept.

Dr. Carl Gassner is credited with constructing the first "dry" cell in 1888. It was similar to the Leclanché system except that ferric hydroxide and manganese dioxide were used as the cathode. The "dry" cell concept grew from the desire to make the cell unbreakable and spill-proof. His cell provided an unbreakable container by forming the anode from zinc sheet into a cup, replacing the glass jar. He then immobilized the electrolyte by using a paste containing plaster of Paris and ammonium chloride. The cylindrical block of cathode mix (called a bobbin) was wrapped in cloth and was saturated with a zinc chloride-ammonium chloride electrolyte. This reduced local chemical action and improved the shelf life. Gassner, as did others, replaced the plaster of Paris with wheat flour as an electrolyte-gelatinizing agent and demonstrated such a battery as a portable lighting power source at the 1900 World's Fair in Paris. These advances were instrumental in establishing industrial production and commercialization of the "zinc-carbon dry cell" and led to the evolution of "dry-cell" portable power.

From the early 1900s through the 1990s, the portable power industry was driven to meet the needs of the electric and electronic industries. In the early part of the 20th century, battery-operated telephones, electric doorbells, toys, lighting devices, and countless other applications placed increasing demands on "dry battery" manufacturers. Through the middle of the century, the advent of radio broadcasting and World War II military applications further increased that demand significantly. In the latter part of the century, demands for an inexpensive battery to power flashlights, portable transistor radios, electric clocks, cameras, electronic toys, and other convenience applications maintained the demand.

Zinc-carbon technology has continued to evolve. During much of the 20th century, the system was continually improved. Manganese dioxide materials, electrolytic and chemical, with higher capacity and substantially higher activity than the natural manganese ores, had been developed. The use of acetylene black carbon as a substitute for graphite has not only provided a more conductive cathode structure, but the higher absorption properties have enhanced the handling characteristics of the cathode powder. Improved manufacturing techniques were implemented that resulted in the production of an improved product at lower costs. A better understanding of the reaction mechanisms, improved separators, and venting seal systems have all contributed to the present state of the zinc-carbon art.

A significant portion of the technology effort since the 1960s has been directed toward developing the zinc chloride cell system. This design provided substantially improved performance on heavy-drain applications over that of the Leclanché cell. From the 1980s to the present time, development effort has been focused on environmental concerns, including the elimination of mercury, cadmium, and other heavy metals from the system. The work of the past century has extended the discharge life and storage life of the zinc-carbon battery over 400% compared to the 1910 version.[3-9]

Most of zinc-carbon cell manufacturing and battery assembly is now done outside of the United States. Manufacturers have opted to consolidate and relocate plants and equipment to achieve cost reductions through the use of economies of scale, low-cost labor, and materials. Regional plants are coming of age rather than local country manufacturing facilities. This has occurred because of the improved conditions in global trade, which in many areas has reduced tariffs and duties. As a direct result, cell prices have generally been maintained at steady levels and business opportunities for zinc-carbon batteries have increased globally.

The advantages and disadvantages of zinc-carbon batteries, compared with other primary battery systems, are summarized in Table 9.2. A comparison of the more popular primary cell systems is given in Chap. 8.

TABLE 9.2 Major Advantage and Disadvantages of Leclanché and Zinc Chloride Batteries

Standard Leclanché battery		
Advantages	Disadvantages	General comments
Low cell cost	Low energy density	Good shelf life if refrigerated
Low cost per watt-hour	Poor low-temperature service	For best capacity, the discharge should be intermittent
Large variety of shapes, sizes, voltages, and capacities	Poor leakage resistance under abusive conditions	Capacity decreases as the discharge drain increases
Various formulations	Low efficiency under high current drains	Steadily falling voltage is useful if early warning of end of life is important
Wide distribution and availability	Comparatively poor shelf life	
Long tradition of reliability	Voltage falls steadily with discharge	

Standard zinc chloride battery		
Advantages	Disadvantages	General comments
Higher energy density	Requires excellent sealing system due to increased oxygen sensitivity	Steadily falling voltage with discharge
Better low-temperature service		Good shock resistance
Good leak resistance		Low to medium initial cost
High efficiency under heavy discharge loads		

9.2 CHEMISTRY

The zinc-carbon cell uses a zinc anode, a manganese dioxide cathode, and an electrolyte of ammonium chloride and/or zinc chloride dissolved in water. Carbon (acetylene black) is mixed with the manganese dioxide to improve conductivity and retain moisture. As the cell is discharged, the zinc is oxidized and the manganese dioxide is reduced. A simplified overall cell reaction is

$$Zn + 2MnO_2 \rightarrow ZnO \cdot Mn_2O_3$$

In actual practice, the chemical processes that occur in the Leclanché cell are significantly more complicated. Despite the 125 years of its existence, controversy over the details of the electrode reactions continues.[7] A chemical "recuperation" reaction may operate simultaneously with the discharge reactions.[5]

This could result in several intermediate states that confuse the reaction mechanisms. Furthermore, the chemistry is complex because it is a non-stoichiometric oxide and is more accurately represented as MnO_x, where x typically equals 1.9+. The efficiency of the chemical reaction depends on such things as electrolyte concentration, cell geometry, discharge rate, discharge temperature, depth of discharge, diffusion rates, and type of MnO_2 used. A more comprehensive description of the cell reaction is as follows:[4]

1. For cells with ammonium chloride as the primary electrolyte:
   Light discharge: $Zn + 2MnO_2 + 2NH_4Cl \rightarrow 2MnOOH + Zn(NH_3)_2Cl_2$
   Heavy discharge: $Zn + 2MnO_2 + NH_4Cl + H_2O \rightarrow 2MnOOH + NH_3 + Zn(OH)Cl$
   Prolonged discharge: $Zn + 6MnOOH \rightarrow 2Mn_3O_4 + ZnO + 3H_2O$

2. For cells with zinc chloride as the primary electrolyte:*
   Light or heavy discharge: $Zn + 2MnO_2 + 2H_2O + ZnCl_2 \rightarrow 2MnOOH + 2Zn(OH)Cl$
   or: $4Zn + 8MnO_2 + 9H_2O + ZnCl_2 \rightarrow 8MnOOH + ZnCl_2 \cdot 4ZnO \cdot 5H_2O$
   Prolonged discharge: $Zn + 6MnOOH + 2Zn(OH)Cl \rightarrow 2Mn_3O_4 + ZnCl_2 \cdot 2ZnO \cdot 4H_2O$

In the theoretical case, as discussed in Chap. 1, the specific capacity calculates to 224 Ah/kg, based on Zn and MnO_2 and the simplified cell reaction. On a more practical basis, the electrolyte, carbon black, and water are ingredients that cannot be omitted from the system. If typical quantities of these materials are added to the "theoretical" cell, a specific capacity of 96 Ah/kg is calculated. This is the highest specific capacity a general-purpose cell can have and is, in fact, approached by some of the larger Leclanché cells under certain discharge conditions. The actual specific capacity of a practical cell, considering all the cell components and the efficiency of discharge, can range from 75 Ah/kg on very light loads to 35 Ah/kg on heavy-duty, intermittent discharge conditions.

9.3 TYPES OF CELLS AND BATTERIES

During the last 125 years, the development of the zinc-carbon battery has been marked by gradual change in the approach to improve its performance. It now appears that zinc-carbon batteries are entering a transitional phase. While miniaturization in the electrical and electronic industries has reduced power demands, it has been offset by the addition of new features requiring high power, such as motors to drive compact disc players or cassette recorders, halogen bulbs in lighting devices, etc. This has increased the need for a battery that can meet heavy discharge requirements. For this reason,

---

*Note: 2MnOOH is sometimes written as $Mn_2O_3 \cdot H_2O$ and Mn_3O_4 as $MnO \cdot Mn_2O_3$. Electrochemical discharge of MnOOH vs. zinc (prolonged discharge) does not provide a useful operating voltage for typical applications.

as well as competition from the alkaline zinc-manganese battery system for heavy-drain applications, many manufacturers are no longer investing capital to improve the Leclanché or zinc-carbon technology. The traditional Leclanché cell construction, which utilizes a starch paste separator, is being gradually phased out and replaced by zinc chloride batteries utilizing paper separators. This results in increased volume available for active materials and increased capacity. In spite of these conversion efforts by manufacturers, a number of developing countries still continue the demand for pasted Leclanché products because of their low cost. The size of that market has prevented a complete conversion. It appears that this situation will continue for the near future.

During this transitional phase, the zinc-carbon batteries can be classified into two types, Leclanché and zinc chloride. These can, in turn, be subdivided into separate general-purpose and premium battery grades, in both pasted and paper-lined constructions.

9.3.1 Leclanché Batteries

General Purpose. Application: Intermittent low-rate discharges; low cost. The traditional, regular battery, which is not too different from the one introduced in the late 19th century, uses zinc as the anode, ammonium chloride (NH_4Cl) as the main electrolyte component along with zinc chloride, a starch paste separator, and natural manganese dioxide (MnO_2) ore as the cathode. Batteries of this formulation and design are the least expensive and are recommended for general-purpose use and when cost is more important than superior service or performance.

Industrial Heavy-Duty. Application: Intermittent medium- to heavy-rate discharges; low to moderate cost. The industrial heavy-duty zinc-carbon battery generally has been converted to the zinc chloride system. However, some types continue to include ammonium chloride and zinc chloride ($ZnCl_2$) as the electrolyte and synthetic electrolytic or chemical manganese dioxide (EMD or CMD) alone or in combination with natural ore as the cathode. Its separator may be of starch paste, but it is typically a paste-coated paper liner type. This grade is suitable for heavy intermittent service, industrial applications, or medium-rate continuous discharge.

9.3.2 Zinc Chloride Batteries

General Purpose. Application: Low-rate discharges both intermittent and continuous; low cost. This battery has replaced the Leclanché general-purpose battery in all Western countries. It is a true zinc chloride battery and possesses some of the heavy-duty characteristics of the premium type. The electrolyte is zinc chloride; however, some manufacturers may add small amounts of ammonium chloride. Natural manganese dioxide ore is used as the cathode. Batteries of this formulation and design are competitive in cost to the Leclanché general-purpose batteries. They are recommended for general-purpose use on both continuous and intermittent discharges and when cost is an important consideration. This battery also exhibits a low leakage characteristic.

Industrial Heavy-Duty. Application: Low to intermediate continuous and intermittent heavy-rate discharges; low to moderate cost. This battery has generally replaced the industrial Leclanché heavy-duty battery. It is a true zinc chloride cell and possesses the heavy-duty characteristics of the premium zinc chloride type. The cell electrolyte is zinc chloride; however, some manufacturers may add small amounts of ammonium chloride. Natural manganese dioxide ore is used along with electrolytic manganese dioxide as the cathode. These cells use paper separators coated with cross-linked or modified starches, which enhance their stability in the electrolyte. Batteries of this formulation and design are competitive in cost to the Leclanché heavy-duty industrial batteries. They are recommended for heavy-duty applications where cost is an important consideration. This battery also exhibits a low leakage characteristic.

Extra/Super Heavy-Duty. Application: Medium and heavy continuous, and heavy intermittent discharges; higher cost than other zinc chloride types. The extra/super heavy-duty type of battery

9.6 PRIMARY BATTERIES

is the premium grade of the zinc chloride line. This cell is composed mainly of an electrolyte of zinc chloride with perhaps a small amount of ammonium chloride, usually not exceeding 1% of the cathode weight. The cathode uses electrolytic manganese dioxide (EMD) exclusively. These cells use paper separators coated with cross-linked or modified starches, which enhance their stability in the electrolyte. Many manufacturers use proprietary separators in almost all their zinc-carbon type batteries. This battery type is recommended when good performance is desired but at higher cost. It also has improved low-temperature characteristics and reduced electrolyte leakage.

In general, the higher the grade or class of zinc-carbon batteries, the lower the cost per minute of service. The price difference between classes is about 10 to 25%, but the performance difference can be from 30 to 100% in favor of the higher grades depending upon the application drain.

9.4 CONSTRUCTION

The zinc-carbon battery is made in many sizes and a number of designs but in two basic constructions: cylindrical and flat. Similar chemical ingredients are used in both constructions.

9.4.1 Cylindrical Configuration

In the common Leclanché cylindrical battery (Figs. 9.2 and 9.3), the zinc can serves as the cell container and anode. The manganese dioxide is mixed with acetylene black, wet with electrolyte, and compressed under pressure to form a bobbin. A wax impregnated carbon rod is inserted into the

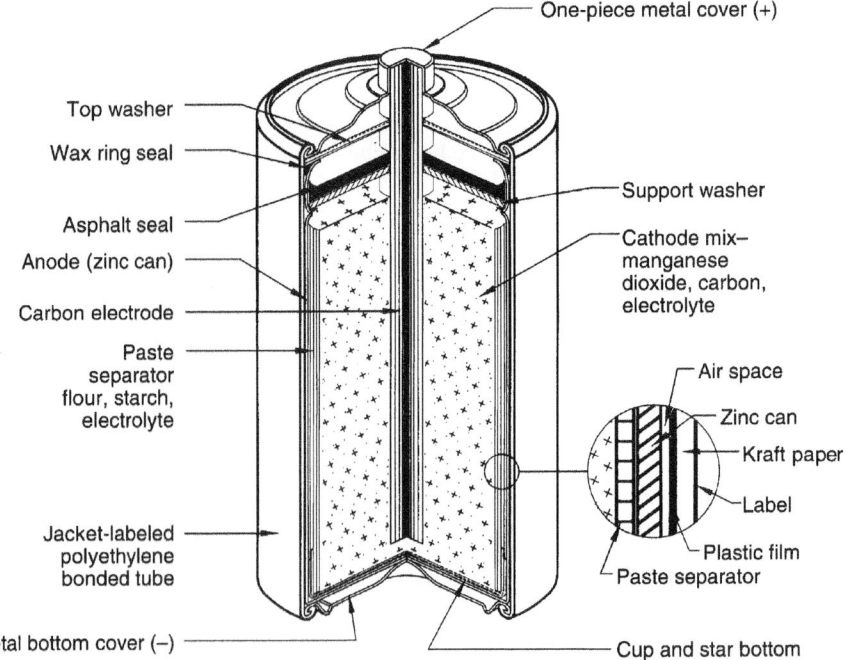

FIGURE 9.2 Typical cutaway view of cylindrical Leclanché battery ("Eveready") paste separator, asphalt seals.

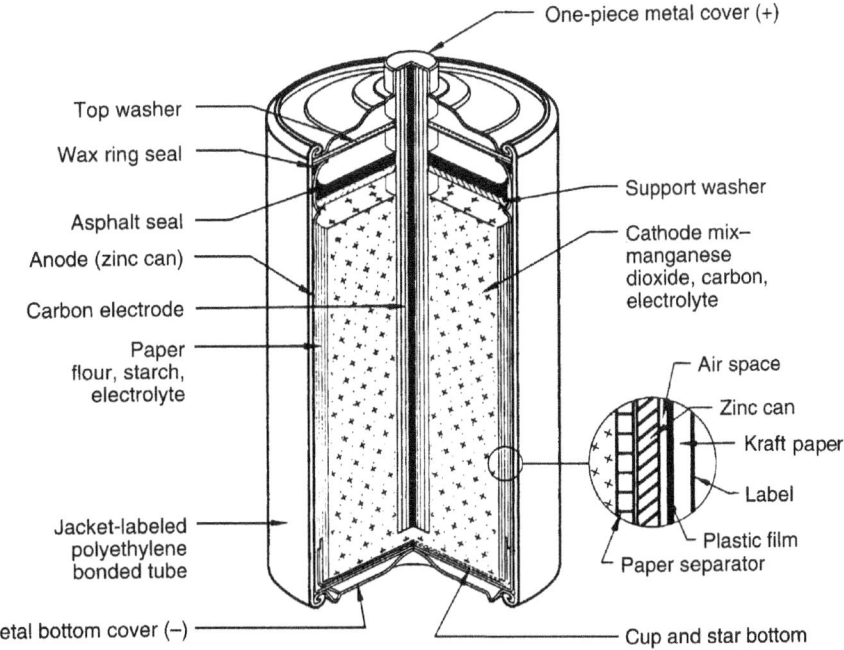

FIGURE 9.3 Typical cutaway view of cylindrical Leclanché battery ("Eveready") paper liner separator, asphalt seal.

bobbin. The rod serves as the current collector for the positive electrode. It also provides structural strength and is porous enough to permit the escape of gases, which accumulate in the cell, without allowing leakage of electrolyte. The separator, which physically separates the two electrodes and provides the means for ion transfer through the electrolyte, can be a cereal paste wet with electrolyte (Fig. 9.2) or a starch or polymer coated absorbent kraft paper in the paper-lined cell (Fig. 9.3). This provides thinner separator spacing, lower internal resistance, and increased active materials volume. Single cells are covered with metal, cardboard, plastic, or paper jackets for aesthetic purposes and to minimize the effect of electrolyte leakage through containment.

Construction of the zinc chloride cylindrical battery (Fig. 9.4) differs from that of the Leclanché battery in that it usually possesses a resealable, venting seal. The carbon rod serving as the current collector is sealed with wax to plug any vent paths (necessary for Leclanché types). Venting is then restricted to only the seal path. This prevents the cell from drying out and limits oxygen ingress into the cell during shelf storage. Hydrogen gas evolved from corrosion of the zinc is safely vented as well. In general, the assembly and finishing processes resemble that of the earlier cylindrical batteries.

9.4.2 Inside-Out Cylindrical Construction

Another cylindrical cell is the "inside-out" construction shown in Fig. 9.5. This construction does not use the zinc anode as the container. This version resulted in more efficient zinc utilization and improved leakage, but has not been manufactured since the late 1960s. In this cell, an impact-molded impervious inert carbon wall serves as the container of the cell and as the cathode current collector. The zinc anode (covered with thin separator) in the shape of vanes coated with separator is located inside the cell and is surrounded by the cathode mix.

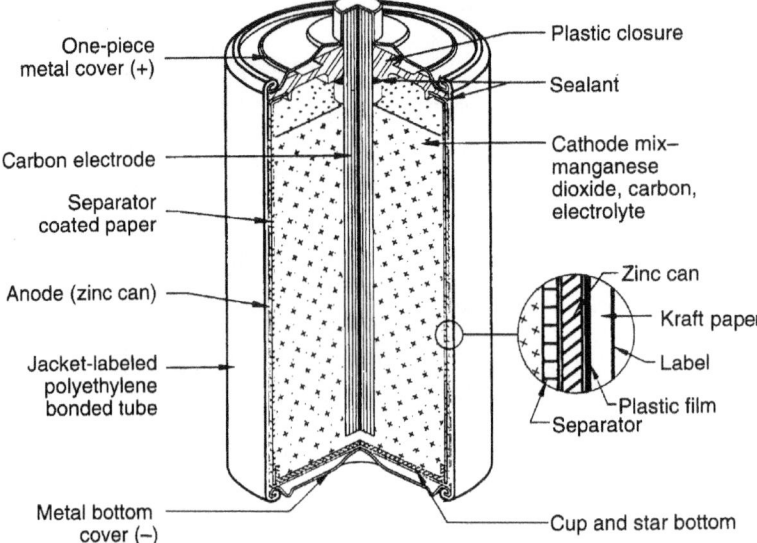

FIGURE 9.4 Typical cutaway view of cylindrical zinc chloride battery ("Eveready") paste separator, plastic seal.

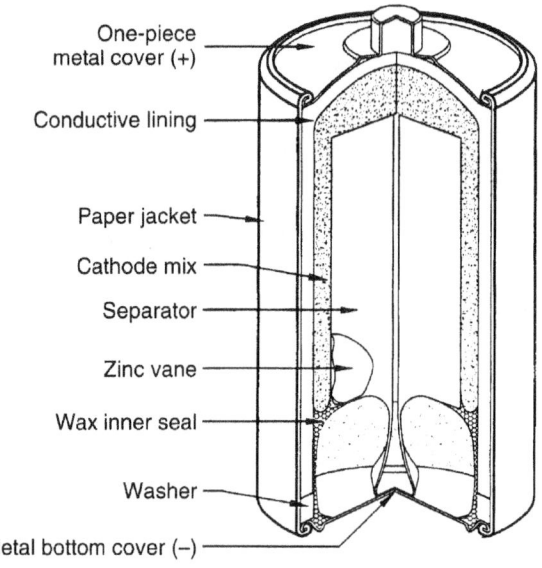

FIGURE 9.5 Typical cutaway view of cylindrical Leclanché battery ("Eveready") inside-out construction.

9.4.3 Flat Cell and Battery

The flat cell is illustrated in Fig. 9.6. In this construction, a duplex electrode is formed by coating a zinc plate with either a carbon-filled conductive paint or laminating it to a carbon-filled conductive plastic film. Either coating provides electrical contact to the zinc anode, isolates the zinc from the cathode of the next cell, and performs the function of cathode collector. The collector function is the same as that performed by the carbon rod in cylindrical cells. When the conductive paint method is used, an adhesive must be placed onto the painted side of the zinc prior to assembly to effectively seal the painted surface directly to the vinyl band to encapsulate the cell. No expansion chamber or carbon rod is used as in the cylindrical cell. The use of conductive polyisobutylene film laminated to the zinc instead of the conductive paint and adhesive usually results in improved sealing to the vinyl; however, the film typically occupies more volume than the paint and adhesive design. These methods of construction readily lend themselves to the assembly of multicell batteries.

Flat cell designs increase the available space for the cathode mix because the package and electrical contacts are minimized, thereby increasing the energy density. In addition, a rectangular construction reduces wasted space in multicell assemblies (which is currently the only application for the flat cell). The volumetric energy density of an assembled battery using flat cells is nearly twice that of cylindrical cell assemblies.

Metal contact strips are used to attach the ends of the assembled battery to the battery terminals; (e.g., 9 V transistor battery). The orientation of the stack subassembly (cathode up or anode up) is only important for each manufacturer's method of assembly. The use of contact strips allows either design mode. The entire assembly is usually encapsulated in wax or plastic. Some manufacturers also sleeve the assembly in shrink film after waxing. This aids the assembly process cleanliness and provides additional insurance against leakage. Cost, ease of assembly, and process efficiencies usually dictate the orientation during the assembly process.

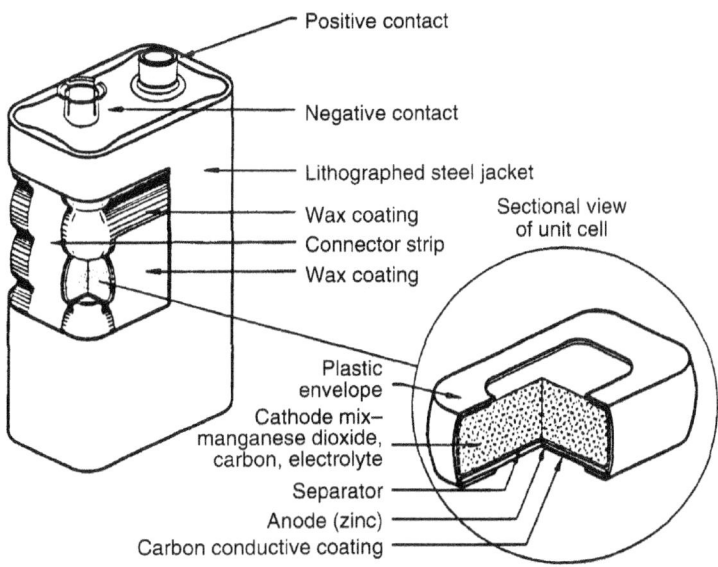

FIGURE 9.6 Typical cutaway view of Leclanché flat cell and battery (e.g., "Eveready" No. 216).

9.4.4 Special Designs

Designs for special applications are also currently in use. These designs demonstrate the levels of innovation that can be applied to unusual application and design problems. These are covered in Sec. 9.7.

9.5 CELL COMPONENTS

9.5.1 Zinc

Battery-grade zinc is 99.99% pure. Classical zinc can alloys contained 0.3% cadmium and 0.6% lead. Modern lubrication and forming techniques have reduced these amounts. Currently, zinc can alloys with cadmium contain 0.03 to 0.06% cadmium and 0.2 to 0.4% lead. The content of these metals varies according to the method used in the forming process. Lead, while insoluble in the zinc alloy, contributes to the forming qualities of the can, although too much lead softens the zinc. Lead also acts as a corrosion inhibitor by increasing the hydrogen overvoltage of the zinc in much the same manner as does mercury. Cadmium aids the corrosion-resistance of zinc to ordinary dry-cell electrolytes and adds stiffening strength to the alloy. For cans made by the drawing process, less than 0.1% of cadmium is used because more would make the zinc difficult to draw. Zinc cans are commonly made by three different processes:

1. Zinc is rolled into a sheet, formed into a cylinder, and, with the use of a punched-out zinc disk for the bottom, soldered together. This method is obsolete except for the most primitive of assemblies. Last use of this method in the United States was during the 1980s in No. 6 cells.
2. Zinc is deep-drawn into a can shape. Rolled zinc sheet is shaped into a can by forming through a number of steps. This method was used primarily in cell manufacturing in the United States prior to the relocation and consolidation of U.S. zinc-carbon manufacturing overseas.
3. Zinc is impact extruded from a thick, flat disk or calot. This is now the method of choice. Used globally, this method reshapes the zinc by forcing it to flow under pressure from the calot shape into the can shape. Calots are either cast from molten zinc alloy or punched from a zinc sheet of the desired alloy.

Metallic impurities such as copper, nickel, iron, and cobalt cause corrosive reactions with the zinc in the battery electrolyte and must be avoided particularly in "zero" mercury constructions. In addition, iron in the alloy makes zinc harder and less workable. Tin, arsenic, antimony, magnesium, etc., make the zinc brittle and prone to perforation.[4,6]

U.S. federal environmental legislation prohibits the land disposal of items containing cadmium and mercury when these components exceed specified leachable levels. Some states and municipalities have banned land disposal of batteries, require collection programs, and prohibit sale of batteries containing added cadmium or mercury. Some European countries have similarly prohibited the sale and disposal of batteries containing these materials. For these reasons, levels of both of these heavy metals have been reduced to near zero. This impacts directly upon global zinc can manufacture due to importation of battery products to the United States and Europe. Manganese is a satisfactory substitute for cadmium and has been included in the alloy at levels similar to that of cadmium to provide stiffening. The handling properties of zinc alloyed with manganese or cadmium are equivalent; however, no corrosion resistance is imparted to the alloy with manganese as is the case with cadmium.

9.5.2 Bobbin

The bobbin is the positive electrode and is also called the black mix, depolarizer, or cathode. It is a wet powder mixture of MnO_2, powdered carbon black, and electrolyte (NH_4Cl and/or $ZnCl_2$, and water). The powdered carbon serves the dual purpose of adding electrical conductivity to the MnO_2, which itself has high electrical resistance. It also acts as a means of holding the electrolyte. The cathode mixing and forming processes are also important since they determine the homogeneity of the cathode mix and the compaction characteristics associated with the different methods of manufacture. This becomes more critical in the case of the zinc chloride cell, where the cathode contains proportions of liquid that range between 60 and 75% by volume.

Of the various forming methods available, mix extrusion and compaction-then-insertion are the two used most widely. On the other hand, there is a wide variety of techniques for mixing. The most popular methods are cement-style mixers, mash mixers, and rotary mullor mixers. Both techniques offer the ability to manufacture large quantities of mix in relatively short times and minimize the shearing effect upon the carbon black, which reduces its ability to hold solution. The bobbin usually contains ratios of manganese dioxide to powdered carbon from 3:1 to as much as 11:1 by weight. Also, 1:1 ratios have been used in batteries for photoflash applications where high pulses of current are more important than capacity.

9.5.3 Manganese Dioxide (MnO_2)

The types of manganese dioxide used in dry cells are generally categorized as natural manganese dioxide (NMD), activated manganese dioxide (AMD), chemically synthesized manganese dioxide (CMD), and electrolytically deposited manganese dioxide (EMD). EMD is a more expensive material that has a gamma-phase crystal structure. CMD has a delta-phase structure, and NMDs have the alpha, beta, and gamma phases of MnO_2. EMD, while more expensive, results in a higher cell capacity with improved rate capability and is used in heavy or industrial applications. As shown in Figs. 9.7a and 9.7b, polarization can be significantly reduced by choice of natural ore. (As noted above, polarization is also reduced by adding EMD or CMD as a substitute for part of the natural ore.)

Naturally occurring ores (in Gabon Africa, Brazil, Greece, and Mexico), high in battery-grade material (70 to 85% MnO_2), and synthetic forms (90 to 95% MnO_2) generally provide electrode potentials and capacities proportional to their manganese dioxide content. Manganese dioxide potentials are also affected by the pH of the electrolyte. Performance characteristics depend upon the crystalline state, the state of hydration, and the activity of the MnO_2. The efficiency of operation

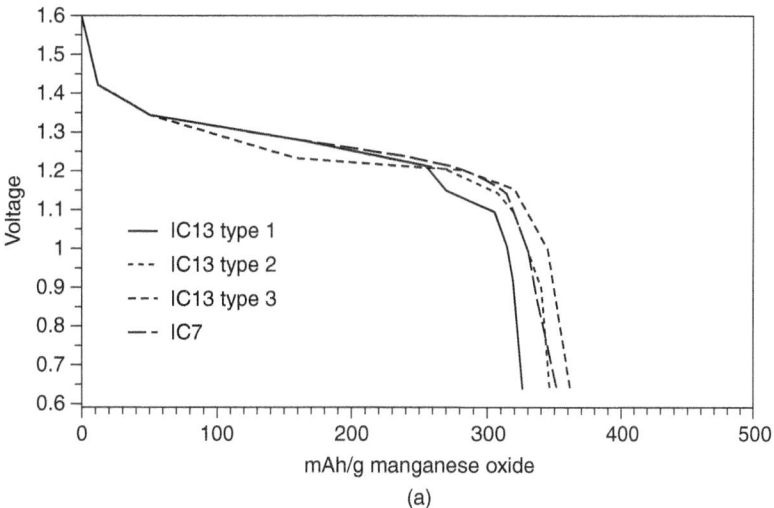

FIGURE 9.7a Ore sample performance, Leclanché 6.38% ore mix (13 mA/g ore)

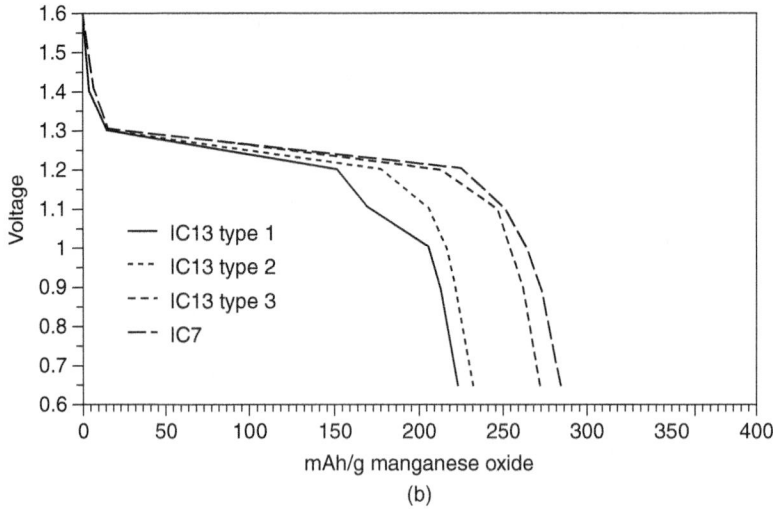

FIGURE 9.7b Ore sample performance, zinc chloride 6.71% ore mix (13 mA/g ore)

under load depends heavily upon the electrolyte, the separator characteristics, the internal resistance, and the overall construction of the cell.[4,5]

9.5.4 Carbon Black

Because manganese dioxide is a poor electrical conductor, chemically inert carbon or carbon black is added to the cathode mix to improve its conductivity. This is achieved by coating the manganese dioxide particles with carbon during the mixing process. It provides electrical conductivity to the particle surface and also serves the important functions of holding the electrolyte and providing compressibility and elasticity to the cathode mix during processing.

Graphite was once used as the principle conductive media and is still used to some extent. Acetylene black, by virtue of its properties, has displaced graphite in this role for both Leclanché and zinc chloride cells. One great advantage of acetylene black is its ability to hold more electrolyte in the cathode mix. Caution must be used during the mixing process so as to prevent intense shearing of the black particles as this reduces their ability to hold electrolyte. This is critical for zinc chloride cells, which contain much higher electrolyte levels than the Leclanché cell. Cells containing acetylene black usually give superior intermittent service, which is the way most zinc-carbon batteries are used. Graphite, on the other hand, serves well for high flash currents or for continuous drains.[4,9]

9.5.5 Electrolyte

The ordinary Leclanché cell uses an aqueous mixture of ammonium chloride and zinc chloride with the former predominating. Zinc chloride cells typically use only $ZnCl_2$, but can contain a small amount of NH_4Cl to ensure high rate performance. Examples of typical electrolyte formulation for the zinc-carbon battery systems are listed in Table 9.3.

Generally some zinc oxide is included in the electrolyte in order to prevent excess corrosion of the zinc.

TABLE 9.3 Electrolyte Formulations*

Constituent	Weight %
Electrolyte I	
NH_4Cl	26.0
$ZnCl_2$	8.8
H_2O	65.2
Zinc-corrosion inhibitor	0.25–1.0
Electrolyte II	
$ZnCl_2$	15–40
H_2O	60–85
Zinc-corrosion inhibitor	0.02–1.0

*Electrolyte I based on Kozawa and Powers.[7]
Electrolyte II based on Cahoon.[5]

9.5.6 Corrosion Inhibitor

The classical zinc corrosion inhibitor has been mercuric or mercurous chloride, which forms an amalgam with the zinc. Cadmium and lead, which reside in the zinc alloy, also provide zinc anode corrosion protection. Other materials like potassium chromate or dichromate, used successfully in the past, form oxide films on the zinc and protect via passivation. Surface-active organic compounds, which coat the zinc, usually from solution, improve the wetting characteristic of the surface unifying the potential. Inhibitors are usually introduced into the cell via the electrolyte or as part of the coating on the paper separator. Zinc cans could be pretreated; however, this is ordinarily not practical.

Environmental concerns have generally eliminated the use of mercury and cadmium in these batteries. These restrictions are posing problems for battery manufacturers in the areas of sealing, shelf storage reliability, and leakage. This is critical for zinc chloride cells in that the lower pH electrolyte can result in the formation of excessive hydrogen gas due to zinc dissolution. Certain classes of materials considered for use to supplant mercury include gallium, indium, lead, tin, and bismuth, either alloyed into the zinc or added to the electrolyte from their soluble salts. Other organic materials, like glycols and silicates, offer protection alternatives. Additional restrictions on lead use, which are already stringent, may also be imposed in the future.

9.5.7 Carbon Rod

The carbon rod used in round cells is inserted into the bobbin and performs the functions of current collector. It also performs as a seal vent in systems without a positive venting seal. It is typically made of compressed carbon, graphite, and binder, formed by extrusion, and cured by baking. It has, by design, a very low electrical resistance. In Leclanché and zinc chloride cells with asphalt seals, it provides a vent path for hydrogen and carbon dioxide gases that might build up in and above the cathode during heavy discharge or elevated temperature storage. Raw carbon rods are initially porous, but are treated with enough oils or waxes to prevent water loss (very harmful to cell shelf life) and electrolyte leakage. A specific level of porosity is maintained to allow passage of the evolved gases. Ideally, the treated carbon should pass internally evolved gases, but not pass oxygen into the cell, which could add to zinc corrosion during storage. Typically this method of venting gases is variable and less reliable then the use of venting seals.[4,6]

Zinc chloride cells using plastic, resealable, venting seals utilize plugged, nonporous electrodes. Their use restricts the venting of internal gas to only the designed seal path. This prevents the cell from drying out and limits oxygen ingress into the cell during shelf storage. Hydrogen gas evolved from wasteful corrosion of the zinc is safely vented as well.

9.5.8 Separator

The separator physically separates and electrically insulates the zinc (negative) from the bobbin (positive), but permits electrolytic or ionic conduction to occur via the electrolyte. The two major separator types in use are either the gelled paste or paper coated with cereal or other gelling agents such as methycellulose.

In the paste type, the paste is dispensed into the zinc can. The preformed bobbin (with the carbon rod) is inserted, pushing the paste up the can walls between the zinc and the bobbin by displacement. After a short time, the paste sets or gels. Some paste formulations need to be stored at low temperatures in two parts. The parts are then mixed; they must be used immediately as they can get at room temperature. Other paste formulations need elevated temperatures (60 to 96°C) to gel. The gelatinization time and temperature depend upon the concentration of the electrolyte constituents. A typical paste electrolyte uses zinc chloride, ammonium chloride, water, and starch and/or flour as the gelling agent.

The coated-paper type uses a special paper coated with cereal or other gelling agent on one or both sides. The paper, cut to the proper length and width, is shaped into a cylinder and, with the addition of a bottom paper, is inserted into the cell against the zinc wall. The cathode mix is then metered into the can forming the bobbin, or, if the bobbin is preformed in a die, it is pushed into the can. At this time, the carbon rod is inserted into the center of the bobbin and the bobbin is tamped or compressed, pushing against the paper liner and carbon rod. The compression releases some electrolyte from the cathode mix, wetting the paper liner to complete the operation.

By virtue of the fact that a paste separator is relatively thick compared with the paper liner, about 10% or more manganese dioxide can be accommodated in a paper-lined cell, resulting in a disproportional increase in capacity, because of the lower load per gram.[4,6,10]

9.5.9 Seal

The seal used to enclose the active ingredients can be asphalt pitch, wax and resin, or plastic (polyethylene or polypropylene). The seal is important to prevent the evaporation of moisture and the phenomenon of "air line" corrosion from oxygen ingress.[4,5]

Leclanché cells typically utilize thermoplastic materials for sealing. These methods are inexpensive and easily implemented. A washer is usually inserted into the zinc can and placed above the cathode bobbin. This provides an air space between the seal and the top of the bobbin to allow for expansion. Melted asphalt pitch is then dispensed onto the washer and is heated until it flows and bonds to the zinc can. One drawback to this method of sealing is that it occupies space that could be used for active materials. A second fault is that this type of seal is easily ruptured by excessive generation of evolved gases and is not suitable for elevated temperature applications.

Premium Leclanché and almost all zinc chloride cells use injection molded plastic seals. This type of seal lends itself to the design of a positive venting seal and is more reliable. Molded seals are mechanically placed onto a swaged zinc can. Many manufacturers have designed locking mechanisms into the seal, void spaces for various sealants, and resealable vents. Several wrap the seal and can in shrink-wrap or tape to prevent leakage through zinc can perforations. It is very important to prevent moisture loss in the zinc chloride system, and to vent the evolved gases generated during discharge and storage. The formation of these gases disrupts the separator surface layer significantly and affects cell performance after storage. Use of molded seals in the zinc chloride cell construction has resulted in the good shelf storage characteristics evidenced by this design.

9.5.10 Jacket

The battery jacket can be made of various components: metal, paper, plastic, polymer films, plain or asphalt-lined cardboard, or foil in combination or alone. The jacket provides strength, protection, leakage prevention, electrical isolation, decoration, and a site for the manufacturer's label. In many manufacturers' designs, the jacket is an integral part of the sealing system. It locks some seals in place, provides a vent path for the escape of gases, or acts as a supporting member to allow seals to flex under internal gas pressures. In the inside-out construction, the jacket was the container in which a carbon-wax collector was impact-molded (Fig. 9.5).

9.5.11 Electrical Contacts

The top and bottom of most batteries are capped with shiny, tin-plated steel (or brass) terminals to aid conductivity, prevent exposure of any zinc, and in many designs enhance the appearance of the cell. Some of the bottom covers are swaged onto the zinc can, others are locked into paper jackets or captured under the jacket crimp. Top covers are almost always fitted onto the carbon electrode with interference. All of the designs try to minimize the electrical contact resistance.

9.6 PERFORMANCE CHARACTERISTICS

9.6.1 Voltage

Open-Circuit Voltage. The open-circuit voltage (OCV) of the zinc-carbon battery is derived from the potentials of the active anode and cathode materials, zinc and manganese dioxide, respectively. As most zinc-carbon batteries use similar anode alloys, the open circuit voltage usually depends upon the type or mixture of manganese dioxide used in the cathode and the composition and pH of the electrolyte system. Manganese dioxides, like EMDs, are of greater purity than the NMDs, which contain a significant quantity of manganite (MnOOH), and thus have lower voltage. Figure 9.8 shows the open circuit voltage for fresh Leclanché and zinc chloride batteries containing various mixtures of natural and electrolytic manganese dioxide.

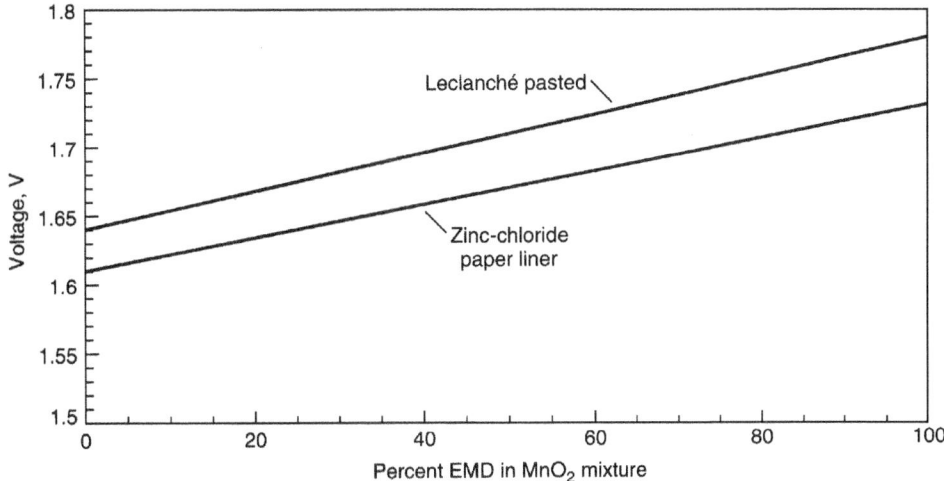

FIGURE 9.8 Comparison of open-circuit voltage for batteries using mixtures of natural and electrolytic manganese dioxide.

TABLE 9.4 Initial Closed-Circuit Voltage of a Typical D-size Zinc-Carbon Battery as a Function of Load Resistance at 20°C

Voltage (V)		Load resistance	Initial current (mA)	
ZC*	LC*	Ω	ZC	LC
1.61	1.56	∞	0	0
1.59	1.52	100	16	15
1.57	1.51	50	31	30
1.54	1.49	25	62	60
1.48	1.47	10	148	147
1.45	1.37	4	362	343
1.43	1.27	2	715	635

*ZC: Zinc chloride battery; LC: Leclanché battery.

Closed-Circuit Voltage. The closed-circuit voltage (CCV), or working voltage, of the zinc-carbon battery is a function of the load or current drain the cell is required to deliver. The heavier the load or the smaller the circuit resistance, the lower the closed-circuit voltage. Table 9.4 illustrates the effect of load resistance on the closed-circuit voltage for D-size batteries in both the Leclanché and zinc chloride systems.

The exact value of the CCV is determined mainly by the internal resistance of the battery as compared with the circuit or load resistance. It is, in fact, proportional to $R_1/(R_1 + R_{in})$ where R_1 is the load resistance and R_{in} is the battery's internal resistance. Another factor, important to the battery's ability to sustain the CCV, is the transport characteristic of the cell component—that is, the ability to transport ionic and solid reaction products, and water, to and from the reaction sites. The physical geometry of the cell, its solution volume, electrode porosity, and solute materials are critical characteristics that affect the diffusion coefficient. Transport is enhanced by use of highly mobile ions, high solution volumes, high electrode porosity, and high surface area. Transport characteristics are diminished by slow ionic transport, low solution volumes, and barriers of precipitated reaction product which block diffusion paths. (This topic is discussed in greater detail in Chap. 2.) Temperature, age, and depth of discharge greatly affect the internal resistance and transport factors as well.

As zinc-carbon batteries are discharged, the CCV and, to a lesser extent, the OCV drop in magnitude. The drop in OCV is attributable to the decrease in the active material manganese dioxide and the increase in the product of the reaction, manganite. Reduction of the CCV is the result of increased electrical resistance and a decrease in transport characteristic. The discharge curve is a graphic representation of the closed-circuit voltage as a function of time and is neither flat nor linearly decreasing but, as seen in Fig. 9.9, has the character of a single- or double-S curve depending upon the depth of discharge. Figure 9.10 illustrates the shape of typical discharge curves for D-size, general purpose, Leclanché and zinc chloride batteries.

End Voltage. The end voltage, or cutoff voltage (COV), is defined as a point along the discharge curve below which no usable energy can be drawn for the specified application. Typically 0.9 V has been found to be the COV for a 1.5 V cell when used in a flashlight. Some radio applications can utilize the cell down to 0.75 V or lower, while other electronic devices may tolerate a drop to only 1.2 V. Obviously, the lower the end voltage, the greater the amount of energy that can be delivered by the battery. The lower voltage will impact certain applications, like flashlights, resulting in a dimmer light and lower volume and/or range for radios. Devices that can operate only within a narrow voltage range would do better with a battery system noted for a flat discharge curve. Although a closed-circuit voltage that steadily decreases may present a disadvantage in some applications, it is advantageous where sufficient warning of the end of battery life is required, as in a flashlight.

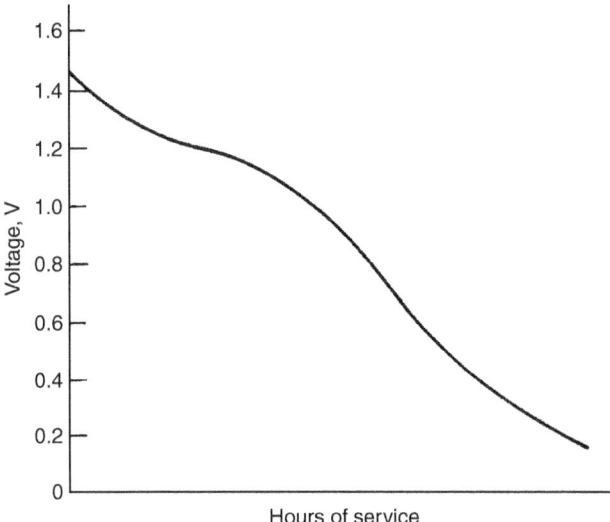

FIGURE 9.9 Typical discharge curve of a Leclanché zinc-carbon battery.

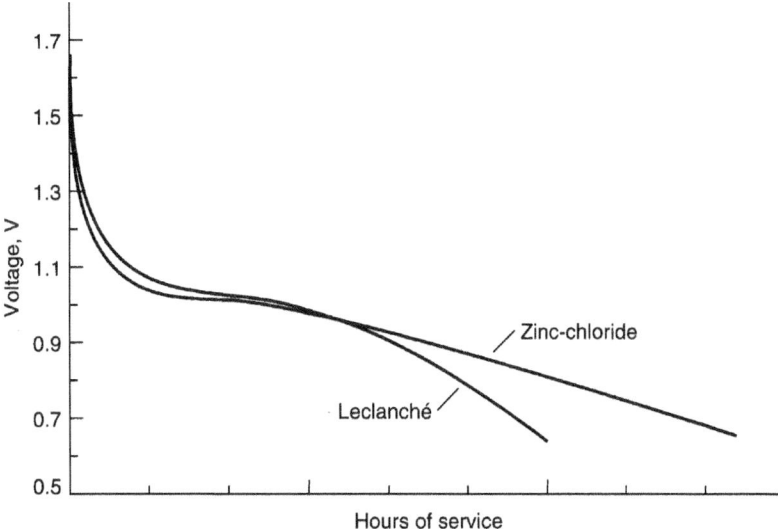

FIGURE 9.10 A comparison of typical discharge curves between Leclanché and zinc chloride batteries.

9.6.2 Discharge Characteristics

Both the Leclanché and zinc chloride batteries have performance characteristics that show advantages in specific applications, but poor performance in others. A variety of factors influence battery performance (see Chap. 3). It is necessary to evaluate specifics about the application (discharge conditions, cost, weight, etc.) in order to make a proper selection of a battery. Many manufacturers provide data for this purpose.

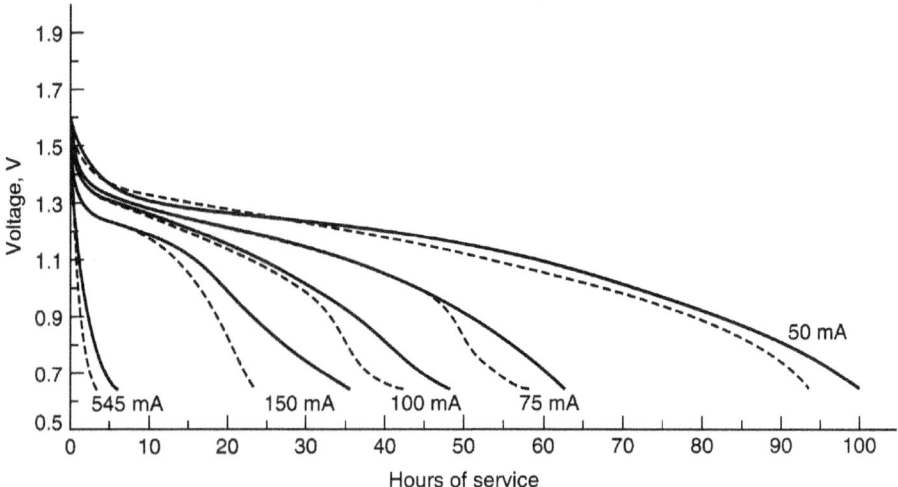

FIGURE 9.11 Typical discharge curves for general-purpose Leclanché and zinc chloride D-size batteries, discharged 2 h/day at 20°C. Solid line: Zinc chloride; broken line: Leclanché.

Typical discharge curves for general-purpose D-size Leclanché and zinc chloride batteries, of equivalent capacity, discharged 2 h per day at 20°C, are shown in Fig. 9.11. These curves are characterized by a sloping discharge and a substantial reduction in voltage with increasing current. The zinc chloride construction shows a higher voltage characteristic and more service at the higher current levels. On the 50 mA drain, both constructions provide nearly equivalent performance. This is the result of the depletion of manganese dioxide at the low discharge rates, as most zinc-carbon batteries are cathode limited.

9.6.3 Effect of Intermittent Discharge

Performance of zinc-carbon batteries varies depending upon the type of discharge. The performance of Leclanché batteries is significantly better when used under intermittent compared to continuous discharge conditions, because (1) a chemical recuperation reaction replaces a small portion of active ingredients during the rest periods, and (2) transport phenomena redistribute reaction products.[5]

Zinc chloride batteries can support heavier drains and respond to intermittent discharges with longer discharge cycles. This system relies upon its improved transport mechanism to support heavier drains and to redistribute reaction product. (Fig. 9.12) illustrates the general effects of intermittency and discharge rate on the capacity of a general-purpose D-size battery. On extremely low-current discharges, the benefit of intermittent rest and discharge is minimal for both systems. It is likely that the reaction rate proceeds more slowly than the diffusion rate and results in a balanced condition even during discharge. Under conditions of extremely low rate of discharge, factors such as age will reduce the total delivered capacity. Most applications fall in the moderate- (radio) to high-rate (flashlight) categories, and for these the energy delivered can more than triple when the cell is used intermittently as compared with continuous usage.

The standard flashlight current drains are 300 mA (3.9 Ω per cell) and 500 mA (2.2 Ω per cell), which correspond to two-cell flashlights using PR2 and PR6 lamps, respectively, or three-cell flashlights using PR3 and PR7 lamps, respectively. The beneficial effects of intermittent discharge are clearly shown in Figs. 9.13 and 9.14 which compare Leclanché general-purpose D-size batteries on four different discharge regimens: continuous, light intermittent flashlight, heavy intermittent flashlight, and a 1 h/day cassette simulation test. Table 9.5 lists the ANSI application tests currently being used to evaluate both cell systems.

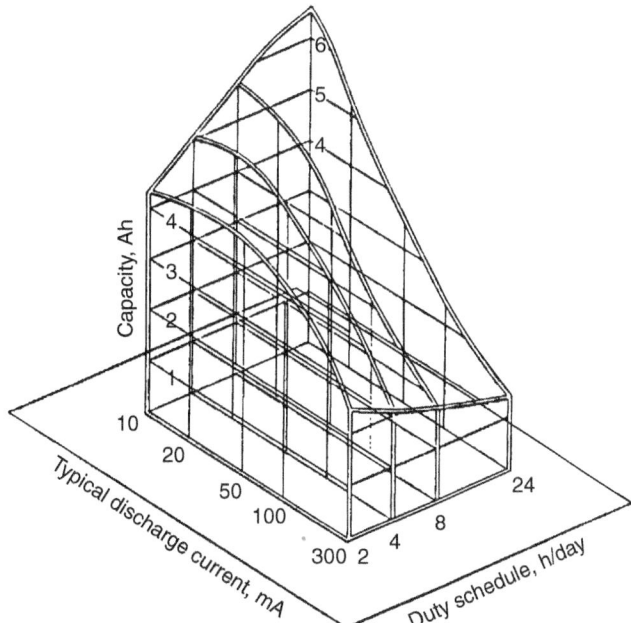

FIGURE 9.12 Battery performance (capacity to 0.9 V) as a function of discharge, load, and duty schedule for a general-purpose D-size zinc-carbon battery at 20°C. (*Source:* Eveready Battery Energy Data.)[12]

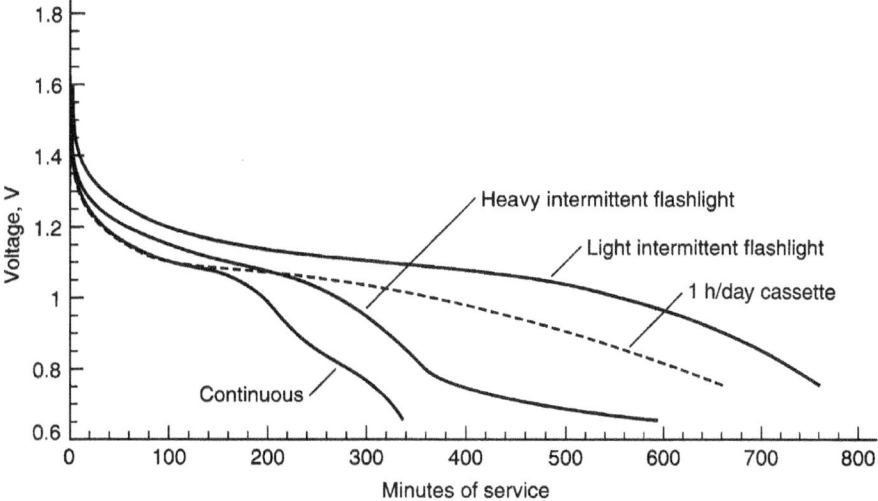

FIGURE 9.13 General-purpose D-size zinc-carbon battery discharged through 3.9 ohm at 20°C, under various discharge conditions.

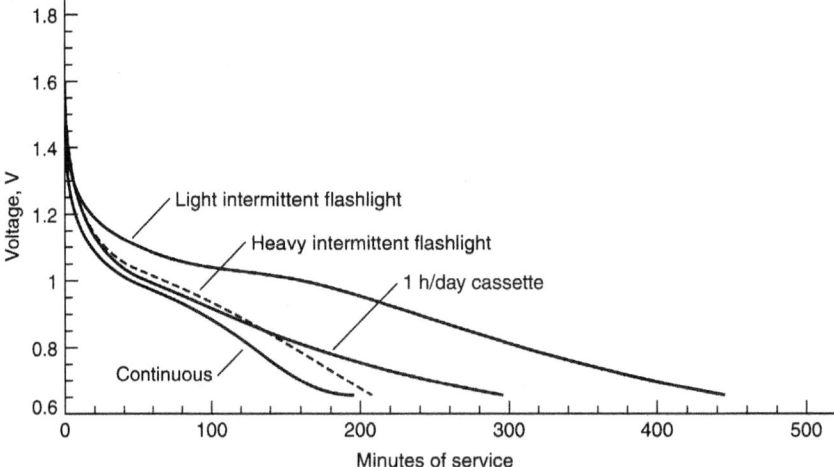

FIGURE 9.14 General-purpose D-size zinc-carbon battery discharged through 2.2 ohm at 20°C, under various discharge conditions.

TABLE 9.5 Standard Application Tests Specified in ANSI Battery Specifications

Typical use or test	Discharge schedule
Pulse test (PHOTO)	15 s ON/min × 24 h/day
Portable lighting (GPI)	5 min ON/day
Portable lighting (LIP)	4 min ON/h × 8 h/day
Portable lighting (LANTERN)	0.5 h ON/h × 8 h/day
Transistor radios	4 h ON/day
Transistor radio (small 9 V)	2 h ON/day
Personal tape recorder, cassette	1 h ON/day
Toys and motors	1 h ON/day
Pocket calculator	0.5 h ON/day
Hearing aid	12 h ON/day
Electronic	24 h ON/day

Source: Based on ANSI C18.1M-2009[11]

9.6.4 Comparative Discharge Curves: Size Effect upon Heavy-Duty Zinc Chloride Batteries

The performance of batteries of different sizes, AAA, AA, C, and D (see Table 9.9 for a list of cell sizes) are given in Figs. 9.15 and 9.16. Note that the AAA- through D-size batteries contain increasing amounts of active materials (zinc and manganese dioxide) with the increase in size. Increasing the size also results in proportionally larger electrode surface areas in the cell and therefore the voltage is maintained at higher levels at the same discharge loads.

Figure 9.15 illustrates a discharge through a relatively high resistance of 150 ohms (about 10 mA) continuously at 20°C. The performance of the D- and C-size batteries on an intermittent discharge would not be too different from that of the continuous discharge, because, for batteries of these sizes, the current drain is low. For the smaller AA and AAA sizes, however, discharge through a 150 ohm load is a heavier load. Under these conditions some benefit could be gained, for Leclanché batteries,

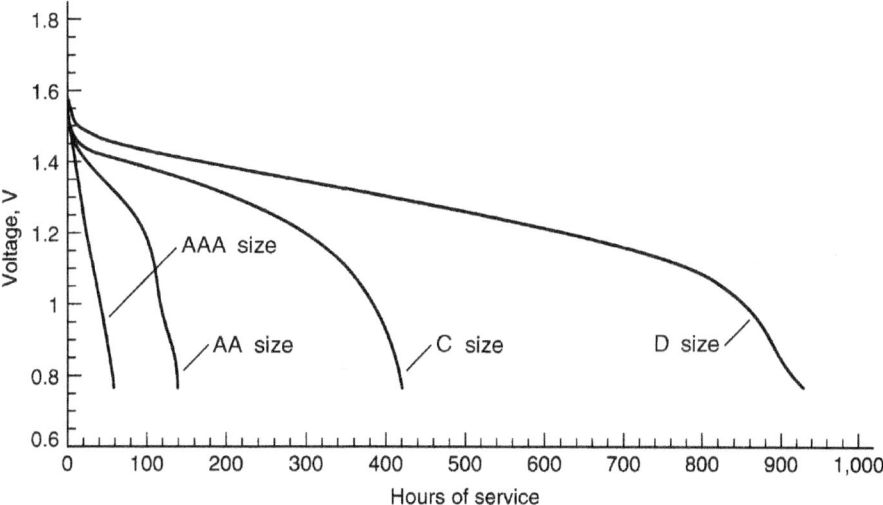

FIGURE 9.15 Zinc-carbon batteries, continuous discharge through 150 ohm at 20°C.

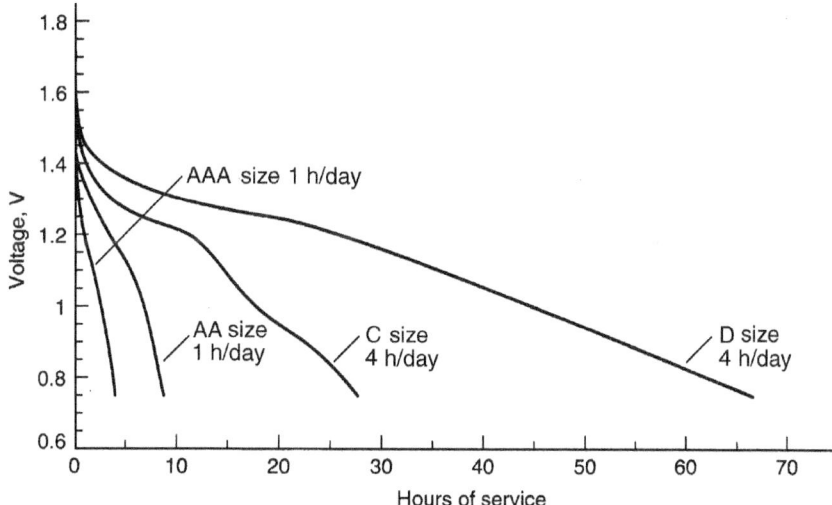

FIGURE 9.16 Zinc-carbon batteries, discharged under simulated cassette application (10 ohm intermittent load) at 20°C.

using an intermittent discharge since it assists in dissipating reaction product barriers and increases service life. Zinc chloride batteries would show less evidence of an effect because of their improved transport characteristic.

Figure 9.16 shows the same four battery sizes discharged through a relatively low resistance of 10 ohms (about 150 mA) on a simulated intermittent cassette application. AAA- and AA-size batteries deliver about 30% less service when continuously discharged at this load.

The relative performance of both Leclanché and zinc chloride AAA, AA, C, and D batteries roughly follows a 1:2:8:16 proportion to the 0.9 V cutoff for the low rate and a 1:2:12:24 proportion

9.22 PRIMARY BATTERIES

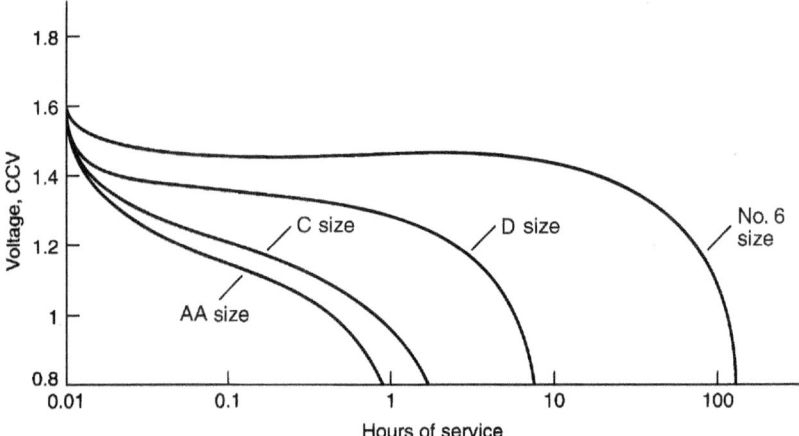

FIGURE 9.17 Zinc-carbon batteries, continuous discharge through 3.9 ohm at 20°C.

for the high-rate drain, illustrating the advantage of the lower current density discharge for the larger batteries. The high discharge rate for the general-purpose C- and D-size batteries at 300 mA (3.9 ohms) is shown in Fig. 9.17 compared with the performance of the general-purpose larger No. 6 battery, for which this discharge rate is low.

9.6.5 Comparative Discharge Curves: Different Battery Grades

Figure 9.18 compares both Leclanché and zinc chloride general-purpose (GP), heavy-duty (HD), and the extra/super heavy-duty (EHD) D-size batteries (as defined in Sec. 9.3) discharged continuously through a 2.2 ohm load at 20°C. A performance ratio, to the 0.9 V cutoff, of 1.0:1.3 between the Leclanché and zinc chloride GP batteries was observed. The same ratio for the HD batteries was

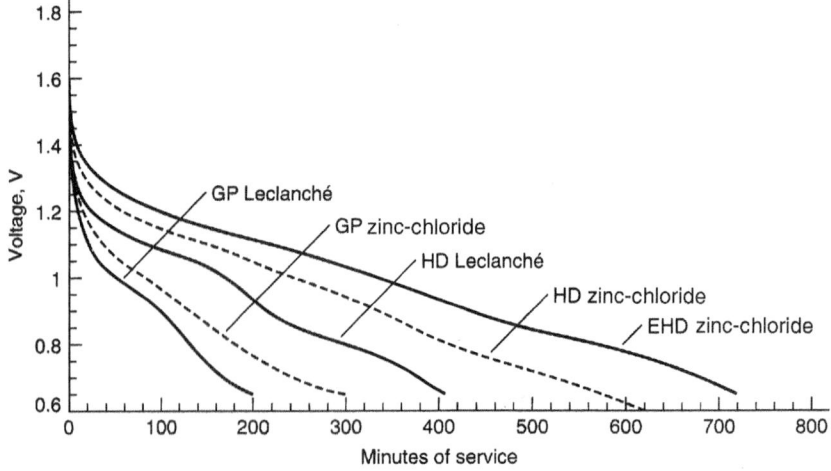

FIGURE 8.18 Comparison of Leclanché and zinc chloride D-size batteries of various grades, continuously discharged through 2.2 ohm at 20°C. GP: General purpose; HD: Heavy duty; EHD: Extra heavy duty.

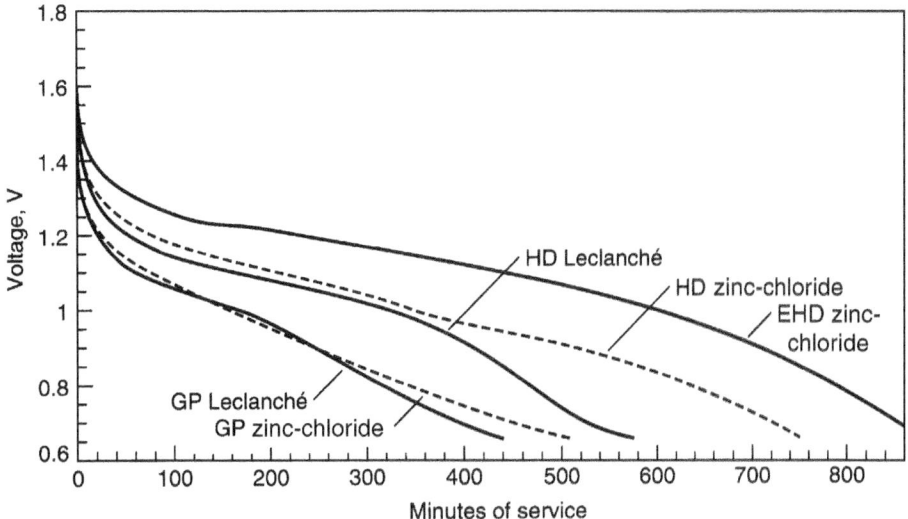

FIGURE 9.19 Comparison of Leclanché and zinc chloride D-size cells of various grades discharged on the ANSI LIF test (4 min/h, 8 h/day) through 2.2 ohm at 20°C. GP: General purpose; HD: Heavy duty; EHD: Extra heavy duty.

1.0:1.5, respectively. Comparison between the Leclanché (LC) and zinc chloride (ZC) GP, HD, and EHD batteries showed ratios of 1:1.3:2.2:3.4:4.4 for LC, GP:ZC, GP:LC, HD:ZC, HD:ZC, EHD respectively.

Figure 9.19 shows a comparison of the same battery grades discharged intermittently through a 2.2 ohm load on the American National Standards Institute (ANSI) light intermittent flashlight (LIF) test. On this regimen, the performance ratio to the 0.9 V cutoff is 1:1 for the GP batteries, 1:1.3 for the HD, and for the LC, GP:ZC, GP:LC, HD:ZC, HD:ZC, EHD, 1:1:1.7:2.1:2.9, respectively. Testing on the intermittent discharge, which allows for a rest period for recovery, results in increased performance for all batteries and evidences a decreased difference in performance between the grades.

Figure 9.20 illustrates the same battery grades discharged continuously through the lighter 3.9 ohm load. The following ratios were obtained; 1:1.3 GP, 1:1.4 HD, and 1:1.3:2.0:2.8:3.5 between all grades to the 0.9 V cutoff. Less of a difference was observed than that obtained at the heavier 2.2 ohm discharge rate. The slower reaction rate at the lighter drain is evident because of the higher battery voltages. A comparison of an intermittent discharge at 3.9 ohm for 1 h/day on a simulated cassette test is shown in Fig. 9.21. On this regimen, the performance ratio for the same grouping of battery grades drops to 1:1.1:1.5:2.4:2.7. This reflects the increase in service and tighter grouping for all grades.

These battery grades are compared once again in Fig. 9.22 on a moderate discharge through a 24 ohm resistor for 4 h continuously with 20 h of rest on the ANSI transistor radio and electronic equipment battery tests. At this more moderate discharge load, the performance ratio is even closer, 1:1.4:1.6:1.9:2.0 to a 0.9 V cutoff.

Continuous discharge tends to increase the difference in performance between the different grades of batteries of the same size. The differences between the Leclanché and zinc chloride systems are evident when tested continuously. Intermittent discharges tend to reduce the differences between systems and grades. Similarly, higher discharge currents tend to increase the performance difference.

Figure 9.23a summarizes the performance of the Leclanché general-purpose D-size battery grade discharged continuously to different end voltages. The performance of the zinc chloride general-purpose

9.24 PRIMARY BATTERIES

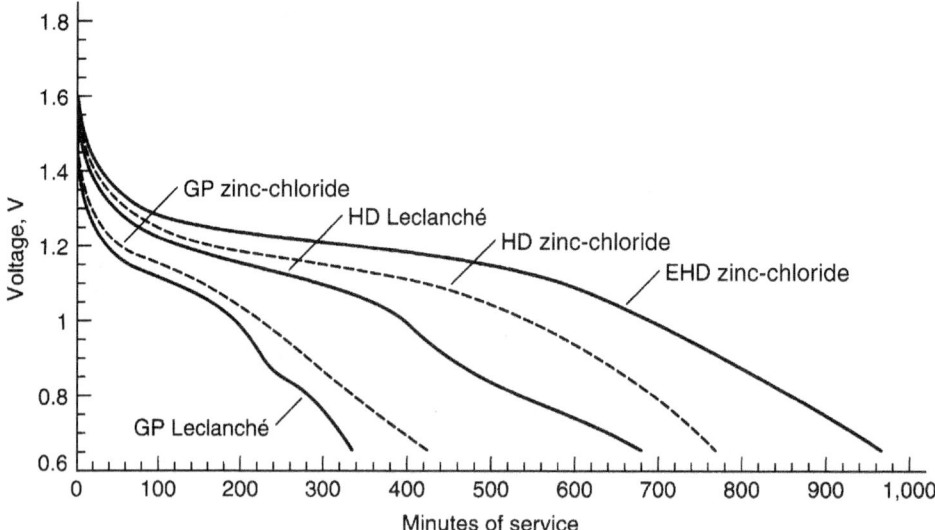

FIGURE 9.20 Comparison of Leclanché and zinc chloride D-size batteries of various grades continuously discharged through 3.9 ohm at 20°C. GP: General purpose; HD: Heavy duty; EHD: Extra heavy duty.

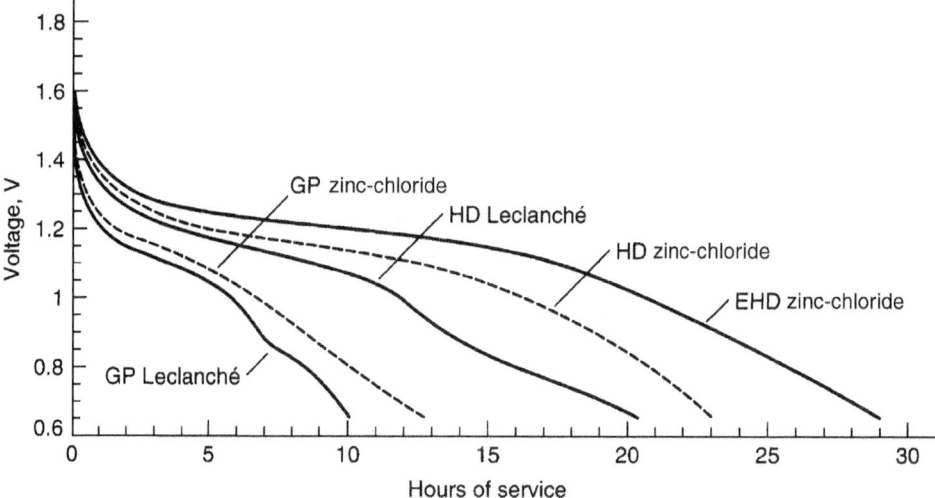

FIGURE 9.21 Comparison of Leclanché and zinc chloride D-size batteries of various grades discharged on a simulated cassette application, 3.9 ohms, 1 h/day at 20°C. GP: General purpose; HD: Heavy duty; EHD: Extra heavy duty.

battery, for the same testing, is shown in Fig. 9.23b. Figures 9.24a and b present the same performance relationships except for both the Leclanché and zinc chloride batteries in the heavy-duty D-size battery and Fig. 9.25 for the zinc chloride extra/super heavy-duty battery.

Performance differences between the batteries of the same grade but from different manufacturers are shown in Fig. 9.26. There is a difference of about 25% between the best and poorest battery to the 0.9 V cutoff.

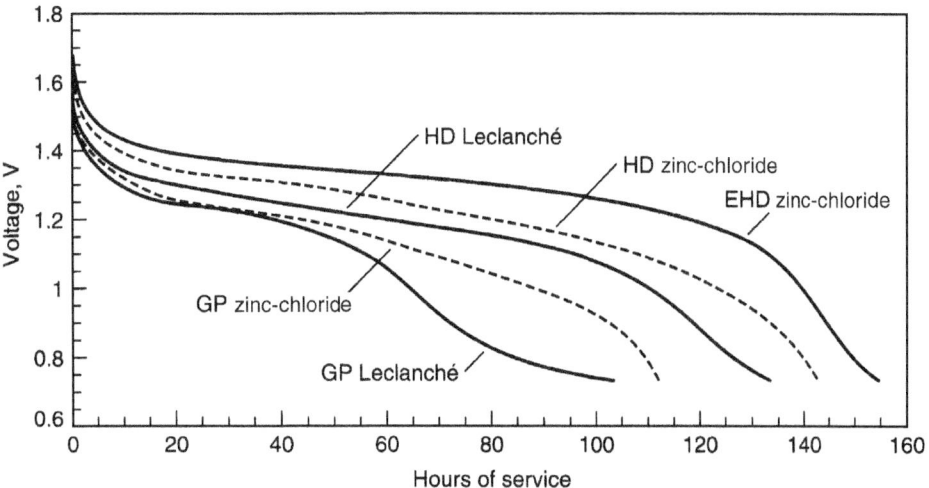

FIGURE 9.22 Comparison of Leclanché and zinc chloride D-size batteries of various grades discharged through 24 ohm for 4h/day at 20°C. GP: General purpose; HD: Heavy duty; EHD: Extra heavy duty.

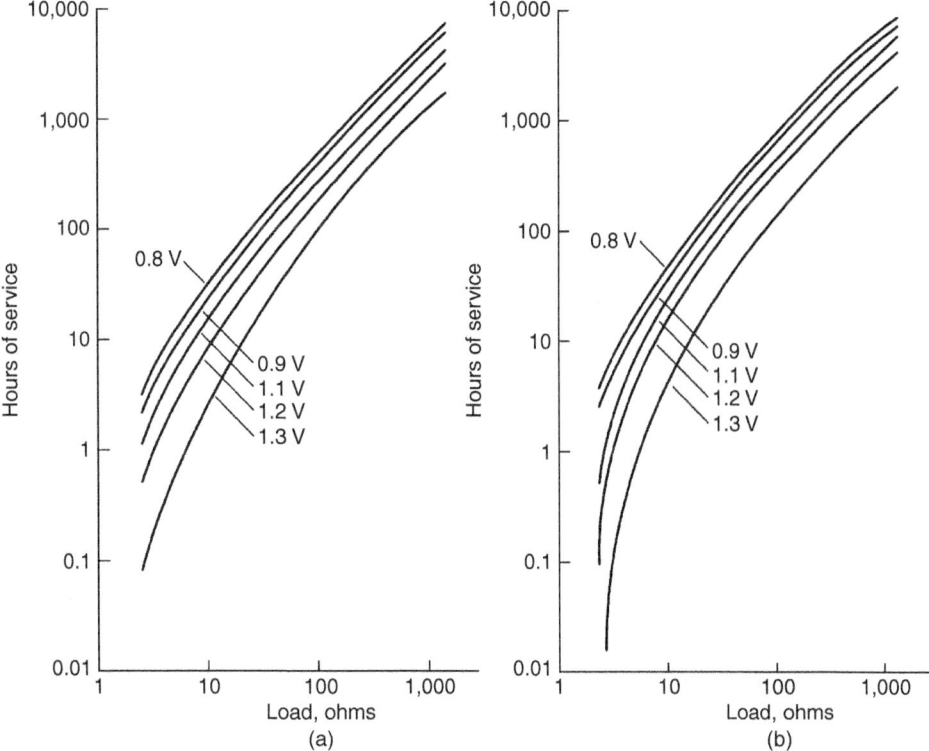

FIGURE 9.23 Discharge load vs. hours of service to different cutoff (end) voltages at 20°C, continuous discharge. (a) General-purpose D-size Leclanché battery construction. (b) General-purpose D-size zinc chloride battery.

9.26 PRIMARY BATTERIES

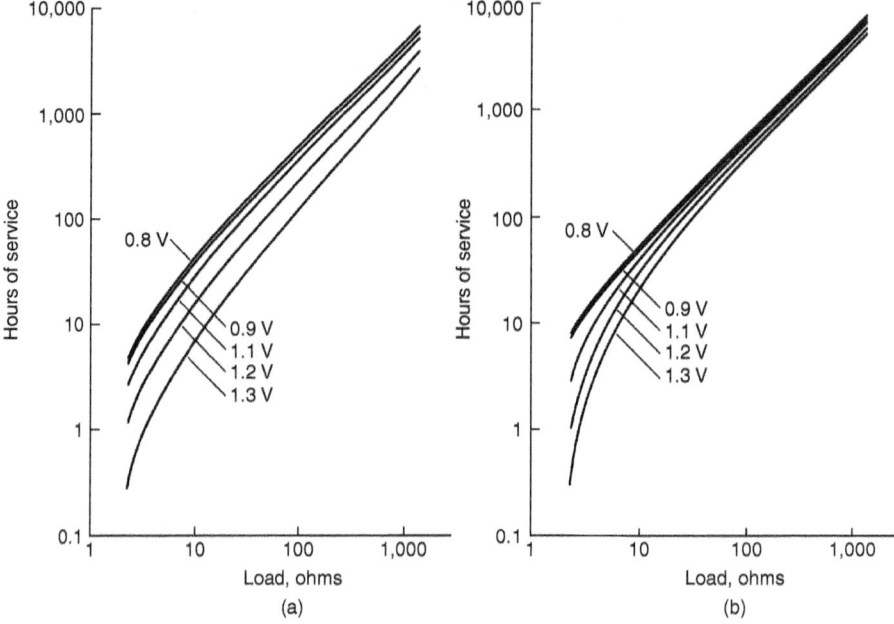

FIGURE 9.24 Discharge load vs. hours of service to different cutoff (end) voltages at 20°C, continuous discharge. (a) Heavy-duty D-size Leclanché battery. (b) Heavy-duty D-size zinc chloride battery.

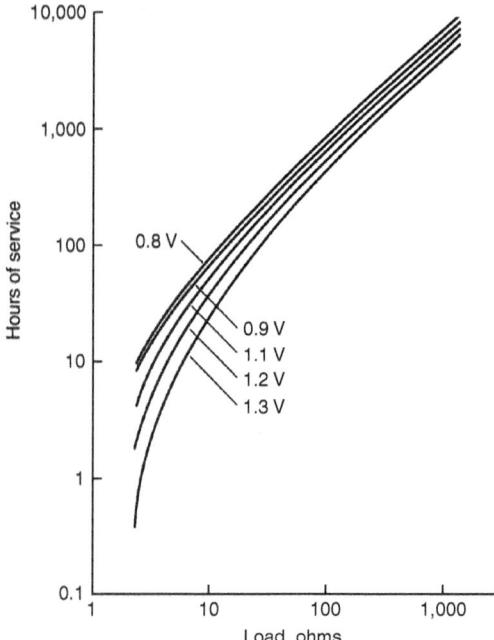

FIGURE 9.25 Hours of service vs. discharge load to different cutoff (end) voltages at 20°C, continuous discharge. Extra heavy-duty D-size zinc chloride battery.

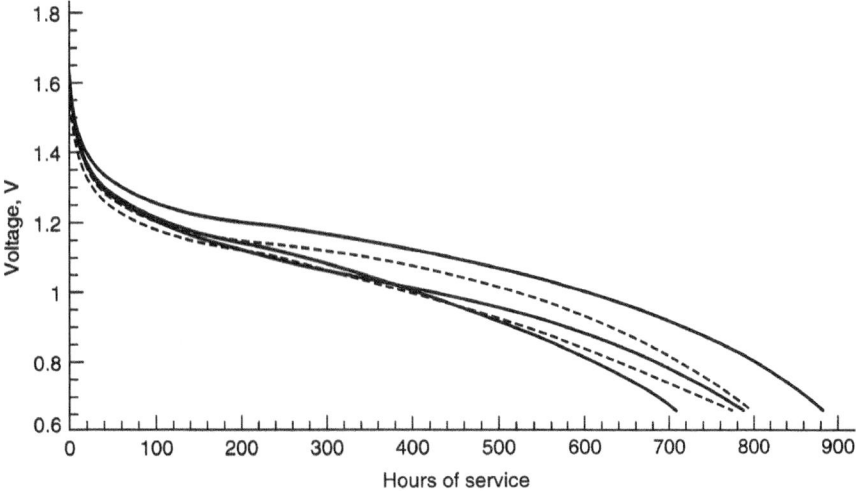

FIGURE 9.26 Comparison of general-purpose D-size zinc-carbon batteries from five different manufacturers, discharged on the ANSI LIF test through 2.2 ohm load at 20°C.

9.6.6 Internal Resistance

Internal resistance (R_{in}) is defined as the opposition or resistance to the flow of an electric current within a cell or battery, i.e., the sum of the ionic and electronic resistances of the cell components. Other considerations include battery size and construction as well as temperature, age, and depth of discharge.

9.6.6.1 Electronic Resistance. Electronic resistance includes the resistance of the materials of construction: metal covers, carbon rods, conductive cathode components, and so on. An approximation of the internal electronic resistance of a battery can be made by determining the OCV and the peak flash current (I), using very low resistance meters. The ammeter resistance should be low enough that the total circuit resistance does not exceed 0.01 ohm and is no more than 10% of the cell's internal resistance. The internal electronic resistance is expressed as

$$R_{in} = OCV/I$$

where R_{in} = internal resistance expressed in ohm
  OCV = open circuit voltage
  I = peak flash current (A)

A more accurate method of calculation is made using the voltage-drop method. In this method, a small initial load is applied on the battery to stabilize the voltage. A load approximating the application load is then applied. The internal resistance is calculated by

$$R_{in} = (V_1 - V_2)R_L/V_2$$

where R_{in} = internal resistance
  V_1 = initial stabilized closed-circuit voltage, V
  V_2 = closed circuit voltage reading at the application load, VO
  R_L = application load, ohms

The application load time should be kept to a pulse of 5 to 50 ms to minimize effects due to polarization. These methods measure the voltage loss due to the electrical resistance component but do not take into account voltage losses due to polarization.

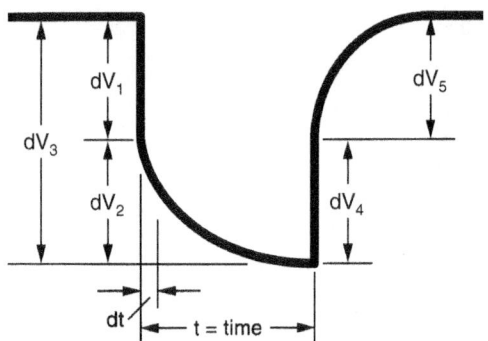

FIGURE 9.27 Voltage pulse/time profile illustration of curve shape and voltage components to calculate internal resistance.

9.6.6.2 Ionic Resistance. Ionic resistance encompasses factors resulting from the movement of ions within the cell. These include electrolyte conductivity, ionic mobility, electrode porosity, electrode surface area, secondary reactions, etc. The polarization effect is best illustrated by a trace of the pulse/time profile as shown in Fig. 9.27. The total resistance (R_T) is expressed using Ohm's law by

$$R_T = dR = dV/dI$$

which also equals

$$(V_1 - V_2)/(I_1 - I_2)$$

where V_1 and I_1 = the voltage and current just prior to pulsing
V_2 and I_2 = the voltage and current just prior to the pulse load removal
dV_3 = total voltage drop shown

The internal resistance of the battery component is expressed as dV_1 and the polarization effect component is the voltage drop dV_2. Since some energy was removed by the pulse, a more correct expression for the battery resistance is the voltage drop expressed by dV_4.

Measurement of the battery voltage drop (dV_4) is very difficult to capture, therefore the pulse duration (dt) is minimized to reduce the polarization effect voltage drop (dV_2). The pulse duration is generally kept in the range of 5 to 50 ms. For accurate and repetitive results, it is recommended that duration times be kept constant by "read and hold" voltage measurements.

Since dV_2 is slightly greater than dV_1, one can see that the resistance due to polarization (R_p) is greater than the internal resistance of the battery (R_{ir}) by the formula

$$R_T = R_{ir} + R_p$$

Partial, light discharge or a light background load prior to the pulse and internal resistance measurements provides equilibration for consistent measurements.

Table 9.6 shows the general relationship of flash current and internal resistance of the more popular cell sizes.

Zinc-carbon batteries perform better on intermittent drains than continuous drains, largely because of their ability to dissipate the effects of polarization. Factors that affect polarization are identified earlier in this section. Resting between discharges allows the zinc surface to "depolarize." One such effect is the dissipation of concentration polarization at the anode surface. This effect is more pronounced as heavier drains and longer duty schedules are applied. The internal resistance of the zinc chloride batteries is slightly lower than that of the Leclanché batteries. This results in a smaller voltage drop for a given battery size.

TABLE 9.6 Flash Current and Internal Resistance for Various Battery Sizes

Size	Typical maximum flash current, A		Approximate internal resistance, ohms	
	LC*	ZC*	LC	ZC
N	2.5	...	0.6	...
AAA	3	4	0.4	0.35
AA	4	5	0.30	0.28
C	5	7	0.39	0.23
D	6	9	0.27	0.18
F	9	11	0.17	0.13
9 V (battery)	0.6	0.8	5	4.5

*LC: Leclanché, ZC: Zinc chloride.
Source: Eveready Battery Engineering Data.[12]

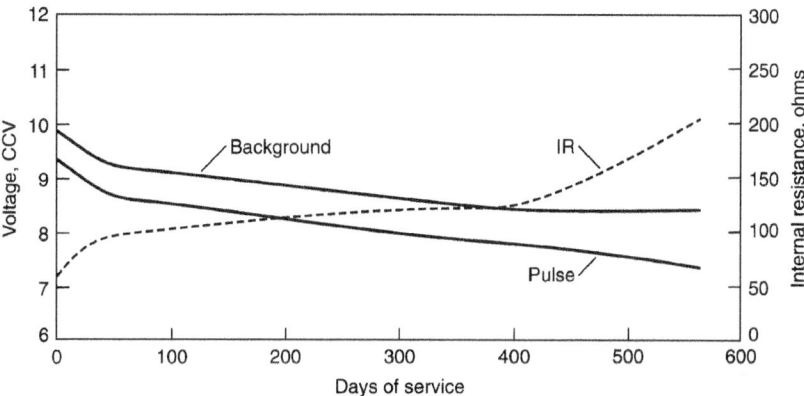

FIGURE 9.28 Comparison of voltage and internal resistance during discharge of a 9 V battery on smoke detector test. Background load = 620,000 ohms continuous; pulse load = 1,500 ohms × 10 ms every 40 s.

The internal resistance of zinc-carbon batteries increases with the depth of discharge. Some applications use this feature to establish low-battery alarms to predict near end of battery life situations (such as in the smoke detector). Figure 9.28 shows the relative battery internal resistance versus depth of discharge of a 9 V Leclanché battery.

One of the reasons for this increase in internal resistance is the cathode discharge reaction. The porous cathode becomes progressively blocked with reaction products. In the case of the Leclanché system, it is in the form of diammine-zinc chloride crystals; in the case of the zinc chloride system, it is in the form of zinc oxychloride crystals. Also the conductivity of the manganese dioxide decreases.

9.6.7 Effect of Temperature

Zinc-carbon batteries operate best in a temperature range of 20 to 30°C. The energy output of the battery increases with higher operating temperatures, but prolonged exposure to high temperatures (50°C and higher) will cause rapid deterioration. The capacity of the Leclanché battery falls off rapidly with decreasing temperatures, yielding no more than about 65% capacity at 0°C, and is essentially inoperative below −20°C. Zinc chloride batteries provide an additional 15% capacity at 0°C or 80%

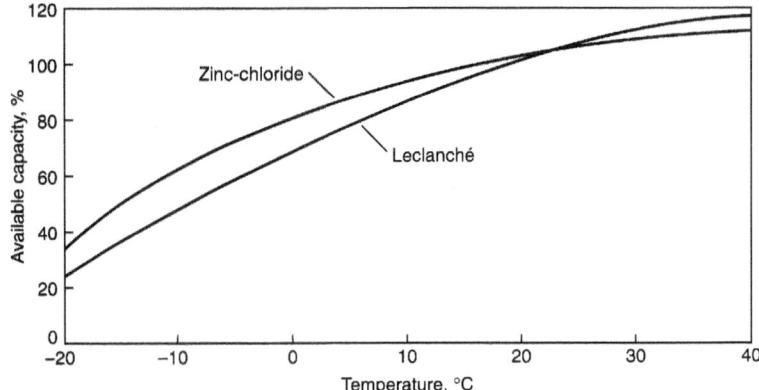

FIGURE 9.29 Percentage of capacity available as a function of temperature, moderate-drain radio-type discharge.

of room temperature capacity. The effects are more pronounced at heavier current drains; a low current drain would tend to result in a higher capacity at lower temperatures than a higher current drain (except for a beneficial heating effect that may occur at the higher current drains).

The effect of temperature on the available capacity of zinc-carbon (Leclanché and zinc chloride systems) batteries is shown graphically in Fig. 9.29 for both general-purpose (ammonium chloride electrolyte) and heavy-duty (zinc chloride electrolyte) batteries. At −20°C, typical zinc chloride electrolytes (25 to 30% zinc chloride by weight) turn to slush. Below −25°C, ice formation is likely. Under these conditions, it is not surprising that performance is dramatically reduced. These data represent performance at flashlight-type current drains (300 mA for a D-size cell). A lower current drain would result in a higher capacity than shown. Additional characteristics of this D-size battery at various temperatures are listed in Table 9.7.

TABLE 9.7 Temperature Effect on Internal Resistance

Battery size	System*	Resistance, ohms			
		−20°C	0°C	20°C	45°C
Single cell batteries					
AAA	ZC	10	0.7	0.6	0.5
AA	LC	5	0.8	0.5	0.4
AA	ZC	5	0.8	0.5	0.4
C	LC	2	0.8	0.5	0.4
C	ZC	3	0.5	0.4	0.3
D	LC	2	0.6	0.5	0.4
D	ZC	2	0.4	0.3	0.2
Flat cell batteries					
9 V	LC	100	45.0	35.0	30.0
9 V	ZC	100	45.0	35.0	30.0
Lantern batteries					
6 V	LC	10	1.0	0.9	0.7
6 V	ZC	10	1.0	0.8	0.7

*LC: Leclanché, ZC: Zinc chloride.
Source: Eveready Battery Engineering Data.[12]

Special low-temperature batteries were developed using low freezing-point electrolytes and a design that minimizes internal cell resistance, but they did not achieve popularity due to the superior overall performance of other types of primary batteries. For best operation, at low ambient temperatures, the Leclanché battery should be kept warm by some appropriate means. A vest battery worn under the user's clothing, employing body heat to maintain it at a satisfactory operating temperature was once used by the military to achieve reliable operation at low temperatures. Addition of other salts or gum karaya can boost low-temperature performance at the expense of high-temperature (>40°C) shelf life.

9.6.8 Service Life

The service life of the Leclanché battery is summarized in Figs. 9.30 and 9.31, which plot the service life at various loads and temperatures normalized for unit weight (Amperes per kilogram) and unit volume (Amperes per liter). These curves are based on the performance of a general-purpose battery at the average discharge current under several discharge modes. These data can be used to approximate the service life of a given battery under particular discharge conditions or to estimate the size and weight of a battery required to meet a specific service requirement.

Manufacturers' catalogs should be consulted for specific performance data in view of the many cell formulations and discharge conditions. Table 9.8 presents typical data from a manufacturer of two formulations of the AA-size battery.

9.6.9 Shelf Life

Zinc-carbon batteries gradually lose capacity while idle. This deterioration is greater for partially discharged batteries than for unused batteries and results from parasitic reactions such as wasteful zinc corrosion, chemical side reactions, and moisture loss. The shelf life or rate of capacity loss is affected by the storage temperature. High temperatures accelerate the loss; low temperatures retard

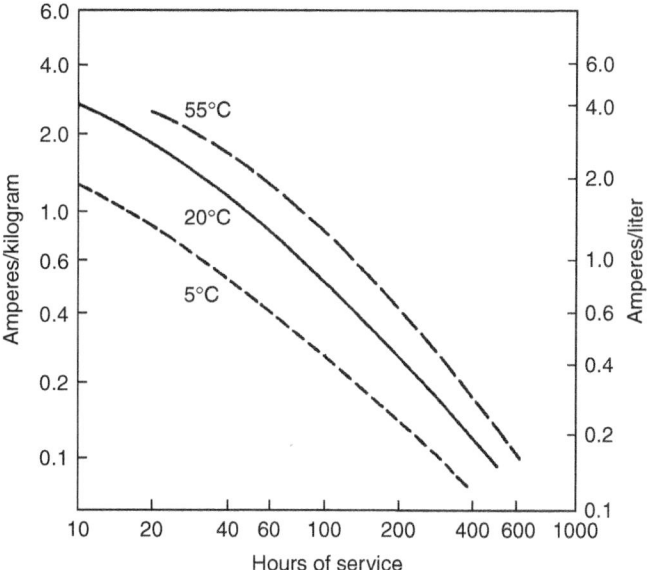

FIGURE 9.30 Service hours for general-purpose zinc-carbon battery, discharged 2 h/day to 0.9 V at three temperatures.

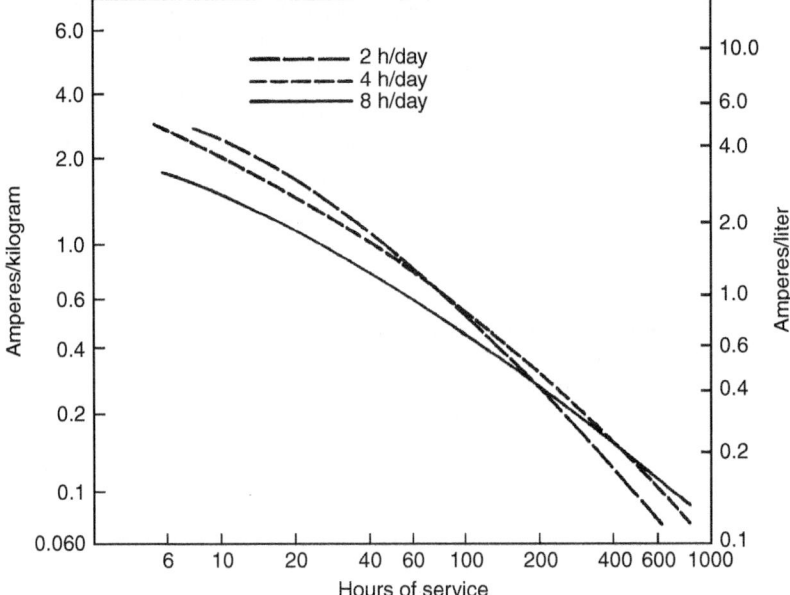

FIGURE 9.31 Service hours for a general-purpose zinc-carbon battery discharged intermittently to 0.9 V at 20°C.

TABLE 9.8 Manufacturer's Data for AA-Size Zinc-Carbon Batteries

Schedule	Drain @ 1.2 V, mA	Load Ω	Cutoff Voltage V					
			1.3	1.2	1.1	1.0	0.9	0.8
Typical service of Eveready No. 1015 general-purpose battery								
			Hours					
4 h/day	28 mA	43	2	5	12	20	24	27
1 h/day	120 mA	10	0.1	0.4	1.2	2.6	3.9	4.5
1 h/day	308 mA	3.9	0.09	0.2	0.4	0.7	0.9	1.0
			Pulses					
15 s/min/24 h/day (pulse)	667 mA	1.8	6	14	30	51	68	73
Typical service of Eveready No. 1215 super heavy-duty battery								
			Hours					
4 h /day	28 mA	43	4	10	21	31	36	39
1 h /day	120 mA	10	0.2	0.4	2.5	5.2	6.4	7.0
1 h /day	308 mA	3.9	0.1	0.3	0.5	1.2	1.7	1.9
			Pulses					
15 s/min/24 h/day (pulse)	667 mA	1.8	7	14	30	89	139	160

Source: Eveready Battery Engineering Data.[12]

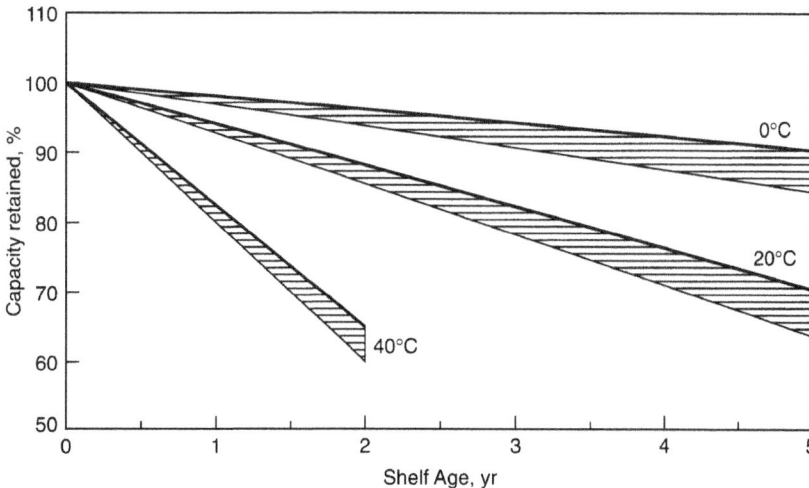

FIGURE 9.32 Capacity retention after storage at 40, 20, and 0°C for paper-lined plastic seal zinc chloride batteries.

the loss. Refrigerated storage will increase the shelf life. Figure 9.32 shows the retention of capacity of a zinc-carbon battery after storage at 40, 20, and 0°C. The capacity retention of a zinc chloride battery is higher than that of the Leclanché type because of the improved separators (coated paper separator types), sealing systems, and other materials used in that design.

Leclanché-type batteries, using the asphalt or pitch-type seals in conjunction with paste-type separators, have the poorest capacity retention. Zinc chloride batteries, using highly crosslinked starch-coated paper separators in conjunction with molded polypropylene or polyethylene seals, provide the best retention.

Batteries stored at −20°C are expected to retain approximately 80 to 90% of their initial capacity after 10 years. Since low temperatures retard deterioration, storage at low temperatures is an advantageous method for preserving battery capacity. A storage temperature of 0°C is very effective.

Freezing usually may not be harmful as long as there is no repeated cycling from high to low temperatures. Use of case materials or seals with widely different coefficients of expansion may lead to cracking. When batteries are removed from cold storage, they should be allowed to reach room temperature in order to provide satisfactory performance. Moisture condensation during warm-up should be prevented as this may cause electrical leakage or short-circuiting.

9.7 SPECIAL DESIGNS

The zinc-carbon system is used in special designs to enhance particular performance characteristics or for new or unique applications.

9.7.1 Flat-Pack Zinc/Manganese Dioxide P-80 Battery

In the early 1970s, Polaroid introduced a new instant camera-film system, the SX-70. A major innovation in that system was the inclusion of a battery in the film pack rather than in the camera. The film pack contained a battery designed to provide enough energy for the pictures in the pack. The concept was that the photographer would not have to be concerned about the freshness of the battery as it was changed with each change of film.

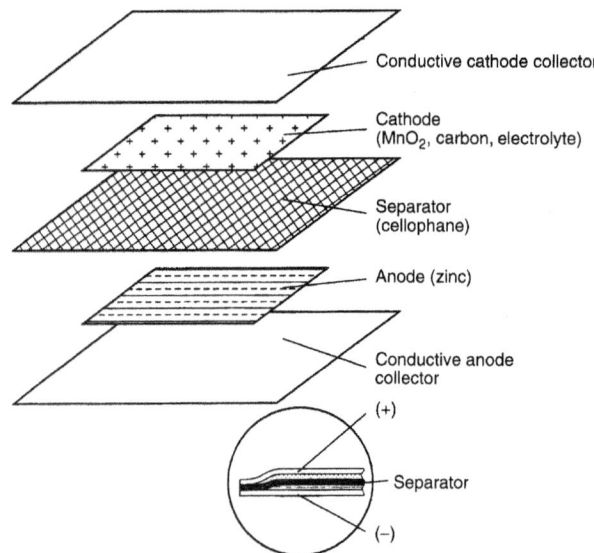

FIGURE 9.33 Exploded view of a single cell from Polaroid P-80 battery pack.

Battery Construction The P-80 battery uses chemistry quite similar to Leclanché round cells, although the shape is unique. Figure 9.33 details one cell (1.5 V). The electrode area is approximately 5.1 cm × 5.1 cm. The anode of zinc dust and binder (without mercury) is coated on a conductive vinyl web.

The manganese dioxide is mixed in a thick paste that contains the electrolyte salts. The electrolyte is mainly zinc chloride with some ammonium chloride. The anode and the cathode are separated by a thin film of cellophane. The complete 6 V battery has four cells. The four identical cells are connected by vinyl frames to each other and the aluminum collector plates. A special conductive coating allows the aluminum to bond to the plastic materials.

Battery Parameters The key battery parameters of the flat battery are similar to those of the cylindrical one. The flat configuration provides low resistance by virtue of the geometry. The thin layers need to stay in intimate contact to maintain the low resistance, and gassing effects have to be minimized.

- Open-circuit (no-load) voltage: The open-circuit voltage in this battery is dependent on the manganese dioxide activity and the system pH. The cathode slurry is adjusted to a constant pH to minimize battery-to-battery voltage variation. For example, the P-80 battery is adjusted so the voltage is 6.40 V at 56 days and 6.30 V after 12 months of shelf storage. The on-load voltage is sensitive to the synthetic manganese dioxide used.
- Closed-circuit (on-load) voltage: The closed-circuit voltage is used as an indicator of the battery's capability to deliver energy at high currents. In the case of the P-80 battery, a 1.63 A load is used since that is one of the operating requirements for the camera. The closed-circuit voltage is measured at 55 ms to minimize polarization effects. The normal closed-circuit voltage is 5.58 V at 56 days and 5.35 V after 12 months of shelf storage. The on-load voltage is sensitive to the synthetic MnO_2 used.
- Internal resistance and voltage drop (delta V): The battery's internal resistance is measured by using the voltage drop or delta V at a given load for a specified pulse period. A major

contributor which effects delta V is the activity at the zinc surface, which is dependent on both zinc particle size and amount of reaction product present. The polarization effect occurs when the load is held for some time period, as in the case of charging the camera's strobe circuit. Total resistance is then a summation of the two resistances—that is, the internal resistance of the battery and the resistance due to polarization effects, the latter being very time sensitive. To minimize polarization effect resistance, the pulse period for delta V measurements was minimized.

The 56-day point is of interest, as that is the normal age when the battery is released for assembly into film packs. At that time, every battery is measured for electrical characteristics and defective ones are screened.

The total internal resistance is expressed by the following:

$$R_t = R_i + R_p$$

where R_i = battery internal resistance
R_p = polarization resistance effect

For the P-80 battery, R_i was 0.50 ohms and the R_p was 0.12 ohms.

- Capacity: The capacity simulator mimics the energy used to charge the camera strobe. The pulse consists of an open-circuit voltage at rest, followed by a pulse at a 2 A load to result in a 50 watt-second (50 Ws) pulse. The 50 Ws cycle test is maintained until the final CCV reaches the 3.7 cutoff voltage, where the number of cycles is determined.

During the time while the 50 Ws load is maintained, the polarization drop occurs. The time to produce the 50 Ws increases with each cycle as the battery is "consumed." A 30 s rest between cycles is used. Initially, the voltage drop is fairly constant; however, near the end of the test, the resistance increases. The test is maintained to 3.7 V to indicate the cutoff point of the camera.

Figure 9.34 illustrates the voltage profile at different discharge loads, and Figure 9.35 shows rate sensitivity of the batteries described above versus capacity.

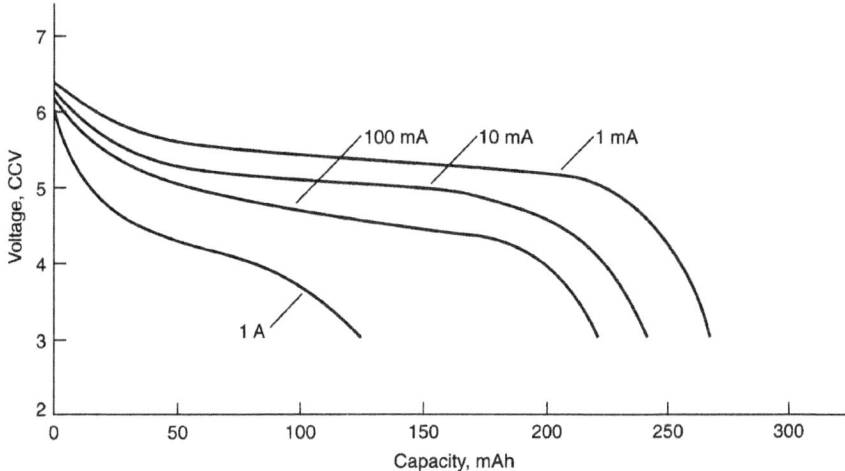

FIGURE 9.34 Polaroid P-80; battery voltage profile at various discharge loads.

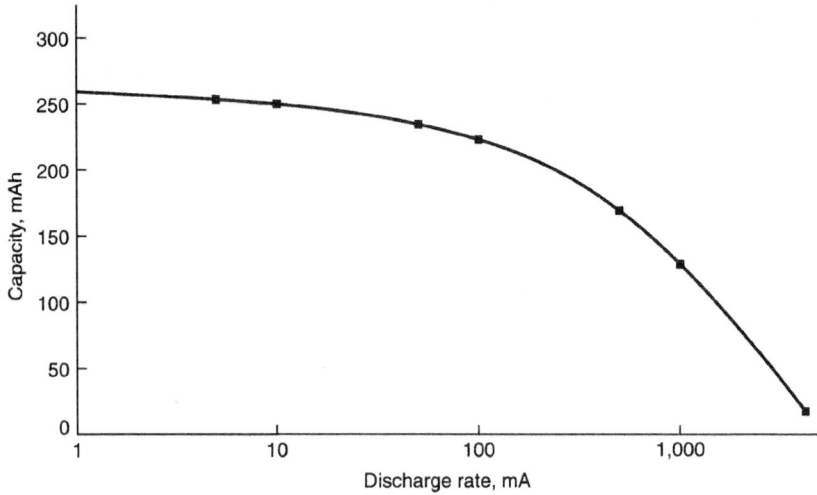

FIGURE 9.35 Polaroid P-80 battery; rate sensitivity vs. capacity (to 3.0 V cutoff).

9.8 TYPES AND SIZES OF AVAILABLE CELLS AND BATTERIES

Zinc-carbon batteries are made in a number of different sizes with different formulations to meet a variety of applications. The single-cell and multicell batteries are classified by electrochemical system, either Leclanché or zinc chloride, and by grade: general purpose, heavy-duty, extra heavy-duty, photoflash, and so on. These grades are assigned according to their output performance under specific discharge conditions.

Table 9.9 lists the more popular battery sizes with typical performance at various loads under a 2 h/day intermittent discharge, except for the continuous toy battery test. The performance of these batteries under several intermittent discharge conditions is given in Table 9.10.

The AA-size battery is becoming the predominant one and is used in penlights, photoflash, and electronic applications. The smaller AAA-size is used in remote control devices and other small electronic applications. The C-and D-size batteries are used mainly in flashlight applications, and the F-size is usually assembled into multicell batteries for lanterns and other applications requiring these large batteries. Flat cells are used in battery assemblies, in particular, the 9 V battery used in smoke detectors and electronic applications such as transistor radios.

Table 9.11 lists some of the major multicell zinc-carbon batteries that are available commercially. The performance of these batteries can be estimated by using the IEC designation to determine the cell compliment (e.g., NEDA 6, IEC 4R25 battery consists of four F-size cells connected in series). Table 9.12 gives cross-references to the zinc-carbon batteries and manufacturers' designations. The most recent manufacturers' catalogs and Web sites should be consulted for specific performance data to determine the suitability of their products for a particular application.

TABLE 9.9 Characteristics of Zinc-Carbon Batteries

| Size | IEC | ANSI, NEDA | Weight g | Maximum dimensions, mm | | Typical service, 2 h/day* | | | |
| | | | | | | Leclanché | | Zinc chloride | |
				Diameter	Height	Drain mA	Service h	Drain mA	Service h
N	R1	910	6.2	12	30.2	1	480		
						10	45		
						15	20		
AAA	R03	24	8.5	10.5	44.5	1	—	1	520
						10	—	10	55
						20	—	20	26
AA	R6	15	15	14.5	50.5	1	950	1	1200
						10	80	10	110
						100	4	100	8
						300	0.6	300	1
C	R14	14	41	26.2	50	5	380	5	800
						25	75	20	150
						100	6	100	20
						300	1.7	300	5.5
D	R20	13	90	34.2	61.5	10	400	10	700
						50	70	50	135
						100	25	100	55
						500	3	500	6
F	R25	60	160	34†	92†	25	300	25	400
						100	60	100	85
						500	5.5	500	9
G	R26	—	180	32†	105†	—			
No. 6	R40	905	900	67	170.7	5	8000		
						50	700		
						100	350		
						500	70		

*Typical values of service to 0.9 V cutoff.
†Typical values.

TABLE 9.10 ANSI Standards for Zinc-Carbon and Alkaline-Manganese Dioxide Batteries

| | | | | | Specifications requirements | |
| | | | | | Zinc-carbon batteries | Alkaline-manganese dioxide batteries |
Size	Use	Ohms	Schedule	Cutoff voltage	Initial*	Initial*
N					910D	910A
	Portable lighting	5.1	5 min/day	0.9	NA	100 min
	Pager	(10.0 then 3000.0	5 s/h 3595 s/h)	0.9	NA	888 h
AAA					24D	24A
	Pulse test	3.6	15 s/min 24 h/day	0.9	150 pulses	450 pulses
	Portable lighting	5.1	4 min/h 8 h/day	0.9	48.0 min	130.0 min
	Recorder	10.0	1 h/day	0.9	1.5 h	5.5 h
	Radio	75.0	4 h/day	0.9	24.0 h	48.0 h
AA					15D	15A
	Pulse test	1.8	15 s/min 24 h/day	0.9	100 pulses	450 pulses
	Motor/toy	3.9	1 h/day	0.8	1.2 h	5 h
	Recorder	10.0	1 h/day	0.9	5.0 h	13.5 h
	Radio	43.0	4 h/day	0.9	27.0 h	60 h
C					14D	14A
	Portable lighting	3.9	4 min/h 8 h/day	0.9	350 min	830 min
	Toy	3.9	1 h/day	0.8	5.5 h	14.5 h
	Recorder	6.8	1 h/day	0.9	10.0 h	24.0 h
	Radio	20.0	4 h/day	0.9	30 h	60.0 h
D					13D	13A
	Portable lighting	1.5	4 min/15 min 8 h/day	0.9	150 min	540 min
	Portable lighting	2.2	4 min/h 8 h/day	0.9	120 min	950 min
	Motor/toy	2.2	1 h/day	0.8	5.5 h	17.5 h
	Recorder	3.9	1 h/day	0.9	10 h	26.0 h
	Radio	10.0	4 h/day	0.9	33 h	90.0 h
9 V					1604D	1604A
	Calculator	180	30 min/day	4.8	380 min	630 min
	Toy	270	1 h/day	5.4	7 h	14 h
	Radio	620	2 h/day	5.4	23 h	38 h
	Electronic	1300	24 h/day	6.0	NA	NA
	Smoke detector	Currently under consideration.				
6 V					908D	908A
	Portable lighting	3.9	4 min/h 8 h/day	3.6	5 h	21 h
	Portable lighting	3.9	1 h/day	3.6	50 h	80 h
	Barricade	6.8	1 h/day	3.6	165 h	300 h

*Performance after 12 month storage:
  zinc-carbon batteries: 80% of initial requirement
  alkaline-manganese dioxide batteries: 90% of initial requirement
Source: ANSI C18.1M-2009.[11]

TABLE 9.11 ANSI/NEDA Dimensions of Zinc-Carbon Batteries

ANSI	IEC	Diameter, mm		Overall height, mm		Length, mm		Width, mm	
		Max	Min	Max	Min	Max	Min	Max	Min
13C	R20S	34.2	32.3	61.5	59.5				
13CD	R20C	34.2	32.3	61.5	59.5				
13D	R20C	34.2	32.3	61.5	59.5				
13F	R20S	34.2	32.3	61.5	59.5				
14C	R14S	26.2	24.9	50.0	48.5				
14CD	R14C	26.2	24.9	50.0	48.5				
14D	R14C	26.2	24.9	50.0	48.5				
14F	R14S	26.2	24.9	50.0	48.5				
15C	R6S	14.5	13.5	50.5	49.2				
15CD	R6C	14.5	13.5	50.5	49.2				
15D	R6C	14.5	13.5	50.5	49.2				
15F	R6S	14.5	13.5	50.5	49.2				
24D	R03	10.5	9.5	44.5	43.3				
903	—			163.5	158.8	185.7	181.0	103.2	100.0
904	—			163.5	158.8	217.9	214.7	103.2	100.0
908	4R25X			115.0	107.0	68.2	65.0	68.2	65.0
908C	4R25X			115.0	107.0	68.2	65.0	68.2	65.0
908CD	4R25X			115.0	107.0	68.2	65.0	68.2	65.0
908D	4R25X			115.0	107.0	68.2	65.0	68.2	65.0
915	4R25Y			112.0	107.0	68.2	65.0	68.2	65.0
915C	4R25Y			112.0	107.0	68.2	65.0	68.2	65.0
915D	4R25Y			112.0	107.0	68.2	65.0	68.2	65.0
918	4R25-2			127.0	—	136.5	132.5	73.0	69.0
918D	4R25-2			127.0	—	136.5	132.5	73.0	69.0
926	—			125.4	122.2	136.5	132.5	73.0	69.0
1604	6F22			48.5	46.5	26.5	24.5	17.5	15.5
1604C	6F22			48.5	46.5	26.5	24.5	17.5	15.5
1604CD	6F22			48.5	46.5	26.5	24.5	17.5	15.5
1604D	6F22			48.5	46.5	26.5	24.5	17.5	15.5

Source: ANSI C18.1M-2009.[11]

TABLE 9.12 Cross-Reference of Zinc-Carbon Batteries

ANSI	IEC	Duracell	Everyday	Rayovac	Panasonic	Toshiba	Varta	Military
13C	R20	M13SHD	EV50	GP-D	—	—	—	—
13CD	R20	M13SHD	EV150	HD-D	UM1D	—	—	—
13D	R20	M13SHD	1250	6D	UMIN	R20U	3020	—
13F	R20	—	950	2D	UM1	R20S	2020	BA-30/U
14C	R14	M14SHD	EV35	GP-C	—	—	—	—
14CD	R14	M14SHD	EV135	HD-C	UM2D	—	—	—
14D	R14	—	1235	4C	UM2N	R14U	3014	—
14F	R14	—	935	1C	UM2	R14S	2014	BA-42/U
15C	R6	M15SHD	EV15	GP-AA	—	—	—	—
15CD	R6	M15SHD	EV115	HD-AA	UM3D	—	—	—
15D	R6	M15SHD	1215	5AA	UM3N	R6U	3006	—
15F	R6	—	1015	7AA	UM3	R6S	2006	BA-58/U
24D	R03	—	1212	—	UM4N	—	—	—
24F	R03	—	—	—	—	—	—	—
210	20F20	—	413	—	—	—	—	BA-305/U
215	15F20	—	412	—	15	—	V72PX	BA-261/U
220	10F15	—	504	—	W10E	—	V74PX	BA-332/U
221	15F15	—	505	—	MV15E	—	—	—
900	R25-4	—	735	900	—	—	—	—
903	5R25-4	—	715	903	—	—	—	BA-804/U
904	6R25-4	—	716	904	—	—	—	BA-207/U
905	R40	—	EV6	—	—	—	—	BA-23
906	R40	—	EV6	—	—	—	—	BA-23
907	4R25-4	—	1461	641	—	—	—	BA-44/U
908	4R25	M908	509	941	4F	—	—	BA-200/U
908C	4R25	M908SHD	EV90	GP-6V	—	—	430	—
908CD	4R25	M908SHD	EV90HP	—	—	—	431	—
908D	4R25	M908SHD	1209	944	—	—	430	—
915	4R25	M915	510S	942	—	—	—	BA-803/U
915C	4R25	M915SHD	EV10S	—	—	—	—	—
915D	4R25	M915SHD	—	945	—	—	—	—
918	4R25-2	—	731	918	—	—	—	—
918C	4R25-2	—	EV31	—	—	—	—	—
918D	4R25-2	—	1231	928	—	—	—	—
922	—	—	1463	922	—	—	—	—
926	8R25-2	—	732	926	—	—	—	—
1604	6F22	—	216	1604	006P	—	2022	BA-90/U
1604C	6F22	M9VSHD	EV22	GP-9V	—	—	—	—
1604CD	6F22	M9VSHD	EV122	HD-9V	006PD	—	—	—
1604D	6F22	M9VSHD	1222	D1604	006PN	6F22U	3022	—

Source: Manufacturers' catalogs.

REFERENCES

1. Frost and Sullivan Inc., Global Battery Market, New York, 2009.
   V. Sapru "Analyzing the Global Battery Market," *Battery Power*, **14**(4)(2010).
2. The Freedonia Group, Inc., Industry Study #1193, Primary and Secondary Batteries, Cleveland, Ohio, December 1999. I. Buchmann, Battery University.com, after Freedonia (2005) and the Freedonia Group, Inc.
3. Samuel Rubin, *The Evolution of Electric Batteries in Response to Industrial Needs*, Chap. 5, Dorrance, Philadelphia, 1978.
4. George Vinal, *Primary Batteries*, Wiley, New York, 1950.
5. N.C. Cahoon, in N. C. Cahoon and G. W. Heise (Eds.), *The Primary Battery*, Vol. 2, Chap. 1, Wiley, New York, 1976.
6. Richard Huber, in K. V. Kordesh (Ed.), *Batteries*, Vol. 1, Chap. 1, Decker, New York, 1974.
7. D. Glover, A. Kozawa, and B. Schumm, Jr., (Eds.), *Handbook of Manganese Dioxides, Battery Grade*, International Battery Material Association (IBA, Inc.), IC Sample Office, 1989.
8. R. J. Brodd, A. Kozawa, and K. V. Kordesh "Primary Batteries 1951–1976," *J. Electrochem. Soc.* **125**(7) (1978).
9. B Schumm, Jr., in *Modern Battery Technology,* C. D. S. Tuck (Ed.), Ellis Horwood, Ltd., London, 1991, pp. 87–111.
10. C. L. Mantell, *Batteries and Energy Systems*, 2d ed., McGraw-Hill, New York, 1983.
11. "American National Standards Specification for Dry Cells and Batteries," ANSI C18.1M-2009, American National Standards Institute, Inc., 2009.
12. Eveready Battery Engineering Data: information is available at www.Energizer.com; technical information website. These data are frequently updated and current.
13. M. Dentch and A. Hillier, Polaroid Corp., *Progress in Batteries and Solar Cells*, Vol. 9 (1990).

CHAPTER 10
MAGNESIUM AND ALUMINUM BATTERIES

Patrick J. Spellman

10.1 GENERAL CHARACTERISTICS

Magnesium and aluminum are attractive candidates for use as anode materials in primary batteries.[*] As shown in Table 1.1, Chap. 1, they have a high standard potential. Their low atomic weight and multivalence change result in a high electrochemical equivalence on both a gravimetric and a volumetric basis. Further, they are both abundant and relatively inexpensive.

Magnesium has been used successfully in a magnesium/manganese dioxide (Mg/MnO_2) battery. This battery has two main advantages over the zinc-carbon battery, namely, twice the service life or capacity of the zinc battery of equivalent size and the ability to retain this capacity during storage, even at elevated temperatures (Table 10.1). This excellent storability is due to a protective film that forms on the surface of the magnesium anode.

Several disadvantages of the magnesium battery are its "voltage delay" and the parasitic corrosion of magnesium that occurs during the discharge once the protective film has been removed, generating hydrogen and heat. The magnesium battery also loses its excellent storability after being partially discharged and, hence, is unsatisfactory for long-term intermittent use. For these reasons, the active (nonreserve) magnesium battery, while used successfully in military applications, such as radio transceivers and emergency or standby equipment, has not found wide commercial acceptance.

Furthermore, the use of the magnesium battery by the U.S. military has ceased due to upgrades in military equipment that require higher rate capability than the magnesium system can provide. The lithium primary and lithium-ion rechargeable batteries are now used.

Aluminum has not been used successfully in an active primary battery despite its potential advantages. Like magnesium, a protective film forms on the aluminum, which is detrimental to battery performance, resulting in a battery voltage that is considerably below theoretical and causing a voltage delay that can be significant for partially discharged batteries or those that have been stored. While the protective oxide film can be removed by using suitable electrolytes or by amalgamation, gains by such means are accompanied by accelerated corrosion and poor shelf life. Aluminum, however, has been more successfully used as an anode in aluminum/air batteries. (See Chap. 33.)

---

[*]The use of magnesium and aluminum in reserve and mechanically rechargeable batteries is covered in Chapters 34 and 33, respectively.

TABLE 10.1 Major Advantages and Disadvantages of Magnesium Batteries

Advantages	Disadvantages
Good capacity retention, even under high-temperature storage	Delayed action (voltage delay)
Twice the capacity of corresponding Leclanché batteries	Evolution of hydrogen during discharge
Higher battery voltage than zinc-carbon batteries	Heat generated during use
Competitive cost	Poor storage after partial discharge

10.2 CHEMISTRY

The magnesium primary battery uses a magnesium alloy for the anode, manganese dioxide as the active cathode material but mixed with acetylene black to provide conductivity, and an aqueous electrolyte consisting of magnesium perchlorate, with barium and lithium chromate as corrosion inhibitors and magnesium hydroxide as a buffering agent to improve storability (pH of about 8.5). The amount of water is critical, as water participates in the anode reaction and is consumed during the discharge.[1]

The discharge reactions of the magnesium/manganese dioxide battery are

$$\text{Anode} \quad Mg + 2OH^- = Mg(OH)_2 + 2e$$
$$\text{Cathode} \quad 2MnO_2 + H_2O + 2e = Mn_2O_3 + 2OH^-$$
$$\text{Overall} \quad Mg + 2MnO_2 + H_2O = Mn_2O_3 + Mg(OH)_2$$

The theoretical potential of the battery is greater than 2.8 V, but this voltage is not realized in practice. The observed values are decreased by about 1.1 V, giving an open-circuit voltage of 1.9 to 2.0 V, still higher than for the zinc-carbon battery.

The rest potential of magnesium in neutral and alkaline electrolytes is a mixed potential, determined by the anodic oxidation of magnesium and the cathodic evolution of hydrogen. The kinetics of both of these reactions are strongly modified by the properties of the passive film, its history of formation, prior anodic (and to a limited extent cathodic) reactions, the electrolyte environment, and magnesium alloying additions. The key to a full appreciation of the magnesium electrode lies in an understanding of the predominantly $Mg(OH)_2$ film,[2] the factors that govern its formation and dissolution, as well as the physical and chemical properties of the film.

The corrosion of magnesium under storage conditions is slight. A film of $Mg(OH)_2$ that forms on the magnesium provides good protection, and treatment with chromate inhibitors increases this protection. As a result of the formation of this tightly adherent and passivating oxide or hydroxide film on the electrode surface, magnesium is one of the most electropositive metals to find use in aqueous primary batteries. However, when the protective film is broken or removed during discharge, corrosion occurs with the generation of hydrogen

$$Mg + 2H_2O \rightarrow Mg(OH)_2 + H_2$$

During the anodic oxidation of magnesium, the rate of hydrogen evolution increases with increasing current density due to destruction of the passive film, which exposes more (cathodic) sites on the bared magnesium surface. This phenomenon has often been referred to as the "negative difference effect." An appreciable rate of anodic oxidation of magnesium can only take place on the bare metal surface. Magnesium salts generally exhibit low levels of anion conductivity, and one could theoretically invoke a mechanism wherein OH^- ions migrate through the film to form reaction product $Mg(OH)_2$ at the magnesium-film interface. In practice this does not occur at a sufficiently rapid rate and instead the film becomes disrupted, in all likelihood mechanically, as the result of anodic current flow.[3] A theoretical model for the breakdown of the passive film has been proposed.[4–7] This model involves, successively, metal dissolution at the metal-film interface, film dilatation, and film breakdown. This wasteful reaction is a problem, not only because of the need to vent the hydrogen from the battery and to prevent it from accumulating, but also because it uses water that is critical to the battery operation, produces heat, and reduces the efficiency of the anode.

The efficiency of the magnesium anode is about 60 to 70% during a typical continuous discharge and is influenced by such factors as the composition of the magnesium alloy, battery components, discharge rate, and temperature. On low drains and intermittent service, the anode efficiency can drop to 40 to 50% or less. The anode efficiency also is reduced with decreasing temperature.

Considerable heat is generated during the discharge of a magnesium battery, particularly at high discharge rates, due to the exothermic corrosion reaction (about 82 kcal per gram-mole of magnesium) and the losses resulting from the difference between the theoretical and operating voltage. Proper battery design must allow for the dissipation of this heat to prevent overheating and shortened life. On the other hand, this heat can be used to advantage at low ambient temperatures to maintain the battery at higher and more efficient operating temperatures.

A consequence of the passive film on these metals is the occurrence of a voltage delay—a delay in the battery's ability to deliver full output voltage after it has been placed under load—which occurs when the protective film on the surface of the metal becomes disrupted by the flow of current, exposing bare metal to the electrolyte (see Fig. 10.1). When the current is interrupted, the passive film does indeed reform, but never to the original degree of passivity. Thus both the magnesium and the aluminum batteries are at a significant disadvantage in very low or intermittent service applications.[3] This delay, as shown in Fig. 10.2, is usually less than 1 s, but can be longer (up to a minute or more) for discharges at low temperatures and after prolonged storage at high temperatures.

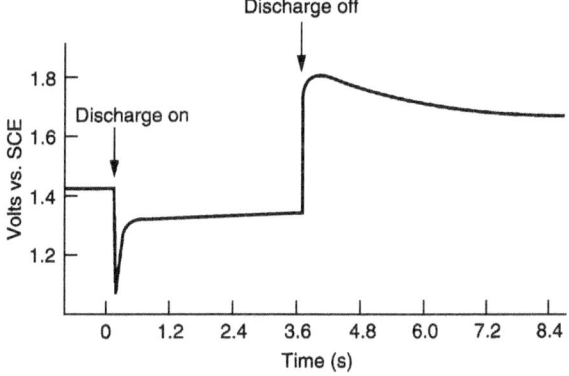

FIGURE 10.1 Voltage profile of magnesium primary battery at 20°C.

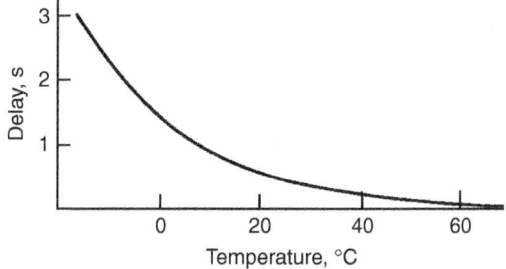

FIGURE 10.2 Voltage delay vs. temperature, Mg/MnO$_2$ battery.

10.2.1 Aluminum

The standard potential for aluminum in the anode reaction,

$$Al \rightarrow Al^{3+} + 3e$$

is reported as -1.7 V. A battery with an aluminum anode should have a voltage about 0.9 V higher than the corresponding zinc battery. However, this voltage is not attained, and the voltage of an Al/MnO$_2$ battery is only about 0.1 to 0.2 V higher than that of a zinc battery. The Al/MnO$_2$ battery never progressed beyond the experimental stage because of the problems with the oxide film, excessive corrosion when the film was broken, voltage delay, and the tendency for aluminum to corrode unevenly. The experimental batteries that were fabricated used a two-layer aluminum anode (to minimize premature failure due to can perforation), an electrolyte of aluminum or chromium chloride, and a manganese dioxide-acetylene black cathode similar to the conventional zinc/manganese dioxide battery. The reaction mechanism is

$$Al + 3MnO_2 + 3H_2O \rightarrow 3MnO \cdot OH + Al(OH)_3$$

10.3 CONSTRUCTION OF Mg/MnO$_2$ BATTERIES

Magnesium/manganese dioxide (nonreserve) primary batteries are generally constructed in a cylindrical configuration.

10.3.1 Standard Construction

The construction of the magnesium battery is similar to the cylindrical zinc-carbon battery. A cross section of a typical battery is shown in Fig. 10.3. A magnesium alloy can, containing small amounts of aluminum and zinc, is used in place of the zinc can. The cathode consists of

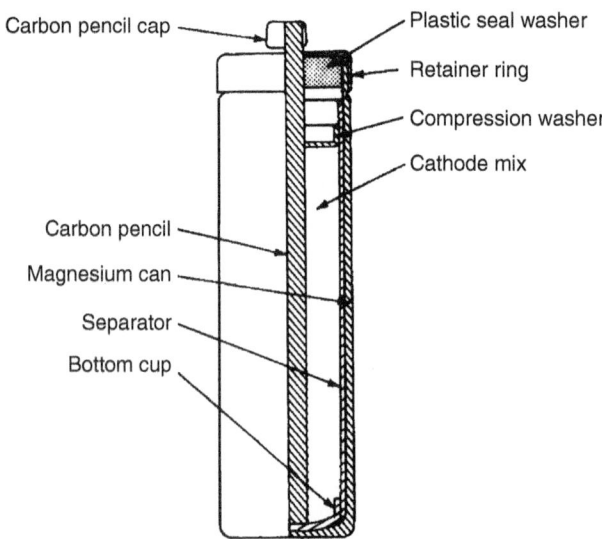

FIGURE 10.3 Cylindrical construction of magnesium primary battery.

an extruded mix of manganese dioxide, acetylene black for conductivity and moisture retention, barium chromate (an inhibitor), and magnesium hydroxide (a pH buffer). The electrolyte is an aqueous solution of magnesium perchlorate with lithium chromate. A carbon rod serves as the cathode current collector. The separator is an absorbent kraft paper as in the paper-lined zinc battery structure. Sealing of the magnesium battery is critical, as it must be tight to retain battery moisture during storage but provide a means for the escape of hydrogen gas which forms as the result of the corrosion reaction during the discharge. This is accomplished by a mechanical vent—a small hole in the plastic top seal washer under the retainer ring which is deformed under pressure, releasing the excess gas.[8]

10.3.2 Inside-Out Construction

The basis of the inside-out design (Fig. 10.4) is a highly conductive carbon structure which can be molded readily into complex shapes. The carbon structure is formed in the shape of a cup which serves as the battery container; an integral center rod is incorporated to reduce current paths. The cups are structurally strong, homogeneous, and impervious to the passage of liquids and gases and the corrosive effects of the electrolyte. The battery consists of the carbon cup, a cylindrical magnesium anode, a paper separator, and a cathode mix consisting of manganese dioxide, carbon black, and inhibitors with aqueous magnesium bromide or perchlorate as the electrolyte. The cathode mix is packed into the spaces on both sides of the anode and is in intimate contact with the inside and outside surfaces of the anode, the center rod, and the inside surfaces of the cup. This configuration provides larger electrode surface areas. External contacts are made by two metallic end pieces. The positive terminal is bonded during the forming process to the closed end of the carbon cup. The negative terminal, to which the anode is attached, together with a plastic ring forms the insulated closure and seal for the open end of the cup. The entire battery assembly is enclosed in a crimped tin-plated steel jacket.[9-11]

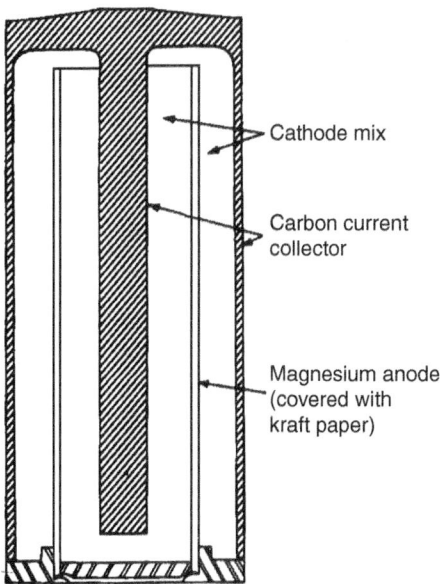

FIGURE 10.4 Inside-out construction of magnesium primary battery. (*Courtesy of ACR Electronics, Inc.*)

10.6 PRIMARY BATTERIES

10.4 PERFORMANCE CHARACTERISTICS OF Mg/MnO$_2$ BATTERIES

10.4.1 Discharge Performance

Typical discharge curves for the cylindrical magnesium/manganese dioxide primary battery are shown in Fig. 10.5. The discharge profile is generally flatter than for the zinc-carbon batteries; the magnesium battery also is less sensitive to changes in the discharge rate. The average discharge voltage is on the order of 1.6 to 1.8 V, about 0.4 to 0.5 V above that of the zinc-carbon battery; the typical end voltage is 1.2 V. The performance characteristics of the cylindrical magnesium battery, type 1LM, on continuous and intermittent discharge are summarized in Figs. 10.6 to 10.8 and Table 10.2. The batteries were discharged under a constant-resistance load at 20°C.

Figure 10.6 provides a summary of the battery's performance under continuous load to 1.1 V end voltage.

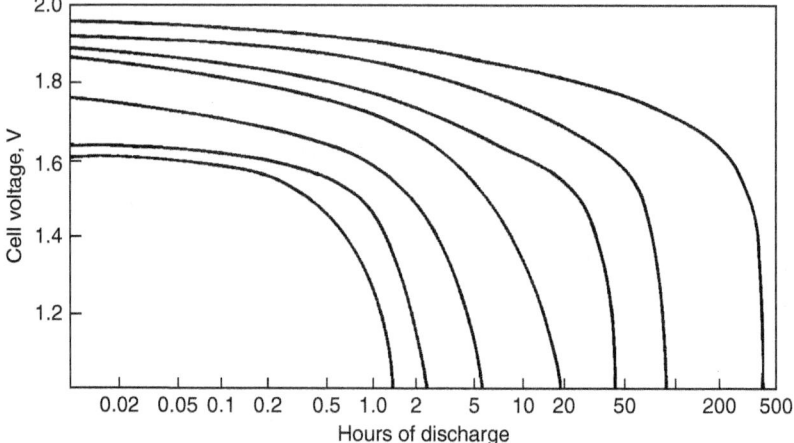

FIGURE 10.5 Typical discharge curves of magnesium/manganese dioxide cylindrical battery at 20°C.

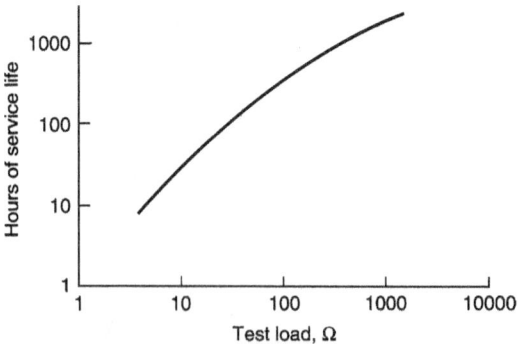

FIGURE 10.6 1LM service life (hours to 1.1 V) vs. test load at room temperature. (*Courtesy of Rayovac Corporation.*)

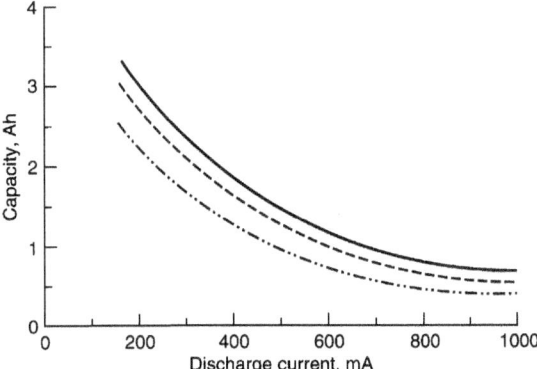

FIGURE 10.7 1LM service life (ampere-hours) vs. constant-current discharge. Dotted line—1.4V end voltage; dashed line—1.2 V end voltage; solid line—1.0 V end voltage. (*Courtesy of Rayovac Corporation.*)

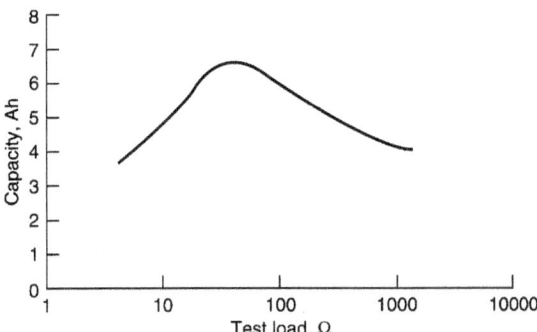

FIGURE 10.8 1LM service life (ampere-hours to 0.8 V) vs. test load. (*Courtesy of Rayovac Corporation.*)

Figure 10.7 shows the relationship of discharge current to delivered ampere-hour capacity of the battery on continuous constant-current discharge to several end voltages. The intermittent discharge characteristics are illustrated in Table 10.2. The sizable reduction in performance under low-rate or long-term discharge is attributed to the corrosion reaction between the discharging magnesium anode and the cell electrolyte. The reaction, which results in the evolution of hydrogen and the concomitant reduction of water, causes a loss of total cell efficiency. This phenomenon is illustrated in Fig. 10.8, which summarizes the ampere-hour output under continuous discharge of the 1LM cell to an 0.8 V end voltage. This loss of capacity on the low-rate (high-resistance) discharges is evident.

The performance of the magnesium primary battery at low temperatures is also superior to that of the zinc-carbon battery, operating to temperatures of −20°C and below. Figure 10.9 shows the performance of the magnesium battery at different temperatures based on the 20-h discharge rate. The low temperature performance is influenced by the heat generated during discharge and is dependent on the discharge rate, battery size, battery configuration, and other such factors. Actual discharge tests should be performed if precise performance data are needed.

On extended low-rate discharges, the magnesium battery may split open. This rupture is due to the formation of magnesium hydroxide, which occupies about one and one-half times the volume of the magnesium. It expands and presses against the cathode mix, which has hardened appreciably

TABLE 10.2 Performance, in Hours, of 1LM Batteries on Continuous and Intermittent Discharge

	End voltage	
Type of discharge	1.1 V	0.8 V
4 ohms, continuous	8.9	9.9
4 ohms, LIFT*	10.7	11.6
4 ohms, HIFT†	11	12
4 ohms, 30 min/h, 8 h/day	9.72	10.60
25 ohms constant resistance		
Continuous	100	104
4 h/day	84.2	88.4
500 ohms constant resistance		
Continuous	1265	1312
4 h/day	752	776

*Light industrial flashlight test, 4 min/h, 8 h/day.
†Heavy industrial flashlight test, 4 min/15 mm, 8 h/day.
Source: Rayovac Corporation.

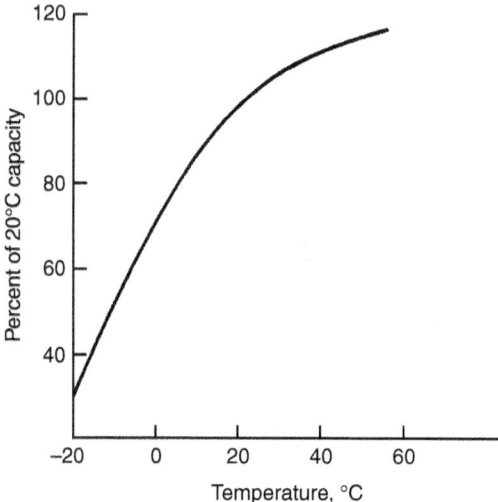

FIGURE 10.9 Performance vs. temperature of magnesium/manganese dioxide cylindrical battery.

from the loss of water during the discharge. This opening of the can causes the voltage to rise about 0.1 V, also increasing capacity due to the air that can enter into the reaction.

The service life of the magnesium/manganese dioxide primary battery, normalized to unit weight (kilogram) and volume (liter), at various discharge rates and temperatures is summarized in Fig. 10.10. The data are based on a rated performance of 60 Ah/kg and 120 Ah/L.

10.4.2 Shelf Life

The shelf life of the magnesium/manganese dioxide primary battery at various storage temperatures is compared with the shelf life of the zinc-carbon battery in Fig. 10.11. The magnesium battery is noted for its excellent shelf life. The battery can be stored for periods of 5 years or longer at 20°C with a total capacity loss of 10 to 20% and at temperatures as high as 55°C with losses of about 20%/year.

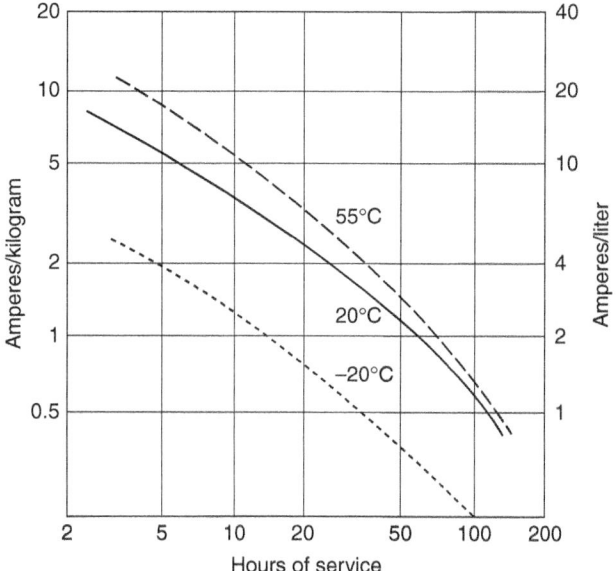

FIGURE 10.10 Service life of magnesium/manganese dioxide primary battery at various discharge rates and temperatures (to 1.2 V/cell end voltage).

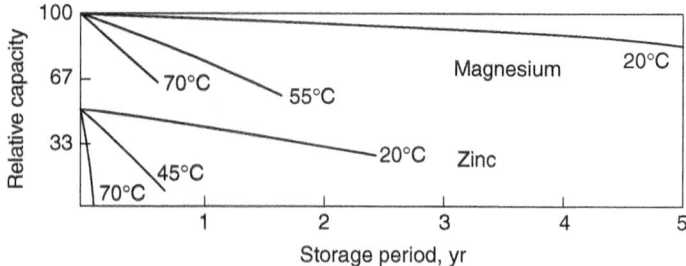

FIGURE 10.11 Comparison of service vs. storage of magnesium/manganese dioxide and zinc-carbon batteries.

10.4.3 Inside-Out Cells

The discharge characteristics of the cylindrical inside-out magnesium primary batteries are shown in Fig. 10.12 for various discharge rates and at 20°C and −20°C. This structure has better high-rate and low-temperature performance than the conventional structure. These batteries can be discharged at temperatures as low as −40°C, although at lighter discharge loads at the lower temperatures. Discharge curves are characteristically flat. They also have good and reproducible low-drain, long-term discharge characteristics as they do not split under these discharge conditions. Discharges for a 2½-year duration are realized with a D-size battery at a 270-μA drain at 20°C.

10.4.4 Battery Design

Battery configuration has an important influence on the performance of the magnesium/manganese dioxide battery because of the heat generated during the discharge. As discussed in Sec. 10.2, proper

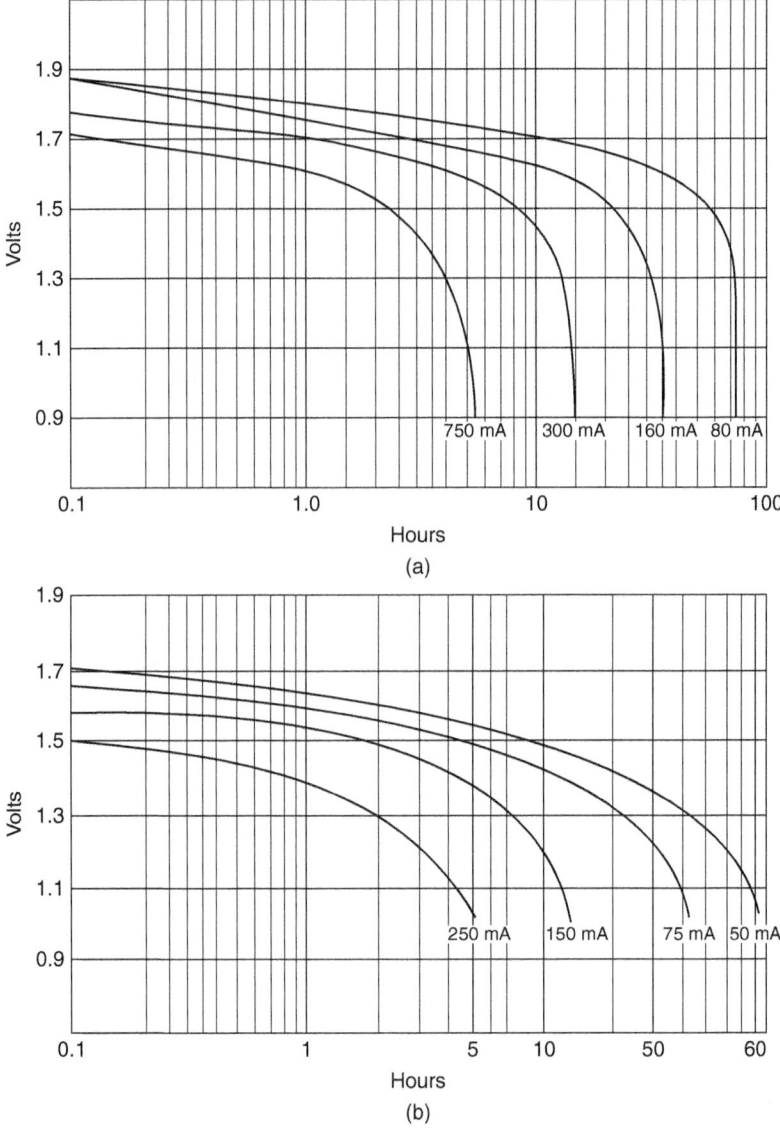

FIGURE 10.12 Typical discharge curves of magnesium inside-out primary battery, D size. (a) 20°C. (b) −20°C. (*Courtesy of ACR Electronics, Inc.*)

battery design must allow for the dissipation of this heat to prevent overheating, premature dry-out, and shortened performance—or for using this heat to improve performance at low ambient temperatures. In some low-temperature applications it is advantageous to insulate the battery against heat loss. Actual discharge tests will be required to obtain precise performance data under a variety of possible conditions and battery designs.

The battery and equipment design must also consider the hydrogen that is generated during discharge. The hydrogen must be vented and kept from accumulating because hydrogen-air mixtures are flammable above 4.1% hydrogen and explosive above 18% hydrogen.

TABLE 10.3 Cylindrical Magnesium Prmary Batteries

Battery type	Diameter, mm	Height, mm	Weight, g	Capacity, Ah[*]	
				Conventional structure[†]	Inside-out cell[‡]
N	11.0	31.0	5	0.5	
B	19.2	53.0	26.5	2.0	
C	25.4	49.7	45	—	3.0
1LM	22.8	84.2	59	4.5	
D	33.6	60.5	105	—	7.0
FD	41.7	49.1	125	—	8.0
No. 6	63.5	159.0	1000	—	65

[*]50-h discharge rate.
[†]Manufacturer: Rayovac Corp.
[‡]Manufacturer: ACR Electronics, Inc., Hollywood, FL (no longer manufactured).

10.5 SIZES AND TYPES OF Mg/MnO$_2$ BATTERIES

The cylindrical magnesium/manganese dioxide batteries were manufactured in several of the popular standard ANSI sizes, as summarized in Table 10.3. Most of the production of the conventional battery was used for military radio transceiver applications, and mainly in the 1LM size.[12,13] The batteries are no longer available commercially. Inside-out batteries are no longer manufactured.

10.6 OTHER TYPES OF MAGNESIUM BATTERIES

Magnesium primary batteries have been developed in other structures and with other cathode materials, but these designs have not achieved commercial success. Flat cells, using a plastic-film envelope, were designed but were never produced commercially.

The use of organic depolarizers, such as *meta*-dinitrobenzene (*m*-DNB), in place of manganese dioxide was of interest because of the high capacity that could be realized with the complete reduction of *m*-DNB to *n*-phenylenediamine (2 Ah/g). The discharge of actual batteries, while having a flat voltage profile and a higher ampere-hour capacity than the manganese dioxide battery, had a low operating voltage of 1.1 to 1.2 V per cell. Watthour energies were not significantly higher than for the magnesium/manganese dioxide batteries. The *m*-DNB battery also was inferior at low temperatures and high current drains. Commercial development of these batteries never materialized.

Magnesium/air batteries were studied, again because of the higher operating voltage than with zinc (see Chap. 33). These batteries, too, were never commercialized. Magnesium, however, is a very useful anode in reserve batteries. Its application in these types of batteries is covered in Chap. 34.

Magnesium rechargeable batteries, although not yet in commercial production, are undergoing development and characterization.[14,15] One approach that is being evaluated uses a pure magnesium anode, a suitable organic electrolyte such as tetrahydrofuran (THF), a suitable active salt such as Mg(butylAlCl$_3$)$_2$, and an intercalating cathode such as Mg$_x$Mo$_6$S$_8$ (x = 0-2). This battery system has a theoretical capacity of 122 mAh/g (based on cathode weight), an operating voltage of 1.1 V, and is reported to be capable of thousands of recharges with little capacity fade. This appears to rival the specific energy of the lead-acid and Ni-Cd systems (~60 Wh/kg).[16,17] The rate capability of the system at present is low, but would support applications such as load leveling or solar support.[18]

Another approach being evaluated utilizes a pure magnesium anode, a suitable salt/electrolyte combination (such as 1 M C$_2$H$_5$MgF in diethyl ether), and a cathode capable of intercalating the magnesium ion easily (such as AgO).[19] This formulation, along with many others, provided

420 mAh/g and >90% capacity retention after 50 1 mA (0.42 mA/cm^2 of cathode area) constant-current discharge/charge cycles from 1 to 3 V.

10.7 ALUMINUM PRIMARY BATTERIES

Experimental work on Al/MnO$_2$ primary or dry batteries was concentrated on the D-size cylindrical battery using a construction similar to the one used for the Mg/MnO$_2$ battery (Fig. 10.3). The most successful anodes were made of a duplex metal sheet consisting of two different aluminum alloys. The inner, thicker layer was more electrochemically active, leaving the outer layer intact in the event of pitting of the inner layer. The cathode bobbin consisted of manganese dioxide and acetylene black, wetted with the electrolyte. Aqueous solutions of aluminum or chromium chloride, containing a chromate inhibitor, were the most satisfactory electrolytes.

Aluminum active primary batteries were never produced commercially. While the experimental aluminum batteries delivered a higher energy output than conventional zinc batteries, anode corrosion, causing problems on intermittent and long-term discharges and irregularities in shelf life, and the voltage-delay problem restrained commercial acceptance. Aluminum/air batteries are covered in Chap. 33.

REFERENCES

1. J. L. Robinson, "Magnesium Cells," in N. C. Cahoon and G. W. Heise (eds.), *The Primary Battery,* Vol. 2, Wiley-Interscience, New York, 1976, Chap. 2.
2. G. R. Hoey and M. Cohen, "Corrosion of Anodically and Cathodically Polarized Magnesium in Aqueous Media," *J. Electrochem. Soc.* **105**:245 (1958).
3. J. E. Oxley, R. J. Ekern, K. L. Dittberner, P. J. Spellman, and D. M. Larsen, "Magnesium Dry Cells," in *Proc. 35th Power Sources Symp.,* IEEE, New York, 1992, p. 18–21.
4. B. V. Ratnakumar and S. Sathyanarayana, "The Delayed Action of Magnesium Anodes in Primary Batteries. Part I: Experimental Studies," *J. Power Sources* **10**:219 (1983).
5. S. Sathyanarayana and B. V. Ratnakumar, "The Delayed Action of Magnesium Anodes in Primary Batteries. Part II: Theoretical Studies," *J. Power Sources* **10**:243 (1983).
6. S. R. Narayanan and S. Sathyanarayana, "Electrochemical Determination of the Anode Film Resistance and Double Layer Capacitance in Magnesium-Manganese Dioxide Cells," *J. Power Sources* **15**:27 (1985).
7. B. V. Ratnakumar, "Passive Films on Magnesium Anodes in Primary Batteries," *J. Appl. Electrochem.* **18**:268 (1988).
8. D. B. Wood, "Magnesium Batteries," in K. V. Kordesch (ed.), *Batteries,* Vol. 1: *Manganese Dioxide,* Marcel Dekker, New York, 1974, Chap. 4.
9. R. R. Balaguer and F. P. Schiro, "New Magnesium Dry Battery Structure," in *Proc. 20th Power Sources Symp.,* Atlantic City, NJ, 1966, p. 90.
10. R. R. Balaguer, "Low Temperature Battery (New Magnesium Anode Structure)," Report: ECOM-03369-F, 1966.
11. R. R. Balaguer, "Method of Forming a Battery Cup," U.S. Patent 3,405,013, 1968.
12. D. M. Larsen, K. L. Dittberner, R. J. Ekern, P. J. Spellman, and J. E. Oxley, "Magnesium Battery Characterization," in *Proc. 35th Power Sources Symp.,* IEEE, New York, 1992, p. 22.
13. L. Jarvis, "Low Cost, Improved Magnesium Battery, in *Proc. 35th Power Sources Symp.,* New York, 1992, p. 26.
14. P. Novak, R. Imhof, and O. Haas, "Magnesium Insertion Electrodes for Rechargeable Nonaqueous Batteries—a Competitive Alternative to Lithium?" *Electrochimica Acta* **45**, (September 1999).
15. D. Aurbach, Y. Gofer, Z. Lu, A. Schechter, O. Chusid, H. Gizbar, Y. Cohen, V. Ashkenazi, M. Moshkovich, R. Turgeman, and E. Levi, "A Short Review on the Comparison between Li Battery Systems and Rechargeable Magnesium Battery Technology," *J. Power Sources* **97–98**:119 (July 2001).

16. D. Aurbach, Y. Gofer, A. Schechter, L. Zhohdghua, and C. Gizbar, "High Energy, Rechargeable Electrochemical Cells with Nonaqueous Electrolytes," U.S. Patent 6,316,141, November 13, 2001.
17. N. Amir, Y. Vestfrid, O. Chusid, Y. Gofer, and D. Aurbach, "Progress in Nonaqueous Magnesium Electrochemistry," *J. Power Sources* **174**:1234–1240 (December 2007).
18. D Aurbach, "Advances in R&D of Electrolyte Solutions for Recharging Batteries," Twenty-Sixth International Battery Seminar, Fort Lauderdale, FL, March 2009.
19. S. Ito, O. Yamamoto, T. Kanbara, and H. Matsuda, "Nonaqueous Electrolyte Secondary Battery with an Organic Magnesium Electrolyte Compund," U.S. Patent 6,713,213 B2, March 30, 2004.

CHAPTER 11
ALKALINE-MANGANESE DIOXIDE BATTERIES

John C. Nardi and Ralph J. Brodd

11.1 GENERAL CHARACTERISTICS

Within the primary battery category—that is, batteries that are used once and then discarded—the alkaline cell has grown to be the dominant battery system for use in portable devices. As the portable device market has increased, so has the market share of this battery system. Alkaline batteries have become the battery of choice in the United States and most other developed nations. While the alkaline cell, named for its use of a basic or alkaline electrolyte, was commercially introduced in 1959, it wasn't until the 1980s that it became widely recognized as being superior to the carbon-zinc type primary battery. A group led by Karl Kordesch and Lew Urry, employees of the Eveready Battery Co. (now Energizer®), is credited with being the inventors of the alkaline round cell, which Lew Urry patented. Because of its important contribution to society, in 1999, Lew Urry's original battery was placed in the Smithsonian National History Museum of American History in the same room as Edison's light bulb.

This superior performance is significantly manifested in the higher-drain devices like electronic toys, CD players, cameras, and remotes. Some of its advantages are listed in Table 11.1. However, these advantages have come at a price, i.e., a higher cost compared to the carbon-zinc battery.

The alkaline cell, otherwise known as the alkaline-manganese dioxide or zinc-manganese dioxide battery ($Zn/KOH/MnO_2$), is produced in two different cell types, the cylindrical type and the miniature button style. In addition, multiple alkaline cells are made into multiple-cell configurations such as the 9 V battery. Demand for the alkaline-MnO_2 cell was about $12 billion worldwide in 2008 and is expected to increase annually over the next several years due to the demand for battery-powered devices, especially for smaller, thinner, and lighter portable devices. In addition, the major battery companies, such as Energizer®, Duracell®, and Rayovac® in the United States have developed a product mix within the alkaline cell primary category shifting toward more powerful batteries for high-tech devices. Some of these companies also provide alkaline brand grades that include their economy or value alkaline, designed for long-lasting service in devices with low to moderate drain rates. Such applications include radios, remote controls, and clocks. A historical trend in the alkaline low-to-moderate performance is shown in Fig. 11.1.

The standard alkaline cell is designed for the widest range of device applications, which include photoflash, games, CDs, tape players, lighting, toys, remote controls, and clocks. Lastly, their premium cell will deliver superior performance for high-tech devices, such as digital cameras, photoflash, games, CDs, and tape players. This high-tech performance increase over a 7-year period is shown in Fig. 11.2.

11.2 PRIMARY BATTERIES

TABLE 11.1 Major Advantages of Alkaline-Manganese Dioxide Battery Compared to Carbon-Zinc Battery

Higher energy density
Superior service performance at all drain rates, higher capacity
Superior cold-temperature performance
Lower internal resistance
Longer shelf life
Less leakage

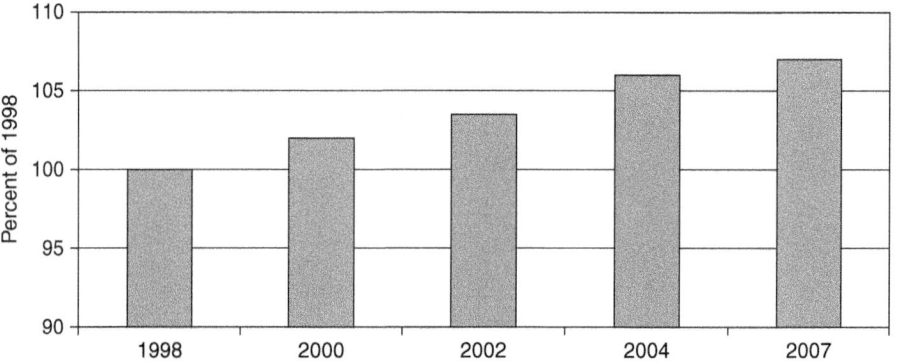

FIGURE 11.1 Historical trend in alkaline low-to-moderate rate performance. (*Courtesy of Energizer, Inc.*)

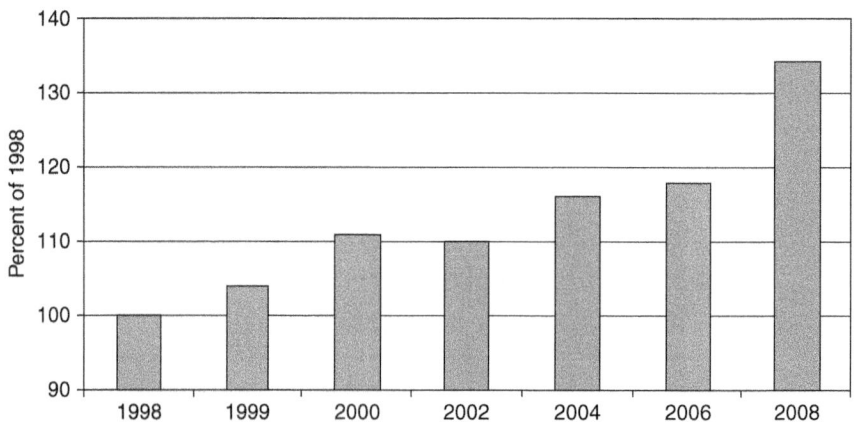

FIGURE 11.2 Historical trend in alkaline high-tech performance (*Courtesy of Energizer, Inc.*)

However, in recent years, the alkaline cell system has new primary competition in powering portable devices, including the Panasonic® Oxyride battery, launched in 2005, and the Duracell® Powerpix battery, which are both based on the nickel oxyhydroxide addition to the cathode formulation. These batteries are claimed to last twice as long as the typical alkaline cell and provide sufficient power for the higher-drain devices. Details about this new system will be discussed in a later section. In addition,

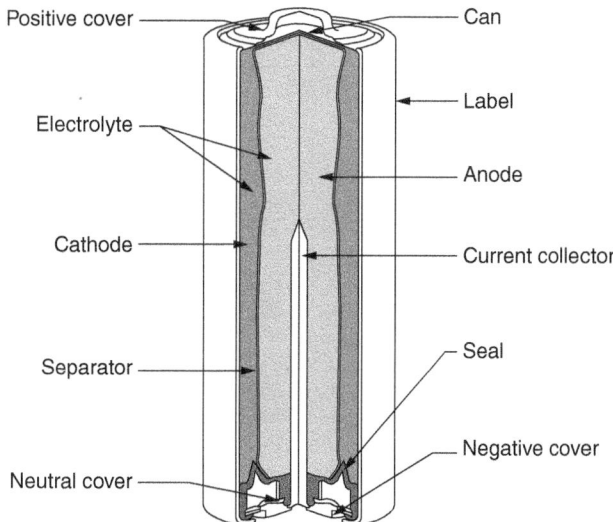

FIGURE 11.3 Schematic view of typical alkaline cell construction. (*Courtesy of Energizer, Inc.*)

the demand for primary batteries could be limited by competition from rechargeable batteries, such as the nickel-metal hydride, and the primary 1.5 V lithium chemistries advertised to last up to 7 times longer than regular alkaline batteries.

Improvements in performance (Fig. 11.2) have resulted in the increased sales shown in Fig. 11.1 due to a combination of improved materials, design, and chemistry. The battery manufacturers have responded to the higher power and constant current drains required by the new portable devices being commercialized over the years. When compared to the carbon-zinc system, the alkaline cell is built inside-out and upside-down, as shown in Fig. 11.3. The positive electrode mix, electrolytic manganese dioxide, graphite, and potassium hydroxide electrolyte, is molded into a steel can whose bottom is at the top of the pictured cutaway. A paper separator basket or two strips are inserted, and a potassium hydroxide gel containing powdered zinc is dispensed into the basket. The electrolyte also includes an inhibitor to mitigate zinc corrosion and ensure long shelf life. A negative collector assembly consisting of a brass nail and a plastic seal are inserted, making contact with the zinc gel. A flat cover is then placed over the open part of the can and crimped shut, becoming the negative end of the cell. The steel can bottom, the positive contact, is also provided with a cover, sometimes having a center dimple, which forms the positive end of the finished cell.

During the evolution of the alkaline cell over the past 50 years, many improvements have been made to this cell design. After the initial concept of using a gelled zinc powder anode and the use of a vented plastic seal, the first major advance was the butt-seam metal finish, which provided an increased internal volume. This was followed by the discovery that the addition of organic inhibitors to the anode could reduce the rate of gassing caused by impurities or contaminants in the zinc anode, which resulted in a product with bulge and leakage problems. Another major development was the introduction of a plastic label and lower profile seal, which even further increased the cell's internal volume that allowed the addition of more active materials and thus a greater discharge capacity. One of the most important developments of the alkaline cell in the 1980s was the gradual reduction in the amount of added mercury in the anode. Most early alkaline cells contained up to 6% mercury in the zinc anode, but with the development of cathode materials with lower impurities and better processing techniques, the level of added mercury was gradually reduced to zero. This objective of removing all added mercury was driven by the worldwide concern over the environmental impact of the cell components after their disposal. Today, most countries have banned batteries that contain mercury.

11.4 PRIMARY BATTERIES

TABLE 11.2 Comparison of Advantages and Disadvantages of Miniature Alkaline-Manganese Dioxide Cell to Other Miniature Systems

Advantages	Disadvantages
Lower cost	Sloping discharge
Lower internal resistance	Lower energy density
Good low-temperature performance	Shorter shelf life
Equivalent leakage resistance	

These improvements in materials and construction have allowed the alkaline-MnO$_2$ battery to gain as much as a 60 to 70% increase in specific energy output since its first introduction. These improvements have allowed alkaline batteries to keep pace with the needs of consumers and their demand for smaller and higher energy devices. The continuing research by the major battery companies will provide further technological improvements to ensure their leadership roles in the market.

As already mentioned, the miniature alkaline cell uses the same zinc/alkaline electrolyte/manganese dioxide configuration as the cylindrical cells. It competes with the other miniature cell battery systems, such as silver oxide and zinc/air as well as lithium-based chemistries. The major uses of this cell configuration are in watches, hearing aids, and specialty items. This cell consists of a shallow steel can that holds the cathode and serves as the positive contact and a copper-clad steel cover containing a potassium hydroxide gel with zinc powder that is the negative contact. Table 11.2 lists the advantages and disadvantages of the alkaline-manganese dioxide miniature battery compared to other lithium-based miniature systems.

11.2 CHEMISTRY

The active components of the alkaline-manganese dioxide cell include powdered zinc, an aqueous KOH electrolyte, and electrolytically produced manganese dioxide. The electrolytic MnO$_2$ or EMD is used instead of either chemical MnO$_2$ or the natural ore because of its higher manganese content, increased activity, and higher purity. The KOH electrolyte is a high-purity concentrated solution typically in the range of 35 to 52%, which provides a high conductivity and reduced gassing rate for the sealed alkaline cells for the various device applications and storage conditions. The zinc powder anode provides a high surface area for high-rate capability, i.e, low local current density, and facilitates the homogeneous distribution of the solid and liquid phases in the anode compartment to minimize concentration polarization of the reactants and products.

During discharge, the manganese dioxide cathode first undergoes a one-electron reduction to the oxyhydroxide with expansion and distortion of its lattice in concentrated alkaline electrolytes.

$$MnO_2 + H_2O + e \rightarrow MnOOH + OH^- \tag{11.1}$$

The MnOOH product forms a solid solution with the reactant, which produces the characteristic sloping discharge.[1] Of the many structural forms of MnO$_2$ that exist, only the gamma form has the best alkaline discharge characteristics, because its surface is not prone to being blocked by the reaction product. Manganese dioxide has been identified to have at least nine crystal structures, one of which is the gamma form, known in nature as nsutite, an intergrowth of the beta or pyrolusite form and ramsdellite. It is this structurally disordered form of manganese dioxide that is found in alkaline cells. It is composed of the 1 × 1 tunnel structure of pyrolusite and the 1 × 2 tunnel form of the ramsdellite, as depicted in Fig. 11.4.[2] Table 11.3 shows the different manganese oxide structures.[3]

The cathode expands about 17% in volume when forming the MnOOH reaction product. MnOOH can also undergo some undesirable chemical side-reactions, depending on the conditions and extent of the discharge. In the presence of the zincate ion, MnOOH, based on its equilibrium with soluble Mn(III), can form the complex compound hetaerolite, ZnMn$_2$O$_4$. Although electroactive, hetaerolite is not as easily discharged as MnOOH, and results in an increased cell impedance. In addition, the

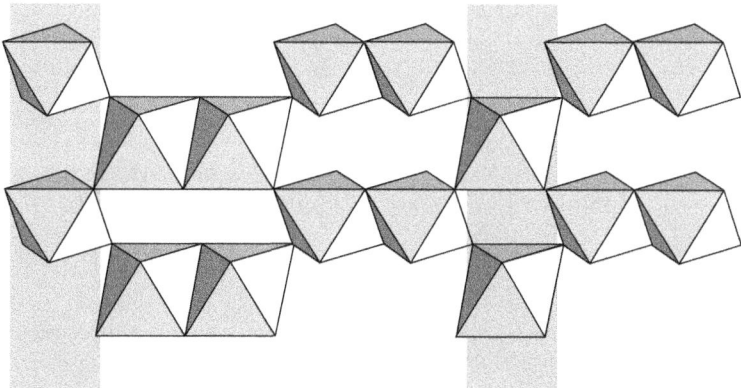

FIGURE 11.4 Schematic representation of intergrowth between pyrolusite and ramsdellite lattices.

TABLE 11.3 Different Manganese Oxide Structures

Mineral	Space Group	Z	a(Å)	b(Å)	c(Å)	β, γ(°)	Reference
Pyrolusite (β)	P4$_2$/mnm	2	4.398	—	2.873	90	Baur, 1976
Ramsdellite	Pbnm	4	4.533	9.27	2.866	90	Byström, 1949
Nsutite (γ)	[intergrowth]	4	4.45	9.305	2.85	90	De Wolff, 1959
Birnessite	P ml	1	2.84	—	7.27	120	Giovanoli et al., 1970
ε-MnO$_2$	P6$_3$/mmc	1	2.80	—	4.45	120	De Wolff et al., 1978
Spinel (λ)	Fd3m	16	8.029	—	—	90	Mosbah et al., 1983
Hollandite (α)	I2/m	2	10.026	2.8782	9.729	91.03	Post et al., 1982
Psilomelane	C2/m	2	13.929	2.8459	9.678	92.39	Turner and Post, 1988
Todorokite	P2/m	8	9.764	2.8416	9.551	94.06	Post and Bish, 1988
Manganite (γ)	B2$_1$/d	8	8.88	5.25	5.71	90	Dachs, 1963, 1973
Groutite (α)	Pbnm	4	4.560	10.7	2.87	90	Glasser and Ingram, 1968
Groutellite	[Pbnm]	4	4.7	9.531	2.864	90	JCPDS 42-1316
Feitkneichtite (β)	P ml	1	3.32	—	4.71	120	Feitnecht et al., 1962
Pyrochroite	P ml	1	3.322	—	4.734	120	Christensen, 1965

MnOOH/MnO$_2$ solid solution can undergo recrystallization into a less active form, resulting in a noticeable loss of cell voltage under certain very slow discharge conditions.[4]

Overall, the first half or more of the MnO$_2$ discharge reaction is a simple proton-electron insertion reaction with no structural change except for expansion and distortion of its lattice. Toward the end of the first electron discharge, however, the reaction proceeds through a soluble Mn^{3+} species along with a variety of Mn^{3+} and Mn^{2+} intermediate products, depending on the discharge conditions.

During discharge at the lower voltages, the MnOOH can be further discharged, as depicted by the following equation:

$$3MnOOH + e \rightarrow Mn_3O_4 + OH^- + H_2O \tag{11.2}$$

This reaction produces a flat discharge curve, but it only occurs under low-rate discharge conditions. No additional volume change occurs during this reaction in the cathode. This step only provides about one-third of the capacity of the first MnO$_2$ reaction. During deep discharge, a further reduction to Mn(OH)$_2$ is possible but seldom occurs.

The following reactions provide a more detailed alkaline cathode reaction scenario.

$$MnO_2 + xH_2O + xe \rightarrow MnOOH_x + xOH^- \qquad (0<x<\sim0.6) \qquad (11.3)$$

With other products, for $x>\sim0.6$ via the soluble Mn^{3+} species of MnOOH, $Mn(OH)_2$, Mn_3O_4, and $ZnMn_2O_4$ are produced.

During the early stage of the cell's discharge, the anode reaction in KOH produces the soluble zinc ion [Reaction (11.4)] which can be found in the separator and cathode

$$Zn + 4OH^- \rightarrow Zn(OH)_4^{-2} + 2e \qquad (11.4)$$

At a certain point in the discharge, depending on the composition of the anode and the rate and depth of discharge, the electrolyte will become saturated with zincate that then causes the reaction product to change to the insoluble $Zn(OH)_2$. Eventually, the anode will become depleted of water and the zinc hydroxide dehydrates to ZnO by the following two reactions:

$$Zn + 2OH^- \rightarrow Zn(OH)_2 + 2e \qquad (11.5)$$

$$Zn(OH)_2 \rightarrow ZnO + H_2O \qquad (11.6)$$

These changes in the different zinc discharge products cannot be easily noted in the discharge curve since the standard reaction potentials for Reactions (11.5) and (11.6) are very similar. However, under certain conditions, the formation of the oxide can be sufficiently high that it passivates any undischarged zinc. Such conditions would include high-rate, low-temperature, and poor electrolyte conductivity. These concerns are typically mitigated by the use of high surface area zincs in order to minimize any cell impedance increase by the anode.

The overall total one-electron reaction of the alkaline cell during a continuous discharge is as follows:

$$2MnO_2 + Zn + 2H_2O \rightarrow 2MnOOH + Zn(OH)_2 \qquad (11.7)$$

Since water is a reactant in Reaction (11.7), the amount of water in a cell is quite important, especially in high-rate discharge applications. Therefore, the total water management in a cell is an important variable that battery manufacturers must control in order to provide good performance over a wide range of discharge conditions. Some battery manufacturers have included additives to the cell, such as TiO_2 and $BaSO_4$, in order to better manage this important characteristic. Also, there appear to be many different ZnO morphologies that could affect the anode's performance.

However, at the low- or intermittent-drain rates, the total cell reaction for 1.33 electrons per mole is

$$3MnO_2 + 2Zn \rightarrow Mn_3O_4 + 2ZnO \qquad (11.8)$$

This reaction clearly indicates that under such conditions there is no water management concern.

The open-circuit voltage of an undischarged alkaline cell is typically between 1.55 and 1.65 V, depending on the purity and activity of the cathode components, the ZnO content of the anode, and the storage temperature of the cell.

Due to the natural corrosion activity of zinc metal in a basic solution, it can reduce water and form hydrogen gas. Such a reaction does occur in the alkaline cell and reduces the overall cell capacity (zinc corrosion) if allowed to become significant. Hydrogen gas evolution can occur during long-term storage of undischarged cells or after partial discharge. The amount of gas formed during the latter event depends on the discharge rate, delivered capacity, and storage temperature. This gas buildup in a cell can cause the cell to bulge and eventually leak. In addition, the formed hydrogen gas can reduce the manganese dioxide, even further reducing the cell's available capacity.

While the hydrogen evolution rate of pure zinc is low, the inevitable presence of impurities, i.e., traces of ppm levels, promotes the rate of gassing by acting as cathodic sites on the zinc. This gassing can be reduced or minimized in several of the following ways: (1) alloying the zinc with known gassing inhibiting elements, (2) reducing the impurity levels of the cell components, (3) adding ZnO to the anode, and (4) adding inorganic or organic gassing inhibitors (e.g., PEG, polyethylene glycol) to the anode. Mercury is the best and most efficient inhibitor, but due to its toxic nature and the international drive for green chemistry, has been banned for use in alkaline cells worldwide.

Alloying elements are incorporated into the zinc to both inhibit gassing and improve performance. The main group of such elements includes lead, bismuth, thallium, and indium. The levels of these elements have been empirically determined based on the performance required by the battery manufacturer.

As already mentioned, the impurity levels in the zinc should be as low as possible, with most of them coming from the natural ore or during processing. Their levels are typically determined by the efficiency of the zinc powder process.

Another important aspect of the zinc reaction in alkaline solution is its equilibrium with its ion in solution. The zinc metal anode is in equilibrium with the zinc ions in the KOH solution. Therefore, on stand, the zinc electrode reaction is continually dissolving and redepositing in the anode compartment of the cell. It has been shown by detailed SEM studies that both type I and type II zinc oxides are present in the discharged zinc particles in the porous anode. The morphology of these two different zinc types appears to depend on the drain rate, i.e., the current density, of the zinc during discharge in a cell and is produced by a solution-precipitation reaction. While most of the zinc particles are discharged, the core of undischarged zinc is covered with these two oxides. At low drain rates, the distribution of ZnO is uniformly distributed within the anode, while at high drain rates, the ZnO mainly forms near the separator.[5,6]

11.3 CELL COMPONENTS AND MATERIALS

11.3.1 Cathode Components

The typical composition of an alkaline cathode and its functions are listed in Table 11.4. The cathode is basically a mixture of manganese dioxide and carbon (typically graphite), binder (typically Portland cement or polymers), along with cell electrolyte. However, other materials may be also present, such as binders, additional water, and/or more electrolyte solution.

Manganese Dioxide. The manganese dioxide is the positive electrode of the cell or the oxidizing component. It must be highly active and very pure as it basically determines the battery's OCV and shape of the discharge curve during use.

The production of high-quality EMD basically involves multiple steps over a period of time. The natural manganese dioxide dug out of the ground is first calcined to form MnO. The MnO is then dissolved in sulfuric acid to produce a manganese sulfate solution. This solution then goes through an electrolyte purification step to remove most of the deleterious heavy metal impurities, e.g., Fe, Cu, Co, Ni, Mo, Cr. This purified solution is then placed in an electrolysis cell and heated

TABLE 11.4 Composition of Typical Alkaline Cell Cathode

Component	Percent of Cathode (%)	Function
Manganese Dioxide	80–90	Active material
Carbon	2–10	Electronic conductor
KOH	7–10	Ionic conductor
Binder	0–1	Maintain cathode integrity

to almost boiling. The typical electrolysis cell of most EMD manufacturers consists of a titanium anode and copper cathode. However, in the past, lead and graphite cathodes have also been employed. The anode reaction to deposit solid MnO_2 alkaline cell product is as follows:

$$Mn^{2+} + 2H_2O \rightarrow MnO_2 + 4H^+ + 2e \tag{11.9}$$

Hydrogen is formed at the cathode according to Reaction (11.10):

$$2H^+ + 2e \rightarrow H_2\uparrow \tag{11.10}$$

Thus the overall reaction for the plating of EMD is

$$MnSO_4 + 2H_2O \rightarrow MnO_2 + H_2SO_4 + H_2 \tag{11.11}$$

The important plating variables to produce an EMD suitable for use in alkaline cells require precise control of the bath's temperature, current density, and component concentrations. Once the EMD is removed from the anode, it is crushed, washed, ground, and dried. Each battery manufacturer has its own EMD specification, so one type does not fill all requirements.

The analysis of a typical EMD is shown in Table 11.5. The low levels of impurities are essential in order to minimize the hydrogen gassing at the anode if these elements become soluble and diffuse to the anode. The other listed parameters are also important, and their listed ranges all go toward providing an EMD suitable for alkaline cell use.

Other important EMD characteristics include its surface area and hardness. The surface area, dictated by the porosity and particle size distribution, determines the current density in the cathode, which is especially important for high-rate discharge applications. EMD is typically a very hard material, and this hardness affects the milling of the EMD and tool wear of the equipment used to make the cathode mixes and molding equipment. Premature tool and mill wear can introduce iron impurities into the pure EMD and add cost to the overall battery manufacturing process.

Carbon. EMD is a relatively poor conductor in its undischarged state and even worse when partially discharged. To overcome this problem, carbon, typically in the form of graphite, is added to the cathode mix to enhance its overall electronic conductivity. The graphite provides a conductive matrix so the electrons can be evenly distributed throughout the cathode, thus lowering the overall current density in the cathode. However, one must strike a balance between the amount of added carbon and the EMD level. Additional carbon provides a more conductive cathode matrix, but it reduces the amount of active material in the cathode. Therefore, the ratio of carbon to EMD in the cathode needs to be optimized for the required applications of the battery. Over the years, many changes

TABLE 11.5 Typical Analysis of Electrolytic Manganese Dioxide (EMD)

Component	Typical Value*	Component	Typical Value*
MnO_2 content	>91%	Ti	<5 ppm
Mn	>60%	Cr	<7 ppm
Peroxidation	>95%	Ni	<4 ppm
H_2O, 120C	<1.50%	Co	<2 ppm
H_2O, 120–400C	>3.0%	Cu	<4 ppm
Real density	4.45 g/cm³	V	<2 ppm
K	<300 ppm	Mo	<1 ppm
Na	<4000 ppm	As	<1 ppm
Mg	<500 ppm	Sb	<1 ppm
Fe	<100 ppm	Pb	<100 ppm
C	0.07%	SO_4^{2-}	≤0.85%

*Based on analyses of typical alkaline-grade EMD.

in the type of carbon added to the alkaline cell cathode have occurred. Natural graphites, synthetic graphites, acetylene black, and, most recently, expanded graphites have been used to improve the cathode conductivity. In all cases, this conductor must be pure so as not to add any more impurities to the cell. The expanded graphite allows less carbon to be used as its synthesis expands the graphite planes, while maintaining its conductivity within the carbon planes.[7] This graphite has a higher liquid absorption value, and the particle size can be optimized for the required cathode formulation.

Other Components. KOH and water are used to form the cathode electrolyte. They are added during the mixing of the cathode ingredients to form a moist paste. This makes the cathode mix easier to handle and mold. Depending on the battery manufacturer, other ingredients, such as binders and additives, are used to produce a dense and stable cathode with a good electronic and ionic conductivity. The battery must also perform efficiently under a variety of discharge conditions, including low and high continuous and intermittent discharge over a wide range of temperatures.

11.3.2 Anode Components

The anode is composed of a mixture of ingredients that allow for good cell performance and provide for easy manufacturing. The typical composition of an alkaline anode is listed in Table 11.6.

TABLE 11.6 Typical Composition of Alkaline Cell Anode

Component	Range (%)	Function
Zinc powder	60–70	Negative electrode material
Aqueous KOH (25–50%)	25–35	Ionic conductor
Gelling agent	0.4–1.0	Control viscosity
ZnO	0–2	Zinc plating; gassing suppressor
Surfactant/gassing inhibitor	0–0.1	Gassing suppressor; improves performance

Zinc Powder. Zinc is the electrochemically active component of the alkaline cell's negative electrode. The pure zinc that is acceptable for use in the alkaline cell is commercially obtained by either a thermal distillation process, i.e., thermal zinc, or by electrolytic deposition from an aqueous solution, i.e., electrolytic zinc. This zinc is converted to a powder by atomizing a thin stream of the molten metal by high-pressure compressed air. Depending on the setup and requirements, the particle shape of the obtained zinc can range from "potatoes" to "dog bones." Improvements in the process have allowed zinc manufacturers to better control the size and shape of the final zinc powder in order to meet the increasing demands of performance improvements and cost savings. Typical battery-grade zinc ranges in particle size from 20 to 500 microns in a log-normal distribution. This zinc is very pure, but alloying elements are added to better control the normal gassing that does occur in a basic electrolyte. Such metallic additives can include indium, lead, bismuth, and aluminum in varying ratios. Such additives have become very important since the intentional addition of mercury to the anode has been banned. A typical analysis of battery-grade zinc is listed in Table 11.7. While typical levels are shown, some battery-grade zincs have lower levels of impurities.

TABLE 11.7 Typical Impurity Analysis of Battery-Grade Zinc Powder

Element	Typical level (ppm)*
Ag	1.56
Al	.14
As	.01
Ca	.20
Cd	4.2
Co	.05
Cu	1.5
Cr	.10
Fe	4.0
Ni	.20
Mg	.03
Mo	.035
Sb	.09
Si	.20
Sn	.10
V	.001

*Based on analyses of typical alkaline-grade zinc powder.

Recent research in developing a zinc powder for high-rate discharge applications has resulted in the patenting of a blended zinc powder.[8] This blended powder contains selected portions of two different particle size powder distributions. Advantageously, it allows battery manufacturers to maximize an alkaline cell's performance while minimizing the cost of the zinc.

Anode Gel. The anode gel serves to suspend the zinc particles and to maintain them in contact with one another. Starch or cellulosic derivatives, polyacrylates, and/or ethylene maleic anhydride copolymers continue to be used as anode gelling agents. Common gelling agents can include sodium carboxymethyl cellulose or the sodium salt of an acrylic acid copolymer. Typically the selected gel is well mixed with the zinc powder and any other additives prior to dispensing into the anode cavity of the cell. As with the other cell components, these materials must also be of high purity in order to minimize gassing. This is especially true of the carbonate, chloride, and iron levels. Depending on the cell's primary application, the volume fraction range of the gelling agent can vary. The lower limit is based on maintaining good electronic conductivity in the anode, while the upper limit is defined by limiting the accumulation of reaction products that could eventually passivate the undischarged zinc and hinder ionic diffusion within the anode. A recent patent has suggested the use of the crosslinked polymer polyvinylbenzyltri-methylammoniumhydroxide.[9] It is claimed that the use of this gelling agent allows better high-rate discharge performance.

Anode Collector. The anode collector used in the alkaline cell has typically been a high-purity cartridge brass, but silicon bronzes have also been used. The collector in most current designs has a pin or nail shape, but strip type collectors have been used in the past. The collector is part of the collector assembly that also consists of the seal and cover. Once the collector is inserted into the anode gel, its surface becomes rapidly coated with zinc, thus acting more like a zinc electrode than brass. This provides good electronic contact with the zinc particles and suppresses gassing in the anode due to any impurities in the brass that could be gassing promoters. In order to provide a rapid zinc plating, the brass collector can undergo a special cleaning or surface coating. One such patented method involves the electroplating of the collector wire with indium that forms an indium-plated wire with a thickness of about 0.1 to 10 microns.[10] This coating reduces the amount of gassing that may occur in the cell, especially in mercury-free alkaline cells, prior to the collector becoming fully coated with zinc.

Separators. The separator insulates the cathode electronically from the anode. However, it must be ionically conductive along with being chemically stable in the concentrated alkaline electrolyte under both oxidizing and reducing conditions while also being strong, flexible, uniform, impurity-free, and very absorptive. There are many ways to produce such a material, but the more frequently used material is a nonwoven or felt-like material. The typical separator material can consist of cellulose, vinyl polymers, polyolefins, or any combination. Depending on the battery manufacturer, the separator can consist of two inserted cross strips or a preformed "convolute" separator basket. Other types of separators that have not proven as successful include gelled, inorganic, and radiation-grafted separators. A cellophane separator has also been used typically if there is a concern about zinc dendrite growth through the separator. A recent patent claimed the use of a reinforced separator that can withstand the forces applied during manufacture and contain any fragmented electrode particles formed when the cell is dropped.[11]

Containers, Seals, and Finish. The can or external container of the alkaline cell, unlike the carbon-zinc can, does not take part in the discharge reaction. It is merely an inert container that provides an external contact for the positive electrode. The can is typically made of mild steel that is thick enough to maintain its shape during discharge as the cathode is known to expand and hydrogen gas can form during storage or discharge, creating internal pressure. Over the years, the can thickness has been reduced in order to provide more internal space for the active battery materials.[12] The can is formed by the deep drawing of a steel strip.

The can materials must also be of high purity as the can does contact the cathode. Depending on the cell construction, the interior contact can be the steel itself or it can be treated to improve the cathode-to-can contact. Such a treatment of the steel could be a nickel plating or coating with a conductive carbon paint. Such treatments are typically for those cells used specifically for high-rate applications.

The seal is typically a plastic material, such as nylon or polypropylene. It is combined with several metal parts that include the brass collector and cover to form the collector assembly. It seals off the open end of the cylindrical can. After it is crimped in place, it prevents leakage of the electrolyte from the cell and provides electrical insulation between the can (positive electrode) and the anode (negative electrode).

The remaining components of the alkaline cell include the label and metal disks at either end of the cell to provide the negative and positive electrode contacts. The label on most current alkaline cell designs is a plastic label that is heat-shrunk onto the can. This has been another recent design feature that allows the can to be larger in diameter, which results in a greater input capacity.

Miniature Cell. The components that make up the miniature cell are essentially the same as those of the cylindrical cell, but scaled down. The can that contains the molded cathode pellet is typically made of mild steel plated on both sides with nickel. Some can designs are even made of a triclad metal material. The seal is a thin plastic gasket, and the separator is the typical nonwoven material. The anode cup, which contains the anode gel, is then pressed into the seal to complete the cell. The outside of the cell contains the manufacturer's logo and cell-size identification.

11.4 CONSTRUCTION

11.4.1 Cylindrical Design

Figure 11.3 is a partial cutaway drawing of a typical alkaline battery that is representative of most alkaline batteries currently being produced. Figure 11.5 shows one method for assembling the battery. The cylindrical steel can is the container for the cell and serves as the current collector for

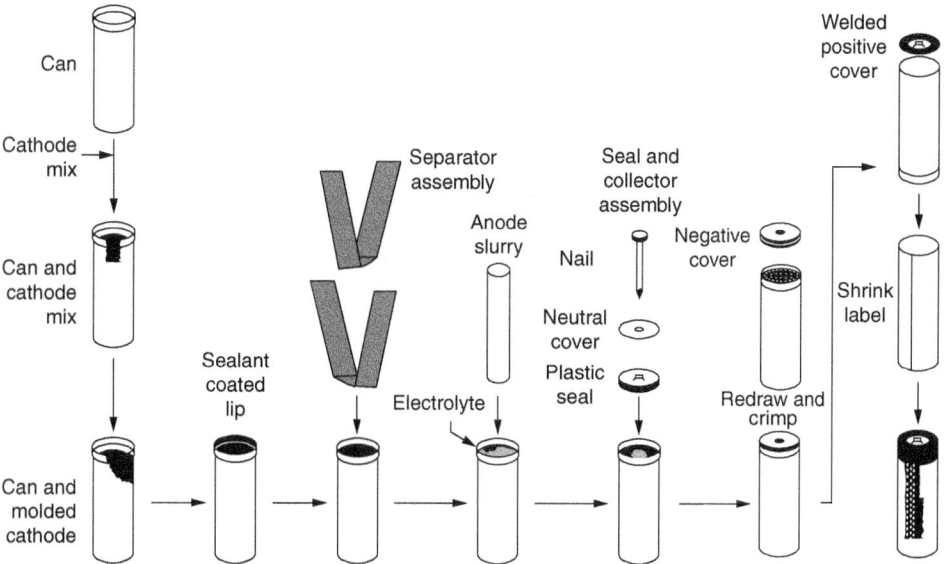

FIGURE 11.5 Assembly process for cylindrical alkaline-manganese dioxide cell. (*Courtesy of Energizer, Inc.*)

the cathode. The cathode mix can be added to the can in two ways, depending on the battery manufacturer's equipment setup. In one process, a prescribed amount of compacted mixture of manganese dioxide, carbon, and other additives are added to the can. The cathode is then directly molded into the can by inserting a ram down the center of the mix. This compacts the cathode mix to the required solids packing and provides a good can-to-cathode contact. The second method, and currently the most common one, is that the cathode mix is first formed into cylindrical rings outside of the cell, then three or four such rings are inserted into the can. Next, separator strips are inserted into the hollow cavity of the rings using either the two-strip method or a convolute separator. Once the separator is in place, the required amount of anode gel is added based on precise calculations, whereas the input capacity of the anode is close to or slightly greater than that of the cathodes. This required ratio is to prevent excessive gassing if and when the cell is deeply discharged. Next, the collector assembly is added and the can crimped over it to provide a leakproof seal. The brass collector pin now provides the external contact for the negative electrode. The cell now has top and bottom covers added to it and then the plastic label is applied. The covers serve a twofold purpose. In addition to providing a label with the manufacturer's distinctive artwork, it also indicates the polarity of the cell. This ensures that the customer will properly position the battery in a device for proper operation. This becomes increasingly important for a device that takes multiple batteries as one inserted backwards could be charged by the others, causing the cell to prematurely leak. In order to prevent this, some manufacturers have also added a reversal protection, which consists of an insulating ring on the positive end of the cell.

Figure 11.6 shows the more typical process of alkaline cell cathode production. Molded cathode rings are inserted into the can, a ram compacts them against the can, and the cell is finished in the normal way.

With research always trying to increase the performance of the alkaline cell, some novel innovations regarding the anode and cathode have recently been patented. One involves the use of zinc ribbons, which allows for a significantly increased high-rate performance.[13] Another involves the formation of a cathode having lobes or a sinusoidal cathode design.[14] This increases the surface area of the cathode, thus decreasing the current density, which allows for a better high-rate performance. Another recent patent, regarding the manganese dioxide itself, discloses better performance using a particulate MnO_2 simultaneously having a micropore surface area of greater than 8.0 m^2/g and BET surface area between 20 and 31 m^2/g.[15]

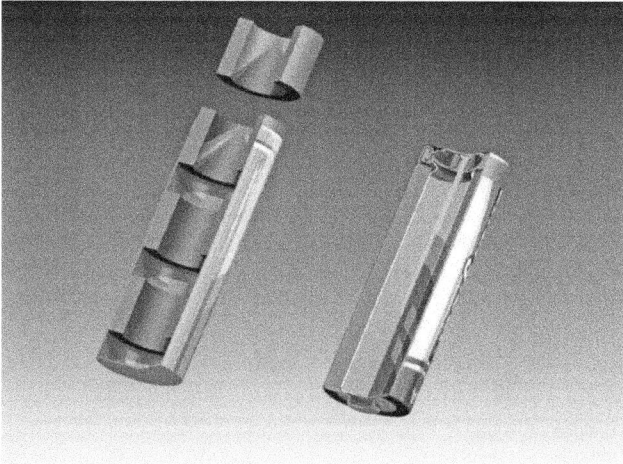

FIGURE 11.6 Illustration showing insertion of four cathode rings in can and finished cell. (*Courtesy of Energizer, Inc.*)

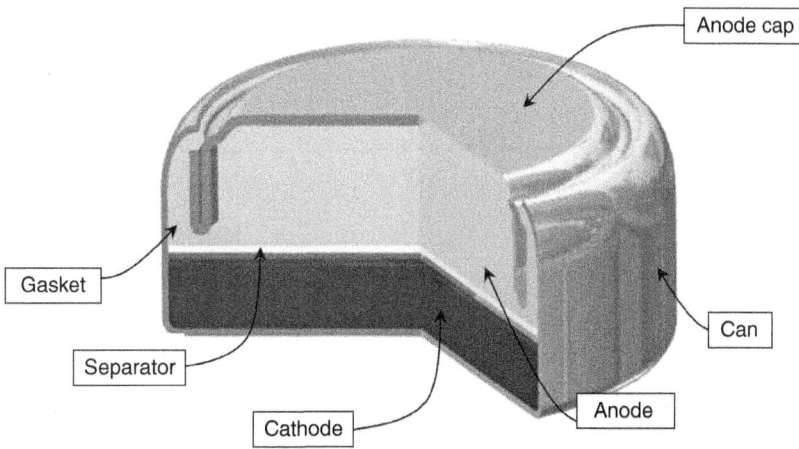

FIGURE 11.7 Cross-section illustration of miniature alkaline-manganese dioxide battery. (*Courtesy of Energizer, Inc.*)

11.4.2 Miniature Cell Configuration

A cross-sectional view of a miniature alkaline-manganese dioxide cell is shown in Fig. 11.7. Its construction is similar to that of the cylindrical cell. There is a bottom cup that holds the cathode pellet, followed by round disks of separator paper, the plastic seal, and the top cover containing the anode mix. The top cover is pressed into the plastic seal, providing a leak-proof seal to prevent leakage.

There are five common sizes of miniature alkaline cells; their dimensions are listed in Table 11.8.

TABLE 11.8 Dimensions of Miniature Alkaline Cells

		Dimensions (mm)					
		A/B		M	N	Φ	
Designation	Voltage	Max.	Min.	Min.	Min.	Max.	Min.
LR41	1.5	3.6	3.3	3.0	3.8	7.9	7.55
LR55	1.5	2.1	1.85	3.8	3.8	11.6	11.25
LR54	1.5	3.05	2.75	3.8	3.8	11.6	11.25
LR43	1.5	4.2	3.8	3.8	3.8	11.6	11.25
LR44	1.5	5.4	5.0	3.8	3.8	11.6	11.25

A/B = cell height; M = diameter of flat negative contact; N = diameter of flat positive contact; Φ = diameter of the cell.
Source: IEC Int'l. Standard Part 2: Primary Batteries, 11th edition, 2006.

11.4.3 Battery Types and Sizes

The alkaline-manganese dioxide chemistry comes in a variety of cell sizes dictated by the many different portable devices and their discharge characteristics. Figure 11.8 shows the cylindrical cell sizes currently using the alkaline-manganese dioxide cell chemistry. While most cans are quite common and easily recognized, there are several that are not so common, e.g., the A, B, F, and G cell

11.14 PRIMARY BATTERIES

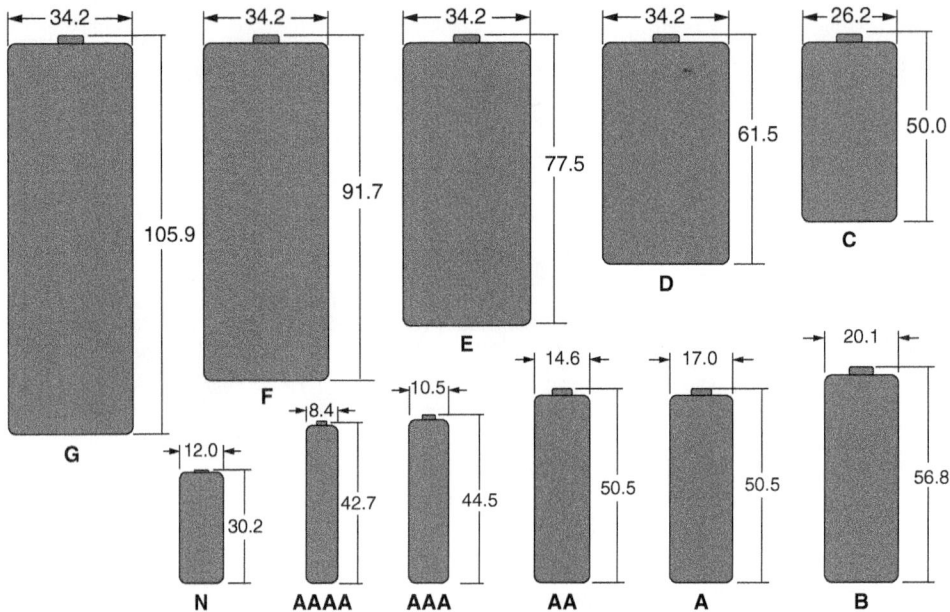

FIGURE 11.8 Typical alkaline cell sizes and their dimensions in millimeters.

sizes. As an example, the B size is commonly used in bicycle lights in Europe, while the F has been used in lantern batteries.

Recognizing the need for smaller, thinner, and lighter device applications, the AAAA(4A) premium alkaline battery has been recently commercially introduced. While initially having limited distribution, it is now becoming more popular with the sales of smaller, more portable electronic devices. It is basically thinner than either the AA or AAA cells. Its current applications include Bluetooth™ devices, flash audio players, remote controls, and noise-canceling headsets.

11.4.4 Testing Standards

With the many different battery sizes and the significant number of battery manufacturing companies worldwide, it is important to establish standard tests and designation codes for cylindrical and miniature primary cells so they can be easily compared to actual devices that they would power. This has been accomplished by the American National Standards Institute (ANSI). The current testing standards are outlined in their current publication, *ANSI C18.1M, Part 1-2008, American National Standard for Portable Primary Cells and Batteries with Aqueous Electrolyte–General and Specifications*. For a brief history of the standardization of these cells, refer to Ref. 16. As the number and type of new devices become commercialized, this organization regularly meets to determine if their current testing requirements need to be updated to reflect the latest device requirements. The designations for the different battery sizes are generally known by letters, i.e., D, C, AA, AAA, AAAA, etc. In addition, battery manufacturers also use their own designations that can be found on the battery and its packaging. More information about the individual manufacturer's battery specifications and performance is available on their individual websites.[17] The IEC (International Electrotechnical Commission) and ANSI also have their own designations. Table 11.9 lists some of the current tests that several of the more common cylindrical alkaline batteries undergo for advertising purposes, along with their different codes.

TABLE 11.9 Designations and Typical ANSI Tests for Cylindrical Alkaline Cells

Size	IEC Designation	ANSI Designation	Test	Load	Cycle	End Voltage (V)	Min. Avg. Duration
D	LR20	13A	Portable stereo	600 ohms	2 h on, 22 h off	0.9	11 h
			Portable lighting	2.2 ohms	4 min on, 56 min off; 8 h on, 16 h off cycle	0.9	15.8 h
			Toy	2.2 ohms	1 h on, 23 h off	0.9	17.5 h
			Radio	10 ohms	4 h on, 20 h off	0.9	90 h
C	LR14	14A	Portable stereo	400 mA	2 h on, 22 h off	0.9	8 h
			Portable lighting	3.9 ohms	4 min on, 56 min off; 8 h on, 16 h off cycle	0.9	13.8 h
			Toy	3.9 ohms	1 h on, 23 h off	0.8	14.5
			Radio	20 ohms	4 h on, 20 h off	0.9	85 h
AA	LR6	15A	Digital camera	1500 mW, 650 mW	1st load for 2 s, then 2nd load for 28 s; 5 min on, 55 min off	1.05	50 pulses
			Toothbrush	500 mA	2 min on, 13 min off; 24 hours	0.8	2.5 h
			CD	250 mA	1 h on, 23 h off	0.9	6 h
			Toy	3.9 ohms	1 h on, 23 h off	0.8	5 h
			Remote control	24 ohms	15 s on, 45 s off; 8 h on, 16 h off	1.0	33 h
			Radio/clock	43 ohms	4 h on, 20 h off	0.9	60 h
AAA	LR03	24A	Photoflash	600 mA	10 s on, 50 s off; 1 h on, 23 h off	0.9	170 pulses
			Portable lighting	5.1 ohms	4 min on, 56 min off; 8 h on, 16 h off cycle	0.9	2.2 h
			Digital audio	100 mA	1 h on, 23 h off	0.9	7.5 h
			Remote control	24 ohms	15 s on, 45 s off; 8 h on, 16 h off	1.0	14.5 h
AAAA	LR8	25A	Lighting	5.1 ohms	5 min on, 23 h and 55 min off	0.9	1.3 h
			Laser pointer	75 ohms	1 h on, 23 h off	1.1	22 h
N	LR1	910A	Portable lighting	5.1 ohms	5 min on, 23 h and 55 min off	0.9	1.6 h
			Pager	10 then 3K ohms	1st load for 5 s, then 2nd load for 3595 s; 24 h	0.9	888 h

11.4.5 Cell Leakage

Commercial alkaline cells are of a sealed construction using a compressible polymer grommet (seal) between the can and the top current collector. Leakage can occur by two different mechanisms: (1) nonelectrochemical leakage resulting from manufacturing defects and/or poor cell design which can occur at either electrode (called either positive or negative leakage, depending on whether it occurs between the seal and can or at the collector, respectively), and (2) electrochemical-related leakage associated with leakage that occurs only at the negative electrode.

Most manufacturers use a surface active coating such as asphalt, polyimide, etc., in the seal area to smooth out surface imperfections in the cell parts to provide a better seal. The amount of seal compression should not exceed the elastic limit of the grommet. Details of the cell design, choice of grommet material, as well as manufacturing processes play a role in cell leakage. Factors that influence leakage include the material of the can, e.g., nickel, stainless steel, or gold, the choice of

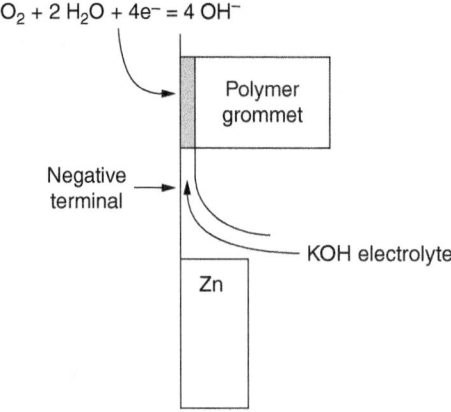

FIGURE 11.9 Illustration of possible leakage path in a cylindrical alkaline cell.

polymer gasket material, as well as the choice of the anode gelling agent and possible damage to the seal area during cell manufacture.

Electrochemical-related leakage occurs only at the negative electrode. Here, leakage, or creepage of the electrolyte, occurs on the negative electrode of the cell and is evident by "salting" or the appearance of white crystals at the negative terminal only. As shown in Fig. 11.9, oxygen ingress into the seal area is reduced to OH^- at the air-electrolyte interface, increasing the local concentration of OH^- in the seal area. The reaction of oxygen with the zinc anode has the same effect. The increased concentration between the creepage film compared to the concentration of the bulk cell electrolyte gives rise to an osmotic pressure difference. This osmotic force can reach several atmospheres and will force electrolyte into the seal area under high pressure. Grommet materials with a high water absorption, such as nylon, readily transmit water into the reaction zone in the seal area as opposed to a vinylidine chloride-acrylonitrile (Saran) which has a low water transmission rate. The choice of gelling agent in the zinc electrode may also play a role by improving the wetting, which facilitates movement of water into the reaction zone. Cells stored in dry nitrogen have lower leakage rates than do cells with polished can walls.

The moist environment between the can wall and the grommet forms a reaction zone for the reduction of oxygen to produce OH^- ions. This increases the hydroxide concentration locally, causing a difference in osmotic pressure between the bulk electrolyte and the electrolyte in the reaction zone. This drives the cell electrolyte out through the seal area where it reacts with CO_2 in the atmosphere to produce the white precipitate, potassium carbonate, e.g., K_2CO_3.[18,19] This situation is particularly disconcerting as it provides the customer with an immediate visual clue that such a cell is defective even prior to purchase.

11.5 EVOLTA™ AND OXYRIDE™ BATTERIES

Competition in the battery business has always been very keen. Most improvements by alkaline battery manufacturers have come in the form of improved active materials, an increased internal volume, and/or reduced internal resistance. These improvements in Ah capacity, longer storage life, and high-rate discharge have not been noticed as the outside dimensions have not changed over the years. Significantly different alkaline-based batteries are seldom introduced. However, in 2004 when Panasonic® first introduced their Oxyride™ battery to the Japanese market, they claimed an impressive improvement over the current alkaline-manganese dioxide battery for use in high-power applications. Panasonic® literature states that this cell includes a finer-grained graphite and manganese dioxide, allowing for a denser fill of material. The positive electrode also contains nickel oxyhydroxide, allowing the battery to maintain a higher operating voltage. These cells also utilize a "vacuum-pouring" technology during the production process, allowing more electrolyte to be packed into each battery for increased durability. This same chemistry is used in the Duracell® PowerPix™ battery. These batteries are for use in digital devices that require more power, however, they do cost more than the normal alkaline-manganese dioxide cells. These cells have a higher open-circuit voltage of 1.7 V compared to the alkaline-manganese dioxide cell's typical open-circuit voltage of 1.60 to 1.65 V. This higher voltage could cause problems, especially with devices containing an incandescent light bulb or no voltage regulator.

Panasonic® launched their new AA Evolta™ product in Japan in April 2008 and in the United States in May 2008. The name Evolta is derived from the words "evolution" and "voltage," and this cell has laid claim to being the first of its kind to have the *Guinness Book of World Records* officially certify it as the "world's longest-lasting AA alkaline battery cell" along with having a reported shelf life of 10 years. It derives its increased performance from the addition of a titanium oxyhydroxide to the cathode, a thinner can, thinner seal, and increased collector length. The discharge performance of this cell can be seen by visiting Panasonic's® website.[17]

REFERENCES

1. A. Kozawa and R.A. Powers, *J. Chem. Educ.* **49**, 587, 1972.
2. R. Burns and V. Burns, *Manganese Dioxide Symposium,* Vol. 1, Cleveland, p. 306, 1975.
3. Y. Chabre and J. Pannetier, *Prog. Solid St. Chem.* **23**, 12, 1995.
4. D. M. Holton, et al., in *Proc. 14th International Power Sources Symposium*, Brighton, England, Pergamon, NY, 1984.
5. Q. C. Horn and Y. Shao-Horn, *J. Electrochem. Soc.*, **150**(5), A652, 2003.
6. R. W. Powers and M. Brieter, *J. Electrochem. Soc.*, **116**, 1652, 1952.
7. J. C. Nardi, U.S. Patent 6,828,064, Dec. 7, 2004.
8. D. Fan, U.S. Patent 7,364,819, April 29, 2008.
9. C. Robert, U.S. Patent 6,916,577, July 14, 2005.
10. D. Mihara, U.S. Patent 5,622,612, April 22, 1997.
11. R. Janmey, U.S. Patent 6,828,061, Dec. 7, 2004.
12. R. Ray, U.S. Patent 6,855,454, Feb. 15, 2005.
13. N. C. Tang, U.S. Patent 6,221,527, April 24, 2001.
14. P. J. Slezak, U.S. Patent 6,869,727, March 22, 2005.
15. S. Davis, U.S. Patent 6,863,876, March 8, 2005.
16. ANSIC18 Committee Doc. 18/382/DOC/, Nov. 21, 2002.
17. www.energizer.com; www.duracell.com; www.rayovac.com; www.sanyo.com; www.panasonic.com; www.varta.com.
18. M. N. Hull and H. I. James, *J. Electrochem. Soc.*, **124**, 332, 1977.
19. S. M. Davis and M. N. Hull, *J. Electrochem. Soc.*, **125**, 1918, 1978.

CHAPTER 12
MERCURIC OXIDE BATTERIES

Nathan D. (Ned) Isaacs

12.1 GENERAL CHARACTERISTICS

The alkaline zinc/mercuric oxide battery is noted for its high capacity per unit volume, constant voltage output, and good storage characteristics. The system has been known for over a century, but it was not until World War II that a practical battery was developed by Samuel Ruben in response to a requirement for a battery with a high capacity-to-volume ratio that would withstand storage under tropical conditions.[1,2]

Since that time, the zinc/mercuric oxide battery has been used in many applications where stable voltage, long storage life, or high energy-to-volume ratios were required. The characteristics of this battery system were particularly advantageous in applications such as hearing aids, watches, cameras, some early pacemakers, and small electronic equipment, where it was widely used. The battery has also been used as a voltage reference source and in electrical instruments and electronic equipment, such as sonobuoys, emergency beacons, rescue transceivers, radio and surveillance sets, small scatterable mines, and early satellites. These applications, however, did not become widespread, except for military and special uses, because of the relatively higher cost of the mercuric oxide system.

The use of cadmium in place of zinc results in a very stable battery with excellent storage and performance at extreme temperatures due to the low solubility of cadmium in caustic alkali over a wide range of temperatures. However, the cost of the material is high and the cell voltage is low, less than 1.0 V. Hence, the cadmium/mercuric batteries were used, but to a lesser degree, in special applications requiring the particular performance capabilities of the system. These include gas and oil well logging, telemetry from engines and other heat sources, alarm systems, and for operation of remote equipment, such as data-monitoring, surveillance buoys, weather stations, and emergency equipment.[3]

The market for mercuric oxide batteries has evaporated due to environmental problems associated with mercury and cadmium. The 1996 Mercury-Containing and Rechargeable Battery Management Act (P.L. 104-142) prohibits the sale of mercuric oxide batteries in the United States unless manufacturers provide for recycling and disposal. Furthermore, these batteries have been removed from the International Electrotechnical Commission (IEC) and the American National Standards Institute (ANSI) standards. In applications, they have been replaced by alkaline-manganese dioxide, zinc/air, zinc/silver oxide, and lithium batteries.

The major characteristics of these two battery systems are summarized in Table 12.1.

12.2 PRIMARY BATTERIES

TABLE 12.1 Characteristics of the Zinc/Mercuric Oxide and Cadmium/Mercuric Oxide Batteries

Advantages	Disadvantages
Zinc/mercuric oxide battery	
High energy-to-volume ratio, 450 Wh/L	Batteries were expensive; although widely used in miniature sizes, but only for special applications in the larger sizes
Long shelf life under adverse storage conditions	
Over a wide range of current drains, recuperative periods are not necessary to obtain a high capacity from the battery	After long periods of storage, cell electrolyte tends to seep out of seal, which is evidenced by white carbonate deposit at seal insulation
High electrochemical efficiency	Moderate energy-to-weight ratio
High resistance to impact, acceleration, and vibration	Poor low-temperature performance
Very stable open-circuit voltage, 1.35 V	Batteries for disposal are considered hazardous wastes under environmental regulations
Flat discharge curve over wide range of current drains	
Cadmium/mercuric oxide battery	
Long shelf life under adverse storage conditions	Batteries are more expensive than zinc/mercuric oxide batteries due to high cost of cadmium
Flat discharge curve over wide range of current drains	
Ability to operate efficiently over wide temperature range, even at extreme high and low temperatures	System has low output voltage (open-circuit voltage = 0.90 V)
Can be hermetically sealed because of inherently low gas evolution level	Moderate energy-to-volume ratio
	Low energy-to-weight ratio
	Batteries for disposal are considered hazardous wastes under environmental regulations

12.2 CHEMISTRY

It is generally accepted that the basic cell reaction for the zinc/mercuric oxide cell is

$$Zn + HgO \rightarrow ZnO + Hg$$

For the overall reaction, $\Delta G^0 = 259.7$ kJ. This gives a thermodynamic value for E^0 at 25°C of 1.35 V, which is in good agreement with the observed values of 1.34 to 1.36 V for the open-circuit voltage of commercial cells.[4] From the basic reaction equation it can be calculated that 1 g of zinc provides 819 mAh and 1 g of mercuric oxide provides 247 mAh.

Some types of zinc/mercuric oxide cells exhibit open-circuit voltages between 1.40 and 1.55 V. These cells contain a small percentage of manganese dioxide in the cathode and are used where voltage stability is not of major importance for the application.

The basic cell reaction for the cadmium/mercuric oxide cell is

$$Cd + HgO + H_2O \rightarrow Cd(OH)_2 + Hg$$

For the overall reaction, $\Delta G^0 = -174.8$ kJ. This gives a thermodynamic value for E^0 at 25°C of 0.91 V, which is in good agreement with the observed values of 0.89 to 0.93 V. From the basic reaction it can be calculated that 1 g of cadmium should provide 477 mAh.

12.3 CELL COMPONENTS

12.3.1 Electrolyte

Two types of alkaline electrolyte were used in the zinc/mercuric oxide cell, one based on potassium hydroxide and one on sodium hydroxide. Both of these bases are very soluble in water and highly concentrated solutions were used; zinc oxide was also dissolved in varying amounts in the solution to suppress hydrogen generation.

Potassium hydroxide electrolytes generally contain between 30 and 45% w/w KOH and up to 7% w/w zinc oxide. They were more widely used than the sodium hydroxide electrolytes because of their greater operating temperature range and ability to support heavier current drains. For low temperature operation, both the potassium hydroxide and the zinc oxide contents were reduced, and this introduced some instability at higher temperatures with respect to hydrogen generation in the cell.

Sodium hydroxide electrolytes were prepared in similar concentration ranges and were used in cells where low temperature operations or high current drains were not required. These electrolytes were suitable for long-term discharge applications because of the reduced tendency of the electrolyte to seep out of the cell seal after long periods of storage.

Generally only potassium-based alkaline electrolytes were used in the cadmium/mercuric oxide cell. As cadmium is practically insoluble in all concentrations of aqueous potassium hydroxide solutions, the electrolyte could be optimized for low-temperature operation.

The freezing-point curve for caustic potash solutions is shown in Fig. 12.1. It shows that the eutectic with a freezing point below −60°C is 31% w/w KOH, which was the electrolyte most frequently used. Improvements in low-temperature performance have been made in some cases by the addition of a small percentage of cesium hydroxide to the electrolyte.

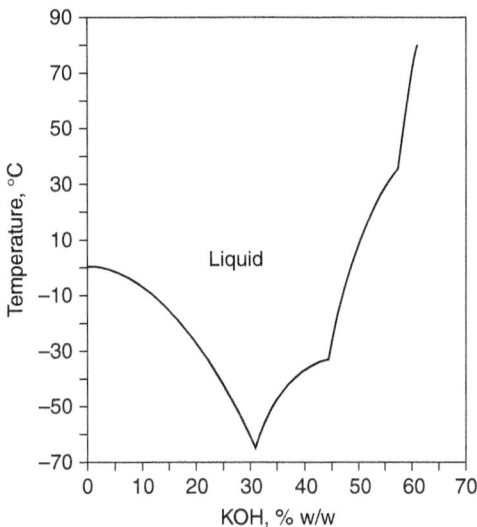

FIGURE 12.1 Freezing-point curve for aqueous caustic potash solutions.

12.3.2 Zinc Anode

Alkaline electrolytes act as ionic carriers in the cell reactions. The reaction at the zinc negative electrode may be written

$$Zn + 4OH^- \rightarrow Zn(OH)_4^{2-} + 2e$$

$$Zn(OH)_4^{2-} \rightarrow ZnO + 2OH^- + H_2O$$

These reactions imply the dissolution of the zinc electrode, with the crystallization of zinc oxide from the electrolyte. The reaction at the anode can be simplified to

$$Zn + 2OH^- \rightarrow ZnO + H_2O + 2e$$

12.4 PRIMARY BATTERIES

Direct solution of the zinc electrode in the alkaline solution on open circuit is minimized by dissolving zinc oxide in the electrolyte and amalgamating the zinc in the electrode. Mercury levels used in zinc electrodes were usually in the range of 5 to 15% w/w. Great attention was also paid to the impurity levels in the zinc, since minor cathodic inclusions in the electrode can drive the hydrogen generation reaction despite the precautions indicated.[5,6]

12.3.3 Cadmium Anode

The reaction at the anode is

$$Cd + 2OH^- \rightarrow Cd(OH)_2 + 2e$$

This implies the removal of water from the electrolyte during discharge, necessitating an adequate quantity of electrolyte in the cell and the desirability of a high percentage of water in the electrolyte. Cadmium has a high hydrogen overvoltage in the electrolyte, and so amalgamation is neither necessary nor desirable, since the electrode potential is some 400 mV less electropositive than zinc.

Cadmium metal powders as produced conventionally were unsuitable for use as electrode materials. Activated cadmium anodes were produced by (1) electroforming the anode, (2) electroforming powder by a special process followed by pelleting, or (3) precipitating by a special process as a low-nickel alloy and pelleting. All of these processes have been used by different manufacturers to give cells with various performance parameters.[7]

12.3.4 Mercuric Oxide Cathode

At the cathode, the overall reaction may be written

$$HgO + H_2O + 2e \rightarrow Hg + 2(OH)^-$$

Mercuric oxide is stable in alkaline electrolytes and has a very low solubility. It is also a nonconductor, and adding graphite is necessary to provide a conductive matrix. As the discharge proceeds, the ohmic resistance of the cathode falls and the graphite assists in the prevention of mass agglomeration of mercury droplets. Other additives that have been used to prevent agglomeration of the mercury are manganese dioxide, which increases the cell voltage to 1.4 to 1.55 V, lower manganese oxides, and silver powder, which forms a solid-phase amalgam with the cathode product.

Graphite levels usually range from 3 to 10% and manganese dioxide from 2 to 30%. Silver powder was used only in special-purpose cells because of cost considerations, but may be up to 20% of the cathode weight. Again, great care was taken to obtain high-purity materials for use in the cathode. Trace impurities soluble in the electrolyte are liable to migrate to the anode and initiate hydrogen evolution. An excess of mercuric oxide capacity of 5 to 10% was usually maintained in the cathode to "balance" the cell and prevent hydrogen generation in the cathode at the end of discharge.

12.3.5 Materials of Construction

Materials of construction for the zinc/mercuric oxide cells were limited not only by their ability to survive continuous contact with strong caustic alkali, but also by their electrochemical compatibility with the electrode materials. External contacts were chosen for their corrosion resistance, galvanic compatibility with the equipment interface and, to some degree, cosmetic appearance. Metal parts may have been homogeneous, plated metal, or clad metal. Insulating parts may have been injection-, compression-, or transfer-molded polymers or rubbers.

With the exception of the anode contact (where slight modification of the top/anode interface was necessary), materials for the cadmium/mercuric oxide cell were generally the same as for the zinc/mercuric oxide cell. However, because of the wide range of storage and operating conditions of most applications, cellulose and its derivatives were not used, and low-melting-point polymers were also avoided. Nickel was usually used on the anode side of the cell and also, conveniently, at the cathode.

12.4 CONSTRUCTION

The mercuric oxide batteries were manufactured in three basic structures—button, flat, and cylindrical configurations. There were several design variations within each configuration.

12.4.1 Button Configuration

The button configuration of the zinc/mercuric oxide battery is shown in Fig. 12.2. The top is copper or copper alloy on the inner face and nickel or stainless steel on the outer face. This part may also be gold plated, depending on the application. Within the top is a dispersed mass of amalgamated zinc powder ("gelled anode"), and the top is insulated from the can by a nylon grommet. The whole top-grommet-anode assembly presses down onto an absorbent that contains most of the electrolyte, the remainder being dispersed in the anode and cathode. Below the absorbent is a permeable barrier, which prevents any cathode material from migrating to the anode. The cathode of mercuric oxide and graphite is consolidated into the can, and a sleeve support of nickel-plated steel prevents collapse of the cathode mass as the battery discharges. The can is made of nickel-plated steel, and the whole cell is tightly held together by crimping the top edge of the can as shown.

The cadmium/mercuric oxide button battery uses a similar configuration.

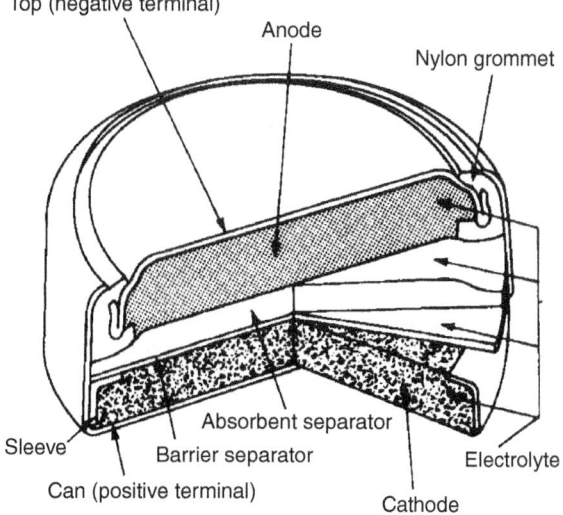

FIGURE 12.2 Zinc/mercuric oxide battery—button configuration. (*Courtesy of Duracell, Inc.*)

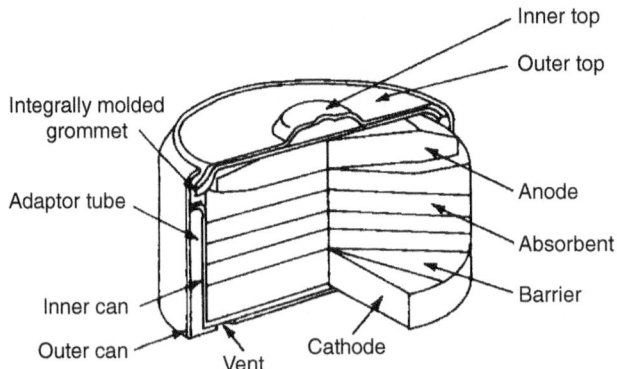

FIGURE 12.3 Zinc/mercuric oxide battery—flat-pellet configuration. (*Courtesy of Duracell, Inc.*)

12.4.2 Flat-Pellet Configuration

A form of a larger-sized zinc/mercuric oxide battery is shown in Fig. 12.3. In these cells, the zinc powder is amalgamated and pressed into a pellet with sufficient porosity to allow electrolyte impregnation. A double top is used, with an integrally molded polymer grommet as a safeguard to relieve excessive gas pressures and maintain a leak-resistant structure. The outer top is of nickel-plated steel, and the inner top is nickel-plated steel but tin plated on its inner face. This cell also uses two nickel-plated steel cans with an adaptor tube between the two, the seal being effected by pressing the top-grommet assembly against the inner can and crimping over the outer can. A vent hole is pierced into the outer can so that if gas is generated within the cell, it can escape between the inner and outer cans, any entrained electrolyte being absorbed by the paper adaptor tube.

12.4.3 Cylindrical Configuration

The larger cylindrical zinc/mercuric oxide battery is constructed from annular pressings, as shown in Fig. 12.4. The anode pellets are rigid and pressed against the cell top by the neoprene insulator slug. A number of variations of the cylindrical cell were used with dispersed anodes, where contact with the anode is made either by a nail welded to the inner top or a spring extending from the base insulator to the top.

12.4.4 Wound-Anode Configuration

Another design of the zinc/mercuric oxide battery that operates better at low temperatures is the wound-anode or jelly-roll structure shown in Fig. 12.5. Structurally the cell is similar to the flat cell shown in Fig. 12.3, but the anode and absorbent have been replaced by a wound anode, which consists of a long strip of corrugated zinc interleaved with a strip of absorbent paper. The paper edge protrudes at one side and the zinc strip at the other. This provides a large surface area anode. The roll is held in a plastic sleeve and the zinc is amalgamated in situ. The paper swells in the electrolyte and forms a tight structure, which is compressed in the cell at the assembly stage with the zinc edge in contact with the top.

Electrolyte formulations can be adjusted for low-temperature operation, long storage life at elevated temperature, or a compromise between the two. The performance is optimized by careful adjustment of the anode geometry.

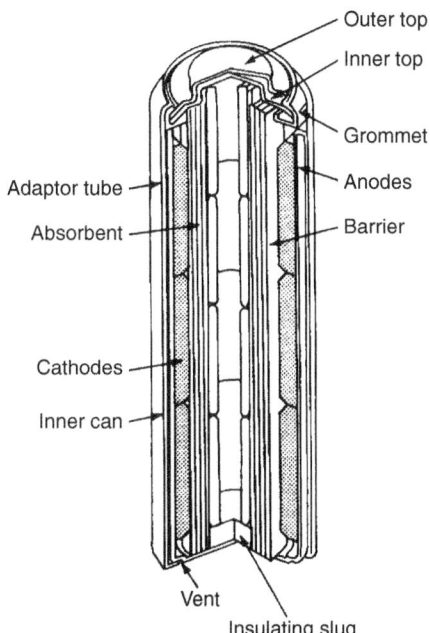
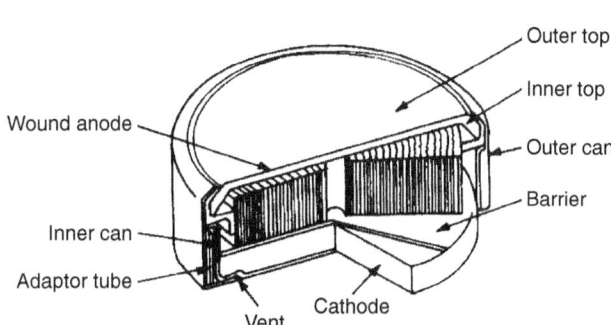

FIGURE 12.4 Zinc/mercuric oxide battery—cylindrical configuration. (*Courtesy of Duracell, Inc.*)

FIGURE 12.5 Zinc/mercuric oxide battery—wound-anode configuration. (*Courtesy of Duracell, Inc.*)

12.4.5 Low-Current-Drain Structures

Batteries designed for operation at low current drain require modification of the structure to prevent internal electrical discharge paths from forming from the conductive materials in both anode and cathode. After partial discharge, metallic mercury globules are particularly troublesome in this respect. The problem can be minimized by the use of silver powder in the cathode.

All available passages through which material could form an electrical track need to be blocked if long-term discharge is to be realized. A typical button battery for watch applications used multiple barrier layers and a polymer insulator washer, effectively sealing off the anode from the cathode by compressing these layers against the support ring. These batteries are discharged at the 1- to 2-year rate.[8]

12.5 PERFORMANCE CHARACTERISTICS OF ZINC/MERCURIC OXIDE BATTERIES

12.5.1 Voltage

The open-circuit voltage of the zinc/mercuric oxide battery is 1.35 V. Its voltage stability under open-circuit or no-load conditions is excellent, and these batteries have been widely used for voltage reference purposes. The no-load voltage is nonlinear with respect to both time and temperature. A voltage-time curve is shown in Fig. 12.6. The no-load voltage will remain within 1% of its initial value for several years. A voltage-temperature curve is shown in Fig. 12.7. Temperature stability is even better than age stability. From −20 to +50°C, the total no-load voltage range is in the region of 2.5 mV.

Batteries containing manganese dioxide as a cathode additive to the mercuric oxide do not show the no-load voltage stability illustrated in Figs. 12.6 and 12.7.

12.8 PRIMARY BATTERIES

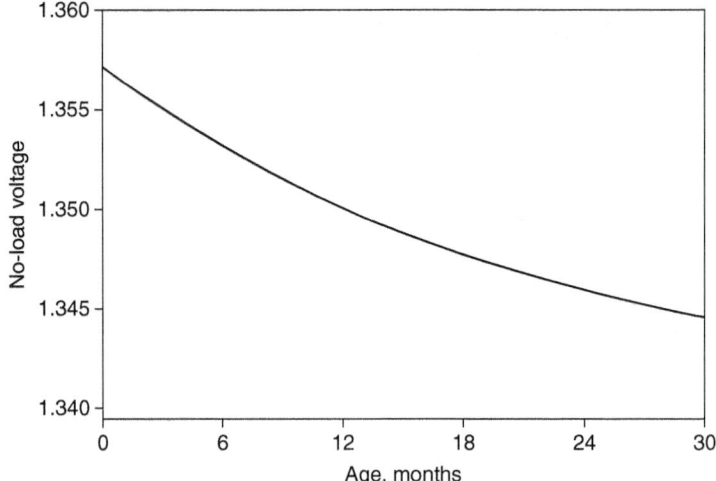

FIGURE 12.6 No-load voltage vs. time, zinc/mercuric oxide battery, 20°C.

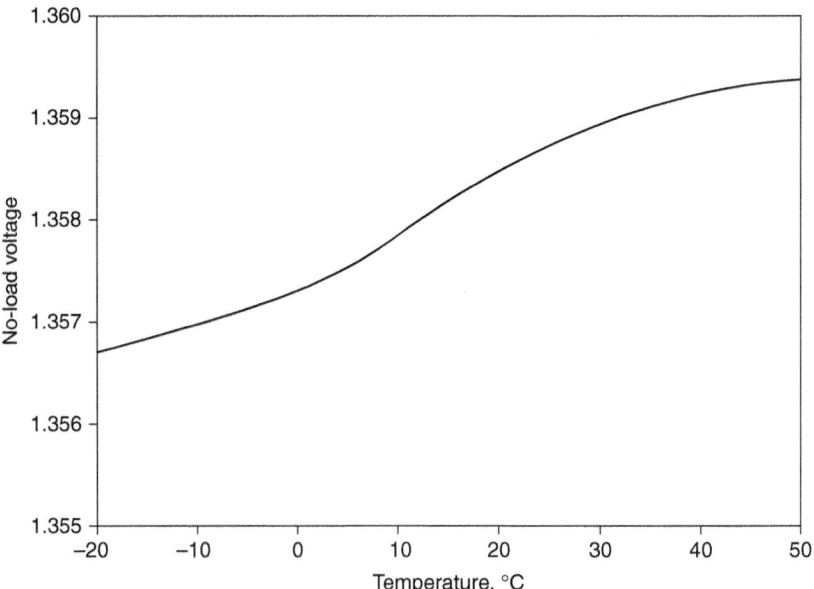

FIGURE 12.7 No-load voltage vs. temperature, zinc/mercuric oxide battery.

12.5.2 Discharge Performance

A flat discharge curve, characteristic of the zinc/mercuric oxide battery, is shown in Fig. 12.8 for a pressed-powder anode battery at 20°C. The end-point voltage is generally considered to be 0.9 V, although at higher current drains the batteries may discharge usefully below this voltage. At low current drains, the discharge profile is very flat and the curve is almost "squared off."

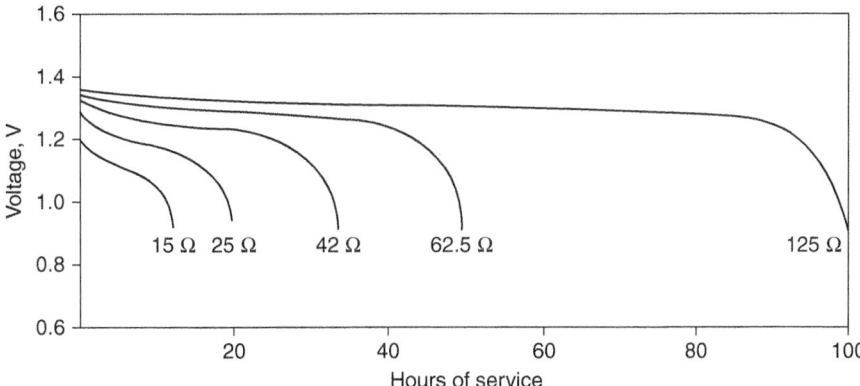

FIGURE 12.8 Discharge curves, zinc/mercuric oxide battery, 1000 mAh size, 20°C.

The capacity or service of the zinc/mercuric oxide battery is about the same on either continuous or intermittent discharge regimes over the recommended current drain range, irrespective of the duty cycle.

Under overload conditions, however, a considerable shift in available capacity can be realized by the use of "rest" periods, which may increase service life considerably.

Problems are not encountered at low rates of discharge with batteries designed for the purpose unless a high-current-drain pulse is superimposed on a continuous low-drain base current; special designs are necessary to cope with this situation.

12.5.3 Effect of Temperature

The zinc/mercuric oxide battery is best suited for use at normal and elevated temperatures from 15 to 45°C. Discharging batteries at temperatures up to 70°C is also possible if the discharge period is relatively short. The zinc/mercuric oxide battery generally does not perform well at low temperatures. Below 0°C, discharge efficiency is poor unless the current drain is low. Figure 12.9 shows the effect of temperature on the performance of two types of zinc/mercuric oxide batteries at nominal discharge drains.

The wound-anode or "dispersed"-powder anode structures are better suited to high rates and low temperatures than the pressed-powder anode.

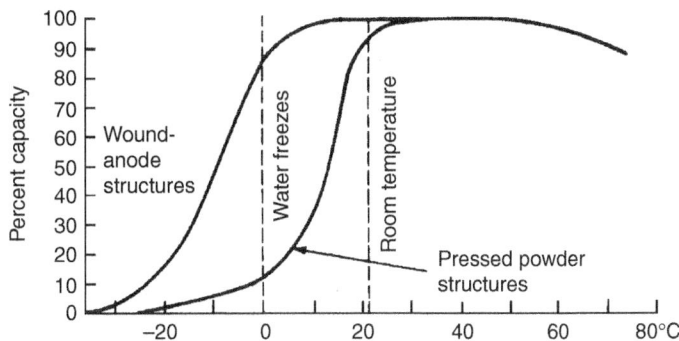

FIGURE 12.9 Effect of temperature on performance of zinc/mercuric oxide batteries.

12.5.4 Impedance

Impedance for the zinc/mercuric oxide button batteries was usually measured at a frequency of 1 kHz because of their use in hearing-aids.[9]

The impedance curve is almost a mirror image of the voltage discharge curve, rising very steeply at the end of the useful discharge life, as illustrated in Fig. 12.10. The value obtained is frequency-dependent to some degree, particularly above 1 MHz, and a fixed frequency has to be specified. A frequency versus impedance curve under no-load conditions is shown in Fig. 12.11.

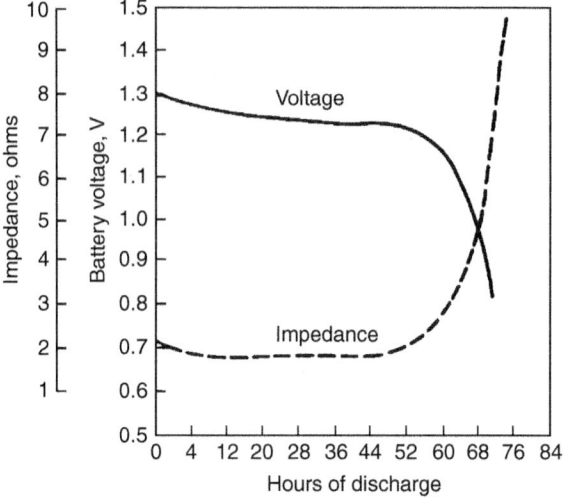

FIGURE 12.10 Internal impedance, zinc/mercuric oxide battery, 350 mAh size, 20°C, 1 kHz, 250 ohms load.

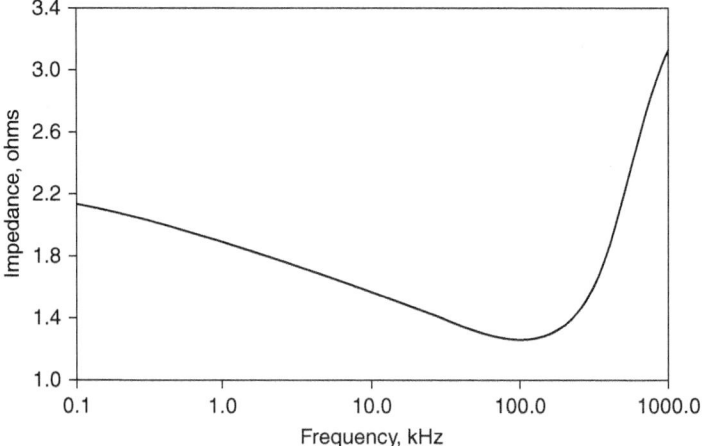

FIGURE 12.11 Variation of impedance with frequency, zinc/mercuric oxide battery, 210 mAh size, 20°C.

12.5.5 Storage

Zinc/mercuric oxide batteries have good storage characteristics. In general they will store for over 2 years at 20°C with a capacity loss of 10 to 20% and 1 year at 45°C with about a 20% loss. Storage at lower temperatures, such as down to −20°C, will, as with other battery systems, increase storage life.

The storability will depend on the discharge load and also on the cell structure. Failure in storage is usually due to the breakdown of cellulosic compounds within the cell which, at first, results in a reduction of the limiting-current density at the anode. Further breakdown produces low-drain internal electrical paths and a real loss of capacity due to self-discharge. Eventually, complete self-discharge can occur, but at 20°C and below these processes take many years.

Long storage lives are within the capabilities of the mercuric oxide system. For example, a wound-anode cell with a noncellulosic barrier has a capacity loss in the region of only 15% over 6 years. With cells designed for long-term storage, dissolution of mercuric oxide from the cathode and its transfer to the anode become a significant factor in capacity loss.

12.5.6 Service Life

The performance of the zinc/mercuric oxide cell at various temperatures and loads is summarized in Figs. 12.12 and 12.13 on a weight and volume basis. These data, based on the performance of a 800 mAh battery with a dispersed anode, can be used to approximate the performance of a zinc/mercuric oxide battery.

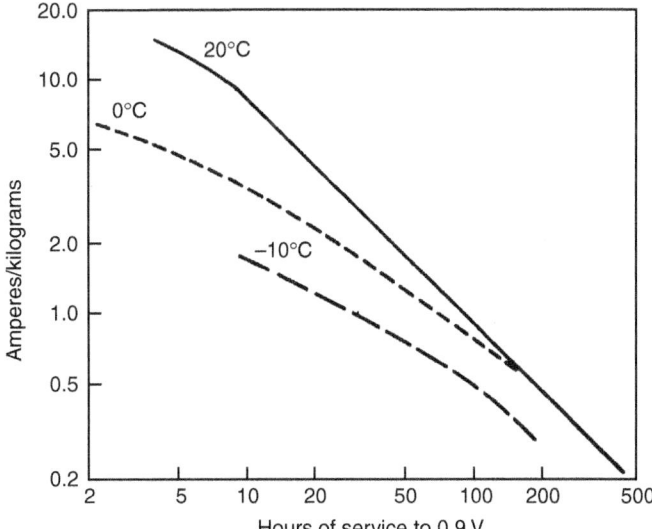

FIGURE 12.12 Service life of typical zinc/mercuric oxide battery (dispersed anode) on a weight basis.

12.12 PRIMARY BATTERIES

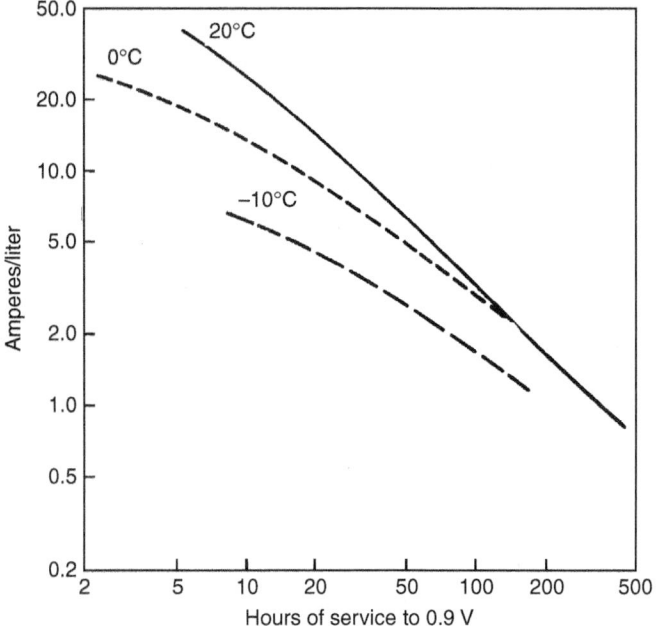

FIGURE 12.13 Service life of typical zinc/mercuric oxide battery (dispersed anode) on a volume basis.

12.6 PERFORMANCE CHARACTERISTICS OF CADMIUM/MERCURIC OXIDE BATTERIES

12.6.1 Discharge

An outstanding feature of the cadmium/mercuric oxide battery is its ability to operate over a wide temperature range. The usual operating range is from −55 to +80°C, but with the low gassing rate and thermal stability of the cell, operating temperatures to 180°C have been achieved with special designs.

Figure 12.14 shows the discharge curves for a typical button battery at various temperatures. Excellent voltage stability and flat discharge curves are characteristic of these but at a low operating voltage (open-circuit voltage is only 0.9 V). Figure 12.15 shows the effect of temperature on the

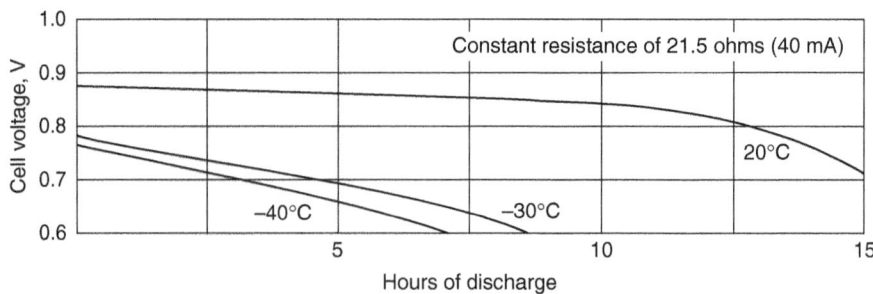

FIGURE 12.14 Discharge curves—cadmium/mercuric oxide button battery (500 mAh size).

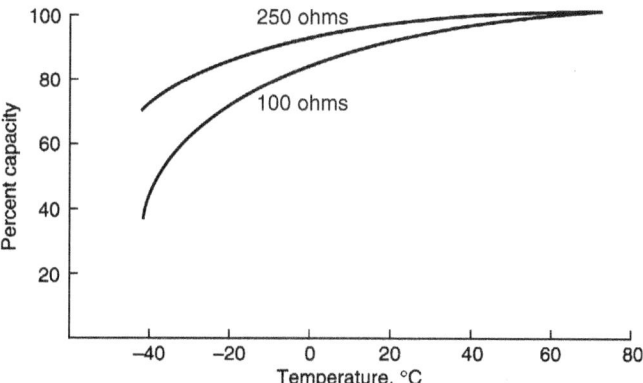

FIGURE 12.15 Effect of temperature on capacity of a cadmium/mercuric oxide button battery (3000 mAh size).

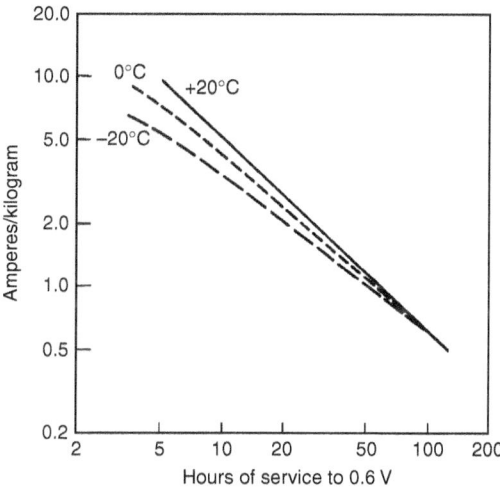

FIGURE 12.16 Service life of typical cadmium/mercuric oxide batteries on a weight basis.

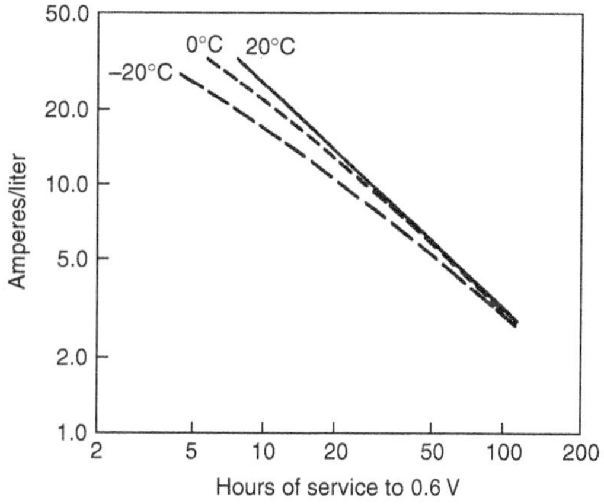

FIGURE 12.17 Service life of typical cadmium/mercuric oxide batteries on a volume basis.

capacity at various discharge loads. A high percentage of the 20°C capacity is available at the lower temperatures. The end-point voltage is usually taken as 0.6 V, although at higher current densities and lower temperatures more useful life can be obtained to lower end voltages.

The performance of the cadmium/mercuric oxide battery is summarized in Figs. 12.16 and 12.17 on a weight and volume basis, respectively. The data were derived from the performance of typical button batteries.

12.6.2 Storage

Storage life over the temperature range of −55 to +80°C is remarkably good, and if the barrier-absorbent system is designed to withstand elevated-temperature storage, the major self-discharge mechanism is by dissolution of the mercuric oxide and its transfer to the anode. A shelf life of 10 years at ambient temperatures with less than 20% capacity loss is within the capabilities of the system. Elevated-temperature storage is exceptionally good (approximately 15% loss per year at 80°C), and since neither electrode should generate hydrogen, the cells can be hermetically sealed with minimal risk of electrolyte leakage or cell distortion.[8]

REFERENCES

1. C. L. Clarke, U.S. Patent 298,175 (1884).
2. S. Ruben, "Balanced Alkaline Dry Cells," *Proc. Electrochem. Soc. Gen. Meeting,* Boston, Oct. 1947.
3. B. Berguss, "Cadmium-Mercuric Oxide Alkaline Cell," *Proc. Electrochem. Soc. Meeting,* Chicago, Oct. 1965.
4. P. Ruetschi, "The Electrochemical Reactions in Mercuric Oxide-Zinc Cell," in D. H. Collins (ed.), *Power Sources,* vol. 4, Oriel Press, Newcastle-upon-Tyne, England, 1973, p. 381.
5. D. P. Gregory, P. C. Jones, and D. P. Redfearn, "The Corrosion of Zinc Anodes in Aqueous Alkaline Electrolytes," *J. Electrochem. Soc.* **119:**1288 (1972).
6. T. P. Dirkse, "Passivation Studies on the Zinc Electrode," in D. H. Collins (ed.), *Power Sources,* vol. 3, Oriel Press, Newcastle-upon-Tyne, England, 1971, p. 485.
7. D. Weiss and G. Pearlman, "Characteristics of Prismatic and Button Mercuric Oxide-Cadmium Cells," *Proc. Electrochem. Soc. Meeting,* New York, Oct. 1974.
8. P. Ruetschi, "Longest Life Alkaline Primary Cells," in J. Thompson (ed.), *Power Sources,* vol. 7, Academic, London, 1979, p. 533.
9. S. A. G. Karunathilaka, N. A. Hampson, T. P. Haas, R. Leek, and T. J. Sinclair. "The Impedance of the Alkaline Zinc-Mercuric Oxide Cell. I. Behaviour and Interpretation of Impedance Spectra." *J. Appl. Electrochem.* **11** (1981).

CHAPTER 13
BUTTON CELL BATTERIES: SILVER OXIDE–ZINC AND ZINC-AIR SYSTEMS

Joseph Passaniti, Denis Carpenter, and Rodney McKenzie

SECTION A
SILVER OXIDE–ZINC BATTERIES

13.1 GENERAL CHARACTERISTICS OF THE SILVER OXIDE-ZINC BATTERY

The zinc/silver oxide system (zinc/alkaline electrolyte/silver oxide) offers several advantages for an aqueous battery system: high capacity, a steady discharge voltage, and good storage capacity retention. The theoretical capacity of monovalent silver oxide (Ag_2O) is 231 mAh per gram. The zinc/silver oxide battery will discharge at a flat, constant voltage typically between 1.5 and 1.6 V at both high and low discharge rates. The battery has long storage life, retaining more than 95% of its initial capacity after 1 year of room-temperature storage. It also has good low-temperature discharge capabilities, delivering about 70% of its nominal capacity at 0°C and 35% at –20°C. These features have enabled the zinc/silver oxide battery to be an important micropower source for electronic devices and equipment, such as watches, calculators, electronic thermometers, glucometers, cameras, and other applications that require small, thin, high-capacity, long-service-life batteries that discharge at a constant voltage. The commercial primary zinc/silver oxide batteries are manufactured mainly in the button cell configuration, with the sizes ranging from 5 to 250 mAh. There are a few applications for this battery system in larger sizes, such as for the military, but its use is limited by the high cost of silver (see Chap. 34).

Of the three oxidation states of silver as an oxide, the monovalent state or silver (I) oxide (Ag_2O) is most commonly used for commercial button cells. The divalent silver oxide or silver (II) oxide (AgO) has a higher theoretical capacity (432 mAh per gram) but also has the disadvantages of a dual voltage discharge and greater instability in alkaline solutions. The divalent silver oxide button cell was sold commercially as "Ditronic" or "Plumbate" batteries. These formulations were discontinued approximately two decades ago; the availability of divalent silver oxide button cells is limited. The trivalent silver oxide or silver (III) oxide (Ag_2O_3) is very unstable and is not used in batteries.

The major advantages and disadvantages of the zinc/monovalent silver oxide battery are summarized in Table 13.1.

13.2 PRIMARY BATTERIES

TABLE 13.1 Major Advantages and Disadvantages of Zinc/Silver Oxide Primary Batteries

Advantages	Disadvantages
High energy density	Use limited to button and miniature cells because of high cost
Good voltage regulation, high rate capability	
Flat discharge curve can be used as a reference voltage	
Comparatively good low-temperature performance	
Leakage and salting negligible	
Good shock and vibration resistance	
Good shelf life	

13.2 BATTERY CHEMISTRY AND COMPONENTS

The zinc/silver oxide cell consists of three main electrochemical components: fine powdered zinc metal as the anode, an aqueous alkaline electrolyte with dissolved zincates, and a compressed silver oxide cathode pellet. The active components are contained in an anode top, cathode can, separated by an ionic conductive barrier membrane and sealed with a nylon gasket.

The overall electrochemical reaction of the zinc/monovalent silver oxide cell is

$$Zn + Ag_2O \rightarrow 2Ag + ZnO \quad (1.59 \text{ V})$$

The zinc/divalent silver oxide cell has a two-step electrochemical reaction

$$\text{Step 1:} \quad Zn + AgO \rightarrow Ag_2O + ZnO \quad (1.86 \text{ V})$$
$$\text{Step 2:} \quad Zn + Ag_2O \rightarrow 2Ag + ZnO \quad (1.59 \text{ V})$$

13.2.1 Zinc Anode

Zinc is used for the negative electrode in aqueous alkaline batteries because of its high half-cell potential, low polarization, and high limiting current density (up to 40 mA/cm^2 in a cast electrode). Its equivalent weight is relatively low, thus resulting in a high theoretical capacity of 820 mAh/g. The low polarization of zinc allows for a high discharge efficiency of 85 to 95% (the ratio of useful capacity to theoretical capacity).

The zinc is prepared as a powder by air or gas atomization of the molten zinc. As in other alkaline zinc anode cells, care is taken during all processes to avoid contamination of the zinc with other metals, especially iron, as the purity of the zinc is critical to the performance and leakage resistance of the finished cells.

Zinc metal is thermodynamically unstable in aqueous alkali. Pure zinc will very slowly reduce water to hydrogen gas and zinc oxide

$$Zn + H_2O \rightarrow ZnO + H_2$$

Commercial zinc often contains trace heavy metal impurities that act as catalytic sites that rapidly increase the rate of hydrogen generation. The generation of hydrogen within a tightly sealed battery may lead to cell distortion, leakage, or, if the pressure is sufficient, rupture. Zinc alloys containing copper, iron, antimony, arsenic, or tin are known to increase the zinc corrosion rate, while zinc alloyed with mercury, cadmium, aluminum, bismuth, or lead will reduce the corrosion rate.[1,2] In commercial applications, anode gassing is brought to tolerable limits by the selection of zinc alloys, by the use of organic gassing inhibitors, and by the addition of mercury to the high surface area zinc powder. While the amount of mercury per cell is very small (typically about 3% of the anode weight), the battery industry in general is eliminating the use of mercury in all batteries in favor of organic gassing inhibitors and the use of lower gassing zinc alloys.

The oxidation of zinc during discharge is a complex phenomenon. The reactions can be written simplistically as[3,4]

$$Zn + 2OH^- \rightarrow Zn(OH)_2 + 2e \qquad E° = +1.249 \text{ V}$$

$$Zn + 4OH^- \rightarrow ZnO_2^{-2} + 2H_2O + 2e \qquad E° = +1.215 \text{ V}$$

As the electrolyte becomes saturated, zinc oxide will precipitate, releasing the bound water

$$Zn(OH)_2 \rightarrow ZnO + H_2O$$

13.2.2 Silver Oxide Cathode

Silver oxide is known to have three oxidation states:[2] monovalent (Ag_2O), divalent (AgO), and trivalent (Ag_2O_3). The trivalent silver oxide is very unstable and is not used for batteries. The divalent form had been used in button cells, generally mixed with other metal oxides. The monovalent silver oxide is the most stable and is the one primarily used for commercial button cell batteries.

The reaction product of the discharge of the monovalent silver oxide cathode is highly conductive silver metal.

$$Ag_2O + H_2O + 2e \rightarrow 2Ag + 2OH^- \qquad E° = +0.342 \text{ V}$$

Prior to discharge, however, the monovalent silver oxide is a very poor conductor of electricity. Without any additives, a monovalent silver oxide cell would initially exhibit a very high cell impedance and an unacceptably low closed-circuit voltage (CCV). To improve the initial CCV, the monovalent silver oxide is generally blended with 1 to 5% powdered graphite. As the cathode discharges, the silver metal produced maintains a low internal cell resistance and a high CCV. The theoretical capacity of the monovalent silver oxide is 231 mAh/g by weight or 1640 Ah/L by volume. The addition of graphite reduces the cathode capacity due to lower packing density and lower silver oxide content.

Unlike the other silver oxides, the monovalent silver oxide is stable to decomposition in alkaline solutions. Some decomposition to silver metal may occur due to the impurities brought into the cathode material by the graphite. The decomposition rate is dependent upon the quality of the graphite blended into the cathode, the amount of graphite, and the cell storage temperature. Greater graphite impurities and higher cell storage temperatures result in greater silver oxide decomposition rates.[5]

Due to the high cost of silver bullion, some manufacturers may reduce the amount of silver in the cathode by the use of other cathode active additives. One common additive is manganese dioxide (MnO_2). With increasing amounts of MnO_2 added to the cathode, the voltage curve changes from a constant voltage throughout the discharge to a curve where the voltage gradually decreases as the cathode is depleted (Fig. 13.1). This gradual drop in voltage has been considered an indicator of the state of cell depletion; the decreasing voltage indicates the cell is nearing its end of useful life.

Another additive that can serve a dual function is silver nickel oxide ($AgNiO_2$). Silver nickel oxide is produced by the reaction of nickel oxyhydroxide (NiOOH) with monovalent silver oxide in hot aqueous alkaline solution[6,7]

$$Ag_2O + 2NiOOH \rightarrow 2AgNiO_2 + H_2O$$

The dual feature of silver nickel oxide is that it is electrically conductive, like graphite, as well as cathode active, like MnO_2. The coulometric capacity of silver nickel oxide (263 mAh/g) is higher than Ag_2O and discharges at 1.5 V against zinc (Figs. 13.2 and 13.3). Silver nickel oxide can replace both the graphite and part of the monovalent silver oxide, reducing the cost of the cell.

13.4 PRIMARY BATTERIES

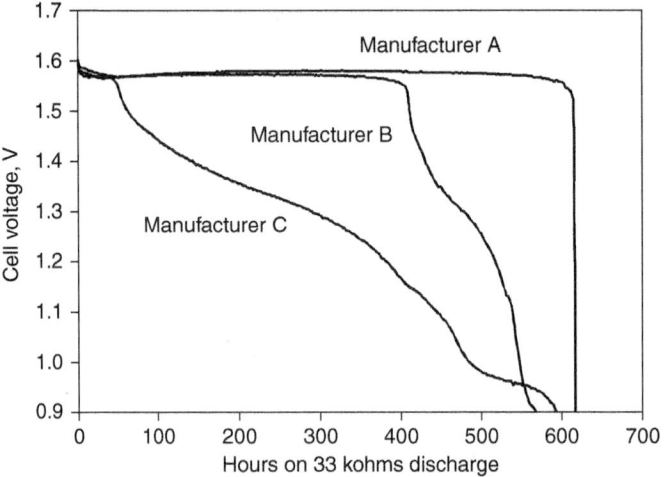

FIGURE 13.1 Voltage profiles of zinc/silver oxide 377 cells from three different manufacturers displaying an increasing dilution of the silver oxide cathode with lower cost manganese dioxide. Discharge on 33 kohms, 21°C. Data from a competitive audit. (*Courtesy of Rayovac*)

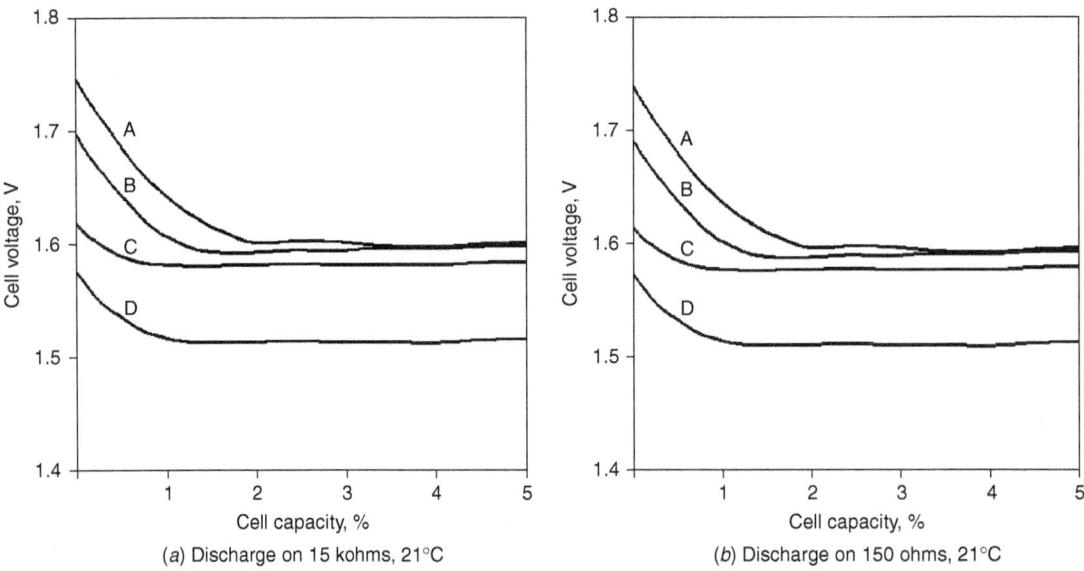

FIGURE 13.2 Closed-circuit voltage of various zinc/silver oxide chemistries, type 392 button cell, 7.8 × 3.6 mm. (A) Zn/"double-treatment" AgO; (B) Zn/Ag$_2$O; (C) Zn/AgO-silver plumbate; (D) Zn/AgNiO$_2$. (*a*) Discharge on 15 kohms, 21°C. (*b*) Discharge on 150 ohms, 21°C.

Divalent silver oxide is unstable in alkaline solutions, decomposing to monovalent silver oxide and oxygen gas[8]

$$4AgO \rightarrow 2Ag_2O + O_2$$

This instability can be improved by the addition of lead or cadmium compounds[9–12] or by the addition of gold to the divalent silver oxide.[13] The zinc/divalent silver oxide battery exhibits a two-step

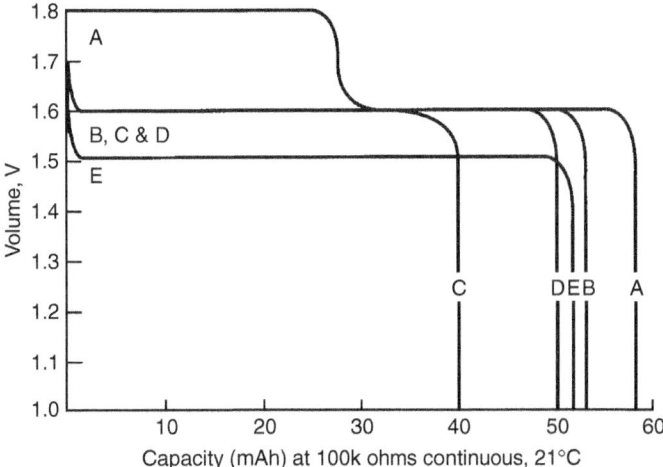

FIGURE 13.3 Comparative performances of various zinc/silver oxide chemistries, type 392 button cell, 7.8 × 3.6 mm. (A) Zn/AgO; (B) Zn/"double-treatment" AgO; (C) Zn/Ag$_2$O; (D) Zn/AgO-silver plumbate; (E) Zn/AgNiO$_2$. Discharge on 100 kohms, 21°C.

discharge curve. The first occurs at 1.8 V, corresponding to the reduction of divalent silver oxide to monovalent

$$2AgO + H_2O + 2e \rightarrow Ag_2O + 2OH^- \qquad E° = +0.607 \text{ V}$$

As the discharge continues, the voltage drops to 1.6 V, corresponding to the reduction of monovalent silver oxide to silver metal

$$Ag_2O + H_2O + 2e \rightarrow 2Ag + 2OH^- \qquad E° + 0.342 \text{ V}$$

The overall electrochemical reaction of the Zn/AgO cell is

$$Zn + AgO \rightarrow Ag + ZnO$$

This two-step discharge is not desirable for many electronic applications where tight voltage regulation is required.

The elimination of the two-step discharge has been resolved by several methods.[11,14–16] One commercial approach, shown schematically in Fig. 13.4, was to treat a compressed pellet of AgO with a mild reducing agent such as methanol. The treatment forms a thin outer layer of Ag$_2$O around a core of AgO. The treated pellet is consolidated into a can and then reacted with a stronger reducing agent such as hydrazine. The hydrazine reduces a thin layer of silver metal across the pellet's exposed surface. The cathode produced by this process has only silver metal and Ag$_2$O in contact with the cathode terminal. The layer of Ag$_2$O masks the higher potential of the AgO, while the thin,

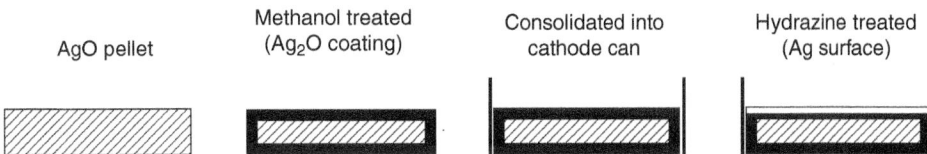

FIGURE 13.4 The Ditronic "double-treatment" process for divalent silver oxide.

TABLE 13.2 Effect of the Coating Thickness for the "Double-Treatment" Method of Divalent Silver Oxide

Ag_2O thickness on pellet, (mm)	Ag thickness on consolidation, (mm)	Final cathode capacity, (mAh/g)	Voltage level with storage		
			1 month	3 months	6 months
0.2	0.12	372	1.73	1.77	1.80
0.6	0.12	360	1.61	1.63	1.71
1.0	0.12	326	1.60	1.59	1.59
0.2	0.24	360	1.60	1.59	1.59
0.6	0.24	348	1.60	1.59	1.59
1.0	0.24	315	1.60	1.59	1.59

conductive silver layer reduces the cell impedance. In use, only the monovalent silver oxide voltage is observed yet the cell delivers the greater capacity of the divalent silver oxide. Even with this double surface treatment, the cells delivered, with the same weight of silver, 20 to 40% more hours of service than the standard monovalent silver oxide cells.

Cells produced by this "double-treatment" process are termed Ditronic™ cells. Figure 13.4 shows the benefit of the Ditronic design in a button cell. The treated cell has about a 30% capacity advantage over a conventional Ag_2O cathode at the same operating voltage.

This "double-treatment" method has the disadvantage that the control of the treatment process is critical to the shelf life of the cell. Reducing the outer surface of the divalent silver oxide pellet to either only monovalent silver oxide or to only silver metal does not have the advantages of the dual process. The same is true if either coating is not sufficiently thick (Table 13.2).

During storage, the cells eventually exhibited the phenomenon referred to as "voltage up" and "impedance up" (Fig. 13.5). The divalent silver oxide slowly oxidizes the conductive silver layer back to the resistive monovalent silver oxide.

$$Ag + AgO \rightarrow Ag_2O$$

As the metallic silver layer is depleted, the cell demonstrates an increase in open-circuit voltage and impedance. The high impedance, due to high internal resistance, results in a low closed-circuit voltage and a nonfunctional cell.

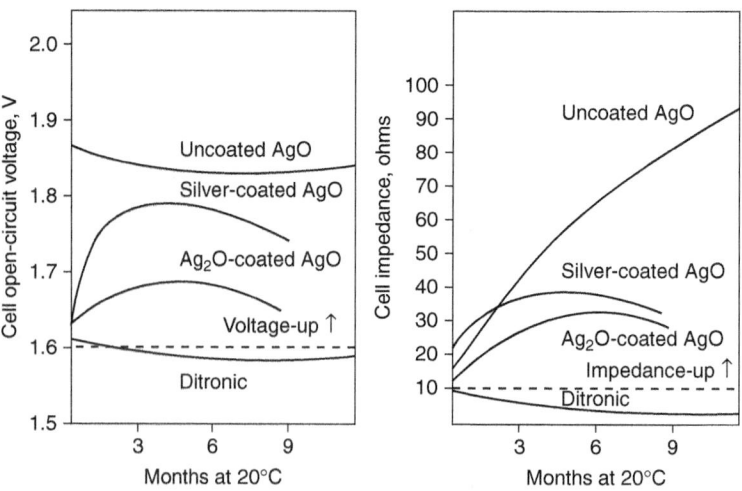

FIGURE 13.5 "Voltage-up" and "impedance-up" of Zn/AgO battery during 1 year storage at 21°C, type 392 button cell, 7.8 × 3.6 mm.

A second approach to eliminate the two-step discharge was through the use of silver plumbate as a cathode additive material.[17] Silver plumbate cathode material was prepared by reacting an excess of divalent silver oxide as a coarse powder with lead sulfide (PbS) in a hot alkaline solution. The product of the reaction is a mixture of remaining divalent silver oxide (AgO), monovalent silver oxide (Ag_2O), and silver plumbate ($Ag_5Pb_2O_6$). The sulfur is oxidized to the sulfate and is washed from the reaction product

$$2PbS + 19AgO + 4NaOH \rightarrow Ag_5Pb_2O_6 + 7Ag_2O + 2Na_2SO_4 + 2H_2O$$

The AgO particles after the reaction retain a core of AgO but have an outer coating of monovalent silver oxide and silver plumbate. The silver plumbate compound is conductive, stable, and cathode active. The Ag_2O serves to mask the AgO, while the conductive $Ag_5Pb_2O_6$ improves the cell impedance. Unlike silver metal, the conductive silver plumbate is not oxidized by the AgO and the cathode impedance remains low throughout the cell life.

Cathode material prepared by the silver plumbate process will discharge through four reaction steps (Fig. 13.6):

Rxn I	$2AgO + H_2O + 2e$	$\rightarrow Ag_2O + 2OH^-$	$E° = +0.607$ V
Rxn II	$Ag_2O + H_2O + 2e$	$\rightarrow 2Ag + 2OH^-$	$E° = +0.342$ V
Rxn III	$Ag_5Pb_2O_6 + 4H_2O + 8e \rightarrow 5Ag + 2PbO + 8OH^-$		$E° = +0.2$ V
Rxn IV	$PbO + H_2O + 2e$	$\rightarrow Pb + 2OH^-$	$E° = +0.580$ V

The open-circuit voltages (OCV) of the silver plumbate cells are found to be stable at about 1.75 V. However, once placed on discharge, the cell voltage quickly drops to the monovalent silver oxide operating voltage of about 1.6 V, eliminating the AgO plateau (Fig. 13.2). As button cells are anode limited, the cells may be depleted in zinc capacity before the $Ag_5Pb_2O_6$ and PbO reduction reactions can be observed.

The silver plumbate approach has advantages over the dual-treatment process in that the treatment is simpler while still retaining a capacity advantage over monovalent silver oxide. The product

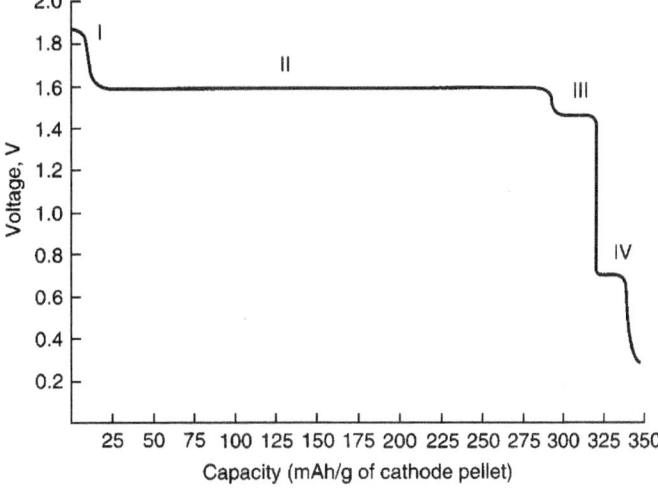

FIGURE 13.6 Cathode limited discharge of the zinc/AgO-silver plumbate system; 300ohms continuous in flooded beaker cell at 21°C. Cathode pellet weight = 0.12 g. Roman numerals indicate the reaction steps.

from the reaction of divalent silver oxide with 8% lead sulfide has a coulometric capacity of 345 to 360 mAh/g.

The silver plumbate process has the disadvantage that the button cells do contain a small amount of lead, 1 to 4% of the cell weight. An alternate approach was developed to use bismuth sulfide in place of the lead sulfide in the material preparation reaction.[18] The reaction product retains the advantages of the silver plumbate material but without the toxicity of lead. Bismuth is not considered toxic and is used in medical and cosmetic applications, both externally and internally within the body.[19] The product of the reaction of bismuth sulfide with divalent silver oxide is believed to be silver bismuthate ($AgBiO_3$)

$$Bi_2S_3 + 28AgO + 6NaOH \rightarrow 2AgBiO_3 + 13Ag_2O + 3Na_2SO_4 + 3H_2O$$

Like the silver plumbate compound, the silver bismuth compound is conductive and cathode active. The monovalent silver oxide produced by the reaction coats the divalent silver oxide particles, while the conductive silver bismuthate reduces the cell impedance, allowing for a high cell CCV. The silver bismuthate will discharge against zinc in alkaline solutions at about 1.5 V. Therefore, in anode-limited button cells only the monovalent silver oxide voltage is observed.

Unlike monovalent silver oxide systems, additives such as graphite or manganese dioxide cannot be added to the divalent silver oxide. Graphite enhances the decomposition of AgO to Ag_2O and oxygen. Manganese dioxide is readily oxidized by AgO to alkali-soluble manganate compounds.

Although the divalent silver oxide has a higher theoretical capacity (432 mAh/g by weight or 3200 Ah/L by volume) than the monovalent silver oxide, the use of the divalent form in button batteries was limited and is no longer marketed commercially.[2,8] This is due primarily to the difficulty in eliminating the two-step discharge and declining prices as the zinc/silver oxide button cells became a commodity.

13.2.3 Alkaline Electrolyte

The electrolytes used for zinc/silver oxide cells are based upon 20 to 45% aqueous solutions of potassium hydroxide (KOH) or sodium hydroxide (NaOH). Zinc oxide (ZnO) is dissolved in the electrolyte as the zincate salt to help control zinc gassing. The zinc oxide concentration varies from a few percent to a saturated solution.

The preferred electrolyte for button cells is potassium hydroxide (KOH). Its higher electrical conductivity[20,21] allows cells to discharge over a wider range of current demands (Fig.13.7). Sodium hydroxide (NaOH) is used mainly for long life cells not requiring a high-rate discharge (Fig 13.8). The sodium hydroxide exhibits less creep, and such cells are less apt to leak than the potassium hydroxide cells. Leakage is evidenced as frosting or salting around the seal. However, leakage issues with potassium hydroxide cells have been resolved by most manufacturers through improvements in seal technology.

Electrolyte gelling agents such as polyacrylic acid, potassium or sodium polyacrylate, sodium carboxymethyl cellulose, or various gums are generally blended into the zinc powder to improve electrolyte accessibility during discharge.

13.2.4 Barriers and Separators

A physical barrier is required to keep the zinc anode and silver cathode apart in the tight volume constraints of a button cell. Failure of the barrier will result in internal cell shorting and cell failure. A silver oxide cell requires a barrier with the following properties:

1. Permeable to water and hydroxyl ions
2. Stable in strong alkaline solutions

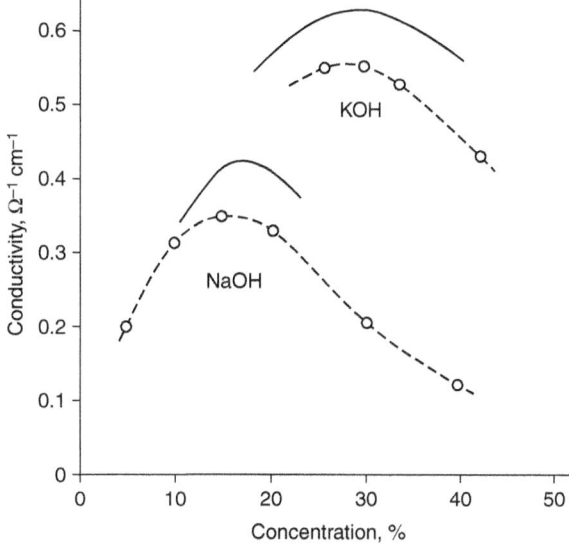

FIGURE 13.7 Specific conductivity of alkaline hydroxide solutions. Solid line: 25°C, Ref. 5; dotted line: 15°C, Ref. 6.

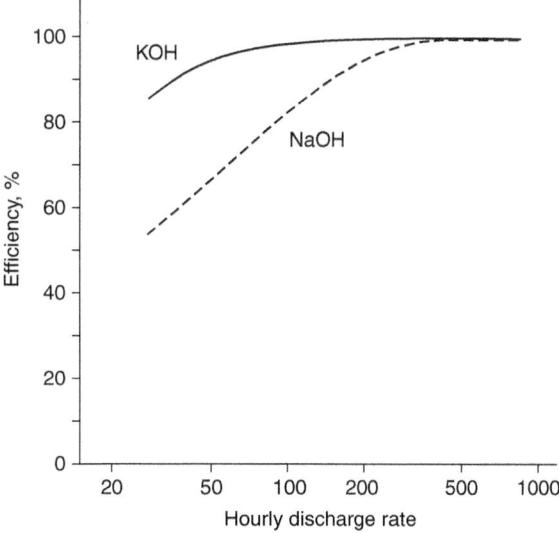

FIGURE 13.8 Dependence of discharge efficiency on discharge rate of zinc/silver oxide button battery at 20°C.

3. Not oxidized by the solid silver oxide or dissolved silver ions
4. Retards the migration of dissolved silver ions to the anode

Because of the slight solubility of silver (I) oxides in alkaline electrolyte, little work was done with zinc/silver oxide cells until 1941 when André[22] suggested the use of a cellophane barrier. Cellophane prevents migrating silver ions from reaching the anode[23,24] by reducing them to insoluble silver metal. The cellophane is oxidized and destroyed in the process, making it less effective for long-life cells.

Many types of laminated membranes are presently available. A commonly used alternate barrier material is prepared from a radiation graft of methacrylic acid onto a polyethylene membrane.[23,24] The graft makes the film wettable and permeable to the electrolyte. Studies have shown that a lower resistance polyethylene barrier membrane is suitable for high-rate KOH cells, while higher resistance polyethylene is suitable for low-rate NaOH cells. Cellophane is used in conjunction with the grafted membrane as a sacrificial barrier. The lamination of cellophane to either side of the polyethylene membrane results in a synergistic action for stopping silver migration.[15]

A separator is commonly used in conjunction with a barrier membrane layer as added protection to the barrier. It is located between the barrier and anode cavity and is multifunctional both during cell manufacture and in performance. Separators in zinc/silver cells are typically fibrous woven or nonwoven polymers such as polyvinyl alcohol (PVA). The fibrous nature of the separator gives it stability and strength that protects the more fragile barrier layers from compression failure during cell closure or through penetration of zinc particles. The separator also acts to moderate the effects of dimensional stresses in the barrier layers developed during the lamination processes. These stresses are relieved as the barrier membranes wet up.

13.3 CONSTRUCTION

Figure 13.9 is a cross-sectional view of a typical zinc/silver oxide button type battery. Zinc/silver oxide button cells are designed to be anode limited; the cell has 5 to 10% more cathode capacity than anode capacity. If the cell were cathode limited, a zinc-nickel or zinc-iron couple could form between the anode and the cathode can, resulting in the generation of hydrogen gas.

The cathode material for zinc/silver oxide cells is monovalent silver oxide (Ag_2O) mixed with 1 to 5% graphite to improve the electrical conductivity. The Ag_2O cathode material may also contain manganese dioxide (MnO_2) or silver nickel oxide ($AgNiO_2$) as cathode extenders. A small amount of polytetrafluoroethylene (Teflon™) may be added to the mix as a binder and to aid pelleting.

The anode is a high surface area, amalgamated, gelled zinc metal powder housed in a top cup, which serves as the external negative terminal for the cell. The top cup is pressed from a triclad metal sheet: the outer surface is a protective layer of nickel over a core of steel. The inner surface that is in direct contact with the zinc is high-purity copper or tin.

The cathode pellet is consolidated into the positive cup, which is formed from nickel-plated steel and serves as the positive terminal for the cell. To keep the anode and cathode separated, a barrier disk of cellophane or a grafted polymeric membrane is placed over the consolidated cathode. The entire system is wetted with potassium or sodium hydroxide electrolyte.

The gasket serves to seal the cell against electrolyte loss and to insulate the top and bottom cups from contact. The gasket material is made from an elastic, electrolyte-resistant plastic such as nylon. The seal may be improved by coating the gasket with a sealant such as polyamide or bitumen to prevent electrolyte leakage at the seal surfaces.

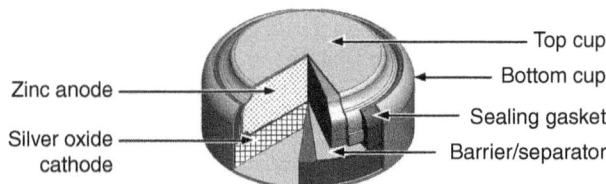

FIGURE 13.9 Cutaway view of typical zinc/silver oxide button type battery.

13.4 PERFORMANCE CHARACTERISTICS

13.4.1 Open-Circuit Voltage

The open-circuit voltage (OCV) of the Zn/Ag_2O battery is about 1.60 V, but will vary slightly (1.595 to 1.605 V) with electrolyte concentration, concentration of zincate in the electrolyte, and temperature exposure.[25] The reaction of the silver oxide with carbon dioxide during battery manufacturing can raise the OCV to 1.65 V due to the formation of silver carbonate. The increase in voltage, however, is temporary and will drop to the operating voltage level of 1.58 V within seconds, for example, in a watch. The depth of discharge has little effect on the OCV of a monovalent silver battery; a partially used battery has the same OCV as a new one.

The OCV of the zinc/divalent silver oxide battery will vary from 1.58 to 1.86 V, depending on the ratios of Ag to Ag_2O to AgO in the cathode. The OCV will decrease with greater Ag_2O to AgO ratios and with the presence of silver metal in the cathode. With divalent silver oxide batteries, the depth of discharge does have an effect on the OCV; a partially used battery will have more Ag_2O and silver metal than a new one and may have a lower OCV.

13.4.2 Discharge Characteristics

Figure 13.10 exhibits the typical curves for a type 389, 11.6 × 3.0 mm monovalent silver oxide battery on constant resistance discharge. These are typical voltage curves; the discharge voltage profiles of other size batteries would be similar. The service life will vary depending upon the size of the battery and the applied load.

The discharge characteristics of the various types of silver oxide batteries are also covered in Sec. 13.2.2.

Figure 13.11 shows the initial closed-circuit voltage of representative sizes of zinc/silver oxide batteries at various loads and temperatures. The zinc/silver oxide button battery is capable of operating over a wide temperature range. The battery can deliver more than 70% of its 20°C capacity at 0°C and 35% at −20°C, at the more moderate loads. At heavier loads, the loss is greater. Higher temperatures tend to accelerate capacity deterioration, but temperatures as high as 60°C can be tolerated for several days with no serious effect.

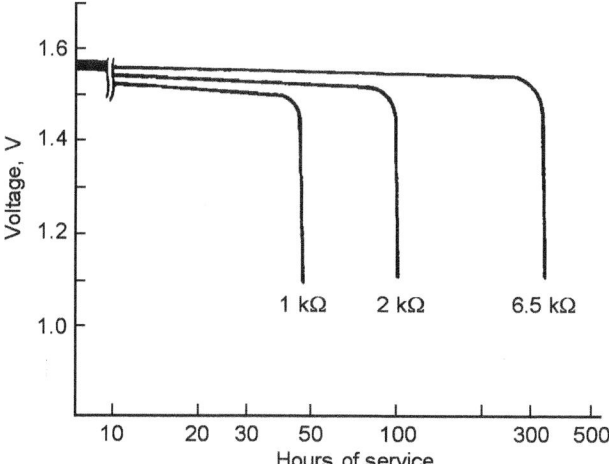

FIGURE 13.10 Typical discharge curves of zinc/silver oxide battery at 20°C, type 389 button cell 11.6 × 3.0 mm.

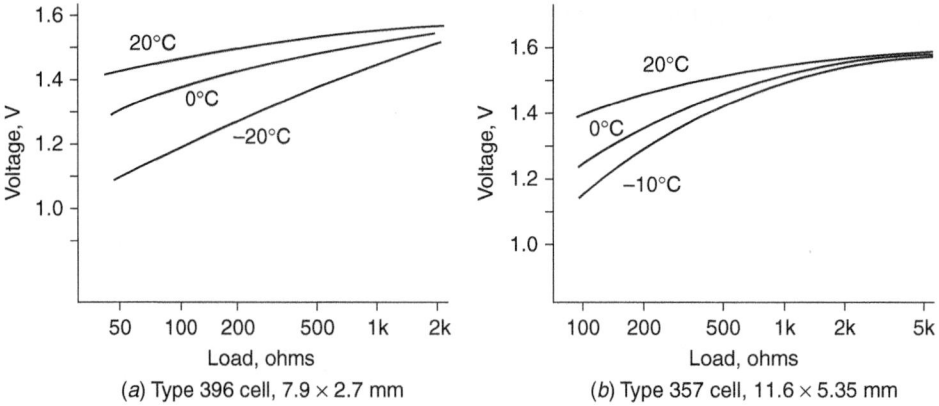

FIGURE 13.11 Closed-circuit voltage of zinc/silver oxide batteries various resistances. (*a*) Type 396 cell, 7.9 × 2.7 mm and (*b*) Type 357 cell, 11.6 × 5.35 mm.

Figure 13.12 shows the pulse performance of treated divalent silver oxide batteries using sodium hydroxide and potassium hydroxide as electrolytes. Potassium hydroxide electrolyte yields a cell that will discharge at a higher operating voltage than the sodium hydroxide electrolyte.

The manufacturers of these two types of batteries do not distinguish them by service life tests. In fact, similar mAh output is obtained at loads lighter than the 500 hr rate. The industry uses pulse CCV tests to differentiate the higher rate KOH version from the low rate NaOH version.

The impedance of a zinc/silver oxide battery is influenced primarily by the conductive diluents in the cathode, the barrier resistivity, and the electrolyte type and concentration.

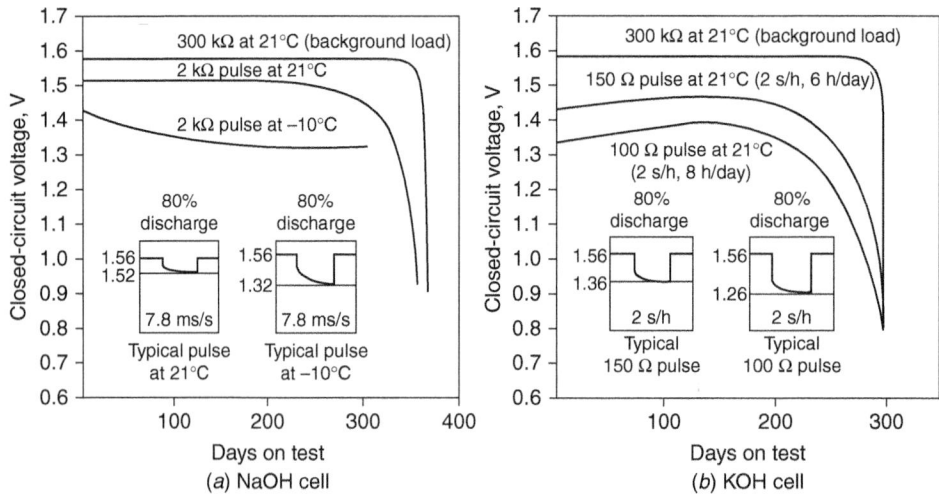

FIGURE 13.12 Closed-circuit voltage curves for Zn-AgO cells with NaOH and KOH as electrolytes, type 392 cell, 7.8 × 3.6 mm. (*a*) NaOH cell, analog watch test. (*b*) KOH cell, LCD watch with backlight test.

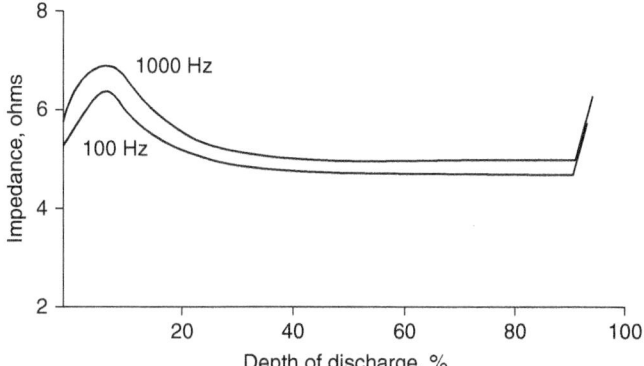

FIGURE 13.13 Impedance of a Zn/Ag$_2$O battery at 100 and 1000 Hz during discharge, type 357 cell, 11.6 × 5.35 mm.

These factors are balanced by battery manufacturers to obtain the desired values required to meet the applications. As the cell is discharged, the impedance will decline as the resistive silver oxide is reduced to conductive metallic silver (Fig. 13.13).

13.4.3 Shelf Life

Major improvements in seal technology and in cell stability have extended the shelf life of watch batteries to more than five years. The effect of temperature and humidity on leakage of button batteries was reported by Hull.[26] Leakage was caused by mechanical means (improper seal, fibers in the seal, scratches) or electrochemical means (high oxygen content or high humidity). Batteries are now designed to operate watches for 5 years without leakage.

Stability of these batteries after high temperature storage or prolonged storage at room temperature is influenced by cathode stability and barrier selection. With the monovalent silver oxide cathode, gassing in aqueous potassium or sodium hydroxide at 74°C is not a problem.

With modified cathodes, however, such as divalent silver oxide, silver plumbate or silver nickel oxide, gas suppression is necessary. CdS, HgS, SnS$_2$, or WS$_2$ were found to reduce oxygen evolution, while BaS, NiS, MnS, and CuS increased oxygen evolution from AgO.[11] Failure on shelf of these zinc/silver oxide batteries is closely connected with barrier selection. Cellulosic membranes were used for many years in Zn/Ag$_2$O cells, but their use in Zn/AgO cells was unsuccessful because of massive silver diffusion. While solubility of AgO and Ag$_2$O was reported[24] to be the same (4.4 × 10^{-4} mol/L in 10N NaOH), AgO decomposition to Ag$_2$O occurred spontaneously, resulting in more silver diffusion with Zn/AgO cells than with Zn/Ag$_2$O cells. The small amount of soluble silver reaching the zinc caused accelerated corrosion and hydrogen evolution. In addition, silver was plated in the barrier, forming electronic shorts that internally discharge the cell. Laminated Permion membranes have been used to stop silver migration to the zinc. Figure 13.14 shows an Arrhenius plot of the storage characteristics of various low- and high-rate zinc/silver oxide systems. The data shows that 10 years of storage at 21°C is possible.

13.4.4 Service Life

Figure 13.15 is a monograph that can be used to calculate the service life of the various sized batteries at various current drains at 20°C.

13.14 PRIMARY BATTERIES

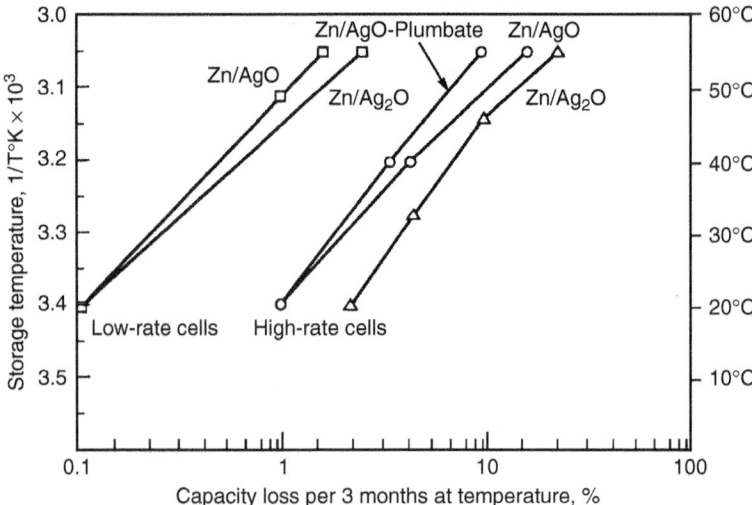

FIGURE 13.14 Arrhenius plot of various zinc/silver oxide chemistries. Type 357 cell, 11.6 × 5.35 mm, 6500 ohms continuous discharge to 0.9 V at 21°C. Projected 10% loss: □ > 10 years, ○ > 5 years, △ > 3 years.

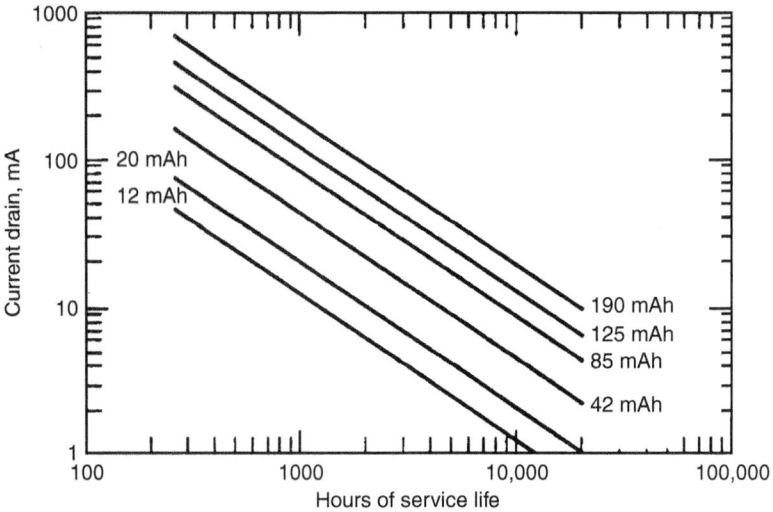

FIGURE 13.15 Service life of zinc/silver oxide batteries at 20°C.

13.5 CELL SIZES AND TYPES

The characteristics of commercially available zinc/monovalent silver oxide button batteries are summarized in Table 13.3.

TABLE 13.3 Characteristics of Commercially Available Zinc/Silver Oxide Batteries

Rayovac model number	ANSI*	IEC†	Drain rate	Rated load kΩ	Nominal capacity (mAh)	Approximate volume cm^2	Maximum dimensions dia. × ht. (millimeters)	Approximate weight (grams)
376	1196SO	SR626	high	47	26	0.09	6.8 × 2.6	0.4
361	1173SO	SR58	high	30	22	0.1	7.9 × 2.1	0.44
396	1163SO	SR59	high	45	35	0.13	7.9 × 2.7	0.56
392	1135SO	SR41	high	15	35	0.17	7.8 × 3.6	0.61
393	1137SO	SR48	high	15	90	0.26	7.8 × 5.4	1.04
370	1188SO	SR69	high	45	35	0.15	9.5 × 2.1	0.6
399	1165SO	SR57	high	20	53	0.19	9.5 × 2.7	0.79
391	1160SO	SR55	high	15	43	0.22	11.6 × 2.1	0.83
389	1138SO	SR54	high	15	85	0.32	11.6 × 3.0	1.21
386	1133SO	SF43	high	6.5	120	0.44	11.6 × 4.2	1.56
357	1131SO	SR44	high	6.5	190	0.57	11.6 × 5.35	2.22
337	NA	SR416	low	100	8.3	0.02	4.8 × 1.65	0.13
335	1193SO	SR512	low	150	6	0.03	5.8 × 1.25	0.13
317	1185SO	NA	low	70	11	0.04	5.8 × 1.65	0.18
379	1191SO	NA	low	70	14	0.06	5.8 × 2.15	0.23
319	1186SO	NA	low	70	16	0.07	5.8 × 2.7	0.26
321	1174SO	SR65	low	70	14	0.06	6.8 × 1.65	0.24
364	1175SO	SR60	low	70	19	0.08	6.8 × 2.15	0.33
377	1176SO	SR66	low	45	26	0.09	6.8 × 2.6	0.4
346	1164SO	SR721	low	100	9.5	0.06	7.9 × 1.25	0.23
341	1192SO	SR714	low	68	13	0.07	7.9 × 1.45	0.3
315	1187SO	SR67	low	70	16	0.08	7.9 × 1.65	0.32
362	1158SO	SR58	low	70	22	0.1	7.9 × 2.1	0.44
397	1164SO	SR59	low	45	35	0.13	7.9 × 2.7	0.56
329	NA	NA	low	20	36	0.15	7.9 × 3.1	0.6
384	1134SO	SR41	low	15	35	0.17	7.8 × 3.6	0.61
373	1172SO	SR68	low	45	24	0.12	9.5 × 1.65	0.44
371	1171SO	SR69	low	45	35	0.15	9.5 × 2.1	0.61
395	1162SO	SR57	low	20	53	0.19	9.5 × 2.7	0.81
394	1161SO	SR45	low	15	64	0.26	9.5 × 3.6	0.96
366	1177SO	SR1116	low	30	30	0.17	11.6 × 1.65	0.7
381	1170SO	SR55	low	20	43	0.22	11.6 × 2.1	0.8
390	1159SO	SR54	low	15	85	0.32	11.6 × 3.0	1.21
344	1139SO	SR42	low	15	105	0.38	11.6 × 3.6	1.35
301	1132SO	SR43	low	6.8	110	0.44	11.6 × 4.2	1.68

*ANSI: American National Standards Institute.
†IEC: International Electrotechnical Commission.
Source: Rayovac Corporation.

SECTION B
ZINC-AIR BATTERIES

13.6 GENERAL CHARACTERISTICS OF BUTTON CELL ZINC/AIR BATTERIES

A zinc/air battery is remarkable because it contains so little oxygen when it is first made. Unlike other electrochemical systems that we know as batteries, zinc/air batteries rely on the oxygen available from the environment to replenish the supply that is used during discharge. The thin layer of cathode contains just enough oxygen to support the next instant of discharge, and that sets up a gradient of oxygen concentration—depleted at the site of use, and promoting the diffusion of more oxygen into the cell, anticipating the next instant of use. While the cathode is constantly being refreshed, it acts catalytically and is essentially unchanged throughout the life of the battery. The anode is zinc powder dispersed in an alkaline electrolyte. During discharge, the zinc gives up electrons, and eventually zinc oxide is formed, building up amid the undischarged zinc. The capacity of a zinc/air cell is dictated by the amount of zinc that is available for discharge. On a volumetric basis, zinc/air cells offer one of the highest energy densities of any commercially available electrochemical system.

Is the zinc/air cell a battery or is it a fuel cell? Of course, the answer is both. The zinc makes a one-way trip to becoming zinc oxide during discharge, and capacity is limited by the material contained within the container when the cell is manufactured. On the other side, the zinc/air system enjoys the fact that oxygen enters the cell only as it is needed. It can do this because the cell is open to the environment. Unfortunately, this is the reason that zinc/air suffers from some inherent limitations. Since both water vapor and oxygen have similar molecular size, it is hard to allow diffusion of one without having roughly equivalent diffusion of the other. Gain or loss of water vapor leads to the most immediate limitations we find in the use of zinc/air cells. If the zinc/air battery is designed to deliver power quickly, and operate in a stable continuous mode, then it will respond quickly to changes in the external environment from which the oxygen comes. The alkaline electrolyte needs to be of a high molarity in order to support high currents, but at this concentration of caustic the electrolyte readily picks up water in high humidity conditions and will lose it in low humidity. When water is lost, the rate capability of the cell is impaired, as is its overall life. A gain of water engorges the cell, and capacity is cut short due to a loss of the available volume for zinc oxide being formed during discharge. Cells may bulge, and electrolyte may be expressed through the vent holes when internal pressure forces it through the cathode structure.

Before a zinc/air cell is needed for use, it must be protected from the external environment. It still needs a little oxygen in order to maintain a good open circuit voltage. The adhesive tab covering the vent holes prevents most of the oxygen in the outside air from freely entering the cell.[27] However, a small amount of air, roughly comparable to a current of a few nanoamperes, is needed to offset the slow oxidation of the zinc that is always occurring in a zinc/air cell. This is the result of side reactions that are minimized by excluding contaminants to the greatest extent possible, but never eliminated completely. If this parasitic oxidation of the zinc is excessive, then the tabbed voltage will drop because insufficient oxygen is able to replenish the active sites on the cathode. Table 13.4 lists the major strengths and weaknesses of zinc/air batteries.

The use of zinc/air batteries has been explored in telecommunications, military, and industrial applications. Backup power and portable power systems are examples of the attempts to apply a high-rate zinc/air system for use over a limited period of time. Certain commercial and medical uses that are not intermittent applications have also used zinc/air as a power source. The best use of zinc/air chemistry is in persistent and repetitive applications. Miniature hearing instruments have found zinc/air batteries to be the ideal power source for almost 30 years, at this writing. Capacity, as compared with the predecessor mercury oxide cells, was almost twice at the outset, and now stands at over three times the comparable mercury cells. The desire for cosmetically discrete hearing instruments caused

TABLE 13.4 Major Strengths and Weaknesses of Zinc/Air Batteries

Strengths	Weaknesses
High energy per unit volume	Responds to environmental conditions
Stable voltage curve	Limited shelf life after open to air
Environmentally friendly	Flooding in high RH
Economical	Poor on intermittent use
Convenient	Tape must be removed from air holes to activate cell

a migration to smaller sizes, and a demand for greater power output as circuitry became more sophisticated. New work was demanded of the battery as digital signal processing was added to the chain of events that takes place between the reception of sound at the microphone and the eventual delivery of amplified sound at the receiver. The advent of cochlear implants raised the bar for zinc/air in the need for a carrier signal to inductively power the implanted electronics. This is used to stimulate the nerves of the cochlea with a superimposed message signal that produces the sensation of sound.

Very low-power applications, where infrequent maintenance and long service life are needed, can also benefit from the high capacity of zinc/air cells. In these applications, the cells are much more isolated from their environment. Remote railroad signaling, marine navigation systems, and electric fence applications are examples of this legacy from the earlier incarnations of commercially available zinc/air products. These batteries are often multicell and prismatic in configuration. Large zinc/air batteries are currently finding use in military applications (see chap. 33).

13.7 CHEMISTRY

The zinc/air battery consists of three main electrochemical components: zinc metal in a fine powder form, aqueous alkaline electrolyte, and a catalytic cathode structure that makes oxygen available for conversion to OH^- ions when an external circuit is established. The zinc is separated from the cathode physically by a wettable microporous membrane, so that only ionic conduction can take place within the body of the cell. The cathode layer is compressed against an insulating seal gasket to contain the liquid constituent of the cell and prevent leakage. One or more holes in the cathode can admit oxygen by diffusion as the internally contained oxygen is consumed during discharge.

The anode reaction as understood from an elementary viewpoint goes as follows:

$$Zn + 2OH^- \rightarrow Zn(OH)_2 + 2e^-$$

$$Zn(OH)_2 \rightarrow ZnO + H_2O$$

The cathode reaction, occurring first at the catalytic sites and then in the electrolyte where the ionic conduction takes place, is as follows:

$$O_2 + 2H_2O + 4e^- \rightarrow 4OH^-$$

The oxygen reduction process is predicated on the peroxide–free radical (O_2H) formation, followed by a subsequent decomposition of the peroxide. This effectively restricts performance of the cell in practice, as it is the rate limiting electrochemical step

$$O_2 + H_2O + 2e^- \rightarrow O_2H^- + OH^-$$

$$O_2H^- \rightarrow OH^- + \tfrac{1}{2}O_2$$

Peroxide decomposition can be accelerated in order to increase overall system performance. This is usually accomplished by catalysts that promote the reaction in the last equation. While one of many transition metal compounds or a rare earth metal may be appropriate, most commonly it is an oxide of manganese that finds use in commercially available products.[28–30] Care must be given to the solubility of the additives placed in the cathode, and the degree to which they will promote parasitic reactions in the anode if they dissolve into the electrolyte over time.

13.8 CONSTRUCTION

Button and coin cells are the most prevalent form for zinc/air batteries to take. Development of other formats, such as cylindrical and prismatic forms, has been somewhat limited due to economics, but also by the frustrations of air management and the poor performance of zinc/air when use is intermittent. Moisture sensitivity in large cells becomes a greater problem when rate capability demands a largely open cathode to air interface. Unless the cathode can be isolated from the environment during the idle periods, the cells will suffer the adverse effects of water gain or loss, as well as the degradation caused by the accumulation of carbonates at the air-electrolyte interface.

Similar in many ways to their predecessors, the zinc/mercury oxide and zinc/silver oxide button cells, the zinc/air button cell is designed with an eye to ultimate electrical capacity within a defined volume. As illustrated in Fig. 13.16, the zinc/air button cell can pack more zinc in the same volume because the system doesn't require a thick metal oxide cathode to complete the cell reaction. Each cell size has an analogue size in the two older chemistries, with the exception of the size 10 cell (IEC PR70), which was developed by Rayovac in the mid-1980s. It was introduced to the hearing instrument industry as an enabling technology for in the ear (ITE) and completely in the canal (CIC) devices.

As seen Fig. 13.17 there are no spare parts in the construction of the zinc/air cell. Each part performs multiple functions and must do so while occupying a minimum of volume of its own. The resulting assembly contains the maximum amount of zinc that can be discharged by the user, delivering the longest number of hours or days of battery performance. The negative top of the cell is an impervious container for the zinc, acts as the current collector, and resists the pressure of closure that seals it against the gasket. The top also makes the electrical contact that brings the electronic current out of the cell to an external device or circuit. The positive can acts as the outer container of the cell, connecting electrically to the cathode disk, and applies the compressive forces of closure against the gasket that stands between the can and the top. Vent holes in the can admit air to the cathode, and the outer surface provides electrical contact to the external circuit. Finally, the gasket between the can and the top provides a seal against electrolyte leakage and prevents electrical shorting between the top and the can. The gasket is usually made of a hard polymeric material, often an engineering plastic such as polyamide, and may be coated with additional sealant material to enhance its resistance to leakage.

Within the cell container, the cathode rests flat (see Fig. 13.18), occupying the inside bottom of the can. It is a multilayer structure consisting of a gas diffusion membrane.[31,32] The first layer is a hydrophobic, microporous membrane that is permeable to gases such as oxygen and water vapor, but

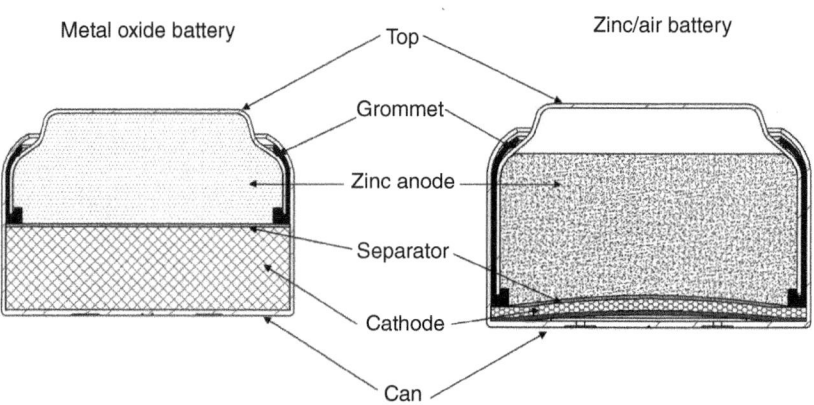

FIGURE 13.16 Cross-sections of metal oxide and zinc/air button batteries.

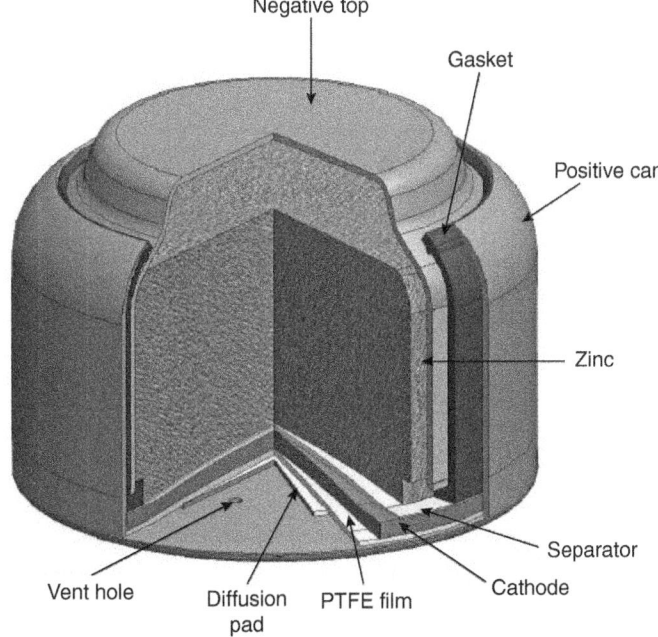

FIGURE 13.17 A typical zinc/air button battery (components not to scale).

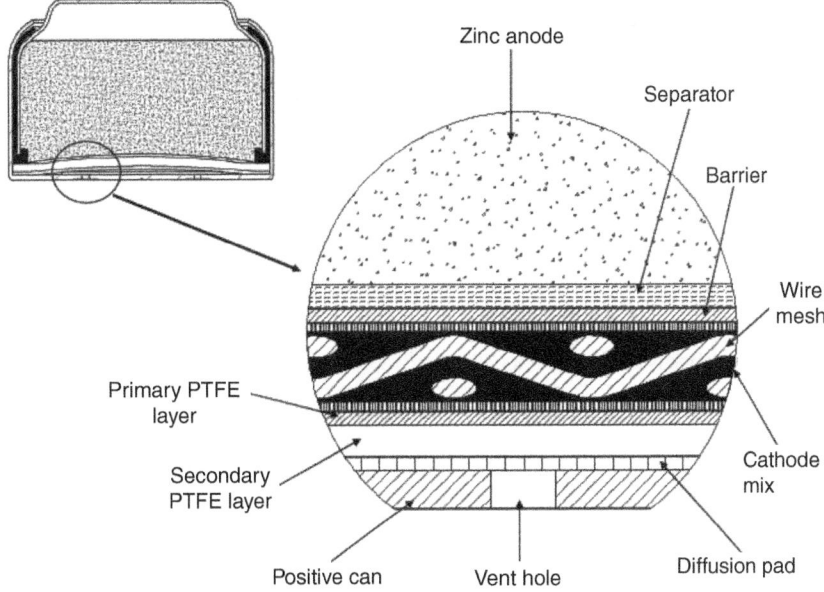

FIGURE 13.18 Key constructional features of the zinc/air cathode.

too fine in pore size to allow liquid to pass from inside the cell to the outside. In most commercial zinc/air batteries, this layer is made up of expanded PTFE film.

The next layer is the conductive mix of carbon and catalysts that are wetted with electrolyte. This creates the conditions for oxygen to be held at activated sites and thus made available for the electrochemical reaction which produces hydroxide ions in the electrolyte, and water is consumed from the electrolyte. To support this reaction, there needs to be a source of electrons. They come in from the external circuit, driven by the potential of the zinc in the anode. The electronic path for the current is through the current collector, usually a wire mesh screen or expanded metal layer impressed into the cathode active material. In the case of most button or coin cells, this is a disk blanked from a larger sheet which fits snugly into the inside diameter of the cathode can.

Air enters the cathode through vent holes that are punched into the flat bottom of the can. The access of air is controlled by a combination of size and number of the vent holes, as well as the degree to which the porosity of the PTFE sheet adjacent to the cathode material has been occluded or compressed during manufacture of that subassembly. Depending on the intended use of the cell, and the electrical current needed to meet that use, the parameters of vent size, their number, and the cathode rate capability might all be increased or decreased to suit the application.

On the anode side of the cathode structure are the separator/barrier layers, wettable to the electrolyte, but capable of preventing zinc or zinc oxide from directly contacting the cathode. If this were to happen, a direct electron path would be established, and the cell would self-discharge. The barrier is most commonly a microporous layer of polymer film that does not break down in the caustic environment of the cell and remains a good ionic conductor throughout discharge. The separator is most commonly a nonwoven cellulosic that is highly absorbent to electrolyte and helps to prevent zinc dendrite shorting. If the separator/barrier system interferes with the conductivity of the cell, then the rate capability, the capacity of the cell, or both will be adversely affected.

The negative top of the button cell contains all of the zinc and the majority of the electrolyte. Only a small portion of the electrolyte wets into the cathode, stopping at the hydrophobic layer. In order to allow for the growth of the anode, it is not possible to initially fill the entire top. Zinc metal has a density of approximately 7.14 g/cc, whereas the density of ZnO is about 5.47 g/cc. Zinc going to zinc oxide gains in mass by a factor of 1.25, with a corresponding volume increase due to the decrease in density of 1.63 g/cc. Space for that growth must be accounted for in the initial fill of anode material into the cell. Under equilibrium conditions, the electrolyte mass does not change during discharge, and occupies essentially the same volume at the end of discharge as it did when first assembled.

High-purity zinc is the active constituent of the anode. It is normally a distribution of finally divided, atomized particles. In some instances, it is alloyed to reduce the tendency for catalytic corrosion of the metal. Historically, as in virtually all button cells, it was also amalgamated with a small amount of mercury typically less than 25 mg/cell to reduce hydrogen overpotential, largely due to the segregation of trace metals to the grain boundaries of the zinc during the solidification that takes place as the zinc is atomized. Mercury would preferentially concentrate at the grain boundaries. Insoluble in the zinc oxide, the mercury would further concentrate into the remaining zinc, and eventually be released as beads of liquid metal suspended in the discharged anode. The impending introduction in 2011 of zero mercury zinc/air cells will eliminate the use of mercury in most commercial products.

The electrolyte used in zinc/air cells is normally a caustic solution of KOH at a concentration of about 30%. The electrolyte is highly conductive and wets through the cathode structure readily, providing excellent ionic access to both zinc and cathode active sites. At this concentration, the electrolyte is at equilibrium with water vapor at roughly 50% RH for ambient temperatures. Performance of the zinc/air system will be compromised if the rate of ingress or egress of water is significant. Gain of water will dilute the electrolyte slightly, but more significant is the increase in the volume of the anode and the reduction of the space that was initially left for the discharge product. The gain in water does not have a large effect on rate capability, but capacity of the battery will be cut short as the internal materials push the cathode against the inside bottom of the cell and reduce the effective cathode active area. When the environmental conditions promote drying out of the cell, there is a more immediate detrimental effect on the battery's electrical performance. Good ionic contact within the cell may be affected as gaps develop at the electrode interface and sufficient water is not available in the cathode to support the destruction of peroxide and subsequent hydroxyl formation.

Dry conditions in the anode promote early cementation of the partially discharged anode, and there is usually an increase in the battery's internal impedance.

There are four main electrical measurements for zinc/air cells:

- Open-circuit voltage
- Closed-circuit or operating voltage
- Internal cell impedance
- Limiting current

The first and second are easily measured. Open-circuit voltage represents the potential for current to flow and is usually 1.4 to 1.5 V for commercial zinc/air cells. Closed-circuit is the actual voltage that the cell can support when a defined load has been applied that is demanding a current to flow through a test circuit or through an actual device under test. The power being delivered is the product of voltage times amps, which is generally in the range of a few milliwatts for a zinc/air button cell.

The internal impedance of the battery is usually measured with an LCR meter and accounts for the AC components of resistance, capacitance, and inductance. For fresh zinc/air button cells, the impedance rises as one goes to low frequency levels. The impedance drops as cells are discharged and remains low until the end of discharge as the last of the available zinc is converted to zinc oxide.

The most dynamic and perhaps interesting electrical property of the zinc/air system is the limiting current, so named because it occurs when an increasing external load is applied, the cell has reached a point where there is no corresponding increase in current, and the voltage of the cell drops precipitously. Below the limiting current value, the cell is not internally rate limited; it is the external circuit that is limiting the flow of electrons. At the limiting current, there is a rate limiting process that has been maximized. At the beginning of discharge, the rate limiting process of the fresh cell is usually oxygen diffusion within the cathode, PTFE diffusion rate, or the vent hole configuration of the can. As discharge continues, the anode active electrode area (metallic zinc surface) decreases and becomes the rate limiting process.

13.9 PERFORMANCE CHARACTERISTICS

13.9.1 Cell Sizes

Zinc/air button and coin cells are available is a variety of sizes. Capacities range from 35 to 1000 mAh. Table 13.5 lists the physical and electrical characteristics of currently available zinc/air button and coin cells. The button cells are primarily used for hearing aid applications while the coin cell has been used in pager applications. With the advances in cochlear implants, a special higher power PR44 (675 Cochlear) has been designed for that application.

TABLE 13.5 Characteristics of Zinc/Air Button and Coin Batteries

Generic type	IEC no.	ANSI no.	Max. height (mm)	Max. diameter (mm)	Average weight (g)	Standard drain (mA)	High power drain (mA)	Rated capacity (mA)
5	–	–	2.15	5.8	.2	.4	–	33
10	PR70	7005ZD	3.6	5.8	.3	.7	1.0	75–105
312	PR41	7002ZD	3.6	7.9	.5	1.2	2.0	145–180
13	PR48	7000ZD	5.4	7.9	.8	2.0	3.0	265–310
675	PR44	7003ZD	5.4	11.6	1.8	5.0	8.0	600–650
675 Cochlear	–	–	5.4	11.6	1.8	10	20	550–570
2330	–	–	3.0	23.2	4	4.0	–	950

Source: Rayovac, Duracell, Energizer, Power One, Zeni, Panasonic.

Zinc/air batteries for hearing aid applications have been improved and refined over the years to meet stringent device and user requirements. With the advent of digital aids in the late 1990's, the battery design has had to adapt to higher current and pulse requirements. Today's zinc/air hearing aid batteries offer up to twice the capacity of the original designs of the late 1970s.

These improvements have been achieved by optimizing the internal anode volume without exceeding the standard external cell dimensions.[33,34] The zinc content is maximized without compromising the internal free volume needed for expansion as zinc metal is converted to zinc oxide. If the free internal volume is not adequate, the cell could have excessive expansion, leakage, or a premature end of life failure.

Recently, zinc/air prismatic batteries have been offered for OEM applications. The prismatic battery offers a 5 mm thin, prismatic construction designed to work with a variety of applications.[35] These batteries can last up to 3 times longer than alkaline batteries of a similar volume. As with all zinc/air batteries, the application needs to consume the energy quickly to achieve the full capacity advantage. Table 13.6 summarizes the characteristics of these prismatic batteries.

TABLE 13.6 Characteristics of Zinc/Air Prismatic Batteries

Size	PP425	PP355	PP255
Length (mm)	36.0	32.2	22.6
Width (mm)	22.0	14.7	10.3
Thickness (mm)	5.0	5.0	5.0
Volume (cm^3)	3.96	2.37	1.16
Cell weight (g)	11.7	6.8	3.4
Continuous rate capability	< 200 mW	< 100 mW	< 50 mW
Capacity rated (mAh)	3600	1800	720

Source: Energizer Zinc/Air Prismatic Handbook.

13.9.2 Voltage

The nominal open-circuit voltage for a zinc/air battery is 1.4 V. This value can vary from manufacturer to manufacturer because of differences in anode and cathode chemistry. Typically, the open-circuit value can range from 1.4 to 1.5 V. The initial closed-circuit voltage at 20°C ranges from 1.15 to 1.35 V depending on discharge load. The discharge is relatively flat, with the typical end voltages falling between 0.9 and 1.1 V.

In order to ensure freshness and long-term shelf life, the zinc/air battery's air holes are covered with a tape tab. The tab is designed to mute air ingress to the point that it lowers the open-circuit voltage (OCV). This lower tabbed voltage helps to determine if the tape tab is properly attached to the battery.

If the battery tape tab does not properly adhere, allowing excess air access, the OCV will be above 1.40 V. This would be the same OCV as if the zinc/air battery had been left untabbed for a couple of hours. If this condition occurs when the battery is in storage, the battery may dry out and not function when used.

It is important that the tape tab does not lower the voltage of the battery too much. A tabbed cell OCV of less than 1.0 V may not rise fast enough when untabbed to properly start the device the battery is powering.

Figure 13.19 illustrates the time it takes to achieve a functional voltage based on the initial tabbed voltage. The lower the tabbed voltage, the longer is takes to reach the functional voltage. Rise time can be influenced by changes in the air cathode chemistry, cell limiting current, or air hole design.

13.9.3 Energy Density

Zinc/air batteries have the highest volumetric energy density of any other primary button or coin cell chemistry system. The common hearing aid batteries range from 1300 Wh/L in the PR70 (size 10) to 1400 Wh/L in the PR44 (size 675).

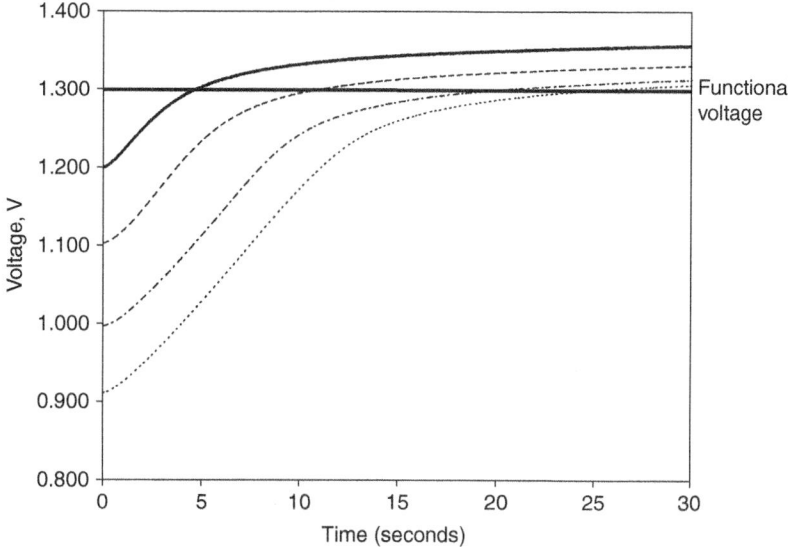

FIGURE 13.19 Typical rise time for zinc/air batteries of selected tabbed OCV.

13.9.4 Cell Internal Impedance

The effects of discharge level and signal frequency on the internal impedance characteristics of a PR48 (13) button battery are presented in Fig. 13.20. Fresh cells have the highest impedance at low frequencies. This low frequency impedance decreases with depth of discharge.

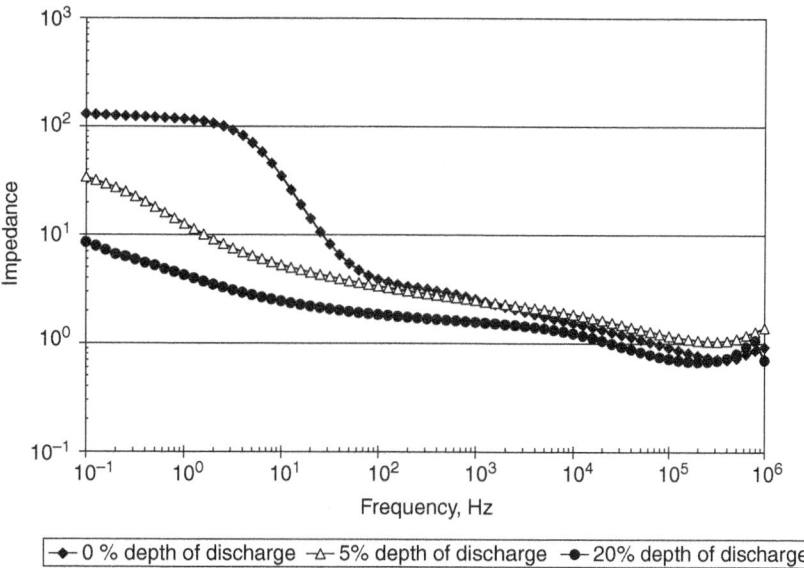

FIGURE 13.20 Impedance vs. frequency and depth of discharge for a PR48 (13) zinc/air battery.

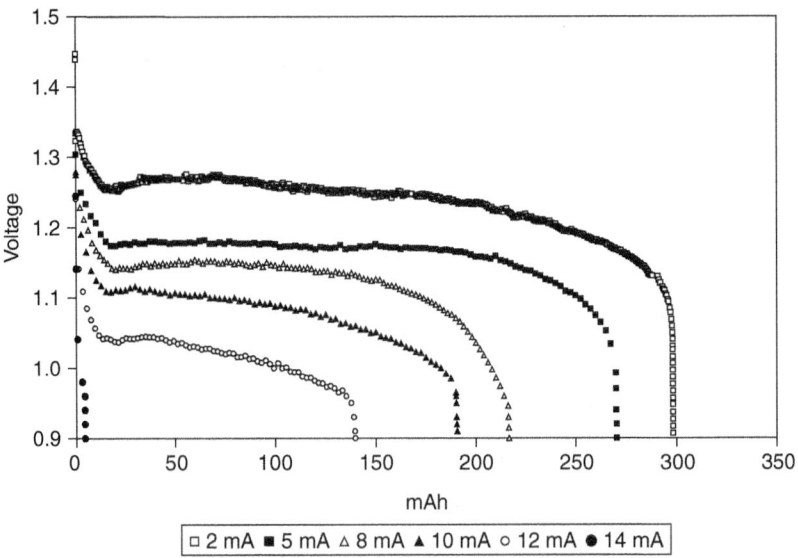

FIGURE 13.21 Discharge curves for a PR48 (13) zinc/air battery at six different rates of discharge

13.9.5 Discharge Characteristics

A set of discharge curves, typical of zinc/air button batteries at 20°C, is presented in Fig. 13.21. At low to moderate current levels, the discharge curves are relatively flat. As the discharge current goes up, the discharge voltage goes down. When the discharge rate approaches the cell's limiting current, the delivered capacity is decreased and cell polarization increases.

13.9.6 Voltage-Current Performance

The catalytic activity of the cathode, along with the amount of oxygen access to the cathode, generally defines the voltage-current profile of a zinc/air battery. Oxygen access is defined by the degree of air access to the battery's cathode. Improved oxygen access to the cathode increases the cell's power output. Oxygen access can be improved by increasing the number or size of the air access holes in the battery case. If the number and size of the air holes are kept constant, the power capability of the battery can be increased by raising the limiting current of the cathode. To minimize the detrimental effects of water vapor transport, air access needs to be properly balanced.[36,37] In a low humidity environment, a high rate of water vapor loss accelerates dry-out, lowering the battery's overall capacity. In a high humidity environment, a high rate of water vapor gain takes up the free volume designed for the zinc anode discharge expansion, reducing the battery's overall capacity along with causing possible swelling or leakage. Understanding the maximum current requirements of the application using the zinc/air battery is important to minimizing the detrimental effects of water vapor transport.

Figure 13.22 illustrates the effect of increasing the total air hole circumference on the cathode limiting current for a PR48 (size 13) zinc/air battery. Limiting current is the maximum current a zinc/air battery or cathode can deliver for a given set of conditions. The limiting current is measured by potentiostatically setting the battery or cathode to a voltage of 0.9 V and then measuring the current output after 1 to 5 minutes from the initiation of the test.

Since the limiting current is a measure of air access to the cathode, this measurement can also be used to determine water vapor transport rates (see Fig. 13.23). Since internal electrolyte evaporation is

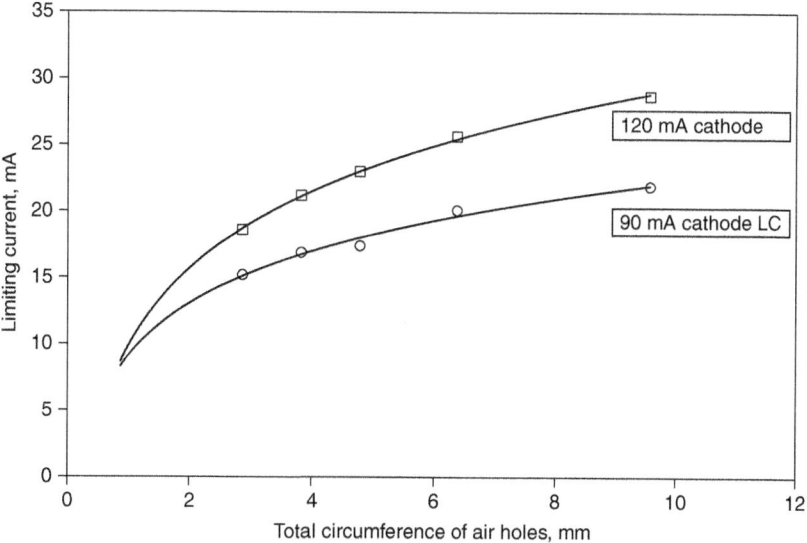

FIGURE 13.22 Battery limiting current as a function of air hole circumference and cathode limiting current in a PR48 (13) battery.

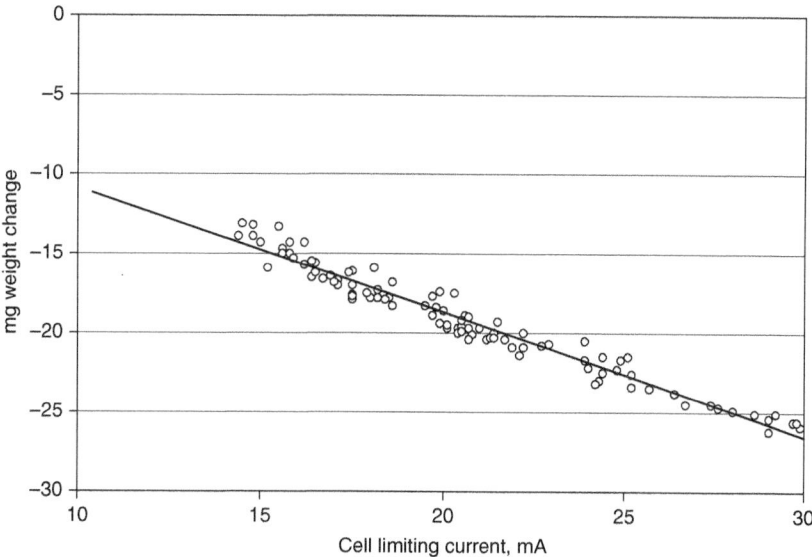

FIGURE 13.23 PR48 (13) weight loss vs. cell limiting current after 6 days at 20°C, 20% RH.

the primary source of cell weight loss, the water vapor transport rate is directly related to cell limiting current in a low-humidity environment.

In Fig. 13.24, a cell limiting current versus water vapor transport relationship also exists in high-humidity conditions. A high cell limiting current increases the water vapor transport rate, increasing the cell weight gain.

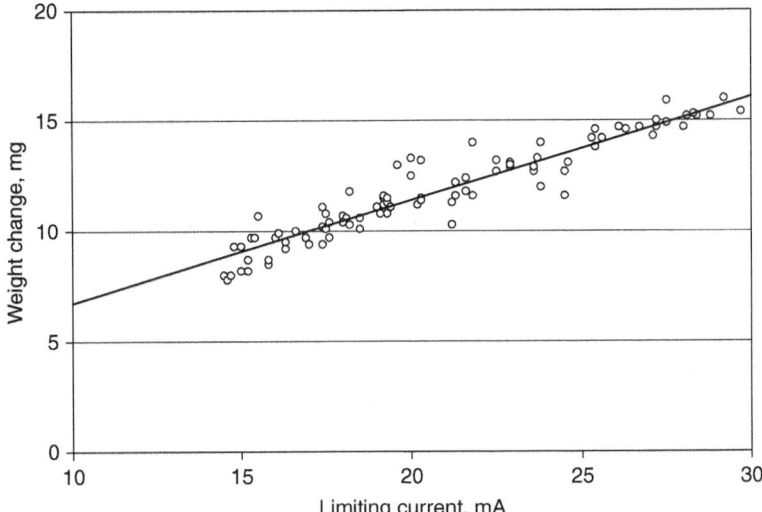

FIGURE 13.24 PR48 (size 13) weight gain vs. cell limiting current after 6 days at 20°C, 80% RH.

The maximum power output of the zinc/air battery can be improved by increasing the catalytic activity of the cathode. Catalytic additives are commonly mixed into the carbon mix component of the cathode.[38,39] Various forms of MnO_x are typically used as catalysts for the zinc/air cathode. Figure 13.25 compares the Tafel plots of a carbon cathode with and without a manganese oxide catalyst. The use of a manganese oxide catalyzed cathode instead of the carbon cathode in a cell will produce a higher operating voltage.

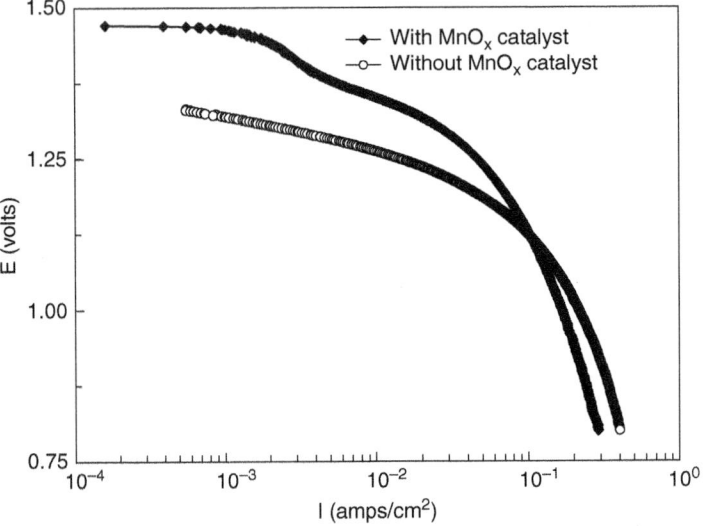

FIGURE 13.25 Voltage as a function of current for an activated carbon cathode with and without a manganese oxide catalyst.

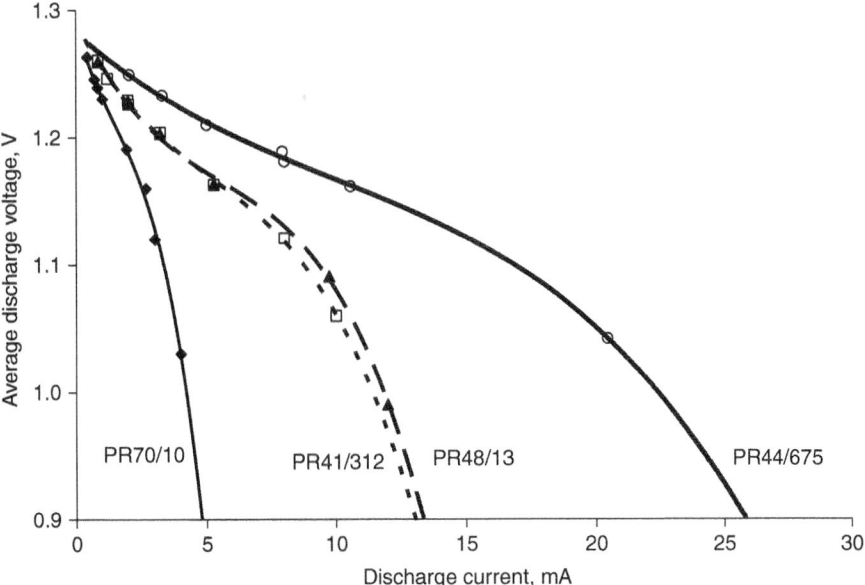

FIGURE 13.26 Average voltage-current profiles for various zinc/air button batteries at 20°C.

The average voltage-current profiles for various sized zinc/air button batteries are shown in Fig. 13.26. The average voltage of the battery falls as the current increases, until the battery becomes oxygen starved. Once the battery is oxygen starved, it can't support the load. Increasing the diameter of the battery will increase its constant current capability.

13.9.7 Pulse Load Performance

Zinc/air batteries can handle pulse currents much higher than the limiting current (I_L) of the battery. How high a pulse current level the battery can handle depends on the nature of the pulse. This capability results from the reservoir of oxygen that builds up within the cell when the current load is below the cell's limiting current.

As long as the average current (I_{ave}) of the pulse load does not exceed the cell limiting current (I_L), the zinc/air battery is able to sustain the pulse load. Figure 13.27 illustrates the resulting voltage profile of a series of PR41(312) batteries with a I_L of 12 mA that had been subjected to an ever-increasing series of 1 s, 15 mA pulses over a 5 mA background current drain. In this illustration, the duty cycle of the 15 mA pulse was increased from 10 to 50%. As the duty cycle increases, the average current is increased, reducing the overall running voltage of the cell. Since the average current of the pulse regime never exceeded the battery's limiting current, the battery was able to sustain the pulse regime. If the pulse regime's average current exceeds the battery's limiting current, the battery will become oxygen starved, and the cell's voltage level will collapse.

Although the battery may not be oxygen starved, the voltage level achieved during the pulse regime must be considered. In our example, if the device using the batteries doesn't work below 1.1 V, increasing the pulse duty cycle will cause premature failure in the device. If, however, the device works down to 0.9 V, increasing the duty cycle will have no effect on the function of the device. Very high, short-duration pulses can be achieved by the zinc/air cell as long as the average current of the pulse and the background current don't exceed the battery's limiting current and the voltage level of the pulse doesn't go lower than the functional voltage level of the device.

13.28 PRIMARY BATTERIES

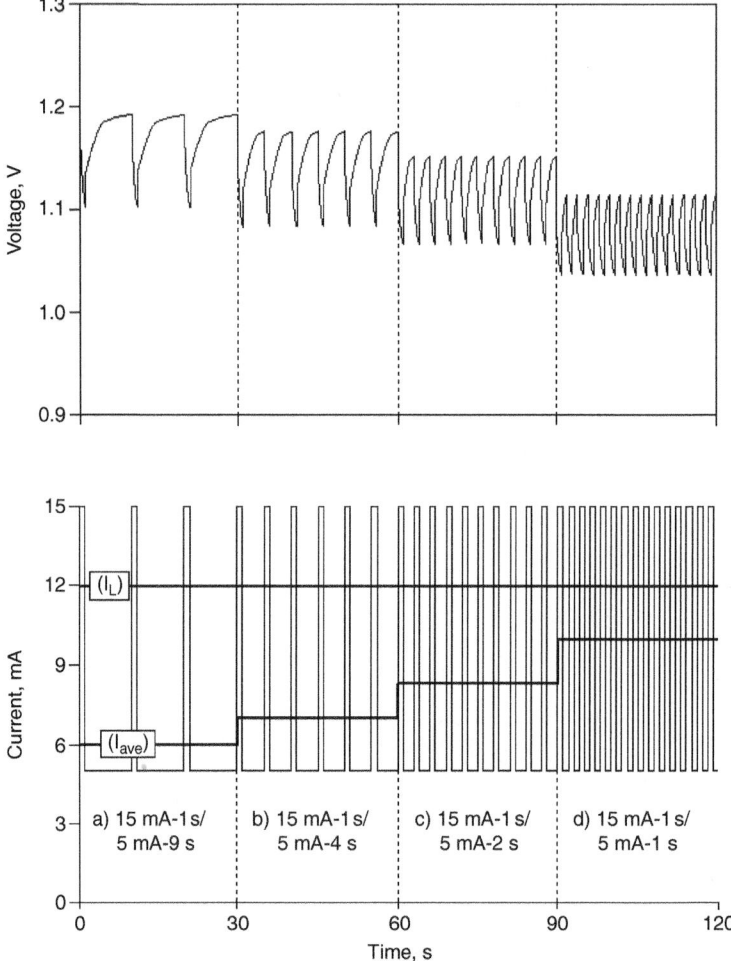

FIGURE 13.27 Pulse loads profiles of a PR41 (312) zinc/air battery after 20 mAh of discharge under the pulsing conditions.

13.9.8 Effect of Temperature

The effect of temperature at various discharge rates is illustrated in Fig. 13.28. The decrease in voltage levels as the temperature decreases is primarily due to electrolyte effects. Adding potassium hydroxide to water lowers the freezing point, and at concentrations typical for zinc/air cells, the electrolyte will freeze at lower than −40°C. Also, as the temperature lowers, the conductivity of the electrolyte is decreased, slowing the discharge reaction. An application running at low discharge rate can function at a lower temperature than an application requiring a higher rate of discharge.

13.9.9 Effect of Altitude

The cathode of a zinc/air cell relies on oxygen from the external environment being available at a given partial pressure (21% of 760 mm at sea level or 160 mm). In low current demand applications, the rate

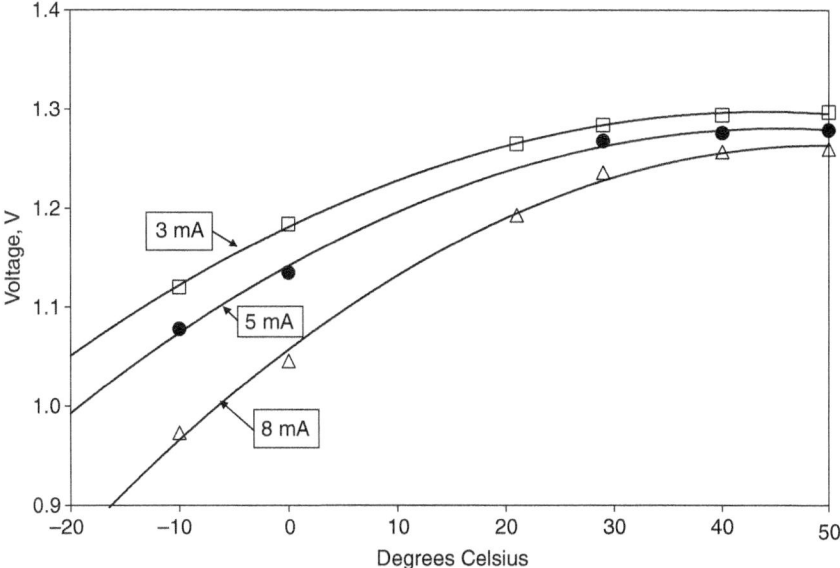

FIGURE 13.28 Discharge voltage level of a PR44 (675) zinc/air battery as a function of discharge current and temperature

of oxygen diffusion is limited by the actual oxygen used in discharge, but the situation changes when the demand is oxygen limited. At increased elevation, the barometric pressure goes down, as does the partial pressure of each gas present, including oxygen. As a result, the limiting current of a cell decreases as it operates at higher altitudes where the concentration of oxygen is lower than it is at sea level. The effect has been noted by active hearing aid users who have hiked to greater elevations. Airliners are pressurized to a "pressure altitude" around 8,000 ft. This is enough to decrease limiting current by about 25%. The relative partial pressure of oxygen at various points on interest, expressed as a percent of the oxygen available at sea level (760 mm barometric pressure) is displayed in Table 13.7.

Tests of cells in a chamber that had been evacuated to simulate the pressure at altitude declined in limiting current (i.e., the current that can be made to flow if the CCV is held to 0.9 V) and in

TABLE 13.7 Partial Pressure of Oxygen at Altitude

Elevation (ft)	Barometric pressure (mm Hg)	Partial pressure of oxygen (mm Hg)	Percent of sea level pressure	Points on earth
−1500	802	168.0	105.5	Dead Sea, Israel-Jordan (−1317 ft)
−500	774	162.5	101.8	Death Valley, CA (−282 ft)
0 (sea level)	760	160.0	100.0	London, England
500	746	156.7	98.2	Montmartre, Paris, France (423 ft)
1,000	733	153.9	96.4	Vaalserberg, Netherlands (1,053 ft)
2,000	707	148.4	93.0	High Willhays, Cumbria, UK (2,037 ft)
5,000	633	132.8	83.2	Denver, CO (5,280 ft)
10,000	523	109.8	77.2	Cascade Mountain, Canadian Rockies (9,836 ft)
20,000	349	73.4	46.0	Mt. McKinley (20,320 ft)
30,000	226	47.48	29.8	Mt. Everest (29,028 ft)

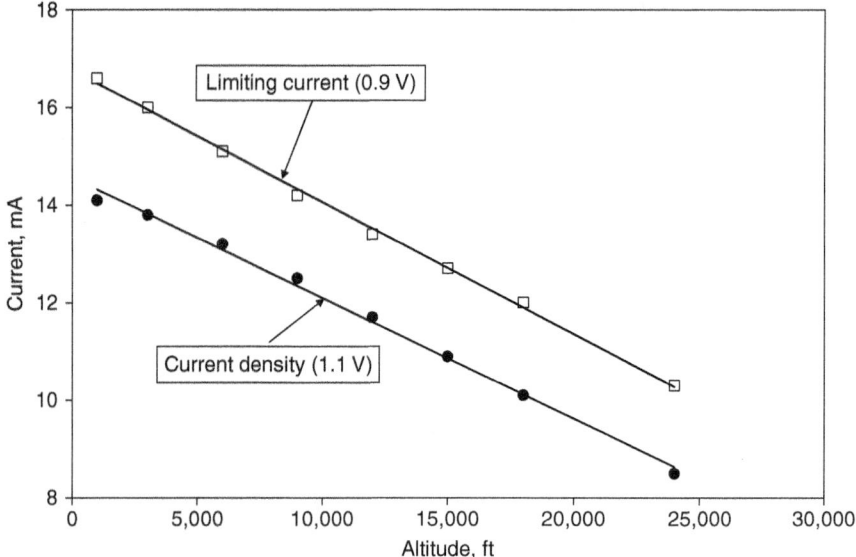

FIGURE 13.29 Change in limiting current (0.9 V) and current density (1.1 V) for a PR48 (13) size cell operated at a pressure equivalent of altitude.

current density (i.e., the current that can be made to flow if the CCV is held to 1.1 V). The current delivered exceeded slightly the predicted levels of the theoretical model. An example of a PR48 (13) size cell is given in Fig. 13.29.

13.9.10 Storage Life

Four principle mechanisms affect the capacity of zinc/air batteries during storage and operating service. One mechanism, self-discharge of the zinc (corrosion), is an internal reaction: the other three are caused by gas transfer. The gas transfer mechanisms are direct oxidation of the zinc anode, carbonation of the electrolyte, and electrolyte water gain or loss.

During storage, the air access holes of the zinc/air battery are sealed to prevent gas transfer decay. Only enough oxygen is allowed into the cell to give a sealed open-circuit voltage of greater than 1 V. Oxygen, one of the cell's reactants, is severely restricted from entering the cell during storage. Limiting air access gives zinc/air batteries excellent shelf-life performance.

The primary mechanism affecting the shelf life of a zinc/air battery is the self-discharge reaction. Zinc is thermodynamically unstable in alkaline electrolyte and reacts to form zinc oxide and hydrogen gas. To control this reaction, additives are used in the anode. Mercury historically has been one of the additives used to control this self-discharge reaction. Environmental concerns will force mercury's removal from the cell anode chemistry, resulting in new additives to control the self-discharge of the zinc/air system.

Capacity retention results of PR41 (312) and PR48 (13) cells are presented in Fig. 13.30. Under low rate conditions, the batteries lose about 3% a year, while increasing the discharge rate 2 to 3 times increases the rate of loss to 7 to 8% a year. Improving self-discharge storage retention can result in trade-offs with other cell performance parameters such as the discharge voltage level.

Elevated temperature will increase the rate of the self-discharge reaction and is used as an analytical tool to accelerate the effects of additives on the self-discharge performance of zinc/air batteries.

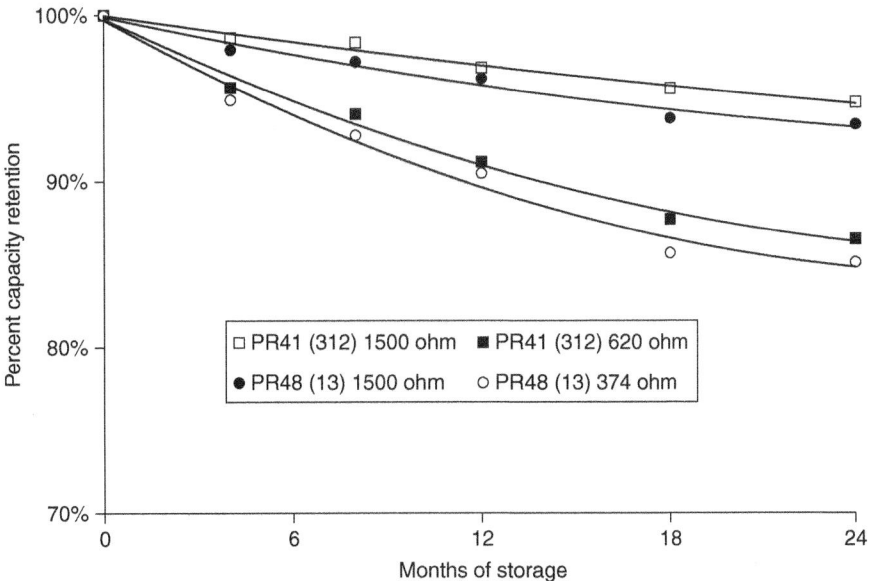

FIGURE 13.30 Capacity retention of PR41 (312) and PR48 (13) batteries at different rates of discharge.

13.9.11 Factors Affecting Service Life

The zinc/air system is open to the atmosphere, even when the tab is in place restricting gas transmission into the cell. The most immediate effect that the external environment can have on the cell is that of relative humidity. Other effects that are generally of less concern, but are well documented, include carbonation of electrolyte, direct oxidation, and the impact that high-altitude use can have on rate capability. In the use profile of a hearing aid, the most common use for zinc/air batteries, these effects are difficult, if not impossible, to notice.

Carbonation of Electrolyte. While alkaline electrolytes have significant solubility for carbon dioxide, most zinc/air cells are used within weeks of being opened to the external environment when the tab is removed. As a result, carbonation will not be a factor in the use of the product. Extremely light drain or intermittent duty that would extend the use of the product beyond a month can challenge the zinc/air cell's service life, first in response to relative humidity, and then to possible carbonate crystals forming in the electrolyte by the gas diffusion membrane of the cathode. Crystallization can produce a pathway for direct electrolyte leakage if this occurs.

Direct Oxidation. Direct oxidation is not a significant factor in the consumption of the zinc in the zinc/air cell as long as the tab is kept in place and oxygen access to the cell is properly restricted. Any alkaline zinc cell needs oxygen to discharge the zinc and release the electrons that will flow back to the positive terminal through the external circuit. Normal discharge occurs when the hydroxyl ions interact with the zinc, producing zinc hydroxide species and eventually ZnO. There is a significant solubility in the electrolyte for zinc, zinc oxide, and the zincates (hydroxyl species). Oxygen is also soluble in the aqueous KOH electrolyte, leading to a secondary means to oxidize the zinc that is present in metallic form. The source of this oxygen is from the gas-liquid interface found by the gas diffusion layer of the cathode.

Effect of Water Vapor Transfer on Service Life. The primary cause of service life reduction in a zinc/air cell is water vapor transfer. As illustrated in Fig. 13.31, water vapor transfer occurs when a

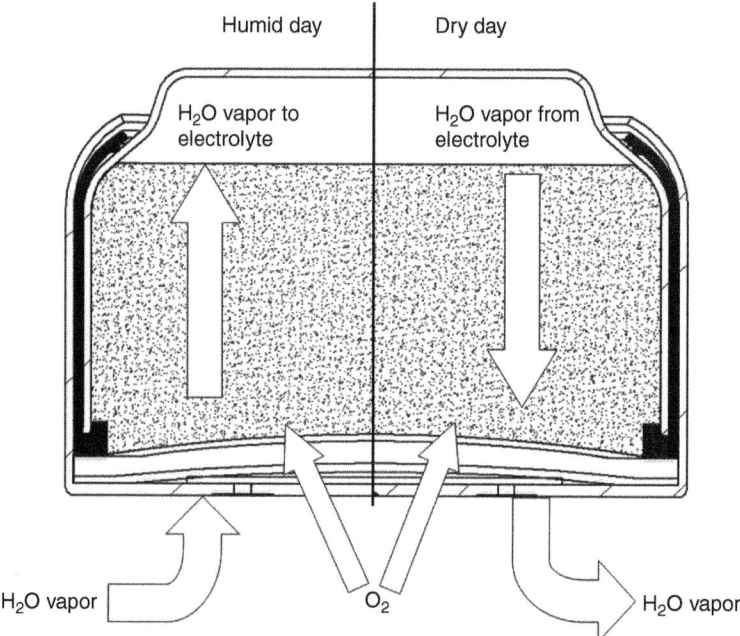

FIGURE 13.31 Water vapor transfer mechanism in a zinc/air cell.

partial pressure difference exists the between the vapor pressure of the internal cell electrolyte and that of the outside environment. The internal cell vapor pressure is determined by the cell's electrolyte at a given temperature. If the external humidity is lower than the cell's internal relative humidity (dry day), the cell will lose water. If the external humidity is higher than the cell's internal relative humidity (humid day), the cell will gain water. Excessive water loss causes the electrolyte to concentrate, increasing cell impedance and promoting carbonation. Eventually the loss of water will cause the cell to dry out to the point where direct oxidation can occur. Excessive water gain dilutes the cell electrolyte, reducing conductivity. The addition of water vapor to the cell can flood the cathode and fill up the anode free space cavity that is designed for zinc oxide expansion, eventually causing loss of rate capability, battery swelling, or leakage.

Figure 13.32 illustrates the relationship between KOH concentration and relative humidity at room temperature. Based on the cell design requirements, the desired electrolyte concentration can range from 25 to 40% by weight. At a given temperature, lowering the electrolyte concentration will raise the internal cell relative humidity. The lowering of the electrolyte concentration will slow the rate of water vapor transport in a high-humidity environment but increase the rate in a low-humidity environment. The opposite effect occurs with vapor transport if the concentration of electrolyte is raised.

A thorough understanding of the intended application is required to properly design a zinc/air battery. Knowing the rate requirements and functional voltages of the application along with environmental conditions of use can determine the trade-offs that can be made to optimize cell performance.

The effect of water vapor transport is demonstrated in the evaluation described below. Table 13.8 compares the average limiting currents and open-stand weight changes of three commercially available PR41 (312) zinc/air batteries, typically used in hearing aids. The batteries, with their seal tabs removed, were initially weighed and placed in three different 20°C relative-humidity environments. After 7 days untabbed, the batteries were weighed again and the change in battery weight was determined. It is assumed that cell weight change is due to the water vapor exchange with the environment, as illustrated in Fig. 13.31.

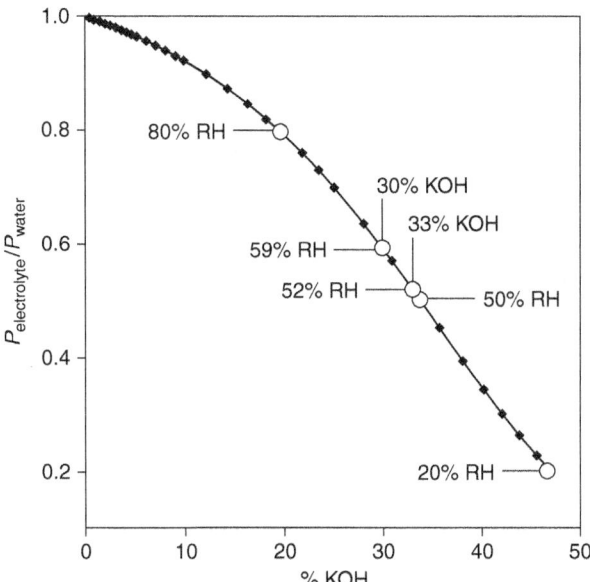

FIGURE 13.32 Closed container relative humidity as a function of KOH concentration.

TABLE 13.8 Average cell weight change after 7 days at 20°C on open stand in various relative humidity conditions for three commercially available size PR41 (312) zinc/air batteries

Design	Cell limiting current (mA)	Cell Weight Change (mg)			
		20% RH	50% RH	80% RH	Total range
A	7.5	−7.3	0	5.8	13.1
B	10.4	−10.4	−1.9	6.7	17.1
C	13.9	−11.7	−1.4	9.7	23.6

With the lowest average limiting current, design A had the lowest total range of cell weight change after 7 days. Design C's average limiting current was 85% higher and had 80% more total weight change than design A. Figure 13.33 compares the average 7-day cell weight changes of the three designs in three different humidity environments. The graph illustrates how the designs are targeted for relative humidity environment. Design A crosses the no-weight-change line at 50% relative humidity, while design B crosses the no-weight-change line at between 55 and 60% relative humidity. If design A and design B had the same limiting current, design A would have less weight change in the low-humidity environment but more weight change in the high-humidity environment.

Initially, the three designs were submitted to two test loads at various relative humidities. The first evaluation was on the lower drain 1500 ohm, 12 h/day test. The typical duration for this test at 50% relative humidity is about 16 to 18 days. Figure 13.33 compares the performance of the three designs on this test in a 20 and 50% relative humidity environment. To normalize the different capacities, the results of each capacity test are plotted in Fig. 13.34 as a percentage of each design's average 1500 ohm, 12 h/day result compared to the 50% RH test result.

The second evaluation was at the moderate drain 620 ohm, 16 h/day test. The typical duration for this test at 50% relative humidity is about 5 to 6 days. Figure 13.35 compares the performance of

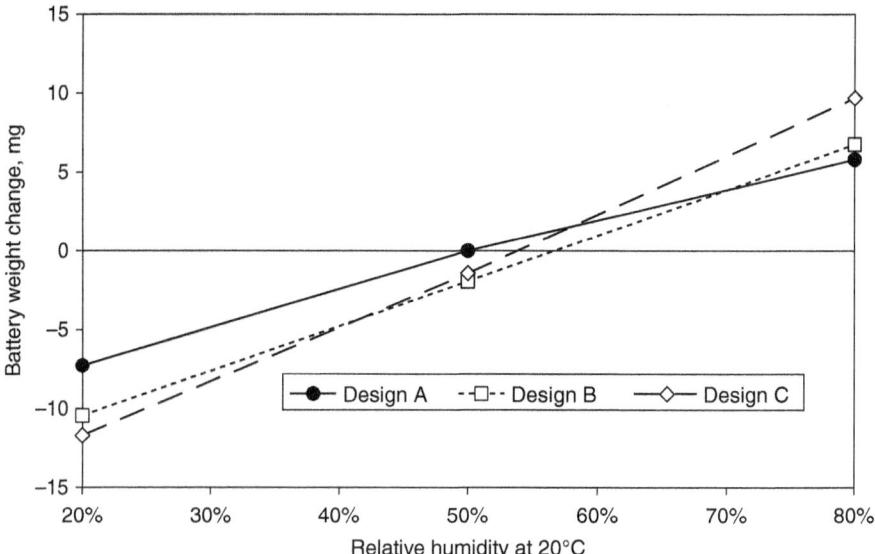

FIGURE 13.33 Change in battery weight of three different manufactures of PR41 (312) zinc/air batteries after 7 days open stand at 20, 50, and 80% relative humidities.

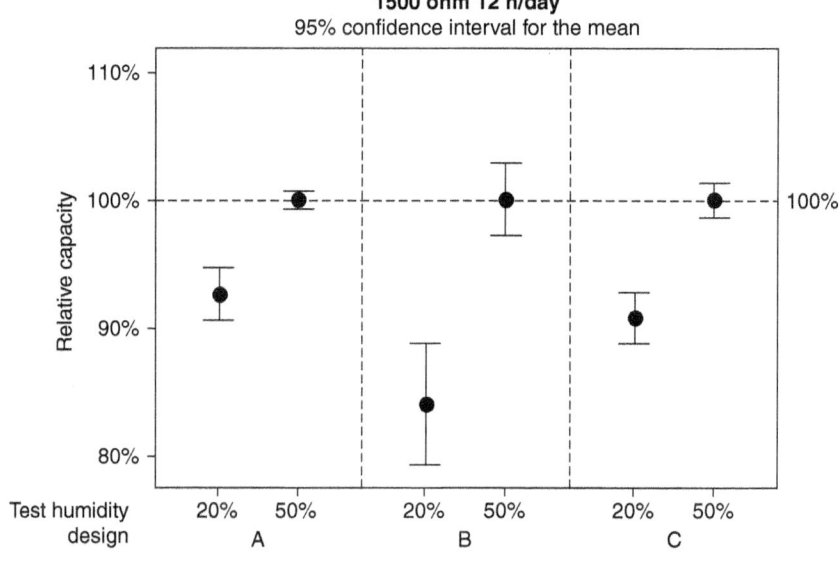

FIGURE 13.34 The initial 20% relative humidity capacity performance as a percentage of each design's 50% relative humidity result on the 1500 ohm, 12 h/day test at 20°C.

the three designs on this test in 20, 50, and 80% relative humidity. To normalize the different capacities of the designs, the results of each capacity test are plotted in Fig. 13.36 as a percentage of each design's average 620 ohm, 16 h/day result at 50% relative humidity.

For the short duration moderate 620 ohm test, none of the designs experienced a significant loss of performance in high- and low-humidity conditions. However, when the typical test time frame

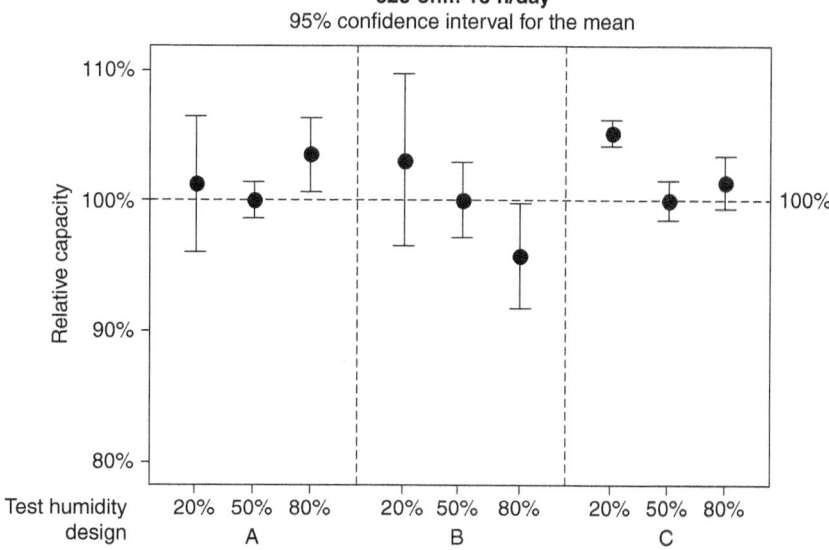

FIGURE 13.35 The initial 20 and 80% relative humidity capacity performance as a percentage of each design's 50% relative humidity performance on the 620 ohm, 16 h/day test at 20°C.

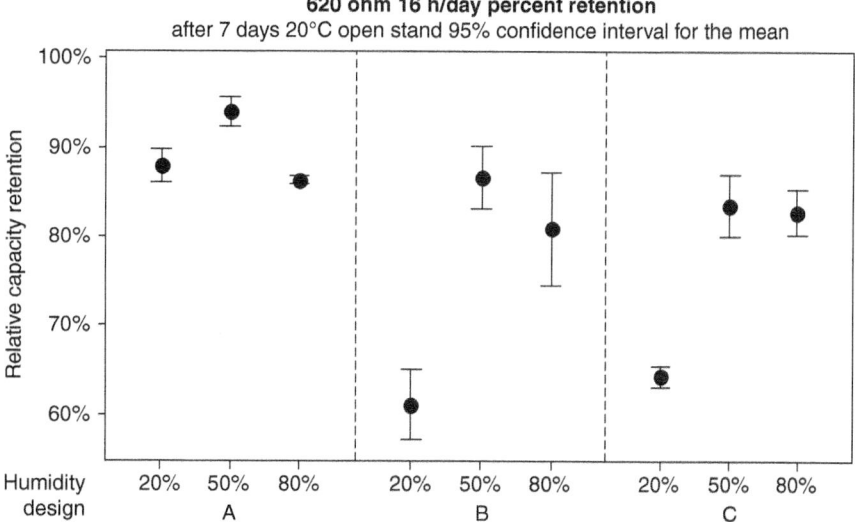

FIGURE 13.36 The percentage retention of the initial capacity on the 620 ohm, 16 h/day test after 7 days open stand in 20, 50, and 80% relative humidity open stand at 20°C.

was changed to 16 to 18 days, a significant loss to performance occurred at 20% relative humidity. Design B with the middle limiting current range and higher internal battery relative humidity had a 15% performance loss in the low-humidity test environment.

Figure 13.36 summarizes the percent of the initial performance the various designs obtained on the moderate drain 620 ohm test after 7 days open stand. For instance, a sample of design C was left open for 7 days in a 20°C, 20% relative humidity environment. In that time the cells lost an average of 11.7 mg water vapor and then when tested on the 620 ohm test in the same 20% relative humidity environment performed at 65% of the typical initial 20% RH results. Design A, with the lowest limiting current, had the best retention after open stand, since it restricts water vapor transport up to 85% better than the other designs.

REFERENCES

1. F. Kober and H. West, "The Anodic Oxidation of Zinc in Alkaline Solutions," Extended Abstracts, The Electrochemical Society, Battery Division 12, 66–69 (1967).
2. A. Fleischer and J. Lander (eds.), *Zinc-Silver Oxide Batteries*, Wiley, New York, 1971.
3. W. M. Latimer, *Oxidation Potentials*, Prentice Hall, Englewood Cliffs, NJ, 1952.
4. D. R. Lide (editor-in-chief), *Handbook of Chemistry and Physics*, 73rd ed., CRC Press, Boca Raton, FL, 1992.
5. A. Shimizu and Y. Uetani, "The Institute of Electronics and Communication Engineers of Japan," Tech. Paper CPM79-55, 1979.
6. T. Nagaura and T. Aita, U.S. Patent 4,370,395 (1981).
7. T. Nagaura, "New Material $AgNiO_2$ for Miniature Alkaline Batteries," *Progress in Batteries and Solar Cells*, **4:**105–107 (1982).
8. E. A. Megahed, "Small Batteries for Conventional and Specialized Applications," *The Power Electronics Show and Conference*, San Jose, CA, pp. 261–272 (1986).
9. B. C. Cahan, U.S. Patent 3,017,448 (1959).
10. P. Ruetschi, in *Zinc-Silver Oxide Batteries*, A. Fleischer and J. J. Lander, eds., Wiley, New York, p. 117 (1971).
11. E. A. Megahed and C. R. Buelow, U.S. Patent 4,078,127 (1978).
12. A. Tvarusko, *J. Electrochem. Soc.* **116:**1070A (1969).
13. S. M. Davis, U.S. Patent 3,853,623 (1974).
14. E. A. Megahed, C. R. Buelow, and P. J. Spellman, U.S. Patent 4,009,056 (1977).
15. E. A. Megahed and D. C. Davig, "Long Life Divalent Silver Oxide-Zinc Primary Cells for Electronic Applications," in *Power Sources*, Vol. 8, Academic, London, 1981.
16. E. A. Megahed and D. C. Davig, "Rayovac's Divalent Silver Oxide-Zinc Batteries," *Progress in Batteries and Solar Cells*. **4:**83–86 (1982).
17. E. A. Megahed and A. K. Fung, U.S. Patent 4,835,077 (1989).
18. J. L. Passaniti, E. A., Megahed, and N. Zreiba, U.S. Patent 5,389,469 (1994).
19. "Bismuth," in *Minerals, Facts, and Problems*, Bureau of Mines Bulletin 675, U.S. Department of the Interior (1985).
20. E. J. Rubin and R. Babaoian, "A Correlation of the Solution Properties and the Electrochemical Behavior of the Nickel Hydroxide Electrode in Binary Aqueous Alkali Hydroxides," *J. Electrochem. Soc.* **118:**428 (1971).
21. "Kagaku Benran," Maruzen, Tokyo, 1966.
22. H. André, *Bull. Soc. Franc. Elect.* **6:**1, 132 (1941).
23. V. D'Agostino, J. Lee, and G. Orban, "Grafted Membranes," in *Zinc-Silver Oxide Batteries*, A. Fleischer and J. J. Lander, eds., Wiley, New York, 1971, pp. 271–281.
24. R. Thornton, "Diffusion of Soluble Silver-Oxide Species in Membrane Separators," General Electric Final Report, Schenectady, NY (1973).
25. S. Hills, "Thermal Coefficients of EMF of the Silver (I) and the Silver (II) Oxide-Zinc-45% Potassium Hydroxide Systems," *J. Electrochem. Soc.* **108:**810 (1961).
26. M. N. Hull and H. I. James, "Why Alkaline Cells Leak," *J. Electrochem. Soc.* **124:**332–339 (1977).

27. J. Oltman, R. Dopp, and D. Carpenter, U.S. Patent 4,649,090.
28. E. Yeager, "Electrochemical Catalysis for Oxygen Electrodes," Rep. LBL-25817, Lawrence Berkeley Lab., CA 1988.
29. C. Warde and A.D. Glasser, U.S. Patent 3935027.
30. B. Szczesniak et al., Abstract Number 280, *Joint International Meeting of ECS and ISE*, Paris, 1997.
31. G. W. Elmore and H. A. Tanner, U.S. Patent 3,419,900.
32. A.M. Moos, U.S. Patent 3,267,909.
33. J. Oltman, B. Dopp, and J. Burns, U.S. Patent 5,567,538.
34. J. Oltman, U.S. Patent 6,245,452 B1.
35. Energizer Zinc Air Prismatic Handbook, www/energizer.com, Winter 2009.
36. A. Ohta, A. Hanafusa, H. Yoshizawa, and Z. Ogumi, "Design of Air Holes on Button Type Zinc-Air Batteries. I. New Evaluation Method of Both Water Vapor and Oxygen Premeabilities," *Denki Kagaku (Electrochemistry)*, Vol. 65, No. 5, 1997.
37. A. Ohta, H. Yoshizawa, A. Hanafusa, and Z. Ogumi, "Design of Air Holes on Button Type Zinc-Air Batteries. II. Simulation of Gas Flow Though Air Holes," *Denki Kagaku (Electrochemistry)*, Vol. 66, No. 4, 1998.
38. J. Passanti and R. Dopp, U.S. Patent 5,308,711.
39. A. Ohta, Y. Morita et al., "Manganese Oxide as a Catalyst for Zinc-Air Cells", *Proc. Battery Material Symp.*, 1985.

BIBLIOGRAPHY

Steven F. Bender, John W. Cretzmeyer, and Terrence F. Riese, "Zinc/Air Batteries-Button Configuration," *Handbook of Batteries*, 3rd ed., McGraw-Hill Companies, New York, 2002, Chapter 13.

CHAPTER 14
LITHIUM PRIMARY BATTERIES

Thomas B. Reddy

14.1 GENERAL CHARACTERISTICS

Lithium metal is attractive as a battery anode material because of its light weight, high voltage, high electrochemical equivalence, and good conductivity. Because of these outstanding features, the use of lithium has dominated the development of high-performance primary batteries during the last three decades. (Chapters 26 and 27 cover lithium secondary batteries.)[1]

Serious development of high-energy-density battery systems was started in the 1960s and concentrated on nonaqueous primary batteries using lithium as the anode. The lithium batteries were first used in the early 1970s in selected military applications, but their use was limited as suitable cell designs, formulations, and safety considerations had to be resolved. Lithium primary cells and batteries have since been designed, using a number of different chemistries, in a variety of sizes and configurations. Sizes range from less than 5 mAh to 10,000 Ah; configurations range from small coin and cylindrical cells for memory backup and portable applications to large prismatic cells for standby power.

Lithium primary batteries, with their outstanding performance and characteristics, are being used in increasing quantities in a variety of applications, including cameras, memory backup circuits, security devices, calculators, watches, etc. Nevertheless, lithium primary batteries have not attained a major share of the market as was anticipated, because of their high initial cost, concerns with safety, the advances made with competitive systems, and the cost-effectiveness of the alkaline/manganese battery. Worldwide sales of lithium primary batteries for 2009 have been estimated at $1.3 billion.[2]

14.1.1 Advantages of Lithium Cells

Primary cells using lithium anodes have many advantages over conventional batteries. The advantageous features include the following:

1. *High voltage.* Lithium batteries have voltages up to about 4 V, depending on the cathode material, compared with 1.5 V for most other primary battery systems. The higher voltage reduces the number of cells in a battery pack by a factor of at least 2.

2. *High specific energy and energy density.* The energy output of a lithium battery (up to 870 Wh/kg and 1180 Wh/L) is 2 to 5 times better than that of conventional zinc anode batteries.

3. *Operation over a wide temperature range.* Many of the lithium batteries will perform over a temperature range from about 70 to −40°C, with some capable of performance to 150°C or as low as −80°C.

14.2 PRIMARY BATTERIES

4. *Good power density.* Some of the lithium batteries are designed with the capability to deliver their energy at high current and power levels.
5. *Flat discharge characteristics.* A flat discharge curve (constant voltage and resistance through most of the discharge) is typical for many lithium batteries.
6. *Superior shelf life.* Lithium batteries can be stored for long periods, even at elevated temperatures. Storage of up to 10 years at room temperature has been achieved and storage of 1 year at 70°C has also been demonstrated. Shelf lives over 20 years have been projected from reliability studies.

The performance advantages of several types of lithium batteries compared with conventional primary and secondary batteries, are shown in Section 8.3. The advantage of the lithium cell is shown graphically in Figs. 8.2 to 8.10 which compare the performance of the various primary cells. Only the zinc/air, zinc/mercuric oxide, and zinc/silver oxide cells, which are noted for their high energy density, approach the capability of the lithium systems at 20°C. The zinc/air cell, however, is very sensitive to atmospheric conditions; the others do not compare as favorably on a specific energy basis nor at lower temperatures.

14.1.2 Classification of Lithium Primary Cells

Lithium batteries use nonaqueous solvents for the electrolyte because of the reactivity of lithium in aqueous solutions. Organic solvents such as acetonitrile, propylene carbonate, and dimethoxyethane and inorganic solvents such as thionyl chloride are typically employed. A compatible solute is added to provide the necessary electrolyte conductivity. (Solid-state and molten-salt electrolytes are also used in some other primary and reserve lithium cells; see Chaps. 27, 31, 33, and 36.) Many different materials were considered for the active cathode material; sulfur dioxide, thionyl chloride, manganese dioxide, iron disulfide, and carbon monofluoride are now in common use. The term "lithium battery," therefore, applies to many different types of chemistries, each using lithium as the anode but differing in cathode material, electrolyte, and chemistry as well as in design and other physical and mechanical features.

Lithium primary batteries can be classified into several categories, based on the type of electrolyte (or solvent) and cathode material used. These classifications, examples of materials that were considered or used, and the major characteristics of each are listed in Table 14.1.

Soluble-Cathode Cells. These use liquid or gaseous cathode materials, such as sulfur dioxide (SO_2) or thionyl chloride ($SOCl_2$), that dissolve in the electrolyte or are the electrolyte solvent. Their operation depends on the formation of a passive layer on the lithium anode resulting from a reaction between the lithium and the cathode material. This prevents further chemical reaction (self-discharge) between anode and cathode or reduces it to a very low rate. These cells are manufactured in many different configurations and designs (such as high and low rate) and with a very wide range of capacities. They are generally fabricated in a cylindrical configuration in the smaller sizes, up to about 35 Ah, using a bobbin construction for the low-rate cells and a spirally wound (jelly-roll) structure for the high-rate designs. Prismatic containers, having flat parallel plates, are generally used for the larger cells up to 10,000 Ah in size. Flat or "pancake-shaped" configurations have also been designed. These soluble cathode lithium cells are used for low to high discharge rates. The high-rate designs, using large electrode surface areas, are noted for their high power density and are capable of delivering the highest current densities of any active primary cell.

Solid-Cathode Cells. The second type of lithium anode primary cell uses solid rather than soluble gaseous or liquid materials for the cathode. With these solid cathode materials, the cells have the advantage of not being pressurized or necessarily requiring a hermetic-type seal, but they do not have the high-rate capability of the soluble-cathode systems. They are designed, generally, for low- to medium-rate applications such as memory backup, security devices, portable electronic equipment, digital cameras, watches, calculators, and small lights. Button, flat, and cylindrical-shape cells are

TABLE 14.1 Classification of Lithium Primary Batteries*

Cell classification	Typical electrolyte	Power capability	Size, Ah	Operating range, °C	Shelf life, years	Typical cathodes	Nominal cell voltage, V	Key characteristics
Soluble cathode (liquid or gas)	Organic or inorganic (w/solute)	Moderate to high power, W	0.5 to 10,000	−80 to 70	5–20	SO_2 $SOCl_2$ SO_2Cl_2	3.0 3.6 3.9	High energy output, high power output, low-temperature operation, long shelf life
Solid cathode	Organic (w/solute)	Low to moderate power, mW-W	0.03 to 1200	−40 to 50	5–8	V_2O_5 $AgV_2O_{5.5}$ MnO_2 CFx CuS FeS_2 FeS	3.3 3.2 3.0 2.6 1.7 1.5 1.5	High energy output for moderate power requirements, non-pressurized cells
Solid electrolyte (see Chap. 31)	Solid state	Very low power, μW	0.003 to 2.4	0 to 100	10–25	$PbI_2/PbS/Pb$ $I_2(P2VP)$	1.9 2.8	Excellent shelf life, solid state—no leakage, long-term microampere discharge

*For reserve lithium batteries, see Chap. 35.

available in low-rate and the moderate-rate jelly-roll configurations. A number of different solid cathodes are being used in lithium primary cells, as listed in Table 14.1. The discharge of the solid-cathode cells is not as flat as that of the soluble-cathode cells, but at the lower discharge rates and ambient temperature, their capacity (energy density) may be higher than that of the lithium/sulfur dioxide cell.

Solid-Electrolyte Cells. These cells are noted for their extremely long storage life, in excess of 20 years, but are capable of only low-rate discharge in the microampere range. They are used in applications such as memory backup, cardiac pacemakers, and similar equipment where current requirements are low, but long life is critical (see Chap. 31).

In Fig. 14.1 the size or capacity of these three types of lithium cells (up to the 30 Ah size) is plotted against the current levels at which they are typically discharged. The approximate weight of lithium in each of these cells is also shown.

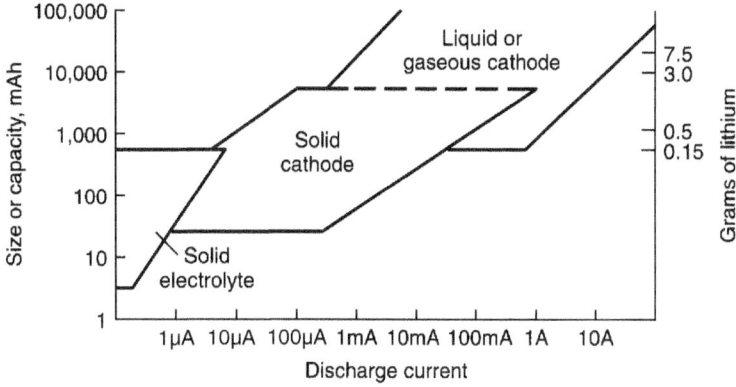

FIGURE 14.1 Classification of lithium primary cell types.

14.2 CHEMISTRY

14.2.1 Lithium

The main requirements for electrode materials used for high-performance (high specific energy and energy density) batteries are a high electrochemical equivalence (high coulombic output for a given weight of material) and a high electrode potential. It is apparent from Table 14.2, which lists the characteristics of metals used as battery anodes, that lithium is an outstanding candidate. Its standard potential and electrochemical equivalence are the highest of the metals; it excels in theoretical gravimetric energy density; and, with its high potential, it is inferior only to aluminum and magnesium on a volumetric energy basis (watthours per liter). Aluminum, however, has not been used successfully as an anode except in reserve systems, and magnesium has a low practical operating voltage. Furthermore, lithium is preferred to the other alkali metals because of its better mechanical characteristics and lower reactivity. Calcium has been investigated as an anode, in place of lithium, because of its higher melting point (838°C compared with 180.5°C for lithium). To date, practical cells using calcium have not been produced.

TABLE 14.2 Characteristics of Anode Materials

Material	Atomic weight, g	Standard potential at 25°C, V	Density, g/cm^3	Melting point, °C	Valence change	Electrochemical equivalence		
						Ah/g	g/Ah	Ah/cm^3
Li	6.94	−3.05	0.534	180	1	3.86	0.259	2.08
Na	23.0	−2.7	0.97	97.8	1	1.16	0.858	1.12
Mg	24.3	−2.4	1.74	650	2	2.20	0.454	3.8
Al	26.9	−1.7	2.7	659	3	2.98	0.335	8.1
Ca	40.1	−2.87	1.54	851	2	1.34	0.748	2.06
Fe	55.8	−0.44	7.85	1528	2	0.96	1.04	7.5
Zn	65.4	−0.76	7.1	419	2	0.82	1.22	5.8
Cd	112	−0.40	8.65	321	2	0.48	2.10	4.1
Pb	207	−0.13	11.3	327	2	0.26	3.87	2.9

Lithium is one of the alkali metals, and it is the lightest of all the metallic elements, with a density about half that of water. When first made or freshly cut, lithium has the luster and color of bright silver, but it tarnishes rapidly in moist air. It is soft and malleable, can be readily extruded into thin foils, and is a good conductor of electricity. Table 14.3 lists some of the physical properties of lithium.[3,4]

Lithium reacts vigorously with water, releasing hydrogen and forming lithium hydroxide

$$2Li + 2H_2O \rightarrow 2LiOH + H_2$$

This reaction is not as vigorous as that of sodium and water, probably due to the fairly low solubility and the adherence of LiOH to the metal surface under some conditions, however, the heat generated

TABLE 14.3 Physical Properties of Lithium

Melting point	180.5°C
Boiling point	1347°C
Density	0.534 g/cm^3 (25°C)
Specific heat	0.852 cal/g (25°C)
Specific resistance	9.35 × 10^6 Ω·cm (20°C)
Hardness	0.6 (Mohs scale)

by this reaction may ignite the hydrogen that is formed and the lithium will then also burn. Because of this reactivity, however, lithium must be handled in a dry atmosphere and, in a battery, be used with nonaqueous electrolytes. (The lithium/water and lithium/air batteries are described in Chap. 33.)

14.2.2 Cathode Materials

A number of inorganic and organic materials have been examined for use as the cathode in primary lithium batteries.[1,5] The critical requirements for this material to achieve high performance are high battery voltage, high energy density, and compatibility with the electrolyte (that is, being essentially nonreactive or insoluble in the electrolyte). Preferably the cathode material should be conductive, although there are few such materials available and solid cathode materials are usually mixed with a conducting material, such as carbon, and applied to a conductive grid to provide the needed conductivity. If the cathode reaction products are a metal and a soluble salt (of the anode metal), this feature can improve cathode conductivity as the discharge proceeds. Other desirable properties of the cathode material are low cost, availability (noncritical material), and favorable physical properties, such as nontoxicity and nonflammability. Table 14.4 lists some of the cathode materials that have been studied for primary lithium batteries and gives their cell reaction mechanisms and the theoretical cell voltages and capacities.

14.2.3 Electrolytes

The reactivity of lithium in aqueous solutions requires the use of nonaqueous electrolytes for lithium anode batteries.[5] Polar organic liquids are the most common electrolyte solvents for the active primary cells, except for the thionyl chloride ($SOCl_2$) and sulfuryl chloride (SO_2Cl_2) cells, where these inorganic compounds serve as both the solvent and the active cathode material. The important properties of the electrolyte are:

1. It must be aprotic, that is, have no reactive protons or hydrogen atoms, although hydrogen atoms may be in the molecule.
2. It must have low reactivity with lithium (or form a protective coating on the lithium surface to prevent further reaction) and the cathode.
3. It must be capable of forming an electrolyte of good ionic conductivity.
4. It should be liquid over a broad temperature range.
5. It should have favorable physical characteristics, such as low vapor pressure, stability, nontoxicity, and nonflammability.

A listing of the organic solvents commonly used in lithium batteries is given in Table 14.5. These solvents are typically employed in binary or ternary combination. These organic electrolytes, as well as thionyl chloride (mp −105°C, bp 78.8°C) and sulfuryl chloride (mp −54°C, bp 69.1°C), are liquid over a wide temperature range with low freezing points. This characteristic provides the potential for operation over a wide temperature range, particularly low temperatures.

The Jet Propulsion Laboratory (Pasadena, CA) has evaluated several types of lithium primary batteries to determine their ability to operate planetary probes at temperatures of −80°C and below.[6] Individual cells were evaluated by discharge tests and Electrochemical Impedance Spectroscopy. Of the five types considered (Li/$SOCl_2$, Li/SO_2, Li/MnO_2, Li-BCX, and Li-CFx), lithium-thionyl chloride and lithium-sulfur dioxide were found to provide the best performance at −80°C. Lowering the electrolyte salt to ca. 0.5 molar was found to improve performance with these systems at very low temperatures. In the case of D-size Li/$SOCl_2$ batteries, lowering the $LiAlCl_4$ concentration from 1.5 to 0.5 molar led to a 60% increase in capacity on a baseline load of 118 ohms with periodic 1-min pulses at 5.1 ohms at −85°C.

Lithium salts, such as $LiClO_4$, LiBr, $LiCF_3SO_3$, LiI, and $LiAlCl_4$, are the electrolyte solutes most commonly used to provide ionic conductivity. The solute must be able to form a stable electrolyte

TABLE 14.4 Cathode Materials Currently or Previously Used in Lithium Primary Batteries

Cathode material	Molecular weight	Valence change	Density, g/cm³	Theoretical faradic capacity (cathode only)			Cell reaction mechanism (with lithium anode)	Theoretical cell	
				Ah/g	Ah/cm³	g/Ah		Voltage, V	Specific Energy Wh/kg
SO_2	64	1	1.37	0.419	—	2.39	$2Li + 2SO_2 \rightarrow 2Li_2S_2O_4$	3.1	1170
$SOCl_2$	119	2	1.63	0.450	—	2.22	$4Li + 2SOCl_2 \rightarrow 4LiCl + S + SO_2$	3.65	1470
SO_2Cl_2	135	2	1.66	0.397	—	2.52	$2Li + SO_2Cl_2 \rightarrow 2LiCl + SO_2$	3.91	1405
Bi_2O_3	466	6	8.5	0.35	2.97	2.86	$6Li + Bi_2O_3 \rightarrow 3Li_2O + 2Bi$	2.0	640
$Bi_2Pb_2O_5$	912	10	9.0	0.29	2.64	2.41	$10Li + Bi_2Pb_2O_5 \rightarrow 5Li_2O + 2Bi + 2Pb$	2.0	544
$(CF)_x$	31	1	2.7	0.86	2.32	1.16	$xLi + (CF)_x \rightarrow xLiF + xC$	3.1	2180
$CuCl_2$	134.5	2	3.1	0.40	1.22	2.50	$2Li + CuCl_2 \rightarrow 2LiCl + Cu$	3.1	1125
CuF_2	101.6	2	2.9	0.53	1.52	1.87	$2Li + CuF_2 \rightarrow 2LiF + Cu$	3.54	1650
CuO	79.6	2	6.4	0.67	4.26	1.49	$2Li + CuO \rightarrow Li_2O + Cu$	2.24	1280
$Cu_4O(PO_4)_2$	458.3	8	—	0.468	—	2.1	$8Li + Cu_4O(PO_4)_2 \rightarrow Li_2O + 2Li_3PO_4 + Cu$	2.7	—
CuS	95.6	2	4.6	0.56	2.57	1.79	$2Li + CuS \rightarrow Li_2S + Cu$	2.15	1050
FeS	87.9	2	4.8	0.61	2.95	1.64	$2Li + FeS \rightarrow Li_2S + Fe$	1.75	920
FeS_2	119.9	4	4.9	0.89	4.35	1.12	$4Li + FeS_2 \rightarrow 2Li_2S + Fe$	1.8	1304
MnO_2	86.9	1	5.0	0.31	1.54	3.22	$Li + Mn^{IV}O_2 \rightarrow Mn^{III}O_2(Li^+)$	3.5	1005
MoO_3	143	1	4.5	0.19	0.84	5.26	$2Li + MoO_3 \rightarrow Li_2O + MoO_5$	2.9	525
Ni_3S_2	240	4	—	0.47	—	2.12	$4Li + Ni_3S_2 \rightarrow 2Li_2S + 3Ni$	1.8	755
$AgCl$	143.3	1	5.6	0.19	1.04	5.26	$Li + AgCl \rightarrow LiCl + Ag$	3.267	583
Ag_2CrO_4	331.8	2	5.6	0.16	0.90	6.25	$2Li + Ag_2CrO_4 \rightarrow Li_2CrO_4 + 2Ag$	3.35	515
$AgV_2O_{5.5}$*	297.7	3.5	—	0.282	—	—	$3.5Li + AgV_2O_{5.5} \rightarrow Li_{3.5}AgV_2O_{5.5}$	3.24	655
V_2O_5	181.9	1	3.6	0.15	0.53	6.66	$Li + V_2O_5 \rightarrow LiV_2O_5$	3.4	490

*Multiple-step discharge; see Ref. 11 (Experimental values to +1.5 V cutoff).

TABLE 14.5 Properties of Organic Electrolyte Solvents for Lithium Primary Batteries

Solvent	Structure	Boiling point at 10^5 Pa, °C	Melting point, °C	Flash point, °C	Density at 25°C, g/cm^3	Specific conductivity with 1M LiClO$_4$, S/cm^{-1}
Acetonitrile (AN)	H$_3$C—C≡N	81	−45	5	0.78	3.6×10^{-2}
γ-Butyrolactone (BL)	H$_2$C—CH$_2$ | | O CH$_2$ \\ / C ‖ O	204	−44	99	1.1	1.1×10^{-2}
Dimethylsulfoxide (DMSO)	H$_3$C—S—CH$_3$ ‖ O	189	18.5	95	1.1	1.4×10^{-2}
Dimethylsulfite (DMSI)	O=S(OCH$_3$)(OCH$_3$)	126	−141		1.2	
1,2-Dimethoxyethane (DME)	H$_2$C—O—CH$_3$ | H$_2$C—O—CH$_3$	83	−60	1	0.87	
Dioxolane (1,3-D)	H$_2$C—O \\ | CH$_2$ H$_2$C—O /	75	−26	2	1.07	
Methyl formate (MF)	H—C—O—CH$_3$ ‖ O	32	−100	−19	0.98	3.2×10^{-2}
Propylene carbonate (PC)	H$_3$C—CH—CH$_2$ | | O O \\ / C ‖ O	242	−49	135	1.2	7.3×10^{-3}
Tetrahydrofuran (THF)	H$_2$C—CH$_2$—CH$_2$—CH$_2$ _____O_____/	65	−109	−15	0.89	

that does not react with the active electrode materials. It must be soluble in the organic solvent and dissociate to form a conductive electrolyte solution. Maximum conductivity with organic solvents at room temperature is normally obtained with a 1-Molar solute concentration, but generally the conductivity of these electrolytes is about one-tenth that of aqueous systems. To accommodate this lower conductivity, close electrode spacing and cells designed to minimize impedance and provide good power density are used.

14.2.4 Cell Couples and Reaction Mechanisms

The overall discharge reaction mechanism for various lithium primary batteries is shown in Table 14.4, which also lists the theoretical cell voltage. The mechanism for the discharge of the lithium anode is the oxidation of lithium to form lithium ions (Li$^+$) with the release of an electron.

$$Li \rightarrow Li^+ + e$$

14.8 PRIMARY BATTERIES

The electron moves through the external circuit to the cathode, where it reacts with the cathode material, which is reduced. At the same time, the Li$^+$ ion, which is small (0.06 nm in radius) and mobile in both liquid and solid-state electrolytes, moves through the electrolyte to the cathode, where it reacts to form a lithium compound.

A more detailed description of the cell reaction mechanism for the different lithium primary batteries is given in the sections on those battery systems.[1,7]

14.3 CHARACTERISTICS OF LITHIUM PRIMARY BATTERIES

14.3.1 Summary of Design and Performance Characteristics

A listing of the major lithium primary batteries now in production or advanced development and a summary of their constructional features, key electrical characteristics, and available sizes are presented in Table 14.6. The types of batteries, their sizes, and some characteristics are subject to change depending on design, standardization, and market development. Manufacturers' data should be obtained for specific characteristics. The performance characteristics of these systems, under theoretical conditions, are given in Table 14.4. Comparisons of the performance of the lithium batteries with comparably sized conventional primary batteries are covered in Sec. 8.3. Detailed characteristics of some of these batteries are covered in Secs. 14.5 to 14.11 and Sec. 31.5.4.

14.3.2 Soluble-Cathode Lithium Primary Batteries

Two types of soluble-cathode lithium primary batteries are currently available (Table 14.1). One uses SO_2 as the active cathode dissolved in an organic electrolyte solvent. The second type uses an inorganic solvent, such as the oxychlorides $SOCl_2$ and SO_2Cl_2, which serves as both the active cathode and the electrolyte solvent. These materials form a passivating layer or protective film of reaction products on the lithium surface, which inhibits further reaction. Even though the active cathode material is in contact with the lithium anode, self-discharge is inhibited by the protective film, which proceeds at very low rates, and the shelf life of these batteries is excellent. This film, however, may cause a voltage delay to occur, i.e., a time delay to break down the film and for the cell voltage to reach the operating level when the discharge load is applied. These lithium batteries have a high specific energy and, with proper design, such as the use of high-surface-area electrodes, are capable of delivering high specific energy at high specific power.

These cells generally require a hermetic-type seal. Sulfur dioxide is a gas at 20°C (bp −10°C), and the undischarged cell has an internal pressure of 3 to 4 × 10^5 Pa at 20°C. The oxychlorides are liquid at 20°C, but with boiling points of 78.8°C for $SOCl_2$ and 69.1°C for SO_2Cl_2, a moderate pressure can develop at high operating temperatures. In addition, as SO_2 is a discharge product in the oxychloride cells, the internal cell pressure increases as the cell is discharged.

The lithium/sulfur dioxide (Li/SO_2) battery is the most advanced of these lithium primary batteries. These batteries are typically manufactured in cylindrical configurations in capacities up to 34 Ah. They are noted for their high specific power (about the highest of the lithium primary batteries), high energy density, and good low-temperature performance. They are used in military and specialized industrial, space, and commercial applications where these performance characteristics are required.

The lithium/thionyl chloride (Li/$SOCl_2$) battery has one of the highest specific energies of all the practical battery systems. Figures 8.8 and 8.9 illustrate the advantages of the Li/$SOCl_2$ battery over a wide temperature range at moderate discharge rates. Figure 14.2 compares typical discharge profile of the Li/$SOCl_2$ cell with the Li/SO_2 cell. At 20°C, at moderate discharge rates, the Li/$SOCl_2$ cell has a higher working voltage and about a 50% advantage in service life. The Li/SO_2 cell, however, does have better performance at low temperatures and high discharge rates and a lower voltage delay after

TABLE 14.6 Characteristics of Lithium Primary Batteries

Soluble cathode batteries

System	Cathode	Electrolyte Solvent	Electrolyte Solute	Separator	Construction	Voltage, V Nominal	Voltage, V Working* (20°C)	Specific energy† Wh/kg	Energy density† Wh/L	Power density	Discharge profile	Available sizes
Lithium/sulfur dioxide (Li/SO_2)	SO_2 with carbon and binder on Al screen	AN	LiBr	Microporous polypropylene	Spiral "jelly-roll" cylindrical construction; glass-to-metal seal	3.0	2.9–2.7	260	415	High	Very flat	Cylindrical batteries up to 34 Ah
Lithium/thionyl chloride (Li/$SOCl_2$) Low rate	$SOCl_2$ with carbon and binder on Ni or SS	$SOCl_2$	$LiAlCl_4$	Glass non-woven	Wafer construction	3.6	3.6–3.4	275	630	Low	Flat	0.4–1.7 Ah
					"Bobbin" in cylindrical construction	3.6	3.5–3.3	590	1100	Medium	Flat	Cylindrical batteries 1.2–35 Ah
High capacity					Prismatic with flat plates	3.6	3.5–3.3	480	950	Medium	Flat	12–10,000 Ah
High rate					Spiral "jelly-roll" cylindrical construction or flat disk	3.6	3.5–3.2	495	970	Medium to high	Flat	Cylindrical: 1.2–14 Ah Flat disk: up to 2300 Ah
		$SOCl_2$ with halogen additives	$LiAlCl_4$	Glass mat	Spiral "jelly-roll" cylindrical construction	3.9	3.8–3.3	485	1070	Medium	Flat	2–30 Ah
Lithium/sulfuryl chloride (Li/SO_2Cl_2)	SO_2Cl_2 with carbon and binder SS screen	SO_2Cl_2 (some with additives)	$LiAlCl_4$	Glass	Spiral "jelly-roll" cylindrical construction; glass-to-metal seal	3.95	3.5–3.1	480	1040	Medium to high	Flat	7–30 Ah

(*Continued*)

TABLE 14.6 Characteristics of Lithium Primary Batteries (*Continued*)

Solid cathode batteries

| System | Electrolyte | | | Cathode | Construction | Voltage, V | | Specific energy[†] Wh/kg | Energy density[†] Wh/L | Power density | Discharge profile | Available sizes |
	Solvent	Solute	Separator			Nominal	Working* (20°C)					
Lithium/carbon mono-fluoride Li/CFx	PC + DME or BL	LiBF$_4$ or LiAsF$_6$	Polypropylene	CFx with carbon and binder on nickel collector	"Coin" construction; crimped seal Pin type	3.0	2.7–2.5	215	550	Low to medium Low	Moderately flat Humped	Coin batteries to 500 mAh Small cylinders 25–50 mAh
					Spiral "jelly-roll" cylindrical construction; crimped or glass-to-metal seal			350 (commercial) 800 (military)	560 1160			Cylindrical batteries to 5 Ah (commercial) and 1200 Ah (military)
					Rectangular with flat plates			440 (biomedical)	900			Rectangular batteries to 40 Ah
Lithium/ copper oxide (Li/CuO)	1,3D	LiClO$_4$	Nonwoven glass	CuO pressed in cell can	"Bobbin" inside-out cylindrical construction	1.5	1.5–1.4	280	650	Low	High initial voltage drop, then moderatley flat	Cylindrical batteries 500–3500 mAh
Lithium/iron disulfide (LiFeS$_2$)	1–3D + DME	LiI	Microporous polyethylene	FeS$_2$ with carbon and binders	"Jelly-roll" cylindrical construction; crimped seal	1.5	1.6–1.4	310	562	Medium to high	High initial drop, then flat	AAA and AA sizes to 3.1 Ah
Lithium/ manganese dioxide (Li/MnO$_2$)	PC + DME	Li salt	Polypropylene	MnO$_2$ with carbon and binder on supporting grid	"Coin construction with flat electrodes	3.0	3.0–2.7	230	545	Low to medium	Moderately flat	Coin batteries 25–1000 mAh
	Organic solvent	Li salt	Polypropylene		"Jelly-roll" cylindrical construction; crimped and hermetic seals	3.0	2.8–2.5	261	546	Medium to high	Moderately flat	2/3 A Cylindrical batteries typical, larger cells available to 11 Ah
	Organic solvent	Li salt	Polypropylene		"Bobbin" cylindrical construction	3.0	3.0–2.8	270	620	Low to medium	Moderately flat	Cylindrical batteries to 2.5 Ah
Lithium/silver vanadium oxide (Li/ AgV$_4$O$_{11}$)	PC, DME	LiAsF$_6$	Microporous polypropylene	AgV$_2$O$_{5.5}$ with graphite and carbon	Rounded prismatic and D-shaped cross section	3.2	3.2–1.5	270	780	Low to medium	Multiple plateaus	Special sizes for implantable medical devices

*Working voltages are typical for discharges at favorable loads.
[†]Energy densities are for 20°C, under favorable discharge conditions. See details in appropriate sections.

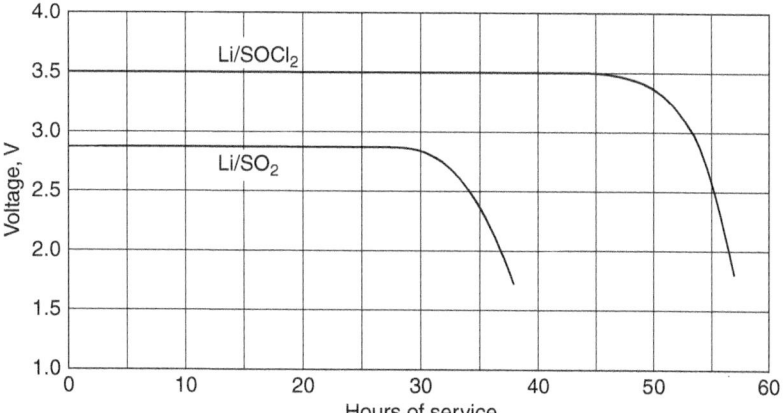

FIGURE 14.2 Comparison of performance of Li/SO$_2$ and Li/SOCl$_2$ C-size batteries; 100-mA discharge load at 20°C.

storage. Li/SOCl$_2$ cells have been fabricated in many sizes and designs ranging from small button and cylindrical cells with capacities below 1 Ah, to large prismatic cells with capacities as high as 10,000 Ah. Low-rate cells have been used successfully in many applications, especially as memory backup, for many years; high-rate cells are used in special applications.

The lithium/sulfuryl chloride (Li/SO$_2$Cl$_2$) battery has potential advantages because of its higher voltage (3.9 open-circuit voltage) and resultant higher specific energy. Suitable cathode electrode formulations and cell designs have been investigated to achieve the full capability of this electrochemical system. Figure 14.3 shows a comparison of the cathode polarization for Li/SO$_2$Cl$_2$ and Li/SOCl$_2$ batteries. Halogen additives such as chlorine have been used as a means to improve performance. Halogen additives are also used in some cells.

Calcium has been investigated as an anode material in place of lithium in thionyl chloride cells. Safer operation was anticipated with calcium since its melting temperature of 838°C is not likely to be reached by any internally driven cell condition. While the discharge voltage is about 0.4 V lower than for the Li/SOCl$_2$ cell (open-circuit voltage is 3.25 V), the Ca/SOCl$_2$ cell has a flat discharge profile and about the same volumetric ampere-hour capacity. Shelf-life characteristics are also similar to those of the lithium anode cell.[9,10] However, calcium is significantly more difficult to process than lithium and passivation is a more significant factor. To date, no calcium-thionyl chloride batteries have been commercialized.

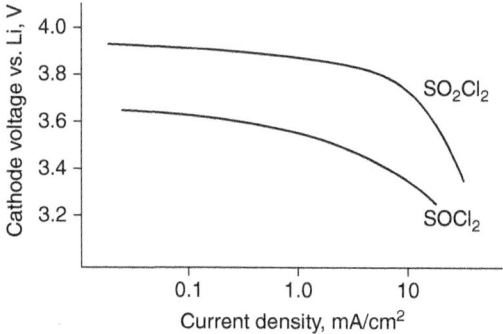

FIGURE 14.3 Comparison of cathode polarization curves, Li/SO$_2$Cl$_2$ vs. Li/SOCl$_2$ batteries.[8]

14.3.3 Solid-Cathode Lithium Primary Cells

The solid-cathode lithium batteries are generally used in low- to moderate-drain applications and are manufactured mainly in small flat or cylindrical sizes ranging in capacity from 25 mAh to about 11 Ah, depending on the particular electrochemical system. Larger batteries have been produced in cylindrical and prismatic configurations. A comparison of the performance of solid-cathode lithium batteries and conventional batteries is presented in Chap. 8.

The solid-cathode batteries have the advantage, compared with the soluble-cathode lithium primary batteries, of being nonpressurized and thus not requiring a hermetic-type seal. A mechanically crimped seal with a polymeric gasket is satisfactory for most applications. On light discharge loads, the energy density of some of the solid-cathode systems is comparable to that of the soluble-cathode systems, and in smaller battery sizes it may be greater. Their disadvantages, again compared with the soluble-cathode batteries, are a lower rate capability, poorer low-temperature performance, and a more sloping discharge profile.

To maximize their high-rate performance and compensate for the lower conductivity of the organic electrolytes, designs are used for these lithium cells to increase electrode area, such as a larger-diameter coin cell instead of button cells, or the spirally wound jelly-roll construction for the cylindrical cells.

A number of different cathode materials have been used in the solid-cathode lithium cells. These are listed in Tables 14.4 and 14.6, which present some of the theoretical and practical performance data of these cells. The major features of the solid-cathode lithium cells are compared in Table 14.7. Many of the characteristics are similar, such as high specific energy and energy density and good shelf life. An important property is the 3 V cell voltage obtained with several of these cathodes. Some cathode materials have been used mainly in the coin or button cell designs while others, such as the manganese dioxide cathode, have been used in coin, cylindrical, and prismatic cells, as well as in both high (spirally wound) and low (bobbin) rate designs.

Although a number of different solid-cathode lithium batteries have been developed and even manufactured, more recently the trend is toward reducing the number of different chemistries that are manufactured. The lithium/manganese dioxide (Li/MnO_2) battery was one of the first to be used commercially and is still the most popular system. It is relatively inexpensive, has excellent shelf life, good high-rate and low-temperature performance, and is available in coin and cylindrical cells. The lithium/

TABLE 14.7 Characteristics of Typical Lithium/Solid-Cathode Batteries

Type of battery	Operating voltage, V	Characteristics
Li/MnO_2	3.0	High specific energy and energy density; wide operating temperature range (−40 to +85°C); performance at relatively high discharge rates; minimal voltage delay; relatively low cost; available in flat (coin) and cylindrical batteries (high and low rates).
Li/CFx	2.8	Highest theoretical specific energy, low- to moderate-rate capability; wide operating temperature range (−20 to 85°C); flat discharge profile; available in flat (coin), cylindrical and prismatic designs.
Li/CuO	1.5	Highest theoretical volumetric coulombic capacity (Ah/L); long storage life; low- to moderate-rate capability; operating temperature range up to 125 to 150°C; no apparent voltage delay. Potential replacement for alkaline-manganese but not currently available.
Li/FeS_2	1.5	Replacement for conventional zinc-carbon and alkaline-manganese dioxide batteries; higher power capability than conventional batteries and better low-temperature performance and storability. Currently available in AA and AAA sizes as a direct replacement for alkaline-manganese. Finding increasing use in digital cameras.
$Li/AgV_2O_{5.5}$	3.2	High specific energy and energy density multiple-step discharge; good rate capability; used in implantable and other medical devices. See Sec. 31.5.4.
Li/V_2O_5	3.3	High energy density; two-step discharge; used in reserve cells (See Chap. 35).

carbon monofluoride (Li/CFx) battery is another of the early solid-cathode batteries and is attractive because of its high theoretical capacity and flat discharge characteristics. It is also manufactured in coin, cylindrical, and prismatic configurations. The higher cost of polycarbon monofluoride has affected the commercial potential for this system, but it is finding use in biomedical, military, and space applications where cost is not a factor. This system is also finding increased use in digital cameras.

The lithium/vanadium pentoxide (Li/V_2O_5) battery has a high volumetric energy density, but with a two-step discharge profile. Its main application has been in reserve batteries (Chap. 35). The lithium/silver vanadium oxide (Li/Ag$V_2O_{5.5}$) battery is used in medical applications, such as defibrillators, which have pulse load requirements as this battery is capable of relatively high-rate discharge.[11] The other solid-cathode lithium batteries operate in the range of 1.5 V and were developed to replace conventional 1.5 V button or cylindrical cells. The lithium/copper oxide (Li/CuO) cell is noted for its high coulombic energy density and has the advantage of higher capacity or lighter weight when compared with conventional cylindrical cells. It is capable of performance at high temperatures and, has a long shelf life under adverse conditions. It is not currently available commercially. The iron disulfide (Li/FeS$_2$) cell has similar advantages over the conventional cells, plus the advantage of high-rate performance. Once available in a button cell configuration, it is now being marketed commercially in high-rate cylindrical AA and AAA sizes as a replacement for alkaline-manganese dioxide batteries.

The remaining solid-cathode systems listed in Tables 14.1 and 14.4 are not currently commercially available.

Typical discharge curves for the major solid-cathode batteries are shown in Fig. 14.4. The discharge curves of the Li/SO$_2$ and Li/SOCl$_2$ batteries showing their flatter discharge profile are also plotted for comparison purposes.

A comparison of the performance of several of the solid-cathode batteries in a low-rate button configuration and the higher-rate cylindrical configuration is presented in Sec. 8.3. In the button configuration the lithium batteries have an advantage in specific energy (Wh/kg) over many of the conventional batteries. This advantage may not be too important in these small battery sizes, but the lithium batteries have an advantage of lower cost, particularly when compared with the silver cells, and longer shelf life. In addition, the zinc/mercuric oxide battery, which once dominated the

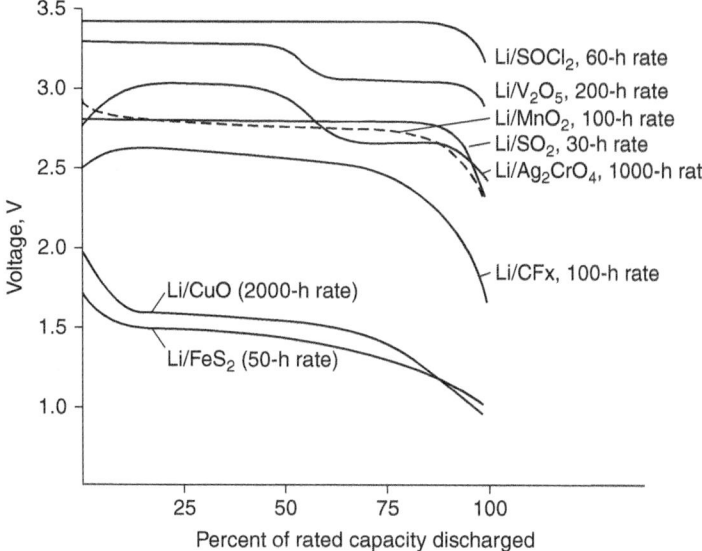

FIGURE 14.4 Typical discharge curves of lithium/solid-cathode batteries.

14.14 PRIMARY BATTERIES

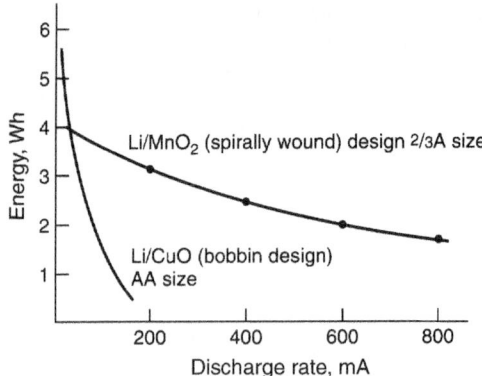

FIGURE 14.5 Comparison of Li/CuO and Li/MnO$_2$ batteries at 20°C. Batteries are equivalently sized.

button battery market, and the cadmium/mercuric oxide battery are not currently available due to environmental factors.

In the larger cylindrical sizes (Table 8.5), the lithium cells have an advantage in both volumetric and gravimetric energy density. In some designs, this advantage is even more significant at higher discharge loads. Figure 8.3 shows another comparison of the performance of solid-cathode and soluble-cathode lithium cells with aqueous cells.

It is important when making these comparisons to identify the specific discharge conditions of the application since the comparative performance of each battery system can vary depending on the discharge conditions. For example, as shown in Fig. 14.5, the lithium/copper oxide battery, designed for optimum performance at low discharge rates, has a comparatively high energy output when discharged at these light discharge loads, but the output drops off considerably at high rates. The similarly sized high-rate spirally wound configuration for the lithium/manganese dioxide cell has a lower energy output at the low discharge rates, but can maintain this performance as the discharge rate is increased. The performance of each of the battery systems, under various discharge conditions, is presented in the sections discussing each specific system.

The selection of a lithium versus a conventional cell thus becomes a trade-off between the lower initial cost of most of the conventional cells, the performance advantages of the lithium cells, and the key requirements of the specific application.

14.4 SAFETY AND HANDLING OF LITHIUM BATTERIES

14.4.1 Factors Affecting Safety and Handling

Attention must be given to the design and use of lithium cells and batteries to ensure safe and reliable operation. As with most battery systems, precautions must be taken to avoid physical and electrical abuse because some batteries can be hazardous if not used properly. This is important in the case of lithium cells since some of the components are toxic or flammable,[12] and the relatively low melting point of lithium (180.5°C) indicates that cells must be prevented from reaching high internal temperatures.

Because of the variety of lithium cell chemistries, designs, sizes, and so on, the procedures for their use and handling are not the same for all cells and batteries and depend on a number of factors such as the following:

1. *Electrochemical system.* The characteristics of the specific chemicals and cell components influence operational safety.

2. *Size and capacity of cell and battery.* Safety is directly related to the size of the cell and the number of cells in a battery. Small cells and batteries, containing less material and, therefore, less total energy, are "safer" than larger cells of the same design and chemistry.
3. *Amount of lithium used.* The less lithium that is used, implying less energetic cells, the safer they should be.
4. *Cell design.* High-rate designs, capable of high discharge rates, versus low-power designs where discharge rate is limited, use of "balanced" cell chemistry, adequate intra- and intercell electrical connections, and other features affect cell performance and operating characteristics.
5. *Safety features.* The safety features incorporated in the cell and battery will obviously influence handling procedures. These features include cell-venting mechanisms to prevent excessive internal cell pressure, thermal cutoff devices to prevent excessive temperatures, electrical fuses, PTC devices, and diode protection. Cells are hermetically or mechanically crimped-sealed, depending on the electrochemical system, to effectively contain cell contents if cell integrity is to be maintained.
6. *Cell and battery containers.* These should be designed so that cells and batteries will meet the mechanical and environmental conditions to which they will be exposed. High shock, vibration, extremes of temperature, or other adverse conditions may be encountered in use and handling, and the cell and battery integrity must be maintained. Container materials should also be chosen with regard to their flammability and the toxicity of combustion products in the event of fire. Container designs should also be optimized to dissipate the heat generated during discharge and to release pressure in the event of cell venting.

14.4.2 Safety Considerations

The electrical and physical abuses that may arise during the use of lithium cells are listed in Table 14.8 together with some generalized comments on corrective action. The behavior of specific cells is covered in the other sections of this chapter. The manufacturer's data should be consulted for more details on the performance of individual cells. Material safety data sheets (MSDSs) should also be obtained.

High-Rate Discharges or Short-Circuiting. Low-capacity batteries, or those designed as low-rate batteries, may be self-limiting and not capable of high-rate discharge. The temperature rise will thus

TABLE 14.8 Considerations for Use and Handling of Lithium Primary Batteries

Abusive condition	Corrective procedure
High-rate discharging or short-circuiting	Low-capacity or low-rate batteries may be self-limiting
	Electrical fusing, thermal protection
	Limit current drain; apply battery properly
Forced discharge (cell reversal)	Voltage cutoff
	Use low-voltage batteries
	Limit current drain
	Special designs ("balanced" cell)
	Use of diode in parallel across cells to bypass current
Charging	Prohibit charging
	Diode protection to prevent or limit charging current
Overheating	Limit current drain
	Fusing, thermal cutoff, PTC devices
	Design battery properly
	Do not incinerate
Physical abuse	Avoid opening, puncturing, or mutilating cells
	Maintain battery integrity

14.16 PRIMARY BATTERIES

be minimal and there will be no safety problems. Larger or high-rate cells can develop high internal temperatures if short-circuited or operated at excessively high rates. These cells are generally equipped with safety vent mechanisms to avoid more serious hazards. Such cells or batteries should be fuse-protected (to limit the discharge current). Thermal fuses or thermal switches should also be used to limit the maximum temperature rise. Positive temperature coefficient (PTC) devices are used in cells and batteries to provide this protection.

Forced Discharge or Voltage Reversal. Voltage reversal can occur in a multicell series-connected battery when the better performing cells can drive the poorer cell below 0 V, into reversal, as the battery is discharged toward 0 V. In some types of lithium cells, this forced discharge can result in cell venting or, in more extreme cases, cell rupture. Precautionary measures include the use of voltage cutoff circuits to prevent a battery from reaching a low voltage, the use of low-voltage batteries (since this phenomenon is unlikely to occur with a battery containing only a few cells in series), and limiting the current drain, since the effect of forced discharge is more pronounced on high-rate discharges. Special designs, such as the "balanced" Li/SO_2 cell (see Sec. 14.5), also have been developed that are capable of withstanding this discharge condition. The use of a current collector in the anode maintains lithium integrity and may provide an internal shorting mechanism to limit the voltage in reversal.

Charging. Lithium primary batteries, as well as the other primary batteries, are not designed to be recharged. If they are, they may vent or explode. Batteries that are connected in parallel or that may be exposed to a charging source (as in battery-backup CMOS memory circuits) should be diode-protected to prevent charging (see chap. 5).

Overheating. As discussed, overheating should be avoided. This can be accomplished by limiting the current drain, using safety devices such as fusing and thermal cutoffs, and designing the battery to provide necessary heat dissipation.

Incineration. Lithium cells are either hermetically or mechanically sealed. They should not be incinerated without proper protection because they may rupture or explode at high temperatures.

Currently special procedures govern the transportation and shipment of lithium batteries, and procedures for the use, storage, and handling of lithium batteries have been recommended.[13-14] Disposal of some types of lithium cells also is regulated. The latest issue of these regulations should be consulted for the most recent procedures (see Sec. 4.10 for details.) The U.S. Federal Aviation Agency has adapted technical standard order TSO-C142-Lithium Batteries, governing the installation and use of lithium primary batteries on commercial aircraft.[15] U.S. DOT, IATA, ICAO, and other governmental agencies issue regulations governing the shipment of lithium batteries.

14.5 LITHIUM/SULFUR DIOXIDE (Li/SO_2) BATTERIES

One of the more advanced lithium primary batteries, used mainly in military and in some industrial and space applications, is the lithium/sulfur dioxide (Li/SO_2) system. The battery has specific energy and energy density of up to 300 Wh/kg and 415 Wh/L, respectively, in large sizes. The Li/SO_2 battery is particularly noted for its capability to handle high current and high power requirements, excellent low-temperature performance, and long shelf life.

14.5.1 Chemistry

The Li/SO_2 cell uses lithium as the anode and a porous carbon cathode electrode with sulfur dioxide as the active cathode material. The cell reaction mechanism is

$$2Li + 2SO_2 \rightarrow Li_2S_2O_4 \downarrow \text{ (lithium dithionite)}$$

As lithium reacts readily with water, a nonaqueous electrolyte consisting of sulfur dioxide and an organic solvent, typically acetonitrile, with dissolved lithium bromide is used. The specific conductivity of this electrolyte is relatively high and decreases only moderately with decreasing temperature (Fig. 14.6), thus providing a basis for good high-rate and low-temperature performance. About 70% of the weight of the electrolyte/depolarizer is SO_2. The internal cell pressure, in an undischarged cell, due to the vapor pressure of the liquid SO_2, is 3–4×10^5 Pa at 20°C. The pressure at various temperatures is shown in Fig. 14.7. The mechanical features of the cell are designed to contain this pressure safely without leaking and to vent the electrolyte if excessively high temperatures and the resulting high internal pressures are encountered.

During discharge, the SO_2 is used up and the cell pressure reduced somewhat. The discharge is generally terminated by the full use of available lithium, in designs where the lithium is the limiting electrode, or by blocking of the

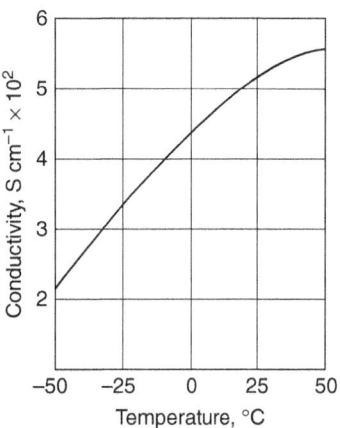

FIGURE 14.6 Conductivity of acetonitrile/lithium bromide/sulfur dioxide electrolyte (70% SO_2).

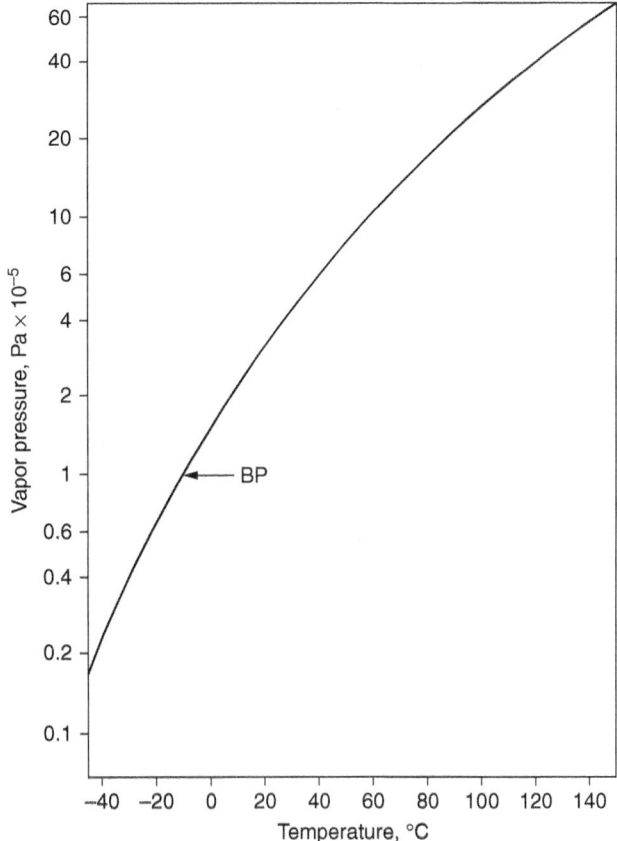

FIGURE 14.7 Vapor pressure of sulfur dioxide at various temperatures.

cathode by precipitation of the discharge product (cathode limited). Current designs are typically limited by the cathode so that some lithium remains at the end of discharge. The good shelf life of the Li/SO_2 cell results from the protective lithium dithionite film on the anode formed by the initial reaction of lithium and SO_2. It prevents further reaction and loss of capacity during storage.

Most Li/SO_2 cells are now fabricated in a balanced construction where the lithium:sulfur dioxide stoichiometric ratio is in the range of $Li:SO_2 = 0.9 - 1.05:1$. With the earlier designs, where the ratio was on the order of $Li:SO_2 = 1.5:1$, high temperatures, cell venting, or rupture and fires due to an exothermic reaction between residual lithium and acetonitrile, in the absence of SO_2, could occur on deep or forced discharge. Lithium cyanide, methane and other organic products can also be generated through this reaction. In the balanced cell, the anode is protected by residual SO_2 and remains passivated. The conditions for the hazardous reaction are minimized since some protective SO_2 remains in the electrolyte.[16] A higher negative cell voltage, in reversal, of the balanced cell is also beneficial when using diode protection, which is used in some designs to bypass the current through the cell and minimize the adverse effects of reversal.

The use of a current collector, typically an inlayed stripe of copper metal, also helps to maintain the integrity of the anode and leads to formation of a short-circuit mechanism since copper dissolution on cell reversal causes plated copper on the cathode to form an internal ohmic bridge.

14.5.2 Construction

The Li/SO_2 cell is typically fabricated in a cylindrical structure, as shown in Fig. 14.8. A jelly-roll construction is used, made by spirally winding rectangular strips of lithium foil, a microporous polypropylene separator, the cathode electrode (a Teflon-acetylene black mix pressed on an expanded aluminum screen), and a second separator layer. This design provides the high surface area and low cell resistance to obtain high-current and low-temperature performance. The roll is inserted in a nickel-plated steel can, with the positive cathode tab welded to the pin of a glass-to-metal seal and the anode tab welded to the cell case, the top is welded in place, and the electrolyte/depolarizer is added. The safety vent releases when the internal cell pressure reaches excessive levels, typically 2.41 MPa (350 psi) caused by inadvertent abusive use, such as overheating or short-circuiting, and prevents cell rupture or explosion. The vent activates at approximately 95°C, well above the upper temperature limit for operation and storage, safely relieving the excess pressure and preventing possible cell rupture. Additional construction details have been previously described.[16] It is important to employ a corrosion-resistant glass to prevent lithiation of the glass due to the potential difference between the cell case and the pin of the glass-to-metal seal.

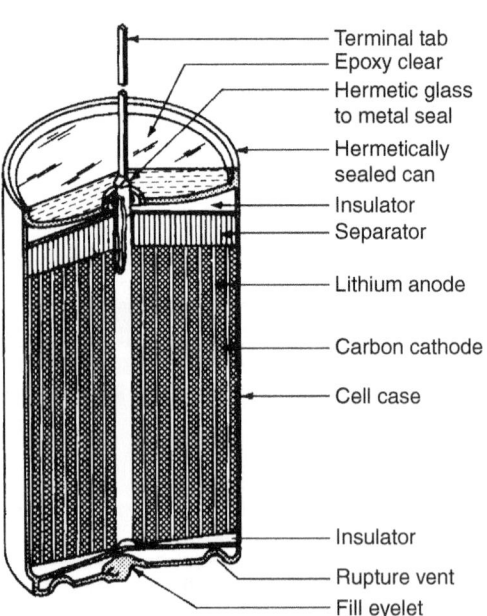

FIGURE 14.8 Lithium/sulfur dioxide batteries.

14.5.3 Performance

Voltage. The open-circuit voltage of the Li/SO$_2$ battery is 2.95 V. The nominal voltage is usually specified as 3 V. The specific voltage on discharge is dependent on the discharge rate, discharge temperature, and state-of-charge; typical working voltages range between 2.7 and 2.9 V (see Figs. 14.9, 14.10, and 14.12). The cutoff voltage, the voltage by which most of the battery capacity has been exhausted, is typically 2 V.

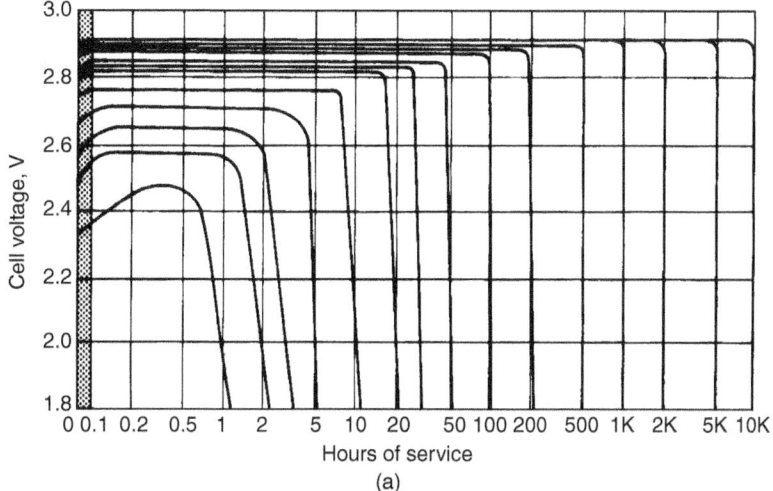

FIGURE 14.9a Typical discharge characteristics of standard-rate Li/SO$_2$ battery at various loads at 20°C.

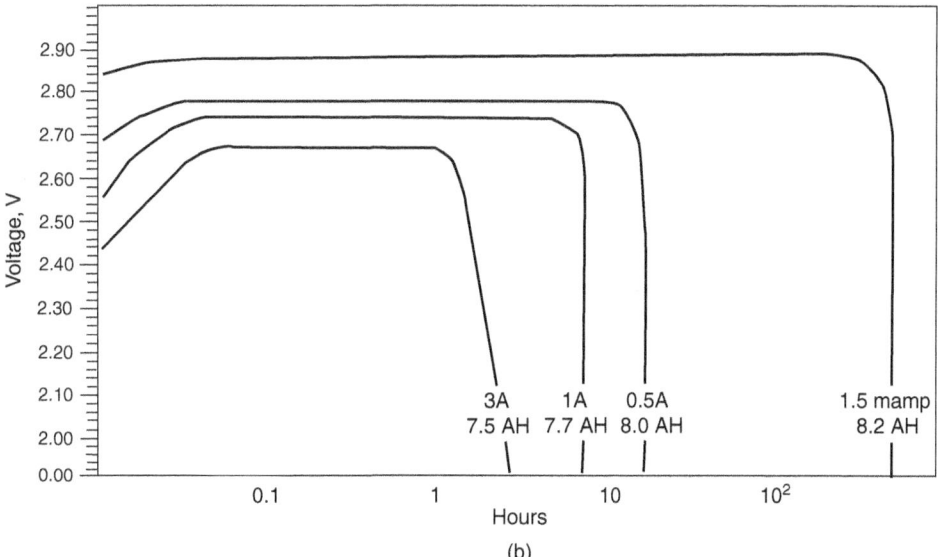

FIGURE 14.9b Discharge characteristics of high-rate Li/SO$_2$ D-size battery at four rates at 23°C.

Discharge. Typical discharge curves for the standard-rate Li/SO$_2$ battery at 20°C are given in Fig. 14.9a. The high cell voltages and the flat discharge profile are characteristic of the Li/SO$_2$ battery. Another unique feature is the ability of the Li/SO$_2$ battery to be efficiently discharged over a wide range of current or power levels, from high-rate short-term or pulse loads to low-drain continuous discharges for periods of 5 years or longer. At least 90% of the battery's rated capacity may be expected on the long-term discharges. Figure 14.9b shows the discharge curves for a high-rate D-size battery at four rates up to 3 A.

The Li/SO$_2$ battery is capable of higher-rate discharges on pulse loads. For example, a squat D cell designed in a high-rate construction can deliver pulse loads as high as 37.5 A, producing 59 watts of power.[17] For high-rate designs, extended discharges, however, at rates above the 2 h rate may cause overheating. The actual heat rise depends on the battery design, type of discharge, temperature, and voltage. As discussed in Sec. 14.4, the design and use of the battery should be controlled to avoid overheating.

A study[17] has shown that the high-rate pulse output of the lithium/sulfur dioxide battery may be enhanced by a variety of design variables. Multiple tabbing (1 to 3) of both anode and cathode, optimizing the composition of the cathode mix and reducing the aspect ration (length/width) of the electrodes were all found to reduce polarization during high-rate, 10 s pulse discharge. D-size cells and thin D-size cells (1.1 in diameter × 2.20 in high) with anodes and cathodes containing 2 tabs using an optimized cathode mix were found capable of producing 99 and 97 watts, respectively, under 50 A, 10 s pulses. Ultimately, a 5/4 C-size cell without multiple tabbing but using the optimized cathode mix was selected for reasons of volumetric efficiency to produce a 74-cell, 110 V battery capable of providing 5500 watt, 10 s pulses for a U.S. Navy application.

A similar design optimization study[18] has resulted in the production of a Li/SO$_2$ D-cell with a room-temperature capacity of 9.1 Ah at 250 mA and 8.8 Ah at 2 A. This compares to 7.75 Ah for the standard design and was achieved through an optimization study in which the aspect ratios of both anode and cathode were varied along with the use of three types of carbon in the cathode and the use of a central cathode tab. When discharged between 2.0 and 0.0 V, these cells were found to generate less heat than the standard cells. The high-capacity cells were used to construct U.S. Military BA-5590 batteries, which were tested to the requirements of MIL-PRF-49471. These batteries met the specification requirements for performance and safety.

Effect of Temperature. The Li/SO$_2$ battery is noted for its ability to perform over a wide temperature range, from –40 to 55°C. Discharge curves for a standard-rate Li/SO$_2$ battery at various temperatures are shown in Fig. 14.10. Significant, again, are the flat discharge curves over a wide temperature range, the good voltage regulation, and the high percentage of the 20°C performance available at the temperature extremes. As with all battery systems, the relative performance of the Li/SO$_2$ battery is dependent on the rate of discharge. In Fig. 14.11, the discharge performance of a standard-rate cell is plotted as a function of load and battery temperature.

Internal Resistance and Discharge Voltage. The Li/SO$_2$ battery has a relatively low internal resistance (about one-tenth that of conventional primary batteries) and good voltage regulation over a wide range of discharge loads and temperatures. The midprint voltage of the discharge of a standard-rate Li/SO$_2$ battery (to an end voltage of 2 V) at various discharge rates and temperatures is plotted in Fig. 14.12.

Service Life. The capacity or service life of the Li/SO$_2$ battery at various discharge rates is given in Fig. 14.13. The data are normalized for a 1-kg or 1-L size battery and presented in terms of hours of service at various discharge rates. The linear shape of this curve is again indicative of the capability of the Li/SO$_2$ battery to be efficiently discharged at these extreme conditions. This data can be used in several ways to calculate the approximate performance of a given battery or to select a Li/SO$_2$ battery of suitable size for a particular application, recognizing that the specific energy of the larger-size batteries is higher than that of the smaller ones.

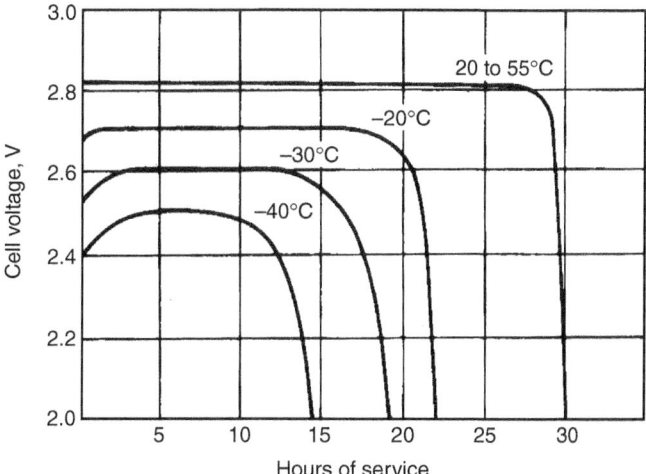

FIGURE 14.10 Typical discharge characteristics of Li/SO$_2$ battery at various temperatures, $C/30$ discharge rate.

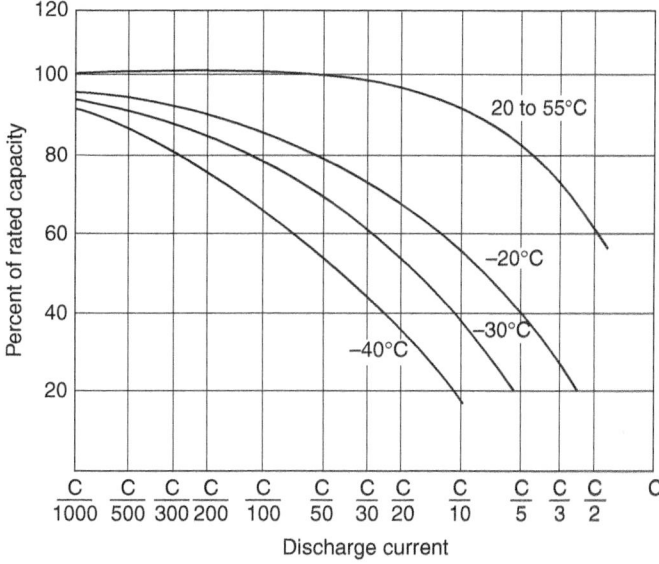

FIGURE 14.11 Performance of Li/SO$_2$ batteries as a function of discharge temperature and load.

The service life of a battery at a given current load can be estimated by dividing the current (in amperes) by the weight or volume of the battery. This value is located on the ordinate, and the service life, at a specific current and temperature, is read on the abscissa.

The weight or volume of a battery needed to deliver a required number of hours of service at a specified current load can be estimated by locating a point on the curve corresponding to the required

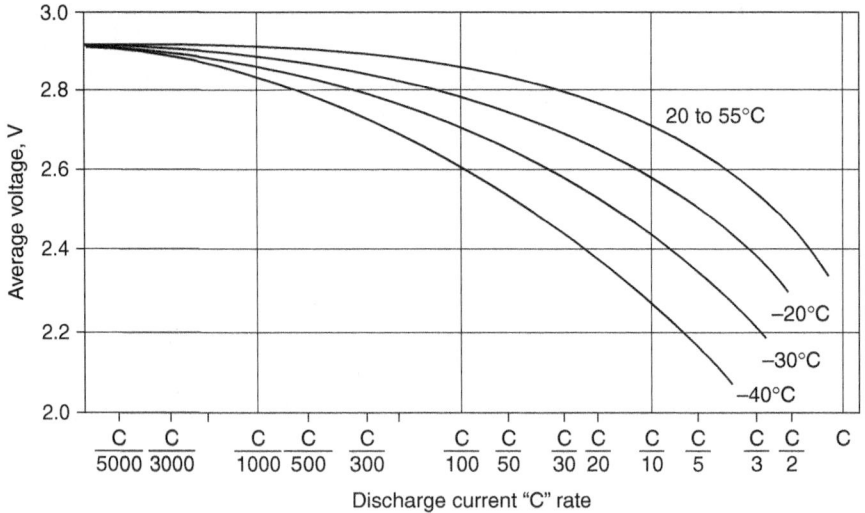

FIGURE 14.12 Midpoint voltage of Li/SO$_2$ batteries during discharge.

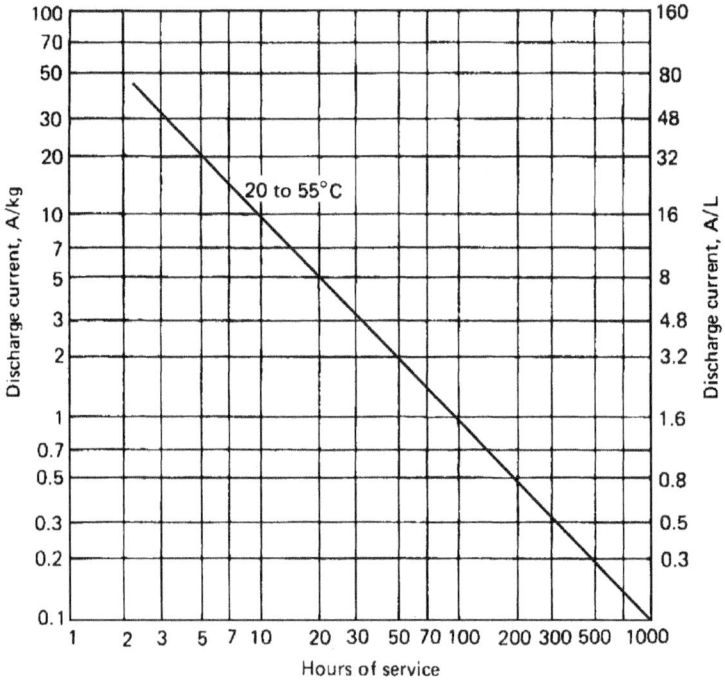

FIGURE 14.13 Service life of high-rate Li/SO$_2$ batteries; 2.0 V end voltage.

service hours and discharge temperature. The battery weight or volume is calculated by dividing the value of the specified current (in amperes) by the value of amperes per kilogram or amperes per liter obtained from the ordinate.

Shelf Life. The Li/SO$_2$ battery is noted for its excellent storage characteristics, even at temperatures as high as 70°C. Most primary batteries lose capacity while idle or on standby due to anode corrosion, side chemical reactions, or moisture loss. With the exception of the magnesium battery, most of the conventional primary batteries cannot withstand temperatures in excess of 50°C and should be refrigerated if stored for long periods. The Li/SO$_2$ battery, however, is hermetically sealed and protected during storage by the formation of a film on the anode surface. Capacity losses during stand are minimal. If cells are partially discharged and then stored, the self-discharge rate is accelerated.

Data[19] on 2-year old BA-5590 batteries consisting of 10 Li/SO$_2$ D-size cells discharged in series at 2 A at +21 and −30°C showed a 6.5% capacity loss at the higher temperature but no loss at the lower temperature. Fourteen-year storage data was also obtained on BA-5598 batteries consisting of five "squat" D-size cells in series. These batteries showed only an 8% capacity loss when discharged at room temperature at 2 A, but virtually no loss at cold temperature. In both cases, a lower operating voltage was observed after storage. Using multiple groups of batteries, stored for 4, 6 and 14 years under ambient conditions, the data shown in Fig. 14.14 were obtained. The capacity loss for the first two years is approximately 3%/yr, but the rate of loss decreases significantly after that period. High-temperature storage of batteries was also carried out at +70 and +85°C, as shown in Fig. 14.15. At 70°C, these batteries showed 92% capacity retention after 1 month and 77% capacity retention after five months. At 85°C, 82% capacity retention was observed after one month's storage. This study concluded that there was no obvious benefit to making long-term storage predictions based on accelerated aging tests at high temperature.

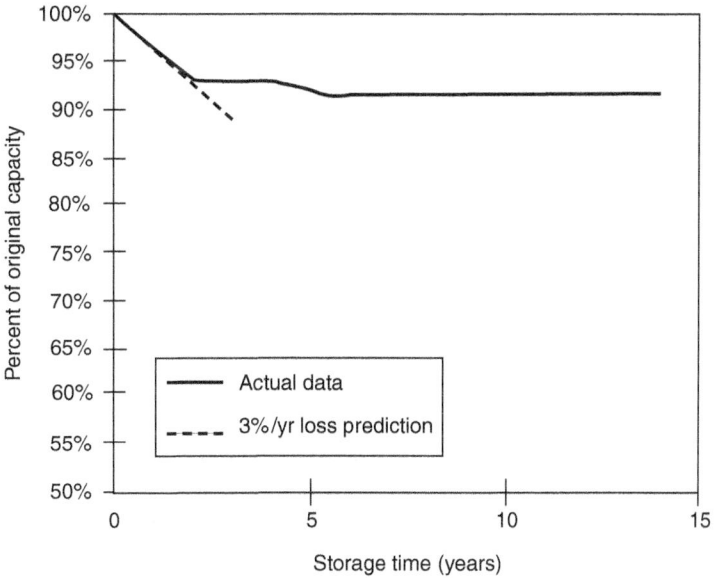

FIGURE 14.14 Capacity retention of Li/SO$_2$ batteries after ambient storage and discharge at the 2 A rate.

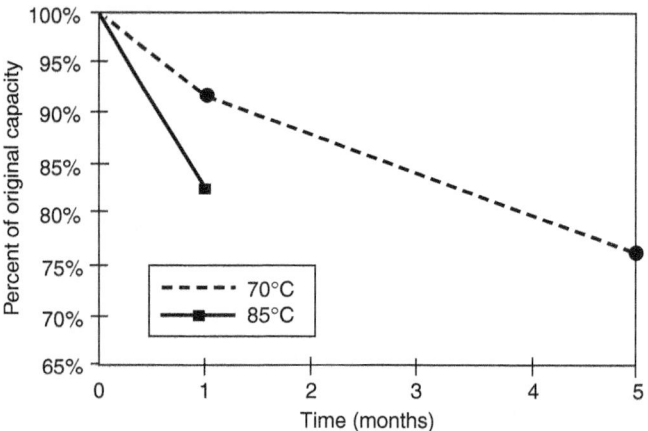

FIGURE 14.15 Effect of storage time/temperature on capacity of Li/SO$_2$ batteries.

Voltage Delay. After extended long-term storage at elevated temperatures, the Li/SO$_2$ battery may exhibit a delay in reaching its normal operating voltage when placed on discharge, especially at high current loads and low temperatures. This start-up or voltage delay is caused by the protective film formed on the lithium anode, the characteristic responsible for the excellent shelf life of the cell. The specific delay time for a battery depends on such factors as the history of the battery, the specific cell design and components, the storage time and temperature, discharge load and temperature. Typically, the voltage delay is minimal or nonexistent for discharges at moderate to low rates at temperatures above −20°C. No delay is evident on discharge at 20°C, even after storage at 70°C for 1 year. On discharge at −30°C, the delay time is less than 200 ms after 8 weeks of storage at 70°C on discharges lower than the 40 h rate. At higher rates, the voltage delay increases with increasing storage temperature and time. At the 2 h discharge rate, for example, the maximum start-up time is about 80 s after 8 weeks of storage at 70°C; it is 7 s after 2 weeks of storage.[20] The start-up voltage delay can be eliminated by preconditioning with a short discharge at a higher rate to depassivate the anode until the operating voltage is reached since the delay will return only after another extended storage period.

14.5.4 Cell and Battery Types and Sizes

Li/SO$_2$ batteries are manufactured in a number of cylindrical cell sizes, ranging in capacity to 34 Ah. Some of the cells are manufactured in standard ANSI cell sizes in dimensions of popular conventional zinc primary batteries. While these single batteries may be physically interchangeable, they are not electrically interchangeable because of the higher cell voltage of the lithium cell (3.0 V for lithium, 1.5 V for the conventional zinc cells). Table 14.9 lists some of the sizes and rated capacities of Li/SO$_2$ batteries that are currently manufactured.

14.5.5 Use and Handling of Li/SO$_2$ Cells and Batteries: Safety Considerations

The Li/SO$_2$ battery is designed as a high-performance system and is capable of delivering a high capacity at high discharge rates. The cell should not be physically or electrically abused, safety features should not be bypassed, and manufacturers' instructions should be followed.

TABLE 14.9 Typical Lithium/Sulfur Dioxide Cylindrical Cells

Size	Open-circuit voltage	Nominal voltage	Nominal capacity (drain)	Max. recommended continuous current	Outside diameter max.	Height max.	Weight	Transport status
½ AA	3 V	2.8 V	450 mAh (50 mA)	250 mA	14.2 mm	27.9 mm	8 g	Nonrestricted
AA	3 V	2.8 V	950 mAh (80 mA)	500 mA	14.2 mm	50.3 mm	15 g	Nonrestricted
2/3 A	3 V	2.8 V	800 mAh (80 mA)	750 mA	16.3 mm	34.5 mm	12 g	Nonrestricted
"Long" A	3 V	2.8 V	1700 mAh (80 mA)	1500 mA	16.3 mm	57.7 mm	18 g	Nonrestricted
1/3 C	3 V	2.9 V	860 mAh (80 mA)	1000 mA	25.9 mm	20.3 mm	18 g	Nonrestricted
2/3 C	3 V	2.8 V	2200 mAh (650 mA)	2000 mA	25.9 mm	35.9 mm	30 g	Nonrestricted
C	3 V	2.8 V	3200 mAh (1000 mA)	2500 mA	25.6 mm	49.5 mm	47 g	Class 9
C	3 V	2.8 V	3750 mAh (250 mA)	2500 mA	25.9 mm	50.4 mm	40 g	Class 9
5/4 C	3 V	2.8 V	5000 mAh (200 mA)	2500 mA	25.6 mm	60.2 mm	58 g	Class 9
5/4 C	3 V	2.8 V	5000 mAh (200 mA)	2500 mA	25.9 mm	59.3 mm	53 g	Class 9
2/3 "Thin" D	3 V	2.8 V	3500 mAh (120 mA)	2000 mA	28.95 mm	42.29 mm	40 g	Class 9
"Thin" D	3 V	2.8 V	5750 mAh (200 mA)	2500 mA	29.1 mm	59.9 mm	63 g	Class 9
D	3 V	2.8 V	7750 mAh (250 mA)	2500 mA	34.5 mm	59.8 mm	85 g	Class 9
D	3 V	2.8 V	7750 mAh (250 mA)	2500 mA	34.2 mm	59.3 mm	85 g	Class 9
D	3 V	2.8 V	9200 mAh (250 mA)	2500 mA	34.2 mm	59.3 mm	85 g	Class 9
D	3 V	2.8 V	7500 mAh (250 mA)	4000 mA	34.2 mm	59.3 mm	85 g	Class 9
"Fat" D	3 V	2.8 V	8000 mAh (270 mA)	2500 mA	39.5 mm	50.3 mm	96 g	Class 9
F	3 V	2.8 V	11500 mAh (1000 mA)	3000 mA	31.9 mm	100.3 mm	125 g	Class 9
DD	3 V	2.8 V	16500 mAh (500 mA)	3000 mA	33.3 mm	120.6 mm	175 g	Class 9
"Long Fat DD"	3 V	2.8 V	34000 mAh (1000 mA)	3000 mA	41.7 mm	141.0 mm	300 g	Class 9

Source: SAFT Batteries.
Cells leakproof up to +95°C. Most cells are UL recognized.
Operating temperature range: −60/+70°C.

Abusive conditions could adversely affect the performance of the Li/SO$_2$ battery and result in cell venting, rupture, explosion, or fire. Preventive measures are discussed in Sec. 14.4.

The Li/SO$_2$ battery is pressurized and contains materials that are toxic or flammable. Properly designed batteries are hermetically sealed so there will be no leakage or out-gassing, and they are equipped with safety vents that release if the batteries reach excessively high temperatures and pressures, thus preventing an explosive condition.

The Li/SO$_2$ batteries can deliver very high currents. Because high internal temperatures can develop from continuous high current drain and short circuit, batteries must be protected by electrical fusing and thermal cutoffs. Charging of Li/SO$_2$ batteries may result in venting, rupture, or even explosion and should never be attempted. Cells or groups of cells connected in parallel should be diode-protected to prevent one group from charging another. The balanced Li/SO$_2$ cell is designed to handle forced discharges or cell reversal and will perform safely within the specified bounds, but design limits should not be exceeded in any application.

Proper battery design, using the Li/SO$_2$ cell, should follow these guidelines:

1. Use electrical fusing and/or current-limiting devices to prevent high currents or short-circuits.
2. Protect with diodes if cells are paralleled or connected to a possible charging source.
3. Minimize heat buildup by adequate heat dissipation and protect with thermal cutoff devices.

4. Do not inhibit cell vents in battery construction.
5. Do not use flammable materials in the construction of batteries.
6. Allow for release of vented gases.
7. Incorporate resistor and switch to activate it to ensure complete depletion of active materials after normal discharge. This allows disposal as nonhazardous waste.
8. In certain cases, a diode is placed in parallel with the cell to limit the voltage excursion in reversal.

Currently special procedures govern the transportation, shipment, and disposal of Li/SO_2 batteries as well as other lithium batteries.[12–15] Procedures for the use, storage, and handling of these batteries also have been recommended. The latest issue of these regulations should be consulted for the most recent procedures.

14.5.6 Applications

The desirable characteristics of the Li/SO_2 battery and its ability to deliver a high energy output and operate over a wide range of temperatures, discharge loads, and storage conditions have opened up applications for this primary battery that heretofore were beyond the capability of primary battery systems.

Major applications for the Li/SO_2 battery are in military equipment, such as night-vision devices, radio transceivers, and portable surveillance devices, taking advantage of its light weight and wide-temperature operation. Table 14.10 lists the most common types of military Li/SO_2 and Li/MnO_2 batteries, their characteristics and applications. These batteries are constructed to meet the requirements of MIL-PRF-49471 B (CR) and the applicable specification sheets for the particular battery type. Other military applications, such as sonobuoys and munitions, have long shelf-life requirements, and the active Li/SO_2 primary battery can replace reserve batteries used earlier. Some industrial applications have developed, particularly to replace secondary batteries and eliminate the need for recharging. Consumer applications have been limited to date because of restrictions in shipment and transportation and concern with its hazardous components.[21]

14.6 LITHIUM/THIONYL CHLORIDE (Li/SOCL$_2$) BATTERIES

The lithium/thionyl chloride ($Li/SOCl_2$) battery has one of the highest cell voltages (nominal voltage 3.6 V) and energy densities of the practical battery systems. Specific energy and energy densities range up to about 590 Wh/kg and 1100 Wh/L, the highest values being achieved with the low-rate batteries. Figures 8.8, 8.9, and 14.2 illustrate some of the advantageous characteristics of the $Li/SOCl_2$ cell.

$Li/SOCl_2$ batteries have been fabricated in a variety of sizes and designs, ranging from wafer or coin cells with capacities as low as 420 mAh, cylindrical cells in bobbin and spirally wound electrode structures, to large 10,000 Ah prismatic cells, plus a number of special sizes and configurations to meet particular requirements. The thionyl chloride system originally suffered from safety problems, especially on high-rate discharges and overdischarge, and a voltage delay that was most evident on low-temperature discharges after high-temperature storage.[22]

Low-rate batteries have been used commercially for many years for memory backup and other applications requiring a long operating life, such as toll tags and RF transponders. The large prismatic batteries have been used in military applications as an emergency backup power source. Medium- and high-rate batteries have also been developed as power sources for a variety of electric and electronic devices. Some of these batteries contain additives to the thionyl chloride and other oxyhalide electrolytes to enhance certain performance characteristics. These are covered in Sec. 14.7.

TABLE 14.10 U.S. Military Lithium Nonrechargeable Batteries (MIL-PRF-49471*)

	Lithium/sulfur dioxide batteries				
Type designation	Open-circuit voltage (series/parallel)(V)	Nominal voltage (series/parallel)(V)	Nominal energy† (Wh)	Weight (g)	Typical/applications
BA-5093/U	27	23.4	77.2	635	Respirators
BA-5557A/U	30/15	16/13	54	410	Digital message devices
BA-5588A/U	15	13	35	290	PRC-68 and PRC-126 radios; respirators
BA-5590A/U‡	30/15	26/13	185	1,021	SINCGARS radios; chemical agent detectors; satellite radios; jammers; loudspeakers; range finders; countermeasures
BA-5590B/U‡	30/15	26/13	185	1,021	Same as BA-5590A/U
BA-5598A/U	15	13	87	650	PRC-77 radios; direction finders; sensors
BA-5599A/U	9	7.8	50	450	Test sets; sensors
	Lithium/manganese dioxide batteries				
Type designation	Open-circuit voltage (series/parallel)(V)	Nominal voltage (series/parallel)(V)	Nominal energy† (Wh)	Weight (g)	Typical applications
BA-5312/U	13.2	10.8	41	275	PRC-112G survival radio
BA-5347/U	6.6	5.4	40	290	Thermal weapons sights; test sets
BA-5360/U	9.9	8.1	65	320	Digital communications devices
BA-5367/U	3.3	2.7	3.25	20	Night vision devices
BA-5368/U	13.2	10.8	12	140	PRC-90 survival radios
BA-5372/U	6.6	5.4	2.3	20	Memory hold function; encoding devices
BA-5380/U	6.6	5.4	45	230	Ground navigation sets; chemical agent Monitors; respirators
BA-5388/U	16.5	13.5	49	500	PRC-68 and PRC-126 radios; respirators
BA-5390/U‡	33/16.5	27/13.5	250	1,350	SINCGARS radios; chemical agent detectors; satellite radios; jammers; loudspeakers; range finders; countermeasures
BA-5390A/U‡	33/16.5	27/13.5	250	1,350	Same as BA-5390/U

Notes:
*MIL-PRF-49471 will be replaced by MIL-PRF-32271 in DOD procurements.
†Nominal energy rating for temperature range of 25 ± 10°C (77 ± 18°F).
‡The BA-5590A/U and BA-5390A/U have built-in state-of-charge indicators (SOCIs); the BA-5590B/U and BA-5390/U do not.
Contributed by Mr. Patrick Lyman, U.S. Army Material Command.

14.6.1 Chemistry

The Li/SOCl$_2$ cell consists of a lithium anode, a porous carbon cathode, and a nonaqueous SOCl$_2$:LiAlCl$_4$ electrolyte. Other electrolyte salts, such as LiGaCl$_4$ have been employed for specialized applications. Thionyl chloride is both the electrolyte solvent and the active cathode material. There are considerable differences in electrolyte formulations and electrode characteristics. The proportions of anode, cathode, and thionyl chloride will vary depending on the manufacturer and the desired performance characteristics. Significant controversy exists as to the relative safety of anode-limited vs. cathode-limited designs.[23] Some cells have one or more electrolyte additives. Catalysts, metallic powders, or other substances have been used in the carbon cathode or in the electrolyte to enhance performance.

The generally accepted overall reaction mechanism is

$$4Li + 2SOCl_2 \rightarrow 4LiCl \downarrow + S + SO_2$$

The sulfur and sulfur dioxide are initially soluble in the excess thionyl chloride electrolyte, and there is a moderate buildup of pressure due to the generation of sulfur dioxide during the discharge. The lithium chloride, however, is not soluble and precipitates within the porous carbon cathode as it is formed. Sulfur may precipitate in the cathode at the end of discharge. In most cell designs and discharge conditions, this blocking of the cathode is the factor that limits the cell's service or capacity. Formation of sulfur as a discharge product can also present a problem because of a possible reaction with lithium, which may result in a thermal runaway condition.

The lithium anode is protected by reacting with the thionyl chloride electrolyte during stand, forming a protective LiCl film on the anode as soon as it contacts the electrolyte. This passivating film, while contributing to the excellent shelf life of the cell, can cause a voltage delay at the start of a discharge, particularly on low-temperature discharges after long stands at elevated temperatures. The presence of trace qualities of moisture leads to the formation of HCl, which increases passivation, as does the presence of ppm levels of iron. Some products have special anode treatments or electrolyte additives to overcome or lower this voltage delay.

The low freezing point of thionyl chloride (below −110°C) and its relatively high boiling point (78.8°C) enable the cell to operate over a wide range of temperature. The electrical conductivity of the electrolyte decreases only slightly with decreasing temperature. Some of the components of the Li/SOCl$_2$ systems are toxic and flammable; thus exposure to open or vented cells or cell components should be avoided.

14.6.2 Bobbin-Type Cylindrical Batteries

Li/SOCl$_2$ bobbin batteries are manufactured in a cylindrical configuration, most in sizes conforming to ANSI standards. These batteries are designed for low- to moderate-rate discharge and are not typically subjected to continuous discharge at rates higher than the $C/50$ rate. They have a high energy density. For example, the D-size cell delivers 19.0 Ah at 3.4 V, compared with 15 Ah at 1.5 V for the conventional zinc-alkaline cells (see Tables 8.5 and 14.11).

Construction. Figure 14.16 shows the constructional features of the cylindrical Li/SOCl$_2$ cell, which is built as a bobbin-type construction. The anode is made of lithium foil which is swaged against the inner wall of a stainless or nickel-plated steel can; the separator is made of nonwoven glass fibers. The cylindrical, highly porous cathode, which takes up most of the cell volume, is made of Teflon-bonded acetylene black. The cathode also incorporates a current collector, which is a metal cylinder in the case of the larger cells and a pin in the case of smaller cells that do not have an annular cavity.

TABLE 14.11 Characteristics of Extended Life Wafer-Type and Cylindrical Bobbin-Type Li/SOCl$_2$ Cells

Size	½ AA	AA	C	1/10 D	1/6 D	D	DD
Rated capacity (Ah)	1.2	2.4	8.5	.1	1.7	19	35
Rated voltage (V)	3.6	3.6	3.6	3.6	3.6	3.6	3.6
Dimensions (max)							
Diameter (mm)	14.5	14.5	26.2	32.9	32.9	32.9	32.9
Height (mm)	25.2	50.5	50	6.5	10.2	61.5	124.5
Volume (cm^3)	4.16	8.34	27.0	5.2	8.2	52.3	105.8
Weight (g)	9.6	18	49.5	16.2	21	93	190
Maximum current for continuous use (mA)	20	60	75	10	10	100	450
Specific energy (Wh/kg)	438	467	610	216	283	695	645
Energy density (Wh/L)	1010	1007	1117	673	726	1235	1158

Operating temperature range: −55 to +85 C. UL Component Recognition: MH12193.
Source: Tadiran Lithium Batteries.

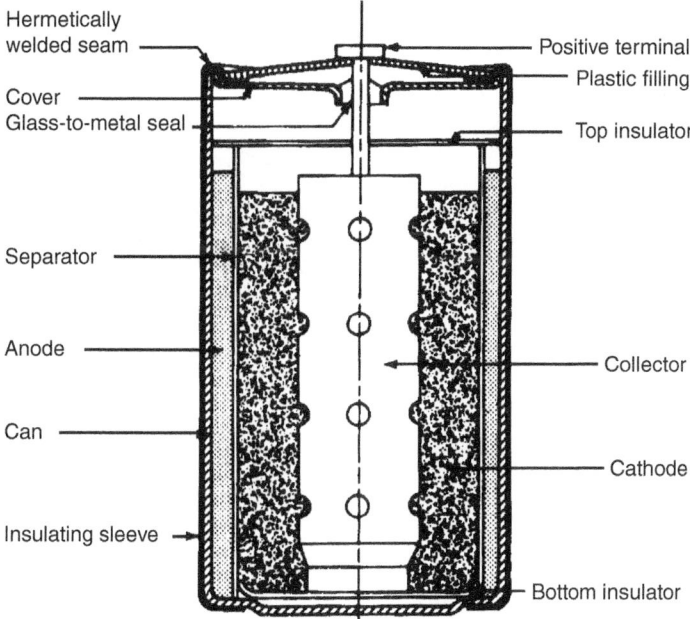

FIGURE 14.16 Cross section of bobbin-type Li/SOCl$_2$ battery.[24]

Performance. The open-circuit voltage of the Li/SOCl$_2$ cell is 3.65 V; typical operating voltages range between 3.3 and 3.6 V with an end voltage of 3.0 V. Typical discharge curves for the Li/SOCl$_2$ battery are shown in Fig. 14.17*a* for the D-size cell. The Li/SOCl$_2$ cell discharges are characterized by a flat profile with good performance over a wide range of temperatures and low- to moderate-rate discharges. Figure 14.17*b* shows the operating voltage of the bobbin D-cell at various drain

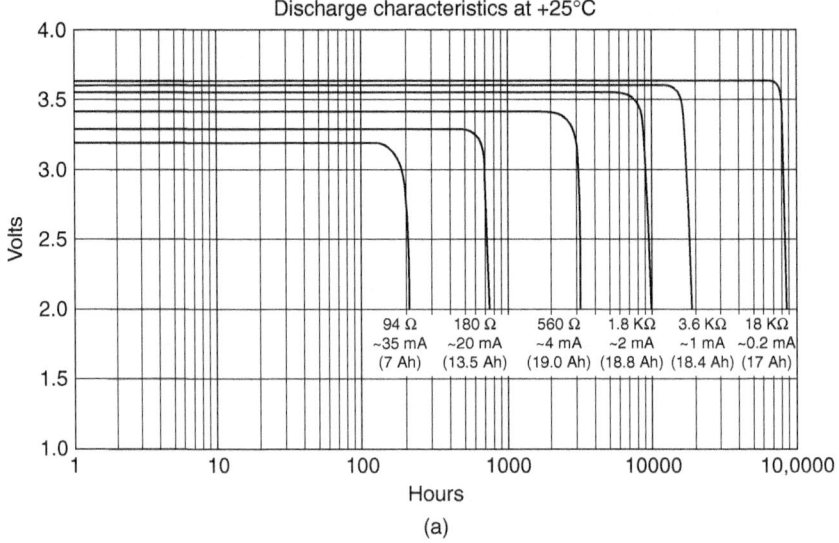

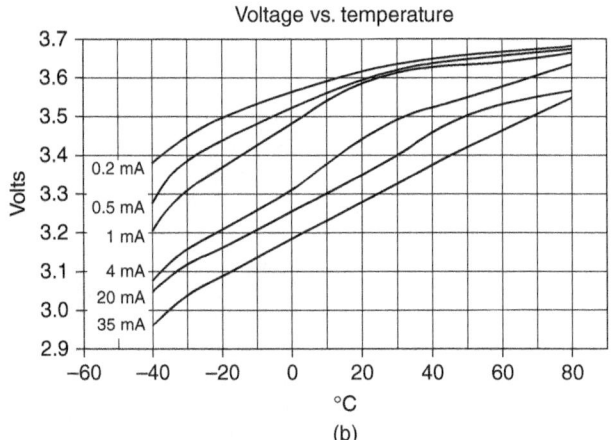

FIGURE 14.17 (*a*) Discharge characteristics of high-capacity Li/SOCl$_2$ cylindrical D-size bobbin battery at +25°C. (*b*) Operating (plateau) voltage of the same battery as a function of temperature at various drain rates.[24]

rates and temperatures. The relationship of capacity with current is given in Fig. 14.18, showing the performance from −40 to 80°C. The Li/SOCl$_2$ cell is capable of performance at unusually high temperatures. At 145°C, (Fig. 14.19) the cells deliver most of their capacity at high rates and up to 70% at low discharge rates (20 days of discharge).[25] Li/SOCl$_2$ cells are used to build battery packs that are employed in oil exploration and most withstand temperatures to 150°C as well as high levels of shock and vibration.

Figure 14.20 shows the behavior of AA cells on continuous low-rate discharge at 25°C. The discharge curve is very flat at these low-current drains, but capacity loss below the 2.4 Ah rating occurs below the 1000-hour rate due to parasitic self-discharge.

The capacity or service life of the high-capacity bobbin-type Li/SOCl$_2$ cell, normalized for a 1-kg and 1-L size cell, at various discharge temperatures and loads, is summarized in Fig. 14.21.

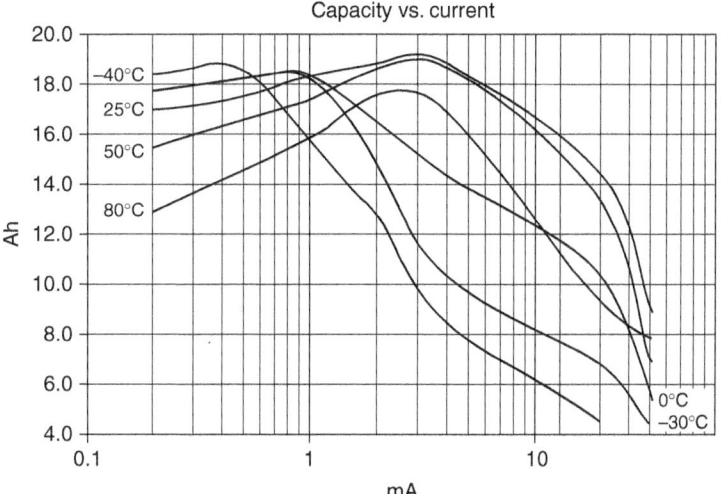

FIGURE 14.18 Performance characteristics of high-capacity cylindrical bobbin D-size batteries as a function of drain rate at various temperatures.[24]

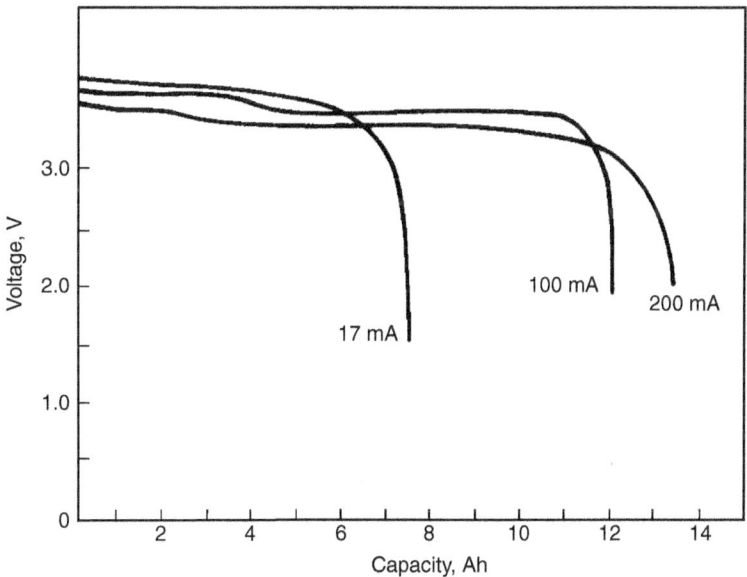

FIGURE 14.19 Discharge characteristics of Li/SOCl$_2$ cylindrical D-size bobbin battery at 145°C.[25]

The long shelf life of the Li/SOCl$_2$ battery is due to the stability of the lithium anode in contact with the electrolyte, as a result of a protective LiCl film that forms on the lithium surface. The long shelf life can also be attributed to the stability of other cell components. For example, the can and cover are cathodically protected by the lithium, and the carbon, stainless-steel collector, and glass separator are all inert in the electrolyte. Figure 14.22 shows the loss of capacity after 3 years at 20°C, a loss of about 1 to 2% per year. Storage at 70°C results in a capacity loss of about 5% per year. Cells should preferably be stored in an upright position; storage on the side or upside-down may result in higher capacity loss.

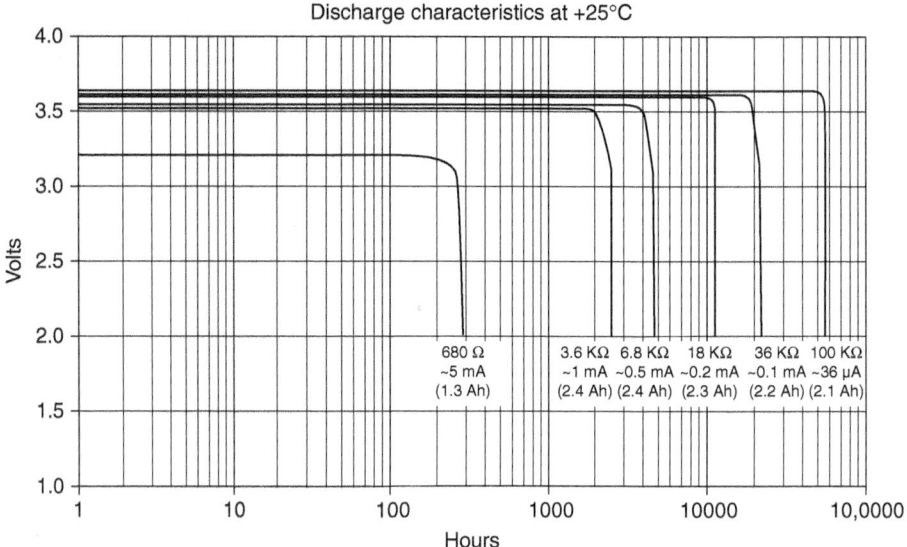

FIGURE 14.20 Discharge characteristics of high-capacity Li/SOCl$_2$ cylindrical AA-size bobbin battery on low-rate discharge at +25°C.[24]

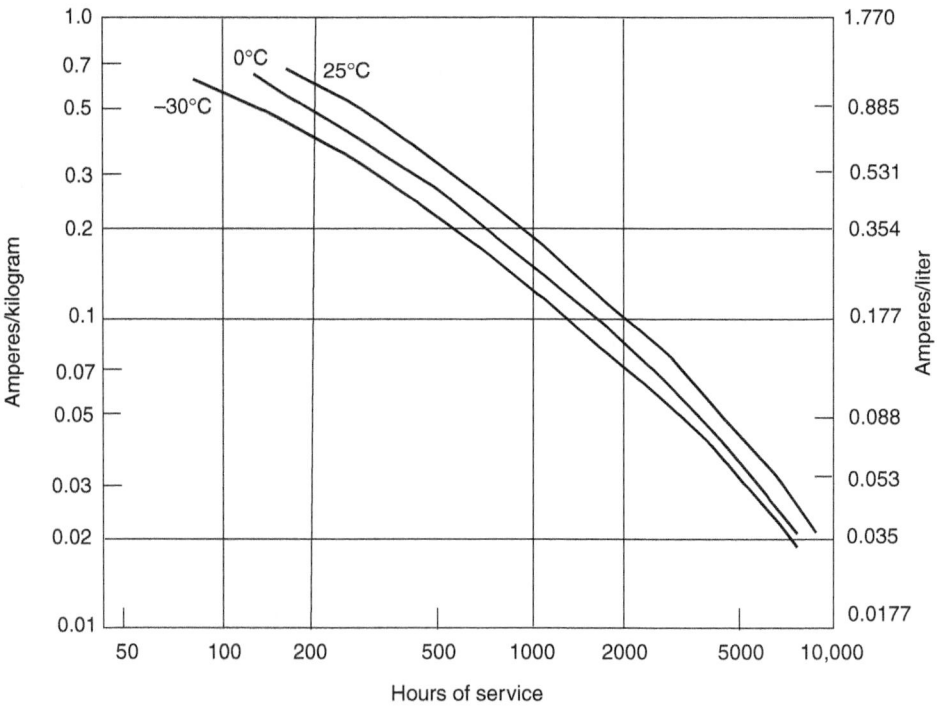

FIGURE 14.21 Service life of Li/SOCl$_2$ cylindrical high-capacity bobbin batteries to 2.0 V cutoff.

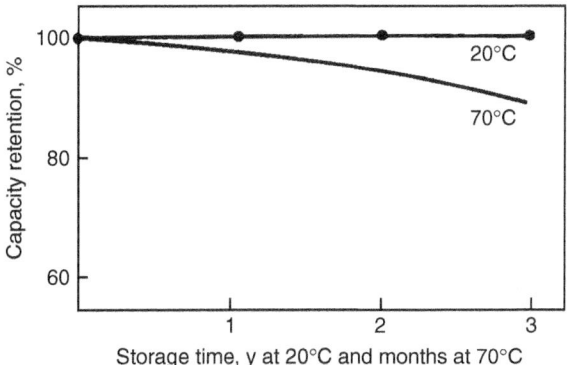

FIGURE 14.22 Capacity retention of Li/SOCl$_2$ cylindrical bobbin battery.[24]

After storage, the Li/SOCl$_2$ battery may exhibit a delay in reaching its operating voltage because of the formation of the LiCl film on the lithium surface. The voltage delay becomes more pronounced with a heavier discharge load and lower discharge temperature. The voltage delay of the Li/SOCl$_2$ cells can be improved by an in situ coating of the lithium anode with an ionic conductor-solid-electrolyte interface. The improvement is shown in Fig. 14.23, which compares the minimum voltage and the load after 2 years of storage for both the standard construction and the coated one. It shows the dependence of the closed-circuit voltage on the discharge current of AA cells after 2 years of storage at 25°C. Once the discharge is started, the passivation film is dissipated gradually, the internal resistance returns to its normal value, and the plateau voltage is reached. The passivation film may be removed more rapidly by the application of high-current pulses for a short period or, alternatively, by short-circuiting the batteries momentarily several times until the cell is activated. The use of a pulse provides more reproducible results.

Special Characteristics. The bobbin batteries are designed to limit the possibility of hazardous operation and to eliminate the need (in some designs) for a safety vent. This is achieved by minimizing the reactive surface area and increasing the heat dissipation, thus limiting the short-circuit current and a hazardous temperature rise, respectively. These cells also are cathode-limited, a feature that was found safer than anode-limited cells for this design.[26] The batteries have withstood short

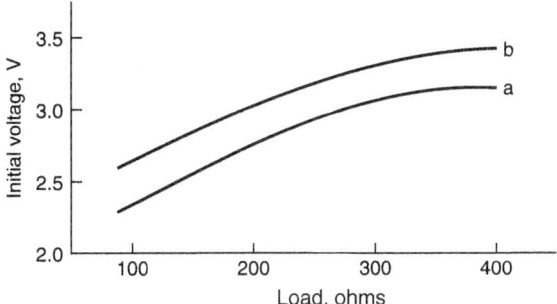

FIGURE 14.23 Li/SOCl$_2$ cylindrical AA-size bobbin batteries—minimum voltage vs. load after 2-year storage at 25°C; (*a*) standard construction; (*b*) with loading on lithium anode.

14.34 PRIMARY BATTERIES

circuits, forced discharge, and charging under certain conditions with no hazardous condition.[24,25,27] Batteries should not be disposed of in fire or subjected to long-term exposure at temperatures near 180°C because they may explode.

Battery Sizes. The bobbin-type Li/SOCl$_2$ batteries are manufactured in the standard ANSI cell sizes as well as in special cell and battery configurations. Although some of these batteries may be physically interchangeable with conventional zinc batteries, they are not electrically interchangeable because of their higher voltages.

Table 14.11 lists the properties of some of the typical bobbin-type batteries that are manufactured. These characteristics may vary with the manufacturer. Manufacturer's data should be consulted for specific data as well as for the characteristics of their other batteries.

14.6.3 Spirally Wound Cylindrical Batteries

Medium- to moderately high–power Li/SOCl$_2$ batteries that are designed with a spirally wound electrode structure are also available. These batteries were developed primarily to meet military specifications where high drains and low-temperature operation are required. They are also used in selected industrial applications where these features are also needed.

A typical construction is shown in Fig. 14.24. The cell container is made of stainless steel, a corrosion-resistant glass-to-metal feed-through is used for the positive terminal, and the cell cover

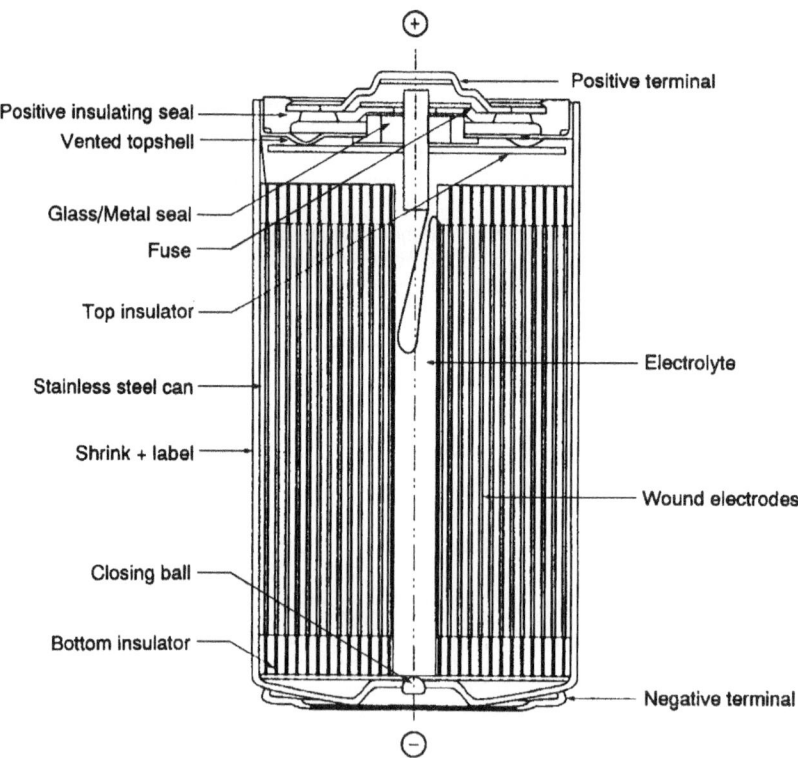

FIGURE 14.24 Cutaway view of lithium/thionyl chloride spirally wound electrode battery. (*Courtesy of SAFT Batteries.*)

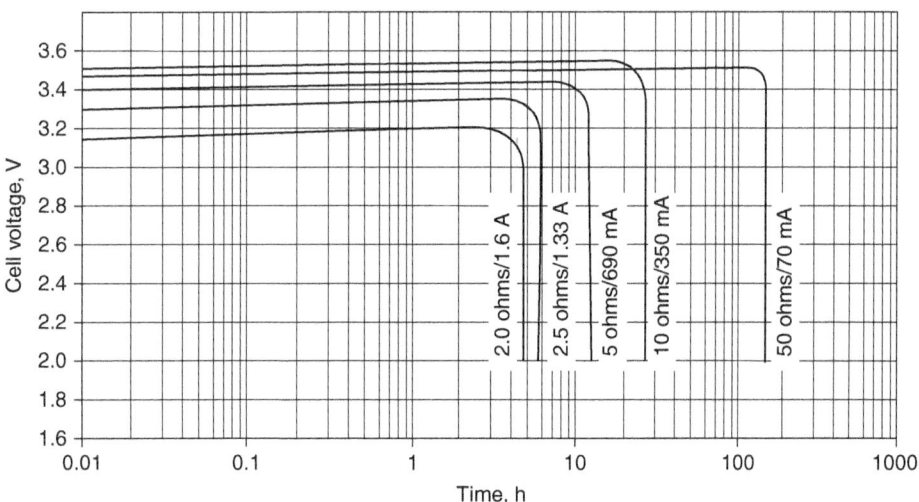

FIGURE 14.25 Discharge characteristics of spirally wound Li/SOCl$_2$ D-size battery, medium discharge rate at 20°C. (*Courtesy of SAFT Batteries.*)

is laser sealed or welded to provide an hermetic closure. Safety devices, such as a vent and a fuse or a PTC device, are incorporated in the cell to protect against buildup of internal pressure or external short circuits.

The discharge curves for a D-size battery are plotted in Fig. 14.25, showing the higher performance at moderate drains compared to the bobbin cell (see Fig. 14.17).

Figure 14.26 shows the performance characteristics of the D-size battery, providing the relationship of voltage and capacity with current drain at several temperatures.

Like the other Li/SOCl$_2$ batteries, these batteries have an excellent storage capability over a wide temperature range due to the buildup of a protective lithium chloride layer on the lithium. Capacity loss on storage at ambient conditions is stated to be less than 3% per year. These products are said to be resistant to passivation. Table 14.12 lists the characteristics of typical cylindrical spirally wound Li/SOCl$_2$ batteries.

14.6.4 Flat or Disk-Type Li/SOCl$_2$ Cells

The Li/SOCl$_2$ system was also designed in a flat or disk-shaped cell configuration with a moderate to high discharge rate capability. These batteries are hermetically sealed and incorporate a number of features to safely handle abusive conditions, such as short circuit, reversal, and overheating, within design limits.

The battery shown in Fig. 14.27 consists of a single or multiple assembly of disk-shaped lithium anodes, separators, and carbon cathodes sealed in a stainless-steel case containing a ceramic feedthrough for the anode and insulation between the positive and negative terminals of the cell.[28]

The batteries were originally manufactured in small and large diameter sizes by Altus Corp., and are currently being produced in large sizes only for U.S. Navy applications by HED Battery Corp., Santa Clara, CA. The characteristics of these batteries that are currently available are summarized in Table 14.13. Discharge curves for large batteries are shown in Fig. 14.28. Typically the cells have a high energy density, flat discharge profiles, and the capability of performance over the temperature range of −40 to 70°C. On storage they can retain 90% of the capacity after storage of 5 years at 20°C, 6 months at 45°C, or 1 month at 70°C.

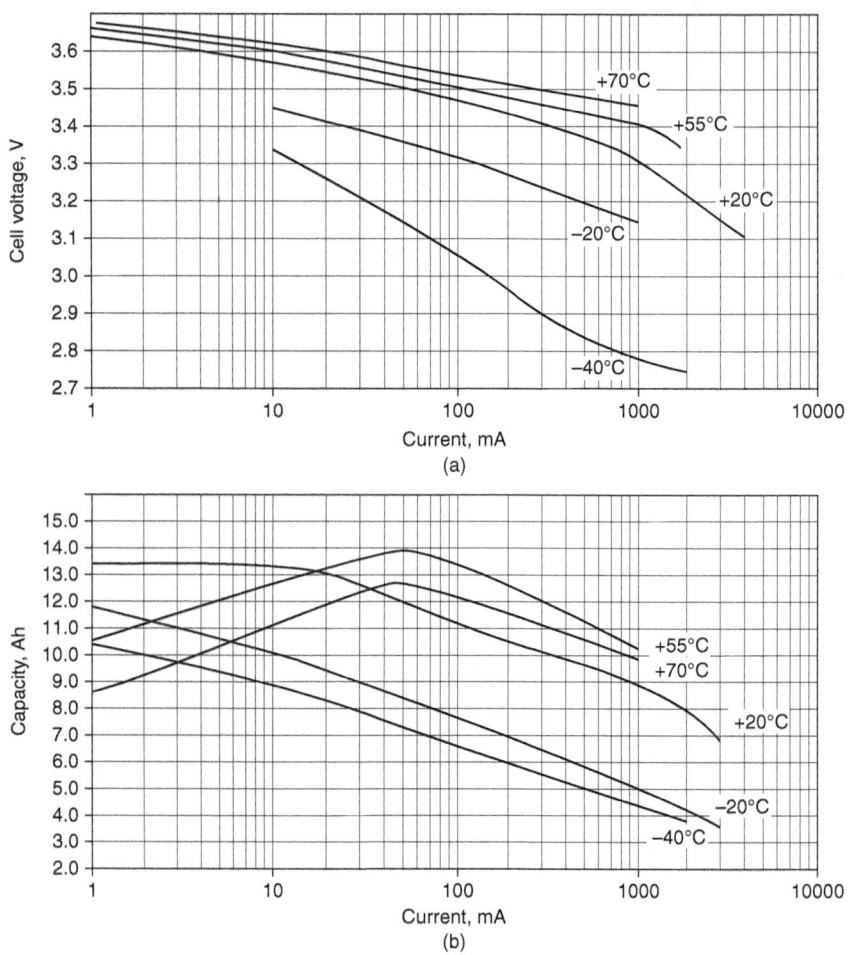

FIGURE 14.26 Discharge characteristics of spirally wound Li/SOCl$_2$ D-size battery, at various temperatures. (*a*) Voltage vs. current. (*b*) Capacity vs. current. (*Courtesy of SAFT Batteries.*)

TABLE 14.12 Characteristics of Typical Cylindrical Spirally Wound Li/SOCl$_2$ Batteries

Size	1/3 C	C	C (light)	D	D	D
Capacity at 20° (Ah)	1.2	5.8	3.6	13.0	12.0	14.0
Rated current (mA)	10	15	15	15	50	300
Nominal voltage	3.6	3.6	3.6	3.6	3.6	3.6
Dimensions (max)						
Diameter (mm)	26.2	26.0	26.0	33.4	33.4	32.05
Height (mm)	18.6	50.4	50.4	61.6	61.6	61.7
Maximum current for continuous use (A)	0.4	1.3	1.3	1.8	1.0	Not specified
Weight (g)	24	51	51	100	100	104.5
Operating temperature range (C)	−60/+85	−60/+85	−60/+85	−65/+85	−65/+120	−40/+150
Transport	Non-restricted	Class 9	Non-restricted	Class 9	Class 9	Class 9

Source: SAFT Batteries.
Open circuit voltage = 3.67 V. Individual cells fitted with non-resettable 5 Amp fuse protection.

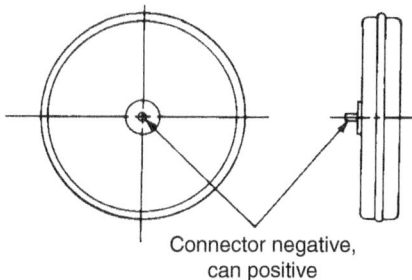

FIGURE 14.27 Disk-type Li/SOCl$_2$ cell.

TABLE 14.13 Characteristics of Disk-Type Batteries Currently Available

Nominal capacity (Ah)	Diameter (cm)	Height (cm)	Weight (kg)	Test current (A)	Average voltage (V)	Actual capacity (Ah)	Specific energy Wh/kg	Energy density (Wh/L)
1200	20.32	12.7	7.63	20	3.34	1170	510	947
2400	40.64	5.84	15.1	8	3.42	2300	523	1043
2400	40.64	5.84	15.1	50	3.28	2000	434	871

Source: HED Battery Corp.

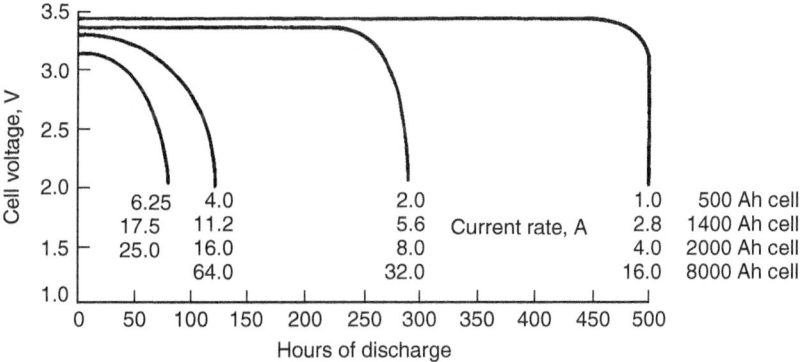

FIGURE 14.28 Performance characteristics of disk-type Li/SOCl$_2$ cells. High-capacity cell; typical performance at 0 to 25°C range to 2.5 V cutoff.

The cell design includes the following features:

1. *Short-circuit protection.* Structure of interconnects fuses at high currents, providing an open circuit.
2. *Reverse-voltage chemical switch.* Upon cell reversal, it allows cell to endure 100% capacity reversal, up to 10 h rate, without venting or pressure increase.
3. *Antipassivation (precoat lithium anode).* Reduces voltage delay by retarding growth of LiCl film; large cells stored for 2 years reach operating voltage within 20 s.
4. *Self-venting.* Ceramic seal is designed to vent cell at predetermined pressures.[28]

These cells are used as multicell batteries in naval applications.

Studies of these designs[29,30] have involved 1000 and 1200 Ah cells for application in a U.S. Navy Long-Range Mine Reconnaissance System (LMRS). These are scaled-down versions of 2350 Ah cells, which had shown the ability to operate at the $C/40$ rate, providing a power density of 2.3 W/kg. Both 1000 and 1200 Ah cells were 20.3 cm in diameter with an annular cavity in the center of the disk. The former unit was 9.53 cm high, while the latter was 12.07 cm high. Both designs incorporate a ceramic-to-metal seal capable of carrying 60 A, and both were limited by the capacity of the carbon cathode with Li/SOCl$_2$ capacity ratio balanced. The 1000 Ah units were tested individually and as 4 and 12-cell batteries with 0.5 cm intercell insulators and compressed between 1.59 cm aluminum end-plates by tie-rods. The 12-cell battery consisted of three stacks of four cells with a diameter of 45.3 cm designed to fit within the hull of LMRS. Test data are summarized in Table 14.14. Based on the results of this testing, a 30-cell battery weighing about 205 kg would deliver 100 kWh at 100 V for operational power up to 5 kW. Subsequently, the cell capacity was increased to 1200 Ah by increasing the cell height.[30] These cells were subjected to a series of safety tests as defined by NAVSEA INST 9310.1B (June 13, 1992) and U.S. Navy Technical Manual S9310-AQ-SAF-010. The 1200 Ah units were subjected to intermittent and sustained short circuits, forced discharge into voltage reversal, charging tolerance high-temperature discharge and high-temperature exposure after low temperature (0°C) discharge. No cells produced venting, loss of material, or case breach of any kind during these tests, nor were there indications of internal shorts or potentially violent conditions. The pulsed and sustained soft-shorts produced significant heating and pressure, but these were within the capability of the battery to operate safely. At sustained currents in excess of 110 A, the cathode appears to clog rapidly, limiting capacity. The exothermic response obtained when the battery was quickly heated to 75°C after cold discharge at 40 A at 0°C is a result of accelerated anode repassivation. The subsequent 55°C short-circuit behavior confirms this hypothesis. There was an indication that this response would have led to a thermal runaway. The 40 A, 55°C discharge demonstrated that the battery could operate safely in the absence of cooling for an extended period of time in a simulated vehicle structure. The tolerance to a moderate charging voltage indicated a margin level in potential failures of diodes. The forced reversal test demonstrated a moderate tolerance to these conditions. A fuse in the negative terminal assembly is being considered to withstand high-rate short circuits. This test program demonstrated the feasibility of using a large lithium/thionyl chloride propulsion battery for LMRS and other similar undersea applications.

TABLE 14.14 Performance Characteristics of 1000 Ah LMRS Lithium/Thionyl Chloride Cells and Batteries (Number of Cells Tested Indicated in Parenthesis After Each Test)

Configuration	Rate	Ah	kWh	Wh/kg
Single (1)	C/22–C/67	931	3.12	108
Single (5)	C/25–C/67	913	3.00	105
Single (2)	C/40	927	3.09	111
4-cell	C/25–C/50	1053	3.58	125
4-cell	C/40	1075	3.67	126
4-cell	C/60	1004	3.41	119
12-cell	C/20–C/40	896	3.03	106
12-cell	C/20–C/40	1016	3.44	121

14.6.5 Large Prismatic Li/SOCl$_2$ Cells

These large, high-capacity Li/SOCl$_2$ batteries were developed mainly as a standby power source for those military applications requiring a power source that is independent of AC line power and the need for recharging.[31–33] They generally were built in a prismatic configuration, as shown schematically in Fig. 14.29. The lithium anodes and Teflon-bonded carbon electrodes were made as rectangular plates with a supporting grid structure, separated by nonwoven glass separators, and housed in an hermetically sealed stainless-steel container. The terminals were brought to the outside by glass-to-metal

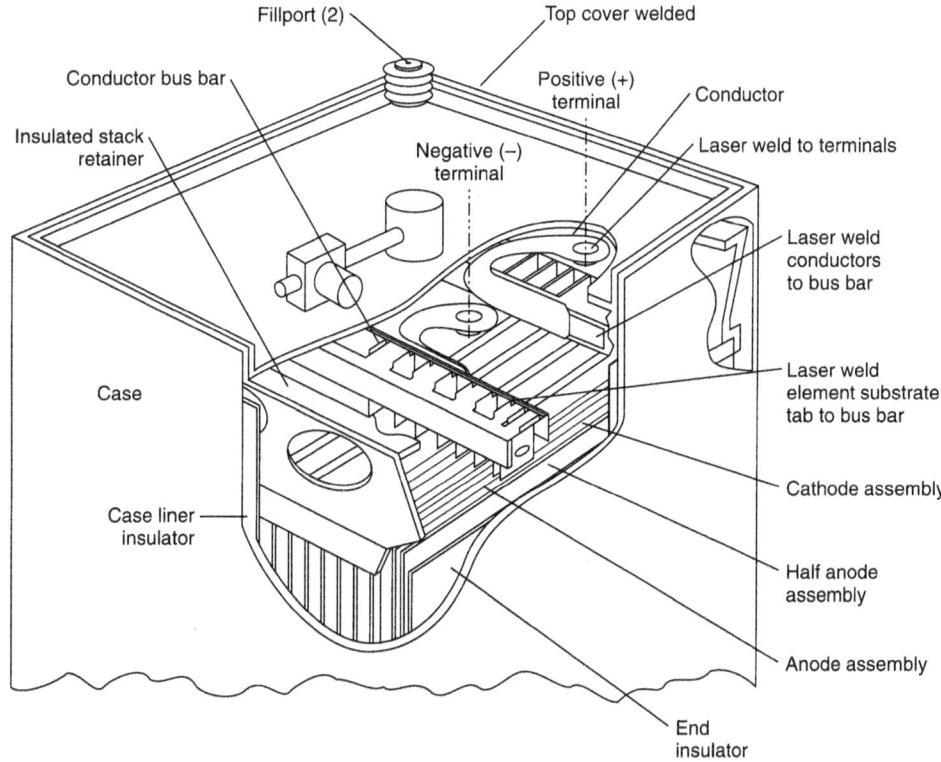

FIGURE 14.29 Cutaway view of 10,000-Ah Li/SOCl$_2$ battery.[33]

feed-through or by a single feed-through isolated from the positive steel case. The cells were filled through an electrolyte filling tube.

The characteristics of several prismatic batteries are summarized in Table 14.15. These cells had a very high energy density. They were generally discharged continuously at relatively low rates (200–300 h rate), but were capable of heavier discharge loads. A typical discharge curve is shown in Fig. 14.30. The voltage profile was flat, and the cell operated just slightly above ambient temperature at this discharge load. During the course of the discharge there was a slight buildup of pressure, reaching a value of about 2×10^5 Pa at the end of the discharge. A higher-rate pulse discharge is shown in Fig. 14.31. The 2000 Ah cell was discharged continuously at a 5 A load, with 40 A pulses, 16 s in duration, superimposed once every day. A steady discharge voltage was obtained throughout most of the discharge, with only a slight reduction in voltage during the pulse. The batteries were capable of performance from −40 to 50°C; shelf-life losses were estimated at 1% per year.[33] These batteries have been decommissioned and are no longer in use but remain the largest lithium batteries ever built.

TABLE 14.15 Characteristics of Large Prismatic Li/SOCl$_2$ Batteries

Capacity, Ah	Height, mm	Length, mm	Width, mm	Weight, kg	Specific energy Wh/kg	Energy density Wh/L
2,000	448	316	53	15	460	910
10,000	448	316	255	71	480	950
16,500	387	387	387	113	495	970

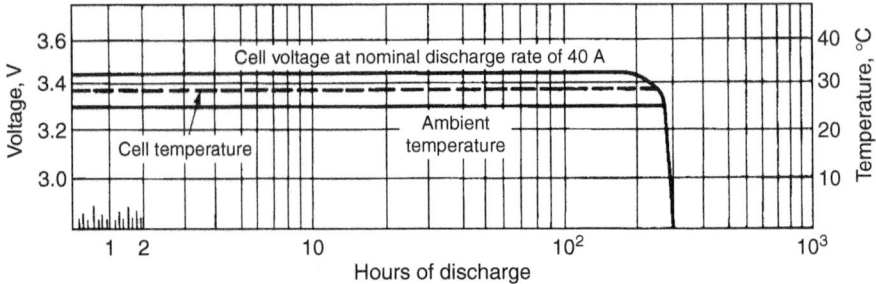

FIGURE 14.30 Discharge curves for 10,000 Ah Li/SOCl$_2$ battery.

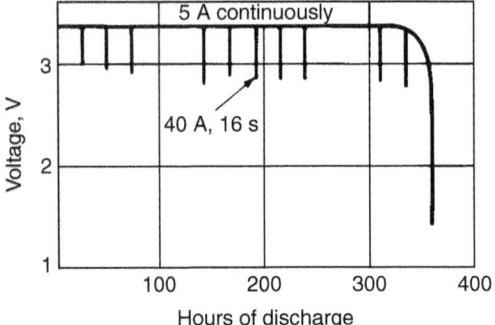

FIGURE 14.31 Discharge of high-capacity 2000 Ah Li/SOCl$_2$ battery.

14.6.6 Applications

The applications of the Li/SOCl$_2$ system take advantage of the high energy density and long shelf life of this battery system. The low-drain cylindrical batteries are used as a power source for CMOS memories, utility meters, and RFID tags such as the EZ Pass Toll collection system, programmable logic controllers, and wireless security alarm system. Wide application in consumer-oriented applications is limited because of the relatively high cost and concern with the safety and handling of these types of lithium batteries.

The higher-rate cylindrical and the larger prismatic Li/SOCl$_2$ batteries are used mainly in military applications where high specific energy is needed to fulfill important mission requirements. A significant application for the large 10,000 Ah batteries was as standby power source as 9-cell batteries for the Missile Extended System Power in the event of loss of commercial or other power. These batteries have been decommissioned.

A lithium/thionyl chloride battery was developed for use on the Mars Microprobe Mission, a secondary payload on the Mars 98 Lander Mission, which disappeared on entry into the Martian atmosphere in December, 1999.[34] The Microprobe power source is a 4-cell lithium/thionyl chloride battery with a second redundant battery in parallel. The eight 2 Ah cells are arranged in a single-layer configuration in the aft-body of the microprobe. The lithium primary cells (and battery configuration) have been designed to survive the maximum landing impact which may reach 80,000 G, and then be operational on the Martian surface to −80°C. Primary lithium-thionyl chloride batteries

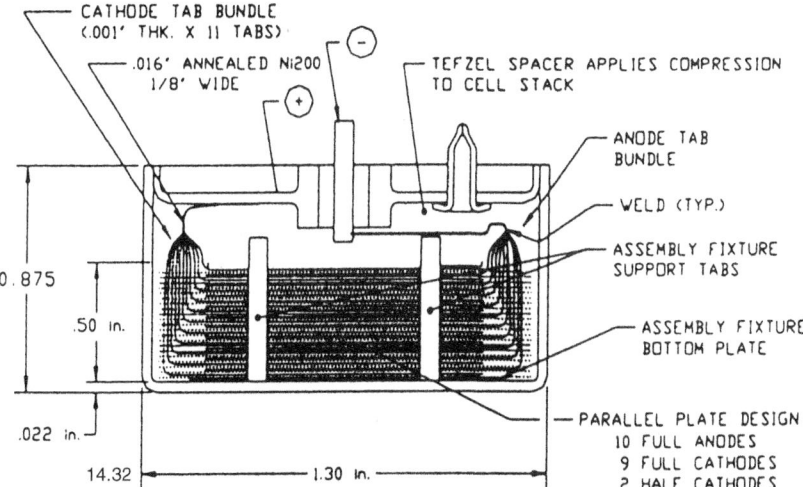

FIGURE 14.32 Vertical cross section of the final 2 Ah cell design. (*Courtesy of Yardney Technical Products, Inc.*)

were selected for the Microprobes based on high specific energy and low-temperature performance. A parallel plate design was selected as the best electrode configuration for surviving the impact without shorting during impact. A cross section of the final 2 Ah Mars cell design showing the parallel plate electrode arrangement is shown in Fig. 14.32. For this cell, the cathodes are blanked from sheets of a Teflon-bonded carbon composition attached to nickel-disc current collectors and connected in parallel. The 10 full disc-anodes are also connected in parallel and are electrically isolated from the case and cover. The assembly fixture helps with component alignment and handling during stack assembly and during connection of the cathode and anode substrate tabs to the cover and the glass-to-metal (GTM) seal anode terminal pin, respectively. The D-size diameter case is 2.22 cm in height. The cover was redesigned after initial tests to minimize the chance of a GTM seal fracture during impact. The Tefzel spacer, located between the cover and stack, helps in handling the stack during substrate tab connections and provides for the proper degree of cathode and separator compression once the cover is TIG welded to the case. The Mars cells were required to deliver 0.55 Ah of capacity at −80°C. A low-temperature thionyl chloride electrolyte consisting of a 0.5 M $LiGaCl_4$ in $SOCl_2$ was developed during the initial phase of the program. As the result of this effort, the battery was able to operate at −80°C on a 1 A discharge as shown in Fig. 14.33. The battery provided over 0.70 Ah at this extremely low temperature. The ability to withstand the 80,000 G impact was demonstrated by Air Gun tests performed at −40°C into frozen desert sand followed by a simulated mission profile at −60°C in an environmental chamber. The battery supplied power for a drill that provided a core sample of subsurface soil for water analysis. In addition, the power requirement for the 20 min water experiment was increased from 2.5 to ~6 W, and increased power levels were required for telemetry at both −60 and −80°C. The total low temperature capacity and major tasks are listed in Table 14.16. The drill operation required an initial current of 1 A for 25 ms, after which the current was in the range of 75 to 85 mA for the duration of the task. The soil sample heating operation lasted for 20 min with power in excess of 6 W required. The high rate transmission started out at 10.4 W (9.7 V), but the power level dropped off toward the end to 6.4 W (7.6 V) after 9 min. The cell delivered a total of 0.724 Ah of low temperature capacity. Although the fate of the Microprobes is unknown, this program extended the state of the art for lithium/thionyl chloride battery technology by demonstrating its ability to withstand 80,000 G impact and then operate at temperatures down to −80°C.

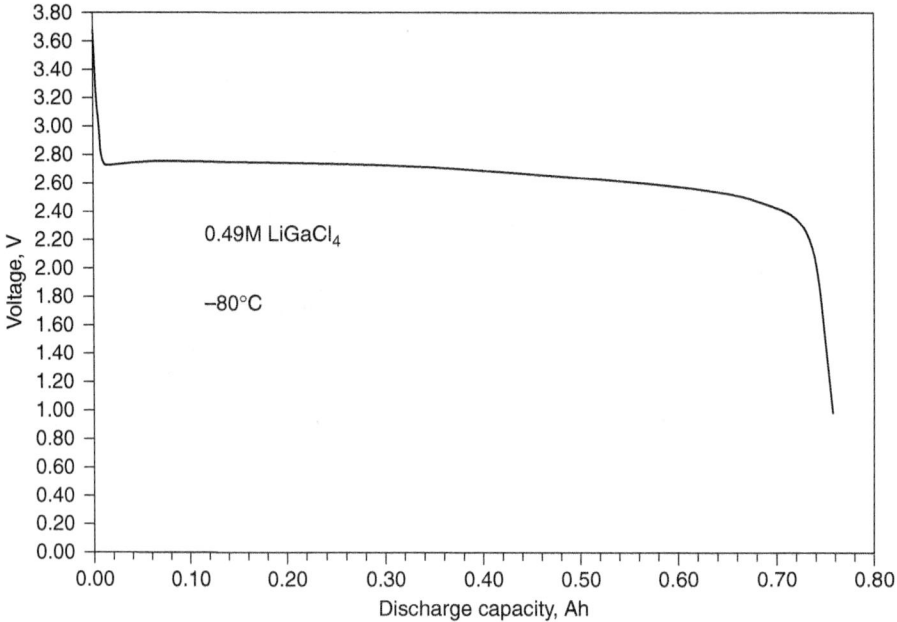

FIGURE 14.33 Capacity in ampere-hours for a 1 A discharge at −80°C for Mars Microprobe battery. (*Courtesy of Yardney Technical Products, Inc.*)

TABLE 14.16 Results of Air Gun Test on Mars Microprobe Battery

Post-impact battery discharge	
Output on profile	0.515 Ah
Additional output at −80°C	0.157 Ah
Total output	0.724 Ah
Major tasks	
Calib. 9Ω −60°C	9.5 V
Drill 136Ω, −60°C	11.7 V
H$_2$O 16Ω, −60°C	10.5 V
High X-mit 9Ω, −60°C	9.7 V
X-mit 59Ω, −80°C	7.6 V

14.7 LITHIUM/OXYCHLORIDE BATTERIES

The lithium/sulfuryl chloride (Li/SO$_2$Cl$_2$) battery is the other oxychloride that has been used for primary lithium batteries. The Li/SO$_2$Cl$_2$ battery has three potential advantages over the Li/SOCl$_2$ battery:

1. A higher energy density as a result of a higher operating voltage (3.9 V open-circuit voltage) as shown in Fig. 14.3 and less solid product formation (which may block the cathode) during the discharge.

2. Inherently greater safety because sulfur, which is a possible cause of thermal runaway in the Li/SOCl$_2$ battery, is not formed during the discharge of the Li/SO$_2$Cl$_2$ battery.
3. A higher rate capability than the thionyl chloride battery as, during the discharge, more SO$_2$ is formed per mole of lithium, leading to a higher conductivity.

Nevertheless, the Li/SO$_2$Cl$_2$ system is not as widely used as the Li/SOCl$_2$ system because of several drawbacks:

1. Cell voltage is sensitive to temperature variations.
2. It has a higher self-discharge rate.
3. It has lower rate capability at low temperatures.

Another type of lithium/oxychloride battery involves the use of halogen additives to both the SOCl$_2$ and SO$_2$Cl$_2$ electrolytes. These additives give an increase in the cell voltage (3.9 V for the Li/BrCl in the SOCl$_2$ system; 3.95 V for the Li/Cl$_2$ in the SO$_2$Cl$_2$ system), energy density and specific energy up to 1070 Wh/L and 485 Wh/kg, and safer operation under abusive conditions.

14.7.1 Lithium/Sulfuryl Chloride (Li/SO$_2$Cl$_2$) Batteries

The Li/SO$_2$Cl$_2$ battery is similar to the thionyl chloride battery using a lithium anode, a carbon cathode, and the electrolyte/depolarizer of LiAlCl$_4$ in SO$_2$Cl$_2$. The discharge mechanism is

$$
\begin{array}{ll}
\text{Anode} & \text{Li} \rightarrow \text{Li}^+ + e \\
\text{Cathode} & \underline{\text{SO}_2\text{Cl}_2 + 2e \rightarrow 2\text{Cl}^- + \text{SO}_2} \\
\text{Overall} & 2\text{Li} + \text{SO}_2\text{Cl}_2 \rightarrow 2\text{LiCl}\downarrow + \text{SO}_2
\end{array}
$$

The open-circuit voltage is 3.909 V.

Cylindrical, spirally wound Li/SO$_2$Cl$_2$ cells were developed experimentally but were never commercialized because of limitations with performance and storage. Bobbin-type cylindrical cells, using a sulfuryl chloride/LiAlCl$_4$ electrolyte and constructed similar to the design illustrated in Fig. 14.16, also showed a variation of voltage with temperature and a decrease of the voltage during storage. This may be attributed to reaction of chlorine, which is present in the electrolyte and formed by the dissociation of sulfuryl chloride into Cl$_2$ and SO$_2$. This condition can be ameliorated by including additives in the electrolyte. Bobbin cells, made with the improved electrolyte, gave significantly higher capacities at moderate discharge currents, compared to the thionyl chloride cells.[35] This system has been employed for reserve lithium/sulfuryl chloride batteries, as well[36] (see Chap. 35).

14.7.2 Halogen-Additive Lithium/Oxychloride Cells

Another variation of the lithium/oxyhalide cell involves the use of halogen additives in both the SOCl$_2$ and the SO$_2$Cl$_2$ electrolytes to enhance the battery performance. These additives result in: (1) an increase in the cell voltage (3.9 V for BrCl in the SOCl$_2$ system [BCX], 3.95 V for Cl$_2$ in the SO$_2$Cl$_2$ system [CSC]), and (2) an increase in energy density and specific energy to about 1054 Wh/L and 486 Wh/kg for the CSC system.

The lithium/oxyhalide cells with halogen additives offer among the highest energy density of primary battery systems. They can operate over a wide temperature range, including high temperatures, and have excellent shelf lives. They are used in a number of special applications—oceanographic and space applications, memory backup, and other communication and electronic equipment.

These lithium/oxychloride batteries are available in hermetically sealed, spirally wound electrode cylindrical configurations, ranging from AA to DD size in capacities up to 30 Ah. These batteries are

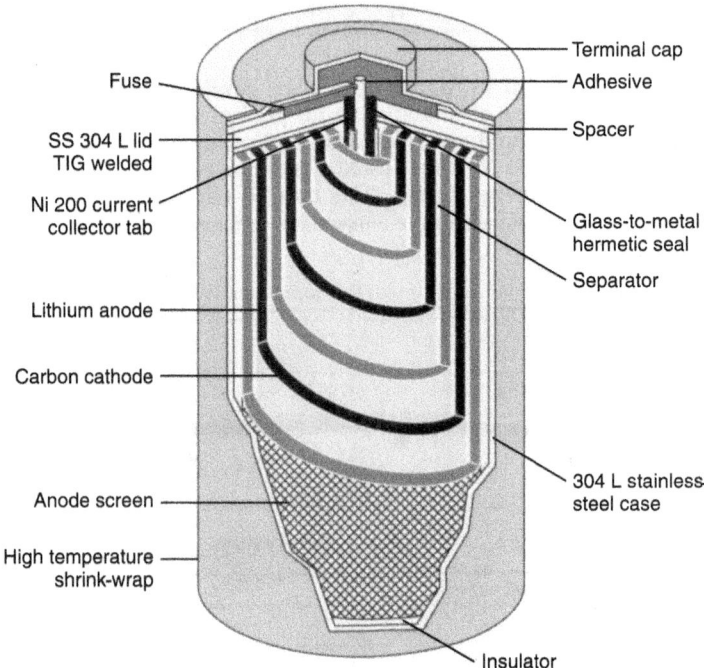

FIGURE 14.34 Cross section of lithium/oxychloride cell. (*Courtesy of Electrochem Solutions Div. Greatbatch Ltd.*)

also available in the AA size containing 0.5 g of Li and in flat disk-shaped cells. Figure 14.34 shows a partial cutaway view of a typical cell. Table 14.17 lists several different lithium-oxychloride batteries manufactured and their key characteristics. Two types of halogen-additive lithium/oxychloride batteries have been developed, as follows:

Li/SOCl$_2$ System with BrCl Additive (BCX). This battery has an open-circuit voltage of 3.9 V and an energy density of up to 1070 Wh/L at 20°C. The BrCl additive is used to enhance the performance. The cells are fabricated by winding the lithium anode, the carbon cathode, and two layers of a separator of nonwoven glass into a cylindrical roll and packaging them in an hermetically sealed can with a glass-to-metal feed-through. The performance of the D-size battery at various temperatures and discharge rates is shown in Fig. 14.35. The discharge curves are relatively flat with a working voltage of about 3.5 V. The batteries are capable of performance over the temperature range of −55 to 72°C. The capacity loss is rated at 3%/yr at 25°C. Capacity loss on storage is higher than with lithium systems using thionyl chloride only.

The addition of BrCl to the depolarizer may also prevent the formation of sulfur as a discharge product, at least in the early stage of the discharge, and minimize the hazards of the Li/SOCl$_2$ battery attributable to sulfur or discharge intermediates. The cells show abuse resistance when subjected to the typical tests, such as short circuit, forced discharge, and exposure to high temperatures.[37]

Li/SO$_2$Cl$_2$ with Cl$_2$ Additive (CSC). This battery has an open-circuit voltage of 3.9 V and an energy density of up to 1050 Wh/L. The additive is used to decrease the voltage-delay characteristic of the lithium/oxyhalide cells. The typical operating temperature of these cells is −20 to 93°C. The cylindrical cells are designed in the same structure as those shown in Fig. 14.34.

TABLE 14.17 Typical Halogen Additive Oxychloride Batteries

	BrCl in $SOCl_2$				Cl_2 in SO_2Cl_2		
	AA	C	D	DD	C	D	DD
Voltage, V							
Open-circuit			3.9			3.9	
Average operating			3.4			3.3 to 3.5	
Rated capacity, 100 h rate, Ah	2.0	7.0	15.0	30.0	7.0	15.0	30.0
Dimensions							
Diameter, mm	13.7	25.6	33.5	33.5	25.6	33.5	33.5
Height, mm	49.2	48.4	59.3	111.5	48.4	59.3	111.4
Volume, cm³	7.25	24.9	52.3	98.3	24.9	52.3	98.2
Weight, g	16	55	115	216	52	116	213
Maximum current capability, mA	100	500	1000	3000	1000	2000	4000
Specific energy/energy density at 100 h rate							
Wh/kg	453	445	433	486	478	452	486
Wh/L	965	984	975	1068	998	990	1054
Operating temperature range, °C			−55 to +85			−20 to +93	

Self-discharge rated at 3%/yr. at 25°C for both types.
Source: Electrochem Solutions Div., Greatbatch Inc.

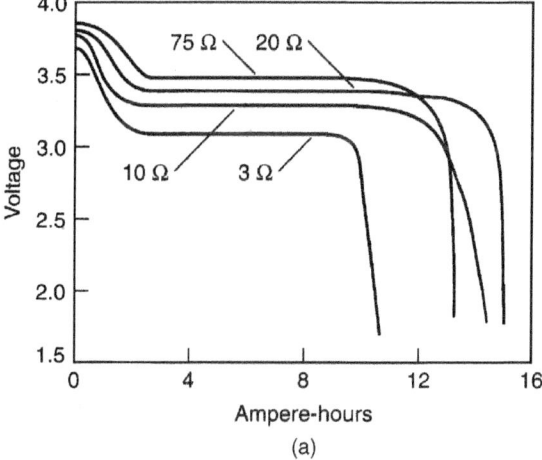

FIGURE 14.35 Performance characteristics of Li/$SOCl_2$ with BrCl-additive. D-size batteries. (*a*) Discharge characteristics at 20°C. (*Courtesy of Electrochem Solutions Div., Greatbatch, Inc.*)

Typical performance characteristics for this battery type are shown in Fig. 14.36. The cells show abuse resistance similar to the Li/BrCl in $SOCl_2$ cells when subjected to abuse tests. Capacity loss is also rated at 3%/yr at 25°C.

Another study[38] has evaluated the effect of ambient temperature storage for up to 6 years. The interrelation of voltage stability, capacity retention, self-discharge, and voltage delay has been delineated. This source should be consulted to obtain detailed data on this system.

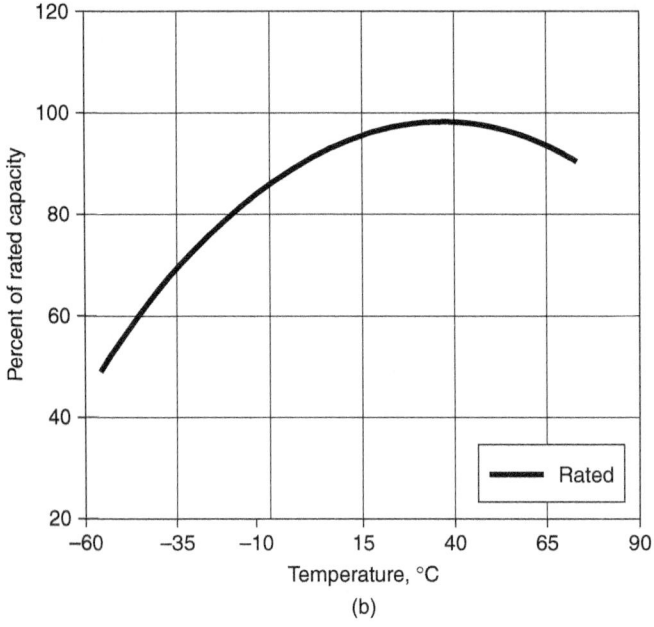

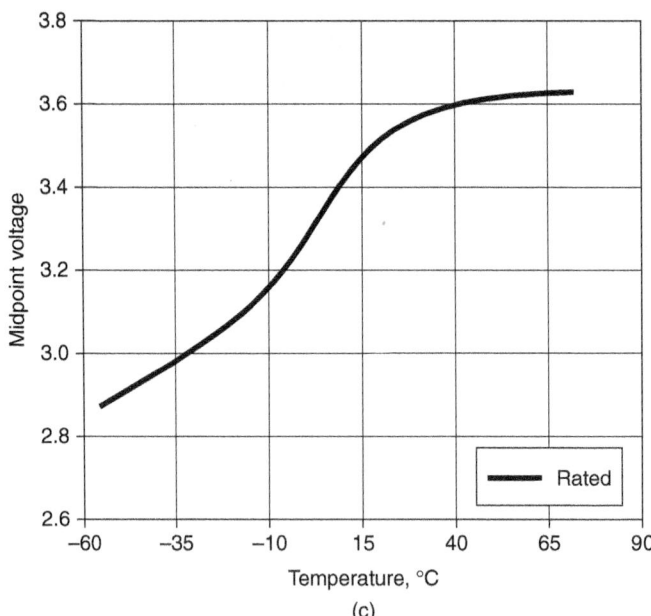

FIGURE 14.35 (*Continued*) (*b*) Capacity as a function of discharge temperature (100% represents rated capacity at room temperature). (*c*) Loaded voltage as a function of temperature.

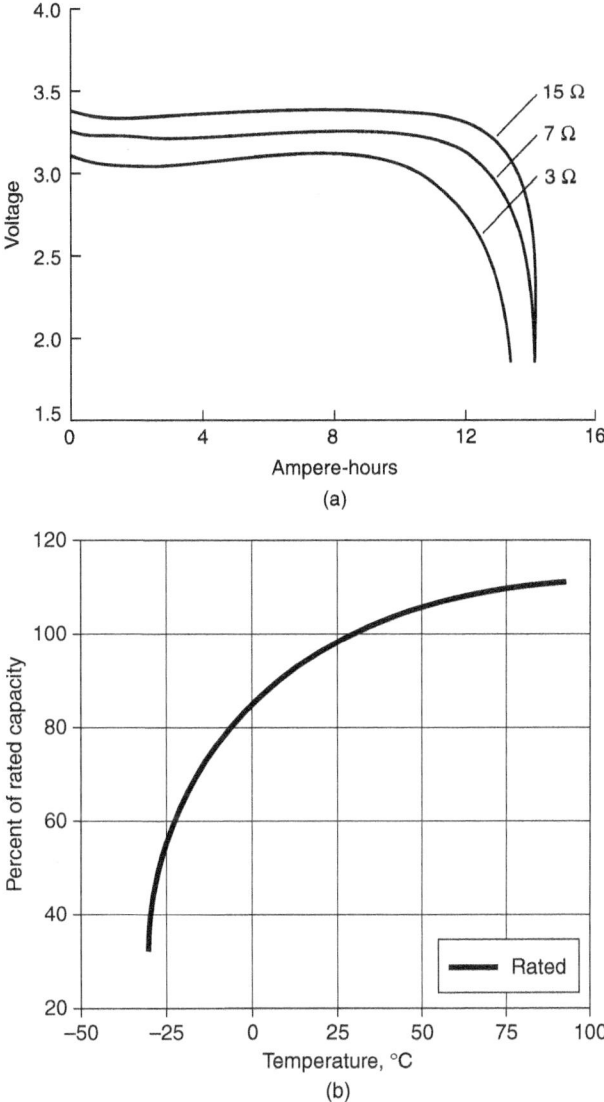

FIGURE 14.36 Performance characteristics of Li/SO$_2$Cl$_2$ with Cl$_2^-$ additive in D-size batteries. (*a*) Discharge at 20°C. (*b*) Capacity vs. discharge temperature; 100% capacity delivered at 20°C. (*Courtesy of Electrochem Power Solutions Div., Greatbatch Inc.*)

14.8 LITHIUM/MANGANESE DIOXIDE (Li/MNO$_2$) BATTERIES

The lithium/manganese dioxide (Li/MnO$_2$) battery was one of the first lithium/solid-cathode systems to be used commercially and is now the most widely used primary lithium battery. It is available in many configurations, including coin, bobbin, spirally wound cylindrical, and prismatic configurations in multicell batteries, and in designs for low, moderate, and moderately high drain applications.

14.48 PRIMARY BATTERIES

The capacity of batteries available commercially ranges up to 11.1 Ah. Larger sized batteries are available for special applications and have been introduced commercially. Its attractive properties include a high cell voltage (nominal voltage 3 V), specific energy about 280 Wh/kg and an energy density above 588 Wh/L, depending on design and application, good performance over a wide temperature range, long shelf life, storability even at elevated temperatures, and low cost.

The Li/MnO$_2$ battery is used in a wide variety of applications such as long-term memory backup, safety and security devices, cameras, many consumer devices, and in military electronics. It has gained an excellent safety record during the period since it was introduced.

The performance of a Li/MnO$_2$ battery is compared with comparable mercury, silver oxide, and zinc batteries in Sec. 8.3, illustrating the higher energy output of the Li/MnO$_2$ battery.

14.8.1 Chemistry

The Li/MnO$_2$ cell uses lithium for the anode, and an electrolyte containing lithium salts in a mixed organic solvent, such as propylene carbonate and 1,2-dimethoxyethane, and a specially prepared heat-treated form of MnO$_2$ for the active cathode material.

The cell reactions for this system are

$$\begin{array}{ll} \text{Anode} & x\text{Li} \rightarrow \text{Li}^+ + e \\ \text{Cathode} & \text{Mn}^{IV}\text{O}_2 + x\text{Li}^+ + e \rightarrow \text{Li}_x\text{Mn}^{III}\text{O}_2 \\ \hline \text{Overall} & x\text{Li} + \text{Mn}^{IV}\text{O}_2 \rightarrow \text{Mn}^{III}\text{O}_2 \end{array}$$

Manganese dioxide, an intercalation compound, is reduced from the tetravalent to the trivalent state, producing Li$_x$MnO$_2$ as the Li$^+$ ion enters into the MnO$_2$ crystal lattice.[1,39]

The theoretical voltage of the total cell reaction is about 3.5 V, but an open-circuit voltage of a new cell is typically 3.3 V. Cells are typically predischarged to lower the open-circuit voltage to reduce corrosion.

14.8.2 Construction

The Li/MnO$_2$ electrochemical system is manufactured in several different designs and configurations to meet the range of requirements for small, lightweight, portable power sources.

Coin Cells. Figure 14.37 shows a cutaway illustration of a typical coin cell. The manganese dioxide pellet faces the lithium anode disk and is separated by a nonwoven polypropylene separator impregnated with the electrolyte. The cell is crimped-sealed, with the can serving as the positive terminal and the cap as the negative terminal.

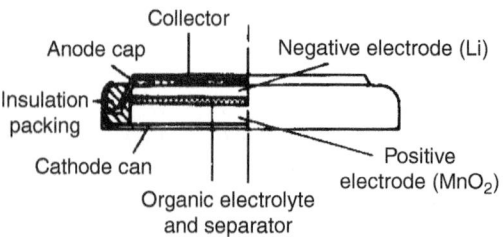

FIGURE 14.37 Cross-sectional view of Li/MnO$_2$ coin-type battery. (*Courtesy of Duracell, Inc.*)

Bobbin-Type Cylindrical Cells. The bobbin-type cell is one of the two Li/MnO$_2$ cylindrical cells. The bobbin design maximizes the energy density due to the use of thick electrodes and the maximum amount of active materials, but at the expense of electrode surface area. This limits the rate capability of the cell and restricts its use to low-drain applications.

A cross section of a typical cell is shown in Fig. 14.38. The cells contain a central lithium anode core surrounded by the manganese dioxide cathode, separated by a polypropylene separator impregnated with the electrolyte. The cell top contains a safety vent to relieve pressure in the event of mechanical or electrical abuse. Welded-sealed cells are manufactured in addition to the crimped-seal design. These cells, which have a 10-year life, are used for memory backup and other low-rate applications.

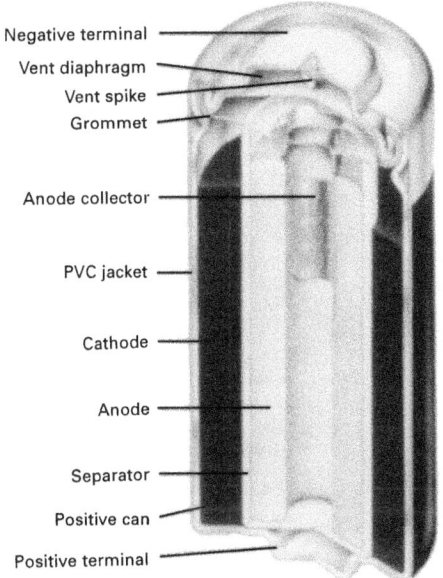

FIGURE 14.38 Cross-sectional view of Li/MnO$_2$ bobbin battery. (*Courtesy of Duracell, Inc.*)

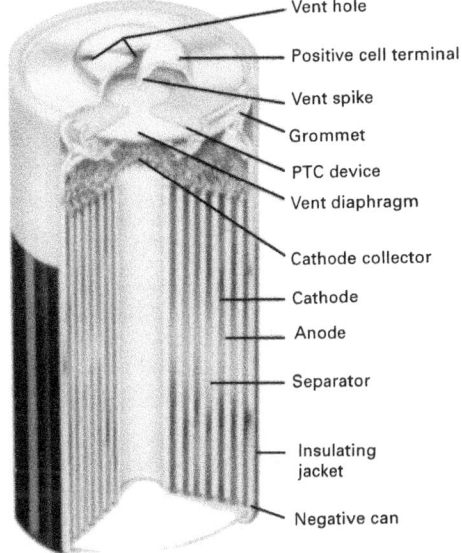

FIGURE 14.39 Cross-sectional view of Li/MnO$_2$ spirally wound electrode battery. (*Courtesy of Duracell, Inc.*)

Spirally Wound Cylindrical Cells. The spirally wound cell, illustrated in Fig. 14.39, is designed for high-current pulse applications as well as continuous moderate-rate operation. The lithium anode and the cathode (a thin, pasted electrode on a supporting grid structure) are wound together with a microporous separator interspaced between the two thin electrodes to form the jelly-roll construction. With this design a high electrode surface area is achieved and the rate capability increased.

High-rate spirally wound cells contain a safety vent to relieve internal pressure in the event the cell is abused. Many of these cells also contain a resettable positive temperature coefficient (PTC) device which limits the current and prevents the cell from overheating if short-circuited accidentally (see also Sec. 14.8.5). Some manufacturers produce these cells with a peripheral laser-welded seal.

Multicell 9 V Battery. The Li/MnO$_2$ system has also been designed in a 9 V battery with 1200 mAh capacity in the ANSI 1604 configuration as a replacement for the conventional alkaline zinc battery. The battery contains three prismatic cells, using an electrode design that utilizes the entire interior volume, as shown in Fig. 14.40. An ultrasonically sealed plastic housing is used for the battery case.

Foil Cell Designs. Other cell design concepts are being used to reduce the weight and cost of batteries by using lightweight cell packaging. One of these approaches is the use of heat-sealable thin

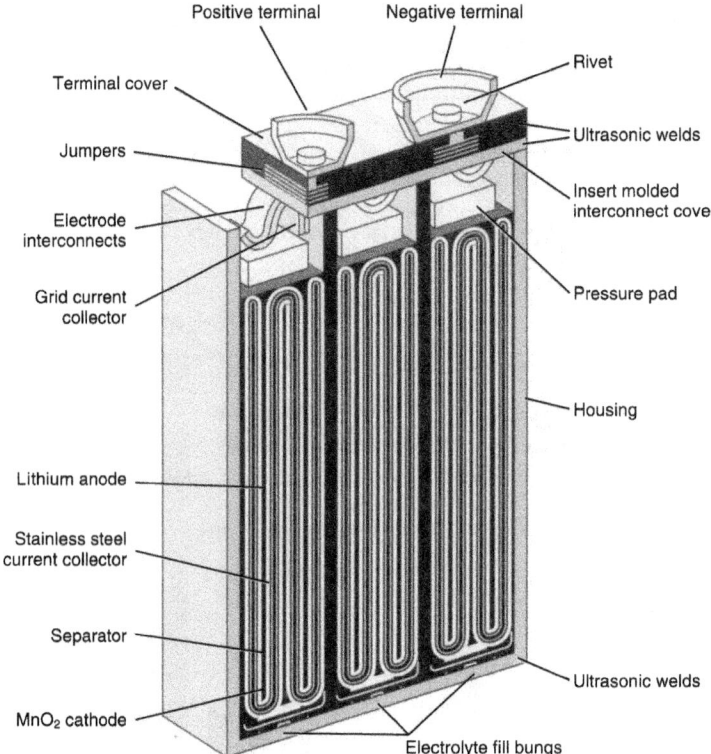

FIGURE 14.40 Cross-sectional view of 3-cell 9 V Li/MnO$_2$ battery. (*Courtesy of Ultralife Batteries, Inc.*)

foil laminates in a prismatic cell configuration in place of metal containers. The design of a cell with a capacity of about 16 Ah is illustrated in Fig. 14.41. The cell contains 10 anode and 11 cathode plates in a parallel plate array.[40]

14.8.3 Performance

Voltage. The open-circuit voltage of the Li/MnO$_2$ battery is typically 3.1 to 3.3 V after predischarge. The nominal voltage is 3.0 V. The operating voltage during discharge ranges from about 3.1 to 2.0 V and is dependent on the cell design, state-of-charge, and the other discharge conditions. The end or cutoff voltage, the voltage by which most of the capacity has been expended, is 2.0 V, except under high-rate, low-temperature discharges, when a lower end voltage may be specified.

Discharge Characteristics of Coin-Type Batteries. Typical discharge curves for the Li/MnO$_2$ coin cells are presented in Fig. 14.42. The discharge profile is fairly flat at these low to moderate discharge rates throughout most of the discharge, with a gradual drop near the end of life. This gradual drop in voltage can serve as a state-of-charge indicator to show when the battery is approaching the end of its useful life.

Some applications (such as an LED watch with backlight) require a high pulse load superimposed on a low background current. The performance of a coin-type battery under these conditions is shown in Fig. 14.43.

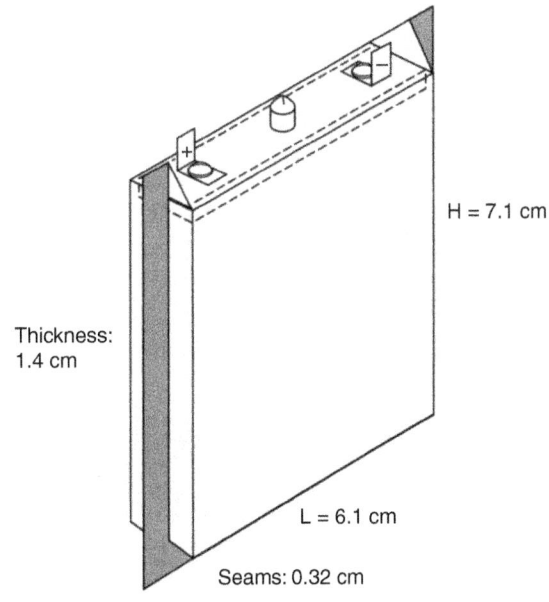

FIGURE 14.41 Foil-cell design. (*From Ref. 40.*)

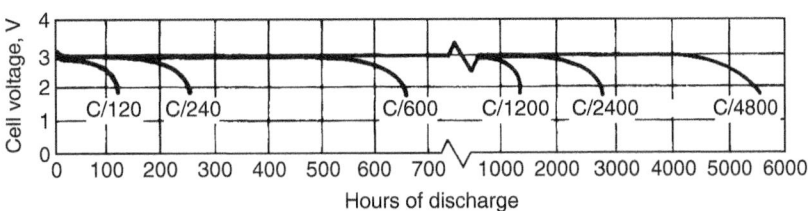

FIGURE 14.42 Typical discharge curves of Li/MnO$_2$ coin-type batteries. (*Courtesy of Duracell, Inc.*)

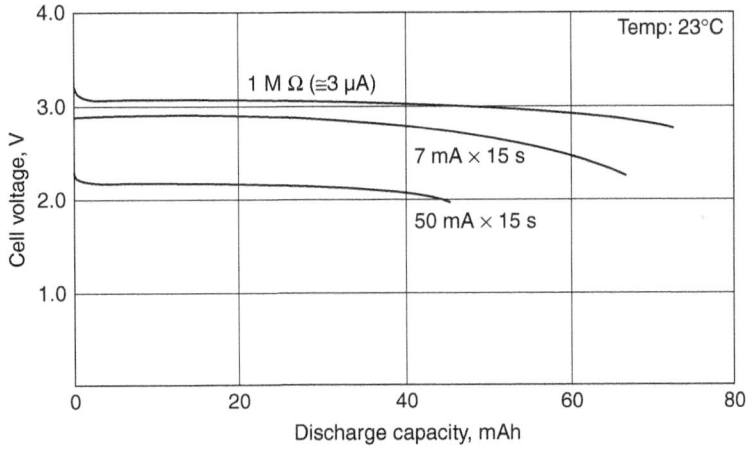

FIGURE 14.43 Pulse characteristics of Li/MnO$_2$ coin-type battery (80 mAh size) at 23°C. Test conditions: continuous load—1 MΩ ≈ 3 μA; pulse load—7 mA × 15 s and 50 mA × 15 s. (*Courtesy of Sanyo Electric Co., Ltd.*)

14.52 PRIMARY BATTERIES

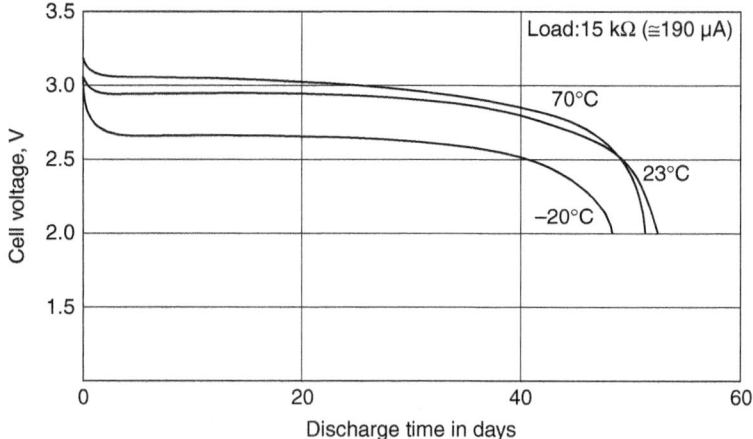

FIGURE 14.44 Typical discharge performance of Li/MnO$_2$ coin-type battery (230 mAh) at various temperatures. (*Courtesy of Sanyo Electric Co., Ltd.*)

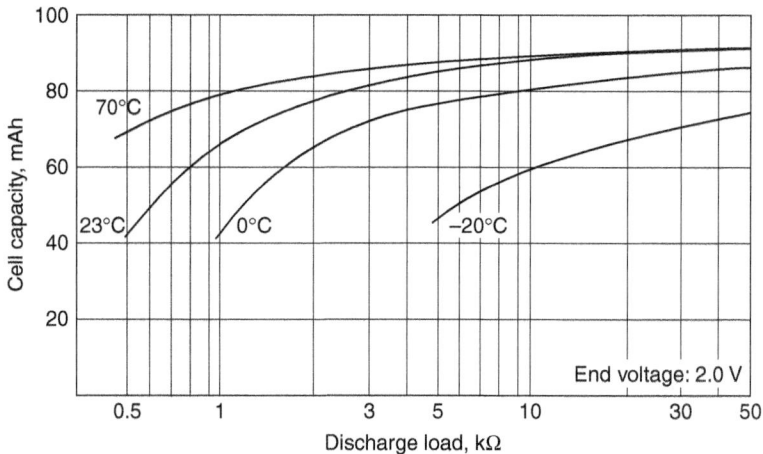

FIGURE 14.45 Delivered capacity of Li/MnO$_2$ coin-type (80 mAh size) at various temperatures and loads. (*Courtesy of Sanyo Electric Co., Ltd.*)

The Li/MnO$_2$ coin-type battery is capable of performing over a wide temperature range, from about −20 to 70°C, as shown in Fig. 14.44.

The discharge characteristics of the Li/MnO$_2$ battery are summarized in Fig. 14.45, which shows the percent capacity delivered at various temperatures and discharge loads.

Discharge Characteristics of Cylindrical Bobbin Batteries. Typical discharge curves for the Li/MnO$_2$ cylindrical bobbin batteries are given in Fig. 14.46. These bobbin electrode batteries are designed for use at low to moderate discharge rates, delivering higher capacities at these discharge rates than the spirally wound electrode batteries of the same size (see Table 14.18). The discharge profile is fairly flat at these low rates throughout most of the discharge, with the typical gradual slope near

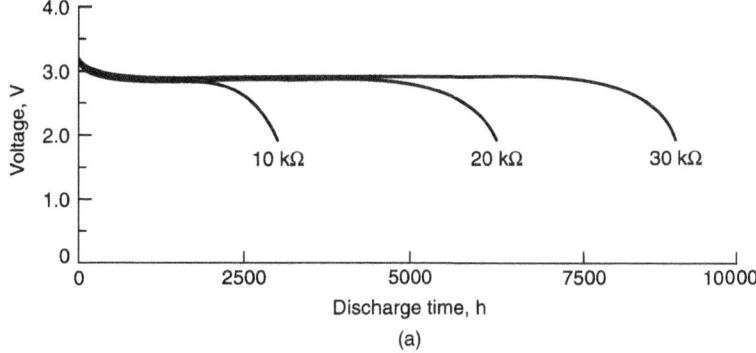

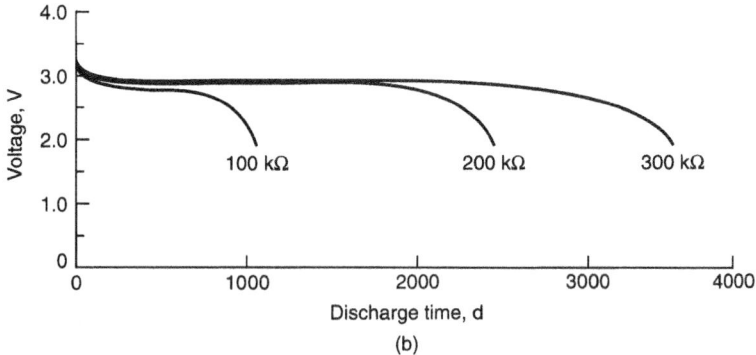

FIGURE 14.46 Discharge characteristics of Li/MnO$_2$ cylindrical bobbin battery (850 mAh size) at 20°C. (*a*) Discharge time in hours. (*b*) Discharge time in days. (*Courtesy of Duracell, Inc.*)

the end of the discharge. The effect of a high pulse load superimposed on a low background current is shown in Fig. 14.47.

The performance of the Li/MnO$_2$ cylindrical bobbin battery at temperatures from −20 to 60°C is shown in Fig. 14.48. Operation of the coin-type and cylindrical bobbin electrode batteries at the lower temperatures is limited to the lower discharge rates.

Discharge Characteristics of Cylindrical Spirally Wound Batteries. Typical discharge curves for Li/MnO$_2$ cylindrical spirally wound batteries at various constant-current discharge loads and temperatures are given in Fig. 14.49. These batteries are designed for operation at fairly high rates and low temperatures. Their discharge profile is flat under most of these discharge conditions. The midpoint voltage when discharged at various loads and temperatures, is plotted in Fig. 14.50.

The characteristics of the batteries under constant power discharge are shown in Fig. 14.51. These data are expressed in terms of *E*-rate, which is calculated in a manner similar to calculating the *C* rate, but based on the rated watt-hour capacity. For example, the *E*/5 rate for a cell rated at 4 Wh is 800 mW.

The discharge characteristics of the cylindrical spirally wound Li/MnO$_2$ battery at various temperatures and loads are summarized in Fig. 14.52. Fig. 14.52(*a*) shows the percent capacity delivered on constant-resistance loads, Fig. 14.52(*b*) the percent capacity delivered on constant-current loads.

TABLE 14.18 Typical Li/MnO$_2$ Batteries

(a) Low-rate coin cell batteries

Model no.	Electrical characteristics (20°C)			Dimensions (mm)		
	Nominal voltage (V)	*Nominal capacity (mAh)	Continuous drain (mA)	Diameter	Height	Weight (g)
CR1025	3	30	0.1	10.0	2.5	0.7
CR1216	3	25	0.1	12.5	1.6	0.7
CR1220	3	35	0.1	12.5	2.0	1.2
CR1612	3	41	0.1	16.0	1.2	0.8
CR1616	3	55	0.1	16.0	1.6	1.2
CR1620	3	75	0.1	16.0	2.0	1.3
CR1632	3	140	0.1	16.0	3.2	1.8
CR2012	3	55	0.1	20.0	1.2	1.4
CR2016	3	90	0.1	20.0	1.6	1.6
CR2025	3	165	0.2	20.0	2.5	2.3
CR2032	3	225	0.2	20.0	3.2	2.9
CR2330	3	265	0.2	23.0	3.0	3.8
CR2354	3	560	0.2	23.0	5.4	5.8
CR2412	3	100	0.2	24.5	1.2	2.0
CR2450	3	620	0.2	24.5	5.0	6.3
CR2477	3	1000	0.2	24.5	7.7	10.5
CR3032	3	500	0.2	30.0	3.2	6.8

*Nominal capacity shown above is based on standard drain and cut off voltage down to 2.0 V at 20°C.

(b) Specialized high-power, cylindrical-type batteries (spiral structure, laser-sealing)

IEC type	Nominal voltage (V)	Nominal capacity (mAh)	Dimensions (mm)		Weight (g)
			Diameter	Height	
CR17335	3	1600	17.0	33.5	17
CR17335	3	1350	17.0	33.5	16
CR17450	3	2400	17.0	45.0	23
CR17450	3	2600	17.0	45.0	23

Operational temperature range: −40 to +85°C.

(c) Standard high-power, cylindrical batteries (spiral-wound, hermetic)

Size	Nominal voltage (V)	Nominal capacity (Ah)	Dimensions (mm)			Continuous current (A)
			Diameter	Height	Weight (g)	
C	3.0	4.8	25.8	50.0	61	2.0
5/4C	3.0	6.1	25.8	60.5	71	2.5
D	3.0	11.1	34.0	60.5	115	3.3

(d) Specialized low-power, cylindrical-type batteries

Model	Nominal voltage (V)	Nominal capacity (mAh)	Dimensions (mm)		Weight (g)
			Diameter	Height	
CR14250	3	850	14.5	25.0	9
CR12600	3	1500	12.0	60.5	15
CR17335	3	1800	17.0	33.5	17
CR17450	3	2500	17.0	45.0	22

Operational temperature range: −40 to +85°C.

TABLE 14.18 Typical Li/MnO$_2$ Batteries (*Continued*)

(e) Specialized cylindrical type primary lithium batteries (spiral structure, crimp-sealing)

IEC type	Nominal voltage (V)	Nominal capacity (mAh)	Dimensions (mm)		Weight (g)
			Diameter	Height	
CR-1/3	3	160	11.6	10.8	3.3
2CR-1/3N	6	160	13.0	25.2	9.1
CR2	3	850	15.6	27.0	11
CR123A	3	1400	17.0	34.5	17
CR-V3	3	3300	28.6 (L) × 14.6 (W) × 52.2 (H)		38
CR-P2	6	1400	34.8 (L) × 19.5 (W) × 35.8 (H)		37
2CR5	6	1400	34 (L) × 17 (W) × 45 (H)		40

Operational temperature range: –40 to + 60°C.

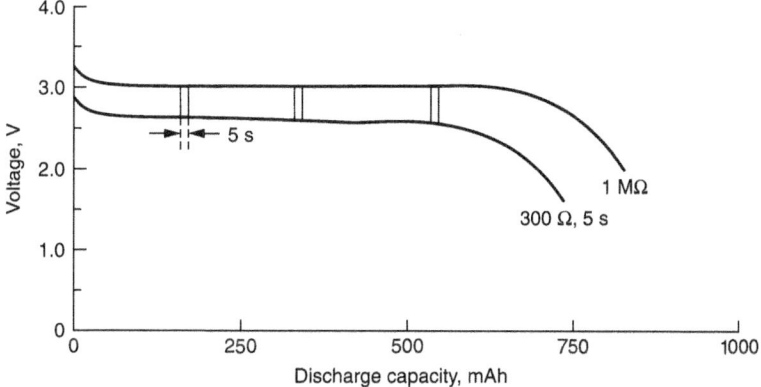

FIGURE 14.47 Pulse discharge characteristics of Li/MnO$_2$ cylindrical bobbin cell (850 mAh size) at 20°C. Test conditions: continuous load—1 MΩ ≈ 2.9 µA; pulse load—300 Ω ≈ 10 mA; duration—5 s; pulses—3; time between pulses—3 h. (*Courtesy of Duracell, Inc.*)

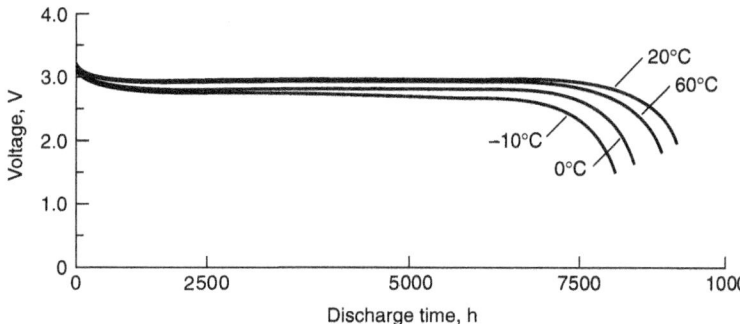

FIGURE 14.48 Discharge performance of Li/MnO$_2$ cylindrical bobbin cell (850 mAh size) at various temperature; 30 kΩ discharge rate. (*Courtesy of Duracell, Inc.*)

14.56 PRIMARY BATTERIES

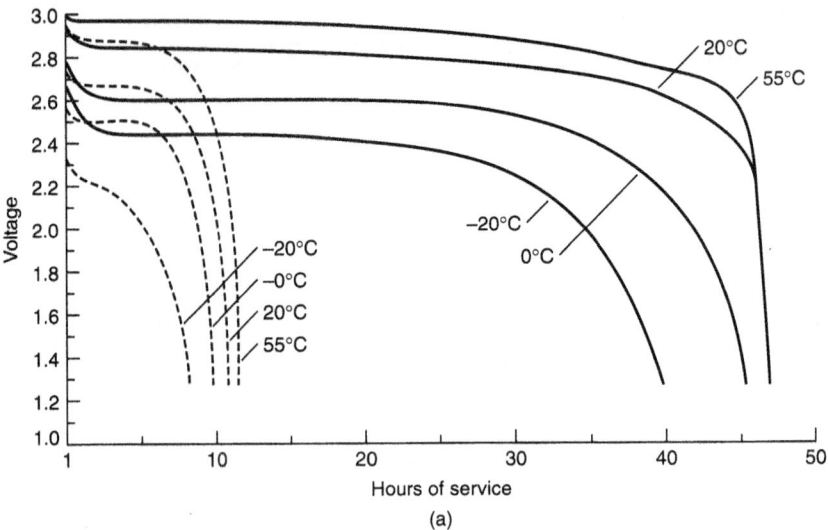

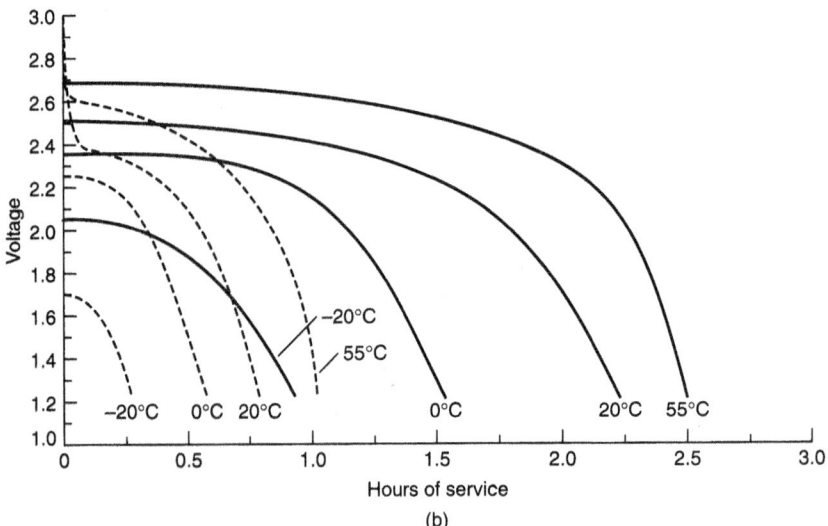

FIGURE 14.49 Discharge characteristics of cylindrical (spirally wound electrode) Li/MnO$_2$ battery (CR123A-size). (*a*) Discharge at 30 and 125 mA. Broken line—125 mA; solid line—30 mA. (*b*) Discharge at 500 and 1000 mA. Broken line—1 A; solid line—0.5 A.

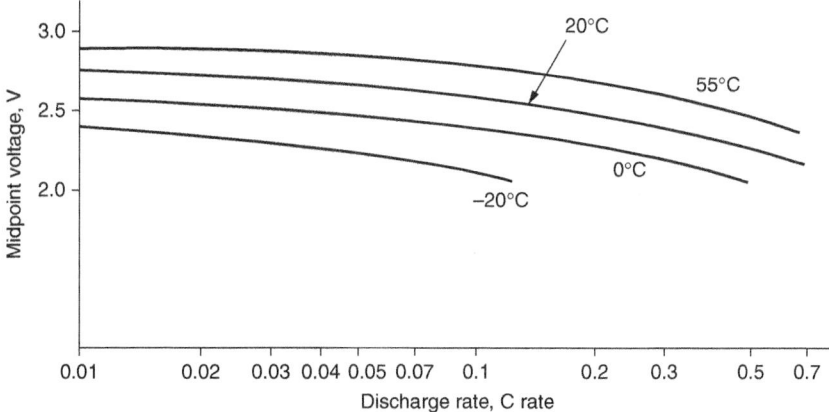

FIGURE 14.50 Midpoint voltage of cylindrical (spirally wound) Li/MnO$_2$ batteries during discharge; 2 V end voltage.

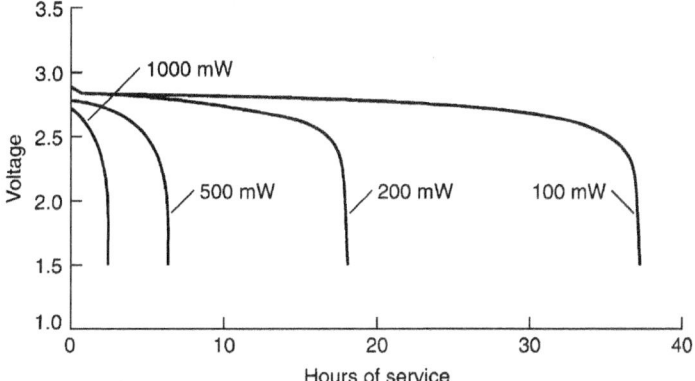

FIGURE 14.51 Discharge characteristics of cylindrical (spirally wound electrode) Li/MnO$_2$ cells (CR123A-size) under constant-power mode at 20°C.

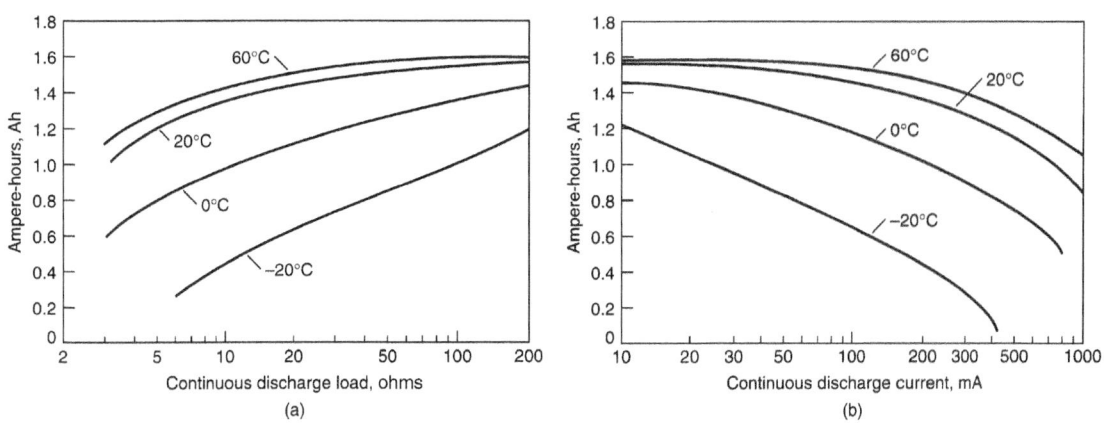

FIGURE 14.52 Summary of discharge characteristics of cylindrical (spirally wound) Li/MnO$_2$ battery (CR123A-size); capacity vs. discharge load to 2.0 V per cell. (*a*) Constant-resistance loads. (*b*) Constant-current loads.

14.58 PRIMARY BATTERIES

The good performance of the Li/MnO$_2$ battery at the lower-rate discharges is evident, and it still delivers a higher percentage of its capacity at relatively high discharge rates compared to conventional aqueous primary cells.

The discharge characteristics of a larger (D-size) spiral-wound Li/MnO$_2$ battery are shown in Fig. 14.53. These figures show the discharge curves at three rates (250 mA, 2.0 A, and 3.0 A) at temperatures from −40 to +72°C. The discharge characteristics indicate the fall-off in performance at lower temperatures.

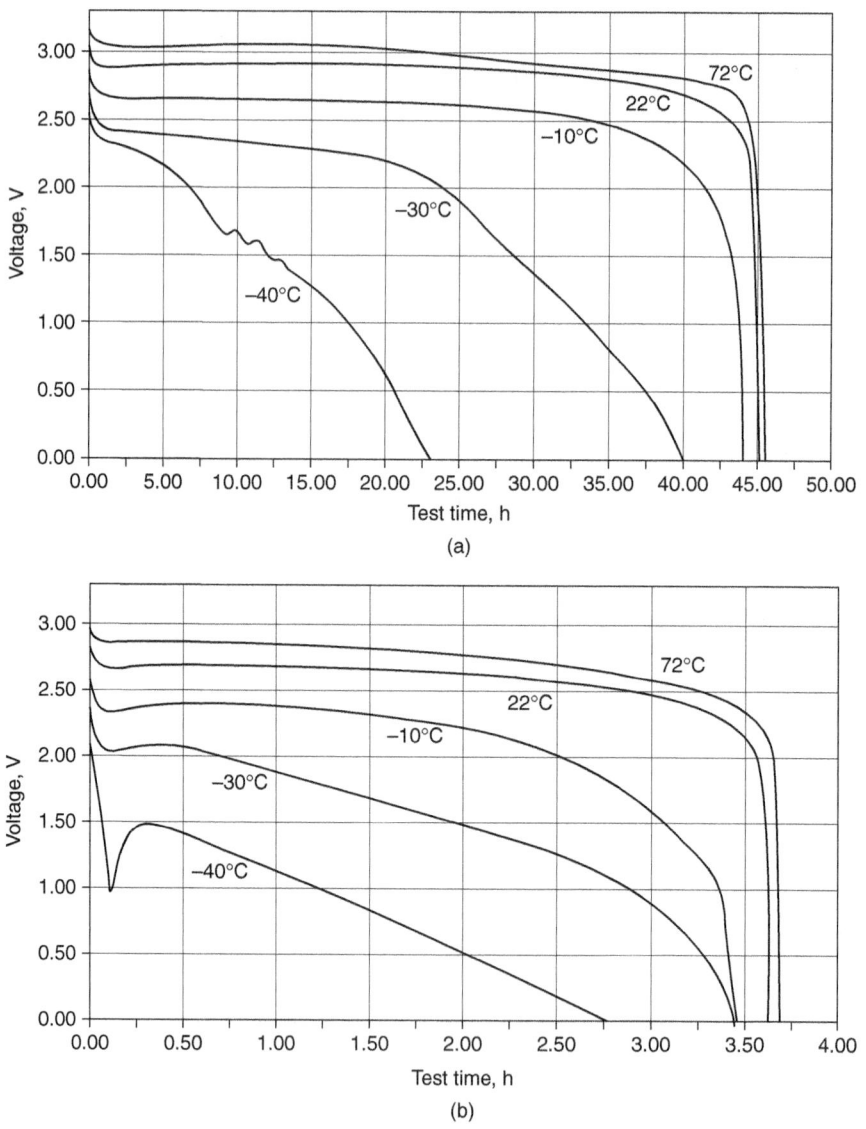

FIGURE 14.53 Discharge characteristics of spiral-wound Li/MnO$_2$ D-size battery. (*a*) Discharge curve at 250 mA amp rate at 5 temperatures. (*b*) Discharge curves at 2.0 A and 5 temperatures. (*c*) Discharge curves at 3.0 A and 5 temperatures. Temperatures for all discharges are: +72, +22, −10, −30, and −40°C. (*Courtesy of Ultralife Batteries, Inc.*)

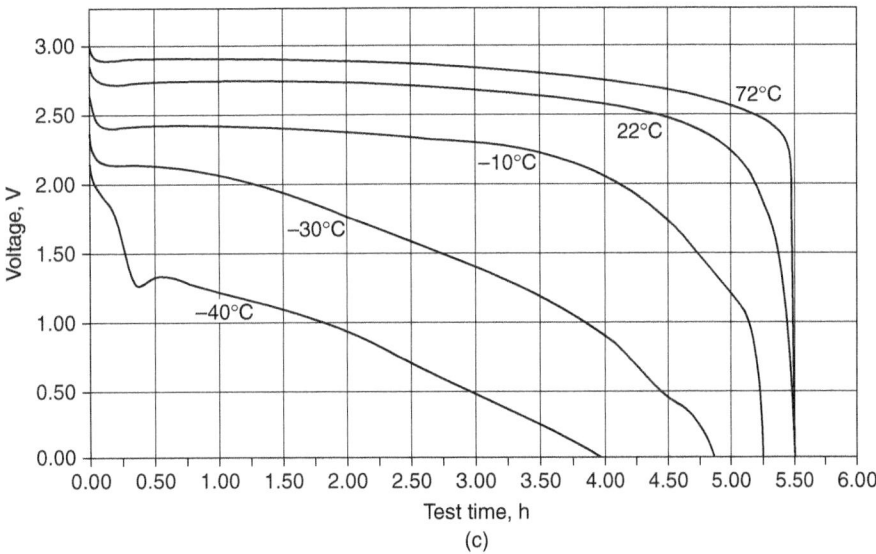

FIGURE 14.53 (*Continued*)

Discharge Characteristics of 3-Cell 9 V Li/MnO$_2$ Battery. The performance of the 9 V, 1.2 Ah Li/MnO$_2$ battery is shown in Fig. 8.6. Typically discharge curves on a 900 ohm load at temperatures from −20 to +23°C are shown in Fig. 14.54(a). Figure 14.54(b) shows the realized capacity in Ah for loads from 60 to 900 ohms at room temperature. The lithium battery has a higher voltage and delivers significantly more service than the comparable zinc–alkaline and carbon–zinc batteries as shown in Fig. 8.6.

Internal Resistance. The internal resistance of the Li/MnO$_2$ battery, as with most battery systems, is dependent on the cell size, design, electrode, separator, as well as the chemistry. Inherently, the conductivity of the organic solvent-based electrolytes is lower than that of the aqueous electrolytes, and the Li/MnO$_2$ system, therefore, has a higher impedance than conventional cells of the same size and construction. Designs that increase electrode area and decrease electrode spacing, such as coin-shaped flat cells and spirally wound jelly-roll configurations, are used to reduce the resistance. Further, the lithium cells will perform relatively more efficiently at the lower temperatures because the conductivity of the organic solvents is less sensitive to temperature changes than it is for the aqueous solvents.

Figure 14.55 shows the change in internal resistance of a 280 mAh coin-type battery during a low-rate discharge at 20°C. Typically, the resistance is a mirror image of the voltage profile. It remains fairly constant for most of the discharge and increases at the end of life.

Service Life. The capacity or service life of the different types of Li/MnO$_2$ cells, normalized for a 1 g and 1 cm^3 cell, at various discharge loads and temperatures, is summarized in Fig. 14.56. These data can be used to approximate the performance of a given cell or to determine the size and weight of a cell for a particular application.

Shelf Life. The storage characteristics of two Li/MnO$_2$ cells are shown in Fig. 14.57. This system is very stable in all of the configurations, with a loss of capacity of less than 0.5%/yr for hermetic and laser-sealed cells. Coin-cells show capacity loss of less than 1% annually at room temperature. The cells do not have any noticeable voltage delay at the start of most discharges, except at low temperature on high discharge rates.

14.60 PRIMARY BATTERIES

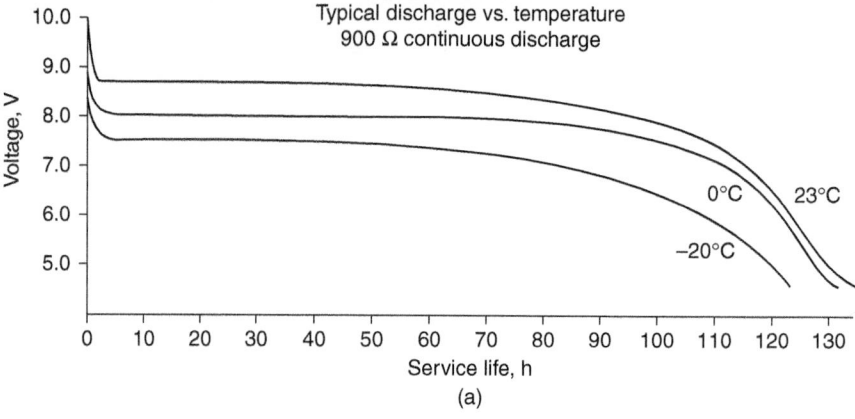

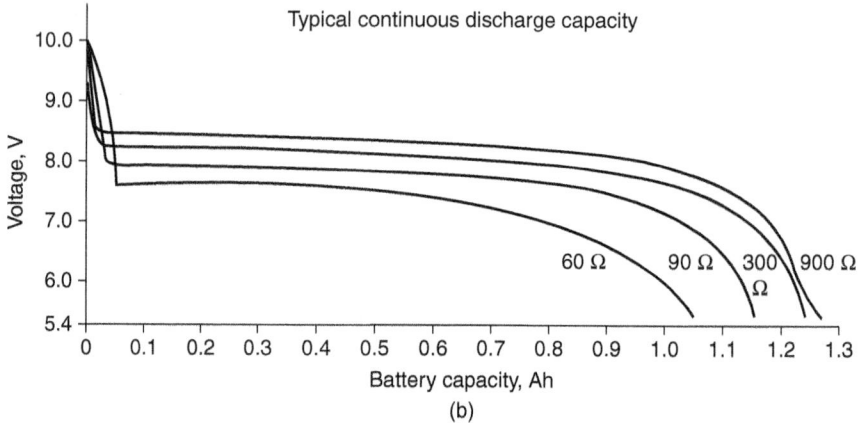

FIGURE 14.54 Discharge characteristics of 9 V Li/MnO$_2$ battery. (*a*) Discharge vs. temperature; 900 ohms continuous discharge. (*b*) Continuous discharge at room temperature on loads from 60 to 900 ohms. (*Courtesy of Ultralife Batteries, Inc.*)

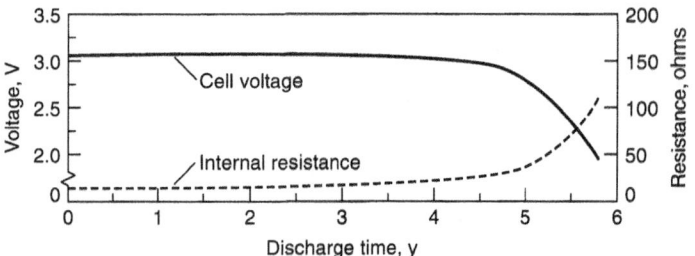

FIGURE 14.55 Internal resistance of Li/MnO$_2$ coin-type battery (280 mAh size), 5-µA drain, at 20°C.

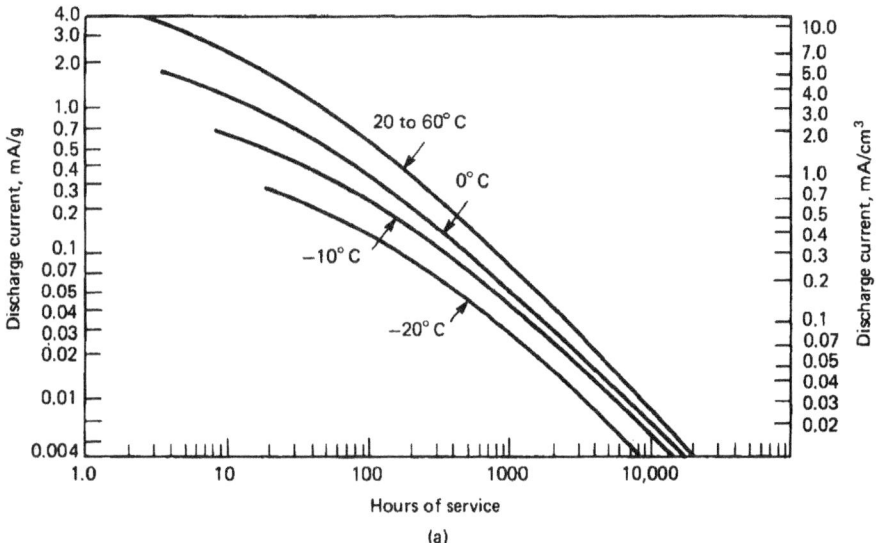

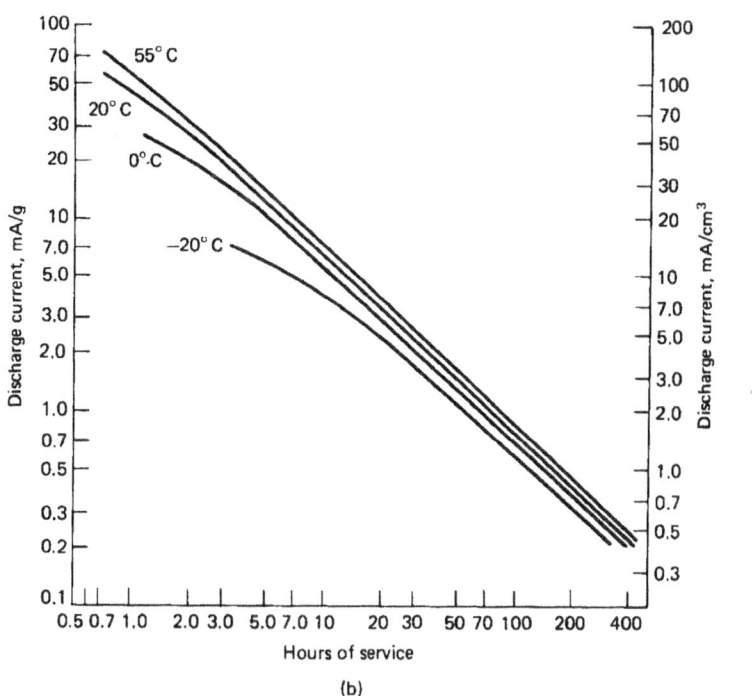

FIGURE 14.56 Service life of Li/MnO$_2$ cells to 2 V end voltage. (*a*) Low-rate coin-type batteries. (*b*) Small cylindrical batteries.

14.62 PRIMARY BATTERIES

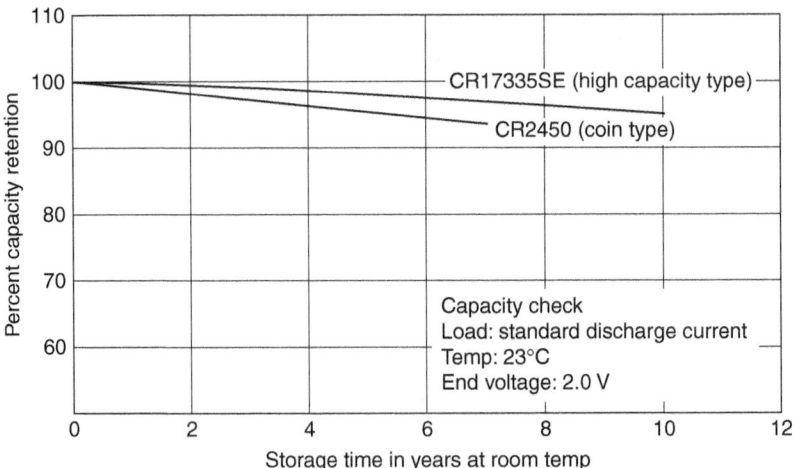

FIGURE 14.57 Storage characteristics of a typical spiral-wound, laser-sealed cylindrical cell and a typical coin cell.

14.8.4 Cell and Battery Sizes

The Li/MnO$_2$ cells are manufactured and commercially available in a number of flat and cylindrical batteries ranging in capacity from about 30 mAh to 11.1 Ah. The physical and electrical characteristics of some of these are summarized in Table 14.18. In some instances interchangeability with other battery systems is provided by doubling the size of the battery to accommodate the 3 V output of the Li/MnO$_2$ cell compared with 1.5 V of the conventional primary batteries, for example, battery type CR-V3. Table 14.19 shows the characteristics of two commercially available thin cells in foil-laminate packaging.

TABLE 14.19 Characteristics of Commercially Available Li/MnO$_2$ Cells in Thin, Foil-Laminate Packaging

Dimensions (mm) Thickness × width × length	Average voltage (V)	Nominal capacity to 1.5 V	Maximum discharge (mA-continuous)	Weight (g)	Pulse capacity (mA)
5.00 × 44.45 × 54.61	3.0	1.5 Ah	250	15.0	Up to 500
2.16 × 32.16 × 40.36	3.0	400 mAh	25	3.5	Up to 130

Operating/temperature: 0 to 71°C.
Capacity of 1.5 Ah cell determined at 10 mA and of the 400 mAh cell at 6 mA.
(Courtesy of Ultralife Batteries, Inc.)

14.8.5 Applications and Handling

The main applications of the Li/MnO$_2$ system currently range up to several ampere-hours in capacity, taking advantage of its higher energy density, better high-rate capability, and longer shelf life compared with the conventional primary batteries. The Li/MnO$_2$ batteries are used in memory applications, watches, calculators, cameras, and radio frequency identification (RFID) tags. At the higher drain rates, motor drives, automatic cameras, toys, personal digital assistants (PDAs), digital cameras, and utility meters are excellent applications.

The low-capacity Li/MnO$_2$ batteries can generally be handled without hazard, but, as with the conventional primary battery systems, charging and incineration should be avoided as these conditions could cause a cell to explode.

The higher-capacity cylindrical batteries are generally equipped with a venting mechanism to prevent explosion, but the batteries, nevertheless, should be protected to avoid short circuits and cell reversal, as well as charging and incineration. Most of the high-rate batteries are also equipped with an internal resettable current and thermal protective system called a positive temperature coefficient (PTC) device. When a cell is short-circuited or discharged above design limits and the cell temperature increases, the resistance of the PTC device quickly increases significantly. This limits the amount of current that can be drawn from the cell and keeps the internal temperature of the cell within safe limits. Figure 14.58 shows the operation of the PTC device when a cell is short-circuited. After a short-circuit peak of about 10 A the current is abruptly limited and maintained at the depressed level. When the short circuit is removed, the cell reverts to its normal operating condition. The short delay of several minutes before the PTC operates permits the cell to deliver pulse currents at higher values than the maximum permitted under continuous drain.

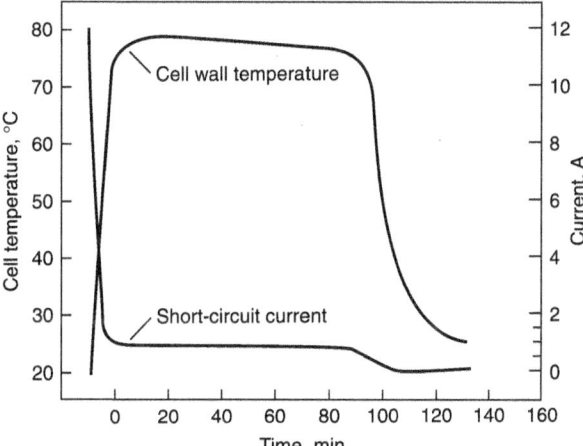

FIGURE 14.58 Short circuit of Duracell XL™ CR123A battery.

Military applications of lithium/manganese dioxide batteries are increasing.[41] At room temperature, they provide higher energy density and slightly higher specific energy than the Li/SO$_2$ batteries commonly employed by the U.S. military. A recently developed Li/MnO$_2$ D-cell provides 14.0 Ah at the 250 mA rate and 13.0 Ah at the 2.0 A rate at room temperature. These cells were produced using a specially heat-treated manganese oxide which is more highly active. When employed in cathodes in a standard cell design with a LiClO$_4$-DME-PC electrolyte, this material provides a specific energy of 339 Wh/kg and an energy density of 742 Wh/L on 250 mA discharge. At –40°C, these cells also provide 3.39 Ah at 250 mA and 0.46 Ah at 2.0 A discharge rates, more than twice the capacity of standard Li/MnO$_2$ cells under these conditions. The higher capacity cell is being utilized in military batteries.

A lithium/manganese dioxide pouch cell in a foil-laminate package[42] has also been developed for use in the BA-7847 battery, which powers the Thermal Weapons Sight and other U.S. military electronics. These cells have dimensions of 8.25 mm × 61 mm × 72 mm and are employed in a 2p2s configuration within the prismatic BA-7847 case.

TABLE 14.20 Performance Data at Room Temperature for Li/MnO$_2$ Pouch Cells Designed for BA-7847 Batteries

Discharge rate (mA)	Capacity (Ah)	Energy (Wh)	Specific energy (Wh/kg)	Energy density (Wh/L)
250	9.94	26.68	402	737
500	9.80	25.77	384	712
1000	9.27	23.91	356	661
2000	9.00	22.68	339	627

(Courtesy of Ultralife Batteries, Inc.)

Performance characteristics of these cells are summarized in Table 14.20. When used in the BA-7847 battery, they provide capacities greater than 19.5 Ah on 250 mA discharge at room temperature. This corresponds to a specific energy of about 300 Wh/kg. These batteries also passed the applicable UN/IATA shipping tests. When tested on the military L-test (8 W for 2 min followed by 5 W to a 4.0 V cutoff), these batteries ran for 9.5 hrs at −10°C but only 0.5 hrs at −20°C. Further improvement in low-temperature performance is being sought. These batteries must comply with the requirements of MIL-PRF-49471 for the particular battery type. A list of BA-type lithium/manganese dioxide batteries currently qualified by the U.S. Army is given in Table 14.10.

Battery packs are also being employed for Emergency Positioning Indicating Radio Beacons (EPIRBs) and pipeline test vehicles. Smaller batteries are also available commercially in foil-laminate packages for use in specialized applications such as toll collection transponders, RFID tags for shipping and inventory control, and smart security tags.

The specific conditions for the use and handling of Li/MnO$_2$ batteries are dependent on the size as well as the specific design features. Manufacturers' recommendations should be consulted.

14.9 LITHIUM/CARBON MONOFLUORIDE (Li/CFx) BATTERIES

The lithium/carbon monofluoride Li/(CFx) battery was one of the first lithium/solid-cathode systems to be used commercially. It is attractive as its theoretical specific energy (about 2190 Wh/kg) is among the highest of the solid-cathode systems. Its open-circuit voltage is 3.2 V, with an operating voltage of about 2.5 to 2.7 V. Its practical specific energy and energy density ranges up to 250 Wh/kg and 635 Wh/L in smaller sizes and 820 Wh/kg and 1180 Wh/L in larger sizes. The system is used primarily at low to medium discharge rates.

14.9.1 Chemistry

The active components of the cell are lithium for the anode and polycarbon monofluoride (CFx) for the cathode. The value of x is typically 0.9 to 1.2. Carbon monofluoride is an interstitial compound, formed by the reaction between carbon powder and fluorine gas. While electrochemically active, the material is chemically stable in the organic electrolyte and does not thermally decompose up to 400°C, resulting in a long storage life. Different electrolytes have been used; 1 Molar lithium tetrafluoro borate (LiBF$_4$) in δ-butyrolactone (GBL) for cylindrical cells and LiBF$_4$ in a mixture of GBL and dimethoxy ethane (DME) or a mixture of propylene carbonate (PC) and DME for coin cells.

The simplified discharge reactions of the cell are

Anode: $xLi \rightarrow xLi^+ + xe$

Cathode: $CFx + xe \rightarrow xC + xF^-$

Overall: $xLi + CFx \rightarrow xLiF + xC$

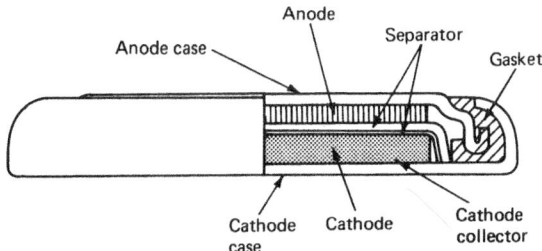

FIGURE 14.59 Cross-sectional view of Li/CFx coin-type battery. (*Courtesy of Panasonic Corp. of North America.*)

The polycarbon monofluoride changes into carbon, which is more conductive as the discharge progresses, thereby increasing the cell's conductivity, improving the regulation of the discharge voltage and increasing the discharge efficiency. The crystalline LiF precipitates in the cathode structure.[1,43,44]

14.9.2 Construction

The Li/CFx system is adaptable to a variety of sizes and configurations. Batteries are available in flat coin or button, cylindrical, and rectangular shapes, ranging in capacity from 0.020 to 25 Ah; larger-sized batteries have been developed for specialized applications.

Figure 14.59 shows the construction of a coin-type battery. The Li/CFx cells are typically constructed with an anode of lithium foil rolled onto a collector and a cathode of Teflon-bonded polycarbon monofluoride and acetylene black on a nickel collector. Nickel-plated steel or stainless steel is used for the case material. The coin cells are crimped-sealed using a polypropylene gasket.

The pin-type batteries use an inside-out design with a cylindrical cathode and a central anode in an aluminum case, as shown in Fig. 14.60.

The cylindrical batteries use a spirally wound (jelly-roll) electrode construction, and the batteries are either crimped or hermetically sealed. Their construction is similar to the cylindrical spiral-wound electrode design of the Li/MnO$_2$ battery shown in Fig. 14.39. The larger cells are provided with low-pressure safety vents.

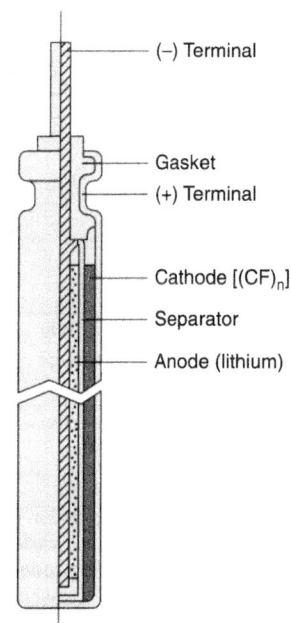

FIGURE 14.60 Cross-sectional view of Li/CFx pin-type battery. (*Courtesy of Panasonic Corp. of North America.*)

14.9.3 Performance

Coin-Type Batteries. Figure 14.61 presents the discharge curves at 20°C for a typical Li/CFx coin-type battery rated at 165 mAh. The voltage is constant throughout most of the discharge, and the coulombic utilization is close to 100% under low-rate discharge. Figure 14.62 presents the discharge curves for the same battery at various discharge temperatures. The behavior of the battery on a pulse discharge at 20°C is shown in Fig. 14.63.

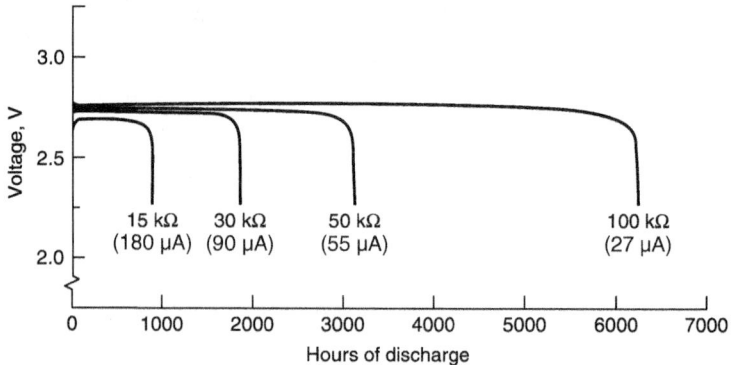

FIGURE 14.61 Typical discharge curves of Li/CFx coin-type battery at 20°C; rated capacity 165 mAh. (*Courtesy of Panasonic Corp. of North America.*)

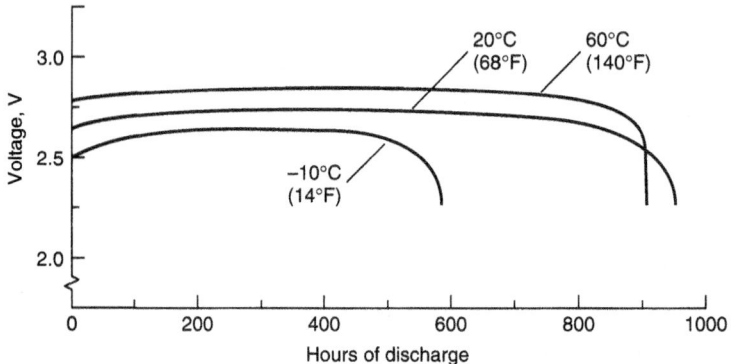

FIGURE 14.62 Typical discharge curves of Li/CFx 165 mAh coin-type battery at various temperatures; 15-kΩ discharge load; 180 µA. (*Courtesy of Panasonic Corp. of North America.*)

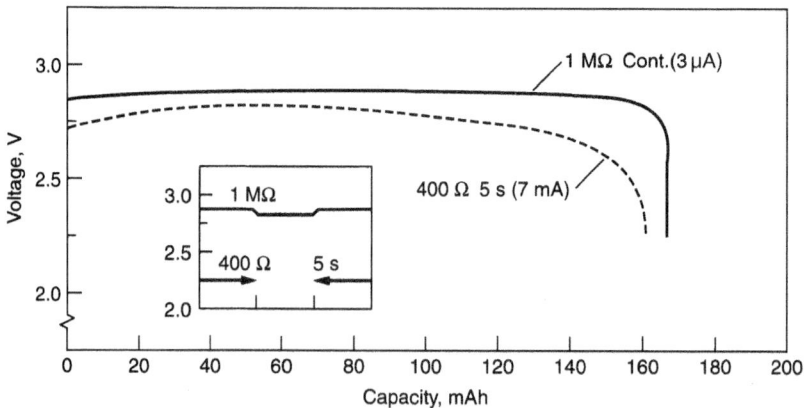

FIGURE 14.63 Pulse discharge characteristics of Li/CFx coin-type (165-mAh size) at 20°C. (*Courtesy of Panasonic Corp. of North America.*)

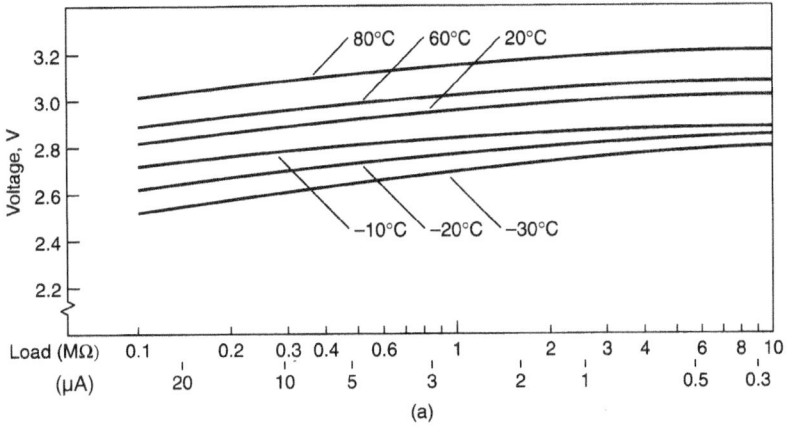

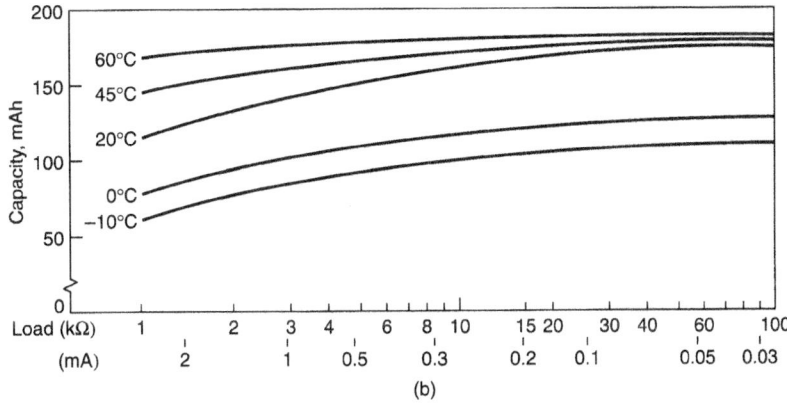

FIGURE 14.64 Discharge characteristics of Li/CFx coin-type battery (165 mAh size). (*a*) Operating voltage vs. discharge load; voltage at 50% discharge. (*b*) Capacity vs. discharge load; cutoff at 2.0 V. (*Courtesy of Panasonic Corp. of North America.*)

The performance of the coin battery (165 mAh capacity) is summarized in Fig. 14.64. Figure 14.64(a) shows the average load voltage (plateau voltage during discharge) and Fig. 14.64(b) shows the capacity for discharges at various loads and temperatures.

Figure 14.65 summarizes the discharge performance data for Li/CFx coin-type batteries normalized for a 1 g and 1 mL battery. These data can be used to approximate the size or performance of a battery for a particular application.

Cylindrical Cells. The cylindrical batteries are designed to operate at higher discharge rates than the coin batteries. Figure 14.66 presents the discharge curves on a 1 kohm load at several temperatures. In some cases, an initial low voltage is observed with the Li/CFx battery; that is, the voltage drops initially below the operating level on load and recovers gradually as the discharge progresses. This is attributed to the fact that CFx is an insulator, but the resistance of the cathode decreases during the discharge as conductive carbon is produced.

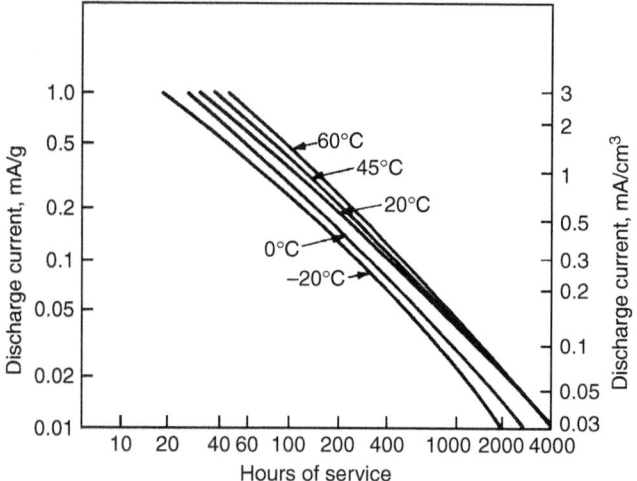

FIGURE 14.65 Service life of Li/CFx coin-type batteries at various discharge rates and temperatures; 2.0 V end voltage.

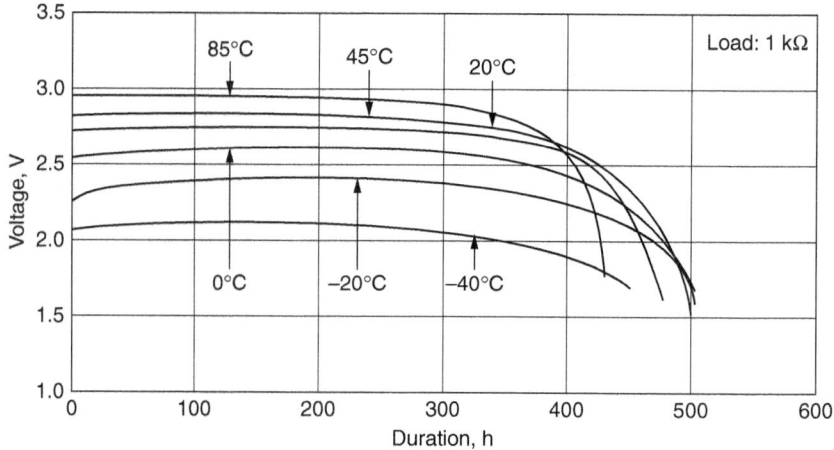

FIGURE 14.66 Discharge curves for Li/CFx 2/3 A-size cylindrical battery on a 1 kiloohm load at temperatures from −40 to +85°C.

The average load voltages for 2/3 A-size cylindrical batteries at various temperatures and rates are given in Fig. 14.67. The performance data are then summarized in Fig. 14.68, which shows the effect of temperature and load on the service life of a battery normalized to unit weight (grams) and volume (cubic centimeters).

Shelf Life. The Li/CFx cells have extremely good storage characteristics. Tests over more than 10 years of storage show a self-discharge rate of about 0.5%/yr at 20°C for coin cells and 1.0%/yr for cylindrical batteries. These rates decrease on longer-term storage.[45] This cell is also well-suited for applications requiring low current drain over an extended period of time. This is illustrated in

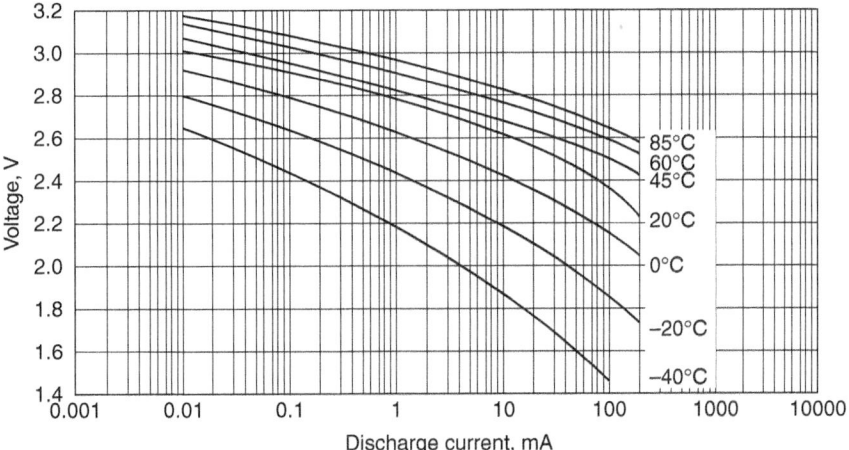

FIGURE 14.67 Mid-point voltage for Li/CFx 2/3 A-size cylindrical battery as a function of temperature and discharge current.

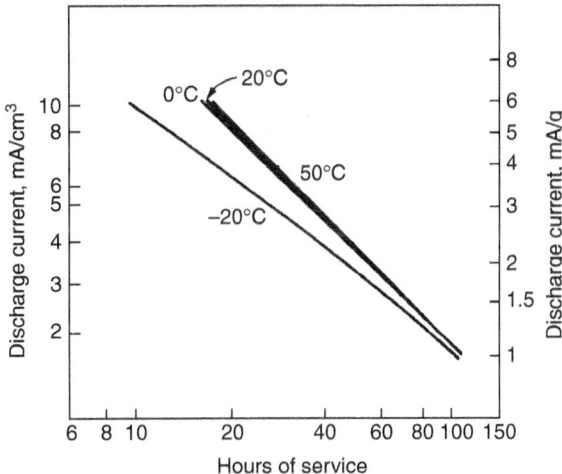

FIGURE 14.68 Service life of Li/(CF)$_n$ cylindrical battery at various discharge rates and temperatures; 1.8 V end voltage. (*Courtesy of Panasonic Corp. of North America.*)

Fig. 14.69, which shows the discharge characteristics of the $\frac{2}{3}$ A-size cell at a 20-μA discharge rate over a period of 7 years. Voltage delay after storage is not apparent with these cells, except under severe discharge conditions.

14.9.4 Cell and Battery Types

The Li/(CF)$_n$ batteries are manufactured in a number of coin, cylindrical, and pin configurations. The major electrical and physical characteristics of some of these batteries are listed in Table 14.21a and b.

14.70 PRIMARY BATTERIES

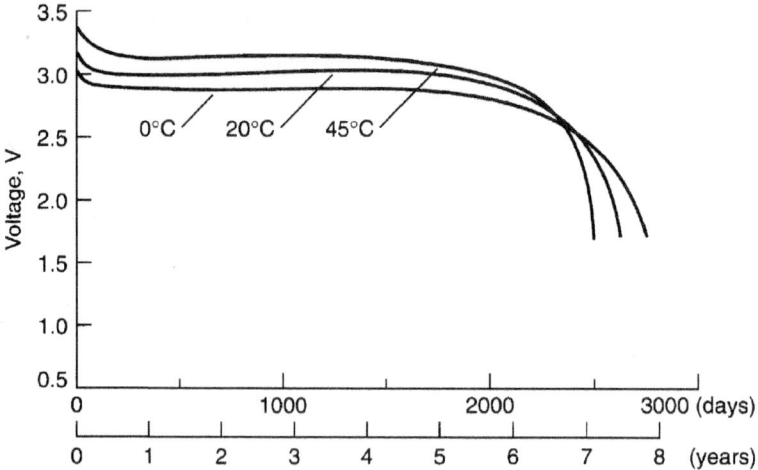

FIGURE 14.69 Long-term discharge of Li/CFx cylindrical battery. BR 2/3 A-size; 150-kΩ discharge. (*Courtesy of Panasonic Corp. of North America.*)

TABLE 14.21(a) Characteristics of Lithium/Carbon Monofluoride (Li/CFx) Batteries

		Coin batteries, 3 V			
	*Nominal capacity	Standard load	Dimensions and weight		
Model No.	(mAh)	Continuous drain (mA)	Diameter (mm)	Height (mm)	Weight (g)
BR1220	35	0.03	12.5	2.00	0.7
BR1225	48	0.03	12.5	2.50	0.8
BR1632	120	0.03	16.0	3.20	1.5
BR2032	190	0.03	20.0	3.20	2.5
BR2325	165	0.03	23.0	2.50	3.2
BR2330	255	0.03	23.0	3.00	3.2
BR3032	500	0.03	30.0	3.20	5.5

*Nominal capacity shown is based on standard drain and cutoff voltage down to 2.0 V at 20°C.

		Pin type, 3 V			
	Nominal capacity	Dimensions (mm)		Basic battery	Continuous drain
Model no.	(mAh)	External diameter	Height	weight (g)	(mA)
BR425	25	4.2	25.9	0.55	0.5
BR435	50	4.2	35.9	0.85	1.0

		Cylindrical type, 3 V				
	*Nominal capacity	Dimensions (mm)		Basic battery	Continuous	Operating temp
Model no.	(mAh)	External diameter	Height	weight (g)	drain (mA)	(0°C)
BR-C	5,000	26.0	50.5	42.0	5.0	−40 ~ +85
BR-A	1,800	17.0	45.5	18.0	2.5	−40 ~ +85
BR-1/2AA	1,000	14.5	25.5	8.0	2.5	−40 ~ +100
BR-2/3A	1,200	17.0	33.5	13.5	2.5	−40 ~ +85
BR-AG	2,200	17.0	45.5	18.0	2.5	−40 ~ +85
BR-2/3AG	1,450	17.0	33.5	13.5	2.5	−40 ~ +85

*Nominal capacity is based on standard drain rate and cutoff voltage of 2.0 V at 20°C.

TABLE 14.21(b) Characteristics of Large Lithium/Carbon Monofluoride (Li/CFx) Batteries

		Single-cell batteries		
		Dimensions (cm)		
Part number	Capacity (Ah)	Diameter	Height	Weight (grams)
LCF-111	240	6.62	16.51	880
LCF-112	35	3.02	13.84	170
LCF-117	1200	11.43	26.67	3950
LCF-119	400	11.43	9.53	1575
LCF-122	18	3.37	6.06	—
LCF-123	35	3.37	11.72	—
LCF-313	40	6.45(L) × 3.43(W) × 7.09(H)		230

			Multicell batteries			
	Capacity	Nominal	Dimensions (cm)			Weight
Part number	(Ah)	voltage	H	L	W	(grams)	Comments
MAP-9036	23.5	39	17.1	20.3	14.0	4586	Former shuttle range safety system
MAP-9046	3.74 (×2)	30 (×2)	15.9	17.3	7.6	3405	2 independent voltage sections
MAP-9225	240	15	24.9	30.7	6.5	6000	
MAP-9257	80	18	12.4	18.5	14.8	—	
MAP-9319	240	21	42.9	29.7	9.7	—	
MAP-9325	120/7.2	12/15	17.1	18.6	9.2	—	Optional casing
MAP-9334	80	6	16.8	7.6	4.8	—	Minuteman III GRP batteries
MAP-9381	70	39	31.3	20.0	9.7	—	Integrated capacitor bank
MAP-9382	80/70	33/12	20.1	17.6	14.1	—	2 independent voltage sections
MAP-9389	280	15	23.6	33.8	11		
MAP-9392	40	39	17.1	20.3	14.0		X-33 Range safety system

Source: Eagle-Picher Technologies.

Manufacturers' specifications should be consulted for the most recent listings of commercially available cells.

Larger sizes of cells and batteries[46] have also been developed for military, governmental and space applications, as given in Table 14.21b. Spiral-wound and prismatic cells are used to build the multicell batteries given in this table. The smaller cylindrical cells employ a 0.030 cm thick steel case, but larger units, such as the 1200 Ah cell, are reinforced with an epoxy-fiberglass cylinder to provide additional strength, with an increase in weight about half that of increasing the steel wall thickness. All these cells employ a Zeigler-type compression seal, a unique cutter vent mechanism and two layers of separator. The first is a microporous layer to prevent particulate migration, and the second is a nonwoven polyphenylene sulfide material to provide high-strength, high-temperature stability and good electrolyte wicking action. These low-rate designs provide a specific energy of 600 Wh/kg and an energy density of 1000 Wh/L in the DD size and higher values for larger units. Capacity to a 2.0 V cutoff as a function of temperature at four rates from 0.04 to 1.00 A is shown in Fig. 14.70 for the DD design. The capacity of this battery is relatively independent of temperature

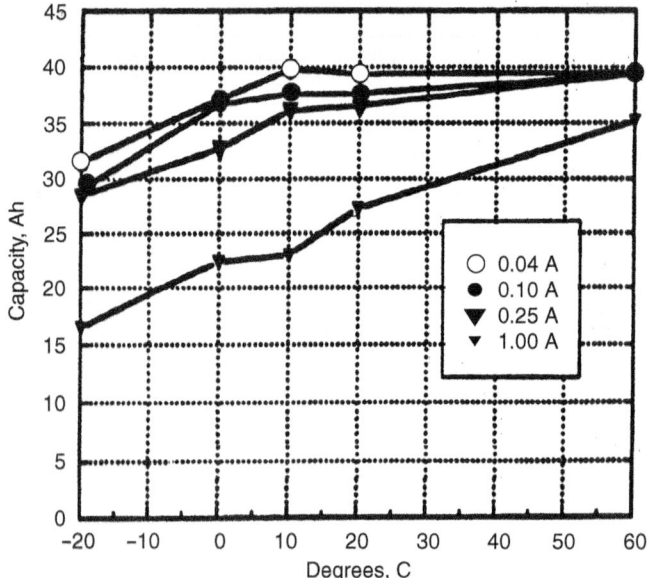

FIGURE 14.70 DD Li/CFx discharge performance as a function of rate and temperature.

at the three lower currents and above 10°C, but decreases at the higher rate and lower temperatures. Discharge curves for the 1200 Ah reinforced cylindrical battery at the 2000-hour (ca. 500 mA) and the 1000-hour rates (ca. 1.0 A) are shown in Fig. 14.71. The trailing knee in these discharge curves has been attributed to electrolyte starvation at the end of the discharge. These batteries effectively demonstrate the ability of the lithium/carbon monofluoride system to provide very high specific energies and energy densities, in these low rate designs.

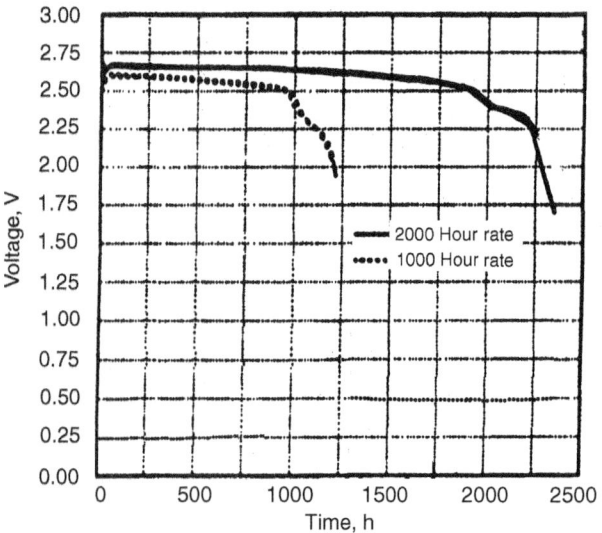

FIGURE 14.71 Typical discharge curves for 1200 Ah Li/CFx battery.

14.9.5 Applications and Handling

The applications of the Li/CFx battery are similar to those of the other lithium/solid-cathode batteries, again taking advantage of the high specific energy and energy density and long shelf life of these batteries. The Li/CFx coin batteries are used as a power source for watches, portable calculators, memory applications, and electronic translators. The low-capacity miniature pin-type batteries have been used as an energy source for LEDs and for fishing lights and microphones. The cylindrical batteries can also be used in memory applications, but their higher drain capability also covers use in cameras, electrical locks, emergency signal lights, and utility meters. The very large cells (Table 14.21b) are used for military and space applications.

Handling considerations for the Li/CFx systems, too, are similar to those for the other lithium/solid-cathode systems. The limited current capability of the coin and low-capacity batteries restricts temperature rise during short circuit and reversal. These batteries can generally withstand this abusive use even though they are not provided with a safety vent mechanism. The larger batteries are provided with a venting device, but short circuit, high discharge rates, and reversal should be avoided as these conditions could cause the cell to vent. Charging and incineration likewise should be avoided for all batteries. The manufacturer's recommendations should be obtained for handling specific battery types.

14.9.6 Recent Advances in Lithium Carbon Monofluoride Technology

Use of Mixtures of Carbon Monofluoride and Manganese Dioxide. The use of both CFx and MnO_2 in the cathode of a lithium primary battery was first described in a U.S. patent issued in 1982.[47] This patent claims the use of mixtures of CFx and MnO_2 but also claims a cell in which a layer of CFx is disposed on top of a manganese dioxide layer. Little data is presented in this patent to support the claims.

The high cost of carbon monofluoride relative to manganese dioxide has limited its use in many applications. A recent report[48] describes a study in which a mixture of CFx with $x = 1$ and heat-treated MnO_2 was used to construct a lithium primary D-cell. The proportions of the mixture were stated to be a 50/50 blend, and the discharge curve shows two plateaus of approximately equal duration. The cells were stated to have a balanced design. The electrolyte was only described as an inorganic lithium salt in an organic electrolyte mixture. A co-polymer film separator was also used. The D cells were discharged at 0.050, 0.250 and 2.0 A at 21°C and at 2.0 A at −30°C. Results are summarized Table 14.22. All three room-temperature discharge curves show two plateaus of approximately equal duration. The 2.0 A discharge shows running voltages of 2.64 and 2.41 V, which are ascribed to MnO_2 and CFx, respectively. On 0.250 A discharge at 21°C, these cells exhibit a specific energy of 380 Wh/kg and an energy density of 923 Wh/L. These parameters represent increases of 35 and 57% in specific energy and energy density compared to standard D cells using manganese dioxide only. On 2.0 A discharge at −30°C, the cells with hybrid cathodes provide a capacity of 12.0 Ah, which is 79% of that obtained at 21°C at the same rate. This corresponds to a specific energy of 227 Wh/kg

TABLE 14.22 Performance Data for High-Capacity Li/CFx-MnO_2 D-Cells at Different Rates and Temperatures to a 2.0 V Cutoff

Temperature °C	Rate (A)	Capacity (Ah)	Specific energy (Wh/kg)	Energy density (Wh/L)
+21	0.050	16.6	407	990
+21	0.250	16.2	380	923
+21	2.0	15.2	338	823
−30	2.0	12.0	227	552

and an energy density of 552 Wh/L, both figures being 67% of the values obtained at 21°C. These results indicate that the hybrid cathode cells provide superior performance to cells using manganese dioxide alone, particularly at low temperature.

Mixtures of carbon monofluoride and silver vanadium oxide have also been employed for biomedical applications. See Sec. 31.5.6.

Subfluorinated and Semi-Ionic Carbon Fluoride Materials. Recent studies have shown that subfluorinated carbon fluorides (SFCFs) provide enhanced performance at low temperatures to −40°C. An initial study,[49] compared partially fluorinated natural graphites with x values of 0.33. 0.48. 0.52, 0.63 to a commercial CFx with x = 1.08. Structural studies[50] showed that the SFCF material consisted of domains of fluorinated carbon intimately mixed with the graphite precursor particles. The partially fluorinated CFx materials exhibit higher power capability at room temperature and superior low-temperature performance compared to $CF_{1.08}$. Figure 14.72 shows discharge curves at room temperature for a Li/CFx coin cell with x = 0.52 at different rates to 2.5C. The cathode mix contained 80% CFx, 10% acetylene black, and 10% binder and was used with a 1.2 M $LiBF_4$ in PC/DME (7/3) electrolyte. A Ragone plot[49] shows better high-rate power capability above 6.4 kW/kg for the SFCFs compared to the standard commercial cells. Figure 14.73 shows discharge curves Li/CFx coin cells with x = 0.65 at −40°C with and without a 3% predischarge at a C/40 rate. These cells used an electrolyte of 1 M $LiBF_4$ in PC/DME (20/80). Predischarged cells with $CF_{0.65}$ provided a specific capacity of 610 mAh/g to a 1.5 V cutoff compared to 200 mAh/g for control cells with $CF_{1.08}$ tested under the same conditions.

Another study[51] to optimize the electrolyte composition for SFCFs found that 0.5 M $LiBF_4$ in PC/DME (20/80) provided improved low-temperature performance compared to higher salt concentrations and eliminated the need for the predischarge step prior to low-temperature discharge.

The use of CFx materials with semi-ionic character[52] prepared in a two-step fluorination process and SFCF materials prepared from multiwalled carbon nanotubes (MWCNs) were also found to have superior properties in terms of rate capability and low-temperature performance compared to conventional CFx materials.

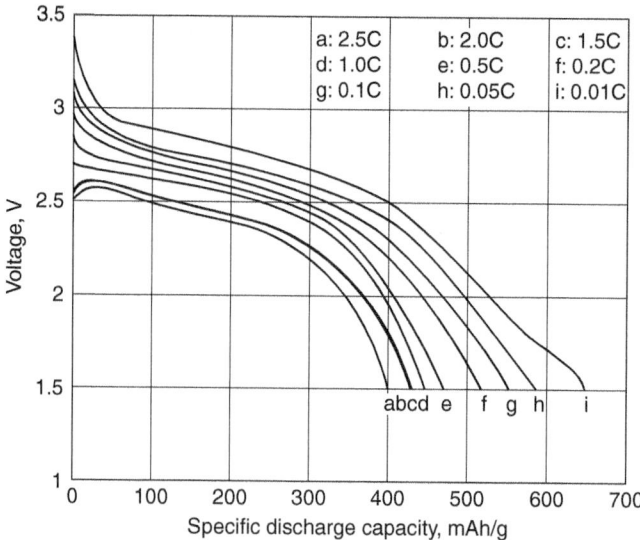

FIGURE 14.72 Discharge profiles for Li/$CF_{0.52}$ coin cells discharged at different rates.

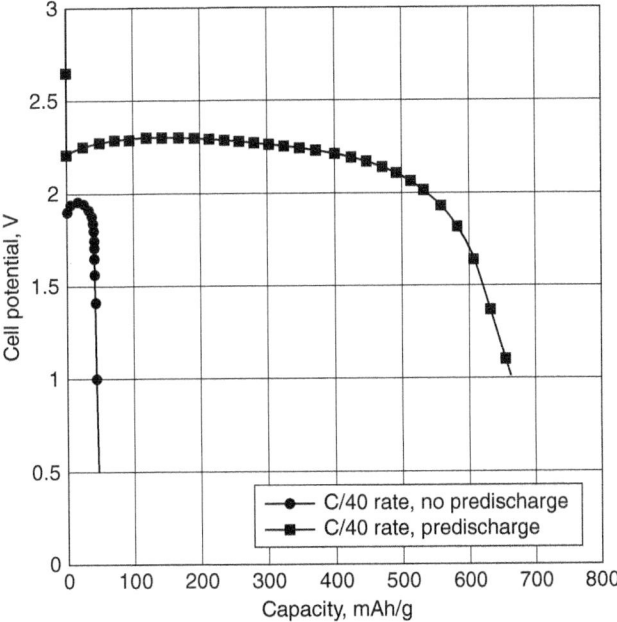

FIGURE 14.73 Discharge data from Li/CF$_{0.65}$ test cells at −40°C with and without a room-temperature predischarge of 3% of the total cell capacity.

14.10 LITHIUM/IRON DISULFIDE (Li/FeS$_2$) BATTERIES

Iron sulfide, in both the monosulfide (FeS) and the disulfide (FeS$_2$) forms, has been considered for use in solid-cathode lithium batteries. Only the disulfide battery has been commercialized because of its performance advantage due to its higher sulfur content and higher voltage. The monosulfide electrode has the advantage of reduced corrosion, longer life, and a single voltage plateau compared to the disulfide electrode, which discharges in two steps.

These batteries have a nominal voltage of 1.5 V* and can therefore be used as replacements for aqueous batteries having a similar voltage. Button-type Li/FeS$_2$ batteries were manufactured as a replacement for zinc/silver oxide batteries but are no longer marketed. They had a higher impedance and a slightly lower power capability but were lower in cost and had better low-temperature performance and storability.

Li/FeS$_2$ batteries are now manufactured in a cylindrical configuration. These batteries have better high-drain low-temperature performance than the zinc/alkaline-manganese dioxide batteries. The capacity of these two systems on constant-current discharge at four rates is compared in Fig. 14.74 for the AA-size cells.

14.10.1 Chemistry

These cells[53] employ a cathode of FeS$_2$ mixed with carbon and a mixed Teflon® organic binder coated on an aluminum foil, an anode of lithium alloyed with 0.5% aluminum, and a 20 micron high-porosity polyethylene separator. The electrolyte is a 0.75 M solution of LiI in a 65:35 (V/V)

---

*ANSI Standard C18.3M, Part 1-2009.

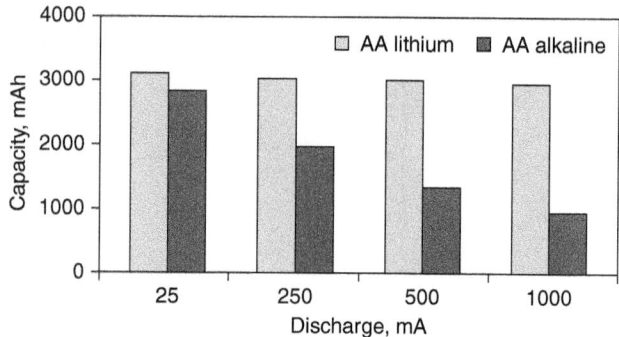

FIGURE 14.74 Comparison of capacity (mAh) of AA Li/FeS$_2$ and alkaline-manganese cells on constant-current discharge at four rates (25, 250, 500, and 1000 mA) at 21°C. (*Courtesy Energizer Battery Co.*)

mixture of 1,3 dioxolane and 1,2 dimethoxyethane, which is reported to increase in conductivity as the temperature decreases. The cell reactions at high rate and ambient temperature are

$$
\begin{array}{ll}
\text{Anode} & 4\text{Li} \rightarrow 4\text{LI}^+ + 4e \\
\text{Cathode} & \underline{\text{FeS}_2 + 4e \rightarrow \text{Fe} + 2\text{S}^{-2}} \\
\text{Overall} & 4\text{Li} + \text{FeS}_2 \rightarrow \text{Fe} + 2\text{Li}_2\text{S}
\end{array}
$$

At low rate and/or high temperature, a two-step discharge process occurs as seen in Fig. 14.75. The cell reactions are then given by

$$2\text{Li} + \text{FeS}_2 \rightarrow \text{Li}_2\text{FeS}_2$$
$$2\text{Li} + \text{Li}_2\text{FeS}_2 \rightarrow 2\text{Li}_2\text{S} + \text{Fe}$$

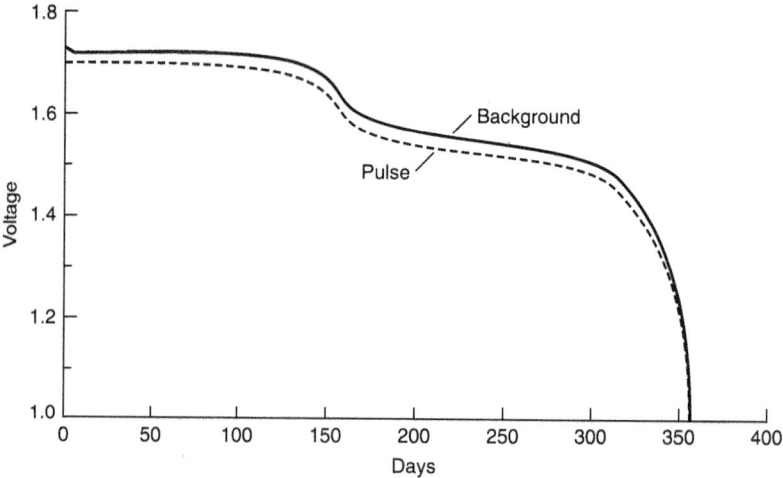

FIGURE 14.75 Stepped discharge curve of Li/FeS$_2$ AA-size batteries on light drain at 21°C. 5000 ohm background with 25 ohm, 1 s/week pulse. (*Courtesy of Energizer Battery Co., Inc.*)

14.10.2 Construction

Li/FeS$_2$ batteries may be manufactured in a variety of designs, including the button and both bobbin and spiral-wound-electrode cylindrical cells. A bobbin construction is most suitable for light-drain applications. The spiral-wound-electrode construction is needed for the heavier-drain applications, and it is this design that has been commercialized.

The construction of the spiral-wound cylindrical battery is shown in Fig. 14.76. These batteries typically have several safety devices incorporated in their design to provide protection against such abusive conditions such as short circuit, charging, forced discharge, and overheating. Two safety devices are shown in the figure—a pressure relief vent and a resettable thermal switch, called a positive thermal coefficient (PTC) device. The safety relief vent is designed to release excessive internal pressure to prevent violent rupture if the battery is heated or abused electrically.

The primary purpose of the PTC is to protect against external short circuits, though it also offers protection under certain other electrical abuse conditions. It does so by limiting the current flow when the cell temperature reaches the PTC's designed activation temperature. When the PTC

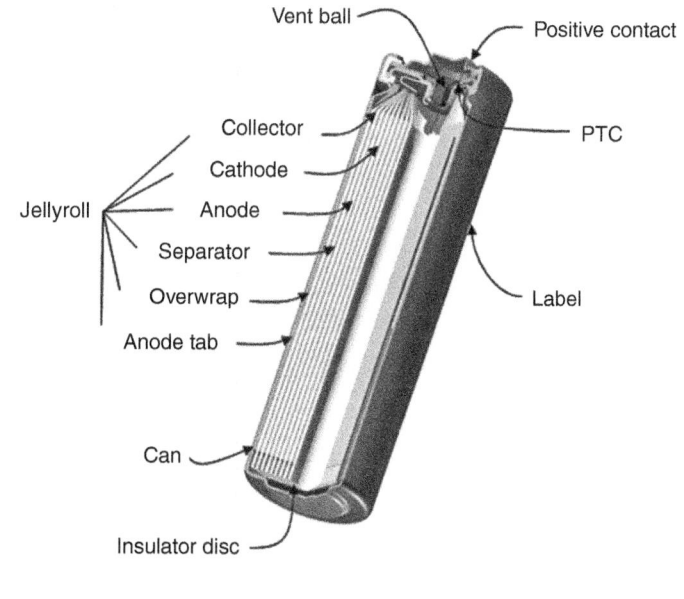

*Additional view (1)
positive end*

*Additional view (2)
negative end*

FIGURE 14.76 Partial cutaway view of spiral-wound Li/FeS$_2$ cell with additional details of positive (1) and negative (2) ends. (*Courtesy of Energizer Battery Co.*)

activates, its resistance increases sharply, with a corresponding reduction in the flow of current and, consequently, internal heat generation. When the battery (and the PTC) cools, the PTC resistance drops, allowing the battery to discharge again. The PTC will continue to operate in this manner for many cycles if an abusive condition continues or recurs. The PTC will not "reset" indefinitely, but when it ceases to do so, it will be in the high-resistance condition. The characteristics of PTCs (or any other current-limiting devices in the battery) may place some limitations on performance. These are discussed in more detail in Sec. 14.10.3.

14.10.3 Performance

Voltage. The nominal voltage of the Li/FeS$_2$ system is given as 1.5 V and the open-circuit voltage of undischarged cells is approximately 1.78 V. The voltage on load drops within milliseconds, as shown in Fig. 14.77.

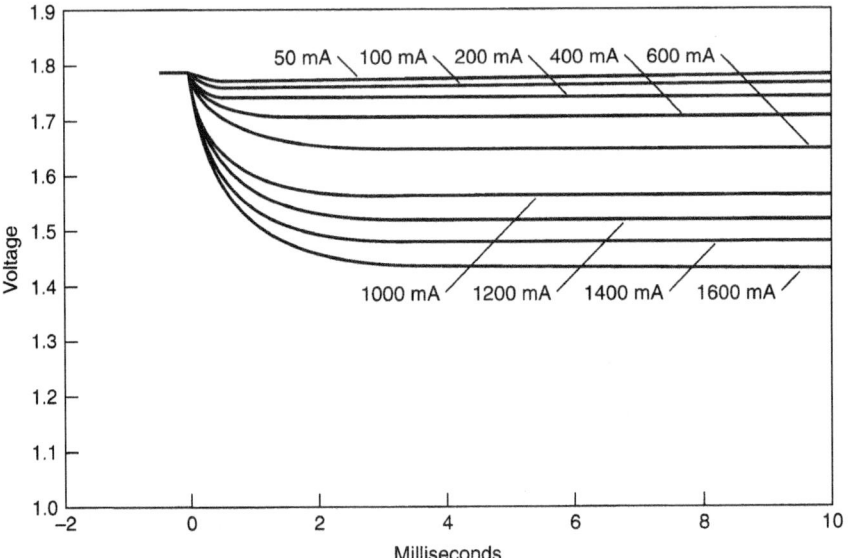

FIGURE 14.77 On-load voltage of Li/FeS$_2$ AA-size cells. (*Courtesy of Energizer Battery Co., Inc.*)

Discharge. Li/FeS$_2$ batteries typically have a higher operating voltage and a flatter discharge profile than aqueous zinc/alkaline manganese dioxide 1.5 V batteries. This is illustrated in Fig. 14.78 which compares the performance of these two battery systems at relatively light and heavy constant-current discharge rates. These characteristics of the Li/FeS$_2$ battery result in higher energy and power output, especially on heavier drains where the operating voltage differences are greatest.

Performance characteristics of AA Li/FeS$_2$ batteries under constant-current and constant-power discharge modes are Figs. 14.79 and 14.80. The improvement to performance with time on an ANSI digital still camera (DSC) test is shown in Fig. 14.81.

Operating Temperature. Li/FeS$_2$ batteries are also suitable for use over a broad temperature range, generally −40 to 60°C. Service life is improved at elevated temperatures. In some applications there may be further limits on the maximum discharge temperature due to current limiting, which are part

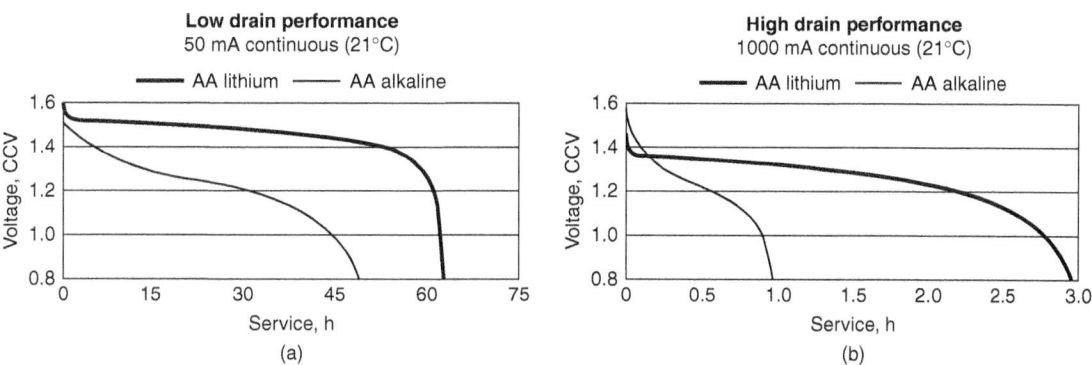

FIGURE 14.78 Comparison of AA Li/FeS$_2$ and alkaline-manganese batteries at two drain rates, (a) 50 and (b) 1000 mA, at 21°C. (*Courtesy Energizer Battery Co.*)

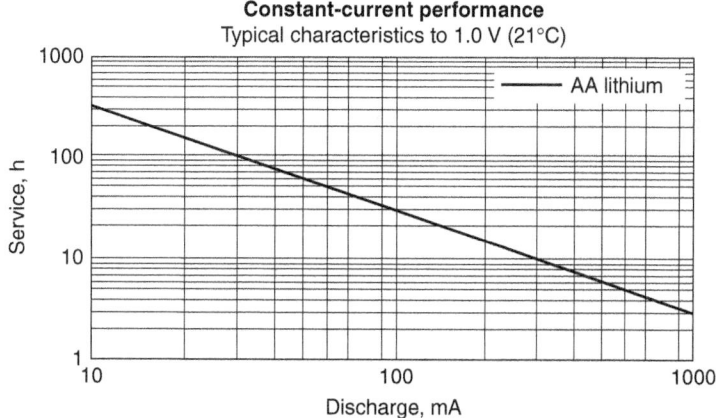

FIGURE 14.79 Constant current performance shows hours of service as a function of discharge rate (mA) for AA Li/FeS$_2$ batteries (*Courtesy Energizer Battery Co.*)

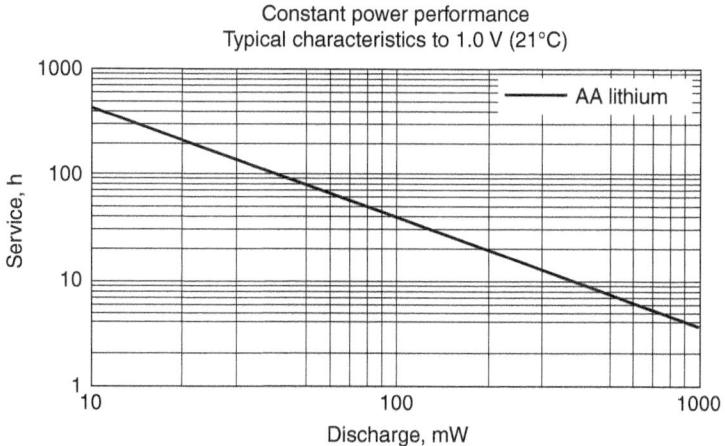

FIGURE 14.80 Constant power performance showing hours of service as a function of constant power discharge rate (mW) for AA Li/FeS$_2$ battery. (*Courtesy Energizer Battery Co.*)

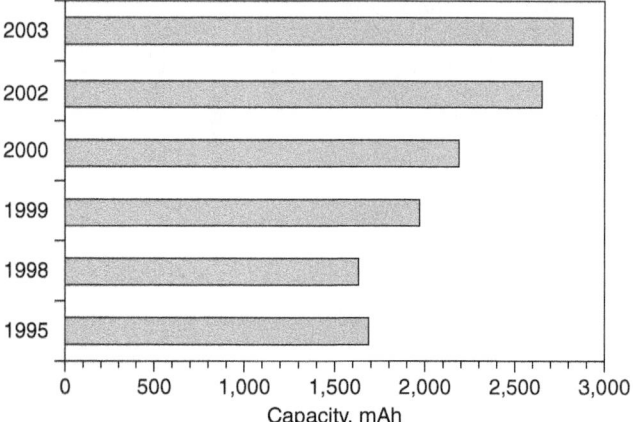

FIGURE 14.81 Improvement in AA high rate performance on ANSI DSC test: 1A continuous test to 1.0 V.

of the cell or battery device designs. Service life is reduced as the discharge temperature is lowered below room temperature, though the performance of the Li/FeS$_2$ battery is affected much less by low temperature than are aqueous systems.

Effects of Current-Limiting Devices. Some current-limiting devices, such as thermal fuses and PTCs, are designed to respond to high temperatures. Both the ambient temperature and internal cell heating can affect these devices, so any of the following factors may play a role:

Surrounding air temperature
Thermal insulating properties of battery container
Heat generated by equipment components during use
Cumulative heating effects of multicell batteries
Discharge rates and durations
Frequency and duration of rest periods

It may be necessary to consult the manufacturer or conduct testing to determine limitations in specific applications.

Impedance. AC impedance is an electrical characteristic that is frequently used as an indicator of performance for aqueous batteries. The correlation is only poor at best with Li/FeS$_2$ batteries. There is a protective film that forms on the surface of the lithium anode. This film is an important factor in the excellent shelf life of the Li/FeS$_2$ cell. As the cell ages, this protective film increases with age. As the film increases, the impedance does as well. However, this film is easily disrupted when the battery is put on load, making impedance inappropriate as an indicator of expected Li/FeS$_2$ battery performance, especially after storage.

Storage Temperature. Storage at high temperature will reduce the service life of Li/FeS$_2$ batteries, as it will with other systems. However, because of the very low levels of impurities in the materials used and the high degree of seal effectiveness required in lithium batteries, service maintenance of Li/FeS$_2$ batteries after high-temperature storage is better than expected with aqueous systems. The typical storage temperature range of Li/FeS$_2$ batteries is −40 to 60°C. Accelerated storage tests at

TABLE 14.23 Characteristics of Commercial Li/FeS$_2$ Batteries

Model no.	Size	Max. diameter (mm)	Max. height (mm)	Typical weight (g)	Typical volume (cc)	Max. continuous current (A)	Max. pulsed current (2 s on/8 s off) (A)	Shelf life at 21°C (yrs)	Capacity at 21°C (mAh)
L92	AAA	10.5	44.5	7.6	3.8	1.5	2.0	15	1200
L91	AA	14.5	50.5	14.5	8.0	2.0	3.0	15	3000
EA92	AAA	10.5	44.5	7.6	3.8	1.0	1.5	10	1200
EA91	AA	14.5	50.5	14.5	8.0	1.5	2.0	10	3000

Storage temperature: −40 to 60°C; operating temperature: −40 to 60°C; typical IR drop: 90–150 milliohms; shelf-life rating to 90% of initial capacity for L92 and L91 models and to 80% of initial capacity for EA92 and EA91 models.
Source: Energizer Battery Co.

temperatures to 85°C have resulted in an estimated shelf life at room temperature between 26 and 40 years to 80% of initial capacity. They are rated for a 10-year shelf life for the EA models and a 15-year shelf life for the L models. See Table 14.23.

14.10.4 Cell Types and Applications

Table 14.23 lists the characteristics of the cylindrical Li/FeS$_2$ batteries that are currently available commercially. These batteries have better high-drain and low-temperature performance than the conventional zinc cells and are intended to be used in applications that have a high current drain requirement, such as cameras, digital audio devices, CD players, portable lighting, toys, and games. In one particular camera test, two Li/FeS$_2$ AA cells provided approximately 1,000 flashes compared to 400 for a high-rate AA alkaline battery.

Button-type Li/FeS$_2$ batteries are no longer manufactured. Multicell batteries using the Li/FeS system have not been manufactured for commercial use.

14.11 LITHIUM/COPPER OXIDE (Li/CuO) CELLS

The lithium/copper oxide (Li/CuO) system is characterized by a high specific energy and energy density (about 280 Wh/kg and 650 Wh/L) as copper oxide has one of the highest volumetric capacities of the practical cathode materials (4.16 Ah/cm^3). The battery has an open-circuit voltage of 2.25 V and an operating voltage of 1.2 to 1.5 V, which makes it interchangeable with some conventional batteries. The battery system also features a long shelf life with a low self-discharge rate and operation over a wide temperature range.

Li/CuO batteries have been designed in button and cylindrical configurations up to about 3.5 Ah in size, mainly for use in low- and medium-drain, long-term applications for electronic devices and memory backup. Higher-rate designs as well as hermetically sealed batteries with glass-to-metal seals have also been manufactured.

Figure 14.82 compares the performance of a Li/CuO AA-size cylindrical battery with the zinc/alkaline/MnO$_2$ battery. The Li/CuO cell has a significant capacity advantage at low discharge rates, but loses this advantage at higher discharge rates.

14.11.1 Chemistry

The discharge reaction of the Li/CuO cell is

$$2Li + CuO \rightarrow Li_2O + Cu$$

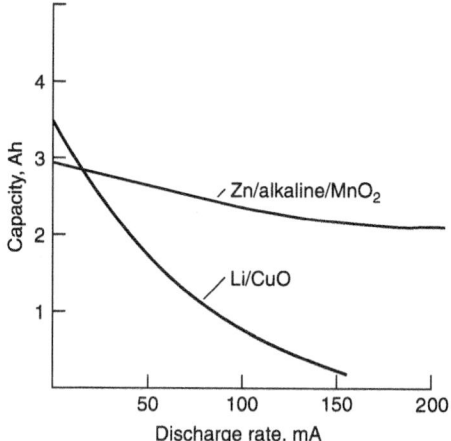

FIGURE 14.82 Comparison of Li/CuO and Zn/alkaline/MnO$_2$ AA-size batteries at 20°C.

The discharge proceeds stepwise, CuO → Cu$_2$O → Cu, but the detailed mechanism has not been clarified.[1,54] A double-plateau discharge has been observed at high-temperature (70°C) discharges at low rates, which blends into a single plateau under more normal discharge conditions.[55]

14.11.2 Construction

The construction of the Li/CuO button-type battery shown in Fig. 14.83a similar to other conventional and lithium/solid-cathode cells. Copper oxide forms the positive electrode and lithium the negative. The electrolyte consists of lithium perchlorate in an organic solvent (dioxolane).

The cylindrical batteries (Fig. 14.83b) use an inside-out bobbin construction. A cylinder of pure porous nonwoven glass is used as the separator, nickel-plated steel for the case, and a polypropylene gasket for the cell seal. The can is connected to the cylindrical copper oxide cathode and the top to the lithium anode.

14.11.3 Performance

Button Battery. The performance of the 60 mAh Li/CuO button cell under various discharge conditions and temperatures is shown in Fig. 14.84.

Cylindrical Bobbin Li/CuO Battery. Typical discharge curves for this system are shown in Fig. 14.85. After a high initial load voltage, the discharge profile is flat at the relatively light loads. The bobbin construction does not lend itself to high-rate discharges, and the battery capacity is significantly lowered with increasing discharge rates. The Li/CuO cylindrical battery operates over a wide temperature range, typically from −20 to 70°C, although the battery can operate outside these limits but with changes in the discharge profile or load capability.

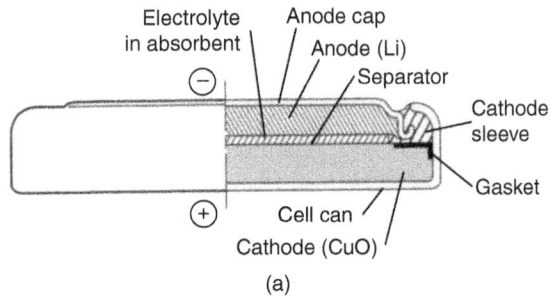

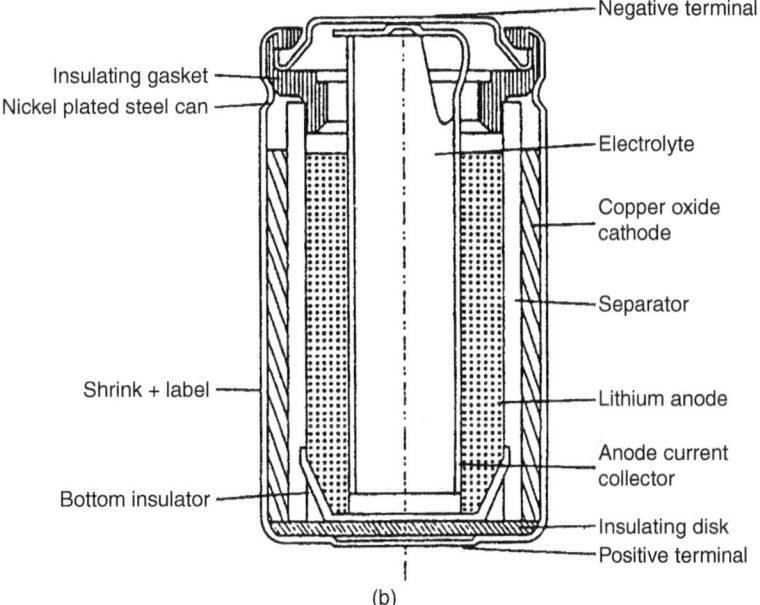

FIGURE 14.83 Lithium/copper oxide batteries (*a*) Button configuration. (*Courtesy of Panasonic Corp. of North America.*) (*b*) Cylindrical battery, bobbin construction. (*Courtesy of SAFT America, Inc.*)

Discharge curves at several different temperatures are shown in Fig. 14.86. The performance of the battery at temperatures from −40 to 70°C and at various loads is summarized in Fig. 14.87. The high capacity of the battery at the lighter loads falls off sharply with increasing load and decreasing temperatures.

The long-term storage capability of these Li/CuO cells is illustrated in Fig. 14.88. Figure 14.88*a* shows that there is only a minimum loss of capacity after 10 years of storage at room temperature, less than 0.5% per year. Performance after storage at high temperatures is plotted in Fig. 14.88*b*. The retention of residual capacity in partially discharged cells is said to be equivalent to that of fully charged cells.

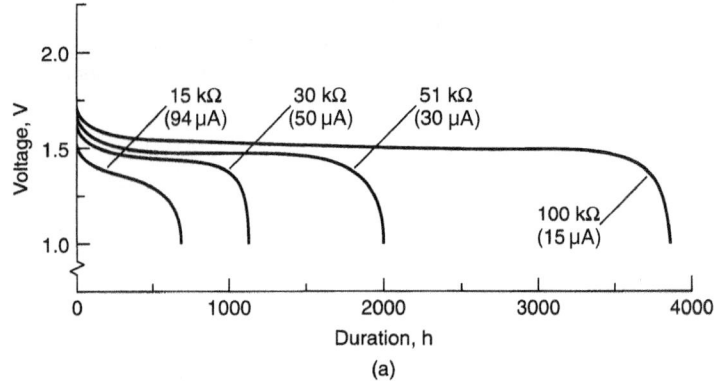

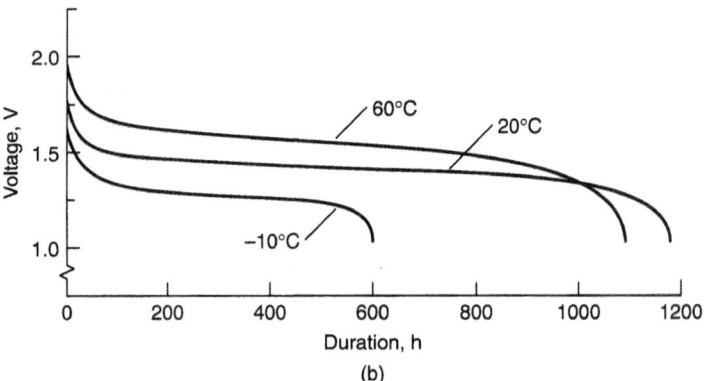

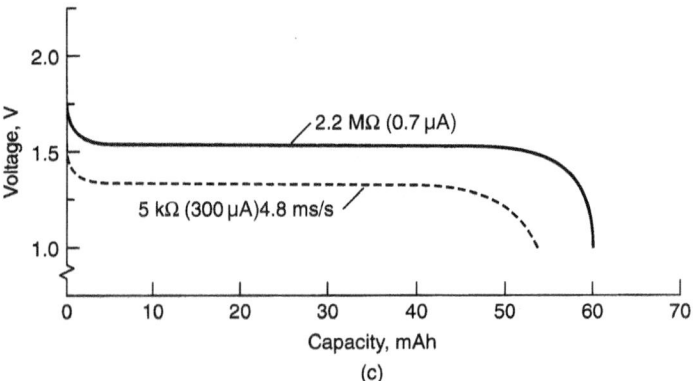

FIGURE 14.84 Discharge characteristics of Li/CuO button-type battery, 60 mAh size. (*a*) Load characteristics. (*b*) Temperature characteristics. (*c*) Pulse discharge characteristics. (*Courtesy of Panasonic Corp. of North America.*)

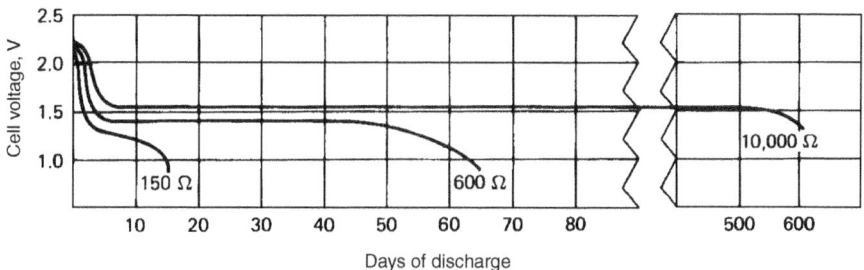

FIGURE 14.85 Typical discharge curves for Li/CuO AA-size battery at 20°C. (*Courtesy of SAFT America, Inc.*)

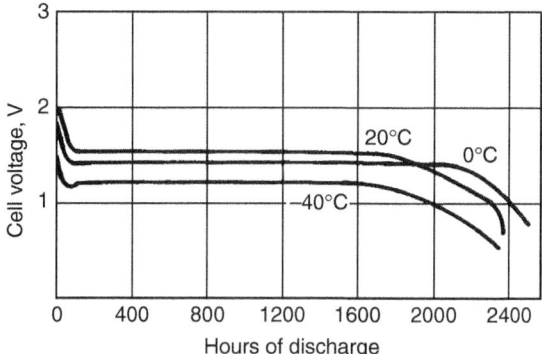

FIGURE 14.86 Effect of temperature on Li/CuO AA-size battery, 1-kΩ load. (*Courtesy of SAFT America, Inc.*)

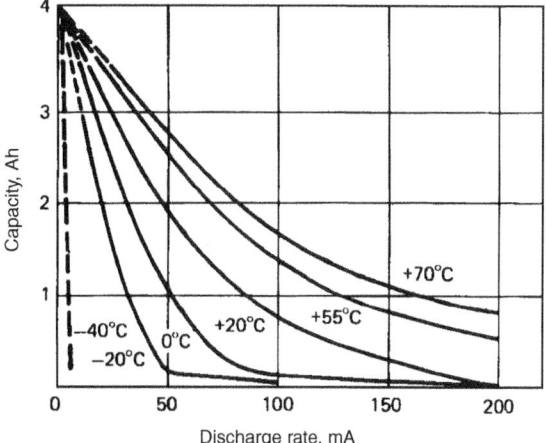

FIGURE 14.87 Capacity of Li/CuO AA-size battery as a function of discharge load and temperature. (*Courtesy of SAFT America, Inc.*)

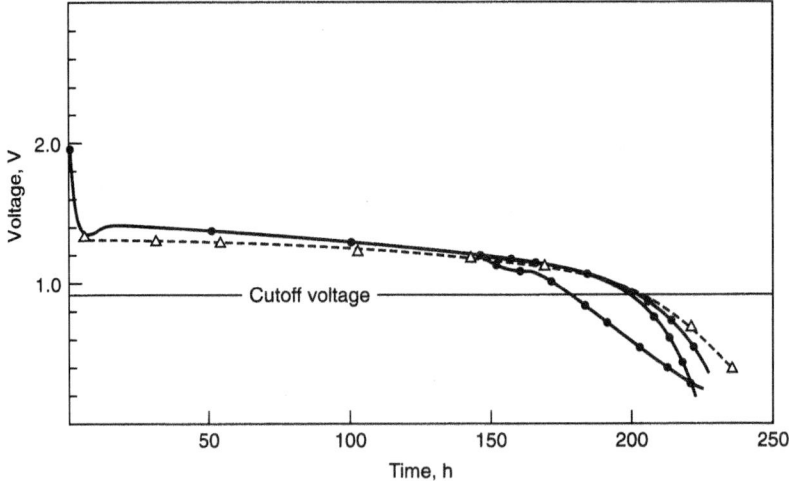

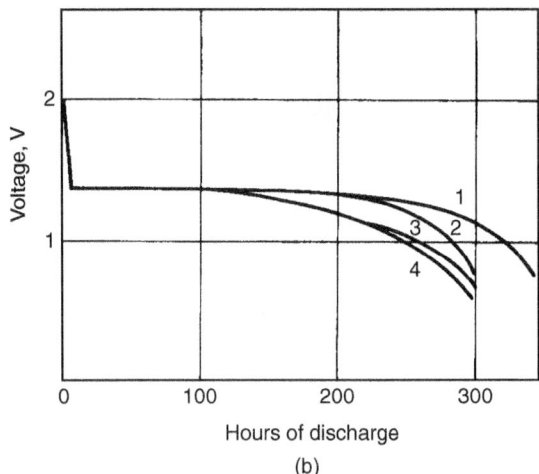

FIGURE 14.88 Effect of storage on performance on Li/CuO cylindrical batteries. (*a*) Discharges at 20°C before and after 10-year storage at 20°C. (*b*) Discharges at 20°C after 70°C storage. Curve 1—No storage; curve 2—6 months' storage; curve 3—12 months' storage; curve 4—18 months' storage. (*Courtesy of SAFT America, Inc.*)

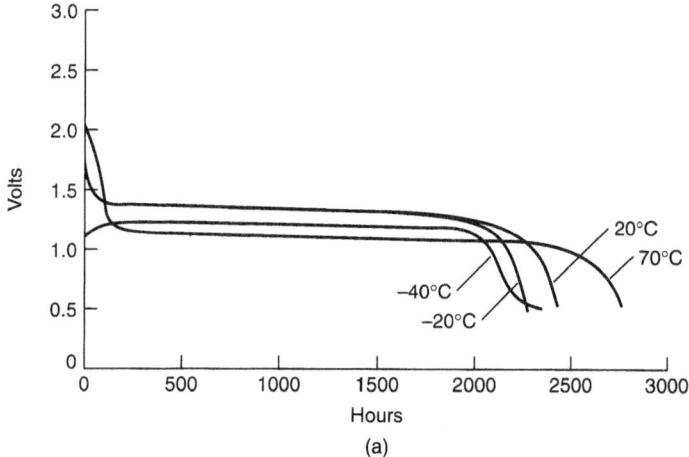

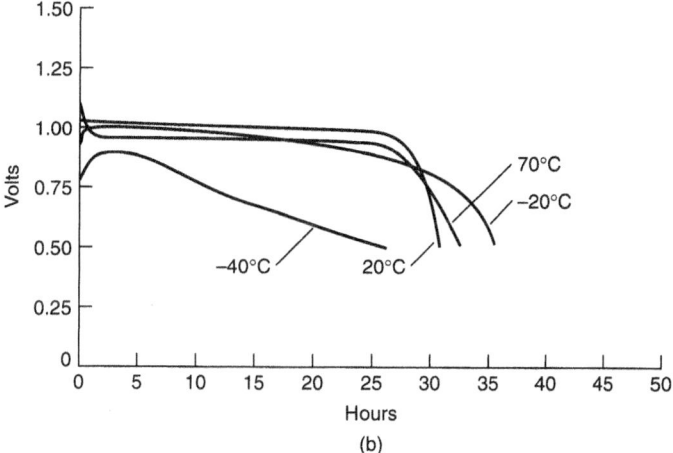

FIGURE 14.89 Performance of high- and low-rate Li/CuO D-size batteries. (*a*) Discharge at 147 ohms. (*b*) Discharge at 1.5 ohms.

High-Temperature Cells. Specially designed hermetically sealed batteries have been developed for use at the high temperatures encountered, for example, by the oil-well logging industry, which uses down-hole tools operating to 150°C, the maximum temperature at which the Li/CuO cells can operate.

Spirally Wound Cells. Cylindrical batteries in C and D sizes have been designed with spirally wound electrodes to meet higher drain requirements. Figure 14.89 shows the discharge performance of a Li/CuO D-size battery at various temperatures and at relatively low and high discharge rates.

14.11.4 Cell Types and Applications

The Li/CuO batteries that have been available in the button and small cylindrical (bobbin) configurations are listed in Table 14.24. Under the low-drain conditions these batteries have a significant

TABLE 14.24 Characteristics of Lithium/Copper Oxide Batteries

	Li/CuO		
	Button	½AA	AA
Nominal voltage, V	1.5	1.5	1.5
Dimensions (max)			
Diameter, mm	9.5	14.5	14.5
Height, mm	2.7	26.0	50.5
Volume, cm^3	0.2	4.3	8.3
Weight, g	0.6	7.3	17.4
Rated capacity, Ah*	0.060	1.4	3.4
Specific energy/Energy density			
Wh/kg	150	285	290
Wh/L	450	485	610
Weight of lithium, g	—	0.4	0.9
Maximum current, mA	0.3	20	40

*At approximately $C/1000$ rate.
Source: SAFT America, Inc. and Panasonic of North America Corp.

capacity advantage over the conventional aqueous batteries. Combined with their excellent storability and operation over a wide temperature range, these batteries provide reliable power sources for applications such as memory backup, clocks, electric meters, and telemetry and, with high-temperature cells, in high-temperature environments. Specially designed units were also manufactured to meet higher drain applications. These batteries are no longer available commercially. Since this technology remains a potential competitor for alkaline-manganese at low-drain rates, information on its properties is retained in this edition.

14.12 LITHIUM/SILVER VANADIUM OXIDE BATTERIES

The lithium/silver vanadium oxide system has been developed for use in biomedical applications, such as cardiac defibrillators, neurostimulators, and drug delivery devices. A description of this system and its applications is found in Sec. 31.5.4.

14.13 LITHIUM/WATER AND LITHIUM/AIR BATTERIES

These two technologies are described in Sec. 33.6.

REFERENCES

1. J. P. Gabano, *Lithium Batteries*, Academic, London, 1983.
2. V. Sapru, *Battery Power*, **14**, pp. 4–8 (2010.)
3. Technical data, Foote Mineral Co., Exton, PA; Lithium Corp. of America, Gastonia, NC.
4. H. R. Grady, "Lithium Metal for the Battery Industry," *J. Power Sources* **5**:127 (1980), Elsevier Sequoia, Lausanne, Switzerland.
5. J. T. Nelson and C. F. Green, "Organic Electrolyte Battery Systems," U.S. Army Material Command Rep. HDL-TR-1588, Washington, DC, Mar. 1972.

J. O. Besenhard and G. Eichinger, "High Energy Density Lithium Cells, pt. I, Electrolytes, and Anodes," *J. Electroanal. Chem.* **68:**1 (1976), Elsevier Sequoia, Lausanne, Switzerland.

G. Eichinger and J. O. Besenhard, "High Energy Density Lithium Cells, pt. II, Cathodes and Complete Cells," *J. Electroanal. Chem.* **72:**1 (1980), Elsevier Sequoia, Lausanne, Switzerland.

6. F. Deligiannis, B. V. Ratnakumar, H. Frank, E. Davies, and S. Surampudi, *Proc. 37th Power Sources Conf.*, pp. 373–377 (1996), Cherry Hill, NJ.

7. A. N. Dey, "Lithium Anode Film and Organic and Inorganic Electrolyte Batteries," in *Thin Solid Films*, Vol. 43, Elsevier Sequoia S. A., Lausanne, Switzerland, 1977, p. 131.

8. S. Gilman and W. Wade, "The Reduction of Sulfuryl Chloride at Teflon-Bonded Carbon Cathodes," *J. Electrochem. Soc.* **127:**1427 (1980).

9. A. Meitav and E. Peled, "Calcium-Ca(AlCl$_4$)$_2$-Thionyl Chloride Cell: Performance and Safety," *J. Electrochem. Soc.* **129:**3 (1982).

10. R. L. Higgins and J. S. Cloyd, "Development of the Calcium-Thionyl Chloride Systems," *Proc. 29th Power Sources Conf.*, Electrochemical Society, Pennington, N.J., June 1980.

    M. Binder, S. Gilman, and W. Wade, "Calcium-Sulfuryl Chloride Primary Cell," *J. Electrochem. Soc.* **129:**4 (1982).

11. E. S. Takeuchi and W. C. Thiebolt, "The Reduction of Silver Vanadium Oxide in Lithium/Silver Vanadium Oxide Cells," *J. Electrochem. Soc.* **135:**11 (1988).

    E. S. Takeuchi, "Lithium/Solid Cathode Cells for Medical Applications," *Proc. Int. Battery Seminar*, Boca Raton, FL, 1993.

    A. Crespi, "The Characterization of Silver Vanadium Cathode Material by High-Resolution Electron Microscopy," *Proc. 7th Int. Meet. Lithium Batteries*, Boston, MA, May 1994.

12. N. I. Sax, *Dangerous Properties of Industrial Materials*, Van Nostrand Reinhold, New York.

13. *Transportation*, Code of Federal Regulations CFR 49, U.S. Government Printing Office, Washington, DC; Exemption DOT-E-7052, Department of Transportation, Washington, DC: "Technical Instructions for the Safe Transport of Dangerous Goods by Air," International Civil Aviation Organization, DOC 9284-AN/905, Montreal, Quebec, Canada.

14. E. H. Reiss, "Considerations in the Use and Handling of Lithium-Sulfur Dioxide Batteries," *Proc. 29th Power Sources Conf.*, Electrochemical Society, Pennington, NJ, June 1980.

15. Technical Standard Order: TSO-C142, Lithium Batteries, U.S. Dept. of Transportation, Federal Aviation Administration, Washington, DC (2000).

16. T. B. Reddy, *Modern Battery Technology*, Sec. 5.2, C. D. S. Tuck, ed., Ellis Horwood, New York (1991).

17. M. Mathews, *Proc. 39th Power Sources Conf.*, pp. 77–80 (2000), Cherry Hill, NJ.

18. S. Charlton, R. Costa, and C. Negrete, *Proc. 41st Power Sources Conf.*, pp. 29–31 (2004), Philadelphia, PA.

19. M. Sink, *Proc. 38th Power Sources Conf.*, pp. 187–190 (1998), Cherry Hill, NJ.

20. H. Taylor, "The Storability of Li/SO$_2$ Cells," *Proc. 12th Intersociety Energy Conversion Engineering Conf.*, American Nuclear Society, LaGrange Park, IL, 1977.

21. D. Linden and B. McDonald, "The Lithium-Sulfur Dioxide Primary Battery—Its Characteristics, Performance and Applications," *J. Power Sources* **5:**35 (1980), Elsevier Sequoia, Lausanne, Switzerland.

22. R. C. McDonald et al., "Investigation of Lithium Thionyl Chloride Battery Safety Hazard," Tech. Rep. N60921-81-C0229, Naval Surface Weapons Center, Silver Spring, MD, Jan. 1983.

23. S. C. Levy and P. Bro, *Battery Hazards and Accident Prevention*, Sec. 10.3.2, Plenum Publishing Corp., New York (1994).

24. Tadiran Batteries, Port Washington, NY, 11050.

25. M. Babai and U. Zak, "Safety Aspects of Low-Rate Li/SOCl$_2$ Batteries," *Proc. 29th Power Sources Conf.*, Electrochemical Society, Pennington, NJ, June 1980.

26. K. M. Abraham and R. M. Mank, "Some Safety Related Chemistry of Li/SOCl$_2$ Cells," *Proc. 29th Power Sources Conf.*, Electrochemical Society, Pennington, NJ, June 1980.

27. R. L. Zupancic, "Performance and Safety Characteristics of Small Cylindrical Li/SOCl$_2$ Cells," *Proc. 29th Power Sources Conf.*, Electrochemical Society, Pennington, NJ, June 1980.

28. J. F. McCartney, A. H. Willis, and W. J. Sturgeon, "Development of a 200 kWh Li/SOCl$_2$ Battery for Undersea Applications," *Proc. 29th Power Sources Conf.*, Electrochemical Society, Pennington, NJ, June 1980.

29. A. Zolla, J. Westernberger, and D. Noll, *Proc. 39th Power Sources Conf.*, pp. 64–68 (2000), Cherry Hill, NJ.

30. C. Winchester, J. Banner, A. Zolla, J. Westenberger, D. Drozd, and S. Drozd, *Proc. 39th Power Sources Conf.*, pp. 5–9 (2000), Cherry Hill, NJ.
31. K. F. Garoutte and D. L. Chua, "Safety Performance of Large Li/SOCl$_2$ Cells," *Proc. 29th Power Sources Conf.*, Electrochemical Society, Pennington, NJ, June 1980.
32. F. Goebel, R. C. McDonald, and N. Marincic, "Performance Characteristics of the Minuteman Lithium Power Source," *Proc. 29th Power Sources Conf.*, Electrochemical Society, Pennington, NJ, June 1980.
33. D. V. Wiberg, "Non-Destructive Test Techniques for Large Scale Li/Thionyl Chloride Cells" *Proc. Int. Battery Seminar*, Boca Raton, FL, 1993.
34. P. G. Russell, D. Carmen, C. Marsh, and T. B. Reddy, *Proc. 38th Power Sources Conf.*, pp. 207–210 (1998), Cherry Hill, NJ.
35. E. Elster, S. Luski, and H. Yamin, "Electrical Performance of Bobbin Type Li/SO$_2$Cl$_2$ Cells," *Proc. 11th Int. Seminar Batteries*, Boca Raton, FL, March 1994.
36. S. McKay, M. Peabody, and J. Brazzell, *Proc. 39th Power Sources Conf.*, pp. 73–76 (2000), Cherry Hill, NJ.
37. C. C. Liang, P. W. Krehl, and D. A. Danner, "Bromine Chloride as a Cathode Component in Lithium Inorganic Cells," *J. Appl. Electrochem* (1981).
38. D. M. Spillman and E. S. Takeuchi, *Proc. 38th Power Sources Conf.*, pp. 199–202 (1998), Cherry Hill, NJ.
39. H. Ikeda, S. Narukawa, and S. Nakaido, "Characteristics of Cylindrical and Rectangular Li/MnO$_2$ Batteries," *Proc. 29th Power Sources Conf.*, Electrochemical Society, Pennington, NJ, 1980.
40. T. B. Reddy and P. Rodriguez, "Lithium/Manganese Dioxide Foil-Cell Battery Development,"*Proc. 36th Power Sources Conf.*, Cherry Hill, NJ, 1994.
41. X. Wang, J. Bennetti, M. Mathews, and X. Zhang, *Proc. 42nd Power Sources Conf.,* pp. 69–72 (2006), Philadelphia, PA.
42. Z. Pi and X. Zhang, *Proc. 42nd Power Sources Conf.,* pp. 65–68 (2006), Philadelphia, PA.
43. A. Morita, T. Iijima, T. Fujii, and H. Ogawa, "Evaluation of Cathode Materials for the Lithium/Carbon Monofluoride Battery," *J. Power Sources* **5:**111 (1980), Elsevier Sequoia, Lausanne, Switzerland, 1980.
44. D. Eyre and C. D. S. Tuck, *Modern Battery Technology*, Sec. 5.3, C. D. S. Tuck, ed., Ellis Horwood, New York (1991).
45. R. L. Higgins and L. R. Erisman, "Applications of the Lithium/Carbon Monofluoride Battery," *Proc. 28th Power Sources Symp.*, Electrochemical Society, Pennington, NJ, June 1978.
46. T. R. Counts, *Proc. 38th Power Sources Conf.*, 143–146 (1998), Cherry Hill, NJ.
47. V. Z. Leger, U. S. Patent No. 4,327,166 (April 1982).
48. X. Wang, J. Mastroangelo, and X. Zhang, *Proc. 43rd Power Sources Conf.*, pp. 541–545 (2008), Philadelphia, PA.
49. P. Lam and R. Yazami, *J. Power Sources*, **153:**354–359 (2006).
50. J. Whitacre et al., *J. Power Sources*, **160:**517 (2006).
51. J. F. Whitacre et al., *Electrochem and Solid-State Letters* **10:**A166–A170 (2007).
52. R. Yazami, 25th *International Florida Battery Seminar*, Ft. Lauderdale, FL, March 2008.
53. A. Webber, *Proc. 41st Power Sources Conf.*, pp. 25–28 (2004), Philadelphia, PA.
54. T. Iijima, Y. Toyoguchi, J. Nishimura, and H. Ogawa, "Button-Type Lithium Battery Using Copper Oxide as a Cathode," *J. Power Sources* **5:**1 (1980), Elsevier Sequoia, Lausanne, Switzerland.
55. J. Tuner et al., "Further Studies on the High Energy Density Li/CuO Organic Electrolyte System," *Proc. 29th Power Sources Conf.*, Electrochemical Society, Pennington, NJ, June 1980.

PART 3

SECONDARY BATTERIES

CHAPTER 15
AN INTRODUCTION TO SECONDARY BATTERIES

Thomas B. Reddy

15.1 GENERAL CHARACTERISTICS AND APPLICATIONS OF SECONDARY BATTERIES

Secondary or rechargeable batteries are widely used in many applications. The most familiar are starting, lighting, and ignition (SLI) automotive applications; industrial truck materials-handling equipment; and emergency and standby power. Small, secondary batteries are also used to power portable devices such as tools, toys, lighting, photographic, radio, and more significantly, consumer electronic devices (laptop computers, camcorders, cellular phones). More recently, secondary batteries are being developed as a power source for electric and hybrid electric vehicles. (See Chap. 29.) Major development programs have been initiated toward improving the performance of existing battery systems and developing new systems to meet the stringent specifications of these new applications.

The applications of secondary batteries fall into two major categories:

1. Those applications in which the secondary battery is used as an energy storage device, being charged by a prime energy source and delivering its energy to the load on demand, when the prime energy source is not available or is inadequate to handle the load requirement. Examples are automotive and aircraft systems, uninterruptible power supplies and standby power sources, and hybrid applications.

2. Those applications in which the secondary battery is discharged (similar in use to a primary battery) and recharged after use, either in the equipment in which it was discharged or separately. Secondary batteries are used in this manner for convenience, for cost savings (as they can be recharged rather than replaced), or for power drains beyond the capability of primary batteries. Most consumer electronics, electric-vehicle, traction, industrial truck, and some stationary battery applications fall in this category.

Conventional aqueous secondary batteries are characterized, in addition to their ability to be recharged, by high power density, flat discharge profiles, and good low-temperature performance. Their energy densities and specific energies, however, are usually lower, and their charge retention is poorer than those of primary battery systems. Rechargeable batteries, such as lithium-ion technologies, however, have higher energy densities, better charge retention, and other performance enhancements characterized by the use of higher energy materials. Power density may be adversely affected because of the use of aprotic solvents in the electrolyte, which have lower conductivity than the aqueous electrolyte. This has been compensated for by using high surface area electrodes and advanced materials. (See Chap. 26.)

Secondary batteries have been in existence for over 150 years. The lead-acid battery was developed in 1859 by Planté. The nickel-iron alkaline battery was introduced by Edison in 1908 as a power source for the early electric automobile. It eventually saw service in industrial trucks, underground work vehicles, railway cars, and stationary applications. Its advantages were durability and long life, but it gradually lost its market share because of its high cost, maintenance requirements, and lower specific energy.[1]

The pocket-plate nickel-cadmium battery has been manufactured since 1909 and was used primarily for heavy-duty industrial applications. The sintered-plate designs, which led to increased power capability and energy density, opened the market for aircraft engine starting and communications applications during the 1950s. Later the development of the sealed nickel-cadmium battery led to its widespread use in portable and other applications. The dominance of this technology in the portable rechargeable market has been surplanted initially by nickel-metal hydride and more recently by lithium-ion batteries, which provide higher specific energy and energy density.

As with the primary battery systems, significant performance improvements have been made with the older secondary battery systems, and a number of newer types, such as silver-zinc, nickel-zinc, nickel-hydrogen, and lithium ion batteries, and the high-temperature system, have been introduced into commercial use or are under advanced development. Much of the development work on new systems has been supported by the need for high-performance batteries for portable consumer electronic applications and electric vehicles.

The specific energy and energy density of portable rechargeable nickel-cadmium batteries have not improved significantly, and now stand at 35 Wh/kg and 100 Wh/L, respectively. Through the use of new hydrogen-storage alloys, improved performance in nickel-metal hydride batteries has been achieved and that system now provides 70 to 100 Wh/kg and up to 430 Wh/L. A major increase in performance of lithium-ion systems was seen in the last decade due to the use of graphitic carbon materials in the negative electrode with much higher specific capacity. These batteries now provide a specific energy of 200 Wh/kg and an energy density of 570 Wh/L in the small cylindrical sizes employed for consumer electronics applications. This improvement was also due to improved cathode materials and cell design. Advanced lithium-ion batteries currently provide 240 Wh/kg and 640 Wh/L. (See Chap. 26.) The lithium metal/lithiated manganese dioxide rechargeable AA cell was withdrawn from the market in the late 1990s and, although significant research and development with lithium metal continue, few products are commercially available at the present time. (See Chap. 27.)

The worldwide secondary battery market was approximately $47.5 billion in 2009.[2] The lead-acid battery is the most popular, with the SLI battery accounting for a major share of the market. This share is declining gradually, due to increasing applications for other types of batteries. In 2008, there were 3,163 billion lithium-ion cells produced, with a market value of about $13 billion.[3] A major growth area has been the non-automotive consumer applications for small secondary batteries. Lithium-ion batteries have emerged in the last decade to capture a 75% share of the market for small, sealed consumer batteries. The typical characteristics and applications of secondary batteries are summarized in Table 15.1.

15.2 TYPES AND CHARACTERISTICS OF SECONDARY BATTERIES

The important characteristics of secondary or rechargeable batteries are that the charge and discharge—the transformation of electric energy to chemical energy and back again to electric energy—should proceed nearly reversibly, should be energy efficient, and should have minimal physical changes that can limit cycle life. Chemical action, which may cause deterioration of the cell's components, loss of life, or loss of energy, should be absent, and the cell should possess the usual characteristics desired of a battery such as high specific energy, low resistance, and good

TABLE 15.1 Major Characteristics and Applications of Secondary Batteries

System	Characteristics	Applications
Lead-acid:		
Automotive	Popular, low-cost secondary battery, low specific-energy, high-rate, and low-temperature performance; maintenance-free designs	Automotive SLI, golf carts, lawn mowers, tractors, aircraft, marine, micro-hybrid vehicles
Traction (motive power)	Designed for deep 6–9 h discharge, cycling service	Industrial trucks, materials handling, electric and hybrid electric vehicles, special types for submarine power
Stationary	Designed for standby float service, long life, VRLA designs	Emergency power, utilities, telephone, UPS, load leveling, energy storage, emergency lighting
Portable	Sealed, maintenance-free, low cost, good float capability, moderate cycle life	Portable tools, small appliances and devices, portable electronic equipment
Nickel-cadmium:		
Industrial and FNC	Good high-rate, low-temperature capability, flat voltage, excellent cycle life	Aircraft batteries, industrial and emergency power applications, communication equipment
Portable	Sealed, maintenance-free, good high-rate low-temperature performance, good cycle life	Consumer electronics, portable tools, pagers, appliances, photographic equipment, standby power, memory backup
Nickel-metal hydride	Sealed, maintenance-free, higher capacity than nickel-cadmium batteries; high energy density and power	Consumer electronics and other portable applications; hybrid electric vehicles
Nickel-iron	Durable, rugged construction, long life, low specific energy	Materials handling, stationary applications, railroad cars
Nickel-zinc	High specific energy, extended cycle life, high-power capability	Bicycles, scooters, consumer electronics such as power tools
Silver-zinc	High specific energy, very good high-rate capability, low cycle life, high cost	Training targets, drones, submarines, other military equipment, launch vehicles and space power
Silver-cadmium	High specific energy, good charge retention, moderate cycle life, high cost	Portable equipment requiring a lightweight, high-capacity battery; space satellites
Nickel-hydrogen	Long cycle life under shallow discharge, long life	Primarily for aerospace applications such as LEO and GEO satellites
Ambient-temperature rechargeable "primary" types (Zn/MnO_2)	Low cost, good capacity retention, sealed and maintenance-free, limited cycle life and rate capability	Cylindrical cell applications, rechargeable replacement for zinc-carbon and alkaline primary batteries, consumer electronics (ambient-temperature systems)
Lithium-ion	High specific energy and energy density, long cycle life; high-power capability	Portable and consumer electronic equipment, electric vehicles (EVs, HEVs, PHEVs), space applications, electrical energy storage

performance over a wide temperature range. These requirements limit the number of materials that can be employed successfully in a rechargeable battery system.

15.2.1 Lead-Acid Batteries

The lead-acid battery system has many of these characteristics. The charge-discharge process is essentially reversible, the system does not suffer from deleterious chemical action, and while its energy density and specific energy are low, the lead-acid battery performs reliably over a wide temperature range. A key factor for its popularity and dominant position is its low cost with good performance and cycle-life.

The lead-acid battery is designed in many configurations, as listed in Table 15.1 (see also Chapters 16 and 17), from small sealed cells with a capacity of 1 Ah to large cells, up to 12,000 Ah. The automotive SLI battery is by far the most popular and the one in widest use. Most significant of the advances in SLI battery design are the use of lighter-weight plastic containers, the improvement in shelf life, the "dry-charge" process, and the "maintenance-free" design. The latter, using calcium-lead or low-antimony grids, has greatly reduced water loss during charging (minimizing the need to add water) and has reduced the self-discharge rate so that batteries can be shipped or stored in a wet, charged state for relatively long periods.

The lead-acid industrial storage batteries are generally larger than the SLI batteries, with a stronger, higher-quality construction. Applications of the industrial batteries fall in several categories. The motive power traction types are used in materials-handling trucks, tractors, mining vehicles, and, to a limited extent, golf carts and personnel carriers, although the majority in use are automotive-type batteries. A second category is diesel locomotive engine starting. Significant advances are the use of lighter-weight plastic containers in place of the hard-rubber containers, better seals, and changes in the tubular positive-plate designs. Another category is stationary service: telecommunications systems, electric utilities for operating power distribution controls, emergency and standby power systems, uninterruptible power supply (UPS), and in railroads, signaling and car power systems.

The industrial batteries use three different types of positive plates: tubular and pasted flat plates for motive power, diesel engine cranking, and stationary applications, and Planté designs, forming the active materials from pure lead, mainly in the stationary batteries. The flat-plate batteries use either lead-antimony or lead-calcium grid alloys. A development for the telephone industry has been the "round cell," designed for trouble-free, long-life service. This battery uses plates, conical in shape with pure lead grids, which are stacked one above the other in a cylindrical cell container, rather than the normal prismatic structure with flat, parallel plates.

An important development in lead-acid battery technology is the valve-regulated lead-acid battery (VRLA) (see Chap. 17). These batteries operate on the principle of oxygen recombination, using a "starved" or immobilized electrolyte. The oxygen generated at the positive electrode during charge can, in these battery designs, diffuse to the negative electrode, where it can react, in the presence of sulfuric acid, with the freshly formed lead. The VRLA design reduces gas emission by over 95% as the generation of hydrogen is also suppressed. Oxygen recombination is facilitated by the use of a pressure-relief valve, which is closed during normal operation. When pressure builds up, the valve opens at a predetermined value, venting the gases. The valve reseals before the cell pressure decreases to atmospheric pressure. The VRLA battery is now used in about 70% of the telecommunication batteries and in about 80% of the uninterrupted power supply (UPS) applications.

Smaller sealed lead-acid cells are used in emergency lighting and similar devices requiring backup power in the event of a utility power failure, portable instruments and tools, and various consumer-type applications. These small sealed lead-acid batteries are constructed in two configurations, prismatic cells with parallel plates, ranging in capacity from 1 to 30 Ah, and cylindrical cells similar in appearance to the popular primary alkaline cells and ranging in capacity up to 25 Ah. The acid electrolyte in these cells is either gelled or absorbed in the plates and in highly porous separators so they can be operated virtually in any position without the danger of leakage.

The grids generally are of lead-calcium-tin alloy; some use grids of pure lead or a lead-tin alloy. The cells also include the features for oxygen recombination and are considered to be VRLA batteries (see Chap. 17).

Lead-acid batteries also are used in other types of applications, such as in submarine service, for reserve power in marine applications, and in areas where engine-generators cannot be used, such as indoors and in mining equipment. New applications, to take advantage of the cost effectiveness of this battery, include load leveling for utilities and solar photo-voltaic systems. These applications will require improvements in the energy and power density of the lead-acid battery.

Newer designs, such as the Ultrabattery® and the extended life flooded (ELF) designs either incorporate carbon plates in parallel with the negative electrodes or use carbon in the negative to provide a capacitive effect to allow use in micro-hybrid vehicles (see Chap. 29).

15.2.2 Alkaline Secondary Batteries

Most of the other conventional types of secondary batteries use an aqueous alkaline solution (KOH or NaOH) as the electrolyte. Electrode materials are less reactive with alkaline electrolytes than with acid electrolytes. Furthermore, the charge-discharge mechanism in the alkaline electrolyte involves only the transport of oxygen or hydroxyl ions from one electrode to the other; hence the composition or concentration of the electrolyte does not change during charge and discharge.

Nickel-Cadmium Batteries. The nickel-cadmium secondary battery is available in several cell designs and in a wide range of sizes. The original cell design used the pocket-plate construction. The vented pocket-type cells are very rugged and can withstand both electrical and mechanical abuse. They have very long lives and require little maintenance beyond occasional topping with water. This type of battery is used in heavy-duty industrial applications, such as materials-handling trucks, mining vehicles, railway signaling, emergency or standby power, and diesel engine starting. The sintered-plate construction is a more recent development, having higher energy density. It gives better performance than the pocket-plate type at high discharge rates and low temperatures but is more expensive. It is used in applications such as aircraft engine starting and communications and electronics equipment, where the lighter weight and superior performance are required. Higher energy and power densities can be obtained by using nickel foam, nickel fiber, or plastic-bonded (pressed-plate) electrodes. The sealed cell is a third design that incorporates excess capacity in the negative electrode. It uses an oxygen-recombination feature similar to the one used in sealed lead-acid batteries to prevent the buildup of pressure caused by gassing during charge. Sealed cells are available in prismatic, button, and cylindrical configurations and are used in consumer and small industrial applications.

Nickel-Iron Batteries. The nickel-iron battery was important from its introduction in 1908 until the 1970s, when it lost its market share to the industrial lead-acid battery. It was used in materials-handling trucks, mining and underground vehicles, railroad and rapid-transit cars, and in stationary applications. The main advantages of the nickel-iron battery, with major cell components of nickel-plated steel, are extremely rugged construction, long life, and durability. Its limitations, namely, low specific energy, poor charge retention, and poor low-temperature performance, and its high cost of manufacture compared with the lead-acid battery led to a decline in usage. Recent research has shown that the use of nanostructured carbon in the iron electrode can significantly improve the capacity of the Fe negative.

Silver Oxide Batteries. The silver-zinc (zinc/silver oxide) battery is noted for its high energy density, low internal resistance desirable for high-rate discharge, and a flat second discharge plateau. This battery system is useful in applications where high energy density is a prime requisite, such as submarine and training target propulsion, and other military and space uses. It is not employed for general storage battery applications because its cost is high, its cycle life and activated life are limited, and its performance at low temperatures falls off more markedly than with other secondary battery systems. Recently, there has been a research effort to develop this system for consumer electronics applications.

The silver-cadmium (cadmium/silver oxide) battery has significantly longer cycle life and better low-temperature performance than the silver-zinc battery but is inferior in these characteristics compared with the nickel-cadmium battery. Its energy density, too, is between that of the nickel-cadmium and the silver-zinc batteries. The battery is also very expensive, using two of the more costly electrode materials. As a result, the silver-cadmium battery was never developed commercially but is used in special applications, such as nonmagnetic batteries and space applications. Other silver battery systems, such as silver-hydrogen and silver-metal hydride couples, have been the subject of development activity but have not reached commercial viability.

Nickel-Zinc Batteries. The nickel-zinc (zinc/nickel oxide) battery has characteristics midway between those of the nickel-cadmium and the silver-zinc battery systems. Its energy density is about twice that of the nickel-cadmium battery, but the cycle life previously has been limited due to the tendency of the zinc electrode toward shape change, which reduces capacity, and dendrite formations, which cause internal short-circuiting.

Recent development work has extended the cycle life of nickel-zinc batteries through the use of additives in the negative electrode in conjunction with the use of a reduced concentration of KOH to repress zinc solubility in the electrolyte. Both of these modifications have extended the cycle life of this system to 900 cycles at 80% DOD, so that it is now being marketed for use in electric bicycles, scooters, and consumer electronics such as power tools in the United States.

Hydrogen Electrode Batteries. Another secondary battery system uses hydrogen for the active negative material (with a fuel-cell-type electrode) and a conventional positive electrode, such as nickel oxide. These batteries are being used exclusively for aerospace programs, which require long cycle life at low depth of discharge in (LEO) or limited deep-discharge during the year (GEO) satellites. The high cost of these batteries is a disadvantage that limits their application. A further extension is the sealed nickel/metal hydride battery where the hydrogen is absorbed, during charge, by a metal alloy forming a metal hydride. This metal alloy is capable of undergoing a reversible hydrogen absorption-desorption reaction as the battery is charged and discharged, respectively. The advantage of this battery is that its specific energy and energy density are significantly higher than those of the nickel-cadmium battery. The sealed nickel-metal hydride battery, manufactured in prismatic and cylindrical cells, is being used for portable electronic applications and has became the battery of choice for use in hybrid electric vehicles during the last decade. Pre-charged consumer cells are now commercially available with a capacity loss of less than 10%/yr.

Zinc/Manganese Dioxide Batteries. Several of the conventional primary battery systems have been manufactured as rechargeable batteries, but the only one currently being manufactured is the cylindrical cell using the zinc/alkaline-manganese dioxide chemistry. Its major advantage is a higher capacity than the conventional secondary batteries and a lower initial cost, but its cycle life and rate capability are limited. This system is currently available in AAA and AA sizes and in multicell packs of these two sizes.

Lithium-Ion Batteries. Lithium-ion batteries have emerged in the last two decades to capture over three-quarter of the sales value of the secondary consumer market, with applications such as laptop computers, cell phones, and E-books. Production capacity has recently been estimated to be 250 million/cells per month.[3] These cells provide high energy density and specific energy and long cycle life, typically greater than 1000 cycles at 80% depth of discharge. When built into batteries, battery management circuitry is required to prevent over charge and over discharge, both of which

15.3 COMPARISON OF PERFORMANCE CHARACTERISTICS FOR SECONDARY BATTERY SYSTEMS

15.3.1 General

The characteristics of the major secondary systems are summarized in Table 15.2. This table is supplemented by Table 1.2, which lists several theoretical and practical electrical characteristics of these secondary battery systems. A graphic comparison of the theoretical and practical performances of various battery systems is given in Fig. 3.3. This shows that up to only about 25 to 35% of the theoretical capacity of a battery system is attained under practical conditions as a result of design and the discharge requirements.

It should be noted, as discussed in detail in Chaps. 3 and 32, that these types of data and comparisons (as well as the performance characteristics shown in this section) are necessarily approximations, with each system being presented under favorable discharge conditions. The specific performance of a battery system is very dependent on the cell design and all the detailed and specific conditions of the use and discharge-charge of the battery.

A qualitative comparison of the various secondary battery systems is presented in Table 15.3. The different ratings given to the various designs of the same electrochemical system are an indication of the effects of the design on the performance characteristics of a battery.

15.3.2 Voltage and Discharge Profiles

The discharge curves of the conventional secondary battery systems, at the $C/5$ rate, are compared in Fig. 15.1. The lead-acid battery has the highest cell voltage of the aqueous systems. The average voltage of the alkaline systems ranges from about 1.65 V for the nickel-zinc system to about 1.1 V. At the $C/5$ discharge rate at 20°C there is relatively little difference in the shape of the discharge curve for the various designs of a given system. However, at higher discharge rates and at lower temperatures, these differences could be significant, depending mainly on the internal resistance of the cell.

Most of the conventional rechargeable battery systems have a flat discharge profile, except for the silver oxide systems, which show the double plateau due to the two-stage discharge of the silver oxide electrode at low rates, and the rechargeable zinc/manganese dioxide battery.

The discharge curve of a lithium-ion battery, the graphite/lithiated cobalt oxide system, is shown for comparison. The cell voltages of the lithium-ion batteries are higher than those of the conventional aqueous cells by a factor of three because of the characteristics of these systems. The discharge profile of the lithium-ion batteries are usually not as flat due to the lower conductivity of the nonaqueous electrolytes that must be used and to the thermodynamics of intercalation electrode reactions (see Chap. 26). The typical discharge voltage for a lithium-ion cell is 3.7 V, which allows one unit to replace three NiCad or NiMH cells in a battery configuration. Lithium-ion batteries operating in the 4 to 5 V range are being developed.

15.3.3 Effect of Discharge Rate on Performance

The effects of the discharge rate on the performance of these secondary battery systems are compared again in Fig. 15.2. This figure is similar to a Ragone plot, except that the abscissa is expressed in hours of service instead of specific energy (Wh/kg). This figure shows that hours of service each battery type (unitized to 1 kg battery weight) will deliver at various power (discharge current ×

TABLE 15.2 Characteristics of the Major Secondary Battery Systems

	Lead-acid				Nickel-cadmium			
Common name	SLI	Traction	Stationary	Portable	Vented pocket plate	Vented sintered plate	Sealed	FNC
Chemistry:								
Anode	Pb	Pb	Pb	Pb	Cd	Cd	Cd	Cd
Cathode	PbO_2	PbO_2	PbO_2	PbO_2	NiOOH	NiOOH	NiOOH	NiOOH
Electrolyte	H_2SO_4 (aqueous solution)	H_2SO_4 (aqueous solution)	H_2SO_4 (aqueous solution)	H_2SO_4 (aqueous solution)	KOH (aqueous solution)	KOH (aqueous solution)	KOH (aqueous solution)	KOH (aqueous solution)
Cell voltage (typical), V:								
Nominal	2.0	2.0	2.0	2.0	1.2	1.2	1.2	1.2
Open-circuit	2.1	2.1	2.1	2.1	1.29	1.29	1.29	1.35
Operating	2.0–1.8	2.0–1.8	2.0–1.8	2.0–1.8	1.25–1.00	1.25–1.00	1.25–1.00	1.25–0.85
End	1.75 (lower operating and end voltage during cranking operation)	1.75	1.75 (except when on float service)	1.75 (where cycled)	1.0	1.0	1.0	1.00–0.65
Operating temperatures, °C	−40 to 55	−20 to 40	−10 to 40[c]	−40 to 60	−20 to 45	−40 to 50	−20 to 70	−50 to 60
Specific energy and energy density (at 20°C)								
Wh/kg	40	25	10–20	30	27	30–37	35	10–40
Wh/L	80	80	50–70	90	55	58–96	100	15–80
Discharge profile (relative)	Flat	Flat	Flat	Flat	Flat	Very flat	Very flat	Flat
Power density	High	Moderately high	Moderately high	High	High	High	Moderate to high	Very high
Self-discharge rate (at 20°C), % loss per month[b]	20–30 (Sb-Pb) 2–3 (maintenance-free)	4–6	—	4–8	5	10	15–20	10–15
Calendar life, years	3–6	6	18–25	2–8	8–25	3–10	5–7	5–20
Cycle life, cycles[c]	200–700	1500	—	250–500	500–2000	500–2000	300–1000	500–10,000 cycles in LEO satellite tests to 35% DOD at 10°C
Advantages	Low cost; ready availability; good high-rate; high- and low-temperature operation (good cranking service); good float service; new maintenance-free designs	Lowest cost of competitive systems (also see SLI)	Designed for "float" service; lowest cost of competitive systems (also see SLI)	Maintenance-free; long life on float service; low- and high-temperature performance; no "memory" effect; operates in any position	Very rugged, can withstand physical and electrical abuse; good charge retention, storage, and cycle life; lowest cost of alkaline batteries	Rugged; excellent storage; good specific energy and high-rate and low-temperature performance	Sealed, no maintenance; good low-temperature and high-rate performance; long life cycle; operates in any position	Sealed, no maintenance; high power capability even at low temperature; long cycle life at low depth of discharge; fast charging
Limitations	Relatively low cycle life; limited energy density; poor charge retention and storability; hydrogen evolution	Low energy density; less rugged than competitive systems; hydrogen evolution	Hydrogen evolution	Cannot be stored in discharged condition; lower cycle life than sealed nickel-cadmium; difficult to manufacture in very small sizes	Low energy density	High cost; "memory" effect; thermal runaway problem	Sealed lead-acid battery better at high temperature and float service; "memory" effect	Lower energy density than sintered plate design
Major battery types available	Prismatic cells: 40–100 Ah at 20 h rate	Based on positive plate design: 45–200 Ah per positive plate	Based on positive plate design: 5–400 Ah per positive to 1440 Ah plate	Sealed cylindrical cells: 2.5–25 Ah; prismatic cells to 1440 Ah	Prismatic cells: 5–1200 Ah	Prismatic cells: 1.5–100 Ah	Button cells to 0.5 Ah; cylindrical cells to 12 Ah	Prismatic designs to 490 Ah

[a]Based on $C/LiCoO_2$ lithium-ion battery (see Chap. 26) (characteristics vary with battery system and design).
[b]Self-discharge rate usually decreases with increasing storage time.
[c]Dependent on depth of discharge.
[d]High rate Zn/AgO battery.

	Nickel-iron (conventional)	Nickel-zinc	Zinc/silver oxide (silver-zinc)	Cadmium/silver oxide (silver-cadmium)	Nickel-hydrogen	Nickel-metal hydride	Rechargeable "primary" types, Zn/MnO_2	Lithium ion systems[a]
Negative	Fe	Zn	Zn	Cd	H_2	MH	Zn	C
Positive	NiOOH	NiOOH	AgO	AgO	NiOOH	NiOOH	MnO_2	$LiCoO_2$
Electrolyte	KOH (aqueous solution)	KOH (aqueous solution)	KOH (aqueous solution)	KOH (aqueous solution)	KOH (aqueous solution)	KOH (aqueous solution)	KOH (aqueous solution)	Organic solvent graphite anode
Nominal voltage	1.2	1.65	1.5	1.1	1.4	1.2	1.5	4.0
OCV	1.37	1.73	1.86	1.41	1.32	1.4	1.5	4.1
Operating V	1.25–1.05	1.6–1.4	1.7–1.3	1.4–1.0	1.3–1.15	1.25–1.10	1.3–1.0	3.7
End V	1.0	1.2	1.0	0.7	1.0	1.0	0.9	3.0
Operating temp (°C)	−10 to 45	−20 to 50	−20 to 60	−25 to 70	0 to 50	−20 to 65	−20 to 40	−20 to 50
		Prismatic / Cylindrical				HEV / Commercial		
Specific energy (Wh/kg)	30	60–100 / 70–110	105[d]	70	64 (CPV)	47 / 90–110	100	203
Energy density (Wh/L)	55	110–200 / 200–360	180	120	105 (CPV)	177 / 430	286	570
Discharge profile	Moderately flat	Flat	Double plateau at low rate	Double plateau	Moderately flat	Flat	Sloping	Sloping
Power capability	Moderate to low	High	Very high (for high-rate designs)	Moderate to high	Moderate	High	Moderate	Moderate (energy cells); high (power cells)
Self-discharge (%/month)	20–40	20	5	5	Very high except at low temp.	15–30		2
Calendar life (years)	8–25	—	6–18 cells (wet)	3 (vented) / 4 (sealed)	—	5–10	5–7	
Cycle life	2000–4000	To 900 at 80% DOD	10–50 (HR)	300–800	1500–6000 / 40,000 at 40% DOD	500–1000 (300,000 for HEV)	15–25	1000+
Advantages	Very rugged, can withstand physical and electrical abuse; long life (cycling or stand)	High energy density; relatively low cost; good low-temperature performance; high-power capability	High energy density; high discharge rate; low self-discharge	High-energy density; low self-discharge; good cycle life	High energy density; long cycle life at low DOD; can tolerate overcharge	High energy density; sealed; good cycle life	Good shelf life; low cost	High specific energy and energy density; low self-discharge; long cycle life; good rate-capability
Limitations	Low power and energy density; high self-discharge; hydrogen evolution; high cost and high maintenance cost	Subject to zinc dendrite shorting	High cost; low cycle life; decreased performances at low temperatures	High cost; decreased performance at low temperatures	High initial cost; self-discharge proportional to H_2 pressure and temperature	Memory effect; must be charged at moderate temperature	Limited cycle life; low drain applications; small sizes only	Requires the use of a battery management system; safety issues
Applications	Decreasing significance in developed countries	Prismatic and cylindrical (AA, sub-C, and D) types available for light vehicles and consumer applications such as power tools and digital cameras	Prismatic cells: <1 to 1000 Ah; special types to 5000 Ah	Prismatic cells: for space application	Aerospace applications (up to 100 Ah)	Button and cylindrical cells to 12 Ah; large prismatics to 250 Ah	AAA and AA cylindrical cells to 2.0 Ah multi-cell bundles	Cylindrical and prismatic cells available in many chemistries $LiCoO_2$/graphite typical for consumer applications

TABLE 15.3 Comparison of Secondary Batteries*

System	Energy density	Power density	Flat discharge profile	Low-temperature operation	Charge retention	Charge acceptance	Efficiency	Life	Mechanical Properties	Cost
Lead-acid:										
Pasted	4	4	3	3	4	3	2	3	5	1
Tubular	4	5	4	3	3	3	2	2	3	2
Planté	5	5	4	3	3	3	2	2	4	2
Sealed	4	3	3	2	3	3	2	3	5	2
Lithium-metal	1	3	3	2	1	3	3	4	3	4
Lithium-ion	1	2	3	2	1	1	1	1	2	2
Nickel-cadmium:										
Pocket	5	3	2	1	2	1	4	2	1	3
Sintered	4	1	1	1	4	1	3	2	1	3
Sealed	4	1	2	1	4	2	3	3	2	2
Nickel-iron	5	5	4	5	5	2	5	1	1	3
Nickel-metal hydride	2	1	2	2	3	1	2	2	2	3
Nickel-zinc	2	1	2	3	4	3	3	4	3	3
Silver-zinc	1	1	4	3	1	3	2	5	2	4
Silver-cadmium	2	3	5	4	1	5	1	4	3	4
Nickel-hydrogen	2	3	3	4	5	3	5	2	3	5
Silver-hydrogen	2	3	4	4	5	3	5	2	3	5
Zinc-manganese dioxide	2	4	5	3	1	4	4	5	4	2

*Rating: 1 to 5, best to poorest.

midpoint voltage) levels. The higher slope is indicative of superior retention of capacity with increasing discharge load. The specific energy can be calculated by the following equation:

$$\text{specific energy} = \text{specific power} \times \text{hours of service}$$

or

$$Wh/kg = W/kg \times h = \frac{A \times V \times h}{kg}$$

Figure 15.3 is a plot comparing the characteristics of the nickel-cadmium, sealed nickel-metal hydride, and lithium-ion batteries. The range of specific energies and energy densities for the three most common rechargeable consumer batteries are compared.

15.3.4 Effect of Temperature

The performance of the various secondary batteries over a wide temperature range is shown in Fig. 15.4 on a gravimetric basis. In this figure, the specific energy for each battery system is plotted from −40 to 60°C at about the $C/5$ discharge rate. The lithium-ion system has the highest energy density to −20°C. The sintered-plate nickel-cadmium and nickel-metal hydride batteries show higher percentage retention. In general, the low-temperature performance of the alkaline batteries is better than the performance of the lead-acid batteries, again with the exception of the nickel-iron system. The lead-acid system shows better characteristics at the higher temperatures. These data are necessarily generalized for the purposes of comparison and present each system under favorable discharge conditions. Performance is strongly influenced by the specific discharge conditions.

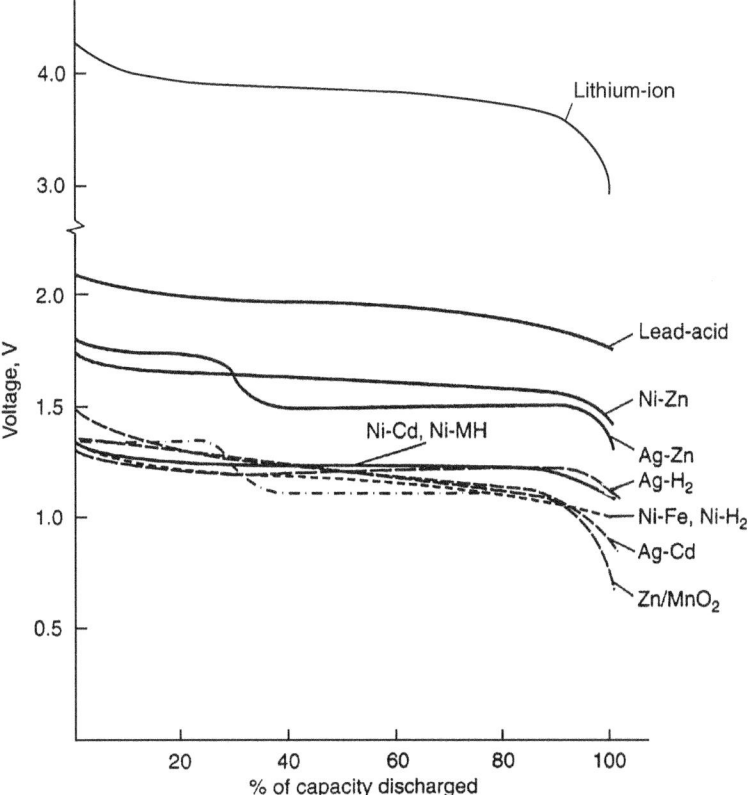

FIGURE 15.1 Discharge profiles of conventional secondary battery systems and rechargeable lithium-ion battery at approximately $C/5$ discharge rate.

15.3.5 Charge Retention

The charge retention of most of the conventional secondary batteries is poor compared with that of primary battery systems (see Fig. 8.10). Normally, secondary batteries are recharged on a periodic basis or maintained on "float" charge if they are to be in a state of readiness. Most alkaline secondary batteries, especially the nickel-oxide batteries, can be stored for long periods of time even in a discharged state without permanent damage and can be recharged when required for use. The lead-acid batteries, however, cannot be stored in a discharged state because sulfation of the plates, which is detrimental to battery performance, will occur. Lithium-ion batteries are best kept at 50% state-of-charge for long-term storage.

Figure 15.5 shows the charge retention properties of several different secondary battery systems. These data are also generalized for the purpose of comparison. There are wide variations of performance depending on design and many other factors, with the variability increasing with increasing storage temperature. Typically, the rate of loss of capacity decreases with increasing storage time.

The silver secondary batteries, the Zn/MnO_2 rechargeable battery, and lithium-ion systems have the best charge retention characteristics of the secondary battery systems. With typical lithium-ion batteries, self-discharge is typically 2% per month at ambient temperature. Low-rate silver cells may lose as little at 10 to 20% per year, but the loss with high-rate cells with large surface areas could be

15.14 SECONDARY BATTERIES

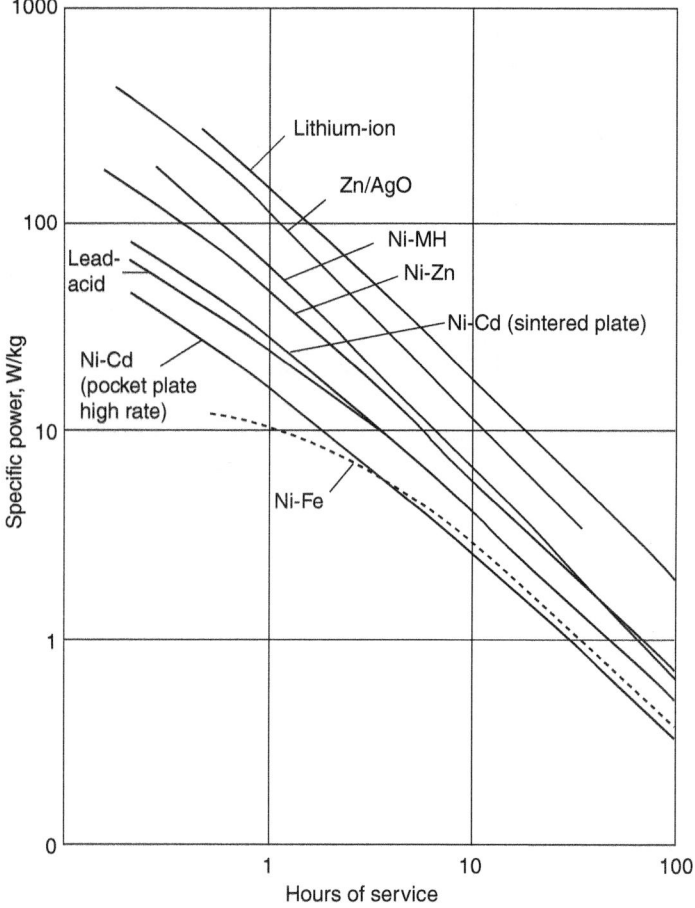

FIGURE 15.2 Comparison of performance of secondary battery systems at 20°C.

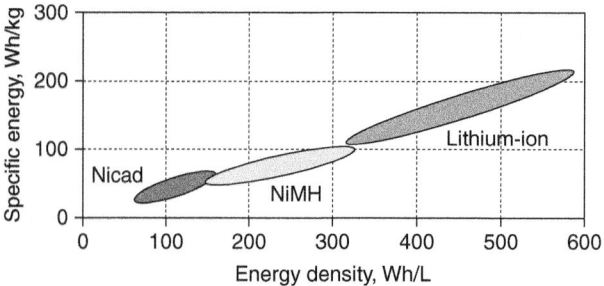

FIGURE 15.3 This figure illustrates the range of specific energies and energy densities for the three most common, commercial rechargeable battery types: nickel-cadmium, nickel metal hydrids, and lithium-ion. (*Courtesy of Prof. Jeff Dahn.*)

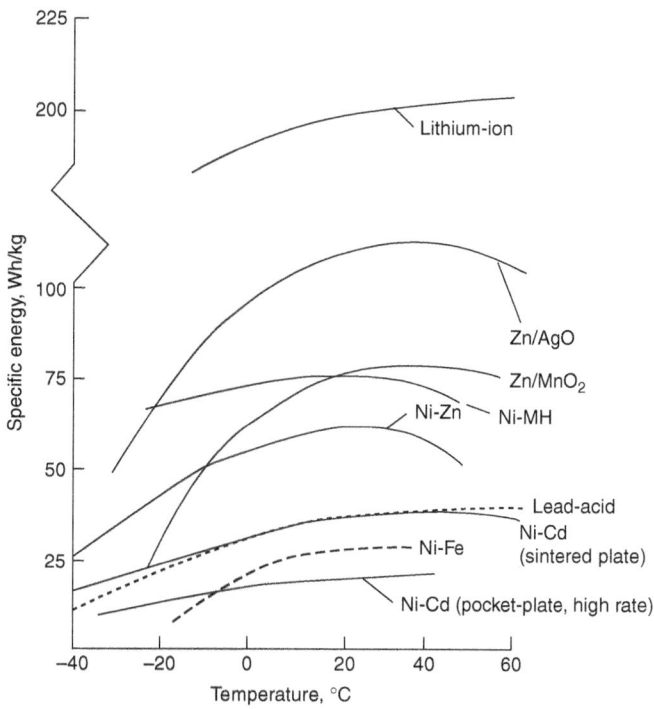

FIGURE 15.4 Effect of temperature on specific energy of secondary battery systems at approximately $C/5$ discharge rate.

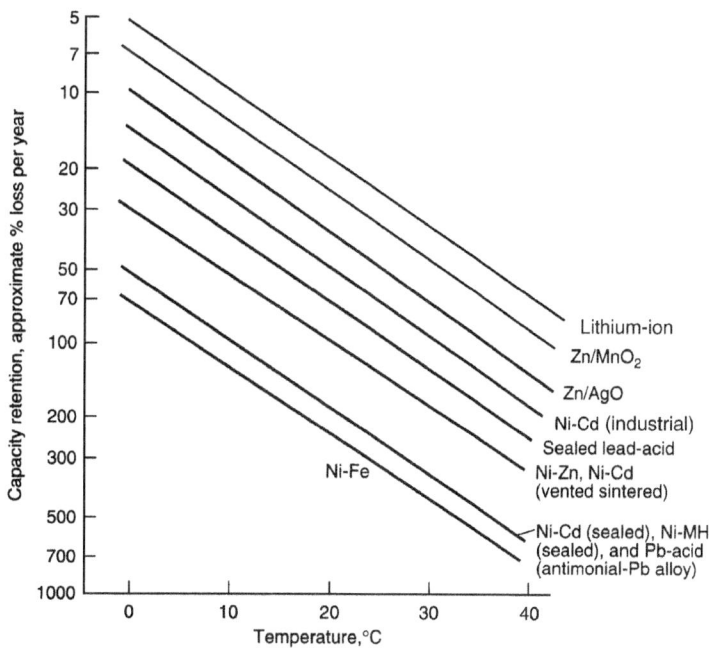

FIGURE 15.5 Capacity retention of secondary battery systems.

5 to 10 times higher. The vented pocket-and sintered-plate nickel-cadmium batteries and the nickel-zinc systems are next; the sealed cells and the nickel-iron batteries have the poorest charge retention properties of the alkaline systems.

The charge retention of the lead-acid batteries is dependent on the design, electrolyte concentration, and formulation of the grid alloy, as well as other factors. The charge retention of the standard automotive SLI batteries, using the standard antimonial-lead grid, is poor, and these batteries have little capacity remaining after 6-months' storage at room temperature. The low antimonial-lead designs and the maintenance-free batteries have much better charge retention, with losses on the order of 20 to 40% per year.

One of the potential advantages of the lithium metal rechargeable batteries is their good charge retention which, in many cases, should be similar to the charge retention characteristics of the lithium primary batteries.

15.3.6 Life

The cycle life and calendar life of the different secondary battery systems are also listed in Table 15.2. Again, these data are approximate because specific performance is dependent on the particular design and the conditions under which the battery is used. The depth of discharge (DOD), for example, as illustrated in Fig. 15.6, and the charging regime strongly influences the battery's life.[4]

Of the conventional secondary systems, the nickel-iron and the vented pocket-type nickel-cadmium batteries are best with regard to cycle life and total lifetime. The nickel-hydrogen battery, developed mainly for aerospace applications, has demonstrated very long cycle life under shallow depth of discharge. The lead-acid batteries do not match the performance of the best alkaline batteries. The pasted cells have the shortest life of the lead-acid cells; the best cycle life is obtained with the tubular design, and the Planté design has the best lifetime.

One of the disadvantages of using zinc, lithium, and other metals with high negative standard potentials in rechargeable batteries is the difficulty of successful recharging and obtaining good cycle and calendar lives. The nickel-zinc battery has recently been improved to provide extended cycle life. The lithium-ion system, however, has also been shown to have good cycle life and has achieved very long cycle life at low DOD by cycling in the middle of the capacity range. (See Fig. 26.59).

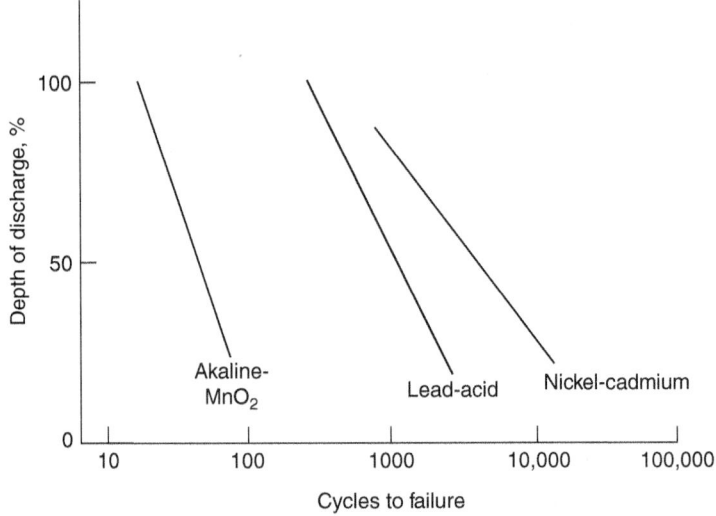

FIGURE 15.6 Effect of depth of discharge on cycle life of secondary battery systems.

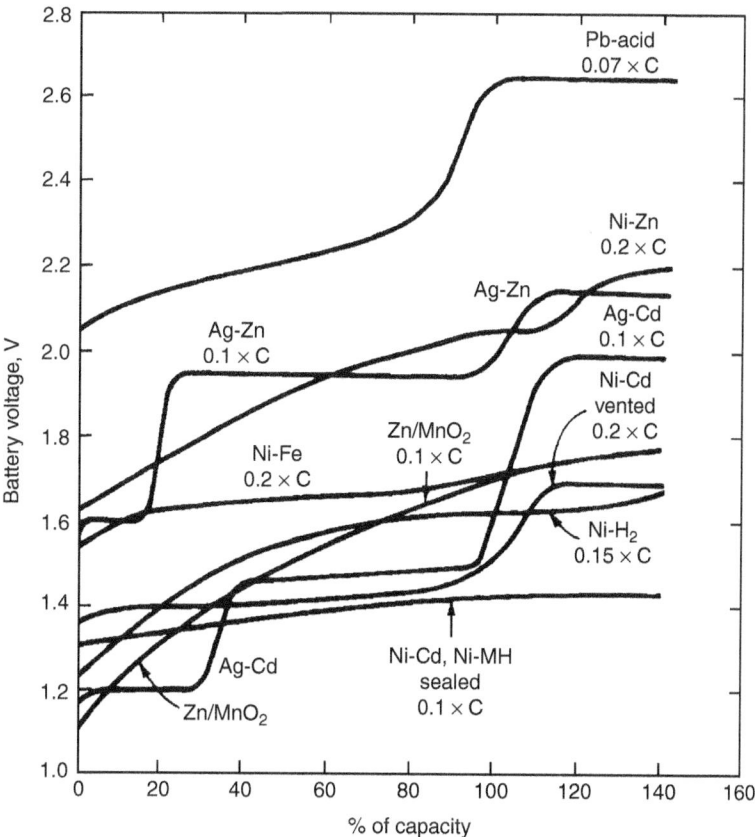

FIGURE 15.7 Typical charge characteristics of secondary battery systems, constant-current charge at 20°C. (*Adapted from Falk and Salkind. Ref. 5.*)

15.3.7 Charge Characteristics

The typical charge curves of the various secondary aqueous-systems at normal constant-current charge rates are shown in Fig. 15.7. Most of the batteries can be charged under constant-current conditions, which is usually the preferred method of charging, although, in practice, constant-voltage or modified constant-voltage methods are used. Some of the sealed batteries, however, may not be charged by constant-voltage methods because of the possibility of thermal runaway. Generally the vented nickel-cadmium battery has the most favorable charge properties and can be charged by a number of methods and in a short time. These batteries can be charged over a wide temperature range and can be overcharged to some degree without damage. Nickel-iron batteries, sealed nickel-cadmium batteries, and sealed nickel/metal hydride batteries have good charge characteristics, but the temperature range is narrower for these systems. The nickel/metal hydride battery is more sensitive to overcharge, and charge control to prevent overheating is advisable. The lead-acid battery also has good charge characteristics, but care must be taken to prevent excessive overcharging. The zinc/manganese dioxide and zinc/silver oxide batteries are most sensitive with regard to charging; overcharging is very detrimental to battery life. Figure 15.8 shows typical constant current–constant voltage charging characteristics of an 18650 lithium-ion battery.

15.18 SECONDARY BATTERIES

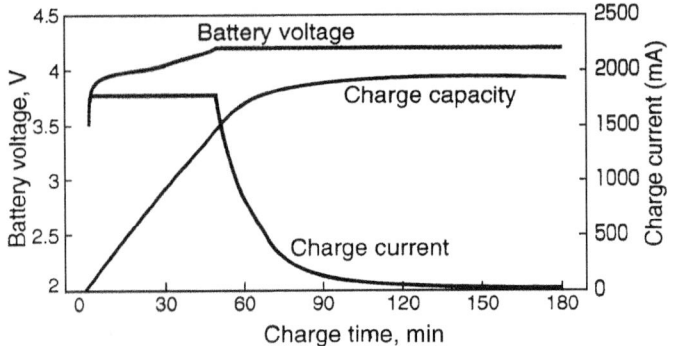

FIGURE 15.8 Charging characteristics of a typical cylindrical 18650 lithium-ion battery at 20°C. Battery is charged at constant current to 4.2 V followed by a taper charge at this voltage to a current limit.

Table 15.4 summarizes the typical conditions for charging the different systems. However, the chapters on individual battery systems and manufacturers' recommendations should be consulted because of the different procedures used.

Many manufacturers are now recommending "fast" charge methods to meet consumer and application demands for recharging in less than 2 to 3 h. These methods require control to cut off the charge before an excessive rise in gassing, pressure, or temperature occurs. These could cause venting or a more serious safety hazard, or they could result in a deleterious effect on the battery's performance or life. Pulse charging is also being employed with some systems to provide higher charge rates.

TABLE 15.4 Charging Characteristics of Secondary Batteries

System	Charged methods* Preferred	Charged methods* Not recommended	Recommended constant-current charge rate, C (A)	Overcharge-ability	Temperature range for charging, °C	Efficiencies† Ah, %	Efficiencies† Wh, %
Lithium-ion	cc, cv		0.20	None	0 to 50	99	95
Lead-acid							
Pasted, Planté	cc, cv		0.07	Fair	−40 to 50	90	75
Tubular	cc, cv		0.07	Fair	−40 to 50	80	70
Nickel-cadmium:							
Industrial vented	cc, cv		0.2	Very good	−50 to 40	70	60
Sintered vented	cc, cv		0.2	Very good	−55 to 75	70–80	60–70
Sealed	cc	cv	0.1–0.3‡	Very good	0 to 40	65–70	55–65
Nickel/metal hydride	cc	cv	0.1‡	Fair	0 to 40	65–70	55–65
Nickel-iron	cc	cv	0.2	Very good	0 to 45	80	60
Nickel-zinc	cc, cv		0.1–0.4	Fair	−20 to 40	85	70
Silver-zinc	cc		0.05–0.1	Poor	0 to 50	90	75
Silver-cadmium	cc		0.01–0.2	Fair	−40 to 50	90	70
Zn/MnO$_2$	cv	cc w/o v. limit		Fair	10–30		55–65

*Constant-current (cc) includes two-rate charging, and constant-voltage (cv) includes modified constant-voltage charging.
†All data are related to normal rates of charge and discharge and room-temperature operation.
‡Fast charge procedures can be used with charge control.
Source: Based on Falk and Salkind, Ref. 5.

In general, control techniques are useful for recharging most secondary batteries. They can be employed in several ways: to prevent overcharging, to facilitate "fast" charging, to sense when a potentially deleterious or unsafe condition may arise and cut off the charge or reduce the charging rate to safe levels. Similarly, discharge controls are also being used to maintain cell balance and to prevent overdischarge.

Another approach is the "smart" battery. These batteries incorporate a battery management system (BMS) with the following features:

1. To control the charge so that the battery can be charged optimally and safely and prevent overcharge or over-discharge
2. For fuel gauging to indicate the remaining charge left in the battery
3. Safety devices to alert the user to unsafe or undesirable operation or to cut off the battery from the circuit when these occur.

See Sec. 5.6 for more information.

15.3.8 Cost

The cost of a secondary battery may be evaluated on several bases, depending on the mode of operation. The initial cost is one of the bases for consideration. Other factors are the number of charge-discharge cycles that are available, or the number delivered in an application, during a battery's lifetime, or the cost determined on a dollar-per-cycle or dollar-per-total-kilowatt-hour basis. The cost of charging, maintenance, and associated equipment may also have to be considered in this evaluation. See Sec. 32.6.10. In an emergency standby service or SLI-type application, the important factors may be the calendar life of the battery (rather than the cycle life), and the cost is evaluated on a dollar-per-operating-year basis.

The lead-acid battery system is by far the least costly of the secondary batteries, particularly the SLI type. The lead-acid traction and stationary batteries, having more expensive constructional features and not as broad a production base, are several times more costly, but are still less expensive than the other secondary batteries. The nickel-cadmium and the rechargeable zinc/manganese dioxide batteries are next lowest in cost, followed by the nickel/metal hydride battery. The cost is very dependent on the cell size or capacity, the smaller button cells being considerably more expensive than the larger cylindrical and prismatic cells. The nickel-iron battery is more expensive and, for this reason among others, lost out to the less expensive battery system.

The most expensive of the conventional-type secondary batteries are the silver batteries. Their higher cost and low cycle life have limited their use to special applications, mostly in the military and space applications, which require their high energy and power densities. The nickel-hydrogen system is more expensive due to its pressurized design and a relatively limited production. However, their excellent cycle life under conditions of shallow discharge make them attractive for aerospace applications. The cost of cylindrical lithium-ion batteries has been decreasing rapidly as production rates have increased and has recently been stated to be $0.20/Wh.[3]

An important objective of the program for the development of secondary batteries for electric vehicles and energy storage is to reduce the cost of these battery systems. Cost factors are discussed in Chap. 29. For electric vehicles (HEV, PHEV, and EV types), the USABC goal for an EV battery is $150/KWh.

REFERENCES

1. A. J. Salkind, D. T. Ferrell, and A. J. Hedges, "Secondary Batteries 1952–1977," *J. Electrochem. Soc.* **125**(8), Aug. 1978.
2. V. Sapru. *Battery Power.* **14**(4):4–8 (2010).

3. H. Takeshita, *Proc. 27th Int. Battery Seminar*, Fort Lauderdale, FL (March 15–18, 2010).
4. L. H. Thaller, "Expected Cycle Life vs. Depth of Discharge Relationships of Well-Behaved Single Cells and Cell Strings," *J. Electrochem. Soc.* **130**(5), May 1983.
5. S. U. Falk and A. J. Salkind, *Alkaline Storage Batteries,* Wiley, New York, 1969.

CHAPTER 16
LEAD-ACID BATTERIES

Alvin Salkind and George Zguris

16.1 GENERAL CHARACTERISTICS

Lead-acid batteries are used extensively in telephone systems, power tools, automotive applications, communication devices, emergency lighting systems, and as the power source for mining and material-handling equipment. The wide use of the lead-acid battery in many designs, sizes, and system voltages is accounted for by the low price and the ease of manufacture on a local geographic basis of this battery system. The lead-acid battery is almost always the least expensive storage battery for any application, while still providing good performance and life characteristics. Lead-acid batteries top the list as highly recycled consumer products. More than 97% of all battery lead is recycled.[2]

The lead-acid battery continues to be the largest segment of the battery market in the world. Its production and use have continued to grow because of new applications for battery power in energy storage, emergency power, and electric and hybrid vehicles (including off-road vehicles) in addition to its traditional usage for automotive engine starting, vehicle lighting, and engine ignition (commonly called SLI).

The lead-acid battery market share of all secondary batteries was about 70% worldwide in 2008. The worldwide sales of rechargeable batteries in 2008 were $36 billion with a projected growth to $51 billion in 2013.[1] These numbers represent manufacturers' pricing levels. The values at retail would be two to three times these amounts. Because of the worldwide recession that started in 2008, these values are somewhat lower than those reported for recent previous years.

However, the lead-acid battery system is facing many competitive challenges in several markets from other battery chemistries. In its traditional markets, such as SLI motive power, and telecom, alternative options such as lithium chemistry designs are being investigated. This is especially true in the SLI sector where a transition in the world automotive market, with the help of new environmental legislation, is moving the car industry toward electric and various degrees of hybrid vehicles. The lead-acid battery has lost its position as the leading chemistry for electric and full hybrid vehicle platforms to lithium-ion and nickel metal hydride battery systems. However, newly reported advanced lead-acid battery designs may provide an opportunity for lead batteries to regain a strong market position. This may not happen in the OEM segment, but as a replacement battery pack, where cost to the consumer may be a greater driver than battery life or weight. These advanced lead-acid designs are mostly of the valve-regulated type of construction (see Chap. 17). They may be of a bipolar design to improve volume and energy density. One such design is from Effpower, shown in Fig. 16.1.[3–7]

Another promising development is the combination of the lead-acid battery with a carbon electrode material as part of the electrode arrangement. The UltraBattery® invented by CSIRO (Australia) and manufactured by Furukawa Battery Co. and East Penn, is one of the advanced lead-acid technologies that may find a market in the medium hybrid or even full plug-in hybrid[8] applications. The UltraBattery is being manufactured as both VRLA and as a flooded version.[9] The flooded version

16.2 SECONDARY BATTERIES

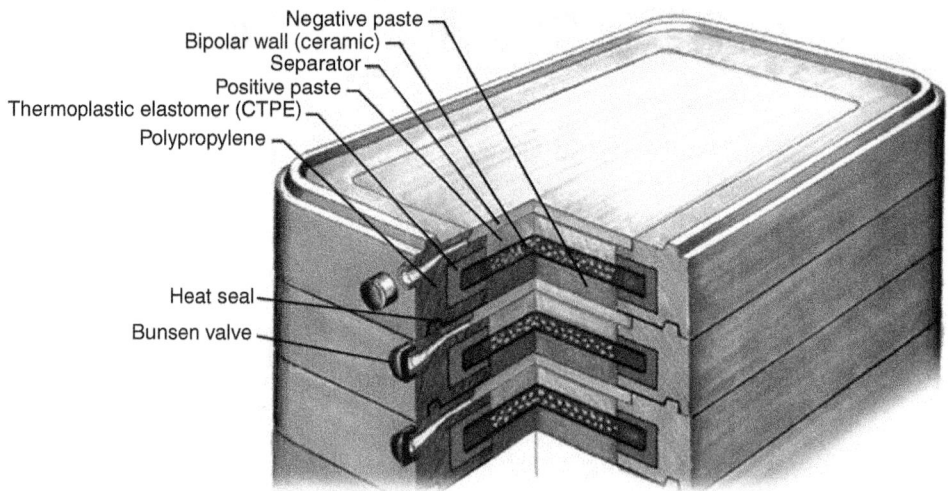

FIGURE 16.1 Effpower bipolar design.

falls into a new SLI category as an extended-life flooded (ELF) design. Extended flooded vehicle lead-acid batteries are being developed for new micro-hybrid vehicle application. This application has start-stop functionality (see Chap. 29). Traditional lead-acid batteries are not capable of meeting this type of duty cycle requirement. The battery must have suitable high-rate partial state-of-charge performance. A standard flooded battery has issues such as stratification and shedding of active material. The ELF battery has additional additives in the negative and positive plates and uses a scrim or separator to apply pressure to the active materials. Different lead alloys may also be used.[10] The ELF has been developed to provide for reduced performance but lower-cost design as compared to a VRLA-AGM (glass mat) battery that provides a longer life and safer battery. The European car industry is leading in this conversion to comply with environmental legislation requiring 120 grams of CO_2/km by 2012. Projections are that by 2013, the 185 million OEM batteries that will be used will be 4.5 million AGM, 4.8 million ELF, and 9.2 million standard flooded.[11]

The UltraBattery technology, along with many of the more significant improvements in the last 15 years, has been sponsored by the Advance Lead Acid Battery Consortium (ALABC). The ALABC is a research consortium originally formed in 1992 to advance the capabilities of the valve-regulated lead-acid battery to help electric vehicles become a reality and since then has adjusted the research focus to the current demands for the battery (Fig. 16.2).

Traditional vertical-plate design batteries are capable of specific energies greater than 40 Wh/kg. Horizontal-plate designs with higher energy and power have found use in traction and fork lift applications. Alternate systems in traditional markets such as fork trucks/material handling sectors are being challenged by other power sources such as fuel cells.[13] In 2008, this was a 717 million dollar market.[14]

The telecom market segment has had significant improvements, especially with the VRLA products, the major type of lead-acid batteries used in this application. With recent changes, made to IEC standards, these batteries for most applications must also meet certain cycle-life requirements (see Chap. 17). Competition is being offered in telecom applications by fuel cells, fly wheels (new), lithium-ion and nickel metal hydride chemistries.

The lead-acid segment is being challenged with change as the other power storage devices attempt to win market share for the secondary battery applications. This technology is moving toward higher performing advanced lead-acid, VRLA, and extended life designs, which would give this battery chemistry a continuing future.

The overall advantages and disadvantages of the lead-acid battery, compared with other systems, are listed in Table 16.1.

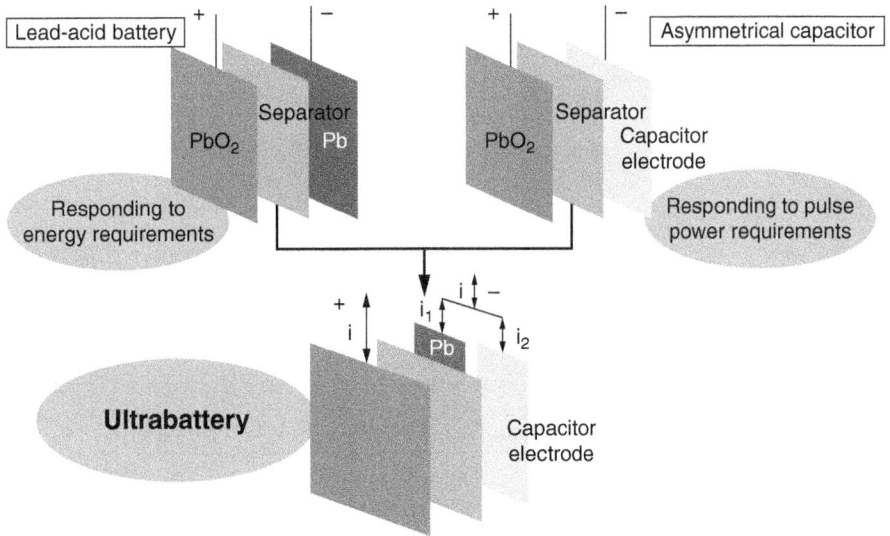

FIGURE 16.2 Diagram of the UltraBattery concept.[12]

TABLE 16.1 Major Advantages and Disadvantages of Lead-Acid Batteries

Advantages	Disadvantages
Popular low-cost secondary battery—capable of manufacture on a local basis, worldwide, from low to high rates of production	Relatively low cycle life (50–500 cycles)*
Available in large quantities and in a variety of sizes and designs—manufactured in sizes from smaller than 1 Ah to several thousand Ah	Limited energy density—typically 30–40 Wh/kg
	Long-term storage in a discharged condition can lead to irreversible polarization of electrodes (sulfation)
	Difficult to manufacture in very small sizes
Good high-rate performance—suitable for engine starting (but outperformed by some nickel-cadmium and nickel metal-hydride batteries)	Hydrogen evolution in some designs can be an explosion hazard (flame arrestors are installed to prevent this hazard)
Moderately good low- and high-temperature performance	Stibene and arsine evolution in designs with antimony and arsenic in grid alloys can be a health hazard
Electrically efficient—turnaround efficiency of over 70%, comparing discharge energy out with charge energy in	Thermal runaway in improperly designed batteries or charging equipment
High cell voltage—open-circuit voltage of >2.0 V is the highest of all aqueous-electrolyte battery systems	Positive post blister corrosion with some designs
Good float service	
Easy state-of-charge indication	
Good charge retention for intermittent charge applications (if grids are made with high-overvoltage alloys)	
Available in maintenance-free designs	
Low cost compared with other secondary batteries	
Cell components are easily recycled	

*Up to 2000 cycles can be attained with special designs.

16.4 SECONDARY BATTERIES

TABLE 16.2 Types and Characteristics of Lead-Acid Batteries

Type	Construction	Typical applications
SLI (starting, lighting, ignition)	Flat-pasted plates (option: maintenance-free construction)	Automotive, marine, aircraft, diesel engines in vehicles and for stationary power
Traction	Flat-pasted plates; tubular and gauntlet plates	Industrial trucks (material handling)
Vehicular propulsion	Flat-pasted plates; tubular and gauntlet plates; also composite construction	Electric vehicles, golf carts, hybrid vehicles, mine cars, personnel carriers
Submarine	Tubular plates; flat-pasted plates	Submarines
Stationary (including energy-storage types such as charge retention, solar photovoltaic, load leveling)	Planté;* Manchester;* tubular and gauntlet plates; flat-pasted plates; circular conical plates	Standby emergency power: telephone exchange, uninterruptible power supply (UPS), load leveling, signaling
Portable (see Chap. 17)	Flat-pasted plates (gelled electrolyte, electrolyte absorbed in separator); spirally wound electrodes; tubular plates	Consumer and instrument applications: portable tools, appliances, lighting, emergency lighting, radio, TV, alarm systems

*Now rarely used.

The lead-acid battery is manufactured in a variety of sizes and designs, ranging from less than 1 to over 10,000 Ah. Table 16.2 lists many of the various types of lead-acid batteries that are available.

16.1.1 History

Practical lead-acid batteries began with the research and inventions of Raymond Gaston Planté in 1860, although batteries containing sulfuric acid or lead components were discussed earlier.[15] Table 16.3 lists the events in the technical development of the lead-acid battery. In Planté's fabrication method, two long strips of lead foil and intermediate layers of coarse cloth were spirally wound and immersed in a solution of about 10% sulfuric acid. The early Planté cells had little capacity, since the amount of stored energy depended on the corrosion of a lead foil to lead dioxide to form the positive active material, and similarly the negative electrode was formed by roughening of another foil (on cycling) to form an extended surface. Primary cells were used as the power sources for this formation. The capacity of Planté cells was increased on repeated cycling as corrosion of the substrate foils created more active material and increased the surface area. In the 1870s, magnetoelectric generators became available to Planté, and about this time the Siemens dynamo began to be installed in central electric plants. Lead-acid batteries found an early market to provide load leveling and to average out the demand peaks. They were charged at night, similar to the procedure now planned for modern load-leveling energy-storage systems.

Subsequent to Planté's first developments, numerous experiments were done on accelerating the formation process and coating lead foil with lead oxides on a lead plate pretreated by the Planté method. Attention then turned to other methods for retaining active material, and two main technological paths evolved.

1. Coating a lead oxide paste on cast or expanded grids, rather than foil, in which the active material developed structural strength and retention properties by a "cementation" process (interlocked crystalline lattice) through the grid and active mass. This is generally referred to as flat-plate design.
2. The tubular electrode design, in which a central conducting wire or rod is surrounded by active material and the assembly encased in an electrolyte porous insulated tube, which can be either square, round, or oval.

TABLE 16.3 Events in Technical Development of Lead-Acid Battery

		Precursor systems
1836	Daniell	Two-fluid cell; copper/copper sulfate/sulfuric acid/zinc
1840	Grove	Two-fluid cell; carbon/fuming nitric acid/sulfuric acid/zinc
1854	Sindesten	Polarized lead electrodes with external source
		Lead-acid battery developments
1860	Planté	First practical lead-acid battery, corroded lead foils to form active material
1881	Faure	Pasted lead foils with lead oxide-sulfuric acid pastes for positive electrode, to increase capacity
1881	Sellon	Lead-antimony alloy grid
1881	Volckmar	Perforated lead plates to provide pockets for support of oxide
1882	Brush	Mechanically bonded lead oxide to lead plates
1882	Gladstone and Tribs	Double sulfate theory of reaction in lead-acid battery: $PbO_2 + Pb + 2H_2SO_4 \rightleftharpoons 2PbSO_4 + 2H_2O$
1883	Tudor	Pasted mixture of lead oxides on grid pretreated by Planté method
1886	Lucas	Formed lead plates in solutions of chlorates and perchlorates
1890	Phillipart	Early tubular construction—individual rings
1890	Woodward	Early tubular construction
1910	Smith	Slotted rubber tube, Exide tubular construction
1920 to present		Materials and equipment research, especially expanders, oxides, and fabrication techniques
1935	Haring and Thomas	Lead-calcium alloy grid
1935	Hamer and Harned	Experimental proof of double sulfate theory of reaction
1956–1960	Bode and Voss Ruetschi and Cahan Burbank Feitknecht	Clarification of properties of two crystalline forms of PbO_2 (alpha and beta)
1970s	McClelland and Devit	Commercial spiral-wound sealed lead acid battery
		Expanded metal grid technology; composite plastic/metal grids; sealed and maintenance-free lead-acid batteries; glass fiber and improved separators; through-the-partition intercell connectors; heat-sealed plastic case-to-cover assemblies; high-energy-density batteries (above 40 Wh/kg); conical grid (round) cell for long-life float service in telecommunications facilities
1980s		Sealed valve-regulated batteries; quasi-bipolar engine starter batteries; improved low-temperature performance; world's largest battery installed (Chino, CA); 40 MWh lead-acid load leveling
1990s		Electric-vehicle interest reemerges; bipolar battery designs for high-power use in uninterruptible power supplies, power tool market, and electronic backup; thin foil cells, small cells for consumer and current road applications
2009		Development of lead-carbon batteries, extended life flooded batteries for micro-hybrids, improved high-rate partial state-of-charge (HRPSOC) VRLA, micro-hybrids for stop-start application, using bipolar battery

Simultaneous with the advances in developing and retaining active material was work in strengthening the grid by casting it from lead alloys such as lead-antimony (e.g., Sellon, 1881) or lead-calcium (e.g., Haring and Thomas, 1935).[16] The technical knowledge for an economical manufacture of reliable lead-acid batteries was in place by the end of the nineteenth century, and subsequent growth of the industry was rapid. Improvements in design, manufacturing equipment and methods, recovery methods, active material utilization and production, supporting structures and components, and nonactive components such as separators, cases, and seals continue to improve the economic and performance characteristics of lead-acid batteries. Battery development continues with the main focus towards the growing hybrid car market. Work sponsored by the AIABC on carbon and other additives in the active materials has improved the battery charge acceptance for improved partial state-of-charge performance. Lead-carbon batteries and bipolar designs continue to demonstrate advanced battery performance.

16.1.2 Manufacturing Data and Battery Usage

The largest use of flooded lead-acid batteries is in automotive vehicle applications for starting, lighting, and ignition purposes (referred to as SLI). Similar usage is prevalent in aircraft, boats, off-road, and farm equipment vehicles. The increased use of electronics in most modern cars has also resulted in increased electrical capacity (Ah). The 12 V systems now commonly used are designed with capacities in the 40 to 100 Ah range, and weigh between 11 and 45 Kg. The high-rate current capability, in the standard cold crank test, can be as high as 900 A.

The number of registered on-road vehicles in the United States has been steadily increasing for decades, as shown in Table 16.4a, and was approximately 260 million vehicles in 2010. There were many million other vehicles (planes, boats, off-road) with similar battery needs. The total number of vehicles in use in the United States in 2010 is estimated to be 300 million. Although in the 1920s the United States represented the bulk of the automotive market, by the 2010s the use of cars had spread throughout the world and the U.S. share was in the 25 to 30% range. The trend toward vehicle use in China, India, Europe, and elsewhere appears to be accelerating, along with technology infrastructure and population. An oddity of the U.S. market is that there are more registered vehicles than licensed drivers. Although influenced by environmental temperature, driving distance, and stop-start driving patterns in a given year, the usage pattern of SLI batteries has been observed to be generally in the range of 38 to 41 SLI units marketed (replacement and new vehicle) for every 100 registered MVS.

TABLE 16.4a Number of Registered Vehicles in the United States: Automobiles, Buses, and Trucks from 1970–2006 in Millions

Year	Autos	Buses	Trucks	Total
1970	89.24	0.38	18.8	108.42
1980	121.6	0.53	33.67	155.8
1990	133.7	0.63	54.47	188.8
2000	133.62	0.75	87.11	221.48
2006	135.4	0.88	107.94	244.16

Source: USDOT, Federal Highway Administration

The growing markets and applications in China and India, the two must populous countries in the world, are extremely important factors. China in 2009 had equivalent OEM sales of cars as the North American Region (U.S. and Canada). China also has a substantial E-bike market with sales in 2007 and 2008 of about 21 million bikes. In 2010, there were about 100 million E-bikes is use in China with usage growing. In order to avoid spills, the batteries in these bikes are generally of the VRLA design. At present, the lead-acid system has about a 95% market share, but lithium-ion and other systems are attempting to provide an appropriate product for this application.

Data on the market growth of lead-acid batteries are given in Table 16.4b. The automobile industry is moving toward the production of greener designs. Various platform technologies are described below:

- *Micro-hybrids:* This application is still expected to utilize a lead-acid product, but the advanced extended-life flooded (ELF) and AGM-VRLA designs may be more significant in the future.
- *Mild-hybrids:* Currently Ni-MH, but lithium-ion batteries are under test.

TABLE 16.4b Market Growth of Lead-Acid Batteries in United States*

	1960	1980	1991	1999*	2010 (Est)
SLI units (original equipment and replacement)	34	62	76	100	120
SLI sales, $	330	1675	2100	2700	3000
Industrial, $	70	380	550	1015	1500
Consumer, $	1	55	100	150	200
Total, $	400	2110	2750	3965	4700

*All units in millions values are at manufacturers' pricing. Battery prices are affected by the price of lead. Lead prices varied from $0.40 to over $3.00/kg between 1978 and 2009 (lead price Dec. 2009 = $2.3 kg.)
Source: London Metal Exchange

- *Full hybrids:* Same situation as mild hybrids.
- *Plug-in hybrids:* Lithium-ion batteries.
- *EV:* Lithium-ion batteries.

See Chap. 29 for more detailed information on hybrid and electric vehicles.

Future lead-acid battery production volumes will also be determined by the penetration of hybrids and EVs. One prediction by an authoritative source is that by 2015 there will be 18 million hybrid vehicles, 78% of them micro-hybrids and 22% all other types. The European market appears to be taking the leading role for the development of micro-hybrids. Governmental legislation in Europe has put a 120 g CO_2/km fleet requirement by 2012, and various European national taxation rules have provided incentives for low CO_2 emitting vehicles.

In addition to automotive use, lead-acid batteries are widely used in industrial and consumer applications. The industrial designs are discussed under traction batteries and standby power in Secs. 16.5 and 16.6. The small consumer design product is discussed in sec. 16.9.2. The industrial designs have a higher cost than the SLI designs per unit of capacity because of more expensive materials, longer warranties, and less automation in production. The overall markets for the industrial battery designs is in the range of 40 to 50% of the market for SLI designs. The use of small sealed lead-acid cells in consumer designs is much smaller.

Overall, it appears that VRLA designs using AGM mats will garner a greater percentage of the total lead-acid market in the future. It should be noted that the lead-acid battery is the easiest battery to recycle of all battery systems.

16.2 CHEMISTRY

16.2.1 General Characteristics

The lead-acid battery uses lead dioxide as the active material of the positive electrode and metallic lead, in a high-surface-area porous structure, as the negative active material. The physical and chemical properties of these materials are listed in Table 16.5.[17] Typically, a charged positive electrode contains both α-PbO_2 (orthorhombic) and β-PbO_2 (tetragonal). The equilibrium potential of the α-PbO_2 is more positive than that of β-PbO_2 by 0.01 V. The α form also has a larger, more compact crystal morphology which is less active electrochemically and slightly lower in capacity per unit weight; it does, however, promote longer cycle life. Neither of the two forms is fully stoichiometric. Their composition can be represented by PbO_x, with x varying between 1.85 and 2.05. The introduction of antimony, even at low concentrations, in the preparation or cycling of these species leads to a considerable increase in their performance. The preparation of the active material precursor consists of a series of mixing and curing operations using leady lead oxide (PbO + Pb), sulfuric acid, and water.

TABLE 16.5 Physical and Chemical Properties of Lead and Lead Oxides (PbO_2)

Property	Lead	α-PbO_2	β-PbO_2
Molecular weight, g/mol	207.2	239.19	239.19
Composition		$PbO_{1.94-2.03}$	$PbO_{1.87-2.03}$
Crystalline form	Face-centered cubic	Rhombic (columbite)	Tetragonal (rutile)
Lattice parameters, nm	$a = 0.4949$	$a = 0.4977$	$a = 0.491-0.497$
		$b = 0.5948$	$c = 0.337-0.340$
		$c = 0.5444$	
X-ray density, g/cm³	11.34	9.80	~9.80
Practical density at 20°C (depends on purity), g/cm³	11.34	9.1–9.4	9.1–9.4
Heat capacity, cal/deg·mol	6.80	14.87	14.87
Specific heat, cal/g	0.0306	0.062	0.062
Electrical resistivity, at 20°C, µΩ/cm	20	~100 × 10³	
Electrochemical potential in 4.4M H_2SO_4 at 31.8°C, V	0.356	~1.709	~1.692
Melting point, °C	327.4		

Source: Ref. 17 and others.

The ratios of the reactants and curing conditions (temperature, humidity, and time) affect the development of crystallinity and pore structure. The cured plate consists of lead sulfate, lead oxide, and some residual lead (<5%). The positive active material, which is formed electrochemically from the cured plate, is a major factor influencing the performance and life of the lead-acid battery. In general the negative, or lead, electrode controls cold-temperature performance (such as engine starting).

The electrolyte is a sulfuric acid solution, typically about 1.28 specific gravity or 37% acid by weight in a fully charged condition.

As the cell discharges, both electrodes are converted to lead sulfate. The process reverses on charge:

Negative electrode
$$Pb \underset{\text{charge}}{\overset{\text{discharge}}{\rightleftarrows}} Pb^{2+} + 2e$$

$$Pb^{2+} + SO_4^{2-} \underset{\text{charge}}{\overset{\text{discharge}}{\rightleftarrows}} PbSO_4$$

Positive electrode
$$PbO_2 + 4H^+ + 2e \underset{\text{charge}}{\overset{\text{discharge}}{\rightleftarrows}} Pb^{2+} + 2H_2O$$

$$Pb^{2+} + SO_4^{2-} \underset{\text{charge}}{\overset{\text{discharge}}{\rightleftarrows}} PbSO_4$$

Overall reaction
$$Pb + PbO_2 + 2H_2SO_4 \underset{\text{charge}}{\overset{\text{discharge}}{\rightleftarrows}} 2PbSO_4 + 2H_2O$$

As shown, the basic electrode processes in the positive and negative electrodes involve a dissolution-precipitation mechanism and not a solid-state ion transport or film formation mechanism.[17] The discharge-charge mechanism, known as double-sulfate reaction, is also shown graphically in Fig. 16.3.[18] As the sulfuric acid in the electrolyte is consumed during discharge, producing water, the electrolyte is an "active" material and in certain battery designs can be the capacity-limiting material. This capacity-limiting effect of the electrolyte is an important design consideration in VRLA batteries.

As the cell approaches full charge and the majority of the $PbSO_4$ has been converted to Pb or PbO_2, the cell voltage on charge becomes greater than the gassing voltage (about 2.39 V per cell)

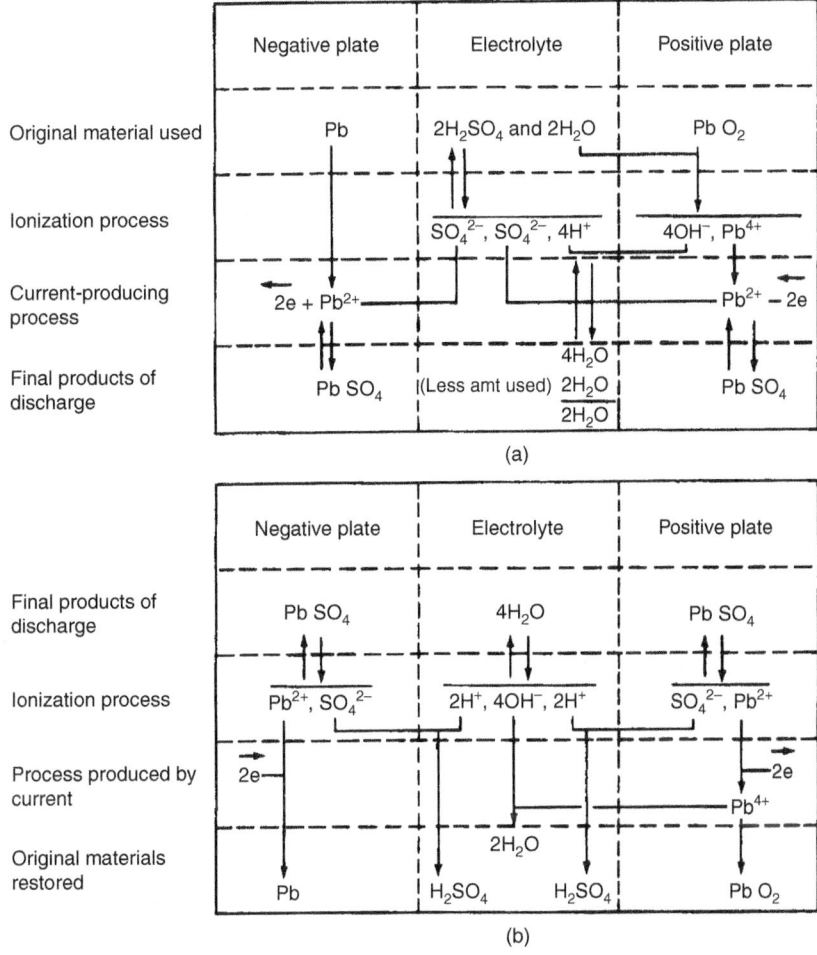

FIGURE 16.3 Discharge and charge reactions of lead-acid cell. (*a*) Discharge reactions. (b) Charge reactions. (*From Ref. 18.*)

and the overcharge reactions begin, resulting in the production of hydrogen and oxygen (gassing) and the resultant loss of water.

Negative electrode	$2H^+ + 2e \rightarrow H_2$
Positive electrode	$H_2O - 2e \rightarrow \frac{1}{2}O_2 + 2H^+$
Overall reaction	$H_2O \rightarrow H_2 + \frac{1}{2}O_2$

In sealed lead-acid cells this reaction is controlled to minimize hydrogen evolution and the loss of water by recombination of the evolved oxygen with the negative plate (see Sec. 17.2).

The general performance characteristics of the lead-acid cell, during charge and discharge, are shown in Fig. 16.4. As the cell is discharged, the voltage decreases due to depletion of material, internal resistance losses, and polarization. If the discharge current is constant, the voltage under

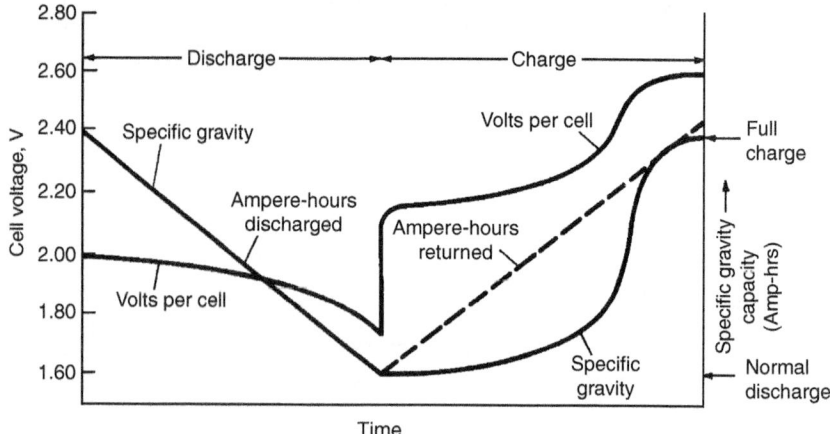

FIGURE 16.4 Typical voltage and specific gravity characteristics of lead-acid cell at constant-rate discharge and charge.

load decreases smoothly to the cutoff voltage and the specific gravity decreases in proportion to the ampere-hours discharged.

An analysis of the behavior of the positive and negative plates can be done by measuring the voltage between each electrode and a reference electrode, the "half-cell" voltage. Figure 16.5 illustrates this analysis, using a cadmium reference electrode. The industry is, however, shifting away from cadmium reference electrodes to more stable materials, as discussed later (Sec. 16.2.3).

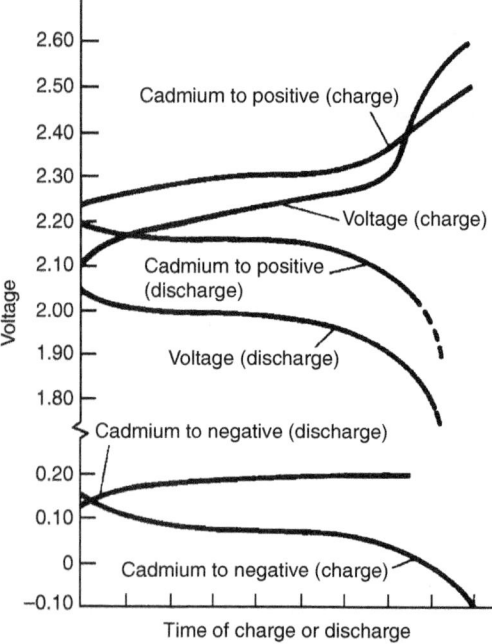

FIGURE 16.5 Typical charge-discharge curves of lead-acid cell. (*From Ref. 18.*)

Voltage. The nominal voltage of the lead-acid cell is 2 V; the voltage on open circuit is a direct function of the electrolyte concentration, ranging from 2.125 V for a cell with 1.28 specific gravity electrolyte to 2.05 V with 1.21 specific gravity (see Sec. 16.2.2). The end or cutoff voltage on moderate-rate discharges is 1.75 V per cell, but may range to as low as 1.0 V per cell at extremely high discharge rates at low temperatures.

Specific Gravity. The selection of specific gravity (relative density) used for the electrolyte depends on the application and service requirements (see Table 16.12). The electrolyte concentration must be high enough for good ionic conductivity and to fulfill electrochemical requirements, but not so high as to cause separator deterioration or corrosion of other parts of the cell, which would shorten life and increase self-discharge. The electrolyte concentration is deliberately reduced in high-temperature climates. During discharge, the specific gravity decreases in proportion to the ampere-hours discharged (Table 16.6). The specific gravity is thus a means for checking the state of charge of the battery. On charge, the change in specific gravity should similarly be proportional to the ampere-hour charge accepted by the cell. However, there is a lag because complete mixing of the electrolyte does not occur until gassing commences near the end of the charge.

TABLE 16.6 Specific Gravity of Lead-Acid Battery Electrolytes at Different States of Charge for Various Designs*

State of charge	Specific gravity			
	A	B	C	D
100% (full charge)	1.330	1.280	1.265	1.225
75%	1.300	1.250	1.225	1.185
50%	1.270	1.220	1.190	1.150
25%	1.240	1.190	1.155	1.115
Discharged	1.210	1.160	1.120	1.0

Assumes flooded cell design.
*Specific gravity may range from 100 to 150 points between full charge and discharge depending on cell design: A—electric vehicle battery; B—traction battery; C—SLI battery; D—stationary battery.

16.2.2 Open-Circuit Voltage Characteristics

The open-circuit voltage for a battery system is a function of temperature and electrolyte concentration as expressed in the Nernst equation for the lead-acid cell (see also Chap. 2).

$$E = 2.047 + \frac{RT}{F} \ln\left(\frac{\alpha H_2SO_4}{\alpha H_2O}\right)$$

Since the concentration of the electrolyte varies, the relative activities of H_2SO_4 and H_2O in the Nernst equation change. A graph of the open-circuit voltage versus electrolyte concentration at 25°C is given in Fig. 16.6. The plot is fairly linear above 1.10 specific gravity, but shows strong deviations at lower concentrations. The open-circuit voltage is also affected by temperature. The temperature coefficient of the open-circuit voltage of the lead-acid battery is shown in Fig. 16.7. Where dE/dT is positive, such as above 0.5 Molar H_2SO_4, the reversible potential of the system increases with increasing temperature. Below 0.5 M, the temperature coefficient is negative. Most lead-acid batteries operate above 2 Molar H_2SO_4 (1.120 specific gravity) and have a thermal coefficient of about +0.2 mV/°C.

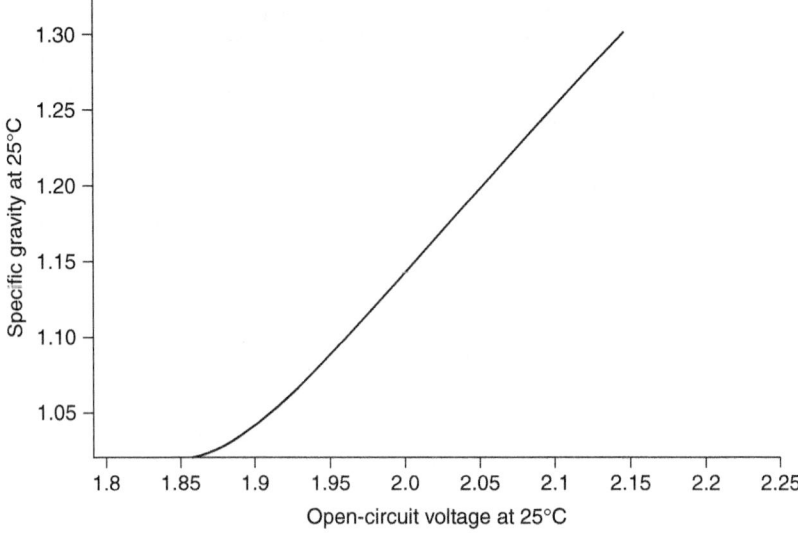

FIGURE 16.6 Open-circuit voltage of lead-acid cell as a function of electrolyte specific gravity.

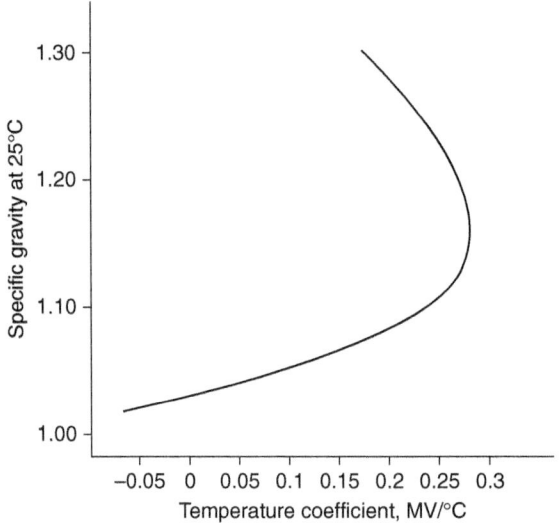

FIGURE 16.7 Temperature coefficient of open-circuit voltage of lead-acid cell as a function of electrolyte specific gravity.

16.2.3 Polarization and Resistive Losses

When a battery is being discharged, the voltage under load is lower than the open-circuit voltage at the same concentrations of H_2SO_4 and H_2O in the electrolyte and Pb or PbO_2 and $PbSO_4$ in the plates. The thermodynamically stable state for batteries is the discharged state. Work (charging) must be done to cause the equilibria of the electrochemical reactions to go toward PbO_2 in the positive and

Pb in the negative. Thus the voltage of the power source for recharging the lead-acid battery must be higher than the Nernst voltage of the battery on open circuit.

These deviations from the open-circuit voltage during charge or discharge are due, in part, to resistive losses in the battery and, in part, to polarization. These losses can be measured by use of an interrupted discharge, where the *IR* losses can be estimated by Ohm's law ($\Delta E/\Delta I = R$) within a few seconds to a few minutes after the discharge is stopped. The effect of polarization can take several hours to measure in order for diffusion to allow the plate interiors to reequilibrate. AC impedance spectroscopy techniques are also of value. Polarization is more easily measured by use of a reference electrode. The standard reference of hydrogen on platinum is not practical for most measurements on lead-acid batteries, and several other sulfate-based reference electrode systems are used. A review of reference electrodes[19] neglects several very practical sulfate electrodes. Still, a commonly used electrode for battery maintenance is the cadmium "stick," but it is not especially stable (± 20 mV/day). Mercury-mercurous sulfate reference electrodes are stable and are available from several vendors. A novel $Pb/H_2SO_4/PbO_2$ reference electrode has been patented.[20] This electrode measures the polarization on charge or discharge directly, without need for a correction of different thermal coefficients of EMF. The change in polarization between the start and the end of discharge is typically 50 to several hundred mV, and the cell capacity is limited by the plate group (positive or negative) that has the largest change in polarization during discharge. When both groups in a cell change about equally, the capacity limitation is more likely depletion of H_2SO_4 in the electrolyte than depletion of Pb or PbO_2 in the plates. On charge, the polarization is a good measure that both positives and negatives have been recharged: the plate polarizations change by more than 60 mV between start and end of recharge. Polarization voltages stabilize at some value when plates are recharged and are gassing freely.

16.2.4 Self-Discharge

The equilibria of the electrode reactions are normally in the discharge direction since, thermodynamically, the discharged state is more stable. The rate of self-discharge (loss of capacity [charge] when no external load is applied) of the lead-acid cell is fairly rapid, but it can be reduced significantly by incorporating certain design features.

The rate of self-discharge depends on several factors. Lead and lead dioxide are thermodynamically unstable in sulfuric acid solutions, and on open circuit, they react with the electrolyte. Oxygen is evolved at the positive electrode and hydrogen at the negative, at a rate dependent on temperature and acid concentration (the gassing rate increases with increasing acid concentration) as follows:

$$PbO_2 + H_2SO_4 \rightarrow PbSO_4 + H_2O + \tfrac{1}{2}O_2$$
$$Pb + H_2SO_4 \rightarrow PbSO_4 + H_2$$

For most positives, the formation of $PbSO_4$ by self-discharge is slow, typically much less than 0.5%/day at 25°C. (Some positives that have been made with nonantimonial grids can fail by a different mechanism on open circuit, namely, the development of a grid-active material barrier layer.) The self-discharge of the negative is generally more rapid, especially if the cell is contaminated with various catalytic metallic ions. For example, antimony lost from the positive grids by corrosion can diffuse to the negative, where it is deposited, resulting in a "local action" discharge cell which converts some lead active material to $PbSO_4$. New batteries with lead-antimony grids lose about 1% of charge per day at 25°C, but the charge loss increases by a factor of 2 to 5 as the battery ages. Batteries with nonantimonial lead grids lose less than 0.5% of charge per day regardless of age. This is illustrated in Fig. 16.8a.[21] Maintenance-free and charge-retention-type batteries, where the self-discharge rate must be minimized, use low-antimony or antimony-free alloy (such as calcium-lead) grids. However, because of other beneficial effects of antimony, its complete elimination may not be desirable, and low-antimony–lead alloys are a useful compromise.

16.14 SECONDARY BATTERIES

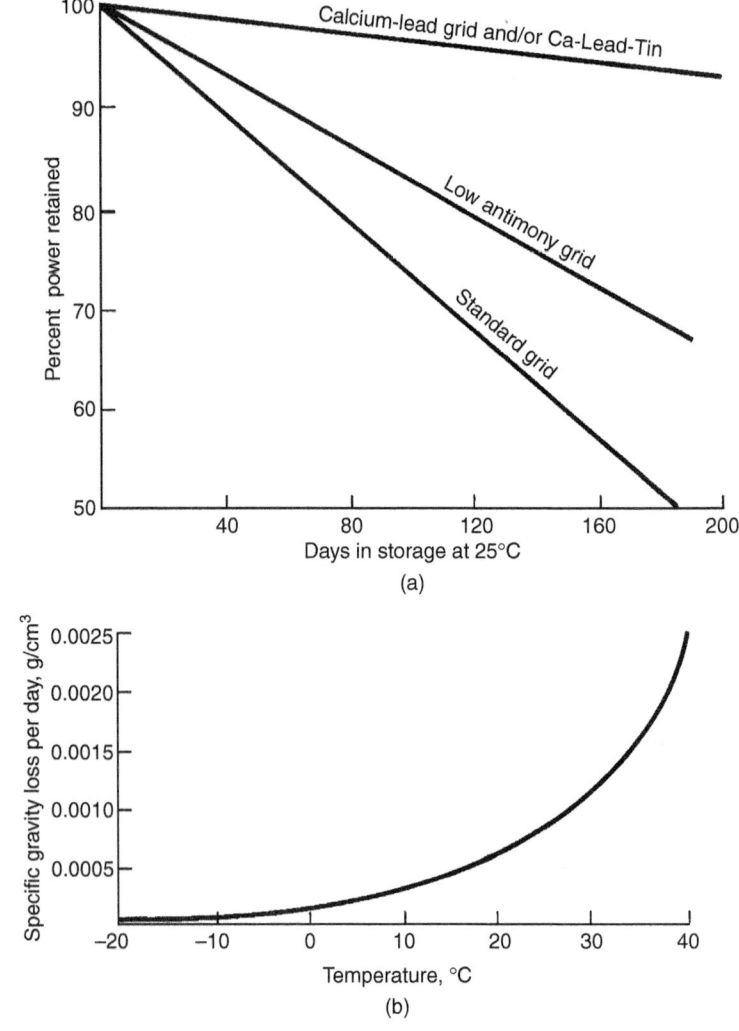

FIGURE 16.8 (*a*) Capacity retention during stand or storage at 25°C. (*From Ref. 21.*) (*b*) Loss of specific gravity per day with temperature of a new, fully charged lead-acid battery with 6% antimonial lead grids. (*From Ref. 22.*)

Self-discharge is temperature-dependent, as shown in Fig. 16.8*b*.[22] The graph shows the fall in specific gravity per day of a new fully charged battery with 6% antimonial lead grids. Self-discharge can thus be minimized by storing batteries at temperatures between 5 and 15°C.

16.2.5 Characteristics and Properties of Sulfuric Acid

The major characteristics and properties of the sulfuric acid electrolyte, as they apply to the operation of the lead-acid battery, are listed in Table 16.7. The freezing points of sulfuric acid solutions

TABLE 16.7 Properties of Sulfuric Acid Solutions*

Specific gravity		Temperature coeff. α	H_2SO_4			Freezing point, °C	Elecro-chemical equivalent (per liter of acid), Ah
At 15°C	At 25°C		Wt., %	Vol., %	Mol/L		
1.00	1.000	—	0	0	0	0	0
1.05	1.049	33	7.3	4.2	0.82	−3.3	22
1.10	1.097	48	14.3	8.5	1.65	−7.7	44
1.15	1.146	60	20.9	13.0	2.51	−15	67
1.20	1.196	68	27.2	17.7	3.39	−27	90
1.25	1.245	72	33.2	22.6	4.31	−52	115
1.30	1.295	75	39.1	27.6	5.26	−70	141
1.35	1.345	77	44.7	32.8	6.23	−49	167
1.40	1.395	79	50.0	38.0	7.21	−36	
1.45	1.445	82	55.0	43.3	8.2	−29	
1.50	1.495	85	59.7	48.7	9.2	−29	

*To calculate the specific gravity for any temperature, °C, SG (t) = SG (15°C) + α × 10^{-5} (15 − t).

at various concentrations are also plotted in Fig. 16.9a. The freezing point of aqueous sulfuric acid solutions varies significantly with concentration. Batteries must therefore be designed so that the electrolyte concentration is above the value at which the electrolyte would freeze when exposed to the anticipated cold. Alternatively, the battery can be insulated or heated so that it remains above the electrolyte freezing temperature.

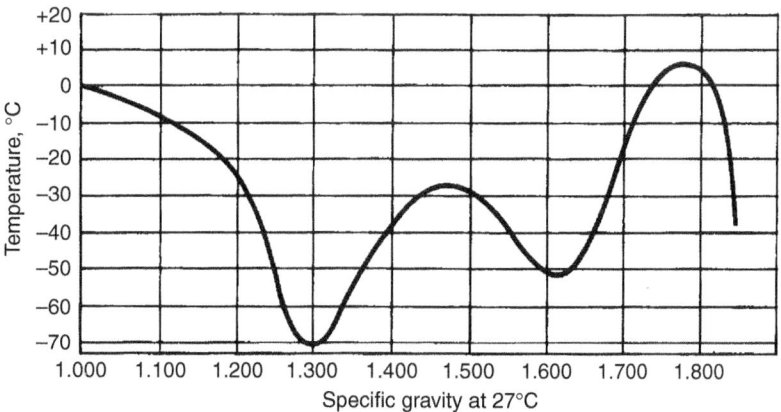

FIGURE 16.9a Freezing points of sulfuric acid solutions at various specific gravities. The inflection points result from the different water to SO_3 hydration ratios.

Figure 16.9b shows the specific resistivity of sulfuric acid solutions at various specific gravities as a function of temperature from −40 to 40°C.

The specific gravities for several types of lead-acid battery designs are given in Table 16.12; the change in specific gravity at different states of charge is shown in Table 16.6. A comparison with freezing-point data will show that battery type A will freeze at −30°C when fully discharged, while battery type D will freeze at about −5°C, a factor which must be considered in the design of the battery and the battery housing. The acid concentration for most lead-acid batteries for use in

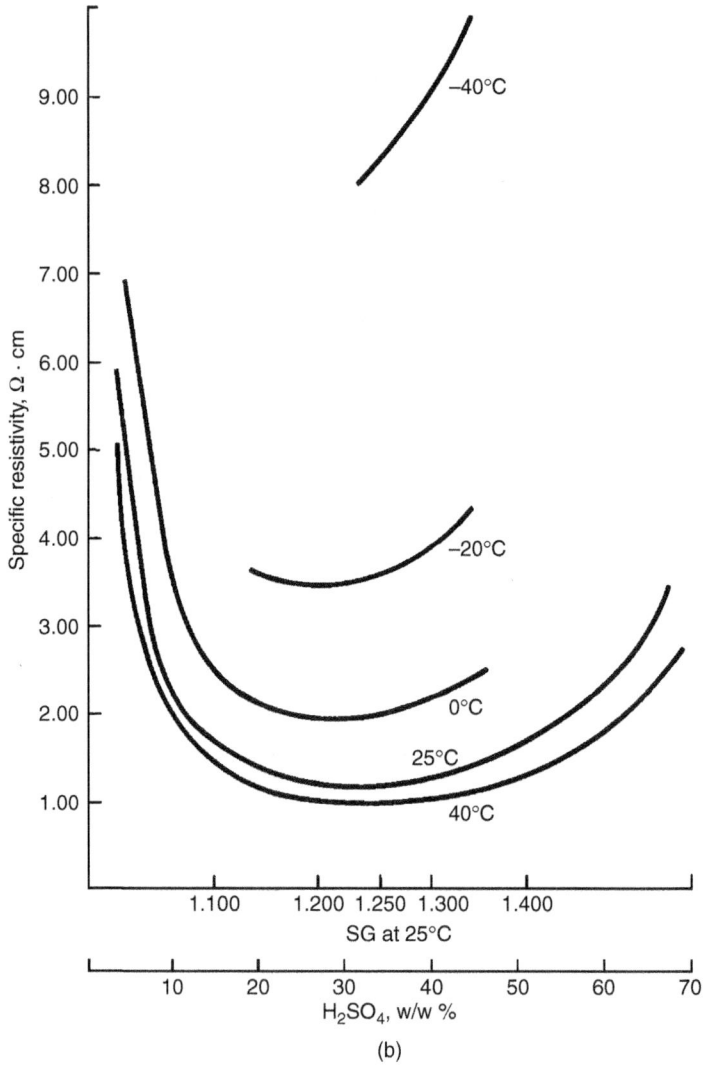

FIGURE 16.9b Specific resistivity of sulfuric acid solutions at various specific gravities and temperatures.

temperate climates is usually between 1.26 and 1.28 specific gravity. Higher-concentration electrolytes tend to attack some separators and other components; lower concentrations tend to be insufficiently conductive in a partially charged cell and freeze at low temperatures. In high-temperature climates, a lower concentration is used, and in stationary cells with larger proportional electrolyte volumes and no high-rate demands, electrolytes with specific gravity as low as 1.21 are used (see Table 16.12).

Figure 16.9c indicates the method of preparing sulfuric acid solutions of any specific gravity from concentrated sulfuric acid.

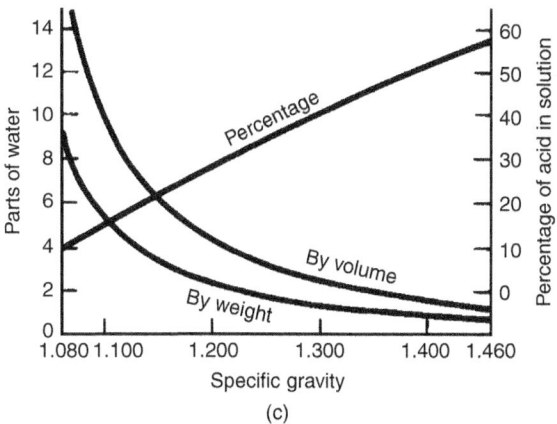

FIGURE 16.9c Preparation of sulfuric acid solutions of any specific gravity from concentrated sulfuric acid. (*From G. W. Vinal, Storage Batteries, 4th ed., Wiley, New York, 1955, p. 129.*)

16.3 CONSTRUCTIONAL FEATURES, MATERIALS, AND MANUFACTURING METHODS

Lead-acid batteries consist of several major components, as shown in a cutaway view in Fig. 16.10. This figure shows the construction of an automotive SLI battery. Batteries for other applications have analogous components, as illustrated and described in Secs. 16.4–16.6. The applications of the various cells and batteries dictate the design, size, quantities, and types of materials that are used.

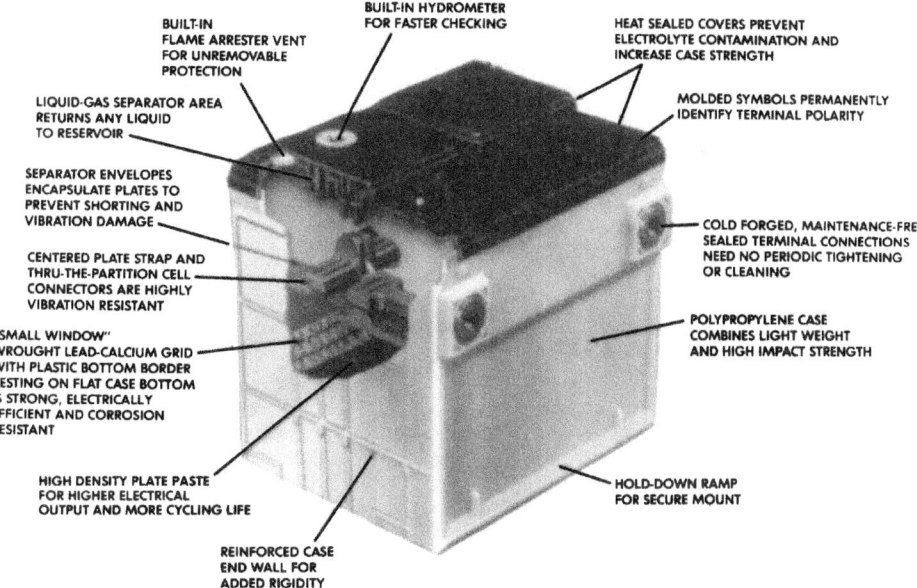

FIGURE 16.10 Typical maintenance-free lead-acid battery overview.

16.18 SECONDARY BATTERIES

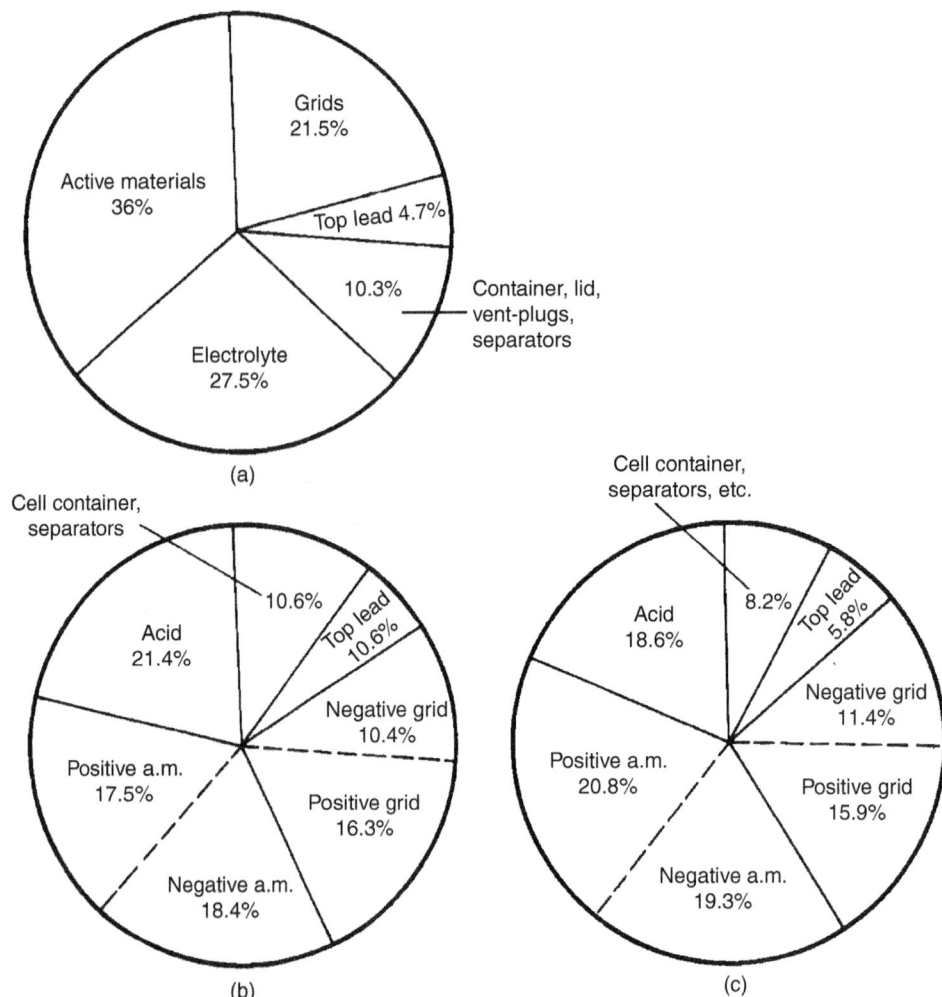

FIGURE 16.11 Weight analysis of typical lead-acid batteries. (*a*) SLI battery. (*b*) Tubular industrial battery. (*c*) Flat-plate traction battery. (*From Ref. 22.*)

The active components of a typical lead-acid battery constitute less than one-half of its total weight. A breakout of the weights of the components of several types of lead-acid batteries is shown in Fig. 16.11.

The battery components are fabricated and processed as shown in the flowsheets of Fig. 16.12. The major starting material is highly purified lead.[23] The lead is used for the production of alloys (for subsequent conversion to grids) and for the production of lead oxides (for subsequent conversion first to paste and ultimately to the lead dioxide positive active material [Fig. 16.12*a*] and the sponge lead negative active material).

Automotive lead-acid batteries (SLI) are produced mainly in high-volume plants with a great deal of automation. Many modern factories are capable of producing quantities on the order of 100,000 batteries per day. On average, an automated facility might require less than 500 employees, including all staffing levels. The automation has been prompted by environmental, reliability, and cost considerations.

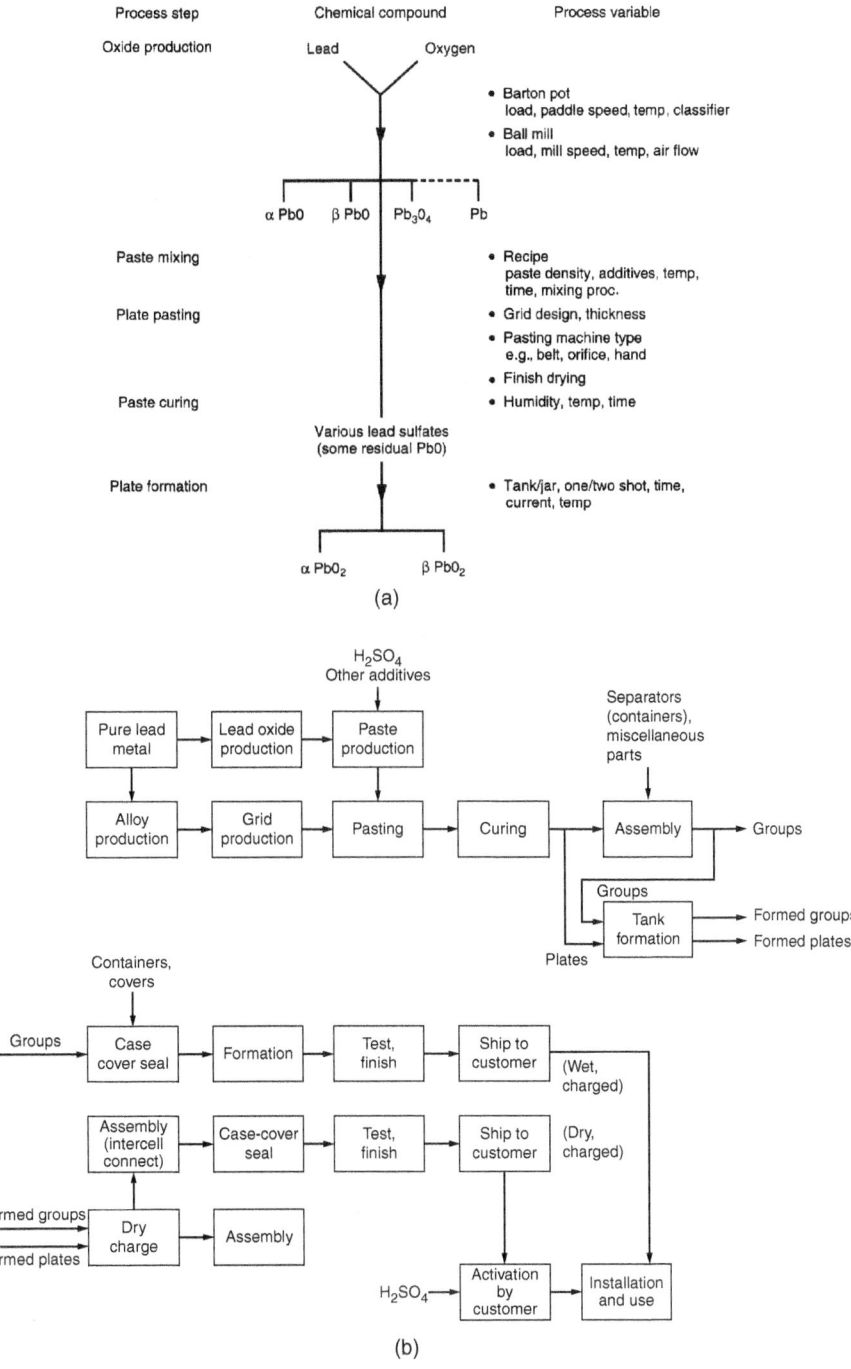

FIGURE 16.12 (*a*) Chemical compounds and process parameters in production of SLI batteries. (*b*) Production flow sheet for lead-acid batteries.

16.3.1 Alloy Production

Pure lead is generally too soft to be used as a grid material. Exceptions that use pure lead plates are some special, very thick plate Planté or pasted-plate batteries, some spiral wound batteries, some valve regulated cells and batteries (see Fig. 16.14c) and a cylindrical cell (see Fig. 16.41).[24]

The lead was hardened in older designs, by the addition of antimony metal. The amount of antimony has varied between 5 and 12% by weight, generally dependent on the availability and cost of antimony. Typical modern alloys, especially for deep-cycling applications, contain 2 to 6% antimony. The trend in grid alloys is to go to even lower antimony contents, in the range of 1.5 to 2% antimony, in order to reduce the maintenance (water addition) that the battery will require. As the antimony content goes below 4%, the addition of small amounts of other elements is necessary to prevent grid fabrication defects and grid brittleness. These elements, such as sulfur, copper, arsenic, selenium, tellurium and various combinations of these elements act as grain refiners to decrease the lead grain size.[25-27]

Some of the alloying elements, not previously described as grain refiners, fall into two broad classes of elements that are beneficial or detrimental to grid production or battery performance. Beneficial elements include tin, which operates synergistically with antimony and arsenic to improve metal fluidity and castability. Silver, cobalt, and selenium are also thought to improve corrosion resistance. Detrimental elements include iron, which increases drossing;[15] nickel, which affects battery operation; and manganese, which attacks paper separators. Cadmium has been used in grid alloys to enhance processability in antimonial alloys to minimize the detrimental effects of antimony. Cadmium, however, is not popular because of its toxicity and difficulty of removal during lead recovery (recycling) operations. Bismuth exists in many lead ore feedstocks and has been reported to both increase and decrease grid corrosion rates.

A second class of lead alloys has been developed which uses calcium or other alkaline earth elements for stiffening. These alloys were developed originally for telephone service applications.[16,28] Antimony from the grids is dissolved during battery operation and migrates to the negative plates where it redeposits, which results in increased hydrogen evolution and water loss. For telephone applications, more stable battery operation and less frequent watering were desired. The composition of the alloy depends somewhat on the grid manufacturing process. Calcium is used in the range of 0.03 to 0.20%, but for corrosion resistance the preferred range is 0.03 to 0.05%. A variation has been to substitute strontium for calcium. Barium has been investigated but is generally felt to be detrimental to performance. Tin has been used to enhance the mechanical and corrosion-resistant properties of the Pb-Ca alloys and is usually used in the range of 0.25 to 2.0% by weight. The trend in nonantimonial alloy development is toward ternary alloys (Pb-Ca-Sn) containing a minimal amount of tin because of the expense of this element. Some batteries are produced with a quaternary alloy—the fourth element being aluminum—to stabilize the drossing loss of the alkaline earth element (calcium or strontium) from the molten alloy. Grain refining is done by the alkaline earth metal, and no other elements (impurities) are desired. The properties of the alloys are summarized in Table 16.8.[25]

16.3.2 Grid Production

Two general classes of grid production methods virtually describe all modern production, but two other classes of production techniques might become widespread in the future. These are listed in Table 16.9.

The purposes of the grid are to hold the active material mechanically and conduct electricity between the active material and the cell terminals. The mechanical support can be provided by nonmetallic materials (polymer, ceramic, rubber, etc.) inside the plate, but these are not electrically conductive. Additional mechanical support is sometimes gained by the construction method or by various wrappings on the outside of the plate. Metals other than lead alloys have been investigated to provide electrical conductivity, and some (copper, aluminum, silver) are more conductive than lead. These alternate conductors are not corrosion-resistant in the sulfuric acid electrolyte and are

TABLE 16.8 Properties of Lead Alloys

	Alloys of the 1970s						
			Cast lead-calcium-tin				Wrought-lead-calcium-tin
Property	Conventional antimony	Low antimony	0.1Ca 0.3Sn	1.1Ca 0.75Sn	Lead-strontium tin-aluminum	Lead-cadmium-antimony	0.065Ca 0.7Sn
Ultimate tensile strength, Pa × 10⁻⁶	38–46	33–40	40–43	47–50	53	33–40	60
Percent elongation	20–25	10–15	25–35	20–30	15	25	10–15

Property	Cast conventional antimony	Cast low-antimony	Cast lead-calcium	Cast lead-calcium-tin	Cast lead-strontium	Cast lead-cadmium antimony	Wrought lead-calcium-tin (1st generation)
Ease of grid manufacture	Good	Fair	Fair	Fair to good	Fair to good	Fair	Good
Mechanical	Good	Fair	Fair to good	Fair to good	Fair to good	Fair	Good
Corrosion	Fair	Fair	Good	Good	Good	Fair	Good
Battery performance	Poor	Fair	Good	Good	Good	Good	Fair to good
Economics	Good	Good	Fair	Fair	Poor	Fair	Fair to good

(*Continued*)

TABLE 16.8 Properties of Lead Alloys (*Continued*)

	Alloys of the 1980s and 1990s							
	Cast alloys				Wrought alloys			Cast and wrought
Property	Lead-calcium-tin 0.1Ca 0.3Sn	Lead-calcium 0.1Ca	Lead-calcium-tin with aluminum	Lead-calcium with aluminum	Lead-calcium-tin 0.065Ca 0.3Sn	Lead-calcium-tin 0.065Ca 0.5Sn	Lead-calcium 0.075Ca	Low antimony 2.5–3.0% Sb	Lead 0.01–1.5Sn
Ultimate tensile strength, Pa × 10^{-6}	40–43	37–39	40–43	37–39	43–47	47	43	37–40	
Percent elongation	25–35	30–45	25–35	30–45	15	15	25	25–40	

	Cast alloys			Wrought alloys		
Property	Low antimony	Lead-calcium	Wrought lead-calcium-tin (2nd generation)	Wrought low antimony		
Ease of grid manufacture	Fair to good	Good (aluminum)	Good	Good		Conductivity and corrosion- equivalent to pure lead
Mechanical	Fair	Fair to good	Good	Good		
Corrosion	Fair	Good	Good	Fair to good		
Battery performance	Fair to good	Good	Fair to good	Fair		
Economics	Good	Good (lower tin)	Good	Good		

NOTE: Alloy constituents given in weight percent.

TABLE 16.9 Grid Production Methods

Book mold cast
   Gravity cast
   Injection molded (die cast)
Mechanically worked (Planté, Manchester)
Continuous cast, drum cast
Continuous cast, wrought-expended, cast-expended
   Casting
   Working
   Expansion
      Progressive die expansion
      Precision expanded
      Rotary expanded
      Rotary expansion
      Diagonal/slit expansion
   Punching
Composite

often more expensive than lead alloys. Titanium has been evaluated as a grid material; it is not corroded after special surface treatments but is very expensive. Copper grids are used in the negatives of some submarine batteries.

The grid design is generally a rectangular framework with a tab or lug for connection to the post strap. For cast grids, the framework features a heavy external frame and a lighter internal structure of horizontal and vertical bars. In some grid designs the frame tapers with the greater width near the lug; the internal bars may also be tapered. A recent advance in grid design is the "radial" grid, with the vertical wires displaced along the frame, pointing directly toward the tab area in order to increase grid conductivity (Fig. 16.13). The radial design has been further refined to a composite of lead alloy radial conductor arrangement cast into a rectangular plastic frame. An example of this composite grid is shown in Fig. 16.14a. The grids used in the cylindrical-cell design (Fig. 16.14b) incorporate both concentric and radial members. This system has been in commercial production since 1970, with most of the original cells still in use. An example of a balanced positive grid design is shown in Fig. 16.14c.[57]

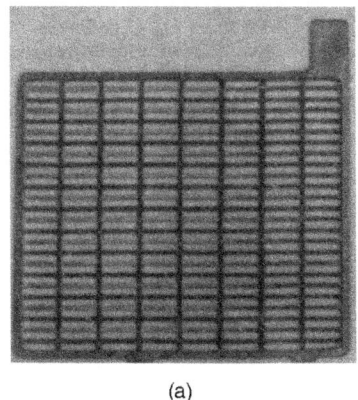 

          (a)   (b)

FIGURE 16.13 Examples of lead-acid battery cast grids. (*a*) Conventional cast flat grid. (*b*) Radial-design grid.

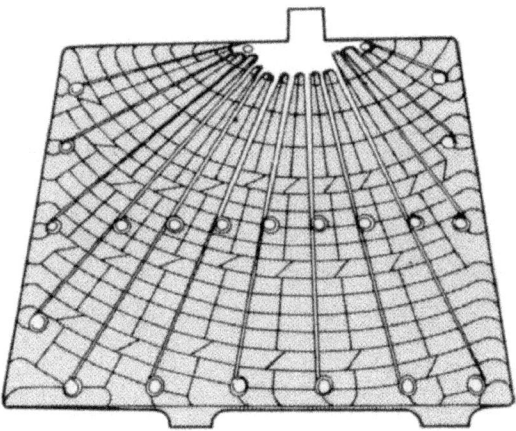

FIGURE 16.14a Composite grid, radial conductor. Grid combines diagonal conducting members with light robust plastic frame.

FIGURE 16.14b Balanced positive design[51] takes into account grid corrosion and growth and promotes the maintenance of contact of the grid with the active material, while maintaining the shape of the plate and its angle with the horizontal. This concept has also been carried into the prismatic grid structure. (*Courtesy of AT&T.*)

"Book-mold" casting historically accounted for most grid production. Permanent molds are made from steel (Meehanite) blocks by machining grooves to form the grid frames and internal lattice structure. The molds are filled when closed with an amount of lead sufficient to form the grid and leave an excess gate or sprue, which is subsequently trimmed off by a cutting or stamping operation. The grid alloy is put into the mold from a ladle in a recirculation lead alloy stream, from a metering valve in a nonrecirculation lead stream or from a hand-filled ladle. A variation on book-mold casting is injection molding or die casting of battery grids. Here the lead alloy is forced into a clamped mold by high injection pressure. Depending on the alloy characteristics, injection molding can be capable of very high production rates suggest adding picture of a Wirtz 220C automatic grid casting machine. (See Figs. 16.15a and 16.15b.)

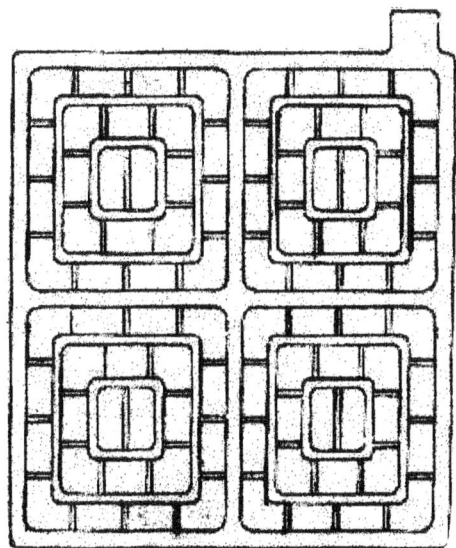

FIGURE 16.14c Balanced rectangular positive grid design. This design promotes active material contact and accounts for grid corrosion and growth in a prismatic cell. (*From Ref. 57.*)

FIGURE 16.15a Wirtz automatic grid casting machine model 220C. (*Courtesy of Wirtz Manufacturing Co.*)

FIGURE 16.15b Wirtz industrial grid casting machine model 450C. (*Courtesy of Wirtz Manufacturing Co.*)

Another method of grid manufacture is via mechanical treatment of a strip or slab of lead alloy. The traditional procedures (Planté-type plate) have been either to cut grooves into a thick lead plate, thereby increasing its surface area, or to crimp and roll up lead strips into rosettes which are inserted into round holes in a cast plate. The resultant plates are formed electrolytically into positives in the classic Planté and Manchester designs (Fig. 16.16).

The third major grid production method is circumferential continuous casting onto a mold cut into the surface of a drum. Successful high-speed production of up to 150 grids per minute has been reported. Continuous-cast grids are not symmetrical about a central planar axis and need to be overpasted to hold the active material in place.

A fourth major grid production method, expansion from wrought or cast lead alloy strip, is rapidly supplanting book-mold casting as the preferred method for the manufacture of SLI battery grids. The advantages of this method are lower grid weight per unit of battery electrical performance, the capability to manufacture a wide variety of sizes with a minimum investment in tooling, a very high-rate production capability (up to 600 plates per minute), and very uniform grid and plate sizes. Most development and commercialization have been done on nonantimonial lead-calcium tin alloys (Fig. 16.17). Strip is produced from cast slabs by a variety of proprietary metal-working processes, and the thin strip is slit to the width specified by the battery manufacturer. The worked metal increases in strength as it decreases in thickness during this processing.

Machinery to produce grids from wrought strip has been developed and put into production by several manufacturers. Four types of machinery are involved: progressive die expansion, precision expansion, rotary expansion, and diagonal slit expansion. Progressive die expansion has been the most extensively utilized of the four methods, but rotary expansion is of increasing importance. Continuous drum casting of automotive grids is also challenging the expansion processes as the dominant manufacturing method for calcium alloy grids used in negative plates in automotive batteries. Positive-plate grids are most often produced of low-antimony lead cast in book molds.

(a)

(b)

FIGURE 16.16 Planté and Manchester plates. (*a*) Planté. (*b*) Manchester.

FIGURE 16.17 Expanded wrought grid for lead-acid batteries.

Whatever grid production method is used, there is often the need for small cast parts for plate and cell interconnections and connection to external equipment. These parts have traditionally been cast in fixed molds, sometimes with mold inserts to allow a variety of similar parts to be made in each mold. Newer battery production methods often produce these various interconnections automatically in the course of battery assembly.

16.3.3 Lead Oxide Production

Lead is used to make the active materials as well as the grids. The lead must be highly refined (usually virgin or primary lead) to preclude contamination of the battery. It is described as corroding-grade lead in ASTM specification B29.[23] Lead is oxidized by either of two processes—the Barton pot or the ball mill.[29] In the Barton pot process, a fine stream of molten lead is swept around inside a heated pot-shaped vessel, and oxygen from the air reacts with fine droplets or particles to produce an oxide coating around each droplet. Typical Barton pot oxides contain 15 to 30% free lead, which usually exists as the core of each fine leady oxide spherically shaped particle. Barton pots are available in a variety of sizes up to 1000 kg/h output.

Ball milling describes a larger variety of processes. Lead pieces are put into a rotary mechanical mill, and the attrition of the pieces causes fine metallic flakes to form. These are oxidized by an airflow, and the airflow also serves to remove the leady oxide particles to collection in a baghouse. The feedstock for ball mills can range from small cast slugs weighing less than 30 g to full pigs of lead weighing approximately 30 kg. Typical ball mill oxides also contain 15 to 30% free lead in the shape of a flattened platelet core surrounded by an oxide coating.

Some battery positive plates use an additive of red lead (Pb_3O_4), which is more conductive than PbO, to facilitate the electrochemical formation of PbO_2. Red lead is produced from leady oxide by roasting this material in an airflow until the desired conversion is complete. Such processing reduces the free lead content and generally increases the oxide particle size. A variety of other oxides and lead-containing materials have been used to produce battery plates but are of only historical interest.[29] Positive plates for the Lucent Technologies batteries (formerly Bell Laboratories) were initially made with tetrabasic lead sulfate ($4PbO \cdot PbSO_4$), which is a precursor for α-PbO_2. These plates now contain up to 25% red lead (Pb_3O_4) in order to facilitate the electrochemical formation process.

16.3.4 Paste Production

Lead oxide is converted to a plastic doughlike material so it can be affixed to the grids. Leady oxide is mixed with water and sulfuric acid in a mechanical mixer. Three types of mixers are commonly used: the change can or pony mixer, the muller, and a vertical muller. The pony mixer is the traditional unit. A preweighed amount of leady oxide is placed into the mixing tub, and this is wetted first with water and then with sulfuric acid solution. Dry paste additives, if any, are premixed into the leady oxide before water addition. These additives can be plastic modified microglass fibers to enhance the mechanical strength and electrical performance of the dried paste, expanders to maintain negative-plate porosity in operation, carbon (especially for extended-life flooded SLI), and various other proprietary additives that ease processing or are believed to improve battery performance.

Muller mixers are usually filled first with the water component, then additives, the oxide, and then the acid. As mixing proceeds, the paste viscosity increases, then decreases, as measured by the amount of power consumed by the mixer motor. The paste becomes hot from the mechanical mixing and from the reaction of H_2SO_4 with the leady oxide. The paste temperature is controlled by cooling jackets on the mixer or by evaporation of water from the paste. The amounts of water and acid for a given amount of oxide will be different for the two mixer types and will also depend on the intended use of the plates: SLI plates are generally made at a low bulking agent—the more acid used, the lower the plate density will be. The total amount of additives, liquids, and the type of mixer used will affect final paste consistency (viscosity). Paste mixing is controlled by the measurement of paste density using a cup with a hemispherical cavity and by the measurement of paste consistency with a penetrometer. In making paste for advance extended-life flooded SLI where the carbon additive is added at high level, the paste density will be lowered due to the low density of the carbon material. When using such additives, a reconsideration of the targeted pasted density should be made.

Another option is to use a continuous paste mixer such as the S-9PM1 Teckominco Continuous Paste Mixer. This type of paste mixer can be used for all types of lead-acid batteries. The mixer has the ability to uniformly distribute fibers and additives (such as carbon) in a paste mix. Moreover, having a uniform distribution that avoids clumping of fibers also eliminates pasting problems that can cause costly line downtime. A diagram of the process can be found in Fig. 16.18.

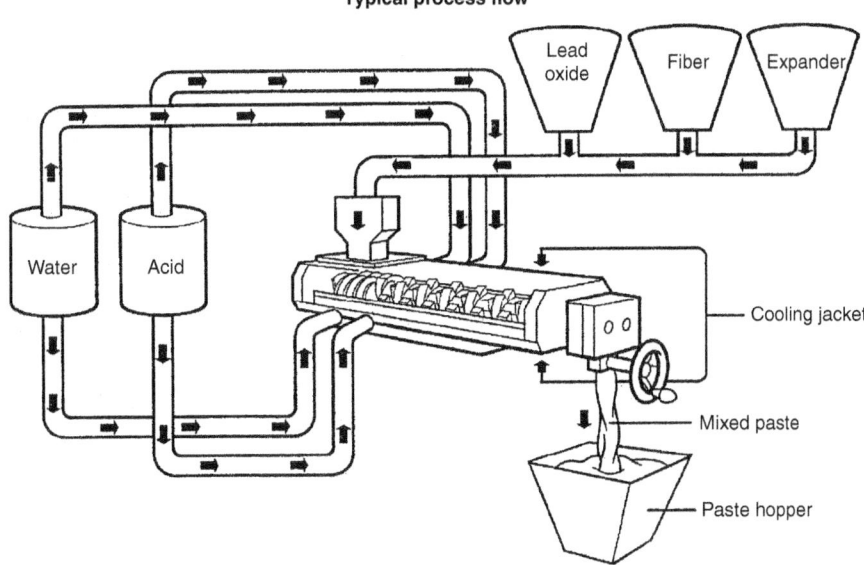

FIGURE 16.18 Typical process flow diagram for paste production.

16.3.5 Pasting

Pasting is the process by which the paste is actually integrated with the grid to produce a battery plate. This process is a form of extrusion, and the paste is pressed by hand trowel or mainly by machine into the grid interstices. Two types of pasting machines are used: a fixed orifice paster that pushes paste into both sides of the plate simultaneously and a belt paster in which paste is pressed into the open side of a grid that is being conveyed past a paste hopper on a porous belt. The amount of paste applied to a plate by a belt paster is regulated by the spacing of the hopper above the grid on the belt and the type of trowel (roller or rubber squeegee) used at the hopper exit. Using identical paste and grids, a trowel roller machine packs the paste both thicker and more densely than a rubber squeegee machine. As plates are pasted on either belt pasting machine, water is forced out of the paste, into the belt, and ultimately to a sump on or near the machine. The sump material can be used in place of some of the liquids for subsequent batches of negative paste.

Grids are automatically or manually placed onto the belt before being moved under the paste hopper. Most smaller-sized plates are made as "doubles" joined at the feet (cast) or at the tab edge (wrought expanded), or as panels of varying number of plates. Typically the belt is 35 to 50 cm wide and can handle such doubles. Industrial stationary or traction plates (being larger) are pasted by lengthwise feed into the machine or are hand-pasted. After pasting, plates are racked or stacked for curing. Stacked plates contain enough moisture to stick together, so before stacking, the plate surfaces are dried somewhat by a rapid passage through a high-temperature drier or over-heated platens. Some carbon dioxide from the combustion process might be absorbed on the surface such that the surface is made "harder." The flash drying process may also help start the curing reactions. Thicker plates are usually placed with the long edge upward in racks rather than being stacked horizontally on pallets after flash drying. Wrought expanded plates and some cast plates are cut into discrete plate portions by a slitter machine in the pasting line. Some manufacturers also have the plate lugs brushed clean of paste and surface oxidation on the same machinery.

In Europe, and less commonly in the United States, many of the heavy-duty battery positive plates are made in porous tubular sheaths. The grid is cast or injection-molded of lead, with long-finned spines attached to a header bar and a connection lug (see Fig. 16.19b). The spines are placed in nonwoven multitube gauntlets and a very fluid paste is added with the gauntlet filtering out some of the fluid. To a lesser extent, woven plastic or glass sheaths are used, with the nonwoven tube being the more common. See Table 16.10 for typical property difference among different gauntlet types. A plastic cap plugs the open sheath ends at the bottom of the plate (Fig. 16.19)

16.3.6 Curing

The curing process is used to make the paste into a cohesive, porous mass and to help produce bonding between the paste and the grid. The curing process oxidizes residue-free lead, it produces the physical and chemical structure of the plate to allow the plate to be handled efficiently through the battery-making process, and it sets the stage for the future electrical performance of the battery. Several different curing processes are used for lead oxides, depending on paste formulation and the intended use of the battery.[30,34]

The typical cure for SLI plates is "hydroset," at low temperature and low humidity for 24 to 72 h. The temperature is preferably between 25 and 40°C; the water content is that contained in the flash-dried plates, typically 8 to 20% H_2O by weight. The plates are usually covered by canvas, plastic, or other material to help retain both temperature and moisture. Some manufacturers use enclosed rooms for the hydroset, and these rooms may be heated where required by climatic conditions. As the plates cure, they reach a peak temperature, and then temperature and humidity decrease. Hydroset typically produces tribasic lead sulfate, which gives high energy density. For the battery manufacture, it is important to obtain a uniform cure in the plates. Additives can impact the consistency of the plate curing, such as adding a microglass paste fiber. Figure 16.20 shows plates that have very uniform cure.[30] The microglass additive will also stop tetrabasic lead sulfate from forming, unless a seed crystal is used.

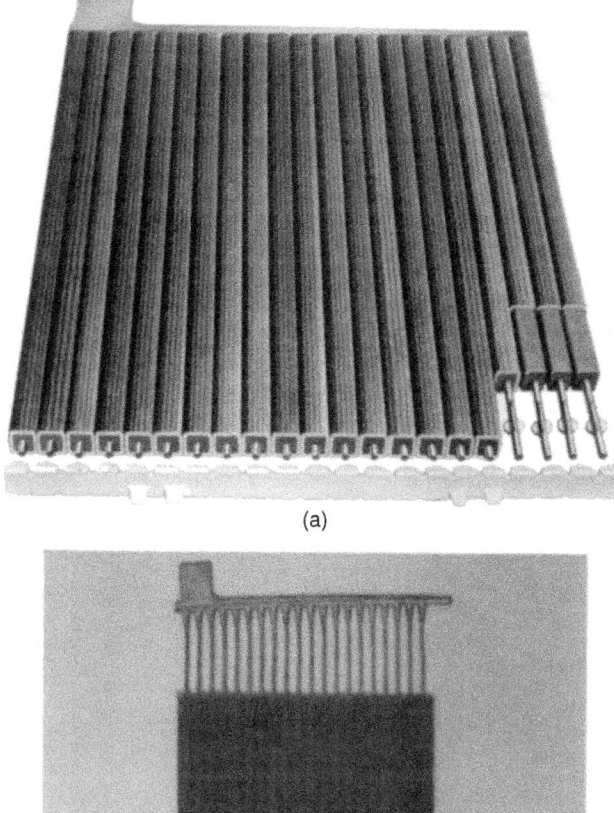

FIGURE 16.19 Tubular and gauntlet plates. (*a*) Tubular. (*b*) Gauntlet.

The type of cure that is obtained can be greatly influenced by additives. The addition of 0.5 to 2.0% of H. and V. modified paste additive (PA-10-6™)[31] prevents a standard tetra basic core process from forming tetra basic lead-oxide. Various seed crystals are being used in lead-acid batteries that provide a likelihood of a tetra basic cure being formed. SureCure[32,33] is used as a seed crystal to promote rapid formation of tetra basic lead sulfate during paste mixing and curing. It is added at 1% to the paste batch with the other dry ingredients. Seed crystals are reported to offer benefits[34] such as:

- Shorter cure time
- Lower material and curing cost
- Better crystal structure that is more uniform
- Improved formation
- Reduced capital investment in curing chambers

TABLE 16.10 Properties of Separator Systems and Materials

Typical properties of industrial PE separators[39]		
Separator properties	Unit	Typical results
Back-web thickness	μm	400–550
Electrical resistance	mΩcm^2	210–270
Porosity	%	55–58
Back-web oil	%	15–17
Total oil	%	19–21
Moisture resistance	%	3

Typical properties of SLI-PE separators[39,41]		
Separator properties	Unit	Typical results
Back-web thickness	μm	60–200
Electrical resistance	mΩcm^2	50–60
Porosity	%	50–60
Total oil	%	10–21
Moisture resistance	%	3–5
Puncture	N	5–13
Elongation XMD	%	200–500

Typical properties of gauntlets[40]				
Gauntlet properties	Unit	Standard	Reinforced	Woven
Electrical resistance	mΩcm^2	180	350	500
Porosity	%	74	60	40
Acid absorbed	g/cm^2	0.12	0.10	0.05
Acid retained	g/g	2.7	2.0	0.8

NOTE: Test methods per Battery Council International, *Flooded Separators*, section 3B.

FIGURE 16.20 Lead-acid plate batch after curing. Note radial grid to tab.

The use of curing ovens, where temperature and humidity can be precisely controlled, ensures that sufficient moisture is available to oxidize the remaining free lead in the paste. Peak temperatures in the range of 65 to 90°C are used. Another process to force completion of curing is to dip the partially cured plates into dilute sulfuric acid. This latter process ("pickling") is also used for cure of powder-filled tubular positive plates. Cured plates are stored until use. Shelf life is not critical, but the high cost of inventory usually makes storage time limited.

16.3.7 Assembly and Separator Materials

The simplest cell consists of one negative, one positive, and one separator between them in electrolyte. Most practical SLI flooded batteries contain about 7 to 30 plates. The positive and negative plates are usually enveloped in a microporous polyethylene (PE) separator. Most battery designs just envelop the positive plate. Individual or leaf separators are generally not used, having been replaced by the sealable microporous polyethylene separator. In hot tropical climates, especially in areas with poor road conditions, a laminate synthetic pulp that is filled with silica particles and glass fibers is still in general use. This separator also has high mechanical strength that resists tearing and cracking, thus allowing automatic assembly. Envelope separators are also used in motive power and standby batteries.

In SLI batteries, a thinner back-web separator has been developed, and these are used to provide for decreased electrical resistivity to improve battery performance such as the cold cranking current. The back-web thicknesses used in SLI are in the 150 to 200 micron range. An industrial application would use a thicker, back-web polyethylene separator of 450 microns. An off-road, industrial, or heavy-duty SLI battery would have a PE separator that has a glass veil attached to the ribs of the separator. This glass veil is used to apply pressure and to hold the positive active material in the plate from falling out due to vibration, etc. The glass veil is a chopped glass fiber mat in the 20 to 60 gsm range. The glass fibers are typically in the 11 to 18 micron range and the fibers are very long, 12 mm or longer. The mat will have a binder content of 15 to 25%. The percentage of the binder tends to decrease as the sheet gets thicker. To help reinforce the positive active materials additives such as glass microfibers could also be added to the active material in the 1 to 2.5% range.[35]

To address the extra cycle requirements needed in micro-hybrid designs, the standard flooded battery has been redesigned to allow the battery to have enhanced cycling under partial state-of-charge conditions. These extended-life flooded (ELF) designs use a combination of changes in the plates, alloy, electrolyte, and separator. These batteries will have a glass mat or synthetic nonwoven material against the positive plate to apply pressure to the positive plate. Having pressure on the active material has been shown to improve cycle life. The positive grid may be an expanded metal, such as Concast® (Tech Cominco Metals Ltd.), Conroll® (Wirtz Manufacturing Co. Inc.), or other continuous strip plate. Instead of using a typical cellulose tissue paper for the pasting paper, companies are investigating alternate materials, such as spunbonded or carded synthetic webs, glass veils, or the conversion of the separator system to an all-glass system. A reference to doing this to extend cycle performance can be found in U.S. patent 4,336,314, issued in 1982.[36] With the increased use of hybrid absorptive mats (HAGM) in VRLA, these very tough AGMs, can have acceptable results in a flooded cell.

For large, industrial flooded cells, the separator construction used has remained basically unchanged. The plate is wrapped with thin, continuous, glass fibers (slyver) that have been laminated to a nonwoven glass mat. The glass fibers assist in allowing the gas bubbles generated during charging to rise to the top of the cell. The glass nonwoven mat is usually a chopped strand glass mat composed of glass fiber diameters in the 10 to 19 micron range, which is bonded together with about 16 to 24% acrylic binder. A plastic outer wrapper (with die cut holes) is used to wrap and seal in place the glass laminate against the plate. Then a plastic boot is usually placed on the bottom of the plate. When the plates are in the case, an industrial grade separator is used to separate the negative and positive plates. Experiments have demonstrated that polyester synthetic nonwovens could be used to replace the slyver-glass-plastic wrapper. However, this has never gained much market penetration due to the extremely long time to test a new separator concept for these cells.

In tubular constructions, gauntlets have moved toward the nonwoven type and away from the woven gauntlet. The nonwoven gauntlet has also been improved, moving away from a carded nonwoven material to a spunbond that provides for greater fabric strength properties.[37,38]

Plates and separators are stacked manually or by a stacking machine. Stacked elements are staged on roller conveyors or carts as input to the interplate welding operations. Welding is done by two general methods: melting of the lugs in a mold with the lugs facing upward, or immersion of lugs facing downward into pools of molten lead alloy contained in a preheated mold. The first method is the traditional assembly method for lead-acid batteries. In this method the plate lugs fit up through slots in a mold "comb;" the shape and the size of the group strap are delineated by the "dam" and "back iron" portions of the tooling. Some battery manufacturers use slotted "crowfoot" posts to fit over the plate lugs to speed the welding process. The second welding method is called the "cast-on strap" process and is typically used for SLI cells. Stacked elements are loaded into slots of the cast-on machine. A mold that has cutouts corresponding to the desired straps and posts is preheated and filled with the appropriate molten lead alloy, making sure not to join lead or lead-calcium alloys with antimony alloys. The mold and the stacked elements are moved until the plate lugs are immersed in the strap cutouts. External cooling solidifies the strap onto and around each lug, and the elements are moved to a point where they can be dropped into a battery case. Visual examination can differentiate between the two welding methods: fixture-welded plate straps are usually thicker and smoother than cast-on straps; cast-on straps also will usually show a convex meniscus of metal between adjacent plate lugs on the underside of the strap if the lug is properly cleaned of paste. A good weld is required between each plate lug and the strap so that high-rate discharge performance is maximized. The resultant assemblage of plates and separators is known as an element, and the welded subelements are known as groups. Electrical testing for short circuits is usually done on elements before further assembly.

Cast-on battery elements are either continuously connected or made in discrete one-cell modules. The first method requires that long intercell connections be used, which travel over the intercell partition and are seated in a slot in this partition; this is known as the loop-over-partition design. In the second cast-on method, tabs on the ends of the plate straps are positioned over holes that have been prepunched into the intercell partitions of a battery case. These tabs are welded together manually with a very small torch or automatically by a resistance welding machine. The latter also squeezes the tabs and the intercell partition to provide a leakproof seal.

Industrial traction cells and old-style SLI cells have been connected into batteries after the cell cases and covers are sealed together. Traction batteries are needed in thousands of different sizes for various applications, and the standard unit of construction is a cell, not a quantity of plates and separators. A heavy steel tray is fabricated and coated with an acid-resistant coating (urethane, epoxy, etc.). Traction cells are placed into the tray and shimmed as necessary, and intercell connections are welded on. Heavy flexible wires (made from welding cable) are welded to the end cells for connection to the external circuit.

16.3.8 Case-to-Cover Seal

Four different processes have been used to seal battery cases and covers together. Enclosed cells are necessary to minimize safety hazards related to the acidic electrolyte, to the potentially explosive gases produced on overcharge, and to electrical shock. Most SLI batteries and many modern traction cells are sealed with fusion of the case and cover. The fusion comes from preheating each on a platen, then forcing the two together mechanically, or from ultrasonic welding of the case and cover. Fusion-sealed batteries are virtually impossible to repair. At best the elements can be salvaged, but the cover and usually the case are discarded and replaced. A few SLI batteries are sealed using an epoxy cement which fills a groove in the cover; the battery is inserted and positioned so that the case and intercell partition lips fit into the epoxy-filled groove. Heat is used to activate the catalyst to set the epoxy.

Some small deep-cycling batteries feature tar (asphalt)-sealed cases and individual cell covers. Here the tar seal allows easy repair to the battery. Traditionally all batteries were made this way before about 1960, but heat seals are typical for SLI batteries today. Molten tar is dispensed from a heated kettle to fill a groove between the cover and the case. The tar must be hot enough to flow easily but cool and viscous enough to solidify before running down into the cell.

Stationary batteries in plastic cases are sealed with epoxy glues, with solvent cement, or (for PVC copolymer cases and covers) with a thermal seal. Terminals are cast or welded on. Some very large stationary and traction cells are made so that coolant can be circulated through the terminals, and others are made with terminals with copper inserts for increased conductivity and mechanical strength.

16.3.9 Tank Formation

Plates or assembled groups can be electrically formed or charged before assembly into the case. When SLI plates are formed, these are usually formed as "doubles," with two to five panels stacked together in a slotted plastic formation tank, spaced an inch or less from stacks of counterelectrode-pasted panels in adjacent slots. The stacks are arranged so that all positive lugs protrude out of one side of the tank top and all negative lugs protrude out of the other side of the tank top. All lugs with the same polarity are connected by welding to a heavy lead bar, and the two bars are connected to a low-voltage, constant-current power supply. The tank is filled with electrolyte, and current is passed until the plates have been formed: the positives are converted to a deep brownish black and the negatives to a soft gray which shows a bright metallic streak when scratched. Industrial plates are usually formed singly. Sometimes these are also formed against dummy plates or grids. A variety of tank materials have been used, but the most common are PVC, polyethylene, or lead. The tanks are arranged so that the acid can be drained and refilled because formation increases the electrolyte concentration.

A variety of formation conditions are used, with variations in electrolyte density, charging rate (current), and temperature. Electrolyte is typically dilute, in the range of 1.050 to 1.150 specific gravity. The charging rate is usually fixed, but some manufacturers use a sequence of two or three different charging rates for different periods of time.

Tank-formed groups or plates are somewhat unstable (negatives will spontaneously oxidize in air) and therefore are "dry charged" before use (see Sec. 16.3.11).

Modern electronic chargers for formation are made to operate in a constant-current mode, either by a saturable reactor control or by use of silicon controlled rectifiers (SCRs). The most recent development in formation chargers is to control the current and time by use of a microcomputer. Some formation schedules include charge at three or more different currents, which start at low currents, go to higher currents, and then revert to lower currents. Current adjustment during formation minimizes damage to the cells by high temperatures and the need for cell cooling by water spray or forced air.

16.3.10 Case Formation

The more usual method of formation is to completely assemble the battery, fill it with electrolyte, and then apply the formation charge. This method is used for SLI and most stationary and traction batteries. A variety of formation conditions are used, similar to those for tank formation. The two major formation processes are the two-shot formation process (used for stationary and traction batteries) and the one-shot formation process (used for most SLI batteries). In the two-shot formation, the electrolyte is dumped to remove the low-density initial electrolyte and refilled with more concentrated electrolyte, chosen so that when this is mixed with the dilute initial acid residue which is absorbed in the elements or trapped in the case, the cell electrolyte will equilibrate at the desired density (Table 16.11). Typical values of the electrolyte specific gravity at full charge after formation are given in Table 16.12.

TABLE 16.11 Formation Processes

	One-shot	Two-shot
Typical application	SLI	All others, some SLI
Electrolyte concentrations, sp gr:		
Initial	1.200	1.005–1.150
Final	1.280	1.150–1.230
Subsequent processing	None	Dump and refill with 1.280–1.330 sp gr electrolyte; continue charge for several hours

TABLE 16.12 Specific Gravity of Electrolytes at Full Charge at 25°C

	Specific gravity	
Type of battery	Temperate climates	Tropical climates
SLI	1.260–1.290	1.210–1.230
Heavy duty	1.260–1.290	1.210–1.240
Golf cart	1.260–1.290	1.240–1.260
Golf cart (electric vehicle)	1.275–1.325	1.240–1.275
Traction	1.275–1.325	1.240–1.275
Stationary	1.210–1.225	1.200–1.220
Diesel starting (raiload)	1.250	
Aircraft	1.260–1.285	1.260–1.285

16.3.11 Dry Charge

The performance of wet batteries degrades with long periods of inactivity, especially when stored at warm temperatures. A loss of 1 to 3% of capacity each day is possible with SLI batteries that contain antimonial lead grids. The loss on stand can be much lower for maintenance-free batteries (0.1 to 3%/day). When lead-acid batteries must be stored for a long time, especially in high ambient temperatures, or when batteries are shipped for export, their performance can be stabilized by removal of the electrolyte by one of several methods.

When the electrolyte is removed, the battery is termed "dry-charged" (that is, charged and dry) or "charged and moist." The first process is done before the battery elements are assembled inside the case and cover. The plates can be tank-formed, water-washed, then dried in an inert gas before the element-welding portion of assembly. Alternately the welded element can be tank-formed, washed, and then dried in an inert gas. The latter process is simpler to carry out, but it is necessary that the separators can be rewetted easily after being washed and dried. The assembly (case, elements, cover) is completed and the battery is sealed. The battery can be stored in this dry-charged state for up to several years before reactivation and use.

Several processing innovations have been commercialized in the past 10 years to convert wet-charged batteries into moist or semidry batteries. In one process, most of the electrolyte is removed by centrifugation. Another process uses an inorganic salt (sodium sulfate) in the electrolyte, which minimizes degradation during storage and assists in an eventual reactivation. A battery is formed, dumped, refilled with electrolyte that contains the additive, high-rate-discharge tested, and then finally dumped. The high-rate electrical discharge (to simulate engine cranking) allows testing of an assembled "damp-dry" battery, but this also probably blocks the plate surfaces with a thin layer of small lead sulfate crystals. These crystals then minimize plate or separator degradation during storage when the battery is sealed.

16.3.12 Testing and Finishing

Electrical tests are used to check the performance of batteries before they are sold and often before they are put into use. The type of test employed depends on the intended use for the battery. SLI batteries are tested by brief discharges at very high currents (200 to 1500 A) to simulate engine-cranking performance. Stationary and traction cells are discharged at a rate specified by the user, usually in the range of 1 to 10 h if stationary and 4 to 8 h for traction. The discharge for SLI batteries is usually done by dissipation through a fixed low-value resistor or by a brief, high-rate electrical discharge driven by a power supply. Heavy-duty batteries are discharged through a resistor, a transistorized load, or an inverter.

The final manufacturing steps consist of improving the battery appearance by washing, drying, painting, installing vent plugs, and labeling as desired. Rubber-case batteries are usually painted; plastic-case batteries are not. A large variety of plaques and labels are available that can describe the battery, its performance, and use. Product liability requirements in many countries mandate that the user be warned of the hazardous nature of the battery, especially that the electrolyte is corrosive and that gases are formed that can be explosive.

Traction batteries are physically sized to fit a myriad of different forklift trucks, and so the final assembly for a traction battery consists of inserting preformed and pretested cells into a sturdy metal box (tray), making intercell connections, making cable connections, and sometimes adding a tar or plastic (urethane) material in the spaces between cell covers.

16.3.13 Shipping

Small batteries (SLI and golf-cart types) are palletized several layers high for long-distance shipment. The batteries are cushioned by five-sided (slipover) or six-sided cardboard boxes with cardboard or wood sheet between layers. The batteries are held laterally by banding or by plastic sheet which is shrink- or stretch-wrapped around a full pallet. Pallets need to be very sturdy to withstand the battery weight and handling abuse. Large batteries are palletized, banded, and cushioned as appropriate.

Batteries have traditionally been shipped only minimal distances because of their fragile nature, their weight, and their corrosive contents. The latter cause common carriers to charge a significant premium for battery shipment and usually preclude shipment by air. As the number of small, localized-sales battery manufacturers continues to decrease in the United States, the remaining large manufacturers now usually ship to their distribution chain on their own trucks.

16.3.14 Activation of Dry-Charged Batteries

When batteries have been dry-charged, they have to be reactivated before use. Activation consists of unpacking the battery, filling the cells with electrolyte (which sometimes is shipped with the battery in a separate package), charging the battery (if time is available), and testing the battery performance. When dry-charged batteries are activated, the materials that were used to seal the vent holes must be removed and discarded.

16.4 SLI (AUTOMOTIVE) BATTERIES: CONSTRUCTION AND PERFORMANCE

16.4.1 General Characteristics

The design of lead-acid batteries is varied in order to maximize the desired type of performance. Trade-offs exist for optimization among such parameters as power density, energy density, cycle life, "float-service" life, and cost.

High power density requires that the internal resistance of the battery be minimal. This affects grid design, the porosity, thickness, and type of separator, and the method of intercell connection. High power and energy densities also require that plates and separators be thin and very porous and, usually, that paste density be very low. High cycle life requires premium separators, high paste density, the presence of α-PbO_2 or another bonding agent, modest depth of discharge, good maintenance, and, usually, the use of high-antimony (5 to 7% Sb) grid alloy. Low cost requires both minimum fixed and variable costs, high-speed automated processing, and no premium materials for the grid, paste, separator, and other cell and battery components.[15,26,42–44]

The automotive industry, to comply with CO_2 fleet emission standards and fleet-mileage requirements, is moving to micro-hybrids, mild hybrids, hybrids, and EVs. These new environmental requirements for car companies have put new load and cycling requirements on the traditional flooded SLI battery. Batteries under micro-hybrid requirements must operate under a start-stop requirement. This requires the battery to have good cycle life especially under high rate partial state-of-charge (HRPSoC). A traditional flooded battery construction cannot meet those requirements. The best technology to address these new load requirements is with a AGM VRLA (see Chapter 17). A new type of flooded lead-acid SLI is also being developed. This technology is identified as advanced flooded, or extended-life flooded, SLI. These batteries are about 1.5 times the cost of standard maintenance-free batteries and use different alloys, much higher additive loading in the active material of carbon, a separator system that applies pressure against and coverage of the surface of the positive plate. This may be that a permanent type of pasting paper is used in place of a standard cellulose tissue. In addition, there are some reports of additives being placed in the electrolyte.

16.4.2 Construction

A traditional SLI battery, whose function is to start an internal combustion engine, discharges briefly but at a high current. Once the engine is running, a generator or alternator system recharges the battery and then maintains it on "float" at full charge or slight overcharge. In recent automobile designs the parasitic electrical load of lights, motors, and electronics causes a gradual discharge of the battery when the engine is not in operation. There is a move toward changing the car fleet over to a micro-hybrid design for stop-start requirements. This factor, coupled with normal self-discharge, introduces a significant cycling component into the normal cranking/floating duty cycles, which has resulted in the traditional SLI battery not to have required cycle capabilities. Studies of SLI battery life and failure modes are presented in Secs. 16.4.3 and 16.8.4.

The cranking ability of an SLI battery is directly proportional to the geometric area of plate surface, with the proportionality factor typically 0.155 to 0.186 cold-crank amperes (CCA) at −17.8°C (0°F) per square centimeter of positive-plate surface. Cranking performance is generally limited by the positive plate at higher temperatures (>18°C) and by the negative plate at lower temperatures (<5°C). The ratio of positive surface to negative surface is fixed by design. To maximize the cranking capacity, SLI battery designs emphasize grids with minimum electrical resistance (using a variety of radial and expanded grid designs), thin plates, and a higher concentration of electrolyte than motive-power or stationary batteries.

Usually an "outside-negative" ($n + 1$ negative plates interspersed with $2n$ separators and n positive plates) design is used. However, in order to balance the cranking rating with the requirement or electrical load, as well as to facilitate automatic assembly, SLI batteries with an even number of plates, or "outside-positive" designs, are widely produced in the United States.

The maintenance-free SLI battery has several characteristics that distinguish it from the non-maintenance free battery. It requires no addition of water during its life; it has significantly improved capacity retention during storage; and it has minimal terminal corrosion. The construction of a typical maintenance-free SLI battery was illustrated in Fig. 16.10. This type of battery relies mainly on charge control to prevent electrolysis of water and dry-out as compared to the AGM SLI designs which rely on oxygen recombination (see Chap. 17).

The SLI maintenance-free battery has a large acid reservoir made possible by the use of smaller plates and placement of the elements directly on the bottom of the container, eliminating the sludge space. The positive plates are usually enveloped in a microporous PE separator that prevents active material from falling to the bottom of the container and creating a short circuit. An important feature of the maintenance-free battery is the use of nonantimonial (such as calcium-lead) or low-antimonial lead grids. The use of these grids reduces the overcharge current significantly, reducing the rate at which water is lost during overcharge, as well as improving the stand characteristics (see Sec. 16.7). The use of the expanded grid, produced from wrought lead-calcium strip, is also shown in the figure. Most SLI batteries, so-called hybrid, are built using lead-calcium-tin grids for the negative and low-antimony lead grids for the positive electrode.

Another refinement of the SLI battery is illustrated in Fig. 16.21. In this design the plates are approximately one-fifth the width of conventional SLI battery plates and are inserted parallel, rather than perpendicular, to the length of the battery case. This design reduces the internal impedance of the cell and gives very high CCA ratings.

Heavy-duty SLI batteries for trucks, buses, and construction equipment are designed similar to the passenger vehicle SLI batteries but use heavier and thicker plates with high-density paste, premium separators often with glass mats, anchor-bonding of the element to the bottom of the case, rubber cases, and other such features to enhance longer life. This is necessary to provide maximum mechanical strength for the physically large (up to 530 × 285 mm) case dimensions. Because the thick plates provide less cranking current than the thinner plates (since fewer can be included in a cell of a given size), series or series-parallel connections of batteries are used. Typically, the 12 V monoblocks are connected in series for cranking at 24 V and in parallel for running and recharging at 12 V. A few sizes of maintenance-free heavy-duty batteries have also been produced.

SLI-type batteries are also used on motorcycles and boats. Batteries for recreational marine use generally have thicker plates (to give more capacity) and higher-density paste. They have the same

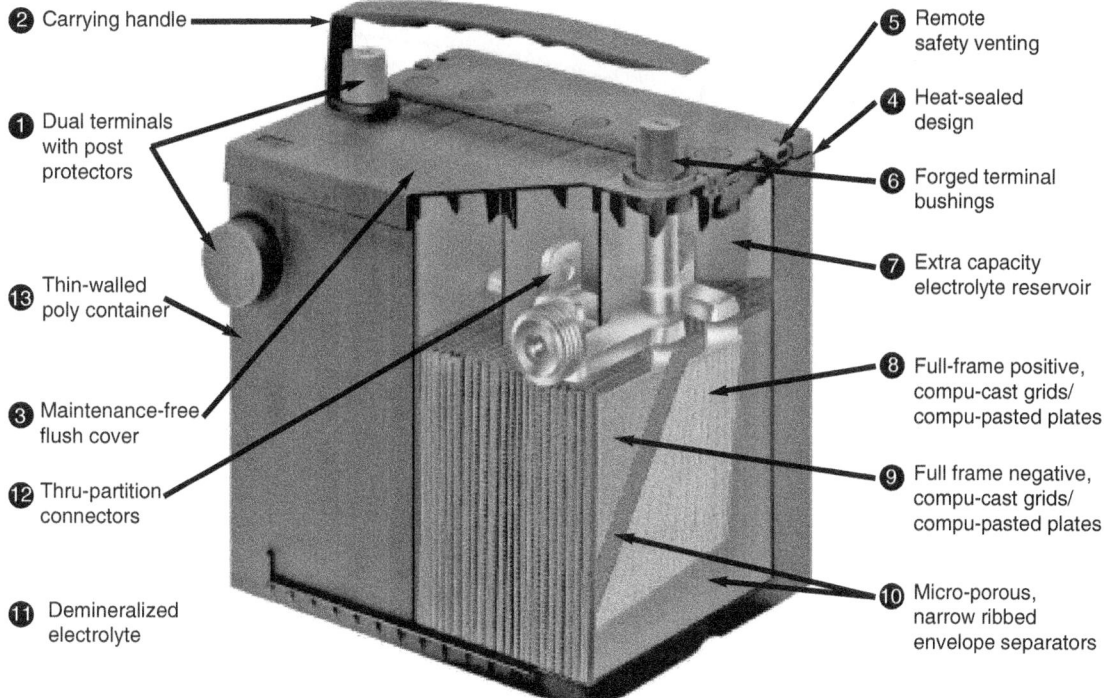

FIGURE 16.21 Cutaway of lead-acid battery. (*Courtesy of East Penn Co.*)

16.40 SECONDARY BATTERIES

Battery Council International type designations as automotive batteries. See Sec. 4.10 for a listing of BCI battery types. Marine batteries are also manufactured in 4-cell 8 V monoblocks. Many of these special applications have been converted to a VRLA design.

Aircraft use SLI-type batteries with special spillproof vent caps which preclude loss of electrolyte during flight, and VRLA designs are well represented in these applications.

16.4.3 Performance Characteristics

Discharge Performance. Discharge curves, showing the discharge profile of SLI-type batteries at several constant-current discharge rates, are presented in Fig. 16.22. The typical final or end voltages at these discharge rates are also shown. Higher service capacity is obtained at the lower discharge rates. At the higher discharge rates, the electrolyte in the pore structure of the plates becomes depleted and the electrolyte cannot diffuse rapidly enough to maintain the cell voltage. Intermittent discharge, which allows time for the electrolyte to recirculate, or forced circulation of the electrolyte

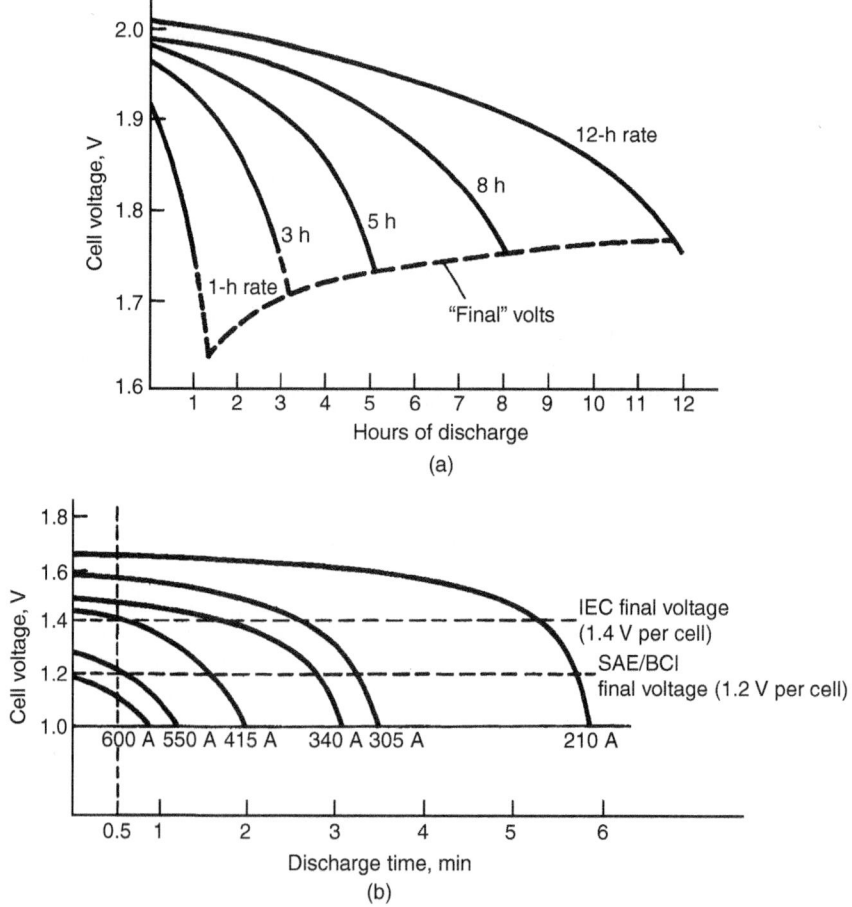

FIGURE 16.22 Discharge curves of lead-acid SLI batteries. (*a*) At various hourly rates and 25°C. (*b*) At various high rates and −17.8°C. Battery rated at 70 Ah, 20-h rate at 25°C.

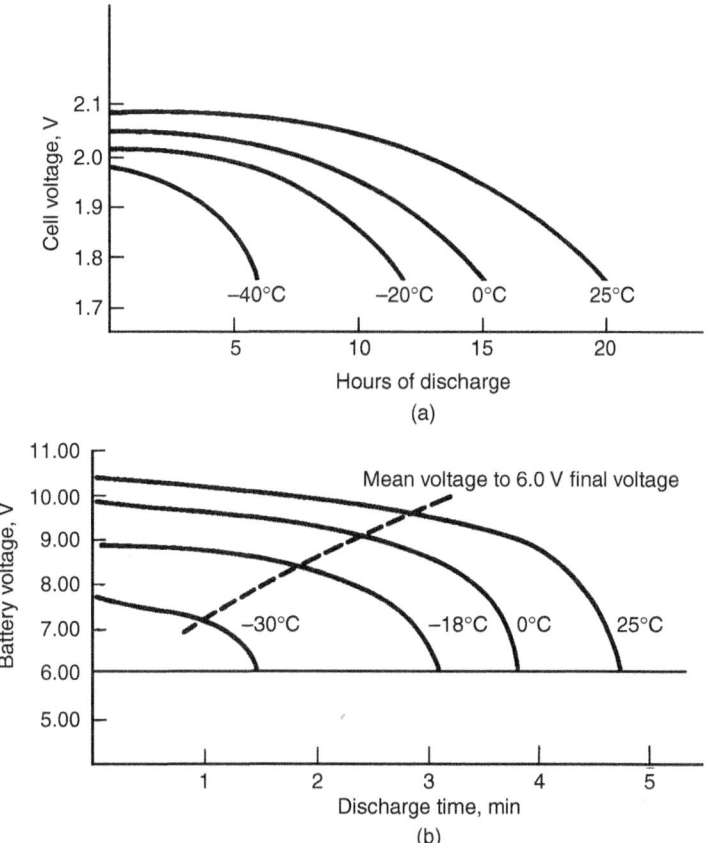

FIGURE 16.23 Discharge curves of lead-acid SLI batteries at various temperatures. (*a*) At C/20 rate. (*b*) At 340 A, 12-V battery, nominal capacity 60 Ah, 20-h rate at 25°C.

will improve high-rate performance. In general, the lead-acid cell may be discharged without harm at any rate of current it will deliver, but the discharge should not be continued beyond the point where the cell approaches exhaustion or where the voltage falls below a useful value.

Effect of Temperature on Performance. Figure 16.23*a* shows typical discharge curves for a lead-acid single-cell battery at several discharge temperatures. Figure 16.23*b* shows the discharge characteristics of a 12 V, 60 Ah battery when discharged at 340 A at temperatures from −30 to 25°C. Higher discharge voltages and capacities are obtained at the higher temperatures and lower discharge rates.

The effect of discharge rate and temperature on the capacity of the lead-acid battery is summarized in Fig. 16.24, which shows the percentage of the 20 h rate capacity delivered under different discharge conditions. Although the battery will operate over a wide temperature range, continuous operation at high temperatures may reduce life as a result of an increase in the rate of corrosion (see Sec. 16.8.1).

The performance of the lead-acid SLI-type cell at different temperatures and loads is given in another form in Fig. 16.25. The logarithm of the current drain is plotted against the logarithm of the service hours, in accordance with Peukert's relationship (Chap. 3, Sec. 3.2.6). The linear relationship is maintained over a wide range, with divergences appearing on the high-rate and low-temperature discharges. In this figure, the data have been normalized to unit cell weight (kilograms) and unit cell

16.42 SECONDARY BATTERIES

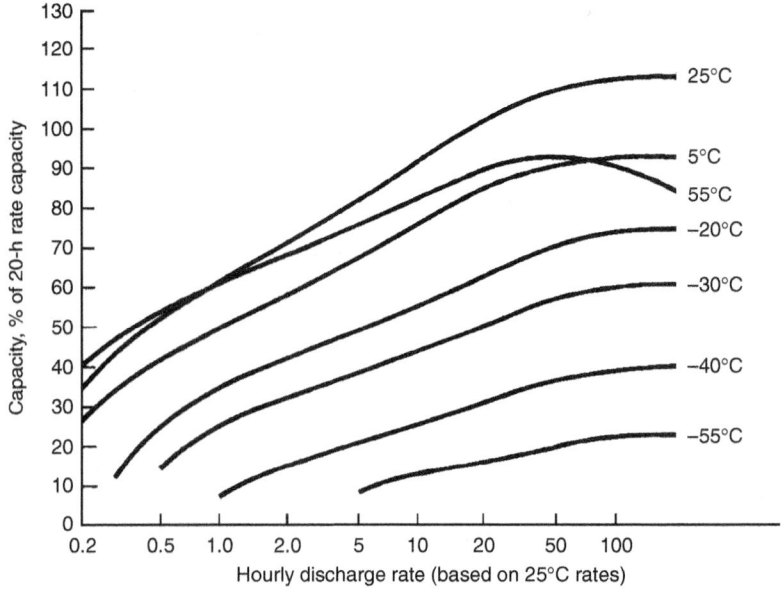

FIGURE 16.24 Performance of lead-acid SLI batteries at various temperatures and discharge rates to 1.75 V per cell end voltage.

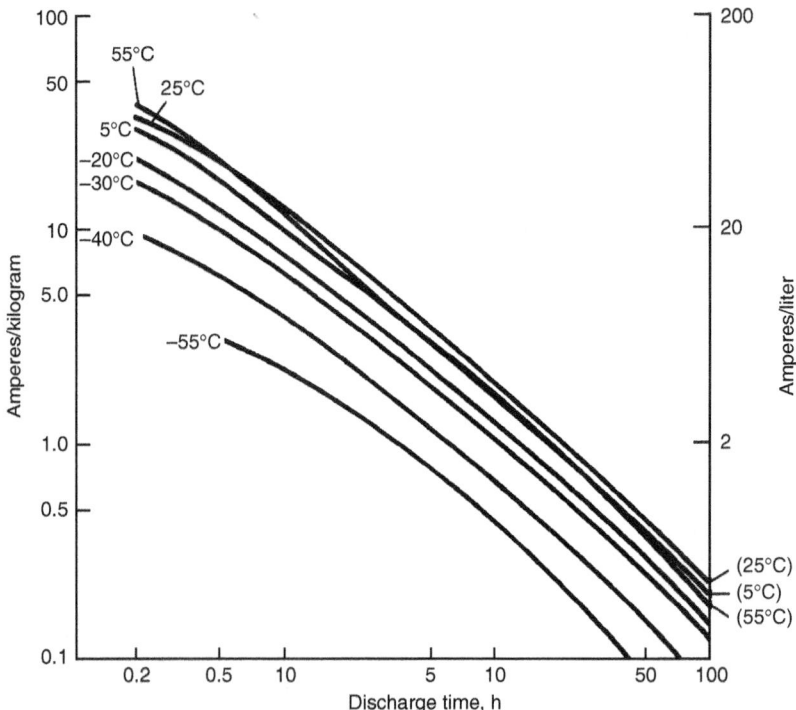

FIGURE 16.25 Service life of lead-acid SLI battery to 1.75 V end voltage per cell.

volume (liters). Figure 16.25 can be used to approximate the performance of various size cells over the operating conditions shown or to determine the size and weight of a battery to meet a particular service requirement.

Internal Impedance. The high current requirement for engine cranking demands that the batteries be designed with low resistance; for example, that conductors have large cross sections and minimal lengths, that separators have maximum porosity and minimum back-web thickness, and that the electrolyte be in the range of low resistance. The relationship between plate surface area and CCA suggests the involvement of the electrochemical double layer of the porous active materials. Low frequencies are generally necessary to evaluate the capacitive reactance component of battery impedance—strictly, the resistance impedance components can be evaluated by Ohm's law by determining the voltage difference at two levels of discharge current. The resistance of a lead-acid battery increases during a discharge almost linearly with the decrease of the specific gravity of the electrolyte. The difference in resistance between full charge and discharge is in the order of 40%. The effect of temperature on the resistance of the battery is shown in Fig. 16.26; the battery resistance increases by about 50% between 30 and −18°C.

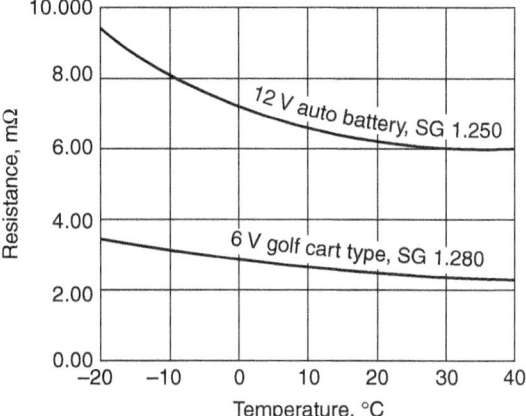

FIGURE 16.26 Comparison of lead-acid battery designs. Effect of temperature on battery resistance.

Self-Discharge. A lead-acid battery loses capacity during open-circuit stand (self-discharge). This loss is more severe with batteries that use antimonial lead grid alloys in the positive plates. A comparison of the open-circuit stand loss of conventional antimonial lead (>4% Sb), low antimonial lead (<3% Sb), and nonantimonial lead grids is shown in Fig. 16.8. This loss is most easily detected by a drop in the terminal voltage of the battery and/or the specific gravity. The sulfuric acid reacts, primarily on the surface of the negative plate, in small local self-discharge "cells" where antimony and lead are in contact, becoming a small particle of lead sulfate. In batteries using calcium-lead nonantimonial lead negative and antimonial lead positive grids, self-discharge loss is minimized until antimony diffuses to the negative. This is especially true for the low-antimonial alloy positives.

Life and Failure Modes. The life of SLI batteries is affected by the design, the processing, and the operational environment of the battery. Because of the automated assembly methods used today, SLI batteries are fairly consistent in life under ideal operating conditions, but the wide variety of operating conditions tends to spread the failure distribution. Warranty coverage for a failed battery is often more dependent on marketing strategy than on the statistical expectations of the failure rate.

SLI battery design, materials, and operation have changed markedly in recent decades; life and failure mode distribution have also changed. In Fig. 16.27a, the average age of failed batteries is plotted. Possible explanations for the shorter life in 1982 may be a reduction in battery size and more demanding performance requirements. These averages include taxis, police cars, and other heavy-duty users, which account for the relatively low age for failed batteries. Figure 16.27b shows the failure modes for these batteries, which are described in more detail in Table 16.13. A higher incidence of short-circuited batteries used in warmer climates suggests that grid corrosion is still a major failure mode. The "worn-out" category includes low electrolyte level, and it should be noted that many maintenance-free SLI batteries are sealed so that water lost to evaporation and electrolysis cannot be replaced.

SLI batteries are not designed for deep-cycling service, and very short lives are generally obtained with such operation. The deep-cycling capability of SLI batteries is covered in Sec. 16.5.2.

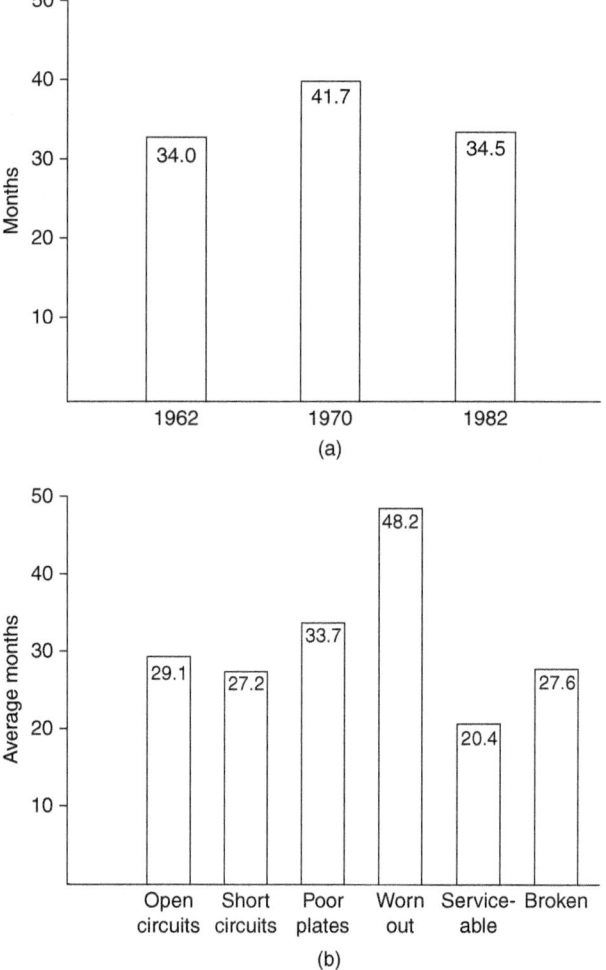

FIGURE 16.27 Failure modes of SLI batteries. (a) Average age of returned batteries. (b) Failure mode of returned batteries. (*From Ref. 2.*)

TABLE 16.13 Summary of Failure Modes of Lead-Acid SLI Batteries

1. Open circuits
   a. Terminal
   b. Cell to terminal
   c. Cell to cell
   d. Broken straps
   e. Plates off
2. Short circuits
   a. Plate to strap
   b. Plate to plate (plate fault)
   c. Plate to plate (separator fault)
   d. Plate to plate (sediment/moss)
   e. Vibration short circuit
3. Poor plates
   a. Overcharge/overheat
   b. Grid corrosion
   c. Paste adhesion
   d. Paste sulfation
   e. Paste undeformed
4. Worn out
   a. Worn out
   b. Undercharge
   c. Low level (electrolyte)
   d. Terminal corrosion
   e. Underformed
5. Serviceable
   a. Serviceable
   b. Discharged only
6. Broken
   a. Broken container
   b. Broken cover
   c. Damaged terminal
   d. Internal damage
   e. Other

Standard Tests for Rating SLI Batteries. Several standard tests have been devised to evaluate and rate the performance of SLI batteries under conditions simulating the major requirements of their applications. The cold-cranking amperes (CCA) test evaluates the capability of the battery to deliver power to crank an engine at cold temperatures. The cranking-test rating is the current that a fully charged battery can deliver at −17.8°C for 30 s to a voltage of 1.2 V per cell. If the measured voltage is above or below this value at 30 s, the CCA value can be calculated by multiplying the discharge current by the correction factor shown in Fig. 16.28. Figure 16.22b illustrates the performance of a 70 Ah cell with a CCA rating of 550 A.

Reserve capacity is measured in a test of the battery's ability to provide power for lights, ignition, and the auxiliaries. The reserve capacity is defined as the number of minutes a fully charged battery can maintain a current of 25 A to 1.75 V per cell at 25°C.

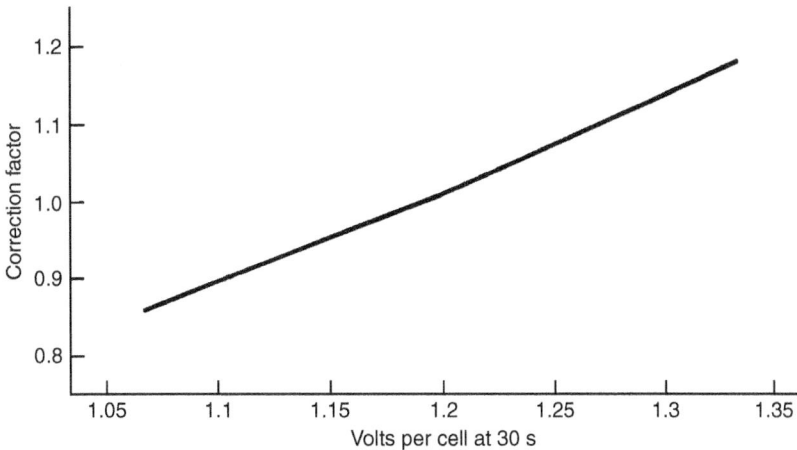

FIGURE 16.28 Correction factor for calculating cold-cranking ampere (CCA) rating.

Other SLI tests are included in the SAE battery test standard J537 on charge rate acceptance, overcharge life, and vibration resistance. A standard SLI life test is specified in SAE J240A. This test consists of a shallow discharge at 25 A followed by a brief charge at voltage and current limits for 10 min.

16.4.4 Cell and Battery Types and Sizes

SLI battery sizes have been standardized by both the automotive industry through the Society of Automotive Engineers (SAE), Warrendale, PA, and the battery industry through the Battery Council International (BCI), Chicago, IL.[45] The BCI nomenclature follows the standards adopted by its predecessor, the American Association of Battery Manufacturers (AABM). The latest standards are published annually by the Battery Council International (BCI).[45,46] Internationally, standardization is handled by the International Electrotechnical Commission (IEC). More detailed information on these standards is found in Chap. 4, Sec. 4.10 and Table 4.10.

16.5 DEEP-CYCLE AND TRACTION BATTERIES: CONSTRUCTION AND PERFORMANCE

16.5.1 Construction

The prime requirement for deep-cycling batteries for traction applications is maximum cycle life, then, if possible, high energy density and low cost. In an electric forklift application, in fact, light weight may not be advantageous because the battery's weight usually is needed to counterbalance the payload. The life of these batteries is improved by the use of thick plates with high paste density, usually a high-temperature and high-humidity cure, low electrolyte density formation, premium separators, and one or more layers of glass fiber matting (to retain the active material in the positive plates). The major modes of failure are disintegration of the PbO_2 positive active mass and corrosion of the positive grids. The deep-cycling battery is usually designed to be capacity-limited when new by the amount of electrolyte and not by the material in the plates. This serves to protect the plates and maximize their life. Both negatives and positives are degraded during use, but at end of life, battery capacity is generally limited by the positive plate. Battery failure, for cycle life rating purposes, is considered to occur when the battery will no longer produce 60 to 80% of its initial or rated discharge capacity.

A typical traction battery, using flat-pasted plates, is shown in Fig. 16.29. Cells are always made with an outside-negative design (e.g., n positive plates, $n + 1$ negative plates). Deep-cycling traction batteries are built as an assemblage of individual cells. If the battery's performance is limited by a catastrophic failure of one (or a few) cell(s), then those cells can be repaired or replaced in a cost-effective manner. Power requirements vary widely with the load, distance traveled, and lifting or climbing requirements. Battery sizes are determined by the forklift truck manufacturer and can be "calibrated" in the actual application by the use of an ampere-hour meter. A rough indication of the suitability of a traction battery for an application is the change in the specific gravity of the electrolyte during use. A larger battery size (or battery replacement or repair) is indicated when full operation cannot be achieved.

Although the flat-pasted (Faure) positive plate is typical for deep-cycling batteries in the United States, some cycling batteries in the United States and most cycling batteries in the rest of the world are built with tubular or gauntlet-type positives (Fig. 16.30). The tubular construction minimizes both grid corrosion and shedding, and long life is characteristic of these designs, but at a higher initial cost. Flat-pasted negative plates are used in conjunction with these positives and the cells are of the outside-negative design.

Small traction batteries (such as for golf carts) are designed to be intermediate between full-sized traction batteries and SLI batteries. Traction design concepts sometimes utilized include high paste

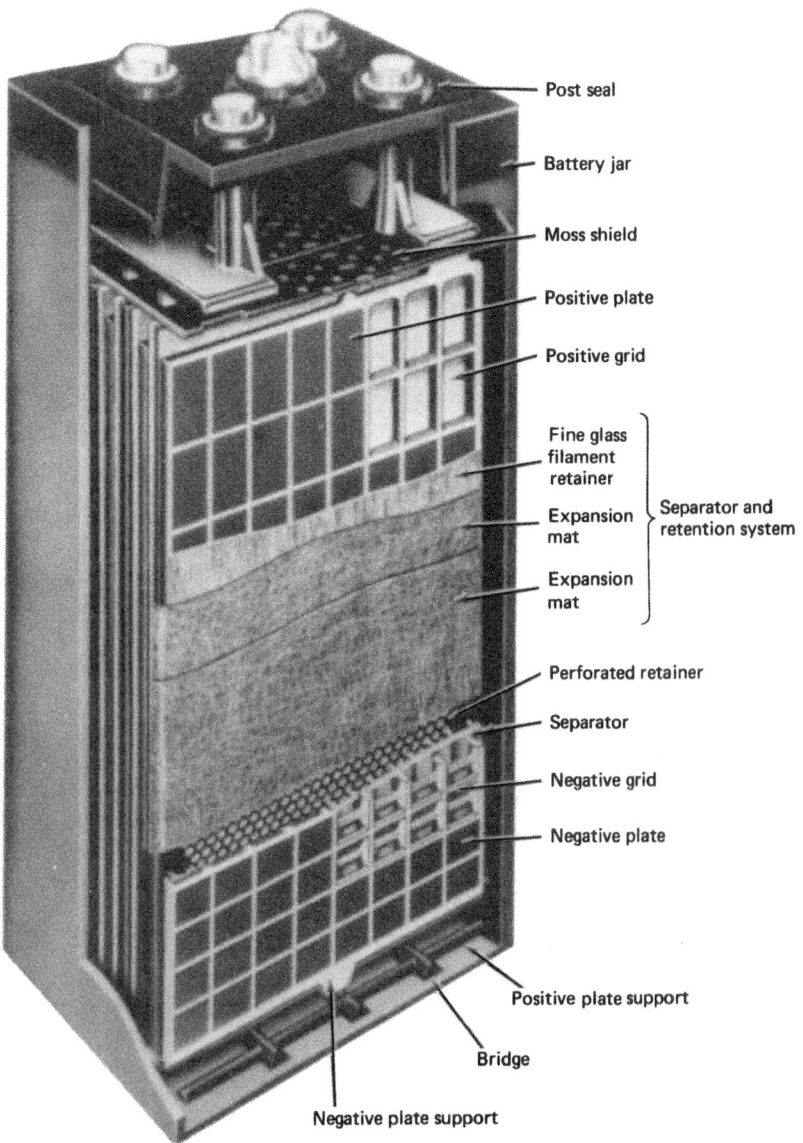

FIGURE 16.29 Flat-pasted-plate lead-acid traction battery. (*Courtesy of C&D Technologies.*)

density, careful control of plate curing and formation, to maximize content of the positive plates, glass matting against the positive, and tubular positives. SLI concepts sometimes utilized for golf-cart and other electric-vehicle batteries include thin cast radial grids, minimum separator resistance, through-the-partition intercell connection, and heat or epoxy-sealed plastic cases and covers. Cost is also an important factor.

For on-the-road electric-vehicle applications, the major criterion has been high energy density, which results in maximum range, and SLI battery design has prevailed over traction

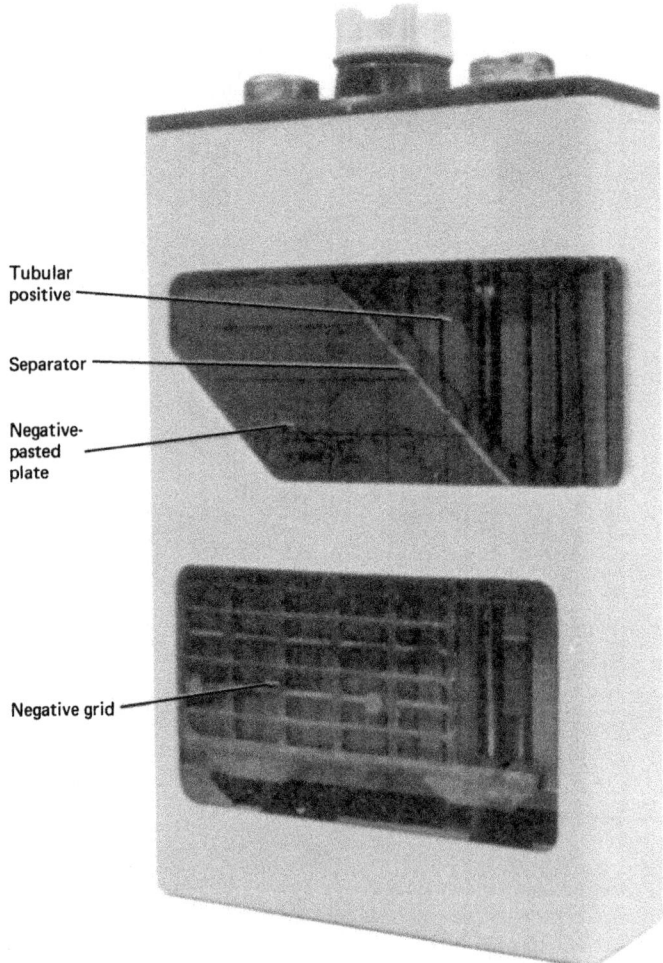

FIGURE 16.30 Lead-acid cell with tubular positive plates. (*Courtesy of Enersys, Inc.*)

battery design. In traditional cycling batteries with a few widely spaced plates, good electrolyte homogeneity occurs by convective flow. When plates are made thinner and more closely spaced for high-discharge-rate applications, such as for electric vehicle propulsion, the electrolyte has been found to become stratified during operation. A variety of electrolyte mixing devices have been designed not only to offset stratification but also to increase the discharge efficiency of the battery.

Military submarines of the diesel electric type require cycling batteries for propulsion. These batteries are made with nonantimonial lead grids because stibine and arsine produced on charge are unacceptable for personnel health in a closed environment. The plates are much larger than most traction cells—up to 600 cm wide and up to 1500 cm tall. Both flat-pasted and tubular positive plates are used.

16.5.2 Performance Characteristics

Batteries for traction and deep-cycle applications use cells with either pasted or tubular positive plates. In general, the performance of the two types of plates is similar, but the tubular or gauntlet plates show lower polarization losses because of the larger active surface area, better retention of the positive active material, good compression of the active material against the spine, and reduced loss on stand. The loss of capacity on stand at room temperature for the two plate structures, as measured by the drop in specific gravity, is shown in Fig. 16.31.

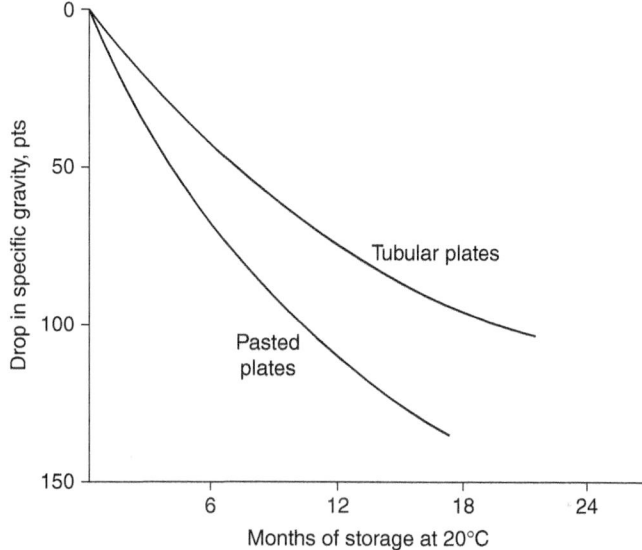

FIGURE 16.31 Retention of charge of pasted- vs. tubular-plate traction batteries.

Typical discharge curves for the two types of traction cells are shown in Fig. 16.32. The relationship of discharge current to ampere-hour capacity, up to various end voltages, is shown in Fig. 16.33. These data are presented on the basis of the positive plate since cell design and performance data of traction batteries are generally based on the number and size of positive plates that are in the cell. As is typical with most batteries, the capacity decreases with increasing discharge load and increasing end voltage.

The same relationship and comparison of the performance of the pasted and tubular plates are plotted in Fig. 16.34. These data show the superiority of the tubular plate as the discharge rate is increased.

Figure 16.35 shows the increase in available service on intermittent discharge, carried out over different periods, as compared with a continuous discharge. The gain is more pronounced at the heavier discharge loads and when the intermittent discharge is spread out over a longer period, thus allowing more time for recovery.

The effect of temperature on the discharge performance of traction-type batteries is illustrated in Fig. 16.36.

The cycle life characteristics of traction batteries are presented in Fig. 16.37. This figure shows the relationship of cycle life to depth of discharge at the 6 h discharge rate, cycle life being defined as the number of cycles of 80% of rated capacity. It is quite evident that the deeper the

16.50 SECONDARY BATTERIES

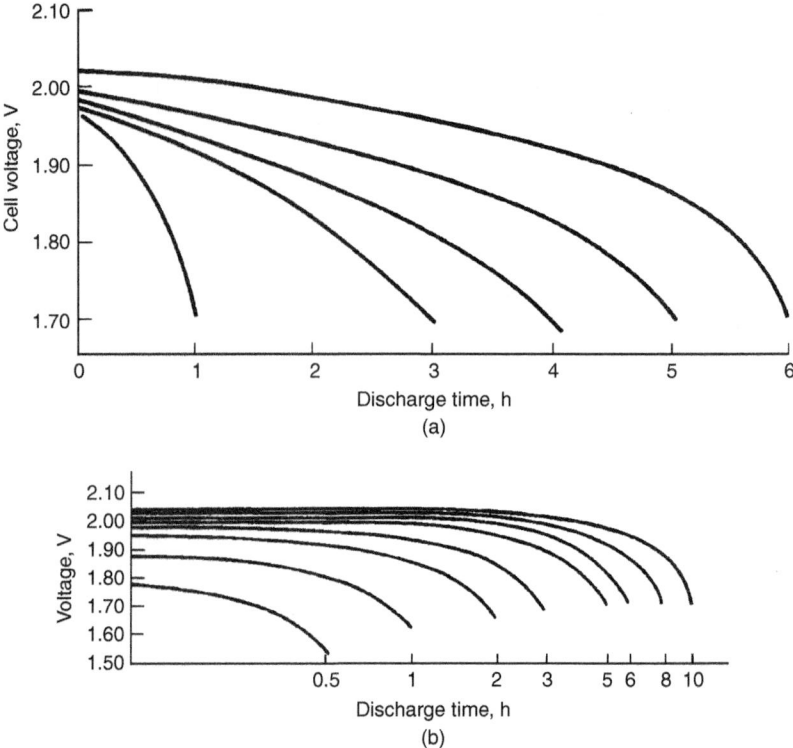

FIGURE 16.32 Discharge characteristics of traction batteries at 25°C. (*a*) Flat-pasted-plate batteries. (*b*) Tubular positive batteries.

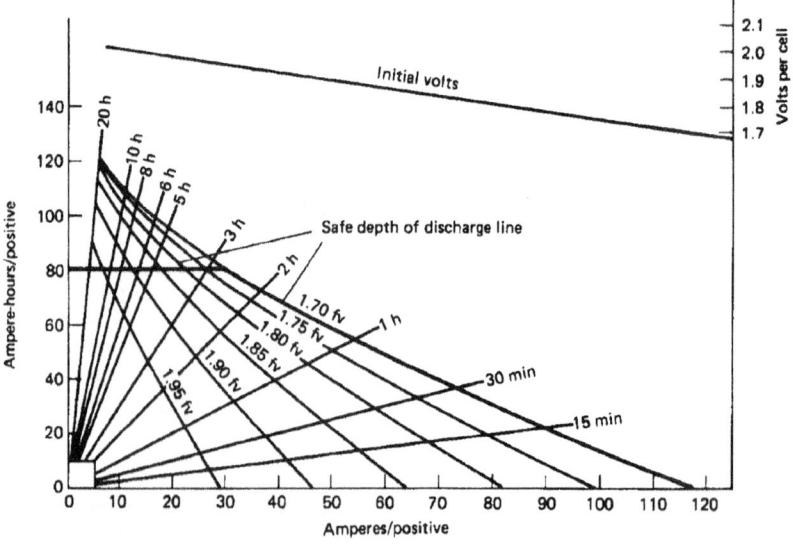

FIGURE 16.33 Performance characteristics of industrial flat-pasted plate traction battery to various final voltages (FV), at 25°C based on positive plate rated at 100 Ah at 6 h rate.

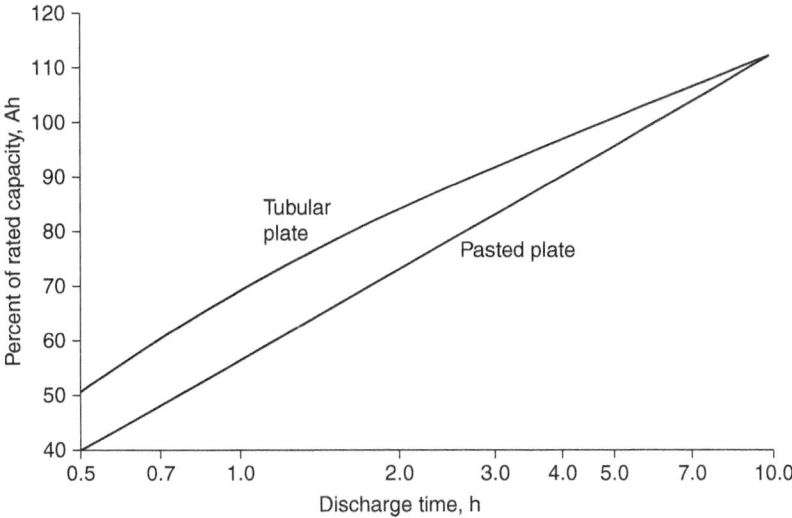

FIGURE 16.34 Effect of discharge rate on capacity of traction batteries at 25°C. Comparison of performance of flat-pasted-plate vs. tubular-plate batteries.

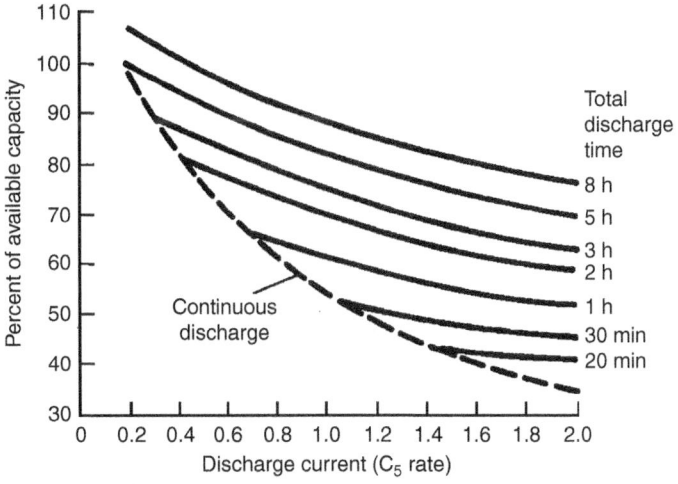

FIGURE 16.35 Capacity available on intermittent discharge of traction batteries at 25°C.

cells are discharged, the shorter their useful life, and that 80% depth of discharge should not be exceeded if full life expectancy is to be attained. Figure 16.33 shows the safe depth of discharge for other discharge rates. At low rates, the discharge should be terminated at the higher end voltages as shown, until the 1.70 V line is intercepted; then the discharges at the higher rates can be run to 1.7 V per cell final voltage. Typical cycle life expectancy is 1500 cycles (approximately 6 years).

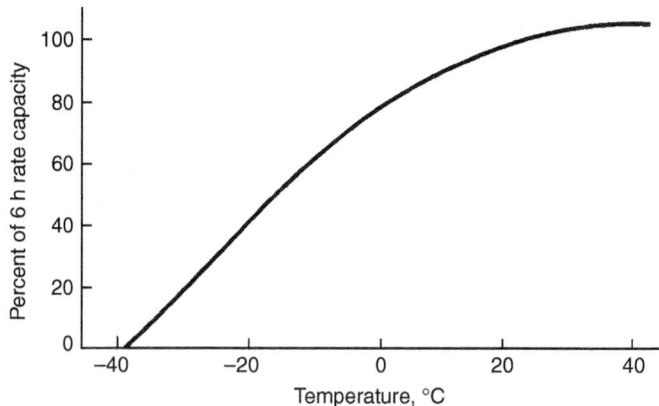

FIGURE 16.36 Effect of temperature on capacity of traction batteries, typical flat-plate design. (*From Ref. 47.*)

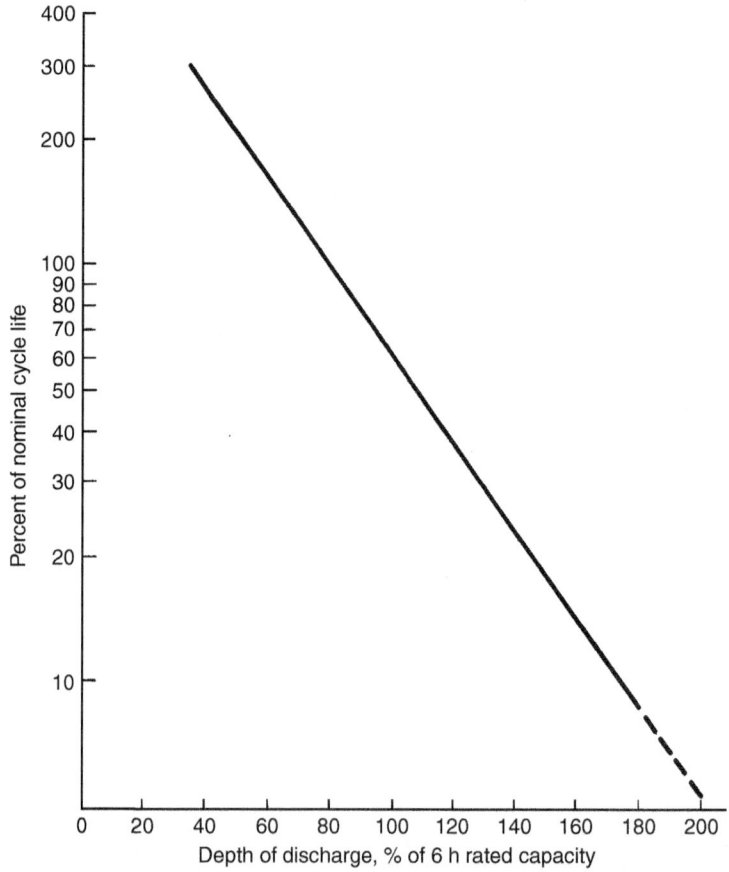

FIGURE 16.37 Cycle life vs. depth of discharge of traction batteries.

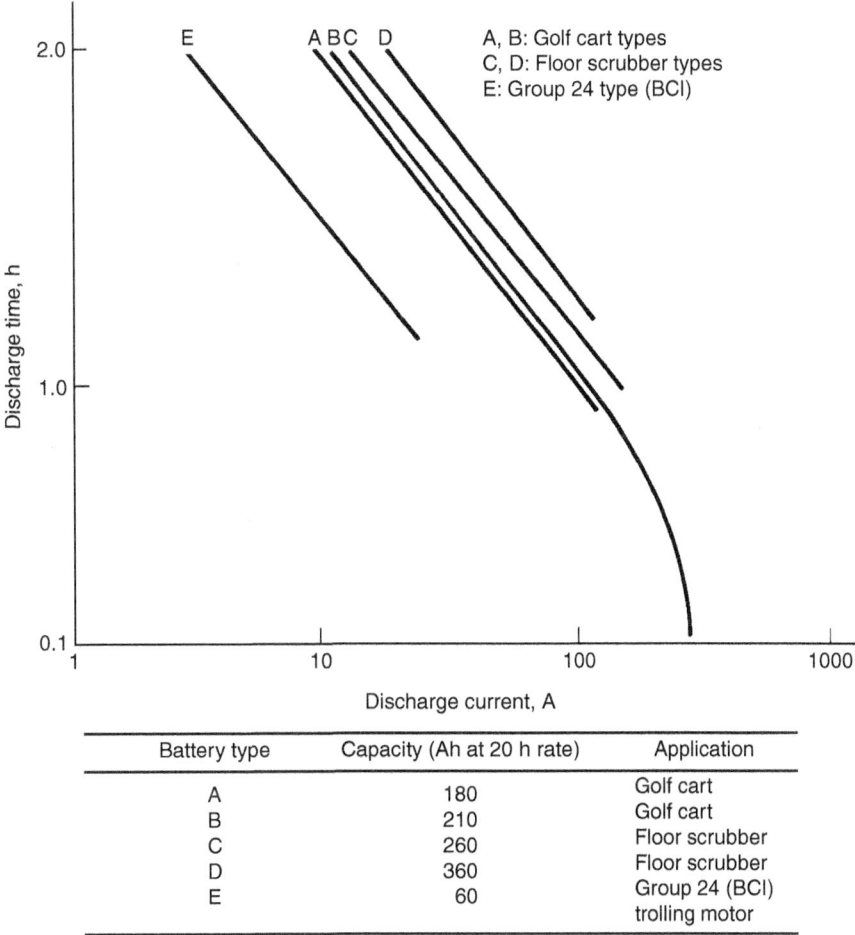

FIGURE 16.38 Performance of electric-vehicle batteries.

The relationship of discharge current and service time for several small deep-cycling batteries is shown in Fig. 16.38. At very high discharge rates, Peukert's relationship does not hold as well as for the SLI types, and the performance deviates at shorter discharge times. Nevertheless, such deep-cycling batteries can be used for cranking service and may be preferred if the battery will be deeply or repeatedly discharged in operation. Conversely, an SLI battery generally makes a poor deep-cycling battery; SLI batteries are usually made with nonantimonial lead grids (U.S. practice), and cycling generally causes the development of a grid-active material barrier layer which shortens cycle life. A comparison of the cycle life at a low discharge rate (25 A) of an SLI battery with a deep-cycle design of the same physical size is shown in Fig. 16.39.

16.5.3 Cell and Battery Types and Sizes

Traction or motive-power batteries are made in many different sizes, limited only by the battery compartment size and the required electrical service. The basic rating unit is the positive-plate capacity,

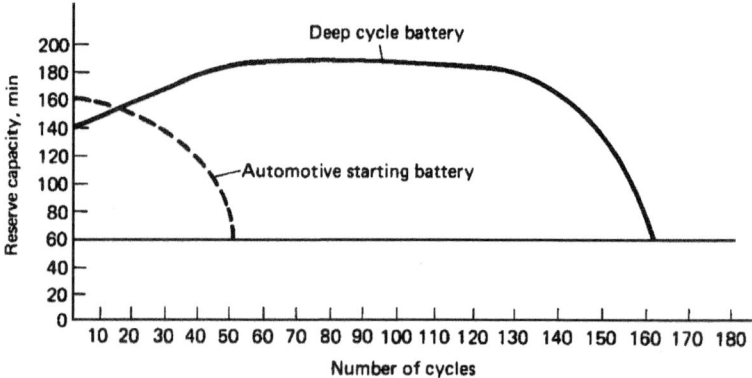

FIGURE 16.39 Cycle life characteristics at low discharge rate (25 A) for deep-cycle vs. SLI-type batteries.

TABLE 16.14a Typical Traction Batteries (United States), Flat-Pasted Plates

Positive-plate capacity, Ah at 6 h rate	Plate dimensions, mm					Cell size,*† (positive plates per cell)
	Height	Width		Thickness		
		Positive	Negative	Positive	Negative	
45	275	143	138	6.5	4.6	5–16
55	311	143	138	6.5	4.6	5–16
60	330	143	138	6.5	4.6	5–16
75	418	143	138	6.5	4.6	2–16
85	438	146	146	7.4	4.6	3–16
90	489	138	143	6.5	4.6	3–16
110	610	143	143	7.4	4.6	4–12
145	599	200	200	6.5	4.7	4–10,12,15
160	610	203	203	7.2	4.7	4–10,12,15

*All cells have n positive plates and $n + 1$ outside negative plates.
†Typical cell characteristics: 6 positive, 85 Ah plates (510-Ah cell); weight: 45 kg; size: length, 127 mm; width, 159 mm; height, 616 mm.
Source: C & D Technologies.

given in ampere-hours at the 5 or 6 h rate. Table 16.14a lists the typical U.S. traction plate sizes using flat-pasted plates; between 5 and 33 plates are used to assemble traction cells, as also shown in the table. Ratings of the cell are the product of the capacity of a single positive plate multiplied by the number of positive plates. The cells, in turn, are assembled in a variety of battery layouts, with typical voltage in 6 V increments (e.g., 6, 12, 18 to 96 V) resulting in almost 1,000 battery sizes. Popular traction battery sizes are the 6-cell, 11-plates-per-cell, 75 Ah positive-plate (375 Ah cell) and the 6-cell, 13 plates-per-cell, 85 Ah positive-plate (510 Ah cell) batteries. Table 16.14b presents similar information on the tubular positive-plate batteries.

Several SLI group sizes have been used for deep-cycling applications, especially taller versions of otherwise SLI lengths and widths. Some of these are listed in Table 16.15.

A variation of the forklift battery design is used for some on-the-road electric vehicles. Table 16.16 lists the characteristics of typical electric-vehicle batteries. Manufacturers' catalogs and data should be consulted for specific information on sizes and performance.

TABLE 16.14b Typical Traction Batteries, Tubular Plates

Positive-plate capacity, Ah at 6 h rate	Dimensions, mm					Cell size, †,‡ (positive plates per cell)
	Height	Width		Thickness		
		Positive	Negative	Positive	Negative	
49	249	147	144	9.1	*	4–10
55	258	147	144	9.1	*	4–10
57	300	147	144	9.1	*	4–10
75	344	147	144	9.1	*	4–10
85	418	147	144	9.1	*	4–10
100	445	147	144	9.1	*	4–10
110	565	147	144	9.1	*	4–10
120	560	147	144	9.1	*	4–10
170	560	204	203	9.1	*	3–8

*Varies from 5 to 8 mm depending on manufacturer.
†All cells have n positive and $n + 1$ outside negative plates. Negatives are flat-pasted plates.
‡Typical cell characteristics; 6 positive, 85 Ah plates (510 Ah cell); weight: 36 kg; size: length, 127 mm; width, 157 mm; height, 549 mm.
Source: Enersys, Inc.

TABLE 16.15 Small Deep-Cycling Batteries

BCI type	Volts	Dimensions, mm			Ratings	Typical operational current, A	Applications
		L	W	H			
U1	12	197	132	186	30–45 Ah at 20 h	25	Trolling motors
24	12	260	173	225	75–90 Ah at 20 h	25	Wheelchairs
27	12	306	173	225	90–105 Ah at 20 h	25	
GC2	6	264	183	260	75 min at 75 A	75 (GC)	Golf carts
(GC2H)	6	264	183	260	95–90 min at 75 A	300 (EV)	Electric vehicles
Not assigned	6	264	183	260	100–100 min at 75 A	300 (EV)	
Not assigned	12	261	181	279	105 Ah at 20 h	150 (EV)	
Not assigned	6	295	178	276	200–230 Ah at 20 h	50–75	Floor maintenance machinery
Not assigned	12	241	166	239	50–70 Ah at 20 h	50–75	
Not assigned	12	518	276	445	350–400 Ah at 20 h	30–50	Mine cars

Source: BCI Technical Committee, Battery Council International.

TABLE 16.16 Typical Electric Vehicle (EV) Batteries

BCI group	Volts	Plates per battery	Maximum overall dimensions, mm			Weight, kg	Ah at 2 h	Ah at 3 h	75 A (min)	Wh/kg at 3 h rate
			Length	Width	Height					
U1	12	54	197	132	186	95	20	22	15	26
24	12	78	260	173	225	22	55	59	39	31
GC2	6	57	264	183	270	26	126	135	100	29
27	12	90	306	173	225	24	62	68	45	32
GC2	6	57	264	183	270	30	150	171	120	33
GC2	6	39	264	183	280	27	158	174	140	37

16.6 STATIONARY BATTERIES: CONSTRUCTION AND PERFORMANCE

16.6.1 Construction

Designs for stationary batteries have changed much more slowly than those for SLI and traction batteries. This is not surprising in light of the much longer service life of the stationary batteries. Heavy, thick plates (including Planté as well as pasted Faure and tubular positives) with high paste density are generally used.[48] Curing is very important, and pasted plates are usually carefully dried to prevent cracks and degradation of the grid-paste interface.

The stationary battery is designed with excess electrolyte (highly flooded) to minimize maintenance and the watering interval and is generally positive-plate-limited in capacity (compared with traction batteries, for example, which are electrolyte- or acid-limited). The stationary batteries are capable of being floated and moderately overcharged.

The thick-plate design of stationary batteries reflects the fact that high energy and power densities are not as necessary as is the case for SLI and traction batteries. The overcharge operation of stationary batteries requires a large electrolyte volume (which can be accommodated as the batteries are mounted in fixed positions) and usually nonantimonial lead grids (to maximize the intervals between watering). The overcharge causes some positive-grid corrosion, and this is manifested as "growth" or expansion of the grid. The dimensions from the positive plate to the inside container are scaled so that the positives can grow by up to 10% before the plates touch the container walls. If the growth is greater than 10%, the active material is sufficiently loose on the grid so that the capacity becomes severely positive-limited and the battery must be replaced.

The positives are usually supported from hanging lugs or nonconductive rods, which are borne by the tops of the negatives. The containers are usually transparent thermoplastics (acrylonitrile-butadiene-styrene, styrene-acrylonitrile resin, polycarbonate, PVC), but some small stationary batteries are built in translucent polyolefin containers similar to those used for SLI batteries. The stationary batteries were the first application for flame-retardant vent caps which are now also standard on most SLI batteries.

The positive plate has the greater influence on the performance and life of the battery. A variety of positives are used, depending as much on tradition and custom as on the performance characteristics. The flat-pasted stationary batteries are popular in the United States because of their lower costs, lower maintenance, and lower generation of hydrogen. Lead-calcium tubular plates are now being introduced. For most of the standby emergency applications the grids are cast in lead-calcium alloy. Planté and tubular designs are popular in Europe because of their longer life. All stationary batteries today use pasted negative plates generally with n positive plates and $n + 1$ negative plates (outside-negative design). This is done because of the need for proper support of the positives, which tend to grow or expand during their life. One design used by some manufacturers is to make the two outermost negative plates thinner than the inside negative plates because the outermost surfaces are not easily recharged. Figure 16.40 is an illustration of a stationary battery system installation.

A significantly different approach to stationary battery design was the cylindrically shaped battery of Lucent, and then Lineage Power (developed by AT&T Bell Laboratories).[49,50] Traditionally, prismatic-shaped stationary batteries had failed after 5 to 20 years of service in telephone systems. The battery, illustrated in Fig. 16.41, incorporates a number of innovations in order to achieve a battery life initially predicted to be 30 years or longer. These include lattice-type circular-shaped pure lead grids (cupped at a 10° angle), plates stacked horizontally one above the other instead of in the conventional vertical construction, chemically produced tetrabasic lead sulfate (TTB) positive paste, positives welded around the external plate circumference, negatives welded to a central conductor core, and heat-sealed copolymer container and cover. The use of pure lead in place of lead-calcium is to reduce positive-plate grid growth; the circular and slightly concave shape of the plates is to counter the effect of growth and ensure good contact of the active material and grid during the life of the battery. An alloy of Pb/Sn[52,56] prevents positive terminal post corrosion (nodular) and postseal leakage for the life of the cell. Twenty-year-old batteries with this alloy have shown no nodular corrosion and no post-seal leakage. Grid growth is caused by the conversion of lead into PbO_2 on the

FIGURE 16.40 Stationary battery system installation.

grid surfaces and at the grain boundaries. The formation of this PbO_2 on the grid adds to the amount of active paste material of the plate and, in the case of concentric grid,[51] increases the plate capacity over time. The Bell Labs researchers[24] found that the growth of the positive-grid members is proportional to their surface area and inversely proportional to their cross-sectional area. Maintaining this ratio of surface area to cross-sectional area for the concentric members in the grid is accomplished by varying the cross section and the surface areas. Thus the shape of the grid remains constant and the only change is caused by the formation of the lead dioxide on the lead grid surfaces.

Accelerated tests projected a capacity increase to accompany the grid growth. In 1988, verification tests were conducted at a site that had cells manufactured in 1973, that is, 15 years earlier. A 24 V string, selected at random out of 11 strings, was discharged at the same 5 h rate used at the time they were manufactured. The capacity behavior on aging is compared to the predicted capacity behavior at 22°C, which was the average yearly temperature of this particular location (Fig. 16.42a). The plate growth of the 54 plates measured is summarized in Fig. 16.42b, where the findings were compared to that predicted from accelerated testing. The expected capacity and plate growth increases at 22°C and 15 years were 0.25% per year or 3.8% total capacity and 0.027% per year or

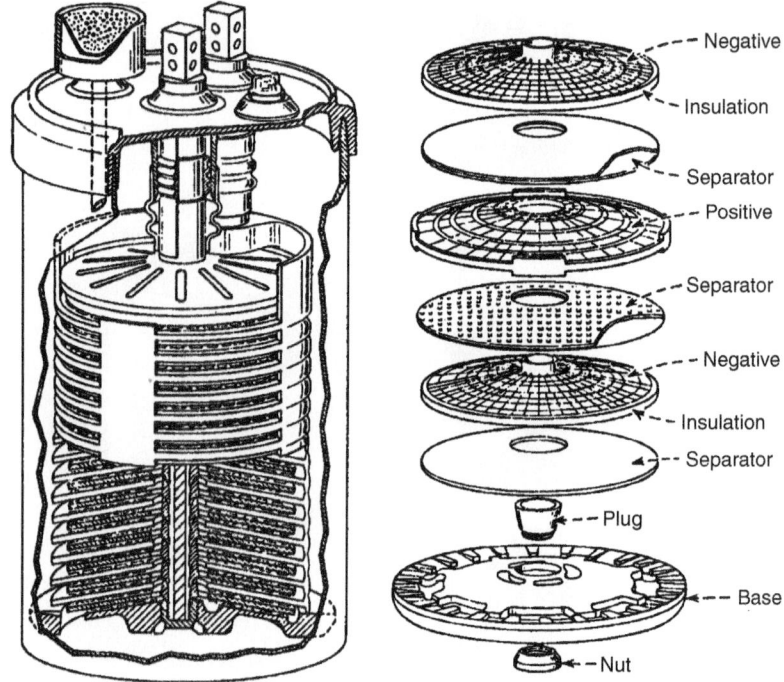

FIGURE 16.41 Cutaway and exploded views of Bell System lead-acid battery. (*From Ref. 49.*)

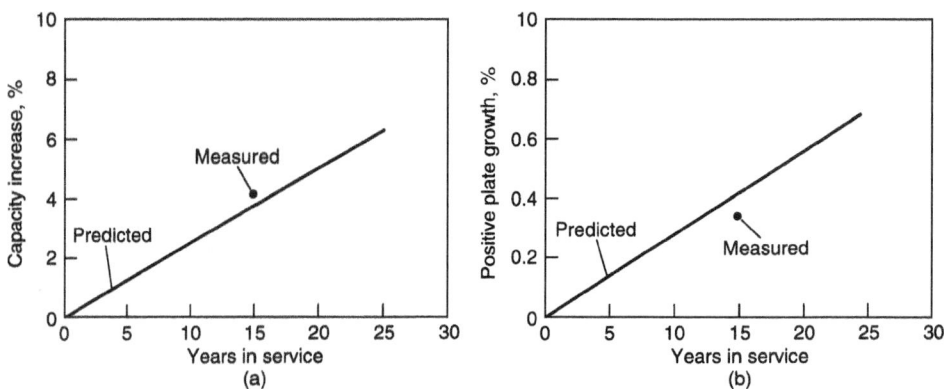

FIGURE 16.42 (*a*) Capacity after 15 years in service vs. prediction from accelerated testing. (*From Ref. 52.*) (*b*) Positive-plate growth after 15 years in service vs. prediction from accelerated testing.

0.4% plate growth, respectively. More recent 23-year corrosion data projects a life of 68 to 69 years for round cell batteries.

Some stationary batteries are designed to cycle rather than "float." For these applications the design criteria of the traction batteries are applicable (see Sec. 16.5.1). Applications of cycling stationary batteries include load leveling, utility peak-power shaving, and photovoltaic energy-storage systems.

16.6.2 Performance Characteristics

Batteries for stationary applications may use cells with flat-pasted, tubular, Planté, or Manchester positive plates. Typical discharge curves for the flat-pasted-type stationary cell at various discharge rates at 25°C are shown in Fig. 16.43, and the effect of the discharge rate on the capacity of the cell is summarized in Fig. 16.44. Generally, the discharge rate for stationary cells is identified as the hourly rate (the current in amperes that the battery will deliver or the rate hours) rather than the C rate used for other types of batteries.

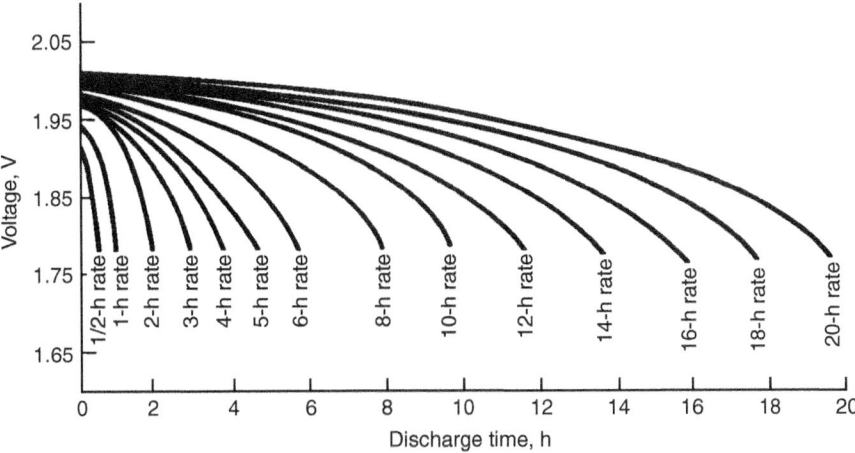

FIGURE 16.43 Discharge curves of flat-pasted lead-acid stationary batteries (specific gravity 1.215) at various discharge rates at 25°C.

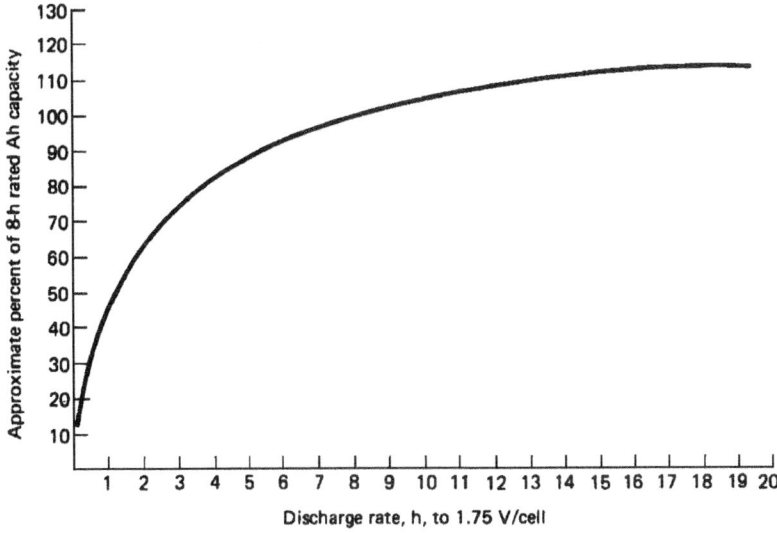

FIGURE 16.44 Effect of discharge rate on cell capacity at 25°C for flat-pasted lead-acid stationary batteries (specific gravity 1.215) to 1.75 V end voltage.

16.60 SECONDARY BATTERIES

Figure 16.45(*a–d*) is a series of curves showing the specific performance characteristics of the four types of stationary batteries at 25°C based on positive-plate design. An electrolyte with a specific gravity of 1.215 is used in all these batteries. The format used in these figures consists of two sections. The lower log-log section shows the capacity (expressed in discharge time) the particular positive plate will deliver at the specified current (expressed in amperes per positive plate) to various voltages including a final voltage. The upper semilog section shows the cell voltage at various stages of the discharge at various discharge rates (also expressed in amperes per positive plate). The final voltage is the voltage at which the cell can no longer supply useful energy.

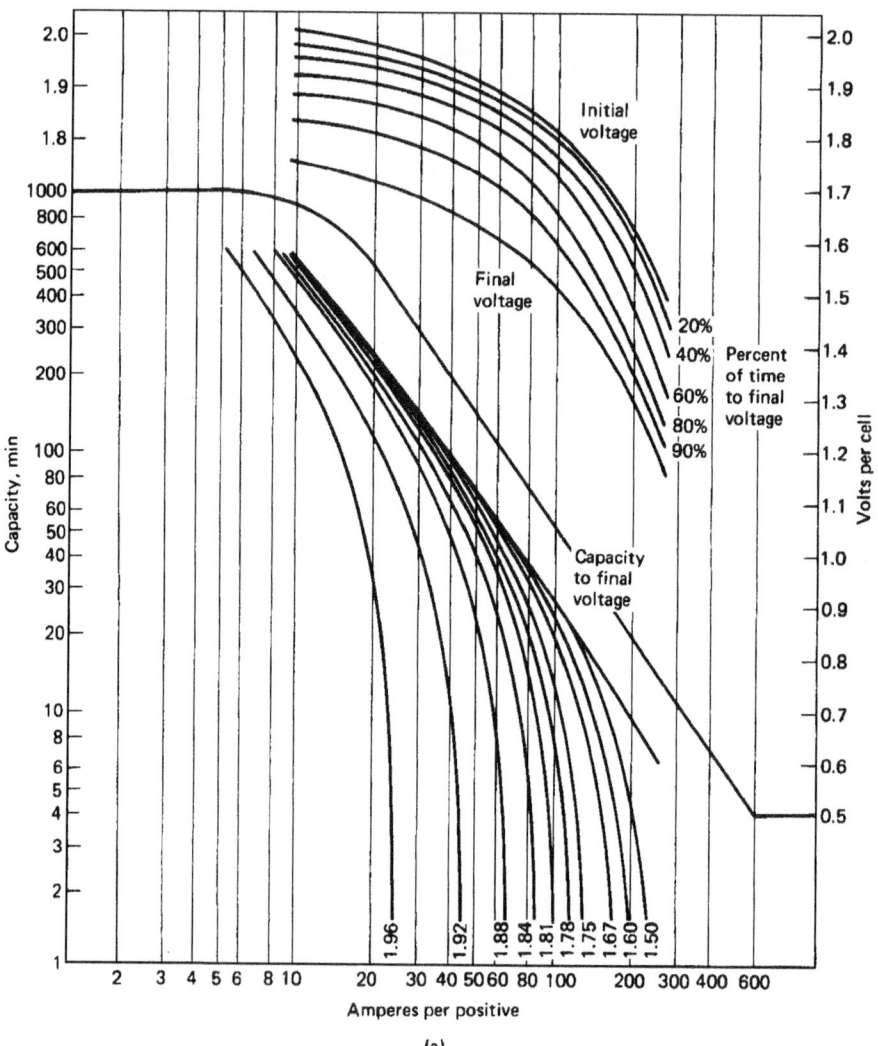

FIGURE 16.45 Performance curves of lead-acid stationary batteries at 25°C (S-shaped curves, based on positive-plate performance). (*a*) Antimony flat-pasted plate, 125 Ah at 8 h rate; 290 mm height, 239 mm width, 8.6 mm thickness. (*Courtesy of Enersys, Inc.*)

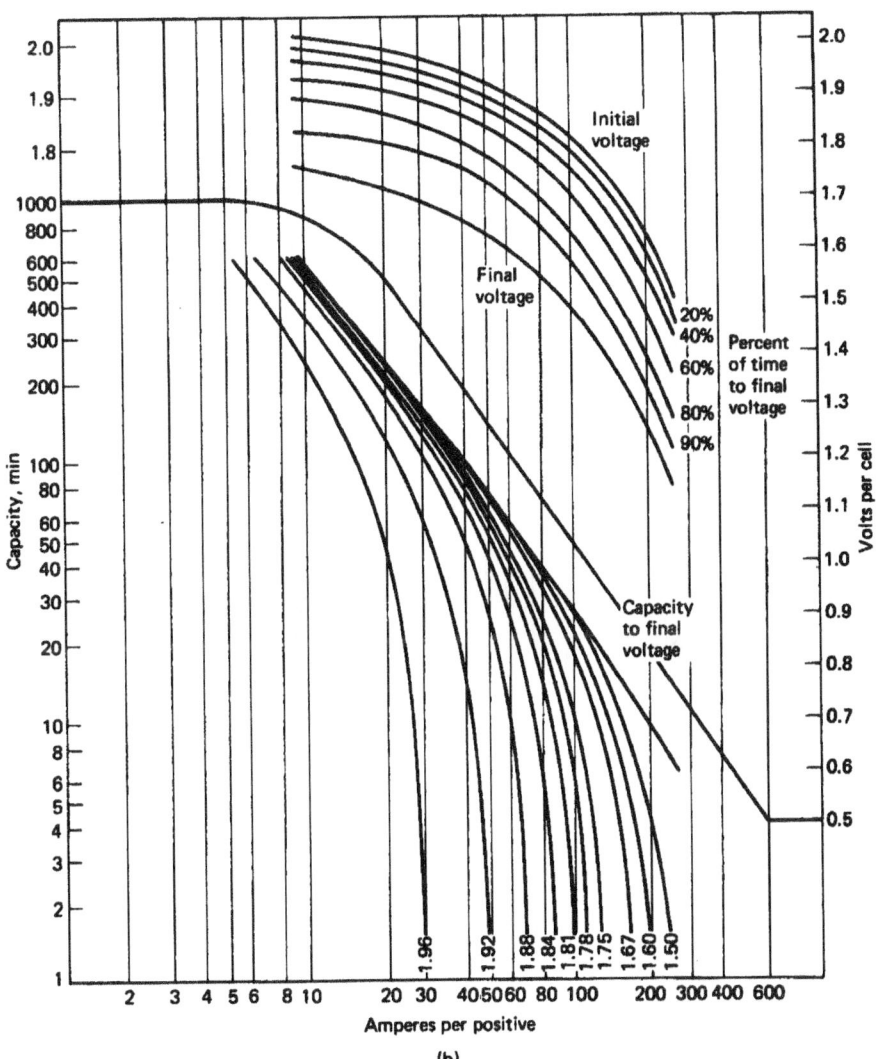

FIGURE 16.45 (*b*) Calcium flat-pasted plate, 125 Ah at 8 h rate; 290 mm height, 239 mm width, 8.6 mm thickness. (*Courtesy of Enersys, Inc.*)

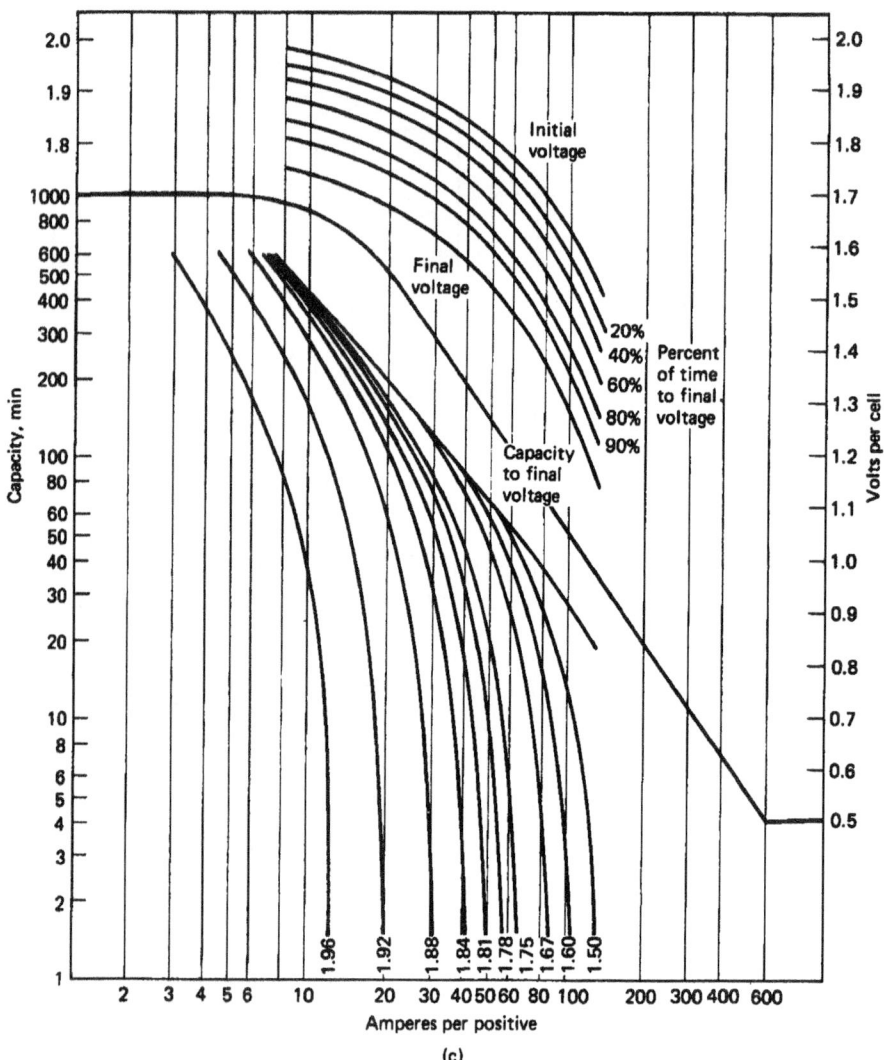

FIGURE 16.45 (c) Ironclad tubular plate, 70 Ah at 8 h rate; 274 mm height, 203 mm width, 8.9 mm thickness. (*Courtesy of Enersys, Inc.*)

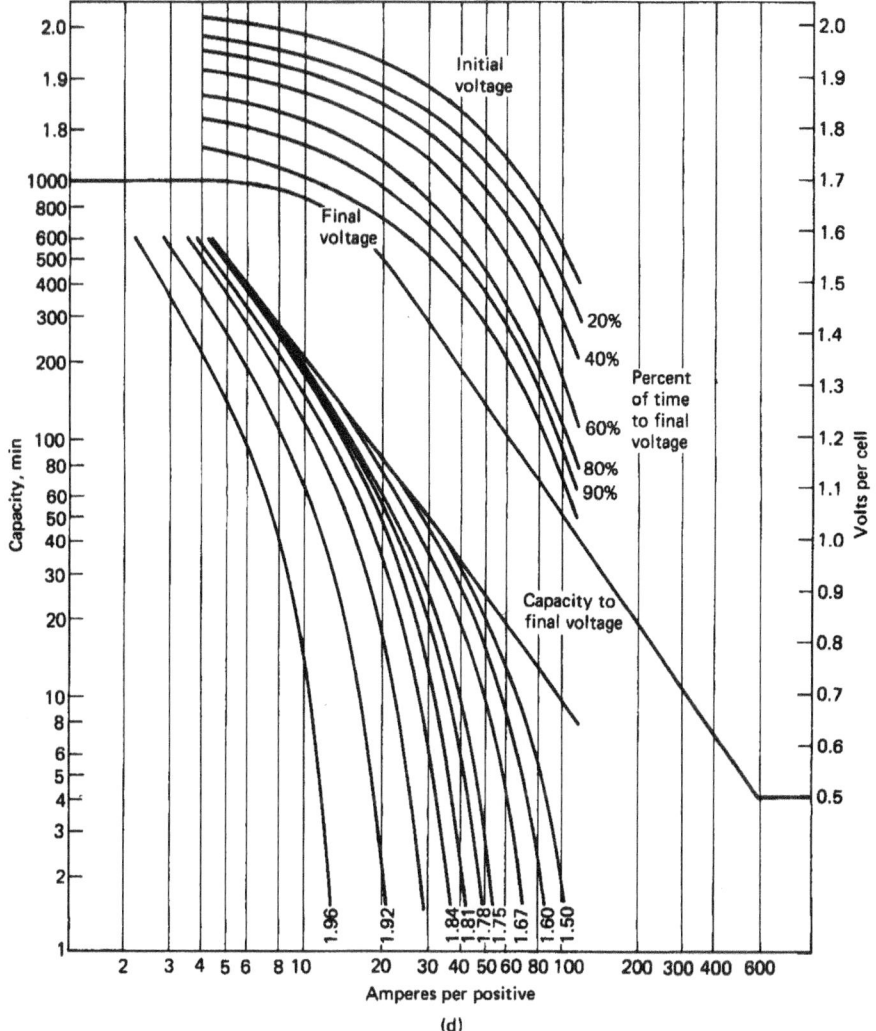

FIGURE 16.45 (*d*) Manchex plate, 40 Ah at 8 h rate; 197 mm height, 197 mm width, 11.2 mm thickness. (*Courtesy of Enersys, Inc.*)

The energy density of the flat-pasted positive-plate and the tubular positive-plate batteries is similar. It is lower for the Planté positive-plate batteries. The high-rate performance of the flat-pasted positive cells is better because these plates can be made thinner than the tubular or Planté plates.

The optimal temperature for the use of stationary batteries ranges from 20 to 30°C, although temperatures from −40 to 50°C can be tolerated. The effect of temperature on the capacity of stationary batteries at different discharge loads is shown in Fig. 16.46. High-temperature operation, however, increases self-discharge, reduces cycle life, and causes other adverse effects, as discussed in Sec. 16.8.1.

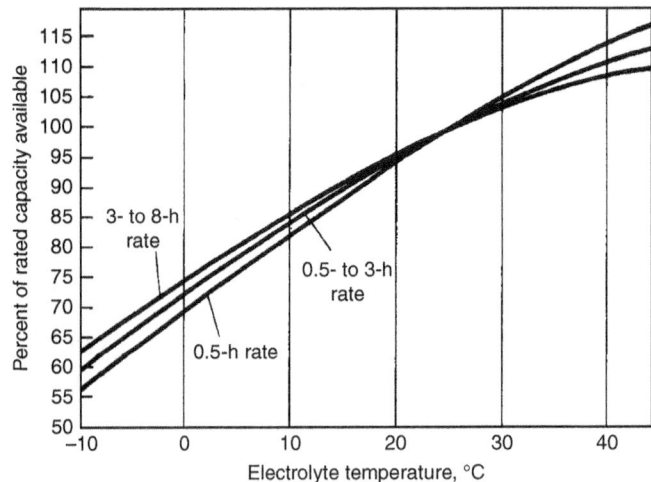

FIGURE 16.46 Performance of flat-pasted lead-acid stationary batteries at various temperatures and discharge rates. (*Courtesy of C&D Technologies.*)

The rates of self-discharge of the various types of stationary batteries are compared in Fig. 16.47, which shows the relative float current at a specified float voltage. The float current under these conditions is a measure of self-discharge or local action. It is lowest for the calcium-lead grid pasted positives and remains low throughout the life. The float current is progressively higher for the tubular antimony-lead positives, the pasted antimony-lead positives, and the Manchester-type positives—at the beginning and throughout the battery's life. If the float current is not increased periodically, the antimonial cells will all become progressively self-discharged and sulfated.

For fully charged batteries, the self-discharge rate at 25°C for the calcium-lead positive-plate cells is about 1% per month, 3% for the Planté, and about 7 to 15% for the antimonial lead positive cells. At higher temperatures, the self-discharge rate increases significantly, doubling for each 10°C rise in temperature.

The float current for the calcium-lead and antimonial lead batteries is shown in Fig. 16.48 under float charge at voltages between 2.15 and 2.40 V per cell. It has been found that more than 50 mV positive and negative overpotential is necessary to prevent self-discharge so that 0.005 A float current per 100 Ah of battery capacity is required for the lead-calcium batteries. Antimonial lead batteries initially require at least 0.06 A per 100 Ah, but this increases to 0.6 A per 100 Ah as the battery ages. The higher float current also increases the rate of water consumption and evolution of hydrogen gas.

Various, and at times conflicting, claims about the life of stationary battery designs are made by the different manufacturers worldwide. Generally, the flat-pasted antimonial lead batteries have the shortest life (5 to 18 years), followed by the flat-pasted calcium-lead batteries (15 to 25 years), the tubular batteries (20 to 25 years), and the Planté batteries (25 years).

Life on float service has been found to be related to temperature (Arrhenius-type behavior), as plotted in Fig. 16.49. The growth rate constant k is plotted for several different types of grid alloys used for the telephone system. At 25°C the time to reach 4% growth, an upper limit before the battery's integrity is impaired, is calculated to be 13.8 years for PbSb, 16.8 years for PbCa, and 82 years for pure lead.[49]

16.6.3 Cell and Battery Types and Sizes

Stationary batteries, like traction batteries, are available in a variety of plate and cell sizes. Stationary battery systems are assembled on insulated metal racks with groups of cells in series, having nominal system voltages of between 12 and 160 V. Some battery installations are made with several such

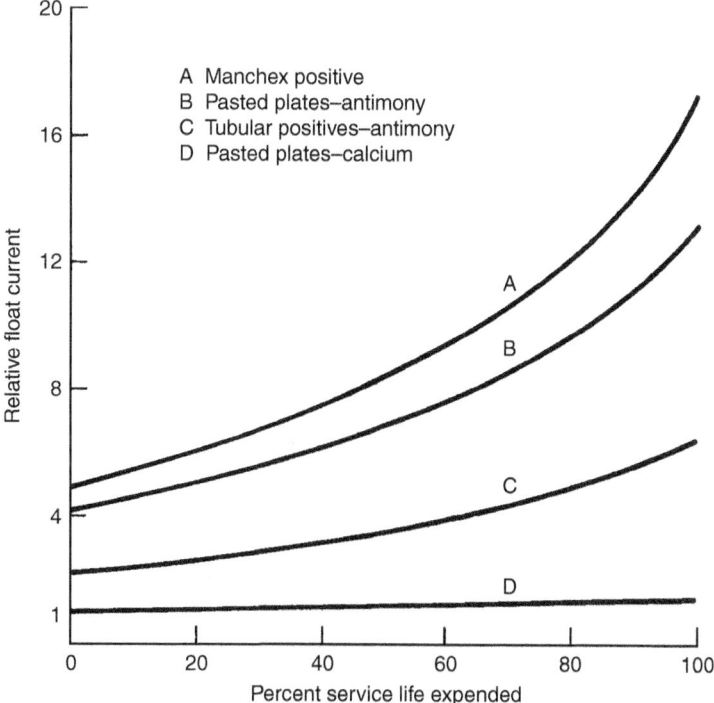

FIGURE 16.47 Relative self-discharge of lead-acid stationary batteries of different construction.

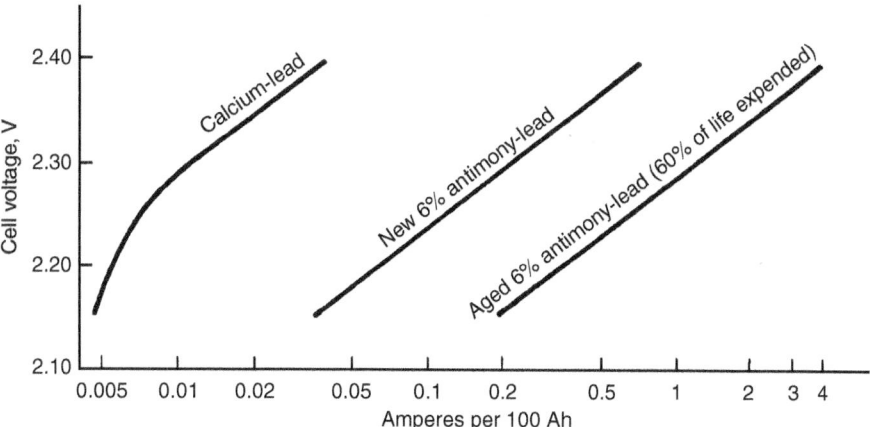

FIGURE 16.48 Float current characteristics of stationary batteries at 25°C, 100 Ah cells, fully charged, 1.210 specific gravity.

16.66 SECONDARY BATTERIES

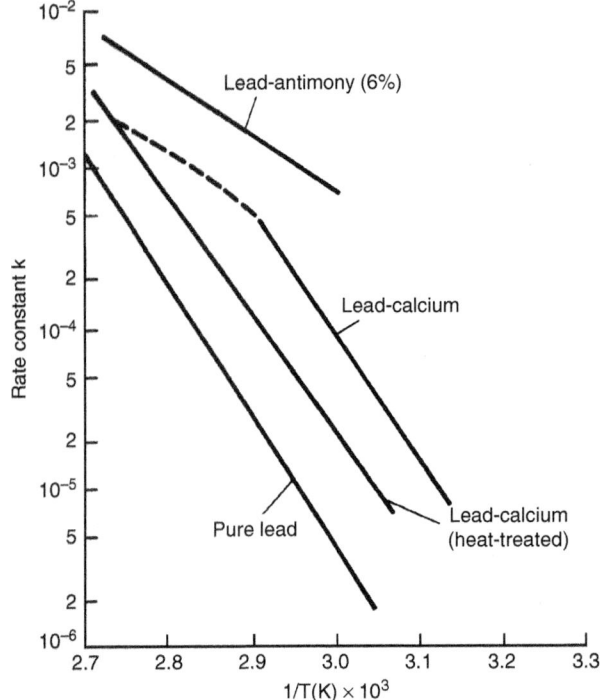

FIGURE 16.49 Corrosion rate constant log k vs. $1/T$ for different lead-alloy grids. (*From Ref. 49.*)

series cell strings connected in parallel for greater storage capacity. Most stationary batteries in the United States are made with pasted positive plates. In Europe, Planté and tubular positives are more popular than in the United States.

Listings of stationary batteries with flat-pasted, tubular, and Planté plates are given in Table 16.17 (*a–c*). The basic rating unit is the positive plate, given in ampere-hours at the 8 h rate

TABLE 16.17a Typical Stationary Batteries, Flat-Pasted Plates

Positive-plate capacity, Ah at 8 h rate	Plate dimensions, mm				Cell size*† (positive plates per cell)
			Thickness		
	Height	Width	Positive	Negative	
5	89	63.5	6.6	4.3	2,4
25	149	143	6.6	4.3	1–8
90–95	290	222	7.9	5.3	2–12
150–155	381	304	6.4	4.6	2–17
166	381	304	7.9	5.3	5–16
195	457	338	6.9	5.3	13–18
412	1816	338	7.6	5.5	17–19

*Typical cell construction: n positive and $n + 1$ outside negative plates per cell. Some smaller cell sizes are assembled in multiples of two, three, or four cells in a monolithic container.
†Typical cell characteristics: 10 positive 168 Ah plates (1680 Ah cell); weight: 140 kg; size: length, 270 mm; width, 359 mm; height, 575 mm.
Source: C & D Technologies, Inc., Blue Bell, PA.

TABLE 16.17b Typical Stationary Batteries, Tubular Plates

Positive-plate capacity, Ah		Plate dimensions, mm				Cell size*† (positive plates per cell)
At 4 h rate	At 8 h rate	Height	Width	Thickness		
				Positive	Negative	
26	31.25	157	203	8.9	5.6	4
76	96	277P	234P	8.9		3–10
		290N	239N		6.1	
88	105	277P	234P	8.9		3–10
		290N	239N		6.1	
124	152	366	307	8.9	4.8	5–14

*Typical cell construction: n positive and $n + 1$ outside negative plates per cell; used with flat-pasted negative plates.
†Typical cell characteristics: 11 positive 152 Ah plates (1672-Ah cell); weight: 128 kg; size: length, 272 mm; width, 368 mm; height, 577 mm.
Source: Enersys, Inc., and Tudor AB.

TABLE 16.17c Typical Stationary Batteries, Planté Plates

Positive-plate capacity, Ah at 8 h rate	Plate dimensions, mm				Cell size* (positive plates per cell)
	Height	Width	Thickness		
			Positive	Negative	
	Planté type†‡				
8	140	140	9.5	4.7	3, 5, 7
20			9.5	4.7	5–17
40			11.1		9–25
80	286	233	9.5	4.7	2–7
83			11.1		13–25
	Manchester type†§				
20	155	149	9.7	4.6	2, 3, 4
40	197	197	11.2	4.6	2–9
83	282	292	11.2	4.6	5–12

*Typical cell construction: n positive and $n + 1$ negative outside plates per cell.
†Used with flat-pasted negative plates.
‡Typical Planté cell characteristics: 2 positive 80 Ah plates (160 Ah cell). Two-cell battery size: L 283 mm, W 159 mm, H 463 mm.
§Typical Manchester cell characteristics: 4 positive 40 Ah plates (160 Ah cell). One-cell battery weight: 40 kg; size: length, 131 mm; width, 257 mm; height, 455 mm.
Source: Enersys, Inc.

unless specific otherwise; ratings of the cell are the product of the capacity of a single positive plate multiplied by the number of positive plates. A popular stationary battery size is the 1680 Ah cell (168 Ah positive, 10 positive or 21 plates per cell) for use in telephone exchanges.

16.7 CHARGING AND CHARGING EQUIPMENT

16.7.1 General Considerations

In the charging process, DC electric power is used to reform the active chemicals of the battery system to their high-energy, charged state. In the case of the lead-acid battery, this involves, as shown in Sec. 16.2, the conversion of lead sulfate in the positive electrodes to lead oxide (PbO_2), the conversion of lead sulfate of the negative electrode to metallic lead (sponge lead), and the restoration of the electrolyte from a low-concentration sulfuric acid solution to the higher concentration of approximately 1.21 to 1.30 specific gravity. Since a change of phase from solid to solution is involved with the sulfate ion, charging lead-acid batteries has special diffusional considerations and is temperature-sensitive. During charge and discharge the solid materials that go into solution as ions are reprecipitated as a different solid compound. This also causes a redistribution of the active material. The rearrangement will tend to make the active material contain a crystal structure with fewer defects, which results in less chemical and electrochemical activity. Therefore the lead-acid battery is not as reversible physically as it is chemically.[53] This physical degradation can be minimized by proper charging, and often batteries discarded as dead can be restored with a long, slow recharge (3 to 4 days at 2 to 3 A for SLI batteries).

A lead-acid battery can generally be charged at any rate that does not produce excessive gassing, overcharging, or high temperatures. The battery can absorb a very high current during the early part of the charge, but there is a limit to the safe current as the battery becomes charged. This is shown in Fig. 16.50, which is a graphic representation of the ampere-hour rule

$$I = Ae^{-t}$$

where I is the charging current, A is the number of ampere-hours previously discharged from the battery, and t equals time. Because there is considerable latitude, there are a number of charging regimes, and the selection of the appropriate method depends on a number of considerations, such as the type and design of the battery, service conditions, time available for charging, number of cells or batteries to be charged, and charging facilities. Figure 16.51 shows the relation of cell voltage to the state of charge and the charging current. The figure shows that a fully discharged battery can absorb high currents with the charging voltage remaining relatively low. However, as the battery becomes charged, the voltage increases to excessively high values if the charge is maintained at the high rate, leading to overcharge and gassing. The charge current should be reduced to reasonable values as the battery reaches full charge.

In automotive, marine, and other vehicle applications, the DC electric power is usually provided by an on-board generator or alternator driven from the prime engine. These devices have a voltage and current limiter to prevent overcharging. The proper limit is dependent on the chemistry and physical construction of the cell or battery. For the traditional automotive batteries that use antimonial lead alloy as grid material, voltage limits in the range of 14.1 to 14.6 V for a nominal 12 V battery are usual. With the newer maintenance-free batteries, which use a calcium-lead alloy grid or

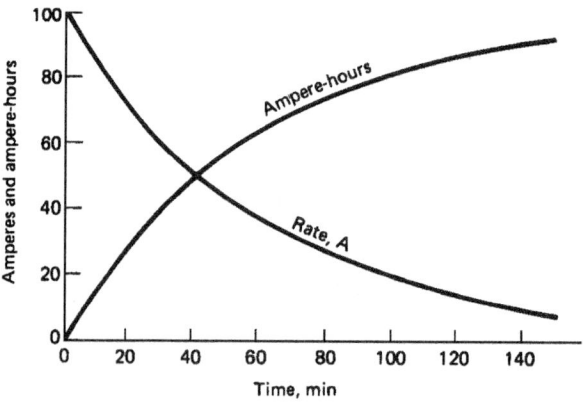

FIGURE 16.50 Graphic illustration of ampere-hour law. (*From Ref. 42.*)

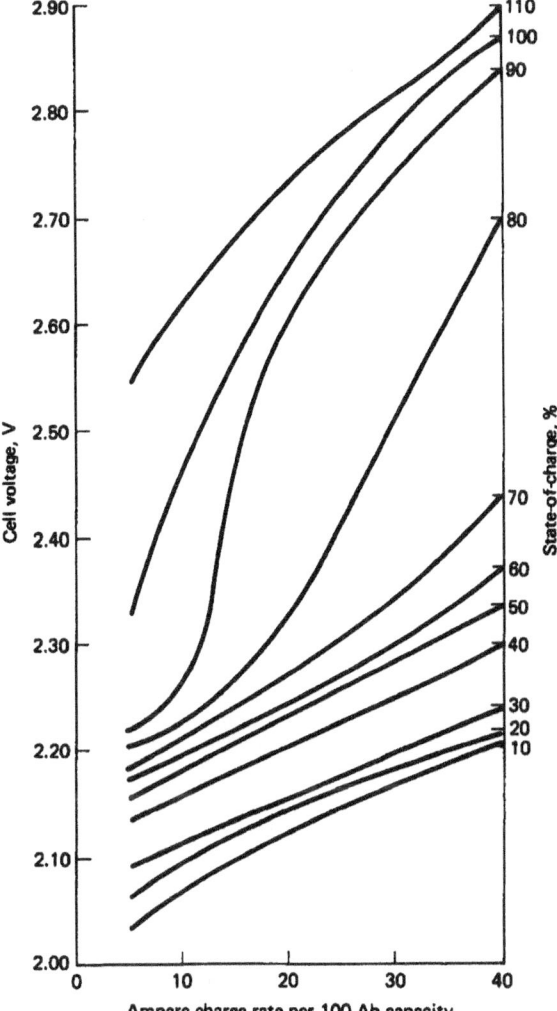

FIGURE 16.51 Charging voltage of lead-acid battery at various states of charge. (*From Ref. 47.*)

other grid material with high hydrogen overvoltage, higher charging voltages, in the range of 14.5 to 15.0 V, can be used without danger of overcharge. Batteries in automobile and similar applications today see what is close to cycling rather than float service, but the charging controls are such that very little gas is evolved on charge. This minimizes the requirement for watering, but makes accurate control of the charge necessary. The charging rates for the different types of SLI batteries are compared in Fig. 16.52.[21] The calcium-lead maintenance-free battery is less affected by high settings in the voltage regulator than the other batteries.

In many nonautomotive applications, charging is done separately from the system using the battery. The direct current necessary for charging is usually obtained by rectifying alternating current. These chargers include wall-hung units and mobile units as well as floor-mounted units. Newer charger designs have microprocessor controls, can sense battery condition, temperature,

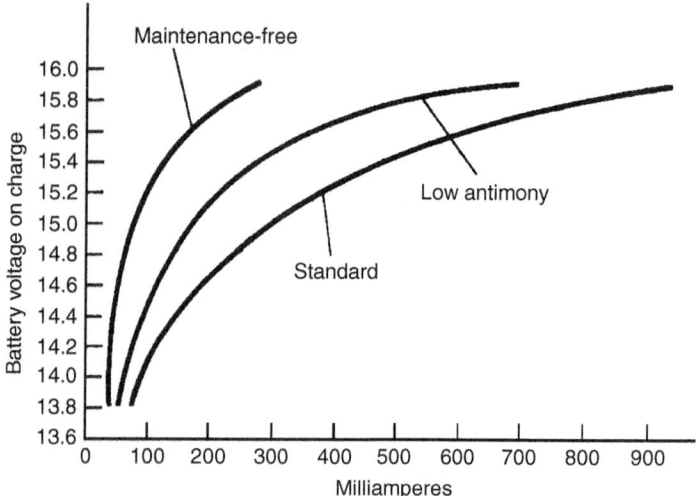

FIGURE 16.52 Charging characteristics of lead-acid SLI batteries at 25°C.

voltage, charge current, and so on, and are capable of changing charging rates during the charge. Most rectifiers produce some AC ripple with the direct current, which causes additional heating of the battery. This should be minimized, especially near the end of the charge when batteries tend to get hot. Pulse charging and the use of asymmetric alternating current have been proposed as a means to overcome this problem, but practical lead-acid batteries have such large capacitances that the pulses are smoothed out and the effects minimized.[53]

16.7.2 Methods of Charging Lead-Acid Batteries

Proper recharging is important to obtain optimum life from any lead-acid battery under any conditions of use. Some of the rules for proper charging are given below and apply to all types of lead-acid batteries.

1. The charge current at the start of recharge can be any value that does not produce an average cell voltage in the battery string greater than the gassing voltage (about 2.4 V per cell).
2. During the recharge and until 100% of the previous discharge capacity has been returned, the current should be controlled to maintain a voltage lower than the gassing voltage. To minimize charge time, this voltage can be just below the gassing voltage.
3. When 100% of the discharged capacity has been returned under this voltage control, the charge rate will have normally decayed to the charge "finishing" rate. The charge should be finished at a constant current no higher than this rate, normally 5 A per 100 Ah of rated capacity (referred to as the 20 h rate).

A number of methods for charging lead-acid batteries have evolved to meet these conditions. These charging methods are commonly known as:

1. Constant-current, one-current rate
2. Constant-current, multiple decreasing-current steps
3. Modified constant current
4. Constant potential

5. Modified constant potential with constant initial current
6. Modified constant potential with a constant finish rate
7. Modified constant potential with a constant start and finish rate
8. Taper charge
9. Pulse charging
10. Trickle charging
11. Float charging
12. Rapid charging

Constant-Current Charging. Constant-current recharging, at one or more current rates, is not widely used for lead-acid batteries. This is because of the need for current adjustment unless the charging current is kept at a low level throughout the charge (ampere-hour rule), which will result in long charge times of 12 h or longer. Typical charger and battery characteristics for the constant-current charge, for single and two-step charging, are shown in Fig. 16.53.

Constant-current charging is used for some small lead-acid batteries (see Chap. 17). The use of a constant-current charge during the initial battery "formation" charge has been described in Sec. 16.3.

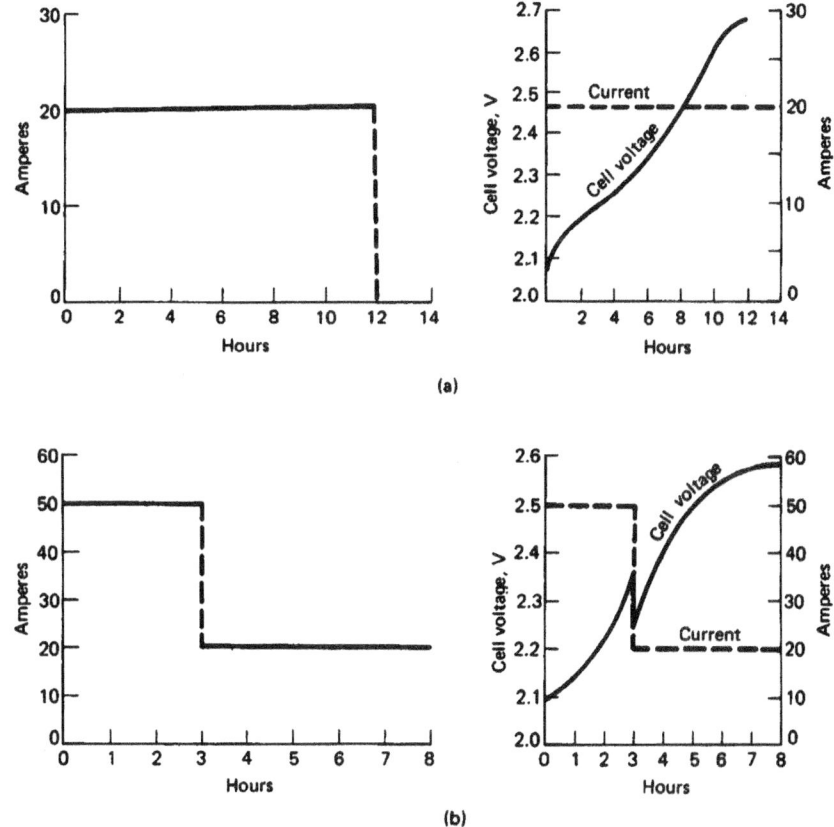

FIGURE 16.53 Typical charger and battery characteristics for constant-current charging of lead-acid batteries. (*a*) Single-step constant-current charging. (*b*) Two-step constant-current charging. (*From Ref. 22.*)

Constant-current charging is also used at times in the laboratory because of the convenience of calculating ampere-hour input and because constant-current charging can be done with simple, inexpensive equipment. Constant-current charging at half the 20 h rate can be used in the field to decrease the sulfation in batteries which have been over-discharged or undercharged. This treatment, however, may diminish battery life and should be used only with the advice of the battery manufacturer.

Constant-Potential Charging. The characteristics of constant-potential and modified constant-potential charging are illustrated in Fig. 16.54. In normal industrial applications, modified constant-potential charging methods are used (methods 5, 7, and 8). Modified constant-potential charging (method 5) is used for on-the-road vehicles and utility, telephone, and uninterruptible power system applications where the charging circuit is tied to the battery. In this case, the charging circuit has a current limit, and this value is maintained until a predetermined voltage is reached. Then the voltage is maintained constant until the battery is called on to discharge. Decisions must be made regarding the current limit and the constant-voltage value. This is influenced by the time interval when the battery is at the constant voltage and in a 100% state of charge. For this "float"-type operation with the battery always on charge, a low charge current is desirable to minimize overcharge, grid corrosion associated with overcharge, water loss by electrolysis of the electrolyte, and maintenance to replace this water. To achieve a full recharge with a low constant potential requires the proper selection of the starting current, which is based on the manufacturer's specifications.

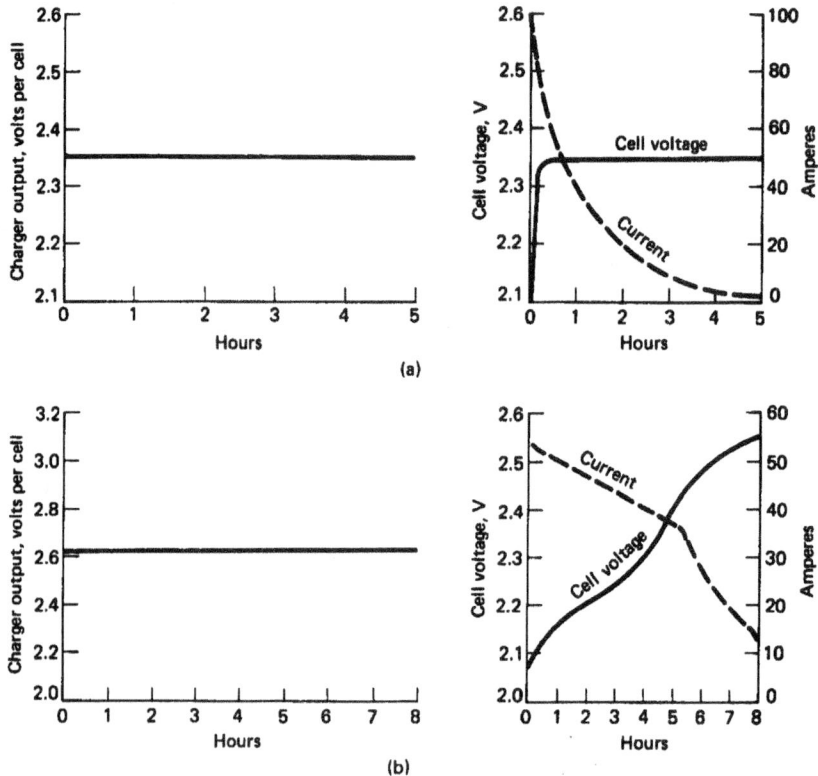

FIGURE 16.54 Typical charger and battery characteristics for constant-potential charging of lead-acid batteries. (*a*) Constant-potential charging. (*b*) Modified constant-potential charging. (*From Ref. 22.*)

The modified constant-potential charge, with constant start and finish rates, is common for deep-cycling batteries which are typically discharged at the 6 h rate to a depth of 80%; the recharge is normally completed in an 8 h period. The charger is set for the constant potential of 2.39 V per cell (the gassing voltage), and the starting current is limited to 16 to 20 A per 100 Ah of the rated 6 h ampere-hour capacity by means of a series resistor in the charger circuit. This initial current is maintained constant until the average cell voltage in the battery reaches 2.39 V. The current decays at constant voltage to the finishing rate of 4.5 to 5 A per 100 Ah, which is then maintained to the end of the charge. Total charge time is controlled by a timer. The time of charge is selected to ensure a recharge input capacity of a predetermined percent of the ampere-hour output of the previous discharge, normally 110 to 120%, or 10 to 20% overcharge. The 8 h charging time can be reduced by increasing the initial current limit rate.

Taper Charging. Taper charging is a variation of the modified constant-potential method, using less sophisticated controls to reduce equipment cost. The characteristics of taper charging are illustrated in Fig. 16.55. The initial rate is limited, but the taper of voltage and current is such that the 2.39 V per cell at 25°C is exceeded prior to the 100% return of the discharge ampere-hours. This method does result in gassing at the critical point of recharge, and the cell temperature is increased. The degree of gassing and temperature rise is a variable depending on the charger design, and battery life can be degraded from excessive battery temperature and overcharge gassing (see Sec. 16.8.3). The gassing voltage decreases with increasing temperature; correction factors given in Table 16.18 provide the voltage correction factors at temperatures other than 25°C.

The end of the charge is often controlled by a fixed voltage rather than a fixed current. Therefore when a new battery has a high counter-EMF, this final charge rate is low and the battery often does not receive sufficient charge within the time period allotted to maintain the optimum charge state. During the latter part of life when the counter-EMF is low, the charging rate is higher than the normal finishing rate, and so the battery receives excessive charge, which degrades life. Thus the taper charge does degrade battery life, which must be justified by the use of less expensive equipment.

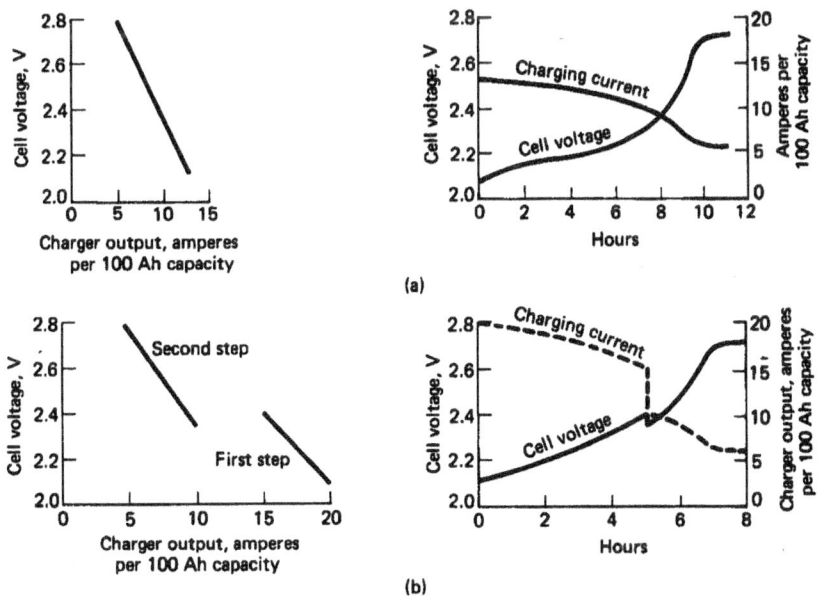

FIGURE 16.55 Typical charger and battery characteristics for taper charging of lead-acid batteries. (*a*) Single-step taper charge. (*b*) Two-step taper charge. (*From Ref. 22.*)

TABLE 16.18 Correction Factors for Cell Gassing Voltage

Electrolyte temperature, °C	Cell gassing voltage, V	Correction factor, V
50	2.300	−0.090
40	2.330	−0.060
30	2.365	−0.025
25	2.390	0
20	2.415	+0.025
10	2.470	+0.080
0	2.540	+0.150
−10	2.650	+0.260
−20	2.970	+0.508

For photovoltaic battery systems and other systems designed for optimum life, charging control and regulation circuits should produce a pattern of voltage and current equivalent to the best industrial circuits. Modified constant-potential charging methods with constant initial current (methods 5 and 7) are preferred. Optimum control to maximize life and energy output from the battery is best achieved when the depth of discharge and the time for recharge are predetermined and repetitive, a condition not always realized in solar photovoltaic applications.

Pulse Charging. Pulse charging is also used for traction applications, particularly in Europe. In this case, the charger is periodically isolated from the battery terminals, and the open-circuit voltage of the battery is automatically measured (an impedance-free measurement of the battery voltage). If the open-circuit voltage is above a preset value, depending on a reference temperature, the charger does not deliver energy. When the open-circuit voltage decays below that limit, the charger delivers a DC pulse for a fixed time period. When the battery state-of-charge is very low, charging current is connected almost 100% of the time because the open-circuit voltage is below the present level or rapidly decays to it. The duration of the open-circuit and the charge pulses are chosen so that when the battery is fully charged, the time for the open-circuit voltage to decay is exactly the same as the pulse duration. When the charger controls sense this condition, the charger is automatically switched over to the finish rate current and short charging pulses are delivered periodically to the battery to maintain it at full charge. In many industrial applications high-voltage batteries may be used and difficulty can be encountered in keeping the cells in a balanced condition. This is particularly true when the cells have long periods of standby use with different rates of self-decay. In these applications, the batteries are completely discharged and recharged periodically (usually semiannually) in what is called an equalizing charge, which brings the whole string of cells back to the complete charge state. On completion of this process, the liquid levels in the cells must be checked and water added to depleted cells as required. With the newer types of maintenance-free cells, which are semisealed, such equalizing charges and differential watering of the cells may not be possible, and special precautions are taken in the charger design to keep the cells at an even state of charge.

Trickle Charging. A trickle charge is a continuous constant-current charge at a low (about $C/100$) rate, which is used to maintain the battery in a fully charged condition, recharging it for losses due to self-discharge as well as to restore the energy discharged during intermittent use of the battery. This method is typically used for SLI and similar type batteries when the battery is removed from the vehicle or its regular source of charging energy for charging.

Float Charging. Float charging is a low-rate constant-potential charge also used to maintain the battery in a fully charged condition. This method is used mainly for stationary batteries which can be charged from a DC bus. The float voltage for a non-antimonial grid battery containing 1.210

specific gravity electrolyte and having an open-circuit voltage of 2.059 V per cell is 2.17 to 2.25 V per cell.

Rapid Charging. In many applications, it is desirable to be able to rapidly recharge the battery within an hour or less. As is the case under any charging condition, it is important to control the charge to maintain the morphology of the electrode, to prevent a rise in the temperature, particularly to a point where deleterious side reactions (corrosion, conversion to nonconducting oxides, high solubility of materials, decomposition) take place, and to limit overcharge and gassing. As these conditions are more prone to occur during high-rate charging, charge control under these conditions is critical.

The availability of small, low-cost but sophisticated semiconductor chips has made effective methods of controlling the charging voltage-current-profile feasible. These devices can be used to either terminate the charge, limit the charge current, or switch between charge regimes when potentially damaging conditions arise during the charge.

A number of different techniques have been developed for effective rapid recharge. In one method, referred to as "reflex" charging, a brief discharge pulse of a fraction of a second, is incorporated into the charging regime. This technique has been found to be effective in preventing an excessive rise in temperature during rapid (15 min) high-rate recharging.

16.8 MAINTENANCE, SAFETY, AND OPERATIONAL FEATURES

16.8.1 Maintenance

It is common for industrial lead-acid batteries to function for periods of 10 years or longer. Proper maintenance can ensure this extended useful life. Five basic rules of proper maintenance are:

1. Match the charger to the battery charging requirements.
2. Avoid overdischarging the battery.
3. Maintain the electrolyte at the proper level (add water as required).
4. Keep the battery clean.
5. Avoid overheating the battery.

In addition to these basic rules, as the battery is made of individual cells connected in series, the cells must be properly balanced periodically.

Charging Practice. Poor charging practice is responsible for short battery life more than any other cause. Fortunately the inherent physical and chemical characteristics of lead-acid batteries make control of charging quite simple. If the battery is supplied with DC energy at the proper charging voltage, the battery will draw only the amount of current that it can accept efficiently, and this current will reduce as the battery approaches full charge. Several types of devices can be used to ensure that the charge will terminate at the proper time. The specific gravity of the electrolyte should also be checked periodically for those batteries that have a removable vent and adjusted to the specified value (see Tables 16.7 and 16.12).

Overdischarge. Overdischarging the battery should be avoided. The capacity of large batteries, such as those used in industrial trucks, is generally rated in ampere-hours at the 6 h discharge rate to a final voltage of 1.75 V per cell. These batteries can usually deliver more than rated capacity, but this should be done only in an emergency and not on a regular basis. Discharging cells below the specified voltage reduces the electrolyte to a low concentration, which has a deleterious effect on the pore structure of the battery. Battery life has been shown to be a direct function of the depth of discharge, as illustrated in Fig. 16.56.[54]

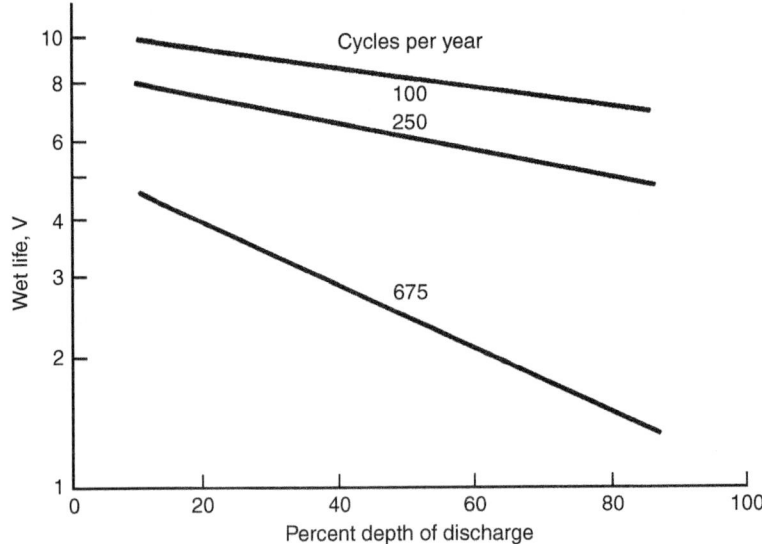

FIGURE 16.56 Effect of depth of discharge and number of cycles per year on wet life at 25°C. (*From Ref. 54.*)

Electrolyte Level. During normal operation, water is lost from a battery as the result of evaporation and electrolysis into hydrogen and oxygen, which escape into the atmosphere. Evaporation is a relatively small part of the loss, except in very hot, dry climates. With a fully charged battery, electrolysis consumes water at a rate of 0.336 cm^3 per ampere-hour overcharge. A 500 Ah cell overcharged 10% can thus lose 16.8 cm^3, or about 0.3% of its water each cycle. It is important that the electrolyte be maintained at the proper level in the battery. The electrolyte not only serves as the conductor of electricity but is a major factor in the transfer of heat from the plates. If the electrolyte is below the plate level, then an area of the plate is not electrochemically active; this causes a concentration of heat in other parts of the cell. Periodic checking of water consumption can also serve as a rough check on charging efficiency and may warn when adjustment of the charger is required.

Since replacing water can be a major maintenance cost, water loss can be reduced by controlling the amount of overcharge and by using hydrogen and oxygen recombining devices in each cell where possible. Addition of water is best accomplished after recharge and before an equalization charge. Water is added at the end of the charge to reach the high acid level line. Gassing during charge will stir the water into the acid uniformly. In freezing weather, water should not be added without mixing, as it may freeze before gassing occurs. Water added must be either distilled water, demineralized water, or local water which has been approved for use in batteries. Automatic watering devices and reliability testing can reduce maintenance labor costs further. Overfilling must be avoided because the resultant overflow of acid electrolyte will cause tray corrosion, ground paths, and loss of cell capacity. A final check of specific gravity should be made after water has been added to ensure correct acid concentration at the end of charge. A helpful approximation is

$$\text{specific gravity} = \text{cell open-circuit voltage} - 0.845$$

which permits electrical monitoring of specific gravity on an occasional basis (see also Fig. 16.6). Although distilled water is no longer specified by most battery manufacturers, good-quality water, low in minerals and heavy-metal ions such as iron, will help prolong battery life.

Cleanliness. Keeping the battery clean will minimize corrosion of cell post connectors and steel trays and avoid expensive repairs. Batteries commonly pick up dry dirt, which can be readily blown off or brushed away. This dirt should be removed before moisture makes it a conductor of stray currents. One problem is that the top of the battery can become wet with electrolyte any time a cell is overfilled. The acid in this electrolyte does not evaporate and should be neutralized by washing the battery with a solution of baking soda and hot water, approximately 1 kg of baking soda to 4 L of water. After application of such a solution, the area should be rinsed thoroughly with water.

High Temperature—Overheating. One of the most detrimental conditions for a battery is high temperature, particularly above 55°C, because the rates of corrosion, solubility of metal components, and self-discharge increase with increasing temperature. High operating temperature during cycle service requires higher charge input to restore discharge capacity and local action (self-discharge) losses. More of the charge input is consumed by the electrolysis reaction because of the reduction in the gassing voltage at the higher temperature (see Table 16.19). While a 10% overcharge per cycle maintains the state of charge at 25 to 35°C, 35 to 40% overcharge may be required to maintain the state of charge at the higher (60 to 70°C) operating temperatures. On float service, float currents increase at the higher temperatures, resulting in reduced life. Eleven days float at 75°C is equivalent in life to 365 days at 25°C.

Batteries intended for high-temperature applications should use a lower initial specific gravity electrolyte than those intended for use at normal temperatures (see Table 16.12). Other design features, such as the use of more expander in the negative plate, are also important to improve operation at high temperatures.

Cell Balancing. During cycling, a high-voltage battery having many cells in a series string can become unbalanced, with certain cells limiting charge and discharge. Limiting cells receive more overcharge than other cells in the string, have greater water consumption, and thus require more maintenance. The equalization charge has the function of balancing cells in the string at the top of charge. In an equalization charge, the normal recharge is extended for 3 to 6 h at the finishing rate of 5 A per 100 Ah, 5 h rated capacity, allowing the battery voltage to rise uncontrolled. The equalization charge should be continued until cell voltages and specific gravities rise to a constant, acceptable value. Frequency of equalization charge is normally a function of the accumulative discharge output and will be specified by the manufacturer for each battery design and application.

16.8.2 Safety

Safety problems associated with lead-acid batteries include spills of sulfuric acid, potential explosions from the generation of hydrogen and oxygen, and the generation of toxic gases such as arsine and stibine. All these problems can be satisfactorily handled with proper precautions. Wearing of face shields and plastic or rubber aprons and gloves when handling acid is recommended to avoid chemical burns from sulfuric acid. Flush immediately and thoroughly with clean water if acid gets into the eyes, skin, or clothing and obtain medical attention when eyes are affected. A bicarbonate of soda solution (100 g per liter of water) is commonly used to neutralize any acid accidentally spilled. After neutralization the area should be rinsed with clear water.

Precautions must be routinely practiced to prevent explosions from ignition of the flammable gas mixture of hydrogen and oxygen formed during overcharge of lead-acid batteries. The maximum rate of formation is 0.42 L of hydrogen and 0.21 L of oxygen per ampere-hour overcharge at standard temperature and pressure. The gas mixture is explosive when hydrogen in air exceeds 4% by volume. A standard practice is to set warning devices to ring alarms at 20 to 25% of this lower explosive limit (LEL). Low-cost hydrogen detectors are available commercially for this purpose.

With good air circulation around a battery, hydrogen accumulation is normally not a problem. However, if relatively large batteries are confined in small rooms, exhaust fans should be installed to vent the room constantly or to be turned on automatically when hydrogen accumulation exceeds 20% of the lower explosive limit. Battery boxes should also be vented to the atmosphere. Sparks or flame can ignite these hydrogen atmospheres above the LEL. To prevent ignition, electrical sources of arcs, sparks, or flame must be mounted in explosion-proof metal boxes. Battery cells can similarly be equipped with flame arrestors in the vents to prevent outside sparks from igniting explosive gases inside the cell cases. It is good practice to refrain from smoking, using open flames, or creating sparks in the vicinity of the battery. A considerable number of the reported explosions of batteries come from uncontrolled charging in nonautomotive applications. Often batteries will be charged, off the vehicle, for long periods of time with an unregulated charger. In spite of the fact that the charge currents can be low, fair volumes of gas can accumulate. When the battery is then moved, this gas vents, and if a spark is present, explosions have been known to occur. The introduction of the calcium alloy grids has minimized this problem, but the possibility of explosion is still present.

Some types of batteries can release small quantities of the toxic gases stibine and arsine. These batteries have positive or negative plates which contain small quantities of the metals antimony and arsenic in the grid alloy to harden the grid and to reduce the rate of corrosion of the grid during cycling. Arsine (AsH_3) and stibine (SbH_3) are generally formed when the arsenic or antimony alloy material comes into contact with nascent hydrogen, usually during overcharge of the battery, which then combines to form these colorless and essentially odorless gases. They are extremely dangerous and can cause serious illness and death. The OSHA 1978 concentration limits for SbH_3 and AsH_3 are 0.1 and 0.05 ppm, respectively, as a maximum allowable weighted average for any 8 h period. Ventilation of the battery area is very important. Indications are that ventilation designed to maintain hydrogen below 20% LEL (approximately 1% hydrogen) will also maintain stibine and arsine below their toxic limits.

The ordinary 12 V SLI automotive battery is a minor shock hazard. The hazard level increases with higher-voltage systems, and systems in the range of 84 to 360 V are being used for electric vehicles. Systems as high as 1000 V are under study for fixed-location energy-storage systems for load leveling. Batteries are electrically alive even in the discharged state, and the following precautions should be practiced:

1. Keep the top of the battery clean and dry to prevent ground short circuits and corrosion.
2. Do not lay metallic objects on the battery. Insulate all tools used in working on batteries.
3. Remove jewelry and any other electrical conductor before inspecting or servicing batteries.
4. When lifting batteries, use insulated lifting tools to avoid risks or short circuits between cell terminals by lifting chains or hooks.
5. Make sure gases do not accumulate in batteries before they are moved.

16.8.3 Effect of Operating Parameters on Battery Life

Operating parameters which have a strong influence on battery life are depth of discharge, number of cycles used each year, charging control, type of storage, and operating temperature. In some cases, the battery design features that increase life tend to decrease the initial capacity, power, and energy output. It is important, therefore, that the design features of the battery be selected to match the operating and life requirements of the application.

1. Increasing the depth of discharge decreases cycle life, as illustrated in Figs. 16.57[54] and 16.37.
2. Increasing the number of cycles performed per year decreases the wet life (Sec. 16.8.1 and Fig. 16.56).
3. Excessive overcharging leads to increasing positive grid corrosion, active material shedding, and shorter wet life.
4. Storing wet cells in a discharged condition promotes sulfation and decreases capacity and life.
5. Proper charging operations with good equipment maintain the desired state of charge with a minimum of overcharge and lead to optimum battery life.

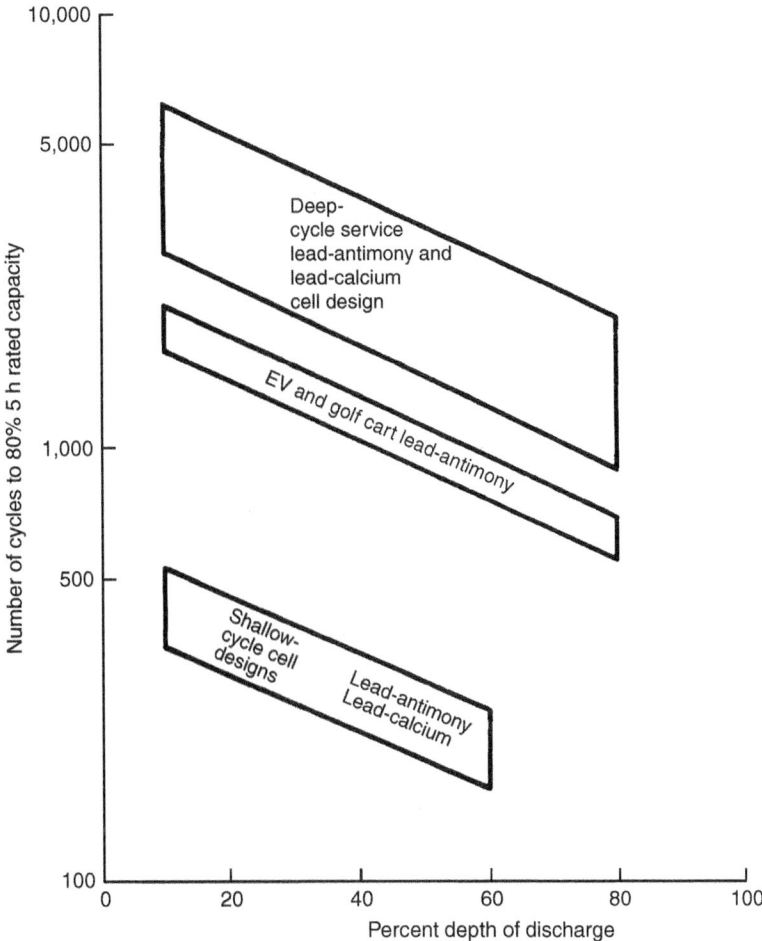

FIGURE 16.57 Effect of cell design and depth of discharge on cycle life of various types of lead-acid batteries at 25°C. (*From Ref. 54.*)

6. Stratification of the electrolyte in large cells into levels of varying concentration can limit charge acceptance, discharge output, and life unless controlled during the charge process. During a recharge, sulfuric acid of higher concentration than the bulk electrolyte forms in the pores of the plates. This higher-density acid settles to the bottom of the cell, giving higher specific gravity acid near the bottom of the plates and lower specific gravity acid near the top of the plates. This stratification accumulates during the nongassing periods of charge. During the gassing periods of overcharge, partial stirring is accomplished by gas bubbles formed at and rising along the surfaces of the plates and in the separator system. During discharge, acid in the pores of the plates and near their surface is diluted; however, concentration gradients set up by longer charge periods are seldom compensated entirely, particularly if the discharge periods are shorter, as is usually the case. Diffusion processes to eliminate these concentration gradients are very slow, and stratification during repetitive cycling can become progressively greater. Two methods for stratification control are by deliberate gassing of the plates during overcharge at the finishing rate and by

stirring of cell electrolyte by pumps (usually airlift pumps). The degree of success in eliminating stratification is a function of cell design, the design of the pump accessory system, and cell operating procedures.

16.8.4 Failure Modes

The failure modes of lead-acid batteries depend on the type of application and the particular battery design. This is the rationale for the manufacture of different batteries since each one is designed to give optimum performance in a specific type of use. The more prevalent failure modes for the different types of lead-acid batteries are listed in Table 16.19.[55] Significantly, if a battery is properly maintained, most of the inherent failures are due to the degradation of the positive plate through either grid corrosion or paste shedding. These failures are irreversible, and when they occur, the battery must be replaced. Details of the failure modes of SLI batteries are given in Sec. 16.4.3.

The failure mode and the time to failure can be modified by changes in the inner parameters (I), such as battery materials, processing, and design, or by the conditions of use, designated as the outer parameters (O). Some of these are listed in Table 16.20.[55]

TABLE 16.19 Failure Modes of Lead-Acid Batteries

Battery type	Normal life	Normal failure mode
SLI	Several years	Grid corrosion
SLI (maintenance-free)	Several years	Lack of water, damage to positive plates
Golf cart	300–600 cycles	Positive shedding and grid corrosion, sulfation
Stationary (Industrial)	6–25 years	Grid corrosion
Traction (Industrial)	Minimum 1500 cycles	Shedding, grid corrosion

TABLE 16.20 Modification of Lead-Acid Battery Failure Rate

Failure mechanism	Rate of failure modification*
Shedding positives	I: active mass structure, battery design
	O: number of cycles, depth of discharge, charge factor
Sulfation/leading of negatives	I: active mass additives
	O: temperature, charge factor, maintenance
Positive grid corrosion (overall, localized, or positive grid growth)	I: grid alloy, casting conditions, active mass
Separators	I: electrolyte concentrations, battery design
	O: temperature, charge factor, maintenance
Case, cover, vents, external battery connections	I: battery materials and design
	O: maintenance, abuse

*I—Inner parameters; O—outer parameters.

16.9 APPLICATIONS AND MARKETS

The lead-acid battery is used in a wide variety of applications, and in the past few years many new applications have arisen. The various types of lead-acid batteries and their applications are listed in Table 16.2. The new uses of lead-acid batteries are mainly associated with the smaller sealed maintenance-free cells used in electronic and portable devices and with the advanced designs for energy storage and electric vehicles.

16.9.1 Automotive Applications

Traditionally the most common use of the lead-acid battery is for starting, lighting, and ignition in automobiles and other vehicles with internal combustion engines. Almost all of these now have 12 V nominal electric systems. Most of the earlier generators have been replaced by alternators and electromechanical regulators by electronic/solid-state controls. High cranking ability at low temperatures is still the major design factor, but SLI batteries today see more cycling-type service (compared with float service) because of the electrical load of the auxiliaries. Size and weight reduction have also become important as well as the battery geometry. Batteries are normally located in the cool air stream ahead of the engine to prevent their overheating; thus their geometry can affect the profile of the vehicle. These factors have led to the redesign of the lead-acid battery for SLI applications. The most important changes were:

- Change from high-antimony (4 to 5%) lead alloy grid to a low-antimony (1 to 2%) or nonantimonial lead alloy grid, thus reducing hydrogen evolution
- Use of thinner electrodes
- Better separators with lower electrical resistance
- Plate tabs located in from corners, and grids redesigned for high conductivity
- Semisealed, maintenance-free construction

Automotive-type batteries are also used on trucks, aircraft, industrial equipment, and motorcycles as well as in many other applications. They are used in off-road vehicles, such as snowmobiles, in boats to crank inboard and outboard engines, and in various farm and construction equipment. Military vehicles in the United States and NATO countries have standardized on a 24 V electric system that is provided by a series connection of two 12 V batteries.

The term "SLI battery" has evolved into something of a misnomer. In addition to starting, lighting and ignition, the automotive SLI battery may provide the power for many other functions. Although these features may not pose much of a burden individually, collectively they add up to a significant drain on the SLI battery. Some of today's automobiles require up to 2 to 3 kW of power. This could more than double in the next few years. Table 16.21 shows some current or anticipated features requiring power exclusive of the SLI functions. The typical SLI battery is not designed to handle the cycling demands prompted by some of these items. The car companies are still investigating higher voltage systems for lead acid, the major advantage of going to the higher voltage being that, at a given power level, the current required would be proportionately less than with the conventional 12 V battery. The higher voltage will result in a substantial saving in weight of the current-carrying distribution system in the car, i.e., less copper will be required, but the higher voltage has created issues of safety regarding the car's wiring insulation and its resistance to the salts, etc., that the wires

TABLE 16.21 Features Requiring Power for Present and Future Automobile Designs (Exclusive of SLI Function)

Alarms (may include flashing LEDs)	Communication devices
Computer	Audio-radio, tape or CD players
Electric suspension	Global positioning features (maps, routing, emergency location)
Automatic start-stop of engine	Electromagnetic valve trains
Air conditioning	Electric heating of catalytic converters
Electric heating of seats	Sensing (e.g., for airbag deployment)
Electric steering	Anti-lock braking
Power windows	Electrochromic mirrors
Rear seat entertainment center	Rear window deicer/defogger
Electric door locks	Cigarette lighter (other functions)
Clock	Cruise control
Regenerative braking	

are exposed to during normal exposure from arch tracking of the insulation. In turn, this saving is projected to translate into a 5 to 10% increase in gas mileage.

Simply scaling up the current SLI lead-acid battery from 6 to 18 cells leads to a number of problems, not the least of which is the significant increase in the weight of the battery. In addition, with all the added drains on the battery and the possibility of automatic start-stop, the battery will experience deeper and more numerous discharges than current SLI batteries are designed to handle. Like VRLA batteries, today's SLI batteries are maintenance free in that they don't require addition of water during their operating life. However, unlike the starved electrolyte VRLA batteries, SLI batteries contain flooded cells. To address the weight problem, VRLA batteries are an obvious choice. Unfortunately, as discussed in Chap. 17, the wide temperature swings encountered under the hood of an automobile present a significant challenge to VRLA technology.

One example of a dual battery now being marketed is the so-called Gemini Twinpower™ battery manufactured in China. This system comprises two 12 V batteries in a single case with an associated "Energy Management Controller" (EMC). One 12 V battery has thin plates for the starter function; the other 12 V battery has thick plates and glass mat separators designed for cycle life. The EMC maintains the starter battery fully charged and, during starting, combines the two batteries for starting, while isolating the batteries when the engine is turned off. This "smart" battery could be considered an intermediate step towards the 36/42 V system.

16.9.2 Small Sealed Lead-Acid Cells

In recent years, there has been a significant increase in the use of battery-operated consumer equipment such as portable tools, lighting devices, instruments, photographic equipment, calculators, radio and television, toys, and appliances. Batteries for these applications are generally of low capacity, up to 25 Ah. Storage batteries are frequently used because of their high power capability and rechargeability, but they have to be sealed or of the nonspill type in order to function in all positions. Vented lead-acid batteries of the electrolyte-retaining (ER) type and cylindrical (see Chap. 17) or prismatic cells are used in competition with sealed nickel-cadmium cells and other technologies. The lead-acid batteries offer lower initial cost, better float service, higher cell voltage, and the absence of memory effect (loss of capacity on shallow cycling). The nickel-cadmium cell has longer life and better cycling service.

The small sealed or semisealed lead-acid cells are available as single 2 V units or as multiple-cell units, usually in 6 V monoblock constructions. They are an outgrowth of the earlier ER-type batteries in which the electrolyte was absorbed in wood pulp separators. The ER-type cells, while spillproof, contained more electrolyte, did not recombine oxygen on overcharge, and were vented. These are mainly VRLA in most applications in today's market.

A related small deep-discharge lead-acid battery is the one used for miner's lamps and similar equipment. These are 4 V units which are vented and can be watered. They are designed to deliver 1 A for 12 h between charges.

16.9.3 Industrial Applications

Applications for lead-acid batteries, other than the SLI and small sealed power units, fall into two categories, as shown in Table 16.22—those based on automotive-type constructions and those based on industrial-type constructions. Often several designs can be used for a single type of application.

16.9.4 Electric Vehicles

The GM EV1 was a limited production vehicle designed from the ground up as an electric vehicle. It had an aerodynamic teardrop shape, regenerative braking to charge the battery, together with an aluminum structure and composite body panels for reduced weight. A battery of twenty-six 12 V

TABLE 16.22 Major Applications of Lead-Acid Batteries (Non-SLI Types)

Automotive and small energy storage designs		Industrial designs		
Traction	Special	Stationary	Traction (motive power)	Special
Golf cart	Emergency lighting	Switch gear	Mine locomotives	Submarines
Off-road vehicles	Alarm signals	Emergency lighting	Industrial trucks	Ocean buoys
On-road vehicles	Photovoltaic	Telecommunication facilities	Large electric vehicles	
	Sealed cells (for tools, instruments, electronic devices, etc.)	Railway signals		
		Uninterrupted power supply		
		Photovoltaics		
		Load leveling and energy storage		

valve-regulated lead-acid batteries powered a 137 horsepower, 3-phase AC induction motor. The estimated driving range was between 55 and 95 miles between charges, depending on driving conditions and driver habits. An optional nickel-metal hydride battery extended the range from 75 to 130 miles. With air conditioning, traction control, cruise control, anti-lock braking, speeds up to 80 miles per hour, and other features, the EV1 was not a stripped-down vehicle. Although such a zero-emissions electric vehicle would appear to be the answer to environmental concerns and the benefits were widely publicized, consumer acceptance of a short range, more expensive vehicle was not widespread. On the other hand, the introduction of low-emissions hybrid electric-internal combustion vehicles, notably the Toyota Prius, generated immediate enthusiasm and sales, with favorable reviews in the automotive press. The Prius employs an Ni-MH battery, as does the Honda Insight. It is clear that mass production of pure on-the-road electric vehicles will require lower cost batteries, increased range between charges, and an infrastructure of charging facilities to support the electric vehicles. These requirements are discussed in Chap. 29.

16.9.5 Energy-Storage Systems

Secondary batteries are used for load leveling in electric utility systems as an alternative to meet peak power demands currently provided with energy-expensive oil- or gas-fueled turbines (see Chap. 30). Large batteries, on the order of 50 MWh at 1000 V, are required. The lead-acid battery, again, is a major candidate for a near-term solution for this application. The goal is to obtain in excess of 2000 cycles or 10 year of operation at a cost of about $90 per kilowatthour.

Smaller-sized batteries are used for energy storage in systems employing renewable but interruptible energy sources, such as wind and solar (photovoltaic) energy. These systems are usually located on the customer side of the utility power grid. The system generally handles the following functions:

1. Converts solar, wind, or other such prime energy source to direct electric power
2. Regulates the electric power output
3. Feeds the electric energy into an external load circuit to perform work
4. Stores the electric energy in a battery subsystem for later use

A block diagram of a typical photovoltaic system is shown in Fig. 16.58.[54]

The selection of the proper battery for these types of applications requires a complete analysis of the battery's charge and discharge requirements, including the load, the output and pattern of the solar or alternative energy source, the operating temperature, and the efficiency of the charger and other

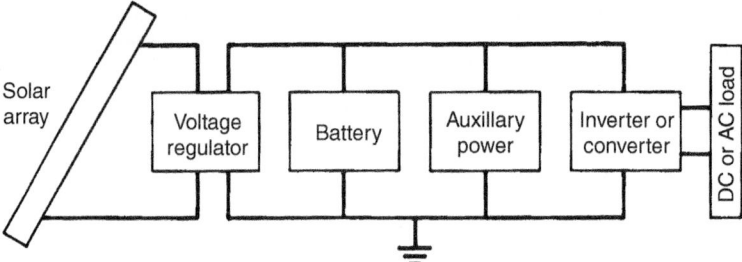

FIGURE 16.58 Components of solar photovoltaic system. (*From Ref. 54.*)

system components. Golf-cart-type lead-acid batteries and modified electric-vehicle designs are widely used in these small stationary energy-storage systems because they are the least expensive design in commercial production. Maintenance-free batteries are also used in these applications. These batteries (100 Ah size) can give maintenance-free unattended operation, a self-discharge rate less than 5% per month, a recharge in less than 8 h, and 1000 to 2000 cycles to an 80% depth of discharge.

16.9.6 Power Conditioning and Uninterrupted Power Supply Systems

DC Power Systems. A new concept in standby power is the DC power system with battery backup. These power systems include a battery charger (rectifier/charger) that has a sufficient capacity to recharge the batteries at the proper voltage while simultaneously supplying power to the DC load. In addition, equipment protection and isolation of the line voltage from the secondary windings of the special power transformer are designed into the system.

Static Uninterruptible AC Power Supply (UPS). In this power system, a storage battery is linked to the utility power to provide a continuity of service in the event of an interruption of the utility power. The continuous UPS system (Fig. 16.59) is illustrative of this type of device. During normal operation, the AC line supplies power to the static battery charger (rectifier-charger) which, in turn, "float" charges the battery and, at the same time, supplies DC power to the static inverter. The inverter, in turn, supplies power to the AC load. A synchronizing signal (if used) from the AC power line can maintain the phase and frequency of the inverter output the same as in the power line. This maintains the accuracy of timing devices such as clocks and recorder charts.

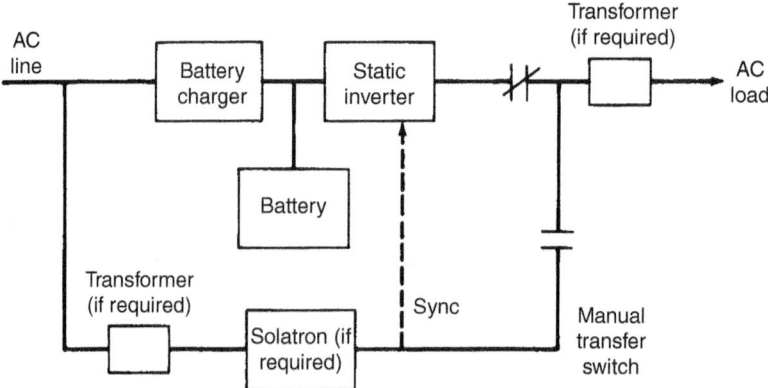

FIGURE 16.59 Schematic of continuous-type uninterruptible power system (UPS).

The voltage regulator within the inverter maintains the AC load voltage constant throughout the load range as well as during periods of equalizing charge on the storage battery. Transient and steady-state power-line variations are isolated from the load by the regulating action of the battery charger in conjunction with the filtering action of the battery and the inverter.

Upon AC line power failure, the battery charger ceases to operate and the battery instantly supplies power to the inverter and sustains the AC load without interruption. The synchronizing signal is also lost during the power failure. The inverter, therefore, continues to operate on its own internal frequency reference (±1% for standard units). The ampere-hour capacity of the battery determines the operating time of the system. The inverter is designed to maintain a constant voltage output as the battery voltage drops during the discharge. When AC power returns, the charger restores the battery energy and at the same time supplies power to the inverter. A more detailed description of other types of UPS systems employing VRLA lead-acid batteries is found in Chap. 17.

16.9.7 Marine Batteries

The marine battery market includes small and large leisure craft used for fishing, sailing, and travel as well as larger commercial vessels engaged in towing, passenger transportation, and workboat activities. In general, lead-acid battery systems are used with system voltages ranging from 6 to 220 V where recharging is accomplished by an engine generator or alternator. Three types of designs are common: conventional flooded cells (mainly in the larger cell designs), absorbed electrolyte designs, and gelled electrolyte designs. (The latter two are VRLA batteries and are discussed in Chap. 17.)

Marine service differs from automotive in several aspects. Lights, refrigeration, blowers, motors, radio, and other electrical equipment results in cycling service, with often a delay between discharge and charge. In marine service, the batteries should have much greater capacity than would normally be specified for the same horsepower equipment in a shore-based application.

The key engineering features utilized by one prominent manufacturer of marine batteries[58] include the following:

- A special grid design with heavier vertical and horizontal members.
- High density active materials, both positive and negative.
- Positive plates double insulated with thick woven glass matting and then sealed in a microporous polyethylene envelope.

In some designs, individual cells can be replaced while outer plastic cases provide high impact and environmental protection.

REFERENCES

1. PRlog press release, "Advance Rechargeable Battery Market: Emerging Technologies and Trends Worldwide," April 6, 2009, www. prlog.org.
2. Battery Council International, www.batterycouncil.org.
3. M. Saakes, R. Woortmeijer, and D. Schmal, "Bipolar Lead Acid Battery for Hybrid Vehicles," *J. Power Sources* **144** (2005).
4. Green Car Congress, "Lead-Acid Battery Developers Targeting Hybrid Applications," May 30, 2007, www.greencarcongress.com/2007/05/leadacid_batter.html.
5. EU Patent Application: WO 2004/0798151 A1, "Partition Wall for Bipolar Battery, Bipolar Electrode, Bipolar Battery and Method for Producing a Partition Wall."
6. EU Patent Application: WO 2004/021478 A1, March 11, 2004, "Separator Battery with Separator and Method Producing a Separator."
7. EU Patent Application: WO 2007/073279 A1, June 28, 2007, "Method and Device for Producing a Battery."

8. L. Lascelles, "CSIRO's UltraBattery to Cut Cost of Hybrid Battery by $2,000 in Two Years," Autoblog Green, January 20, 2008, green.autoblog.com.
9. J. Furukawa et al., "Development of the Flooded Type UltraBattery and Battery Sensor for Micro-HEV Application," 9th AABC, June 11, 2009, Long Beach, CA.
10. M. Suzuki, "The Extended Life Flooded (ELF) Battery of Micro-hybrids," 9th AABC, June 11, 2009, Long Beach, CA.
11. G. Fraser-Bell and D. Prengaman, "The European SLI Battery Market Past, Present, and Future," 9th AABC, June 11, 2009, Long Beach, CA.
12. www.furukawadenchi.co.jp/English/rd/nt_ultra.htm.
13. "This May Be the Turning Point for the PEM-Fuel Cell Manufacturers," June 9, 2009, www.glgroup.com.
14. B. Cullen, "North American Industrial Market Forecast 2009–2011," *121st Battery Council Convention*, Las Vegas, NV, May 3–6, 2009.
15. H. Bode, *Lead-Acid Batteries*, Wiley, New York, 1977.
16. H. E. Haring and U. B. Thomas, *Trans. Electrochem. Soc.* **68**:293 (1935).
17. P. Ruetschi, "Review of the Lead-Acid Battery Science and Technology," *J. Power Sources* **2**:3 (1977/1978).
18. C. Mantell, *Batteries and Energy Systems*, 2nd ed., McGraw-Hill, New York, 1983.
19. D. J. G. Ives and G. J. Janz, *Reference Electrodes*, Academic, New York, 1961.
20. E. A. Willihnganz, U.S. Patent 3,657,639.
21. A. Sabatino, *Maintenance-Free Batteries, Heavy Duty Equipment Maintenance*, Irving-Cloud Publishing, Lincolnwood, IL, 1976.
22. Special Issue on Lead-Acid Batteries, *J. Power Sources* **2**(1) (1977/1978).
23. ASTM Specification B29, "Pig Lead Specifications," American Society for Testing and Materials, Philadelphia, 1959.
24. A. G. Cannone, D. O. Feder, and R. V. Biagetti, *Bell Sys. Tech. J.* **19**:1279 (1970).
25. A. T. Balcerzak, *Alloys for the 1980s*, St. Joe Lead Co., Clayton, MO, 1980.
26. *Grid Metal Manual*, Independent Battery Manufacturers Association (IBMA), Key Largo, FL, 1973.
27. N. E. Hehner, *Storage Battery Manufacturing Manual*, Independent Battery Manufacturers Association (IBMA), Key Largo, FL, 1976.
28. U. B. Thomas, F. T. Foster, and H. E. Haring, *Trans. Electrochem. Soc.* **92**:313 (1947).
29. N. E. Hehner and E. Ritchie, *Lead Oxides*, Independent Battery Manufacturers Association (IBMA), Key Largo, FL.
30. T. Ferreira et al., "Stronger, Cleaner Plates Make Better Batteries," *116th Convention of Battery Council International*, May 4, 2004.
31. Trademark of Hollingsworth & Vose Company, East Walpole, MA.
32. Trademark of the Hammond Group, IN.
33. U.S. Patent 7,118,830, "Battery Paste Additive and Method for Producing Battery Plates," D. Boden, October 10, 2006.
34. D. Boden, "Sure Cure™ Technology, and Applications," *120th Convention of Battery Council International* Tampa, FL, April 2008.
35. T. Ferreira, "Development of an Inorganic Additive to Active Materials of Lead-Acid Batteries," *Long Beach Battery Conference*, 2002.
36. Yonezu, et al, "Pasted Type Lead Acid Battery." U.S. Patent 4,336,314, June 22, 1982.
37. V. Toniazzo, European Patent Application E.P 1,720,210 A1(200), "Non-Woven Gauntlet for Batteries."
38. V. Toniazzo, "New Generation of Non-Woven Gauntlets for Tubular Positive Plate," *J. Power Sources*, **158**(2):1062–1068 (2006).
39. Data from Daramic technical data sheets at www.daramic.com/products/daramic_products.cfm.
40. Data from Amer-sil website at www.amer-sil.com/Frames/Prod-AmerTube.htm.
41. Data from Entek International data sheet at www.entek-international.com/Products/RhinoHide.html.

42. G. W. Vinal, *Storage Batteries*, 4th ed., Wiley, New York, 1955.
43. M. Barak, *Electrochemical Power Sources*, Peter Peregrinus, Stevanage, U.K., 1980.
44. *Battery Service Manual*, 9th ed., Battery Council International, Chicago, 1982.
45. Battery Council International, 401 N. Michigan Ave., Chicago, 60611.
46. *Battery Replacement Data Book*, Battery Council International, 2000.
47. *Gould Battery Handbook*, Gould Inc., Mendota Heights, MN, 1973.
48. E. J. Friedman et al., *Electrotechnology*, vol. 3: *Stationary Lead-Acid Batteries*. Ann Arbor Science Publishers, Ann Arbor, MI, 1980.
49. *Bell Sys. Tech. J.* **49**(7) (Sept. 1970).
50. R. V. Biagetti and H. J. Luer, "A Cylindrical, Pure Lead, Lead-Acid Cell for Float Service," *J. Power Sources* **4** (1979).
51. A. G. Cannone, U.S. Patent 3,556,853, Jan. 19, 1971.
52. R. V. Biagetti, "The AT&T Lineage 2000 Round Cell Revisited: Lessons Learned; Significant Design Changes; Actual Field Performance v. Expectations," *INTELECT—Int. Telecommunications Energy Conf.*, Kyoto, Japan, Nov. 5–8, 1991.
53. E. Ritchie, International Lead-Zinc Research Organization Project LE-82-84, Final Rep., New York, Dec. 1971.
54. "Handbook of Secondary Storage Batteries and Charge Regulators in Photovoltaic Systems." Exide Management and Technology Co., Rep. 1-7135, Sandia National Laboratories, Albuquerque, NM, Aug. 1981.
55. G. E. Mayer, "Critical Review of Battery Cycle Life Testing Methods," *Proc. 5th Int. Electric Vehicle Symp.*, Philadelphia, Oct. 1978.
56. A. G. Cannone, U.S. Patent 4,605,605, Aug. 12, 1986.
57. A. G. Cannone, U.S. Patent 4,980,252, Dec. 25, 1980.
58. Product Literature, Rolls Battery Engineering, Salem, MA, 01970, USA.

CHAPTER 17
VALVE REGULATED LEAD-ACID BATTERIES

Kathryn R. Bullock and Alvin J. Salkind

17.1 GENERAL CHARACTERISTICS

Lead-acid battery designs for many small portable, and some larger fixed applications, have often been referred to as sealed and/or maintenance free. This is an accurate description to the extent that there is no need or opportunity to replace electrolyte. However, virtually no design includes a true hermetic seal, but only a pressure release valve which limits inflow or outflow of gas to the cell. The valves release internal gases pressures ranging from a few tenths of an atmosphere to a few atmospheres. In addition, most lead-acid battery cases are molded of plastics, which are hydrogen permeable.

A more accurate and now commonly accepted term for these designs is "valve-regulated lead-acid battery" or VRLA. A one-way, pressure-relief valve is designed to seal the cell unless its internal pressure exceeds a design maximum. The resealable valves are normally closed to prevent the entrance of oxygen from the outside air. The vent pressure design depends on the manufacturer and predominantly on the case shape and material. VRLA designs have two usual shapes, one with spirally wound electrodes (jelly-roll construction) in a cylindrical container, and the second with flat plates in a prismatic container. The cylindrical containers can maintain higher internal pressures without deformation and can be designed to have a higher release pressure than the prismatic cells. In some designs, an outer metal container is used to prevent deformation of the plastic cases at higher temperatures and internal cell pressures. The range of venting pressures includes a high of 25 to 40 psi for a metal-sheathed, spirally wound cell to 1 to 2 psi for a large prismatic battery.

The electrolyte is commonly immobilized in two ways:

- Gelled electrolyte: The liquid electrolyte is typically immobilized by mixing it with fumed or colloidal silica. The mixture is agitated to keep it in a liquid state until after it is poured into the cells and then hardens into a thixotropic gel. To preserve a gap between the plates for electrolyte filling, separators typically include a ribbed sheet made of microporous materials that are stronger and more rigid than AGM separators. In new batteries, some water is lost from the cell at the end of charge and during overcharge. Dry-out forms a network of cracks that enhance the passage of oxygen from the positive to negative plates. Hydrogen evolution and water loss are reduced, and acid leakage is minimized in gelled electrolyte VRLA batteries.
- Absorptive glass mat (AGM): In AGM batteries, the electrolyte is immobilized by absorption in highly porous AGM separators between the positive and negative electrodes. AGM separators are made primarily from glass microfibers that may not be fully saturated. Oxygen generated at the positive plate during charge can pass to the negative plate through the microporous structure, where it reacts to form water. This reduces hydrogen evolution from the cell and dry-out.

The separators are slightly compressed against the surfaces of the electrodes to facilitate the reaction of the electrolyte with electroactive lead materials in the positive and negative plates. Stronger fibers, hybrid blends of polymeric and glass fibers, and surface treatments are sometimes added to the AGM for strength and/or to enhance performance.

Some manufacturers use the AGM system between the plates but also insert gelled electrolyte into free spaces in the sides, bottom, or top of the cells of the plate stack to increase the amount of acid in the cell. Excess gelled acid can facilitate heat transfer to the case and reduce the rate of dry-out. This is especially advantageous in hot operating environments.

It should be noted that silica reacts with sulfuric acid and the absorption or gelation is chemical as well as physical in nature. The immobilization of the electrolyte allows batteries to operate in different orientations without spillage. In larger industrial applications, the batteries can be installed on their sides, permitting compact installations that use up to 40% less floor space and volume. An orientation in which the plates are horizontal to the ground is referred to as "pancake style."[1] Significant improvements in cycle life have been reported, in some circumstances, in cells cycled with their electrodes parallel to the ground.[2]

The use of VRLA designs is becoming more popular and accounted for over 75% of telecommunication and UPS applications in 1999. The development of advanced charging techniques has also increased the use of VRLA batteries in cycling applications such as forklift service. New market opportunities in portable electronics, power tools, and hybrid electric vehicles have stimulated the development of new designs of lead-acid batteries. However, the VRLA designs do not have the capability of handling certain types of abuse as well as conventional flooded batteries. The electrolyte, which provides the major internal heat sink in cells, is much more limited in VRLA cells. As a result, VRLA designs are more prone to thermal runaway under abusive conditions that pose little hazard for flooded cells. This is particularly true when VRLA batteries are subjected to operations at elevated temperatures. Recent discussion of high temperature (over 40°C) effects as well as a discussion of advantages and disadvantages have been given.[3,4] A detailed review of VRLA applications has been given in a series of articles.[5] To insure long battery life under conditions of higher temperature operations or overheating due to oxygen recombination, precautions must be taken.

A comparison of gel designs with absorbed electrolyte designs in forklift truck applications was presented[6] for one particular set of applications. However, new materials and designs are evolving and the comparisons should be ongoing. Much activity is being reported in the area of special fibers, blends, and surface treatments for the absorbing mat separators. Currently, many innovations and improvements in VRLA batteries are under development. These are discussed in Sec. 17.9.

The major advantages and disadvantages of VRLA batteries are listed in Table 17.1.

TABLE 17.1 Major Advantages and Disadvantages of VRLA Batteries

Advantages	Disadvantages
Maintenance-free	Should not be stored in discharged condition
Moderate life on float service	Relatively low energy density
High-rate capability	Lower cycle life than sealed nickel-cadmium battery
Little "memory" effect (compared to nickel-cadmium battery)	Thermal runaway can occur with incorrect charging or improper thermal management
"State of charge" can usually be determined by measuring voltage	More sensitive to higher temperature environment than conventional lead-acid batteries
Relatively low cost	
Available from small single-cell units (2 V) to large 48 V batteries	
Certain designs can be installed on their side, simplifying maintenance	

17.2 CHEMISTRY

Although the design and the construction of the VRLA battery are different, its chemistry is that of the traditional lead-acid battery. The basic "double sulfate" reaction applies to the overall reaction

$$PbO_2 + Pb + 2H_2SO_4 \underset{\text{charge}}{\overset{\text{discharge}}{\rightleftarrows}} 2PbSO_4 + 2H_2O$$

The reaction at the positive electrode is

$$PbO_2 + 3H^+ + HSO_4^- + 2e \underset{\text{charge}}{\overset{\text{discharge}}{\rightleftarrows}} 2H_2O + PbSO_4$$

and at the negative electrode

$$Pb + HSO_4^- \underset{\text{charge}}{\overset{\text{discharge}}{\rightleftarrows}} PbSO_4 + H^+ + 2e$$

When the cell is recharged, finely divided particles of $PbSO_4$ are electrochemically converted to sponge lead at the negative electrode and PbO_2 at the positive electrode. As the cell approaches complete recharge, when the majority of the $PbSO_4$ has been converted to Pb and PbO_2, the overcharge reactions begin. For conventional flooded lead-acid batteries, the result of these reactions is the production of hydrogen and oxygen gas and subsequent loss of water.

A unique aspect of the VRLA design is that the majority of the oxygen generated within the cells at normal overcharge rates is recombined within the cell. Grids made of high purity lead, often alloyed with tin, are used in cylindrical VRLA cells to collect and transport current from the active materials to the battery terminals. In rectangular battery designs, such as the gelled electrolyte designs, more rigid grids made of lead-calcium-tin alloys are generally used. The lead-antimony alloys that have been used in traditional batteries are avoided. Control of lead purity minimizes hydrogen evolution on overcharge and reduces the rate of self-discharge on stand. The oxygen cycle minimizes hydrogen formation. VRLA batteries must nonetheless be operated in ventilated areas, because small quantities of hydrogen, carbon oxides, and oxygen are released through the pressure release valve or the plastic container.

Oxygen will react with lead at the negative plate in the presence of H_2SO_4 as quickly as it can diffuse to the lead surface

$$Pb + HSO_4^- + H^+ + \tfrac{1}{2}O_2 \underset{\text{charge}}{\overset{\text{discharge}}{\rightleftarrows}} PbSO_4 + H_2O$$

In a flooded lead-acid battery, this diffusion of gases is a slow process, and virtually all the H_2 and O_2 escape from the cell rather than recombine. In the VRLA battery, the closely spaced plates are separated by a glass mat which is composed of fine glass strands in a porous structure. The cell is filled with only enough electrolyte to coat the surfaces of the plates and the individual glass strands in the separator, thus creating the starved-electrolyte condition. This condition allows for the homogeneous gas transfer between the plates, necessary to promote the recombination reactions. Additional discussion of VRLA chemical kinetics can be found in the literature.[4]

The pressure release valve maintains an internal pressure, and this condition aids recombination by retaining the gases long enough within the cell for diffusion to take place. The net result is that water, rather than being released from the cell, is cycled electrochemically to take up the excess overcharge current beyond that used for conversion of active material. Thus the cell can be overcharged sufficiently to convert virtually all the active material without loss of water, particularly at the recommended recharge rates. At continuous high overcharge rates (such as $C/3$ and above), gas buildup becomes so rapid that the recombination process is not as highly efficient, and O_2 as well as H_2 gas is released from the cell as the cell vents. Charging at these higher rates should therefore be avoided.

17.4 SECONDARY BATTERIES

17.3 CELL CONSTRUCTION

17.3.1 VRLA Cylindrical Cells[7]

A cross section of a VRLA cell and a breakdown of the basic components contained in the cell are shown in Figs. 17.1 and 17.2. Both the positive and the negative grids are made from 99.99% pure lead with 0.6% tin added for deep discharge recovery. The lead grid is relatively thin, 0.6 to 0.9 mm, to provide for a high plate surface area and resultant high discharge rates. The plates are pasted with lead oxides, separated by an absorbing glass mat, and spirally wound to form the basic cell element. Lead posts are then welded to the exposed positive- and negative-plate tabs. The terminals are inserted through the polypropylene inner top and are effectively sealed by expansion into the lead posts. The element is then inserted into the liner, and the top and liner are bonded together. At this state of construction, the cell is sealed except for the open vent hole. Sulfuric acid is then added and the relief valve is placed over the vent hole. The sealed element is then inserted into the metal can, the outer plastic top is added, and crimping completes the assembly. The metal case provides mechanical strength and does not affect the operation of the resealable vent. The cell is then formed electrochemically.

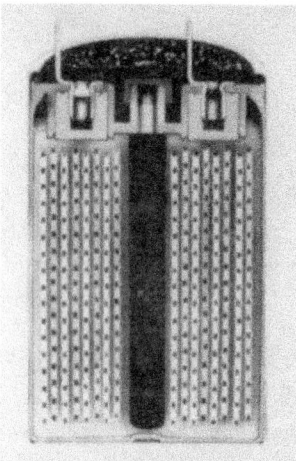

FIGURE 17.1 Cross section of the VRLA cell. Components identified in Fig. 17.2. (*Courtesy of EnerSys Energy Products, Inc., formerly Hawker Energy Products, Inc.*)

Monobloc batteries, using the cylindrical cell, are produced with two to six cells interconnected in a single plastic container. These 4, 6, and 12 V batteries have performance characteristics similar to those of the single cell. The monobloc design is illustrated in Fig. 17.3.

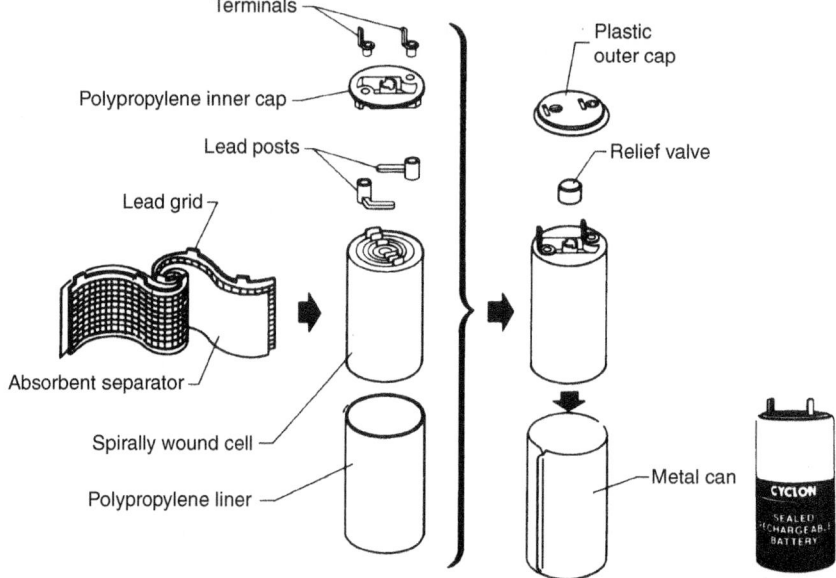

FIGURE 17.2 Components of VRLA cell. (*Courtesy of EnerSys Energy Products, Inc., formerly Hawker Energy Products, Inc.*)

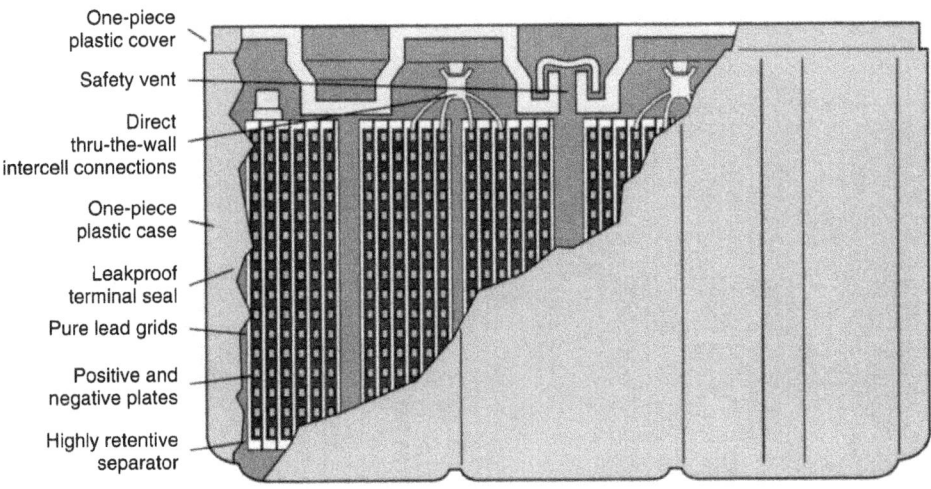

FIGURE 17.3 Monobloc battery. (*Courtesy of EnerSys Energy Products, Inc., formerly Hawker Energy Products, Inc.*)

17.3.2 VRLA Prismatic Cells

A cutaway view of a prismatic lead-acid cell is shown in Fig. 17.4a. An exploded view of a three-cell monobloc battery, using this prismatic cell, is given in Fig. 17.4b.

Nonantimonial calcium-lead alloys are usually used for the grids for the prismatic cell because these grids have a lower self-discharge rate and reduced gassing. However, the nonantimonial lead grids tend to reduce the deep-cyclability of the batteries unless additives are incorporated into the cell. Most of these are proprietary, the most common being phosphoric acid. The balance between cyclability, capacity, and float life is controlled by the ratio of α-PbO_2 to β-PbO_2, the paste density, and the amount and concentration of the electrolyte.

The electrolyte is absorbed in blotter-like glass-fiber separators or is gelled. Many acid-resistant materials, such as burnt clay, pumice, sand, Fuller's earth, plaster of paris, and asbestos, have been used for gelling. Fumed silica is the most common gelling agent. Recently other compounds of silica, such as alkali metal poly-silica[8] and a mixed poly-siloxane gel[9] have been applied to commercial VRLA products.

17.3.3 High-Power Cell Designs

Power, or current times voltage, can be achieved either by increasing the current or by increasing the cell voltage. New designs have been investigated using both of these approaches to increase the gravimetric and volumetric power of VRLA batteries.

Very thin electrodes on metal substrates have been stacked in cells to increase the current density and/or to increase the number of cells in a given volume. These designs have been largely abandoned for two major reasons.

First, the high surface areas between thin sheet-metal current collectors and the active electrode materials accelerate the rate of voltage loss by self-discharge. Although the current increases because of the high electrode surface area, the power drops rapidly due to voltage loss. The result is that the products cannot provide the required shelf life in most applications. This is especially true in portable product markets, where thin VRLA electrode designs have to compete with the very long shelf lives of alkaline batteries.

17.6 SECONDARY BATTERIES

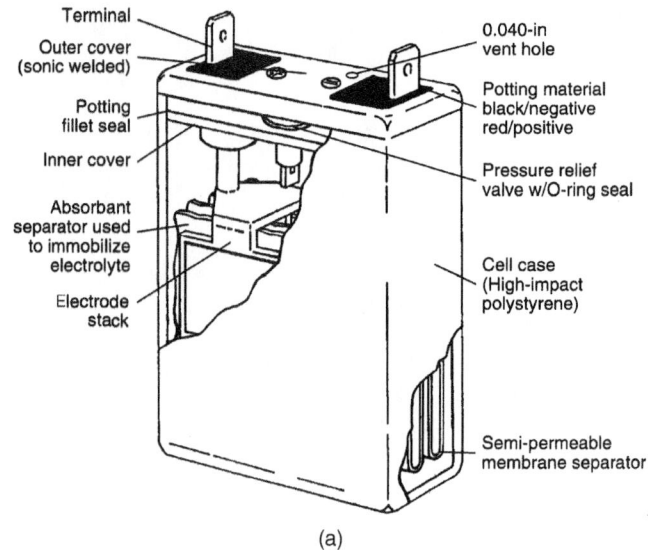

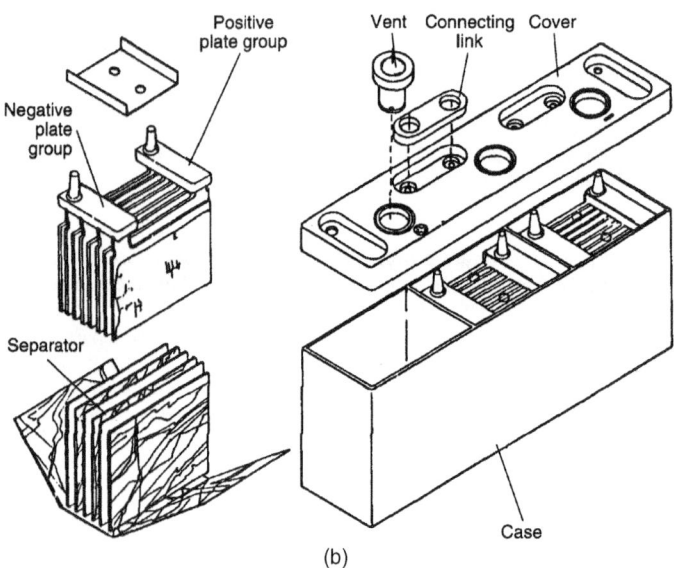

FIGURE 17.4 Typical prismatic lead-acid cell and monobloc battery (*a*) Cutaway view (*Courtesy of Eagle-Picher Industries, Inc.*) (*b*) Exploded view. (*Courtesy of Johnson Controls, Inc.*)

Second, compared to a VRLA battery lithium-ion batteries have higher energy density and higher open-circuit voltages. The lower material costs of the lead-acid system are less of a benefit in small portable battery sizes, where cell production costs are a larger percentage of the overall cost.

Another approach to power is the development of bipolar VRLA batteries. In these designs, the positive and negative electrode materials are placed on the opposite sides of an electrically conductive, bipolar substrate. The electroactive materials must adhere well to the substrate but not penetrate through it. Bipolar electrodes are stacked in a series configuration, as shown in Figure 17.5.

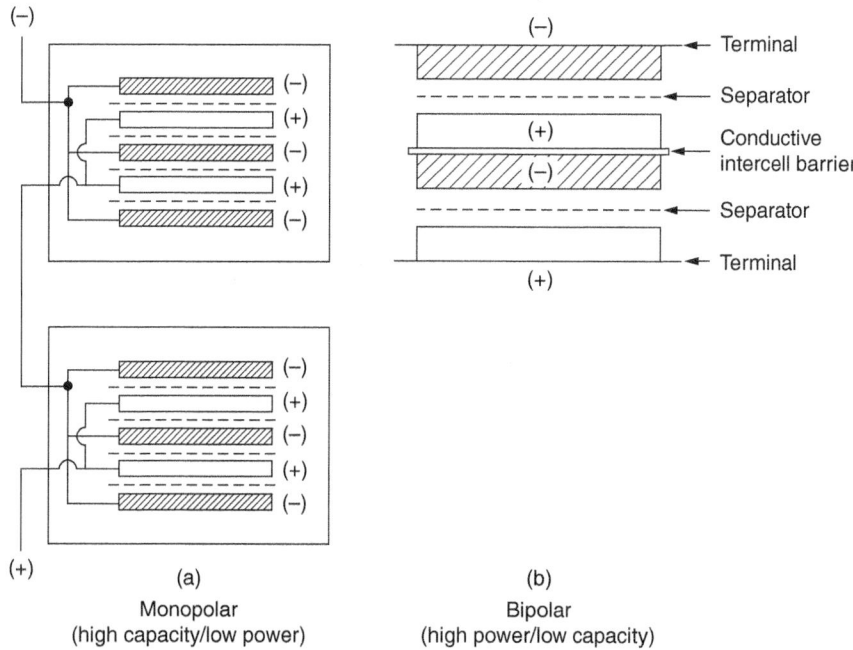

FIGURE 17.5 Schematics comparing monopolar and bipolar cell designs. Reprinted from *J. Electrochem. Soc.*, **142**:1726, (1995). (*Courtesy of the Electrochemical Society.*)

Three major challenges in the development of a bipolar VRLA battery are (1) finding a conductive, stable bipolar substrate, (2) sealing the multiple cells required to achieve a high voltage, and (3) developing a resealable valve that will fit in a two-electrode cell.

Work is ongoing to develop bipolar substrates using conductive materials that are chemically stable inside a lead-acid battery and provide a barrier to ion transport between the positive and negative sides of the bipolar plate. Conductive, stable materials that are compatible with VRLA components include a lead-infiltrated-ceramic plate, as well as barium plumbate ($BaPbO_3$), carbons, or non-stoichiometric titanium dioxides (Ti_4O_7 and Ti_4O_9) formed into solid foams or other structures or mixed with polymers or epoxy resins.[10,11]

In contrast to a monopolar battery design with multiple positive and negative plates per cell, a bipolar cell has only one positive and one negative plate per cell. Bipolar plates can be stacked together to get a high voltage in a small space, but they have low capacities. The voltaic efficiency increases as the number of bipolar plates in the battery increases, but the ampere-hour capacity does not. The current path is directly though the substrate, instead of across the surface area of the plate in monopolar designs. The resistance is lower and the current distribution is more uniform compared to monopolar designs.

Bipolar designs in large VRLA batteries may have improved lives in high-rate applications such as hybrid electric vehicles. The international Advanced Lead Acid Battery Consortium (ALABC) is road-testing a Swedish bipolar battery design in a Honda Civic in the U.K.[11,13]

17.8 SECONDARY BATTERIES

17.4 PERFORMANCE CHARACTERISTICS

17.4.1 VRLA Cylindrical Cells

Voltage. The nominal voltage of a VRLA single-cell battery is 2.2 V but is commonly called 2.0 V. It is typically discharged to 1.75 V per cell under load. The open-circuit voltage depends on the state of charge, as plotted in Fig. 17.6, based on a $C/10$ discharge rate. The open-circuit voltage can therefore be used to approximate the state-of-charge. The curve is accurate to within 20% if the battery has not been charged or discharged within 24 h; it is accurate to within 5% if the battery has not been used for 5 days. The measurement of the open-circuit voltage to determine the state of charge is based on the relationship between the electromotive force (OCV) and the concentration of the sulfuric acid in the battery.

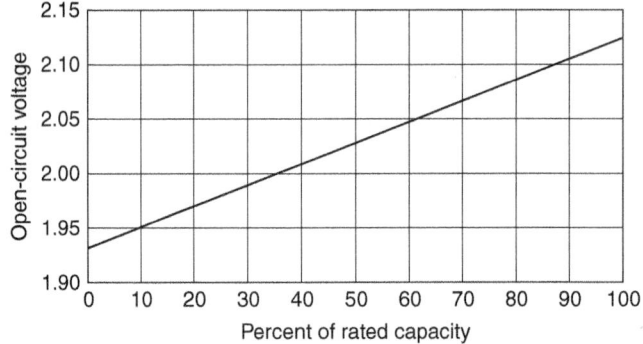

FIGURE 17.6 Open-circuit voltage vs. state-of-charge.

Discharge Characteristics. The discharge voltage profiles of typical VRLA single-cell batteries, at various temperatures ranging from –40 to 65°C for various discharge rates, are shown in Fig. 17.7 (see Table 17.2 for capacity data on various size cells). The discharge voltage curves are relatively flat at medium to low rates. These curves are based on smaller 2.5 and 5 Ah batteries. Discharge curves for the larger 25 Ah battery are slightly different from those of the smaller batteries because of the greater distance from the center of the plate to the external stud. This gives a higher effective internal impedance per unit of capacity and results in a slightly lower performance at higher rates and lower temperatures. Figure 17.8 shows a set of discharge voltage curves for a 2.5 Ah cell at 25°C, which further illustrates the good voltage performance of the cell even at high rates of discharge.

Effect of Temperature and Discharge Rate. The capacity of a VRLA battery, as with most batteries, is dependent on the discharge rate and temperature, the capacity decreasing with decreasing temperature and increasing discharge rate. The effect of temperature at the $C/10$, C, and $5C$ rates is shown in Fig. 17.9 for the cylindrical type D and X for the batteries. The larger 25 Ah cell gives a lower percentage of the 25°C performance at lower temperatures and higher discharge rates.

High-Rate Pulse Discharge. The VRLA battery is effective in applications that require high-rate discharge, such as in engine-starting. The voltage-time curves for the battery at room temperature at the $10C$ discharge rate, both on continuous discharge and for a 16.7% duty cycle (10 s pulse, 50 s rest), are shown in Fig. 17.10 for 25 and –20°C.

It is apparent from these data that the capacity of the VRLA battery is increased greatly when an intermittent pulse discharge is used. This is true because of the phenomenon known as "concentration polarization." As a discharge current is drawn from the cell, the sulfuric acid in the electrolyte

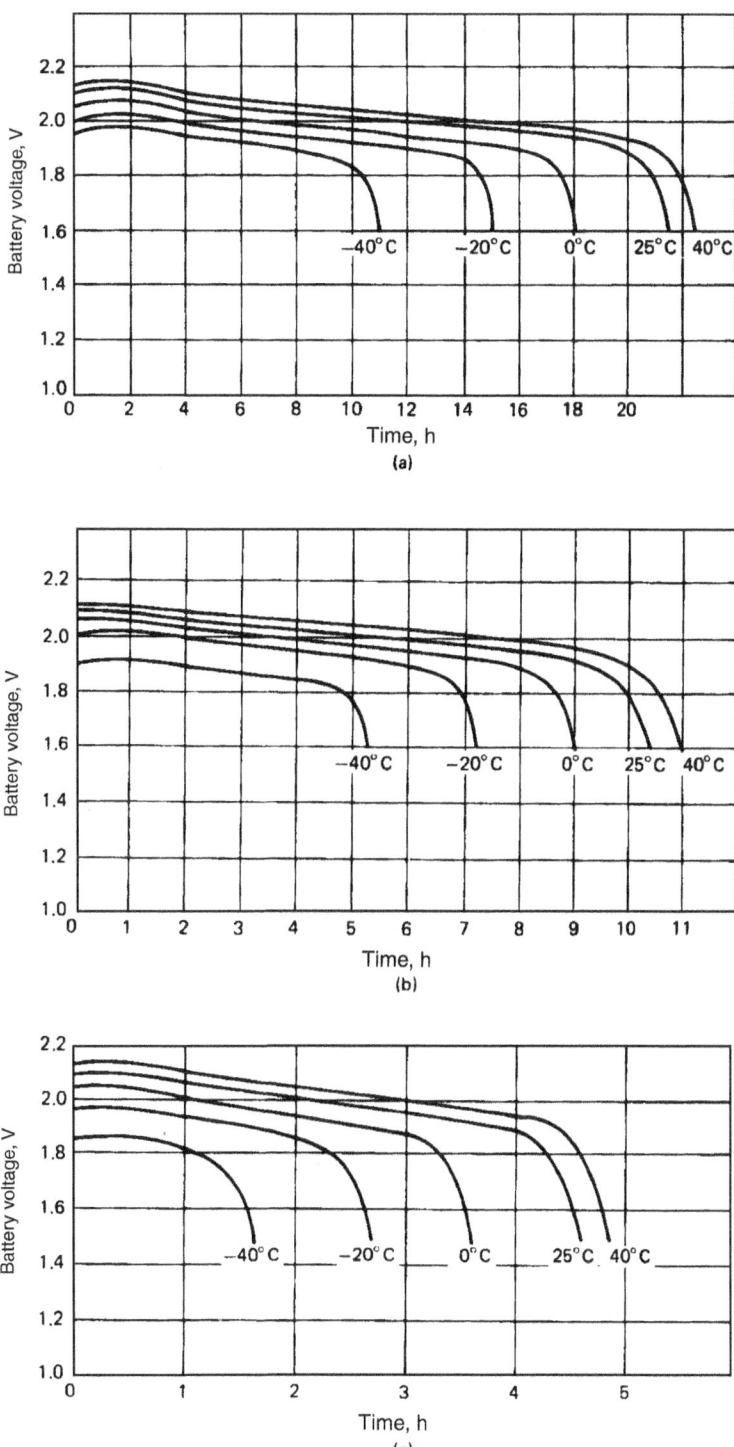

FIGURE 17.7 Discharge curves of cylindrical VRLA D and X single-cell batteries. Discharge rate (*a*) at C/20, (*b*) at C/10, (*c*) at C/5.

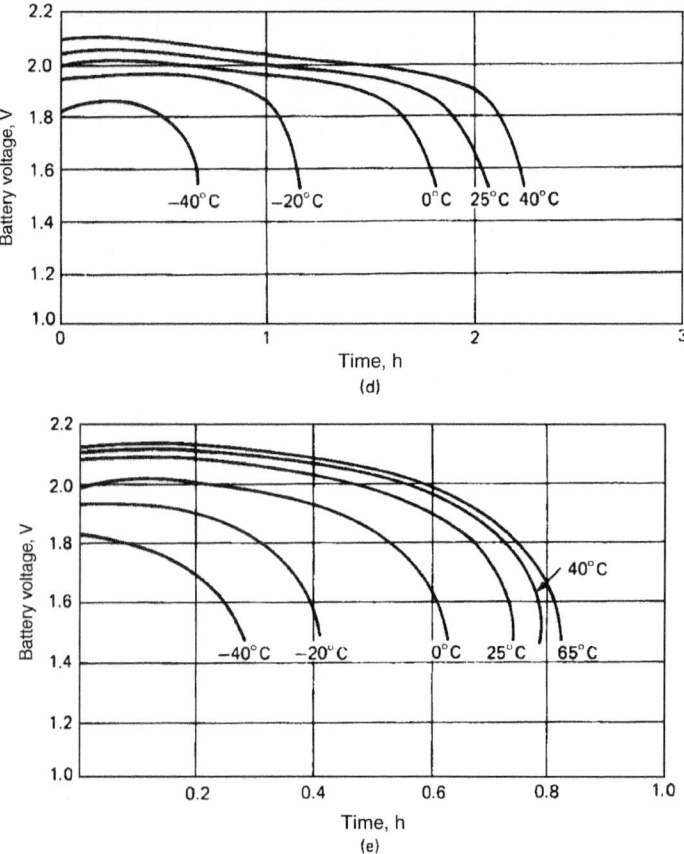

FIGURE 17.7 (d) at C/12.5, (e) at 1C (see Table 17.2 for capacity data) (Continued).

TABLE 17.2 VRLA Cylindrical Batteries

Model	Capacity, Ah			Dimensions, mm			Weight (typical), g	Specific energy @C/20	
	C/10	C/20	1C	Height	Diameter, width	Length		Wh/kg	Maximum discharge, A
Single cells									
D	2.5	2.7	1.8	67.3	34.3	N/A	180	30.0	40
X	5.0	5.4	3.2	80.3	49.5	N/A	390	27.6	40
J	12.5	13.0	9.0	135.7	51.8	N/A	840	30.8	60
BC	25.0	26.0	17.5	172.3	65.3	N/A	1580	32.9	250
DT	4.5	4.8	3.7	102.9	34.3	N/A	272	35.3	40
E	8.1	8.4	6.2	108.7	44.5	N/A	549	30.6	40
Monobloc batteries (preassembled batteries in common sizes)									
D, 4 V	2.5	2.7	1.8	70	45	78	360		40
D, 6 V	2.5	2.7	1.8	70	45	113	540		40
X, 4 V	5.0	5.4	3.2	77	54	96	740		40
X, 6 V	5.0	5.4	3.2	77	54	139	1110		40
E, 4 V	8.0	8.6	5.8	102	54	96	1110		40
E, 6 V	8.0	8.6	5.8	102	54	139	1670		40

Source: EnerSys Energy Products, Inc.

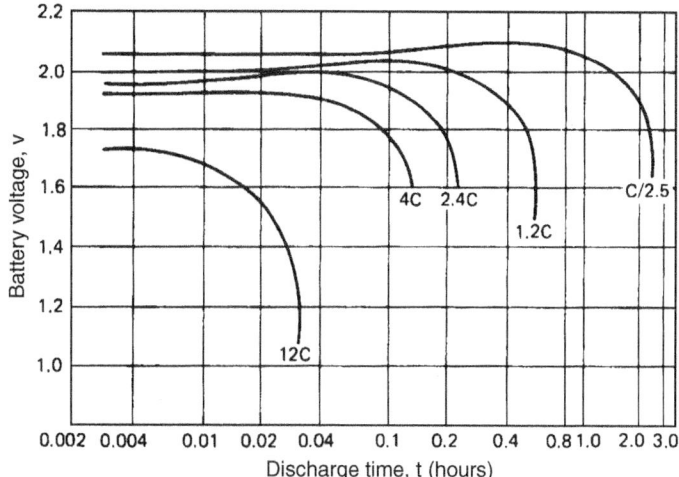

FIGURE 17.8 Discharge curves of VRLA cylindrical D and X batteries at 25°C, high discharge rates.

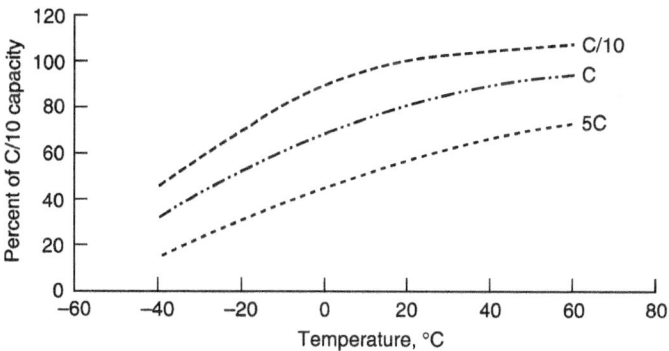

FIGURE 17.9 Effect of temperature on capacity for D and X size cylindrical units.

reacts with the active materials in the electrodes. This reaction reduces the concentration of the acid at the electrode-electrolyte interfaces. Consequently the cell voltage drops. During the rest period, the acid in the bulk of the solution diffuses into the electrode pores to replace the acid that has been used up. The cell voltage then increases as acid equilibrium is established. During a pulse discharge, the acid is allowed to equilibrate between pulses, it is not depleted as quickly, and the total cell capacity is increased.

The ability of the cell to provide high discharge currents and maintain usable voltage is illustrated in Fig. 17.11. The effect of the discharge rate on the midpoint voltage is shown at both 22 and −20°C. (The voltage indicated was measured midway through the discharge.)

17.12 SECONDARY BATTERIES

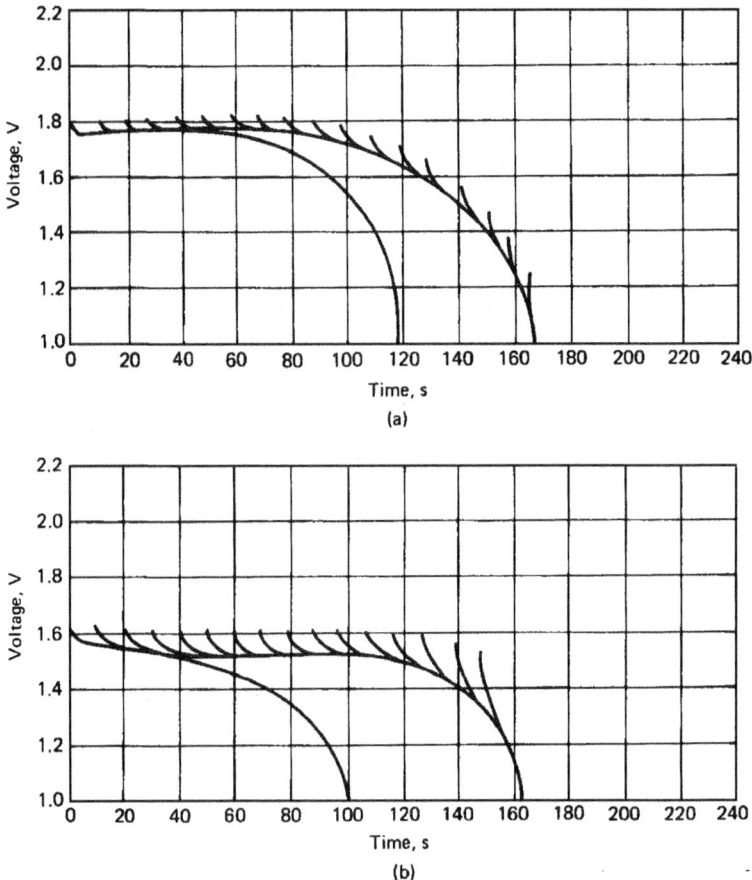

FIGURE 17.10 High-rate pulse performance at 10C discharge rate for VRLA cylindrical units. (*a*) 25°C. (*b*) −20°C. The upper curves are pulsed, the lower curves are continuous discharges.

The curves in Fig. 17.12 illustrate the maximum power that can be delivered as a function of the discharge rate at room temperature and at −20°C. The maximum power obtainable increases as the temperature increases.

Discharge Level. As with all rechargeable batteries, discharging the VRLA battery beyond the point at which 100% of the capacity has been removed can shorten the life of the battery or impair its ability to accept a charge.

The voltage point at which 100% of the usable capacity of the cell has been removed is a function of the discharge rate, as shown in the upper envelope of the curve of Fig. 17.13. The lower curve shows the minimum voltage level to which the battery may be discharged with no effect on recharging capability. For optimum life and charge capability, the cell should be disconnected from the load at the voltages within the gray area between the two curves.

Under these "overdischarge" conditions, the sulfuric acid electrolyte can be depleted of the sulfate ion and become mainly water, which can create several problems. A lack of sulfate ions as charge conductors will cause the cell impedance to appear high and little charge current to flow. Longer charge times or alteration of the charge voltage may be required before normal charging may resume.

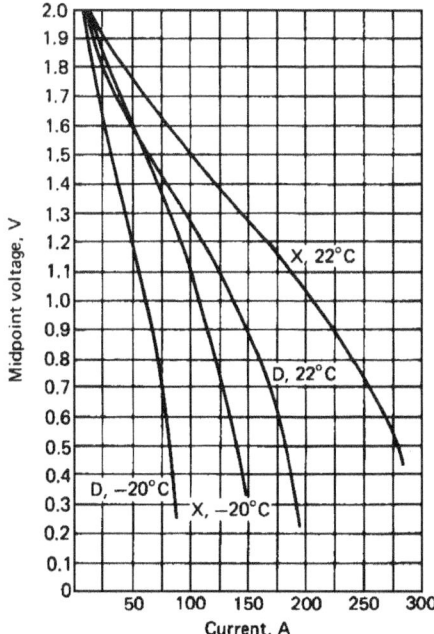

FIGURE 17.11 Effect of discharge rate on cell midpoint voltage of VRLA batteries.

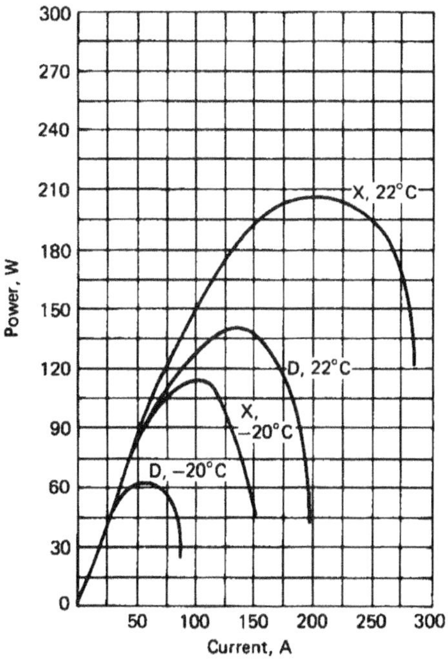

FIGURE 17.12 Instantaneous maximum peak power of VRLA batteries at 22 and −20°C.

Another potential problem is the solubility of lead sulfate in water. In a severe deep-discharge condition, the lead sulfate present at the plate surfaces can go into solution in the aqueous electrolyte. Upon recharge, the water and the sulfate ion in the lead sulfate convert to sulfuric acid, leaving a precipitate of lead metal in the separator. This lead metal can result in dendritic short circuits between the plates and subsequently cell failure.

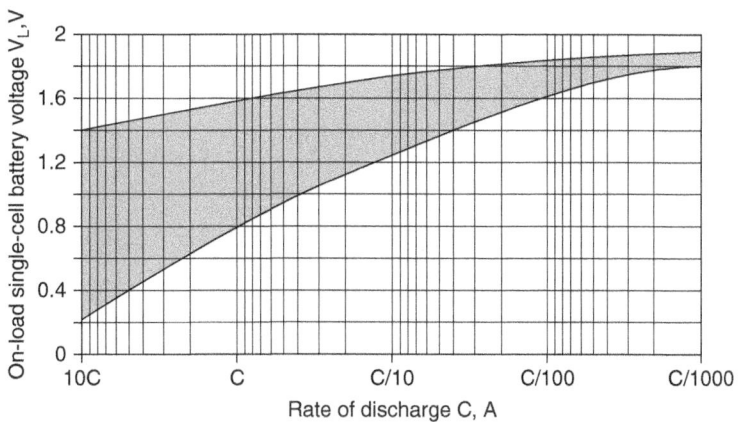

FIGURE 17.13 Acceptable voltage discharge levels of VRLA cells.

17.14 SECONDARY BATTERIES

Storage Characteristics. Most batteries lose their stored energy when allowed to stand on open circuit due to the fact that the active materials are in a thermodynamically unstable state.[12] The rate of self-discharge is dependent on the chemistry of the system and the temperature at which it is stored. The cylindrical VRLA battery is capable of long storage without damage, as can be seen in Fig. 17.14, which plots the maximum storage time against the storage temperature. This curve shows the maximum number of days at any given temperature from 0 to 70°C for the cell to self-discharge from 2.18 down to 1.81 V open circuit.

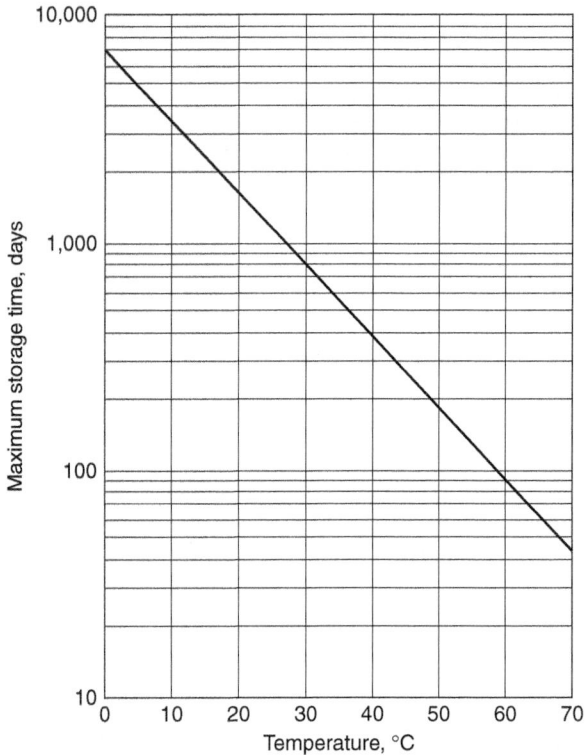

FIGURE 17.14 Storage characteristics of VRLA battery.

The cell should not be allowed to self-discharge below 1.76 V, because the recharge characteristics of the cell change appreciably and the cycle life cannot be predicted accurately. The capacity of these cells can be restored by recharging; however, the first charge on a cell that has been allowed to self-discharge down to 1.76 V will take longer than normal, and the first discharge will generally not deliver rated capacity. Subsequent cycles, however, will show an increase in the cell capacity to rated value.

It is important to recognize that the self-discharge rate of the VRLA battery is nonlinear; thus the rate of self-discharge changes as the state-of-charge of the cell changes. When the cell is in a high state-of-charge, that is, 80% or greater, the self-discharge is very rapid. The cell may discharge from 100 to 90% at room temperature in a matter of a week or two. Conversely, at the same temperature it may take 10 weeks or longer for the same cell to self-discharge from 20% state of charge down to 10% state of charge. Figure 17.15 is a curve of open-circuit voltage versus percent remaining storage time, which shows the non-linearity of the self-discharge reaction.

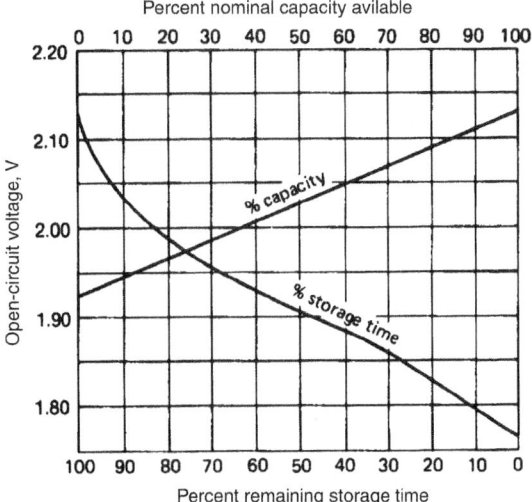

FIGURE 17.15 Open-circuit voltage vs. percent remaining storage time of VRLA batteries.

By the use of Figs. 17.14 and 17.15, the number of days of storage that remain before a battery must be recharged can be calculated. As an example, if a battery has an open circuit voltage of 2.00 V, the state-of-charge, as determined from Fig. 17.15, is 37%. From Fig. 17.15 (again at an open-circuit voltage of 2.00 V), the remaining storage time is 82%. Figure 17.14 shows that at 20°C, the battery can be stored for a total of 1,200 days before it must be recharged. Therefore the remaining storage time is 82% of 1,200 or 984 days. This is the number of days at 2.00 V open-circuit voltage that a battery can be stored before it will reach 1.76 V and must be recharged.

Figure 17.16 is a curve of the remaining usable capacity in a VRLA battery versus months of storage at various temperatures. This curve is convenient in determining the approximate remaining capacity after a given storage time at a particular temperature.

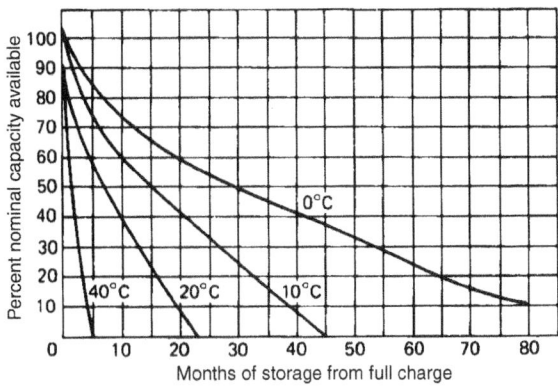

FIGURE 17.16 Remaining usable capacity of VRLA batteries after storage.

Life. The life of all rechargeable battery systems is variable, depending on the type of use, environment, cycling, and charge to which the battery is subjected during its life.

1. *Cycle life:* Figures 17.17 and 17.18 illustrate the effect of several of the factors that control the cycle life. Figure 17.17 shows the effect of the charging voltage and demonstrates the need to select the proper charging voltage for a particular cycle regime. The figure is somewhat misleading, however, in that it would indicate that a low charging voltage, say 2.35 V, would yield a low cycle life, and this is not true. For example, for an application where the cells would be used about three times a week and left on charge the rest of the time, 2.35 V would be quite adequate. Most of the capacity can be returned to the cells within 16 h. The occasional long charge periods would maintain capacity and thus optimize the total cycle life. Generally, 2.45 V per cell is a better charging voltage for regimes of about one cycle per day.

Figure 17.18 shows the general effect of the depth of discharge (DOD) on the cycle life; typically at 100% DOD, 200 cycles are characteristic. It demonstrates that high cycle life can be achieved by slightly oversizing the battery for the application to reduce the depth of discharge. Figure 17.19 is a curve of capacity versus cycles for a 2.5 Ah cell (2.35 Ah at 5 h rate) cycled at 1 cycle per day at a $C/5$ discharge rate to 1.6 V per cell and an 18 h charge at 2.5 V constant voltage. The cell takes from 20 to 25 cycles to achieve rated capacity, exceeds rated capacity, and then begins to fall off slowly. The initial increase in capacity is a function of forming the cell.

2. *Float life:* The expected float life of the VRLA battery is greater than 8 years at room temperature, arrived at by using accelerated testing methods, specifically, at high temperatures.

The primary failure mode of the VRLA battery can be defined as growth of the positive plate. Because this growth is the result of chemical reactions within the cell, the rate of growth increases with increasing temperature. In Fig. 17.20, the float life is plotted against temperature. The solid

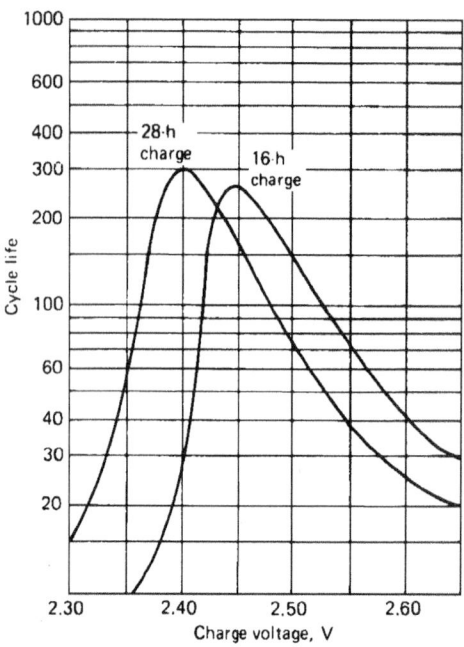

FIGURE 17.17 Effect of charge voltage on cycle life of VRLA batteries at various charge times at 25°C and approximately 100% DOD. End of cycle life—80% of rated capacity; discharge rate—$C/5$ to 1.6 V.

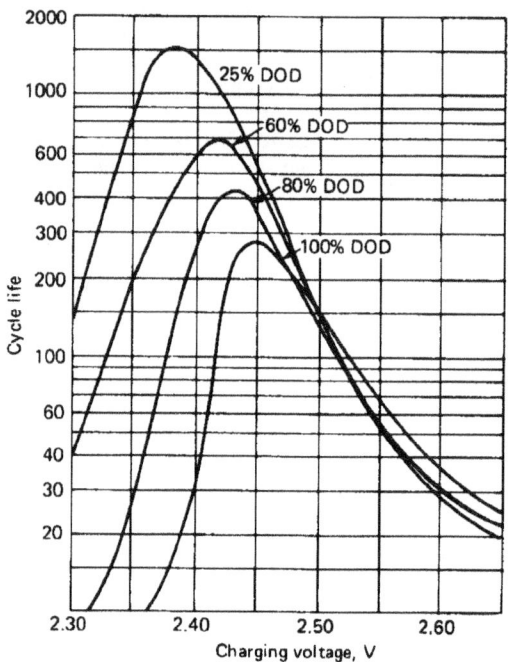

FIGURE 17.18 Effect of depth of discharge on cycle life of VRLA batteries as a function of charging voltage at 25°C, 16 h charge.

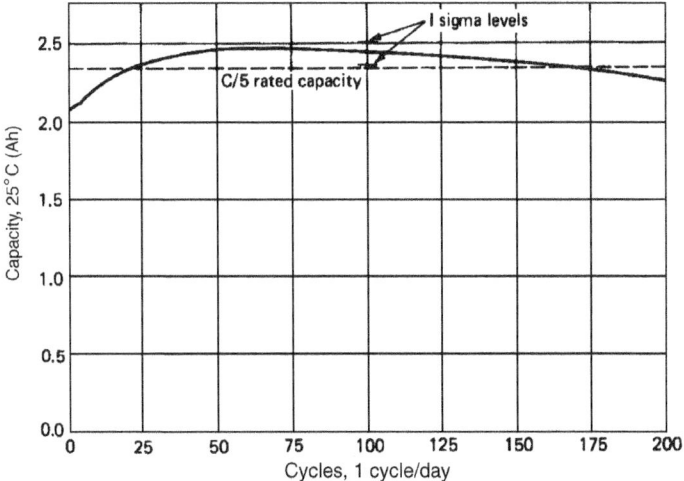

FIGURE 17.19 Effect of cycling on cell capacity, VRLA cell, $C/5$ discharge rate. Rated at 2.35 Ah at $C/5$ rate.

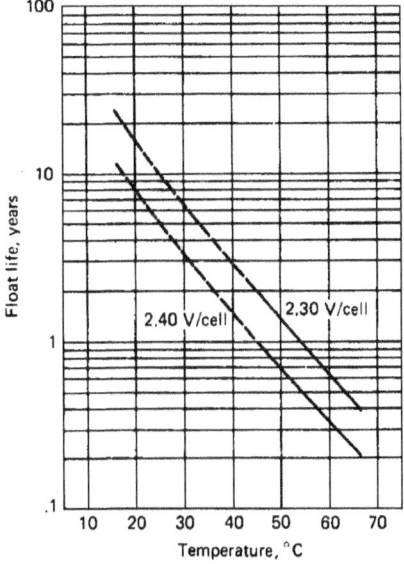

FIGURE 17.20 Float life of VRLA batteries.

lines represent data from float-life tests performed at two float voltages, 2.3 and 2.4 V per cell. This graph can be used to determine the expected float life at various temperatures. End of life is defined as the failure of the cell to deliver 80% of rated capacity.

17.4.2 Performance Characteristics—VRLA Prismatic Batteries

Typical discharge curves of the prismatic lead-acid battery at 20°C are shown in Fig. 17.21, which illustrates the high-rate capability and the flat discharge profile of this battery system. The cells are

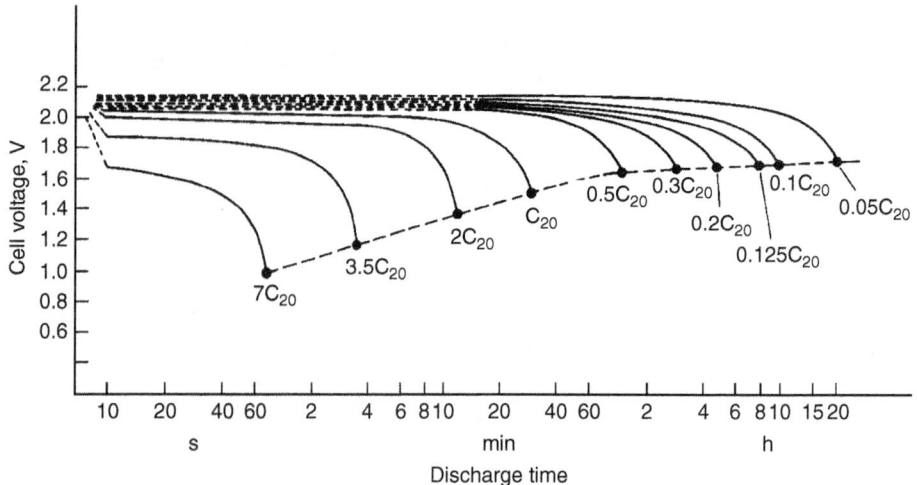

FIGURE 17.21 Typical discharge curves of prismatic lead-acid battery at various $C/20$ discharge rates. 20°C. (*Courtesy of Johnson Controls.*)

usually rated at the 20 h or .05C rate, and the figure shows the performance of the battery at discharge currents, expressed in terms of the $C/20$ rate, to various end voltages. For example, for a cell rated at 5 Ah at the 20 h rate, the $C/20/5$ rate is

$$\frac{C/20}{5} = 0.2\,C/20 = (0.2)(5) = 1\text{ A}$$

The $C/20/5$ rate is 1.0 A. The $C/20$ rate is 0.25 A.

Figure 17.22 shows the effect of discharge rate and temperature on the delivered capacity of the battery. A fully charged battery can operate over a very wide temperature range. These data are

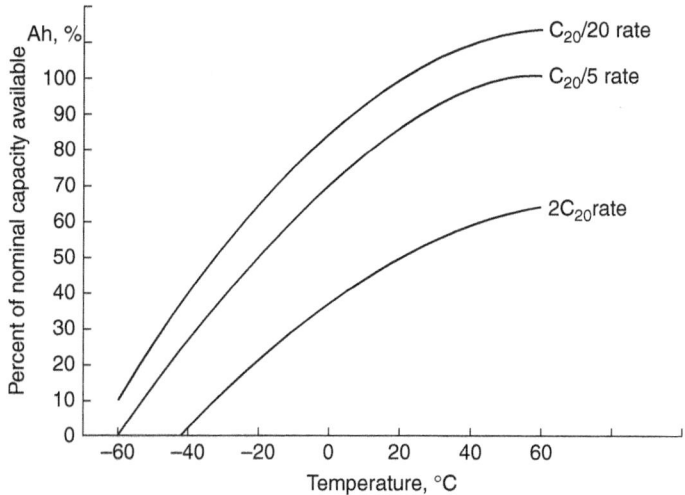

FIGURE 17.22 Effect of temperature and discharge rate on capacity of prismatic lead-acid battery.

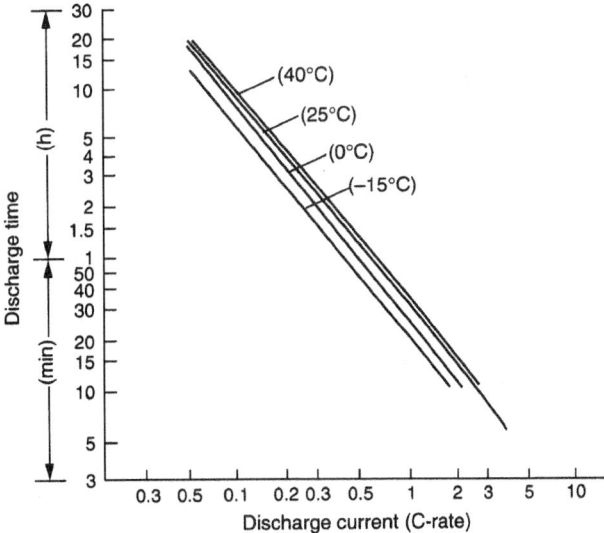

FIGURE 17.23 Service life of sealed prismatic lead-acid batteries.

summarized in Fig. 17.23, which plots the service life of the sealed prismatic lead-acid battery at various temperatures and discharge rates (C-rates). Manufacturers' data should be obtained for specific performance characteristics as the characteristics may vary depending on the size and design of the battery.

The typical self-discharge characteristics of the sealed prismatic lead-acid battery are shown in Fig. 17.24. Figure 17.24a shows the capacity retention after storage for different periods of time at several temperatures. Figure 17.24b shows the time it takes for the capacity to decrease to 50% of the rated capacity throughout the temperature range. It loses relatively little capacity in storage compared with the more conventional antimonial-lead grid battery. The rate of self-discharge is about 4% per month. Once the battery has self-discharged to the level where it has about a 50% state of charge, recharging the battery is advisable. The residual capacity can be estimated by measuring the open-circuit voltage, as shown in Fig. 17.25.

The service-life characteristics of the gelled lead-acid battery on float service are shown in Fig. 17.26. The cycle-life characteristics, to different depths of discharge (DOD), are shown in Fig. 17.27. A characteristic of the battery is that the capacity increases in the early stages of life and reaches a maximum at about 10 to 30 cycles. It will also gain in capacity while on extended charge, as in float service where discharges are infrequent. The battery also does not exhibit a "memory" effect, as is the case with the sealed nickel-cadmium battery, which may become conditioned when used for short periods, and may not be able to deliver full capacity when required.

The following are the recommended temperature ranges for the operation and storage of the prismatic lead-acid battery: discharge, −15 to 50°C; charge, 0 to 40°C; and storage, −15 to 40°C.

17.4.3 New Cell Designs for High-Rate, Partial State-of-Charge Cycling

Over the past decade, new applications have emerged in which VRLA batteries are frequently cycled but are seldom, if ever, fully charged or discharged. Batteries in partial-state-of-charge (PSoC) applications satisfy many needs, including energy for braking and acceleration in hybrid electric vehicles (HEVs), energy storage in photovoltaic and wind generators, and propulsion energy for off-road vehicles such as forklift trucks. Most of these applications require a battery that is able to accept a high-rate charge and deliver a high-rate discharge.

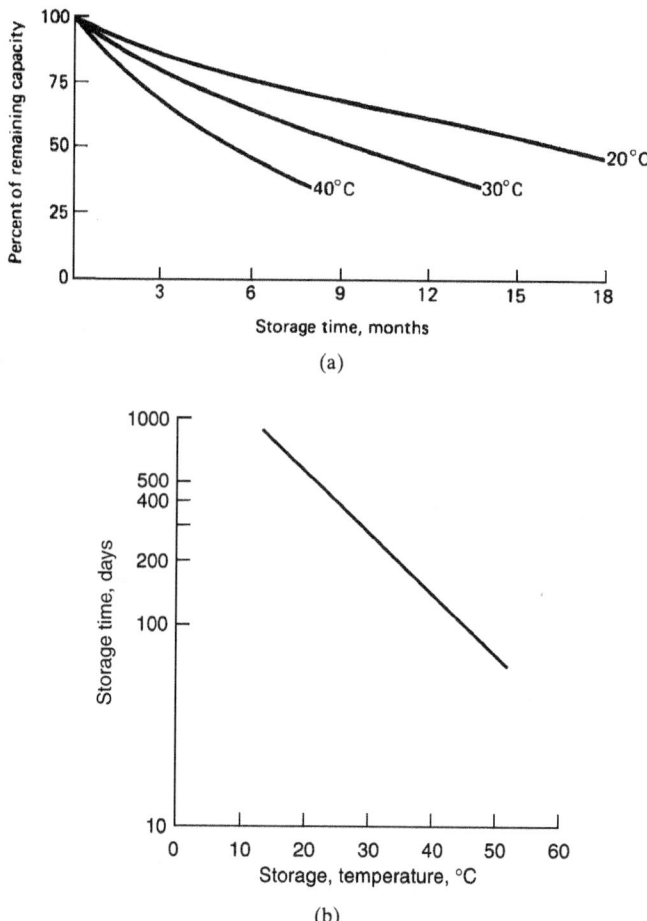

FIGURE 17.24 Characteristics of prismatic lead-acid battery. (*a*) Self-discharge at various temperatures. (*b*) Storage time and temperature to 50% of nominal capacity.

In traditional applications, the lead-acid battery positive electrode has typically been the first to fail. In PSoC cycling at high rates, little or no equalizing charge produces negative plate passivation. Lead sulfate crystals grow progressively larger on the negative plates, increasing the battery's resistance and decreasing its power and charge acceptance.

Manufacturing variables and slight differences in cell temperatures cause a distribution in the cell states-of-charge. When the cells are not regularly charged with a mild overcharge, cell voltages cannot be equalized across the battery string. The reduction in battery performance and life increases with the number of cells in the battery string.

Recent studies have shown that battery life can be greatly extended in HEV PSoC applications by adding higher levels of carbon powders and/or graphites to VRLA negative plates. The carbon may have several effects, including providing some additional capacitance for vehicle acceleration and energy storage during high-current regenerative braking. Other possible functions of the carbon include facilitating the recharge of lead sulfate and providing additional conductivity and porosity in the negative plate.

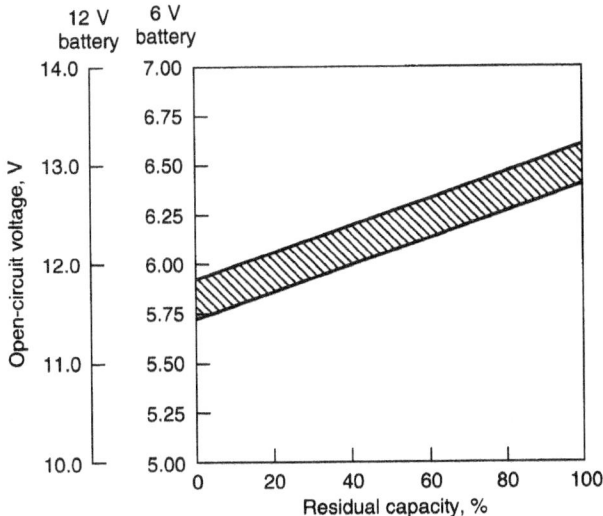

FIGURE 17.25 Open-circuit voltage vs. residual capacity of sealed prismatic lead-acid batteries at 25°C.

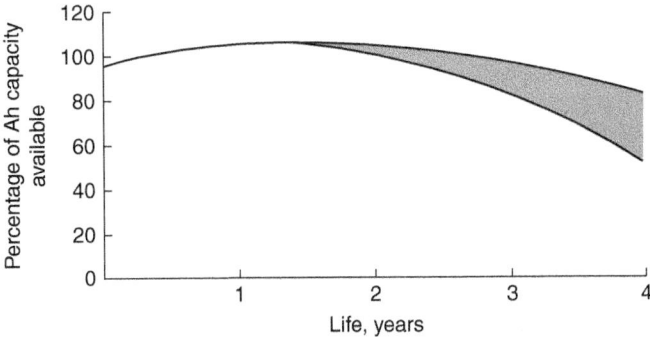

FIGURE 17.26 Performance of prismatic lead-acid battery on float service at 20°C. Float voltage = 2.25 – 2.3 V per cell.

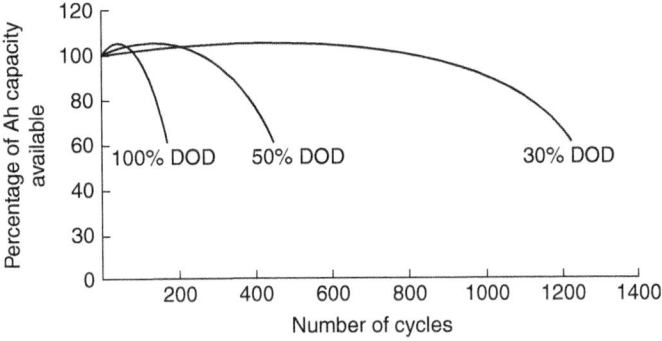

FIGURE 17.27 Cycle service life of prismatic lead-acid battery in relation to depth of discharge (DOD) at 20°C.

With these improvements, the negative plate failure mode can be overcome. The battery lasts longer until a new failure mode causes the end of life. In PSoC testing, the high-carbon batteries appear to fail by water loss. This dry-out is likely increased by the reaction of some of the oxygen generated by the positive plate with carbons, forming small amounts of volatile carbon dioxide that is vented from the battery. The oxygen lost by this mechanism cannot recombine to form water. However, cell operation in a PSoC mode minimizes dry-out so that VRLA batteries can now achieve cycle lives in HEV applications that are comparable to the performance of other battery systems.

New designs called the e3 Supercell and the UltraBattery are combinations of a battery and a capacitor in one case. The e3 Supercell has a lead dioxide positive electrode and an activated carbon negative electrode. In the UltraBattery, the negative electrode is vertically divided in half, with a lead electrode on one half and a carbon double layer capacitor electrode on the other half.[11,12,13] (See Chap. 16.)

The capacitor can store and deliver the high-rate, short-term currents required during regenerative breaking and acceleration in a hybrid electric vehicle. The lead-acid cell provides more stored energy for use in cruising and running electrical equipment. Life is extended because the battery is not strained by high currents. A combination of a battery and a capacitor in separate cases can work the same way, but the parallel combination of a capacitor and battery in a single case requires less electronic control and is more compact.

The UltraBattery with the lead/carbon negative electrode combination has lasted over 100,000 miles in a Honda Insight® HEV in a road test at the GM Millbrook Proving Ground near London, in a test sponsored by the ALABC. The use of components from the capacitor and lead battery industries has made this design relatively easy to manufacture at a low cost. The UltraBattery has been licensed and a manufacturing process developed by Furukawa. East Penn Manufacturing Company has also licensed the design. Cost estimates are about 70% of the cost of the nickel batteries used in HEVs. These new approaches to HEV propulsion may provide an opportunity within the next few years to develop a mass market in hybrid electric vehicles.

17.5 CHARGING CHARACTERISTICS

17.5.1 General Considerations[14]

Charging a VRLA battery, like charging other secondary systems, is a matter of replacing the energy depleted during discharge. Because this process is inefficient, it is necessary to return more than 100% of the energy removed during discharge. The amount of energy necessary for recharge depends on how deeply the battery was discharged, the method of recharge, the recharge time, and the temperature.[15] In high-temperature environments, the charge current or voltage should be controlled by the battery temperature.[14] The overcharge required in the lead-acid battery is associated with the generation of gases and corrosion of the positive-grid materials. In conventional flooded lead-acid batteries the gases generated are released from the system, resulting in a loss of water, which is replenished by maintenance. The VRLA battery incorporates the gas recombination principle, which allows the oxygen generated at normal overcharge rates to be reduced at the negative plate, eliminating oxygen outgassing. Hydrogen gas generation has been substantially reduced by the use of nonantimonial lead grid material. The corrosion of the positive grid has been reduced by the use of pure or special alloy lead. Also, the effects of corrosion of the positive grid have been minimized by the element construction.

Charging can be accomplished by various methods. Constant-voltage charging is the conventional method for lead-acid batteries and is also preferred for VRLA batteries. However, constant current, taper current, and variations thereof can also be used.

VRLA batteries during cycling and float charge can become unbalanced in that the corrosion rate of the positive becomes unequal to the self-discharge rate of the negative. A suggested remedy[16] is to utilize a catalyst in the vent space of the cell to restore balance by recombining hydrogen and oxygen.

17.5.2 Constant-Voltage Charging

Constant-voltage charging is the most efficient and fastest method of charging a VRLA battery. Figure 17.28 shows the recharge times at various charge voltages for a cell discharged to 100% of capacity. The charger required to achieve these times at given voltages must be capable of at least the $2C$ rate. If the constant-voltage charger used has less than the $2C$ rate of charge capability, the charge times should be lengthened by the hourly rate at which the charger is limited; that is, if the charger is limited to the $C/10$ rate, then 10 h should be added to each of the charge voltage-time relationships; if the charger is limited to the $C/5$ rate, then 5 h should be added, and so on. There are no limitations on the maximum current imposed by the charging characteristics of the battery.

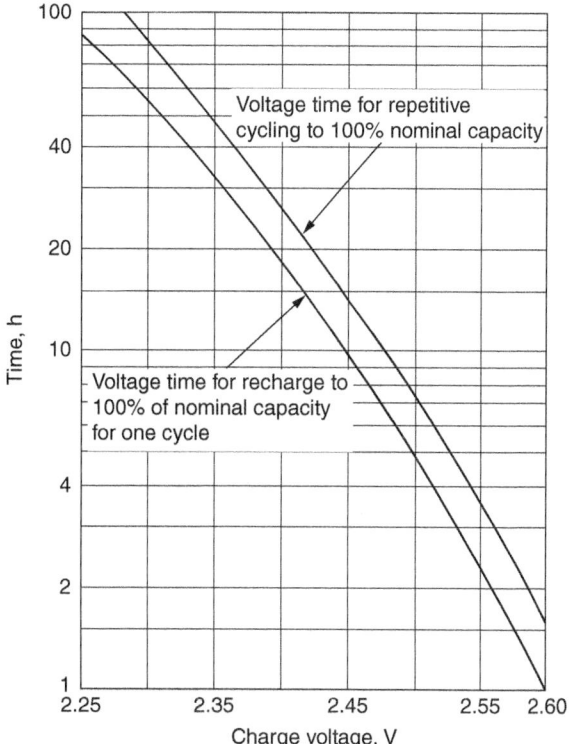

FIGURE 17.28 Charge voltage vs. time on charge at 25°C.

Figure 17.29 is a set of curves of charge current versus time for 2.5 Ah batteries charged by a constant voltage of 2.45 V with chargers limited to 2, 1, and 0.3 A currents. As shown, the only difference in these three charges is the length of time necessary to recharge the battery.

Figure 17.30a shows the charge rate versus time and the percent of previous discharge capacity returned versus time at 2.35 V constant voltage for a battery discharged at the $C/2.5$ rate to 80% DOD. The time necessary to recharge the battery to 100% of the previous discharge capacity is 1.5 h. The charger used was capable of an output in the $4C$ range and of voltage regulation at the terminals of better than 0.1%. The initial high inrush of current caused internal cell heating, which enhanced charge acceptance. If the battery had been more deeply discharged, the length of time to reach 100% capacity would have increased. The current decays exponentially with time, and after 3 h on charge,

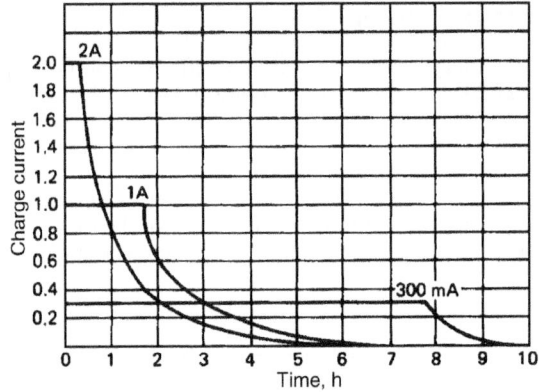

FIGURE 17.29 Charge current vs. time at 2.45 V constant voltage with various current limits (2.5 Ah battery, $C/10$ rate).

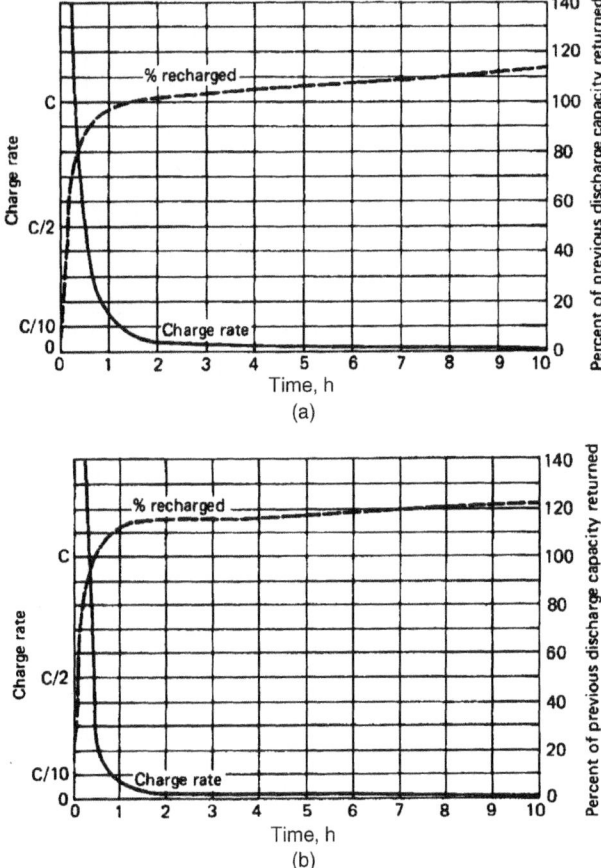

FIGURE 17.30 Charge rate and percent recharged vs. time during constant-voltage charge at 25°C. (*a*) 2.35 V. (*b*) 2.50 V.

the current has decayed to a low level. At 2.35 V, the cell will accept whatever current is necessary to maintain capacity.

Figure 17.30b is a similar plot but for a battery charged at 2.50 V constant voltage. All of the capacity taken out on the previous discharge is returned in approximately one-half hour.

17.5.3 Fast Charging[17]

A fast charge is defined as a method of charge that will return the full capacity in less than 4 h. However, many applications require 1 h or less.

Unlike conventional flooded lead-acid batteries, the VRLA design uses a starved electrolyte system where most of the electrolyte is contained within a highly retentive separator, which then creates the starved plates necessary for homogeneous gas-phase transfer. The gassing problem inherent in conventional lead-acid cells is not evident with this system, as the oxygen given off on overcharge is able to recombine within the VRLA battery. The large surface area of the thin plates used in some VRLA batteries reduces the current density to a level far lower than normally seen in the fast charge of conventional lead-acid batteries, thereby enhancing the fast-charge capabilities.

Figure 17.31 shows the charge rate or the current the VRLA battery can accept for a 1 h charge at three different voltages. The charger has a capability of delivering up to a $5C$ charge rate. The battery has a high charge acceptance at the beginning of the charge time; in fact, in the case of the 2.55 V per cell charge, the cell accepted the full current capability of the charger for the first 3 to 4 min. In the case of the 2.7 V per cell charge, there was a considerable amount of overcharging starting at 30 min, which caused internal heating and a consequent increase in charge current.

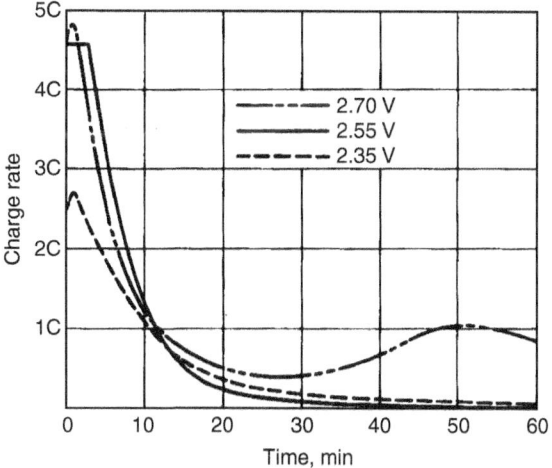

FIGURE 17.31 Charge rate vs. time for three charge voltages.

Figure 17.32 is a set of curves of normalized charge efficiency versus time in minutes for the three different voltages. This efficiency figure was calculated by dividing the total ampere-hour capacity returned by the previous discharge capacity removed. On the 2.55 V charge, 100% of the capacity removed on the previous cycle was returned in 15 min. With the 2.7 V charge, a 60% overcharge was returned at the end of the 60 min charge.

Figure 17.33 plots the discharge time in minutes versus cycle number for the three charge voltages. Also, a set of reference batteries was charged at 2.5 V constant voltage for 16 h and discharged at the $1C$ rate. This reference curve is displayed to show the expected capacity at the $1C$ rate. It can

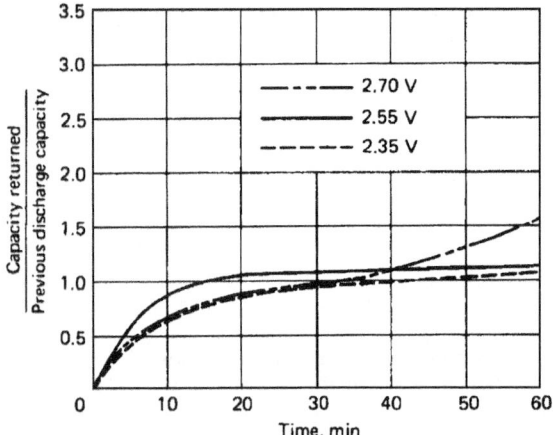

FIGURE 17.32 Charge efficiency vs. time for three charge voltages.

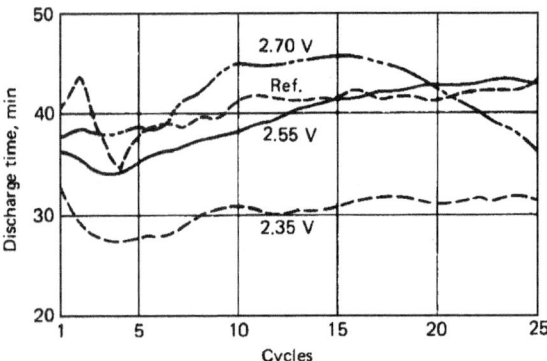

FIGURE 17.33 Effect of cycling on discharge time for three charge voltages.

be seen from these data that the 2.55 V per cell curve most closely approximates the reference line. The battery charged at 2.7 V per cell received too much overcharge and, therefore, the capacity degraded after 15 cycles. The battery charged at 2.35 V achieved a value of approximately 75% of the reference and continued to cycle at that level.

These data show that the thin-plate VRLA battery can be fast-charged to 100% of the rated capacity in less than 1 h. A constant-voltage charger set at 2.5 to 2.55 V per cell and capable of the 3 C to 4 C rate of charge is preferred. It should be noted, as discussed, that charging at 2.7 V per cell for prolonged periods will damage the battery.

17.5.4 Float Charging

When the VRLA battery is to be float-charged as in a standby application, the constant-voltage charger should be maintained between 2.2 and 2.3 V for maximum life. Continuous charging at greater than 2.4 V per cell is not recommended because of accelerated grid corrosion. Figure 17.34

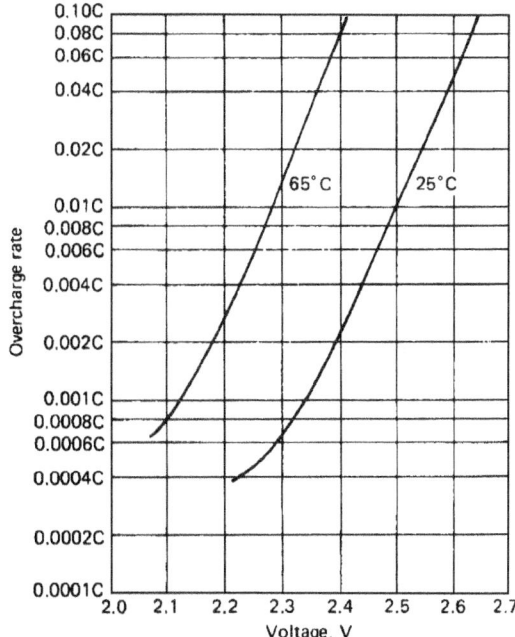

FIGURE 17.34 Overcharge current and voltage.

gives the approximate values of voltage a battery will attain when float-charged at 25 and 65°C, or the charge rate a battery will accept if it has been charged for a sufficient period of time so that it is in a state of overcharge equilibrium. These curves can also be used to determine the approximate value of continuous constant current (trickle charge) that will maintain the proper float voltage. As an example, if a battery were trickle-charged at the $0.001C$ rate, its average voltage per cell on overcharge would be 2.35 V at 25°C. Conversely, if a cell were constant-voltage-charged at 2.35 V, its overcharge rate would be $0.001C$.

High temperatures accelerate the rate of the reactions that reduce the life of a battery. At increased temperatures, the voltage necessary for returning full capacity in a given time is reduced because of the increased reaction rates within the battery. To maximize life, a negative charging temperature coefficient of approximately -2.5 mV/°C per cell is used at temperatures significantly different from 25°C. Figure 17.35 shows the recommended charging voltage at various temperatures for a sealed battery float-charged at 2.35 V per cell at 25°C. It is obvious from this curve that at extremely low temperatures, a significantly greater temperature coefficient than -2.5 mV/°C is required to achieve full recharge of the cell. Figure 17.35 also shows the voltage compensation under cycling

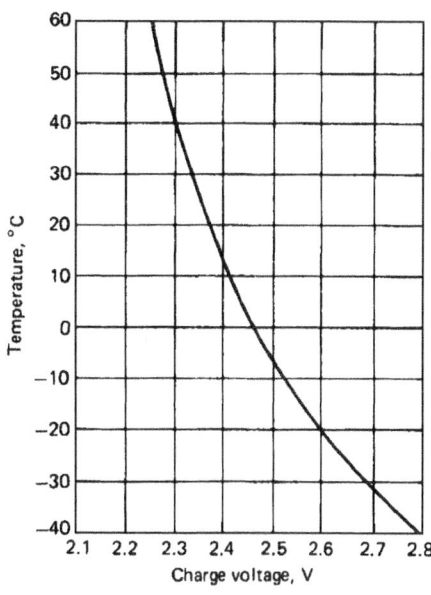

FIGURE 17.35 Recommended charge voltage at various temperatures (temperature compensation).

17.28 SECONDARY BATTERIES

service. The voltage compensation keeps the charging current at about the same value that it would be at 25°C when the battery temperature is different. Temperature compensation of the charging voltage prevents thermal runaway of the batteries when they are used at high temperatures and ensures adequate charging if the battery temperature is low.

When trickle-charging, it may be necessary to increase the charge rate at higher temperatures to maintain the proper float voltage. From Fig. 17.34, it can be seen that for a battery trickle-charged at the $0.001C$ rate at 25°C, the float voltage would be 2.34 V. However, at the same rate at 65°C, the float voltage would be approximately 2.12 V, which is below the open-circuit voltage of the cell. At 65°C, the trickle-charge current would need to be increased to approximately $0.01C$ to maintain the proper float voltage.

17.5.5 Constant-Current Charging

Constant current is another efficient method of charging the VRLA battery. Constant-current charging is accomplished by the application of a nonvarying constant-current source. This charge method is especially effective when several cells are charged in series, since it tends to eliminate any charge imbalance in a battery. Constant-current charging charges all cells equally because it is independent of the charging voltage of each cell in the battery. Figure 17.36 shows a family of curves of battery voltage versus percent capacity of previous discharges returned at different constant-current charging rates. As shown by these curves at different charge rates, the voltage increases sharply as the full charge state is approached. This increase in voltage is caused by the plates going into overcharge when most of the active material on the plates has been converted from lead sulfate to lead on the negative plate and to lead dioxide on the positive plate. The voltage increase will occur at lower states of charge when the cell is being charged at higher rates. This is because at the higher constant-current charge rates the charging efficiency is reduced. The voltage curves in Fig. 17.36 are somewhat different from those for a conventional lead-acid battery due to the effect of the recombination of gases on overcharge within the system. The VRLA battery is capable of recombining the oxygen produced on overcharge up to the $C/3$ rate of constant-current charge. At higher rates the recombination reaction is exceeded by the rate of gas generation.

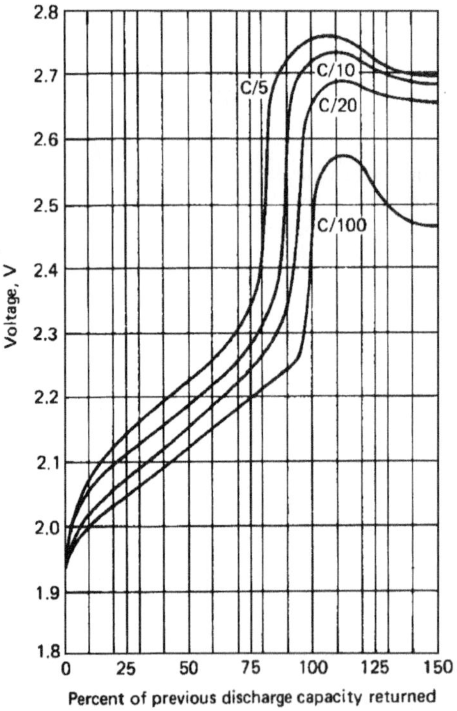

While constant-current charging is an efficient method, continued application at rates above $C/500$ after the battery is fully charged can be detrimental to life. At overnight charge rates ($C/10$ to $C/20$), the large increase in voltage at the nearly fully charged state is a useful indicator for terminating or reducing the rates for a constant-current charger. If the rate is reduced to $C/500$, the battery can be left connected continuously and give 8 to 10 years of life at 25°C. Figure 17.37 is a plot of voltage versus time for a battery charging at the $C/15$ rate of constant current at 25°C. This battery had previously been discharged to 100% depth of discharge at the $C/5$ rate. This curve shows that the battery is not fully charged at the time the voltage increase occurs and must

FIGURE 17.36 Voltage curves for batteries charged at various constant-current rates at 25°C.

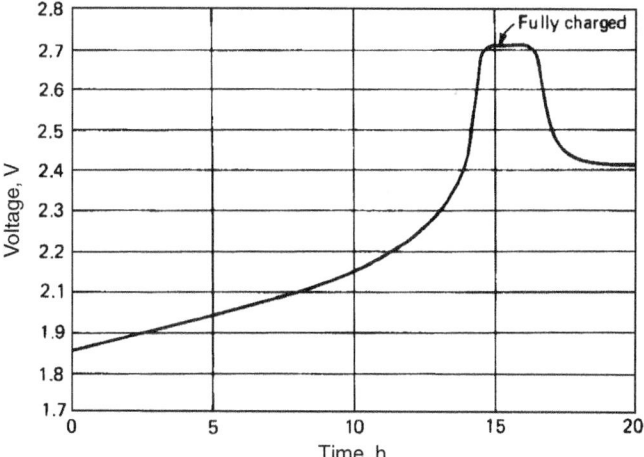

FIGURE 17.37 Constant-current charge at $C/15$ rate, 25°C.

receive additional charging. If a battery is to be charged with constant current at higher than room temperature, then some temperature compensation must be built into the voltage-sensing network. As explained in Sec. 17.5.4, at higher temperatures and given charging rates, the battery voltage on overcharge is reduced. Therefore the rise in voltage at close to full charge will be somewhat depressed.

17.5.6 Taper-Current Charging

Although taper-current chargers are among the least expensive types of chargers, their lack of voltage regulation can be detrimental to the cycle life of any type of battery. The VRLA battery can withstand charge voltage variations, but some caution in using taper-current chargers is recommended. A taper-current charger contains a transformer for voltage reduction and a half- or full-wave rectifier for converting from alternating to direct current. The output characteristics are such that as the voltage of the battery increases during charge, the charging current decreases. This effect is achieved by use of the proper wire size and the turns ratio. Basically, the turns ratio from primary to secondary determines the output voltage at no load, and the wire size in the secondary determines the current at a given voltage. The transformer is essentially a constant-voltage transformer which depends entirely on the AC line voltage regulation for its output-voltage regulation. Because of this method of voltage regulation, any changes in input line voltage directly affect the output of the charger. Depending on the charger design, the output-to-input voltage change can be more than a direct ratio; for example, a 10% line-voltage change can produce a 13% output-voltage change.

When considering the cost advantage of using a half-wave rectifier versus a full-wave rectifier in a taper-current charger, it should be noted that the half-wave rectifier supplies a 50% higher peak-to-average-voltage ratio than the full-wave rectifier. Therefore the total life of the battery for a given average charge voltage can be reduced for the half-wave type of charger because of the higher peak voltages. A DC ripple can lead in time to decreased performance through degradation of the active material and the grid. An AC ripple can be a more significant factor in premature battery failure, especially in float or uninterruptible power systems. The repeated charging and discharging of the battery shortens the battery life through heat generation and corrosion.

There are several charging parameters that must be met. The parameter of main concern is the recharge time to 100% nominal capacity for cycling applications. This parameter can primarily be defined as the charge rate available to the battery when each cell is at 2.2 and 2.5 V. The charge voltage at which approximately 50% of the charge has been returned to the battery at normal charge rates of $C/10$ to $C/20$ is 2.2 V per cell; the 2.5 V per cell point represents the voltage at which the battery is in overcharge. Given the charge rate at 2.2 V, the recharge time for a taper-current charger can be defined by

$$\text{Recharge time} = \frac{1.2 \times \text{capacity discharged previously}}{\text{charge rate at 2.2 V}}$$

It is recommended that the charge rate at 2.5 V be between $C/50$ maximum and $C/100$ minimum to ensure that the battery will be recharged at normal rates and will not be severely overcharged if the charger is left connected for extended time periods.

Figure 17.38 is a set of output voltage versus current curves for a typical 2.5 Ah battery taper-current charger. The three curves show the change in output with a variation in input voltage from 105 to 130 V AC. This particular charger, at 120 V AC input, will charge a three-cell D-sized battery (rated at 2.5 Ah) which had been previously discharged to 100% depth of discharge in 30 h by the following equation:

$$\text{Recharge time} = \frac{1.2 \times 2.5}{0.100 \text{ A}} = 30 \text{ h}$$

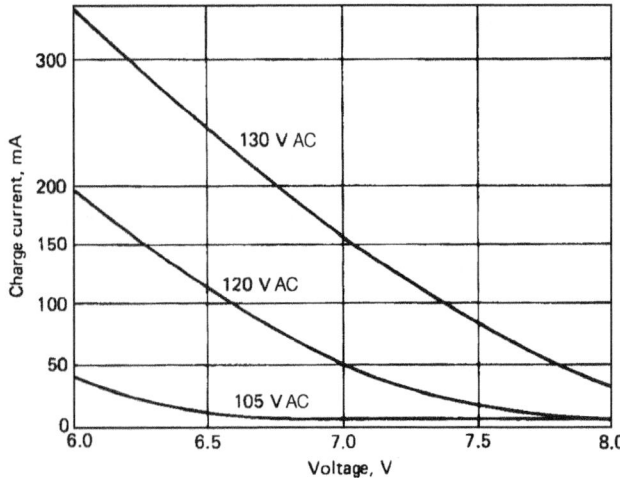

FIGURE 17.38 Taper-current charger characteristics for 2.5 Ah cylindrical three-cell VRLA battery.

17.5.7 Parallel/Series Charging

VRLA batteries can be charged or discharged in parallel. When more than four strings of cells are paralleled, it is advisable to use diodes in both the charge and the discharge path in the circuit. The discharge diodes prevent a battery from discharging into a paralleled battery should a cell short-circuit in the battery. The charge diodes, in conjunction with the fuse, will prevent a battery with a short-circuited cell from accepting all the charge current from the charger and subsequently prevent the other paralleled batteries from being fully charged. The fuse should be sized by dividing the maximum charge current by the number of batteries in parallel and multiplying this value times 2. This should result in the fuse opening on charge in a parallel string which has a short-circuited cell.

When float-charging many cells in series, 12 or more, for example, it is advantageous to use a trickle charge of $C/500$ maximum in parallel with the float charger. This trickle charge will tend to balance all cells in the battery by driving a continuous trickle charge equally through all cells.

17.5.8 Charge-Current Efficiency

Charge-current efficiency is the ratio of the current that is actually used for electrochemical conversion of the active material from lead sulfate to lead and lead dioxide to the total current supplied to the cell on recharge. The current that is not used for charging is consumed in parasitic reactions within the cell such as corrosion and gas production.

The charging efficiency is high for a VRLA battery. The distinctly high ratio of plate surface area to ampere-hour capacity allows for higher charging rates and, therefore, efficient charging.

Charge-current efficiency is a direct function of the state of charge. The charge efficiency of a battery is high until it approaches full charge, at which time the overcharge reactions begin and the charge efficiency decreases. Obviously, past the point of full recharge, the efficiency falls to zero.

Figure 17.39 is a curve of charge-current efficiency versus voltage at various constant voltages. Increasing voltage decreases the efficiency because of increased parasitic currents. The efficiency shows a marked decrease below the open-circuit voltage, typically 2.15 to 2.18 V, because the charge voltage is not high enough to support the charging reaction.

Figure 17.40 is a curve of efficiency versus log rate at various constant-current charge rates. As can be seen from the curve, at rates up to $C/10$ the efficiency approaches 100%. At higher rates, there is a decrease in efficiency because as the cell approaches the fully charged state, the surfaces of the plates become fully charged. This increases the charging reaction rates and results in increased voltages and gassing. At low charge rates, the efficiency drops because the charge current is equivalent to the parasitic currents and the battery voltage approaches the open-circuit value.

Figure 17.41 shows the charge acceptance during charge at various temperatures and charge rates.

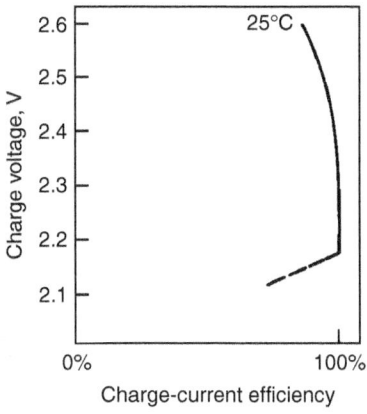

FIGURE 17.39 Constant-voltage charge efficiency.

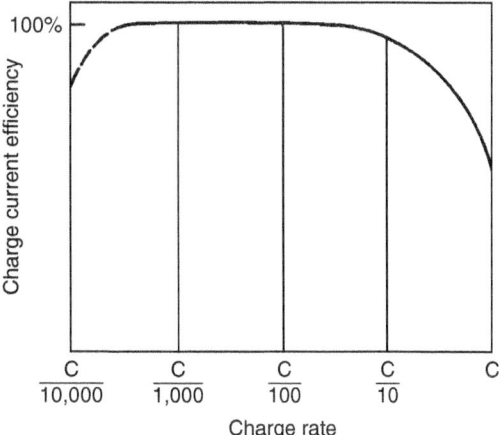

FIGURE 17.40 Constant-current charge efficiency.

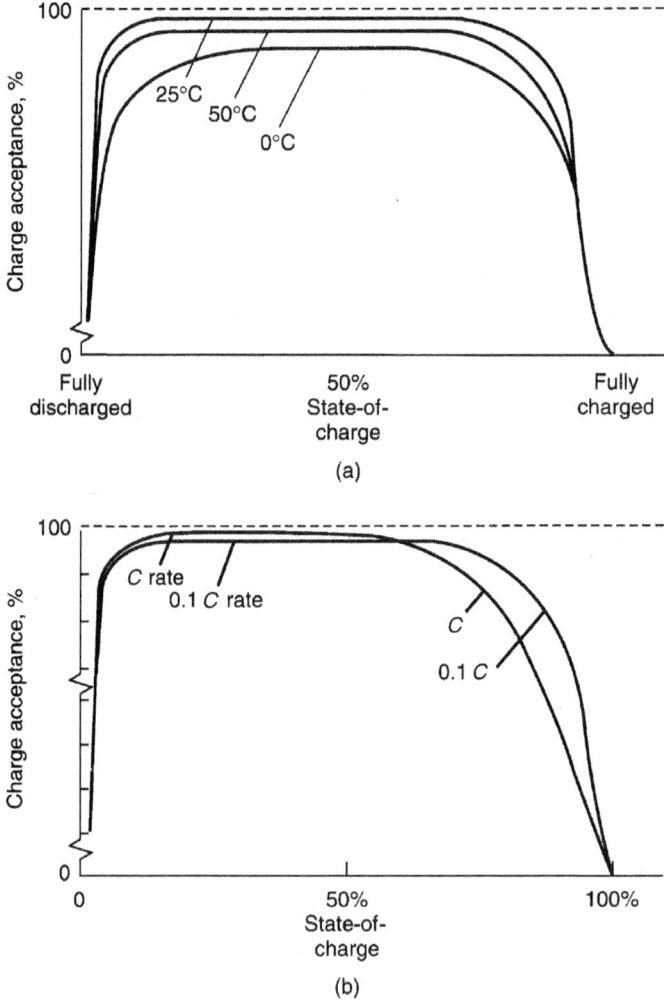

FIGURE 17.41 Charge acceptance of sealed lead-acid batteries. (*a*) At various temperatures. (*b*) At various charge rates.

17.6 SAFETY AND HANDLING

Two primary considerations relative to the application of VRLA batteries should be recognized to ensure that their usage is safe and proper: gassing and short-circuiting.

17.6.1 Gassing

Lead-acid batteries produce hydrogen and oxygen gases internally during charging and overcharging. These gases are released in an explosive mixture from conventional lead-acid batteries and therefore must not be allowed to accumulate in a confined space. An explosion could occur if a spark were introduced.

The VRLA battery, however, operates on 100% recombination of the oxygen gas produced at recommended rates of charging and overcharging, and so there is no oxygen outgassing. During normal operation, some hydrogen gas and also some carbon dioxide gas are given off. The hydrogen outgassing is essential with each cycle to ensure continued internal chemical balance. The lead grid construction of the VRLA battery minimizes the amount of hydrogen gas produced. Carbon dioxide is produced by oxidation of organic compounds in the cell.

The minute quantities of gases that are released from the VRLA battery with recommended rates of charge and overcharge will normally dissipate rapidly into the atmosphere. Hydrogen gas is difficult to contain in anything but a metal or glass enclosure, that is, it can permeate a plastic container at a relatively rapid rate. Because of the characteristics of gases and the relative difficulty in containing them, most applications will allow for their release into the atmosphere. However, if the VRLA battery is being designed into a gastight container, precautions must be taken so that the gases produced can be released to the atmosphere. If hydrogen is allowed to accumulate and mix with the atmosphere at a concentration of between 4 and 79% (by volume at standard temperature and pressure), an explosive mixture would be present which would be ignited in the presence of a spark or flame.

Another consideration is the potential failure of the charger. If the charger malfunctions, causing higher-than-recommended charge rates, substantial volumes of hydrogen and oxygen will be vented from the battery. This mixture is explosive and should not be allowed to accumulate. Adequate ventilation is required. Therefore the VRLA battery should never be operated in a gastight container. The batteries should never be totally encased in a potting compound since this prevents the proper operation of the venting mechanism and free release of gas. Furthermore, considerable pressure can build up in a gastight container. This can occur during storage because of the continuous generation of carbon dioxide gas. Such pressure is further compounded during charging by the generation of hydrogen.

17.6.2 Short-Circuiting

These batteries have low internal impedance and thus are capable of delivering high currents if externally short-circuited. The resultant heat can cause severe burns and is a potential fire hazard. Particular caution should be used when any person working near the open terminals of cells or batteries is wearing metal rings or watchbands. Inadvertently placing these metal articles or tools across the terminals could result in severe skin burns.

17.7 *BATTERY TYPES AND SIZES*

A listing of VRLA cylindrical batteries is given in Table 17.2. The performance of these units at various current drains at 25°C is given in Fig. 17.42. A number of multicell batteries are available that use these cells in various series/parallel configurations.

Table 17.3 and 17.4 list some of the typical VRLA prismatic lead-acid batteries that are manufactured. Information on other manufacturers' products can be obtained by consulting their websites. In alphabetical order, the principal suppliers include; C and D Technologies, East Penn-Deka, EnerSys, Exide, HBL (India), NorthStar, Panasonic, Power Battery, and Yuasa. Unlike some of the other types of lead acid batteries, there is no standardized list of sizes. Hence sizes, weights, and capacity ratings may vary from manufacturer to manufacturer.

Note that in telecom applications, there are numerous national and international standards that are applicable. These include:

1. IEC 6096-21/22. Global Standard for Stationary Valve-Regulated Lead-Acid Batteries (2003).
2. Telecordia SR-4228 (Bellcore TR-NWT-000766). VRLA Battery String Certification Based on Requirements for Safety and Performance.

17.34 SECONDARY BATTERIES

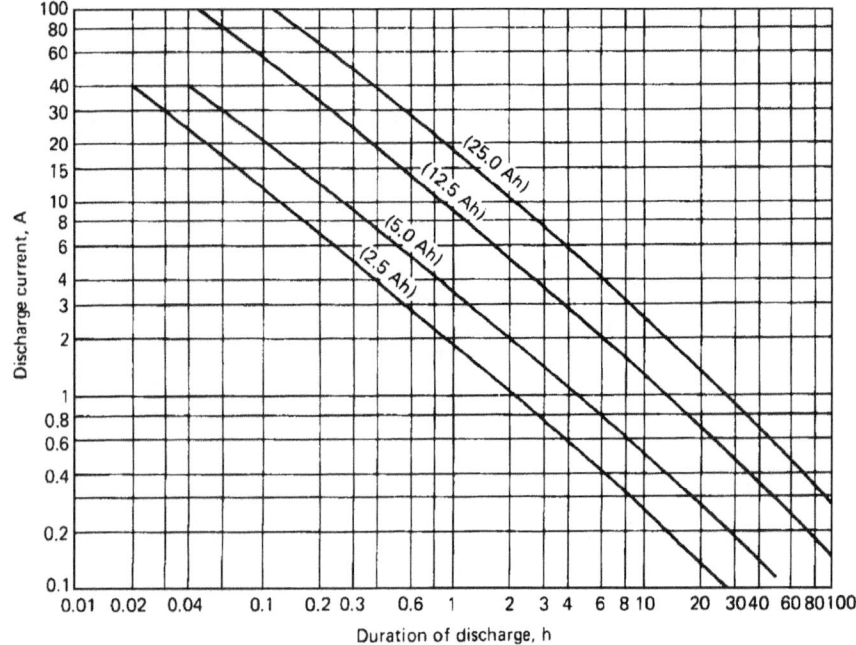

FIGURE 17.42 Discharge times of four types of VRLA cylindrical batteries at 25°C.

TABLE 17.3 Typical VRLA Batteries for Telecom Applications

	NSB40	NSB70	NSB75	NSB90	NSB125
Height	176 mm 6.93 in	176 mm 6.93 in	200 mm 7.87 in	213 mm 8.39 in	275 mm 10.81 in
Length	197 mm 7.76 in	331 mm 13.02 in	261 mm 10.27 in	341 mm 13.42 in	345 mm 13.57 in
Width	165 mm 6.50 in	165 mm 6.50 in	173 mm 6.80 in	173 mm 6.80 in	173 mm 6.80 in
Weight	16.0 kg 35.3 lbs	27.3 kg 60.0 lbs	27.3 kg 60.0 lbs	37.8 kg 83.1 lbs	54.0 kg 119 lbs
Terminal	M6 × 1.25	M6 × 1.25	M6 × 1.25	M6 × 1.25	M6 × 1.25
C/10 Cap	40 Ah	66 Ah	69 Ah	96 Ah	129 Ah
Impedance (1 kHz)	4.5 mΩ	2.7 mΩ	2.6 mΩ	2.0 mΩ	2.0 mΩ
Conductance @ 25°C (77°F)	1052 S	1589 S	1398 S	1806 S	2103 S
Short-circuit current	2000 A	3200 A	3200 A	4300 A	5000 A

(Courtesy of NorthStar Battery Company.)

TABLE 17.4 Specifications for Typical Pure Lead-Tin VRLA Batteries

Product (capacity)	Internal res. of fully charged battery mΩ @ 25°C	Nominal short-circuit current for charged battery	Dimensions			Weight lb. (kg)
			Length in. (mm)	Width in. (mm)	Height in. (mm)	
G13EP (13 Ah)	8.5	1400 A	6.910 (175.51)	3.282 (83.36)	5.113 (129.87)	10.8 (4.9)
G13EPX (13 Ah)	8.5	1400 A	6.998 (177.75)	3.368 (85.55)	5.165 (131.19)	12.0 (5.4)
G16EP (16 Ah)	7.5	1600 A	7.150 (181.61)	3.005 (76.33)	6.605 (167.77)	35.5 (6.1)
G16EPX (16 Ah)	7.5	1600 A	7.267 (184.58)	3.107 (78.92)	6.666 (169.32)	14.7 (6.7)
G26EP (26 Ah)	5.0	2400 A	6.565 (166.75)	6.920 (175.77)	4.957 (125.91)	22.3 (10.1)
G26EPX (26 Ah)	5.0	2400 A	6.636 (168.55)	7.049 (179.04)	5.040 (128.02)	23.8 (10.8)
G42EP (42 Ah)	4.5	2600 A	7.775 (197.49)	6.525 (165.74)	6.715 (170.56)	32.9 (14.9)
G42EPX (42 Ah)	4.5	2600 A	7.866 (199.80)	6.659 (169.14)	6.803 (172.80)	35.1 (15.9)
G70EP (70 Ah)	3.5	3500 A	13.020 (330.71)	6.620 (168.15)	6.930 (176.02)	53.5 (24.3)
G70EPX (70 Ah)	3.5	3500 A	13.020 (330.71)	6.620 (168.15)	6.930 (176.02)	56.0 (25.4)

Source: EnerSys Energy Products, Inc.

3. Bellcore GR-63-Core, Compliance Test Program.
4. DOT 49CFR 173.159(d) (i) and (ii). U.S. Non-Hazardous Shipping Regulations.
5. UL Approval. Requirements for Flame-Retardancy, UL V-0, and Proper Vent Operation.
6. EuroBatt. Requirement for Design Life Greater than Fifteen Years at 20°C.
7. British Standard BS 6920: Part 4: 1997. Specification for Classifying Valve-Regulated Lead-Acid Stationary Cells and Batteries.
8. Deutsche Telekom TL 4423-06.

The products described above, and similar products from other manufacturers, have been specifically engineered for use in indoor/outdoor cabinets, have a flame-retarded case and cover, and can be installed in any orientation other than inverted, which is not recommended. The effect of temperature on OCV and state-of-charge (SoC) is shown in Fig. 17.43 at three levels along with the rate of decay of OCV at these temperatures.

17.36 SECONDARY BATTERIES

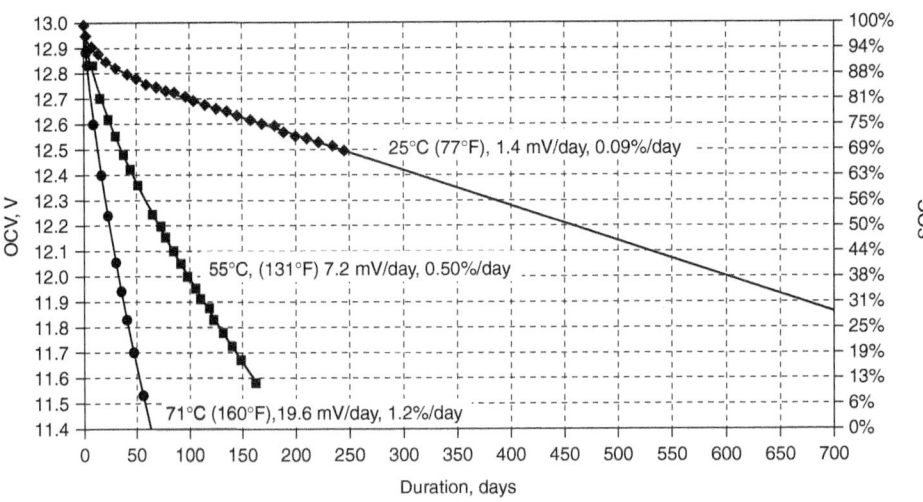

FIGURE 17.43 Effect of temperature on OCV and SOC at three levels. (*Courtesy of NorthStar Battery Co.*)

17.8 APPLICATIONS OF VRLA BATTERIES TO UNINTERRUPTIBLE POWER SUPPLIES

The major application of the VRLA battery is in the standby power market, ranging from low-power (generally less than 5 KVA) applications such as emergency lighting or uninterruptible power supplies (UPSs) for individual computers or work stations to high-power UPSs in telecommunications facilities. A continuous supply of power is also critical in areas such as banking, stock exchanges, hospitals, air traffic control centers, etc., where brief interruptions pose the risk of loss of critical data or hazards to health and safety. The low-power UPS systems are generally used where a power loss is acceptable as long as there is sufficient power to allow time for a safe shutdown of equipment. In a high-power application, the UPS is typically required to provide power until a generator can be brought on line.

UPS systems are generally one of three basic designs: (1) standby or off-line; (2) on-line; or (3) line interactive/hybrid. In most UPSs, AC power is fed to a battery charger/rectifier to provide the DC power to float charge the battery. The output of the battery is connected to an inverter that converts the DC output of the battery and/or the battery charger/rectifier to provide AC power needed to run the load. In the standby mode off-line UPS system, the battery and inverter only come into play when the normal AC power fails. In the on-line UPS system, the battery and inverter are always in the circuit. When the AC power fails, the battery is already on-line supplying power to the inverter and no voltage dropout occurs, as opposed to the off-line case where the voltage may drop out, typically for milliseconds before the battery/inverter duo is switched into service. The on-line UPS system serves also to smooth out any voltage fluctuations with the battery/inverter, a continuously active element in the circuit (see Sec. 16.9.6). The line interactive hybrid UPS system utilizes an automatic voltage regulator and a special transformer to smooth any under- or overvoltages and to ease the transition to complete battery backup only during outages of the input AC power. A comprehensive treatment of UPS systems, markets and alternative UPS systems (e.g., flywheels) is given in Ref. 18.

The experiences with VRLA batteries in the high-power UPS arena have served to demonstrate the complex problems associated with these batteries compared to their flooded counterparts. For most of the 20th century, lead-acid batteries for high-power UPS systems, notably those employed in telecommunications facilities, were of the flooded type. Flooded batteries have relatively large footprints, can spill acid, and require periodic maintenance in the form of watering, a costly operation for a UPS facility involving a large number of cells. In the mid-1980s, with the advent of the VRLA battery and its maintenance-free feature, there was immediate interest in replacing flooded with VRLA batteries. Initially, expectations were that VRLA batteries would provide the roughly 20-year life found with flooded batteries. Failures of VRLA batteries occurred after only a few years of service in many cases. One unanticipated problem, negative plate self-discharge, is considered below. Currently, more realistic claims of 5–10 year battery life are the rule and there is a trend back to flooded lead-acid for the high-power UPS applications. Some switching from lead-acid to nickel-cadmium or nickel-metal hydride batteries for low-power UPS applications is also taking place or being considered.

The life of a VRLA battery in a UPS application depends not only on the design of the battery and the quality of its manufacture, but also very strongly on the usage. Most of the performance figures quoted in the manufacturers' specifications are for operating temperatures of 25°C. Any significant deviation from that temperature, higher or lower, can result in poor performance, especially in the hands of a customer without knowledge of the proper handling of VRLA batteries. For example, the optimal charging regime for a VRLA battery is quite dependent on the temperature and must be modified for either higher or lower temperatures (see Sec. 16.5.4). VRLA batteries that perform perfectly well in a constant temperature environment may perform quite poorly, even exploding or catching fire, in an outdoor cabinet in a telecommunications application in a variable climate.

However, a considerable amount of research has gone into various ways to improve VRLA technology to overcome some of its deficiencies. These deficiencies are related generally to the oxygen recombination feature of the VRLA battery. One problem is negative plate self-discharge, a problem which has been found even in batteries that have been operated under conditions conforming to the battery manufacturers' recommendations. This negative self-discharge leads to the negative being in a significantly reduced state of charge, and a reduced battery capacity results. In flooded batteries, the float current is more than ample to keep the negative plate fully charged. In VRLA batteries, this is not necessarily the case, and at least three approaches have been suggested for improving the situation. One is to increase the purity of the negative plate. Certain impurities tend to lower the overvoltage for hydrogen evolution. A second approach is to increase the corrosion rate of the positive grid, not an attractive alternative for long-life batteries. Another approach is the addition of a catalyst to promote the recombination of hydrogen and oxygen.[19] This catalytic approach should compensate for impure active materials as well as for air leaks into the cell. The addition of a catalyst may also increase the negative polarization, thus lowering the positive polarization, which in turn would lessen positive grid corrosion. A combination of higher purity negative plates and a catalyst could be an even better solution to negative plate polarization, but the purity problem is complicated by the desire to use recycled lead, with attendant needs for improved refining processes. One company is manufacturing catalytic devices for use in new batteries or, in some cases, retrofitting batteries already in service.

Control of charging current as a function of battery temperature is a good way to minimize problems with VRLA batteries in hot environments. Current can be controlled directly in some applications. But since telecommunications systems control the charge voltage, battery current must be controlled by reducing the float voltage as the temperature increases.[14]

Data on two examples of VRLA batteries are given in Tables 17.5a, b and c. These cells employ lead-calcium-tin positive grids and lead-calcium negative grids, a self-resealing safety vent releasing at 2 psi. Lead-calcium alloy grids are also used by other manufacturers. Batteries are floated at 2.23 to 2.35 V per cell at 25°C.

TABLE 17.5a Prismatic VRLA Single-Cell Battery Characteristics at 25°C

Voltage	Capacity, Ah	Height, mm	Width, mm	Length, mm	Weight, kg	Max. amps*
2 V/cell	500 (C/8) 346 (C/2)	368	182	228	32	1000
2 V/cell	1440 (C/8) 968 (C/2)	580	328	183	88	2880

*Maximum current for 1 minute duration.
Source: Panasonic website: www.panasonic.com.

17.38 SECONDARY BATTERIES

TABLE 17.5b Self-Discharge Data for Cells in Table 17.5(a) at 25°C

Storage time	3 months	6 months	12 months	18 months
% Initial capacity	91	82	64	50

TABLE 17.5c Discharge Rate (A) for 1440 Ah Battery in Table 17.5(a)

Hours	20	10	8	4	1
Cutoff voltage:					
1.60	76	151	187	319	894
1.75	72	143	180	300	731
1.90	64	128	151	259	540

The discharge rate data for a small 6-cell 12 V, 25 Ah (at $C/8$ to 10.5 V) VRLA battery are presented in Table 17.6. The dimensions of this battery are: height, 181 mm; width, 132 mm; length, 194 mm, with a weight of 10 kg.

TABLE 17.6 Discharge Rate (A) for 6-Cell, 25 Ah Battery

Time	20 h	10 h	30 min	10 min	1 min
Cutoff voltage:					
1.75 V	1.4	2.6	28.4	57.2	113.1
1.90 V	1.2	2.2	22.4	39.8	57.9

Source: C&D Technologies website: www.cdtechno.com.

The discharge rate data are presented in Table 17.7 for a 8 V, 427 kg, 1360 Ah battery model #4DDV85-33 manufactured by EnerSys, Inc.

TABLE 17.7 Discharge Rate (A) for 8 V, 1360 Ah Battery

Time	24 h	10 h	1 h	15 min	1 min
Cutoff voltage:					
1.75 V/cell	61	145	672	1248	1472

Source: EnerSys, Inc.

The above data are for randomly selected batteries out of many marketed for UPS applications, and are believed to be representative of the products currently on the market for UPS applications.

17.9 CURRENT DEVELOPMENTS AND FUTURE OPPORTUNITIES FOR VRLA BATTERIES

Significant improvements in the rate capability and cycle life of VRLA designs, as well as reduced maintenance, have provided opportunities for their use in several new applications. Recent studies include the use of improved glass and hybrid glass-polymeric separators, high-capacitance cells using carbon electrodes in parallel with the negative plates (UltraBattery), improved alloys and structures, and providing higher charge-rate capability. High-speed automation in manufacturing facilities has led to a more uniform product, which allows the use of series-connected modules for high-voltage applications. Some of the new applications that these designs and improvements facilitate include microhybrid electric vehicles (see Chap. 29) and forklift trucks for exterior use. The ability to stack VRLA cells on their side provides better use of floor space and reduced maintenance for interior applications, as well.

REFERENCES

1. S. Takahashi, K. Hirakawa, M. Morimitsu, Y. Yamagachi, and Y. Nakayama, "Development of a Long Life VRLA Battery for Load Leveling-2," *Yuasa-JIHO*, **88**:34–38, 2000.
2. A. G. Cannone, A. J. Salkind, and F. A. Trumbore. *Proc. 13th Annual Battery Conf.* Long Beach, CA, Jan. 1998.
3. M. Pavlov, *Conference on Oxygen Cycle in Lead-Acid Batteries*, 7th ELBC, Dublin, Ireland, September, 2000.
4. D. Berndt, "Valve-Regulated Lead-Acid Battery" and "Lead-Acid Batteries," *Conference on Oxygen Cycle in Lead and Batteries*, 7th ELBC, Dublin, Ireland, Sept. 2000.
5. P. Moseley, "Improving the Valve Regulated Lead-Acid Battery," *Proc. 1999 IBMA Conf., Battery Man*, p. 16–29, Feb. 2000.
6. W. W. McGill III, "Gel vs. VRLA Lift Truck Batteries," *Battery Man*, Feb. 2000, p. 34–36.
7. D. H. McClelland et al., U.S. Patent 3,704,173 and U.S. Patent 3,862,861.
8. S. S. Misra and T. M. Noveske, U.S. Patent 4,889,778, Dec. 26, 1989.
9. Z. Tang et al., *J. Appl. Electrochem.*, **37**:1163–1169, 2007.
10. K. R. Bullock, *J. Electrochem. Soc.*, **142**:1726–1731, 1995.
11. M. Weighall, *BEST* Magazine, Spring:67–75 (2008).
12. K. R. Bullock and D. H. McClelland, "The Kinetics of the Self-Discharge Reaction in a Sealed Lead-Acid Cell." *J. Electrochem. Soc.*, **123**:327, 1976.
13. R. Putnam, *BEST* Magazine, Spring:47–52 (2008).
14. R. O. Hammel, "Charging Sealed Lead Acid Batteries," *Proc. 27th Annual Power Sources Symp,* 1976.
15. K. R. Bullock, D. Fent, and P Ng, *Proc. 17th International Telecommunications Energy Conference*, pp. 8–13. (1995), IEEE 0-7803-2750-0/95.
16. W. Jones and D. O. Feder, *Batteries International*, pp. 77–83 (1997).
17. R. O. Hammel, "Fast Charging Sealed Lead Acid Batteries," extended abstracts, pp. 34–36, *Electrochem. Soc. Meeting*, Las Vegas, NV, 1976.
18. J. Plante, *Power 2000*, pp. 30–94, Supplement to *EE Times* (2000).
19. E. Jones, INTELEC 2000, paper 23.3; IEEE 10.1109/INTLEC.2000.884288.

CHAPTER 18
IRON ELECTRODE BATTERIES

Gary A. Bayles

18.1 GENERAL CHARACTERISTICS

Iron electrodes have been used as anodes in rechargeable battery systems since the introduction of the nickel-iron rechargeable battery at the turn of the century by Junger in Europe and Edison in the United States.[1] Even today the batteries are produced in a fashion similar to the original construction. New constructions have been developed which give better high-rate performance and have lower manufacturing costs. Today the nickel-iron battery is the most common rechargeable system using iron electrodes. Iron-silver batteries have been tested in special electronic applications, and iron/air batteries have shown promise as motive power systems. The characteristics of the iron battery systems are summarized in Tables 18.1 and 18.2.

As designed by Edison, the nickel-iron battery was and is almost indestructible. It has a very rugged physical structure and can withstand electrical abuse such as overcharge, over-discharge, discharged stand for extended periods, and short-circuiting. The battery is best applied where high cycle life at repeated deep discharges is required (such as traction applications) and as a standby power source with a 10- to 20-year life. Its limitations are low power density, poor low-temperature performance, poor charge retention, and gas evolution on stand. The cost of the nickel-iron battery lies between the lower-cost lead-acid and the higher-cost nickel-cadmium battery in most applications, with the exception of limited use applications in electric vehicles and mobile industrial equipment.

Most recently, iron electrodes have been considered and tested as cathodes too. Based upon high valence state iron, Fe(VI), these cathodes have shown promise in experimental cells when coupled with zinc or metal hydride anodes for portable primary and secondary batteries. The results are discussed in Sec. 18.8.

18.2 CHEMISTRY OF NICKEL-IRON BATTERIES

The active materials of the nickel-iron battery are metallic iron for the negative electrode, nickel oxide for the positive, and a potassium hydroxide solution with lithium hydroxide for the electrolyte. The nickel-iron battery is unique in many respects. The overall electrode reactions result in the transfer of oxygen from one electrode to the other. The exact details of the reaction can be very complex and include many species of transitory existence.[2–4] The electrolyte apparently plays no part in the overall reaction, as noted in the following reactions:

$$Fe + 2NiOOH + 2H_2O \xrightleftharpoons[\text{charge}]{\text{discharge}} 2Ni(OH)_2 + Fe(OH)_2 \quad \text{(first plateau)}$$

$$3Fe(OH)_2 + 2NiOOH \xrightleftharpoons[\text{charge}]{\text{discharge}} 2Ni(OH)_2 + Fe_3O_4 + 2H_2O \quad \text{(second plateau)}$$

18.2 SECONDARY BATTERIES

TABLE 18.1 Iron Electrode Battery Systems

System	Uses	Advantages	Disadvantages
Iron/nickel oxide (tubular)	Material handling vehicles, underground mining vehicles, miners' lamps, railway cars and signal systems, emergency lighting	Physically almost indestructible Not damaged by discharged stand Long life, cycling or stand Withstands electrical abuse: overcharge, overdischarged, short-circuiting	High self-discharge Hydrogen evolution on charge and discharge Low power density Lower energy density than competitive systems Poor low-temperature performance Damaged by high temperatures Higher cost than lead-acid Low cell voltage
Iron/air	Motive power	Good energy density Uses readily available materials Low self-discharge	Low efficiency Hydrogen evolution on charge Poor low-temperature performance Low cell voltage
Iron/silver oxide	Electronics	High energy density High cycle life	High cost Hydrogen evolution on charge

TABLE 18.2 System Characteristics

System	Nominal voltage, V		Specific energy Wh/kg	Energy density Wh/L	Specific power W/kg	Cycle life, 100% DOD
	Open-circuit	Discharge				
Iron/nickel oxide						
Tubular	1.4	1.2	30	60	25	4000
Developmental	1.4	1.2	55	110	110	>1200
Iron/air	1.2	0.75	80	—	60	1000
Iron/silver oxide	1.48	1.1	105	160	—	>300

The overall reaction is

$$3Fe + 8NiOOH + 4H_2O \underset{\text{charge}}{\overset{\text{discharge}}{\rightleftharpoons}} 8Ni(OH)_2 + Fe_3O_4$$

The electrolyte remains essentially invariant during charge and discharge. It is not possible to use the specific gravity of the electrolyte to determine the state of charge as for the lead-acid battery. However, the individual electrode reactions do involve an intimate reaction with the electrolyte.

A typical charge-discharge curve of an iron electrode is shown in Fig. 18.1. The two plateaus on charge correspond to the formation of the stable +2 and +3 valent states of the iron reaction products. The reaction of the iron electrode can be written as

$$Fe + nOH^- \rightarrow FE(OH)_n^{2-n} + 2e \quad \text{(first plateau)}$$

and

$$Fe(OH)_n^{2-n} \rightarrow Fe(OH)_2 + nOH^-$$

$$Fe(OH)_2 + OH^- \rightarrow Fe(OH)_3 + e \quad \text{(second plateau)}$$

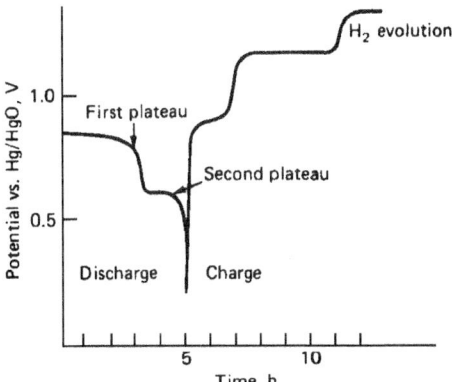

FIGURE 18.1 Discharge-charge curve of an iron electrode. (*From Ref. 5.*)

Then

$$2Fe(OH)_3 + Fe(OH)_2 \rightarrow Fe_3O_4 + 4H_2O$$

Iron dissolved initially as the +2 species in alkaline media. The divalent iron complexes with the electrolyte to form the $Fe(OH)_n^{2-n}$ complex of low solubility. The tendency to supersaturate plays an important role in the operation of the electrode and accounts for many important aspects of the electrode performance characteristics. Continued charge forms the +3 valent iron which, in turn, interacts with +2 valent iron to form Fe_3O_4.

The superior life-cycling characteristics of the iron electrode result from the low solubility of the reaction intermediates and oxidized species. The supersaturation on discharge results in the oxidized material forming small crystallites near the reaction site. On charge, the low solubility also slows the crystal growth of the iron, thereby helping to ensure formation of the original active high-surface-area structure. The low solubility also accounts for poor high-rate and low-temperature performance as the discharged (oxidized) species precipitate at or near the reaction site and block the active surface. The performance characteristics are substantially improved, however, in the advanced nickel-iron batteries by the use of a superior electrode grid structure, such as fiber-metal, which provides intimate contact with the iron active material throughout the volume of porous structure.

Sulfide addition to the iron electrode radically changes the electrocrystallization kinetics. It increases the supersaturation and makes reaction more reversible. Sulfide also absorbs on the surface to block crystallization sites and raises the hydrogen evolution reaction on charge. The self-discharge rate of the iron electrode is correspondingly reduced by the sulfide as well, with studies showing that PbS outperforms FeS in this regard.[6] Lithium salt additions seem to make the electrode perform more reversibly, perhaps by enhancing the solubility of the reaction intermediates.

Nickel electrode reactions[7,8] are generally thought to be solid-state-type reactions wherein a proton is injected or rejected from the lattice reversibly on discharge and on charge, respectively.

$$\beta\text{-}Ni(OH)_2 \xleftarrow[\text{in KOH}]{\text{transformation}} \alpha\text{-}Ni(OH)_2$$

$$\text{reduction} \updownarrow \text{oxidation} \qquad \text{oxidation} \updownarrow \text{reduction}$$
$$\text{(discharge)} \quad \text{(charge)} \qquad \text{(charge)} \quad \text{(discharge)}$$

$$\beta\text{-}NiOOH \xrightarrow[\text{in KOH}]{\text{overcharge}} \gamma\text{-}NiOOH$$

The oxidation (charge) voltage for the α and β materials is more positive than the discharge voltage by 60 mV and 100 mV, respectively. The β-$Ni(OH)_2$ is the usual electrode material. It is converted on charge to β-NiOOH with about the same molar volume. On overcharge, the γ structure can form. This form also incorporates water and potassium (and lithium) into the structure. Its molar volume is about 1.5 times the β form. This is thought to be responsible in large part for the volume expansion (swelling) which occurs on charging the battery. The α form then results on discharge of the γ form. Its molar volume is about 1.8 times the β form, and the electrode can swell further on discharge. On discharge stand in concentrated electrolyte, the α form converts to the β form. Cobalt additions (2 to 5%) improve the charge acceptance (reversibility) of the nickel electrode.

18.3 CONVENTIONAL NICKEL-IRON BATTERIES

18.3.1 Construction

The construction of a tubular or pocket plate nickel-iron cell is shown in Fig. 18.2. The active materials are filled in nickel-plate perforated steel tubes or pockets. The tubes are fastened into plates of desired dimensions and assembled into cells by interleaving the positive and negative plates. The container is fabricated from nickel-plated sheet steel. The cells may be assembled into batteries in molded nylon cases or mounted into wooden traps. The steel cases may be coated with plastic or rubber for insulation or spaced by insulating buttons.

The manufacturing process has remained relatively unchanged for over 50 years. The processes are designed to produce materials of highest purity and with special particle characteristics for good electrochemical performance.

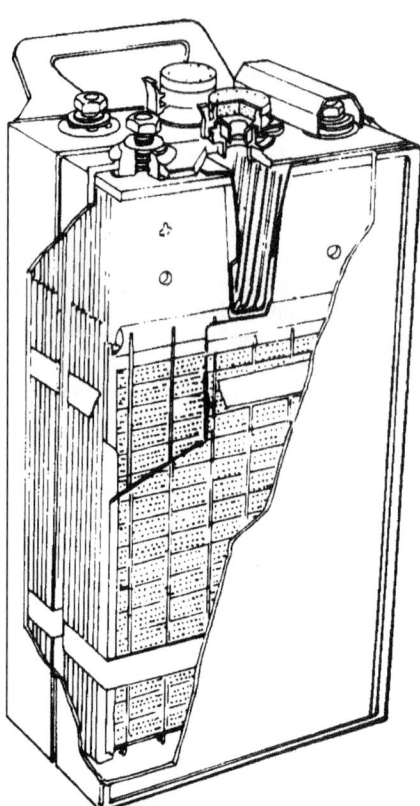

FIGURE 18.2 Cross section of typical nickel-iron battery. (*Courtesy of SAFT America, Inc.*)

Negative Electrode. To produce the anode active material, pure iron is dissolved in sulfuric acid. The $FeSO_4$ is recrystallized, dried, and roasted (815 to 915°C) to Fe_2O_3. The material is washed free of sulfate, dried, and partially reduced in hydrogen. The resulting material (Fe_3O_4 and Fe) is partially oxidized, dried, ground, and blended. Small amounts of additives, such as sulfur, FeS, and HgO, are blended in to increase battery life by acting as depassivators, reducing gas evolution, or improving conductivity.

To make the anode current collector, steel strips or ribbon are perforated and nickel-plated. After drying and annealing, the strip is formed into a pocket, about 13 mm wide and 7.6 mm long. One end is left open and filled with the iron active material. A machine automatically introduces the active material and tamps it into the pockets. After filling, the negative pockets are crimped and pressed into openings in a nickel-plated steel frame.

Positive Electrode. The positive active material consists of nickel hydroxide in alternate layers with nickel flake. High-purity nickel powder or shot is dissolved in sulfuric acid. The hydrogen evolved is used in making the iron active material. The acidity of the resulting solution is adjusted to pH 3 to 4 and filtered to remove ferric iron and other insoluble materials. If needed, the solution may be further purified to remove traces of ferrous iron and copper. Cobalt sulfate may be added in the proportion of 1.5% to improve nickel electrode performance. The nickel sulfate solution is sprayed into hot 25 to 50% NaOH solution. The resulting slurry is filtered, washed, dried, crushed, and screened to yield particles that pass 20- but not 200-mesh screens.

Special nickel flake (1.6 × 0.01 mm) is produced by electrodepositing alternate layers of nickel and copper on stainless steel. The electroplate is stripped and cut into squares. The copper is dissolved out in hot sulfuric acid, and the resulting nickel flakes are washed free of copper and dried at low temperature to prevent nickel oxide formation. With a modified process[9] the flakes of proper shape and size can be produced as a single layer, eliminating the need for deposition of alternate copper layers. As in the negative electrode, the positive-electrode process starts with perforated steel ribbon which is nickel-plated and annealed. The ribbon is wound into tubes with an interlocked seam. Two types, right- and left-wound tubes, are produced, typically of 6.3 mm diameter. The tubes are filled with alternate layers of nickel hydroxide and nickel flakes. Each layer is tamped (144 kg/cm^2) to ensure good contact. There are 32 layers of flake per centimeter. To prevent the seam from opening during the rigors of charge and discharge, rings are placed around the tubes at uniform intervals of about 1 cm. The tubes are enclosed, and the pinched ends are locked into the nickel-plated steel grid frame. The "rights" and "lefts" are alternated so that any tendency to distort on the part of one tube is counteracted by the next one. The positive electrode can also be made in the pocket plate construction, as described above under "Negative Electrode."

Cell Assembly. The configuration and size of the tubes and pockets determine the capacity for each plate. The plates are then assembled into electrodes to meet the capacity requirements of each cell.

Each plate group is assembled by bolting a terminal pole and a selected number of plates, depending on capacity, to a steel rod which passes through the grid at the top of the plates. Groups of positive and negative plates are intermeshed to form the element. A cell usually contains one or more negative than positive plates. The cells are made positive limiting for best cycle life.

The positive and negative plates are separated by hard rubber or plastic pins called "hair pins" or "hook pins," which fit into spaces formed by the tubular positive and flat negative electrodes.

Electrolyte. The electrolyte is a 25 to 30% KOH solution with up to 50 g/L of LiOH added. The composition of the replacement electrolyte to compensate for losses due to spray from the vent cap is about 23% caustic with about 25 g/L LiOH. Occasionally the electrolyte is replaced completely to rejuvenate the cell performance. The renewal electrolyte is about 30% KOH with 15 g/L LiOH.

Lithium additions to the electrolyte are important but not completely understood. Recent work on the mechanism of the lithium interaction suggests that Li^+ is reduced within the iron oxide lattice to produce intermediate $Li_xFe_yO_z$ intercalation-compounds, which are subsequently reduced to metallic iron and lithium hydroxide.[10] Lithium hydroxide improves cell capacity and prevents capacity loss on cycling and also seems to facilitate nickel electrode kinetics. It expands the working plateau on charge and delays oxygen evolution. Some evidence exists for the formation of Ni^{4+}, which improves electrode capacity. Lithium also decreases the carbonate content in the electrolyte since Li_2CO_3 is not very soluble. It also decreases the tendency for swelling of the positive active material but increases the resistivity of the cell electrolyte.

Shortly after initiation of charge, hydrogen evolution begins on the iron electrode. The considerable hydrogen evolution on charge presumably helps counteract iron passivation in alkaline solution. Mercury additions also have a similar effect, but only in the early formation cycles.

18.3.2 Performance Characteristics of Nickel-Iron Battery

Voltage. A typical discharge-charge curve of a commercial iron/nickel oxide battery is shown in Fig. 18.3. The battery's open-circuit voltage is 1.4 V; its nominal voltage is 1.2 V. On charge, at rates most commonly used, the maximum voltage is 1.7 to 1.8 V.

18.6 SECONDARY BATTERIES

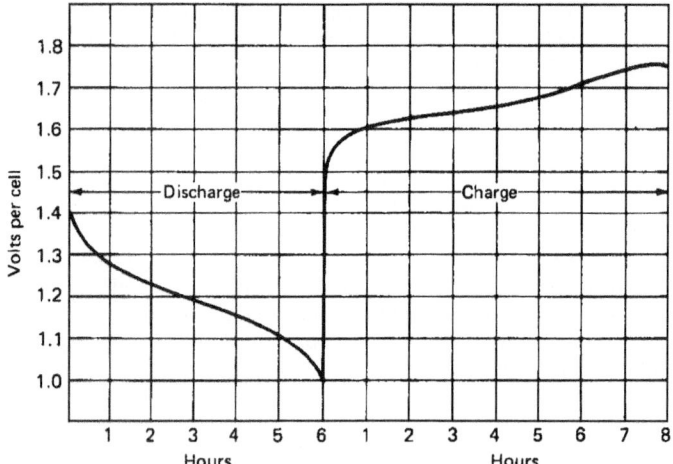

FIGURE 18.3 Typical voltage characteristics during constant-rate discharge and recharge. (*From Ref. 11.*)

Capacity. The capacity of the nickel-iron battery is limited by the capacity of the positive electrode and, hence, is determined by the length and number of positive tubes in each plate. The diameter of the tubes generally is held constant by each manufacturer. The 5 h discharge rate is commonly used as the reference for rating its capacity.

The conventional nickel-iron battery has moderate power and energy density and is designed primarily for moderate to low discharge rates. It is not recommended for high-rate applications such as engine starting. The high internal resistance of the battery lowers the terminal voltage significantly when high rates are required. The relationship between capacity and rate of discharge is shown in Fig. 18.4.

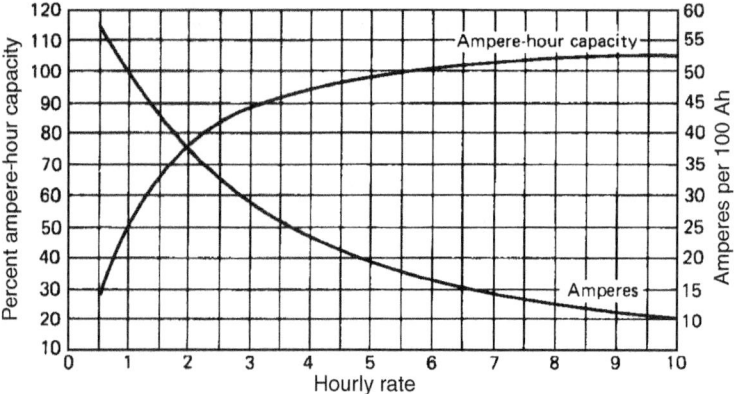

FIGURE 18.4 Curves of capacity vs. discharge rate at 25°C; end voltage 1.0 V per cell. (*From Ref. 11.*)

If a battery is discharged at a high rate and then at a lower rate, the sum of the capacities delivered at the high and low rates nearly equals the capacity that would have been obtained at the single discharge rate. This is illustrated in Fig. 18.5.

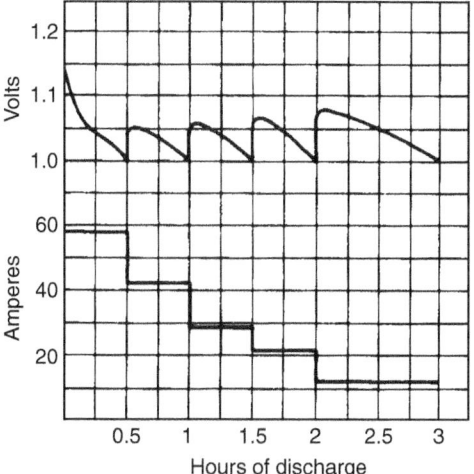

FIGURE 18.5 Effect of decreasing rate on battery voltage of nickel-iron cell.

Discharge Characteristics. The nickel-iron battery may be discharged at any current rate it will deliver, but the discharge should not be continued beyond the point where the battery nears exhaustion. It is best adapted to low or moderate rates of discharge (1 to 8 h rate). Figure 18.6 shows the discharge curves at different rates of discharge at 25°C.

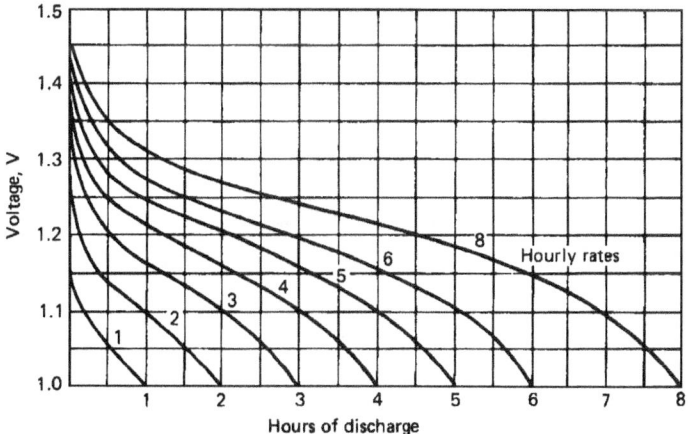

FIGURE 18.6 Time-voltage discharge curves of nickel-iron battery; end voltage 1.0 V per cell. (*From Ref. 11.*)

Effect of Temperature. Figure 18.7 shows the effects of temperature on the discharge. The capacity at 25°C is normally taken as the standard reference value. The decrease in performance is generally attributed to passivity of the iron electrode and decreased solubility of the reaction intermediate. At low temperature, increased resistivity and viscosity of the electrolyte along with slower nickel

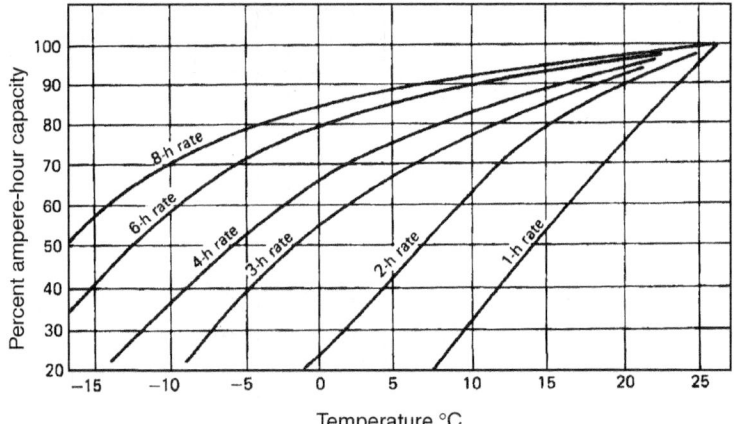

FIGURE 18.7 Effect of temperature on capacity at various rates. (*From Ref. 11.*)

electrode kinetics contribute to the fall-off of capacity. Care must be exercised to keep the temperature from exceeding about 50°C as the self-discharge of the nickel positive electrode is accelerated. Also, the increased solubility of iron at high temperature can adversely affect operation of the nickel electrode by incorporating soluble iron into the nickel hydroxide crystal lattice. The battery is seldom used below −15°C.

Hours of Service. The hours of service on discharge that a typical nickel-iron battery, normalized to unit weight (kilograms) and volume (liters), will deliver at various discharge rates and temperatures are summarized in Fig. 18.8.

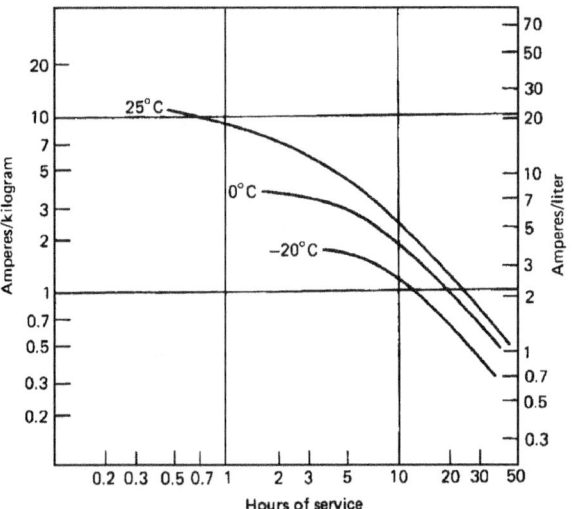

FIGURE 18.8 Hours of service of nickel-iron battery at various discharge rates and temperatures; end voltage 1.0 V cell.

Self-Discharge. The self-discharge rate, charge retention, or stand characteristic of the nickel-iron battery is poor. At 25°C a cell can lose 15% of its capacity in the first 10 days and 20 to 40% in a month. At lower temperatures the self-discharge rate is lower. For example, at 0°C the losses are less than one-half of those experienced at 25°C.

Internal Resistance. To a rough approximation, the internal resistance R_i can be estimated for tubular Ni-Fe from the equation

$$R_i \times C = 0.4$$

where R_i = internal resistance, Ω
C = battery capacity, Ah

For example, $R_i = 0.004\ \Omega$ for a 100 Ah battery. The value of R_i remains constant through the first half of the discharge, then increases about 50% during the latter half of the discharge.

Life. The main advantages of the tubular-type nickel-iron battery are its extremely long life and rugged construction. Battery life varies with the type of service but ranges from 8 years for heavy duty to 25 years or more for standby or float service. With moderate care, 2000 cycles can be expected; with good care, for example, by limiting temperatures to below 35°C, cycle life of 3000 to 4000 cycles has been achieved.

The battery is less damaged by repeated deep discharge than any other battery system. In practice, an operator will drive a battery-operated vehicle until it stalls, at which point the battery voltage is a fraction of a volt per cell (some cells may be in reverse). This has a minimal effect on the nickel-iron battery in comparison with other systems.

Charging. Charging of the batteries can be accomplished by a variety of schemes. As long as the charging current does not produce either excessive gassing (spray out of the vent cap) or temperature rise (above 45°C), any current can be used. Excessive gassing will require more frequent addition of water. If the cell voltage is limited to 1.7 V, these conditions should not be a consideration. Typical charging curves are given in Fig. 18.9. The ampere-hour input should return 25 to 40% excess of the previous discharge to ensure complete charging. The suggested charge rate is normally between 15 and 20 A per 100 Ah of battery capacity. This rate would return the capacity in the 6 to 8 h time frame. The effect of temperature on charging is shown in Fig. 18.10.

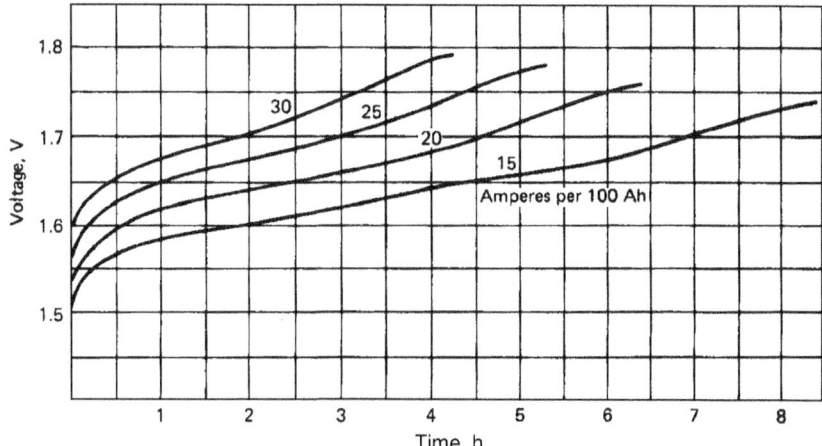

FIGURE 18.9 Typical charging voltage for nickel-iron battery at various rates. (*From Ref. 11.*)

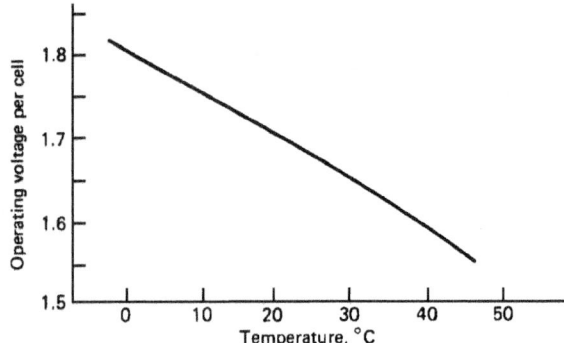

FIGURE 18.10 Variation of relay operating voltage with temperature. (*From Ref. 11.*)

Constant current and modified constant potential (taper), shown in Fig. 18.11, are common recharging techniques. The charging circuit should contain a current-limiting device to avoid thermal runaway on charge. Recharging each night after use (cycle charging) is the normal procedure. The batteries can be trickle-charged to maintain them at full capacity for emergency use. A trickle charge rate of 0.004 to 0.006 A/Ah of battery capacity overcomes the internal self-discharge and maintains the battery at full charge. Following an emergency discharge, a separate recharge is needed. For applications such as railroad signals, charging at a continuous average current may be the most economic method. Here a modest drain is required when no trains are passing but quite a heavy drain when a train passes, yet the total ampere-hours over a period of 24 h remains fairly constant. For this situation, a constant current equal to that required to maintain the battery can be used.

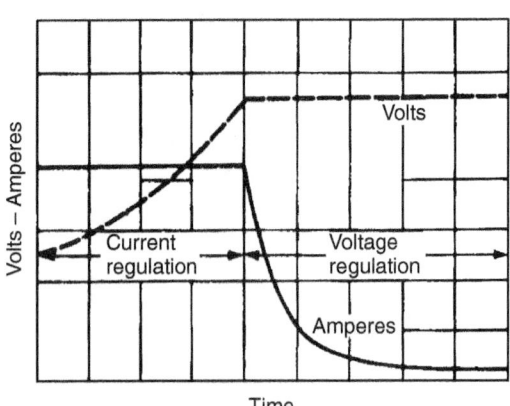

FIGURE 18.11 Effects of "regulators" with voltage and current regulation. (*From Ref. 11.*)

18.3.3 Sizes of Nickel-Iron Batteries

Nickel-iron batteries have been available in sizes ranging from about 5 to 1250 Ah. In recent years they have become less popular, giving way to the lead-acid and nickel-cadmium batteries, and are no longer manufactured by many of the original manufacturers. Table 18.3 lists the physical and electrical characteristics of typical nickel-iron batteries.

18.3.4 Special Handling and Use of Nickel-Iron Batteries

The battery should be operated in a well-ventilated area to prevent the accumulation of hydrogen. Under certain circumstances, hydrogen can be ignited by a spark to cause an explosion with a resulting fire. In multicell batteries, the usual precautions in dealing with high voltages should be taken.

TABLE 18.3 Typical Nickel-Iron Batteries

Nominal capacity, Ah	169	225	280	337	395	450	560	675
Nominal current, A: 5 h discharge	34	45	56	67	79	90	112	135
Cell weight, filled, kg	8.8	10.8	12.9	15.3	17.4	19.5	24.3	28.6
Installed weight, kg	9.8	12.0	14.3	16.9	19.3	21.7	26.5	31.2
Electrolyte (1.17 kg/L), kg	1.8	2.2	2.6	3.0	3.4	3.8	4.9	5.9
Cell dimensions, mm*								
Length	52	66	82	96	111	125	156	186
Width	130	130	130	130	130	130	135	135
Height	534	534	534	534	534	534	534	534
Battery dimensions, mm†								
Length:								
2 cells	—	—	—	265	295	321	343	343
3 cells	—	—	—	376	421	460		
4 cells	284	367	431	487				
5 cells	346	448	545					
6 cells	408	546						
Width	161	161	161	161	161	161	197	228
Height	568	573	582	582	582	582	590	590

*See drawing (a).
†See drawing (b).
Source: Varta Batteries AG, Hanover, Germany.

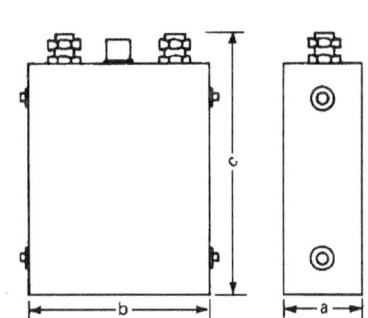

(a) Cell showing dimensions used in Table 18.3.

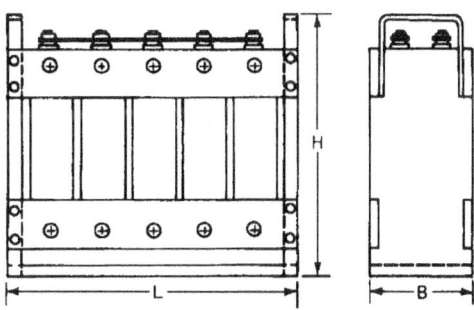

(b) Multicell battery showing dimensions used in Table 18.3. Tolerances are 5, 3, and 3 mm for dimensions L, B, and H, respectively.

If the battery is to be out of service for more than a month, it should be stored in the discharged condition. It should be discharged and short-circuited, then left in that condition for the storage period. Filling caps must be kept closed. If this procedure is not followed, several cycles are required to restore the capacity upon reactivation.

Constant-voltage charging is not recommended for conventional nickel-iron batteries. It may lead to a thermal runaway condition which results in dangerous conditions and can severely damage the battery. When the battery nears full charge, the gassing reactions produce heat and the temperature rises, lowering the internal resistance and the cell EMF. Accordingly, the charge current increases under constant-voltage charge. This increased current further increases the temperature, and a vicious cycle is started. A modified constant-voltage charging with current limiting is, however, acceptable.

18.12 SECONDARY BATTERIES

18.4 ADVANCED NICKEL-IRON BATTERIES

The desire to use the attractive features of the nickel-iron couple, such as ruggedness and long life, in applications requiring high-rate performance and low manufacturing costs has led to the development of advanced nickel-iron batteries with performance characteristics suitable for electric automobiles and other mobile traction applications. The capability of these batteries permits an electric vehicle a range of at least 150 km between charges, acceleration rapid enough to merge into highway traffic, and a cycling life equivalent to 10 or more years of on-the-road service. The advanced battery utilizes sintered-fiber metal (steel wool) plaques, impregnated with active material, for both the positive and the negative electrodes. Nonwoven polypropylene sheets are used as separators between electrodes. The techniques for making plaques, impregnation and activation, stacking, and assembly are all amenable to high-volume production methods similar to those used in lead-acid battery manufacture.

The battery system design incorporates an electrolyte management system to minimize the maintenance problems associated with its widespread deployment in the public sector. This system, shown schematically in Fig. 18.12, provides for semiautomatic watering of the cells by utilizing a single-point watering port. The flow of electrolyte through the cells during the charging cycle permits heat removal and effective management of gas evolved during the charge. Uniform specific gravity for all cells is ensured, and specific gravity maintenance is easily achieved by use of the system.

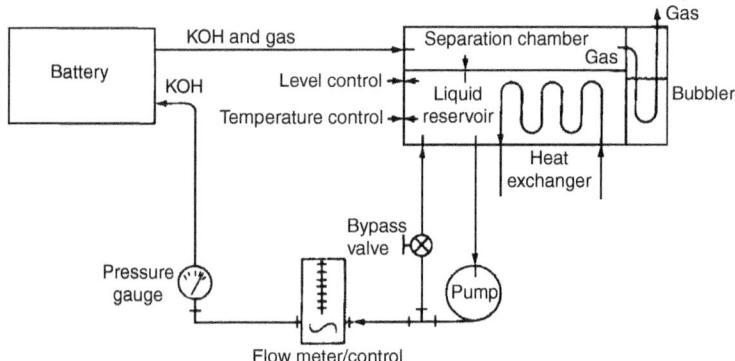

FIGURE 18.12 Schematic of electrolyte circulation system.

Both the positive and the negative plates of the Westinghouse nickel-iron battery used fiber metal plaques as the substrate. Two methods of active nickel impregnation were developed and used in demonstration batteries. An electroprecipitation process (EPP), developed in the mid-1960s, demonstrated good performance, ruggedness, and long cycle life. The EPP deposits nickel hydroxide electrochemically into the porous substrate. Efficient use of the nickel material is achieved, with active material utilization of 0.14 Ah/g of total electrode. An alternate nickel electrode manufacturing process was also developed which entails the preparation of a nickel hydroxide paste that is then loaded into the fiber metal substrate by roll pasting methods. Pasted nickel electrodes demonstrated performance equivalent to EPP electrodes (0.14 Ah/g of total electrode) while demonstrating a less expensive manufacturing process. The iron electrodes were also produced by a pasting process. Iron oxide, Fe_2O_3, was paste-loaded into the fiber metal electrode substrate and then furnace-reduced in a hydrogen atmosphere. These electrodes demonstrated 0.26 Ah/g of total electrode or, better, at $C/3$ discharge rates.

The performance characteristics of the advanced nickel-iron batteries, as typified by the Westinghouse system, are summarized in Table 18.4. Figure 18.13 shows a typical discharge curve for a 90-cell electric-vehicle battery at the $C/3$ rate. The battery capacity and energy as a function of discharge rate are shown in Fig. 18.14. Cell power and voltage characteristics, as a function of discharge rate and state

TABLE 18.4 Advanced Nickel-Iron Battery Performance Characteristics* Demonstrated as of December 1991

Capacity,[†] Ah	210
Specific energy,[†] Wh/kg	55
Energy density,[†] Wh/L	110
Specific power,[‡] W/kg	100
Cycle life[§]	>900
Urban range, km	
With regenerative braking	154
Without regenerative braking	125
Projected production cost, $/kWh (1990 $)	200–250

*Based on the Westinghouse nickel-iron battery.
[†]At the $C/3$ rate.
[‡]30 s average at 50% state of charge.
[§]Cycle to 100% depth of discharge; life to 75% of rated energy.

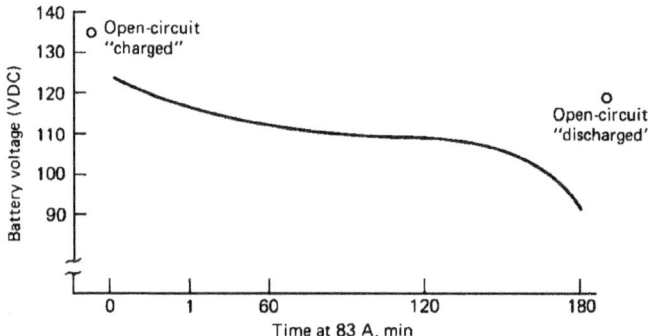

FIGURE 18.13 Battery voltage at $C/3$ (83 A) discharge rate.

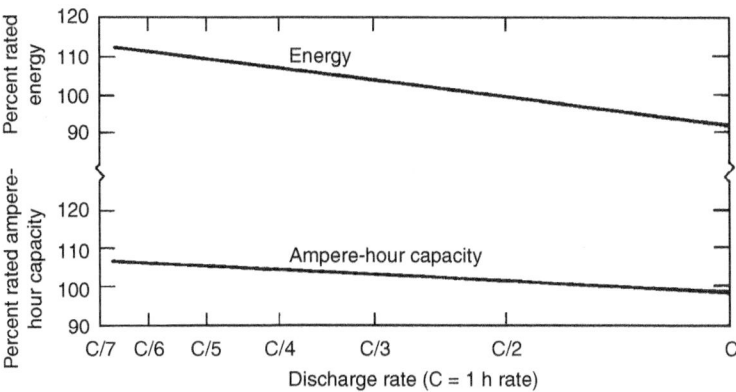

FIGURE 18.14 Capacity as a function of discharge rate. (*Courtesy of Westinghouse Electric Corp.*)

of discharge, are presented in Fig. 18.15. The variation in battery capability with temperature, based on tests on 5-cell modules, is shown in Fig. 18.16.[12–14] The Eagle-Picher Company also developed a nickel-iron battery using sintered-nickel electrodes similar to those described in Chap. 20. The iron electrode is similar to the Swedish National Development Corporation iron electrode discussed in Sec. 18.5. The performance of this battery is similar to that given in Figs. 18.13 to 18.16.[15]

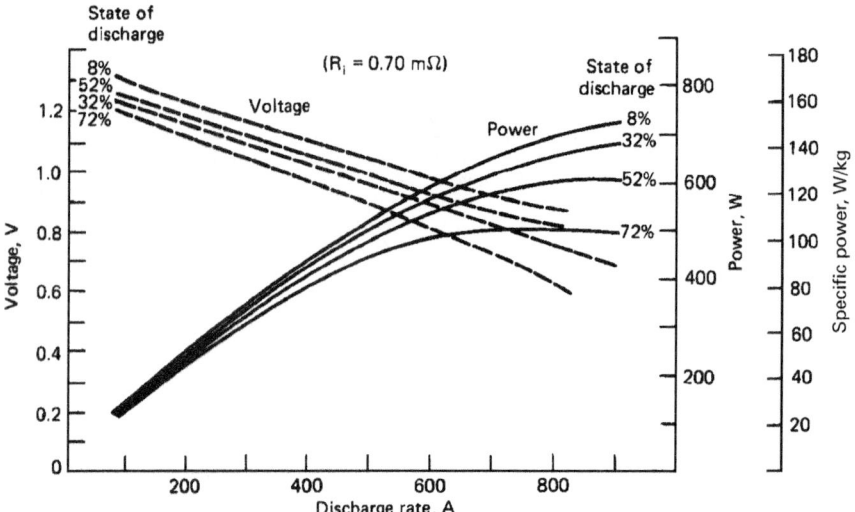

FIGURE 18.15 Power characteristics of 210 Ah nickel-iron battery. (*Courtesy of Westinghouse Electric Corp.*)

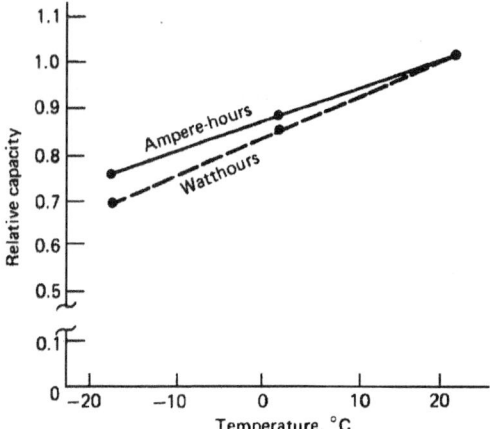

FIGURE 18.16 Effect of temperature on capacity and energy of nickel-iron battery ($C/3$ rate).

As noted earlier, interest in nickel-iron batteries has declined in recent years, in favor of other emerging electrochemical couples. Nonetheless, the rugged and economical nature of this couple continues to sustain a level of interest. Efforts to develop advanced iron electrodes, as described in Section 18.7 below, and to create a sealed maintenance-free design[16] are examples. If successful, a renewed thrust in nickel-iron batteries could result.

18.5 IRON/AIR BATTERIES

Rechargeable metal/air batteries have an advantage over conventional systems as only one reactant (the anode material) need be contained within the battery. The electrically rechargeable iron/air cell has a lower specific energy than the mechanically rechargeable cells (see Chap. 33) but has the advantage of potentially lower life-cycle costs. Unlike zinc, the iron electrodes do not suffer a severe redistribution of active materials or gross shape change upon prolonged electrical cycling. The iron/air cell is another candidate as a motive power source, especially for electric vehicles. The cell reactions are

$$O_2 + 2Fe + 2H_2O \underset{\text{charge}}{\overset{\text{discharge}}{\rightleftharpoons}} 2Fe(OH)_2 \quad \text{(first plateau)}$$

$$3Fe(OH)_2 + \tfrac{1}{2}O_2 \underset{\text{charge}}{\overset{\text{discharge}}{\rightleftharpoons}} Fe_3O_4 + 3H_2O \quad \text{(second plateau)}$$

The iron electrode kinetics are covered in Sec. 18.2. The oxygen electrode reactions follow the kinetic path with peroxide as an intermediate. The oxygen electrode reactions in simple form are

$$O_2 + 2H_2O + 2e \rightarrow H_2O_2 + 2OH^-$$

$$H_2O_2 + 2e \rightarrow 2OH^-$$

The single most important life-limiting factor in this battery system is the stability of the air electrode, which loses its ability to function reversibly as it undergoes repeated charges and discharges. The oxygen and peroxide evolved on charge and discharge may attack the substrate, alter the activity of the catalyst, and delaminate the wetproofing film. Separate air (oxygen) electrodes and circuits can be employed in the charge and discharge modes; however, considerations of system weight and volume favor the use of a bifunctional electrode, that is, a single electrode capable of sustaining either oxygen reduction or evolution. These electrodes must be stable over the potential range of both reactions, a fact which poses constraints on material stability and electrode design.

Several designs were developed, and research continues on the iron anode for use in iron/air batteries.[17–19] Most work on iron/air has been discontinued in favor of zinc/air.

The Swedish National Development Corporation's iron/air cell used the sintered-iron-mesh anode construction.[5,20,21] A pore-forming material could be included to control the development of the optimum electrode structure. The resulting pressed matrix was treated in H_2 at 650°C. The pore-forming material could be leached out after treatment. The active material utilization approached 65%. The air electrode was a porous-nickel double-layer structure (0.6 mm thick) composed of sintered nickel of coarse and fine porosities. The coarse layer on the electrolyte side was catalyzed with silver and impregnated with hydrophobic agents. The electrodes were welded into a polymer frame and formed into cells, as shown in Fig. 18.17. There were two air electrodes for each iron electrode. Air was forced past the electrode at about 2 times the stoichiometric requirement during operation. A schematic and a photo of a 30 kWh battery are shown in Fig. 18.18 and Fig. 18.19, respectively. Electrolyte circulation was used to control heat balance and remove gases generated during operation. Carbon dioxide was scrubbed from the incoming air using NaOH. The air was then humidified to minimize electrolyte loss. Overall the auxiliary systems require less than 10% of the system output.

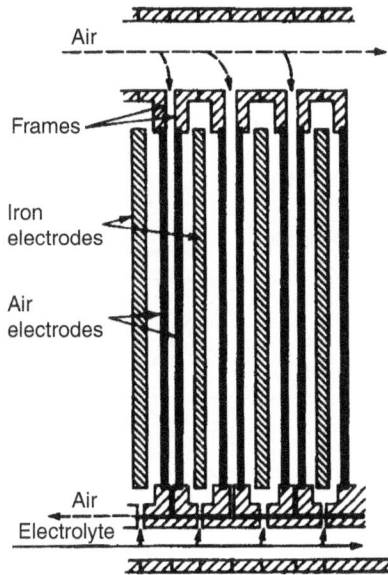

FIGURE 18.17 Cross section of Swedish National Development Corporation's iron/air battery pile. *(From Ref. 5.)*

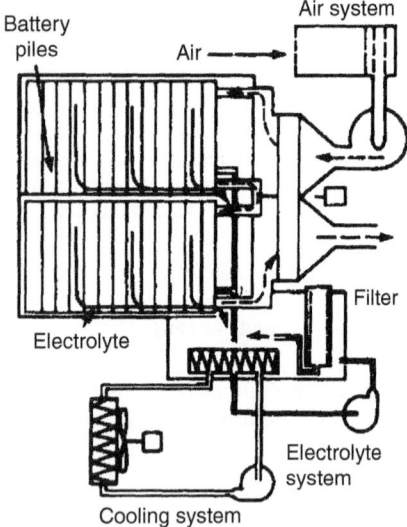

FIGURE 18.18 Schematic cross section of Swedish National Development Corporation's iron/air battery, including auxiliary system. (*From Ref. 5.*)

FIGURE 18.19 Swedish National Development Corporation's 30 kWh iron/air battery system. (*Courtesy of Swedish National Development Corp.*)

Typical charge-discharge curves for an average battery in the iron/air battery are shown in Fig. 18.20. The marked difference in charge and discharge voltages accounts largely for the low overall system efficiency. Figure 18.21 shows the power-producing characteristics. The system is capable of over 1000 cycles, limited by the gradual deterioration of the air electrode.

The Westinghouse iron/air battery used a construction similar to that described for the Swedish National Development Corporation's system.[22] The sintered-iron electrode was somewhat similar to

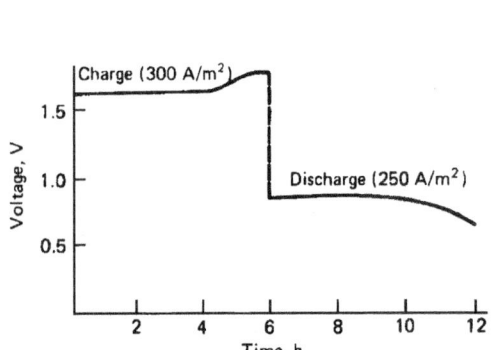

FIGURE 18.20 Charge-discharge voltages for battery in Swedish National Development Corporation's iron/air battery. (*From Ref. 5.*)

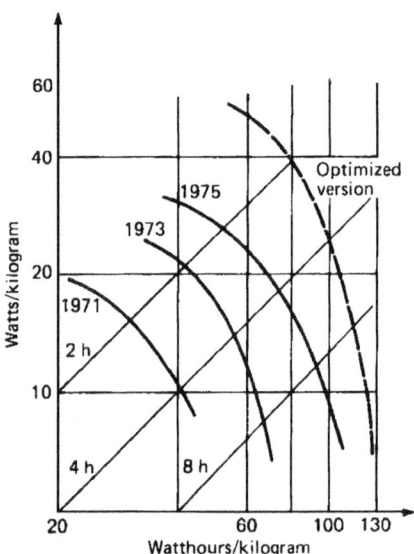

FIGURE 18.21 Performance of Swedish National Development Corporation's iron/air batteries. (*From Ref. 5.*)

TABLE 18.5 Characteristics of Westinghouse Iron/Air Electric-Vehicle Battery

Electric vehicle:	
Weight	900 kg curb weight
Range	240 km
Battery:	
Energy	40 kWh
Power	10 kW continuous power
Weight	530 kg
Volume	0.04 m^3
Cost	$150/kWh

that described before. The iron electrode for this iron/air battery had a high active iron content and lower cycle life compared with the electrode described previously. Particles of iron powder were sintered to form a structure without the steel fiber substrate. Electrodes with this construction demonstrated up to 0.44 Ah/g. The air electrode was bifunctional and used a Teflon-bonded carbon-based structure with complex silver catalysts (silver content was less than 2 mg/cm^2) supported on a silver-plated nickel screen. The Westinghouse system used a horizontal flow concept to improve performance and control gas and heat. Good life was demonstrated for over 300 cycles with an air electrode of potentially very low cost. A summary of the projected characteristics of the 40 kWh battery is given in Table 18.5.

The Siemens cell was similar except that the air electrode was fabricated with two layers: a hydrophilic layer of porous nickel on the electrolyte side for oxygen evolution and a hydrophobic layer (carbon black bonded with Teflon® [PTFE] and catalyzed with silver) on the air side for oxygen reduction. The dual porosity helped to shield the silver catalyst from oxidation. As many as 200 cycles were achieved.[23]

18.6 SILVER-IRON BATTERY

The silver-iron battery has been limited in use because of its high cost. Its theoretical energy density is essentially equal to that of the more popular silver-zinc system. The silver-iron battery has good cycle life compared with the silver-zinc and provides a battery of high reliability, long life, and better durability where high specific energy content is essential.[24–29] Figure 18.22 shows a 3.5 kWh battery designed for telecommunication use, whereas Fig. 18.23 shows a 9.5 kWh battery designed for use in a submersible.

FIGURE 18.22 A 3.5 kWh telecommunications iron/silver oxide battery. (*Courtesy of Westinghouse Electric Corp.*)

FIGURE 18.23 A 9.5 kWh iron/silver oxide battery for a submersible vehicle. (*Courtesy of Westinghouse Electric Corp.*)

The cell reactions are

$$\mathrm{Fe} + 2\mathrm{AgO} + \mathrm{H_2O} \underset{\mathrm{charge}}{\overset{\mathrm{discharge}}{\rightleftharpoons}} \mathrm{Fe(OH)_2} + \mathrm{Ag_2O} \qquad \text{(first plateau)}$$

$$\mathrm{Fe} + \mathrm{Ag_2O} + \mathrm{H_2O} \underset{\mathrm{charge}}{\overset{\mathrm{discharge}}{\rightleftharpoons}} \mathrm{Fe(OH)_2} + 2\mathrm{Ag} \qquad \text{(second plateau)}$$

$$3\mathrm{Fe(OH)_2} + \mathrm{Ag_2O} \underset{\mathrm{charge}}{\overset{\mathrm{discharge}}{\rightleftharpoons}} \mathrm{Fe_3O_4} + 3\mathrm{H_2O} + 2\mathrm{Ag} \qquad \text{(third plateau)}$$

In practice, only the first and second discharge plateaus are used. The overall reaction is

$$2\mathrm{Fe} + 2\mathrm{AgO} + 2\mathrm{H_2O} \underset{\mathrm{charge}}{\overset{\mathrm{discharge}}{\rightleftharpoons}} 2\mathrm{Fe(OH)_2} + 2\mathrm{Ag} \qquad E^0 = 1.34 \text{ V}$$

Charge/Discharge Characteristics. Typical charge-discharge curves for a silver-iron battery of the type shown in Fig. 18.22 are given in Fig. 18.24. The electrolyte is KOH of 1.31 specific gravity with 15 g/L LiOH added. The batteries can withstand several complete reversals without appreciable adverse effect on capacity.

Separator System and Cycle Life. Multilayer microporous polyethylene, nonwoven felt polypropylene, and cellophanes are some of the materials typically used as separators in this system. It is important to note that the choice of separator has very little to do with the iron electrode, which is extremely stable in KOH and does not react with the separator system. Rather, the separator system must be selected to retard the migration of silver to the iron electrode and to withstand the oxidative effects of the silver electrode itself. The particular separator system chosen will therefore determine the cycle life, the shelf life, and the power capabilities. Consequently, the separator system is usually application-specific. The typical cycle life at 100% depth of discharge and 10% overcharge is shown in Fig. 18.25.

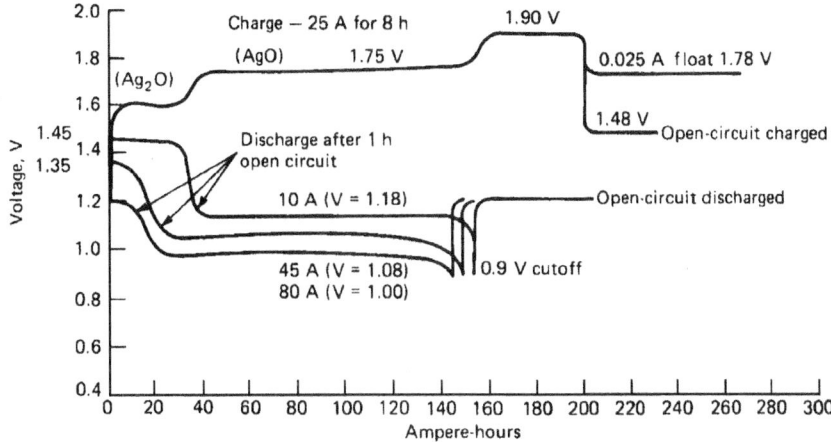

FIGURE 18.24 Charge-discharge characteristics of nominal 140 Ah iron/silver oxide battery. (*From Ref. 26.*)

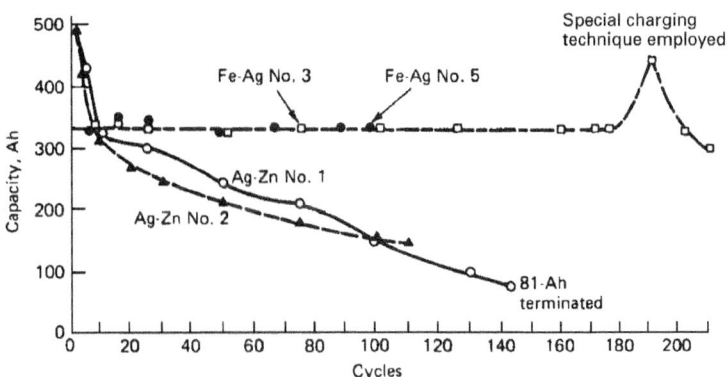

FIGURE 18.25 Cyclic life performance of zinc/silver oxide and iron/silver oxide prototype batteries. (*From Ref. 26.*)

Temperature Effects. As with other alkaline battery systems, silver-iron performance can depend on the operating temperature. Cells designed for long-life and low-rate applications normally have a higher internal resistance than those designed for high rate and shorter life. The two designs therefore behave differently when the discharge temperature is decreased, as shown in Figs. 18.26 and 18.27.

Experimental Designs. Experimental tests were conducted on designs other than the monopolar, prismatic designs discussed so far. Hand-assembled bipolar and jelly-roll cells were tested at the laboratory demonstration level. Results of voltage polarization tests on those designs are compared to the prismatic design in Fig. 18.28. The voltage characteristics of the jelly-roll design would be expected to improve considerably as assembly is refined and automated. Designs such as these would be suited for use in smaller portable systems such as communication devices that may require high power and energy density.

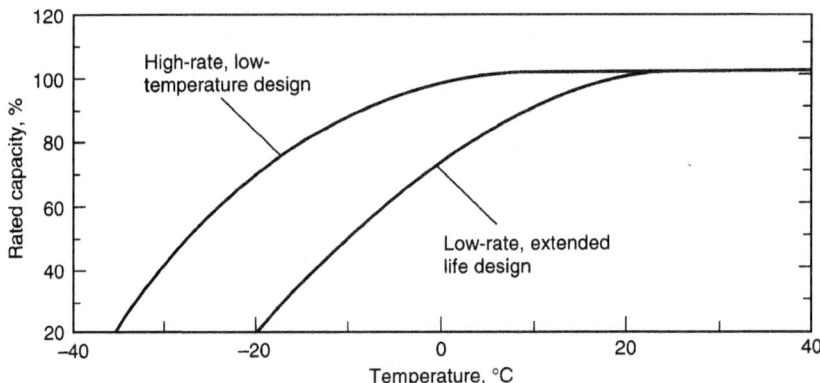

FIGURE 18.26 Effect of temperature on discharge capacity for different battery designs, $C/10$ discharge rate. (*Courtesy of Westinghouse Electric Corp.*)

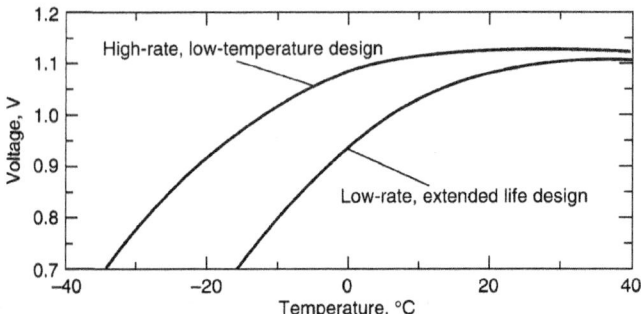

FIGURE 18.27 Effect of temperature on discharge voltage for different battery designs; $C/10$ discharge rate. (*Courtesy of Westinghouse Electric Corp.*)

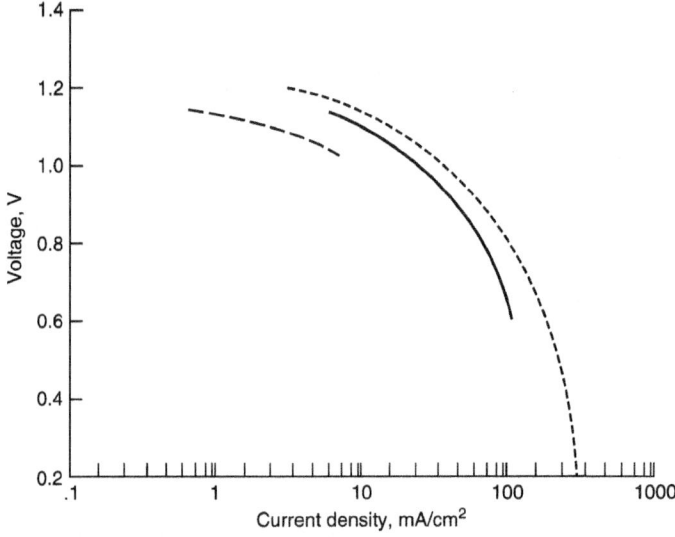

FIGURE 18.28 Voltage polarization characteristics for silver-iron system in experimental tests on three conventional types of cell designs. - - -, bipolar; ——, prismatic; – – –, jelly roll. (*Courtesy of Westinghouse Electric Corp.*)

18.7 RECENT ADVANCES IN IRON ANODE MATERIALS

The increased use of carbon in battery electrodes, coupled with advances in the development of nano-materials in the last 10 years, has generated renewed interest in iron electrode research. The ability to disperse iron in carbon nano-structures offers the potential for better active material utilization and cyclic efficiency. Iron active material utilizations as high as 510 mAh/g have been reported,[30] which is a significant improvement over the commonly achieved levels of 350 mAh/g, and a respectable approach toward the 962 mAh/g theoretical limit. The challenge in this work has been to maintain the nano-scale character, since the iron particle size has been shown to increase with repeated cycling, accompanied by a decrease in the surface area, which in turn produces a linear decline in the active material utilization of about 30 mAh/g for every m²/g of surface area lost.

Several methods of fabricating iron electrodes with nano-scale properties have been studied. A slow addition of a ferric chloride solution to a chilled sodium borohydride solution produces iron particles with sizes in the range of 30 to 70 nm.[30] Another approach[31] uses iron carbide particles in the 20 micron range as a starting material, which after cycling, ultimately produces particles with a size of 100 nm or less, presumably through repeated dissolution and redeposition of Fe and Fe(OH)$_2$. Yet another approach[17] uses carbon substrates with nano-scale features upon which iron is deposited. Carbon nano-fibers, nano-tubes, or platelets having particle sizes in the 100 nm range are impregnated with an aqueous iron nitrate solution and then dried and calcined to produce an iron-carbon composite with finely dispersed Fe$_2$O$_3$. In an attempt to keep the iron particles from growing during cycling (which degrades capacity) the particles have been preferentially deposited within the nano-tubes, rather than on the surface.[18] This approach has shown only limited success, since the penetration of the electrolyte into the nano-tubes then becomes the limiting factor, and this limitation is only partly overcome by treating the nano-tubes to create pores in the walls of the tubes.

As with sintered porous iron electrodes, adding sulfide compounds to the electrolyte or the electrode material appears to have the beneficial effects of increasing hydrogen over-potential, improving electrode capacity, and enhancing charge efficiency of these iron-carbon composite nano-scale variants.[19]

18.8 IRON MATERIALS AS CATHODES

Iron has conventionally been used as the anode or negative active material in batteries, but iron compounds have also been used as the cathode or positive active material. The use of iron sulfides (FeS and FeS$_2$) in lithium primary is covered in Chap. 14.

In the late 1990s an iron oxide, having a high valence state, was reported for use as a cathode active material.[32] Iron normally exists as a metal or in the valence states of Fe(II) and Fe(III). The new cathode material is an Fe(VI)-containing compound which has a high specific capacity due to a 3-electron change in its reduction reaction, as follows:

$$FeO_4^{2-} + 3H_2O + 3e \rightarrow FeOOH + 5OH^- \qquad E^0 = \sim 0.9\,\text{Volt}$$

The theoretical capacity of several of these Fe(VI) compounds is listed in Table 18.6. These values can be compared with the values given in Table 1.1 for the more conventional cathode materials. Recent work has extended the list of Fe(VI) salts to include Cs$_2$FeO$_4$, Rb$_2$FeO$_4$, K$_x$Na$_{(2-x)}$FeO$_4$, and SrFeO$_4$ as well as a transition metal Fe(VI) salt (Ag$_2$FeO$_4$).[33]

TABLE 18.6 Theoretical Capacities of Fe(VI) Compounds

Material	Molecular weight (g.)	Valence change	Electrochemical Equivalence	
			(mAh/g)	(g/Ah)
Li$_2$FeO$_4$	133.7	3	601	1.66
Na$_2$FeO$_4$	165.8	3	485	2.06
K$_2$FeO$_4$	198.1	3	406	2.46
BaFeO$_4$	257.2	3	313	3.19

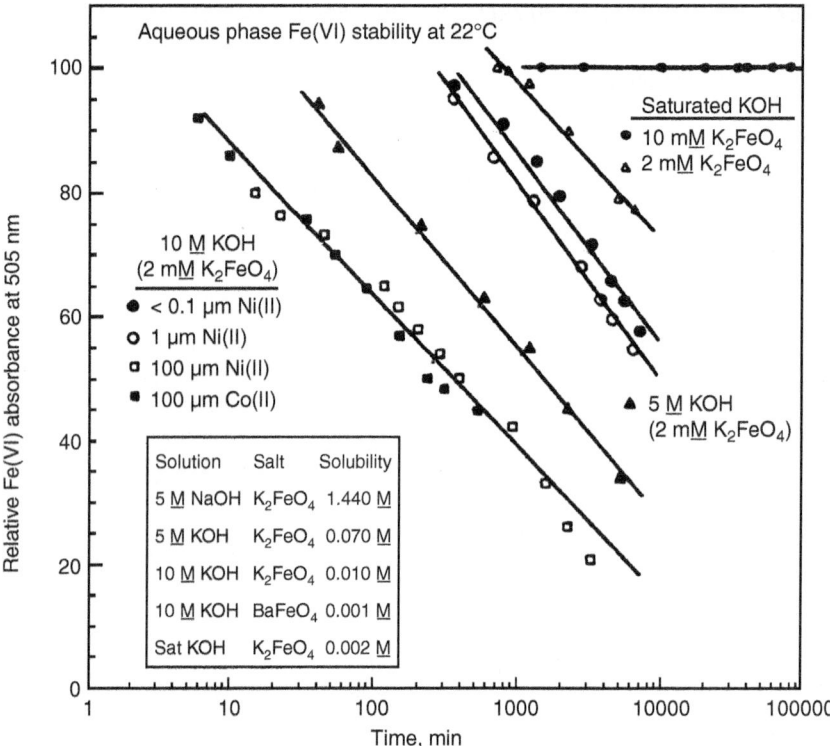

FIGURE 18.29 Stability of Fe(VI) in alkaline electrolyte with various concentrations of OH⁻, K_2FeO_4 salts, and Co(II) and Ni(II) impurities. (*From Ref. 32.*)

The characteristics of Fe(VI) compounds have not been studied extensively in the past, mainly because of the perception that these materials are highly unstable. While Li_2FeO_4 and Na_2FeO_4 are soluble in alkaline hydroxide, $BaFeO_4$ and K_2FeO_4 show evidence of low alkaline solubility and high stability as shown in Fig. 18.29. Further, their stability is greater in the more concentrated alkaline solutions that are used as battery electrolytes. These data have been extrapolated to suggest that, over a 10-year period, there will be less than a 10% loss of Fe(VI) in concentrated potassium hydroxide solutions using highly purified materials.

Electrochemically, the FeO_4^{2-} species have a high reduction potential, on the order of 0.9 V. Against an anode of zinc, the open circuit potential was found to be 1.75 V and 1.85 V for the K_2FeO_4 and $BaFeO_4$ cell, respectively. The proposed discharge reaction mechanism is as follows:

$$MFe(VI)O_4 + \tfrac{3}{2} Zn \rightarrow \tfrac{1}{2} Fe(III)_2O_3 + \tfrac{1}{2} ZnO + MZnO_2$$

where M = K_2 or Ba

The theoretical capacities and specific energy for the two batteries are given in Table 18.7. These values can be compared with those of other batteries listed in Table 1.2. The values for the zinc/iron oxide cells are higher than most of the other batteries with the exception of the lithium and air-breathing systems.

The discharge characteristics and specific energy of experimental primary alkaline button cells, with zinc anodes and Fe(VI) compound cathodes, were measured and compared to those with MnO_2

TABLE 18.7 Theoretical Capacity and Specific Energy for $MFeO_4$ Batteries

Couple	Open circuit voltage (V)	Theoretical specific capacity (g/Ah)	Theoretical specific capacity (Ah/g)	Theoretical specific energy (Wh/kg)
Zn/K_2FeO_4	1.75	3.68	0.271	475
$Zn/BaFeO_4$	1.85	4.41	0.226	419

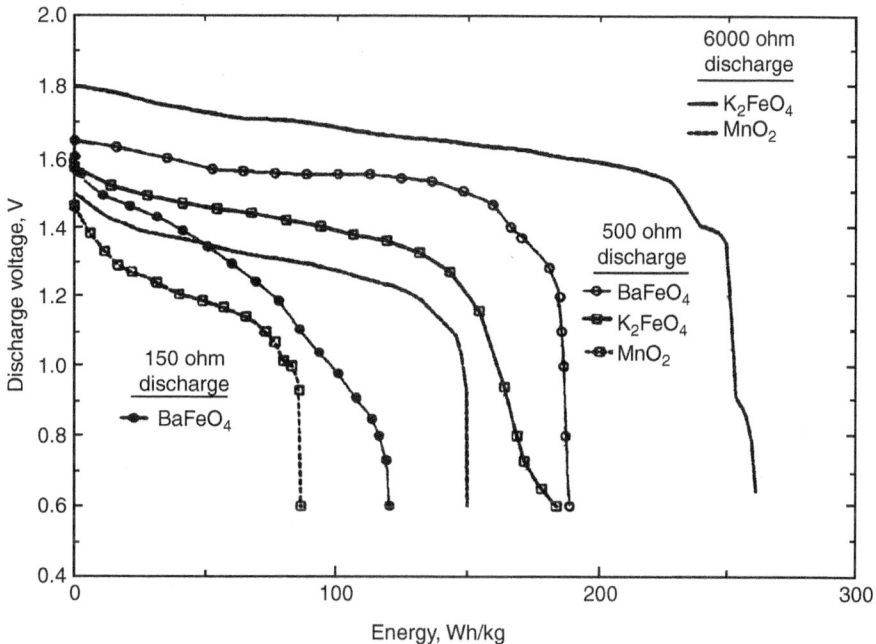

FIGURE 18.30 Capacity of several experimental button cells using Fe(VI) compounds as cathodes and Zn as anodes, compared to conventional Zn/MnO_2 button cells. (*From Ref. 32.*)

cathodes. These data are plotted in Figure 18.30 and illustrate the higher energy output of the cells fabricated with the Fe(VI) cathodes. Similar results were obtained with cells fabricated in the conventional cylindrical construction.

The Fe(VI) compounds were also shown to be rechargeable. A button cell, using a metal hydride anode and a capacity limited K_2FeO_4 cathode, was cycled for several cycles to a 75% depth of discharge and for more than 400 cycles to a 30% depth of discharge. The open circuit voltage of the cell was 1.3 V and the midpoint voltage was 1.1 V, similar to the voltage characteristics of the nickel-metal hydride cell. Efforts to increase the film thickness and prevent the buildup of a passivating Fe(III) layer have been the focus of the most recent studies, with the objective of improving practical capacity and long-term rechargeability of this couple.[34]

The Fe(VI) compounds are promising candidates for cathode materials for both primary and rechargeable alkaline batteries. The reported results demonstrate their higher specific energy compared to other cathode materials now used in alkaline batteries. The questions of long-term stability, shelf life and other critical performance characteristics, large scale manufacture, materials cost, and so on still have to be resolved and are subject to further evaluation.

REFERENCES

1. S. U. Falk and A. J. Salkind, *Alkaline Storage Batteries*, Wiley, New York, 1969.
2. A. J. Salkind, C. J. Venuto, and S. U. Falk, "The Reaction at the Iron Alkaline Electrode," *J. Electrochem. Soc.*, **111**:493 (1964).
3. R. Bonnaterre, R. Doisneau, M. C. Petit, and J. P. Stervinou, in J. H. Thompson (ed.), *Power Sources*, vol. 7, Academic, London, 1979, p. 249.
4. L. Ojefors, "SEM Studies of Discharge Products from Alkaline Iron Electrodes," *J. Electrochem. Soc.*, **123**:1691 (1976).
5. B. Anderson and L. Ojefors, in J. H. Thompson (ed.), *Power Sources*, vol. 7, Academic, London, 1979, p. 329.
6. C. A. C. Souza, I. A. Carlos, M. Lopes, G. A. Finazzi, and M. R. H de Almeida, "Self-Discharge of Fe-Ni Alkaline Batteries," *J. Power Sources*, **132**:288–290 (2004).
7. J. L. Weininger, in R. G. Gunther and S. Gross (eds.), *The Nickel Electrode*, vol. 82-84, Electrochemical Society, Pennington, NJ, 1982, pp. 1–19.
8. D. Tuomi, "The Forming Process in Nickel Positive Electrodes," *J. Electrochem. Soc.*, **123**:1691 (1976).
9. INCO ElectroEnergy Corp. (formerly ESB, Inc.), Philadelphia.
10. U. Casellato, N. Comisso, and G. Mengoli, "Effect of Li Ions on Reduction of Fe Oxides in Aqueous Alkaline Medium," *Electrochimica Acta*, **51**:5669–5681 (2006).
11. "Nickel Iron Industrial Storage Batteries," Exide Industrial Marketing Divisions of ESB, Inc., 1966.
12. F. E. Hill, R. Rosey, and R. E. Vaill, "Performance Characteristics of Iron Nickel Batteries," *Proc. 28th Power Sources Symp.*, Electrochemical Society, Pennington, NJ, 1978, p. 149.
13. R. Rosey and B. E. Tabor, "Westinghouse Nickel-Iron Battery Design and Performance," EV Expo 80, EVC #8030, May 1980.
14. W. Feduska and R. Rosey, "An Advanced Technology Iron-Nickel Battery for Electric Vehicle Propulsion," *Proc. 15th IECEC*, Seattle, Aug. 1980, p. 1192.
15. R. Hudson and E. Broglio, "Development of the Nickel-Iron Battery System for Electric Vehicle Propulsion," *Proc. 29th Power Sources Conf.*, Electrochemical Society, Pennington, NJ, 1980.
16. B. Hariprakash, S. K. Martha, M. S. Hegde, and A. K. Shukla, "A Sealed, Starved-Electrolyte Nickel-Iron Battery," *J. Applied Electrochemistry*, **35**:27–32, (2005).
17. B.T. Hang, T. Watanabe, M. Egashira, S. Okadab, J. Yamaki, S. Hata, S-H. Yoon, and I. Mochida, "The Electrochemical Properties of Fe_2O_3-Loaded Carbon Electrodes for Iron-Air Battery Anodes." *J. Power Sources*, **150**:261–271 (2005).
18. B. T. Hang, H. Hayashi, S. H. Yoon, S. Okada, and J. Yamaki, "Fe_2O_3-Filled Carbon Nano-tubes as a Negative Electrode for an Fe-Air Battery," *J. Power Sources*, **178**:393–401 (2008).
19. B. T. Hang, T. Watanabe, M. Egashira, I. Watanabe, S. Okada, and J. Yamaki, "The Effect of Additives on the Electrochemical Properties of Fe/C Composite for Fe/Air Battery Anode," *J. Power Sources*, **155**:461–469 (2006).
20. L. Carlsson and L. Ojefors, "Bifunctional Air Electrode for Metal-Air Batteries," *J. Electrochem. Soc.*, **127**:525 (1980).
21. L. Ojefors and L. Carlson, "An Iron-Air Vehicle Battery," *J. Power Sources*, **2**:287 (1977/78).
22. J. F. Jackovitz and C. T. Liu, *Extended Abstracts: 9th Battery and Electrochemical Contractors' Conf.*, USDOE, Alexandria, Va., Nov. 12–16, 1989, pp. 319–324.
23. H. Cnoblock, D. Groppel, D. Kahl, W. Nippe, and G. Siemsen, in D. H. Collins (ed.), *Power Sources*, vol. 5, Academic, London, 1975, p. 261.
24. O. Lindstrom, in D. H. Collins (ed.), *Power Sources*, vol. 5, Academic, London, 1975, p. 283.
25. *The Silver Institute Letter*, vol. 7, no. 3 (1977).
26. J. T. Brown, Extended Abstract No. 28, Battery Div., the Electrochemical Society, Las Vegas, NV, pp. 76–77 (1977).
27. E. Buzzelli, "Silver-Iron Battery Performance Characteristics," *Proc. 28th Power Sources Symp.*, Electrochemical Society, Pennington, NJ, 1978, p. 160.

28. G. A. Bayles, E. S. Buzzelli, and J. S. Lauer, "Progress in the Development of a Silver-Iron Communications Battery," *Proc. 34th Int. Power Sources Symp.*, Cherry Hill, NJ, June 1990.
29. G. A. Bayles, J. S. Lauer, E. S. Buzzelli, and J. F. Jackovitz, "Silver-Iron Batteries for Submersible Applications," *Proc. 3rd Annual Underwater Vehicle Conf.*, Baltimore, June 1989.
30. K. C. Huang and K. S. Chou, "Microstructure Changes to Iron Nanoparticles During Discharge/Charge Cycles," *Electrochemistry Communications*, **9:**1907–1912 (2007).
31. K. Ujimine and A. Tsutsumi, "Electrochemical Characteristics of Iron Carbide as an Active Material in Alkaline Batteries," *J. Power Sources*, **160:**1431–1435 (2006).
32. S. Licht, B. Wang, and S. Ghosh, *Science,* **128:**1039–1042 (1999).
33. X. Yu and S. Licht, "Advances in Fe(VI) Charge Storage Part I. Primary Alkaline Super-Iron Batteries," *J. Power Sources*, **171:**966–980 (2007).
34. X. Yu and S. Licht, "Advances in Fe(VI) Charge Storage Part II. Reversible Alkaline Super-Iron Batteries and Nonaqueous Super-Iron Batteries," *J. Power Sources*, **171:**1010–1022 (2007).

CHAPTER 19
INDUSTRIAL AND AEROSPACE NICKEL-CADMIUM BATTERIES

John K. Erbacher

19.1 INTRODUCTION

The vented pocket-plate battery is the oldest and most mature of the various designs of nickel-cadmium batteries available. It is a very reliable, sturdy, long-life battery, which can be operated effectively at relatively high discharge rates and over a wide temperature range. It has very good charge retention properties, and it can be stored for long periods of time in any condition without deterioration. The pocket-plate battery can stand both severe mechanical abuse and electrical maltreatment such as overcharging, reversal, and short-circuiting. Little maintenance is needed on this battery. The cost is lower than for any other kind of alkaline storage battery; still, it is higher than that of a lead-acid battery on a per watthour basis. The major advantages and disadvantages of this type of battery are listed in Table 19.1.

The pocket-plate battery is manufactured in a wide capacity range, 5 to more than 1200 Ah, and it is used in a number of applications. Most of these are of an industrial nature, such as railroad service, switchgear operation, telecommunications, uninterruptible power supplies (UPSs), and emergency lighting. The pocket-plate battery was also used in military and space applications.

Pocket-plate batteries are available in three plate thicknesses to suit the variety of applications. The high-rate designs use thin plates for maximum exposed plate surface per volume of active material. They are used for the highest-rate discharge. The low-rate designs use thick plates to obtain maximum volume of active material per exposed plate surface. These types are used for long-term discharge. The medium-rate designs use plates of medium thickness and are suited for applications between, or combinations of, high-rate and long-term discharge.

Developmental work has been conducted almost continuously since the introduction of the pocket-plate nickel-cadmium battery to improve the performance characteristics of the battery and reduce weight. The sintered plate, which can be constructed in a thinner form than the pocket plate, was developed during the 1940s and has a lower internal resistance and gives superior high-rate and low-temperature performance compared to the pocket plate. It is used in high-power applications, such as engine starting, and in low-temperature environments. The sintered-plate battery is covered in Chap. 20. Further development of the sintered plate led to the design of smaller batteries for portable equipment and subsequently to the sealed, maintenance-free nickel-cadmium battery covered in Chap. 21.

The sintered-plate battery was found to be too expensive and complex to manufacture. It used a large amount of nickel and was impractical for medium-rate thick electrodes or for cells larger than 100 Ah. The pocket-plate battery was too heavy for many applications. Recent developmental work has been directed to more effectively use the costly materials—nickel and cadmium—and toward

TABLE 19.1 Major Advantages and Disadvantages of Industrial and Aerospace Nickel-Cadmium Batteries

Advantages	Disadvantages
Long cycle life	Low energy density
Rugged; can withstand electrical and physical abuse	Higher cost than lead-acid batteries
Reliable; no sudden death	Contains cadmium
Good charge retention	Caustic alkaline electrolyte
Excellent long-term storage	Memory effect
Low maintenance	Temperature-controlled charging system required to extend life
Flat discharge profile	

simplified manufacturing processes. The design philosophy was to develop a high-surface-area, conductive-plate structure that would be light, easy to manufacture, and inexpensive, and to eliminate the troublesome aspects of sintered-plate technology, namely, the sintering process and the chemical impregnation of the active materials. Taking advantage of new polymer materials and plating techniques, this work has resulted in a new electrode structure, the fiber-structured electrode (the Fiber Nickel Cadmium Battery—FNC) developed by Deutsche Automobilgesellschaft GmbH (DAUG).

The fiber plates are manufactured from either a mat of pure nickel fibers or, more commonly, nickel-plated plastic fibers. To make the plastic fiber conductive, a thin layer of nickel is applied by electroless plating and thereafter a sufficiently thick layer of nickel for good conductivity is applied by electroplating. The plastic is then burned off, leaving a mat of hollow nickel fibers. The nickel-fiber plaque is welded to a nickel-plated steel tab. Figure 19.1a shows the structure of a nickel-fiber plaque before impregnation, and Fig. 19.1b shows an unformed positive electrode.

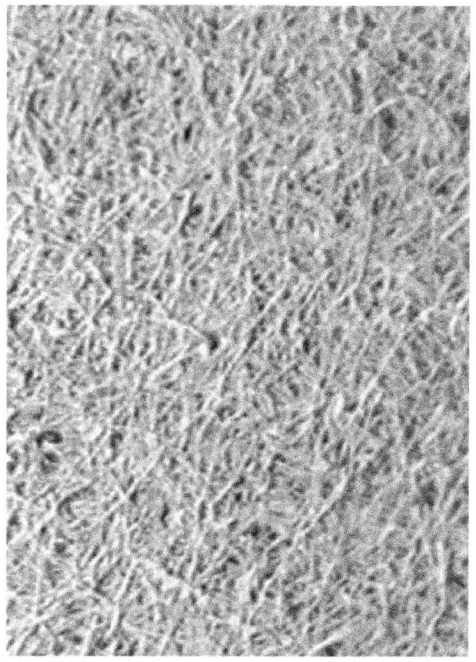

FIGURE 19.1a Nickel-fiber electrode structure before impregnation. (*Courtesy of Acme Electric Corp.*)

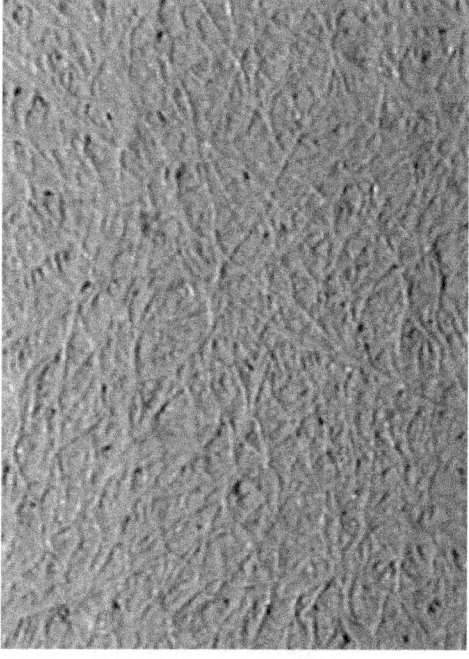

FIGURE 19.1b Unformed, pasted nickel positive electrode. (*Courtesy of Acme Electric Corp.*)

This fiber electrode technology, while originally developed for EV applications, was first used for industrial low- and medium-rate vented cells. It is now being used in all types of nickel-cadmium as well as nickel-metal hydride batteries, including high-rate batteries for engine-cranking and sealed cells with oxygen recombination. Details of this technology are covered in Sec. 19.7.

A more recent design that has shown significantly improved performance characteristics is the plastic-bonded or pressed-plate electrode. This new development of electrode materials in industrial batteries is a spin-off from the development of electrode materials for use in aircraft and sealed portable consumer batteries. In the plastic-bonded plate, which is mainly used in the cadmium electrode, the active material cadmium oxide is mixed with a plastic powder, normally PTFE, and a solvent to produce a paste. The paste is isotropic, and the materials are manufactured at the final density for the active material. As a result, dust problems are eliminated during manufacturing. The paste is extruded, rolled, or pasted onto a center current collector normally made of nickel-plated perforated steel. The plate structure is welded to nickel-plated steel tabs.

19.2 CHEMISTRY

The basic electrochemistry is the same for the vented pocket-plate, sintered-plate, fiber and plastic-bonded plate types, as well as for other variations of the nickel-cadmium system. The reactions of charge and discharge can be illustrated by the following simplified equation:

$$2NiOOH + 2H_2O + Cd \underset{\text{charge}}{\overset{\text{discharge}}{\rightleftharpoons}} 2Ni(OH)_2 + Cd(OH)_2$$

On discharge, trivalent nickel oxy-hydroxide is reduced to divalent nickel hydroxide with consumption of water. Metallic cadmium is oxidized to form cadmium hydroxide. On charge, the opposite reactions take place. The electromotive force (EMF) is 1.29 V.

The potassium hydroxide electrolyte is not significantly changed with regard to density or composition during charge and discharge, in contrast to the sulfuric acid in lead-acid batteries. The electrolyte density is generally approximately 1.2 g/mL. Lithium hydroxide is often added to the electrolyte for improved cycle life and high-temperature operation. (See Chap. 7.) A more detailed description of the cell reaction on overcharge is found in Sec. 19.7.

19.3 CONSTRUCTION

A cutaway view of a modern pocket-plate cell is shown in Fig. 19.2. The active material for the positive electrodes consists of nickel hydroxide mixed with graphite for conductivity and additives such as barium or cobalt compounds for improved life and capacity. The active material for the negative electrodes is prepared from cadmium hydroxide or cadmium oxide mixed with iron or iron compounds and sometimes also with nickel. The iron and nickel materials are added to stabilize the cadmium, prevent crystal growth and agglomeration, and improve conductivity. Typical active material compositions are shown in Table 19.2.

The positive and negative electrodes of pocket-plate nickel-cadmium batteries are made using the same basic design to hold the active materials. The pocket plates are built up of flat pockets of perforated steel strips holding the active materials. The thin steel strips are perforated by hardened steel needles or by a technique using profiled roller dies. The specific hole area is between 15 and 30%. The strips are nickel-plated to prevent "iron poisoning" of the positive active material.

The active mass is either pressed into briquettes, which are fed into the preshaped perforated strip, or fed into the preshaped strip as a powder. The upper and lower steel strips are folded together by rollers. A number of these folded strips are arranged to interlock with each other to form long electrode sheets, which are then cut to electrode blanks. Electrodes are made from these blanks by providing them with steel frames for mechanical stability and for current takeoff.

19.4 SECONDARY BATTERIES

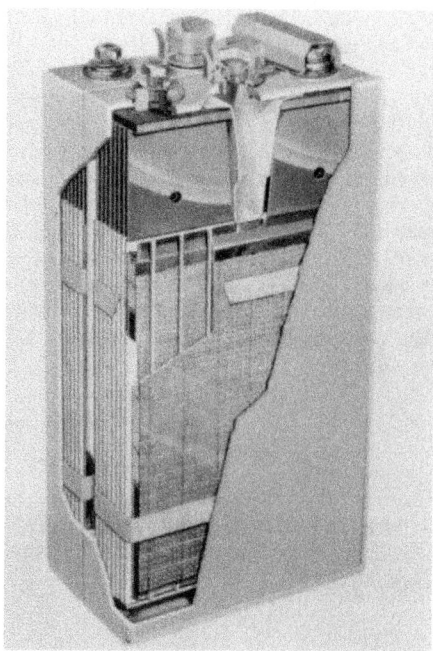

FIGURE 19.2 Pocket plate cell.

TABLE 19.2 Typical Composition of Active Materials for Pocket-Plate Cells in the Discharged State

Positive active material		Negative active material	
Substance	Weight %	Substance	Weight %
Nickel (II) hydroxide	80	Cadmium hydroxide	78
Cobalt (II) hydroxide	2	Iron	18
Graphite	18	Nickel	1
		Graphite	3

The electrodes are made with different thicknesses (1.5 to 5 mm) to provide cells for high-, medium-, and low-rate discharge rates. The negative plate is always thinner (30 to 40%) than the positive.

The electrodes are bolted or welded to electrode groups. Plate groups of opposite polarity are intermeshed and electrically separated from each other by plastic pins and plate edge insulators. Sometimes separators or perforated plastic sheets or plastic ladders are used between the electrodes. The distance between plates of different polarity in an element may vary from less than 1 mm for high-rate cells to 3 mm for low-rate cells.

The elements are inserted into cell containers of plastic or stainless steel. Plastic containers are made from polystyrene, polypropylene, or flame-retarded plastics. Important advantages of plastic containers over steel containers are that they allow visual control of the electrolyte level and require no protection against corrosion. Also, they have lower weight and can be more closely packed in the battery. The main drawbacks are that they are more sensitive to high temperatures and require somewhat more space than steel containers. A plastic-bonded plate cell in a plastic container is shown in Fig. 19.3.

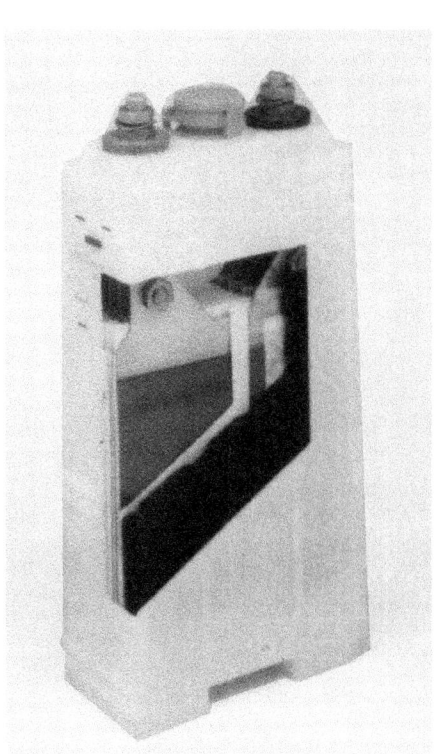

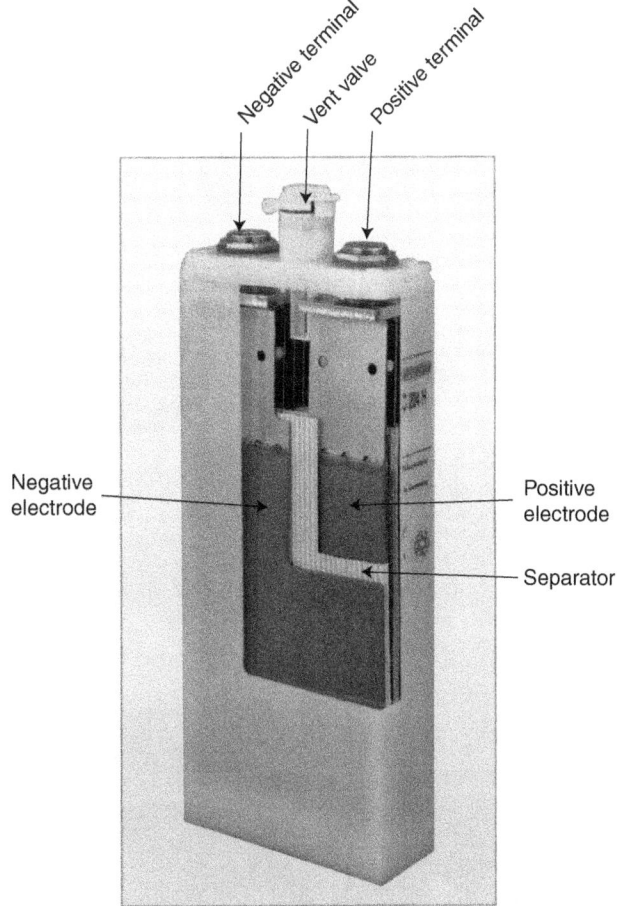

FIGURE 19.3 Plastic-bonded plate cell.

FIGURE 19.4 Partial cutaway view of fiber nickel-cadmium (FNC) cell. *Source: Hoppecke Batteries.*

Figure 19.4 shows a partial cutaway view of a fiber nickel-cadmium (FNC) cell. The case and cover are polypropylene and are welded together. The electrode assembly shows a negative electrode, a corrugated separator, and a positive electrode. O-ring seals are employed in the bushings for the terminals to ensure gas retention, and a vent valve is seen in the case cover between the terminals. A catalytic gas-recombination plug in the vent valve is employed for some applications. The terminals are nickel-plated copper, and a 1.19 kg/L KOH electrolyte is typically employed.

Cells are assembled into batteries in many different ways. Often 2 to 10 cells are mounted in a separate battery unit, several of which may be used to form the complete battery. A typical battery is shown in Fig. 19.5. Cells in plastic containers are also assembled into batteries by putting the single cells close together on a rack or a stand and connecting them with intercell connectors. This is especially the case for stationary applications (Fig. 19.6). Cells in steel containers can be assembled in a similar way; however, here the cells must be spaced from one another and insulated from the rack.

Aerospace batteries are assembled with 19 to 21 cells per battery in a configuration similar to that shown in Fig. 19.7. In many cases, voltage monitoring at the half or quarter battery configuration is used to monitor cell balance and battery state-of-charge.

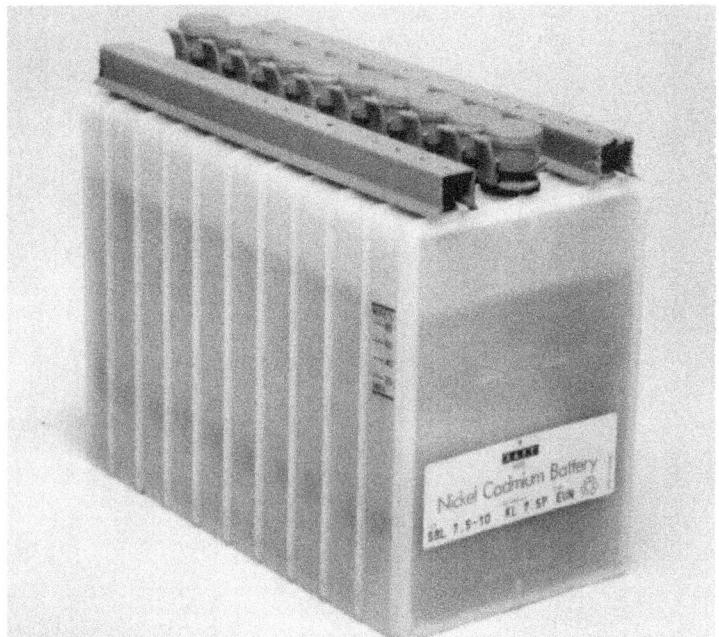

FIGURE 19.5 Ten-cell welded polypropylene unit. (*Courtesy of SAFT America, Inc.*)

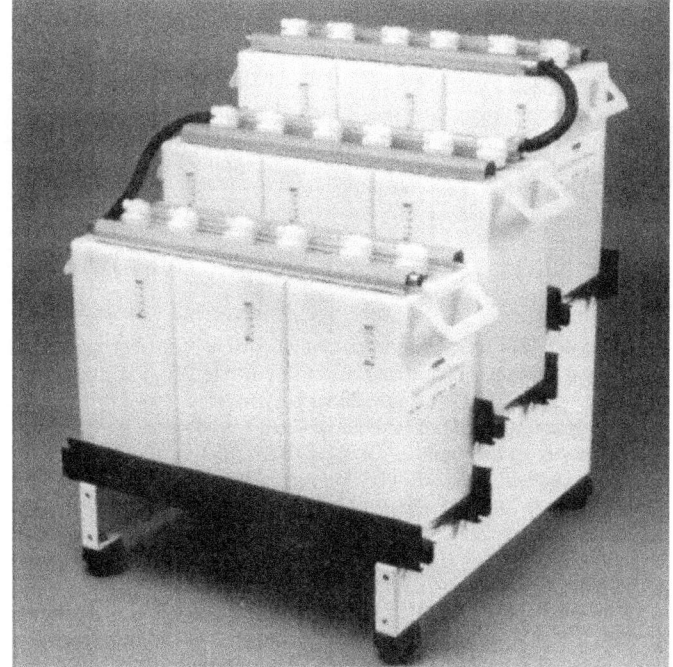

FIGURE 19.6 Typical rack assembly of cells in plastic containers. (*Courtesy of SAFT America, Inc.*)

FIGURE 19.7 Typical vented Ni-Cd aviation battery and top view of cell assembly (*Courtesy of SAFT America, Inc.*)

19.4 PERFORMANCE CHARACTERISTICS

19.4.1 Energy Density and Specific Energy

Typical specific energy and energy density values for pocket-plate, single-cell batteries are 20 Wh/kg and 40 Wh/L, with the best values for commercially available units reaching 27 Wh/kg and 55 Wh/L. The corresponding values for complete pocket-plate batteries are 19 Wh/kg and 32 Wh/L and 27 Wh/kg and 44 Wh/L, respectively. These data are based on the nominal capacity and the average discharge voltage at the 5 h rate. The specific energy and energy density of larger fiber plate batteries approach 40 Wh/kg and 80 Wh/L. Batteries with plastic-bonded plates approach 56 Wh/kg and 110 Wh/L. This compares to a specific energy of 30 to 37 Wh/kg and an energy density of 58 to 96 Wh/L for sintered-plate designs. (See Chap. 20.)

19.4.2 Discharge Properties

The nominal voltage of a nickel-cadmium battery is 1.2 V. Although discharge rate and temperature are of importance for the discharge characteristics of all electrochemical systems, these parameters have a much smaller effect on the nickel-cadmium battery than on, for instance, the lead-acid battery. Thus pocket-plate nickel-cadmium batteries can be effectively discharged at high discharge rates without losing much of the rated capacity. They can also be operated over a wide temperature range.

Typical discharge curves at room temperature for pocket-plate and plastic-bonded plate batteries at various constant discharge rates are shown in Fig. 19.8. Even at a discharge current as high as $5C$ (where C is the numerical value of the capacity in Ah), a high-rate pocket-plate battery can deliver 60% of the rated capacity and a plastic-bonded battery as much as 80%. Battery capacities as a function of discharge rate and cutoff voltage are given in Fig. 19.9.

Pocket-plate nickel-cadmium batteries can be used at temperatures down to −20°C with the standard electrolyte. Cells filled with a more concentrated electrolyte can be used down to −50°C. Figure 19.10 shows the effect of temperature on the relative performance of a nickel-cadmium medium-rate battery with standard electrolyte.

19.8 SECONDARY BATTERIES

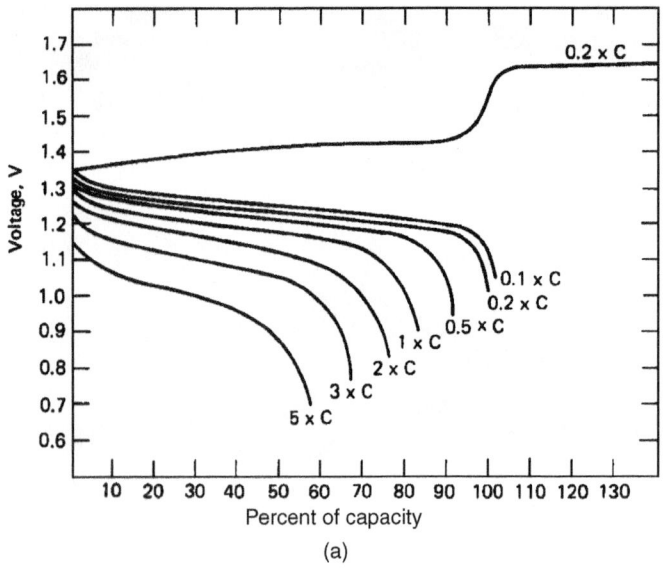

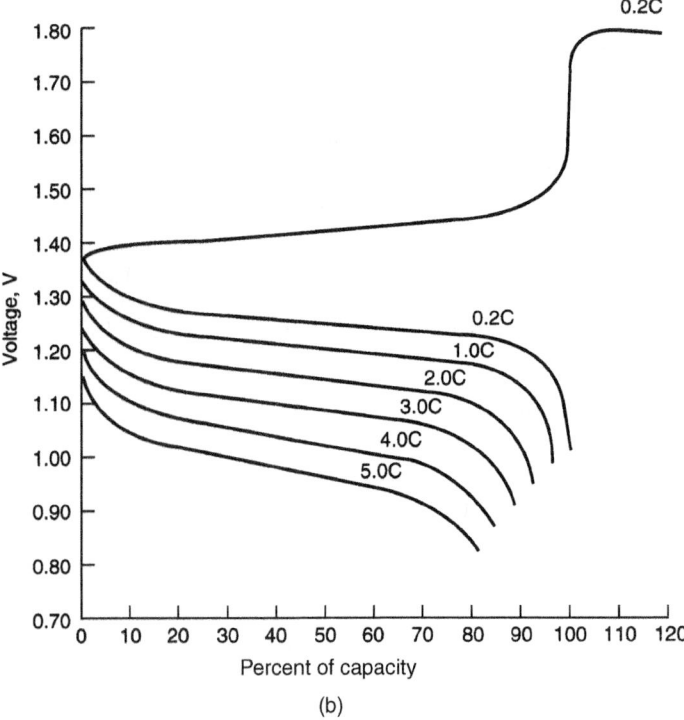

FIGURE 19.8 Charge and discharge characteristics of nickel-cadmium batteries at 25°C. (*a*) Pocket-plate battery, high rate. (*b*) Plastic-bonded plate battery, high rate.

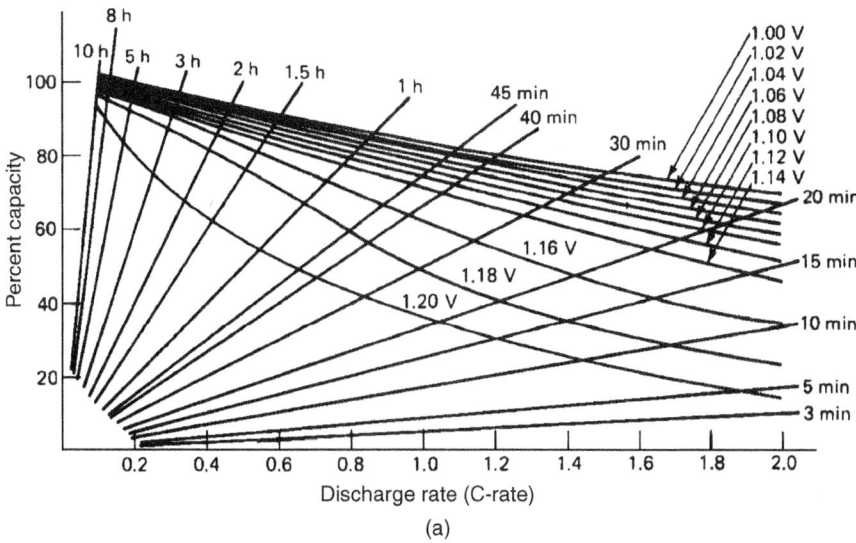

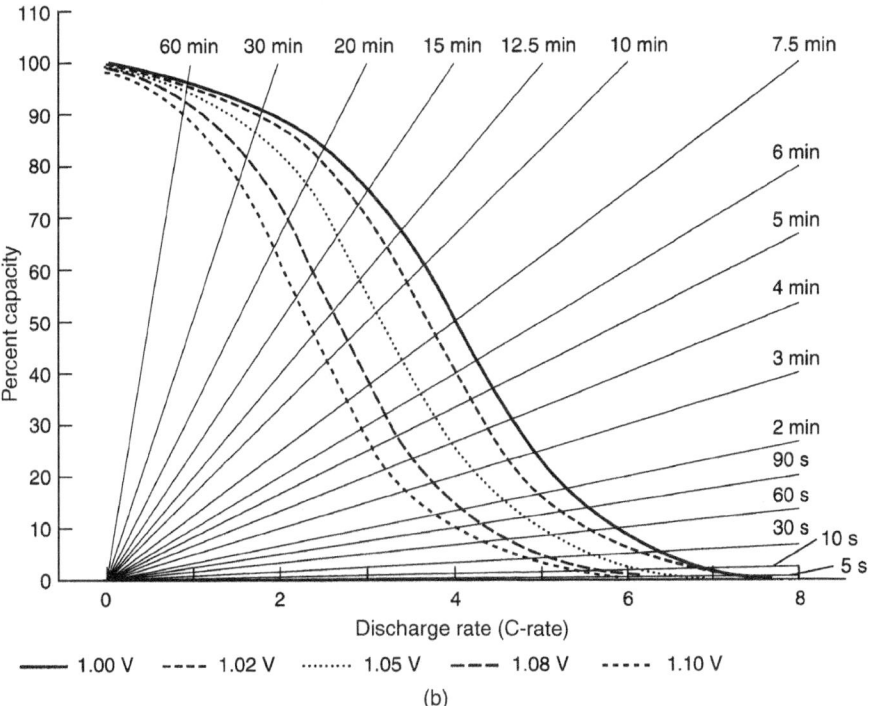

FIGURE 19.9 Discharge characteristics of nickel-cadmium batteries at 25°C; capacity as a function of discharge rate and cutoff voltage. (*a*) Pocket-plate battery, high rate. (*b*) Plastic-bonded plate battery, high rate.

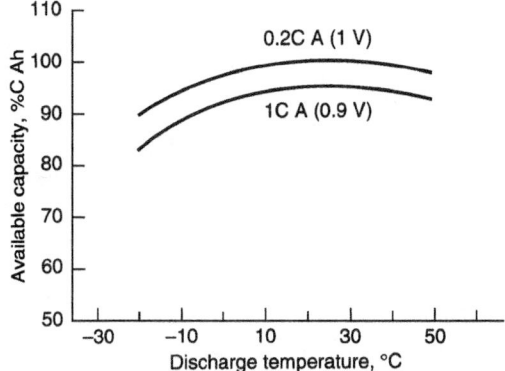

FIGURE 19.10 Typical available capacity at different temperatures for nickel-cadmium medium-rate batteries with standard electrolyte, fully charged at 25°c.

Batteries can also be used at elevated temperatures. Although occasional operation at very high temperatures is not detrimental, 45 to 50°C is generally considered as the maximum permissible temperature for extended periods of operation. Recent tests on aviation batteries exposed to the higher temperature regime of Southwest Asia has led to changes in the high-temperature limits for operation and maintenance of these batteries, which can be operated as high as 70°C.

Figure 19.11 shows typical so-called starter curves for high-rate batteries. The batteries can deliver as much current as $20C$ A during 1 s to a final voltage of 0.6 V.

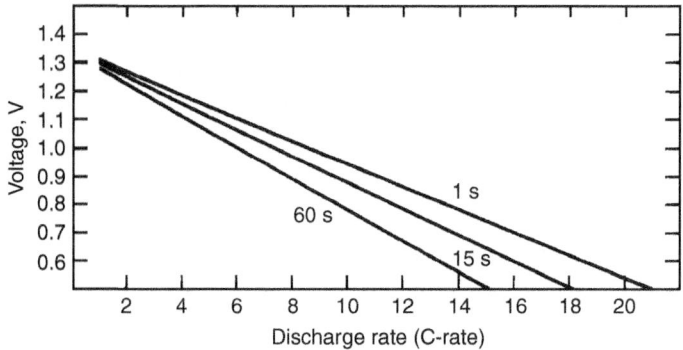

FIGURE 19.11 Voltage-current curves for high-rate pocket-plate batteries at 25°C.

Occasional overdischarge or reversal of nickel-cadmium batteries is not detrimental nor is complete freezing of the cells. After warming up, they will function normally again.

19.4.3 Internal Resistance

Nickel-cadmium batteries generally have a low internal resistance. Typical DC resistance values are 0.4, 1, and 2 mΩ, respectively, for a charged 100 Ah high-, medium-, and low-rate pocket-plate single-cell battery. The internal resistance is largely inversely proportional to the battery size in a given series. Decreasing temperature and decreasing state-of-charge of a battery will result in an increase of the internal resistance. The internal resistance of fiber-plate batteries is 0.3 mΩ for a high-rate design and 0.9 mΩ for a low-rate design. Plastic-bonded plate batteries have an internal resistance as low as 0.15 mΩ.

19.4.4 Charge Retention

Charge retention characteristics of vented pocket-plate batteries at 25°C are shown in Fig. 19.12. Charge retention is temperature-dependent, the capacity loss at 45°C being about three times higher than at 25°C. There is virtually no self-discharge at temperatures lower than −20°C. Charge retention for fiber and plastic-bonded plate batteries has similar characteristics; their charge retention corresponds to that shown in Fig. 19.12 for high-rate batteries.

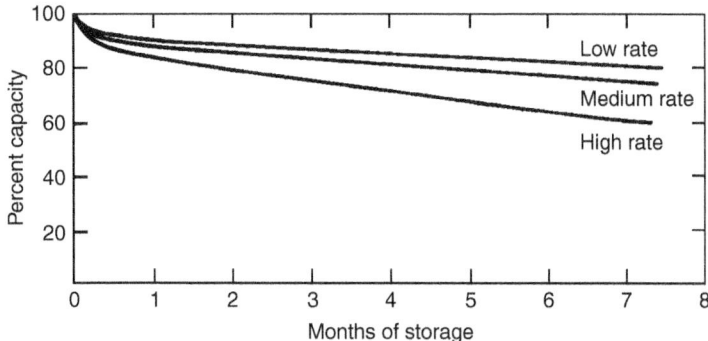

FIGURE 19.12 Charge retention of pocket-plate batteries at 25°C.

19.4.5 Life

The life of a battery can be given either as the number of charge and discharge cycles that can be delivered or as the total lifetime in years. Under normal conditions, a nickel-cadmium battery can reach more than 2000 cycles. The total lifetime may vary between 8 and 25 years or more, depending on the design and application and on the operating conditions. Batteries for diesel engine cranking normally last about 15 years, batteries for train lighting have normal lives of 10 to 15 years, stationary standby batteries have lives of 15 to 25 years, and aircraft batteries have lives of 3 to 5 years.

Factors that affect battery life are the operating temperature, the discharge depth, and the charging regime. Low or moderate operating temperatures should always be preferred. Batteries operating at elevated temperatures or in cycling applications should be filled with electrolyte to which lithium hydroxide has been added.

The factors behind the excellent reliability and very long life of the nickel-cadmium batteries are the mechanically strong design, the absence of corrosive attack of the electrolyte on the electrodes and other components in the cell, and, furthermore, the ability of the battery to withstand electrical abuse, such as reversal or overcharging, and to stand long-time storage in any state-of-charge.

19.4.6 Mechanical and Thermal Stability

Nickel-cadmium cells and batteries are mechanically very robust and can withstand severe mechanical abuse and rough handling in general. The electrode groups are carefully bolted or, in more recent designs such as FNC, welded together. The cell containers are made of steel or high-impact plastics.

19.12 SECONDARY BATTERIES

The electrolyte does not attack any of the components in the cell, and, accordingly, there is no risk of decreased strength during the lifetime of the battery. Cases of so-called sudden death due to corroded lugs or terminals cannot occur.

The thermal resistance of the nickel-cadmium batteries is also very good. These batteries can withstand temperatures up to 85°C or more without mechanical damage. Cells in polypropylene or steel containers are the best in this respect. Saline or corrosive environments present no problems for cells in plastic containers.

19.4.7 Memory Effect

The memory effect—the tendency of a battery to adjust its electrical properties to a certain duty cycle to which it has been subjected for an extended period of time—has been a problem with nickel-cadmium batteries in some applications. Pocket, fiber, and plastic-bonded plate cells do not show this tendency. See Sec. 20.7.2 for a description of the memory effect with sintered-plate nickel-cadmium batteries.

19.5 CHARGING CHARACTERISTICS

Pocket-plate nickel-cadmium batteries may be charged at constant current, constant voltage, or modified constant voltage. Constant-current charge characteristics are shown in Fig. 19.8. Charging is normally carried out at the 5 h rate for 7 h for a fully discharged battery. Overcharging is not detrimental but should be avoided as it leads to increased gassing and decomposition of water. Charging can be carried out in the temperature range of −50 to 45°C. However, at the extreme temperatures, the charging efficiency is lower.

Constant-voltage charging characteristics with current limitations are shown in Fig. 19.13. The current is often limited to 0.1 to 0.4C A, and charging is normally carried out in the voltage range of 1.50 to 1.65 V per cell. The charging time may vary from 5 to more than 25 h, depending on current limitation value and cell type.

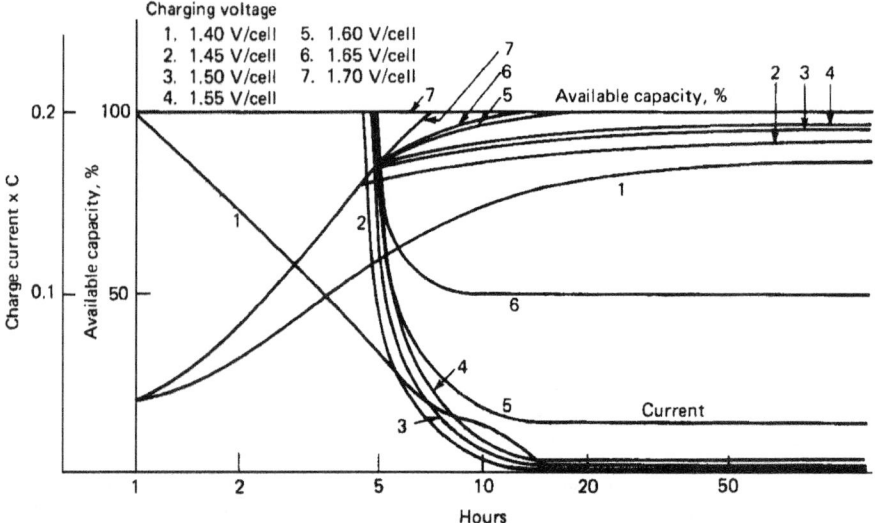

FIGURE 19.13 Constant-voltage charging with current limitation 0.2C of medium-rate pocket-plate batteries at 25°C.

In some applications such as emergency and standby, it is necessary to keep the battery in a high state-of-charge. A convenient way is to connect the battery in parallel with the ordinary current source and the load and to float the battery at 1.40 to 1.45 V per cell. The floating may be combined with a supplementary charge at fixed intervals or after each discharge.

The ampere-hour efficiency of the pocket plate battery is 72% when going from the discharged to the fully charged state. The watthour efficiency is approximately 60%. The best plastic-bonded plate batteries have an ampere-hour efficiency of 85% and a watthour efficiency of 73%.

19.6 SEALED NICKEL-CADMIUM (SNC) BATTERY TECHNOLOGY

19.6.1 Development of the SNC Cell Technology

This work for aviation batteries started in the 1970s but was not very successful due to lack of quality control in materials and chemistry. However, advances in space nickel-cadmium technologies indicated that quality control of materials, manufacturing, and assembly were needed to develop a low-maintenance, long-life battery for aviation applications. Programs to improve these aspects were initiated at Wright-Patterson AFB in the '70s and '80s and resulted in development of the advanced, maintenance-free battery system (AMFABS) which was tested on several AF aircraft in the early '90s and eventually placed in service on the B-52 aircraft. This success led to use of an SNC battery with low maintenance on other military aircraft, and eventually it was adopted for commercial aircraft by the Boeing Aircraft Company. Unfortunately, the SNC technology with its more sophisticated charging requirement and increased quality control was more costly in the acquisition phase of aircraft, so implementation was slower than anticipated. Simultaneously other commercial battery companies developed a low-cost version of this technology, such as the Micro-Maintenance and Ultra Low Maintenance battery concepts from Marathon and SAFT, respectively, along with the fiber nickel-cadmium (FNC) battery described in Section 19.7. Details of the SNC technology are sufficiently different from industrial nickel-cadmium technology and are really a derivative of vented, sintered-plate nickel-cadmium batteries, so the chemistry and technology development are covered separately in Chap. 20.

19.7 FIBER NICKEL-CADMIUM (FNC) BATTERY TECHNOLOGY

19.7.1 FNC Electrode Technology

An ideal electrode will feature the following characteristics:

- Provide a high surface area conductive matrix to contact the active material
- Provide sufficient porosity for high active material loading and an open structure for good electrolyte penetration
- Have sufficient electrical conductivity to carry the current to the tab with minimal voltage drop, yet will still be light
- Fully contain the active material
- Able to accommodate the dimensional changes during battery charge and discharge without fatigue cracking
- Tolerate mechanical shock and vibration
- Chemically and thermally inert to the battery environment; will not introduce any undesirable impurities into the cell
- Utilize a simple process for the loading of the active material

- Strong enough to tolerate the cell manufacturing processes
- Versatile enough to allow manufacturing of various sizes, thickness, conductivity, and porosity
- Economical

It is in this area of plate design that the FNC technology has made a significant step forward in comparison to older technologies. The core of the FNC technology is the three-dimensional nickel-plated fiber matrix. The nickel coating is optimized to the expected current density of the battery. Thus, there is no excess nickel. Electrodes of thickness ranging from 0.5 to 10 mm targeted at ultra-high (XX), high (X and H), medium (M), and low (L) rate designs are fabricated in a common process. The nickel fibers are very compact with one cubic centimeter of electrode volume nominally containing 300 meters of conducting filament. This current collecting matrix is 90% porous, allowing excellent utilization of the active material. The result is improved low-temperature performance, a lower charge coefficient, and significantly higher power capability (see Fig. 19.1a).

The structure is highly porous and thus allows for high loading of active materials as well as excellent penetration of electrolyte. No conductive diluents such as graphite or iron are required. Yet, due to the very high surface area of the fibers, the contact between the current carrying fiber matrix and the paste is very good. Because of this, losses are low, resulting in improved efficiency. The paste is loaded into the electrodes mechanically, and no impurities are introduced in the process. Active material (nickel hydroxide in the positive plate and cadmium hydroxide in the negative) is mechanically imbedded directly into the fiber plate. The pure active material contributes to longer life, lower self-discharge, and a more consistent and reliable product (see Fig. 19.1b). Consequently, a high surface area plate capable of high-current loads and very long life has been realized.

The processes and cell design associated with FNC technology have resulted in improved battery performance. Improved charging efficiency has reduced the gassing on overcharge, and with it, the frequency of water topping for the vented cells. The design of the FNC plate allows elastic expansion and contraction during charge and discharge, eliminating one of the main causes of nickel-cadmium plate degradation. This plate flexibility also provides increased shock and vibration tolerances. The flexibility of the electrode structure eliminates mechanical cracks and associated plate degradation, resulting in increased battery life.

19.7.2 Manufacturing Flexibility

The power capability of a battery will affect its potential applications. A high-power battery will be capable of delivering most of its capacity in a few minutes. To maximize battery power, it is necessary to minimize cell resistance. To that end, high-power cells are designed with high surface areas, thin electrodes, and high metallic contents. However, the above measures will increase battery weight, volume, and cost. Given these trade-offs, it is desirable to optimize the cell design for the application.

Practical manufacturing constraints inhibited the development of high-capacity low-rate sintered-plate batteries. In contrast, the FNC technology covers a wide range of power capabilities. The thickness of the fiber electrode and the amount of conductive metallic nickel on it are varied within an order of magnitude. This results in the capability to produce high-capacity, low-weight, low-cost batteries or ultra-high-power, higher-weight, and higher-cost cells using the same processes and equipment. For the user, this means that the characteristics of sinter foils and the various types of pocket- or foam-plate batteries no longer have to be considered separately. The FNC system has the same properties and basic characteristics over the entire range of applications.

19.7.3 Sealed versus Vented Designs

The charging of aqueous-nickel batteries always occurs in competition with water electrolysis. Toward the end of the charging cycle, oxygen is typically evolved at the positive electrode, and hydrogen may be evolved at the negative electrode. The way that the cell deals with these evolved gases will determine whether the cell can be sealed. In sealed cells, the gases are recombined internally. In an open cell, the gases are allowed to vent, hence the name "vented cell."

The reactions that produce the gases are called the overcharge reactions. These reactions differ depending on whether one deals with a sealed or vented cell.

Overcharge reactions:

Vented (open) cell:

Positive: $\quad 4OH^- \rightarrow 2H_2O + O_2 + 4e^-$
Negative: $\quad 4H_2O + 4e^- \rightarrow 2H_2 + 4OH^-$

Net: $\quad 2H_2O \rightarrow 2H_2 + O_2$

The net result here is the electrolysis of water to hydrogen and oxygen.

Sealed cell:

Positive: $\quad 4OH^- \rightarrow 2H_2O + O_2 + 4e^-$
Negative: $\quad 2Cd(OH)_2 + 4e^- \rightarrow 2Cd + 4OH^-$

Net electrochemical: $\quad 2Cd(OH)_2 \rightarrow 2Cd + 2H_2O + O_2$
Chemical recombination on negative: $\quad 2Cd + O_2 + 2H_2O \rightarrow 2Cd(OH)_2$

The result in a sealed cell is that electricity is converted into heat without any net chemical change in the cell. The overcharge reaction is exothermic, particularly the chemical recombination reaction in the sealed cell.

The overcharge reaction on the positive plate starts before the cell is fully charged, so that some oxygen evolution on charging is unavoidable. At higher temperatures, the oxygen evolution starts at a lower voltage. This results in lower charging efficiencies at higher temperatures and, in the case of the vented cell, an increased need for the addition of water. Additionally, this leaves the positive plate undercharged while the negative plate continues toward a full charged state. The resulting plate imbalance reduces battery capacity. To regain the lost capacity, vented batteries require deep discharge conditioning with each cell being clipped out (shorted).

19.7.4 Sealed FNC Maintenance-Free Batteries

With the development of sealed FNC technology, the first maintenance-free high-rate prismatic nickel-cadmium battery was introduced. In the sealed cell, an unfilled nickel-coated fiber plate is placed between two cadmium-filled negative plates. This effectively results in a split negative with an unfilled central region. This inactivated section serves as a catalytic site for rapid oxygen reduction. The main oxygen pathway to the recombination site is through the plate pores, which, in the FNC plate construction, are relatively large. This provides the oxygen with easy access to the large recombination reaction surfaces, as shown in Fig. 19.14.

Because rapid oxygen recombination eliminates the pressure buildup normally associated with sealed nickel-cadmium cells, high charging rates can be sustained even in the overcharge mode. Also, it is possible to use conventional nylon cell construction to produce prismatic sealed cells rather than the cylindrical design required for high pressure cells. The sealed FNC prismatic cell case is made of either polyamide (nylon) or stainless steel. The negative pressure within the cell (approximately 0.1 bar absolute) allows for pressure change due to oxygen generation during overcharge without causing the cell walls to expand.

All nickel-cadmium batteries must be overcharged to achieve a 100% state-of-charge. During the overcharge stage, excessively charged portions evolve oxygen and hydrogen. In vented batteries, these gases, along with water vapor, are vented outside the cell. The lost liquid must be replaced. The sealed FNC battery completely eliminates the loss of any gases from within the cell. Oxygen

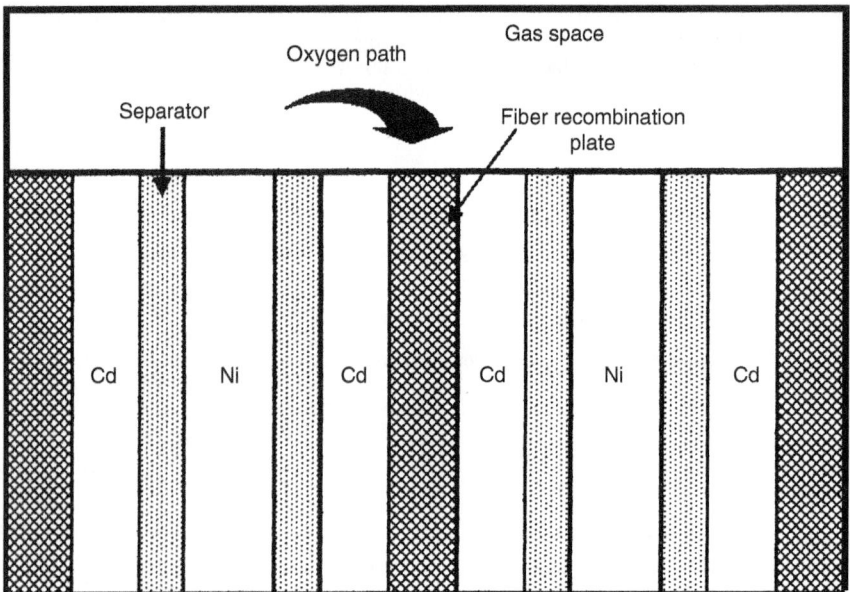

FIGURE 19.14 Electrode structure of fiber nickel-cadmium cell.

generated is rapidly recombined on the negative electrode. Hydrogen evolution is avoided by excess discharged cadmium on the negative electrode. This recombination process also keeps the plates in balance, eliminating the capacity loss that would otherwise result. Electrolyte spillage and corrosion is completely eliminated.

In the event of cell reversal or a failure to control charging voltage, hydrogen gas will be produced. A recombination plate of Pt/Pd-catalyzed plaque located within the cell provides for hydrogen recombination. The oxygen source for hydrogen recombination is provided by the self-discharge reaction on the positive electrode or the overcharge reaction on the following charge.

A safety valve at the top of the cell is provided to allow excessive pressure to escape should the battery be abused to the point that the electrolyte boils. This condition might be caused by a severe overcharge where adequate heat dissipation is not provided. Under high temperature abuse (+100°C and above), the safety valve will open at approximately 45 psia over pressure, allowing water vapor to escape. Electrolyte will not be expelled, even with the cell in an inverted position. When the cell is allowed to cool, the valve will reseal and the negative pressure cell will return to a normal operating condition. A reduction in cell capacity may be anticipated due to the loss of water from within the cell.

Positive and negative plates are connected to their respective terminal posts by nickel tabs. The tabs are attached directly to the fiber plates by a patented welding process and then fastened directly to nickel-plated, solid-copper terminal posts. The electrical path of each cell type is designed for maximum electrical performance.

Plate stacking is the same as previously discussed. Single positive plates are separated from the negative cadmium electrode by an electrolyte wet separator. The cadmium electrode is in three parts: two fiber frameworks carrying the negative active material, and an unfilled fiber recombination electrode placed between them. The large recombination surface area is sufficient to handle a 2 C rate charge on a fully charged battery. With the unfilled recombination plate being the primary gas path for oxygen recombination, a small pore size separator can be used. The separator is designed to be completely filled with electrolyte, thus contributing to improved high-rate performance. Additionally, the recombination plate acts as a reservoir for electrolyte, allowing for volumes in excess of 4 ml/Ah. This prevents stack dry-out as a possibility for premature cell failure.

Sealed FNC batteries are fail-safe. Even if subjected to extreme overcharge to the point at which the electrolyte boils, the battery will not go into thermal runaway. Instead, the hot cells dry out, with the loss of water vapor causing the battery impedance to increase. As the impedance increases, the current will decrease. After a time, the battery will no longer accept the charge current and it will cool down.

19.7.5 Performance

The high current performance capability of the sealed FNC battery design is exemplified by a short-circuit test performed on a model KCF XX47 battery, which produced currents approaching 4000 A (see Fig. 19.15). The KCF XX47 unit was designed to meet the requirements for large auxiliary power unit (APU) and direct engine starting. The constant voltage discharge of 12 V demonstrates the extraordinary high power capability of the battery (see Fig. 19.16). The KCF XX47 battery's high-power performance is also demonstrated with start curves for a large, wide-body aircraft APU. The first two discharge sequences represent unsuccessful start attempts with the third showing a successful start. The minimum voltage required for this particular specification is 13 V, while the FNC battery provides over 16 V (see Fig. 19.17).

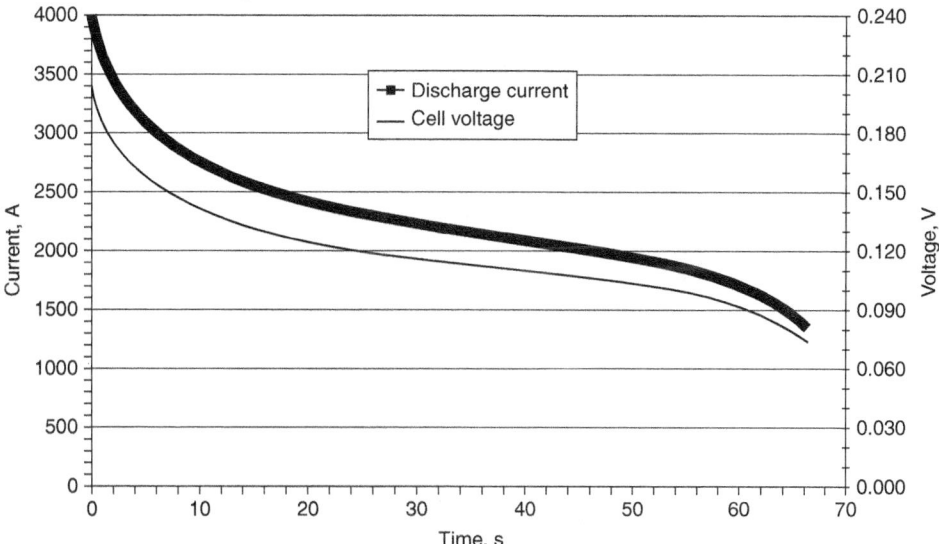

FIGURE 19.15 Short-circuit current. Model KCF XX47 FNC cell (47 Ah rated).

Cold-temperature performance available with the sealed FNC cells is also impressive. Figure 19.18 shows the capacity of a 28 V, 47 Ah battery for four different discharge rates with the battery soaked at a temperature of −18°C.

Sealed FNC batteries have shown outstanding cycle life at both low and high rates of discharge. Cycle life data for low earth orbit (LEO) cycle testing has demonstrated a cycle life in excess of 10,000 cycles for 35% DOD, 10°C, C/2 cycling. By maintaining the stable end of discharge voltage, the low recharge coefficient (approximately 3%) demonstrates the superior charge efficiency of the sealed FNC cell.

Deep discharge cycling is not necessary, eliminating the necessity for battery removal from the aircraft during maintenance. Capacity checks, if desired, can be accomplished with the battery installed by using a portable discharge/charger unit because the sealed FNC design does not exhibit the memory effect typically found in other NiCds.

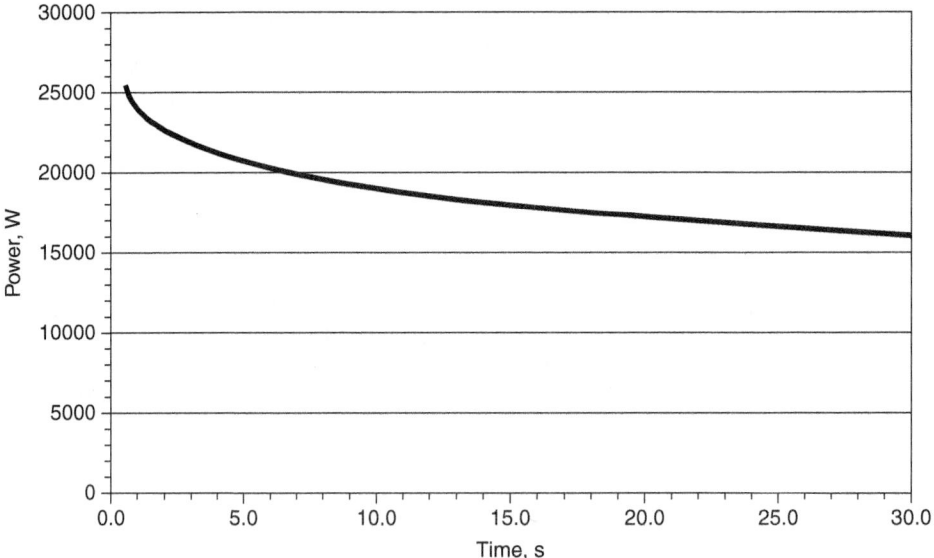

FIGURE 19.16 Constant-voltage (12.0 Volts) discharge. Room temperature. Model XX47 FNC battery (47 Ah rated).

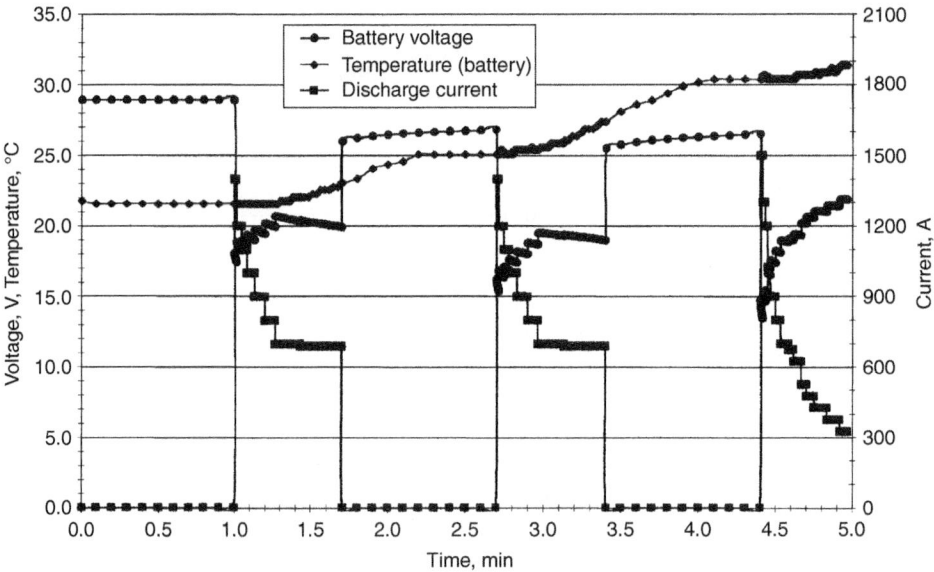

FIGURE 19.17 APU starts: 2 unsuccessful, 1 successful. Model XX47 FNC battery (47 Ah rated).

Charging characteristics of the sealed FNC cells are simple, yet different from vented NiCds. Because of the recombination that takes place during overcharge, the normal dV/dt behavior is not always observed. In addition, heat is generated during the overcharge from the recombination reaction, providing a reliable parameter for charge control. Changing from main mode to topping mode and charge termination is determined by battery temperature rise (ΔT).

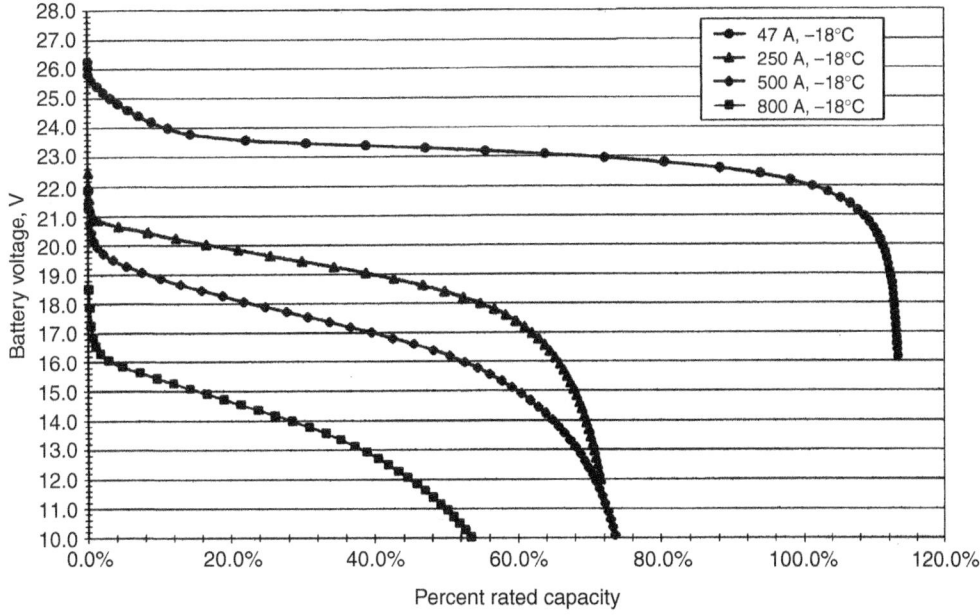

FIGURE 19.18 Constant current discharge. Battery charged at 25°C. FNC battery discharged at −18°C. Model XX47 battery (47 Ah rated).

The preferred charge is at constant current with a voltage clip (maximum voltage) of 1.55 V per cell. This voltage is also sufficient to charge the battery at temperatures as low as −40°C. For many applications, this means that a heater blanket is not required.

A completely charged FNC battery has sufficient recombination to continue to accommodate a 2 C overcharge rate.

19.8 MANUFACTURERS AND MARKET SEGMENTS

Table 19.3 contains data regarding prominent manufacturers of industrial nickel-cadmium batteries, Table 19.4 lists the market segments and applications for these batteries.

TABLE 19.3 Major Manufacturers of Industrial and Aerospace Nickel-Cadmium Batteries (Does Not Include Sintered-Plate Designs—See Chap. 20)

Manufacturer/country	Trademark	Product range		
		Pocket-plate	Fiber plate	Plastic-bonded plate
Acme Electric, U.S.A	Acme		X	
Alcad Ltd., U.K.	Alcad	X		
HBL Power Systems, India	HBL	X		
Hoppecke Batterien, Germany	Hoppecke		X	
Japan Storage Battery, U.S.A.	GS	X		X
Marathon Battery, U.S.A.	Marathon	X		X
SAFT, S.A., France	SAFT	X		X
Tudor S.A., Spain	Tudor	X	X	
Varta, Germany	Varta	X		
Yuasa, Japan	Yuasa	X		

TABLE 19.4 Market Segments and Applications for Vented Industrial Nickel-Cadmium Batteries

	Pocket plate				Fiber plate			Plastic-bonded plate	
Cell range*	H	M	L	XX	H or X	M	L	H	M
Capacity, Ah	10–1000	10–1250	10–1450	23–47	10–220	20–450	20–490	11–190	20–200
Applications	UPS, starting, switchgear	UPS, switchgear, auxiliary power, emergency power	Lighting, alarms, signaling, communications, standby power	Aircraft	UPS, satellites, starting, switchgear, traction, power stations and substations	UPS, switchgear, auxiliary power, emergency power	Lighting, UPS, alarms, signaling, telecommunications, standby power	UPS, starting, switchgear, traction, aircraft	Lighting, auxiliary power, traction
Railroad	X	X	X		X	X	X	X	X
Mass transit	X	X	X		X	X	X	X	X
Industry	X	X	X		X	X	X	X	
Buildings	X	X	X		X	X	X	X	
Hospitals	X	X	X		X	X	X	X	
Oil and gas	X	X	X		X	X	X	X	
Airports	X	X	X		X	X	X	X	
Marine	X	X	X		X	X	X	X	X
Military	X	X	X		X	X	X	X	X
Telecommunications	X	X	X		X	X	X	X	
Photovoltaics			X				X		
AGV/hybrid vehicles					X			X	

*H or X—high-rate; M—medium-rate; L—low-rate.
XX—ultra-high-rate.

FIGURE 19.19 FNC airborne battery system. Battery (left); charger (right). 28 V, 47 Ah battery. (Courtesy of Acme Electric, USA)

Figure 19.19 is a picture of a 28 V, 47 Ah airborne battery system with battery and dedicated charger which uses the model XX47 Acme FNC cell.

19.9 APPLICATIONS

Because of their favorable electrical properties, excellent reliability, low maintenance, rugged design, and long life, nickel-cadmium batteries are used in a large variety of applications, as indicated in Table 19.4. Most of these are of an industrial nature, but this type of battery is also used in many commercial, military, and space applications.

The nickel-cadmium battery was originally developed for traction applications, and since the early years of the 20th century, it has been used extensively in railroad applications. Today the nickel-cadmium battery is the system of choice in a variety of railroad and mass-transit installations around the world. Approximately 40% of all industrial nickel-cadmium batteries produced are used in train lighting and air-conditioning for rail cars, emergency and standby systems such as emergency brakes, door openers, and lighting in mass-transit and subway cars, diesel-engine cranking in locomotives and commuter cars, railroad signaling, communication along tracks, as well as standby power in rail stations and traffic control systems. The pocket-plate battery has traditionally dominated this market segment, but in recent years, with demands for higher energy per unit weight and volume, plastic-bonded and fiber-plate batteries have penetrated this market, particularly for high-speed trains, mass-transit cars, subway cars, and light rail vehicles. Where ruggedness and long durability are the main requirements, the pocket-plate battery still maintains its position.

In stationary applications where reliability is a must, nickel-cadmium batteries are used in standby and emergency installations where life and great economic values would be endangered by a power failure. Examples of such installations are emergency power in hospital operating theaters, standby power for all vital functions on offshore oil rigs, uninterruptible power supplies (UPSs) for large computer systems in banks and insurance companies, standby power in process industries, and emergency lighting and landing systems in airports.

The nickel-cadmium battery is also used in power-generating stations and power distribution networks where power supplies must not break down. The batteries are used in switch-gear applications and for control and monitoring functions. In centralized emergency lighting systems in hospitals, public buildings, sports arenas, and schools, nickel-cadmium batteries are often specified in building codes and by consultants in many industrialized countries.

In case of failure of the primary power supply, diesel generators or gas turbines are installed to take over the power supply. For reliable and fast-acting startup of these engines, nickel-cadmium batteries have proven to be the best emergency power source.

In portable applications where batteries are exposed to temperature extremes or rough handling, nickel-cadmium batteries are used for signal lamps, hand lamps, search lights, and portable instruments. Vented spillproof batteries are used in large devices, whereas sealed nickel-cadmium batteries dominate the smaller ones (see Chap. 20).

The industrial battery market is dominated by the lead-acid battery, and the nickel-cadmium battery is a niche-market product. The reason for this is the higher capital cost for nickel-cadmium batteries compared to lead-acid batteries. Where only energy is required, the lead-acid battery is the least expensive, as its cost per watthour is lower than that for nickel-cadmium batteries. However, in cost per watt or life cycle cost, nickel-cadmium batteries can compete with lead-acid batteries due to much better high-rate performance and longer life combined with low maintenance costs. A typical example is locomotive diesel-engine cranking, where a nickel-cadmium battery with only one-third of the ampere-hour capacity and a life four times that of the lead-acid battery can do the job. In applications with short-duration discharges—standby and emergency equipment are usually used for less than a half-hour—the rated capacity of a battery is of little importance. The size of the battery is chiefly determined by the power need. The nickel-cadmium battery is unmatched in industrial applications when reliability and durability are considered in a life-cycle cost calculation.

Fiber nickel-cadmium batteries of the ultra-high-rate (XX) and high-rate (X) design are employed primarily in aircraft, military, and space applications. Because of the variety of applications for nickel-cadmium batteries, it is important to select the best technology for the application. The characteristics are somewhat different for the three technologies available today for industrial use.

The pocket-plate battery has the lowest cost of the three technologies and is known for high reliability and fail-safe operation. However, the energy and power density limit its use in some areas. The fiber-plate battery has lower internal resistance than the pocket-plate battery and is also available in ultra-high-, high-, medium- and low-rate cells. Where very high energy and power density are required, the plastic-bonded plate may be the choice. The plastic-bonded and fiber plate batteries are the only technologies possible for use in automated guided vehicles (AGVs). They have also cost and performance advantages in some traditional pocket-plate battery applications such as engine cranking, switchgear, and uninterruptible power supplies (UPSs) where only very short duration discharge is required.

BIBLIOGRAPHY

General:

Barak, M. (ed.), *Electrochemical Power Sources*, Peter Peregrinus, London, 1980.

Brunamonti, P., *Life Cycling at Elevated Temperatures Battery Types M81757/8-5 and M81757/15: Marathon Power Tech. Co., EDD 99–127*, Nov. 30, 1999, Crane Div., Naval Surface Warfare Center, Crane, IN 47522-5001.

Falk, S. U., and A. J. Salkind, *Alkaline Storage Batteries*, Wiley, New York, 1969.

Jacksch, H.-D., *Batterie Lexikon*, pp. 348–394, Pflaum Verlag, Munich, 1993.

Kinzelbach, R., *Stahlakkumulatoren*, Varta, Hannover, Germany, 1968.

Miyake, Y., and A. Kozawa, *Rechargeable Batteries in Japan*, JEC Press, Cleveland, OH, 1977.

Newman, B., *Life Cycling at Elevated Temperatures Battery Types M81757/15: SAFT America, Inc., EDD 99–122*, Nov. 17, 1999, Crane Div., Naval Surface Warfare Center, Crane, IN 47522-5001.

Plastic-Bonded Electrode Technology:

McRae, B., and D. Nary, *Proceedings of the 38th Power Sources Conference,* pp. 123–126 (1998).

FNC Technology:

Anderman, M., C. Baker, and F. Cohen, *Proceedings of the 32nd Intersociety Energy Conversion Conference,* Vol. 1, p. 97465 (1997).

Baker, C., *Proceedings of the SAE Power Systems Conference*, Williamsbury, VA (1997). See *Advanced Battery Technology*, April 1997.

Baker, C., and M. Barekatien, *Proceedings of the SAE Power Systems Conference,* San Diego, CA (2000).

FNC Vented Nickel-Cadmium Batteries, Hoppecke Batterien.

CHAPTER 20
VENTED SINTERED-PLATE NICKEL-CADMIUM BATTERIES

R. David Lucero

20.1 GENERAL CHARACTERISTICS

The sintered-plate nickel-cadmium battery is a mature development of the nickel-cadmium system, having a higher energy density, up to 50% greater than its predecessor, the pocket-type construction. The sintered plate can be constructed in a much thinner form than the pocket plate, and the cell has a much lower internal resistance and gives superior high-rate and low-temperature performance. A flat discharge curve is characteristic of the battery, and its performance is less sensitive than other battery systems to changes in discharge load and temperature. The sintered-plate battery has most of the favorable characteristics of the pocket-type battery, although it is generally more expensive. It is electrically and mechanically rugged, is very reliable, requires little maintenance, can be stored for long periods of time in a charged or uncharged condition, and has good charge retention. Batteries losing capacity through self-discharge can be restored to full service with a normal charge. The major advantages and disadvantages of this battery type are given in Table 20.1.

For these reasons, vented sintered-plate nickel-cadmium batteries are used in applications requiring high-power discharge service such as aircraft turbine engine and diesel engine starting as well as other mobile and military equipment. The battery provides outstanding performance where high peak power and fast recharging are required. In many applications the vented sintered-plate battery is used because it leads to a reduction in size, weight, and maintenance as compared to other battery systems. This is particularly true in systems subject to low-temperature operation. The rise in terminal voltage of the vented cell at the end of charge also provides a useful characteristic for controlling the charge.

20.2 CHEMISTRY

Vented sintered-plate nickel-cadmium cells, in the discharged state, consist of flat positive nickel hydroxide and negative cadmium hydroxide plates, separated by materials that act as a gas barrier and electrical separator. The electrolyte, normally a 31% potassium hydroxide solution, completely covers the plates and separators: for this reason vented cells are referred to as "flooded cells."

In the sintered-plate design, the active materials are held within the pores of a sintered-nickel structure. Nickel hydroxide with 3 to 10% cobalt hydroxide is the active material of the positive plate, while cadmium hydroxide is the active material of the negative plate.

The electrochemistry of the charge and discharge of the positive electrode is quite complex and not well understood,[1] especially the role that cobalt plays in the active material.[2] For simplicity, let's consider the role of nickel hydroxide in the charge-discharge reaction.

20.2 SECONDARY BATTERIES

TABLE 20.1 Major Advantages and Disadvantages of Vented Sintered-Plate Nickel-Cadmium Batteries

Advantages	Disadvantages
Flat discharge profile	Higher cost
Higher energy density (50% greater than pocket plate)	Memory effect (voltage depression)
Superior high-rate and low-temperature performance	Temperature controlled charging system required to enhance life
Excellent long-term storage	
Good capacity retention; capacity can be restored by recharge	

During charge, the nickel hydroxide in the positive electrode is oxidized to nickel oxyhydroxide (NiOOH) and higher valence states of nickel. Potassium and water are also incorporated into the active material as potassium hydroxide according to the following equation:[3]

$$Ni(OH)_2 + xK^+ + (1+x)OH^- \rightleftharpoons NiOOH \cdot xKOH \cdot (H_2O) + e$$

The fraction of potassium that is bonded into the nickel oxyhydroxide lattice is represented by x. The value is small (much less than 1.0) and varies according to manufacturing process.

The cadmium hydroxide in the negative electrode is reduced to metallic cadmium during charge

$$Cd(OH)_2 + 2e \rightleftharpoons Cd + 2OH^-$$

The overall charge-discharge reaction is thus

$$2Ni(OH)_2 + 2xKOH + Cd(OH)_2 \underset{\text{discharge}}{\overset{\text{charge}}{\rightleftharpoons}} 2NiOOH \cdot xKOH \cdot (H_2O) + Cd$$

According to the above equation, one might think that the change in the electrolyte concentration might offer a means of state-of-charge determination by measuring the specific gravity of the electrolyte. Unfortunately, the complication of potassium in the active material, accumulation of carbonates, along with the large volume of electrolyte make this type of measurement unreliable and impractical.

The positive electrode does not accept charge, and converts nickel hydroxide to nickel oxyhydroxide, at the thermodynamically reversible potential.[3] In fact, with a low enough charge rate, gassing according to the equation below occurs:

$$4OH^- \rightarrow 2H_2O + O_2 + 4e$$

If the rate is increased appreciably, this will result in an oxygen overvoltage sufficiently high to allow the preferred conversion of the nickel hydroxide to the nickel oxyhydroxide instead of oxygen gassing. However, when about 80% conversion of nickel hydroxide to nickel oxyhydroxide is achieved, the competing oxygen generating reaction occurs gradually and remains until 100% state-of-charge is achieved, and then the only reaction occurring is oxygen evolution.

The negative electrode accepts charge until it is essentially 100% charged, at which time the favored reaction is hydrogen gassing as shown in the following equation:

$$2H_2O + 2e \rightarrow H_2 + 2OH^-$$

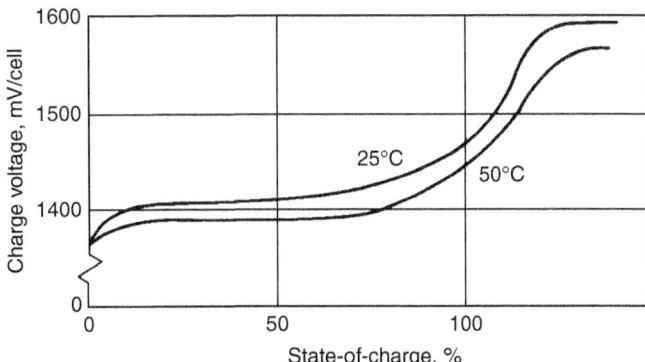

FIGURE 20.1 Constant-current charge voltage of vented sintered-plate nickel-cadmium cell; $C/10$ charge rate.

The hydrogen gassing reaction, with a $C/10$ charge rate, occurs at a cell voltage close to 1600 mV as shown in Fig. 20.1.

The hydrogen overvoltage on the cadmium electrode is quite high, about 110 mV at the $C/10$ rate. Consequently there is a sharp rise in voltage as the negative electrode goes into overcharge. This rise in voltage is used in various charging schemes to control or terminate charging.

During overcharge, all the current is used to electrolyze water to hydrogen and oxygen, as shown in the overall reaction

$$2H_2O \rightarrow 2H_2 + O_2$$

This overcharge reaction consumes water and thereby decreases the level of electrolyte in the cell. The water loss can be limited by controlling the amount of overcharge so as to maximize the interval between needed water replenishments.

Cells are constructed with 50% excess capacity in the negative electrodes and thus are positive limited.

20.3 CONSTRUCTION

Vented cells are designed so that both electrodes reach full charge at about the same time. The positive electrode, as noted, will begin to evolve oxygen before it is fully charged. If this gas is allowed to reach the negative electrode due to failure of the gas barrier, it will recombine and generate heat. This will not only prevent the negative from reaching a full state-of-charge, but it will also result in reduced voltage due to depolarization of the cadmium electrode. To maintain the fullest capability, adequate precautions must be taken to prevent oxygen recombination at the negative plate. This is accomplished by providing a gas barrier between the positive and negative plates and by flooding the plates with excess electrolyte.

Figure 20.2 shows details of a typical vented sintered-plate nickel-cadmium cell.

20.3.1 Plates and Processes

A variety of plate formulations are used in vented sintered-plate nickel-cadmium cells produced by different manufacturers. The plates differ according to the nature of the substrate, method of sintering, impregnation process, formation, and termination techniques. The predominate

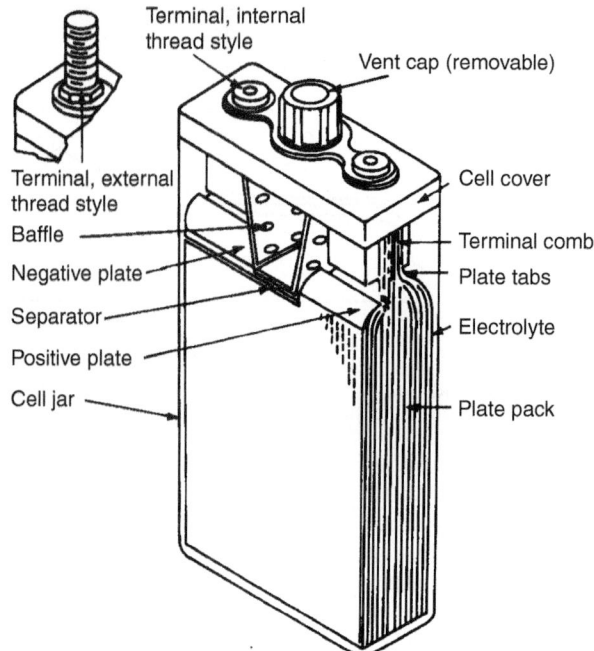

FIGURE 20.2 Cross section of vented sintered-plate nickel-cadmium cell.

plate-fabrication process used for vented sintered plates over the years has been described by Fleischer.[4] There are several reviews on electrode fabrication processes that have been used in flooded vented cells.[3,5,6]

Substrate. The substrate serves as a mechanical support for the sintered structure and as a current collector for the electrochemical reactions that occur throughout the porous sintered portion of the plate. It also provides mechanical strength and continuity during the manufacturing processes. Two types of substrate are typically used: (1) perforated nickel-plated steel or pure nickel strip in continuous lengths, and (2) woven screens of nickel or nickel-clad steel wire. A common perforated type may be 0.1 mm thick with 2 mm holes and a void area of about 40%. A typical screen may use 0.18 mm diameter wire with 1.0 mm openings.

Plaque. The sintered structure before impregnation is generally referred to as "plaque." It usually has a porosity of 80 to 85% and ranges in thickness from 0.40 to 1.0 mm. Two generic sintering processes are used: (1) the slurry coating process and (2) the dry-powder process. Both processes employ special low-density battery grades of carbonyl nickel powder.

In the slurry coating process, the nickel powder is suspended in a viscous, aqueous solution containing a low percentage of a thixotropic agent. The nickel-plated strip with the desired perforated pattern is pulled through the suspension. The thickness is controlled by passing it through doctor blades, while wiping the edges free of slurry. The continuous strip is then dried before sintering in a reducing atmosphere at about 1000°C.

The dry-powder processes generally employ wire screen precut to the so-called master plaque dimension. The screens are placed in molds with loose powder on each side. They are then typically sintered in a belt furnace in a reducing atmosphere at 800 to 1000°C.

Impregnation. A review of various impregnation processes, used to load the porous sintered structure of the positive with nickel hydroxide and of the negative with cadmium hydroxide, has been given by Pickett.[6] The plaque is impregnated with a concentrated solution of nitrate that is then converted to hydroxide by chemical precipitation[4] or electrochemical precipitation.[7,8,9] The most widely used process for vented cells involves a chemical precipitation, and with minor variations, follows, in principle, the process described in 1948.[4] The plaque is impregnated with a concentrated solution of the nitrate, briefly rinsed, and the nitrate salts are precipitated as hydroxide with caustic. Following the addition of caustic, the plaque is made cathodic. This is called polarization. The polarization cycle usually consists of a high current charge (C rate or higher) for approximately 1 h. The plaque is then rinsed and this sequence of steps is repeated a number of times so as to fill about 40 to 60% of the sintered pore volume (or until a targeted weight gain is achieved).

Plate Formation. Following impregnation, the plates are mechanically brushed and electrochemically cleaned and formed by charging and discharging the electrode. In the master-plaque process they are formed against inert counterelectrodes (typically stainless steel or nickel) and can be performed in a loose pack or tight pack configuration. Formation is essential for properly converting hydroxide into the pores of the sinter structure, as well as the reduction of nitrates in the plates. Typical formation cycles for chemical plates consist of high-current cycling. This regime or time may vary for plaque type and capacity. In the case of the continuous-strip process, the formation is done on a machine similar in appearance to a continuous-strip-electroplating machine. Plates blanked from the continuous strip have a clean, wiped area at the top that serves as attachment points for nickel or nickel-plated steel current-collector tabs. In the case of the master-plaque process, a coined or densified area is provided for attachment of these collector tabs.

20.3.2 Separator

The separator system is a thin, multilayered combination. It consists of a cloth that electrically separates the positive and negative plates and an ion-permeable plastic membrane that serves as the gas barrier.

Electrical and mechanical separation of the plates is typically provided by either woven or felted nylon material. This material is relatively porous in order to provide a good ionic conduction path through the electrolyte with a microporous polypropylene separator.

The microporous polypropylene membrane, typically Celgard® (Celgard 3400, manufactured by Celgard LLC, Charlotte, NC, 28273),[10] is utilized as the gas barrier while at the same time it offers minimum ionic resistance. This thin gas barrier, which becomes relatively soft when wetted, is frequently placed between two layers of the cloth separator and receives significant mechanical support from them. Substantial improvements have been made to the toughness of the plastic membrane gas barrier.

20.3.3 Plate-Pack Cell Assembly

Plate packs are assembled by alternately stacking positive and negative plates with the separator-gas barrier system interleaved between them. The cell terminals are bolted, riveted, or welded to the current-collector plate tabs. In the case of cells with many plates, the tabs from the outermost plates may need to be bent quite significantly inward to reach the cell terminals. Spacers at the terminals are sometimes used in these situations to keep the angle of the tabs at a minimum.

20.3.4 Electrolyte

Potassium hydroxide electrolyte is used in a concentration of approximately 31% at full charge (specific gravity 1.30). Performance of the cell, particularly at low temperatures, is significantly dependent on this concentration (see Sec. 20.4.2).

Electrolyte purity can also have significant effects on cell performance. The level of potassium carbonate in the cell relates directly to the cell's performance. Increasing carbonate concentration changes the characteristics of the electrolyte, reducing the high-rate charge and discharge capability of the cell. Fresh electrolyte contains very low levels of carbonate. However, organic components in the cell are slowly oxidized in the presence of the electrolyte and oxygen, forming small amounts of carbonate. The carbonates accumulate as the cell ages and eventually reduce cell performance. Carbonate levels at the time of cell activation are on the order of 80 to 90 g/L due to reaction of the electrolyte with residues from the impregnation process. High-quality cells are designed with components that do not degrade in KOH. In addition, at least one manufacturer flushes new cells repeatedly with fresh electrolyte, lowering final carbonate levels to 6 to 8 g/L.

20.3.5 Cell Container

The plate pack is placed into the cell container with the cell terminals extending through the cover. The cell container is usually made of a low-moisture-absorbent nylon and consists of the cell jar and matching cover that are permanently joined together at assembly by solvent sealing, thermal fusion, or ultrasonic bonding. The container is designed to provide a sealed enclosure for the cell, thus preventing electrolyte leakage or contamination, as well as providing physical support for the cell components. The terminal seal is generally provided by means of O-rings with Belleville washers and retaining clips.

20.3.6 Vent Cap and Check Valve

The vent cap serves as a removable cap to provide the access required for replenishment of water to the electrolyte and also to function as a check valve to release gases generated when water is consumed during overcharge. The check valve prevents atmospheric contamination of the electrolyte. It consists of a nylon body with a hollow center post, through which a cross-hole is drilled and around which an elastomeric sleeve is placed. This functions as a Bunsen valve to allow gas to escape from the cell but not to enter. Typical sleeves used for this application have developed significantly over time, and ethylene-propylene rubber seems to have the best characteristics for vented cells. Neoprene vent sleeves were previously used, but the neoprene is attacked by potassium hydroxide and can soften, swell, and split. It also frequently erodes at the interface between the vent cap and sleeve until the neoprene no longer seals. Before erosion occurs, a sleeve surface can soften due to electrolyte at the interface between the sleeve and vent cap, dry during a subsequent storage, and literally glue itself to the vent. When this occurs, the pressure will build up in the cell during charge until the sleeve breaks free or ruptures or the cell explodes.[10]

20.4 PERFORMANCE CHARACTERISTICS

20.4.1 Discharge Properties

The discharge curves for a typical vented sintered-plate nickel-cadmium battery at various constant-discharge loads are shown in Fig. 20.3. The discharge curves for a typical battery at various temperatures are shown in Fig. 20.4. The curves for this battery are characterized by a flat voltage profile, even at relatively high discharge rates and low temperatures. Voltages at various constant-current discharge loads and states of discharge are given in Fig. 20.5.

The battery, because of its low internal resistance, is capable of delivering pulse currents as high as the 20 to 40 C rate. For this reason it can be used successfully for very high-power applications, such as engine starting (see Sec. 20.4.3).

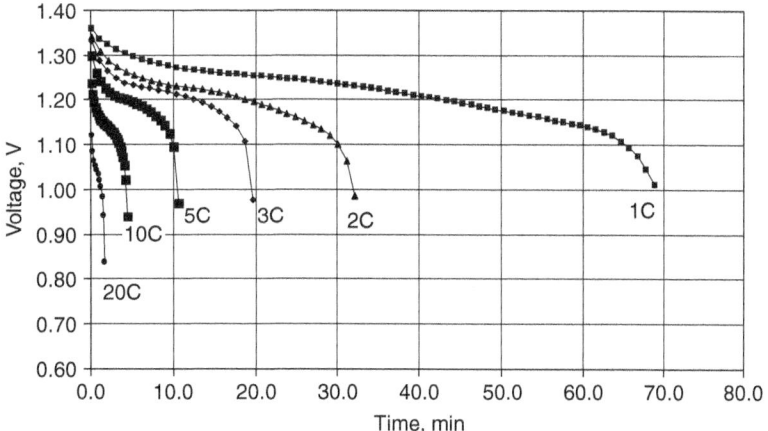

FIGURE 20.3 Typical discharge curves at various C rates, 25°C.

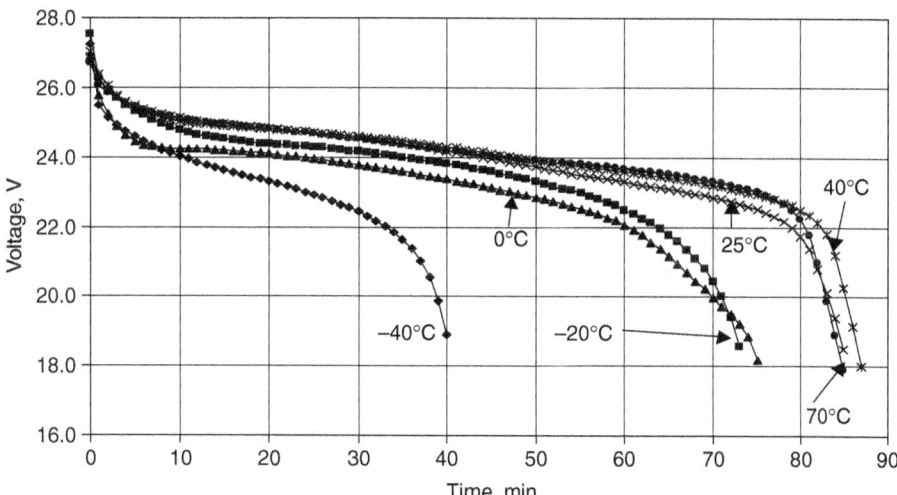

FIGURE 20.4 Typical discharge curves at various temperatures 1 C rate, 20-cell battery.

20.4.2 Factors Affecting Capacity

The total capacity that the fully charged sintered-plate vented battery is capable of delivering is dependent on both discharge rate and temperature, although the sintered-plate battery is less sensitive to these variables than most other battery systems. The relationships of capacity to discharge load and temperature are shown in Figs. 20.6 and 20.7, respectively.

Low-temperature performance is enhanced by the use of eutectic 31% KOH (1.30 specific gravity) electrolyte, which freezes at −66°C. Higher or lower concentrations will freeze at higher temperatures; for example, 26% KOH freezes at −42°C. As shown in Fig. 20.7, more than 60% of the 25°C capacity is available at −35°C, with the temperature having an increasingly significant effect as it is lowered toward −50°C. At high discharge rates, heat that is generated may cause the battery

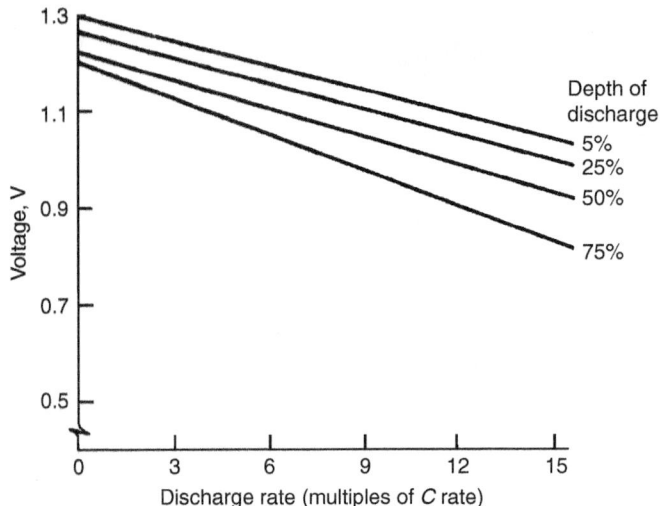

FIGURE 20.5 Voltage as a function of discharge load and at various states of charge at 25°C.

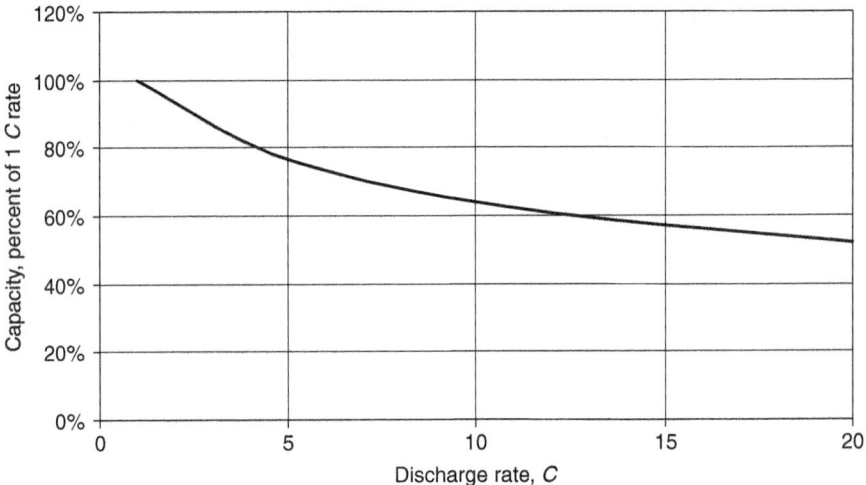

FIGURE 20.6 Capacity derating as a function of discharge rate at 25°C.

to warm up, giving improved performance on immediate or subsequent discharges than would be expected under ambient conditions.

Vented sintered-plate batteries can also be discharged at elevated temperatures. Strict control is required, however, when charging at high temperature. As with most chemically based devices, exposure to high temperatures for extended periods of time will shorten the life of the battery (see Sec. 20.7.3).

The combined effects of increased discharge rate and low temperature may be approximated by multiplying the two derating factors.

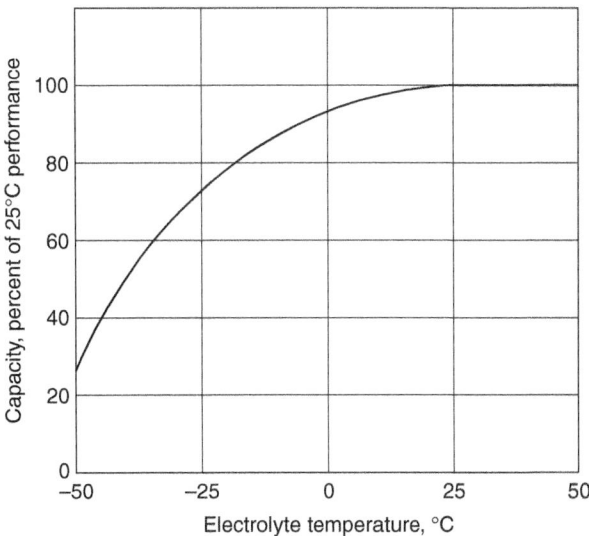

FIGURE 20.7 Capacity derating as a function of discharge temperature at 1 C rate discharge.

20.4.3 Variable-Load Engine-Start Power

The most common and demanding use of the vented sintered-plate nickel-cadmium battery is as the power source for starting turbine engines onboard aircraft. The discharge in this application occurs at relatively high rates for periods of 15 to 45 s. Typically the load resistance when the start is initiated, particularly in low temperature in a marginal-start situation, is of the same order of magnitude as the effective internal resistance of the battery, R_e. The apparent load resistance increases as the engine rotor gradually comes up to speed. This results in a typical discharge current, which slowly decreases from some high initial value while the battery voltage recovers from an initial drop of perhaps 50% or more, back toward 1.2 V per cell, the effective zero load voltage. A representative graph of the battery-starter voltage and current, expressed as a function of time, is shown in Fig. 20.8.

A common and useful measure of battery performance is the maximum power current. This property is generally defined as the load current at which the battery voltage would be $0.6N \times$ V, or one-half of the effective open-circuit voltage (1.2 V/cell) and where N is the number of cells in the battery. The instantaneous maximum power current decreases with decreasing state of charge due to rising internal resistance. Its value versus state-of-charge tends to behave exponentially, as shown in Fig. 20.9. An approximation of I_{mp} may also be measured by performing a "constant-potential" discharge at $0.6N \times$ V for 15 to 120 s. A typical discharge is shown in Fig. 20.10.

The maximum power delivery P_{mp} and the effective internal resistance R_e are related to the value of I_{mp} as follows:

$$P_{mp} = 0.6N \, I_{mp}$$

and

$$R_e = \frac{0.6N}{I_{mp}}$$

20.10 SECONDARY BATTERIES

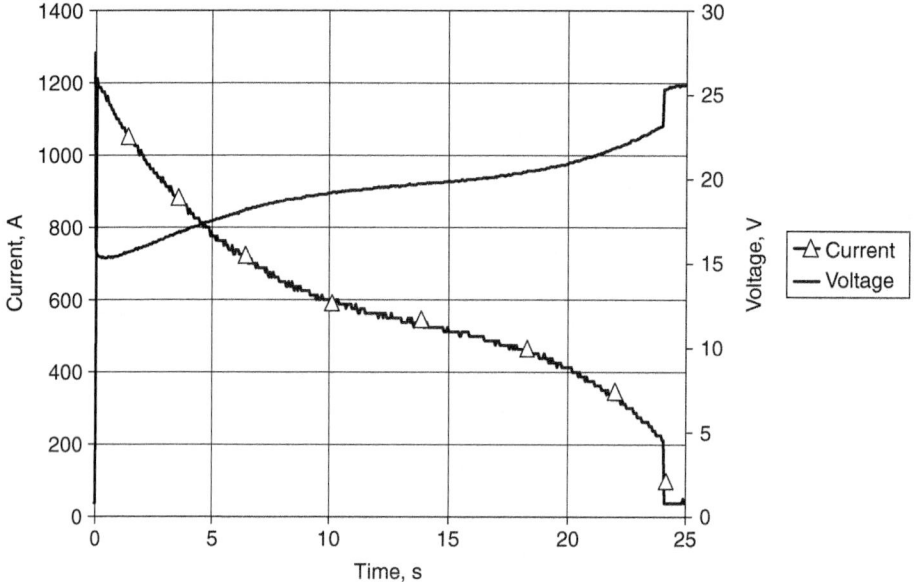

FIGURE 20.8 Battery voltage and current as a function of time for a typical turbine engine start (20-cell battery).

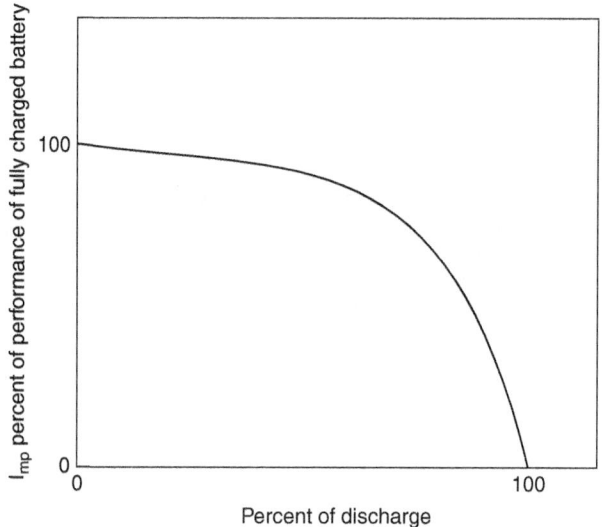

FIGURE 20.9 Maximum power current derating as a function of state-of-charge at 25°C.

20.4.4 Factors Affecting Maximum Power Current

The value of I_{mp}, which a battery is capable of delivering, is maximum at full charge and at 25°C electrolyte temperature. Derating effects due to state-of-charge and electrolyte temperature factors are shown in Figs. 20.9 and 20.11, respectively. It will be noted that both relationships are nonlinear in that the effects on maximum power delivery, per unit of change, increase with decreasing state-of-charge and with decreasing temperature. As with capacity, the approximate effect of combined

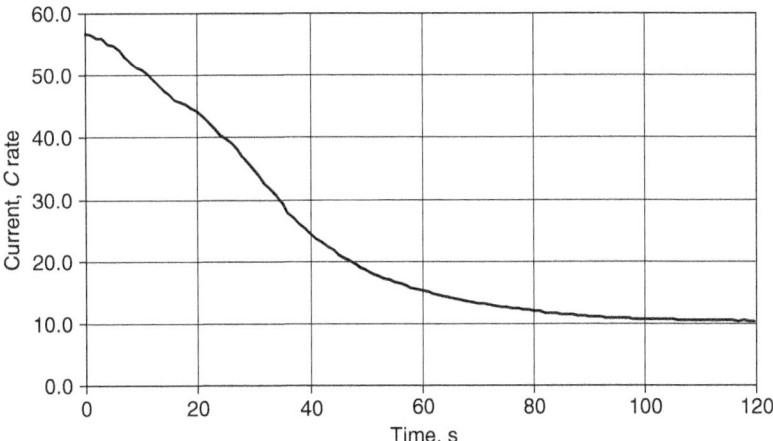

FIGURE 20.10 Representative 0.6 V constant potential discharge at 25°C.

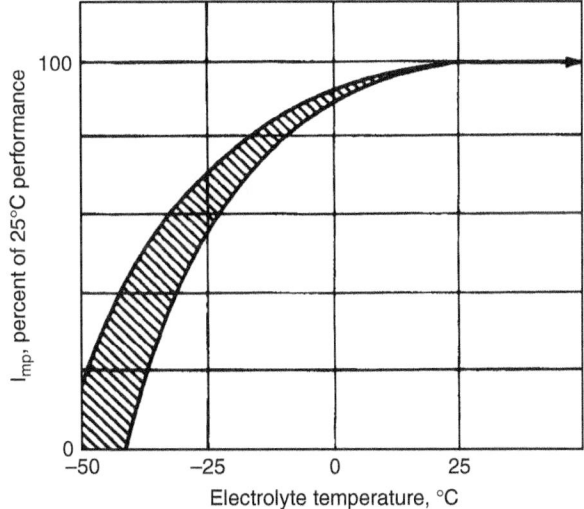

FIGURE 20.11 Maximum power current derating as a function of battery temperature (fully charged).

low electrolyte temperature and decreased state-of-charge may be determined by multiplying the individual factors. It should be noted, however, that high-rate discharge at low temperature may increase the battery temperature. This self-heating must be accounted for when determining the combined derating factors for a subsequent discharge. A negligible effect on I_{mp} occurs with increases in electrolyte temperature above 25°C.

20.4.5 Energy/Power Density

Typical average values for the energy and power densities of the vented sintered-plate nickel-cadmium battery at 25°C are shown in Table 20.2.

TABLE 20.2 Energy and Power Characteristics of Vented Sintered-Plate Nickel-Cadmium Battery (Single Cell Basis)

Specific capacity (single cell, C rate)	25–31 Ah/kg
Capacity density	48–80 Ah/L
Specific energy (C rate)	30–37 Wh/kg
Energy density	58–96 Wh/L
Specific power (at maximum power)	330–460 W/kg
Power density	730–1250 W/L

20.4.6 Service Life

The service life (discharge time) of the vented sintered-plate nickel-cadmium cell, normalized to unit weight (kilogram) and volume (liter) at various discharge rates at 25°C, is approximated in Figs. 20.12 and 20.13.

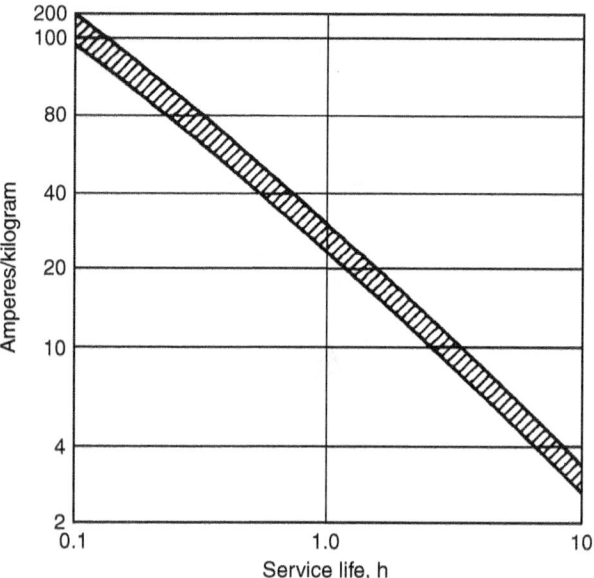

FIGURE 20.12 Service life of typical vented sintered-plate nickel-cadmium battery (gravimetric) at 25°C.

20.4.7 Charge Retention

Charge retention or capacity retention refers to the amount of dischargeable capacity remaining in a battery following prolonged storage under open-circuit conditions. Two mechanisms are responsible for the loss of charge, namely, self-discharge and electrical leakage between cells.

Self-discharge rates are an intrinsic property of cells. Typically, experimental results for the capacity retained as a function of open-circuit storage time best fit a semilogarithmic relationship such as that shown in Fig. 20.14. The self-discharge rate of a cell is affected by impurity levels and the electrochemical stability of the electrodes.

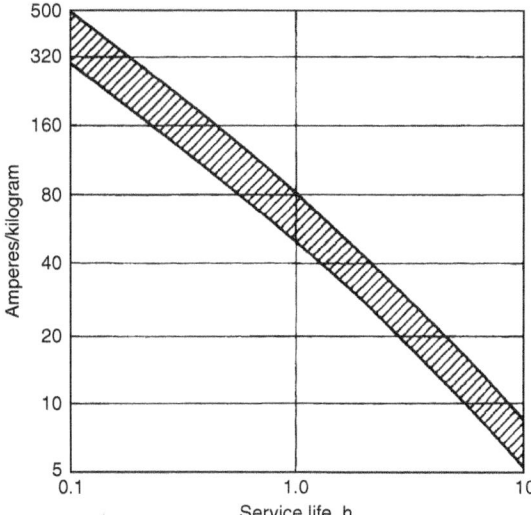

FIGURE 20.13 Service life of typical vented sintered-plate nickel-cadmium battery (volumetric) at 25°C.

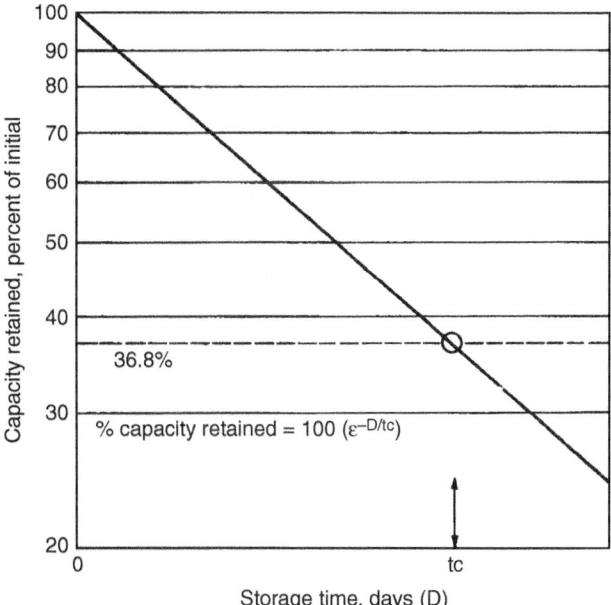

FIGURE 20.14 Capacity retention as a function of storage time.

The effect of temperature is shown in Fig. 20.15, where the exponential time constant (t_c), the time to retention of 36.8% of initial capacity, is plotted against temperature. Storage temperature is the most important factor affecting the self-discharge rate.

The second mechanism, the loss of charge due to electrical leakage, is influenced by the history of the battery's use and maintenance. Charge retention usually improves with cycling of the battery,

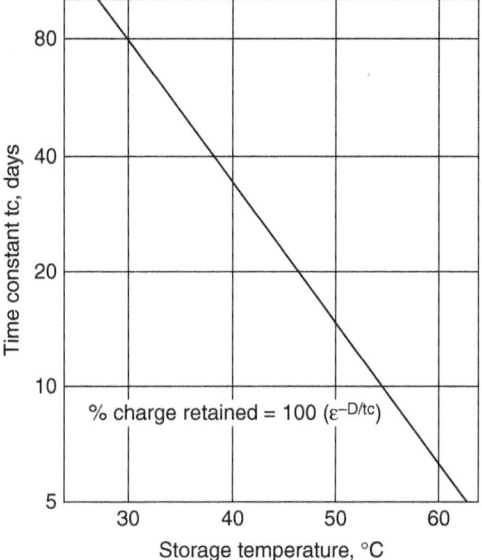

FIGURE 20.15 Charge retention time constant as a function of storage temperature.

and this will be true unless this cycling history has been abusive. The maintenance factor influencing charge retention is primarily battery cleanliness. Battery charge can leak from the terminals of one cell to the terminals of other cells across the cell tops if they are wet with potassium hydroxide. Loss of charge from this cause is relatively unpredictable, but it can usually be prevented by good housekeeping practices. Although surface leakage may affect only a portion of the cells in the battery, it is important since the capacity of the battery is limited to that of the lowest-capacity cell. Additionally it unbalances the cells in the timing of the onset of overcharge voltage response. (See Fig. 20.1.)

It should be noted that loss of charge through these mechanisms is not permanent since the battery capability can be completely restored through comprehensive maintenance practices and recharging.

20.4.8 Storage

The sintered-plate cell can be stored in any state-of-charge and over a very broad temperature range (−60 to 60°C) for an unlimited period. The battery should be clean and dry before placing it in storage. Intercell hardware may have a light coating of petroleum jelly to prevent corrosion. It should be fully discharged and shorted prior to storage periods greater than 30 days. Fully discharged batteries that have been stored longer than 30 days should be charged by a "slow charge" method. The "slow charge" method typically consists of incremental charge rates to voltage cutoff (i.e., C rate to 1.57 V, $C/2$ rate to 1.6 V), with a final charge ($C/10$ or lower) for 2 h. It is the best practice to store the cells shorted and upright with proper electrolyte level at a temperature between 0 and 30°C. The preferable storage method is to allow the battery to discharge through a resistor until the battery voltage is close to zero. Because a vented Ni-Cd still has considerable power available even at very low states-of-charge, failure to completely discharge the battery prior to applying a shorting device can create a hazardous situation.

20.4.9 Life

The life of the battery is strongly influenced by factors such as the design, the care with which it is maintained and reconditioned as well as the way it is used; hence it is difficult to predict battery life. Best life performance is obtained with operation at normal temperatures, temperature-controlled charging, and minimum reconditioning. Some design features that improve the life expectancy of a battery are: the use of modern separator materials and gas barriers, the elimination of materials that degrade in KOH (e.g., O-rings), the reduction of electrolyte impurity levels in manufacturing (by electrolyte flushing and replacement), and the use of pure nickel components versus nickel-plated steel.

20.5 CHARGING CHARACTERISTICS

The functional design of the vented cell battery differs from that of the sealed cell battery primarily by the inclusion of a gas barrier between the positive and negative electrodes. This gas barrier has one principal function, which is to prevent, as discussed in Sec. 20.3, the cross-plate migration and recombination of generated gases within the cell. Preventing this recombination allows both positive and negative plates to return to full charge. This results in an overvoltage during onset of overcharge, which is used as the feedback signal to control the charging device. Because the gas is driven out of the cell, however, the vented cell consumes water, which must be replenished.

Charging of the vented sintered-plate nickel-cadmium battery following its discharge in cyclic use, has four significant objectives. These may be stated as follows:

1. Restore the charge used during discharge as quickly as possible.
2. Maintain the fully charged capacity as high as possible during the use intervals between removals for maintenance.
3. Minimize the amount of water usage during overcharge.
4. Minimize the damaging effects of overcharge.

Fulfillment of the first objective is the principal reason for the design and use of vented cells, since the gas barrier provides the voltage signal, which may be utilized in several different ways to terminate the fast recharge. The charge may thus be accomplished at the desired high rate, without compromising the battery, by continuing that rate in overcharge. Objective **2** must inherently be balanced against objectives **3** and **4** in the design and control of the charging method. Generally, a continued good capacity between reconditionings is enhanced by providing more overcharge, while more overcharge inherently utilizes more water and, if sufficiently high, may result in damage to the battery. A compromise must therefore be struck. Usually about 101 to 105% of the ampere-hours removed on discharge are replaced on the subsequent charge.

Charging techniques that are used in onboard systems utilize the signal provided by the overvoltage of the vented cell in overcharge. This overvoltage signal is shown in Fig. 20.1. This significant voltage rise is present at all charge rates, and its sharpness actually improves as the cell is cycled in high-rate discharge and recharge. The corollary to this curve, which shows voltage response at constant current, is the current response at constant voltage, which may be expected to be somewhat the reverse of Fig. 20.1 and is illustrated in Fig. 20.16.

20.5.1 Constant-Potential Recharging

Constant-potential (CP) charging, the oldest of the methods still in use, is typically utilized in general aviation aircraft. Similar to an automobile battery charging system, CP charging utilizes a regulated voltage output from the aircraft DC generator, which is mechanically coupled to the engine. The voltage is typically regulated at 1.40 to 1.50 V per cell. Figure 20.16 illustrates the

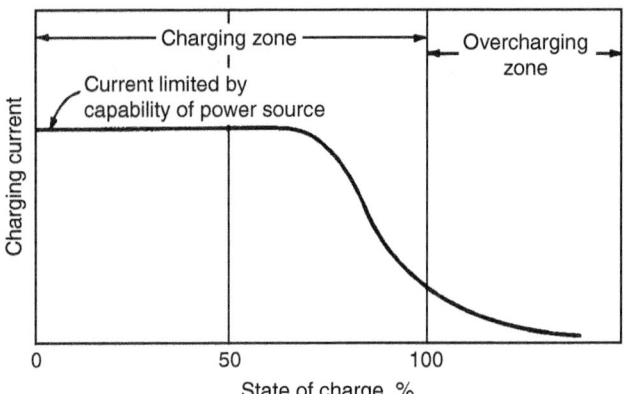

FIGURE 20.16 Constant-potential charge current.

form of the charge current as a function of the state of charge of the vented cell battery during CP recharging. Although the initial current could be quite high if limited only by the voltage response of the battery, it most frequently is limited by the capability of the source as shown. As the battery approaches full charge, however, the "back EMF" of the battery, illustrated as the charge voltage for the constant-current case in Fig. 20.1, reduces the current to that required by the battery to provide an overvoltage equal to the regulated voltage of the charging source. CP charging requires very careful consideration of the selection of charge voltage and its proper maintenance in order to achieve the balance between objectives **2** and **3**, stated in Sec. 20.5. This is particularly difficult to achieve when the battery temperature experiences significant variation, since overvoltage is dependent on battery temperature. This balance may be made essentially independent of battery temperature effects by means of temperature compensation of the CP voltage, as discussed in Sec. 20.5.4.

20.5.2 Constant-Current, Voltage-Controlled Recharging

A number of commercially available chargers based in general on constant-current charging with voltage cutoff control are utilized in modern aircraft. One of the simplest and most effective of these chargers applies an approximately *C*-rate constant-current charge to the battery and then terminates it when the battery voltage reaches a predetermined voltage cutoff (VCO) value such as 1.50 V per cell. The control also reinitiates the constant-charge current whenever the open-circuit battery voltage falls to a predetermined lower level, such as 1.36 V per cell. The net result is that the charger will recharge the capacity used during an engine start, typically 10% of the battery's total, in approximately 6 min and then cut the charger off due to the sharp rise in voltage as the cells go into overcharge, as illustrated in Fig. 20.1. Shortly thereafter, as the voltage falls below the turn-on voltage, the charge is reinitiated for a short period of time until the battery voltage again reaches the cutoff voltage. This simple on-off action continues at decreasing frequency and decreasing lengths of on-time, thereby maintaining the battery in a float condition at a completely full state-of-charge.

The battery voltage reduction, due to a discharge, inherently initiates the recharging of the battery without additional controls, thus automatically providing the recharge signal function regardless of the discharge rate or the reason. Adjustment of the cutoff and turn-on voltages as a function of battery temperature, which is described in Sec. 20.5.4, matches this mode of charging to the temperature characteristics of the vented sintered-plate nickel-cadmium battery, thereby maintaining the desired balance of objectives. Both cutoff and turn-on voltages are compensated at the same rate, thereby maintaining a constant differential between turn-on and cutoff. A diagram of the function of this simple basic charge control scheme is shown in Fig. 20.17.

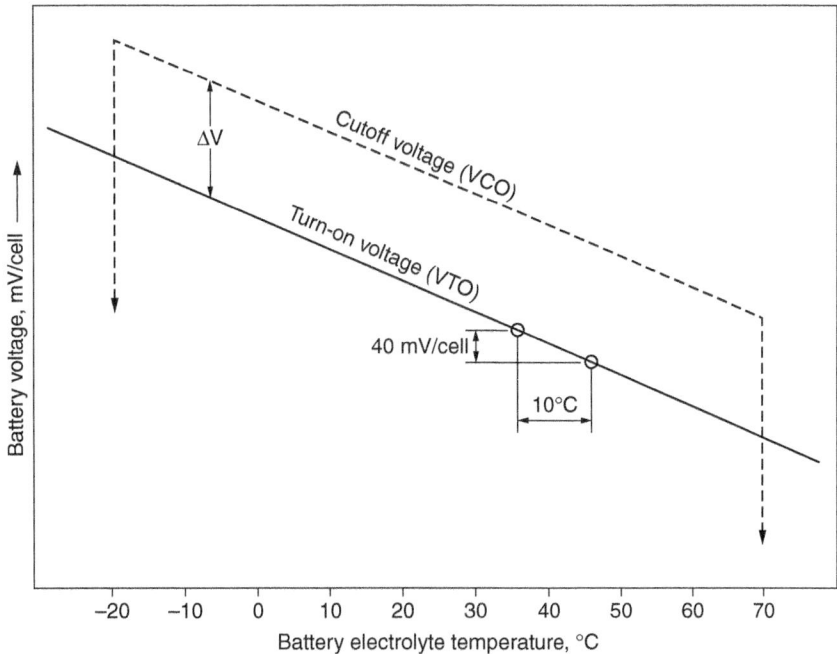

FIGURE 20.17 Charger control voltages as a function of battery temperature, *C*-rate charge (nominal values).

Several other proprietary chargers, based in part on the simple techniques outlined here, may also be found in commercial use. Many of these chargers also provide auxiliary functions such as upper and lower battery temperature charge discontinuation, detection of malfunctioning cells in the battery by detecting half-battery voltage imbalance, and signaling the user in the event of either of these conditions.

20.5.3 Other Charging Methods

The charging methods outlined in the preceding sections are those used in order to achieve fast recharging of a battery which has been discharged in normal use. Periodic maintenance of the vented sintered-plate nickel-cadmium battery, however, requires a full and complete discharge of each cell followed by a thorough recharge well into overcharge. This places both positive and negative plates in each cell of the battery in full and complete overcharge. The battery may then be returned to service with all plates of all cells in the same full charged condition, thus enabling the battery to work from the top down.

The simplest maintenance charge method, requiring the least complex equipment to ensure this fully balanced, fully overcharged condition, is the low-rate charge. This technique utilizes a constant-current, approximately $C/10$ charge-overcharge current without voltage feedback control. At this low rate, the charge may be continued safely into overcharge without compromising the physical integrity of the components of the cell. This charge current should be maintained until at least twice the rated capacity of the battery has been replaced. Since this will inherently result in water usage, water level replenishment is best performed on a fully charged battery just prior to placing the battery back into service at the conclusion of this maintenance charging routine.

Batteries in standby service can be maintained in a fully charged condition by a float or trickle charge similar to pocket-plate batteries. The float voltage for vented sintered-plate batteries is 1.36 to 1.38 V per cell.

20.5.4 Temperature Compensation of Charge Voltage

In both the constant-potential and the constant-current VCO charging methods, it has been pointed out that the selection of voltage is a compromise between the minimization of water usage and the maintenance of a high state of charge. The inherent change of overcharge voltage as a function of battery temperature increases the difficulty of this compromise significantly. This voltage effect is shown as the Tafel curves in Fig. 20.18. The relationship between the Tafel curves at various temperatures indicates a temperature coefficient of −4 mV/°C at constant-current conditions. In other words, overcharge voltage, at constant current, decreases by approximately 4 mV per cell for each 1°C rise in cell temperature.

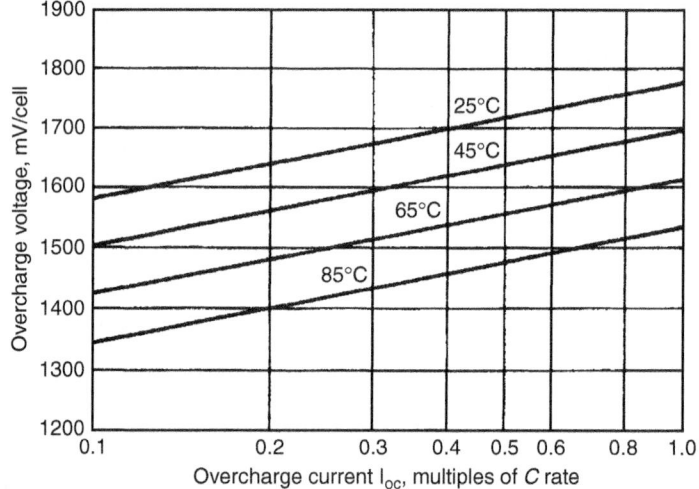

FIGURE 20.18 Overcharge voltage as a function of current and temperature (Tafel curves).

As shown by the slope of these Tafel curves, overcharge voltage is also a linear function of the logarithm of overcharge current. That slope for vented sintered-plate nickel-cadmium batteries is approximately 200 mV per cell per decade of change in overcharge current. Thus with constant-potential charging without temperature compensation, the overcharge current, and therefore both water usage and the overcharge damaging effects, will increase approximately 60% for each 10°C increase in electrolyte temperature.

A convenient technique for avoiding the detrimental effects of an increasing electrolyte temperature is to compensate the constant-potential voltage, or the constant-current cutoff/turn-on voltages, for this change in battery temperature at the rate of −4 mV/°C per cell. This may be accurately accomplished through the use of thermistors or other temperature-sensitive electric devices installed in the battery case to signal cell temperature. It is important that cell temperature and not ambient temperature be sensed by this device. There may be significant differences between the two. This function for the constant-current charging system is also shown in Fig. 20.17. The selection and use of the proper value of temperature compensation permits the battery charging system to function as though the battery were maintained at a constant temperature. Care must be exercised in the design and manufacture of these devices, since they must operate in an environment of potassium hydroxide, which is electrically conductive and also has a propensity to wet and creep on most surfaces. High-grade potting processes must therefore be used to insulate and protect all auxiliary electronic components and wiring placed inside the battery case.

20.6 MAINTENANCE PROCEDURES

20.6.1 Electrical Reconditioning

The periodic maintenance of the vented sintered-plate nickel-cadmium battery has six specific objectives:

1. Assess the timeliness of the preselected maintenance period schedule.
2. Restore the electrical performance, both capacity and power.
3. Detect and isolate cell failures and facilitate their replacement.
4. Physically clean the battery.
5. Replenish the water in the electrolyte.
6. Maintain the charging-system voltage calibration.

A relatively simple electrical procedure fulfills the first objective, namely a single discharge, initially at a relatively high rate to simulate engine start and second at a relatively low rate, approximating that of the 1 h value. This split-rate discharge of the battery as removed from the aircraft serves as a measure of its performance readiness while it was in the aircraft. The 15 s high-rate portion, at approximately the I_{mp} rate of discharge, while measuring the voltage with the capacity removed, determines the relative engine start power capability. The capacity removed during the low-rate portion of the discharge at approximately the C rate, when added to the ampere-hour capacity removed during the high rate test, determines the status of the available emergency energy capacity. The battery prior to performing this discharge should be in the same fully charged condition that would typically be encountered when it is in the normal installation. Comparison of this power delivery, and the total available capacity as removed from the aircraft, with the requirements of the application will allow the user to determine whether the maintenance schedule interval may be increased or whether it needs to be decreased.

The restoration of the electrical characteristics of the battery, known as electrical reconditioning or deep cycle maintenance, objective **2**, may be accomplished in two additional simple steps. The first is a thorough and complete discharge of each cell in the battery in order to discharge all the active material. The second step is the complete recharge of each plate of each cell into full gassing overcharge. The first step consists of a C-rate discharge to approximately 0.7 V. Once all cells have reached voltage, then resistors short cells to 0.010 V per cell or 16 h, whichever occurs first. The second step is accomplished by charging the battery with constant current at a value of one-tenth its ampere-hour rating ($C/10$) for at least 20 h. Since the capacity for some of the cells in the battery may be as much as 30 to 40% greater than the rated value, the total charge of 2.0 C Ah is sufficient to ensure that both plates of all cells reach the full overcharge required. Adjust electrolyte levels accordingly. The battery should have deep cycle maintenance performed every 1000 flight hours or 500 starts, whichever occurs first or when any abnormal operation of the battery is observed.

There are other procedures used and recommended by the manufacturers of specialized proprietary equipment to recondition cells in a shorter period of time. Periodic maintenance in a qualified service center is necessary for optimum performance of the battery. The efficacy of each should be verified, and the added expense and complexity entailed in the use of these proprietary reconditioning devices justified in specific applications. Care must always be exercised in their use, however, to avoid sustained high-rate overcharge, which may damage the gas barrier material.

Evaluation of cell-to-battery case leakage current, part of objective **3**, when the battery is first received for maintenance will determine the electrical need for cell cleanup as well as the presence of cracked or leaking cell cases. This procedure may be conveniently carried out by the simple expedient of completing a circuit from each cell terminal to the battery case with a fused ammeter. A significant amount of leakage current through the ammeter to the case from anywhere in the cell electric circuit indicates the presence of a conductive path, through potassium hydroxide, on the external surfaces of the cell cases. Such a conductive path may result from spewing of the electrolyte during overcharge, which may indicate either overfilling or the existence of a cracked or leaky cell case. Isolation of the exact cause can be accomplished by determining the leakage nodal point in the cell string by repeating the measurements after physical cleanup of the battery, objective **4**.

Detection of the failure of the gas barrier, a very important part of objective **3**, may be reliably and conveniently accomplished by extending the $C/10$ charge to 24 h. This overcharge will indicate accurately the failure of the gas barrier by either or both of two principal measurements near the end of that overcharge. First, the overcharge gassing rate is extremely sensitive to gas barrier condition and gas recombination. When measured with a simple ball flowmeter, the 24 h gas rate will be less than 80% of normal if the barrier has failed significantly in the cell. The normal value is 11 ml/min for each ampere of the $C/10$ overcharge rate. The second indicator of gas barrier failure will be a 24 h overcharge voltage of less than 1.5 V if the cell is being charged at a 23°C ambient. Some downward adjustment of this voltage criterion may be made at the rate of 4 mV/°C if the battery is being charged at a higher ambient temperature.

20.6.2 Mechanical Maintenance

The replenishment of water in the electrolyte to return the electrolyte to the level recommended, objective **5**, is the most important routine mechanical procedure employed during battery reconditioning. It is best accomplished near the end of the 24 hours of the $C/10$ rate charge by replenishing with deionized water until the electrolyte reaches the recommended level for a battery in overcharge. A record of the amount of water usage in each cell should be maintained and compared with the battery manufacturer's statement of reserve electrolyte level in each cell. If the total water usage between maintenance fillings, after deducting the amount used during the maintenance procedure, exceeds the reserve available in that cell design, the maintenance interval must be shortened to prevent plate dry-out during use and resultant cell failure. Note that the 24 h $C/10$ reconditioning procedure will itself use approximately 0.4 ml of water for each ampere-hour of rated capacity during the 24 h reconditioning period. For example, 12 mL of water would be used during the reconditioning period for a 30 Ah rated cell, and this must be subtracted from the amount added to determine the amount actually used in service. It should also be noted that a cell with a damaged gas barrier may use less water.

That point in the maintenance procedure, following the thorough short-circuiting of each cell, may be utilized to perform physical maintenance. Cells may only be replaced while in the discharged state. Cleanup generally consists of a thorough rinsing with clear water followed by warm-air drying of the battery. This will dissolve and remove any accumulation of potassium hydroxide and carbonates from the outside of the cell jars. Vent caps should also be washed in warm deionized water, warm water forced through the vent, and then dried with warm air. This is safely accomplished only with the cells in the completely discharged state. Replacement of any cells not found defective until the conclusion of the $C/10$ overcharge requires discharging the cells a second time.

Other typical hardware problems include:

- Loose terminal nuts—indicated by burns or arcing on intercell links.
- Terminal seal failure—various heavy deposits around cell terminal; remove all hardware and the O-ring.
- Vent failure—various heavy deposits on or around the vent valve; the valve may have been installed improperly or the vent sleeve or O-ring has failed.
- Also inspect the vent sleeve to insure it is not torn or broken. Power delivery of the battery may be enhanced by removing intercell link hardware, buffing contact surfaces, then replacing and retorquing all connectors.

20.6.3 System Inspection Criteria

Reinstallation of the reconditioned and fully charged battery into the aircraft system presents the opportunity for performing the system voltage calibration check in fulfillment of objective **6**. The only measurement required for this on a CP charging system is to record the value of the battery voltage after reactivating the system and following stabilization of the voltage. This stable value is the regulated float voltage to which the battery will be subjected during extended overcharge in use. This voltage measurement should be made at the battery with the engine running at a sufficiently high speed to produce a representative and stable value.

The battery voltage measurement on a constant-current VCO systems should be made just as the system reaches cutoff voltage. The regulated voltage on either of the two systems should then be adjusted to the manufacturer's recommended value if necessary. These adjustments must consider the effects of any automatic temperature compensation of voltage present in the system.

20.7 RELIABILITY

20.7.1 Failure Modes

The sintered plate construction is very robust and capable of operating in both high and low temperature extremes. The cell is capable of withstanding substantial abuse and still performing as intended. It can be discharged into reversal and given a substantial amount of overcharge, as long as the temperature is controlled. The gas barrier that prevents recombination of the oxygen on the cadmium electrode aids control of the temperature during charge. In the past, cellophane was mainly used as the gas barrier. It had a tendency to hydrolyze in the electrolyte and eventually decompose to carbonate and derivatives of the cellophane structure. In recent times, the cellophane has been replaced with Celgard® 3400 or other similar materials.[10] These materials are polyethylene and polypropylene based, and do not degrade in the 30% KOH electrolyte. Should the cell's gas barrier fail, continuing in operation for enough cycles will result in the battery losing capacity and maximum power capability. Continued temperature increase in the cell can result in fusing, or melting, of the nylon separator and result in an internal short circuit of the cell.

Although the cellophane replacement used for the gas barrier alleviates the above failure mechanism, it introduces another problem if the cell is not manufactured and maintained properly. Without proper additives in the electrolyte, which find their way into the cadmium electrode during cycling, the cell can lose capacity in the negative electrode. The role of supplying an oxidized cellophane expander for the cadmium electrode needs to be replaced by cellulose derivatives and other additives to maintain the negative capacity.[11]

Several other failure modes which account for a small portion of cell and battery failures include the following:

- Internally short-circuited cells can result from cut-through of the electrical separator by burrs and other plate irregularities, aggravated by cell interplate pressures and vibration.
- Cracked and leaking cell cases may result from abusive handling of the cells during cell replacement procedures and maintenance, or from defective manufacturing or sealing procedures.
- Burned terminal contacts may result from faulty cleaning and buffing procedures during maintenance, insufficient link assembly torquing, terminal screw failure, or conductive articles being dropped on the internal connectors of a charged battery.

20.7.2 Memory Effect

In addition to these permanent failures, there is a reversible effect that may result in a gradual reduction of both power and capacity with cycling. This effect, sometimes referred to as "memory effect," "fading," or "voltage depression," results from charging following repetitive shallow discharges where some portion of the active materials in the cell is not used or discharged, such as in a typical engine-start use. This effect is noticed when the previously undischarged material is eventually discharged. The terminal voltage during the latter part of that full discharge may be lower by approximately 120 mV (hence, "voltage depression"). The total capacity is not reduced, however, if the discharge is continued to the lower voltages, as, for example, to the "knee" of the curve.

This effect is completely reversible by a maintenance cycle consisting of a thorough discharge followed by a full and complete charge-overcharge as described in Sec. 20.6.1.

20.7.3 Factors Influencing Gas Barrier Failure

Gas barrier failure is generally acknowledged to be caused or aggravated by excessive overcharge current, excessive overcharge temperatures, and discharge at high rates with low electrolyte levels. Gas barrier failure may not be detectable during reconditioning with other than the low-rate constant-current procedures. Barrier failures may actually occur during poorly structured maintenance and then manifest themselves at a later time after reinstallation in the aircraft. This possibility emphasizes the importance of an accurate assessment of the condition of the gas barrier at the end of the reconditioning period just prior to reinstallation. This assessment is accurately made by the measurement of overcharge gas flow following the extended $C/10$ overcharge, as described in Sec. 20.6.1.

One indication of the significant importance of the two factors of (1) temperature compensation of charger voltage and (2) effective maintenance practices is the existence of large-scale field data that document real-time failure differences of up to 100:1. These life differences exist between well-maintained batteries in temperature-compensated systems on the one hand and identical battery designs in uncompensated CP systems with frequent and poorly managed maintenance procedures on the other. Recent improvements in gas barrier materials have significantly reduced these failures.

20.7.4 Thermal Runaway

The loss of the gas barrier in one or more cells of a vented nickel-cadmium battery can lead to thermal runaway. Loss of this function allows oxygen, generated in overcharge, to reach the negative plate and recombine on it. This generates heat. The temperature increase that follows causes the internal cell voltage to decrease. Charge current then increases exponentially to increase cell voltage to match the charger voltage.

Thermal runaway occurs with the use of a voltage-regulated (CP) charge source on a battery containing cells with a failed gas barrier. Thermal runaway begins when the failed cells approach overcharge following recharge. The (over)charge current may reach a minimum and then gradually increase. Voltage inequities may exist at this point unless all cells are experiencing similar recombination (gas barrier damage). Oxygen recombination heats, and begins to increase the temperature of the failed cell or cells and thus their neighboring cells unless the battery is effectively air cooled. The resulting temperature increase, however, proceeds slowly due to the large thermal mass involved. It may take 2 to 4 hours of (near) consecutive overcharging for a cell to reach boiling temperature.

If the boiling phase continues long enough, or is repeated, and the failed cell becomes dry, large inequities in cell voltage will appear. The voltage across the cell that has boiled dry will increase, thereby decreasing the charge current and the voltage across the cells that are still wet with electrolyte. The next event probably will be internal short-circuiting of the dried-out cell due to very high temperatures and voltage at the last remaining damp spots with consequent burning of the electrical separator insulation. The (over)charge current then increases sharply due to cell loss, and the process repeats itself with the next cell to go dry.

Because of extensive heating and boil-away times, thermal runaway may go undetected for many flight hours following the onset of gas barrier failure if the use of the system is not consecutive or continuous. This can confuse the perceived connection between the cause of the gas barrier damage and the resultant thermal runaway.

20.7.5 Potential Hazards

Potential hazards that may be present during the use and maintenance of the vented sintered-plate nickel-cadmium battery fall into five general categories, as described in this section.

Gas Fire and/or Explosion. Since all functional vented batteries generate a stoichiometric mixture of hydrogen and oxygen gases during overcharge and expel them normally from the cell into the battery container, a potential for explosion of these gases is always present. Two conditions are necessary for such an explosion, however, and both are recognized and accounted for in the design of a typical system. The first condition is the accumulation of a sufficient quantity of this gas mixture. This condition is minimized in all system designs by the incorporation of adequate battery case ventilation. Some designs rely on supplying a modest quantity of air to purging tubes on the battery from overboard vents in the aircraft. Others incorporate natural convection of the gases from a louvered battery case into a ventilated compartment. Air-cooled designs inherently accomplish the required ventilation due to the large volume of air used.

Unusual circumstances, however, may defeat any of these gas-purging techniques. It should also be remembered that batteries generate a significant amount of explosive gas during the maintenance procedure, and therefore maintenance should always be performed in a well-ventilated shop.

The second condition necessary for explosion of the gases is the presence of a source of ignition. Although normally there are no ignition sources inside the battery case, several abnormal possibilities do exist. One is the internal short-circuiting of a relatively dry cell in overcharge, resulting in an explosion inside the cell with a subsequent ejection of flames into the battery case. A second and more likely source of ignition may exist at an improperly maintained cell terminal due to the high temperatures generated during high-rate discharge. A third source of ignition may occur at the site of stray leakage currents.

Although the coincidence of both a sufficient amount of gas accumulation and a source of ignition is quite rare, it can happen and has happened. Because of this possibility, many batteries are also designed to be physically capable of managing a hydrogen or oxygen explosion and containing the effects entirely within the battery case. Typically these batteries will also be electrically functional for at least one C-rate discharge following such an explosion.

Arcing and Burning. This potential hazard concerns excessive leakage currents through electrolyte paths outside the cells. Such currents can occur either between cells that are physically adjacent but with a wide voltage separation in the cell circuit or, more probably, from a cell to a grounded metallic case. Arcing is more likely to occur, however, in the circuit of an inappropriately protected auxiliary device located inside the battery case in the environment containing potassium hydroxide. Some examples of these devices are battery heaters, thermal detectors, and voltage sensors. The proper design of these auxiliary appliances must recognize the conductivity of KOH and the ability of that electrolyte to creep along wires and even into mechanically "sealed" insulation. Devices of this type should be tested with high dielectric voltages while submerged in a water-detergent mix before they are installed in the battery case environment.

The result of a sustained leakage current through relatively localized KOH conducting paths may be the ignition of the explosive environment by arcing, as discussed previously. The result might also be the carbonization of adjacent insulating materials and the subsequent burning of those materials within the battery case.

Electrical Power. One of the essential functional capabilities required of the vented nickel-cadmium battery is the ability to deliver high-power rates for engine starting. This very capability, however, presents a potential risk in the form of hot spots on improperly torqued cell terminals during high-rate discharge. It is also a potential hazard to the unwary maintenance technician operating with metallic tools or other objects, such as jewelry, in a careless manner in the vicinity of charged batteries. Since the short-circuit current of these cells (batteries) may exceed 1000 to 4000 A, it should be obvious that the exposed conductors of a charged battery should be treated with respectful caution. Very severe burns may occur if, for example, a ring should accidentally make contact between two adjacent cell terminals. Although one of the most obvious, this hazard is one of the most frequently encountered. Insulated cell hardware does provide a partial solution, however, caution and respect for the available power must always be exercised.

Corrosive KOH. Because of the corrosive nature of the KOH used as an electrolyte, all material employed in the construction of the battery and its accessory appliances must be KOH immune. Materials such as nylon, polypropylene, nickel-plated steel, steel, and stainless steel are therefore used. The potential hazard of KOH corrosiveness, however, is primarily encountered during maintenance. The use of safety glasses and safety face shields should be mandatory while performing maintenance on these batteries. A very small amount of KOH in the eye, for example, without prompt, continued, and adequate flushing followed by medical treatment can result in the loss of eyesight. KOH is also corrosive to the skin, and the affected area should be thoroughly washed and rinsed with water, thereby minimizing the detrimental effect.

Electric Shock. Most vented sintered-plate nickel-cadmium batteries are arrayed in groups of 10 to 30 cells, presenting maximum voltages of 15 to 45 V. There are applications, however, in which batteries are connected electrically in series strings of 90 to 200 cells or more. It should be apparent that the voltages presented by this number of cells in series may be lethal to anyone exposed to them. Personnel should also be cautioned, because of the high probability of electrolyte being present between the cell circuit and the conductive external surface of the battery, to exercise care by disconnecting series-connected batteries prior to personal exposure. Significant shock currents may be carried by relatively small amounts of KOH.

20.8 CELL AND BATTERY DESIGNS

20.8.1 Typical Vented Sintered-Plate Nickel-Cadmium Cells

A listing of several typical vented sintered-plate nickel-cadmium cells is given in Table 20.3. The 10, 20, and 36 Ah sizes are those typically employed in aircraft batteries. Other cells are available in sizes up to about 350 Ah. The larger cells are generally constructed in steel containers rather than the plastic containers now used for the aircraft-size cells.

TABLE 20.3 Typical Vented Sintered-Plate Nickel-Cadmium Cell Properties

Rated capacity (Ah)	Height (cm)	Width (cm)	Thickness (cm)	Weight (grams)
1.5	10.16	2.92	1.70	86
2	8.74	3.81	1.83	95
5.5	10.31	5.51	2.39	236
5.5	10.36	5.51	2.39	272
7	18.85	6.65	1.29	299
6	11.60	5.89	2.69	354
13	11.93	7.95	3.02	486
10	14.48	5.89	2.69	445
12	14.38	5.86	2.69	422
20	17.42	7.95	3.53	1067
28	17.27	7.95	3.53	1149
23	20.57	8.08	2.72	903
36	17.42	7.95	5.08	1562
40	23.31	7.95	3.53	1453
42	23.31	7.95	3.53	1453
100	24.48	12.7	3.83	2860

Source: Courtesy of EaglePicher Technologies, LLC.

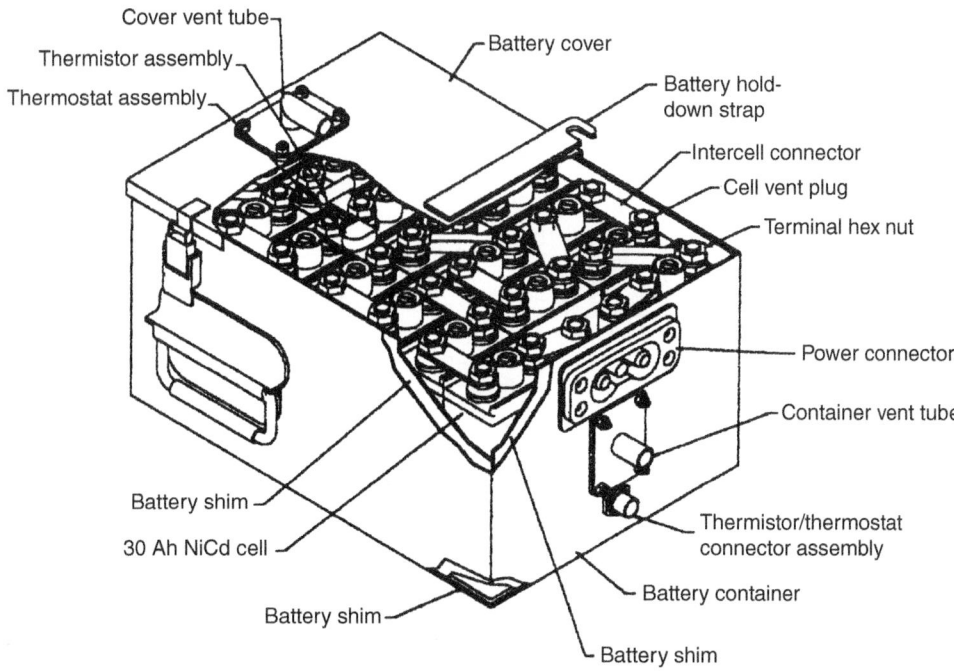

FIGURE 20.19 Vented sintered-plate nickel-cadmium battery. (*Courtesy of EaglePicher Technologies, LLC.*)

20.8.2 Typical Battery Designs

A typical arrangement of vented sintered-plate nickel-cadmium cells into a battery configuration is the conventional aircraft battery. An example of this use is shown in Fig. 20.19 and in detail in Fig. 20.20. This arrangement generally consists of a completely enclosing battery case and cover made of either stainless steel or steel with a KOH-resistant finish of epoxy or paint. The cover is typically secured with over-center-type latches. The battery case is provided with gas-purging vents or with freely convective gas-diffusion openings for dilution. The cells are encased in nylon-molded cell cases with terminals extending through a nylon cover sealed to the case. The cells are electrically connected in series with nickel-plated copper links from cell terminal to cell terminal and from the first and last cell to the battery termination and disconnect device. This battery termination extends through the battery case wall and is present on the outside surface of the battery case as a recessed double-male, polarized, high-current receptacle. Functional requirements of aircraft batteries are specified in SAE standard AS 8033A.

20.8.3 Air Cooling/Heating

The major battery manufacturers produce some battery designs with provision for forced-air cooling. These designs generally take the form of plenum spaces below and above the cells, with cooling passages between the cells connecting the two plenums. The construction provides a means for the external connection of a high-volume, low-pressure air source. Supplying 23°C air will not only effectively cool an overheating battery, it will also rapidly warm a cold battery. The heat transfer coefficient from the battery core is improved up to 10 times by this technique. The thermal time constant for a battery with this feature may be as short as 10 to 20% of that of a standard non-air-cooled battery.

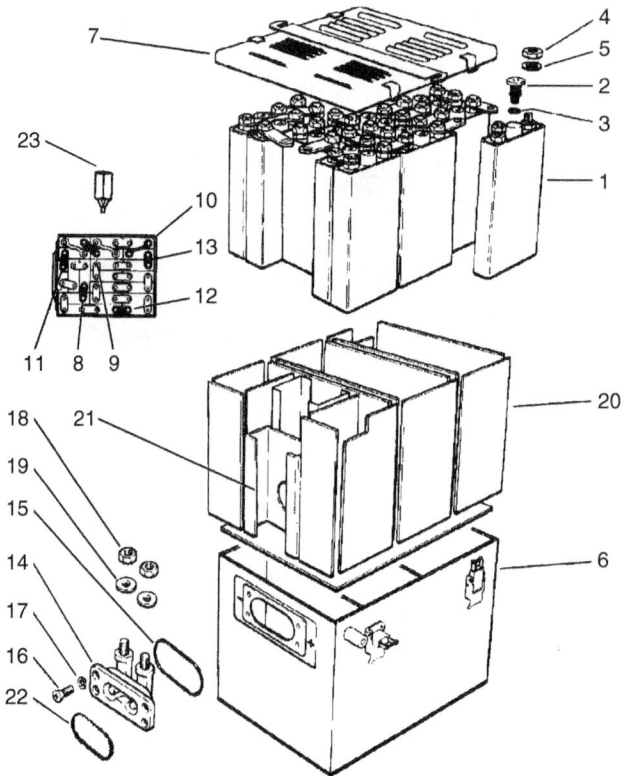

Ref. no.	Description	Qty. req.	Ref. no.	Description	Qty. req.
1	Cell assembly	20	13	Link, terminal–flat	7
2	Valve, vent	1	14	Receptacle, battery connector	1
3	O-ring	1	15	Gasket, connector receptacle	1
4	Nut, terminal	2	16	Screw, connector receptacle mtg.	4
5	Washer, terminal	2	17	Washer, connector receptacle mtg.	4
6	Case, battery	1	18	Nut, terminal adapter	2
7	Cover assembly battery case	1	19	Washer, terminal adapter	2
8	Link, terminal–flat	2	20	Liner–spacer kit	1
9	Link, terminal–flat	1	21	Bracket, spacer	1
10	Link, terminal–curved	3	22	Spring, shorting	1
11	Link, terminal–flat	1	23	Wrench, vent	1
12	Link, terminal–flat	7			

FIGURE 20.20 Typical assembly of vented sintered-plate nickel-cadmium aircraft battery.

20.8.4 Temperature Sensors

Sensors may be provided inside the battery case that sense either typical or average cell temperatures. They are equipped with a provision for external electrical connections. These devices may be of the on-off type, such as thermostats, or they may be the continuous type, such as thermistor assemblies. Continuous types have the capability of providing continuous modulation of, for example, the regulated charging voltage of a CP charging source or of the cutoff/turn-on voltages of a constant-current VCO charging system.

20.8.5 Battery Cases

Although the corrosive effect of KOH on bare steel is minimal and cosmetic in effect, additional KOH resistance in the case material is desirable. In addition to stainless steel and steel with a protective finish, some special applications use KOH-resistant plastic materials. It must be emphasized, however, that battery cases may require withstanding significant rough treatment in shock and vibration without losing their KOH-containment capability.

20.8.6 Battery Electrical Termination

Aircraft batteries are normally terminated on the front case surface by a connector of the type shown in Fig. 20.19. Special applications, however, may utilize direct cable connection to the first and last cell terminals, as well as various other special configurations capable of handling the high-current rates available in the event of a short circuit of the external battery circuit.

20.8.7 Battery Heaters

Alternative to the airflow heating discussed in Sec. 20.8.3, heater blankets are sometimes employed on the inside or outside of the battery case. These blankets may be energized with any available electric energy source. Primarily, this source will be either the DC bus of the same voltage as the battery or an aircraft AC bus of higher potential. Heaters may also be energized from an auxiliary ground supply. Heater blankets have the inherent poor thermal time constant of non-air-cooled batteries.

20.8.8 Extensions of Vented Sintered-Plate Nickel-Cadmium Designs

To avoid costly maintenance procedures associated with vented nickel-cadmium battery designs and to improve general battery reliability, the battery industry, in conjunction with the U.S. government, has incorporated lessons learned from vented sintered-plate nickel-cadmium batteries. There are two alternate versions of the standard nickel-cadmium battery that are being used in both commercial and military aircraft service, the vented low-maintenance nickel-cadmium battery and the maintenance-free nickel-cadmium aircraft battery.

The vented low-maintenance nickel-cadmium battery was derived from the standard vented sintered-plate nickel-cadmium cell design with the exception of a few internal components. As in the standard configuration, the positive electrodes are sintered nickel and chemically impregnated. The standard negative electrode has been replaced with a pasted cadmium oxide electrode, and the electrolyte composition has been modified to increase ionic condition. This modified cell configuration allows for an increase in cell voltage and, on the aircraft bus, reduces the degree of cell overcharge the battery is exposed to and therefore reduces the cell gassing. This results in significantly lower water consumption and subsequently leads to longer maintenance intervals for this battery. The maintenance period can be extended in application from 2 to 3 months to 12 months or longer with the same battery performance and reliability.

The maintenance-free nickel-cadmium aircraft battery uses the same geometric cell shape as the vented nickel-cadmium design, but allows gasses to recombine inside the cell. This cell design uses all sintered electrodes but a modified electrolyte composition and has a number of features found in aerospace cell technology. It is not strictly a sealed design since the cell will vent at a high pressure, then reseal itself well above the ambient pressure. To avoid excessive overcharge and associated thermal runaway, the cell is charged and controlled by its own integrated charger and associated electronics. This battery design has been supplied to several military aircraft programs by EaglePicher Technologies, LLC.[12]

Additional information on industrial and aerospace nickel-cadmium batteries is found in Chap. 19.

REFERENCES

1. J. McBreen, *The Nickel Oxide Electrode*, Modern Aspects of Electrochemistry, No. 21, Ralph E. White, J. O'M. Bockris, and B. E. Conway, Eds., Plenum Press, 1990, New York, p. 29.
2. D. F. Pickett and J. T. Maloy, *J. Electrochem. Soc.*, **12:**1026 (1978).
3. S. U. Falk and A. J. Salkind, *Alkaline Storage Batteries*, Wiley, New York, 1969.
4. A. Fleischer, *J. Electrochem. Soc.* **94:**289 (1948).
5. G. Halpert, *J. Power Sources* **12:**117 (1984).
6. D. F. Pickett, in *Proceedings of the Symposium on Porous Electrodes, Theory and Practice*, H. C. Maru, T. Katan, and M. G. Klein, Eds., The Electrochemical Society, Pennington, NJ, 1982, p. 12.
7. M. B. Pell and R. W. Blossom, U.S. Patent 3,507,699 (1970).
8. R. L. Beauchamp, U.S. Patent 3,573,101 (1971); U.S. Patent 3,653,967 (1972).
9. D. F. Pickett, U.S. Patent 3,827,911 (1974); U.S. Patent 3,873,368 (1975).
10. Mil-B-81757, Performance Specification, Batteries and Cells, Storage, Nickel Cadmium, Aircraft General Specification, Crane Division, NSWC, July 1, 1984.
11. J. J. Lander, personal communication.
12. T. M. Kulin, *33rd Intersociety Engineering Conference on Energy Conversion* IECEC-98-145, Colorado Springs, CO, August 2–6, 1998.

CHAPTER 21
PORTABLE SEALED NICKEL-CADMIUM BATTERIES

Joseph A. Carcone

21.1 GENERAL CHARACTERISTICS

Sealed nickel-cadmium batteries incorporate specific battery design features to prevent a buildup of pressure in the battery caused by gassing during overcharge. As a result, batteries can be sealed and require no servicing or maintenance other than recharging. These unique characteristics for a secondary battery have created wide acceptance for use in a variety of applications, ranging from lightweight portable power (photography, toys, housewares) to high-rate, high-capacity power (electronic devices such as phones, computers, camcorders, power tools) and standby power (emergency lighting, alarm, memory backup). Some nickel-cadmium batteries are now incorporating smart battery control circuitry to give state-of-charge indication and to control overcharge and overdischarge.

The major advantages and disadvantages of the sealed nickel-cadmium battery are summarized in Table 21.1. The important characteristics are described in this section.

Maintenance-free operation: The batteries are sealed, contain no free electrolyte, and require no servicing or maintenance other than recharging.

High-rate charging: Sealed nickel-cadmium batteries are capable of recharge at high rates within 30 min under controlled conditions. Many batteries can be charged in 3 to 5 h without special controls, and all can be recharged within 14 h.

High-rate discharge: Low internal resistance and constant-discharge voltage make the nickel-cadmium battery especially suited for high-rate discharge or pulse-current applications.

Wide temperature range: Sealed nickel-cadmium batteries can operate over the range from about −20 to +70°C and are particularly noted for their low-temperature performance.

Long service life: Over 500 cycles of discharge or up to 5 to 7 years of standby power are common for sealed nickel-cadmium batteries.

21.2 CHEMISTRY

The active materials of the sealed nickel-cadmium battery are the same as for other types of nickel-cadmium batteries, namely, in the charged state, cadmium for the negative electrode, nickel oxyhydroxide for the positive, and a solution of potassium hydroxide for the electrolyte. In the discharged

21.2 SECONDARY BATTERIES

TABLE 21.1 Major Advantages and Disadvantages of Sealed Nickel-Cadmium Batteries

Advantages	Disadvantages
Batteries are sealed; no maintenance required	Voltage depression or memory effect in certain applications[†]
Long cycle life	
Good low-temperature and high-rate performance capability	Higher cost than sealed lead-acid battery
	Poor charge retention
Long shelf life in any state of charge	Sealed lead-acid battery better at high temperature and float service[*]
Rapid recharge capability	
Safety vent system	Environmental concern with the use of cadmium
Low internal impedance	Lower capacity than other competitive batteries

[*]High-temperature nickel-cadmium batteries are available (see Sec. 21.6.3).
[†]See Sec. 21.4.11.

state, nickel hydroxide is the active material of the positive electrode, and cadmium hydroxide that of the negative.

During charge, nickel hydroxide, $Ni(OH)_2$, is converted to a higher-valence oxide

$$Ni(OH)_2 + OH^- \rightarrow NiOOH + HO + e$$

At the negative electrode, cadmium hydroxide, $Cd(OH)_2$, is reduced to cadmium

$$Cd(OH)_2 + 2e \rightarrow Cd + 2OH$$

The overall discharge/charge reaction is

$$Cd + 2NiOOH + 2H_2O \underset{charge}{\overset{discharge}{\rightleftarrows}} Cd(OH)_2 + 2Ni(OH)_2$$

During operation, the active materials undergo changes in their oxidation states but little change in their physical states. Similarly, there is little if any change in the electrolyte concentration. The active materials of both electrodes, in both charged and discharged states, are relatively insoluble in the alkaline electrolyte, remain as solids, and do not dissolve while undergoing changes in their oxidation states. Because of these and other properties, nickel-cadmium batteries are characterized by long life in both cyclic and standby operation and by a relatively flat voltage profile over a wide range of discharge currents.

The operation of the sealed battery is based on the use of a negative electrode having a higher effective capacity than the positive. During charge, the positive plate reaches full charge before the negative and begins to evolve oxygen. The oxygen migrates to the negative electrode, where it reacts with and oxidizes or discharges the cadmium to produce cadmium hydroxide

$$Cd + \frac{1}{2}O_2 + H_2O \rightarrow Cd(OH)_2$$

A separator permeable to oxygen is used so that oxygen can pass through the separator to the negative electrode. Also, a limited amount of electrolyte is used (starved electrolyte system) as this facilitates the transfer of oxygen. This process is illustrated in Fig. 21.1.

At a steady state, the recombination reaction rate during overcharge must be no lower than the rate of oxygen generation to prevent buildup of pressure. The internal pressure is related to charge current, the reactivity of the negative electrode, the electrolyte level, and the temperature. Solid

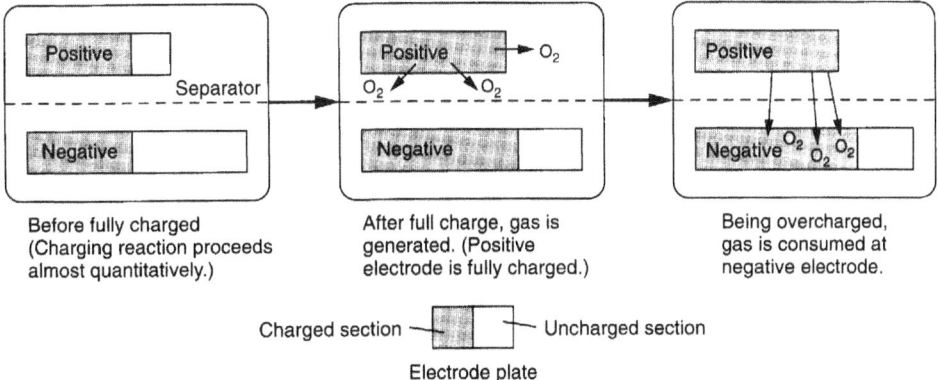

FIGURE 21.1 Oxygen recombination process. (*Courtesy of Sanyo Mobile Energy Co.*)

cadmium, gaseous oxygen, and liquid water must coexist in mutual contact for the recombination reaction to occur. If, for example, the electrolyte level is too high (the electrodes are in a flooded state), the oxygen is prevented from contacting the electrode, and the reaction rate at a given temperature and pressure is substantially lowered.

A safety venting mechanism is used in the battery design to prevent rupture in case of excessive pressure buildup due to a malfunction, high charge rate, or abuse.

21.3 CONSTRUCTION

Sealed nickel-cadmium cells and batteries are available in several constructions. The most common types are the cylindrical shaped batteries (see Table 21.3). Smaller button batteries and rectangular batteries are also manufactured.

21.3.1 Cylindrical Batteries

The cylindrical battery is the most widely used type because the cylindrical design lends itself readily to mass production and because excellent mechanical and electrical characteristics are achieved with this design. Figure 21.2 shows a cross section of the cylindrical battery.

The positive electrode is a highly porous sintered, foam, or fibered nickel structure into which the active material is introduced by embedding or impregnation with a molten nickel salt, followed by the precipitation of nickel hydroxide by immersion or electrochemical deposition in an alkaline solution. The negative electrodes are made by several methods: using a sintered-nickel substrate similar to the positive, by pasting or pressing the negative cadmium active material onto a substrate, or by a continuous electrodeposition process.

After processing, the positive and negative electrodes are cut to size and wound together in a jelly-roll fashion with a separator material between them. The separator material, usually unwoven nylon or polypropylene, is highly absorbent to the potassium hydroxide electrolyte and permeable to oxygen. The roll is inserted into a rugged nickel-plated steel can, and the electrical connections are made. The negative electrode is welded or press connected to the can and the positive is usually welded to the top cover. The very small amount of the electrolyte, enough for efficient operation, is absorbed by the separator. There is no free liquid electrolyte. The cover assembly incorporates

21.4 SECONDARY BATTERIES

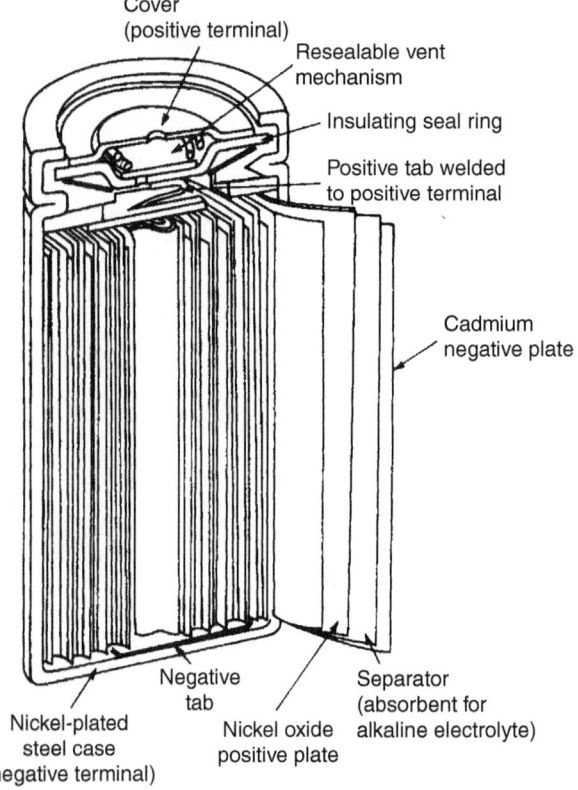

FIGURE 21.2 Construction of sealed nickel-cadmium cylindrical battery.

a fail-safe one-time or resealable vent mechanism to prevent rupture in case of excessive pressure buildup, which could result from extreme overcharge or discharge rates.

21.3.2 Button Batteries

Nickel-cadmium button cells and batteries usually have electrodes made from "pressed" plates. The active materials are compressed in molds into disks or plates, and the electrodes are assembled in a sandwich configuration, as shown in Fig. 21.3.

In some cases, the electrodes are backed with expanded metal or screen to enhance electrical conductivity and mechanical strength. The button battery does not have a resealable fail-safe device, but its construction allows the battery to expand, either breaking the electrical continuity or opening the seal to relieve excess pressure caused by an abnormal circumstance. Button batteries are very suitable for low-current, low-overcharge-rate applications.

21.3.3 Rectangular Batteries

The flat or rectangular batteries are designed to meet the needs of lightweight and compact equipments. The rectangular shape permits more efficient battery assembly, eliminating the voids that

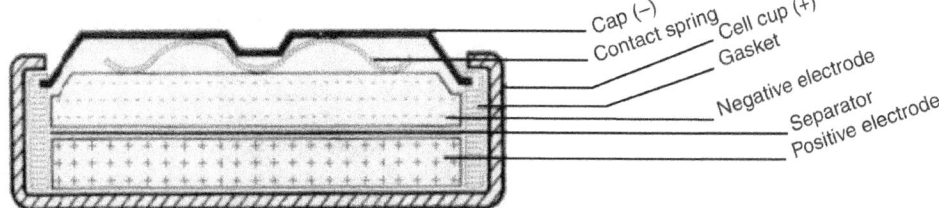

FIGURE 21.3 Nickel-cadmium battery, button configuration.

occur with the assembly or cylindrical batteries. The volumetric energy density of batteries can be increased by a factor of about 20%.

Figure 21.4 shows the structure of the rectangular battery. The plates are manufactured as described in Sec. 21.3.1, but the finished electrodes are cut to predetermined dimensions and placed in the metal casing. These are then hermetically sealed in place to the cover plate. All sides of the casing and cover-plate assembly are laser-welded together to prevent electrolyte leakage. A resealable safety venting system is built into these batteries similar to the ones used in the cylindrical designs. These batteries are no longer available from Sanyo Mobile Energy Co.

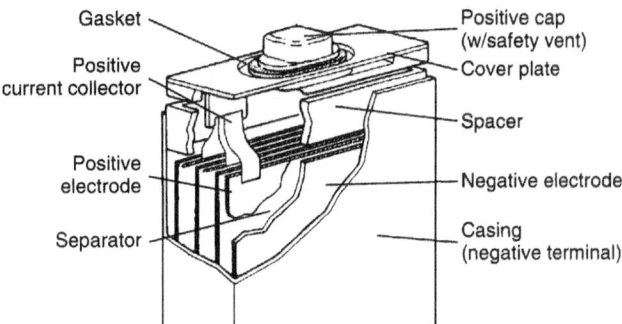

FIGURE 21.4 Construction of sealed rectangular battery. (*Courtesy of Sanyo Mobile Energy Co.*)

21.3.4 Rectangular Batteries

The rectangular battery is housed in a nickel-plated steel can using a construction similar to the vented battery, but incorporating the features needed for sealed-battery operation. This construction is particularly suited for high discharge rates because of the large electrode area. Figure 21.5 is an illustration of a sealed rectangular battery, in this case using fiber-structured electrodes (see Sec. 19.7).

21.6 SECONDARY BATTERIES

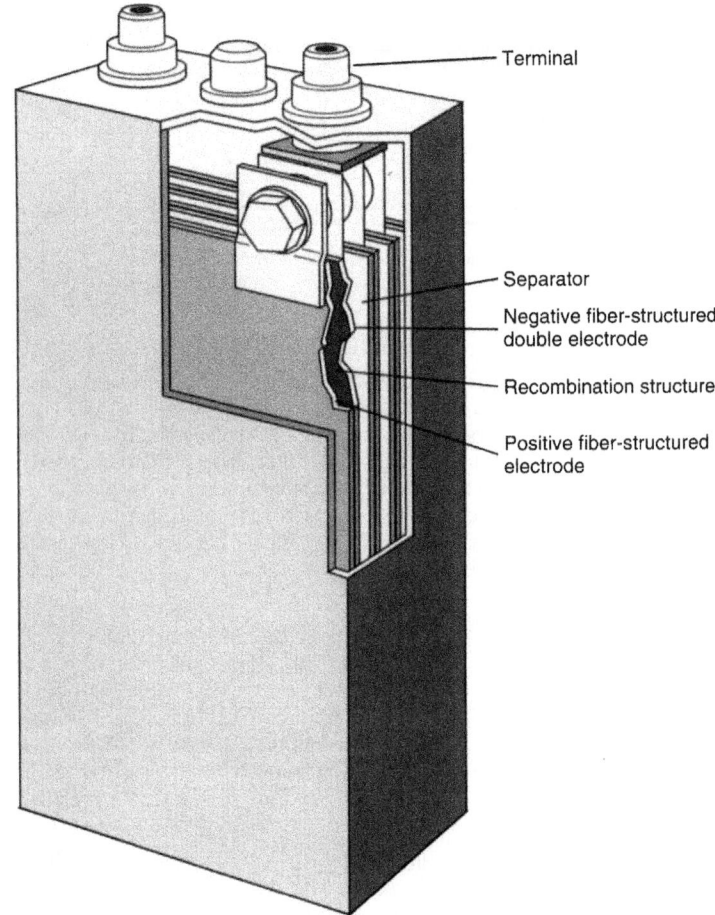

FIGURE 21.5 Sealed rectangular nickel-cadmium battery using fiber-structured electrodes. (*Courtesy of Hoppecke Batteries.*)

21.4 PERFORMANCE CHARACTERISTICS

21.4.1 General Characteristics

A typical charge-discharge cycle for the sealed nickel-cadmium cylindrical battery, at 20°C, is shown in Fig. 21.6. The voltage increases slowly but steadily at the $C/10$ charge rate to a steady-state condition, decays slightly during the 1 h rest, and is relatively flat during the 1 h discharge to 1.0 V. The voltage recovers rapidly over the next hour, while at rest, to near 1.2 V.

21.4.2 Discharge Characteristics

Typical discharge curves for the cylindrical battery at 20°C at various discharge loads are shown in Fig. 21.7. The flat voltage profile, after the initial voltage drop, is characteristic.

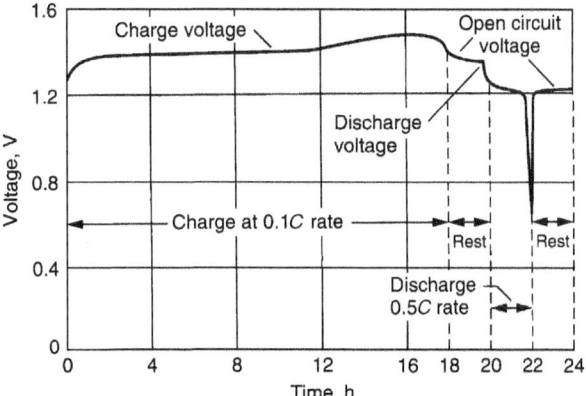

FIGURE 21.6 Voltage profile of nickel-cadmium battery in a typical charge/discharge cycle.

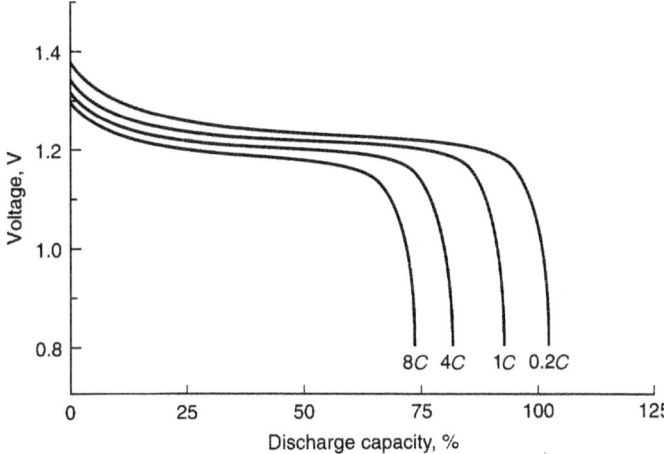

FIGURE 21.7 Constant-current discharge curves for sealed nickel-cadmium batteries at 20°C, charge 0.1C, 16 h. (*Courtesy of Sanyo Mobile Energy Co.*)

The capacity that can be obtained from a battery is dependent on the rate of discharge, the voltage at which discharge is terminated, the discharge temperature, and the previous history of the battery. Figure 21.8 shows the percentage of rated capacity delivered during discharges at various rates and temperatures. The midpoint voltage during discharge decreases as the discharge rate increases (see Fig. 21.11). If the battery were allowed to discharge to a lower cutoff voltage, a greater percentage of the $C/5$ rate capacity will be obtained. However, batteries should not be discharged to below the specified cutoff voltage as the battery or individual cells may be damaged (see Sec. 21.4.6).

21.4.3 Effect of Temperature

The sealed nickel-cadmium battery is capable of good performance over a wide temperature range. Best operation is between −20 and +30°C, although usable performance can be obtained beyond this range. The low-temperature performance, particularly at high rates, is generally better than that of

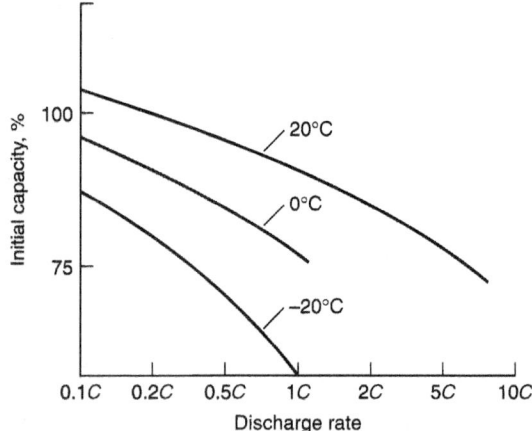

FIGURE 21.8 Percent of $C/5$-rate capacity vs. discharge rate to 1.0 V cutoff for typical sealed nickel-cadmium battery.

the lead-acid battery but usually inferior to that of the vented sintered-plate battery. The reduction in performance at low temperatures is due to an increase in the internal resistance. At high temperatures, the loss can be due to a depressed operating voltage or to self-discharge.

Figure 21.9 shows some typical discharge curves of the sealed battery at various temperatures at the $0.2C$ and $8C$ rates; Fig. 21.10 shows typical discharge curves at $-20°C$. A flat discharge profile is still characteristic, but at a lower operating voltage than at room temperature. Figure 21.11 shows the effect of temperature on the midpoint voltage. Ambient temperatures significantly above or below 20 to 25°C have a depressing effect on the average discharge voltage.

The effect of temperature and discharge rate on the capacity of a sealed nickel-cadmium battery is shown in Fig. 21.12. These data are typical of standard batteries. Manufacturers should be contacted to obtain performance characteristics of specific batteries.

21.4.4 Internal Impedance

The internal impedance of a battery is dependent on several factors, including ohmic resistance (due to conductivity, the structure of the current collector, the electrode plates, separator, electrolyte, or other features of the battery design), resistance due to activation and concentration polarization, and capacitive reactance. In most cases, the effects of capacitive reactance can be ignored. Polarization effects are dependent in a complicated way on current, temperature, and time; they decrease with increasing temperature (see Chap. 2). The effect may be negligible for pulses of short duration, that is, less than a few milliseconds.

The nickel-cadmium battery is noted for its low internal resistance due to the use of thin and large-surface-area plates with good electrical conductivity, a thin separator with good electrolyte retention, and an electrolyte having a high ionic conductivity. During discharge, the activation and concentration polarization effects are negligible, at least at low and moderate rates, and the internal resistance of the battery and the discharge voltage remain relatively constant from the state of full charge to the point where almost 90% of the capacity has been discharged. At that point, the resistance increases due to the conversion of active materials in the electrode plates, which tends to lower electrical conductivity. Figure 21.13 shows the change in internal resistance with the depth of discharge for two batteries of different size and capacity. Figure 21.14 illustrates the effect of temperature. The internal resistance increases as the temperature drops because the conductivity of the electrolyte and other components is lower at the lower temperatures.

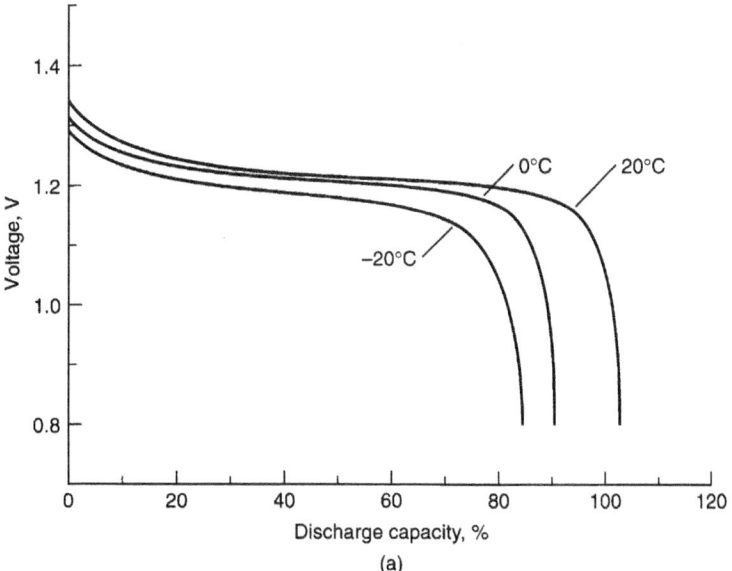

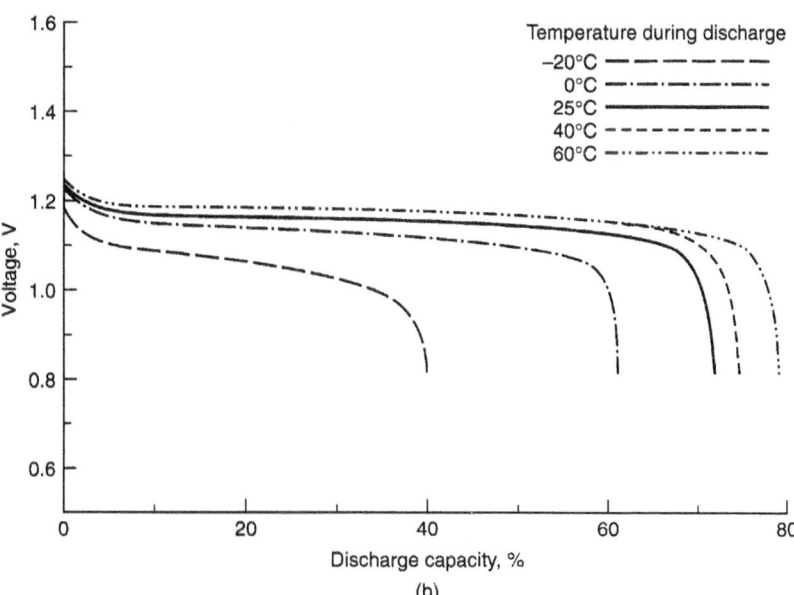

FIGURE 21.9 Constant-current discharge curves of sealed nickel-cadmium batteries at various temperatures. (*a*) 0.2*C* discharge rate. (*b*) 8*C* discharge rate.

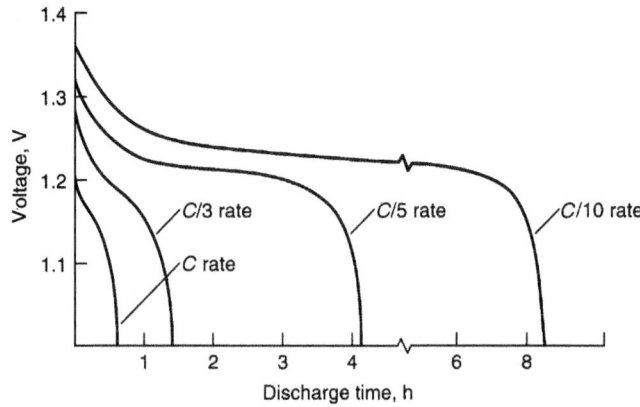

FIGURE 21.10 Constant-current discharge curves of sealed nickel-cadmium batteries at −20°C.

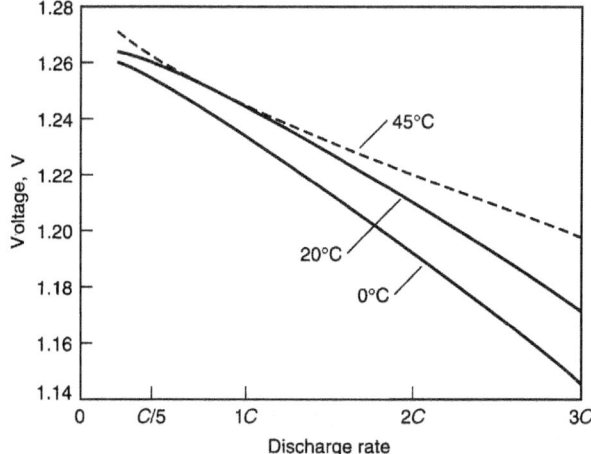

FIGURE 21.11 Midpoint discharge voltage vs. rate at various temperatures for sealed nickel-cadmium batteries, 1 V cutoff.

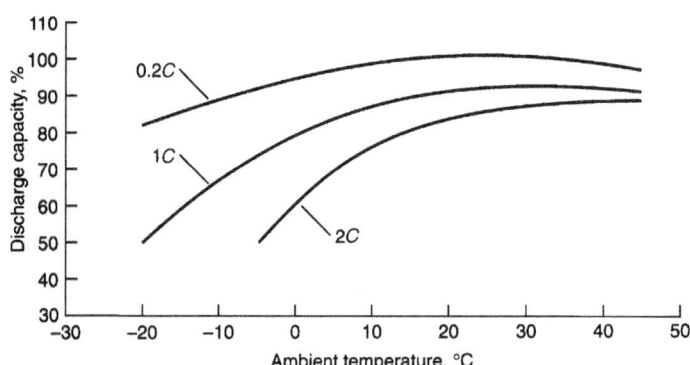

FIGURE 21.12 Percentage of rated capacity vs. temperature at different discharge rates for typical sealed nickel-cadmium batteries, 1.0 V cutoff.

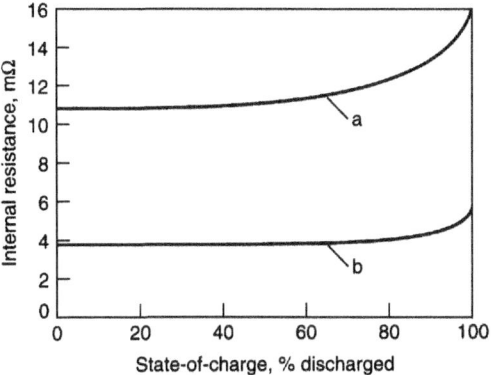

FIGURE 21.13 Resistance vs. state of charge at 20°C, discharged at 0.2C rate, for sealed nickel-cadmium batteries. *a*—AA-size battery, *b*—sub-C-size battery. (Typical for sintered-plate electrode type batteries.)

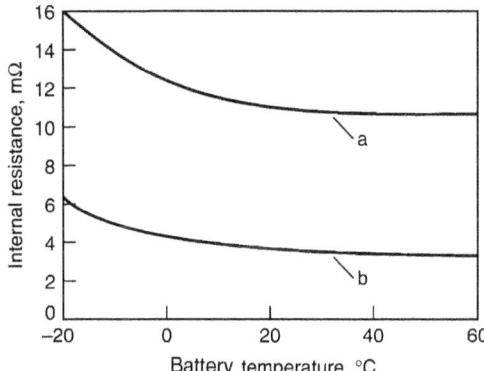

FIGURE 21.14 Resistance vs. temperature for fully charged sealed nickel-cadmium batteries. *a*—AA-size battery, *b*—sub-C-size battery. (Typical for sintered-plate electrode type batteries.)

With use over time, a nickel-cadmium battery gradually loses capacity, resulting in a gradual increase in internal resistance. This is caused by gradual deterioration of the separator and electrodes and by loss of liquid through the seals, which changes the electrolyte concentration and level. The net effect is an increase in internal impedance.

21.4.5 Service Life

The service life of a sealed nickel-cadmium battery, normalized to unit weight (kilograms) and size (liters), at various discharge rates and temperatures is summarized in Fig. 21.15. The curves are based on a capacity, at the *C*/5 discharge rate at 20°C, of 30 Ah/kg and 85 Ah/L, reflecting the performance of standard-type sealed cylindrical batteries. Manufacturers should be contacted for performance characteristics of specific batteries.

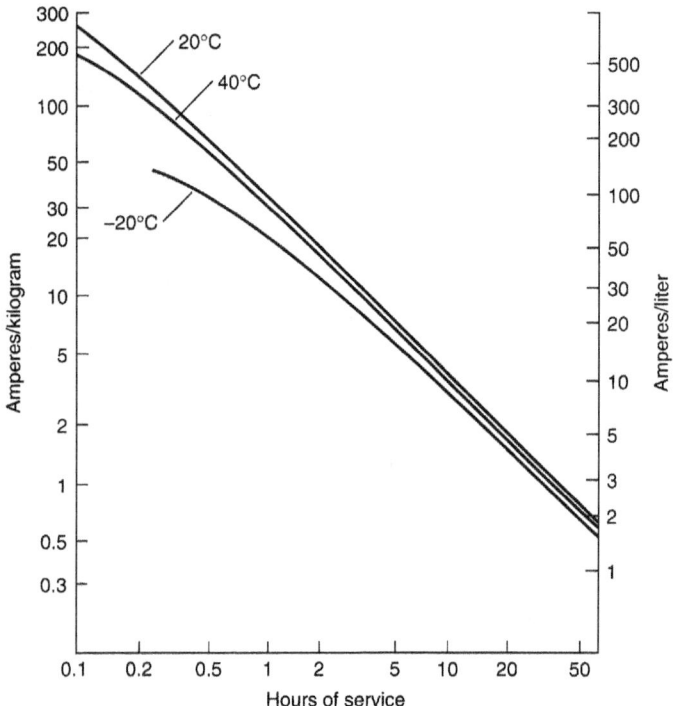

FIGURE 21.15 Service life of sealed nickel-cadmium battery at various constant-current discharge rates and temperatures; end voltage at 1.0 V.

21.4.6 Reversal of Voltage Polarity

When three or more batteries are series-connected, the lowest-capacity battery can be driven into voltage reversal by the others. The larger the number of batteries in series, the greater the possibility of this occurring. During reversal, hydrogen may evolve from the positive electrode and oxygen from the negative. Figure 21.16 shows the complete discharge curve of a battery, including polarity reversal. Section 1 is the normal period of discharge with active materials remaining on both electrodes. Section 2 represents the period when the discharge has been extended and all of the active material on the positive electrode has been discharged and hydrogen gas is generated at this electrode. Active material still remains on the negative electrode and its normal discharge reaction continues. The battery voltage varies with the discharge current, but stays at about −0.2 to −0.4 V. In Section 3, the negative active material has been discharged and oxygen gas is generated at this electrode.

Continued discharging during polarity reversal will lead to high battery pressure and opening of the safety vent. This then results in a loss of gas and electrolyte and breakdown of the capacity balance of the positive and negative electrodes.

Some battery designs provide a limited amount of built-in protection against deep reversal by adding a small amount of cadmium hydroxide to the positive electrode. The term used for the material added to the positive electrode for reversal protection is "antipolar mass" (APM). When the positive electrode is completely discharged, the cadmium hydroxide in that electrode is converted to cadmium, which, combining with the oxygen generated from the negative electrode, depolarizes the positive, preventing hydrogen generation for a time. This reaction occurs at about −0.2 V. This reaction can sustain for only a limited time, after which hydrogen is evolved from the positive electrode. Because hydrogen combines with the battery materials to only a limited extent, repetitive battery reversal will gradually increase a battery's internal pressure, ultimately causing the battery to vent.

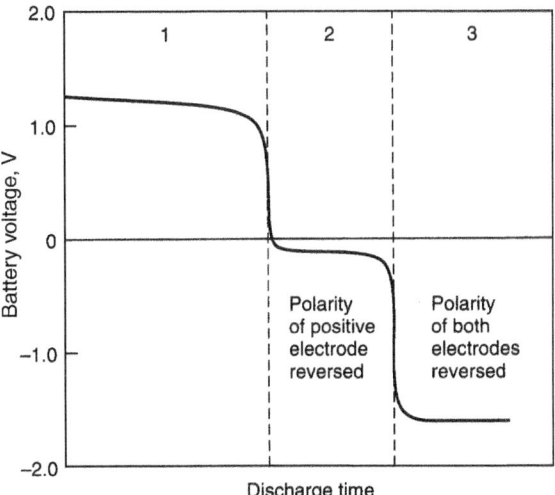

FIGURE 21.16 Discharge of sealed nickel-cadmium battery showing polarity reversal.

Discharging to the point of reversal should be avoided. In order to prevent voltage reversal in any of the cells of a multicell battery, particularly with a series string of more than four cells, the battery should not be discharged to a voltage below 0.8 V per cell. In applications of multicell batteries where it is likely that the battery will be frequently discharged below 1.0 V per cell, a voltage-limiting device is recommended to avoid cell reversal.

21.4.7 Types of Discharge

As discussed in Sec. 3.2.3, a battery may be discharged under different modes (such as constant resistance, constant current, or constant power), depending on the characteristics of the equipment load. The type of discharge mode selected has a significant impact on the service life delivered by a battery in a specified application. The voltage profiles of a nickel-cadmium battery discharged under the three different modes are plotted in Fig. 21.17. The data are based on a discharge of a 650 mAh battery so that, at the end of the discharge (1.0 V per cell), the power output is the same for all modes of discharge. In this example, the power output is 130 mW. To discharge at 130 mW at 1.0 V, the constant-current discharge is 130 mA ($C/5$ rate) and the constant-resistance discharge is 7.7 Ω. As shown, the longest service life is obtained under the constant-power mode, as the average current is the lowest under this mode of discharge.

21.4.8 Constant-Power Discharge

The discharge characteristics of the nickel-cadmium battery under the constant-power mode, at several different power levels, are shown in Fig. 21.18. These are similar to the data presented in Fig. 21.7 for constant-current discharges, except that the performance is presented in hours of service instead of percent discharge capacity. The power levels are shown based on the E-rate. The E-rate is calculated in a manner similar to calculating the C-rate, but based on the rated watthour capacity. For example, for the $E/5$ power level, the power for a battery rated at 780 mWh is 156 mW.

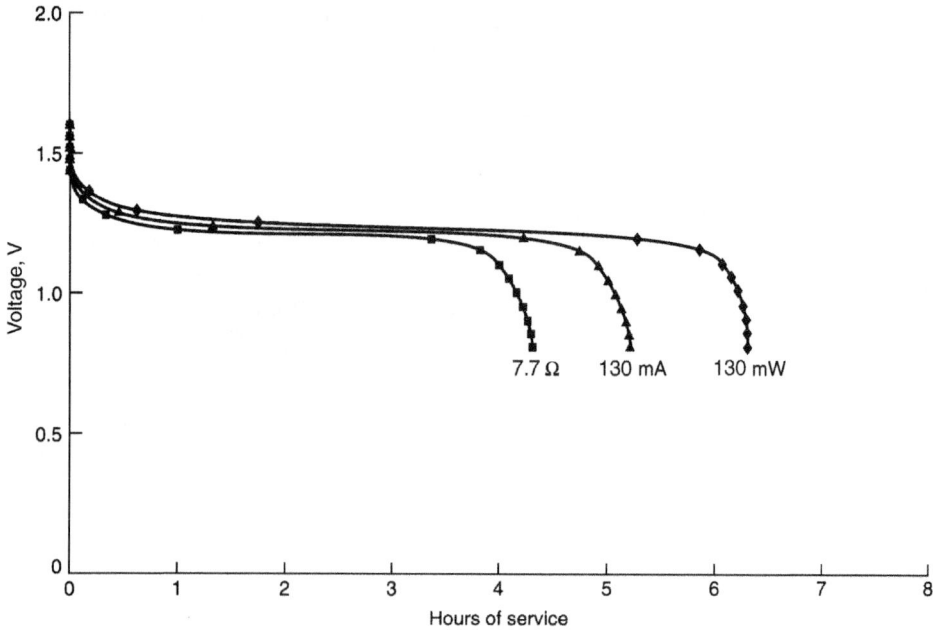

FIGURE 21.17 Discharge curves of AA-size (650 mAh) nickel-cadmium battery—constant power (♦) vs. constant current (▲) vs. constant resistance (■).

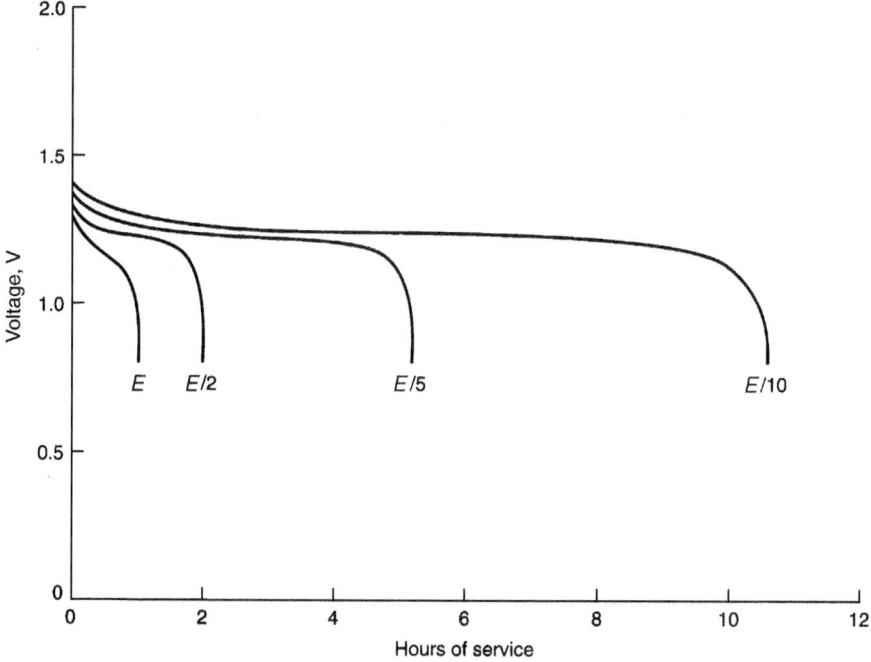

FIGURE 21.18 Constant-power discharge curves at various E rates for sealed nickel-cadmium batteries at 20°C.

21.4.9 Shelf Life (Capacity or Charge Retention)

Nickel-cadmium batteries lose capacity during storage. The rate of this self-discharge is a function of storage temperature and battery design. Figure 21.19 can serve as a guide for the shelf life (capacity or charge retention) at several temperatures for typical standard type nickel-cadmium sealed batteries. Specifically designed batteries, as discussed in Sec. 21.6, may have considerably different charge retention characteristics. For example, the button batteries designed for memory-backup application have significantly better charge retention characteristics than the standard lower-resistance higher-discharge-rate cylindrical batteries.

Sealed nickel-cadmium batteries can be stored in a charged or a discharged condition. Except for extended storage at high temperatures, they can be restored to full capacity after storage by recharging (two or three charge-discharge cycles). Figure 21.20 illustrates the capacity recovery after prolonged storage at several temperatures. The recovery time may be longer after high-temperature storage.

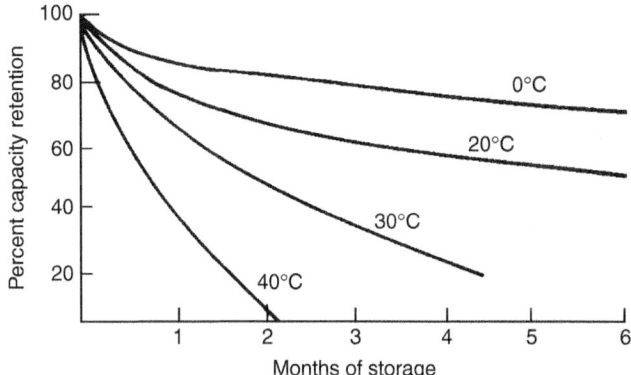

FIGURE 21.19 Capacity retention (shelf life) of standard type sealed nickel-cadmium batteries.

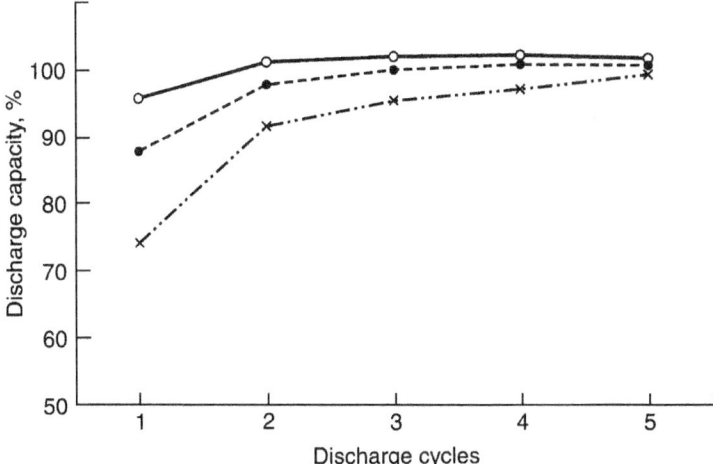

FIGURE 21.20 Capacity recovery of standard type sealed nickel-cadmium batteries, discharge at $0.2C$ rate after 2-year storage. Storage temperatures: at 20°C (○); 35°C (●); and 45°C (×).

21.4.10 Cycle Life

The cycle life is usually measured to the point when the battery will not deliver more than a given percentage (usually 60 to 80%) of its rated capacity. Sealed nickel-cadmium batteries have long cycle lives. Under controlled conditions over 500 cycles can be expected on a full discharge, as illustrated in Fig. 21.21. On shallow discharges considerably higher cycle life can be obtained, as shown in Fig. 21.22. Cycle life is also very dependent on the many conditions to which the battery has been exposed, including charge, overcharge, and discharge rates, frequency of cycling, the temperatures to which the battery has been exposed, and battery age, as well as battery design and battery components. Specially designed batteries, such as those using alkali-resistant materials, are also manufactured, which have longer life, particularly at the higher temperatures (see Sec. 21.6).

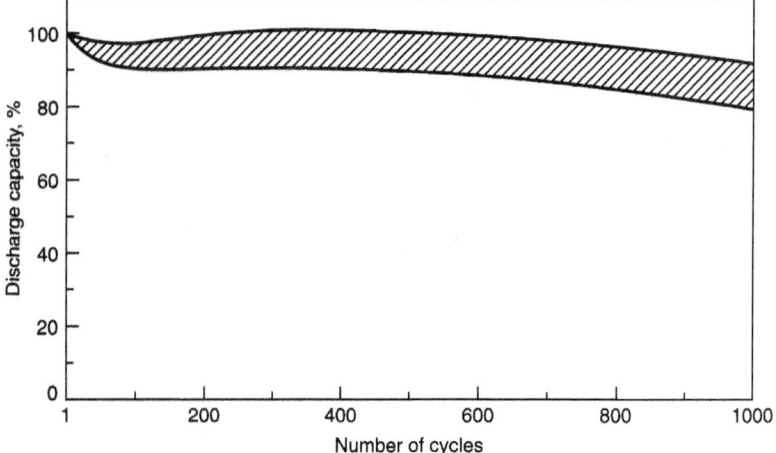

FIGURE 21.21 Cycle life of sealed nickel-cadmium batteries at 20°C. Cycle conditions: charge—$0.1C \times 11$ h: discharge—$0.7C \times 1$ h. Capacity-measuring conditions: charge—$0.1C \times 16$ h; discharge—$0.2C$, end voltage—1 V.

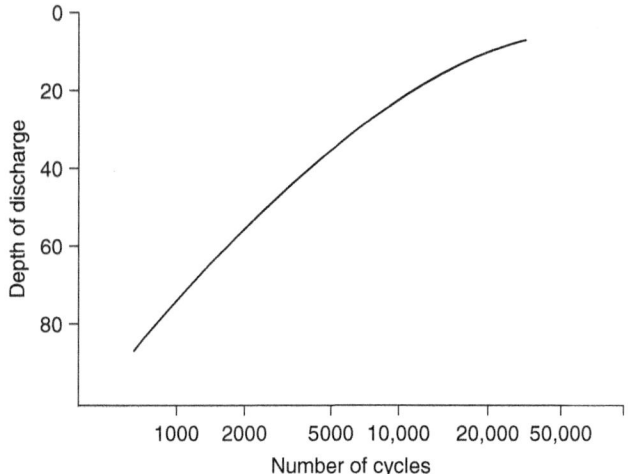

FIGURE 21.22 Cycle life of sealed nickel-cadmium batteries at shallow discharge.

21.4.11 Life Expectancy and Failure Mechanisms

The useful life of a nickel-cadmium battery can be measured either in terms of the number of cycles before failure or in units of time. It is virtually impossible to know all the detailed information necessary to make any kind of accurate prediction of battery life in a given application. The best that can be provided is an estimate based on laboratory test data and field experience or extrapolation of accelerated test data.

Basically, failure of a battery occurs when it ceases to operate the device, for whatever reason, at the prescribed performance level, despite the possibility that the battery may still be useful in another application with less demanding requirements.

Failure of a nickel-cadmium battery can be classified into two general categories: reversible and irreversible failure. When a battery fails to meet the specified performance requirements but can, by appropriate reconditioning, be brought back to an acceptable condition, it is considered to have suffered a reversible failure. Permanent or irreversible failure occurs when the battery cannot be returned to an acceptable performance level by reconditioning or any other means.

Reversible Failures.

Voltage Depression (Memory Effect). A sealed nickel-cadmium battery may suffer a reversible loss of capacity when it is cycled repetitively on shallow discharges (discharge terminated before its full capacity is delivered) and recharged. For example, as shown in Fig. 21.23, if a battery is cycled repetitively, but only partially discharged and then recharged, the voltage and delivered capacity will gradually decrease with cycling (curves 2 representing repetitive cycling). If the battery is then fully discharged (curve 3), the discharge voltage is depressed compared to the original full discharge (curve 1). The discharge profile may show two steps and the battery may not deliver the full capacity to the original cutoff voltage. This condition is known as "voltage depression." It also is referred to as "memory effect," as the battery appears to "remember" the lower capacity of the shallow discharge. Operation at higher temperatures accelerates this type of loss.

The battery can be restored to full capacity with a few full discharge-charge reconditioning cycles. The discharge characteristics on the reconditioning cycles are illustrated in Fig. 21.23 (curves 4 and 5).

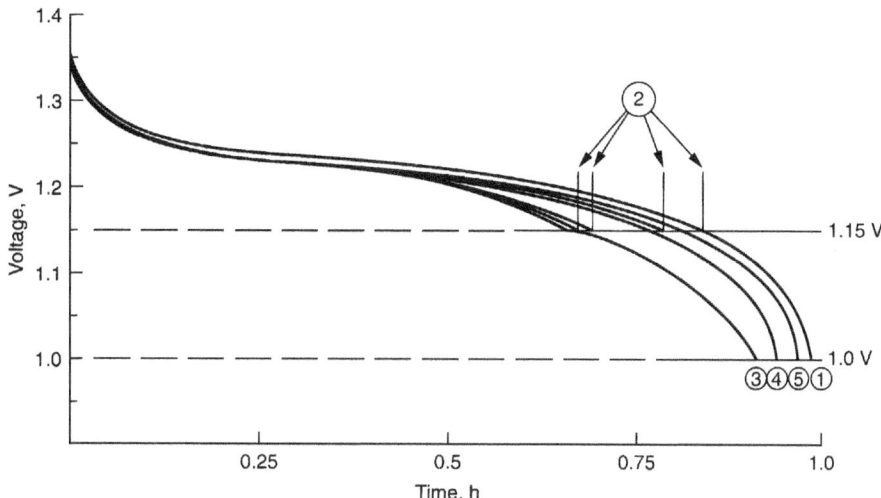

FIGURE 21.23 Voltage depression and subsequent recovery.

The voltage drop occurs because only a portion of the active materials is discharged and recharged during shallow or partial discharging. The active materials that have not been cycled, particularly the cadmium electrode, may undergo a change in physical characteristics and an increase in resistance. The effect has also been ascribed to structural changes in the nickel electrode.[1] Subsequent cycling restores the active materials to their original state.

The extent of voltage depression depends on the depth of discharge and can be avoided or minimized by the selection of an appropriate end voltage. Too high an end voltage, such as 1.16 V per cell, terminates the discharge prematurely. (A high end voltage should be used only if an extended cycle life is desired and the lower capacity can be tolerated, as in some satellite applications.) A small voltage depression may be observed if the discharge is terminated between 1.16 and 1.10 V per cell. The extent of the depression is dependent on the depth of discharge, which is also rate-dependent. Discharging to an end voltage below 1.1 V per cell should not result in a subsequent voltage depression. Discharging to too low an end voltage, however, should be avoided, as discussed in Sec. 21.4.6.

This condition varies with the design and formulation of the electrode and may not be evident with all sealed nickel-cadmium batteries. Modern nickel-cadmium batteries use electrode structures and formation processes that reduce the susceptibility to voltage depression, and most users may never experience low performance due to memory effect. However, the use of the term "memory effect" persists, since it is often used to explain low battery capacity that is attributable to other problems, such as ineffective charging, overcharge, battery aging, or exposure to high temperatures.

Overcharging. A similar reversible failure can occur with long-term overcharging, particularly at elevated temperatures. Figure 21.24 shows the voltage "step" near the end of discharge that can be induced by long-term overcharging. The capacity is still available, but at a lower voltage than when it was freshly cycled. Again, this is a reversible failure; a few charge and discharge cycles will restore normal voltage and expected capacity.

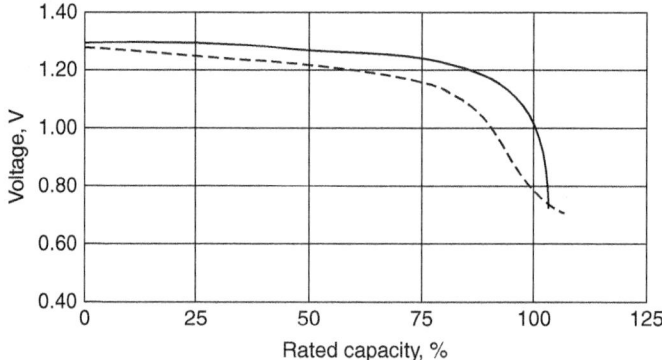

FIGURE 21.24 Typical discharge voltage profile of sealed nickel-cadmium batteries after long-term overcharge (dotted line) vs. 16 h charge, both at $C/10$ rate.

Irreversible Failures. Permanent failure in nickel-cadmium batteries results from essentially two causes: short-circuiting and loss of electrolyte. An internal short circuit may be of relatively high resistance and will be evidenced by an abnormally low on-charge voltage and by a drop of voltage as the battery's energy is dissipated through the internal short circuit. A short circuit may also be of such a low resistance that virtually all the charge current travels through it or the battery electrodes are totally shorted internally.

Any loss of electrolyte will cause degradation in capacity. Charging at high rates, repeated voltage reversal, and direct short-circuiting are ways that can cause loss of electrolyte through the pressure relief device. Electrolyte can also be lost over a long period of time through the battery seals, and capacity is lost in proportion to the reduction in electrolyte volume. Capacity degradation caused by electrolyte losses is more pronounced at high discharge rates.

High temperature degrades battery performance and life. A nickel-cadmium battery gives optimum performance and life at temperatures between 18 and 30°C. Higher temperatures reduce life by promoting separator deterioration and increasing the probability of short-circuiting. Higher temperatures also cause more rapid evaporation of moisture through the seals. These effects are all long-term; but the higher the temperature, the more rapid the deterioration. Table 21.2 lists the recommended temperature limits for sealed nickel-cadmium batteries.

TABLE 21.2 Operating and Storage Limits for Sealed Nickel-Cadmium Batteries

	Temperature, °C	
Type	Storage	Operating
Button	−40 to 50	−20 to 50
Standard cylindrical	−40 to 50	−40 to 70
Premium cylindrical	−40 to 70	−40 to 70

21.5 CHARGING CHARACTERISTICS

21.5.1 General Characteristics

Sealed nickel-cadmium batteries are usually charged by means of the constant-current method. The $0.1C$ rate can be used and the battery is charged for 12 to 16 h (140%). At this rate, the battery can withstand overcharging without harm, although most sealed nickel-cadmium batteries can be safely charged at the $C/100$ to $C/3$ rate. At higher charge rates, care must be taken not to overcharge the battery excessively or develop high battery temperatures and pressures.

The voltage profile of a sealed nickel-cadmium battery during charge at the $C/10$ and $C/3$ rates is shown in Fig. 21.25. A sharp rise in voltage to a peak near the end of the charge is evident.

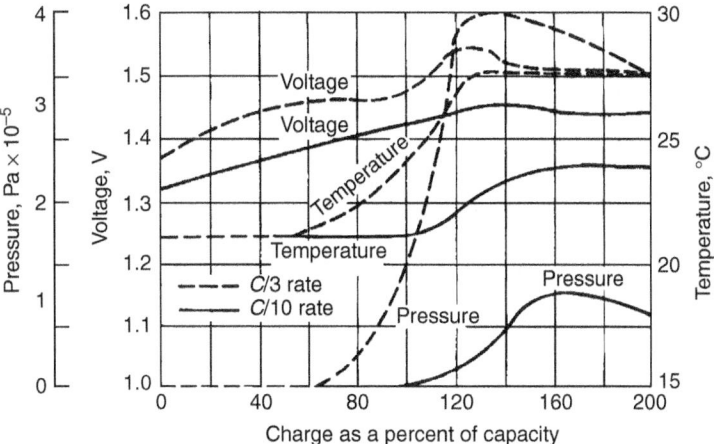

FIGURE 21.25 Typical pressure, temperature, and voltage relationships of sealed nickel-cadmium battery during charge at constant current.

21.20 SECONDARY BATTERIES

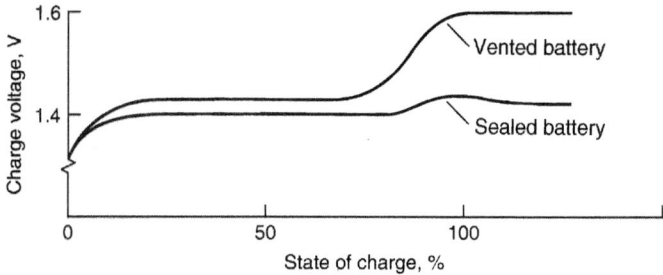

FIGURE 21.26 Charge vs. voltage for sealed and vented nickel-cadmium batteries at 25°C; 0.1C charge rate.

The voltage profile of a sealed nickel-cadmium batteries is different from the one for a vented one, as illustrated in Fig. 21.26 The end-of-charge voltage for the sealed battery is lower. The negative plate does not reach as high a state of charge as it does in the vented construction because of the oxygen recombination reaction.

Constant-potential charging is not recommended for sealed nickel-cadmium batteries as it can lead to thermal runaway. It can, however, be used if precautions are taken to limit the current toward the end of charge.

21.5.2 The Charge Process

The charge process is summarized in Fig. 21.27. Figure 21.27a plots the charge efficiency

$$\text{Charge efficiency} = \frac{\text{discharge energy (on subsequent discharge)}}{\text{charge input energy}}$$

against total input energy. At the start of the discharge (area 1), the charge energy is consumed by the conversion of the active materials into a chargeable form, and the charge efficiency is low. In area 2, charging is most efficient and almost all of the input energy is used to convert the discharged active material into the charged state. As the battery approaches the full charge state (area 3), most of the energy goes to the generation of oxygen, and the charge efficiency is low.

Figure 21.27b presents the relationship of charge efficiency to charge rate. It shows that the charge efficiency, as well as the output capacity, is lower at a lower charge rate.

The charge efficiency also depends on the ambient temperature during charge. This relationship is shown in Fig. 21.27c. There is a decrease in capacity in the high-temperature range due to a fall in potential for oxygen gas generation at the positive electrode.

The principle of the sealed battery is based on the ability of the negative electrode to recombine this oxygen gas and prevent the buildup of internal gas pressure. The capacity for this recombination is limited. Hence the maximum charge rate that can be tolerated is the one that keeps the rate of oxygen generation below the gas recombination rate so that the internal gas pressure does not build up excessively.

Overcharging at rates beyond the ability of oxygen recombination or heat dissipation can result in failure. "Fast" charging methods can be used successfully, but a means must be provided for monitoring and terminating the charge before excessive overcharging occurs. Temperature rise, voltage, or pressure can be monitored and used effectively as a cutoff.

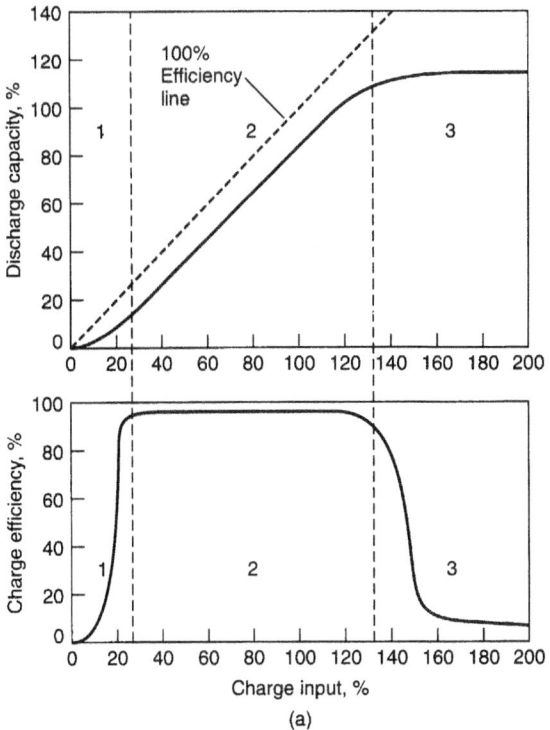

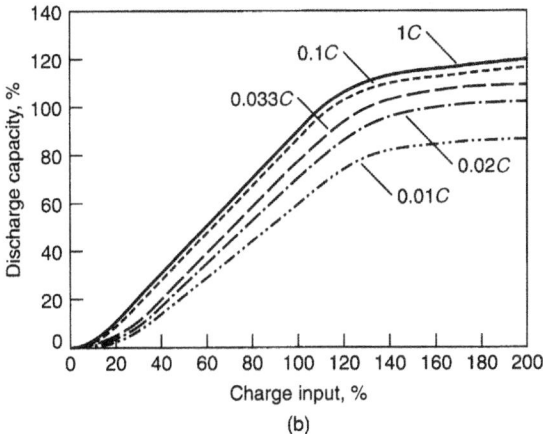

FIGURE 21.27 Charge process of sealed nickel-cadmium batteries. (*a*) Charge efficiency at 20°C. Charge—$0.1C \times 16$ h; discharge—$0.2C$; end voltage—1 V. (*b*) Charge efficiency vs. charge rate at 20°C.

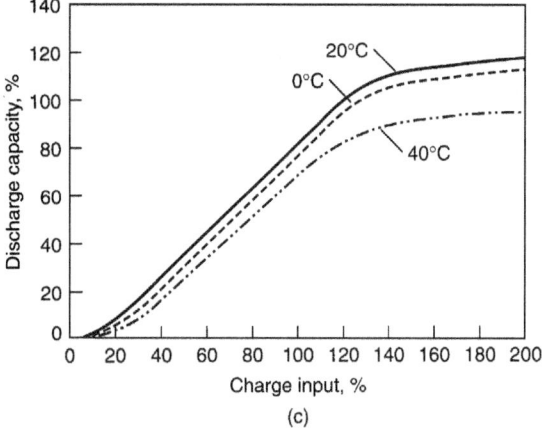

FIGURE 21.27 (*Continued*) (*c*) Charge efficiency vs. ambient temperature during charge at $0.1C$ rate.

21.5.3 Voltage, Temperature, and Pressure Relationships

Figure 21.25 also shows the relationship of voltage, temperature, and pressure of a typical sealed battery during charging at the $C/10$ and $C/3$ rates. The voltage increases gradually during the charge until the battery is about 75 to 80% charged. The voltage then rises more sharply due to the generation of oxygen at the positive electrode. The temperature remains relatively constant during the early part of the charge, as the charge reaction is endothermic. It then rises as the battery reaches the overcharge state due to the heat generated by the oxygen recombination reaction. Similarly, the internal pressure remains low until the battery goes into the overcharge condition, when most of the current is used to produce oxygen and the pressure rises. Finally, as the battery reaches full charge, the voltage drops because of a decrease in the battery's internal resistance due to the increase in the battery temperature. This drop in voltage can be used effectively in a control circuit to terminate the charge.

As shown, when the battery is overcharged at acceptable rates, the pressure and temperature tend to stabilize. These steady-state conditions are governed by such factors as ambient temperature, overcharge rate, heat transfer characteristics of the cell and battery, cell design and components such as the separator, recombination capability of the negative electrode, and the resistance of the cell and battery. Charging at higher rates, such as the $C/3$ rate compared to the $C/10$ rate, results in higher temperatures and internal pressures. At higher charge rates, temperature and pressure will rise more significantly, particularly if the oxygen recombination rate is exceeded. Because of the possibility of venting and other deleterious effects on battery performance due to these high temperatures and pressures, it is necessary to terminate the charge before these conditions are reached.

21.5.4 Voltage Characteristics during Charge

The voltage profile of a sealed nickel-cadmium battery during charge at various charge rates at 20°C is shown in Fig. 21.28. The charge voltage also depends on temperature, as shown in Fig. 21.29. The voltage and voltage peak decrease with a rise in temperature. Charging at temperatures between 0 and 30°C is best for sealed batteries. At lower temperatures, the voltage increases, recombination of oxygen is slower, and the internal gas pressure tends to increase. Charging rates must be reduced. Above 40°C, the charging efficiency is low, and higher temperatures cause battery deterioration.

generation of oxygen is below the recombination rate. This rate also limits the temperature rise. Excessive overcharge should be avoided. The battery should be charged to about 140 to 150% charge input.

Timer Control. (Fig. 21.30*b*). For moderate charge rates, a timer can be used to cut off the charge or reduce the charge current to the trickle charge level. This is a relatively inexpensive control device and suitable for applications where the battery is usually fully discharged before charging. It is not suitable for applications where the battery is frequently charged without prior deep discharging as this could result in excessive overcharge. A thermal cutoff control should be used when charging at rates higher than the $C/5$ rate or without deep discharging to prevent the battery from reaching high temperatures.

Temperature Detection. (Fig. 21.30*c*). This control system uses a sensor to detect the temperature rise of the battery and terminate the charge. A thermostat or thermistor is used as the detection device, and the detecting temperature is usually set at 45°C. It is important that the sensor be located so that it can accurately determine the battery's temperature. Charging in high ambient temperatures can result in an insufficient charge, while low ambient temperatures may result in overcharge. The cycle life with this method may be shorter than with the $-\Delta V$ method or peak voltage methods as the battery could be subjected to more overcharge.

Negative Delta V ($-\Delta V$). (Fig. 21.30*d*). This is one of the preferred charge control systems for sealed nickel-cadmium batteries. The drop in voltage of the battery is detected after the battery voltage has reached its peak during charge. The signal can be used to terminate the charge or reduce the charge current to a trickle charge. The method provides a complete charge regardless of ambient temperature or residual capacity from the previous charge. A value of 10 to 20 mV per cell is usually used for the control. The method is not suitable for charging below the $0.5C$ as the $-\Delta V$ value is too low to be detectable.

Trickle and Float Charge. (Fig. 21.30*e*). Trickle-charge systems are used in two different situations (1) in a standby power application where the battery is on continual charge to maintain it in a state of full charge (compensating for self-discharge) until it is connected to the load when the prime power fails, and (2) as a supplementary charge after the termination of rapid charging. Charging is at the 0.02 to $0.05C$ rate, depending on the frequency and depth of discharge. A periodic discharge every six months followed by a charge is advisable to ensure optimum performance.

21.6 SPECIAL-PURPOSE BATTERIES

Special-purpose sealed nickel-cadmium batteries are manufactured with specifically designed characteristics, overcoming some of the limitations of standard batteries to meet the requirements for certain applications. Manufacturers' recommendations should be followed because of the specific performance characteristics of these batteries.

21.6.1 High-Capacity Batteries

These batteries incorporate design features, such as nickel foam substrate positive plates, pasted negative electrodes, thin-walled battery containers, and increased amounts of active material. These changes result in a 20 to 40% increase in capacity. These batteries are also designed with improved oxygen recombination capability and are capable of being charged at the $0.2C$ rate or less without control. They are capable of fast 1 h charging using $-\Delta V$ charge control. Figure 21.31 compares the discharge characteristics of the high-capacity battery with those of a standard battery. Figure 21.32 shows the relationship of battery capacity and discharge current for the two designs.

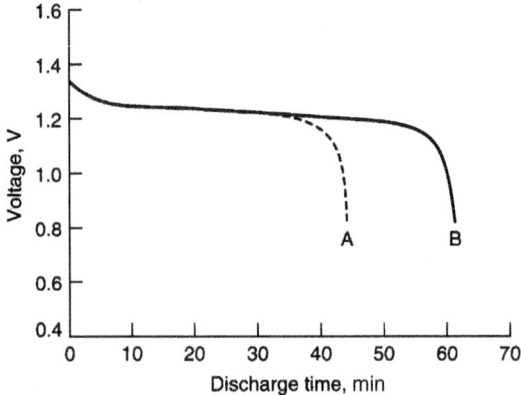

FIGURE 21.31 Comparison of discharge characteristics of sub-C size standard battery (A) vs. high-capacity battery (B) on discharge at 20°C. Discharge at C rate, charge at $0.1C$ rate for 16 h.

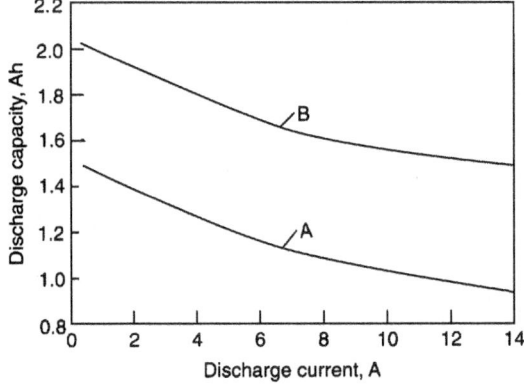

FIGURE 21.32 Comparison of performance of standard battery (A) vs. high-capacity battery (B) (sub-C size), at 20°C.

21.6.2 Fast-Charge Batteries

These batteries have electrode structures and electrolyte distribution designs to enhance oxygen recombination. They can be charged at the fast 1 h rate with charge control (such as temperature-sensing and $-\Delta V$ techniques) and at the $C/3$ rate without charge control because of their ability to withstand this level of overcharge. They are also capable of performance at high discharge rates, though this is achieved at the expense of a slightly reduced battery capacity. These batteries have improved internal heat conductivity, which results in a faster increase in surface temperature. This feature can be used advantageously in a temperature-sensing fast-charge system. Figure 21.33 shows the charge characteristics of a fast-charge battery compared to a standard one. The internal gas pressure of the standard battery increases quickly during charging, whereas that of a fast-charge battery stabilizes.

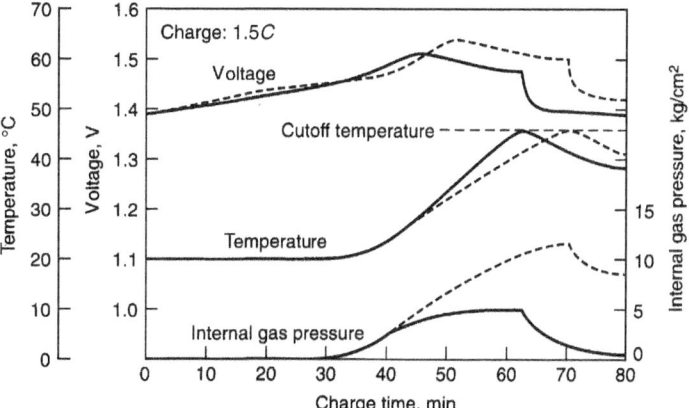

FIGURE 21.33 Comparison of charge characteristics of fast-charge battery (solid line) vs. standard battery (broken line).

21.6.3 High-Temperature Batteries

These batteries are designed to operate at high temperatures without the service life deterioration and charging inefficiencies experienced with conventional designs. Figure 21.34 compares the performance of the high-temperature battery with the standard battery as a function of ambient temperature during charge. This type of battery is capable of charge-discharge cycling at temperatures as high as 35 to 45°C and is particularly designed for trickle charging ($C/20$ to $C/50$ rate) at these high temperatures. The charge voltage of these batteries is slightly higher than that of the standard battery due to the designed-in control of the oxygen-generating potential.

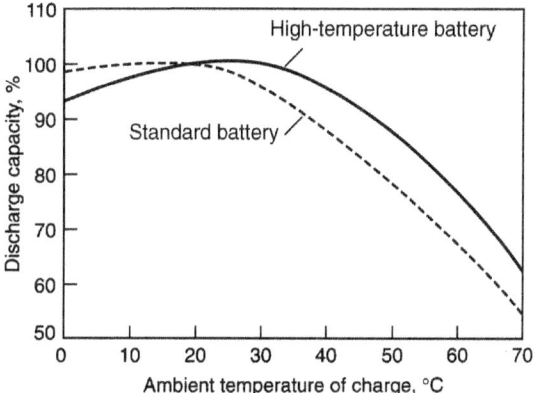

FIGURE 21.34 Comparison of performance of high-temperature battery vs. standard battery. Charge—$C/30$ rate; discharge—$1C$ rate at 20°C.

21.6.4 Heat-Resistant Batteries

These batteries are designed for fast charging at high temperatures. For example, charging at the $0.3C$ rate is possible even at temperatures as high as 45 to 70°C. Their performance characteristics

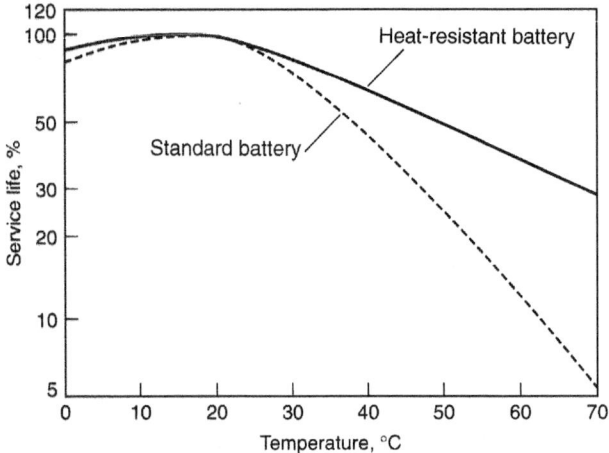

FIGURE 21.35 Comparison of performance of heat-resistant battery vs. standard battery.

are similar to those of the standard battery. However, they have a superior service life when used at high temperatures because of the use of specially selected materials with minimum deterioration at high temperatures. Figure 21.35 compares the service life for standard and heat-resistant batteries throughout the temperature range.

21.6.5 Memory-Backup Batteries

These batteries are used to provide battery backup for volatile semiconductor memory devices. The key requirements for this type of battery are long life (up to 10 years in certain applications), low self-discharge, and good performance at low discharge rates. Figure 21.36 shows the storage characteristics of the memory-backup battery. (This can be compared to the characteristics of the standard battery shown in Fig. 21.19.) The low-rate discharge characteristics of the battery are plotted in Fig. 21.37. As the backup battery is designed for low-rate use, its internal resistance is higher than that of the standard battery and its high-rate discharge characteristics are not as good.

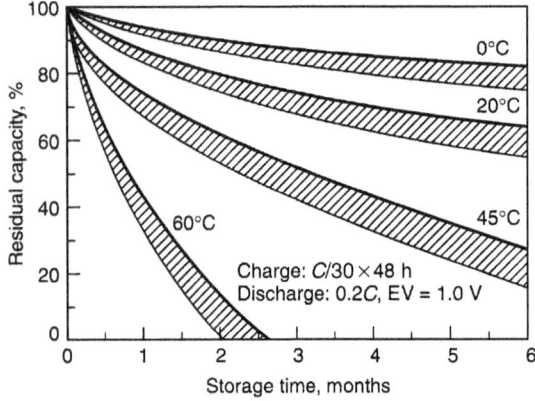

FIGURE 21.36 Storage characteristics of memory-backup batteries.

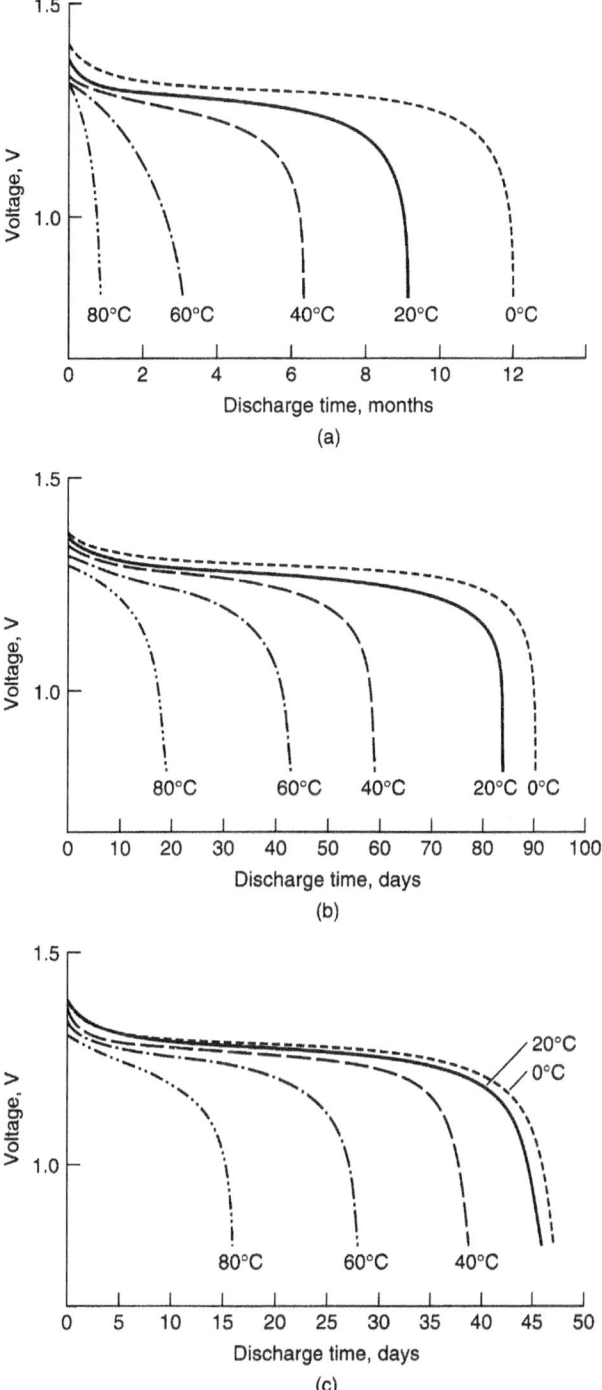

FIGURE 21.37 Performance of memory-backup batteries. Charge—$C/30$ for 48 h at 20°C. Discharge rate: (*a*) $C/10,000$; (*b*) $C/2000$; (*c*) $C/1000$.

21.6.6 Rectangular Batteries

The constructional features of the flat rectangular battery are described in Sec. 21.3.3. The advantage of the rectangular battery is that it permits more efficient battery design, eliminating the voids that occur with the assembly of cylindrical batteries. The volumetric energy density of these batteries can be about 20% higher than a battery using a cylindrical design.

Most of the performance characteristics are similar to those of the standard cylindrical battery, except that it also incorporates some of the features of the high-capacity battery. Gas recombination has been improved to permit charging at the $0.2C$ rate or less and 1 h charging with charge control, preferably with $-\Delta V$ sensing. This is illustrated in Fig. 21.38. The voltage profile on discharge is flat, as with the cylindrical battery, as shown in Fig. 21.39. However, because the resistance of the rectangular battery is higher, performance at rates greater than $4C$ is not as good as with the cylindrical battery. Storage characteristics and cycle life are similar to those of the cylindrical battery.

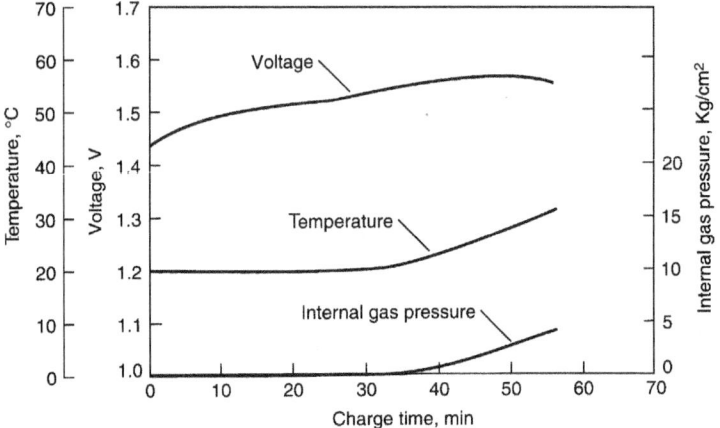

FIGURE 21.38 Charge characteristics of rectangular batteries at 20°C. Charge—$1.5C$ rate; $-\Delta V = 10$ mV.

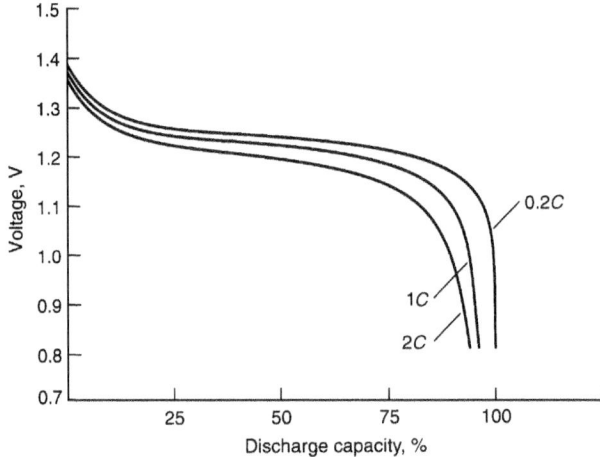

FIGURE 21.39 Discharge characteristics of rectangular batteries at 20°C. Charge—$0.1C$ rate for 16 h.

21.7 BATTERY TYPES AND SIZES

Table 21.3 lists some of the types of sealed nickel-cadmium single-cell batteries that are manufactured and some of their physical and electrical specifications. Multicell batteries are also manufactured, using these cells, in a variety of output voltages and configurations.

Figure 21.40 is a guide to determining the approximate battery size required for a given performance requirement or application. These data are based on the performance of a standard battery at 20 to 25°C. Allowance must be factored into the estimate to determine the performance under other discharge conditions.

Manufacturers' data should be consulted for specific details on dimensions, ratings, and performance characteristics as they may be different from those shown.

TABLE 21.3 Specifications of Typical Sealed Nickel-Cadmium Single-cell Batteries

Battery size	Typical capacity at 0.2C rate (mAh)	Dimensions, Max., mm		Weight, g
		Diameter	Height	
Cylindrical batteries				
Standard batteries: Charging: 0–45°C. Discharge: –20–+60°C				
F	7,000	33.2	91.0	224
M	12,000	43.1	91.0	395
Extended service life batteries: Charging—standard: 0–45°C; quick charge: 10–45°C. Discharge: –20–+60°C				
AA	600	14.3	50.3	22
AA	700	14.3	50.3	23
AA	600	14.3	48.9	22
AA	700	14.3	48.9	23
SC	1,200	22.9	43.0	52
Fast charge batteries: Charging—standard: 0–45°C; quick charge: 10–45°C; fast 1 h: 5–45°C. Discharge: –20–+60°C				
4/5 SC	1,200	22.9	34.0	43
SC	1,300	22.9	43.0	51
SC	1,700	22.9	43.0	55
C	3,000	26.0	50.0	86
High-temperature batteries: Charging—standard: 0–70°C. Discharge: –20–+70°C				
AA	600	14.3	48.9	23
SC	1,600	22.9	43.0	49
C	2,900	26.0	50.5	78
F	7,000	33.2	91.0	224
M	10,000	43.1	91.0	395
Heart-resistant batteries: Charging—standard: 0–70°C; quick: 10–70°C. Discharge: –20–+70°C				
AA	600	14.3	50.2	22
SC	1,200	22.9	43.0	52

Courtesy of Sanyo Mobile Energy Co.

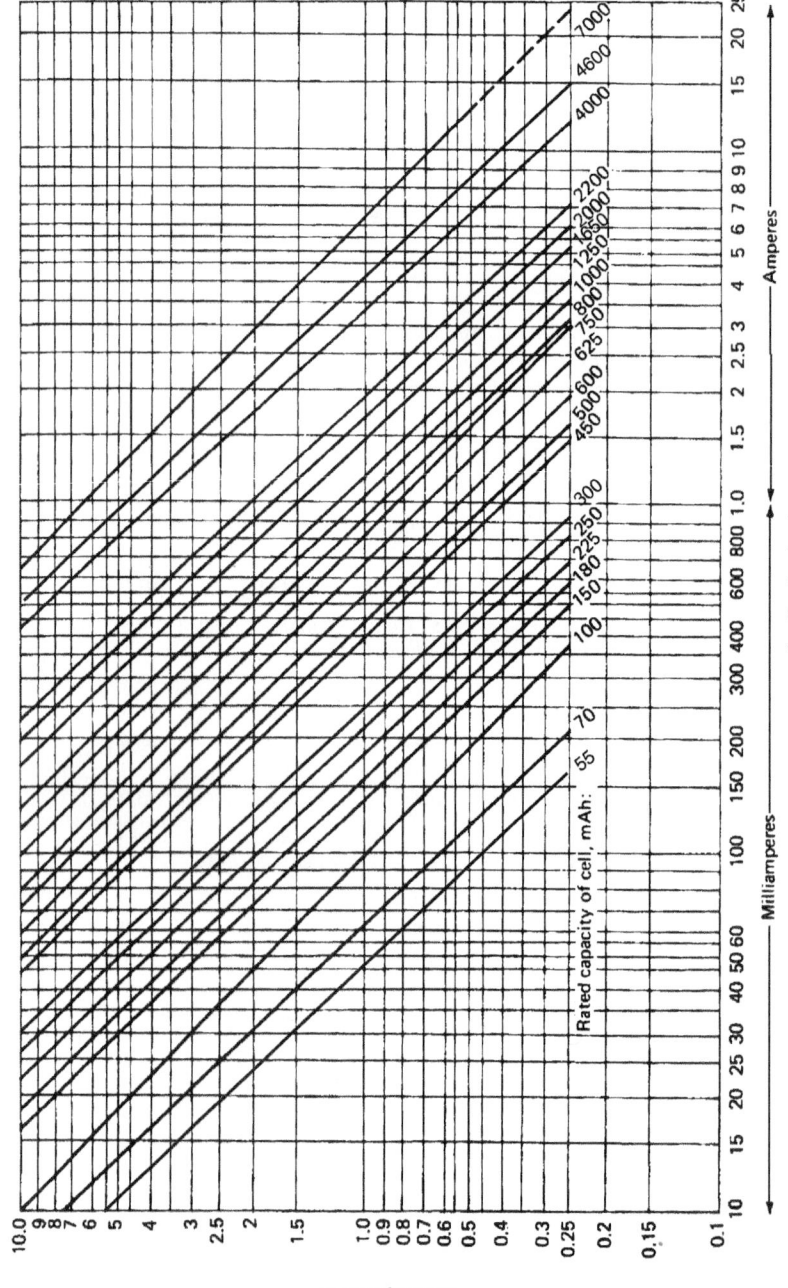

FIGURE 21.40 Selector guide for sealed nickel-cadmium cylindrical batteries. Guide can be used to determine approximate required battery size, given the load and desired run (service) time. Data based on fully charged battery and 20°C operating temperature.

21.8 BATTERY SIZES AND AVAILABILITY

Currently, nickel-cadmium cells and batteries are available from suppliers worldwide. However, manufacturing of sealed cells is predominantly conducted in Asia. In selecting a cell or battery for your application, you can refer to the power data chart shown in Fig. 21.40 to determine ampere-hour capacity requirement.

It is recommended that you then refer to a manufacturer's data sheet, which can address your requirements. Ultimately conversation with a cell manufacturer or capable added-value battery assembler regarding your application and design is necessary to facilitate finalizing design for production.

REFERENCE

1. Y. Sato, K. Ito, T. Arakawa, and K. Kobaya Kawa "Possible Causes of the Memory Effect Observed in Nickel-Cadmium Secondary Batteries, *J. Electrochemical Society,* **143:**L225 (October 1996).

BIBLIOGRAPHY

Cadnica Sealed Type Nickel-Cadmium Batteries Engineering Handbook, Sanyo Electric Co., Osaka, Japan.

Ford, Floyd E., *Handbook for Handling and Storage of Nickel-Cadmium Batteries: Lessons Learned,* NASA Ref. Publ. 1326, Feb. 1994.

Nickel-Cadmium Batteries, Charge System Guide, Panasonic Industrial Co., Secaucus, NJ.

Nickel-Cadmium Batteries, Technical Handbook, Panasonic Industrial Co., Secaucus, NJ.

Sealed NiCad Handbook, SAFT America Inc., Valdosta, GA.

CHAPTER 22
NICKEL-METAL HYDRIDE BATTERIES

Michael Fetcenko and John Koch

22.1 INTRODUCTION

Nickel-metal hydride (NiMH) batteries have become a commercially important rechargeable battery system in high volume production for multiple consumer and vehicle propulsion applications. From commercial introduction in 1989 for portable PCs, today NiMH has enabled widespread commercialization of hybrid vehicles and is increasingly important in many consumer applications. In 2008, the NiMH market became a $1.2B business, gaining a 10% share of the total rechargeable battery industry. The key driving forces for the rapid growth of NiMH have been the growth of HEVs and the development of NiMH batteries as direct replacements for alkaline primary batteries.

NiMH was first studied in the 1960s as a derivative of both The NiCd and The NiH_2 batteries used in satellites. The key motivations for the NiMH studies were the environmental advantages associated with higher energy, lower pressure, and cost of NiMH compared to NiCd. Implementation issues included corrosion and catalytic stability of the hydrogen storage alloy in a hostile alkaline environment. Key research organizations during the 1970s included Battelle (Beccu et al.) and Anwar (Percheron et al.), where progress was reported. Accelerated R&D efforts in the 1980s by Philips (Willems et al.) and Ovonic (Ovshinsky et al.) targeted advanced materials to overcome technical limitations that hindered commercial introduction.

In the mid-1970s, electric vehicles emerged due to what turned out to be temporary gasoline shortages. In the early 1990s, vehicle electrification reemerged due to environmental air quality concerns in urban areas. Today, NiMH is the rechargeable battery chemistry of choice for commercial HEVs, providing overall excellent performance, reliability, and cost. NiMH dominates HEV applications through a combination of desirable performance attributes such as high energy and power, an excellent range of operating temperatures, and low cost. Further, NiMH has demonstrated excellent safety, abuse resistance, and cycle life, which have translated into superior field reliability for advanced vehicular applications and overall performance in consumer applications since 1991.

The early 1990s also saw the parallel introduction of rechargeable lithium technologies, which featured still higher gravimetric energy density. Markets that were once dominated by NiCd and NiMH, laptop computers and cell phones, have been replaced with Li-ion. Yet, NiMH technology has expanded into new markets that were until recently the exclusive domain of alkaline primary cells. The recent introduction of NiMH rechargeable batteries, with their "ready to use" capability and equal or better performance than alkaline, has achieved wide market acceptance. The key to this NiMH technology advancement is charge retention, which has significantly improved to a remarkable 85% for 1 year. Further NiMH market growth based on its other performance attributes, such as cost, safety, life, and versatility of cell sizes and geometry, is expected to continue.

22.2 GENERAL CHARACTERISTICS

Table 22.1 summarizes the key advantages and disadvantages of the sealed NiMH battery. In addition to the essential characteristics of low cost, reliability, cycle life, and operating temperature, the following features of NiMH[1,2] have helped establish the technology's preeminence:

- Flexible cell sizes from 0.06 to 250 Ah
- Safe operation at high voltage (320+ V)
- Excellent volumetric energy and power, flexible vehicle packaging
- Easy application to series and series/parallel strings
- Choice of cylindrical or prismatic cells
- Safety in charge and discharge, including tolerance to abusive overcharge and overdischarge
- Maintenance free
- Excellent thermal properties (–30 to 70° C)
- Capability to utilize regenerative braking energy
- Simple and inexpensive charging and electronic control circuits
- Environmentally acceptable and recyclable materials

TABLE 22.1 Major Advantages and Disadvantages of Sealed Nickel-Metal Hydride Batteries

Advantages	Disadvantages
Higher energy density and specific energy compared to lead-acid and nickel-cadmium	Higher cost than lead-acid
Good high-temperature and high-rate capability	Lower specific energy and specific power compared to Li-ion
Good charge retention	Decreased performance at low temperature
Long cycle life	
Rapid recharge capability	
Long shelf life	
Sealed maintenance-free design	
Safe	

22.3 NiMH BATTERY CHEMISTRY

22.3.1 Chemical Reactions

During discharge, nickel oxyhydroxide is reduced to nickel hydroxide

$$NiOOH + H_2O + e \rightarrow Ni(OH)_2 + OH^- \quad E_o = 0.52 \text{ V}$$

and the metal hydride (MH) is oxidized to the metal alloy (M).

$$MH + OH^- \rightarrow M + H_2O + e \quad E_o = 0.83 \text{ V}$$

The overall reaction on discharge is

$$MH + NiOOH \rightarrow M + Ni(OH)_2 \quad E_o = 1.35 \text{ V}$$

The process is reversed during charge.

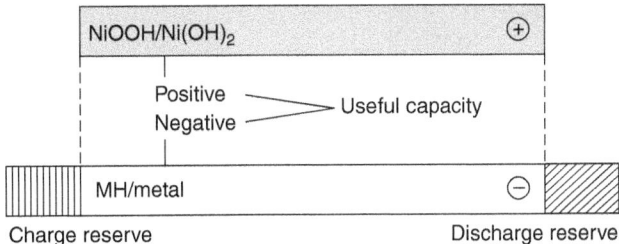

FIGURE 22.1 Schematic representation of electrodes of sealed nickel-metal hydride cell, divided into useful capacity, charge reserve, and discharge reserve.

The sealed nickel-metal hydride cell uses an oxygen-recombination mechanism to prevent the buildup of pressure that may result from the generation of gases toward the end of the charge and in overcharge. This is shown schematically in Fig. 22.1. During charge the positive electrode reaches full charge before the negative and begins to evolve oxygen.

$$2OH^- \rightarrow H_2O + \tfrac{1}{2}O_2 + 2e^-$$

The oxygen gas diffuses through the separator to the negative electrode facilitated by the starved-electrolyte design and the use of an appropriate separator.

At the negative electrode, the oxygen reacts with the hydrogen electrode to produce water. This results in stabilizing the internal pressure of the battery.

$$4MH + O_2 \rightarrow 4M + 2H_2O$$

Furthermore, the negative electrode does not become fully charged, which prevents the generation of hydrogen. This is true for the early stages of cycling where the only gas found inside the cell is oxygen. However, continued cycling of the cell results in hydrogen gas beginning to be evolved, and a noted increase in the proportional hydrogen within is observed. The reason for this behavior is the result of oxygen recombination on the surface of the metal hydride negative electrode, an exothermic reaction that causes localized heating of the metal hydride alloy. This heating in turn affects the equilibrium pressure of the alloy resulting in the release of hydrogen and increased pressure during charging. The charge current must then be controlled at the end of charge and during overcharge to limit the generation of oxygen to below the rate of recombination to prevent the buildup of gases and pressure.

A factor in the design of NiMH batteries is the negative to positive ratio (N/P). It is based on the use of a negative electrode (the metal hydride electrode) that has a higher effective capacity than the positive (or nickel oxyhydroxide electrode). Overall, as shown in Fig. 22.1, having an excess of metal hydride (MH) capacity allows for gas recombination reactions to occur during overcharge (oxygen recombination) and overdischarge (hydrogen recombination, see Sec. 22.10.11). In addition, excess MH electrode capacity is provided to inhibit oxidation and corrosion of the MH alloy. N/P ratios vary by cell design and for a given manufacturer but are typically within the range of 1.3 to 2.0. The lower N/P values are used to maximize energy while higher values are used in power and cycle life designs. The useful capacity of the battery is thus determined by the positive electrode.

22.3.2 Metal Hydride Alloy

NiMH batteries are an unusual battery technology in that the metal hydride active material is an engineered alloy made up of many different elements, and the MH alloy formulas vary to a significant degree.[3,4] The negative electrode contains either an AB_5 (LaCePrNdNiCoMnAl), an A_2B_7 (LaCePrNdMgNiCoMnAlZr), or an AB_2 (VTiZrNiCrCoMnAlSn) disordered type metal hydride active material.[5,6,7,8] AB_5 type alloys are more common, despite significantly lower hydrogen storage capacity as compared to A_2B_7 (320 vs. 380 mAh/g) and AB_2 (320 vs. 440 mAh/g). The advantages of the AB_5 alloys include low raw-material cost, easy activation and formation, flexibility in electrode processing, and high discharge-rate capability. On the other hand, there has been significant ongoing development to improve the properties of A_2B_7 and AB_2 materials to take advantage of their inherently higher energy, which is especially important for cost reduction.

Acceptable electrochemical utilization of metal hydride materials as anodes in NiMH batteries requires that those materials meet a demanding list of performance attributes, including hydrogen storage capacity, suitable metal-to-hydrogen bond strength, acceptable catalytic activity, discharge kinetics, and acceptable oxidation/corrosion resistance. Multi-element, multiphase, disordered alloys of the $LaNi_5$, LaMgNi, and VTiZrNiCr types are attractive development candidates for atomic engineering due to a broad range of elemental addition and substitution possibilities, availability of alternative crystallographic phases that form the matrix for chemical modification, and a tolerance for non-stoichiometric formulas. Through the introduction of modifier elements, easy activation and formation has been achieved. Special processing steps (such as alloy melting and size reduction) suitable for these metallurgically challenging materials have been developed.

The metal hydride active materials also have similar special design options. The active materials may be adjusted to obtain added capacity, power, or cycle life.

AB_5 Metal Hydride Alloys. For the AB_5 system, typical formulas include:

- $La_{5.7}Ce_{8.0}Pr_{0.8}Nd_{2.3}Ni_{59.2}Co_{12.2}Mn_{6.8}Al_{5.2}$ (atomic percent a/o)
- $La_{10.5}Ce_{4.3}Pr_{0.5}Nd_{1.3}Ni_{60.1}Co_{12.7}Mn_{5.9}Al_{4.7}$

While the capacity of the various AB_5 alloys is typically around 290 to 320 mAh/g, other overall performance attributes can be greatly influenced by changing the ratio of A to B elements in the alloy. It is common for the ratio of La/Ce to be reversed to emphasize cycle life and power. The total amount of Co, Mn, and Al significantly affect the ease of activation and formation, but increased cobalt has higher-cost implications. After production of the AB_5 alloy ingot, it is common to further refine the microstructure of the material by a post-anneal treatment at higher temperature for several hours. The annealing treatment can have a significant effect on capacity, discharge rate, and cycle life by adjusting crystallite size and grain boundaries, as well as by eliminating unwanted crystal structures that may be formed during ingot melting and casting. After annealing, the ingot is first crushed and then ground to the desired final particle size range. Additional special processing methods such as melt spinning and other rapid solidification techniques can also promote higher cycle life, although these processing methods are more expensive and there may be other trade-offs, such as discharge-rate capability.[9]

Commercial AB_5 alloys have a predominantly $CaCu_5$ crystal structure. However, within that structure, there are a range of lattice constants brought about by compositional disorder[10] within the material that are important to catalysis, storage capacity, and stability to the alkaline environment and embrittlement. These materials also precipitate a nickel-cobalt crystal structure, which is important to high-rate discharge.[11,12]

A_2B_7 Metal Hydride Alloys. A_2B_7 alloys also provide for compositional and processing choices. Popular A_2B_7 alloy formulas (a/o) include:

- $La_{4.8}Ce_{0.4}Pr_{9.1}Nd_{5.4}Mg_{1.7}Ni_{68.8}Co_{3.0}Mn_{0.2}Al_{5.5}Zr_{0.2}$
- $Nd_{18.7}Mg_{2.5}Ni_{74.7}Co_{0.1}Al_{3.6}Zr_{0.2}$

A_2B_7 alloy capacity may range from 335 to 400 mAh/g. Ce is often excluded from the alloy in an effort to reduce the tendency to form AB_5 crystal structures. The addition of Co helps reduce lattice expansion while Al forms a protective dense surface oxide, and together both improve the cycleability of the alloy. The addition of Mn has been shown to play a role in regulating the amount of the various crystal structures present, while the addition of Zr has been shown to improve the high-rate discharge capability of the alloys.

Alloy preparation using conventional methods results in a combination of PuNi$_3$ and CaCu$_5$ crystal structures within the alloy, but Mg is present only in the PuNi$_3$ crystal structure. After production of the A$_2$B$_7$ alloy ingot, further refinement of the material microstructure is required using a post-annealing treatment at a high temperature for several hours. The annealing treatment has been shown to be critical in reducing or eliminating the amount of AB$_5$ or other unwanted crystal structures that may be formed during ingot melting and casting. After annealing, the ingot is first crushed and then ground to the desired final particle size range. Additional, special processing methods such as melt spinning and other rapid solidification techniques may be required to completely eliminate unwanted AB$_5$ crystal structures depending on the alloy formulation.

AB$_2$ Metal Hydride Alloys. AB$_2$ alloys also provide for compositional and processing choices. Popular AB$_2$ alloy formulas (a/o) include:

- $V_{18}Ti_{15}Zr_{18}Ni_{29}Cr_5Co_7Mn_8$
- $V_5Ti_9Zr_{26.7}Ni_{38}Cr_5Mn_{16}Sn_{0.3}$
- $V_5Ti_9Zr_{26.2}Ni_{38}Cr_{3.5}Co_{1.5}Mn_{15.6}Al_{0.4}Sn_{0.8}$

AB$_2$ alloy capacity may range from 385 to 450 mAh/g. Higher vanadium-content alloys may suffer from higher rates of self-discharge due to the solubility of vanadium oxide and its consequent ability to form a special type of undesirable redox shuttle. The concentration of Co, Mn, Al, and Sn is important for easy activation, formation, and long cycle life. The ratio of hexagonal C$_{14}$ to cubic C$_{15}$ crystal structure is important for improving capacity or power.

Alloy preparation is done using conventional melting and casting methods to produce an ingot. After casting, the ingot is processed using a hydride/dehydride sequence to reduce the ingot into smaller pieces. These smaller pieces are then ground to the desired final particle size range. After grinding, oxygen is slowly introduced to allow for the growth of a protective oxide layer.

For all metal hydride alloys, the metal/electrolyte surface oxide interface is a crucial factor in discharge rate capability and cycle life stability. Original LaNi$_5$ and TiNi[13] alloys extensively studied in the 1970s and 1980s for NiMH battery applications were never commercialized due to poor discharge rate and cycle life capability.[14,15] Lack of catalytic activity at the surface oxide limits discharge, and lack of sufficient oxidation/corrosion resistance is a critical obstacle to long cycle life. The complicated chemical formulas and microstructures of present disordered AB$_5$, A$_2$B$_7$, and AB$_2$ alloys extend to the surface oxide. For the surface oxide, important factors include thickness, microporosity, and catalytic activity. Of importance to a high discharge rate was the discovery that ultrafine metallic nickel particles having a size on the order of 50 to 70 Å or less and dispersed within the oxide are excellent for catalyzing the reaction of hydrogen and hydroxyl ions.[16]

The other critical design factor within the surface oxide is to achieve a balance between surface oxide passivation and corrosion. Porosity with the oxide is important to allow ionic access to the metallic catalysts and therefore promote high-rate discharge. While passivation of the surface oxide is problematic for high-rate discharge and cycle life, unrestrained corrosion is equally destructive. Oxidation and corrosion of the anode metals consume electrolyte, change the state of charge balance, and create corrosion products that are sometimes soluble and capable of poisoning the positive electrode. Establishing a balance between passivation and corrosion for stability is a primary function of the compositional and structural disorder.

22.3.3 Nickel Hydroxide

Positive electrodes for use in NiMH batteries, whether cylindrical or prismatic, can be of the sintered or pasted type, which are also common to NiCd batteries. The nickel hydroxide for use in NiMH batteries is fundamentally the same as that used in NiCd and NiFe, and from a simple viewpoint, the basic compound is the same as that used by Edison and Junger 100 years ago. However, today's high-performance nickel hydroxide is more advanced and continues to improve in capacity, utilization, power and discharge rate capability, cycle life, high temperature charging efficiency, and cost.

Spherical Nickel Hydroxide. As mentioned previously, one type of nickel hydroxide is by far the most common—a high-density spherical type used in pasted electrodes which became commercial around 1990.[17,18] High-density spherical nickel hydroxide is made in a precipitation reactor where metal salts such as nickel sulfate are reacted with a caustic such as NaOH in the presence of ammonia. The nickel source may have additives such as cobalt and zinc to enhance performance. The important physical parameters within this type of nickel hydroxide are:

- **Chemical formula:** The nickel hydroxide active material itself is most commonly a NiCoZn tri-precipitate; a common composition is $Ni_{94}Co_3Zn_3$. However, the amounts of cobalt and zinc, which are usually about 1 to 5% each, can be adjusted for conductivity, oxygen overvoltage, and microstructure refinement with some design trade-offs in terms of active material capacity and cost. Other more complicated multi-elemental precipitates such as NiCoZnCaMg offer higher capacity, cycle life, and high-temperature performance, but cannot be manufactured by conventional precipitation processes.
- **Tap density:** Usually around 2.2 g/cc, tap density is a measure of the dry nickel hydroxide powder packing efficiency and influences the amount of active material that can be loaded into the pores of the nickel foam current collector.
- **Particle size:** An average particle size of about 10 microns.
- **Surface area:** Measured by the BET method, surface area refers not to the geometric area of each nickel hydroxide sphere, but rather to the total surface area of each particle, which contributes to the charge-discharge reactions and can thus affect utilization and high-rate discharge capability. Typical BET surface area for high-density spherical nickel hydroxide is about 10 to 20 m^2/g.
- **Crystallinity:** Each nickel hydroxide sphere has an extremely high surface area corresponding to the nickel hydroxide crystallites themselves. Crystallinity is measured by x-ray diffraction, where the full width at half maximum (FWHM) of a reflection, such as the <101> plane, may yield a typical crystallite size of about 110 Å.

A variety of other factors contribute to performance, such as impurities from processing like residual sulfates, nitrates, sodium sulfate, and others.

The more common pasted nickel hydroxide positive electrode is typically produced by mechanically pasting high-density spherical nickel hydroxide into the pores of a foam metal substrate (Fig. 22.2). The foam metal substrate is typically produced by coating polyurethane foam with a layer of nickel either

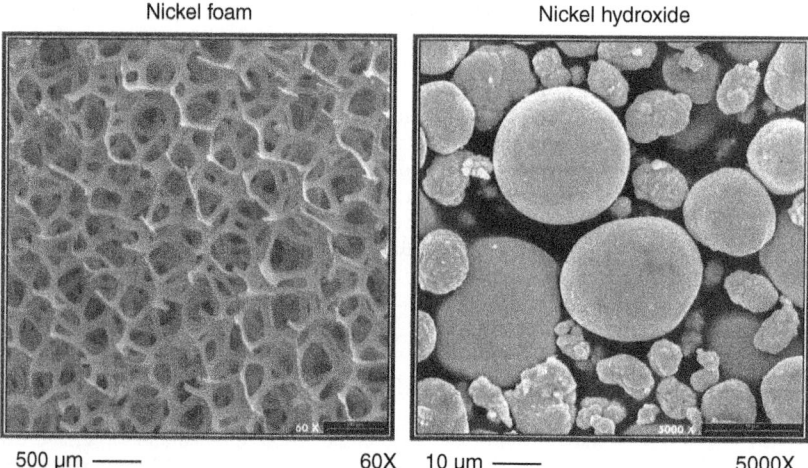

FIGURE 22.2 SEM micrographs of positive electrode nickel foam substrate and high-density spherical nickel hydroxide.

by electroplating or by chemical vapor deposition, followed by a heat treatment process to remove the base polyurethane. The pore size may be decreased from about 400 microns to 200 microns for better conductivity. The density of the foam may also be adjusted to promote conductivity and power versus capacity and utilization.

The foam[19] is then physically loaded with nickel hydroxide having an average diameter of about 10 microns in a paste containing conductive cobalt oxides, which form a conductive network to compensate for the large distance from the nickel hydroxide to the metal current collector and for the fact that the nickel hydroxide itself is relatively low in conductivity. A cross-sectional comparison of sintered and pasted positive electrodes is presented in Fig. 22.3 under backscattered electron imaging, where the bright areas indicate metallic nickel current collection.

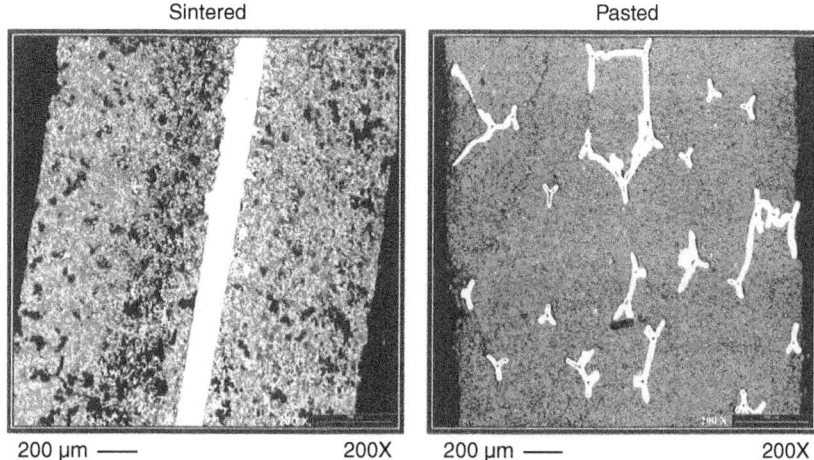

FIGURE 22.3 SEM micrographs under BEI imaging where bright areas indicate nickel metal current collection, illustrating the difference in active material distance to current collector for pasted and sintered positive electrodes.

The nickel hydroxide active material and electrode formula can be specially formulated for specific applications. For operation at temperatures above 35°C, some manufacturers may use other additives to the paste formula to inhibit premature oxygen evolution on charge (see "High-Temperature Nickel Hydroxide" below).[20] In addition, paste formula modifications may adjust the type and quantity of conductive network additives such as cobalt metal and cobalt monoxide.[21,22,23] Usually, the paste additives are finely divided cobalt metal and cobalt monoxide that will dissolve and reprecipitate on the surface of the nickel hydroxide active material. However, the coating may not be uniform and complete in coverage.

For ultra-high-power discharge, it is possible to add metallic nickel fibers to the paste formula to enhance conductivity. In the case of calcium additives, the additive may cause a loss in power and/or cycle life. Cobalt metal and cobalt monoxide are relatively expensive battery materials, and increased use has cost implications. Addition of metallic nickel fibers lowers the amount of active material, resulting in reduced capacity and specific energy.

High-Temperature Nickel Hydroxide. To combat the premature oxygen evolution mechanism resulting from temperatures above ambient, manufacturers often introduce oxygen evolution suppressants such as $Ca(OH)_2$, CaF_2, or Y_2O_3. As shown in Fig. 22.4, introduction of these additives can reduce capacity loss under 65°C charging from about 50% to below 20%. NiMH manufacturers must carefully select the oxygen suppressant type, amount, and location to avoid deleterious effects such as power loss and cycle life reduction due to the nonconductive nature of many of these materials.

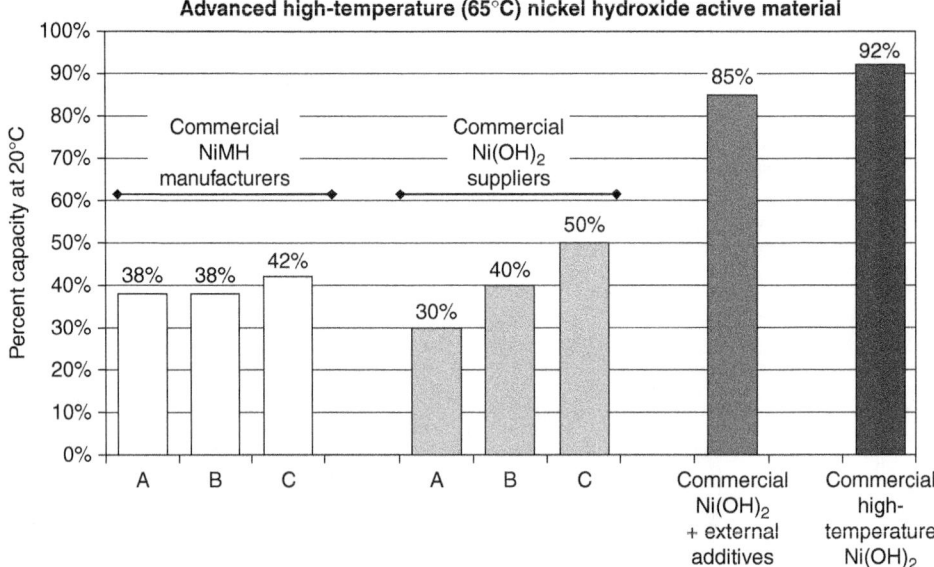

- Use of external additives may significantly degrade power and cycle life of commercial Ni(OH)$_2$ due to conductivity loss.
- High-temperature Ni(OH)$_2$ overcomes this by modifying the base Ni(OH)$_2$ formulation to increase charge acceptance.

FIGURE 22.4 Temperature performance of cylindrical C-size batteries using commercial nickel hydroxide.

Another method is to modify the formula of the nickel hydroxide itself. The most common NiMH positive active material is NiCoZn. To improve high-temperature performance, multi-element formulas such as NiCoZnCaMg have been developed, as shown in Fig. 22.5.

Sintered Nickel Hydroxide. Sintered electrodes have the best rate and power capability,[24] but sacrifice capacity on a weight and volume basis and are more expensive to manufacture. Sintered

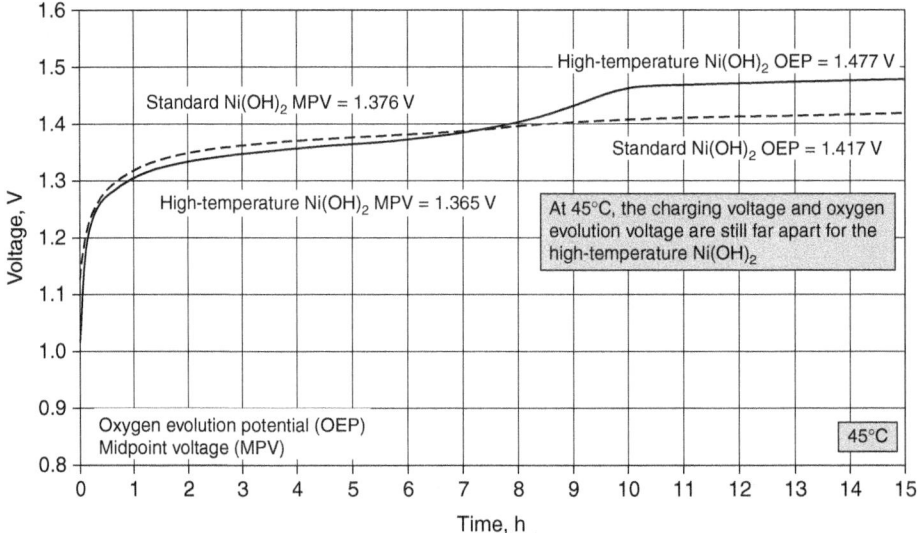

FIGURE 22.5 Charge characteristics of nickel hydroxide as a function of chemical formulation at 45°C.

electrodes involve an expensive and complicated sequence of manufacturing steps, and consequently it is typical that only companies with an existing capital investment for this kind of process would manufacture sintered electrodes. Sintered positives begin with the pasting of filamentary nickel[25] onto a substrate such as perforated foil, where the nickel fibers are then sintered under a high-temperature annealing furnace in a nitrogen/hydrogen atmosphere, and binders from the pasting process are burned away to leave a conductive skeleton of nickel having a typical average pore size of about 30 microns.

Nickel hydroxide is then precipitated into the pores of the sinter skeleton using either a chemical or electrochemical impregnation process. The impregnated electrode is then formed or preactivated in an electrochemical charge/discharge process. Important design variables in the manufacture of sintered nickel hydroxide electrodes include:

- Strength and pore diameter of the filamentary nickel skeleton
- Chemical composition of the nickel hydroxide active material
- Active material loading
- Amount of harmful impurities (e.g., nitrate, carbonate)

One aspect of sintered versus pasted nickel hydroxide technology is that sintered electrodes require the battery manufacturer to make a significant capital investment in facilities and equipment, and to have a great deal of internal expertise in processing. Pasted electrodes, conversely, place a great deal of emphasis on the expertise of the suppliers for both the nickel foam substrate and for the high-density spherical nickel hydroxide. Recent development in pasted electrode technology has brought about exceptional improvements in power and high-rate discharge capability to a level comparable to that of the sintered electrode technology.

22.3.4 Electrolyte

The electrolyte in NiMH batteries of all types is routinely a mixture of about 30% potassium hydroxide in water, providing high conductivity over a wide temperature range. It is most common for the electrolyte to have a lithium hydroxide additive at a concentration of about 17 g/L to promote improved charging efficiency at the nickel hydroxide electrode by suppressing oxygen evolution, the competing reaction to charge acceptance.

An important feature of the electrolyte is related to fill fraction. Essentially all NiMH batteries are of the sealed, starved electrolyte design. Similar to NiCd, the electrodes are nearly saturated with electrolyte, while the separator is only partially saturated to allow for rapid gas transport and recombination.

Special electrolytes are also used in NiMH batteries to enhance high-temperature operation. Instead of binary KOH/LiOH electrolytes, it is also possible to substitute a portion of the KOH with NaOH. The ternary KOH/NaOH/LiOH electrolyte is still at a high concentration of about 7M, but the contribution of NaOH promotes high-temperature charging-efficiency although it is typical for this electrolyte to decrease cycle life through increased corrosion of the metal hydride active materials.

22.3.5 Separator

The primary function of the separator is to prevent electrical contact between the positive and negative electrodes, while holding electrolyte necessary for ionic transport. Original separators for NiMH batteries were standard NiCd and NiH_2 separator materials. However, NiMH batteries proved to be more sensitive to self-discharge, especially when conventional nylon separators were used.[26] The presence of oxygen and hydrogen gas causes the polyamide materials in the nylon separator to decompose. The corrosion products from this decomposition allowed for poisoning of the nickel hydroxide, promoting premature oxygen evolution and also forming compounds capable of redox shuttle between the two electrodes, which further increases the rate of self-discharge.

In addition, the separator plays a crucial role relative to cycle life. In the starved electrolyte design, it is a common design principle to essentially saturate the electrodes with electrolyte at the assembly stage. The separator is designed to have a high electrolyte fill fraction in order to hold as much electrolyte as possible but not be overfilled so as to inhibit gas recombination. To the battery manufacturer, this means that during the first few charge/discharge cycles ("formation"), when the electrodes have not yet absorbed their full amount of intended electrolyte, charging must be initiated carefully to avoid venting.

The electrolyte design concept relates to capillary theory so that the electrolyte will migrate to the smallest pores. In the NiMH battery, this translates to the nickel hydroxide positive electrode having the smallest pores, followed by the metal hydride negative electrode, and finally the separator. At cell assembly, it is common for the separator to be about 90% filled, and then reduced to about 70% during the cell formation process of the first few charge/discharge cycles as both the positive and negative electrodes expand and contract, opening interior regions for electrolyte absorption. This process continues to some degree over many hundreds of charge/discharge cycles, where NiMH cell failure is common when the separator fill fraction has been reduced to about 10 to 15% of its original level. As a result, separators that can absorb larger quantities of electrolyte at cell assembly, have small pores able to compete for electrolyte, and retain surface wettability are highly desirable. Inspection of the separator in failed batteries shows even these types of separators undergo some degradation and loss of electrolyte absorption ability, although not nearly to the same degree as earlier generation separators.

Consequently, there was a need for a more stable NiMH separator material to reduce self-discharge while still retaining electrolyte crucial for maintaining cycle life. In NiMH batteries, there is now widespread use of what is termed "permanently wettable polypropylene." In fact, this separator is a composite of polypropylene and polyethylene where the base composite fibers require special surface treatments to make them wettable to the electrolyte. Currently there are two major types of surface treatments:

- **Acrylic acid:** This process involves the grafting of a chemical such as acrylic acid to the base fibers to impart wettability and is accomplished using a variety of techniques such as UV or cobalt radiation.[27]
- **Sulfonation:** This process imparts wettability to the polypropylene by exposing the base fiber material to fuming sulfuric acid. The separator surface is designed to be made hydrophilic to the electrolyte after completion of the treatment. Use of sulfonated separator materials is a key component in enabling "precharged" NiMH battery technology.

22.4 CELL CONSTRUCTION TYPES

Sealed nickel-metal hydride cells and batteries are constructed in cylindrical, button, and prismatic (both large and small) configurations, similar to those used for sealed nickel-cadmium batteries.

The electrodes are designed with highly porous structures having a large surface area to provide a low internal resistance and a capability for high-rate performance. The positive electrode in the cylindrical NiMH cell is a highly porous sintered or foam-nickel substrate into which the nickel compounds are impregnated, or pasted, and converted into the active material by electro-deposition. Foams have generally replaced sintered plaque electrodes. Expanded metals and perforated sheets are lower cost, but they have poor high-rate capability. Sintered structures are much more expensive. The negative electrode, similarly, is a highly porous structure using a perforated nickel foil or grid onto which the plastic bonded active hydrogen storage alloy is coated. The electrodes are separated with a synthetic nonwoven material, which serves as an insulator between the two electrodes and as a medium for absorbing the electrolyte.

22.4.1 Cylindrical Configuration

The assembly of the cylindrical unit is shown in Fig. 22.6. The electrodes are spirally wound, and the assembly is inserted into a cylindrical nickel-plated steel can. The electrolyte is added and contained within the pores of the electrodes and separator.

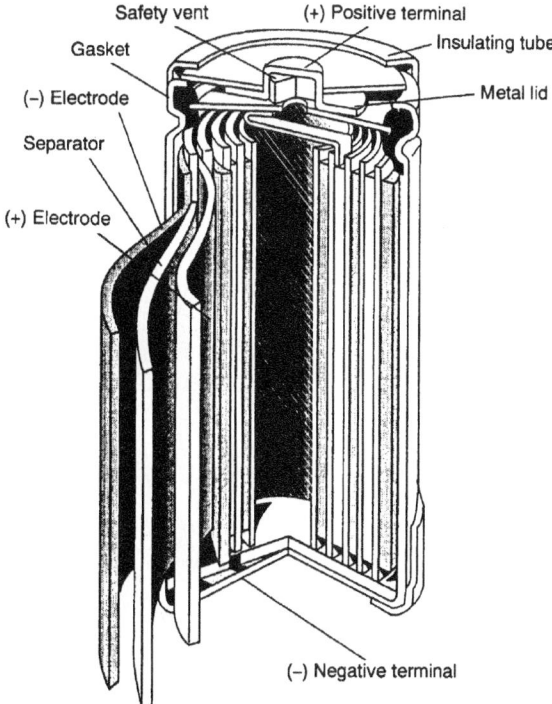

FIGURE 22.6 Construction of a sealed, cylindrical nickel-metal hydride battery.

The cell is sealed by crimping the top assembly to the can. The top assembly consists of a lid, which includes a resealable safety vent, a terminal cap, and a plastic gasket. The can serves as the negative terminal and the lid as the positive terminal, both insulated from each other by the gasket. The vent provides additional safety by releasing any excessive pressure that may build up if the battery is subjected to abuse.

22.4.2 Button Configuration

The button configuration is illustrated in Fig. 22.7. It is similar in construction to the nickel-cadmium button cell, except that the cadmium is replaced by the hydrogen storage alloy.

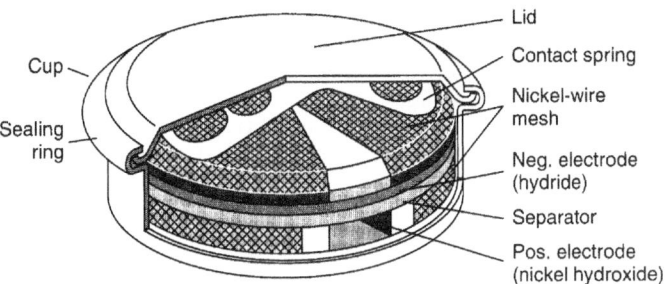

FIGURE 22.7 Construction of a sealed, nickel-metal hydride button cell.

22.4.3 Small Prismatic Configuration

The thin prismatic batteries are designed to meet the needs of compact equipment. The rectangular shape permits more efficient battery assembly, eliminating the voids that occur within the assembly of cylindrical cells. The volumetric energy density of the battery can be increased by a factor of about 22%. The prismatic cells also offer more flexibility in the design of batteries, since the battery footprint is not controlled by the diameter of the cylindrical cell.

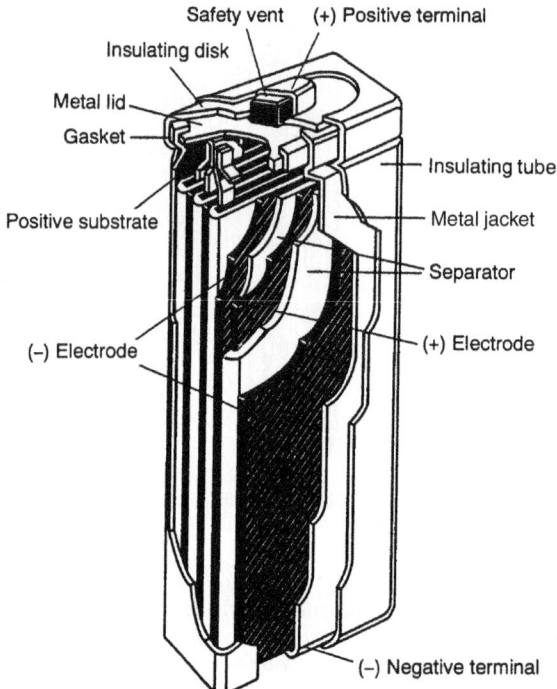

FIGURE 22.8 Construction of a sealed, thin, prismatic nickel-metal hydride button cell.

Figure 22.8 shows the structure of a prismatic battery. The electrodes are manufactured in a similar manner as the electrodes for the cylindrical cell, except that the finished electrodes are flat and rectangular in shape. The flat electrodes are then assembled, with the positive and negative electrodes interspaced by separator sheets, and welded to the cover plate. The assembly is then placed in the nickel-plated steel can and the electrolyte is added. The cell is sealed by crimping the top assembly to the can. The top assembly is a lid which incorporates a resealable safety vent, a terminal cap, and a plastic gasket, similar to the one used on the cylindrical cell. An insulating heat shrink tube is placed over the metal can (jacket). The bottom of the metal can serves as the negative terminal and the top lid as the positive terminal. The gasket insulates the terminal from each other.

22.4.4 9 V Multicell Configuration

The construction of a typical 9-Volt multicell battery is shown in Fig. 22.9.

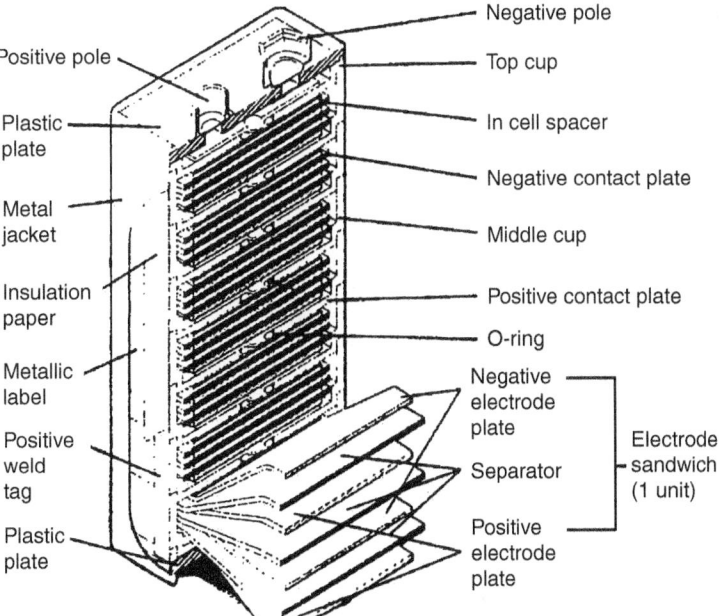

FIGURE 22.9 Cutaway view of a sealed 9-Volt NiMH cell.

22.4.5 Large Prismatic Configuration

The large prismatic configuration is similar in design to that of the small prismatic but with some exceptions, noted below.

Figure 22.10 shows the structure of a large prismatic battery. The electrodes are manufactured in a similar manner as the electrodes for the small prismatic cell. The electrodes are assembled with the positive and negative electrodes interspaced by separator sheets and welded to the individual terminals. The assembly is then placed into the can and the electrolyte is added. The cell is sealed by welding the top assembly to the can. The top assembly is a lid which incorporates a resealable safety vent, terminals, and plastic gaskets, similar to the one used on the cylindrical and small prismatic cell. The metal can (jacket) can be powder coated or taped with an insulating material to prevent cell to cell shorting. Two terminals projecting through the can lid serve as connections to the negative and positive electrodes, with a gasket insulating the terminals from the can.

22.4.6 Monoblock Construction

NiMH battery chemistry is especially well suited for HEV monoblock construction due to its high tolerance to overcharge and overdischarge, which pose problems for other battery chemistries. The monoblock design reduces cost by having a common pressure vessel construction and far fewer cell parts; a single vent assembly and shared hardware can be used in multicell modules. Further attributes of the monoblock construction include reduced volume, since interior cell walls can be shared between cells, and flexible choices of liquid or air cooling. An air-cooled 7.2 V 6.5 Ah plastic monoblock HEV NiMH battery is shown in Fig. 22.11.

Issues of monoblock construction include selection of the plastic casing material to avoid gas permeation and the need for individual cells to have well-matched capacity and impedance to avoid

22.14 SECONDARY BATTERIES

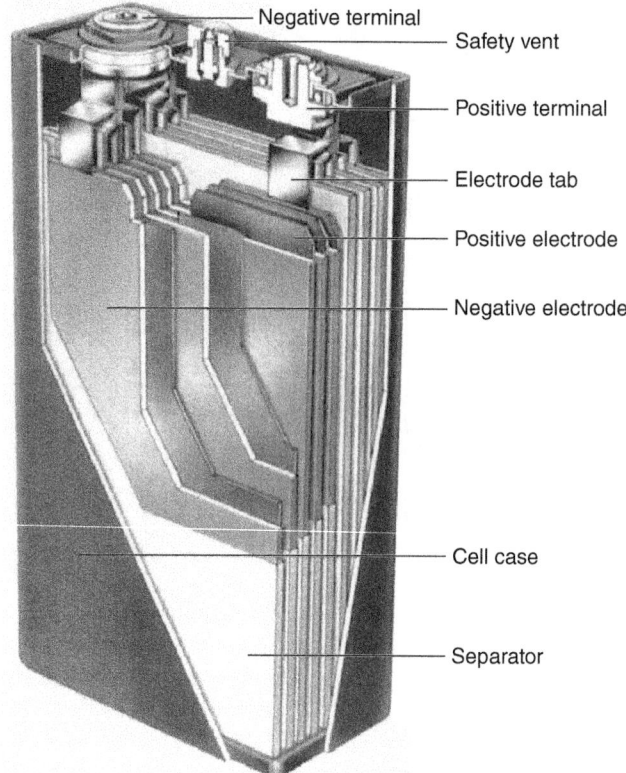

FIGURE 22.10 Cutaway view of a prismatic NiMH cell.

FIGURE 22.11 HEV plastic case prismatic module used in Toyota Prius.

cell-to-cell imbalance. Further, monoblock construction must recognize that electrodes within each cell expand and contract during charge and discharge and that the ensuing swelling of the electrode stack must be compensated for in the mechanical design and loading of the monoblock.

22.5 CELL DESIGN ISSUES

22.5.1 Cylindrical versus Prismatic Configuration

NiMH batteries are versatile in that both cylindrical and prismatic constructions can be utilized. Each type of construction has advantages and disadvantages, and a particular end-use can predetermine the most suitable configuration. For NiMH applications below about 10 Ah, cylindrical construction dominates due to the lower cost and higher speed of manufacture. Above 20 Ah, cylindrical construction is extremely difficult, and the prismatic configuration dominates. In the 10 to 20 Ah cell size range, manufacturers are offering both cylindrical and prismatic designs, although prismatic designs are most common.

Cylindrical cells for industrial and propulsion applications are similar to high-volume production consumer batteries in that the well-known "jelly-roll" construction is used. However, most small portable cylindrical cells require only low to moderate discharge rate capability, and electrode terminal connections are usually quite simple. Conversely, since industrial and propulsion NiMH applications require high to ultra-high discharge rate capability and low internal resistance, multiple tab current collection is used. This type of construction is termed "edge welding" and requires that each electrode has a current collecting strip on one of its sides. The current collecting strips (both for the positive and negative electrodes) are disposed on opposite sides of the jelly-roll. After coil winding of the jelly-roll, the edge current collector is welded in multiple locations to each electrode. Other aspects of cell assembly are virtually identical to small consumer batteries. The net result of the enhanced current collection is a reduction in specific energy due to added weight, but an increase in specific power due to decreased cell AC impedance (usually around 8 to 12 mΩ for small portable batteries and around 1 to 2 mΩ for industrial cylindrical cells). For industrial applications such as HEV, motorcycle, and bicycle applications, the most popular cylindrical cell sizes are standard C and D sizes, although a multitude of height changes within that diameter are also used. Some work on larger size cells such as F size has also been reported.[28]

Prismatic construction is conventional in that electrode stacks of alternating positive and negative electrodes with intermediate separators are used (as shown in Sec. 22.4.5). The main design alternatives involve the thickness and number of each of the electrodes and the aspect ratio (relative proportions of cell height to width to thickness). Key design variations include the ratio of active materials to inactive components (such as the cell can and terminal, and current collectors). In all cases, the cell designer has the objective of emphasizing one or more properties of performance characteristics such as energy versus power, while maintaining a minimum threshold of other performance parameter such as cycle life. One example is that for EV NiMH prismatic cells,[29] a specific power of about 200 W/kg is acceptable for most vehicles. Consequently, relatively thick positive and negative electrodes can be used to increase the ratio of active material to inactive cell components, thereby providing specific energy in the 63 to 80 Wh/kg range. Alternately, HEV prismatic NiMH batteries typically must deliver a specific power greater than 1300 W/kg, and therefore electrode thickness is reduced as compared to the electrode thickness used for EV applications. NiMH batteries for HEV use have a typical specific energy ranging from about 42 to 68 Wh/kg.

22.5.2 Metal versus Plastic Cell Cases

NiMH cylindrical cells exclusively use metal cell cases. One important reason for this is that the can itself is electrically connected to the metal hydride negative electrode and serves as the negative terminal. Another important reason is that many applications require fast charge in which gas recombination can cause considerable internal pressure and only the strength of a metal container would suffice without significant volumetric penalties. Finally, the metal crimp seal to the cover plate assembly with a polysulfone seal ring is inexpensive, fast, and reliable.

Both metallic and plastic cell cases are common for the prismatic NiMH batteries used in automotive applications such as electric vehicles. Unlike cylindrical cells where the can itself is the negative terminal, prismatic designs have both a positive and negative terminal at the top cover plate. Key decision criteria for selecting metal cell cases include excellent thermal conductivity, inexpensive prototyping costs for changing cell dimensions, and small volumetric penalties.

The primary advantage of plastic cell cases are cost and electrical isolation in the ≥ 320 V battery packs common to today's electric vehicles, where even high resistance leakage currents must be considered. Further design considerations for plastic cell cases include development costs for permanent molds, gas permeation, thermal conductivity, and sufficient plastic thickness for gas pressure containment without wall bulging. NiMH battery modules using plastic cell cases are shown in Fig. 22.12.

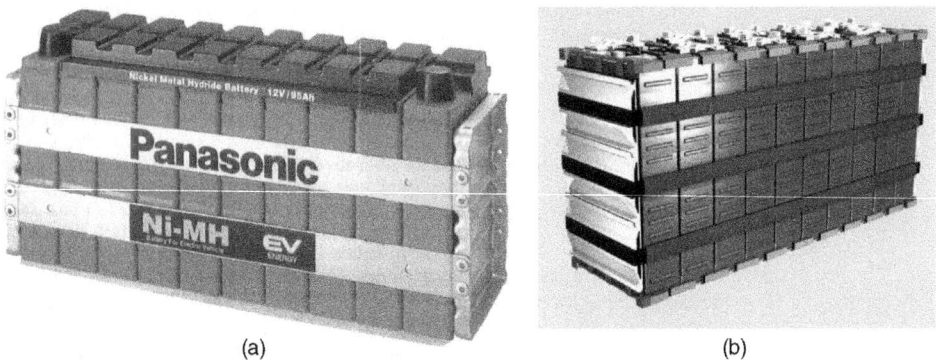

FIGURE 22.12 NiMH battery modules using (a) plastic cases; (b) metal cases.

22.5.3 Energy versus Power Design Trade-offs

Similar to other rechargeable battery technologies, NiMH batteries can be designed to emphasize energy, power, or some combination of the two. The application itself may dictate the choice, but perhaps not for obvious reasons. In electric vehicle applications, a certain threshold power such as 200 W/kg is required for adequate vehicle performance, and once the power requirement is met, competing designs usually tout specific energy in the range of 62 to 80 Wh/kg. The motivation for the higher energy is of course longer vehicle range, but can be attained only with added cost. The typical figure of merit for electric vehicle battery cost is $/kWh, where the USABC/PNGV goal of $150 per kWh has proven to be one of the most challenging development targets. NiMH cost is mainly controlled by the amount, cost, and utilization of raw materials rather than processing labor, assembly, packaging, etc. Consequently, an important development activity for many NiMH manufacturers is to obtain higher utilization of less expensive active materials on a mAh/g basis and therefore reduce battery costs.

For HEV applications, specific energy is of far less importance since the battery has a vastly different function than in pure electric vehicles. For HEVs, the main purpose of the battery is to accept and utilize the energy from regenerative braking and assist in acceleration at startup. The battery is commonly exposed to very-high-current pulses during both charge and discharge, but energy extraction is usually limited to relatively small depths of discharge. The crucial requirement for the NiMH HEV battery is therefore specific power, and the USABC/PNGV goal for battery developers is greater than 1000 W/kg. NiMH batteries available today are in the 1000 to 1300 W/kg specific power range, with development reports approaching 2000 W/kg. Besides performance required for HEV operation, an important figure of merit for NiMH battery developers is power cost in $/kW, and the PNGV goal is below $12 per kW. In this case, the specific energy for a typical HEV battery may be in the 32 to 56 Wh/kg range combined with specific power in the 1000 to 1300 W/kg range.

22.5.4 Cell, Module, and Pack Design

A typical 5.0 Ah C-size NiMH portable cell has an AC impedance on the order of 8 to 15 mΩ, and a $\Delta V/\Delta I$ resistance of 15 to 30 mΩ. In contrast, a 100 Ah NiMH EV cell has an impedance on the order of 0.4 mΩ, and a $\Delta V/\Delta I$ resistance of 0.9 mΩ. Despite the low resistance, heating is still a concern due to the extremely high current pulses resulting from regenerative braking and during vehicle acceleration. Even the I^2R heating effects at these high currents are small in comparison to heating due to overcharge. Consequently, an essential aspect of NiMH EV and HEV battery design involves thermal management.

Proper thermal management begins at the metal hydride negative electrode, understanding that overcharge heat is generated at the surface of the hydride electrode where oxygen is recombined. Heat must migrate from the negative electrode to the cell case and therefore good thermal conductivity within the electrode and electrode stack bundle is important. Cells are usually packaged into 12 V modules, with cells bound together in a back-to-back arrangement. An important design feature is that the cells at each end will have a much higher exposed surface area available for convective cooling. This raises concerns of thermal imbalance within a module and a resultant state-of-charge imbalance which will lead to premature failure. Therefore, proper module design must include endplate and heat sink considerations.

Modules are packaged in a variety of configurations. Within the various packaging arrangements, important design considerations are distance between modules, air or water flow channels, and battery tray heat sink characteristics to equalize cooling from module to module.

22.5.5 Thermal Management—Water Cooling versus Air Cooling

A first step in NiMH EV, PHEV, and HEV thermal management is to decide whether to use water cooling or air cooling. There has not yet been a clear-cut consensus on which approach will dominate. At present, all hybrid electric vehicles use air-cooled batteries, including the Toyota Prius, Honda Insight, Ford Escape, and Fusion. Currently, demonstration models using other battery technologies for plug-in hybrids and pure electric vehicles require water-cooled batteries to more closely control temperature variations. The merits of each approach are highlighted in Table 22.2.

TABLE 22.2 Summary of the Merits for Thermal Management of HEV, PHEV, and EV Batteries

	Air cooling	Water cooling
Advantages	• Lightweight • Simple	• More effective heat transfer • Average fluid temperature more consistent • Integrated with vehicle cooling
Disadvantages	• Complicated distribution of air within pack • Incoming air must be free of dirt and water from road • Variable ambient temperature	• Increased weight • Elaborate module design • Higher average fluid temperature

22.6 EV BATTERY PACKS

22.6.1 USABC Performance Targets

It is interesting to review the USABC technical performance requirements, (see Table 22.3) and then evaluate why NiMH batteries led in virtually all worldwide EV battery demonstration and development programs before most manufacturers abandoned EV work in favor of HEVs.

TABLE 22.3 Primary USABC Midterm Performance Goals for the EV Battery and Actual Performance of the Commercial and Advanced NiMH Battery

Property	USABC	NiMH Commercial	NiMH Prototype
Specific energy (Wh/kg)	80 (100 desired)	63–75	85–90
Energy density (Wh/L)	135	220	250
Power density (W/L)	250	850	1000
Specific power (W/kg) (80% DOD in 30 s)	150 (>200 desired)	220	240
Cycle life (cycles) (80% DOD)	600	600–1200	600–1200
Life (years)	5	10	10
Environmental operating temperature	−30° to 65°C	−30° to 65°C	−30° to 65°C
Recharge time	< 6 hours	15 min (60%) <1 h (100%)	15 min (60%) <1 h (100%)
Self-discharge	<15% in 48 h	<10% in 48 h	<10% in 48 h
Ultimate projected price ($/kWh) (10,000 units at 40 kWh)	<$150	$220–400	$150

Despite the complicated appearance of the USABC requirements, the actual development targets were quite simple. In 1991, there were only two types of automotive-size rechargeable batteries that could be used for electric vehicles: lead-acid and nickel-cadmium. Lead-acid batteries simply did not have sufficient specific energy and cycle life to be viable for EV use. NiCd batteries, while somewhat higher in specific energy and with good cycle life, were not high enough in energy and still faced the environmental issues inherent in the use of cadmium, operation in a flooded electrolyte configuration requiring maintenance, and with the memory effect.

NiMH dominated early EV applications over Na-S, Ni-Zn, and other advanced systems due to excellent overall performance, environmental friendliness, and safety factors. Essentially, an advanced battery technology for EV applications had to satisfy all minimum performance requirements and then compete on energy and cost. NiMH EV batteries demonstrated:

- **High energy:** Commercial electric vehicles (such as GM's EV1) achieved a range of over 200 mile when using NiMH batteries with a specific energy of 70 Wh/kg. Prototype vehicles (such as Solectria Force) using NiMH batteries demonstrated a 350 mile range.
- **High power:** Commercial electric vehicles using 220 W/kg specific power NiMH batteries had demonstrated acceleration comparable to ICE powered vehicles. Advanced EVs using NiMH routinely set race and demonstration range and speed records.
- **Flexible packaging:** EV NiMH batteries operated well in 320 V AC propulsion or 180 V DC propulsion systems and had a wide variety of cell capacities. The volumetric performance of NiMH was exceptional, offering vehicle designers a small size battery package.
- Long life (600 to 1200 cycles to 80% DOD).
- Wide operating temperature range (−30°C to + 65°C).
- Fast charge and simple normal charging.
- Maintenance-free operation.

The main developmental activity for NiMH EV batteries was to reduce cost. Because only prototype production volumes existed, the initial cost of manufacture for NiMH batteries was relatively high. Significant investment was made to decrease cost, with highlights including:

- **Pilot manufacturing capability:** Increased volume production allowed reduced overhead and labor, and the incorporation of automated and semi-automated manufacturing.
- **Increased specific energy:** By raising specific energy from 56 to 80 Wh/kg through increased utilization of active materials, the battery cost in $/kWh had decreased significantly (>$1000/kWh to <$350/kWh, even at relatively low production volumes).
- **Lower cost nickel hydroxide:** From 1995, where high-density spherical nickel hydroxide prices were around $30 per kg, new lower cost suppliers and manufacturing processes allowed nickel hydroxide prices to drop to about $20 per kg.
- **Lower cost metal hydride materials:** Improved efficiency melt/cast processes, scrap recycle, elimination of costly sintering, versatile use of inexpensive substrate materials and significantly reduced battery formation.

22.7 HYBRID BATTERY PACKS

22.7.1 Hybrid Electric Vehicle Types

The crucial targets for HEV, as compared to PHEV or EV goals, include a much greater emphasis on power, both on charge and on discharge, and less emphasis on energy. For HEV applications, typical NiMH batteries have a specific energy in the 45 Wh/kg range and have attained a specific power of about 1000 to 1300 W/kg. HEV cycle life is measured at low depth of discharge (approximately 2 to 5% DOD) compared to 50 to 80% for PHEV and 80% DOD for EV. Even cost is calculated on available power in $/kW. Car makers essentially want to use the smallest energy battery capable of supplying the required power. An important figure of merit target for the HEV car makers is a power to energy ratio (P/E) of 40. Present NiMH HEV batteries have attained P/E ratios of about 25.

In this context, the Partnership for the Next Generation Vehicles (PNGV) was established to enable automakers and the government to meet an equivalent fuel economy of 80 miles per gallon of gasoline. For many car manufacturers, development efforts focused on a variety of different hybrid electric vehicles of a variety of concepts.[30] For example, several types of HEVs were based on how the internal combustion engine (ICE) and electric drive are cooperatively incorporated into the system and, in a parallel design, how the ICE and the electric motor drove the wheels.

In all cases, the electric motor can also be used to recharge the battery. There are several different types of hybrids on the market today. The various operating modes for each type of vehicle make the size of the battery extremely wide ranging, anywhere from 0.5 to 20 kWh. The main difference between such operating modes is the degree of electrification (purely electric drive) involved. Below is a brief breakdown of these operating modes based on increasing levels of electric motor usage. More detailed information is found in Chap. 29.

Micro Hybrid. A micro hybrid uses start/stop technology where the ICE shuts down upon braking thereby reducing fuel consumption and emitting less pollution while stopped. The electric motors are incapable of providing the vehicle with any level of propulsion without ICE help and can recapture some energy through regenerative breaking. The system consists of a 42 V battery and electric motor, with most manufacturers opting for lead-acid technology due to cost concerns. This type of hybrid operates in a charge-sustaining mode and is widely used by European automotive manufacturers.

Mild Hybrids. A mild hybrid, like the micro hybrid, uses stop/start technology, shutting down the engine for stops; is incapable of providing the vehicle with any level of propulsion without ICE help; and can recapture some energy through regenerative breaking. Unlike the micro hybrid, a mild hybrid can provide some level of power assist to the ICE and operate at a higher voltage—140 to 160 V. This type of hybrid operates in a charge-sustaining mode and is used by automotive manufacturers such as GM, Mercedes, and BMW.

Full Hybrids (HEV). A full hybrid incorporates the same features as the micro and mild hybrids, i.e., start/stop and regenerative braking, and adds a limited range of pure electric propulsion—typically less than 5 miles at speeds under 35 mph. These full hybrids operate in a charge-sustaining mode and are used by automotive manufacturers such as Toyota, Honda, and Ford. The systems vary from automaker to automaker, but the batteries operate in the 200 to 300 V range with less than 2.0 kWh of energy storage.

Plug-in Hybrids (PHEV). A plug-in hybrid is a full hybrid but with a larger pure electric driving range—typically 10 to 40 miles at speeds up to 35 mph. The batteries for PHEVs operate in the same voltage range as the full hybrid and contain more energy, 5 to 20 kWh depending on vehicle size. They are equipped with a power cord and are designed to be recharged either at home or at public charging stations. This type of hybrid operates both in a charge-sustained mode and a charge-depletion mode, as shown in Fig. 22.13. This dual mode operation places a high degree of stress on the battery, making cycle life a determining factor in the choice of battery chemistry. Lithium-based PHEVs are currently in prototype production and testing by most major automotive manufacturers, such as Toyota, GM, Mercedes, BMW, and Ford, while Plug-In Conversions Corporation (PICC) is presently providing a 6.4 kWh NiMH battery for the aftermarket conversion of HEVs to PHEVs.

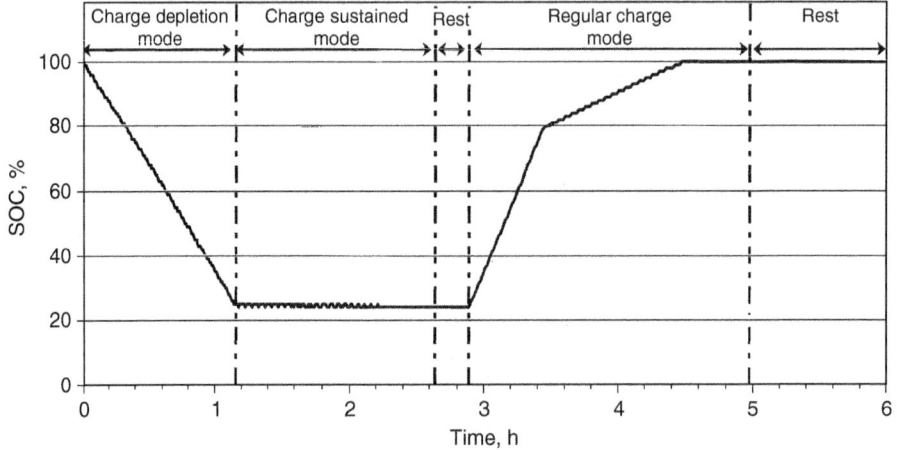

FIGURE 22.13 A simplified representation of a PHEV duty cycle.

22.7.2 Charge Depletion

Charge depletion concept hybrid vehicles are designed to have a certain vehicle range when driven solely in the electric vehicle mode. The motivation is that in heavily congested urban areas where air quality is a vital concern, the vehicle operates with zero emissions. Therefore, in the charge depletion mode the specific energy of the battery remains important, and a specific energy above 60 Wh/kg may be required. For example, a Toyota Prius HEV uses a 1.3 kWh NiMH prismatic comprised of 28 6-cell modules and has a pure electric vehicle range of less than 6 miles. However, when replaced with a 6.4 kWh, 30 Ah NiMH battery pack, the vehicle range on electric drive can be extended to over 25 miles.

22.7.3 Charge Sustained

The more common HEV concept is termed charge sustained, where the vehicle is designed to be operated under the various modes giving the greatest overall fuel economy. In this case, a primary function of the battery is to utilize the otherwise wasted energy of braking, and another main function is to assist the vehicle during acceleration. This concept essentially embraces the fact that ICE engines are very efficient on the highway at constant speed but very inefficient under start-and-stop city traffic conditions. It is a common strategy for the engine to be shut off while the vehicle is stopped at a traffic light, and for the battery to supply the energy needed for the initial vehicle acceleration while the ICE is restarted.

22.8 FUEL CELL STARTUP AND POWER ASSIST

Fuel cells are a significant developmental focus for many companies worldwide, offering the potential for fuel economy exceeding the equivalent of 100 miles per gallon. The most commonly discussed fuel cell is the proton exchange membrane (PEM) type.

The relevance to NiMH batteries is that fuel cell vehicles will still require a substantial (approximately 1 to 10 kWh) battery for several reasons:

- **Startup:** The PEM fuel cell and reformer may operate at a temperature of about 100°C, and the battery must supply sufficient energy for the vehicle to operate until the system comes up to that operating temperature.
- **Acceleration:** A critical obstacle for a PEM fuel cell will be cost, due to its use of expensive catalysts and an expensive membrane. The fuel cell will be sized for an average required vehicle power, not for peak power conditions where the battery will dominate. Essentially, the fuel cell in a fuel cell hybrid electric vehicle is used to charge the battery and operate the vehicle, while the battery is used for power load-leveling.
- **Efficiency:** The fuel cell can be operated under close to steady state conditions in an optimum region of fuel utilization.
- **Regenerative braking:** Similar to ICE powered hybrid vehicles; the fuel cell has no mechanism to take advantage of the otherwise wasted energy of braking without the battery.

22.9 CONSUMER BATTERIES—PRECHARGED NiMH

NiMH technology continues to expand into new markets. Consumer products such as digital cameras and RC toys that were previously the exclusive domain of alkaline primary cells are now beginning to use rechargeable NiMH batteries with "ready to use" capability that provides the consumer a choice with equal or better performance than alkaline in some applications (see Chap. 32). Hence, "precharged" NiMH batteries are achieving wide market acceptance.

The key to this NiMH technology advancement has been the improvement in charge retention to a remarkable 85% for 1 year (comparable to alkaline). Much work has gone into reducing self-discharge mechanisms, resulting in two distinct classes of NiMH technology:

- **Conventional NiMH:** The rate of self-discharge is dependent on storage temperature and time—the higher the temperature, the greater the rate of self-discharge. This is illustrated in Fig. 22.14, which shows charge retention for conventional sealed cylindrical NiMH batteries following storage at different temperatures for varying periods of time. The comparison is for a battery discharged at the rated discharge load (approximately the $1C$ rate) at the stated temperature. Note that conventional, NiMH technology has a charge loss in the range of 20% for 30 days.
- **Advanced precharged NiMH:** This class of NiMH technology is often referred to as "ready to use." These batteries are sold in the precharged state similar to alkaline primary cells and can be stored for up to 1 year with less than 20% loss in capacity. In order to achieve this low level of self-discharge, manufacturers typically use sulfonated separators, cobalt encapsulated $Ni(OH)_2$ with little added cobalt to the positive paste, and low corrosion AB_5 or A_2B_7 metal hydride alloys. This is illustrated in Fig. 22.15.

22.22 SECONDARY BATTERIES

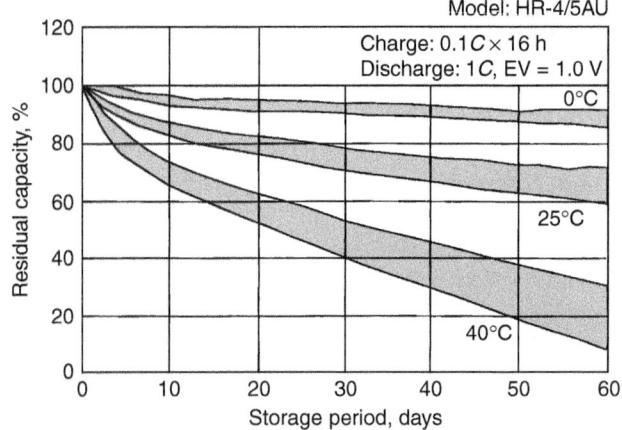

FIGURE 22.14 Charge retention characteristics of conventional NiMH cylindrical batteries at various temperatures.

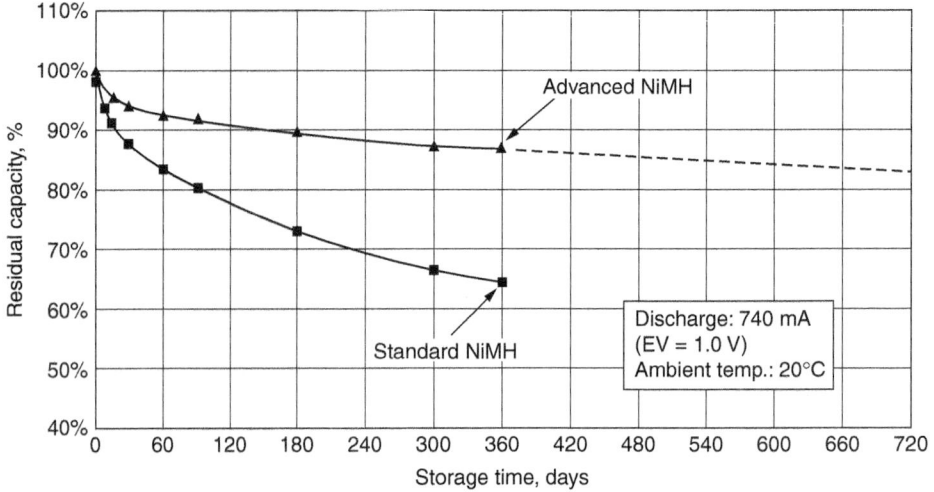

FIGURE 22.15 Charge retention characteristics of advanced vs. conventional NiMH cylindrical batteries.

Storage of the NiMH cell in a charged condition has no permanent effect on its capacity. Any capacity loss due to self-discharge is reversible, and batteries can recover to full capacity by simply recharging. However, storage at high temperatures, similar to operation at elevated temperatures, may deteriorate seals, MH alloys, and separators, leading to permanent damage, such as loss of capacity, cycle life, and overall battery lifetime. Hence, it is recommended that the temperature range for storage of NiMH cells be maintained at 20 to 30°C.

22.10 DISCHARGE PERFORMANCE

22.10.1 General Characteristics

The discharge characteristics of sealed NiMH batteries are very similar to those of sealed nickel-cadmium batteries. Several comparisons are illustrated in Chap. 15. The open-circuit voltage of the batteries of both systems ranges from 1.25 to 1.35 V, the nominal voltage is 1.2 V, and the typical end voltage is 1.0 V.

(Note: All batteries were recharged between 20° and 25°C under conditions shown, unless specified otherwise.)

22.10.2 Discharge Characteristics

Cylindrical Batteries. Typical discharge curves for a cylindrical sealed NiMH battery under various constant current loads and temperatures are shown in Fig. 22.16. (The data are based on the rate performance at 20°C at the 0.2C discharge rate to 1.0 V.) A flat discharge profile is a characteristic. The discharge voltage, as expected, is dependent on the current and temperature. Typically, with higher currents or lower temperatures, one will observe a lower operating voltage in the battery. This is due to a higher IR drop resulting from increasing current or increasing resistance from the lower temperatures. However, because of the relatively low resistance of NiMH batteries (as well as nickel-cadmium batteries), this drop in voltage is less than experienced with other types of portable primary and rechargeable batteries.

Button Batteries. Typical discharge curves for button-type sealed NiMH batteries at room and other temperatures are shown in Fig. 22.17.

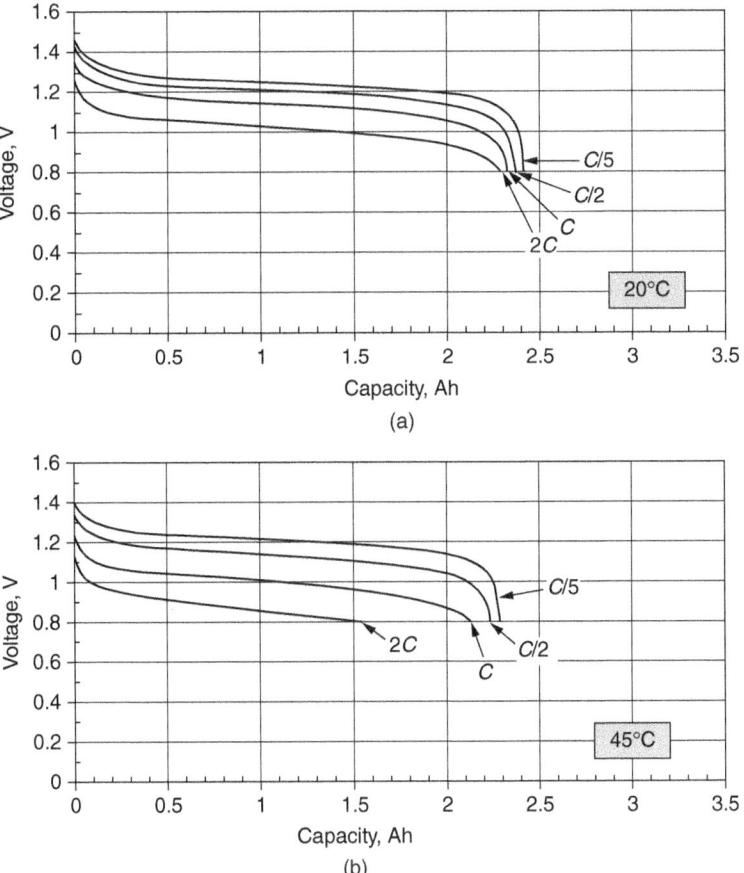

FIGURE 22.16 Discharge characteristics of NiMH cylindrical batteries. (a) Discharged at 20°C; (b) discharged at 45°C; (c) discharged at 0°C.

22.24 SECONDARY BATTERIES

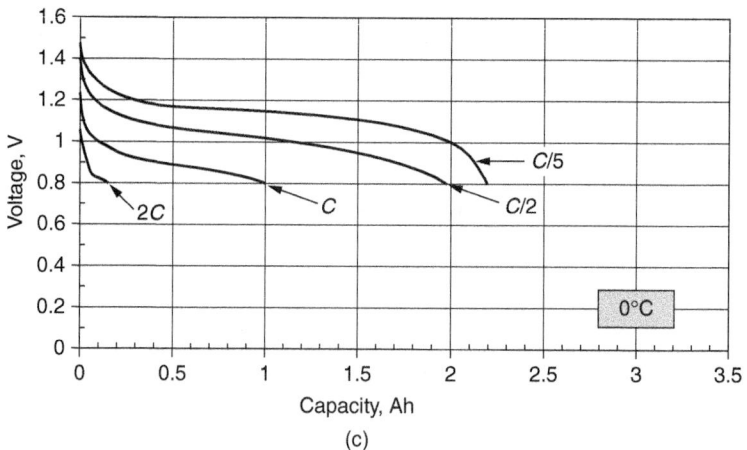

FIGURE 22.16 (*Continued*)

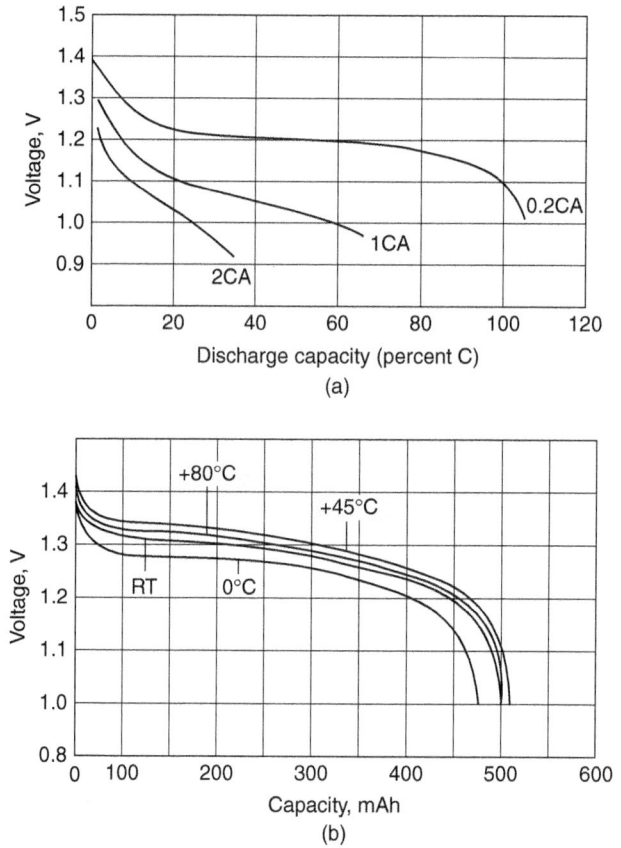

FIGURE 22.17 Discharge characteristics of NiMH button batteries. (a) Discharged at 20°C; (b) discharged at different temperatures and at a 0.2 C rate.

Prismatic Batteries. Typical discharge curves for the prismatic sealed NiMH batteries at room and other temperatures are shown in Fig. 22.18.

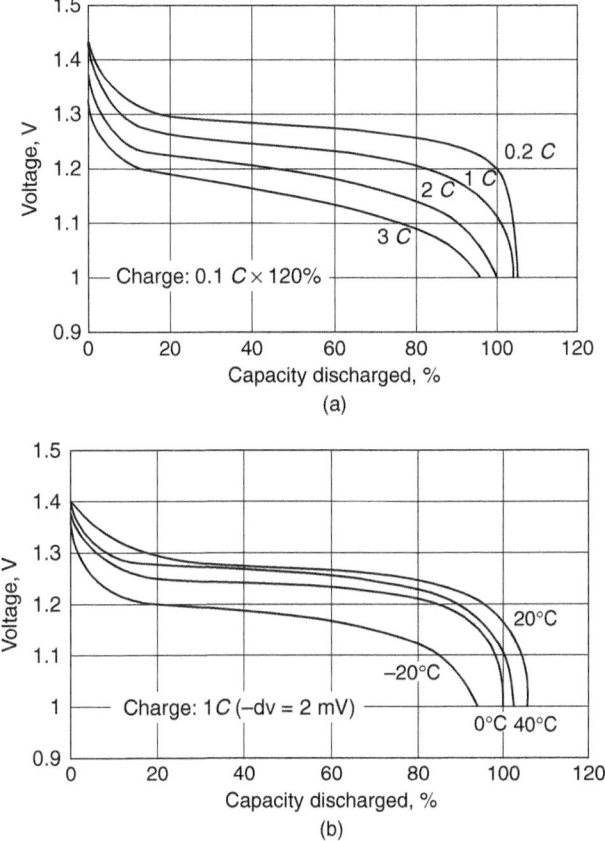

FIGURE 22.18 Discharge characteristics of NiMH prismatic batteries. (a) Discharged at 20°C; (b) discharged at different temperatures and at a 0.2 C rate.

9-Volt Battery. Typical discharge curves for the 9 V sealed NiMH batteries at room and other temperatures are shown in Fig. 22.19.

22.10.3 Specific Energy

The specific energy of NiMH batteries can vary anywhere from 42 to 110 Wh/kg depending on particular application requirements. For laptop computers, where run-time is of paramount importance, NiMH batteries do not need high-power capability or ultra-long cycle life. On the other hand, for extremely high-power charge and discharge, extra current collection, high N/P ratios, and other cell design and construction decisions can additively reduce specific energy. Figure 22.20 presents NiMH specific energy improvements over the last 18 years in portable cylindrical cells. For the most common small portable NiMH batteries, specific energy is usually about 90 to 110 Wh/kg, for EV batteries usually about 65 to 75 Wh/kg, and for HEV batteries and other high-power applications about 45 to 60 Wh/kg.

While gravimetric energy usually receives the attention for advanced battery technologies, in many cases volumetric energy density (in watthours per liter) is more important. NiMH has exceptionally high energy density, having achieved 430 Wh/L.

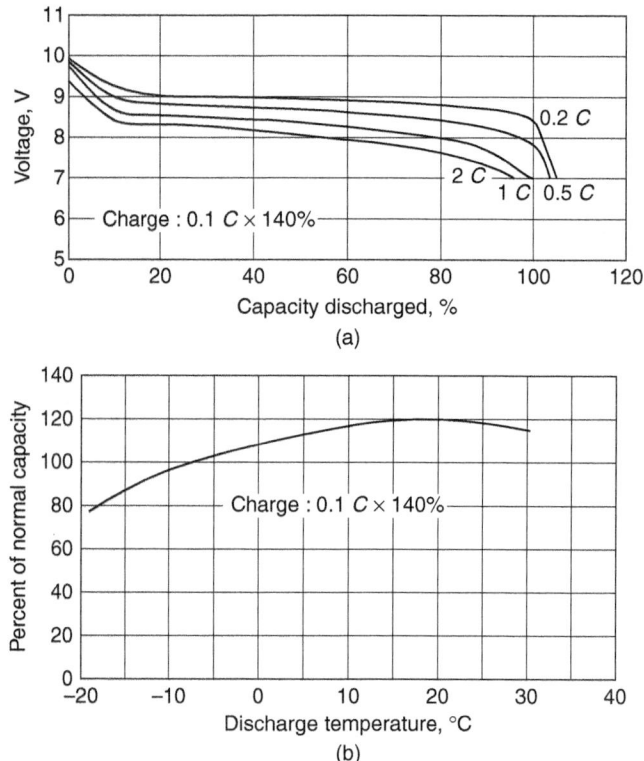

FIGURE 22.19 Discharge characteristics of NiMH 9 Volt batteries. (a) Discharged at 20°C; (b) discharged at different temperatures and at a 0.2 C rate.

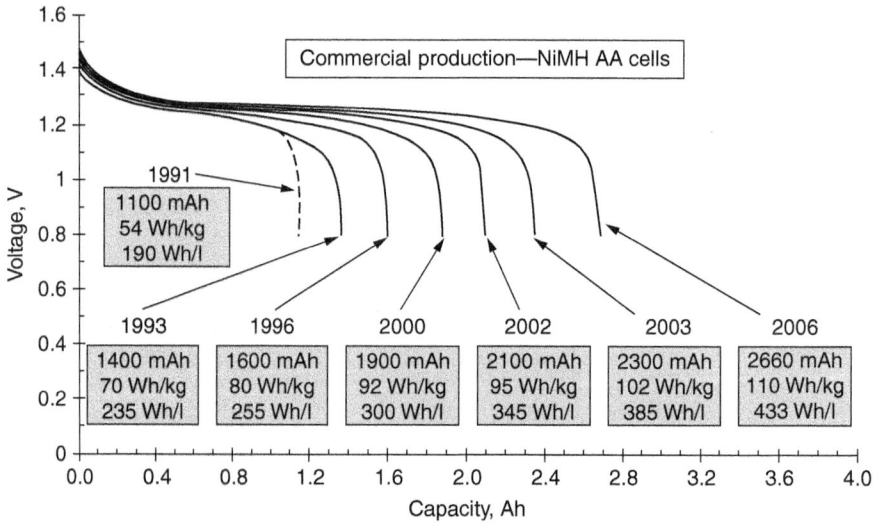

FIGURE 22.20 Evolution of specific energy for NiMH cylindrical AA-size batteries.

22.10.4 Specific Power

The power capability is a main strength of NiMH batteries relative to other advanced battery chemistries. For many years, it was widely believed that NiMH chemistry would never achieve a sufficiently high-rate discharge capability to replace NiCd. In fact, NiMH chemistry is now quickly replacing NiCd in these applications due to NiMH's higher energy and environmental concerns.

Voltage profiles up to the 10 C discharge rate for high-power cylindrical NiMH cells are presented in Fig. 22.21, attaining a specific power of 865 W/kg (Fig. 22.22). NiMH HEV module power shown in Fig. 22.23 is in excess of 1300 W/kg under both charge and discharge conditions.

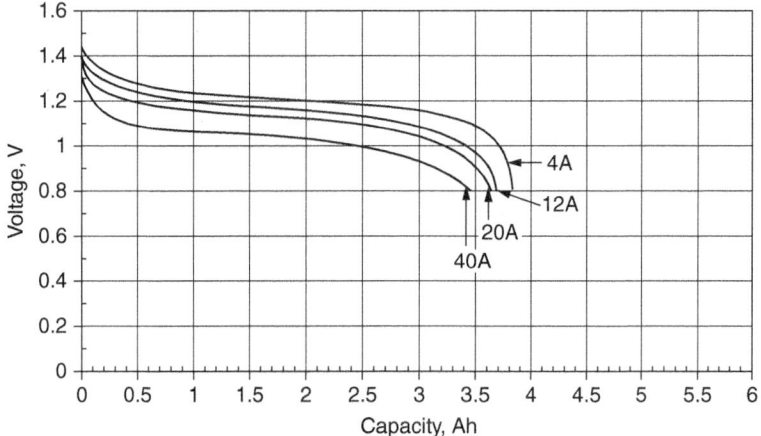

FIGURE 22.21 Voltage-capacity profiles for high-power NiMH cylindrical 3.5 Ah C-size batteries at different rates on continuous discharge.

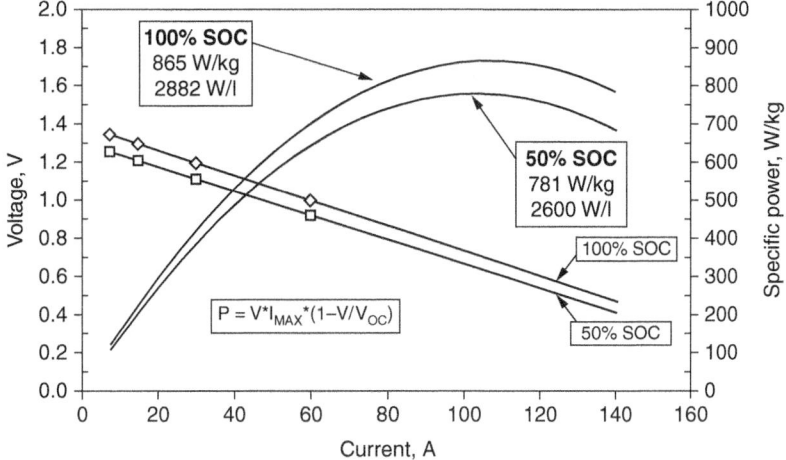

FIGURE 22.22 Specific power for ultra-high power NiMH cylindrical 3.5 Ah C-size batteries.

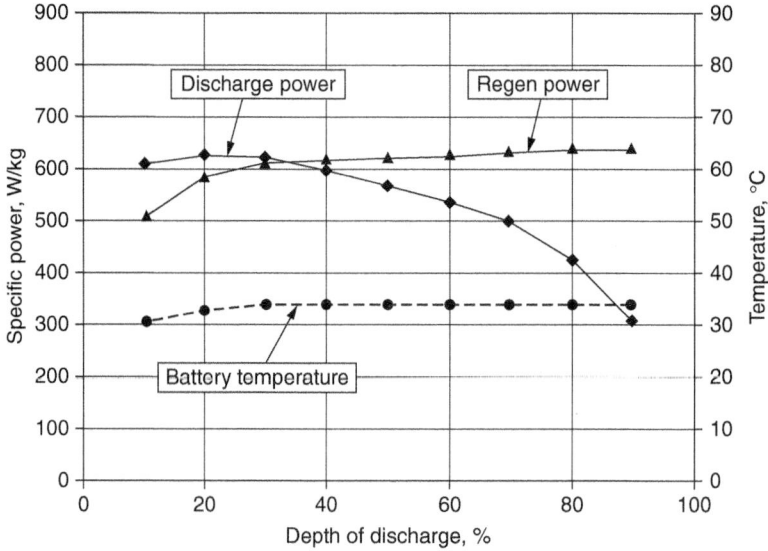

FIGURE 22.23 Specific power of a 12 V, 20 Ah NiMH HEV battery module as a function of depth of discharge.

22.10.5 Effect of Discharge Rate and Temperature on Capacity

The capacity of a battery is dependent on several factors, including discharge current, temperature, and the cutoff or end voltage. The capacity can be increased by continuing the discharge to the lower end voltages, particularly at the higher current drain rates and lower temperatures, where the voltage drops off more rapidly than at the lighter drain rates. However, it should be noted there is a risk to the battery by discharging to too low a cutoff voltage as the cells may be permanently damaged (see Sec. 22.10.11). Hence, the cutoff or end voltage for the NiMH battery is typically 1.0 V per cell.

The relationship between the capacity (expressed as a percentage of the capacity at 20°C on discharge at 0.2 C rate) of the sealed NiMH battery and the discharge temperature and current (expressed in C rate) is summarized in Fig. 22.24.

Typically, the best performance for NiMH batteries is obtained at temperatures between 0 and 40°C. Outside of this range, while the discharge performance characteristics of the battery are affected moderately at higher temperatures, the effects at lower temperatures are more pronounced. Similarly, the effect of a change in temperature becomes more significant at higher discharge rates.

Reduced discharge performance over the life of the battery is mainly due to an increase in internal resistance resulting from the polarization of the metal hydride electrode due to the generation of water during discharge. Capacity dependence as a function of temperature is strongly influenced by the selection of active metal hydride material, active nickel hydroxide material, and the formulation of the electrolyte.

High-temperature charge efficiency is a matter of significant importance for propulsion applications. Electric and hybrid electric vehicles are intended for use in warm weather climates, where vehicle range in the summer on a hot blacktop road is of critical importance. First-generation NiMH batteries for EV applications would lose almost 50% of room temperature capacity when charged at 60°C. The problem with high-temperature charge acceptance is the oxygen evolution characteristics of the nickel hydroxide positive electrode. Normally, at room temperature, the nickel hydroxide electrode has almost complete charge acceptance until about 80% charge input when the competing reaction of oxygen evolution begins. At full charge, continued charging causes 100% oxygen evolution at the nickel hydroxide electrode, and the oxygen migration to the metal hydride or cadmium electrode forms the well-known oxygen recombination overcharge mechanism. At elevated temperatures, the

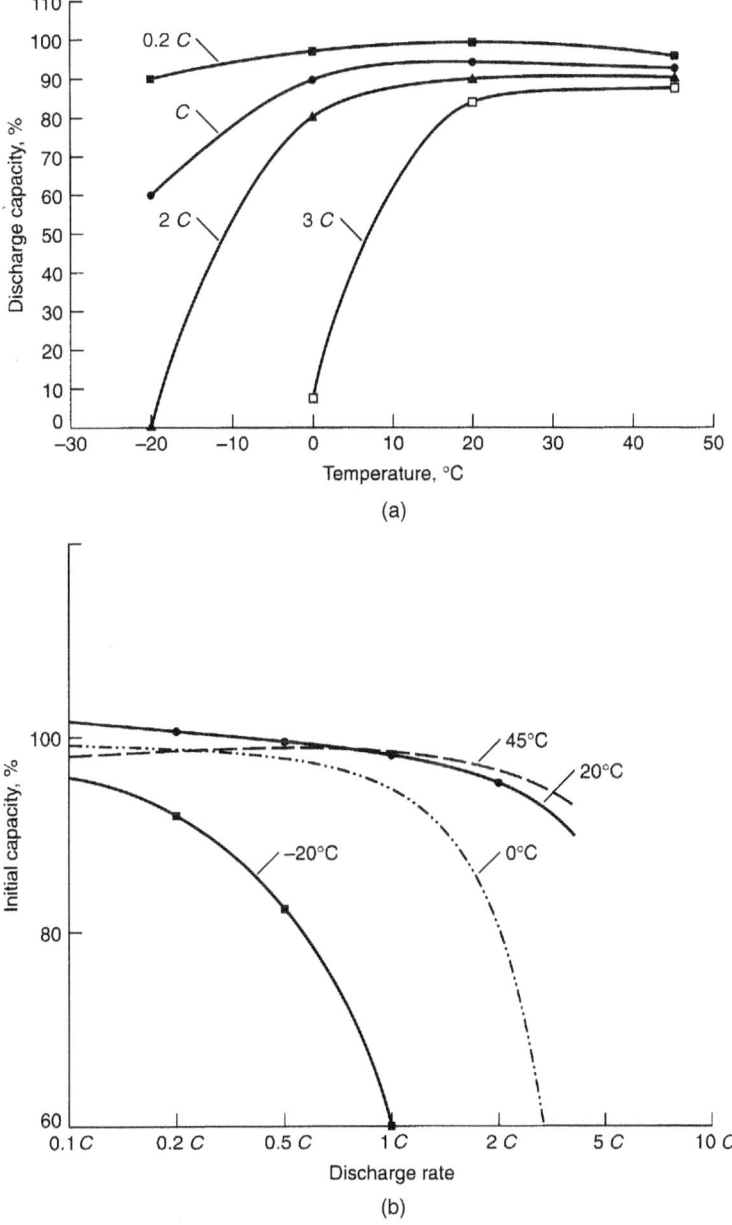

FIGURE 22.24 (a) Discharge capacity vs. ambient temperature for sealed cylindrical NiMH batteries at various discharge rates; end voltage 1.0 V/cell. (b) Discharge capacity (% of 0.2 C rate) vs. discharge rate (C rate) for NiMH cylindrical batteries at various temperatures; end voltage 1.0 V/cell.

22.30 SECONDARY BATTERIES

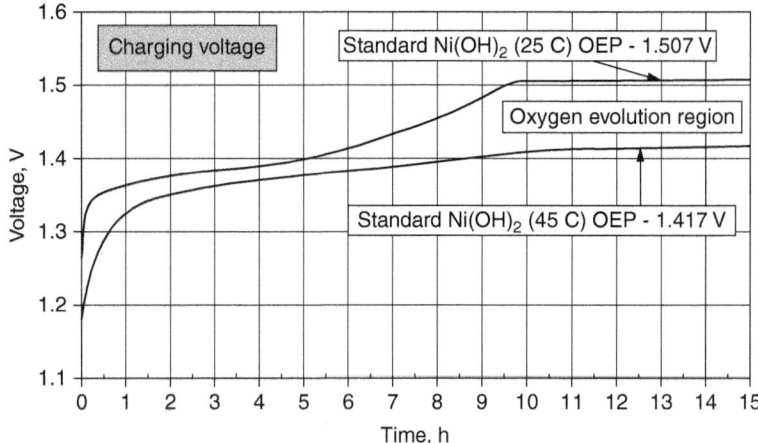

FIGURE 22.25 Charge characteristics of commercial nickel hydroxide as a function of temperature.

issue is therefore that oxygen evolution occurs at much earlier states of charge, and total charge acceptance is reduced, as seen in Fig. 22.25.

Another temperature-related property of NiMH chemistry is the storage effect on life (discussed in Sec. 22.10.9). Extended storage over temperatures above about 45°C can reduce life due to degradation of the separator, oxidation and corrosion of the metal hydride alloy, and disruption of the cobalt conductive network in the positive electrode. Each of these mechanisms is highly dependent on the manufacturer's choices of the active electrode materials.

22.10.6 Service Life (Hours of Service)

Figure 22.26 can be used to approximate the capacity and the service life of standard cylindrical NiMH cells at various discharge rates and temperatures if the rated capacity (at 20°C and the 0.2 C discharge rate) is known. The percentage of the rated capacity delivered under other conditions can be determined directly from this figure. The approximation is valid if the cells are of similar construction and behave similarly to the standard cell on which the data are based. The specific data for a given cell type obtained from the manufacturer should be used to more precisely estimate the performance.

Another form for presenting these data is shown in Fig. 22.26. The service life, in hours, of a cylindrical NiMH battery is plotted against the discharge current, normalized to the unit weight (Ah/kg) and size (Ah/L), based on a rated capacity at the 0.2 C rate of 60 Ah/kg and 200 Ah/L, reflecting the performance of a standard type battery. As discussed in Chap. 3 "Factors Affecting Battery Performance," this figure provides a convenient monograph to determine the approximate performance, in service hours, of a battery or to estimate the size of a battery that will deliver the desired performance under specified discharge conditions, with the caveat that the battery has similar construction and characteristics to the "standard" battery on which the data are based and an energy density close to the one specified.

22.10.7 Charge Retention

Charge retention is an area where NiMH manufacturers compete and the end-user must be careful to compare all performance properties for a given design. It is understood that the state-of-charge of a NiMH battery decreases during storage due to self-discharge and is highly dependent on

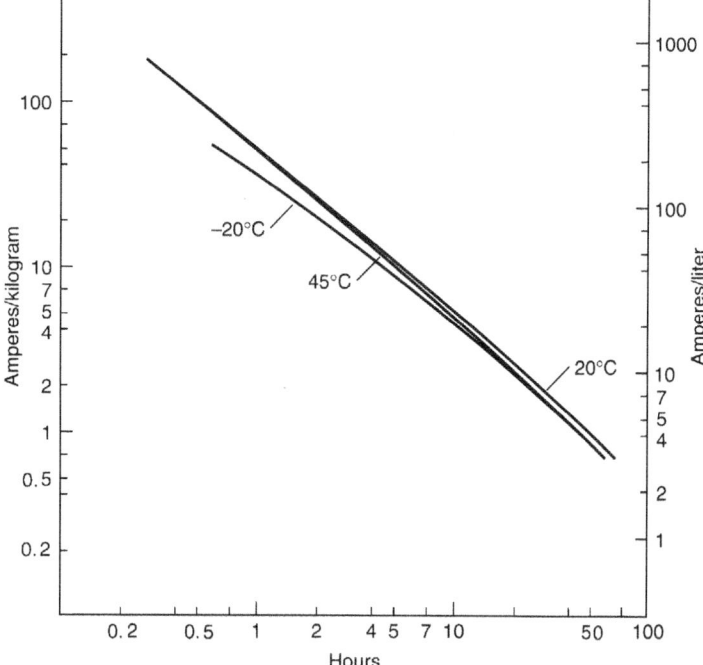

FIGURE 22.26 Service life of a standard NiMH cylindrical battery at various discharge rates and temperatures based on a specific capacity and capacity density under rated conditions at 20°C of 60 Ah/kg and 200 Ah/L; end voltage 1.0 V/cell.

temperature as well as the separator material, active positive and negative electrode materials employed. The specific formula of the metal hydride active material, separator material, and the quality of the nickel hydroxide active material all play important roles in either decreasing or increasing the self-discharge rate. Because there is such a large selection of materials for use in NiMH batteries, charge retention performance can vary.

For example, advanced separator materials have the ability to reduce self-discharge from about a 30% loss in one month at room temperature to about a 15% loss over the same time period. However, these same materials may reduce cycle life from 15 to 50%. The chemical mechanisms for how the separator reduces self-discharge are complicated, involving chemical grafting agents that can bind and thereby inactivate chemical species which may promote self-discharge. However, these separator treatments may have deleterious effects on electrolyte absorption and the ability of the separator to retain electrolyte during cycling. Separator dry-out is a very common mechanism of battery failure.

The choice of the metal hydride alloy can provide similar design trade-offs. Higher capacity AB_2 alloys also have higher rates of self-discharge compared to lower capacity AB_5 or A_2B_7 alloys discussed previously. The mechanisms for the effect of the MH alloy on self-discharge are twofold. First, corrosion products from the hydride alloy may migrate to the nickel hydroxide positive electrode and promote oxygen evolution during storage. Second, other corrosion products such as vanadium with its multivalent oxides may form redox shuttle mechanisms similar to nitrate ions. Likewise, the quality of the nickel hydroxide material, the use of encapsulated nickel hydroxides as well as residual impurities such as nitrates and carbonates can influence the above-cited self-discharge mechanisms. However, ultra-low impurity levels and added encapsulation may add to processing costs.

The positive electrode plays the most significant role in reducing self-discharge. Losses associated with self-discharge leave the cobalt conductive network open to degradation. As the cell gradually loses its charge, the cobalt conductive network can be reduced to Co (2+) or Co metal, allowing the cobalt to go into solution and to migrate elsewhere in the cell. Currently, there are two competing methods to combat this problem. One method, when long-term storage over several years is required, is to increase the levels of cobalt additives to combat breakdown of the conductive network and isolation of the active material. Another, increasingly more popular, way to avoid cobalt loss and uniformity issues is to use active materials that are "cobalt coated" or "cobalt encapsulated." Although more expensive, nickel hydroxide coated with cobalt hydroxide by the active material manufacturer has proven the cobalt to be in a more stable form. Further, it has been reported to offer increased utilization, high-rate discharge performance,[31] and significantly improved charge retention.

22.10.8 Cycle Life

Cycle life for industrial-sized NiMH batteries has both similarities and differences to that of small portable NiMH batteries. Cycle life for small portable NiMH batteries can vary from manufacturer to manufacturer, but usually falls in the range of 500 to 1000 cycles (100% DOD under 2 h charge/discharge). Design and chemistry factors affecting cycle life that are common to both large and small NiMH batteries include:

1. Metal hydride electrode
   - Active alloy formula (oxidation/corrosion properties)
   - Alloy processing effect on microstructure (particle disintegration)
   - Electrode construction (swelling in x-y-z directions and stability of conduction pathways)
2. Nickel hydroxide electrode
   - Active material formula (swelling and poisoning resistance)
   - Conductive network stability (amount and type of cobalt oxides)
   - Substrate (pore size, strength, and resistance to fracture)
3. Cell design
   - N/P ratio (amount of excess negative electrode capacity to influence cell pressure, MH corrosion, disintegration)
   - MH discharge reserve (overdischarge protection)
   - Separator (stability to corrosion, electrolyte absorption and retention, thickness, and resistance to short circuit)
   - Electrolyte (composition, amount, and fill fraction)
   - Vent pressure (weight loss, charge imbalance)
   - Electrode stack design (compression, electrode thickness, aspect ratio of height to width)

Factors affecting cycle life in industrial-sized NiMH batteries that differ from those factors affecting cycle life in portable NiMH batteries include:

- Significantly higher battery voltages (42 to 320 V versus 12 V) increases risk of abusive overcharge and overdischarge due to capacity or state-of-charge mismatch.
- Overall higher energy (0.1 kWh versus 33 kWh) increases heat generation and therefore the criticality of thermal management (which can be further influenced by battery pack enclosure heat transfer limitations).
- Typically higher operating temperature: Air-cooled and water-cooled vehicle batteries usually operate at a temperature of 35°C or higher, whereas operating temperatures for portable batteries may experience transient high temperatures, but on average are at or near room temperature.
- Series/parallel strings: In portable batteries, there are a large number of cell sizes available, ranging from 100 mAh button cells to 7 to 12 Ah D and F cells. There is a much smaller availability of cell sizes for HEV and EV NiMH batteries. Consequently, vehicle propulsion applications may

have to use series/parallel combinations of NiMH cells to meet the required pack voltage and energy demands, thereby increasing the risk of pack imbalances.
- End of life definition for portable batteries is usually based on capacity loss. In contrast, in EV and HEV NiMH batteries, end of life is due to power limitations.
- Qualification and operation testing: The emphasis on power greatly influences testing and methodology. In portable battery cycle life testing, it is most common to use 1 or 2 h constant current charge and discharge, usually to 100% DOD each cycle. For EV cycle life testing, discharge is usually a variable current/time profile to simulate driving conditions—the so-called "DST" drive profile. The significance of pulsed discharge cycle life testing is that high current pulses dominate the test. On the other hand, most EV cycle life tests are done at 80% DOD and typical NiMH module cycle life is from 600 to 1200 cycles. HEV testing emphasizes power capability even more, and de-emphasizes depth of discharge further still. Typical HEV mode cycle life testing is under a high current pulse profile with a 2 to 10% state of charge swing. Typical NiMH cycle life under these conditions is over 300,000 cycles, which corresponds to nearly 150,000 miles in a vehicle. As can be seen if Fig. 22.27, the difference in the depth of discharge (EV versus HEV cycling) can play a significant role in cycle life.

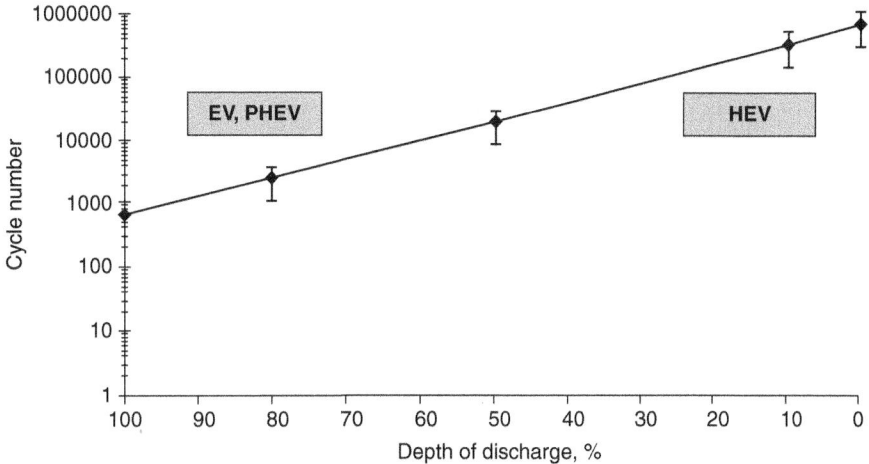

FIGURE 22.27 Relationship between depth of discharge (DoD) and cycle life for NiMH batteries.

Failure modes for EV, PHEV, and HEV batteries included short circuit due to mechanical penetration through the separator. The frequency of such events is usually small based on sound cell and electrode engineering and if manufacturing quality control is sufficient. Another failure mode may be due to abusive overcharge where excessive venting results in insufficient electrolyte within the separator. Abusive overcharge may result from charge imbalances caused by thermal differences from one part of the large battery to another. The problem may be compounded by the sophistication of the charger, where voltage and temperature sensing may not be on an individual module or cell basis. Another failure mode stems from overdischarge, where a cell or module within a high-energy PHEV or EV battery is discharged below the minimum recommended cell voltage of 1.0 V. Overdischarge is likely caused by state-of-charge imbalance within a high voltage string due to thermal gradients within a battery. Another source of abusive overcharge and overdischarge is the "weak cell or weak module" concept. This involves the statistical predictability within a large number of cells as to the decay rate of capacity, power, and resistance as a function of cycle.

A common feature of the above-cited EV/PHEV/HEV failure modes is the importance of maintaining state-of-charge balance within a battery pack that may contain several hundred individual cells. The method used to maintain equalized state-of-charge within an industrial NiMH battery is in effect to routinely bring all the cells to the same state-of-charge. This method of using overcharge to equalize state-of-charge is corrupted if the cell temperature within a battery pack is extreme or if cell-to-cell temperature gradients are too large. One of the biggest factors in replicating the excellent cycle life of small portable NiMH batteries in EV, PHEV, and HEV applications is proper thermal management, which was discussed in Sec. 22.12.2.

If premature failure due to short circuit and abusive overcharge/overdischarge is prevented, the principal failure mode in industrial EV, PHEV, and HEV NiMH batteries is increasing internal resistance with cycling. The EV or PHEV user will observe that top acceleration will diminish after long-term use, or that vehicle range will gradually decrease. To the HEV end-user, battery failure due to increasing internal resistance and resultant power loss will be observed as an inability of the battery to assist acceleration during drive and the inability of the battery to utilize regenerative braking energy due to excessive heating caused by the high charging currents being supplied during braking.

This primary failure mode of increasing internal resistance and power loss during cycling is caused by the same mechanisms as observed in small portable NiMH batteries; namely, separator dry-out as a result of electrolyte redistribution due to swelling of the metal hydride and nickel hydroxide electrodes, consumption of electrolyte due to oxidation of the separator, metal hydride active material, and positive electrode materials, and loss of electrolyte through venting.[32] These mechanisms may be exaggerated for large NiMH batteries due to their prismatic construction. Cylindrical cells have one positive electrode, one negative electrode, and one separator. NiMH prismatic EV or PHEV batteries may have 20 positives, 21 negatives, and a corresponding number of separator sheets. The cylindrical can is more effective in pressure containment than a rectangular container, both in terms of gas pressure and the mechanical forces applied to the can from the electrode stack. Therefore, another critical factor for large NiMH batteries is the management of compressive forces within a module. Typically, restraining bands are used to secure a 10- or 11-cell module, which has an endplate to equalize lateral forces on side wall of the can. Failure to adequately equalize compression within each cell in a module and within the internal cell stack itself will lead to premature failure due to unequal electrolyte distribution.

22.10.9 Shelf Life

Shelf life, also termed "calendar life," for large NiMH batteries is from 5 to 10 years based on a variety of factors, including temperature, charge equalization, electrolyte compensation, and gas permeation. Over perhaps 6 months to a year of storage, NiMH batteries may completely self-discharge. Further storage, under open circuit, may cause the cell voltage to gradually decline to 0 to 0.4 V, which can cause a breakdown of the cobalt conductive network in the positive electrode and/or increased surface oxidation of the metal hydride active material.[33] The length of time the battery is stored under this low voltage condition, the temperature of low voltage storage, and cell design influence the ease and degree of recovery of the battery. For example, a few cycles of low rate charge and discharge may be needed to recover cell capacity and power. If the degree of low voltage degradation is severe, the battery may not be recoverable.

Design factors that must considered for good shelf life are the oxidation and corrosion resistance of the metal hydride alloy, the amount of precharge on the metal hydride electrode, the formula of the nickel hydroxide active material, and the quality of the cobalt conductive network in the positive electrode.

Considering the cost of industrial NiMH batteries, most users leave the battery on low-current trickle charge if the battery will not be used for extended periods. Alternately, the battery may receive a periodic top-off charge designed to compensate for normal self-discharge capacity losses.

22.10.10 Coulombic/Voltaic Efficiency and Internal Resistance

The NiMH battery has a low internal resistance because of the use of thin plates with large surface areas and low resistance and an electrolyte having a high conductivity. Figure 22.28 shows the change in internal resistance with the depth of discharge. The resistance remains relatively constant during

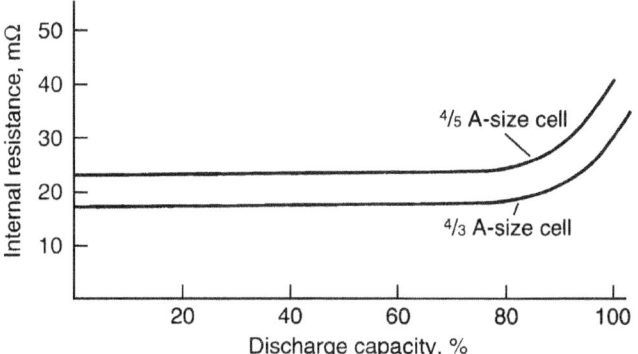

FIGURE 22.28 Internal resistance vs. discharge capacity of NiMH cylindrical batteries.

most of the discharge cycle. Toward the end of the discharge cycle, the resistance also increases due to conversion of the active materials. The internal resistance also increases as the temperature drops because the resistance of the electrolyte and other components are higher at lower temperatures. The resistance of the NiMH battery increases with use and cycling.

A key strength of NiMH technology is high efficiency due to low internal resistance. As shown in Fig. 22.29, over 90% voltaic efficiency was observed for 60 Ah prismatic NiMH EV cells at 100 A; over 75% efficiency at 300 A.

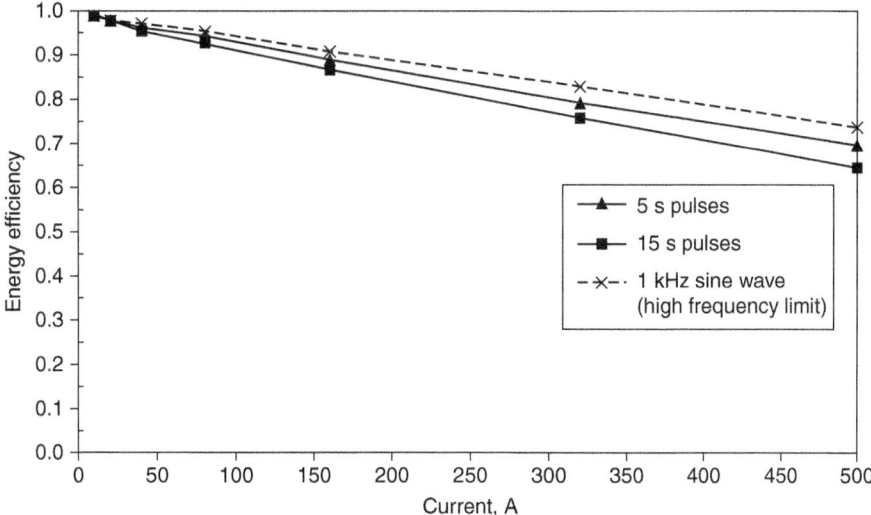

FIGURE 22.29 Energy efficiency of a 60 Ah NiMH HEV battery module as a function of discharge rate.

Voltaic efficiency is largely determined by the linear resistance components in the cell, the electronic and ionic resistances, which can be lowered by further engineering changes. Coulombic efficiency was determined to be 99% at 50% state of charge under an aggressive simulated HEV driving cycle, which is a typical operating point for charge-sustaining HEVs. Under the EPA combined city-highway FTP driving schedule, energy efficiency is about 93 to 95%.

22.10.11 Polarity Reversal During Overdischarge

When a multicell, series-connected battery is discharged, the lowest capacity cell will reach full discharge before others. If the discharge is continued, this lower capacity cell can be driven into an overdischarge condition through 0 V and its polarity (voltage) reversed. This is illustrated in Fig. 22.30.

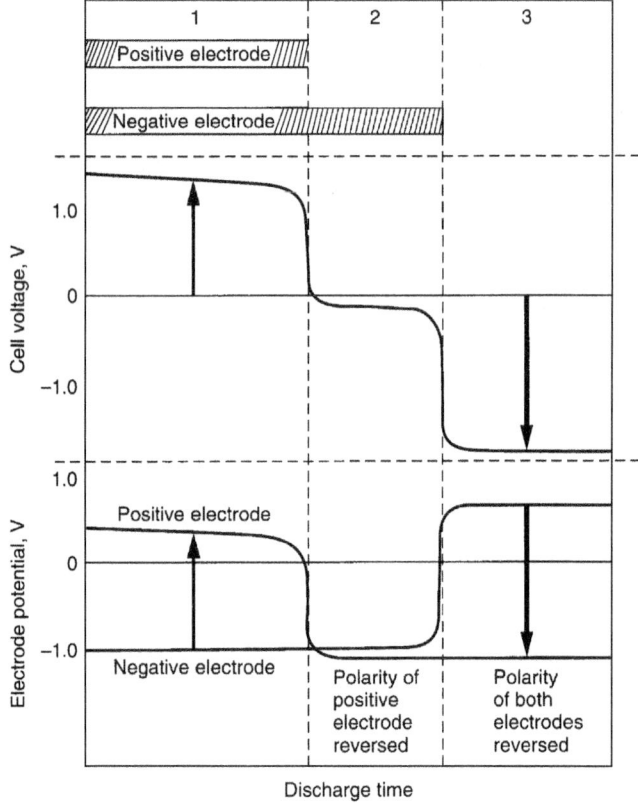

FIGURE 22.30 Charge characteristics of commercial nickel hydroxide as a function of temperature.

Phase 1 of the figure is the normal phase of the discharge with active material remaining on both the positive and the negative electrodes.

During phase 2, the active material on the positive electrode has been fully discharged and continued discharge causes the generation of hydrogen gas. Some of this gas may be absorbed by the hydrogen storage metal alloy in the negative electrode and the remainder builds up in the cell. Active material, however, still remains on the negative electrode and the discharge continues. The cell voltage is dependent on the discharge current, but remains within -0.2 to about -0.4 V.

In phase 3, the active materials on both electrodes have been depleted and oxygen is produced at the negative electrode. Prolonged overdischarge leads to gassing, higher internal cell pressure, venting, and deterioration of the cell.

The greater the number of cells in series (in a multicell battery), the greater the possibility a reversal of polarity will occur. To minimize the effect, whenever three or more cells are connected in series, the cell selection method employed should group cell capacities within a narrow range of $+/-5\%$.

The process of selecting cells of similar capacity is called "matching." Further, a cutoff voltage of 1.0 V per cell or higher should be used for discharge rates up to 1 C to prevent the possibility that any cell goes into reversal. Higher cutoff voltages should be used for batteries containing more than 10 cells in series and for discharge rates exceeding 1 C.

22.10.12 Type of Discharge

As discussed in the Chap. 3 ("Factors Affecting Battery Performance") a battery may be discharged under different modes (such as constant resistance, constant current, or constant power), depending on the characteristics of the load. The type of discharge mode selected has a significant impact on the service life delivered by a battery in a specified application. The discharge profiles of a NiMH battery under the three different modes are plotted in Fig. 22.31. Figure 22.31a shows the voltage profile, Fig. 22.31b shows the current profile, and Fig. 22.31c shows the power profile during discharge of the battery. As an example, the data are based on the discharge of a 1000 mAh battery such that, at the end of the discharge (1.0 V per cell), the power output is the same for all modes of discharge. In this example the power output at the 1.0 V cutoff is 100 mW. To discharge at the 100 mW at 1.0 V, the constant current discharge is 100 mA (C/10 rate) and the constant resistance discharge is 10 ohms. As shown, the longest service life is obtained under the constant power mode as the average current is the lowest under this mode of discharge.

22.10.13 Constant Power Discharge Characteristics

The discharge characteristics of the NiMH battery under constant power mode, at several different power levels, are shown in Fig. 22.32. These are similar to the data presented in Fig. 22.31 for the constant current discharges. The power levels are shown on the basis of E-rate. The E-rate is calculated in a manner similar to calculating the C-rate but based on rated watthour capacity rather than ampere-hour capacity. For example, for the E/2 power level, the power for a battery rated at 1200 mWh (rated at 1000 mAh at the C/5 rate at 1.2 V) is 600 mW.

22.10.14 Voltage Depression (Memory Effect)

A reversible drop in voltage and loss of capacity may occur when a sealed NiMH battery is partially discharged and recharged repetitively without the benefit of a full discharge. This is illustrated in Fig. 22.33. After an initial full discharge (cycle 1) and charge, the battery is partially discharged (in this example to 1.15 V) and recharged for a number of cycles. During cycling, the discharge voltage and the capacity drop gradually (cycles 2 to 18). On a subsequent full discharge (cycle 19), the discharge voltage is depressed compared to the original full discharge (cycle 1). The discharge profile may show two steps, and the cell does not deliver the full capacity to the original cutoff voltage.

This phenomenon is known as "voltage depression" or "memory effect," as the battery appears to "remember" the lower capacity. The battery can be restored to full capacity within a few full discharge-charge cycles, as illustrated in Fig. 22.33 (cycle 20 and 21).

The voltage drop occurs because only a portion of the active materials are discharged and recharged during shallow or partial discharging. The active material that has not been cycled changes in physical characteristics and increases in resistance. The active materials are restored to their original state by subsequent full discharge-charge cycling.

The extent of voltage depression and capacity loss depends on the depth of discharge and can be avoided or minimized by discharging to an appropriate end voltage. The effect is most apparent when the discharge is terminated at higher end voltages, such as 1.2 V per cell. A smaller loss occurs if the discharge is cut off between 1.15 and 1.10 V per cell. Discharging to an end voltage

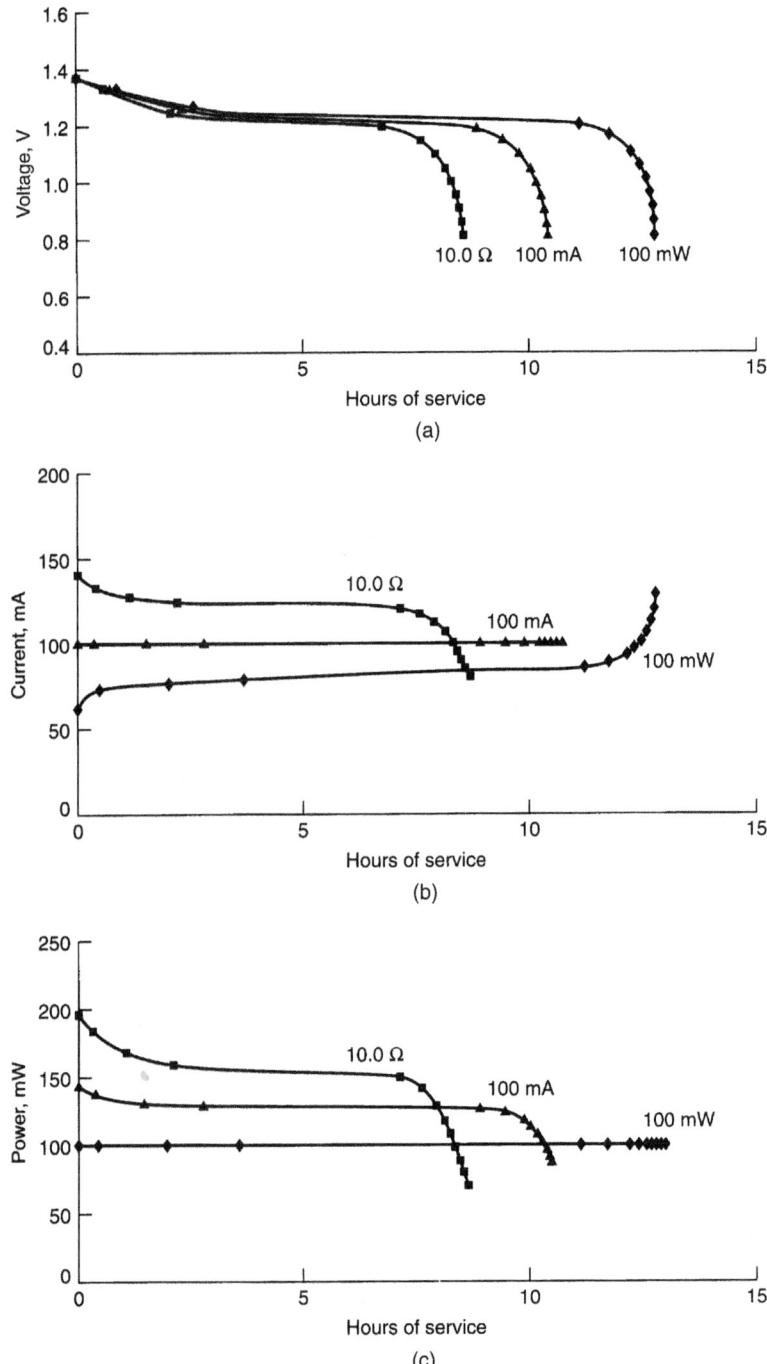

FIGURE 22.31 Discharge of NiMH cylindrical batteries—constant power vs. constant current vs. constant resistance: (a) voltage profile; (b) current profile; (c) power profile. All based on a battery rated at 1000 mAh.

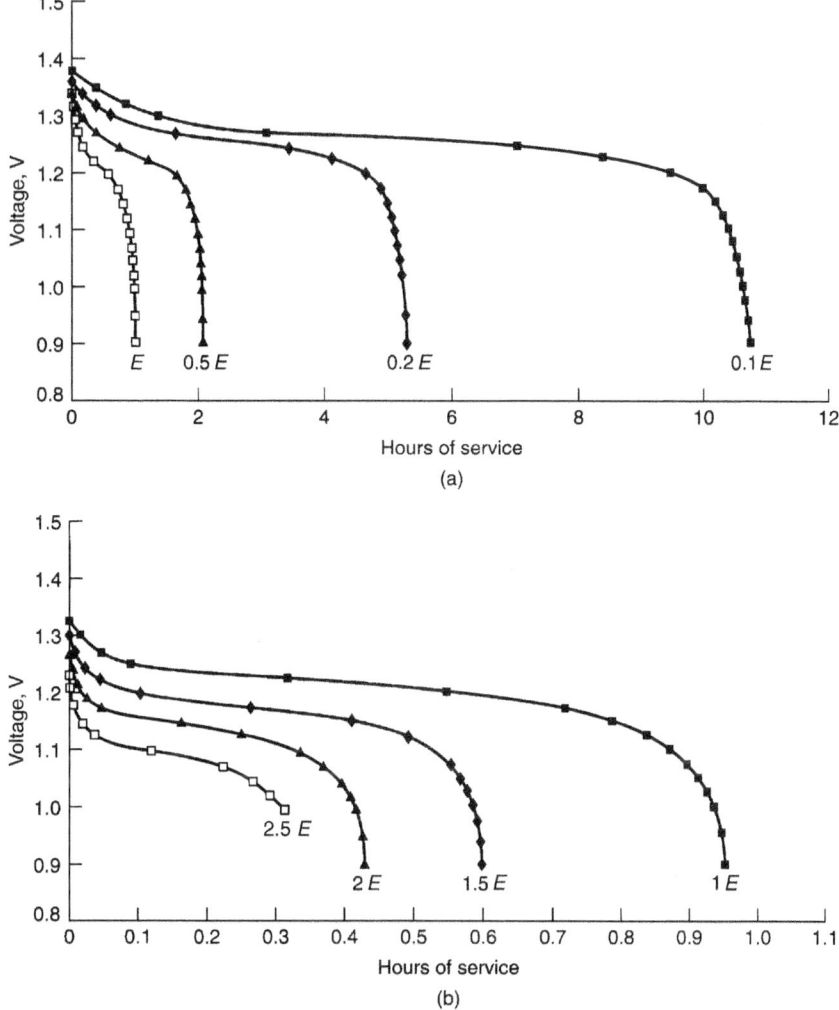

FIGURE 22.32 Constant-power discharge curves for NiMH cylindrical batteries at 20°C. (a) 0.1 E to 1 E discharge rate; (b) 1 E to 2.5 E discharge rate.

below 1.1 V per cell should not result in a significant voltage depression or apparent capacity loss on subsequent discharges. Discharging to too low an end voltage, however, should be avoided, as discussed in Sec. 22.10.11.

The effect is also dependent on the discharge rate. To a given end voltage, the depth of discharge will be less on discharges at the higher rates. This will increase the capacity loss as less of the active material is cycled.

While the memory effect may result in reduced battery performance, the actual voltage depression and capacity loss are only a small fraction of the battery's total capacity. Most users may never experience low performance due to this behavior. Often, the memory effect is used incorrectly to explain a low battery capacity that should be attributed to other problems, such as inadequate charging, overcharge, or exposure to high temperatures.

22.40 SECONDARY BATTERIES

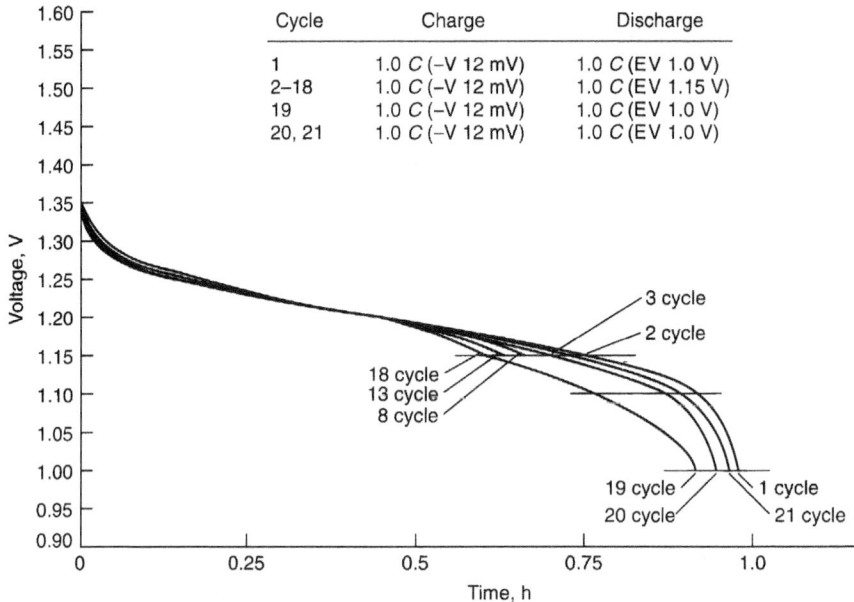

FIGURE 22.33 Voltage depression during cycling of sealed cylindrical NiMH batteries, 20°C.

22.11 CHARGE METHODS

22.11.1 General Principles

Recharging is the process of replacing energy that has been discharged from the battery. Subsequent performance of a battery (as well as its overall life) is dependent on effective charging. The main criteria for effective charging are to:

- Recharge the battery to its full capacity
- Limit the extent of overcharge
- Avoid high temperatures and excessive temperature fluctuations

Recharging characteristics of the NiMH battery are generally similar to those of the sealed nickel cadmium battery. However, there are some distinct differences, particularly on the requirements for charge control, as the NiMH battery is more sensitive to overcharge. Caution should be exercised before interchangeably using the same battery charger.

The most common charging method for the sealed NiMH battery is a constant current charge, but with the current limited to avoid an excessive rise of temperature or exceeding the rate of the oxygen recombination reaction.

The voltage profiles of NiMH and nickel cadmium batteries during charge at a moderate constant current rate are compared in Fig. 22.34. The voltage of both battery systems rises as the battery accepts the charge. During the first phase of the charge, the temperature of both batteries increases slightly due to Joule heating, which stems from the internal resistance of the batteries. Both battery chemistries are associated with endothermic charging processes ($Q_r = T\Delta S$ values are 27 KJ and 40 KJ for NiCd and NiMH batteries, respectively, where Q_r is the thermodynamic heat effect of the process related to the entropy change ΔS). An increase in the temperature of the battery indicates that the heating resulting from the Joule effect is larger than the cooling effect due to reversible heat ($T\Delta S$)

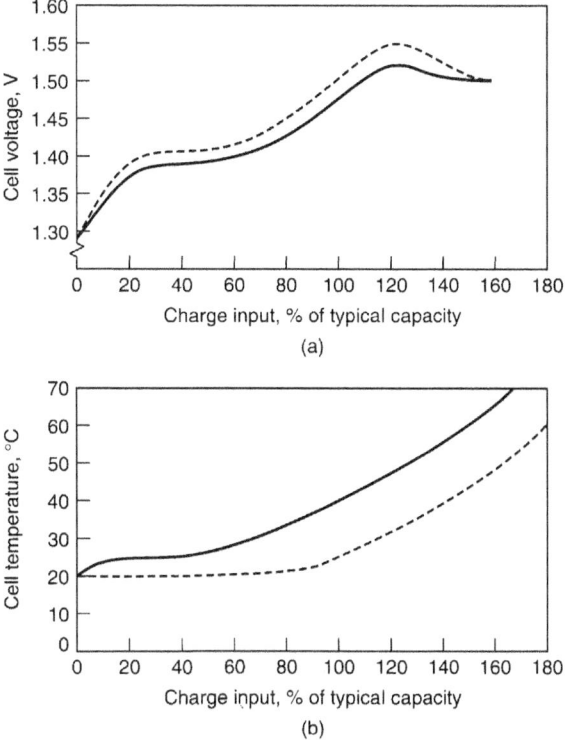

FIGURE 22.34 Comparison of typical charge characteristics of Ni-Cad and NiMH batteries: (a) voltage characteristics; (b) temperature characteristics. Solid line – NiMH; broken line – Ni-Cad.

remains relatively constant because its charge reaction is endothermic. As the batteries approach 75 to 80% recharge, the voltage rises more sharply due to the starting of the oxygen evolution reaction at the positive electrode. The temperature increases sharply at this stage due to the high Joule heating associated with the oxygen evolution and recombination at the negative electrode. The increase in cell temperature causes the voltage to drop as the battery reaches full charge and goes into overcharge due to the endothermic reversible heat effect (TΔS) of the charging process.

The voltage profile of the NiMH battery does not show as prominent a peak as that of the NiCd battery. The shallower drop in a NiMH battery may be due to a somewhat higher exothermic recombination reaction ($\Delta H = -572$ KJ/mol O_2 for NiMH versus $\Delta H = -550$ KJ/mol O_2 for NiCd), which compensates for the reversible endothermic effect of the charging reactions. Both the voltage drop after peaking ($-\Delta V$) and the temperature rise can be used to terminate the charge. However, while similar charge techniques can be used for both types of batteries, the conditions to terminate the charge may differ because of the different behavior of the two battery systems during charge.

The voltage of the sealed NiMH battery during charge depends on a number of conditions, including charge current and temperature. Figure 22.35 shows the voltage profile of the NiMH battery at different charge rates and temperatures. The voltage rises with an increase in charge current due to higher IR and overpotential during the electrode reaction. The voltage decreases with increasing temperature as the internal resistance and the overpotential during electrode reaction decreases. The voltage peak is not as evident at the low charge rates and at the higher temperatures.

The increase in battery temperature during charge at various charge rates is shown in Fig. 22.36. The internal cell pressure increases similarly. This rise in temperature and pressure at the higher

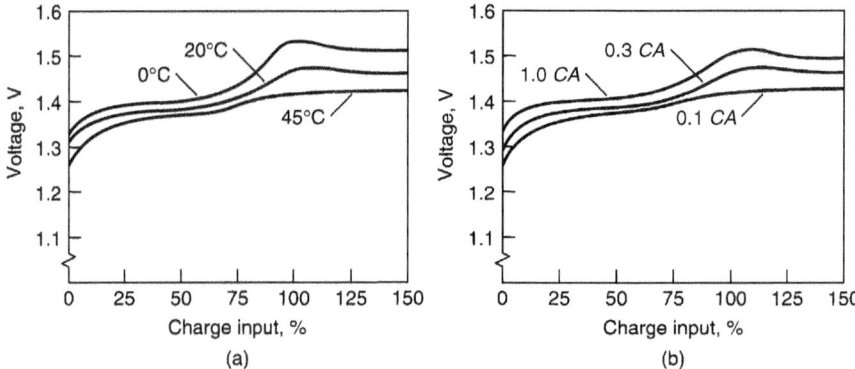

FIGURE 22.35 Cell voltage vs. charge input for NiMH cylindrical batteries: (a) at various temperatures (charge rate 0.3 C); (b) at various charge rates at 20°C.

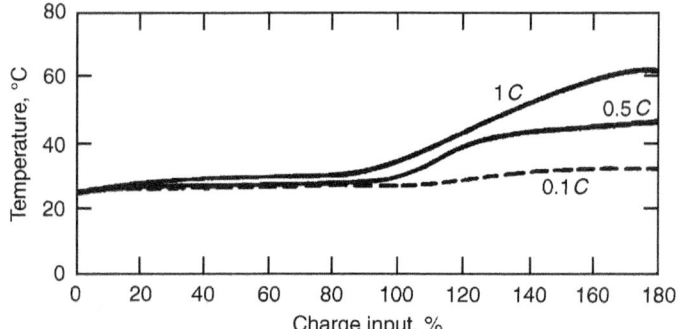

FIGURE 22.36 Battery temperature during charge of NiMH cylindrical batteries.

charge rates emphasizes the need for proper charge control and effective charge termination when "fast charging" to avoid venting and other deleterious effects.

The charge efficiency is also dependent on temperature. It decreases at the higher temperatures due to increasing evolution of oxygen at the positive electrode. At lower temperatures, oxygen recombination is slowed and a rise in internal cell pressure may occur depending on the charge rate. Figure 22.37 shows the available discharge capacity following charging at various temperatures and several charge rates. As shown, the battery capacity is reduced after high temperature charging. The extent of this effect is also dependent on the conditions of the discharge following the charge, as well as on other charge conditions.

Proper recharging is critical not only to obtain maximum capacity on subsequent discharges, but also to avoid high temperatures, overcharge, and other conditions that could adversely affect battery life.

22.11.2 Techniques for Charge Control

Characteristics of the NiMH battery define the need for charge control to terminate the charge, thereby preventing the battery from being overcharged or exposed to high temperatures. The advantage of employing proper charge control to maximize the life of the battery is illustrated in Fig. 22.38. The highest capacity levels are achieved with the 150% charge input, but at the expense of cycle life. The longest cycle life is attained with the 120% charge input, but with lower capacity due to

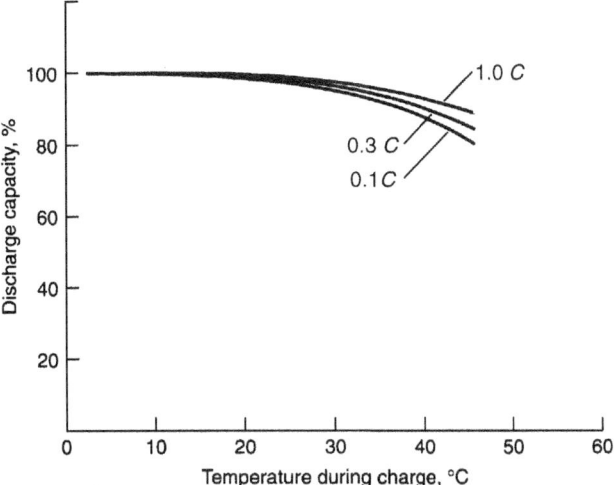

FIGURE 22.37 Charge efficiency vs. charge temperature at various charge rates for NiMH cylindrical batteries. Discharge at 0.2 C rate to 1.0 V.

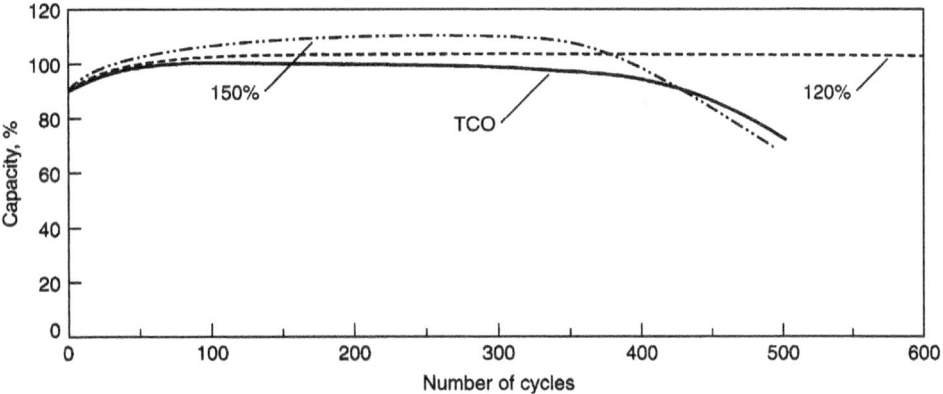

FIGURE 22.38 Effect of charge control on cycle life of NiMH cylindrical batteries. 1 C charge rate, discharge at 1 C to 1.0 V. TCO-charge termination at 40°C; 120%-charge termination at 120% charge input; 150%-charge termination at 150% charge input.

insufficient charge input. Thermal cutoff charge control may reduce cycle life because the battery is usually allowed to reach higher temperatures during charge. This method, however, is useful as a backup control in the event that the maximum temperature is exceeded during charge.

Some popular methods for charge control are summarized hereafter. The characteristics of these methods are illustrated in Fig. 22.39. In many cases, several methods are used during a single charge, particularly to control high-rate charging.

Timed Charge. Under this charge control method, charge is terminated after the battery has been charged for a predetermined amount of time. This method should only be used for charging at low rates to avoid excessive overcharge because the state-of-charge of the battery, prior to charging, cannot always be determined. This procedure is also used as a "topping" charge to other charge termination methods to ensure a complete recharge.

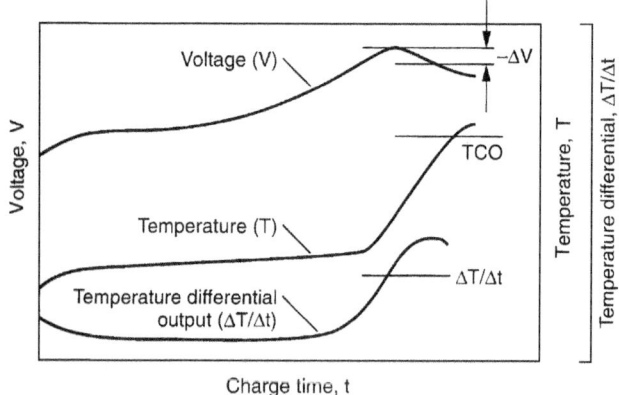

FIGURE 22.39 Comparison of charge termination methods: TCO, $\Delta T/\Delta t$, and $-\Delta V$.

Voltage Drop ($-\Delta V$). With this technique, widely used with sealed nickel cadmium batteries, the voltage during charge is monitored and charge terminated when the voltage begins to decrease. This approach can be used with the NiMH battery, but as noted in Sec. 22.10.1, the peak voltage with the NiMH cell is not as prominent and may be absent in charge currents below the 0.3 C rate, particularly at elevated temperatures. The voltage signal must be sensitive enough to terminate charge when the voltage drops, but not so sensitive that it will terminate charge prematurely due to noise or other normal voltage fluctuations. A 10 mV per cell drop is generally used for the NiMH battery.

Voltage Plateau ($0\ \Delta V$). As the sealed NiMH battery does not always show an adequate voltage drop, an alternate method is to terminate charge when the voltage peaks and the slope is zero rather than waiting for the voltage to drop. The risk of overcharge is reduced as compared to the $-\Delta V$ method. A topping charge can follow to ensure a full recharge.

Temperature Cutoff (TCO). Another technique for charge control is to monitor the temperature rise of the battery and terminate charge when the battery has reached a temperature that indicates the beginning of overcharge. It is difficult to determine this point precisely as it is influenced by ambient temperature, cell and battery design, charge rate, and other factors. For example, a cold battery may be overcharged before reaching the cutoff temperature, while a warm battery may be undercharged. Usually this method is used in conjunction with other charge control techniques and mainly to terminate the charge in the event the battery reaches excessive temperatures before the other charge controls activate.

Delta Temperature Cutoff (ΔT). This technique measures the battery temperature rise above the ambient temperature during charging and terminates charge when this rise exceeds a predetermined value. In this way, the influence of ambient temperature is minimized. The cutoff value is dependent on several factors, including cell size, configuration and number of cells in the battery, and the heat capacity of the battery. Therefore, the cutoff value must be determined for each type of battery.

Rate of Temperature Increase ($\Delta T/\Delta t$). In this method, the change in temperature with time is monitored and charge is terminated when a predetermined incremental temperature rise is reached. The influence of ambient temperature is virtually eliminated. A $\Delta T/\Delta t$ cutoff is a preferred charge control method for NiMH batteries because it provides long cycle life.

Note: Details on the design of batteries using protective devices and a description of the thermal protective devices that can be used for charge control are discussed in Chap. 5, "Battery Design."

22.11.3 Charging Methods

NiMH batteries are extremely flexible and accept diverse charge methods. The principal factor in designing a charging algorithm is to prevent excessive overcharge, especially at high rates, in order to avoid heat buildup and electrolyte venting losses. Several methods of sensing overcharge are common, including time, absolute temperature, ΔT, $\Delta T/\Delta t$, $-\Delta V$, and pressure rise. In all cases, the oxygen evolution/recombination mechanism to create heat is the basis for each charge termination approach. The method employed to terminate charge will depend on the charging rate (slow to fast charging) and should best limit the amount of heat generation in the NiMH battery in order to prevent damage.

Low Rate Charge (12 to 15 h). A convenient method to fully charge sealed NiMH batteries is to charge at a constant current at about 0.1 C rate with time limited charge termination. At this current level, the generation of gas will not exceed the oxygen recombination rate. The charge should be terminated after 150% capacity input (approximately 15 h for a fully discharged battery). Excessive overcharge should be avoided as this can damage the battery. The temperature range for this charge method is 5 to 45°C, with the best performance obtained between 15 and 30°C.

Quick Charge (4 to 5 h). NiMH batteries can be recharged efficiently and safely at higher rates. Charge control is required in order to terminate the charge when the rate of oxygen recombination is exceeded or the battery temperature rises excessively. The fully discharged battery can be charged at a 0.3 C rate for a charge time equivalent to a 150% charge input (approximately 4.5 to 5 h). In addition to the timer control, a thermal cutoff device should be used as a backup control to terminate charge at about 55 to 60°C to avoid exposing the battery to excessively high temperatures. This charging method may be used in an ambient temperature range of 10 to 45°C.

As a further precaution, the decrease in voltage ($-\Delta V$) should also be sensed to ensure that charge is terminated early enough to minimize overcharge. This is particularly advisable if the battery being charged was not fully discharged. A "topping" charge at the 0.1 C rate, as described previously, may then be used to assure 100% recharge.

Charging sealed NiMH batteries at rates between 0.1 and 0.3 C generally is not recommended. At these rates, the voltage and temperature charge profiles may not exhibit characteristics suitable for voltage-based cutoff control and the batteries may otherwise be exposed to harmful overcharge.

Fast Charge (1 h). Another method of charging NiMH batteries in an even shorter time is to charge at constant current at the 0.5 to 1 C rates. At these high charge rates, it is essential that the charge be terminated early during overcharge. Timer control is inadequate as the time needed for charge cannot be predicted. A partially charged battery could easily be overcharged, while a fully discharged one could be undercharged, depending on how the timer control is set.

With fast charging, the decrease in voltage ($-\Delta V$) and the increase in temperature (ΔT) can be used to terminate the charge. For better results, termination of fast charge can be controlled by sensing the rate of temperature increase ($\Delta T/\Delta t$) with a thermal cutoff (TCO) backup.

Figure 22.40 shows the advantage of using a $\Delta T/\Delta t$ method compared to $-\Delta V$ in terminating a fast charge. The $\Delta T/\Delta t$ method can sense the start of the overcharge earlier than the $-\Delta V$ method. The battery is exposed to less overcharge and overheating, resulting in less loss of cycle life. A temperature increase of 1°C/min should be used for $\Delta T/\Delta t$; a temperature of 60°C is recommended for the TCO.

In the case of multicell batteries of three cells or more, $-\Delta V$ termination with TCO backup may be adequate. The $-\Delta V$ value usually is 10 to 15 mV per cell and 60°C for the TCO.

While the infrastructure to fast charge EV batteries is not fully in place, a strength of NiMH battery technology is the ability to accept fast charge when it is required and the proper power is available. As an example, a typical EV cell capacity may be approximately 100 Ah. Therefore, if 15 min charging is desired, currents of approximately 400 A (at 360+ V) must be available.

NiMH batteries can accept from 60 to 80% charge within 15 min, at which point current must be reduced. Temperature rise due to internal resistance heating is small, on the order of 15°C for

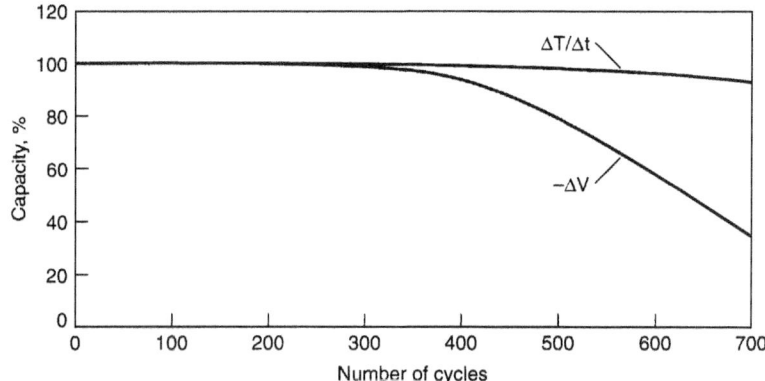

FIGURE 22.40 Cycle life and capacity as a function of charge termination method for NiMH cylindrical batteries. Charge 1 C, rest 30 min, discharge 1 C to 1.0 V, rest for 2 h.

a 33 kWh battery, whereas heating due to oxygen recombination is very large. Again, the crucial aspect of fast charge is proper sensing of the onset of overcharge.

Trickle Charge. A number of applications require the use of batteries that are maintained in a fully charged condition. This is done by trickle charging—charging at a rate that will replace the capacity loss due to self-discharge. A trickle charge at a current of between 0.03 and 0.05 C rates is recommended. The preferred temperature range for trickle charging is between 10 and 35°C. Trickle charge may be used following any of the previously discussed charging methods.

Three-Step Charge Procedure. A three-step procedure provides a means of rapidly charging a sealed NiMH battery to full charge without excessive overcharge or exposure to high temperatures.

1. The first step is a charge at a 1 C rate terminated by using the $\Delta T/\Delta t$ method or the $-\Delta V$ method.
2. This is followed by a 0.1 C topping charge, terminated by a timer after ½ to 1 h of charge.
3. The third step is a maintenance charge of indefinite duration at a current of between the 0.05 and 0.02 C rates. The battery should also be protected with a thermal cutoff device to terminate the charge so that the temperature does not exceed 60°C.

22.11.4 Regenerative Braking

Regenerative braking is a vehicle characteristic that helps NiMH dominate EV and HEV applications. Vehicle range is an area of competition for EV and HEV manufacturers, and a method of extending vehicle range from 5 to 20% is to utilize the energy lost during conventional braking to charge the battery. Regenerative braking energy is available to the battery at extremely high power (approximately 500 W/kg), and many rechargeable battery technologies cannot efficiently utilize the energy supplied at such a high power. NiMH batteries are able to accept regenerative braking energy over a wide state of charge range and over a wide temperature range.

Figure 22.41 shows a plot of voltage-current measurements on an Ovonic NiMH 13-HEV-60 (13 V, 60 Ah) module during an aggressive simulated HEV driving cycle. In contrast to results of a typical high-power lead-acid battery, also shown, the V-I slope is relatively constant. The lead-acid battery shows a significantly higher resistance for charge. For the Ovonic 13-HEV-60 module, the effective resistance on charge and discharge were the same, about 6 mΩ. This provides a substantially higher capability for acceptance of regenerative braking energy.

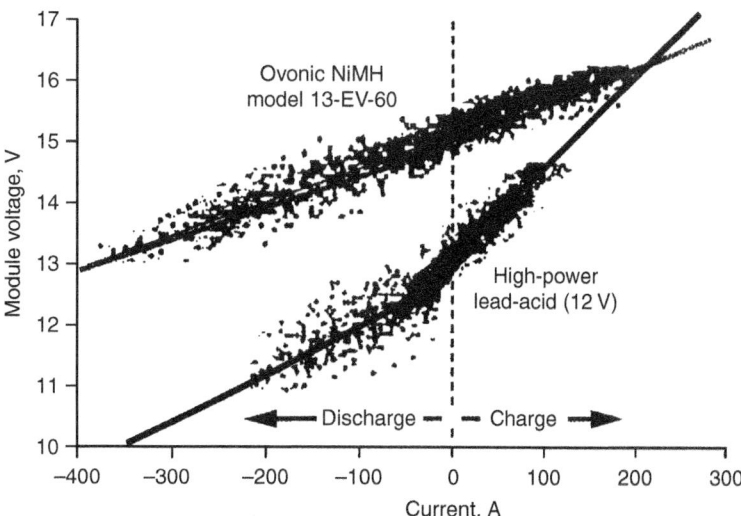

FIGURE 22.41 A comparison of the high-rate charge-discharge performance for HEV applications between NiMH and lead-acid batteries.

22.11.5 Charge Algorithms

Charge algorithms refer to the programming used to charge the NiMH HEV battery. Conceptually, the battery can accept charge input at extremely high rates until the battery is about 80% fully charged, at which point the charge current must be reduced. In overcharge, the charge current should not exceed the 10 h charge rate, and generally, the total charge input should be below about 110 to 120%.

An inherent virtue of NiMH batteries is that simple charging methodology can be used, with inexpensive equipment, and that it is not necessary to monitor the voltage of each cell in the high voltage string.

One method used to reliably charge NiMH EV batteries utilizes constant current steps to a temperature compensated voltage limit. Practically, this means that charging is done at high current until a predetermined voltage is reached which is an indication of a certain state-of-charge. At this voltage, the current is reduced or stepped down, until another predetermined voltage is reached. An important aspect of the charge algorithm is that the prescribed voltage set points are variable, based on temperature and current.

22.12 ELECTRICAL ISOLATION

Isolation of the EV and HEV battery from the vehicle is essential due to the high voltages involved. Isolation begins at the cell level, where plastic cell cases are preferred from an electrical isolation perspective, and metal cased cells must have some kind of insulative coating that is stable and free of pinholes.

The battery module (usually 12 V) must also be isolated from the battery tray, usually done with plastic isolation mounts. It is important that the cell and module interconnecting straps also be isolated from the battery tray.

In air-cooled battery packs, pack-to-vehicle resistance is strongly influenced by road contaminants such as dirt and salt, and pack case design must take this into consideration. For water-cooled designs, the issue is the resistance of the plastic casing materials to the coolant (e.g., ethylene glycol).

There is a strong dependence of the plastic resistance to temperature where a pack isolation resistance of 500 MΩ at 20°C may be reduced to 5 MΩ at 65°C. A figure of merit for HEV and EV battery pack isolation is a resistance of 1 to 10 MΩ.

22.13 NEXT GENERATION NiMH

Despite the huge commercial production of portable NiMH batteries, the technology is still being developed.[34,35] Since the initial introduction of portable NiMH in 1987, great strides have been made in terms of increasing specific energy (from 52 to 80–110 Wh/kg), specific power (from 180 to 850–2000 W/kg), cycle life (from 300 to 1000+), charge retention (from 70% loss in one month to less than 15% in one year), and cell sizes (from 1–4 Ah to 30 mAh–250 Ah). Beginning with only small, low-power cylindrical cells, NiMH is now available in high-power cylindrical cells and large high-power prismatic designs. Beginning with only metal case cells, NiMH is now available in metallic and plastic prismatic cells and even in plastic monoblock configurations.

Intensive development of NiMH battery technology continues in many diverse locations worldwide. Widespread activity has been reported on magnesium-based metal hydride alloys for capacity increases and cost reductions, evaluation of bipolar NiMH designs, satellite NiMH battery development, and a myriad of technical goals too lengthy to list. However, the most intensive NiMH development at this time involves efforts to achieve "half the weight, half the cost" by continuing to reduce cost along with raising specific energy and specific power. R&D activity also focuses on increasing the depth of discharge under a given duty cycle while maintaining cycle life, thereby allowing for a smaller, less expensive battery to be employed. In addition, NiMH technology is looking at expanding into markets dominated by other battery chemistries.

22.13.1 Cost Reduction

Cost reduction continues to be at the forefront of NiMH development. High-volume portable battery production has seen NiMH cost reach or come below $800/kWh, previously thought to be unobtainable. Key achievements in NiMH which have allowed this significant cost reduction include:

- Nickel hydroxide cost reduction from $30 per kg to about $20 per kg
- Nickel foam substrate cost reduction of about 50%
- Increased cell capacity and power by about 50% with virtually no additional cost
- Replacement of pure nickel wire mesh metal hydride electrode substrates with copper and nickel-plated steel on expanded metal and perforated sheet
- Use of pasted metal hydride electrodes
- Improved NiMH battery activation and formation

Despite these dramatic achievements, EV, PHEV, and HEV applications continue to demand even further cost reductions. Initial NiMH battery costs were above $1500 per kWh at a prototype basis, and has steadily progressed to the $800 to 1000 per kWh range (at projected production volumes of about 7,000 to 20,000 vehicles per year). At the $800 per kWh level, and with a 1.6 kWh Prius HEV battery as an example, the vehicle battery cost is $1280. Battery costs associated with PHEV and EV applications have most vehicle manufacturers placing a greater emphasis on hybrid electric vehicles where the battery may be only 10 to 25% of the EV battery size. However, there remains a tremendous development push for reduced NiMH cost to a target of $150 per kWh.

Since NiMH costs are primarily driven by materials, important development activities to meet this goal include:

- Raising NiMH HEV specific energy from 45 Wh/kg to 65 Wh/kg
- Replacement of expensive inactive components such as nickel foam substrate and cobalt additives

- Higher specific capacity metal hydride and nickel hydroxide materials
- Use of inexpensive plastic cell cases
- Monoblock construction to reduce the number of parts

Efforts to raise specific energy involve development of metal hydride alloys with higher hydrogen storage capacity (from 320–385 mAh/g active material to 450 mAh/g) and higher utilization of nickel hydroxide (from 240 mAh/g active material to 280–300 mAh/g).[36,37] Each of these higher utilization active materials requires innovative materials research involving modified alloy formulas and advanced processing techniques.

Reduction of expensive inactive cell components is focused on the positive electrode, including the nickel foam substrate and cobalt metal and cobalt monoxide used to form the conductive network. Several approaches are being studied: replacement of the cobalt compounds by metallic nickel fibers, use of reduced quantity, and less expensive cobalt compounds. An innovative approach being studied is the atomical engineering of inherently more conductive nickel hydroxide accomplished by multielement modification, and by heterogeneous nickel hydroxide powder particles where filamentary metallic fibers have been embedded into the base nickel hydroxide to ultimately connect to the outside conductive network. To reduce the cost of the foam substrate, cost reduction development includes less expensive nickel coating processes and elimination of the foam entirely by enhancing the conductivity of the nickel hydroxide and the conductive network.

Cell construction cost reduction involves development of novel plastic monoblock assemblies, reduction of parts through monoblock construction, lower cost plastic materials where possible, shared terminals and vents, and standardized sizes.

22.13.2 Ultra-High-Power Designs

For many years, it was thought that NiMH would never be able to completely replace NiCd in portable battery applications in which extremely high rates of discharge were required. In particular, power tools such as cordless drills require almost continuous discharge capability of 10 C rate. Today's NiMH cylindrical cells have exceeded the power of NiCd cells. Such high rate discharge capability has been accomplished through application of low-resistance current collection as in NiCd and through the use of improved surface catalytic activity metal hydride materials. High-power 3.5 to 7.0 Ah NiMH cylindrical cells for power tool and HEV applications have an AC impedance of about 1.7 mΩ and a $\Delta V/\Delta I$ internal resistance of about 4 mΩ. NiMH products in a variety of sizes (most commonly sub C, C, and D) are rapidly capturing portable power tool, scooter, and electric bicycle applications.

While NiMH chemistries for HEV applications have already demonstrated specific power in excess of 1300 W/kg, there is a worldwide development effort to achieve 2000 W/kg. NiMH cylindrical and prismatic prototype batteries have been announced with room temperature peak power in excess of 2000 W/kg and power over 2400 W/kg at 35°C. In some cases, power has been increased by sacrificing energy to about 44 Wh/kg, while in other designs inherently higher power active materials and casing construction have provided extremely high power without energy trade-offs. Figure 22.23 illustrates advanced HEV cell performance at 1050 W/kg. The concept is to reduce battery cost further by recognizing that for power-assist HEVs, the energy of the battery is not crucial, but rather the size of the battery is set to provide a predetermined power. Battery cost is primarily determined by the cost of battery components, and higher power utilization materials reduce the amount of material used.

A crucial factor in high-power NiMH batteries continues to be the development of highly catalytic metal hydride active materials. In particular, the interface between the bulk metal hydride and the electrolyte has been identified as essential for low voltage loss under pulse discharge. The surface oxide thickness and microporosity influence the reaction of H^+ and OH^-, and a critical observation was that within the high-power surface oxide metallic catalysts of enriched nickel regions having sizes less than 70 Å are especially important for reducing activation polarization.[38,39]

22.13.3 Stationary Power

In North America, stationary power applications represent nearly a $1B market. Today's existing applications—telecommunications, emergency lighting, and UPS systems—are dominated by lead-acid technology. Energy storage is considered by utilities as a means to reduce the economic and environmental impact associated with new plant construction. In addition, energy independence and energy security are driving increased investments in advanced forms of energy storage and the "Smart" grid. The adoption of renewables (solar and wind) that are intermittent in nature require new energy storage technologies in order to be integrated into the grid. Of the various energy storage technologies currently available, battery technology is considered to be the most proven and advanced form of energy storage to enable optimized power transmission and distribution from renewables.

Energy storage has long been dominated by lead-acid batteries of various different construction methods. Lead-acid at present dominates bulk energy storage due to its comparatively low initial cost. Routine maintenance can lead to high overall costs, including checking electrolyte level and adding water as necessary to each cell; keeping track of charge voltage; inspecting individual cases for leaks; periodic cleaning; and performing routine discharge tests. In addition, a growing need to add capacity to existing installations and a desire to utilize more "green" technology (wind and solar) are opening the door for other battery technologies.

For PV, telecom, and UPS (data centers, hospitals) systems integration, NiMH is well positioned against lead-acid due to its high energy, environmentally friendly, high durability, and long-life characteristics. This market will require high reliability, long life, and low maintenance to meet the demands of advanced energy storage technologies. NiMH battery technology can satisfy the needs of this application with proven high energy and power, cycle life, high temperature, float charge capability, reliability, and cost. Considering total cost of ownership (initial cost, operating costs, maintenance and life-cycle costs), NiMH may provide advantages over lead-acid.

REFERENCES

1. S. R. Ovshinsky, S. K. Dhar, M. A. Fetcenko, K. Young, B. Reichman, C. Fierro, J. Koch, F. Martin, W. Mays, B. Sommers, T. Ouchi, A. Zallen, and R. Young, *17th International Seminar and Exhibit on Primary and Secondary Batteries,* Ft. Lauderdale, FL, March 6–9, 2000.
2. R. C. Stempel, S. R. Ovshinsky, P. R. Gifford, and D. A. Corrigan, *IEEE Spectrum,* Vol. 35, No. 11, November 1998.
3. S. R. Ovshinsky, *Materials Research Society Fall Meeting,* Boston, November 1998.
4. K. Sapru, B. Reichman, A. Reger, and S. R. Ovshinsky, United States Patent 4,623,597 (1986).
5. S. R. Ovshinsky, M. Fetcenko, and J. Ross, *Science,* 260:176 (1993).
6. S. R. Ovshinsky in *"Disordered Materials: Science and Technology,"* D. Adler, B. Schwartz, and M. Silver, eds., Institute for Amorphous Studies Series, Plenum Publishing Corporation, New York, 1991.
7. J. R. van Beek, H. C. Donkersloot, and J. J. G. Willems, *Proceedings of the 14th International Power Sources Symposium,* 1984.
8. R. Kirchheim, F. Sommer, and G. Schluckebier, *Acta metall,* 30:1059–1068 (1982).
9. R. Mishima, H. Miyamura, T. Sakai, N. Kuriyama, H. Ishikawa, and I. Uehara, *Journal of Alloys and Compounds,* 192 (1993).
10. T. Weizhong and S. Guangfei, *Journal of Alloy and Compounds,* 203:195–198 (1994).
11. P. H. L. Notten and P. Hokkeling, *Journal of the Electrochemical Society,* 138(7), July 1991.
12. P. H. L. Notten, J. L. C. Daams, and R. E. F. Einerhand, *Ber. Bunsenges. Phys. Chem.* 96:5 (1992).
13. K. Beccu, United States Patent 3,669,745 (1972).
14. M. H. J. van Rijswick, *Proceedings of the International Symposium on Hydrides for Energy Storage,* (Pergamon, Oxford, 1978), p. 261.

15. M. A. Gutjahr, H. Buchner, K. D. Beccu, and H. Saufferer, *Power Sources 4,* D.H. Collins, ed. (Oriel, Newcastle upon Tyne, United Kingdom, 1973), p. 79.
16. M. A. Fetcenko, S. R. Ovshinsky, B. Chao, and B. Reichman, United States Patent 5,536,591 (1996).
17. M. Oshitani, H. Yufu, K. Takashima, S. Tsuji, and Y. Matsumaru, *Journal of the Electrochemical Society,* 136:6, June 1989.
18. M. Oshitani and H. Yufu, United States Patent 4,844,999 (1989).
19. V. Ettel, J. Ambrose, K. Cushnie, J. A. E. Bell, V. Paserin, and P. J. Kalal, United States Patent 5,700,363 (1997).
20. K. Ohta, H. Matsuda, M. Ikoma, N. Morishita, and Y. Toyoguchi, United States Patent 5,571,636 (1996).
21. I. Matsumoto, H. Ogawa, T. Iwaki, and M. Ikeyama, *16th International Power Sources Symposium,* 1988.
22. S. Takagi and T. Minohara, *Society of Automotive Engineers,* 2000-01-1060, March 2000.
23. K. Watanabe, M. Koseki, and N. Kumagai, *Journal of Power Sources,* 58:23–28 (1996).
24. V. Puglisi, *17th International Seminar and Exhibit on Primary and Secondary Batteries,* Ft. Lauderdale, FL, March 6–9, 2000.
25. G. Halpert, *Proceedings of the Symposium on Nickel Hydroxide Electrodes,* Electrochemical Society, Hollywood, FL, October 1989 (Electrochemical Society, Pennington, NJ, 1990), pp. 3–17.
26. M. A. Fetcenko, S. Venkatesan, and S. Ovshinsky, *Proceedings of the Symposium on Hydrogen Storage Materials, Batteries, and Electrochemistry* (Electrochemical Society, Pennington, NJ, 1992), p. 141.
27. J. Cook, *"Separator—Hidden Talent,"* Electric and Hybrid Vehicle Technology, 1999.
28. F. J. Kruger, *15th International Seminar on Primary and Secondary Batteries,* Ft. Lauderdale, FL, March 1998.
29. D. A. Corrigan, S. Venkatesan, P. Gifford, A. Holland, M. A. Fetcenko, S. K. Dhar, and S. R. Ovshinsky, *Proceedings of the 14th International Electric Vehicle Symposium,* Orlando, FL, 1997.
30. R. Elder, R. Moy, and M. Mohammed, *16th International Seminar on Primary and Secondary Batteries,* Ft. Lauderdale, FL, March 1999.
31. I. Kanagawa, *15th International Seminar on Primary and Secondary Batteries,* Ft. Lauderdale, FL, March 1998.
32. M. A. Fetcenko, S. Venkatesan, K. C. Hong, and B. Reichman, in *Proceedings of the 16th International Power Sources Symposium* (International Power Sources Committee, Surrey, United Kingdom, 1988), p. 411.
33. D. Singh, T. Wu, M. Wendling, P. Bendale, J. Ware, D. Ritter, and L. Zhang, *Materials Research Society Proceedings,* 496:25–36 (1998).
34. T. Doan, "Nickel Metal Hydride for Power Tools," *16th International Seminar on Primary and Secondary Batteries,* Ft. Lauderdale, FL, March 1999.
35. K. Ishiwa, T. Ito, K. Miyamoto, K. Takano, and S. Suzuki, "Evolution and Extension of NiMH Technology," *16th International Seminar on Primary and Secondary Batteries,* Ft. Lauderdale, FL, March 1999.
36. D. A. Corrigan and S. K. Knight, *Journal of the Electrochemical Society,* 143(5), May 1996.
37. S. R. Ovshinsky, D. A. Corrigan, S. Venkatesan, R. Young, C. Fierro, and M. Fetcenko, United States Patent 5,348,822, April 14, 1994.
38. B. Reichman, W. Mays, M. A. Fetcenko, and S. R. Ovshinsky, *Electrochemical Society Proceedings,* Vol. 97–16, October 1999.
39. K. Young, M. A. Fetcenko, B. Reichman, W. Mays, and S. R. Ovshinsky, *Proceedings of the 197th Electrochemical Society Meeting,* May 2000.

CHAPTER 23
NICKEL-ZINC BATTERIES

Jeffrey Phillips and Samaresh Mohanta

23.1 INTRODUCTION

Nickel-zinc batteries are the ideal choice when there is a need for a small, lightweight power source with excellent high-rate discharge capability at a cost point significantly lower than lithium-ion. It is an abuse-tolerant technology also capable of replacing both nickel-cadmium and nickel-metal hydride batteries in most high-volume applications. Portable applications with high current drains are well suited to nickel-zinc batteries because of their exceptional power density and the high specific energy values, currently ranging between 70 Wh/kg and 110 Wh/kg. Cycle lives that exceed 900 cycles at 80% depth of discharge are competitive with those of the best nickel-cadmium and lithium-ion technologies.

Nickel-zinc batteries are a member of a family of alkaline aqueous systems based on nickel positive electrodes that include nickel-iron, nickel-cadmium, and nickel-metal hydride. The common nickel electrode can achieve high cycle life under a wide variety of charge and discharge rates and is designed in all cases to be the ampere-hour (Ah) capacity limiting electrode under normal discharge conditions. The different counter electrodes establish the operating voltage, the size and weight of the battery, and influence the system cost and performance characteristics, such as rate capability, cycle life, and shelf life at various temperatures.

The distinctive characteristic of the nickel-zinc chemistry is the high open-circuit voltage of 1.73 V that enhances the energy density and reduces the number of cells required in higher voltage batteries. The latter factor reduces the cost of the battery with respect to other alkaline systems and proportionately lowers the internal impedance. Since the zinc electrode also demonstrates fast electrochemical kinetics, the battery is ideally suited to high discharge rate applications such as power tools and hybrid electric vehicles. Other applications include commercial AA cells that, under load, demonstrate close voltage compatibility with alkaline primary cells while, delivering superior runtimes at higher power levels.

23.1.1 History

Realization that the nickel-zinc electrochemical couple could be exploited as a rechargeable battery is documented in the patents of de Michalowski,[1] Junger,[2] and Edison[3] in 1899 and 1901. The first successful commercial demonstration of the technology did not occur until 30 years later in Ireland when Dr. James Drumm pioneered the electrification of the Bray to Dublin passenger rail cars in 1931 using large flooded electrolyte 440 V, 600 Ah batteries.[4,5] A total of four rail cars accumulated more than 700,000 miles during a service life that culminated in 1949, ostensibly due to the post–World War II availability of low-cost diesel fuel. Nickel-zinc batteries were used in a niche application as the power source in Russian mines,[6] but there was little further development until interest in electric vehicles was stimulated again during the gasoline shortages of the 1970s. At that time, the United States Department of Energy supported efforts from a number of companies to develop higher-energy alternatives to lead-acid electric vehicle batteries. The nickel-zinc battery was targeted as a high-energy-density battery that had the potential to meet aggressive cost goals provided that the cycle life of the zinc electrode could be significantly improved. General Motors Corporation engaged in one of the more ambitious programs to develop the battery for a small electric car based on the Chevrolet Chevette, but abandoned the program as enthusiasm for electric transportation abated in phase with the decrease in the price of oil.

Further development of the nickel-zinc couple was undertaken mainly in Japan between 1970 and 1990. However, despite substantial development efforts, the only product commercialized was a sealed, prismatic, maintenance-free battery of less than 10 Ah capacity from Yuasa. While the product was 60% lighter than an equivalent lead-acid battery, the cycle life at 100% depth of discharge was only 200 cycles at 2 h discharge rate.[7] Target markets were electric lawn mowers and light electric vehicles in direct competition with low-cost, maintenance-free lead-acid batteries.

After 2000, Evercel introduced a variety of larger prismatic cells and batteries[8] that achieved up to 600 cycles. The Evercel zinc electrode technology followed the development path of General Motors, but a customized electrode fabrication method ultimately proved difficult to commercialize. The batteries employed a starved electrolyte design and incorporated a gas recombination catalyst to help reduce pressure and minimize gas escape through the vent mechanism. While the technology appeared viable for the light electric vehicle and trolling-motor markets, sealed lead-acid batteries provided extremely cost effective competition.

In 2008, breaking the tradition of introducing prismatic nickel-zinc batteries in direct competition with lead-acid batteries, PowerGenix introduced the first sealed cylindrical nickel-zinc batteries for the power tool market.[9] These batteries were designed to directly compete with nickel-cadmium, nickel-metal hydride, and lithium-ion batteries. The company strategy adopted the manufacturing techniques and material development efforts of other alkaline battery manufacturers together with separator improvements from the lithium-ion industry. The combination delivers energy and power densities that have kindled renewed interest in this higher-voltage nickel battery as the competition between alkaline aqueous systems and lithium-ion batteries intensifies for market share.

23.2 NICKEL-ZINC CHEMISTRY

The chemistry of the nickel-zinc battery system is similar to that of nickel-cadmium, except that cadmium is replaced by zinc. The electrolyte is typically a 3 to 8 $\underline{M}$ alkali metal hydroxide solution in which specific additives are sometimes added to "manage" the charge–discharge behavior of the zinc electrode. The active material of the positive electrode is nickel (III) oxy-hydroxide and that for the negative electrode is high surface area metallic zinc. When discharged, the nickel (III) oxy-hydroxide is reduced to nickel (II) hydroxide and zinc is oxidized to zinc (II) species. The electrochemistry and mechanism of the reactions are quite complex[10] and for illustrative purposes, the simplified forms of the reactions are given below. Depending on the conditions, zinc hydroxide, zincate, or zinc oxide and their mixtures may constitute the final reaction product at the negative electrode on discharge. All may be reconverted to zinc during the charging process. The positive electrode reactions are provided below in a simplified form with more details found elsewhere.[11]

Negative electrode discharge reactions:

$$Zn + 2OH^- \rightarrow Zn(OH)_2 + 2e^- \quad E^0 = -1.24 \text{ V} \tag{23.1}$$

$$Zn(OH)_2 + 2OH^- \rightarrow ZnO_2^{2-} + 2H_2O \tag{23.2}$$

$$ZnO_2^{2-} + H_2O \rightarrow ZnO + 2OH^- \tag{23.3}$$

have focused on modifications to the negative electrode and the electrolyte that often accelerate zinc passivation and inhibit high-rate capability and low-temperature performance.

There have been a large variety of chemicals added to the negative electrode to improve cycle life and aid electrode manufacturing. The more common additives and their function are listed below. A more complete listing has been published previously.[12]

- Reduction of shape change, agglomeration, and dendrite formation: calcium hydroxide, calcium zincate, fluoride, phosphate, calcium titanate
- Corrosion reduction: bismuth, indium, lead, cadmium added as oxide
- Conductivity enhancement: bismuth, lead, conductive ceramic, metallic zinc powder
- Internal wicking of electrode: cellulose, chopped newsprint, inorganic fibers
- Electrode binder for integrity and handling: PTFE

Although lead and cadmium additions may help zinc cycle life, they are no longer environmentally acceptable and have been replaced with less hazardous materials.

The alkaline electrolyte used in the nickel-zinc battery is usually a combination of hydroxides of alkali metals such as potassium, sodium, and lithium. Unfortunately, the conflicting electrolyte requirement of the nickel and the zinc electrodes usually necessitates compromise. Nickel electrode capacities are maximized by higher hydroxide concentrations, but this can significantly increase zinc solubility and reduce cycle life.[13] Acids with buffering capability have been used in the electrolyte to reduce free alkalinity and improve zinc cycle life while maintaining nickel electrode efficiency.[14,15] Other anions such as fluoride and silicate have been used in the electrolyte and in the negative electrode to reduce the zinc solubility.

Examples of negative electrode and electrolyte compositions used by various organizations are given in Table 23.1

TABLE 23.1 Examples of Critical Components in Various Nickel-Zinc Designs

Organization	Typical negative electrode composition	Typical electrolyte composition	Reference
PowerGenix	ZnO 85 to 95% Bi_2O_3 1 to 10% KF 0.05 to 4.5%	Total excess OH^- 2.5 $\underline{M}$ to 5 $\underline{M}$ Borate 0.6 $\underline{M}$ to 1.3 $\underline{M}$	21
S.C.P.S.	Contains $(M_2O)_n (TiO_2)_m \cdot xH_2O$; M is an alkali metal, where 0.5<n<2 and 1<m<10, 0<x<10, typical Zn = 5% TiN = 10% CaTitanate = 1.25% CaAluminate = 3% Bi_2O_3 = 5% $Ni(OH)_2$ = 5% Rest is ZnO Plasticizer = PTFE	OH^- = 7 $\underline{M}$ to 15 $\underline{M}$; typically 7 $\underline{M}$ with saturated ZnO	18
Massey University	Ratio of Znstearate to ZnO = 0.075 to 0.25 Ratio of Castearate to ZnO = 0.03 to 0.15 Graphite = 15 to 35% Some amounts of lead and/or copper salt	KOH solution with saturated ZnO and tetrabutyl ammonium hydroxide	23
Evercell or Energy Research Corporation	CaZincate = 30 to 60% PTFE = 1 to 4% ZnO, $Ca(OH)_2$, PbO	KOH = 31 to 35%	25
Lawrence Berkeley Laboratory	ZnO = 93% Lead oxide = 2% Newsprint = 1% PTFE = 4%	KOH = 3.2 $\underline{M}$ KF = 1.8 $\underline{M}$ K_2CO_3 = 1.8 $\underline{M}$	24, 15

23.2.2 Important Considerations of Pairing Nickel with Zinc Electrodes

The beneficial effects of the inclusion of zinc ions into the rechargeable positive nickel hydroxide electrode lattice structure have been established.[16] A corresponding benefit may be derived from the zinc ions present in the nickel-zinc cell electrolyte. However, a secondary effect of the presence of zinc ions close to the positive electrode is the precipitation of zinc oxide in the pore structure, with a resulting restriction of ionic flow. More severe plugging can result in capacity loss that may happen at the early stage of cycling,[12] but the severity is highly dependent on the conditions that control the zinc solubility in the electrolyte. These include the temperature and hydroxyl ion concentration as well as the rates of charge and discharge of the battery. During the charging process, hydroxyl ions are consumed at the positive electrode and generated at the negative electrode. These changes in hydroxyl ion concentration can precipitate zinc oxide directly within the pores of the positive electrode.

A second problem with nickel-zinc cells is the disparity in charge efficiency of the positive and negative electrodes. At the positive electrode, oxygen evolution (reaction 23.9) competes with the nickel hydroxide charging reaction such that an additional 20% overcharge may be required before the electrode is fully charged. Techniques to improve charge efficiency include the addition of oxides, such as calcium oxide or zinc oxide to the nickel hydroxide, or simply charging to less than full capacity. The latter approach unfortunately reduces the cost advantage of the technology over competitive batteries. In contrast, the charging process at the zinc electrode is 99% efficient with small amounts of hydrogen being produced as in reaction 23.8. In vented cell configurations, the disparity in electrode charge efficiencies results in the rapid conversion of all of the zinc oxide in the negative electrode to zinc metal. This promotes the evolution of hydrogen during charge with resulting dry-out and premature failure. One solution to this problem is to operate the cell with a starved electrolyte to encourage the reaction of gaseous oxygen with the negative electrode. The immediate effect of "oxygen recombination" is to both proportionately reduce the charging efficiency of the negative electrode and to equalize charging efficiencies at both electrodes. While the microporous separator may block the access of oxygen to the central area of the negative electrode, the edges can often provide sufficient recombination for charge efficiency equalization. If differences in charging efficiencies can be moderated, then gas recombination catalysts can be used effectively to maintain low cell pressure and allow oxygen recombination to occur at a lower rate at the negative electrode.

23.2.3 The Separator System

The choice of the separator system is a key element in the life of a nickel-zinc battery. The separator is present to electronically insulate the nickel and the zinc electrodes while minimally restricting passage of ions to allow efficient electrochemical discharge and charge of the active materials. Achieving these goals requires a robust separator that resists mechanical penetration while providing a sufficiently open structure for low electrical resistance. In nickel-zinc cells, the separator also inhibits the growth of metallic zinc dendrites that may bridge the positive and negative electrodes. This requirement normally defines a pore size that is both small and tortuous but necessarily at odds with a requirement for ease of ion flow associated with low resistance. Achieving the balance between these requirements is particularly difficult for higher discharge rate applications where there is a greater need to minimize cell resistance for efficient high-power discharge without excessive temperature buildup. Combining a highly conductive electrolyte with less conducting, dendrite inhibiting separators may deliver good high-rate performance. However, the highly conducting electrolytes that provide low resistive losses are the most effective in dissolving the zinc oxide to promote failure modes such as shape change, densification, and dendrite growth. It is therefore very important to address the zinc failure modes from the perspective of separator, electrolyte, and the negative electrode composition so that high-rate charge and discharge are enabled together with high cycle life. High-energy and high-power delivery from a nickel-zinc cell provides an incentive for the use of thin separator materials, but this must be achieved without compromising the safety of the cells. Polyolefin materials are favored as separators since they can be fabricated in high-strength thin films with small, tortuous pores. Unfortunately, these materials are not easily wetted by an aqueous alkaline solution and must be treated either with a surfactant or by chemical grafting of surface groups. Even with such treatment, the wicking ability of the polyolefin separator alone is insufficient to promote good uniform electrolyte distribution without the help of an additional separator such as a high-porosity nonwoven polyamide, a wetting capable polyolefin, or an electrolyte absorbent PVA/cellulose mix.

Parameters normally used to characterize separators are porosity, ionic resistance, air permeability, pore size, mechanical strength, chemical stability, dimensional stability, and wetting ability. Most out-of-cell separator evaluations are based on characterization of resistance and air permeability since these properties are determined by the aggregate of porosity, pore size, pore tortuosity, and wetting ability. The other physical properties such as mechanical strength and thermal stability govern the issues of manufacturability, reliability, and safety. Even using these criteria as guidelines, it is often difficult to predict the suitability of various separator combinations in the nickel-zinc system without extensive in-cell testing.

23.2.4 The Positive Electrode

The positive electrode is very similar to that used in nickel-metal hydride and nickel-cadmium systems. The most widely used nickel electrode is a high porosity nickel foam pasted with spherical nickel hydroxide particles with additions of binder and conductivity enhancing material such as nickel powder and cobalt oxide. Commercially available nickel hydroxide contains 0 to 3% cobalt hydroxide and 0 to 3% zinc hydroxide as co-precipitated material. These additives improve charge acceptance and stabilize the preferred beta crystal structure of nickel hydroxide.

The pasted alkaline nickel electrode demonstrates a mild memory effect or voltage depression if repetitively discharged to the same state of partial charge. However, one or more full discharges to 1.1 V in the case of nickel-zinc batteries will restore the discharge voltage curve to its original condition.

23.3 CELL CONSTRUCTION

Nickel-zinc batteries have traditionally followed the construction guidelines set up for nickel-cadmium and nickel-metal hydride batteries in prismatic and cylindrical formats. The most significant differences lie in the separator system used and the choice of negative electrode substrate. Normally a high cycle life nickel-zinc cell will employ a multilayer separator consisting of one or more layers of a microporous membrane together with a single more highly porous "wicking" separator that promotes good electrolyte distribution and is capable of buffering the electrolyte volume changes that occur during charge and discharge of the cell. The microporous membrane separator plays an important role in preventing electrical short conditions resulting from zinc forming a conducting bridge between the positive and negative electrodes. A secondary consequence of using a less open separator system is a significant reduction in the oxygen and hydrogen gas recombination rates as compared to other alkaline systems. Particular attention must be given to the charging procedures to prevent undue gas pressure buildup in overcharge with resulting vent operation and shortening of cycle life.

The choice between cylindrical and prismatic designs depends on the application. Smaller capacity cells below around 20 Ah favor the cylindrical format because there are fewer individual parts and ease of construction on a jelly-roll winder. This format becomes more challenging for higher capacity cells as plate widths and lengths increase and manufacturing becomes more difficult. Under higher-rate discharges, there are also thermal management issues associated with the difficulty in heat removal from larger diameter jelly-rolls. At lower discharge rates and high ampere-hour requirements, the prismatic cells gain favor on the basis of the easier assembly of thicker plates and the superior packing density of this configuration.

23.3.1 Prismatic Construction

Whether in single cell or monoblock configuration, the prismatic construction has been the most common format for nickel-zinc batteries. The zinc electrodes are individually tabbed before being bagged in one, two, or even three layers of microporous polyolefin film. A manual or automatic stack and wrap

23.8 SECONDARY BATTERIES

machine can alternate these zinc electrodes with tabbed positive electrodes between a web of nonwoven wicking separator that may be based on nylon or grafted polyolefin. The negative and positive tabs are individually grouped and bolted, or resistance welded to the terminals of the cell. Sealing through the lid of the cell is generally accomplished using O-ring seals that resist the creeping of alkaline electrolyte. The cell stack is compressed and inserted into the case before the lid is welded or glued in position. After cell sealing, the electrolyte is introduced under vacuum and is allowed to soak for several hours prior to applying the first charge to the cell. Figure 23.1 shows a typical prismatic construction for a 20 Ah cell. Energy densities range from 60 to 100 watthours per kilogram and 110 to 200 watthours per liter. Most designs are maintenance-free and use sufficient electrolyte to wet the separators without saturation and without electrolyte outside of the cell stack. Because of the low pressure tolerance of the prismatic design, safety vents normally do not exceed 100 psi. Recombination has less driving force at lower ambient pressures, and for long life, it is commonplace to include free-standing recombination catalysts that can promote the production of water from gaseous hydrogen and oxygen present in the cell.

23.3.2 Sealed Cylindrical Construction

A jelly-roll construction can provide very uniform current density if current collection take place at either ends of the jelly-roll. This configuration is thus well suited to high-power applications such as power

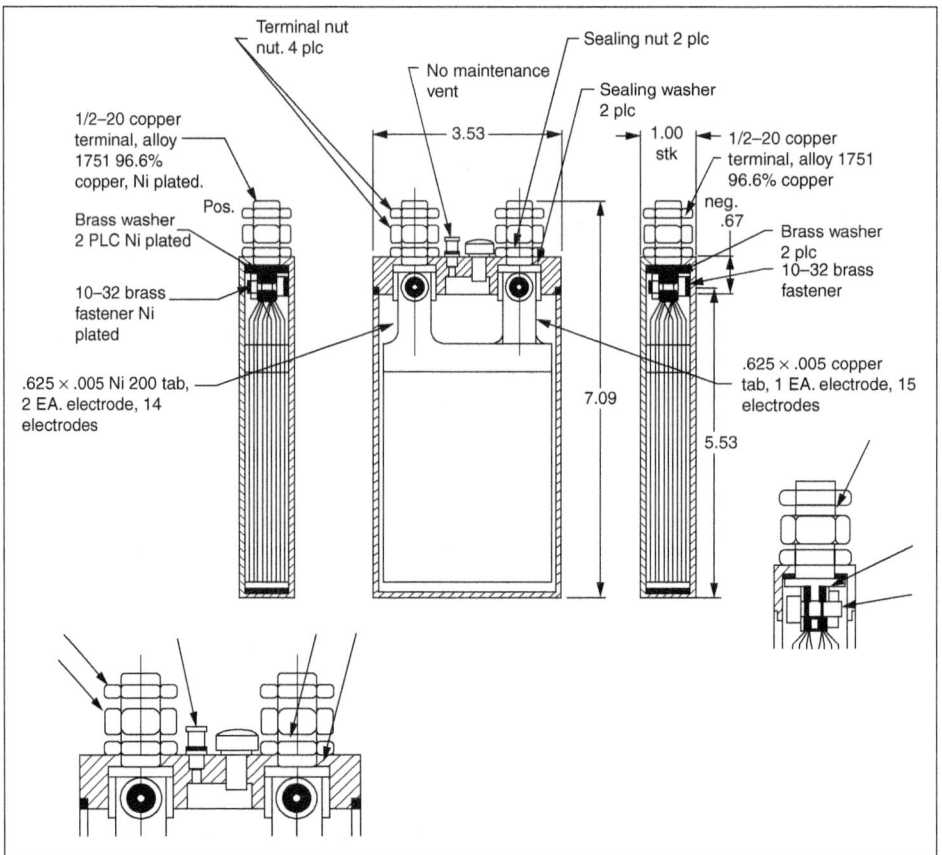

FIGURE 23.1 A prismatic cell construction.

tool and hybrid electric vehicle batteries. A rapid winding process can achieve 5 to 8 jelly-rolls per minute from one machine and can guarantee cost competitive production and throughput. After the jelly-roll is wound, one or more contacts may be attached to the electrodes before it is inserted into the can where welds may be made between the can surface and tabs or discs attached to the jelly-roll. The can containing the jelly-roll is then filled under vacuum with an appropriate amount of electrolyte, and the cell is capped and allowed to soak before the application of the first charge. Figure 23.2 is an exploded view of a typical nickel-zinc spiral-wound construction where a spring that is welded to the cap provides contact to the zinc current collector and a nickel "spiked" disc welded to the can base provides a similar function for the positive electrode.

The design differs from the prismatic cell in several key respects. The microporous separator is wound such that the edges are open with free access to the electrodes from either end of the jelly-roll. This allows contacts to be made either by welding or simple pressure contact. Unlike the prismatic design, the vent should not open unless there is an abnormal pressure buildup. Excessive opening of the vent may result in electrolyte fouling and ineffective subsequent closure with oxygen ingress eventually discharging the zinc electrode. Vent opening pressures must be high enough to remain closed at all operating conditions yet low enough to guarantee safety under abuse. The polarity of the can may be either positive or negative, but hydrogen catalysts such as iron should not be exposed to electrolyte at the zinc potential. General practice is to coat any metal in contact with the zinc electrode with material that can inhibit hydrogen release.

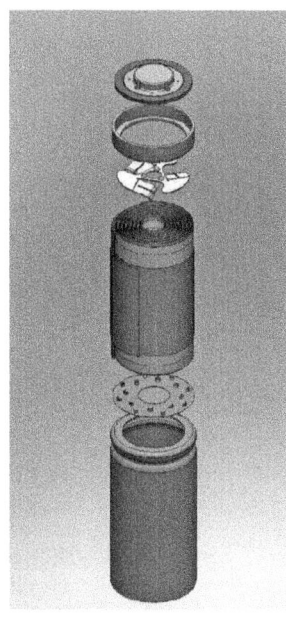

FIGURE 23.2 Exploded view of sealed sub C nickel zinc cell (*Courtesy of PowerGenix, Inc.*)

Small sealed cylindrical nickel-zinc cells exhibit higher specific energy and energy density (70 to 110 Wh/kg and 200 to 360 Wh/L) than the prismatic cell. This is primarily a result of an efficient bussing design and minimal electrolyte amounts in the totally sealed configuration. In multicell batteries, the less efficient volumetric packing capability of cylindrical designs sometimes may eliminate this advantage.

23.3.3 Nickel Electrode

The active material of the positive electrode, nickel hydroxide, has little electronic conductivity in either the charged or discharged state. Electronic conductivity to the active material is provided by the supporting electrode substrate and other conducting additives that provide "micro" conductive elements within the bulk active material.

Nickel electrode technologies[11] have been incorporated with minimal changes from nickel-cadmium and nickel-metal hydride cells. The most robust nickel hydroxide electrode incorporates a porous, sintered nickel substrate as the current carrying backbone. The highly porous nature allows nickel hydroxide active material to be chemically or electrochemically impregnated into the structure along with small amounts of cobalt hydroxide that help active material conductivity and promote efficient charging. While the sintered nickel electrode is capable of thousands of cycles, it is an expensive electrode to manufacture with a requirement for strict environmental controls during the solution processing and washing. The high nickel-metal content provides good overdischarge tolerance, good current distribution, and excellent high-power performance. However, such robustness exacts a gravimetric and volumetric energy penalty.

An alternate pasted-positive electrode does offer significant cost and energy density advantages that enable nickel-zinc technology to compete effectively with existing commercial lithium-ion batteries. The electrode substrate is a highly porous nickel foam that functions as a three-dimensional electronic conductor. The basis weight may vary from 300 to 600 g/m^2 with the choice mainly being determined by the current carrying requirement of the plate. The active nickel hydroxide is manufactured as 10 micron diameter spherical particles containing approximately 2% co-precipitated cobalt hydroxide. Additions of nickel and cobalt metal to the positive mix improves electronic conductivity, and promotes high active material utilization particularly at high discharge currents. Unfortunately, the cobalt additions to the positive electrode can be extremely detrimental to the negative zinc electrode if the ions are allowed to diffuse or migrate across the separator barrier. The offending material is generally soluble cobalt (II) that is reduced in contact with zinc metal to produce cobalt metal sites capable of catalyzing hydrogen evolution. The more recent use of spherical nickel hydroxide particles with a conductive cobalt (III) layer has proved more beneficial to the nickel-zinc cell. Without the need for excessive additions of cobalt metal or cobalt oxide, the contamination of the negative electrode has been minimized. Specific treatments of oxidized cobalt-treated nickel hydroxide have proved even more beneficial to the nickel-zinc battery by optimizing conditions of high cobalt stability with high positive electrode efficiency and minimal expansion over the cycle life of the cell.[17] These continuous improvements have gradually enhanced the reliability and abuse tolerance of the nickel foam/spherical nickel hydroxide based electrode technology and allowed a continuous production pasting process to be easily incorporated into the nickel-zinc manufacturing process.

Other nonsintered positive electrodes such as the roll-bonded positive method have been described as lower-cost alternatives to the pasted foam technology. In the former technique, nonaqueous fluids are used to fibrillate dry Teflon particles in the creation of rubberized active material sheets that are pressed onto either side of a nickel-plated steel substrate to create the three-dimensional electrode. As a further cost reduction step, graphite or carbon is employed as a low-cost substitute for nickel-metal conductive diluents, but invariably there is gradual oxidation of the carbon to create carbonates that slowly destroy the alkalinity of the electrolyte over time with a resulting degradation of the midpoint discharge voltage and power capability. A major disadvantage of this type of electrode is the loss in volumetric energy density from the use of such a low-density conductive material.

The most widely adopted positive electrode for the nickel-zinc system has proved to be the pasted nickel-foam variety. The electrode used readily available starting materials and delivered high energy density and good current carrying capability. In addition, the pasting technique provided a clean continuous electrode processing method at low capital cost for high volume production.

23.3.4 The Zinc Electrode

The rechargeable zinc electrode normally contains 50 to 200% ampere-hour excess of zinc and zinc oxide as compared to the positive nickel hydroxide electrode. The electrode is prepared in a fully discharged state although zinc metal or alloy may be present for enhanced electrode conductivity and additional active discharge material. Methods of preparation of the electrode include slurry coating and pasting as well as less widely used methods such as roll bonding and simple pressing. The most common current collector substrates are perforated or expanded sheets of copper or brass, although substantial advantage is claimed from the use of a copper foam that creates a three-dimensional conducting matrix.[18] Substrates have generally incorporated a surface layer that inhibits hydrogen evolution at the zinc potential and is stable under conditions where the negative electrode undergoes polarization. This may occur during high-rate or low-temperature discharge. The necessity for more "green" batteries has eliminated lead plating as the preferred surface layer in favor of silver or tin. A more recent approach created a corrosion resistant alloy[19] by heat treating a tin-plated copper current collector.

The composition of the negative electrode precursor mix includes conductive elements, anticorrosive additives as well as the fine particle (<1 micron) zinc oxide active material. Other additives such as cellulose[20] and aluminum fiber[21] can provide electrolyte irrigation, while other ceramic materials such as aluminum oxide are useful in trapping soluble zinc hydroxides before they migrate away from their original location. As with most battery electrodes, there is always the need for one or more binders to provide electrode integrity and prevent powder generation during processing. In aqueous pasting or slurry coating techniques, additional processing additives such as carboxymethyl cellulose and various oxide dispersants may be added as an aid to paste stability and flow characteristics. The conductive elements may include metal oxides with redox potentials positive of the zinc potential or zinc metal itself. Other electronically conductive materials such as carbon, conductive polymers, and ceramic materials have also been used.

Perhaps one of the most extensively used oxide additives has been calcium oxide. This material has the benefit of combining with zinc oxide to create calcium zincate that is a much less soluble discharge product than the normal zinc hydroxides. The weight of calcium oxide added may be as much as 25 wt % of the active zinc. This significantly increases the weight of the negative electrode and decreases cell energy density. There may also be electrode kinetic factors that restrict the effectiveness of this additive to lower rate discharges and charges. Studies have determined the kinetics of calcium zincate formation is relatively slow, indicating that as discharge rates increase, there may be less benefit to the calcium addition.[22]

From an electrode fabrication perspective, calcium oxide additions normally restrict electrode processing to nonaqueous techniques such as roll bonding where rubberized sheets of the zinc electrode are pressed into or onto the substrate. Attempts to form the zincate during aqueous paste mixing invariably failed as the exothermic reaction between zinc and calcium oxides resulted in an unworkable mixture. Pasting or slurry coating a zinc electrode containing calcium oxide from an aqueous medium required the preformation of calcium zincate.

In summary, the application and current drain requirements may determine the type of negative zinc electrode that may be included in the cell. High-current drain favors copper foam current collection and zinc oxide containing compositions. Lower-current drains may favor the copper foil current collector with calcium zincate additives. Both types of electrode must work in tandem with specific electrolytes and separator systems to reduce dendrite formation and electrode active material redistribution.

23.3.5 Separator and Electrolyte Design Considerations

Most nickel-zinc cell designs employ a multicomponent separator that includes a thin microporous polyolefin dendrite penetration barrier combined with a more highly porous material capable of absorbing and retaining electrolyte. Traditionally, at least one layer of microporous polyolefin is wrapped around the zinc electrode to limit electrolyte access and reduce the dissolution of zinc oxide. Since this separator is a major cost element of the cell, it is prudent to minimize the number of layers necessary to achieve the desired cycle life. Although one layer is often all that is necessary to provide several hundred charge-discharge cycles, two layers can provide the necessary insurance against separator defects or mechanical damage during the manufacturing process. Of particular concern in cylindrical cells is the winding process, where the thin polyolefin separator sheets may be damaged as components slide against one another during jelly-roll assembly. Some simplification in the assembly process can be achieved by laminating the separators, particularly when using more than two individual layers. High-power and high-energy applications will seek to minimize these separator layers and reduce resistive losses by the choice of a highly conductive electrolyte.

To meet the demand for even higher-energy-density nickel-zinc batteries in the future, one of the most probable approaches will be to use thinner microporous separator materials that have been developed for high-power, high-energy-density lithium-ion cells.

23.4 PERFORMANCE CHARACTERISTICS

The general characteristics of nickel-zinc batteries are given in Table 23.2. Relevant parameters are given value ranges that encompass both cylindrical and prismatic versions of the technology. The higher volumetric energy density and power density figures correspond to small, cylindrical cells optimized for small size and high power. Cycle life figures are highly variable and dependent upon the design, discharge and charge conditions, as well as the defined capacity level at which the test is terminated. Nickel-zinc batteries generally show a gradual loss in Ah capacity during cycling, with end of life being determined when the value reaches 60 to 80% of the initial rated capacity. At higher discharge rates, original equipment manufacturers frequently specify an end of life capacity that is less than the normally accepted 80%.

One of the most important advantages of the nickel-zinc battery is the low impedance, which is a direct consequence of the reduced cell count necessary to achieve battery voltage. A 14.4 V battery will require approximately 30% fewer cells than an equivalent nickel-cadmium or nickel-metal hydride version. While this has a significant impact on the weight and size, there is an immediate 30% reduction in the impedance if the individual cells have comparable impedance values. In general, the nickel-zinc cell will have lower impedance than similar size alkaline cells because of the use of a

TABLE 23.2 The Range of Characteristics of Nickel-Zinc Batteries

Parameters	
Cathode electrochemistry	$Ni(OH)_2$ /NiOOH
Anode electrochemistry	ZnO/Zn
Electrolyte (% potassium hydroxide)	20 to 35
Separator	Microporous + wicking
Nominal cell voltage (volts)	1.65
Operating temperature range (°C)	–20 to 50
Theoretical specific energy (watthour per kilogram)	334
Specific energy (watthour per kilogram)	70 to 110
Energy density (watthour per liter)	130 to 350
Specific power (watts per kilogram)	280 to 2500
Power density (watts per liter)	420 to 7000
Charge retention (percent loss per month at 25°C)	20
Cell cycle life (cycles at 100% DOD)	300 to 900

low resistivity copper current collector and the inherently fast electrochemical kinetics of the zinc electrode. The room-temperature voltage profile of a PowerGenix 48 g sub C size cylindrical cell is shown in Fig. 23.3 for differing discharge rates as a function of the delivered capacity. The AC impedance of this cell measured at 1 kHz is under 4 mΩ and the direct-current impedance calculated from the midpoint voltage is less than 8 mΩ. Even at a 20 A discharge current, the midpoint voltage is 1.55 V and the difference in delivered capacity from cells discharged at 400 mA is less than 100 mAh. The temperature rise during the discharge as measured by a thermocouple located on the side of the cell case is below 50°C when the cell is completely discharged at the 15 C rate.

Extremely high currents can be delivered by a nickel-zinc sub C size cell under constant voltage load. At the 1 V level, currents in excess of 140 A can be delivered with calculated power densities greater than 2.5 kW/kg over a 20 s time period (Fig. 23.4).

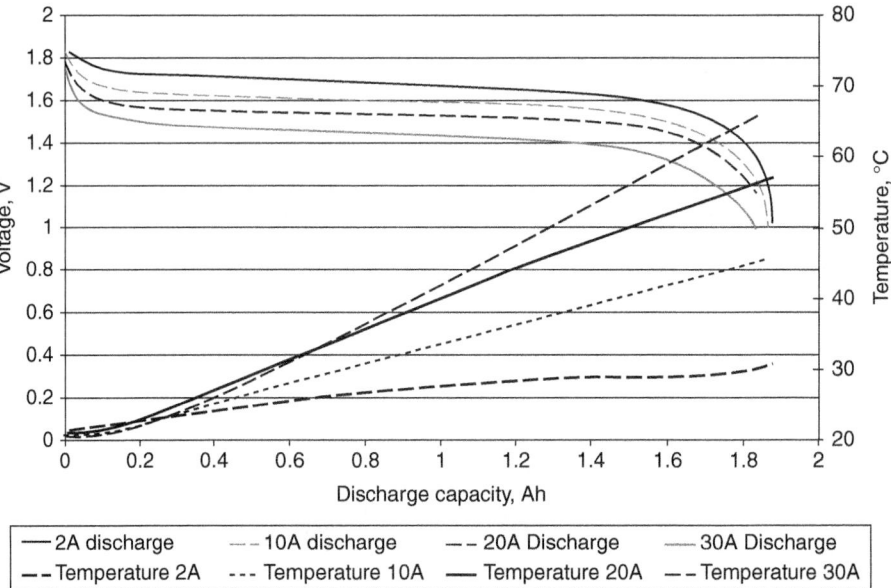

FIGURE 23.3 The voltage-capacity curves for a 1.9 Ah sub C cell and associated temperature rise. (*Courtesy of PowerGenix, Inc.*)

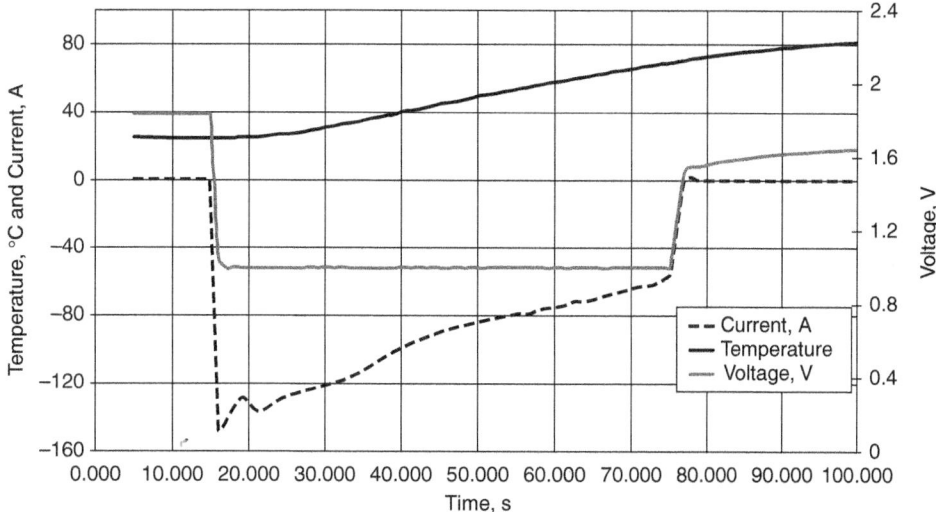

FIGURE 23.4 Current and temperature response of sub *C* size cell held at 1 V. (*Courtesy of PowerGenix, Inc.*)

Ambient temperature affects both the delivered capacity and the discharge voltage level of the cell. The discharge curves of the high-rate sub *C* size cell at the 5 *C* discharge rate and various temperatures are shown in Fig. 23.5 after charging was completed at room temperature.

The high-rate, room-temperature cycle life of both the single cell and a 19.2 V battery encased in plastic is shown in Fig. 23.6. The battery consists of 12 sub *C* size cells connected in series. The discharge current is maintained at 10 A to 1.1 V per cell with a 20 A current being applied periodically every 50 cycles as a check of power degradation. In the case of the single cell and the battery pack, power degradation was minimal over the period of the test and no soft shorts were detected by periodic open-circuit voltage decay checks. The single cells achieved approximately 900 cycles to

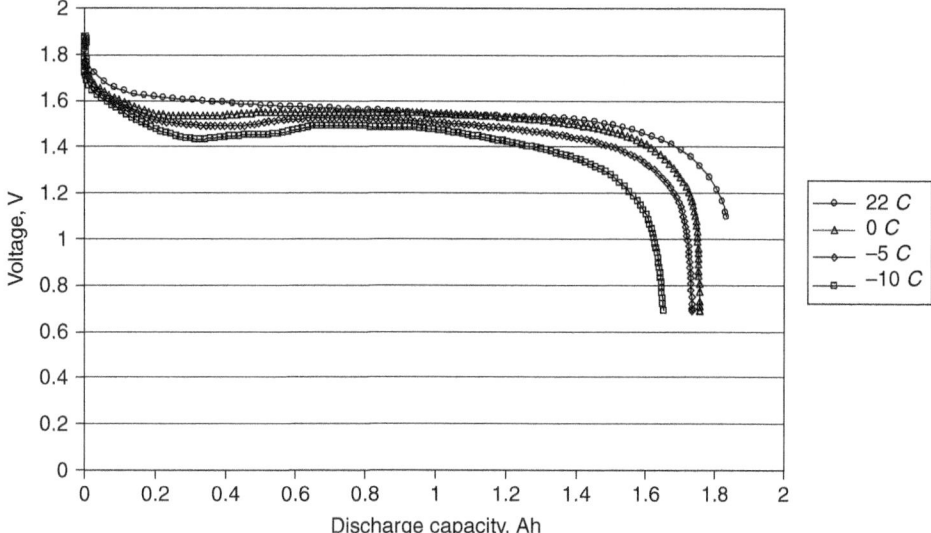

FIGURE 23.5 The effect of temperature on the voltage response of a nickel-zinc sub *C* size cell discharged at the 5 *C* rate. (*Courtesy of PowerGenix, Inc.*)

23.14 SECONDARY BATTERIES

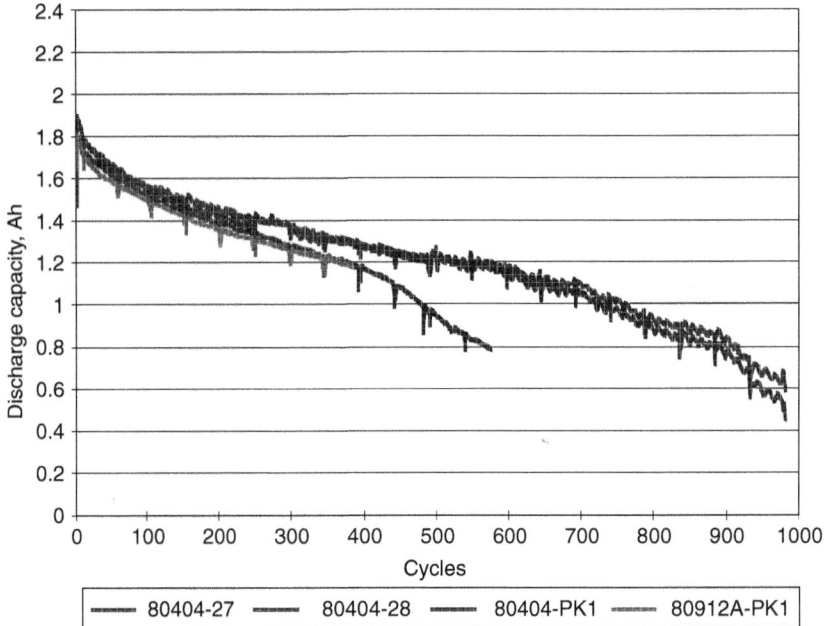

FIGURE 23.6 10 A cycling of single sub C size cells and the 19.2 V plastic encased battery. (*Courtesy of PowerGenix, Inc.*)

a capacity value of 0.8 Ah. There is a faster degradation in the pack capacity versus the single cell, probably as a result of the increased temperature within the pack and the inevitable overdischarge and overcharge of some of the closely packed cells with differing thermal environments.

Figure 23.7 shows the voltage curves at different discharge rates for a 1500 mAh AA-size cell and Fig. 23.8 illustrates how the discharge curves are affected by temperature at a fixed discharge rate of 1 C.

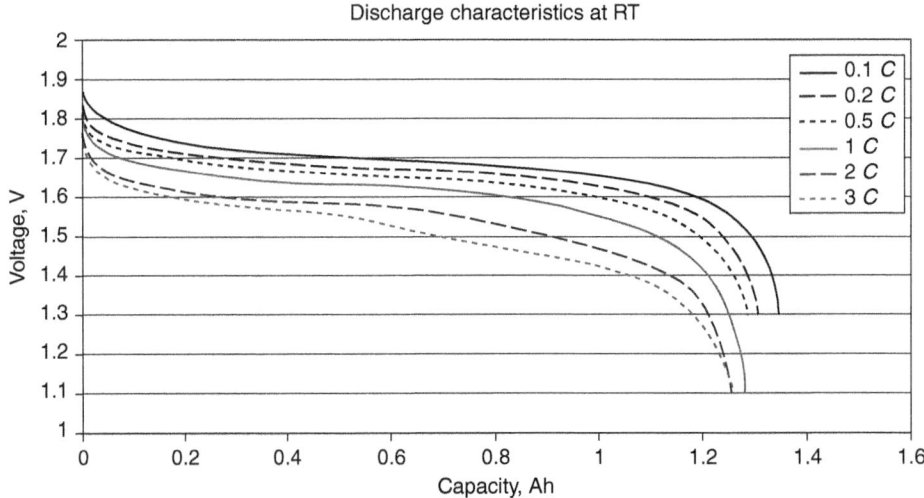

FIGURE 23.7 The voltage-capacity curves of the 1500 mAh AA-size cell. (*Courtesy of PowerGenix, Inc.*)

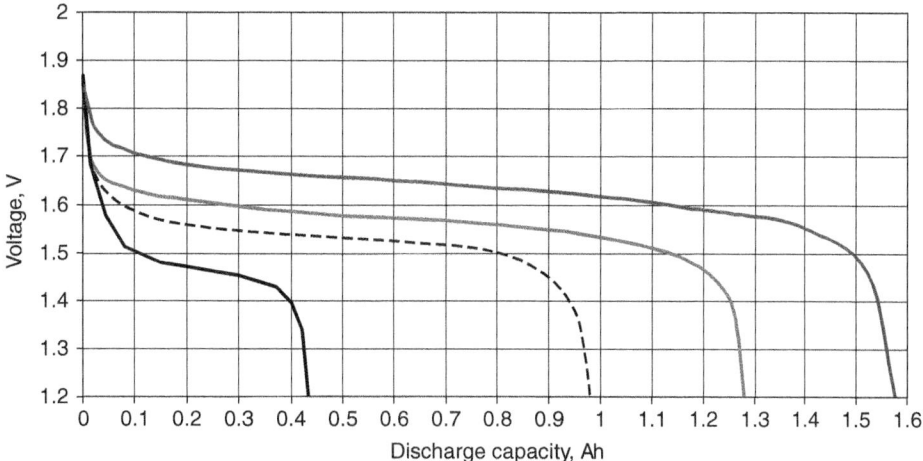

FIGURE 23.8 The 1 C discharge behavior of 1500 mAh AA-size cells at 25 C, 0 C, −10 C, and −18 C after charging at 0.5 C rate at room temperature. (*Courtesy of PowerGenix, Inc.*)

The 1500 mAh AA-size cell is a general-purpose, medium-rate cell with an internal impedance of 12 mΩ for target applications such as digital cameras and electric toothbrushes. Figure 23.9 shows the cycle life when charged within 2 h and discharged completely over a 1 h period. For higher-power applications, a 1300 mAh AA-size cell is available with an AC impedance of 6 mΩ. Figure 23.10 shows a relatively flat voltage response during discharges between 0.25 A and 10 A. The cycle life of a high-power AA-size 10.4 volt pack discharged at 10 A is shown in Fig. 23.11.

The largest cylindrical cells currently available are D size with capacities of 6.5 and 8 Ah. This cell size is specifically designed for light electric vehicles where periodic high-pulse currents are required for acceleration and hill climbing. Figure 23.12 shows the voltage response to various current levels up to 80 A.

Larger capacity prismatic nickel-zinc cells have been produced by Evercel and S.C.P.S. in France. The latter company has demonstrated 30 Ah cells with an excellent cycle life of over 900 cycles to

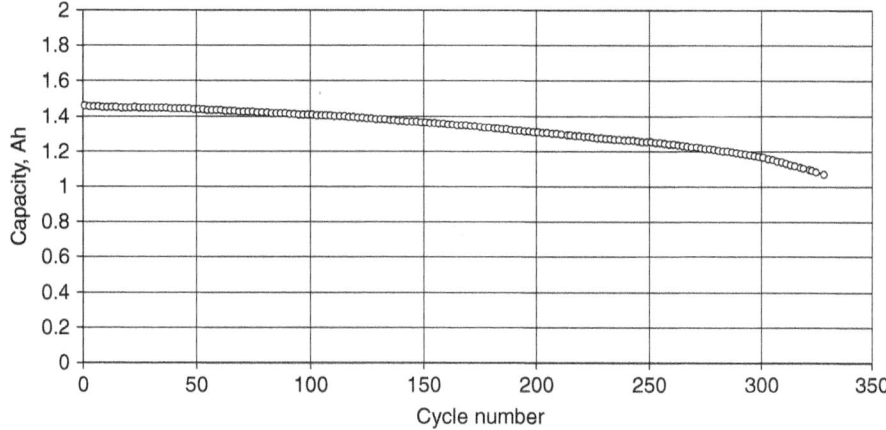

FIGURE 23.9 The capacity-cycle number graph for the 1500 mAh AA-size cell discharged at the 1 C rate and charged at the 0.5 C rate. (*Courtesy of PowerGenix, Inc.*)

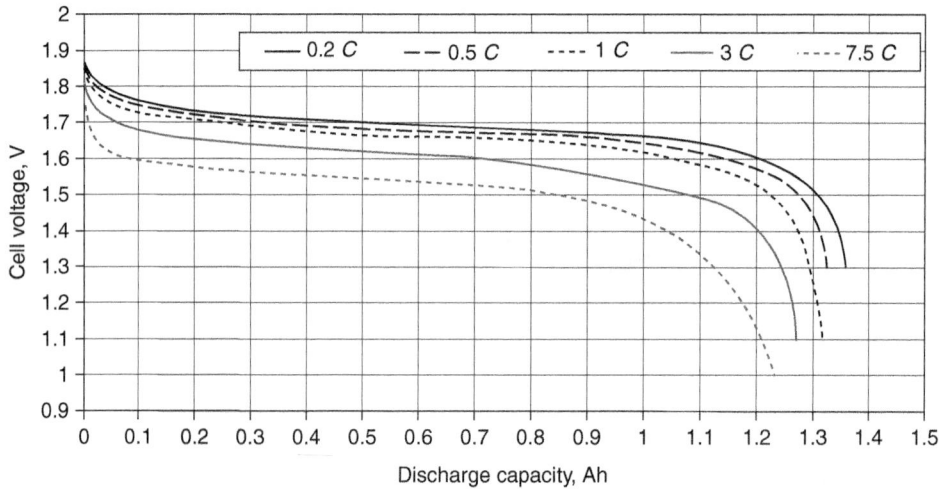

FIGURE 23.10 The cell voltage vs. discharge capacity for the 1300 mAh high rate AA-cell at various discharge rates. (*Courtesy of PowerGenix, Inc.*)

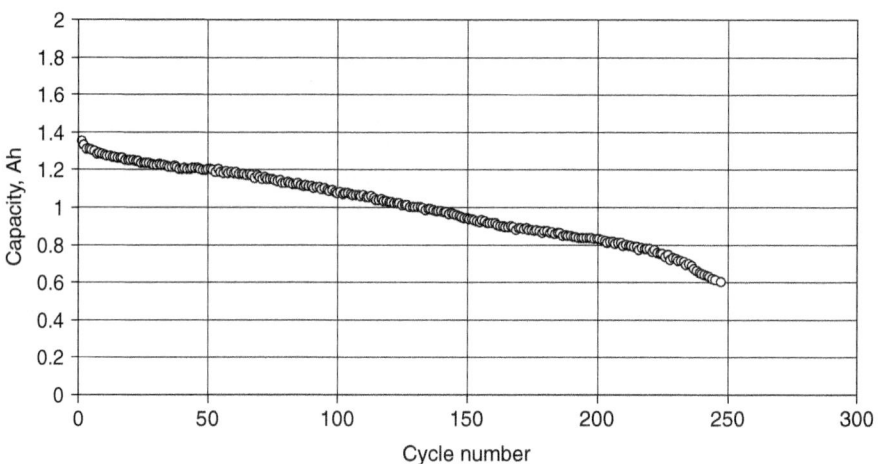

FIGURE 23.11 The capacity-cycle number graph for a 6-cell 10.4 V pack of 1300 mAh high-power AA-size cell discharged at 10 A and charged at 600 mA. (*Courtesy of PowerGenix, Inc.*)

80% depth of discharge at the $C/3$ discharge rate with specific energy values between 70 and 80 Wh/kg and energy density values of 130 Wh/L. Figures 23.13 and 23.14 illustrate the discharge rate dependence of the capacity and the cell cycling performance. Special design features focus on improving the zinc electrode conductivity by inclusion of copper foam as the negative current collector together with the use of conductive ceramic additives to suppress dendrite formation. These cells extend service life by including gas recombination electrodes and a charging method that limits the oxygen generated by the positive electrode.

A decrease in the depth of discharge during cycling will increase the number of cycles to a given capacity; however, for nickel-zinc cells, cycling to 100% depth of discharge may only decrease the cycle life by 20%. For multicell batteries, this factor will be greater and dependent on discharge rate, thermal environment, and the extent to which the cells have been capacity matched initially.

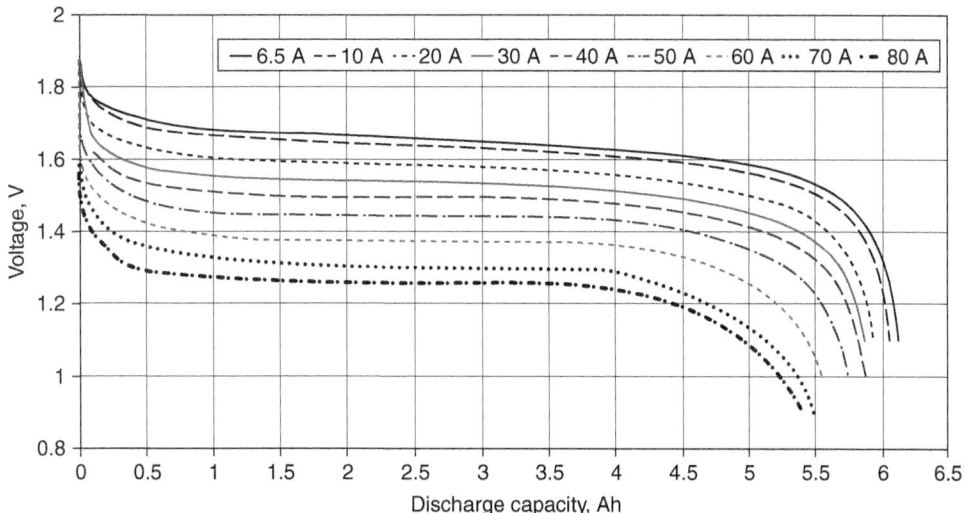

FIGURE 23.12 The voltage vs. discharge capacity for a 6.5 Ah D-size cell at various discharge rates. (*Courtesy of PowerGenix, Inc.*)

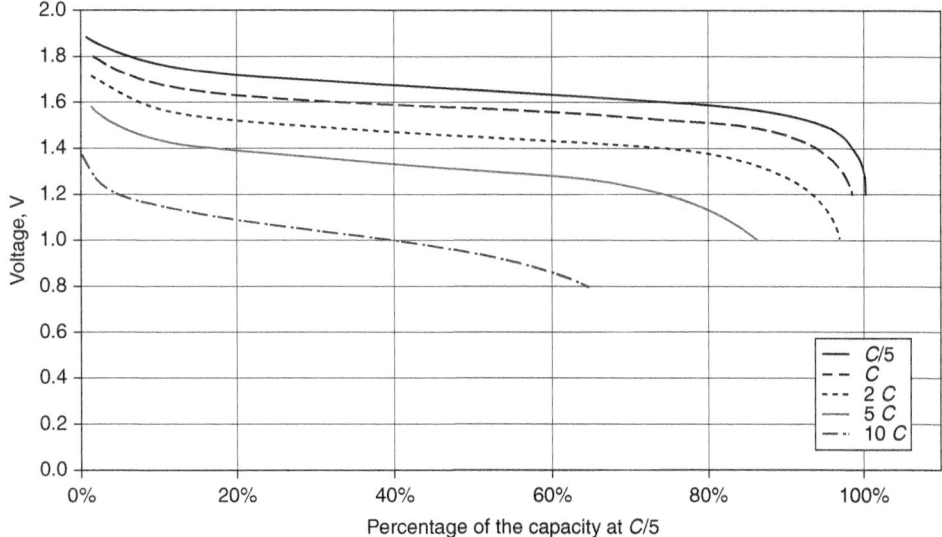

FIGURE 23.13 Discharge curves at various discharge rates for 30 Ah prismatic cell. (*Courtesy of S.C.P.S. France.*)

23.4.1 Storage Characteristics

The room temperature self-discharge rate of both the prismatic and cylindrical cells is approximately 1% of the ampere-hour capacity per day. As with all alkaline nickel batteries, long-term storage above 50°C is not recommended, however, accelerated testing is often performed at 60°C to establish the cells tolerance to extreme temperature conditions likely to be encountered during transportation. Fully charged cylindrical cells subjected to 60°C temperatures for a period of 30 days usually show capacity loss values of 20 to 30% under 5 C discharge conditions but do

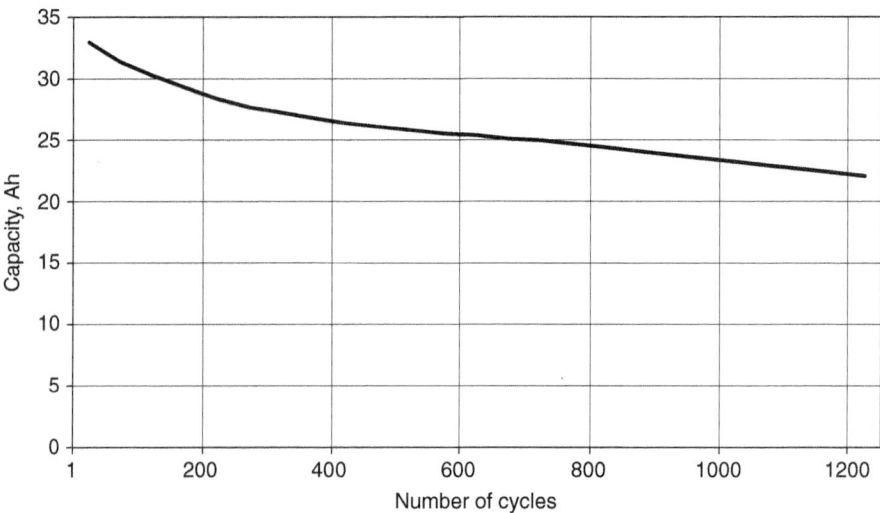

FIGURE 23.14 The cycle life of a 30 Ah prismatic cell under C/3 discharge and 80% depth of discharge. (*Courtesy of S.C.P.S. France.*)

experience approximately 5% loss in initial delivered capacity after recharging. Cycling performance is unaffected by the storage. For prismatic cells with incorporated recombination catalysts, the capacity retention may be higher since the evolved gases that recombine at electrode surfaces in cylindrical cells can contribute to the self-discharge of the electrode.

23.4.2 Safety Considerations

All commercially available cells and batteries should meet the minimum safety standards outlined in Underwriters Laboratories Inc. Document UL 2054 for Household and Commercial Batteries. This safety standard includes a series of mechanical tests that include shock and vibration and more severe crush tests. Various electrical tests that range from simple short-circuit tests to abnormal and abusive charge and discharge are defined with the purpose of creating standards that will reduce the risk of fire, explosion, and personal injury. All the PowerGenix cylindrical nickel-zinc cells described here pass these minimum standards and closely resemble the behavior of nickel-cadmium batteries. Any dangerous pressure buildup within the cell is released through a resealable vent designed to remain closed under all normal operating conditions. Testing above and beyond the UL standard has established nickel-zinc as exceptionally safe under the most abusive 5 C continuous overcharging and overdischarging. In both cases, the passage of current pressurizes the cell by breaking down the electrolyte into hydrogen and oxygen. Eventually the vent opens to release the pressure with resulting dry-out in less than 20 min (Fig. 23.15). Other extreme abuse tests such as a vertical crush test and screw and nail penetration tests have equally benign results to demonstrate a safety level that is necessary in the more abuse-prone power-tool industry.

23.4.3 Charging Nickel-Zinc Cells and Batteries

Figure 23.16 shows the voltage response at room temperature when a cylindrical sub C cell is charged continuously at 1 A. The internal pressure is also monitored using a strain gauge attached to the side of the cell. The voltage rises quickly above 1.8 V and maintains an efficient charge plateau with a midpoint

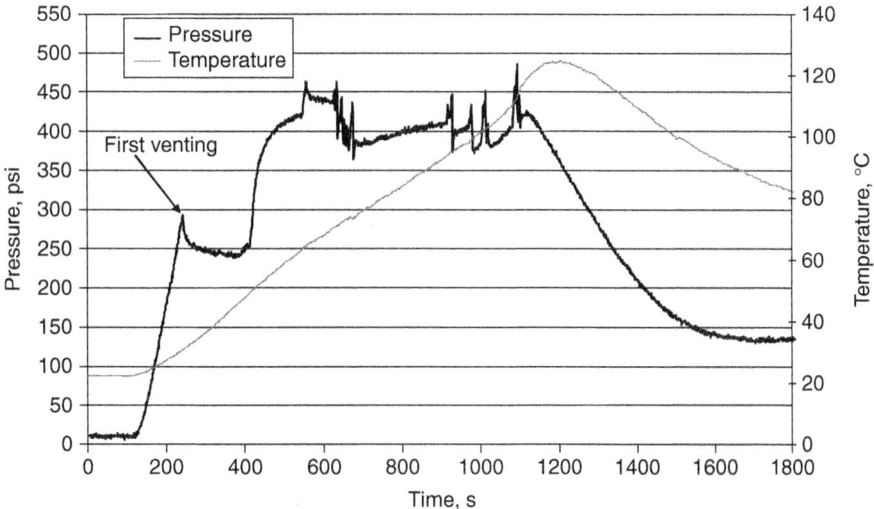

FIGURE 23.15 The 10 A overcharge test on sub C cells without internal fuse. (*Courtesy of PowerGenix, Inc.*)

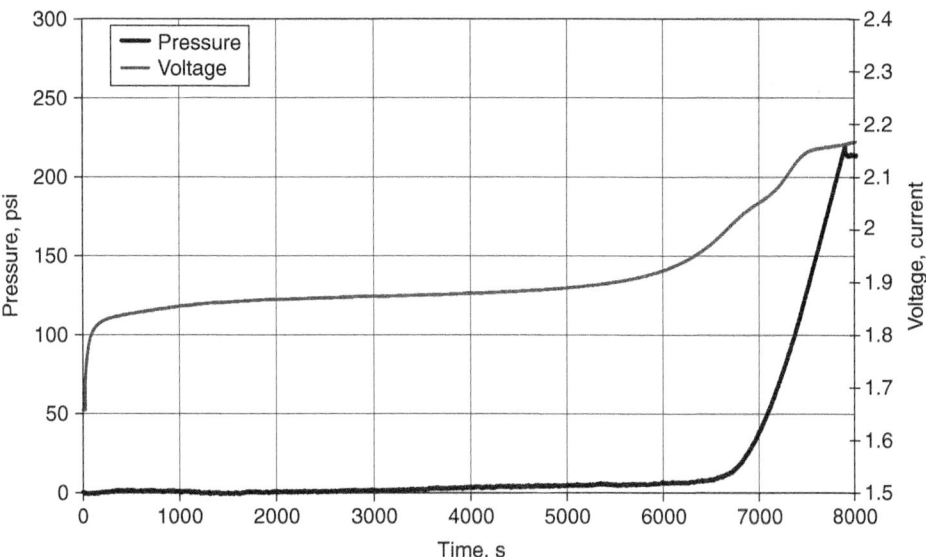

FIGURE 23.16 The gas pressure and voltage variation during charge a sub C of cell at 1 A. (*Courtesy of PowerGenix, Inc.*)

voltage of 1.875 V. The shape of the voltage–time charge curve resembles that of nickel-cadmium batteries and reflects the acceptance of charge by the positive nickel-hydroxide electrode. Significant oxygen gas pressure buildup occurs when the voltage of the cell increases above 1.95 V and the rate of generation of oxygen outstrips the rate of recombination. It is the relatively low oxygen recombination rate that prevents the use of constant current charging techniques at higher charge currents. An alternative is the constant current–constant voltage technique that resembles the taper current charge used in lead-acid and lithium-ion batteries. The cell is charged at a constant current until a predetermined voltage is reached and held as the current tapers to a given value or a fixed time elapses. The constant-voltage value must be manufacturer determined since it is affected by many cell variables including the composition of the positive electrode, electrolyte concentration, cell impedance, gas recombination rate, as well as the charge current and cell temperature. For the example shown in Fig. 23.16, the optimal voltage is 1.9 V per cell at 22°C with a temperature variation of –3.5 mV per degree C.

23.4.4 Failure Mechanisms of Nickel-Zinc Cells and Batteries

The failure modes of nickel-zinc batteries and cells are intimately linked to the application, the rate of charge and discharge, and the mechanical and thermal characteristics of the battery pack. Many of the historical early failure modes associated with nickel-zinc cells have been eliminated, to be replaced in later life by the more generic modes experienced by other systems. The predominant failure mode of multicell batteries is the dry-out associated with continuous overcharge and overdischarge of specific cells that have deviated from the average capacity. This can occur naturally over hundreds of cycles despite efforts to match their initial capacities, or it may be gradually precipitated by variations in the thermal environment as a result of suboptimal pack design. The service life benefits of good thermal management in the design of high-rate nickel-zinc battery packs can be substantial, particularly in higher voltage packs. Dry-out can also eventually be a problem for prismatic single cells with relatively low pressure vents where gas recombination may not be high enough to prevent gradual water loss over hundreds of cycles. This does not appear to be an issue for the sealed cylindrical cells even when subjected to high-rate discharges. The major failure mode at high charge and discharge rates is the gradual pore blockage of the separator as zinc deposits build up, a result of the rapid changes in the hydroxide concentration. However, this is rarely seen below 600 cycles. At low discharge rates ($< C/2$), cells may fail more often by the formation of soft shorts after several hundred cycles. These failure mechanisms are not unique to nickel-zinc batteries, and may be exacerbated by manufacturing defects that can promote shorts due to misalignment of components or damage to key elements such as the microporous separator. The elimination of cell manufacturing defects follows industry standards with focus on active process monitoring and the implementation of preshipment cycling and storage conditioning with close inspection of delivered capacities, open-circuit voltages, and impedance measurements.

23.5 APPLICATIONS

Nickel-zinc batteries are the low-cost, small, lightweight alternative to nickel-metal hydride, nickel-cadmium, and lithium-ion technologies. Unlike lithium-ion, they are abuse-tolerant, do not require extensive battery management electronics, and have the ability to deliver high currents that exceeds all three competitive technologies. They are an ideal candidate to replace nickel-cadmium in all applications because of their performance and environmental advantages.

23.5.1 Power Tools

Excellent abuse tolerance and high-power delivery are prerequisites for this application. The most common cells used in the power tool application are the 4/5 sub *C* and the sub *C* sizes. The direct replacement of nickel-cadmium or nickel-metal hydride by nickel-zinc technology results in an almost 30% reduction in the number of cells required and a corresponding reduction in the size and weight of the battery. The result is that an unwieldy 18 V nickel-cadmium pack can be replaced with an equivalent nickel-zinc pack in a lighter, smaller, 14.4 V form factor. A significant benefit of the reduction in cell count is the lowering of the pack impedance, which improves the power delivery

TABLE 23.3 A Comparison of Battery Performance for Professional Drills Using a 1 Inch Auger Bit in 2 Inch by 4 Inch Douglas Fir

Professional drill model	Technology	Cell count	Capacity (Ah)	Energy content (Wh)	No. of 1 in auger bit holes	Holes per watthour
Brand A 19.2 V	Ni-Zn	12	1.7	33	25	0.8
Brand B 18 V	Ni-Cd (sintered)	15	2.4	43	18	0.4
Brand D 36 V	Li-ion	10	3.0	102	49	0.5

and torque of the motor. Table 23.3 compares the number of 1 in auger bit holes that can be drilled using professional-grade cordless drills powered by three different battery technologies. Despite having less watthour energy content than the sintered nickel-cadmium powered drill, the nickel-zinc unit was capable of drilling 40% more holes. The nickel-zinc powered drill was also able to drill half as many holes with one third of the energy content as a 36 V lithium-ion powered drill. Significantly, the lithium-ion pack was approximately twice the size of the nickel-zinc unit. The advantage of the nickel-zinc pack over competitive packs is maintained for all high-power operations such as drilling holes with larger diameter spade bits and cutting wood with circular and reciprocating saws. In this respect, the nickel-zinc technology demonstrates a clear advantage in performing the more demanding tasks that might normally favor the use of a corded AC tool. Lower-power applications such as dry wall screw insertion are more likely to deliver performance levels that accurately reflect the energy content of the battery.

23.5.2 Lawn and Garden Tools

Power tools in this market segment tend to be larger and higher voltage than electric drills and circular saws. Typical tools include hedge trimmers, string trimmers, and electric lawn mowers that tend to be more seasonal in their use. Many of these tools use a lead-acid battery to maintain competitive pricing with AC-driven or gasoline versions, but this can lead to early problems with sulfation if the tool is not used for prolonged periods or the battery is not fully charged prior to storage. Overall, nickel-zinc technology provides a significant gain in the total cost of ownership combined with a 50% reduction in the size and weight of the power source. A high-rate nickel-zinc cell can offer as much as double the service life of a standard lead battery with less catastrophic long-term storage failures.

23.5.3 Light Electric Vehicles (LEVs)

Nickel-zinc batteries can offer electric scooters or electric bicycles a 50% smaller, lighter alternative to the lead-acid packs available now without sacrificing any range or hill-climbing capability. The relatively low cost per watthour of energy compared to other alkaline systems can provide the option of increasing battery capacity to more than double the range of the lead-acid pack with a similar size and weight. The battery-pack voltage of these vehicles is normally 24 or 48 V with a capacity requirement exceeding 15 Ah. This requirement may be met with a series-string 12 V nickel-zinc prismatic monoblocks or single cells. Alternatively, larger form factor cylindrical cells such as F or double D-size can provide battery capacities of 15 Ah without the need for parallel strings. These larger and more expensive packs can justify the inclusion of BMS modules that extend pack cycle life by limiting the overdischarge and by monitoring smaller series strings during the charge operation to prevent individual cell overcharge.

23.5.4 Hybrid Electric Vehicles (HEVs)

Hybrid electric vehicle battery packs operate in a power-assist mode where the most important characteristic of the battery is the ability to absorb high-rate regenerative braking charge and provide the high-rate discharge necessary for acceleration. The superior power density of the nickel-zinc alkaline system allows for a smaller and lighter battery pack compared to the nickel-metal hydride system without any sacrifice in performance. Since nickel-zinc is approximately 25% less costly than nickel-metal hydride, there is considerable opportunity for the development of lower-cost hybrid electric vehicle.

Figure 23.17 shows a size comparison of the nickel-zinc pack (top) against the standard nickel-metal hydride pack for a Toyota Prius. Table 23.4 outlines in more detail the differences between the two packs. While the pack energies were equivalent, the nickel-zinc pack was 30% smaller and lighter and delivered 30% more peak power.

FIGURE 23.17 A nickel-zinc D cell Prius battery pack compared to the existing prismatic nickel-metal hydride pack. (*Courtesy of PowerGenix, Inc.*)

TABLE 23.4 Comparison of Nickel-Metal Hydride and Nickel-Zinc Prius Packs

Parameter	PEVE NiMH	PGX Ni-Zn
Form factor	Prismatic	Cylindrical
Number of cells	168	128
Nominal voltage	201.6 V	204.8 V
Nominal capacity	6.5 Ah	6.5 Ah
Pack energy	1338 Wh	1357 Wh
Pack peak power	20 kW	26 kW
Gravimetric energy density	46 Wh/kg	69.3 Wh/kg
Bare pack weight	29.1 kg	19.2 kg

Evaluation criteria for hybrid electric vehicle batteries are given in the Partnership for a New Generation of Vehicles Battery Test Manual (see Chap. 29). Nickel-zinc batteries appear capable of meeting all key criteria of the power assist specification, including the 300,000 cycles required to be carried out over a specified state-of-charge range.

23.5.5 AA Consumer Cells

More than 90% of AA consumer cells are primary cells that ultimately end up in landfills. Both nickel-metal hydride and nickel-zinc cells provide a rechargeable option with overall cost benefits to the consumer. Unfortunately, the substitution of nickel-metal hydride for the primary 1.5 V zinc–manganese dioxide cell results in reduced performance and premature device shutdown as a result of having a 1.3 V operating voltage. Nickel-zinc technology provides a number of advantages that promote its adoption as a rechargeable alternative to the alkaline primary battery. Its operating voltage is compatible with the existing primary cells coupled with a low enough impedance to meet the demands of high-current draw devices such as digital cameras with built-in photoflash. The performance of the commercially available cells (Fig. 23.18) compares extremely favorably to other primary and rechargeable AA cells as shown in Fig. 23.19. The 1500 mAh nickel-zinc cell is capable of taking more than 5 times the number of pictures as the primary alkaline cell and has equivalent performance to the nickel-metal hydride cell with a much larger 2500 mAh capacity. In addition, the nickel-zinc cell is environmentally friendly and fully compliant with the European Union regulation for the Reduction of Hazardous Substances (RoHS).

FIGURE 23.18 Commercially available AA-size nickel-zinc cells. (*Courtesy of PowerGenix Inc.*)

23.24 SECONDARY BATTERIES

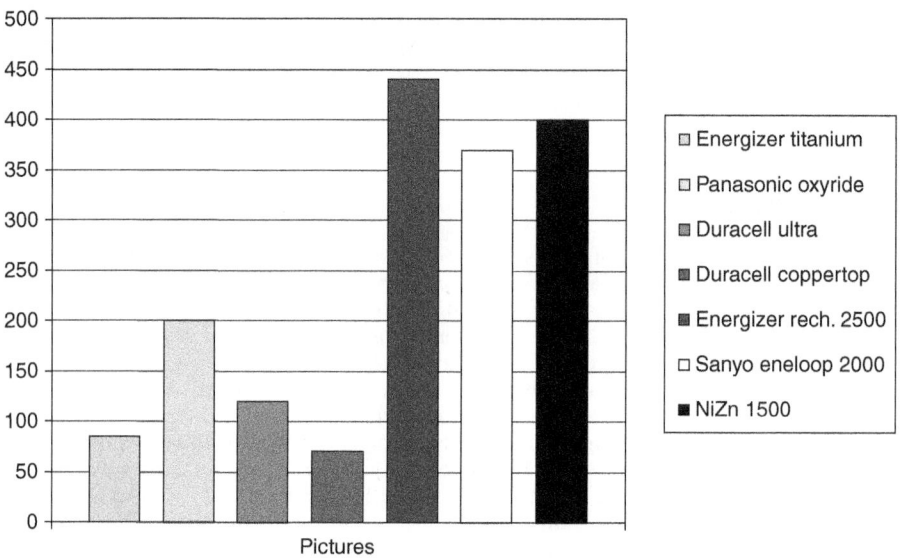

FIGURE 23.19 The number of pictures per cycle of various AA cells. (*Courtesy of PowerGenix, Inc.*)

23.6 ENVIRONMENTAL ASPECTS OF THE NICKEL-ZINC BATTERY

The nickel-zinc battery offers compelling environmental reasons for its adoption in a progressively "green" world. Nickel-zinc batteries are recognized and listed in the Rechargeable Battery Recycling Corporation (RBRC) recycle program, and may be recycled at more than 50,000 collection sites in North America.

The European Union RoHS regulation of 2006 required manufacturers to meet concentration limits on specific heavy metals and organic compounds. An equivalent directive was adopted in California, and similar requirements are likely to be imposed in China and Korea. Although batteries are exempt from the general RoHS provisions, a 2003 battery directive exists in Europe that tightens the limits for lead, mercury, and cadmium. Furthermore, unlike the general RoHS provisions, the concentration limits apply to the entire cell rather than specific components.

The 2003 battery directive requires the following maximum levels by weight:

- Cadmium –0.002%
- Mercury –0.0005%
- Lead –0.004%

Most nickel-zinc batteries meet these requirements. In general, nickel-zinc batteries are much less hazardous than both nickel-cadmium and nickel-metal hydride batteries and represent a greater opportunity for more complete recycling of the component chemicals.

REFERENCES

1. T. de Michalowski, Brit. Patent 15,370 (1899).
2. W. Junger, Swed. Patent 15,567 (1901).
3. T. A. Edison, Brit. Patent 20,072 (1901).

4. J. J. Drumm U.S. Patent 1,955,155 (1934).
5. J. J. Drumm, T. Hagyard, and R. H. D. Burklie, Brit. Patent 407,074 (1934).
6. K. Fujii, H. Yufu, and C. Kawamura, *Yuasa Jiho*, **57:**32 (1984).
7. N. A. Zhulidov and F.I. Yefremov, "The New Nickel-Zinc Storage Battery," Air Force Systems Command, WPAFB, Translation FTD-TT, 64-605.
8. D. Coates and A. Charkey, "Nickel-Zinc Batteries for Commercial Applications," Intersociety Energy Conversion Engineering Conference (IECEC), Vancouver, BC, Canada, August 1999.
9. J. Phillips, S. Mohanta, M. M. Geng, J. Barton, B. McKinney, and J. Wu, "Environmentally Friendly Nickel-Zinc Battery for High Rate Applications with Higher Specific Energy," *ECS Trans*. 16 (16), 11 (2009).
10. X. G. Zhang, *Corrosion and Electrochemistry of Zinc*, pp. 35–36, Plenum Press, New York, 1996.
11. A. K. Shukla, S. Venugopalan, and B. Hariprakash, *J. Power Sources*, **100:**125–148 (2001).
12. R. Jain, T. C. Adler, F. R. McLarnon, and E. J. Cairns, *J. Applied Electrochem.*, **22:**1039–1048 (1992).
13. E. G. Gagnon, *J. Electrochem. Soc.*, **133:**1989 (1986).
14. M. Eisenberg, U.S. Patent 5,215,836 (1993).
15. T. C. Adler, F. R. McLarnon, and E. J. Cairns, U.S. Patent 5,453,336 (1995).
16. Y. Seyama, K. Shichimoto, H. Sasaki, and T. Murata, Jpn. Patent JP 411,297,352A.
17. M. M. Geng, S. Mohanta, J. Phillips, Z. Muntasser, and J. Barton, U.S. Patent application number 2009-0202904. Filed February 4, 2009.
18. B. Bugnet, D. Doniat, and R. Rouget, U.S. Patent 7,300,721 B2 (2007).
19. F. Feng, J. Phillips, S. Mohanta, J. Barton, and Z. Muntasser, Publication No. 2009/0090636 (Published April 9, 2009).
20. R. A. Jones, U.S. Patent 4,358,517 (1979).
21. J. Phillips, U.S. Patent 6,818,350 B2 (2004).
22. Y. Wang and G. Wainwright, *J. Electrochem. Soc.*, **133:**1869 (1986).
23. S. B. Hall and J. Liu, International Patent WO 02/075830 A1 (2002).
24. T. C. Adler, F. R. McLarnon, and E. J. Cairns, U.S. Patent 5,302,475 (1994).
25. A. Charkey, U.S. Patent 5,863,676 (1999).

CHAPTER 24
NICKEL-HYDROGEN BATTERIES

Jack N. Brill

24.1 GENERAL CHARACTERISTICS

A sealed nickel-hydrogen (Ni-H$_2$) secondary battery is a hybrid, combining battery and fuel-cell technologies.[1] The nickel oxide positive electrode comes from the nickel-cadmium cell, and the hydrogen negative electrode comes from the hydrogen-oxygen fuel cell. Major advantages and disadvantages are listed in Table 24.1.

Salient features of this hybrid Ni-H$_2$ battery are a long cycle life that exceeds any other maintenance-free secondary battery system; high specific energy (gravimetric energy density) compared to other aqueous batteries; high power density (pulse or peak power capability); and a tolerance to overcharge and reversal. It is these features that make the Ni-H$_2$ battery system the energy storage subsystem currently employed in many aerospace applications, such as geosynchronous earth-orbit (GEO) commercial communications satellites, and low earth-orbit (LEO) satellites, such as the Hubble space telescope.

Application of the Ni-H$_2$ battery has mainly been directed toward the aerospace field. Recently, however, programs have been started for terrestrial applications, such as long-life stand-alone photovoltaic systems.

24.2 CHEMISTRY

The electrochemical reactions of the Ni-H$_2$ cell for normal operation, overcharge, and reversal are

Normal operation:

Nickel electrode $\quad NiOOH + H_2O + e \underset{charge}{\overset{discharge}{\rightleftarrows}} Ni(OH)_2 + OH^-$

Hydrogen electrode $\quad \tfrac{1}{2}H_2 + OH^- \underset{charge}{\overset{discharge}{\rightleftarrows}} H_2O + e$

Net reaction $\quad \tfrac{1}{2}H_2 + NiOOH \underset{charge}{\overset{discharge}{\rightleftarrows}} Ni(OH)_2$

Overcharge:

Nickel electrode $\quad 2OH^- \rightarrow 2e + \tfrac{1}{2}O_2 + H_2O$

Hydrogen electrode $\quad \tfrac{1}{2}O_2 + H_2O + 2e \rightarrow 2OH^-$

TABLE 24.1 Major Advantages and Disadvantages of the Nickel-Hydrogen Battery

Advantages	Disadvantages
High specific energy (60 Wh/kg)	High initial cost
Long cycle life, 40,000 cycles at 40% DOD for LEO applications	Self-discharge proportional to hydrogen pressure
Long lifetime in orbit, over 15 years for GEO applications	Low volumetric energy density: 50–90 Wh/L (IPV cell) 20–40 Wh/L (battery)
Cell can tolerate moderate overcharge and reversal	
Hydrogen pressure gives an indication of state-of-charge	

Reversal:

  Hydrogen (negative) precharge

  Nickel electrode $\quad H_2O + e \rightarrow OH + \frac{1}{2}H_2$

  Hydrogen electrode $\quad \frac{1}{2}H_2 + OH^- \rightarrow H_2O + e$

 Positive precharge

  Nickel electrode $\quad 2NiOOH + 2H_2O + 2e^- \rightarrow 2Ni(OH)_2 + 2(OH)^-$

  Hydrogen electrode $\quad 2(OH)^- \rightarrow 2e^- + \frac{1}{2}O_2 + H_2O$

  Net reaction $\quad 2NiOOH + 2H_2O \rightarrow 2Ni(OH_2) + \frac{1}{2}O_2$

24.2.1 Normal Operation

Electrochemically, the half-cell reactions at the positive nickel oxide electrode are similar to those occurring in the nickel-cadmium system. At the negative electrode, hydrogen gas is oxidized to water during discharge and is reformed, during charge, from the water by electrolysis. The net reaction shows hydrogen reduction of nickel oxyhydroxide to nickelous hydroxide on discharge with no net change in KOH concentration or in the amount of water within the cell.

24.2.2 Overcharge

During overcharge, oxygen is generated at the positive electrode. An equivalent amount of oxygen is recombined electrochemically at the catalytic platinum electrode. Again, there is no change in KOH concentration or the amount of water in the cell with continuous overcharge. The oxygen recombination rate at the negative platinum electrode is very rapid, sustaining moderate rates of continuous overcharge, provided there is adequate heat transfer away from the cell to avoid thermal runaway. This is one of the operational advantages of the Ni-H_2 cell.

24.2.3 Reversal

Two types of precharge are used with the nickel-hydrogen system. With hydrogen (negative) precharge during cell reversal, hydrogen is generated at the positive electrode and consumed at the negative electrode at the same rate. Therefore the cell can be operated continuously in the cell reversal mode without pressure buildup or net change in electrolyte concentration. This is a unique feature of the system. If nickel (positive) precharge is used, oxygen gas is generated during reversal until the nickel positive material is consumed. The oxygen is consumed again during charge. Once the positive material is consumed, hydrogen is generated and consumed at the same rates. It is possible that the hydrogen oxygen mixture could be in the combustible range for this mode of operation and that a rapid recombination could occur causing damage to the cell stack.

24.2.4 Self-Discharge

The electrode stack is surrounded by hydrogen under pressure. A salient feature is that the hydrogen reacts electrochemically but not chemically to reduce the nickel oxyhydroxide. Actually, the nickel oxyhydroxide is reduced chemically, but at such an extremely low rate that performance for aerospace applications is not affected.

24.3 CELL AND ELECTRODE-STACK COMPONENTS

Ni-H_2 cell stacks are assembled in three distinct configurations. These include the COMSAT back-to-back, the Air Force recirculating, and the hybrid, Mantech back-to-back designs. This section describes the electrode-stack components used for the fabrication of aerospace Ni-H_2 cells in these configurations. Figure 24.1 shows the truncated disk electrode-stack components used in the COMSAT design.

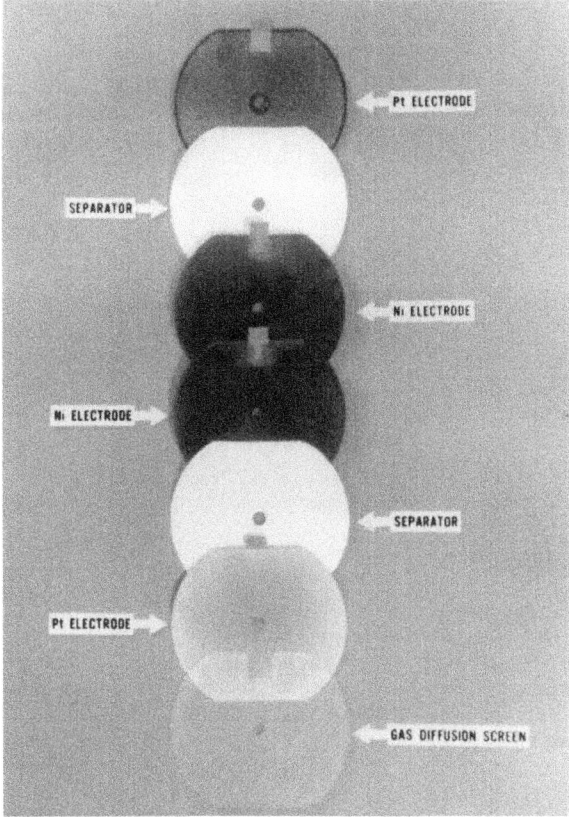

FIGURE 24.1 COMSAT bus-bar-configuration electrode-stack components.

24.4 SECONDARY BATTERIES

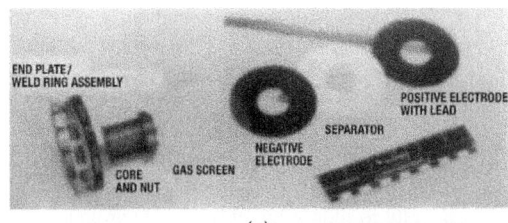

(a)

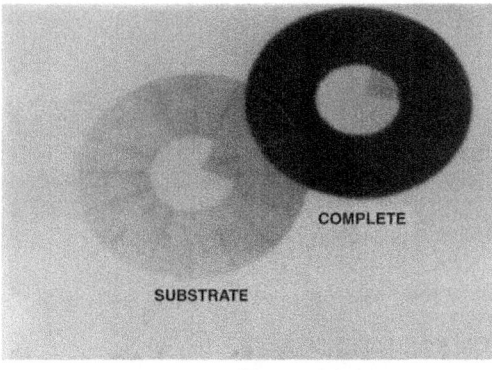

(b)

(c)

FIGURE 24.2 Air Force pineapple-slice configuration. (*a*) Stack components. (*b*) Negative electrode. (*c*) Pressure-vessel cylinder and dome.

Figures 24.2*a* and *b* show the circular components for the Air Force recirculating and hybrid, Mantech back-to-back designs.

24.3.1 Positive Electrodes (Sintered)

The sintered positive electrode consists of a sintered porous nickel plaque that is impregnated with nickel hydroxide active material. The porous sintered plaque serves to retain the active nickel hydroxide material within its pores and conduct the electric current to and from the active material. Essential features of the sintered plaque are high porosity, large surface area, and electrical conductivity in combination with good mechanical strength.[2]

Active material is impregnated into the sintered plaque by an electrochemical impregnation process. There are two electrochemical impregnation processes used—aqueous and alcoholic. The aqueous impregnation process (Bell Laboratories process)[3] uses an aqueous-based nickel nitrate solution for the impregnation bath. The alcoholic impregnation process (Air Force process)[4] uses an alcohol-based nickel nitrate solution for the impregnation bath. Both processes provide the following advantages:

1. *Loading of active material:* Electrochemical impregnation gives very uniform loading of the active material within the pores of the nickel sinter.
2. *Loading level:* The loading level of active material can be accurately controlled by the electrochemical impregnation process. Typical loading values are

   1.67 ± 0.1 g/cm^3 void volume for GEO applications

   1.65 ± 0.1 g/cm^3 void volume for LEO applications

24.3.2 Hydrogen Electrode

The hydrogen electrodes consist of a Teflon-bonded platinum black catalyst supported on a photo-etched nickel substrate with Teflon bonding. The sintered Teflon-bonded platinum electrodes were originally developed at Tyco Laboratories for the fuel-cell industry.[5] For Ni-H$_2$ cells, a hydrophobic Teflon backing was added to these platinum electrodes to stop water or electrolyte loss from the back side of the negative platinum electrode during charge and overcharge while readily allowing diffusion of hydrogen and oxygen gas. The use of Gortex® as the microporous Teflon membrane resulted from a development contract with HAC (Hughes Aircraft Corporation).[6] The platinum content is normally specified as 7.0 ± 1.0 mg/cm^2. The physical properties of this hydrogen electrode provide the right interface for the electrochemical reactions to occur without flooding or drying out the electrode at the separator interface.

24.3.3 Separator Materials

Two types of separator materials have seen use in aerospace Ni-H_2 cells: (1) asbestos (fuel-cell-grade asbestos paper) and (2) Zircar (untreated knit ZYK-15 Zircar cloth). The predominant use in cells manufactured today is Zircar.

Fuel-cell-grade asbestos is a nonwoven fabric with a thickness of 10 to 15 mil. The asbestos fibers are made into a long roll of nonwoven cloth by a paper-making process. As an added precaution, the asbestos can be reconstituted in a blender and then reformed into a cloth to avoid any nonuniformity in the original structure that would allow oxygen to bubble through. The fuel-cell-grade asbestos has a high bubble pressure for oxygen gas; a pressure difference of more than 1.7×10^5 Pa is required across the separator cloth (250 µm thick) to force oxygen bubbles through the material. During overcharge, oxygen gas is forced off the backside of the positive electrode. The oxygen cannot channel or bubble through the separator to cause rapid recombination at the negative electrode. Zircar fibrous ceramic separators are available in textile product forms (Zircar Products, Inc.). These textiles are composed of zirconia fibers stabilized with yttria. These materials offer the extreme temperature and chemical resistance of the ceramic zirconia. They are constructed of essentially continuous individual filaments fabricated in flexible textile forms. Even with the fibrous structure, the inherently brittle nature of the ceramic material zirconia makes these separators fragile and susceptible to breaking. They must be handled with care. Untreated knit ZYK-15 Zircar cloth material is the tensile form used for Ni-H_2 cells.[7] Either one of two 250 to 380 µm-thick layers of this separator material can be used. The second ZYK-15 layer is normally used as a backup to prevent oxygen channeling in the event of assembly damage to the first layer. The knit Zircar cloth has a very low oxygen bubble-through pressure, and during charge and overcharge, oxygen gas readily permeates through the separator to recombine at the hydrogen platinum electrode to form water.

Both the asbestos and the Zircar separators serve the following functions:

1. They act as separators between positive and negative electrodes.
2. They serve as reservoirs for KOH electrolyte and remain stable in the electrolyte, allowing long-term storage and cycling.
3. They serve as media for charge and discharge current through the separator via ionic conduction of hydroxyl ions in the electrolyte.

24.3.4 Gas Screen

A polypropylene gas diffusion screen is placed behind the hydrogen electrode to allow hydrogen gas and oxygen gas to diffuse to the back side of the negative electrode with the Teflon backing.

24.4 Ni-H_2 CELL CONSTRUCTION

Sealed Ni-H_2 cells contain hydrogen gas under pressure within a cylindrical pressure vessel (see Fig. 24.2c). They are referred to as individual pressure-vessel (IPV) cells because each individual cell is contained within its own pressure vessel. IPV cells are assembled using either single or dual electrode stack configurations inside the pressure vessel. An extension of IPV is the two cell (2.5 V) CPV design made by connecting the dual electrode stacks in series within a single pressure vessel. IPV designs include cells having diameters of 6.35, 8.89, 11.43, and 13.97 cm.

Descriptions of the various cell designs follow. These designs represent the first generation of Ni-H_2 cell technology, which was developed in the 1970s and utilized in the 1980s along with the baseline designs currently in use.

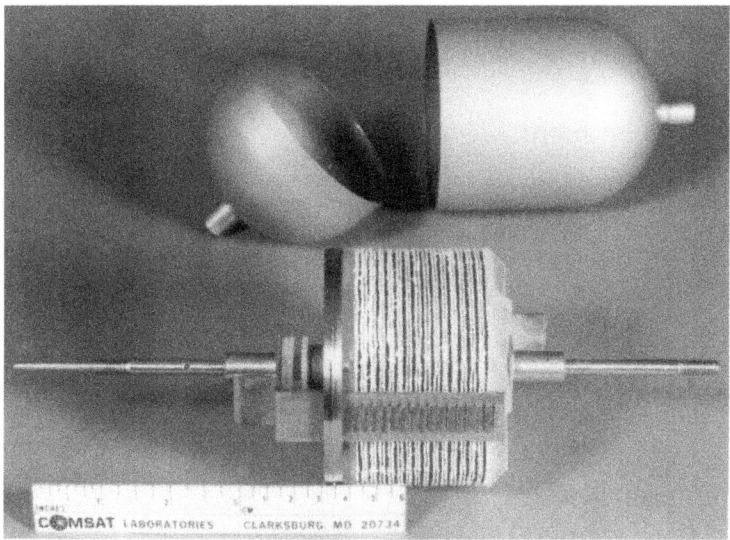

FIGURE 24.3 COMSAT NTS-2 Ni-H$_2$ cell components.

24.4.1 COMSAT Ni-H$_2$ Cell

Components of the COMSAT NTS-2 Ni-H$_2$ cell are shown in Fig. 24.3 with the electrode-stack assembly and weld ring positioned in front of the pressure shells.[8,9] These cells were built by Eagle-Picher Technologies (EPT) under an INTELSAT/COMSAT licensing agreement. The U.S. Navy's Navigation Technology Satellite-2 (NTS-2), launched on June 23, 1977, was the first flight demonstration of the Ni-H$_2$ battery.

Electrode Stack. Figure 24.1 shows the basic arrangement of the electrode-stack components for the COMSAT back-to-back design. Two positive nickel oxide electrodes are positioned back-to-back. A separator is placed on each side of the positive electrodes. The negative platinum electrodes are placed with the platinum black surface next to the separator material. A plastic diffusion screen is placed on the back side of each negative electrode to facilitate gas diffusion to the back side of this electrode. These components constitute one module of the electrode stack. This arrangement is repeated until the number of modules is reached to provide the required capacity. A complete stack can be seen in Fig. 24.3. The bus bars for the positive and negative electrodes are located along the outside of the electrode stack.

During charge and overcharge, oxygen gas that evolves at the nickel electrodes is forced out between the back-to-back positive electrodes. The oxygen diffuses into the gas space between the electrode stack and the pressure vessel wall into the region of the gas diffusion screens on the back of the negative electrode, and through the porous backing of the negative electrode, where it combines with hydrogen to form water. The partial pressure of oxygen is dependent on this diffusion process. The limiting step is the oxygen diffusion in the gas-phase pores of the Teflon-bonded electrode, not in the Teflon backing.[6] The fraction of oxygen gas should be less than 0.5% in the surrounding hydrogen gas when the cell is continuously overcharged at a $C/2$ rate.

Pressure Vessel. The pressure vessel (dome and cylinder), terminal bosses, and weld rings are all fabricated from Inconel alloy 718. The weld ring is manufactured by one of two methods. The first is manufactured using an investment casting process and is then machined to final dimensions. The second is manufactured by machining to final dimensions from an extruded or wrought Inconel metal.

The outside diameter of the weld ring is machined as a T section to position the pressure shells on the weld ring and provide a backup support for an electron beam girth weld. The Inconel 718 pressure vessel shells are manufactured to a near uniform thickness using either a hydroforming or drawn process and then cut to length. The thickness is determined by the operating pressure and cycle requirements for the particular use. The pressure shells are "age hardened" using a standard heat treatment process. The terminal bosses for the compression seals are machined from Inconel 718 material and electron beam welded into the domes of the pressure vessel shells. Nylon plastic is injection-molded into these barrels. The Ziegler compression seal[10] is made by crimping the bosses. Cell designs commonly operate under maximum operating pressures between 4.1×10^6 Pa and 8.3×10^6 Pa. Depending on the particular use, the vessels are designed to provide a safety factor between 2:1 to over 4:1.

24.4.2 Air Force Ni-H$_2$ Cell

Typical components for the Air Force NiH$_2$ cell are shown in Fig. 24.2 including the electrode stack components, the negative electrodes with the chem-milled substrate, and the pressure vessel cylinder and dome with the plasma sprayed zirconium oxide wall wick for the electrolyte. These components are typically assembled in one of two configurations. Separators used in these designs are asbestos and Zircar alone or in combination with each other.

The first is commonly referred to as the recirculating electrode stack design. It is made of a number of modules comprised of a gas screen, hydrogen electrode (negative), separator(s), and nickel electrode. The capacity of the cell is determined by the number of modules used to assemble the stack. A single gas screen and hydrogen electrode are used as the final module to maintain a hydrogen electrode for recombination on both sides of the positive throughout the stack. A wall wick is used on the interior of the pressure vessel to return electrolyte to the stack, hence the name recirculating design.

In this design, using asbestos or a combination of asbestos and Zircar separators, the oxygen generated on overcharge diffuses off the back of the nickel electrode. The diffusion path is very short; the oxygen gas simply travels through the gas screen to recombine at the next hydrogen electrode. Oxygen comes off the back side of the positive electrode of one module and recombines to form water at the next module. During overcharge, this transfer of water to the next module occurs throughout the electrode stack. The last module in the stack is simply a negative electrode and separator reservoir. The water formed at this electrode-separator combination goes to the wall wick and recirculated back to balance the electrolyte throughout the stack. With this recirculating design, the oxygen concentration is kept very low (below 0.2% in the surrounding hydrogen gas) during continuous overcharge at the C rate.

The second configuration is the Air Force Mantech back-to-back design. It is made in the same manner as the COMSAT design. A separator is placed on each side of the positive electrodes. The negative platinum electrodes are placed with the platinum black surface next to the separator material. A plastic diffusion screen is placed on the back side of each negative electrode to facilitate gas diffusion to the back side of this electrode. These components constitute one module of the electrode stack. This arrangement is repeated until the number of modules is reached to provide the required capacity. This design also uses the wall wick for return of electrolyte to the stack.

Designs using Zircar alone normally extend the separator to the pressure vessel wall. The Zircar contains large enough pores for oxygen gas to permeate through the pores of this separator to the negative electrode, where it recombines to form water. Oxygen can, of course, also emerge off the back side of the nickel electrode and diffuse through the gas screens as before. However, most of the oxygen permeates through the separator and there is little or no recirculation of water in cells with Zircar separators. For this design, the concentration of oxygen in the surrounding hydrogen gas is negligible.

The electrode-stack components are shaped like a pineapple slice (see Fig. 24.2b) with provisions for the tab in the center hole. These electrode-stack components are assembled onto a polysulfone central core (see Fig. 24.2a). The electrode tabs are brought out through this central core. The positive and negative tabs can be in opposite directions or in the same direction depending on the terminal configuration. This center core serves to align the electrode-stack components, provide a conduit for the positive and negative tabs, and insulate the positive and negative tabs from each other and from the electrode-stack components.

The cell capacity in a particular diameter is limited by the ability to manufacture a pressure vessel of sufficient length. In designing cells of larger capacity without increasing the diameter of the pressure vessel, two approaches are used. The first involves the use of a dual stack design to increase the capacities of cells for the individual cell diameters.[15] This is accomplished by using two stacks assembled on a single core as described above. The two stacks are separated by end plates and a weld ring and are connected electrically in parallel to attain a 1.25 V cell. The second approach utilizes a three-piece pressure vessel assembly.[16] A single electrode-stack is made with weld rings at each end. A cylinder is placed over the stack and joined at each end with the weld rings and two dome assemblies.

Heat transfer is better with the pineapple-slice configurations than with the COMSAT back-to-back configuration. Heat is transferred uniformly from the entire circumference of the pineapple-slice electrodes, whereas sections are removed from the circumference of the COMSAT back-to-back electrodes.

Pressure Vessel. The pressure vessels used for the Air Force designs are essentially the same as those for the COMSAT designs. Certain designs utilize chemical milling to remove material from lower stress areas for weight reduction. Typically the weld areas are not reduced to compensate for the strength reduction of the age-hardened, Inconel 718 material in the heat-affected zone of the weld areas. The chemical milling is done prior to heat treating (age hardening) of the pressure vessel. The operating pressures and design margins are similar to those discussed for the COMSAT designs.

Two terminal designs have been used. One involves the terminal design using the compression seal described for the COMSAT designs. The second utilizes a hydraulic seal design. When this is used, the seal area is hydroformed as an integral part of the dome and cylinder. The terminal seals are hydraulic cold-flow Teflon seals.[14]

Electrolyte Management. There are three mechanisms for the loss of electrolyte from the electrode stack: (1) by entrainment in the hydrogen and oxygen gases evolved during charge and overcharge, (2) by weeping of the negative electrode, and (3) by electrolyte displacement, that is, the electrolyte being pressed out of the positive electrodes in the cell stack by oxygen gas evolved during charge and overcharge.

Electrolyte loss by both entrainment and weeping of the negative electrode was determined to be negligible for negative electrodes with Gortex backing for both back-to-back and recirculating electrode-stack configurations. The major electrolyte loss mechanism is by displacement. When electrolyte is added to the cell, the void volume of the positive electrode is completely saturated with electrolyte. During activation, approximately 25% of the electrolyte in the positive electrodes is displaced by oxygen gas during charge and overcharge of the cell.[6] It was found that electrolyte loss by displacement occurred during initial cycling (activation) but eventually decreased to zero, leaving enough electrolyte to operate the cell efficiently.[11,12]

Water Loss. Water loss from the electrode stack can result from evaporation and condensation of water vapor from the cell stack to the pressure-vessel wall when a large enough temperature difference exists between stack and wall (approximately 10°C difference). The plasma-sprayed zirconium oxide wall wick[13] shown in Fig. 24.2c provides a return path for any water loss from the cell stack independent of the mechanism.

24.4.3 Specific Energy and Energy Density

Figures 24.4 and 24.5 depict the specific energy and energy density that could be projected for the different nickel-hydrogen cell designs. The actual values may vary depending on manufacturer.

1. In general, the specific energy increases as the capacity increases.
2. The choice and number of separators affect the weight (quantity of electrolyte) and thus the specific energy of the cell.
3. The energy density is primarily a function of the pressure range or free volume in the cell. Cells operating at higher maximum operating pressures have higher energy densities. This increase is a result of the reduction in weight of the pressure vessel at the higher pressures.

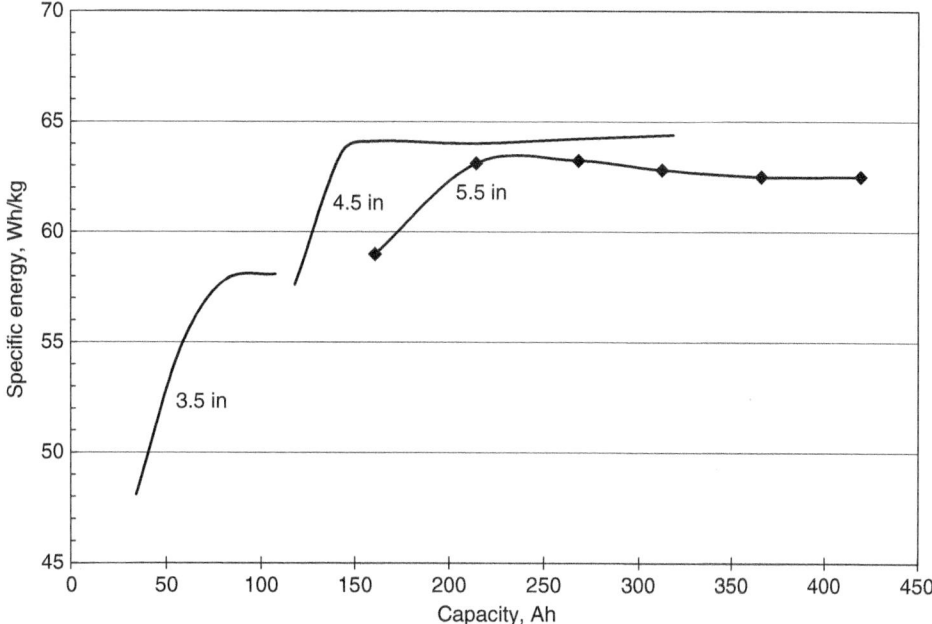

FIGURE 24.4 Specific energy of Ni-H$_2$ cells.

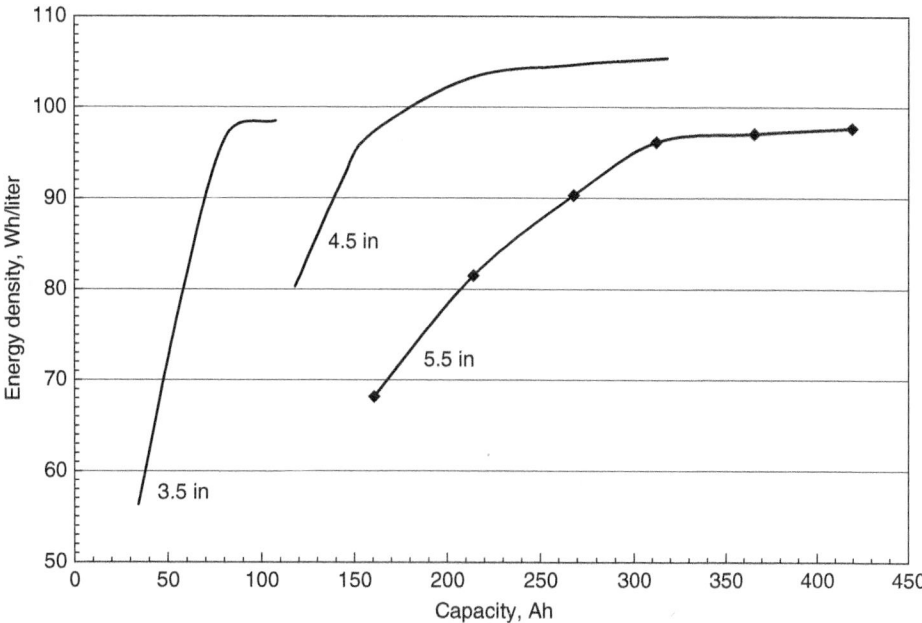

FIGURE 24.5 Energy density of Ni-H$_2$ cells.

24.5 Ni-H₂ BATTERY DESIGN

Various Ni-H$_2$ battery designs have evolved through the years. They are tailored to the specific application and interface with the particular satellite. Mechanical and thermal requirements are the primary drivers for the configuration and interface of each battery. The nickel-hydrogen system is sensitive to temperature and performs best between –10°C and +10°C. Thus, thermal control of the battery is important to minimize size and weight.

Several features have been integrated into the battery designs which enhance the performance and reliability. These include: pressure monitoring of cells through strain gauge or transducers, strain gauge voltage amplification circuits, individual cell voltage monitors, temperature monitoring, redundant individual cell heaters, and individual cell diode bypass protection. Bypass diodes on each cell protect the battery against a failure from an open-circuited cell. Protection in the charge direction is provided by three silicon diodes in series, while protection in the discharge direction is provided by one Schottky barrier diode. The diodes are mounted on heat sinks on the thermal sleeves near the base of the cells or on a separate panel attached to the battery base plate.

Several different battery configurations can be seen in Figs. 24.6, through 24.10. One of the earliest Ni-H$_2$ battery designs was flown on the INTELSAT V program. Two 27-cell, 30 Ah Ni-H$_2$ batteries were used to provide the electric energy during launch, transfer orbit, and solar eclipses.[17]

The first battery using the dual electrode stack configuration was made by Eagle-Picher Technologies, LLC, for the EUTELSAT II Program. The first of two flight sets was delivered in February 1990 and launched in August 1990.[18] This battery is shown in Fig. 24.7.

The photovoltaic power subsystem for the International Space Station uses Ni-H$_2$ batteries for energy storage to support eclipse and contingency operations. These batteries are designed for LEO operation with a 6.5-year design life expectancy and are configured as orbital replacement units (ORU), permitting replacement of worn-out batteries over the anticipated 30-year station life.[19] The baseline energy storage

FIGURE 24.6 DMPS 100 Ah Ni-H$_2$ battery assembly.

FIGURE 24.7 EUTELSAT II 58 Ah Ni-H$_2$ battery.

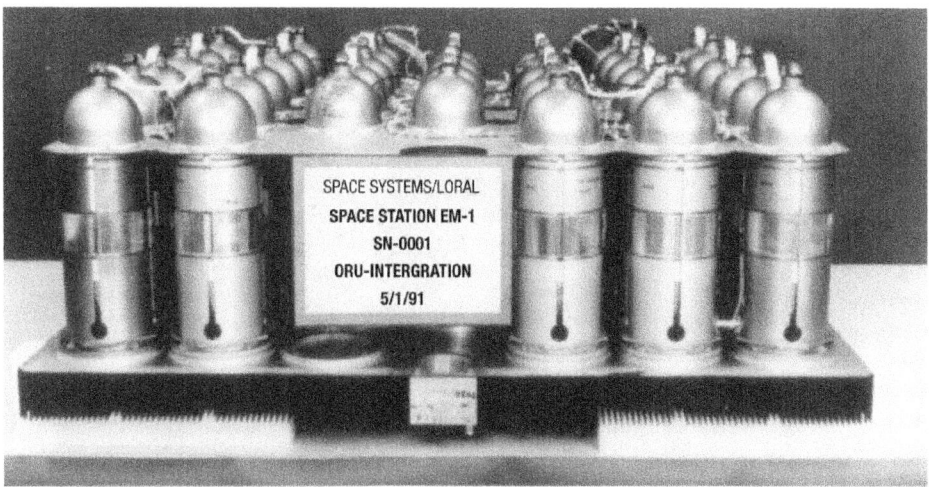

FIGURE 24.8 International Space Station 81 Ah Ni-H$_2$ 38-cell assembly.

system design contains 2 batteries of 76 cells of 81 Ah capacity, packaged as 38-cell assemblies (Fig. 24.8), or approximately 184.7 kWh of stored energy. The initial batteries were placed in service on the space station in November 2000. Six batteries were placed in service in July, 2009 during the visit of STS 127 to the ISS, and another six batteries were replaced in May 2010 during the servicing mission of STS 132.

FIGURE 24.9 TRW 81 Ah Ni-H$_2$ battery for a flight program. (*Photograph provided courtesy of TRW.*)

FIGURE 24.10 MIDEX 23 Ah CPV Ni-H$_2$ battery assembly.

the capacity removed on discharge at a high charge rate, and then switch to a low trickle charge rate for the remainder of the 24 h eclipse day to maintain the batteries at the full (100%) state-of-charge. For the 135 days between eclipse seasons, the batteries are maintained on trickle charge in the fully charged condition.

Reconditioning. Batteries are typically reconditioned prior to each eclipse season.

24.6.2 LEO Applications

Battery Requirements. A 96 min orbit is typically used to characterize LEO satellite applications. The time it takes a satellite to orbit the earth at 555 km is 96 min. The satellite orbits the earth 15 times in one day. The orbital duration remains fixed, but the sunlight and eclipse periods vary with each orbit. For example, a LEO satellite orbiting 555 km above the earth, at an inclination of 28.3°, has an orbital period that is constant at 96 min; but during a given month, such as December 1991, the eclipse durations vary from a maximum of 35.58 min on December 1 to a minimum of 26.97 min on December 30.

Charge Control. The battery is charged during the sunlight period and discharged during the eclipse period. With this high duty cycle, it is essential to minimize overcharge (heat dissipation) and maximize overall watthour efficiency for the battery. A charge method is needed to compensate for the variation in the depth of discharge (variation in eclipse duration) and to minimize overcharge. If the battery can be maintained at a low temperature on charge between 0 and 10°C, the ampere-hour charge efficiency approaches 100%, and the watthour efficiency approaches 85%.

NASA's Marshall Space Flight Center (MSFC) and Lockheed Missile and Space Company (LMSC) selected a temperature-compensated voltage-limit charging method for charging the 88 Ah Ni-H$_2$ batteries that replaced the Ni-Cd batteries on board the Hubble Space Telescope (HST) satellite.[22,23] Figure 24.13 shows the battery voltage limits used for the HST program as a function of temperature. At the beginning of life, the batteries were charged to the K1-L3 and K2-L3 voltage limits at the high rate, then switched to trickle charge, which is approximately a $C/100$ rate. The K1-L3 setting has a cell voltage limit of 1.513 V at 0°C, or a battery voltage limit of 33.28 V for the 22-cell battery. The battery is not fully charged at this voltage limit but rather charged to about 73 Ah (83% of

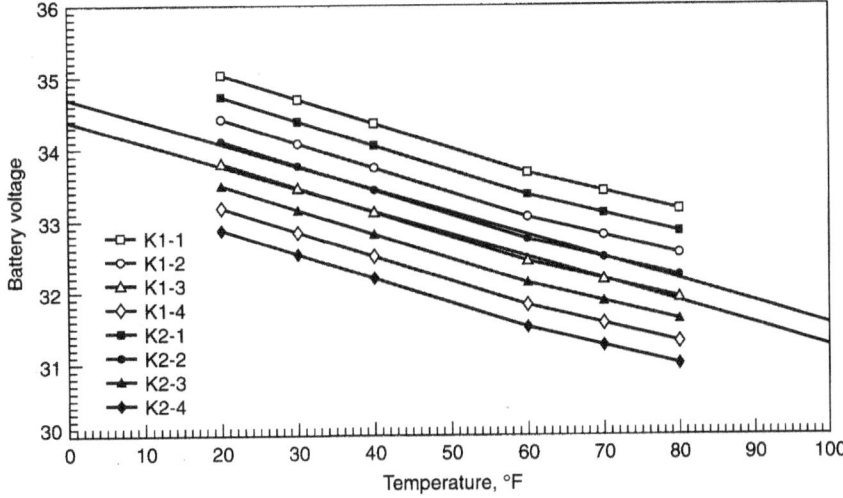

FIGURE 24.13 Charge control *V-T* limits for HST 22-cell Ni-H$_2$ battery, $y = 34.7412 - 0.0329x$, $R = 1.00$; $y = 34.316 - 0.0318x$, $R = 1.00$.

24.16 SECONDARY BATTERIES

its rated capacity). The battery is thermally stable with this charging method. The overall battery watthour efficiency is 80 to 85% during cycling with these level 3 control limits. Charging the battery to a higher voltage limit would decrease the coulombic efficiency, reduce the overall energy efficiency, increase the heat dissipated internally within the battery, and possibly exceed the constraints for thermal control of the battery. The cell pressure serves as an indication of the state-of-charge of the battery and is very useful in this type of application where the battery is not fully charged.

24.6.3 Terrestrial Applications

The advantages offered by Ni-H_2 batteries, including long calendar and cycle life, low maintenance, and high reliability, make them very attractive for terrestrial applications such as stand-alone photovoltaic systems or standby power for emergency or remote site use. The major drawback to the wider use of the Ni-H_2 battery is its high initial cost. The following are two examples of terrestrial applications addressing Ni-H_2 batteries.

Starting in 1983, Sandia National Laboratories sponsored a cost-sharing program with COMSAT Laboratories and Johnson Controls, Inc. for the design and development of a sealed Ni-H_2 battery for deep-discharge terrestrial applications that would be cost-competitive with lead-acid batteries in a system designed for a 20-year life. The main thrust of this program was to reduce the cost of the aerospace technology without comprising the desirable features of the Ni-H_2 system. Figure 24.14 shows a 5-cell, 6 V 100 Ah Ni-H_2 battery assembled in a pressure vessel. Assemblies have been

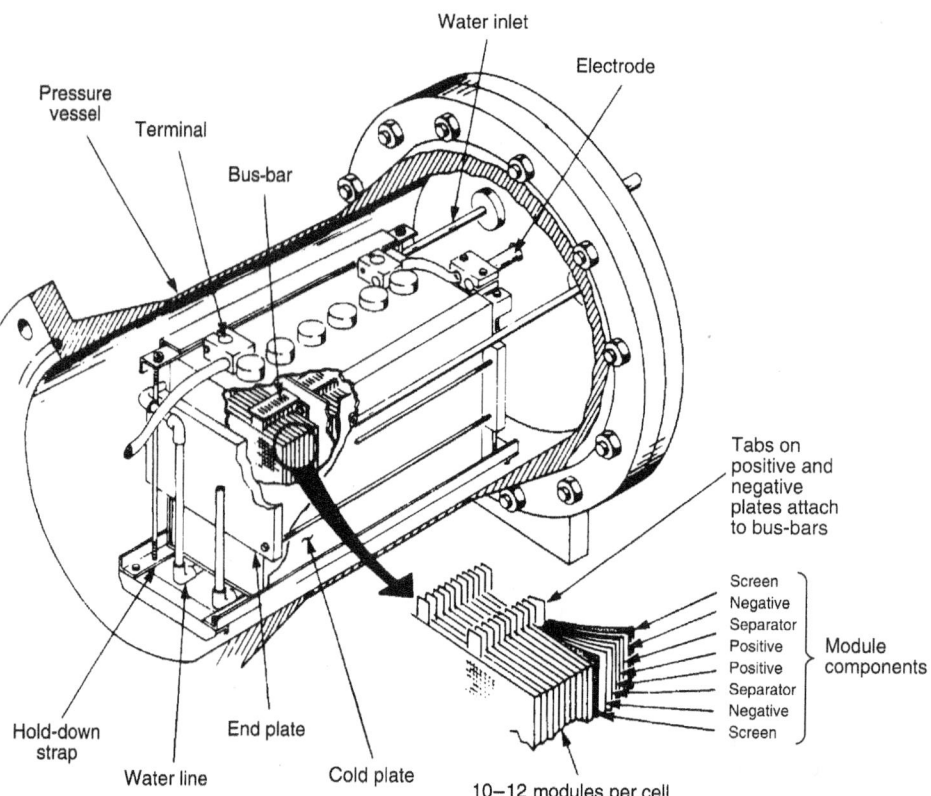

FIGURE 24.14 6 V 100 Ah terrestrial Ni-H_2 battery. (*Courtesy of COMSAT Laboratories.*)

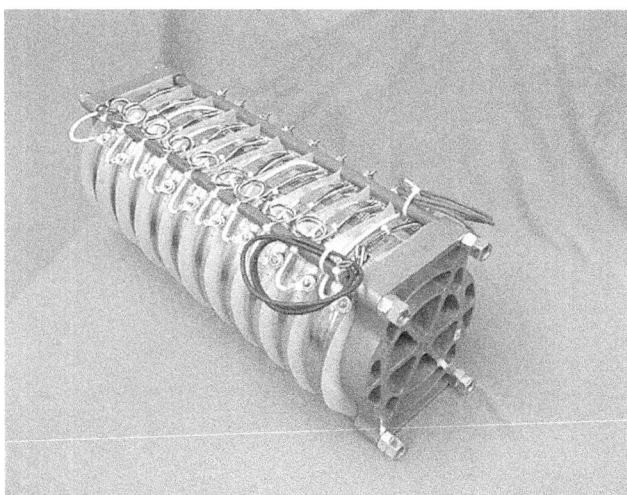

FIGURE 24.15 DPV battery assembly.

tested by Sandia National Laboratories in combination with photovoltaic arrays. The conclusion was expressed that the cells "continue to perform well, reinforcing the projection of a 20-year life, matching that of photovoltaic panels."[21]

Eagle-Picher Technologies, LLC, has developed another design for use as standby power for a remote site in a terrestrial application. Again the primary effort in the design was to create a more cost-effective, reliable battery for terrestrial use. This design utilizes a combination of the dependent pressure vessel (DPV) and 2-cell common pressure vessel (CPV) technologies. Five 2-cell DCPV units are assembled, creating a 12 V battery. The nominal capacity of the battery is 44 Ah at 10°C. This battery is shown in Fig. 24.15.

24.7 PERFORMANCE CHARACTERISTICS

24.7.1 Voltage Performance

Electrochemically impregnated nickel oxide electrodes are used in Ni-H$_2$ cells because of their excellent cyclic performance capabilities.[9,24] The capacity of these electrodes increases as the temperature decreases. The measured capacity of an electrochemically impregnated electrode is about 20% greater at 10°C than at 20°C. The capacity of the NTS-2 35 Ah cell at different temperatures is presented in Fig. 24.16 to show the variation in capacity with temperature. These NTS-2 cells were discharged at the $C/1.67$ rate. Note that the mid-discharge voltage is between 1.2 and 1.25 V.

The high-rate discharge of an INTELSAT V cell at 200 A (12 min rate) is shown in Fig. 24.17. The discharge profile is almost flat at 0.6 V; the potential drop of approximately 600 mV is due to the 3 mΩ terminal impedance of the cell (3 m$\Omega \times$ 200 A = 600 mV). As previously seen in Figure 24.16, the mid-discharge voltage is between 1.2 and 1.25 V for a cell discharged at the $C/1.67$ rate. The aerospace IPV Ni-H$_2$ cells are not optimized for high-rate discharge but are optimized for maximum specific energy at discharge rates of between $C/2$ and $C/1.5$. At higher rates, the usable energy drops off because of the I^2R losses in the terminals. For example, the INTELSAT V cell is capable of delivering about 50 Wh/kg up to the C rate (1 h rate) of discharge at 0°C. Above the C rate (30 A rate), however, the specific energy starts to drop off, as shown in Fig. 24.18.

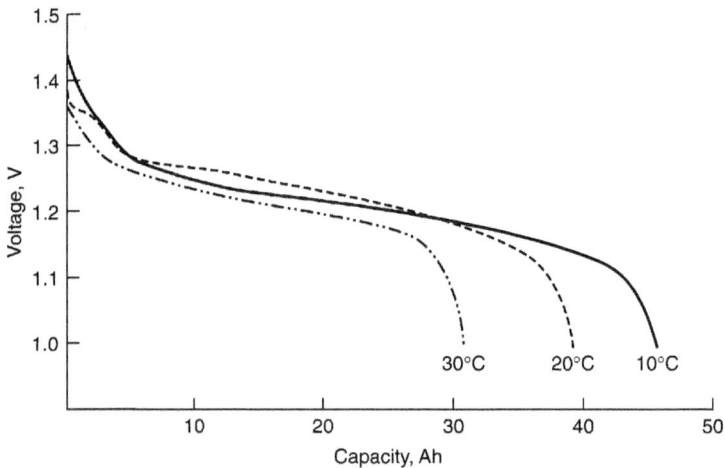

FIGURE 24.16 Capacity of NTS-2 35 Ah cell at different temperatures. Discharge rate, $C/1.67$.

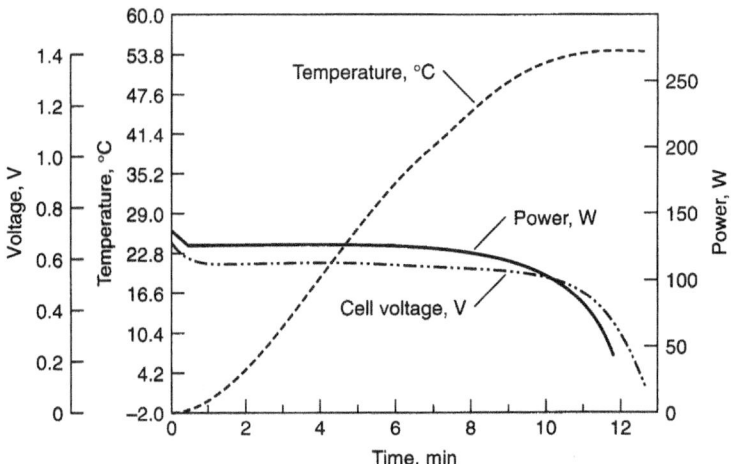

FIGURE 24.17 Discharge of INTELSAT V 30 Ah cell at 200 A rate or 6.7 C-rate discharge.

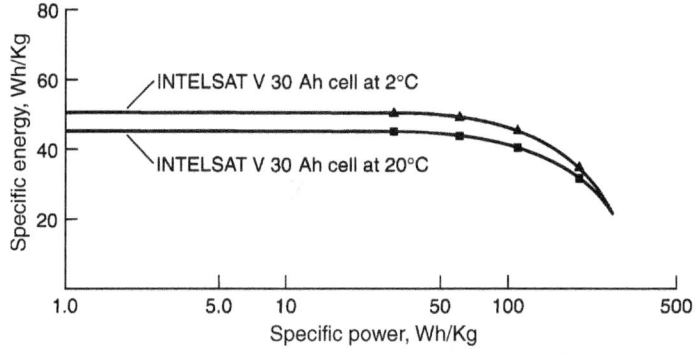

FIGURE 24.18 Specific energy vs. specific power of INTELSAT V Ni-H_2 cell.

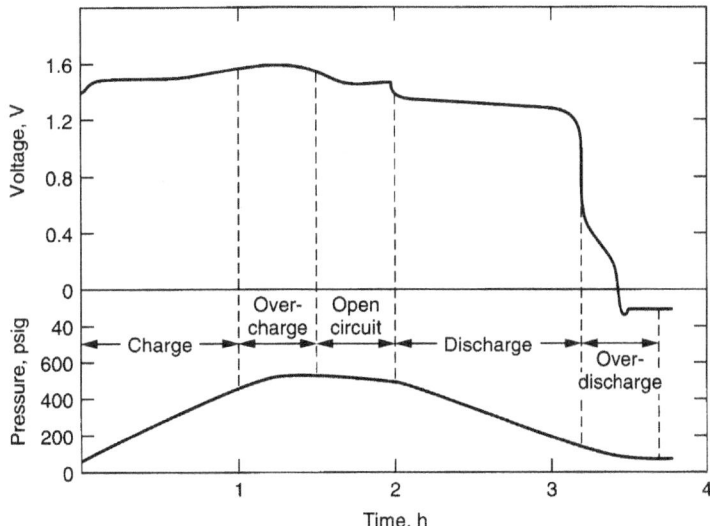

FIGURE 24.19 Pressure and voltage characteristics of NTS-2 Ni-H$_2$ cell at 23°C.

A salient feature of the Ni-H$_2$ cell is that the pressure is a direct indication of the state-of-charge of a cell (Fig. 24.19). On charge, the hydrogen pressure increases linearly until the nickel oxide electrode approaches the fully charged condition. During overcharge, oxygen evolved at the positive electrode recombines at the negative electrode, and the pressure stabilizes. On discharge, the hydrogen pressure decreases linearly until the nickel oxide electrode is fully discharged. If the cell is reversed by overdischarging, hydrogen generated at the positive electrode is consumed at the negative electrode, and again the pressure is constant.

The effects on cell voltage and capacity of discharge at different rates for a 90 Ah Hubble Space Telescope cell can be seen in Fig. 24.20. The data were gathered from cells stabilized at 10°C. As the current increases, the capacity to 1.00 V decreases. Changes in the cell voltage and capacity can be attributed to the impedance of the cell (0.9 mΩ).

24.7.2 Self-Discharge Characteristics of Ni-H$_2$ Cells

The self-discharge rate as a function of temperature was determined experimentally for Air Force 50 Ah cells used in the INTELSAT VI program.[25] Figure 24.21 gives the data for the self-discharge of these 50-Ah cells at 10, 20, and 30 C. Figure 24.22 shows the Arrhenius plot for these three temperatures. The slope of the straight-line regression fit to these three data points indicates an activation energy of 13.6 kcal /mol.[25]

24.7.3 Capacity as a Function of Electrolyte Concentration

The effects of electrolyte concentration on capacity were determined experimentally. Air Force positive electrodes and the Air Force 50 Ah Ni-H$_2$ cells from the INTELSAT VI program were used for this investigation. The Air Force positive electrodes were impregnated with active material by the alcoholic electrochemical impregnation process. The plaque was manufactured using the dry-powder process.

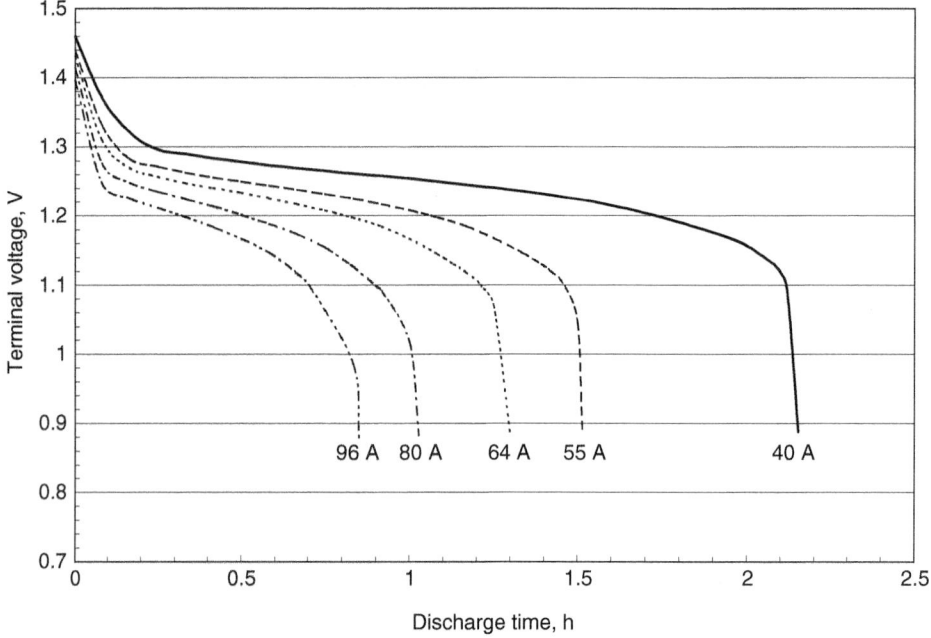

FIGURE 24.20 Discharge of Hubble Space Telescope cell at different rates.

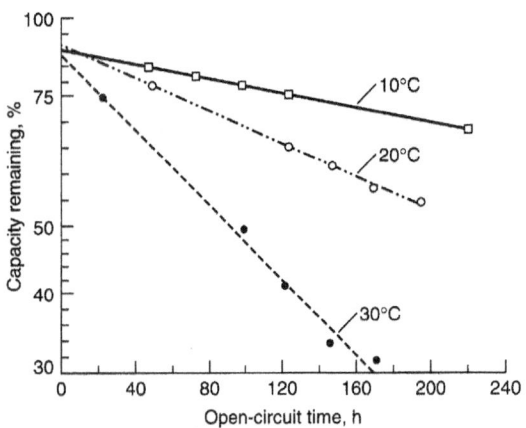

FIGURE 24.21 Self-discharge rates vs. temperature for 50 Ah Ni-H$_2$ cell. (*Courtesy of COMSAT Laboratories.*)

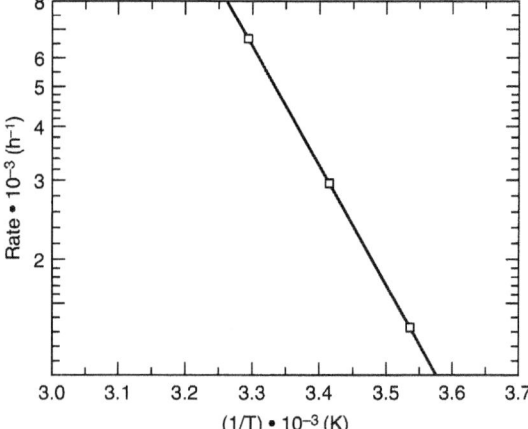

FIGURE 24.22 Arrhenius plot, self-discharge rates vs. temperature. (*Courtesy of COMSAT Laboratories.*)

For the Air Force standard 50 Ah Ni-H$_2$ cell, the electrolyte concentration was determined to be 26% KOH in the charged state and 31% KOH in the discharged state.[25] Cells were activated with three different levels of electrolyte concentration: 25, 31, and 38 wt% concentration of KOH. The electrolyte concentration in these cells was determined by analyses in both the charged and the discharged conditions. Table 24.2 presents cell capacity, electrolyte concentration, and average discharge voltage.

TABLE 24.2 Cell Capacity and Voltage vs. Electrolyte Concentration at 10°C

	Electrolyte concentration		
Parameter	38%	31%	25%
Cell capacity, Ah	64	56	43
Number of positive plates	40	40	40
Capacity per plate, Ah	1.60	1.40	1.08
Electrolyte concentration:			
Charged, wt% KOH	32*	26	21*
Discharged, wt% KOH	38	31	25
Average discharge voltage, V	1.247	1.268	1.290

*Estimated.

24.7.4 GEO Performance

The Ni-H$_2$ batteries on the INTELSAT V satellites have completed up to 9 years in orbit.

Voltage Performance in Orbit. In-orbit performance of the INTELSAT V batteries is judged by the minimum end-of-discharge voltage observed during an eclipse season. The minimum battery voltage requirement is 28.6 V, or 1.10 V per cell average which allows for one cell to fail via a short-circuit. The actual minimum battery voltages and the corresponding load currents and depths of discharge for each of the 14 batteries on the F6 through F15 satellites are presented in Table 24.3 for the Fall 1990 eclipse season.[26] Also presented are the minimum cell voltages within each battery and the corresponding average cell voltage per battery. After up to 7 years in orbit, the minimum end-of-discharge battery voltages on the longest eclipse day ranged from 31.2 to 32.4 V for all 14 batteries. The cells within the batteries were well matched at their minimum end-of-discharge voltages during the Fall 1990 eclipse season (maximum deviation is ±20 mV between cells within the same battery). The one exception is cell 22 in battery 1 on the F-6 spacecraft. This cell was 40 mV below the average cell voltage. These battery voltages are well above the minimum voltage requirement.

Pressure Data. INTELSAT V batteries are reconditioned prior to each eclipse season. The reconditioning capacity and pressure data for INTELSAT V F-6 battery 2 are presented in Table 24.4.[26] The pressure data were measured during the reconditioning discharge. The EOC pressure and the EOD pressure are the pressure at the start and the pressure at the end of the reconditioning discharge, respectively. The pressure constant is ΔP per measured capacity.

TABLE 24.3 Battery Loads and Minimum Voltages for Fall 1990 Eclipse Season

	DOD, %		Current, A		Voltage, V		Cell voltage, V			
	Battery 1	Battery 2	Battery 1	Battery 2	Battery 1	Battery 2	Batt. 1 av.	Batt. 1 min.	Batt. 2 av.	Batt. 2 min.
F-6	55.8	53.1	14.2	13.5	32.0	32.4	1.20	1.16	1.20	1.19
F-8	54.0	54.4	13.7	13.8	32.0	32.0	1.20	1.18	1.20	1.18
F-10	56.9	55.7	14.4	14.3	31.8	32.0	1.19	1.18	1.20	1.18
F-11	55.3	60.0	14.1	15.4	32.0	32.0	1.20	1.18	1.20	1.19
F-12	53.5	58.0	13.6	14.8	32.0	31.8	1.20	1.18	1.18	1.18
F-13	67.0	59.0	16.9	15.0	31.2	31.8	1.17	1.15	1.19	1.17
F-15	67.0	62.3	16.9	15.8	31.2	31.8	1.17	1.16	1.18	1.16

TABLE 24.4 Reconditioning Capacity and Pressure Data for INTELSAT V F-6 Battery 2

Eclipse season	Measured capacity, Ah	Max EOC pressure, lb/in²	Min EOD pressure, lb/in²	ΔP, lb/in²	Pressure constant ΔP/measured capacity, lb/in²/Ah*
F83	38.1	No pressure data in database			
S84	35.4	No pressure data in database			
F84	37.7	516.39	13.87	502.62	13.33
S85	37.6	518.49	17.90	500.59	13.31
F85	37.5	515.14	17.23	497.9	13.27
S86	37.9	519.34	15.32	504.02	13.29
F86	37.6	519.73	22.03	497.70	13.23
S87	38.3	514.34	13.87	505.47	13.19
F87	37.2	519.73	22.03	497.7	13.37
S88	38.3	525.78	16.20	509.58	13.30
F88	37.8	521.86	17.90	503.96	13.33
S89	36.9	526.91	18.67	508.24	13.77
F89	40.2	534.22	−0.57	534.79	13.30
S90	38.6	551.73	19.22	532.51	13.79
F90	36.0	530.87	38.04	492.83	13.68
S91	39.5	546.52	17.23	529.29	13.39
F91	39.0	545.30	17.90	527.40	13.52
					Average 13.37

*1 lb/in² = 6895 pa.

The data in Table 24.4 show the following:

1. The strain-gauge bridge circuit provides useful pressure data.
2. No change occurred in the reconditioning EOD pressure with time for these INTELSAT V battery cells. The EOD pressure in Fall 1991 was almost the same as the EOD pressure at the beginning of life in orbit.
3. The significance of these data is that no oxidation or corrosion has occurred within these cells. Any oxidation of the cell components would result in a pressure increase at the end of the reconditioning discharge.

24.7.5 LEO Performance Data

The HST was launched on April 24, 1990, with six 88 Ah Ni-H$_2$ batteries as the primary energy storage subsystem. This was the first reported nonexperimental mission to use Ni-H$_2$ batteries in an LEO application.[27] The batteries are being charged to a temperature-compensated voltage limit as described in Sec. 24.6.2. The batteries are discharged to 7 to 10% of depth of discharge. As reported at the 1991 IECEC, "To date (April 1991) the performance of the batteries has been flawless."[27] Orbital data had shown an expected normal slow loss in useful capacity which could eventually limit support of the HST. During servicing mission STS 125 in May 2009, after 18 years of service (13 years beyond design orbital life), the 6 batteries were replaced to extend the life of the HST.

24.8 ADVANCED DESIGNS

24.8.1 Advanced Designs for IPV Ni-H$_2$ Cells

A number of advanced concepts for IPV Ni-H$_2$ cells are being used to improve cycle life at deep DOD[28] mitigating failure modes commonly found in Ni-H$_2$ cells. They include: (1) the use of alternative methods for recombination (that is, the catalyzed wall wick), (2) the use of serrated-edge separators to facilitate the movement of gas within the stack while maintaining physical contact with the cell wall, (3) the incorporation of Belleville washers to yield an expandable stack capable of accommodating some of the nickel electrode expansion that is known to occur with cycling, and (4) the use of lower KOH concentrations to improve cycle life.

Cells utilizing the catalyzed wall wick are now in use. This concept offers an improved thermal design with recombination taking place on the pressure vessel wall. The heat from recombination is removed immediately through the pressure vessel wall to the cell thermal sleeve. This design also mitigates damage from internal stack popping since the recombination site is outside the cell stack.

The use of serrated edges for the separator is usually employed in designs utilizing asbestos. The irregular edge allows the unrestricted passage of oxygen along the edge of the cell stack while maintaining contact with the wall wick of the pressure vessel for recovery of electrolyte and removal of heat from the stack.

The Belleville washer acts as a spring, allowing further compression to accommodate any plate expansion during cycling.

The effects of KOH concentration on the cycle life of Ni-H$_2$ were investigated.[29] A breakthrough in the LEO cycle life of individual pressure-vessels cells was reported and cell cycle life was improved by greater than a factor of 10 when KOH concentration was reduced from 31 to 26% in the fully discharged state. The lower concentrations, while enhancing the cycle life, result in a slightly higher mid-discharge operating voltage with a lower available capacity to the 1.00 V per cell limit.

24.8.2 Advanced Battery Design Concepts

The common pressure vessel (CPV) Ni-H$_2$ battery and the bipolar Ni-H$_2$ battery are two advanced battery design concepts investigated to improve the gravimetric and volumetric energy densities as compared to the IPV cell and battery Ni-H$_2$ technology.

Common Pressure Vessel. Conceptually a CPV Ni-H$_2$ battery consists of a number of individual cells connected together in series and contained within one common pressure vessel.[30] For the IPV cell, each individual Ni-H$_2$ cell is contained within its own pressure vessel. Potential advantages for the CPV Ni-H$_2$ batteries include a significant increase in volumetric energy density (a decrease in volume), a decrease in manufacturing cost, a reduction in the complexity associated with the wiring and interconnection of IPV cells, an increase in specific energy, a reduction in the internal impedance of the battery, and improved heat transfer between the electrode stack and the pressure-vessel wall.

Several dual-cell CPV designs have been developed and tested. This design utilizes the dual-stack configuration used for IPV cells. For the CPV cell, the two stacks are connected in series, as shown in Fig. 24.23. This dual cell CPV battery offers a 30% reduction in volume and a 7 to 14% reduction in mass compared to an equivalent battery with IPV cells.[31]

Batteries utilizing these cells have been used in several flight programs including LEO and interplanetary missions. Two batteries used on the Mars Global Surveyor and Mars Polar Lander flight programs can be seen in Figs. 24.24 and 24.25. These are 28 V batteries having capacities of 23 Ah and 16 Ah, respectively.

A lightweight CPV Ni-H$_2$ battery was designed and developed jointly by COMSAT and Johnson Controls, Inc.[32] A prototype 10 in diameter 26-cell, 24 Ah CPV battery was fabricated and tested to

FIGURE 24.23 EPT CPV design (2.5 v.) (*Courtesy of Power Subsystems Group, Eagle-Picher Technologies, LLC.*)

FIGURE 24.24 Mars Global Surveyor 23 Ah CPV battery assembly.

FIGURE 24.25 Mars Polar Lander 16 Ah CPV battery assembly.

FIGURE 24.26 COMSAT/JCI CPV Ni-H$_2$ battery (10 in diameter).

demonstrate the feasibility of this lightweight design for LEO applications. This battery used two 13-cell half-stacks connected in series within the single common pressure vessel to provide a nominal 32 V battery. The 10 in aerospace design used a semicircular cell with a double-tap design to enhance current distribution. The components of the prototype CPV battery are shown in Fig. 24.26 with the fixed heat-fin cavity and lightweight pressure vessel.

Johnson Controls developed a new 5 in diameter 9.6 Ah CPV battery with loose heat fins (Fig. 24.27). The loose heat fin design was designed to overcome the problems encountered with the insertion of the cells into the heat fin cavity for the 10 in CPV cell described.[33]

A 5 in diameter, 28 V, 15 Ah CPV battery with the loose fin design was flown on the Clementine Program. This battery was manufactured by Johnson Controls under contract with the Naval Research Laboratory. This flight was launched and flown successfully in January 1994.

24.26 SECONDARY BATTERIES

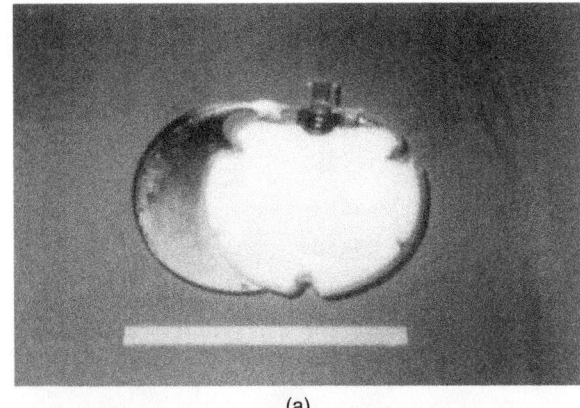

FIGURE 24.27 JCI CPV Ni-H$_2$ battery (5 in diameter). (*a*) Circular cell and loose heat fin. (*b*) 10-cell stack.

The advantages of the CPV Ni-H$_2$ battery make it a candidate for use in large multikilo-watthour LEO energy storage applications, such as the International Space Station or constellation systems such as Iridium®. It also appeals to the other end of the spectrum—the small 100 to 400 Wh applications that need low-cost lightweight batteries.[34]

Eagle-Picher Technologies, LLC, supplied 10 in diameter 28 V, 50 and 60 Ah CPV batteries for the Iridium® program. Over 80 satellites using these CPV batteries have been launched. A 28 V, 60 Ah CPV battery manufactured for the Iridium® program can be seen in Fig. 24.28. This design offered impedances less than 25 milliohm, a specific energy of 55 Wh/kg, and an energy density of 68 Wh/L.

24.8.3 Bipolar Ni-H$_2$ Batteries

Studies have shown that the bipolar batteries promise savings in weight and volume as compared to IPV batteries.[35] The research has been directed toward large energy storage requirements for LEO applications such as the International Space Station program. Several bipolar Ni-H$_2$ batteries were

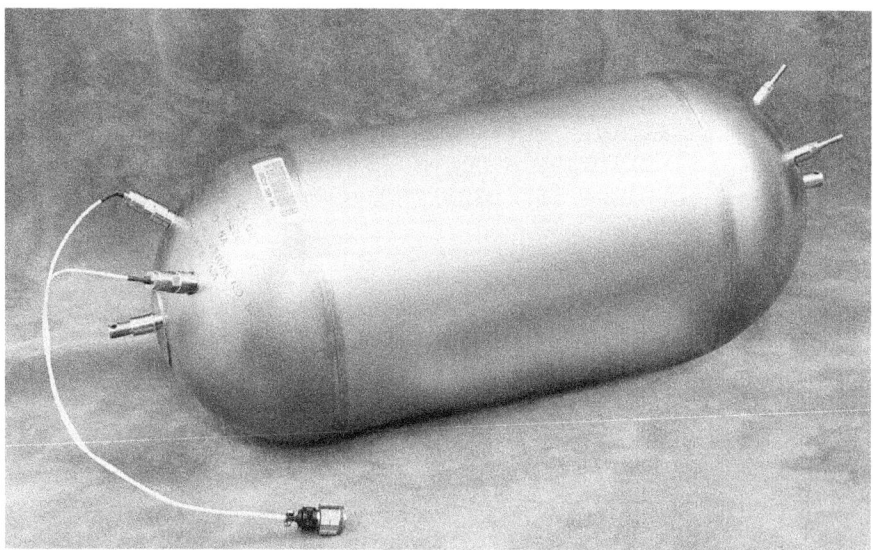

FIGURE 24.28 IRIDIUM® 60 Ah CPV battery assembly.

designed, fabricated, and tested at NASA Lewis Research Center. The second one, assembled in 1983, was a 10-cell 6.5 Ah bipolar Ni-H$_2$ battery. Useful data were generated from tests of this 10-cell bipolar battery and results should aid the development work needed to improve performance.[35]

REFERENCES

1. J. Dunlop, J. Giner, G. van Ommering, and J. Stockel, "Nickel-Hydrogen Cell," U.S. Patent 3,867,199, 1975.
2. S. U. Falk and A. J. Salkind, *Alkaline Storage Batteries,* Wiley, New York, 1969, sec. 2.5
3. R. L. Beauchamp, "Positive Electrodes for Use in Nickel Cadmium Cells and the Method for Producing Same and Products Utilizing Same," U.S. Patent 3,653,967, Apr. 4, 1972.
4. D. F. Pickett, H. H. Rogers, L. A. Tinker, C. Bleser, J. M. Hill, and J. Meador, "Establishment of Parameters for Production of Long Life Nickel Oxide Electrodes for Nickel-Hydrogen Cells," *Proc. 15th IECEC,* Seattle, WA, 1980, p. 1918.
5. L. W. Niedrach and H. R. Alford, *J. Electrochem. Soc.* **112:**117–124 (1965).
6. G. Holleck, "Failure Mechanisms in Nickel-Hydrogen Cells," *Proc. 1976 Goddard Space Flight Center Battery Workshop,* pp. 279–315.
7. E. Adler, S. Stadnick, and H. Rogers, "Nickel-Hydrogen Battery Advanced Development Program Status Report," *Proc. 15th IECEC,* Seattle, WA, 1980, p. 189.
8. G. van Ommering and J. F. Stockel, "Characteristics of Nickel-Hydrogen Flight Cells," *Proc. 27th Power Sources Conf.,* June 1976.
9. J. Dunlop, J. Stockel, and G. van Ommering, "Sealed Metal Oxide-Hydrogen Secondary Cells," *Proc. 9th Int. Symp. on Power Sources,* 1974; in D. H. Collins (ed.), *Power Sources,* Academic, New York, Vol. 5, 1975, pp. 315–329.
10. E. McHenry and P. Hubbauer, "Hermetic Compression Seals for Alkaline Batteries," *J. Electrochem. Soc.* **119:**564–568 (May 1972).

11. H. H. Rogers, S. J. Krause, and E. Levy, Jr., "Design of Long Life Nickel-Hydrogen Cells," *Proc. 28th Power Sources Conf.,* June 1978.
12. G. L. Holleck, M. J. Turchan, and D. DeBiccari, "Improvement and Cycle Testing of Ni/H$_2$ Cells," *Proc. 28th Power Sources Symp.,* June 1978, pp. 139–141.
13. H. H. Rogers, U.S. Patent 4,177,325, Dec. 4, 1979.
14. S. J. Stadnick, U.S. Patent 4,224,388, Sept. 23, 1980.
15. L. Miller, J. Brill, and G. Dodson, "Multi-Mission Ni-H$_2$ Battery Cells for the 1990s," *Proc. 24th IECEC,* Washington, DC, 1989, p. 1387.
16. T. M. Yang, C. W. Koehler, and A. Z. Applewhite, "An 83-Ah Ni-H$_2$ Battery for Geosynchronous Satellite Applications," *Proc. 24th IECEC,* Washington, DC, 1989, p. 1375.
17. G. van Ommering, C. W. Koehler, and D. C. Briggs, "Nickel-Hydrogen Batteries for INTELSAT V," *Proc. 15th IECEC,* Seattle, WA, 1980, p. 1885.
18. P. Duff, "EUTELSAT II Nickel-Hydrogen Storage Battery System Design and Performance," *Proc. 25th IECEC,* Reno, NV, 1990, Vol. 6, p. 79.
19. R. J. Hass, A. K. Chawathe, and G. van Ommering, "Space Station Battery System Design and Development," *Proc. 23d IECEC,* 1988, Vol. 3, pp. 577–582.
20. D. Bush, "Evaluation of Terrestrial Nickel/Hydrogen Cells and Batteries," SAND88-0435, May 1988.
21. D. Bush, "Terrestrial Nickel/Hydrogen Battery Evaluation," SAND90-0390, July 1990.
22. D. E. Nawrocki, J. D. Armantrout, et al., "The Hubble Space Telescope Nickel-Hydrogen Battery Design," *Proc. 25th IECEC,* Reno, NV, 1990, Vol. 3, pp. 1–6.
23. J. E. Lowery, J. R. Lanier Jr., C. I. Hall, and T. H. Whitt, "Ongoing Nickel-Hydrogen Energy Storage Device Testing at George C. Marshall Space Flight Center," *Proc. 25th IECEC,* Reno, NV, 1990, pp. 28–32.
24. M. P. Bernhardt and D. W. Mauer, "Results of a Study on Rate of Thickening of Nickel Electrodes," *Proc. 29th Power Sources Conf.,* Electrochemical Society, Pennington, NJ, 1980.
25. J. F. Stockel, "Self-Discharge Performance and Effects of Electrolyte Concentration on Capacity of Nickel-Hydrogen (Ni/H$_2$) Cells," *Proc. 20th IECEC,* 1986, Vol. 1, p. 1171.
26. J. D. Dunlop, A. Dunnet, and A. Cooper, "Performance of INTELSAT V Ni-H$_2$ Batteries in Orbit (1983–1991)," *Proc. 27th IECEC,* 1992.
27. J. C. Brewer, T. H. Whitt, and J. R. Lanier, Jr., "Hubble Space Telescope Nickel-Hydrogen Batteries Testing and Flight Performance," *Proc. 26th IECEC,* 1991.
28. J. J. Smithrick, M. A. Manzo, and O. Gonzalez-Sanabria, "Advanced Designs for IPV Nickel-Hydrogen Cells," *Proc. 19th IECEC,* San Francisco, CA, 1984, p. 631.
29. H. S. Lim and S. A. Verzwyvelt, "KOH Concentration Effects on the Cycle Life of Nickel-Hydrogen Cells," *Proc. 20th IECEC,* Miami Beach, FL, 1985, p. 1165.
30. D. Warnock, U.S. Patent 2,975,210, 1976.
31. T. Harvey and L. Miller, private communication on EPI handout.
32. M. Earl, J. Dunlop, R. Beauchamp, J. Sindorf, and K. Jones, "Design and Development of an Aerospace CPV Ni-H$_2$ Battery," *Proc. 24th IECEC,* 1989, Vol. 3, pp. 1395–1400.
33. J. Zagrodnik and K. Jones, "Development of Common Pressure Vessel Nickel-Hydrogen Batteries," *Proc. 25th IECEC*, Reno, NV, 1990.
34. J. Dunlop and R. Beauchamp, "Making Space Nickel-Hydrogen Batteries Lighter and Less Expensive," *AIAA/DARPA Meeting on Lightweight Satellite Systems*, Monterey, CA, Aug. 1987, NTIS N88-13530.
35. R. L. Cataldo, "Life Cycle Test Results of a Bipolar Nickel-Hydrogen Battery," *Proc. 20th IECEC,* 1985, Vol. 1, pp. 1346–1351.

BIBLIOGRAPHY

NASA Handbook for Nickel-Hydrogen Batteries, NASA Reference Publ. 1314, September 1993.

CHAPTER 25
SILVER OXIDE BATTERIES

Alexander P. Karpinski

25.1 GENERAL CHARACTERISTICS

The rechargeable silver oxide batteries are noted for their high specific energy and power-per unit weight and volume. The high cost of the silver electrode, however, has limited their use to applications where these qualities are prime requirements, such as lightweight medical and electronic equipment, submersibles, torpedoes, and space applications. The characteristics of the silver oxide secondary batteries are summarized in Table 25.1.

The first recorded use of a "silver battery" was by Volta with his now historic silver-zinc pile battery, which he introduced to the world in 1800.[1] This battery dominated the scene in the early 19th century, and during the next 100 years many experiments were made with cells containing silver and zinc electrodes. All these cells, however, were of the primary (nonrechargeable) type.

The first scientist to report a workable secondary silver battery was Jungner in Sweden in the late 1880s.[2] Although he experimented in the early stages with iron/silver oxide and copper/silver oxide batteries (which reportedly delivered as much as 40 Wh/kg), he settled on the cadmium/silver oxide battery for his experiments with electric car propulsion. The short cycle life and high cost of these batteries, however, made them commercially unattractive. During the next 40 years, other scientists (including Edison) experimented with various electrode formulations and separators, but without much practical success. It was the French professor Henri André who provided the key to the practical rechargeable zinc/silver oxide (silver-zinc) battery in 1941.[3,4] He described the use of a semipermeable membrane, cellophane, as a separator which would retard the migration of the soluble silver oxide to the negative plate and also impede the formation of zinc "trees," or dendrites, from the negative to the positive plate, the two major causes of cell short circuits.

In the 1950s, interest was revived in the silver-cadmium battery using the then newly available silver-zinc and nickel-cadmium technologies. This provided improved cycle life over the silver-zinc system. These batteries were first commercialized by Yardney International Corporation. Later, Westinghouse Corporation reported the commercial application of a silver-iron battery (see Chap. 18) in which they sought to "eliminate the zinc plate problems with a trouble-free iron plate, ease the separator materials and life problem and shift the deep discharge capacity stability to that limited by the silver plate."[5] The goal now, as for the past two centuries, is to provide the high energy content and power capability of the silver electrode in an improved-life, lower-cost, commercially viable secondary battery.

Zinc/silver oxide batteries provide the highest energy per unit weight and volume of any commercially available aqueous secondary batteries. They can operate efficiently at extremely high discharge rates, and they exhibit good charge acceptance at moderate rates and low self-discharge. The disadvantages are low cycle life (ranging from 10 up to 150** deep cycles, depending on design and use),

---

**Up to 250 cycles are possible with improved negative electrode or separators (see Sec. 25.9).

TABLE 25.1 Advantages and Disadvantages of Silver Oxide Secondary Cells

Advantages	Disadvantages
Silver-zinc (zinc/silver oxide)*	
Highest energy and power per unit weight and volume	High cost
High discharge rate capability	Relatively low cycle life
Moderate charge rate capability	Sensitivity to overcharge
Good charge retention	
Flat discharge voltage curve	
Low maintenance	
Safety	
Silver-cadmium (cadmium/silver oxide)*	
High energy and power per unit weight and volume (approx. 60% of silver-zinc)	High cost
Cycle life (up to 250 cycles vented, 100 cycles sealed)	Decreased performance at low temperatures
Flat discharge voltage curve	
Low maintenance	
Nonmagnetic construction possible	
Safety	
Silver-iron (iron/silver oxide)*	
High energy and power capability	High cost
Good capacity maintenance	Water and gas management requirements
Overcharge capability	Not yet proven in field use

*The corrected designation for these battery systems is shown in parentheses. However, the initial designation (silver-zinc, etc.) is more popular and generally used.

decreased performance at low temperatures, sensitivity to overcharge, and high cost. Rates as high as 20 times the nominal capacity (20C rate) can be obtained from specially designed silver-zinc batteries because of their low internal impedances. These high rates, however, must often be limited in duration to avoid a potentially damaging temperature rise within the cells.

Cadmium/silver oxide batteries have been viewed as a compromise between the high energy density but short life of the silver-zinc system and the long cycle life but low energy density of the nickel-cadmium system. Their energy density is roughly 2 to 3 times higher than that of nickel-cadmium, nickel-iron, or lead-acid batteries, with a relatively long cycle life, especially during shallow cycling. Charge retention is excellent. In addition, the ability to fabricate the cells without use of magnetic materials has made them the battery of choice for several scientific satellite programs. The major disadvantage of the silver-cadmium system is cost; the cost per unit energy is even higher than for the silver-zinc battery.

Iron/silver oxide batteries provide high energy and power capability with long service life under deep-discharge use. They are capable of withstanding overcharge and over-discharge without damage and can provide good capacity maintenance with cycling. Disadvantages are, once again, cost and also the need for gas and water management in overcharge applications. Their nominal load voltage of 1.1 V is comparable to that of the silver-cadmium system, but lower than the 1.5 V level for silver-zinc. Sufficient data have not been published for these batteries to date to permit complete characterization of their properties.

All three systems also offer the advantages of long dry shelf life and of providing a flat discharge voltage during the major portion of their discharge. This latter characteristic is related to the fact that as the silver oxide is reduced to metallic silver during discharge, the conductivity of the silver electrode increases and serves to counteract polarization effects. It is noteworthy that other couples successfully used the silver oxide electrode. These include metal hydride/silver oxide cells, hydrogen/silver oxide cells and aluminum/silver oxide pile batteries, with the latter successfully used in torpedo applications.

25.2 CHEMISTRY

25.2.1 Cell Reactions

The overall electrochemical cell reactions, all of which use aqueous solutions of potassium hydroxide (KOH) for electrolyte, can be summarized as follows:

Silver-zinc: $\quad AgO + Zn + H_2O \underset{\text{charge}}{\overset{\text{discharge}}{\rightleftarrows}} Zn(OH)_2 + Ag$

Silver-cadmium: $\quad AgO + Cd + H_2O \underset{\text{charge}}{\overset{\text{discharge}}{\rightleftarrows}} Cd(OH)_2 + Ag$

Silver-iron: $\quad 4AgO + 3Fe + 4H_2O \underset{\text{charge}}{\overset{\text{discharge}}{\rightleftarrows}} Fe_3O_4 \cdot 4H_2O + 4Ag$

Silver metal-hydride: $\quad AgO + 2MH \underset{\text{charge}}{\overset{\text{discharge}}{\rightleftarrows}} Ag + 2M + H_2O$

Silver-hydrogen: $\quad AgO + H_2 \underset{\text{charge}}{\overset{\text{discharge}}{\rightleftarrows}} Ag + H_2O$

Silver-aluminum: $\quad 3AgO + 2Al \xrightarrow{\text{discharge}} 3Ag + Al_2O_3$

These are simplified equations since there is still no general agreement on the detailed mechanisms of these reactions or on the exact form of all the reaction products.

25.2.2 Positive-Electrode Reactions

The charge and discharge processes of the silver electrode in alkaline systems are of special interest because they are characterized by two discrete steps which manifest themselves as two plateaus in the charge and discharge curves. The reaction occurring at the silver electrode at the higher (peroxide) voltage plateau is shown as

$$2AgO + H_2O + 2e \underset{\text{charge}}{\overset{\text{discharge}}{\rightleftarrows}} Ag_2O + 2OH^-$$

and at the lower, (monoxide) voltage plateau as

$$Ag_2O + H_2O + 2e \underset{\text{charge}}{\overset{\text{discharge}}{\rightleftarrows}} 2Ag + 2OH^-$$

As shown, these reactions are reversible.

25.3 CELL CONSTRUCTION AND COMPONENTS

Secondary silver cells have been produced in prismatic, spirally wound cylindrical, and button shape configurations. The most common shape is prismatic. The construction of a typical prismatic cell is shown in Fig. 25.1. This cell contains flat electrodes which are wrapped with multiple layers of separator to provide mechanical separation and inhibit migration of the silver to the zinc plate and the growth of zinc dendrites toward the positive plate. The plate groups are intermeshed, and the pack is placed in a tightly fitting case (Fig. 25.2). Because of the relatively short shelf life of the activated silver cells, they are usually supplied by the manufacturers in the dry charged or dry unformed condition with filling kits and instructions. The cells are filled with electrolyte and activated just prior to use. They may also be supplied in the filled and ready-to-use condition if required by the user.

The mechanical strength of these cells is usually excellent. The electrodes are generally strong and are fitted tightly into the containers. The cell containers are made of high-impact plastics.

25.4 SECONDARY BATTERIES

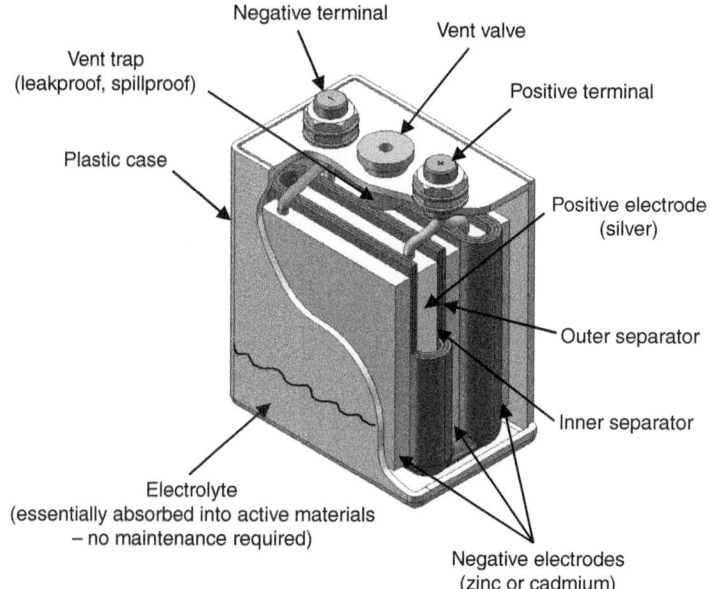

FIGURE 25.1 Cutaway view of typical prismatic zinc/silver oxide or cadmium/silver oxide secondary cell.

FIGURE 25.2 Cell stack being assembled into cell case; model LR-190, 210 Ah silver-zinc battery. (*Courtesy of Yardney Technical Products, Inc.*)

Specific designs of these cells, when properly packaged, have met the high-shock, vibration, and acceleration requirements of launch vehicles, missiles, and torpedoes with no degradation.

25.3.1 Silver Electrodes

The most common fabrication technique for silver electrodes is by sintering silver powder onto a supporting silver grid. The electrodes are manufactured either in molds (as individual plates or as master plates which are later cut to size) or by continuous rolling techniques. They are then sintered in a furnace at approximately 700°C.

Alternate techniques include dry processing and pressing, as well as slurry pasting of AgO or Ag_2O onto a grid. If pasted, the plates are often sintered, converting the silver oxide into metallic silver and burning off the organic additives. The grid may be a woven or expanded metal form of silver, or silver-plated copper.

After being cut to size and having wires or tabs hot-forged onto an appropriately coined (compressed) area to carry current to the cell terminals, the electrodes are either electro-formed (charged in tanks against inert counterelectrodes) before assembly into cells or assembled into the cells in the metallic state and later charged in the cell.

Grid material, density and thickness, electrical lead type and size, and final electrode size, thickness, and density are all design variables that depend on the intended application for the cells. The silver powder particle size may be varied, with the finer powders approaching the theoretical silver utilization of 2.0 g/Ah. The use of very fine powder, however, results in an initial voltage dip (typically less than 120 ms) at medium (C) to high discharge rates.

25.3.2 Zinc Electrodes

Zinc electrodes are most widely made by dry pressing, by a slurry or paste method, or by electrodeposition. In the dry-pressing method, a mixture of metallic zinc or zinc oxide, binder, and additives is compressed around a metal grid; this is normally done in a mold. The grid usually has the current-carrying leads prewelded in place. As the unformed powder electrodes have little strength, one component of the separator system, the negative interseparator, is usually assembled around the electrode as part of the fabricating operation. Rolling techniques have also been developed to permit continuous fabrication of dry-powder electrodes.[6]

In the paste or slurry method, a mixture of zinc oxide, binder, and additives is combined with water and applied continuously to a carrier paper or directly to an appropriate metal grid. Again, the negative interseparator is usually integral to provide needed physical strength. After drying, multiple layers of these pasted slabs may be pressed together about a pretabbed grid to form the final electrode. These plates may be assembled unformed into the cell, or they may be electroformed in a tank against inert counterelectrodes.

Electrodeposited negative electrodes are manufactured by plating zinc in tanks onto metallic grids. The plates must then be amalgamated and pressed or rolled to the desired thickness and density, followed by drying.

The zinc electrode is acknowledged as the life-limiting component in both the silver-zinc and the nickel-zinc systems. Accordingly, much work has been done in the area of additives for these electrodes, both to reduce hydrogen evolution and to improve cycle life. The common additive to reduce hydrogen evolution has traditionally been mercury (1 to 4% of the total mix), but this is being replaced, for personnel safety and environmental reasons, by small amounts or mixtures of the oxides of lead, cadmium, indium, thallium, gallium,[7-11] and bismuth.[23] Many other (proprietary) additives have been introduced into the zinc electrode by various manufacturers in attempts to increase life.

Zinc electrodes also suffer capacity loss, which results from "shape change," or the migration of materials from the sides and top to the center and bottom of the electrode.

Several approaches have been taken to improve the stability of the zinc electrode, including (1) an excess of zinc to compensate for losses during cycling, (2) oversized electrodes on the basis that shape change starts on the electrode edges where current densities are higher, (3) binders such as PTFE, potassium titanate, neoprene latex or other polymers to hold the active materials together, and (4) electrolyte additives.[12-14]

As is the case for the silver electrodes, the grid material, additives, and final electrode size, thickness, and density are all design variables that depend on the final application.

25.3.3 Cadmium Electrodes

Most silver-cadmium cells contain cadmium electrodes that are manufactured by pressed-power or pasting techniques. Although other methods have been used, such as impregnating nickel plaque with cadmium salts, as is done for nickel-cadmium cells, the most common method in silver-cadmium cells is to press or paste a mixture of cadmium oxide with a binder onto a silver or nickel grid. These processes are similar to those used for the pressed and pasted zinc electrodes.

25.3.4 Iron Electrodes

The iron electrodes used here are generally manufactured by powder-metallurgy techniques (see Chap. 18).

25.3.5 Separators

The separators in the silver cells must meet the following major requirements:

1. Provide a physical barrier between positive and negative electrodes
2. Have minimum resistance to the flow of electrolyte and ions
3. Prevent migration of particles and dissolved silver compounds between positive and negative electrodes
4. Be stable in the electrolyte and cell operating environment

In general, secondary silver-zinc and silver-cadmium cells require up to three different separators, as shown in Fig. 25.1. The inner separator, or positive interseparator, serves both as an electrolyte reservoir and as a barrier to minimize oxidation of the main separator by the highly oxidative silver electrode. This separator is usually made of an inert fiber such as nylon or polypropylene, usually with an added wetting agent.

The outer separator, or negative interseparator, also serves as an electrolyte reservoir and can also, ideally, stabilize the zinc electrode and retard zinc penetration of the main separator, and dendritic growth. Much work has been done to develop improved inorganic positive or negative electrode interseparators utilizing such materials as asbestos and potassium titanate fibers. Improvements in life have been reported as a result of this work.[7-10,11,15] However, many of these separators are no longer commercially available because of health hazard considerations.

The main, or ion exchange, separator remains the key to the wet life of the secondary silver cell. It was André's[3] use of cellophane as a main separator that first made the secondary silver cells feasible. The cellulosics (cellophane, treated cellophane, and fibrous sausage casing) are usually employed in multiple layers as the main separators for these cells. Again, much work has been done in recent years to develop improved separators utilizing such materials as radiation-grafted polyethylene,[16] inorganic separators,[10,11,15,17,24] and other synthetic polymer membranes. Improved cell life has been reported through use of these new membranes either alone or in combination with cellulosics. Some of these have yet to be applied extensively to commercial silver cells, however, because of drawbacks sometimes involving high impedance, limited availability, and high cost.

25.3.6 Cell Cases

The cell cases must be chemically resistant to attack by the corrosive concentrated potassium hydroxide electrolyte and to oxidizing effects of the silver electrodes. They must also be strong enough to contain any internal pressure generated in the cells and to maintain structural integrity throughout the anticipated range of environmental conditions that will be experienced by the cells.

The majority of secondary silver cells are assembled in plastic cases. The plastic most commonly used is an acrylonitrile-styrene copolymer (SAN). This material is relatively transparent and can be sealed easily by solvent cement or epoxy. However, its relatively low softening temperature (80°C) precludes it from use in some applications. A wide variety of other plastics have been used for cell cases. Table 25.2 lists some of these materials and gives their characteristics. Metal cases have been used for some sealed cell applications; however, these present problems in sealing and in electrically isolating the electrodes from the cases and are not used widely, except for button cells.

TABLE 25.2 Properties of Cell Case and Cover Materials

Material	MABS Methyl methacrylate-acrylonitrile-butadiene-styrene	ABS acrylonitrile-butadiene-styrene		Polysulfone	Modified Polyphenylene Oxide (PPO)		Nylon[3]
Trade name	Terlux 2802 HD	MG37EP	Lustran 448	Bakelite polysulfone P-1700	Noryl 731	Noryl SE1X	Zytel 151 or 151L
Specific gravity	1.08	1.05	1.05	1.24	1.04–1.09	1.06–1.10	1.05–1.07
Minimum tensile strength (psi)	6,960	4,900	6,100	10,200	8,000	9,800	8,850
Min. impact strength izod, notched (ft-lb/in)	1.31 @ 73°F; 0.37 @ −22°F	6.50 @ 73°F	6.2 @ 73°F; 1.20 @ −22°F	6.50 @ 73°F	3.0, min	3.9, min	1.29
Conductivity (Btu/hr/ft^2/°F/in)			1.3–2.3	1.8	—	—	1.5
Flexural modulus (psi)		355,000	348,000	380,000	351,000	363,000	247,000
Specific heat (Btu/lb/°F)[1]			0.30–0.40	0.31	—	—	0.3–0.4
Heat defl. temp.							
°F @ 66 psi load	273	210	up to 252	358	274	262	275
°F @ 264 psi load	194	185	221, min	345	240, min	244, min	131
Use temp. (°F)[2]	167	140–210	190–230	320			180–250
Hardness (Rockwell)		R75–105	R109	R120	R119	—	R110
Transparency	Yes	No	No	Yes	No	No	Translucent
Annealing temp (°F)	185 ± 5	180 ± 5		333 ± 5	214 ± 5		
Manufacturer	BASF	SABIC	BAYER	AMOCO Performance Products	SABIC	SABIC	E.I. DuPont

Notes:
1. Or cal/g/°C.
2. Maximum continuous, at no load.
3. Data are from ASTM D4066.

TABLE 25.3 Physical and Electrical Characteristics of KOH Solutions

% KOH	Specific gravity at 15.6°C	Conductivity at 18°C, $\Omega^{-1} \cdot cm^{-1}$	Specific heat at 18°C, cal/g·°C	Freezing point, °C	Boiling point, °C at 760 mm Hg	Boiling point, °C at 100 mm Hg	Vapor pressure, mm Hg at 20°C	Vapor pressure, mm Hg at 80°C	Viscosity, cP at 20°C	Viscosity, cP at 40°C
0	1.0000		0.999	0	100	52	17.5	355	1.00	0.66
5	1.0452	0.170	0.928	−3	101	52.5	17.0	342	1.10	0.74
10	1.0918	0.310	0.861	−8	102	53	16.1	327	1.23	0.83
15	1.1396	0.420	0.801	−14	104	54	15.1	306	1.40	0.95
20	1.1884	0.500	0.768	−23	106	56	13.8	280	1.63	1.10
25	1.2387	0.545	0.742	−36	109	59	11.9	250	1.95	1.31
30	1.2905	0.505	0.723	−58	113	62	10.1	215	2.42	1.61
35	1.3440	0.450	0.707	−48	118	66	8.2	178	3.09	1.99
38	1.3769	0.415	0.699	−40	122	69	7.0	156	3.70	2.35
40	1.3991	0.395	0.694	−36	124	71	6.2	140	4.16	2.59
45	1.4558	0.340	0.678	−31	134	80	4.5	106	5.84	3.49
50	1.5143	0.285	0.660	+6	145	89	2.6	70	8.67	4.85

25.3.7 Electrolyte and Other Components

The electrolyte used in secondary silver cells is generally an aqueous solution (35 to 45% concentration) of potassium hydroxide (KOH). Lower concentrations of electrolyte provide lower resistivity and thus a higher voltage output under load as well as a lower freezing point. Concentrations below 45% KOH, however, are more corrosive to the cellulosic separators typically used in silver-based batteries and are not used for extended wet-life applications. Table 25.3 depicts the critical parameters of KOH solutions. Various additives such as zinc oxide, lithium hydroxide, potassium fluoride, potassium borate, tin, and lead have been used to reduce the solubility of the zinc electrode.[14]

Since potassium hydroxide readily combines with carbon dioxide in the air to form potassium carbonate, thus reducing conductivity, cell vents are usually covered with a vent cap or a low-pressure relief valve.

Cell terminals are typically made of steel or brass and are almost always silver- or nickel-plated to improve corrosion resistance.

25.4 PERFORMANCE CHARACTERISTICS

25.4.1 Performance and Design Tradeoffs

The secondary silver batteries provide high energy capability combined with minimum weight and volume. The advantages and disadvantages of the various systems have been described earlier in this chapter. The performance of the batteries for specific applications will depend on the internal designs of the cells. It is rare that one can select an "off-the-shelf" battery that will meet all the requirements of a specific application.

Starting with the basic parameters, the cell design involves a series of compromises to obtain the most favorable combination of voltage, electrical capacity, and cycle life characteristics within the allowable battery weight and volume.

Assuming, for example, a nominal 1.5 V per silver-zinc battery at low current densities (0.01 to 0.03 A/cm^2) and lower voltages at higher currents, the designer selects the number of cells for the application. The problem is increased if high current pulse loads are required and the battery must provide voltage above the minimum allowable at the high rate, while not exceeding the maximum allowable voltage at initial low rates. The size of the cell is then chosen by dividing the allowable volume by the calculated required number of cells.

The voltage, current, electrical capacity, and cycle life requirements must then be reviewed in conjunction with the allowable weight and the environmental conditions that the battery must be able to withstand. Each of these will be a factor in determining the choice of separator material for the cell. The stability and number of layers of separator must be sufficient to provide the desired wet life under these conditions while having a resistance low enough to prevent undue voltage drop at the high current load. Each of these requirements is also a factor in choosing the number of electrodes within the cell. As the number of electrodes (and thus the active electrode area) is increased, the current density during any discharge (amperes per square centimeter) is decreased, raising the output voltage. It should be noted that a cell design optimized for high discharge rates will, by nature of the design, have a reduced capacity under low discharge rates. This is a result of having many electrodes each having to be wrapped with the required number of layers of separators. Given a fixed internal volume, it follows that in such a high-rate cell, less space is available for active electrode material.

The cell must also be designed to contain enough active silver and zinc to supply the required electrical capacity for the desired number of cycles. Theoretically, 2.01 g of silver and 1.22 g of zinc are required in the cell for each ampere-hour of electrical capacity desired. Since these values are the theoretical capability of the pure material, and since some of the active materials will go into solution with each charge-discharge cycle, the designer must work with higher values—of the order of 3.5 g of silver and 3.0 g of zinc per nominal ampere-hour for long cycle-life cells. Other design variables, such as silver powder particle size, will also ultimately affect cell performance.

Because of these considerations, the performance curves shown in the following sections must be viewed as general characteristics of the systems and not necessarily of specific batteries for a particular application.

25.4.2 Discharge Performance for Zinc/Silver Oxide Batteries

The open-circuit voltage of a fully charged zinc/silver oxide cell is 1.84 to 1.86 V. The discharge is characterized by two discrete steps, the first corresponding to the divalent oxide and the second to the monovalent oxide, as shown in Fig. 25.3. The flat portion of the curves is referred to as "plateau voltage." This voltage is rate-dependent; at high rates the voltage steps may be obscured.

The performance of the cells at various discharge rates and temperatures can be seen in Figs. 25.4 to 25.6, which show the effect on plateau voltage and capacity. The high-rate capability of the zinc/silver oxide battery is a complex process which can be characterized as the result of the electrical conductivity of the silver grid and the conductivity of the positive electrode as it is discharged, as well as the thin multiplate design of the cell. The performance of the battery falls off with decreasing temperature, particularly below −20°C. Allowing the battery to warm up or by retaining the heat generated during the discharge can improve the performance at low ambient temperatures, over the range of values shown in Fig. 25.5.

The performance characteristics of the zinc/silver oxide battery are summarized in Figs. 25.7 and 25.8, which can be used to determine the capacity, service life, and voltage under a variety of discharge conditions. These figures present typical performance data. Performance differences can occur for each specific design and even for each battery, depending on cycling history, state of charge, storage time, temperature, and other conditions of use.

Figures 25.3 to 25.8 are specifically for high-rate (HR) designs. For many applications, trade-offs can be made to provide longer life at the expense of somewhat lower energy density. Alternative low-rate (LR) designs contain additional layers of separator, meaning, of necessity, fewer electrodes with higher impedance and lower capacity within a given volume. Typically, the LR battery cannot be discharged at higher than the 1 h rate and will provide about 3 to 5% lower average voltage and capacity than its HR counterpart at the 1 h rate. However, the LR batteries do provide substantial wet shelf life and cycle life advantages (see Table 25.4).

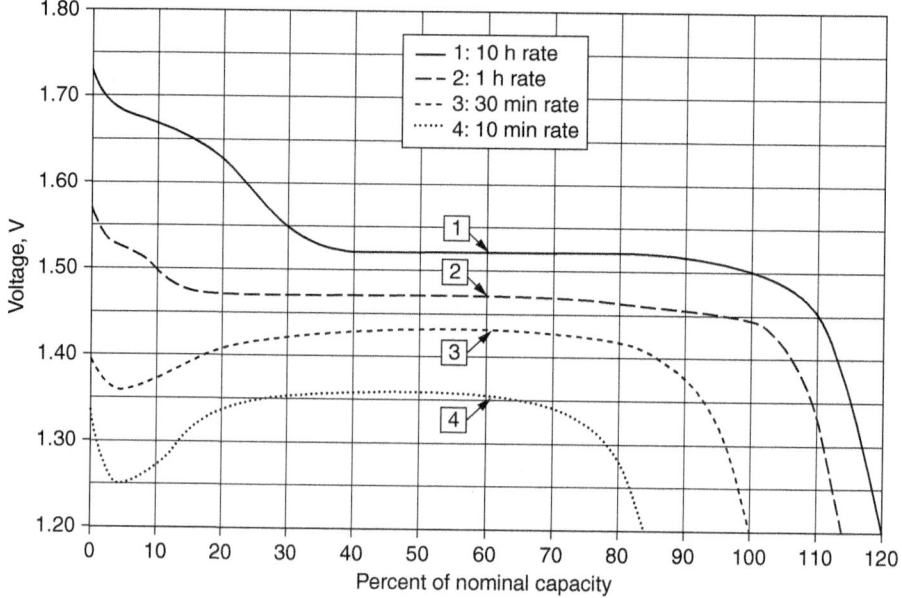

FIGURE 25.3 Typical discharge curves of silver-zinc battery at various rates, at 20°C.

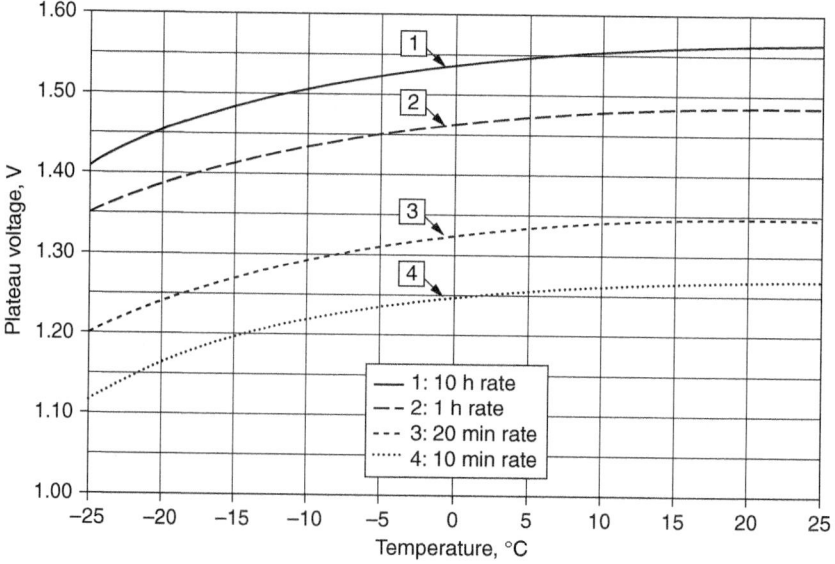

FIGURE 25.4 Typical effect of temperature on plateau voltage for high-rate silver-zinc cells.

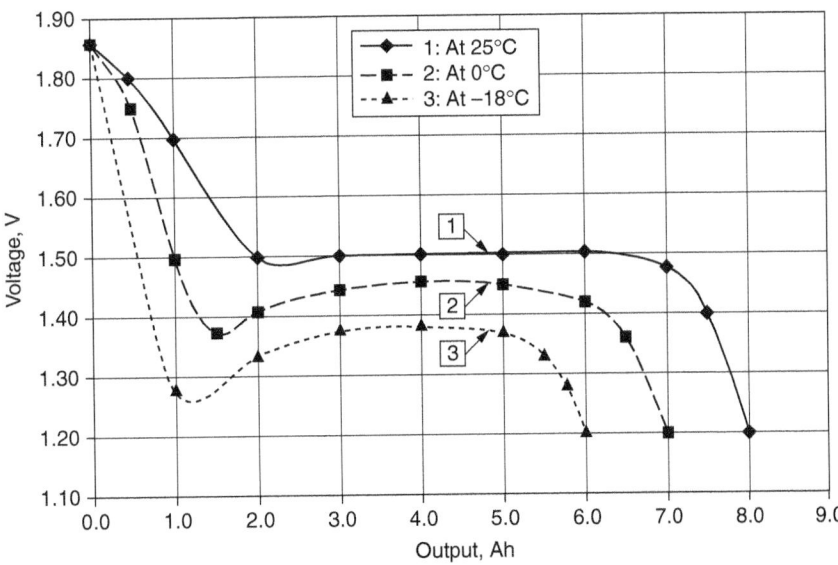

FIGURE 25.5 Typical discharge curve of HR5 silver-zinc cell at various temperatures, with no insulation.

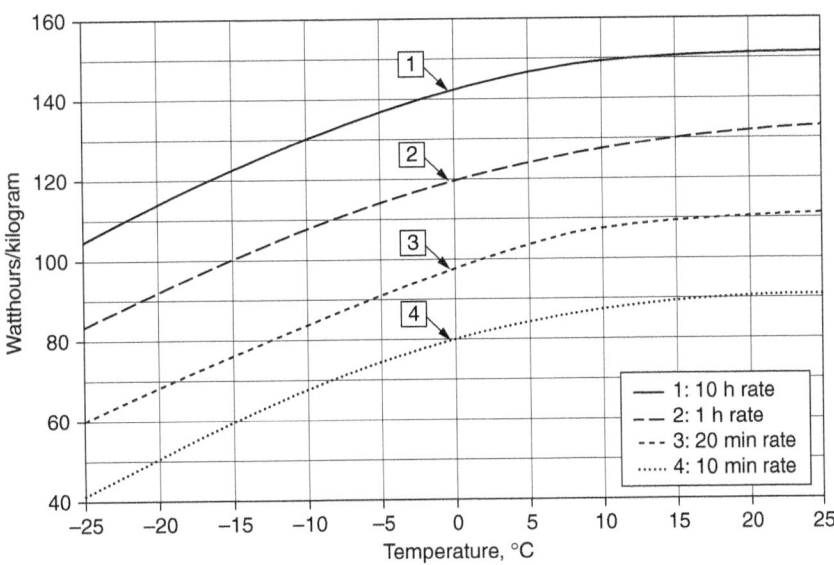

FIGURE 25.6 Typical effect of temperature on specific energy of silver-zinc cells.

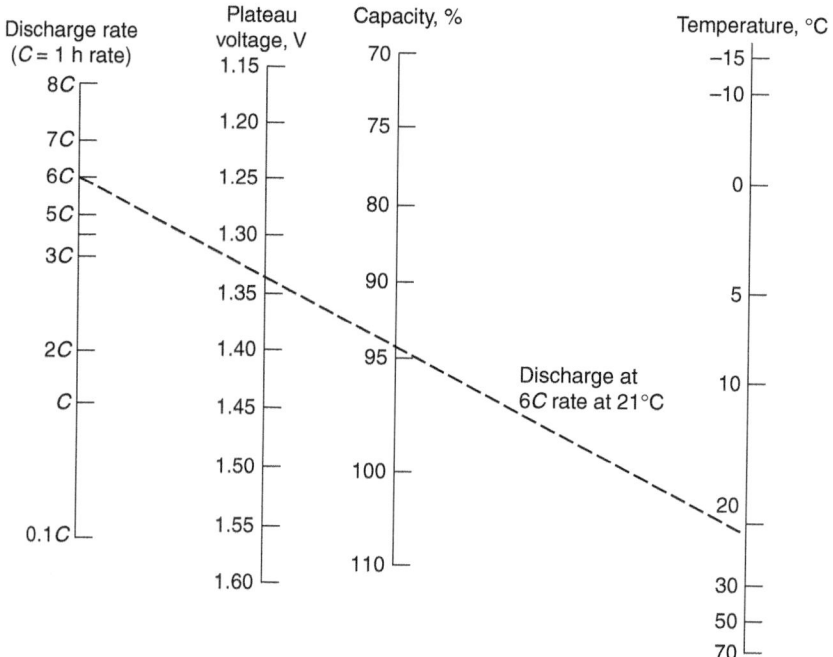

FIGURE 25.7 Performance characteristics of silver-zinc batteries under various conditions. (To find the capacity and the plateau voltage of a silver-zinc battery, draw a straight line between the discharge rate and the ambient temperature at which the battery is stabilized.)

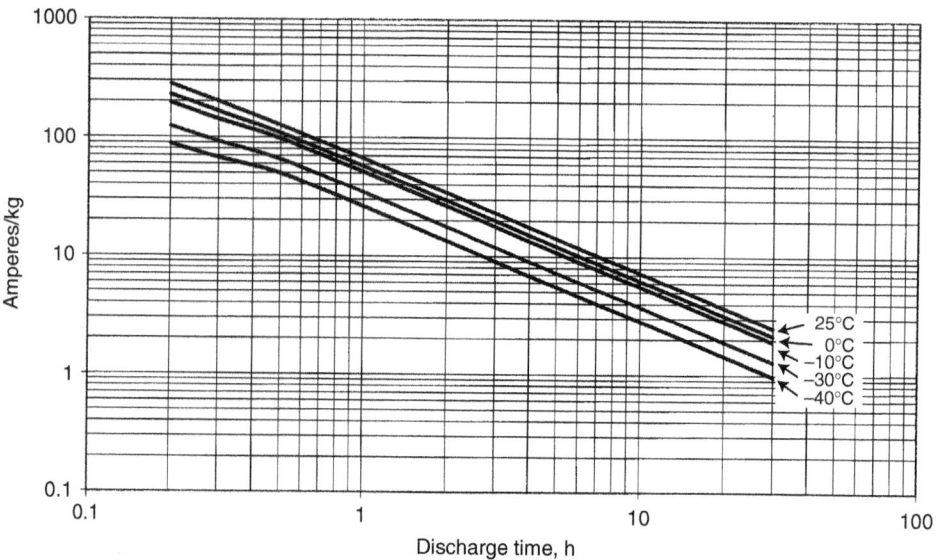

FIGURE 25.8a Service life of silver-zinc batteries at various discharge rates and temperatures. (Amperes/kg vs. discharge time in hours.)

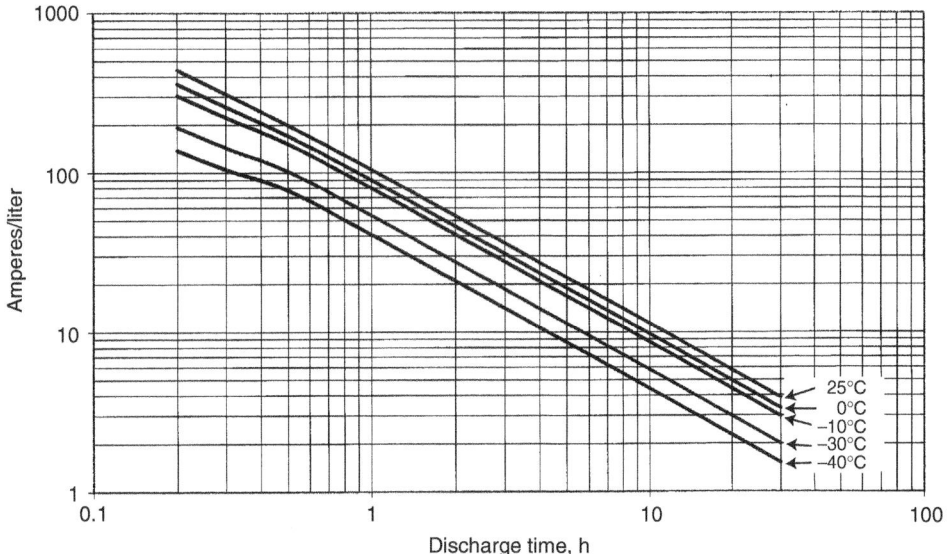

FIGURE 25.8b Service life of silver-zinc batteries at various rates and temperature. (Amperes/liter vs. discharge time in hours.)

25.4.3 Discharge Performance for Cadmium/Silver Oxide Batteries

The open-circuit voltage of the cadmium/silver oxide battery is 1.38 to 1.42 V. Typical discharge curves at 20°C are given in Fig. 25.9, showing the two-step discharge typical of the silver oxide electrode. The discharge characteristics are similar to those of the zinc/silver oxide battery except for the lower operating voltage; ampere-hour capacities per gram of silver are about the same.

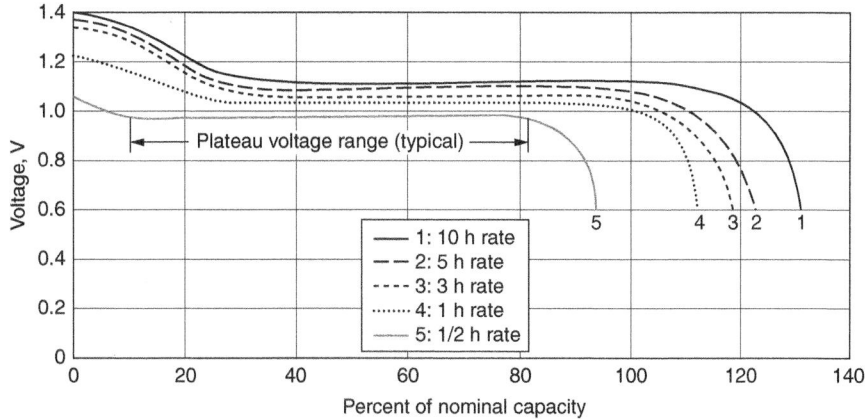

FIGURE 25.9 Typical discharge curves of silver-cadmium battery at various rates at 20°C.

The capacity and the discharge voltage of the battery are temperature-dependent, again similar to the zinc/silver oxide battery. The effects of temperature and discharge rate on voltage and capacity are shown in Figs. 25.10 and 25.11, respectively. The recommended operational temperature range is −25 to 70°C, with the optimum performance obtained between 10 and 55°C. With external heating, the temperature range can be lowered to −60°C.

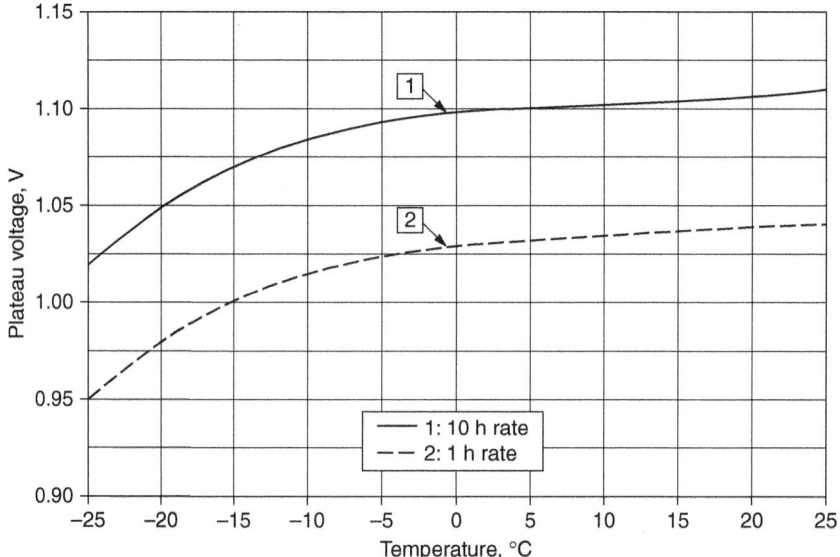

FIGURE 25.10 Typical effect of temperature on plateau voltage of silver-cadmium cells (discharged without heaters).

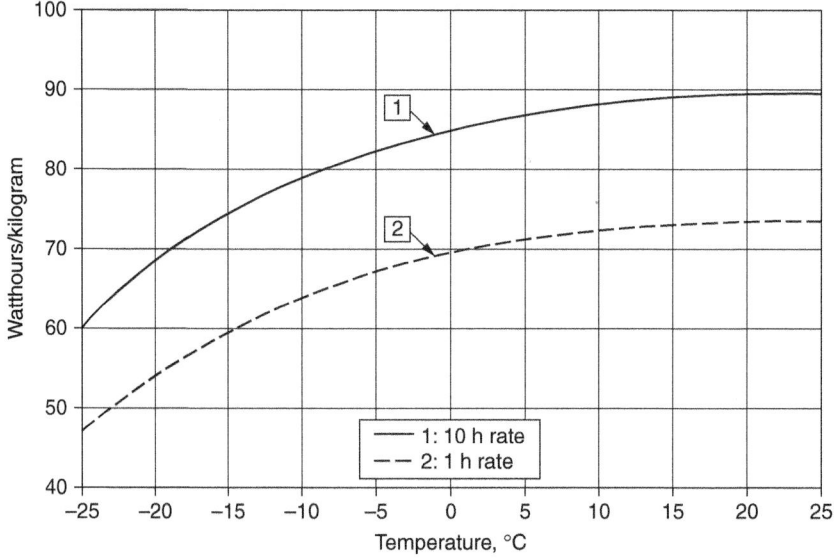

FIGURE 25.11 Typical effect of temperature on specific energy of silver-cadmium cells.

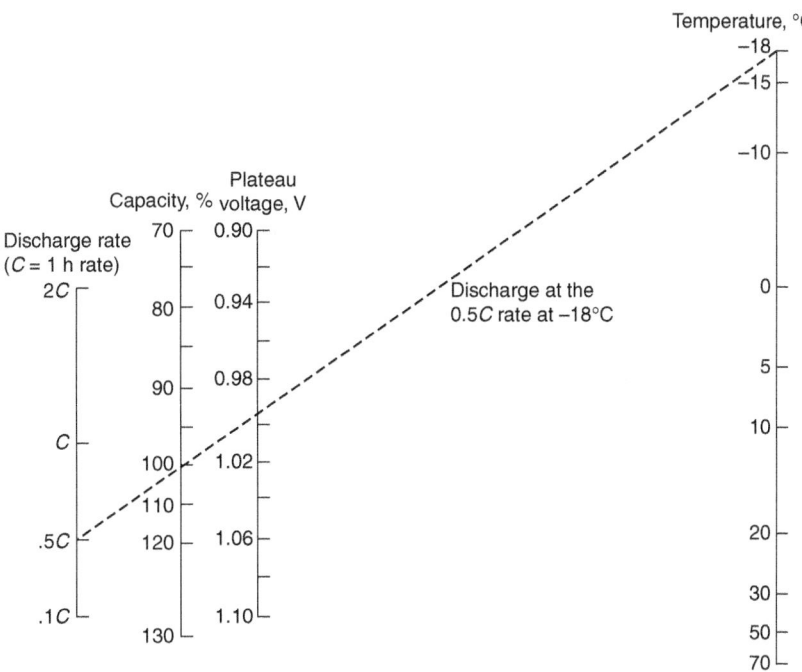

FIGURE 25.12 Performance characteristics of silver-cadmium batteries under various conditions. (To find the capacity and the plateau voltage of a silver-cadmium battery, draw a straight line between the discharge rate and the ambient temperature at which the cell is stabilized.)

The performance characteristics of the cadmium/silver oxide battery are summarized in Figs. 25.12 and 25.13a and b, which can be used to determine the capacity, service life, and voltage levels under a variety of discharge conditions.

25.4.4 Impedance

The impedance of the silver oxide cells is normally low but can vary considerably with many factors, including the separator system, current density, state-of-charge, stand time, cell age, temperature, and, importantly, cell size. In a study of the effect of storage time on the impedance of silver-zinc cells, initial values of 5 to 11 mΩ for partially charged cells were reported,[18] with the values rising to as much as 3 Ω following 8 months' storage at 21°C and 9 to 15 Ω following 8 months at 38°C. Cells stored in the fully discharged condition retained their low impedance (ranging from 2 to 10 mΩ) throughout the entire test period. The high-impedance cells returned to normal low values, however, within several seconds of the start of discharge.

The AC impedance of the silver oxide batteries is highly dependent on the frequency of the load, with the impedance rising sharply above 5 kHz. The impedance of a 6-cell, 350 Ah silver-zinc battery, discharged to approximately 50% DOD, is shown in Fig. 25.14, and the corresponding phase angle is shown in Fig. 25.15.[19]

25.4.5 Charge Retention

The charge retention of the activated and charged silver oxide cells is better than that of most secondary batteries, with retention of more than 85% of charge after 3 months' storage at 20°C.

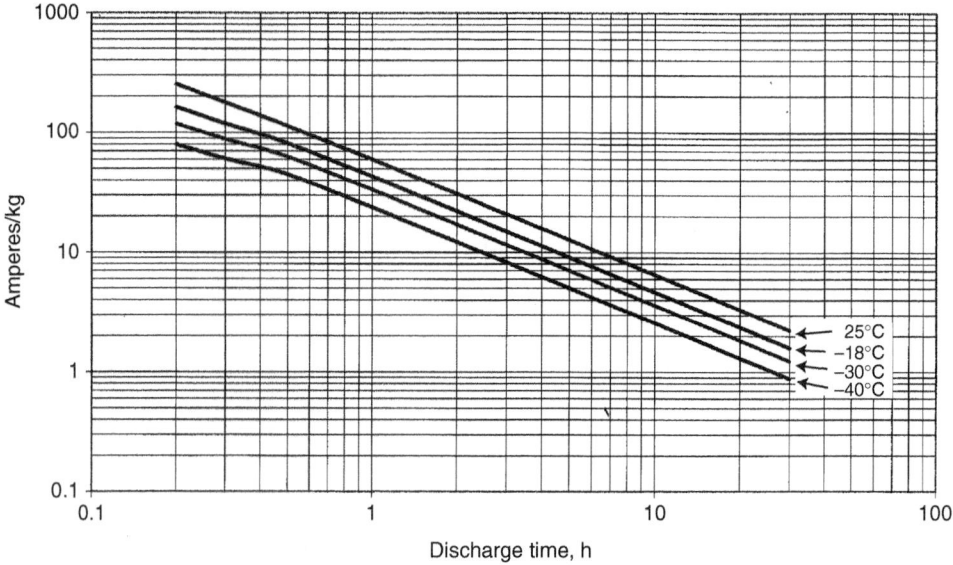

FIGURE 25.13a Service life of silver-cadmium batteries at various discharge rates and temperatures. (Amperes/kg vs. discharge time in hours.)

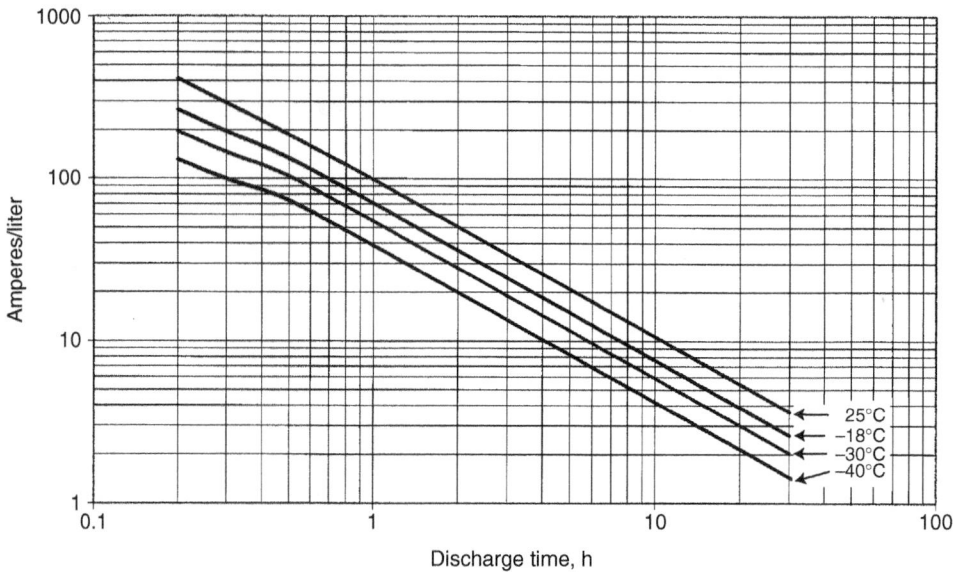

FIGURE 25.13b Service life of silver-cadmium batteries at various discharge rates and temperatures. (Amperes/liter vs. discharge time in hours.)

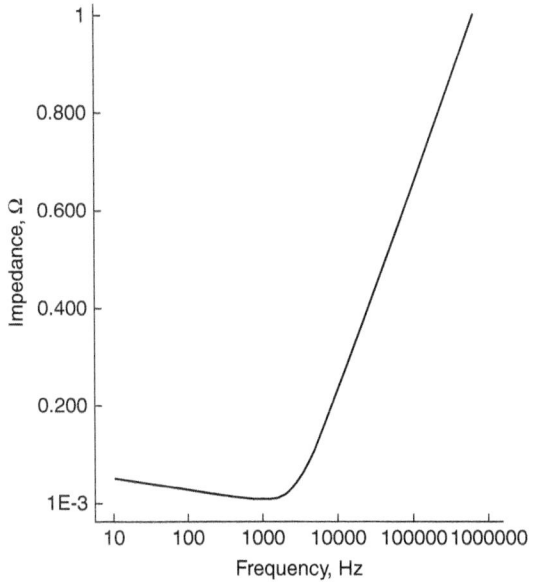

FIGURE 25.14 Impedance magnitude vs. frequency.

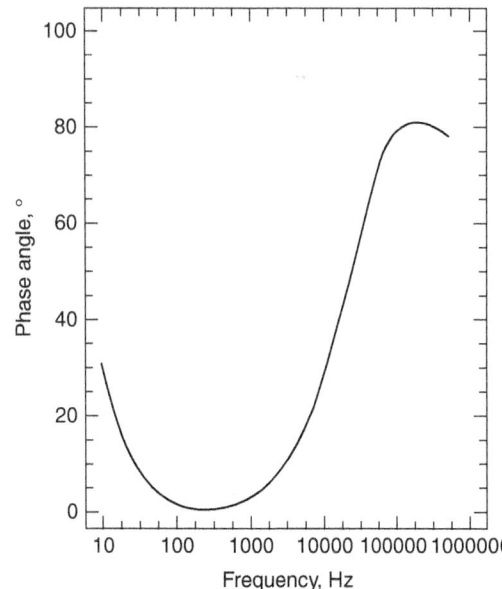

FIGURE 25.15 Impedance phase angle vs. frequency.

As with other chemical reactions, the rate of loss of charge is dependent on the storage temperature (see Fig. 25.16). Storage at −20 to 0°C is recommended for maximizing charge retention. In the dry and charged condition, properly sealed and stored cells will retain their charge for over 10 years. Here again, low temperature storage is highly recommended.

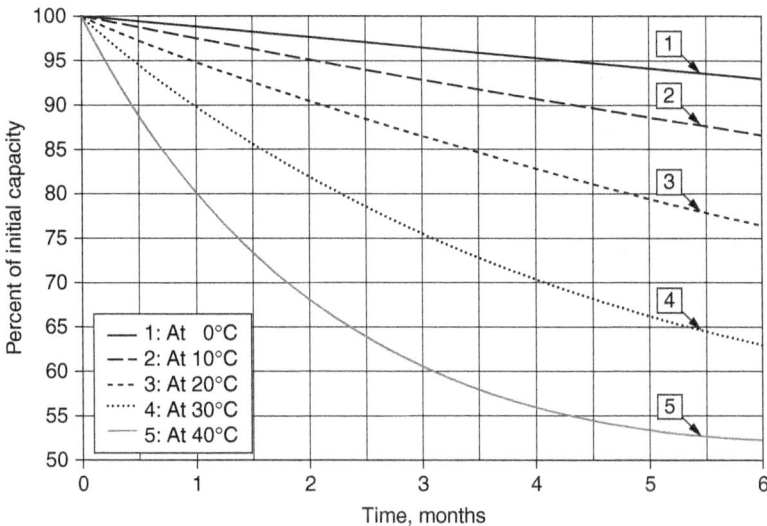

FIGURE 25.16 Capacity retention of silver-zinc and silver-cadmium cells at various temperatures.

25.4.6 Cycle Life and Wet Life

The separator system and the solubility of the active materials play critical roles in determining the wet and cycle lives of the silver-based cells. The separator must have a low electrolytic resistance for discharges at high rates, yet it must have high resistance to chemical oxidation by the silver species as well as low permeability to colloidal silver, zinc, cadmium, or iron.

Since cadmium and iron are relatively insoluble in concentrated alkaline electrolytes, the life of the silver-cadmium and silver-iron cells is therefore limited by the rate of silver migration through the various layers of the separator system. Failure (internal short circuit) occurs when a metallic bridge is established through the separator between positive and negative electrodes. Multiple layers of separator are used to extend the life capability, however, at the expense of higher internal resistance.

The life of the silver-zinc battery is further hindered by the high solubility of zinc in alkaline electrolytes. The problem manifests itself in two failure mechanisms: shape change and growth of dendrites. Shape change is the migration of zinc from the electrode tops and edges where it becomes depleted, to the center and bottom where it densifies, resulting in capacity loss. Dendrites, sharp, needlelike structures of the metal, are formed during overcharge. They may perforate the separators and cause internal short circuits. The capacity degradation in silver-zinc batteries due to shape change is illustrated in Fig. 25.17. The decline in capacity of silver-cadmium batteries is at a much slower rate, as depicted in Fig. 25.18.

Aside from normal capacity loss due to extended wet or cycle life, dry-charged zinc-silver oxide cells may exhibit a one-time deviation in capacity (typically less than 20% of initial capacity) during the second cycle discharge, called "second cycle syndrome." This deviation, which manifests itself as the inability to accept a full charge during the second cycle and has never been fully explained, is well known to the user community. It may be addressed in two ways:

1. If the reduced capacity is acceptable for the application, do nothing; the capacity returns to normal during the next cycle.
2. If not acceptable, the capacity can be increased by a partial discharge followed by vacuuming and recharging, preferably at a lower rate. Note that overcharging will not improve the capacity and may harm the cell.

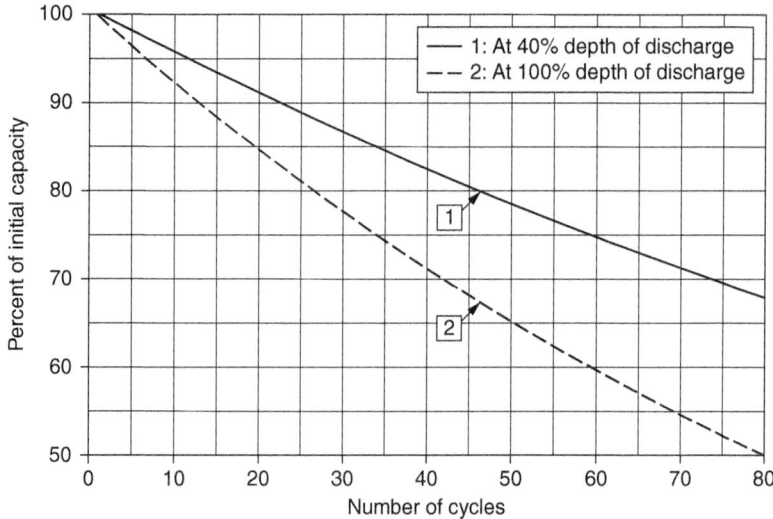

FIGURE 25.17 Typical capacity degradation of silver-zinc cells discharged at low rate.

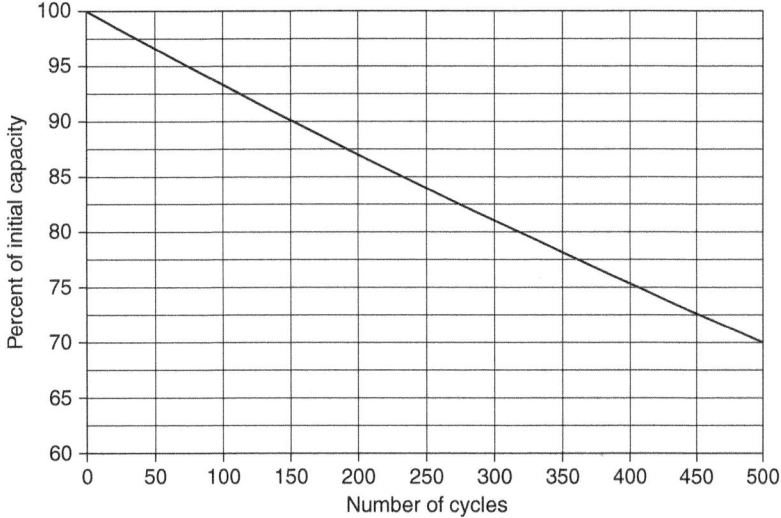

FIGURE 25.18 Typical capacity degradation of sealed silver-cadmium cells discharged at low rate to 100% depth of discharge.

The nominal life ratings for the silver-zinc, silver-cadmium, and silver-iron batteries are given in Table 25.4. The life of the silver oxide cells will also vary greatly with operating and storage conditions. High rates of discharge to 100% depth of discharge and high-temperature exposure (for more than 30 days) will significantly reduce the wet and cycle lives of the batteries. Cold-temperature storage (at less than −10°C), when not in use, on the other hand, will greatly increase the life of the cells. The cycle and wet lives will also increase with a decreasing depth of discharge.

TABLE 25.4 Nominal Life Characteristics of Secondary Silver Batteries*

	High rate (HR) Ag-Zn	Low rate (LR) Ag-Zn	Ag-Cd	Ag-Fe
Wet shelf life	6–18 months	1–2.5 years	2–3 years	2–4 years
Cycle life†	10–50 cycles	50–150 cycles	150–1000 cycles	100–300 cycles

*These characteristics are nominal and vary with operating conditions and design of individual models.
†Cycle life characteristics are for deep (80–100% of full capacity) discharge cycles. Cycle life improves considerably with partial discharges.

In a study to evaluate the capabilities of silver-cadmium batteries for satellite applications, an extensive test program was run on 3-cell, 3 Ah silver-cadmium batteries at various depths of discharge.[20] These results are summarized in Table 25.5, showing the increase in cycle life with decreasing depth of discharge.[20] Another study on 250 ampere-hour silver-zinc cells, cycled at less than 1% depth of discharge, with 14 full-capacity cycles, resulted in a cycle life of 7280 cycles over a 38-month period.[21]

TABLE 25.5 Cycle Life vs. Depth of Discharge for 3 V, 3 Ah Sealed Cadmium/Silver Oxide Batteries

Depth of discharge, %	Cycle life at first cell failure
65	1800
50	3979
50	>5400 (375 days)
35	>5400 (375 days)
25	>5400 (375 days)

Source: Ref. 20.

25.5 CHARGING CHARACTERISTICS

25.5.1 Efficiency

The *ampere-hour* efficiency (ampere-hour output/ampere-hour input) of the silver-zinc and silver-cadmium systems under normal operating conditions is high—greater than 99%—because practically no side reactions occur when charging at normal rates. The *watthour* efficiency (watthour output/watthour input) is about 70% under normal conditions because of the difference between charge and discharge voltages.

25.5.2 Zinc/Silver Oxide Batteries

The manufacturers of these batteries recommend constant-current charging at the 10 to 20 h rate for most applications. However, constant-potential and pulsed charging techniques have also been applied.

A typical charge curve at constant current is shown in Fig. 25.19. The two plateaus reflect the two levels of oxidation of the silver electrode: the first from silver to monovalent silver oxide (Ag_2O),

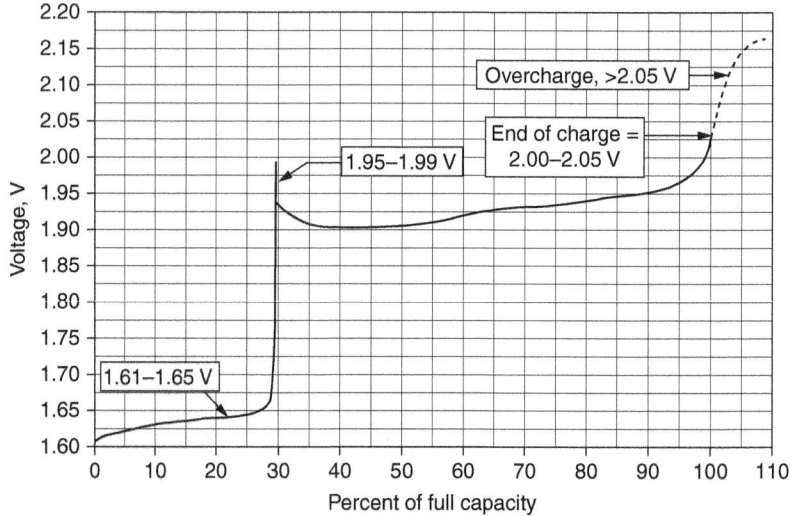

FIGURE 25.19 Typical charge curve of silver-zinc cell at the 10 to 20 h rate.

which occurs at a potential of approximately 1.6 V; the second from the monovalent to the divalent silver oxide (AgO), which occurs at approximately 1.9 V. It should be noted that during this transition from the monovalent to the divalent state of charge, a momentary spike in the charge voltage, of up to 2.0 V, may occur prior to stabilizing at the 1.90 to 1.95 V plateau. To ensure a full charge, the charging system must be designed to ignore this temporary rise in voltage.

Charging is normally terminated when the voltage during charge rises to 2.0 V. Above 2.1 V the cell begins to generate oxygen at the silver electrode and/or hydrogen at the zinc electrode, decomposing water from the electrolyte. Overcharge is also detrimental to cell life in that it promotes the growth of zinc dendrites and subsequent short circuits.

The importance of proper charging to the life of these batteries cannot be overemphasized.

25.5.3 Cadmium/Silver Oxide Batteries

Except for lower voltages on each of the plateaus (1.2 V on the lower level, 1.5 V on the upper level), the charging characteristics of the silver-cadmium battery are similar to those of silver-zinc. A typical charge curve is shown in Fig. 25.20.

As with silver-zinc, silver-cadmium batteries are usually charged at constant current at the 10 to 20 h rates. The recommended cutoff voltage during charge is normally 1.6 V per cell.

The silver-cadmium battery, however, is less sensitive to overcharge than the silver-zinc battery. Other charge methods can be and have been adapted to specific applications.

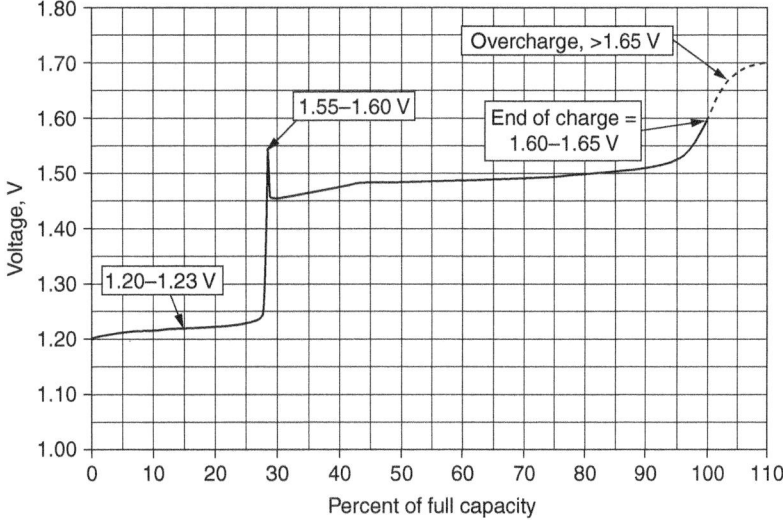

FIGURE 25.20 Typical charge curve of silver-cadmium cell at the 10 to 20 h rate, at room temperature.

25.6 CELL TYPES AND SIZES

Tables 25.6 to 25.8 present examples of the products from two major silver battery manufacturers—Yardney Technical Products, Inc. and Eagle-Picher Technologies, Inc. They are intended as a guide only, since the design parameters can be varied to meet specific customer requirements. Many applications for the high-energy silver batteries, in fact, require special designs, which often dictate new cell case and cover designs and tooling. These then become the "available" models for future applications.

TABLE 25.6 Nominal Characteristics of Typical Vented Zinc/Silver Oxide Batteries

Cell type	Capacity, Ah	Cell dimensions, mm (including terminals)			Cell weight, g (including electrolyte)	Maximum continuous rate, A
		Length	Width	Height		
High-rate types						
HR-02	0.2	5.6	16.0	49.3	6.5	2.0
HR-05	1.3	13.7	27.4	39.6	21.3	4.0
HR-1	2.0	13.7	27.4	51.3	31.2	6.0
HR-2	4.5	15.0	43.7	63.5	68.0	20.0
PMV-2(4.5)*	5.0	15.2	43.7	64.3	72.6	100
HR-5	8.5	20.3	52.8	73.7	127.6	60.0
PM-15*	21.8	20.3	58.9	125.5	295	200
HR-21	35.9	20.6	58.4	191.5	439	160
PM-30*	44.0	25.4	77.7	166.4	607	400
HR-40	46.0	25.1	82.6	180.3	646	200
HR-105	121	35.2	96.9	137.4	950	120
HR-140	190	72.4	82.5	183.4	1721	600
PML-170*	221	35.3	97.0	184.4	1520	120
HR-190	238	39.4	152.6	165.4	2217	400
PML-2500*	2750	107.2	107.2	479.0	18,150	600
MR-200	250	53.5	101.6	206	2156	200
Low-rate types						
LR-1	2.1	13.7	27.4	51.3	30.1	4.5
LR-4	7.5	15.0	43.7	85.3	99.2	16.0
LR-8	10.0	16.3	29.9	120.1	116.3	16.0
LR-12	16.0	19.1	47.2	100.1	163.0	20.0
LR-40	64.0	25.1	82.6	180.3	638	64.0
LR-70	100	36.1	92.5	155.4	1055	160
LR-90	155	54.9	82.9	179.3	1588	150
LR-190	220	39.1	151.6	162.6	2048	200
LR-350	560	107.4	107.4	222.3	5615	350
LR-360	570	69.9	147.3	162.6	4391	300
LR-660	840	79.2	161.3	177.8	6183	180
Special deep submersible types						
LR625	692	161	80	187	5,470	125
LR-700(DS)†	1060	107	107	486	11,200	900
LR-750(DS)†	1075	142	97	513	12,500	750
LR-850	1200	119	114	479	13,200	800
LR-875	1050	160	79.6	183	7,000	125
LR-1000(DS)	1072	137	137	513	18,500	1250

*Primary, manually activated.
†Pressure compensated
Source: Yardney Technical Products, Inc.

TABLE 25.7 Nominal Characteristics of Typical Vented Zinc/Silver Oxide Cells

	High rate						Low rate							
	Rated capacity	Nominal capacity Ah rates			Weight, g		Rated capacity	Nominal capacity Ah rates			Weight, g	Physical dimensions, mm		
Cell type	Ah	15 min	30 min	60 min		Cell type	Ah	4 h	10 h	20 h		L	W	H
SZHR 0.8	0.8	0.7	0.7	0.8	22.7	SZLR 0.8	0.8	0.8	0.8	0.8	22.7	10.9	26.9	51.6
1.5	1.5	1.4	1.5	1.5	39.7	1.5	1.5	1.5	1.5	1.5	42.6	12.4	30.7	57.2
2.8	2.8	2.6	2.7	2.8	53.9	3.0	3.0	3.0	3.0	3.0	56.7	14.2	35.1	63.2
5.0	5.0	4.8	5.0	5.0	76.6	5.3	5.3	5.3	5.3	5.3	85.1	16.3	40.1	70.9
6.5	6.5	6.2	6.4	6.5	119.1	7.5	7.5	7.4	7.5	7.5	124.8	14.9	43.7	90.2
10.5	10.5	10.0	10.3	10.5	170.2	11.5	11.5	11.5	11.4	11.5	184.4	20.1	49.5	84.8
15	15	12	14	15	210.0	16.5	16.5	15.5	16.5	16.5	215.6	21.3	41.1	120.7
26	26	20	24	26	312.1	30	30	28.0	30.0	30.0	326.3	25.4	62.7	103.9
48	48	*	40	48	595.9	51	51	48	51	51	624.2	18.5	89.9	167.9
65	65	*	50	65	737.8	70	70	65	70	70	780.3	26.9	83.1	155.4
100	100	*	80	100	1107	115	115	100	110	115	1220	37.3	92.7	150.9
140	140	*	*	140	1944	160	160	*	150	160	2049	74.17	75.7	161.8

*Not applicable at this rate.
Source: Eagle-Picher Technologies.

25.24 SECONDARY BATTERIES

TABLE 25.8 Nominal Characteristics of Typical Cadmium-Silver Oxide Batteries

Cell type	Capacity, Ah	Cell dimensions, mm (including terminals)			Cell weight, g (including electrolyte)	Maximum continuous rate, A
		Length	Width	Height		
YS-1	1.5	13.7	27.4	51.3	31.2	5.0
YS-3	4.2	15.2	43.7	72.6	82.2	12.0
YS-5	7.8	19.1	51.1	73.9	130.5	25.0
YS-5 (sealed)	6.8	20.1	52.8	73.9	141.8	15.0
YS-10	14.5	18.8	58.9	122.2	246.7	30.0
YS-16 (sealed)	21.0	20.6	58.4	146.1	348.8	50.0
YS-20	32	43.9	52.1	108.7	450.9	40.0
YS-40	54	25.1	82.6	179.8	745.9	100
YS-100	132	70.6	87.4	122.2	1503	150
YS-150	240	45.2	106.4	272.0	2978	150
YS-300	420	45.2	106.4	444.5	5190	150

Source: Yardney Technical Product, Inc.

25.7 SPECIAL FEATURES AND HANDLING

Silver cells are capable of providing extremely high currents if short-circuited. Accordingly, provisions must be made to insulate all tools used with the batteries and to protect the cells against grounding in their application.

The electrolyte is a caustic solution of potassium hydroxide. Precautions such as the use of gloves and safety goggles are required when handling the electrolyte. In most applications, addition of electrolyte or water is not required. However, the manufacturer's recommendations for periodic maintenance and electrolyte checks should be followed closely.

Proper ventilation of these as well as other vented batteries, although not as much a problem here as with other battery systems, is required to avoid the accumulation of hazardous hydrogen, especially during charge. For larger installations, forced air or fans may also be required to prevent undesirable temperature buildups. When close voltage regulation is required at cold temperatures, thermostatically controlled heaters are often used with the batteries.

Because of the sheer size and power of the batteries used, and because of critical personnel safety requirements (for example, the U.S. Navy's NR-1 submersible had a 240 V, 850 Ah silver-zinc backup battery installed under the deck that vented into the operator's quarters), a whole new body of engineering technology has been developed for the application of these batteries for underwater power. Some of the special features developed include removal of all mercury, provision for fire walls inside cells, and provision for pressure compensation for those batteries that are external to the vessels' pressure hulls.[8] Electronic systems have also been developed to permit continuous scanning of individual cell voltages to maximize battery life. Special applications such as these can be successfully developed only if the battery designers, manufacturers, and users work closely together during the design of the product.

25.8 APPLICATIONS

Because of the fluctuating cost of silver, the major applications for these batteries have historically been, and continue to be, governmental. However, because of their high power and energy density, these batteries have found many varied uses where space and weight limitations are critical. In addition, the cost of silver for most applications can be offset by reclaiming the metal after the battery completes its useful life.

FIGURE 25.21 ADMATT medium performance propulsion battery. (*Courtesy of Yardney Technical Products, Inc.*)

One of the original applications for the silver-zinc battery was for use in torpedoes.[22] Much of the original development work was sponsored by the U.S. Navy. Later, development expanded to other underwater applications, including mines, buoys, special test vehicles, swimmer aids, and manual submersibles such as the deep submergence rescue vehicles (DSRV) and Advanced Seal Delivery System (ASDS), such as exploratory underwater vehicles as UUV and NR-1, and various antisubmarine warfare (ASW) applications. The MK40 ADMATT target torpedo battery is illustrated in Fig. 25.21. The ADMATT Propulsion Battery has two configurations, a 60 × HR300DC/58 × HR300DC medium performance battery and a 120 × HR215DC/116 × HR215DC high performance battery. The discharge rate for the nominal 147 V medium performance battery is 325 A while the nominal 290 V high performance battery is 650 A at room temperature conditions; operating time is 35 min for the medium performance battery and 6 min for the high performance battery. The deep submergence and rescue vehicle (DSRV) battery illustrated in Fig. 25.22 is a pressure compensated design using mineral oil to fill the cell and battery void during ocean descent. It is a 115 V, 700 Ah rated battery and the pressure compensation allows it to be mounted outside of the pressurized vessel.

Silver-zinc batteries have found wide use in numerous space applications, including launch-vehicle guidance and control, telemetry, and flight termination power; Apollo lunar spacecraft, lunar and Mars rovers, and lunar drill power; space shuttle payload launch; and getaway special batteries, as well as power for the life-support equipment used by the U.S. astronauts during extra-vehicular activities (EVA). Figure 25.23 shows a typical aerospace battery (Model 20 × HR2DC) consisting of twenty 2 Ah silver-zinc cells housed in a cast aluminum case, equipped with a pressure relief valve, a pressurizing valve, and a battery connector.

Silver-cadmium batteries have been used in a number of space applications requiring nonmagnetic properties. One such battery provided the main power for the *Giotto Halley Comet* intercept spacecraft. Another group of batteries provides backup power for the solar cells (for the periods when the earth eclipses the sun) on board the Cluster II scientific spacecraft launched by the European Space Agency in the year 2000. Although only required to operate for 2 years, those batteries were still operating as of May 2009.

Ground applications include communications equipment, portable television cameras, portable lights, camera drives, medical equipment, vehicle motive power, and similar uses requiring high-energy-density rechargeable batteries.

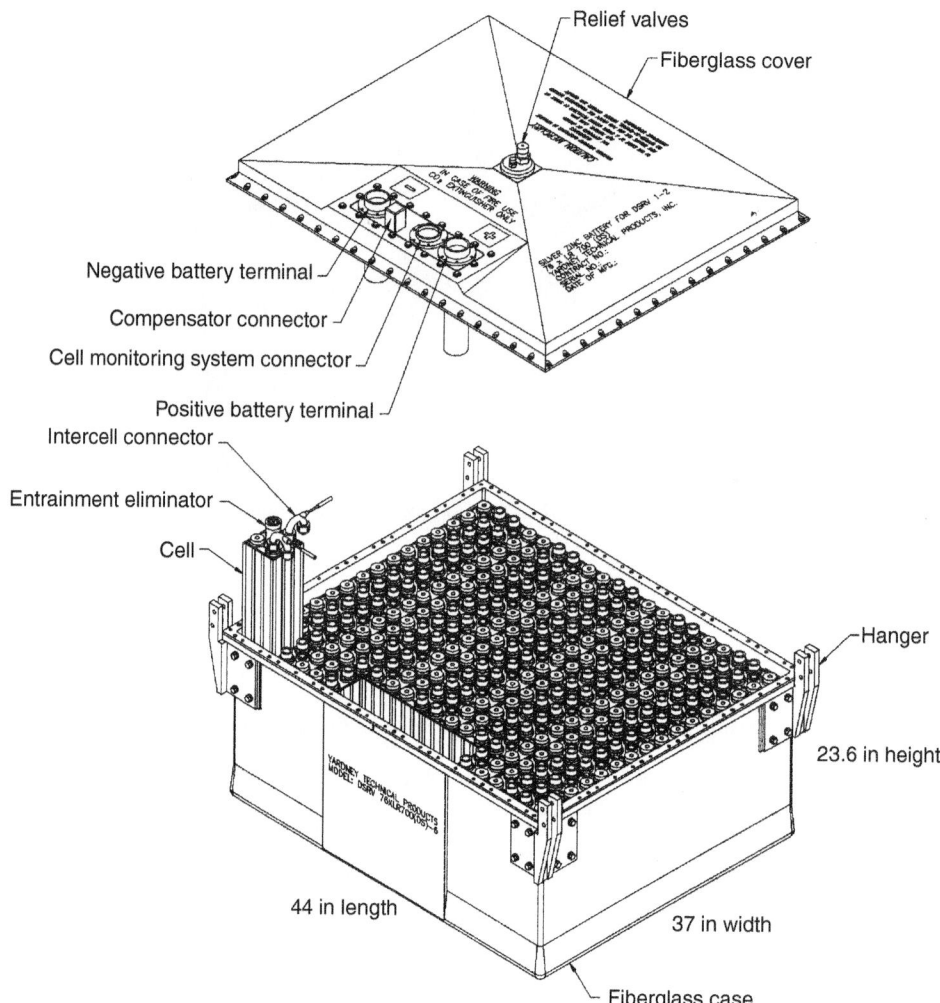

FIGURE 25.22 DSRV pressure compensated battery. Model 76 × LR700(DS) (*Courtesy of Yardney Technical Products, Inc.*)

Silver-zinc batteries are making a resurgence in the portable electronics market as a purportedly safer alternative for lithium-ion batteries used in notebook computers, cell phones, and consumer electronics. The user should keep in mind that no one type of battery is suitable for all applications. Optimum performance of a battery in an application can usually be achieved only by meeting the critical needs of the application and subordinating the others. The best approach for battery selection is to work with the battery manufacturers during the early stages of equipment design, rather than asking the battery designers, as is often done, to "design a battery that meets all my requirements and fits into this remaining cavity in my equipment."

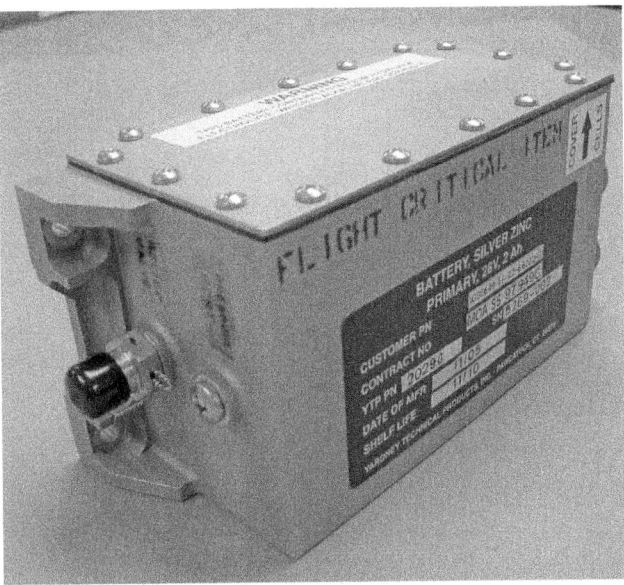

FIGURE 25.23 Silver-zinc aerospace battery, 28 V, 2 Ah. (*Courtesy of Yardney Technical Products, Inc.*)

25.9 RECENT DEVELOPMENTS

Since the days of André, many improvements to the silver-zinc system have been attempted to enhance its performance and manufacturability. Without those improvements, listed in Table 25.9, the cells would have very limited high-rate capability and their cost would be prohibitive.

However, not all of those improvements were adopted, primarily due to an unfavorable cost-benefit analysis. Others found only limited applications, generally those requiring very long cycle and/or wet lives, or unusually high levels of safety.

Some of the most recent work, not listed in Table 25.9, aimed at the negative electrodes and the separators has been quite successful and is briefly described below.

- Use of bismuth oxide as an additive to the zinc electrode slows down "shape change," thus improving capacity retention and cycle life of the cells.[25] This relatively inexpensive additive has been tried in combination with lead and cadmium oxides, in concentrations from 1 to 10%, and the best results were obtained in the range of 2.5 to 5%, with improvements in cycle life of about 60%.

- Presently, the preferred method of manufacturing zinc electrodes is the paste or slurry method, followed by electroforming, as described in sec. 25.3.2. This process is complex and requires large expenditures of labor and energy. It also requires constant surveillance to prevent problems such as the imperfect mixing of the additives and/or binders and the reoxidation of the zinc.

    A method has been developed to overcome some of the above problems, consisting of: (1) blending the dry powders (zinc + additives) or preferably purchasing them prealloyed or premixed, (2) adding a bonding agent, (3) drying (the amount of water to be removed is but a small fraction of that used during pasting and electroforming), and (4) milling to the appropriate particle size. The zinc-powder was tested in 12 Ah cells, with excellent results, including a cycle life improvement of 200%. These were scaled up to 190 Ah cells used in the Mk 30 Mod1 torpedo target. As shown in Fig. 25.24, these cells outperformed the standard cells made with electroformed plates by 100%.

TABLE 25.9 Summary of Recent Developments in Silver Oxide Secondary Batteries and Components

Development area	Advantages	Disadvantages
Zinc electrode		
Increased zinc-to-silver-weight ratio	Delays onset of capacity losses	Gains not proportional to extra material used; reduces energy density
Oversized negatives	Reduce current density at plate edges where shape changes start	Reduce energy density
"Contoured" negatives	Reinforce areas of maximum erosion	Reduce energy density; high cost
PTFE binder	Reduces shape changes and dendrite growth; improves low-temperature performance	High cost; difficult to disperse evenly; may interfere with normal electrode reactions
Potassium titanate fibers	Reduce shape changes and dendritic growth; reduce probability of "hot" short circuits	Somewhat higher cost; hazardous during fabrication; restricted availability
Lead, lead/cadmium, bismuth additives as mercury substitutes	Reduce hazard to health and equipment in cases of "hot" short circuits; improve capacity maintenance	Smaller performance database
Main separator		
Inorganic separators	Resist temperature above 150°C; resist attack by silver oxides and electrolyte	High electrolytic resistance; bulky, difficult to handle; high cost
Cellulosics with metallic groups in molecule	Improved resistance to attack by silver oxides and electrolyte; extend cycle life	Somewhat higher cost
Microporous polypropylene	Proven resistance to attack by the electrolyte	Higher cost
Polyolefin film/inorganic fillers	Major improvement in cycle life	High cost; needs additional development work
Interseparators		
Positive: asbestos	Prevents or reduces magnitude of short circuits; acts as silver stopper	Bulky; reacts with silver oxides; may contaminate cell with iron; hazardous during fabrication
Zirconium oxide	Reduce hazard to health during cell manufacture	Higher cost
Negative: potassium titanate mat	Reduces zinc shape changes; reduces incidence and magnitude of short circuits	Bulky; reduces energy density; somewhat costly; hazardous during fabrication
Silver electrode		
Close control of particle-size distribution	Improves control of cell voltage and capacity	
Hardware		
New plastics for cases and covers (e.g., modified PPO, polysulfone)	High-temperature operation; better mechanical properties	Higher cost
New adhesives	More efficient at high temperatures; comply with EPA requirements	

- The use of novel polymers, nano-technology, and so called matrix materials[27] that act on gelling agents has been applied to the zinc electrode. These materials have also demonstrated a stabilized electrode by reducing shape changes and gas evolution.
- Over the years, many new separators have been evaluated with the aim of replacing cellophane as the main separator for silver cells to improve wet life and cycle life.
  One promising material, developed by Advanced Membrane Systems (AMS), designated as a Flexible Alkaline Separator (FAS), consists of a microporous polyolefin with a titanium dioxide filler. It has excellent resistance to KOH solutions, even at elevated temperatures, dimensional stability, and low electrolytic resistance. Limited testing on 12 Ah cells yielded encouraging results.

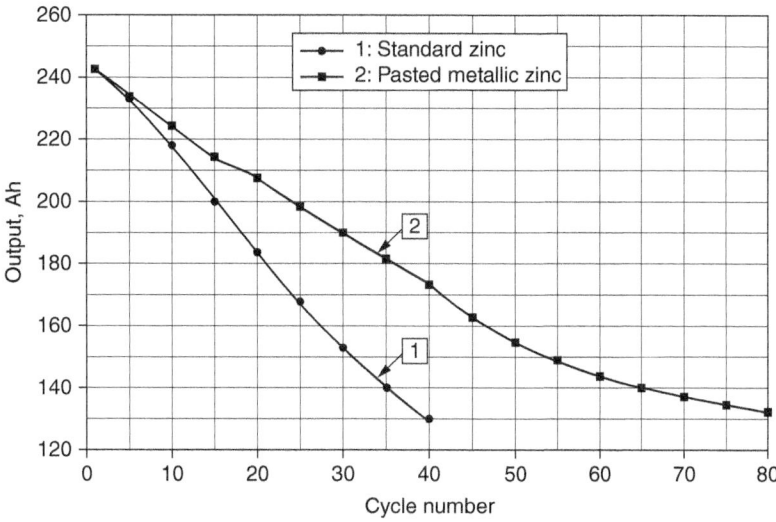

FIGURE 25.24 LR190DC cells with electroformed and pasted metallic zinc electrodes.

REFERENCES

1. A. Volta, *Phil. Trans. R. Soc. London* **90**:403 (1800).
2. S. U. Falk and A. J. Salkind, *Alkaline Storage Batteries*, Wiley, New York, 1967.
3. H. André, *Bull. Soc. Fr. Electrochem.* (6th ser.) **1**:132 (1941).
4. H. André, U.S. Patent 2,317,711 (1943).
5. J. T. Brown, "Iron-Silver Battery—A New High Energy Density Power Source," Westinghouse Corp., Rep. 77-5E6-SILEL-RI, 1977.
6. "Design & Cost Study, Zinc/Nickel Oxide Battery for Electric Vehicle Propulsion, Yardney Electric Corp., Final Rep., Contract 31-109-38-3543, Oct. 1976.
7. R. Serenyi, "Recent Developments in Silver-Zinc Batteries," Yardney Electric Corp., Internal Rep. 2449–79, Oct. 1979.
8. G. W. Work and P. A. Karpinski, "Energy Systems for Underwater Use," *Marine Tech. Expo. Int. Conf*, New Orleans, LA, Oct. 1979.
9. A. Himy, "Substitutes for Mercury in Alkaline Zinc Batteries," *Proc. 28th Annual Power Sources Symp.*, 1978, pp. 167–169.
10. R. Serenyi and P. Karpinski, "Final Report on Silver-Zinc Battery Development," Yardney Electric Corp., Contract N00140-76-C-6726, Nov. 1978.
11. "Medium Rate Rechargeable Silver-Zinc 850 Ah Cell," Eagle-Picher Industries, Final Rep., USN Conract N00140-76-C-6729, Mar. 1978.
12. R. Einerhand, W. Visscher, J. de Goeij, and E. Barendrecht, "Zinc Electrode Shape Change," *J. Electrochem. Soc.* **138**:7–17 (Jan. 1991).
13. K. Choi, D. Bennion, and J. Newman, "Engineering Analysis of Shape Change in Zinc Secondary Electrodes," *J. Electrochem. Soc.* **123**:1616–1627 (Nov. 1976).
14. K. Bass, P. J. Mitchell, G. D. Wilcox, and J. Smith, "Methods for the Reduction of Shape Change and Dendritic Growth in Zinc-Based Secondary Cells," *J. Power Sources* **35**:333–351 (1991).
15. A. Charkey, "Long Life Zinc-Silver Oxide Cells," *Proc. 26th Ann. Power Sources Symp.*, 1976, pp. 87–89.
16. V. D'Agostino, J. Lee, and G. Orban, "Grafted Membranes," in A. Fleischer and J. Lander (eds.), *Zinc-Silver Oxide Batteries*, Wiley, New York, 1971, Chap. 19, pp. 271–281.

17. C. P. Donnel, "Evaluation of Inorganic/Organic Separators," Yardney Electric Corp., Contract NAS3-18530, Oct. 1976.
18. H. A. Frank, W. L. Long, and A. A. Uchiyama, "Impedance of Silver Oxide-Zinc Cells," J. *Electrochem. Soc.* **123**(1):1–9 (Jan. 1976).
19. J. C. Brewer, R. Doreswamy, and L. G. Jackson, "Life Testing of Secondary Silver-Zinc Cells for the Orbital Maneuvering Vehicle," *Proc. 25th IECEC*, Reno, NV, Aug. 1990.
20. "Evaluation of Silver-Cadmium Batteries for Satellite Applications," Boeing Co., Test D2-90023, Feb. 1962.
21. A. P. Karpinski and J. A. Patten, "Performance Characteristics of Silver-Zinc Cells for Orbiting Spacecraft," *Proc. 25th IECEC*, Reno, NV, Aug. 1990.
22. A. Fleischer and J. Lander (eds.), *Zinc-Silver Oxide Batteries*. Wiley, New York, 1971.
23. R. Serenyi, U.S. Patent 5,773,176.
24. A. P. Karpinski, B. Makovetski, S. J. Russell, J. R. Serenyi, and D. C. Williams, "Silver-Zinc: Status of Technology and Applications," *J. of Power Sources* **80**:53–60 (1999).
25. *Proceedings of the 5th Workshop for Battery Exploratory Development*, Burlington, VT, July 1997, pp. 153–157.
26. *Proceedings of the 38th Power Sources Conference*, Cherry Hill, NJ, June 1998, pp. 175–178.
27. M. Cheiky et al., U.S. Patent 6,582,851B2.

CHAPTER 26
LITHIUM-ION BATTERIES

Jeff Dahn and Grant M. Ehrlich

26.1 GENERAL CHARACTERISTICS

Lithium-ion (Li-ion) batteries employ lithium storage compounds as the positive and negative electrode materials. As a battery is cycled, lithium ions (Li$^+$) exchange between the positive and negative electrodes. Li-ion batteries have been referred to as rocking chair batteries because the lithium ions "rock" back and forth between the positive and negative electrodes as the cell is charged and discharged. The positive electrode material is typically a metal oxide with a layered structure, such as lithium cobalt oxide (LiCoO$_2$), or a material having a tunneled structure, such as lithium manganese oxide (LiMn$_2$O$_4$), on an aluminum current collector. The negative electrode material is typically a graphitic carbon, also a layered material, on a copper current collector. In the charge-discharge process, lithium ions are inserted or extracted from interstitial space between atomic layers of the active materials. The first batteries to be marketed by Sony in 1991, and the majority of those currently available utilize LiCoO$_2$ (LCO) as the positive electrode material. LCO provides good electrical performance, is easily prepared, has good safety properties, and is relatively insensitive to process variation and moisture. More recently, other positive electrode materials have been introduced, such as LiFePO$_4$ (LFP), LiMn$_2$O$_4$ (spinel), Li(NiMnCo)O$_2$ (NMC), and Li(NiCoAl)O$_2$ (NCA). These materials each offer different advantages, such as high-rate capability, low cost, high thermal stability, long cycle life, and/or high capacity, which are useful for different applications.

The batteries that were first commercialized employed cells with coke negative electrode materials. As improved graphites became available, the industry shifted to graphitic carbon negative electrode materials because graphitic carbon offers better specific capacity, cycle life, and rate capability than coke materials. More recently, Li$_{4/3}$Ti$_{5/4}$O$_4$ spinel (LTO) has been introduced as an anode material. Although LTO has lower specific capacity and capacity density than graphite, it offers longer cycle life and good thermal stability characteristics. In 2005, commercial cells with negative electrodes containing nanostructured Sn-Co-C alloys were introduced by Sony. Presently, Si-based negative electrodes are poised to enter the marketplace. Sn and Si-based negative electrodes hold the promise of Li-ion cells with higher specific energy and energy density.

The Li-ion battery market has grown in two decades from an R&D interest to sales of over 4 billion units in 2009. By 2015, the market is expected to grow to over 8 billion units with much of the increase in the automotive sector. This technology has rapidly become the standard power source in a broad array of markets, and battery performance continues to improve as Li-ion batteries are applied to an increasingly diverse range of products. To meet market demand, an array of designs and form factors have been developed, including spiral-wound cylindrical, wound prismatic, flat plate or "stacked" prismatic, and "pouch" cell designs in small (0.1 Ah) to large (160 Ah) sizes. In each form factor, cells using a liquid electrolyte or a polymer or polymer gel electrolyte have been developed. Applications now addressed with Li-ion batteries include consumer electronics, such as cell

26.2 SECONDARY BATTERIES

TABLE 26.1 Advantages and Disadvantages of Li-ion Batteries

Advantages	Disadvantages
Sealed cells; no maintenance required	Moderate initial cost
Long cycle life	Degrades at high temperature
Broad temperature range of operation	Need for protective circuitry
Long shelf life	Capacity loss and potential for thermal runaway when overcharged.
Low self-discharge rate	
Rapid charge capability	Possible venting and possible thermal runaway when crushed
High-rate and high-power discharge capability	
High coulombic and energy efficiency	May become unsafe if rapidly charged at low temperatures (< 0°C)
High specific energy and energy density	
No memory effect	
Many possible chemistries offer design flexibility	
Can be made in aluminized plastic cases as "pouch" or polymer cells	

phones, laptop computers, and digital cameras, power tools, electric bikes and scooters, as well as military devices. The big new application for Li-ion cells between 2010 and 2020 will be electrified vehicles. Even in 2009, Tesla Motors produced an all-electric roadster with a 52 kWhr Li-ion battery. The battery is based on 6801 18650-sized Li-ion cells. One of the most promoted extended-range electric vehicles, or plug-in hybrids, the Chevrolet Volt, is expected to use a 16 kWhr Li-ion battery pack. Electrified vehicles with Li-ion batteries from Nissan, Toyota, Volkswagen, and BYD will be introduced in the next few years. It is an exciting time in the industry!

The major advantages and disadvantages of Li-ion cells, relative to other types of cells, are summarized in Table 26.1. The high specific energy (up to 240 Wh/kg) and energy density (up to 640 Wh/L) of commercial products makes them attractive for weight or volume sensitive applications. Li-ion batteries offer a low self-discharge rate (2 to 8% per month), long cycle life (greater than 1000 cycles), and a broad temperature range of operation (commercially available cells may be charged at 0 to 45°C and discharged at −40 to 65°C), enabling their use in a wide variety of applications. A wide array of sizes and shapes is now available from a variety of manufacturers. Single cells typically operate between 2.5 and 4.3 V, which is approximately three times that of NiCd or NiMH cells, and thus fewer cells are required for a battery pack of a given voltage than if a cell using an aqueous electrolyte were used. Li-ion batteries can also offer high rate capability. Discharge at $30C$ continuous, or $100C$ pulse, has been demonstrated and is available in commercial cells. It is common for manufacturers to offer "energy cells," where specific energy and energy density are maximized, and "power cells," where specific power and power density are maximized while still retaining energy density greater than competitive technologies. The combination of these qualities within a cost effective, hermetic package has enabled diverse application of the technology.

A disadvantage of some Li-ion batteries is that they degrade when discharged below 2 V and may vent when overcharged because they do not inherently have a chemical mechanism to manage overcharge, unlike most aqueous cell chemistries. Therefore Li-ion batteries typically employ management circuitry and mechanical disconnect devices to provide protection from overdischarge, overcharge, or over temperature conditions. In addition, redox shuttles, such as the 3M product 2,5-di-tert-butyl-1,4-dimethoxybenzene, can protect $LiFePO_4$-based Li-ion cells from overcharge.[1,2] Other disadvantages of Li-ion products are that they may permanently lose capacity at elevated temperatures (65°C), albeit at a lower rate than most NiCd or NiMH products, and may become unsafe if rapidly charged at low temperatures (<0°C).

26.1.1 Designation and Markings

The International Electrotechnical Commission (IEC) has developed standards for the designation, marking, electrical testing, and safety testing of Li-ion cells and batteries. The IEC designation and marking system for Li-ion cells utilizes five numbers in the case of cylindrical cells and six numbers in the case of prismatic cells. For cylindrical cells, the first two digits designate the diameter in millimeters and the next three digits designate the length in tenths of millimeters. For example, an 18650 cell is 18 mm in diameter and 65.0 mm in length. For prismatic cells, the first two digits designate the thickness in tenths of millimeters, the next two designate the width in millimeters, and the last two designate the length of the cell in millimeters. For example, a 564656P prismatic cell is 5.6 mm thick × 46 mm wide × 56 mm long.

There are other markings on the cells that suggest the chemistry in the cell. However, based on the publications of various manufacturers, it is clear that these nomenclatures are not yet fully standardized. This is presumably due to the large numbers of cell chemistries, which can include graphite, Sn-Co-C, or LTO negative electrodes and LCO, spinel, LFP, NMC, or NCA positive electrodes. Blends of these electrode materials are also often used. For example, E-One Moli Energy lists an IMR26650 cell, which has intercalation chemistry (I), a manganese-based positive electrode (M) and is round (R). A China BAK Battery lists an 18650-Fe product, which is a cylindrical cell having a $LiFePO_4$-based positive electrode. In addition, Samsung SDI[3] lists a variety of 18650-size cells. For example, the ICR18650-30A is a 3.0 Ah cell having an unspecified positive electrode (thought to be $LiCoO_2$), the ICR18650-26D is a 2.6 Ah cell having an NMC positive electrode, and the ICR18650-20F is a 2.0 Ah cell having an $NMC/LiMn_2O_4$ blended positive electrode.

The latest IEC standards should be obtained for detailed information on nomenclature, performance, and safety guidelines. Also, ANSI standard C18.2M "Standard for Portable Rechargeable Cells and Batteries" includes standards for portable Li-ion batteries.

26.2 CHEMISTRY

The electrochemically active electrode materials in Li-ion batteries are a lithium metal oxide or a lithium metal phosphate for the positive electrode material and typically a lithiated graphite for the negative electrode material. These materials are adhered to a metal foil current collector with a binder, typically polyvinylidene fluoride (PVDF), carboxymethylcellulose, and/or styrene-butadiene rubber, and a conductive diluent, typically a carbon black or graphite. In some designs, thin layers (having a thickness of approximately 2 µm) of ceramic particles, such as Al_2O_3, are coated on one electrode surface to increase stability by reducing the likelihood of internal shorting.[4] The positive and negative electrodes are electrically isolated by a microporous polyethylene or polypropylene separator film. Since the commercialization of Li-ion batteries by Sony in 1991, a broad array of variants has been introduced. For example, in some polymer Li-ion cells, the separator is coated with a polymer layer, which aids in bonding the separator to the electrodes. Alternatively, Li-ion polymer cells are made by adding a monomer to the electrolyte and thermally polymerizing after cell assembly. In such polymer cells, the positive, separator, and negative layers are bound by the polymer and are laminated together to form a monolithic device normally having a thin form-factor, as illustrated in Figure 26.1. Despite these differences, the active cell chemistry in "polymer" cells usually is identical to that in cylindrical or prismatic Li-ion batteries.

26.2.1 Intercalation Processes

The active materials in conventional Li-ion cells operate by reversibly incorporating lithium in an intercalation process, a topotactic reaction where lithium ions (guests) are reversibly removed or inserted into a host without a significant structural change to the host. The positive material in a Li-ion cell is a metal oxide, and the metal oxide has either a layered or a tunneled structure. The graphitic carbon negative materials have a layered structure similar to graphite. Thus the metal oxide, graphite, and other materials act as hosts, reversibly incorporating the lithium ion guests to form sandwich-like structures.[5]

Intercalation materials, originally discovered by the Chinese 2700 years ago,[6] have been the subject of contemporary chemical research for only the last half-century. Today, intercalation

26.4 SECONDARY BATTERIES

FIGURE 26.1 A typical polymer Li-ion cell. These cells are available in a broad array of sizes and normally have flexible packaging like that shown here.

compounds form the basis for a variety of technologies, ranging from superconductors to catalysis. Intercalation materials in common use in various industries include graphite,[7,8] layered silicates such as talc $[Mg_3(OH)_2(Si_4O_{10})]$,[9] clays, and layered transition metal dichalcogenides, such as TiS_2.[10] The intercalation of a variety of electron donors, including lithium, and electron acceptors, such as halogens, into graphite has been studied.[8,9] The field of graphite intercalation compounds is especially rich,[11] both in the diversity of the chemistry and the depth of study.

Of particular interest to the field of Li-ion batteries is the work on alkali metal intercalation of graphite and related carbons, in particular Li_xC_6 ($0 \leq x \leq 1$).[12,13] When a Li-ion cell is charged, the active positive electrode material is oxidized and the active negative electrode material is reduced. In this process, lithium ions are deintercalated from the positive material and intercalated into the negative material, as illustrated in Fig. 26.2. In this scheme, $LiMO_2$ represents the metal oxide positive material, such as $LiCoO_2$, and C represents the carbonaceous negative material, such as

$$\text{Positive:} \quad LiMO_2 \underset{\text{Charge}}{\overset{\text{Discharge}}{\rightleftarrows}} Li_{1-x}MO_2 + x\ Li^+ + x\ e^-$$

$$\text{Negative:} \quad C + y\ Li^+ + y\ e^- \underset{\text{Charge}}{\overset{\text{Discharge}}{\rightleftarrows}} Li_yC$$

$$\text{Overall:} \quad LiMO_2 + x/y\ C \underset{\text{Charge}}{\overset{\text{Discharge}}{\rightleftarrows}} x/y\ Li_yC + Li_{1-x}MO_2$$

FIGURE 26.2 Electrode reactions in the most common Li-ion cell.

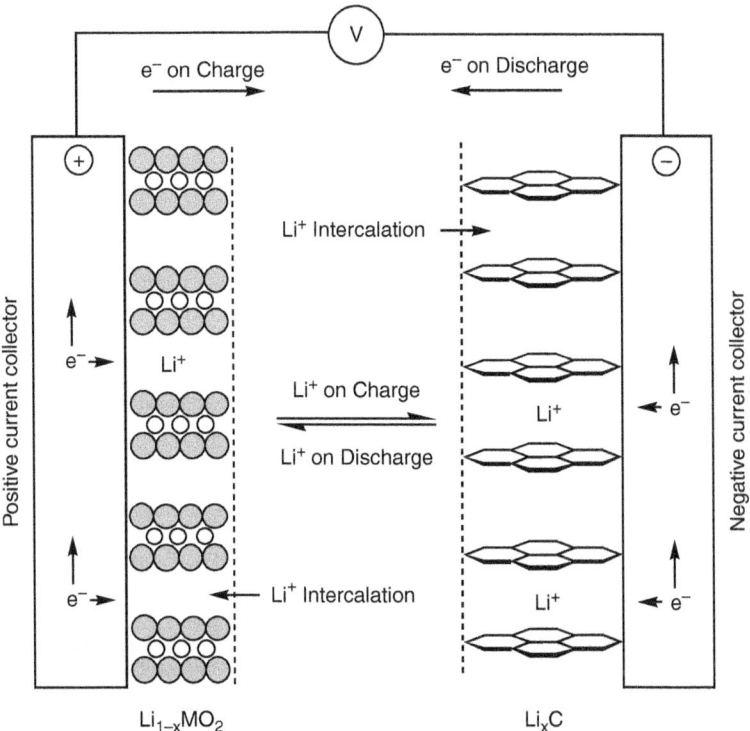

FIGURE 26.3 Schematic of the electrochemical process in a Li-ion cell.

graphite. In Fig. 26.2, x and y are selected based on the molar capacities of the electrode materials for lithium. Normally x is about 0.5 and y is about 0.16, therefore x/y is about 3. The reverse occurs on discharge. Because metallic lithium is not present in the cell, Li-ion batteries are chemically less reactive, safer, and offer longer cycle life than is possible with rechargeable lithium batteries that employ lithium metal as the negative electrode material. The charge-discharge process in a Li-ion cell is further illustrated graphically in Fig. 26.3. In this figure, the layered active materials are shown on metallic current collectors.

26.2.2 Positive Electrode Materials

Positive electrode materials in commercially available Li-ion batteries utilize a lithiated metal oxide or lithiated metal phosphate as the active material. The first Li-ion products marketed by Sony used $LiCoO_2$. Goodenough and Mizushima developed this material, as described in a series of patents.[14] Recently, cells have been developed that utilize less costly materials, such as $LiMn_2O_4$ (spinel), $LiNi_{1-x-y}Mn_xCo_yO_2$ (NMC), $LiFePO_4$, or materials with higher specific capacity, such as $LiNi_{0.8}Co_{0.15}Al_{0.05}O_2$ (NCA). $LiNi_{1/3}Mn_{1/3}Co_{1/3}O_2$, $LiNi_{0.5}Mn_{0.3}Co_{0.2}O_2$, and $LiNi_{0.42}Mn_{0.42}Co_{0.16}O_2$ are some of the most commonly commercialized types of NMC materials. Commercial interest in $LiNiO_2$ and $LiNi_{1-x}Co_xO_2$ has waned because of their instability, driven by the energetic formation of NiO and oxygen, which has been shown to contribute to safety issues.[15]

TABLE 26.2 Requirements for Li-ion Positive Electrode Materials

---
High free energy of reaction with lithium (high voltage vs. lithium metal potential)
Can incorporate large quantities of lithium
Reversibly incorporates lithium without structural change
High lithium ion diffusivity
Good electronic conductivity
Insoluble in the electrolyte
Prepared from inexpensive materials
Low-cost synthesis

Viable positive electrode materials must satisfy a number of requirements, which are summarized in Table 26.2. These factors guide materials selection and development. The positive electrode material is the source of all the active lithium ions in a lithium-ion cell. Therefore to provide high capacity, these materials must incorporate a large amount of lithium as made. Further, the materials must reversibly exchange that lithium with little structural change to permit long cycle life, high coulombic efficiency, and high energy efficiency. To achieve high cell voltage and high energy density, the lithium exchange reaction must occur at a high potential relative to lithium. When a cell is charged or discharged, an electron is removed from or returned to the positive material, respectively. In order that the charge and discharge processes occur at a high rate, the electronic conductivity and lithium ion mobility in the material must be high. $LiFePO_4$ is an exception to this rule. In $LiFePO_4$, adequate lithium ion transport is achieved by use of electrode particles having a nanometer particle size. The positive electrode material must also be compatible with the other materials in the cell. In particular, the positive electrode material must not be soluble in the electrolyte. In addition, the positive electrode material must have an acceptable cost. To minimize cost, preparation from inexpensive materials in a low-cost process is preferred.

FIGURE 26.4 The idealized structure of layered MnO_2. (*Courtesy of CISR.*)

Characteristics of Positive Electrode Materials. A variety of positive electrode materials have been developed, many of which are commercially available. These materials can have one of three structure types: an ordered rock salt-type structure, a spinel-type structure, or an olivine-type structure. The ordered rock-salt structure is a layered structure in which the lithium atoms, transition metal atoms, and oxygen atoms occupy octahedral sites in alternate layers. Exemplary layered materials include LCO, NMC, and NCA. The ideal layered structure of MnO_2 is shown in Fig. 26.4.

The spinel and olivine structures are both three-dimensional "framework" structures. The term *spinel* formally refers to the mineral $MgAl_2O_4$, although the term is used for materials, such as $LiMn_2O_4$, that have an equivalent structure. Likewise, the term *olivine* formally refers to the mineral $(Mg,Fe)_2SiO_4$, although the term is used for materials having an equivalent structure, such as $LiFePO_4$ and $LiMnPO_4$. The three-dimensional framework or tunnel structure of $LiMn_2O_4$ (spinel), which is based on λ-MnO_2, is illustrated in Fig. 26.5. In spinel $LiMn_2O_4$, lithium fills one-eighth of the tetrahedral sites within the λ-MnO_2 structure and Mn-centered oxygen octahedra fill one-half of the octahedral sites.

Materials having the olivine structure have a three-dimensional framework structure based on PO_4 tetrahedra and FeO_6 octahedra, as shown in Fig. 26.6. In $LiFePO_4$, the Li atoms move along one-dimensional tunnels. It is thought that defects and imperfections in these tunnels can lead to poor rate capability.

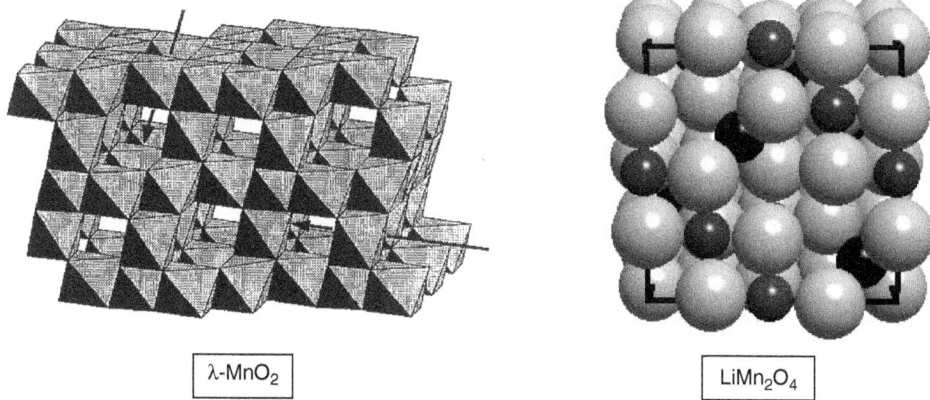

FIGURE 26.5 The idealized structure of λ-MnO$_2$ and LiMn$_2$O$_4$ spinel. The model on the left shows the manganese centered oxygen octahedra of λ-MnO$_2$. In the model on the right of LiMn$_2$O$_4$, oxygen is gray and lithium is black. (*Courtesy of CISR and Michael Tucker.*)

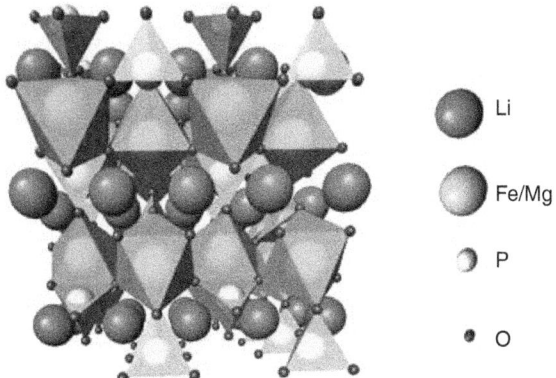

FIGURE 26.6 The crystal structure of LiFe$_{1-x}$Mg$_x$PO$_4$, which is derived from olivine. The FeO$_6$ octahedra and PO$_4$ tetrahedra are shown.

A truly amazing accomplishment is that John Goodenough's research group is responsible for the first clear demonstrations that each of these three structure types could be used as positive electrode materials. Layered materials are described in "Li$_x$CoO$_2$ (0 ≤ x ≤ 1.0): A New Cathode Material for Batteries of High Energy Density," which was published in the Materials Research Bulletin in 1980.[16] Spinel materials are described in "Lithium Insertion in Manganese Spinels," which was also published in the Materials Research Bulletin.[17] Olivine materials are described in "Phospho-olivines as Positive-Electrode Materials for Rechargeable Lithium Batteries."[18] These three papers had been cited 843, 653, and 1044 times, respectively, as of December 20, 2009. John Goodenough has been recognized by numerous major awards for these achievements, including the Japan Prize in 2000 and the Enrico Fermi award in 2009, which is fitting given that his research group is responsible for all major classes of lithium-ion positive electrode materials.

TABLE 26.3 Characteristics of Some Positive Electrode Materials

Material	Specific capacity mAh/g	Midpoint V vs. Li at C/20	Comments
$LiCoO_2$	155	3.9	Still the most common. Co is expensive.
$LiNi_{1-x-y}Mn_xCo_yO_2$ (NMC)	140–180	≈3.8	Safer and less expensive than $LiCoO_2$. Capacity depends on upper voltage cutoff.
$LiNi_{0.8}Co_{0.15}Al_{0.05}O_2$	200	3.73	About as safe as $LiCoO_2$, high capacity.
$LiMn_2O_4$	100–120	4.05	Inexpensive, safer than $LiCoO_2$, poor high temperature stability (but improving with R&D).
$LiFePO_4$	160	3.45	Synthesis in inert gas leads to process cost. Very safe. Low volumetric energy.
$Li[Li_{1/9}Ni_{1/3}Mn_{5/9}]O_2$	275	3.8	High specific capacity, R&D scale, low rate capability.
$LiNi_{0.5}Mn_{1.5}O_4$	130	4.6	Requires an electrolyte that is stable at a high voltage.

The voltage and capacity characteristics of common positive electrode materials are summarized in Table 26.3. The most commonly used positive electrode material, $LiCoO_2$ (Fig. 26.7), offers good capacity, 155 mAh/g, and high average voltage, 3.9 V versus Li. NMC materials have the same structure and provide performance substantially equivalent to that of $LiCoO_2$, and have the benefits of lower raw materials costs and improved thermal stability during abuse.[19] The $LiNi_{0.8}Co_{0.15}Al_{0.05}O_2$ (NCA) materials offer higher capacity, up to 200 mAh/g, albeit at a voltage which is about 0.2 V lower than that of $LiCoO_2$ or $LiMn_2O_4$. Recently, special coatings have been used to increase the performance of layered positive electrode materials by making them more tolerant to charging to higher voltages. Some commercially available cells employ coated positive electrode materials having performance that exceeds that listed in Table 26.3. $LiMn_2O_4$ is also used commercially, particularly in applications that are cost sensitive or require exceptional stability upon abuse. $LiMn_2O_4$ has lower capacity, 100 to 120 mAh/g, slightly higher voltage, 4.0 V versus lithium, but has higher capacity loss on storage or cycling, especially at elevated temperature, relative to cells that use $LiCoO_2$, or NCA. $LiFePO_4$ has a specific capacity of about 160 mAh/g and an average voltage of 3.45 V versus lithium. $LiFePO_4$ is virtually unreactive with electrolytes in the charged or discharged state up to at least 350°C. A disadvantage of $LiFePO_4$ is that it has both low capacity density and a low packing efficiency (tap density), making it difficult to produce $LiFePO_4$ cells having high energy density.

There are several potential new positive electrode materials that could see widespread application during the next decade. One of these, $Li[Li_{1/3-2x/3}Ni_xMn_{2/3-x/3}]O_2$ ($0 \leq x \leq 0.5$)[20] shows stable specific capacity, up to 275 mAh/g for materials with $x \approx 1/3$. Also being developed are "5 V" materials, such as $LiNi_{0.5}Mn_{1.5}O_4$. Recent improvements to this class of materials include reduction of the irreversible capacity loss and improvement of the rate capability.[21]

Physical Properties of Positive Electrode Materials. The particle-size distribution, particle shape, specific surface area, and tap density of positive electrode materials all play important roles in determining the properties of Li-ion batteries that incorporate them. Because the particle size determines the solid-state diffusion path

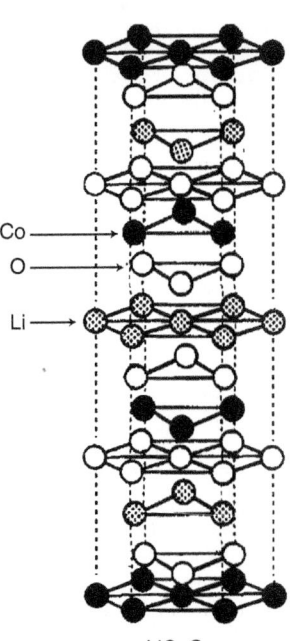

FIGURE 26.7 The idealized layered structure of $LiCoO_2$. Li is speckled, O is white, and Co is black.

length, the particle-size distribution controls aspects of the rate capability. Also, the particle-size distribution and specific surface area control the properties of slurries used for coating electrodes. For lithium transition metal oxides, the specific surface area also controls the reactivity of the charged material with the electrolyte at elevated temperature. This is because interactions between the particles of positive electrode material and the electrolyte occur at the surfaces of the particles, thus surface area should be minimized to make the safest materials. Finally, the tap density is a useful indicator of the ultimate density of a calendered electrode: a material having a high tap density normally provides a high density electrode. In this section, the physical properties of some commercially available positive electrode materials will be reviewed.

Wang et al.[22] reviewed the properties of $LiCoO_2$, $LiNi_{1/3}Mn_{1/3}Co_{1/3}O_2$, $LiNi_{0.42}Mn_{0.42}Co_{0.16}O_2$, and $LiNi_{0.80}Co_{0.15}Al_{0.05}O_2$. Figure 26.8 shows SEM micrographs of these materials. The two samples of $LiNi_{0.8}Co_{0.15}Al_{0.05}O_2$ were obtained from Toda Kogyo Corp. of Japan, the $LiNi_{1/3}Mn_{1/3}Co_{1/3}O_2$ and $LiNi_{0.42}Mn_{0.42}Co_{0.16}O_2$ material was obtained from 3M, and the $LiCoO_2$ was obtained from E-One/Moli Energy of Canada.

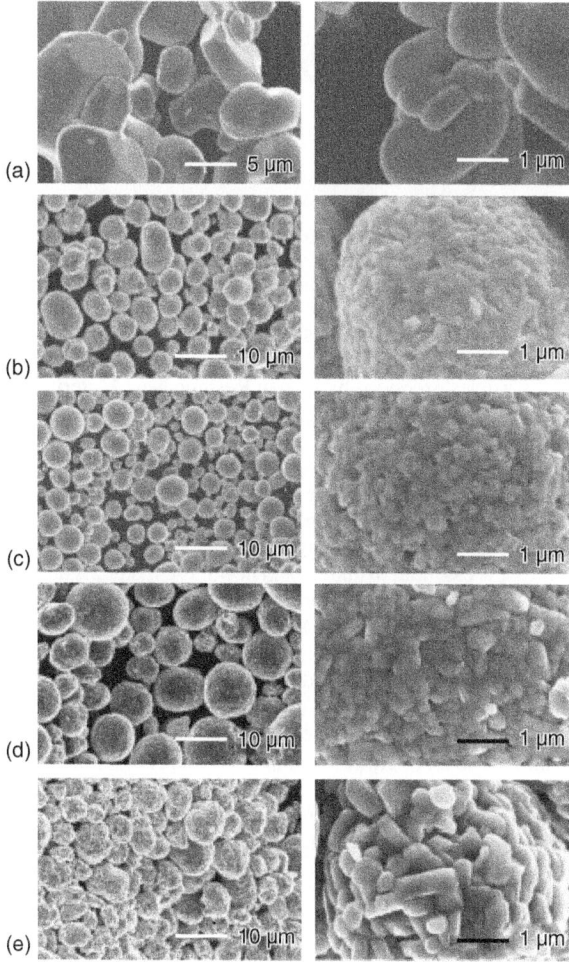

FIGURE 26.8 SEM micrographs at two magnifications of (a) $LiCoO_2$, (b) $LiNi_{0.80}Co_{0.15}Al_{0.05}O_2$ (sample 1), (c) $LiNi_{0.80}Co_{0.15}Al_{0.05}O_2$ (sample 2), (d) $LiNi_{1/3}Mn_{1/3}Co_{1/3}O_2$ and (e) $LiNi_{0.42}Mn_{0.42}Co_{0.16}O_2$.[23] (Reproduced with permission from *Electrochemistry Communications* **9**, 2534–2540 [2007]).

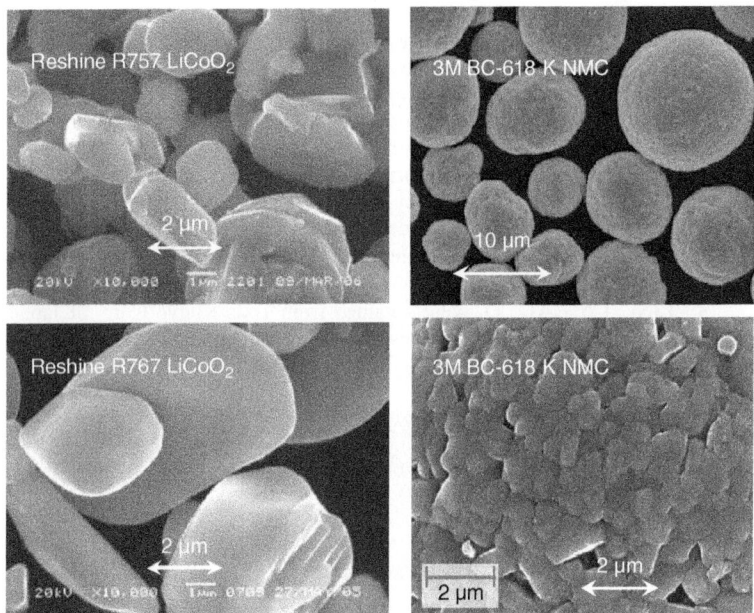

FIGURE 26.9 SEM images of LiCoO$_2$ and NMC. (*Courtesy of Hunan Reshine New Material Co. and 3M, respectively.*)

LiCoO$_2$ typically has a very smooth surface, while NMC and NCA materials typically have rough surfaces, as shown in Figs. 26.8 and 26.9. The left panels of Fig. 26.9 show high fidelity images of LiCoO$_2$ particles produced by Hunan Reshine New Material Co., while the right panels show NMC materials produced by 3M Co.

Figure 26.9 shows SEM images of LiFePO$_4$ and LiMn$_2$O$_4$ produced by Hunan Reshine New Material Company. The LiFePO$_4$ has a very small particle size, to accommodate the slow rate of Li diffusion in this material, while the LiMn$_2$O$_4$ can support large particle sizes due to the relatively fast rate of Li diffusion in LiMn$_2$O$_4$.

In order to understand the particle shapes and sizes of LiCoO$_2$, NMC, NCA, LiFePO$_4$, and LiMn$_2$O$_4$, it is essential to understand how these materials are synthesized. LiCoO$_2$ is easy to make by mixing cobalt oxide or cobalt carbonate with lithium carbonate or lithium hydroxide and sintering the mixture at 700 to 1000°C in air. Various methods of producing LiCoO$_2$ are known.[24] Normally, materials are made with a Li:Co stoichiometric ratio of 1:1 or with a slight excess (0 to 5%) of lithium. LiCoO$_2$ sinters very quickly and particles comprising large single crystals with smooth surfaces (as shown for LiCoO$_2$ in Figs. 26.8 and 26.9) grow in times as short as 1 h at the synthesis temperature.

LiMn$_2$O$_4$ materials can have the generic formula Li$_{1+x}$Mn$_{2-x-y}$M$_y$O$_4$, where M is a substituent atom and can be Al or Ni, for example. Normally x is about 0.07, and y is greater than or equal to 0. A typical preparation used by academic researchers has been published by the Ohzuku group.[25] The Ohzuku group prepared lithium aluminum manganese oxide (LAMO) from electrolytic manganese dioxide (EMD), Li$_2$CO$_3$, AlOH$_3$, and boric acid (H$_3$BO$_3$), where boric acid was added to improve particle growth and sintering. Appropriate amounts of starting materials were mixed with water to make a slurry having a particle size below 1 μm, and the slurry was spray-dried at 250°C. The resulting powder was heated at 900°C for 12 h and then at 650°C for 24 h in air. The reaction product was washed with hot water at 95°C to remove boron oxide and then dried at 200°C for 16 h in air. Industrial scale syntheses are believed to be similar, except for changes to increase simplicity and reduce cost. If EMD is used as the Mn-containing precursor, the angular shape of the EMD particles is normally retained by the product as shown in Fig. 26.10.

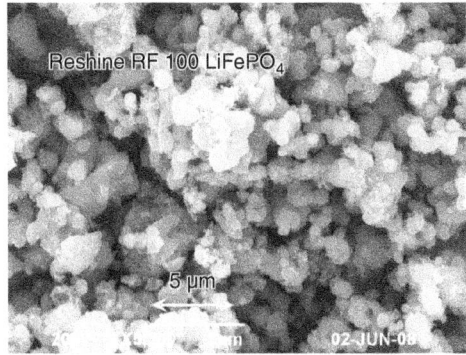 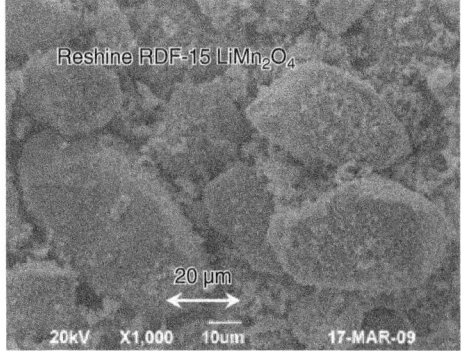

FIGURE 26.10 SEM images of $LiFePO_4$ and $LiMn_2O_4$ products from Hunan Reshine New Material Company. (*Courtesy of Hunan Reshine New Material Company.*)

NMC and NCA materials rely on a uniform and homogeneous distribution of cations in the transition metal layers of the structure (see Fig. 26.7). The most common way to ensure this is to use a mixed transition metal hydroxide or carbonate precursor that has the cations perfectly mixed on the atomic scale. Such precursors are normally prepared by coprecipitation of mixed metal sulphate solutions in a base or in a carbonate. The precursors made by this route form spherical agglomerates of many primary particles, resulting in a rough surface morphology. The spherical particle size and tap density of the precursors can be controlled by the pH, temperature, and ammonia concentration during precipitation.[26,27] Oxides are produced by sintering the mixed transition metal hydroxide precursors with lithium carbonate. The spherical shape and rough surface of the hydroxide precursors are maintained in the resulting NMC and NCA oxides, as shown in Figs. 26.8 and 26.9.

$LiFePO_4$, unlike the layered oxides and the spinel materials, must be synthesized in an inert atmosphere to prevent oxidation of the desired divalent Fe to trivalent Fe. The requirement for an inert atmosphere increases process complexity and hence the cost of the resulting material. In addition, the primary $LiFePO_4$ particles have a particle size of less than about 500 nm to provide sufficient rate capability. As shown in Fig. 26.11, a commercially available $LiFePO_4$ material (Phostech Lithium Life Power® P2 grade "C-$LiFePO_4$," a composite of pyrolytic carbon on $LiFePO_4$) has a particle size of about 100 to about 500 nm.

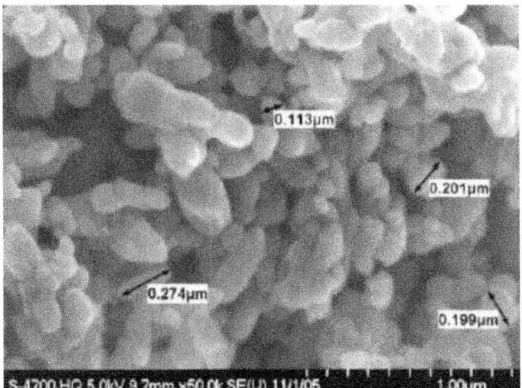 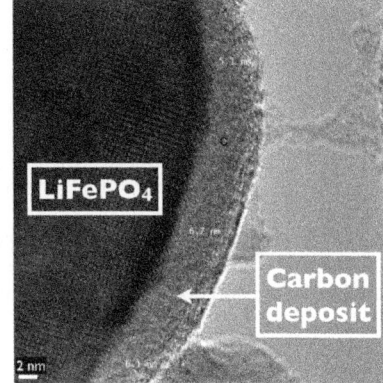

FIGURE 26.11 SEM images of Phostech Lithium "Life Power® P2 grade" $LiFePO_4$ ("C-$LiFePO_4$"). The left image shows the 100 to 300 nm primary particle size. The right image shows the pyrolytic carbon coating (i.e., "deposit") on the $LiFePO_4$ particles. (*Courtesy of Phostech Lithium Inc./Süd-Chemie AG.*)

Inexpensive $LiFePO_4$ is normally made from an Fe^{3+} precursor, which must be reduced during synthesis to Fe^{2+}. This reduction can be accomplished by an elegant method using carbothermal reduction[28] because $LiFePO_4$ is stable at elevated temperatures in contact with carbon (The same is not true for the layered and spinel oxides, which are reduced by carbon at elevated temperature.) The carbothermal reduction synthesis conveniently results in a thin layer of carbon formed on the $LiFePO_4$ particle surface. Ravet et al. showed that a thin carbon coating on the surfaces of the $LiFePO_4$ particles enhances electronic conductivity and improves rate capability.[29] Commercial $LiFePO_4$ materials, such as the Phostech "Life Power® P2 grade" shown above, are apparently produced by pyrolysis of an organic precursor in a similar but different process.

Table 26.4 compares some physical properties of various positive electrode materials from several manufacturers. The data in this table are typical for each of the material classes listed. All oxide materials have low specific surface areas, generally less than 1.0 m²/g and preferably less than 0.5 m²/g, to minimize reaction with the electrolyte under abuse conditions. $LiFePO_4$ is more stable to thermal abuse, and therefore materials having a larger surface area can be used in order to increase rate performance. However, the very low tap density and large surface area of typical $LiFePO_4$ materials also result in the low energy density of $LiFePO_4$-based cells.

TABLE 26.4 Physical Properties of Various Positive Electrode Materials[30]

Material	Manufacturer	Specific surface area (m²/g)	Tap density (g/mL)	Mean particle size (d_{50}) (μm)
$LiCoO_2$ (R757)	Reshine	0.4–0.75	2.1–2.9	6.5–9.5
$LiCoO_2$ (R767)	Reshine	<0.45	>2.5	9.0–14.0
NMC 532	Reshine	0.1–0.4	>2.3	9.0–15.0
NMC 111 BC-618	3M	0.26	2.69	10.5
NMC 442 BC-723	3M	0.39	2.29	7.8
NCA-01	Toda Kogyo	0.47[30]	N/A	N/A
NCA-02	Toda Kogyo	0.34[30]	N/A	N/A
$LiMn_2O_4$ TR-LMO-300	Tronox	0.88–1.0	1.5–1.7	6–15
$LiMn_2O_4$ TR-LMO-100	Tronox	0.5–0.7	2.0–2.4	10–30
$LiFePO_4$ (RF-100)	Reshine	<16	>0.8	2.5-5 (agglomerates)
$LiFePO_4$ (P2)	Phostech Lithium	12–18	N/A	0.5–1.0 (agglomerates)
NCA (N95)	BTR New Materials Co.	0.4–0.9	>2.2	8–12

NMC 111 = $LiNi_{1/3}Mn_{1/3}Co_{1/3}O_2$
NMC 442 = $LiNi_{0.42}Mn_{0.42}Co_{0.16}O_2$
NCA = $LiNi_{0.8}Co_{0.15}Al_{0.05}O_2$

Electrochemical Properties of Positive Electrode Materials. Figure 26.12 shows potential versus specific capacity of a $Li/LiCoO_2$ cell and a Li/graphite cell in which the specific capacity axis of the Li/graphite cell has been scaled by dividing by 2. Normally, researchers test Li-ion electrodes versus metallic lithium to determine specific capacity, differential capacity, and charge-discharge cycle life. If such preliminary results are encouraging, full Li-ion cells with electrodes having active materials in appropriate ratios are constructed.

For the example in Fig. 26.12, the specific capacity of graphite is about 350 mAh/g and the specific capacity of $LiCoO_2$ (to 4.2 V) is about 140 mAh/g, thus the graphite would be properly included in a Li-ion cell at a mass per unit electrode area of electrode, which is about 50% of the $LiCoO_2$ mass per unit area of electrode. In addition, about 10% excess graphite is typically added to the negative electrode to avoid lithium plating at a full state-of-charge, as illustrated in Fig. 26.12. The voltage of the Li-ion cell is calculated by taking the difference between the curves in Fig. 26.12 at a selected of state-of-charge. $LiCoO_2$ can deliver more capacity if it is charged to higher potentials, e.g., about 155 mAh/g at 4.3 V, but may need to be coated and/or doped with other elements to allow excursions to high voltages without detrimental impacts to cycle life or thermal stability.

Figure 26.13 shows potential versus specific capacity for $LiNi_{0.5}Mn_{1.5}O_4$ spinel, a $LiMn_2O_4$ spinel-based lithium aluminum manganese oxide (LAMO, with Al and excess Li), $LiNi_{1/3}Mn_{1/3}Co_{1/3}O_2$

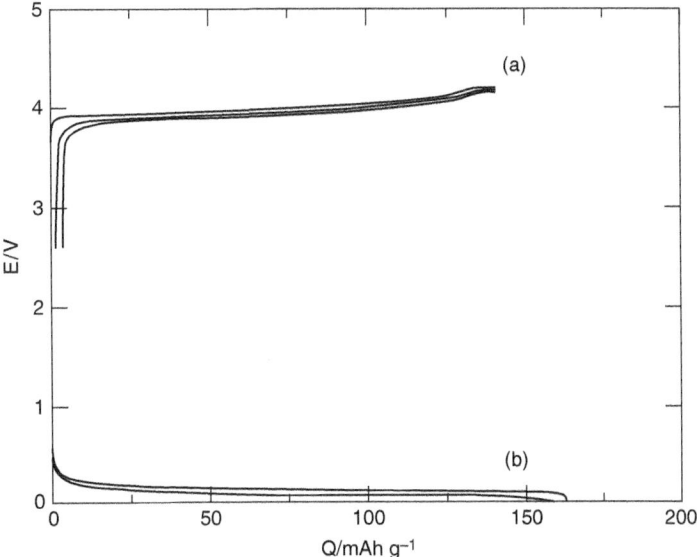

FIGURE 26.12 Potential vs. specific capacity for Li/LiCoO$_2$ and Li/graphite cells. The specific capacity axis for the Li/graphite cell has been divided by 2.[31] (Reproduced with permission from *Journal of Power Sources* **174**, 449–456 [2007].)

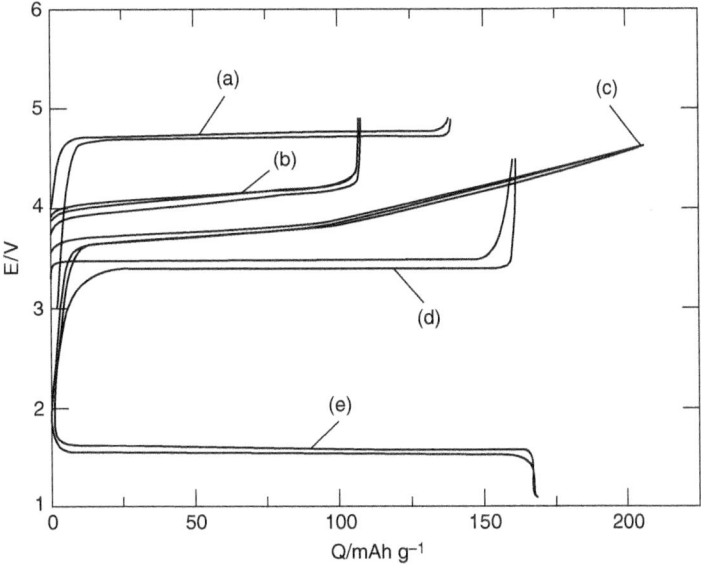

FIGURE 26.13 Potential vs. specific capacity for (a) LiNi$_{0.5}$Mn$_{1.5}$O$_4$ spinel, (b) a LiMn$_2$O$_4$ spinel-based lithium aluminum manganese oxide (LAMO, with Al and excess Li), (c) LiNi$_{1/3}$Mn$_{1/3}$Co$_{1/3}$O$_2$ (NMC 111), (d) LiFePO$_4$, and (e) Li[Li$_{1/3}$Ti$_{5/3}$]O$_4$ spinel (a potential negative electrode material).[31] (Reproduced with permission from *Journal of Power Sources* **174**, 449–456 [2007].)

(NMC 111), LiFePO$_4$, and Li$_{4/3}$Ti$_{5/3}$O$_4$ spinel (a potential negative electrode material also written as Li$_4$Ti$_5$O$_{12}$). Figure 26.13 shows that LiNi$_{0.5}$Mn$_{1.5}$O$_4$ has the highest potential and LiFePO$_4$ has the lowest potential among these positive electrode materials. LAMO has the smallest specific capacity. NMC 111 provides a very high specific capacity, 200 mAh/g if it is charged to 4.6 V as in Fig. 26.13. However, due to a compromise between capacity and cycle life, NMC materials are typically charged to only about 4.3 V where their capacities are similar to LiCoO$_2$.

Many of these electrode materials display spectacular physics and chemistry associated with metal-insulator transitions, stacking rearrangements, and ordering of the intercalated lithium within the available sites. Perhaps the most famous is the feature that appears in the differential capacity of Li/Li$_x$CoO$_2$ cells at x = ½ which is caused by the ordering of lithium atoms along rows in the structure.[32,33]

Figure 26.14 shows differential capacity versus potential for a Li/LiCoO$_2$ cell. The local minimum near 4.14 V (x = ½ in Li$_x$CoO$_2$) corresponds to a structure in which Li is ordered along every second row of available sites in the Li layers of Fig. 26.7. The peaks on either side of the minimum correspond to order-disorder phase transitions. Disruptions to the Co layer, by the substitution of Ni or by the inclusion of an excess of Li, destroys the order-disorder transition, even with a few percent of substituent atoms, and the feature in the differential capacity plot (dQ/dV versus V) disappears. Some commercial LiCoO$_2$ grades do not show this order-disorder feature and some do.

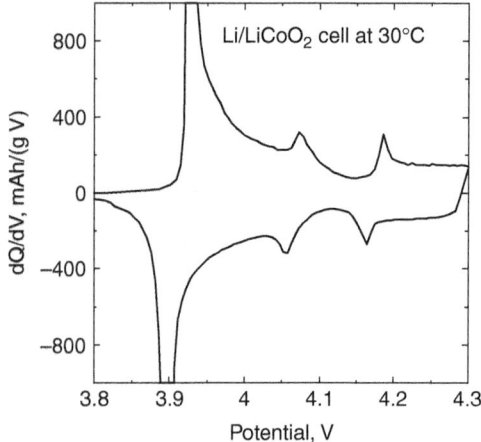

FIGURE 26.14 Differential capacity (dQ/dV) vs. cell potential for a Li/LiCoO$_2$ cell charged and discharged at a 0.1C rate at 30°C. (*Courtesy of Wenbin Luo.*)

A huge emphasis has been placed on the electrochemical performance of LiMn$_2$O$_4$ spinel. It has been shown that excess Li in Li$_{1+x}$Mn$_{2-x}$O$_4$, and other substituents, notably Al and/or Ni, improve the capacity retention of spinel, especially at elevated temperature. Figure 26.15 shows data from a review article by Whittingham.[34] The results shown in Fig. 26.15 clearly illustrate the importance of excess Li on reducing the capacity-loss rate. However, the specific capacity of the material is reduced as excess Li is added. Ohzuku et al.[35] showed the importance of Al additions in improving the charge-discharge cycling of spinel. Their preferred LAMO material has a lattice constant of a = 8.221 Å and a specific capacity near 110 mAh/g. Figure 26.16 shows a comparison between the potential-capacity curves of LAMO and stoichiometric LiMn$_2$O$_4$ spinel.

Figure 26.16 shows that properly prepared spinel materials can provide excellent potential-capacity profiles. In addition, such materials can provide a long charge-discharge cycle life at room temperature. Capacity retention of spinel materials at elevated temperature (above 45°C) is still not

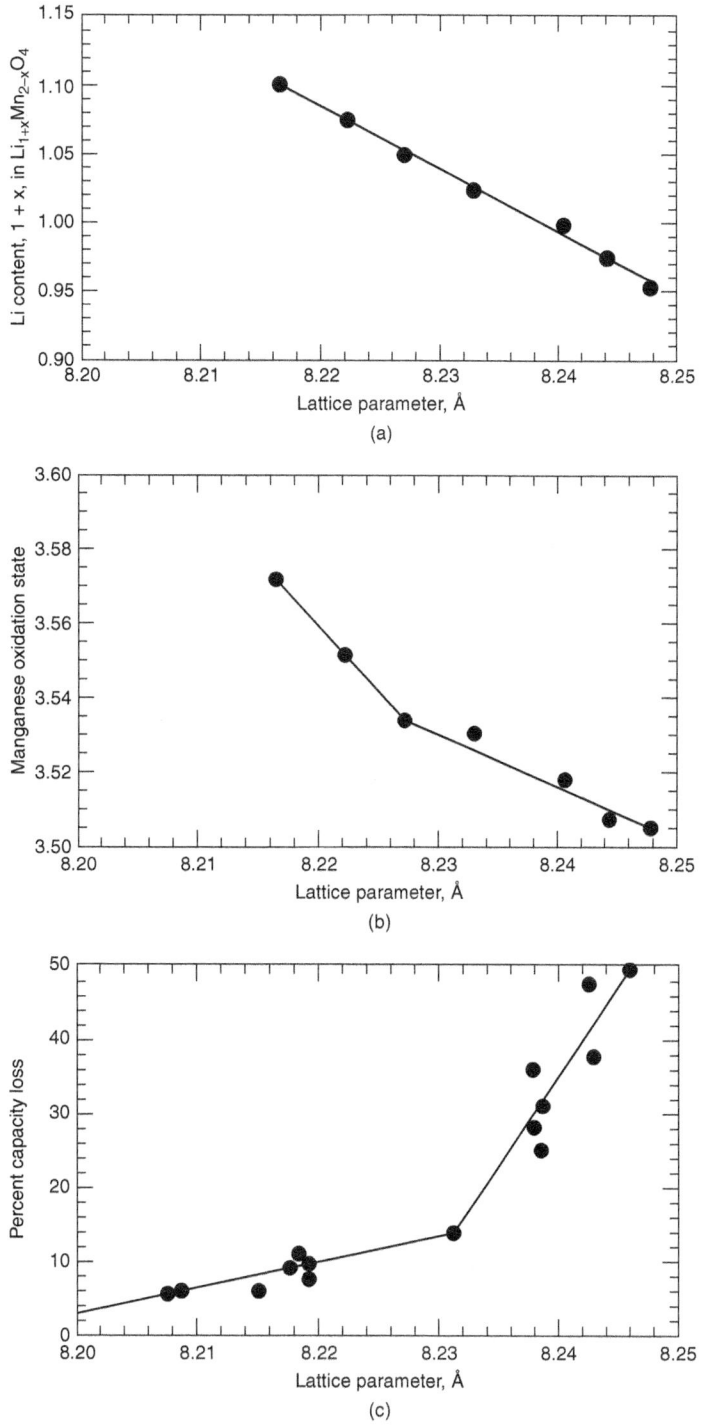

FIGURE 26.15 Correlation of the lattice parameter of the spinel $Li_{1+x}Mn_{2-x}O_4$ with (a) the lithium content, (b) manganese oxidation state, and (c) capacity loss of the cell over the first 120 cycles.[36] (Reproduced with permission from *Chem. Rev.* **104**, 4271–4301 [2004].)

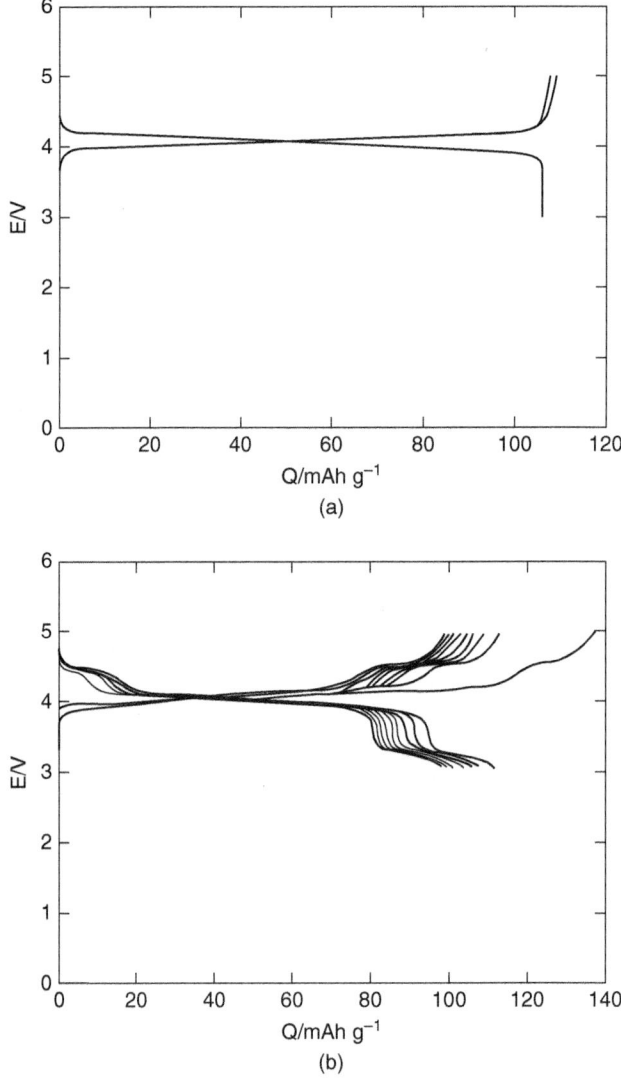

FIGURE 26.16 Charge and discharge curves of (a) Li/LAMO and (b) Li/LiMn$_2$O$_4$ cells operated between 3 and 5 V. The current densities applied to the cells were 0.20 mA/cm^2 for LAMO and 0.17 mA/cm^2 for LiMn$_2$O$_4$. In both cells, the electrolyte was 1 M LiPF$_6$ in EC:DMC 3:7 by volume.[37] (Reproduced with permission from *Electrochemical and Solid-State Letters*, **9**, A557–A560 [2006].)

as good as LiCoO$_2$, LiFePO$_4$, or NMC, for example. The reasons for the capacity loss of spinel at elevated temperature are the subject of hundreds of reports. The review by Whittingham and a recent review by Fergus[38] provide numerous references, many of which are often contradictory.

The consensus seems to be that capacity loss in spinel at elevated temperature is caused by Mn dissolution and/or instability of the electrolyte at the spinel surface. Mn dissolution and electrolyte instability can be mitigated to some extent by electrolyte additives that reduce the HF and water content of the electrolyte and by coatings on the electrode material that apparently slow the rate of Mn dissolution and the rate of the reaction of spinel with electrolyte. As an example, SEM/EDX studies of $LiMn_2O_4$ spinel produced by Tronox show improvement using rare earth atoms (primarily La) and fluorine at the particle surface. This type of approach is popular in academia[39] and industry.[40]

Considerable effort has been expended by academic researchers and battery manufacturers to understand capacity loss mechanisms in Li-ion positive electrodes. Figure 26.17 summarizes those mechanisms thought to be the most influential as they apply to the active and inactive phases in the positive electrode material. Electrolyte additives and positive electrode material surface coatings are thought to be effective ways to mitigate these issues.

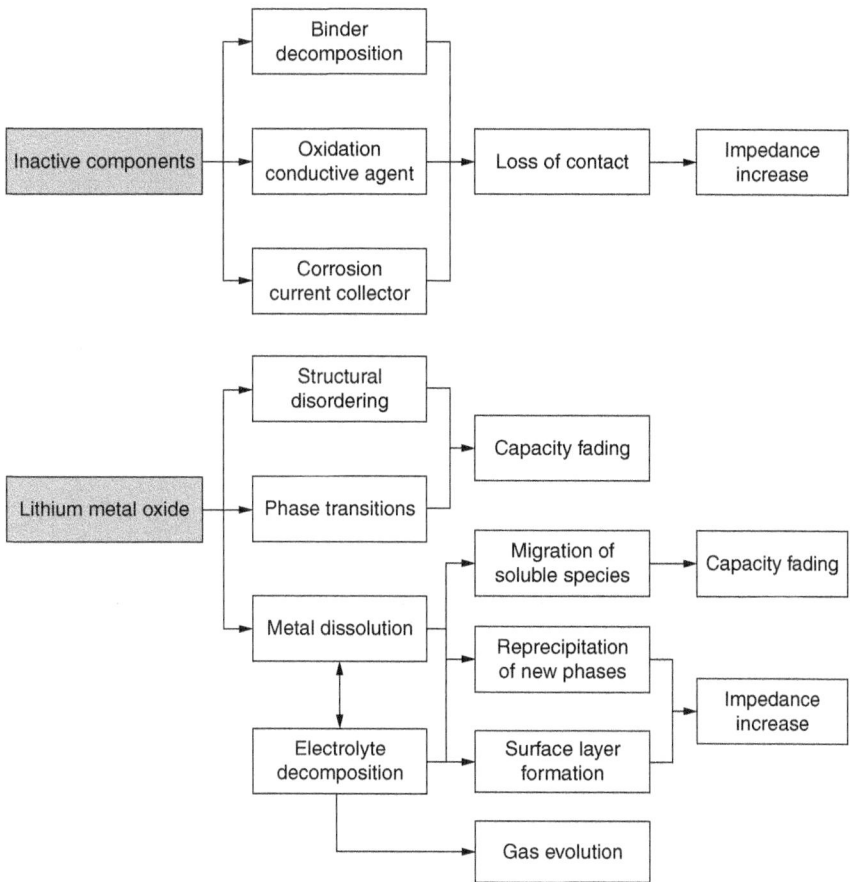

FIGURE 26.17 Schematic overview of capacity loss mechanisms for Li-ion positive electrodes.[41] (Reproduced with permission from *Journal of Power Sources* **127**, 58–64 [2004].)

26.2.3 Negative Electrode Materials

Historical Overview. Since the early 1970s, intercalation compounds have been considered as electrode materials for secondary lithium batteries. However, secondary lithium battery development effort throughout the 1970s and early 1980s focused on the use of lithium metal as the negative electrode because of the high specific capacity of the metal. Cells with impressive performance were developed and some were commercialized, however, safety issues with lithium metal batteries[42,43] caused the industry to concentrate on using lithium intercalation into carbon at the negative electrode instead of lithium metal.[44] The safety issues with lithium metal have been attributed to the changing morphology of lithium as a cell is cycled. As described in Chap. 27, the safety properties of negative electrodes may be correlated to their surface area. Thus while the properties of lithium metal negative electrodes change with use, carbon electrodes offer stable surface morphology, resulting in consistent safety properties over their useful life.[44] By utilizing low-surface-area carbons, electrodes with acceptable self-heating rates may be fabricated.

The first Li-ion batteries marketed by Sony utilized petroleum coke at the negative electrode. Coke-based materials offer good capacity, 180 mAh/g, and are stable in the presence of propylene carbonate (PC), in contrast to graphitic materials, which can cointercalate PC and exfoliate unless stabilizing additives are used. The disorder in coke materials is thought to pin the graphene layers, inhibiting reaction or exfoliation in the presence of propylene carbonate.[44] In the mid-1990s, most Li-ion cells utilized electrodes employing graphitic spheres, in particular a mesocarbon microbead (MCMB) carbon. MCMB carbon offers higher specific capacity, 300 to 350 mAh/g, and low surface area, thus providing low irreversible capacity and good safety properties. Recently, a wider variety of carbon types has been used in negative electrodes. Many commercial cells utilize synthetic or natural graphite, which is available at very low cost, and is usually highly graphitized to provide the highest specific capacity and excellent packing efficiency.

Types of Carbon. Many types of carbon materials are commercially available. The structure of the carbon greatly influences its electrochemical properties, including lithium intercalation capacity and potential. The basic building block for carbon materials is graphene, which is a planar sheet of carbon atoms arranged in a hexagonal array, as shown in Fig. 26.18. These sheets are stacked in a registered fashion to form graphite. In Bernal graphite, the most common type, ABABAB stacking occurs, resulting in hexagonal or 2H graphite. In a less common polymorph, ABCABC stacking occurs, termed rhombohedral or 3R graphite.[5]

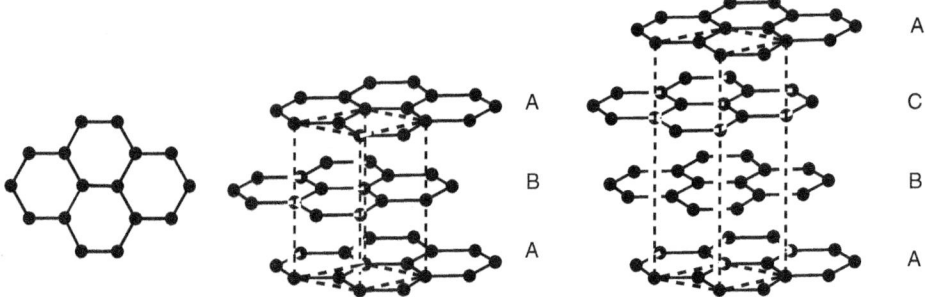

FIGURE 26.18 The hexagonal structure of a carbon layer and the structures of hexagonal (2H) and rhomboherdal (3R) graphite.

Most real materials contain disorder, including the 2H and 3R stacking orders as well as random stacking, thus a more precise way to identify a graphite is to specify the relative fractions of 2H, 3R, and random stacking. Various forms of carbon have been developed with a range of stacking disorders and different morphologies. Stacking disorders include those where the graphitic planes are parallel but shifted or rotated, termed turbostratic disorder,[45] or those in which the planes are not parallel, termed unorganized carbon.[44] Particle morphologies include flat plates, which are present in natural graphites, carbon fibers, and spheres.

Carbon materials can be considered as different aggregations of a basic structural unit (BSU) consisting of two or three parallel planes with a diameter of 2 nm.[46] The BSUs may be oriented randomly, resulting in carbon black, or oriented to a plane, axis, or point, resulting in a planar graphite, a whisker,

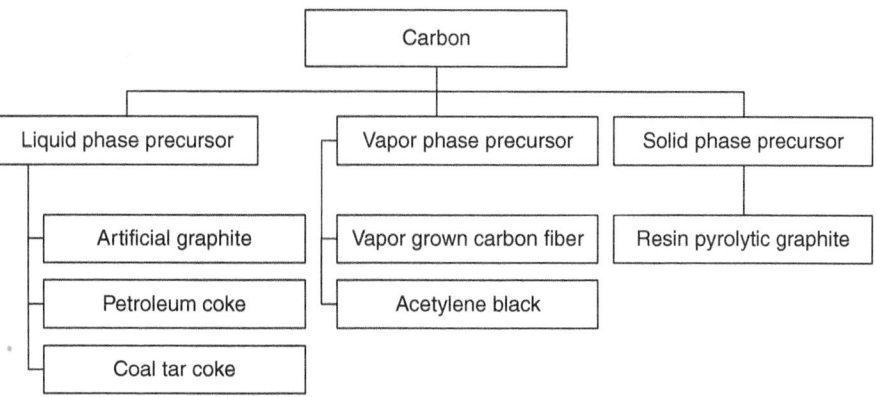

FIGURE 26.19 Carbons classified by the precursor phase.

or a spherule. The types of carbon may alternatively be organized based on the type of precursor material, as illustrated in Fig. 26.19 because the precursor material, and the processing parameters determine the nature of carbon produced. Materials that can be graphitized by treatment at high temperature (2000 to 3000°C) are termed soft carbons. Upon graphitization, the turbostratic disorder is removed progressively with increasing temperature, and strain in the material is relieved.[47] Hard carbons, such as those prepared from phenolic resin, cannot be readily graphitized, even when treated at 3000°C. Coke-type materials are prepared at about 1000°C, typically from an aromatic petroleum precursor.

Staging and Electrochemical Intercalation into Carbon. When lithium is intercalated into graphite, the ABAB structure transforms to an AAAA structure and distinct voltage plateaus are observed as distinct phases (stages) are formed.[48] This is illustrated in Fig. 26.20, which shows the voltage of a Li/graphite cell over one cycle at a low rate for a highly ordered graphite. A classical model of

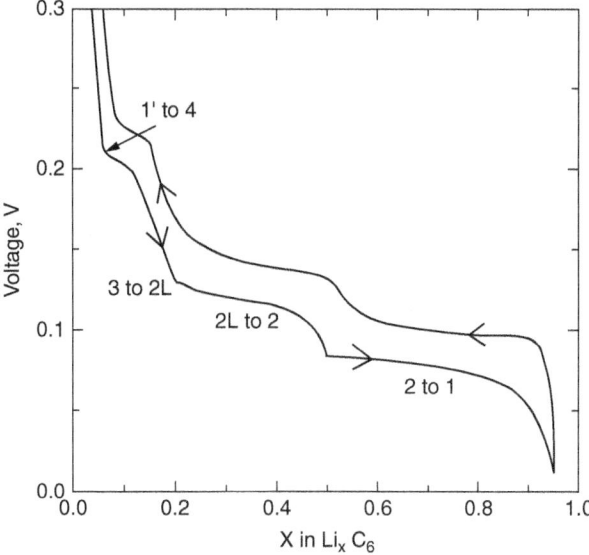

FIGURE 26.20 The voltage of a Li/graphite cell illustrating Li staging upon intercalation of graphite.[47]

FIGURE 26.21 Schematic diagram of lithium staging in graphite.[49] (Reproduced with permission from *Phys. Rev. B.* **51**, 734 [1995].)

lithium staging is illustrated in Fig. 26.21. As shown in Fig. 26.21, lithium forms "islands" within graphite instead of distributing homogeneously. The most lithium-rich stage, LiC_6, is termed stage 1 and is formed at the lowest voltage, as indicated in Fig. 26.20. As lithium is removed from the graphite, higher stages are formed, as indicated in Figs. 26.20 and 26.21.

The specific capacity of a graphitic carbon is determined by the fraction of adjacent graphene layers that are in turbostratic misalignment. It has been found that lithium cannot be inserted between parallel graphene sheets that are turbostratically misaligned. The specific capacity, Q, of graphite is then simply calculated as

$$Q = 372 (1 - P) \text{ mAh/g}$$

where P is the fraction of adjacent misaligned layers. Alternatively, the capacity (x_{max}) in Li_xC_6, can be written as

$$x_{max} = 1 - P$$

The fraction of layers in turbostratic misalignment can be determined by careful studies of the x-ray diffraction pattern of graphite samples. Hang Shi et al. wrote a software package that is used to determine P, by comparing calculated patterns to measured patterns, as illustrated in Fig. 26.22.[50] As P decreases, the diffraction peaks become sharper and more peaks appear.

Figure 26.23 shows how the potential-capacity profile of lithiated graphite depends on the heat-treatment temperature and on P. Note that the curves have been sequentially offset by 0.1 V for clarity. As P approaches zero, the capacity increases. This increase in capacity is understood to result from relief of turbostratic misalignment as the heat treatment temperature is increased.

Table 26.5 provides structural parameters of the graphitic carbons described in Figs. 26.22, 26.23, and 26.24.

Figure 26.24 shows the capacity of the graphite samples listed in Table 26.5 as well as other samples plotted versus P. Figure 26.24 shows a clear linear correlation between the fraction of turbostratically misaligned layers and the reduction in capacity. Evidently, in order to obtain the highest specific capacity, it is important to use graphitic carbons with the highest degree of order. In addition, particle size, specific surface area, tap density, impurity content, and surface treatments, are other significant parameters that influence the lithium intercalation performance of graphites. As further described below, a small specific surface area is desirable.

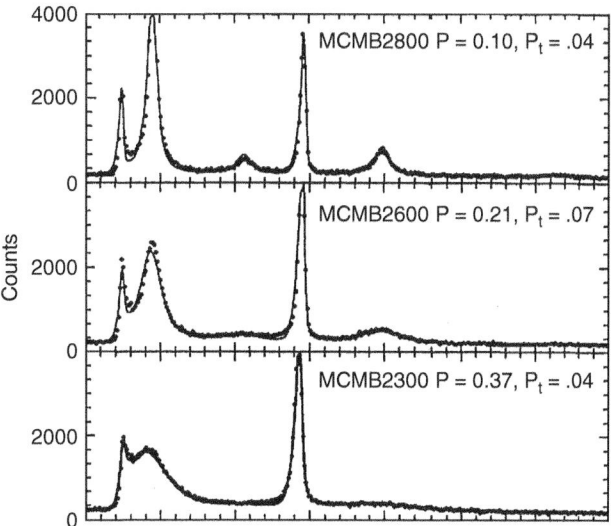

FIGURE 26.22 X-ray diffraction patterns of materials listed in Table 26.5. The solid lines are calculated best-fit patterns used to extract the value of P.[49] (Reproduced with permission from *Phys. Rev. B.* **51**, 734 [1995].) Note: P_t is the probability of finding layers in rhombohedral (ABC) stacking instead of hexagonal (ABAB) stacking.

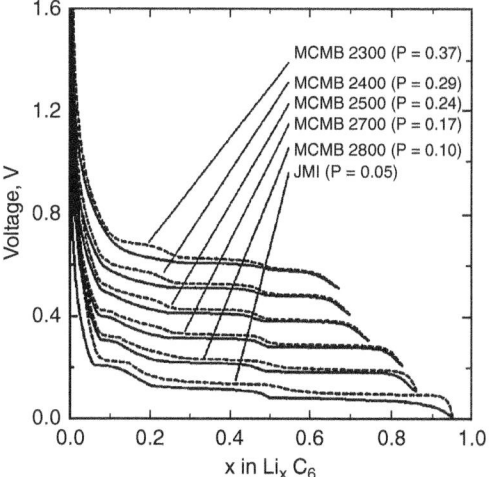

FIGURE 26.23 The potential-capacity profiles for the second discharge-charge cycle of lithiated synthetic graphite samples that were initially heated to different temperatures, as listed in the legend. Note that in Fig. 26.23 the curves have been sequentially offset by 0.1 V for clarity.[49] (Reproduced with permission from *Phys. Rev. B.* **51**, 734 [1995].)

TABLE 26.5 Structural Parameters of the Graphitic Carbons Described by Figs. 26.22, 26.23, and 26.24

Sample	Heat-treatment temperature (°C)	$d_{(002)}$ (Å)	P
JMI	?	3.356	0.05
MCMB2800	2800	3.352	0.10
MCMB2700	2700	3.357	0.17
MCMB2600	2600	3.358	0.21
MCMB2500	2500	3.359	0.24
MCMB2400	2400	3.363	0.29
MCMB2300	2300	3.369	0.37

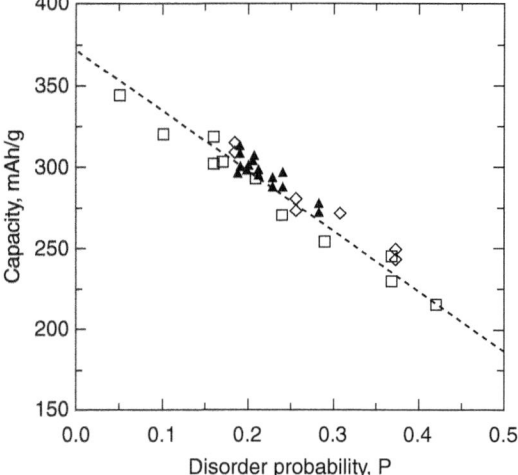

FIGURE 26.24 Reversible specific capacity of Li/graphite cells plotted vs. the fraction P of adjacent graphene layers in turbostratic misalignment.[49] (Reproduced with permission from *Phys. Rev. B*. **51**, 734 [1995].)

During the first electrochemical reaction of lithium with graphite, some of the lithium transferred to the graphite reacts with electrolyte to form a passivation layer at the electrode-electrolyte interface, which is commonly referred to as the solid electrolyte interphase (SEI). The passivation layer contains lithium that is no longer electrochemically active, thus the formation of the SEI results in irreversible capacity (IRC). The capacity difference between the charge and discharge curves in Fig. 26.25 results from irreversible capacity. This initial capacity loss is an undesirable property of all current materials and occurs largely on the first cycle. However, once formed, the SEI layer protects the graphite surface from further reaction with the electrolyte.

Properties of Commercial Graphites for Li-ion Battery Negative Electrodes. The most significant selection factors for a graphite powder to be used in a negative electrode material for a Li-ion battery include the fraction of the layers in turbostratic misalignment, the specific surface area, the content of impurities, the particle size, and the tap density.

The fraction of layers in turbostratic misalignment is characterized by the parameter *P*, which represents the probability of turbostratic misalignment in adjacent parallel layers. Because *P* is

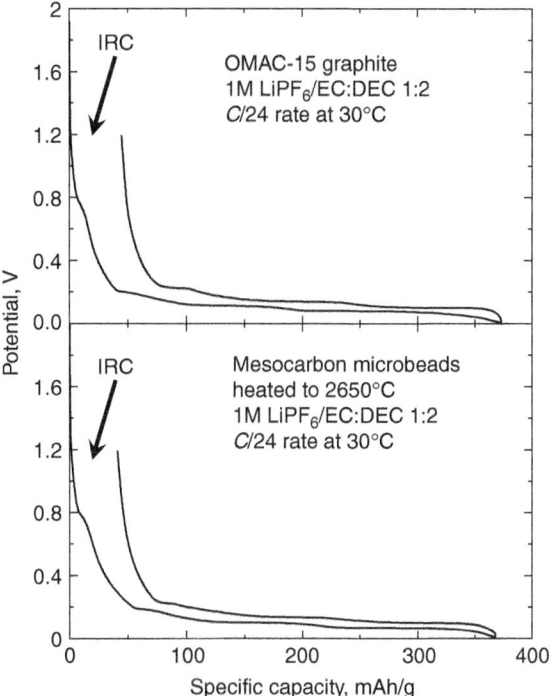

FIGURE 26.25 First lithiation and delithiation of graphite electrodes in Li/graphite cells illustrating the irreversible capacity (IRC) associated with the formation of the SEI layer on the graphite surface. *(Courtesy of Aaron Smith.)*

correlated to the heat-treatment temperature and to the d(002) spacing (the spacing between graphene sheets, see Table 26.5), the heat-treatment temperature and the lattice parameters can often be used to characterize the graphite. As shown in Fig. 26.24, the reversible specific capacity of a graphite is directly correlated to P. P can be reduced by increasing heating time and/or temperature, however this contributes to increased product cost.

The reactions that form the SEI and contribute to the irreversible capacity are proportional to the specific surface area. Fong et al. showed that a linear relationship exists between specific surface area and irreversible capacity.[51] In addition, the reactivity of the negative electrode with electrolyte under abuse conditions also scales with the surface area,[52] thus suggesting that surface area should be minimized to avoid irreversible capacity and improve abuse performance. However, to obtain a reasonable rate capability, the specific surface area cannot be too small because this would require use of undesirably large particles. Although graphite is normally a flaky material, mesophase graphites and spherical or "potato-shaped" graphites are commonly found in Li-ion cells because a spherical geometry minimizes surface area and also minimizes Li-diffusion path lengths in the material.

Impurities normally found in graphite are not detrimental to cell performance but contribute dead weight to the electrode. Natural graphites undergo a stringent purification process to reduce impurity content.

The particle-size distribution and tap density are not always independent of the specific surface area and control the ability to make well-controlled electrode coatings and electrodes having high density after compression. High-density electrodes are important for maximizing Li-ion cell energy density.

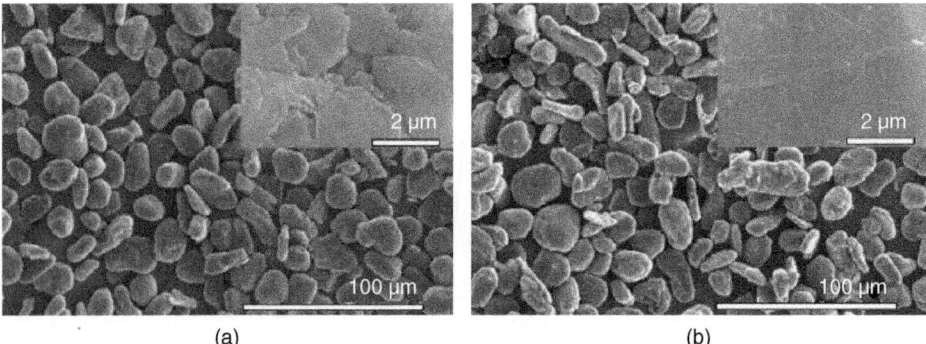

FIGURE 26.26 SEM images of (a) uncoated and (b) coated natural graphite.[54] (Reproduced with permission from *Journal of Power Sources* **190**, 553–557 [2009].)

Recent advances in graphitic materials for Li-ion batteries include the application of carbon coatings on the surface of graphites to reduce surface area and inhibit exfoliation. This reduces irreversible capacity and improves the thermal stability of the charged electrode materials in electrolyte. Nozaki et al.[53] and Park et al.[54] of the graphite producers Toyo Tanso (Japan) and Carbonix Inc. (Korea), respectively, have discussed the application of coatings on natural graphite. Figure 26.26 shows SEM images of uncoated and coated natural graphite.

The insets in Fig. 26.26 shows that the "craggy" surface of the natural graphite is "smoothed" by the coating, while the lower magnification sections of the images show that the coating does not change the overall particle size or shape.

The application of the coating improves the electrochemical behavior of the materials described by Park et al.[54] as follows:

The initial discharge capacity (intercalation) of the uncoated NG [natural graphite] was 415 mAh/g and the charge capacity (deintercalation) was 362 mAh/g, which results in a coulombic efficiency of 87.2% on the first cycle. The carbon-coated NG delivered 374 mAhg of discharge capacity and 348 mAh/g of charge capacity with a coulombic efficiency of 93.0%. The reversible capacity of the carbon-coated NG slightly decreased, compared to that of the unmodified NG (362 mAh/g). However, the irreversible capacity of the carbon-coated NG drastically decreased from 54 to 26 mAhg with improved coulombic efficiency. The enhanced coulombic efficiency and the reduced irreversible capacity of the carbon-coated NG on the first cycle can be attributed to a decrease in BET specific surface area and the existence of non-graphitic carbon on the surface structure. The measurement of the BET specific surface area for the unmodified NG gave 5.67m^2/g, whereas that for the carbon-coated NG was only 0.6 m^2/g.[54]

The same materials were tested for their reactivity with electrolyte under abuse conditions by DSC methods. In such experiments, charged LiC_6 is placed with electrolyte in a sealed DSC capsule and the sample is heated while the heat flow from the sample is monitored. Figure 26.27 shows the cumulative heat generated up to a temperature, T, plotted versus T for the uncoated and coated samples described in Fig. 26.26. The coated sample shows far less heat generation between 200 and 350°C. This is due to the lower specific surface area, which slows reaction kinetics by creating a thicker film of reaction products on the particle surfaces. This slows the reaction between the intercalated lithium and the electrolyte.

Table 26.6 summarizes the properties of commercially available graphites from a number of manufacturers around the world. As pointed out above, those materials with the smallest specific surface area generally have the lowest irreversible capacity or the highest first cycle efficiency. Most of these materials provide capacity near the theoretical capacity of graphite, which is 372 mAh/g, indicating that P in these materials is about zero.

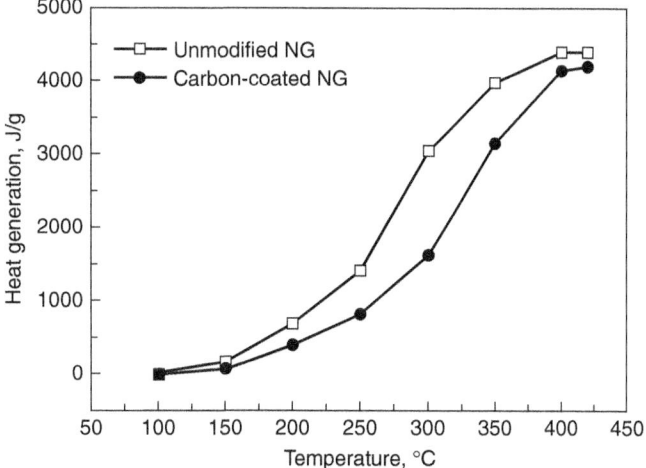

FIGURE 26.27 Overall heat evolution of uncoated and carbon-coated natural graphite (NG) electrodes in a fully-lithiated state as function of temperature.[54] (Reproduced with permission from *J. of Power Sources* **190**, 553–557 [2009].)

TABLE 26.6 Properties of Graphitic Carbons for Li-ion Battery Negative Electrodes

Manufacturer	Grade	Tap density (g/mL)	BET surface area (m²/g)	Particle size (d_{50}) μm	Reversible capacity (mAh/g)	First cycle efficiency (%)
BTR New Energy Materials Inc.	SAG artificial graphite	>1.0	4–5	18–22	>310	>89
BTR New Energy Materials Inc.	818 natural graphite	>1.1	1–2	17–19	>360	>95
China Steel	MGP	1.4	<1.2	22	>330	>90
China Steel	CMG_M	>1	<2	16–26	>345	>90
Gelon (China)	MCMB high cap.	>1.29	<1.0	22–26	>330	>91
Henglide (China)	HSG-20	>1	3.5–7.5	20±1	360	>86
ShanShan Tech. (China)	3H	0.76	3.2	19.5	>355	92.7
Timcal	SLP-50 Potato shape	N/A	6.5	21–25	N/A	N/A
Conoco (USA)	G-15	1.1	<1.4	12–5	330	95
Osaka Gas	OMAC 1.5	1.18	1.3	21.4	356	93.8
Carbonix (Korea)	Coated NG	N/A	0.6	N/A	348	93

Lithium Titanate Negative Electrode Material, $Li_{4/3}Ti_{5/3}O_4$.

Overview. Lithium titanate, $Li_{4/3}Ti_{5/3}O_4$ (i.e., $Li_4Ti_5O_{12}$, but commonly written $Li_{4/3}Ti_{5/3}O_4$ in view of its spinel structure), is an alternative to graphite for Li-ion cells that require extremely long cycle life and better safety characteristics. This is because the potential of lithium titanate (LTO) is about 1.55 V versus Li, therefore the reactivity of LTO with electrolyte is less than LiC_6. However, this penalty of about 1.5 V in the terminal voltage of Li-ion cells using LTO and LTO's low packing efficiency compared to graphite leads to Li-ion cells of lower energy density and specific energy than cells using graphite. However, such cells may find application in stationary applications where extremely long cycle life is required, such as grid-energy storage.

Some of the first studies of LTO were by the Murphy,[55] Dahn,[56] and Ohzuku groups.[57] LTO has a spinel structure like that of $LiMn_2O_4$ but with some Li occupying 16d metal sites. LTO is the end member of the solid solution series $Li_{1+x}Ti_{2-x}O_4$ with x = 1/3.[56]

The electrochemical behavior of LTO is shown by curve (e) in Fig. 26.13. The charge-discharge curve is a plateau that represents the coexistence between two phases, the $Li_{4/3}Ti_{5/3}O_4$ starting material and the $Li_{7/3}Ti_{5/3}O_4$ fully lithiated end phase. Ohzuku et al.[57] showed that these two phases have exactly the same lattice constant, thus the intercalation-deintercalation reaction proceeds topotatically, i.e., without any volume change. Ohzuku called this material a "zero-strain" insertion electrode material. Ohzuku and others believe that this is one reason that LTO shows excellent charge-discharge cycle performance.

Characteristics of Commercially Available $Li_{4/3}Ti_{5/3}O_4$. LTO ($Li_{4/3}Ti_{5/3}O_4$) is now commercially available from a number of suppliers, including Süd-Chemie, NEI Corporation, and BTR New Energy Materials Inc., among others. Figure 26.28 compares SEM images of LTO produced by BTR with that produced by Sud-Chemie. The SEM of the Süd-Chemie material is a high magnification image of agglomerates with a particle size of about 9 μm. Typical materials produced today achieve capacities of about 160 mAh/g, which is very close to the theoretical limit of 170 mAh/.

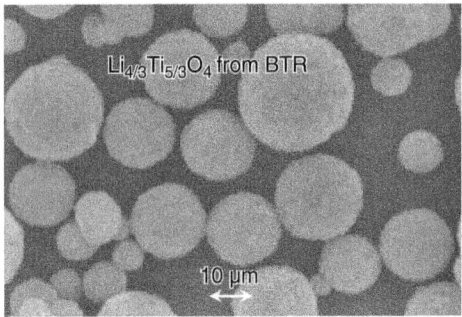

FIGURE 26.28 SEM micrographs of LTO. Notice that the BTR material is aggregated into spherical particles. (*Courtesy of BTR and Phostech Lithium Inc./Süd-Chemie AG.*)

Table 26.7 compares the properties of several LTO grades from several manufacturers. A comparison of the two Süd-Chemie grades shows the importance of high specific surface area in obtaining high-rate capability for this material, as further illustrated in the half-cell data shown in Fig. 26.29. In addition, LTO materials are exceptionally stable and can provide thousands of cycles, as illustrated in Fig. 26.30.

TABLE 26.7 Properties of $Li_{4/3}Ti_{5/3}O_4$ from Several Manufacturers

Manufacturer	Grade	d(50) (μm)	Surface area (m²/g)	Tap density (g/mL)	Specific capacity @ C-rate (mAh/g)	Specific capacity (mAh/g) @ rate
Süd-Chemie	EXM 1979	9	10	0.65	160	150 @ 20C
Süd-Chemie	EXM 1037	2.3	3	1.25	150	60–80 @ 20C
BTR New Energy Materials Inc. (China)	LTO anode material	≈10	N/A	N/A	≈165	N/A
Gelon (China)	LTO	0.7	11	1.03	>160	N/A

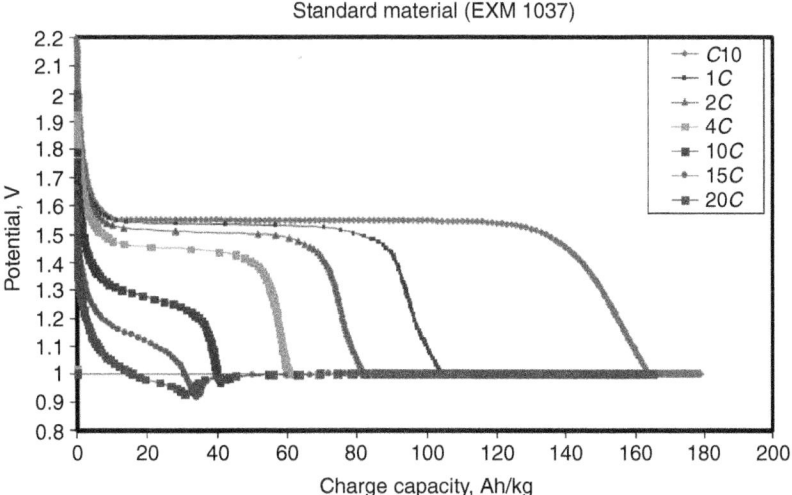

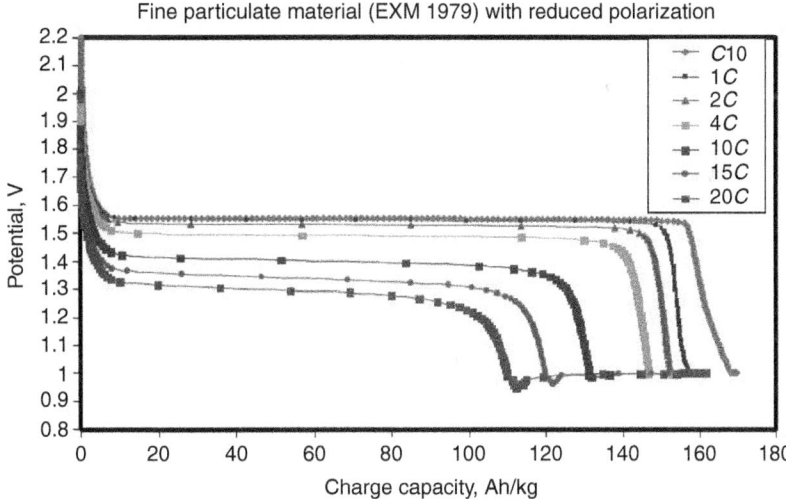

FIGURE 26.29 Charge capacity of Süd-Chemie EXM 1037 LTO, which has primary particle size of about 1 μm, and EXM 1979, which has a primary particle size of 100 to 200 nm, at various rates. (*Courtesy Phostech Lithium Inc./Süd-Chemie AG.*)

Sn and Si-Containing Negative Electrode Materials.

Overview. During the last decade there has been an enormous amount of work on Si and Sn-containing negative electrode materials for Li-ion batteries. This is because Si and Sn offer significant volumetric and gravimetric energy advantages over graphite. However, at this point only Sony is mass-producing large quantities of Li-ion cells with Sn-based negative electrodes. No manufacturer is currently producing cells with a Si-based negative electrode on a commercial scale. It is likely that more cells using these materials will appear between 2010 and 2020, so it is worth spending time briefly discussing these materials here. For example, Panasonic has announced a 4.0 Ah 18650-size cell for introduction in 2013 which uses a NCA/Si-based alloy cell chemistry.

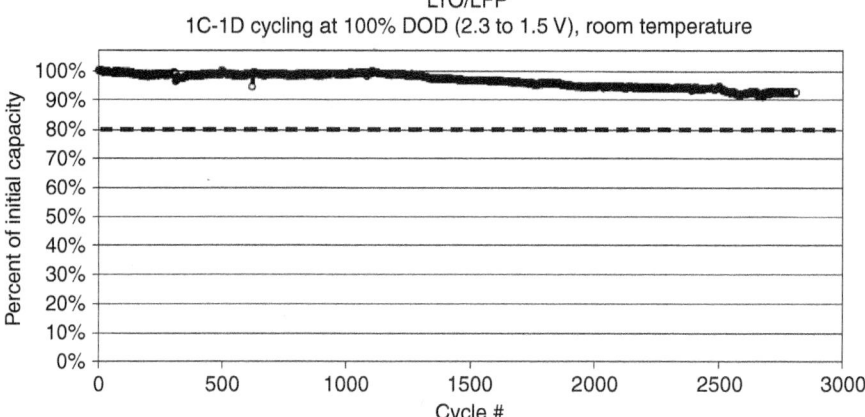

FIGURE 26.30 Cycle life of a LFP/LTO cell using Süd-Chemie EXM 1037 LTO when cycled at a 1C rate at 100% DOD. (*Courtesy of Süd-Chemie AG.*)

Figure 26.31 shows the specific and volumetric capacities of various Li-metal alloys compared to graphite. This figure clearly illustrates why the switch to alloy-based negative electrode materials would be attractive. However, all the alloys undergo large volume changes (up to 280%) when they react with lithium. This presents numerous challenges in electrode material design and electrode design, which have been partly overcome.

It is beyond the scope of this chapter to describe the path to Si and Sn-based materials that are now almost commercially viable. Instead, a recent review by Todd et al.[58] and recent papers by Obrovac et al.[59] describe strategies to make successful electrode materials and electrodes from alloys of Sn and Si, respectively. Silicon suboxide (SiO_x, where $x \sim 1$) is another material being widely considered for high capacity negative electrodes. Si-O is composed of nanosilicon grains in a matrix of SiO_2. According to recent patent applications by Shin-Etsu Corporation, SiO_x can be made by vapor deposition from mixtures of silicon and SiO_2 powders heated to high temperatures. Although SiO_x has similar volumetric and specific capacity as alloy materials, it suffers from high irreversible capacity loss (~50%) due to the formation of lithium oxide or lithium silicates during initial lithiation.[60] In order to compensate for the high irreversible capacity loss from materials like SiO_x,

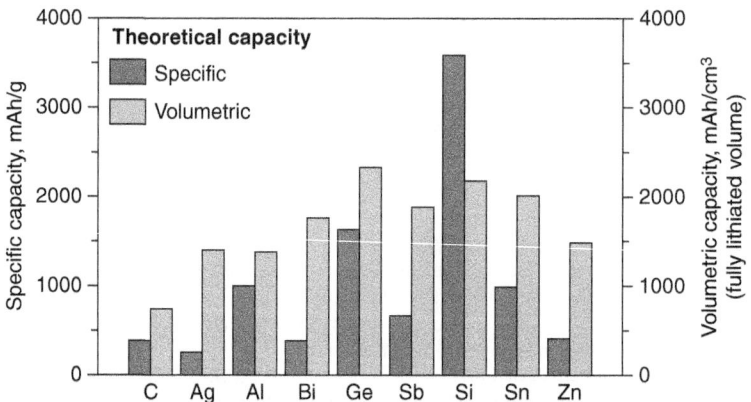

FIGURE 26.31 Specific and volumetric capacities of Li alloys compared to LiC_6. The volumetric capacities were calculated based on the fully lithiated volume.

FMC Corporation has developed an air-stable form of lithium metal powder (SLMP) that may be incorporated in the anode during cell assembly. It remains to be seen if the use of this powder will be accepted by major battery manufacturers.

Although there has been much recent fanfare about the importance of nanomaterials for batteries,[61] nanometer-sized Si and Sn based particles are unlikely to be useful for Li-ion cell negative electrodes. This is because negative electrode materials with high specific surface area show large irreversible capacity and poor thermal stability under cell abuse conditions. On the other hand, nanostructured or amorphous materials, such as alloys or SiO_x, which comprise micron-size particles made up of nanometer-size grains of electrochemically active and inactive materials, are very useful. The material used in the Sony Nexelion cell is a nanostructured material.

Characteristics of the Sony Nexelion Sn-Co-C Negative Electrode Material. Sony has released some information about the Sn-based negative electrode material in their Nexelion cells in a press release, at conferences, and in Japanese language journals.[62] Figure 26.32 shows the proposed nanostructure of the Sn-Co-C negative electrode material, which consists of nanometer-size Co-Sn grains in a carbon matrix.

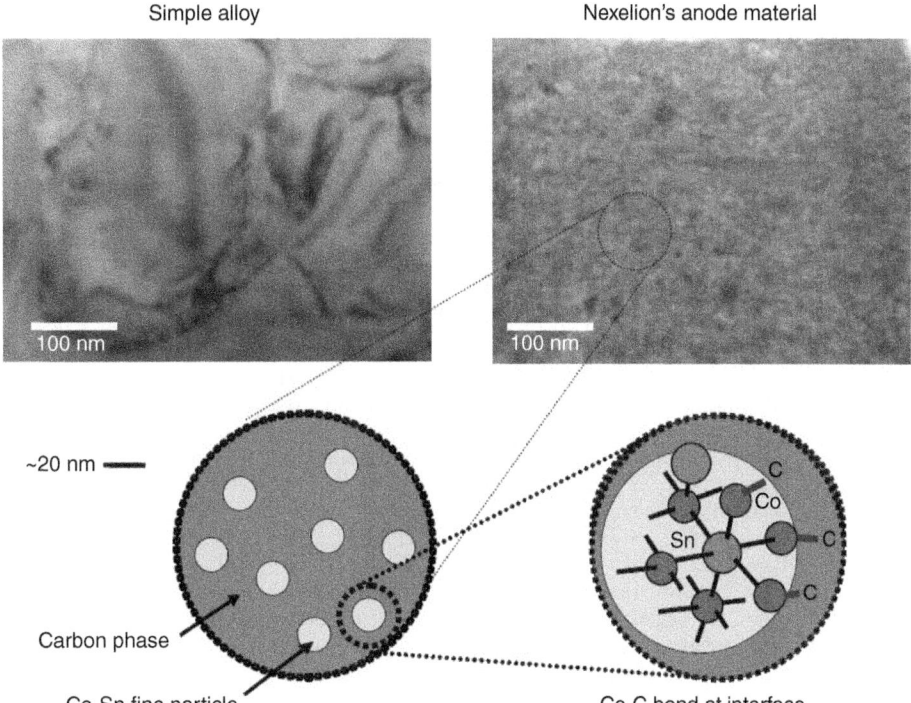

FIGURE 26.32 Fine structure of the Sn-Co-C negative electrode material. Top: TEM images; bottom: structural model.[63] (*Courtesy of Sony Energy Devices.*)

The proposed nanostructure shown in Fig. 26.32 has been confirmed by TEM studies, by the Whittingham group,[64] and by small angle neutron scattering studies by Todd et al.[65]

To our knowledge, Sony has never published the exact composition nor the electrochemical characteristics of their materials as measured in half cells. However, work by Todd et al.[58] appears to recreate the characteristics of Sony's Sn-Co-C negative electrode material. Todd et al. used magnetron sputtering or mechanical attrition to prepare $Sn_{30}Co_{30}C_{40}$ which had a nanostructure much like that shown in Fig. 26.32. Figure 26.33 includes a graph of specific capacity versus cycle number for three

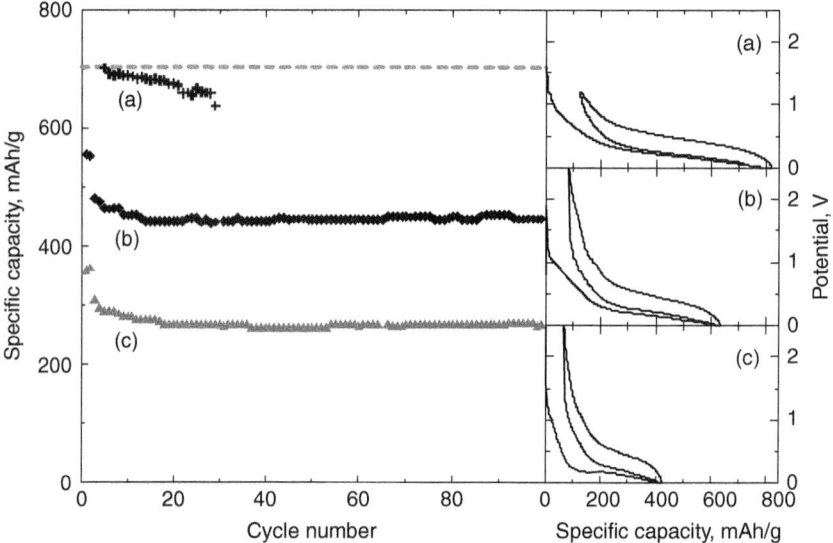

FIGURE 26.33 Specific capacity vs. cycle number and potential vs. specific capacity for Li/$Sn_{30}Co_{30}C_{40}$ cells: (a) material prepared by sputtering; (b) material prepared by mechanical attrition starting from $CoSn_2$, Co, and C; and (c) material prepared by mechanical attrition starting from CoSn and C.[58]

Li/$Sn_{30}Co_{30}C_{40}$ coin-type test cells and graphs of potential versus specific capacity of the same cells. Material prepared by sputtering provides the theoretical capacity of about 700 mAh/g, but the sputtering process would not be expected to be economical on a commercial scale. Material prepared by mechanical methods can have specific capacities of about 400 to about 500 mAh/g and good capacity retention, depending on the preparation method, as shown in Fig. 26.33. The attritted materials have densities (measured by He pychnometry) of about 6.5 g/mL and provide high capacity density.

$Sn_{30}Co_{30}C_{40}$ has a reversible volume change of about 150% if it reaches the theoretical specific capacity as shown by Tian et al.[66] In order to buffer this large volume change, it is believed that Sony adds about 50% by weight of graphite to the negative electrodes in the Nexelion cell. Furthermore, special electrolyte additives, like fluoroethylene carbonate (FEC), and special binder materials are required to form an SEI that is stable on the $Sn_{30}Co_{30}C_{40}$ particle surfaces under such large volume changes.[67]

Co is a very expensive raw material. Therefore, it is important to reduce or eliminate Co from the Sn-Co-C negative electrode material to reduce cost. Ferguson et al.[68] showed that it is possible to replace about 50% of the Co by Fe in these materials without causing substantial performance degradation. Disassembly and analysis of recent Nexelion-type cells suggests this substitution has been made.[69] Considering that it took about a decade for all Li-ion battery researchers and manufacturers to realize that the best carbonaceous material to use for negative electrodes was highly graphitized (P near zero) graphite, it may take several decades for the "best" alloy negative electrode material to be identified because there are so many possible elements that can be combined with Si and/or Sn to produce alloys of the appropriate nanostructure. The challenge for all these alloys will be to reduce cost to be competitive with graphite.

26.2.4 Nonaqueous Lithium Electrolytes

Two main types of electrolytes are used in Li-ion batteries: liquid electrolytes and gel electrolytes. Liquid electrolytes are solutions of a lithium salt in one or more organic solvents, typically carbonates.

As described in a monograph,[70] a gel electrolyte is an ionically conductive material wherein a salt and a solvent are dissolved or mixed with a high molecular weight polymer. Gel electrolytes developed for Li-ion batteries are typically films of PVDF-HFP, LiPF$_6$, and a carbonate solvent(s). Fumed silica may be added to the PVDF-HFP film for additional structural integrity. Gel electrolytes can also be formed by including a polymerizable monomer in the electrolyte which is thermally or otherwise crosslinked after cell assembly (see Sec. 26.3.3). This method helps bind the cell components together, which is especially important in prismatic-type cells. Another possible advantage of gel electrolytes is that the liquid phase is absorbed within the polymer, and thus is less likely to leak from a battery. However, in a typical Li-ion battery employing a liquid electrolyte, the electrolyte is almost completely absorbed into the electrode and separator materials. In the marketplace and the literature, gel electrolytes are often termed gel-polymer electrolytes, and cells that employ a gel (or gel-polymer) electrolyte are often termed gel-polymer or simply polymer cells.

Most Li-ion electrolytes in current use utilize LiPF$_6$ as the salt because LiPF$_6$ solutions offer high ionic conductivity, 10^{-2} S/cm, have a high lithium-ion transference number (~0.35), and provide acceptable safety properties. As reviewed below, many other salts have attracted industrial interest, notably LiBF$_4$, LiN(CF$_3$SO$_2$)$_2$, and lithium bis-oxalato borate (LiBOB). Electrolytes in current use are formulated almost exclusively with carbonate solvents. Carbonates are aprotic, polar, and have a high dielectric constant, thus can solvate lithium salts to high concentration ($\geq$1 M). They also provide compatibility with cell electrode materials over a broad range of potential. While the industry focused initially on propylene carbonate (PC)-based solutions, current formulations utilize ethylene carbonate (EC), dimethyl carbonate (DMC), ethyl methyl carbonate (EMC), and diethyl carbonate (DEC) as well as some PC. PC used alone, without EC or small additions of LiBOB, can cause degradation in graphite electrodes as it co-intercalates with lithium, resulting in exfoliation. The choice of solvents for a Li-ion electrolyte is also influenced by any low-temperature requirements of the application. Low-temperature electrolytes utilize low-viscosity solutions with low freezing points.

Salts. Salts commonly used in Li-ion cells are listed in Table 26.8. Most cells currently marketed use LiPF$_6$, as their solutions have high conductivity and good safety properties. However, the salt is relatively costly, hygroscopic, and LiPF$_6$ yields hydrofluoric acid (HF) upon reaction with water, thus must be handled in a dry environment. Organic salts have also been developed. They are more stable to water, thus easier to handle. In particular, lithium bistrifluoromethanesulfonimide (LiN(CF$_3$SO$_2$)$_2$) has received significant attention as its use as an additive in conventional electrolytes results in improved high-temperature cell performance and reduced gas formation.

TABLE 26.8 Salts Commonly Used in Li-ion Electrolytes

Name	Formula	Mol. weight (g/mol)	Typical impurities	Comments
Lithium hexafluorophosphate	LiPF$_6$	151.9	H$_2$O, HF	Most common salt
Lithium tetrafluoroborate	LiBF$_4$	93.74	H$_2$O, HF	Less hygroscopic than LiPF$_6$
Lithium bisoxalatoborate	LiB(C$_2$O$_4$)$_2$	193.7	H$_2$O	Can help SEI
Lithium bistrifluoromethane sulfonimide	LiN(CF$_3$SO$_2$)$_2$	286.9	H$_2$O	Can reduce gassing and improve high-temperature cycle life

Solvents. A wide variety of solvents, including carbonates, ethers, and acetates, have been evaluated for nonaqueous electrolytes. The industry has now focused on the carbonates as they offer excellent stability, good safety properties, and compatibility with electrode materials. Neat carbonate solvents typically have an intrinsic solution conductivity of less than 10^{-7} S/cm, dielectric constant $\geq$3, and solvate lithium salts to a high concentration. Table 26.9 presents the properties of some commonly used solvents.

Electrolyte formulations in current Li-ion cells typically utilize three to five solvents (excluding additives, which will be addressed later). Formulations with multiple solvents can provide better cell

TABLE 26.9 Characteristics of Organic Solvents*

Characteristic	EC	PC	DMC	EMC	DEC	1,2-DME	AN	THF	γ-BL
Structure									
BP (°C)	248	242	90	109	126	84	81	66	206
MP (°C)	39	−48	4	−55	−43	−58	−46	−108	−43
Density (g/ml)	1.41	1.21	1.07	1.0	0.97	0.87	0.78	0.89	1.13
Viscosity (cP)	1.86 (40°C)	2.5	0.59	0.65	0.75	0.455	0.34	0.48	1.75
Dielectric constant	89.6 (40°C)	64.4	3.12	2.9	2.82	7.2	38.8	7.75	39
Donor number	16.4	15	8.7[70]	6.5[70]	8[70]	—	14	—	—
Mol. wt.	88.1	102.1	90.1	104.1	118.1	90.1	41.0	72.1	86.1

*EC = ethylene carbonate, PC = propylene carbonate, DMC = dimethyl carbonate, EMC = ethyl methyl carbonate, DEC = diethyl carbonate, DME = dimethylether, AN = acetonitrile, THF = tetrahydrofuran, γ-BL = γ-butyrolactone. (*From Refs. 71, 73*)

performance, higher conductivity, and a broader temperature range than is possible with a single solvent electrolyte. For example, as will be explained below, ethylene carbonate (EC) is associated with low irreversible capacity and low capacity fade when used in conjunction with graphitic negative electrodes. Ethylene carbonate is found in many commercial electrolyte formulations, but is a solid at room temperature. Multiple solvent formulations often include EC, thereby incorporating its desirable properties, while using other solvents to lower the freezing point and viscosity of the mixture. Further properties of electrolyte solvents and strategies for electrolyte formulation are included in several nice review articles.[71,72]

Conductivity of Electrolytes. The conductivity of 1 M $LiPF_6$ solutions in solvents commonly used in Li-ion electrolytes is provided in Table 26.10 and plotted in Fig. 26.34 at temperatures from −40 to 80°C. In general, these solutions offer high conductivity, 10^{-2} S/cm, and a few solvents, such as PC and EMC, offer good conductivity at low temperature and a high boiling point. MA and MF offer high conductivity, but cell performance is poor if either is present in an amount greater than about 25% by weight.

TABLE 26.10 Conductivity, in mS/cm, of 1M $LiPF_6$ Solutions in Various Solvents

Solvent*	−40.0°C	−20.0°C	0.0°C	20.0°C	40.0°C	60.0°C	80.0°C
DEC	—	1.4	2.1	2.9	3.6	4.3	4.9
EMC	1.1	2.2	3.2	4.3	5.2	6.2	7.1
PC	0.2	1.1	2.8	5.2	8.4	12.2	16.3
DMC	—	1.4	4.7	6.5	7.9	9.1	10.0
EC	—	—	—	6.9	10.6	15.5	20.6
MA	8.3	12.0	14.9	17.1	18.7	20.0	—
MF	15.8	20.8	25.0	28.3	—	—	—

*DEC = diethyl carbonate, EMC = ethyl methyl carbonate, PC = propylene carbonate, DMC = dimethyl carbonate, EC = ethylene carbonate, MA = methyl acetate, MF = methyl formate.
(*Courtesy of Merck KGaA, Darmstadt, Germany.*)

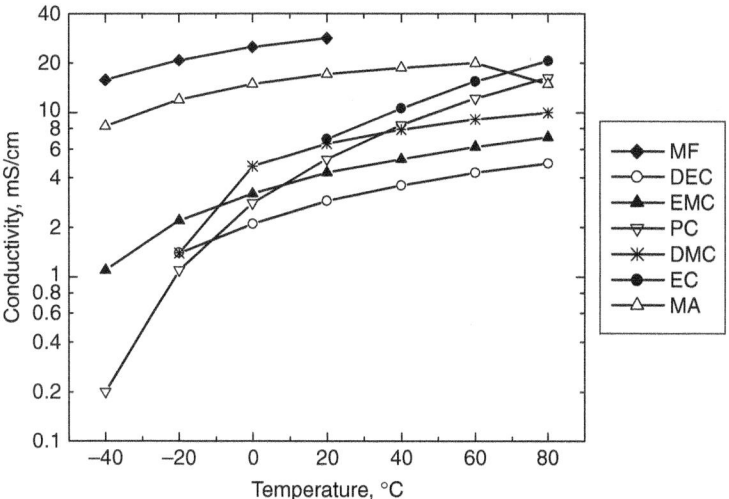

FIGURE 26.34 Conductivity (mS/cm) of 1M $LiPF_6$ solutions in various solvents. (*Courtesy of Merck KGaA, Darmstadt, Germany.*)

TABLE 26.11 Conductivity, in mS/cm, of LiPF$_6$ Solutions in Binary Mixtures, 1:1 by Weight. C = partially crystallized, S = saturated.

Solvents*	Concentration	–40°C	–20°C	0°C	20°C	40°C	60°C	80°C
EC:DEC	0.25 M	—	—	1.7 (C)	4.2	5.8	7.3	8.8
	0.50 M	—	2.5 (C)	3.0	6.4	8.7	11.1	13.6
	1.00 M	0.7	2.2	4.2	7.0	10.3	13.9	17.5
	1.25 M	0.4	1.7	3.6	6.4	9.7	13.5	17.4
	1.50 M	—	—	—	5.6	—	—	—
	1.75 M	—	—	—	4.8 (S)	—	—	—
EC:DMC	0.25 M	—	—	4.2	5.8	7.8	9.7	11.5
	0.50 M	—	—	6.5	9.3	12.8	16.0	19.1
	0.75 M	—	3.8	6.9	10.3	14.0	17.9	21.6
	1.00 M	—	3.7	7.0		15.0	19.5	24.0
	1.25 M	0.7	2.7	5.6	9.3	13.7	18.4	23.3
	1.50 M	—	2.2	5.4	9.3	14.1	19.2	24.7
	1.75 M	—	—	—	7.5	—	—	—
	2.00 M	—	—	—	6.7	—	—	—
	2.25 M	—	—	—	0.9 (S)	—	—	—
EC:EMC	0.25 M	—	—	3.7	5.3	7.2	9.1	10.9
	0.50 M	—	3.0	5.1	7.5	10.2	12.8	15.4
	1.00 M	0.9	2.7	5.3	8.5	12.2	16.3	20.3
	1.25 M	0.6	2.3	4.7	8.0	12.0	16.2	20.6
	3.50 M	—	—	—	0.9 (S)	—	—	—
EC:MP	0.25 M	2.4 (C)	4.6	6.3	8.3	10.4	12.4	—
	0.50 M	3.1 (C)	6.7	9.8	13.1	16.0	19.3	—
	1.00 M	3.8	7.8	12.2	17.1	22.3	27.3	—
	1.25 M	—	7.1	11.8	17.2	22.7	28.4	—
	3.0 M	—	0.5	2.1	5.2	—	15.4	21.8
	3.5 M	—	—	—	3.4 (S)	—	—	—
EC:MPC	1.00 M	C	1.5	3.6	6.3	9.5	12.9	16.8

*DEC = diethyl carbonate, EMC = ethyl methyl carbonate, DMC = dimethyl carbonate, EC = ethylene carbonate, MA = methyl acetate.
(*Courtesy of Merck KGaA, Darmstadt, Germany.*)

The conductivity of binary 1:1 mixtures of EC with common Li-ion electrolyte solvents over a range of salt concentrations and temperatures is given in Table 26.11. For many solvent pairs, conductivity is highest with 1 M LiPF$_6$, and these formulations are liquid from –40 to 80°C. The conductivity of 1 M LiPF$_6$ binary mixtures with EC is plotted in Fig. 26.35. As shown, the EC:MA mixture offers the highest conductivity although such levels of MA are associated with high capacity fade.[74] Other mixtures, including EC:DEC, EC:DMC, and EC:EMC, offer good conductivity and lower capacity fade. In particular, EC:EMC mixtures offer 0.9 mS/cm conductivity at –40°C and low capacity fade. The conductivity of LiPF$_6$ and a variety of organic salts in a mixture of PC:DME 1:1 at concentrations from 0.25 to 2 M is illustrated in Fig. 26.36. As shown in this figure, of the salts included, LiPF$_6$ offers the highest conductivity, 13 mS/cm at 1.2 M, although solutions with the organic salts offer comparable conductivity, up to 11 mS/cm.

The conductivity of 1 M LiPF$_6$ in selected ternary solvent mixtures is provided in Table 26.12 and plotted in Fig. 26.37. These mixtures contain 33% EC, as is common in some Li-ion electrolytes, and provide high conductivity and a broad temperature range, illustrating the utility of multiple component mixtures. For example, four of these mixtures provide at least 1 mS/cm at –40°C, of which three are liquid at 80°C. Quaternary solvent mixtures have also been developed to provide electrolytes with better low-temperature performance. The conductivity of 1.0 M LiPF$_6$ in various

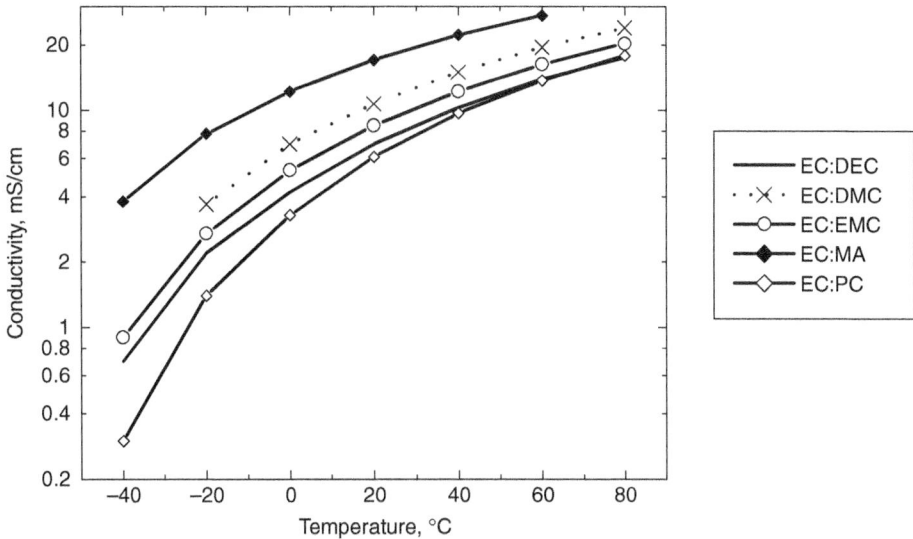

FIGURE 26.35 The conductivity of 1M $LiPF_6$ in various binary solvent mixtures, 1:1 by weight. (*Courtesy of Merck KGaA, Darmstadt, Germany.*)

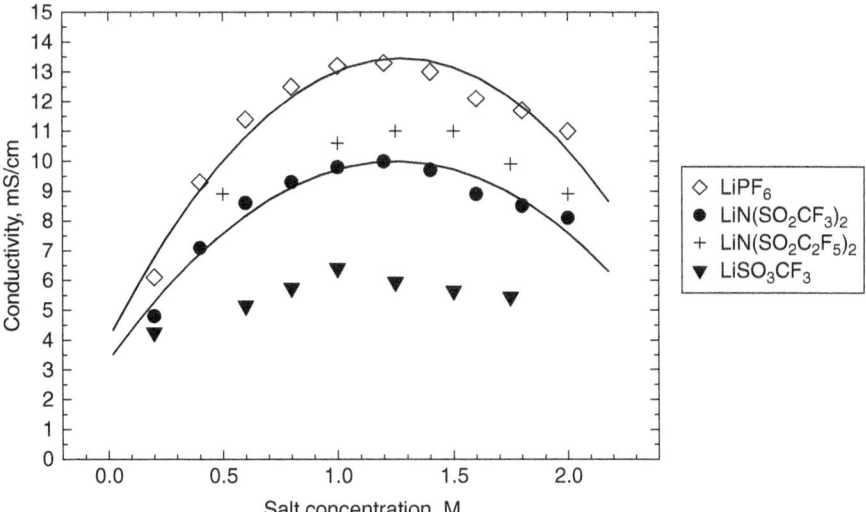

FIGURE 26.36 The conductivity of $LiPF_6$ and various organic salts in a mixture of PC:DME (1:1, V/V) at 20°C. (*Courtesy of 3M Co., St. Paul, Minnesota.*)

quaternary mixtures is shown in Fig. 26.38. As illustrated, these solutions offer over 1 mS/cm at −40°C and up to 0.6 mS/cm at −60°C.

A survey of the conductivity and solvent properties of electrolytes, including $LiAsF_6$, $LiPF_6$, $LiSO_3CF_3$, and $LiN(SO_2CF_3)_2$, and a wide variety of solvents, including EC, DME, ethyldiglyme, triglyme, tetraglyme, sulfolane, Freon, and methylene chloride, has been published.[75]

TABLE 26.12 The Conductivity, in mS/cm, of 1 M LiPF$_6$ in Various Ternary Solvent Mixtures.

Solvent*	Wt. ratio	−40°C	−20°C	0°C	20°C	40°C	60°C	80°C
EC:PC:DMC	20:20:60	—	—	6.9	10.6	14.5	18.4	22.2
EC:PC:EA	15:25:60	3	6.2	9.8	13.7	17.8	21.6	25.1
EC:PC:EMC	15:25:60	1	2.8	5.3	8.1	11.5	14.6	17.8
EC:PC:MA	15:25:60	4.1	8.1	12.9	17.8	22.8	27.6	boils
EC:PC:MPC	15:25:60	0.5	1.4	3.3	5.6	8.2	10.9	13.9
EC:DMC:EMC	15:25:60	1.4	3.2	5.3	7.6	10	12.1	14.1
EC:DMC:MPC	15:25:60	0.7	1.8	3.4	5.3	7.2	9	10.9

*DEC = diethyl carbonate, EMC = ethyl methyl carbonate, PC = propylene carbonate, DMC = dimethyl carbonate, EC = ethylene carbonate, MA = methyl acetate, MPC = methyl propyl carbonate, EA = ethyl acetate. (*Courtesy of Merck KGaA, Darmstadt, Germany.*)

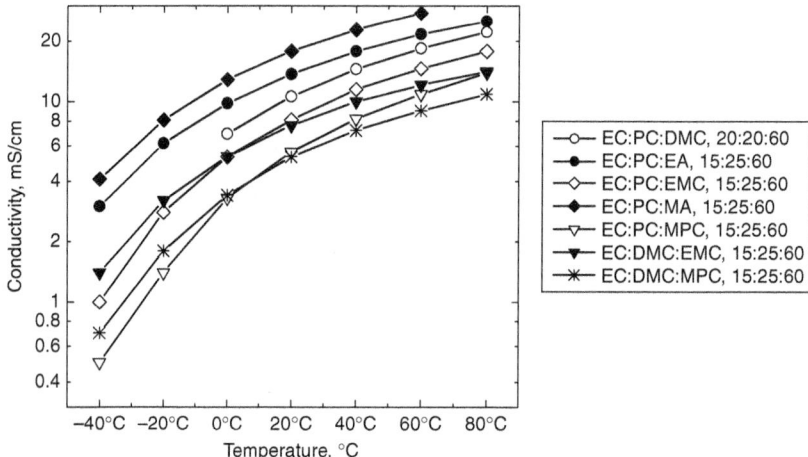

FIGURE 26.37 The conductivity of 1M LiPF$_6$ in ternary solvent mixtures. (*Courtesy of Merck KGaA, Damstadt, Germany.*)

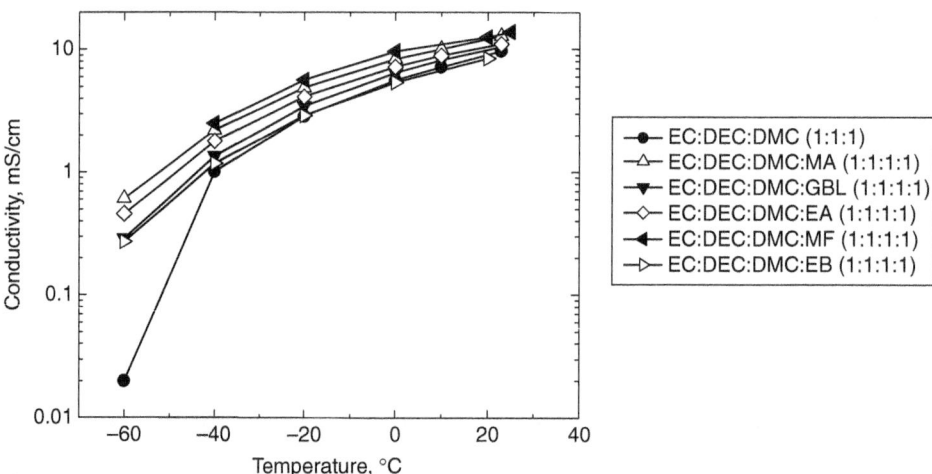

FIGURE 26.38 The conductivity of 1M LiPF$_6$ in quarternary solvent mixtures. (*Courtesy of Merck KGaA, Damstadt, Germany.*)

Electrolyte Formulation, Irreversible Capacity, and the SEI. For lithium-ion cells to function, the electrolyte must be stable at both the anodic and cathodic potentials found in Li-ion cells, which are about 0 V to about 4.4 V versus Li. No practical solvents are thermodynamically stable with lithium or Li_xC_6 near 0 V versus Li, but many solvents undergo a limited reaction to form a passivation film on the electrode surface. The resulting film spatially separates the electrolyte from the electrode, yet is ionically conductive, and thus allows passage of lithium ions. The passivation film, termed the solid electrolyte interphase (SEI), imparts extrinsic stability to the system, allowing the fabrication of cells that are stable for years without significant degradation.[74]

When the SEI is formed, lithium is incorporated into the passivation film. This process is irreversible and is observed as a loss of capacity, primarily on a cell's first cycle. The amount of irreversible capacity is dependent on the electrolyte formulation and the electrode materials, in particular the type of carbon used in the negative electrode. Because the reaction occurs at the surface of a particle, materials with low specific surface area typically offer lower irreversible capacity.

A poorly formed SEI layer can result in capacity fade and/or poor capacity retention due to the consumption of lithium by continuous reaction with the solvent at the electrolyte surface. Such cells also can suffer from high impedance, due to electrolyte depletion, and thick deposits of electrolyte decomposition products. Under normal circumstances, even a stable SEI layer can take many cycles to form completely. In order to establish a stable SEI layer on the anode surface, commercial cells typically undergo a regime of charging, floating, and discharging, sometimes at elevated temperatures. This cell "formation" step can take many weeks to complete.

Cells with electrolyte formulations that contain alkyl carbonates, in particular EC, have been shown to offer lower capacity fade, low irreversible capacity, and high capacity.[76] In EC containing electrolytes, the passivation film formed on the surface of Li-ion electrodes is formed with a minimum amount of lithium. This SEI has been shown to consist primarily of $Li_2(OCO_2(CH_2)_2OCO_2)_2$,[77] and related reaction products, such as Li_2CO_3 and $LiOCH_3$,[78] which are derived from the electrolyte solvent and either lithium or a lithiated species such as Li_xC_6. While solvents other than EC, typically esters or alkyl carbonates such as EMC or MPC,[79] also form stable passivation films, most solvents do not. If an ester or alkyl carbonate is not used, graphite can be cycled in a solvent that does not form a stable passivation film if an additive, such as a crown ether[80] or CO_2,[81] is added to the electrolyte. The use of additives can significantly change the chemistry of the SEI layer and enhance its ability to protect electrode surfaces from reaction with the electrolytes. The use of additives is discussed in detail in the next section.

26.2.5 Electrolyte Additives

Undesired reactions in Li-ion batteries, which lead to capacity loss, can occur at either the positive or the negative electrode. For example, as discussed in the previous section, the formation and repair of the SEI consumes lithium at the negative electrode, resulting in irreversible capacity loss.[82,83] Electrolyte oxidation at the positive electrode,[83] and the presence of impurities such as water or HF, can also lead to capacity loss. These reactions can also result in the production of gas, which is particularly serious in prismatic cell designs, where cell bulging and the loss of stack pressure can result. The use of high-purity electrolytes, electrode coatings,[84] and special electrode materials[85] can be used to improve cycle life and reduce gassing. In particular, the use of various electrolyte additives in the form of organic molecules, salts, inorganic compounds, or gases[91–97] can significantly enhance performance by improving SEI stability and scavenging HF or water.[86] Additives can also significantly improve cell safety characteristics, as will be discussed below. Xu[83] describes many aspects of Li-ion electrolytes and additives.

Additives are typically present in electrolyte formulations at less than about 10%. After cell formation by the manufacturer, many additives are completely consumed in the formation of the SEI layer and are difficult or impossible to detect in commercially available cells. Because additives can be retained as a trade secret, the identity and amount of additives used is highly proprietary, and as a result there is relatively little information regarding additives in the open literature in comparison to the knowledge of commercial cell makers. Nevertheless, a small number of additives are known to be in common use. For example, vinylene carbonate (VC)[87] was found to be highly beneficial and this additive is used today in many commercial Li-ion cells. Aurbach et al.[88] report that:

> VC is a reactive additive that reacts on both the anode and the cathode surfaces. The influence of this additive on the behavior of Li-graphite anodes is very positive since it improves their cyclability, especially at elevated temperatures, and reduces the irreversible capacity. Spectroscopic studies indicate that VC polymerizes on the lithiated graphite surfaces, thus forming polyalkyl Li-carbonate species that suppress both solvent and salt anion reduction. The presence of VC in solutions reduces the impedance of the $LiMn_2O_4$ and $LiNiO_2$ cathodes at room temperature. However, we have not yet found any pronounced impact of VC on the cycling behavior of the cathodes, either at room temperature or at elevated temperatures. Thus, VC can be considered as a desirable additive for the anode side in Li-ion batteries, one which has no adverse effect on the cathode side.

Broussely et al.[82] demonstrated the effectiveness of VC as an additive to promote the formation of a stable SEI that consumes little lithium after many cycles.

Lithium bis(oxalato)borate (LiBOB) has been found to dramatically affect the SEI that forms on the graphite electrode. The effect of LiBOB is pronounced in Li/graphite cells that use a 1 M $LiPF_6$/PC or a 0.8 M LiBOB/PC electrolyte.[89] In cells using a 1 M $LiPF_6$/PC electrolyte, Xu et al. report that exfoliation of graphite occurred and effectively no lithium intercalation took place, while in cells that also included 0.8 M LiBOB, exfoliation of graphite was effectively eliminated and lithium intercalation to form LiC_6 occurred. According to Xu et al.:

> Due to the presence of the BOB anion, the content of semicarbonate-like components in the graphite/electrolyte interface increases significantly, as indicated by the conspicuous peak located at 289 eV. These components, believed to originate from the oxalate moiety of the anion, are mainly responsible for the protection of graphitic anodes, either at elevated temperatures or in the presence of PC.[90]

Yamane et al.[86] demonstrate the effectiveness of adding hexamethyldisilazane to remove water from electrolytes and improve capacity retention of $LiMn_2O_4$-based cells. Abe et al.,[91] in a discussion of "functional electrolytes," describe how various additives create an "electro-conductive membrane" (ECM) on the positive electrode surface, leading to improved cycling of $LiCoO_2$/graphite cells. Li et al.[92] describe the use of heptamethyldisilazane for removing HF from $LiPF_6$-containing electrolytes, leading to improved storage properties of Mn-spinel-based cells. Patoux et al.[93] describe the use of a number of additives, including 1,3-propane sultone, to reduce the self-discharge of cells with the 4.7 V positive electrode $LiNi_{0.5}Mn_{1.5}O_4$. In an excellent review, Zhang[94] describes hundreds of additives that can be used to stabilize the SEI on graphite, protect the positive electrode, remove HF, and mitigate the effects of overcharge. El-Ouatani et al.[95] describe the impact of VC in $LiCoO_2$/graphite, $LiFePO_4$/graphite, and $LiCoO_2/Li_{4/3}Ti_{5/3}O_4$ cells. Abe et al.[96] show an unexpected synergistic effect between the additives propargyl methanesulfonate and vinylene carbonate. Wrodnigg et al.[97] report beneficial effects of ethylene sulfite on the cycle life of $LiMn_2O_4$ positive electrodes.

In addition to electrolyte additives that ultimately modify surface films, coatings on electrode particles can directly modify the surfaces in contact with electrolyte. Lee et al.[98] show the usefulness of a BiOF coating on LAMO electrodes in protecting against HF attack. Sun et al.[99] show the positive benefits of $(NH_4)_3AlF_6$ and AlF_3 coatings on NMC for improving capacity retention during high potential cycling at 55°C. Sun et al.[100] used a coating of a Co-poor NMC on $LiNi_{0.8}Mn_{0.1}Co_{0.1}O_2$ to improve cycle life and thermal stability. Li et al.[101] show that an $FePO_4$ coating on $LiCoO_2$ can improve capacity retention and thermal stability. Chen et al.[102] show that simple thermal treatments of "aged" $LiCoO_2$ improves its capacity retention when charged to 4.5 V. Patoux et al.[103] describe the impact of a number of coatings on the capacity retention of the 4.7 V spinel $LiNi_{0.5}Mn_{1.5}O_4$. Biphenyl is an additive that is often used in commercial cells as an overcharge protection additive. When exposed to high potential in an overcharged cell, biphenyl polymerizes, significantly decreasing the ionic conductivity of the electrolyte and effectively shutting down the cell. Other additives known to be in common use include ethylene sulfite (ES), fluoroethylene carbonate (FEC), and lithium bistrifluoromethanesulfonimide. A review of electrolyte additives can be found in Ref. 83.

There can be no doubt that electrolyte additives and electrode material coatings are beneficial and that Li-ion battery makers use them. However, because many additives are consumed during cell formation and because manufacturers do not disclose the additives they use, it is very difficult to determine what additives or coatings are employed within a particular Li-ion cell. It is reasonable to expect that a typical electrolyte will contain an HF scavenger, a water scavenger, an SEI modifier (e.g., VC), and an overcharge protection additive (e.g., biphenyl). As of this writing, the academic and research communities are far behind the cell manufacturers in developing a detailed understanding of additives.

A relatively recent development is the discovery of the redox shuttle electrolyte additive 2,5-di-tert-butyl-1,4-dimethoxybenzene (DDB), which can prevent the overcharge of $LiFePO_4$-based Li-ion cells.[1,2] Figure 26.39 shows the operation of two $LiFePO_4$/graphite 18650 cells that contain the DDB redox shuttle additive. These cells were charged at a constant current corresponding to a 10 h charging rate for a time of 20 h. After the normal 10 h charge period, the voltage of the cells rises to about 3.9 V and is held at about 3.9 V by the oxidation potential of the DDB molecule.[104] The lower panel of Fig. 26.39 shows that when the shuttle-protected overcharge occurs, the cell temperature rises. This is because the excess charge energy is dissipated as heat within the cell; no chemical changes to the electrode materials are taking place. The DDB shuttle has been shown to provide overcharge protection for many hundreds of overcharges of $LiFePO_4$-based cells. Such an additive may enable the introduction of lower cost $LiFePO_4$-based Li-ion batteries in retail blister packs, because the cells could be continuously charged with an inexpensive charger, provided that a 10 h charging rate is used.

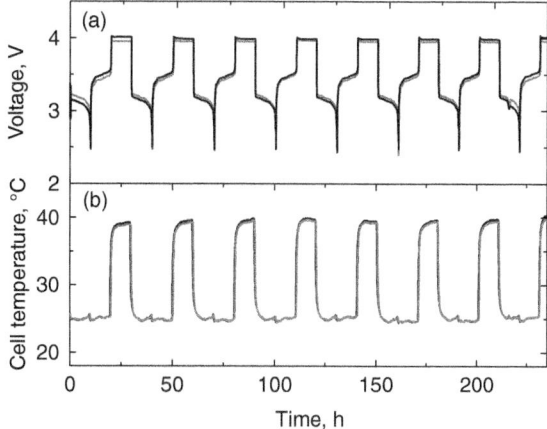

FIGURE 26.39 Cell terminal voltage (a) and cell surface temperature (b) vs. time for two 18650-size $LiFePO_4$/graphite cells charged and discharged at 140 mA (a 10 h rate). These cells include the redox shuttle electrolyte additive 2,5-di-tert-butyl-1,4-dimethoxybenzene at a concentration of 0.15 M. During shuttle-protected overcharge, the energy supplied to the cell is evolved entirely as heat and the cell temperature rises by about 14°C.

26.2.6 Separator Materials

Li-ion cells use thin (16 to 40 μm) microporous films to electrically isolate the positive and negative electrodes. To date, all commercially available liquid electrolyte cells use microporous polyolefin materials because polyolefins provide excellent mechanical properties, chemical stability, and acceptable cost. Nonwoven materials have also been developed but have not been widely adopted, in part because it is difficult to fabricate a thin and uniform nonwoven material having high strength.[105] Requirements for Li-ion separators include:

- High machine direction strength to permit automated winding
- Does not yield or shrink in width

- Resistant to puncture by electrode materials
- Effective pore size less than 1 μm
- Easily wetted by electrolyte
- Compatible and stable in contact with electrolyte and electrode materials

Microporous polyolefin materials in current use are made of polyethylene, polypropylene, or laminates of polyethylene and polypropylene. Also available are surfactant coated materials, designed to offer improved wetting by the electrolyte. These materials are fabricated by either a dry, extrusion-type process or a wet, solvent-based process.[106] The properties of commercial materials, including pore dimensions, porosity, and permeability, have been reported[107] and are available on the manufacturers' websites (see, for example, www.celgard.com/products/default.asp). Commercial materials offer pore size of 0.03 to 0.1 μm, and 30 to 50% porosity, as illustrated by the SEM micrographs of commercial separators in Fig. 26.40.

The low melting point of polyethylene (PE) materials enables their use as a thermal fuse. As the temperature approaches the melting point of the polymer, 135°C for polyethylene (PE) and 155°C for polypropylene (PP), porosity is lost.[108] Trilayer materials (PP/PE/PP) have been developed in which the polypropylene layers are designed to maintain the integrity of the film, while the lower melting point of the polyethylene layer is intended to shut down the cell if an over-temperature condition occurs. Shutdown occurs when the polymer melts and the pores close, stopping the transport of Li$^+$ ions from one electrode to the other. Obviously, such multi-component separators are very useful for helping to ensure Li-ion battery safety. Ultimately, what is needed is a separator that has a shutdown component and a second component that does not melt at any temperature. According to a recent press release, DuPont is developing a high-temperature battery separator for electric/hybrid vehicles to improve safety and enable higher power devices.

Some manufacturers have implemented high-temperature separators in commercially available cells. For example, Panasonic has disclosed improved safety and capacity using its "Heat Resistance Layer" (HRL) technology[109] that forms an insulating metal oxide layer between the positive and negative electrodes. The layer prevents the battery from overheating even if an internal short circuit occurs. Such layers have been incorporated by several manufacturers and are generally a thin

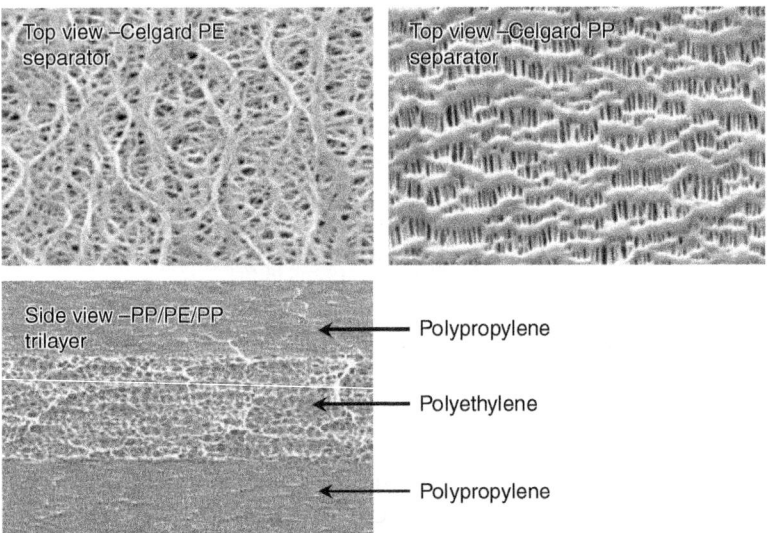

FIGURE 26.40 SEM micrographs of Celgard polyethylene (top left view), polypropylene (top right view) and PP/PE/PP trilayer (side view) separators. (*Courtesy of Celgard.*)

(several micrometers thick) coated layer of a refractory metal oxide (e.g., Al_2O_3) on the top surface of the negative electrode. This layer prevents direct electrical contact between the positive and negative electrodes should the separator melt and flow.

26.3 CONSTRUCTION

Cylindrical, prismatic, and so-called "polymer" Li-ion batteries are presently being mass-produced by more than 100 manufacturers worldwide. Wound (cylindrical or prismatic) designs are typical in small cells (< 4 Ah), however, in larger cells prismatic configurations having a flat-plate or stacked construction are commercially available. For prismatic and "polymer" Li-ion cells, two cell design types are commonly practiced: flat-mandrel wound pseudo-prismatic designs and stacked, true-prismatic designs.

Because Li-ion cells are fabricated in the discharged state, they must be charged before use. As previously mentioned, the first few charge and discharge cycles of a lithium-ion cell are done by the manufacturer in a controlled procedure termed formation. During formation, the cell capacity and voltage profile can be carefully monitored as a quality control check. Cells are normally discharged to about 50% capacity and stored for several weeks while the open-circuit voltage is monitored. This procedure allows any self-discharging cells (normally less than 0.5% for automated manufacturers) to be identified and removed.

26.3.1 Construction of Wound Li-ion Cells

The construction of a wound cylindrical Li-ion cell is illustrated in Fig. 26.41. The construction of a wound prismatic cell is similar except that a flat mandrel is used instead of a cylindrical mandrel. A schematic diagram of a wound prismatic cell is shown in Figure 26.42. The construction consists of a

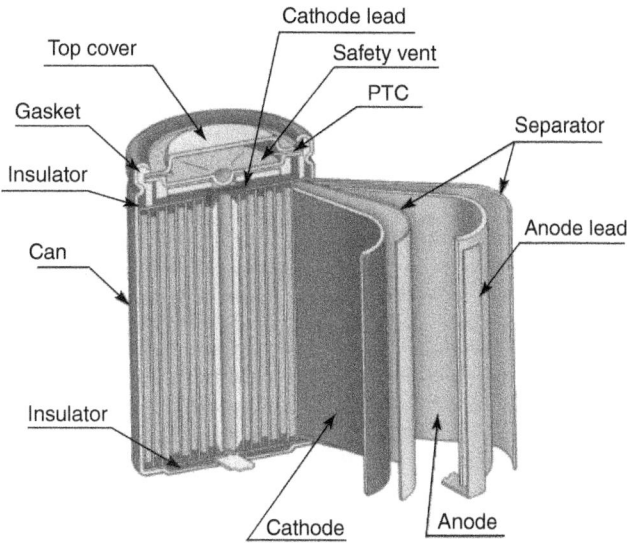

FIGURE 26.41 Schematic diagram of a wound cylindrical Li-ion cell. (*Courtesy of the University of South Carolina. Reproduced with permission from J. Power Sources.*)

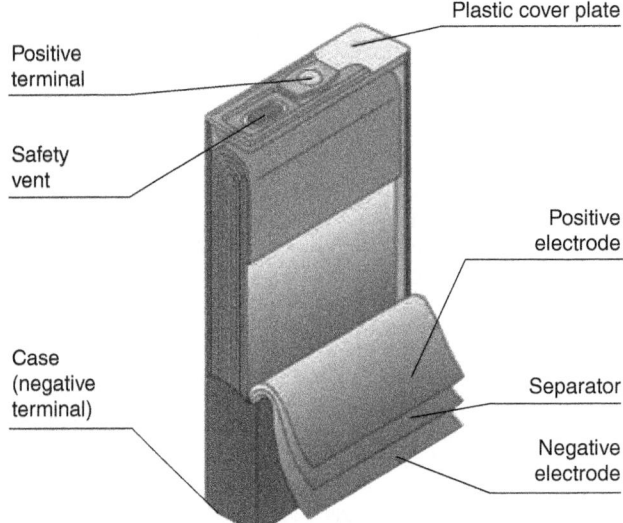

FIGURE 26.42 Schematic diagram of a wound prismatic cell. (*Courtesy of Japan Storage Battery Co., Ltd.*)

positive and negative electrode separated by a 16 to 25 μm thick microporous separator as described in Sec. 26.2.6. Positive electrodes consist of 10 to 20 μm Al foil coated with the active material (on both sides) to a total thickness of typically 100 to 250 μm. Negative electrodes are typically 8 to 15 μm Cu foil coated with a graphite-type active material to a total thickness of 100 to 250 μm. In cells designed for power applications (where power density is maximized), the coatings are porous and thickness is limited by the low conductivity of nonaqueous electrolytes, which is about 10 mS/cm,[110] and slow Li$^+$ diffusion in the positive and negative electrode materials, which is about 10^{-9} cm^2/s. Typical coating thicknesses for power cells are about 50 μm on each side of the foil. Coatings for cells designed for energy applications (where energy density and specific energy are maximized) are typically highly densified. For energy cells, the coating thickness is limited by the ability of the coating to be handled during the manufacturing process without damage. This is especially true for wound cells, where coatings with excessive thickness can easily crack and flake off the current collector when wound. Coatings for energy cells are typically about 125 μm thick on each side of the foil. Cells designed for intermediate applications requiring both power and energy have intermediate electrode thicknesses. In energy cells, a single tab at the end of the wind is used to connect the current collectors to their respective terminals. Multiple tabs spaced along the current collector are a common feature in power cells. The case, commonly used as the negative terminal, is typically nickel-plated steel. When used as the positive terminal, as in many designs, the case is typically aluminum. Most commercially available cylindrical or prismatic cells utilize a header that incorporates one or more disconnect devices, which are activated by temperature or pressure, such as a PTC device or a safety vent. (See App. H.) One design is illustrated in Fig. 26.43. The header-can seal is typically formed through a crimp.

26.3.2 Construction of Stacked Li-ion Batteries

The construction of a stacked prismatic cell is illustrated in Fig. 26.44. As in a wound cell, a microporous separator separates the positive and negative electrodes. Typically, each plate in the cell has a tab, and the tabs are bundled and welded to their respective terminals or to the cell case. Cell cases of either aluminum, nickel-plated steel, or 304L stainless steel have been used. As shown, the cover typically incorporates one or two terminals, a fill port, and a rupture disk. The terminal may be a

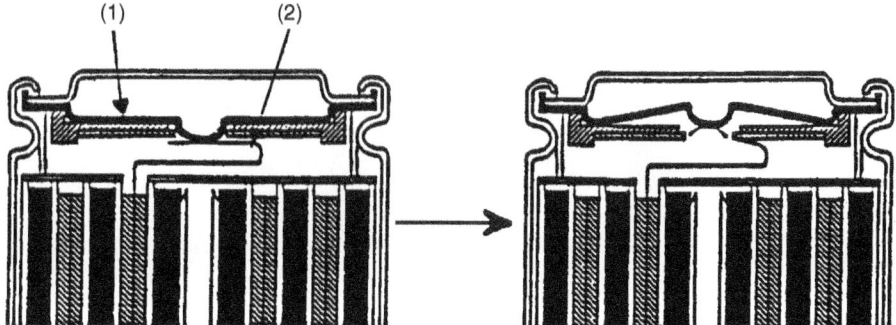

FIGURE 26.43 Detail of the construction of a cell header with a breaker and vent mechanism which can be activated by an abnormal rise in internal pressure, (1) aluminum burst disk, (2) aluminum lead current interrupter device (CID). (*Courtesy of Sony Corp.*)

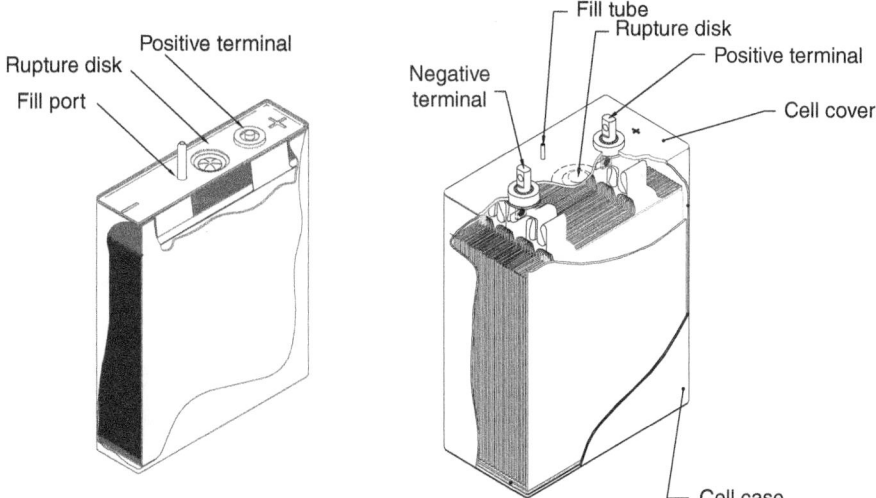

FIGURE 26.44 Schematic view showing the header and electrodes of a 7 Ah case negative and a 40 Ah case neutral flat plate prismatic Li-ion cell. (*Courtesy of Yardney Technical Products, Inc.*)

glass-to-metal seal, although for low-cost applications, compression-type seals have been used, or the terminal may incorporate devices similar to those found in the header of cylindrical products to provide pressure, temperature, and over-current interrupt in one component. The case-to-cover seal is typically formed either by TIG or laser welding.

Prismatic batteries are attractive in retrofit and volume-sensitive applications because the dimensions of the battery can be selected to efficiently use the available space. For example, prismatic lithium-ion batteries have been used to replace NiCd and lead-acid batteries in aircraft and marine applications. Shown in Fig. 26.45 is the first lithium-ion battery used on a U.S. military aircraft, a 24 V, 50 Ah battery developed for the B-2 Spirit.

Prismatic lithium-ion batteries have also been used in extraterrestrial applications. Pictured in Fig. 26.46 is a Mars Exploration Rover (MER) battery, which includes two 28 V, 10 Ah prismatic batteries. As of June 2010, one MER (Opportunity) is still operational after 7 years while the second MER (Spirit) was in a somnabulent state.

26.44 SECONDARY BATTERIES

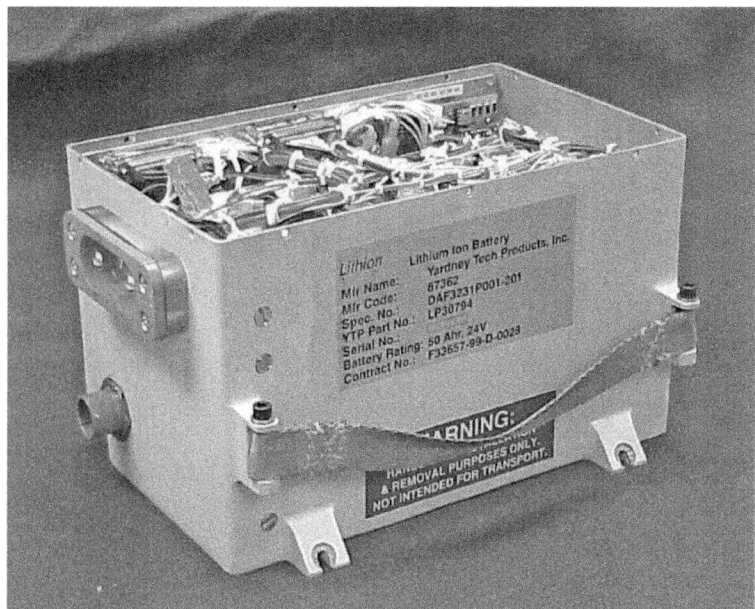

FIGURE 26.45 The first lithium-ion battery used on a U.S. military aircraft, a 24 V, 50 Ah battery developed for the B-2 Spirit. (*Courtesy of Yardney Technical Products.*)

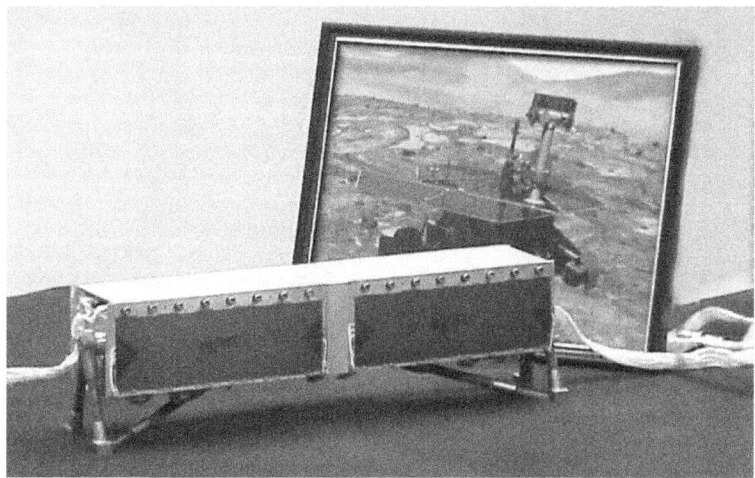

FIGURE 26.46 The Mars Exploration Rover battery, which comprised two 28 V, 10 Ah batteries. (*Courtesy of Yardney Technical Products.*)

26.3.3 Construction of "Polymer" Li-ion Cells

There is no common definition of a "polymer" Li-ion cell. Originally, such cells had a polymer or gel electrolyte. However, cells marketed today as "polymer" lithium-ion cells are no more than a regular Li-ion cell in a flexible aluminized polymer package, thus may be more accurately termed "pouch" cells. The flexible package is made from a heat-sealable aluminized plastic. Typical packaging consists of a trilayer of polypropylene/Al foil/polypropylene that has been thermally laminated together. Other materials are also used.

Figures 26.47 and 26.48 illustrate the construction of a "polymer" Li-ion cell using a flat-mandrel pseudo-prismatic jelly-roll. The jelly-roll has flat tabs attached to each of the positive and negative electrodes, and the tabs protrude from the packaging after the plastic case is sealed. The packaging can be sealed using heat or ultrasonic welding. The electrolyte is added to the case under vacuum, and after the cell is sealed, exterior air pressure then applies about 100 kPa of pressure to the electrode stack.

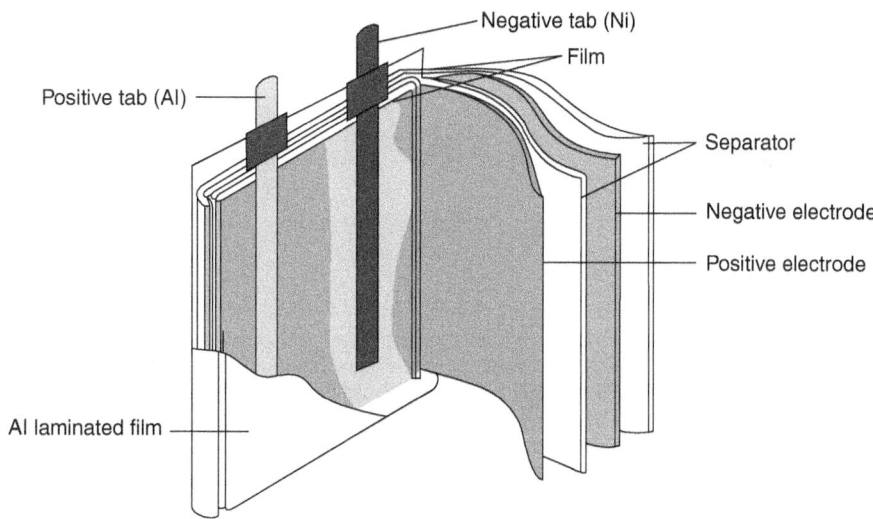

FIGURE 26.47 Schematic diagram of a Li-ion "polymer" cell. (*Courtesy of BYD.*)

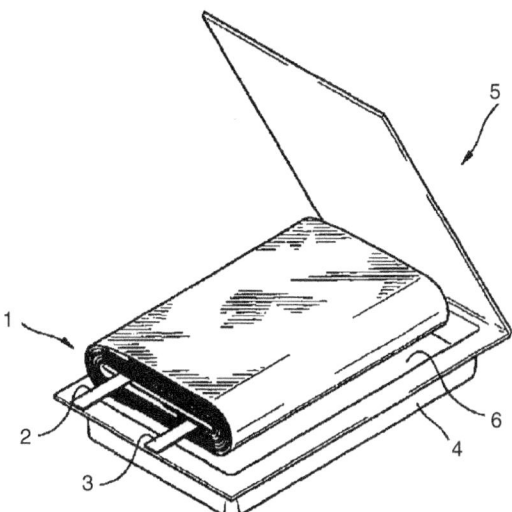

FIGURE 26.48 Schematic of the construction of a "polymer" Li-ion cell. The jelly-roll (1) has anode (2) and cathode (3) tabs that protrude. The aluminized film case (4) has been formed with a "pocket" to accommodate the jelly-roll. The top (5) is heat sealed or ultrasonically welded to the case (6).[111]

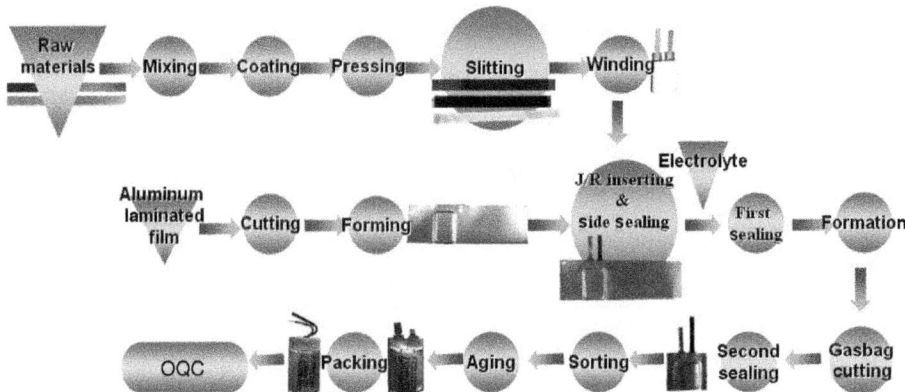

FIGURE 26.49 Schematic of the polymer Li-ion battery construction process. (*Courtesy of Hangzhou Future Power Technology Ltd.*)

A flow diagram describing the construction of a Li-ion polymer cell is provided in Fig. 26.49. The top row describes the electrode coating, slitting, and winding steps to produce a jelly-roll. The middle row shows, from the left, the steps of cutting the aluminized film, forming the pocket for the jelly-roll, and leaving excess film which is used to create the "gas bag." Then the jelly-roll is inserted and the sides of the pouch are sealed. After the electrolyte is added, the oversized pouch is completely sealed. During the first charge of the Li-ion cell after assembly (the "formation" step), the creation of the SEI can lead to the production of gases. The gases expand into the gas bag, which is later cut off, and the cell is then sealed a second time, packaged, and subjected to outgoing QC procedures.

Many lithium-ion polymer cells have the electrodes "bonded" to the separator. Two methods of bonding the electrodes to the separator are commercially practiced. The first uses an electrolyte that contains an oligomer that can be polymerized at a relatively low temperature to form a gel electrolyte throughout the negative/separator/positive electrode stack, thereby providing bonding. This general procedure is shown in Fig. 26.50.

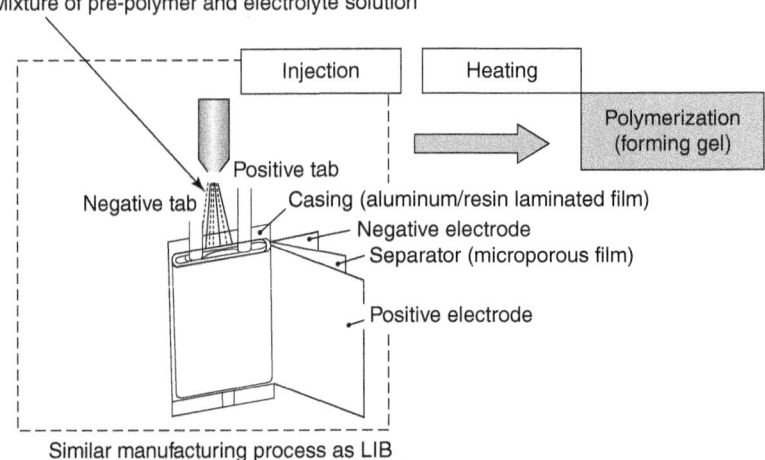

FIGURE 26.50 Illustration of the formation of a gel-polymer electrolyte within a Li-ion polymer battery by in situ polymerization during a heating step. (*Courtesy of Sanyo.*)

The second method of bonding involves coating either the separator or the electrodes with a polymer, such as PVDF. Some manufacturers also coat their separator with a ceramic prior to the PVDF for enhanced safety. According to some experts, in commercial Li-ion cells the separator is the most expensive component. Depending upon the coating used, in particular the molecular weight of the polymer, a temperature of about 70 to 90°C and pressure is sufficient to laminate and bond the separators to the electrodes. After bonding, the cell is ready for electrolyte filling.

26.4 CHARACTERISTICS AND PERFORMANCE

The general performance characteristics of Li-ion batteries are outlined in Table 26.13. As indicated in the table, Li-ion batteries provide a high operating voltage, typically 2.5 to 4.2 V, which is approximately three times that of NiCd or NiMH cells. Accordingly, fewer cells are required for a battery of a given voltage. Li-ion batteries also provide high specific energy and energy density. Cells having a specific energy up to 240 Wh/kg and energy density up to 640 Wh/L are commercially available. Li-ion batteries designed for power applications also offer high-rate capability, over 30C continuous or 100C pulse, thus Li-ion cells provide high power density. Further, Li-ion cells provide a low self-discharge rate, years of calendar life, no memory effect, and a broad temperature range of operation. Common Li-ion batteries can be charged from 0 to 45°C and discharged from −40 to 60°C. Specialized cells can offer wider temperature ranges. The combination of these qualities within a variety of cost-effective and hermetic packages has enabled the diverse applicability of this technology.

TABLE 26.13 General Performance Characteristics of Li-ion Cells (Cylindrical, Prismatic, and "Polymer") Using Common Cell Chemistries

Characteristic	$LiCoO_2$/graphite NMC/graphite NCA/graphite Energy cells	NMC/graphite LMO/graphite Power cells	$LiFePO_4$/graphite Power cells	LMO/$Li_{4/3}Ti_{5/3}O_4$
Voltage range (V)	2.5–4.2 typ. 2.5–4.35 for some cells	2.5–4.2	2.5–3.6	2.8–1.5
Avg. Voltage	3.7	3.7	3.3	2.3
Specific energy (Wh/kg)	175–240 cylinder 130–200 polymer	100–150	60–110	70
Energy density (Wh/L)	400–640 cylinder 250–450 polymer	250–350	125–250	120
Continuous rate capability (C)	2–3	Over 30	10–125	10
Pulse-rate capability (C)	5	Over 100	Up to 250	20
Cycle life at 100% DOD (to 80% capacity)	500+	500+	1000+	4000+
Calendar life (yr)	>5	>5	>5	>5
Self-discharge rate (%/month)	2–10 %/mo	2–10 %/mo	2–10 %/mo	2–10 %/mo
Charge temperature range (°C)	0–45 Some cells have wider range	0–45 Some cells have wider range	0–45 Some cells have wider range	−20–45 Some cells have wider range
Discharge temperature range (°C)	−20–60	−30–60	−30–60	−30–60
Memory effect	None	None	None	None
Power density (W/L) (pulse)	~2000	~10000	~10000	~2000
Specific power (W/kg) (pulse)	~1000	~4000	~4000	~1100

NMC = $LiNi_{1/3}Mn_{1/3}Co_{1/3}O_2$, $LiNi_{0.5}Mn_{0.3}Co_{0.2}O_2$ or $LiNi_{0.42}Mn_{0.42}Co_{0.16}$, etc.
NCA = $LiNi_{0.8}Co_{0.15}Al_{0.05}O_2$, etc.
LMO = $Li_{1+x}Mn_{2-x}O_4$, etc.

26.4.1 Characteristics of Li-ion Cells

As illustrated in Table 26.14, Li-ion cells are available in cylindrical, prismatic, or "polymer" form factors using liquid or polymer gel electrolytes. Manufacturers' websites can be consulted for the latest specifications, and Table 26.14 was current as of December 30, 2009. Although the cylindrical cells in Table 26.14 have been split into "energy cells" and "power cells," many manufacturers offer a range of cells. For example, the Samsung 18650 size cells in Table 26.14 have a range of power and energy characteristics. The most common sizes for cylindrical cells are the 18650 and 26650 sizes, although other sizes (e.g., 14500) are available. Numerous Chinese battery manufacturers offer a huge range of polymer Li-ion battery sizes; for example, Harbin Coslight Power and BYD each offer over 100 polymer Li-ion cell sizes.

Table 26.14 shows that the cylindrical cells having the highest specific energy and energy density are now being charged to voltages as high as 4.35 V, instead of the typical 4.2 V for $LiCoO_2$, NMC, and NCA positive electrodes. Each of these electrode materials delivers more specific capacity in the 4.2 to 4.4 V range, and even more if charged to 4.6 V. Manufacturers have apparently determined how to obtain acceptable cycle life for cells charged to 4.35 V through a combination of electrolyte additives, electrode material treatments, or coatings. Increased voltage is the easiest way to increase the energy density of a Li-ion battery, therefore it is likely that further increases in upper charging voltage will occur over the next years. This will lead to energy increases in power cells as well.

Integration and Impedance Characteristics. Lithium-ion cells are currently the power source of choice for portable electronics and many power tools. They are being used in electric cars by a variety of manufacturers (e.g., BYD, Nissan Tesla Motors) and have been deployed in hundreds of intracity buses beginning in 2008. Additional introductions are planned, and application of Li-ion batteries in extended-range electric vehicles (or plug-in hybrids), such as the Chevy Volt, is expected by the end of 2010. (See Chap. 29.) Li-ion batteries are also being proposed for grid energy storage. (See Chap. 30.) Thus lithium-ion batteries are being applied in an increasingly diverse array of applications, many of which were previously supported by other ambient-temperature rechargeable battery technologies.

Power is frequently a factor when determining how many or what kind of cells are required for a given application. For example, one may want to know whether a particular battery can power a cell phone, or how many cells are required to make a pack that can power a car to provide the desired acceleration. The power capability of a battery is determined by its cell voltage and its impedance. The impedance of batteries is complex; it can vary with the time of discharge (or charge), current, state-of-charge, and temperature. Thus, when sizing a battery for a desired application, it is important to specify the operating temperature, length of time the power will be needed, and minimum (or maximum) voltage cutoff.

Batteries typically show an Arrhenius dependence of impedance on temperature, i.e., a plot of log Z (real) versus 1/T absolute is linear. There may be a discontinuity at the freezing point of the electrolyte, which is typically between −20 and −40°C, although lithium-ion cells with freezing points below −20°C have been demonstrated for specialty applications. At the freezing point of the electrolyte, the impedance dramatically increases.

The impedance of batteries increases with increasing time of passing current. The increase arises at short times from the double-layer capacitance at the surface of the electrodes, and at long times from diffusion of lithium ions within the electrolyte and from lithium within the intercalation materials that serve as the positive and negative electrodes. In some chemistries, impedance strongly changes with state-of-charge, and thus never reaches a constant value. In other chemistries, the impedance may "saturate" at some relatively constant value after current has been passed for a duration of seconds to minutes, with the time to saturation depending on the design of the battery. The dependence of impedance on time is often characterized using AC impedance, as shown in the Nyquist plot in Fig. 26.51. Frequency (inverse time) decreases from left to right.

The shape of the profile in the Nyquist plot contains three features that relate to the three components of impedance in batteries: the intercept with the real axis is the ohmic component that

TABLE 26.14 Specifications of Various Commercially Available Li-ion Cells from Several Manufacturers

Manufacturer	Cell	Average voltage (V)	Charge endpoint (V)	Capacity (mAh)	Diameter (mm)	Length (mm)	Width (mm)	Volume (mL)	Mass (g)	Specific energy (Wh/kg)	Energy density (Wh/L)	Positive electrode
Energy cells												
Panasonic	NCR18650	3.6	4.2	2900	18.6	65.2		17.71	N/A	N/A	589.31	NCA
Panasonic	CGR18650E	3.7	4.2	2550	18.6	65.2		17.71	46.5	202.90	532.58	
Panasonic	CGR18650CG	3.6	4.2	2250	18.6	65.2		17.71	45	180	457.22	
LG Chem	ICR18650C1	3.75	4.35	2800	18.29	65.02		17.08	48	218.75	614.66	
Samsung	ICR18650-30A	3.78	4.35	3000	18.6	65.2		17.71	48	236.25	640.12	
Samsung	ICR18650-28A	3.75	4.3	2800	18.6	65.2		17.71	48	218.75	592.70	
Samsung	ICR18650-26F	3.7	4.2	2600	18.6	65.2		17.71	46	209.13	543.03	LCO/NMC
Samsung	ICR18650-24F	3.7	4.2	2400	18.6	65.2		17.71	45	197.33	501.25	NMC
Samsung	ICR18650-22F	3.7	4.2	2250	18.6	65.2		17.71	44.2	188.34	469.93	NMC
Sanyo	UR18650-ZT	3.7	4.3	2800	18.24	65.1		17.01	48	215.83	609.04	Hybrid(?)
Sanyo	UR18650-F	3.7	4.2	2600	18.1	64.8		16.67	47	204.68	576.98	NMC
ATL	18650E	3.7	4.2	2150	18.4	65		17.28	45	176.77	460.27	LCO
Boston Power	Sonata 4400	3.7	4.2	4400	18.5	65.2	37.1	44.75	92	176.95	363.79	NMC
E-One Moli	IHR18650B	3.6	4.2	2250	18.4	65.2		17.3	47.5	166	457	NMC
Power cells												
A123	APR18650M1 AHR18700M1	3.3	3.6	1100	18.4	65		17.3	39	93.07	210.02	LFP
A123	Ultra	3.3	3.6	700	18.4	70.0		18.6	38	61	124	LFP
Samsung	IFR18650-11P	3.2	3.6	1100	18.6	65.2		17.71	43	81.86	198.69	LFP
ATL	18650P	3.7	4.2	1380	18.4	65		17.28	45	113.46	295.43	NMC
E-One Moli	IMR18650E	3.8	4.2	1400	18.24	65		16.98	42	126.66	313.23	LMO
E-One Moli	IMR18650D	3.8	4.2	1530	18.4	65.3		17.7	44.5	133	344	LMO
E-One Moli	IBR18650B	3.6	4.2	1500	18	65		16.6	42	129	327	LMO/NMC

(*Continued*)

TABLE 26.14 Specifications of Various Commercially Available Li-ion Cells from Several Manufacturers (*Continued*)

Manufacturer	Cell	Average voltage (V)	Charge endpoint (V)	Capacity (mAh)	Thickness (mm)	Width (mm)	Length (mm)	Volume (mL)	Mass (g)	Specific energy (Wh/kg)	Energy density (Wh/L)	Positive electrode
Polymer cells												
LISUN	IMP225058S	3.7	4.2	4000	22	50	58	63.8	160	92.5	231.97	LMO
Harbin Coslight	CA401230	3.7	4.2	100	4	12.5	30.5	1.52	3	123.33	242.62	LCO
Harbin Coslight	CA582237	3.7	4.2	390	5.8	22.5	37.5	4.89	10	144.3	294.86	LCO
Harbin Coslight	CA463946	3.7	4.2	900	4.6	39.5	46.5	8.44	20	166.5	394.12	LCO
Harbin Coslight	CA103450	3.7	4.2	1800	10	34.5	50.5	17.42	35.5	187.60	382.26	LCO
BYD	SL755850	3.7	4.2	2100	7.5	58	50	21.75	41	189.51	357.24	LCO
BYD	SL685183	3.7	4.2	3000	6.1	51	83	25.82	57	194.73	429.87	LCO
LG CHEM	E2	3.8	4.2	6200	4.7	93.6	201.6	88.68	160	147.25	265.65	LMO:NMC
LG CHEM	E1	3.85	4.2	10000	7.2	94	201.5	136.37	245	157.14	282.30	LMO:NMC
Altair Nano*	11 Ah polymer	2.3	2.8	11000	8	129	207	213.624	366	69.12	118.43	LAMO/LTO
Electrovaya	35 Ah polymer	3.7	4.2	35000	12.5	142	210	373	710	183	347	NMC(?)
Prismatic cells												
BYD	LP443446ARU	3.7	4.2	1020	5.5	34	46	8.602	19	198.63	438.73	LCO
BYD	LP103450ARU	3.7	4.2	1650	10.5	34	50	17.85	35	174.42	342.01	LCO
Panasonic	CGA103450A	3.7	4.2	1950	10.6	34	50	18.02	39	185	400.38	LCO

*The Altair Nano polymer Li-ion cell uses a $Li_{4/3}Ti_{5/3}O_4$ negative electrode. All other cells in this table use graphite negative electrodes.

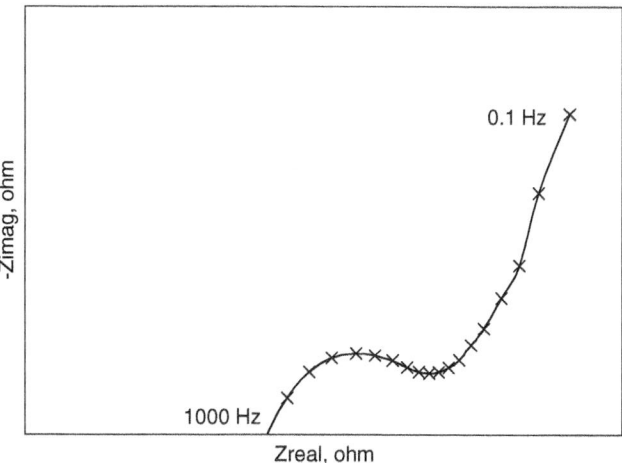

FIGURE 26.51 A schematic Nyquist plot for a lithium-ion battery. (*Courtesy of Karen Thomas-Alyea.*)

arises from resistance to electron and ion migration; the semicircle arises from charge-transfer at the interfacial surface between the electrolyte and the electrode materials, and the tail at low frequencies arises from diffusion of lithium ions in the electrolyte and from lithium within the negative and positive electrode materials. These three components of impedance have different dependencies on current. Ohmic impedance is independent of current. Charge-transfer impedance can *decrease* with increasing current at high current (a phenomenon known as Tafel kinetics). Diffusion impedance *increases* with increasing current. Also, the components of impedance all have differing dependencies on temperature. Thus, the dependence of impedance on current may be different at different temperatures. Typically, the dependence of impedance on current is small above 0°C, where ohmic components may dominate cell impedance, and impedance may vary substantially with current at low temperature. Also the direction of the dependence will depend on whether charge-transfer or diffusion dominates the impedance of the particular cell design.

Other significant challenges to the widespread adoption of the Li-ion technology relate to either stability at high temperature or safety. While Li-ion batteries may be exposed to temperatures as high as 70°C for short periods, the rate of degradation of current Li-ion batteries is significant above 60°C. Li-ion batteries are generally safe, although venting can occur if they are overcharged or crushed. To vent from overcharge, batteries must typically be charged to greater than 250% of their rated capacity at high current. Protective devices are employed to prevent venting under abusive conditions.

In current Li-ion batteries, the overcharge, overdischarge, and over-temperature issues have been largely addressed by the incorporation of electronic circuits called a Battery Management System (BMS) into batteries to provide protection from overcharge, overdischarge, or over-temperature. In addition, the BMS may also perform state-of-charge or "fuel gauge" functions and can record battery history. The controller in the BMS typically monitors the voltage of each cell or string of cells in a battery. In addition, circuits typically include a thermistor or other thermostat, for restorable over-temperature control, and a thermal fuse for nonrestorable over-temperature control. Additional information on battery management systems is found in Sec. 5.6.

26.4.2 Performance of Commercial Li-ion cells

As shown in Tables 26.13 and 26.14, there are numerous electrode material choices and cell construction formats available for Li-ion cells. To a first approximation, there is not a large difference between the performance characteristics of a particular chemistry in cylindrical, prismatic, or "polymer" Li-ion formats. Therefore, this section will not provide examples of every possible cell configuration and instead, selected examples, which normally focus on 18650-size cells, will be provided. The following chemistries will be discussed: $LiCoO_2$/graphite, NMC/graphite, $LiMn_2O_4$/graphite, ($LiMn_2O_4$:NMC)/graphite, $LiFePO_4$/graphite, $LiMn_2O_4$/$Li_{4/3}Ti_{5/3}O_4$, and the Sony Nexelion cell, thought to be ($LiCoO_2$:NMC)/(Sn-Co-C:graphite).

$LiCoO_2$/Graphite Cells. Figure 26.52 shows the discharge voltage profiles at 20°C of a Sanyo UR18650F cell at several currents up to the 2C rate. The cell provides about 2550 mAh at 0.2C and about 2380 mAh at a 2C rate. The midpoint cell voltage drops about 0.4 V when the current is increased from 0.5 A (0.2C) to 5.0 A (2C). This corresponds to an effective cell resistance of about 0.09 Ω (0.4 V/4.5 A). This cell was designed for notebook PC use where high-rate discharge is not a requirement. The 2450 mAh Panasonic CGR18650E cell also shows a voltage drop of 0.4 V when the current is increased from 490 mA to 2.45 A, thus also has a resistance of about 0.09 Ω, which is typical for energy cells for laptop applications.

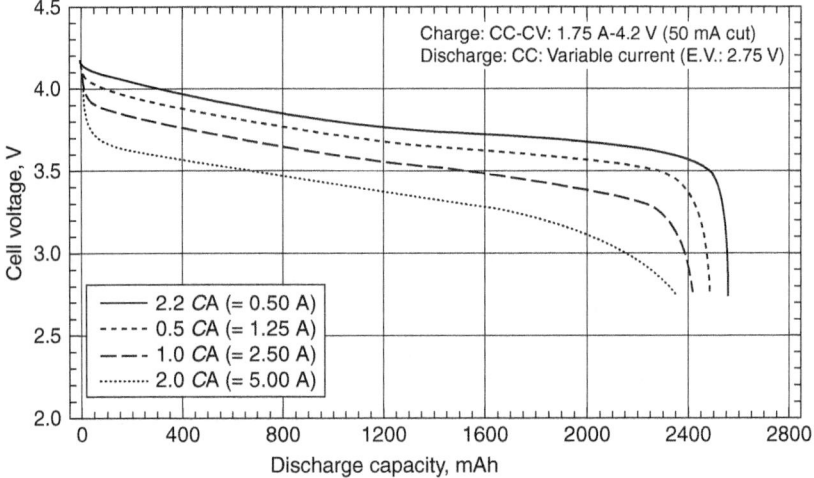

FIGURE 26.52 Discharge rate characteristics at 20°C of a Sanyo UR18650E $LiCoO_2$/graphite cell. Currents are indicated in the legend. (*Courtesy of Sanyo.*)

Figure 26.53 shows the discharge-temperature characteristics at a 1C rate of a Sanyo UR18650E $LiCoO_2$/graphite cell. The cell delivers substantially less capacity at subambient temperatures. However, this cell is designed for laptop use, where normally work is done at room temperature. Figure 26.54 shows the charging characteristics of the same cell under a CCCV protocol. The current is set to a constant 1.75 A until the voltage reaches 4.2 V, then the current decays while the voltage is held fixed. Even at 0°C, over 75% of the cell capacity is charged in under 2 h.

Figure 26.55 shows the percent capacity versus cycle number for an E-One Moli $LiCoO_2$/graphite ICR18650H cell. This cell has a nominal capacity of 2200 mAh. The cell retains about 87% of its capacity after 300 cycles at 23°C and about 84% at 45°C. Such results are typical for laptop PC Li-ion batteries.

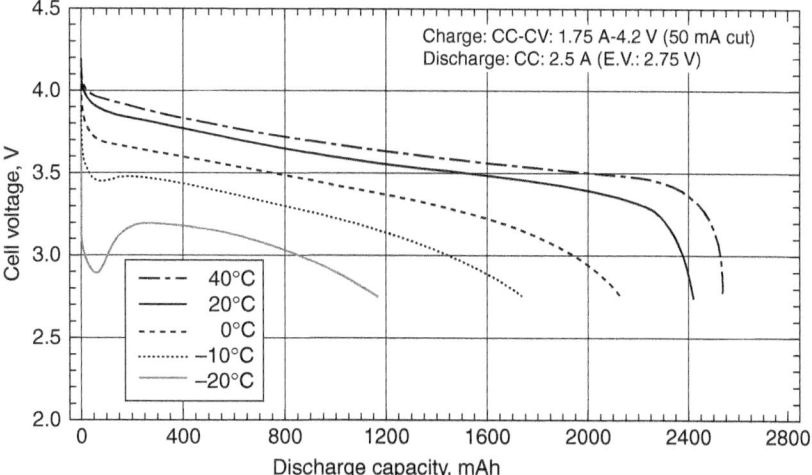

FIGURE 26.53 Discharge temperature characteristics at a 1C rate of a Sanyo UR18650E LiCoO$_2$/graphite cell. Temperatures are indicated in the legend. (*Courtesy of Sanyo.*)

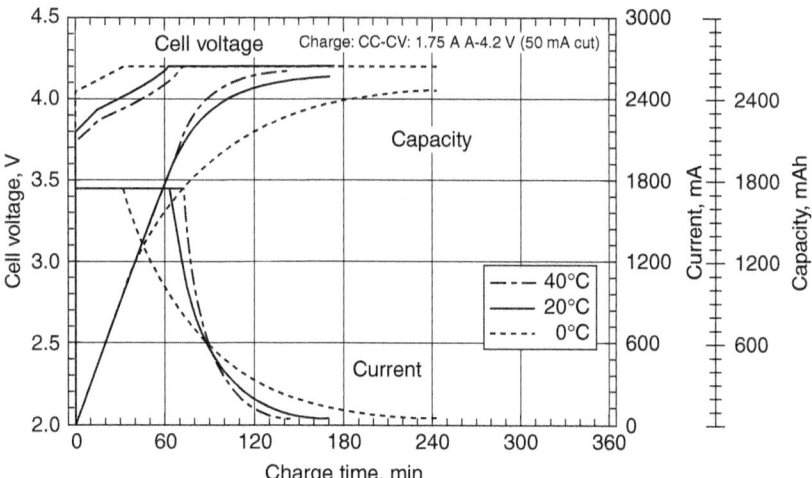

FIGURE 26.54 Charge characteristics (1.75 A, 4.2 V CCCV) of a Sanyo UR18650E LiCoO$_2$/graphite cell. Temperatures are indicated in the legend. (*Courtesy of Sanyo.*)

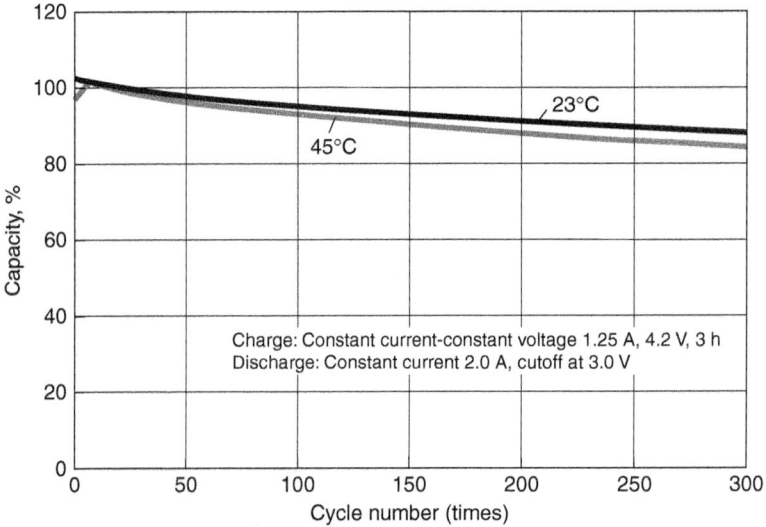

FIGURE 26.55 Percent capacity vs. cycle number for an E-One Moli LiCoO$_2$/graphite ICR18650H cell. This cell has a nominal capacity of 2200 mAh. (*Courtesy of E-One Moli, Ltd.*)

NMC/Graphite Energy and Power Cells. NMC is being used in cells designed for high energy and in cells designed for high power. Cells comprising NMC can provide lower cost and better inherent safety compared to LiCoO$_2$/graphite cells.

Energy Cells. The characteristics of a typical energy cell will be examined first. Figure 26.56 shows the discharge voltage profiles at 23°C of an E-One Moli Energy IHR18650B cell at several

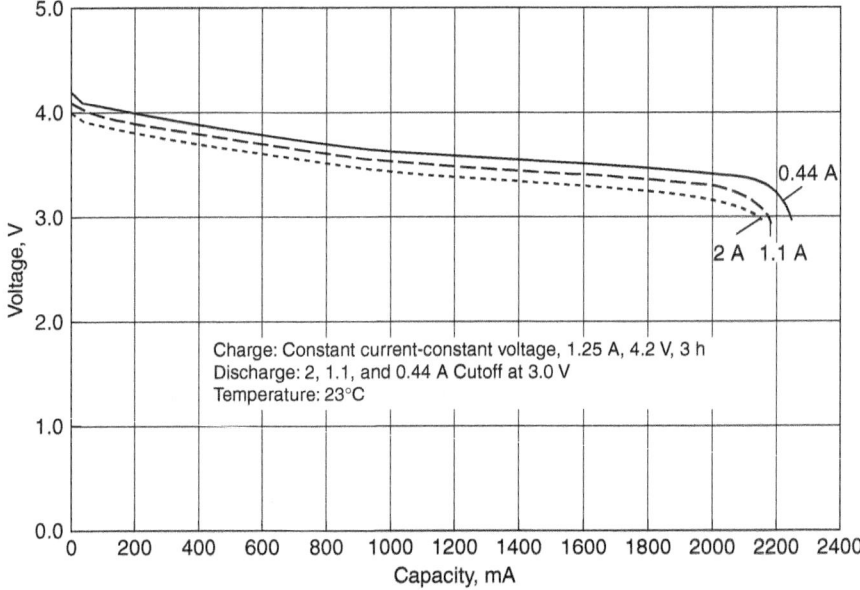

FIGURE 26.56 The discharge voltage profiles at 23°C of an E-One Moli Energy IHR18650B cell at several currents up to 2 A. (*Courtesy of E-One Moli Energy.*)

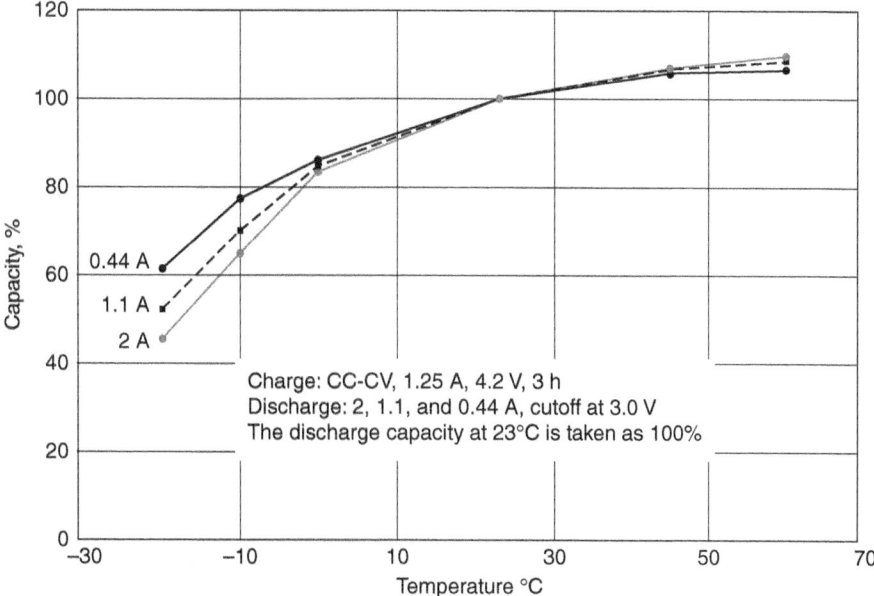

FIGURE 26.57 Percent capacity as a function of temperature at several discharge currents for the E-One Moli Energy IHR18650B cell, which uses an NMC positive electrode. (*Courtesy of E-One Moli Energy.*)

currents up to 2 A. The cell provides about 2250 mAh at 0.2C and about 2150 mAh at 2 A. The midpoint cell voltage drops about 0.22 V when the current increases from 0.44 A (0.2C) to 2.0 A (about 1C). This corresponds to an effective cell resistance of about 0.14 Ω (0.22 V/1.56 A), which is adequate for laptop applications.

Figure 26.57 shows the percent capacity as a function of temperature at several discharge currents for the E-One Moli Energy IHR18650B cell. The cell can deliver 50% of its capacity at −20°C at a 2 A current. This type of performance is typical for energy cells using almost any positive electrode material.

Figure 26.58 shows the capacity versus cycle number for the E-One Moli Energy IHR18650B cell, which uses an NMC positive electrode. The cell was tested at 23°C using a 1.1 A discharge current. The cell provides about 87% of its capacity after 300 cycles, which is similar to the $LiCoO_2$/graphite cell shown in Fig. 26.55.

Much higher cycle life has been demonstrated in cells developed for use in satellites. Flat-plate prismatic cells developed for low earth orbit (LEO) applications have demonstrated over 39,000 cycles over 7 years as of this writing. They employ a layered lithiated nickel-cobalt oxide positive electrode material. The LEO cycle includes 40% depth of discharge (DOD) cycles with periodic 100% DOD cycles to monitor the capacity of the cells. Shown in Fig. 26.59 are the results of LEO testing of cells charged to 3.9 or 3.7 V. The cells charged to 3.9 V demonstrated better stability than the cells charged to 3.7 V. Figure 26.60 is a photograph of a 28 V, 43 Ah satellite battery.

Power Cells. NMC/graphite cells have also been developed for high-power applications. Cylindrical power cells can deliver high discharge rate, up to 30C, for hundreds of cycles. NMC/graphite power cells typically have higher energy densities than $LiMn_2O_4$/graphite or $LiFePO_4$/graphite cells. Fig. 26.61 shows the discharge voltage profiles of a CHAM NMC/graphite power cell at rates of about 1C to about 24C. The right panel of Fig. 26.61 shows discharge capacity versus current. Although the average voltage decreases with increasing current, the capacity increases due to cell heating and an increase in electrolyte conductivity. As a result, the energy provided by the cell is nearly the same at all rates tested. In Fig. 26.61, the midpoint cell voltage on discharge drops from about 3.7 V at 1.3 A to about 3.1 V at 32.5 A. This corresponds to an effective DC resistance of about 0.02 Ω (0.6 V/31.2 A). Type 18650 power cells typically have an effective DC resistance of about 0.02 Ω.

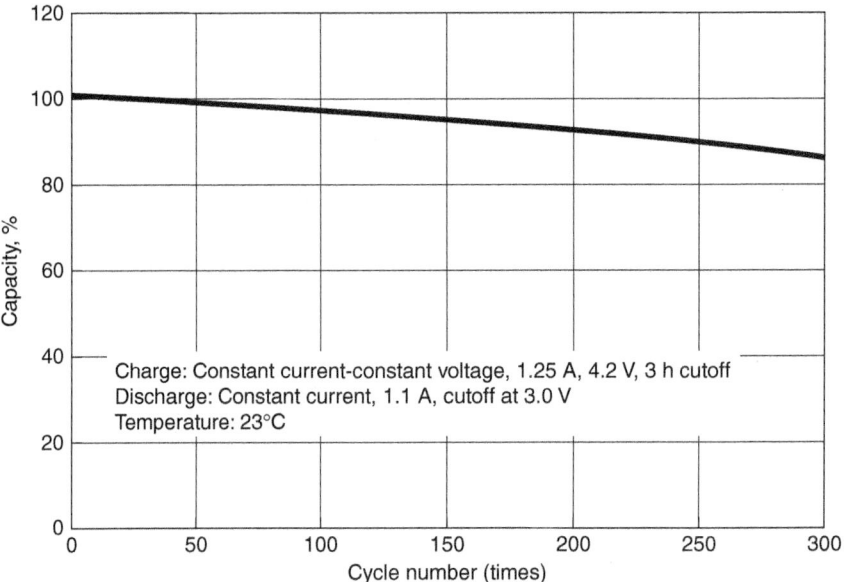

FIGURE 26.58 The capacity vs. cycle number for the E-One Moli Energy IHR18650B cell, which uses an NMC positive electrode. (*Courtesy of E-One Moli Energy.*)

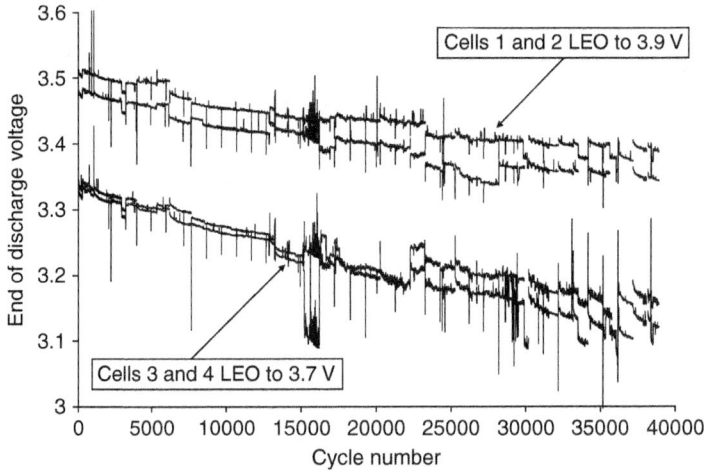

FIGURE 26.59 End of discharge voltage for 20 Ah flat-plate prismatic lithium-ion cells cycled under LEO conditions to 40% DOD with periodic 100% DOD discharges over 7 years. The upper data represent cells charged to 3.9 V, the lower data for cells discharged to 3.7 V. (*Courtesy of Yardney Technical Products.*)

FIGURE 26.60 A 28 V, 43 Ah satellite battery. (*Courtesy of Yardney Technical Products.*)

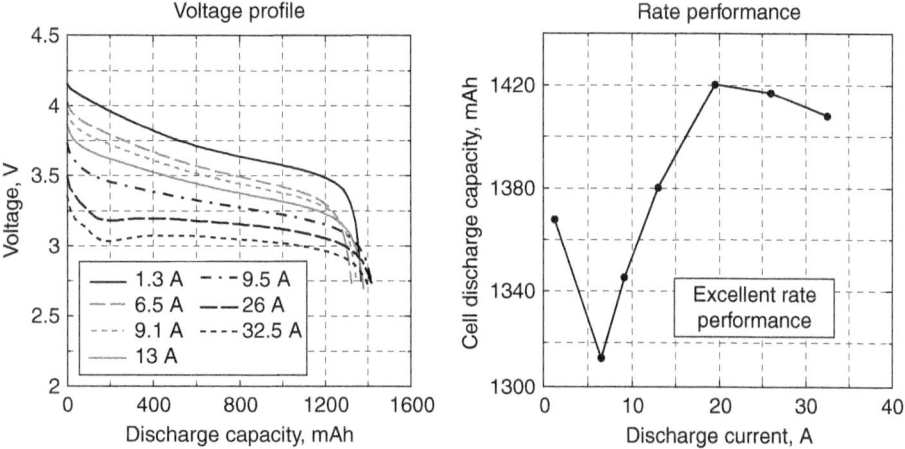

FIGURE 26.61 The left panel shows the discharge potential profiles of a CHAM NMC/graphite power cell at different currents. The right panel shows discharge capacity versus current. At high rates, the increase in cell temperature improves the discharge capacity. (*Courtesy of J. Jiang.*)[112] (Note: CHAM is a custom battery manufacturer.)

Figure 26.62 shows the capacity versus cycle number for the CHAM 18650 NMC graphite power cell. The cell was discharged at a $10C$ rate and charged to 4.2 V using a CCCV protocol. The cell retains almost 90% of its capacity after 500 charge-discharge cycles.

Flat-plate prismatic cells have demonstrated exceptional power capability, providing 6 kW/kg continuous and 15 kW/kg peak specific power. Shown in Fig. 26.63 is pulse-discharge performance of a 7 Ah cell using a duty cycle having a 6 s, $60C$ (450 A) discharge and a 30 s rest to simulate use in a directed energy weapon. The DC cell resistance is about 0.7 milliohms (0.3 V/450 A).

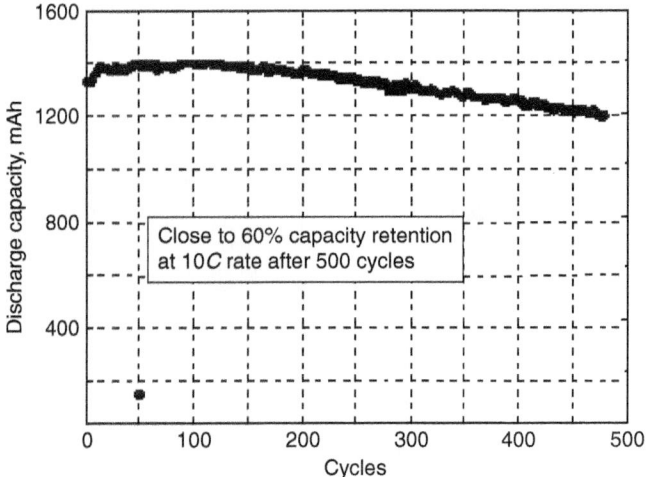

FIGURE 26.62 Capacity vs. cycle number of the CHAM 18650 NMC/graphite power cell using a 10C discharge and CCCV charge. (*Courtesy of J. Jiang.*)[112]

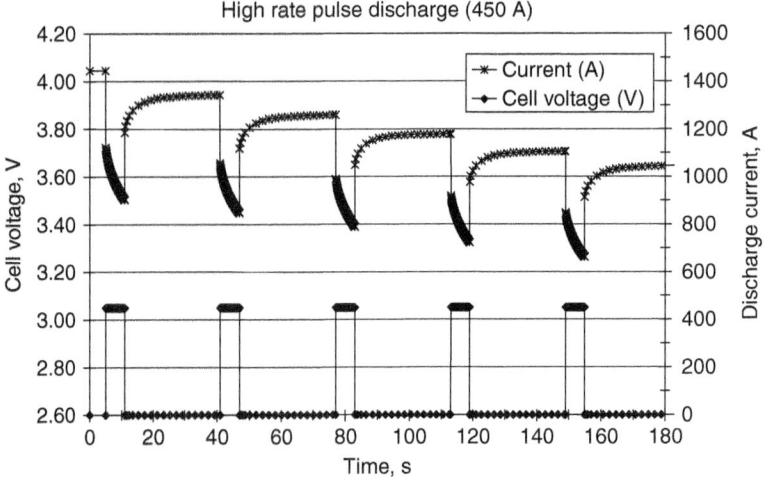

FIGURE 26.63 Pulse discharge of a 7 Ah flat-plate primatic NMC/graphite cell. The cell provided 15 kW/kg. (*Courtesy of Yardney Technical Products.*)

Continuous 60C (450 A) discharge of the 7 Ah cell is shown in Figure 26.64. The cell provided 76% of the rated capacity in 46 s when discharged to 3.0 V. The specific energy of this cell is 100 Wh/kg, which is about half the 205 Wh/kg of a comparable energy cell.

The performance of a 9 Ah prismatic cell when discharged at rates from 0.2C to 200C is shown in Fig. 26.65. At 200C, the cell provided 40% of the capacity delivered at 0.2C. At rates above about 75C, it is believed that the capacity is limited by bulk diffusion.

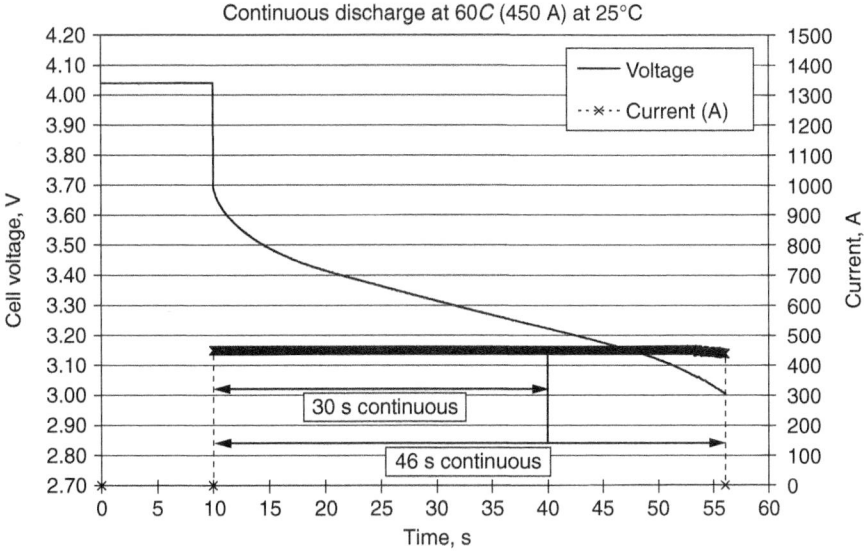

FIGURE 26.64 Continuous 60C discharge of a 7 Ah flat-plate prismatic NMC/graphite cell. The cell delivered 76% of the rated capacity above 3.0 V in 46 s. (*Courtesy of Yardney Technical Products.*)

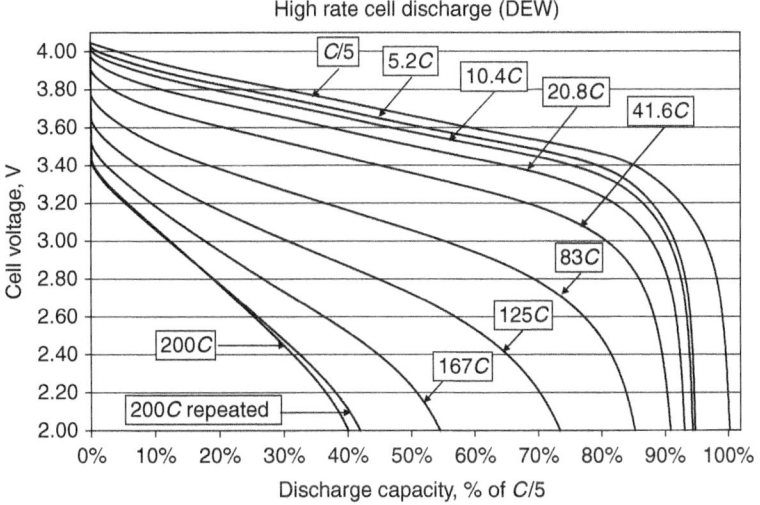

FIGURE 26.65 Continuous discharge of a 9 Ah, flat-plate primatic NMC/graphite cell at rates up to 200C. At 200C, the cell delivered 40% of the capacity delivered at 0.2C. (*Courtesy of Yardney Technical Products.*)

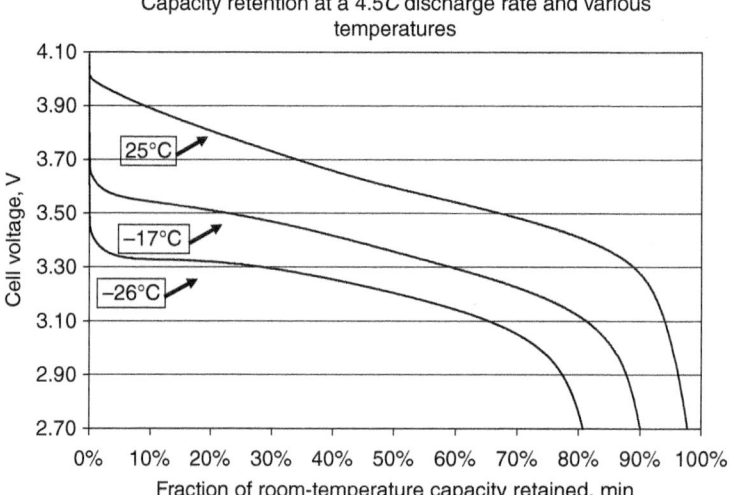

FIGURE 26.66 Discharge of a 5 Ah, flat-plate prismatic NMC/graphite cell at 4.5C at 25, −17, and −26°C. (*Courtesy of Yardney Technical Products.*)

Flat-plate prismatic power cells can also provide high power at low temperatures. Figure 26.66 shows 4.5C discharges of a 5 Ah cell at 25, −17, and −26°C. The cell provided about 82% of the rated capacity when discharged at 4.5C at −26°C.

To illustrate the effect of the discharge rate, Fig. 26.67 illustrates the discharge of the 5 Ah, flat-plate prismatic NMC/graphite cell at −20°C at rates from 0.2C to 5C. Note the increase in cell voltage in the 5C curve, which results from an increase in internal temperature.

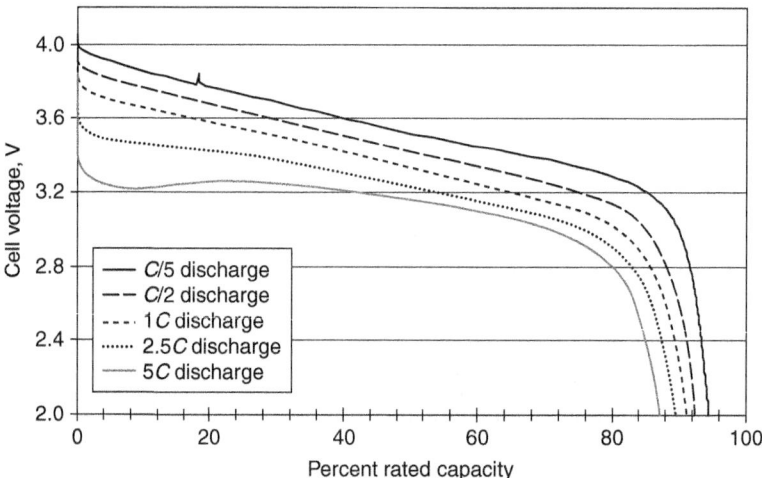

FIGURE 26.67 Discharge of a 5 Ah, flat-plate prismatic NMC/graphite cell at −20°C at rates from 0.2C to 5C. (*Courtesy of Yardney Technical Products.*)

LiMn₂O₄/Graphite and LiMn₂O₄:NMC/Graphite Power Cells. $LiMn_2O_4$ is a very attractive positive electrode material because it can be prepared at low cost, provides good safety properties, and can deliver high power and moderate energy density. Numerous companies produce power cells based on positive electrodes of $LiMn_2O_4$ and also on blends of $LiMn_2O_4$ and NMC. The first company to mass produce $LiMn_2O_4$/graphite Li-ion cells was NEC/Moli Energy in the late 1990s. E-One/Moli Energy (Taiwan) currently produces $LiMn_2O_4$/graphite power cells. It is believed that the LG Chemical cells destined for the Chevy Volt (similar to the E1 and E2 cells in Table 26.14) have $LiMn_2O_4$:NMC blended positive electrodes. Other manufacturers are doing the same. There is a synergistic effect between NMC and $LiMn_2O_4$ that contributes to enhanced storage properties and better cycle life.[113] Additions of NMC also improve energy density but increase cost.

Figure 26.68 shows the discharge voltage profiles at various currents for an IMR18650E $LiMn_2O_4$/graphite cell operating at 23°C. The cell is rated at 1.4 Ah capacity and provides about 1.3 Ah when discharged at 20 A. The midpoint voltage drops by about 0.35 V when the current is increased from 5 to 20 A, thus the effective DC resistance is about 0.023 Ω (0.35 V/15 A), which is typical of such cells.

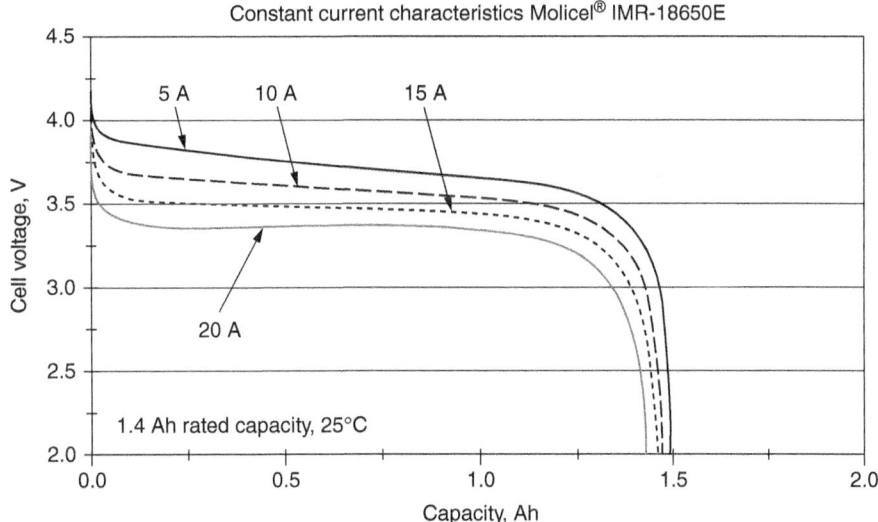

FIGURE 26.68 Discharge voltage profiles at various currents for an IMR18650E $LiMn_2O_4$/graphite cell operating at 23°C. (*Courtesy of E-One/Moli Energy.*)

Figure 26.69 shows the discharge voltage profiles at various currents for an IMR18650E $LiMn_2O_4$/graphite cell operating at −20°C. The cell is not heat sunk and warms significantly as the current flows, leading to the increase in cell voltage with discharge time.

Figure 26.70 shows the discharge voltage profiles at various currents for an IBR18650B ($LiMn_2O_4$:NMC)/graphite cell operating at 23°C. The positive electrode of this cell is a blend of $LiMn_2O_4$ and NMC. The cell is rated at 1.5 Ah and provides over 1.4 Ah at a 30 A (20*C*) discharge rate. The midpoint voltage drops by about 0.6 V when the current increases from 1.5 to 30 A, thus the effective DC resistance is about 0.021 Ω (0.6 V/28.5 A), which is typical of such cells. At 30 A and 3.2 V, this cell continuously provides a specific power of 2.3 kW/kg and a power density of 5.6 kW/L. Figure 26.71 shows the charge discharge cycle life of this cell. The cell provides over 84% capacity retention after 1000 cycles. This is quite impressive!

Further information about ($LiMn_2O_4$:NMC)/graphite cells is available on the Electric Mobility Canada website.[115]

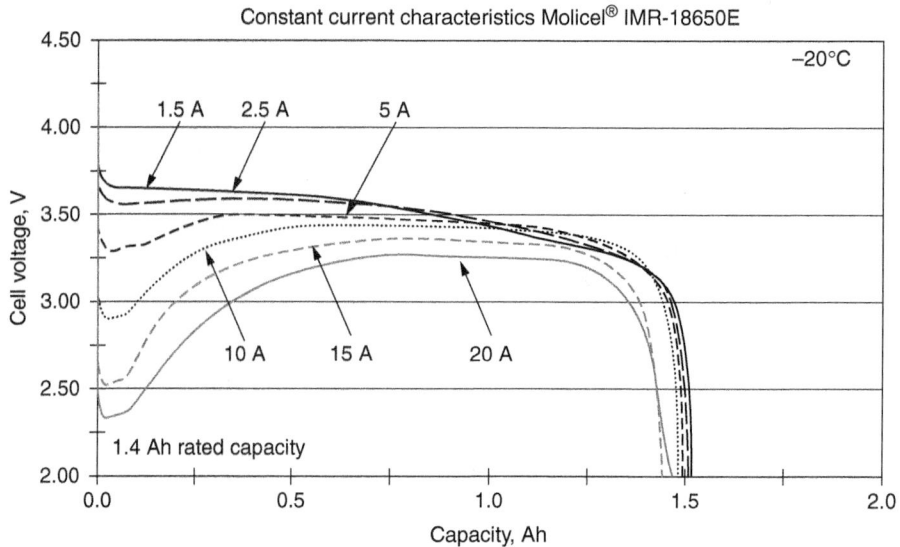

FIGURE 26.69 Discharge voltage at various currents for an IMR18650E LiMn$_2$O$_4$/graphite cell operating at −20°C. (*Courtesy of E-One/Moli Energy.*)

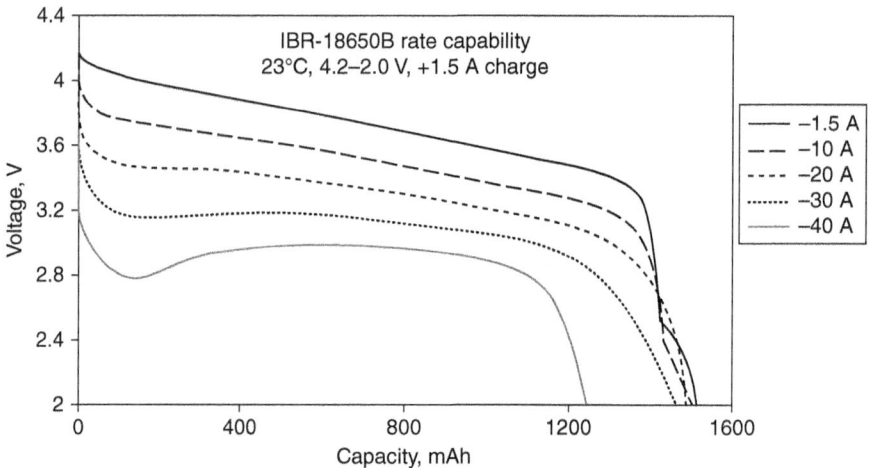

FIGURE 26.70 The discharge voltage profiles at various currents for an IBR18650B (LiMn$_2$O$_4$:NMC)/graphite cell operating at 23°C.[114] (*Courtesy of E-One/Moli Energy.*)

Performance of LiFePO$_4$/Graphite Power Cells. LiFePO$_4$-based Li-ion cells have impressive power capabilities and provide long cycle life. Additionally, they are the most thermally stable of all the Li-ion cells that use a graphite negative electrode. The drawback of LiFePO$_4$/graphite cells is their relatively low specific energy and energy density, as shown in Tables 35.13 and 35.14.

Figure 26.72 shows the discharge voltage versus capacity for an A123 Systems ANR26650M1A LiFePO$_4$/graphite cell when discharged at rates from 1 to 40 A. The midpoint cell voltage drops about 0.7 V between 1 and 40 A discharge, thus the effective DC resistance is about 0.017 Ω (0.7 V/ 39 A).

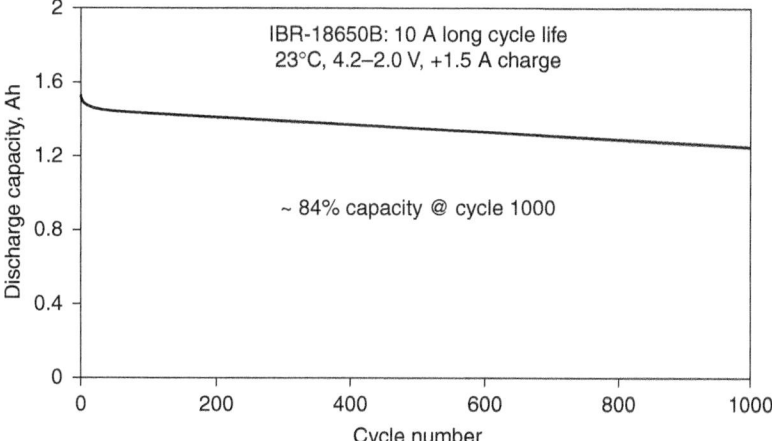

FIGURE 26.71 The charge-discharge capacity vs. cycle number for an IBR18650B (LiMn$_2$O$_4$:NMC)/graphite cell operating at 23°C. The discharge current was 10 A and a 1.5 A CCCV charge to 4.2 V was used.[114] (*Courtesy of E-One/Moli Energy.*)

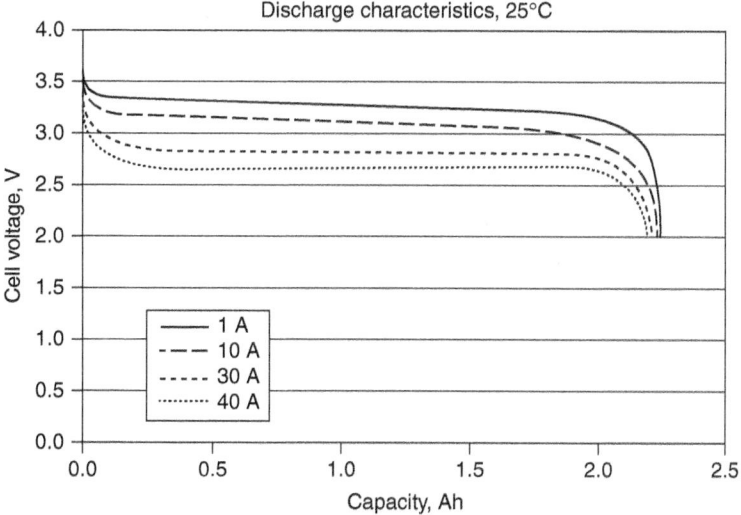

FIGURE 26.72 Discharge voltage vs. capacity for an A123 Systems ANR26650M1A LiFePO$_4$/graphite cell discharged using various currents. (*Courtesy of Yet-Ming Chiang.*)

Figure 26.73 shows the capacity versus cycle number for the ANR26650M1A cell. The cell provides over 95% of its capacity at *C*-rate cycling at 25°C after 1000 cycles. Capacity retention at 60°C is over 75% after 1000 cycles.

A123 Systems worked with a Formula 1 race team to introduce hybrid power to the racetrack. A special, extremely high-power, graphite/LiFePO$_4$ cell, the AHR18700M1 *Ultra*, has been produced. The cell was specifically designed for Formula 1 racing and was successfully used by the Vodafone-McLaren-Mercedes team throughout the 2009 racing season. This high-power cell is designed for use at operating temperatures up to 100°C, can deliver specific power of over 20 kW/kg, and provides specific energy of 60 Wh/kg. In the Formula 1 application, the Kinetic Energy Recovery

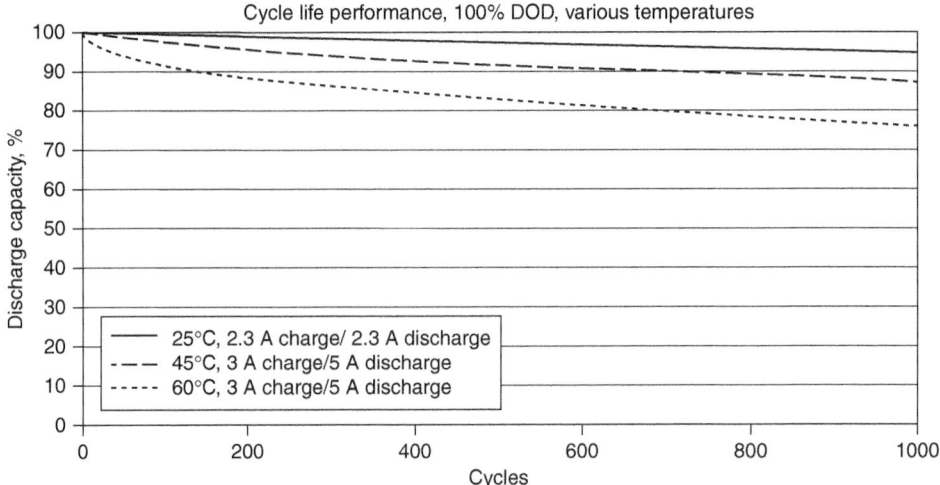

FIGURE 26.73 The capacity vs. cycle number for the ANR26650M1A cell. (*Courtesy of Yet-Ming Chiang.*)

System (KERS) using these cells is limited by race regulations to delivering 400 kJ total energy per lap. The electric boost is delivered at the driver's discretion by the push of a button on the steering wheel for purposes of overtaking and defending against such. Under race conditions, the discharge C-rate is about $250C$, corresponding to about 80% of the stored energy being delivered in 6 to 8 s. Figure 26.74 shows the race car believed to be provided with the KERS system. Figure 26.75 shows the discharge voltage versus state-of-charge for discharges of the cell at various C-rates up to $150C$. At $125C$ (87.5 A for this cell) the midpoint cell voltage is 2.5 V, and at 0.7 A the midpoint voltage is 3.3 V. Thus, the effective DC resistance of this cell is only about 0.01 Ω (0.8 V/86.8 A). The cell continuously provides specific power at the midpoint of the $125C$ discharge of 5.6 kW/kg.

FIGURE 26.74 Photograph of the Vodafone-McLaren-Mercedes Formula-One race car believed to have a kinetic energy recovery system. The batteries were from A123 Systems.

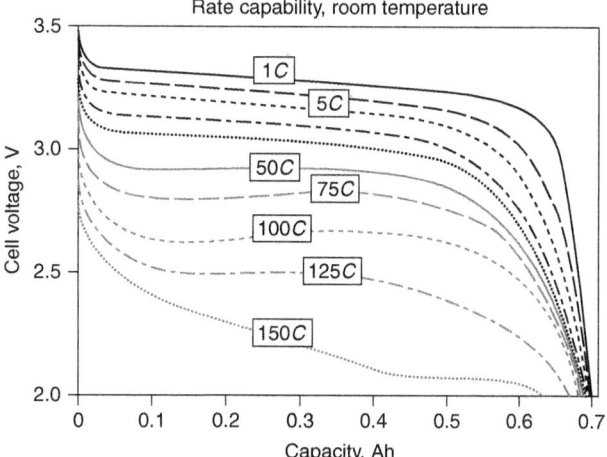

FIGURE 26.75 Discharge voltage vs. capacity at various *C*-rates as indicated for the AHR18700M1*Ultra* graphite/LiFePO$_4$ cell from A123 Systems. (*Courtesy of Yet-Ming Chiang.*)

According to the manufacturer's specification sheet, the cell can supply 12 kW/kg and can sink (during regenerative braking) 8 kW/kg at 60°C for 10 s periods throughout most of the state-of-charge of the cell.

Figure 26.76 shows capacity retention versus cycle number for the AHR18700M1*Ultra* graphite/LiFePO$_4$ cell from A123 Systems when the cell is charged and discharged at a 10*C* rate at room temperature. More than 93% capacity remains after 1400 cycles. Figure 26.77 shows capacity retention versus cycle number for AHR18700M1*Ultra* cells cycled at various temperatures up to 100°C. The capacity retention at 100°C is very impressive, given that many Li-ion cells do not function well above 60°C.

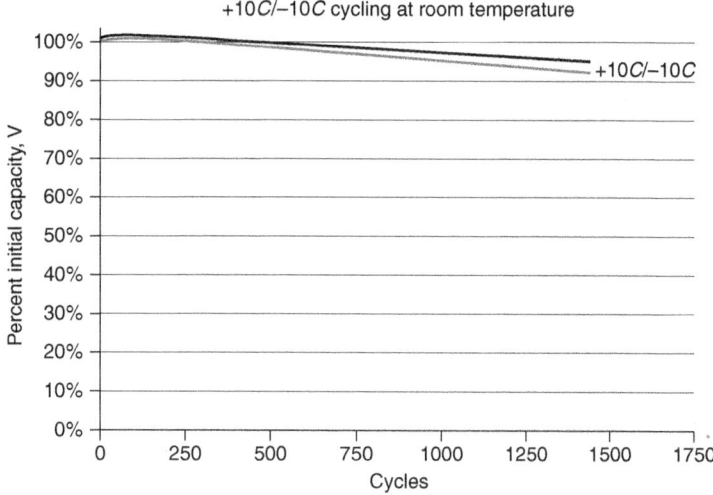

FIGURE 26.76 Capacity retention versus cycle number for the AHR18700M1*Ultra* graphite/LiFePO$_4$ cell from A123 Systems. The cell was charged and discharged at a 10*C* rate at room temperature. (*Courtesy of Yet-Ming Chiang.*)

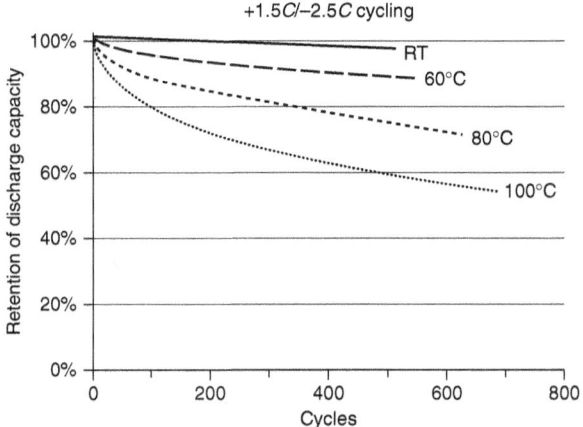

FIGURE 26.77 Capacity retention vs. cycle number for the AHR18700M1*Ultra* graphite/LiFePO$_4$ cell from A123 Systems. The cell was charged at a 1.5C rate and discharged at a 2.5C rate at the temperatures indicated. (*Courtesy of Yet-Ming Chiang.*)

Performance of LiMn$_2$O$_4$/Li$_{4/3}$Ti$_{5/3}$O$_4$ Cells. LiMn$_2$O$_4$/Li$_{4/3}$Ti$_{5/3}$O$_4$ cells have been developed by a few companies. As shown in Tables 26.13 and 26.14, these cells have low specific energy and energy density. On the other hand, they have impressive safety characteristics and extremely long charge-discharge cycle life. Such cells may have a role in stationary applications such as grid energy storage. Figure 26.78 shows the discharge voltage versus capacity for an 11 Ah LiMn$_2$O$_4$/Li$_{4/3}$Ti$_{5/3}$O$_4$ Li-ion polymer cell from Altair Nano, at various currents, up to 10C. The midpoint cell voltage drops about 0.3 V as the current is increased from 11 to 110 A, yielding a DC resistance of about 0.003 Ω (0.3 V/99 A). Therefore a 1.1 Ah cell (~18650 size for this technology) would have a resistance of about 0.03 Ω, which is similar to other typical 18650-size power cells.

A careful examination of the shape of the voltage-capacity cure of Figure 26.78 suggests that the positive electrode in this cell includes a blend of LiMn$_2$O$_4$ and another material, likely NMC.

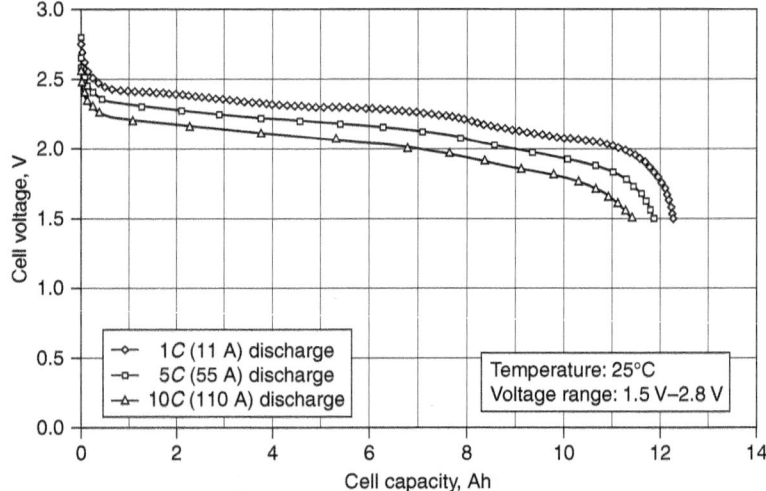

FIGURE 26.78 The discharge voltage vs. capacity for an 11 Ah LiMn$_2$O$_4$/Li$_{4/3}$Ti$_{5/3}$O$_4$ cell from Altair Nano at various currents, up to 10C, at 25°C. (*Courtesy of Altair Nano.*)

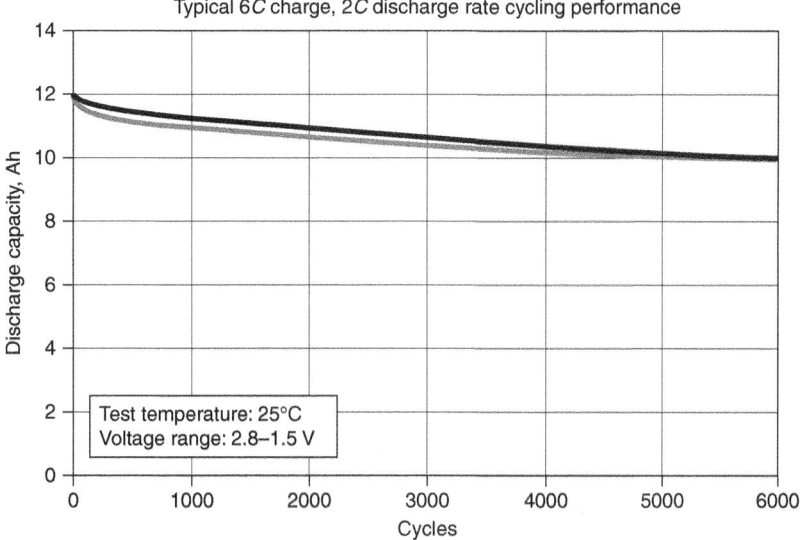

FIGURE 26.79 The discharge capacity vs. cycle-number performance of the Altair Nano 11 Ah LiMn$_2$O$_4$/Li$_{4/3}$Ti$_{5/3}$O$_4$ cell. (*Courtesy of Altair Nano.*)

Figure 26.79 shows discharge capacity versus cycle number for the Altair Nano 11 Ah LiMn$_2$O$_4$/Li$_{4/3}$Ti$_{5/3}$O$_4$ cell. The capacity retention is extremely impressive; the cell provided over 85% retention after 6000 high-rate cycles.

Performance of the Sony Nexelion Cell (Sn-Co-C Negative Electrode). To our knowledge, the Nexelion cell has, so far, only been incorporated in battery packs made by Sony. Therefore performance data for the Nexelion cell is limited. Figure 26.80 shows discharge voltage versus capacity for a Sony 14430 cylindrical Nexelion cell compared to the same size cell using a graphite negative electrode. The Nexelion delivers a greater capacity, albeit an lower voltage than the graphite cell.

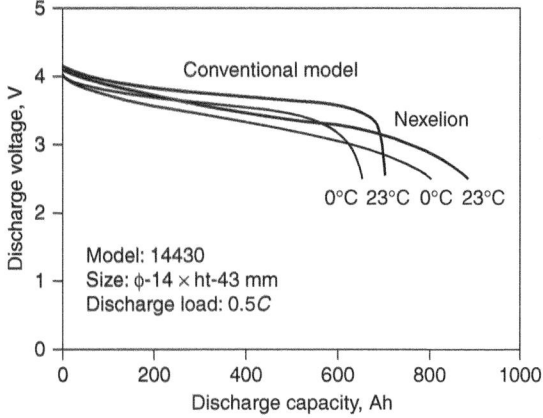

FIGURE 26.80 Discharge profiles of the Sony Nexelion cell measured at 0.5C at 23 and 0°C. The results are compared to those from a cell with a graphite negative electrode.[63] (*Courtesy of Sony Energy Devices Corporation.*)

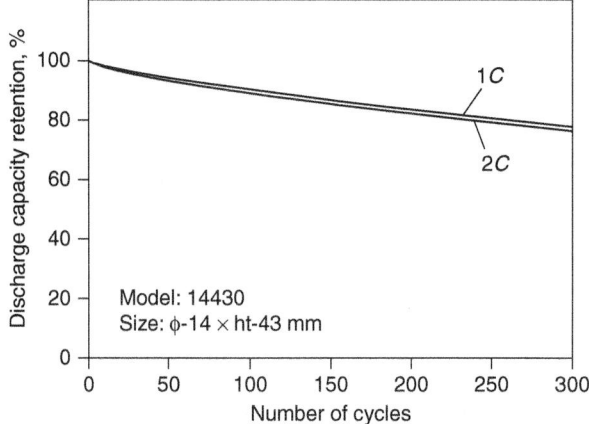

FIGURE 26.81 Capacity vs. cycle number for a 14430 size Sony Nexelion cell. The testing was performed at 23°C with C-rate and $2C$-rate discharges.[63] (*Courtesy of Sony Energy Devices Corporation.*)

Figure 26.81 shows capacity versus cycle number for the 14430-size Nexelion cell as presented by Inoue et al. The cell retains over 75% capacity after 300 cycles at both C-rate and $2C$-rate discharges. Such performance is comparable to that of many cylindrical $LiCoO_2$/graphite cells. As stated earlier, it is believed that the composition of the Sony negative electrode has changed since the original introduction. The data in Figs. 26.80 and 26.81 are for the original Nexelion cells.

26.5 SAFETY PROPERTIES

26.5.1 Temperature Dependence of the Reactions Between Charged Electrode Materials and Electrolyte

There have been numerous safety events documented involving lithium-ion batteries. Many of these have led to recalls involving significant numbers of batteries. Nevertheless, the safety record of Li-ion batteries is extremely good and is improving due to the worldwide efforts of Li-ion battery manufacturers, OEM equipment designers, and regulatory agencies. If one assumes about 100 safety incidents per year from a global production of 4 billion cells, this translates to a safety incident rate of only 1 per 40 million cells, which is impressive.

All safety issues with Li-ion batteries ultimately stem from the fact that the charged positive and negative electrode materials react with the cell electrolyte at elevated temperature. Figure 26.82 shows Accelerating Rate Calorimeter (ARC) data for the reaction of various lithiated carbon negative electrode samples with 1M $LiPF_6$-containing electrolyte.

Figure 26.82 shows that all of these materials begin reacting with electrolyte near 80°C at a low rate. (This is fundamentally why Li-ion batteries begin to lose capacity when cycled above 60°C—reactions between the lithiated graphite and electrolyte do not just "turn on," but are exponentially activated by temperature.) As the temperature increases, the reaction rate increases strongly for the high-surface area KS and SFG samples. The MCMB and fiber samples have low specific surface area, and the reaction rate reaches a plateau after about 100°C and increases dramatically above 200°C.

A common misconception is that lithium in $Li_{7/3}Ti_{5/3}O_4$ (charged lithium titanate) is not reactive with electrolyte at elevated temperature and therefore Li-ion cells with LTO negative electrodes would show no negative electrode reactivity at elevated temperature. Jiang et al.[117] showed that the heat of reaction between lithiated negative electrode materials and electrolyte decreases linearly with

1. Underwriters Laboratories UL1642, "Standard for Li-ion Batteries"
2. International Electrotechnical Commission IEC 62133, "Secondary Cells and Batteries Containing Alkaline or Other Non-acid Electrolytes"
3. Institute of Electrical and Electronics Engineers IEEE 1625, "Standard for Rechargeable Batteries for Multi-Cell Mobile Computing Devices"
4. IEEE 1725, "Standard for Rechargeable Batteries for Cellular Telephones"

These standards have been selected because they include the essential safety tests and design rules that Li-ion batteries must pass or comply with to obtain these and other certifications. Other Li-ion battery standards exist—for example, the Battery Association of Japan (BAJ) and the United Nations have numerous standards. A UN committee of experts recommends tests for shipping by air which are then adopted by the International Air Transport Association (IATA).

Table 26.15 lists and defines the safety tests and required outcomes to obtain UL1642 certification. These tests require that both fresh cells (fully charged) and cells that have been cycled to 25% of the specified cycle life (or cycled for 90 days, whichever is shorter) and then are fully charged, meet the test requirements. The tests cover the three main areas of electrical, mechanical, and environmental

TABLE 26.15 Safety Tests for Lithium-Ion Batteries as Outlined by UL1642

Test	Description	Number of cells	Required outcome
Electrical tests			
Short circuit (23°C)	< 0.1 Ω to 0.1 V, monitor until T returns to 33°C.	5 fresh charged, 5 cycled charged	No explosion, no fire, $T_{cell} < 150°C$
Short circuit (55°C)	< 0.1 Ω to 0.1 V, monitor until T returns to 65°C.	5 fresh charged, 5 cycled charged	No explosion, no fire, $T_{cell} < 150°C$
Abnormal charge	Charge at 3 times manufacturers' recommended rate for 7 h.	5 fresh charged, 5 cycled charged	No explosion, no fire
Forced discharge	One discharged cell in series with the number of series-connected cells (charged) used in the device. Discharge the series assembly through a resistance < 0.1 Ω to 0.1 V. Monitor until T returns to 10°C above ambient.	5 fresh charged, 5 cycled charged	No explosion, no fire
Mechanical tests			
Flat plate crush	Between flat surfaces to 13 kN.	5 fresh charged, 5 cycled charged	No explosion, no fire
Impact test	15.8 mm diameter bar placed across cell or battery. 9.1 kg weight dropped onto the bar from a height of 61 cm. Prismatic cells to be tested in both directions.	5 fresh charged, 5 cycled charged	No explosion, no fire
Shock test	3 axis, minimum 75 g, peak 125 to 175 g.	5 fresh charged, 5 cycled charged	No explosion, no fire, no leaking, no venting
Vibration test	0.8 mm amplitude, 10 to 55 Hz at a rate of 1 Hz/min and back again.	5 fresh charged, 5 cycled charged	No explosion, no fire, no leaking, no venting
Environmental tests			
Heating test	Heat to 130°C at 5°C/min, and hold at 130°C for 10 min. Return to room temperature and examine.	5 fresh charged, 5 cycled charged	No explosion, no fire
Temperature cycling test	Room T: 4 h, 70°C: 4 h, room T: 4 h, –40°C: 4 h, room T: 4 h; repeat cycle 10 times.	5 fresh charged, 5 cycled charged	No explosion, no fire, no leaking, no venting
Altitude test	11.6 kPa for 6 h.	5 fresh charged, 5 cycled charged	No explosion, no fire, no leaking, no venting
Projectile test	Cells are incinerated.	5 fresh charged	Cell parts cannot penetrate the wire screen used in the test

Note: The projectile test is one in which the cells are heated on a screen over a burner. When they explode or vent and burn, the cells must not puncture the screen on which they rest by projectiles produced by the event.

abuse. Virtually all Li-ion cells that reach consumers for power tool and portable electronics applications pass these tests.

The International Electrotechnical Commission IEC 62133 standard is similar to the UL-1642 standard. However, there are some very important differences. First, the IEC standard includes a free-fall test where charged batteries and devices containing the batteries must be dropped three times from a 1 m height. No fire or explosion can occur. The IEC standard contains an additional overcharge test. Cells are charged at 2C continuously with a thermocouple attached to the cell casing. The test continues until the temperature reaches steady-state conditions or returns to room temperature. No fire or explosion can occur.

The most significant difference between the IEC and UL standards is the proposed inclusion of a forced internal short-circuit test in the latter standard. The forced internal short-circuit test is extremely important because most of the safety events in the field are thought to have been caused by internal shorts. The proposed test procedure in the IEC guidelines is detailed and essentially consists of placing a small "L" shaped Ni particle in between the negative electrode and separator of a charged cell, followed by crushing the cell under a specified pressure. Figure 26.84 shows a diagram of the Ni particle, and Fig. 26.85 shows the Ni particle being inserted into the jelly-roll of a prismatic cell. After the jelly-roll is reassembled, it is pressed between flat plates to create an internal short. Once the short is initiated, the cell is monitored and cannot catch fire to pass the test. The Battery Association of Japan (BAJ) uses a similar test in which a nickel particle is placed in a cell's jelly-roll electrode configuration during cell assembly to simulate an internal short circuit.

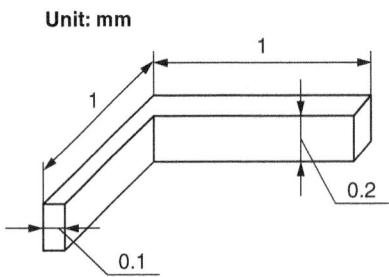

FIGURE 26.84 Shape of the Ni particle placed between the positive electrode and separator in the IEC internal short test.[121]

The IEEE standards 1625 and 1725 define the design, assembly, and safety response of Li-ion cells for computing and cellular phone applications, respectively. In order to obtain these certifications, cells must meet the UL1642 or IEC 62133 safety standards and meet a significant number of design and quality assurance guidelines. Table 26.16 is an overview of these guidelines. Many of these guidelines are designed to prevent the occurrence of cell defects that might ultimately cause an internal short circuit during the use of the cell. The IEEE 1625 standards also include protocols for the safe operation of multicell battery packs.

The IEEE 1625 and 1725 standards were prepared by the IEEE Portable Battery Working Group, which at the time of the submission of the document to the IEEE-SA standards board for approval had the following members: Amperex Technology Ltd. (ATL), Apple Inc., Asahi KASEI Chemicals Corporation, Battery Association of Japan, BYD, Celgard, Compal Electronics, Inc., Dell Inc., Gateway, Hewlett Packard Company, IBM Corporation, Intel, Intersil, Lenovo, LG Chem, Panasonic,

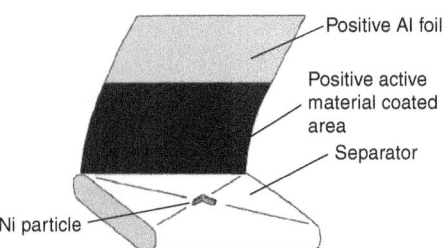

FIGURE 26.85 Ni particle inserted between the positive electrode and the separator for the proposed IEC internal short test.[121]

TABLE 26.16. Criteria that Must Be Met to Obtain IEEE 1725 Certification (*Italics Added By Authors*)

Topic	Sub-topic	Sub-sub-topic	Comments
Design requirements			
	Separator selection		
		Stability	Separator needs sufficient chemical, electrochemical, mechanical, and thermal stability.
		Shutdown performance	Separator must have at least two orders of magnitude resistance increase during shutdown at a minimum rate of 2000 Ω cm^2/s.
		Strength and thickness	Must provide sufficient strength and thickness to prevent puncture.
		Shrinkage allowance	Separator must extend beyond the negative and positive electrodes in all cases (*to prevent electrode contact*).
	Electrode design		
		Capacity balance	Reversible charge capacity of negative electrode must be greater than positive electrode (*to prevent Li plating*).
		Electrode geometry	The active area of the negative electrode must completely cover that of the positive electrode (*to prevent Li plating*).
	Electrode tabs		Optimal tab length required
		Tab insulation	Tab with opposite polarity of case must be insulated (*to prevent shorts during shock and vibration*).
		Insulation adherence	Insulation needs to be permanently adhered.
		Insulation characteristics	Insulation needs sufficient chemical, electrochemical, mechanical, and thermal stability (*for long-term stability*).
	Cell vent mechanism		Cell must incorporate a reliable pressure-vent mechanism to avoid dangerous pressure buildup. Pouch cells do not require a vent.
		Retention of contents	Vent mechanism must retain cell contents during venting.
		Projectiles	Cell must pass projectile test of UL1642.
	Overcurrent protection		Cell may incorporate an overcurrent device such as a positive temperature coefficient resistor to limit current in case of short circuit.
	Overvoltage protection		Manufacturer must provide recommended charging current for the cell and upper-voltage limit to the purchaser.
Manufacturing considerations			
	Materials specifications		Materials specifications must be developed to limit impurities below critical limits.
	Impurity avoidance		All known likely impurities in materials should be identified and controlled.
	Cleanliness		Temperature range, humidity, and dust levels must meet specifications. Metal contamination from equipment or process shall be prevented (*to prevent subsequent internal shorts*).
	Traceability		All cells must be marked in a way to ensure traceability even during an exothermic event (*to find out what went wrong*).

(*Continued*)

TABLE 26.16. Criteria that Must Be Met to Obtain IEEE 1725 Certification (*Italics Added By Authors*) (*Continued*)

Topic	Sub-topic	Sub-sub-topic	Comments
Winding or stacking process	Electrode production		
		Uniform coating	Coating density, thickness, and surface roughness of the electrodes must meet specifications (*to ensure cell balance and prevent internal shorts*).
		Burr control	Burrs cannot exceed 50% of separator thickness. Burrs on electrodes must be measured once per day at minimum (*to prevent internal shorts*).
	Prevent damage to electrodes		Wrinkling, tearing, and/or deformation of electrodes should be prevented. Manufacturer must have a way to detect this.
	Manufacturing equipment		Must prevent damage or modification to the cell.
	Defective electrodes		Must be scrapped.
	Preventive maintenance plan		Must have an effective preventive maintenance plan (i.e., replacement of electrode slitter knives).
	Cell teardown		Cell teardown to ensure specifications are met must be done at least once per machine shift.
	Care during winding or stacking		
		Tension and damage	Avoid excessive tension or damage by twisting or bending of electrodes.
		Loose material	Must have an effective method to collect all loose material produced in various manufacturing steps (*to prevent internal shorts*).
		Damaged cells	Must have methods to detect damaged cell cores. Voltage test, resistance test, x-ray check (*to prevent internal shorts*).
	Electrode spacing		Winding spindle removal process cannot damage the jelly-roll.
	Winding pressure		Pressure selected to avoid cell damage and avoid introduction of burrs, etc.
Assembly precautions	No contaminants		No dust, flakes, etc., can be introduced (*to prevent internal shorts*).
	No internal shorts		Assembly method and location of insulating material shall provide reliable protection against internal shorts over the lifetime of the cell.
	Tab positioning		
		Staggered	Positive and negative tabs should be staggered so they do not overlap with each other.
		Integrity of cell core	Resistance check to be used to ensure core is undamaged after assembly.
	Insulators		Insulators must be in the proper positions.
	Electrode alignment		Electrode alignment is critical to prevent hazards. Positive electrode must be fully overlapped by negative electrode. 100% of cells must be checked by a vision system to ensure alignment.
Cell ageing			Aging tests must be carried out.
Cell safety			Tests according to UL1642 or IEC 62133 must be carried out and cells must pass all tests.

PC TEST Engineering, Laboratory, Inc., Samsung SDI, Sanyo Electric Co., Ltd., Shenzhen BAK Battery Co., Sony Corporation, Texas Instruments, Inc., Tianjin Lishen Battery Jt-Stock Co., Ltd., Tonen Chemical Nasu Co., Ltd., Tyco Electronics, and Underwriters Laboratories Inc. These are industry leaders committed to ensuring battery and product safety.

26.6 CONCLUSIONS AND FUTURE TRENDS

Li-ion batteries are used in virtually all types of rechargeable, portable electronic devices, including laptop computers, cellular phones, and digital cameras. They are now used in a large percentage of commercially available battery-powered tools. Lithium-ion batteries are being used in E-bikes and scooters in Asia and Europe. New markets and emerging markets for Li-ion batteries include electrified vehicles (HEVs, PHEVs, EREVs, and EVs) and grid-energy storage. The acceptance of the technology has been driven by its unique ability to offer a high level of performance in many aspects, including energy density, specific energy, specific power, cycle life, storage life, and temperature range, in a safe, low-cost product. The various common choices of positive electrode materials enable trade-offs among the relative importance of cost, power, energy, and thermal stability.

Further improvements in mechanical design and materials will provide additional improvement in cell performance. Li-ion materials are currently the subject of huge interest in the research and development community. Improved positive electrode materials that offer higher capacity and improved safety properties are in development, as are negative electrode materials, such as the tin- and silicon-based materials, that offer further improvement in specific energy, energy density, rate capability, and longevity. Electrolyte additives and electrode material coatings are being employed to improve both cycle life and calendar life.

As an example, a recent Panasonic press release[122] announced the introduction of an 18650-size Li-ion cell with an "enhanced Ni-based" positive electrode and a silicon negative electrode. Panasonic reports that the cell has a capacity of 4.0 Ah, an average voltage of 3.4 V, and a mass of 54 g, thus this cell provides a specific energy of 251 Wh/kg and an energy density of 800 Wh/L! Release of these cells is scheduled for 2013.

There has been some discussion recently that there may not be enough lithium available to satisfy the requirements of the automotive market. However, recent opinion suggests that there is enough lithium available in the world's reserves, but perhaps not enough short-term production capacity to meet demand in the 2011–2015 time frame. Nevertheless, alternatives to Li-ion technology having improved sustainability, such as Na-ion technology, are being explored by a number of research groups.

ACKNOWLEDGMENTS

The authors acknowledge the assistance of numerous experts in preparing this chapter. Special thanks to Junwei Jiang (3M Co.), Mark Obrovac (3M Co.), Yong-Shou Lin (E-One/Moli Energy Canada), Ulrich von Sacken (BAK Battery Canada), Qiming Zhong (BAK Battery Canada), Mark Shoesmith (E-One/Moli Energy Canada), Karim Zaghib (Hydro-Quebec), Ken Rudisuela (Mobilogy), Sherman Hon (Electrovaya), Yet-Ming Chiang (A123 Systems and MIT), Ralph Brodd (Broddarp of Nevada), John Zhang (Celgard), Hiroshi Inoue (Sony), Karen Thomas-Alyea (A123), Frank Puglia and Maggie Gulbinska (both of Yardney Technical Products).

REFERENCES

1. J. Chen, C. Buhrmester, and J. R. Dahn, *Electrochem. Solid-State Lett.* **8**, A59 (2005).
2. J. R. Dahn, J. Jiang M. D. Fleischauer, C. Buhrmester, and L. J. Krause, *J. Electrochem. Soc.* **152**, A1283–A1291 (2005).

3. www.samsungsdi.com/storage/battery/circle/cylindrical-rechargeable-battery.jsp, accessed Dec. 20, 2009.
4. www.ubergizmo.com/15/archives/2009/12/panasonic_lithium_ion_battery_enters_production.html, "Panasonic managed to achieve this milestone by relying on its unique Heat Resistance Layer (HRL) technology which forms an insulating metal oxide layer between the positive and negative electrodes, resulting in the battery not overheating should a short circuit occur," accessed Dec. 20, 2009.
5. M. S. Whittingham and M. B. Dines, *Surv. Prog. Chem.* **9**, 55 (1980).
6. A. Weiss, *Angew. Chem.* **75**, 755–761 (1963).
7. A. Herold, *Bull. Soc. Chim. Fr.*, 999 (1955).
8. A. Herold, in *Intercalated Materials,* F. Levy (ed), D. Reidel Publishing, Dordrecht, the Netherlands, 1979, p. 323.
9. D. M. Adams, *Inorganic Solids,* Wiley, New York, 1974.
10. A. R. West, *Solid State Chemistry,* Wiley, New York, 1984, pp. 25–29.
11. H. Selig and L. B. Ebert, *Advances in Inorganic Chemistry and Radiochemistry* **23**, 281 (1980).
12. J. R. Dahn, A. K. Sleigh, H. Shi, B. M. Way, W. J. Weydanz, J. N. Reimers, Q. Zhong, and U. von Sacken in "Lithium Batteries—New Materials, Developments and Perspectives," G. Pistoia (ed.) 1994, pp. 1–97.
13. T. Zheng, J. N. Reimers, and J. R. Dahn, *Phys. Rev. B* **51**, 734–741 (1995).
14. U.S. Patent 4,357,215. U.S. Patent 4,302,518.
15. M. R. Palacin, D. Larcher, A. Audemer, N. Sac-Epee, G. G. Amatucci, and J.-M. Tarascon, *J. Electrochem. Soc.* **144**, 4226 (1997).
16. K. Mizushima, P. C. Jones, P. J. Wiseman, and J. B. Goodenough, *Mat. Res. Bull.* **15**, 783 (1980).
17. M. M. Thackeray, W. I. F. David, P. G. Bruce, and J. B. Goodenough, *Mater. Res. Bull.* **18**, 461–472 (1983).
18. A. K. Padhi, K. S. Nanjundaswamy, and J. B. Goodenough, *J. Electrochem. Soc.* **144**, 1188 (1997).
19. Z. Lu, D. D. MacNeil, and J. R. Dahn, "Layered Li[Ni$_x$Co$_{1-2x}$Mn$_x$]O$_2$ Cathode Materials for Lithium Ion Batteries," *Electrochemical and Solid State Letters* **4**, A200–A203 (2001).
20. Z. Lu, D. D. MacNeil, and J. R. Dahn, "New Layered Cathode Materials Li[Ni$_x$Li$_{(1/3-2x/3)}$Mn$_{(2/3-x/3)}$]O$_2$ for Lithium Ion Batteries," *Electrochemical and Solid State Letters* **4**, A191–A194 (2001).
21. Q. Y. Wang, J. Liu, A. V. Murugan, and A. Manthiram, *Journal of Materials Chemistry* **19**, 4965 (2009).
22. Y. Wang, J. Jiang, and J. R. Dahn, *Electrochemistry Communications* **9**, 2534–2540 (2007).
23. Reproduced with permission from Y. Wang, J. Jiang, and J. R. Dahn, *Electrochemistry Communications* **9**, 2534–2540 (2007).
24. M. Yoshio, H. Tanaka, K. Tominaga, and H. Noguchi, *J. Power Sources* **40**, 347–353 (1992); E. K. Mizushima, P. C. Jones, P. J. Wiseman, and J. B. Goodenough, *Mat. Res. Bull.* **15**, 783–789 (1980); W. D. Johnson, R. R. Heikes, and D. Sestrich, *Phys. Chem. Solids* **7**, 1–13 (1958); E. Jeong, M. Won, and Y. Shim, *J. Power Sources* **70**, 70–77 (1998); Zhecheva, R. Stoyanova, M. Gorova, R. Alcantra, J. Moales, and J. L. Tirado, *Chem. Mater.* **8**, 1429–1440 (1996); B. Garcia, J. Farcy, J. P. Pereira-Ramos, J. Perichon, and N. Baffier, *J. Power Sources* **54**, 373–377 (1995); P. N. Kumta, D. Gallet, A. Waghray, G. E. Blomgren, and M. P. Setter, *J. Power Sources* **72**, 91–98 (1998); Y. Chiang, Y. Jang, H. Wang, B. Huang, D. Sadoway, and P. Ye, *J. Electrochem. Soc.* **145**, 887 (1998); Y. Chiang, Y. Jang, H. Wang, B. Huang, D. Sadoway, and P. Ye, *J. Electrochem. Soc.* **145**, 887 (1998); T. J. Boyle, D. Ingersoll, T. M. Alam, C. J. Tafoya, M. A. Rodriguez, K. Vanheusden, and D. H. Doughty, *Chem. Mater.* **10**, 2270–2276 (1998); G. G. Amatucci, J. M. Tarascon, D. Larcher, and L. C. Klein, *Solid State Ionics* **84**, 169–180 (1996); D. Larcher, M. R. Palacin, G. G. Amatucci, and J.-M. Tarascon, *J. Electrochem. Soc.* **144**, 408 (1997); M. Antaya, J. R. Dahn, J. S. Preston, E. Rossen, and J. N. Reimers, *J. Electrochem. Soc.* **140**, 575 (1993); M. Antaya, K. Cearns, J. S. Preston, J. N. Reimers, and J. R. Dahn, *J. Appl. Phys.* **75**, 2799 (1994); P. Frajnaud, R. Nagarajan, D. M. Schleich, and D. Vujic, *J. Power Sources* **54**, 362–366 (1995); E. Antolini, *J. Eur. Ceram. Soc.* **18**, (10), 1405–1411 (1998).
25. K. Ariyoshi, E. Iwata, M. Kuniyoshi, H. Wakabayashi, and T. Ohzuku, *Electrochemical and Solid-State Letters* **9**, A557 (2006).
26. A. van Bommel and J. R. Dahn, *J. Electrochem. Soc.* **156**, A362–A366 (2009).
27. A. van Bommel and J. R. Dahn, *Chemistry of Materials* **21**, 1500–1503 (2009).
28. J. Barker, M. Y. Saidi, and J. L. Swoyer, *Electrochem. and Solid State Letters* **6**, A53–A55 (2003).

29. N. Ravet, S. Besner, M. Simoneau, A. Vallee, M. Armand and J. F. Magnan, U.S. Patent No. 6,855,273 (2005).
30. Y. Wang, J. Jiang, and J. R. Dahn, *Electrochemistry Communications* **9**, 2534–2540 (2007).
31. T. Ohzuku and R. J. Brodd, *J. Power Sources* **174**, 449–456 (2007).
32. J. N. Reimers and J. R. Dahn, *J. Electrochem. Soc.* **139**, 2091 (1992).
33. A. Van der Ven, M. K. Aydinol, G. Ceder, G. Kresse, and J. Hafner, *Physical Review B* **58**, 2975 (1998).
34. M. S. Whittingham, *Chem. Rev.* **104**, 4271–4301 (2004).
35. K. Ariyoshi, E. Iwata, M. Kuniyoshi, H. Wakabayashi, and T. Ohzuku, *Electrochemical and Solid-State Letters* **9**, A557–A560 (2006).
36. M. S. Whittingham, *Chem. Rev.* **104**, 4271–4301 (2004).
37. K. Ariyoshi, E. Iwata, M. Kuniyoshi, H. Wakabayashi, and T. Ohzuku, *Electrochemical and Solid-State Letters* **9**, A557–A560 (2006).
38. J.W. Fergus, *J. Power Sources* **195**, 939–954 (2010).
39. C. Feng, H. Li, P. Zhang, Z. Guob, and H. Liu, "Synthesis and modification of non-stoichiometric spinel ($Li_{1.02}Mn_{1.90}Y_{0.02}O_{4-y}F_{0.08}$) for lithium-ion batteries," *Materials Chemistry and Physics* **119**, 82–85 (2010).
40. Y. Wang, M. Zhang, U. von Sacken, and B. M. Way, U.S. Patent No. 6,045,948 (2000).
41. M. Wohlfahrt-Mehrens, C. Vogler, and J. Garche, *Journal of Power Sources* **127**, 58–64 (2004).
42. "Cellular Phone Recall May Cause Setback for Moli," *Toronto Globe and Mail*, August 15, 1989 (Toronto, Canada).
43. *Adv. Batt. Technology,* **25**(10), 4 (1989)
44. J. R. Dahn, A. K. Sleigh, H. Shi, B. M. Way, W. J. Weydanz, J. N. Reimers, Q. Zong, and U. von Sacken, in "Lithium Batteries—New Materials, Developments and Perspectives," G. Pistoia (ed.), pp. 1–97.
45. R. E. Franklin, *Proc. Roy. Soc.* (London) **A209**, 196 (1951).
46. M. Inagaki, *Solid State Ionics* **86–88**, 833–839 (1996).
47. T. Zheng, J. N. Reimers, and J. R. Dahn, *Phys. Rev. B* **51**, 734 (1995).
48. H. Selig and L. B. Ebert, *Advances in Inorganic Chemistry and Radiochemistry,* **23**, 281 (1980).
49. T. Zheng, J. N. Reimers, and J. R. Dahn, *Phys. Rev. B* **51**, 734 (1995).
50. H. Shi, J. N. Reimers, and J. R. Dahn, *J. Appl. Crystallography* **26**, 827–836 (1993).
51. R. Fong, U. von Sacken, and J. R. Dahn, *J. Electrochem. Soc.* **137**, 2009–2013 (1990).
52. D. D. MacNeil, D. Larcher, and J. R. Dahn, *J. Electrochem. Soc.* **146**, 3596–3602 (1999).
53. H. Nozaki, K. Nagaoka, K. Hoshi, N. Ohta, and M. Inagakic, *Journal of Power Sources* **194**, 486–493 (2009); Y.-S. Park, H. J. Bang, S.-M. Oh, Y.-K. Sun, S.-M. Lee, Journal of Power Sources **190**, 553–557 (2009).
54. Y.-S. Park, H. J. Bang, S.-M. Oh, Y.-K. Sun, and S.-M. Lee, *Journal of Power Sources* **190**, 553–557 (2009).
55. D. W. Murphy, R. J. Cava, S. M. Zahurak, and A. Santaro, *Solid State Ionics* **9–10**, 413 (1983).
56. K. M. Colbow, R. R. Haering, and J. R. Dahn, *J. Power Sources* **26**, 397–402 (1989).
57. T. Ohzuku, A. Ueda, and N. Yamamoto, *J. Electrochem. Soc.* **142**, 1431 (1995).
58. A. D. W. Todd, P. P. Ferguson, and J. R. Dahn, accepted for publication in *International Journal of Energy Research*, 2010.
59. M. N. Obrovac, L. Christensen, D. B. Le, and J. R. Dahn, *J. Electrochem. Soc.* **154**, A849 (2007); M. N. Obrovac and L. J. Krause, *J. Electrochem.Soc.* **154**, A103 (2007).
60. A. B. McEwen, H. L. Ngo, K. LeCompte, and J. L. Goldman, *J. Electrochem. Soc.* **146**, 1687–1695; M. Miyachi, H. Yamamoto, H. Kawai, T. Ohta, and M. Shirakata, *J. Electrochem. Soc.* **152**, A2089 (2005).
61. C. K. Chan, H. Peng, G. Liu, K. McIlwrath, X. F. Zhang, R. A. Huggins, and Y. Cui, *Nature Nanotechnology* **3**, 31 (2008).
62. www.sony.co.jp/SonyInfo/News/Press/200502/05-006/index.html; H. Inoue, in *International Meeting on Lithium Batteries, IMLB2006, Abstr.# 228* (2006); H. Inoue, S. Mizutani, H. Ishihara, and S. Hatake, *214th ECS Meeting, Abstr., #1160*, (2008); H. Inoue, S. Mizutani, H. Ishihara, and Y. Fukushima, *Denchi Gijutsu* **19**, 86 (2007); H. Inoue, T. Takada, and Y. Kudo, *Electrochemistry* **76**, 358 (2008) [in Japanese].

63. H. Inoue, S. Mizutani, H. Ishihara, and Y. Fukushima, *Denchi Gijutsu* **19**, 86 (2007).
64. Q. Fan, P. J. Chupas, and M. S. Whittingham, *Electrochemical and Solid State Letters* **10**, A274 (2007).
65. A. D. W. Todd, P. P. Ferguson, J. G. Barker, M. D. Fleischauer, and J. R. Dahn, *J. Electrochem. Soc.* **156**, A1034–A1040 (2009).
66. Y. Tian, A. Timmons, and J. R. Dahn, "In-Situ AFM Measurements of the Expansion of Nanostructured Sn-Co-C Films Reacting with Lithium," *J. Electrochem. Soc.* **156** A187–A192 (2009).
67. J. Li, P. P. Ferguson, D.-B. Le, and J. R. Dahn, accepted for publication in *Electrochimica Acta*.
68. P. P. Ferguson, P. Liao, R. A. Dunlap, and J. R. Dahn, *J. Electrochem. Soc.* **156**, A13-A17 (2009).
69. P. P. Ferguson, A. D. W. Todd, and J. R. Dahn, submitted to *J. Electrochem. Soc.*
70. F. M. Gray, *Polymer Electrolytes,* The Royal Society of Chemistry, 1997.
71. A. B. McEwen, H. L. Ngo, K. LeCompte, and J. L. Goldman, *J. Electrochem. Soc.* **146**, 1687–1695 (1999); A. B. McEwen, S. F. McDevitt, and V. R. Koch, *J. Electrochem. Soc.* **144**, L84 (1997).
72. K. Xu, *Chem. Rev.* **104**, 4303 (2004).
73. H. Nakamura, H. Komatsu, and M. Yoshio, *J. Power Sources* **62**, 219–222 (1996); B. Scrosati and S. Megahed, Electrochemical Society Short Course, New Orleans, Oct. 10, 1993; D. Linden (ed.), *The Handbook of Batteries*, 2nd ed., McGraw-Hill, New York, 1995, p. 36.14.
74. S. T. Mayer, H. C. Yoon, C. Bragg, and J. H. Lee, "Low Temperature Ethylene Carbonate Based Electrolyte for Lithium-Ion Batteries," Polystor Corporation, Dublin, CA, 1997.
75. J. T. Dudley, D. P. Wilkinson, G. Thomas, R. LaVae, S. Woo, H. Blom, C. Horvath, M. W. Juzkow, B. Denis, P. Juric, P. Aghakian, and J. R. Dahn, *J. Power Sources* **35**, 59–82 (1991).
76. D. Guyomard and J. M. Tarascon, *J. Electrochem. Soc.* **54**, 92 (1995); T. Zheng, Y. Liu, E. W. Fuller, U. von Sacken, and J. R. Dahn, *J. Electrochem. Soc.* **142**, 2581 (1995); D. Aurbach, B. Markovsky, A. Schechter, Y. Ein-Eli, and H. Cohen, *J. Electrochem. Soc.* **143**, 3809 (1996).
77. D. Aurbach, Y. Ein-Eli, B. Markovsky, A. Zaban, S. Luski, Y. Carmeli, and H. Yamin, *J. Electrochem. Soc.* **142**, 2882 (1995).
78. H. Yoshida, T. Fukunaga, T. Hazama, M. Terasaki, M. Mizutani, and M. Yamachi, *J. Power Sources* **68**, 311–315 (1997).
79. Y. Ein-Eli, S. F. McDevitt, D. Aurbach, B. Markovsky, and A. Schechter, *J. Electrochem. Soc.* **144**, L180 (1997).
80. Z. X. Shu, R. S. McMillian, and J. J. Murray, *J. Electrochem. Soc.* **140**, 922 (1993).
81. D. Aurbach, Y. Ein-Eli, B. Markovsky, A. Zaban, S. Luski, Y. Carmeli, and H. Yamin, *J. Electrochem. Soc.* **142**, 2882 (1995); O. Chusid, Y. Ein-Eli, M. Babai, Y. Carmeli, and D. Aurbach, *J. Power Sources,* **43–44**, 47 (1993).
82. M. Broussely, Ph. Biensan, F. Bonhomme, Ph. Blanchard, S. Herreyre, K. Nechev, and R.J. Staniewicz, *J. Power Sources* **146**, 90–96 (2006).
83. K. Xu, *Chem. Rev.* **104**, 4303 (2004).
84. K.-S. Lee, S.-T. Myung, K. Amine, H. Yashiro, and Y.-K. Sun, *J. Materials Chemistry* **19**, 1995 (2009); Y.-K. Sun, S.-T. Myung, C.S. Yoon, and D.-W. Kim, *Electrochemical and Solid State Letters* **12**, A163 (2009); Y.-K. Sun, S.-W. Cho, S.-W. Lee, C. S. Yoon, and K. Amine, *J. Electrochem. Soc.* **154**, A168 (2007); Y.-K. Sun, S.-T. Myung, B.-C. Park, J. Prakash, I. Belharouk, and K. Amine, *Nature Materials* **8**, 320 (2009); G. Li, Z. Yang, and W. Yang, *J. Power Sources* **183**, 741 (2008); Z. H. Chen and J. R. Dahn, *Electrocimica Acta* **49**, 1079 (2004).
85. S. Patoux, F. Le Cras, C. Bourbon, and S. Jouanneau, U.S. Patent Application Publication No. 2008/0107968 A1 (2008).
86. H. Yamane, T. Inoue, M. Fujita, and M. Sano, *J. Power Sources* **99**, 60 (2001).
87. D. Aurbach, K. Gamolsky, B. Markovsky, Y. Gofer, M. Schmidt, and U. Heider, *Electrochimica Acta* **47**, 1423 (2002).
88. D. Aurbach, K. Gamolsky, B. Markovsky, Y. Gofer, M. Schmidt, and U. Heider, *Electrochimica Acta* **47**, 1423 (2002).
89. K. Xu, U. Lee, S. Zhang, M. Wood, and T. R. Jow, *Electrochemical and Solid-State Letters* **6**, A144 (2003).

90. K. Xu, U. Lee, S. Zhang, M. Wood, and T. R. Jow, *Electrochemical and Solid-State Letters* **6**, A144 (2003).
91. K. Abe, Y. Ushigoe, H. Yoshitake, and M. Yoshio, *J. Power Sources* **153**, 328 (2006).
92. Y. Li, R. Zhang, J. Liu, and C. Yang, *J. Power Sources* **189**, 685 (2009).
93. S. Patoux, L. Daniel, C. Bourbon, H. Lignier, C. Pagano, F. Le Cras, S. Jouanneau, and S. Martinet, *J. Power Sources* **189**, 344 (2009).
94. S. S. Zhang, *J. Power Sources* **162**, 1379 (2006).
95. L. El-Ouatani, R. Dedryvere, C. Siret, P. Biensan, and D. Gonbeau, *J. Electrochem. Soc.* **156**, A468 (2009).
96. K. Abe, K. Miyoshi, T. Hattori, Y. Ushigoe, and H. Yoshitake, *J. Power Sources* **184**, 449 (2008).
97. G. H. Wrodnigg, J. O. Besenhard, and M. Winter, *J. Electrochem. Soc.* **146**, 470 (1999).
98. K.-S. Lee, S.-T. Myung, K. Amine, H. Yashiro, and Y.-K. Sun, *J. Materials Chemistry* **19**, 1995 (2009).
99. Y.-K. Sun, S.-T. Myung, C. S. Yoon, and D,-W. Kim, *Electrochemical and Solid State Letters* **12**, A163 (2009); Y.-K. Sun, S.-W. Cho, S.-W. Lee, C. S. Yoon, and K. Amine, *J. Electrochem. Soc.* **154**, A168 (2007).
100. Y.-K. Sun, S.-T. Myung, B.-C. Park, J. Prakash, I. Belharouk, and K. Amine, *Nature Materials* **8**, 320 (2009).
101. G. Li, Z. Yang, and W. Yang, *J. Power Sources* **183**, 741 (2008).
102. Z. H. Chen and J. R. Dahn, *Electrocimica Acta* **49**, 1079 (2004).
103. S. Patoux, F. Le Cras, C. Bourbon, and S. Jouanneau, U.S. Patent Application Publication 2008/0107968 A1 (2008).
104. J. Dahn, J. Jiang, C. Buhrmester, and L. Moshurchak, "Studies of the 2,5-ditertbutyl-1,4-dimethoxybenzene Overcharge Shuttle in 18650-sized LiFePO4/graphite cells," *Meet. Abstr. Electrochem. Soc.* **502**, 217 (2006).
105. R. Spotnitz, in *Handbook of Battery Materials*, J. O. Besenhard (ed.), VCH Wiley, Amsterdam and New York, 1999.
106. H. S. Bierenbaum, R. B. Isaacson, M. L. Druin, and S. G. Plovan, *Ind. Eng. Chem. Prod. Res. Dev.*, **13**, 2 (1974).
107. G. Venugopal, J. Moore, J. Howard, and S. Pendalwar, *J. Power Sources* **77**, 34–41 (1999).
108. R. P. Quirk and M. A. A. Alsamarraie, in *Polymer Handbook*, J. Brandrup and E. H. Immergut (eds.), Wiley, New York, 1989.
109. Y. Fukumoto, T. Hayashi, and K. Kubota, World Intellectual Property Organization Publication No. WO/2008/010423 (2008).
110. M. C. Smart, B. V. Ratnakumar, and S. Surampudi, *J. Electrochem. Soc.* **146**, 486–492 (1999).
111. U.S. Patent 7,501,200 (2009).
112. J. Jiang, Z. Lu, and M. Triemert, Presentation at the 25th International Battery Seminar and Exhibit, Fort Lauderdale, Florida, March 17–20, 2008.
113. H. Kitao, T. Fujihara, K. Takeda, N. Nakanishi, and T. Nohma, *Electrochemical and Solid-State Letters* **8**, A87 (2005).
114. Y.-S. Lin and L. Feng, IBA Meeting and PPSS 2010, Waikoloa, Hawaii, USA, January 11–15, 2010.
115. www.emc-mec.ca/phev/en/Proceedings.html.
116. D. D. MacNeil, D. Larcher, and J. R. Dahn, *J. Electrochem. Soc.* **146**, 3596 (1999).
117. J. Jiang and J. R. Dahn, *J. Electrochem. Soc.* **153**, A310 (2006).
118. D. D. MacNeil, Zhonghua Lu, Zhaohui Chen, and J. R. Dahn, *J. Power Sources* **108**, 8 (2002).
119. Y. Wang, J. Jiang, and J.R. Dahn, *Electrochemistry Communications* **9**, 2534 (2007).
120. Jim Mc Dowell, Philippe Biensan, and Michel Broussely, "Industrial Lithium Ion Battery Safety—What Are the Tradeoffs?" IEEE document # 978-1-4244-1628-8/07 (2007).
121. IEC 61233.
122. http://panasonic.co.jp/corp/news/official.data/data.dir/jn091225-1/jn091225-1.html.

CHAPTER 27
RECHARGEABLE LITHIUM METAL BATTERIES (AMBIENT TEMPERATURE)

Daniel H. Doughty

27.1 GENERAL CHARACTERISTICS

Lithium metal rechargeable batteries operating at or near room temperature offer a very desirable advantage compared to lithium-ion battery technologies because of their much higher specific energy. There are currently not many manufacturers of cells, but development work is increasing. The highest energy cells available are rated at 350 Wh/kg and 350 Wh/L, with credible projections of 30 to 50% higher energy within the next few years.

However, in many other aspects, Li-ion batteries have better performance metrics. Over the last several years, Li-ion batteries have met with substantial technical and market success, which has influenced several companies to terminate research and development on lithium metal rechargeable battery activities. However, with the renewed interest and increasing demand for higher energy rechargeable batteries, there has been a revival of interest in lithium metal rechargeable batteries. In addition to improving energy content, lithium metal rechargeable battery development activities focus on improving performance aspects such as longer cycle life, wider operational temperature limits, and safety. These objectives are being met by employing innovative strategies for eliminating dendritic growth and the formation of mossy Li on the Li anode during cycling by developing protective layers on the Li surface as well as polymeric and inorganic electrolyte development.

Commercial sources for lithium metal rechargeable batteries provide cells with capacities ranging from < 0.1 mAh to 2.7 Ah, with many more manufacturers supplying small cells in the mAh size. Avestor, now Bolloré, had produced much larger cells (up to 80 Ah) in their batteries, but these are no longer commercially available. Lithium metal rechargeable batteries fall into three categories:

- Liquid electrolyte with cathodes based on sulfur or transition metal oxides and phosphates typically used in Li-ion batteries
- Polymer electrolyte with cathodes typically used in Li-ion batteries
- Inorganic electrolytes with air cathodes or cathodes typically used in Li-ion batteries

Some cell designs may include combinations of these three categories. Cathode materials that are used in Li-ion batteries are also used in lithium metal rechargeable batteries, and since these materials are discussed in Chap. 26, they will not be reviewed here.

27.1.1 History

The lithium-sulfur (Li-S) rechargeable battery, which has a theoretical specific energy of 2500 Wh/kg, the highest of any sealed rechargeable system, was first reported in 1979 by Rauh and coworkers[1] at EIC (Norwood, MA, USA). They observed good electrochemical reversibility at low current and elevated temperature (50°C). It was recognized that the solubility of polysulfide reaction products of sulfur reduction (Li_2S_x) strongly influence the performance of this rechargeable liquid cathode system. The final discharge product, Li_2S, was insoluble and electronically insulating, which was a challenge for this system. The open-circuit voltage is 2.5 V and operating voltage is 2.3 to 1.7 V. The redox behavior of polysulfides has been studied.[2] Dioxolane[3] and glymes[4] (which have a chemical structure similar to poly[ethylene oxide]) were identified as preferred electrolyte solvents for practical Li-S cells. Work has continued at a few companies and universities, and the cell performance has improved dramatically.

In the late 1970s, Armand proposed ionically conducting polymer electrolyte materials for use in solid-state battery designs[5] and a considerable development effort has ensued. The unique aspect of these batteries is that the electrolyte is a solid flexible film comprised of a polymer matrix and an ionic salt complexed into the matrix. Thin-film solid-polymer electrolyte batteries offer the possibility of improved safety as well as good high-rate capability. Poly(ethylene oxide) (PEO) was the first material utilized as a true (dry) solid polymer electrolyte.[6] Development met with substantial success in terms of cycleability that relies on stability of the lithium-polymer electrolyte interface[7] as well as the whole battery system.[8] Avestor (previously Argotech in Canada) batteries were the only systems to reach commercialization. The Avestor battery, originally designed as an electric vehicle (EV) battery module, was redesigned for telecommunication applications and was an outcome of a development program sponsored by the U.S. Advanced Battery Consortium (USABC) and U.S. Department of Energy (DOE). Avestor closed in 2006 and their assets were purchased by Bolloré in France and that company has announced plans to commercialize the lithium metal polymer (LMP) battery technology for BlueCar EVs.[9]

Lithium metal rechargeable batteries with liquid electrolyte and solid cathodes were developed in the early 1980s at EIC and were comprised of a lithium metal anode, a variety of cathodes such as vanadium oxide,[10] TiS_2,[11] and $MoO_xS_{(3-x)}$[12] with an electrolyte solution containing tetrahydrofuran (THF), 2Me-THF, methyl furan (a stabilizer), and $LiAsF_6$. These battery systems were the first practical Li metal organic electrolyte rechargeable cells with more than 100 cycles. They operated from −10 to 50°C and had reasonable stability to 70°C. They were not found to be safe enough for commercial uses but could be used for space and military applications.

Another lithium metal rechargeable battery that was developed and sold commercially was lithium/molybdenum disulfide (Li/MoS_2) batteries by Moli Energy (Vancouver, Canada). The Li/MoS_2 system was introduced in the mid-1980s in a cylindrical AA-size.[13] The cell used thin lithium metal anodes (125 μm), with a stoichiometric excess of about three times first discharge capacity. The MoS_2 cathode slurry was coated on a thin aluminum foil (150 μm). The open-circuit voltage was 2.3 V and the operating voltage was 2.2 to 1.4 V. A spirally wound construction was used. The electrolyte was 1 M $LiAsF_6$ dissolved in a 50:50 mixture of propylene carbonate and ethylene carbonate. These batteries were withdrawn from the market after several safety incidents occurred in the late 1980s,[14] and in 1990 the company was acquired by a Japanese consortium led by NEC and also included Mitsui and Yuasa.

A rechargeable Li metal battery system that became a commercial success during the mid-1990s was developed and produced by Tadiran, Ltd. (Israel). The battery was AA-sized and comprised of a Li_xMnO_2 (0.3 < x < 1), a 3 V cathode, an electrolyte solution containing 1,3-dioxolane, $LiAsF_6$, and tributylamine (TBA) in trace amount as a stabilizer. The operating voltage was between 3.4 and 2.0 V. This battery system had several attractive features such as an energy density >140 Wh/kg, a wide temperature range of operation (−30°C < T < 60°C), excellent shelf life, a reasonable cycle life (more than 300 cycles at 100% depth of discharge [DOD]), and internal safety mechanisms.[15] These safety mechanisms, which made this battery safe for use and of commercial value, were based on a shutdown of the battery in abuse cases, such as short-circuiting, heating above 130°C, and overcharge by a fast polymerization of the solvent before dangerous thermal runaway processes can take place.[16] However, this battery had a limited cycle life because the charging rates have to be very low, < C/9, (corresponding to current densities less than 0.5 mA/cm^2) to avoid formation of mossy Li. The high surface area Li deposits react with electrolyte components, consuming the electrolyte and causing premature end of life.[17] Because Li-ion cells became available with similar energy and longer cycle life, Tadiran discontinued marketing this cell and has not announced plans to resume work on lithium metal rechargeable batteries.

Lithium/air (Li-O$_2$) rechargeable batteries will be covered in Chap. 33, "Metal/Air Batteries," but a few comments will be made here. This electrochemical system also has exceedingly high theoretical energy, 5200 Wh/kg (including oxygen). Working cells with high temperature ceramic electrolytes (operating at 650°C) were reported in 1987,[18] and room temperature organic electrolyte cells were first demonstrated by workers at EIC in 1996.[19] (Cells with aqueous electrolyte suffer from corrosion of the Li electrode by water.[20]) Open circuit voltages approached 3.0 V, and working voltage was between 2.8 and 2.0 V. The cell uses a polyacrylonitrile (PAN) polymer electrolyte membrane into which propylene carbonate (PC) and salts were added and a cobalt compound catalyzed the air electrode. Other polymers can be used as well.[21] Cycle life is limited due to adsorption and reaction of other gaseous species (H_2O and CO_2). This is an active area of research and development, with recent reports of rechargeable lithium cells with ceramic[22,23] as well as glass-ceramic[24] and polymer-ceramic[25] electrolytes.

Thin-film, solid-state batteries are a specialized type of lithium metal rechargeable battery developed for low current semiconductor and microelectronic applications. These microbatteries, which employ a metallic Li negative, solid electrolyte, and transition metal oxide cathode materials, can be fabricated by high volume manufacturing techniques on silicon wafers and are viable as on-chip or on-board power sources for microelectronics. The first lithium battery of this type was enabled by fabrication of lithium phosphorus oxynitride (LiPON) glassy electrolyte by Bates.[26] LiPON exhibits a single Li-ion conducting phase between −26 and 140°C, with an average conductivity of 2.3×10^{-6} S/cm at 25°C and an average activation energy of 0.55 eV. LiPON is mechanically stable and has a 5 V stability window. It acts as a rigid barrier against the possible growth of lithium dendrites, yet it does not crack as the cathode volume changes during cycling as evidenced by the long cycle lives of these thin-film batteries. Li/LiCoO$_2$ batteries have been cycled over 40,000 times at 25°C to ~96% depth of discharge with total capacity losses of less than 5%. The possibility of using LiPON as a protective layer between the lithium anode and electrolyte in order to improve the cycle life of conventional lithium metal batteries has spawned several companies and substantial commercial interest.

Additional details and information on other lithium metal rechargeable batteries that were developed but not commercialized in the 1990s and earlier can be found in previous editions of this Handbook and will not be discussed here.

27.2 CHEMISTRY

The objective of the rechargeable lithium metal battery development is to produce batteries that have high energy density, high power density, good cycle life, charge retention, and to provide this performance reliably and safely. The judicious selection of cell components and designs is necessarily a compromise to achieve the optimum balance. Many of the characteristics and criteria for selection are similar to those for primary Li batteries covered in Chap. 14 and Li-ion rechargeable batteries covered in Chap. 26. The process, however, is even more complex for lithium metal rechargeable batteries since the reactions that occur during recharge (lithium plating) affect all of the characteristics and the performance on subsequent cycling.

27.2.1 Negative Electrodes

The search for high-energy-density primary and rechargeable batteries has inevitably led to the use of lithium, as the electrochemical characteristics of this metal are unique. Lithium is the lightest and most electropositive metal, and it has a high specific capacity, 3862 mAh/g,[27] compared to 372 mAh/g for

LiC_6, which is the typical anode in Li-ion batteries. The volumetric comparison is still favorable for Li metal (2061 Ah/L compared to 837 Ah/L for LiC_6), but the low density of Li metal (0.534 g/cm^3) means the advantage is not as great. Lithium is also more easily handled than the other alkali metals. Metallic lithium is more reactive than lithium aluminum and other lithium alloys that have been used as anodes. Alloys have been used mainly in small flat or coin cells, as it is difficult to scale up to larger and spirally wound designs because most lithium alloys are brittle and cannot be extruded into thin foils.

A number of batteries, both primary and rechargeable, using a lithium anode in conjunction with intercalation cathodes, were developed which had attractive energy densities, excellent storage characteristics, and, for rechargeable cells with liquid electrolytes, a limited but reasonable cycle life. In contrast to Li primary batteries, commercial success has been a challenge for all but very small capacity rechargeable Li cells due to persistent safety problems. Dendrite formation is a primary failure mechanism in lithium metal rechargeable batteries. Specifically, control of the lithium/electrolyte interface is essential to avoid dendritic growth[28] and formation of mossy lithium deposits at the Li surface on cycling. While such deposits were first hypothesized to occur in lithium/organic systems in 1974[29] and were first directly observed in 1980,[30] dendrites were directly linked to cell failure in 1988.[31] While the thermal stability of lithium metal foil in many organic electrolytes is sufficiently good early in life, with minimal exothermic reaction occurring up to temperatures near the melting point of lithium (180.5°C), after cycling the surface area of the lithium increases significantly with a corresponding increase in the reactivity, which degrades the thermal stability of the system, with the result that cells become increasingly sensitive to abuse as they are cycled.

The physics and chemistry of dendrite formation are being analytically modeled by Monroe and Newman,[32] which is providing understanding and directions for future materials development in lithium metal rechargeable batteries. The model of Li anode and PEO polymer electrolyte with lithium trifluro sulfonimide (LiTFSI) salt provides trends and predictions that are observed in practical cells. For example, dendrite accelerates across cells under all conditions, and dendrite growth is always slowed by lowering the current density. Cell shorting occurs during charges at current densities above 75% of the limiting current. Increased interelectrode distance slows failure, but the advantages decrease as distance lengthens. Further studies have identified the effect of stack pressure on stabilizing the Li surface[33] and suppressing or eliminating dendrite formation.[34] Stack pressures applied to the cell just over the yield strength of lithium (>25 to 35 psi) were found to physically confine the anode between the flat solid polymer electrolyte (SPE) and greatly diminish adverse morphology development and increase cycle life more than five times in certain cases.

The model predicts that by introducing a highly rigid electrolyte between the two electrodes (the elastic modulus of the polymer electrolyte should exceed 1 GPa) to transmit pressure to the lithium interface, smooth Li interfaces will remain after each cycle. PEO has an elastic modulus of ~3 MPa, so new materials are clearly needed. Further analysis of the elastic deformation of the polymer electrolyte[35] analyzes the stress and surface-tension forces. The incorporation of elastic effects into a kinetic model demonstrates regimes of electrolyte mechanical properties where amplification of surface roughness can be inhibited. This theoretical study provides qualitative justification for the improved resistance to dendrites exhibited by the current Li/polymer system.

The difficulties associated with the use of metallic lithium also are a consequence of its reactivity with the electrolyte and the changes that occur after repetitive charge-discharge cycling. Most solvents are unstable in the presence of Li metal and are reduced at the surface. The reaction of electrolyte at the lithium surface results in irreversible capacity loss associated with formation of a solid electrolyte interphase (SEI), which in some cases may be part of a thicker passivating layer.[36] The major role of the SEI is to separate the negative electrode from electrolyte and to eliminate (or drastically reduce) the transfer of electrons from the electrode surface to solvent molecules. When the film is sufficiently thick to prevent electron tunneling, the electrolyte reduction is suppressed.

Each electrode/electrolyte combination has its own unique features and problems, but there are some general phenomena common to lithium metal systems. The morphology of the SEI is very complex and changes with time, electrolyte composition, and other factors. The morphology is a heterogeneous, polyparticle (in some cases polycrystalline) thin layer upon which a thick porous layer may be formed. Cationic transport is through polymeric sections or on the surface of particles. When lithium is electroplated during recharge, a fresh lithium surface is created which will form a new SEI and passivation layer. Electroplated Li may form a mossy and in some cases a dendritic deposit with a larger surface area than the original metal. The nature of Li deposits in organic electrolytes has been studied by impedance and electrochemical techniques,[37] infrared spectroscopy (IR),[38] atomic force microscopy (AFM),[39] electrochemical quartz crystal microbalance (EQCM) techniques,[40] as well as scanning electron microscopy (SEM).[41] While these studies are complex and the results depend on the nature of electrolyte solvents and salts, current density, additives and temperature, the trends are clear. The important causes of premature failure of lithium metal rechargeable batteries are internal shorting due to Li dendrites and depletion of solvent.[41]

Another contributing effect is the inability of attaining 100% lithium cycling efficiency. This happens because lithium is not thermodynamically stable in the organic electrolytes and the surface of the lithium is covered with a film of the reaction products between the lithium and the electrolyte. Every time the lithium is stripped and replated during discharge and charge, a new lithium surface is exposed and then passivated with a new film, consuming electrochemically active lithium. Also, as a consequence of a mossy Li deposit, some lithium may become electrochemically isolated from the lithium electrode and becomes "dead lithium" on repeated cycling.[31] In order to obtain a reasonable cycle life, a two- to threefold excess of lithium may be required.

A recent report by SION Power Corp. describes a way of substantially reducing the development of high surface mossy lithium deposits in the lithium-sulfur (Li-S) rechargeable system.[42] Figure 27.1 shows SEM photomicrographs of lithium anode morphology after 50 cycles after using conventional cycling conditions that developed porous Li deposit with a thickness of 60 μm compared to "improved" cycling conditions that produced dense Li deposit with a thickness of 47 μm, nearly the same as the original thickness of the Li foil anode. Eliminating the generation of highly porous lithium over the life of the cell would substantially improve cycle life and safety. Protective layers on lithium are also being pursued and can be helpful, but the challenge is to make them durable during many charge/discharge cycles.

The failure to control the growth of surface area of the lithium anode remains a problem in many electrochemical systems, limiting the commercialization of lithium metal cells with liquid organic electrolytes. Several approaches are being investigated to overcome this problem, such as a solid inorganic or organic polymer electrolyte, which are less reactive with lithium.

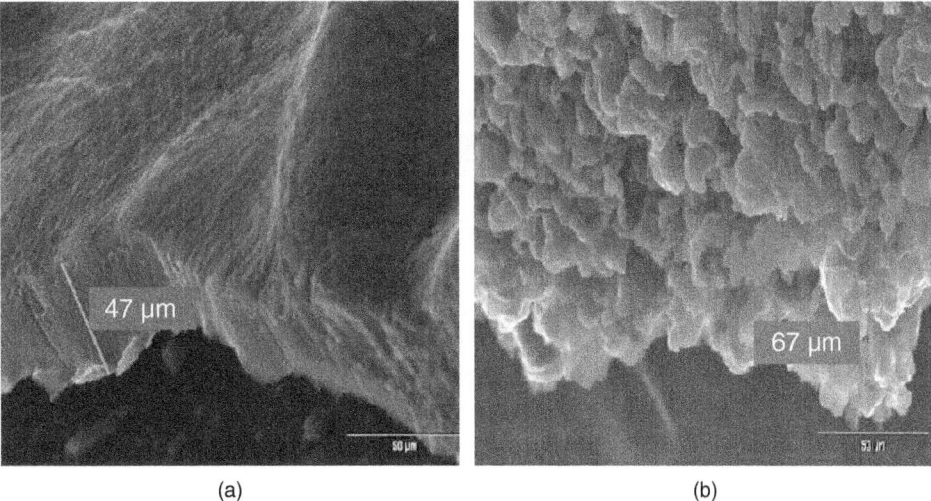

FIGURE 27.1 SEM photomicrograph of fracture edge of lithium anode showing morphology after 50 cycles revealing dense Li deposits with new cycling protocol (a) and porous Li deposits with conventional cycling (b). (*From Ref. 42.*)

Finally, within a given type of lithium rechargeable cells, there is a trade-off between anode depth of discharge and cycle life (measured by cumulative capacity attained during all cycles). The general trend observed is the less lithium that is plated per cycle (microns plated or mAh/cm² per cycle), the greater cumulative capacity is attained and longer cycle life is measured.

27.2.2 Positive Electrodes

There is a wide choice of materials that can be selected for the positive electrodes of lithium metal rechargeable batteries. Cathode materials that are used in Li-ion batteries are also suitable for lithium metal rechargeable batteries. In addition, other cathodes are possible, such as sulfur or oxygen (Li/air battery). Li-ion battery cathode materials will not be discussed except where specific information relates to lithium metal rechargeable batteries.

Sulfur Cathode. The sulfur cathode is an attractive choice because of its high energy. The reduction of elemental sulfur (S_8) to sulfide ion (S^{2-}) has a theoretical capacity of 1675 mAh/g, about an order of magnitude higher than lithiated transition metal oxides used in Li-ion rechargeable batteries.

Two approaches have been explored, incorporating elemental sulfur in the cathode and preparing an electrolyte in which sulfur is dissolved in the form of polysulfides Li_2S_x. The early work[1] showed solvents with high basicity could dissolve large amounts of lithium polysulfides. In dimethyl sulfoxide or ethers like tetrahydrofuran,[43] the sulfur solubility as Li_2S_x can exceed 10 *M*. Spectroscopic and electrochemical studies of polysulfides in nonaqueous solutions suggest that their dynamic equilibrium, redox chemistry, and kinetics are strongly affected by solvent complexation.[1] High polysulfide solubility electrolytes enable the Li-S battery to operate as a liquid cathode system no matter how the starting sulfur active material was incorporated into the cell, solid as S_8 or liquid as dissolved polysulfides.

Cathodes are typically prepared from coating a slurry that contains elemental sulfur, acetylene black, graphite, and a binder. A variety of current collectors may be used, but Al foil is common. Sulfur is reduced in a stepwise fashion and a series of lithiated polysulfides are formed. The first ambient temperature discharge of a Li-S cell consists of two plateaus as shown in Figure 27.2. The regions 1 to 4 identify areas where different polysulfide species predominate. It is important to note that the sulfur species are in equilibrium with each other so a mixture of species will be expected to exist in the electrolyte at all times.

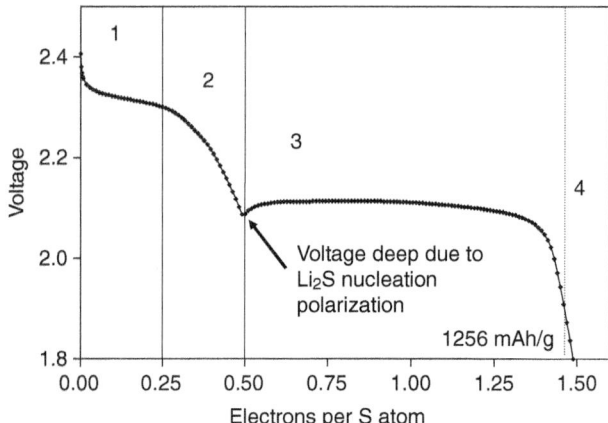

FIGURE 27.2 First discharge curve of a Li-S cell. Sulfur utilization up to 1256 mAh/g sulfur was achieved at low rate (*C*/30). (*Courtesy of SION Power Corp., Tucson AZ.*)

In region 1, sulfur is reduced to form Li_2S_8 according to the reaction

$$S_8 + 2e^- + 2Li^+ \leftrightarrow Li_2S_8$$

The reduction continues with the formation of Li_2S_6 in region 2. Region 3 contains the lower order polysulfides, Li_2S_4, Li_2S_2, and Li_2S. Li_2S_4 is soluble, but the lower order polysulfides are only sparingly soluble in dioxolane (DOL) and dimethoxyethane (DME), solvents that are commonly used in Li-S cells. Thus, full reduction to Li_2S is achieved only at low discharge rates because of the high polarization (region 4) which is caused by exhaustion of soluble Li_2S_4 and cathode porosity blocking by precipitation of solid Li_2S reaction product. Li_2S has been shown to be electrochemically reversible,[44] thus solubility is the primary factor that limits full sulfur utilization. Therefore, in a well-functioning cell discharged at low rate, 1256 mAh/g sulfur (¾ of the theoretical capacity of the sulfur) can be achieved.

The Li-S electrochemical performance is strongly influenced by the polysulfide shuttle (see Fig. 27.3) which has been described and modeled.[45] The shuttle reaction affects many cell properties, including self-discharge, charge-discharge efficiency, charge profile, and overcharge protection. The recharge of a Li-S cells produces high order polysulfides (e.g., Li_2S_8) rather than elemental sulfur.[46] The higher-order polysulfides, which are generated at the "sulfur" electrode during the latter stages of the charge, diffuse to the lithium electrode where they react directly with the lithium in a parasitic reaction to recreate the lower-order polysulfides. These species diffuse back to the sulfur electrode to generate the higher forms of polysulfide again, thus creating a shuttle mechanism.

The shuttle mechanism is a powerful overcharge protector,[47] but it reduces charge efficiency and is an impediment to attain higher specific capacity. An additional advantage is that the shuttle mechanism will react with and remove any Li dendrites that may form during charge, eliminating the potential for internal short circuit due to dendrite formation.

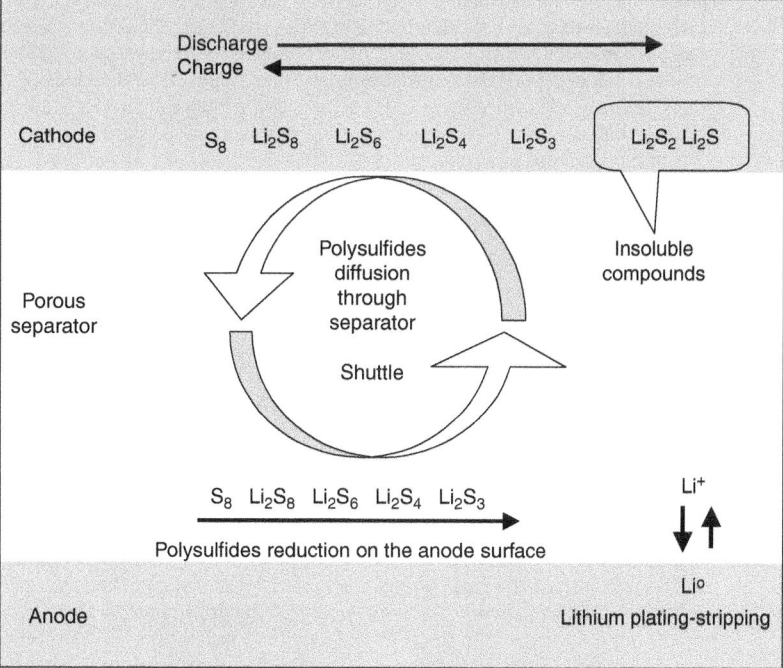

FIGURE 27.3 Shuttle mechanism for Li-S batteries (*from Ref. 46.*)

Polymeric Cathode. Electronically conductive redox polymers may also be used as cathode materials in rechargeable lithium batteries. Organosulfur polymers for sodium batteries were first described by Visco and coworkers,[48] and lithium batteries soon followed.[49] The discharge-charge process is a redox reaction in the polymer. The most popular polymers are polyacetylene, polypyrrole,[50] polyaniline, and polythiophene, which are made conductive by doping with suitable anions. The relatively high energy[51] and low cost make them attractive, but their instability on cycling, sluggish kinetics of the redox reactions at room temperature, and lack of electronic conductivity have been barriers to commercialization. Additives are useful in increasing cycle life[52] and new materials are being explored that have better reversibility,[53] but in most of the recent work polymer cathodes are combined with lithium intercalation anodes and thus are outside the scope of a discussion on lithium metal rechargeable batteries.

27.2.3 Electrolytes

The choice of electrolyte for rechargeable lithium batteries is also critical. The liquid electrolyte should have the following characteristics:

1. Good ionic conductivity ($\sim 10^{-3}$ S/cm from -40 to $70°C$) to minimize internal resistance
2. Lithium-ion transference number approaching unity (to limit concentration polarization)
3. Wide electrochemical voltage window (0 to 5 V)
4. Thermal stability (up to $70°C$)
5. Compatibility with other cell components

Liquid Electrolytes.

Aprotic Organic Electrolytes. Aprotic liquid organic electrolyte solvents, such as dioxolane (DOL), propylene carbonate (PC), ethylene carbonate (EC), diethyl carbonate (DEC), ethylmethyl carbonate (EMC), and dimethyl ether (DME) are the most common electrolyte solvents because of their low reactivity with lithium. A list of the electrolyte solvents used in rechargeable lithium batteries with their major characteristics is given in Table 27.1.[54] Choices for the electrolyte solute and their ionic conductivities in various solvents at different temperatures are listed in Table 27.2. These organic liquid electrolytes generally have conductivities that are about two orders of magnitude lower than aqueous electrolytes.

Room Temperature Ionic Liquids. Room temperature ionic liquids (RTILs) possess important attributes that make them attractive for use as solvents in electrochemical applications,[55] and lithium metal rechargeable batteries have potential to benefit from the advantages they promise. These include a wide electrochemical window, high conductivity, high thermal stability, low safety hazards (nonflammable, nonvolatile), and low toxicity. The limitations to be overcome include high viscosity, poor wetting of battery electrodes and separators, as well as low ionic conductivity at room temperature, not to mention purity and cost. The advantages of RTIL electrolytes have been recently reviewed.[56] This article highlights the growing interest in these materials. RTILs are seen as a means to improve the safe performance of lithium-based cells, particularly large-scale batteries, because of their nonvolatile properties. High cycling efficiency and uniform, non-dendritic lithium anode plating deposits have been obtained[57] at room temperature. The lithium morphology depends on deposition rate, temperature, substrate (copper versus lithium or platinum), and cycling history, with low rates and platinum metal substrate producing the most uniform and highest efficiency (>99%) deposits. The solid electrolyte interphase (SEI) formation in the presence of RTILs has different characteristics. A pronounced difference in composition was observed between the SEI formed on the lithium surface and that formed in situ during deposition. The surface film formed on the lithium deposit did not contain species associated with the lithium native film. Instead, significant quantities of species associated with the RTIL cation were observed.[58] RTILs offer an electrolyte system that has the potential to allow researchers a degree of control over the SEI composition that has not previously existed, which could be of substantial benefit to lithium metal rechargeable batteries.

TABLE 27.1 Characteristics of Organic Solvents*

Characteristic	γ-BL	THF	1,2-DME	PC	EC	DMC	DEC	DEE	Dioxolane (DN)
Structural formula	CH₂—CH₂ \ CH₂ C=O \ O	CH₂—CH₂ \ CH₂ CH₂ \ O	CH₂—O—CH₃ \ CH₂—O—CH₃	O=C \ O—CH₂—CH \ CH₃	O=C \ O—CH₂—CH₂	O=C \ O—CH₃ \ O—CH₃	O=C \ O—CH₂—CH₃ \ O—CH₂—CH₃	CH₂—O—C₂H₅ \ CH₂—O—C₂H₅	CH₂ O \ O CH₂ \ CH₂—CH₂
Boiling point, °C	202–204	65–67	85	240	248	91	126	121	78
Melting point, °C	−43	−109	−58	−49	−39 to −40	4.6	−43	−74	−95
Density, g/cm³	1.13	0.887	0.866	1.198	1.322	1.071	0.98	0.842	1.060
Viscosity at 25°C, cP	1.75	0.48	0.455	2.5	1.86 (at 40°C)	0.59	0.75	0.65	0.58
Dielectric constant at 20°C	39	7.75	7.20	64.4	89.6 (at 40°C)	3.12	2.82	5.1	6.79
Molecular weight	86.09	72.10	90.12	102.0	88.1	90.08	118.13	118.18	74.1
Typical H₂O content, ppm	<10	<10	<10	<10	<10	<10	<10	<10	<10
Electrolytic conductivity at 20°C, 1M LiAsF₆, mS/cm	10.62	12.87	19.40	5.28	6.97	11.00 (1.9 M)	5.00 (1.5 M)	~10.00[†]	~11.20[†]

*γ-BL = γ-butyrolactone; THF = tetrahydrofuran; 1,2-DME = 1,2-dimethoxyethane; PC = propylene carbonate; EC = ethylene carbonate; DMC = dimethyl carbonate; DEC = diethyl carbonate; DEE = diethoxyethane.
[†] Estimation based on Walden's rule.
Source: From Ref. 54.

TABLE 27.2 Ionic Conductivity of Some 1 Molar Organic Liquid Electrolytes Used in Secondary Lithium Battery Systems

Salt	Solvents	Solvent, vol %	Conductivities at °C, mS/cm							References
			-40	-20	-0	20	40	60	80	
$LiPF_6$	EC/PC	50/50	0.23	1.36	3.45	6.56	10.34	14.63	19.35	*
	2-MeTHF/EC/PC	75/12.5/12.5	2.43	4.46	6.75	9.24	11.64	14.00	16.22	*
	EC/DMC	33/67	—	1.2	5.0	10.0	—	20.0	—	†
	EC/DME	33/67	—	8.0	13.6	18.1	25.2	31.9	—	‡
	EC/DEC	33/67	—	2.5	4.4	7.0	9.7	12.9	—	‡
$LiAsF_6$	EC/DME	50/50	Freeze	5.27	9.50	14.52	20.64	26.65	32.57	*
	PC/DME	50/50	Freeze	4.43	8.37	13.15	18.46	23.92	28.18	*
	2-MeTHF/EC/PC	75/12.5/12.5	2.54	4.67	6.91	9.90	12.76	15.52	18.18	*
$LiCF_3SO_3$	EC/PC	50/50	0.02	0.55	1.24	2.22	3.45	4.88	6.43	*
	DME/PC	50/50	—	2.61	4.17	5.88	7.46	9.07	10.61	*
	DME/PC	50/50	—	Freeze	5.32	7.41	9.43	11.44	13.20	*
	2-MeTHF/EC/PC	75/12.5/12.5	0.50	0.93	1.34	1.78	2.31	2.81	3.30	*
$LiN(CF_3SO_2)_2$	EC/PC	50/50	0.28	1.21	2.80	5.12	7.69	10.70	13.86	*
	EC/DME	50/50	—	Freeze	7.87	12.08	16.58	21.25	25.97	*
	PC/DME	50/50	—	3.92	7.19	11.23	15.51	19.88	24.30	*
	2-MeTHF/EC/PC	75/12.5/12.5	2.07	3.40	5.12	7.06	8.71	10.41	12.02	*
$LiBF_6$	EC/PC	50/50	0.19	1.11	2.41	4.25	6.27	8.51	10.79	*
	2-MeTHF/EC/PC	75/12.5/12.5	—	0.38	0.92	1.64	2.53	3.43	4.29	*
	EC/DMC	33/67	—	1.3	3.5	4.9	6.4	7.8	—	‡
	EC/DEC	33/67	—	1.2	2.0	3.2	4.4	5.5	—	‡
	EC/DME	33/67	—	6.7	9.9	12.7	15.6	18.5	—	‡
$LiClO_4$	EC/DMC	33/67	—	1.0	5.7	8.4	11.0	13.9	—	‡
	EC/DEC	33/67	—	1.8	3.5	5.2	7.3	9.4	—	‡
	EC/DME	33/67	—	8.4	12.3	16.5	20.3	23.9	—	‡

*J. T. Dudley et al., *J. Power Sources* **35**(59), 82 (1991).
†D. Guyomard and J. M. Tarascon. *J. Electrochem. Soc.* **140**, 3071–3081 (1993).
‡S. Sosnowski and S. Hossain, unpublished results, Yardney Technical Products, Inc.

Solid Inorganic Electrolytes. Solid inorganic electrolytes can be comprised of glass or ceramic (crystalline) phases. Useful materials have been produced by thin film deposition techniques such as RF magnetron sputtering. Powders and pellets exhibit interesting properties, but are difficult to fabricate into commercially attractive electrochemical cells.

Lithium Phosphorus Oxynitride Electrolyte (LiPON). The thin film glassy solid electrolyte invented at Oak Ridge National Laboratory in the early 1990s is the most widely used solid electrolyte for thin film batteries.[59] The key insight by J. B. Bates was that addition of nitrogen to the glass structure might enhance the chemical and thermal stability of a lithium glass, as it does for sodium phosphate and sodium silicate glasses. The lithium phosphorus oxynitride electrolyte, known as LiPON, is deposited by RF magnetron sputtering from a ceramic target of Li_3PO_4 using a nitrogen process gas to form the plasma.[60] The films are amorphous and free of any columnar microstructure or boundaries. With a nitrogen/oxygen ratio as small as 0.1, the ionic conductivity is 1 to 2 μS/cm, which is about 40-fold higher than glassy films of nitrogen-free Li_3PO_4. More importantly, the cationic transport number of Li^+ is unity, the electrochemical stability extends to 5.5 V versus Li/Li^+, and LiPON is stable at both elevated temperatures and in contact with metallic lithium.[26] The lithium-ion conductivity, although 100 times lower than for many liquid electrolytes, is sufficient because 1 μm-thick films are adequate to create a pinhole-free barrier over most thin film electrodes. Furthermore, the electronic resistivity of LiPON is very high, greater than 10^{14} ohm-cm.

Lithium Sulfide Glasses. Sulfide glasses have sufficient lithium-ion conductivity to be of scientific interest for Li rechargeable batteries. Arsenic sulfide glasses have room temperature conductivity of 2.9×10^{-5} S/cm.[61] However, toxicity, stability, and processing issues prevented commercialization of this material. Other sulfide glasses such as Li_3PO_4-Li_2S-SiS_2,[62] and Li_4SiO_4-Li_2S-SiS_2,[63] glasses, crystalline Li_2S-P_2S_5,[64] and thio-LISICON type ($Li_{3.25}Ge_{0.25}P_{0.75}S_4$)[65] compounds have been studied in all-solid-state lithium cells. Processing produces a pellet that has reasonable room temperature conductivity as high as 1×10^{-4} S/cm,[66] but processing of thin films has been problematic. An additional complication is the reactivity of Li with the sulfide glasses. A black film has been observed at the interface between the lithium metal electrode and the sulfide glass electrolyte.[67] Deposition of Li_3N film can help stabilize the lithium/electrolyte interface,[68] but these barriers remain a substantial challenge.

A lithium sulfur oxynitride (LiSON) amorphous film[69] has been reported that is also stable to 5.5 V versus Li/Li^+. RF magnetron sputtering produced films that had a room temperature conductivity as high as 2×10^{-5} S/cm. Reactivity with lithium was not reported with this electrolyte.

A variety of other inorganic glassy electrolytes are being evaluated, but none has been as widely tested as the LiPON electrolyte in thin film Li metal rechargeable batteries. Inorganic glasses may include binary and ternary mixes of lithium borate, phosphate, silicate, and vanadate, but most compositions do not appear to meet conductivity or stability requirements.

Glass-Ceramic Electrolytes. Glass-ceramics with higher ionic conductivities in the range of 10^{-4} S/cm at room temperature are available in 1 in^2 plates as thin as 0.3 mm.[70] These materials might be attractive as a self-supporting electrolyte if fabricated in much thinner sheets and stability with the lithium interface could be demonstrated.

Solid Polymer Electrolytes. An alternative to the liquid electrolytes is a solid polymer electrolyte (SPE) formed by incorporating lithium salts into polymer matrices and casting into thin films. These films can function as both the electrolyte and separator. SPE electrolytes have lower ionic conductivities and the same or higher lithium-ion transport numbers compared to the liquid electrolytes, and they are less reactive with lithium, which should enhance the safety of the battery. Cells with thin polymer films operate at higher temperatures (60 to 100°C) to compensate in part for the lower conductivity of the polymer film. The solid polymers also offer design advantages of a "nonliquid" battery, allowing flexibility to manufacture thin batteries in a variety of configurations.

Initially, high-molecular-weight polymers such as polyethylene oxide (PEO) and lithium salts such as $LiClO_4$ and $LiN(CF_3SO_2)_2$ (Li Imide) were used.[71] These PEO-lithium salt electrolytes have good mechanical properties but low conductivities, which are on the order of 10^{-8} S/cm at 20°C. A significant improvement in conductivity to approximately 10^{-5} S/cm has been achieved with the combination of modified comb-shaped PEO structures with lithium salts,[72] but these solid polymer electrolytes have poor mechanical properties and their conductivity is still two orders of magnitude lower than that of most organic liquid electrolytes. Further improvement in conductivity was obtained with the addition of liquid plasticizers such as propylene carbonate.[73,74] The amount of plasticizer may be as high as 70%, resulting in limited chemical and mechanical stability.

Composite polymer electrolytes, prepared by the addition of particulate metal oxide to the polymer film,[75] gave improved interfacial stability and improved cycle life. Orientation of polymer chains[76] perpendicular to the face of the electrode has improved ionic conductivity and increased the transference number to 0.6. The transference number was improved further, achieving $t_+ = 1$ in a PEO system that contained an anion-trapping supermolecule additive.

Another class of polymer electrolytes called "gelled" electrolytes has been developed by trapping liquid solutions of lithium salts in aprotic organic solvents (for example, $LiClO_4$ in propylene carbonate-ethylene carbonate [PC/EC] solvent) into a solid polymer matrix such as poly(vinylidene difluoride) (PVdF)[77] and poly(acrylonitrile) (PAN).[78,79] The "gel" electrolytes are made by adding liquid electrolyte solutions into polymer porosity with an immobilization procedure such as cross-linking, gellification, and casting. Crosslinking may also be carried out by ultraviolet, electron-beam, or gamma-ray irradiation. Conductivities as high as 10^{-3} S/cm at 20°C and transference number around 0.6 have been obtained. However, these plasticized and gelled electrolytes are more reactive with lithium than true solid polymers. The conductivities of the various classes of polymer electrolytes and their variations with temperature are presented in Fig. 27.4.

A combined polymer approach (PEO-PVdF) has been applied to lithium metal cells[80] with intercalation cathodes. A longer cycle life (up to 200+ cycles at room temperature at C/2 discharge and C/10 charge rate) and a more stable lithium interface have been achieved. Modified PEO-based electrolytes with poly(ethylene glycol)-borate ester gave stable cycling at 60°C, reaching 150 cycles with 90% of original capacity.[81] The polymer was formed by UV irradiation and $AlPO_4$ was added to stabilize the $LiFePO_4$ cathode surface.

Single-ion conducting polymeric electrolytes designed for use with lithium metal anodes have been prepared using graft copolymers[82] and comb-branched polyepoxide ethers.[83] In the latter case, the conductivity of the polymer was increased to the point that a Li/V_6O_{13} cell was cycled at room temperature, showing the promise of this approach. The flexibility of tailored mechanical properties of graft copolymers, while maintaining a Li transference number of unity,[84] is an attractive property. Moreover, the polymer voltage stability window extends to 4.5 V versus Li/Li^+, more than 1 V higher than PEO-based polymer electrolytes.

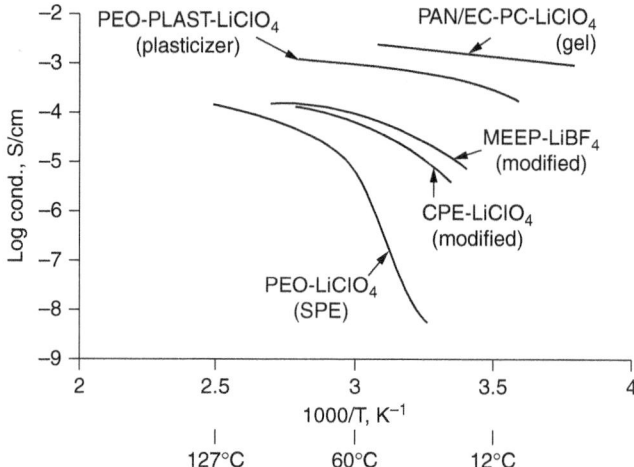

FIGURE 27.4 Variation of conductivity with temperature for different classes of polymer electrolytes. PEO = poly(ethylene oxide). CPE = cross-linked poly (vinyl ether). MEEP = poly(bis[methoxy ethoxy ethoxide]). PAN = poly(acrylonitrile).
Source: Chapter 34, 3rd ed. Handbook of Batteries.

27.3 CHARACTERISTICS OF LITHIUM RECHARGEABLE BATTERIES

A number of different battery systems have been investigated for the development of lithium metal rechargeable batteries. The efforts reviewed here achieve the high specific energy that lithium metal batteries offer without sacrificing other important characteristics, such as specific power and cycle life, while maintaining safe and reliable operation.

27.3.1 Electrochemical Systems

The different types of ambient-temperature lithium metal rechargeable batteries can be classified into three design categories.

- Liquid organic electrolyte cells
- Polymer electrolyte cells
- Inorganic electrolyte cells

The components, chemical reactions, and performance characteristics of typical examples of the three types are summarized and compared below.

27.3.2 Liquid Organic Electrolyte Cells

Lithium metal rechargeable batteries with liquid electrolyte are of high interest because of their advantage in specific energy and energy density compared to Li-ion rechargeable batteries. There are several types, which can be classified as liquid cathode (Li-S) and solid cathode (Li metal anode with rechargeable intercalation cathode, that has been developed for Li-ion rechargeable batteries).

Liquid Cathode: Lithium Sulfur Rechargeable Cells. Lithium sulfur rechargeable cells are the highest energy liquid electrolyte lithium metal sealed rechargeable cell available today. Sulfur is inexpensive and nontoxic. Li-S has the advantage of combining the highest capacity anode (lithium metal) and cathode (sulfur) with exceedingly high theoretical specific energy (2500 Wh/kg). Cells are available with a specific energy of 350 Wh/kg and an energy density of 350 Wh/L, with credible projections of 30 to 50% higher energy and power (>3 kW/kg) within the next few years. Li-S chemistry has benefited from sustained development over the last 15 years. Cycle life and safety are areas where improvements are required. The cycle life of today's Li-S cells is generally not much above 100 cycles. The first practical application for Li-S cells was in a solar-electric high altitude unmanned aerial vehicle (UAV).[85] These aircraft, designed to fly continuously for weeks to months at over 60,000 feet altitude, use photovoltaic (PV) panels on top of the wings to generate electricity for daytime flight as well as charging a battery pack that provides propulsion during the night. The very high specific energy rechargeable Li-S battery is a key enabler of this UAV technology.

The cell is a wound prismatic construction. The cathode is typically prepared from coating a slurry that contains elemental sulfur, acetylene black, graphite, and a binder coated on aluminized poly(ethylene terephthalate) (PET) or aluminum foil current collectors. The electrolyte is a solution of lithium bis(trifluoromethylsulfonyl)imide $LiN(CF_3SO_2)_2$ in a 45:55 volume ratio mixture of 1,3-dioxolane (DOL) and 1,2-dimethoxyethane (DME). Using a Li foil anode thickness of 50 μm and a polyolefin separator, the cell is wound together into a "wound prismatic" construction. After connection of tabs, the cell is sealed into an aluminized polypropylene pouch. Cell construction details are given in Fig. 27.5.

Li-S cells are typically thought of as low-rate cells due to relatively high cell resistance. However, power capability has been improved so that the cell delivers over 60% capacity at 6 C discharge at room temperature as shown in Fig. 27.6.

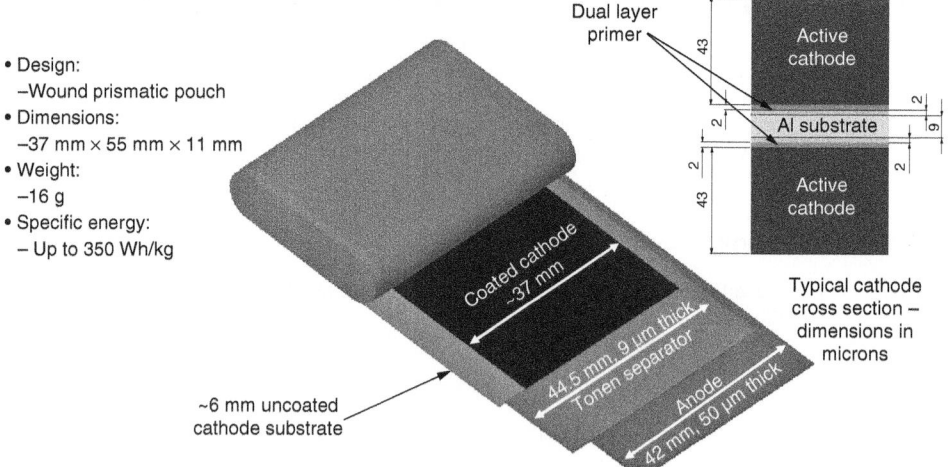

FIGURE 27.5 Li-S cell construction details. (*Courtesy of SION Power Corp., Tucson, AZ.*)

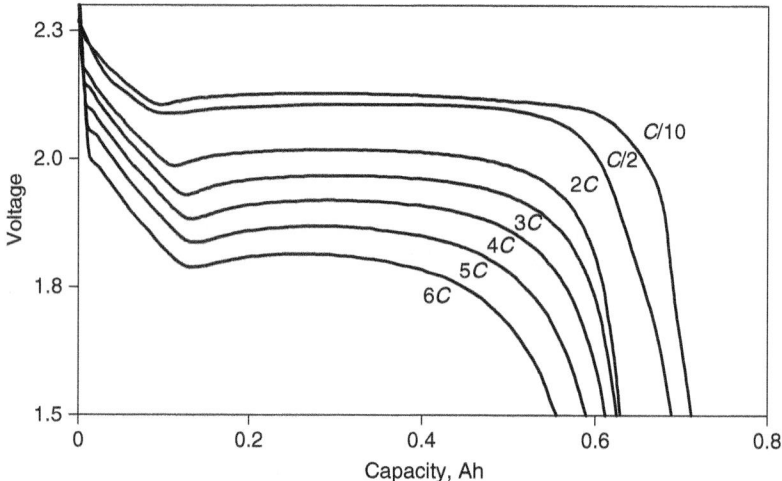

FIGURE 27.6 Discharge voltage vs. discharge capacity at room temp (+25°C) at different discharge rates. (*From Ref. 47.*)

Low-temperature performance of Li-S cells depends on current levels and is quite good. When discharged at $C/10$ rate at −40°C, the cell delivers 80% of room temperature capacity, as shown in Fig. 27.7, and the 5 C discharge capacity at −40°C is still over 50% of the room temperature value.[47]

Analysis of cell area specific resistance has resulted in improved cell design and higher rate capability, with improved cells providing 3000 W/kg at 350 Wh/kg.[86] Figure 27.8 shows specific power as a function of depth of discharge (DOD). The power delivery of Li-S cells is relatively independent of state-of-charge. Thus, it is clear that Li-S cells can be designed to provide high discharge rates that are competitive with high-rate Li-ion rechargeable batteries.

Increasing sulfur utilization will improve rechargeable capacity and increase cycle life. Additives have been developed[87] that provide higher discharge capacity and help stabilize cycle life. Systematic

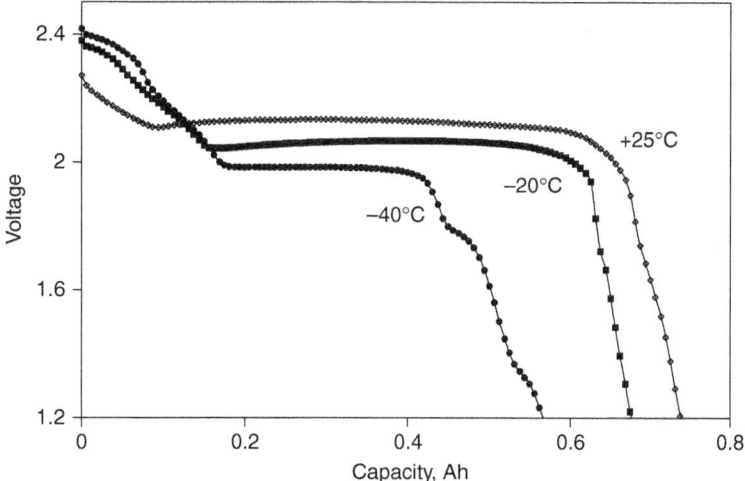

FIGURE 27.7 Discharge profiles for Li-S cells at different temperatures at 10 h discharge rate (70 mA = C/10). −40°C capacity retention is 80% of room temperature value. (*From Ref. 47.*)

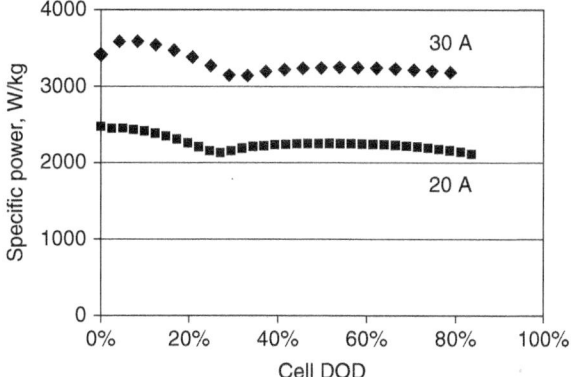

FIGURE 27.8 Specific power vs. cell depth of discharge (DOD) for Li-S cells at 20 A and 30 A pulses. Pulse duration 10 s. (*From Ref. 86.*)

studies have related capacity fade to structural change in the sulfur cathode.[88] These studies highlighted the importance of a stable microstructure of the carbon framework in the cathode, since an insoluble discharge product (Li_2S) precipitates on the electrode surface, reducing the sulfur utilization. Multiwalled carbon nanotubes have been added to the cathode[89] to provide controlled surface area and structure. While these modifications stabilized the rechargeable capacity, sulfur utilization remained under 50% of the theoretical value.

At SION Power Corp., improved cathode microstructures with engineered porosity were developed and the stability of the cathode improved during cycling. These changes, as well as changes in electrolyte composition, have resulted in a 50% increase in sulfur utilization (from 800 mAh/g S to over 1200 mAh/g S) in the last few years.[46,90] New cell designs have reached 400 Wh/kg at room temperature (and 450 Wh/kg at 60°C) in experimental cells,[91] which also delivered ~50% of room temperature capacity at −70°C. However, cycle life remained at ~55 cycles.

A highly ordered nanostructured carbon-sulfur cathode has been reported with good sulfur utilization.[92] The tailored composite was a combination of sulfur and mesoporous carbon which was further modified by polyethylene glycol (PEG). The maximum sulfur utilization was 1320 mAh/g, nearly 80% of theoretical. The capacity retention on cycling was reasonably good, with sulfur utilization over 1100 mAh/g after 20 cycles.

When the shuttle mechanism is active, capacity that resides in the upper voltage plateau is unavailable since the high order polysulfides rapidly discharge. This phenomenon also decreases sulfur utilization. Thus, the shuttle mechanism, which occurs on the Li anode surface, must be prevented to retain high capacity over long cycle life. Surface analytical studies of lithium anodes in active Li-S cells explored the reactivity of electrolyte on freshly deposited Li.[93] These studies show that additives[94] have a positive effect in preventing the shuttle mechanism. In addition to chemical mechanisms, physical coatings are being pursued to stabilize the lithium surface.[42] These multilayer anode assemblies with physical protection have improved cycle life and increased sulfur utilization to the highest level to date—over 1400 mAh/g sulfur after 80 cycles.

The solution chemistry and electrochemistry for the Li-S system is more complicated than Li-ion rechargeable batteries.[95] Interconversion between the polysulfide species is in dynamic equilibrium, which influences the redox chemistry and kinetics. Moreover, solvent complexation[1] plays a major role in cell performance. An electrochemical model has been developed[96] that describes the complete Li-S cell performance. Sulfur speciation and solubility on both high voltage and low voltage plateaus provide a quantitative description of cell response and concentration gradients of the polysulfide species. The new understanding provided by the model of the electrochemistry of the Li-S system should provide insights that will allow significant advances in cell design and performance.

Ionic liquids are being explored as solvents in Li-S electrolytes. Cairns[97] has shown that ionic liquids are compatible with the Li-S system. The electrolyte was a mixture of low molecular weight poly(ethylene glycol) dimethyl ether (PEGDME, MW = 250) or tetra(ethylene glycol)dimethyl ether (TEGDME)[98] with N-methyl-N-butyl pyrrolidinium bis(trifluoromethanesulfonyl)imide (PYR14TFSI) and LiTFSI. Sulfur utilization was only average on the first cycle (887 mAh/g sulfur) and declined to half that value after 20 cycles. The thermal stability and nonflammability are important advantages, but the higher viscosity and lower ionic conductivity resulted in higher overvoltages at room temperature. However, the potential usefulness of ionic liquids in Li-S chemistry has been demonstrated and holds promise for development of cells with improved high temperature stability and safety.

Another factor that has a strong influence on achieving higher cycle life is the nature of lithium deposits during recharge. Electroplated lithium should not form a mossy or dendritic deposit with a larger surface area than the original metal. Additionally, the safety issues with lithium metal cells have been attributed to the changing morphology of lithium as a cell is cycled. If the properties of lithium metal negative electrodes change with use, additional testing will be required to ensure that cell performance will be consistent and predictable over the lifetime of the cell.

The safety performance of Li-S cells has been reported.[99] Tests at the cell level have shown that short circuit, overcharge, and overdischarge result in benign failures. Gas generation during the test typically caused the cell to fail in an open-circuit condition. Thermal stability was measured by a thermal ramp test.[100] Results were strongly influenced by the condition of the physical protection coatings—when the protective films applied to the anode were intact, the runaway temperature was delayed until near the melting point of Li (175 to 180°C).

A variant of Li-S, termed "lithium-sulfide" is being developed but very little is published about this cell.[101] It is claimed to be different from Li-S, but the details have not been divulged. A recent report[102] shows the development of cells with Li metal as well as Li-ion anodes, with sulfur utilization ~600 mAh/g for over 250 cycles.

The improved performance and safety results that have been achieved with the Li-S system over the last few years are very encouraging and there will likely be a resurgence in commercial sources of Li-S cells and batteries in the future.

Solid Cathode: Lithium Intercalation Cathode Rechargeable Cells. Intercalation materials that are useful in Li-ion rechargeable batteries may also be used in lithium metal rechargeable batteries. Cathode choices are discussed in Chap. 26 and will not be reviewed here. The barrier to development of practical commercial cells is controlling the surface morphology of the lithium anode on cycling.

There are a number of studies of the influence of liquid electrolyte composition on the morphology of deposited lithium.[103] The lithium cycling efficiency ($Q_{dissolution}/Q_{deposition}$) on discharge and charge is related to the lithium morphology and is a measure of how reversibly the anode cycles. Studies cover a variety of charge/discharge rates and various solvents and salts that make up the electrolyte. Li efficiency was seldom over 90% and in some cases was as low as 60%. A porous lithium anode structure has been documented after the first charge/discharge cycle. The best Li morphology and efficiency were achieved when the electrolyte contained cyclic ethers such as tetrahydrofuran (THF) and tetrahydropyran (THP) and $LiN(SO_2C_2F_5)_2$ (Li BETI) or $LiAsF_6$ salts. In the best cases, capacity degraded by 35% after 120 cycles, and in the worst cases, capacity degraded by 95% after 100 cycles.

Conditions that favor dendritic morphology development form thicker surface film by repeated deposition/dissolution cycles of lithium. Composition of surface films and gases generated during cycling has been studied,[104] and decomposition products of reaction of the Li with both the electrolyte and salt are observed. Small amounts of aluminum alloyed with lithium (0.1 wt.% Al) increased lithium cycling efficiency above 95%, but the cycle life of the cell was only about 30 cycles.[105] The cell used a PVDF gel electrolyte and a $LiCoO_2$ cathode.

Addition of vinylene carbonate, a common additive in Li-ion rechargeable batteries, produced a film on the anode similar in composition to that observed in Li-ion cells.[106] The cycling behavior was marginally better, and lithium cycling efficiency was slightly above 90% for the first 50 cycles. A larger improvement was observed using the solid electrolyte interface additive, triacetoxyvinylsilane at 2 wt% in the electrolyte.[107] After precycling, no lithium dendrites on the lithium metal surface were observed, and low interfacial resistance of the unit cell ($Li/LiCoO_2$) was achieved. Cells could maintain about 80% of the initial capacity after 200 cycles at a high charge/discharge current density ($C/2 = 1.25$ mA/cm^2).

Indeed, when lithium metal anodes can be cycled to produce dense, dendrite-free lithium metal after every charge, a substantial increase in energy content of rechargeable batteries can be expected. Recent reports of cells that use a lithium anode and high capacity cathodes such as cobalt trifluoride (CoF_3),[108] iron difluoride (FeF_2),[109] and lithium manganese bismuth oxide ($Li_yMnO_2(Bi_2O_3)_x$ where x = 0.0 to 0.125 and y ~0.5)[110] had very promising energy content but poor cycle life, often due to instability and dendrite formation on the lithium anode.

27.3.3 Polymer Electrolyte Cells

Polymer electrolyte lithium batteries contain all solid-state components: lithium as the anode material, a thin polymer film as a solid electrolyte and separator, and a transition metal chalcogenide, oxide or phosphate, or a sulfur-based polymer as the cathode material. These features offer the potential for (a) improved safety because of the reduced activity of lithium with the solid electrolyte, (b) flexibility in design as the cell can be fabricated in various sizes and shapes, and (c) high energy density.

The cathode and electrolyte are coated onto a current collector to form a thin sheet, called the cathode laminate. The lithium metal foil is applied to the cathode laminate to form a layered structure, with the solid polymer separating the lithium from cathode. These cells use extremely thin components with high surface areas to minimize the internal resistance and compensate for the lower conductivity of the polymer electrolyte. The thickness depends on the specific cell design and required capacity. A thicker laminate delivers a higher capacity per unit area of electrode but with lower efficiency at the higher current drains.

Lithium Batteries Using Solid Polymer Electrolytes. Lithium solid polymer electrolyte (SPE) batteries show two major advantages with respect to standard batteries that use liquid electrolyte and inert porous polymeric separators:

- The electrode and electrolyte layers are laminated (usually by heating and pressing the layers), thus allowing more varied battery shapes without loss of contact.
- Even if a low molecular weight (liquid) plasticizer is added to obtain high conductivity at ambient and subambient temperatures, there is no free liquid present in the battery thus preventing any leakage problems.

They can be classified into two categories:

1. Dry polymer electrolyte (PEO-based).
2. Gel polymer separator layer filled with a liquid electrolyte (e.g., PVdF- or PAN-based).

SPE cells that are fabricated from Li-ion rechargeable battery electrode materials are covered in Chap. 26 and will not be covered here. In this chapter, the attention is focused on the lithium metal polymer electrolyte systems. Gel polymer separator membranes used with lithium metal anodes have not met with success in the marketplace because of the reactivity of lithium with liquid electrolytes (the gel usually contains the same liquid electrolyte as found in Li-ion rechargeable batteries with polymeric separator).

Dry Polymer PEO Electrolyte Battery Using V_3O_8. Avestor, Inc. was engaged in the development, manufacture, and commercialization of a family of SPE batteries with lithium metal anodes which were being developed for stationary telecommunications market and EV applications. The lithium-metal-polymer (LMP) cell was a laminate of four thin materials:

- A metallic lithium foil anode. The ultra-thin lithium foil (less than 50 μm thick) acted as both a lithium source and a current collector.
- A solid polymeric electrolyte. This lithium-ion carrier was obtained by dissolving a lithium salt in a solvating copolymer.
- A metallic oxide cathode based on a reversible intercalation compound of vanadium oxide, blended with lithium salt and polymer to form a plastic composite.
- An aluminum foil current collector.

The solid, dry, lithium-ion conducting PEO polymer membrane served both as the electrolyte and as the separator between the anode and cathode foils. The membrane was a solid polymer PEO matrix with an ionic lithium salt (LiTFSI) complexed into the matrix. The elastic behavior of the polymer assures a low impedance interface to the surfaces of the two electrodes. The result was a totally solid-state electrochemical cell, having neither liquid nor gel components.[111] The operating voltage was 2.0 to 3.1 V per cell. The minimum operating temperature was 60°C, and the battery module contained a thermal management control system.

The modules were equipped with cell equalization and balancing to maintain all cells at a uniform float voltage of 3.1 V. Safety systems, such as charge control and disconnect switching, were integrated to protect against operation under excessive, abnormal conditions. The module also had a mechanical pressure subsystem to ensure the stability of the interface between the anode and the electrolyte. The polymer electrolyte transmits pressure to the surface of the lithium anode to resist the formation of Li dendrites. (See sec. 27.2.1.) The mechanical pressure system (50 to 100 psi) can only maintain the uniformity of the lithium foil surface if the rate of plating is low. Consequently, the charging current must be limited to a maximum of $C/8$ to prevent battery deterioration over its life.[111]

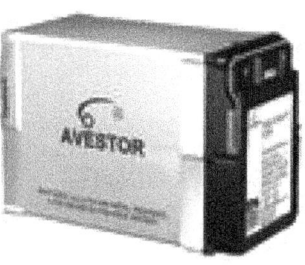

FIGURE 27.9 Avestor 80 Ah lithium-metal-polymer battery.

The primary Avestor product was a 80 Ah SE48S80 battery that operated at 48 V (Fig. 27.9). In August 2006, the company had produced and shipped its 20,000th battery. At that time, Avestor said it had signed multimillion dollar, multiyear contracts with major telecommunications service providers in North America. By October 2006, the company closed. In March 2007, Avestor, Inc., was acquired by Bolloré S.A., who has announced that the LMP will be developed and commercialized to provide power to the EU electric vehicle program BlueCar.[9] Properties of the Bolloré module are given in Fig. 27.10.

Dry Copolymers Based on PEO Electrolyte. The insight gained by Newman's analysis of experimental factors that will suppress dendrite formation[34,35] spawned a search for a new class of polymer electrolytes that could provide the high modulus SPE (~1 GPa) that the calculations predict

Bolloré LMP battery for BlueCar	
General characteristics	
Volume (L)	300
Mass (kg)	300
Communication bus	CAN
Thermal characteristics	
Internal temperature	60 to 80°C
Operating temperature	−20 to +60°C
Electrical characteristics	
Power rating	30 kWh
Nominal voltage	410 V
Peak power output	45 kW(30s)
Min./max. battery voltage	300/435 V
Capacity at $C/4$	75 Ah
Energy density per unit mass	100 Wh/kg
Energy density per unit volume	100 Wh/L

FIGURE 27.10 Bolloré LMP battery for European BlueCar EV program. (*From Ref. 9.*)

will suppress or eliminate dendrite formation. Because segmental motion of polymer chains is very important for ionic mobility, high conductivity and high modulus are almost mutually exclusive goals in homopolymer polymer electrolytes. For example, PEO has an elastic modulus of less than 1 MPa. Glassy polymers such as polystyrene offer very high modulus (~3 GPa) but are poor ion conductors.

Balsara and coworkers[112] are synthesizing dry block copolymer electrolytes for lithium metal batteries. They have chosen polystyrene/PEO block copolymers[113] because it may allow polymer electrolytes to meet both demanding material requirements. In their nanostructured electrolytes, the PEO phase is the ionic conducting part of the copolymer, and the polystyrene provides the high elastic modulus. Block copolymers self-assemble into well-defined structures such as lamellae or cylinders with domain spacings on the order of tens of nanometers, with specifics depending on the molecular weight and volume fraction of each block.[114] The conductivity depends on the molecular weight of PEO.[115]

A small startup company, SEEO Inc.,[116] is trying to commercialize these developments. While there are many hurdles, such as cycle life, safety, etc., to be addressed, this approach may lead to a greater availability of lithium metal rechargeable batteries with higher capacity.

27.3.4 Inorganic Electrolyte Cells

A number of small format, thin film rechargeable lithium metal rechargeable batteries have been developed for portable applications such as a power source for electronic devices, memory backup, and other types of auxiliary power sources. Thin film, all solid state batteries are based on the technology developed by Bates et al.[117] and fabricated by a sequential series of physical vapor deposition processes. The key similarity is the use of a lithium phosphorous oxynitride (LiPON) glassy electrolyte which is applied onto a variety of cathode thin films. The main problems are low ionic conductivity of solid electrolytes and a large charge-transfer resistance at the electrode/solid electrolyte interface.

A schematic drawing of a thin film, solid state Li cell[118] is shown in Fig. 27.11. Each of the component layers is 0.1 μm to several micrometers thick. Ideally, the substrate is also a component of the device; otherwise even with very thin substrates, the weight and volume of the battery is largely determined by the inactive support. Thin film batteries are being developed using a variety of supports, including silicon, quartz, mica, alumina, polymers, soda-lime glass, and metal foils. As shown in Fig. 27.12, many of the batteries are not only thin, but also quite flexible.[119]

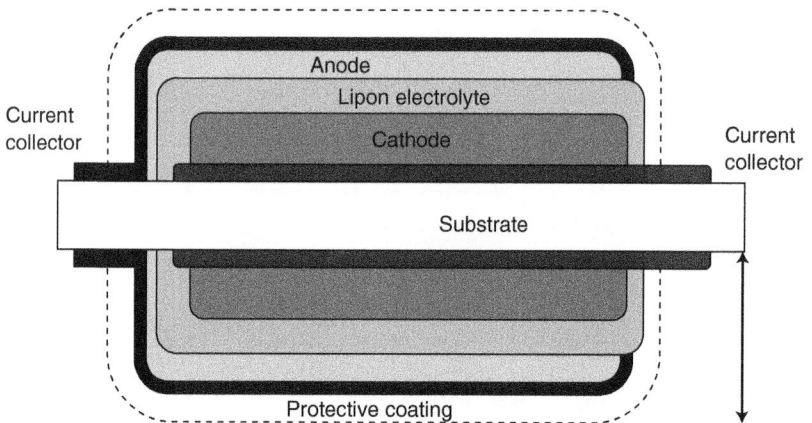

FIGURE 27.11 Schematic cross section of thin film microbattery based on LiPON electrolyte. (*From Ref. 118.*)

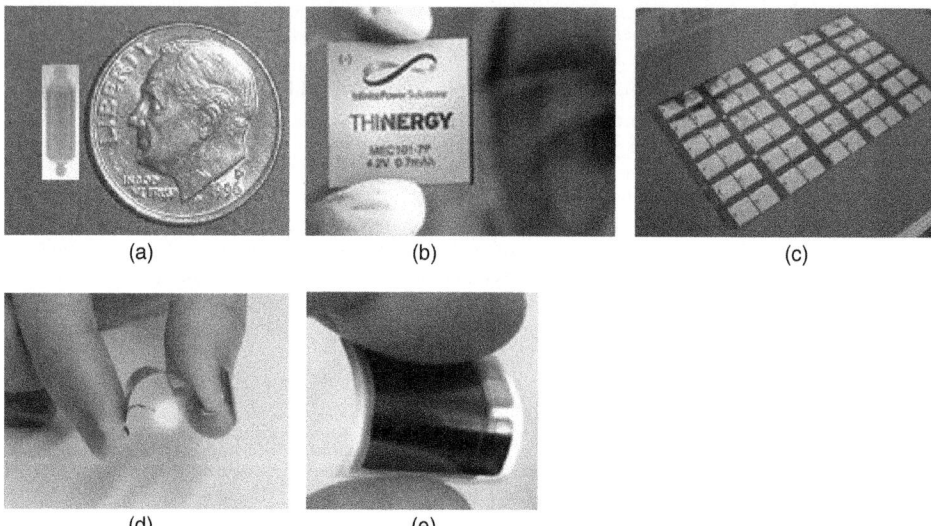

FIGURE 27.12 Examples of prototype batteries manufactured by (a) Oak Ridge Micro-Energy (www.oakridgemicro.com/), (b) Infinite Power Solutions (www.infinitepowersolutions.com/), (c) Excellatron (www.excellatron.com/), (d) Front Edge Technology (www.frontedgetechnology.com/), and (e) Cymbet Corporation (www.cymbet.com/). (*From Ref. 119.*)

These cells are fabricated by sequential layer deposition of the cell components using RF magnetron sputtering, except for evaporated lithium and for the metallic current collector components that are deposited by DC magnetron sputtering. The deposition conditions for $LiCoO_2$[117] and lithium phosphorous oxynitride (LiPON) electrolyte are reported in the literature.[120] Positive current collectors of gold or platinum (0.1 to 0.3 μm), over a layer of cobalt (0.01 to 0.05 μm, to improve adhesion), have been used. Cells using either $LiCoO_2$ or $LiMn_2O_4$ positive electrode materials have been fabricated. The positive electrode layer of laboratory test cells is typically 0.05 to 5 μm thick and 0.04 to 25 cm² in area, depending on the capacity required by the application. Figure 27.13 is a SEM photomicrograph that shows the columnar growth of the $LiCoO_2$ cathode on an alumina substrate after high temperature annealing.

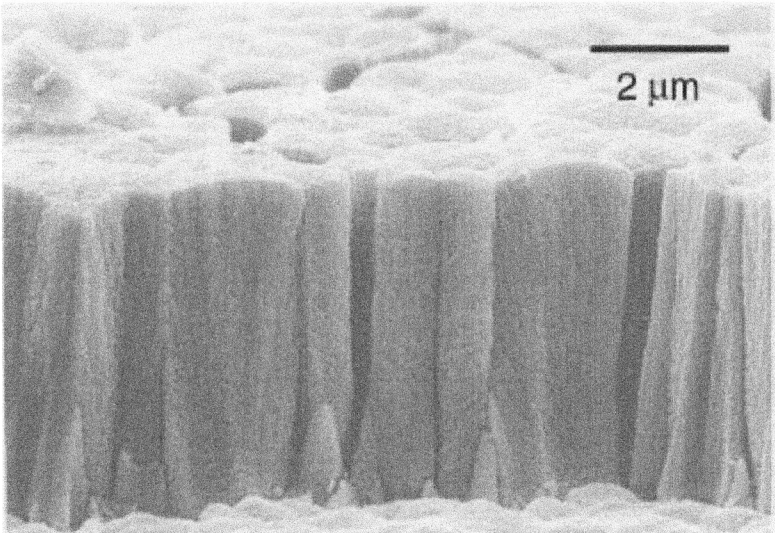

FIGURE 27.13 A fracture edge of a thin film, annealed $LiCoO_2$ cathode film on an alumina substrate. (*From Ref. 118.*)

Negative current collectors of copper, titanium, or titanium nitride (0.1 to 0.3 μm) are typical. To enhance the hermeticity of the cell, protective overlayers of LiPON (1 μm) or parylene[121] (6 μm) and titanium or aluminum (0.1 μm) have been used.

Capacity of cells depends on cathode choice, coating thickness, degree of crystallization of the cathode, and other processing conditions. Typical results are shown in Fig. 27.14. Thick $LiCoO_2$ cathodes have the highest capacity. Capacity is typically rated by active cell area, but studies have shown[122] that cells can achieve 100 Wh/kg at 1 kW/kg and 100 Wh/L at 1 kW/L (excluding substrate weight and volume).

The rate capability of thin film batteries using LiPON film strongly depends on the cathode,[123] and $LiCoO_2$ is attractive for high-power applications. Additionally, the charge transfer at the electrode/electrolyte interface can influence the charge and discharge rate.[124] The use of alternate electrode materials such as $Li_4Ti_5O_{12}$, a material with zero strain insertion on lithiation,[125] has been shown to provide more rapid kinetics. Interestingly, $Li_4Ti_5O_{12}$ was used (in separate cell configurations) both as cathode (with lithium anode) and as anode (with $LiFePO_4$ cathode) in thin film batteries. A $Li_4Ti_5O_{12}$ anode can be imbedded in the LiPON, and the cell operates in air without any additional protection.

Commercialization has proceeded at a rapid pace from the early developments at Oak Ridge National Laboratory to spawn at least 6 commercial endeavors. Cells are available from these

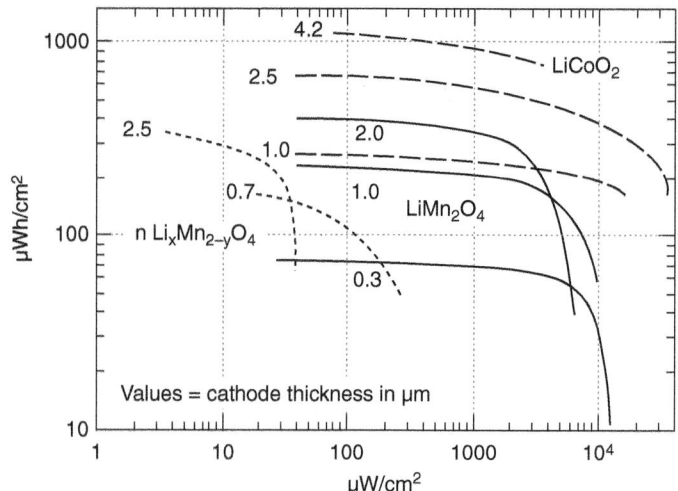

FIGURE 27.14 Ragone plot for thin film lithium batteries comparing the energy and power delivered by constant current discharge. As indicated, the batteries have various crystalline and nanocrystalline (n) cathode materials with different film thicknesses. The energy and power are normalized by the active battery area. (*From Ref. 118.*)

companies. An example is Infinite Power Solutions.[126] Figure 27.15 shows manufacturer's data from the "Thin Energy" cell line that describe the performance parameters of cells from 0.1 to 2.5 mAh capacity.

The lithium plating-stripping process is efficient with LiPON electrolyte technology, which provides long cycle life. Figure 27.16 shows the discharge curves at the 1st, 500th, and 1000th cycle of 0.9 mAh NanoEnergy® cell manufactured by Front Edge Technology Inc.[127] The charging rate is lower at the 1000th cycle than that of the first cycle. The charging time required to obtain 95% of the rated capacity is 4 min at the first cycle and increases to 6 min at the 1000th cycle.

High rates are achievable. For example, NanoEnergy® cells can be continuously discharged at rates of more than 10 C, and more than 20 C in pulsed discharge. Figure 27.17 shows the discharge characteristics of a 0.9 mAh NanoEnergy® cell at rates from 0.5 C to 10 C.

27.4 CONCLUSIONS

In conclusion, the development of very high-energy storage devices will enable solutions to society's pressing needs for more efficient generation and utilization. Li metal rechargeable batteries are commercially available in small sizes with good-to-excellent performance. Limitations in temperature stability, cycle life, and safety are being addressed. Larger cell sizes (>10 Ah) are still in development. The future will likely see a range of high-capacity lithium metal rechargeable batteries become commercialized with performance that meets market demands.

ACKNOWLEDGMENTS

Thanks to Yuriy V. Mikhaylik, Nancy J. Dudney, K. M. Abraham, and Martin Simoneau for providing information and figures for this chapter.

Device	Voltage	Capacity	Current	Size
 MEC125	4.1 V	0.1 mAh 0.2 mAh	7.5 mA	.5 in × .5 in × 0.0067 in 12.7 mm × 12.7 mm × 0.17 mm
 MEC120	4.1 V	0.2 mAh 0.3 mAh 0.4 mAh	15 mA	1.0 in × 0.5 in. × 0.007 in. 25.4 mm × 12.7 mm × 0.17 mm
 MEC101	4.1 V	0.5 mAh 0.7 mAh 1.0 mAh	40 mA	1.0 in × 1.0 in × 0.007 in 25.4 mm × 25.4 mm × 0.17 mm
 MEC102	4.1 V	1.2 mAh 1.7 mAh 2.5 mAh	100 mA	1.0 in × 2.0 in × 0.0067 in 25.4 mm × 50.8 mm × 0.17 mm

FIGURE 27.15 Examples of thin film lithium metal rechargeable battery products from Infinite Power Solutions. (*Courtesy of Infinite Power Solutions.*)

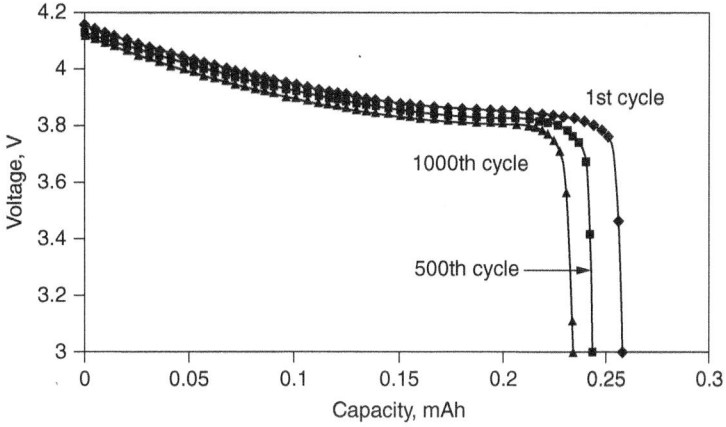

FIGURE 27.16 Capacity of 0.9 mAh NanoEnergy® cell manufactured by Front Edge Technology Inc. at cycles 1, 500, and 1000. (*From Ref. 127.*)

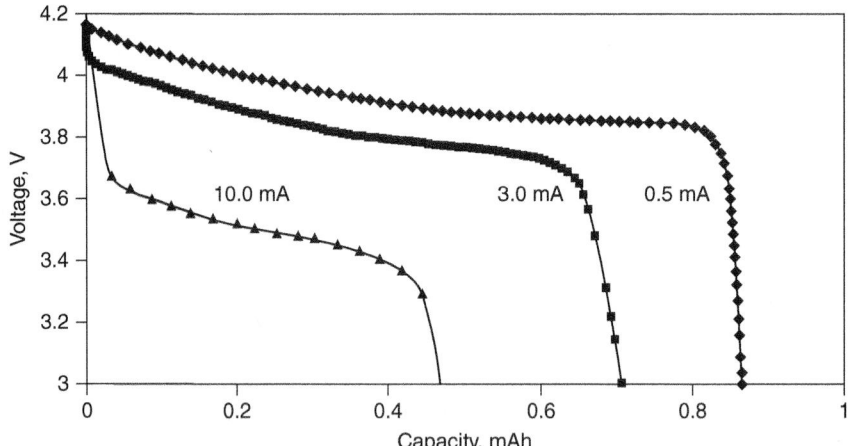

FIGURE 27.17 Discharge characteristics of a 0.9 mAh NanoEnergy® at rates from 0.5 C to 11 C. (*From Ref. 127.*)

REFERENCES

1. R. D. Rauh, K. M. Abraham, G. F. Pearson, J. K. Surprenant, and S. B. Brummer, *J. Electrochem. Soc.* **126**, 523 (1979).
2. H. Yamin, J. Penciner, A. Gorenshtein, M. Elam, and E. Peled, *J. Power Sources* **14**, 129 (1985); H. Yamin, A. Gorenshtein, J. Penciner, M. Segal, and Y. Sternberg, *J. Electrochem. Soc.* **135**, 1045 (1988).
3. E. Peled, A. Gorenshtein, and M. Elam, *J. Power Sources* **26**, 269 (1989).
4. J. Shim, K. A. Striebel, and E. J. Cairns, *J. Electrochem. Soc.* **149**, A1321 (2002).
5. M. B. Armand, J. M. Chabagno, and M. Duclot, "Extended Abstracts," *2nd Int. Meeting on Solid Electrolytes*, St. Andrews, Scotland, Sept. 1978; M. B. Armand, J. M. Chabagno, and M. Duclot, in *Fast Ion Transfer in Solids*, P. Vashishta, ed., p. 131, North Holland, New York, 1979.
6. M. Gauthier et al., *J. Electrochem. Soc.* **132**, 1333 (1985).
7. P. P. Prosini, S. Passerini, R. Vellone, and W. H. Smyrl, *J. Power Sources* **75**, 73–83 (1998).
8. B. B. Owens and S. Passerini, "International Development Trends of Energy Storage Technology for EV/HEV" *4th Symposium of Advanced Technology of Energy Storage for EV*, Tokyo, Japan, Nov. 1999.
9. www.bluecar.fr/en/pages-innovation/batterie-lmp.aspx.
10. K. M. Abraham, J. L. Goldman, and M. D. Dempsey, *J. Electrochem. Soc.* **128**, 2493 (1981).
11. M. W. Rupich, L. Pitts, and K. M. Abraham, *J. Electrochem. Soc.* **129**, 1857 (1982); K. M. Abraham, J. S. Foos, and J. L. Goldman, *J. Electrochem.* Soc. **131**, 2197 (1984); U.S. Patent 4,911,996 (1990).
12. K. M. Abraham, D. M. Pasquariello, and E. B. Willstaedt, *J. Electrochem. Soc.* **136**, 576 (1989).
13. D. Fouchard, in *Proc. 33rd Power Sources Symp.*, the Electrochemical Society, Pennington, NJ, 1988; J. A. R. Stilb, *J. Power Sources* **26**, 233 (1989).
14. L. Dominey, in *Non-Aqueous Electrochemistry*, D. Aurbach, ed, Chap. 8, pp. 437–460, Marcel Dekker, New York, 1999. Also see "Cellular Phone Recall May Cause Setback for Moli," *Toronto Globe and Mail*, August 15, 1989, and *Adv. Batt. Technology* **25**(10), 4 (1989).
15. P. Dan, E. Mengeritsky, Y. Geronov, D. Aurbach, and I. Weissman, *J. Power Sources* **54**, 143 (1995).
16. D. Aurbach, I. Weissman, A. Zaban, Y. Ein-Eli, E. Mengeritsky, and P. Dan, *J Electrochem. Soc.* **143**, 2110 (1996).
17. D. Aurbach, E. Zinigrad, H. Teller, Y. Cohen, G. Salitra, H. Yamin, P. Dan, and E. Elster, *J. Electrochem. Soc.* **149**, A1267 (2002).

18. K. W. Semkow and A. F. Sammells, *J. Electrochem. Soc.* **134**, 2084 (1987).
19. K. M. Abraham and Z. Jiang, *J. Electrochem. Soc.* **143**, 1 (1996); K. M. Abraham, *ECS Trans.* **3**(42), 67 (2008).
20. E. L. Littauer and K. C. Tsai, *J. Electrochem. Soc.* **124**, 850 (1977).
21. J. Read, *J. Electrochem. Soc.* **153**, A96 (2006).
22. S. J. Visco, E. Nimon, and B. Katz, *Meet. Abstr.-Electrochem. Soc.* **602**, 389 (2006).
23. N. Imanishi, T. Zhang, Y. Shimonishi, S. Hasegawa, A. Hirano, Y. Takeda, and O. Yamamoto, *Meet. Abstr.-Electrochem. Soc. Fall 2009*, Vienna, Austria, Abstract 215.
24. B. Kumar, N. Gupta, J. Kumar, J. P. Fellner, and S. J. Rodrigues, *Proc. of 43rd Power Sources Conference*, June 7–10, 2008, p. 35.
25. B. Kumar, J. Kumar, R. Leese, and K. M. Abraham, *Meet. Abstr.-Electrochem. Soc. Fall 2009*, Vienna, Austria, Abstract 210.
26. X. Yu, J. B. Bates, G. E. Jellison, Jr., and F. X. Hart, *J. Electrochem. Soc.* **144**, 524 (1997).
27. D. Linden, *Handbook of Batteries,* 2nd ed., McGraw-Hill, Inc., New York, 1995, p. 36.9.
28. M. Dollé, L. Sannier, B. Beaudoin, M. Trentin, and J.-M. Tarascon, *Electrochem. Solid-State Lett.* **5**, A286 (2002).
29. R. Selim and P. Bro, *J. Electrochem. Soc.* **121**, 1457 (1974).
30. I. Epelboin, *J. Electrochem. Soc.* **127**, 2100 (1980).
31. I. Yoshimatsu, T. Hirai, and J. I. Yamaki, *J. Electrochem. Soc.* **135**, 2422 (1988).
32. C. Monroe and J. Newman, *J. Electrochem. Soc.* **150**, A1377 (2003).
33. M. Gauthier, A. Belanger, and A. Vallee, U.S. Patent 6,007,935 (1999).
34. C. Monroe and *J. Newman, J. Electrochem. Soc.* **151**, A880 (2004).
35. C. Monroe and J. Newman, *J. Electrochem. Soc.* **152**, A396 (2005).
36. E. Peled, D. Golodnitsky, G. Ardel, C. Menachem, D. Bar Tow, and V. Eshkenazy, *Mat. Res. Soc. Proc.* **393**, D. H. Doughty et al., eds., p. 209 (1995).
37. D. Aurbach, A. Zaban, Y. Gofer, O. Abramson, and M. Ben-Zion, *J. Electrochem. Soc.* **142**, 687 (1995).
38. D. Aurbach, Y. Ein-Eli, and A. Zaban, *J. Electrochem. Soc.* **141**, L1 (1994).
39. D. Aurbach and Y. Cohen, *J. Electrochem. Soc.* **144**, 3355 (1997).
40. D. Aurbach and M. Moshkovich, *J. Electrochem. Soc.* **145**, 2629 (1998).
41. E. Zinigrad, E. Levi, H. Teller, G. Salitra, D. Aurbach, and P. Dan, *J. Electrochem. Soc.* **151**, A111 (2004).
42. Y. V. Mikhaylik, I. Kovalev, R. Schock, K. Kumaresan, J. Xu, and J. Affinito, *Meet. Abstr.-Electrochem. Soc. Fall 2009*, Vienna, Austria, Abstract 216.
43. R. D. Rauh, F. S. Shuker, J. M. Marston, and S. B. Brummer, *J. Inorg. Nucl. Chem.* **39**, 1761 (1977).
44. G. Roberts, D. H. Doughty, Y. Gerenov, M. Simoneau, and V. Puglisi, *Meet. Abstr.-Electrochem. Soc.* **602**, 164 (2006).
45. Y. V. Mikhaylik and J. R. Akridge, *J. Electrochem. Soc.* **151**, A1969 (2004) and references therein.
46. J. R. Akridge, Y. V. Mikhaylik, and N. White, *Solid State Ionics* **175**, 243–245 (2004).
47. Y. V. Mikhaylik and J. R. Akridge, *J. Electrochem. Soc.* **150**, A306 (2003).
48. S. J. Visco, C. C. Mailhe, L. C. De Jonghe, and M. B. Armand, *J. Electrochem. Soc.* **136**, 661 (1989).
49. S. J. Visco, M. Liu, and L. C. De Jonghe, *J. Electrochem. Soc.* **137**, 1191 (1990); M. Liu, S. J. Visco, and L. C. De Jonghe, *J. Electrochem. Soc.* **138**, 1896 (1991).
50. S. Kakuda, T. Momma, T. Osaka, G. B. Appetecchi, and B. Scrosati, *J. Electrochem. Soc.* **142**, L1 (1995).
51. K. Naoi, K-I. Kawase, M. Mori, and M. Komiyama, *J. Electrochem. Soc.* **144**, L173 (1997).
52. N. Oyama, J. M. Pope, and T. Sotomura, *J. Electrochem. Soc.* **144**, L47 (1997).
53. Y. Kiya, J. C. Henderson, and H. D. Abruña, *J. Electrochem. Soc.* **154**, A844 (2007).
54. B. Scrosati and S. Megahed, *Electrochemical Society Short Course*, New Orleans, Oct. 10, 1993.
55. S. Forsyth, J. Golding, D. R. MacFarlane, and M. Forsyth, *Electrochim. Acta*, **46**, 1753 (2001).

56. A. Webber and G. E. Blomgren, in *Advances in Lithium-Ion Batteries, Ionic Liquids for Lithium Ion and Related Batteries*, W. A. van Schalkwijk and B. Scrosati, eds., p. 185, Kluwer Academic/Plenum Publ., New York, 2002.
57. P. C. Howlett, D. R. MacFarlane, and A. F. Hollenkamp, *Electrochem. Solid-State Lett.* **7**, A97 (2004).
58. P. C. Howlett, N. Brack, A. F. Hollenkamp, M. Forsyth, and D. R. MacFarlane, *J. Electrochem. Soc.* **153**, A595 (2006).
59. N. J. Dudney, "Thin Film Micro-Batteries," *The Electrochemical Society Interface* **17**(3), 44 (2008).
60. J. B. Bates, N. J. Dudney, G. R. Gruzalski, R. A. Zuhr, A. Choudhury, C. F. Luck, and J. D. Robertson, *Solid State Ionics* **53–56**, 647 (1992); J. B. Bates, N. J. Dudney, G. R. Gruzalski, R. A. Zuhr, A. Choudhury, C. F. Luck, and J. D. Robertson, *J. Power Sources* **43–44**, 103 (1993).
61. S. J. Visco, P. J. Spillane, and J. H. Kennedy, *J. Electrochem. Soc.* **132**, 1766 (1985).
62. K. Takada, N. Aotani, K. Iwamoto, and S. Kondo, *Solid State Ionics* **86–88**, 877 (1996).
63. R. Komiya, A. Hayashi, H. Morimoto, M. Tatsumisago, and T. Minami, *Solid State Ionics* **140**, 83 (2001).
64. Y. Seino, K. Takada, B. Kim, L. Zhang, N. Ohta, H. Wada, M. Osada, and T. Sasaki, *Solid State Ionics* **176**, 2389 (2005); M. Tatsumisago, *Solid State Ionics* **175**, 13 (2004).
65. R. Kanno and M. Murayama, *J. Electrochem. Soc.* **148**, A742 (2001).
66. H. Okamoto, S. Hikazudani, C. Inazumi, T. Takeuchi, M. Tabuchi, and K. Tatsumi, *Electrochem. Solid-State Lett.* **11**, A97 (2008).
67. J. H. Kennedy and Z. Zhang, *Solid State Ionics* **28–30**, 726 (1988).
68. H. Takahara, M. Tabuchi, T. Takeuchi, H. Kageyama, J. Ide, K. Handa, Y. Kobayashi, Y. Kurisu, S. Kondo, and R. Kanno, *J. Electrochem. Soc.* **151**, A1309 (2004).
69. K.-H. Joo, H.-J. Sohn, P. Vinatier, B. Pecquenard, and A. Levasseur, *Electrochem. Solid-State Lett.* **7**, A256 (2004).
70. www.ohara-inc.co.jp/en/product/electronics/licgc.html.
71. M. B. Armand, J. M. Chubagno, and M. Duclot, in *Fast Ion Transport in Solid*, P. Vashista, J. M. Mundy, G. K. Sherroy, eds., North-Holland, Amsterdam, 1979; M. B. Armand, *Solid State Ionics* **9810**, 745 (1979).
72. M. B. Armand, in *Polymer Electrolyte Reviews-1*, J. R. MacCallum and C. A. Vincent, eds., Elsevier Applied Science, New York, 1987.
73. K. M. Abraham and M. Alamgir, *J. Electrochem. Soc.* **136**, 1657 (1990).
74. R. Koksbang, M. Gauthier, A. Belanger, in *Proc. Symp. Primary and Secondary Lithium Batteries*, K. M. Abraham and M. Salomon, eds., the Electrochemical Society, Pennington, NJ, 1991.
75. G. B. Appetecchi, F. Croce, G. Dautzenberg, M. Mastragostino, F. Ronci, B. Scrosati, F. Soavi, A. Zanelli, F. Alessandrini, and P. P. Prosini, *J. Electrochem. Soc.* **145**, 4126 (1998); G. B. Appetecchi, F. Croce, M. Mastragostino, B. Scrosati, F. Soavi, and A. Zanelli, *J. Electrochem. Soc.* **145**, 4133 (1998).
76. A. Blazejczyk, W. Wieczorek, R. Kovarsky, D. Golodnitsky, E. Peled, L. G. Scanlon, G. B. Appetecchi, and B. Scrosati, *J. Electrochem. Soc.* **151**, A1762 (2004).
77. A. S. Gozdz, C. N. Schmutz, J.-M. Tarascon, and P. C. Warren, U.S. Patent 5,456,000 (1995).
78. K. M. Abraham, in *Applications of Electroactive Polymers*, B. Scrosati, ed., Chapman and Hall, London, 1993.
79. D. H. Shen, G. Nagasubramanian, C. K. Huang, S. Surampudi, and G. Halpert, in *Proc. 36th Power Sources Conf.*, pp. 261–263, Cherry Hill, NJ, 1994.
80. L. Sannier, R. Bouchet, L. Santinacci, S. Grugeon, and J.-M. Tarascon, *J. Electrochem. Soc.* **151**, A873 (2004).
81. Z. Bakenov, M. Nakayama, and M. Wakihara, *Electrochem. Solid-State Lett.* **10**, A208 (2007).
82. P. E. Trapa, Y.-Y. Won, S. C. Mui, E. A. Olivetti, B. Huang, D. R. Sadoway, A. M. Mayes, and S. Dallek, *J. Electrochem. Soc.* **152**, A1 (2005).
83. X.-G. Sun and J. B. Kerr, *Macromolecules* **39**, 362 (2006).
84. P. E. Trapa, M. H. Acar, D. R. Sadoway, and A. M. Mayes, *J. Electrochem. Soc.* **152**, A2281 (2005).
85. news.bbc.co.uk/2/hi/science/nature/7577493.stm.

86. Y. Mikhaylik, I. Kovalev, J. Xu, and R. Schock, *ECS Trans.* **13**:19, 53 (2008).
87. Y.-G. Ryu et al., U.S. Patent 7,517,612 (2009).
88. S.-E. Cheon, K.-S. Ko, J.-H. Cho, S.-W. Kim, E.-Y. Chin, and H.-T. Kim, *J. Electrochem. Soc.* **150**, A796 (2003); S.-E. Cheon, K.-S. Ko, J.-H. Cho, S.-W. Kim, E.-Y. Chin, and H.-T. Kim, *J. Electrochem. Soc.* **150**, A800 (2003).
89. S.-C. Han, M.-S. Song, H. Lee, H.-S. Kim, H.-J. Ahn, and J.-Y. Lee, *J. Electrochem. Soc.* **150**, A889 (2003).
90. F. B. Tudron, J. R. Akridge, and V. J. Puglisi, *Proc. of 41st Power Sources Conference*, June 14–17, 2004, p. 341.
91. Y. Mikhaylik, I. Kovalev, and C. Burgess, *Meet. Abstr.-Electrochem. Soc.* **702**, 753 (2007).
92. X. Ji, K. T. Lee, and L. F. Nazar, *Nature Materials* **8**, 500–506 (2009).
93. D. Aurbach, E. Pollak, R. Elazari, G. Salitra, C. Scordilis Kelley, and J. Affinito, *J. Electrochem. Soc.* **156**, A694 (2009).
94. Y. Mikhaylik, U.S. Patent 7,352,680 (2008).
95. S.-I. Tobishima, H. Yamamoto, and M. Matsuda, *Electrochim. Acta.* **42**, 1019 (1997).
96. K. Kumaresan, Y. Mikhaylik, and R. E. White, *J. Electrochem. Soc.* **155**, A576 (2008).
97. J. H. Shin and E. J. Cairns, *J. Electrochem. Soc.* **155**, A368 (2008).
98. J. H. Shin, P. Basak, J. B. Kerr, and E. Cairns, *Meet. Abstr.-Electrochem. Soc.* **802**, 1265 (2008).
99. D. H. Doughty, D. L Coleman, and M. J. Berry, *Proc. of 43rd Power Sources Conference*, June 7–10, 2008, p. 39.
100. D. H. Doughty, E. P. Roth, C. C. Crafts, G. Nagasubramanian, G. Henriksen, and K. Amine, *J. Power Sources* **146**, 116–120 (2005).
101. www.oxisenergy.com.
102. G. Ivanov, V. Kolosnitsyn, and K. Pelton, *Adv. Auto. Battery Conf. 2009 Proceedings*, Long Beach, CA, June 8–12, 2009, Poster #40.
103. H. Ota, X. Wang, and E. Yasukawa, *J. Electrochem. Soc.* **151**, A427 (2004) and references therein.
104. H. Ota, Y. Sakata, X. Wang, J. Sasahara, and E. Yasukawa, *J. Electrochem. Soc.* **151**, A437 (2004) and references therein.
105. F. Ding, Y. Liu, and X. Hu, *Electrochem. Solid-State Lett.* **9**, A72 (2006).
106. H. Ota, Y. Sakata, Y. Otake, K. Shima, M. Ue, and J.-I. Yamaki, *J. Electrochem. Soc.* **151**, A1778 (2004).
107. Y. M. Lee, J. E. Seo, Y.-G. Lee, S. H. Lee, K. Y. Cho, and J.-K. Park, *Electrochem. Solid-State Lett.* **10**, A216 (2007).
108. J. Read and W. Behl, *Proc. of 43rd Power Sources Conference*, June 7–10, 2008, p. 165.
109. S. Cordova, Z. Johnson, N. Pereira, F. Badway, G. G. Amatucci, and K. M. Abraham, *Proc. of 43rd Power Sources Conference*, June 7–10, 2008, p. 369.
110. T. B. Atwater and A. J. Salkind, *Proc. of 43rd Power Sources Conference*, June 7–10, 2008, p. 577.
111. V. Dorval, C. St-Pierre, and A. Vallee, *Proc. of 2004 BATCON Conf.*, p. 19-1; available at www.battcon.com/PapersFinal2004/ValleePaper2004.pdf.
112. N. Balsara, M. Singh, and L. Odusanya, *Meet. Abstr.-Electrochem. Soc.* **501**, 1690 (2006).
113. N. P. Balsara, M. Singh, V. Chen, and E. D. Gomez, *Meet. Abstr.-Electrochem. Soc.* **701**, 293 (2007).
114. S. A. Mullin, A. Panday, N. Wanakule, and N. Balsara, *Meet. Abstr.-Electrochem. Soc.* **802**, 1269 (2008).
115. M. Singh et al., *Macromolecules* **40**, 4578–4585 (2007).
116. www.seeo.com/.
117. J. B. Bates, N. J. Dudney, B. J. Neudecker, F. X. Hart, H. P. Jun, and S. A. Hackney, *J. Electrochem. Soc.* **147**, 59–70 (2000).
118. N. J. Dudney, *The Electrochemical Society Interface* **17**(3), 44 (2008).
119. N. J. Dudney, "Thin Film Batteries for Energy Harvesting," in *Energy Harvesting Technologies*, S. Priya and D. J. Inman, eds., pp. 349–357, Springer Publisher, Dec. 2008.

120. B. J. Neudecker, R. A. Zhur, and J. B. Bates, *J. Power Sources* **81–82**, 27–32 (1999).
121. www.vp-scientific.com/parylene_properties.htm.
122. N. J. Dudney, "Solid-State Thin-Film Rechargeable Lithium Batteries," *Mat. Sci. Eng. B.* **116**, 245–249 (2005).
123. N. J. Dudney, and Y. I. Jang, *J. Power Sources* **119**, 300 (2003).
124. Y. Origami, D. Shimizu, T. Abe, M. Sodom, and Z. Ogumi, *ECS Transactions* **16**(26), 45–52 (2009).
125. T. Ohzuku, A. Ueda, and N. Yamamoto, *J. Electrochem. Soc.* **142**, 1431 (1995).
126. www.infinitepowersolutions.com/.
127. www.frontedgetechnology.com/.

CHAPTER 28
RECHARGEABLE ZINC/ALKALINE/ MANGANESE DIOXIDE BATTERIES

Josef Daniel-Ivad and Karl Kordesch

28.1 GENERAL CHARACTERISTICS

The rechargeable zinc/alkaline/manganese dioxide battery is an outgrowth of the primary battery. Zinc is used as the negative active material (the anode during discharge), manganese dioxide for the positive active material (the cathode during discharge), and a potassium hydroxide solution for the electrolyte.

The original design of this rechargeable battery closely followed the cylindrical inside-out design of the alkaline primary battery and retained its advantages of long shelf life, good current density, and safety.[1] This battery was marketed in the mid-1970s, but only briefly, for 6 V lanterns and portable TV sets. Its advantages were its lower cost compared to other rechargeable batteries and that it was manufactured in a fully charged state. The problems that limited the commercialization of this design were that the cells were not strictly zinc-limited and would lose their ability to be recharged due to the expansion of the cathode if the discharge continued below the voltage level corresponding to a one-electron discharge of the manganese dioxide ($MnO_{1.5}$). Thus voltage control was needed to limit the discharge to 1.1 to 1.0 V per cell, depending on the load and age of the battery, and the capacity was reduced because of this higher end voltage. Further, the capability of catalytic hydrogen-gas recombination was not included in the cell design.

A way to control the cathodic discharge is to limit the capacity of the zinc electrode. This can, however, cause poor rechargeability of the zinc electrode. Further study of this battery system led to the development of reliable techniques for limiting the capacity of the zinc electrode.[2-4] Present-day batteries can be discharged to lower end voltages and have much improved cycle life over original designs.[5]

The major advantages and disadvantages of the rechargeable zinc/alkaline/MnO_2 battery are listed in Table 28.1.

28.2 CHEMISTRY

The discharge mechanism of electrolytic manganese dioxide, which is essentially γ-MnO_2, has been studied extensively.[1] It is generally assumed that the first electron discharge step proceeds in a homogeneous reaction by the movement of protons and electrons into the lattice, resulting in a

TABLE 28.1 Major Advantages and Disadvantages of Rechargeable Zinc/Alkaline/Manganese Dioxide Batteries

Advantages	Disadvantage
Low initial cost (and possible lower operating cost than other rechargeable batteries)	Useful capacity about two-thirds of primary battery but higher than most rechargeable batteries
Manufactured in a fully charged state	Limited cycle life
Good retention of capacity (compared to other rechargeable batteries)	Available energy decreases rapidly with cycling and depth of discharge
Completely sealed and maintenance-free	Higher internal resistance than NiCd and NiMH
No "memory effect" problem	
No toxic materials, green certified	

gradually decreasing value of x in MnO_x from $x = 2$ to $x = 1.5$. The reaction is a conversion of one solid structure (MnO_2) into another ($MnOOH$), with manganese (formally) in the trivalent state[6]

$$MnO_2 + H_2O + e \rightarrow MnOOH + OH$$

Soluble manganese species begin to appear if the discharge is continued, especially when the lower voltage second-electron range is approached. The manganese ions find their way to the zinc anode, increasing the corrosion reaction and reducing the shelf life characteristics.

When electrolytic manganese dioxide is recharged, the process is reversed. The number of discharge and charge cycles obtainable depends on the depth of discharge, indicating that irreversible electrode processes are occurring. The relationship between cycle number and depth of discharge was found to be essentially logarithmic. Coulometric studies indicate that the recharge efficiency reaches nearly 100% after a few cycles. The initial losses are probably related to "formation problems," as manganese dioxide is a nonconductor and an interface with the graphite structure must be formed.

The cathode-dominated cycle characteristic of the rechargeable batteries and the logarithmic relationship of loss of capacity with cycling are shown in Fig. 28.1. A zinc capacity of about 2 Ah is usually provided in an AA-size battery to maintain a high first-cycle discharge capacity. This still prevents the MnO_2 from discharging beyond the one-electron capacity. In this special experiment, the discharge is limited to predetermined capacity at each cycle until the cutoff voltage of 0.9 V is reached. The estimated cycle life number and cumulative capacity can be determined from Fig. 28.1.

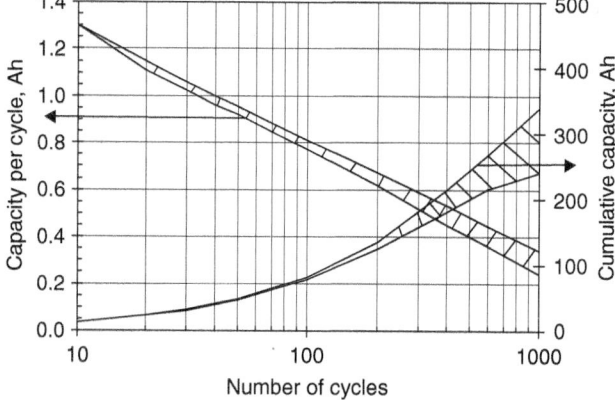

FIGURE 28.1 Performance of rechargeable zinc/alkaline/manganese dioxide battery on cycling at 20°C, recharging after each discharge. (*Courtesy of Pure Energy Visions, Inc.*)

Charging of the zinc/alkaline/manganese dioxide battery must be controlled to limit the charge voltage to about 1.65 V. Charging to higher voltages produces a hexavalent manganate and oxygen gas. The soluble manganate disproportionates into tetravalent MnO_2 and a nonrechargeable divalent manganese species, which results in a loss in cycle capacity. The oxygen gas reacts with zinc to produce ZnO.

The use of special catalysts was investigated to suppress the manganate formation and overcome the poor reversibility of the MnO_2 discharge. One approach is to use cobalt-spinel type catalysts.[7–11] Another is to replace the electrolytic manganese dioxide (EMD) with a specially prepared bismuth-doped form of manganese dioxide (BMD). This modified cathode permits a deep two-electron $MnO_2 \rightarrow Mn(OH)_2$ reversible discharge.

Bismuth is incorporated as Bi_2O_3 at about 10 weight-percent of the MnO_2. A cathode of BMD formulated with high surface area, oxidation-resistant carbon can give several hundred charge-discharge cycles with a depth of discharge equivalent to over 80% of the theoretical two-electron capacity of the MnO_2. The bismuth appears to act as a redox catalyst, extending the heterogeneous discharge regime as well as blocking the formation of the nonrechargeable manganese compounds. This produces a markedly flatter voltage discharge profile relative to EMD and, in addition, a higher capacity as illustrated in Fig. 28.2. The presence of bismuth is essential in order for the charging to proceed through the reaction sequence:

1. $Bi \rightarrow Bi^{3+} + 3e^-$ (electrochemical)
2. $Bi^{3+} + 3Mn^{2+} \rightarrow 3Mn^{3+} + Bi$ (chemical)
3. $3Mn^{3+} \rightarrow 1.5Mn^{2+} + 1.5Mn^{4+}$ (chemical)

Mn^{2+} in the last disproportionation step recycles to step 2 until charging is complete. Soluble Mn^{3+} and Bi^{3+} species must be retained at the cathode. Special chemically functionalized separators have been developed for the BMD cathode.[12]

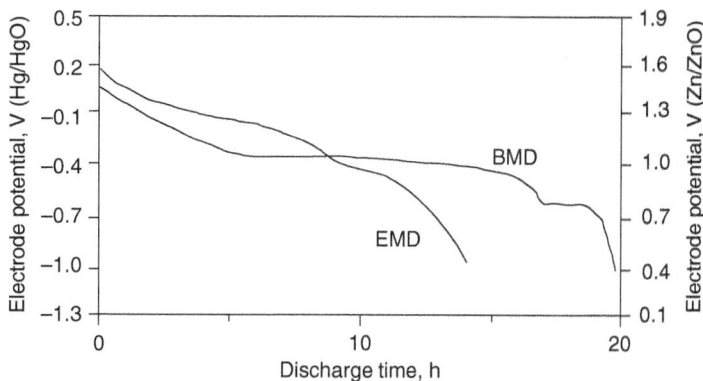

FIGURE 28.2 Discharge curve for electrolytic Manganese Dioxide (EMD) and Bismuth modified Manganese Dioxide (BMD) in KOH. (*Courtesy of Pure Energy Visions, Inc.*)

28.3 CONSTRUCTION

The construction of the cylindrical rechargeable alkaline cell is shown in Fig. 28.3. The construction is similar to the primary cell, using an inside-out design. The positive electrode consists of three or four cathodic rings, which are formed under high pressure to a slightly oversized diameter and then inserted into the steel can. The cathodic mix formulation uses electrolytic MnO_2 and up to 10% graphite. The negative electrode, consisting of a powder zinc mass containing gelled KOH, is in the center. A nail, located in the center of the gel, serves as the negative current collector. The cell is crimped-sealed and contains a vent mechanism.

28.4 SECONDARY BATTERIES

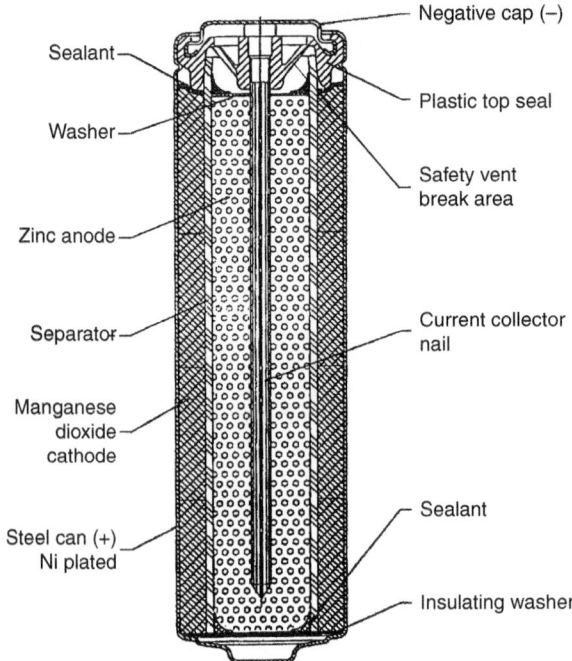

FIGURE 28.3 AA-size rechargeable zinc/alkaline/manganese dioxide battery. (*Courtesy of Pure Energy Visions, Inc.*)

The following features distinguish the rechargeable cell. The cathode contains additives, such as $BaSO_4$ or other alkaline earth compounds which increase the cathode capacity and improve cycling.[13] The cathode also contains a catalyst, such as silver on acetylene black or carbon, to recombine any hydrogen that may form. The limiting zinc powder anode contains KOH and a gelling agent. The amount of zinc determines the depth of discharge and thereby the capacity of the cell. Mercury is not added to the anode, and special zinc alloys and/or organic inhibitors in combination with a special anode preparation process are used to reduce the zinc corrosion and control dendritic zinc growth on recharge.[14] However, for rechargeable cells, this is not sufficient as the zinc deposited after the first charge is much finer than the original granulated zinc powder. The silver catalysts added to the cathode recombine the hydrogen that results from the zinc corrosion. Zinc oxide is dissolved into the KOH to ensure that on charge (or overcharge) only oxygen, not hydrogen, can be formed by electrolysis. That is, ZnO reduction occurs instead of the generation of hydrogen. The separator, which is multilayered, contains regenerated cellulose with a high oxidation resistance to caustics. It also prevents internal short-circuiting due to zinc dendrite formation.[15a,b]

28.4 PERFORMANCE

28.4.1 First Cycle Discharge

The rechargeable batteries are manufactured and shipped in a charged state, as are the primary cells. Because of their good shelf life, they can retain most of this capacity (depending on storage prior to use) and do not need to be recharged before first use. The discharge characteristics of the rechargeable zinc/alkaline/manganese dioxide batteries are similar to those of the primary batteries. However, due to the zinc-limited design of the rechargeable cell, its terminal voltage on a medium- or high-load discharge drops rapidly once 0.8 V is reached. On a low-load regime, a slope to about 0.6 to 0.7 V is sometimes noticeable before the voltage drops practically to zero. Figure 28.4 shows the discharge curves of fresh AA-size rechargeable batteries on the first cycle of discharge at several constant-current discharge loads. Figure 28.5 shows similar discharge curves for discharges at constant-resistance loads.

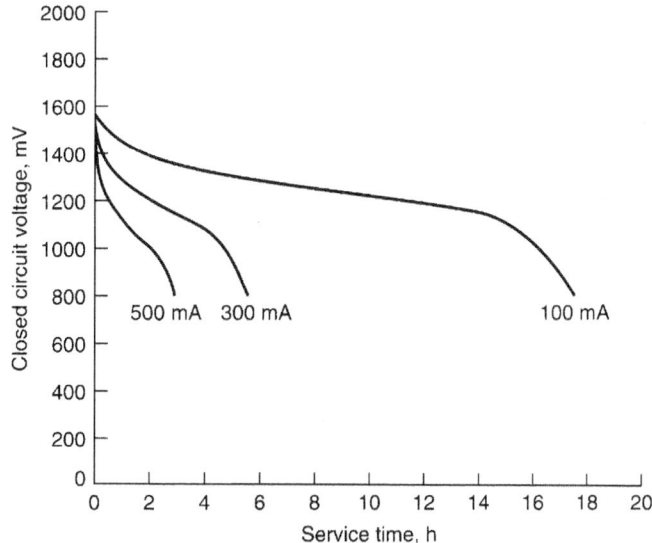

FIGURE 28.4 First-cycle discharge characteristics of rechargeable zinc/alkaline/manganese dioxide AA-size batteries discharged continuously at different constant-current loads at 22°C. (*Courtesy of Pure Energy Visions, Inc.*)

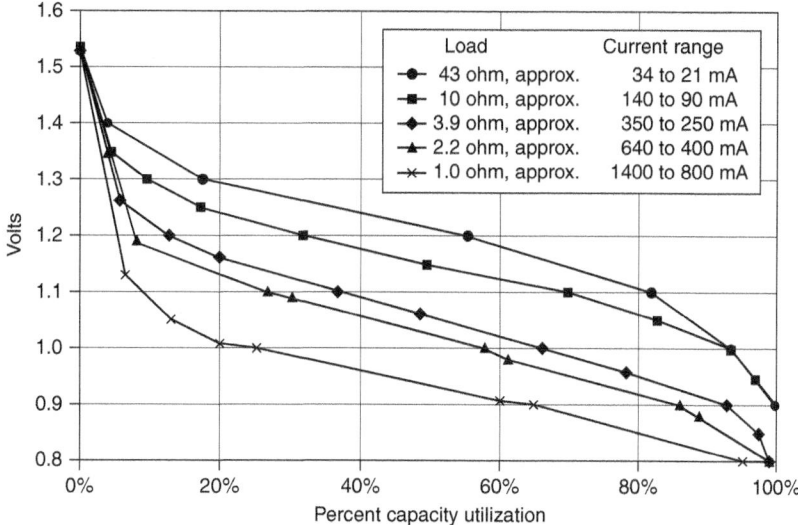

FIGURE 28.5 First-cycle discharge characteristics of rechargeable zinc/alkaline/manganese dioxide AA-size batteries discharged continuously at different constant-resistance loads at 22°C. (*Courtesy of Pure Energy Visions, Inc.* [*From Ref. 15a.*])

28.4.2 Cycling

The rechargeable battery has its highest capacity on the first cycle, and that value at 20°C is about 70 to 80% of the capacity of the primary cell. On subsequent charge-discharge cycles, if the cells are completely discharged before being recharged, 20 discharge cycles can be obtained until the capacity drops to about 50% of the initial capacity. The shape of the discharge curve changes slightly during cycling, but the voltage level drops with cycling, as shown in Fig. 28.6. Additional deep discharge cycles can be obtained, but at reduced capacity, if the cycling is continued.

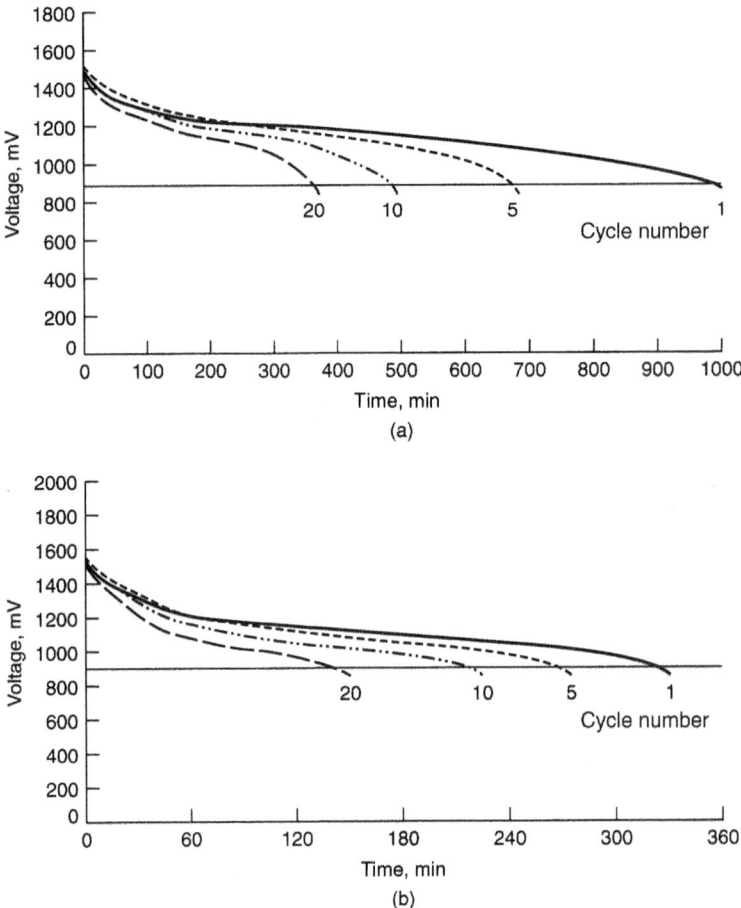

FIGURE 28.6 Continuous-discharge characteristics of rechargeable zinc/alkaline/manganese dioxide AA-size batteries after cycling at 20°C. (a) 10 Ω cycling. (b) 4 Ω cycling. (*Courtesy of Pure Energy Visions, Inc.*)

The number of useful cycles and the cycle capacity increase when the batteries are only partially discharged and recharged after use. Figure 28.7a shows the increased cycle life obtained on an intermittent discharge of a AA-size battery on a 10 ohm load, applied daily for 4 h (about a 25% depth of discharge). The cells are charged overnight on a constant voltage charger set at about 1.65 V. Although the same drop in voltage level with cycling as shown in Fig. 28.6 is present, more than 200 cycles can be obtained with the terminal voltage about 0.9 V.

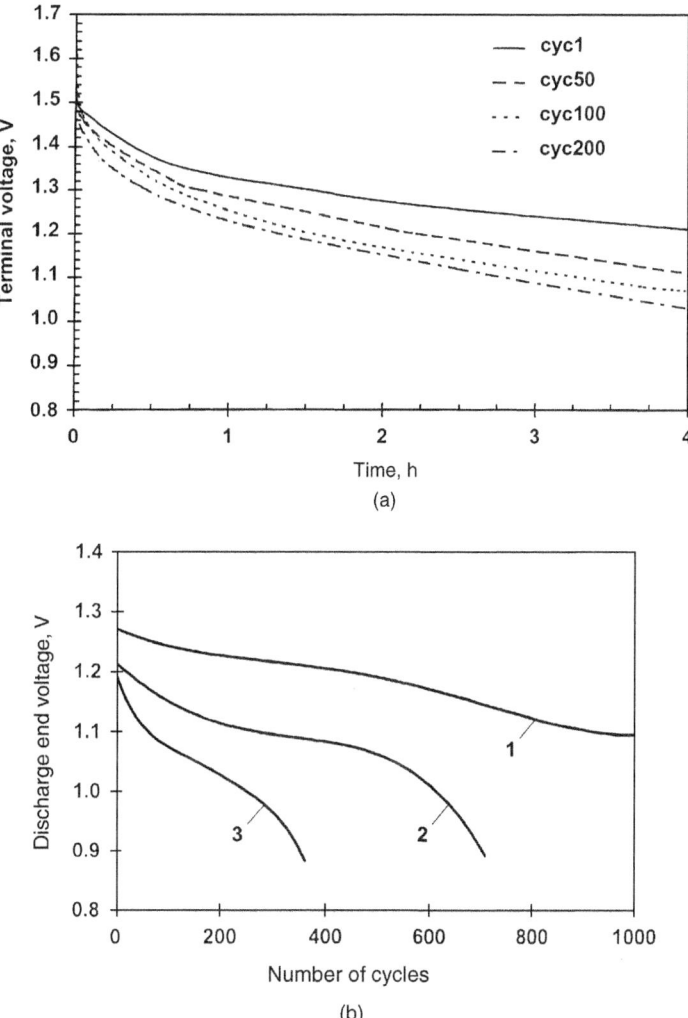

FIGURE 28.7 (a) Discharge characteristics of rechargeable zinc/alkaline/manganese dioxide AA-size batteries after cycling at 20°C. Cells discharged 4 h/day at 10 Ω; full recharge after each discharge. (b) The cycle life of AA-size batteries on a 10 Ω load. Curve 1 discharged for 300 mAh, then recharged; curve 2 discharged for 400 mAh, then recharged; curve 3 discharged for 500 mAh, then recharged. (c) Performance of rechargeable zinc/alkaline/manganese dioxide batteries on very shallow discharge cycling. (*Courtesy of Pure Energy Visions, Inc.*)

The cycle life, when the battery is discharged to other depth of discharge, is shown in Fig. 28.7b. This figure shows the results of repeated discharge of the rechargeable AA-size battery to approximately 15, 20 and 25% depth of discharge, followed by recharge. The cycle life increases with reduced depth of discharge and the voltage drop decreases with lowering the depth of discharge. If the depth of discharge is very shallow, several thousand cycles have been demonstrated in lab testing, which is illustrated in Fig. 28.7c, showing the cycling performance for the shallow discharge–frequent recharge mode, achieving up to 5000 cycles without reaching the cutoff criteria.

28.8 SECONDARY BATTERIES

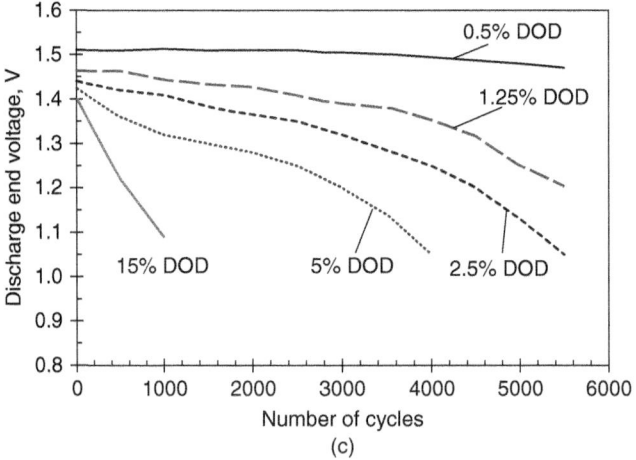

FIGURE 28.7 *(Continued)*

28.4.3 Performance of Different Sizes of Batteries

The rechargeable alkaline-manganese dioxide batteries are available in other sizes. Figures 28.8 and 28.9 show the performance of "true" C- and D-size batteries and Fig. 28.10 that of AAA-size.

Note that the AA-size and AAA-size batteries, with their thinner positive electrodes, give relatively better performance than the larger diameter C-size and D-size batteries. The efficiency of utilization of the manganese dioxide and the deep cycling performance of the rechargeable battery are related to the thickness of this electrode. This is further illustrated in Fig. 28.11, which shows the utilization of the manganese dioxide and the ampere-hour capacity delivered from the batteries as a

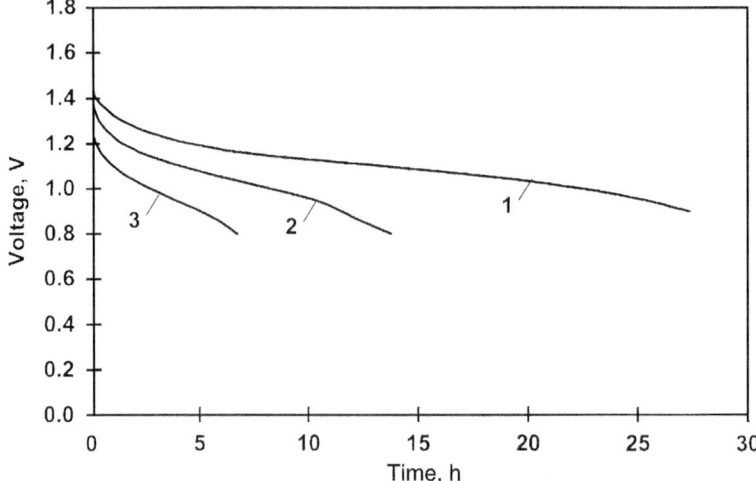

FIGURE 28.8 First cycle discharge characteristics of rechargeable zinc/alkaline/manganese dioxide C-size batteries discharged continuously at different constant resistance loads at 20°C. Curve 1 – 6.8 Ω, 160 mA (approx.); curve 2 – 3.9 Ω, 270 mA (approx.); curve 3 – 2.2 Ω, 450 mA (approx.). (*Courtesy of Pure Energy Visions, Inc.*)

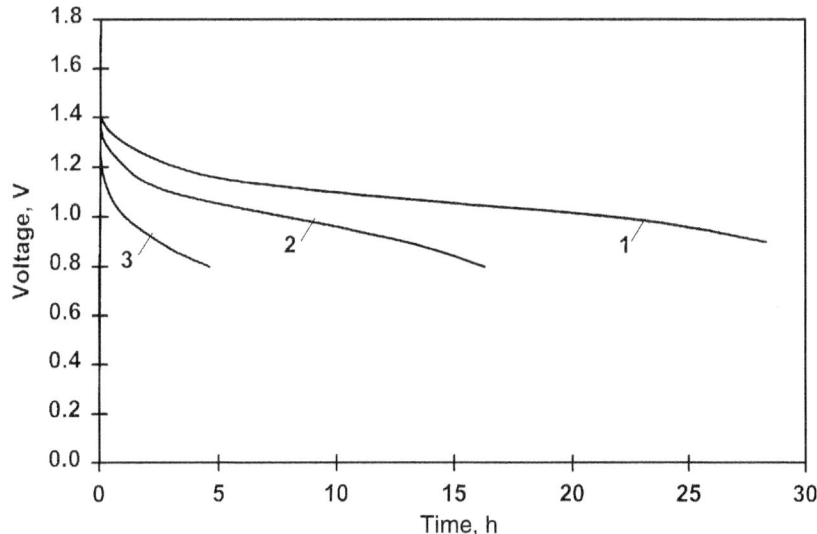

FIGURE 28.9 First cycle discharge characteristics of rechargeable zinc/alkaline/manganese dioxide D-size batteries discharged continuously at different constant resistance loads at 20°C. Curve 1 – 3.9 Ω, 280 mA (approx.); curve 2 – 2.2 Ω, 460 mA (approx.); curve 3 – 1.0 Ω, 1 A (approx.). (*Courtesy of Pure Energy Visions, Inc.*)

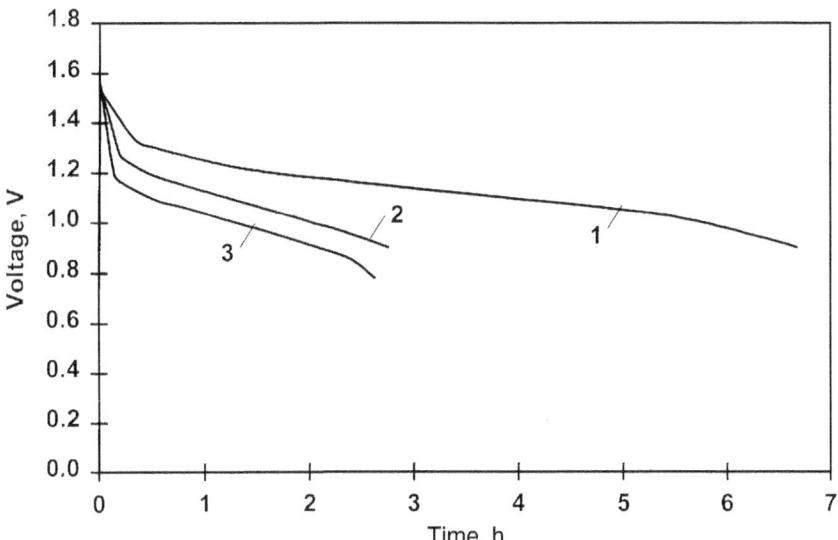

FIGURE 28.10 First cycle discharge characteristics of rechargeable zinc/alkaline/manganese dioxide AAA-size cells discharged continuously at different constant resistance loads at 20°C. Curve 1 – 10 Ω, 110 mA (approx.); curve 2 – 5.1 Ω, 190 mA (approx.); curve 3 – 3.9 Ω, 260 mA (approx.). (*Courtesy of Pure Energy Visions, Inc.*)

28.10 SECONDARY BATTERIES

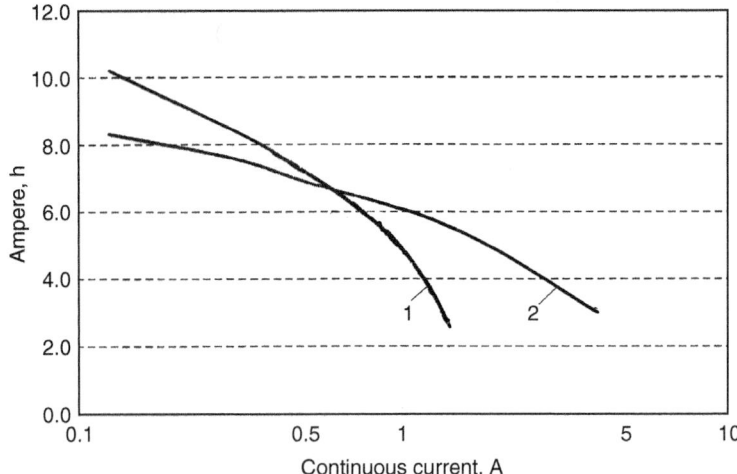

FIGURE 28.11 Comparison of 20°C performance of rechargeable zinc/alkaline/manganese dioxide D-size battery (curve 1) to output of 4 AA-cells connected in parallel (in D-size can) to 0.9 V end-voltage (curve 2). (*Courtesy of Pure Energy Visions, Inc.*)

function of load current. The thinner batteries deliver a higher percentage of their capacity than the larger diameter batteries when discharged at the higher discharge currents.[16–18]

28.4.4 Bundle (Parallel) Multicell Batteries

Figure 28.11 also illustrates the advantage of using a multicell battery design, with smaller cells in parallel, than a larger single-cell battery. The battery with four AA-size cells in a container having the same dimensions as a D-size battery weighs about 90 g, compared to a single D-size battery, which weighs about 125 g, but will outperform it at the higher discharge currents. The beneficial effect of using several small-diameter cylindrical cells in parallel in a multicell battery instead of a single larger-diameter battery also results in a gain in total internal electrode interface, thereby decreasing the current density at a given load and improving performance.[19]

28.4.5 Effect of Temperature

The performance of the rechargeable zinc/manganese dioxide batteries at various temperatures is shown in Fig. 28.12. At low temperatures down to −30°C, the batteries function, but performance is decreased. The decrease is more severe for moderate and higher rates. At higher temperatures up to 50°C, low-rate performance is unchanged, but performance at moderate and higher rates is improved. The performance of the AA-size cell at 45°C and 65°C is shown in Fig. 28.13. It should also be noted that the capacity and high current drain capability are higher at the higher temperatures due to better diffusion and higher MnO_2 utilization.[20,21]

28.4.6 Shelf Life

The shelf life of fresh, unused (charged) rechargeable alkaline-manganese dioxide batteries is about the same as that of the primary batteries (20 to 25% loss after 5 to 7 years when stored at room temperature). The data for open-circuit voltage during high-temperature storage are shown in

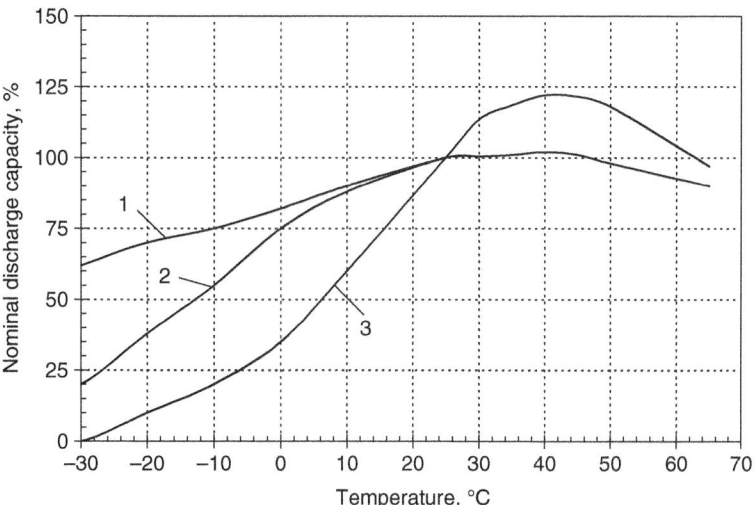

FIGURE 28.12 Effect of temperature on capacity of rechargeable zinc/alkaline/manganese dioxide batteries for different rates of discharge, 1 = very low rate: 1–5 mA; 2 = low rate: 50–100 mA; 3 = moderate rate: 250–300 mA. (*Courtesy of Pure Energy Visions, Inc.*)

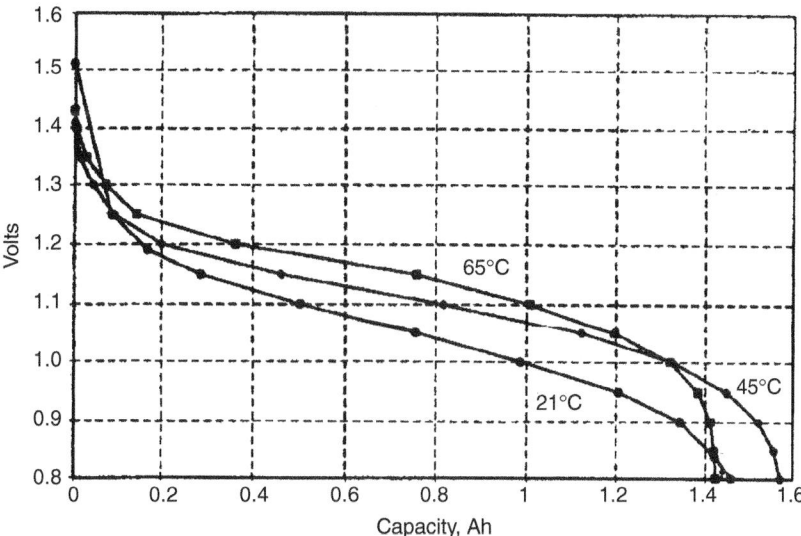

FIGURE 28.13 Discharge of rechargeable AA-size zinc/alkaline/manganese dioxide batteries at different temperatures at a 3.9 ohm load. (*From Ref. 20.*)

Fig. 28.14. Over a storage period of 12 weeks at 71°C, which represents more than 7 years at 21°C, the open-circuit voltage drops only 6%, demonstrating that no self-discharge reaction of significance takes place.

The shelf life of cycled cells depends on whether they are stored in a charged or a discharged condition. Batteries stored in a charged state after cycling show about the same losses as a fresh

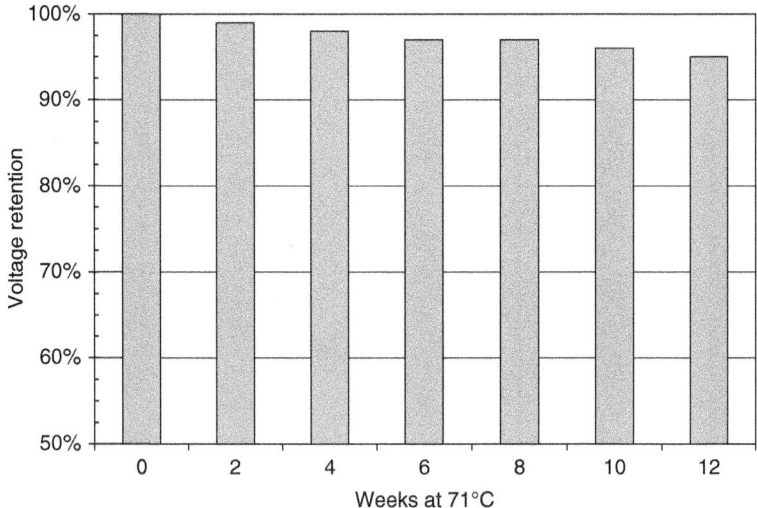

FIGURE 28.14 Open-circuit voltage stability of mercury-free rechargeable zinc/alkaline/manganese dioxide AA-size batteries at 71°C. (*Courtesy of Pure Energy Visions, Inc.*)

uncycled battery. Storage of batteries in a discharged condition, particularly at elevated temperatures (65°C) may be detrimental to the anode performance on subsequent cycles. However, under normal usage, batteries can be recharged close to the capacity level of the previous cycle.

28.5 CHARGE METHODS

In the charging process for the zinc/alkaline/manganese dioxide cell, the discharged positive active material, manganese oxyhydroxide (MnOOH), is oxidized to manganese dioxide (MnO_2) and the zinc oxide (ZnO) in the negative is reduced to metallic zinc. Manganese dioxide can be further oxidized to higher oxides (Mn^{+6} compounds) which are soluble, resulting in loss of rechargeability. Therefore proper recharging is important to obtain optimum life. Charging over 1.75 V per cell for days or over 1.70 V per cell for weeks can damage the battery. Batteries should not be charged after 105% of the ampere-hours removed have been replaced. Batteries can be float-charged for extended periods at 1.65 V per cell.[21]

28.5.1 Constant-Potential Charging

Constant-potential charging is the preferred method. This is equivalent to a taper current charge method. The voltage on charge should not exceed 1.65 to 1.68 V. If charging is continued at higher voltages, current will continue to flow and some damage to the cell can be expected due to increased anode corrosion by the soluble Mn^{+6} species. If the end voltage is set to below 1.65 V, the charging takes longer and the cell may not be fully charged overnight, but the cycle life of the battery is improved. Figure 28.15 shows the voltage and current profiles during the charge. A constant potential charger using a voltage regulator such as an LM317 device is shown in Fig. 28.16.

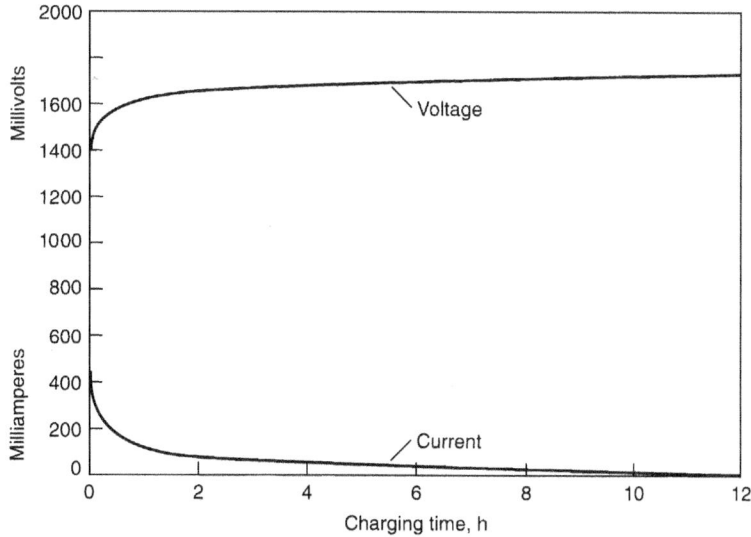

FIGURE 28.15 Constant-potential charging of AA-size rechargeable zinc/alkaline/manganese battery at 20°C. (*Courtesy of Pure Energy Visions, Inc.*)

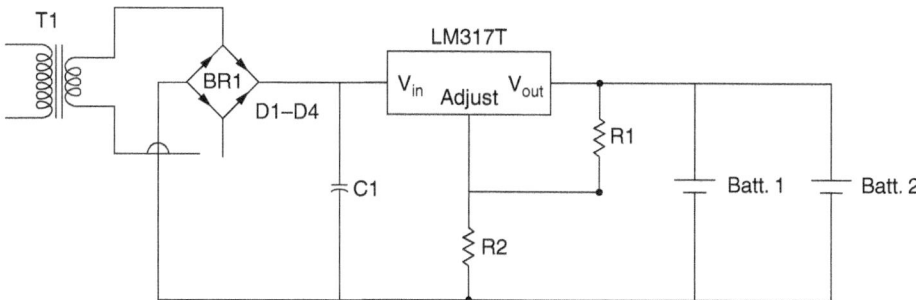

FIGURE 28.16 Principle circuit diagram for a constant potential charger utilizing an LM317T voltage regulator. (*Courtesy of Pure Energy Visions, Inc.*)

28.5.2 Constant-Current Charging

Uncontrolled constant-current charging over an extended period of time leads to electrolysis of the electrolyte, which causes a buildup of internal gas pressure and results in the rupturing of the safety vent, allowing the release of the gases. Constant-current charging is feasible if the charge voltage is limited to 1.65 V per cell (resistance-free) and a shutoff control is incorporated in the charge circuit.

28.5.3 Pulse Charging

Pulse charging can be used to permit quick charging of the rechargeable zinc-alkaline batteries. The pulse-charging method utilizes halfwave rectified 60 Hz alternating current. During pulse pause, the circuit measures the cell voltage. Because no charge current flows through the battery during the pause, the true electrochemical voltage, without any ohmic resistance, is measured. This voltage is often called the "resistance-free" voltage. The pulse-charger circuit regulates the time period when the charge is on by comparing the actual resistance-free cell voltage to the preset cutoff voltage. As long as the resistance-free voltage is lower than the cutoff voltage, charge current passes into the battery. If the resistance-free voltage is equal to or higher than the charge cutoff voltage, the charge current is cut off. The charge voltage can be much higher than the specified charge cutoff voltage as long as the resistance-free voltage does not exceed the charge cutoff voltage. This causes the initial charge current to be much higher, making fast charging possible. Pulse charging has also been shown to increase the cycle life of the battery due to improved replating of zinc.[22] Figure 28.17 shows the current profile for pulse charging of one to four AA-size cells in parallel.

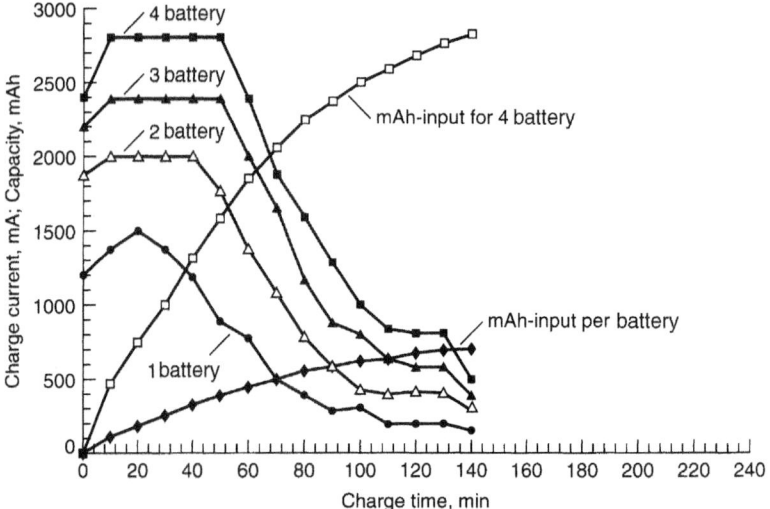

FIGURE 28.17 Pulse charging of AA-size rechargeable zinc/alkaline/manganese dioxide batteries. One to four batteries are charged in parallel to 1.68 V end voltage at 20°C. (Batteries discharged at 100% DOD at 1 Ω.) The charge current flow is determined by a regulating circuit that reads the battery voltage between current pulses. (*Courtesy of Pure Energy Visions, Inc.*)

It was found that the total charging time can be lowered if the electrodes are given time to equalize their internal charge polarization gradients by "recovery periods" in the order of minutes. After the recovery period, the charge current, which had dropped to low values, will start and continue at the higher values until the polarization gradients reach the previous high level. Overall, the charging can be carried out with a higher average current over a shorter time period. These "charge and recovery periods" in the time range of minutes act differently from the usual pulse charge methods, which interrupt for only fractions of a second. Diffusion and interface-concentration gradients in the porous manganese dioxide graphite electrodes (with a slow proton transport) change gradually.

28.5.4 Overflow Charging

The term "overflow charging" describes a process for charge control using electronic devices that become conductive at a given voltage and then divert the charging current from the battery as it becomes fully charged. Precision voltage shunt regulators, voltage detectors, LEDs, Zener diodes, and/or other diodes can be used to provide this overcharge protection.[23]

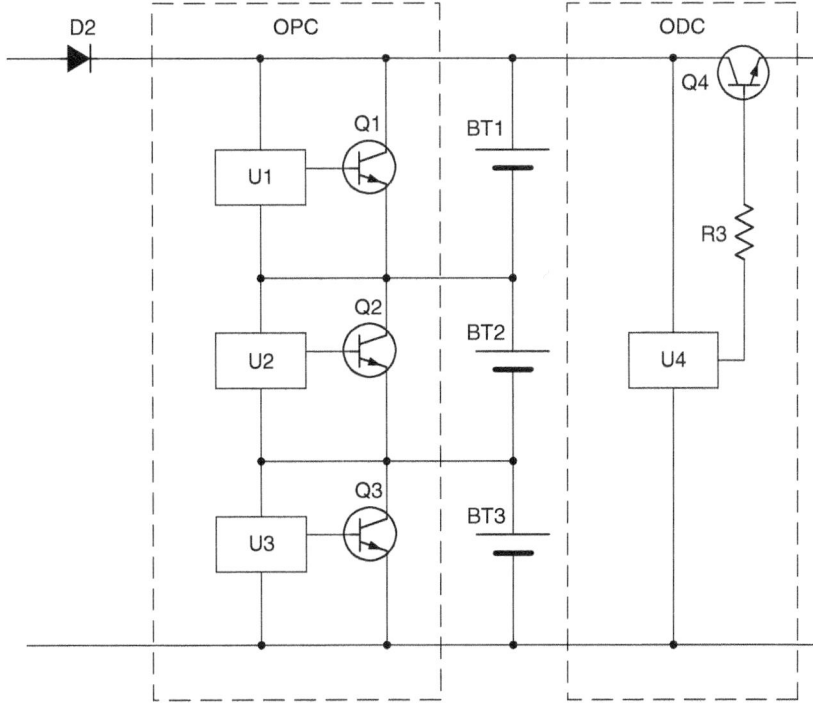

FIGURE 28.18 Charger circuit for a 3-cell battery with voltage detector/transistor overflow and protection against overdischarge. (*Courtesy of Pure Energy Visions, Inc.*)

Figure 28.18 illustrates a circuit for charging three cells in a series connection equipped with voltage detector overcharge protection. In addition, a voltage detector activated transistor switch is provided across the output terminal to protect against overdischarge. As long as the cell voltage is low, all charge current will flow through the cells. As the cell voltage increases and approaches the fully charged state, the voltage detector will switch the transistor into an ON state by-passing charge current. Meanwhile, the cell is not receiving a charge and the cell voltage will drop, causing the voltage detector to turn the transistor into an OFF state again. This way the cells receive a pulsating DC charge when they approach the fully charged state. The ON/OFF hysteresis voltage window can be adjusted, depending on application needs.

For small, welded battery packs, when the individual cells are not replaceable, a pack overvoltage protection is suitable. An example for a 2-cell pack is shown in Fig. 28.19. For further reduced electronic back leakage, voltage detector circuits can be used.

28.6 TYPES OF CELLS AND BATTERIES

The characteristics of commercially available rechargeable zinc/alkaline/manganese dioxide batteries are listed in Table 28.2, Table 28.3 shows the electrical performance of rechargeable zinc/alkaline/manganese dioxide batteries on international standard tests for primary alkaline batteries, according to IEC 60086-2 for AA-size. Table 28.4 shows it for AAA-size.

28.16 SECONDARY BATTERIES

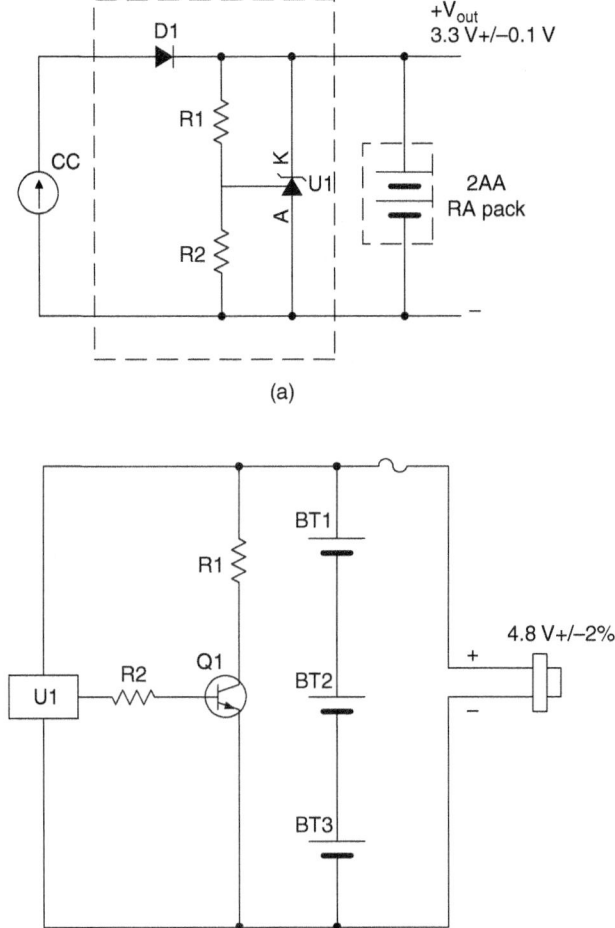

FIGURE 28.19 Charger circuit for 2- and 3-cell batteries using a shunt regulator (a) and a voltage detector (b) for overcharge protection of pack. (*Courtesy of Pure Energy Visions, Inc.*)

TABLE 28.2 Typical Rechargeable Zinc/Alkaline/Manganese Dioxide Batteries

Cell type	Dimensions, (mm)		Weight, G	Rated Capacity, Ah* (initial discharge)
	Height	Diameter		
AAA	44	10	11	0.90 at 75 Ω
AA	50	14	22	2.00 at 43 Ω
Bundle-C (3AAA)	50	26	58	2.40 at 10 Ω
Bundle-D (3AA)	60	34	104	6.00 at 10 Ω

*Based on discharge, through specified resistance, to 0.9 V per cell.
Source: Pure Energy Visions, Inc.

TABLE 28.3 Performance of Rechargeable Zinc/Alkaline/Manganese Dioxide AA-Size Batteries on International Standard Tests for Primary Alkaline Batteries According to Standard IEC 60086-2.

Application test	Load	Duty cycle	End voltage	IEC minimum average duration	Typical service duration
Radio	43 ohms	4 h/d	0.9 V	60 h	75 h
Toy	3.9 ohms	1 h/d	0.8 V	4.0 h	6.0 h
Cassette	10 ohms	1 h/d	0.9 V	11.5 h	16 h
CD/MD/Games	250 mA	1 h/d	0.9 V	4.5 h	6.4 h
Photoflash	1000 mA	10 s/m, 1 h/d	0.9 V	200 pulses	315 pulses
Remote	24 ohms	15 s/min, 8 h/d	1.0 V	31 h	40 h

TABLE 28.4 Performance of Rechargeable Zinc/Alkaline/Manganese Dioxide AAA-Size Batteries on International Standard Tests for Primary Alkaline Batteries According to Standard IEC 60086-2.

Application test	Load	Duty cycle	End voltage	IEC minimum average duration	Typical service duration
Radio	75 ohms	4 h/d	0.9 V	44 h	65 h
Cassette	10 ohms	1 h/d	0.9 V	5 h	6.8 h
Lighting	5.1 ohms	4 min/h, 8 h/d	0.9 V	130 min	190 min
Photoflash	600 mA	10 s/m, 1 h/d	0.9 V	140 pulses	250 pulses
Remote	24 ohms	15 s/min, 8 h/d	1.0 V	14.5 h	17 h

REFERENCES

1. K. Kordesch (ed.), *Batteries*, vol. 1, "Manganese Dioxide," Dekker, New York, 1974.
2. K. Kordesch, J. Gsellmann, R. Chemelli, M. Peri, and K. Tomantschger, *Electrochim. Acta* **26**:1495–1504 (1981).
3. K. Kordesch et al., "Rechargeable Alkaline Zinc Manganese Dioxide Batteries," *33d Int. Power Sources Symp.*, Cherry Hill, NJ, June 13–16, 1988.
4. Environmental Battery Systems, Richmond Hill, Ont., L4B 1C3, Canada.
5. J. Daniel-Ivad, *23rd International Seminar and Exhibit on Primary and Secondary Batteries*. Ft. Lauderdale, FL, March 13–16, 2006.
6. A. Kozawa, "Electrochemistry of Manganese Oxide," in K. Kordesch (ed.), *Batteries*, vol. 1, Dekker, New York, 1974, Chap. 3.
7. K. Kordesch and J. Gsellmann, German Patent DE 3,337,568 (1989).
8. M. A. Dzieciuch, N. Gupta, and H. S. Wroblowa, *J. Electrochem. Soc.* **135**:2415 (1988), also U.S. Patents 4,451,543 (1984), 4,520,005 (1985).
9. D. Y. Qu, B. E. Conway, and L. Bai, *Proc. Fall Meet. of the Electrochemical Soc.*, Toronto, Ont., Canada, Oct. 1992, abstract 8; Y. H. Zhou and W. Adams, *ibid.*, abstract 9.
10. B. E. Conway et al., "Role of Dissolution of Mn(III) Species in Discharge and Recharge of Chemically Modified MnO_2 Battery Cathode Materials," *J. Electrochem. Soc.* **140** (1993).
11. E. Kahraman, L. Binder, and K. Kordesch, *J. Power Sources* **36**:45–56 (1991).
12. Private communication, B. Coffey, Rechargeable Battery Corp., College Station, TX.
13. J. Daniel-Ivad, "Rechargeable Alkaline Cell Having Reduced Capacity Fade and Improved Cycle Life," U.S. Patent Application Publication 2007/0122704 (2007).
14. J. Daniel-Ivad, R. J. Book, and E. Daniel-Ivad, "Method of Manufacturing Anode Compositions for Use in Rechargeable Electrochemical Cells." U.S. Patent 7,008,723 (2006).

15a. K. Kordesch et al., "Rechargeable Alkaline Zinc-Manganese Dioxide Batteries" *36th Power Sources Conference*, Palisades Institute for Research Services, Inc., New York, 1994.

15b. T. Messing et al., "Improved Components for Rechargeable Alkaline Manganese-Zinc Batteries," *36th Power Sources Conference*, Palisades Institute for Research Services, Inc., New York, 1994.

16. J. Daniel-Ivad, K. Kordesch, and E. Daniel-Ivad, "An Update on Rechargeable Alkaline Manganese RAM™ Batteries," *Proc. 39th Power Sources, Conf.*, Cherry Hill, NJ, 2000, pp. 330–333.

17. K. Kordesch et al., *Proc. 26th JECEC*, Boston, 1991, vol. 3, pp. 463–468.

18. K. Kordesch, L. Binder, W. Taucher, J. Daniel-Ivad, and Ch. Faistauer, *18th Int. Power Sources Symp.*, Stratford-on-Avon, England, Apr. 19–21, 1993.

19. K. Kordesch, J. Daniel-Ivad, and Ch. Faistauer, "High Power Rechargeable Alkaline Manganese Dioxide-Zinc Batteries," *Proc. 182th Meet. of the Electrochemical Soc.*, Toronto, Ont., Canada, Oct. 11–16, 1992, abstract 10.

20. J. Daniel-Ivad, K. Kordesch, and E. Daniel-Ivad, "Performance Improvements of Low-Cost RAM™ Batteries," *Proc. 38th Power Sources Conf.*, Cherry Hill, NJ, pp. 155–158, (1998).

21. J. Daniel-Ivad, K. Kordesch and E. Daniel-Ivad, "High-Rate Performance Improvements of Rechargeable Alkaline (RAM™) Batteries," *Proc. Vol. 98-15 Aqueous Batteries of the 194th Electrochem. Soc. Meeting*, Boston, Nov. 1–6, 1998.

22. K. V. Kordesch, "Charging Methods for Batteries Using the Resistance-Free Voltage as Endpoint Indication," *J. Electrochem. Soc.* **119**:1053–1055 (1972).

23. J. Daniel-Ivad, "Rechargeable Alkaline Battery with Overcharge Protection." U.S. Patent application no. 2007/0122704 A1 (May 31, 2007).

PART · 4

SPECIALIZED BATTERY SYSTEMS

CHAPTER 29
BATTERIES FOR ELECTRIC AND HYBRID VEHICLES

Dennis A. Corrigan and Alvaro Masias

29.1 INTRODUCTION

In the 21st century, electric drive vehicles, including electric and hybrid vehicles, are viewed by the United States and the world with great promise to provide major societal benefits:

- To eliminate or reduce toxic exhaust emissions from automobiles, especially in urban areas of high air pollution
- To provide strategic flexibility in national energy needs by reducing dependence on foreign oil for transportation
- To reduce carbon dioxide greenhouse gas emissions, addressing concerns over global climate change

Hybrid electric vehicles are now commercially available and growing in market share. Additionally, there is increased interest internationally in the development and commercialization of modern battery-powered electric vehicles. Batteries are the enabling technology for electric drive vehicles, and the recent advances in battery technology are driving this new industry. In this chapter, we describe electric drive applications for batteries, the performance requirements for batteries, and discuss promising battery technologies for these applications.

Batteries for the pure battery electric vehicle (BEV), which is more concisely referred to as the electric vehicle (EV), are discussed first. A historical summary of the EV is provided, followed by a description of vehicle requirements and performance targets for batteries. We discuss the performance of batteries in recent electric vehicle applications and the prospects of batteries and electric vehicles in coming years.

Second, we discuss batteries for the hybrid electric vehicle (HEV), which utilizes battery energy storage in conjunction with an internal combustion engine (ICE) and/or other power sources. A historical summary culminates with the recent commercialization of hybrid vehicles, with a description of the variety of types of hybrid vehicles under consideration, together with a discussion of performance requirements for various types. We discuss the performance of batteries in recent hybrid electric vehicle applications and the prospects of batteries and hybrid electric vehicles in coming years. Alternative energy storage devices for hybrid and electric vehicle applications are also discussed.

29.1.1 Electric Vehicles

Battery electric vehicles utilize rechargeable batteries to power electric motors for propulsion.[1-6] Electric vehicles typically utilize electric motors that can also be operated in reverse as a generator to partially recharge the batteries through regenerative braking. However, the batteries are principally recharged from external power sources, such as the electrical grid using a charger than can be either on-board or off-board the electric vehicle. The elegant simplicity of electric vehicle design is illustrated in the schematic in Fig. 29.1. Electric vehicles feature highly efficient utilization of electrical energy in the battery. However, the size, weight, and cost of the battery are the major barriers to widespread commercialization of electric vehicles.

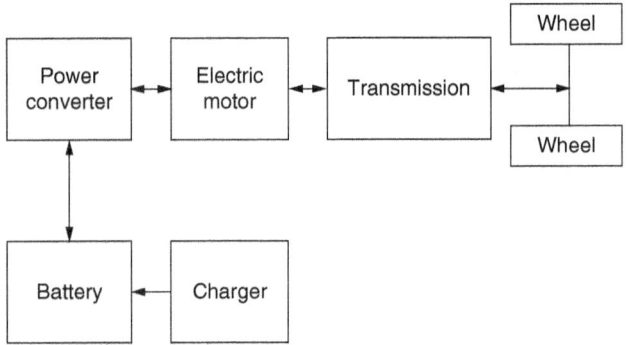

FIGURE 29.1 Schematic diagram for battery electric vehicle.

Electric vehicles date all the way back to 1837, shortly after the invention of the primary battery and the electric motor. After the invention of rechargeable lead-acid batteries in 1860, they became a practical device, even before the development of the internal combustion engine. EVs were highly competitive with vehicles powered by steam engines and internal combustion engines for the rest of the 19th century. In the golden age of electric vehicles (1900–1912), over 30,000 electric vehicles were introduced into the United States, and several times that worldwide. Electric vehicles were especially popular in cities, to overcome the pollution issues of the day by replacing horse-drawn vehicles. Electric vehicles also did not share a principal disadvantage of gasoline-powered cars: the difficulty and dangers of hand-cranking internal combustion engines. An early electric vehicle is shown in Fig. 29.2.

It was ironically the development of another battery-powered device, the electric starter motor, that allowed the gasoline-powered car to prevail. Once the disadvantage of the hand-crank issue was resolved, the advantages of gasoline-powered cars, including longer range and lower cost, allowed them to fully dominate the automotive industry for the rest of the 20th century. Then and thereafter, electric vehicles cost several times as much as gasoline-powered vehicles due to the high cost of batteries. Throughout the rest of the 20th century, electric vehicles were relegated to niche applications such as milk trucks in England, golf carts, and forklift trucks. However, automotive companies continued periodic cycles of R&D into EVs in response to environmental issues and episodes of high fuel costs such as the OPEC oil embargo in the 1970s.

At the end of the 20th century, electric vehicles were developed vigorously due to the 1990 California Zero Emission Vehicle (ZEV) mandate, which initially required automotive OEMs to make available for sale EVs equivalent to 2 and 10% of their fleets in 1998 and 2003, respectively. This mandate was enacted by the California Air Resources Board (CARB) in response to the display of the GM Impact EV prototype at the Los Angeles Auto Show in 1989. Subsequently, GM developed this prototype into a remarkable and famous electric vehicle known as the EV1, which was introduced in limited numbers at the end of the last century.[7] Other car companies developed limited production EVs as well, including the Chrysler EPIC minivan, Ford Ranger Truck EV, Toyota RAV4

FIGURE 29.2 Nickel-iron battery inventor Thomas Edison with 1912 Detroit Electric EV. (*Courtesy of the National Museum of American History.*)

electric, and the Honda EV Plus. Several electric vehicles introduced in the 1990s in response to the California ZEV mandate are shown in Fig. 29.3. These vehicles were powered mainly by nickel-metal hydride (NiMH) batteries. However, lead-acid batteries and other advanced batteries were also utilized. The Ford Ecostar was developed to use high-temperature sodium-sulfur batteries, and the Nissan Altra utilized lithium-ion batteries.

The electric vehicles developed in response to the ZEV mandate were only offered in limited quantity because they could not be manufactured and sold with a profit. Disadvantages to the customer were limited range, slow refueling, and, most importantly, the high cost of this new technology. In particular, the battery cost was the key issue. High battery costs resulted in electric vehicles costing more than double that of comparable gasoline-powered vehicles.

The California mandate also created great conflict and controversy with disputes between stakeholders, including government, the auto industry, the oil industry, electric utilities, EV proponents,

FIGURE 29.3 Electric vehicles developed in response to the California ZEV mandate include (upper row, left to right) GM EV1, Chrysler EPIC Minivan, and Ford Ecostar, and (lower row, left to right) Toyota RAV4 EV, Honda EV Plus, and Chevy S10 Truck EV.

and EV antagonists. The spirit of these times has been captured in the 2006 documentary *Who Killed the Electric Car?* which views the demise of the California mandate as a great conspiracy. In reality, many socioeconomic factors played a role, but the biggest issue preventing commercialization was the high cost of batteries.

Despite the rollback of the ZEV mandate, societal motivations for electric vehicles remain strong, especially including environmental concerns with increased carbon dioxide emissions and global warming, and also including strategic issues related to the limited supply of petroleum and the geopolitical distribution of existing supplies. A resurgence of interest in electric vehicles has resulted from recent peaks in the price of gasoline, as well as continuous improvement in electric vehicle technology, especially batteries.

Additionally, the situation in certain rapidly developing countries such as China and India may greatly favor the introduction of electric vehicles. These most populated countries on earth have much more serious issues than the United States in terms of petroleum supply and air pollution. The population density may not allow the same penetration of polluting gasoline-powered vehicles in major cities. Additionally, these overcrowded cities do not have the same acceleration and range requirements that have driven auto development in the industrialized world. Even with range limitations, electric vehicles may reach much wider acceptance in developing countries.

Some electric vehicles under recent commercial development throughout the world are shown in Fig. 29.4. In 2008, U.S.-based Tesla Motors introduced the high-performance, all-electric Tesla Roadster. Thousands have been ordered and a thousand were delivered by the end of 2009. Ford announced plans to introduce a battery EV version of the Ford Focus in 2011. EV development at GM and Chrysler continue with prototypes on display at 2009 auto shows. Mitsubishi has developed the iMiEV for initial introduction in Japan in 2010. Nissan has developed the Leaf EV, which it intends to introduce into the United States and Canada in a demonstration fleet program starting in 2010. The Think EV is being commercialized in Norway. Several Chinese automakers are developing battery electric vehicles, including BYD. The REVA electric vehicle is being commercialized in India. Notably, all of these vehicles have lithium-ion (Li-ion) battery packs. A list of EV developers of interest is shown in Table 29.1.

29.1.2 Power and Energy for Electric Vehicle Propulsion

Propulsion power and energy for electric vehicles are provided by the battery.[6,8] The most basic requirement for electric vehicle batteries is to provide the electrical power to propel electric vehicles,

FIGURE 29.4 Notable electric vehicles under commercial development in the first decade of the 21st century, including (upper row, left to right) Tesla Roadster EV, BYD e6, and REVA NXR, and (lower row, left to right) Think EV, Nissan Leaf EV, and Mitsubishi iMiEV.

TABLE 29.1 Selected Developers of 21st Century Electric Vehicles

Developer	Location	Current EV model	Battery technology
BMW	Munich, Germany	Mini E (2012)	Li-ion
BYD	Shenzhen, China	e6	Li-ion
Chrysler	Auburn Hills, Michigan	GEM	Lead-acid
Daimler Benz	Stuttgard, Germany	Smart EV (2010)	Li-ion
Ford	Dearborn, Michigan	Focus EV (2011)	Li-ion
Mitsubishi	Tokyo, Japan	iMiEV (2010)	Li-ion
Nissan	Yokahama, Japan	Leaf EV (2010)	Li-ion
REVA	Banglaore, India	REVA NXR	Lead-acid, lithium-ion
Tesla	San Carlos, California	Tesla Roadster (2009)	Li-ion
Think	Norway	Think EV	Li-ion, sodium/metal-chloride

to accelerate them, and to sustain speed. The force required to propel a vehicle is greater than the force to accelerate the vehicle mass due to additional forces for rolling resistance and aerodynamic drag:

$$F = ma + mgC_{rr} + \tfrac{1}{2}\rho C_D A v^2$$

where F = force required at the wheels of the vehicle
 m = mass of the vehicle
 a = acceleration of the vehicle
 g = acceleration of free fall due to gravity
 C_{rr} = coefficient of rolling resistance between tires and road surface
 ρ = density of the ambient air
 C_D = coefficient of drag of the vehicle in the direction of travel
 A = cross-sectional area of the vehicle
 v = speed in the direction of travel

Strictly speaking, there are two additional terms. There is a small force required related to rotational acceleration of rotating parts, especially the motor. There is also a force related to propelling the vehicle up a grade (negative for a descending grade) and this force can be very large for steep grades.

The power required to propel the vehicle depends on this force and the vehicle velocity according to

$$P = Fv = mav + mgC_{rr}v + \tfrac{1}{2}\rho C_D A v^3$$

Note that this power is somewhat more than directly proportional to velocity, due to higher velocity dependence of the aerodynamic drag force term. In uniform acceleration, the power demand increases nearly linearly with time, peaking at the end of the acceleration period.

For an electric vehicle, the power capability must be sufficient to meet acceleration requirements, typically specified in terms of acceleration time from 0 to 60 mi/h. Acceleration from 0 to 60 mi/h within the 12 to 15 s typical for a 3000 lb U.S. midsize passenger car requires about 80 to 100 kW. Most of the power is required for acceleration, but around 10% is required to overcome aerodynamic drag and 3 to 5% can be required to overcome rolling resistance.

Electric vehicles can also brake using the electric propulsion motor in such a way as to store the braking energy in the battery. This electric braking is known as regenerative braking since it provides charge input to regenerate stored energy in the battery. The braking force is provided through the transfer of kinetic energy of vehicle motion into electrical energy that is then stored as chemical energy in the battery. In this case, the braking force required is less than the force needed to decelerate the vehicle mass since the rolling resistance and aerodynamic drag forces also act to slow the vehicle:

$$F = ma - mgC_{rr} - \tfrac{1}{2}\rho C_D A v^2$$

The power required to brake the vehicle is the product of this force and the vehicle velocity:

$$P = Fv = mav - mgC_{rr}v - \tfrac{1}{2}\rho C_D A v^3$$

Taking into account again the approximate proportionality of power to velocity, the braking power required declines about linearly with time in a uniform deceleration. Thus, for braking, the peak regen braking power requirement occurs at the beginning of the braking event. Braking is partially assisted by the aerodynamic drag and rolling resistance forces. Thus, moderate decelerations comparable to acceleration performance require moderately less power than acceleration. However, deceleration during braking may be up to several times quicker than acceleration, thus the maximum power required for braking may be several times higher than that required for acceleration. Mechanical braking is usually used to supplement regenerative braking in electric vehicles.

High-power batteries are now capable of providing excellent power performance for a wide range of electric vehicles, including high-speed sports cars. Energy, the capability to provide power over time, is a related but separate key requirement to allow a practical range for electric vehicles. It is the energy requirement that is challenging for batteries. Conventional ICE-powered vehicles typically provide a range of 300 to 400 mi powered by gasoline with a theoretical specific energy of 13,000 Wh/kg. This is nearly 400 times more than the 35 Wh/kg specific energy of typical lead-acid batteries. A superficial conclusion would be that an electric vehicle version of a conventional gasoline-powered car with a 300 mi range would have a range of less than 1 mi.

While still extremely challenging, practical opportunities for battery electric vehicles arise from the efficiencies of electric drive and a more detailed comparison of the relative drivetrains as well as the opportunity for higher energy batteries. First, internal combustion engines are heat engines that are subject to Carnot cycle inefficiencies. The operating efficiency of an internal combustion engine can be as low as 20% or less in actual operation. The efficiency of battery discharge can exceed 85%. Even including losses in the electric motor and power electronics, an EV drivetrain efficiency in excess of 80% is feasible. Thus, there can be a 4:1 efficiency advantage from electric drive in using on-board energy. Additionally, electric vehicles can capture regenerative braking energy by the operation of motors in reverse where they function as generators. Especially in stop-and-go city driving, this can provide for substantial energy efficiency benefits. Finally, electric motors and power electronics do not weigh as much as automotive engine systems and transmissions. Thus, the battery weight in an electric vehicle can be several times the weight of the gasoline tank in a conventional gasoline-powered vehicle. By utilizing advanced batteries, a practical vehicle range of over 100 mi is feasible.

The energy consumption during driving depends strongly on the size, weight, and type of vehicle, as well as the characteristics of driving to which the vehicle is subjected. Representative energy consumption results are usually obtained through measurements during dynamometer testing of vehicles subjected to standard driving schedules, such as those established by the EPA for fuel economy standards. Standard driving schedules include the Urban Dynamometer Driving Schedule (UDDS), which represents city driving, and the Highway Fuel Economy Test (HWFET), which represents highway driving. The velocity profile over time for these driving schedules is shown in Fig. 29.5. The energy requirement can be determined from the average power draw under these conditions.

Typical energy consumption results for a midsize electric vehicle are 225 Wh/mi for city driving and 275 Wh/mi for highway driving (according to the HWFET and UDDS schedules, respectively).[9] Under the US06 schedule, commonly used to represent the more aggressive highway driving now typical in the United States, the energy consumption is typically higher, at about 400 Wh/mi. In contrast to results with conventional gasoline-powered vehicles, electric vehicles are more efficient under city driving conditions where regenerative braking can provide substantial energy efficiencies in stop-and-go driving. The energy input from regenerative braking is subtracted from the energy

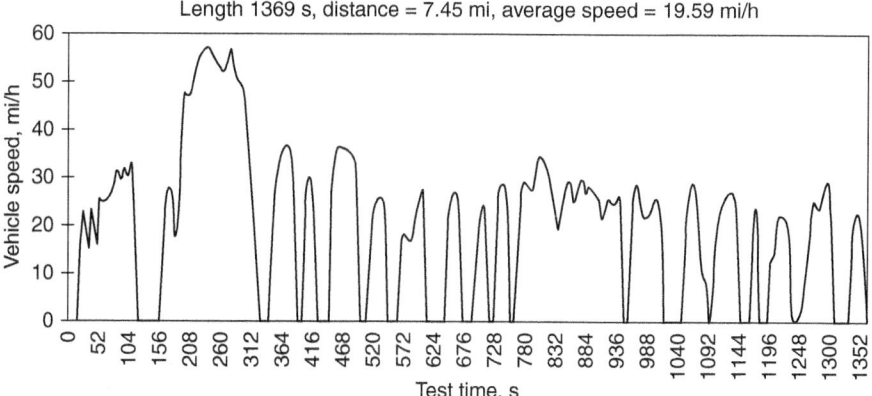

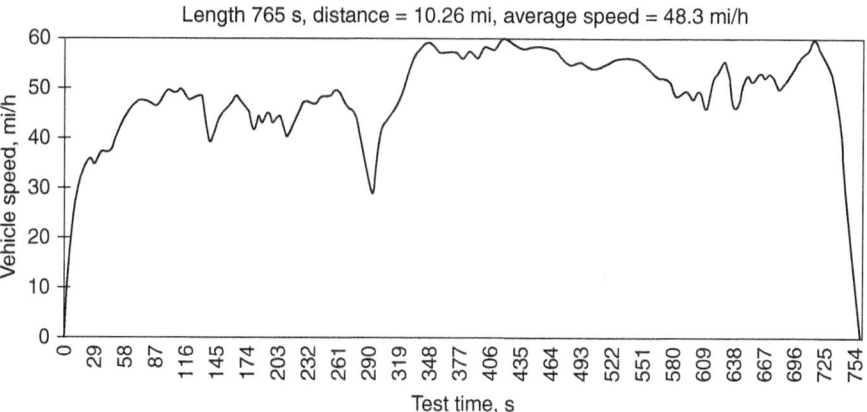

FIGURE 29.5 EPA standard UDDS driving schedule representing city driving (top) and HWFET schedule representing highway driving (bottom). (*Courtesy of U.S. EPA.*)

required to propel the vehicle to determine the net energy consumption. Using the average of these driving cycle results yields 300 Wh/mi as a typical energy consumption, so that a 100 mi range would require a battery pack of 30 kWh. A 40 kWh battery pack would yield a range of 133 mi.

However, we are used to conventional vehicles with a range of 300 to 400 mi. This would require a battery pack energy of 90 to 120 kWh. With commercially available batteries, such a battery pack would be very large and weigh 2000 to 2600 lb even using 100 Wh/kg technology. Some lowering of these energy requirements can be achieved by improved vehicle design for higher efficiency. However, the weight, size, and cost of these large battery packs are the key challenges for the EV industry.

Another challenge for battery electric vehicles relates to recharge of the battery at the end of a trip. Rapid recharge, comparable with gasoline refueling of a conventional vehicle at the gas station, is not feasible due to practical charge power limitations. Recharge in 2 to 3 min would require megawatt power capability. A more practical solution is charging at home or at work locations, with full charge expected in 8 to 10 h. Even this would require installation of 220 V electrical power to provide around 6 kW of power needed for overnight charging of a 60 kWh battery pack.

29.10 SPECIALIZED BATTERY SYSTEMS

29.1.3 EV Battery Pack Systems

Batteries consist of one or more electrochemical cells. Electric vehicles are powered by battery packs, which contain multiple battery modules that in turn are comprised of several battery cells, as illustrated in Fig. 29.6.

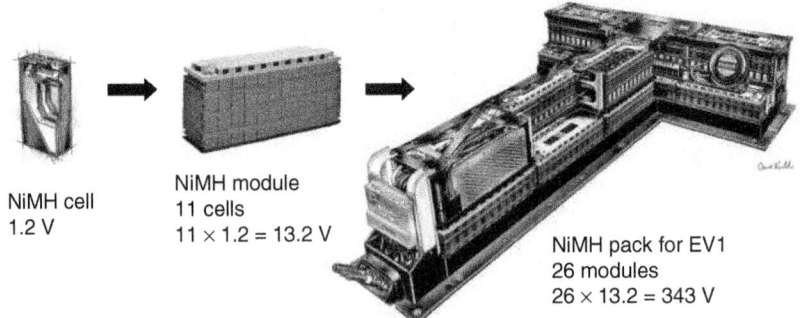

NiMH cell
1.2 V

NiMH module
11 cells
11 × 1.2 = 13.2 V

NiMH pack for EV1
26 modules
26 × 13.2 = 343 V

FIGURE 29.6 Battery modules comprised of cells are the building blocks of battery packs. (*Courtesy of General Motors.*)

Battery cells operate typically in the range of 1 to 4 V. The nominal discharge voltage for nickel-metal hydride battery cells is 1.2 V. Lead-acid batteries operate at 2 V per cell. Lithium-ion batteries typically operate in the range of 3 to 4 V depending on the chemistry. This low-voltage range is insufficient to provide the tens of kilowatts required for electric vehicle applications since practical considerations with electric motors and power electronics limit current flow to less than 500 A. Electric vehicles operate generally in the range of 100 to 500 V. A group or series-parallel arrangement of dozens or hundreds of cells is used to power electric vehicles.

A battery cell is a single electrochemical cell. Multiple cells are combined electrically and mechanically to provide higher voltage units called battery modules, which are the building blocks used to assemble EV battery packs. Battery modules typically contain 5 to 25 cells arranged so the voltage is limited to less than 50 V (safe against electrocution hazards) with a weight of less than 50 lb for easy handling. Means for thermal management, cell voltage monitoring, and control electronics may also be incorporated into the module depending on the specific technology and product.

Battery modules are combined electrically and mechanically into battery packs to provide the full power and energy needed for electric vehicles. Battery pack systems incorporate a series-connected string of modules together with power interconnections, mechanical packaging, electronic sensors and controls, and thermal management components necessary for full electric vehicle battery operation.

29.1.4 Electronic Controls for EV Batteries

The electronic controls assist the battery pack in providing required propulsion power while maintaining the battery pack within the normal operating conditions. The battery pack voltage and current are measured as well as battery module and/or cell voltages. Multiple temperature sensors are also utilized to determine the battery pack average operating temperature. The battery management system (BMS) utilizes these inputs to control the battery pack operation so that cell voltages are maintained within normal operating limits. On discharge, cell voltage excursions below specified low-voltage limits are avoided by reduction and/or suspension of discharge power. The cell voltage limits are usually compensated for temperature and discharge current based on the electrical response of the battery cells. Similarly, the BMS acts to keep the battery pack from exceeding high-temperature limits. It also provides for battery pack ground fault detection. The BMS also serves to monitor the state-of-charge of the electric vehicle battery pack during discharge by amp-hour counting of the net discharge capacity.

Electric vehicles can be charged with on-board and/or off-board chargers that interact with the battery management system as well. Together the charger and BMS act to provide for full charge within normal cell voltage and temperature conditions. On charge, cell voltage excursions above upper voltage limits are avoided by reduction and/or suspension of charge power. The upper voltage limits are compensated for temperature and charge current. The battery pack temperature must also be controlled to provide for efficient charging as well as to ensure adequate life.

In the case of lithium-ion batteries, which cannot be overcharged without safety issues and/or compromise of battery life, electronic controls are essential to keep cells within proper operational voltage limits. These electronic controls can be incorporated at both the battery module and battery pack level. They must be capable of detecting voltage deviations between cells arising from variations in self-discharge rates. These differences are usually due to thermal inhomogeneities within battery packs. Electronic circuits to balance the state-of-charge between individual cells are also included. These controls add significantly to the complexity and cost of lithium-ion battery packs. (See also Sec. 5.6.)

29.1.5 Thermal Management for EV Batteries

Electric vehicle battery packs employ a thermal management system designed to maintain the battery temperature within the normal operating range. Ambient temperature batteries such as lead-acid, nickel-metal hydride, and lithium-ion batteries operate best in the range of 20 to 40°C, coincidentally temperatures at which human beings are comfortable. The power performance suffers as we approach freezing temperatures. Elevated temperatures over 40°C can result in reduced charging efficiency as well as acceleration of failure modes leading to reduced life. Extreme elevated temperatures can also lead to safety issues.

Electric vehicle operation generates significant waste heat within a battery pack that must be rejected to keep the battery within normal operating temperatures. While entropic heating and cooling can play a role in low-performance electric vehicles, more frequently the heat generation is dominated by Joule heating of the battery pack and can be on the order of 10% or more of the traction power output. Thus, passive cooling utilizing natural convection is not generally adequate, and forced-air or forced liquid cooling is the usual approach.

Electric vehicle battery packs typically incorporate a carefully designed thermal management system that keeps the average battery temperatures within the normal operating range and provides for uniform temperatures across the entire battery pack. Uniform battery cell temperatures are needed to avoid operating cell voltage variations caused by variations in resistance, which can have a significant temperature dependence. It is equally important to avoid cell-to-cell temperature variations to maintain a uniform state-of-charge since self-discharge has a significant temperature dependence.

Forced-air cooling has been the most common approach utilized for electric vehicle battery packs. It has the advantage of simplicity, and it also typically provides a lighter and cheaper solution than can be achieved with liquid cooling. However, air has a limited heat capacity and it can be difficult to achieve uniform battery temperatures. Air flow channels are bulky and require symmetrical geometries that may not be available in some vehicle designs. Fans consume energy and can be noisy. Finally, dirt and water must be removed from intake air to avoid ground-fault issues.

Utilizing the much higher heat capacity and thermal conductivity of liquid cooling media, significantly higher heat-rejection rates can be obtained with liquid cooling. Liquid cooling also provides for a more uniform distribution of battery cell temperatures. A more compact battery pack design can be achieved since flow channels can be smaller. Automotive coolants comprised of water-glycol formulations provide for high heat capacity and good flow characteristics under vehicle operation. Liquid cooling is especially useful in providing for uniform cooling of complex and asymmetric battery packs. However, liquid-cooled thermal management also carries disadvantages in terms of complexity, weight, and cost. The liquid coolant can add complexity by introducing another conduction path, and additional discipline is required to avoid reliability issues with respect to leaks. Liquid cooling also requires a second heat exchanger, such as a radiator, which is then ultimately air cooled. Both air cooling and liquid cooling can also utilize active refrigeration systems to further accelerate heat rejection from the battery pack. For example, air inlets from an air conditioned cabin have provided for refrigerated air to cool propulsion battery packs.

29.1.6 Vehicle Integration of EV Batteries

Battery packs must also be contained physically within the vehicle in such a way as to be safe in the event of crashes and remain isolated electrically from the vehicle. This structure and containment also provide significant weight and volume burden to the battery pack. Due to the restrictive nature of volume limitations on the battery pack, the engineering of the battery packaging envelope is always a critical feature in the design of an electric vehicle. As a rule of thumb, the battery pack is typically less than one-third of the vehicle weight and even a smaller fraction of the vehicle volume.

Electric vehicles utilize battery pack systems for propulsion, so the EV battery requirements are based on the power, energy, life, weight, volume, and cost of full battery pack systems. System components other than the battery cells and modules add no power or energy, but can contribute significantly to the weight, volume, and cost of the battery pack system. Thus, the specific performance of the battery pack system is always less than that of the battery cells and modules. Similarly, the cost per unit energy ($/kWh) is always higher. EV battery performance targets are thus rightfully set at the battery system level.

29.2 EV BATTERY PERFORMANCE TARGETS

The United States Advanced Battery Consortium (USABC) is part of the U.S. Council for Automotive Research (USCAR), an umbrella organization for precompetitive collaborative research between Chrysler, Ford Motor Company, and General Motors with support from the U.S. Department of Energy (DOE). The mission of USABC is to develop electrochemical energy storage technologies that support commercialization of fuel cell, hybrid, and electric vehicles. An important contribution of USABC has been the development of performance targets and detailed test procedures for electric vehicle batteries.[10] The USABC established quantitative performance targets for batteries for electric vehicle applications starting in the early 1990s.

Midterm goals, including 80 Wh/kg specific energy, 200 W/kg specific power, and 1000 lifetime charge-discharge cycles, have been largely achieved by nickel-metal hydride EV batteries developed in the 1990s, at least on a battery module basis. This enabled the demonstration EV programs in response to the CARB ZEV mandate. However, range limitations of this midterm technology limited its commercial appeal. More significantly, battery costs were several times the $150/kWh cost target, and commercialization did not proceed.

The development focus is now on the long-term goals aimed at more widespread commercialization of electric vehicles. The long-term goals for performance and cost are provided in Table 29.2. Minimum goals for commercialization are also included in Table 29.2.

These criteria for EV batteries were developed through a consensus of various viewpoints of the USABC partners GM, Ford, and Chrysler, with additional inputs from the U.S. DOE. While they do not represent battery product targets for any particular electric vehicle, they were developed to guide development of batteries for midsize electric vehicle passenger cars in North American markets. They are generally accepted in the EV community as useful targets to guide development of batteries for electric vehicle applications. They do capture the complexity of the variety of performance criteria that must be simultaneously met at an affordable cost. The USABC performance and cost goals are especially challenging in that the quantitative targets are for fully integrated battery packs at the systems level.

The specific power and power density requirements are based on the peak discharge power at the end of a 30 s discharge pulse at 80% depth of discharge. The USABC defines peak battery power as

TABLE 29.2 USABC Goals for Advanced Batteries for Electric Vehicles

Parameter (for fully burdened system)	Condition	Units	Minimum goals for commercialization	Long-term goals
Power density	30 s discharge pulse at 80% DOD	W/L	460	600
Specific power	30 s discharge pulse at 80% DOD	W/kg	300	400
Specific regen power	10 s regen (charge) pulse at 20% DOD	W/kg	150	200
Energy density	$C/3$ discharge rate	Wh/L	230	300
Specific energy	$C/3$ discharge rate	Wh/kg	150	200
Power/energy ratio		1/h	2:1	2:1
Total pack energy	(implied) $C/3$ discharge rate	kWh	40	40
Life	<20% power and energy performance degradation	years	10	10
Cycle life	DST cycles to 80% DOD, <20% performance degradation	cycles	1,000	1,000
Selling price	40 kWh packs at production volume of 25,000 units	$/kWh	< 150	100
Operating environment	<20% performance degradation throughout range	°C	−40 to +50	−40 to +85
Normal recharge time		h	6	3 to 6
High rate charge			20–70% SOC in 30 min	40–80% SOC in 15 min
High rate discharge	Fraction rated energy discharged in 1 h with no failure	%	75	75

the peak power delivered at a voltage no lower than two-thirds of the battery open-circuit voltage. Although 11% higher power is theoretically available at half of the open-circuit voltage, it is not deemed practical to utilize this power in an EV where power electronics typically are not operated at such low voltage levels. In a well-behaved battery characterized by an ohmic resistance, the USABC expression for peak power is given by

$$\text{Peak power} = 2/9 \times V_{oc}^2/R$$

where V_{oc} is the open circuit voltage and R is the effective resistance of the battery in ohms. The specific power and power density are the peak power of the battery pack divided by the battery pack weight and volume, respectively.

The energy requirements are based on the energy available in a constant-current discharge at the $C/3$ rate. The specific energy and energy density are the discharge energy of the battery pack divided by the battery pack weight and volume, respectively. Measurements of discharge energy can be obtained in one charge-discharge cycle that usually incorporates pulse-power tests to simultaneously determine the peak power. Experimentally, duplicate measurements of specific power, power density, specific energy, and energy density can be determined in 1 day since one full charge-discharge cycle typically takes about 10 h or less.

While the USABC goals are written in intrinsic variables of specific power, power density, specific energy, and energy density, there was an implicit assumption of a 40 kWh battery pack with a pulse-power discharge power of 80 kW yielding a power-to-energy ratio of 2:1 (in units of reciprocal hours). This battery pack would weigh 200 kg or 440 lb and occupy a volume of 133 L or 35 gal so that the density would be 1.5 kg/L. The regen power target corresponds to 40 kW.

The life requirements for the battery pack are based on the expectation that the battery will last the life of the vehicle, generally 10 years and/or 100,000 mi. This is especially necessary due to the high cost of the batteries. For a 100-mi range vehicle, 1000 charge-discharge cycles would be required to meet this target. The cycle life targets are based on deep depth cycles under Dynamic

Stress Test (DST) simulated driving profile conditions. The DST profile was developed to simulate a more realistic and aggressive version of the Federal Urban Driving Schedule (FUDS) in a limited number of constant-power steps, as shown in Fig. 29.7. USABC cycle life tests subject batteries to repeated charge-discharge cycles to 80% of the rated capacity until the power and energy drop to 80% of their rated values. The cycle life target of 1000 deep cycles is designed to provide for the battery to last a vehicle life of 100,000 mi, implicitly assuming a range of at least 100 mi can be achieved with one 80% DOD discharge of the battery pack. Cycle life testing can be time consuming. Assuming 6 h charge times and 3 h discharge times, it would take about 1 year to perform 1000 charge-discharge cycles. Even accelerated testing at higher rates takes at least several months.

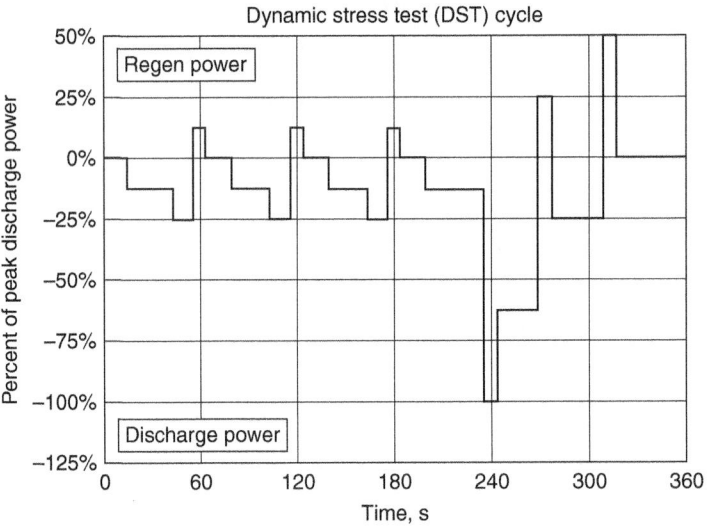

FIGURE 29.7 USABC Dynamic Stress Test power profile. (*Courtesy of USABC.*[10])

Calendar life is a separate issue from cycle life.[11] Some failure modes are independent of charge-discharge cycling and are driven by time-dependent chemical processes such as corrosion. Thus, it is more difficult to reliably accelerate calendar-life testing than it is to accelerate cycle-life testing. A general approach is to accelerate known stress conditions. For example, failure due to temperature-dependent failure mode processes can be accelerated by life testing at elevated temperatures. Results at several temperatures can then be extrapolated back to estimate the calendar life at room temperature. However, accelerated calendar-life testing to predict a battery life meeting USABC goals of 10 years or more is a somewhat speculative art since failure modes may change with stress conditions. There is an inherent risk in basing business decisions on these unproven projections.

The selling price for a 40 kWh battery pack produced at relatively low volumes of 25,000 units per year is targeted at $6000 for the minimum goal for EV commercialization. The long-term target of $100/kWh corresponds to $4000. The battery pack plus the electric motor replace the gas tank and internal combustion engine in a conventional vehicle that cost on the order of $2000 or less. However, savings in fuel cost in comparison to less expensive energy in the form of electricity may be used to justify a higher battery cost. Cost is the single most challenging target for EV batteries. This is the key issue holding back the commercialization of electric vehicles.

Other USABC goals for advanced batteries for electric vehicles in Table 29.2 include an operational temperature range between −40 and +50°C that is to be achieved on a system level. Capability for rapid recharge on the order of 40% in 15 min is desired to supplement overnight charging at the

6 h rate. The high-rate discharge requirement to enable nearly complete discharge in 1 h simulates high-speed highway operation and/or severe grade conditions.

In addition to the primary criteria of Table 29.2, USABC has established secondary criteria for energy efficiency, charge retention, maintenance-free operation, and abuse tolerance. The long-term goal for energy efficiency is 80% based on a 6 h charge followed by discharge at the 3 h rate. The charge retention target is 85% of the rated capacity after 1 month of open-circuit stand. Battery systems should be fully maintenance-free. Finally and importantly, battery packs must be tolerant of electrical and mechanical abuse on a system level. A series of electrical and mechanical abuse tests have been developed that battery packs must pass to ensure safe operation in electric vehicles.[12]

29.3 BATTERIES FOR ELECTRIC VEHICLES

With the introduction of electric vehicles in the 19th century, almost all electric vehicles were powered by lead-acid batteries. Around the turn of the 20th century, Edison developed the nickel-iron battery to provide a battery solution with a somewhat higher energy density that was much more durable as well as very robust to electrical abuse. While it provided a significantly improved performance over lead-acid batteries, its high cost precluded widespread commercial success.

In the development of electric vehicles in the 20th century, a variety of other batteries have also been utilized including:[8]

- Nickel-cadmium
- Nickel-zinc
- Sodium-sulfur
- Sodium-metal chloride
- Zinc-bromine
- Zinc-chlorine

However, until near the end of the 20th century, none of these eclipsed the lead-acid battery in terms of overall commercial viability. The other battery types above all provided higher specific energy and energy density as well as longer life. However, several battery types raised significant toxicity and/or safety issues. All of the other batteries were significantly more expensive than lead-acid batteries by several-fold. It is noteworthy that despite many years of advanced battery development in the 20th century, when GM introduced its revolutionary EV1, it was initially powered by deep-cycle (VRLA) lead-acid batteries. However, the practical range of less than 100 mi was not satisfactory.

Midterm EV battery development in the 1990s focused largely on nickel-metal hydride batteries by developers including Ovonic Battery Company (a subsidiary of Energy Conversion Devices), its manufacturing operation GM Ovonics (joint venture with GM and predecessor to Cobasys), SAFT, and Panasonic EV Energy or PEVE (joint venture of Panasonic and Toyota). GM Ovonic, SAFT, and PEVE all produced NiMH battery packs for EVs introduced by major OEMs in California around 2000. Table 29.3 shows representative performance achieved by the GM Ovonics NiMH battery module and the GM EV1 battery pack.[13,14]

In this NiMH battery pack for the EV1, the battery modules were arranged in a T-shape, as illustrated in Fig. 29.6. Thermal management was achieved by air cooling, which was complicated by the asymmetry of the battery pack shape. The battery pack also included a battery management system that operated effectively with module-level voltage and selected temperature sensors. Safety disconnects, fuses, and power contactors were also included. The battery pack hardware added nearly 15% to the battery pack weight.

PEVE developed 12 V NiMH battery modules with a capacity of 100 Ah, which were used in air-cooled battery packs for the Toyota RAV-4 EV and the Honda EV-Plus. SAFT developed similar

TABLE 29.3 Performance Specifications for GM Ovonic NiMH Battery Module and EV1 Battery Pack

	GM Ovonic battery module (11 cells)	EV1 battery pack (26 modules)
Nominal voltage	13.2 V	343 V
Discharge capacity	90 Ah	90 Ah
Energy	1.2 kWh	30 kWh
Power	3.6 kW	94 kW
Weight	18 kg	535 kg
Specific power	200 W/kg	175 W/kg
Specific energy	66 Wh/kg	56 Wh/kg

NiMH battery modules that were utilized in liquid-cooled battery packs for the Chrysler EPIC minivan. NiMH batteries provided typically 65 Wh/kg and 200 W/kg at the battery module level, which was adequate to provide an EV range in excess of 100 mi. Additionally, a cycle life of 1000 cycles was achieved, which provided for excellent durability and several times the life of lead-acid battery packs. The NiMH electric vehicle performance was generally quite good, and these vehicles—stimulated by ZEV mandate—were quite popular. NiMH technology was further proven as a robust technology for automotive propulsion as the enabling technology for power-assist hybrid electric vehicles.

It was the cost of nickel-metal hydride batteries per unit energy that was not commercially sustainable for EV applications. At the low volumes, the nickel-metal hydride battery packs cost well in excess of $1000/kWh. Such costs were affordable for some small, high-power battery packs for hybrid electric vehicle applications. However, for large EV battery packs required to provide a vehicle range of more than 100 mi, this amounted to costs of over $25,000 per vehicle for the battery alone. This was not a viable proposition. NiMH battery developers projected production costs to drop below $300/kWh at mature volumes of 20,000 battery packs per year. Actual volumes achieved were more than an order of magnitude, lower, and the lower-cost potential was not realized. This commercial issue resulted in modification of the California ZEV mandate, and attention was subsequently refocused on fuel-cell electric vehicles.

More recently, renewed interest in battery electric vehicles has been spurred by great progress in the development of high energy density lithium-ion cells. Strong commercial development in lithium-ion batteries for portable electronics applications such as laptop computers and cell phones has resulted in $5B market with strong competition for high performance and low cost. The specific energy and energy density of small 18650 Li-ion cells (cylindrical cells for laptop computers with a standard size of 18 mm diameter by 65 mm high) have improved from 100 Wh/kg and 200 Wh/L in 1990 to around 200 Wh/kg and 570 Wh/L in 2010. Figure 29.8 shows the steady increase in cell performance, accompanied by an even more dramatic decrease in cell cost.[15]

Large lithium-ion batteries were developed by Sony and Hitachi for electric vehicle applications in the early 1990s due to their strong promise of high energy density. In those early days, the promise of high energy density was tempered by low power performance and serious safety issues. However, not only has the promise of high energy density been realized, but the thin electrode designs have evolved to provide very high-power lithium-ion batteries. The serious safety issues associated with earlier products have been greatly mitigated by improved manufacturing techniques and cell designs, as well as new chemistries with greatly reduced volatility.

Lithium-ion cells encompass a wide variety of chemistries and cell designs that now offer a variety of technology options for electric vehicle developers.[15-17] The cobalt oxide cathode chemistry provides very high energy density, but many EV developers have avoided this chemistry due to concerns about its safety. Mixed layered-oxide cathode chemistries such as the nickel-cobalt-aluminum oxide and the nickel-cobalt-manganese oxide have relatively high energy density with improved stability with respect to safety issues. The most benign chemistries from a safety point of view include the manganese oxide spinel cathode and the lithium iron phosphate olivine-phase cathode.

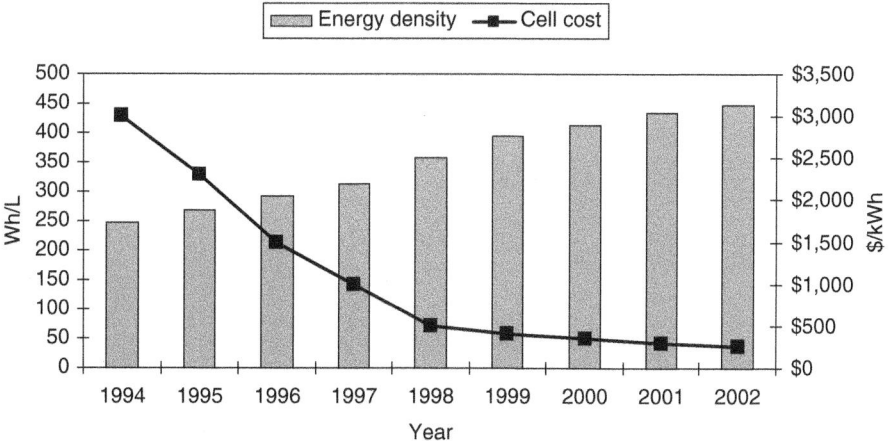

FIGURE 29.8 Progression of improvements in energy density and cost of 18650 lithium-ion cells.

However, these latter cathode materials offer a lower battery energy density. On the other hand, all of these alternative chemistries are attractive in mitigating materials costs issues with cobalt. All of the cell chemistries are offered in a variety of cell designs, including cylindrical-wound cells, elliptical-wound cells, traditional large-format prismatic cells, and laminated pouch cells. The great commercial promise of lithium-ion batteries has spurred intense international competition resulting in a diverse multitude of suppliers and developers as shown in Table 29.4.

In one case, 18650 lithium-ion battery cells developed for portable electronics are being directly used in EV battery packs.[18] Tesla Motors has developed a novel battery pack for its Tesla Roadster sports car that incorporates 6831 of these cells in a series-parallel arrangement to provide 53 kWh of energy with a peak power of about 200 kW. High specific energy cobalt oxide chemistry is used to provide over 2 Ah of capacity at about 3.6 V in a cell weighing less than 50 g yielding nearly 200 Wh/kg. The parts count and manufacturing complexity of a battery pack with over 6000 cells is challenging from a cost point of view. However, the use of the high volume 18650 cell provides very competitive cell costs with the full effect of high volume materials production. Additionally, a variety of alternative suppliers are available to provide high quality and reliable cells from this mature market.

The potential volatility of this high-energy chemistry is mitigated by the small cell size and internal cell safety features, such as a PTC (positive temperature coefficient) current limiting device and a current interrupt device (CID) incorporated into each cell. The series-parallel network provides for inherent redundancy, and the pack is designed with cell-level fuses to survive the failure of multiple individual cells. Sophisticated electronics at the battery module and pack level are designed to provide for safe control of this high-energy battery pack. Thermal management of the battery pack is achieved with liquid cooling. The entire battery pack system weighs 450 kg. The battery pack level specific energy is an impressive 120 Wh/kg, and the specific power is over 400 W/kg. Battery packs ready for installation in the Tesla Roadster are shown in Fig. 29.9.

More typically, larger prismatic lithium-ion battery cells are being utilized in battery packs developed for EV applications. An example is the battery pack being developed for the Mitsubishi iMiEV that utilizes a single series string of 88 prismatic lithium-ion batteries developed by GS Yuasa.[19] This pack, shown in Fig. 29.10, consists of 22 series-connected modules, each consisting of 4 series-connected cells. The cells deliver 50 Ah at a nominal cell voltage of 3.7 V, yielding a specific energy of 109 Wh/kg. Thermal management of this pack is achieved by forced-air cooling. The battery pack, including batteries, thermal management, and control electronics, weighs 200 kg, yielding a specific energy of 82 Wh/kg on a battery pack level. Performance specifications for the cells, modules, and battery packs for the iMiEV are summarized in Table 29.5.

TABLE 29.4 Some Developers of Lithium-Ion EV Batteries

Li-ion EV battery developer	Location	Products
A123 Systems	USA	Cells, modules, packs
Automotive Energy Supply Corp. (AESC)	Japan	Cells, modules, packs
AltairNano	USA	Cells, modules, packs
Compact Power Inc. (LG Chem)	USA	Packs
Electrovaya	Canada	Cells, modules
Delphi	USA	Packs
Dow-Kokam	USA/Korea	Cells, modules
EnerDel	USA	Cells, modules, packs
Hitachi	Japan	Cells, modules, packs
Johnson Controls—SAFT (JCS)	USA/France	Cells, modules, packs
International Battery, Inc.	USA	Cells, modules
K2 Energy Solutions, Inc.	USA	Cells, modules
LG Chem	Korea	Cells, modules, packs
Lishen	China	Cells, modules, packs
Lithium Technology Corporation (GAIA)	USA/Germany	Cells, modules, packs
Magna	Canada	Packs
Panasonic	Japan	Cells, modules, packs
SAFT	France	Cells, modules, packs
Sanyo	Sanyo	Cells, modules, packs
Shin-Kobe	Japan	Cells, modules, packs
SK Energy	Korea	Cells, modules, packs
Tesla	USA	Packs
Thunder Sky Batteries	China	Cells, modules, packs
Toshiba	Japan	Cells, modules, packs
Valence	USA	Cells, modules
Yuasa	Japan	Cells, modules, packs

FIGURE 29.9 Tesla Roadster battery packs in production. (*Courtesy of Tesla Motors.*)

FIGURE 29.10 Lithium-ion battery cells, module, and battery pack for the Mitsubishi iMiEV (*Courtesy of Mitsubishi.*)

TABLE 29.5 Performance Specifications for GS Yuasa LEV50 Cells and Modules and the Mitsubishi iMiEV Battery Pack

	Yuasa LEV50 battery cell	Yuasa LEV50 battery module (4 cells)	Mitsubishi iMiEV battery pack (22 modules)
Nominal voltage	3.7 V	14.8 V	326 V
Discharge capacity	50 Ah	50 Ah	50 Ah
Energy	185 Wh	740 Wh	16.3 kWh
Power	935 W	3740 W	60 kW
Weight	1.7 kg	7.5 kg	200 kg
Length	171 mm	175 mm	1400 mm
Width	44 mm	194 mm	700 mm
Height	114 mm	116 mm	200 mm
Volume	0.85 L	3.9 L	196 L
Specific power	550 W/kg	500 W/kg	300 W/kg
Power density	1100 W/L	960 W/L	306 W/L
Specific energy	109 Wh/kg	100 Wh/kg	82 Wh/kg
Energy density	218 Wh/L	190 Wh/L	83 Wh/L

Another approach utilizes a series-parallel arrangement of laminated type lithium-ion pouch cells for EV battery packs, as illustrated in Fig. 29.11 with an EnerDel EV battery pack.[20] In this example, each module consists of a series-parallel arrangement of 48 pouch cells each with a capacity of 20 Ah delivered at 3.7 V. The module capacity of 120 Ah and voltage of 30 V is achieved by arranging 6 cells in parallel and 8 cells in series. A series connection of 8 modules provides for a battery pack voltage of 240 V and total energy of 28 kWh. High-energy pouch cells with a specific energy of 150 Wh/kg can be assembled into a full battery pack system with a specific energy of about 90 Wh/kg using this approach.

In 2009, USABC presented the spider chart in Fig. 29.12 showing the technology status of lithium-ion EV batteries relative to the USABC minimum goals for commercialization.[21] There is strong optimism that the specific power and power density goals have been achieved as well as the

29.20 SPECIALIZED BATTERY SYSTEMS

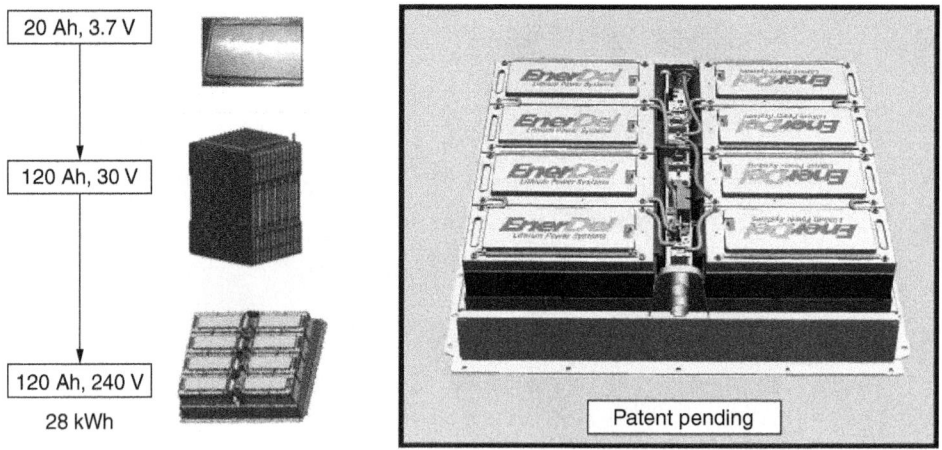

FIGURE 29.11 EnerDel lithium-ion EV battery pack consisting of modules of pouch cells. (*Courtesy of EnerDel.*)

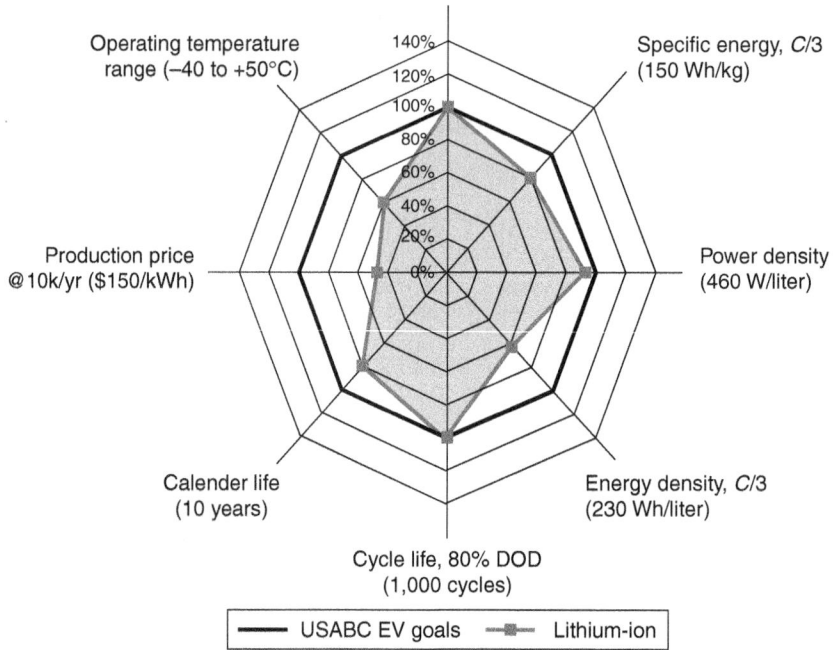

FIGURE 29.12 Status of lithium-ion battery technology in comparison to USABC minimum goals for commercialization for EV applications (presented by USABC in May 2009).

cycle life goal of 1000 deep cycles. The specific energy status is short of the target of 150 Wh/kg on a battery pack level. There is also a lack of confidence about the calendar life reaching 10 years. An area of significant concern is the operating temperature range, particularly due to concerns about charging lithium-ion batteries at temperature below freezing. With graphite anodes, kinetic limitations on lithium insertion reactions result in the plating of lithium metal at low temperatures. This can result in the formation of lithium dendrites that can form shorting paths through the separator with significant impacts on safety and/or life. However, the most serious shortcoming is in the area of cost. System-level costs for lithium ion battery packs are still several times the cost target.

There remains optimism about the long-term prospects to meet EV battery cost targets with lithium-ion batteries. The cost of lithium-ion cells is now lower than that of comparable NiMH battery cells that are more burdened with high materials costs. The high cell voltage (which is about triple that of NiMH batteries) provides triple the energy for the same amount of active materials, assuming comparable specific capacities. Additionally, at least some cell chemistries involve relatively inexpensive active materials. This has not immediately been translated into a significant cost advantage on an EV battery pack system level due to the higher complexity of lithium-ion battery packs resulting from safety concerns. The need to avoid overcharge requires cell-level voltage monitoring and state-of-charge balancing circuits. Systems development to simplify and reduce the cost of electronic controls is under way. Additionally, research and development approaches to enable overcharge through the use of redox shuttles have shown some promise.[22]

Lithium-ion battery technology has profited from being the principal focus of funded EV battery R&D in the United States in the last two decades. This funding is expected to continue to exploit the promise of this high-energy technology. Further improvements in energy density can drive the cost down by lowering materials costs. One promising approach is the development of higher-voltage cathode materials. A second approach is the development of higher specific capacity cathodes and anodes. Both are being vigorously pursued in dozens of university, industrial, and government labs.

Additionally, development of other battery technologies exceeding the performance of lithium-ion is now being encouraged. A recent U.S. DOE ARPA-E solicitation has encouraged the development of higher specific energy batteries.[23] This solicitation targeted development of batteries with a specific energy of 400 Wh/kg at the cell level to enable battery packs to meet the 200 Wh/kg long-term specific energy goal. Specific technologies encouraged include metal/air batteries (see Chap. 33), lithium batteries that utilize lithium metal anodes and/or displacement reaction cathodes, and lithium-sulfur batteries that utilize multielectron transfer cathode materials (see Chap. 27). These advanced batteries have promise for significant improvements in energy density, if cycle life and safety issues can be resolved.

Metal/air batteries, including aluminum/air, zinc/air, and lithium/air, offer some of the highest theoretical energy densities available ranging from 1300 to 13,000 Wh/kg. Challenges include the development of high cycle life rechargeable anodes and the development of efficient bifunctional air cathodes. The air cathodes are formally fuel cell electrodes, and perhaps the ultimate high-energy density battery is the "hydrogen/air battery," more commonly known as the hydrogen fuel cell, which has a theoretical specific energy of 32,000 Wh/kg.

29.4 OTHER ENERGY STORAGE TECHNOLOGIES FOR ELECTRIC VEHICLES

Fuel cell systems could conceivably replace batteries entirely in electric vehicles. Utilizing the high specific energy of hydrogen, fuel cell systems offer the promise of longer range. Additionally, fast refueling comparable with that in conventional gasoline-powered vehicles may be possible. In the last decade of the 20th century and the first decade of the 21st century, there has been tremendous progress in the development of PEM (proton exchange membrane) fuel cell power plants that meet the power requirements for automotive propulsion. However, some significant technical and commercial barriers remain, including durability under practical driving conditions, hydrogen storage system weight and size, and the cost of hydrogen fuel. The most serious commercial barrier is the cost of platinum-based PEM fuel cells that utilize noble metal catalysts and expensive proprietary membrane materials. Fuel cell vehicles also utilize high-power batteries to enable them to capture and utilize regenerative braking energy as well as provide for peak acceleration power, which allows utilization of smaller and less expensive fuel cell stacks. Thus, the proper term for the modern fuel cell vehicle is fuel cell hybrid electric vehicle (FCHEV). Though the timing for commercial introduction now appears to be 2020 or beyond, optimism remains for this promising technology within the automotive community.

In principle, other energy storage technologies could be used for electric vehicles, including supercapacitors, mechanical flywheels, and hydraulic accumulators. However, though these technologies have excellent specific power capability, more than sufficient to propel an electric vehicle, their energy density is more limited than that of batteries. Hybridized electrical energy sources have been explored to utilize these devices synergistically to enhance the power capability of high-energy-density batteries with inadequate pulse power capability. A recent example is the study of high energy density lithium-ion batteries coupled with high-power supercapacitors.[24] The supercapacitor responds to the high-rate charge and discharge pulses, reducing heat generation in the high-energy lithium-ion battery to provide for increased life.

The incumbent vehicle technology, internal combustion engines, can also be co-opted to extend the range of electric vehicles through the use of a small generator. Electric vehicle batteries can be directly used for hybrid electric vehicles, particularly plug-in hybrid vehicles or extended-range electric vehicles. However, smaller batteries can generally be used for hybrid vehicle applications with advantages of lower weight, size, and cost. The quantitative requirements depend profoundly on the type and operating mode of the hybrid vehicle, which are addressed in the next section.

29.5 HYBRID ELECTRIC VEHICLES

The hybrid electric vehicle (HEV) is a vehicle in which propulsion energy is available from two or more kinds of power sources or converters, at least one of them being electric.[2–6,25–26] Recent HEV development and commercialization have focused on vehicles powered by an internal combustion engine (ICE) combined with a battery electric drivetrain. The HEV battery serves as an electrochemical "flywheel" that improves the efficiency of operation by allowing the internal combustion engine to operate closer to its optimal efficiency. Additionally, the storage and utilization of braking energy through regenerative braking is even more important for hybrid vehicles than for electric vehicles. In regenerative braking, the HEV electric motor operates in reverse as a generator and charges the battery, providing energy for subsequent accelerations.

Hybrid electric vehicles are more complex than either conventional vehicles powered by internal combustion engines or battery electric vehicles, since they involve the combination of two drivetrains. Additionally, there is a great diversity of HEV designs, arising from engineering choices on the relative size of the drivetrains and how they are combined. For example, the ICE and electric drivetrains can be combined in series or parallel, as shown in Fig. 29.13. In the series configuration, the electric motor drives the wheels, utilizing power from the battery that is recharged by a generator driven by the ICE. The series hybrid vehicle can be thought of as an electric vehicle that uses a gasoline-powered ICE and generator as a range extender. In the parallel configuration, the electric and ICE drivetrains both power the wheels through the transmission connection. In this case, either drivetrain or both can propel the vehicle. Additionally, there are more complex hybrid vehicle designs that combine series and parallel hybrid features using power-split transmissions.

The relative power and energy delivered from the electric and ICE drivetrains can vary by an order of magnitude in hybrid electric vehicles, and this greatly affects the size and performance requirements of the HEV battery. Small batteries with a few kilowatts and a fraction of a kilowatt-hour may be sufficient for micro-hybrid vehicles limited to stop-start functionality. Large battery packs similar in size to EV battery packs may be required for plug-in hybrid vehicles with extended all-electric range.

Regardless of the HEV design, the objective is to improve energy efficiency. The hybrid electric drive to supplement acceleration power enables the use of more efficient engine technologies

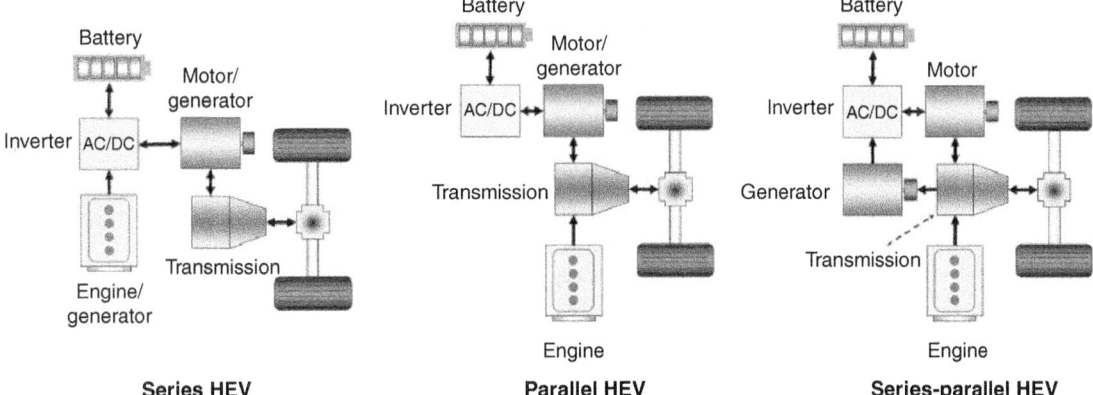

FIGURE 29.13 Schematic diagram for series and parallel hybrid electric vehicle configurations.

such as Atkinson cycle engines in place of the higher-power but less efficient Otto cycle ICE. In both series and parallel architectures, control flexibility of the electric drive can be utilized to operate the internal combustion engine closer to its peak efficiency operating point, which also generally coincides with minimal emissions. The HEV electric power also enables the engine to be idled during deceleration and stop periods.

While hybrid vehicles do consume petroleum-based fuel and do produce emissions of toxic pollutants as well as greenhouse gases, very substantial benefits in both fuel economy improvements and emissions reduction have been achieved in commercially viable vehicles. In contrast to the performance limitations of pure battery electric vehicles, hybrid electric vehicles achieve these benefits with little or no sacrifice in performance or range. Furthermore, hybrid electric vehicles are less expensive than pure battery electric vehicles because they utilize smaller batteries, which are generally the most expensive component in electric drivetrains. The cost premium for hybrid electric vehicles over conventional ICE vehicles can often be recouped in the lower cost of fuel over several years.

29.5.1 Commercialization of Hybrid Electric Vehicles

Hybrid electric vehicles date back to the end of the 19th century. Ferdinand Porche developed a gasoline-electric hybrid vehicle in 1899. Several hybrid electric vehicles were exhibited at the Paris Auto Show in 1906. However, though engineering concepts were developed, hybrid electric vehicles did not attain early commercial success. Internal combustion engine power plants were hybridized to use electric motor-generator sets in some special applications such as submarines and diesel locomotives. It is also true that, in some sense, the conventional automobile can be considered a hybrid electric vehicle since it incorporates electric power in the starter, which is powered by a battery that is recharged by the alternator or generator.

Hybrid electric vehicles remained in the R&D realm of the automotive industry throughout most of the 20th century. Serious development of hybrid vehicles accelerated around 1970 in response to various strategic oil-supply issues. More intensive development was initiated with the Partnership for a New Generation of Vehicles (PNGV) initiated in 1993 by the U.S. Department of Energy (DOE). The PNGV program was a research program created by the Clinton Administration between U.S. OEMs (Chrysler, Ford, and GM) and government agencies with the goal of producing highly fuel-efficient automobiles in 10 years' time. The DOE FreedomCAR program and USABC work closely together on the development of batteries for hybrid electric vehicles and the establishment of performance and cost targets they have mutually adopted.

By 2000, the U.S. OEMs participating in the PNGV program had created three 5-passenger concept vehicles that achieved greater then 72 mi/gal: the GM Precept, the Ford Prodigy, and the

Chrysler ESX-3. Development of these prototype vehicles provided the necessary technology for the subsequent commercialization of hybrid vehicles. At the change of the U.S. presidential administration in 2001, the DOE PNGV program was incorporated into a new program called FreedomCAR, which emphasized development of hydrogen fuel cell vehicles, but continued development of hybrid electric vehicle technology as well.

The first mass-produced hybrid car was the Toyota Prius, which was introduced into the Japanese domestic market in 1997. The Prius was a distinctively styled sedan that utilized an electric drivetrain powered by a small 288 V nickel-metal hydride battery pack consisting of high-power cylindrical cells. The electric motor delivered 21 kW of traction power from a high-power NiMH battery pack with less than 2 kWh of energy. The electric drivetrain was coupled to a high-efficiency 43 kW Atkinson cycle engine through a planetary gear train in a series-parallel architecture. This full hybrid vehicle achieved a fuel economy in excess of 41 mi/gal. It utilized regenerative braking energy to enable electric launch as well as the power-assist function.

A Prius with upgraded technology was successfully introduced in the United States in 2001. It utilized a prismatic, 274 V NiMH battery pack produced by Panasonic EV Energy (PEVE), a joint venture battery manufacturer formed by Toyota and Panasonic. A new model in 2004 achieved improved fuel economy with a yet smaller 202 V NiMH battery. This vehicle generation also utilized a boost power converter to provide 500 V power to the electric drivetrain. The latest Prius generation introduced in 2009 provides improved performance with improved fuel economy of about 50 mi/gal. This updated Prius, with electric drivetrain and battery arrangement, is shown in Fig. 29.14. The Prius is the most successful hybrid vehicle yet commercialized by far, with over 1.5 million units sold globally by 2009. In fact, the Prius has become one of the top-selling vehicles of all types for Toyota. More than 158,000 were sold in the United States in 2008.

The first hybrid electric vehicle sold into the U.S. market was the vividly aerodynamic yet sporty Honda Insight, a 2-seater, 3-door hatchback hybrid vehicle introduced in 1999. This mild-hybrid vehicle utilized a small 144 V nickel-metal hydride battery pack and a less powerful electric drivetrain that delivered 10 kW of traction power. The Insight utilized a parallel drive architecture that coupled the electric drivetrain to the 43 kW ICE drivetrain directly through the crankshaft. This vehicle utilized regenerative braking energy for power-assist during acceleration, but had insufficient electric drive to provide electric launch.

Despite a relatively low degree of electrification, very high fuel economy was achieved utilizing a very efficient 3-cylinder engine in this hybrid configuration. Additionally, Honda utilized a very lightweight design with extensive use of plastic and aluminum parts and highly optimized aerodynamics with a coefficient of drag of only 0.25. In the initial manual transmission version, a fuel economy of 49 mi/gal in the city and 61 mi/gal on the highway was achieved. This vehicle also met stringent ULEV (ultra-low-emission vehicle) emissions standards. Honda also introduced a version of the Insight that utilized a continuously variable transmission (CVT) that achieved comparable fuel economy and met the more stringent SULEV (super-ultra-low-emission vehicle) emissions standards. This vehicle convincingly demonstrated the dual advantages of increased fuel economy and reduced emissions.

Early in the 21st century, the Prius and the Insight were joined in the U.S. market by a variety of other hybrid vehicles as shown in Fig. 29.15. In 2003, Honda introduced the parallel mild-hybrid version of the Honda Civic that ramped to sales volumes of about 30,000 per year. The Ford Escape

FIGURE 29.14 2009 Toyota Prius (left) and layout of hybrid power train components (right) with battery located behind rear seat. (*Courtesy of Toyota Motors.*)

FIGURE 29.15 Hybrid electric vehicles of Honda (Insight, Accord Hybrid, and Civic Hybrid in row 1), General Motors (Saturn VUE Hybrid, Saturn Aura Hybrid, and Cadillac Escalade Hybrid in row 2), and Ford (Ford Escape Hybrid, Mercury Mariner Hybrid, and Ford Fusion Hybrid in row 3).

was introduced as the first hybrid SUV in 2004, using a power-split transmission architecture enabling full hybridization. Ford sold about 20,000 units per year of this vehicle, advertised as the most fuel-efficient SUV in America. Excellent durability has been demonstrated with this robust vehicle that exceeded 200,000 mi in a fleet of New York taxi cabs. Ford built on this success with the introduction of the Ford Fusion Hybrid, which they billed as the most efficient midsize sedan in America at the 2009 North American International Auto Show. GM introduced multiple HEV modules including stop-start and mild-hybrid models such as the Chevy Silverado Truck HEV and the Saturn VUE HEV. Building on success with the Prius, its highest volume HEV, Toyota produced several other full hybrids, as illustrated in Fig. 29.16. Toyota increased its hybrid share of their U.S.

FIGURE 29.16 Hybrid vehicle lineup of Toyota (Prius, Highlander Hybrid, and Camry Hybrid in top row) and Lexus (GS 450h, RX 400h, and LS 600h in bottom row) models.

sales to over 10% in 2008. Overall, 1.3 million HEVs have been sold in the United States from 1999 to 2008, and the market share approached 3% in 2009. The steady market growth and proliferation of models are also illustrated in Fig. 29.17.

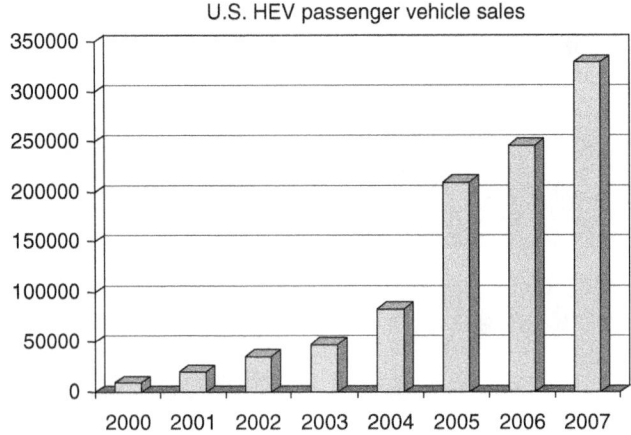

FIGURE 29.17 Growth of HEV passenger vehicle sales 2000–2007. (*Courtesy of Electric Drive Transportation Association.*)

The principal advantage of hybrid electric vehicles is fuel economy. Table 29.6 compares the fuel economy of hybrid vehicles to their comparable conventional ICE models, where available. The city and highway mileage results are taken from the 2008 EPA data based on the UDDS and HWFET driving schedules (see Fig. 29.5). For this mix of strong and mild hybrids, a fuel economy increase of about 50% was observed in city driving where the regenerative braking, power-assist, and engine

TABLE 29.6 City and Highway Fuel Economy Mileage for Hybrid Electric Vehicles in Comparison to Conventional ICE Model (From U.S. EPA 2008 data)

Vehicle model	Hybrid vehicle		ICE vehicle		Percent increase	
	City	Highway	City	Highway	City	Highway
Honda Civic	40	45	26	34	54	32
Nissan Altima	35	33	23	32	52	3
Ford Escape FWD	34	30	22	28	55	7
Mazda Tribute 2WD	34	30	22	28	55	7
Mercury Mariner FWD	34	30	22	28	55	7
Toyota Camry	33	34	21	31	57	10
Ford Escape 4WD	29	27	19	24	53	13
Mazda Tribute 4WD	29	27	19	24	53	13
Mercury Mariner 4WD	29	27	19	24	53	13
Toyota Highlander 4WD	27	25	17	23	59	9
Saturn Vue	25	32	19	26	32	23
Chevrolet Malibu	24	32	22	30	9	7
Saturn Aura	24	32	22	30	9	7
Chevrolet Tahoe 2WD	21	22	14	20	50	10
GMC Yukon 1500 2WD	21	22	14	20	50	10
Chevrolet Tahoe 4WD	20	20	14	19	43	5
GMC Yukon 1500 4WD	20	20	14	19	43	5

FIGURE 29.18 Selected recently developed HEVs, including new 2010 Honda Insight, Mercedes-Benz S400 HEV, and Hymotion CS Prius PHEV conversion.

idle during deceleration and stop can be highly utilized to provide maximum fuel economy benefits. A more modest fuel economy saving of about 10% is achieved on average in highway driving, which involves more constant-speed driving. As a consequence, the city fuel economy generally was comparable to the highway fuel economy for HEV models. HEVs feature smaller gas tanks and a significantly extended urban range in comparison to conventional ICE vehicles.

The introduction of new and improved HEV models continues, as illustrated in Fig. 29.18. In addition to the 2009 Ford Fusion strong HEV, new developments include the all-new 2010 Honda Insight mild hybrid, priced below the Toyota Prius with a fuel economy of over 40 mi/gal. Introduced first into Japan, this 5-passenger, 5-door compact hatchback was the first hybrid vehicle to top the automotive sales charts for all classes of vehicles in April 2009. In 2010, Toyota offers its latest generation of the Prius with improved performance and superlative fuel economy at 51 and 48 mi/gal in city and highway driving, respectively. The model year 2010 Mercedes-Benz S400 HEV luxury car is the first mass-production HEV to use lithium-ion batteries. This mild-hybrid vehicle, first introduced in Europe in 2009, utilizes a 126 V lithium-ion battery pack for stop-start and power-assist functions, improving the fuel economy 20% to 21 mi/gal. Another development is the introduction of small volumes of plug-in HEVs in the form of conversions of power-assist HEVS, particularly the high-volume Prius. In addition to the Hymotion conversion shown in Fig. 29.18, Energy CS and Hybrids-Plus have offered small numbers of Prius HEVs where the small NiMH battery pack has been replaced with high-energy Li-ion battery packs. Toyota also displayed a plug-in hybrid version of the Toyota Prius in the 2009 North America International Auto Show.

In 2007, GM announced their intention to develop a high-performance plug-in HEV known as the Chevy Volt on an aggressive schedule for production launch in late 2010. This vehicle, illustrated in Fig. 29.19, utilizes a 16 kWh lithium-ion battery pack to enable an all-electric range of 40 mi. GM describes the series HEV design as an extended-range electric vehicle (E-REV) and indeed it is expected to operate mostly as an electric vehicle since almost all trips in the United States are within the 40 mi range. GM has raised the awareness of plug-in hybrid vehicles and this opportunity to revolutionize the automotive industry. Other automotive companies, including Toyota, Ford, and BYD, have announced plans to introduce plug-in hybrid vehicles in the 2010–2012 time frame. Smaller battery packs are more frequently used, but this lower-cost approach results in blended ICE operation with electric drive operation during most trips. The proposed timing of future PHEV/EV vehicles aligns well with the implementation schedule for CARB mandates and credits.

FIGURE 29.19 GM Chevy Volt range-extended electric vehicle at the 2008 North American International Auto Show (left), together with 16 kWh battery pack (middle), and as pictured in the 2010 Chevy Volt website (right). (*Courtesy of GM.*)

29.6 TYPES OF HYBRID ELECTRIC VEHICLES

Battery requirements vary substantially across the wide variety of types of hybrid electric vehicles under development, including micro hybrids, mild hybrids, strong or full hybrids, and plug-in hybrids.[27-30] While the definitions of HEV types are not entirely consistent, a general consensus is evolving. Particularly from the point of view of understanding battery requirements, it is useful to classify hybrid electric vehicles according to their performance features and degree of electrification, as shown in Fig. 29.20. In the most minimal hybrid design, the stop-start micro HEV, the internal combustion engine is turned off during idle and/or deceleration. In the mild-hybrid electric vehicle, the electric motor additionally provides assist during acceleration. In the strong-hybrid or full HEV, engine-off electric drive operates at least in some speed ranges. In the plug-in HEV (PHEV), a significant all-electric range is added with the capability to recharge the battery from the utility grid. In all cases, regenerative braking energy can be captured to power the various electric performance features. All designs provide for improved fuel economy, and the benefit generally increases with degree of electrification. The size of the batteries also increases with degree of electrification as well as the overall system's complexity and HEV cost.

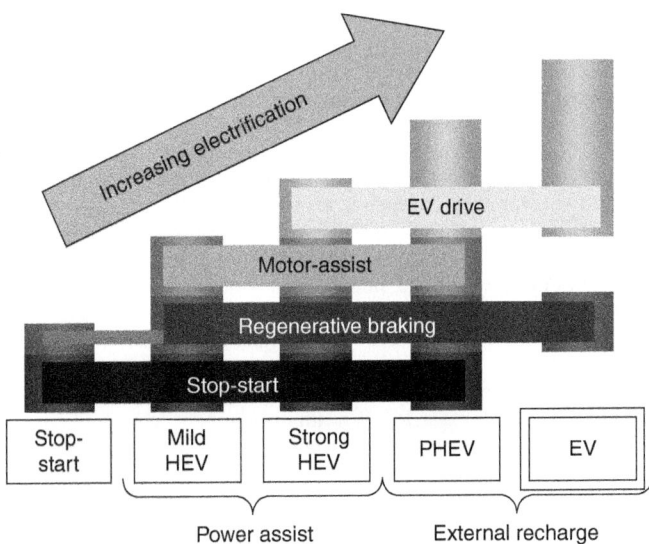

FIGURE 29.20 Types of hybrid electric vehicles based on performance features.

29.6.1 Stop-Start (Micro) Hybrid Electric Vehicles

Stop-start HEVs provide for improved vehicle efficiency by turning off the internal combustion engine at idle.[27-30] During the engine-off time, the vehicle's electrical system is powered by the vehicle starter-lighting-ignition (SLI) battery, which is typically enhanced for this hybrid application. At idle, the battery must provide power for auxiliary loads, including, most significantly, the heating/ventilation/air conditioning (HVAC) system. When the driver's throttle command prompts a shift into drive, the HEV's starter/generator is able to restart the internal combustion engine to provide the required traction power. Extended idle periods can also trigger engine starts, particularly during HVAC operation, which can draw several kilowatts of power.

Stop-start HEV systems aim to provide seamless engine to motor transitions and provide their greatest benefits in urban driving with its frequent stops. Stop-start hybrids, which are also referred to as micro hybrids, feature a small and relatively inexpensive hybrid electric drivetrain, but offer a minimal fuel economy improvement of ~5% depending on design and driving conditions. Although the fuel economy benefit is not large, stop-start micro hybrids offer the least expensive hybridization strategy. Micro HEVs are being pursued particularly in the European market where high fuel efficiency diesel engines are predominant. Examples include the GMC Silverado and Sierra parallel hybrid trucks of 2005–2006. The Mazda3 compact car offers the stop-start option on cars sold in Japan. However, the opportunity appears greatest in Europe, with companies such as BMW planning to offer this feature on many models.

The simplest approach to the micro hybrid utilizes a 14 V electrical system (alternator voltage) with an enhanced 12 V lead-acid battery. Similar to the conventional SLI operation, the battery is charged whenever the engine is operating so it is maintained at a high state-of-charge. Although this limits the opportunity to accept regenerative braking energy, it avoids premature sulfation failure of lead-acid batteries in partial state-of-charge operation.

The power requirement for stop-start or micro HEVs is defined by the cold-start requirement, which is typically in the range of 6 to 8 kW at –30°C. SLI batteries are generally sized to meet the cold-start requirement and provide an operational life of at least 3 to 5 years. The micro-hybrid application requires an enhanced battery because both the number of charge-discharge cycles and the capacity throughput are increased by an order of magnitude in comparison to the operation of an SLI battery in conventional vehicles. The multiple engine starts per mile of operation expected will require more than 100,000 shallow cycles in 4 to 5 years. Operation during idle periods has been estimated to average from 1 to 5 Wh to as much as 50 Wh with HVAC operation. Thus, the total energy throughput requirement is on the order of 1 MWh or more.

Given the limitation of conventional flooded lead-acid SLI batteries to provide an accumulated Ah turnover of about 150 times, the nominal capacity in shallow discharge cycles, they are not expected to provide even 10,000 miles of operation in this more challenging application. Thus, larger valve-regulated lead-acid (VRLA) batteries which can provide a major improvement in life are being introduced in these applications, particularly in Europe and Japan.

Micro hybrids have also been developed using 42 V electrical systems (42 V charge voltages with batteries having a nominal voltage of around 36 V). These systems have higher capabilities approaching those of the mild hybrid and may accommodate some acceptance of regenerative braking energy. Additionally, an objective of these more sophisticated and more expensive systems is to provide for a vehicle lifetime durability of 150,000 mi over 15 years. Thus, battery performance and durability requirements are more challenging. In Table 29.7, detailed requirements for this application as developed by DOE under the FreedomCAR are summarized.[31]

TABLE 29.7 FreedomCAR/USABC Goals for Batteries for 42 V Stop-Start HEVs

Parameter (for fully burdened system)	Condition	Units	Goal
Discharge pulse power	2 s @ >27 V	kW	6
Regenerative pulse power	Not required	kW	N/A
Cold cranking power	3 pulses of 2 s >21 V at –30°C	kW	8
Engine-off accessory load	5 min	kW	3
Available energy	@ 3 kW discharge >27 V	Wh	250
Recharge rate		kW	2.4
Energy efficiency	Zero pwr asst load profile	%	90
Cycle life	Zero pwr asst load profile	starts	450,000
Mileage	Vehicle application	miles	150,000
Calendar life	Vehicle application	years	15
Maximum system weight		kg	10
Maximum system volume		L	9
System selling price	@ 100,000 production volume	$	150
Maximum open-circuit voltage	after 1 s	V	48
Self-discharge		Wh/day	<20
Operating temperature range		°C	–30 to +52
Survival temperature range		°C	–46 to +66

29.30 SPECIALIZED BATTERY SYSTEMS

Superficially, the power and energy requirements for the 42 V stop-start HEV are no more challenging than for the conventional SLI application. The specific power and specific energy requirements of 600 W/kg and 25 Wh/kg, respectively, can be met with lead-acid battery technology. Additionally, because of the minimal power and energy requirements, passive thermal management can be utilized. In comparison to conventional SLI battery operation, a more elaborate battery management and monitoring system is required to provide top-off cycles and state of health monitoring, but the control electronics are simple in comparison to other hybrid vehicle types. It is the life requirement of several hundred thousand cycles and several megawatt-hours of total energy throughput that make this application very difficult for lead-acid batteries. This requires the use of advanced batteries, making it difficult to meet full system cost targets at $25/kW on a cost per unit power or $600/kWh on a cost per unit energy basis.

29.6.2 Power-Assist Hybrid Electric Vehicles

In power-assist hybrid electric vehicles, supplementary traction power is provided by the on-board electric power train. This motor-assist is accomplished by using battery energy to power the HEV's electric motor and enabling it to contribute power to the driveshaft. Many different hybrid vehicle designs are under development and vary in their integration into the existing mechanical drivetrain, such as through simple clutches or more complicated power-split devices. The extent of electric power available for motor-assist varies. At the low-power end are mild hybrids, which provide motor-assist but no electric launch or all-electric propulsion. More substantial power-assist HEVs feature electric launch and engine-off electric drive at least at low speeds. These higher power HEVs with electric drive are called full hybrids or strong hybrids and have also been termed dual-mode hybrids since they feature the electric drive mode. All power-assist HEVS, both mild and strong hybrids, utilize regenerative braking.

Table 29.8 summarizes FreedomCAR energy storage system performance goals for power-assist hybrid vehicles.[32] The most important criteria for HEV applications is power, both pulse-discharge power for acceleration and pulse-charge power for regenerative braking. The discharge and regen

TABLE 29.8 FreedomCAR/USABC Goals for Batteries for Power-Assist HEVs

Parameter (for fully burdened system)	Condition	Units	Power-assist miniumum goal	Power-assist maximum goal
Pulse discharge power	10 s pulse	kW	25	40
Peak regenerative pulse power	10 s pulse	kW	20	35
Cold cranking power	3 pulses of 2 s at –30°C	kW	5	7
Total available energy	1C discharge over DOD range where power goals are met	kWh	0.3	0.5
Round-trip energy efficiency	25 Wh cycle for minimum goal 50 Wh cycle for maximum goal	%	90	90
HEV cycle life	25 Wh cycle for minimum goal 50 Wh cycle for maximum goal	cycles	300,000	300,000
HEV cycle life energy throughput		MWh	7.5	15
Calendar life		Years	15	15
Maximum weight		kg	40	60
Maximum volume		L	32	45
Operating voltage limits		V	max ≤ 400, min ≥ 0.55 × Vmax	max ≤ 400, min ≥ 0.55 × Vmax
Self-discharge	Maximum allowable rate	Wh/day	50	50
Temperature range	Equipment operation	°C	–30 to +52	–30 to +52
Temperature range	Equipment survival	°C	–46 to +66	–46 to +66
System selling price	@ 100,000 production volume	$	500	800

power must also be delivered at voltages within the operating range of the power electronics for the electric motor. So, the minimum voltage on discharge must be no lower than 55% of the maximum voltage on charge. (This is roughly equivalent to the USABC practice of specifying the peak discharge power for an EV at two-thirds of the open-circuit voltage.) The discharge and regen pulse-power performance are specified for 10 s pulses. FreedomCAR provides goals for two representative cases. The minimum power-assist HEV targets call for 25 kW for discharge and 20 kW for regen charge. The maximum power-assist HEV targets call for 40 kW discharge and 35 kW regen charge. With the battery system weight targets of 40 and 60 kg and volume targets of 32 and 45 L for the minimum and maximum power-assist HEV cases, respectively, this represents over 600 W/kg and around 800 W/L for the specific power and power density requirements at a battery-pack system level.

The power-assist HEV application target for energy is specified as the total available energy over the state-of-charge range at which the battery system will provide the specified pulse-discharge power and pulse-regen charge power. This available energy, also termed usable energy, is specified as energy that can be delivered on the C-rate discharge from the highest state-of-charge where the regen-power goals are met to the lowest state-of-charge where the discharge power goals are met. This is illustrated in Fig. 29.21. The power and regen performance are determined from the Hybrid Pulse Power Characterization (HPPC) test, which utilizes constant-current pulses, 10 s in duration at 10% depth of discharge increments in a C-rate discharge. The available energy is the discharge energy corresponding to the state-of-charge region where the discharge-power and regen-power targets are simultaneously met.

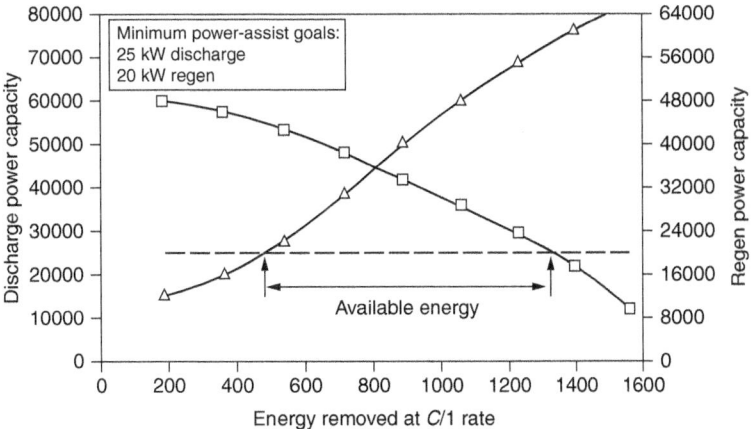

FIGURE 29.21 Illustration of USABC available energy determined from pulse power and regen performance as a function of state-of-charge. (*Courtesy of USABC.*)

The 90% round-trip energy efficiency target is based on a simple series of discharge and charge pulses with peak discharge pulses at 15 and 24 kW and peak charge pulses at 12 and 21 kW for the minimum and maximum power-assist HEV cases, respectively. The energy-efficiency profile for the maximum power-assist HEV is provided in Fig. 29.22. This profile delivers 50 Wh of discharge energy per cycle. The similar but lower-power profile for the minimum power-assist HEV delivers 25 Wh of discharge energy per cycle.

The 25 Wh and 50 Wh energy-efficiency cycles are also utilized for HEV cycle life testing of the minimum and maximum power-assist HEVs, respectively. Battery pack systems for the maximum and minimum power-assist HEVs must be capable of delivering 300,000 HEV cycles. This is equivalent to an energy throughput of 7.5 and 15 MWh for the minimum and maximum power-assist HEVs, respectively. These goals are designed to provide the capability for the battery to last the life of a car or 150,000 mi and 15 years. A calendar life of 15 years is also separately targeted, since energy throughput and time can stress the battery differently for some failure modes.

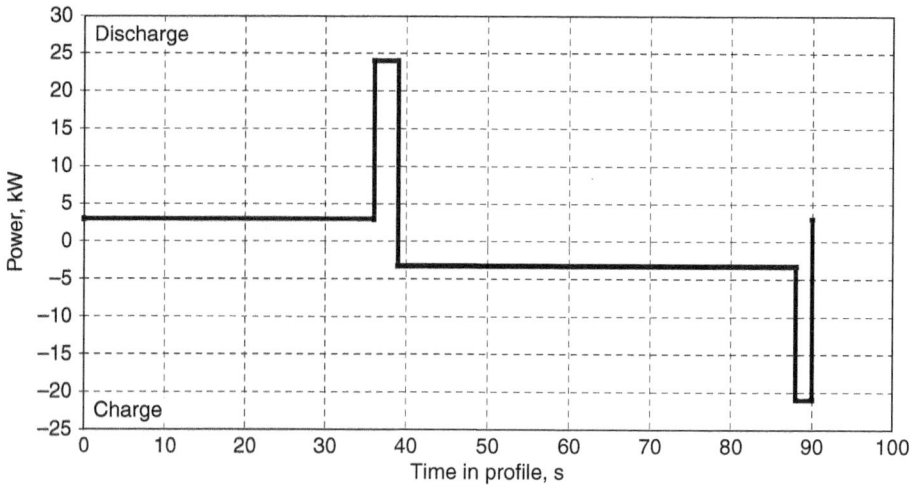

FIGURE 29.22 Energy-efficiency and baseline cycle-life test profile for power-assist HEV. (*Courtesy of USABC.*)

A key point for FreedomCAR/USABC life targets for hybrid vehicles is that end of life is defined as failure to meet either the power or energy performance goals. This differs from the USABC EV life targets, in which end of life was defined as the point where the energy or power performance dropped to less than 80% of the performance goals. Thus, to meet the USABC HEV life targets, HEV batteries need to be designed with excess power and energy, providing a buffer so that performance targets are still met at the end of life. This is illustrated in Fig. 29.23.

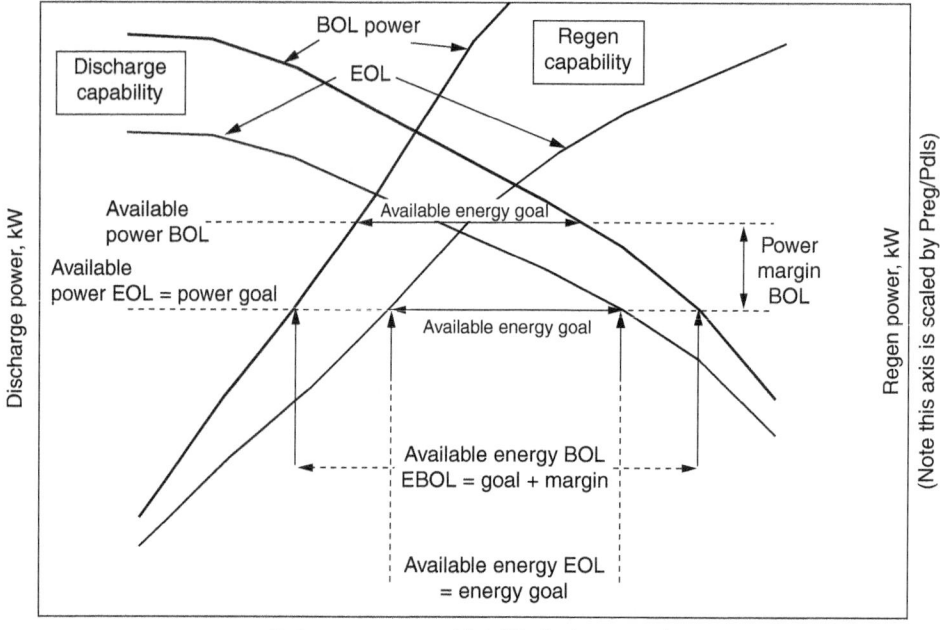

FIGURE 29.23 Illustration of available energy and power margins over battery system life. BOL = Beginning of Life, EOL = End of Life. (*Courtesy of USABC.*)

29.6.3 Strong HEVs

Strong-hybrid vehicles, or full HEVs as they are sometimes called, are a subset of power-assist hybrid vehicles which are capable not only of motor-assist but also all-electric propulsion that may include electric launch. The torque and power capability of the electric drivetrain is comparable to that of the internal combustion engine. The 25–50 kW power capability is provided at voltages of over 200 V to maintain electrical currents at practical levels. These vehicles require very high specific power batteries, approaching 1000 W/kg or more, with capability to accept high-power regen charge pulses as well as deliver high-power discharge pulses. The FreedomCAR goals for power-assist HEVs are aimed principally at strong HEVs, especially the maximum power-assist HEV case. Strong HEVs utilize petroleum-based fuels in internal combustion engines as the source of energy, but use batteries to improve the efficiency of operation by maximizing the operation of the internal combustion engine at its maximum efficiency operational conditions, which also provides for reduced emissions of noxious pollutants and greenhouse gases. Thus, the strong or full hybrid provides for the high fuel-economy improvements of 50% or more in city driving and large emission reduction opportunities. Good examples of the technical and engineering success of this approach are the Toyota Prius and the Ford Escape Hybrid.

29.6.4 Mild HEVs

Mild-hybrid vehicles are the lower-power subset of power-assist hybrid vehicles that are capable of motor-assist, but not all-electric propulsion. This design approach addresses the cost challenges of the full hybrid and provides an alternative lower-cost design that still provides substantial fuel economy and emissions reduction advantages. The torque and power capability of the electric drivetrain is significantly less than that of the internal combustion engine. The 10 to 20 kW power capability is provided at a lower voltage that still may exceed 100 V, but some mild-hybrid systems have been designed with 42 V systems. These vehicles still require very high specific power batteries, approaching 1000 W/kg or more, but smaller batteries are used, which lowers the system cost. Examples of mild-hybrid vehicles produced commercially include the Honda Civic Hybrid, which utilized a 144 V NiMH battery with a 10 kW power capability, and the Saturn Vue Greenline mild hybrid, which utilized a 42 V NiMH battery pack with a power capability of less than 10 kW. Battery packs for mild-hybrid vehicles thus span a rather wide range from moderate power performance, similar to that in the minimum power-assist HEV given in Table 29.8 down to 42 V systems similar to those with the stop-go hybrid, but with regen capability. Table 29.9 summarizes FreedomCAR performance targets for 42 V systems for mild-hybrid applications.[31]

29.6.5 Plug-In Hybrids

Plug-in hybrid vehicles have the potential to provide most of the benefits of electric vehicles at a lower cost and without the key performance disadvantages in terms of range and refueling time. The very substantial fraction of ZEV operation provides high fuel economy through the efficient use of grid electric power and reduced use of petroleum-based fuel. This vehicle type, which was not a target of the original PNGV program, was developed and promoted in the 1990s by university vehicle competition teams.[33-34] More recently, it has been championed by Plug-In America and other groups for its potential to strategically reduce U.S. dependence on foreign oil.

The key feature of plug-in hybrid vehicles (PHEVs) is the ability to recharge the battery by plugging into the electrical grid. These vehicles utilize both grid electricity and on-board petroleum-based fuel as energy sources, with the electrical energy fraction depending on the size of the battery and the operating strategy. The usual operating strategy is to first utilize HEV charge-depletion operation after recharge of the battery pack from the grid connection. This consists of all-electric operation with the electric drivetrain providing full-traction power unless the ICE drivetrain is needed to meet power demands. Charge-depletion operation is continued until the battery pack is substantially depleted, at which point charge-sustaining HEV operation

TABLE 29.9 FreedomCAR/USABC Goals for Batteries for 42 V Mild HEVs

Parameter (for fully burdened system)	Condition	Units	42 V M-HEV goal	42 V P-HEV goal
Discharge pulse power	2 s @ >27 V for M-HEV 10 s @ >27 V for P-HEV	kW	13	18
Regenerative pulse power	2 s	kW	8	18
Cold-cranking power	3 pulses of 2 s >21 V at –30°C	kW	8	8
Engine-off accessory load	5 min	kW	3	3
Available energy	@ 3 kW discharge	Wh	300	700
Recharge rate		kW	2.6	4.5
Energy efficiency	Partial/full pwr asst for M-HEV/P-HEV	%	90	90
Cycle life	Partial/full pwr asst for M-HEV/P-HEV	profiles	450,000	450,000
Life	Vehicle application	mi	150,000	150,000
Calendar life	Vehicle application	years	15	15
Maximum system weight		kg	25	35
Maximum system volume		L	20	28
System selling price	@ 100,000 production volume	$	260	360
Maximum open-circuit voltage	after 1 s	V	48	48
Self-discharge		Wh/day	<20	<20
Operating temperature range		°C	–30 to +52	–30 to +52
Survival temperature range		°C	–46 to +66	–46 to +66

is commenced. In charge-sustaining operation, the state-of-charge is maintained at a target level by balancing the discharge of the battery with charge from regenerative braking as in the power-assist hybrid. Charge-sustaining operation is essentially the same as the operational mode for the power-assist hybrid vehicle, but typically at a lower state-of-charge. The PHEV operational modes are illustrated by the plot of state-of-charge in Fig. 29.24.

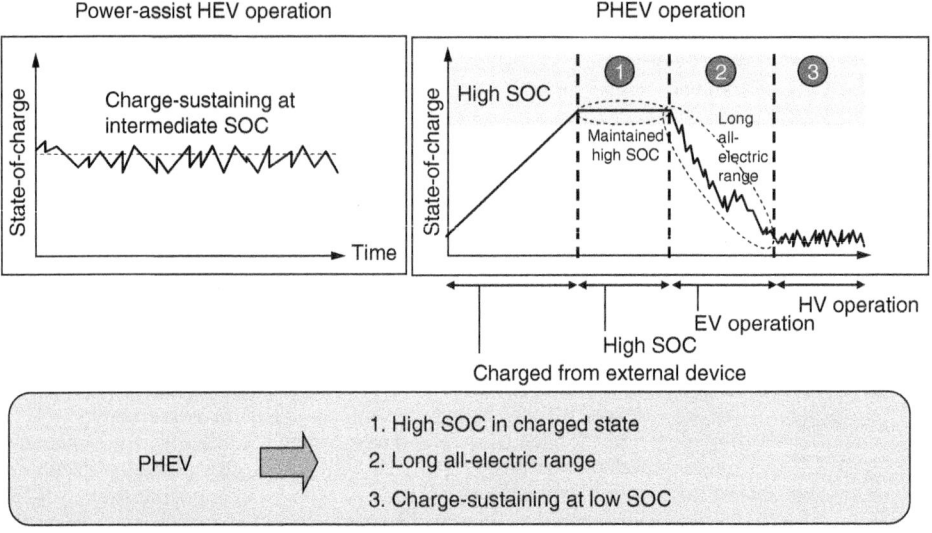

FIGURE 29.24 Plug-in hybrid electric vehicle battery state-of-charge during operational modes, including charge, standby, charge-depletion operation, and charge-sustaining operation.

PHEVs mix features of strong hybrids and electric vehicles. They can be thought of as strong hybrids with extended EV drive capability. Alternatively, they can be thought of as EVs with extended range from an ICE-driven generator. An important variable in PHEV design is the degree of electrification. Generally, the electric drivetrain of the PHEV is designed to provide full-acceleration power capability without use of the ICE drivetrain during some significant all-electric range. The all-electric range of the plug-in hybrid is an important design criterion. Most development has focused on designs with 10 to 40 mi range. The larger the all-electric range, the larger the benefit in terms of improved fuel economy and reduced emissions. However, this requires larger batteries, which increase the weight and cost of the vehicle substantially.

Another key aspect of PHEV design is the control strategy. With large battery packs providing substantial all-electric range, the electric drivetrain may be designed to provide all traction power during the charge-depletion phase of operation. In this way, the PHEV may operate totally as a battery electric vehicle unless the all-electric range is exceeded. These PHEVs may be properly described as extended-range electric vehicles. There are advantages to utilizing series HEV architectures with E-REVs such as the Chevy Volt. Alternatively, to avoid the cost of large battery packs and full-power electric drivetrains, some PHEVs are being designed with blended-mode charge-depletion operation. In this case, the ICE drivetrain will turn on at certain power-demand or vehicle speed thresholds prior to reaching charge-sustaining operation. These PHEVs typically utilize a parallel HEV architecture. For comparison purposes, measurements on PHEVs with blended electric and ICE operation in the charge-depletion mode are still analyzed to provide an equivalent electric range, the range provided by the battery energy by subtracting out the range provided by the ICE drivetrain in the charge-depletion mode.

Another aspect to PHEV control strategy is the state-of-charge set points for full charge and for charge-sustaining operation. To maximize all-electric range for the greatest energy efficiency and emissions benefits, it is desirable to fully charge the battery pack and drive the battery discharge almost to completion in the charge-depletion mode. However, fully charging a lithium-ion PHEV battery pack accelerates failure modes and may expose the battery pack to overcharge abuse during regen in subsequent driving. The life of the battery pack is enhanced and control is also simplified by limiting charge to about 80% of the rated capacity. Similarly, the efficiency in charge-sustaining mode is reduced if the battery pack state-of-charge is too low. Additionally, there is risk of overdischarge under high-power demand conditions. So, the HEV efficiency and control is simplified by limiting discharge to operate in charge-sustaining mode at about 30% state-of-charge. This has the effect of reducing the usable energy to about half of the total rated discharge energy. As a result, a vehicle with an energy consumption of about 200 Wh/mi would require a 16 kWh battery pack to achieve a 40 mi range.

Table 29.10 summarizes the FreedomCAR energy storage system performance goals for plug-in hybrid vehicles.[35] Two sets of goals are listed: high-power battery goals and high-energy battery goals. The FreedomCar test PHEV battery test manual identifies the intended vehicle platform for the high-power battery to be a 4400 lb sport-utility vehicle with an equivalent electric range of 10 mi. The intended platform for the high-energy battery is a 3300 lb midsize car with an equivalent electric range of 40 mi. The biggest difference between these goals and those for the power-assist HEV is the requirement of a large charge-depletion available energy of 3.4 and 11.6 kWh for the high-power and high-energy goals, respectively. This is an order of magnitude larger than the available energy goals for charge-depletion operation. The goals for charge sustaining operation are similar to those for the power-assist HEV. In addition to goals for charge-sustaining cycle life (similar to the power-assist goals), challenging goals are added for the charge-depletion cycle life. However, the cost targets amounting to $500/kWh and $293/kWh for the high-power and high-energy PHEV, respectively, are challenging but more achievable than the EV cost goal of $100/kWh. Thus, it is considered that the PHEV is closer to commercial reality than the pure EV.

29.7 COMPARISON OF HEV BATTERY PERFORMANCE REQUIREMENTS

The HEV specific performance targets derived from the USABC goals in Tables 29.7, 29.8, 29.9, and 29.10 are compared to the specific performance targets for EV batteries in Table 29.11. The EV specific performance targets are from Table 29.2. The USABC and FreedomCAR requirements were

TABLE 29.10 FreedomCAR/USABC Goals for Batteries for Plug-In HEVs

Parameter (for fully burdened system)	Condition	Units	High-power battery goal	High-energy battery goal
Reference equivalent electric range		mi	10	40
Peak pulse discharge power	10 s pulse	kW	45	38
Peak regen pulse power	10 s pulse	kW	30	25
Cold-cranking power at –30°C, 2 s –3 pulses	3 pulses of 2 s at –30°C	kW	7	7
Available energy for charge depleting mode	10 kW rate	kWh	3.4	11.6
Available energy for charge sustaining mode	See power-assist HEV available energy	kWh	0.5	0.3
Minimum round-trip energy efficiency	See power-assist HEV energy efficiency	%	90	90
Charge depleting cycle life	10 kW rate	cycles	5000	5000
Charge depleting energy throughput	10 kW rate	MWh	17	58
Charge sustaining HEV cycle life	50 Wh HEV cycle	cycles	300,000	300,000
Charge sustaining HEV energy throughput	50 Wh HEV cycle	MWh	15	15
Calendar life	40°C temperature	year	15	15
Maximum system weight		kg	60	120
Maximum system volume		L	40	80
Maximum operating voltage		V	400	400
Minimum operating voltage		V	>0.55 × Vmax	>0.55 × Vmax
Maximum self-discharge		Wh/day	50	50
System recharge rate	30°C temperature	kW	1.4	1.4
Operating temperature range:		°C	–30 to +52	–30 to +52
Equipment survival temperature range		°C	–46 to +66	–46 to +66
Maximum current (10 s pulse)	10 s pulse	A	300	300
System selling price	@ 100,000 production volume	$	$1700	$3400

developed at different points in time and the assumptions were not constant. However, it is still useful to compare these requirements and draw some general conclusions. Specific performance criteria listed for the hybrid electric vehicle applications in order of increasing energy were specific energy, specific power, power-to-energy ratio (units of W/Wh = 1/h), and cost in $/kWh and $/kW.

The specific energy targets for stop-start, mild hybrids, and power-assist hybrids are all quite modest and within the capability of lead-acid batteries. The specific energy requirements for plug-in hybrids are more challenging, but could be met by lithium-ion batteries. The power capability required by most HEV batteries is around 600 W/kg, well within the capability of high-power nickel-metal hydride and lithium-ion batteries. The high-energy PHEV and EVs application require moderately less power. High-power nickel-metal hydride and lithium-ion batteries have been developed that can easily meet these power targets.

TABLE 29.11 Comparison of Specific Performance Targets for HEVs and EVs

	Specific energy (Wh/kg)	Specific power (W/kg)	P/E ratio	Energy cost ($/kWh)	Power cost ($/kW)
Stop-start 42 V micro hybrid	25	600	24	$600	$25
42 V mild hybrid (lower power)	12	520	43	$867	$20
42 V mild hybrid (higher power)	20	514	26	$514	$20
Power-assist (lower power)	8	625	83	$1667	$20
Power-assist (higher power)	8	667	80	$1600	$20
Plug-in hybrid (higher power)	57	750	13	$500	$38
Plug-in hybrid (higher energy)	97	317	3	$293	$89
Electric vehicle	200	400	2	$100	$50

The key difference between the specific performance requirements for HEV and EV applications is the power-to-energy ratio. Hybrid vehicles need high power, but have modest energy requirements, except for PHEVs. Consequently, their specific energy requirements are lower. Electric vehicles need high energy, resulting in larger batteries. While the power requirement is also higher, the specific power requirement is lower because the weight is higher and the ratio of power to weight is lower. As a result, the power-to-energy ratio is more than an order of magnitude higher for HEV applications except for PHEV applications, which more closely resemble EVs in their requirements.

Batteries can be designed for high power or high energy. High-energy designs tend to feature a smaller number of thicker electrodes and have a higher fraction of active materials. High-power batteries tend to feature a larger number of thinner electrodes and utilize a higher fraction of conductive components at the expense of the active material fraction. An example can be provided in a comparison of nickel-metal hydride batteries for HEV and EV applications. High-energy NiMH EV batteries have been developed with a specific energy of 80 Wh/kg. They have a specific power of about 200 W/kg. High-power NiMH HEV batteries have been developed with a specific power of 1300 W/kg. However, this was achieved through design trade-offs that resulted in a specific energy of 45 Wh/kg in the HEV battery.

The cost requirements and criteria are different for HEV and EV applications. The cost target for EV batteries is $100/kWh. In HEV batteries, cost per unit power is the key variable. For power-assist HEVs, the cost target is $20/kW. On a cost per unit energy basis, the power-assist HEV cost target is about $1600/kWh, an order of magnitude higher than the EV battery target. Even taking into account the higher cost of the high-power HEV designs, this cost target is much more achievable. It is the principal reason why HEVs are commercially available and EVs are not yet, except for initial trials. One somewhat oversimplified way to look at this is that it is easier for HEV batteries to compete with engines on the cost of providing power than it is for EV batteries to compete with gasoline tanks on the cost of providing energy.

The EV battery safety requirements for tolerance to electrical and mechanical abuse on a system level are also inherent requirements for all types of HEV battery packs. Safety requirements and abuse test procedures have been updated for hybrid electric vehicle batteries.[36] Additionally, USABC has developed a battery hazard mode and risk mitigation analysis.[37] This is based on following a similar methodology to the failure modes and effects analysis (FMEA) commonly used in the international automotive community. This disciplined systems approach to battery pack safety is important in the design of battery packs, their control systems, and their integration into vehicles.

29.8 VEHICLE INTEGRATION OF HEV BATTERIES

Vehicle integration issues for hybrid electric vehicles are similar to those for electric vehicles. While the power, energy, and size of HEV battery packs are highly dependent on the type and design of the hybrid vehicles. Some generalizations are useful. For HEVs, a smaller and lighter battery pack is utilized, but due to the size and weight of the ICE drivetrain in hybrid vehicles, the size and weight constraints for the battery pack are equally challenging. Particularly in the development of hybrid versions of conventional ICE vehicles, it is often challenging to find a suitable location for the battery pack that can simultaneously meet requirements of facile integration with the drivetrain, avoidance of crash safety issues, and environmental issues, including exposure to water or extreme temperatures.

Thermal management issues are more moderate, especially for HEVs with lower degrees of electrification. Stop-start micro hybrids typically utilizing lead-acid batteries are often designed with only passive or natural convection to provide heat rejection. Mild-hybrid vehicles powered by nickel-metal hydride batteries generally use forced air for thermal management, often utilizing cabin air that is preconditioned for passenger comfort. Strong-hybrid vehicles with larger battery packs typically require higher heat rejection capability with more attention to temperature uniformity in the battery pack through detailed thermal modeling and design. The Ford Escape Hybrid successfully utilized dedicated HVAC systems to control the input temperature in the forced air thermal management system, providing excellent temperature uniformity within conservative temperature limits to substantially improve the NiMH HEV battery life.

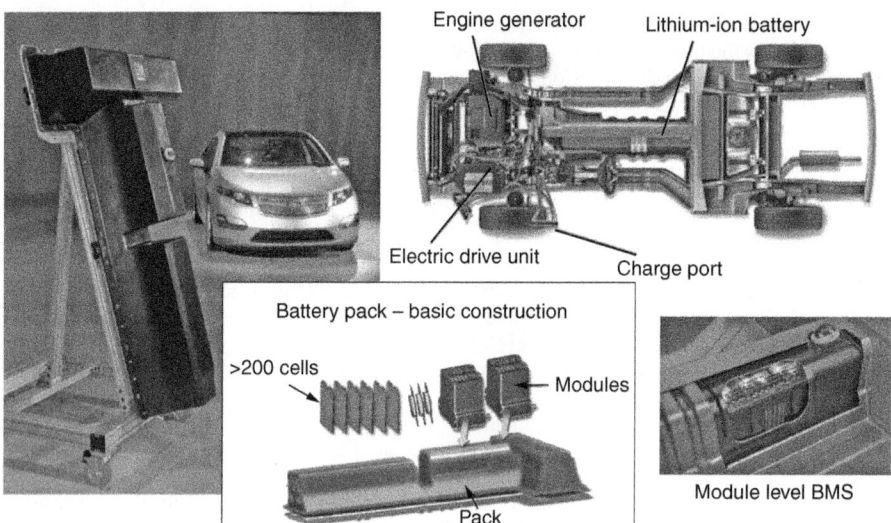

FIGURE 29.25 Lithium-ion battery pack for Chevy Volt E-REV. (*Courtesy of GM.*)

The largest HEV battery packs, those utilized for PHEV applications, are also the most challenging from a thermal management perspective. In size, energy capacity, and heat rejection requirements, PHEV battery packs are similar to EV battery packs. It can be difficult to keep these large battery packs within the recommended operating range and even more difficult to keep the battery cells at uniform temperatures. Particularly for battery pack designs lacking in symmetry, liquid cooling schemes offer significant advantages. This approach is being pursued for the T-shaped battery pack for the Chevy Volt PHEV, shown in Fig. 29.25.

Electronic controls for HEV batteries depend on the type of HEV, but are generally more complex than controls for EV applications. Battery module voltages are monitored at the battery pack, module, and cell level. In comparison to EV applications, it is even more important to keep the battery within proper operational voltage limits because of the higher discharge and regen rates involved. The charge-discharge rates to which power-assist HEV batteries are typically subjected are an order of magnitude higher than those to which EV batteries are subjected in their respective applications. Thus, the degree of electrical abuse to which an HEV battery could be subjected is about an order of magnitude greater. This is particularly true for high-energy-density lithium-ion batteries which are rather intolerant of overcharge abuse. In this case, Li-ion HEV batteries must be integrated with cell-level monitoring and balancing circuits. Typically, the electronic controls may reside on the battery module level as well as the battery pack level, as shown in Fig. 29.25 for the Chevy Volt Li-ion pack.

Additionally, efficient HEV charge-sustaining operation relies on control of the regen charging to maintain the battery pack at a target state-of-charge with capability for high-power discharge and regen. With NiMH batteries, the power and regen capability are not highly dependent on state-of-charge, resulting in a relatively large available energy as a fraction of the total energy. However, this is a consequence of a weak dependence of open-circuit voltage on state-of-charge. Thus, the state-of-charge cannot be accurately estimated from the battery voltage except at extreme high and low state-of-charge. Consequently, state-of-charge algorithms typically utilizing amp-hour counting mechanisms with corrections for self-discharge can be quite complex. Lithium-ion HEV batteries are typically more tightly controlled because the power and regen capability is much more strongly dependent on state-of-charge. However, the state-of-charge of lithium-ion batteries can usually be monitored directly from cell voltage because of the strong dependence of open-circuit voltage on state-of-charge for most Li-ion chemistries. The HEV battery management system (BMS) also typically provides state-of-health monitoring based on time and temperature dependent resistance measurements.

29.8.1 Batteries for Hybrid Electric Vehicles

Lead-acid batteries were initially utilized for many HEV development projects in the period from 1970 to 1990. For power-assist HEV applications, with both mild and strong hybrids, lead-acid batteries did not offer adequate durability. This is principally due to premature sulfation failure during operation at partial states-of-charge. Long life can be obtained with lead-acid batteries maintained at the fully charged state, but this is not feasible in HEV applications utilizing regenerative braking. The power-assist HEV battery must be operated at intermediate states-of-charge to provide capability to accept regenerative-braking charge current.

For stop-start micro-hybrid applications, lead-acid batteries remain the battery of choice because they perform the necessary functions at the lowest cost. As discussed in the section on micro hybrids, the substantially increased number of engine starts required for this application necessitates a larger and more robust battery. Generally, VRLA technology will be preferred over the conventional flooded maintenance-free batteries used for SLI applications.

Additionally, micro-hybrid developers are exploring ways to accept regen braking energy while still utilizing relatively inexpensive lead-acid batteries. One approach under consideration is hybridization with supercapacitors, which can capture the regenerative braking energy at very high rates and transfer it to the battery as needed. Battery companies are also developing lead-acid batteries that have improved durability in partial state-of-charge operation. One promising development is the Firefly Energy lead-acid battery that utilizes carbon foam for the lead negative electrode.[38] The firefly energy battery is no longer available. Another is the asymmetric supercapacitor type lead-acid-carbon hybrid battery from Axion Power.[39] A third is the UltraBattery developed by the Australian Commonwealth Scientific and Industrial Research Organization (CSIRO). The UltraBattery utilizes supercapacitor carbon materials in the anode that enable partial state-of-charge operation.[40] This battery may also have potential for use on power-assist hybrid applications as promoted by CSIRO in a 100,000 mi demonstration in a Honda Insight HEV. The UltraBattery served as a drop-in replacement to the original NiMH battery pack in the Insight. Cooperative development is under way with the Furukawa Battery Company in Japan as well as East Penn in the United States.

For power-assist hybrid vehicles, the application of advanced batteries has been demonstrated starting with NiCd batteries utilized in a number of development and demonstration projects in the 1980s. In the 1990s, there was significant development of high-powered batteries aimed at hybrid electric vehicle applications. The Electrochemical Energy Storage Technical Team of PNGV (now FreedomCAR) was formed in 1993. High-power NiMH batteries for HEV applications were developed by a variety of battery companies, including ECD Ovonics, Panasonic EV Energy (PEVE), Sanyo, Varta, and SAFT. High-power NiMH batteries were first introduced commercially in the original Toyota Prius in 1997. The high specific power and high regen capability enabled the Prius to achieve excellent fuel economy under power-assist HEV operation. Nickel-metal hydride battery technology has the capability to operate in intermediate states-of-charge and deliver hundreds of thousands of shallow charge-discharge cycles with a 1 to 2% state-of-charge swing. This has enabled a battery pack capable of operating for 10 years and/or 100,000 miles with appropriate control strategies.

These first NiMH HEV batteries evolved from high-power cylindrical cells produced at high volume for high-power consumer applications.[41] They featured the usual jelly-roll electrode winding of long, thin electrodes with circular nickel end plates welded to the electrode edges to provide higher specific power. Cylindrical NiMH HEV batteries supplied by PEVE were used to power the original Prius. These batteries were also utilized in the Honda Insight in 1999 in an air-cooled battery pack.[42,43] This battery pack, illustrated in Fig. 29.26, consisted of 120 cells packaged in 20 modules each with 6 cells in a string. The pack provided slightly less than 1 kWh of energy but 10 kW of power, enabling the Insight to achieve exceptional fuel economy. The specifications for the PEVE battery modules and Honda Insight pack are given in Table 29.12. At a module level, a specific power of 600 W/kg was achieved. At the battery pack systems level, a substantial hardware burden reduced the specific power to less than 400 W/kg. The specific-power performance of the battery approached the USABC targets for mild HEV applications. Due to inherent inefficiencies in

29.40 SPECIALIZED BATTERY SYSTEMS

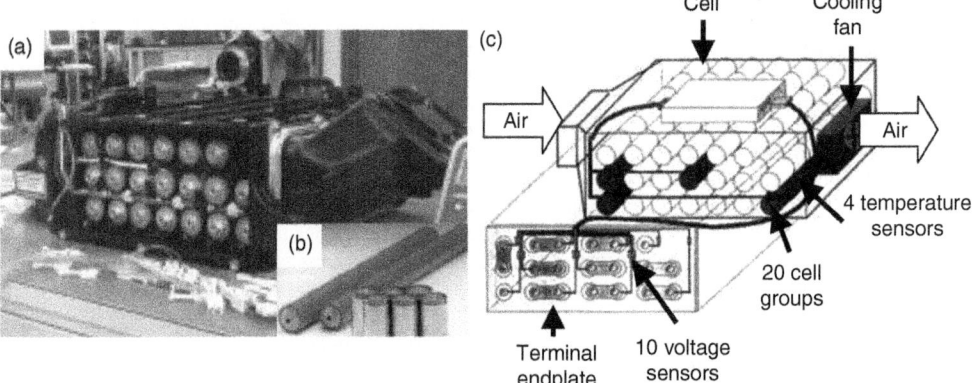

FIGURE 29.26 Photos of HEV battery pack (a) and modules (b) with drawing (c) showing air flow in NiMH HEV battery pack for 1999 Honda Insight. (*Courtesy of NREL.*)

packaging cylindrical cells, the volumetric power density of the pack was several times lower than the module power density, which increased the challenge for packaging on a vehicle level. The total energy for this pack was more than sufficient to meet the available energy requirements for this mild-hybrid application.

Panasonic EV Energy later developed higher-power prismatic NiMH batteries that were utilized first in the 2001 Toyota Prius when it was introduced in North America.[44–46] This module also consisted of 6 series-connected cells, each 6.5 Ah in capacity, but featured a unique prismatic design. Cells in the module were placed edge to edge in a long, thin package that provided for the thin cells

TABLE 29.12 Specifications for PEVE Battery Module and Honda Insight Battery Pack

	PEVE HEV battery module (6 cells)	1999 Honda Insight battery pack (20 modules)
Nominal voltage	7.2 V	144 V
Discharge capacity	6.5 Ah	6.5 Ah
Energy	47 Wh	0.94 kWh
Discharge power	654 W	13 kW
Regen power	545 W	11 kW
Power-to-energy ratio	14 (per h)	14 (per h)
Weight	1.09 kg	29 kg
Length	384 mm	495 mm
Diameter	35 mm	
Width		372 mm
Height		174 mm
Volume	0.37 L	32 L
Specific power	600 W/kg	451 W/kg
Power density	1771 W/L	408 W/L
Regen specific power	500 W/kg	376 W/kg
Specific energy	43 Wh/kg	32 Wh/kg
Energy density	127 Wh/L	29 Wh/L

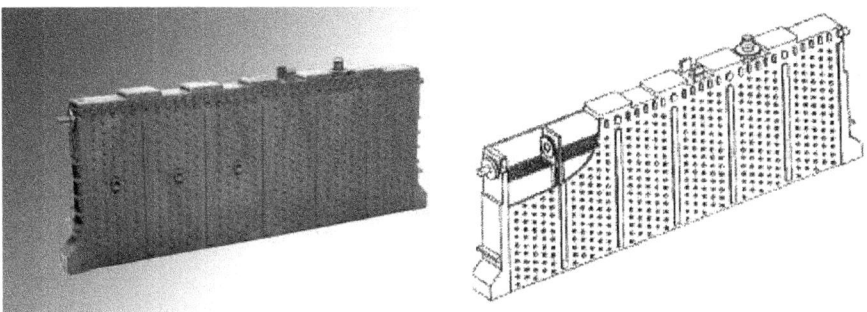

FIGURE 29.27 Photo of PEVE 2000 prismatic NiMH HEV battery module delivering 6.5 Ah at 7.2 V nominal voltage and having a power capability of 1000 W/kg (left); cutaway diagram (right). (*Courtesy of Panasonic EV Energy.*)

to have a high-area cooling surface, as shown in Fig. 29.27. The prismatic PEVE NiMH HEV battery modules utilized thin electrodes and separators with current collection on the long edge of the electrodes. The extra electrode substrates and long current collectors substantially lowered the resistance, but also resulted in a higher fraction of inactive weight and volume in the cell, in comparison to high-energy batteries designed for EV applications. NiMH HEV battery cells are typically engineered for higher power at the expense of moderately lower capacity and specific energy. The specific power of 1000 W/kg for the PEVE NiMH battery module in Fig. 29.27 thus substantially exceeds the 200 W/kg power capability of the 95 Ah NiMH EV battery developed by PEVE for the Toyota RAV-4 EV. Although the 45 Wh/kg specific energy of the HEV battery module is significantly lower than the 63 Wh/kg achieved in the EV-95 battery, it is more than sufficient to satisfy the available energy requirements for power-assist applications such as the Prius. HEV cycle life tests of these prismatic HEV NiMH battery modules have exceeded USABC goals of 300,000 HEV cycles.

The power and energy performance of the prismatic NiMH HEV batteries developed by PEVE are compared to the performance of the cylindrical cell modules in Table 29.13. The 2003 PEVE module battery achieved a specific power of 1300 W/kg in large part by improving electrical power interconnections between cells.[46] The 2000 battery module featured cell-to-cell interconnections at the top of the cells. The 2003 battery module featured an additional through-the-wall electrical

TABLE 29.13 Specifications for Three Generations of Panasonic NiMH EV Battery Modules

	1997 Cylindrical battery module	2000 Prismaic battery module	2003 Prismatic battery module
Nominal voltage	7.2 V	7.2 V	7.2 V
Discharge capacity	6.5 Ah	6.5 Ah	6.5 Ah
Energy	47 Wh	47 Wh	47 Wh
Discharge power	654 W	1050 W	1352 W
Power-to-energy ratio	14 (per h)	22 (per h)	29 (per h)
Weight	1.09 kg	1.05 kg	1.04 kg
Length	384 mm	275 mm	285 mm
Diameter	35 mm		
Width		19.6 mm	19.6 mm
Height		106 mm	114 mm
Volume	0.37 L	0.57 L	0.64 L
Specific power	600 W/kg	1000 W/kg	1300 W/kg
Power density	1771 W/L	1838 W/L	2123 W/L
Specific energy	43 Wh/kg	45 Wh/kg	45 Wh/kg
Energy density	127 Wh/L	82 Wh/L	73 Wh/L

power interconnect between cells that provided the battery with a pseudo-bipolar characteristic. The high-power capability of these batteries enabled the USABC specific power targets to be met at the battery pack level. These prismatic modules could also be packaged more efficiency than cylindrical cells, leading to more substantial gains in volumetric power density at the battery pack level.

Toyota Prius HEVs introduced to the United States utilized PEVE prismatic battery modules that provided for highly efficient packaging, as illustrated in Fig. 29.28. The prismatic battery modules are stacked together with interlocking plastic circular nubs and ribs that hold the batteries in place and space them for optimal flow of cooling air. A cutaway view shows the module stack encased in a steel case that provides for an upper and lower plenum directing air flow between the modules. This arrangement provides for more uniform cooling than achieved in the previous cylindrical-cell battery pack for the Prius, where air flow sequentially cooled a series of battery modules. In this design, all battery modules and cells are cooled with parallel air flow.

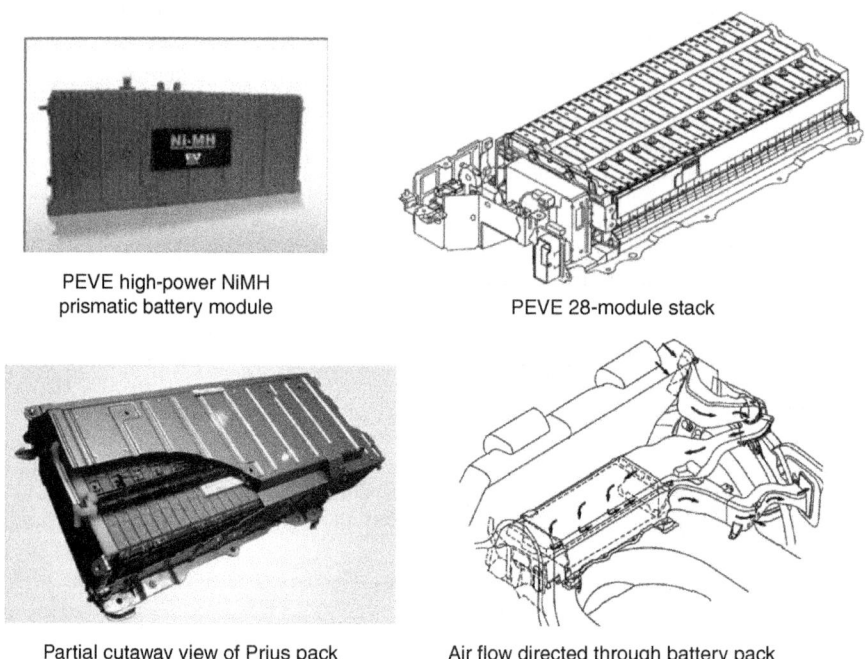

FIGURE 29.28 PEVE 2003 prismatic NiMH HEV battery module and 2004 Prius battery pack. (*Courtesy of Panasonic EV Energy.*)

Higher-power prismatic NiMH batteries enabled substantial size and weight reduction of the battery pack in comparison to the original 1997 Prius battery packs that utilized cylindrical NiMH cells, as shown in Fig. 29.29. The higher-power 2000 Prius battery pack with 38 prismatic batteries was more than 50% smaller and 30% lighter than the 1997 Prius pack with 40 cylindrical modules. Further gains in specific power with the 2003 batteries and Toyota's introduction of a DC/DC boost converter into the 2004 Prius enabled the removal of 10 more battery modules and further reductions in size and weight. The total energy at 1.3 kWh is still more than sufficient for this application and the power-capability significantly exceeds the vehicle power specification. In fact, on the battery pack level, the specific power and power density exceed the target of the USABC for power-assist hybrid vehicles. Consistent with the bench cycle-life results of the PEVE HEV batteries, the Prius battery packs have demonstrated excellent life in the field.

Sanyo has developed a 1000 W/kg NiMH HEV battery with excellent life performance. It is based on high-power D-sized cylindrical cells and has been utilized in various Ford and Honda

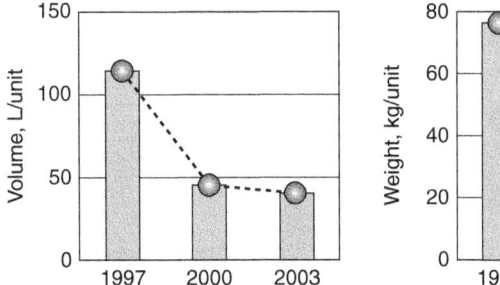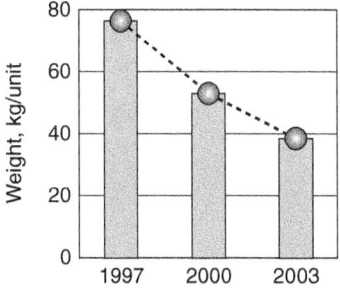

FIGURE 29.29 Development of smaller and lighter weight NiMH battery packs for successive generations of the Toyota Prius HEV.

hybrid vehicles. This battery has demonstrated excellent field performance in these strong-hybrid applications. In a New York City taxi fleet application, excellent durability has been demonstrated in Ford Escape Hybrids with nearly 300,000 miles of aggressive driving.[47]

Cobasys has developed a high-power prismatic battery with 1000 W/kg power capability that has been integrated into mild-hybrid applications for a number of automotive OEMs, including GM in their Saturn Vue Greenline mild hybrid. This battery has capability for applications requiring liquid cooling for thermal management.

Great efforts have been expended in the development of lithium-ion batteries for HEV applications with strong funding from the USABC and DOE for more than a decade. This has been complemented by excellent work in U.S. national labs and globally in government and private sector labs throughout the world. Lithium-ion batteries have achieved exceptional specific-power performance, exceeding the USABC HEV battery goals several-fold. Cost reduction in lithium-ion cells for portable electronics has generated interest in lithium-ion batteries as a potentially lower cost alternative to nickel-metal hydride HEV batteries currently in commercial production. Lithium-ion batteries show strong potential for lower-cost cells than currently available nickel-metal hydride HEV batteries.[47] The diversity of lithium-ion technologies with a variety of cathode and anode compositions available provides opportunities for relatively low material costs. However, at the battery pack level, lithium-ion batteries require more extensive controls and thermal management to ensure safety and durability. At the battery pack level, it is more challenging to develop lithium-ion HEV batteries that are cost competitive with the incumbent NiMH HEV batteries. Additionally, there is concern about the durability Li-ion batteries in the HEV application, particularly with respect to calendar life, which is difficult to ascertain without more field experience in these new battery products. For power-assist HEV applications, operation at intermediate states-of-charge avoids the fully charged state where degradation mechanisms are known to be accelerated. For plug-in hybrid applications, some developers are reducing the charge level even at the expense of energy density to avoid the detrimental effects of standing at very high states-of-charge.

After several years of anticipation, Mercedes-Benz was the first to go into production with a hybrid vehicle utilizing a lithium-ion battery.[48] This mild-hybrid version of the Mercedes S-class luxury sedan, the S400, is illustrated in Fig. 29.30. The battery pack is supplied by the Johnson Controls-SAFT (JCS) joint venture and leverages many years in the development of high-power lithium-ion batteries at SAFT. The battery pack comprises 35 high-power cylindrical lithium-ion cells, each with a capacity of 6.5 Ah delivered at a nominal voltage of 3.6 V. The 126 V battery pack is capable of 19 kW, corresponding to about 750 W/kg on a battery pack level. Mercedes utilized the high specific power and power density of the JCS lithium-ion battery to locate this battery under the hood within the same general packaging envelope of the lead-acid SLI battery. To maintain the temperature of the battery cells below 50°C in the engine compartment, the battery pack is liquid-cooled and utilizes the vehicle air conditioning system as needed.

A variety of battery companies, including SAFT, LG Chem, SK Energy, Hitachi, EnerDel, AESC, Panasonic, and A123, have now developed very high-power lithium-ion HEV batteries in the range of 2000 to 5000 W/kg on a cell level. This represents tremendous progress over the relatively low-power

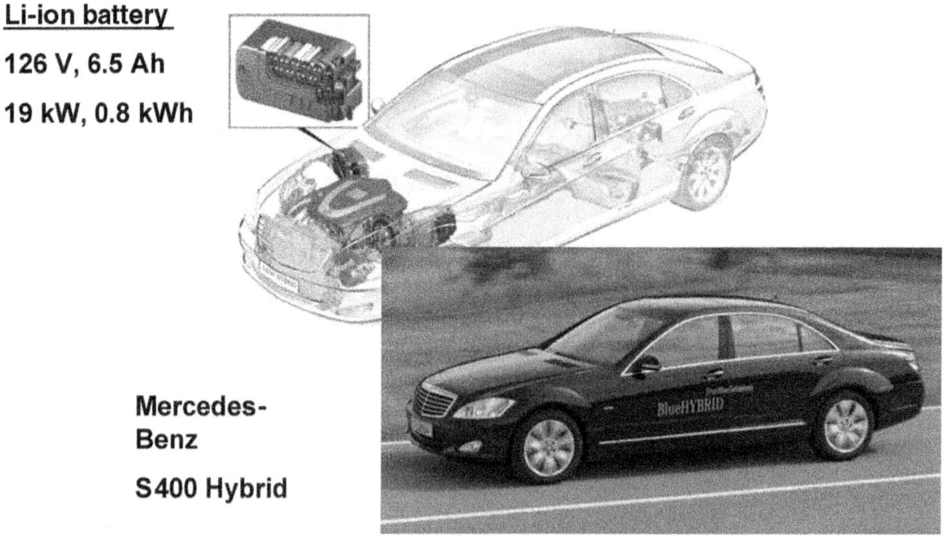

29.44 SPECIALIZED BATTERY SYSTEMS

Li-ion battery

126 V, 6.5 Ah

19 kW, 0.8 kWh

Mercedes-Benz S400 Hybrid

FIGURE 29.30 JCS lithium-ion battery for Mercedes S400 mild-hybrid luxury car. (*Courtesy of Daimler-Benz.*)

performance of early Li-ion battery prototypes. Lithium-ion batteries utilize organic electrolytes with an ionic conductivity two orders of magnitude less than the traditional aqueous battery electrolytes. A key challenge initially facing lithium-ion batteries two decades ago was to overcome the high ionic resistance in the electrolyte that was only partially offset by the high cell voltage. The general approach was to use thin electrodes and separators in a very high surface area configuration. High-power cylindrical cells utilized substantially thinner and longer electrodes. With this design utilizing very long electrodes, current collection became a critical issue. Current collectors were welded directly to the entire length of the electrode edge. High-power cylindrical Li-ion HEV battery cells have similar construction to cylindrical NiMH HEV battery cells, but with electrode active materials typically an order of magnitude thinner. More recently, even higher power cells have been developed in the prismatic soft-package format (also called laminate or pouch cells). These generally provide for smaller and lighter weight modules and easier thermal management designs. In Fig. 29.31, some examples of high-power HEV battery cells of the cylindrical and pouch cell formats are illustrated.[49]

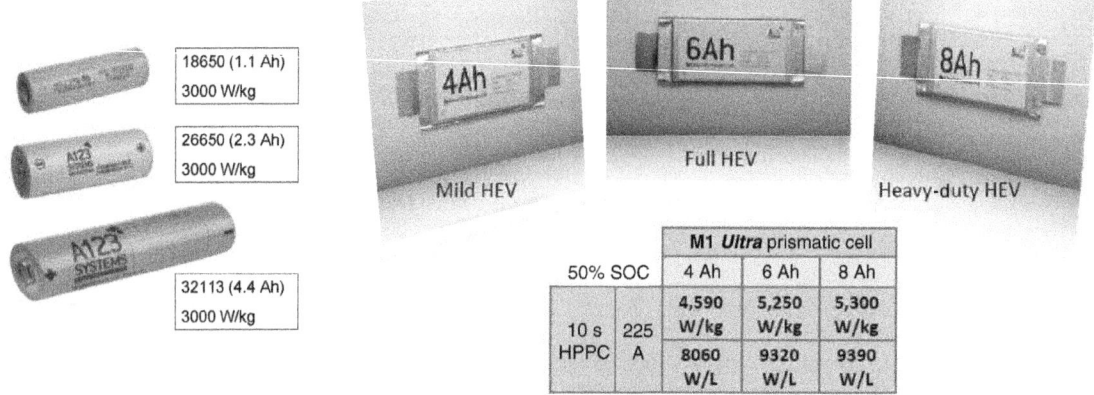

FIGURE 29.31 High-power HEV Li-ion battery cells in cylindrical and pouch cell formats. (*Courtesy of A123.*)

In addition to the impressive progress in developing high-power Li-ion batteries for HEV applications, there has been very significant progress in modeling thermal behavior that has aided the development of effective thermal management systems. There has also been substantial progress in developing more robust battery chemistries with respect to their ability to safely withstand electrical abuse. In particular, batteries utilizing the lithium manganese oxide spinel and lithium iron phosphate cathode materials provide more benign responses when voltage or temperature limits are exceeded. Additionally, the lithium titanate anode materials can provide for more safe operation for low-temperature charging. U.S. national labs, including NREL[42,45] and Sandia,[12,36] have provided invaluable support to development activities through modeling and safety testing activities.

In 2009, USABC presented the spider chart in Fig. 29.32 showing the technology status of high-power lithium-ion batteries relative to the USABC performance goals for the power-assist HEV application.[21] The progress toward performance goals has been excellent, and there is confidence that the specific power, power density, energy, and cycle life targets can be met with several lithium-ion technologies. The areas of concern are cost on a system level, especially due to the additional electrical and thermal management features and controls required to operate the pack safely. Even with sophisticated management, there remains some level of concern about potential safety issues. The other area of doubt is calendar life, which is difficult to ascertain with limited experience with this new technology. However, the potential for smaller and lighter battery packs and, eventually, lower cost has prompted continued enthusiasm, and automakers continue to announce plans to utilize lithium-ion batteries in HEV programs within the next several years.

Nickel-metal hydride batteries remain the incumbent technology for power-assist HEV applications. In principle, higher specific-power nickel-metal hydride batteries could also be developed utilizing similar approaches that have been successful with lithium-ion batteries. NiMH HEV batteries with increased specific power could also be less expensive per unit power ($/kW). Additionally, NiMH battery packs have an intrinsic advantage over their lithium-ion counterparts because simpler electronics and pack designs could be utilized. Nickel-metal hydride batteries continue to be developed in Japan and China. Despite successful USABC-sponsored development work in the 1990s, there has been relatively little recent development work directed at NiMH batteries in the United States.

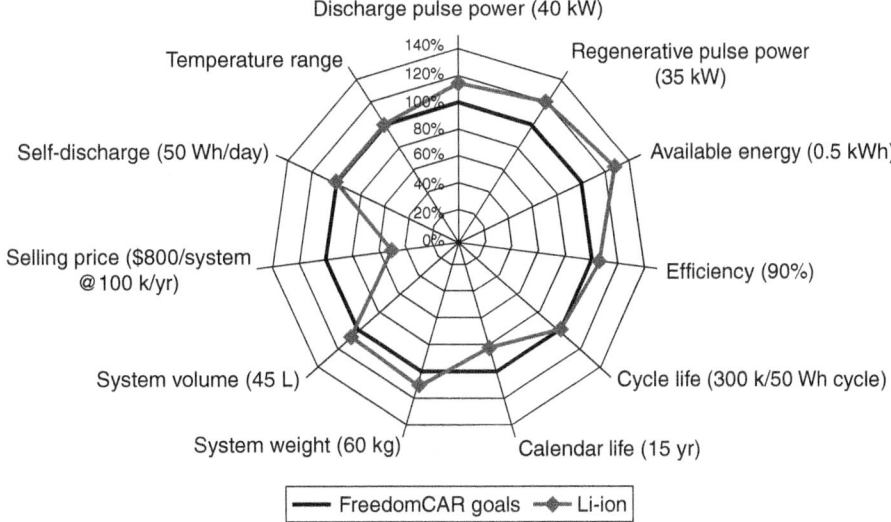

FIGURE 29.32 Status of lithium-ion battery technology in comparison to USABC minimum goals for commercialization for power-assist HEV applications (presented by USABC in May 2009).

29.9 OTHER ENERGY STORAGE TECHNOLOGIES FOR HYBRID ELECTRIC VEHICLES

Some other technologies can provide high power comparable to that of batteries without the complications of the electrochemical charge-storage reactions with respect to cycle life durability. Mechanical flywheels, hydraulic accumulators, and supercapacitors have all been incorporated in developmental hybrid electric vehicles with some promising results. Flywheels store energy in rotational momentum of very high speed rotating disks and can provide very high specific power. Flywheels were pursued at a variety of automotive companies in the last two decades of the 20th century. There are some issues with packaging and cost, but the most serious concern is safety. There were some serious accidents that discouraged development for automotive applications. Hydraulic accumulators have been considered for truck applications which often already have auxiliary hydraulic functions on board that could be utilized.

Of the alternatives to battery technology, supercapacitors, also known as ultracapacitors or electrochemical capacitors, have shown the most promise for HEV applications. They have been developed by a variety of companies throughout the world, including Maxwell, Ness Cap, and Batscap. These devices, discussed in Chap. 39 of this Handbook, have been commercially introduced in heavy-duty HEV applications, including several hundred HEV buses by ISE Corporation powered by Maxwell ultracapacitors.

In addition to high specific power, supercapacitors offer advantages in HEV cycle life with the capability to deliver hundreds of thousands of deep charge-discharge cycles. A major advantage is also the excellent low-temperature performance, particularly of supercapacitors with little attenuation in battery power, down to –30°C in some devices. The most significant performance drawback is specific energy. The 5 Wh/kg commercially available in 2010 is insufficient to meet the 8 Wh/kg specific energy requirements for available energy in power-assist HEV applications. However, there has been reconsideration of this target in recent studies. Incorporation of supercapacitors into the Saturn Vue mild-hybrid vehicle has provided excellent and promising results.[50] Modeling studies have also shown effective HEV operation with available energy storage at 4 Wh/kg levels and below. The main deterrents to the development of power-assist hybrid cars utilizing supercapacitors currently appears to be cost, in part due to low production volume, and modifications required in the power electronics to adapt to the sloping discharge curve of supercapacitors. In the future, there may also be more promise in asymmetric supercapacitors, which incorporate battery and supercapacitor electrodes in the same device.

REFERENCES

1. E. H. Wakefield, "History of the Electric Automobile: Battery-Only Powered Cars," Society of Automotive Engineers, Warrendale, PA (1994).
2. E. H. Wakefield, "History of the Electric Automobile: Hybrid Electric Vehicles," Society of Automotive Engineers, Warrendale, PA (1998).
3. M. H. Westbrook, "The Electric and Hybrid Electric Car," Society of Automotive Engineers, Warrendale, PA (2001).
4. C. C. Chan and K. T. Chau, "Modern Electric Vehicle Technology," Oxford University Press, New York (2002).
5. I. Husain, "Electric and Hybrid Vehicles," CRC Press, New York (2003).
6. J. Larminie and J. Lowry, "Electric Vehicle Technology Explained," John Wiley, Hoboken, NJ, 2003.
7. M. Shnayerson, "The Car that Could: The Inside Story of GM's Revolutionary Electric Vehicle," Random House, New York, 1996.
8. D. A. J. Rand, R. Woods, and R. M. Dell, "Batteries for Electric Vehicles," Society of Automotive Engineers, Warrendale, PA (1998).
9. P. Savagian, "Driving the Volt," *SAE Hybrid Vehicle Technology Conference*, San Diego, CA, February 2008.

10. "USABC Electric Vehicle Battery Test Procedures Manual," U.S. Department of Energy Contract DE-AC07-94ID13223, Report Number DOE/ID-10479, Revision 2, January 1996.
11. "Battery Technology Life Verification Manual, Advanced Technology Development Program for Lithium-Ion Batteries," FreedomCAR Vehicle and Technologies Program, Idaho National Laboratory, INEEL/EXT-04-01986, February 2005.
12. T. Unkelhaeuser and D. Smallwood, "USABC Electrochemical Storage System Abuse Test Procedure Manual," Sandia National Laboratories, SAND99-0497, July 1999.
13. GM Ovonic Application Manual, "Nickel-Metal Hydride Battery Electric Vehicle Battery Model GMO-0900," GM Ovonic, Troy, MI, August 1999.
14. R. S. Stempel, S. R. Ovshinsky, P. R. Gifford, and D.A. Corrigan, "Lithium-Ion: Ready to Serve," *IEEE Spectrum*, **35**, 29 (1998).
15. R. Spotnitz, "Large LiIon Battery Design Principles," Tutorial A, *8th International Advanced Automotive Battery Conference*, Tampa, FL, May 2008.
16. M. S. Whittingham, "Lithium Batteries and Cathode Materials," *Chem. Rev.*, **104**, 4271 (2004).
17. G. Nazri and G. Pistoria, "Lithium Batteries: Science and Technology," Kluwer Academic Publishers, New York (2004).
18. G. Berdichevsky, K. Kelty, J. B. Straubel, and E. Toomre, "The Tesla Roadster Battery System," Tesla Motors, December 2007.
19. T. Miyashita and Y. Tominga, "Development of High Energy Lithium-Ion Battery Pack for Pure EV Applications," Large Lithium-Ion Battery Technology and Application Symposium, *8th International Advanced Automotive Battery Conference*, Tampa, FL, May 2008.
20. S. Hendrix and D. Buck, "Lithium-Ion Battery System Architecture for HEV and EV Applications," Large Lithium-Ion Battery Technology and Application Symposium, *8th International Advanced Automotive Battery Conference*, Tampa, FL, May 2008.
21. K. Snyder, "U.S. Advanced Battery Consortium," *2009 DOE Hydrogen Program and Vehicle Technologies Program Annual Merit Review and Peer Evaluation Meeting*, U.S. Department of Energy, Arlington, VA, May 2009.
22. L. M. Moshurchak, W. M. Lamanna, M. Bulinsky, R. L. Wang, R. R. Garsuch, J. Jiang, D. Magnuson, M. Triemert, and J. R. Dahn, "High Potential Redox Shuttle for Use in Lithium-Ion Batteries," *J. Electrochem. Soc*, **156**, A309 (2009).
23. DOE ARPA-E Funding Opportunity Announcement, "Batteries for Electrical Energy Storage in Transportation," U.S. Department of Energy, Funding Opportunity Number: DE-FOA-0000207, December 2009.
24. M. Verbrugge and R. Matthe, "Energy Storage Progress and Concepts for Plug-In Hybrid and Extended Range Electric Vehicles," *8th International Advanced Automotive Battery Conference*, Tampa, FL, May 2008.
25. M. Ehsani, Y. Gao, and A. Emadi, "Modern Electric, Hybrid, and Fuel Cell Vehicles: Fundamentals, Theory, and Design," 2nd ed., CRC Press, Boca Raton, FL (2009).
26. C. C. Chan, "The State of the Art of Electric, Hybrid, and Fuel Cell Vehicles," *Proceedings of the IEEE*, **95**, 704–718 (2007).
27. M. Anderman, "The Challenge to Fulfill Electrical Power Requirements of Advanced Vehicles," *J. Power Sources*, **127**, 2–7 (2004).
28. O. Bitsche and G. Gutman, "Systems for Hybrid Cars," *J. Power Sources*, **127**, 8–15 (2004).
29. E. Karden, P. Shinn, P. Bostock, J. Cunningham, E. Schoultz, and D. Kok, "Requirements for Future Automotive Batteries—a Snapshot," *J. Power Sources*, **144**, 505–512 (2005).
30. E. Karden, S. Ploumen, B. Fricke, T. Miller, and K. Snyder, "Energy Storage Devices for Future Hybrid Electric Vehicles," *J. Power Sources*, **168**, 2–11 (2007).
31. "FreedomCAR 42 V Battery Test Manual," U.S. Department of Energy Contract DE-AC07-99ID13727, Report Number DOE/ID-11070, April 2003.
32. "FreedomCAR Battery Test Manual for Power Assist Hybrid Electric Vehicles," U.S. Department of Energy Contract DE-AC07-99ID13727, Report Number DOE/ID-11069, October 2003.
33. A. A. Frank, "Charge Depletion Control Method and Apparatus for Hybrid Powered Vehicles," U.S. Patent 5,842,534, December 1, 1998.

34. B. Johnston, T. McGoldrick, D. Funtson, H. Kwan, M. Alexander, F. Aliato, N. Culaud, O. Lang, H. A. Mergen, R. Carlson, A. Frank, and A. Burke, University of California, Davis, PNGV FutureCar Technical Report, SP-1359 SAE, June 1997.
35. "Battery Test Manual for Plug-In Hybrid Electric Vehicles," U.S. Department of Energy Contract DE-AC07-05ID14517, Report Number INL/EXT-07-12536, March 2008.
36. D. H. Doughty and C. C. Craft, "FreedomCAR Electrical Energy Storage System Abuse Test Manual for Electric and Hybrid Vehicle Applications," Sandia National Laboratories, SAND 2005-3123, June 2005.
37. C. N. Ashtiani, "Battery Hazard Modes and Risk Mitigation Analysis," United States Advanced Battery Consortium Manual, August 2007.
38. K. C. Kelley and J. J. Votoupal, "Battery Including Carbon Foam Current Collectors," U.S. Patent 6,979,513, December 27, 2005.
39. W. Buiel, "Axion Power's Asymmetric Ultracapacitor/Lead-Acid Technology Applied to High-Rate Partial State of Charge HEV Cycling," Large EC Capacitor Technology and Application Symposium, *9th International Advanced Automotive Battery Conference*, Long Beach, CA, June 2009.
40. A. Cooper, J. Furakawa, L. Lam, and M. Kellaway, "The UltraBattery—A New Battery Design for a New Beginning in Hybrid Electric Vehicle Energy Storage," *J. Power Sources*, **188**, 642–649 (2009).
41. A. Taniguchi, N. Fujioka, M. Ikoma, and A, Ohta, "Development of Nickel/Metal-Hydride Batteries for EVs and HEVs," *J. Power Sources*, **100**, 117–124 (2001).
42. M. Zolot, K. Kelly, M. Keyser, M. Mihalic, A. Pesaran, and A. Hieronymus, "Thermal Evaluation of the Honda Insight Battery Pack," *Proceedings of the 36th Intersociety Energy Conversion Engineering Conference (IECEC'01)*, Savannah, GA, July 2001.
43. N. Sato, "Overview of Progress in Ni-MH and Li-ion Automotive Batteries," *3rd International Advanced Automotive Battery Conference*, Nice, France, June 2003.
44. B. G. Potter, T. Q. Duong, and I. Bloom, "Performance and Cycle Life Test Results of a PEVE First-Generation Prismatic Nickel/Metal Hydride Battery Pack," *J. Power Sources*, **158**, 760–764 (2006).
45. M. Zolot, A. A. Pesaran, and M. Mihalic, "Thermal Evaluation of Toyota Prius Battery Pack," SAE Technical Paper No. 2002-01-1962, Society of Automotive Engineers, Warrendale, PA (2002).
46. M. Ohnishi, K. Ito, S. Yuasa, N. Fujioka, T. Asahina, S. Hmada, and T. Eto, "Development of Prismatic Type Nickel/Metal-Hydride Battery for HEV," *3rd International Advanced Automotive Battery Conference*, Nice, France, June 2003.
47. K. Snyder, X. G. Yan, and T. J. Miller, "Hybrid Vehicle Battery Technology—The Transition from NiMH to Li-Ion," SAE Technical Paper No. 2009-01-1385, Society of Automotive Engineers, Warrendale, PA (2009).
48. W. Wiedemann, O. Vollrath, N. Armstrong, J. Schenk, O. Bitsche, and A. Lamm, "Advanced Energy Storage Systems for Hybrids," *9th International Advanced Automotive Battery Conference*, Long Beach, California, June 2009.
49. A. Fulop, "A123 Program Review, Vehicle Technologies Program Annual Merit Review," U.S. Department of Energy, Arlington, VA, May 2009.
50. J. Gonder, A. Pesaran, J. Lustbader, and H. Tataria, "Fuel Economy and Performance of Mild Hybrids with Ultracapacitors: Simulations and Vehicle Test Results," Large EC Capacitor Technology and Application Symposium, *9th International Advanced Automotive Battery Conference*, Long Beach, CA, June 2009.

CHAPTER 30
BATTERIES FOR ELECTRICAL ENERGY STORAGE APPLICATIONS

Abbas A. Akhil, John D. Boyes, Paul C. Butler, and Daniel H. Doughty

30.1 INTRODUCTION: ENERGY STORAGE ON THE ELECTRIC GRID

The electricity supply grid has been described as the largest Just-In-Time (JIT) supply system in the world because the electricity generated at power plants is instantly used by the loads connected to it. Today, there is very little capacity to store the energy generated on the electric grid. Without the ability to store excess energy, grid operators are constantly balancing the energy needs of consumers with the generation resources they control. Thus, the generation of electricity must equal the demand for electricity at any given time, and the entire electric grid operates without any inventory of the product it supplies.

The ability to store significant amounts of electricity would change this JIT mode of grid operation and offer major benefits that could improve not only the reliability of the grid, but also decrease its carbon footprint by reducing fuel consumption. More than 17 utility, end-user, and renewable applications for energy storage systems have been identified.[1-6] These applications fall into two broad categories: energy applications and power applications. Energy applications involve continuous storage system discharges over periods of hours with correspondingly long charging periods. Energy applications typically involve only one charge-discharge cycle per day. Power applications involve comparatively short periods of discharge (seconds to minutes), and short recharging periods, and often require many cycles per day. Most storage technologies are better suited to one application category than the other; few technologies can meet all requirements. Energy applications include peak shaving, load-leveling, transmission and distribution upgrade deferral, customer demand charge and energy charge reduction, renewable generation shifting and energy arbitrage or commodity storage. Power applications include frequency and voltage regulation, power quality, renewable generation smoothing and ramp-rate control, and trackside regulation for electric rail operations.

The demand for electricity varies from moment to moment, throughout the day, and seasonally. The moment-to-moment variations are relatively small, the variation between a daily peak and minimum is typically larger, and seasonal variations between summer peak demand and spring or fall minimum demand can be very large. Typical daily utility load curves are shown in Fig. 30.1. Utility systems (generation, transmission, and distribution) are designed to deliver power during the peak demand period of the year. At nonpeak times, this leaves a significant portion of that capacity unutilized. Transmission and distribution subsystems also are designed to meet peak load. A utility load duration curve, Fig. 30.2, shows that the peak 20% of the demand will occur for less than about 250 h during the year.

30.2 SPECIALIZED BATTERY SYSTEMS

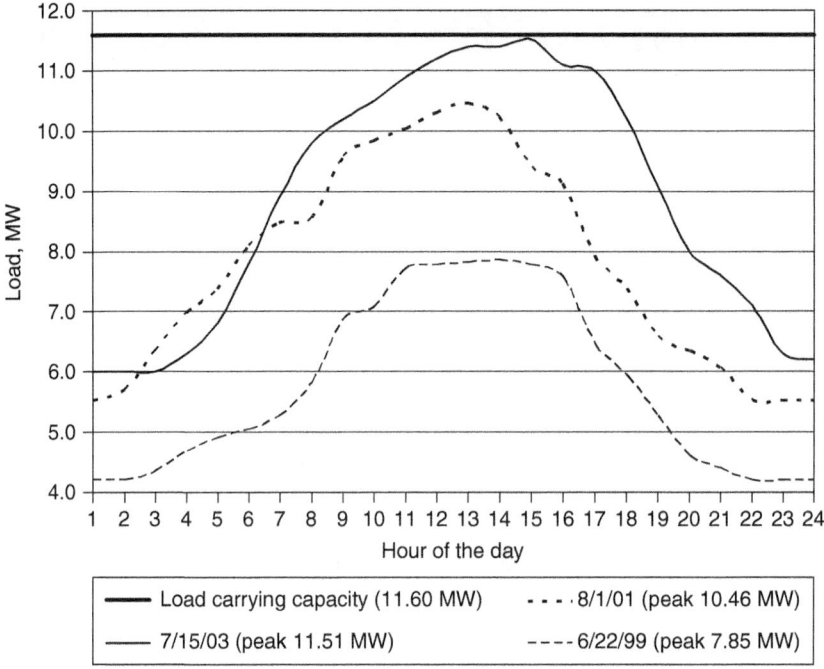

FIGURE 30.1 Hourly load profiles for three days with broad peaks.[5]

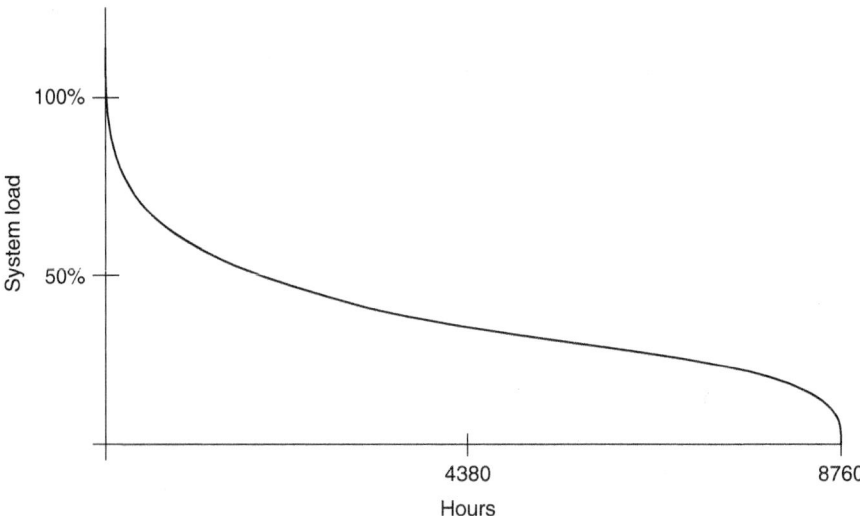

FIGURE 30.2 Typical utility load duration curve.

Variations between generation and demand result in variations in frequency and/or voltage. Frequency and voltage variations must be kept within strict limits or power quality will suffer; the system can become unstable, equipment can be damaged, and blackouts can occur. Energy storage systems can supply that energy and power and act as a buffer between the constantly changing demand for electricity and the capacity to generate that electricity. Placing energy storage near load centers will allow a load-leveling of the demand and allow systems with lower peak capacity to be installed initially, or can allow deferral of upgrades by the utility as the load demand grows.

Without energy storage, the electric grid is constantly balancing supply with the instantaneous demand of all its connected loads. The balancing act is necessary to maintain the stability of the generators and, eventually, control of the grid within the tolerance limits of frequency and voltage. This is currently accomplished by dedicating part or all of some fossil-fueled generators to "regulation" service. This requires that a generator change its output every few minutes (either raise or lower) to balance the system. Energy storage can be the "buffer" that decouples the generators and their loads by absorbing or discharging energy in response to the needs of the grid, which can be very fast, with time constants measured in fractions of a second, or relatively slow in minutes to hours. Energy storage that absorbs or discharges in these very fast time frames will maintain grid stability, whereas slower acting storage creates arbitrage opportunities and other operational benefits to the grid. The demand for electricity also varies, to a lesser extent, over the period of seconds and minutes, shifting both up and down.

In addition, the demand for electricity varies slowly over the course of a day, as well as seasonally. Typically, the minimum demand for electricity occurs between 12 AM and 5 AM. Peak demands typically occur in the afternoon and early evenings, often driven by air-conditioning loads. This variation forces systems that are designed to meet peak requirements to be underutilized during off-peak times. Pumped hydro storage systems, for example, are typically used to allow large central generators in a grid to operate optimally during nonpeak periods. As peak loads increase, new generation must be added, transmission and distribution systems must be upgraded to handle the increased power, and the overall load factor of the system worsens. However, these generators have an optimum load range at which operation is most efficient, and emissions per unit output are minimized. Operating them at low loading levels away from the optimum loading range, and particularly ramping them up and down, increases fuel use and emissions.

Energy storage for grid operations covers several orders of magnitude both in time scale and in power and energy requirements. The rapid response requirements for the fast-acting energy storage can only be met by batteries, electrochemical capacitors, or flywheels, and the slower acting, very large energy storage needs can be met by pumped hydro or compressed air energy storage systems due to their larger storage capacities and slower response times.

To meet the potentially large storage requirements of some energy applications, multiple battery energy storage systems (BESS) can be distributed around the grid. If energy storage is located near load centers, the storage can be charged during off-peak times and discharged during on-peak times. This can delay or prevent upgrading the system as demand grows. Distributed storage can also increase power quality, buffer variable, distributed solar or wind generators, and provide additional flexibility to grid operators.

Renewable generation is rapidly being added to electric grids around the world. Both wind generation and solar photovoltaic (PV) generation varies as the wind rises and falls, or as clouds pass over the solar field, respectively. These variations can be large and occur over short time scales. Wind farms can increase or decrease output by 90% over tens of minutes, and PV systems vary in a matter of seconds. The utility affected must accommodate these changes to maintain grid stability. Energy storage is one option that can be used to absorb and moderate these fluctuations. Further, renewable generation also does not always coincide with peak demand. For example, in some locations, the wind often blows hardest at night. In areas of large wind penetration, utilities are forced to either spilling (wasting) the wind-generated energy, or turning down base-load coal or nuclear generation plants that are not designed to cycle. Energy storage would allow this "excess" generation to be saved and used on-peak when demand (and value) is highest.

30.2 HISTORICAL PERSPECTIVE

30.2.1 Pumped Hydro Storage

Electric utilities added small amounts of energy storage in the United States starting in 1929[7] with water as the storage medium. This technology, commonly called "pumped hydro storage" or "pumped storage," stores and recovers energy as water is pumped between upper and lower reservoirs, as shown in the schematic in Fig. 30.3.

The inherent nature of this technology requires water storage reservoirs that are very large, and yields energy that is typically in the thousands of megawatthours. The ability to store such large quantities of energy made pumped-storage a preferred technology to complement nuclear and coal power plants by allowing them to operate at constant-power levels during peak and off-peak periods. Large nuclear- or coal-fueled conventional steam-turbine generators, as mentioned above, are not designed to adjust their output up or down in diurnal cycles as the grid load changes, and such fluctuations affect both the life and reliability of nuclear- and coal-powered generation turbines. Pumped storage provided a means of absorbing the excess generation capacity of these units during the nighttime off-peak periods, and discharging the stored energy during the daytime peak periods. This strategy allowed pumped storage to cushion the impact of the diurnal load changes of the electric grid on nuclear and coal units.

This minimal energy storage technology has served the electric utility industry needs reliably since its introduction, and many pumped storage projects have been built by electric utilities around the world. Europe and Japan added the largest numbers of pumped-storage facilities, and they comprise 13 to 19% of installed generation capacity. By contrast, the United States has only 3% of pumped storage capacity in its national generation portfolio. Deployment of pumped storage projects stopped in the United States in the mid-1980s due to environmental concerns about its large land and water requirements.

30.2.2 Legacy, Regulated Electric Utility Structure

The vertically integrated utility structure and regulated markets that prevailed until the 1980s provided limited opportunities for energy storage in electric grids around the world. The role of energy

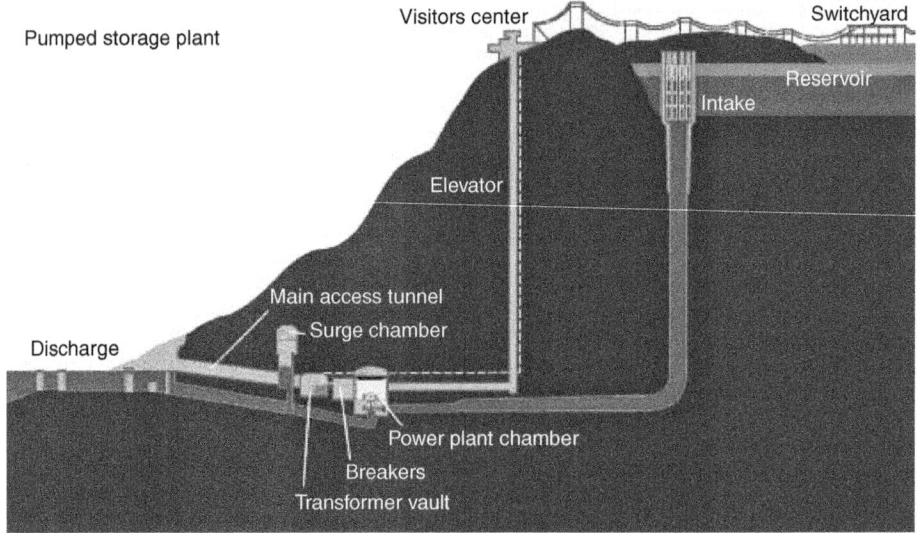

FIGURE 30.3 Schematic of a pumped storage plant. (*Courtesy of Tennessee Valley Authority, Raccoon Mountain Pumped Hydro.*)

storage was largely relegated to peak shifting and load-leveling of large coal and nuclear power plants, as described earlier. During this period, battery energy storage could not offer a technically or economically viable energy storage solution, because the requirements were tied to 8 to 14 h of energy storage, typical of a pumped storage unit. During this period, lead-acid batteries were the only commercially feasible available battery technology, and it was practically impossible for this technology to match the performance of a pumped-storage facility.

It is instructive to note the physical characteristic of one demonstration battery storage facility that was built during this time to appreciate the limitations of lead-acid batteries compared to a pumped-storage unit. Southern California Edison, the Electric Power Research Institute (EPRI), and the U.S. Department of Energy (DOE) collaborated to build a 10 MW/40 MWh lead-acid battery energy storage facility at Chino, California, that was commissioned in 1988. All such facilities contain three major components: the battery storage system, the power conversion system (PCS), and the balance of plant, which includes the system controller, enclosure or shelter, temperature control, etc. The Chino system used 8256 cells to achieve its energy storage capacity. This facility was a landmark achievement in the application of battery energy storage in electric utility grid applications and provided valuable learning experiences that had long-term impact on the future evolution of battery storage system design for electric utility applications. However, even this large battery system only provided 10 MW of power for a duration of 4 h—a limited energy storage resource for electric utilities that visualized energy storage needs only in the size ranges that pumped-hydro plants provided, hundreds of megawatts for 10 h or more. The Chino battery system is described in more detail later in this chapter.

30.2.3 Unregulated Environment

In the legacy, regulated electric utility structure, all assets and operations of the electric grid system were owned by a vertically integrated utility. Energy storage was rare, and approximately 3% of the U.S. generation capability came from pumped-hydro storage facilities. This small energy storage base did not provide widespread benefits to electric utilities, and energy storage was generally used only to provide load-leveling service. Deregulation of the electric industry separated the generation and the transmission and distribution (T&D) sides of the vertically integrated utility structure, and created markets for ancillary and other grid-support services in which energy storage could be more valuable. The prior, limited role of energy storage for load-leveling could now be expanded to a range of services. Around the same time, the lessons of building very large battery facilities such as the Chino battery system played a key role in pointing toward designing smaller, and even semiportable, battery storage systems. Battery energy storage was no longer competing with the capability of pumped-storage plants, but could find lower power and shorter duration applications that offered value in the deregulated electric marketplace.

The transmission and distribution assets of regulated utilities were retained by the electric utility companies and continued to operate as a regulated monopoly, while generation was opened to competition. Independent System Operators (ISOs) were created to oversee the transmission of electricity from generators to distribution companies. A market was created to sell the electricity and to sell various services (such as regulation) necessary to the operation of the grid.

The regulatory environment is still evolving, but several ISOs have created market rules that enable energy storage systems to participate in the marketplace, not just as a "load-leveling" resource, but also by providing operational benefits in all three major components of the electric grid: generation, transmission, and distribution. This chapter discusses these benefits, the role of battery energy storage, and the battery technologies that meet these needs, now and in the future.

30.3 BATTERY ENERGY STORAGE FOR ELECTRICITY APPLICATIONS: HOW STORAGE SYSTEMS CREATE VALUE

The shift toward an expanded role for battery energy storage in the deregulated electricity market became evident by the late 1980s and early 1990s. Two studies by Sandia National Laboratories in 1993 and 1994 identified specific opportunities for battery energy storage in the generation as well as on the T&D side of the electric grid. The first report, titled "Battery Energy Storage: A Preliminary Assessment of National Benefits (The Gateway Benefits Study),"[1] identified potential battery storage application requirements and a preliminary estimate of potential costs and benefits of these applications for the entire U.S. electric grid. These potential applications included:

- Spinning reserve
- Capacity deferral
- Generation dispatch
- Transmission line deferral
- Distribution substation facility deferral
- Demand-side management

This report advocated the concept of "distributed" energy storage within the grid, in contrast to the "centralized" approach for large pumped hydro installations. The distributed approach reduced the storage requirements from 8+ h to less than 4 h for the applications studied.

A subsequent report in 1994, "Battery Energy Storage for Utility Applications: Phase I–Opportunities Analysis,"[2] and the follow-up Phase II study,[3] expanded the Gateway study and identified more than 10 battery applications within the electric grid, and for the first time matched preferred battery technologies to application requirements. The applications identified in this study dispelled the prevailing notion that battery energy storage could only be used in a load-leveling mode within the electric grid. The applications described in the Phase II study are listed in Table 30.1, and as shown in Table 30.2, the required battery size in MW, approximate storage time, and the number of duty cycles per year of operation were estimated. Since publication of these reports, additional applications (for both on- and off-grid) have been identified, although some studies use different terminologies for the same application.

Table 30.2 indicates that battery storage systems in the 1 to 10 MW size range, with storage capacities of 15 min to 4 h, could satisfy the majority of the application requirements. The study concluded that with these requirements, all three battery technologies that were prevalent at that time—lead-acid, sodium/sulfur, and zinc/bromine—could meet the requirements of almost all these utility applications.

Further, the Opportunities Analyses presented illustrative scenarios for each of the applications showing the specific response of the battery to each state or condition, as described below.[3]

30.3.1 Rapid Reserve

The National Electricity Reliability Council (NERC) requires utilities to avoid interruption of service to customers, even if an electrical generating unit fails. The reserve power supply must have instantaneous response to comply with NERC Policy 10 requirements. Satisfying this requirement can represent a significant cost to power producers.

Because cold thermal power plants require hours and combustion turbines require a half-hour to get generators ready to accept load, utilities operate thermal plants and combustion turbines at less-than-full capacity to keep generators hot and spinning and ready to provide reserve power. Energy storage can help utilities maintain rapid reserve, reduce or eliminate the need for supplemental power from combustion turbines, and free thermal plants to generate at full capacity (for greater efficiency and economy). Storage systems designed for rapid reserve can replace generation units that fail and provide power until the utility brings other sources of power on-line or repairs the failed unit.

Since the power plants that they would temporarily replace have power ratings on the order of 10 to 100 MW, storage systems for rapid reserve must have power capacities in this same range. Generation outages that require rapid reserve typically occur about 20 to 50 times per year. These outages occur randomly. Therefore, storage facilities for rapid reserve must be able to address up to 50 significant discharges that occur randomly through the year.

TABLE 30.1 Definitions and Categories of Electric Power Applications of Energy Storage[3]

Category	Application name and definition
Generation	**Rapid reserve** Generation capacity that a utility holds in reserve to meet North American Electric Reliability Council (NERC) Policy 10* requirements to prevent interruption of service to customers in the event of a failure of an operating generating station. **Area control and frequency responsive reserve** The ability for grid-connected utilities to prevent unplanned transfer of power between themselves and neighboring utilities (area control) and the ability of isolated utilities to instantaneously respond to frequency deviations (frequency responsive reserve). Both applications stem from NERC Policy 10 requirements. **Commodity storage** Storage of inexpensive off-peak power for dispatch during relatively expensive on-peak hours. In this report, commodity storage refers to applications that require less than 4 h of storage.
Transmission and distribution	**Transmission system stability** Ability to keep all components on a transmission line in sync with each other and prevent system collapse. **Transmission voltage regulation** Ability to maintain the voltages at the generation and load ends of a transmission line within 5% of each other. **Transmission facility deferral** Ability of a utility to postpone installation of new transmission lines and transformers by supplementing the existing facilities with another resource. **Distribution facility deferral** Ability of a utility to postpone installation of new distribution lines and transformers by supplementing the existing facilities with another resource.
Customer service	**Customer energy management** Dispatching energy stored during off-peak or low cost times to manage demand on utility-sourced power. **Renewable energy management** Applications through which renewable power is available during peak utility demand (coincident peak) and available at a consistent level. **Power quality and reliability** Ability to prevent voltage spikes, voltage sags, and power outages that last for a few cycles (less than 1 s) to minutes from causing data and production loss for customers.

*Available for download at www.nerc.com/~oc/

Figure 30.4a illustrates the generation capacity of a utility for a typical week in which a significant failure occurs; the balloon shows the detail of the capacity loss and recovery with appropriate resources. Figure 30.4b shows how storage would respond to the demand to maintain the utility's ability to satisfy the load.

30.3.2 Area Control and Frequency Responsive Reserve

NERC requires that electric power producers deliver power to and draw power from their neighbors according to prearranged power transfers. This requirement stems from the fact that large changes in electrical load affect the operating speed of generators at power plants. The frequency of the electricity that the generators produce depends on the operating speeds of the generators. When the electrical frequency differs significantly from the 60 cycles per second (Hz) for which electrical equipment in the United States is designed, both the customers' equipment and the utilities' generators can be

30.8 SPECIALIZED BATTERY SYSTEMS

TABLE 30.2 Summary of Applications Requirements[3]

Application	Power*	Storage (min)	AC voltage (kV_{RMS})	Floor space (importance)	Portability (importance)	Duty cycle requirements	Special demands
Rapid reserve	10^1–10^2 real	10^1–10^2	10^1–10^2	Medium	Low	10^1/year, random, discharge only	None
Area control and frequency responsive reserve	10^1–10^2 real	Charge-discharge cycles of <10^1	10^1–10^2	Low	Low	Random, continuous charge-discharge cycles clustered in 2 h blocks daily	None
Commodity storage	10^0–10^2 real	10^2–10^3	10^1–10^3	Medium	Negligible	10^2/year, regular, periodic, weekday block discharge, increased use in shoulder months	Harmonics are more important than in other generation applications
Transmission system stability	10^1–10^2 complex	10^{-3}–10^{-1}	10^1–10^3	Medium	Low	10^2/year, random, charge and discharge cycles	None
Transmission voltage regulation	10^0–10^1 reactive	10^1–10^2	10^1–10^2	Medium	High	10^2/year, random charge and discharge cycles typically weekdays, seasonal by region, at least 6–7 months	Safety concerns are important
Transmission facility deferral	10^{-1}–10^1 complex	10^2	10^1–10^2	High	High	10^2/year, most likely during weekday peaks, charge and discharge	Safety concerns are important
Distribution facility deferral	10^{-1}–10^0 real	10^2	10^0–10^1	High	High	10^2/year, most likely during weekday peaks, charge and discharge	Safety concerns are important
Customer energy management	10^{-2}–10^1 complex	10^1–10^2	10^{-1}–10^1	High	Varies	10^2–10^3/year, regular periods	Safety concerns are important
Renewable energy management	10^{-2}–10^2 complex	10^{-3}–10^3	Variable	High	High	10^2–10^3/year, regular periods, discharge only, unpredictable source	Hostile environments, including extreme heat and cold, particulates and corrosive atmospheres
Power quality and reliability	10^{-2}–10^1 complex	10^{-3}–10^0	10^{-1}–10^1	High	Varies	10^2–10^3/year, irregular periods, charge and discharge	Safety concerns are important

*Real (MW), Reactive (MVAR), or Complex (MVA)

damaged. To regulate frequency, utilities can install storage systems that discharge to meet rising load, and charge when loads fall-off. In this way, the storage system protects the generator from the fluctuation in load, and prevents subsequent frequency variations.

Isolated utilities are not subject to neighbors' power fluctuations, but these utilities with no connection to a large stabilizing grid are very vulnerable to customers' load-switching and failures of small generation plants. Isolated utilities have no neighbors from which they can draw or to

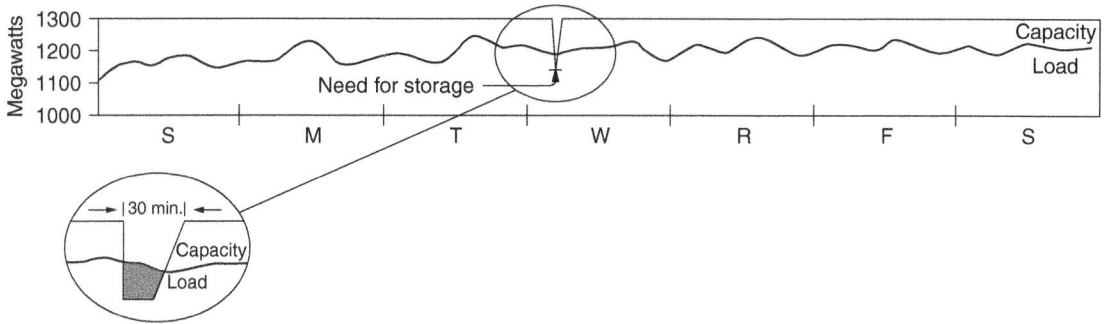

FIGURE 30.4a System need for rapid reserve power.

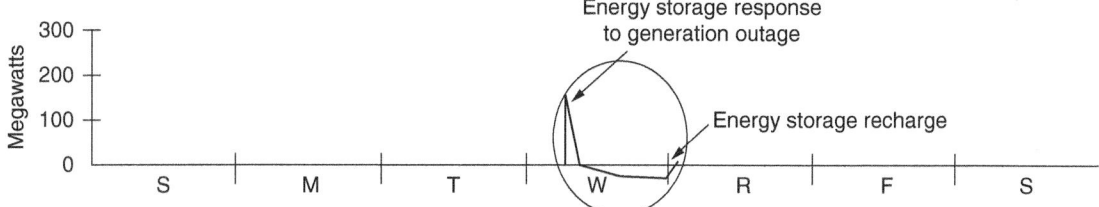

FIGURE 30.4b Storage response to provide rapid reserve power.

which they can feed power. They must balance the generation and load without outside resources. To achieve such area regulation and frequency control, both interconnected and island utilities can install storage systems to accept unwanted power during customer load-drop and deliver additional power during customer load-rise or during an outage of a small generating station. Such storage systems would have to deliver on the order of 10 to 100 MW to absorb and deliver power as it fluctuates. The system would have to be able to dispatch continuously, especially during peak load times, in frequent, shallow charging and discharging that would occur. Peak loading may occur up to 250 weekdays each year for most utilities, and the fluctuations are numerous during those periods, but have total durations of 10 min or less. During low demand periods, when conventional equipment provides frequency and area control, the storage system would be inactive.

Figure 30.5a shows unscheduled power imbalances between one utility's power output and the power level of neighboring utilities on the grid. Figure 30.5b shows how storage would respond to help maintain a scheduled transfer of power.

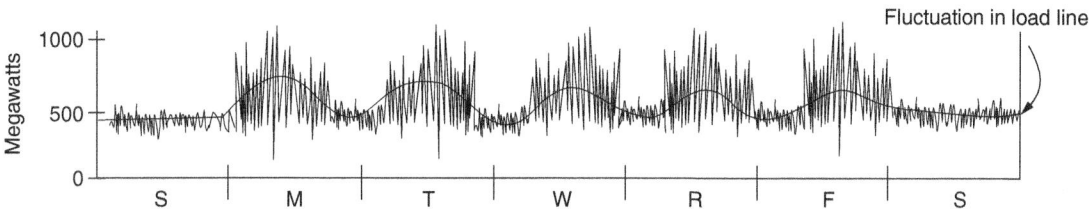

FIGURE 30.5a System need for area control and frequency responsive reserve.

30.10 SPECIALIZED BATTERY SYSTEMS

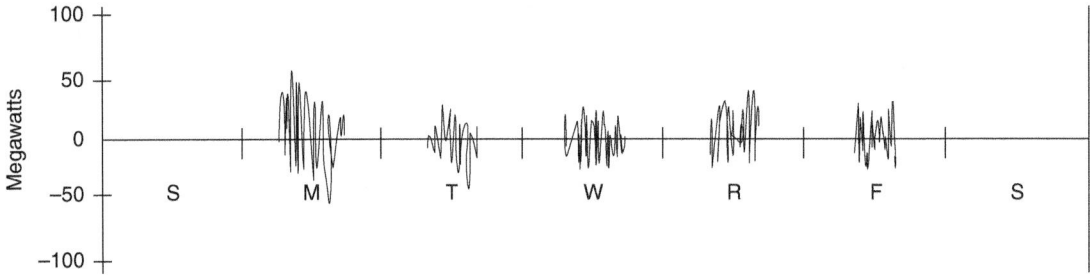

FIGURE 30.5b Storage response to provide area control or frequency responsive reserve.

30.3.3 Commodity Storage

During peak load times, utilities often need to operate costly combustion-turbine units to meet customer demands. With energy storage, utilities can store electricity produced by inexpensive base-loaded units during off-peak hours and discharge power during peak demand times. Leveling out the load demand in this way allows utilities to improve profitability by selling power produced during off-peak times at premium on-peak rates. Although commodity storage (previously referred to as "load-leveling") was the first application that utilities recognized for energy storage, the differences in the marginal cost of generation during peak and off-peak periods for many utilities are quite small. Therefore, commodity storage is generally a secondary benefit that utilities derive from an energy storage system installed for other applications that offer greater economic benefits.

Commodity storage applications require energy storage systems that are on the order of at least 1 MW and up to hundreds of MW. The systems must have several hours of storage capacity (between 2 and 8 h). For utilities without a seasonal demand variation, a system used for commodity storage would operate on weekdays (250 days per year). For utilities that experience seasonal peaking, commodity energy storage systems might operate much less frequently. Operation would be clustered during seasonal peaking months.

Figure 30.6a shows a typical utility load shape and the amount of peaking reduction that an energy storage system used for this application would have to supply. Figure 30.6b shows how storage could offer electricity on demand, using low-cost power as a high-priced commodity.

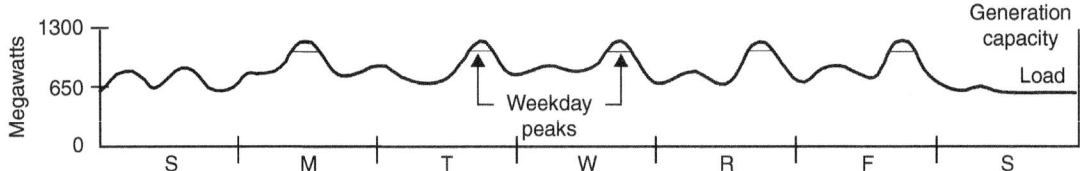

FIGURE 30.6a Typical load profile for an aggregation of customers. Peak has potential to approach capacity as load grows.

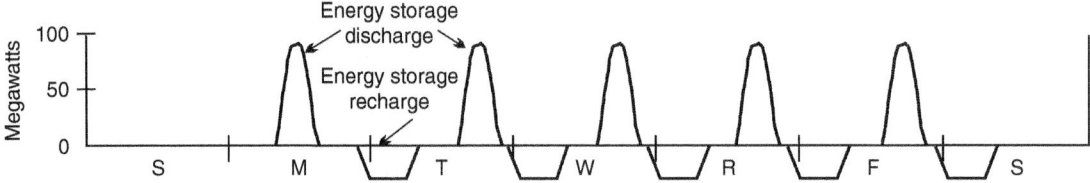

FIGURE 30.6b Energy storage response to commodity storage application demands.

30.3.4 Transmission System Stability

NERC requires numerous safeguards regarding the stability of the nation's transmission system. These safeguards are the topic of debate as the country moves to a competitive electricity industry and power providers seek ways in which they can avoid the cost of maintaining stability. This problem is especially difficult since many events in routine utility operation can cause instability in transmission systems. Events as common as customers switching loads, lightning strikes, and generators going on- or off-line cause generators in the system to fall out of sync with the rest of the system. The difference between the phase-angle of a generator and the phase-angle of the load-end of the transmission line measures the synchronization and stability of the system. If the difference between those angles is too large and the utility cannot quickly (within a few cycles) damp unstable oscillations, the power system can collapse. In this very undesirable circumstance, the utility must shut down and restart its equipment to resynchronize the system.

Energy storage systems can help utilities maintain synchronous operation of their systems by discharging to provide power and charging to absorb power as system loading conditions change. Energy storage systems for transmission line stability require power in the hundreds of MW, have a self-commutated converter (to provide real and reactive power), and have enough storage capacity to discharge at full power from a minute up to hours. Energy storage system operation could typically occur about 100 times annually.

Figure 30.7a illustrates two instances of transmission line stability; both events take the generator away from synchronous operation with the system, and toward an angle difference that could cause system collapse. The balloon shows an expanded time-scale of the first transient event, and the generator's return to stable operation. Figure 30.7b shows the energy storage system discharging and charging multi-megawatt pulses into the system to counter instabilities.

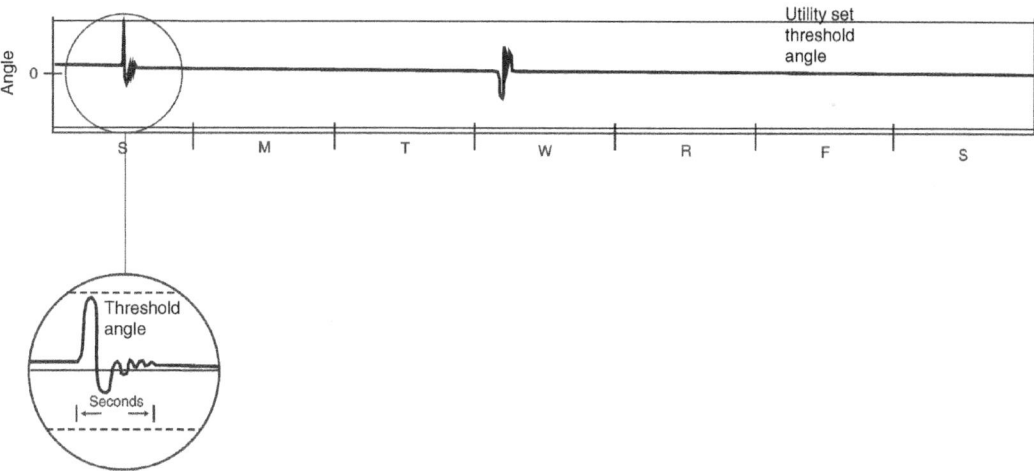

FIGURE 30.7a Storage system needs to address transients to achieve stability.

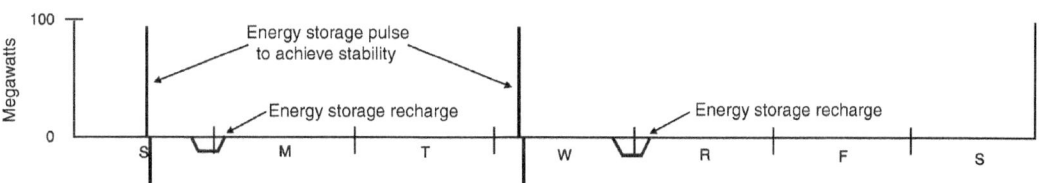

FIGURE 30.7b Storage response to address transients.

30.12 SPECIALIZED BATTERY SYSTEMS

30.3.5 Transmission Voltage Regulation

Without corrective measures, impedance in transmission lines causes the voltage at the generation-end of a line to be greater than voltage at a load location at the other end of the line. To offset this effect, utilities inject reactive power and maintain the same voltage at all locations on the line. Traditionally, fixed and switched capacitors have provided the reactive power (VARs) necessary for voltage regulation. An energy storage system that a utility has installed for some other primary application can provide VARs to the system to augment existing capacitors and replace capacitors planned for future installation.

An energy storage system can provide VARs during discharge, charge, or inactivity. For this reason, utilities can use energy storage systems in megawatt sizes (that have other primary functions in the utility) to achieve voltage regulation on the order of 1 to 10 MVARs. The energy storage system for voltage regulation must provide MVARs for 15 min to an hour during daily load peaks (250 times per year). Peaks might not happen as frequently in regions where loading is seasonal. The power conversion system must be self-commutated to provide reactive power.

Figure 30.8a illustrates high-demand times that might require voltage regulation. Figure 30.8b shows an energy storage system operating to provide VARs during discharge, charge, and inactivity. The circular plots associated with voltage regulation periods show real and reactive power the system provides. As inferred by the plots, the relative magnitudes of VARs and watts that the system provides are not independent, and the inverter must be large enough to provide VARs while discharging at full real power levels.

30.3.6 Transmission Facility Deferral

When growing demand for electricity approaches the capacity of the transmission system, utilities add new lines and transformers. Because load grows gradually, new facilities are designed to be larger than necessary at the time of their installation, and utilities underutilize them during their first

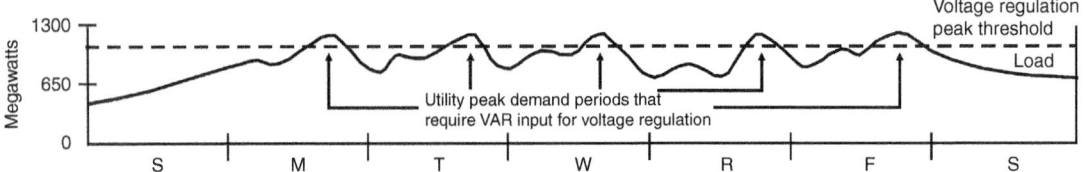

FIGURE 30.8a System need for VARS to correct voltage loss due to impedance.

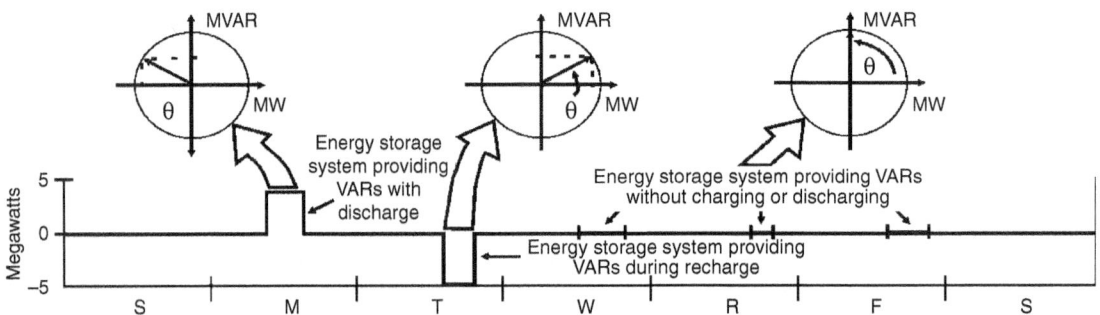

FIGURE 30.8b Storage injecting and absorbing VARS during discharge, charge, and idle.

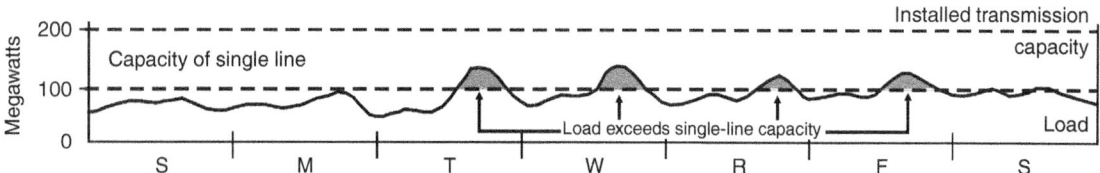

FIGURE 30.9a Load exceeds acceptable percentage of transmission capacity.

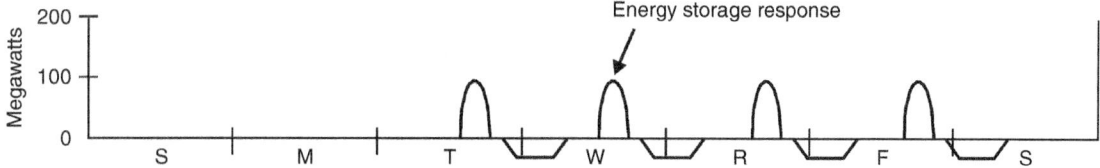

FIGURE 30.9b Storage response to increase transmission capacity until new facilities are cost-justified.

several years of operation. To defer a line or transformer purchase, a utility can employ an energy storage system until load demand will better utilize a new transformer.

Utilities sometimes define the demand at which they need to add transmission facilities as the load at which the transmission system can continue full operation in the event of the loss of one line or transformer. In Fig. 30.9a, the utility has applied this evaluation technique to two 100 MW transmission lines. One power line can carry the entire load during a period of low demand. However, during a high-demand time, a single line cannot provide the power that is needed. Although the transmission capacity does not satisfy the evaluation criterion, existing demand would not fully utilize a third line. The utility could meet the load demand with an energy storage system and defer an expensive facility upgrade.

Figure 30.9b shows an energy storage system operation to help a single transmission line to meet peak demand. Operation would occur hundreds of times per year, mostly during seasonal peaks (when heavy load demand on the lines is more likely). The power requirement for this application would be on the order of hundreds of kW to several hundred MW. The energy storage system would need to provide 1 to 3 h of storage to provide support to the constrained transmission facility.

30.3.7 Distribution Facility Deferral

As load demand approaches the capacity of distribution facilities, utilities add new lines and transformers. Figure 30.10a shows a distribution load that allows an insufficient margin between its peaks and the system capacity. Because demand will continue to grow, utilities install facilities that exceed existing load demands. Therefore, utilities underutilize expensive distribution facilities during their first several years in service. With energy storage systems, utilities can meet current load demands

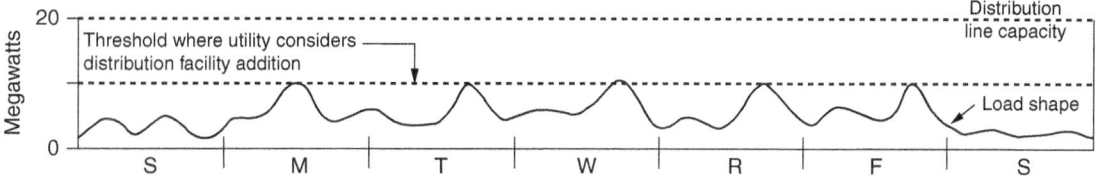

FIGURE 30.10a Distribution load exceeds acceptable percentage of capacity.

30.14   SPECIALIZED BATTERY SYSTEMS

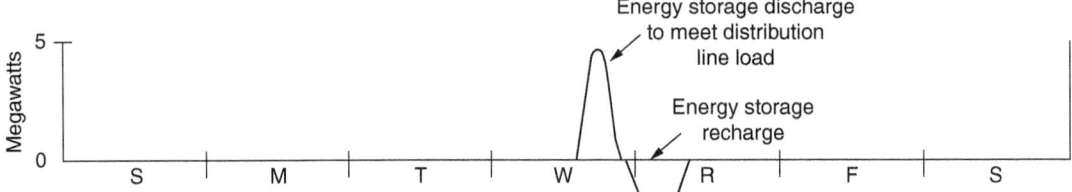

FIGURE 30.10b   Storage provides peak power to temporarily defer purchase of new equipment until it is cost-justified.

with existing distribution facilities, and defer the purchase and installation until the demand better justifies new facilities.

An energy storage system to defer installation of new distribution capacity requires power on the order of tens of kW to a few MW, and must provide 1 to 3 h of storage. In a typical distribution facility, the battery system would operate most frequently during daily high-load periods that occur during seasonal peaks. Figure 30.10b shows the energy storage discharging to meet demand.

30.3.8 Customer Energy Management

Utilities typically charge commercial and industrial customers a monthly fee (peak-demand charge) based on the highest power drawn during the month. By reducing peak demand or by "peak shaving," customers can significantly reduce peak demand charges. Figure 30.11a illustrates the way that customers typically try to reduce monthly demand peaks. At the beginning of the month, the energy storage system shaves the first peak and notes the reduced peak-power level. Then the energy storage system remains idle until power demand exceeds the reference value noted during the previous peak-shaving event. When load exceeds the reference value, the system discharges the battery to shave this peak, and again notes the maximum power that the utility provided to the customer. This process continues until the end of the month, when the system resets.

In Fig. 30.11a, peak shaving occurs twice to the customer's load during the first week in a billing period. The energy management system shaves the first monthly peak, stores the maximum load value (represented by the lowest dashed line) and waits for load to exceed the stored value to operate a second time. Figure 30.11b shows the energy storage operation with discharge for peak

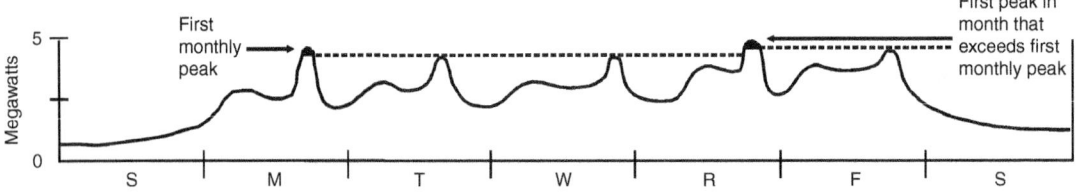

FIGURE 30.11a   Customer demand that creates economic need to reduce peaks.

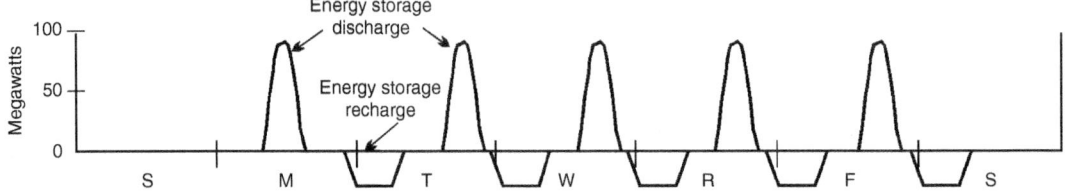

FIGURE 30.11b   Energy storage providing power to manage customer load profile.

shaving and recharge during off-peak hours. In this application, the energy storage system would discharge 7 or 8 times per month, or about 100 times per year. The system size would be in the 10 kW to 1 MW range. The energy storage system would need 1 to 2 hours of storage capacity.

30.3.9 Renewable Energy Management

Energy storage systems have several potential applications for renewable systems. In one near-term application, a storage system can help deliver renewable energy when it is most needed. By storing power from renewable energy systems that produce power at times that do not coincide with the utility system demand peak, the owner of the renewable resource can deliver power at peak times, and create "coincident peaks" between utility demand and the renewable supply. Because utilities will pay a higher rate for renewable energy delivered on peak, renewable power delivered during the utility peak has greater economic value. The "value" of the electricity produced by renewables will be a growing driver in our nation's inclusion of green resources in the generation mix.

Another near-term renewable system application for energy storage takes energy from a source with variable power and delivers reliable, constant power on demand. Because utilities must guarantee the amount of power they have available, such power "firming" makes variable renewable sources more viable and adds to their economic value.

The storage system for either application would need to provide from 10 kW to 100 MW. The storage system would need storage capacity in the fractions of seconds to address transient fluctuations and 1 to 10 h for diurnal storage or coincident peaking. For coincident peaking, the storage system would discharge about 250 times per year during weekday utility peaks. For power firming, the storage system would charge and discharge randomly, as renewable sources wax and wane.

In the long term, a utility with a significant percentage of renewable power may require storage capacity of days to weeks to ride through periods with cloudy skies or windless days. However, this application is still on the horizon of energy storage development.

Figure 30.12a shows the utility load shape with daily peaks in the afternoon and early evening. Figure 30.12b shows the storage response to make energy available coincident with demand peaks.

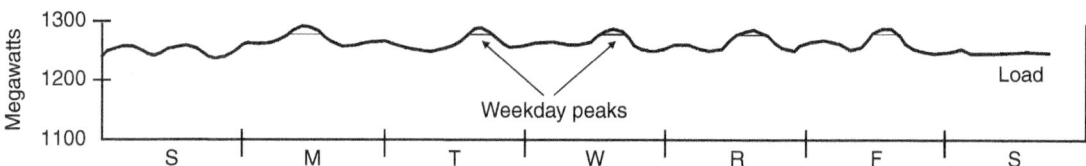

FIGURE 30.12a Aggregate load peaks during a typical week that do not coincide with renewable production peaks.

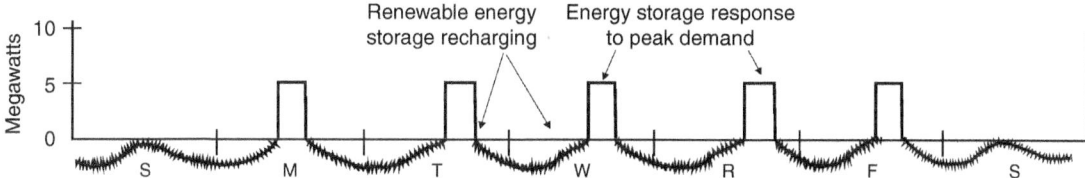

FIGURE 30.12b Off-peak storage of renewable energy for on-peak dispatch to increase capacity credit and economic value of renewables.

30.3.10 Power Quality and Reliability

Small industrial and commercial customers often operate sensitive electronic systems that cannot tolerate voltage sags, spikes, or loss of power. The duration of a power sag may be only one or two cycles (~1/60th of a second) but the effects can be costly. Microprocessors on assembly lines may shut down, and production and data processing suffer. Figure 30.13a illustrates a momentary voltage spike that might cause such production loss.

To protect these electronic devices, customers can install energy storage systems to prevent power sags, spikes, and failures from ever reaching their equipment. If an energy storage system operates in parallel with the load, the battery system disconnects load from a faulted power supply, and provides power until normal utility voltage returns. If an energy system operates in series with the load, the power conversion system always operates. However, energy storage provides power only when voltage sags and interruptions occur. The energy storage system would require hundreds of kilowatts and 15 min of storage.

Voltage sags, spikes, and power loss typically occur about 10 times a year. A self-commutated converter is necessary to reform 60 Hz voltage.

Figure 30.13b shows a storage system installed in parallel with the load where it operates all of the time and provides or absorbs backup power as needed.

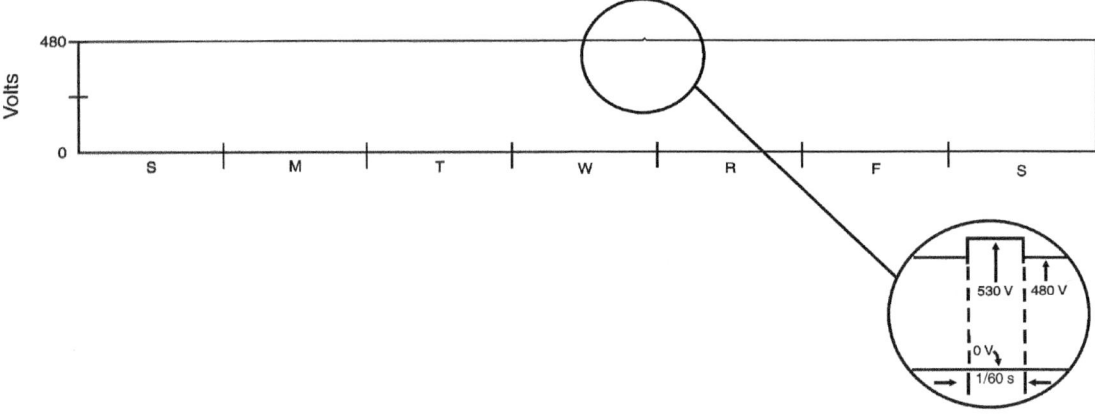

FIGURE 30.13a A 50 V transient spike on a 480 V line that could cause an outage at a production facility.

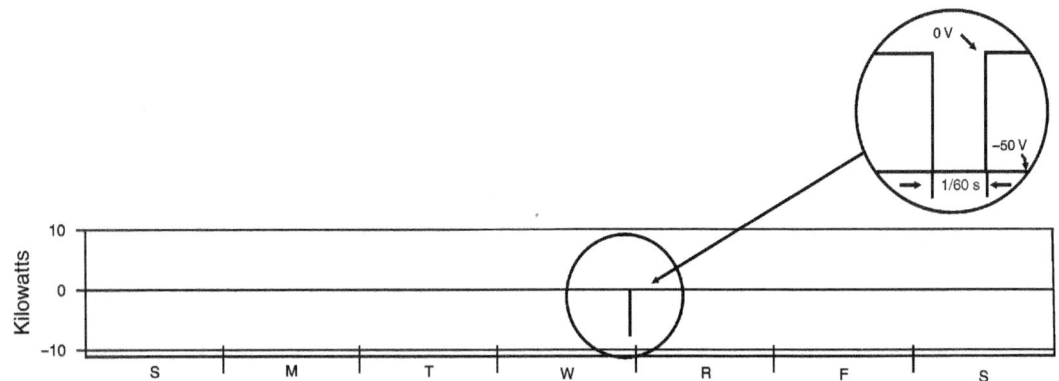

FIGURE 30.13b Storage delivers and absorbs power fluctuations as needed.

These examples of the many uses of battery energy storage were postulated in these studies, and since then most of these applications have been successfully used by electric utilities nationwide, and there is a better understanding of how battery systems can be used effectively to support the electric grid. Notable examples of early uses include:

- Puerto Rico Electric Power Authority battery system at Sabana Llana substation, San Juan, installed to provide spinning reserve and frequency control for the island's electric grid.
- Fairbanks, Alaska, battery owned by Golden Valley Electric Association, installed to provide transmission line stability and voltage regulation to the Fairbanks local grid.

Details of both of these installations are described later in this chapter.

These studies also introduced another important concept that has been instrumental in creating a better appreciation of the economic and functional value of battery energy storage on the electric grid. This was the concept of "clustered" or multiple benefits, such that the same battery system could offer several benefits to the utility grid. Thus, a single battery system could perform more than one application to capture multiple benefits, beyond that which could accrue from a single investment, by either the utility or its customers. Three groups of applications were proposed,[2] each consisting of several mutually compatible applications for certain sizes and types of battery systems, as shown in Table 30.3.

A cost/benefit calculation at today's price points shows that it is usually necessary to combine multiple applications for a single energy storage system to achieve a favorable return on investment.

During the same period, the mid-1980s to late 1990s, there were significant battery energy storage developments in Japan, most notably the Moonlight Project,[8] which was a multiyear undertaking by the Japanese Ministry of International Trade and Industry (MITI). One of the goals of this national initiative was to develop a portfolio of battery technologies, including lead-acid and sodium/sulfur. Two large lead-acid battery demonstration projects were built in Japan, but the major development focus soon shifted to sodium/sulfur as the preferred technology due to its performance characteristics. These efforts are described in more detail in the section on sodium-beta batteries below.

TABLE 30.3 Mutually Compatible, Combined Applications for Energy Storage[2]

Group	Applications	Power* (MW)	Discharge duration (h)	Discharge depth	Discharge frequency (in a 24 h period)
I	Spinning reserve, area/frequency regulation, load-leveling, generation capacity deferral	10–100	1–3	Shallow (A/F regulation)	Continuous charge and discharge, 250 weekdays
				Medium (load leveling/gen. deferral)	One discharge-charge cycle, 250 weekdays
				Deep (spinning reserve)	One discharge-charge cycle, 20–50 days per year
II	Distribution facility deferral, voltage regulation	1	1–3	Shallow (voltage regulation)	Minimal storage for VAR injection, 250 weekdays
				Medium (distribution facility deferral)	One discharge-charge cycle, less than 30 days per year
				Deep (distribution facility deferral)	One discharge-charge cycle, less than 30 days per year
III	Reliability (UPS), power quality, peak shaving	0.1–1	1–2	Shallow (power quality and reliability)	One discharge-charge cycle, less than 20 days per year
				Medium (reliability)	One discharge-charge cycle, less than 20 days per year
				Deep (peak shaving)	One or two charge-discharge cycles, 250 days per year

*Values in this column reflect order of magnitude of power that combined applications require.

The deregulation of the electric utility companies and the increasingly widespread deployment of renewables continue to open greater opportunities for battery energy storage. Other studies that examined the specific role of battery energy storage in a deregulated environment were performed in the 2000 decade.[9–11] Further information is available from the Electricity Storage Association (ESA) website, at www.electricitystorage.org/. In addition, the Sandia National Laboratories, Energy Storage Systems website is a valuable resource, at www.sandia.gov/ess. Finally, U.S. government stimulus funds under the Advanced Research Projects Agency-Energy (ARPA-E) have been invested in energy storage technologies, and a number of technologies and demonstrations were initiated in early 2010. Please see the ARPA-E website for current information: arpa-e.energy.gov/FundedProjects.aspx.

30.4 LANDMARK BATTERY ENERGY STORAGE SYSTEMS

Many battery systems have been installed in a variety of innovative and groundbreaking electric supply systems. The following systems are viewed by many as the most significant, in terms of size, capabilities, or novel applications. Most of these are in the United States, although many others have been installed in Europe and in the far east.[12] The following summaries reflect the major attributes of these systems and their impacts on the continuing use of electrical energy storage.

30.4.1 Crescent Electric Membership Cooperative (now EnergyUnited) BESS, Statesville, NC[12,13]

Application: Peak shaving

Operational dates: 1987 to May 2002

Power: 500 kW

Energy: 500 kWh

Battery type: Lead-acid, flooded cell

Cell size: 2080 Ah at C/5; 324 cells

Battery manufacturer: GNB Industrial Battery, now Exide Battery

Power conversion system manufacturer: Firing Circuits, line-commutated, 12-pulse thyristor converter

The Crescent battery was obtained for a peak shaving application, and represents the first application of battery energy storage for electric utility applications on the U.S. electric grid. The battery was originally installed in 1983 at the Battery Energy Test Facility (BEST) that was built in the Public Service Electric and Gas (PSE&G) NJ service area as a joint venture of EPRI and the U.S. Department of Energy (with PSE&G as the host utility), to establish a laboratory for testing large stationary batteries for electric utility applications. After completing an initial test phase, during which this battery was used for a few hundred charge-discharge cycles, it was purchased by the North Carolina Electric Membership Cooperative (EMC) and moved to the Crescent EMC location in Statesville, NC, in 1987.

Crescent EMC was a rural cooperative electric utility that purchased bulk power from Duke Energy and supplied it to end-use customers in its service area. Crescent EMC and Duke had an arrangement whereby the 500 kW battery was used to reduce peak demand during peak load conditions that offered mutual benefits to Crescent EMC and Duke. The battery controls allowed it to perform this peak shaving function in two modes: a 500 kW constant power discharge for 1 h, or a lower, 200 kW constant discharge for 3 h, as determined by Duke Energy's system load requirements for that specific day. The battery was required only on those days when Duke Energy needed to limit the peaks, and this occurred infrequently, on either hot summer afternoons or very cold winter days.

The battery performed its required duty cycle reliably from the time of its installation in 1987 and proved to be very robust. There are no specific records of the discharge cycles the battery performed during its operational life, but it far outlived its original warranty of 8 years, and it is believed that it exceeded its design life of 2000 cycles. As of 1995, the battery had performed its intended function 95% of the time, and saved Crescent EMC over $223,000. Capacity tests performed around 1999 confirmed that the cells still retained their original capacity of over 2000 Ah. It is also pertinent to note that the battery was installed in a modest, metal, prefabricated building, was cooled only with ambient air ventilation, and required minimal routine maintenance during its operational life. However, in May 2002, the battery charger apparently malfunctioned, causing a significant battery overcharge and likely damage. EnergyUnited performed an analysis of the system, considering its age, recent performance (the battery's capacity had begun to decline noticeably), and the cost to repair/replace the PCS and the battery, and made a business decision to decommission the BESS.

30.4.2 Berliner Kraft- und Licht (BEWAG) Battery System, Berlin, Germany[14]

Application: Frequency regulation and spinning reserve

Operational dates: 1987 to 1995

Power: 8.5 MW in 60 min of frequency regulation; 17 MW for 20 min of spinning reserve

Energy: 14 MWh

Battery type: Lead-acid, flooded cell

Battery: 7080 cells in 12 parallel strings of 590 cells each; cell size: 1000 Ah

Battery manufacturer: Hagen OCSM cells

Power conversion system: AEG 8.5 MW, line-commutated, 12 pulse, thyristor-based inverters

The BEWAG battery was the largest battery for utility applications in the world at the time it was commissioned. It was installed to provide frequency regulation and spinning reserve to the West Berlin grid when East and West Berlin were still divided, which technically created an electrical "island" grid since West Berlin was not connected to the grid in Western Europe. The battery system operated in the multifunction frequency regulation and spinning reserve mode until 1993, when the frequency regulation mode was no longer needed after the reunification, and West Berlin was connected to the European grid. However, the battery remained in operation until 1995, in the spinning reserve mode, when it reached its end of life.

There were no major technical problems with this battery system, and it operated near or at its design specification during its entire use period, despite severe operating requirements of the frequency regulation application requirement, as shown in Figs. 30.5a and 30.5b. In this mode, the battery experienced rapid cycling and continuous charge-discharge cycles that are in sync with the frequency deviations of the electric grid. At the time Figs. 30.5a and 30.5b were conceived, it was anticipated that there would be rest periods for the battery when the grid load was low, on weekends or during off-peak periods, when the grid would not have continuous frequency swings. However, experience at the BEWAG facility, and subsequently with the battery system at PREPA, showed that frequency deviations occur continuously and at all times. Consequently, the battery experienced a very high throughput of ampere-hours during its operational life, more so than many other utility applications. It is estimated that the BEWAG battery had an energy throughput almost 7000 times its rated capacity, or approximately 100 GWh during its 9-year life.

30.4.3 Southern California Edison Chino Battery Storage Project[14]

Application: Several demonstration modes, including load-leveling, transmission line stability, T&D deferral, local VAR control, black start

Operational dates: 1988 to 1997

30.20 SPECIALIZED BATTERY SYSTEMS

Power: 10 MW

Energy: 40 MWh

Battery type: Lead-acid, flooded cells

Battery: 8256 cells in 8 parallel strings of 1032 cells each; cell size: 2600 Ah

Battery manufacturer: Exide Batteries GL-35 cells

Power conversion system: GE bidirectional, 18-pulse, stepped wave gate turn-off (GTO), thyristor-based converter

The Chino battery system was a jointly sponsored project by EPRI, the U.S. DOE, and the International Lead Zinc Research Organization (ILZRO), with Southern California Edison (SCE) as the host utility. It was an early demonstration of a utility-scale battery energy storage system for multiple applications, as outlined in Table 30.3 above. At the time of its completion, this became the largest utility battery energy storage system in the world. It was not until the PREPA battery and later the Fairbanks battery systems were commissioned that this distinction was earned by these projects, in 1994 and 1999, respectively.

The power conversion system that controlled the Chino battery and connected it to the SCE grid was built by GE using GTO thyristors. The battery cells had an automatic, compressed-air bubbling agitation system to prevent stratification of the electrolyte, an individual cell watering system, flame arrestors, and arsine/stibine filters mounted on each cell. The battery was controlled through an elaborate Supervisory Control and Data Acquisition (SCADA) system. Some of these were experimental designs that were installed to observe their performance for use in future large-battery systems.

The battery system operated with minor problems, including failures of the automatic watering system, a rain-induced (roof leak) short circuit, cell-case expansion and resulting leaks, and some PCS failures. Overall, the system performed as expected and provided valuable insights for operating battery energy storage in multiple applications. The overall system efficiency (AC to AC), including all losses, was calculated to be 72%.

30.4.4 Puerto Rico Electric Power Authority (PREPA) Battery System[15,16]

Application: Frequency control and spinning reserve

Operational dates: 11/1994 to 12/1999

Power: 20 MW

Energy: 14 MWh

Battery type: Lead-acid, flooded cell

Original battery: 6000 cells in 6 parallel strings of 1000 cells each; cell size: 1600 Ah

Battery manufacturer: C&D Battery

Power conversion system: GE bidirectional, 18-pulse, stepped wave GTO thyristor-based converter

The PREPA was a fully commercial battery system that was acquired for daily operation in a frequency control and spinning reserve mode for the island grid of Puerto Rico. The grid had ongoing stability issues, and routinely had frequency and voltage excursions that could only be controlled by aggressive load-shedding, unless new generation was added to provide regulation and stability. The choices for new generation included fast-acting combustion turbines or battery energy storage. PREPA's analysis showed that battery energy storage systems offered superior operational benefits due to their faster reaction times, both for frequency regulation and spinning reserve requirements. The comparable slower response times of combustion turbines required more installed capacity, whereas the faster reaction time of a battery system meant that a much smaller battery could offer the same functionality as larger sizes of combustion turbines. Typically, battery systems can reach full operating power in less than 1 s, whereas mechanical systems such as combustion turbines need several seconds to minutes to reach their full power output. The seemingly small difference in reaction times translates into very large consequences for the stability of the electric grid, where events that lead to outages propagate within cycles, and a difference of 1 min translates to a complete blackout under some conditions. Thus, it was shown that battery systems were a more cost effective option compared to combustion turbines because a smaller battery could outperform a much larger block of combustion turbines. This is particularly applicable to an "island" system, where there is no interconnection with a neighboring system to balance the momentary shortage of generation resources during an operational contingency.

The PREPA battery was patterned after the BEWAG battery in application as well as battery type. Valve-regulated lead-acid (VRLA) batteries were commercially available by the time the PREPA battery project was started, but PREPA chose a flooded, flat-plate cell because it had a proven track record at BEWAG. However, once utility operations began in 1994, the battery was cycled more frequently than planned, which caused the battery to age more rapidly than expected. This use led to positive-plate growth, which caused cell/jar cracks, leaks, short circuits, and ultimately early battery failure.

PREPA made the decision in 2001 to repower, or replace, the battery. A tubular positive plate, flooded battery was selected, and the new battery was installed in mid-2004. Several problems occurred, however, and the system was taken out of service.

30.4.5 Metlakatla Power and Light (MP&L) Battery System[17,18]

Application: Voltage regulation and displacing diesel generation

Operational dates: 1997 to the present

Power: 1 MW

Energy: 1.4 MWh

Battery type: Valve-regulated lead-acid (VRLA) Absolyte IIP

Battery: 1134 cells/378 each, 100A75 modules in 1 string

Manufacturer: GNB Industrial Battery, now Exide Technologies, and General Electric

MP&L is a rural cooperative that serves the community of Metlakatla on Annette Island, about 25 miles southwest of Ketchikan, Alaska. MP&L generates almost all of its power from hydroelectric generators. At the time the battery system was planned, the MP&L load included an industrial mill that processed lumber and used large 400 and 600 hp motors for debarking and chipping logs. The intermittent operation of these motors produced 600 to 900 kW spikes on the island grid, which could not be controlled by the hydro generation units due to their slow response times and low ramp rates. MP&L acquired a 3.3 MW diesel genset in 1986 to control the power spikes caused by the mill operations. However, the diesel fuel consumption and maintenance requirements were expensive on this remote island community. A battery energy storage system (BESS) was proposed and installed by the manufacturers with the assistance of the Department of Energy and Sandia Labs to reduce the use of the diesel genset. The battery system became operational in February 1997. It operated continuously for 11½ years until September 2008, when the battery was replaced with a new VRLA product. The VRLA battery exceeded its 8-year design life by over 3 years. The BESS is estimated to have saved MP&L over $6.5 million over its life in reduced fuel consumption and maintenance, in addition to reduced greenhouse emissions. A new battery was installed with the support of the Alaska Commerce, Community, and Economic Development Agency, so the BESS will continue to provide benefits to the community for many years to come.

30.4.6 PQ2000 Installation at the Brockway Standard Lithography Plant in Homerville, Georgia[14,19–21]

Application: Power quality, uninterruptable power supply

Operational dates: 1996 to 2001

Power: 2 MW

Energy: 10 s (55 kWh)

Battery type: Lead-acid

Battery: Delco 2000 low-maintenance, truck-starting batteries, 48 per 250 kW module, 8 modules per 2 MW PQ2000 system

Manufacturer: AC Battery, acquired by Omnion Power Engineering in 1997, in turn acquired by S&C Electric in 1999

The PQ2000 battery system was designed to be a factory-assembled, transportable unit that would protect an entire facility from power disturbances. Until this system was introduced, uninterruptable power supply (UPS) systems protected only a single piece of equipment or small group of items. The PQ2000 system detected variations in supply voltage at the utility interface, and should the utility voltage exceed the voltage limits (either high or low), the system disconnected from the utility and supplied up to 2MW of electrical power from the battery. Transfer from utility power to battery backup typically occurred in ½ to 1 ½ cycles, or 8 to 25 ms. The 15 s duration of battery power is sufficient to correct over 80% of the voltage sags occurring on a typical utility. After the utility returned to normal, the PQ2000 resynchronized with the utility AC-wave form and returned the facility to utility power. A diesel generator backup was added as an option to allow for extended outage protection.

A number of prototype tests were conducted, first by Pacific Gas & Electric's (PG&E) system test facility in San Ramon, California, then later at a Salt River Project facility near Phoenix, Arizona, with the support of EPRI. Through a series of acquisitions, S&C Electric continues to offer this system under the Pure Wave trade name.

30.4.7 Golden Valley Electric Association (GVEA) Fairbanks Battery System[22,23]

Application: VAR support, spinning reserve, power system stabilization

Operational dates: 9/19/2003 to the present

Manufacturer; ABB and SAFT

Power: 27 MW for 15 min

Energy: 14.6 MWh

Battery type: Nickel/cadmium type SBH920 cells

Battery: 4 strings of 3440 cells each, for a total of 13,760 cells

The GVEA battery energy storage system (BESS) is one of the largest utility battery systems in operation, and one of the few that utilizes a nickel/cadmium battery. The system was built and installed by a consortium of ABB and SAFT in 2003. It is being used for voltage (VAR) support, spinning reserve, and overall grid system stabilization. The system cost $30.3 million. It is also unusual in that it operates at over 5 kV DC. It has prevented over 200 outages on the grid since it was commissioned. During the first 16 months of operation, the availability of the BESS was over 98%.

30.5 ADVANCED BATTERY TECHNOLOGIES FOR STATIONARY APPLICATIONS

While many battery technologies have been proposed and developed for electrical-energy storage applications, only a handful have actually been used in fielded systems. Some, such as lead-acid batteries and the many variations thereof, and nickel/cadmium, are described above in fielded systems, and in other sections of this Handbook in terms of technical detail. Other technologies, such as the many lithium-ion battery variations, are in their infancy for these applications,[24] and the current state-of-the-art is described in more detail in Chaps. 26 and 27 of this Handbook. The technologies described below are those that are focused on these applications and not described in more detail elsewhere in this Handbook.

30.5.1 Sodium-Beta High-Temperature Batteries[25]

Rechargeable high-temperature battery technologies that utilize metallic sodium offer attractive solutions for many large-scale, electric utility energy storage applications. Candidate uses include many of the applications identified in Table 30.2, including load-leveling, power quality and peak shaving, renewable energy management and integration, and T&D deferral.

A number of sodium-based battery options have been proposed over the years, but the variants that have been developed the furthest are referred to as sodium-beta batteries. This designation is used because of two common and important features: liquid sodium is the active material in the negative electrode, and the ceramic beta-alumina functions as the electrolyte. Sodium/sulfur technology was introduced in the mid-1970s.[26] Its advancement has been pursued in a variety of designs since that time. The attractive properties and the primary limitations of the sodium/sulfur battery are summarized in Table 30.4.

A decade later, the development of the sodium/metal-chloride system was launched.[27] This technology offered potentially easier solutions to some of the development issues that sodium/sulfur was experiencing at the time. A feature comparison between the sodium/nickel-chloride and the sodium/sulfur technologies is presented in Table 30.5.

From the time of their invention through the mid-1990s, these two technologies were among the leading candidates believed to be capable of satisfying the needs of a number of emerging battery energy storage applications. The one application that generated the most interest centered on powering electric vehicles (EVs). However, economic analyses performed by the funding organizations showed that the market was expected to be very slow to develop, and support for most of the developers of the sodium-beta battery technologies was terminated.

TABLE 30.4 Advantages and Limitations of Sodium/Sulfur Battery Technology

Characteristic	Comments
Advantages	
Potential low cost relative to other advanced batteries	Inexpensive raw materials, sealed, no-maintenance configuration
High cycle life	Liquid electrodes
High energy and good power density	Low-density active materials, high cell voltage
Flexible operation	Cells functional over wide range of conditions (rate, depth of discharge, temperature)
High energy efficiency	80+% due to 100% coulombic efficiency and reasonable resistance
Insensitivity to ambient conditions	Sealed high-temperature systems
State-of-charge identification	High resistance at top of charge and straightforward current integration due to 100% coulombic operation
Limitations	
Thermal management	Effective enclosure required to maintain energy efficiency and provide adequate stand time
Safety	Reaction with molten active materials must be controlled
Durable seals	Cell hermeticity required in a corrosive environment
Freeze/thaw durability	Due to the use of a ceramic electrolyte with limited fracture toughness that can be subjected to high levels of thermally driven mechanical stress

TABLE 30.5 Characteristic Comparison of the Sodium/Nickel-Chloride Technology with the Sodium/Sulfur Technology

---

Higher cell open-circuit voltage—2.59 V open-circuit (2.076 V for sodium/sulfur).

Wider operating temperature range—sodium/nickel-chloride can function at temperatures from as low as 220 to 450°C, whereas sodium/sulfur is limited to a range from 290°C to approximately 390°C. However, the range over which practical power levels and long service life have been established is between 270 to 350°C versus 310 to 350°C for sodium/sulfur.

Safer products of reaction—the exothermic heats of reaction are lower and the vapor pressure of the reactants less than atmospheric up to a temperature level of 900°C.

Less metallic component corrosion—the chemistry of the positive electrode is nonaggressive compared to molten Na_2S_x.

Assembly in the fully discharged state without the handling of metallic sodium—a discharged positive electrode can be used.

Reliable failure mode—if the electrolyte fails, sodium will react with the secondary electrolyte to short-circuit the cell.

No freeze/thaw limitation—the thermally induced mechanical stress on the electrolyte is lower due to (a) the positive electrode is located inside of the electrolyte; (b) a smaller difference exists between the solidification temperature of the positive electrode and ambient, and (c) there is less mismatch in thermal expansion between the secondary and primary electrolyte.

Easier reclamation—primarily because of the value of the nickel in spent batteries (<2 kg/kWh), reclamation is an economic necessity. Due to the cell configuration, recycling is a straightforward process. Recovery of the nickel pays for the recycling.

The relatively low power density observed in early 1990s cells that incorporated a tubular electrolyte configuration has been overcome. The lower power was caused by higher cell resistance, especially near the end of discharge. The improved cell design uses a cruciform shaped electrolyte and incorporates a doping addition in the positive electrode material.

Slightly lower energy density.

As described earlier, development of the sodium/metal-chloride technology in Europe along with sodium/sulfur development for stationary applications continued in Japan. For historical purposes, the major sodium-beta battery developers are listed in Table 30.6, along with their primary applications.

More detailed technical information on these battery technologies is included below since they are not covered elsewhere in the Handbook.

TABLE 30.6 Principal Sodium-Beta Battery Developers Since the 1970s

Company	Abbreviation	Country	Primary application	Status
Sodium/Sulfur				
NGK Insulator, Ltd.—Tokyo Electric Power Co.	NGK	Japan	Stationary	Active
Yuasa Corp.	YU	Japan	Stationary	Terminated
Hitachi Ltd.	HIT	Japan	Stationary	Terminated
Silent Power Ltd.	SPL	U.K. U.S.	Motive Stationary	Terminated
Asea Brown Boveri	ABB	Germany	Motive	Terminated
Eagle-Picher Technologies	EPT	U.S.	Stationary	Resuming
Sodium/Nickel-Chloride				
MES-DEA SA	MES	Switzerland	Motive	Active

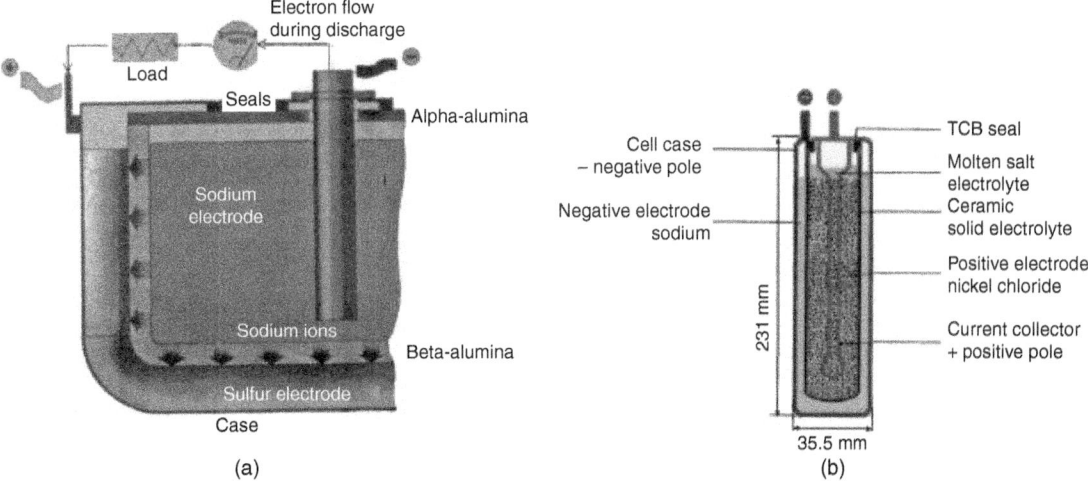

FIGURE 30.14 Diagrams showing the basic functionality of the two types of sodium-beta cells: (a) sodium/sulfur, and (b) sodium/nickel-chloride. (Diagram *b* is courtesy of MES-DEA SA.)

30.5.2 Description of the Electrochemical Systems

The basic cell structure and associated electrochemistry of the two sodium-beta technologies are depicted in Fig. 30.14. As introduced in the previous section, both sodium-beta cells use the solid, sodium ion-conducting β''-Al_2O_3 electrolyte. Cells must be operated at a sufficiently high temperature (270 to 350°C) to keep all (Na/S) or portions (Na/MeCl$_2$) of the active electrode materials in a molten state and to ensure adequate ionic conductivity through the β''-Al_2O_3 electrolyte.

30.5.3 Sodium/Sulfur Electrochemistry

During discharge, sodium (negative electrode) is oxidized at the sodium/β''-Al_2O_3 interface, forming Na$^+$ ions. These ions migrate through the electrolyte and combine with sulfur that is being reduced in the positive electrode compartment to form sodium pentasulfide (Na$_2$S$_5$). The sodium pentasulfide is immiscible with the remaining sulfur, thus forming a two-phase liquid mixture. After all of the free sulfur phase is consumed, the Na$_2$S$_5$ is progressively converted into single-phase sodium polysulfides with progressively higher sulfur content (Na$_2$S$_{5-x}$). During charge, these chemical reactions are reversed. The two half-cell and full-cell reactions are as follows:

Negative electrode: $\quad 2Na \underset{\text{charge}}{\overset{\text{discharge}}{\rightleftarrows}} 2Na^+ + 2e^-$

Positive electrode: $\quad xS + 2e^- \underset{\text{charge}}{\overset{\text{discharge}}{\rightleftarrows}} S_x^{-2}$

Overall cell: $\quad 2Na + xS \underset{\text{charge}}{\overset{\text{discharge}}{\rightleftarrows}} Na_2S_x \quad (x = 5-3) \quad E_{ocv} = 2.076 - 1.78 \text{ V}$

Although the actual electrical characteristics of sodium/sulfur cells are design dependent, the general voltage behavior follows that predicted by thermodynamics. Typical cell performance is shown in Fig. 30.15. This figure is a plot of the equilibrium potential, or open-circuit voltage, and the working voltages (charge and discharge) as a function of depth of discharge. The open-circuit

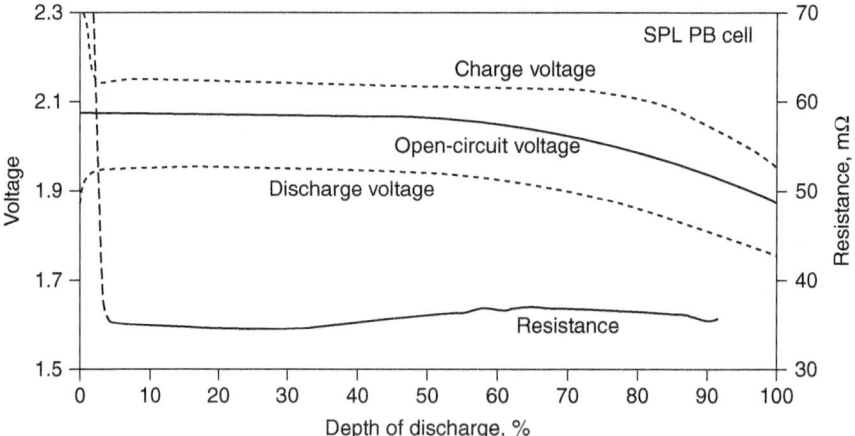

FIGURE 30.15 Cell voltage and resistance as a function of depth of discharge for SPL PB cell. Discharge rate $C/3$; charge rate $C/5$.

voltage is constant (at 2.076 V) during the 60 to 75% of the discharge when the two-phase mixture of sulfur and Na_2S_5 is present. The voltage then linearly decreases in the single phase Na_2S_x region to the selected end-of-discharge point. End of discharge is normally defined at open-circuit voltages of 1.78 to 1.9 V. The approximate sodium polysulfide composition corresponding to 1.9 V per cell is Na_2S_4; for 1.78 V per cell, it is Na_2S_3. Many developers choose to limit the discharge to less than 100% of theoretical (such as to 1.9 V) for two reasons: (1) the corrosivity of Na_2S_x increases as x decreases, and (2) to prevent local cell overdischarge due to possible non-uniformities within the battery (e.g., temperature or depth of discharge). If the discharge is continued past Na_2S_3, another two-phase mixture again forms, except that now the second phase is solid Na_2S_2. The formation of Na_2S_2 is undesirable in the cell because high internal resistance, very poor rechargeability, and structural damage to the electrolyte can result.

Several other important characteristics of the sodium/sulfur electrochemical couple are evident from Fig. 30.15. At high states-of-charge, the working voltage during charge increases dramatically due to the insulating nature of pure sulfur (shown also by the higher cell resistance). This same factor also causes a slight decrease in cell voltage at the start of discharge. At the $C/3$ discharge rate, the average cell working voltage is approximately 1.9 V. The theoretical specific energy of the electrochemical couple is 755 Wh/kg (to 1.76 V open-circuit). Although not all of the sodium is recovered during the initial charge, cells subsequently deliver 85 to 90% of their theoretical ampere-hour capacity. Finally, the existence of wholly molten reactants and products eliminates the classical morphology-based electrode aging mechanisms, thus yielding an intrinsically long cycle life.

30.5.4 Sodium/Metal-Chloride Electrochemistry

The primary electrochemical difference between the two sodium-beta technologies is the metal-chloride positive electrode. This component contains a molten secondary electrolyte ($NaAlCl_4$) and an insoluble and electrochemically active metal-chloride phase (Fig. 30.14b). The secondary electrolyte is needed to conduct sodium ions from the primary β''-Al_2O_3 electrolyte to the solid metal-chloride electrode. Cells using positive electrodes with two transition metal-chlorides, nickel and iron, have been developed. These specific metals were selected based on their insolubility in the molten $NaAlCl_4$ secondary electrolyte.[27,28] During discharge, the solid metal-chloride is converted to the parent metal and sodium chloride crystals. The overall cell reactions for these two chemistries are as follows:

Nickel-based cell: $NiCl_2 + 2Na \underset{charge}{\overset{discharge}{\rightleftharpoons}} Ni + 2NaCl$ $E_{ocv} = 2.58$ V

Iron-based cell: $FeCl_2 + 2Na \underset{charge}{\overset{discharge}{\rightleftharpoons}} Fe + 2NaCl$ $E_{ocb} = 2.35$ V

The voltage behavior as a function of rate and depth of discharge for an early 1990s vintage sodium/nickel-chloride cell is shown in Fig. 30.16. This figure also includes the thermodynamic potentials for the overall cell reaction and two additional cell reactions that only become active during overcharge and overdischarge, respectively. Past the end of normal discharge, the working voltage quickly drops when all of the nickel chloride is consumed. At this point, the reduction of the $NaAlCl_4$ to aluminum begins to occur according to the following reaction:

$$3Na + NaAlCl_4 \rightleftharpoons 4NaCl + Al \quad E_{ocv} = 1.58 \text{ V}$$

If discharge is continued to the point of complete sodium depletion, electrolyte fracture will occur. However, because of the presence of the metallic aluminum, the cell will remain electrically conductive. This characteristic permits batteries to be configured with long series strings. The quick decrease in cell voltage functions as a reliable indicator for the end of discharge and is used to provide overdischarge protection.

If the cell is overcharged, excess nickel chloride will be produced by the decomposition of the secondary electrolyte according to this reaction

$$Ni + 2NaAlCl_4 \rightleftharpoons 2Na + 2AlCl_3 + NiCl_2 \quad E_{ocv} = 3.05 \text{ V}$$

Although degradation of the positive electrode will occur during excessive overcharge, this reaction will prevent voltage-induced fracturing of the $\beta''\text{-}Al_2O_3$ electrolyte. In practice, cells and batteries can be safely overcharged by more than 50%.

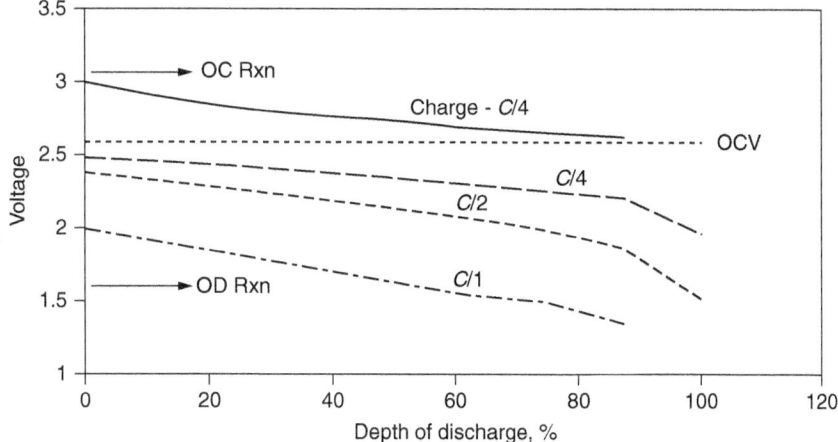

FIGURE 30.16 Cell voltage as a function of depth of discharge for a 1990 vintage sodium/nickel-chloride cell. Lower curves are for three discharge rates and the two arrows indicate the voltage at which the intrinsic overcharge (OC) and overdischarge (OD) reactions occur.

30.5.5 Sodium/Sulfur Cell Technology

The majority of the work on the sodium/sulfur technology has been directed to electric vehicle and stationary energy storage applications. Cells intended for aerospace and defense-related applications (e.g., satellites, submarines, and tanks) have primarily been based on electric vehicle designs. As noted above, the only active development of this technology is currently being performed in Japan for stationary applications.

The dominant organization that is presently developing and commercializing the sodium/sulfur technology is a Japanese collaboration between NGK Insulator, Ltd. and Tokyo Electric Power Company (TEPCO) that started in 1984.[29] Their goal was to develop cells with sufficient energy capacity for use in utility-based load-leveling and peak-shaving applications (e.g., TEPCO's) that require up to an 8 h discharge period. The critical technology for such cells involved the manufacture of large diameter beta-alumina tubes of very high quality and precise dimensions.

The other Japanese developers included Yuasa Corporation and Hitachi Ltd. These organizations originally targeted large-scale utility load-leveling as their prime intended use. Yuasa's initial effort was part of the national Moonlight Project that resulted in the design and fabrication of a large number of 300 Ah cells.[30–32] Hitachi developed several sodium/sulfur cell designs since 1983. They also targeted other applications, including renewable energy storage.[33,34]

Before the discontinuation of their programs in the mid-1980s, two U.S. companies, Ford Aerospace and Communications Corporation (FACC) and General Electric, had also been developing large central-sodium cells for utility load-leveling applications. Silent Power Ltd. (SPL), a company that operated in the U.K. and the U.S., developed a cell specifically for utility applications that had a nominal capacity of 30 Ah.[35] This cell (the XPB) was essentially an extended version of their electric-vehicle cell.

30.5.6 Sodium/Nickel-Chloride Cell Technology

A schematic diagram of a sodium/nickel-chloride cell was shown previously in Fig. 30.14b and a photograph of a representative cell is presented in Fig. 30.17. In this standard configuration, the sodium is located on the outside of the β''-Al_2O_3 electrolyte (outside sodium). An inside sodium configuration would require the use of an expensive nickel container. Another advantage of the outside-sodium cell configuration is that an external cell case with a square cross section can be used. This cell geometry permits maximum volumetric packing efficiency within the battery enclosure to be attained. A third advantage of this configuration is the cell behavior during thermal "freeze/thaw" cycling. Here, detrimental tensile stresses on the electrolyte do not develop, thus effectively eliminating this potential failure mode. The positive electrode itself is contained within the electrolyte. In a fully charged cell, this electrode is a porous nickel matrix that is partially chlorinated to nickel dichloride. The remaining nickel backbone serves as part of the positive-electrode current collector. About 30% of the nickel is used in the cell reactions. The matrix is impregnated with the $NaAlCl_4$ molten salt. The sodium compartment is less complex than that of the sodium/sulfur cell because safety features are not needed. The approach to primary containment for both the outer container and the electrode seals is similar to that with sodium/sulfur. As discharge proceeds, the reactions in the positive electrode occur from the outside of the solid nickel structure and proceed inward through an ever-increasing thickness of reduced nickel. This shrinking-core process results in an increasing electrical resistance as the cell discharges because the effective area of the reacting nickel chloride is constantly being reduced. The chemistry of the discharge process provides another significant advantage relative to sodium/sulfur: cells can be safely assembled with the discharge products (nickel metal and salt) and then charged.

Also in contrast to the sodium/sulfur technology, the development of the sodium/metal chloride system was pursued by a successive progression of single, integrated organizations for one primary application—electric vehicles. The prime developer is the Swiss company, MES-DEA SA. They purchased the technology from AEG Anglo Battery Holdings (AABH), an organization formed by the German company AEG in cooperation with the original developers (Zebra Power Systems

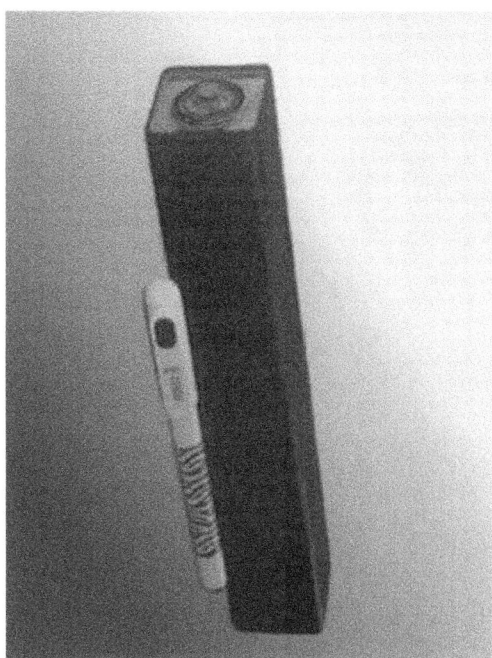

FIGURE 30.17 An MES-DEA sodium/nickel-chloride cell (ML3). For reference, the dimensions of the MES ML3 cell are 36 mm square × 232 mm long. (*Photograph courtesy of MES-DEA SA.*)

and Beta R&D Ltd.). This technology is often referred to as ZEBRA because of its origins. The acronym ZEBRA stands for Zero Emission Battery Research Activities. To date, development has focused almost exclusively on the higher voltage nickel variant, but using iron as a doping addition to the positive electrode. As such, the pure iron-based system will not be discussed further in this chapter.

The cell designs that have been developed to date have capacities ranging from 20 to 200 Ah. Advancements resulted in an improved pulse-power capability (especially near the end of discharge) and energy content.[36] At 80% depth of discharge, the power of an ML3 type cell is 2.5 times that of a 1992 vintage cell (SL09). This enhanced power performance can be determined by comparing the cell data previously presented in Fig. 30.16 with modern performance data shown in Fig. 30.18 for a full-sized battery. The most important modifications involved: (1) the use of a fluted or cruciform-shaped electrolyte that minimized the thickness of the positive electrode and increased the area of the β''-Al_2O_3 electrode; and (2) the introduction of iron as a dopant to the nickel positive electrode. Optimization of the design and improvements in the chemistry of the positive electrode also resulted in a 20 to 40% increase in energy content. The demonstrated reliability of the early 1990s ZEBRA cells was outstanding, with cell failures virtually nonexistent. More recently, further development and testing of stationary ZEBRA energy storage systems are continuing in Canada.[37] A recent article describes the current ZEBRA technology and notes significant interest in stationary applications.[38]

30.5.7 Sodium/Sulfur Battery Design Considerations

Battery-level components include mechanical supports for the cells, a thermal management system (incorporating the thermal enclosure) to ensure that each cell is maintained at a relatively high

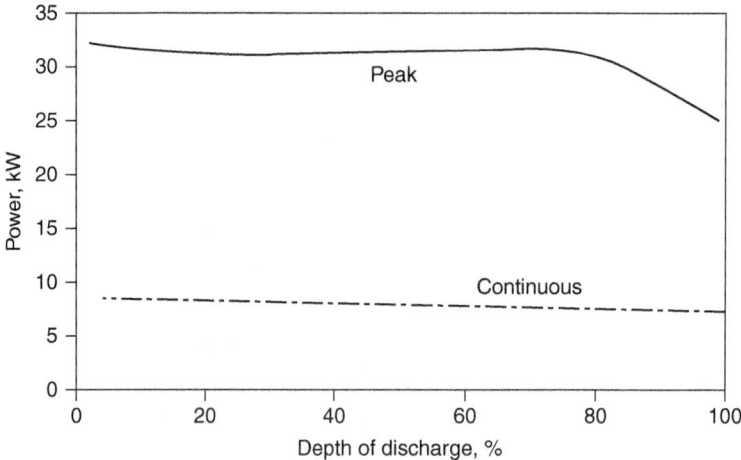

FIGURE 30.18 Power capability (peak and continuous) as a function of depth of discharge for a modern sodium/nickel-chloride battery (Z5C). (*Courtesy of MES-DEA SA.*)

temperature (e.g., for Na/S from 300 to 350°C), electrical interconnects (cell-cell, cell-module, module-battery), possibly cell-failure devices, and safety-related hardware (such as thermal fuses). Batteries are assembled by connecting cells into series and series-parallel arrays to produce the required battery voltage, energy, and power. Electrical heaters are installed within the enclosures to initially warm the cells and then to offset heat loss during periods while the battery is at temperature, but idle. Normally, extra heat is not required during regular discharging and charging due to ohmic heating and chemical reaction effects.

As discussed above, many stationary applications represent a very promising use for the sodium/sulfur technology primarily because of its small relative footprint (high energy density), excellent electrical efficiency if routinely used, ease of thermal management, lack of required maintenance, and cycling flexibility. The developers have adopted a similar design approach for their battery systems that involves the use of self-contained modules, each with 10 to 50 kW of power and 50 to 400 kWh of energy. That is, independent battery-level modules are manufactured that consist of various series-parallel configurations of cells within a thermal enclosure. An analogy to these modules is an integrated electric vehicle battery. The battery itself is then constructed by connecting these modules in a series-parallel arrangement (often in a common structure) to give the desired voltage, energy, and power. The resultant battery is combined with a power-conversion system (PCS) to form an integrated facility that can be connected to an electrical system (either utility or customer side).[39]

Yuasa built and operated the first large sodium/sulfur battery, an integrated 1 MW 8 MWh system, in the early 1990s that contained 26,880 cells.[30-32] As noted earlier, this effort was part of the Japanese national research program called the Moonlight Project.[8] Battery performance was demonstrated in a utility load-leveling application. The basis for this system was a 50 kW, 400 kWh module, and the battery itself consisted of two 10-module strings connected in parallel.[31,32] Similar to the other Japanese developers, Yuasa had been advancing the modular concept using smaller (~25 kW) modules.[33] Hitachi had been pursuing renewable applications for their Na/S batteries. These were hybrid power applications and needed batteries to level power fluctuations from energy sources such as wind and solar. Their program reached the proof-of-concept stage, but was terminated.[34]

The NGK modular concept and its implementation are shown in Fig. 30.19 which shows one of the NGK modules along with an integrated system that contains these same 50 kW modules.

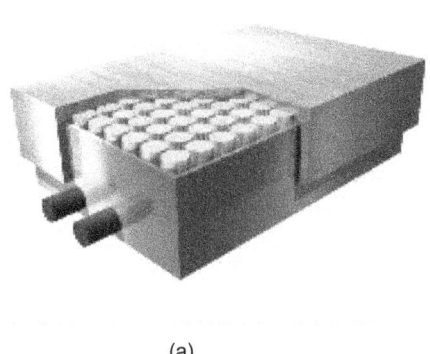 

(a)   (b)

FIGURE 30.19 NGK stationary energy storage batteries: (a) the 50 kW modular battery component, and (b) an integrated 500 kW/4 MWh demonstration battery system that uses 10 of these modular batteries, operating since June 1998 and still in use as of January 2010 at NGK Head Office. (*Courtesy of Tokyo Electric Power Company and NGK Insulators, Ltd.*)

30.5.8 Sodium-Beta Battery System Applications

As noted earlier, sodium-beta battery technologies have been developed for use in relatively large-scale energy storage applications (i.e., those requiring tens to thousands of kWh). The prime attributes that make these technologies attractive candidates for such uses include high energy density, lack of required maintenance, performance independent of ambient temperature, 100% coulombic efficiency, potential for low cost (relative to other advanced batteries), and operating flexibility compared to existing conventional rechargeable systems.

A number of storage applications involving electric power generation, distribution, and consumption are emerging that require a battery that remains stationary during use. Examples include some located at generation utilities (e.g., load-leveling, spinning reserve, area regulation), at renewable generation facilities (e.g., solar, wind), at distribution facilities (e.g., line stability, voltage regulation), or at a customer site (e.g., demand peak reduction, power quality). An assessment of the opportunities for battery-based energy storage in these types of applications in the United States was described earlier.[2,3]

A detailed assessment using this information was performed for the projected U.S. stationary energy-storage market. This study concluded that the sodium-beta technologies would be very attractive candidates for use in renewable energy systems, customer-side system peak reduction (shaving), and to defer the need for distribution facilities.[35] Another important finding of this study was that to be economically attractive to U.S. customers, the battery system would probably need the capability to simultaneously satisfy two or more individual applications. In fact, recent changes in the marketplace have created higher value opportunities for protection from short-duration power losses than has traditionally been recognized for UPS markets. Specifically, the broad dependence of commercial and industrial businesses on computer-based systems has made much of the world economy vulnerable to power disturbances that may last only a few seconds, but can disrupt expensive automated data handling and manufacturing processes.

NGK is in the commercialization stage with sodium/sulfur batteries for stationary applications, and the product is designated NAS. NGK has constructed a highly automated production facility and has built and is currently operating many significant-sized integrated battery systems (see Table 30.7).

From the beginning of the commercialization of sodium/sulfur batteries, they have been installed at substations, factories, or large scale commercial facilities for load-leveling. Emergency power supply systems (EPS) and stand-by power supply systems (SPS) have been realized by allocating partial battery capacity for load-leveling operation and keeping the remaining capacity of energy

TABLE 30.7 NAS Battery Projects as of December 2009

Name of developer	Country	Location	KW	Start of operation/status
TEPCO (Tokyo Electric Power Company)	Japan	Many locations around Tokyo	200,000 (approx.)	Tests ongoing; accumulated installed kW as of 12/2008
HEPCO (Hokkaidou Electric Power Company)	Japan	Wakkanai City, Hokkaido	1,500	Feb. 2008
Other Japanese Electric Companies	Japan	Many locations other than Tokyo area	60,000 (approx.)	Tests ongoing; accumulated installed kW as of 12/2008
JWD (Japan Wind Development Co., Ltd.)	Japan	Rokkasho Village, Aomori	34,000	Aug. 2008
AEP (American Electric Power)	USA	Charleston, WV; Bluffton, OH; Milton, WV; Churubusco, IN; Presidio, TX	11,000	4 sites except for Presidio: July 2006–Jan. 2009; Presidio: shipped in Nov. 2009
NYPA (New York Power Authority)	USA	Long Island, NY	1,000	April 2008
PG&E (Pacific Gas and Electric Company)	USA	Not decided	6,000	Shipped in 2008
Xcel	USA	Luveme, MN	1,000	Nov. 2008
Younicos	Germany	Berlin	1,000	July 2009
Enercon	Germany	Emden, Lower Saxony	800	July 2009
EDF	France	Reunion Island	1,000	Dec. 2009
ADWEA (Abu Dhabi Water and Electricity Authority)	UAE	Abu Dhabi	48,000	Partially operated
Total			365,300	

(Courtesy of NGK)

for the power source during outage. Recently, installations of sodium/sulfur batteries for renewable energy applications have been increasing. Output of wind-power generation and photovoltaic power generation can be intermittent and unstable. Sodium/sulfur batteries absorb these fluctuations of power and mitigate the influence of them on the power grid.

The Rokkasho project in northern Japan is an example of a sodium/sulfur battery with a wind-power generation system (Fig. 30.20). This project was developed by Japan Wind Development Co., Ltd. It includes 34 MW (17 units of the 2 MW battery system) of sodium/sulfur battery and was installed with 51 MW of wind-power generation. The commercial operation at Rokkasho was started in August 2008. In this project, the sodium/sulfur battery is used not only for absorbing the power fluctuations of the wind-power generation, but also for supplying the required constant power to the electrical power market (Fig. 30.21). Besides Japan Wind Development Co., Ltd., Xcel Energy (USA) and Enercon GmBH (Germany) have installed sodium/sulfur batteries to operate with wind-power generation systems.

The Wakkanai project in the northern part of Japan is an example of a sodium/sulfur battery with a photovoltaic generation system (Fig. 30.22). This project was developed by Hokkaido Electric Power Company supported by NEDO (New Energy and Industrial Technology Development Organization). The installation of a 1.5 MW sodium/sulfur battery with 5 MW of photovoltaic power generation system was completed in 2008, and the system is operational. In this project, the sodium/sulfur battery is used not only for absorbing the power fluctuations of the photovoltaic power generation, but also for shifting the power for peak demand times. Besides Hokkaido Electric Power Company, Younicos (Germany) installed a sodium/sulfur battery with a photovoltaic power generation system.

34 MW NAS alongside 51 MW Wind Farm

FIGURE 30.20 34 MW NGK sodium/sulfur battery that is installed with 51 MW wind-power generation system at Rokkasho, Japan. (*Courtesy of NGK.*)

Planned applications for the sodium/nickel-chloride technology include the General Electric (GE) hybrid locomotive, and other heavy equipment and stationary storage uses. In mid-2009, GE announced plans to build a manufacturing plant for this technology in upstate New York capable of producing approximately 10 million "sodium-metal halide" cells annually, equivalent to 900 MWh of energy storage, and enough for 1000 GE hybrid locomotives[R]. GE is planning to commercialize the hybrid locomotive sometime in 2010. The specific sodium/nickel-chloride cell and battery system design details have not been disclosed. GE also is considering the possibility of a dual-battery system for the locomotive, teaming the sodium/nickel-chloride battery, for its energy capabilities, with a lithium-ion battery, for its power capabilities. (See http://greencarcongress.com/2009/05/ge-halide-20090512.html, May 12, 2009.)

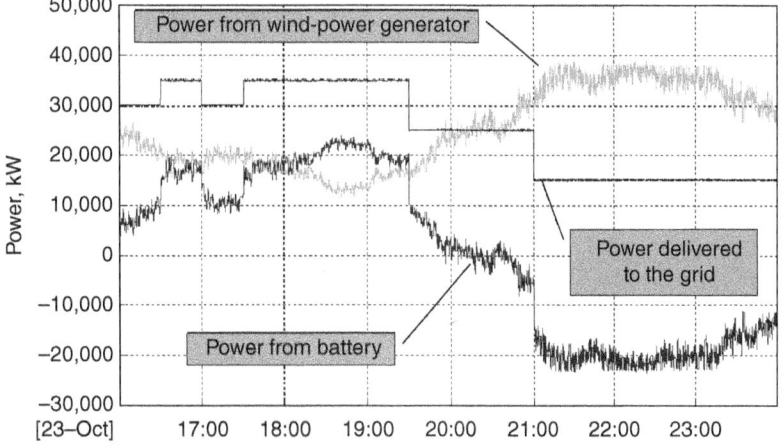

FIGURE 30.21 Operation data from the Rokkasho project showing that constant power is delivered to the grid from the wind-power generator by the NGK sodium/sulfur battery. (*Courtesy of NGK.*)

FIGURE 30.22 1.5 MW NGK sodium/sulfur battery that is installed with a 5 MW photovoltaic power generation system at Wakkanai, Japan. (*Courtesy of NGK.*)

AEP NAS Distributed Energy Storage System at Chemical Station, N. Charleston, WV[39]

Application: Substation upgrade deferral

Operational dates: 2006 to the present

Power: 1.0 MW

Energy: 7.2 MWh

Battery type: Sodium/sulfur

Battery: 50 kW NAS battery modules, 20 each

System manufacturer: NGK Insulators LTD (battery)/S & C Electric Co. (balance of system)

The chemical substation on the American Electric Power (AEP) grid includes a combination of transmission (138 kV) and distribution (12 kV) facilities. The 20 MVA, 46 kV/12 kV distribution transformer and the voltage regulator that supply the three 12 kV feeders out of this substation were very close to their rating limits during the 2005 summer peak (June through August), and were very likely to surpass them during that period the next year. AEP decided to install a 1.2 MW distributed energy storage system (DESS) with a sodium/sulfur battery to mitigate this problem for a few years, until a new substation could be justified. The installation was successful (see Fig. 30.23) in deferring the substation upgrade, and was operational as of January 2010.

Long Island Bus Terminal Energy Storage System

Application: Load shifting

Operational Dates: 2008 to the present

Power: 1.2 MW

Energy: 6.5 MWh

Battery type: Sodium/sulfur

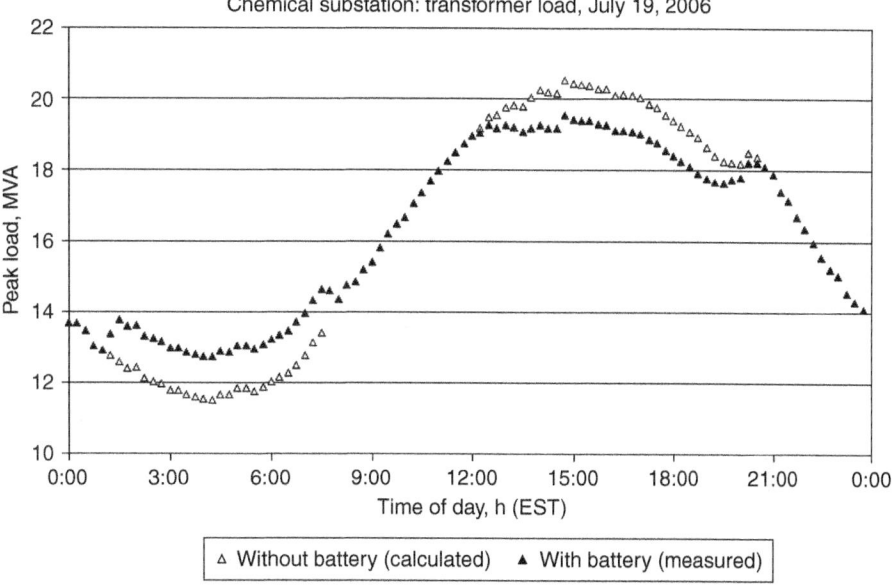

FIGURE 30.23 AEP chemical substation load showing effect of the DESS battery.

Battery: (20) 50 kW (60kW peak) NAS battery modules

System manufacturer: NGK Insulators LTD (battery)/ABB Inc. (integration and balance of system)

The Long Island Transit Authority (LITA) has begun operating a new fleet of natural gas–powered buses. Refueling these buses requires compressing natural gas to fill the bus tanks. Three 600 HP electricity-powered compressors are used to accomplish this. Refueling the buses is done during a third night shift to take advantage of the lower nighttime electricity rates and to reduce transmission congestion during summer peak periods. The LITA installed a NAS battery system to store the off-peak priced electricity, and store that electricity until the next day to power the compressors for daytime refueling of the bus fleet. This installation eliminated the third shift for the terminal, added flexibility, permitting daytime refueling on demand, and has reduced the operating costs of the bus terminal. This project was sponsored by the New York State Energy Research and Development Authority (NYSERDA), DOE, and EPRI.

Issues encountered during this project included the negotiation of interconnection agreements between three utilities and the LITA, and dealing with damage incurred by the NAS battery during an extended storage period.

30.6 FLOWING ELECTROLYTE BATTERIES

Flowing electrolyte batteries are advanced, aqueous-electrolyte battery systems and have the advantage of operating at close to ambient temperature. Nevertheless, complex system design and circulation of electrolyte are needed to meet performance objectives. Work on developing flow batteries started with the invention of the zinc/chlorine hydrate battery in 1968.[40] This system was the subject of development for EV and electric utility storage applications[41] from the early 1970s to the late-1980s in the United States, and from 1980 to 1992 in Japan.[42] Most work on zinc/chlorine battery stopped at that time, but recently has been restarted. Currently there are two main types of flowing electrolyte batteries that are under development: zinc/bromine and vanadium-redox. A comparison of flowing electrolyte batteries for utility applications has been published.[43] More detailed technical information on these battery types is included here since they are not covered elsewhere in the Handbook.

30.6.1 Flowing Electrolyte Batteries: Zinc/Bromine

The zinc/bromine battery technology is being developed for stationary energy storage applications. The system offers good specific energy and design flexibility, and battery stacks can be made from low-cost and readily available materials using conventional manufacturing processes. Bromine is stored remotely as a second-phase, polybromide complex that is circulated during discharge. Remote storage limits self-discharge during standby periods. An added safety benefit of the complexed polybromide is greatly reduced bromine vapor pressure compared to that of pure bromine.

The major advantages and disadvantages of this battery technology are listed in Table 30.8. The concept of a battery based on the zinc/bromine couple was patented over 100 years ago,[44] but development of a commercial battery was blocked by two inherent properties: (1) the tendency of zinc to form dendrites upon deposition, and (2) the high solubility of bromine in the aqueous zinc bromide electrolyte. Dendritic zinc deposits could easily short-circuit the cell, and the high solubility of bromine allows diffusion and direct reaction with the zinc electrode, resulting in self-discharge of the cell. Development programs at Exxon and Gould in the mid-1970s to early 1980s resulted in designs that overcame these problems, however, and allowed development to proceed.[45] The Gould technology was developed further by the Energy Research Corporation, but a high level of activity was not maintained.[46–48] In the mid-1980s, Exxon licensed its zinc/bromine technology to Johnson Controls, Inc., JCI (Americas), Studiengesellschaft fur Energiespeicher und Antriebssysteme, SEA (Europe), Toyota Motor Corporation and Meidensha Corporation (Japan), and Sherwood Industries (Australia). Johnson Controls sold their interest in zinc/bromine technology in 1994 to ZBB Energy Corporation. Powercell Corporation was formed in 1993 and included SEA, and is now known as Premium Power Company.

TABLE 30.8 Major Advantages and Disadvantages of Zinc/Bromine Battery Technology

Advantages	Disadvantages
Circulating electrolyte allows for ease of thermal management and uniformity of reactant supply to each cell.	Auxiliary systems are required for circulation and temperature control.
Good specific energy.	System design must ensure safety as for all batteries.
Good energy efficiency.	Initially high self-discharge rate when shut down while being charged.
Made of low-cost and readily available materials.	Improvements to moderate power capability may be needed.
Low-environmental-impact recyclable/reusable components made using conventional manufacturing processes.	
Flexibility in total system design.	
Ambient-temperature operation.	
Adequate power density for most applications.	
Capable of rapid charge.	
100% depth of discharge does not damage battery but improves it.	

30.6.2 Description of the Electrochemical System

The electrochemical reactions that store and release energy take place in a system whose principal components include bipolar electrodes, separators, aqueous electrolyte, and electrolyte storage reservoirs. Figure 30.24 shows a schematic of a three-cell, zinc/bromine battery system. The electrolyte is an aqueous solution of zinc bromide, which is circulated with pumps past both electrode surfaces. The electrode surfaces are in turn separated by a microporous plastic film. Thus two electrolyte flow streams are present—one on the positive side and one on the negative side. The directions of the flow streams may differ depending on the designs of different companies.

The electrochemical reactions can be represented as follows:

		(25°C)	(50°C)
Negative electrode:	$Zn^{2+} + 2e \underset{\text{discharge}}{\overset{\text{charge}}{\rightleftarrows}} Zn^0$	$E^0 = 0.763$ V	0.760 V
Positive electrode:	$2Br^- \underset{\text{discharge}}{\overset{\text{charge}}{\rightleftarrows}} Br_2(aq) + 2e$	$E^0 = 1.087$ V	1.056 V
Net cell reaction:	$\underset{\text{discharged state}}{ZnBr_2(aq)} \underset{\text{discharge}}{\overset{\text{charge}}{\rightleftarrows}} Zn^0 + Br_2(aq)$	$E^0_{cell} = 1.85$ V	

During charge, zinc is deposited at the negative electrode, and bromine is produced at the positive electrode. During discharge, zinc and bromide ions are formed at the respective electrodes. The microporous separator between the electrode surfaces impedes diffusion of bromine to the zinc deposit, which reduces direct chemical reaction and the associated self-discharge of the cell.

The chemical species present in the electrolyte are actually more complicated than that described. In solution, elemental bromine exists in equilibrium with bromide ions to form polybromide ions,

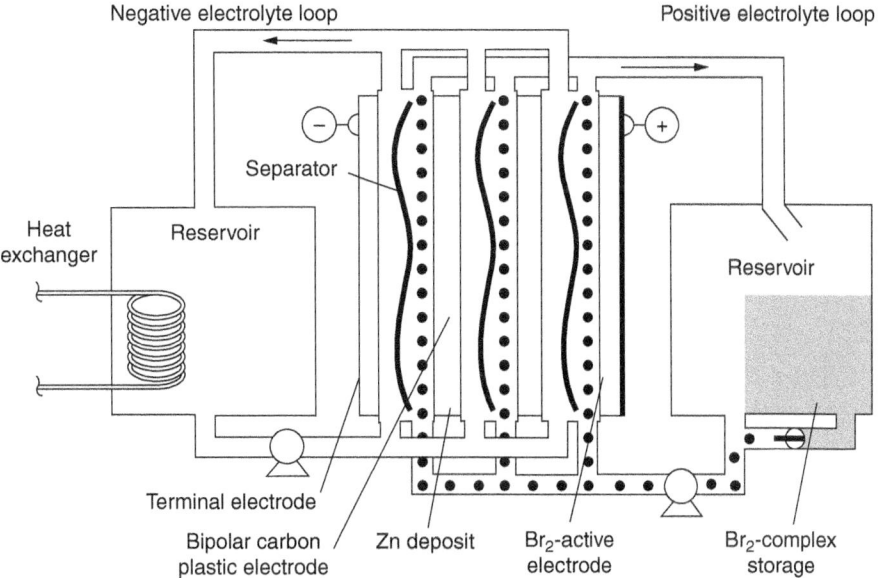

FIGURE 30.24 Schematic of a 3-cell zinc/bromine battery module.

Br_n^-, where n = 3, 5, 7.[49] Aqueous zinc bromide is ionized, and zinc ions exist as various complex ions and ion pairs. The electrolyte also contains complexing agents which associate with polybromide ions to form a low-solubility, second liquid phase. The complex reduces the amount of bromine contained in the aqueous phase 10- to 100-fold, which, in addition to the separator, also reduces the amount of bromine available in the cell for the self-discharge reaction. The complex also provides a way to store bromine at a site remote from the zinc deposit and is discussed further in the next section. Salts with organic cations such as N-methyl-N-ethylmorpholinium bromide (MEMBr) are commonly used as the complexing agents.

The electrodes are bipolar and are typically composed of carbon plastic. The presence of bromine precludes the use of metal electrodes—even titanium can corrode in this environment.[50] A high-surface-area carbon layer is added to the positive side of the electrode to increase the area for reaction. On charge, circulation of the electrolyte removes the complexed polybromide as it is formed, and on discharge complexed polybromide is delivered to the electrode surface. Circulation of the electrolyte also reduces the tendency for zinc dendrites to form and simplifies thermal management of the battery. Thermal management will be needed in many applications of present and advanced batteries.

In a system where the cells are connected electrically in series and hydraulically in parallel, an alternate pathway for the current exists through the common electrolyte channels and manifolds during charge, discharge, and at open-circuit. These currents are called shunt currents, and cause uneven distribution of zinc between end cells and middle cells. This uneven distribution causes a loss of available capacity because the stack will reach the voltage cutoff upon discharge sooner than if the zinc were evenly distributed. In addition, shunt currents can lead to uneven plating on individual electrodes, especially the terminal electrodes. This uneven plating can in turn lead to zinc deposits that divert or even block the electrolyte flow.

Shunt currents can be minimized by designing the cells to make the conductive path through the electrolyte as resistive as possible. This is done by making the feed channels to each cell long and narrow to increase the electrical resistance. This, however, also increases the hydraulic resistance and thus the pump energy. Good battery design balances these factors.[51]

30.6.3 Performance

Zinc/bromine batteries are typically charged and discharged using rates of 15 to 30 mA/cm². A charge-discharge profile for a 50-cell stack is shown in Fig. 30.25. The amount of charge is based on the zinc loading that is defined as 100% state of charge. This amount is always less than the total zinc ion dissolved in the electrolyte. Thus, rate of charge, time duration of charge, and charge efficiency are used to determine the end of charge. The voltage rises at the end of charge, and severe overcharge will electrolyze water. Discharge is usually terminated at about 1 V per cell since the voltage is decreasing rapidly at this point.

The capacity of a battery is directly related to the amount of zinc that can be deposited on the negative electrodes, and zinc loadings can range from 60 to 150 mAh/cm². One hundred percent state-of-charge is defined as a specific zinc loading and can vary depending on the battery. Considerable effort has been expended to ensure good-quality, dense zinc plating. It is important to control the pH to avoid undesirable mossy zinc deposits. Circulation of the electrolyte reduces the occurrence of dendritic deposits. Studies have shown that current density, zinc bromide concentration, electrolyte additives, and operating temperature also affect the quality of the zinc deposit.[47,52] With these studies and improvements, the problems are being managed or have been eliminated.

In a battery system, a portion of the energy will be diverted to auxiliary systems such as thermal management, pumps, valves, controls, and shunt current protection as required. The energy needed for auxiliaries depends on a number of factors, including the efficiency of pumps and motors, pump run-time, and system design. Energy will also be lost during stand time. This was measured in one zinc/bromine battery system to be about 1%/h (watthour capacity lost) over an 8 h period.[53] During the test, electrolyte, which did not contain the complexed bromine phase, was circulated periodically to remove heat. The self-discharge reaction ceases once bromine in the stacks has been depleted.

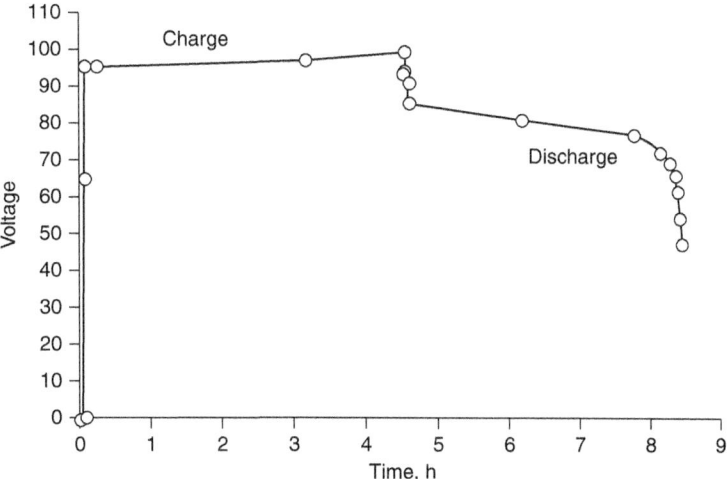

FIGURE 30.25 Charge-discharge profile for 50-cell stack. 80% electrolyte utilization; 30°C; 90 mAh/cm² zinc loading; 20 mA/cm² or C/4.5 charge rate; 20 mA/cm² or C/4 discharge rate. (*Courtesy of Sandia National Laboratories.*)

Zinc/bromine batteries normally operate between 20 and 50°C. Typically, the operating temperature has little effect on energy efficiency, as shown in Fig. 30.26. At low temperature, the electrolyte resistivity increases, resulting in lower voltaic efficiency. This is offset by slowed bromine transport, which results in higher coulombic efficiency. At high temperature, the resistance decreases and the bromine transport increases, again partially compensating for each other. Temperature control is accomplished with a heat exchanger and the circulating electrolyte. The optimum temperature will vary depending on the individual battery design and electrolyte used.

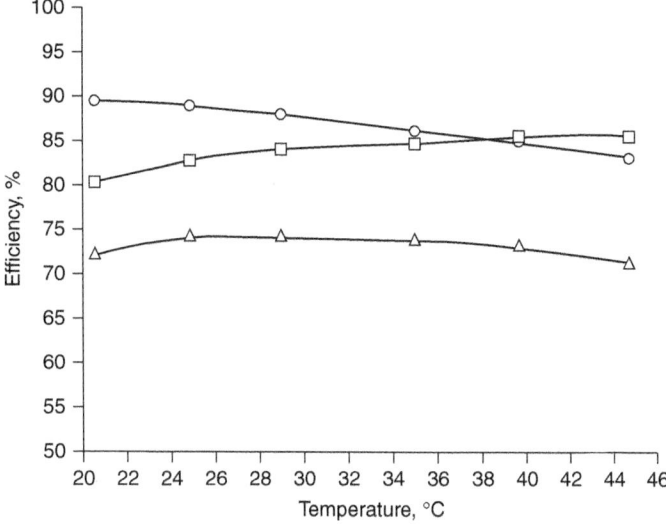

FIGURE 30.26 Efficiencies vs. operating temperature for battery with load-leveling electrolyte. ○—coulombic; □—voltaic; △—energy. (*Courtesy of Johnson Controls Group, Inc.*)

For applications that require high-power discharges, such as electric vehicles, the conductivity of the electrolyte can be enhanced by using additives such as KCl or NH_4Cl. In this way, internal ohmic energy losses are decreased.

30.6.4 Energy Storage Applications of Zinc/Bromine Batteries

The use of zinc/bromine batteries in energy storage applications has been demonstrated. A study by Sandia National Laboratories rated the zinc/bromine battery as excellent for these four utility applications: storage of energy generated by renewable sources, transmission facility deferral, distribution facility deferral, and demand peak reduction.[2,54] In the United States, the most likely near-term market for zinc/bromine batteries is electricity storage applications. Sales of millions of kWh of capacity per year may be possible at an estimated cost of $150/kWh or less.

ZBB Energy Corporation designed and manufactured a 50 kWh battery module in the late 1990s that served as a building block for larger systems. Each module was made up of three 60-cell stacks connected in parallel, an anolyte reservoir, a catholyte reservoir, and an electrolyte circulation system, as shown in Fig. 30.27.[55] These modules offered flexibility in building larger batteries because they

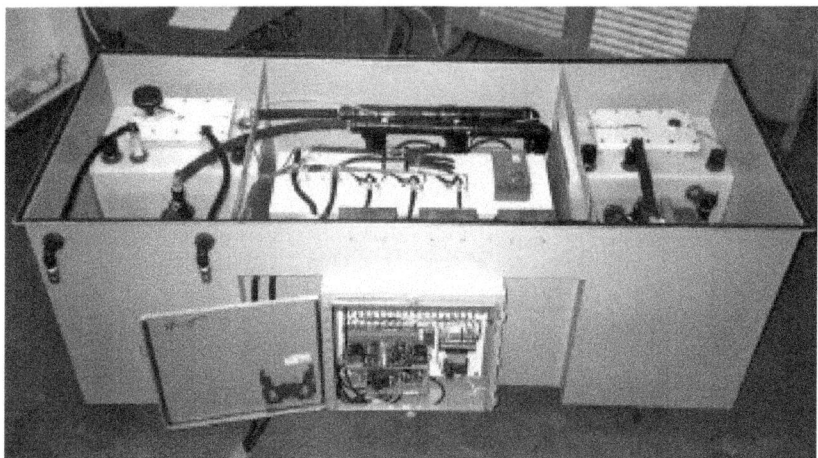

FIGURE 30.27 Photograph of ZBB Energy Corporation's 50 kWh battery module. (*Courtesy of ZBB Energy Corporation.*)

could be placed in a variety of series and parallel arrangements. A 200 kW, 400 kWh battery was designed and built for utility demonstrations and consisted of two strings connected in parallel, with each string composed of four 50 kWh modules in series. See Fig. 30.28 for a photograph of the system installed on a trailer for ease of transport and for the ability to redeploy the system as needed.

This system was tested at two Detroit Edison sites in 2000–2001,[56] and performed successfully although it was not sized optimally for the applications. The first test was for voltage regulation at a grain dryer in Akron, Michigan, and the battery system was intended to reduce voltage spikes on the local grid caused by large motors turning on and off. The battery power capacity was not sufficient to fully eliminate the voltage spikes, but it did moderate the spikes significantly, within its power rating limits. In the second test, the battery system was used for peak shaving/T&D upgrade deferral and power quality improvement for a substation dealing with a summer air conditioning peak load. The system performed as designed and experienced several power level tests.

Subsequently, ZBB Energy Corp. was awarded a contract in 2004 from the California Energy Commission (CEC) to demonstrate a 2 MW/2 MWh zinc/bromine battery system in a peak shaving application.[57] ZBB Energy designed a 500 kW/500 kWh module, designated Z-BESS, which was also

FIGURE 30.28 Photograph of a transportable 200 kW/400 kWh ZBB Energy Corporation Zn/Br battery system installed on a 40 × 8 ft trailer.[56]

installed on a trailer. This module was to be the building block for the full system, and it was installed and tested at the Distributed Utility Integration Test (DUIT) substation in the Pacific Gas & Electric (PG&E) service area in San Ramon, California, in 2005–2006.[58] The module was tested extensively, but due to thermal management limitations, was not able to provide full output. The CEC project was subsequently terminated as the intended substation was upgraded by conventional means.

More recently, ZBB Energy Corporation developed a ZESS 50 module and announced plans to demonstrate it as part of the Future House USA program at the 2008 Beijing Summer Olympics.[59] No further information is available on its status. Another Zn/Br battery system developed by Premium Power has been investigated by First Energy.[60] Plans were described for testing a 100 kW/150 kWh Power Block™ 150 unit in a utility application with support from EPRI. Premium Power also developed a smaller product, designated the Zinc-Flow 45, which is a 30 kW/45 kWh unit for UPS applications.[61]

30.6.5 Flowing Electrolyte Batteries: Vanadium-Redox

Another type of aqueous flowing electrolyte system is the redox flow technology. There are several systems of this type, only one of which, the vanadium-redox battery (VRB), has any significant development continuing. Work on this category of flow battery started with a development program at NASA[62] on a system using $FeCl_3$ as the oxidizing agent (positive) and $CrCl_2$ as the reducing agent (negative). The aim of this work was to develop redox flow batteries for stationary energy storage applications. The term "redox" is obtained from a contraction of the words "reduction" and "oxidation." Although reduction and oxidation occur in all battery systems, the term "redox battery" is used for those electrochemical systems where the oxidation and reduction involves only ionic species in solution and the reactions take place on inert electrodes. This means that the active materials must be mostly stored externally from the cells of the battery. Although redox systems are capable of long life, their energy density is low because of the limited solubility of the active materials typically involved.[63]

In Japan, development of iron/chromium redox flow battery technology was included as part of the Moonlight Project[8] in the 1980s. The goal of this work was electric utility energy storage. Improvements made in the course of the Moonlight Project included new electrode materials and a reduction in the requirement for pumping power.[64] A 60kW battery was tested[65] and 1-MW system was designed,[66] but the redox flow technology was not chosen to advance to the 1-MW pilot plant stage.[67] More recently, iron/chromium redox battery development has been resumed under an ARPA-E project.

Other redox systems have been proposed, such as the zinc/alkaline sodium ferricyanide $[Na_3Fe(CN)_6 \cdot H_2O]$ couple, and initial development work was performed.[68] However, none of these efforts proved successful, mainly because of difficulties resulting from the efficacy and resistance of the ionic exchange membranes, until the development of the VRB by the University of New South Wales, Australia, in the late 1980s.[69] Almost concurrently with this, development work started on VRBs at Sumitomo Electric Industries (SEI) of Osaka, Japan.[70] Starting in the mid-1990s, VRB development was also conducted at Mitsubishi Chemical's Kashima-Kita facility, although at a lower level of effort than at SEI.

The electrolytes in the positive and negative electrode compartments of VRBs are different valence states of vanadium sulfate. Both solutions are 2 Molar in concentration and contain sulfuric acid as a supporting electrolyte. The electrode reactions occur in solution, with the reaction at the negative electrode in discharge being

$$V^{2+} \rightarrow V^{3+} + e$$

and at the positive electrode

$$V^{5+} + e \rightarrow V^{4+}$$

Both reactions are reversible on the carbon felt electrodes that are used. An ion-selective membrane is used to separate the electrolytes in the positive and negative compartments of the cells. Cross-mixing of the reactants would result in a permanent loss in energy storage capacity for the system because of the resulting dilution of the active materials. Migration of other ions (mainly H^+) to maintain electroneutrality, however, must be permitted. Thus, ion-selective membranes are required.

A schematic of a VRB system is shown in Fig. 30.29.[70] The construction of the cell stacks is bipolar. The electrolyte solutions are stored remotely in tanks and are pumped through the cells when needed. The capacity of the redox flow system depends on the size of the storage tanks. The volume of electrolyte needed is large and results in a low energy density for this technology.

The VRB technology is continuing to be developed.[71,72] Efforts are focused on improved efficiency by reducing self-discharge losses and on lower cost electrode structures. Self-discharge is being addressed by only pumping electrolyte through the electrochemical stacks when necessary due to the magnitude of the load. These efforts are continuing.

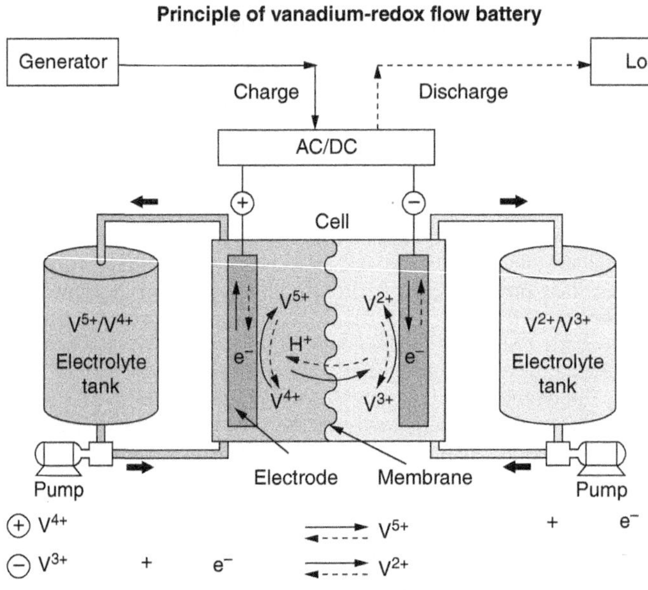

FIGURE 30.29 Block diagram of vanadium-redox system.[70]

30.6.6 Energy Storage Applications of Vanadium-Redox Batteries

Several multi-kW systems have been built and tested by several organizations around the world. A photograph of an SEI 100 KW, 8 h VRB system installed in 2000 is shown in Fig. 30.30. The tanks for the two electrolytes are not in view in the photograph since they are installed in a subbasement below the level of the battery stacks and the AC-DC-AC converter that are shown.

SEI built and successfully tested a VRB system at a liquid crystal plant beginning in 2001, shown in Fig. 30.31. The system is capable of either voltage regulation of 3 MW for 1.5 s, or load-leveling of 1.5 MW for one h.[73,74] More recently, in 2005 SEI initiated the Subaru Project, which tested a 4 MW for 90 min. VRB system connected to a 30.6 MW wind farm.[75] Testing was concluded in 2008 and over 270,000 short cycles were successfully performed by the battery system. (See www.pdenergy.com/applications-solutions/projects installations/.)

Another SEI VRB system was tested in South Africa in a UPS application. The system was 250 kW/520 kWh and was installed in about 2001.[76] A 6 to 12 month test was planned, and initial results confirmed the design expectations. This was the first VRB tested outside of Japan. Subsequently, other VRB systems were installed and tested in Austria,[77] and in other locations in Europe,[78] and in the United States.[79] VRB Power Systems of Vancouver, Canada, has provided the VRB technology in the United States and in other sites around the world.[80] By far, the most attention has been on the VRB system that has been tested in Utah.

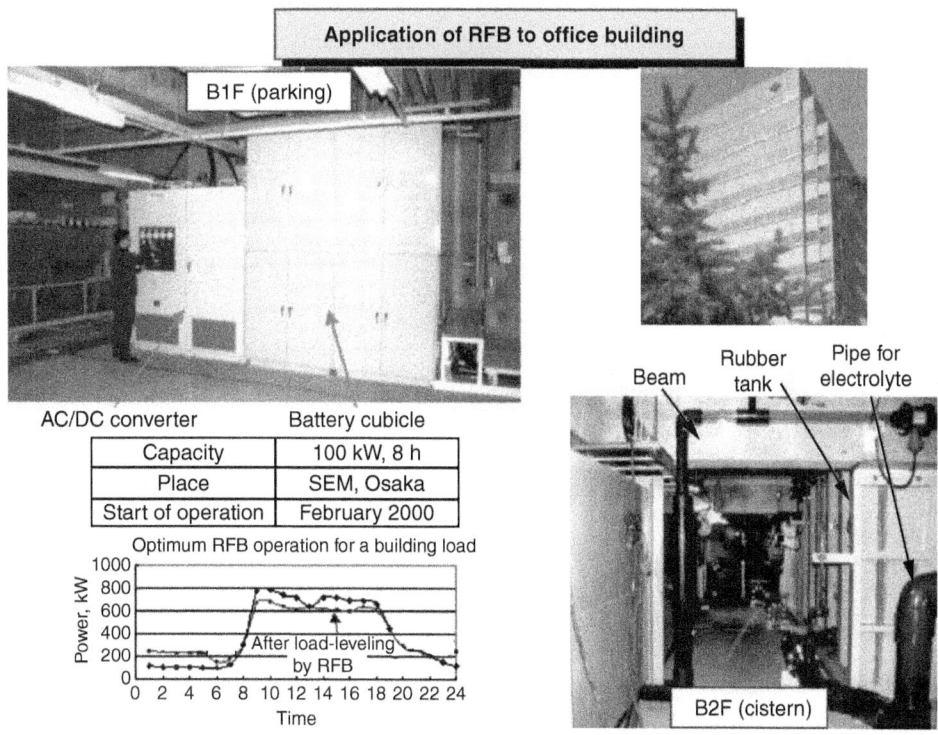

FIGURE 30.30 100 kW, 8 h vanadium-redox battery at the Sumitomo Densetsu office building for peak shaving applications.

30.44 SPECIALIZED BATTERY SYSTEMS

FIGURE 30.31 3 MW for 1.5 s VRB system at Tottori Sanyo Electric Co. Ltd. liquid crystal plant.[74]

30.6.7 PacifiCorp, Castle Valley, Utah Vanadium-Redox Battery (VRB) System[81,82]

Application: Distribution line upgrade deferral, voltage support

Operational dates: March 2004 to March 2009

Power: 250 kW

Energy: 2 MWh

Battery type: Vanadium-redox

Battery: 50 kW Sumitomo battery modules, 250 kW for 8 h

Manufacturer: VRB Power Systems (purchased by Prudent Energy Co., Beijing, China, in 2009)

The PacifiCorp Vanadium-Redox Battery (VRB) in Castle Valley, Utah, is a 250 kW, 2 MWh system installed to defer for at least 4 years the installation of a major new 69 kV transmission line and substation in an environmentally sensitive area. The system cost $1.3 million to build and install, while the new transmission facilities are expected to cost over $5M. The system was installed in late 2003 and became operational in March 2004. See Fig. 30.32. The battery system is expected to be

FIGURE 30.32 Exterior and interior views of the 2 MWh VRB system at Castle Valley, Utah.[81]

relocated to another site and continue operation in another application after the new transmission facilities are constructed in Castle Valley. This was the first VRB system installed in North America. Prudent Energy Inc., based in Beijing, China, purchased VRB Power Systems in early 2009 and continues development of these systems.

30.7 CONCLUSION

Electrical energy storage using batteries is continuing to evolve technologically, and is increasingly being accepted as a viable and potentially revolutionary resource which could fundamentally change the way electricity is supplied and used. More and more, battery storage is being considered for integration with renewable systems to increase the availability and value of those resources. The use of a single storage facility in multiple applications for utility-grid systems is also expanding and becoming recognized for its value. As the cost of energy storage systems decreases and their reliability increases with improved technologies, these trends are likely to accelerate and make storage an essential part of the electricity delivery system.

ACKNOWLEDGMENTS

The authors of this chapter acknowledge the contributions and assistance of Tak Eguchi of NGK Insulators for his timely updates of the information on sodium/sulfur batteries. In addition, we acknowledge Joe Leach of EnergyUnited for his update on the Crescent BESS. We also acknowledge Nancy Clark of Sandia for her helpful comments and reference materials.

REFERENCES

1. A. Akhil, H. Zaininger, J. Hurwitch, and J. Baden, *Battery Energy Storage: A Preliminary Assessment of National Benefits (The Gateway Benefits Study),* Sandia National Laboratories, SAND93-3900, 1993.
2. P. C. Butler, *Battery Energy Storage for Utility Applications: Phase I—Opportunities Analysis,* Sandia National Laboratories, SAND94-2605, 1994.
3. P. Butler, J. L. Miller, and P. A. Taylor, *Energy Storage Opportunities Analysis Phase II Final Report, A Study for the DOE Energy Storage Systems Program,* Sandia National Laboratories, SAND2002-1314, 2002.
4. S. M. Schoenung, *Characteristics and Technologies for Long- versus Short-Term Energy Storage,* Sandia National Laboratories, SAND2001-0765, March 2001.
5. J. J. Iannucci, J. M. Eyer, and W. Erdman, *Innovative Applications of Energy Storage in a Restructured Electricity Marketplace—Phase III Final Report,* Sandia National Laboratories, SAND2003-2546, April 2005.
6. J. M. Eyer, J. J. Iannucci, and G. P. Corey, *Energy Storage Benefits and Market Analysis Handbook, A Study for the DOE Energy Storage Systems Program,* Sandia National Laboratories, SAND2004-6177, December 2004.
7. P. Donalek and W. Hassenzahl, "Pumped Storage Hydroelectric," *2005 Energy Storage Association Annual Meeting,* May 25–26, Toronto, Canada.
8. S. Furuta, "NEDO's Research and Development on Battery Energy Storage System," *Utility Battery Group Meeting,* Valley Forge, PA, November 1992.
9. J. J. Iannucci and S. M. Schoenung, *Energy Storage Concepts for a Restructured Electric Utility Industry, Phase 1,* Sandia National Laboratories, SAND2000-1550, June 1999.
10. J. J. Iannucci, J. M. Eyer, and P. C. Butler, *Innovative Business Cases for Energy Storage in a Restructured Electricity Marketplace, Phase 2,* Sandia National Laboratories, SAND2003-0362, February 2003.

11. S. M. Schoenung and W. M. Hassenzahl, *Long- versus Short-Term Energy Storage Technologies Analysis, A Life-Cycle Cost Study,* Sandia National Laboratories, SAND2003-2783, August 2003.
12. D. A. J. Rand, P. Moseley, J. Garche, and C. Parker, *Valve-Regulated Lead-Acid Batteries,* Elsevier, Amsterdam, 2004.
13. R. B. Sloan, "Crescent Electric Membership Cooperative," *Proc. 10th Meeting of the Utility Battery Group,* Charlotte, NC, November 1995, pp. 145–147.
14. *EPRI-DOE Handbook of Energy Storage for Transmission and Distribution Applications, EPRI Report #1001834,* EPRI, Palo Alto, CA, December 2003. Available at www.epri.com/OrderableitemDesc.asp?product_id=000000000001001834.
15. M. Farber-DeAnda, J. Boyes, and W. Torres, *Lessons Learned form the Puerto Rico Battery Energy Storage System,* Sandia National Laboratories, SAND99-2232, September 1999.
16. J. E. Pueyo Font, C. J. Castro Montalvo, and A. R. Fernandez, "Repowering the Sabana Llana BESS," *2005 Energy Storage Association Annual Meeting,* May 25–26, Toronto, Canada.
17. J. Szymborski, G. Hunt, and R. Jungst, "Examination of the VRLA Battery Cells Sampled from the Metlakatla Battery Energy Storage System," *EESAT 2000,* September 18–20, Orlando, FL.
18. G. Hunt and J. Szymborski, "Achievements of an ABSOLYE® Valve-Regulated Lead-Acid Battery Operating in a Utility Battery Energy Storage (BESS) for 12 Years," *EESAT 2009,* October 4–7, Seattle, WA.
19. G. Corey, W. Nerbun, and D. Porter, *Final Report on the Development of a 250-kW Modular, Factory-Assembled Battery Energy Storage System,* Sandia National Laboratories, SAND97-1276, August 1998.
20. B. Norris and G. J. Ball, *Performance and Design Analysis of a 250-kW, Grid-Connected Battery Energy Storage System,* Sandia National Laboratories, SAND99-1483, June 1999.
21. M. Farber-DeAnda, L. L. Bush, J. Philip, and P. Taylor, *Data Management of Grid-Connected Systems: PQ2000 at Brockway Standard Lithography Plant,* Final Report to Sandia National Laboratories, May 25, 2000.
22. T. DeVries, "Keeping the Lights on in Fairbanks, Alaska," *2005 Energy Storage Association Annual Meeting,* May 25–26, Toronto, Canada.
23. J. McDowall, "The GVEA BESS," *2008 Energy Storage Association Annual Meeting,* May 20–21, Orange, CA.
24. J. McDowell, "Lithium-Ion Technologies—Gaining Ground in Electricity Storage," *2008 Energy Storage Association Annual Meeting,* May 20–21, Orange, CA.
25. D. Linden and T. Reddy, *Handbook of Batteries,* 3rd ed., McGraw-Hill, 2002.
26. J. Sudworth and R. Tilley, *The Sodium/Sulfur Battery,* Chapman and Hall, London, 1985.
27. J. Coetzer, "A New High-Energy-Density Battery System," *J. Power Sources,* **18:**377–380 (1986).
28. J. Prakash, L. Redy, P. Nelson, and D. Vissers, "High Temperature Sodium Nickel Chloride Battery for Electric Vehicles," *Electrochemical Society Proceedings,* **96:**14 (1996).
29. T. Oshima and H. Abe, "Development of Compact Sodium Sulfur Batteries," *Proc. 6th Int. Conf. on Batteries for Utility Energy Storage,* Wissenschaftspark Gelsenkirchen Energiepark Herne, Germany, September 1999.
30. A. Kita, "An Overview of Research and Development of a Sodium Sulfur Battery," *Proc. DOE/ EPRI Beta Battery Workshop VI,* pp. 3.23–3.27, May 1985.
31. K. Takashima et al., "The Sodium Sulfur Battery for a 1 MW/8 MWh Load-Leveling System," *Proc. Int. Conf. on Batteries for Utility Energy Storage,* March 1991, pp. 333–349.
32. E. Nomura et al., "Final Report on the Development and Operation of a 1 MW/8 MWh Na/S Battery Energy Storage Plant," *Proc. 27th IECEC Conf. 3,* pp. 3.63–3.69, 1992.
33. R. Okuyama and E. Nomura, "Relationship Between the Total Energy Efficiency of a Sodium-Sulfur Battery System and the Heat Dissipation of the Battery Case," *J. Power Sources,* **77:**164–169 (1999).
34. A. Araki and H. Suzuki, "Leveling of Power Fluctuations of Wind Power Generation Using Sodium Sulfur Battery," *Proc. 6th Int. Conf. on Batteries for Utility Energy Storage,* Wissenschaftspark Gelsenkirchen Energiepark Herne, Germany, September 1999.
35. A. Koenig, *Sodium/Sulfur Battery Engineering for Stationary Energy Storage, Final Report,* Sandia National Laboratories, SAND Rep. 96-1062, April 1996.
36. R. Galloway and S. Haslam, "The ZEBRA Electric Vehicle Battery: Power and Energy Improvements," *J. Power Sources,* **80:**164–170 (1999).

37. D, Guatto, "Electricity On-Demand: Load Shifting Using Sodium-Nickel Chloride Technology," *2007 Energy Storage Association Annual Meeting*, May 23–24, Boston, MA.
38. R. Manzoni, M. Metzger, and G. Crugnola, "ZEBRA Electric Energy Storage System: From R&D to Market," *HTE hi.tech.expo*, November 25–28, 2008, Milan, IT.
39. A. Nourai, *Installation of the First Distributed Energy Storage System (DESS) at American Electric Power (AEP), A Study for the DOE Energy Storage Systems Program*, Sandia National Laboratories, SAND2007-3580.
40. P. C. Symons, "Process for Electrical Energy Using Solid Halogen Hydrate," U.S. Patent 3,713,888, 1973.
41. Energy Development Associates, "Development of the Zinc Chloride Battery for Utility Applications," Electric Power Research Institute, EPRI AP-5018, January 1987; C. C. Whittlesey, B. S. Singh, and T. H. Hacha, "The FLEXPOWER TM Zinc-Chloride Battery: 1986 Update," *Proc. 21st IECEC*, San Diego, CA, 1986, pp. 978–985.
42. T. Horie, H. Ogino, K. Fujiwara, Y. Watakabe, T. Hiramatsu, and S. Kondo, "Development of a 10kW (80 kWh) Zinc-Chloride Battery for Electric Power Storage Using Solvent Absorption Chlorine Storage System (Solvent Method)," *Proc. 21st IECEC*, San Diego, CA, 1986, vol. 2, pp. 986–991; Y. Misawa, A. Suzuki, A. Shimizu, H. Sato, K. Ashizawa, T. Sumii, and M. Kondo, "Demonstration Test of a 60kW-Class Zinc/Chloride Battery as a Power Storage System," *Proc. 24th IECEC*, Washington, D.C., 1989, vol. 3, pp. 1325–1329; H. Horie, K. Fujiwara, Y. Watakabe, T. Yabumoto, K. Ashizawa, T. Hiramatsu, and S. Kondo, "Development of a Zinc/Chloride Battery for Electric Energy Storage Applications," *Proc. 22nd IECEC*, Philadelphia, 1987, vol. 2, pp. 1051–1055.
43. C. Lotspeich, "A Comparative Assessment of Flow Battery Technologies," *EESAT 2002*, April 15–17, San Francisco.
44. C. S. Bradley, U.S. Patent 312,802, 1885.
45. R. A. Putt and A. Attia, "Development of Zinc Bromide Batteries for Stationary Energy Storage," Gould, Inc., for Electric Power Research Institute, Project 635-2, EM-2497, July 1982.
46. L. Richards, W. Van Schalwijk, G. Albert, M. Tarjanyi, A. Leo, and S. Lott, *Zinc-Bromine Battery Development*, Final Report, Sandia Contract 48-8838, Energy Research Corporation, Sandia National Laboratories, SAND90-7016, May 1990.
47. A. Leo, "Zinc Bromide Battery Development," Energy Research Corporation for Electric Power Research Institute, Project 635-3, EM-4425, January 1986.
48. P. C. Butler, D. W. Miller, C. E. Robinson, and A. Leo, *Final Battery Evaluation Report: Energy Research Corporation Zinc/Bromine Battery*, Sandia National Laboratories, SAND84-0799, March 1984.
49. D. J. Eustace, "Bromine Complexation in Zinc-Bromine Circulating Batteries," *J. Electrochem. Soc.*, **528** (March 1980).
50. R. Bellows, H. Einstein, P. Grimes, E. Kantner, P. Malachesky, K. Newby, H. Tsien, and A. Young, *Development of a Circulating Zinc-Bromine Battery Phase II*, Final Report, Exxon Research and Engineering Company, Sandia National Laboratories, SAND83-7108, October 1983.
51. R. Bellows, H. Einstein, P. Grimes, E. Kantner, P. Malachesky, K. Newby, and H. Tsien, *Development of a Circulating Zinc-Bromine Battery Phase I*, Final Report, Exxon Research and Engineering Company, Sandia National Laboratories, SAND82-7022, January 1983.
52. J. Bolsted, P. Eidler, R. Miles, R. Petersen, K. Yaccarino, and S. Lott, *Proof-of-Concept Zinc/Bromine Electric Vehicle Battery*, Johnson Controls, Inc., Advanced Battery Engineering, Sandia National Laboratories, SAND91-7029, April 1991.
53. N. J. Magnani, P. C. Butler, A. A. Akhil, J. W. Braithwaite, N. H. Clark, and J. M. Freese, *Utility Battery Exploratory Technology Development Program Report for FY91*, Sandia National Laboratories, SAND91-2694, December 1991.
54. N. Clark, P. Eidler, and P. Lex, *Development of Zinc/Bromine Batteries for Load-Leveling Applications: Phase 2 Final Report*, Sandia National Laboratories, SAND99-2691, October 1999.
55. P. Lex and B. Jonshagen, "The Zinc/Bromine Battery System for Utility and Remote Area Applications," ZBB Energy Corporation, *Proc. of the Electrical Energy Storage Systems Applications & Technologies (EESAT) Conference*, Chester, UK, June 16–18, 1998. Also found in *Power Engineering Journal*, June 1999, pp. 142–148.
56. V. Scaini, P. J. Lex, T. W. Rhea, and N. H. Clark, "Battery Energy Storage for Grid Support Applications," *EESAT 2002*, April 15–17, San Francisco.

57. P. Lex, "Demonstration of a 2-MWh Peak Shaving Z-BESS," *EESAT 2005*, October 17–19, San Francisco.
58. B. Norris and R. Winter, "Test and Demonstration of a 2 MWh Zinc-Bromine Battery System," *2006 ESA Annual Meeting*, May 16–18, 2006, Knoxville, TN.
59. "Future House USA: The Complete Package," *2008 Energy Storage Association Annual Meeting*, May 20–21, 2008, Orange, CA.
60. E. Gardow, "Electricity Storage: The Utility Application," *2007 ESA Annual Meeting*, May 23–25, Boston.
61. J. Capes, "Zinc-Bromide Regenerative Energy Storage," *2006 ESA Annual Meeting*, May 16–18, Knoxville, TN.
62. L. H. Thaller, "Recent Advances in Redox Flow Cell Storage Systems," DOE/NASA/1002-79/4, NASA TM 79186, August 1979; N. Hagedorn, "NASA Redox Storage System Development Project," U.S. Dept. of Energy, DOE/NASA/12726-24, October 1984.
63. M. Bartolozzi, "Development of Redox Flow Batteries. A Historical Bibliography," *J. Power Sources*, **27**:219–234 (1989).
64. Z. Kamio, T. Hiramatsu, and S. Kondo, "Research and Development of 10-kW Redox Flow Battery," *Proc. 22nd IECEC*, Philadelphia, 1987, vol. 2, pp. 1056–1059.
65. T. Tanaka, T. Sakamoto, N. Mori, T. Shigematsu, and F. Sonoda, "Development of a 60-kW Class Redox Flow Battery System," *Proc. 3rd Int. Conf. on Batteries for Utility Energy Storage*, Kobe, Japan, 1991, pp. 411–423.
66. H. Izawa, T. Hiramatsu, and S. Kondo, "Research and Development of 10 kW Class Redox Flow Battery," *Proc. 21st IECEC*, San Diego, 1986, vol. 2, pp. 1018–1021.
67. T. Hirabayashi, S. Furuta, and H. Satoh, "Status of the 'Moonlight Project' on Advanced Battery Energy Storage System," *Proc. 26th IECEC*, Boston, 1991, vol. 6, pp. 88–93.
68. R. P. Hollandsworth, "Zinc-Redox Battery, A Technology Update," *Electrochemical Society, Fall Meeting*, October, 1987.
69. M. Skyllas-Kazacos et al., AU Patent 575247 (1986).
70. N. Tokuda et al, "Vanadium Redox Flow Battery for Use in Office Buildings," *Proc. of Conference on Electric Energy Storage Applications and Technologies*, Orlando, FL, September 2000.
71. M. Schreiber, "Vanadium Redox Flow Battery Layout for Improved Efficiency," *EESAT 2007*, September 23–26, San Francisco.
72. M Schreurs and J. Timpert, "Third Generation Redox Flow Battery: A Development Update," *EESAT 2009*, October 4–7, Seattle, WA.
73. T. Shinzato et al., "Vanadium Redox-Flow Battery for Voltage Sag," *EESAT 2002*, April 15–17, San Francisco.
74. K. Emura et al., "Recent Tendency of VRB Technology and Experience," *2002 ESA Annual Meeting*, October 10–11, Milwaukee, WI.
75. G. Koshimizu et al., "Subaru Project: Analysis of Field Test Results for Stabilization of 30.6 MW Wind Farm with Energy Storage," *EESAT 2007*, September 23–26, San Francisco.
76. J. Hawkins and T. Robbins, "A Vanadium Energy Storage System Field Trial," *EESAT 2002*, April 15–17, San Francisco.
77. A. Whitehead, M. Harrer, and M. Schreiber, "Field Test Results for a 1 kW, 50 kWh Vanadium Redox Flow Battery," *EESAT 2005*, October 17–19, San Francisco.
78. M. Schreiber et al., "Vanadium Redox Flow Battery for Remote Area Power Supply," *EESAT 2009*, October 4–7, Seattle, WA.
79. T. Hennessy and J. Davis, "Permanent Load Shifting and UPS Functionality at a Telecommunications Site Using the VRB-ESS™—A Case Study," *EESAT 2007*, September 23–26, San Francisco.
80. B. Beck, "Vanadium-Redox Flow Cell Product," *2008 Energy Storage Association Annual Meeting*, May 20–21, Orange, CA.
81. B. Williams, "Operational Update of a 2 MWh VRB Energy Storage System (VRB-ESS) at PacifiCorp," *2005 Energy Storage Association Annual Meeting*, May 25–26, Toronto, Canada.
82. S. Lathrop, T. Hennessy, B. Steeley, and H. Kamath, "Progress with Flow Batteries: The Vanadium Redox Battery at Castle Valley," *2006 ESA Annual Meeting*, May 16–18, Knoxville, TN.

CHAPTER 31
BATTERIES FOR BIOMEDICAL APPLICATIONS

Randolph A. Leising, Nancy R. Gleason, Barry C. Muffoletto, and Curtis F. Holmes

31.1 APPLICATIONS AND REQUIREMENTS FOR IMPLANTABLE BATTERIES

Batteries for biomedical applications have unique need sets which vary widely according to the large range of devices now used to treat many different medical conditions. Considerable research has been devoted to the materials, chemistry, and design of batteries for these applications over the past 50 years. For example, the first lithium anode battery used in a cardiac pacemaker was implanted in 1972, and this application was one of the first successful commercial uses of primary lithium batteries. Since then, the number of biomedical devices and the conditions being treated by these devices have grown rapidly and continue to grow every year. A representative sample of applications for batteries used in implantable medical devices is presented in this section. Table 31.1 also provides a summary of some of these applications.

31.1.1 Implantable Cardiac Pacemakers

Cardiac pacemakers were the first implantable device. In fact, the year 2008 marked the 50th anniversary of the first implant of a cardiac pacemaker, signaling the birth of the implantable medical device industry. This first pacemaker was implanted in Arne Larsson, a Swedish engineer, on October 8, 1958, by heart surgeon Dr. Ake Senning working with electrical engineer Rune Elmquist in Stockholm. This first pacemaker, only lasted a few hours, and its replacement lasted just a few weeks.[1] At the same time in the United States, Wilson Greatbatch was developing an implantable cardiac pacemaker, and the Chardack-Greatbatch pacemakers were implanted in animals in 1958 and in 10 patients in 1960. These pacemakers had greater longevity, but still suffered from poor performance due to the mercury-zinc batteries that were used. This led Greatbatch to seek a more robust power source for the pacemaker, which led to the development of the Li/Iodine battery for this application.[2-4] It is estimated that more than 600,000 pacemakers are implanted each year worldwide.[1]

As an inspiring footnote, Larsson (the first recipient of a pacemaker) lived another 43 years to the age of 86 after the first pacemaker implant. During that time, he received 26 different implantable pacemakers.

A pacemaker treats brady cardia, a slow heartbeat. The pulse generator (as shown in Fig. 31.1, and a cutaway view in Fig. 31.2) contains the battery and the circuits that generate the pacing pulses. Pacing leads connect to the pulse generator and transmit the electrical pulses to the heart.

31.2 SPECIALIZED BATTERY SYSTEMS

TABLE 31.1 Examples of Battery Powered Implantable Medical Devices

Condition	Device	Typical battery discharge rate
Brady cardia	Pacemaker	Low
Tachy cardia	Cardioverter defibrillator (ICD)	Low and high
Congestive heart failure	Cardiac resynchronization therapy defibrillator (CRT-D)	Low and high
Syncope	Implantable cardiac monitor	Medium
End-stage heart failure	Left ventricular assist devices (LVAD) Total artificial heart (TAH)	High
Chronic pain	Neurostimulator	Medium
Epilepsy	Neurostimulator	Medium
Hearing loss	Neurostimulator	Medium

The pacemaker monitors the heart rate of the patient and provides the electrical pulses to maintain a healthy rate. The size of the pulse generator is typically in the 8 to 15 cc range. The device illustrated in Fig. 31.1 is less than 6 cc in volume, designed for pediatric patients.

Battery longevity is determined by a number of factors, including the amount of pacing required by the patient, as well as size of the battery designed into the device. The typical power consumption for a conventional cardiac pacemaker is on the order of 10 to 30 microwatts. Because of this low power requirement, Li/Iodine batteries have been well suited to the pacemaker application. However, pacemakers are increasingly incorporating wireless telemetry features, which greatly improve patient monitoring functions, but also increase the power consumption of the device. In many cases, a medium-rate type battery, providing mA currents on a periodic basis, is needed to power these advanced devices.

31.1.2 Implantable Cardioverter-Defibrillators

The implantable cardioverter-defibrillator (ICD) was invented by Dr. Michel Mirowski in 1979. Mirowski took his idea for an implantable device from external defibrillators (first developed in the 1940s), where these devices were found to be successful in converting ventricular tachycardia into a

FIGURE 31.1 Compact implantable cardiac pacemaker designed for pediatric patients. (*Courtesy of St Jude Medical.*)

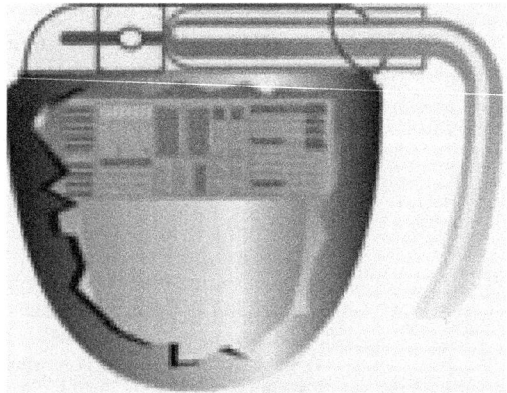

FIGURE 31.2 Cutaway graphic of implantable cardiac pacemaker. (*Courtesy of ELA Medical.*)

normal heart rhythm. Mirowski witnessed repeated fainting spells, which led to the eventual death of a colleague who suffered from ventricular tachycardia. After this experience, Mirowski collaborated with Dr. Martin Mower and Dr. Stephen Heilman to develop the first ICD; this first prototype was built using two Mallory mercuric oxide cells.[5]

An ICD treats ventricular tachycardia, a rapid heartbeat which can be in the 160 to 240 beats per minute range. This is a potentially lethal disruption of the normal heart rhythm which can cause the heart to be unable to pump blood and can lead to ventricular fibrillation. Ventricular fibrillation is a condition where the heart's electrical activity becomes totally disordered. The ventricles contract in a rapid, unsynchronized way and the heart pumps little or no blood. The ICD detects this condition and applies a high voltage shock to the heart to temporarily stop the heart and interrupt the arrhythmia. The heart then recovers and can regain a normal sinus rhythm. The efficacy of ICDs has been demonstrated in clinical studies. In a comparison of over 1000 patients, ICDs were found to be superior to antiarrhythmic drugs for increasing overall survival.[6]

Batteries are low-voltage, high-energy-density electrochemical devices that are unable to meet the high-voltage, high-power requirements of the ICD to defibrillate the heart. The ICD batteries charge capacitors, which then provide a high voltage used to shock the heart. The additional components needed for the ICD lead to a larger device size compared to pacemakers. The ICDs first used in the early 1980s were quite large, over 160 cc in volume.[7] The large size of the device required abdominal implantation. As the technology continued to improve, the size of the device became smaller, reaching ~60 cc in the mid-1990s, while typical modern ICDs are in the ~30 to 40 cc volume range. The ICD illustrated in Fig. 31.3 is 29 cc in volume. This decrease in size led to the pectoral implant of the ICD, which is a more desirable position for device placement.

The battery chemistries of choice for ICDs have been lithium/silver vanadium oxide and lithium/manganese dioxide, both primary battery systems. A new hybrid lithium/silver vanadium oxide + carbon monofluoride primary battery system has also been utilized in ICDs. These batteries have been able to meet the diversified power requirements for ICDs:

- Low current draw for monitoring and pacing functions
- Low self-discharge
- High pulse current on demand to charge capacitors
- State-of-charge indication through voltage differentiation
- Safety and reliability

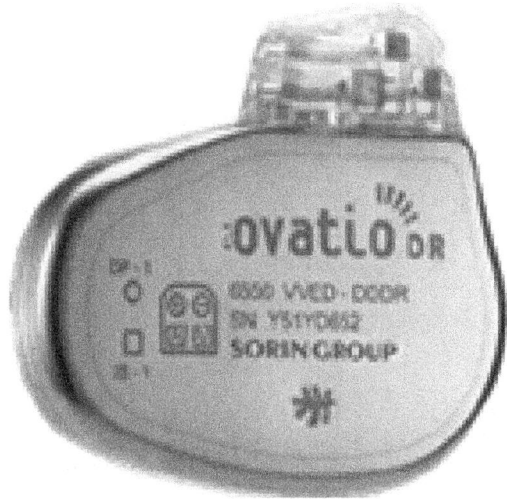

FIGURE 31.3 Implantable cardioverter-defibrillator. (*Courtesy of ELA Medical.*)

31.4 SPECIALIZED BATTERY SYSTEMS

The pacing/monitoring functions of the ICD typically consume low power, at < 20 μA. The high pulse currents, on the other hand, can be in the range of 2 to 4 A, with pulse durations ranging from 5 to 15 s, with several pulses in a row (as many as needed to restore the patient's normal heart rhythm, although there are typically much fewer than 20 pulses applied in any particular episode). Some devices are programmed to provide an audible alarm as a patient alert when the battery needs to be replaced. This elective replacement indicator (ERI) is typically designed to allow three months for a device explant prior to the device coming to end of life (EOL).[8]

31.1.3 Implantable Cardiac Resynchronization Therapy Defibrillators

Modern ICDs, in addition to supplying defibrillation therapy, also function as pacemakers.[9] This includes single and dual chamber pacing, where pacing leads are placed in either the right atrium or right ventricle for single chamber, and in both the right atrium and right ventricle for a dual chamber pacemaker. A dual chamber defibrillator has a pacing lead in the right atrium and a pacing/defibrillation lead placed in the right ventricle. A tri-chamber pacing system utilizes a third pacing lead, which is placed in the left ventricle.[9] Implantable cardiac resynchronization therapy defibrillators (CRT-Ds) (Fig. 31.4) utilize this ventricular resynchronization to treat congestive heart failure in certain patients. Congestive heart failure (CHF) is a condition where the heart cannot pump enough blood to the body's other organs. The weakened, often enlarged, heart muscle cannot pump all the blood it receives, which causes congestion (fluid in the lungs) from which the term congestive heart failure is derived. CHF can result from a number of factors:[10]

- Coronary artery disease
- Scar tissue from past heart attack
- High blood pressure
- Heart valve disease
- Disease of the heart muscle (cardiomyopathy)

FIGURE 31.4 Implantable cardiac resynchronization therapy defibrillator. (*Courtesy of Boston Scientific.*)

- Congenital heart defects
- Infection of the heart valves and/or muscle

While the battery requirements for CRT devices are similar to ICDs in the necessity to provide high power pulse currents for defibrillation, the lower power pacing requirements are significantly (up to 5 times) higher for the CRT versus a single or dual chamber pacing ICD. Thus, a CRT device may require pacing currents in the 50 to 100 µA range.

31.1.4 Implantable Cardiac Monitors

Implantable cardiac monitors (Fig. 31.5) are devices that are implanted under the skin in the upper chest area. These devices continuously monitor heart rhythms by recording electrocardiograms (ECG). The application for these devices is to monitor patients who have displayed syncope, or fainting spells. Syncope results from a temporary reduction in blood flow producing a shortage of oxygen to the brain. However, there can be different causes for this condition (including many noncardiac causes) so continuous cardiac monitoring, especially during a syncopal episode, can be critical in determining a correct diagnosis of the condition.

These devices are small in size, typically on the order of 6 to 10 cc, so a small battery is required for the device. The implantable device is used in conjunction with an external handheld device that activates the internal monitor. Communication between the devices is performed by telemetry, requiring the battery to produce mA currents on a periodic basis. The typical lifetime of the implanted device, which defines the necessary longevity of the battery, is 3 years.

FIGURE 31.5 Implantable cardiac monitor. (*Courtesy of Medtronic.*)

31.1.5 Heart Assist and Total Artificial Heart Devices

Left ventricular assist devices (LVAD) are implantable pumps that pull blood from the left ventricle and send it to the aorta, effectively assisting the left ventricle in pumping blood to the body.[10] These devices are often used as a "bridge to transplant" for patients who require a heart transplant and are waiting for a suitable donor heart. Interestingly, published studies have shown that the use of an LVAD can actually improve the heart of a patient over time, where after the LVAD is removed, the patient shows cardiac recovery compared to pre-implant.[11] Figure 31.6 shows an LVAD system, including the implantable pump, and an external rechargeable battery pack.

Total artificial hearts (TAH) are also used as bridge-to-transplant devices, where patients in end-stage heart failure have their diseased heart replaced by an implanted artificial heart. In end-stage heart failure, the heart slowly loses its ability to efficiently pump blood and results in the death of the patient. A famous example of a TAH is the Jarvik-7, designed by Dr. Robert Jarvik, which was first implanted in Dr. Barney Clark in 1982. Clark, a retired dentist, suffered from severe congestive heart failure. The experimental device succeeded in operating as a replacement heart for 112 days, and provided an initial step in the development of the TAH. While the use of artificial hearts is still not widespread today, several companies and medical centers have continued to develop these devices. Some of these devices are pneumatically powered, while others utilize electrically powered pumps.

The total energy needs for an electrically powered LVAD or TAH cannot be met by reasonably sized primary batteries but require the use of rechargeable batteries. In one version of the TAH, the

31.6 SPECIALIZED BATTERY SYSTEMS

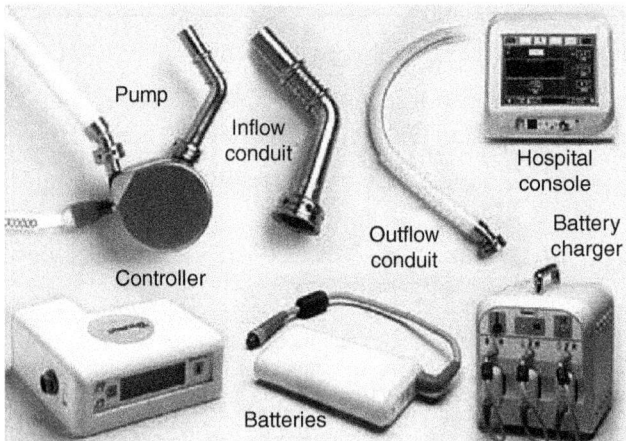

FIGURE 31.6 Left ventricular assist device. (*Courtesy of Terumo Heart.*)

system includes both external and implantable battery packs.[12] The internal batteries provide ~30 min of power for the TAH, allowing the patient to conduct activities such as taking a shower, while disconnected from the external battery pack. The external battery pack powers the TAH for ~4 h. While the size of the external battery pack can be significantly larger than the implantable battery, it still must be small enough to be worn by the recipient of the TAH device. General requirements for powering an LVAD or TAH include cycling the battery at high rates in the range of 0.5 to 3 A, with an average current ~1 A, as well as a high working voltage of 20 to 30 V.[13]

31.1.6 Neurostimulators

Neurostimulators are implanted devices that provide electrical stimulation to treat a wide variety of disorders, from movement disorders to bladder control to pain management. This field continues to expand and offers alternatives to surgery or drug therapy. Conditions that are potentially treated by neurostimulation are listed in Table 31.2.

TABLE 31.2 Examples of Potential Applications of Neurostimulation

Stimulated area	Condition treated
Bone	Bone fractures
Cochlea	Severe to profound hearing loss
Deep brain	Parkinson's disease, essential tremor, dystonia, Alzheimer's disease, obsessive compulsive disorder
Gastric nerve	Obesity
Gracilus muscle	Fecal incontinence
Middle ear	Mild to severe sensorineural hearing loss
Occipital nerve	Chronic migraine headaches
Optic nerve	Blindness
Phrenic nerve	Diaphragm pacing to restore breathing
Sacral nerve	Incontinence
Spinal cord	Chronic pain
Stomach muscle	Gastroparesis
Vagus nerve	Epilepsy, depression

Implantable neurostimulators are typically between the size of implantable pacemakers and defibrillators. Power requirements for neurostimulators are typically in the medium-rate category, with pulses currents in the mA range. Devices can utilize either a primary or secondary battery, depending on size and overall energy required.

31.1.7 Clinical Trials

A wide variety of battery chemistries have been used in implantable biomedical device clinical trials.[14] A list of battery chemistries is summarized in Table 31.3. This list provides a historical perspective of the type of batteries utilized in prototype devices.

TABLE 31.3 Battery Chemistries Used in Clinical Trials for Implantable Devices

Battery chemistry	Device application
Zinc/Mercury	Pacemaker
Li/Iodine	Pacemaker
Li/Copper Sulfide	Pacemaker
Nickel/Cadmium	Pacemaker, Biotelemeter, TICA (hearing prosthesis), Middle Ear Implant
Li/CFx	Pacemaker, Neurostimulator
Li/SOCl$_2$	Pacemaker, Cardiac Monitor
Li/V$_2$O$_5$	ICD
Li/SVO	Neurostimulator, ICD
Li/Ag$_2$CrO$_4$	ICD
Li/MnO$_2$	Middle Ear Implant, ICD
Li-ion	Neurostimulator, Ventricular Assist, Artificial Heart
Li-polymer rechargeable	Spinal Cord Oscillatory Field Stimulator

31.2 APPLICATIONS AND REQUIREMENTS FOR BATTERIES FOR EXTERNALLY POWERED MEDICAL DEVICES

In addition to batteries used to power implantable medical devices, there are many different primary and secondary batteries used to power externally worn or nonimplanted medical devices. One important difference with these systems is that the power source can be replaced without having to explant the medical device. A few examples of the many external medical devices in use today are listed below.

31.2.1 External Drug Pumps

Insulin pump therapy is used to treat diabetes. According to the Centers for Disease Control and Prevention, 24 million people in the United States have diabetes, and 57 million Americans are prediabetic. An insulin pump is a device that delivers insulin continuously to the body to keep blood sugar levels at a steady state. This device is worn outside the body and delivers insulin through a soft tube connected to a cannula which is inserted under the skin. The use of an insulin pump has been found to be superior to insulin injections because pumps continuously supply small amounts of fast-acting insulin compared to larger doses of long-acting insulin supplied by injections.[15] Among other things, this allows the user greater freedom in deciding when to eat meals.

Batteries for insulin pumps are typically consumer-type, such as silver oxide button cells or alkaline AA or AAA cells. These batteries typically last on the order of several weeks in this application.

31.8 SPECIALIZED BATTERY SYSTEMS

31.2.2 Hearing Assist Devices

Hearing assist devices, such as traditional hearing aids but also including implantable cochlear devices, typically utilize small button cell batteries that are external to the body. Zinc/air is a leading battery chemistry used for hearing aids. In addition, rechargeable nickel-metal hydride and lithium-ion cells are also designed for use in hearing assist devices. In one application, the NiMH batteries in the device can be recharged by using a portable card charger powered by a lithium polymer battery.[16]

Cochlear implants are very different from hearing aids, which amplify sounds for damaged ears. Cochlear implants bypass the damaged portion of the ear and stimulate the auditory nerve. A cochlear implant system (shown in Fig. 31.7) contains[17] (1) a microphone, which picks up sound from the environment, (2) a speech processor, (3) a transmitter and receiver/stimulator, which receives signals from the speech processor and converts them into electrical pulses, and (4) an electrode array, which stimulates the auditory nerve. The electrode array is partially implanted, but the microphone, processor, and transmitter are worn externally.

Cochlear implants typically utilize zinc/air battery chemistry. A commonly used battery size is the 675 button cell. These cells are 0.6 cc in volume, 1.8 g, and can provide ~570 mAh capacity with a nominal voltage of 1.4 V.[16] High-rate tests of the zinc/air cells at 27 to 18 mA continuous (51 ohm) discharge provides ~400 mAh capacity.[18] Rechargeable lithium-ion cells are also used for cochlear implant devices, where a prismatic 120 mAh Li-ion cell is contained in the externally worn speech processor, as illustrated in Fig. 31.8.

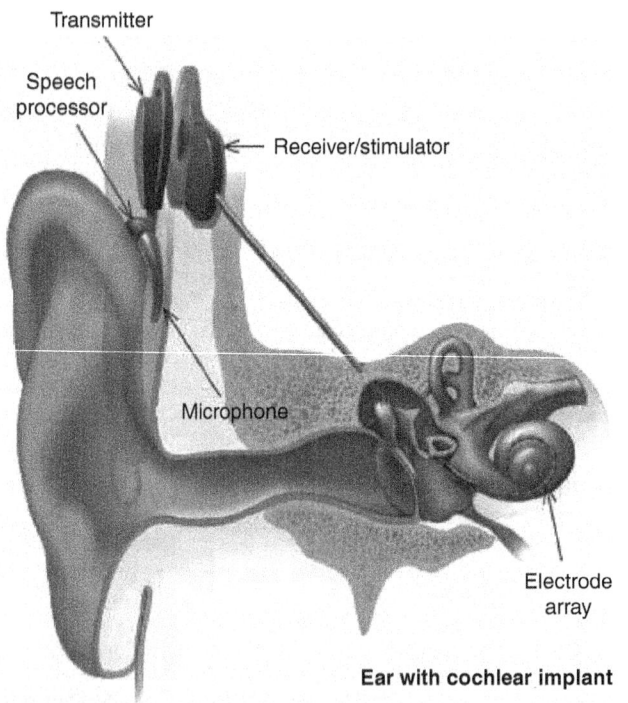

FIGURE 31.7 Illustration of cochlear implant (*Ref. 17*).

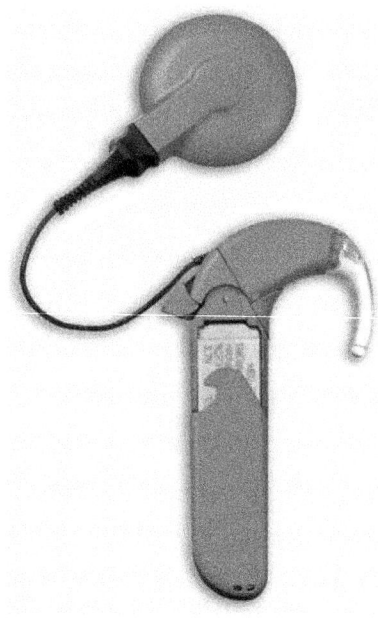

FIGURE 31.8 Transmitter and speech processor of cochlear implant containing a rechargeable Li-ion battery. (*Courtesy of MED-EL.*)

31.2.3 Automated External Defibrillators

Automated external defibrillators (AEDs) are portable devices that are used to automatically diagnose and treat someone who has gone into a life-threatening arrhythmia. The devices are automated to the point that they can be used by ordinary individuals with a small amount of training. The AED checks a person's heart rhythm and can automatically apply a shock when needed. It uses voice commands and other messages to guide the rescuer, and can also aid the rescuer in applying CPR to the patient. Initially these devices were used by police and rescue squads, but have since been located in many public places, such as schools and churches, airports and aboard airplanes, and even on board the international space station.

While all of the AEDs need to be small enough to be portable, size of the batteries or battery packs are not a limiting issue. Batteries used in the devices vary widely based on the several brands of devices that are commercially available. Some devices use primary batteries, typically lithium-based, such as Li/MnO_2 and Li/SO_2, while others use rechargeable batteries such as Li-ion or Ni/Cd. All of the batteries require high-rate pulsing capability (in the ampere range) to provide adequate power needed for shocks. AEDs typically apply two sequential shocks of 120 to 200 joules each. A photo of AEDs is presented in Figure 31.9.

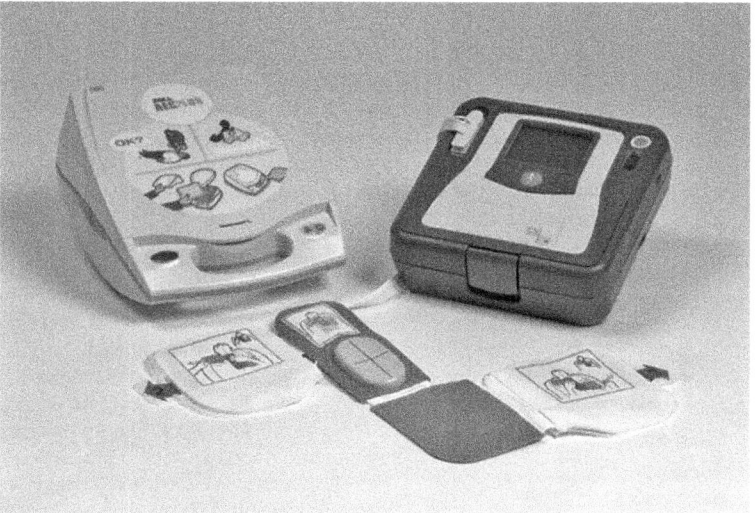

FIGURE 31.9 Automated external defibrillators. (*Courtesy of Zoll Medical.*)

31.3 SAFETY CONSIDERATIONS

Safety and reliability of batteries for biomedical devices is of extreme importance and the number one consideration in the selection, design, and manufacture of a battery chemistry for these applications. The incorporation of redundant safety features, hermetically sealed enclosures, and individual serialization and traceability typify the batteries that have been used for implantable devices. During the more than 30 years that lithium primary batteries have been extensively utilized in implantable devices, there is a proven track record for the safety and reliability of these systems.

In defining safety of medical devices, the World Health Organization has presented the following key elements:[19]

- Absolute safety cannot be guaranteed.
- It is a risk management issue.

- It is closely aligned with device effectiveness/performance.
- It must be considered throughout the life span of the device.
- It requires shared responsibility among the stakeholders.

As a component of the medical device, the battery is part of the system where the risk must be managed. Clearly, safety of the battery must be considered throughout the life of the device, and all necessary precautions must be taken to ensure safety from manufacturing to distribution and end-user.

31.3.1 Safety of Primary Batteries

A majority of primary batteries used in biomedical devices utilize lithium metal anodes due to the high energy density of the lithium battery systems. While these batteries have been safely used for several decades, the high reactivity of lithium does require that significant precautions are taken. The melting point of lithium is 180°C; if this temperature is exceeded for a lithium containing battery, the melted lithium may produce a violent reaction which can result in the explosion of the battery. This situation is highly relevant to implanted devices containing lithium batteries, where it is necessary to explant the device prior to the cremation of a deceased patient. A survey of crematoria in the U.K. indicated that about half had experienced explosions due to devices which were not explanted, where the explosions could cause structural damage and injury, and most of the crematoria staff surveyed were unaware of the explosive potential of the implanted devices.[20]

Localized heating in a lithium battery, such as caused by an internal or external short circuit can also result in the venting or explosion of the battery, if the heat cannot be dissipated quickly enough from the affected site. The chemistry and design of the battery can have a significant impact on the ability of the cell to adequately dissipate heat. In one example related to lithium batteries for biomedical devices, safety incidents were reported for automated external defibrillators that utilize primary Li/SO_2 cells where, in a few cases, the lithium cells vented, causing injury to firefighters.[21] During this event, the venting of the D-size cell was also accompanied by release of sulfur dioxide, which is a hazardous material.

To ensure the safety of each particular battery design to be used in a biomedical device, standardized tests must be conducted and passed. Safety standards for batteries are issued by a number of regulatory agencies or laboratories. A list of safety specifications are given in Table 31.4. UL 1642 is the Underwriters Laboratory safety specification for lithium batteries; it covers both primary and secondary lithium batteries, including lithium-ion batteries.[22] This specification defines construction,

TABLE 31.4 Standards Concerning Battery Safety

Standard	Description
IEC 60086-4	Primary Batteries—Part 4: Safety of Lithium Batteries
IEC 61960	Secondary Cells and Batteries Containing Alkaline or Other Non-Acid Electrolytes—Secondary Lithium Cells and Batteries for Portable Applications
IEC 62281	Safety of Primary and Secondary Lithium Cells and Batteries During Transport
ANSI C18.1, Part 2	American National Standard for Portable Primary Cells and Batteries with Aqueous Electrolyte—Safety Standard
ANSI C18.2, Part 2	American National Standard for Portable Rechargeable Cells and Batteries—Safety Standard
ANSI C18.3, Part 2	American National Standard for Portable Lithium Primary Cells and Batteries—Safety Standard
UL 1642	Lithium Batteries

performance, tests, and marking for batteries. These tests are further delineated into subcategories of electrical, mechanical, and environmental, as follows:

- Electrical tests
  - Short-circuit test
  - Abnormal charging test
  - Forced-discharge test
- Mechanical tests
  - Crush test
  - Impact test
  - Shock test
  - Vibration test
- Environmental tests
  - Heating test
  - Temperature cycling test
  - Low-pressure test
- User-replaceable lithium batteries
  - Projectile test

ANSI standard C18.3M, Part 2 is the safety standard for portable lithium primary cells and batteries. This standard defines tests in categories of Intended Use Simulation (altitude simulation, thermal shock, vibration, and shock) and Reasonably Foreseeable Misuse (external short circuit, forced discharge, incorrect installation, free fall, and crush). In the international standard, IEC 60086-4, safety of primary lithium batteries, the Reasonably Foreseeable Misuse category also contains impact, abnormal charging, thermal abuse, and overdischarge tests.

31.3.2 Safety of Rechargeable Batteries

In addition to the safety considerations for primary batteries, secondary batteries have the added complication of charging. In the case of lithium-containing batteries, early research in the development of lithium metal anode rechargeable cells were plagued by the formation of high surface area metal and dendrites at the lithium anode after recharging which resulted in internal short circuits of the cells. In certain cases, these short circuits resulted in venting or explosion of the cells and the associated safety hazards. More recent work in this area has focused on investigating the influence of specific surface layers on the lithium anode to prevent the formation of dendrites during charging.[23,24]

The lithium-ion cell, containing a carbon-based or lithium alloy-based anode, was developed as a safe rechargeable lithium-containing cell. The use of an insertion electrode as the anode avoids the formation of lithium dendrites during charging under normal operation. However, overcharging of lithium-ion batteries is a significant safety concern and needs to be avoided.[25] Charging control circuits must be utilized to avoid overcharging of these systems. In addition, there is considerable materials research in the area of safety enhancement of cathodes, anodes,[26] electrolytes,[27] and the interaction of these components as a function of temperature. This safety research is targeted at improving lithium-ion technology for consumer and electric vehicle applications but is also of particular use for biomedical applications.

While billions of lithium-ion batteries have been produced and used safely for small portable consumer devices, a few well-publicized incidents involving fires with lithium-ion laptop batteries have highlighted the potential for safety issues with this system. Most of these have been ascribed to a manufacturing defect. The combination of high energy density with flammable electrolytes makes these systems potent power sources that need to be treated with respect.[28] Lithium-ion battery incidents on aircraft have led to the issuance of new packaging requirements.[29] The National Transportation Safety Board (NTSB) investigated one of these incidents and made recommendations to the U.S. Pipeline and Hazardous Material Safety Administration on modifying the regulations for the transport of lithium-containing batteries.[30]

31.3.3 Shipping Regulations

The shipment of lithium and lithium-ion batteries is regulated by a number of different governing agencies, including the U.S. Department of Transportation (DOT), the International Air Transport Association (IATA), International Civil Aviation Organization (ICAO), the International Maritime Organization (IMO), and the United Nations (UN). Table 31.5 lists shipping regulations concerning lithium and lithium-ion batteries. In general, the shipment of these batteries will be included under Class 9 of hazardous materials, unless the cells or batteries contain less than a certain weight of lithium metal or equivalent amount of lithium for lithium-ion cells.

TABLE 31.5 Regulations Concerning Lithium Battery Shipment

Organization	Regulation
IATA	International Air Transport Association Dangerous Goods Regulations
ICAO	International Civil Aviation Organization Technical Instructions
IMO	International Maritime Dangerous Goods Code
U.S. DOT	Part 49 of the Code of Federal Regulations of U.S. Hazardous Materials Regulations
UN	United Nations Recommendations on the Transport of Dangerous Goods

It should be noted that the regulations for shipment of lithium batteries are updated often, so the most current documents should be consulted prior to shipping these materials. For example, the 50th edition of the IATA Dangerous Goods Regulations was published in 2009 and contains updated regulations for the shipment of lithium metal and lithium-ion batteries.[29] International Standard IEC 62281, "Safety of primary and secondary lithium cells and batteries during transport," lists UN-derived transport and misuse tests T1 through T8: (T1) altitude, (T2) thermal cycling, (T3) vibration, (T4) shock, (T5) external short circuit, (T6) impact, (T7) overcharge, and (T8) forced discharge.

31.4 RELIABILITY CONSIDERATIONS

Levy and coauthors have written a number of publications dealing with the subject of reliability of primary batteries with a special focus on lithium anode battery systems.[31–34] These authors divided reliability considerations into three parts: lot reliability, individual cell reliability, and root cause analysis of failed cells. This relates to the reliability based on manufacturing of the batteries (lot analysis), the design of the batteries (individual cell reliability), and postmortem analysis of a failed cell to identify root cause. All of these topics have particular relevance to battery systems for biomedical applications. Methods used to estimate the reliability and root cause analysis for failed batteries are summarized below.

31.4.1 Failure Modes and Fault Tree Analysis

Both failure modes and effects analysis and fault tree analysis have been used in industries that require high reliability, such as the nuclear energy and aerospace industries. These analysis techniques are also directly applicable to the design and manufacture of batteries for biomedical devices. The use of failure mode and effects analysis (FMEA) in the battery design process is aimed at identifying potential problems in a battery early in the design process so these issues can be addressed to improve the overall reliability of the battery.[35] The FMEA process uses a standardized approach to evaluating the battery design by evaluating the risk of failure of a particular component or system and assigning a risk priority number. This results in a numerical output that can be used in assessing risk and assigning priority to items to be addressed. Process FMEAs, or PFMEAs, are also conducted to analyze all of the processes that go into the manufacturing of the battery.

Fault tree analysis is another technique that has been used to evaluate the design of batteries, although this technique typically starts from a failed cell and works backward to determine the events

31.4.2 Qualification of Battery Designs

Once a high reliability battery design has been completed, it must pass a qualification based on the particular application. The qualification procedure for batteries designed for implantable device applications may vary depending on the manufacturer, but a sample procedure was presented by Visbisky et al. for the qualification of ICD batteries:[36]

- Thermal cycling from 70 to –40°C with 1 min transition time
- High-pressure testing at 90 and 120 psi
- Low-pressure testing under vacuum equivalent to 4500, 12,030, and 15,000 meters above sea level
- Low and high temperature exposure
- Short circuit at room temperature and at 37°C
- Forced overdischarge at $C/10$ rate
- Forced overdischarge of a depleted cell by a fresh cell
- Shock testing with a 1000 g force
- Vibration testing at frequencies ranging from 5 to 5000 Hz, peak acceleration 5 g

The protocol is displayed graphically in Fig. 31.10.

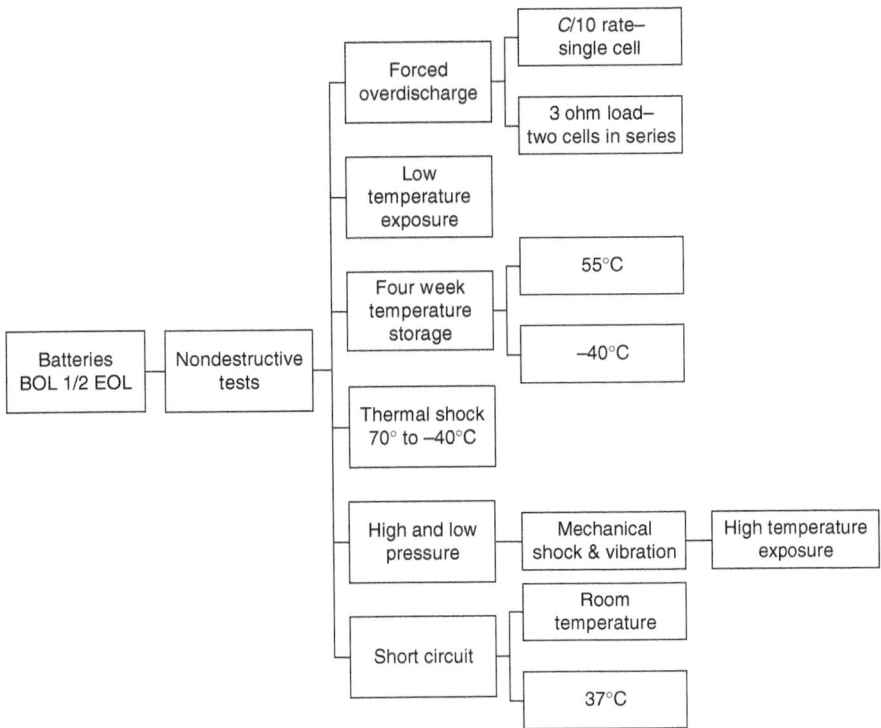

FIGURE 31.10 Standard qualification test protocol for an ICD battery (*Ref. 36*). BOL = Beginning of life; EOL = End of life.

31.4.3 Nondestructive Testing

There are several nondestructive analysis techniques that can be used to estimate the reliability of a given battery system. Regression trees have been developed to utilize nondestructive measurements as a screening tool to identify cells with poor performance. An example of this type of analysis is the use of complex impedance data in conjunction with microcalorimetry to screen Li/SO_2 cells.[37]

The use of microcalorimetry in the analysis of battery self-discharge was first developed in the late 1970s.[38–40] A schematic diagram of a microcalorimeter designed for the analysis of pacemaker batteries is displayed in Fig. 31.11 This describes a twin-cell differential heat flow instrument. The sensitivity of this type of instrument is typically down to the microwatt level with absolute heat rate measurement accuracies from 1 to 0.1% traceable to U.S. NIST.[41] Understanding the self-discharge properties of battery chemistries and cell design is important in determining the deliverable capacity for cells that are intended to be implanted over several years. The heat dissipation of a cell resulting from internal parasitic reactions is quantified by the microcalorimeter, and can be related to the total capacity of the cell in terms of a percent self-discharge, as long as the battery chemistry is well understood. This technique is also an important nondestructive test that can be used to assess failed batteries. The presence of a small internal "soft" short can be identified by elevated heat dissipation from a cell. Electrochemical impedance spectroscopy (EIS) can be used to effectively assess the condition of a cell in a nondestructive manner.[42] This technique is widely used to characterize battery systems and identify the internal impedance of the cathode, anode, and separator/electrolyte components.

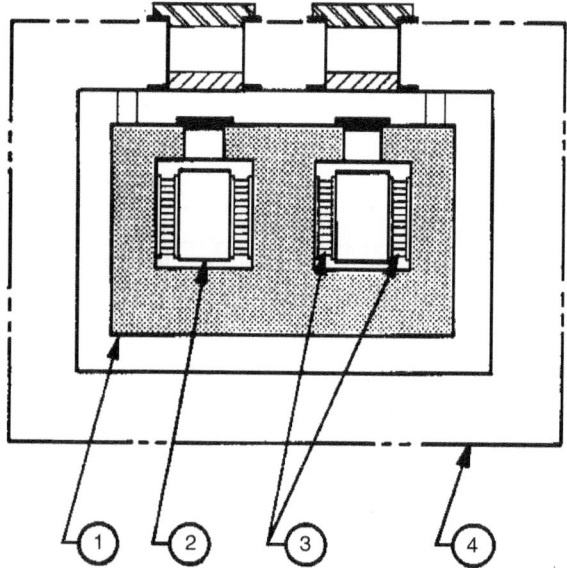

FIGURE 31.11 A schematic diagram of the design of a twin cell microcalorimeter used for analysis of batteries: (1) heat sink, (2) one of the twin cells, (3) thermoelectric sensors, and (4) stable temperature environment (*Ref. 41*).

In conjunction with the other nondestructive techniques, open-circuit voltage (OCV) measurements and limited pulsing of a battery can also be used to assess the lot reliability of cells. Open-circuit voltage is a quick and simple measurement that can provide important information about the state of the battery. A low value can indicate an internal short or corrosion reaction, while a high value may indicate the presence of a foreign contaminant in the battery.[33] Pulsing that removes 1% or less of the capacity of the cell, while still depleting a small amount of the finite capacity of a primary battery, can be quite valuable in identifying batteries with undesirable high internal resistance. It is typical that batteries for implantable medical applications are tested on a 100% sample basis in this fashion. Additional nondestructive analysis techniques include dimensional and weight measurements, as well as x-ray and leak test of hermetically sealed batteries.[33]

31.4.4 Destructive Testing

Destructive physical analysis (DPA) refers to the physical dismantling of a battery for the purpose of analyzing the components. This process can be used as part of failure analysis for a particular failed unit or it can be part of a systematic lot analysis scheme. In a typical application, a certain percentage of cells manufactured for implantable applications are subjected to accelerated discharge followed by destructive analysis. During the DPA, the cell is cut open and all of the cell components are closely examined for any evidence of defects. This destructive analysis procedure may be part of the lot acceptance criteria which must be passed prior to releasing a lot of batteries.

Life test refers to a test where battery samples are subjected to electrochemical discharge that utilizes all of the available capacity of the cell. These tests are typically performed on a lot sample of batteries where the test conditions approximate the real-time use cycle. One example of a life test for Li/Iodine pacemaker batteries involves a discharge under a 100 kΩ constant resistance load at 37°C. Voltage and impedance measurements are taken bimonthly, and a computer program selects and prints out data for any cell whose voltage falls below a predetermined standard.[36] The life test program was reported to serve three purposes. First, it provides a verification in real time of accelerated projections made for battery performance. Second, it permits a determination of cell-to-cell variability and failure rates. Finally, it provides a warning of any major defect or unexpected cell behavior.[43]

31.5 CHARACTERISTICS OF BATTERIES FOR BIOMEDICAL APPLICATIONS

Batteries for biomedical applications are typically classified into three groups based on the discharge rate needed by the application: low rate, medium rate, and high rate. The low-rate applications are limited to 100 µA of current draw or less, while medium rate are normally in the milliamp range, and high rate can reach as high as several amps of pulsatile current draw. The Li/Iodine battery is prototypical for low-rate batteries, as it has a long history of powering low-rate devices such as the implantable cardiac pacemaker. Likewise, the Li/SVO battery has typified the high-rate battery with its use as an ICD power source. While both of these battery chemistries have been used almost exclusively for implantable medical applications, there are many other battery chemistries that have been utilized in biomedical devices and have seen more extensive usage in military, commercial, and other areas. This section describes the characteristics of the main battery chemistries utilized in this area with an emphasis on design and usage tailored to the biomedical applications.

31.5.1 Lithium/Iodine Batteries

The first implant of a lithium/iodine powered pacemaker occurred in 1972 in Italy,[44] and since then, millions of these batteries have been implanted, demonstrating their high degree of safety, reliability, and predictable performance.[45] Following the initial use of mercury/zinc and nickel/cadmium cells in the first pacemakers, several other battery chemistries were implanted in pacemakers starting in the 1970s. These included lithium battery chemistries, such as lithium/silver chromate

$$Ag_2CrO_4 + xLi^+ + xe^- \rightarrow Li_xAg_2CrO_4$$

The Li/Ag$_2$CrO$_4$ system displayed high voltage (~3 V) and a stepped discharge curve due to multiple electrochemical reactions linked to the rate of discharge.[46] Another lithium battery chemistry used in implantable pacemakers was lithium/copper sulfide

$$2Li^+ + 2e^- + 2CuS \rightarrow Li_2S + Cu_2S \quad \text{(1st plateau)}$$

$$2Li^+ + 2e^- + Cu_2S \rightarrow Li_2S + 2Cu \quad \text{(2nd plateau)}$$

At low drain rates, as used in a pacemaker, the Li/CuS cell system displayed a stepped discharge, where the discharge equations outlined above describe the 1st and 2nd voltage plateaux. These voltage steps occur at ~2.2 V and 1.7 V, respectively, and were used for end-of-life indication.[47] A quite different power source that saw limited use in implantable pacemakers was the nuclear battery. These batteries were based on the radioactive decay of nuclear isotopes.[48] In one case, electrons from a beta emitter were caused to impinge on a semiconductor, giving off electrons and converted to electric current. In another design, kinetic energy of emitted particles from plutonium-238 was converted into thermal energy, which was in turn converted into electrical energy through use of the Seebeck effect. Multiple companies produced nuclear batteries for pacemaker applications in the 1970s, and several thousand of these pacemakers were implanted. The nuclear batteries offered a very long lifetime, on the order of several decades. In one case, it was reported that the pacemaker was still operating 34 years later for a young patient implanted with a nuclear pacemaker in 1973.[49] However, stringent safety testing and device tracking was required for these systems to ensure that there would be no possible release of the radioactive fuel under extreme conditions and this included the accidental cremation of a patient with an implanted nuclear pacemaker. This test required that the pacemaker be placed in a furnace at 1300°C for 2 h with no release of the radioactive fuel.[48] By the early 1980s, the Li/iodine battery chemistry supplanted these other technologies and became the clear choice for implantable pacemaker devices.

The lithium/iodine battery is based on the chemical reaction

$$Li + 1/2I_2 \rightarrow LiI$$

TABLE 31.6 Thermodynamic Functions for Formation of Lithium Iodide (300 K)

Parameter	Value
ΔH (kcal/mol)	−64.551
$T\Delta S$ (kcal/mol)	−0.101
ΔG (kcal/mol)	−64.450
OCV (V)	2.795

The thermodynamic parameters for this reaction are listed in Table 31.6.[50]

The kinetics of the reaction are improved through the use of charge-transfer complexes between iodine and organic donors. The first report of the use of an iodine charge transfer complex was made by Gutman et al. in 1967.[2] This was followed by a 1972 patent that described a solid-state primary cell having a lithium anode, an iodine cathode, and a solid lithium halide electrolyte.[3]

The thermal reaction between iodine and poly-2-vinylpyridine (PVP) to form a I$_2$/PVP cathode material was described in a 1973 patent,[51] and I$_2$/PVP charge transfer complex materials continue to be in use in the manufacture of Li/iodine pacemaker batteries today.

The reaction product, lithium iodide, forms *in situ* as the cell is discharged. The lithium iodide acts as both the separator and the solid electrolyte for the battery, and as the cell is discharged, the LiI layer between the cathode and anode increases in thickness, causing a gradual rise in the internal impedance of the battery. The discharge voltage curve for a Li/iodine cell along with the battery resistance curve is plotted as a function of capacity in Fig. 31.12. The log of the increase in resistance was linear with respect to capacity. It is also notable that the resistance varied between 100 to 8000 ohms during discharge, illustrating the large impact of the formation of the LiI layer. The formation of the solid electrolyte interface between I$_2$/PVP and the lithium anode during discharge also translates into a separator system that is self-healing; an internal short between cathode material and anode generates LiI at the site of the short and thus forms a layer of solid electrolyte/separator at that point, eliminating the short circuit.

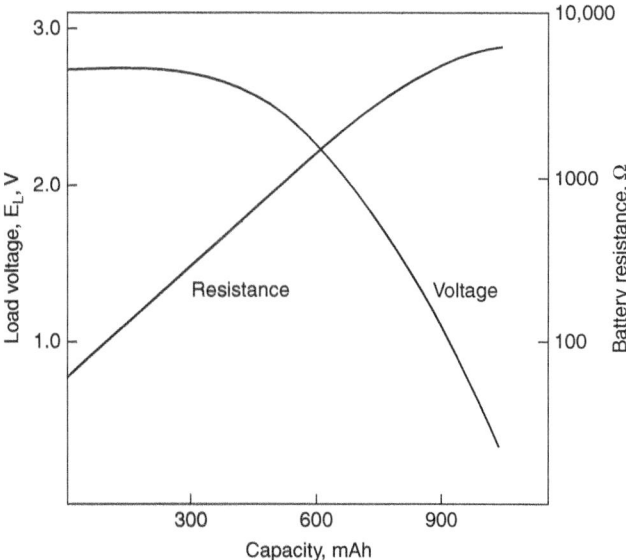

FIGURE 31.12 Loaded voltage and battery resistance for coated-anode Li/iodine battery discharged at 100 μA at 37°C. (*Courtesy of Medtronic.*)

A cutaway view of a Li/iodine cell is displayed in Fig. 31.13. As illustrated in this figure, the standard design for a Li/iodine cell includes a central lithium anode that is corrugated to increase the surface area of this electrode. The cell is filled with I_2/PVP, which surrounds the lithium anode, and is in contact with the stainless steel case, making this cell a case-positive design. In addition to the PVP thermally reacted with the I_2 depolarizer material, PVP is also coated onto the lithium anode as

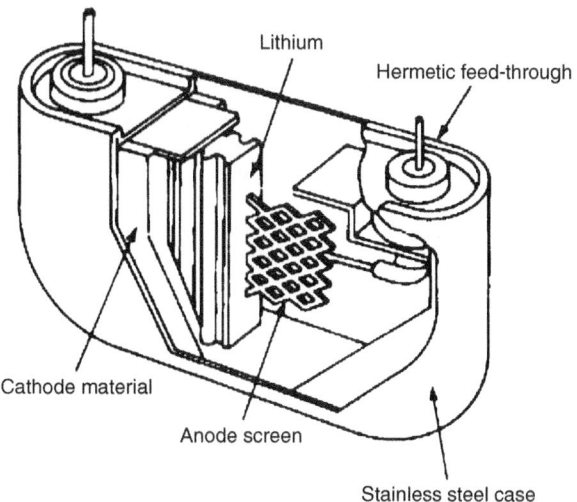

FIGURE 31.13 Cutaway view of typical case-positive Li/iodine cell. (*Courtesy of Greatbatch, Inc.*)

31.18 SPECIALIZED BATTERY SYSTEMS

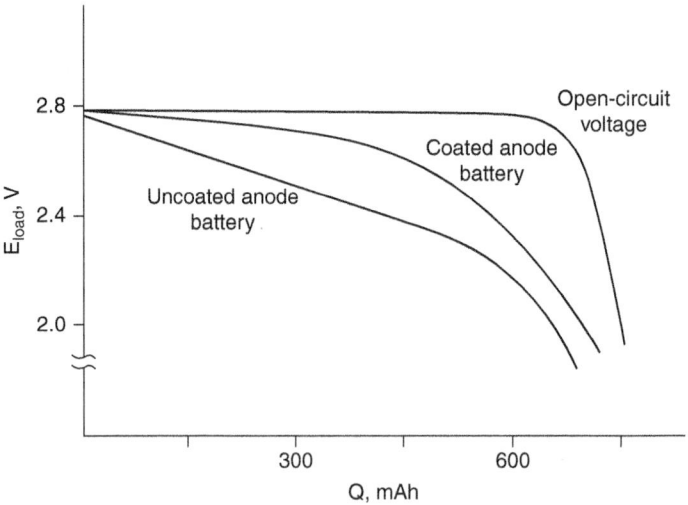

FIGURE 31.14 Loaded voltage vs. discharge state for uncoated and PVP-coated anode Li/iodine batteries discharged at 6.7 µA/cm² at 37°C. (*Courtesy of Medtronic.*)

a thin film. This anode PVP-coating technique, developed in the 1970s,[52] significantly lowers the internal impedance of the cell throughout discharge, as illustrated in Fig. 31.14.

Self-discharge of the Li/iodine battery system occurs by the direct combination of lithium and iodine that has diffused through the lithium iodide layer. The amount of iodine that is able to diffuse through the separator layer is dependent on the thickness of that separator film and thus, self-discharge is larger early in the discharge reaction of the battery, when the LiI layer is thinnest. A plot of heat dissipation data, collected by microcalorimeter measurements made on open-circuit, as a function of cell capacity is illustrated in Fig. 31.15. From this figure it is clear that the great majority of the battery self-discharge process occurs during the first 25% of the battery discharge.

31.5.2 Lithium/Thionyl Chloride Batteries

Medical devices such as neurostimulators, drug pumps, and cardiac monitors require a battery that can supply a constant microamp current for patient monitoring and powering the internal circuitry of the device but also supply short pulses in the milliamp range at intervals to delivery therapy.[53] Implantable-grade lithium/thionyl chloride batteries were developed starting in the 1980s as a medium-rate cell for use in these devices because of its high operating voltage, high energy density, and good rate capability for pulsing applications.[54]

This primary lithium anode battery is a liquid cathode type, where the thionyl chloride ($SOCl_2$) serves as both the cathode and the electrolyte with the addition of the electrolytic salt lithium tetrachloroaluminate ($LiAlCl_4$). The overall cell reaction is defined in the equation

$$4Li + 2SOCl_2 \rightarrow 4LiCl \downarrow + S + SO_2$$

The nominal open-circuit voltage for the system is 3.65 V, and the nominal running voltage is 3.5 to 3.6 V. During the discharge, the $SOCl_2$ is reduced at the carbon black surface which acts as a cathode matrix. The LiCl discharge product is insoluble in the $SOCl_2$ electrolyte and deposits in the carbon matrix. More detailed information on this system for other applications is found in Sec. 14.6. A representative discharge curve is shown in Fig. 31.16, where the 49.9 kΩ resistance load discharged the cell in 4 years' time.

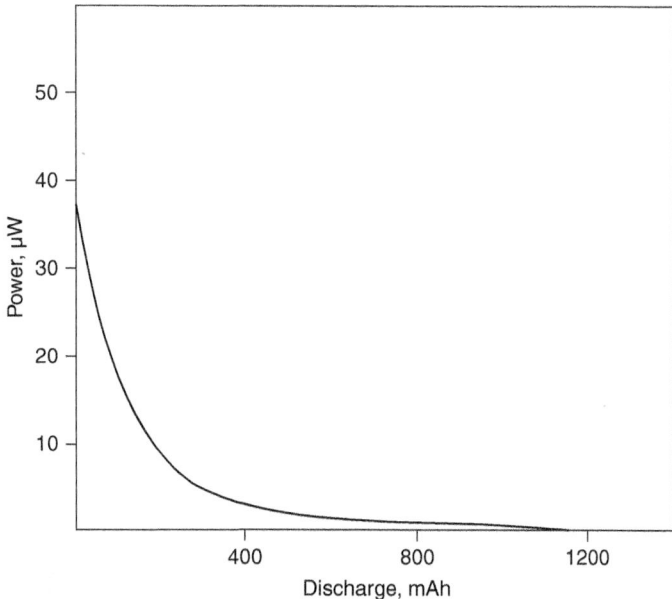

FIGURE 31.15 Power (heat) loss due to self-discharge vs. discharge state for typical Li/iodine battery (calorimetric measurements made at open-circuit voltage). (*Courtesy of Medtronic.*)

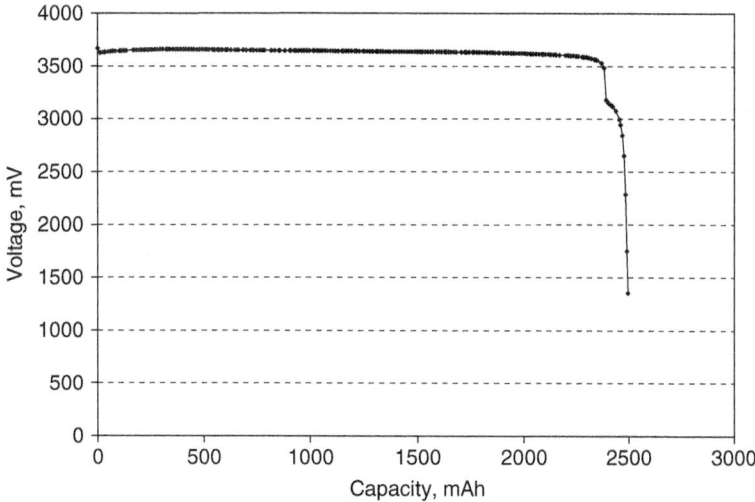

FIGURE 31.16 Discharge curve for an implantable grade Li/SOCl$_2$ battery discharged under a 49.9 kΩ constant resistance load at 37°C. (*Courtesy of Greatbatch, Inc.*)

31.20 SPECIALIZED BATTERY SYSTEMS

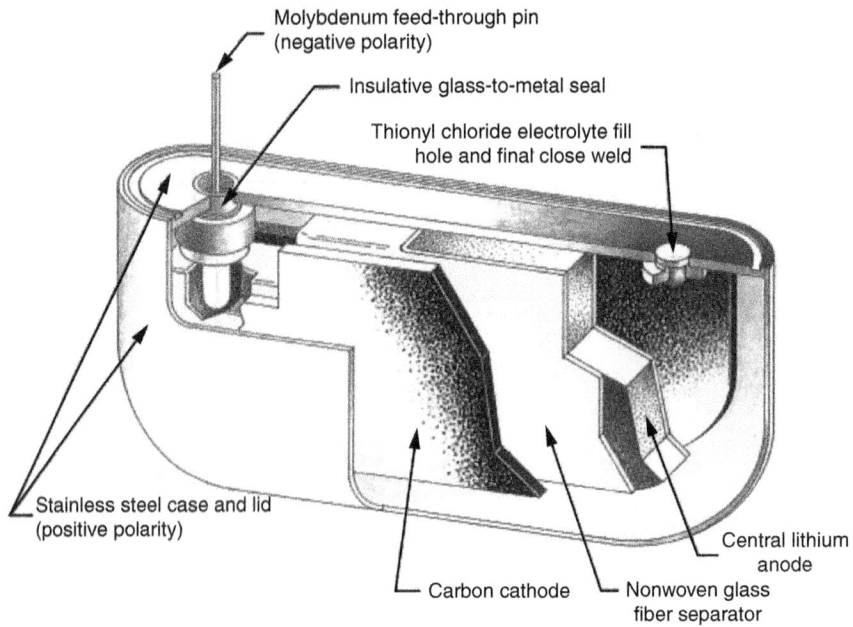

FIGURE 31.17 A cutaway view of a Li/SOCl$_2$ battery designed for implantable biomedical applications. *(Courtesy of Greatbatch, Inc.)*

A cell designed for implantable medical applications is illustrated in the cutaway view in Fig. 31.17. This design features a central lithium anode where lithium metal is pressed to both sides of a screen and covered with a porous separator material. The anode assembly is connected to the feed-through pin coming out of the header. Two cathode current collectors surround the anode and are comprised of a high surface area carbon/polymer binder mixture pressed onto screens that are connected to the battery case. At the time the cell is filled, there is some reaction between the lithium anode and the SOCl$_2$ liquid cathode material, forming a Solid Electrolyte Interface (SEI) layer on the lithium anode, primarily formed of LiCl.[55] Undischarged Li/SOCl$_2$ cells will show an increase in OCV due to the formation of S$_2$Cl$_2$ and SO$_2$ products of the reaction of SOCl$_2$ and lithium, where the S$_2$Cl$_2$ byproduct has a higher potential than SOCl$_2$.[56]

A neurostimulator that operated on a Li/SOCl$_2$ battery was described for diaphragm pacing to restore breathing and for dynamic gracilioplasty to treat fecal incontinence.[57] An implantable grade Li/SOCl$_2$ cell was contained in the neurostimulator device. The system was designed such that two skeletal muscles could be activated simultaneously, with up to 4 electrodes for each nerve. The life span reached by the device was between 4.1 and 7.2 years for diaphragm pacing and 6.4 years for dynamic gracilioplasty, based on the primary battery. The authors of the study remarked that the disadvantage of having to exchange implants by surgical procedure was balanced by the better quality of life achieved from the elimination of permanent external components.[57]

Li/SOCl$_2$ cells suffer from voltage delay, due to the formation of an anode surface film layer, and display higher self-discharge during storage than other low-to medium-rate battery systems (such as Li/iodine and Li/CFx).

31.5.3 Lithium/CFx Batteries

Battery systems containing a carbon monofluoride (CFx) cathode material were among the first lithium anode cells commercialized. A Li/CFx battery was first patented by Watanabe and Fukuda of Matsushita Electric Industrial Co. in 1970.[58] The cell was described as being of high energy density, with nearly 100% utility of the cathode active material, having a flat discharge voltage curve and a long shelf life. The discharge reaction for a Li/CFx cell is described below, where the overall reaction of lithium with CFx results in the formation of lithium fluoride and carbon.[59]

$$nx\text{Li} + (\text{CFx})n \rightarrow nx\text{LiF} + n\text{C}$$

Carbon monofluoride is a gray powder that can be prepared by the direct reaction of carbon powder with fluorine gas at an elevated temperature, following the reaction below:[59]

$$x\text{C (solid)} + x/2\ \text{F}_2\ \text{(gas)} \rightarrow (\text{CF})x\ \text{(solid)}$$

The carbon material used to make CFx can be chosen from a variety of different feedstocks, as illustrated in Table 31.7. As seen in the table, the type of carbon feedstock impacts the physical properties of the resulting fluorinated material.

TABLE 31.7 Properties of CFx Materials Prepared from Different Carbon Feedstocks[59]

Carbon feedstock	Atomic ratio (C:F)	True specific gravity (g/cc)	Surface area (m²/g)	Decomposition temperature (°C)
Graphite fiber	1:0.96	2.52	340	390
Petroleum coke	1:0.98	2.50	290	380
Wood charcoal	1:0.91	2.35	176	333
Binder carbon	1:0.92	2.34	180	320
Carbon black	1:0.97	2.52	297	392
Natural graphite	1:0.98	2.68	293	485

The theoretical capacity of CFx cathode material also directly depends on the level of fluorination, as shown in Table 31.8. While most CFx used as cathode materials for lithium battery systems have utilized x ~ 1 to maximize the capacity of the system, investigation into subfluorinated (x < 1) CFx for lithium batteries has been made with the goal of increasing the inherent rate capability of the material.[60]

TABLE 31.8 Theoretical Capacity of CFx Materials with Different C:F Ratios[59]

CFx material	CFx capacity (Ah/g)	Specific energy (Wh/kg)
$CF_{1.0}$	0.864	2190
$CF_{0.8}$	0.788	1960
$CF_{0.6}$	0.687	1740
$CF_{0.4}$	0.547	1430

Implantable grade Li/CFx batteries have been designed and used in pacemakers and neurostimulators. Such a battery was described by Greatbatch and coworkers.[61] The battery used CFx prepared by the direct fluorination of petroleum coke, with a lithium anode and gamma butyrolactone electrolyte containing the salt lithium tetrafluoroborate ($LiBF_4$). The system was compatible with a titanium casing, which allowed for a 50% reduction in weight over the same size Li/iodine cell. The discharge curves in Fig. 31.18 were collected for implantable grade Li/CFx cells using a range of 2000 to 100,000 ohm constant resistance loads. The 100,000 ohm load data was collected over 10 years of a real-time test at 37°C. Additional information on Li/CFx batteries for other applications may be found in Sec. 14.9.

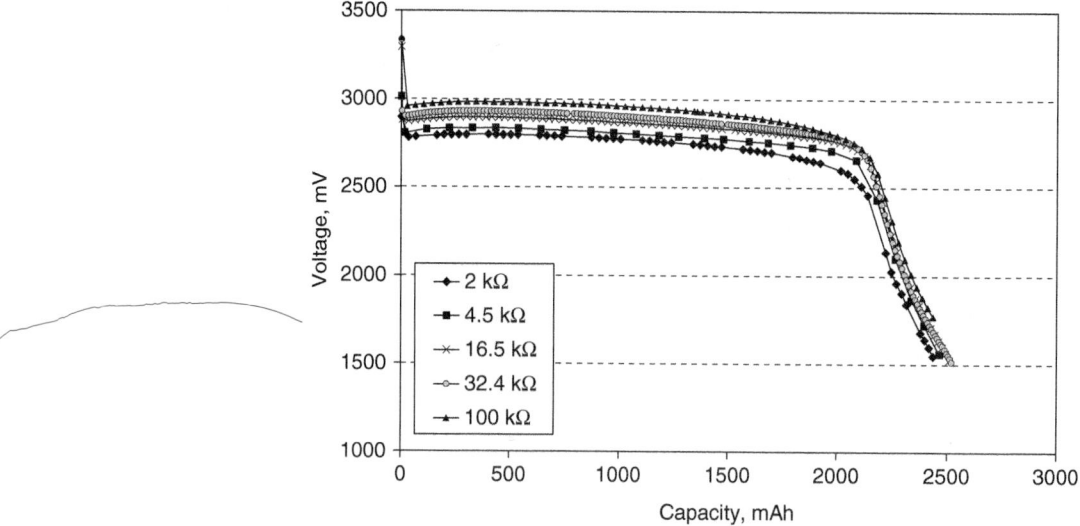

FIGURE 31.18 Discharge of implantable grade Li/CFx batteries under constant resistance loads (2.0, 4.5, 16.5, 32.4, and 100.0 kΩ loads) at 37°C. (*Courtesy of Greatbatch, Inc.*)

31.5.4 Lithium/Silver Vanadium Oxide Batteries

Silver vanadium oxide (SVO) was first investigated as a cathode material for lithium batteries for high-temperature commercial applications.[62] It wasn't until the implantable cardioverter defibrillator device emerged that the application of this battery system to implantable medical devices was identified. The first battery system used in ICDs was the lithium/vanadium oxide (Li/V$_2$O$_5$) cell designed by the Honeywell Corporation.[63] This battery chemistry was used in ICDs until the late 1980s. Since the first implant of a Li/SVO battery in 1987, the great majority of ICDs have utilized this power source, and it was this battery system that enabled the tremendous growth of the ICD market.

The active cathode of a Li/SVO battery utilizes a solid-state material with a stoichiometry of Ag$_2$V$_4$O$_{11}$ (ε-phase). The structure of SVO is composed of edge-shared distorted-octahedra in V$_4$O$_{11}$ clusters which then form corner-shared vanadium oxide strings in a layered structure.[64] Silver is located between the V$_4$O$_{11}$ layers. During the discharge of the cathode, lithium ions replace silver between the vanadium oxide layers in a displacement reaction.[65] The silver is displaced from the structure and can be seen to extrude from the SVO particles, as shown in the scanning electron microscope image of discharged SVO cathode material in Fig. 31.19.[66]

The anode for the Li/SVO battery is lithium metal, and the electrolyte is based on organic solvents (propylene carbonate and dimethoxyethane) to which a lithium salt has been added. The cell reactions are defined in the following equations:

Anode reaction: $\qquad\qquad\qquad Li \rightarrow Li^+ + e^-$

Cathode reaction: $\qquad Ag_2V_4O_{11} + 7Li^+ + 7e^- \rightarrow Li_7Ag_2V_4O_{11}$

Overall cell reaction: $\qquad 7Li + Ag_2V_4O_{11} \rightarrow Li_7Ag_2V_4O_{11}$

The discharge reaction of the Li/SVO battery has been studied in detail by a combination of physical, chemical, and electrochemical methods.[67–69] The first reduction step of SVO primarily forms

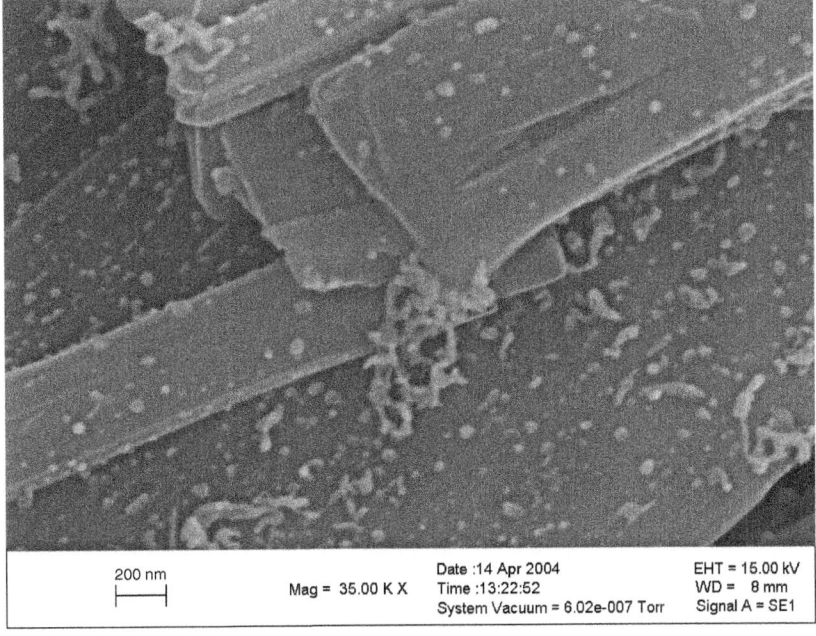

FIGURE 31.19 Scanning electron microscope (SEM) image of discharged SVO, $Li_{1.4}Ag_{0.6}V_4O_{11}$, after the insertion of lithium. (*Courtesy of Greatbatch, Inc.*)

silver metal (reduction of Ag^+ to Ag^0). This metallic silver greatly increases the conductivity of the cathode material, which contributes to the high current-carrying capability of the Li/SVO system. However, some vanadium is also reduced during the first plateau region of discharge, as identified by NMR spectroscopy of the discharged cell materials.[70] As the discharge reaction continues, the vanadium(V) is reduced to vanadium(IV) and vanadium(III). The discharge of the SVO cathode provides a stepped discharge curve due to the multiple oxidation states of vanadium. Figure 31.20 provides a typical voltage curve for a Li/SVO cell discharged under a light load of 100 kΩ. This stepwise change in voltage is predictable and provides state-of-charge indication for the battery, which is important in determining the remaining lifetime in the implantable device.

During the middle-of-life discharge zone, Li/SVO batteries typically display an increase in internal cell resistance. This increased resistance is evident in a voltage delay observed during high current draw on the battery and has been shown to be the result of film formation on the lithium anode.[71] Figure 31.21 displays the discharge of a Li/SVO battery under a 7-year test regime. The cell was discharged under a 80 kΩ constant resistance load, with pulse loads applied every 180 days. The pulse load was applied for 10 s, in groups of 4 pulses for each pulse train, with a current density of 24.5 mA/cm^2. Voltage delay (pulse minimum voltage lower than pulse end voltage) can be seen in the 40 to 50% DOD region of the discharge. Under ordinary conditions, the anode film is thin enough that it is removed by exercising the cell under high current pulses, making this a reversible condition. If cells are not pulsed at a sufficient level, the film on the anode is no longer effectively removed and a permanent increase in cell resistance results. A number of studies have been done to explore the effectiveness of electrolyte additives on reducing voltage delay by modifying the film formed on the lithium anode.[71–73] The addition of CO_2 or substances such as dibenzyl carbonate (DBC) or benzyl succinimidyl carbonate (BSC) have been reported to reduce anode passivation. These materials are believed to operate by lowering the impedance of the SEI layer on the surface of the lithium anode.

31.24 SPECIALIZED BATTERY SYSTEMS

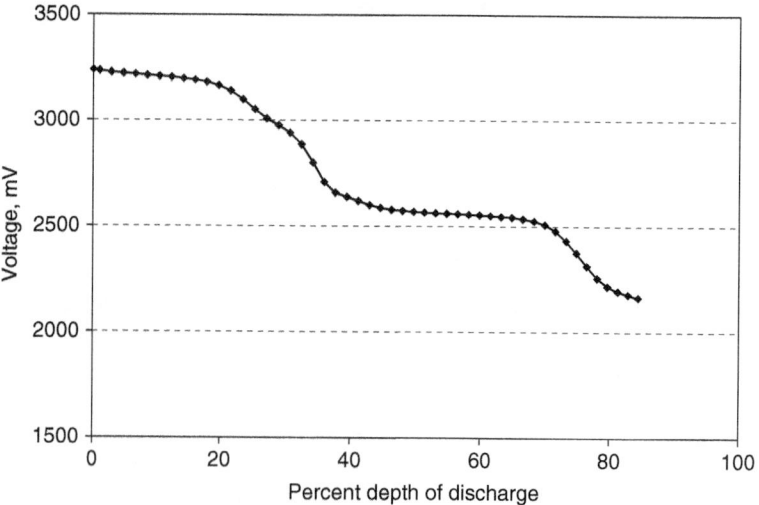

FIGURE 31.20 Discharge of a Li/SVO battery under a 100 kΩ constant resistance load at 37°C, with pulse loads applied every 30 days. The total test time for the discharge was 4 years. (*Courtesy of Greatbatch, Inc.*)

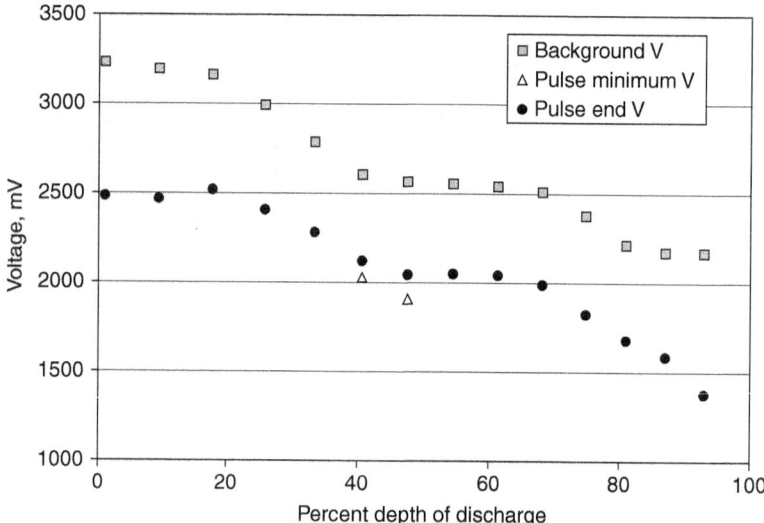

FIGURE 31.21 Discharge of a Li/SVO battery under a 7-year test regime at 37°C. The cell was discharged under a 80 kΩ constant resistance load, with pulse loads applied every 180 days. (*Courtesy of Greatbatch, Inc.*)

Typical construction of a prismatic implantable Li/SVO cell designed for use in an ICD where high power capability is required is shown in Fig. 31.22. The case and lid are typically made of 304 L stainless steel and are the negative cell terminal. A glass-to-metal seal (GTMS) employs TA-23 glass and a molybdenum or niobium pin.[74] Although this GTMS is intrinsically corrosion resistant in lithium primary cells, an elastomeric material and an insulating cap are added to the underside

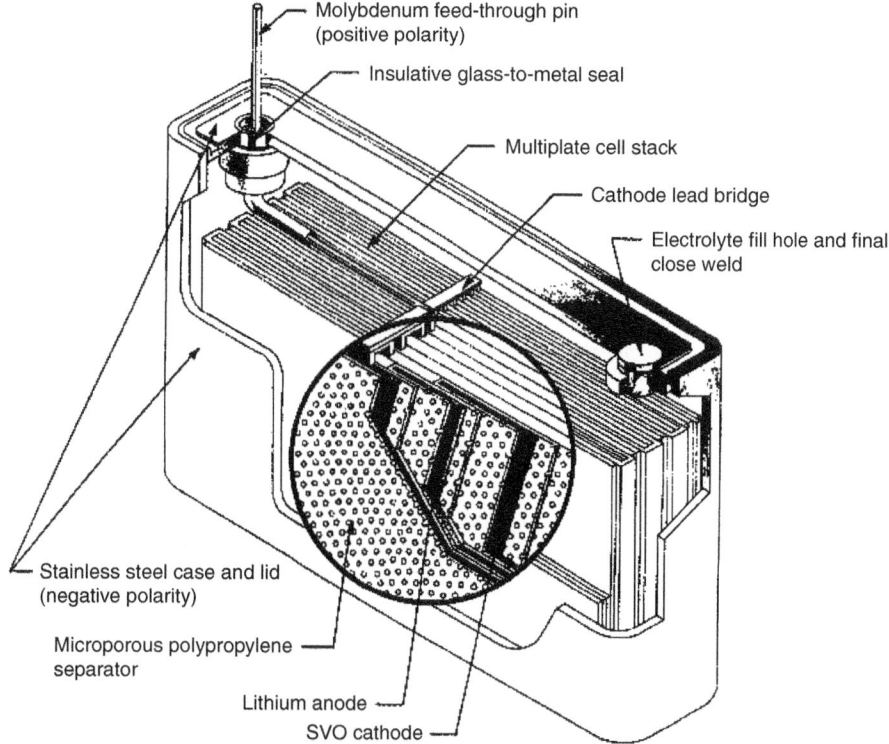

FIGURE 31.22 Construction of a Li/SVO cell designed for an ICD. (*Courtesy of Greatbatch, Inc.*)

of the seal to provide an additional mechanical barrier and enhanced protection for the GTMS. Insulating straps are added to the underside of the lid to prevent contact of the cathode lead to the lid. The anode is constructed of two layers of lithium foil pressed on a nickel screen and heat sealed in a microporous polypropylene or polyethylene separator. The cathode consists of SVO with added carbon, graphite, and binder, pressed on a finely meshed titanium screen under high pressure and heat sealed in a microporous separator. One anode assembly is wound around the individual cathode plates, as shown in Fig. 31.22, and the anode tab is welded to the case. Multiple cathode plates are welded to a cathode lead bridge, and an insulated tab connects the bridge to the Mo pin of the GTMS. Following welding of the lid to the case, the cell is filled with electrolyte through the fill hole and a stainless steel ball inserted in the fill port, and an additional plug is then laser welded over the fill hole, providing a hermetic seal.

Capacity requirements for a Li/SVO battery used to power an ICD are typically in the range of 1 to 3 Ah. The battery is required to deliver 10 to 30 µA continuously for the sensing and/or pacing functions for many years. It must also be capable of providing the energy required to charge the capacitors in the defibrillator. This energy is typically 2 to 4 A for 10 s.

31.5.5 Lithium/Manganese Oxide Batteries

Lithium/manganese dioxide (Li/MnO_2) batteries were first developed by Sanyo in 1975, and were originally designed and used for low-power applications, such as watches, calculators, and memory backup.[75] Higher-power Li/MnO_2 cells were then developed for commercial applications, with the

2/3A camera battery being a prime example. See Sec. 14.8 for more detailed information of lithium/manganese dioxide batteries for other applications. Li/MnO$_2$ batteries have typically been applied to medical devices that require high power, such as implantable cardiac defibrillators and automated external cardiac defibrillators. Advantages of the Li/MnO$_2$ battery system have been defined as high voltage (3 V) and high energy density, excellent discharge, storage stability and safety.[75]

The manganese dioxide used in lithium cells is heat treated electrolytic MnO$_2$. The lithium cell discharge reaction is described by

$$Li + Mn^{4+}O_2 \rightarrow LiMn^{3+}O_2$$

Figure 31.23 presents the discharge curve for an example Li/MnO$_2$ battery designed for an implantable medical device, such as a defibrillator.[76] This battery has a capacity of ~2 Ah, with a total cell volume of 8.6 cc. For the example given in Fig. 31.23, the battery is pulsed with a 3 A current. The cell utilized a high overall cathode surface area of 180 cm^2, giving a cathode current density of 17 mA/cm^2.

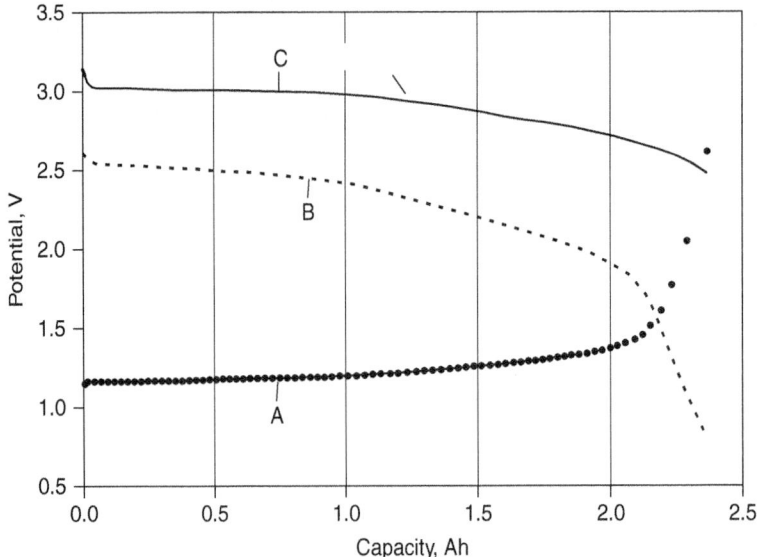

FIGURE 31.23 Discharge curve for an implantable Li/MnO$_2$ cell designed for ICD applications. Curve A represents ICD charge time, curve B is the pulse 1 average voltage, and curve C is the background voltage (*Ref. 76*).

The discharge curve for the Li/MnO$_2$ cell displayed in Fig. 31.23 is similar to those found for Li/MnO$_2$ cells used for commercial applications. Another example of a Li/MnO$_2$ cell that has been utilized for biomedical applications is a dual battery which utilizes two Li/MnO$_2$ cells separated in the battery can and connected in series to produce a 6 V battery.[77,78] A diagram of this type of battery design is presented in Fig. 31.24. These cells have been developed for use in ICDs to provide high rate pulse currents. A plot of the pulse discharge of a Li/MnO$_2$ double cell from Reference 78 is presented in Figure 31.25. In this test, the double cell was pulsed with a current of 1.4 A for a 9 s pulse duration followed by 15 s of open-circuit rest, for a total of 4 pulses in each group. The cell was then allowed to relax on open circuit for 30 min before applying the next pulse train. The higher nominal voltage of the 6 V cell allows the cell to be discharged at high pulse currents and still maintain device function, allowing for a decreased charge time of the output capacitors. A decrease in charge time is desired, as this can translate into more rapid delivery of therapy. The charge time

BATTERIES FOR BIOMEDICAL APPLICATIONS **31.27**

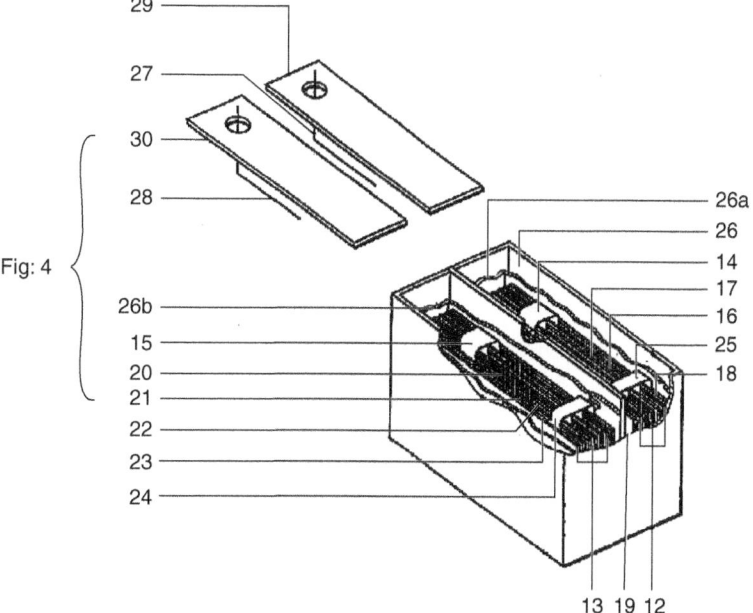

FIGURE 31.24 Schematic representation of an implantable 6 volt Li/MnO$_2$ double-cell used for ICD applications. (*See Reference 78 to identify all components.*)

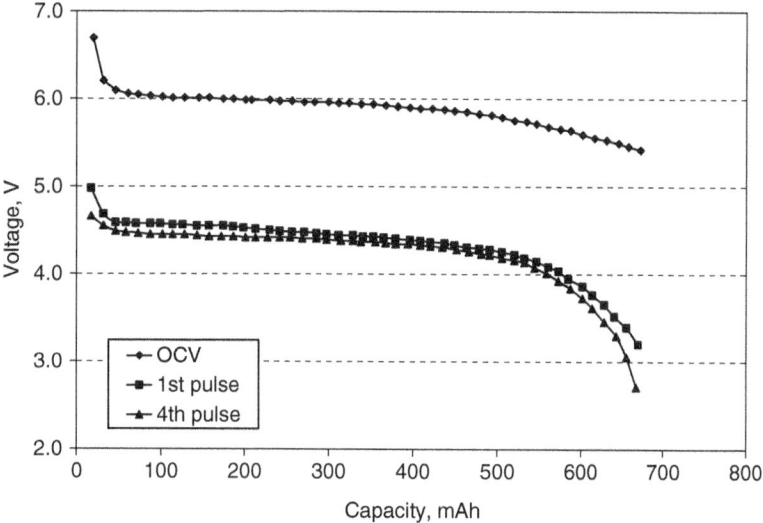

FIGURE 31.25 Pulse discharge results for an implantable Li/MnO$_2$ double cell (*Ref. 77*).

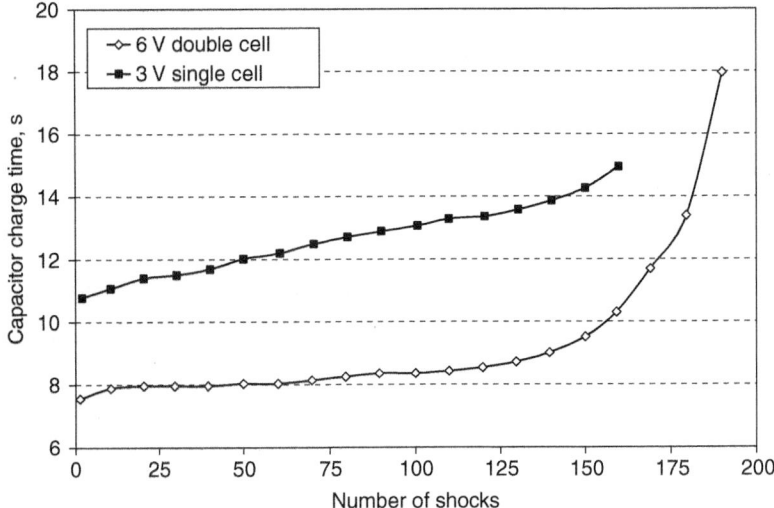

FIGURE 31.26 Time needed to charge capacitors using implantable Li/MnO$_2$ single and double cells (*Ref. 77*).

for a test circuit designed to mimic the output stage of an ICD was presented for a double Li/MnO$_2$ cell compared to a single Li/MnO$_2$ cell in Reference 77, and is displayed in Fig. 31.26.

The addition of lead compounds (PbO, PbCrO$_4$, and PbMoO$_4$) to the manganese dioxide cathode material was described to provide a sloping voltage curve for battery end-of-life indication.[78]

A Li/MnO$_2$ cell was developed for implantable medium-rate power applications (advanced pacer, neurostimulator, micromechanical devices, and drug pumps).[79] This cell utilized a stainless steel case with a case negative design. One central anode was utilized, with a cathode on each side. The cell delivered 2.5 Ah capacity when discharged under a 3 kΩ load, yielding 0.588 Wh/cc and 0.149 Wh/g energy density and specific energy. Pulsing capability up to 400 mA was also demonstrated for this cell.

31.5.6 Lithium/Silver Vanadium Oxide plus CFx Batteries

Lithium/silver vanadium oxide and lithium/carbon monofluoride batteries have been utilized in implantable medical applications successfully for more than two decades. A comparison of Li/SVO and Li/CFx battery technologies is shown in Table 31.9.[80] As can be seen in the table, the Li/SVO system has much higher rate capability and lower internal resistance, while the Li/CFx system has much higher energy density (nearly 300 Wh/L higher than Li/SVO) and is very stable.

TABLE 31.9 Comparison of Li/SVO and Li/CFx Implantable Battery Technologies

Category	Li/SVO	Li/CFx
Typical running voltage (V)	2.7	2.9
Energy density (Wh/L)	730	1000
Typical current density (mA/cm^2)	35	1
Typical internal resistance (BOL) (ohms)	0.250	40
Time-dependent internal resistance increase (%)	40–60 DOD	none
Self discharge (% per year)	1	<1
Stepped discharge curve	Yes	No

BOL = Beginning of life

More recently, the combination of these two materials into one cathode system has been introduced to improve the rate and capacity of implantable batteries. The SVO material serves to improve the rate of the cathode, while the CFx serves to improve the overall capacity of the cathode. One form of this system uses a mixture of SVO with CFx to form a hybrid cathode,[81] while another form uses a distinct layer of SVO on top of a CFx layer, with a current collector between.[82] The use of two different battery chemistries within a single implantable device has historical precedence, where dual energy sources, both a high-rate Li/SVO and a low-rate Li/iodine battery, have been used to power ICDs.[83] Here the Li/SVO cell was used to provide power for the defibrillation shocks and the Li/iodine cell was used to provide power for the monitoring functions.

A physical mixture of SVO with CFx in a hybrid cathode in a primary lithium battery has been used to power a number of implantable devices, including pacemakers, hemodynamic monitors, drug-delivery devices, and pulse generators to treat atrial fibrillation and to provide cardiac resynchronization therapy.[83] The ratio of SVO to CFx can be varied in these batteries, where a greater proportion of CFx enhances the capacity, and a greater proportion of SVO improves the power capability and end-of-service characteristics. For low- to medium-rate applications, designs with 85 to 90% CFx were chosen, and discharged over a range of current densities from 3.8 to 30.1 mA/cm^2. These batteries matched Li/iodine cells in energy density at about 1 Wh/cc.[83]

A cross-sectional view of a primary lithium battery using a SVO/CFx hybrid cathode in a medium-rate cell design is displayed in Fig. 31.27.[84] In this diagram, the cell case (labeled 10) can be made of stainless steel or titanium, and the conducting pin (25) is insulated from the case via a glass-to-metal-seal (GTMS). The anode current collector is labeled 45, and the cathode current collector is labeled 60. In this design, the cathode material is pressed into the D-shaped cathode current collector, with the current collector surrounding the cathode, so that swelling of the cathode during discharge does not result in impedance variability.

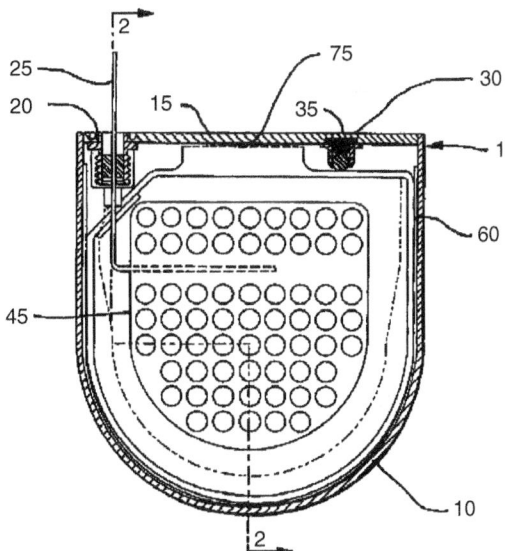

FIGURE 31.27 Cross-sectional view of medium-rate implantable battery utilizing SVO-CFx hybrid cathode. (*Ref. 84.*)

Extensive modeling studies have been conducted on the Li/SVO-CFx hybrid cathode batteries.[85] The physically based models showed good agreement with data collected for hybrid batteries that varied in cathode thickness, geometric area, and SVO-CFx mix ratio. Moderate rates of discharge were modeled in these studies, ranging from 7.5 to 188 μA/cm^2.

31.30 SPECIALIZED BATTERY SYSTEMS

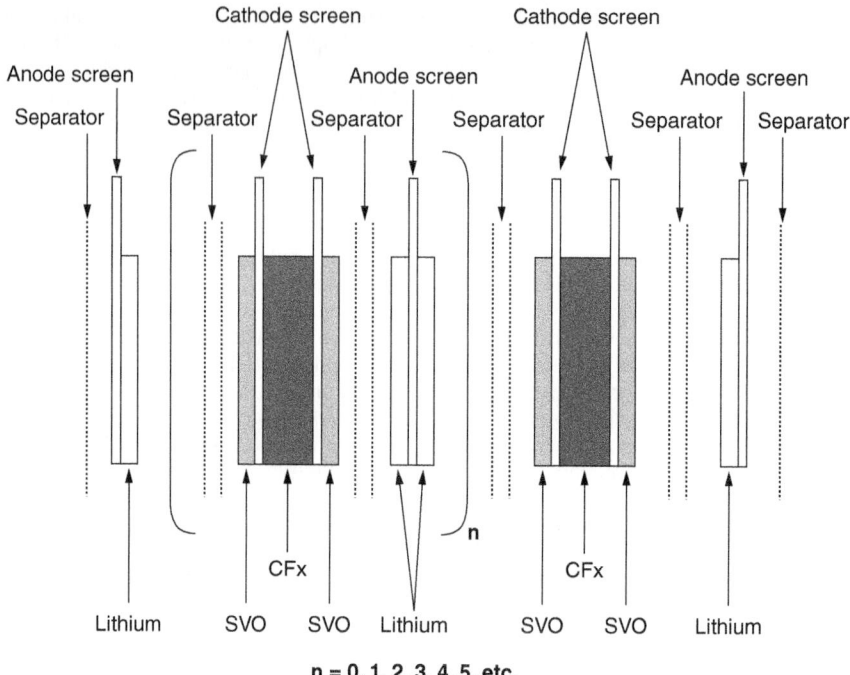

FIGURE 31.28 Schematic diagram for a high-rate Li/SVO-CFx sandwich cell. (*Courtesy of Greatbatch, Inc.*)

Another application of a cathode for implantable medical batteries that contains both SVO and CFx utilizes distinct layers of material in a sandwich configuration, as displayed in Fig. 31.28. In this configuration, the high-rate SVO cathode material is placed on the outside of the cathode, while the lower-rate CFx material is sandwiched between current collectors that are also in contact with the SVO layers.[86] By maintaining a distinct SVO layer in contact with the current collector, this configuration allows the system to maintain the high-rate capability of a SVO cathode, while increasing the energy density and stability of the cathode through the incorporation of CFx.

The sandwich cathode design also allows for flexibility in the end-of-service point for the battery by adjusting the SVO:CFx ratio. The SVO/CFx system provides two replacement indication points: one between the CFx plateau and the second SVO plateau, and another after the second SVO plateau.[80] Figure 31.29 displays a discharge profile for an implantable grade high-rate Li/SVO-CFx sandwich cell, pulsed with 3 A current. Figure 31.30 displays a discharge profile for an implantable-grade, medium-rate Li/SVO-CFx sandwich cell, pulsed with 19 mA current.

31.5.7 Lithium-Ion Batteries

Lithium-ion batteries have seen widespread use in consumer electronics due to their high energy density, long cycle life, and lack of a memory effect. These batteries have also been introduced into medical devices where rechargeability is needed to satisfy the energy requirements of the system. Inductive coupling is used to charge the implanted batteries.

The first commercialized Li-ion cells utilized $LiCoO_2$ cathodes coupled with carbonaceous anodes for reversible intercalation of lithium. Over the past several years, the materials used for cathodes and anodes for Li-ion cells have been the subject of countless studies, and there are many choices, ranging from mixed metal oxides and phosphates for the cathode, to varied carbons and lithium alloys for the anode. However, the use of Li-ion cells for implantable medical applications

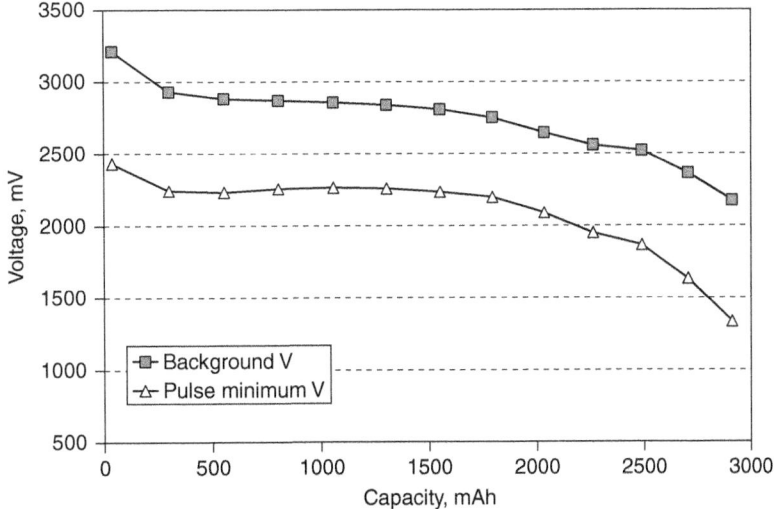

FIGURE 31.29 Discharge profile for an implantable high-rate Li/SVO-CFx sandwich cell. Cell pulsed with 3000 mA current for 10 s every 60 days, with a background load of 20 kΩ, at 37°C. (*Courtesy of Greatbatch, Inc.*)

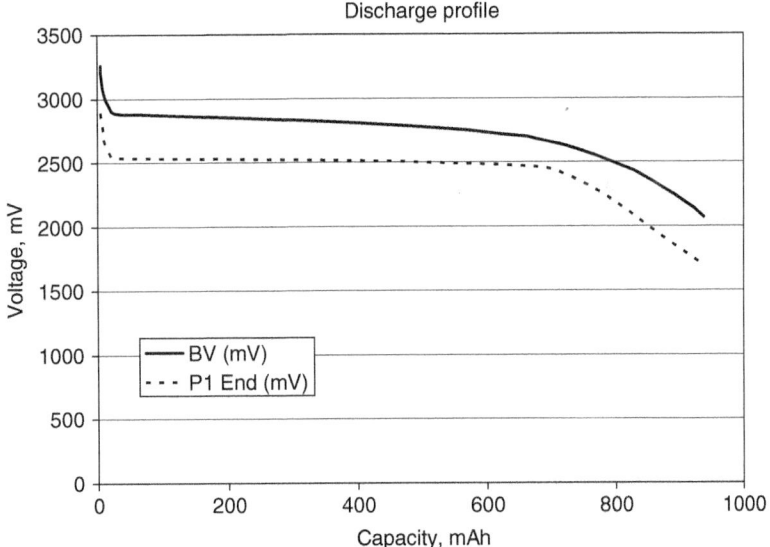

FIGURE 31.30 Discharge profile for an implantable medium-rate Li/SVO-CFx sandwich cell. Cell pulsed with 19 mA current for 900 s every 28 days, with a background load of 25 kΩ, at 37°C. BV = Background voltage; P1 End = Voltage at end of pulse 1.

has thus far been dominated by the LiCoO$_2$/carbon anode chemical system, since this system is well characterized and known to be very stable.[81] This electrochemical system is described by the following equations:

Charge reaction:
$$LiCoO_2 + Li_{1-x}CoO_2 + xLi^+ + xe^-$$
$$C_6 + xLi^+ + xe^- \rightarrow Li_xC_6$$

Discharge reaction:
$$Li_{1-x}CoO_2 + xLi^+ + xe^- \rightarrow LiCoO_2$$
$$LiC_6 \rightarrow Li_xC_6 + xLi^+ + xe^-$$

The theoretical capacity of LiCoO$_2$ is 270 mAh/g, with a practical capacity of ~150 mAh/g at $x \sim 0.5$. The theoretical capacity for graphite anode material is 372 mAh/g with a practical capacity of ~335 mAh/g at $x \sim 0.9$. Typical discharge curves for a LiCoO$_2$/graphite Li-ion cell at a variety of C rates are displayed in Fig. 31.31. While the volumetric energy density of an implantable Li-ion cell is lower than a comparable primary implantable cell, the overall energy delivered by the system is very high when the number of recharge cycles is factored into the equation. Fig. 31.32 displays a graph of the cycle life curve for an implantable-grade Li-ion cell. Here the cell can be cycled at least 1000 times with more than 70% of initial capacity.[87]

Several examples of lithium-ion cells used for medical applications have been reported in the literature. Li-ion batteries are being used as the implantable battery in ventricular assist and total artificial heart devices. The large energy draw of these devices requires that the battery be recharged often (typically daily) and rules out the use of a primary battery for these applications. Because these devices place a significant load on the internal battery, the heat rise during use is a significant concern. Excessive heat generated by the battery/device can lead to tissue damage in the patient. A study of the heat generation of lithium-ion cells used to power a ventricular assist device was reported, where a 16.5 cc (1800 mAh) lithium-ion cell was compared to a 7.5 cc (730 mAh) polymer lithium-ion cell.[88] The cells were discharged using a 1 A current at an ambient

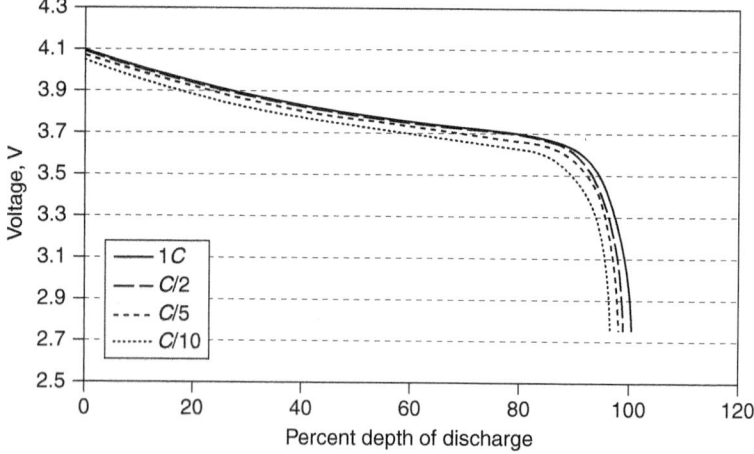

FIGURE 31.31 Discharge curves for an implantable Li-ion cell at rates from $C/10$ to $1C$. (*Courtesy of Greatbatch, Inc.*)

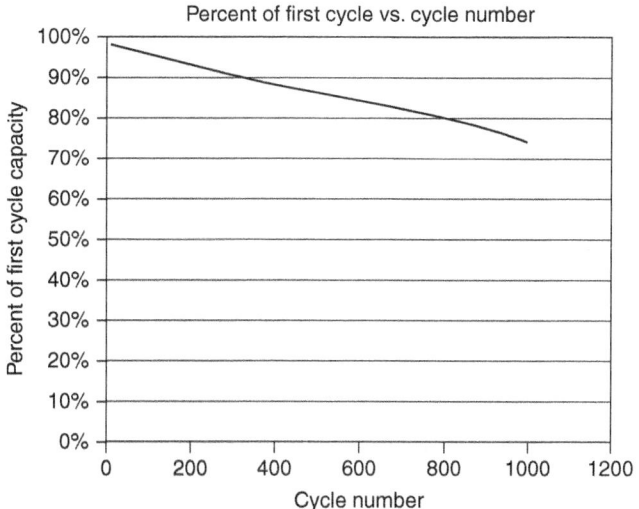

FIGURE 31.32 Cycle life curve for an implantable Li-ion cell (*Ref. 87*).

temperature of 37°C. The temperature of the batteries was found to be determined by ohmic loss due to internal resistance, chemical loss due to chemical reactions, and release of heat to the surroundings. The internal resistance was found to be relatively constant as a function of state-of-charge (SOC) of the battery, with the polymer cell having a lower resistance than the lithium-ion cell. However, the polymer cell displayed a slightly larger entropy change (ΔS) compared to the lithium-ion cell, and both were strongly dependent on state-of-charge, where at <50% SOC the ΔS significantly increased. Overall, the cell skin temperature increased <2°C between 100 and 50% SOC for both cells, while a 3 to 4°C temperature rise was observed when the cells were discharged to 10 to 30% SOC.

In another study of rechargeable batteries for total artificial heart devices, Li-ion batteries were compared to nickel/metal hydride and nickel/cadmium batteries.[89] All of the batteries were charged using a 1C rate followed by full discharge to mimic device operation. The Li-ion batteries met the 1500 cycle life requirement set by the researchers and showed a significantly lower temperature increase (3°C) during usage compared to the NiMH and NiCd cells (15 to 25°C).

A battery pack designed to be implantable for LVAD and TAH applications contained 7 Li-ion cells, 700 mAh each, connected in series to achieve a pack voltage in the range of 20.3 to 28.7 V.[13] The pack was tested at discharge rates from $C/5$ to $2C$ and displayed good rate capability. Cycling tests were completed on individual 700 mAh size cells at $C/2$ and 1.5C rates and were found to give 2000 cycles with 75% capacity retention, and 1000 cycles with 81% capacity retention, respectively.

Small Li-ion cells have been developed for hearing aid devices.[90] These rechargeable cells displayed an average discharge voltage of 3.6 V and a capacity of ~10 mAh, yielding an average energy density of 200 Wh/L and a specific energy of 75 Wh/kg. The cells maintained a voltage above 2.8 V when pulsed using a 100 mA pulse current for a duration of 10 s. Charging time was 2 h or less using a 3 mA charge current. Cycle life testing of the batteries demonstrated 2700 cycles at 60% of full charge for each cycle, corresponding to a capability of powering a hearing aid for 7 years on a daily use cycle. The construction of the cells was a standard button cell, with an external cell volume of 0.18 cc, and a weight of ~0.45 g. The cells used graphite anodes and $LiCoO_2$ cathodes, with 1 M LiAsF6-EC/DMC electrolyte.

Li-ion cells have also received FDA approval for use in cochlear implants.[91] A 120 mAh cell is contained in the external speech processor (see Fig. 31.8) worn behind the ear for a cochlear implant system. The Li-ion cell used in this system is prismatic in shape and has a nominal running voltage of 3.7 V. Charging time is defined as <4 h, and the cell is rated for 500 cycles or greater. With a 10 h run time, charged once a day, a cell would last slightly less than 2 years before being replaced with a new Li-ion cell. A comprehensive treatment of lithium-ion technology is given in Chap. 26.

31.5.8 Zinc/Air Batteries

Zinc/air batteries, in the small button cell format, are extensively used for hearing aid applications. These cells have exceptionally high specific energy (since one active cathode material is taken from air), and high rate capability for a button cell. Zinc/air initially replaced mercuric oxide batteries for hearing aids based on environmental reasons, but the high-rate capability and energy density of the zinc/air chemistry produced greater battery life and improved sound quality relative to the mercuric oxide batteries.[92] The chemistry for the discharge reaction for zinc/air batteries is described in the following equations:

Cathode reaction: $O_2 + 2H_2O + 4e^- \rightarrow 4OH^-$

Anode reaction: $Zn + 2OH^- \rightarrow ZnO + H_2O + 2e^-$

Overall cell reaction: $2Zn + O_2 \rightarrow 2ZnO$

The design of a Zn/air button cell is illustrated in Fig. 31.33. The cathode consists of a thin catalyzed carbon electrode which utilizes O_2 as the active cathode material. Holes must be included in the cell case to allow for air access to the gelled zinc powder anode. These holes are covered until the cell is ready to use to prevent the cell from depleting prior to the intended usage.

Depending on the rate of discharge, the zinc/air battery has a mainly flat discharge curve. Figure 31.34 displays discharge curves collected at three different rates for 675-size button cells.

Since the hearing aid application typically requires the battery to be placed in a small device behind or in the ear, contamination of the battery by human sweat is a distinct possibility. The

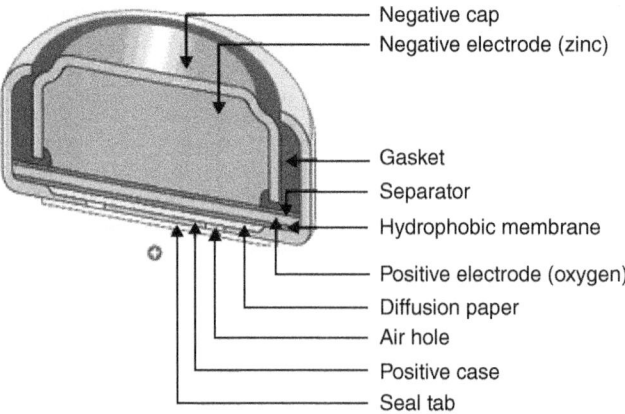

FIGURE 31.33 Cutaway view of zinc/air button-cell battery. (*Courtesy of Japanese Battery Association.*)

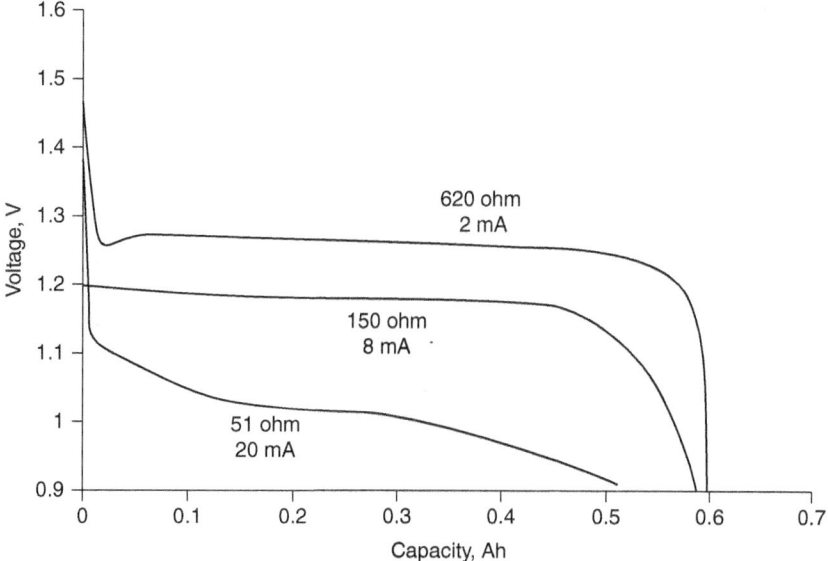

FIGURE 31.34 Discharge curves for a 675-size zinc/air button-cell battery discharged at three loads at 20°C. (*Courtesy of Duracell, Inc.*)

potential for corrosion of the battery components under these conditions needs to be considered in the materials selection aspects of the design of these cells.[93] In addition, the loss of capacity and performance in zinc/air cells was assigned to four basic mechanisms:[94]

- Gassing: The reaction of zinc with the aqueous electrolyte to form zinc oxide and hydrogen
- Water transfer: The gain or loss of water by the cell due to differences between the partial pressure of water vapor in the electrolyte and in the surrounding atmosphere
- Direct oxidation: The chemical oxidation of the zinc by atmospheric oxygen
- Carbonation: The reaction of carbon dioxide in the air with the alkaline (KOH) electrolyte

More detailed information on zinc/air button-cell batteries is given in Chap. 13.

31.5.9 Biological Fuel Cells

As power sources for biomedical devices, biological fuel cells are at present experimental devices that convert biological substances to electrical energy. In one embodiment, these devices utilize biocatalysts to catalyze the oxidation of the biomass material, coupled with the reduction of oxygen to water, to generate electricity. They could potentially be used in implantable applications, such as powering implantable devices and sensors, or utilizing the biofuel cell itself as a sensor.[95,96] While the first glucose biofuel cell was reported in 1962, there are still issues to be resolved with these systems, including short active lifetimes, low power densities, and low efficiency due to partial oxidation of the biofuel.[97]

Biofuel cells are similar to other fuel cells in containing an anode, cathode, and separator, but also include redox enzymes to complete the oxidation and reduction at the anode and cathode, respectively. Figure 31.35 displays a schematic diagram for a biofuel cell. The anode catalysts used to oxidize the biofuel can be biocatalysts, like glucose oxidase, or even living microbes. However, complete oxidation of a complex biofuel (such as glucose) to CO_2 and water in a fuel cell is typically

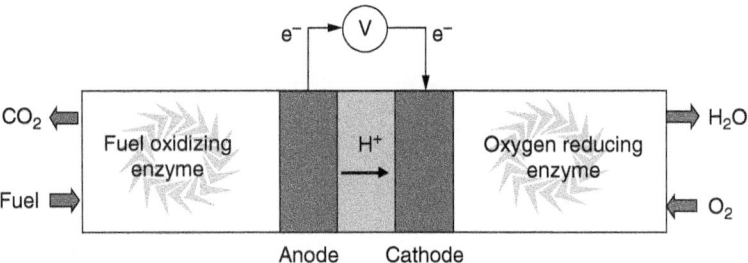

FIGURE 31.35 Schematic diagram of a biofuel cell utilizing enzymatic catalysts.

not feasible. On the cathode side, the reduction of oxygen to water can be catalyzed by enzymes such as lactase, bilirubin oxidase, or ascorbate oxidase. The theoretical cell reactions for the anode, cathode, and overall cell are defined below:

Anode: $C_6H_{12}O_6 + 24OH^- \rightarrow 6CO_2 + 18H_2O + 24e^-$

Cathode: $6O_2 + 12H_2O + 24e^- \rightarrow 24OH^-$

Overall: $C_6H_{12}O_6 + 6O_2 \rightarrow 6CO_2 + 18H_2O$

The Gibbs free energy of the above reaction is given in Reference 98 as $\Delta G° = -2.870 \times 10^6$ Jmol^{-1}. The thermodynamic potential of glucose oxidation is -0.63 V versus Ag/AgCl, while the potential for oxygen reduction is $+0.59$ V, yielding an overall maximum theoretical potential for a glucose-based biofuel cell of 1.2 V. In practice, the combination of the appropriate enzymes in a biofuel cell can produce an open-circuit voltage as high as ~1.0 V.[97,98] One important challenge with enzymatically catalyzed biofuel cells is optimizing the electron transfer between the biocatalyst and the electrode. Mediators are often employed to electrically connect the enzyme to the electrode and improve the catalytic efficiency of the system.[99]

Another embodiment of the biofuel cell is the abiotically catalyzed fuel cell, where abiotic catalysts, such as noble metals or activated carbons, are used instead of enzymatic catalysts.[98] These biofuel cells, using glucose and oxygen as the fuels, were under development in the late 1960s for use to power implantable cardiac pacemakers. The successful development and implementation of the Li/iodine battery for pacemakers resulted in the abandonment of the biofuel cell for this purpose. While the abiotically catalyzed biofuel cells are limited by low power density (microwatt range), the catalysts are very stable and capable of long-term performance. By comparison, the enzymatically catalyzed biofuel cells can offer higher power density but suffer from limited enzyme stability. The long-term stability of the abiotically catalyzed biofuel cells has led to a resurgence in research in this area for low-power implantable microelectromechanical systems (MEMS) applications.

REFERENCES

1. V. S. Mallela, V. Ilankumaran, and N. S. Rao, "Trends in Cardiac Pacemaker Batteries," *Indian Pacing ElectroPhys. J.* **4**:201 (2004).
2. F. Gutmann, A. M. Hermann, and A. Rembaum, "Solid-State Electrochemical Cells Based on Charge Transfer Complexes," *J. Electrochem. Soc.* **114**:323 (1967).
3. J. R. Moser, "Solid State Lithium-Iodine Primary Battery," U.S. Patent 3,660,163, May 2, 1972.
4. R. T. Mead, C. F. Holmes, and W. Greatbatch, "Design Evolution of the Lithium Iodine Pacemaker Battery," *Proc. Electrochem. Soc.* **79-1**:327 (1979).

5. H. F. Clemo and K. A Ellenbogen, Chapter 4 in *Implantable Cardiac Pacemakers and Defibrillators*, A. W. C. Chow and A. E. Buxton, Eds., Blackwell Publishing, Malden, MA, 2006.
6. D. P. Zipes, D. G. Wyse, P. L. Friedman, A. E. Epstein, A. P. Hallstrom, H. L. Greene, E. B. Schron, and M. Domanski, "A Comparison of Antiarrhythmic-Drug Therapy with Implantable Defibrillators in Patients Resuscitated from Near-Fatal Ventricular Arrhythmias," *N. Eng. J. Med.* **337**:1576 (1997).
7. R. S. Nelson, Chapter 12 in *Implantable Cardioverter Defibrillator Therapy: The Engineering-Clinical Interface*, M. W. Kroll and M. H. Lehmann, Eds., Kluwer Academic Publishers, Norwell, MA, 1996.
8. K. E. Ellison, Chapter 6 in *Implantable Cardiac Pacemakers and Defibrillators*, A. W. C. Chow and A. E. Buxton, Eds., Blackwell Publishing, Malden, MA, 2006.
9. M. Kirk, Chapter 1 in *Implantable Cardiac Pacemakers and Defibrillators*, A. W. C. Chow and A. E. Buxton, Eds., Blackwell Publishing, Malden, MA, 2006.
10. American Heart Association website, www.americanheart.org, accessed 3/17/09.
11. M. Dandel, Y. Weng, H. Siniawski, E. Potapov, T. Drews, H. B. Lehmkuhl, C. Knosalla, and R. Hetzer, "Prediction of Cardiac Stability after Weaning from Left Ventricular Assist Devices in Patients with Idiopathic Dilated Cardiomyopathy," *Circulation* **118**:S94 (2008).
12. ABIOMED website, www.abiomed.com, accessed 3/17/09.
13. J. Dodd, C. Kishiyama, H. Mukainakano, M. Nagata, and H. Tsukamoto, "Performance and Management of Implantable Lithium Battery Systems for Left Ventricular Assist Devices and Total Artificial Hearts," *J. Power Sources* **146**:784 (2005).
14. F. Albano, M. D. Chung, D. Blaauw, D. M. Sylvester, K. D. Wise, and A. M. Sastry, "Design of an Implantable Power Supply for an Intraocular Sensor, Using POWER (Power Optimization for Wireless Energy Requirements)," *J. Power Sources* **170**:216 (2007).
15. Medtronic Minimed website, www.minimed.com, accessed 2/6/09.
16. PowerOne website, www.powerone-batteries.com, accessed 3/17/09.
17. National Institute on Deafness and Other Communication Disorders website, www.nidcd.nih.gov, accessed 3/17/09.
18. 675CP Zinc Air Cochlear datasheet S6600399, Rayovac website, www.rayovac.com, accessed 3/17/09.
19. World Health Organization, "Medical Device Regulations. Global Overview and Guiding Principles," Geneva (2003).
20. C. P. Gale and G. P. Mulley, "Pacemaker Explosions in Crematoria: Problems and Possible Solutions," *J. Royal Soc. Med.* **95**:353 (2002).
21. N. Lozare and D. J. Iannone, "Real Danger or Freak Incident," from www.firehouse.com, accessed 3/10/09, originally posted 11/5/99.
22. Underwriters Laboratories website, www.ul.com, accessed 3/16/09.
23. J.-I. Yamaki, S.-I. Tobishima, K. Hayashi, K. Saito, Y. Nemoto, and M. Arakawa, "A Consideration of the Morphology of Electrochemically Deposited Lithium in an Organic Electrolyte," *J. Power Sources* **74**:219 (1998).
24. D. Aurbach, E. Zinigrad, Y. Cohen, and H. Teller, "A Short Review of Failure Mechanisms of Lithium Metal and Lithiated Graphite Anodes in Liquid Electrolyte Solutions," *Solid State Ionics* **148**:405 (2002).
25. R. A. Leising, M. J. Palazzo, E. S. Takeuchi, and K. J. Takeuchi, "Abuse Testing of Lithium-Ion Batteries: Characterization of the Overcharge Reaction of $LiCoO_2$/Graphite Cells," *J. Electrochem. Soc.* **148**:A838 (2001).
26. D. D. MacNeil, D. Larcher, and J. R. Dahn, "Comparison of the Reactivity of Various Carbon Electrode Materials with Electrolyte at Elevated Temperature," *J. Electrochem. Soc.* **146**:3596 (1999).
27. J. S. Gnanaraj, E. Zinigrad, L. Asraf, H. E. Gottlieb, M. Sprecher, D. Aurbach, and M. Schmidt, "The Use of Accelerating Rate Calorimetry (ARC) for the Study of the Thermal Reactions of Li-Ion Battery Electrolyte Solutions," *J. Power Sources* **119–121**:794 (2003).
28. S. R. Alavi-Soltani, T. S. Ravigururajan, and M. Rezac, "Thermal Issues in Lithium Ion Batteries," *Proceedings of the Materials Division, the ASME Non-Destructive Evaluation Division, and the ASME Pressure Vessels and Piping Division*, 383 (2006).
29. New Lithium Battery Packing Instructions, "Lithium Batteries – Significant Changes on the Way," www.iata.org, accessed 3/16/09.

30. NTSB press release, December 4, 2007, "NTSB Recommends Fire Suppression Systems on All Cargo Airplanes," www.ntsb.gov, accessed 3/16/09.
31. K. Fester and S. C. Levy, Chapter 4 in *Batteries for Implantable Biomedical Devices*, B. B. Owens, Ed., Plenum Press, New York, 1986.
32. S. C. Levy and P. Bro, "Reliability Analysis of Lithium Cells," *J. Power Sources* **26**:223 (1989).
33. P. Bro and S. C. Levy, *Quality and Reliability Methods for Primary Batteries*, John Wiley & Sons, New York, 1990.
34. S. C. Levy, "Safety and Reliability Considerations for Lithium Batteries," *J. Power Sources* **68**:75 (1997).
35. R. E. McDermott, R. J. Mikulak, and M. R. Beauregard, *The Basics of FMEA*, 2nd Ed., CRC Press, New York, 2009.
36. M. Visbisky, R. C. Stinebring, and C. F. Holmes, "An Approach to the Reliability of Implantable Lithium Batteries," *J. Power Sources* **26**:185 (1989).
37. E. V. Thomas, C. D. Jaeger, and S. C. Levy, "Improving Battery Reliability by Using Regression Trees," *Proc. 3rd Annual Battery Conference on Applications and Advances*, California State University, Long Beach, 1988.
38. L. D. Hansen and R. M. Hart, "The Characterization of Internal Power Losses in Pacemaker Batteries by Calorimetry," *J. Electrochem. Soc.* **125**:842 (1978).
39. D. F. Untereker, "The Use of a Microcalorimeter for Analysis of Load-Dependent Processes Occurring in a Primary Battery," *J. Electrochem. Soc.* **125**:1907 (1978).
40. W. Greatbatch, R. McLean, W. Holmes, and C. Holmes, "A Microcalorimeter for Nondestructive Analysis of Pacemakers and Pacemaker Batteries," *IEEE Trans. Biomed. Eng.* **26**:309 (1979).
41. R. M. Hart, E. A. Lewis, and L. D. Hansen, "Theory and Application of Battery Calorimetry," *Proc. 4th Annual Battery Conference on Applications and Advances*, California State University, Long Beach, 1989.
42. M. E. Orazem and B. Tribollet, *Electrochemical Impedance Spectroscopy*, Wiley, Hoboken, NJ, 2008.
43. M. Visbisky, R. C. Stinebring, and C. F. Holmes, "The Reliability Evaluation of Medical Implantable Batteries," *Proc. 3rd Annual Battery Conference on Applications and Advances*, California State University, Long Beach, 1988.
44. G. Antonioli, F. Baggioni, F. Consiglio, G. Grassi, R. LeBrun, and F. Sanardi, "Stimulatore Cardiaco Impiantabile con Nuova Battaria a Stato Solido al Litio," *Minerva Med.* **64**:2298 (1973).
45. C. F. Holmes, The Lithium/Iodine-Polyvinylpyridine Pacemaker Battery—35 Years of Successful Clinical Use, *211th Meeting of the Electrochemical Society*, Chicago, May 6, 2007.
46. J. P. Rivault and M. Broussely, Chapter 10 in *Lithium Batteries*, J.-P. Gabano, Ed., Academic Press, London, 1983.
47. N. Margalit, Chapter 7 in *Lithium Batteries*, J.-P. Gabano, Ed., Academic Press, London, 1983.
48. D. L. Purdy, Chapter 11 in *Batteries for Implantable Biomedical Devices*, B. B. Owens, Ed., Plenum Press, New York, 1986.
49. G. Emery, "Nuclear Pacemaker Still Energized After 34 Years," Dec. 19, 2007, www.reuters.com, accessed 4/7/09.
50. C. F. Holmes, Chapter 6 in *Batteries for Implantable Biomedical Devices*, B. B. Owens, Ed., Plenum Press, New York, 1986.
51. R. T. Mead, "Solid State Battery," U.S. Patent 3,773,557, Nov. 20, 1973.
52. R. T. Mead, W. Greatbatch, and F. W. Rudolph "Lithium-Iodine Battery Having Coated Anode," U.S. Patent 3,957,533, May 18, 1976.
53. P. M. Skarstad, Chapter 8 in *Batteries for Implantable Biomedical Devices*, B. B. Owens, Ed., Plenum Press, New York, 1986.
54. D. R. Berberick, R. C. Buchman, B. F. Heller, W. G. Howard, M. Jain, D. R. Merritt, P. S. Skarstad, and E. R. Scott, "A Twenty-Five Year Perspective on Lithium-Thionyl Chloride Batteries for Implantable Medical Applications," *211th Meeting of the Electrochemical Society*, Chicago, May 6, 2007.
55. T. I. Evans, T. V. Nguyen, and R. E. White, "A Mathematical Model of a Lithium/Thionyl Chloride Primary Cell," *J. Electrochem. Soc.* **136**:328 (1989).
56. J. B. Bailey, "Investigation of Thionyl Chloride Decomposition and Open-Circuit Potential in Lithium-Thionyl Chloride Cells," *J. Electrochem. Soc.* **136**:2794 (1989).

57. H. Lanmuller, S. Sauermann, E. Unger, G. Schnetz, W. Mayr, M. Bijak, D. Rafolt, and W. Girsch, "Battery-Powered Implantable Nerve Stimulator for Chronic Activation of Two Skeletal Muscles Using Multichannel Techniques," *Artificial Organs* **23**:399 (1999).
58. N. Watanabe and M. Fukuda, "Primary Cell for Electric Batteries," U.S. Patent 3,536,532, Oct. 27, 1970.
59. M. Fukuda and T. Iijima, Chapter 9 in *Lithium Batteries*, J.-P. Gabano, Ed., Academic Press, London, 1983.
60. R. Yazami, Y. Ozawa, S. Miao, A. Harmwi, J. Whitacre, M. Smart, W. West, and R. Bugga, "The Kinetics of Sub-Fluorinated Carbon Fluoride Cathodes for Lithium Batteries," *ECS Trans.* **3**:199 (2006).
61. W. Greatbatch, C. F. Holmes, E. S. Takeuchi, and S. J. Ebel, "Lithium/Carbon Monofluoride (Li/CFx): A New Pacemaker Battery," *PACE* **19**:1836 (1996).
62. C. C. Liang, M. E. Bolster, and R. M. Murphy, "Metal Oxide Composite Cathode Material for High Energy Density Batteries," U.S. Patent 4,391,729, Jul. 5, 1983.
63. C. F. Holmes, Chapter 10 in *Implantable Cardioverter Defibrillator Therapy: The Engineering-Clinical Interface*, M. W. Kroll and M. H. Lehmann, Eds., Kluwer Academic Publishers, Norwell, MA, 1996.
64. M. Onoda and K. Kanbe, "Crystal Structure and Electronic Properties of the $Ag_2V_4O_{11}$ Insertion Electrode," *J. Phys.: Condens. Matter* **13**:6675 (2001).
65. M. Morcrette, P. Martin, P. Rozier, H. Vezin, F. Chevallier, L. Laffont, P. Poizot, and J.-M. Tarascon, "$Cu_{1.1}V_4O_{11}$: A New Positive Electrode Material for Rechargeable Li Batteries," *Chem. Mater.* **17**:418 (2005).
66. N. R. Gleason, R. A. Leising, M. Palazzo, E. S. Takeuchi, and K. J. Takeuchi, "Microscopic Study of the First Voltage Plateau in the Discharge of SVO and the Consequences on Electrical Conductivity," *208th Meeting of the Electrochemical Society*, Los Angeles, Oct. 21, 2005.
67. R. A. Leising, W. C. Thiebolt, and E. S. Takeuchi, "Solid-State Characterization of Reduced Silver Vanadium Oxide from the Li/SVO Discharge Reaction," *Inorg. Chem.* **33**:5733 (1994).
68. P. M. Skarstad, "Lithium/Silver Vanadium Oxide Batteries for Implantable Cardioverter-Defibrillators," *Proceedings of the Twelfth Annual Battery Conference on Applications and Advances (IEEE 97th 8226)*, pg. 151, IEEE (1997).
69. R. P. Ramasamy, C. Feger, T. Strange, and B. N. Popov, "Discharge Characteristics of Silver Vanadium Oxide Cathodes," *J. Appl. Electrochem.* **36**:487 (2006).
70. N. D. Leifer, A. Colon, K. Martocci, S. G. Greenbaum, F. M. Alamgir, T. B. Reddy, N. R. Gleason, R. A. Leising, and E. S. Takeuchi, "Nuclear Magnetic Resonance and X-Ray Absorption Spectroscopic Studies of Lithium Insertion in Silver Vanadium Oxide Cathodes," *J. Electrochem. Soc.* **154**:A500 (2007).
71. H. Gan and E. S. Takeuchi, "Lithium Electrodes With and Without CO_2 Treatment: Electrochemical Behavior and Effect on High Rate Lithium Battery Performance," *J. Power Sources* **62**:45 (1996).
72. H. Gan and E.S. Takeuchi, "Correlation of Anode Surface Film Chemical Composition and Voltage Delay in Silver Vanadium Oxide Cell System," *198th Meeting of the Electrochemical Society*, Phoenix, Oct. 22, 2000.
73. H. Gan and E.S. Takeuchi, U.S. Patent *5,753,389*, 1998.
74. A. M. Crespi, F. J. Berkowitz, R. C. Buchman, M. B. Ebner, W. G. Howard, R. E. Kraska, and P. M. Skarstad, "The Design of Batteries for Implantable Cardioverter Defibrillators," Chapter 26 in *Power Sources 15*, A. Attewell and T. Keily, Eds., p. 349 (1995).
75. T. Nohma, S. Yoshimura, K. Nishio, and T. Saito, "Commercial Cells Based on MnO_2 and MnO_2-Related Cathodes," in *Lithium Batteries: New Materials, Developments and Perspectives*, G. Pistoia, Ed., p. 417 (1994).
76. M. J. O'Phelan, T. G. Victor, B. J. Haasl, L. D. Swanson, R. J. Kavanagh, A. G. Barr, and R. M. Dillon, "Batteries Including a Flat Plate Design," U.S. Patent 7,479,349 B2, Jan. 20, 2009.
77. J. Drews, R. Wolf, G. Fehrmann, and R. Staub, "High-Rate Lithium/Manganese Dioxide Batteries; the Double Cell Concept," *J. Power Sources* **65**:129 (1997).
78. R. Staub, G. Fehrmann, R. Wolf, T. Fischer, and H. Heimer, "Implantable Medical Device with End-of-Life Battery Detection Circuit," U.S. Patent 5,713,936, Feb. 3, 1998.
79. J. Drews, G. Fehrmann, R. Staub, and R. Wolf, "Primary Batteries for Implantable Pacemakers and Defibrillators," *J. Power Sources* **97–98**:747 (2001).
80. H. Gan, R. Rubino, and E. Takeuchi, "Dual-Chemistry Cathode System for High-Rate Pulse Applications," *J. Power Sources* **146**:101 (2005).

81. C. L. Schmidt and P. M. Skarstad, "The Future of Lithium and Lithium-Ion Batteries in Implantable Medical Devices," *J. Power Sources* **97–98**:742 (2001).
82. H. Gan and E. Takeuchi, "Novel Electrode Design for High Rate Implantable Medical Cell Application," Abst. 219, *204th Meeting of the Electrochemical Society*, Oct. 12–16, 2003.
83. K. Chen, D. R. Merritt, W. G. Howard, C. L. Schmidt, and P. M. Skarstad, "Hybrid Cathode Lithium Batteries for Implantable Medical Applications," *J. Power Sources* **162**:837 (2006).
84. W. C. Sunderland, A. W. Rorvick, D. C. Merritt, C. L. Schmidt, and D. P. Haas, "Electrochemical Cell," U.S. Patent 5,716,729, Feb. 10, 1998.
85. P. M. Gomadam, D. R. Merritt, E. R. Scott, C. L. Schmidt, P. M. Skarstad, and J. W. Weidner, "Modeling Lithium/Hybrid-Cathode Batteries," *J. Power Sources* **174**:872 (2007).
86. H. Gan, "Sandwich Cathode Design for Alkali Metal Electrochemical Cell with High Discharge Rate Capability," U.S. Patent 6,551,747, Apr. 22, 2003.
87. Greatbatch website,www.greatbatch.com, accessed 3/18/09.
88. E. Okamoto, M. Nakamura, Y. Akasaka, Y. Inoue, Y. Abe, T. Chinzei, I. Saito, T. Isoyama, S. Mochizuki, K. Imachi, and Y. Mitamura, "Analysis of Heat Generation of Lithium Ion Rechargeable Batteries Used in Implantable Battery Systems for Driving Undulation Pump Ventricular Assist Device," *Artificial Organs* **31**:538 (2007).
89. E. Okamoto, K. Watanabe, K. Hashiba, T. Inoue, E. Iwazawa, M. Momoi, T. Hashimoto, and Y. Mitamura, "Optimum Selection of an Implantable Secondary Battery for an Artificial Heart by Examination of the Cycle Life Test," *ASAIO Journal* **48**:495 (2002).
90. S. Passerini and B. B. Owens, "Medical Batteries for External Medical Devices," *J. Power Sources* **97–98**:750 (2001).
91. MedEL website, www.medel.com, accessed 4/4/09.
92. C. Sparkes, "A Study of Mercuric Oxide and Zinc-Air Battery Life in Hearing Aids," *J. Laryngology & Otology* **111**:814 (1997).
93. M. Valente, J. H. Cadieux, L. Flowers, J. G. Newman, J. Scherer, and G. Gephart, "Differences in Rust in Hearing Aid Batteries Across Four Manufacturers, Four Battery Sizes, and Five Durations of Exposure," *J. Am. Acad. Audiology* **18**:846 (2007).
94. H. F. Gibbard, H. R. Espig, J. C. Hall, J. W. Cretzmeyer, and R. S. Melrose, "Mechanisms of Operation of the Zinc-Air Battery," *Proceed. Electrochem. Soc.* **79-1**:232 (1979).
95. A. Heller, "Potentially Implantable Miniature Batteries," *Anal. Bioanal. Chem.* **385**:469 (2006).
96. N. Kakehi, T. Yamazaki, W. Tsugawa, and K. Sode, "A Novel Wireless Glucose Sensor Employing Direct Electron Transfer Principle Based Enzyme Fuel Cell," *Biosensors and Bioelectronics* **22**:2250 (2007).
97. P. Atanassov, C. Apblett, S. Banta, S. Brozik, S. C. Barton, B. Cooney, B. Y. Liaw, S. Mukerjee, and S. D. Minteer, "Enzymatic Biofuel Cell," *Electrochem. Soc. Interface* **X**:28 (2007).
98. S. Kerzenmacher, J. Ducree, R. Zengerle, and F. von Stetten, "Energy Harvesting by Implantable Abiotically Catalyzed Glucose Fuel Cells," *J. Power Sources* **182**:1 (2008).
99. E. Nazaruk, S. Smolinski, M. Swatko-Ossor, G. Ginalska, J. Fiedurek, J. Rogalski, and R. Bilewicz, "Enzymatic Biofuel Cell Based on Electrodes Modified with Lipid Liquid-Crystalline Cubic Phases," *J. Power Sources* **183**:533 (2008).

CHAPTER 32
BATTERY SELECTION FOR CONSUMER ELECTRONICS

John A. Wozniak

32.1 INTRODUCTION

A proliferation of consumer electronic applications in recent years has pushed battery technology to the forefront of concerns when designing a new product. Run-time, talk time, standby time, and shelf life are all important buzz words in consumer electronics that contribute directly to how well a product sells, and the values used for these variables depend on the choice of battery. This chapter will cover the needs of typical consumer electronic devices, common battery chemistries, and key selection criteria.

There have been many advances in battery technology in recent years: new chemistries, higher energy densities, new form factors, and new coatings for improved reliability. Yet in the end, there is still no single perfect battery that performs optimally under all electrical and environmental conditions. This ideal battery would have unlimited energy and power capabilities, operate well under all environmental conditions, be inexpensive, have unlimited shelf life and be completely safe and consumer-proof. One may find two or three of these characteristics that a single battery may approach, but other characteristics will suffer.

The component materials of batteries exhibit a wide range of electrochemical properties, and the continued push to achieve higher energy densities requires caution on the part of the designer. As consumer electronic applications seek smaller and more powerful batteries, there runs a parallel risk of keeping this power under control. Selection of the proper battery for a specific application is a study in trade-offs. In considering these trade-offs, one must also keep in mind the education of the customer in the proper use and care of the battery in their device.

32.2 KEY CONSIDERATIONS IN SELECTING A BATTERY

Many factors must be considered when selecting the battery that best meets the needs of a specific application. The characteristics of available batteries must be compared with the needs of the electronic device. It is critical to consider these needs early in the development of the device since the battery will have a direct impact on the size and weight of the device. Addressing the various trade-offs from the beginning of development is the most effective means of ensuring reasonable compromises resulting in a good product design.

32.2 SPECIALIZED BATTERY SYSTEMS

Key considerations include:

- *Type of battery:* Primary (single use) or secondary (rechargeable)
- *Voltage:* Nominal or operating voltage, profile of the discharge curve, maximum and minimum permissible voltages
- *Physical size:* Weight, shape, size, and terminal requirements
- *Capacity:* Required Ah or Wh to achieve run, talk, or standby times
- *Load current and profile:* Constant power, constant current, constant impedance, or other; value of load current or profile; constant, variable, or pulsed load and duty cycle requirements
- *Temperature requirements:* Operating and storage temperature ranges
- *Shelf life:* State-of-charge during storage; storage time as a function of temperature, humidity and other environmental factors; active/standby/sleep modes
- *Charging (if rechargeable):* Float or charge cycling; cycle life requirements, simplicity and availability of charging source, charging efficiency
- *Safety and reliability:* Permissible variability and failure rates; use of potentially hazardous or toxic materials; operation under severe, hazardous, or abusive conditions; failure mode (out gassing, leakage, swelling)
- *Cost:* Initial cost; operating cost or life-cycle cost; use of exotic or critical materials with potentially volatile pricing; cost of charging circuit or charger, if rechargeable
- *Regulatory requirements:* Country of origin and country of delivery concerns; special shipping concerns; recycling requirements and labeling
- *Environmental conditions:* Shock and vibration, acceleration or other mechanical demands and forces; atmospheric conditions (pressure, humidity, altitude, etc.)

32.3 TYPICAL PORTABLE APPLICATIONS

The demand for portability in the consumer electronics world has driven the demand for a wide range of electrochemical battery technologies. The wide range of power and environmental requirements for these devices requires an equivalent wide range of battery technologies to best match these needs. Portable consumer electronics is a rapidly expanding area as an increasing number of portable devices are being introduced that are designed to operate solely with batteries or, in some cases such as notebook computers, to operate with either batteries or AC line power.

Consumer electronics typically include electronic devices that are intended for everyday use. Communication, entertainment, and office applications dominate this product class and the Consumer Electronics Association (CEA) estimated 2008 U.S. revenue to be nearly $180 billion.[1] Devices classified as consumer electronics include personal computers, cell phones, DVD/CD/video players, MP3 players, Bluetooth headsets, GPS navigational systems, televisions, digital cameras, camcorders, electronic toys, and calculators. Even simple devices such as laser pointers and hearing aids are considered in this category. The trend in the industry is convergent devices where a single device provides multiple functions. One example is how PDAs (personal digital assistants) have been virtually eliminated from the marketplace through the incorporation of office organizational functions, such as digital business cards and scheduling, into cell phones. Table 32.1 lists typical current drains of common portable electronic devices. This serves to illustrate the wide range of requirements from microamperes to well over an amp. Although the values are specified in amps, many devices present a variable load to the battery.

---

[1]CEA: Industry Statistics, www.ce.org/Research/Sales_Stats/default.asp

TABLE 32.1 Current Drain in Battery-Operated Portable Electronics

Device	Current drain (mA)
CD players	100–350
Cell phones (talking)	300–600
Digital camera	500–1200
Camcorders	500–1000
Notebook computer	200–2000
Memory backup	Microamperes
Radio	20–50
Radio-controlled toys	600–1500
TV (portable)	300–700
Travel shaver	300–500
Remote controls	10–50
Watches (LED)	10–40

TABLE 32.2 Application of Batteries: Primary vs. Secondary

Application	Primary or secondary battery
Portable tools	Secondary
Notebook computer	↑
Cordless telephone	
Camcorders	
Video player	
Portable shaver	
Hand vacuum	
Cell phone	
Audio players	
Digital camera	
Toys	
Hearing aid	
Remote control	
Watches	↓
Smoke detector	
Memory backup	Primary

Battery solutions for consumer electronics can be divided into two main groups: Primary single-use batteries and secondary rechargeable batteries. Primary batteries are typically used in low to medium power applications, whereas secondary batteries are used in nearly all other applications. The following sections will cover typical applications for both primary and secondary batteries and conclude with some specific examples of the trade-offs to consider when making a final choice. Table 32.2 lists some common applications and their suitability for primary and secondary battery solutions. There is significant overlap and, in the end, comparison of key characteristics of the battery choices to the device requirements will result in an acceptable solution.

32.4 PRIMARY BATTERY TYPES AND APPLICATIONS

Primary batteries are appropriate when power demands are relatively low. Devices such as garage door openers and remote controls that are only active for extremely short periods of time as well as very few times per day are often well suited for primary batteries. Watches and digital clocks are other examples where continuous, extremely low power demands may allow a primary battery to be used for at least a year without replacement. Backup memory power is also a typical application where a device such as an Internet-enabled phone or notebook computer may have a secondary battery for power, but may also incorporate a primary battery to keep volatile memory alive (e.g., RTC, or real-time clock).

Primary batteries come in a variety of shapes and sizes from coin cells to large cylindrical and prismatic cells (lantern batteries). Along with these diverse form factors, there is a wide range of chemistries, including carbon-zinc, alkaline, lithium metal, and others.

Primary batteries are usually less expensive to produce, with a smaller physical size than secondary batteries capable of supporting the same power demands. However, with more global awareness of the effects of electronic waste on the environment, many devices that once were powered by primary batteries (digital cameras, MP3 players) are now designed to accept secondary batteries instead of, or in addition to, primary batteries.

The most common primary battery chemistries used in consumer electronics are zinc/air, alkaline (Zn/MnO_2), lithium/manganese dioxide (Li/MnO_2), lithium/iron disulfide (Li/FeS_2), and lithium/sulfur dioxide (Li/SO_2). Lithium sulfur dioxide batteries are largely used in military electronics, but serve as a reference system for consumer applications. There are several other lithium primary

32.4 SPECIALIZED BATTERY SYSTEMS

chemistries, as covered in Chap. 14, but either cost or safety concerns preclude their use in consumer electronic applications.

Specific details regarding zinc/air batteries can be found in Chaps. 13 and 33. These batteries have a high energy density and stable voltage curve in addition to being environmentally friendly. However, they are extremely sensitive to relative humidity and the oxygen level in the air. They also have limited shelf life when exposed to air and do not perform well with intermittent loads. A principal application is hearing aids where compact size and constant low power demands are required.

Alkaline primary batteries are arguably the most ubiquitous cells available. Ease of replacement and low initial cost make them attractive for inexpensive consumer products such as toys, clocks, radios, and remote controls. Their relative insensitivity to discharge rate and duty cycle, a wide operating temperature range, and variety of shapes and sizes make them suitable for most consumer electronic applications. The key drawback for alkaline battery use is the need for regular replacement in medium- to high-drain rate applications such as audio and video players. They also exhibit a sloping discharge curve that can limit their use where constant voltage is required.

Lithium-manganese dioxide batteries are commonly used in button or coin cell form for watches, garage door openers, and memory backup. Good shelf life, pulse capability, and energy density make these batteries the most commonly used of the lithium primary cells. Reduced volatility compared to Li/SO_2 and relatively low cost make them attractive for many consumer electronic applications. They are, however, subject to the same regulatory restrictions as other lithium primary chemistries.

Lithium iron disulfide batteries are manufactured in AAA and AA sizes for consumers. They have a lower voltage (1.5 V), but similar energy density as Li/MnO_2. Performance characteristics are also similar, but the lower operating voltage makes them useful as a higher-power replacement for alkaline batteries. This makes them preferable to alkaline cells in higher pulse power applications such as digital cameras with photo flash capability. They are also available in two grades, "ultimate" and "advanced." The former is an energy product while the latter provides higher power.

Positive characteristics of lithium sulfur dioxide batteries are high energy density, good low-temperature performance, superior rate capability, and long shelf life. Relatively high cost and safety and environmental concerns have limited their use in consumer applications to higher priced products like security systems and some telecommunications systems. They are more appropriate for military, aerospace, and some biomedical applications, although they are included in subsequent figures for reference purposes.

32.4.1 Comparing Primary Battery Characteristics

Characteristics of conventional primary batteries based on theoretical limits are summarized in Table 1.2. This table also lists characteristics based on the actual performance of a practical battery under near optimal conditions for each specific type. It is important to note the following:

- The actual capacity available from a battery is *significantly* less than the theoretical capacity of the active materials.
- The actual capacity is also less than the theoretical capacity of a practical battery because an actual battery has weight and volume attributed to non-energy-producing materials used in construction of the battery in addition to the active materials.
- The capacity of a battery can vary greatly from the values listed in Table 1.2. The table values are based on optimum conditions for that battery, and the real-world conditions for use are rarely optimal. Measurements should be made under actual real-world conditions for the product before a final judgment is made.

The following figures compare some key characteristics of the above described primary batteries. More detailed characteristics can be found in the appropriate chapter of this Handbook. Those data, rather than the generalized data in this section, should be used to evaluate the specific performance of each battery.

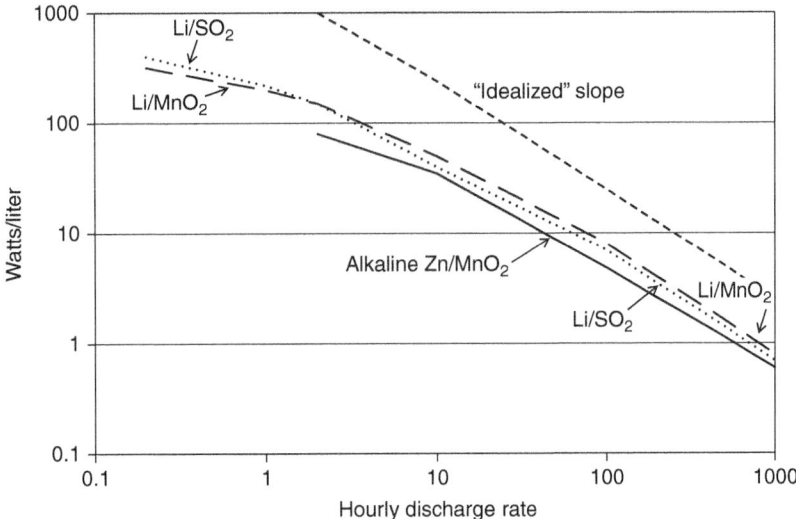

FIGURE 32.1 Performance characteristics of primary batteries on a volumetric basis, 20°C.

Figure 32.1 is a comparison on a volumetric basis, while Fig. 32.2 is a Ragone plot presenting similar data on a gravimetric basis. Lithium batteries have a considerable advantage in weight over their alkaline counterparts.

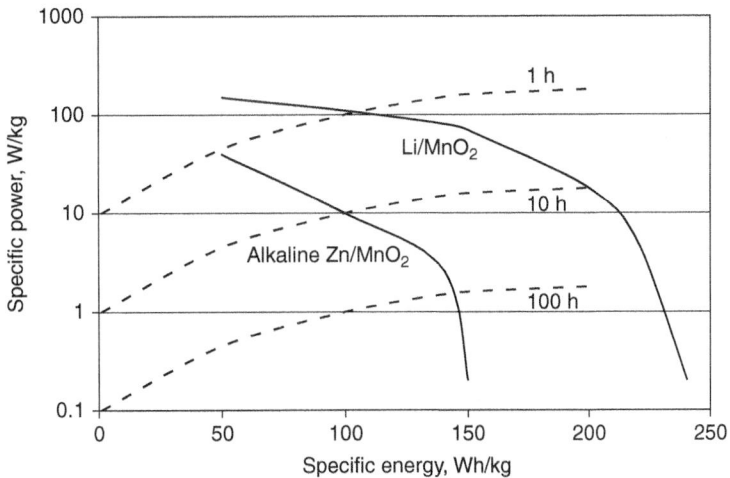

FIGURE 32.2 Performance capabilities of primary batteries—specific energy vs. specific power.

The comparative performances as a function of temperature on a gravimetric basis (Wh/kg) and on a volumetric basis (Wh/L) of the various primary batteries are shown in Figs. 32.3 and 32.4, respectively. These data are based on a moderate 20 h rate. One can see that primary battery performance drops off rather quickly at low-temperatures. Again, lithium batteries have better low-temperature characteristics than the zinc-based chemistries.

32.6 SPECIALIZED BATTERY SYSTEMS

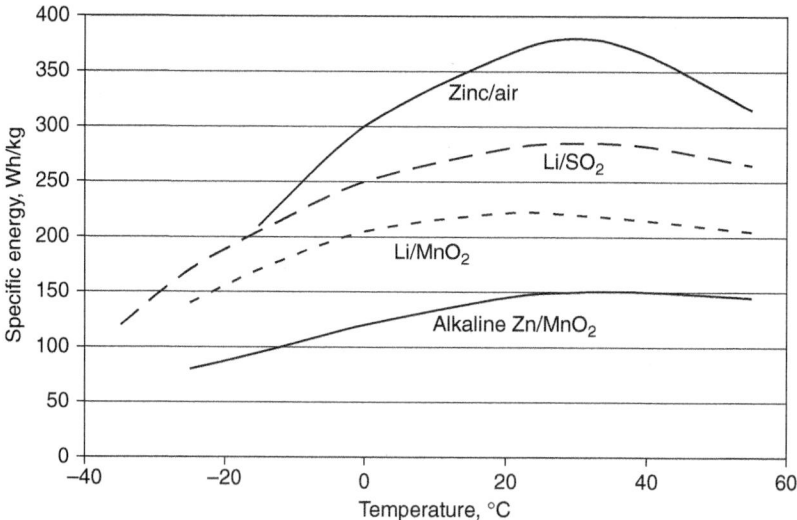

FIGURE 32.3 Effects of temperature on specific energy of primary cells.

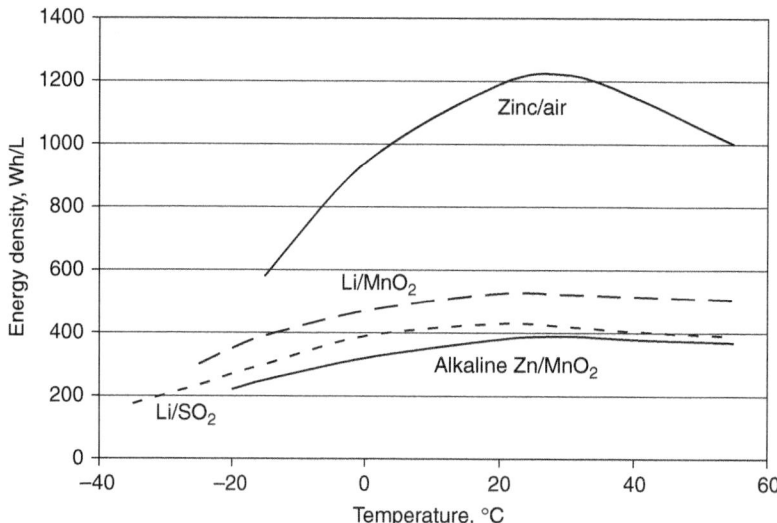

FIGURE 32.4 Effects of temperature on volumetric energy density of primary cells.

32.5 SECONDARY BATTERY TYPES AND APPLICATIONS

Rechargeable batteries for consumer electronics have become a $600 billion per year industry that is continuing to grow at a double-digit rate. The environmental "friendliness" of reusing the battery for months or years is a key selling point in the marketplace. Although these batteries are somewhat bulkier than primary batteries of equivalent power, consumers demand the convenience of not having to regularly replace the battery. Of course, a key factor in designing a secondary battery into an application is the charger. Decisions need to be made early in development as to whether the charger is embedded in the device or a stand-alone option, or both. There may be a higher initial cost of ownership, but the *lifetime* cost of ownership is typically much lower.

Cell phones, notebook computers, music and video players and recorders, cordless phones, and navigational devices are all heavily dependent on secondary batteries for power. These are devices that are used many hours per day, and the device is expected to last for years. Secondary batteries can be used in nearly every consumer electronics application. The eco-friendly consumer will even use secondary batteries in simple devices such as remote controls and electronic toys that have commonly used primary batteries in the past.

Common secondary battery chemistries for consumer electronics include nickel-cadmium (NiCd), nickel-metal hydride (NiMH), and a variety of lithium-ion chemistries. These lithium-ion batteries include lithium cobalt oxide (Li/CoO_2), lithium manganese oxide spinel (Li/Mn_2O_4), lithium iron phosphate (Li/$FePO_4$), and a variety of blended oxides Li/(Ni-Co-Mn)O_2 where manganese, nickel, and/or aluminum may replace some of the cobalt in Li/CoO_2. Nearly all of the lithium-ion chemistries have carbon/graphite anodes, so the key differences lie in the cathode materials.

Nickel-cadmium is quickly being replaced by NiMH as the environmental concerns for cadmium have become more prevalent. NiCd batteries are typically lower cost, but regulatory restrictions and cost of disposal can make them unattractive in many countries. One key advantage that NiCd maintains over NiMH is high-temperature performance. Both are good for high-drain applications. NiCd batteries are typically used in low cost consumer devices such as cordless phones. NiMH batteries can be retrofitted into many NiCd applications with proper design to get more capacity in the same space. NiMH batteries can still be found in some low cost cameras and audio players and recorders, but these applications are typically where lithium-ion batteries are a better fit. Both NiCd and NiMH batteries come in a wide variety of form factors ranging from button cells to F-size cylindrical.

With a high operating voltage, high energy density, good cycle life, and reasonable operating temperature range, lithium-ion has become the workhorse of the portable consumer electronics industry. There are few electronic applications that cannot take advantage of the improvements lithium-ion has to offer over other secondary batteries. Cost has become the key factor that could exclude lithium-ion from some low cost devices. The cost of using lithium-ion batteries comes not only from the cells, but from the need for protection devices and more complex chargers. Notebook computers, cell phones, video and audio recorders and players as well as GPS devices have all become more portable and ubiquitous due to lithium-ion battery technology. Lithium-ion cells are typically manufactured in cylindrical, prismatic, or "pouch" forms.

It is important to note that the term "polymer cells" is often misused to describe "pouch" cells. These are lithium-ion cells that are built into a relatively thin sealed pouch rather than a metallic can. Many of these cells use a polymerized gel electrolyte, thus the term "polymer." However, there are many that use liquid electrolyte just like their cylindrical or prismatic counterparts. These are typically referred to as "starved liquid" electrolyte cells because they have little free electrolyte in the cell. The pouch form factor allows for thinner cells and a wide variety of x-y dimensional variability. They tend to have better cycle life than those cells in metallic cans, because the pouch form allows the electrodes to swell some. This swelling, however, is also a design challenge when using pouch cells in batteries.

Chapter 26 covers details of the various lithium-ion chemistries, but the next section will compare some of the key characteristics of the secondary battery chemistries.

32.5.1 Comparing Secondary Battery Characteristics

Characteristics of conventional secondary batteries based on theoretical limits are summarized in Table 1.2. This table also lists characteristics based on the actual performance of a practical battery

TABLE 32.3 General Secondary Battery Comparison for Consumer Applications

Characteristic	NiCd	NiMH	Lithium-Ion
Cell voltage	1.2	1.2	3.6–3.7
Cycle life at 80% DOD	1000+	500+	400–500
Temperature range	−40 to 70°C	−40 to 50°C	−20 to 60°C
Memory effect	Yes	Yes	No
High rate discharge	$10C+$	Up to $5C$	Up to $2C$ (typical)
Fast charge time	<1 h	2 h	2 h
Capacity after 1 year storage @ 25°C	<30%	<20%	80%
Energy density @ 10 h discharge rate	60 Wh/kg	90 Wh/kg	230 Wh/kg

under near optimal conditions for each specific type. The same factors must be noted as in Sec. 32.4.1 for primary batteries.

Table 32.3 compares some key characteristics of the most common secondary battery types. These data are typical of what was commercially available at the time of publication. Exotic versions of these chemistries are not included. For example, there are now some lithium-ion chemistries that offer fast charge at the expense of slightly reduced energy density. Figure 32.5 is a Ragone plot of these common secondary battery chemistries. One can see the huge weight savings that can be achieved by using lithium-ion instead of the nickel chemistries. This weight savings is often critical in how well a consumer electronic product sells in the marketplace.

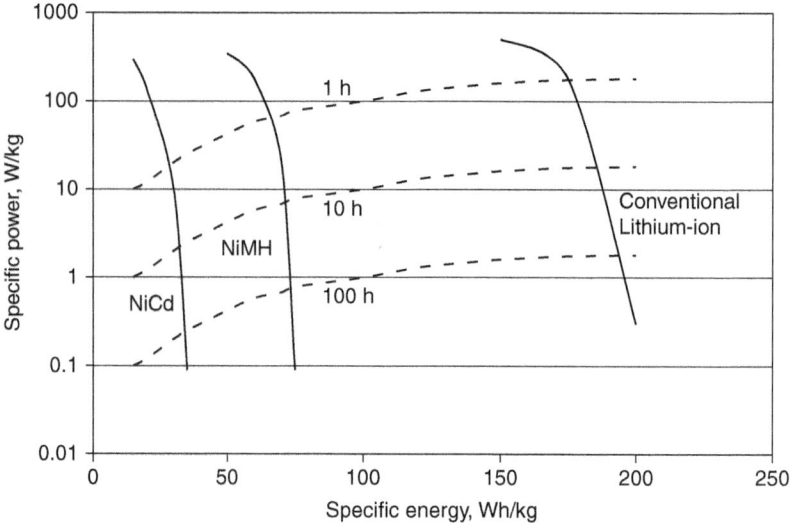

FIGURE 32.5 Performance capabilities of secondary batteries—energy vs. specific power.

Further comparisons of some of the key differences between the different lithium-ion chemistries that are available in the market today are presented in Table 32.4. Li/FePO$_4$ has a lower operating voltage, but generally higher power capabilities than the other lithium-ion chemistries. Li/FePO$_4$ also requires less safety circuitry and exhibits a greater resistance to thermal runaway.

The next section will discuss specific criteria for battery selection and begin with the first big question: primary or secondary?

TABLE 32.4 General Rechargeable Lithium-Ion Battery Comparisons

Characteristic	Li/CoO$_2$	Li/Mn$_2$O$_4$	Li/(Ni-Co-Mn)O$_2$	Li/FePO$_4$
Cell voltage	3.6+	3.6	3.5–3.8	3.2
Cycle life	400–500	400–500	400–500	1000+
Temperature range	−20 to 60°C	−20 to 60°C	−20 to 50°C	−20 to 60°C
Discharge rate	Up to 2C	Up to 5C	Up to 1.5C	10C+
Charge rate	2 h	1–2 h	2–3 h	<1 h
Energy density @ $C/5$ discharge	200 Wh/kg	150 Wh/kg	230 Wh/kg	120 Wh/kg

32.6 SPECIFIC CRITERIA FOR BATTERY SELECTION

32.6.1 Primary versus Secondary

Is a single use or rechargeable battery more appropriate? This appears to be a simple question, but the answer is more complex. Primary batteries are typically used in low to medium power applications, whereas secondary batteries can be used in nearly all consumer electronic applications. The exception is the Li/FeS$_2$ primary, which offers high power capability at increased cost compared to alkaline manganese. Primary batteries for consumer electronics offer the convenience of simple replacement rather than recharge, but the long-term cost of operation can quickly escalate in medium to high power devices. This is the principal trade-off when considering which battery is appropriate for a device. The development of standard consumer sizes (AAA, AA, C) maintenance-free rechargeable batteries has allowed electronic designers of low and moderately priced devices to forego this question completely. Since NiCd and NiMH batteries can provide a voltage similar to primary alkaline batteries, the question becomes irrelevant. The only thing a consumer must do is choose between the low initial cost of primary batteries and the higher initial cost of rechargeables, along with a suitable charger. This means that end-user preferences must also be considered. In the end, other factors may come into play that will sway the decision.

32.6.2 Voltage Concerns

A critical element in selecting a battery is the operating voltage range required for a specific application. Figure 32.6 is a voltage comparison chart. Many electronic components have a minimum voltage requirements near 3 V. The higher operating voltage of the lithium chemistries, both primary and secondary, can simplify battery design by requiring only a single cell, instead of two or three in series. One must also note that in order to utilize the full capacity of NiCd or NiMH in place of alkaline batteries, the device must be capable of operating down closer to 1.0 V instead of 1.2 V for single-cell applications. Basically, the nominal operating voltage for the rechargeable chemistries is at the bottom end of the voltage range for the primary battery. In addition to the *range* of voltage, the actual discharge profile of that voltage is important if the entire functional capacity of the battery is to be used. Figure 32.7 illustrates the flatter discharge curves of conventional secondary batteries.

32.6.3 Physical Size

Battery sizes will be classified into four general cell groups for this section: button/coin cells, cylindrical cells, prismatic cells, and pouch cells.

32.10 SPECIALIZED BATTERY SYSTEMS

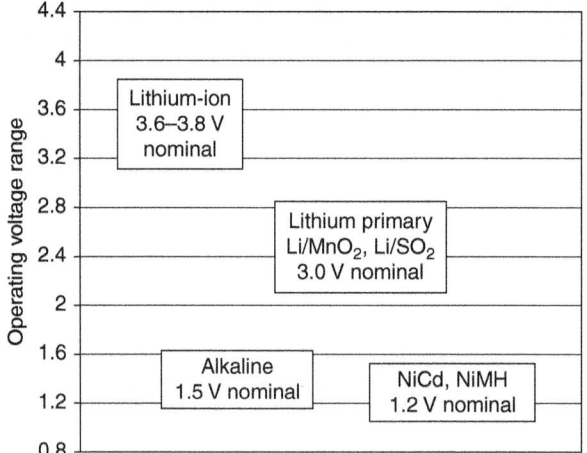

FIGURE 32.6 Voltage comparison chart for primary and secondary cells.

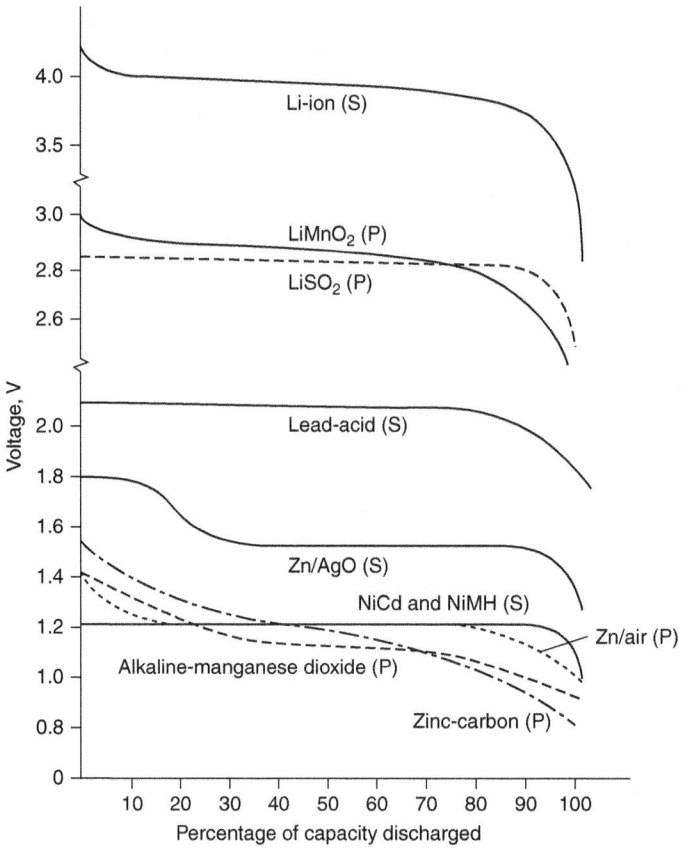

FIGURE 32.7 Discharge profiles of primary (P) and secondary (S) batteries.

Coin cell part numbers usually describe the cell dimension and chemistry. For example, a BR2032 coin cell is decoded as follows:

- BR = chemistry/manufacturer designation
- 20 = 20 mm diameter
- 32 = thickness/10 = 3.2 mm

Some coin cells come socketed, some with tabs for welding or soldering, and some with wires attached. Check specifications from the manufacturer.

Cylindrical cells tend to follow ANSI guidelines as shown in Table 32.5, but are often available in fractions of the standard (4/3A, 1/2AA, 2/3D). The standard lithium-ion cell is actually a "short fat A" cell or 18650 (18 mm diameter by 65 mm long). The original standard NiMH cells were 17670 cells (17 mm diameter by 67 mm long). Settling on a lithium-ion cell size that would not fit into the same space as NiMH was a conscious decision within the industry to prevent accidental overcharging of lithium-ion cells.

TABLE 32.5 Standard Cylindrical Cell Sizes (mm)

Full size	Diameter	Length
N	12	30
AAA	10.5	44.5
AA	14.5	50
A	17	50
Af (fat A)	18	67
SC (sub-C)	23	43
C	25.8	50
D	33	61

Prismatic cells also come in a wide variety of sizes, and part numbers often designate the actual size in millimeters. For example, an ICP103450 prismatic cell can be decoded as follows:

- ICP = chemistry/manufacturer designation
- 10 = 10 mm nominal thickness
- 34 = 34 mm width
- 50 = 50 mm length

It is important to note that the *nominal* thickness is *not* what should be designed into a product. Prismatic cells will swell over cycle life and at higher temperatures. It is good engineering to allow 10% above the nominal thickness for swelling over the life of the product. Ask the manufacturer for a *maximum* thickness specification and the factors that influence it. This is especially important if cells are to be stacked on top of one another and welded together in the battery pack. Figure 32.8 illustrates

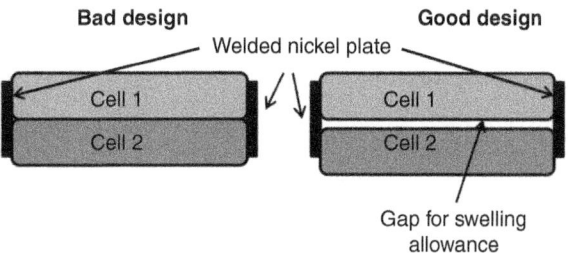

FIGURE 32.8 Proper stacking assemblies of prismatic cells.

how vertical strapping between cells must allow for this swelling. The risk of assembling a battery as shown in the "bad design" is that the nickel strapping can actually pull away from the cell when swelling occurs. The end result can be as simple as a broken connection and a nonfunctional battery pack, or as dangerous as creating short circuits within the battery pack. It is also important to note that most prismatic cells greater than 10 mm in thickness have a built-in safety device such as a thermal fuse or bimetallic switch. Smaller size prismatic cells do not typically incorporate this feature.

Pouch batteries also come in a wide variety of sizes. They can be manufactured less than 2 mm thin and typically up to 7 or 8 mm thick. The maximum thickness depends on the manufacturing technique. Pouch batteries can use a wound, folded, or stacked electrode design. This is covered in Chap. 26. Again, when designing into a product it is important to allow for the maximum thickness after swelling. Early pouch batteries were susceptible to swelling because of insufficient drying, resulting in moisture in the cell. As manufacturing has improved, the main reason for swelling beyond the maximum specification is damage to the cell that allows environmental moisture to intrude. Always check the manufacturer's specification for maximum swelling and the conditions under which this takes place.

32.6.4 Capacity

The rated capacity of a battery in milliamp-hours or milliwatt-hours is a significant factor in determining the number, size, and type of cell needed to meet run-time, talk time, or standby time expectations for a particular device. This capacity is directly related to the energy density of the materials used in the cell. Table 32.6 is a comparison of primary and secondary battery chemistries in AA size (14.5 mm × 50 mm). The capacity rating that a manufacturer gives a cell is based on a specific drain rate at room temperature (20 to 25°C). This drain rate is typically $C/5$ for secondary cells and $C/100$ or more for primary cells. It is important to note that very little development work has been done in the AA-size lithium-ion batteries. As a result, the capacity in this size does not reflect the increased energy density over other secondary chemistries.

TABLE 32.6 Capacity Comparisons for AA-Size Cells

Chemistry	Voltage	Capacity (mAh)	Drain rate	Energy (mWh)
Primary				
Alkaline	1.2	2850	25 mA	3420
Li/FeS$_2$	1.5	3000	500 mA	4500
Li/SO$_2$	3.0	2450	2 mA	7350
Li/MnO$_2$	3.0	2000	10 mA	6000
Secondary				
NiCd	1.2	700	140 mA	840
NiMH	1.2	2100	420 mA	2520
Lithium-ion	3.6	800	160 mA	2880

32.6.5 Load Current and Profile

Table 32.6 is based on particular drain rates. Battery chemistries perform differently under different load conditions. This means that a battery that may appear to have better capacity than another according to a specification sheet may actually perform more poorly if the load in the application is much higher.

Figure 32.9 compares the performance of several AA-size primary and rechargeable batteries. The actual capacity drawn from the batteries is shown over a range of discharge currents. Typically, the primary batteries perform better under low-drain conditions, but they lose this advantage as the

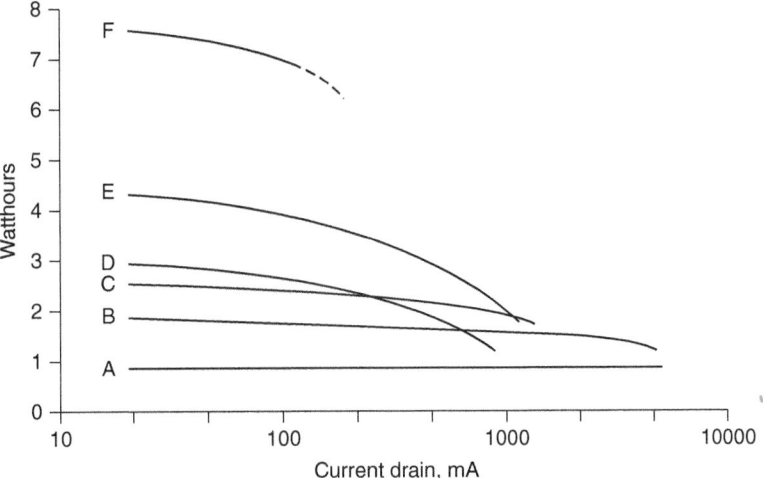

FIGURE 32.9 Performance of AA-size (or equivalent) batteries at various current drains at 20°C. A: NiCd; B: NiMH; C: Lithium-ion; D: zinc-alkaline; E: Li/MnO$_2$ (2/3A size); F: zinc/air (button type). A–C are secondary batteries. D–F are primary batteries.

drain rate increases. Again, it is not completely fair to plot AA lithium-ion, since little work has been done in this size. The current generation of 18650 lithium-ion cells provides 2400 mAh of capacity and has twice the volume of AA cells. Thus an optimized AA lithium-ion cell should provide a capacity of 1200 mAh.

32.6.6 Temperature Requirements

Operating temperature of a portable device is another key factor in selecting a battery. Some consumer electronic applications are designed to be used indoors and have much more conservative thermal demands. Others must be used outside, sometimes in extreme conditions, and this can be problematic for a battery. It was shown in Figs. 32.3 and 32.4 how temperature affects the gravimetric and volumetric energy density of primary cells. The effect of temperature on secondary batteries is much less severe, but the overall operating range of secondary cells is more restricted. This is shown in Fig. 32.10.

An operating characteristic that must also be considered is the effect of temperature on charging the secondary batteries. Both NiCd and NiMH batteries can use the cell temperature to determine the fully charged status of the battery. Operating in a high ambient environment can mask this termination or cause charging to terminate early.

Lithium-ion cells present a different problem. Typically, they can only be charged at the maximum rate within a fairly narrow temperature window (~20 to 45°C). Below or above this temperature, much lower currents and/or voltages must be used. This can result in extremely long charge times in certain thermal environments. In fact, the self-heating during high-rate discharge can elevate the battery temperature of conventional lithium-ion batteries to the point where it must cool down before charging can be initiated. This is evident in the operating specifications of the cell where the "operational temperature" may be specified as –20°C to 60°C, but the normal *charging* temperature range may be only –10°C to 45°C. At temperatures below 5 or 10°C, a trickle charge may be required to prevent lithium plating. At temperatures above 45°C, the voltage may be limited, as well.

32.14 SPECIALIZED BATTERY SYSTEMS

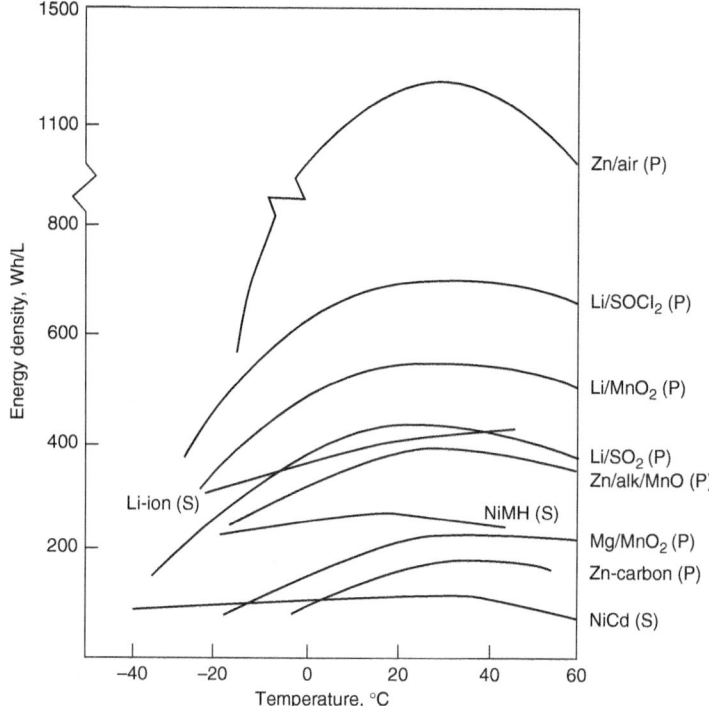

FIGURE 32.10 Effect of temperature on primary (P) and secondary (S) batteries.

32.6.7 Shelf Life

How long will a battery sit on a shelf or in a storeroom before it is put into use? The self-discharge rate, or charge retention, of a particular battery determines this. Primary batteries have self-discharge rates that are typically an order of magnitude or more *lower* than secondary batteries. This allows them to have a shelf life measured in years, rather than months for secondary batteries. An exception is the new Eneloop NiMH battery from Sanyo, which boasts capacity retention of 70 to 85% after 1 year at room temperature compared to 20 to 30% for conventional NiMH. Temperature also has a profound effect on shelf life, with a general rule being that the rate doubles for every 10°C increase in temperature. This means that the shelf life of *all* batteries can be maximized by keeping them cool, as seen in Fig. 32.11.

32.6.8 Charging

The charging of secondary batteries varies considerably from simple float charging over days to rapid charging in less than an hour. Consumer electronic applications typically require something in between, although there are some rapid charge applications. The *rate* of charge can have a direct impact on cycle life of the battery. Higher charge rates typically mean lower charge efficiency and elevated temperatures, resulting in accelerated capacity loss with lithium-ion batteries. With NiCd and NiMH, high charge rates can cause premature termination of charge and a resulting loss of capacity. However, there are secondary cells that are specifically designed for high-power applications that can tolerate higher rates without loss of performance. The trade-off is that these cells typically have much lower energy densities.

Figure 32.12 illustrates the effect of charge rate on the capacity of typical lithium-ion cells charged with a conventional CC-CV method.

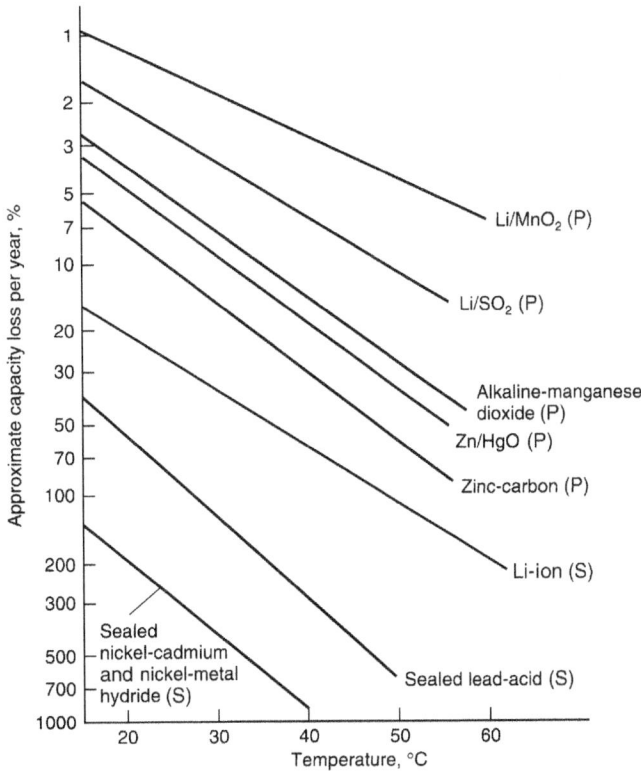

FIGURE 32.11 Shelf life (charge retention) characteristic of a variety of batteries—primary (P) and secondary (S).

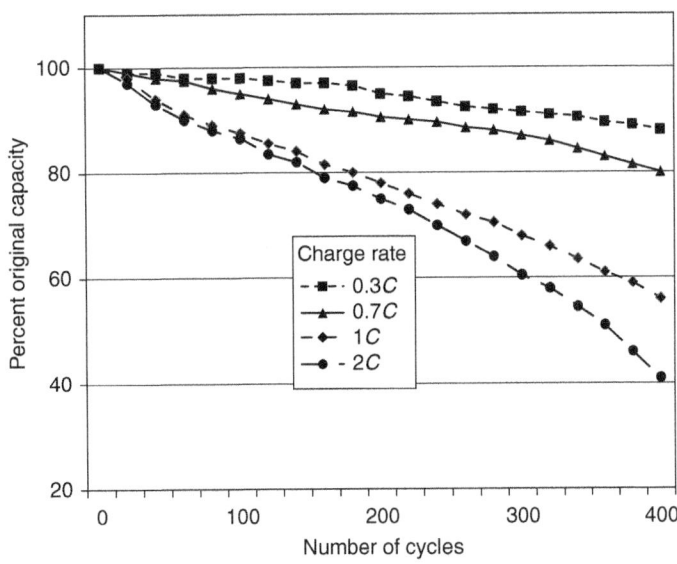

FIGURE 32.12 Effect of charge rate on typical lithium-ion battery capacity during cycling.

32.16 SPECIALIZED BATTERY SYSTEMS

This effect of charge rate on cycle life can be exacerbated by increased or decreased temperature, depending on what additives may be used in the electrolyte. It is critical to follow the cell manufacturers' recommendations for charging as closely as possible.

32.6.9 Safety and Regulatory Concerns

Batteries are energy storage devices that contain relatively large amounts of energy in a small space. If mishandled, abused, or used outside of certain environmental specifications, they can release this energy in a short period of time. Rechargeable batteries pose an additional risk in that they are connected to an essentially unlimited power source when charging. Using rechargeable batteries requires multiple levels of safety to protect against misuse by the consumer.

The first critical layer is the charger. The charger should have protection to avoid excessive current and to disable charging when certain environmental conditions exist (e.g., high temperature).

The second layer is the battery pack itself. Good battery pack design protects terminals from short circuit and protects the cells from damage if a removal battery is dropped. The electronics in a battery pack protect against common conditions that can cause harm to the cells in the battery pack. Typical protection devices include PTCs, thermostats, chemical fuses, and dedicated safety integrated circuits. These protect against conditions such as overvoltage, undervoltage, over temperature, high current, and more (see Chap. 5).

The final layer of protection is the cell itself. A reliable vent mechanism to relieve excessive internal pressures is essential in batteries built into metal cans. Some cells have internal current interrupt devices or thermal fuses to permanently disable the cell.

It is good practice to perform a Failure Modes and Effects Analysis (FMEA) prior to designing the battery. This type of study examines potential failure modes and ranks them based on the severity of the effect of the failure. Potential catastrophic failures can be mitigated through independent, redundant safety mechanisms that address the potential fault conditions. A FMEA analysis should ideally be done on the cells themselves, the battery pack, and the entire battery pack/host device combination. This concept is addressed in IEEE 1625 and 1725 documentation.

Regulatory requirements must also be considered, particularly for a product that will be shipped between countries. There are transportation regulations that define packaging, labeling, and documentation, and these vary from country to country and are constantly changing. Certain markings are also required on the batteries from government agencies. Again, these vary from country to country. Additionally, there are environmental regulations that govern recycling or disposal of the batteries. Planning for end of life is essential in the consumer electronics industry (see Chap. 4).

32.6.10 Cost

Cost effectiveness of primary and secondary battery solutions versus life cycle analysis can be used to evaluate which is more appropriate for a particular application. Table 32.7 is an example of such an analysis, comparing zinc-alkaline primary batteries to rechargeable nickel-metal hydride batteries for a typical low-power, portable electronic device application. A low usage rate makes the primary battery solution much more cost effective in addition to offering the convenience of not having to be recharged periodically. The payback computation is based on little degradation in capacity of the NiMH batteries over the time in question. Another factor to keep in mind is that NiMH capacity is based on discharge down to 1.0 V, so the device must be able to operate from 3 V down to 2 V with the 2 AA cells. Otherwise, the effective capacity of the NiMH solution is reduced and the payback time increased.

Another example is shown in Table 32.8, where power demands are higher. A 1 A load is typical of some digital video cameras and motorized remote-controlled toys. In this example, a high powered primary lithium battery is compared to the rechargeable NiMH battery. Conventional alkaline primary cells simply do not perform well at these higher loads. It is clear that even with moderate to low usage, the payback time for the rechargeable solution is rather short. This is due principally to the higher cost of the more advanced lithium primary batteries.

TABLE 32.7 Cost Effectiveness of Primary (Zinc/Alkaline/MnO$_2$) vs. Secondary (Nickel-Metal Hydride) Batteries

Assumptions
Nominal voltage: 3 V
Drain rate: 150 mA
2 AA Zn/MnO$_2$ batteries (2500 mAh): $0.64
2 AA NiMH batteries (2300 mAh): $6.20
NiMH charger: $4.00

Usage (h/day)	Change NiMH (days)	Change Zn/MnO$_2$ (days)	NiMH battery payback (days)
0.5	30.7	33.3	530
1.0	15.3	16.7	266
2.0	7.7	8.3	133
4.0	3.8	4.2	66
6.0	2.6	2.8	44
8.0	1.9	2.1	33

TABLE 32.8 Cost Effectiveness of Primary (Lithium Iron Disulfide) vs. Secondary (Nickel-Metal Hydride) Batteries

Assumptions
Nominal voltage: 3 V
Drain rate: 1000 mA
2 AA Li/FeS$_2$ batteries (3000 mAh): $4.10
2 AA NiMH batteries (2300 mAh): $6.20
NiMH charger: $4.00

Usage (h/week)	Change NiMH (days)	Change Li/FeS$_2$ (days)	NiMH battery payback (days)
1.0	14	21	53
3.0	4.7	7	17.5
7.0	2	3	7.5
14.0	1	1.5	4
21.0	0.7	1	2.5

If a decision is made to use a rechargeable battery solution, the cost of a lithium-ion solution should be considered where weight and size are critical. In general, cylindrical lithium-ion batteries have a lower dollar per watthour cost than prismatic or pouch batteries. This is due to the high level of automation and the huge volumes that the key manufacturers are producing. A typical lower capacity (2200 mAh) 18650 cell can cost as little as $0.90/Wh, with premium cells (2600 mAh) commanding as much as twice that amount. Cost varies widely between vendors, with Japanese and Korean sources being more expensive than Chinese manufacturers. Of course, one must also consider the cost of quality. Vendors should be thoroughly audited, and it is wise to select a battery with multiple sourcing to ensure good supply and competitive pricing.

Lithium-ion pouch batteries can cost 1.2 to 1.5 times more than an equivalent capacity cylindrical cell. The manufacturing technique for pouch batteries is slower and less cost effective. This, combined with somewhat lower factory yields, has kept lithium-ion pouch batteries more expensive than other form factors.

Lithium-ion batteries require more sophisticated charging circuits and protection circuits compared to NiCd or NiMH. A complete protection circuit for a single-cell lithium-ion application can cost $1.25, an expense not needed for NiCd or NiMH. In addition, a lithium-ion charger can cost twice that of a NiCd/NiMH charger. Although "universal" lithium-ion chargers are available, it is typically good practice to design a charger for the specific application. This usually ends up being more cost effective as the flexibility in voltage and current that a "universal" charger incorporates is eliminated.

To estimate the total cost of ownership, one must consider much more than the cost of the battery itself. Protection circuitry, charge circuitry, regulatory certification, disposal cost and responsibility, and shipping costs all need to be factored into the total cost. One additional consideration is any special assembly techniques that may add cost to the battery in the factory.

32.7 DECISION MAKING AND TRADE-OFFS

Selecting a battery solution for a consumer electronics application involves narrowing down the available options, then weighing critical criteria to come up with the best solution. Keep in mind, there may not be one best solution. Consider the following five things to help reduce the available number of options: function/application, performance, cost, integrity, and safety. Questions to be asked when considering these factors will be covered next.

32.7.1 Narrowing Choices

A particular application may make it immediately clear whether rechargeability is required. Once this initial question is answered, many other application-dependent answers should be gathered.

- Primary or secondary?
- Will a charger be required in the device or as a stand-alone option?
- Resting state-of-charge can be an important factor. Will the device be required to power up at the time of purchase, or will charging be necessary?
- How will the battery be packaged to be compatible with the application?
- What overall size is acceptable, and how will the battery connect to the device?
- Will the battery be user-replaceable or embedded in the device?
- What type of protective circuitry (if any) will be required?

Once a basic size and shape and packaging are decided upon, the performance characteristics must be considered. This is where many of the factors discussed in Sec. 32.6 need to be sorted through to eliminate incompatible chemistries. The following performance characteristics should be included in this process. Further comparisons and trade-offs will be discussed later in this section.

- Energy density
- Voltage range, depth of discharge, shape of curve
- Discharge rates, continuous, pulse
- Temperature, both operating and storage requirements
- Reliability, environmental factors
- Self-discharge, shelf life
- Rechargeability, cycle life, capacity fade, charge rate, trickle charge, protection

Naturally, cost is a key consideration for consumer products, but some devices are less sensitive than others to the cost of the battery. The battery solution may be 10% of the total device cost of a notebook computer, but as much as 50% of the cost for an MP3 player. Cost associated factors include:

- Initial cost of battery
  - Compared to other components
  - Budgetary constraints
- Operating cost of battery—cost per cycle (primary vs. secondary)
- Cost implications for host device; protective devices, charger, special assembly
- Disposal costs and responsibility

Integrity is a broad-ranging topic that is often not considered thoroughly enough. This incorporates topics relating to design, marketing, and manufacturing. The following key points should be considered:

- Reputation: cell vendor, battery pack assembler
- Sourcing
  - Do volumes require multiple sources?
  - Is there a competing alternative technology available (fuel cells, etc.)?
- Perceptions
  - How will the customer perceive the product?
  - How will the product compete in the marketplace?

- Government/regulatory
  - Safety
  - Transportation: strict guidelines are in place for lithium-based products
  - Recycling requirements

The final narrowing factor to be considered, but certainly not the least important, is safety. Safety concerns are not simply whether or not special circuits are required. They also encompass *how* and *where* the device and battery will be used. Anticipating what a consumer will actually do with a device is critical. Will it be left in a hot car? Will it be dropped a lot? Will it be exposed to high humidity or submerged? These are conditions of misuse that must be considered during design. There are also abusive conditions that are difficult to design for, but risk can be mitigated by considering potential abuse. These questions lead to the following safety considerations:

- Mechanical misuse/abuse: what level of shock/vibration tolerance is needed?
- Thermal misuse/abuse: temperature extremes during operation, storage, or charging
- Electrical hazards: battery pack or cell shorting risks
- Disposal/environmental
- Toxicity: special dangers and precautions

32.7.2 Performance Criteria Trade-Offs

All performance characteristics cannot be equally important for a specific application. Capacity becomes a critical parameter in consumer electronic applications where run-time of the device is used as a marketing tool. What can be sacrificed for higher battery capacity? Higher capacity may be achieved by simply using a larger battery. Are increased size and/or weight acceptable for the device in question? Other factors that can be reduced to increase capacity are things like cycle life, charge rate, and high- or low-temperature performance. In *all* cases, safety should *not* be sacrificed.

It is often useful to role play and think like a consumer for a particular portable device. Table 32.9 lists some common applications along with characteristics to be considered. For each application, the relative importance or value of the characteristic is given. It is interesting to note that for a low current drain application, where the battery may be quite small from a capacity standpoint, the size, weight and cost are of little importance because the battery is such a minor part of the device.

This can be taken even further within a particular application, such as notebook computers. These devices come in a wide range of sizes and target many different types of users. These applications range from 17-in displays that function as desktop computer replacements to 10-in screens that function as web browsers (netbooks). Table 32.10 illustrates the relative importance of some key battery characteristics for several classes of portable computers.

TABLE 32.9 Relative Importances of Characteristics for Specific Applications

Application characteristic	Notebook computer	Cell phone	MP3 player	Memory backup
Min. operating voltage	6V	3 V	3 V	3 V
Max. current drain	3–4 A	800 mA	60 mA	100 μA
Run-time	2–3 h	2–4 h talk, 24 h+ standby	6–8 h	24 h
Weight	Somewhat important	Very important	Very important	Not important
Size	Very important	Very important	Very important	Not important
Cost	Somewhat important	Somewhat important	Very important	Not important

32.20 SPECIALIZED BATTERY SYSTEMS

TABLE 32.10 Battery Criteria for Portable Computers

	Display size	Cost	Power	Run-time	Weight/size
Portable workstation	15–17 in		+++		
Desktop replacement	15–20 in	++	++	+	
Mainstream	14–16 in	+++	+	++	+
Thin/light	13–14 in		++	+++	++
Ultraportable	10–12 in			+++	+++
Netbook	< 10 in	+++		++	+++

Battery selection strategies can be summarized as follows:

- Workstations typically have high-powered graphics and CPU/chipsets. They also sell for a premium price and are rarely unplugged from AC power, thus power is king.
- Cost-sensitive programs (mainstream, netbook, desktop replacement) should use the lowest capacity that meets the minimum run-time requirements.
- More mobile devices (thin/light, ultraportable, netbooks) are extremely weight and size sensitive. Performance can also be important, so multiple batteries may be required to reach multiple market segments.
- Netbooks also need to be low cost. This may require a sacrifice in size or weight. The size question can also become an industrial design question. The thinness of a device may be a marketing point. If thinness is more important than cost, perhaps a pouch battery solution is needed. These are all examples of the trade-offs to consider when selecting a battery.

32.8 AVOIDING COMMON PITFALLS IN BATTERY SELECTION

After considering device requirements, examining different battery choices, analyzing battery characteristics, and weighing relative importance of a variety of criteria, you are ready to design your battery solution. Did you forget anything? This final section lists some common pitfalls to avoid.

- Do not neglect shelf life. Some batteries, such as lithium-ion, may require periodic recharging to avoid irrecoverable capacity loss. If a product runs the risk of sitting in a store or inventory pipeline for several months, consider this parameter.
- Leave enough room for the battery.
  - All batteries have tolerances in their size specifications. Designing a battery to use 18.1 mm diameter cells may eliminate several potential sources with maximum diameters of 18.3 or 18.5 mm.
  - Prismatic and pouch batteries also have swelling specifications that must be allowed for.
- Do not disregard temperature. Batteries have a limited operating temperature range. Using a battery beyond the specified range will result in decreased performance and possible loss of capacity over time.
- Consider choices for sourcing. Many batteries come in standard sizes from several manufacturers. If volume deliveries are critical, qualifying multiple sources may be necessary. Once locked in to a unique size, you are at the mercy of the supplier when it comes to pricing and delivery.
- Do not overlook the device power profile. Consider the entire discharge voltage profile of the battery under the expected load to be certain that run-time requirements can be met.
- Whenever possible, use actual performance data to make decisions. Battery specifications are never exact, and cells are typically rated with nominal and minimum capacities. Be certain you know what you are getting.

- Do not neglect cycle life. Different battery chemistries age differently. Trade-offs between cycle life and cell capacity are commonly needed. *Always* expect shorter run-times as the battery ages.
- Fully understand the effects of charging on capacity and cycle life. A particular device may need faster charging, but this usually results in more rapid degradation of capacity and shortened cycle life.
- Do not discount parasitic drain. Some devices have a volatile memory that continues to draw power even when the device is off. Many "smart" batteries require communication with the host device, resulting in some parasitic power drain on the battery itself. Some safety circuits have small drain rates. Be careful!
- Consider unusual battery behavior. Battery voltage drops as the load increases or when the temperature is low. A sudden surge in the power demand, especially in low temperature, can result in a voltage dip below a device's cutoff voltage.

A final thought: ***Design the battery into the system as early as possible***!

CHAPTER 33
METAL/AIR BATTERIES

Terrill B. Atwater and Arthur Dobley

33.1 GENERAL CHARACTERISTICS

The electrochemical coupling of a reactive anode to an air electrode provides a battery with an inexhaustible cathode reactant and, in some cases, very high specific energy and energy density. The capacity limit of such systems is determined by the ampere-hour capacity of the anode and the techniques for handling and storage of the reaction product. As a result of this performance potential, a significant effort has gone into metal/air battery development.[1,2] The major advantages and disadvantages of the metal/air battery system are summarized in Table 33.1.

Primary, reserve, electrically rechargeable, and mechanically rechargeable metal/air battery configurations have been explored and developed. In the mechanically rechargeable designs (that is, replacing the discharged metal electrode), the battery essentially functions as a primary battery and can use relatively simple "unifunctional" air electrodes that need to operate only in a discharge mode. Conventional electrical recharging of metal/air batteries requires either a third electrode (to sustain oxygen evolution on charge) or a "bifunctional" electrode (a single electrode capable of both oxygen reduction and evolution).

Table 33.2 lists the metals that have been considered for use in metal/air batteries with several of their electrical characteristics. Of the potential metal/air battery candidates, zinc has received the most attention because it is the most electropositive metal that is relatively stable in aqueous and alkaline electrolytes without significant corrosion, provided the appropriate inhibitors are used.

Zinc has been used for many years in commercial primary zinc/air batteries. Initially, the products were large batteries using alkaline electrolytes for such applications as railroad signaling, remote communications, and ocean navigational units requiring long-term, low-rate discharge. As thin electrodes were developed, the technology was applied to small (button-type), high-capacity primary cells used in hearing aids, pagers, and similar applications (see Chap. 13).

Zinc is also attractive for electrically rechargeable metal/air systems because of its relative stability in alkaline electrolytes and also because it is the most active metal that can be electrodeposited from an aqueous electrolyte. The development of a practical rechargeable zinc/air battery with an extended cycle life would provide a promising high-capacity power source for many portable applications (computers, communications equipment) as well as, in larger sizes, for electric vehicles. Problems of dendrite formation, nonuniform zinc dissolution and deposition, limited solubility of the reaction product, and unsatisfactory air electrode performance have slowed progress toward the development of a commercial rechargeable battery. However, there is a continued search for a practical system because of the potential of the zinc/air battery.

The lithium/air battery is attractive because lithium has the highest theoretical voltage and electrochemical equivalence (3860 Ah per kilogram of lithium) of any metal anode considered for a

33.2 SPECIALIZED BATTERY SYSTEMS

TABLE 33.1 Major Advantages and Disadvantages of Metal/Air Batteries

Advantages	Disadvantages
High energy density	Dependent on environmental conditions:
Flat discharge voltage	Drying-out limits shelf life once opened to air
Long shelf life (dry storage)	Flooding limits power output
No ecological problems	Limited power output
Low cost (on metal use basis)	Limited operating temperature range
Capacity independent of load and temperature when within operating range	H_2 from anode corrosion
	Carbonation of alkali electrolyte

TABLE 33.2 Characteristics of Metal/Air Cells

Metal anode	Electrochemical equivalent of metal, Ah/g	Theoretical cell voltage,* V	Valence change	Theoretical specific energy (of metal), kWh/kg	Practical operating voltage, V
Li	3.86	3.4	1	13.0	2.4
Ca	1.34	3.4	2	4.6	2.0
Mg	2.20	3.1	2	6.8	1.2–1.4
Al	2.98	2.7	3	8.1	1.1–1.4
Zn	0.82	1.6	2	1.3	1.0–1.2
Fe	0.96	1.3	2	1.2	1.0

*Cell voltage with oxygen cathode.

practical battery system. In the cell discharge reaction, lithium metal, atmospheric oxygen, and water are consumed, and LiOH is generated. The cell can operate at high coulombic efficiencies because of the formation of a protective film on the metal that retards rapid corrosion after formation. On open-circuit and low-drain discharge, the self-discharge of the lithium metal is rapid due to the parasitic corrosion reaction. This reaction degrades the anode's coulombic capacity and must be controlled if the full potential of the lithium anode is to be realized. This self-discharge also necessitates the removal of the electrolyte during stand.

The principal advantage of the lithium/air battery is its higher cell voltage, which translates into higher power and specific energy. However, in view of their availability, cost, and safety advantages, the development of metal/air batteries has previously concentrated on zinc and aluminum.

Other metals have also been investigated as electrode materials for metal/air batteries. Calcium, magnesium, and aluminum have attractive energy densities. Lithium/air,[3,4] calcium/air, and magnesium/air batteries[5,6] have been studied, but cost and problems such as anode polarization or instability, parasitic corrosion, nonuniform dissolution, safety, and practical handling have so far inhibited the development of commercial products. The voltage and the specific energy of the iron/air battery are relatively low. Development on this battery, therefore, has concentrated on an electrically rechargeable system (see Chap. 18), as the iron electrode is long-lived and more adapted to recharging.

Aluminum is attractive for use because of its geological abundance (third most abundant element in the earth's crust), its potentially low cost, and its relative ease of handling.[7-9] However, the aluminum/air battery has too high a charging potential to be electrically recharged in an aqueous system (water is preferentially electrolyzed). Therefore the effort has been directed to reserve and mechanically rechargeable designs.

Table 33.3 summarizes the work on the various types and designs of metal/air batteries.

TABLE 33.3 Metal/Air Batteries

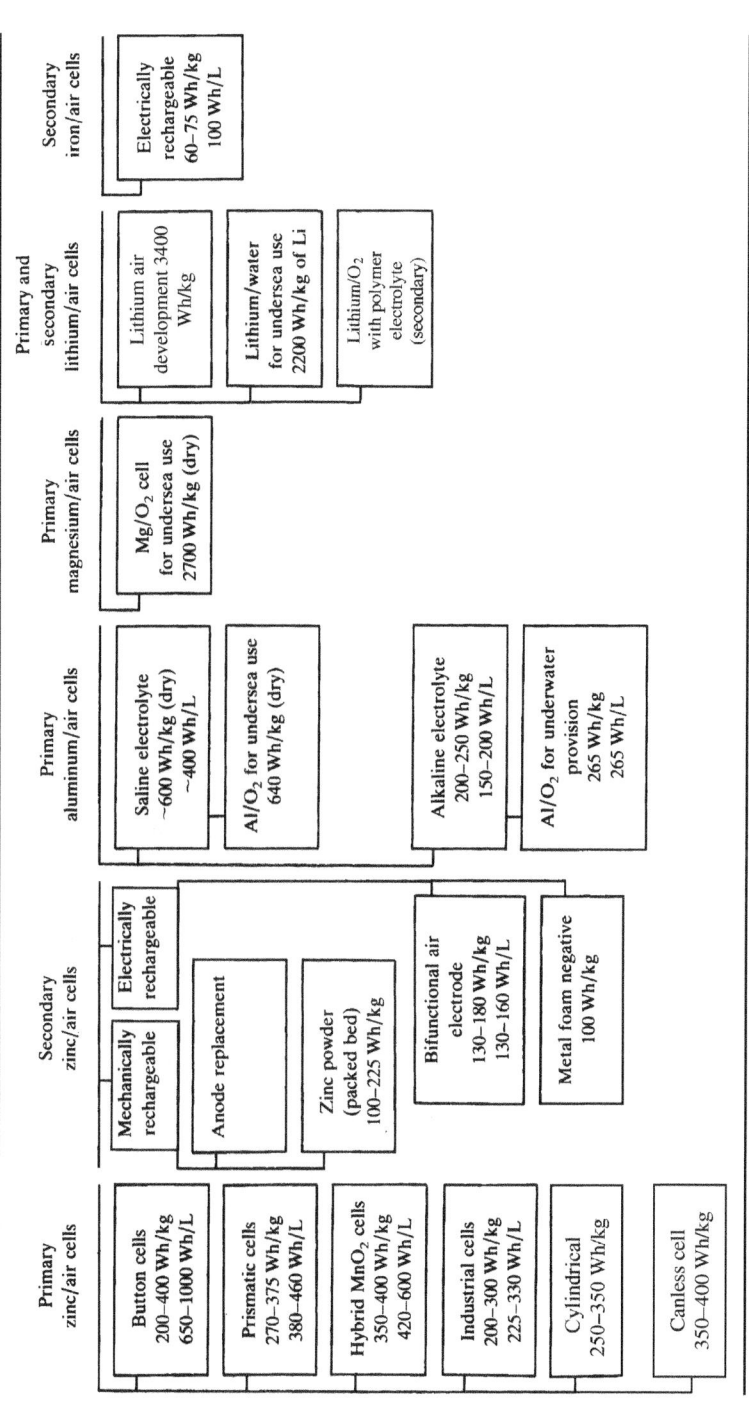

33.2 CHEMISTRY

33.2.1 General

The metal/air batteries being developed use neutral or alkaline electrolytes. The oxygen-reduction half-cell reaction during discharge may be written

$$O_2 + 2H_2O + 4e \rightleftharpoons 4OH^- \qquad E^0 = +0.401 \text{ V}$$

The theoretical cell voltages, the equivalent weights of the metals, and the theoretical specific energies obtained when this oxygen electrode is coupled with various metal anodes are given in Table 33.2. Polarization effects at both electrodes degrade these voltages to those shown in the table at practical operating discharge rates. Note that the theoretical specific energy of metal/air batteries is based only on the negative electrode (anode or fuel electrode during discharge) as this is the only reactant that has to be carried in the battery. The other reactant, oxygen, is introduced into the battery from ambient air during discharge.

The discharge reaction at the negative or metal electrode (anode during discharge) is dependent on the specific metal used, the electrolyte, and other factors in the cell chemistry. The discharge reaction at the negative electrode can be generalized as

$$M \rightarrow M^{n+} + ne$$

The generalized overall discharge reaction may be written

$$4M + nO_2 + 2nH_2O \rightarrow 4M(OH)_n$$

where M is the metal and the value of n depends on the valence change for the oxidation of the metal, as listed in Table 33.2.

Most metals are thermodynamically unstable in an aqueous electrolyte and react with the electrolyte to corrode or oxidize the metal and generate hydrogen as follows:

$$M + nH_2O \rightarrow M(OH)_n + \frac{n}{2}H_2$$

This parasitic corrosion reaction, or self-discharge, degrades the coulombic efficiency of the anode and must be controlled to minimize this loss of capacity.

Other factors that can affect the performance of the metal/air battery are the following:

Polarization. The voltage of a metal/air battery drops off more sharply with increasing current than that of other types of batteries because of diffusion and other limitations in the oxygen or air cathode. This means that these air systems are more suited for low- to moderate-power applications than to high-power ones.

Electrolyte Carbonation. As the cell is open to air, carbon dioxide can be absorbed. This can result in the crystallization of carbonate in the porous air electrode, which may impede air access and cause mechanical damage and a decreasing electrode performance. Potassium carbonate is also less conductive than the KOH electrolyte normally employed in metal/air batteries.

Water Transpiration. Again, as the cell is open to air, water vapor can be transferred if a vapor partial pressure difference exists between the electrolyte and the surrounding environment.

Excessive water loss increases the concentration of the electrolyte and leads to drying out and premature failure. Gain of water can lead to dilution of the electrolyte. This gain can cause flooding of the air electrode pores and electrode polarization due to the inability of the air to reach the reaction sites.

Efficiency. The oxygen electrode at moderate temperatures displays a significant irreversibility during both charge and discharge. As a result there is generally about a 0.2 V difference between the actual charging voltage and the reversible potential, with the same situation on discharge. For example, a zinc/air battery generally discharges at a voltage of about 1.2 V, while the charging voltage is about 1.6 V or higher. This results in a loss of overall energy efficiency even before any other factors are considered.

Charging. Oxidation of catalysts and electrode supports during charging can be a problem for those systems which are recharged electrically, such as zinc/air and iron/air. Approaches to solving this problem generally involve either the use of oxidation-resistant substrates and catalysts, the use of a third electrode for charging, or charging the negative (metal) electrode material external to the cell.

33.2.2 Air Electrode

Successful operation of metal/air batteries depends on an effective air electrode. As a result of the interest in gaseous fuel cells and metal/air batteries over the past 40 years, a significant effort has been aimed at improved high-rate, thin air electrodes, including the development of better catalysts, longer-lived physical structures, and lower-cost fabrication methods for such gas diffusion electrodes.

An alternative approach is to use a low-cost air cathode with more modest performance, but this requires a greater cathode area in each cell. Figure 33.1 shows a type of electrode which is produced by a continuous process using low-cost materials.[10-12] This electrode is composed of two active layers bonded to each side of a current-collecting screen, with a microporous Teflon layer bonded to the air side of the electrode. The active layers are fabricated by passing a nonwoven web of carbon fibers (see Fig. 33.1b) through a slurry containing the catalyst, a dispersing agent, and a binder in a continuous process, with a drying and compacting step built into the process. The active layers, the screen, and the Teflon layer are then bonded in the continuous process. These electrodes are used in the aluminum air reserve standby batteries (see Sec. 33.4.2).

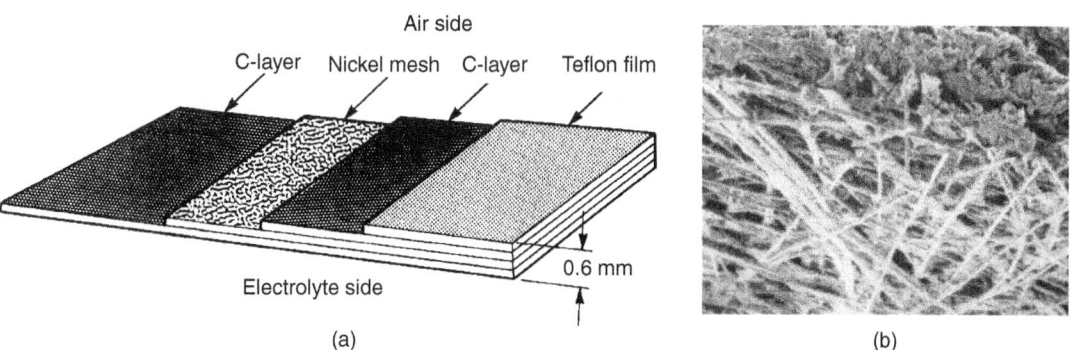

FIGURE 33.1 (a) Laminated air cathode. (b) Carbon fiber substrate. (*Courtesy of Alupower, Inc.*)

33.3 ZINC/AIR BATTERIES

33.3.1 General

Zinc/air batteries are commercially available in primary button-type batteries (see Chap. 13), and in the late 1990s 5 to 30 Ah prismatic batteries as well as larger primary industrial-type batteries were developed. Electrically rechargeable batteries are being considered for both portable and electric-vehicle applications, but the control of the replating of zinc during recharging and the development of an efficient high-rate bifunctional air electrode remain a challenge. In some designs, a third oxygen-evolving electrode is used for recharging, or recharging is done external to the cell to avoid the need for the bifunctional air electrode. Another approach to avoid the difficulties with electrical recharging is the mechanically rechargeable battery, where the spent zinc electrode and/or the discharged products are removed and physically replaced.

The developmental progression of the primary zinc/air system lends itself to be described in four generations.

First generation (GEN 1) zinc/air batteries were introduced in the 1930s, resemble automotive SLI batteries in construction, and are used in remote applications such as buoys and railroad signaling (see Sec. 33.3.3). They are designed for low-rate (< 1 A), multi-year service and have moderate specific energy.

The second generation (GEN 2) zinc/air is the button cell, commercialized in the 1970s for hearing aids (see Chap. 13). With specific energies of more than 400 Wh/kg, the button cell is typically limited to about 10 mW in power and has a service life of one month.

Third generation (GEN 3) zinc/air cells, employ molded plastic cell housings sealed and joined with epoxy adhesives (see Sec. 33.3.2). Batteries using 30 Ah GEN 3 zinc/air cells were first available in 2003 in a 12/24 V, 750 Wh battery. The battery is designed for moderate power (up to 50 W) and a service life of several months. It powers tactical radios for up to a week or more for typical duty cycles.

The fourth generation (GEN 4), began development in the late 1990s and early 2000s (see Sec. 33.3.2 for more detailed information on GEN 4 batteries).

Table 33.3 contains a summary of the different types of zinc/air batteries.

The overall cell reaction for a zinc/air battery on discharge in an alkaline electrolyte may be represented as

$$Zn + \tfrac{1}{2}O_2 + H_2O + 2(OH)^- \rightarrow Zn(OH)_4^{-2} \qquad E° = 1.62 \text{ V}$$

The initial discharge reaction at the zinc electrode can be simplified to

$$Zn + 4(OH)^- \leftrightarrow Zn(OH)_4^{-2} + 2e$$

This reaction occurs as a result of the solubility of the zincate anion in the electrolyte and proceeds until the zincate level reaches the saturation point. There is no well-defined solubility limit, since the degree of supersaturation is time-dependent. After partial discharge, the solubility exceeds the equilibrium solubility level, with subsequent precipitation of zinc oxide, as follows:

$$Zn(OH)_4^{-2} \rightarrow ZnO + H_2O + 2(OH)^-$$

The overall cell reaction then becomes

$$Zn + \tfrac{1}{2}O_2 \leftrightarrow ZnO$$

This transient solubility is one of the main reasons for the difficulty in making a successful rechargeable zinc/air battery. The location of the precipitation of the reaction product cannot be controlled, so that on a subsequent recharge, the amount of zinc deposited on different parts of the electrode area of the cell can vary.

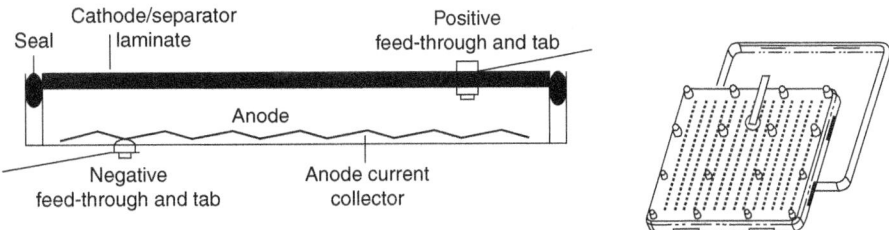

FIGURE 33.2 Design of a prismatic primary zinc/air cell. (*Courtesy of Electric Fuel Corp.*)

33.3.2 Portable Primary Zinc/Air Batteries

Primary zinc/air button-type batteries are described in Chap. 13. This configuration is an effective way to package the zinc/air system in small sizes, but scaling up to larger sizes tends to lead to performance and leakage problems, but these can be overcome with prismatic cell designs. Figure 33.2 shows the basic schematic of a prismatic zinc/air cell. A typical prismatic cell uses a metal or plastic tray, which holds the zinc anode/electrolyte blend while the separator and cathode are bonded onto the rim of the tray. The anode/electrolyte blend is similar to the anode blend used in zinc/alkaline primary cells, containing zinc powder in a gelled aqueous potassium hydroxide electrolyte. The cathode is a thin gas diffusion electrode comprising two layers, an active layer and a barrier layer. The active layer of the cathode, which interfaces with the electrolyte, uses a high surface area carbon and a metal oxide catalyst bonded with Teflon. The high surface area carbon is required for oxygen reduction and the metal oxide catalyst (MnO_2) for peroxide decomposition. The barrier layer, which interfaces with the air, consists of carbon bonded together with Teflon. A high concentration of Teflon prevents electrolyte from weeping from the cell. Prismatic zinc/air cells have been designed with moderately high rate and high capacity. The thickness of the cell determines the anode capacity of the cell, and the cross-sectional surface area determines the maximum rate capability.[13,14]

In addition to prismatic cell designs, cylindrical zinc/air cells (see Fig. 33.3), have been designed.[15–17]

The high specific energy, low cost and safety of the zinc/air primary battery make it an attractive choice for many portable electronics applications. It is particularly advantageous for applications

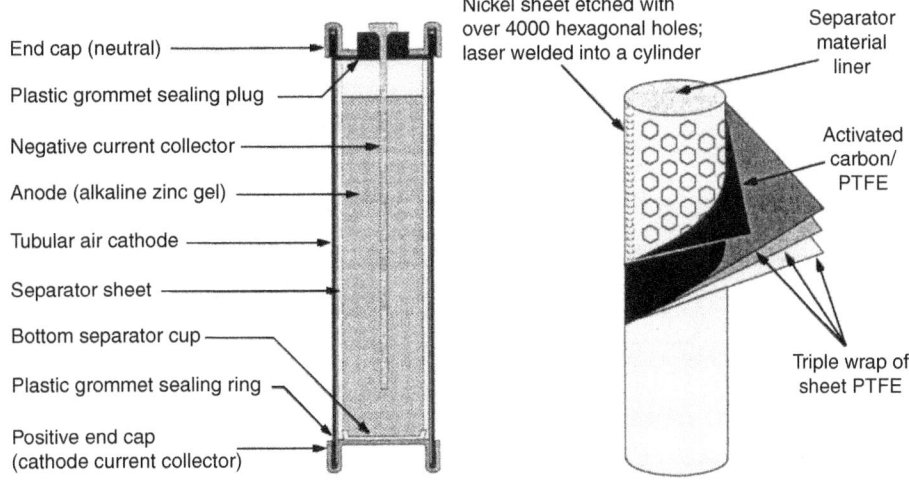

FIGURE 33.3 Design of a cylindrical primary zinc/air cell. (*Courtesy of Rayovac Corp.*)

where the battery energy is consumed within a range of 1 to 14 days, since the high specific energy and energy density of the zinc/air system can be realized and the impact of environmental interactions (dry-out, flooding, and carbonation) is low. Typical cell discharge curves at 25°C are shown in Fig. 33.4. The cell voltage is relatively flat throughout most of the discharge, with little capacity remaining beyond 0.9 volts per cell. Figures 33.5 and 33.6 show the specific energy of prismatic zinc/air as a function of drain rate. Figure 33.5 shows the specific energy for 5 Ah zinc/air cells over

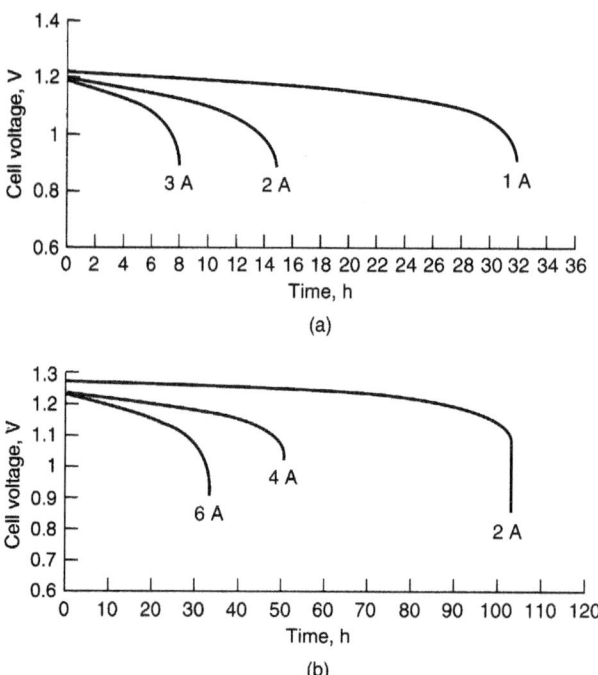

FIGURE 33.4 Discharge curves for prismatic primary zinc/air cells at 25°C. (a) High-rate cell. (b) High-capacity cell. (*Courtesy of Matsi, Inc.*)[18]

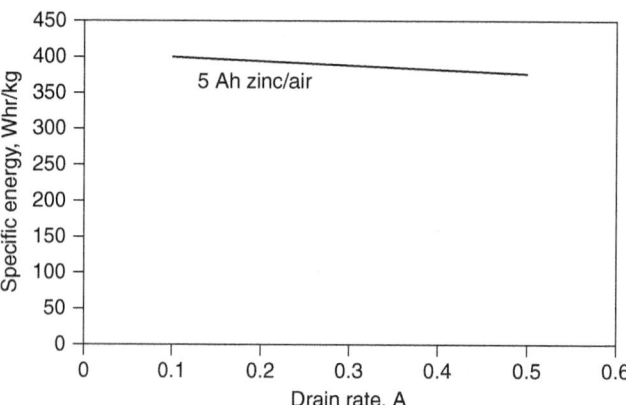

FIGURE 33.5 Specific energy for 5 Ah zinc/air cell as a function of drain rate. (*Courtesy Electric Fuel Corp.*)

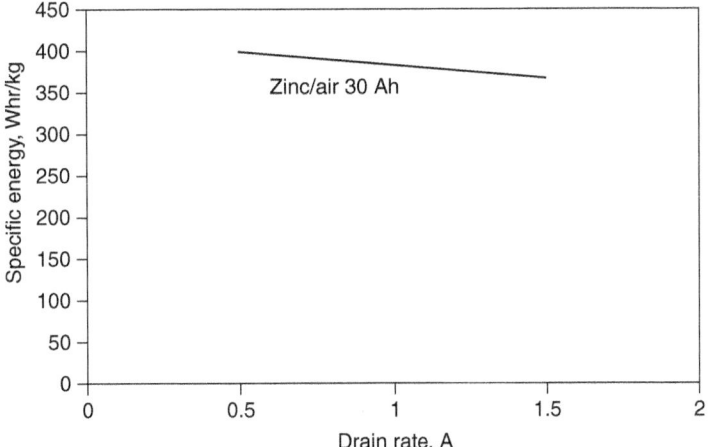

FIGURE 33.6 Specific energy for 30 Ah zinc/air battery as a function of drain rate. (*Courtesy Electric Fuel Corp.*)

TABLE 33.4 Specifications of Prismatic Zinc/Air Cells

Variable	Cellular phone cell	Auxiliary power cell
Facial dimensions, cm (length × width)	4.6 × 2.7	7.6 × 7.6
Height, cm	0.43	0.6
Weight, g	15	87
Capacity, Ah	3.6	30
Specific energy, Wh/kg	300	500
Energy density, Wh/L	800	1250

typical current ranges for portable tape players and analog cellular phones. Figure 33.6 shows the specific energy for 30 Ah zinc/air batteries over typical current ranges for portable stereo systems and camcorders. A summary of discharge characteristics of representative state-of-the-art prismatic zinc/air cells is given in Table 33.4.

Three approaches are being taken to the design of prismatic zinc/air cells for portable batteries (GEN 3). The first is a metal case prismatic cell. This design is essentially an adaptation of button cell technology. In this design, a cathode subassembly, contained in a nickel-plated steel can, is crimp-sealed onto an anode subassembly, contained in a copper lined nickel-plated stainless steel can. A molded plastic insulator seal separates the anode and cathode assemblies. This design has performed well for smaller sizes (5 Ah or less). Figure 33.7 shows a battery designed for cellular telephone applications; the characteristics are listed in Table 33.5.

The second design uses plastic for the case of the prismatic zinc/air cell. This design employs adhesive technology to bond the cell anode and cathode subassemblies. The plastic cell design is preferred for large capacity cell sizes (> 5 Ah) due to technological limitations imposed on the metal cell design. In particular, leak-tight crimp seals become a challenge as cell dimensions increase due to the need for close dimensional tolerances. The key challenges for the plastic cell include the development of the proper designs and materials for the cathode and cell seals and for the current feed-throughs. The latter is required for the plastic cell but not the metal cell, in which the cans serve as terminals for electrical contact. Figures 33.8 and 33.9 show a cell and battery designed for auxiliary power and for remote applications. The characteristics of this auxiliary power battery are listed in Table 33.5.

FIGURE 33.7 Zinc/air battery for cellular phone applications. (*Courtesy of Electric Fuel Corp.*)

TABLE 33.5 Specifications of Prismatic Zinc/Air Batteries

Variable	Cellular phone battery	Auxiliary power battery
Number of cells	4	24
Voltage, V (nominal)	4.8	28
Capacity, Ah	3.6	30
Dimensions, cm		
Length	10.4	31
Width	4.5	18.5
Height	1.5	6
Weight, g	78	2400
Volume, cm^3	70	3500
Energy density, Wh/L	250	300

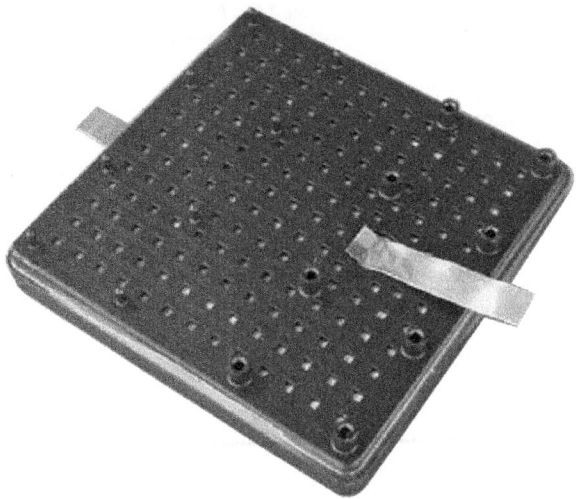

FIGURE 33.8 Zinc/air cell for auxiliary power and for remote applications. (*Courtesy of Electric Fuel Corp.*)

FIGURE 33.9 Zinc/air (BA-8180) battery for auxiliary power and for remote applications. (*Courtesy of Electric Fuel Corp.*)

The third prismatic zinc/air cell design consists of the fourth-generation (GEN 4) zinc/air systems, whose development began in the late 1990s and early 2000s. These cells are designed to use the air cathode as the cell housing, folded in half around the zinc electrode and its edges sealed for leaktight integrity. Figure 33.10 shows a GEN 4 zinc/air cell. The "can-less" design of the GEN 4 zinc/air electrochemical cell allows for increased power density and increased specific power. These increases come at the expense of increased parasitic reactions.

FIGURE 33.10 GEN 4 zinc/air cell. (*Courtesy of Electric Fuel Corp.*)

Prismatic cells are designed so they can be stacked as multicell batteries for use in various portable electronic equipment. Stacking of the cells requires a provision, such as a spacer, to permit air access to the cathode and a fan to provide forced flow of air. The thickness of the spacer is dependent on the dimensions of the cell and the required current density. If the spacer is too thin, the cell can become oxygen starved, while if too thick, it increases the battery weight and volume unnecessarily. Figure 33.11 shows a typical zinc/air cell stack. An alternative approach to dealing with oxygen diffusion is by providing a positive pressure of air by designing a fan and air channels into the battery design (see Fig. 33.12).

33.12 SPECIALIZED BATTERY SYSTEMS

FIGURE 33.11 Typical zinc/air cell stack. Representative stack consists of GEN 4 zinc/air cells. (*Courtesy of Electric Fuel Corp.*)

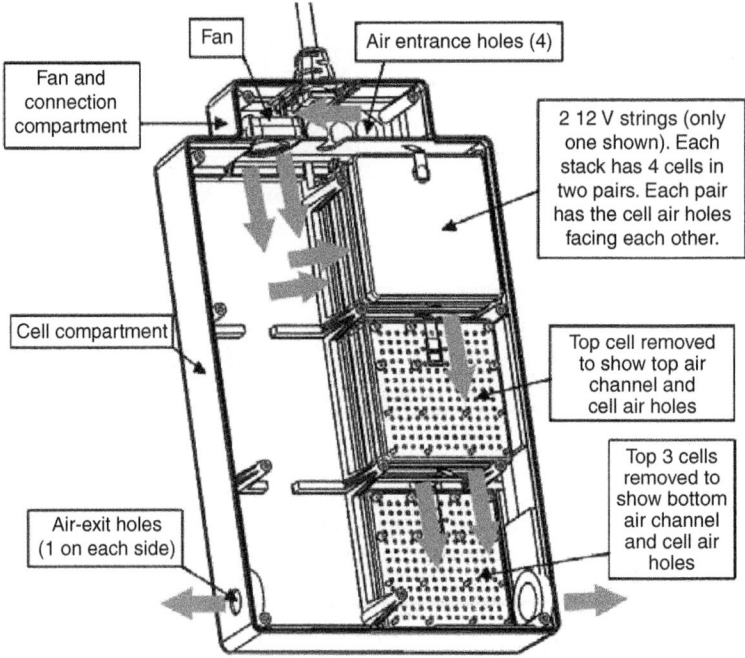

FIGURE 33.12 BA-8180 battery design. (*Courtesy of Electric Fuel Corp.*)

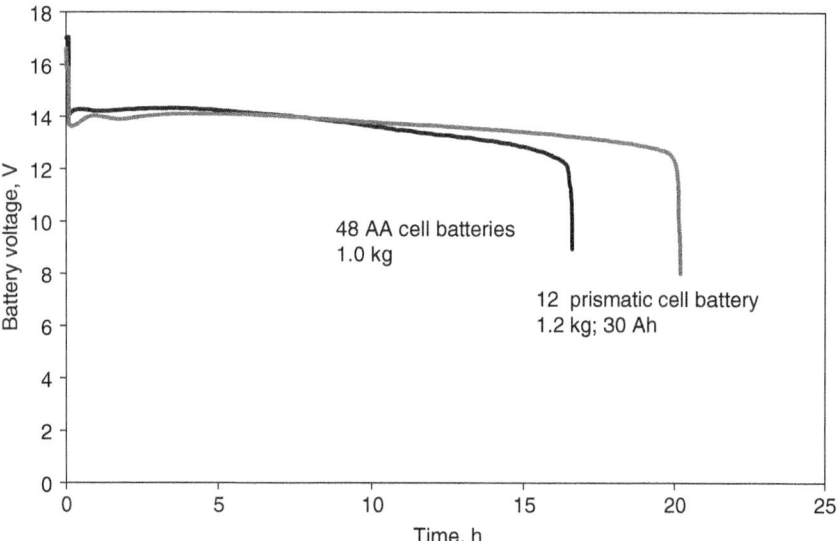

FIGURE 33.13 Discharge profile for 12 V zinc/air batteries discharged at 18 W continuous. Data from U.S. Army tests.

Cylindrical zinc/air cells (Fig. 33.3) have been designed primarily in the AA cell size. These cells allow for a direct replacement of zinc alkaline manganese dioxide cells. The zinc/air technology uses a very thin cathode, allowing for the bulk of the cell to contain the anode/electrolyte mixture. The relatively high surface area of AA cells provides high power discharge rates. Batteries constructed from arrays of these cells do not provide for forced flow of air, but it has been shown that thermal gradients within the battery pack do provide convective flow.

Figure 33.13 shows a typical discharge curve for two 12 V zinc/air battery configurations: (1) twelve 30 Ah prismatic zinc/air cells in series and (2) 48 AA zinc/air cells consisting of four parallel strings of 12 cells in series. Figure 33.14 shows the discharge characteristics for 3 single-cell zinc/air batteries designed for portable electronic equipment.

The advancement of the zinc/air electrochemical cell can be seen in Fig. 33.14. The figure presents the performance characteristics of each generation in the form of specific energy versus specific power. At more than 400 Wh/kg, the button cell (GEN 2) has the highest specific energy, while the power-optimized GEN 4 cell has the highest specific power, exceeding 100 W/kg. However, the GEN 4 cell operates at a reduced specific energy when compared to the GEN 3 cell. The figure shows that, at low rates (<10 W/kg), the specific energy of GEN 3 approaches that of the button cell, but falls off rapidly as the rate increases. This is because the anode is very thick (more than 5 mm, at least twice those of typical GEN 4 cells). For reference, a GEN 3 cell delivering 30 Ah is used in the BA-8180 to power radiotelephones over a period of several days to a week, at low cost.

33.3.3 Industrial Primary Zinc/Air Batteries

Large primary zinc/air batteries have been used for many years to provide low-rate, long-life power for applications such as railroad signaling, seismic telemetry, navigational buoys, and remote communications. They are available in either water-activated (containing dry potassium hydroxide) or preactivated versions.[19] Preactivated versions are also available with a gelled electrolyte to minimize the possibility of leakage. Until recently, the zinc contained a few percent of mercury to minimize self-discharge after activation. The newer batteries use additives and alloys to eliminate the mercury and minimize hydrogen generation and corrosion.

33.14 SPECIALIZED BATTERY SYSTEMS

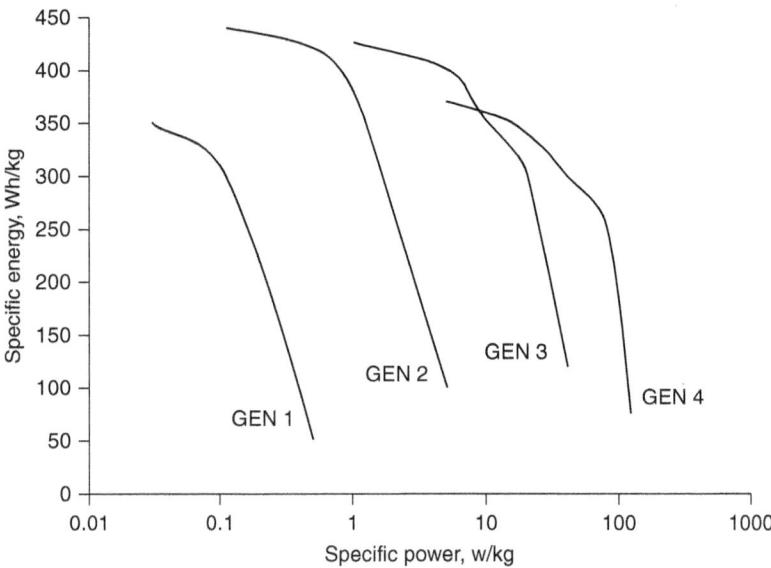

FIGURE 33.14 Discharge characteristics for zinc/air cells designed for portable electronic equipment. Data from U.S. Army tests.

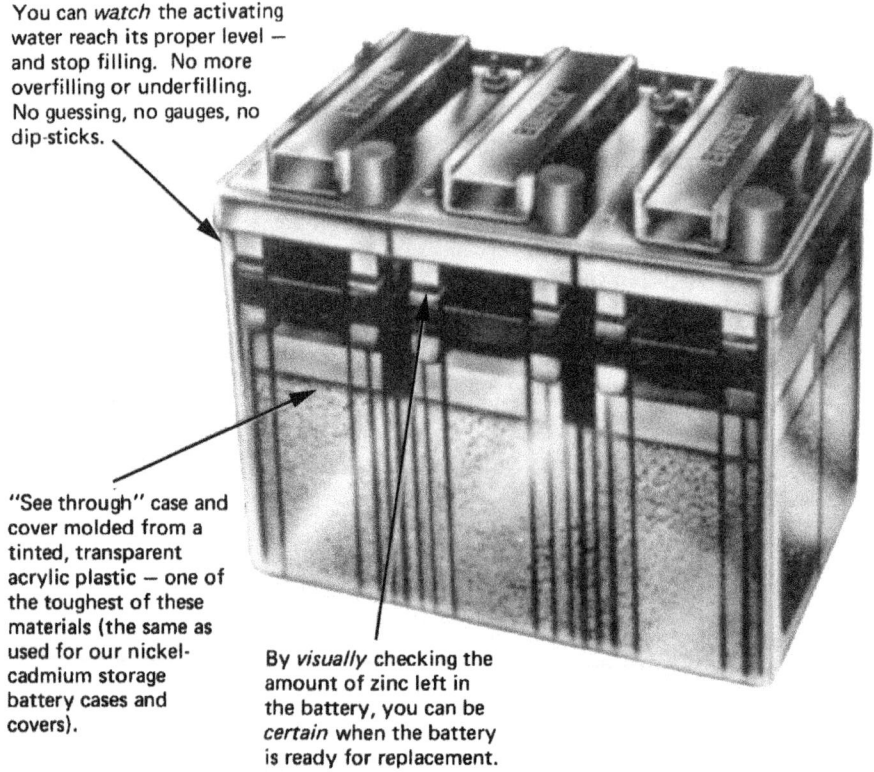

You can *watch* the activating water reach its proper level — and stop filling. No more overfilling or underfilling. No guessing, no gauges, no dip-sticks.

"See through" case and cover molded from a tinted, transparent acrylic plastic — one of the toughest of these materials (the same as used for our nickel-cadmium storage battery cases and covers).

By *visually* checking the amount of zinc left in the battery, you can be *certain* when the battery is ready for replacement.

FIGURE 33.15 Edison Carbonaire zinc/air battery. (*Courtesy of SAFT America, Inc.*)

Preactivated and Water-Activated Types. A typical preactivated industrial-type zinc/air cell, the Edison Carbonaire cell, is manufactured in a 1100 Ah size and is available in two- and three-cell configurations, as illustrated in Fig. 33.15. The cell case and cover are molded from a tinted transparent acrylic plastic. The construction features are shown in Fig. 33.16 identifying the wax-impregnated carbon cathode block, the solid zinc anodes, and the lime-filled reservoir. These cells normally have a bed of lime to absorb carbon dioxide and to remove soluble zinc compounds from solution and precipitate them as calcium zincate. They are made with transparent cases so that the electrolyte

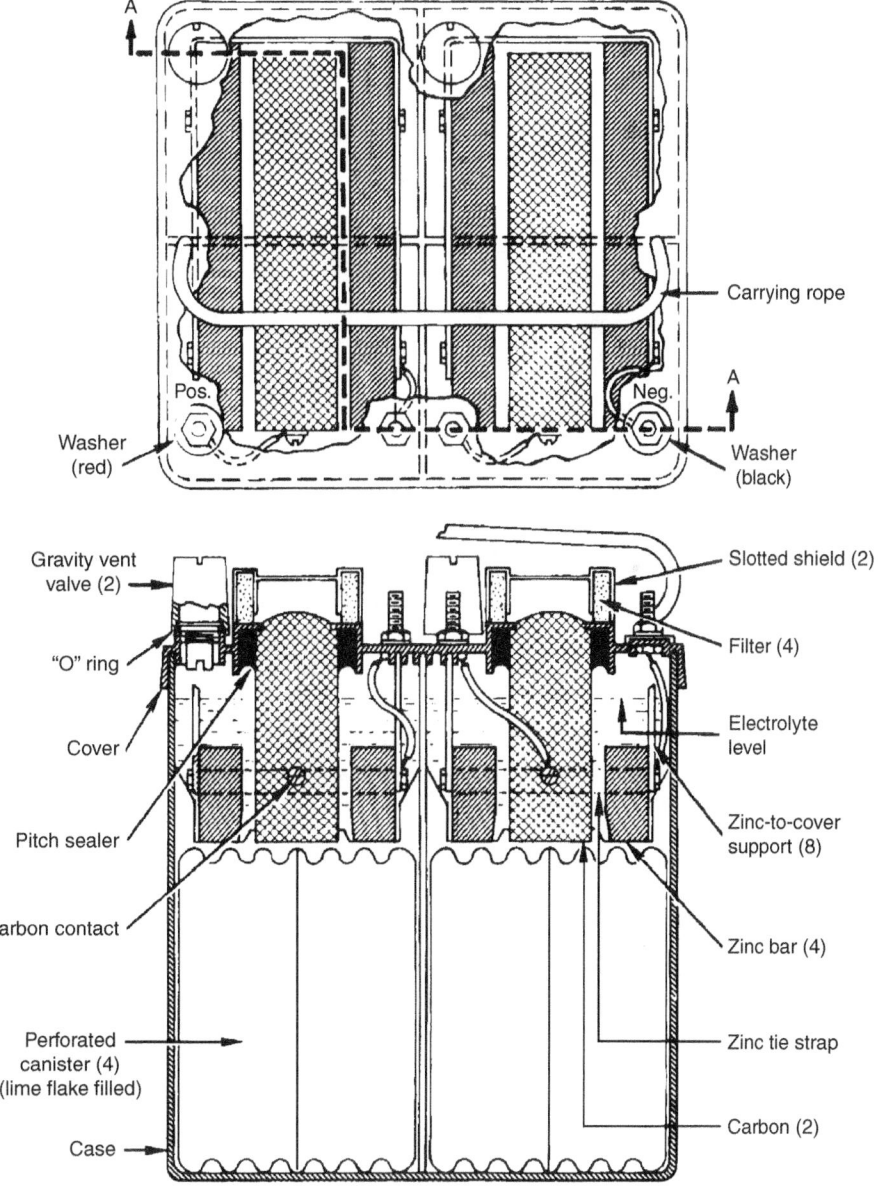

FIGURE 33.16 Top and side views of type ST-2 Carbonaire zinc/air battery. (*Courtesy of SAFT America, Inc.*)

level and the state-of-charge can be monitored visually. The state-of-charge can be monitored by observing the condition of the zinc plates and the condition of the lime bed. The bed turns darker as it is converted to zincate.

Water-activated cells and batteries are supplied sealed. The caustic (potassium hydroxide) electrolyte and the lime flake are present in the dry form. The cell is activated by removing the seals and adding the appropriate amount of water to dissolve the potassium hydroxide. Periodic inspection and addition of water are the only required maintenance.

These cells are manufactured in a 1100 Ah size and are available in multicell batteries, with the cells connected in series or parallel. The maximum continuous discharge rate for this battery at 25°C is 0.75 A. A preactivated 1100 Ah three-cell battery weighs about 2 kg, giving an energy density of about 180 Wh/kg. The physical and electrical characteristics of these batteries are listed in Tables 33.6 and 33.7.

TABLE 33.6 Edison Carbonaire Zinc/Air Batteries

Type	Dimensions, cm			Weight (filled), kg	Connection	Nominal voltage, V	Nominal capacity, Ah
	Length	Width	Height				
Two cells:							
ST-22-1100	21.9	20.0	28.9	14	Series	2.5	1100
ST-22-2200					Parallel	1.25	2200
Three cells:							
ST-33-1100	32.4	20.0	28.9	21	Series	3.75	1100
ST-33-3300					Parallel	1.25	3300

Source: SAFT America, Inc.

TABLE 33.7 Maximum Discharge Rates (Amperes) to 1.0 V per Cell for Edison Carbonaire ST Type Zinc/Air Batteries

	Duty cycle			
	10% on (up to 0.5 s)	20% on (up to 2.0 s)	50% on (up to 1 s)	100% on (continuous)
20°C	3.5	2.8	2.3	1.25
−5°C	2.4	1.9	1.6	0.75

Source: SAFT America, Inc.

Voltage versus time curves plotted in Fig. 33.17 depict the performance obtained at various discharge rates. Capacities obtained are quite consistent over the range of 0.15 to 1.25 A continuous discharge, although at the higher rates, voltage variation with temperature must be considered. If tight voltage control is required, a series-parallel assembly of batteries may be needed to reduce the current drain per cell and thus minimize the voltage fluctuations with temperature.

Gelled-Electrolyte Types. An alternative version uses a gelled electrolyte to eliminate the possibility of leakage during operation. The zinc electrode is composed of zinc powder mixed with a gelling agent and the electrolyte, and the reaction product is zinc oxide rather than calcium zincate. The battery is filled with electrolyte during manufacture. Figure 33.18 shows a cross section of the cell.

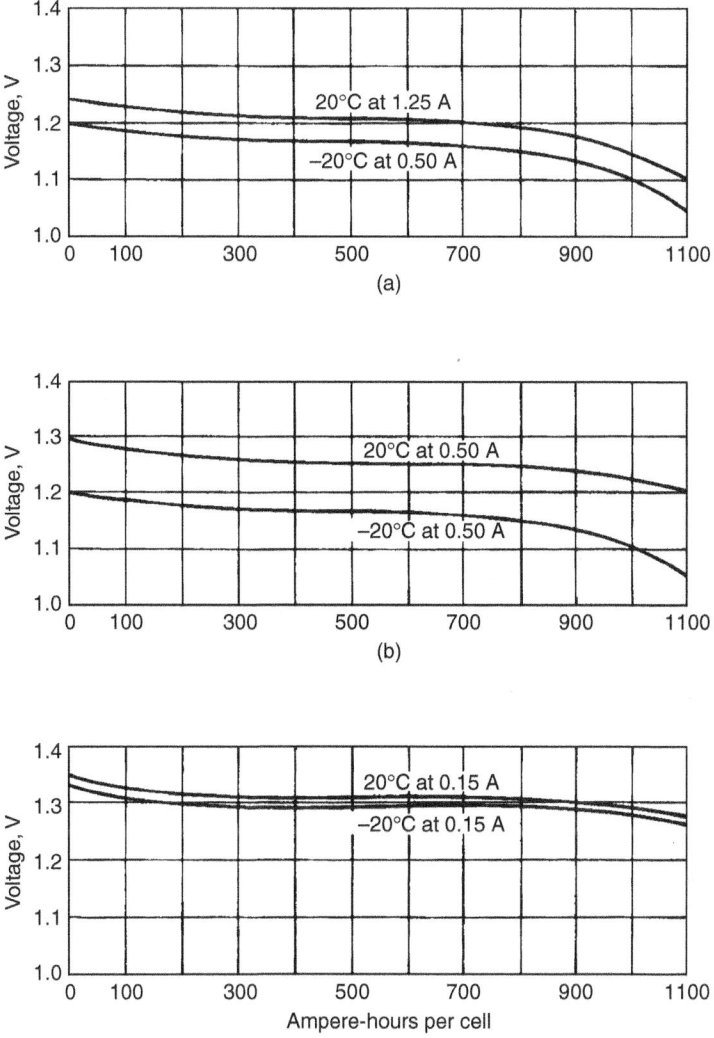

FIGURE 33.17 Typical discharge characteristics of Carbonaire 1100 Ah zinc/air batteries. (a) Maximum continuous rates. (b) Moderate continuous rates. (c) Low continuous rates. (*Courtesy of SAFT America, Inc.*)

The Gelaire battery is manufactured in a nominal 1200 Ah size and is available in multicell batteries with the cells connected in series or parallel. The physical and electrical characteristics of these batteries are listed in Table 33.8.

Figure 33.19 shows the discharge characteristics of the 1200 Ah cell at several discharge loads. As a result of the use of a porous zinc electrode, and the direct formation of zinc oxide as the reaction product, the specific energy of this type of battery is higher than for the water-activated types. At low rates of discharge, it is 285 Wh/kg.

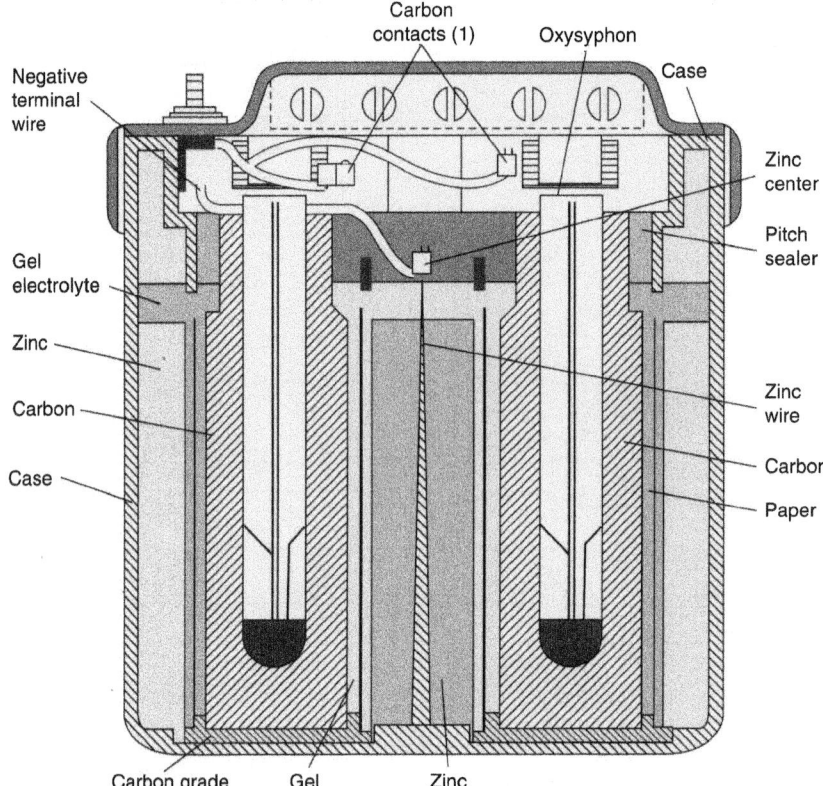

FIGURE 33.18 Cross section of Gelaire cell. (*Courtesy of SAFT America, Inc.*)

TABLE 33.8 Physical and Electrical Characteristics of Gelaire Battery

	NT 1000X (single cell)	2NT 1000X (two-cell block)*	3NT 1000X (large-cell block)*
Dimensions, mm:			
Length	108	216	324
Width	200	200	200
Height	213	213	213
Weight, kg	5.4	10.9	16.4
Nominal voltage, per cell, V	1.3	1.3	1.3
Nominal capacity, per cell, Ah	1200	1200	1200

*Units can be connected either in series or in parallel.
Source: SAFT America, Inc.

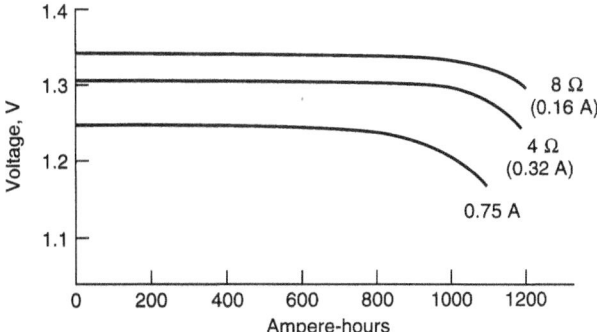

FIGURE 33.19 Discharge performance of Gelaire battery (NT1000X). (*Courtesy of SAFT America, Inc.*)

33.3.4 Primary Hybrid-Air/Manganese Dioxide Batteries

Another approach to primary zinc/air batteries is to use a hybrid cathode that contains a significant amount of manganese dioxide.[20] During low-rate operation, the battery functions as a zinc/air system. At high rates, as the oxygen may be depleted, the discharge function at the cathode is taken over by the manganese dioxide. This means that such a battery should essentially have the capacity of a zinc/air battery when discharged at low rates, but it should have the pulse current capability of a manganese dioxide battery. After the high-current pulse, the manganese dioxide is partially regenerated by air oxidation so that the pulse current capability is restored.

Figure 33.20 is a side view of a flat-pack cell. Figure 33.21 compares the performance of a 6 V four-cell hybrid "lantern" battery with similar alkaline and zinc-carbon batteries at an intermittent low-rate discharge. The specific energy of this battery is about 350 Wh/kg. Single and multicell batteries are available in capacities of 40 to 4800 Ah.

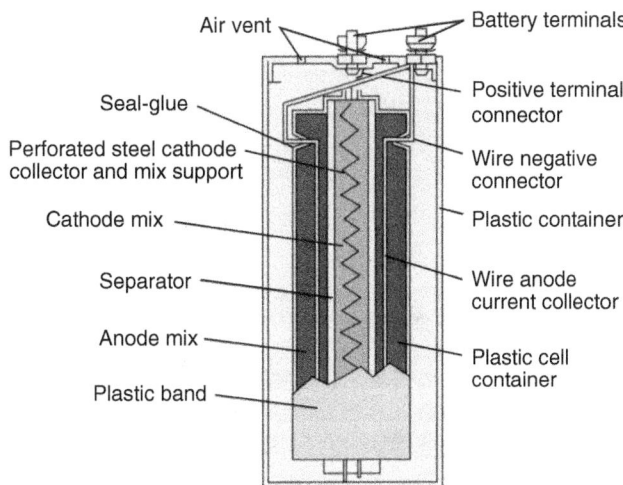

FIGURE 33.20 Side view of hybrid zinc/air-manganese dioxide cell. (*Courtesy of Celair Corp.*)

33.20 SPECIALIZED BATTERY SYSTEMS

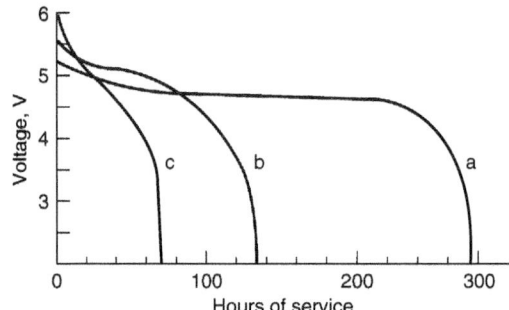

FIGURE 33.21 Comparison of performance of hybrid "lantern" battery with zinc-carbon and alkaline-manganese dioxide battery. Curve *a*—air-alkaline; curve *b*—alkaline-MnO_2; curve *c*—heavy-duty zinc-carbon. (*Courtesy of Celair Corp.*)

33.3.5 Electrically Rechargeable Zinc/Air Batteries

Electrically rechargeable zinc/air batteries use a bifunctional oxygen electrode so that both the charge process and the discharge process take place within the battery structure.

The basic reactions of an electrically rechargeable zinc/air cell using a bifunctional oxygen electrode are shown in Fig. 33.22. Advances in electrically rechargeable zinc/air cells have concentrated on the bifunctional air electrode.[21–23] Electrodes based on La, Sr, Mn, and Ni perovskites have demonstrated good cycle life. Figure 33.23 shows the gains in cycle life for the bifunctional air cathode achieved going from Phase I to Phase II of a research and development program.

Portable Electrically Rechargeable Batteries. An electrically rechargeable zinc/air cell having a bifunctional oxygen electrode, designed for use in computers and other electronic communication

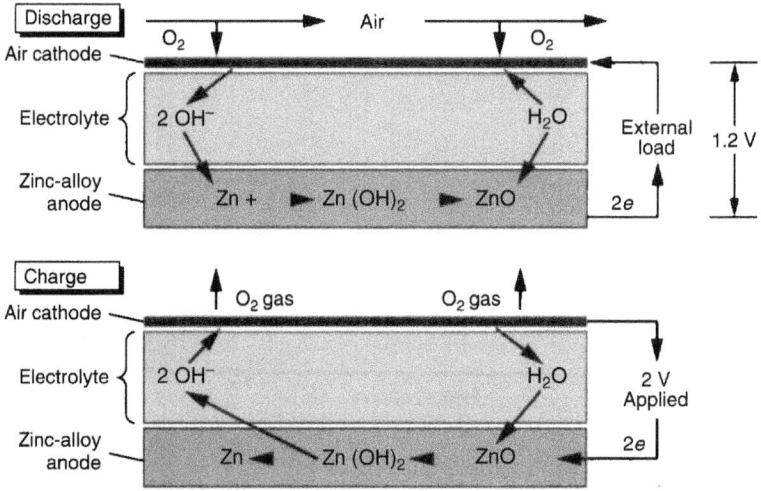

FIGURE 33.22 Basic operation of electrically rechargeable zinc/air cell. (*Courtesy of AER Energy Resources, Inc.*)

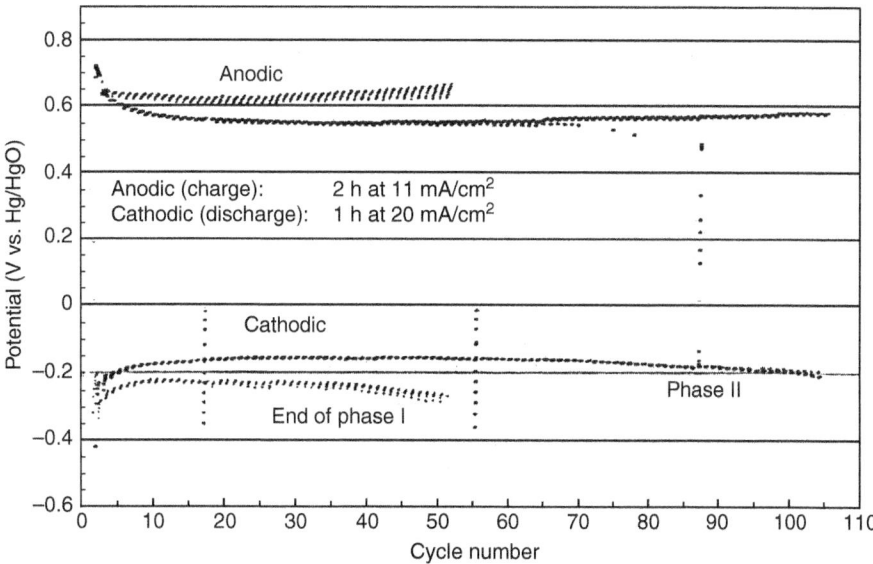

FIGURE 33.23 Advances in bifunctional air electrodes. LSNC perovskite plus Shawinigan black carbon. Area = 25 cm^2. 8 M KOH at room temperature. (*Courtesy of Alupower, Inc.*)

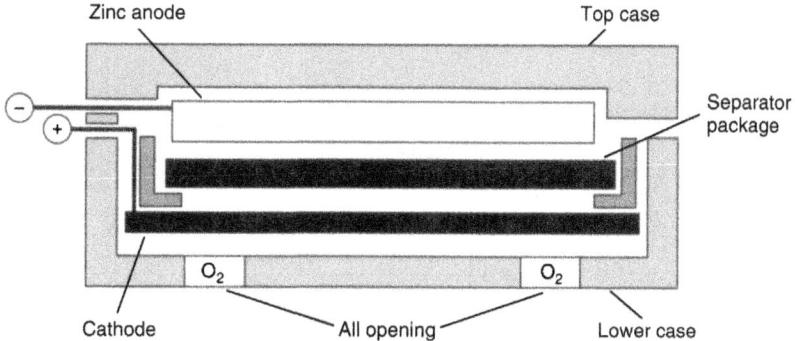

FIGURE 33.24 Cross section of electrically rechargeable zinc/air cell. (*Courtesy of AER Energy Resources, Inc.*)

equipment, is shown in Fig. 33.24. The cell is a prismatic or thin rectangular design. A high-porosity zinc structure, which maintains its integrity and morphology during cycling, is used. The air electrode is a corrosion-resistant carbon structure, containing a large number of small pores and a catalyst. The structure is permeable to oxygen, hydrophobic, and supported by a low-impedance current collector. The flat zinc negative and the air electrode plates face each other, separated by a high-porosity separator with low electrochemical resistance and the ability to absorb and retain the potassium hydroxide electrolyte. The cell case is injection-molded polypropylene with openings to permit the inflow of oxygen during discharge and release of the oxygen generated during charge.

A critical factor in the design of the cell and battery is the means of controlling the flow of air into and out of the cell, which must be matched to the needs of the application. Excessive amounts of air could result in drying out the cell; too little air (oxygen-starved) will result in a drop-off of performance. The stoichiometric quantity of air required is 18.1 cm^3/min per ampere of continuous current. An air manager is used to control the flow of air by opening air access to the cathode during discharge and sealing the battery from the air when it is not in use to minimize self-discharge. A fan, powered by the battery, also is used to assist the airflow.

A discharge and charge profile of a 20 Ah zinc/air battery is shown in Fig. 33.25. The battery is typically discharged at the $C/20$ rate or lower to about 0.9 V. The voltage profile is flat with a sharp drop at the end of the discharge. Deep discharge to 0 V may be detrimental. Discharging at rates higher than the $C/10$ rate is not feasible because of the battery's relatively high internal resistance. This power limitation dictates the minimum size and weight, and the battery cannot be designed efficiently to operate for less than 8 to 10 h. When discharged at the acceptable loads, the battery can deliver about 150 Wh/kg and 160 Wh/L.

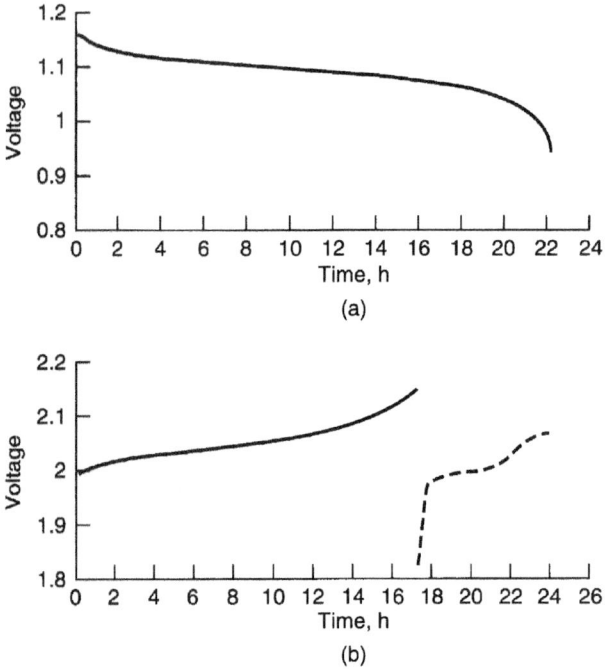

FIGURE 33.25 Electrically rechargeable zinc/air battery. (a) Representative discharge profile, 1 A discharge. (b) Representative charge profile, 1.25 A charge followed by 0.5 A charge. (*Courtesy of AER Energy Resources, Inc.*)

The battery is recharged by constant-current methods, using a two-step process, as shown in Fig. 33.25b. A moderate rate is used initially until the battery is about 85% charged. This is followed by a low rate to complete the charge. Charging a fully discharged battery takes about 24 h. Both charge rate and overcharging must be controlled. Overcharging will result in the generation of hydrogen at the negative electrode. It will damage the cell and shorten life due to the corrosion of the air cathode. Energy efficiency is about 50% due to the large difference between discharge and charge voltages. The overall life of the battery is independent of the number of cycles. About 400 h of operation has been demonstrated, but further development of this battery has been terminated because of its limited cycle life.

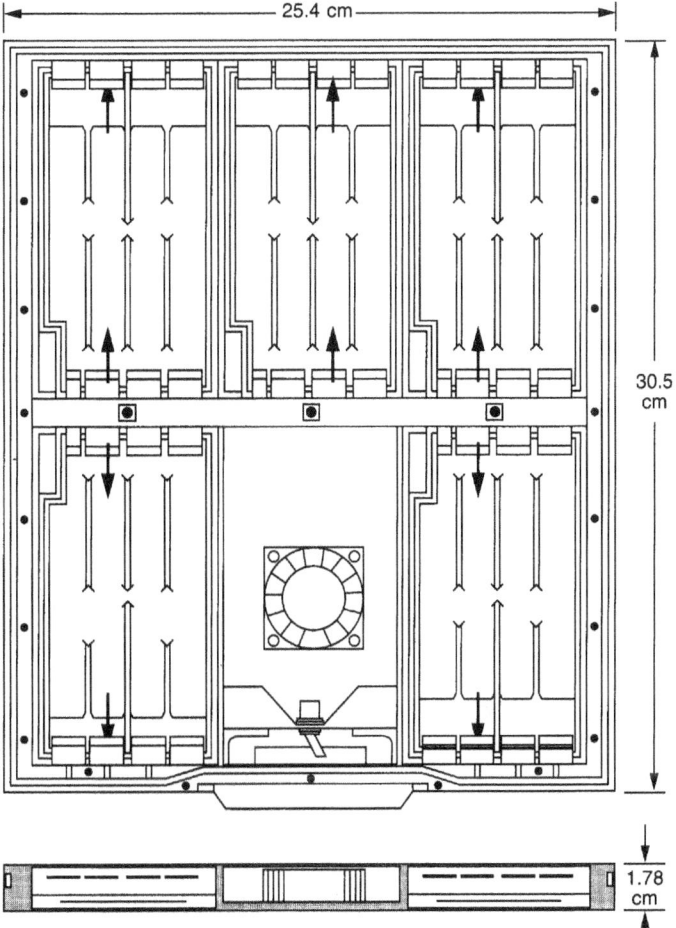

FIGURE 33.26 Prototype electrically rechargeable zinc/air battery for notebook computer. (*Courtesy of AER Energy Resources, Inc.*)

A sketch of a battery design for a notebook computer, fitting into the base of the computer case, is shown in Fig. 33.26. The battery contains five cells and is rated at 5 V and 20 Ah. Table 33.9 provides the physical and electrical characteristics of this battery.[24]

Electrically Rechargeable Systems for Electric Vehicles. A similar rechargeable zinc/air cell, operating at room temperature, was being developed for use in electric vehicles. The cell uses a planar bipolar configuration. The negative electrode consists of zinc particles in a paste form, similar to the electrode used in alkaline-manganese dioxide primary cells. The bifunctional air electrode consists of a membrane of carbon and plastic with appropriate catalysts. The electrolyte is potassium hydroxide with gelling agents and fibrous absorbing materials. A typical cell is rated at 100 Ah with an average operating voltage of 1.2 V.

Specific energies up to 180 Wh/kg at the 5 to 10 h discharge rates and a battery life of about 1500 h have been achieved. Technical limitations are limited power density and a relatively short separator

TABLE 33.9 Physical and Electrical Characteristics of Electrically Rechargeable Zinc/Air Battery

Open-circuit voltage	1.45 V
Nominal operating voltage	1.2–0.9 V (design using 1 V)
Cutoff voltage	0.9 V
Capacity	20 Ah (1 A)
Capacity retention	Capacity loss less than 2% per month when stored sealed at room temperature
Max current:	
Continuous	2 A
Pulse	3 A
Specific energy	130 Wh/kg
Energy density	160 Wh/L
Cycle life	400 h
Charge characteristics	2.0 V/cell at 750 mA
Overcharge sensitivity	Overcharge degrades battery life
Charge termination	dV/dT and maximum voltage
Cathode air rate	100 cm^3/min per cell at 1 A rate
Weight	155 g
Dimensions:	
Length	13.5 cm
Width	7.6 cm
Height	1.22 cm
Temperature:	
Operating	5–35°C
Storage	−20 to 55°C
Relative humidity:	
Operating	20–80%
Storage	5–95%

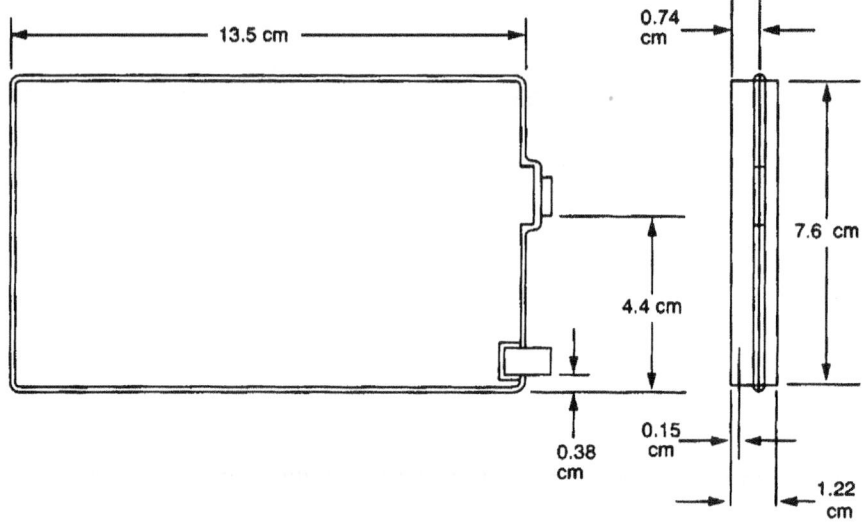

Source: AER Energy Resources, Inc.

TABLE 33.10 Characteristics of Zinc/Air Traction Battery

Physical characteristics:	
Cell size	$33 \times 35 \times 0.75$ cm
Cell weight	1.0 kg (typical)
Cell voltage:	
Open-circuit	1.5 V
Average	1.2 V
High load	1.0 V
Charging	1.9 V
Configurations:	
General purpose	120 Wh/kg, 120 W/kg peak power
High energy	180 Wh/kg at 10 W/kg
High power	200 W/kg peak at 100 Wh/kg

Source: Dreisbach Electromotive, Inc. (DEMI).

life. The air must be managed to remove carbon dioxide, and to provide humidity and thermal management. Table 33.10 provides some of the characteristics of this battery, which is no longer under development.[25,26]

33.3.6 Mechanically Rechargeable Zinc/Air Batteries

Mechanically rechargeable or refuelable batteries are designed with a means to remove and replace the discharged anodes or discharge products. The discharged anode or discharge products can be recharged or reclaimed external to the cell. This avoids the need for a bifunctional air cathode and the shape change problems resulting from the charge-discharge cycling of an in situ zinc electrode.

Mechanically Refueled Systems—Anode Replacement. Mechanically replaceable zinc/air batteries were seriously considered for powering portable military electronic equipment in the late 1960s because of their high specific energy and ease of recharging. This battery contained a number of bicells connected in series to provide the desired voltage. Each bicell, as illustrated in Fig. 33.27, consisted of two air cathodes connected in parallel and supported by a plastic frame which together formed an envelope for the zinc anode. The anode, which was a highly porous zinc structure enclosed in an absorbent separator, was inserted between the cathodes. The electrolyte, KOH, was contained in a dry form in the zinc anode and only water was needed to activate the cell. "Recharging" was accomplished by removing the spent anode, washing the cell, and replacing the anode with a fresh one. These batteries were never deployed because of their short activated life, poor intermittent operation, and the development of new high-performance primary lithium batteries which were superior in rate capability and ease of handling in the field.[27,28]

A design similar to the portable mechanically rechargeable zinc/air battery has been considered for electric vehicle applications. The battery would be refueled "robotically" at a fleet servicing location or at a public service station by removing and replacing the spent anode cassettes. The discharged fuel would be electrochemically regenerated, using a modified zinc electrowinning process, at central facilities that serve regional distribution networks.[29]

This zinc/air battery consisted of modular cell stacks, each containing a series of individual bicells. Each bicell consists of an anode cassette containing a zinc-based electrolyte slurry, contained between air cathodes, and a separator system. The slurry is maintained in a static bed without circulation. In addition, the battery contains subsystems for air provision and heat management and is adapted for fast mechanical replacement of the cassettes.

The technology has been evaluated in a full-size 264 V, 110 kWh battery weighing 650 kg in a van that was converted to electric drive. The battery delivered 230 Wh/kg and 230 Wh/L with a power density of 100 W/kg.

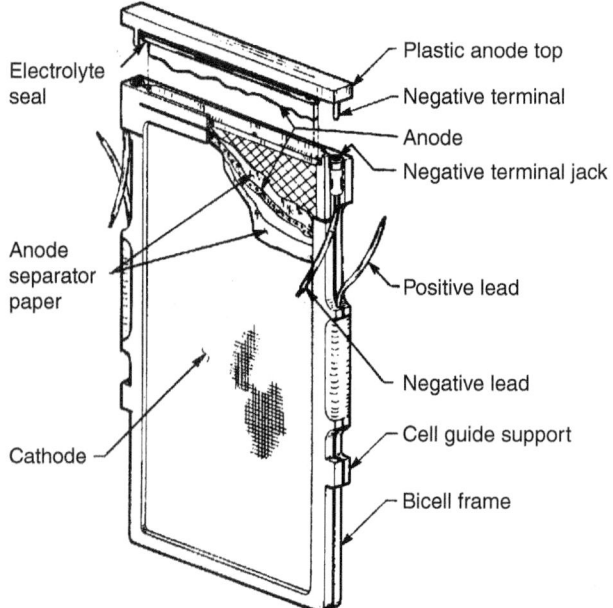

FIGURE 33.27 Zinc/air bicell.

Another approach to powering electric vehicles with mechanically rechargeable zinc/air batteries is a hybrid configuration where the zinc/air battery is combined with a rechargeable battery, such as a high-power lead-acid battery.[30] With this approach, the performance of each battery can be optimized, using the high specific energy zinc/air battery as the energy source with a high specific power rechargeable battery to handle the peak power requirements. The power battery can also be sized to handle the anticipated peak load and duty cycle. In operation, during periods of light load, the zinc/air battery handles the load and recharges the rechargeable battery through a voltage regulator. The load is shared by both batteries during peak load conditions. When fully discharged, the zinc/air battery is recharged by removing and replacing the zinc oxide discharge product, which can be regenerated externally and efficiently in designated facilities. The advantage of this hybrid design is illustrated in Fig. 33.28, a Ragone plot comparing the performance of the hybrid with the performance of the individual batteries. In this example, the hybrid lead-acid battery is one specifically designed for high rate performance.

Mechanically Refueled System—Zinc Powder Replacement.[31-33] Figure 33.29 is a sketch of an 80 cm^2 laboratory cell using a packed bed of zinc powder, which can be replaced when depleted. Natural convection is utilized for electrolyte circulation. During operation, electrolyte flows downward through the zinc bed and upward around the back of the current collector, which is either graphite or copper. Figure 33.30 shows the voltage profile on constant-current discharge for each of these current collectors.

The cell was designed so that the zinc bed and electrolyte could be pumped out at the end of discharge and replaced with a fresh charge of zinc and electrolyte to simulate operation in an electric vehicle. The cell was discharged at 2 A for 4 h. Most of the electrolyte and the residual particles were then sucked out of the anode side of the cell through a tube passing through a hole in the top of the cell and connected to a water jet aspirator. Without rinsing, fresh particles and electrolyte were placed in the cell through the hole and a second discharge was carried out. Following this, about 90% of the particles were removed less carefully, and the cell was refilled and discharged for a third time. The data in Fig. 33.31 shows that the three discharges were essentially the same.

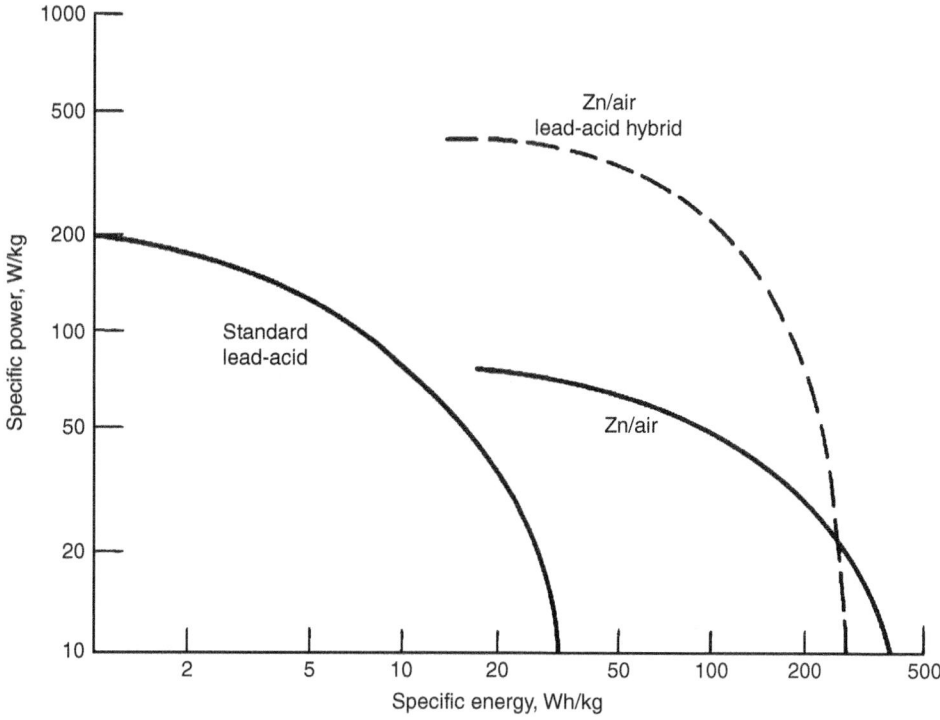

FIGURE 33.28 Comparison of Zn/air lead-acid hybrid battery with individual lead-acid and Zn/air batteries. Lead-acid battery uses special high-rate design.

Based on these experiments, a conceptual design was made for a 55 kW (peak power) electric-vehicle battery. Projected specific energy of the battery was 110 Wh/kg at 97 W/kg under a modified Simplified Federal Urban Driving Schedule (SFUDS). These values were increased to 228 Wh/kg at 100 W/kg when the battery was designed for optimum capacity, and to 100 Wh/kg at 150 W/kg when designed for optimum power output, based on the results of discharge experiments at 45°C.

Efficient regeneration of the zinc particles is required to provide for a practical, efficient system. It is projected that for the practical application of this system, the spent electrolyte and residual particles would be removed at local service centers and the vehicle would be quickly refueled by the addition of regenerated zinc powder and electrolyte. The system under development[33] involves stopping the discharge of the battery described when the voltage falls below a practical value rather than when the voltage becomes zero. Under these conditions, no precipitation has occurred and the electrolyte is clear. The processing of products removed from the cell is then simply one of redeposition of zinc onto the particles.

33.4 ALUMINUM/AIR BATTERIES

Aluminum has long attracted attention as a potential battery anode because of its high theoretical ampere-hour capacity, voltage, and specific energy. While these values are reduced in a practical battery because of the inability to operate aluminum and the air electrodes at their thermodynamic potentials and because water is consumed in the discharge reaction, the practical energy density still exceeds that of most battery systems. The inherent hydrogen generation of the aluminum anode in aqueous electrolytes

33.28 SPECIALIZED BATTERY SYSTEMS

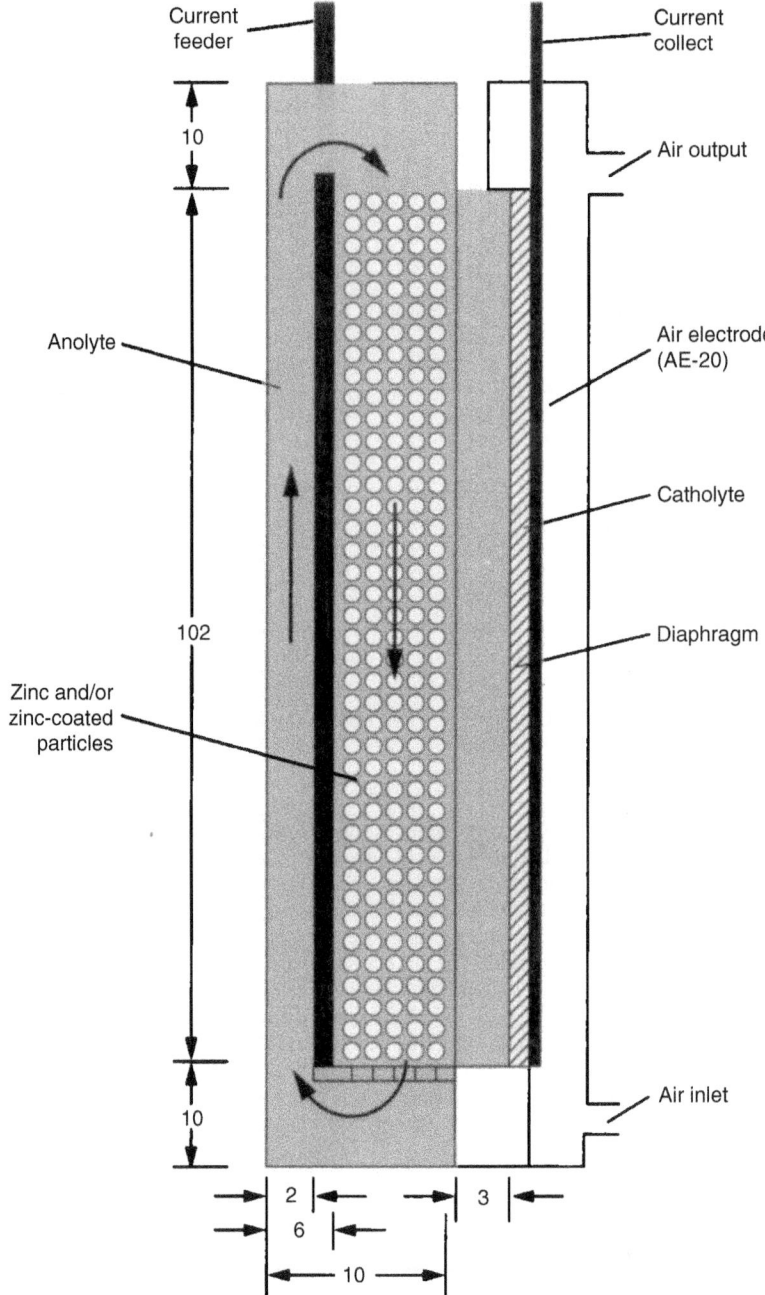

FIGURE 33.29 Schematic of mechanically refueled 80 cm² laboratory zinc/air cell. (*Courtesy of Lawrence Berkeley Laboratory.*)

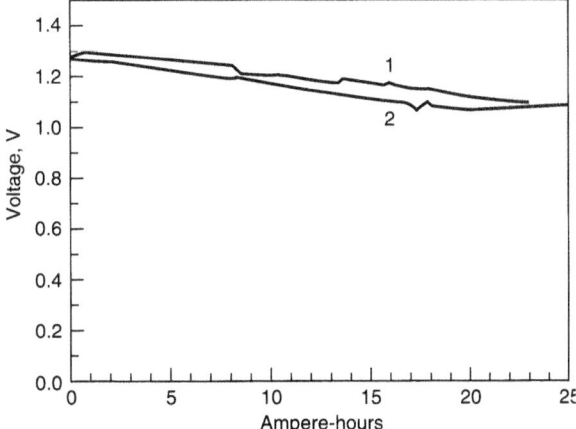

FIGURE 33.30 Constant-current discharges of mechanically refueled zinc/air battery, graphite vs. copper feeders. Anolyte/catholyte—45% KOH; anode—30-mesh zinc; cathode—AE-20 air electrode; I = 2A; A = 78 cm^2. Curve 1—1.5-mm copper current feeder; curve 2—4.0-mm graphite current feeder. (*Courtesy of Lawrence Berkeley Laboratory.*)

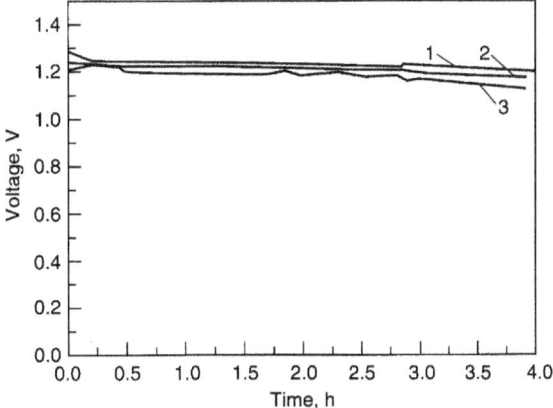

FIGURE 33.31 Voltage vs. time during subsequent mechanical recharging for mechanically refueled zinc/air battery. Anolyte/catholyte—45% KOH; anode—20-mesh zinc particles; cathode—AE-20 air electrode; I = 2 A; A = 76 cm^2. Curve 1—first run; curve 2—100% of anolyte/particles suctioned out, cell refilled with fresh ones, no rinsing; curve 3–90% of anolyte/particles suctioned out, cell refilled with fresh ones, no rinsing. (*Courtesy of Lawrence Berkeley Laboratory.*)

is such that the batteries are designed as reserve systems with the electrolyte added just before use, or as "mechanically" rechargeable batteries with the aluminum anode replaced after each discharge. Electrically rechargeable aluminum/air batteries are not feasible using aqueous electrolytes.

The discharge reactions for the aluminum/air cell are

$$\text{Anode:} \quad \text{Al} \rightarrow \text{Al}^{+3} + 3e$$
$$\text{Cathode:} \quad \text{O}_2 + 2\text{H}_2\text{O} + 4e \rightarrow 4\text{OH}^-$$
$$\text{Overall:} \quad 4\text{Al} + 3\text{O}_2 + 6\text{H}_2\text{O} \rightarrow 4\text{Al(OH)}_3$$

The parasitic hydrogen-generating reaction is

$$Al + 3H_2O \rightarrow Al(OH)_3 + \tfrac{3}{2}H_2$$

Aluminum can be discharged in neutral (saline) solutions as well as in caustic solutions. The neutral electrolytes are attractive because of the relatively low open-circuit corrosion rates and the reduced hazards of these solutions compared with concentrated caustic. Saline systems were under development for relatively low-power applications, such as ocean buoys and portable battery applications, with specific energies of a "dry" battery as high as 800 Wh/kg. Seawater batteries for underwater vehicle propulsion and other applications, using oxygen present in the ocean, rather than air, or operating as corrosion cells, also are of interest because of the potentially high energy output.

Alkaline systems have an advantage over saline systems because the alkaline electrolyte has a higher conductivity and a higher solubility for the reaction product, aluminum hydroxide. Thus the alkaline aluminum/air battery is a candidate for high-power applications such as standby batteries, propulsion power for unmanned underwater vehicles, and has been proposed for electric vehicle propulsion. The specific energy can be as high as 400 Wh/kg. Aluminum/air batteries (as well as zinc/air batteries), because of their high energy densities, can also be used as power sources for recharging lower-energy rechargeable batteries in remote areas where line power is not available.

33.4.1 Aluminum/Air Cells in Neutral Electrolytes

Aluminum/air cells using neutral electrolytes have been developed for portable equipment, stationary power sources, and marine applications. Aluminum alloys are now available for saline cells with low polarization voltages, which can operate with coulombic efficiencies in the range of 50 to 80%. Alloying elements are required to enhance the disruption of the anodic surface film when current is drawn. Interestingly, in neutral electrolytes the corrosion reaction, resulting in the direct evolution of hydrogen, occurs at a rate linearly proportional to the current density, starting from near zero at zero current.[34]

Cathodes, such as those described earlier, are satisfactory. However, there are some extra limitations which apply in a saline solution. Nickel is not a suitable substrate where extensive periods on open-circuit are involved. Under these conditions, the potential of the active material in contact with the screen is high enough to oxidize the screen. One way to minimize this problem is to continue to draw a very low current during no-load periods, which is sufficient to keep the cathode potential from rising to its open-circuit value.

A suitable neutral electrolyte is a 12 wt % solution of sodium chloride, which is near the maximum conductivity. Current densities are limited to 30 to 50 mA/cm^2 as a result of the limitation imposed by the conductivity of the electrolyte. Such batteries may also be operated in seawater, with obvious limitations in current capability as a result of the lower conductivity of seawater.

Electrolyte management is required because of the behavior of the reaction product, aluminum hydroxide. It has a transient high solubility in the electrolyte and tends to become gellike when it first precipitates. In an unstirred system, the electrolyte starts to become "unpourable" when the total charge produced exceeds 0.1 Ah/cm^3. Up to this point, the electrolyte and the reaction product can be poured out of a cell and more saline solution added to continue the discharge until all of the aluminum is consumed. If the discharge is continued without draining the electrolyte, it will proceed satisfactorily until the total discharge reaches approximately 0.2 Ah/cm^3. At this point, the cell contents are nearly solid.

Approaches to minimizing the amount of electrolyte required have been studied.[35] In one approach, the electrolyte was stirred in a reciprocating manner, which minimized gel formation and produced a finely divided product which was dispersed in the electrolyte. A total electrolyte capacity of 0.42 Ah/cm^3 was achieved using reciprocated 20% potassium chloride electrolyte. A similar result was achieved by injecting a pulsed air stream at the bottom of each cell. This has the additional advantage that it sweeps the hydrogen out of each cell in a concentration below the flammability limit. An electrolyte utilization of 0.2 Ah/cm^3 was achieved in a system from which the electrolyte could be easily drained.

Portable Aluminum/Air Batteries. A number of batteries using saline electrolytes have been designed. In general, they are built as reserve batteries and activated by adding the electrolyte to the battery.

A saltwater battery, illustrated in Fig. 33.32, was designed for field recharging of nickel-cadmium and lead-acid storage batteries.[36] Figure 33.33 shows the charge and discharge characteristics of a 2 Ah 24 V sealed nickel-cadmium battery being charged within 4 h. The aluminum/air battery can recharge this size nickel-cadmium battery about seven times before the aluminum is depleted. The specific energy of a dry battery, with enough metal for the anode and salt for the electrolyte to provide for a complete discharge, is about 600 Wh/kg.

FIGURE 33.32 Aluminum/air field recharger. 600 Wh, 6 V. (*Courtesy of Alupower, Inc.*)

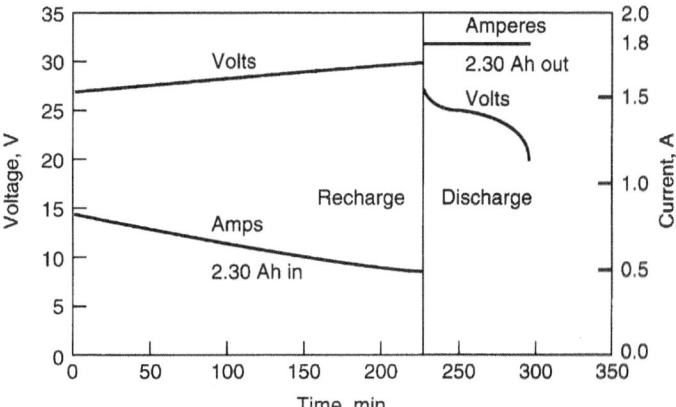

FIGURE 33.33 Charge and discharge of a nickel-cadmium battery, aluminum/air field recharger. (*Courtesy of Alupower, Inc.*)

Ocean Power Supplies. Batteries based on the use of oxygen dissolved in seawater have an advantage over others as all reactants, except for the anode material, are supplied by the seawater. In these batteries a cathode, which is open to the ocean, is spaced around an anode so that the reaction products can fall out into the ocean.[37] Relatively large surface areas are required as there is not much oxygen in seawater. In addition, because of the conductivity of the ocean, there can be no series arrangement of cells. Higher voltages are obtained by the use of a DC-to-DC converter.

Many instruments and devices used in the ocean have to operate over long periods of time, and aluminum is a candidate for the anode for missions requiring months or years of service.

Figure 33.34 shows a flat-plate aluminum/dissolved oxygen battery.[38] The battery, about 1.5 m high, has a dry specific energy of 500 Wh/kg and can operate at power densities of up to 1 W/m^2. This battery can be installed beneath a buoy, as shown in the illustration, and used with a DC-to-DC converter to charge a lead-acid battery.

FIGURE 33.34 Aluminum dissolved oxygen flat-plate battery (attached to Woods Hole Oceanographic Institution buoy.) (*Courtesy of Alupower, Inc.*)

33.4.2 Aluminum/Air Cells in Alkaline Electrolytes

The concept of operating aluminum/air batteries at high energy and power densities was described in the early 1970s, but successful commercialization was impeded because of technological limitations, including the high open-circuit corrosion rate of aluminum alloys in alkaline electrolyte, the nonavailability of thin, large-dimension air cathodes, and the difficulty of handling and removing the cell reaction products (precipitated aluminum hydroxide) to prevent cell clogging.

Significant advances have been made in reducing the corrosion of aluminum alloys in caustic electrolytes.[39,40] An aluminum anode, containing magnesium and tin, has approximately two orders of magnitude reduction in corrosion current at open-circuit and operates at greater than 98% coulombic efficiency over a wide range of current densities. Even at open-circuit, the alloy can remain in a caustic electrolyte with a relatively low rate of self-discharge. Prior to this development, the amount of hydrogen and heat evolved during open-circuit stand was usually so high as to require that the electrolyte be drained from the alloy during this period to prevent it from boiling.

Techniques to handle the aluminum so that it can be continuously fed as chips or pellets into the electrochemical system have also been developed.[41] One design uses aluminum particles having diameters of 1 to 5 mm.[42] The electrode is a pocket whose walls are composed of cadmium-plated expanded steel screen. The electrode is fed by a system that keeps the cell maintained with an aluminum particulate at an optimum level. Figure 33.35 shows the performance of a cell operating at 50°C using 8N KOH-containing stannate. The battery, with an electrode area of 360 cm^2, was able to deliver a current of 56 A at 1.35 V for 110 h, with the automatic addition of aluminum every 20 min.

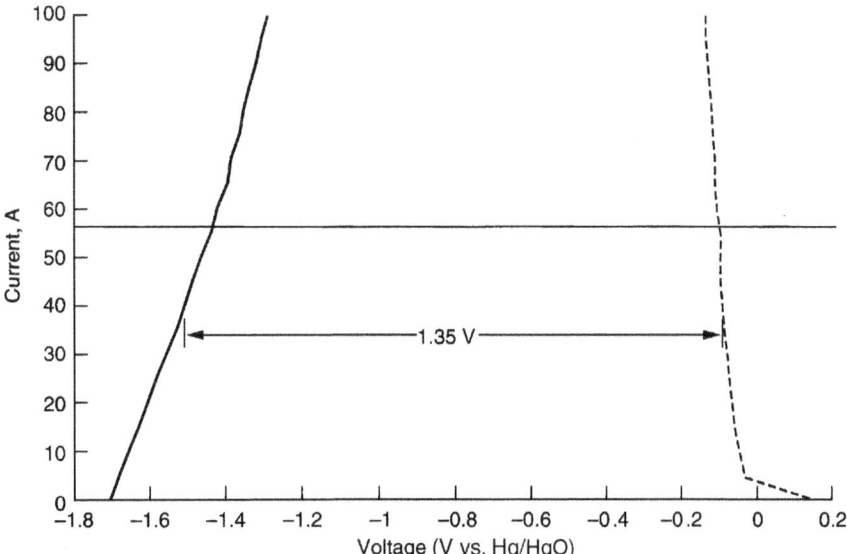

FIGURE 33.35 Polarization curves of particulate-feed aluminum/air battery. ---, positive electrode; ——, negative electrode. (*Courtesy of Sorapec.*)

Management of the electrolyte to remove the reaction product from it is required as the conductivity of the electrolyte decreases with increasing aluminate concentration. As shown in Fig. 33.36, the voltage decreases if the aluminate is not removed. Several techniques have been developed to remove the reaction products and are discussed later in this section.

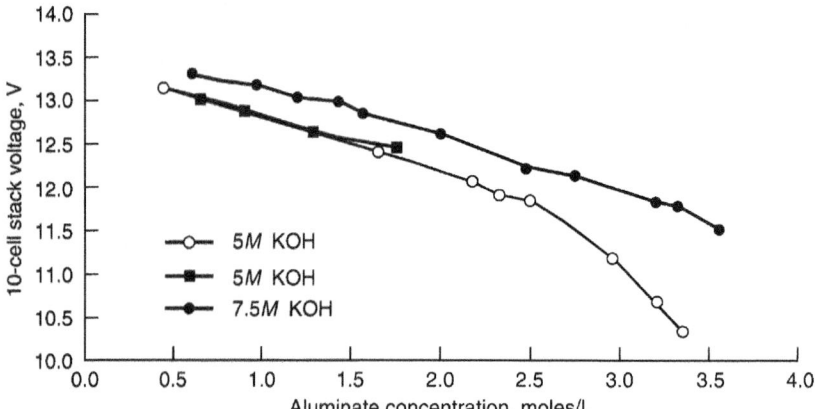

FIGURE 33.36 Voltage vs. aluminate concentration. (*Courtesy of Eltech Systems.*)

Applications of Alkaline Aluminum/Air Batteries. The alkaline aluminum/air batteries being developed cover a wide range of applications from emergency power supplies to field-portable batteries for remote power applications and underwater vehicles. Most of these are designed as reserve batteries, which are activated before use, or "mechanically" recharged by replacing the exhausted aluminum anodes.

Reserve Power Units, Standby Battery.[43,44] This is a reserve battery, used with conventional lead-acid batteries, to provide a standby power supply with extended service life. The aluminum/air battery is about one-tenth the weight and one-seventh the volume of a lead-acid battery containing the same energy. The basic elements of the power supply design are shown in Fig. 33.37. The aluminum/air battery consists of an upper cell stack, a lower electrolyte reservoir (which is isolated from the stack during periods of nonuse), and auxiliary systems to pump and cool the electrolyte and circulate air through the battery. The electrolyte is an 8 M solution of potassium hydroxide containing a stannate additive. During operation, the electrolyte becomes increasingly saturated and then supersaturated with potassium aluminate. Eventually the conductivity of the electrolyte decreases to the point where the battery is unable to sustain the load. At that point, it has reached the end of its capacity based on total electrolyte volume (VLD). The electrolyte can be changed at this point, and the discharge continued to the point where the aluminum in the anode is exhausted (ALD). Figure 33.38 shows the performance for a nominal 1200 W battery discharged in the two modes. Operation in the electrolyte-capacity limited mode will yield a total discharge time of 36 h, while operation to anode exhaustion, incorporating one electrolyte change, will result in a total discharge time of 48 h. The overall energy density and specific energy are greater than 150 Wh/L and 250 Wh/kg.

The control system for this power unit is arranged so that in the event of a power outage, the lead-acid batteries provide the backup for the first 1 to 3 h. As the voltage of the lead-acid battery begins to fall, the aluminum/air battery is activated by a controller which initiates the pumping of electrolyte from a reservoir through the aluminum/air cell stack. Once the aluminum/air battery reaches full power (about 15 min from activation), it provides full power to the load and recharges the depleted lead-acid batteries. The electrical characteristics of the unit are shown in Figure 33.39. The aluminum/air battery has limited restart capability but can be refurbished by replacing the cell stack and the electrolyte.

Battlefield Power Unit. This power source, called the Special Operations Forces Aluminum Air (SOFAL) battery,[45] was developed as a reserve system to support specialized military communications equipment in covert field operations. The SOFAL weighs approximately 7.3 kg after activation and powers 12 and 24 VDC equipment with pulse currents up to 10 A, continuous drains up to 4 A, and has a design capacity of 120 Ah. To minimize weight, this battery is carried to the field dry and can be activated with any source of water.

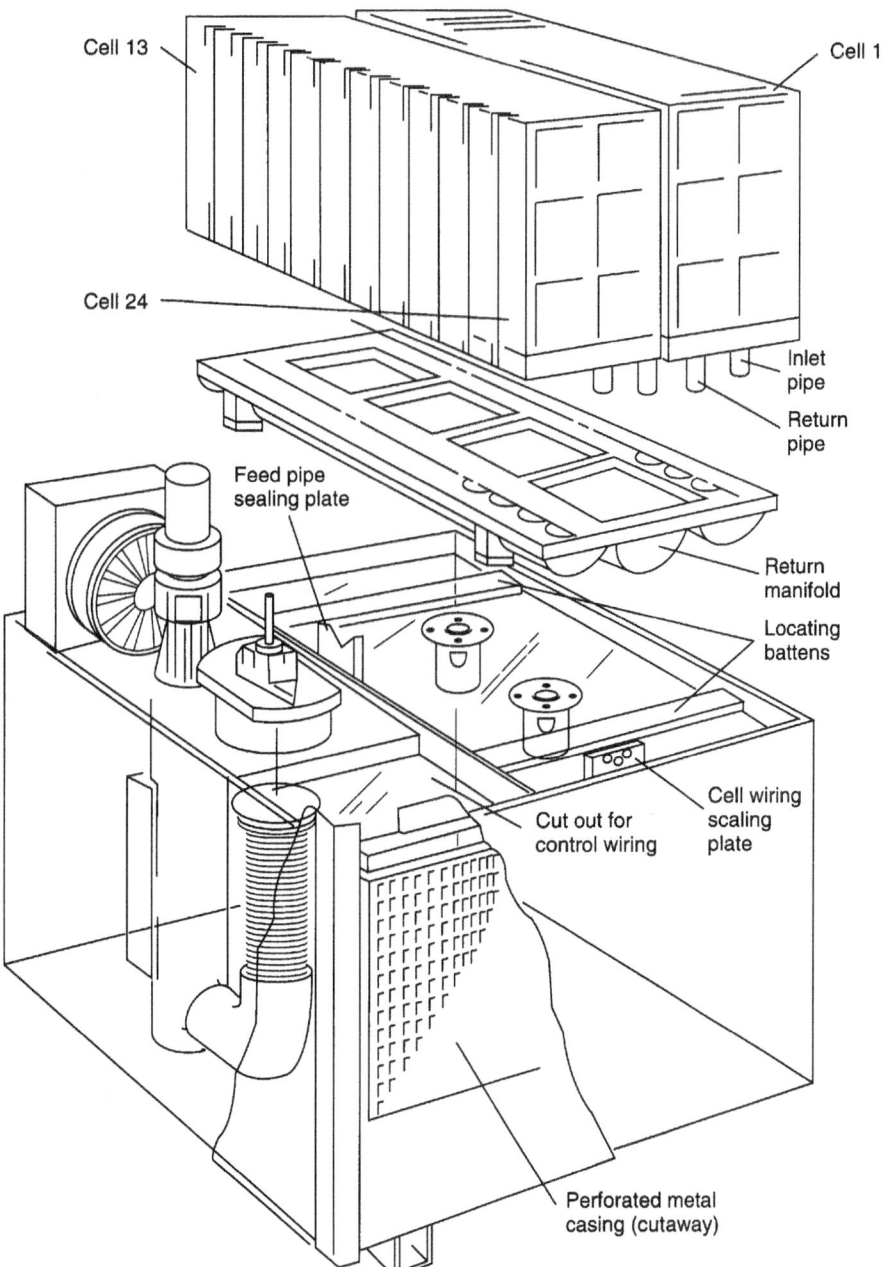

FIGURE 33.37 Aluminum/air reserve power unit, standby battery. (*Courtesy of Alupower, Inc.*)

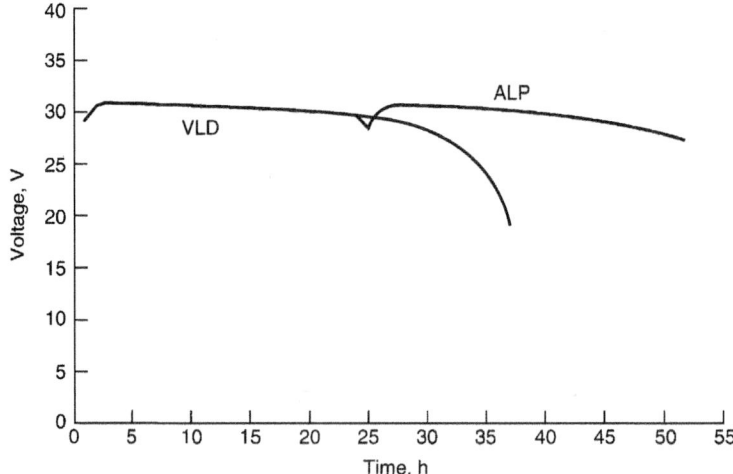

FIGURE 33.38 Comparison of volume (VLD) vs. anode (ALD) limited discharges of aluminum/air reserve power unit. (*Courtesy of Alupower, Inc.*)

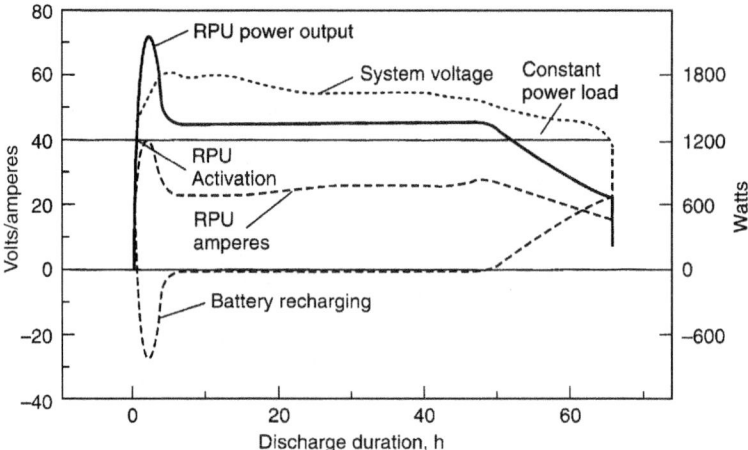

FIGURE 33.39 Discharge profile of 6 kW aluminum/air reserve power unit. (Direct connection to power system.) (*Courtesy of Alupower, Inc.*)

The SOFAL unit consists of 16 series-connected cells (Fig. 33.40a) with intercell connections provided by a printed circuit board. The cell stack, which weighs 3.5 kg dry, is shown in Fig. 33.40b. Activation of the system is accomplished with 2.5 L of water through a manifold system to each cell where it dissolves a cast block of stannated potassium hydroxide giving a 30% (w/w) KOH solution. After activation, each cell has an open-circuit voltage of 1.7 V or 27.2 V for the cell stack. Electrochemistry of the battery is similar to that described earlier. Dissolution of the KOH and corrosion of the aluminum provide heat to operate the system even at low temperature. The unit has been designed to access 1.6 L/min of air, which provides for low-power operation. If ambient air flow is insufficient, a small fan, activated by the battery, will provide the required airflow and will dissipate excess heat during high-power use. The SOFAL unit provides up to two weeks of service after activation.

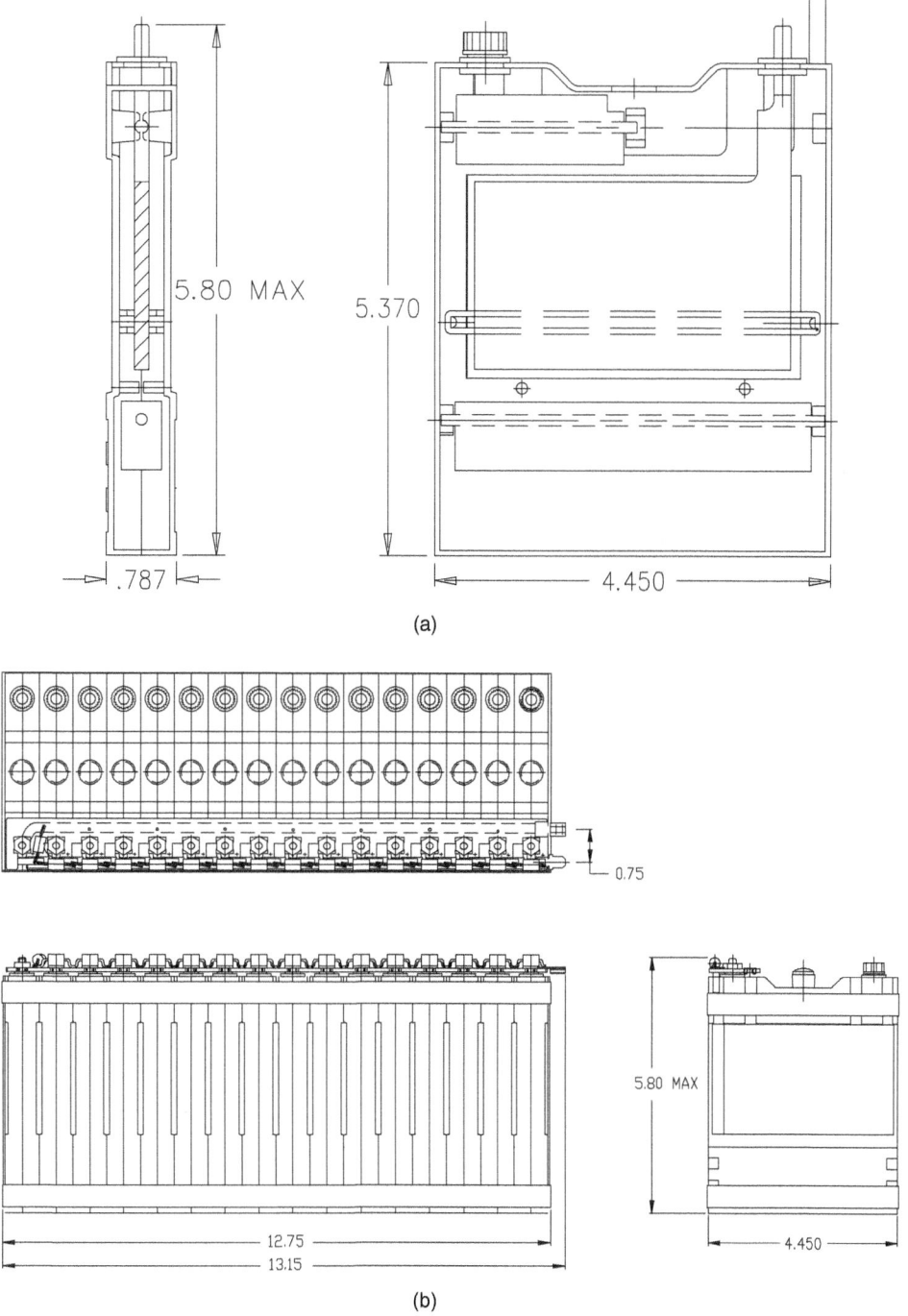

FIGURE 33.40 SOFAL battery. (a) Cell design configuration, (b) Battery block design configuration, (c) Full-scale design layout.

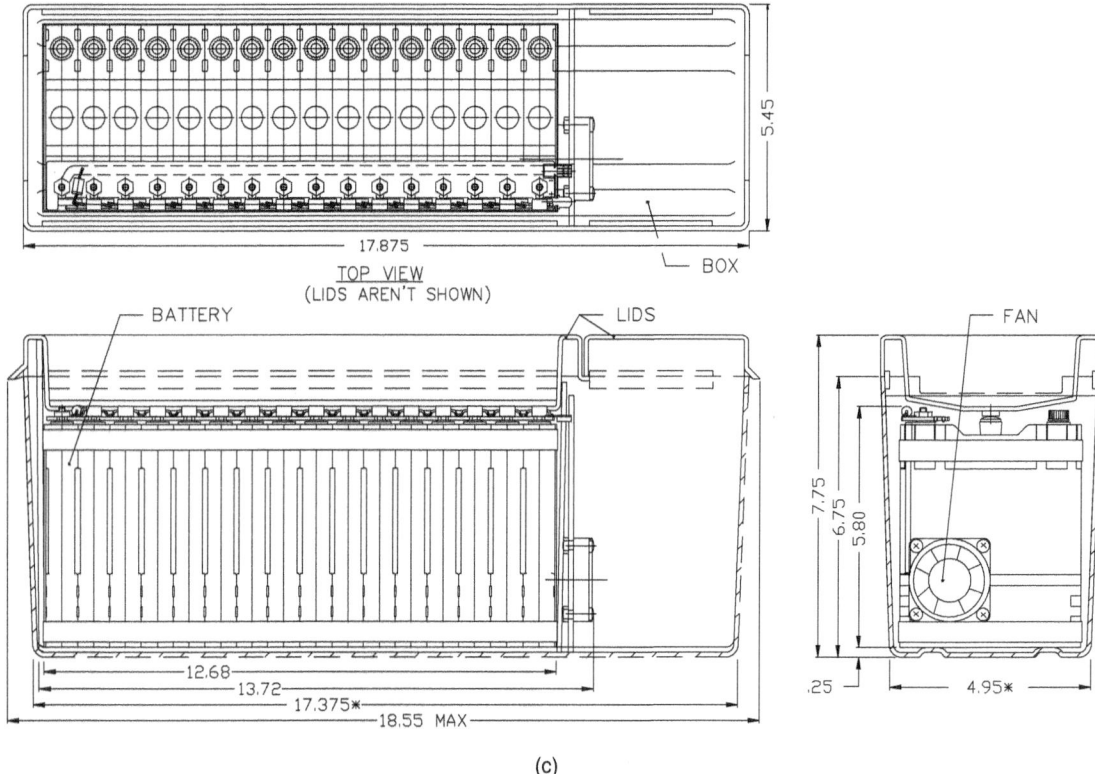

(c)

FIGURE 33.40 (*Continued*)

The Electronics Module Package (EMP) shown in Fig. 33.40c contains the electronic components, provides the mounting for an internal rechargeable battery, and houses the fan. The electronics package contains the power management circuitry to keep the internal secondary battery fully charged, provides both 24 V and a regulated 12 V output, and can be used to power electronics directly or recharge external batteries. Figure 33.41 shows the discharge profile of the SOFAL battery on a 2 A continuous load on 24 V operation. Figure 33.42 shows the discharge curves for two cells from the SOFAL battery. Cell no. 1 was activated with 8 molar KOH, while cell no. 2 employed KOH pellets in the cell and was activated with water. Both cells were discharged at 0.5 A, and provided 135 Ah capacity, but cell no. 1 operated at slightly higher voltage, particularly at the end of discharge.

Following discharge, the cell stack can be replaced to provide a new battlefield power unit.

Underwater Propulsion. Another area of application for alkaline aluminum batteries is in self-contained extended-duration power supplies for underwater vehicles such as unmanned vehicles for submarine and mine surveillance, long-range torpedoes, swimmer delivery vehicles, and submarine auxiliary power.[46,47] In these applications, the oxygen can be carried in pressurized or cryogenic containers, or it can be obtained from the decomposition of hydrogen peroxide or from oxygen candles. The aluminum/oxygen system can produce almost twice as much energy per kilogram of oxygen as a hydrogen/oxygen fuel cell, as the operating voltage of 1.2 to 1.4 V is almost twice that of the fuel cell. One type of aluminum/oxygen battery for the propulsion of underwater vehicles is shown in Fig. 33.43, and its characteristics are listed in Table 33.11.

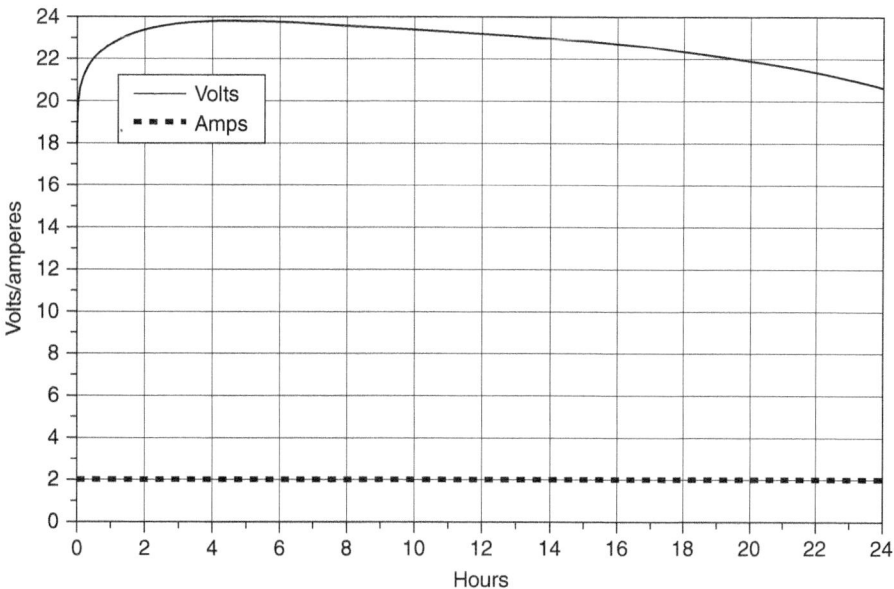

FIGURE 33.41 Sixteen cell SOFAL battery. Constant current 2.0 A discharge.

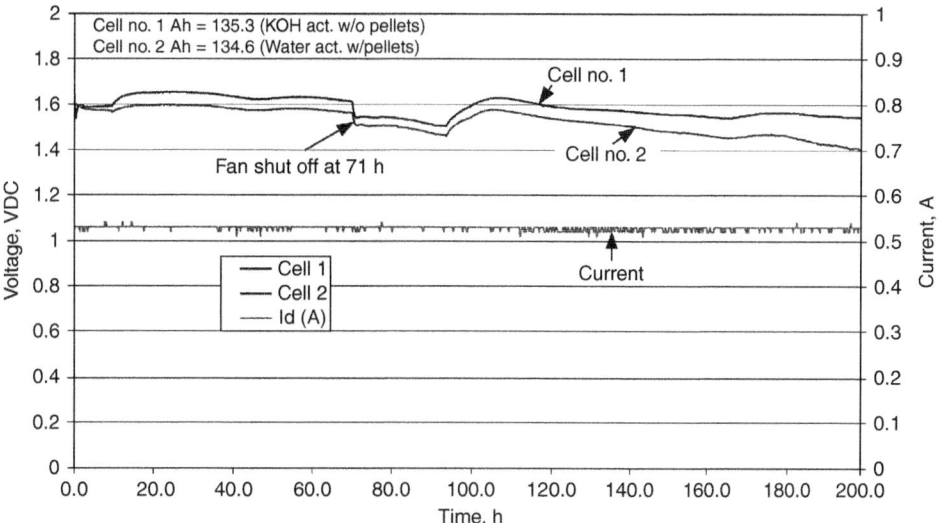

FIGURE 33.42 SOFAL test cells on 0.5 A resistive discharge.

This battery uses a "self-managing" electrolyte system, where the required electrolyte circulation and precipitation take place within the cell chamber, without the requirement for external pumps. This has the advantage of allowing the design of battery systems without electrolyte circulation pumps. There are no shunt currents between cells as each cell is independent, and there are no electrolyte paths between cells. In addition, the cells can be conformal with the system they are designed

33.40 SPECIALIZED BATTERY SYSTEMS

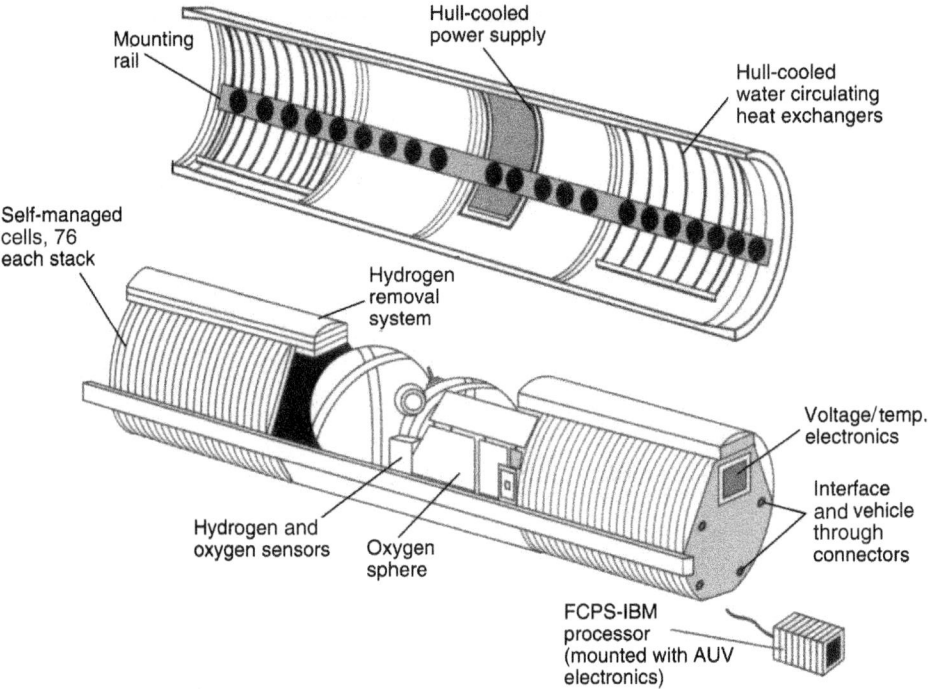

FIGURE 33.43 Aluminum/oxygen power system. (*Courtesy of Alupower, Inc.*)

TABLE 33.11 Characteristics of Aluminum/Oxygen Battery

Performance:	
Power	2.5 kW
Capacity	100 kWh
Voltage	120 V nominal
Endurance	40 h at full power
Fuel	25 kg aluminum anodes
Oxidant	22 kg oxygen at 4000 lb/in^2
Buoyancy	Neutral, including aluminum hull section
Time to refuel	3 h
Dimensions:	
Mass	360 kg
Battery diameter	470 mm
Hull diameter	533 mm
System length	2235 mm
Performance:	
Energy density	265 Wh/L
Specific energy	265 Wh/kg

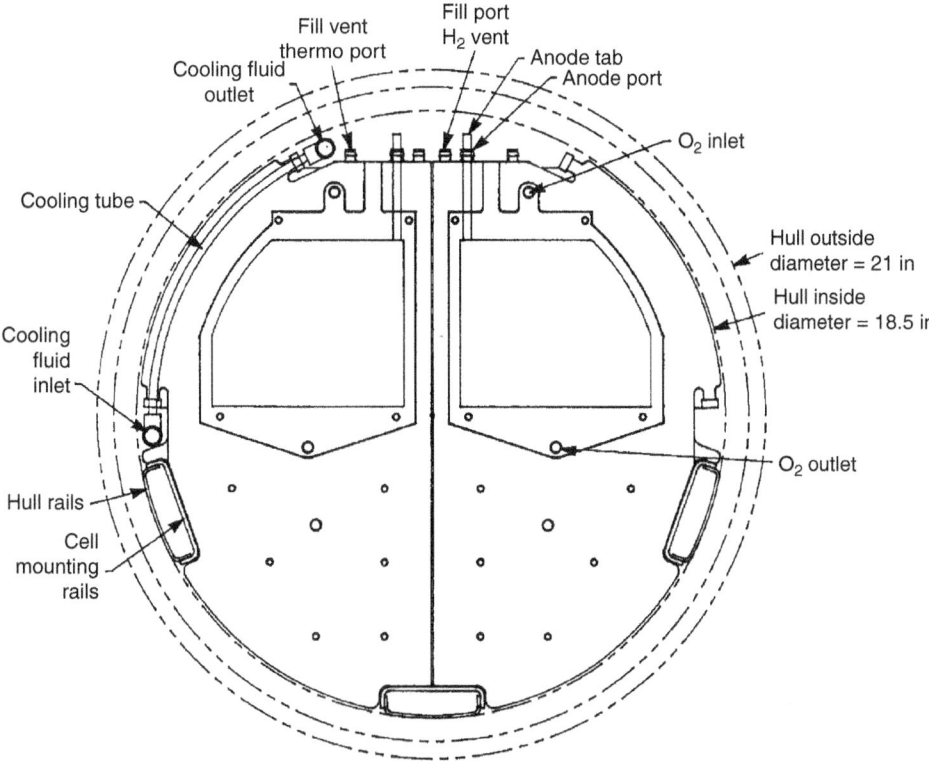

FIGURE 33.44 Self-managing cell system. (*Courtesy of Alupower.*)

to power. Figure 33.44 shows the design of a system using 19-in diameter (48.25 cm) cells. Each cell is about 0.5 in (1.25 cm) thick. Thermal and concentration gradients and the resulting convection currents within the cell precipitate the reaction product to the bottom. With this type of system, it is possible to utilize about 0.8 Ah/cm^3 of electrolyte. Figure 33.45 gives a discharge curve for a cell which is exactly one-half the cell shown in Fig. 33.44. The cell was divided down the center to provide for redundancy. The discharge was carried out at a constant power level of 18 W and a current density of about 50 mA/cm^2. The figure shows that the voltage remains relatively constant, between 1.4 and 1.5 V, for most of the run.

To maximize the capacity, the amounts of aluminum and electrolyte are matched so that at the end of discharge the aluminum is consumed and the electrolyte is completely filled with reaction product. Under this condition, the module is either discarded or rebuilt after use. Alternatively, a higher concentration of electrolyte can be used and discharge stopped before the onset of precipitation. In this mode of operation, the amount of aluminum incorporated into the cell can be sufficient for several runs with only the electrolyte being replaced between runs.

Another requirement of the underwater power system is the hydrogen-removal system, which is needed to safely remove the hydrogen that is generated by corrosion of the anode. Catalytic recombination is especially attractive for applications where space and energy efficiency are needed.[48] The unit shown in Fig. 33.43 uses a hydrogen-removal system, but the amount of hydrogen generated is not excessive since a low-corrosion aluminum alloy is used.

Another approach to removing the reaction product is a filter/precipitator system,[48] as shown in Fig. 33.46. The aluminate concentration is controlled by pumping the electrolyte out of the cell

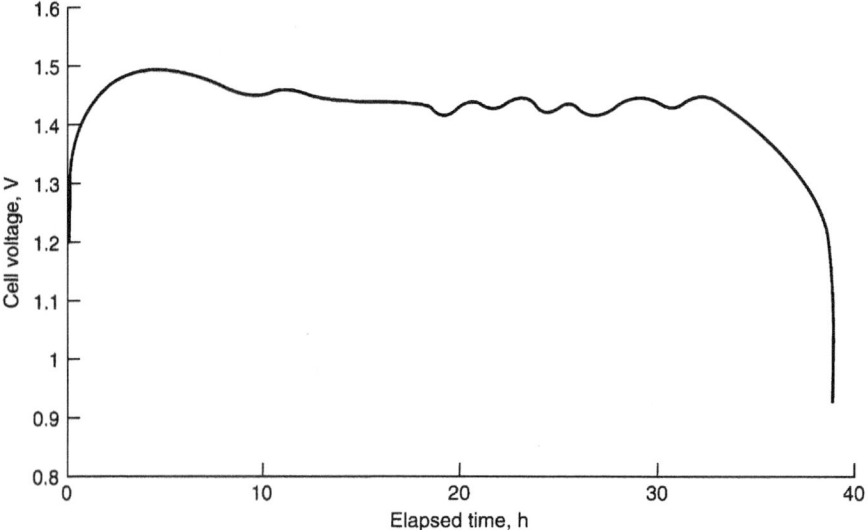

FIGURE 33.45 Discharge curve for self-managing cell. (*Courtesy of Alupower.*)

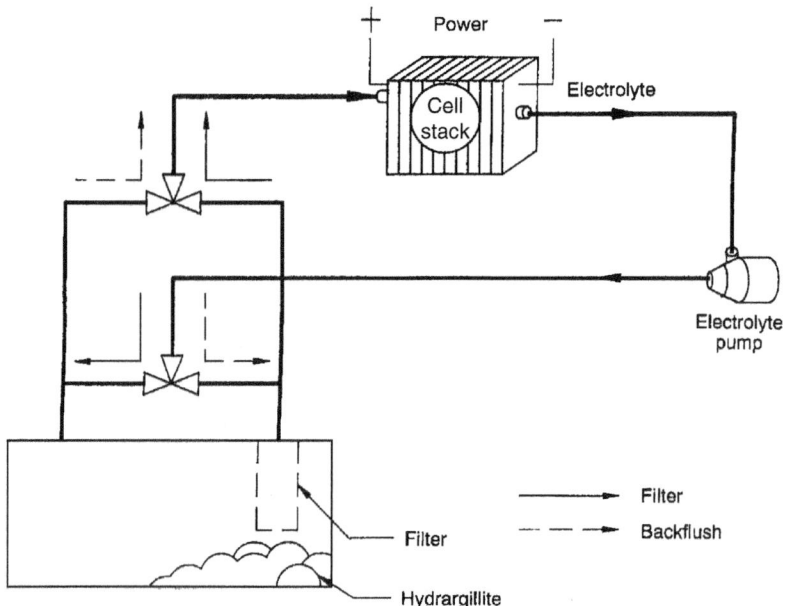

FIGURE 33.46 Conceptual design of filter/precipitator system when integrated with aluminum/oxygen battery. (*Courtesy of Eltech Systems.*)

stack and through the filter/precipitator. The filter promotes the growth of the aluminum trioxide and regeneration of KOH as follows:

$$KAl(OH)_4 \rightarrow KOH + Al(OH)_3$$

The crystal cake gradually increases in thickness with a subsequent increase in the pressure drop across the filter. When the pressure drop reaches a predetermined level, the cake is pulsed off the filter by backflushing (flow reversal) and collected in the bottom of the precipitate tank.

33.5 MAGNESIUM/AIR BATTERIES

The discharge reaction mechanisms of the magnesium/air battery are

$$\begin{aligned}\text{Anode:} \quad & Mg \rightarrow Mg^{+2} + 2e \\ \text{Cathode:} \quad & O_2 + 2H_2O + 4e \rightarrow 4(OH) \\ \text{Overall:} \quad & Mg + \tfrac{1}{2}O_2 + H_2O \rightarrow Mg(OH)_2\end{aligned}$$

The theoretical voltage of this reaction is 3.1 V, but in practice, the open-circuit voltage is about 1.6 V.

Magnesium anodes tend to react directly with the electrolyte with the formation of magnesium hydroxide and the generation of hydrogen

$$Mg + 2H_2O \rightarrow Mg(OH)_2 + H_2$$

This reaction stops in alkaline electrolytes because of the formation of an insoluble film of magnesium hydroxide on the electrode surface which prevents further reaction. Acid tends to dissolve the film. An important consequence of the film on magnesium electrodes (see also Chap. 10) is that there is a delayed response to an increase in the load because of the need to disrupt the film to create new bare surfaces for reaction. "Pure" magnesium anodes usually do not give good cell performance, and several magnesium alloys have been developed for use as anodes tailored to provide the desired characteristics.

Magnesium/air batteries have not been successfully commercialized, and an effort has been directed toward undersea applications using the dissolved oxygen in seawater as the reactant. The battery uses a magnesium alloy anode and a catalytic membrane cathode, and it is activated by the seawater. The main advantage of this system is that, with the exception of the magnesium, all of the reactants are supplied by the seawater. The battery can have a specific energy of about 700 Wh/kg.

The concentration of oxygen in seawater is only 0.3 mol/m^3, corresponding to 28 Ah per ton of seawater. Therefore the cathode must have an open structure to ensure that there is sufficient contact with the seawater. Further, as seawater is highly conductive, it is not feasible to use more than one cell. A DC-to-DC converter is used to increase the low cell voltage to the required voltage range.

Figure 33.47 illustrates a cell design for an undersea mission designed to deliver 3 to 4 W for one year or longer at a total weight of 32 kg. In this design, the oxygen-reduction cathode is positioned on the circumference of a cylinder with a total cathode area of 3 m^2. The anode is a 19 kg cylinder of magnesium, located internal to the air cathode. The weight of the cathode is about 1.8 kg, and the remainder of the weight is for support structure and other necessary hardware. The single-cell battery has a long shelf life in its dry, unactivated state. It is immediately active on deployment when it is immersed in the seawater electrolyte. Figure 33.48 shows the discharge of a test battery with periodic voltage spikes when the load is increased.[49]

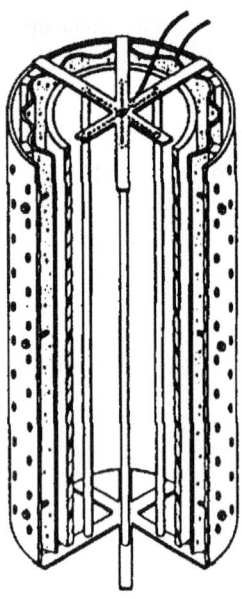

FIGURE 33.47 Schematic of concentric cylinder configuration of seawater cell. The output cylinder is comprised of porous fiberglass coated with an antifoulant. The corrugated structure is the air cathode, which is exterior to the magnesium anode. The entire structure is open to the seawater electrolyte. (*Courtesy of Westinghouse Corp.*)

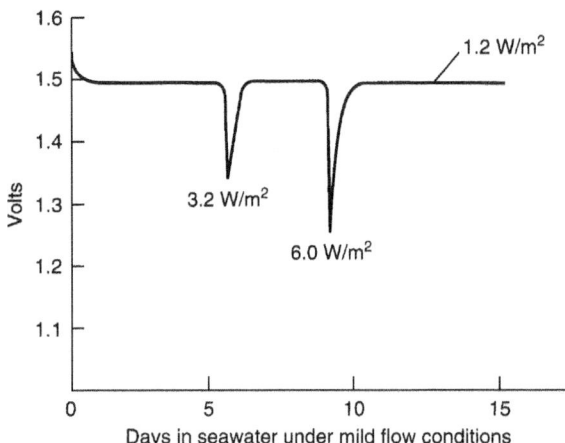

FIGURE 33.48 Discharge profile of seawater power source at 80 μA, 20°C. Current spikes represent charging of a small silver-iron battery. Neutral pH was maintained by periodic addition of hydrochloric acid. (*Courtesy of Westinghouse Corp.*)

33.6 LITHIUM/AIR BATTERIES

33.6.1 Background

Lithium/air batteries consist of lithium anodes electrochemically coupled to atmospheric oxygen through an air cathode. Lithium/air batteries are also referred to as lithium/oxygen batteries. These batteries are related to lithium/water batteries, which will be discussed later in this section. For lithium/air batteries, oxygen gas introduced into the battery through an air cathode is essentially an unlimited cathode reactant source. Theoretically, with oxygen as a cathode reactant, the capacity of the battery is limited by the Li anode. The theoretical specific energy of the Li/oxygen cell is 13.0 kWh/kg, the highest for a metal/air battery. In addition to this very high specific energy, the Li/air battery offers a flat discharge voltage profile, environmental friendliness, and long storage life. Disadvantages, which depend on the cell design, include reliance on the environment, drying out, low discharge rates, and safety concerns. In the past, lithium/air batteries used solely alkali aqueous electrolytes.[50] These alkali aqueous cells suffered from parasitic corrosion of the lithium metal anode and cell failures. Cell designs utilizing a nonaqueous electrolyte and/or a protected lithium anode alleviate the parasitic corrosion reactions of the Li anode that plagued past lithium/air batteries based on aqueous alkaline electrolytes. The nonaqueous electrolyte-based cell design also overcomes safety concerns of the Li/air system. Most advancements over the last few years have focused on nonaqueous systems for both primary and secondary lithium/air chemistries.

Lithium/air batteries consist of lithium (Li) anodes electrochemically coupled to atmospheric oxygen through an air cathode. Oxygen gas (O_2) introduced into the battery through the air cathode is essentially an unlimited cathode reactant source. Because of this, the air cathode is the most important component of the system. In the nonaqueous lithium/air system, the lithium metal reacts with O_2 to produce electricity according to the following reactions:

Discharge:

$$4Li \rightarrow 4Li^+ + 4e^- \quad \text{(lithium electrode) (anode)}$$
$$O_2 + 4e^- \rightarrow 2O^{2-} \quad \text{(gas electrode) (cathode)}$$
$$4Li + O_2 \rightarrow 2Li_2O \quad \text{(cell)} \quad E^\circ = 2.91 \text{ V}$$
$$2Li + O_2 \rightarrow Li_2O_2 \quad \text{(cell)} \quad E^\circ = 3.10 \text{ V}$$

The system and cells are capable of being recharged. Electricity is applied to the cell to convert the lithium oxide (stored in the cathode) back to lithium metal and oxygen gas. The reactions involved in recharging the cells are as follows:

Charge:

$$4Li^+ + 4e^- \rightarrow 4Li \quad \text{(lithium electrode) [cathode (reduction)]}$$
$$2O^{2-} \rightarrow O_2 + 4e^- \quad \text{(gas electrode) [anode (oxidation)]}$$
$$2Li_2O \rightarrow 4Li + O_2 \quad \text{(cell)}$$
$$Li_2O_2 \rightarrow 2Li + O_2 \quad \text{(cell)}$$

Theoretically, with unlimited oxygen, the capacity of the battery is limited by the Li anode. The theoretical specific energy of the Li/oxygen cell, as shown with the above reactions, is 13 kWh/kg (excluding oxygen), the highest for a metal/air battery. In addition the Li/air battery offers lightweight components, rechargeability, and potentially low cost components. This calculation does not include the need to store the reaction product(s) within the cell.

33.6.2 Anodes

The anodes for lithium/air batteries are typically lithium metal on a current collector. This type of electrode construction is simple and works well for many applications. A more sophisticated design is the same, but incorporates a protective layer. This protective layer is often a ceramic or glass lithium-ion conductor. The first patented ceramic and glass protective layer for lithium metal anodes opened a new field within lithium/air cells (Fig. 33.49). Examples of other work that followed included the use of LiSICON,[51] LiPON,[52] LATP,[52] and LiGC as solid electrolytes,[53] Many protected lithium anodes are constructed with a special interlayer between the lithium metal and the ion conductor. One of the first protected lithium electrodes is composed of LMP (a solid ion conducting layer), a special interlayer, and lithium metal (see Figs. 33.49, 33.50, and 33.51).[54] This protected lithium anode is stable in both aqueous and nonaqueous systems and has been used successfully in a variety of cell types. The protected lithium anode was built into a lithium/air cell with aqueous

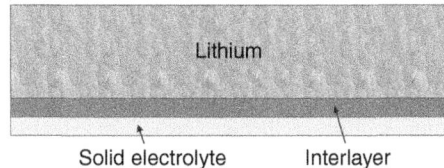

FIGURE 33.49 Cross-sectional illustration of protected lithium electrode: lithium/interlayer/solid electrolyte: U.S. Patents 7,282,295, 7,282,296, 7,282,302, 7,390,591, and 7,491,458. (*Courtesy of PolyPlus Battery Company.*)

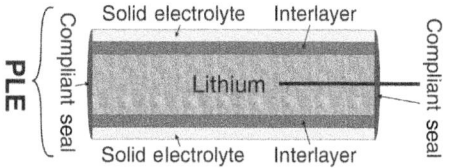

FIGURE 33.50 Cross-sectional view of protected lithium electrode having a compliant seal; patent pending. (*Courtesy of PolyPlus Battery Company.*)

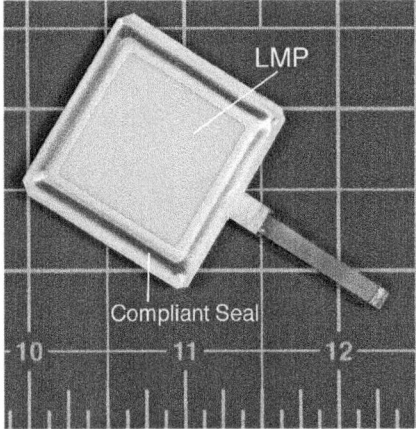

FIGURE 33.51 Fully functional protected lithium electrode; lithium electrode is stable to a broad range of protic and aprotic solvents including water; 2400 Wh/kg with a 2.8 V cathode, patent pending. (*Courtesy of PolyPlus Battery Company.*)

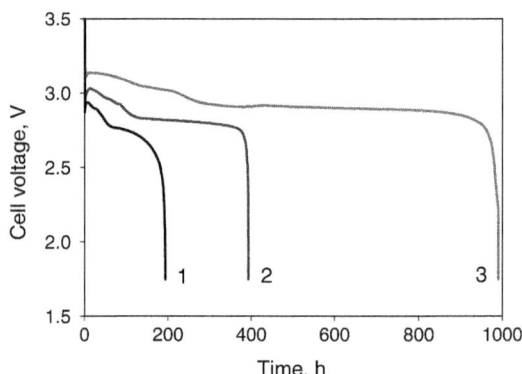

FIGURE 33.52 The protected lithium anode was built into the lithium/air cells and discharged in aqueous electrolyte at three current densities: (1) 1.0 mA/cm^2, (2) 0.5 mA/cm^2, and (3) 0.2 mA/cm^2. The specific energy of these cells is about 800 Wh/kg (includes all components except the external battery case). (*Courtesy of PolyPlus Battery Company.*)

electrolyte and successfully discharged at three different current densities (Fig. 33.52). Many of these protected lithium anodes can be used for the lithium/air chemistry and also lithium/water and even lithium-ion systems.

33.6.3 Electrolytes and Separators

The majority of recent research on electrolytes for lithium/air batteries has advanced the use of nonaqueous electrolytes. This was first done with lithium salts in organic carbonates and conductive polymer membranes.[55] The most popular salts and solvents used to make electrolytes for lithium-air cells are listed in Table 33.12. Many of the electrolytes used for lithium/air are similar to the salts and solvents used in lithium-ion chemistry (see Chaps. 7 and 26). Electrolyte work has also been done recently on aqueous systems,[54,56,57] and ionic liquids.[58,59]

Separators used for lithium/air cells are typically polyolefins such as Setela® and Celgard®. Glass fibers and solid ion conducting membranes are also used.

TABLE 33.12 The Most Commonly Used Salts and Solvents for Lithium-Air Cells

Salts	Solvents
LiPF$_6$	Propylene carbonate (PC)
LiBF$_4$	1,2-Dimethoxyethane (DME)
LiCF$_3$SO$_3$	Ethylene carbonate (EC)
LiN(SO$_2$CF$_3$)$_2$	Diethyl carbonate (DEC)
LiClO$_4$	Dimethylcarbonate (DMC)

33.6.4 Cathodes

The limiting factor in nearly all metal/air batteries is the air cathode (also known as an oxygen cathode).[55] The slow rate of the chemical reaction is the diffusion of oxygen into the cell through the air cathode. The performance of the lithium/air battery has also been limited by the air cathode.

A typical cathode preparation entails depositing carbon, binder, and catalyst on a metal current collector. This process can be done by coating, impregnating, or pressing. Additionally a substrate may be used to increase surface area, or an atmospheric membrane could be placed on top of the cathode to protect it from the environment. This produces an air cathode suitable for laboratory testing or actual field use.

A commercially available air cathode that is carbon based has a double-sided electrode.[60,61,62] This electrode consists of two carbon layers sandwiched around a current collector and then covered with a PTFE film. The carbon layers contain the high surface carbon and the metal catalysts. Catalysts incorporated into the carbon electrode enhance the oxygen reduction kinetics and increase the specific capacity of the cathode. The importance of catalysts can be seen in Fig. 33.53. Several cathodes were constructed with different metal catalysts such as silver, platinum, and ruthenium. Oxides such as manganese, cobalt, and a manganese-cobalt mixture also provide catalytic activity, as seen in Fig. 33.53. The PTFE film acts as an atmospheric water barrier. Preventing water from entering the battery increases safety and performance.

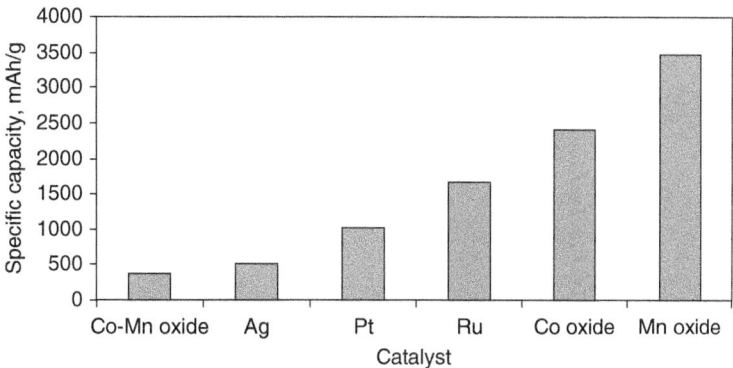

FIGURE 33.53 The specific capacities of lithium/air pouch cells utilizing various catalysts. Specific capacity is based on grams of carbon in the air cathode, which is customary. The discharge current was 1.0 mA and corresponds to 0.1 mA/cm². (*Courtesy of Yardney Technical Products, Inc.*)

33.6.5 Cell Designs and Performance

The first nonaqueous lithium/air battery using polymer electrolytes was reported by Abraham and Jiang.[55] It comprised a lithium-ion conductive polymer electrolyte membrane between a thin Li metal foil anode and a thin carbon composite on which oxygen, the electro-active cathode material, is reduced during discharge. Fig. 33.54 shows the cell structure, which is encapsulated in a metallized plastic envelope with pores on the cathode surface which are covered by tape prior to activation. The cathode is composed of 20 wt % acetylene black (or graphite powder) and 80 wt % polymer electrolyte catalyzed with cobalt phthalocyanine, in some cases, and pressed on a Ni or Al screen current collector. The electrolyte was composed of 12% polyacrylonitrile (PAN), 40% ethylene carbonate, 40% propylene carbonate, and 8% $LiPF_6$, all by weight, formed into a film 75 to 100 microns thick. The lithium electrode was 50 microns thick. Figure 33.55 shows an intermittent discharge curve for the Li/PAN-based electrolyte/O_2 battery at a current density of 0.1 mA/cm, using an acetylene black cathode in an atmosphere of flowing oxygen. The capacity was found to be proportional to the carbon weight. This battery exhibited an open circuit voltage (OCV) of 2.85 V prior to discharge. The OCV remained steady during intermittent discharge, indicating a two-phase equilibrium at the electrode surface. Raman spectra of the reaction product absorbed on the electrode surface showed it was Li_2O_2 and that the following reaction was occurring during discharge:

$$2Li + O_2 \rightarrow Li_2O_2 \qquad E = 3.10 \text{ V}$$

It was also determined that the absorbed lithium peroxide on a catalyzed electrode could be reoxidized to oxygen. Figure 33.56 shows the first discharge and recharge, followed by the second discharge of one battery. Although this system is of considerable technological interest, its active life

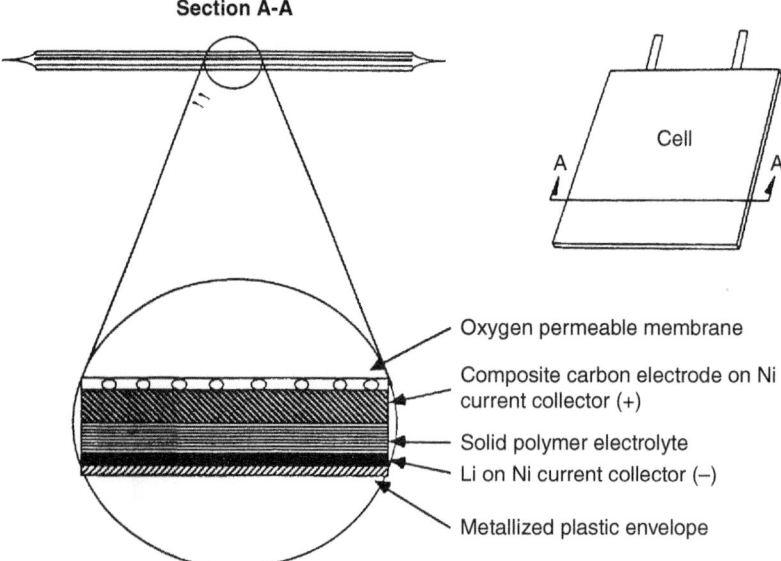

FIGURE 33.54 Lithium/oxygen battery with solid polymer electrolyte in metallized plastic envelope.

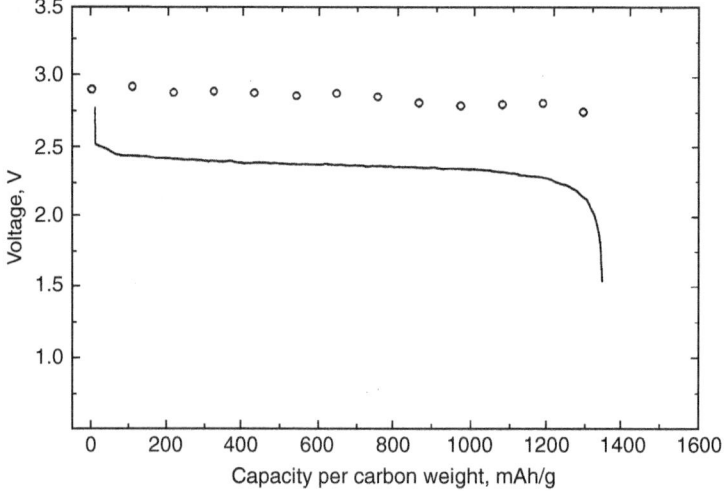

FIGURE 33.55 The intermittent discharge curve and the open-circuit voltages of a Li/PAN-based polymer electrolyte/oxygen battery at a current density of 0.1 mA/cm^2 at room temperature in an atmosphere of oxygen. The cathode contained acetylene black carbon. The cell was discharged in 1.5 h increments with an open-circuit stand of about 15 min between discharges. Open circles: OCV; solid line: load voltage.

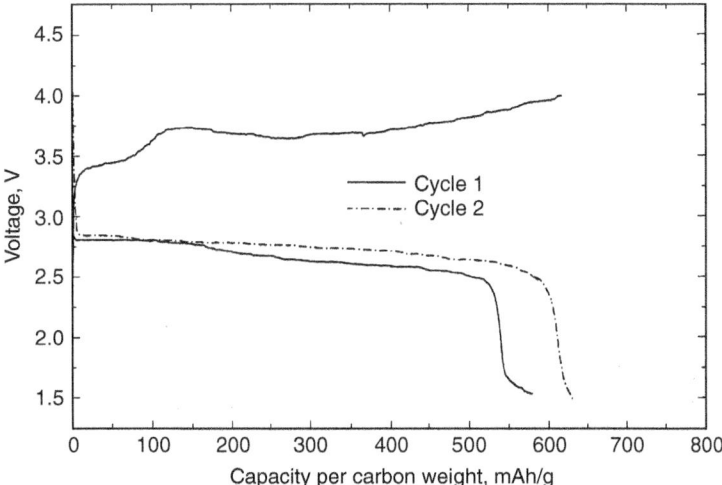

FIGURE 33.56 Cycling data for a Li/PAN-based polymer electrolyte/oxygen battery at room temperature in an atmosphere of oxygen. The cathode contained 20 w/o catalyzed Chevron carbon black and 80 w/o polymer electrolyte. The battery was discharged at 0.1 mA/cm² and charged at 0.05 mA/cm².

was limited by the diffusion of oxygen through the PAN electrolyte, where it reacted chemically with lithium. This started the interest in the nonaqueous lithium/air system and in other cell designs.

Lithium/air cells have been built using several types of cell construction techniques, such as using Swagelok® fittings,[63] pouch cell,[64] coin cell,[65] and plastic cases.[61,64] The most common construction to date is the pouch cell design due to its ease of fabrication and design adaptability. These cells are suitable for testing in different environments. The cell construction involves layering the cell components and sealing them in a plastic case resembling a packet. The lithium metal anode, separator, electrolyte, and carbon air cathodes are sealed inside the metallized plastic packaging material. A pouch cell used for testing is shown in Fig. 33.57. The anode is composed of a lithium metal foil, pressed into a nickel current collector with a nickel tab. The area of the anode is slightly larger than the 10 cm². The separator used was Setela®, which is a microporous polyolefin film. The liquid electrolyte used was 1 M $LiPF_6$ in 1:1:1 EC/ DEC/ DMC. (EC is ethylene carbonate, DEC is diethyl carbonate, and DMC is dimethyl carbonate). The air cathode is a carbon composite made by combining carbon, a metal catalyst, and a binder, deposited on a metal current collector. The binder used was PTFE. The air cathode structure is a layered composite with an increased capacity. A thin PTFE film between the air cathode and the atmosphere provides hydrophobicity to the cathode to repel atmospheric water and creates channels for oxygen diffusion. The cell was run in an oxygen atmosphere, at room temperature, and about 1 atmosphere of pressure. The lithium/air pouch cell had a capacity of 91 mAh with a relatively flat discharge profile (Fig. 33.58). Discharging further to 1.5 V resulted in 100 mAh total capacity. The corresponding energy yield is 246 mWh. With 0.028 g of carbon impregnated into the air cathode current collector, the specific capacity is 3137 mAh/g of carbon. The goal for this lithium/air cell was 3000 Wh/kg for a full cell.[66]

A large lithium/air cell was produced to test assembly procedures and manufacturing processes for nonaqueous lithium/air batteries.[64] The large prototype cell is composed of a two-piece molded plastic case (Fig. 33.59). The case is about 5 in square with "windows" on both sides for access to air. The two large air cathodes are stacked on both sides of the anode. This case was designed with many special features, including the ability to house both primary and secondary lithium/air systems.

Several groups have reported on the rechargeability of the lithium/air system.[54,55,67,68] Each group used a different air cathode, with specific catalysts to recharge the cell. The chemistry of each system was essentially the same, but recharging each of the cells was different due to the specific

FIGURE 33.57 A picture of a lithium/air pouch. The "window" toward the top is the air cathode that allows oxygen into the cell. The cell is about 3 in on each side. (*Courtesy of Yardney Technical Products, Inc.*)

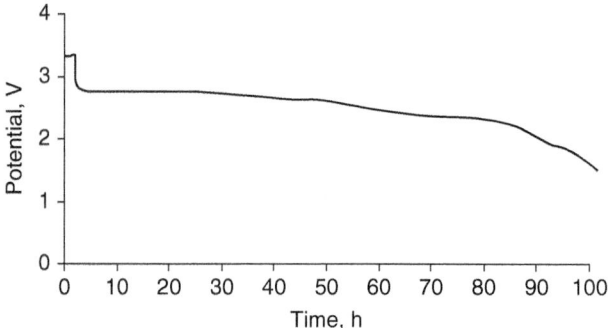

FIGURE 33.58 Discharge profile of Mn oxide catalyzed cathode in a lithium/air pouch cell at 0.1 mA/cm^2 shown with an initial open-circuit rest of 2 h. (*Courtesy of Yardney Technical Products, Inc.*)

nature of the air cathodes, catalysts, and cell designs. All groups were able to cycle the cells. Due to the length of discharge, which may take several days, most groups limited the cycling to a small number. This recent work may create rechargeable lithium/air cells with a high specific capacity and a long cycle life.

33.6.6 Battery Design

For lithium/air batteries, there has only been one publicly released battery design to date.[64] This battery utilizes a large cell design to make a 12 V battery. The cells are capable of fitting together to form a cell stack (Fig. 33.60). The most important design feature is air management. Air needs

FIGURE 33.59 This large lithium/air cell was designed with manufacturing features to allow the construction of primary or secondary cells. It has windows on both sides for two large air cathodes. One air cathode is seen in the image, with the other air cathode on the back side. (*Courtesy of Yardney Technical Products, Inc.*)

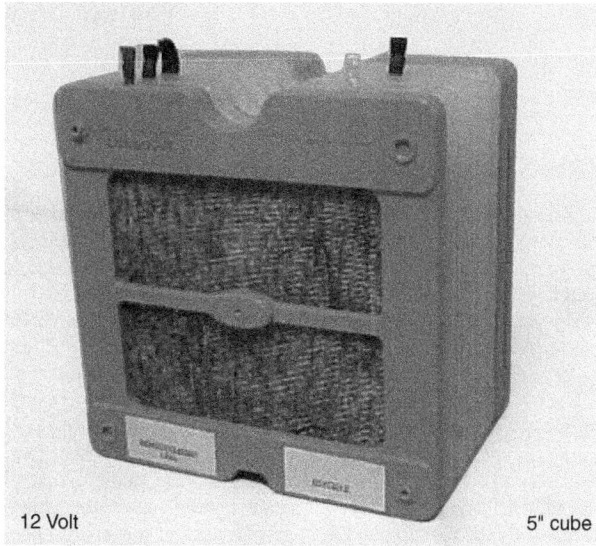

FIGURE 33.60 A 12 V lithium-air battery using a large cell design. This cell stack is cubic, with each side about 5 in. The battery was successfully discharged. (*Courtesy of Yardney Technical Products, Inc.*)

33.6.7 Lithium/Water Batteries

The high energy density of lithium is attractive as a reserve battery for undersea applications using its reaction with water.[50] In general, the combination of lithium with water may be considered hazardous because of the high heat of reaction. However, in the presence of hydroxyl (OH$^-$) ion at concentrations greater than 1.5 M, a protective film is formed which exists in a dynamic steady state. The film is pseudo-insulating, which permits the cathode to be pressed against it without causing a short circuit, thus reducing IR losses and concentration polarization and achieving high current outputs.

The major difficulty in the use of lithium in aqueous solution is the fact that it will not discharge efficiently at current densities less than about 0.2 to 0.4 kA/m^2, and the battery cannot be placed on open-circuit with electrolyte present within it. Also, under certain conditions the lithium will passivate. This feature can be used to advantage, however, to temporarily terminate reaction for standby with an electrolyte-filled battery.

The aqueous lithium systems are in principle quite simple, but in practice the electrolyte management subsystem and internal cell features involve sophisticated design and a level of complexity characteristic of fuel cells, requiring a reservoir, pump, heat exchanger, and controller, and using a flowing electrolyte. The electrodes are held in close juxtaposition, and frequently they are pressed together. A film on the lithium prevents short-circuiting, but permits high flux rates. The rate of discharge is inversely proportional to the concentration of electrolyte, and the output of the cell can be uniquely controlled by adjusting the molarity M of the LiOH produced at the anode. The voltage of the batteries is influenced more by the characteristics of the cathode than by the lithium anode.

The discharge reactions of the lithium-water battery are

Anode: $\quad$ Li $\rightarrow$ Li$^+$ + e

Cathode: $\quad$ H$_2$O + e $\rightarrow$ OH$^-$ + ½H$_2$

Overall: $\quad$ Li + H$_2$O $\rightarrow$ LiOH + ½H$_2$

Parasitic corrosion: $\quad$ Li + H$_2$O $\rightarrow$ LiOH + ½H$_2$

Precipitation of LiOH occurs as the monohydrate crystal,

$$\text{LiOH} + \text{H}_2\text{O} \rightarrow \text{LiOH·H}_2\text{O}$$

The overall electrochemical reaction has a thermodynamic potential of 2.21 V and a theoretical specific energy of 8530 Wh/kg based on lithium, which is the only reactant that has to be supplied with the battery. Dissolved oxygen is not necessary to depolarize the cathode as lithium possesses a high enough voltage to reduce water to hydrogen.

The parasitic corrosion reaction is highly undesirable as it produces no electric energy but consumes lithium. This highly exothermic reaction (−53.3 kcal/g·mol of lithium) can accelerate corrosion detrimentally. Efficient minimum-weight batteries require that this parasitic reaction be minimized.

The overall battery concept is shown in Fig. 33.61.[69] The neoprene bellows enable pressure equalization between the inside of the battery and the surrounding ocean. The bellows can also expand or contract to take up any changes in the cell volume with time. During operation, water is pumped into the battery to satisfy the water requirements. Due to the nature of the water-pumping concept, some hydroxide will be slowly lost to the surrounding seawater. However, hydroxide is continually being generated by the reaction and the rate of hydroxyl-ion loss will be slower than its rate of generation.

The discharge characteristics of a 2-month test of a prototype battery are shown in Fig. 33.62. The operating voltage on a discharge of 2.0 mA/cm^2 (equivalent to a 2 W discharge on a full-size battery) was between 1.4 and 1.43 V.

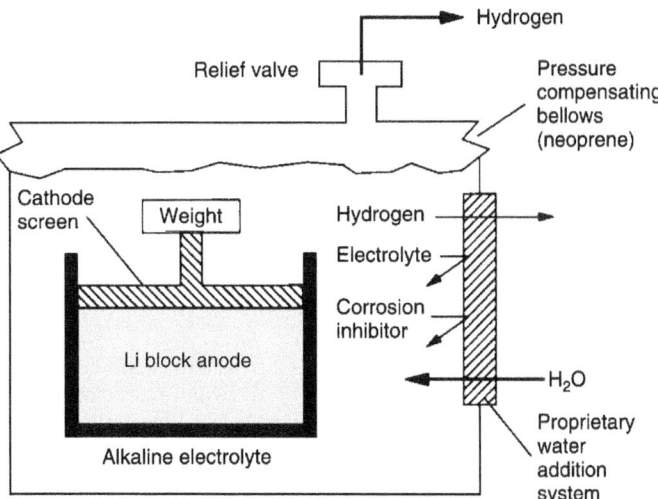

FIGURE 33.61 Low-rate lithium/water battery concept. (*From Shuster.*[69])

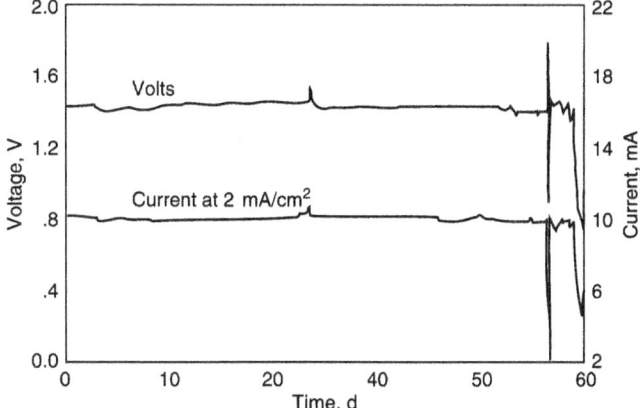

FIGURE 33.62 Typical performance of 28 cm diameter low-rate lithium/water test battery.

A typical lithium/water battery uses a 30 cm diameter, 30 cm thick solid cylindrical anode with a weight of approximately 11.5 kg. This battery is designed to deliver 2 W at 1.4 V for about a year with a specific energy of 1800 to 2400 Wh/kg.

For undersea applications, seawater can be used as a catholyte.[70] In this mode, the battery is immersed in a limitless supply of the electroactive species (water). This results in weight and volume savings and contributes significantly to the energy density and specific energy of this system.

The most significant challenge to the commercialization of Li/water batteries is the high rate and heat generation of the parasitic corrosion of the Li/water reaction. As a result, the development of high efficiency Li metal anodes for aqueous batteries has been the focus of Li/water battery research since the 1970s. Early work explored the use of concentrated aqueous hydroxide catholytes, typically KOH or NaOH.[71] Under the proper operating conditions, LiOH precipitates on the anode surface, forming a porous passivation film. Film thickness and porosity affect the anode corrosion rate and rate performance. Hydroxide ion concentration, Li^+ formation rate (current density), electrolyte temperature, and electrolyte flow rate are the operating parameters that control the properties of the anode film. Fine control of these variables is necessary to maintain the conditions for high Li coulombic efficiency. If the anode film is too thin, the corrosion rate is high, with the potential for thermal runaway. If it is too thick, current cannot be maintained or may stop. Once at the extremes, reacquiring the desired film properties and current density can be problematic. Prototype Li/water batteries designed for high[72] and low rate[73] operation were demonstrated using hydroxide catholytes.

Recently, glass ceramic electrolytes (GCEs) have been developed to provide a protective layer on the Li metal anode (see Sec. 33.6.1). This physically separates the Li metal anode and water while maintaining ionic contact between them. The solid-state Li^+ conductivity (10^{-4} S/cm) of GCEs limits rate performance; however, they are water impermeable, so parasitic Li corrosion is eliminated. In addition, GCE electrical conductivity is low ($< 10^{-11}$ S/cm), which prevents self-discharge after the battery is activated. Anodes for Li/water batteries have been developed that encase the Li anode inside a flexible, hermetic pouch.[54,74] A GCE window provides ionic contact with the aqueous catholyte. Figure 33.63 shows a drawing of this Li/water battery concept with a seawater catholyte. Typically, the metal anode is isolated from seawater by a pouch composed of a flexible laminate film which adheres to the GCE. A Li^+ conductive separator is used between the Li and GCE to prevent a possible reaction between them[75] Gas and hydroxide ion are produced at the cathode during battery discharge. In principle, battery operation is simple and efficient. The key to long service and shelf life, and high Li coulombic efficiency is low pouch permeability. The reported open-circuit potential of Li/GCE anodes in aqueous electrolytes is 3.04 V versus SHE, indicating no mixed potential, therefore no reaction with water. Coulombic efficiencies > 96% have also been reported for Li/CGE anodes in aqueous electrolytes.

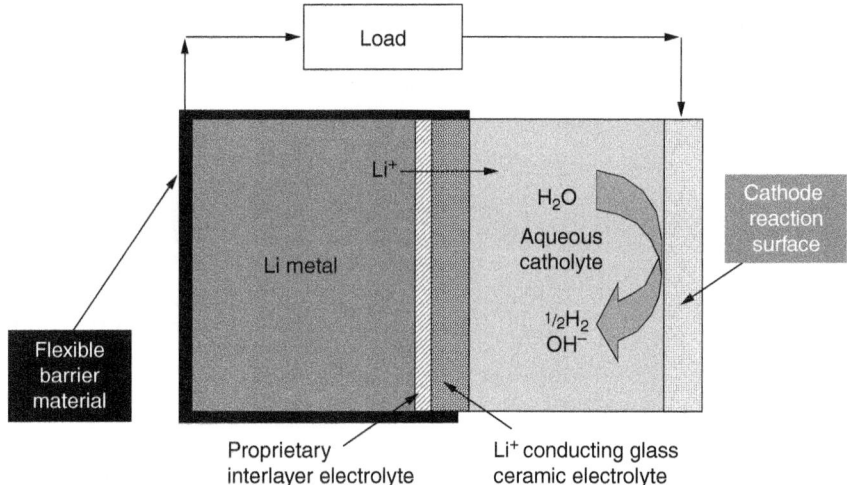

FIGURE 33.63 Schematic drawing of the Li/water battery concept using a glass ceramic electrolyte separator and a seawater catholyte. (*Courtesy of Naval Undersea Warfare Center, Newport, RI.*)

Dissolved oxygen and Mg in seawater are an opportunity and a challenge, respectively. The direct reduction of oxygen gas has a voltage advantage compared to water.

$$O_2 + H_2O + 2e^- \rightarrow HO_2^- + OH^- \qquad E^0 = -0.08 \text{ V}$$

The dissolved oxygen gas concentration in the ocean is low (up to 6 mL per L of seawater) and significantly depends on location, e.g., surface concentration is higher than concentration at depth.

Therefore, the contribution of dissolved O_2 to Li/seawater battery voltage depends on battery location, current density, and seawater mass transport. The high concentration of Mg^{2+} in the ocean (1.4×10^{-3} mg/L) and the low solubility of $Mg(OH)_2$ (Ksp = 2×10^{-11}) are enough to warrant concern about precipitate formation at high current density.

REFERENCES

1. D. A. J. Rand, "Battery Systems for Electric Vehicles: State of Art Review," *J. Power Sources* **4:**101 (1979).
2. K. F. Blurton and A. F. Sammells, "Metal/Air Batteries: Their Status and Potential—A Review," *J. Power Sources* **4:**263 (1979).
3. H. F. Bauman and G. B. Adams, "Lithium-Water-Air Battery for Automotive Propulsion," Lockheed Palo Alto Research Laboratory, Final Rep., COO/1262-1, Oct. 1977.
4. W.P. Moyer and E. L. Littauer, "Development of a Lithium-Water-Air Primary Battery," *Proc. IECEC,* Seattle, WA, Aug. 1980.
5. W. N. Carson and C. E. Kent, "The Magnesium-Air Cell," in D. H. Collins (ed.), *Power Sources,* 1966.
6. R. P. Hamlen, E. C. Jerabek, J. C. Ruzzo, and E. G. Siwek, "Anodes for Refuelable Magnesium-Air Batteries," *J. Electrochem. Soc.* **116:**1588 (1969).
7. J. F. Cooper, "Estimates of the Cost and Energy Consumption of Aluminum-Air Electric Vehicles," *ECS Fall Meeting,* Hollywood, FL, Oct. 1980; Lawrence Livermore, UCRL-84445, June 1980; update UCRL-94445 rev. 1, Aug. 1981.
8. R. P. Hamlen, G. M. Scamans, W. B. O'Callaghan, J. H. Stannard, and N. P. Fitzpatrick. "Progress in Metal-Air Battery Systems," *International Conference on New Materials for Automotive Applications,* Oct. 10–11, 1990.
9. A. S. Homa and E. J. Rudd, "The Development of Aluminum-Air Batteries for Electric Vehicles," *Proc. 24th IECEC,* vol. 3, 1989, pp. 1331–1334.
10. W. H. Hoge, "Air Cathodes and Materials Therefore," U.S. Patent 4,885,217, 1989.
11. W. H. Hoge, "Electrochemical Cathode and Materials Therefore," U.S. Patent 4,906,535, 1990.
12. W. H. Hoge, R. P. Hamlen, J. H. Stannard, N. P. Fitzpatrick, and W. B. O'Callaghan, "Progress in Metal-Air Systems," *Electrochem. Soc.*, Seattle, WA, Oct. 14–19, 1990.
13. T. Atwater, R. Putt, D. Bouland, and B. Bragg, "High-Energy Density Primary Zinc/Air Battery Characterization," *Proc. 36th Power Sources Conf.,* Cherry Hill, NJ, 1994.
14. R. Putt, N. Naimer, B. Koretz, and T. Atwater, "Advanced Zinc-Air Primary Batteries," *Proc. 6th Workshop for Battery Exploratory Development,* Williamsburg, VA, 1999.
15. J. Passanitti, "Development of a High Rate Primary Zinc-Air Cylindrical Cell," *Proc. 5th Workshop for Battery Exploratory Development,* Burlington, VT, 1997.
16. J. Passanitti, "Development of a High Rate Primary Zinc-Air Cylindrical Cell," *Proc. 38th Power Sources Conf.,* Cherry Hill, NJ, 1998.
17. J. Passanitti and T. Haberski, "Development of a High Rate Primary Zinc-Air Battery," *Proc. 6th Workshop for Battery Exploratory Development,* Williamsburg, VA, 1999.
18. R. A. Putt and G. W. Merry, "Zinc-Air Primary Batteries," *Proc. 35th Power Sources Symp.,* IEEE, 1992.
19. Sales literature, SAFT, Greenville, NC.
20. Celair Corp., Lawrenceville, GA.
21. A. Karpinski, "Advanced Development Program for a Lightweight Rechargeable AA Zinc-Air Battery," *Proc. 5th Workshop for Battery Exploratory Development,* Burlington, VT, 1997.
22. A. Karpinski, B. Makovetski, and W. Halliop, "Progress on the Development of a Lightweight Rechargeable Zinc-Air Battery," *Proc. 6th Workshop for Battery Exploratory Development,* Williamsburg, VA, 1999.
23. A. Karpinski, and W. Halliop, "Development of Electrically Rechargeable Zinc/Air Batteries," *Proc. 38th Power Sources Conf.,* Cherry Hill, NJ, 1998.

24. AER Energy Resources, Inc., Atlanta, GA.
25. L. G. Danczyk, R. L. Scheffler, and R. S. Hobbs, "A High Performance Zinc-Air Powered Electric Vehicle," *SAE Future Transportation Technology Conference and Exposition*, Portland, OR, Aug. 5–7, 1991, paper 911633.
26. M. C. Cheiky, L. G. Danczyk, and M. C. Wehrey, "Second Generation Zinc-Air Powered Electric Minivans," *SAE International Congress and Exposition*, Detroit, MI, Feb. 24–28, 1992, paper 920448.
27. S. M. Chodosh et al., "Metal-Air Primary Batteries, Replaceable Zinc Anode Radio Battery," *Proc. 21st Annual Power Sources Conf.*, Electrochemical Society, Pennington, NJ, 1967.
28. D. Linden and H. R. Knapp, "Metal-Air Primary Batteries, Metal-Air Standard Family," *Proc. 21st Annual Power Sources Conf.*, Electrochemical Society, Pennington, NJ, 1967.
29. Electric Fuel, Ltd. Jerusalem, Israel.
30. R. A. Putt," Zinc-Air Batteries for Electric Vehicles," *Zinc/Air Battery Workshop*, Albuquerque, NM, Dec. 1993.
31. H. B. Sierra Alcazar, P. D. Nguyen, G. E. Mason, and A. A. Pinoli, "The Secondary Slurry-Zinc/Air Battery," LBL Rep. 27466, July 1989.
32. G. Savaskan, T. Huh, and J. W. Evans, "Further Studies of a Zinc-Air Cell Intended for Electric Vehicle Applications, Part I: Discharge," *J. Appl. Electrochem.* (Aug. 1991).
33. T. Huh, G. Savaskan, and J. W. Evans, "Further Studies of a Zinc-Air Cell Intended for Electric Vehicle Applications, Part II: Regeneration of Zinc Particles and Electrolyte by Fluidized Bed Electrodeposition," *J. Appl. Electrochem.* (Aug. 1991).
34. A. R. Despic, "The Use of Aluminum in Energy Conversion and Storage," *First European East-West Workshop on Energy Conversion and Storage,* Sintra, Portugal, Mar. 1990.
35. N. P. Fitzpatrick and D. S. Strong, "An Aluminum-Air Battery Hybrid System," *Elec. Vehicle Develop.* **8:**79–81 (July 1989).
36. T. Dougerty, A. Karpinski, J. Stannard, W. Halliop, V. Alminauskas, and J. Billingsley, "Aluminum-Air Battery for Communications Equipment," *Proc. 37th Power Sources Conf.*, Cherry Hill, NJ, 1996.
37. C. L. Opitz, "Salt Water Galvanic Cell With Steel Wool Cathode," U.S. Patent 3,401,063, 1968.
38. D. S. Hosom, R. A. Weller, A. A. Hinton, and B. M. L. Rao, "Seawater Battery for Long-Lived Upper Ocean Systems," *IEEE Ocean Proc.,* vol. 3, Oct. 1–3, 1991.
39. J. A. Hunter, G. M. Scamans, and J. Sykes, "Anode Development for High Energy Density Aluminium Batteries," *Power Sources,* vol. 13 (Bournemouth, England, Apr. 1991).
40. R. P. Hamlen, W. H. Hoge, J. A. Hunter, and W. B. O'Callaghan, "Applications of Aluminum-Air Batteries," *IEEE Aerospace Electron. Mag.* **6:**11–14 (Oct. 1991).
41. S. Zaromb, C. N. Cochran, and R. M. Mazgaj, "Aluminum-Consuming Fluidized Bed Anodes," *J. Electrochem. Soc.* **137:**1851–1856 (June 1990).
42. G. Bronoel, A. Millott, R. Rouget, and N. Tassin, "Aluminum Battery with Automatic Feeding of Aluminium," *Power Sources,* vol. 13 (Bournemouth, England, Apr. 1991); also French Patents 88.15703, 1988; 90.07031, 1990; 90.14797, 1990.
43. W. B. O'Callaghan, N. Fitzpatrick, and K. Peters, "The Aluminum-Air Reserve Battery—A Power Supply for Prolonged Emergencies," *Proc. 11th Int. Telecommunications Energy Conf.,* Florence, Italy, Oct. 15–18, 1989.
44. J. A. O'Conner, "A New Dual Reserve Power System for Small Telephone Exchanges," *Proc. 11th Int. Telecommunications Energy Conf.,* Florence, Italy, Oct. 15–18, 1989.
45. A. P. Karpinski, J. Billingsley, J. H. Stannard, and W. Halliop, *Proc. 33rd IECEC,* 1998.
46. K. Collins et al., "An Aluminum-Oxygen Fuel Cell Power System for Underwater Vehicles," Applied Remote Technology, San Diego, 1992.
47. D. W. Gibbons and E. J. Rudd, "The Development of Aluminum/Air Batteries for Propulsion Applications," *Proc. 28th IECEC,* 1993.
48. D. W. Gibbons and K. J. Gregg, "Closed Cycle Aluminum/Oxygen Fuel Cell with Increased Mission Duration," *Proc. 35th Power Sources Symp.,* IEEE, 1992.
49. J. S. Lauer, J. F. Jackovitz, and E. S. Buzzelli, "Seawater Activated Power Source for Long-Term Missions," *Proc. 35th Power Sources Symp.,* IEEE, 1992.

50. E. L. Littauer and K. C. Tsai, "Anodic Behavior of Lithium in Aqueous Electrolytes, ii. Mechanical Passivation," *J. Electrochem. Soc.* **123**:964 (1976); "Corrosion of Lithium in Aqueous Electrolytes," *ibid.* **124**:850 (1977); "Anodic Behavior of Lithium in Aqueous Electrolytes, iii. Influence of Flow Velocity, Contact Pressure and Concentration," *ibid.* **125**:845 (1978).
51. D. L. Foster, J. R. Read, M. Shichtman, S. Balagopal, J. Watkins, and J. Gordon, "High Energy Lithium-Air Batteries for Soldier Power," http://oai.dtic.mil/oai/oai?verb=getRecord&metadataPrefix=html&identifier=ADA481576. Accessed Oct. 2009. Paper from unspecified conference.
52. N. Imanishi, S. Hasegawa, T. Zhang, A. Hirano, Y. Takeda, and O. Yamamoto, "Lithium Anode for Lithium-Air Secondary Batteries," *J. Power Sources* **185**:1392 (2008).
53. I. Kowalczk, J. Read, and M. Salomon, "Li-Air Batteries: A Classic Example of Limitations Owing to Solubilities," *Pure Appl. Chem.* **79**(5):851 (2007).
54. S. J. Visco, E. Nimon, B. Katz, L. D. Jonghe, and M.-Y. Chu, "The Development of High Energy Density Lithium/Air and Lithium/Water Batteries with No Self-Discharge," *210th Meeting of the Electrochemical Society*, Cancun, Mexico, 2006.
55. K. M. Abraham and Z. Jiang, *J. Electrochem. Soc.* **143**:1 (1996).
56. T. Zhang, N. Imanishi, S. Hasegawa, A. Hirano, J. Xie, Y. Takeda, O. Yamamoto, and N. Sammes, "Water-Stable Lithium Anode with the Three-Layer Construction for Aqueous Lithium-Air Secondary Batteries," *Electrochemical and Solid-State Letters* **12**(7):A132 (2009).
57. M. B. Marx and J. A. Read, "Performance of Carbon/Polyetraflouroethylene (PTFE) Air Cathodes from pH 0 to 14 for Li-Air Batteries," Army Research Laboratory Summary Report ARL-TR-4334 (2007).
58. H. Ye, J. Huang, J. J. Xu, A. Khalfan, and S. G. Greenbaum, "Li Ion Conducting Polymer Gel Electrolytes Based on Ionic Liquid/PVDF-HFP Blends," *J. Electrochem. Soc.* **154**(11):A1048 (2007).
59. T. Kuboki, T. Okuyama, T. Ohsaki, and N. Takami, "Lithium/Air Batteries Using Hydrophobic Room Temperature Ionic Liquid Electrolyte," *J. Power Sources* **146**:766 (2005).
60. A. Dobley, R. Rodriguez, and K. M. Abraham, "High Capacity Cathodes for Lithium-Air Batteries," *Electrochemical Society Conference*, Honolulu, HI, Oct. 2004.
61. A. Dobley, C. Morein, and R. Roark, "Lithium Air Cells with High Capacity Cathodes" *Electrochemical Society 210th Meeting Proceedings*, Cancun, Mexico, 2006.
62. A. Dobley, J. DiCarlo, and K. M. Abraham, "Non-aqueous Lithium-Air Batteries with an Advanced Cathode Structure," *41st Power Sources Conference,* Philadelphia, June 2004.
63. S. D. Beattie, D. M. Manolescu, and S. L. Blair, "High-Capacity Lithium-Air Cathodes," *J. Electrochem. Soc.* **156**(1):A44 (2009).
64. A. Dobley, C. Morein, and R. Roark, "Design Options for Emerging Lithium-Air Technology," *212th Electrochemical Society Conference*, Washington, DC, Oct. 2007.
65. J. Ostroha, "Lithium-Air System Development," *11th Electrochemical Power Sources R&D Symposium*, Baltimore, MD, July 2009.
66. The value of 3000 Wh/kg is the weight of the cell components and its packaging.
67. J. Read, "Characterization of the Lithium/Oxygen Organic Electrolyte Battery," *J. Electrochemical Society* **149**(9):A1190 (2002).
68. T. Ogasawara, A. Debart, M. Holzapfel, P. Novak, and P. G. Bruce, "Rechargeable Li_2O_2 Electrode for Lithium Batteries" *J. Am. Chem. Soc.* **128**(4)1390 (2006).
69. N. Shuster, "Lithium-Water Power Source for Low Power Long Duration Undersea Applications," *Proc. 35th Power Sources Symp.*, IEEE, 1992.
70. The majority of the information and text for the lithium/water section was provided by C.J. Patrissi, Naval Undersea Warfare Center, Newport, RI.
71. E. L. Littauer and K. C. Tsai, *J. Electrochem. Soc.* **845**(1978); E. L. Littauer and K. C. Tsai, *J. Electrochem. Soc.*, **964**(1976); E. L. Littauer and K. C. Tsai, *J. Electrochem. Soc.*, **771**(1976); P. Darby and M. Schmier, "Lithium-Aqueous Electrolyte Battery: Preliminary Studies," TM No. SB322-4326-72, Naval Underwater Systems Center, August 4, 1972.
72. Conceptual Design of a 164-kW Lithium Seawater Power System, U.S. Navy Contract No. N00017-73-C-4311, Lockheed Missiles & Space Company.
73. N. Shuster, *Proc. 34th International Power Sources Symposium*, p. 118, Cherry Hill, NJ, 1990.

74. C. J. Patrissi, C. R. Schumacher, S. P. Tucker, J. H. Fontaine, D. W. Atwater, T. M. Fratus, and C. M. Deschenes, "Electrochemical Performance of Pressure Tolerant Anodes for a Li-Seawater Battery," *215th Meeting of the Electrochemical Society*, San Francisco, 2009.

75. S. J. Visco, B. D. Katz, Y. S. Nimon, and L. C. D. Jonghe, "Protected Active Metal Electrode and Battery Cell Structures with Non-aqueous Interlayer Architecture," U.S. Patent, 7,282,295 B2, Oct. 16, 2007.

CHAPTER 34
RESERVE MAGNESIUM ANODE AND ZINC/SILVER OXIDE BATTERIES

R. David Lucero and Alexander P. Karpinski

SECTION A
MAGNESIUM WATER-ACTIVATED BATTERIES

34.1 GENERAL CHARACTERISTICS OF RESERVE MAGNESIUM BATTERIES

The water-activated battery was first developed in the 1940s to meet a need for a high-energy-density, long-shelf-life battery, with good low-temperature performance, for military applications.

The battery is constructed dry, stored in the dry condition, and activated at the time of use by the addition of water or an aqueous electrolyte. Most of the water-activated batteries use magnesium as the anode material. Several cathode materials have been used successfully in different types of designs and applications.

The magnesium/silver chloride seawater-activated battery was developed by Bell Telephone Laboratories as the power source for electric torpedoes.[1] This work resulted in the development of small high-energy-density batteries readily adaptable for use as power sources for sonobuoys, electric torpedoes, weather balloons, air-sea rescue equipment, pyrotechnic devices, marine markers, and emergency lights.

The magnesium/cuprous chloride system became commercially available in 1949.[2,3] Compared with the magnesium/silver chloride battery, this system has lower energy density, lower rate capability, and less resistance to storage at high humidities, but its cost is significantly lower. Although the magnesium/cuprous chloride system can be used for the same purposes as the magnesium/silver chloride battery, its major application was in airborne meteorological equipment, where the use of the more expensive silver chloride system was not warranted. The cuprous chloride system does not have the physical or electrical characteristics required for use as the power source for electric torpedoes. More recently, the magnesium/cuprous chloride chemistry has been developed for aviation and marine life jacket lights (see Sec. 34.5.4).

Because of the high cost of silver and the impracticality of recovering it after use, other nonsilver water-activated batteries were developed, primarily as the power source for anti-submarine warfare (ASW) equipment.

34.2 SPECIALIZED BATTERY SYSTEMS

TABLE 34.1 Comparison of Silver and Nonsilver Cathode Batteries

Advantages	Disadvantages
Silver chloride cathodes	
Reliable	High raw material costs
Safe	High rate of self-discharge after activation
High power density	
High energy density	
Good response to pulse loading	
Instantaneous activation	
Long unactivated shelf life	
No maintenance	
Nonsilver cathodes	
Abundant domestic supply	Requires supporting conductive grid
Low raw-material cost	Operates at low current densities
Instantaneous activation	Low energy density compared to silver
Reliable, safe	High rate of self-discharge after activation
Long unactivated shelf life	
No maintenance	

The systems that have been developed and used successfully are magnesium/lead chloride,[4] magnesium/cuprous iodide-sulfur,[5-7] magnesium/cuprous thiocyanate-sulfur,[8] and magnesium/manganese dioxide utilizing an aqueous magnesium perchlorate electrolyte.[9-11] None of these systems can compete with the magnesium/silver chloride system in almost every attribute except cost.

Magnesium seawater-activated batteries, using dissolved oxygen in the seawater as the cathode reactant, also have been developed for application in buoys, communications, and underwater propulsion. These batteries, as well as the use of other metals as anodes for water-activated batteries, are covered later in this chapter and in Chap. 33.

Another seawater battery system considered for low-rate, long-duration undersea vehicle applications consists of a magnesium anode, a palladium and iridium catalyzed carbon paper cathode, and a solution-phased catholyte of seawater, acid, and hydrogen peroxide. The magnesium/hydrogen peroxide system has a voltage of 2.12 V and is expected to be capable of more than 500 Wh/kg when configured for large-scale unmanned undersea vehicles.[12]

The advantages and disadvantages of the various water-activated magnesium batteries are given in Table 34.1.

Information on nonreserve primary and secondary magnesium anode batteries is found in Chap. 10.

34.2 CHEMISTRY

The principal overall and current-producing reactions for the water-activated magnesium batteries are as follows:

1. Magnesium/silver chloride

Anode	$Mg - 2e \rightarrow Mg^{2+}$
Cathode	$2AgCl + 2e \rightarrow 2Ag + 2Cl^-$
Overall	$Mg + 2AgCl \rightarrow MgCl_2 + 2Ag$

2. Magnesium/cuprous chloride

   Anode $Mg - 2e \rightarrow Mg^{2+}$
   Cathode $2CuCl + 2e \rightarrow 2Cu + 2Cl^-$
   Overall $Mg + 2CuCl \rightarrow MgCl_2 + 2Cu$

3. Magnesium/lead chloride

   Anode $Mg - 2e \rightarrow Mg^{2+}$
   Cathode $PbCl_2 + 2e \rightarrow Pb + 2Cl^-$
   Overall $Mg + PbCl_2 \rightarrow MgCl_2 + Pb$

4. Magnesium/cuprous iodide, sulfur

   Anode $Mg - 2e \rightarrow Mg^{2+}$
   Cathode $Cu_2I_2 + 2e \rightarrow 2Cu + 2I^-$
   Overall $Mg + Cu_2I_2 \rightarrow MgI_2 + 2Cu$

5. Magnesium/cuprous thiocyanate, sulfur

   Anode $Mg - 2e \rightarrow Mg^{2+}$
   Cathode $2CuSCN + 2e \rightarrow 2Cu + 2SCN^-$
   Overall $Mg + 2CuSCN \rightarrow Mg(SCN)_2 + 2Cu$

6. Magnesium/manganese dioxide

   Anode $Mg - 2e \rightarrow Mg^{2+}$
   Cathode $2MnO_2 + H_2O + 2e \rightarrow Mn_2O_3 + 2OH^-$
   Overall $Mg + 2MnO_2 + H_2O \rightarrow Mn_2O_3 + Mg(OH)_2$

A side reaction also occurs between the magnesium anode and the aqueous electrolyte, resulting in the formation of magnesium hydroxide, hydrogen gas, and heat.

$$Mg + 2H_2O \rightarrow Mg(OH)_2 + H_2$$

In immersion-type batteries, the hydrogen evolved creates a pumping action which helps purge the insoluble magnesium hydroxide from the battery. Magnesium hydroxide remaining within a cell can fill the space between the electrodes, which can become devoid of electrolyte, prevent ionic flow, and cause premature cell and battery failure.

The heat evolved improves the performance of immersion-type batteries; it enables dunk-type batteries to operate at low ambient temperatures and forced-flow batteries to operate at high current densities.

Those cathodes containing sulfur exhibit a higher potential versus magnesium than cathodes possessing only the prime depolarizer. During discharge, the sulfur probably reacts with the highly active copper formed when the prime depolarizer is reduced, producing a copper sulfide, thus accounting for the fact that no copper is observed at end of discharge. This reaction may also prevent copper from plating out on the magnesium, thus deterring premature voltage drop. In those cases where the battery is allowed to discharge past the point where all prime depolarizer is gone and magnesium is present, hydrogen sulfide can be produced. Hydrogen sulfide can also result if the cell is short-circuited.

34.3 TYPES OF WATER-ACTIVATED BATTERIES

Water-activated batteries are manufactured in the following basic types:

1. *Immersion* batteries are designed to be activated by immersion in the electrolyte. They have been constructed in sizes to produce from 1.0 V to several hundred volts at currents up to 50 A. Discharge times can vary from a few seconds to several days. A typical immersion-type water-activated battery is shown in Fig. 34.1.

2. *Forced-flow* batteries are designed for use as the power source for electric torpedoes. The name is derived from the fact that seawater is forced through the battery as the torpedo is driven through the water. Because of the heat generated during discharge and electrolyte recirculation, these systems can perform at current densities up to 500 mA/cm^2 of cathode surface area. Batteries containing from 118 to 460 cells which will produce from 25 to 460 kW of power have been developed. Discharge times are about 10 to 15 min. A diagrammatic representation of a torpedo battery and a torpedo battery with recirculation voltage control is shown in Fig. 34.2.

3. *Dunk-type* batteries are designed with an absorbent separator between the electrodes and are activated by pouring the electrolyte into the battery, where it is absorbed by the separator. Batteries of this type have been designed to produce from 1.5 to 130 V at currents up to about 10 A. Lengths of discharge vary from about 0.5 to 15 h. Figure 34.3 is a diagrammatic representation of a magnesium/cuprous chloride battery used in radiosonde applications. A pile-type construction is used. A sheet of magnesium is separated from the cuprous chloride cathode by a porous separator, which also serves to retain the electrolyte. The cathode is a pasted type made by applying a paste of powdered cuprous chloride and a liquid binder onto a copper grid or screen. The assembly is taped together to form the battery. The batteries are also made in spiral or jelly-roll design. (See Fig. 34.22 for an illustration of this battery.)

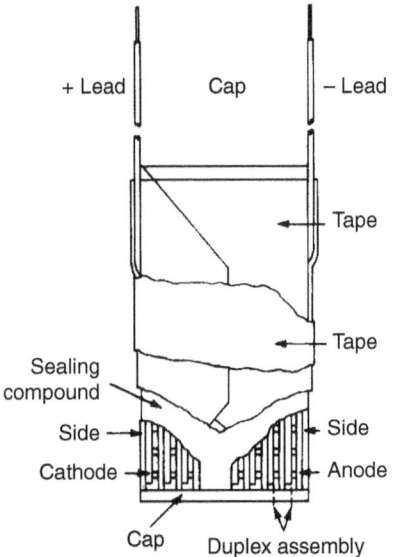 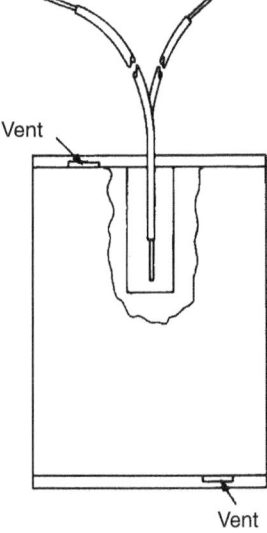

FIGURE 34.1 Seawater battery, immersion type.

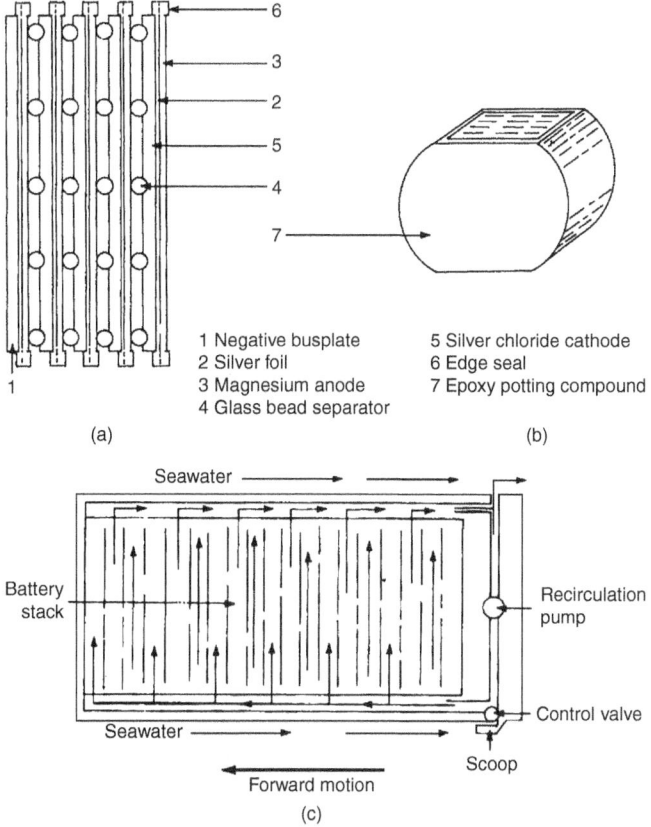

FIGURE 34.2 Diagrammatic representation of torpedo battery construction. (*a*) Cell construction. (*b*) Battery configuration. (*c*) Recirculation voltage control.

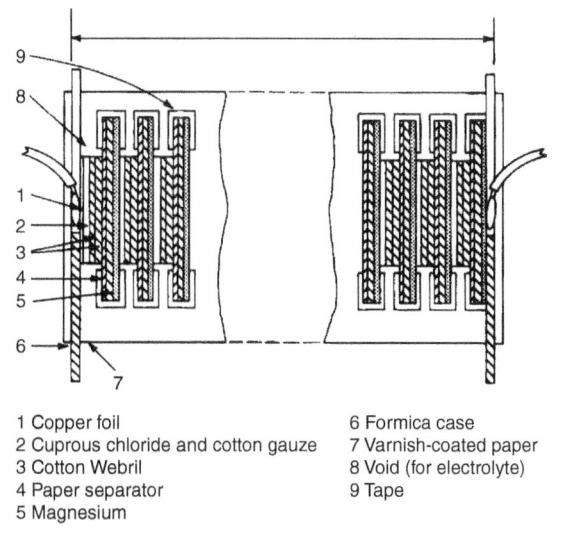

FIGURE 34.3 Diagrammatic representation of magnesium/cuprous chloride dunk-type battery. (*Courtesy of Eagle-Picher Technologies.*)

34.4 CONSTRUCTION

Water-activated cells consist of an anode, a cathode, a separator, terminations, and some form of encasement. A battery consists of a multiplicity of cells connected in series or series-parallel. Such an assembly requires a method to connect the cells in the desired configuration plus a method to control leakage currents. The voltage of a cell depends primarily on the electrochemical system involved. To increase voltage, a number of cells must be connected in series. The capacity of a cell in ampere-hours is primarily dependent on the quantity of active material in the electrodes. The ability of a cell to produce a given current at a usable voltage depends on the area of the electrode. To decrease current density so as to increase load voltage, the electrode area must be increased. Power output depends on the temperature and salinity of the electrolyte. Power output can be increased by increasing the temperature or the salinity of the electrolyte.

The basic components of a single cell, a duplex assembly for connecting cells in series, and a finished battery are illustrated in Figs. 34.4, 34.5, and 34.1, respectively.[12–16] The illustrations represent batteries designed for use by immersion in the electrolyte as contrasted to a dunk-type (radiosonde) battery, which is activated by pouring the electrolyte into the battery, or a forced-flow electric torpedo battery. The construction principles with slight variations are similar in all cases.

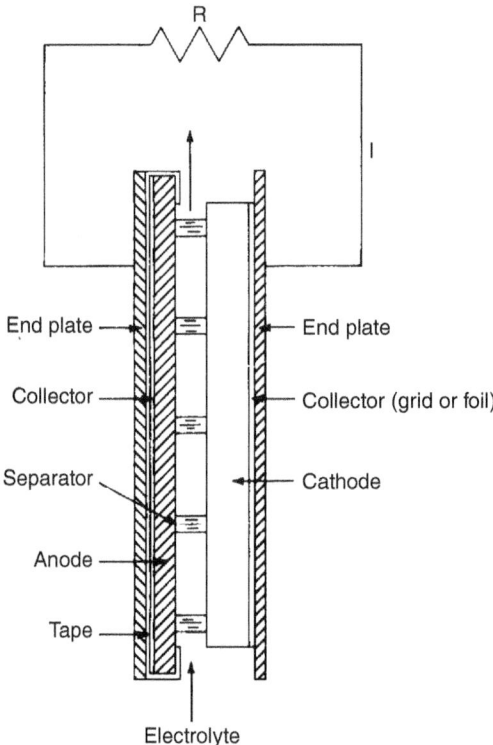

FIGURE 34.4 Basic water-activated cell.

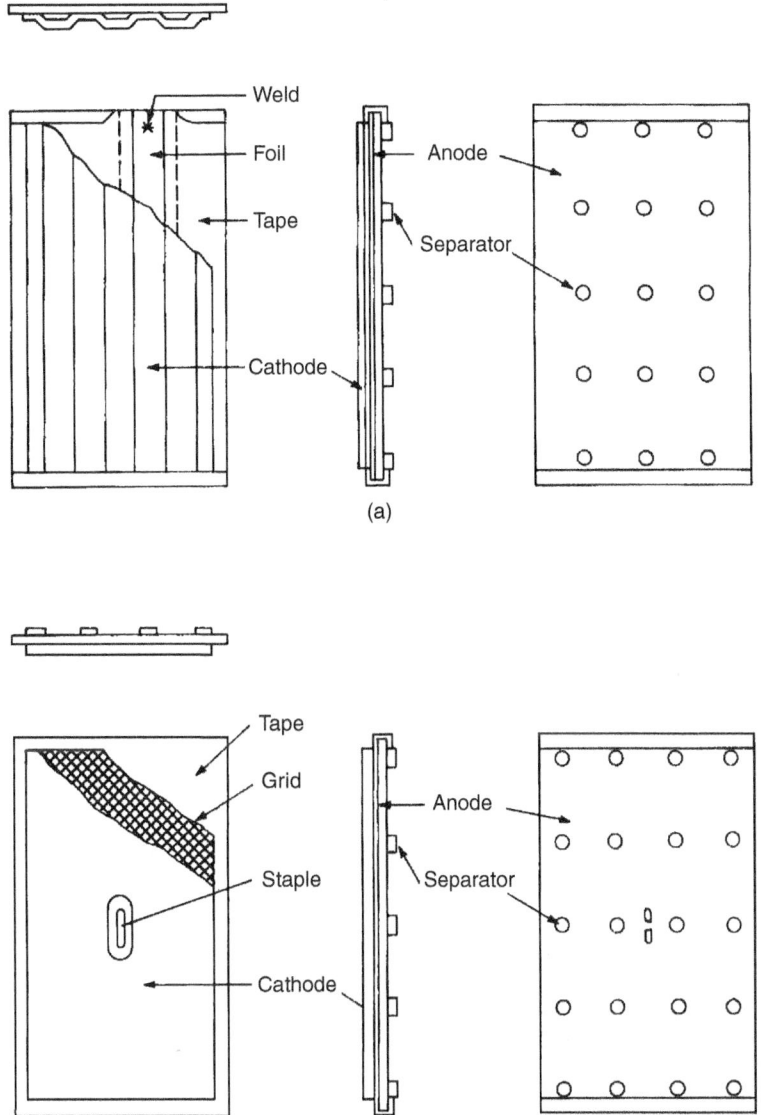

FIGURE 34.5 Duplex electrode assemblies. (*a*) Silver. (*b*) Nonsilver.

34.4.1 Components

A more detailed description of the various cell and battery components and construction elements follows.

Anode (Negative Plate). The anode is made from sheet magnesium. Magnesium AZ61A is preferred because it tends to sludge and polarize less. In some cases AZ31B alloy is used; however, this alloy gives slightly lower voltage, polarizes at high current densities, and sludges more. In recent

TABLE 34.2 Composition Range for Battery Plate Alloys

	AZ31		AZ61		AP65		MELMAG 75	
Element	% Min.	% Max.	% Min.	% Max.	% Min.	% Max.	% Min.	% Max.
Al	2.5	3.5	5.8	7.2	6.0	6.7	4.6	5.6
Zn	0.6	1.4	0.4	1.5	0.5	1.5	—	0.3
Pb	—	—	—	—	4.4	5.0	—	—
Tl	—	—	—	—	—	—	6.6	7.6
Mn	0.15	0.7	0.15	0.25	0.15	0.30	—	0.25
Si	—	0.1	—	0.05	—	0.3	—	0.3
Ca	—	0.04	—	0.3	—	0.3	0.3	—
Cu	—	0.05	0.05	0.05	0.005	—	—	—
Ni	—	0.005	—	0.005	—	0.005	—	0.005
Fe	—	0.006	—	0.006	—	0.010	—	0.006

years, magnesium alloys AP65 and MTA75 have been developed and evaluated. These are high-voltage alloys giving load voltages of 0.1 to 0.3 V higher than AZ61A. MTA75 is a higher-voltage alloy than AP65. These alloys sludge more; under some forced-flow discharge conditions, the sludging problem may be controlled. These alloys are not used extensively in the United States; they are used in the United Kingdom and Europe in electric torpedo batteries. Composition ranges of these alloys are shown in Table 34.2.

Cathode (Positive Plate). The cathode consists of a depolarizer and a current collector. These depolarizers are powders and are nonconductive. In order for the depolarizer to function, a form of carbon is added to impart conductivity; a binder is added for cohesion, and a metal grid is used as a current collector, a base for the cathode, to facilitate intercell connections and battery terminations. Possible cathode formulations are shown in Table 34.3.[1,3–5,8]

TABLE 34.3 Cathode Compositions

	Silver chloride[1]	Cuprous iodide[5,6]	Cuprous thiocyanate[8]	Lead chloride[4]	Cuprous chloride
Depolarizer, %/w	100	73	75–80	80.7–82.5	95–100
Sulfur, %/w	—	20	10–12	—	—
Additive, %/w	—	—	0–4	2.3–4.4	—
Carbon, %/w	—	7	7–10	9.6–9.8	—
Binder, %/w	—	—	0–2	1.5–1.6	0–5
Wax, %/w	—	—	—	3.8	—

Silver chloride is a special case. Silver chloride can be melted, cast into ingots, and rolled into sheet stock in thicknesses from about 0.08 mm up. Since this material is malleable and ductile, it can be used in almost any configuration. Silver chloride is nonconductive and is made conductive by superficially reducing the surface to silver by immersion in a photographic developing solution. No base grid need be used with silver chloride.

Nonsilver cathodes are usually prismatic in shape and are flat. Silver chloride cathodes are used flat and corrugated in many configurations.

Separators. Separators are nonconductive spacers placed between the electrodes of immersion- and forced-flow-type batteries to form a space for free ingress of electrolyte and egress of corrosion products. Separators in the form of disks, rods, glass beads, or woven fabrics may be used.[13–14]

Dunk-type batteries utilize a nonwoven, absorbent, nonconductive material for the dual purpose of separating the electrodes and absorbing the electrolyte.

Intercell Connections. In a series-arranged battery of pile construction, the anode of one cell is connected to the cathode of the adjacent cell. To accomplish this without producing a short-circuited cell, an insulating tape or film is placed between the electrodes on nonsilver batteries. For silver batteries, silver foil is used alone or in conjunction with an insulating tape.

For nonsilver cells, the connection is made by stapling the electrodes together through the insulator.[15] For silver cells, the silver chloride, surface-reduced to silver, is heat-sealed to silver foil, which has been previously welded to the anode. Where large surface areas are involved, contact between silver and silver foil can be made by pressure alone.

Terminations. For silver chloride cathodes, the lead is soldered directly to silver foil, which has been heat-sealed to one surface of the silver chloride. Leads are soldered directly to the collector grid of nonsilver cathodes or soldered to a piece of copper foil, which has been stapled to the collector grid.

The anode connection is made by soldering the lead to silver foil that has been welded to the anode or by welding directly to the anode.

Encasement. The battery encasement must effectively rigidize the battery and provide openings at opposite ends to allow free ingress and egress of electrolyte and corrosion products.

The periphery of the battery must be sealed in such a manner that the cells contact the external electrolyte only at the openings provided at the top and bottom of the battery. The encasement can be accomplished by using premolded pieces, caulking compounds, epoxy resins, an insulating sheet, or hot-melt resins.[13-16] For single batteries, these precautions are not necessary.

34.4.2 Leakage Current

All the cells in the immersion- and forced-flow-type batteries operate in a common electrolyte. Since the electrolyte is conductive and continuous from cell to cell, conductive paths exist from each point in a battery to every other point. Current will flow through these conductive paths to points of different potential. This current is referred to as "leakage current" and is in addition to the current flowing through the load. Electrodes must be designed to compensate for these leakage currents.

Leakage currents for a small number of cells can be reduced by increasing the resistance path from a cell to the common electrolyte or that of the common electrolyte between adjacent cells. Leakage currents for a large number of cells can be reduced by increasing the resistance of the common electrolyte external to the individual cells.

During construction, the conducting paths from cell to cell are made as long as possible. In many instances, the negative or positive of the battery is connected to an external metal surface. Leakage currents flow from the battery to this surface. These leakage currents are controlled by placing a cap containing a slot over the battery openings. If one terminal is connected to an external conductive surface, the slot in the cap is opened to the electrolyte only on that side of the battery. Where neither terminal is connected to an external conductive surface, either end of the cap may be opened, but only on one side of the battery.

The resistance (ohms) of the slot in the cap may be calculated using the formula

$$R = p\frac{l}{a}$$

where R = resistance, Ω
l = length of slot, cm
a = cross-sectional area of slot, cm^2
p = resistance of electrolyte for temperature and salinity in which battery is operating, $\Omega \cdot$ cm

For dunk-type batteries the electrolyte continuity from cell to cell is broken when the electrolyte is absorbed in the separator. The excess is poured off the battery or spun away from the cells by some external force applied to the battery.

34.4.3 Electrolyte

Seawater-activated batteries are designed to operate in an infinite electrolyte, namely, the oceans of the world. However, for design, development, and quality control purposes, it is not practical to use ocean water. Thus it is common practice throughout the industry to use a simulated ocean water. A commercial product, composed of a blend of all the ingredients required, simplifies the manufacture of simulated ocean water test solutions.

Dunk-type batteries, activated by pouring the electrolyte into the battery where it is absorbed by the separator, can utilize water or seawater when the temperature is above freezing. At lower temperatures special electrolytes can be used. The use of a conducting aqueous electrolyte will result in faster voltage buildup. However, the introduction of salts in the electrolyte will increase the rate of self-discharge.

34.5 PERFORMANCE CHARACTERISTICS

34.5.1 General

A summary of the performance characteristics of the major water-activated batteries currently available is given in Table 34.4.

TABLE 34.4 Performance Characteristics of Water-Activated Batteries

Cathode	Silver chloride	Lead chloride	Cuprous iodide	Cuprous thiocyanate	Cuprous chloride[a]
Anode	Magnesium				
Electrolyte	Tap water, seawater, or other conductive aqueous solutions				
Open-circuit voltage, V	1.6–17	1.1–1.2	1.5–1.6	1.5–1.6	1.5–1.6
V per cell at 5 mA/cm^{2b}	1.42–1.52	0.90–1.06	1.33–1.49	1.24–1.43	1.2–1.4
Activation, s:					
35°C[c]	<1	<1	<1	<1	
RT[d]	—	—	—	—	1–10
0°C[e]	45–90	45–90	45–90	45–90	
Internal resistance, Ω[f]	0.1–2	1–4	1–4	1–4	2
Ah/g cath. theor.[g]	0.187	0.193	0.141	0.220	0.271
Usable capacity, % of theoretical	60–75	60–75	60–75	60–75	60–75
Wh/kg	100–150	50–80	50–80	50–80	50–80
Wh/L	180–300	50–120	50–120	50–120	20–200
Operating temperatures, °C[h]			−60 to +65		

[a] All but cuprous chloride are immersion type. Cuprous chloride is dunk type.
[b] See voltage vs. current density curves.
[c] Battery preconditioned at +55°C, then immersed in simulated ocean water of 3.6 wt %.
[d] Electrolyte at room temperature poured into battery and absorbed by separator.
[e] Battery preconditioned at −20°C, then immersed in simulated ocean water of 1.5 wt %.
[f] Depends on battery design.
[g] 100% active material.
[h] Following activation at room temperature.

Voltage versus Current Density. Figures 34.6 and 34.7 are representative of voltage versus current density curves for several water-activated battery systems at 35 and 0°C, respectively, using a simulated ocean water electrolyte.

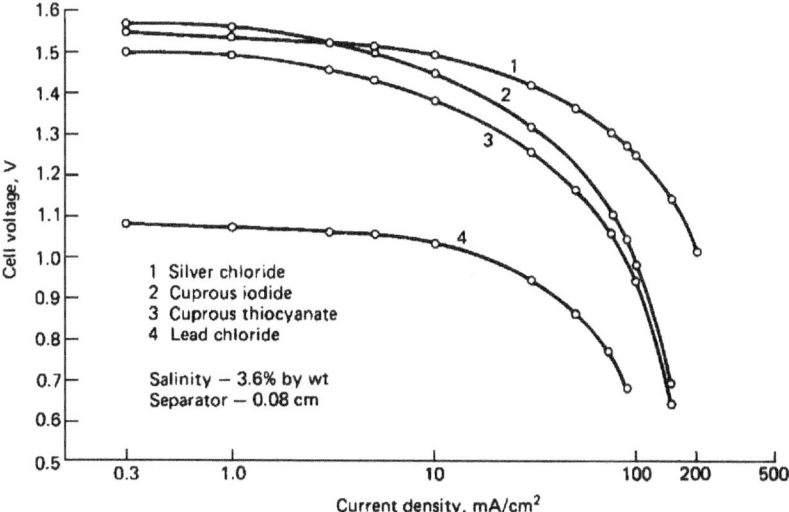

FIGURE 34.6 Representative cell voltages vs. current density at 35°C.

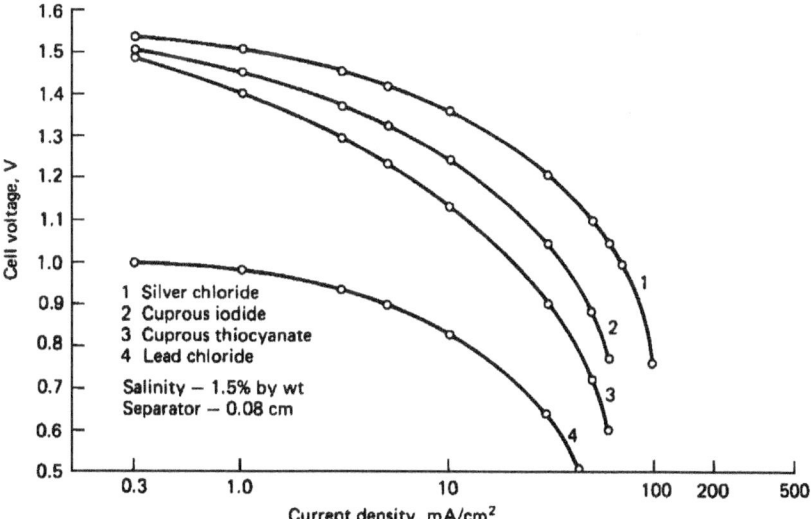

FIGURE 34.7 Representative cell voltages vs. current density at 0°C.

Discharge Curves. Discharge curves of the magnesium/silver chloride, magnesium/cuprous thiocyanate-sulfur, magnesium/cuprous iodide-sulfur, and magnesium/lead chloride electrochemical systems, discharged continuously through various resistances in simulated ocean water at high and low temperatures and salinities, are shown in Figs. 34.8 to 34.15. These data show the advantageous performance of the silver chloride system.

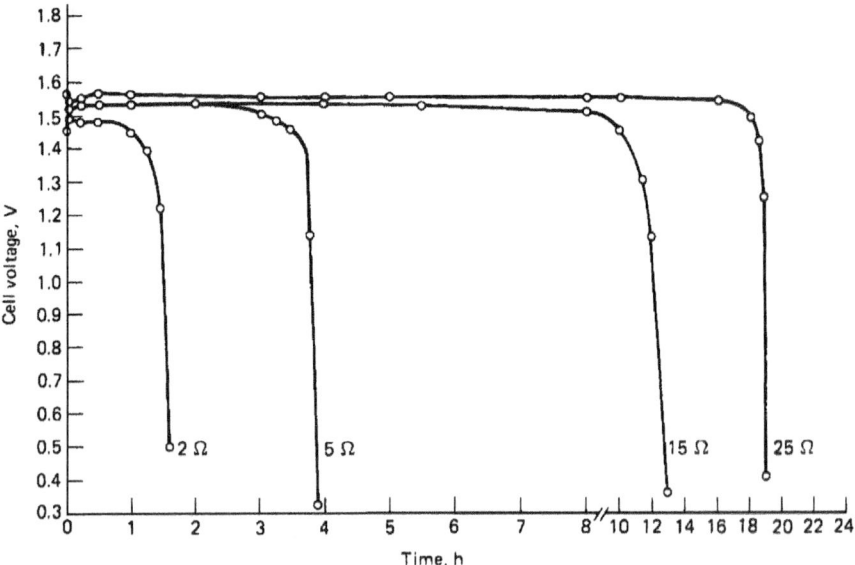

FIGURE 34.8 Magnesium/silver chloride seawater-activated cell discharged continuously at 35°C in simulated ocean water, 3.6% salinity.

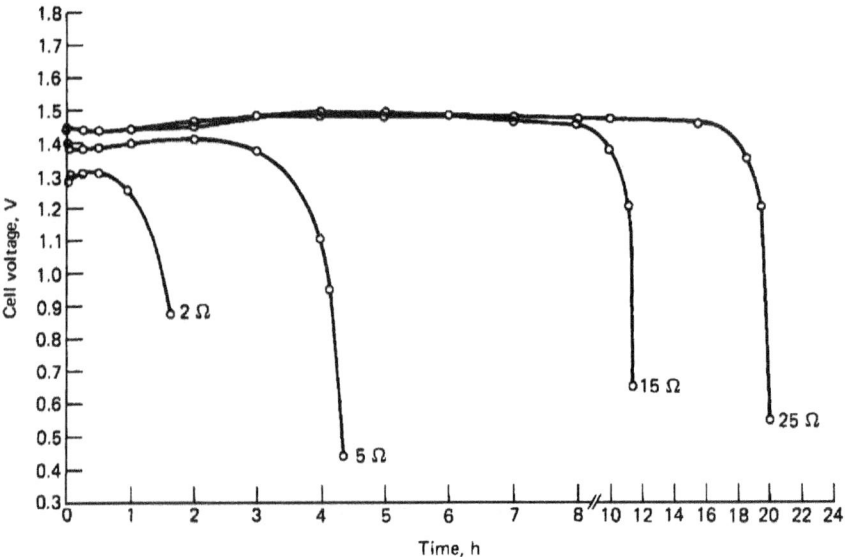

FIGURE 34.9 Magnesium/silver chloride seawater-activated cell discharged continuously at 0°C in simulated ocean water, 1.5% salinity.

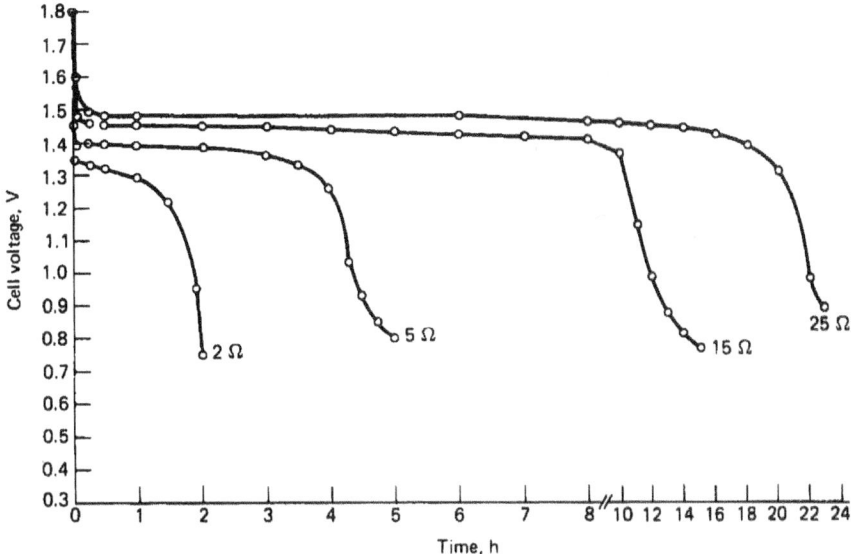

FIGURE 34.10 Magnesium/cuprous thiocyanate seawater-activated cell discharged continuously at 35°C in simulated ocean water, 3.6% salinity.

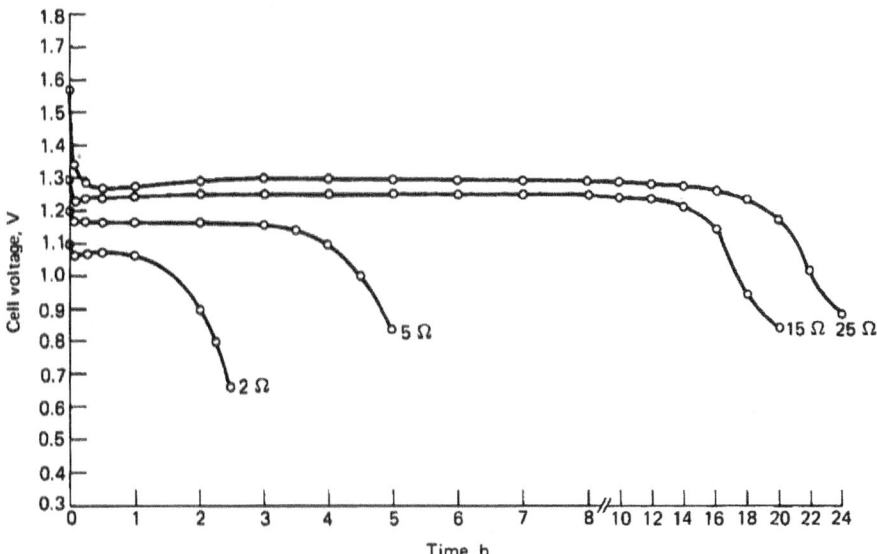

FIGURE 34.11 Magnesium/cuprous thiocyanate seawater-activated cell discharged continuously at 0°C in simulated ocean water, 1.5% salinity.

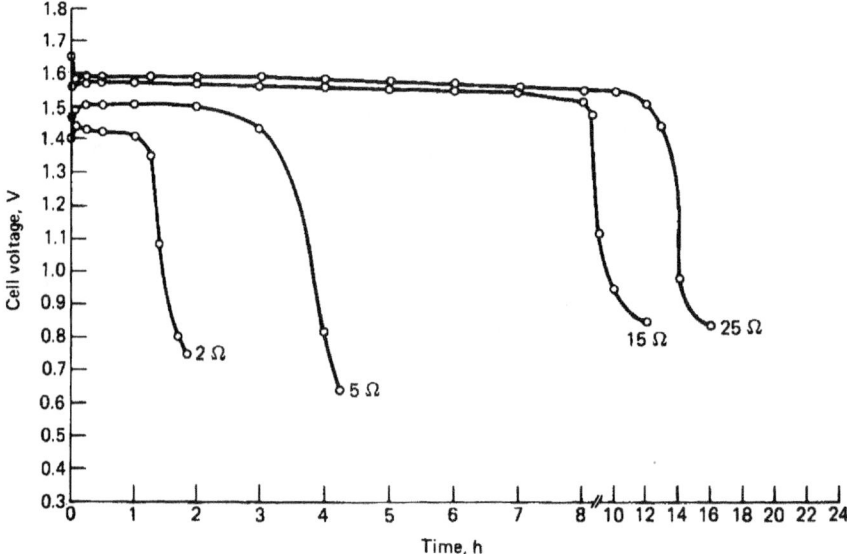

FIGURE 34.12 Magnesium/cuprous iodide seawater-activated cell discharged continuously at 35°C in simulated ocean water, 3.6% salinity.

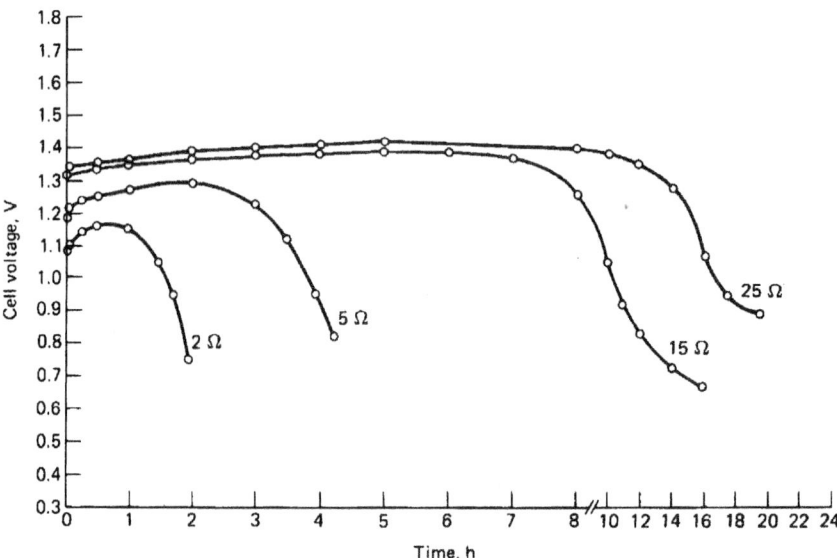

FIGURE 34.13 Magnesium/cuprous iodide seawater-activated cell discharged continuously at 0°C in simulated ocean water, 1.5% salinity.

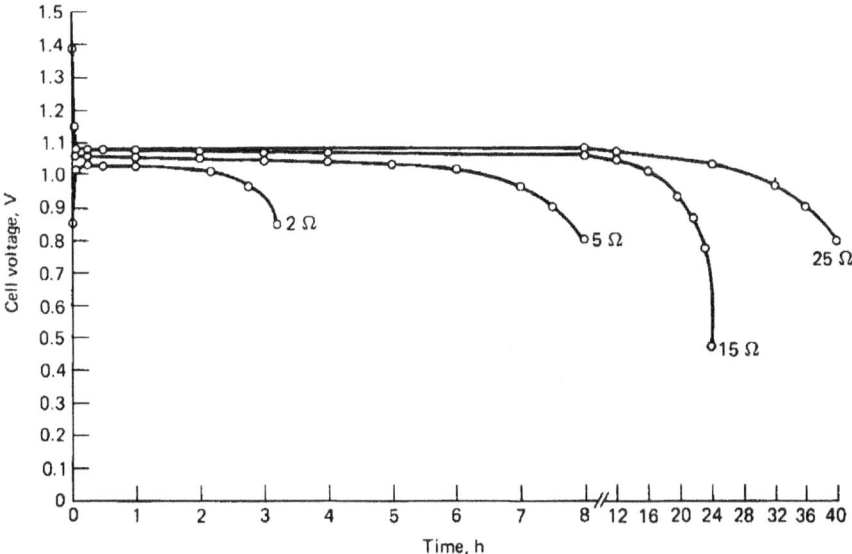

FIGURE 34.14 Magnesium/lead chloride seawater-activated cell discharged continuously at 35°C in simulated ocean water, 3.6% salinity.

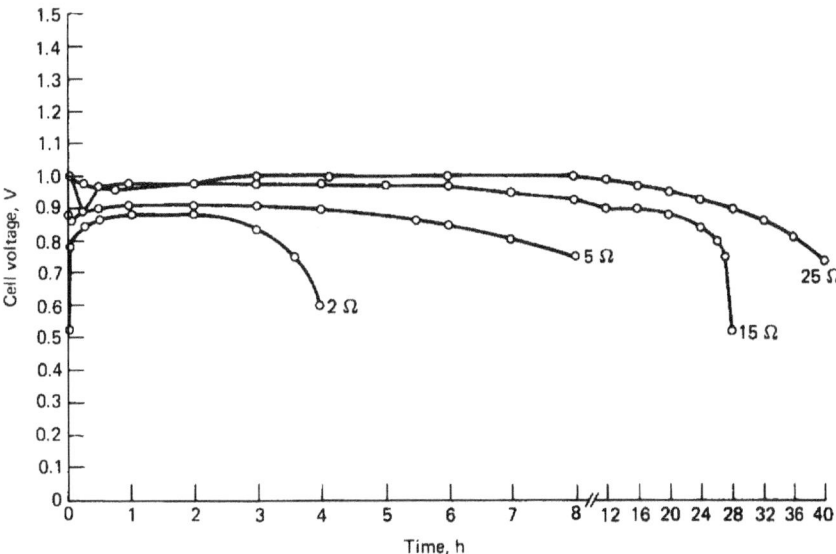

FIGURE 34.15 Magnesium/lead chloride seawater-activated cell discharged continuously at 0°C in simulated ocean water, 1.5% salinity.

Service Life. The capacities per unit of weight versus the average power output of these same electrochemical systems, at high and low temperatures and salinities, are shown in Figs. 34.16 and 34.17.

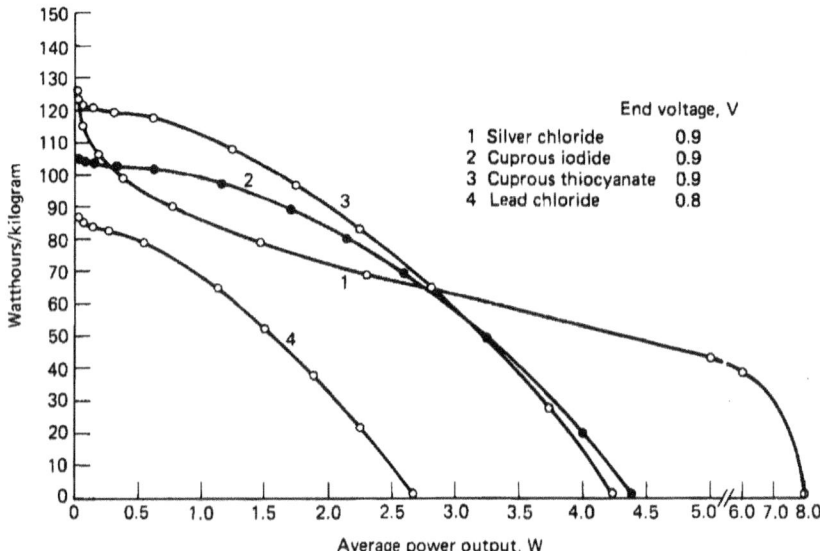

FIGURE 34.16 Capacity vs. power output of seawater-activated cells discharged continuously at 35°C in simulated ocean water, 3.6% salinity.

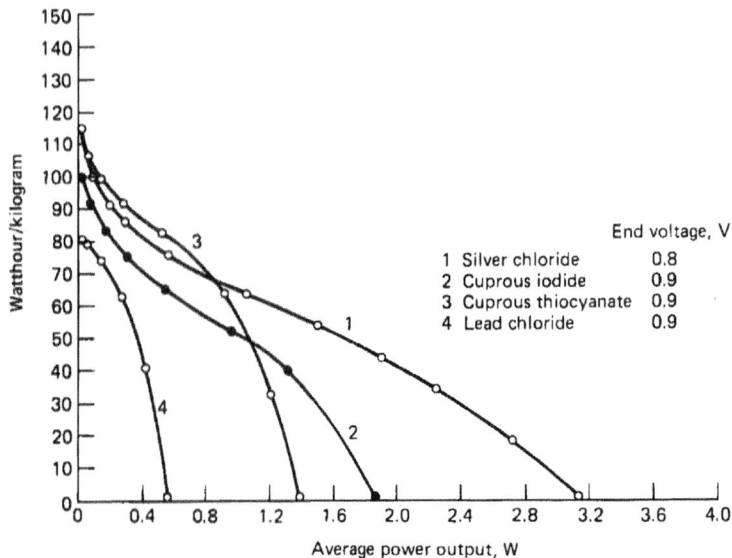

FIGURE 34.17 Capacity vs. power output of seawater-activated cells discharged continuously at 0°C in simulated ocean water, 1.5% salinity.

34.5.2 Immersion-Type Batteries

The performance of these same systems, designed as immersion-type batteries to meet the physical, electrical, and environmental specifications listed in Table 34.5, is shown in Figs. 34.18 to 34.20. The performance characteristics are summarized in Table 34.6.

TABLE 34.5 Performance Specifications for Seawater-Activated Battery

Load	80 ± 2 Ω
Life	9 h
Voltage	15.0 V: Initial from 90 s to 9 h
	19.0 V max.
Activation*	60 s to 13.5 V
	90 s to 15.0 V
Battery size:	Silver Nonsilver
Height, cm	7.7 max. 10.6 max.
Width, cm	5.7 max. 7.6 max.
Thickness, cm	4.2 max 5.7 max.
Weight, g	255 ± 14 482 ± 85
Environmental:	
Storage	From −60 to +70°C for 5 years†
	90 days at −50 to +40°C at 90% RH
	10 days per MIL-T-5422E
Vibration, Hz	5–500
Electrolyte:	
Low temperature	Ocean water of 1.5% salinity by weight at 0 ± 1°C
High temperature	Ocean water of 3.6% salinity by weight at +34 ± 1°C

*Battery preconditioned at −20°C prior to immersion in ocean water of 1.5% salinity by weight at 0 ± 1°C.
†In equipment packed in sealed plastic container with appropriate desiccant.

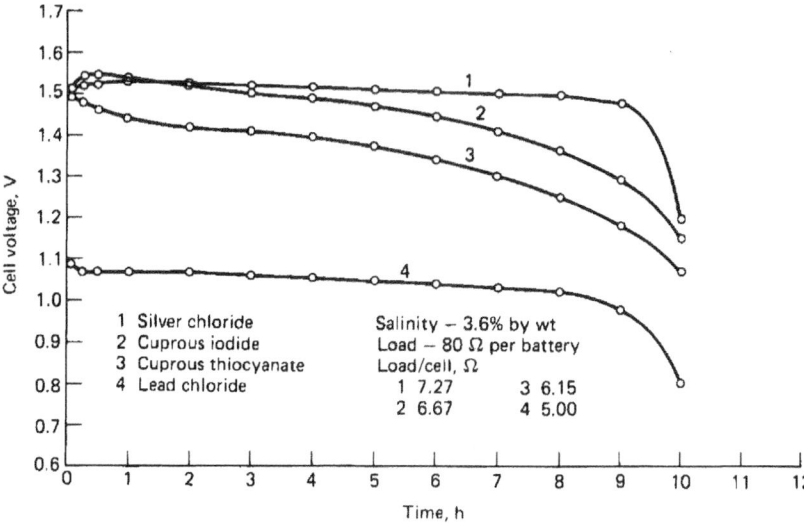

FIGURE 34.18 Discharge curves of seawater-activated batteries at 35°C.

34.18 SPECIALIZED BATTERY SYSTEMS

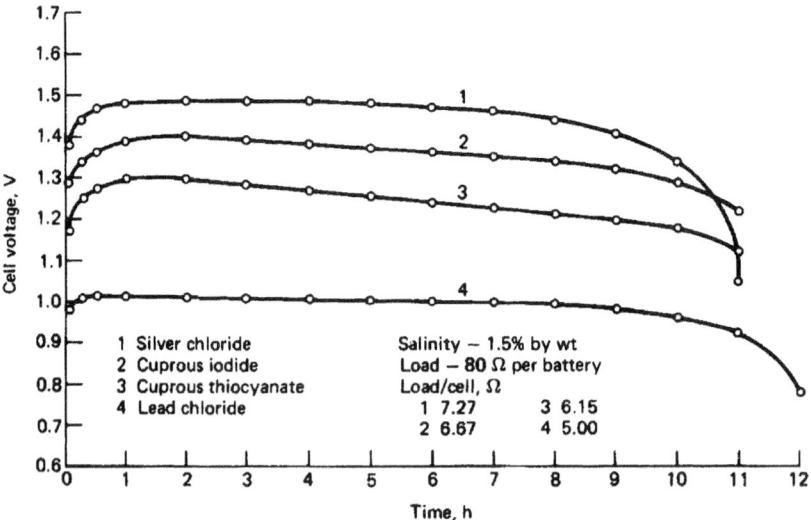

FIGURE 34.19 Discharge curves of seawater-activated batteries at 0°C.

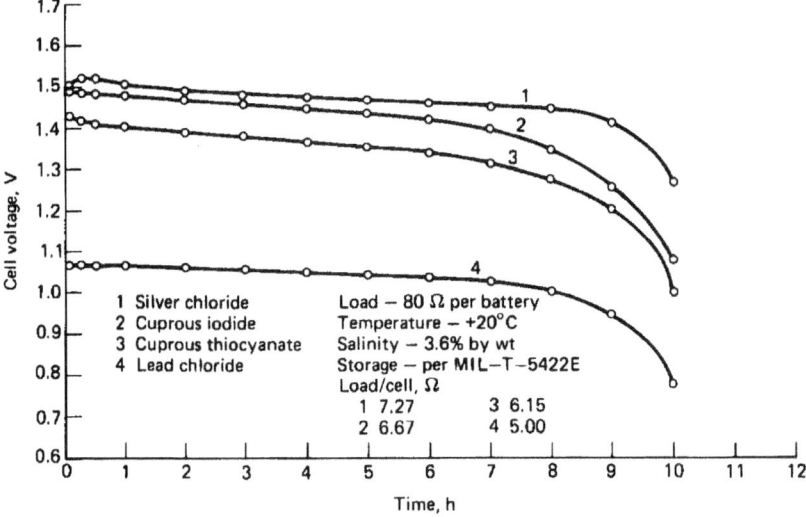

FIGURE 34.20 Discharge curves of seawater-activated batteries, 10-day humidity at 90% RH.

34.5.3 Forced-Flow Batteries

With the development of the recirculation system in which the inflow of fresh electrolyte can be controlled, thereby maintaining the temperature and conductivity of the electrolyte, the performance of electric torpedo batteries has been improved markedly. With recirculation and flow control, a recirculation pump (see Fig. 34.2) and a voltage-sensing mechanism are added to the battery system. By this method the temperature of the battery and the conductivity of the seawater electrolyte increase. Since battery voltage increases directly with temperature and conductivity, it is possible to control the output of the battery by controlling the intake of electrolyte by means of the voltage-sensing mechanism.

The performance of one type of torpedo battery with and without recirculation voltage control is shown in Fig. 34.21.[17] The blocked-in area represents the limits within which an electric torpedo

TABLE 34.6 Performance Summary of Seawater-Activated Batteries

	Silver chloride	Cuprous iodide	Cuprous thiocyanate	Lead chloride
Number of cells	11	12	13	16
Battery dimensions:				
Height, cm	7.5	9.8	10.2	10.5
Width, cm	5.5	7.6	7.4	7.5
Thickness, cm	3.9	4.4	5.7	4.5
Weight, g	252	516	478	458
Activation:				
Low temp.:				
To 13.5 V, s	<15	<15	<15	<15
To 15.0 V, s	60	60	60	15
High temp.:				
To 15.0 V, s	<1	<1	<1	<1
Life:				
High temp., h	9.67	9.4	9.3	9.5
Low temp., h	9.80	10.3	10.3	10.7
Load resistance (per cell), Ω*	7.27	6.67	6.15	5.0
Cutoff voltage (per cell), V*	1.364	1.25	1.154	0.9375
Average current, A	0.206	0.220	0.236	0.219
Average volts per cell, V*	1.497	1.463	1.378	1.048
Wh/L	204	110	90	100
Wh/kg	130	70	79	75

*As each battery system contains a different number of cells, cell load resistances and cell voltages are different for each battery.

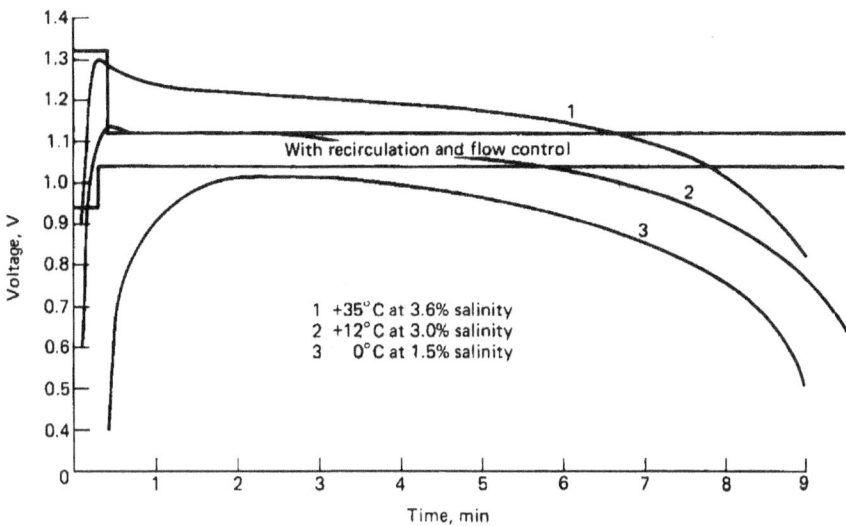

FIGURE 34.21 Discharge curves of torpedo battery—effect of recirculation and flow control.

34.20 SPECIALIZED BATTERY SYSTEMS

battery with recirculation and flow control will perform when discharged under any of the conditions shown by the three individual curves. All voltages pertinent to the start and finish of the battery are shown by the three individual curves.

34.5.4 Dunk-Type Batteries

Magnesium/Cuprous Chloride Batteries. The magnesium/cuprous chloride battery was widely used in applications requiring low-temperature performance, such as radiosondes, having replaced the more expensive magnesium/silver chloride system in applications where weight and volume are not critical. Figure 34.22 illustrates a typical magnesium/cuprous chloride battery. The pile-type construction shown in Fig. 34.3 is used.

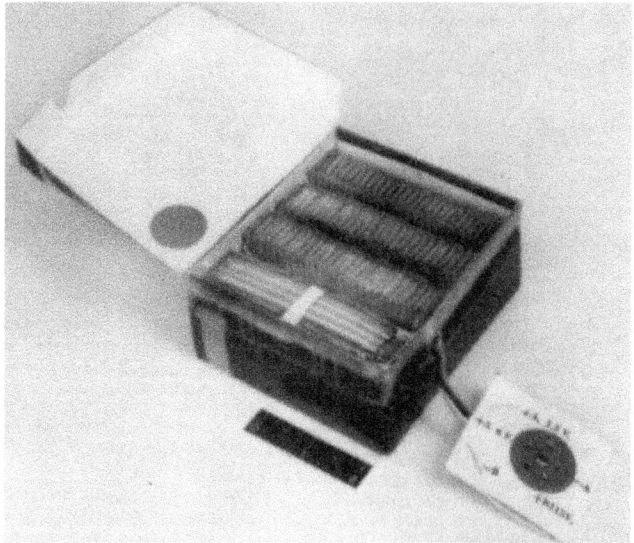

FIGURE 34.22 Magnesium/cuprous chloride radiosonde battery. Size: 10.2 × 11.7 × 1.9 cm; weight: 450 g; rated capacity: A_1 section—1.5 V, 0.3 Ah; A_2 section—6.0 V, 0.4 Ah; B section—115 V, 0.08 Ah.

The battery is activated by filling it with water, and full voltage is reached within 1 to 10 min. The battery is best suited for discharge at about the 1 to 3 h rate at temperatures from +60 to –50°C after activation at room temperature. Overheating and dry-out will occur on high current drains, and self-discharge limits the life after activation. For best service, these batteries should be put into use soon after activation. The heat that is developed during discharge can be used to advantage in batteries that are operated at low temperatures; therefore, the energy output varies little with decreasing temperature. Figure 34.23 shows the discharge curve for this battery at various temperatures. Figure 34.24 gives some typical discharge curves for this type of battery with a similar design at various discharge loads.

Magnesium/Manganese Dioxide Battery. This reserve battery consists of a magnesium anode and a manganese dioxide cathode.[10,18] It is activated by pouring an aqueous magnesium perchlorate electrolyte into the cells of the battery, where it is absorbed by the separators. Electrolyte absorption occurs within a few seconds at 0°C or above, but 3 min or more are required at –40°C due to the viscosity of the electrolyte.

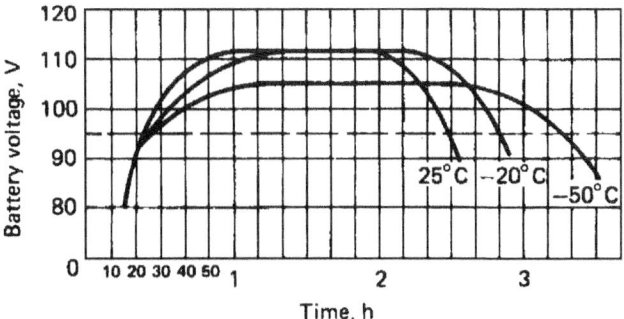

FIGURE 34.23 Discharge curves of magnesium/cuprous chloride radiosonde battery, 115 V section; discharge load: 3050 Ω.

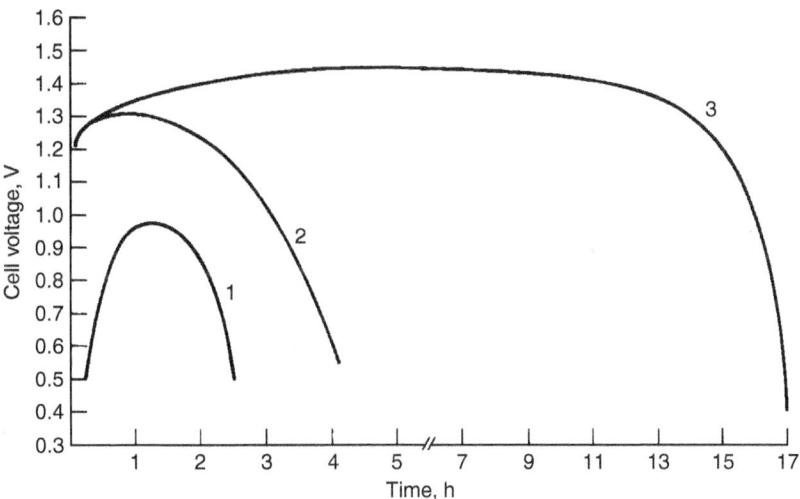

Cell no.	Load, Ω	Dimensions, cm			
		Volume	Length	Height	Thickness
1	2.5	10.2	8.2	2.5	0.5
2	8.0	2.5	2.2	3.8	0.3
3	125	1.3	2.0	2.0	0.3

FIGURE 34.24 Discharge curves of magnesium/cuprous chloride water-activated batteries at 20°C; electrolyte: tap water.

The battery can deliver between 80 and 100 Wh/kg over the temperature range of –40 to +45°C at the 10 to 20 h discharge rate. Over 75% of the battery's fresh capacity is available after 7 days' activated stand at 20°C and 4 days' storage at 45°C. Typical discharge curves are shown in Fig. 34.25 for a five-cell 10 Ah battery, weighing about 1 kg with a volume of 655 cm³.

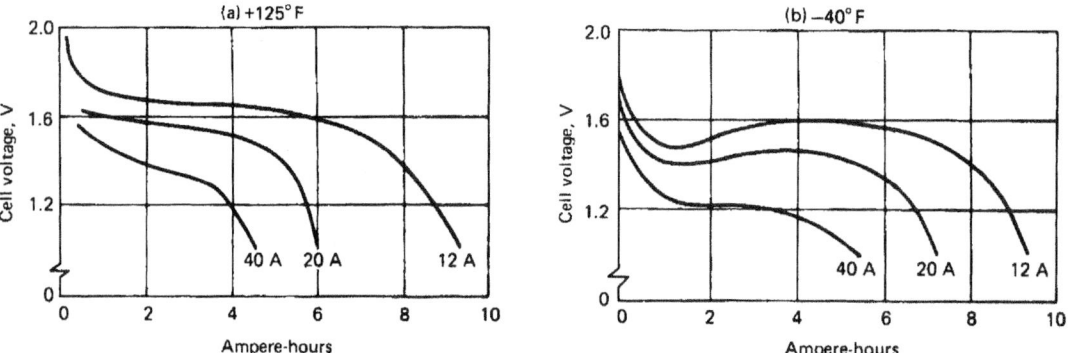

FIGURE 34.25 Typical discharge curves of magnesium/manganese dioxide cell, 10 Ah size. (*Courtesy of Eagle-Picher Industries.*)

34.6 BATTERY APPLICATIONS

Water-activated batteries can be viable candidates as the power source for many types of equipment. The choice of which battery to use becomes one of economics. With proper design, all will perform similarly. Where high current densities are required and cost is secondary, the magnesium/silver chloride system is best. All can be used as immersion or dunk-type batteries; all but the magnesium/cuprous chloride system will withstand long storage times at high temperatures and high humidities. At the present state of the art, only the magnesium/silver chloride system is suitable for use in forced-flow batteries.

34.6.1 Water-Activated Batteries for Aviation and Marine Life Jacket Lights

The magnesium/cuprous chloride water-activated battery system is being used in FAA and U.S. Coast Guard approved aviation and marine life jacket lifts. A typical light is shown in Fig. 34.26.

The single-cell battery has a cathode approximately 5 mm thick with a footprint of 7.25 by 2 mm. Table salt is added to the cathode mix[19] to obtain an adequate voltage in freshwater. (The holes in the battery case are optimized to maintain electrolyte salinity while allowing flushing of discharge products.) After being mixed while heated, and then cooled and rechopped, the powder is pressed and reheated in an automatic hydraulic press. The cathode is pressed with a titanium wire current collector, which is wire brushed before manufacture to remove oxide buildup.

The cell is constructed with two anodes, each with the same footprint as the cathode, connected in parallel and placed on either side of the cathode. The anodes are AZ61 electrochemical magnesium sheet.

Typical cell voltage at a 220 to 240 mA (C/12) discharge (against a miniature incandescent lamp) starts at 1.77 V in saltwater and goes down gradually to about 1.65 V before a sharp voltage drop signaling the end of discharge. Voltages in freshwater are about 0.1 V lower. Total capacity is about 3000 mAh.

A battery, with two cells wired in series for international marine use, uses a AT61 sheet because of the requirement for higher voltage. In saltwater, the cell voltage at a 340 mA (C/8) discharge (against a highly efficient gas-filled miniature lamp) is as high as 1.87 V early in the discharge and drops to about 1.8 V after 8 h. Again, the freshwater voltage is about 100 mV less per cell. This discharge is shown in Fig. 34.27.

Because the salt added to the cathode makes the cathode even more hygroscopic than it would otherwise be, the batteries are preferably stored with a removable pull-plug used to seal the holes in the battery case.

The characteristics of the life jacket lights are given in Table 34.7.

RESERVE MAGNESIUM ANODE AND ZINC/SILVER OXIDE BATTERIES **34.23**

FIGURE 34.26 Life jacket light, using magnesium/cuprous chloride water-activated battery. (*Courtesy of Electric Fuel Ltd.*[20])

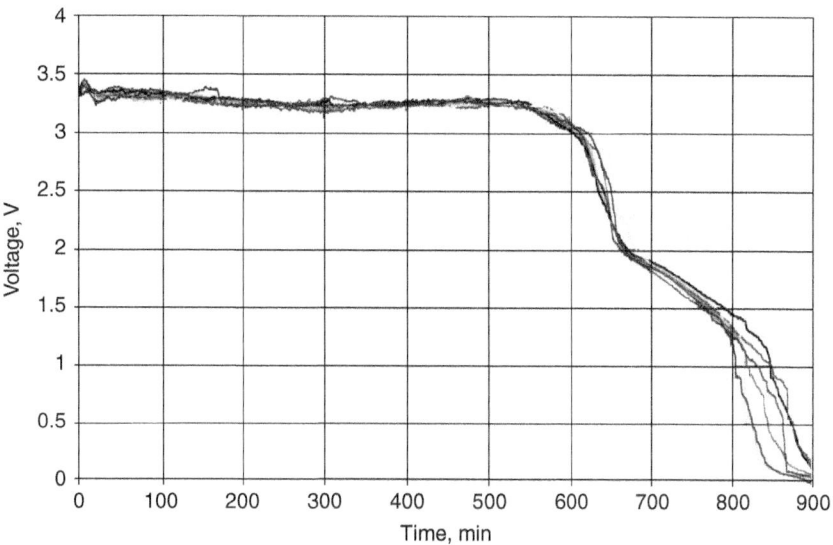

FIGURE 34.27 Typical discharge of 6 WAB-MX8 batteries in fresh tap water at 330 mA. (*Courtesy of Electric Fuel Ltd.*[20])

34.24 SPECIALIZED BATTERY SYSTEMS

TABLE 34.7 Characteristics of Life Jacket Lights

Electric fuel model no.	Nominal voltage, V	Nominal size, cm			Nominal discharge capacity		
		Length	Width	Height	Time	Wh	Normal usage mode
WAB-H12	1.7	2.9	1.6	9.3	12h	4.4	Aviation/marine life jacket light
WAB-H18	1.7	2.9	1.6	9.3	8h+	3.3	Aviation life jacket light
WAB-MX8	3.6	3.1	3.3	9.5	8h+	10.7	Marine life jacket light

Source: Electric Fuel Ltd.[20]

34.6.2 Magnesium/Silver Chloride Batteries

Figure 34.28 illustrates two of the magnesium/silver chloride batteries currently manufactured. These batteries are used in the following types of applications:

Lifeboat emergency equipment on commercial airlines

Sonobuoys

Radio and light beacons

Underwater ordnance

Radiosonde units—balloon transport equipment; high altitude, low ambient temperature operation.

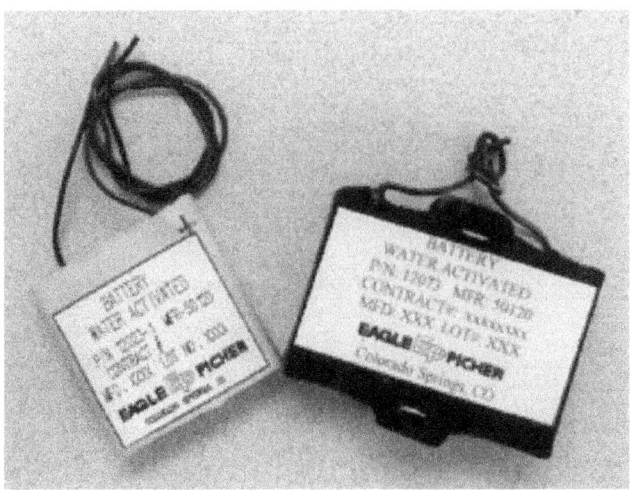

FIGURE 34.28 Magnesium/silver chloride batteries; 12023-1 and 12073.

34.7 BATTERY TYPES AND SIZES

Although "standard lines" of water-activated batteries were once manufactured, most batteries now are designed and manufactured for specific applications. Tables 34.8 and 34.9 list some of the standard and special purpose magnesium/cuprous chloride and magnesium/silver chloride batteries that were manufactured. Of these, only the two batteries illustrated in Fig. 34.28 are currently manufactured.

TABLE 34.8 Magnesium/Cuprous Chloride Water-Activated Batteries

E-P number	Other designation	Nominal voltage/ selection, V	Nominal size, cm			Nominal discharge capacity		Normal usage mode
			Length	Width	Height	Time	Wh	
MAP-12037	PIBAL	3.0	1.3	3.2	5.1	30 min	0.8	Airborne, lighting type
MAP-12051	—	18.0	6.8	3.8	5.7	120 min	2.16	Airborne, radiosonde
MAP-12053	BA-259	A: 1.5 B: 6.0 C: 115.0	11.7	10.2	6.0	A: 90 min B: 90 min C: 90 min	0.34 1.89 650.4	Airborne, radiosonde
MAP-12060	—	18.0	5.1	5.4	5.1	120 min	5.4	Airborne, radiosonde
MAP-12061	—	22.5	5.1	7.0	5.1	90 min	7.59	Airborne, radiosonde
MAP-12064	BA-253	6.0	10.2	3.8	3.8	45 min	2.25	Airborne, lighting type
MAP-12071	—	20.0	6.3	7.6	16.0	8.1 h	53.46	Submerged, buoy system

Source: Eagle-Picher Technologies, LLC.[21]

TABLE 34.9 Magnesium/Silver Chloride Water-Activated Batteries

E-P number	Other designation	Nominal voltage, V	Approximate size, cm			Nominal discharge capacity		
			Length	Width	Height	Time	Wh	Ah
MAP-2023-1	Squib firing battery	5.5	5.1	2.5	5.4	1 min	0.315	0.0572
MAP-12062	—	48	12.1	dia.	33	20 min	400	8.33
MAP-12065	—	4.5	6.3	6.7	13.9	50 h	157.5	35
MAP-12066	—	7.5	5.1	5.1	16.5	14 h	138	18.4
MAP-12067	MK-72 Squib firing battery	0.75	2.8	dia.	2.5	13 s	0.0010	0.0014
MAP-12069	—	10	7.6	2.5	8.9	6 h	1.5	0.15
MAP-12070	—	12	5.1	2.6	10	9 h	53.2	4.44
MAP-12073	—	14.5	7.6	2.8	5.1	15 h	14.55	1
MAP-12074	—	10.5	4.1	5.1	25	48 h	95.35	9

Source: Eagle-Picher Technologies, LLC.[21]

SECTION B
ZINC/SILVER OXIDE RESERVE BATTERIES

34.8 GENERAL CHARACTERISTICS OF ZINC/SILVER OXIDE RESERVE BATTERIES

An important reserve battery, particularly for missile and aerospace applications, is the zinc/silver oxide electrochemical system, which is noted for its high-rate capability and high energy density. The cell is designed with thin plates and large-surface-area electrodes, which augment its high-rate and low-temperature capability and provide a flat discharge characteristic. This design, however, reduces the activated or wet shelf life of the battery, necessitating the use of a reserve-battery design to meet storage requirements.

The zinc/silver oxide electrochemical system was the metallic couple with which Volta demonstrated the possibility of using dissimilar metals in a "pile-type" multicell construction to obtain a substantial electric voltage. The system existed somewhat as a laboratory device until Professor André designed a practical secondary cell early in World War II.

Subsequent to World War II, the U.S. military became interested in a dry-charged primary version for use in airborne electronics and missiles because of its very high energy output per unit weight and volume and high-rate capability. The ultimate result of this interest was the development of lightweight batteries for the aerospace industry, both military and civilian. The entire manned space program was keyed to zinc/silver oxide reserve batteries as the power sources for the various flight vehicles.

Zinc/silver oxide reserve batteries are divided into two classes, manually activated and remotely activated. In general, the manually activated types are used for space systems and accessible terrestrial applications and are usually packaged in more conventional configurations. The remote or automatically activated types are used principally for weapon and missile systems. This use requires a long period of readiness (in storage), a means for rapid remote activation, and an efficient discharge at high discharge rates, typically from about 10 s to over 4 h, inclusive of open-circuit wet stand periods. The specific energy and energy density of manually activated types range from about 60 to 220 Wh/kg and 120 to 550 Wh/L. For remotely activated types, the specific energy and energy density are reduced because of the self-contained activating device and range from about 11 to 88 Wh/kg and 24 to 320 Wh/L.

34.9 CHEMISTRY

The electrochemical reactions associated with the discharge of a zinc/silver oxide battery as a primary system are generally considered to proceed as follows. The cathode or positive electrode is silver oxide and may be either Ag_2O (monovalent), AgO (divalent), or a mixture of the two. The anode or negative electrode is metallic zinc, and the electrolyte is an aqueous solution of potassium hydroxide. The chemical reactions and the associated voltages at standard conditions are

$$Zn + 2AgO + H_2O \rightarrow Zn(OH)_2 + Ag_2O \qquad E^0 = 1.815 \text{ V}$$

$$Zn + Ag_2O + H_2O \rightarrow Zn(OH)_2 + 2Ag \qquad E^0 = 1.589 \text{ V}$$

The total cell reaction with 31% KOH electrolyte at 25°C is

$$Zn + AgO + H_2O \rightarrow Zn(OH)_2 + Ag \qquad E = 1.852 \text{ V}$$

34.10 CONSTRUCTION

A typical assembly of the manually activated reserve zinc/silver oxide cell is shown in Fig. 34.29. These batteries are designed to be filled with electrolyte just before use. The conventional cell design is a prismatic container with positive and negative terminals and a combination fill/vent cap. Batteries are formed by connecting single cells in series and packaging them in a unit container. Batteries used in space programs utilize thin-gauge stainless steel, titanium, magnesium, or composite containers to minimize weight.

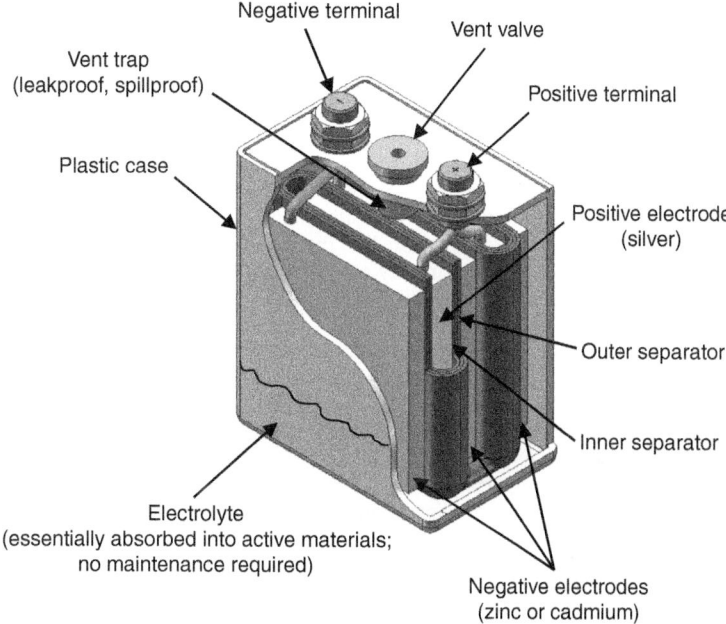

FIGURE 34.29 Typical construction of primary reserve zinc/silver oxide cell. (*Courtesy of Yardney Technical Products, Inc.*)

34.10.1 Cell Components

The components of a reserve zinc/silver oxide cell consist of the positive plates, the negative plates, and the separators. The components are assembled such that each negative plate is protected from direct contact with the adjacent positive plate by a separator. The cell components are assembled and packaged in a container; the plates can be prepared in either a dry and charged condition or dry and uncharged condition.

An alternate construction technique that has been successfully used on various applications is pile batteries. The major incentive for use of bipolar cells in a pile configuration is the elimination of intercell connectors, other heavy current carrying structural elements, and some of the cell containment features. This allows for a substantial weight reduction, increasing the power and energy density of the battery. Some of the disadvantages of the bipolar electrode are that the electrode can only react on one surface and the potential for intercell electrolyte leakage, which could create parasitic currents between some cells.

The construction usually consists of a bipolar or duplex matrix where both positive and negative electrodes are built on a common current collector. These are stacked in a pile configuration with a separator placed between each bipolar electrode. This results in a high voltage, multiple cell battery. Each cell consists of the positive side of a bipolar electrode, the negative side of the next bipolar electrode, the separator in between, and a plastic frame. The common current collector has two main functions. It separates the positive and negative portion of each bipolar electrode and serves as an intercell connector and electrolyte barrier.

Positive Plates. The positive plates are prepared by applying silver or silver oxide powder to a metallic grid. Copper, nickel, and silver have all been used for grid material, with silver the most prevalent for reasons of electrochemical stability and conductivity. After the silver powder is pressed or sintered to the grid, the plates are electroformed in an alkaline solution, then washed thoroughly and air-dried at a moderate temperature (usually 20 to 50°C). The nominally divalent oxide thus formed is relatively stable at ambient temperatures but tends to lose oxygen and degrade to the monovalent state with increasing temperatures and time. Continuous exposure to high temperatures (70°C) causes reduction to the monovalent oxide in a few months.

Negative Plates. The negative plates may be prepared by pasting or pressing zinc powder or zinc oxide onto a grid or by electroplating zinc from an alkaline bath to form a very active spongy zinc deposit.

Both positive and negative electrodes may vary in thickness from 0.12 mm as a practical minimum to 2.5 mm maximum for positives and 3.5 mm maximum for negatives. The extremely thin plates are utilized for very short-life, high-discharge-rate automatically activated batteries; the thick plates are employed in manually activated batteries designed for continuous discharge over several months at very low currents.

Separator Materials. Typical separator materials used in zinc/silver oxide cells include regenerated cellulose films (cellophane, fiber-reinforced, or silver-treated cellophane), and woven nylon or nonwoven synthetic fiber mats of nylon, hemp tissue, polyvinyl alcohol, rayon, or high alpha cellulose content papers. The woven nylon, paper, or synthetic fiber mats are frequently placed adjacent to the positives to protect the cellophane from the highly oxidizing influence of that material. The cellophane, a semipermeable film, prevents buildup of particles between plates (while allowing ionic transfer), thus preventing interplate short circuits. The mats also absorb the electrolyte solution and distribute it over the electrode surfaces. Cells intended for automatic activation normally are not designed with the film separators for they require too long for complete wetting. The open-mat separators provide sufficient protection from interplate short-circuiting for several minutes to hours.

Separator materials are necessary for cell operation because they prevent short circuits, but they also impede current flow, causing an *IR* drop within the cell. Very high-discharge-rate cells must have very low internal impedance, hence a minimum of separator material. As a result, this type of cell is restricted to very short wet-life applications. The semipermeable film is the separator which contributes most to *IR* drop and also to protection against short circuits. Long-life cells may contain five or six layers of cellophane. They are therefore better suited to medium or low discharge rates.

Electrolyte. The electrolyte used for reserve zinc/silver oxide cells is an aqueous solution of potassium hydroxide. High and medium discharge rate cells use a 31% by weight electrolyte solution because this composition has the lowest freezing point and is close to the minimum resistance, which occurs at 28 wt.%. Low-rate cells may use a 40 to 45% solution since lower rates of hydrolysis of cellulosic separators occur with the higher KOH concentrations.

34.10.2 High- and Low-Rate Designs

A battery intended to be discharged at a 5 to 60 min rate is considered a high-rate design. These cells are designed primarily to deliver high current and require a large plate surface area. They therefore contain many very thin plates. The separators also have as low an impedance as possible, that is, one or two layers of cellophane versus five or six layers for low-rate cells. The 31% potassium hydroxide electrolyte has a high conductivity and is therefore employed in high-rate cells.

Low-rate batteries are in a class intended for discharge at rates ranging from 10 to 1000 h with emphasis on high specific energy and energy density. The plates are thick (2 mm), and relatively high impedance separator wraps are used. A higher concentration of electrolyte (40%), which permits a greater ampere-hour capacity, can also be used. This design configuration also gives a substantial improvement in the activated or wet stand capability of the cell.

34.10.3 Automatically Activated Types

The automatically activated type battery is a class of reserve battery intended for quick preparation for use after an undetermined period subsequent to installation. The very high power output of the primary zinc/silver oxide system and the use of an integrally designed system for injecting the electrolyte into the cells combine to provide an efficient power source for weapons and other systems requiring a long-term ready state. Figure 34.30a shows typical automatically activated batteries for military applications. Figure 34.30b shows a typical automatically activated battery for a torpedo application.

Four kinds of activation systems have been utilized in this type of battery for transferring electrolyte from a reservoir to the cells. All the systems depend on pressure above ambient (e.g., 4 to 1 ratio) to move the electrolyte, and the most conventional source of gas is a pyrotechnic device.

The "gas generator" is a small cartridge that contains an ignitable propellant material and an electrically fired ignitor or "match." Figure 34.31 shows the four types of battery designs.

The tubular reservoir (Fig. 34.31a) can assume many forms. It is usually coiled around the battery, as shown in Fig. 34.32 (an assembly of a battery with a tubular reservoir), but it can also be formed with 180° bends into a flat shape, or it can be configured to fit into available nonstandard volumes into which a missile battery is often mounted. The tubular reservoir is fitted with foil diaphragms at each end. For activation, the gas generator located at one end can be electrically ignited; the gas causes the diaphragms to break, and the electrolyte is forced into a manifold which distributes it to the cells of the battery. The piston activator (Fig. 34.31b) operates by pushing the electrolyte out of a cylindrical reservoir when a gas generator is fired behind it. The tank activator (Fig. 34.31c) contains the electrolyte in a variable-geometry tank with a gas generator located at the top. When the gas enters at the top, the electrolyte is forced out through an aperture at the bottom. The system is position-sensitive and will operate properly only when in an upright position relative to the components. The tank-diaphragm activator (Fig. 34.31d) uses a sphere or spheroid tank with a diaphragm attached internally at the major circumference. When the gas generator is fired, the diaphragm moves to the opposite side, forcing the electrolyte out through an aperture in the reservoir side of the tank.

Of the four systems, the tubular system is the most versatile, but in simple battery shapes may be heavier. The piston and diaphragm systems have moving parts and thus can be less reliable; they are also less adaptable to special shapes. The tank is efficient but position-sensitive.

Depending on the battery design and intended applications, the battery can be configured to externally vent or retain any internal gases due to the activation system and normal gassing of the cells.

The operating sequence of an automatically activated battery involves: (1) application of ignition current, (2) gas generator propellant burning and associated gas production, (3) rupture of a diaphragm, (4) movement of electrolyte out of the reservoir into the distribution manifold, and (5) filling of the cells with electrolyte. In a typical operation, the total sequence involves less than 1 s. In many applications the electrical load is wired directly to the battery, and so the battery activates under load. Figure 34.33 shows the rise times of the voltage under load (A) and under no-load (B) condition for a battery not used until 6 h later. The delayed-use battery has film separators, and the slower wetting is reflected by the longer rise time.

In some applications, reserve batteries include multiple power taps and can utilize cells with different capacity ratings within the same battery container.

Automatically activated batteries suffer a weight and volume penalty compared with manually activated types, but the design permits the use of a high-performance battery when there is no time available to activate manually or the unit is inaccessible. In many applications, both conditions exist. The volume penalty is usually about 2 times and the weight penalty about 1.6 times the basic battery.

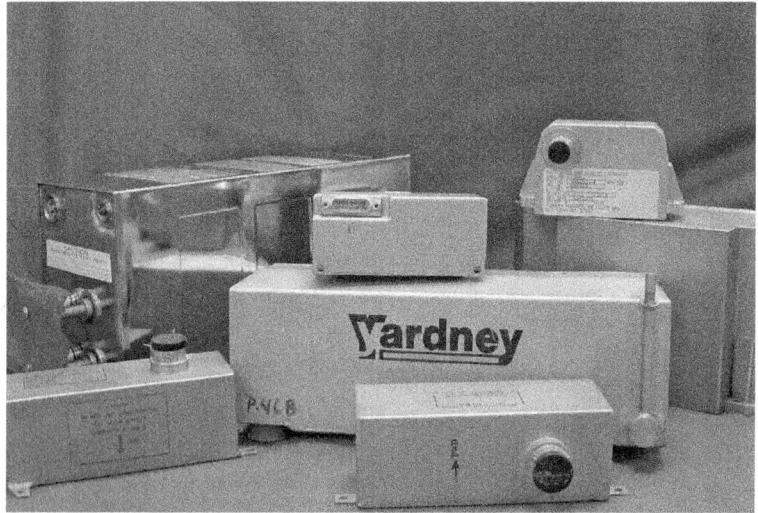

(a)

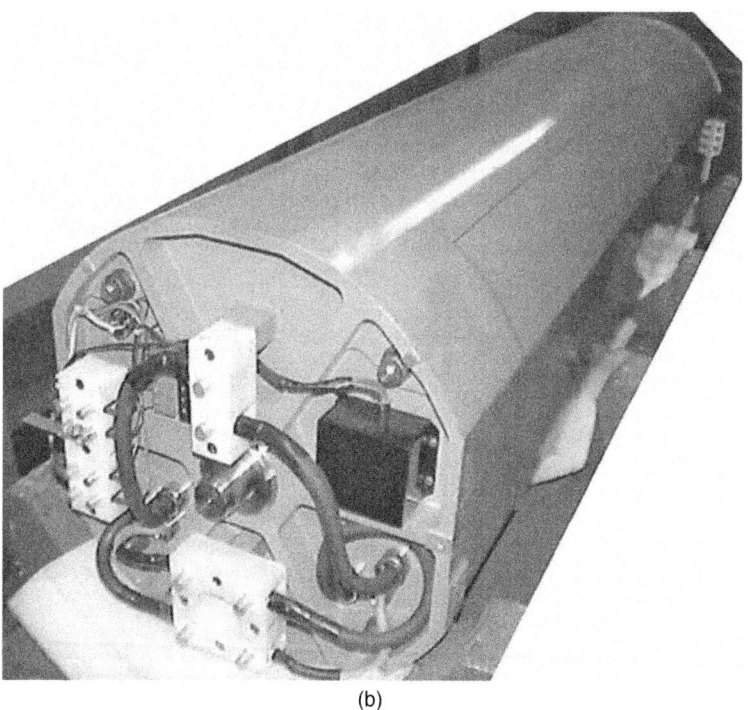

(b)

FIGURE 34.30 (*a*) Zinc/silver oxide primary reserve batteries designed for automatic activation. (*b*) Automatically activated zinc/silver oxide torpedo battery. (*Courtesy of Yardney Technical Products, Inc.*)

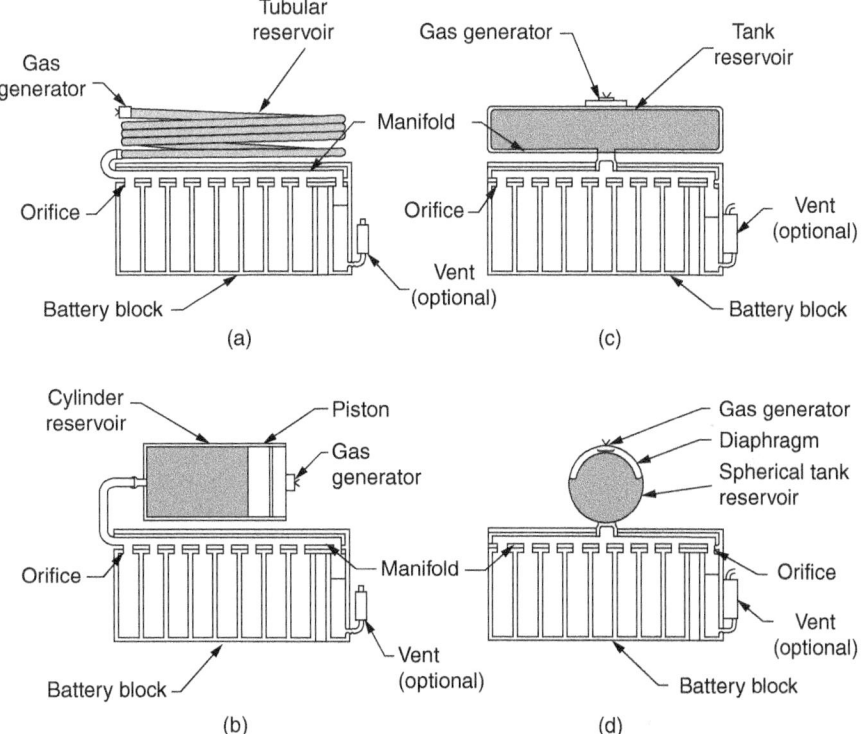

FIGURE 34.31 Schematic drawings of four types of activation systems used in automatically activated batteries. (*a*) Tubular reservoir. (*b*) Piston activator. (*c*) Tank activator. (*d*) Tank-diaphragm activator. (*Courtesy of Yardney Technical Products, Inc.*)

FIGURE 34.32 Assembly of automatically activated zinc/silver oxide primary battery with tubular coil reservoir. (*Courtesy of Eagle-Picher Technologies.*)

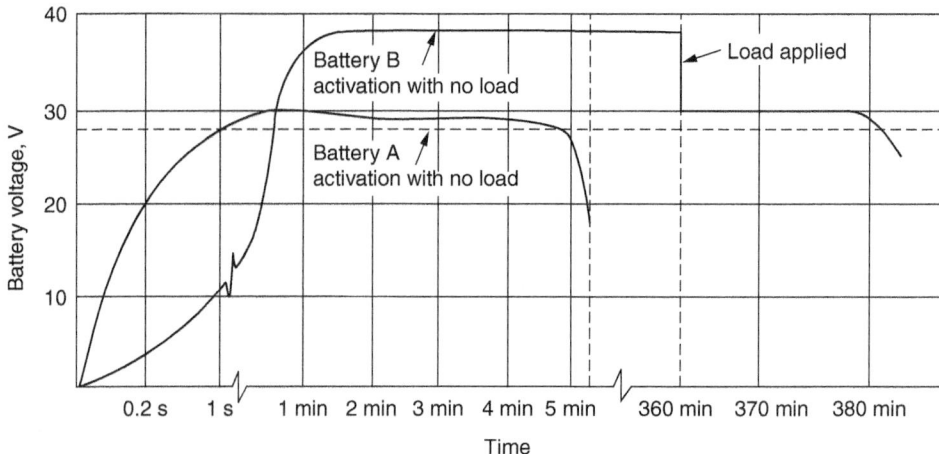

FIGURE 34.33 Voltage rise time for automatically activated zinc/silver oxide batteries at 25°C. (*Courtesy of Eagle-Picher Technologies.*)

Most automatically activated battery designs utilize an integral electric heater. The heater maintains the electrolyte at about 40°C or at a temperature that will raise a cold battery to 40°C when activation occurs. The use of heaters permits the design of batteries that can meet close voltage tolerances, thus improving the capability of the weapons' electric and electronic systems when operating over a wide temperature range.

34.11 PERFORMANCE CHARACTERISTICS

Zinc/silver oxide reserve batteries as a class are somewhat unique in that they are almost entirely committed to specific applications. These applications require the flat voltage profile and the high specific energy and energy density available from this system, and they often demand a special design for each requirement. If a low-temperature environment is involved, battery heaters are used. If the discharge requires a wide range of current with only a small voltage variation, many very thin plates are used. A very high capacity requirement at low rates requires the use of thick plates and more concentrated electrolyte. There is no standard design or size because there is no typical application. The applications always demand the maximum from the battery design in capacity and voltage regulation at a minimum weight and volume. Multiple batteries are commonly packaged as a single unit, providing a range of load current and ampere-hour capacity in one convenient package.

34.11.1 Voltage

The open-circuit voltage of the zinc/silver oxide cell will range from 1.6 to 1.85 V per cell. The nominal load voltage is 1.5 V, and typical end voltages are 1.4 V for low-rate cells and 1.2 V for high-rate cells. At high rates, such as a 5 to 10 min discharge rate, the output voltage would be about 1.3 to 1.4 V per cell, whereas the 2 h rate discharge voltage would be slightly above 1.5 V. Figure 34.34 shows a family of discharge curves at four different current densities. The voltage level is inversely related to the current density (calculated from the area of the active plate surface). Thus, based on 100 cm^2 of positive-plate surface, a 10 A discharge rate would be a 0.1 A/cm^2 current density. If the discharge rate is doubled to 0.2 A/cm^2, the voltage level would decrease, and if the rate is lowered to 0.05 A/cm^2, the voltage level would increase.

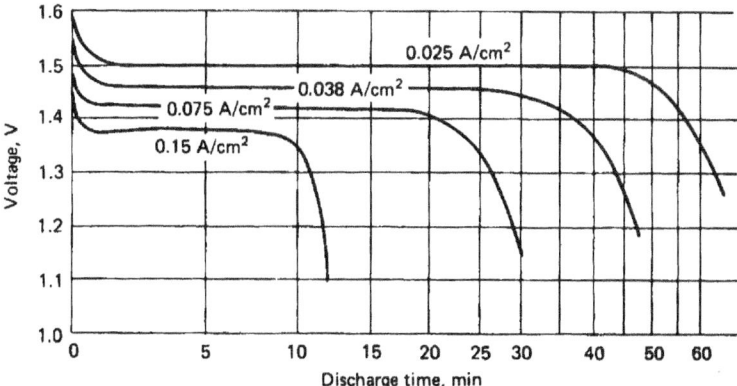

FIGURE 34.34 Effect of changing current density on battery voltage at 25°C. (*Courtesy of Eagle-Picher Technologies.*)

In cell design, the ampere-hour capacity of the cell is determined by the amount of silver oxide active material present (zinc active material is provided in excess because of the cost relationship to silver), but the voltage is determined by the current density. In a fixed volume, higher discharge rates can be obtained without lowering the battery voltage by using thinner plates (thus providing more plates per cell element and lowering the current density), but with a reduction of capacity. The lower the current density at which a battery can operate, the better the voltage regulation with changing rates of discharge.

34.11.2 Discharge Curves

A set of discharge curves for high-rate batteries is shown in Fig. 34.35 and for low-rate batteries in Fig. 34.36. The designs for these two types of batteries are quite different, with the principal difference being the thickness of the plates. The thin plates used in high-rate cells provide more surface area for lower current density, thus better voltage control and also more efficient utilization of the active material. At lower rates of discharge, as in Fig. 34.36, the voltage level is higher and active material utilization is also excellent, both because of lower current density. It will be noted that the

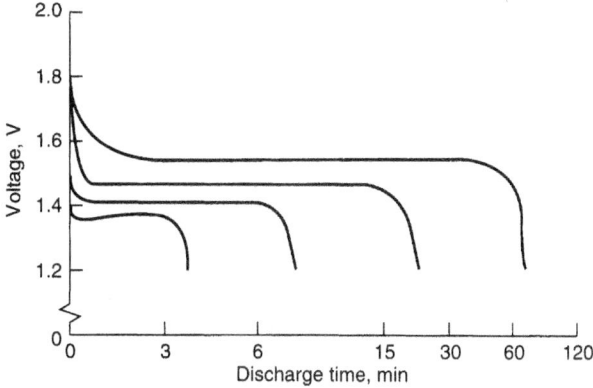

FIGURE 34.35 Discharge curves for high-rate zinc/silver oxide batteries at 25°C. (*Courtesy of Eagle-Picher Technologies.*)

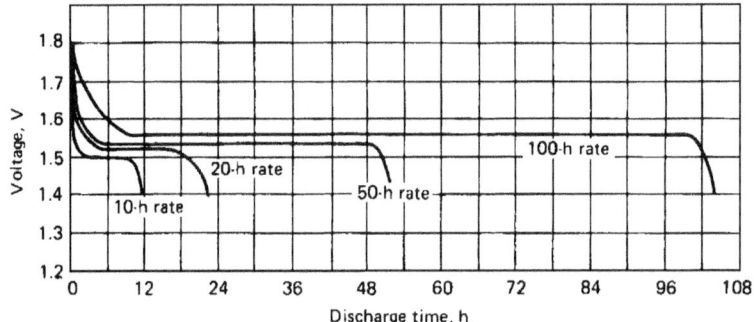

FIGURE 34.36 Discharge curves for low-rate zinc/silver oxide batteries at 25°C. (*Courtesy of Eagle-Picher Technologies.*)

low-rate discharge curves are above 1.6 V for a period of time. This is the effect of the divalent oxide, which affects voltage only at low rates. Most of the divalent capacity is obtained at high rates, but its voltage is decreased by the higher current density imposed.

34.11.3 Effect of Temperature

The family of curves shown in Fig. 34.37 illustrates the performance obtained from a high-rate battery when discharged over a range of temperatures. It should be understood that the change in voltage levels caused by temperature is closely related to the changes caused by current density. Thus the adverse effect of cold temperature can be improved by reducing the current density of the cell, and the voltage and capacity of batteries discharged at high current densities can be improved by increasing their operating temperature. Figure 34.38 shows a family of curves for low-rate batteries discharged at various temperatures. The two sets of curves show that the zinc/silver oxide system is significantly affected at temperatures below 0°C and thus is not recommended for applications in this environment without heaters.

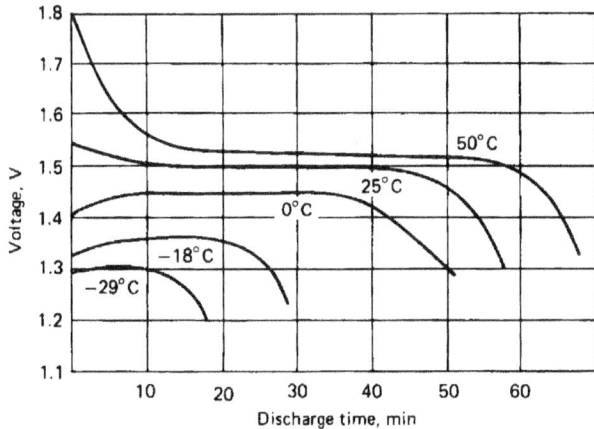

FIGURE 34.37 Effects of temperature on high-rate zinc/silver oxide primary batteries discharged at the 1 h rate. (*Courtesy of Eagle-Picher Technologies.*)

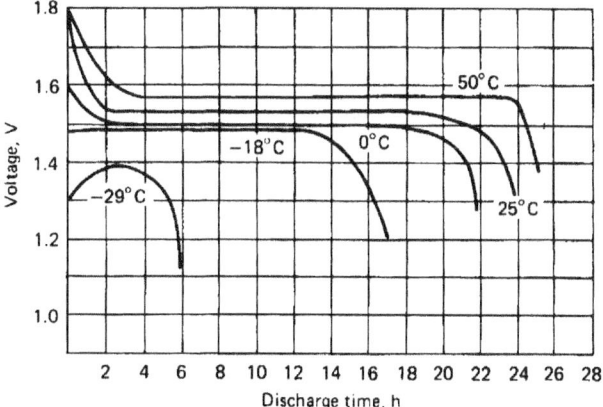

FIGURE 34.38 Effects of temperature on low-rate zinc/silver oxide primary batteries discharged at the 24 h rate. (*Courtesy of Eagle-Picher Technologies.*)

34.11.4 Impedance

Figure 34.39 shows the dynamic internal resistance (DIR) of a high-rate cell at various stages of discharge and temperature. These curves show a declining ($\Delta V/\Delta I$) ratio until the end of discharge, at which time the dynamic resistance rises rapidly. The declining impedance is caused by an improvement in the positive-plate conductivity and a temperature rise during the discharge. This feature can vary considerably, depending on cell design, ambient temperature of the discharge, and the point in time after the change of discharge rate when the voltage change is observed.

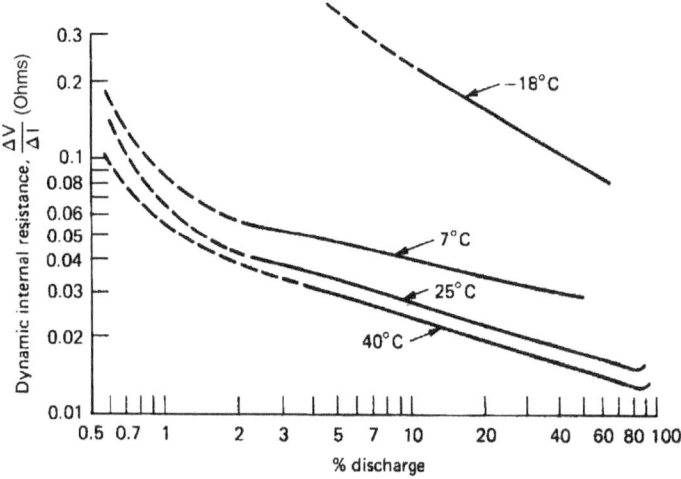

FIGURE 34.39 Dynamic internal resistance of zinc/silver oxide primary batteries. (*Courtesy of Eagle-Picher Technologies.*)

34.11.5 Service

The performance of zinc/silver oxide batteries in amperes per unit weight and volume versus service time is given in Fig. 34.40. It can be noted, again, that this battery system is particularly sensitive to temperatures below 0°C. These data are applicable, within reasonable accuracy, for both high- and low-rate designs.

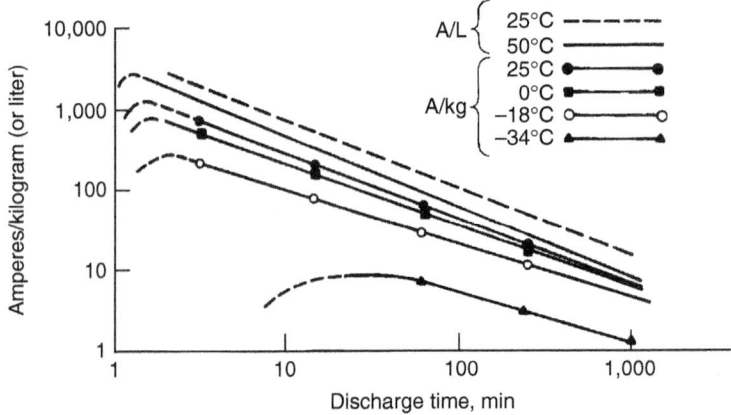

FIGURE 34.40 Service life of zinc/silver oxide primary batteries. (*Courtesy of Eagle-Picher Technologies.*)

34.11.6 Shelf Life

The dry shelf life of the zinc/silver oxide battery is shown in Fig. 34.41, which gives storage data at 25, 50, and 74°C for periods of up to 2 years. The losses shown are based on the assumption that the positive active material is divalent silver oxide, which slowly degrades to monovalent oxide at temperatures above about 20°C. Degradation of the negative plate is minimal. It is expected that the monovalent oxide level would be reached in about 30 months when the storage temperature is 50°C. Experience has shown that batteries stored at average ambient temperatures of 25°C or lower retain capacity at or above the monovalent oxide level for a period of 25 years or longer.

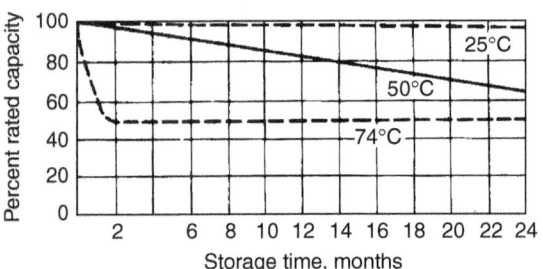

FIGURE 34.41 Dry storage of zinc/silver oxide primary batteries. (*Courtesy of Eagle-Picher Technologies.*)

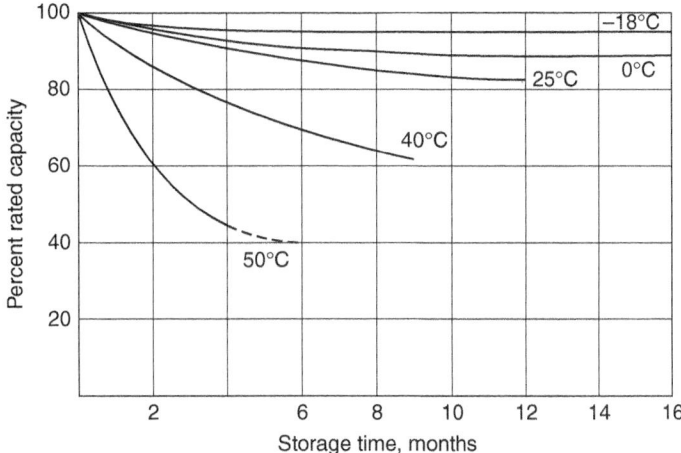

FIGURE 34.42 Wet (activated) storage of zinc/silver oxide primary batteries. (*Courtesy of Eagle-Picher Technologies.*)

The wet shelf life of the zinc/silver oxide battery varies considerably with design and method of manufacture. Figure 34.42 provides a guide to the expected performance of most designs. The wet shelf-life degradation is caused principally by loss of negative-plate capacity (dissolution of the sponge zinc in the electrolyte) or development of short circuits through the cellulosic separators.

34.12 CELL AND BATTERY TYPES AND SIZES

Single-cell units of the reserve zinc/silver oxide type are available in sizes from about 1 Ah as a minimum up to about 775 Ah. Tables 34.10a and 34.10b provide the specifications for a series of high-rate cells ranging in capacity from 1 to 250 Ah and a series of low-rate cells ranging in capacity from about 2 to 2680 Ah. These are all manually activated.

Table 34.11 lists a number of automatically activated batteries which have been designed to meet various specific applications. Most of these batteries are high-rate with a short wet-life. The weight and volume of this type are more a function of the load requirements and the space envelope provided than voltage and capacity.

34.13 SPECIAL FEATURES AND HANDLING

Both manually and automatically activated zinc/silver oxide batteries were developed to meet highly stringent requirements with regard to performance and reliability. The time and temperature of storage prior to use are of importance, and records should be maintained to ensure use within allowable limits. Special care must be exercised to ensure that the proper amount of the specified type of electrolyte is added to each cell of a manual-type battery and that, after activation, the unit is discharged within the shelf-life limitation at the proper temperature. Some battery containers have pressure-relief valves or heaters, or both, and these must be carefully maintained and monitored.

Automatically activated batteries require special preinstallation check out of gas generator ignitor circuits, heater circuits, and vent fittings. For long-term installations, there should be monitoring of the ambient temperature to prevent degradation caused by exposure to high temperatures. Periodic

TABLE 34.10a Zinc/Silver Oxide Manually Activated Batteries

	High-rate cells, 15 min rate				Low-rate cells, 20 h rate					Physical dimensions, cm		
Cell type*	Cap., Ah	Specific energy Wh/kg	Energy density Wh/L	Wt, g.	Cell type*	Cap., Ah	Specific energy Wh/kg	Energy density Wh/L	Wt, g	Length	Width	Height
SZH 1.0	1.0	57	104	25	SZL 1.7	1.7	84	171	30	1.09	2.69	5.16
SZH 1.6	1.6	66	110	35	SZL 2.8	2.8	88	201	50	1.25	3.07	5.72
SZH 2.4	2.4	66	116	55	SZL 4.5	4.5	92	220	75	1.42	3.50	6.32
SZH 4.0	4.0	66	128	90	SZL 7.5	7.5	97	250	120	1.63	4.00	7.09
SZH 7.0	7.0	66	134	160	SZL 16.8	16.8	106	305	240	2.00	4.95	8.48
SZH 16.0	16.0	66	140	370	SZL 43.2	43.2	125	397	520	2.54	6.27	10.39
SZH 68.0	68.0	80	196	1290	SZL 160.0	160.0	187	470	1330	3.73	9.27	15.09
SZH 250.0	250.0	154	410	2450	—	—	—	—	—	4.32	9.45	22.43
—	—	—	—	—	SZL 410.0	410.0	210	560	3000	4.22	13.84	19.35
—	—	—	—	—	SZL 775.0	775.0	276	957	4380	6.96	8.36	21.70

*Eagle-Picher Technologies.

TABLE 34.10b Zinc/Silver Oxide Manually Activated Primary Batteries*

Cell model[†]	Type[‡]	Voltage[§]	Capacity Ah	Specific energy Wh/kg	Energy density Wh/L	Weight g	Dimensions, cm Height	Width	Depth	Volume L
PM1	HR	1.42	2.0	92	147	31	5.13	2.74	1.37	0.019
PMV2	HR	1.48	5.3	103	184	76	6.42	4.37	1.52	0.043
PM3	HR	1.41	6.4	106	187	85	7.26	4.37	1.52	0.048
PML4	HR	1.42	8.3	113	208	104	8.53	4.37	1.52	0.057
PM5	MR	1.49	9.9	119	187	124	7.36	5.28	2.03	0.079
PMC5	MR	1.48	12.3	141	231	129	7.36	5.28	2.03	0.079
PMC10	MR	1.48	28	152	312	272	12.00	5.89	1.88	0.133
PM15	HR	1.42	19	92	180	292	12.55	5.89	2.03	0.150
PMV16	HR	1.47	18	72	141	365	15.57	5.84	2.06	0.187
PM30	HR	1.44	41	98	169	600	16.64	8.28	2.54	0.350
PM58	HR	1.42	56	85	162	938	18.42	8.26	3.23	0.491
PML100	LR	1.50	118	180	376	982	13.74	9.70	3.53	0.470
PML140	LR	1.49	165	197	439	1,250	16.36	9.70	3.53	0.560
PML170	LR	1.48	200	197	469	1,500	18.44	9.70	3.53	0.631
PML400	LR	1.47	375	218	566	2,525	16.10	15.27	3.96	0.974
PML2500	LR	1.48	2680	221	721	17,960	47.90	10.72	10.72	5.505

*These batteries are normally used as primaries. However, they all can be recharged (typically 3 to 10 cycles).
[†]Yardney Technical Products.
[‡]HR = High rate, MR = Medium rate, LR = Low rate.
[§]HR = 15 min rate, MR = 1 h rate, LR = 5 h rate.

TABLE 34.11 Zinc/Silver Oxide Automatically Activated Batteries

Part number*	Application	Weight, kg	Volume, L	Voltage, V	Current, A	Capacity, Ah	Specific energy and energy density Wh/kg	Wh/L
EPI 4331	AIM-7	1.0	0.45	26	10.0	0.8	21	46
EPI 4568	Peacekeeper	3.3	1.89	30	2.0	3.8	35	60
EPI 4500	Patriot	3.6	1.61	51	18.0	1.5	21	48
YTP 15148	Trident I (C-4)	5.0	1.20	28	6.0	12.0	65	284
EPI 4567	Peacekeeper	6.2	3.46	30	11.0	16.0	77	139
EPI 4470	Harpoon	8.6	3.5	28	27, 40	8, 12	65	160
EPI 4445	Torpedo	9.3	4.8	28	30	20	60	117
YTP 15066	Trident I	14.5	3.8	30, 31	15, 23	4, 10	30	112
YTP 5659985	Trident II (D-5)	30.0	31.2	34, 32	113, 10	15, 7	21.7	20.83
YTP P-530	Minuteman	0.77	0.36	30	10.0	0.46	17.9	38.3
YTP P-515	Sparrow	0.99	0.45	24	11.0	0.45	10.9	24.0
YTP P-512	NMD	0.86	0.30	30	13.0	0.30	10.5	30.0
YTP P-468	AGM130	7.03	2.70	28	30	12.08	45.0	117.0
YTP P-471	Peacekeeper	19.5	12.1	76, 31	16.7, 40	5.46, 40.90	86.3	139.3
YTP P-329[†]	Hawk	3.18	1.64	59, 25, 19, 12, −13, −59	2.2, 0.8	0.5	5.7	11.0
YTP 17511	MK37 Torpedo	120	96.6	85, 76	900, 450	79, 22	67.5	83.9
YTP 19580	SST-4 Torpedo	408	467	210, 115	480, 525	29, 110	54.3	47.4
YTP[‡]	Tigerfish	583	367	45/60	1200/750	240	54.3, 47.6	75.61

*EPI—Eagle-Picher Technologies; YTP—Yardney Technical Products, Inc.
[†]6 V taps.
[‡]Consists of one forward and one aft battery.

checks should be made to ensure that the ignitor circuits are intact, as some circuits are sensitive to electromagnetic fields. After activation, if the battery is not discharged within the specified time, it must be replaced.

The proper electrical performance of these batteries is best ensured by operating them at temperatures at or slightly above room temperature. Temperatures below 15°C can adversely affect the voltage regulation of high-rate batteries, and below 0°C there is also considerable loss of capacity for both types.

34.14 COST

The cost of high-performance primary zinc/silver oxide batteries is dependent on the specifications to which they are built and the quantity involved. Manual-type batteries may cost anywhere from $5 to $15 per watthour; remote-activated types will cost about $15 to $20 per watthour. When the price of silver is high, material cost becomes one of the chief disadvantages of these batteries. There are many applications, however, in which no other technology can meet the high energy density of the zinc/silver oxide primary system.

REFERENCES (SECTION A)

1. National Defense Research Committee, *Final Report on Seawater Batteries,* Bell Telephone Laboratories, New York, 1945.
2. L. Pucher, "Cuprous Chloride-Magnesium Reserve Battery," *J. Electrochem. Soc.* **99**:203C (1952).
3. B. N. Adams, "Batteries," U.S. Patent 2,322,210, 1943.
4. H. N. Honer, F. P. Malaspina, and W. J. Martini, "Lead Chloride Electrode for Seawater Batteries," U.S. Patent 3,943,004, 1976.
5. H. N. Honor, "Deferred Action Battery," U.S. Patent 3,205,896, 1965.
6. N. Margalit, "Cathodes for Seawater Activated Cells," *J. Electrochem. Soc.* **122**:1005 (1975).
7. J. Root, "Method of Producing Semi-Conductive Electronegative Element of a Battery," U.S. Patent 3,450,570, 1969.
8. R. F. Koontz and L. E. Klein, "Deferred Action Battery Having an Improved Depolarizer," U.S. Patent 4,192,913, 1980.
9. E. P. Cupp, "Magnesium Perchlorate Batteries for Low Temperature Operation," *Proc. 23d Annual Power Sources Conf.,* Electrochemical Society, Pennington, N.J., 1969, p. 90.
10. N. T. Wilburn, "Magnesium Perchlorate Reserve Battery," *Proc. 21st Annual Power Sources Conf.,* Electrochemical Society, Pennington, N.J., 1967, p. 113.
11. W. A. West-Freeman and J. A. Barnes, "Snake Battery; Power Source Selection Alternatives," NAVSWX TR 90-366, Naval Surface Warfare Center, Carderock Div. 1990.
12. M. G. Medeiros and R. R. Bessette, "Magnesium-Solution Phase Catholyte Seawater Electrochemical System," *Proc. 39th Power Sources Conf.,* Cherry Hill, N.J., June 2000, p. 453.
13. M. E. Wilkie and T. H. Loverude, "Reserve Electric Battery with Combined Electrode and Separator Member," U.S. Patent 3,061,659, 1962.
14. K. R. Jones, J. L. Burant, and D. R. Wolter, "Deferred Action Battery," U.S. Patent 3,451,855, 1969.
15. H. N. Honor, "Seawater Battery," U.S. Patent 3,966,497, 1976.
16. H. N. Honer, "Multicell Seawater Battery," U.S. Patent 2,953,238, 1976.
17. J. F. Donahue and S. D. Pierce, "A Discussion of Silver Chloride Seawater Batteries," Winter Meeting, American Institute of Electrical Engineers, New York, 1963.
18. H. R. Knapp and A. L. Almerini, "Perchlorate Reserve Batteries," *Proc. 17th Annual Power Sources Conf.,* Electrochemical Society, Pennington, N.J., 1963, p. 125.

19. U.S. Patent 5,424,147.
20. Electric Fuel, Ltd., Beit Shemesh, Israel.
21. Eagle-Picher Technologies, LLC.

BIBLIOGRAPHY (SECTION B)

Bauer, P.: *Batteries for Space Power Systems,* U.S. Government Printing Office, Washington, D.C., 1968.
Cahoon, N. C., and G. W. Heise: *The Primary Battery,* Wiley, New York, 1969.
Chubb, M. F., and J. M. Dines: "Electric Battery," U.S. Patent 3,022,364.
Eagle-Picher Technologies, Joplin, MO, website: www.eagle-picher.com.
Fleiseher, A., and J. J. Lander: *Zinc Silver Oxide Batteries,* Wiley, New York, 1971.
Hollman, F. G., et al.: "Silver Peroxide Battery and Method of Making," U.S. Patent 2,727,083.
Jasinski, R.: *High Energy Batteries,* Plenum, New York, 1967.
Yardney Technical Products, Pawcatuck, CT, website: www.yardney.com.

CHAPTER 35
RESERVE MILITARY BATTERIES

David L. Chua, Benjamin M. Meyer, William J. Epply,
Jeffrey A. Swank, and Michael Ding

SECTION A
AMBIENT-TEMPERATURE LITHIUM ANODE RESERVE BATTERIES

35.1 GENERAL CHARACTERISTICS

The use of lithium metal as an anode in reserve batteries provides a significant energy advantage over traditional reserve batteries because of the high potential and low equivalent weight (3.86 Ah/g) of lithium. A lithium reserve battery can operate at a voltage close to twice that of the conventional aqueous types. Due to the reactivity of lithium in aqueous electrolytes, with the exception of the special lithium/water and lithium/air batteries (see Sec. 33.6), lithium batteries must use a nonaqueous electrolyte with which lithium is nonreactive.

The various ambient-temperature active (nonreserve) lithium primary batteries are covered in Chap. 14. Of these systems, the ones demonstrating the higher energy densities and rate capabilities are Li/SO_2, Li/V_2O_5, $Li/SOCl_2$, Li/SO_2Cl_2, and Li/Li_xCoO_2. The discharge characteristics of these batteries are shown in Fig. 35.1. These are the electrochemical systems that are predominately employed in the reserve-type configurations.

In the reserve construction, the electrolyte is physically separated from the electrode active materials until the battery is used, and it is stored in a reservoir prior to activation. This design feature provides a capability of essentially undiminished output even after storage periods in the inactive state of over 20 years. The reserve feature, however, results in an energy density penalty of as much as 50% compared with the active lithium primary batteries. Key contributors to this penalty are the activation device and the electrolyte reservoir.

In the selection of a lithium anode electrochemical system for packaging into a reserve battery, besides such important considerations as physical properties of the electrolyte solution and performance as a function of the discharge conditions, factors such as the stability of the electrolyte and the compatibility of the electrolyte with the materials of construction of the electrolyte reservoir are of special importance. Use of environmentally friendly cell systems plays some importance with respect to manufacturing. Thus, there is an emerging interest in pursuing organic-based electrolyte solutions in the development of new cell systems.

35.2 SPECIALIZED BATTERY SYSTEMS

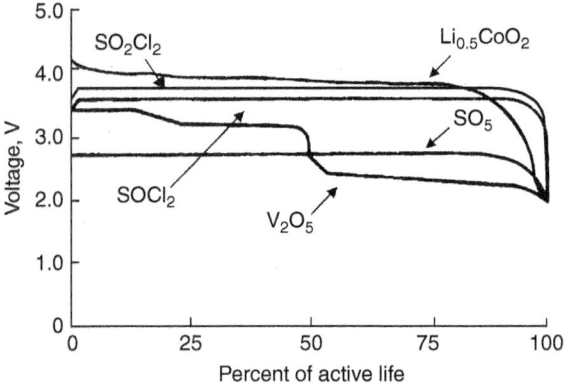

FIGURE 35.1 Performance comparison of lithium anode primary systems at 20°C. Thionyl chloride ($SOCl_2$)—3.6 V; vanadium pentoxide (V_2O_5)—3.4 V; sulfur dioxide (SO_2)—2.9 V; precharged lithiated cobalt oxide (Li_xCoO_2, $x = 0.4 - 0.5$) 4.0 V; sulfuryl chloride (SO_2Cl_2)—3.8 V.

35.2 CHEMISTRY

35.2.1 Lithium/Vanadium Pentoxide (Li/V_2O_5) Cell

The basic cell structure of this system consists of a lithium anode, a microporous polypropylene film separator, and a cathode that is usually composed of 90% V_2O_5 and 10% graphite, on a weight basis. When it is used in a reserve battery, historically the prevalent electrolyte is $2M$ $LiAsF_6 + 0.4M$ $LiBF_4$ in methyl formate (MF) because of its excellent stability during long-term storage.

As shown in Fig. 35.1, the Li/V_2O_5 system has a two-plateau discharge characteristic. A net cell reaction, involving the incorporation of lithium in V_2O_5, has been postulated to account for the first plateau:

$$Li + V_2O_5 \rightarrow LiV_2O_5$$

The initial voltage level ranges from 3.4 to 3.3 V, decreases to 3.3 to 3.2 V for approximately 50% of the active life of the first discharge plateau, at which point the range again decreases to a level of 3.2 to 3.1 V, which is maintained for the balance of the first plateau discharge. After completion of the first plateau, the Li/V_2O system undergoes a rapid change in voltage to the second discharge plateau around a voltage range of 2.4 to 2.3 V. This step involves the formation of reduced forms of V_2O_5, although specific mechanisms remain unclear.[1] This second plateau is relatively more sensitive to temperature and discharge rate, and it is for this reason that most Li/V_2O_5 cells (active and reserve) are designed to operate at only the first discharge plateau level.[2]

The long-term storage capability of Li/V_2O_5 reserve cells is heavily dependent on the stability of the electrolyte solution. $LiAsF_6$ in MF electrolyte is unstable due to the decomposition reactions involving the hydrolysis of methyl formate followed by the dehydration of the hydrolysis product(s).[3] These reactions result in a premature fracture of the glass ampoule used as the electrolyte reservoir. The stability of the LiAsF:MF electrolyte solution was achieved by making the solution either neutral or alkaline. In practice, this is accomplished by using two electrolyte salts ($LiAsF_6$ + $LiBF_4$:MF) and by incorporating lithium metals to scavenge water in the glass ampoule. A $LiBF_4$-based electrolyte solution is one newer electrolyte technology that is being developed as a replacement for the $LiAsF_6$/MF-based electrolyte technology.[4]

35.2.2 Lithium/Thionyl Chloride (Li/SOCl$_2$)

The basic cell structure that is generally used for this system consists of a lithium anode, a nonwoven glass separator, and a Teflon®-bonded carbon cathode, which serves only as the reaction site medium. One unique feature of this chemistry is the fact that thionyl chloride (SOCl$_2$) serves two functions—as the solvent of the commonly used LiAlCl$_4$ in SOCl$_2$ electrolyte solution and as the active cathode material (see Sec. 14.6).

Figure 35.1 shows the marked advantage in discharge performance of the Li/SOCl$_2$ system. The accepted net cell reaction for this system is

$$4Li + 2SOCl_2 \rightarrow 4LiCl + S + SO_2$$

Most of the sulfur dioxide formed during discharge is dissolved in the electrolyte and practically no gas pressure is generated.[5] Depending on the discharge rate and temperature, the Li/SOCl$_2$ system normally exhibits a working voltage range of between 3.0 and 3.6 V with a flat discharge characteristic. These excellent discharge characteristics—high voltage and flat discharge—are best attained in a reserve cell, especially if the discharge current density is high. In an active primary Li/SOCl$_2$ cell, the lithium anode is coated with a passive LiCl film. Under sustained storage coupled with high-temperature exposure, the passivating film will limit the current-handling capability of the anode as well as increase the time required to reach operating voltage.[5]

The conventional LiAlCl$_4$ in SOCl$_2$ electrolyte solution has proved to have excellent stability. Electrolyte glass ampoules exposed to +74°C did not show any sign of apparent degradation up to at least 12 years of storage. Because of this and its overall performance superiority, the Li/SOCl$_2$ reserve cell has become the system of choice for high-energy reserve batteries.

The use of excess AlCl$_3$, a Lewis acid, in the conventional LiAlCl$_4$ in SOCl$_2$ electrolyte solution has been shown to improve the rate capability of the Li/SOCl$_2$ system.[6,7] It should be noted, however, that the inherent stability of this high-rate electrolyte has yet to be established so as to ensure its application in reserve cells.

35.2.3 Lithium/Sulfur Dioxide (Li/SO$_2$)

The Li/SO$_2$ system uses a basic cell structure consisting of a lithium anode, a separator, and a Teflonated carbon cathode, similar to one used in the Li/SOCl$_2$ system, which serves as the reaction site. The electrolyte solution commonly employed contains a mixture of lithium bromide (LiBr), acetonitrile (AN), and sulfur dioxide (SO$_2$), which also serves as the active cathode material.

One serious problem in using the LiBr in AN-SO$_2$ electrolyte solution for reserve cells is its instability during storage. Although this electrolyte solution is commonly employed in active primary cells, it is unsuitable for reserve battery applications because it decomposes to form highly reactive and solid products when stored in the absence of cell components. Replacing LiBr with lithium hexafluoroarsenate (LiAsF$_6$) results in an electrolyte solution with good stability. The functional performance of the LiAsF$_6$ electrolyte is equivalent or superior to that of the LiBr solution for low to moderate rate.[8-10]

The Li/SO$_2$ reserve battery, using the stable electrolyte solution (LiAsF$_6$ in AN-SO$_2$), follows the same net cell reaction

$$2Li + 2SO_2 \rightarrow Li_2S_2O_4$$

as in the active primary cell. It should be noted, however, that LiAsF$_6$ in AN-SO$_2$ electrolyte solution is limited to moderate- or lower-rate applications due to poorer electrolyte conductivity. Because of this, the emphasis on using the Li/SO$_2$ system for reserve applications has shifted to the higher-performance Li/SOCl$_2$ system.

35.2.4 Lithium/Precharged Li$_x$CoO$_2$ (0.5 ≤ x < 1) Cell

Motivated to have higher single cell voltage cell systems, Lin and Burgess[11] have investigated many of the higher voltage cathode-based systems that were popularly being used in the lithium-ion cell technologies; e.g., LiCoO$_2$, LiFePO$_4$, and varied compositions of the cobalt-based mixed oxide chemistries, the LiCoO$_2$-based cathode is the only one to have applicability for reserve battery applications.

The use of precharged Li$_x$CoO$_2$ is a new approach to high-voltage and high-energy cathode systems for reserve battery applications. Key criteria are the ability to process the cathode materials into a stable raw material for the reserve cells. In the case of the Li$_x$CoO$_2$ cathode, lithium can be extracted electrochemically to where $x \geq 0.5$ and then processed successfully as a raw material for primary active cell applications.[11] The Li/precharged Li cathode Li$_x$CoO$_2$ (0.5 ≤ x < 1) cell structure is very similar to the Li/V$_2$O$_5$ cell. Except for the difference in cathode material, all other design features are essentially the same.

Both the Li/V$_2$O$_5$ and Li/precharged Li$_x$CoO$_2$ (0.5 ≤ x < 1) cells offer an unique design feature, permitting the cells to be recharged when demanded by a specific application. One application that used the precharged Li$_x$CoO$_2$ cathode technology is described in Sec. 35.3.2.

35.3 CONSTRUCTION

35.3.1 General Considerations

Lithium anode reserve batteries are basically composed of three major components:

1. Activation and electrolyte delivery system
2. Electrolyte reservoir
3. Cell and/or battery unit

However, the actual design can vary widely, depending on the application. The design can vary from a simple, small, single cell with an ampoule manually activated, to a very large, complex, multicell battery with an automatic electric initiation mechanism to transfer the electrolyte from the reservoir chamber to a high-voltage multicell battery stack. Both the electrodes and the hardware components are essentially the same as the primary active units, but with allowances made for electrolyte storage and electrolyte delivery into the cells at the time of activation. In addition, the electrochemical and hardware components must be constructed of a rugged maintenance-free design to survive severe environmental and performance requirements as most are used in military or special applications. For example, Table 35.1 lists typical requirements of lithium reserve batteries and illustrates the reason for many of their unique construction and design features.

TABLE 35.1 Typical Characteristics of Lithium Anode Reserve Batteries

Operating temperature range −55 to 70°C
10- to 20-year unactivated storage life
Hermetically sealed
High energy density
High reliability
Low electrical noise
Flat discharge voltage profile
Rapid voltage rise after initiation
Mechanical environmental capability:
Acceleration shocks up to 20,000 g
High spin up to 20,000 rpm
Transportation and deployment vibration levels
Operating life from several seconds up to 1 year

Some common construction features are used in the design of lithium reserve batteries. The outer case is generally made of a 300-series stainless steel since it offers the corrosion resistance against both the internal system and the external environment during its long-term use. Various welding techniques such as laser, tungsten inert gas (TIG), resistance, and electron beam can be applied to the 300-series stainless steels. Thus the outer case provides a true 20-year reliability, capable of maintaining the hermeticity required for reserve lithium batteries. The electrical terminals used are generally glass-to-metal types, which also provide the hermeticity required for long-term storage.

35.3.2 Types of Lithium Anode Reserve Batteries

Three basic lithium reserve battery types are being manufactured at the present time:

1. Single-cell battery with electrolyte stored in a glass ampoule
2. Multiple single cells using bellows for the electrolyte storage reservoir
3. Multicells of bipolar construction with either a glass ampoule or a reservoir for electrolyte storage

Ampoule Type. Single-cell reserve types using an ampoule as the electrolyte storage reservoir are the most reliable of the reserve designs due to their simple construction and lack of intercell leakage problems associated with multicell batteries. One group of these cells is sized to the ANSI standard specifications, and the other group consists of those cells built for special-purpose applications which are not sized to the ANSI specifications. Both groups, however, are very similar in construction.

Figure 35.2 shows the cross section of a reserve lithium anode cell in an A-size configuration of about 1 Ah, using the $Li/SOCl_2$ system.[7] The cell consists of concentrically arranged components. A lithium anode is swaged against the inner wall of a stainless-steel cylindrical can. A nonwoven glass separator is located adjacent to the anode. The Teflon®-bonded carbon cathode is inserted against the separator. A cylindrical nickel current collector provides the electrical contact to the positive terminal and houses the hermetically sealed glass ampoule. The ampoule is held firmly in place by upper and lower insulating supports, which protect it from premature breakage while permitting transmission of a direct force at the bottom of the case to shatter the ampoule at the time of activation. The unit is sealed hermetically to ensure long shelf life in the unactivated condition. Activation is achieved by applying a sharply directed force at the bottom of the cell case to shatter the glass ampoule. The electrolyte is absorbed by the porous cathode and the glass separator, thereby activating the battery.

Another design has been developed for mine and fuze applications, using both the Li/V_2O_5 and the $Li/SOCl_2$ systems, in the capacity range of 100 to 500 mAh.[12] The cross sections of these two cells are shown in Figs. 35.3 and 35.4 respectively. Both cells are similar with respect to the external hardware and the internal arrangement of the components. The case and header assembly

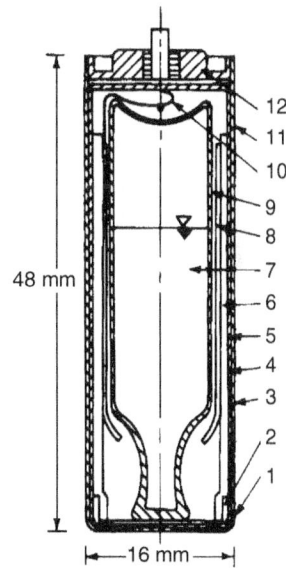

1 Insulator
2 Bottom separator
3 Cell can
4 Lithium anode
5 Separator
6 Carbon cathode
7 Electrolyte
8 Current collector
9 Glass ampoule
10 Positive terminal tab
11 Top spacer
12 Cell cover

FIGURE 35.2 Cross section of $Li/SOCl_2$ A-size reserve cell. (*Courtesy of Tadiran Industries, Ltd.*)

35.6 SPECIALIZED BATTERY SYSTEMS

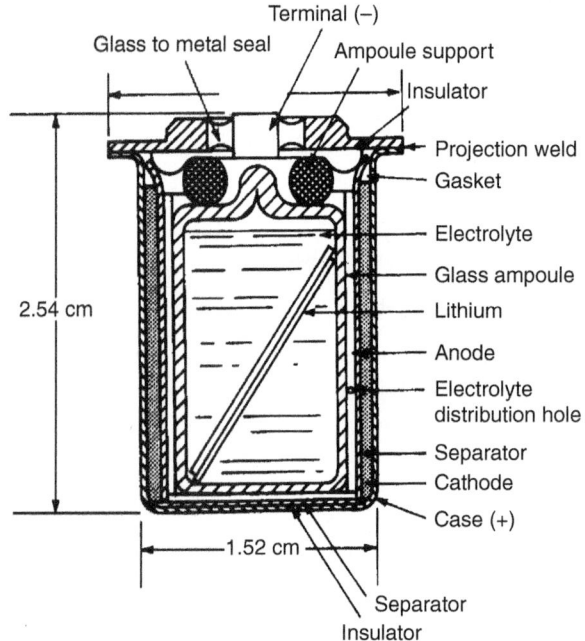

FIGURE 35.3 Cross section of Li/V$_2$O$_5$ reserve cell. Alliant model G2659. (*Courtesy of Alliant Techsystems, Inc.*)

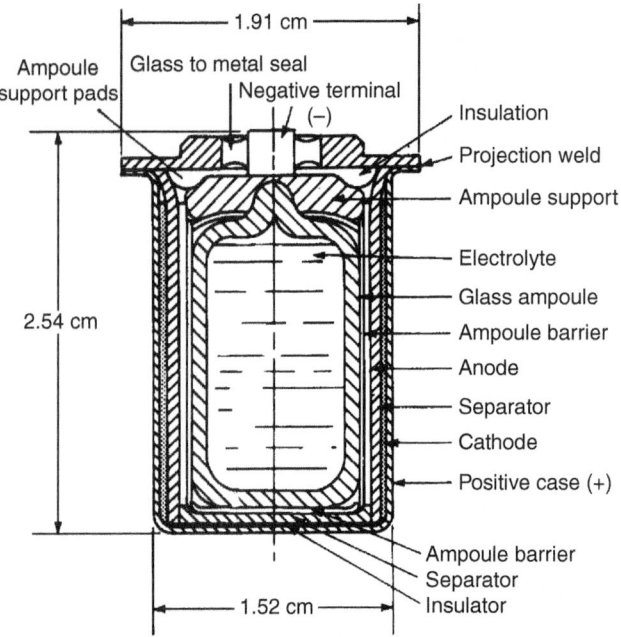

FIGURE 35.4 Cross section of LiSOCl$_2$ reserve cell. Alliant model G2659B1. (*Courtesy of Alliant Techsystems, Inc.*)

are projection-welded together at the case flange. The header serves as the cover for the cells and incorporates a glass-to-metal seal for the center terminal pin made of nickel-iron Alloy 52. The terminal pin has negative polarity (both cell designs), and the balance of the header and case surface have positive polarity. The hermetically sealed hardware in conjunction with the reserve feature of the design makes it possible to achieve storage times in excess of 20 years.

The internal arrangement of the components consists of annularly located electrodes about a central glass ampoule used as the electrolyte solution reservoir. In addition, there are various insulating components in the upper and lower portions of the cell, used to prevent internal short-circuiting.

Several features account for most of the design differences between these two cells. In the Li/$SOCl_2$ reserve cell, the glass ampoule also contains the cathode oxidant, $SOCl_2$, while the cathode oxidant of the Li/V_2O_5 reserve cells is contained in the cathode structure. Directly adjacent to the Li/$SOCl_2$ cell case is the Teflonated carbon, while in the case of the Li/V_2O_5 cell, the cathode is molded from a dry mixture of V_2O_5 and graphite. The Teflonated carbon cathode for the reduction of $SOCl_2$ is made in sheet form and is attached to a metal grid, rolled to shape, and inserted against the inside wall of the case. Another difference is the way the electrical connection is made for the two cathodes. The V_2O_5 connection is made by the direct-pressure contact of the molded cathode, whereas with the $SOCl_2$ system the cathode lead is welded to the case at the time the cover is welded. The lithium anode structure consists of pure lithium metal, which is pressed onto an expanded metal grid of 316L stainless steel. One end of a flat 316L stainless-steel lead is spot-welded to the pin of the glass-to-metal seal. Rolled into a cylinder, the anode is inserted into the cell next to the separator. Both cells are provided with an ampoule support in order to survive the shock environment specified. In the Li/$SOCl_2$ system, Tefzel and glass have been found to be chemically stable for use as insulators, separators, and supports. The Li/V_2O_5 system allows more flexibility because many rubbers and plastics can be used.

Multicell Single-Activator Design. For those applications where higher than single-cell voltages are required, a battery is constructed of two or more cells, depending, of course, on the voltage needed. Typical voltages are 12 and 28 V, and for lithium anode cells with a 2.7 to 3.3 V operating voltage, this would require anywhere from 4 to 10 cells for each battery. This family of batteries is unique with respect to the method of cell activation and the containment of electrolyte in multiple cells initiated from a single self-contained reservoir of electrolyte. Batteries of this design are used in preference to the bipolar type to achieve higher cell capacities and to allow discharge times up to 1 year or more, through the tight control of intercell leakage. The leakage currents are controlled and limited to usually less than several percent of the discharge current. This feature, however, limits these batteries from being miniaturized, which is possible with many bipolar designs.

An example of this design approach is the Li/SO_2 reserve battery illustrated in Fig. 35.5. The battery is cylindrical and contains three main components: (1) the electrolyte storage reservoir section, (2) the electrolyte manifold and activation system, and (3) the reserve cell compartment. About one-half of the internal battery volume contains the electrolyte reservoir. The reservoir section consists primarily of a collapsible bellows in which the electrolyte solution is stored. Surrounding the bellows, between it and the outer battery case, is a space that holds a specific amount of gas/liquid. The gas is selected such that its vapor pressure always exceeds that of the electrolyte, thereby providing the driving force for eventual liquid transfer into the cell chamber section once the battery has been activated.

In the remaining half of the battery volume there is the centrally located electrolyte manifold and activation system housed in a 1.588-cm-diameter tubular structure plus the series stack of four torroidally shaped cells that surround the manifold/activation system.

The manifold and cells are separated from the reservoir by an intermediate bulkhead. In the bulkhead there is a centrally positioned diaphragm of thin section to be pierced by the cutter contained within the manifold. In fabrication, the diaphragm is assembled as part of the tubular manifold which, in turn, is welded as a subassembly to the intermediate bulkhead. Figure 35.6 is a more detailed cross-sectional view of the electrolyte manifold and activation system with the major components identified.

The activating mechanism consists of a cutter that is manually moved into the diaphragm, cutting it and thereby allowing electrolyte to flow. The movement of the cutter is accomplished by the turning

35.8 SPECIALIZED BATTERY SYSTEMS

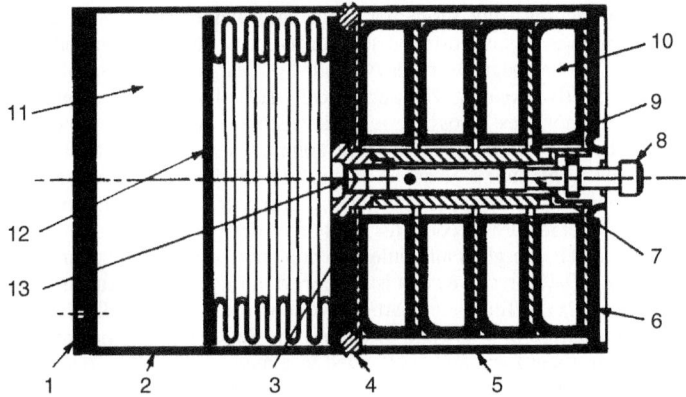

Legend:

1 Battery top bulkhead
2 Upper battery case
3 Bulkhead
4 Intermediate bulkhead ring
5 Lower battery case
6 Battery bottom bulkhead
7 Activation manifold

8 Activation stud
9 Intercell insulation
10 Single 20 Ah cell
11 Freon backfill volume
12 Electrolyte storage bellows
13 Manifold diaphragm

FIGURE 35.5 Cross section of 20 Ah Li/SO$_2$ multicell battery.

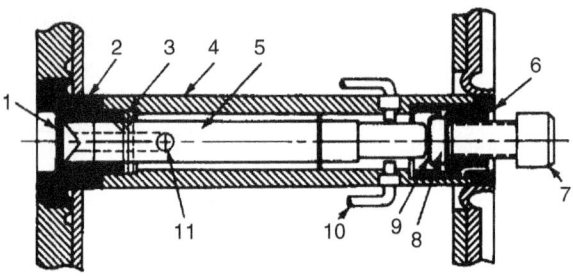

Legend:

1 Diaphragm
2 Top bushing
3 Shear pin
4 Center tube
5 Cutter
6 Bottom bushing

7 Activator stud
8 Drive disk
9 Collapsing cup
10 Electrolyte supply tube
11 Electrolyte entry flow

FIGURE 35.6 Cross section of electrolyte manifold and activation system.

of an external screw that is accessible in the bottom base of the battery. The cutter section and the screw mechanism are isolated from one another by a small collapsible metal cup that is sealed hermetically between the two sections. This prevents external electrolyte leakage. The manifold section is a series of small nonconductive plastic tubes connected to one end of the central cylinder and to each of the individual cells at the other end. The long length and small cross-sectional area of the tubes minimize intercell leakage losses during the period of time that electrolyte is present in the manifold structure.

In this application, four individual cells are required to meet the voltage requirement. (The number of cells is, of course, adjustable with minor modification to meet a wide range of voltage needs.) Each cell contains flat circular anodes and cathodes that are separately wired in parallel to achieve the individual cell capacity and plate area needed for a given set of requirements. To fabricate, the components, with intervening separators, are alternately stacked around the cell center tube, after which the parallel connections are made. The cells are individually welded about the inner tube and outer perimeter to form hermetic units ready for series stacking within the battery. Connections from the cells are made to external terminals, which are located in the bottom bulkhead of the battery.

Figure 35.7 shows the major battery components prior to assembly. The components shown are fabricated primarily from 321 stainless steel, and the construction is accomplished with a series of TIG welds. The hardware shown is designed specifically for use with the lithium/sulfur dioxide electrochemical system: however, it is adaptable, with minor modifications, to other liquid and solid oxidant systems. The battery can also be adapted to electrical rather than manual activation.

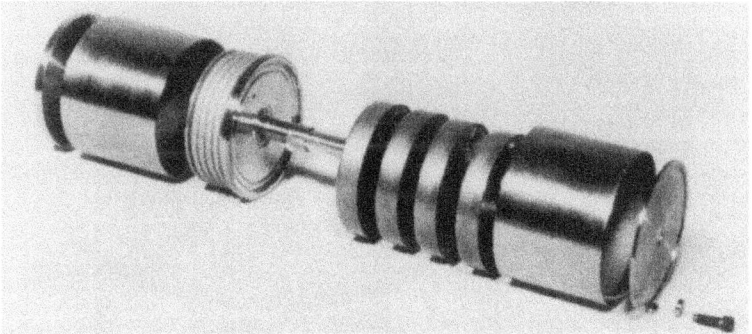

FIGURE 35.7 Pictorial view of 20 Ah Li/SO_2 multicell battery.

An example of this reserve design approach being used with the lithium/thionyl chloride chemistry is shown in Fig. 35.8a. This high-power reserve battery, designated by the U.S. Navy as battery BA-6511 SLQ, was developed to provide electric power for a family of ocean buoys.[13] The reserve battery was selected for this application to eliminate the problems of self-discharge and passivation associated with extended stand of an active battery, but also for safety as the electrolyte is stored separately from the battery until activation.

The battery weighs about 145 lb and is contained in a package that is 29.2 cm in diameter and 43.2 cm long. The battery contains 21 cells; 18 cells compose a 56 V section, delivering 4 kW, and rated at 65 Ah; 3 cells are in a 10 V section, delivering 7 A, and rated at 57 Ah. The electrolyte is stored in a reservoir and is distributed to the 21 cells via a unique manifolding system. Activation is initiated by an explosive squib, and a stored energy system within the reservoir provides the motive power. The cell design used for this battery (Fig. 35.8b) is a circular wafer with a hole through the center to provide a channel for electrical and tubing connections. The two types of cells used in the battery are physically similar, differing only in height and capacity as a result of one less set of electrodes. The cells used in the high-voltage, high-rate section contain five anodes and six cathodes.

35.10 SPECIALIZED BATTERY SYSTEMS

(a)

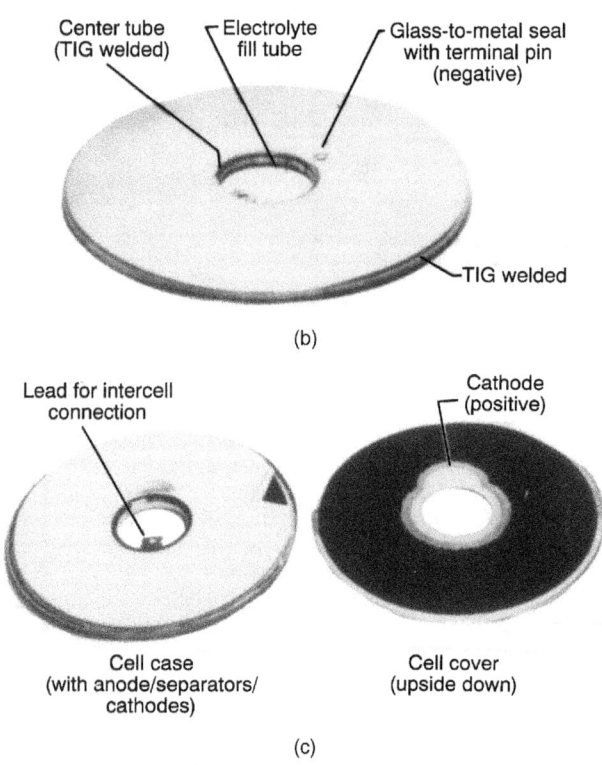

FIGURE 35.8 High-power reserve battery BA-6511/SLQ. (a) Li/SOCl$_2$ reserve battery. (b) High-power cells. (c) High-power cell case and electrode assembly. (*Courtesy of Alliant Techsystems, Inc.*)

TABLE 35.2 Characteristics of Li/SOCl$_2$ Reserve Cells Model G3070A2

	Low-rate reserve cell	High-rate reserve cell
Performance		
Open-circuit voltage (activated)	3.67 V	3.67 V
Voltage under load	3.40 V, 7 A at 20°C	3.10 V, 72 A at 20°C
Rated capacity	57 Ah at 7 A to 2.67 V at 20°C	65 Ah at 72 A to 2.63 V at 20°C
Physical characteristics:		
Max. diameter, OD	28.5 cm	28.5 cm
Max. diameter, ID	6.7 cm	6.7 cm
Max. height	0.89 cm	1.04 cm
Cell weight with electrolyte	1310 g	1485 g
Case material	Stainless steel	Stainless steel

Source: Alliant Techsystems, Inc.

Anodes are single-sided with lithium pressed onto expanded nickel grids. The cathode is cut from coated stock of Teflonated carbon on a nickel screen, as shown in Fig. 35.8c. Nonwoven glass separators are used. The specifications for the two cells are listed in Table 35.2.

Another example of this reserve design using a precharged Li$_x$CoO$_2$ (0.5 ≤ x < 1) chemistry[14] is shown in Fig. 35.9. This reserve battery consists of three hermetically welded cell cases with a central reservoir enclosed in a stainless steel battery housing. Such a battery was developed to power the Hand-emplaced Wide Area Munitions (HWAM).

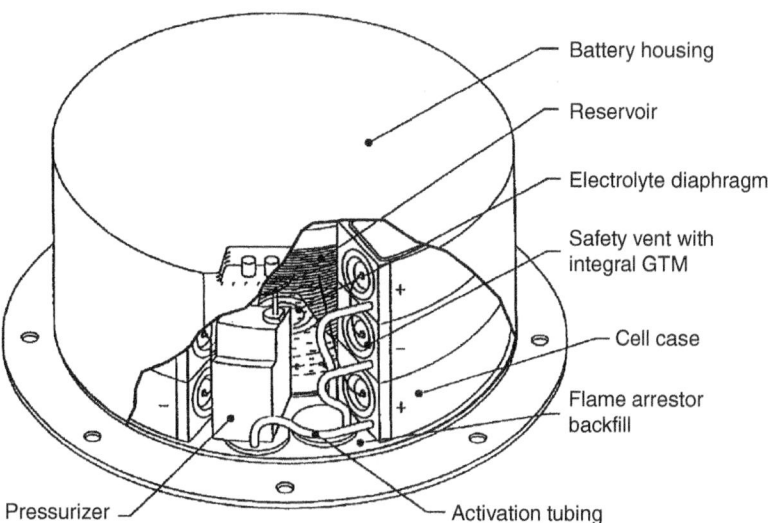

FIGURE 35.9 Design of reserve Li/Li$_x$CoO$_2$ battery for Hand-emplaced Wide Area Munitions (HWAM). (*Courtesy of Alliant Techsystems, Inc.*)

Figure 35.10 shows another multicell battery designed for lightweight missile applications. Based on the Li/oxyhalide technology, it used advanced thin electrode technology that has high energy utilization and low electrical impedance. A composite separator was also used that combines high electrolyte absorption and mechanical integrity. This type of high power design sees applications such as the Theater High Altitude Area Defense (THAAD) in a Kill Vehicle for the Ground

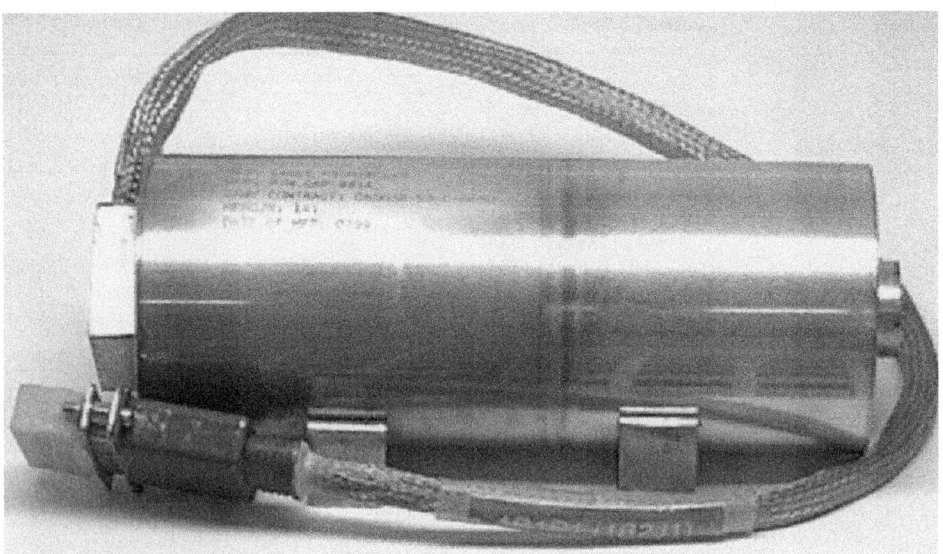

FIGURE 35.10 High-power 1 KW Li/oxychloride reserve battery. (*Courtesy of Eagle-Picher Technologies.*)

Based Interceptor (GBI) program.[15] Other advantages of this lightweight, high-power battery are: (1) reduction in battery weight over the thermal or silver-zinc system; (2) specific energy greater than 250 Wh/kg; (3) gain in weight advantage as the mission time increases or as the energy to power ratio increases; (4) high power delivery even after 10 years of storage at temperature below 32°C; and (5) low operating temperature allowing locations near heat-sensitive electronics.

Multicell Bipolar Construction with Single-Activator Reservoir. Lithium anode reserve batteries, using bipolar construction, are relatively few in number and always developed for specific applications. The bipolar construction—one component used as both the anode collector of one cell and the cathode collector of the next cell in the stack—is not unique to the lithium reserve battery, but an adaptation of techniques used in other types of batteries. There are several advantages of the bipolar construction:

- Very high energy and power density for high-voltage batteries
- Rugged construction to withstand spin and setback forces from artillery firing
- Flexibility to adjust voltages in the cell stack
- Adaptability to varying energy and power requirements

Figure 35.11 is an illustration of a reserve lithium/thionyl chloride battery using a bipolar plate construction. This battery weighs approximately 5.4 kg and has a volume of 2000 cm^3.

Activation of the reserve battery is accomplished by supplying an electric pulse to the battery by firing an electric squib or actuator or by some mechanical means. This type of reserve battery has been used chiefly in artillery shells for electronic fuze power supplies and in missiles for the electronic power supply. Therefore the electric pulse can be supplied prior to firing or at the time of launch. However, for artillery fuze power supplies, the battery is usually activated by the launch acceleration (setback) and/or the spin forces. The acceleration force of the artillery shell releases a firing pin which strikes and fires a primer. The primer can ignite a gas generator or directly release a stored gas by breaking open a metal diaphragm.

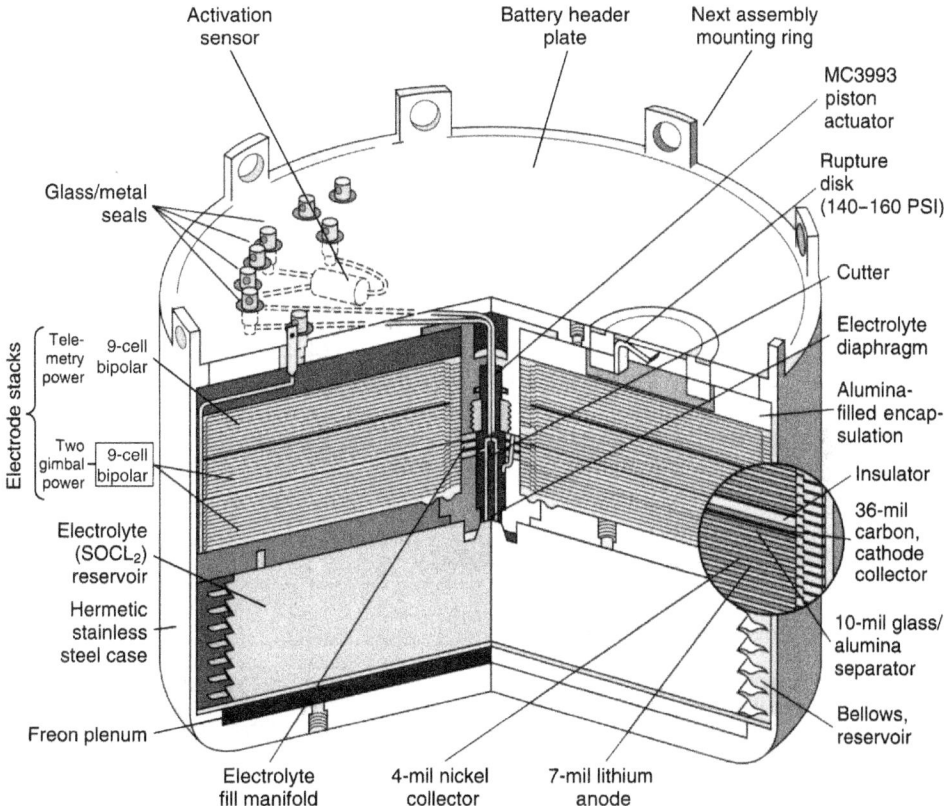

FIGURE 35.11 Sandia National Laboratories Li-SOCl$_2$ reserve battery model MC3945.

Once the battery has been initiated as described, the gas pressure (such as from a gas generator, stored gas/liquid, or CO$_2$) forces the electrolyte into each of the cells through a manifold (electrolyte distribution network).

The electrolyte reservoir is generally made using a collapsible cup, a bellows, or a wound tubing design. These serve to hold the electrolyte during the long inactive storage period and act as the delivery mechanism during activation. Each reservoir has some type of diaphragm, which is broken with high pressure or mechanical means to allow the electrolyte to enter the cell-stack part of the battery hardware.

The bipolar cell stack with the electrolyte distribution manifold in the center comprises the battery section. When electrolyte enters the center manifold, it is distributed to each cell through holes or passageways in the housing encompassing the battery. The design of the manifold is the key to controlling intercell leakage. For bipolar batteries, life requirements are relatively short (seconds to several hours), therefore the manifolding is relatively simple. But when longer life is needed, the parasitic leakage currents are controlled by the length and area of the leakage path.

Another battery of this design was developed as a power source for the Extended Range Guided Missile (ERGM).[16] Ultimately, this design would be similar to the design shown in Fig. 35.10. However, the development test fixture used is shown in Fig. 35.12. It uses the Li/SOCl$_2$ chemistry but with a special lithium tetrachlorogallate electrolyte optimized for performance. The test fixture employs gas pressure to compress the bellows, rupturing the diaphragm and activating the battery. In actual use, setback forces on launch perform this function. The specifications for the ERGM battery are given in Table 35.3.

35.14 SPECIALIZED BATTERY SYSTEMS

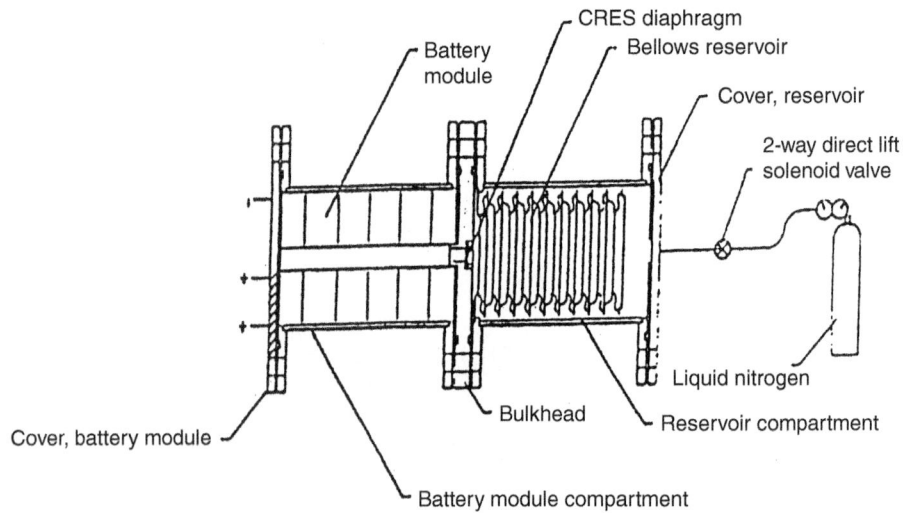

FIGURE 35.12 Laboratory test fixture for activation of Extended Range Guided Missile (ERGM) battery using Li/SOCL$_2$ system. (*Courtesy Yardney Technical Products, Inc.*)

TABLE 35.3 Specifications for ERGM Battery

Description	Specifications
12 V section regulation	• Load voltage range: 9.5 to 16.0 V
	• 60 W applied continuously
28 V section regulation	• Load voltage range: 24 to 40 V
	• 15 W applied continuously
	• 125 pulses, 8 amperes, 0.1 s
	• Duration, evenly distributed
Operating life	• 480 s minimum

35.4 PERFORMANCE CHARACTERISTICS

35.4.1 Ampoule-Type Batteries

Voltage characteristics at the "time of activation" are unique and an important feature of reserve batteries. This is especially true for military applications, where reserve batteries must normally be designed to meet operational voltage in less than 1 s and in many cases even less than ½ s. For nonmilitary use, activation times to operating voltage level are less critical. However, for a given reserve battery design and the electrochemical couple used, the activation time is dependent on the discharge rate and temperature.

In general, the voltage rise times for both Li/SOCl$_2$ and Li/V$_2$O$_5$ have similar characteristics. Figure 35.13 shows the rise-time characteristics for the Li/SOCl$_2$ battery (illustrated in Fig. 35.4) at five temperatures at a current density of 0.1 mA/cm (approximately a $C/500$ rate). Rise times are typically below 20 ms at ambient (24°C) and higher temperatures, but increase up to 500 ms at the lower temperatures. The ability to activate rapidly is primarily due to the cell design, which allows the electrolyte to penetrate and wick into the porous electrode and separator at the instant of ampoule breakage.

The voltage levels of both the Li/V$_2$O$_5$ and the Li/SOCl$_2$ systems (batteries illustrated in Figs. 35.3 and 35.4) under steady-state discharge conditions are shown in Fig. 35.14. These two systems are very close in voltage at the lower temperatures, ranging from 3.3 to 3.0 V at current densities of less

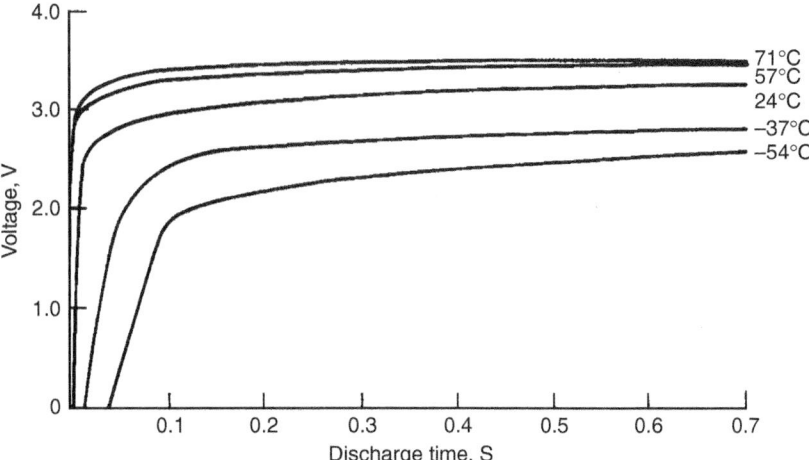

FIGURE 35.13 Rise-time characteristics after activation of Li/SOCl$_2$ reserve battery, Alliant model G2659B1; load = 4.35 kΩ. (*Courtesy of Alliant Techsystems, Inc.*)

than 1 mA/cm^2. At higher temperatures, ambient and above up to 74°C, the Li/SOCl$_2$ battery operates above 3.5 V, whereas the Li/V$_2$O$_5$ normally operate between 3.2 and 3.4 V. The higher voltage and increased capacity account for the significant increase in energy density of the SOCl$_2$ battery over the V$_2$O$_5$ battery. As shown in Fig. 35.14, the V$_2$O$_5$ system has very little change in capacity over the wide temperature range but is still much lower in capacity than the SOCl$_2$ battery when discharged at the same rate, namely 0.1 mA/cm^2. Although the capacity and voltage of the Li/SOCl$_2$ battery are lower at cold temperatures, its output is still higher than that of most other systems and its voltage profile is characterized by a flat single-step plateau. The high-temperature curve is also extremely flat and typically discharges above 3.6 V at a current density of 0.1 mA/cm^2. The voltage characteristics on ambient temperature discharges are similar to those at high temperature except for a slightly lower load voltage when discharged at the same rate, averaging 3.5 V. Table 35.4 compares the output parameters of the two systems with identical hardware and shows the superior

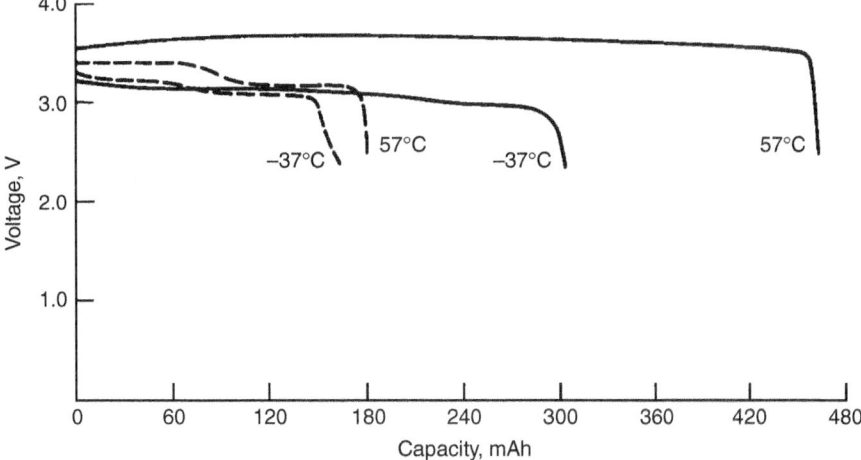

FIGURE 35.14 Comparison of discharge profiles for reserve-type Li/V$_2$O$_5$ (— —) batteries and (——) Li/SOCl$_2$. Current density = 0.1 mA/cm^2.

TABLE 35.4 Performance Comparison between $Li/SOCl_2$ and Li/V_2O_5 Systems

System	Temperature, °C	Cell voltage, V	Capacity, mAh	Cell volume, cm³	Cell weight, g	Specific energy, Wh/kg	Energy density, Wh/L
Li/V_2O_5*	−37	3.15	160	5.1	10	50.4	98.8
	57	3.30	180	5.1	10	59.4	116.5
$Li/SOCl_2$†	−37	3.05	300	5.1	10.5	87.1	179.4
	57	3.60	450	5.1	10.5	154.3	317.6

*Alliant model G2659.
†Alliant model G2659B1.

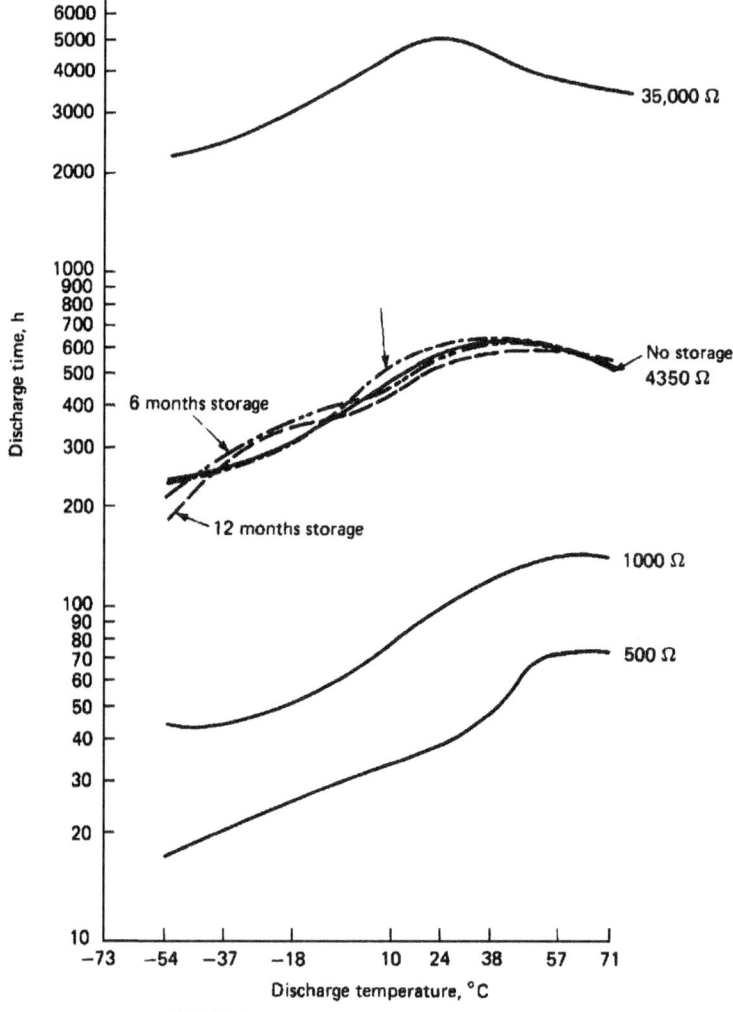

FIGURE 35.15 Effect of discharge rate at various loads and inactive storage on $Li/SOCl_2$ reserve battery, Alliant model G2659B1. (*Courtesy of Alliant Techsystems, Inc.*)

performance of the Li/SOCl$_2$ battery. The similarity in voltage and the fact that the same hardware is used for both systems permits a one-for-one replacement.

Figure 35.15 shows the effect of inactive storage of up to 12 months at 71°C on the Li/SOCl$_2$ battery performance over the temperature range of −54 to 71°C. No significant effect on performance was found as a result of the storage. The slightly lower voltages during discharge or the voltage delays when the load is first applied on active (nonreserve) batteries were not present with the reserve batteries. Figure 35.15 also gives a summary of the performance of fresh batteries at various discharge loads and temperatures.

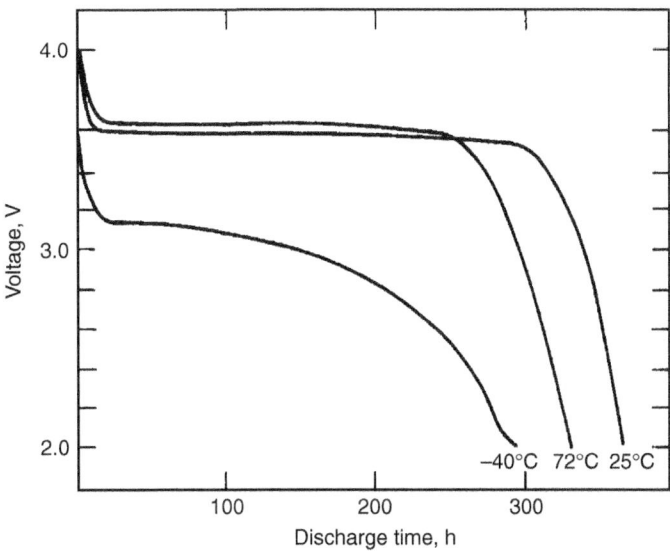

FIGURE 35.16 Typical discharge curves on 1.25 kOhm of Li/SOCl$_2$ reserve battery, Tadiran model TL-5160. (*Courtesy of Tadiran Industries, Ltd.*)

Figure 35.16 shows the discharge curves of the Li/SOCl$_2$ reserve A-size battery (illustrated in Fig. 35.2). The current drain capability of a reserve system significantly exceeds that of the corresponding active primary battery. Figure 35.16 shows the discharge characteristics at 1.25 kΩ (about 3 mA) or 0.15 mA/cm^2. Currents higher than 1.5 A (current density of 100 mA/cm^2) can be obtained at voltages higher than 2.0 V for several minutes at −10°C. The specific energy of the cells, to a cutoff voltage of 2.0 V, as a function of the discharge current at various temperatures is shown in Fig. 35.17. The performance at higher temperatures is close to that for 25°C.

35.4.2 Multicell Battery Design

The performance characteristics of the Li/SO$_2$ multicell single-activation design battery are shown in Fig. 35.18. The activation and discharge profiles for a 12 V, 100 Ah battery using an LiAsF$_6$ in AN-SO$_2$ electrolyte are illustrated. Because the battery is activated manually, the slower voltage rise time is expected because the cutting of the diaphragm in the center bulkhead requires several turns on the activation bolt. The battery could easily be activated with an electric or mechanical input to a piston actuator or squib to cut the diaphragm to improve the voltage rise time. Although the operational life of the battery can be very short, the data illustrate its capability for a long-term discharge at low discharge rates if a stable electrolyte is used.

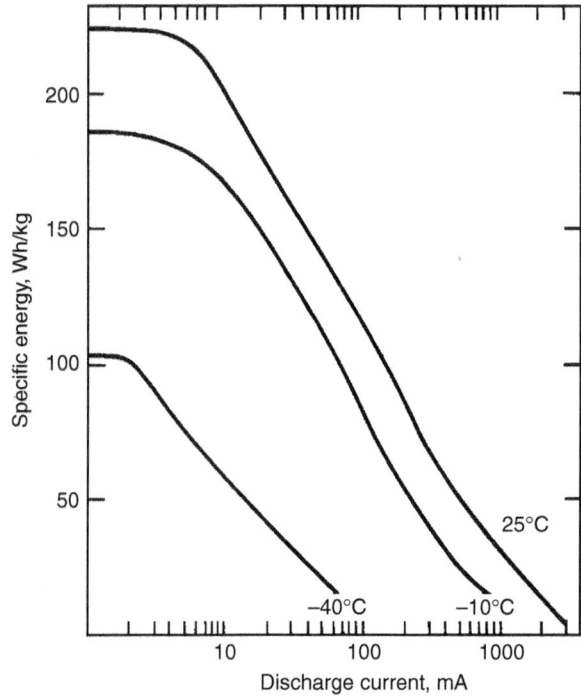

FIGURE 35.17 Specific energy as a function of discharge current and temperature for Li/SOCl$_2$ reserve battery, Tadiran model TL-5160. (*Courtesy of Tadiran Industries, Ltd.*)

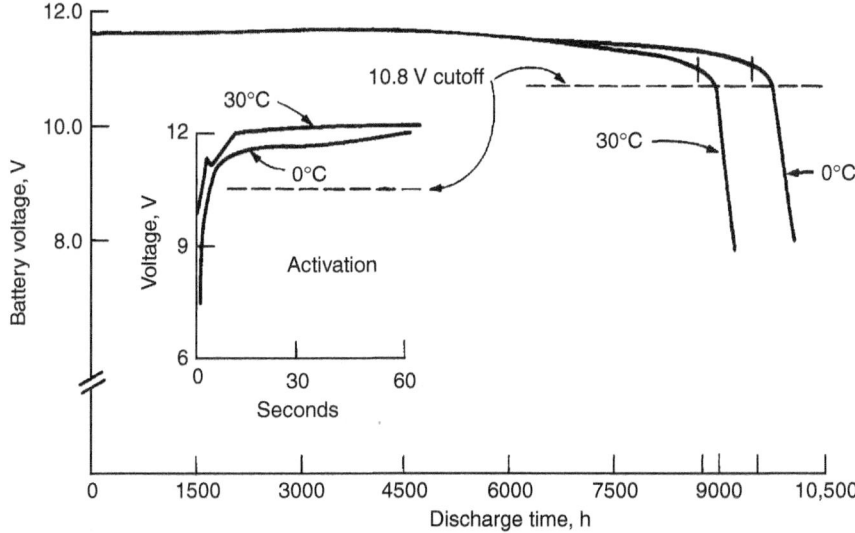

FIGURE 35.18 Activation and discharge voltage profile for 12 V, 100 Ah Li/SO$_2$ battery; electrolyte: LiAsF$_6$ in AN-SO$_2$.

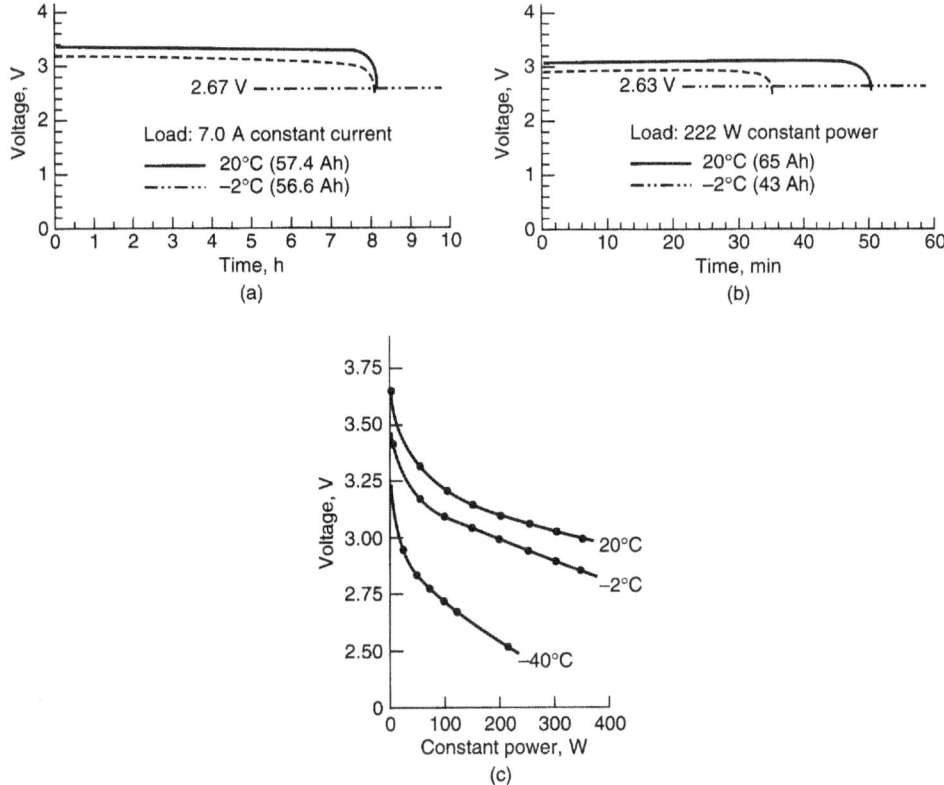

FIGURE 35.19 Discharge characteristics of Li/SOCl$_2$ reserve batteries, Alliant model G3070A2. (a) Low-rate cell discharge profile. (b) High-rate cell discharge profile. (c) Voltage characteristics of high-rate cell at various temperatures and power levels. (*Courtesy of Alliant Techsystems, Inc.*)

Typical discharge curves at several temperatures for the two types of single-cells used in the Li/SOCl$_2$ reserve battery illustrated in Fig. 35.8 are shown in Fig. 35.19.

35.5 APPLICATIONS

35.5.1 Production Batteries

- MOFA for prox-fuzed artillery
- M762/M767 electronic time fuzes
- Excalibur 155 mm round for prelaunch battery
- Self-destruct Fuze (SDF).

35.20 SPECIALIZED BATTERY SYSTEMS

SECTION B
SPIN-DEPENDENT RESERVE BATTERIES

35.6 GENERAL CHARACTERISTICS

Various military, and a few civilian, applications with long shelf-life requirements must turn to reserve batteries for their electric power. This is particularly true when the system requires that the power supply be integrally packaged with the electronics and not replaced throughout the storage life of the system. Typical of such applications are fuzing, control, and arming systems for artillery and other spin-stabilized projectiles.

High spin forces, such as those encountered in artillery projectiles, may produce a difficult environment for many battery designs. However, special designs for liquid-electrolyte reserve batteries have evolved that take advantage of spin to bring about their activation and then keep the electrolyte within the cell structure.

A typical spin-dependent reserve battery is illustrated in Figs. 35.20 and 35.21. The electrode stack consists of electrodes and cell spacers of an annular configuration packaged dry and therefore capable of long-term storage. A metal ampoule, inserted in the center hole of the stack, houses the electrolyte. Upon firing of the gun, the ampoule opens; the electrolyte is released and is then distributed into the annular-shaped cells centrifugally, thereby causing the battery to become active.

FIGURE 35.20 Cross section of lead/fluoboric acid/lead dioxide multicell reserve battery showing "dashpot" cutter for copper ampoule. (*Courtesy of U.S. Department of the Army.*)

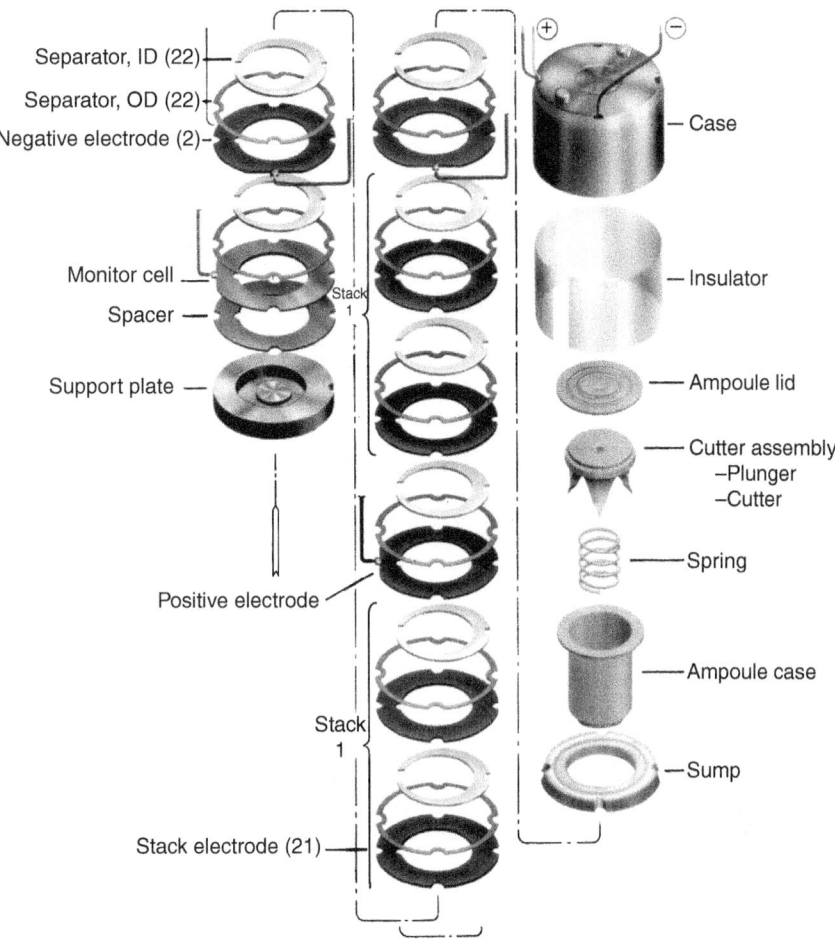

FIGURE 35.21 Component parts of lead/fluoboric acid/lead dioxide multicell reserve battery, PS 416 power supply. (*Courtesy of U.S. Department of the Army.*)

35.7 CHEMISTRY

For several decades, the chemistry most commonly employed in spin-dependent liquid-electrolyte reserve batteries had been the lead/fluoboric acid/lead dioxide cell represented by the following simplified reaction:

$$Pb + PbO_2 + 4HBF_4 \rightarrow 2Pb(BF_4) + 2H_2O$$

Fluoboric acid, rather than the more common sulfuric acid electrolyte, is used because it performs better at the very low temperatures required for these military applications. This low-temperature performance is due in part to the absence of insoluble reaction products as the reserve battery discharges. In the early 1990s, the last two facilities that were capable of producing the specialized electrode material used in these batteries were decommissioned, primarily for business reasons, and the availability of this technology essentially ended. However, fuzes powered by this electrochemistry remain in the U.S. Army's inventory.

To replace the lead/fluoboric acid/lead dioxide system spin-dependent liquid-electrolyte reserve batteries employing lithium anodes have been developed. The most common system today is that in which thionyl chloride serves in the dual role of electrolyte carrier and active cathodic depolarizer (see Chap. 35, Section A). The accepted cell reaction for this system is

$$4Li + 2SOCl_2 \rightarrow 4LiCl + S + SO_2$$

At one time, the zinc/potassium hydroxide/silver oxide system was also employed in spin-dependent reserve batteries. More frequently, this reserve system has been used in non-spin applications, such as missiles, where the electrolyte is driven into place by a gas generator or other activation method (Chap. 34). This system is again finding favor in some applications where the potential hazards of lithium-based systems can create safety problems. The chemistry of the zinc/silver oxide couple can be represented by either of two reactions, depending on the oxidation state of the silver oxide:

$$2AgO + Zn \rightarrow Ag_2O + ZnO$$

$$Ag_2O + Zn \rightarrow 2Ag + ZnO$$

In the early 1970s, thermal batteries using the Ca/LiCl-KCl-CaCro$_4$/Fe system that could operate at high spin rates (300 rps) were developed and successfully demonstrated by Sandia National Laboratories. In the early 1990s, this process was repeated for the now standard lithium (alloy)/iron disulfide couple employed in most thermal batteries (see Chap. 36 for a detailed discussion of these chemistries).

35.8 DESIGN CONSIDERATIONS

35.8.1 Electrode-Stack Arrangement

The electrode stack may be arranged in two ways. One favors a high-voltage output, and the other a high-current output. The former generally uses bipolar electrodes—that is, electrodes wherein anodic and cathodic materials, respectively, are applied to the opposite sides of a metal substrate. Such bipolar electrode plates are stacked in a pile or series configuration, making automatic contact from one cell to the next. The voltage output of such a stack is the sum of all the cells. In the high-current configuration, electrode plates coated with anodic material on both sides of the substrate are stacked alternately with plates coated with cathodic material on both sides. All anodic plates are connected electrically in parallel through tabs. All cathodic plates are similarly connected. The two electrical connections constitute the effective terminals of the battery. This type of parallel stack is, in effect, a single electrochemical cell with a larger electrode area. Where required by the application, multiple series stacks can be connected in parallel, thereby yielding both high-voltage and high-current outputs (Fig. 35.22).

35.8.2 Electrolyte Volume Optimization

The electrolyte capacity of the ampoule must be matched to the composite volume of all the cells in the battery. A parallel construction battery is reasonably tolerant of electrolyte flooding or starvation since it is a single cell. A series configuration, however, can be greatly degraded by flooding since that condition produces intercell short circuits in the electrolyte fill channel or manifold. The opposite condition—that is, insufficient electrolyte—may leave one or more cells empty and therefore fail to provide continuity throughout the cell stack.

Since temperature extremes have a greater effect on the expansion and contraction of the liquid electrolyte than on the volume of the cells, an electrolyte volume that ensures that the cells will be reasonably full at low temperatures usually leads to an excess of electrolyte at higher temperatures.

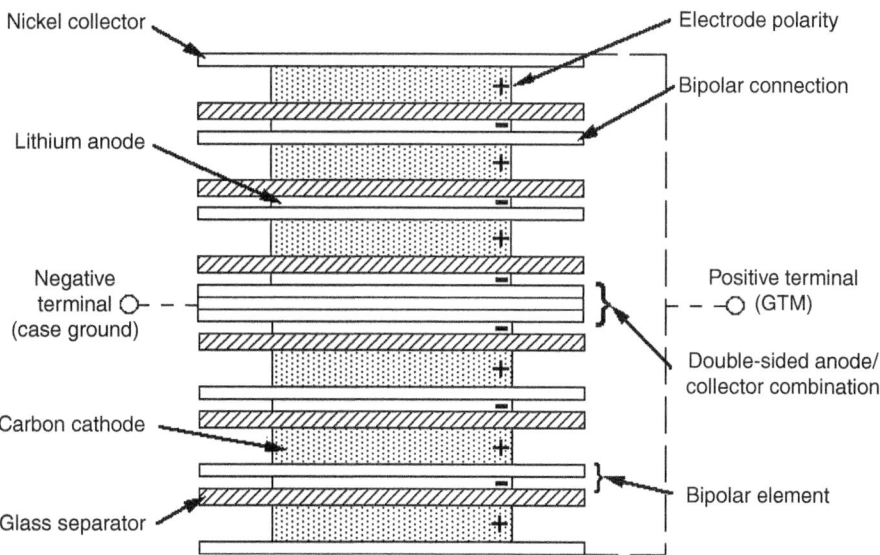

FIGURE 35.22 Quarter cross-sectional view of the cell stack of a lithium/thionyl chloride reserve battery showing an example of series and parallel construction. (*Courtesy of EnerSys Advanced Systems.*)

In the lead/fluoboric acid/lead dioxide system, this excess must be accommodated in the design of the battery by the use of a "sump," which is placed at the end of the path the electrolyte follows to fill the cells. The sump is used to catch not only any excess of electrolyte created by thermal expansion, but also that resulting from eccentric spin conditions. Lithium/thionyl chloride batteries may be more tolerant of cell flooding, depending on the particulars of the individual design. In some short-life batteries, a match is established at high temperature with the recognition that cells will be less than full at lower temperatures. To ensure that some electrolyte enters each cell (so that continuity can be maintained), leveling holes may be provided from cell to cell. Though kept very small to reduce the effect of inevitable intercell short circuits, these holes do dissipate some of the capacity of the battery.

35.8.3 Cell Sealing

Since the individual cells of a spin-dependent liquid-electrolyte reserve battery are generally annular in shape and are filled by centrifugal force, the periphery of the cell must be sealed to keep electrolyte from leaking out. This sealing is typically accomplished by a plastic barrier formed around the outside of the electrode-spacer stack. For lead/fluoboric acid/lead dioxide batteries, this barrier is formed by fish paper (a dense, impervious paper) coated with polyethylene that melts at a relatively low temperature (similar to that used on milk cartons). Cell spacers are punched from the coated fish paper and placed between the electrodes. The stack is then clamped together and heated in an oven at a temperature sufficient to fuse the polyethylene, which then acts as an adhesive and sealer between the electrodes. Cell sealing is critically important in this system because the electrolyte is highly conductive and any leakage that leads to intercell shorting will rapidly deplete the capacity of the battery.

Cell sealing can be somewhat less critical in some lithium/thionyl chloride batteries, as in the moderately powerful batteries typically found in modern radar proximity fuzes used in artillery applications. In a recently produced multicell battery (Fig. 35.23), the cell parts were designed so that the annular electrodes and separators could more easily be inserted into a Tefzel® cell cup to create the cell stack. This was done to simplify automated assembly. The intercell shorting that resulted from the slightly relaxed fit at the outside diameter of the cell stack was reduced by the less conductive (relative to the fluoboric acid used in the previous system) catholyte, and somewhat offset by

35.24 SPECIALIZED BATTERY SYSTEMS

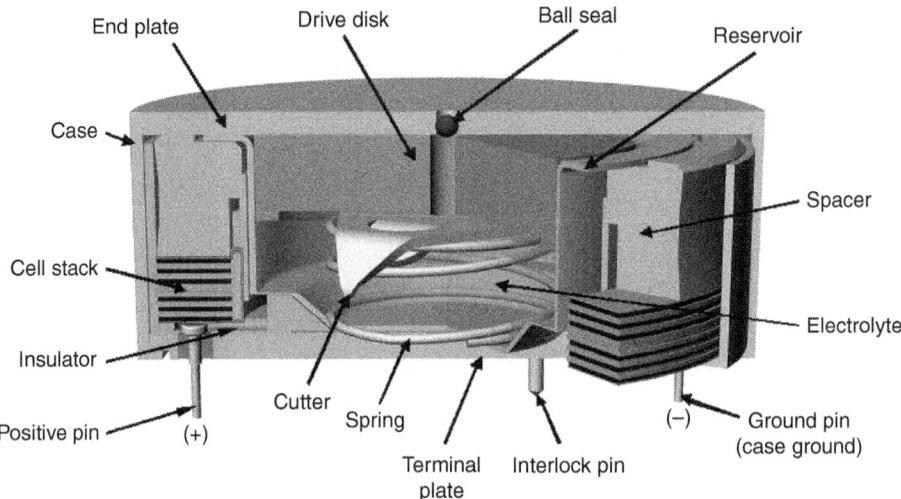

FIGURE 35.23 Cross section of a lithium/thionyl chloride multicell reserve battery. (*Courtesy of EnerSys Advanced Systems.*)

the ability to easily package excess capacity into that particular design. For the higher power and energy batteries required by missile applications (Chap. 35, Section A), cell sealing and isolation remain critically important.

35.8.4 Ampoules

Early designs of liquid-electrolyte reserve batteries used glass ampoules to house the electrolyte, and in fact, some modern batteries still use such ampoules. These ampoules are generally smashed by the acceleration force of gunfire or by the explosive output of a primer or squib. Although these forces are ample, there is also a tendency for rough handling or a drop on a hard surface to cause inadvertent glass ampoule breakage. This destroys the battery due to the premature leakage of electrolyte into the cells.

A major advance in battery ruggedness resulted from the design of metal, usually copper, in lead/fluoboric acid/lead dioxide batteries and stainless steel in the lithium/thionyl chloride designs ampoules (or reservoirs) with internal cutting mechanisms. One version employs a cutter that is activated by a combination of spin and acceleration (Fig. 35.24), both provided by the act of gunfire. Other versions rely on a dashpot cutter mechanism (Figs. 35.20, 35.21, and 35.23). This mechanism requires a sustained acceleration (several milliseconds experienced in gunfire), but will not function when subjected to the much shorter (a portion of a millisecond) shock pulse resulting from being dropped on a hard surface. The use of these "intelligent" ampoules, which are capable of discriminating between the forces of gunfire and those of rough handling, has resulted in a substantial improvement in battery reliability and safety.

35.8.5 Safety in Lithium-Based Batteries

For at least the past 20 years, reserve single-cell batteries that use various lithium-based electrochemistries have been employed in a number of fuzing applications. These cells are normally constructed with an anode-separator-cathode assembly spirally wound around a centrally located glass electrolyte ampoule. The electrolyte ampoule is normally broken upon gunfire or as the result of the bottom of the cell case being struck with a squib- or spring-driven device. Since these are single-cell devices, there is no chance for intercell short-circuiting and subsequent safety problems.

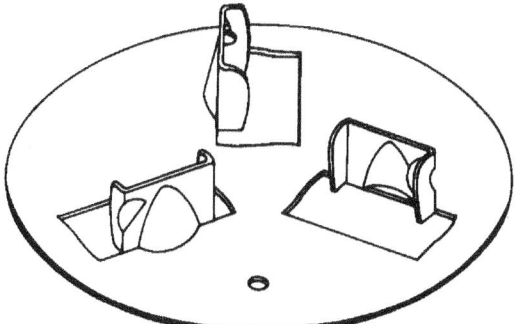

FIGURE 35.24 Three-bladed cutter for copper ampoule requiring spin and acceleration for activation. (*Courtesy of U.S. Department of the Army.*)

However, in multicell pile configuration reserve batteries, there is a considerable chance for intercell short-circuiting in the common electrolyte manifold. This intercell short-circuiting not only dissipates the capacity of the cells, but it can also allow for dendritic growth, which can lead to electronic short-circuiting of the cells with catastrophic results. Experience has shown that such dendritic growth can be minimized or eliminated if all interior metallic surfaces of the battery have a nonconductive (usually Teflon-based) coating.

35.9 PERFORMANCE CHARACTERISTICS

35.9.1 General

Energy and Power Density. Liquid-electrolyte reserve batteries are not normally rated in terms of energy or power per unit weight or per unit volume. Because of the need to provide double the volume for the electrolyte (one volume in an ampoule, the other in the cells themselves), such batteries are not highly space efficient. Space is also consumed by the ampoule-opening mechanism and the cell-sealing material. Furthermore, the cell area is sometimes not exposed to the electrolyte because of the spin eccentricity of the projectile, which houses the battery. Finally, such batteries are generally designed for short-lifetime applications, such as the flight time of an artillery projectile (approximately 3 min).

Operating Temperature Limits. Like most other batteries, the performance of liquid-electrolyte reserve batteries is affected by temperature. Military applications frequently demand battery operations at all temperatures between −40 and 60°C, with storage limits of −55 to 70°C. These requirements are routinely met by the lead/fluoboric acid/lead dioxide systems and, with some difficulty at the low-temperature end, by the lithium/thionyl chloride and zinc/potassium hydroxide/silver oxide systems. Provision is occasionally made to warm the electrolyte prior to the activation of the two latter systems.

Voltage Regulation. Since the voltage sustained by a liquid-electrolyte reserve battery at low temperatures and under heavy electric loading is much lower than that which it delivers at high temperatures, a serious problem of voltage regulation frequently results. In some situations, the ratio of high- to low-temperature voltage may be as much as 2:1. This problem may be avoided by the use of thermal batteries (Chap. 36), which provide their own pyrotechnically induced operating temperature, irrespective of the ambient temperature. Until recently, thermal batteries were extremely ineffective at high spin rates, but progress has been made in this field and thermal batteries capable of withstanding spin rates of 300 rev/s are now available.

Shelf Life. The shelf life of liquid-electrolyte reserve batteries is highly dependent on the storage temperature, with high temperatures being the more deleterious. Zinc/silver oxide cells are probably the most vulnerable of the generally used systems because of the reduction of silver oxide and the passivation of zinc. Ten-year storage life is probably the best that can be expected unless the battery is substantially overdesigned. Lead/fluorboric acid/lead dioxide batteries also degrade with time, in both the loss of capacity and the lengthening of activation time. However, if objectionable organic materials are avoided in battery construction and the battery is designed with some safety factors, 20 to 25 years of shelf life may be realized. Lithium/thionyl chloride reserve systems employing a neutral catholyte formulation stored in a glass ampoule, as shown in Fig. 35.4, have demonstrated excellent shelf life, exceeding 16 years in a real-time study. Other lithium/thionyl chloride batteries (Fig. 35.23), have incorporated formed and/or welded stainless steel (304L or 316L) reservoirs filled with nonneutral catholytes that may also contain additives to improve certain aspects of performance. Predicting shelf life in this circumstance is less straightforward because of the highly corrosive nature of the catholytes and the potential interaction with/of the various additives, but several studies have projected long storage capability for properly (dry) built and sealed batteries.

Linear and Angular Acceleration Limits. Since spin-activated batteries are normally expected to be used in environments where guns are used, they must be built to withstand the forces of gunfire. With the development of the ampoules and the construction methods described, such batteries can withstand linear acceleration to the 20,000 to 30,000 g level and spin rates as great as 30,000 rev/m. The sizes intended for small-caliber (20 to 40 mm) projectiles will withstand linear g levels 2 to 5 times that high.

As an assist in withstanding these forces, the battery assembly is sometimes encapsulated in a supporting plastic. A popular design involved a molded plastic cup to house the stack and ampoule assembly, which was locked in place with an epoxy resin. More recently, the stack and ampoule assembly was encapsulated in situ in a RIM (reaction impingement molding) process using a high-impact polyurethane foam, a process that allows demolding in just minutes. These two types of support are shown in Fig. 35.25.

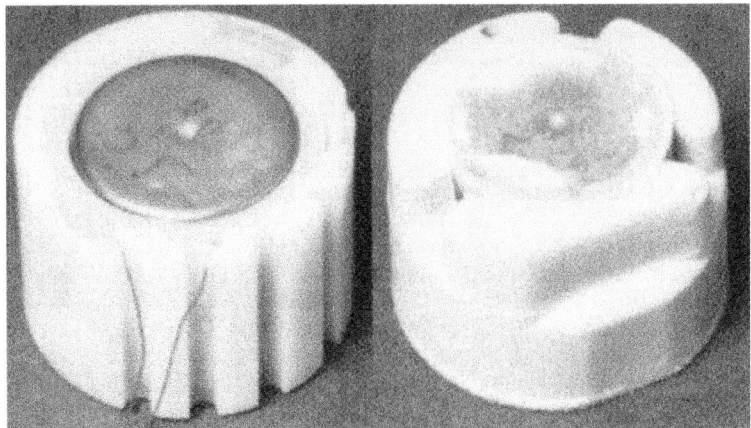

FIGURE 35.25 Stack and ampoule assembly of a lead/fluoboric acid/lead dioxide reserve battery supported by potting in epoxy in a molded case (left) and by in situ molding using a reaction impingement molded polyurethane foam (right). (*Courtesy of U.S. Department of the Army.*)

Activation Time. The time from initiation of the battery to the point at which it delivers and sustains a requisite level of voltage across a specified electric load is defined as the activation time. For a spin-dependent liquid-electrolyte reserve battery, this time would include the times for ampoule opening, electrolyte distribution, clearing of electrolyte short circuits in the filling manifold, depassivation of electrodes, and elimination of any form of polarization. Activation times are usually longest at low temperatures, where increased viscosity of the electrolyte and decreased ion mobility are most significant.

The application normally establishes the maximum allowable activation time, and reserve batteries are frequently designed to reach 75 or 80% of their peak voltage within this required time. A typical application requiring a very short activation time, perhaps less than 100 ms, would be a time fuze for an artillery projectile. Battery power is required to start the timer. Hence a stretch-out or uncertainty of time to reach timer voltage could result in a serious timing error, with a corresponding ineffectiveness of gunfire. In some cases, safety can be adversely affected by a timing error. In less critical situations, 0.5 to 1.0 s is allowed for activation.

35.9.2 Performance of Specific Electrochemical Systems

The physical and electrical characteristics of several typical spin-dependent reserve batteries are presented in Table 35.5.

TABLE 35.5 Typical Spin Dependent Reserve Batteries

Reference	Electrochemical system	Height, cm	Diameter, cm	Weight, g	Nominal voltage, V	Nominal energy, Wh
Fig. 35.20	Pb/HBF$_4$/PbO$_2$	4.1	5.7	280	35	0.5
Fig. 35.23	Li/SOCl$_2$	1.67	3.8	70	9	0.37
	Zn/KOH/AgO	1.3	5.1	80	1.4	0.65

Lead/Fluoboric Acid/Lead Dioxide Battery. Discharge curves for a typical lead/fluoboric acid/lead dioxide liquid-electrolyte reserve battery employed to power the proximity fuze of an artillery shell are given in Fig. 35.26. The slight rise in the low-temperature curve is due to its gradual rise in temperature in a room-temperature spinning tester. Similarly, the high-temperature curve is falling faster than it would in a true isothermal situation.

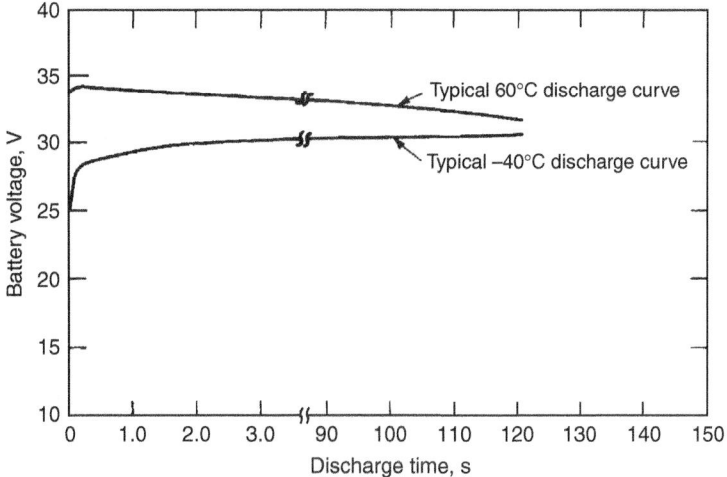

FIGURE 35.26 Discharge curves of a spinning lead/fluoboric acid/lead dioxide series-configuration reserve battery. Current density: 100 mA/cm^2. (*Courtesy of U.S. Department of the Army.*)

Lithium/Thionyl Chloride Battery. Discharge curves for the multicell lithium/thionyl chloride liquid-electrolyte reserve battery shown in Fig. 35.23, which was designed to power the proximity fuze of an artillery shell, are shown in Figs. 35.27 and 35.28 and indicate the effect of operating temperature on the output voltage level, deliverable capacity, and rise time of the battery.

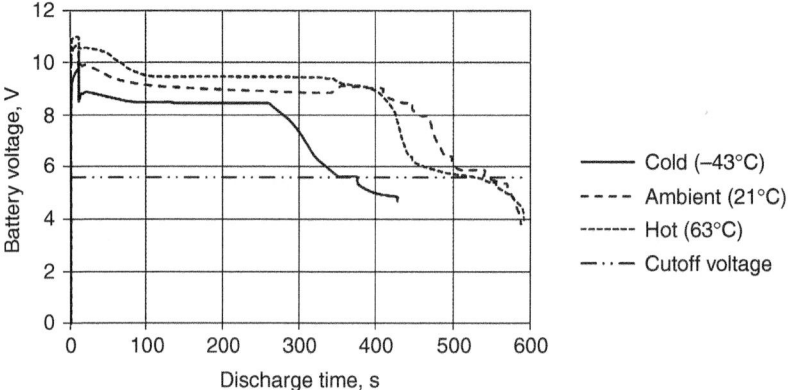

FIGURE 35.27 Discharge curves of a spinning (80 rev/s) lithium/thionyl chloride reserve battery at various temperatures. Discharge current density is 2 mA/cm^2 until 10 s have elapsed, after which it is 35 mA/cm^2. (*Courtesy of U.S. Department of the Army.*)

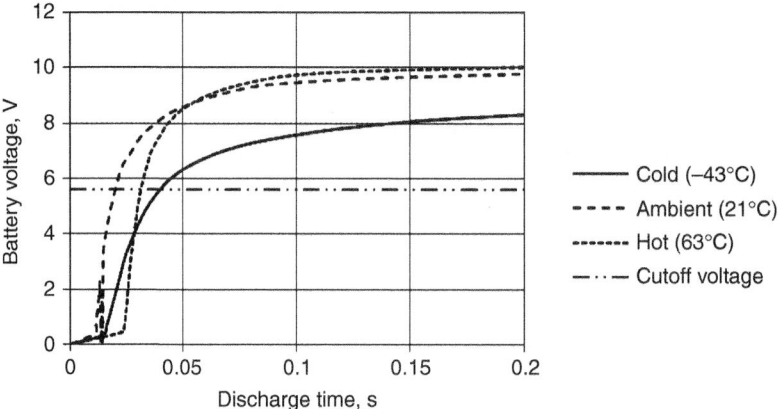

FIGURE 35.28 Rise-time curves of the spinning (80 rev/s) lithium/thionyl chloride reserve battery shown in Fig. 35.27 at various temperatures. Discharge current density is 2 mA/cm^2. (*Courtesy of U.S. Department of the Army.*)

In addition to simply replacing the lead/fluoboric acid/lead dioxide system, the advent of lithium/thionyl chloride batteries introduced some operational advantages. With the former system spin-dependent batteries were expected to function for short periods of time and only under sustained spin (necessary to keep the electrolyte within the cells). New applications have arisen that require a battery capable of withstanding artillery fire and spinning for a short time followed by some substantial operating time in a nonspin mode. Such applications include artillery delivery of mines or communications jammers intended to function after impact with the ground, or projectiles and submunitions that are operative while being slowed down by parachute.

The lithium-based liquid-electrolyte reserve battery is capable of fulfilling this difficult combination of requirements. A typical cell, as illustrated in Fig. 35.29, incorporates an absorbing separator such

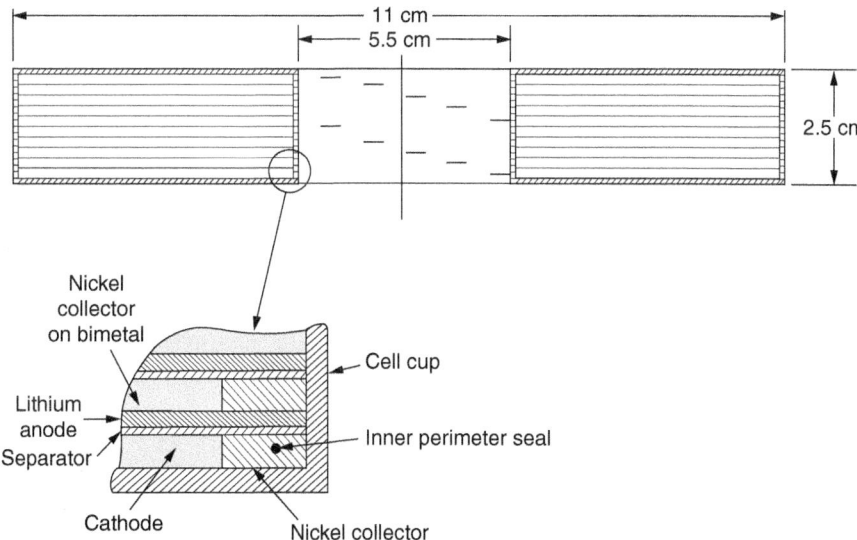

FIGURE 35.29 Lithium anode/carbon cathode cell stack for thionyl chloride liquid-electrolyte reserve battery. (*Courtesy of EnerSys Advanced Systems.*)

as a nonwoven glass mat between the electrodes, and a long, high-resistance electrolyte filling path. After cell filling under spin, the absorbing material causes the electrolyte to be retracted away from the manifold and retained within the cell after cessation of spin. These design features, coupled with the long wet-stand capability of the lithium/thionyl chloride system, have paved the way for reserve batteries in applications that previously had to depend on the use of active batteries with relatively shorter storage capability. The discharge curves for such a multicell, liquid-electrolyte reserve battery that demonstrate this long wet-stand capability are given in Fig. 35.30 (also see Part A of this chapter).

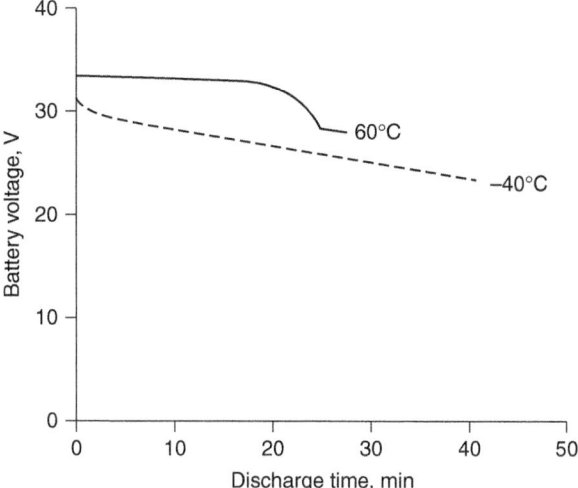

Figure 35.30 Discharge curves of a lithium/thionyl chloride series-configuration reserve battery. Current density is 50 mA/cm^2. (*Courtesy of U.S. Department of the Army.*)

Spin-Capable Thermal Batteries. Because of their lower susceptibility to temperature extremes and known superior shelf life (without degradation), thermal batteries have been desired as an alternative to liquid-electrolyte reserve batteries for some time. The primary failure mode for thermal batteries in a high-spin environment had always been intercell short-circuiting at the cell stack edges due to the leakage of molten conductive materials at battery operating temperatures. Novel construction techniques, electrochemistries that allow for higher electrolyte binder contents, and lithium alloy anodes that prevent migration of the anode material have made spin-capable thermal batteries practical.

REFERENCES (PART A)

1. A. N. Dey, "Lithium Anode Film and Organic and Inorganic Electrolyte Batteries," in *Thin Solid Films*, vol. 43, Elsevier Sequoia, Lausanne, Switzerland, p. 131, 1997.
2. R. J. Horning, "Small Lithium/Vanadium Pentoxide Reserve Cells," *Proc. 10th Intersoc. Energy Convers. Eng. Conf.*, 1975.
3. W. B. Ebner and C. R. Walk, "Stability of $LiAsF_6$-Methyl Formate Electrolyte Solutions," *Proc. 27th Power Sources Conf.*, 1976.
4. "Organic-Based M762 Reserve Battery," ARL Sponsored Program, 2009.
5. B. Ravid, *A Reserve-Type Lithium-Thionyl Chloride Battery*, Tadiran Israel Electronics Industries, 1979.
6. M. J. Domenicomi and F. G. Murphy, "High Discharge Rate Reserve Cell and Electrolyte," U.S. Patent 4,150,198, Apr. 17, 1979.
7. M. Babai, U. Meishar, and B. Ravid, "Modified $Li/SOCl_2$ Reserve Cells with Improved Performance," *Proc. 29th Power Sources Conf.*, June 1980.
8. P. M. Shah, "A Stable Electrolyte for Li/SO_2 Reserve Cells," *Proc. 27th Power Sources Symp.*, 1976.
9. P. M. Shah and W. J. Eppley, "Stability of the $LiAsF_6$: $AN:SO_2$ Electrolyte," *Proc. 28th Power Sources Symp.*, 1978.
10. R. J. Horning and K. F. Garoutte, "Li/SO_2 Multicell Reserve Structure," *Proc. 27th Power Sources Symp.*, 1976.
11. Hsiu-Ping Lin and K. Burgess, "Synthesis of Charged Li_xCoO_2 ($0 < x < 1$) for Primary and Secondary Batteries," U.S. Patent 5,667,660 (1997).
12. W. J. Eppley and R. J. Horning, "Lithium/Thionyl Chloride Reserve Cell Development," *Proc. 28th Power Sources Symp.*, 1978.
13. J. Nolting and N. A. Remer, "Development and Manufacture of a Large Multicell Lithium-Thionyl Chloride Reserve Battery," *Proc. 35th International Power Sources Symp.*, 1992.
14. C. Kelly, "Development of HWAM Li_xCoO_2 Reserve Battery," Report No. NSWCCD-TR-98/005, April 1997.
15. S. McKay, M. Peabody, and J. Brazzell, *Proc. 39th Power Sources Conf.*, pp. 73–76, 2000.
16. P. G. Russell, D. C. Williams, C. Marsh, and T. B. Reddy, *Proc. 6th Workshop for Battery Exploratory Development*, pp. 277–281, 1999.

BIBLIOGRAPHY (PART B)

Benderly, A. A.: "Power for Ordnance Fuzing," *National Defense*, Mar.–Apr. 1974.

Biggar, A. M.: "Reserve Battery Requiring Two Simultaneous Forces for Activation," *Proc. 24th Annual Power Sources Symp.*, Electrochemical Society, Pennington, NJ, pp. 39–41, 1970.

Biggar, A. M., R. C. Proestel, and W. H. Steuernagel: "A 48-Hour Reserve Power Supply for a Scatterable Mine," *Proc. 26th Annual Power Sources Symp.*, Electrochemical Society, Pennington, NJ, pp. 126–129, 1974.

Cieslak, W. R., F. M. Delnick, and C. C. Crafts: "Compatibility Study of 316L Stainless Steel Bellows for XMC3690 Reserve Lithium/Thionyl Chloride Battery," Sandia Report SAND85-1852, February 1986.

Doddapaneni, N., D. L. Chua, and J. Nelson: "Development of a Spin Activated, High Rate, Li/SOCl$_2$ Bipolar Reserve Battery," *Proc. 30th Annual Power Sources Symp.,* Electrochemical Society, Pennington, NJ, pp. 201–204, 1982.

Grothaus, K. R.: "Thermal Battery for Artillery," *Proc. 26th Power Sources Conference*, U.S. Army CECOM/ARL, pp. 141–144, Apr. 29–May 2, May 1974.

Morganstein, M., and A. B. Goldberg: "Reaction Impingement Molding (RIM) Encapsulation of a Fuze Power Supply," *Proc. of the 4th International SAMPE Electronics Conference,* Society for the Advancement of Material and Process Engineering, Covina, CA, pp. 753–764, 1990.

Schisselbauer, P. F., and D. P. Roller, "Reserve g-Activated, Li/SOCl$_2$ Primary Battery for Artillery Applications," *Proc. 37th Annual Power Sources Conference,* U.S. Army CECOM/ARL, Cherry Hill, NJ, pp. 357–360, 1996.

Turrill, F. G., and W. C. Kirchberger: "A One-Dollar Power Supply for Proximity Fuzes," *Proc. 24th Annual Power Sources Symp.,* Electrochemical Society, Pennington, NJ, pp. 36–39, 1970.

CHAPTER 36
THERMAL BATTERIES

Charles M. Lamb

36.1 GENERAL CHARACTERISTICS

Thermal batteries are primary reserve batteries that employ inorganic salt electrolytes. These electrolytes are relatively nonconductive solids at ambient temperatures. Integral to the thermal battery are pyrotechnic materials scaled to supply sufficient thermal energy to melt the electrolyte. The molten electrolyte is highly conductive, and high currents may then be drawn from the cells.

The activated life of a thermal battery depends on several factors involving cell chemistry and construction. Once activated, and as long as the electrolyte remains molten, thermal batteries may supply current, discharging the active materials to the point of functional exhaustion. On the other hand, even with excess active materials present, the batteries will eventually cease functioning due to the loss of internal heat and subsequent re-solidification of the electrolyte. Hence, two of the primary factors behind thermal battery active life are:

1. Compositions and masses of the active cell-stack materials (i.e., anodes and cathodes)
2. Other construction details, including the overall battery shape and the types and amounts of thermal insulation

Depending on the battery design, which is ultimately determined by the specific requirements of the application, the activated thermal battery may supply electric power for only a few seconds or may function for over an hour.

Initiation of a thermal battery is normally provided by an energy impulse from an external source to a built-in initiator. The initiator, typically an electric match, an electro-explosive device (squib), or a percussion primer, ignites the cell-stack pyrotechnics. Rise time, the time interval between the initiation impulse and that time at which the battery can sustain a current at voltage, varies as a function of battery size, design, and chemistry. Rise times of several hundred milliseconds are not uncommon for large units. Small batteries have been designed to reliably achieve operating conditions within 10 to 20 ms.

The shelf life of an unactivated thermal battery is typically 10 to 25 years, depending upon design. Once activated and discharged, though, they are not reusable or rechargeable.

Current developments in extending the activated life capabilities of thermal batteries have widened their suitability and application potential in new military as well as industrial/civilian systems.

Thermal batteries were first developed in Germany in the 1940s, and were used primarily for weapons applications.[1-3] Batteries containing multiple cells and integral pyrotechnic materials have been produced since 1947.[4] Because of their high reliability and long shelf life, thermal batteries are ideally suited for military ordnance purposes. Consequently, they have been widely

FIGURE 36.1 Typical thermal batteries. (*Courtesy of Catalyst Research Corp.*)

used in missiles, bombs, mines, decoys, jammers, torpedoes, space exploration systems, emergency escape systems, and similar applications. Figure 36.1 illustrates typical thermal battery configurations.

Some of the advantages of thermal batteries include:

1. Very long shelf life (up to 25 years) in a ready state without degradation in performance.
2. Almost instant activation; fast start designs can provide useful power in hundredths of a second.
3. Peak-power densities can exceed 11 W/cm^2.
4. Very high demonstrated reliability and ruggedness following long-term storage over a wide temperature range and severe dynamic environments.
5. No maintenance required; they can be permanently installed in equipment.
6. Self-discharge is generally negligible. An unactivated battery can support almost no current.
7. Wide operating temperature range.
8. No outgassing; the batteries are hermetically sealed.
9. Custom designed for specific voltage, start time, current, and physical configuration requirements.

The disadvantages of thermal batteries include:

1. Generally short activated lives (typically less than 10 min), but they can be designed to operate for more than 2 h.
2. Low to moderate energy densities and specific energies.
3. Surface temperatures can typically reach 230°C or higher.
4. Voltage output is nonlinear and decreases with life.
5. One time use. Once activated, they cannot be turned off or reused (recharged).

36.2 DESCRIPTION OF ELECTROCHEMICAL SYSTEMS

A number of electrochemical systems have been used in thermal batteries. As materials and techniques have improved the state-of-the-art (SOA) performance of these batteries, older designs have gradually disappeared. Battery designs with older technologies still exist, however, and continue to be manufactured. In some cases, the continuing production of an "antiquated" system is driven by economics. Redesign and requalification of an existing battery with a newer technology is often economically unacceptable. Table 36.1 lists some of the more common types of electrochemical systems that have been used over the years.

All thermal battery cells consist of an alkali or alkaline earth metal anode, a fusible salt electrolyte, and a metal salt cathode. The pyrotechnic heat source is usually inserted between cells in a series cell-stack configuration.

36.2.1 Anode Materials

Until the 1980s, most thermal battery designs employed a calcium metal anode—with calcium foil generally attached to an iron, stainless steel, or nickel foil current collector or backing. A "bimetal" anode is manufactured by vapor depositing the calcium on the backing material. Here, the calcium anode thickness usually ranges between 0.03 and 0.25 mm. In other designs, calcium foil is either pressed onto a perforated "cheese grater"–type backing sheet or is spot-welded to the backing. Magnesium metal is another anode material that has been widely used, both in "bimetal"-form and in pressed or spot-welded anode configurations.

Introduced in the mid-1970s, lithium has become the most widely used anode material in thermal batteries. There are two major configurations of lithium anodes: lithium alloy and lithium metal. The most commonly used alloys are lithium aluminum, with about 20 wt % lithium and lithium (silicon), with about 44 wt % lithium. Lithium-boron alloy has also been evaluated, but has not been used widely because of its higher cost.

LiAl and Li(Si) alloys are processed into powders, which are cold-pressed into anode wafers or pellets that range in thickness from 0.75 to 2.0 mm. In the cell, the alloy pellet is backed with an iron, stainless steel, or nickel current collector. Lithium alloy anodes function in activated cells as solid anodes, and must be maintained below melt or partial melt temperatures. Forty-four wt % Li(Si) alloy will partially melt at 709°C, while α, β-LiAl will exhibit partial melting at 600°C. If these

TABLE 36.1 Types of Thermal Batteries

Electrochemical system: anode/electrolyte/cathode	Operating cell voltage	Characteristics and/or applications
Ca/LiCl-KCl/$K_2Cr_2O_7$	2.8–3.3	Very fast activation times; short lives; used in "pulse" applications
Ca/LiCl-KCl/WO_3	2.4–2.6	Medium-short lives; low electrical noise; not severe physical environments
Ca/LiCl-KCl/$CaCrO_4$	2.2–2.6	Medium lives; severe dynamic environments
Mg/LiCl-KCl/V_2O_5	2.2–2.7	Medium-short lives; severe physical environments
Ca/LiCl-KCl/$PbCrO_4$	2.0–2.7	Fast activation; short lives
Ca/LiBr-KBr/$K_2Cr_2O_4$	2.0–2.5	Short lives; used in high-voltage, low-current applications
Li(alloy)/LiCl-LiBr-LiF/FeS_2	1.6–2.1	Short to medium lives, high current capacity; severe physical environments
Li(metal)/LiCl-KCl/FeS_2	1.6–2.2	Long lives, high current capacity; severe physical environments
Li(alloy)/LiBr-KBr-LiF/CoS_2	1.6–2.1	Long lives (past 1 h), high current capacity; severe physical environments

36.4 SPECIALIZED BATTERY SYSTEMS

melting temperatures are exceeded, the melted anode may come in contact with cathode material, allowing a direct, highly exothermic chemical reaction and cell short-circuiting.

Lithium metal anodes function in activated cells at temperatures above the melting temperature of lithium, 181°C. To prevent the molten lithium from flowing out of the cells and short-circuiting the battery, it is combined with a high-surface-area binder of metal powder or metal foam. The binder holds the lithium in place by surface tension.

Lithium metal anodes are prepared by combining the binder material with molten lithium, followed by pressing the solidified mixture into thin foil, typically 0.07 to 0.65 mm thick. The foil is then cut into cell-sized parts. The anode foil parts are enclosed in iron-foil cups, which provide added protection against the migration of any free lithium (which can result in cell shorting) and also serve as electron collectors (electrical connections). Such anodes can function at cell temperatures greater than 700°C without significant loss of performance.[5] Each thermal battery designer or manufacturer has developed a number of anode configurations, from which the most suitable may be selected, depending upon specific battery performance requirements.

36.2.2 Electrolytes

Historically, most thermal battery designs have used a molten eutectic mixture of lithium chloride and potassium chloride as the electrolyte (45:55 LiCl:KCl by weight, mp = 352°C). Halide mixtures containing lithium have been preferred because of their high conductivities and general overall compatibility with the anodes and cathodes. Compared to most lower-melting oxygen-containing salts, the halide mixtures are less susceptible to gas generation via thermal decomposition or other side reactions. More recent electrolyte variations, containing bromides, have been developed for thermal batteries to achieve a lower melting point (and thus extend the operating life) or to reduce the internal resistance (and raise the current capability) of the batteries. These include LiBr-KBr-LiF (mp = 320°C), LiCl-LiBr-KBr (mp = 321°C), and the all-Li^+ electrolyte LiCl-LiBr-LiF (mp = 430°C).[6] Electrolytes with mixed-cations (e.g., Li^+ and K^+, instead of all-Li^+) are subject to the establishment of Li^+ concentration gradients during discharge. These concentration gradients can give rise to localized freezing out of salts, especially during high current draw.[7]

At battery operating temperatures, the viscosity of molten salt electrolytes is very low (ca. 1 cP). In order to immobilize the molten electrolytes, binders are added to the salts during compounding. Earlier blends, originally developed for $Ca/CaCrO_4$ systems and the original $LiAl/FeS_2$ batteries, employed clays, such as kaolin, and fumed silica as effective binders for the salts. These siliceous materials will react with Li(Si) and lithium metal anodes, however. High surface area MgO is sufficiently inert for the more reactive anodes, and is presently the binder of choice in most systems.

36.2.3 Cathode Materials

A wide variety of cathode materials has been used for thermal batteries. These include calcium chromate ($CaCrO_4$), potassium dichromate ($K_2Cr_2O_7$), potassium chromate (K_2CrO_4), lead chromate ($PbCrO_4$), metal oxides (V_2O_5, WO_3), and sulfides (CuS, FeS_2, CoS_2). The criteria for suitable cathodes include high voltage against a suitable anode, compatibility with halide melts, and thermal stability to approximately 600°C. Calcium chromate has been most often used with calcium anodes because of its high potential (at 500°C, V = 2.7) and its thermal stability at 600°C. FeS_2 and (more recently) CoS_2 are used with modern lithium-containing anodes (FeS_2 to 550°C and CoS_2 to 650°C).

36.2.4 Pyrotechnic Heat Sources

The two principal heat sources that have been used in thermal batteries are heat paper and heat pellets. Heat paper is a paper-like composition of zirconium and barium chromate powders supported in an inorganic fiber mat. Heat pellets are pressed tablets or pellets consisting of a mixture of iron powder and potassium perchlorate.

The Zr-BaCrO$_4$ heat paper is manufactured from pyrotechnic-grade zirconium powder and BaCrO$_4$, both with particle sizes below 10 microns. Inorganic fibers, such as ceramic and glass fibers, are used as a structure for the mat.[8] The mix, together with water, is formed into a paper—either as individual sheets by use of a mold or continuously through a paper-making process. The resultant sheets are cut into parts and dried. Once dried, the material must be handled very carefully since it is very susceptible to ignition by static charge and friction. Heat paper has a burning rate of 10 to 300 cm/s and a usual heat content of about 1675 J/g (400 cal/g). Heat paper combusts to an inorganic ash with electrical resistivity. If inserted between cells, it must be used in combination with highly conductive intercell connectors. In some battery designs, combusted heat paper serves as an electrical insulator between cells. In those applications it may have an additional layer of ceramic fibers only, known as base sheet, to enhance its dielectric properties. In most modern pellet-type batteries, heat paper is used only as an ignition or fuse train, if at all. In this application, the heat paper fuse, which is ignited by the initiator, in turn ignites the heat pellets, which are the primary heat source in these batteries.

Heat pellets are manufactured by cold-pressing a dry blend of fine iron powder (1 to 10 microns) and potassium perchlorate. The iron content ranges from 80 to 88% by weight and is considerably in excess of stoichiometry. Excess iron provides the combusted pellet with sufficient electronic conductivity, eliminating the need for separate intercell connectors. The heat content of Fe-KClO$_4$ pellets ranges from 920 J/g for 88% iron to 1420 J/g for 82% iron. Burning rates are generally slower than those of heat paper, and the energy required to ignite them is greater. For that reason, the heat pellet is less susceptible to inadvertent ignition during battery manufacture. Heat pellets (and especially unpelletized heat powder) must nevertheless be handled with extreme care and protected from potential ignition sources.

After combustion, the heat pellet is an electronic conductor, simplifying intercell connection and battery design. It also retains its physical shape and is very stable under dynamic environments (such as shock, vibration, and spin). This contributes greatly to the general ruggedness of battery designs that incorporate heat pellets. Another major advantage of heat pellets is that their enthalpy of reaction is much higher than that of heat paper ash. Thus, they serve as heat reservoirs, retaining considerable heat after combustion, and tend to extend the battery active life by virtue of their greater thermal mass.

36.2.5 Methods of Activation

Thermal batteries are activated by applying an external signal to an initiation device that is incorporated in the battery. There are four generally used methods of activation: electric signal to an electric igniter; mechanical impulse to a percussion primer; mechanical shock to an inertial activator; and optical energy (laser) signal to a pyrotechnic material.

Electric igniters typically contain one or more bridge wires and a heat-sensitive pyrotechnic material. Upon application of an electric current, the bridge wire ignites the pyrotechnic, which in turn ignites the heat source in the thermal battery. Igniters generally fall into two categories: squibs and electric matches. A typical squib is enclosed in a sealed metal or ceramic enclosure and contains one or two bridge wires. The most commonly used types require a minimum activation current of 3.5 A and have a maximum no-fire limit of 1 A or 1 W (whichever is greater). Electric matches do not have a sealed enclosure and typically contain only one bridge wire. They require an activation current of 500 mA to 5 A and should not be subjected to a no-fire test current of more than 20 mA. Squibs are 4 to 10 times more expensive than electric matches, but are required for applications that may encounter environments with electromagnetic radiation.

Percussion primers are pyrotechnic devices that are activated by impact from a mechanical striking device. Typically, a primer is activated by an impact of 2016 to 2880 g × cm applied with a 0.6 to 1.1 mm spherical radius firing pin. Primers are installed in primer holders that are integral parts of the battery enclosure.

Inertial or setback activators are devices that are activated by a large-magnitude shock or rapid acceleration, as is generated upon firing of a mortar or artillery round. They are designed to react to a predetermined combination of g force and its duration. Inertial activators are typically firmly mounted inside the battery structure in order to withstand severe dynamic environments.

Optical energy (laser) activation of thermal batteries is accomplished by firing a laser beam through an optical window installed in the outer enclosure of the battery and igniting a suitable pyrotechnic material inside the unit. This method has found utility in applications where severe electromagnetic interference would be disruptive to an electrical firing method.

Thermal batteries can be equipped with more than one activation device. The multiple activators can be of the same type or of any combination required by the application.

36.2.6 Insulation Materials

Thermal batteries are designed to maintain hermeticity throughout their service lives, even though their internal temperatures reach or even exceed 600°C. The thermal insulation used to retard heat loss from the cell stack and minimize peak surface temperatures must be anhydrous and must have high thermal stability. Ceramic fibers, glass fibers, certain high-temperature polymers, and their combinations have been used as thermal insulators. Older battery designs may still have asbestos insulation, which was widely used before the 1980s.

Electrical insulation materials for conductors, terminals, initiators, and other electrically conductive components are typically made of mica, glass or ceramic fiber cloths, and high temperature–resistant polymers.

Thermal insulation is located around the periphery and at both ends of the cell stack. Some designs also incorporate high-temperature epoxy potting materials as insulation and structural support for the initiators and electric conductors on the terminal end (header) of the batteries. Long-life batteries (20+ min) usually incorporate high-efficiency thermal insulation materials such as Min-K (Johns Manville Co.) or Micro-Therm (Constantine Wingate, Ltd.). Extended life batteries (1 h and longer) may incorporate vacuum blankets and/or double cases with vacuum space between them to retard heat loss. Special high-thermal-capacity pellets and extra "dummy" cells are also used at the ends of cell stacks to retard heat loss and thus prolong the activated life of some batteries.[9] Figure 36.2 shows a typical arrangement of thermal insulation and an encapsulated header assembly with initiator (squib).

A very effective method for extending the activated battery life and reducing the effects of heat on thermally sensitive components located near the battery is to use an external thermal blanket. Provided that it is protected from external contamination, a thermal blanket is more effective than

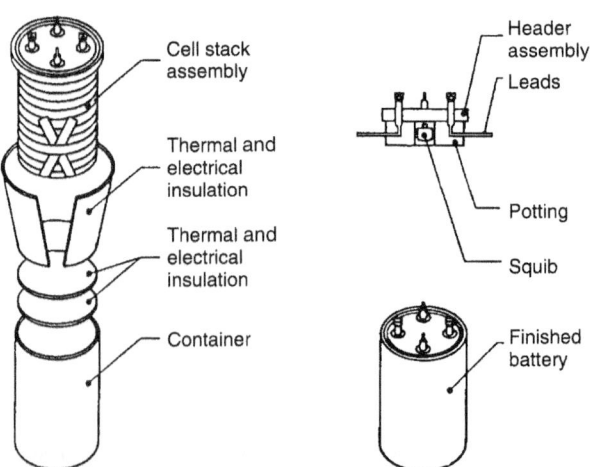

FIGURE 36.2 Typical thermal battery assembly. (*Courtesy of Eagle-Picher Technologies, LLC.*)

internal insulation, primarily because the hot gases that are generated inside the battery during activation cannot penetrate it. External insulation, mounting methods, and the surrounding environment have a significant effect on the heat loss from the battery and all of these must be taken into consideration in the design of thermal batteries.

36.3 CELL CHEMISTRY

A wide variety of different cell chemistries have been developed and used in thermal batteries. At this time, the most widely used chemistry is lithium/iron disulfide (Li/FeS_2), with calcium/calcium chromate ($Ca/CaCrO_4$) as a distant second. There are special applications, though, where one of the other less used chemistries could offer special advantages. As an example, the requirement for a very fast activation time with a relatively short activated life would be provided by the calcium/potassium dichromate ($Ca/LiCl-KCl/K_2Cr_2O_7$) system or the calcium/lead chromate ($Ca/LiCl-KCl/PbCrO_4$) system. For a very general overview, Table 36.2 lists some example-specific performance characteristics of various thermal battery chemistries and designs.

36.3.1 Lithium/Iron Disulfide

There are three common lithium anode configurations: Li(Si) alloy, LiAl alloy, and Li metal in metal matrix, Li(M), where the matrix is usually iron powder. With the difference that the alloy anodes remain solids and the lithium in the Li(Fe) mix is molten in an activated cell, all three anodes participate in the cell reaction similarly. All may be used with the same FeS_2 cathode and the same electrolytes. These electrolytes may be the basic LiCl-KCl eutectic electrolyte, LiCl-LiBr-LiF electrolyte for best ionic conductivity, or a lower-melting-point electrolyte such as LiBr-KBr-LiF for extended activated life. Since the FeS_2 is a good electronic conductor, the electrolyte layer is necessary in order to prevent direct anode-to-cathode contact and cell short-circuiting. When molten,

TABLE 36.2 Characteristics of Various Thermal Batteries

Cell type	Volume, cm^3	Weight, g	Nominal voltage, V	Current, A	Peak power, W	Average life, s	Specific energy (Wh/kg)	Energy density (Wh/L)
Cup/WO_3	450	850	7	5.8	41	70	2.3	4.3
Open cell/tape/WO_3	100	385	50	0.36	15	70	1.3	5.0
Open cell/tape/dichromate	44	148	18	26.0	462	1.2	1.0	3.5
Open cell/tape/dichromate	1	5.5	10	5.0	50	0.15	0.4	2.1
Open cell/tape/bromide	81	225	203	0.02	4	45	0.2	0.7
Pellet/$CaCrO_4$/heat paper	123	310	42	2.9	125	25	2.8	6.8
Pellet/$CaCrO_4$/heat pellet	105	307	28	1.2	34	150	4.6	13.4
Pellet/$CaCrO_4$/heat pellet	105	307	28	2.5	75	60	3.8	11.1
Pellet/LiM*/FeS_2	92.3	320	25	15.0	420	35	11.4	39.0
Pellet/LiM*/FeS_2	170	505	28	12.0	378	140	26.2	82.0
Pellet/LiM*/FeS_2	208	544	138	1.0	138	250	32.2	84.1
Pellet/LiM*/FeS_2	3120	6620	315	10.0	3600	250	33.1	77.0
Pellet/LiM*/FeS_2	334	907	65	7.95	541	320	43.0	116.0
Pellet/LiM*/FeS_2	552	1400	27	12.0	372	600	38.7	111.8
Pellet/LiM*/FeS_2	1177	270	27	17.0	459	900	35.1	97.5

M*—either alloy or metal.

the electrolyte between the anode and the cathode is held in place by capillary action through the use of a chemically compatible (inert) binder material. MgO is the preferred material for this application.[10]

The Li/FeS$_2$ electrochemical system has become the preferred system because it does not contain any parasitic chemical reactions. The extent of self-discharge depends on the type of electrolyte used and the cell temperature.[11] The predominant discharge path for cathodes is

$$3Li + 2FeS_2 \rightarrow Li_3Fe_2S_4 \ (2.1 \text{ V})$$

$$Li_3Fe_2S_4 + Li \rightarrow 2Li_2FeS_2 \ (1.9 \text{ V})$$

$$Li_2FeS_2 + 2Li \rightarrow Fe + 2Li_2S \ (1.6 \text{ V})$$

Most batteries are designed to use only the first and sometimes the second cathode transition to avoid changes in cell voltage.

The transitions that occur at the anode depend on the alloy used. For LiAl:

$$\beta\text{-LiAl} \ (ca.\ 20 \text{ wt \% Li}) \rightarrow \alpha\text{-Al (solid solution)}$$

Below approximately 18.4 wt % lithium (lower limit for all β-LiAl) and above 10 wt % lithium (upper limit for α-Al), the alloy is two-phase α,β-LiAl. This fixes the alloy voltage on a plateau. This plateau is about 300 mV less than the voltage afforded by pure lithium metal.

The composition transitions for Li(Si) are

$$Li_{22}Si_5 \rightarrow Li_{13}Si_4 \rightarrow Li_7Si_3 \rightarrow Li_{12}Si_7$$

An anode voltage plateau is defined for compositions falling between each adjacent pair of alloys. That is, the first plateau occurs between $Li_{22}Si_5$ and $Li_{13}Si_4$. The 44 wt % Li(Si) composition falls here, and begins its discharge approximately 150 mV less than that of pure lithium.

The use of FeS$_2$ as a cathode material can cause a large voltage transient or spike of 0.2 V or more per cell, which is evident immediately after activation and lasts from milliseconds to a few seconds. This phenomenon is related to the impact of temperature, the amounts of electroactive impurities in the raw material (iron oxides and sulfates), elemental sulfur from FeS$_2$ decomposition, and the activity of lithium not being fixed in the cathode. In applications where the voltage has to be well regulated, this spike is not acceptable. The voltage transient can be virtually eliminated by the addition of small amounts of Li$_2$O or Li$_2$S (typically 0.16 mol Li per mol FeS$_2$) to the catholyte (FeS$_2$ and electrolyte blend), a method known as *multiphase lithiation*.[12] The spike can also be reduced (but not eliminated) by thoroughly washing or vacuum treating the FeS$_2$ to remove acid-soluble impurities and elemental sulfur.

The Li/FeS$_2$ electrochemical system has a number of important advantages over other systems, including Ca/CaCrO$_4$. These advantages include:

- Tolerance of a wide range of discharge conditions, from open-circuit to high current densities
- High current capabilities, 3 to 5 times that of Ca/CaCrO$_4$
- Highly predictable performance
- Simplicity of construction
- Tolerance to processing variations
- Stability in extreme dynamic environments

As a result of these advantages, this system has become the predominant choice for a wide range of high-reliability military and space applications.

36.3.2 Lithium/Cobalt Disulfide

As a cathode versus lithium in molten salt electrolyte cells, cobalt disulfide exhibits a slightly lower voltage than iron disulfide. Cobalt disulfide has a greater thermal stability with respect to loss of sulfur, however. The decomposition reactions for cobalt disulfide at elevated temperatures are

$$3CoS_2 \rightarrow Co_3S_4 + S_2 \text{ (g)}$$

$$3Co_3S_4 \rightarrow Co_9S_8 + 2S_2 \text{ (g)}$$

For iron disulfide at elevated temperatures

$$2FeS_2 \rightarrow 2FeS + S_2 \text{ (g)}$$

As a rough indicator of the relative stabilities, FeS_2 will have a sulfur vapor pressure (p_{S2}) of 1 atm in equilibrium with it at 700°C, whereas $p_{S2} = 1$ atm for CoS_2 at 800°C. It is, therefore, no surprise that the substitution of CoS_2 for FeS_2 can yield a cell that is more stable at high temperature, and is therefore useful in batteries with activated lives of over 1 h.[13] In an active battery, the decomposition of FeS_2 to FeS and elemental sulfur becomes significant above approximately 550°C. The free sulfur can combine directly with the Li anode in a highly exothermic reaction. Not only would this reduce available anode capacity, but the extra heat can cause even more thermal decomposition of the cathode. CoS_2, which is stable up to 650°C, allows higher initial stack temperatures to be sustained without excessive degradation of the cathode.[14] It has also been demonstrated that cells with CoS_2 cathodes have a lower internal resistance later in activated life than do FeS_2 cathodes.

36.3.3 Calcium/Calcium Chromate

The reactions that take place in a $Ca/CaCrO_4$ thermal cell during activation must be in critical balance for the cell to function properly. Upon activation (application of heat), the calcium anode reacts with lithium ions in the LiCl-KCl eutectic electrolyte to form liquid beads of Ca-Li alloy. This alloy becomes the operational anode in the subsequent electrochemical reaction. The anodic half-cell reaction is

$$CaLi_x \rightarrow CaLi_{x-y} + yLi^+ + ye^-$$

The Ca-Li alloy beads also react with dissolved $CaCrO_4$, forming a coating of $Ca_5(CrO_4)_3Cl$.[15,16] This Cr(V) compound is the same species that is formed in the cathodic half-cell reaction

$$3CrO_4^{-2} + 5Ca^{+2} + Cl^- + 3e^- \rightarrow Ca_5(CrO_4)_3Cl$$

This product acts as a separator or mass transport barrier between the cathode and the anode to limit electrochemical self-discharge. If the integrity of this separator is breached, the battery can experience a "thermal runaway" condition, whereby the active electrochemical components are chemically consumed with accompanying generation of large amounts of excess heat. At the same time, if battery conditions are such that alloy formation exceeds usage, the excess alloy can cause periodic shorting, the "alloy noise" sometimes seen in cold-stored batteries.

The balance between chemical and electrochemical reactions in this system is dependent on the source of materials (particularly $CaCrO_4$), processing variations, density of compression-formed pellets, operating temperature of the cell, rate of current drain, and other variables.[17] Consequently, this system has been gradually phased out in favor of the more stable and predictable lithium/iron disulfide cell chemistry, which also has a higher energy density.

36.4 CELL CONSTRUCTION

A number of factors, including the cell chemistry used, the operating environments of the battery and the preferences of the designer, determine the choice of cell design. Basically, all cell designs fall into three categories: cup cells, open cells, and pelletized cells. To meet specific performance requirements, some designs may incorporate aspects of more than one cell category. Figure 36.3 illustrates the relative thickness ranges of the different cell designs.

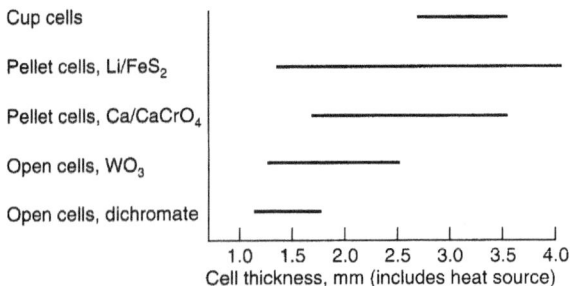

FIGURE 36.3 Thicknesses of thermal battery cells.

36.4.1 Cup Cells

The typical cup cell features a two-layer anode (calcium or magnesium) having active anode material on both sides of a central current collector. On either side of the anode is an electrolyte pad made of glass tape impregnated with eutectic electrolyte. Next to each electrolyte are depolarizer pads consisting of cathode material ($CaCrO_4$ or WO_3) in an inorganic fiber matrix (paper). The cell is enclosed in a nickel foil cup and cover that are tightly crimped (Fig. 36.4a). Some designs also incorporate inorganic fiber mat gaskets and nickel eyelets to help prevent the molten electrolyte from leaking out of the activated cell. Zr/ $BaCrO_4$ heat paper pads located on either side provide heat to the cup cell.

Cup cells have the advantage of large reactive surfaces (they are two-sided or bipolar), and can contain relatively large amounts of reactive materials. Their disadvantages are that they are difficult to seal against electrolyte leakage and they have low heat capacity. The $Ca/CaCrO_4$ cell chemistry is also prone to "alloying" (producing excess molten Ca-Li alloy), which can short-circuit the cells. In order to obtain required short activation times, cup cells typically have to be premelted or prefused prior to assembly into cell stacks. Intercell electrical connections are accomplished by spot-welding the cell output leads between each cell, which presents a potential reliability problem.

Currently, cup cells have limited application and are found primarily in older battery designs.

36.4.2 Open Cells

The open-cell design is similar in construction to the cup cell, except that it is not enclosed in a cup (Fig. 36.4b). Elimination of the cup is possible because the amount of electrolyte is reduced to the extent that practically all of it is bound to the glass fiber cloth matrix by surface tension. Some designs use homogeneous electrolyte-depolarizer pads; others have discrete parts. Open-cell designs typically incorporate a combination anode-electron collector, usually in the shape of a "dumbbell." This combination part has anode material vacuum-deposited on one end (which serves as anode in one cell), while the other end is an electron collector in the next, series-connected cell. A narrow bridge connects the two ends of the dumbbell. The bridge serves as an intercell connector, eliminating the need for spot welds. $Zr/BaCrO_4$ heat paper pads heat the open cells, which are assembled between the folded dumbbells.

The open-cell design is used in relatively short-life applications and in pulse batteries. Their parts can be made very thin to promote very rapid heat transfer and obtain short activation times.

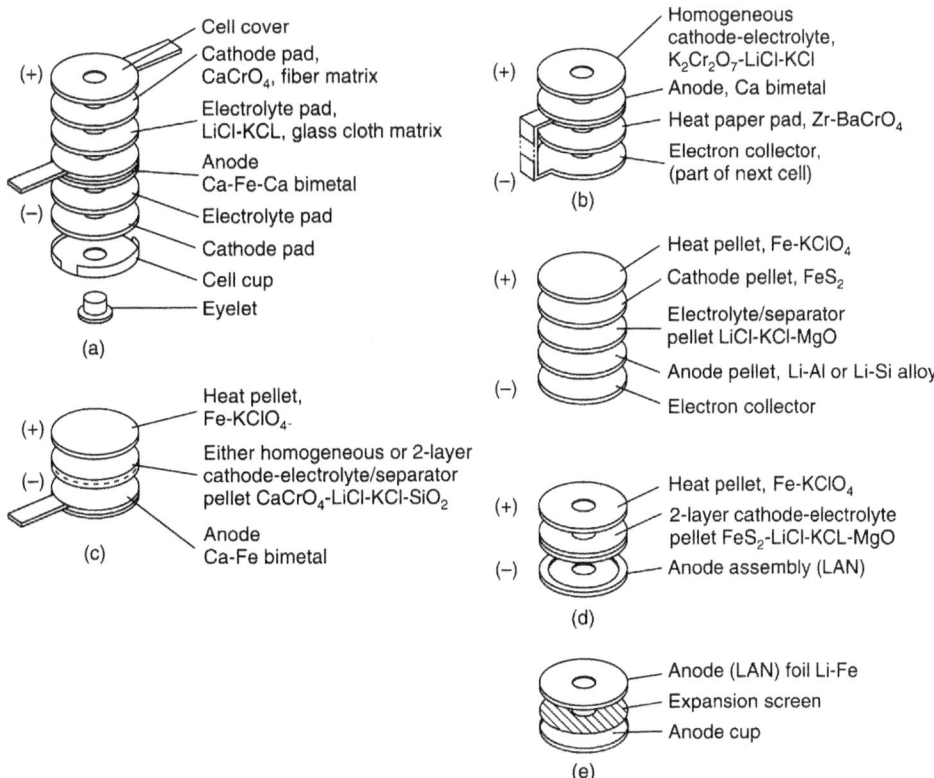

FIGURE 36.4 Variations in cell configurations. (*a*) Cup cell. (*b*) Open cell. (*c*) Ca/CaCrO$_4$ pellet cell. (*d*) Li alloy/FeS$_2$ and Li/FeS$_2$ pellet cell. (*e*) Li metal/FeS$_2$, (LAN) anode assembly.

36.4.3 Pellet Cells

In pellet cells, the electrolyte, cathode, and heat source are in pellet (wafer) form. Anodes can be of different configurations, depending on which electrochemical system is used. For pellet production, the cell component chemicals are processed into powders, and the powders are uniaxially pressed into the parts. Electrolytes, which melt at cell operating temperatures, are combined with inert binders, which hold the molten salts in place by capillary action or surface tension, or both.

A typical pelletized Ca/CaCrO$_4$ cell, as shown in Fig. 36.4*c*, is made up of the following:

1. A calcium anode—either calcium foil (on nickel or iron foil collector) or calcium bimetal (deposited on either iron or nickel collector).
2. A pelletized electrolyte powder blend—consisting of LiCl-KCl eutectic salts and either SiO$_2$ or kaolin as binders.
3. A pelletized cathode powder blend—consisting of CaCrO$_4$, LiCl-KCl eutectic salts, and SiO$_2$ or kaolin binder.
4. A pelletized heat source—a blend of iron powder and KClO$_4$. (Alternatively, this may be a non-pelletized heat source assembly made up of Zr-BaCrO$_4$ heat paper in a nickel or iron foil dumbbell with the anode of the next cell on the outside—similar to the anode and heat source in open-cell designs.)

Variations of this cell design include (1) the use of a two-layer pellet with discrete electrolyte and cathode layers formed into one part, and (2) the use of a homogeneous pellet that has the electrolyte and cathode powders blended together (depolarizer-electrolyte-binder or DEB pellet).[18] A typical Li/FeS$_2$ cell, as illustrated in Fig. 36.4d, is made up of the following:

1. A lithium anode—of either pelletized lithium alloy powder or a lithium metal anode assembly.
2. A pelletized electrolyte powder blend—consisting of a salt mixture and MgO binder. The salts may include mixtures such as LiCl-KCl eutectic, LiBr-KBr-LiF, or LiCl-LiBr-KBr.
3. A pelletized cathode powder blend—of FeS$_2$ and electrolyte with either MgO or SiO$_2$ binder.
4. A pelletized heat powder blend—of pyrotechnic-grade iron powder and KClO$_4$.
5. An electrical collector—of iron or stainless steel foil, located between the heat pellet and the lithium alloy anode pellet. This part is not used with a lithium metal anode assembly, which has an integral metal foil cup. In some cases, especially in longer-life batteries, a second metal foil "collector" is placed between the FeS$_2$ cathode and the heat pellet to buffer or prevent the cathode from exposure to excessive heat.

The pressure used for pelletizing the cell components is critical. In the case of Ca/CaCrO$_4$ designs, the forming pressures, and hence the resultant densities of the electrolyte and cathode pellets, have a profound effect on the reactivities of the cells. The components of the Li/FeS$_2$ systems, except for the heat pellets, are less sensitive to variations in density. Heat pellet ignition sensitivity and burning rate are significantly affected by changes in density, however, with high density decreasing ignition sensitivity and rate. The design parameters of a representative Li(Si)/FeS$_2$ cell are shown in Table 36.3.[19]

The use of pellet-type cell construction has significantly increased the performance capability of thermal batteries. Pellet designs have particular advantages in longer-activated-life, high-current-drain applications. They are structurally very rugged, can operate reliably over wider ambient temperature ranges, and are generally less expensive to manufacture than older designs. There are applications, however, such as those requiring fast activation times and high-voltage pulses, where open-cell designs with Ca/LiCl-KCl/K$_2$Cr$_2$O$_7$ or Ca/LiCl-KCl/PbCrO$_4$ cell chemistries and heat paper are more suitable.

TABLE 36.3 Cell Components of 3400 A/s, Li-Si/FeS$_2$, Thermal Battery Cell[19]

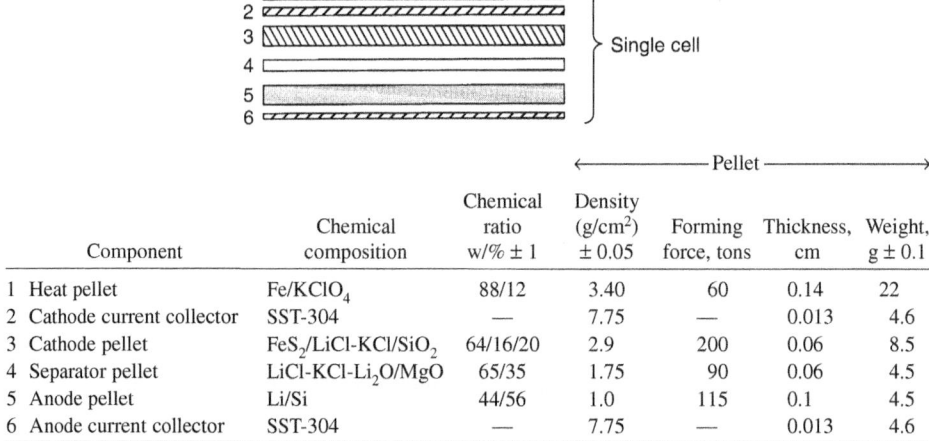

Component	Chemical composition	Chemical ratio w/% ± 1	Density (g/cm^2) ± 0.05	Forming force, tons	Thickness, cm	Weight, g ± 0.1
1 Heat pellet	Fe/KClO$_4$	88/12	3.40	60	0.14	22
2 Cathode current collector	SST-304	—	7.75	—	0.013	4.6
3 Cathode pellet	FeS$_2$/LiCl-KCl/SiO$_2$	64/16/20	2.9	200	0.06	8.5
4 Separator pellet	LiCl-KCl-Li$_2$O/MgO	65/35	1.75	90	0.06	4.5
5 Anode pellet	Li/Si	44/56	1.0	115	0.1	4.5
6 Anode current collector	SST-304	—	7.75	—	0.013	4.6

36.5 CELL-STACK DESIGNS

All thermal batteries are designed to satisfy a specific set of performance requirements, each of which includes output voltage, current drain, and activated life. In designing a battery, the output voltage determines the number of cells that must be connected in series. Since each cell produces a fixed maximum voltage (from 1.6 to 3.3 V on open-circuit, depending on the cell chemistry used), the battery output will be in multiples of discrete cell voltages. Batteries containing over 180 series-connected cells with an overall output voltage near 400 V have been successfully manufactured. Typical batteries contain 14 to 80 cells, and have an output voltage of 28 to 140 V. Figure 36.5 illustrates two different cell-stack configurations, one with cup cells and the other with pellet-type cells.

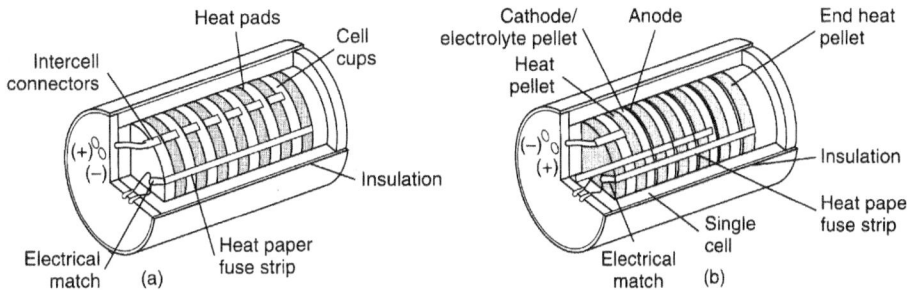

FIGURE 36.5 Typical thermal battery assemblies. (*a*) Cup cells. (*b*) Pellet cells.

The current-carrying capacity of each cell is determined by the reactive surface area of the cell, which is directly related to the cell size (diameter). As with cell voltages, the maximum useful current densities (amperes per unit area) differ greatly among cell chemistries (see Tables 36.4 and 36.5). The effective cell area, and hence the current-carrying capacity of a battery, can be adjusted by electrically connecting any number of cells in parallel.

Thermal batteries can be designed to provide multiple output voltages by electrically connecting the required number of cells in series. The multiple-voltage outputs can be drawn either from cells that are common to more than one output or from isolated cells whose output is not shared. An electrically isolated group of cells must be used for circuits that cannot tolerate "crosstalk" from other circuits in a system. It is also possible to combine cell-stack sections with different cell chemistries in the same battery. Such combinations yield the specific performance characteristics of both chemistries from a common unit. An example of this is a battery that combines a cell stack with a chemistry that has a very short start time with a different cell stack that can provide a high current over a long activated life. Where such combinations are used, the outputs from

TABLE 36.4 Attainable Current Density of Different Cell Designs

Cell design	Current density, mA/cm^2		
	10 s rate	100 s rate	1000 s rate
Cup cell	620	35	
Open cell/dichromate	54		
Pellet cell/two-layer Ca/CaCrO$_4$	790	46	
Pellet cell/DEB Ca/CaCrO$_4$	930	122	
Pellet cell/Li/FeS$_2$	>2500	610	150

36.14 SPECIALIZED BATTERY SYSTEMS

TABLE 36.5 Typical Power and Energy Densities of Li/FeS$_2$ Thermal Batteries

Battery volume, cm^3	Power density, W/cm^2	Energy density, Wh/L	Activated life, s
20	11.25	46.87	15
29	1.44	34.20	85
70	2.59	35.97	50
108	0.65	32.41	180
170	1.98	109.80	200
171	10.64	118.26	40
183	2.29	63.75	100
306	0.51	39.65	280
311	2.25	75.03	700
552	0.15	67.63	1600
1176	0.40	101.19	900
1312	0.17	85.37	1800
3120	1.11	83.30	270

the different cell-stack sections are often diode-isolated to prevent one section from charging the other. Some thermal battery designs combine two or more discrete batteries into an assembly that may have a number of different, mutually isolated voltage outputs with widely varying current capabilities.

Cells comprising a cell stack are typically held in place by the closing compression applied when the battery cover is secured by welding it to the battery case. Some battery designs incorporate an inner case to maintain compression on the cell stack while the outer case and cover combination provides hermetic enclosure for the unit. Figure 36.6 pictures a battery design that employs an inner cell-stack case.

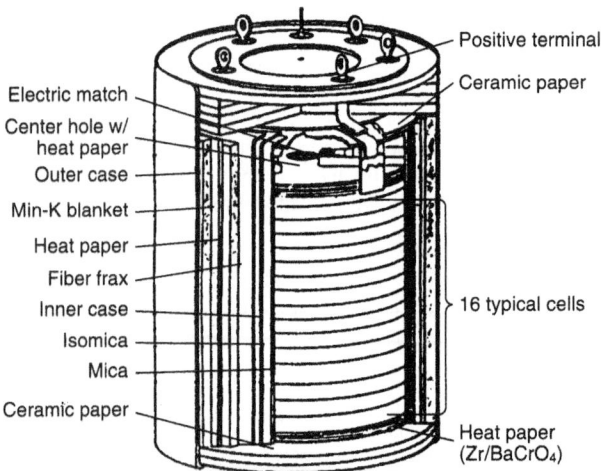

FIGURE 36.6 Typical thermal battery assembly with inner case. (*Courtesy of SAFT Batteries.*)

36.6 PERFORMANCE CHARACTERISTICS

Thermal batteries are custom-designed to satisfy a specific set of performance requirements. These include not only output voltage, current, activated life, and voltage rise time (start), but also storage and activated-life environments, mounting, surface temperature, activation method and energy, and others. For this reason it is very important that the user or systems designer have a close technical interface with the battery designer during the design and development phases of the battery.

36.6.1 Voltage Regulation

Thermal battery output voltages are not linear. After reaching a peak level, typically within 1 s after activation, the voltage starts to decay until it eventually drops below the minimum useful level. Voltage regulation is the range between the specified minimum and maximum limits. Typically, the minimum voltage limit is 75% of the peak voltage. The battery output profile (consisting of the rise time, peak voltage, and rate of decay) depends on the cell chemistry, and is strongly affected by the operating temperature and applied load. Figure 36.7 illustrates the effects of discharge load on a typical battery output profile.

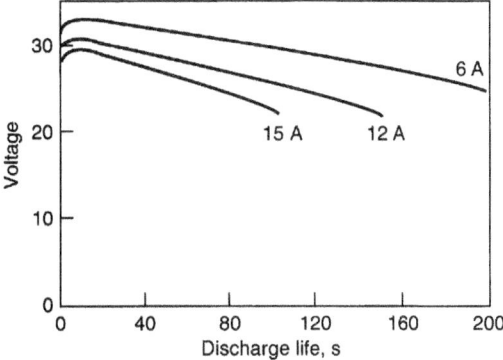

FIGURE 36.7 Discharge voltage curves of typical Li/FeS$_2$ thermal battery at three different current drain rates.

36.6.2 Activation Time

The activation time (rise time) is the time interval from the application of energy to the initiation device until the battery output voltage reaches the minimum specified limit. The activation time is affected by the operating temperature, applied load, and cell chemistry used. Lowering the operating temperature or increasing the load typically increases the activation time. Typical Li/FeS$_2$ batteries have activation times from 0.35 to 1.00 s. Large, high-capacity batteries can have activation times as long as 3 s. (Large diameter heat pellets take longer times to burn.) On the other hand, fast-activating chemistries such as Ca/K$_2$Cr$_2$O$_7$ can yield activation times as short as 12 ms. Figure 36.8 shows activation time ranges for various cell chemistries and Fig. 36.9 illustrates the effects of ambient temperature.

36.6.3 Activated Life

The activated (operating) life is typically specified as the time from the initial application of the activation energy until the battery voltage drops below the minimum specified limit. Activated life

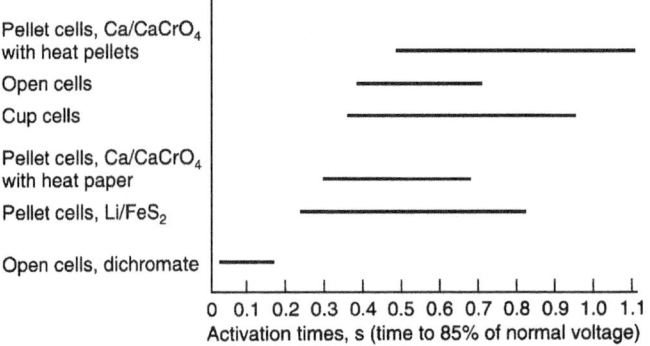

FIGURE 36.8 Activation times of different thermal battery cell designs.

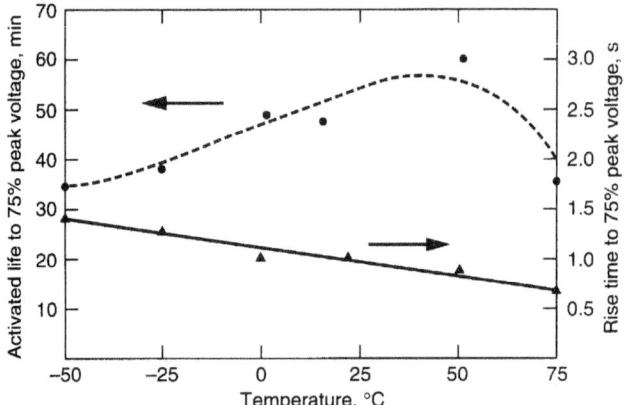

FIGURE 36.9 Activated life and rise time of Li/FeS$_2$ battery.[20]

is affected by the cell chemistry used, the operating temperature environment, and the current drain. Typically, thermal batteries are thermally balanced (total cell mass versus caloric input) to have the longest activated lives between the high and low operating temperature limits, or near room ambient. Lives will get shorter near each temperature limit. At the low limit, the electrolyte will start freezing sooner, whereas at the high limit the thermal degradation of FeS$_2$ occurs at a faster rate, depleting active materials.

36.6.4 Interface Considerations

The following performance and design characteristics must be noted when designing a system that interfaces with a thermal battery:

1. An unactivated battery has a very high internal resistance (megohm). Once activated, an individual cell's resistance is between 0.003 and 0.02 ohm, depending on the cell design. The internal resistance of the battery is equal to the sum of the resistances of all series-connected cells.
2. Some cell chemistries, such as Li/FeS$_2$, are tolerant of backcharging from an external power source. Others, however, such as Ca/CaCrO$_4$, must not be subjected to backcharging at all.

3. Electric actuators contain bridge wires that may not burn through during activation and, if not disconnected, may act as a parasitic load on the external ignition circuit.
4. Leakage paths that can adversely load the battery may develop in an activated battery between electrically live components and the battery case or activator circuits. System requirements, such as case grounding, cell-stack common output, and activator circuit grounding must be specified so that special insulation provisions can be incorporated into the battery design.
5. The surface temperature of an activated battery may reach 400°C. The type of battery mounting, the heat transfer properties of the mounting, the effects of high temperature on the surrounding components, and the proximity of combustible materials must be considered. The battery surface temperature can usually be reduced significantly by incorporating added (or more efficient) thermal insulation. This is achieved, however, at considerable cost and increase in battery volume. Figures 36.10 and 36.11 illustrate typical surface temperatures of thermal batteries.

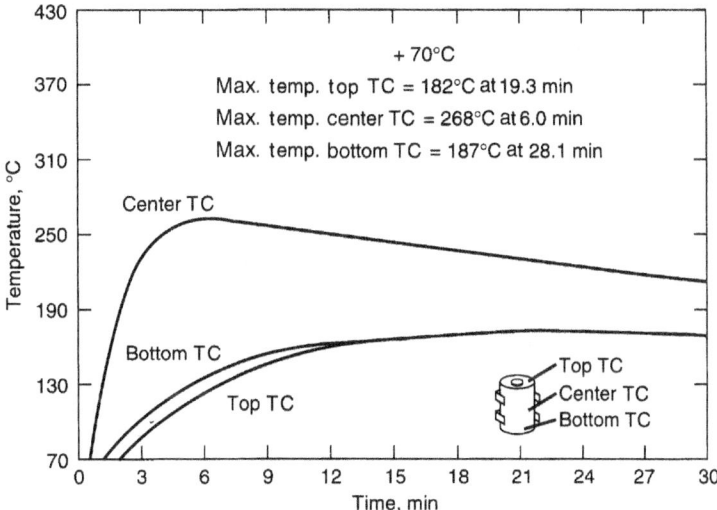

FIGURE 36.10 Surface temperature profiles for long-life thermal battery.[19]

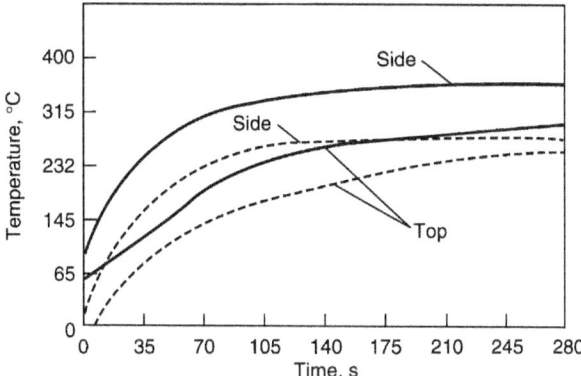

FIGURE 36.11 Surface temperature profiles for a medium-life thermal battery. Solid line—tested at 71°C; broken line—tested at −53°C.

36.7 TESTING AND SURVEILLANCE

The safety and reliability of thermal batteries has been a matter of continuing study since they were first developed. To identify defective units, most designs are 100% tested for hermeticity, polarity, electrical insulation resistance, and activation circuit resistance (if applicable) on manufacture. Most units are also radiographed. Prior to commencement of production, a sample group of 10 to as many as 500 batteries is subjected to qualification tests. This series of tests includes the most severe environmental and discharge conditions to which the particular battery design will be exposed in actual field use. Almost all thermal batteries are fabricated in homogeneous groups or lots, and samples from each lot are discharged to demonstrate compliance with the performance requirements. Usually the samples are discharged at maximum specified loads, often with concurrently imposed environmental forces. By using such test programs, reliability values greater than 99% and safety values greater than 99.9% have been demonstrated innumerable times in the last five decades.

Lithium thermal batteries designed for use in U.S. Navy systems are subject to safety tests per Navy technical manual S9310-AQ-SAF, "Battery, Navy Lithium Safety Program Responsibilities and Procedures." These tests are designed to ensure that the battery design is safe not only in proper storage and use, but also when subjected to inadvertent misuse and conditions caused by accidents, such as backcharging, short circuits, and fires.

36.8 NEW DEVELOPMENTS

The primary aim of new research and development in the thermal battery area has been to increase the energy density and specific energy of the practical unit. Two possible approaches to this goal are to (1) decrease the total volume and mass of the battery, and (2) increase the voltage or the current-carrying capacity per cell volume and mass.

In the area of decreasing battery mass, investigations have been conducted by substituting lighter-weight materials for the currently used stainless steel battery housings. Titanium, aluminum, composites, and other materials have been suggested and tried with varying degrees of success. Titanium cases and headers have been successful but suffer from higher cost.

Efforts to deposit active material films by plasma-spraying powders onto stainless steel substrates have yielded promising results. This has been accomplished both for FeS_2 cathodes[21] and for Li(Si) anodes.[22] This technology could potentially reduce the mass and volume of a thermal battery by virtue of the higher active material densities afforded. More recently, several organizations have been investigating tape casting[23-25] or conventional spraying[24] to achieve thinner components.

Development efforts to produce cells with higher voltage have demonstrated the potential of employing molten nitrate electrolytes with lithium anodes.[26] These systems have the added benefit of lowering the battery operating temperature by more than 200°C.

Efforts have also been directed toward increasing the activated life of batteries past 2 h up to 4 h. These have required the development of more efficient thermal insulation, such as the use of double-walled vacuum enclosures (cases) and multilayered insulating blankets, as well as lower melting point electrolyte compositions.

The long-time use of $Fe\text{-}KClO_4$ heat pellets as combination heat/cathodes has attracted increased interest with recent claims of improvement.[27] This technology, in which the iron oxide produced by the heat source is also the cathode, was employed by Catalyst Research Corporation as early as 1981.[28]

REFERENCES

1. G. O. Erb, "Theory and Practice of Thermal Cells," *Publication BIOS/Gp 2/HEC 182 Part II*, Halstead Exploiting Centre, June 6, 1945.
2. O. G. Bennett et al., U.S. Patent 3,575,714, Apr. 20, 1971.
3. B. H. van Domelen, and R. D. Wehrle, "A Review of Thermal Battery Technology," *Intersoc. Energy Convers. Conf.*, 1974.

4. F. Tepper, "A Survey of Thermal Battery Designs and Their Performance Characteristics," *Intersoc. Energy Convers. Conf.,* 1974.
5. G. C. Bowser, D. E. Harney, and F. Tepper, "A High Energy Density Molten Anode Thermal Battery," *Power Sources* **6** (1976).
6. R. A. Guidotti, and F. W. Reinhardt, "Evaluation of Alternate Electrolytes for Use in Li(Si)/FeS$_2$ Thermal Batteries," *Proc. 33rd Power Sources Symp.,* 1988, pp. 369–376.
7. L. Redey, J. A. Smaga, J. E. Battles, and R. Guidotti, "Investigation of Primary Li-Si/FeS$_2$ Cells," *ANL-87-6,* Argonne National Laboratory, Argonne, IL, June 1987.
8. W. H. Collins, U.S. Patent 4,053,337, Oct. 11, 1977.
9. C. S. Winchester, "The LAN/FeS$_2$ Thermal Battery System," *Power Sources* **13** (1982).
10. Z. Tomczuk, T. Tani, N. C. Otto, M. F. Roche, and D. R. Vissers, *J. Electrochem. Soc.* **129(5):**925–932 (1992).
11. R. A. Guidotti, R. M. Reinhardt, and J. A. Smaga, "Self-Discharge Study of Li-Alloy/FeS$_2$ Thermal Cells," *Proc. 34th Int. Power Sources Symp.,* 1990, pp. 132–135.
12. R. A. Guidotti, "Methods of Achieving the Equilibrium Number of Phases in Mixtures Suitable for Use in Battery Electrodes, e.g., for Lithiating FeS$_2$," U.S. Patent 4,731,307, Mar. 15, 1988.
13. R. A. Guidotti, and F. W. Reinhardt, "The Relative Performance of FeS$_2$ and CoS$_2$ in Long-Life Thermal-Battery Applications," *Proc. 9th Int. Symp. Molten Salts,* 1994.
14. R. A. Guidotti, and F. W. Reinhardt, "Characterization of the Li(Si)/CoS$_2$ Couple for a High-Voltage, High-Power Thermal Battery," *SAND2000-0396,* 2000.
15. R. A. Guidotti, and F. W. Reinhardt, "Anodic Reactions in the Ca/CaCrO$_4$ Thermal Battery," *SAND83-2271,* 1985.
16. R. A. Guidotti, and W. N. Cathey, "Characterization of Cathodic Reaction Products in the Ca/CaCrO$_4$ Thermal Battery," *SAND84-1098,* 1985.
17. R. A. Guidotti, F. W. Reinhardt, D. R. Tallant, and K. L. Higgins, "Dissolution of CaCrO$_4$ in Molten LiCl-KCl Eutectic," *SAND83-2272,* 1984.
18. D. M. Bush et al., U.S. Patent 3,898,101, Aug. 3, 1975.
19. H. K. Street, "Characteristics and Development Report of the MC3573 Thermal Battery," *SAND82-0695,* 1983.
20. R. K. Quinn, and A. R. Baldwin, "Performance Data for Lithium-Silicon/Iron Disulfide Long Life Primary Thermal Battery," *Proc. 29th Power Sources Symp.,* 1980.
21. H. Ye et al, "Novel Design and Fabrication of Thermal Battery Cathodes Using Thermal Spray," Spring Meeting of the Materials Research Society, San Francisco, CA, April 5–9, 1999.
22. C. J. Crowley et al., "Development of Fabricating Processes for Plasma-Sprayed Li-Si Anodes," *Proc. 40th Power Sources Conference,* 2002, pp. 303–306.
23. J. K. Pugh et al., "Tape Cast Technology as Applied to Thermal Batteries," *Proc. 43rd Power Sources Conference,* 2008, pp. 369–372.
24. S. B. Preston et al., "Development of Coating Process for Production of Low-Cost Thermal Batteries," *Proc. 43rd Power Sources Conference,* 2008, pp. 373–376.
25. J. Edington et al., "Development of Thin Components for Thermal Batteries," *Proc. 43rd Power Sources Conference,* 2008, pp. 177–180.
26. M. H. Miles, "Lithium Batteries Using Molten Nitrate Electrolytes," *Proc. 14th Annual Battery Conf.,* Long Beach, 1999.
27. D. R. Dekel and D. Laser, U.S. Patent Appl. 2007/0292748.
28. C. S. Winchester, NSWC Carderock, personal communication.

BIBLIOGRAPHY

Askew, B. A., and R. Holland: "A High Rate Primary Lithium-Sulfur Battery," *Power Sources* **4** (1972).

Baird, M. D., A. J. Clark, C. R. Feltham, and L. H. Pearce: "Recent Advances in High Temperature Primary Lithium Batteries," *Power Sources* **7** (1978).

Birt, D., C. Feltham, G. Hazzard, and L. Pearce: "The Electrochemical Characteristics of Iron Sulfide and Immobilized Salt Electrolytes," *Power Sources* **7** (1978).

Bowser, G. C., et al.: U.S. Patent 3,891,460, June 24, 1975.

Bowser, G. C., et al.: U.S. Patent 3,930,888, Jan. 1976.

Bush, D. M., and D. A. Nissen: "Thermal Cells and Batteries Using the Mg/FeS$_2$ and LiAl/FeS$_2$ Systems," *Proc. 28th Power Sources Symp.*, 1978.

Collins, W. H.: U.S. Patent 1,482,738, Aug. 10, 1977.

De Gruson, J. A.: "Improved Thermal Battery Performance," *AFAPL-TR-79-2042*, Eagle-Picher Industries, 1979.

Delnick, F. M., R. A. Guidotti, and D. K. McCarthy: "Chromium (V) Compounds as Cathode Materials in Electrochemical Power Sources," U.S. Patent 4,508,796, Apr. 2, 1985.

Guidotti, R. A., and F. W. Reinhardt: "Lithiation of FeS$_2$ for Use in Thermal Batteries," *Proc. 2nd Annual Battery Conf. on Applications and Advances,* 1987, paper 87DS-3.

Guidotti, R. A., F. M. Reinhardt, and W. F. Hammeter: "Screening Study of Lithiated Catholyte Mix for a Long-Life Li(Si)/FeS$_2$ Thermal Battery," *SAND 85-1737*, 1988.

Hansen, M.: *Constitution of Binary Alloys*, McGraw-Hill, New York, 1958.

Harney, D. E.: U.S. Patent 4,221,849, Sept. 9, 1980.

Kuper, W. E.: "A Brief History of Thermal Batteries," *Proc. 36th Power Sources Conf.,* Cherry Hill, N.J., June 1994.

Quinn, R. K., et al.: "Development of a Lithium Alloy/Iron Disulfide 60-Minute Primary Thermal Battery," *SAND79-0814*, 1979.

Schneider, A. A., et al.: U.S. Patent 4,119,796, Oct. 10, 1978.

Searcey, J. Q., et al.: "Improvements in Li(Si)/FeS$_2$ Thermal Battery Technology," *SAND82-0565*, 1982.

Szwarc, R.: "Study of Li-β Alloy in LiCl-KCl Eutectic Thermal Cells Utilizing Chromate and Iron Disulfide Depolarizer," Gepp-TM-426, General Electric Co., Neut. Dev. Dept., 1979.

PART 5

FUEL CELLS AND ELECTROCHEMICAL CAPACITORS

CHAPTER 37
INTRODUCTION TO FUEL CELLS

David Linden and H. Frank Gibbard

37.1 GENERAL CHARACTERISTICS

A fuel cell is a galvanic device that continuously converts the chemical energy of a fuel (and oxidant) to electrical energy.[1,2] Like batteries, fuel cells convert this energy electrochemically and are not subject to the Carnot cycle limitation of thermal engines, thus offering the potential for highly efficient conversion. The essential difference between a fuel cell and a battery is the manner of supplying the source of energy. In a fuel cell, the fuel and the oxidant are supplied continuously from an external source when power is desired. The fuel cell can produce electrical energy as long as the active materials are fed to the electrodes. In a battery, the fuel and oxidant (except for metal/air and redox flow batteries) are an integral part of the device. The battery will cease to produce electrical energy when the limiting reactant is consumed. The battery must then be replaced or recharged.

The electrode materials of the fuel cell are inert in that they are not consumed during the cell reaction, but have catalytic properties that enhance the electroreduction or electrooxidation of the reactants (the active materials).

The anode active materials used in fuel cells are generally gaseous or liquid fuels (in contrast to the metal anodes generally used in most batteries), such as hydrogen, methanol, hydrocarbons, or natural gas, which are fed into the anode side of the fuel cell. As these materials are like the conventional fuels used in heat engines, the term "fuel cell" has become popular to describe these devices. Oxygen, most often air, is the predominant oxidant and is fed into the cathode.

Fuel cell technology can be classified into two categories[3]:

1. Direct systems where fuels, such as hydrogen, methanol, and hydrazine, can react directly in the fuel cell (see Sec. 38.5).

2. Indirect systems in which the fuel, such as natural gas or other fossil fuels, is first converted by reforming to a hydrogen-rich gas which is then fed into the fuel cell. (See Secs. 38.6.4 and 38.6.5.)

Fuel cell systems can take a number of configurations depending on the combinations of fuel and oxidant, the type of electrolyte, temperature of operation, application, etc.[4] Table 37.1 is a summary of various types of fuel cells distinguished by the electrolyte and operating temperature. The PEM fuel cell and its variant, the direct methanol fuel cell, are currently the predominant ones for portable and small fuel cells as they are the only ones operating near ambient conditions.*

---

*Large fuel cells are beyond the scope of this Handbook. See Appendix F, Bibliography for references.

TABLE 37.1 Types of Fuel Cells

1. **Solid oxide (SOFC):** These cells use a solid oxygen-ion-conducting metal oxide electrolyte. They operate at about 1000°C, with an efficiency of up to 60%. They are slow to start up, but once running, provide high-grade waste heat, which can be used to heat buildings. They may find application in industrial and large-scale applications. This technology has been heavily supported by the U.S. government under the Solid-State Energy Conversion Alliance program (SECA) since 2005 because of its potential for high efficiency and low cost in central generation facilities.
2. **Molten carbonate (MCFC):** These cells use a mixed alkali-carbonate, molten-salt electrolyte and operate at about 650°C. They are being developed for continuously operating facilities, and can use coal-based or marine diesel fuels.
3. **Phosphoric acid (PAFC):** This is the most commonly used type of fuel cell for stationary commercial sites such as hospitals, hotels, and office buildings. The electrolyte is concentrated phosphoric acid. The fuel cell operates at about 230°C. It is highly efficient and can generate energy at up to 85% (40% as electricity and another 45% if the heat given off is also used).
4. **Alkaline (AFC):** These are used by NASA on the manned space missions, and operate well at about 70°C. They use alkaline potassium hydroxide as the electrolyte and can generate electricity with up to 60% efficiency. A disadvantage of this system is that it is restricted to fuels and oxidants which do not contain or produce carbon dioxide.
5. **Proton exchange membrane (PEM):** These cells use a perfluorinated ionomer polymer membrane electrolyte, which conducts protons from the anode to the cathode. They operate at a relatively low temperature (70 to 85°C), and are especially notable for their rapid startup time. These are being developed for use in transportation and small (low-kW) stationary applications and for portable and small fuel cells. In order to reach higher operating temperatures, porous membranes of polybenzimidazole containing phosphoric acid within the pores have been developed. Such membranes allow operation up to 220°C and enable higher efficiency, ease of cooling, and reduced sensitivity to carbon monoxide than the lower-temperature PEM systems.
6. **Direct methanol (DMFC):** These fuel cells directly oxidize liquid methanol (methyl alcohol) in an aqueous solution at the anode. Like PEMs, these also use a membrane electrolyte, and operate at similar temperatures. This fuel cell has received a great deal of attention during the last decade and is in low-volume commercial production for subkilowatt applications.
7. **Regenerative (RFC):** These are closed-loop generators. A powered electrolyzer separates water into hydrogen and oxygen, which are then used by the fuel cell to produce electricity and exhaust (water). That water can then be recycled into the powered electrolyzer for another cycle. RFC has been considered for storage of energy generated by solar and wind power, but low round-trip efficiency has limited deployment of such systems.

Source: Fuel Cells 2000

A practical fuel cell power plant consists of four basic subsystems:

1. A power section, which consists of one or more fuel cell stacks—each stack containing a number of individual fuel cells, usually connected in series to produce a stack output ranging from a few to several hundred volts (direct current). This section converts the fuel and the oxidant into DC power.

2. A fuel subsystem that manages the fuel supply to the power section. This subsystem can range from simple flow controls to a complex fuel-processing facility. This subsystem processes fuel to the type required for use in the fuel cell (power section).

3. A power conditioner that converts the output from the power section to the type of power and quality required by the application. This subsystem could range from a simple voltage control to a sophisticated device that converts the DC power to an AC power output. In addition, a fuel cell power plant, depending on size, type, and sophistication, may require an oxidant subsystem, thermal and fluid management subsystems, and other ancillary subsystems.

4. Except for the simplest fuel cell systems, a control subsystem that manages the operating parameters of the fuel cell system, e.g., temperatures, mass flows, and power conditioning, and integrates the operations of the other subsystems.

Fuel cells have been of interest for over 170 years as a potentially more efficient and less polluting means for converting hydrogen and carbonaceous or fossil fuels to electricity compared to conventional heat engines. A significant application of the fuel cell has been the use of the hydrogen/oxygen fuel cell by NASA, using cryogenic fuels, in space vehicles for over 40 years, including the new fleet of space shuttles. Use of the air-breathing fuel cell in terrestrial applications, such as for utility power and electric vehicles, has been ongoing for some time but has been developing slowly. Recent advances have now revitalized interest for these and other new applications.

During recent decades, interest in small air-breathing fuel cells has arisen for dispersed or on-site electric generators, remote devices, and other such applications in the subkilowatt power range, replacing engine-generators and larger-sized batteries. At lower power levels, from below 1 to 50 W, historically the domain of batteries, fuel cell technology is seen as an approach to achieve higher specific energy than those delivered by batteries. Progress has been made with fuel cell systems in the sizes above 50 W, especially for extended long-term service (see Chap. 38). However, the development of yet smaller portable fuel cells (which can be "recharged," for example, by replacing a small container of fuel), competitive in size and performance with batteries, remains a challenge (see Sec. 38.3).

37.2 OPERATION OF THE FUEL CELL

37.2.1 Cell Reactions

A simple fuel cell is illustrated in Fig. 37.1a. Two catalyzed electrodes are immersed in an electrolyte (acid in this illustration) and separated by a gas barrier. The fuel, in this case hydrogen, is bubbled across the surface of one electrode while the oxidant, in this case oxygen from ambient air, is bubbled across the other electrode. When the electrodes are electrically connected through an external load, the following events occur:

1. The hydrogen dissociates on the catalytic surface of the fuel electrode, forming hydrogen ions and electrons.
2. The hydrogen ions migrate through the electrolyte (and a gas barrier) to the catalytic surface of the oxygen electrode.
3. Simultaneously, the electrons move through the external circuit, doing useful work, to the same catalytic surface.
4. The oxygen, hydrogen ions, and electrons combine on the oxygen electrode's catalytic surface to form water.

The cell reactions of this fuel cell, in acid and alkaline electrolytes, are shown in Table 37.2. The major differences, electrochemically, are that the ionic conductor in the acid electrolyte is the hydrogen ion (or, more correctly, the hydronium ion, H_3O^+) and the OH^- or hydroxyl ion in the alkaline electrolyte. The only by-product of a hydrogen/oxygen fuel cell is water; in the acid electrolyte, water is produced at the cathode and, in the alkaline electrolyte fuel cell, it is produced at the anode.

The net reaction is that of hydrogen and oxygen producing water and electrical energy. As in the case of batteries, the reaction of one electrochemical equivalent of fuel will theoretically produce 26.8 Ah of DC electricity at a voltage that is a function of the free energy of fuel-oxidant reactions. At ambient conditions, this reversible voltage is 1.23 V DC for a hydrogen/oxygen fuel cell.

Figure 37.1b is a schematic of the proton exchange membrane fuel cell (PEMFC), presently the best candidate for use in small portable fuel cells. Passing through a gas diffuser, hydrogen reacts on a catalyst (small circles) at the anode, sending protons and electrons to the cathode. The protons migrate through the membrane and the electrons through the external circuit. The protons react with the oxygen, supplied at the cathode, to form water. Product water and unused reactants exit through the gas vents.

37.6 FUEL CELLS AND ELECTROCHEMICAL CAPACITORS

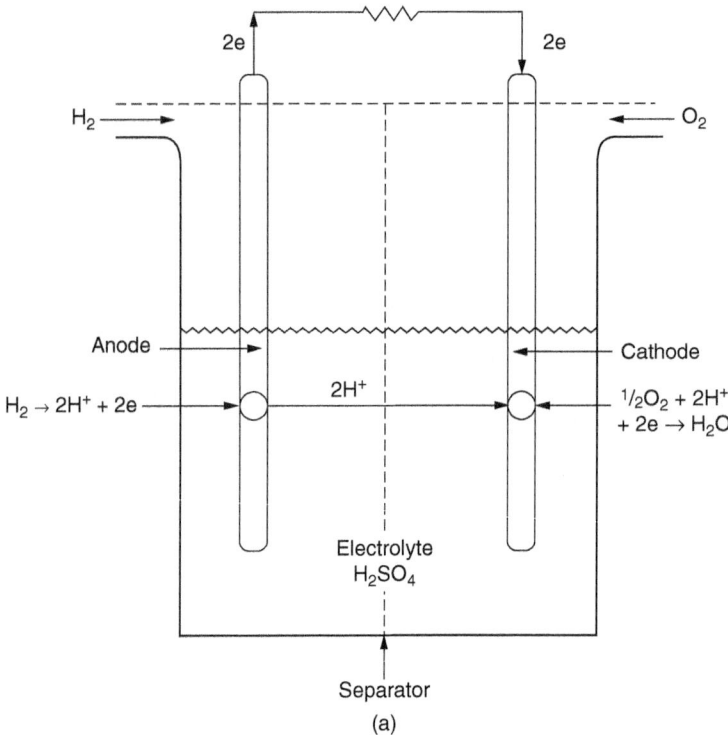

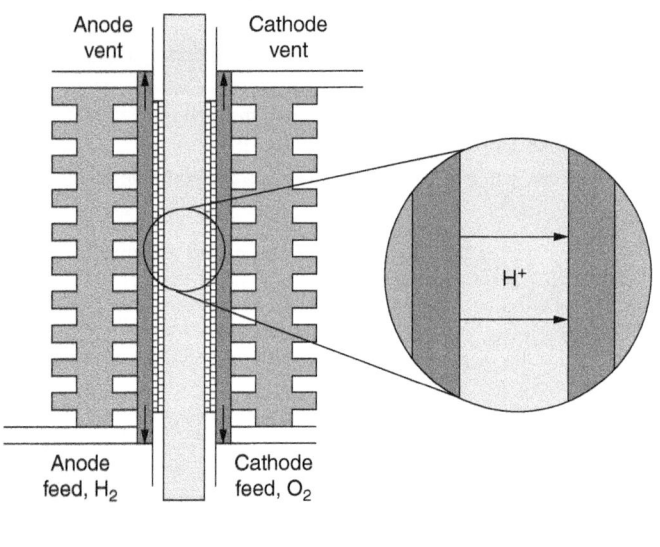

FIGURE 37.1 Operation of the fuel cell. (a) Reactions in an acid electrolyte. (b) Based on a proton exchange membrane (PEM). (*Source: Chemical and Engineering News*, American Chemical Society, Washington, DC, June 14, 1999.)

TABLE 37.2 Cell Reactions of the H_2/O_2 Fuel Cell

	Acid electrolyte	Alkaline electrolyte
Anode	$H_2 \rightarrow 2H^+ + 2e$	$H_2 + 2OH^- \rightarrow 2H_2O + 2e$
Cathode	$1/2 O_2 + 2H^+ + 2e \rightarrow H_2O$	$1/2 O_2 + 2e + H_2O \rightarrow 2OH^-$
Overall	$H_2 + 1/2 O_2 \rightarrow H_2O$	$H_2 + 1/2 O_2 \rightarrow H_2O$

37.2.2 Major Components of the Fuel Cell

The important components of the individual fuel cell are:

1. The *anode* (fuel electrode) must provide a common interface for the fuel and electrolyte, catalyze the fuel oxidation reaction, and conduct electrons from the reaction site to the external circuit (or to a current collector that, in turn, conducts the electrons to the external circuit).

2. The *cathode* (oxygen electrode) must provide a common interface for the oxygen and the electrolyte, catalyze the oxygen reduction reaction, and conduct electrons from the external circuit to the oxygen electrode reaction site.

3. The *electrolyte* must transport one of the ionic species involved in the fuel and oxygen electrode reactions while preventing the conduction of electrons (electron conduction in the electrolyte causes a short circuit). In addition, in practical cells, the role of gas separation is usually provided by the electrolyte system. In systems that employ concentrated aqueous electrolytes, such as phosphoric acid in the PAFC, or potassium hydroxide in the alkaline fuel cell, this is accomplished by retaining the electrolyte in the pores of a matrix. The capillary forces of the electrolyte within the pores allow the matrix to separate the gases, even under some pressure differential. Currently, the technology in use for portable ambient temperature fuel cells is the electrolyte membrane Nafion®.

37.2.3 General Characteristics

The performance of a fuel cell is represented by a plot of cell voltage versus current or current density (Fig. 37.2). An equation describing this curve can be written as follows:

$$E = E^o - \eta_{c,act} - \eta_{a,act} - \eta_{c,conc} - \eta_{a,conc} - iR_{int}$$

where $\eta_{c,conc}$ and $\eta_{a,conc}$ = concentration overpotentials for the cathode and anode, V
$\eta_{c,act}$ and $\eta_{a,act}$ = activation overpotentials for the cathode and anode, V
i = cell current density, A/cm^2
R_{int} = the internal resistance of the cell, $\Omega \cdot$cm^2
E^o = reversible cell potential, V

Ideally, in the absence of all overpotentials, a single H_2/O_2 fuel cell could produce 1.23 V DC at ambient conditions, as shown by the dotted line in Fig. 37.2. In practice, fuel cells produce useful voltage outputs that are less than the ideal and decrease with increasing discharge rate (current density). The losses or reductions in voltage from the ideal are referred to as "polarization," as illustrated in Fig. 37.2 (also see Chap. 2).

These losses include:*

1. Activation polarization represents energy losses that are associated with the electrode reactions. Most chemical reactions involve an energy barrier that must be overcome for the reactions to

---

*Simplifying assumptions are made in the mathematical formulas for the polarization losses, such as the use of planar-electrode theory and simple electrode kinetics. Extensions to porous electrodes and detailed descriptions of the electrode mechanisms, especially for the oxygen electrode, are exceedingly complex, but they preserve the qualitative features of Fig. 37.2.

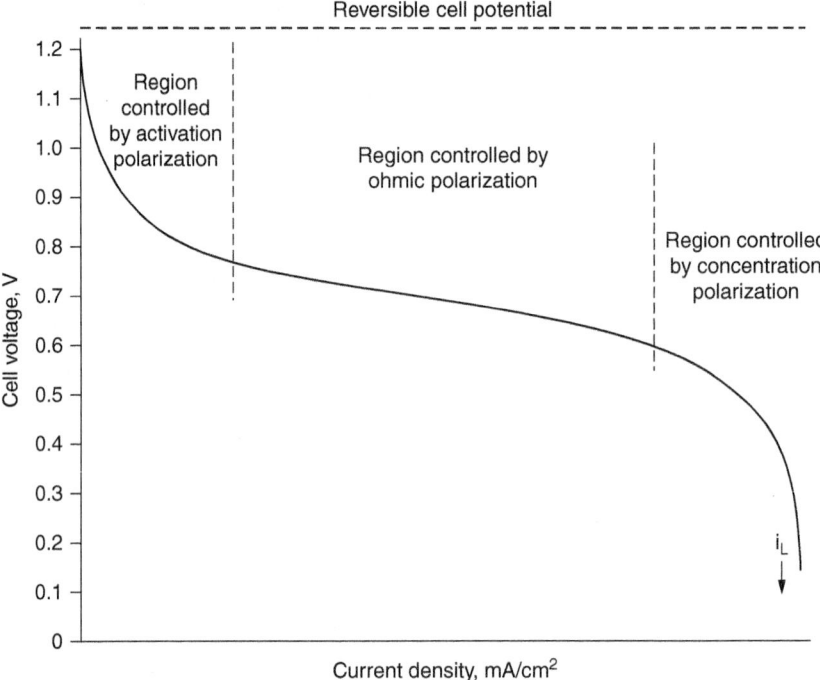

FIGURE 37.2 Fuel cell polarization curve.

proceed. For electrochemical reactions, the activation energy lost in overcoming this barrier takes the form

$$\eta_{act} = a + b \ln i$$

where a, b = constants, and η_{act} = activation polarization = $\eta_{a,act} + \eta_{c,act}$, V

2. Ohmic polarization represents the summation of all the ohmic losses within the cell, including electronic impedances through electrodes, contacts, and current collectors and ionic impedance through the electrolyte. These losses follow Ohm's law

$$\eta_{ohm} = iR_{int}$$

3. Concentration polarization represents the energy losses associated with mass transport effects. For instance, the performance of an electrode reaction may be inhibited by the inability for reactants to diffuse to or products to diffuse away from the reaction site. In fact, at some current, the limiting current density i_L, a situation will be reached wherein the current will be completely limited by the diffusion processes (see Fig. 37.2). Concentration polarization can be represented by

$$\eta_{conc} = \left(\frac{RT}{nF}\right) \ln\left(1 - \frac{i}{i_L}\right)$$

where η_{conc} = concentration polarization = $\eta_{c,conc} + \eta_{a,conc}$
R = gas constant, J/mol·°K
T = temperature, °K
n = number of electrons in cell reaction
i_L = limiting current, A/cm^2

The net result of these polarization effects is that fuel cells generally operate between 0.5 and 0.9 V DC. Fuel cell performance can be increased by increasing cell temperature and reactant partial pressure. However, for small or portable fuel cells, operation near ambient conditions is usually a requirement, particularly when the fuel cell is to be used as a replacement for batteries.

37.3 SUBKILOWATT FUEL CELLS

The fuel cell has many attractive features that have increased interest in its use in small and/or portable power sources below a kilowatt in power output. At the same time, because of the unique requirements of portable devices, there are limitations in fuel cell technology that present challenges for its deployment, particularly in the sizes below 20 W as replacements for batteries. These include:

37.3.1 Hydrogen and Energy-Rich Fuels

Hydrogen and other energy-rich fuels have a higher energy density than the active materials normally used in batteries. Table 37.3 lists the theoretical specific energy and energy density of several of these materials, which are significantly higher than those of batteries and are practical for use in portable fuel cells. Of these, hydrogen stands out, not only because of its high specific energy, but because it can be directly converted to electrical energy in a fuel cell operating at ambient temperatures. Natural gas, propane, gasoline, and other fossil fuels cannot be considered as they cannot be converted directly, except at very high temperatures. Incorporating a fuel processing unit would not be feasible for a small portable device for battery replacement. Methanol is the

TABLE 37.3 Characteristics of Fuels for Use in Portable Fuel Cells

	Theoretical*		Current state-of-art[†]	
	Wh/kg	Wh/L	Wh/kg	Wh/L
Hydrogen				
Hydrogen (gas)	32,705			
Hydrogen (liquid) cryogenic	32,705	2310		
Pressurized H_2 containers				
70 MPa		3925		
Metal hydride				
MH (2%H_2)	655		164	426
MH (7%H_2)	2290	3400		
Chemical hydrides				
LiH + H_2O	2539			
$NaBH_4$ + 2H_2O	3590		592[‡]	
30% $NaBH_4$ solution	2375	2080		
Methanol (MeOH)				
100% MeOH	6088	4810	289–805[§]	141–385[§]
MeOH-H_2O solution (equimolar)	~3900	~3350		

*Based on 1.23 V for H_2/O_2 fuel cell
[†]Based on actual watthour output of a fuel cell running on the specified H_2 source
[‡]Includes container/packaging and required water
[§]Depends on power and run-time—see Fig. 37.5

only liquid fuel that, at this time, shows promise for direct conversion at reasonable temperatures (see Sec. 38.3).

The necessity for containing and supplying hydrogen to the fuel cell, in a practical and safe method, substantially reduces its practical specific energy. A number of methods are being used, including compressed gas cylinders, reversible storage in hydrides, and chemical methods for generating hydrogen that require specific methods for generating and controlling the generation of hydrogen (see Sections 38.6.1 and 38.6.2). Table 37.3 also lists the theoretical values of the various methods for supplying hydrogen and, as applicable, the status of current technology. While these values, albeit much lower than that of hydrogen gas, are still higher than those of most battery systems, a comparison to the specific energy of battery systems, which is often done, is not a correct one. It compares only the fuel supply of the fuel cell (omitting the fuel cell stack and other fuel cell components) to a complete battery system. A more reasonable method of comparison is discussed below and illustrated in Fig. 37.5.

37.3.2 Electrochemical Conversion

The fuel cell converts chemical energy to electrical energy electrochemically at high conversion efficiencies (in the order of 30 to 60%, depending on the output voltage) in small as well as large units and even when it is operating at partial power. However, while conversion efficiency may not be affected, scaling down to the lower power levels may not result in a proportional decrease in the weight and size of the power unit or the auxiliary devices.

37.3.3 Operating Temperature

Most practical portable fuel cells operate at internal temperatures not much different from ambient. Based on the current technology, as summarized in Table 37.1, this has for the most part limited the choice to the proton exchange membrane fuel cell (PEMFC) and direct methanol fuel cell (DMFC). Operation at temperatures near ambient results in somewhat lower fuel-to-electricity efficiency, and the dependence of the electrical conductivity in the membrane electrolyte on the presence of water also limits operation at subfreezing temperatures. Although the membrane can withstand freezing of the water inside it, the PEMFC startup will be slower and may require source of heat, e.g., a battery or combustion heat from a fuel.

A notable exception to the need for a direct fuel cell for small power sources is the subkilowatt reformed methanol fuel cell, which has been developed at power levels down to 25 W at a system weight of 1.24 kg (Fig. 37.3). As shown schematically in Fig. 37.4, this system is based on the use of a miniaturized fuel processing unit that produces hydrogen gas from a mixture of methanol and water and delivers this fuel to a PEMFC to generate electric power.

37.3.4 Modular Features

The modular features of the fuel cell, with separate units for power conversion and fuel storage, facilitate designing the fuel cell system to meet application requirements and equipment footprints. The power conversion unit (the fuel cell stack) can be sized to meet the power requirement, and the fuel container can be sized to contain sufficient fuel to meet the service time requirement.

Figure 37.5 compares the performance of several primary and secondary batteries with that of a fuel cell, showing the total mass of each system designed, in this example, to deliver 20 W, for different times of operational service. The secondary battery systems deliver their rated capacity even at the highest discharge rates shown in the figure; hence, their performance is characterized by a straight sloping line. The slope is equivalent, as shown, to their specific energy, and the mass of the battery is reduced almost proportional to the reduction in service time. At the low operational times, the curve for the fuel cell levels off, reflecting the inactive mass of the system, i.e., the fuel cell stack

INTRODUCTION TO FUEL CELLS **37.11**

FIGURE 37.3 UltraCell XX25™ Fuel Cell System.

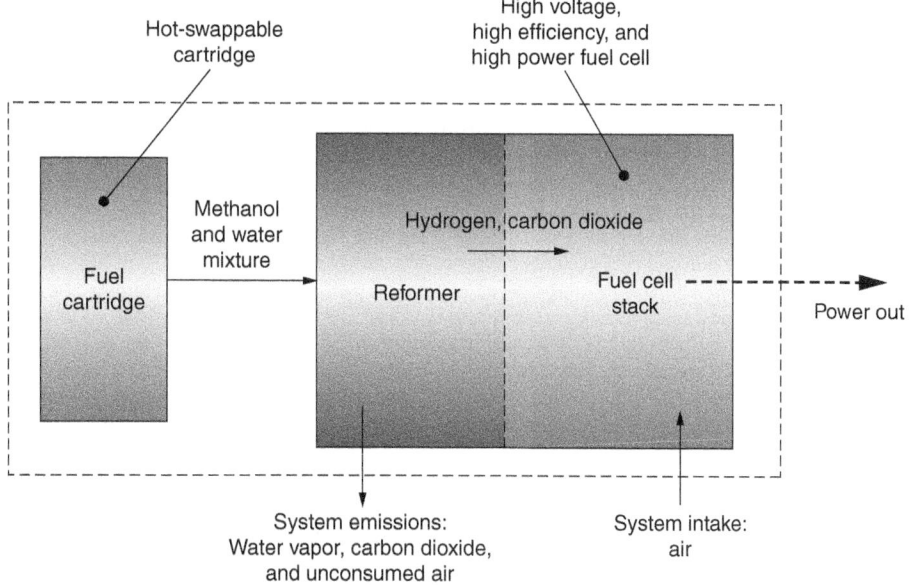

FIGURE 37.4 Reformed methanol fuel cell schematic. (*Source: UltraCell Corp.*)

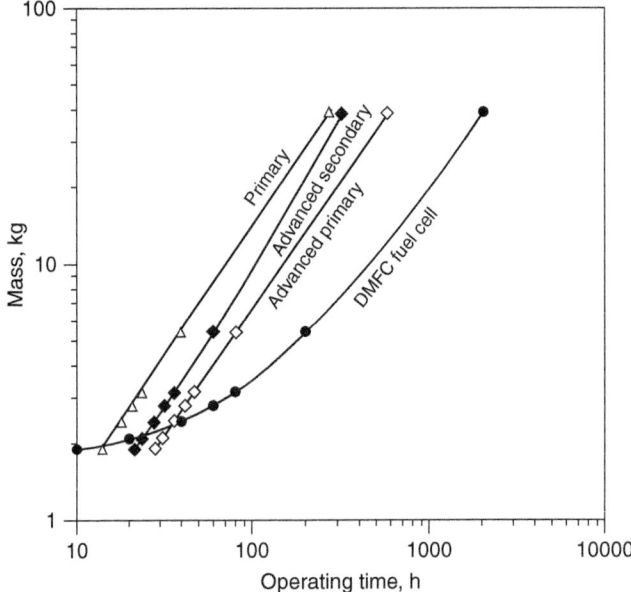

FIGURE 37.5 Comparison of electrochemical systems—mass vs. service life (based on 20 W output and the following specific energies: primary, 145 Wh/kg; advanced primary, 300 Wh/kg; advanced secondary, 225 Wh/kg).

and the "balance of plant" components required to operate the system. At the longer service times, the mass of the power unit becomes insignificant, and the system mass asymptotically approaches the specific energy of the fuel and its containment system. For times less than 10 h, the primary battery systems, which generally do not perform as well at high discharge rates, also tend to show decreasing slopes.

This figure graphically illustrates the respective advantages of batteries and fuel cells. The battery shows its advantage on the relatively short-term applications as the fuel cell is penalized by the mass of the power unit. On longer-term applications, the fuel cell benefits from the higher specific energy of the replacement fuel compared to that of nearly all battery systems. A similar relationship exists if the comparison is made on a volumetric basis.

This figure points out the direction that fuel cell development must take to compete successfully with battery systems for relatively short (say, less than 10 h) mission times in the subkilowatt range. The weight and size of the power unit must be reduced substantially, as the emphasis in the design of portable equipment is toward lower size and weight even at the sacrifice of service time. Unless this is done, the advantage of the fuel cell, the lighter weight fuel replacement, will not be significant. We note the lack of inexpensive, highly efficient miniaturized components such as gas and liquid pumps of the size and capacity needed for small fuel cells. Such components may become available as the need for them is driven by the introduction of small fuel cells for large-market applications.

An interesting consideration is a possible trade-off in the design of the fuel cell component and the fuel source. In the case of the direct methanol fuel cell (DMFC), for example, water is required for the reaction of methanol. The discharge product of the fuel cell, water, can be used if the water management or recovery is incorporated in the fuel cell—a one-time cost of increased size, weight, and complexity of the fuel cell. Alternatively, water can be added to the fuel source at the expense of a recurring lower specific energy of the diluted fuel source.

37.3.5 Air-Breathing Systems

Most terrestrial fuel cells are air-breathing and an oxidant does not have to be stored and carried with the fuel cell, thus keeping the size and weight of the system to a minimum. Depending on the power level, the airflow may be insufficient, necessitating the addition of fans or other means of forced convection for the electrochemical reaction, cooling, and water balance.

37.3.6 Environmentally Friendly

Fuel cells are *environmentally friendly* and, while the large sizes can be complex, much like a chemical plant, in the small and portable sizes they can be quiet and relatively simple in design. While these characteristics are superior to the engine-generators and other heat engines they may replace, for these characteristics they offer no advantage over batteries. Further, the need to provide a method for attaching the fuel supply and an infrastructure to supply the fuel makes it more complex as these components are not required for the battery, which is self-contained.

37.3.7 Cost

Cost will be a major factor for the acceptance of fuel cells as a replacement for batteries. The cost of the fuel cell is determined by its two components: the fuel cell and auxiliaries, and the fuel source. At this time, the cost of the fuel cell is high compared to batteries, not only because it has not attained commercial production status, but also because the polymers, catalysts, and other components are expensive. A potential advantage of the fuel cell again focuses on the fuel supply. If the cost of fuel replacement can be reduced so that it is lower than that of battery replacement, fuel cell deployment may be a cost-effective approach for extended periods of operation.

37.4 INNOVATIVE SUBKILOWATT DESIGNS: SOLID OXIDE FUEL CELLS

At first glance, the solid oxide fuel cell (SOFC), traditionally operating at 800 to 1000°C, does not appear to be a promising candidate for use in small fuel cells. The ratio of surface area to volume for small systems would appear to require too much insulation to prevent excessive heat loss and consequent loss of fuel energy to maintain the system at its operating temperature. Nevertheless, companies such as Adaptive Materials have pursued the development of such systems and have advanced them to the point of significant demonstrations. Systems ranging from 25 to 250 W have been developed using a "microtubular" design based on cells only a few millimeters in diameter and using an oxygen-ion-conducting ceramic electrolyte capable of operating in the temperature range of 600 to 800°C. Such cells can be brought to operating temperature within a few seconds and can withstand the associated thermal shock. Figure 37.6 shows a conceptual diagram of the operation of a single planar SOFC. The SOFC is well suited to the use of hydrocarbon fuels such as propane or butane, as minimal fuel processing is required at the high operating temperature of the system.

Figure 37.7 depicts a 25 W SOFC with a mass of 1.5 kg (without fuel), a volume of 2.0 L, and the remarkable specific energy of 661 Wh/kg, including the mass of the fuel and its container.

As with many other fuel cell systems under development, the initial market for SOFC appears to be for military systems such as unmanned aerial vehicles, field battery chargers, small robotic vehicles, and electronics and sensors. This market, though it requires ruggedization for harsh conditions, can also tolerate a relatively high initial price and a lifetime in the hundreds or low thousands of hours. Success in the military arena may also open the larger markets for industrial and consumer applications. The worldwide availability of propane and butane in canisters from 0.5 to 10 kg would provide a ready fuel infrastructure for consumer applications that is totally lacking for hydrogen gas.

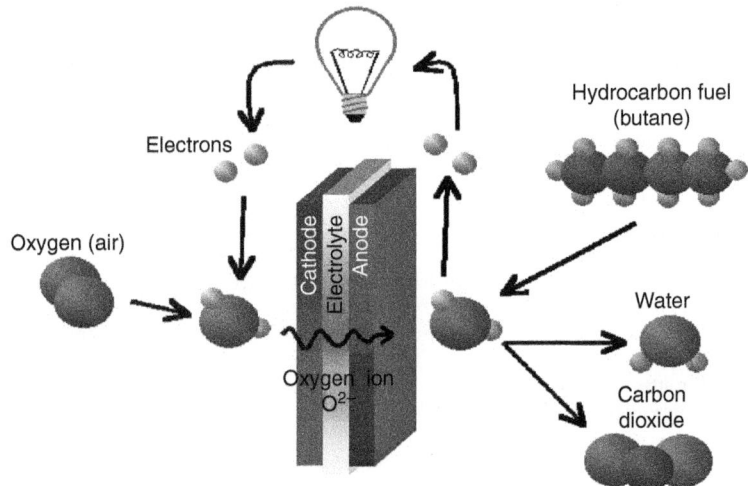

FIGURE 37.6 Operating principles of the solid oxide fuel cell. (*Source: Adaptive Materials, Inc.*)

FIGURE 37.7 A 25 W portable SOFC. (*Source: Adaptive Materials, Inc.*)

REFERENCES

1. K. Kordesch and G. Simader, *Fuel Cells and Their Applications*, VCH Publishers, NY, 1996.
2. B. V. Tilak, R. S. Yeo, and S. Srinivasan, "Electrochemical Energy Conversion—Principles," in *Comprehensive Treatise of Electrochemistry*, Vol. 3, J. O'M. Bockris, B. E. Conway, E. Yeager, and R. E. White, eds., Plenum Press, New York, 1981, pp. 39–122.
3. S. R. Narayan and T. I. Valdez, "High-Energy Portable Fuel Cell Power Sources," the Electrochemical Society, *Interface*, pp. 40–44, Winter, 2008.
4. S. Srinivasan, *Fuel Cells: From Fundamentals to Applications*, Springer, New York, 2006.

CHAPTER 38
SMALL FUEL CELLS

Arthur Kaufman and H. Frank Gibbard

38.1 GENERAL

Small fuel cells that provide power at ratings below 1000 W are being implemented in numerous application areas. Representative examples of such applications are shown in Table 38.1. Descriptions of associated hardware development/commercialization efforts are provided in Sec. 38.8. These existing and candidate applications may include power systems that are fuel-cell-only, fuel-cell/battery hybrids, and fuel-cell/solar/battery hybrids, depending on the nature of the system's requirements.

The interest in small fuel cells stems primarily from their potential to replace batteries with systems having higher specific energy or energy density, and small engine-generators with more portable, efficient, and environmentally friendly conversion systems.

The energy-storage and power-generating elements of a fuel cell system are separate entities—the fuel storage and fuel cell power section (including its auxiliaries), respectively. In a battery, on the other hand, the energy-storage and power-generating elements are the same. Hence, the fuel cell system could be designed to relate optimally to its operating mode—the fuel cell power section to satisfy the power requirements, and the fuel storage to satisfy the energy requirements. This decoupling of energy and power can be particularly advantageous in applications where the energy requirement is great and the power requirement is minimal—that is, in applications of long duration. In such applications, the fuel cell section, with its auxiliaries, becomes relatively insignificant in size and weight within the overall system; and the system's energy density and specific energy approach that of the fuel storage subsystem alone. The mission duration beyond which fuel cells would tend to be favored over batteries, by providing a smaller and/or lighter system, depends on the specific application requirements.

Certain applications are well suited to a fuel-cell/battery hybrid system by nature of their duty cycle. Those that have high peak-to-average load ratios and relatively short-duration peaks are generally attractive candidates. Such a system allows the fuel cell to be rated near the average power, while a relatively small battery provides excess power on demand and is recharged by the fuel cell during normal load operation. Hybrid systems thus exploit the strengths of both batteries and fuel cells—the wide dynamic range of the former and the high energy content per unit weight or volume of the latter.

Solar/battery power systems can also be combined advantageously with fuel cells in various applications. Use of the fuel cell can often allow solar power to be exploited without the need for excessive battery size and weight associated with prolonged or unpredictable lack of availability of solar energy. Small fuel cells are also expected to become an alternative to small engine-generator sets in some applications. Fuel cells systems are expected to demonstrate advantages over engine-generators in the areas of life, reliability, efficiency (fuel consumption), noise, and emissions. While larger engine-generators tend to have size and weight advantages over comparable fuel cell systems

TABLE 38.1 Small Fuel Cell Applications

Application
Remote power, including battery charging
Portable power, including soldier-wearable power
Mobile power, including vehicle-based auxiliary power
Unmanned vehicle and robotic power
Cellular phone power
Power for portable digital devices
Backup power
General purpose power

operating on the same fuel, these advantages are expected to diminish as systems are scaled down in power rating. There are prospects of fuel cells competing effectively in these categories in the low power area that are the subject of this chapter. A key competitive challenge will be in the area of cost.

38.2 APPLICABLE FUEL CELL TECHNOLOGIES

Various types of fuel cell technologies are either in use or under development, and these are generally distinguished on the basis of their electrolyte. These technologies exhibit viable operation in different temperature regimes:

1. *Phosphoric acid fuel cells (PAFC)* use immobilized highly concentrated phosphoric acid electrolyte and generally operate in the 160 to 200°C range. Their electrodes are typically resin-bonded, carbon-supported, platinum-based catalyst layers on wetproofed carbon-fiber substrates.
2. *Molten carbonate fuel cells (MCFC)* use mixed alkali-carbonate molten salt electrolyte and typically operate at about 600°C. Their electrodes are non-precious-metal structures.
3. *Solid oxide fuel cells (SOFC)* use solid oxygen-ion-conducting metal oxide electrolyte in the 600 to 1000°C range. Their electrodes are non-precious-metal structures.
4. *Alkaline fuel cells (AFC)* typically use circulating solutions of potassium hydroxide electrolyte at temperatures ranging from ambient to about 80°C. Their electrodes are typically non-precious-metal mesh structures, although precious metals may be used at the anodes.
5. *Proton-exchange membrane fuel cells (PEMFC)* use solid-polymer proton-conducting membrane electrolyte at temperatures generally ranging from somewhat above ambient to about 80°C. Current power sources (including conventional hydrogen-fueled PEMFCs and direct-methanol-fueled units [DMFCs]) primarily use a trifluoromethanesulfonic-acid-based electrolyte membrane, such as DuPont's Nafion®. However, higher-temperature membranes, such as phosphoric-acid-doped polybenzimidazole, are now seeing service in power sources. Although these electrolytes literally do not conduct protons through the membrane per se (the phosphoric acid phase provides this function), cells utilizing them are commonly grouped into the proton-exchange membrane category. Operating temperatures for these units can be in the 100 to 180°C range.[1] These are commonly referred to as high-temperature proton-exchange membrane fuel cells (HTPEMFC). The electrodes for both the conventional and high-temperature PEMs are typically ionomer-bonded, carbon-supported, platinum-based catalyst layers on wetproofed carbon-fiber substrates.

Small fuel cells can be exploited most effectively if they can stand by and operate at ambient temperatures (and can therefore start rapidly), can operate on ambient air, respond rapidly to load changes, have a nonmigrating (solid) electrolyte, and have a reasonably high power density and specific power. The fuel cell type that best suits these criteria is the PEMFC, despite a drawback related to the fact that liquid water embodied in the solid polymer tends to freeze, and thereby impede proton conduction, when its temperature drops below the freezing point. The PEMFC can stand by under freezing conditions, however, and can generally operate under these conditions as well, taking advantage of self-generated heat; external means, such as power from a battery or an electric grid, are sometimes required to execute a sub-freezing start-up or to prevent freezing.

Direct-methanol fuel cells (DMFCs) and high-temperature PEM fuel cells (HTPEMFCs) are included in the PEMFC category. Both suit the stated criteria to a somewhat lesser degree than does the conventional PEMFC (excepting the freezing issue) but offer advantages for certain system approaches. DMFCs offer the major advantage of utilizing a logistically attractive liquid fuel without the burden of external processing. This comes with a sacrifice in fuel cell voltage-current performance (as compared to that for hydrogen-fueled cells); incremental current inefficiency and cathode polarization caused by methanol crossover from the anode across the electrolyte-membrane to the cathode (often measured to be equivalent to a current density in excess of 100 mA/cm^2, but likely to have been reduced via claimed membrane improvements);[2] and the need for increased precious-metal loading, especially at the anode. HTPEMFCs are not vulnerable to dry operating conditions (unlike conventional PEMFCs) and tolerate carbon monoxide in processed fuel far better than conventional PEMFCs (see Sec. 38.6.5), although their electrochemical activity is inherently lower using today's state-of-the-art electrolyte-membranes.

Proton-exchange membrane fuel cells are indeed the type that has received the predominant share of development and implementation in the small fuel cell arena. Nevertheless, despite their high operating temperature, innovation in solid oxide fuel cells (SOFC) as relevant to small power sources has opened the door to customer examination of hardware utilizing this technology.[3]

38.3 CELL ELECTROCHEMICAL OPERATION

PEM fuel cells use ion-exchange membranes. The electrolyte membrane supports cell current generation via proton transport from the anode at one membrane surface to the cathode at the opposite surface, as illustrated in Fig. 38.1.

Hydrogen ions are produced at the anode-electrolyte interface from hydrogen gas that diffuses through the anode structure. Electrons generated in this process pass through the electronically conductive phase of the anode to an adjacent current collector. Hydrogen ions reaching the cathode-electrolyte interface react with electrons returning from the external load and with oxygen gas (from air) that diffuses through the cathode structure, producing water, either a mixture of the vapor and liquid phases or, at higher temperature, all in the vapor phase.

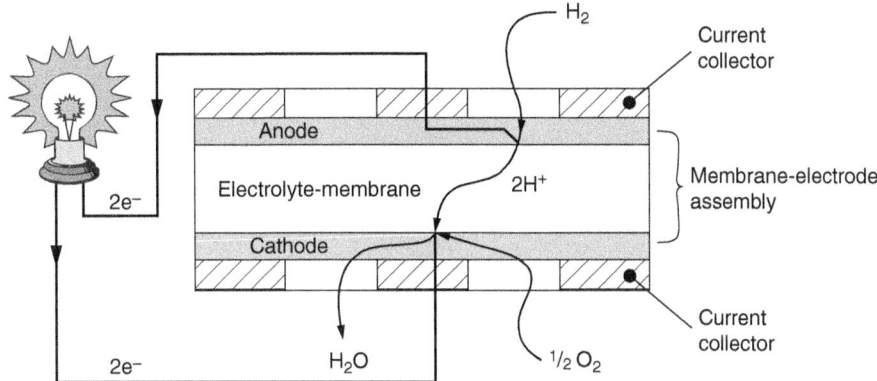

FIGURE 38.1 Schematic representation of an individual proton-exchange membrane fuel cell.

38.4 FUEL CELLS AND ELECTROCHEMICAL CAPACITORS

The cell reactions are:

Anode: $H_2 \rightarrow 2H^+ + 2e^-$

Cathode: $½ O_2 + 2H^+ + 2e^- \rightarrow H_2O$

Overall: $H_2 + ½ O_2 \rightarrow H_2O$

Direct-methanol cells use proton-exchange membranes as well, and rely on the reaction of methanol and water at the anode to generate protons and electrons. The protons migrate across the membrane to the cathode, while the electrons enter the external circuit, as illustrated in Fig. 38.2. Oxygen gas (from air) reacts with the protons and the returning electrons at the cathode to produce excess water.

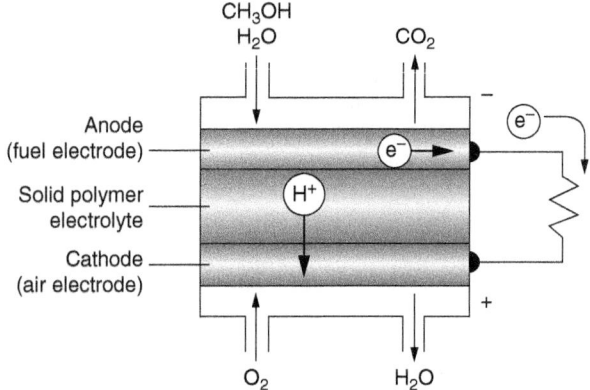

FIGURE 38.2 Schematic representation of an individual direct-methanol fuel cell.

The reactions are:

Anode: $CH_3OH + H_2O \rightarrow 6H^+ + 6e^-$

Cathode: $\frac{3}{2} O_2 + 6H^+ + 6e^- \rightarrow 3H_2O$

Overall: $CH_3OH + \frac{3}{2} O_2 \rightarrow CO_2 + 2H_2O$

Solid oxide cells feature oxygen ion conduction from cathode to anode across a solid-state metal oxide electrolyte. These ions react with hydrogen gas at the anode, producing water vapor and generating electric current as electrons flow into the external circuit, as illustrated in Fig. 38.3. Oxygen gas (from air) reacts with these electrons at the cathode to generate the oxygen ions.
The reactions are:

Anode: $H_2 + O^{2-} \rightarrow H_2O + 2e^-$

Cathode: $½ O_2 + 2e^- \rightarrow O^{2-}$

Overall: $H_2 + ½ O_2 \rightarrow H_2O$

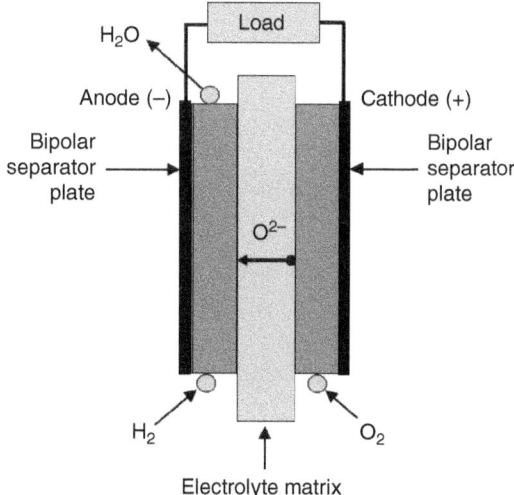

FIGURE 38.3 Schematic representation of an SOFC single cell.

38.4 CELL STACKING CONFIGURATIONS

Since cell voltages generated in the discharge of fuel cells are typically in the 0.6 to 0.8 V range (about 0.4 to 0.6 V for direct-methanol cells), multiple cells are connected in series to obtain practical voltage levels. These are most commonly arrayed as a bipolar stack, as illustrated for a PEMFC in Fig. 38.4. The electrochemically active cell elements shown schematically in Figs. 38.1, 38.2, and 38.4 are commonly referred to as membrane-electrode assemblies (MEAs). The MEAs are interleaved with bipolar plates that have multiple functions: (1) conduction of electronic current from cell to cell; (2) dispersion of the active hydrogen and oxygen (air) gases through flow channels on the two opposing surfaces; (3) prevention of mixing of these gases; and (4) in many cases providing a means for removing heat that is generated during operation of the fuel cell. The voltage or current of the stack shown in Fig. 38.4 can be raised by increasing the number of cells or their active area, respectively.

An alternative to the bipolar stacking arrangement is the so-called monopolar approach, whereby the individual MEAs are nominally arranged in a side-by-side configuration such that the electronic current is edge-collected at one electrode and passed on to the opposite electrode of the adjacent MEA.[4] The voltage of the array is increased by increasing the number of MEAs in the line, and possibly by using multiple series-connected arrays. The current is increased by increasing the cell active area and/or by connecting arrays in parallel. This approach eliminates the need for bipolar plates, thereby providing the opportunity for significant volume and weight reduction. In order to be effective, the design must minimize the tendencies toward nonuniform current distribution and substantial ionic and electronic losses associated with in-plane conduction within the electrolyte-membranes and electrodes (and interconnects), respectively.

In addition to the planar geometry described for fuel cells to this point, the adaptation of SOFC technology for small power sources has focused in part on the use of cells in the form of small-diameter tubes. Each tube in an array has an anode, cathode, electrolyte, and provisions for fuel and air delivery. Cells are arranged in series-parallel combination, so as to obtain the desired voltage-current characteristics. This design is analogous to the commonly used method of constructing batteries of monopolar cells.

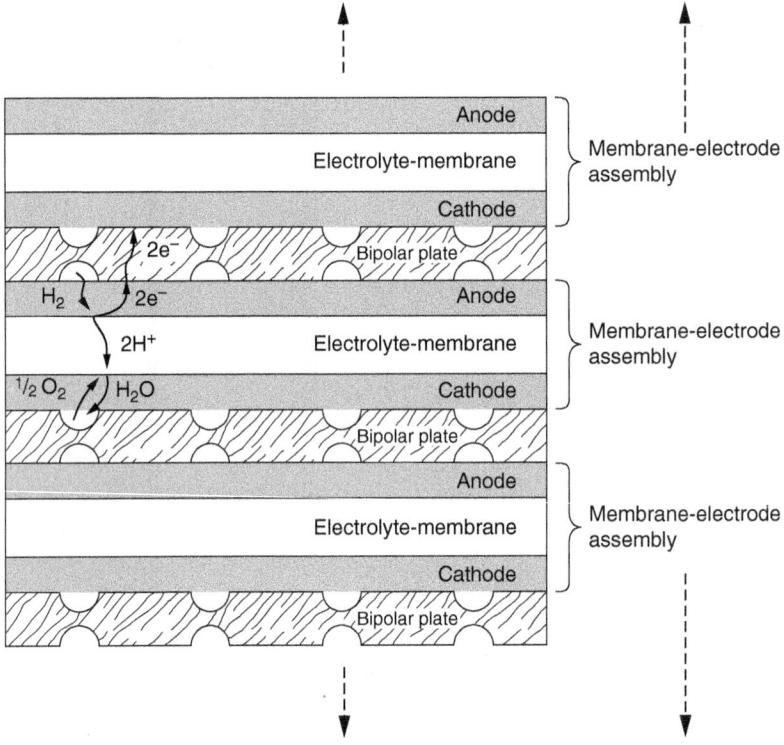

FIGURE 38.4 Schematic representation of proton-exchange membrane fuel cell stack.

38.5 FUEL SELECTION

The viability of small fuel cells in their various applications is heavily dependent on the fuel that is utilized. The predominant fuels implemented in PEM-based small fuel cells to date are hydrogen and methanol. Hydrogen fuel may be obtained directly from storage or indirectly via processing of another chemical or fuel entity. PEM systems utilizing methanol at the anodes (referred to as direct-methanol fuel cells or DMFC) generally store the fuel as neat methanol. Small SOFC systems, operating at much higher temperatures, are far more tolerant of fuel by-products and typically utilize and process compressed/liquefied hydrocarbon gases.[3]

38.6 FUEL PROCESSING AND STORAGE CONFIGURATIONS

Since the advantages of small fuel cells are dependent on their fuel processing and storage characteristics, it is important to examine the relevant options available for fuel cell systems.

38.6.1 Compressed Hydrogen Storage

The simplest form of fuel storage and utilization for fuel cell systems is compressed hydrogen. Such storage is impractical for larger fuel cells, except in cases where operation is limited to brief periods in a backup mode, because of volume and overall weight considerations as well as the higher cost and inferior logistics of transport for hydrogen in comparison with common fuels. With respect to small fuel cells, however, compressed hydrogen is sometimes an acceptable storage option. Examples include operation at low-power level, short-duration, and backup service, as in applications where the relative system simplicity offered by operation on hydrogen (with attendant benefits in system weight, size, and reliability) prevails over the benefits provided by common fuels (weight, size, and cost for the fuel and its storage).

Since the weight of the active material is negligible, compressed hydrogen storage provisions can be exploited where light weight is a priority. This requires the use of lightweight, high-pressure canisters and pressure regulators, and the associated cost factors must be taken into account. This approach is illustrated in Fig. 38.5, where a lightweight energy system developed for military field use is shown. This 50 W system from the late 1990s was designed for 1 kWh over a period of 20 h. It should be noted that small (1.5 L, 1.3 kg) commercially available cylinders operating at 5000 psi can store hydrogen with approximately 1750 Wh of energy content (lower heating value).[5] Larger lightweight cylinders with operating pressure capability greater than 10,000 psi have been developed for potential use in automotive fuel cell systems.[6]

FIGURE 38.5 50 W fuel cell with compressed hydrogen gas.

38.6.2 Indirect Hydrogen Storage

Storage of hydrogen by indirect means for small fuel cell systems has included its generation from metal hydrides and chemical hydrides.

Metal Hydrides. The storage of hydrogen in the form of hydrides of metal-alloy powders is often an attractive and convenient energy storage mode for small fuel cell systems. This is attributable to their simplicity of operation and compactness; the benefits are most realizable in particularly small systems. The energy densities of these materials can range up to 500 to 1000 Wh/L (electric), substantially higher than that of compressed hydrogen. This reflects a hydrogen loading approaching 2% by weight, which is characteristic of the maximum obtainable in alloys (typically AB_2 type, where, for example, A is Zr or a mixture of Zr and Ti, and B is a mixture of transition metals) that have useful hydrogen pressures at ambient temperatures. (Magnesium-based alloys have been formulated to obtain hydrogen loadings on the order of 5% by weight, but these require discharge temperatures in the neighborhood of 300°C, which necessitates combustion means with a percentage of the hydrogen being sacrificed to generate heat.)

Chemical Hydrides. Primary hydride systems of various types have seen service in small fuel cell systems. These are irreversible (throwaway) chemical systems that generate hydrogen on demand. The active reactant is generally an alkali or alkaline earth metal hydride (sometimes in complex form). This typically is caused to react with water to form hydrogen and a metal oxide (or mixed metal oxide). Analogous chemical hydride systems are also being explored.

The potential advantages of chemical hydride systems include high specific energy, since the hydrogen yield by weight can be a far higher percentage of the reactant weight in comparison with reversible metal hydrides. On the other hand, energy densities for these systems are not as attractive because of the relatively low densities of the reactants. The hydrogen-generating reaction and theoretical hydrogen yield are shown in Table 38.2 for representative chemical hydrides. (Note that the mass of water required to generate the hydrogen is included in the calculations of percent hydrogen for each chemical hydride.) The specific energy advantage can be enhanced if product water from the fuel cell can be recovered for use in the chemical hydride reaction, in which case the system could be refueled simply via a stored reserve of reactant powder or granules.

TABLE 38.2 Representative Chemical Hydride Reactions and Theoretical Hydrogen Yields

Reaction	Theoretical hydrogen yield	Theoretical specific energy
$LiH + H_2O \rightarrow LiOH + H_2$	7.8%	2540 Wh/kg
$CaH_2 + 2H_2O \rightarrow Ca(OH)_2 + 2H_2$	5.2%	1700 Wh/kg
$NaBH_4 + 2H_2O \rightarrow NaBO_2 + 4H_2$	10.9%	3590 Wh/kg

Theoretical, including weight of hydride and water, in H_2/air fuel cell based on 1.23 V per cell.

The challenges associated with chemical hydrides include the requirement that the respective reactants be brought into contact such that the rate of reaction just meets the fuel cell's hydrogen needs. Also, since the reaction products are disposed of, as opposed to being regenerated, the cost of the replenishing reactant chemicals must be taken into account in evaluating the economics of system operation. The most likely use of chemical hydrides in small fuel cells is for high-performance military systems.

38.6.3 Fuel Processing

The applicability of small fuel cells could clearly be greatly enhanced if compact systems using conventional fuels are implemented. In most cases, this requires a fuel processor to convert the fuel into a hydrogen-rich gas that would be delivered to the fuel cell. Much of the challenge in such an approach relates to attaining a sufficiently compact and low-cost fuel processor.

A variety of common fuels and chemicals can be considered as candidate fuels for small fuel cells. These include the following.

Ammonia. Ammonia is commonly used in industry and agriculture in the form of a liquid stored at its own modest vapor pressure. Liquid ammonia offers high specific energy and energy density based on a relatively simple thermocatalytic hydrogen-generating dissociation reaction

$$NH_3 \rightleftharpoons 3H_2 + N_2$$

Thus, ammonia can yield hydrogen at about 17% of its own weight. This corresponds to about 3 kWh/kg (elec.) and almost 2 kWh/L (elec.); however, a fraction of the hydrogen formed must be consumed in generating heat to sustain the endothermic dissociation reaction. Liquefied petroleum gas (LPG) is generally preferred as a fuel over ammonia because of its greater distribution infrastructure, even higher energy density, and lower cost per unit of energy content, and because of ammonia's reputation as a toxic chemical. Nevertheless, ammonia could play a role in selected small fuel cell applications as a result of its far simpler hydrogen-generating process.

Methanol. Methanol is a widely available chemical that is relatively easy to handle and store as a liquid at atmospheric pressure. It can also be converted to a hydrogen-rich gas by a process that is the simplest among those for carbon-containing fuels. The endothermic reaction

$$CH_3OH + H_2O \rightleftharpoons CO_2 + 3H_2$$

is carried out at modest temperature (about 200 to 250°C) since methane (favored by thermodynamic equilibrium) cannot be formed when conventional methanol steam-reforming catalysts are used. Accordingly, little or no downstream processing is needed because very little carbon monoxide is formed at these temperatures. Here again, whereas LPG is favored from the points of view of energy content and cost, methanol is an attractive fuel for small fuel cells because its processing burden is greatly eased. Small PEM fuel cell systems using processed methanol are being provided by several companies to a variety of customers.

Ethanol. Ethanol is similar to methanol in its handling and storage. Its reaction with steam can be expressed as

$$C_2H_5OH + H_2O \rightleftharpoons 2CO + 4H_2$$

Unlike methanol, it is considered nontoxic. Its availability and cost as a fuel are irregular; more importantly, ethanol processing requires the rupture of a carbon-carbon bond and cannot be carried out at the low temperatures characteristic of methanol steam reforming. Downstream processing is required, at least for low-temperature (PEM) fuel cells, and its potential attractiveness as a fuel for compact systems of this type is therefore diminished.

Liquefied Petroleum Gases. LPG (principally propane in the United States, but sometimes principally butane, as in Japan) is often the preferred fuel for dispersed fuel cell systems. It has substantially higher specific energy than both ammonia and methanol along with lower cost per unit energy. Indeed, in the absence of pipeline natural gas, it is the fuel of choice for stationary fuel cells. However, small fuel cells generally have a different set of requirement criteria. In a relative sense, fuel cost is less important, and compactness, simplicity, and hardware cost more important. (High-temperature fuel processing is required, as is the case with natural gas and transportation-type fuels as well.) However, this tendency cannot be over generalized; for example, missions with very long durations could be an exception. Also, this type of fuel is attractive for SOFCs because its processing integrates well with high-temperature, byproduct-tolerant fuel cells.

Natural Gas. Stationary fuel cell systems that have access to a natural gas pipeline would typically be fueled by this methane-rich gas. In addition to eliminating the storage burden, pipeline natural gas provides a lower cost per unit energy than LPG, and methane yields more hydrogen per unit weight than propane in a fuel cell's fuel processor. Here again, however, these factors are unlikely to be the prevailing issues in the case of small fuel cell systems. In cases where onboard storage of fuel is required (for portability or mobility), LPG would be preferred based on logistics and compactness.

Transportation-Type Fuels. Aviation-type fuels (such as JP-8) are preferred for military applications since these are in general use, readily available, and the safest (very low vapor pressure). Diesel fuel (widely used in heavy vehicles, to a lesser degree in automobiles, and often as a general-purpose fuel in underdeveloped regions) is less desirable for small fuel cells because of its somewhat higher molecular weight and typically higher sulfur content. Low-sulfur (15 ppm) diesel fuel is now employed in the United States. Indeed, these factors pose a difficult challenge to the processing of aviation fuel as well. This is attributable to the difficulty of vaporizing and breaking down its large molecules without carbon formation and/or catalyst poisoning by sulfur at some point in the process. The compactness and low hardware cost required in most small fuel cell applications exacerbate these issues, but aviation fuel use remains a target for small SOFC systems. Gasoline is more readily processed than aviation fuel because of its lower molecular weight, and concomitant easier vaporization and reactivity, and its relative absence of sulfur. However, this is not a preferred fuel for small fuel cell systems because of handling and safety issues, related to its vapor pressure and flammability.

38.6.4 Methodologies for Fuel Processing

The generation of hydrogen from carbon-containing fuels (with the exception of methanol) requires a high-temperature (usually catalytic) process. The fuel is reacted (1) with steam catalytically (steam reforming, or SR), (2) with substoichiometric oxygen from air (partial oxidation, homogeneous or catalytic, POX), or (3) with steam and oxygen catalytically (autothermal, ATR). SR is an endothermic reaction carried out typically at 730°C or higher. POX is an exothermic process, usually carried out at higher temperatures (perhaps as high as 900 to 1000°C). The ATR process is almost thermally neutral and typically operates at or somewhat above SR temperatures. Representative reactions for these processes using methane as the fuel can be expressed as follows:

$$\text{SR:} \quad CH_4 + H_2O \rightleftharpoons CO + 3H_2$$

$$\text{POX:} \quad CH_4 + 0.5O_2 \rightleftharpoons CO + 2H_2$$

$$\text{ATR:} \quad CH_4 + 0.25O_2 + 0.5H_2O \rightleftharpoons CO + 2.5H_2$$

In simplistic terms, the SR process generally provides the highest hydrogen yield and consequently the highest efficiency. Because of the endothermic reaction, its thermal management tends to be the most complex and bulky. Conversely, POX is typically the least efficient but has the simplest configurations. The ATR process tends to be between the other two in both respects.

The selection of a preferred fuel processor type is decided based on the application requirements. For example, a conventional stationary fuel cell system operating continuously on natural gas might be best suited to the SR processor to minimize fuel cost, while a small, mobile system requiring rapid startup might be best served via a POX or ATR system.

38.6.5 Process Gas Upgrading

The high-temperature processes described above yield a reformate gas that is high in carbon monoxide content (usually greater than 10%). Low-temperature fuel cells, like PEM, would require further processing to maximize the hydrogen yield and minimize the fuel cell anode-catalyst inhibiting effects of CO. Thus the reformate gas is then passed through a catalytic water-gas shift-converter at lower temperature (sometimes in two stages, the second at lower temperature) where the following reaction takes place:

$$CO + H_2O \rightleftharpoons CO_2 + H_2$$

Here again, as in the case of methanol steam-reforming, the formation of methane under these conditions is prevented via the specificity of the shift-converter catalyst. For PEM fuel cells, further reduction in CO concentration is usually necessary (from about 0.5% down to less than 100 ppm). This is often carried out by way of catalytic preferential oxidation of CO in the presence of hydrogen with the addition of air at a flow rate that is a small multiple of the stoichiometric rate required for complete oxidation of the CO. With regard to process gas upgrading, the high-temperature PEM fuel cell (HTPEMFC), operating at about 170°C, has the advantage of tolerating CO to a concentration of about 2% or more. This would allow a more modest shift-converter and eliminate the need for a preferential oxidizer.

From the above discussion, it is evident that steam or water vapor is an essential player in the fuel processor, whether it be in the reforming reaction itself or in the subsequent shift-conversion stage. The source of this water must come from storage, make-up, or condensation and recovery from the fuel cell system. In any event the design and logistics for water management must be provided for the system, and the selected mode must best reflect the requirements for the specific application. Among potential military users of portable fuel cell systems there is a strong preference for systems that do not require the carrying of excess water onboard the system.

The complexity of the overall high-temperature fuel processing system for carbonaceous fuels indicates the challenge associated with adapting conventional fuels for use in small fuel cell systems. Considerable effort is required in designing and optimizing the system to achieve the requisite miniaturization and low cost in these applications. Even in the best case it appears that only the SOFC system, which can tolerate partially processed fuel, would be a potentially viable small power source for fuels requiring high-temperature processing.

38.7 SYSTEM INTEGRATION REQUIREMENTS

The requirements for a small fuel cell system vary from application to application. Among those that are most important for a given rated-power output are size, weight, initial cost, fuel cost, fuel supply infrastructure, energy storage capacity, durability, ambient-temperature operating range, and startup characteristics.

38.7.1 Fuel Delivery

Systems utilizing virtually pure hydrogen typically provide this fuel to the anodes at a modest pressure (generally in the 10 to 50 kPa range) and dead-ended, allowing the anodes to consume the amount of hydrogen required to sustain the electrochemical reaction at any given rate. A momentary "purge" of the exit port is implemented to allow accumulated impurities and water to be discharged from the anode compartment at selected time intervals.

If the power source uses processed fuel to generate a hydrogen-rich gas, the H_2 is utilized at the anodes to the extent practical to balance fuel consumption, processor temperature, and cell performance. Depending on specific system characteristics this rate could be in the 60 to 90% range. The unused fuel is combusted in the fuel processor section to generate heat needed therein.

PEM fuel cells fueled directly by methanol (DMFCs) are typically fed by a solution of methanol in water at the anodes (although the methanol is generally stored in neat form and enriches a circulating methanol-water solution as needed). Carbon dioxide formed at the anodes must be separated from the circulating stream.

38.7.2 Air Supply

Small fuel cells can operate on either diffused or forced-reactant air. Diffused-air units are generally limited in their applicability because of air supply rate issues and the impact on size and weight; also, the requisite openness of the air compartments in such devices tends to render them vulnerable to atmospheric conditions. Hence, diffused-air (static) fuel cells are usually practical only in certain particularly low-power applications, up to perhaps 25 W, where extreme simplicity of operation is essential.[7]

Forced-air fuel cells are practical over the entire output range of small power sources. The reactant air for this type of system is generally delivered to the fuel cell cathodes at whatever pressure is necessary to overcome the pressure drop through that section and associated plumbing. This is typically a small fraction of atmospheric pressure, in the range of 1 to 20 kPa, depending on stack design characteristics. The air-moving devices are usually small pumps (such as rotary-vane, diaphragm, or, less commonly, piezoelectric types) or sometimes blowers. The cathodes' utilization rate of oxygen in the reactant air will vary in accordance with operating conditions, but typical rates are about 40 to 50%. The exit air is generally discharged to the atmosphere; however, DMFC systems often recover a portion of the water in this stream to offset water consumption at the anodes.

38.7.3 Water Management

A key design issue for small PEM fuel cells (and other PEM fuel cells as well) is the management of water with respect to the fuel cell stack. Today's technology primarily utilizes the trifluoromethane-sulfonic-acid-based electrolyte membrane, such as DuPont's Nafion®. This type of membrane requires a certain level of water content in order to conduct protons efficiently, with water molecules effectively serving as carriers for the protons in their migration across the membrane. Accordingly, the system design must provide for a reasonably high relative humidity in the reactant passages that are in fluid communication with the membrane.

The moisture requirements for the reactant streams place significant limitations on operating regimes. Ambient (nonhumidified) reactant air is highly preferred in small fuel cells in order to achieve the simplicity and compactness that are generally sought in these power sources. The use of ambient air, however, requires design measures to prevent the membrane from drying out. The flow rate of air must be limited to reduce the drying effect, and the cell design often must be tailored to take advantage of the product water. The threat of drying clearly becomes far more acute as the ambient temperature increases and/or relative humidity decreases, and as the current density of the stack is increased, thereby increasing the fuel cell's temperature in relation to ambient.

The water management burden is not limited to preventing membrane dry-out. The need to operate at relatively high oxygen utilization rates (relatively low air flow rates) increases the tendency to form water droplets within the cell from the formation of product water at the cathode. This can lead to accumulation of water on the surface of or within the electrode substrate or in the air distribution channels of the cathode flow-field. Such events could result in serious performance losses from the ensuing restriction of air access to impacted regions of the cathode electrocatalysts. Consequently, the cell design approach must also serve to prevent such accumulation of water droplets.

High-temperature PEM fuel cells (HTPEMFCs) tend to have lower electrochemical activity than conventional PEMFCs. However, the HTPEMFC is particularly tolerant of dry operating conditions; consequently, these cells can be operated at relatively high temperatures (about 100 to 180°C) without requiring water management provisions. Higher-temperature operation tends to offset their lower intrinsic activity and serves to diminish the catalyst-poisoning effects of carbon monoxide. Therefore, this type of fuel cell is more readily integrated with fuel processors using methanol or other fuels.

DMFC systems, which use electrolyte-membranes of the same type as those used in conventional PEMFCs, do not have the same water management issues. Dry-out conditions do not exist since the feed to the anodes is a methanol-water liquid mixture, provided that sufficient excess water generated at the cathode is returned to the anode feed. The cell design, however, must take into account the potential impact of water accumulation on the cathode side. SOFC systems, operating at very high temperatures, of course do not have water management issues.

38.7.4 Thermal Management

The thermal management requirements for conventional PEM fuel cells are intimately associated with water management. As discussed above, the level of hydration of the electrolyte membrane must be maintained in order to prevent dry-out and thus loss of proton conduction in the membrane.

The temperatures within the stack must accordingly be restricted. Cooling of the individual cells of the stack is carried out to ensure that temperatures are moderate and rather uniform throughout the stack.

Liquid cooling in larger PEM fuel cells is very effective because of the relatively high thermal conductivity and high volumetric heat capacity of liquids. However, since small fuel cell applications benefit from simple systems, the preferred approach for small fuel cells is usually air cooling. This may be based on the use of forced air delivery through channels spaced appropriately throughout the stack of cells or, alternatively, across external surfaces of the stack along with conduction of heat from interior to exterior sections (enhanced by favorable geometric profiles). The latter approach could employ finned extensions of the bipolar plates to expand the cells' external surface area. Nevertheless, the potential benefits of liquid cooling are significant, and embodiments emphasizing minimized complexity have been implemented in small power sources.[8] These are typically based on a circulating water loop with external air cooling of the water stream.

Thermal management in HTPEMFCs is less demanding because of their tolerance of dry operating conditions; heat rejection to the surroundings is also enhanced because of the greater temperature driving forces. The high operating temperature and the configuration of the cell arrays in SOFC systems allow a simple air-cooled system. The thermal management burden is also eased in the case of DMFC systems since these can use the circulating methanol-water stream to manage fuel cell temperature.

38.7.5 Operational Control

Certain control elements must be imposed to foster stable operation of the fuel cell section. These are most comprehensive for conventional PEM fuel cells. A representative methodology for these is as follows:

1. The reactant-air flow rate must be controlled as a function of load to ensure that neither excessive water buildup (low flow rate) nor cell dry-out (high flow rate) is encountered; this requires measurement of fuel-cell current and corresponding speed adjustment in the air-moving device.

2. Fuel cell temperature must be controlled to prevent operation in a dry-out condition (too hot); this requires that the stack cooling fan (or external heat-rejection fan for liquid cooling) be either turned on or ramped to a higher speed in response to a high-temperature signal from the temperature sensor.

3. Since systems fueled by virtually pure hydrogen generally run dead-ended, a timer (or coulometer) is utilized to impose a brief open-close cycle to a solenoid valve in the exit line to purge the anode compartment of accumulated impurities and water on a regular basis.

4. Other control means are provided on an application-specific basis, as appropriate.

38.8 HARDWARE AND PERFORMANCE

Activity in commercial and pre-commercial small fuel cell hardware has increased markedly over the last several years. Many companies worldwide have participated in these developments. The activity is currently distributed among several fuel cell system types.

38.8.1 PEM Fuel Cells

PEMFCs have been represented in various forms in commercial and pre-commercial hardware focusing on small fuel cell applications. Examples of these types are discussed below.

Direct-Hydrogen-Fueled PEMFCs. Compressed hydrogen has been implemented as a fuel in various small fuel cell systems; see Fig. 38.5 for example. This system (from the late 1990s) is indicative of compressed hydrogen being competitive in short-duration missions (20 h); specific energy approaching 200 Wh/kg (1000 Wh at 50 W at a total system weight of 5.22 kg) with decade-old fuel cell and hydrogen-cylinder technologies. Compressed hydrogen is now a candidate for use in certain systems that are being designed or developed. An example is a third-generation power supply for unmanned aerial vehicles (UAVs) envisioned by Protonex Technology Corporation.[8] The approach, utilizing a compressed-hydrogen cylinder, is illustrated in Figs. 38.6 and 38.7. This further suggests the viability of direct hydrogen as a fuel for small fuel cells in short-duration missions.

FIGURE 38.6 Protonex developmental high-pressure cylinder.

FIGURE 38.7 Protonex developmental 500 W PEMFC power supply fueled by pressurized hydrogen.

Indirect-Hydrogen-Fueled PEMFCs

Chemical Hydrides. Unmanned aerial vehicle (UAV) applications have also used chemical hydrides for fueling small fuel cell power sources. A Protonex UAV system operates on a sodium borohydride–based chemical hydride system. Rated at 250 W continuous power, and using a hybrid battery for peak power, it provides 1500 Wh of energy with an activated (hydrated) cartridge weight of 1.8 kg. The overall system weight is 3.0 kg, yielding a specific energy of 500 Wh/kg. This system is depicted in Fig. 38.8.[8]

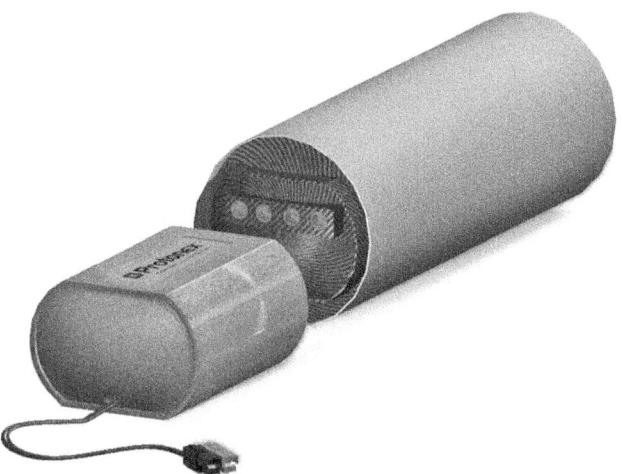

FIGURE 38.8 Protonex 250 W chemical-hydride-fueled power system for unmanned aerial vehicle applications.

Contemplating a much wider range of applications, Horizon Fuel Cell offers a 60 W continuous power system with a water-activated chemical hydride cartridge, for consumer and commercial applications. The energy content of a cartridge is 200 Wh, and its hydrated weight is 1.0 kg. The overall system weight is 3.5 kg, yielding a specific energy of 57 Wh/kg.[9]

Metal Hydrides. Metal-hydride-based hydrogen is being implemented by Angstrom Power Inc. for a fuel-cell-based mini-system to power a cell phone. The system is being applied to an existing Motorola cell phone case. The metal hydride fuel would need to be either recharged with hydrogen when spent or replaced with a fresh fuel cartridge.[10]

Horizon Fuel Cell has developed a metal-hydride-based fuel cell (5 V, 0.4 A output) for charging or extending operation for a variety of portable/mobile electronic devices (see Fig. 38.9). The fuel cartridge for the unit has an energy of 12 Wh. These can be replaced or recharged using a home refilling set that generates hydrogen from electricity and water. The overall weight of the system is 155 grams.[9]

Reformed Methanol. An UltraCell Corporation methanol-reforming fuel cell system (XX25™) delivers a continuous maximum power of 25 W. This system, designed for a variety of military and commercial portable missions, can be worn by soldiers in the field. It utilizes a methanol-water mix to feed its integrated fuel processor.[11] For a mission operating at a net power output of 20 W the system provides 900 Wh of energy in 45 h. The system weight is 1.24 kg; and, using a 900 Wh fuel cartridge weighing 1.2 kg, the overall system weight becomes 2.44 kg. The corresponding specific energy is 370 Wh/kg.

A Protonex fuel cell system (M250-CX) operating on reformed methanol with the fuel stored as a methanol-water mix delivers a continuous output power of 250 W. It serves as a battery charger as

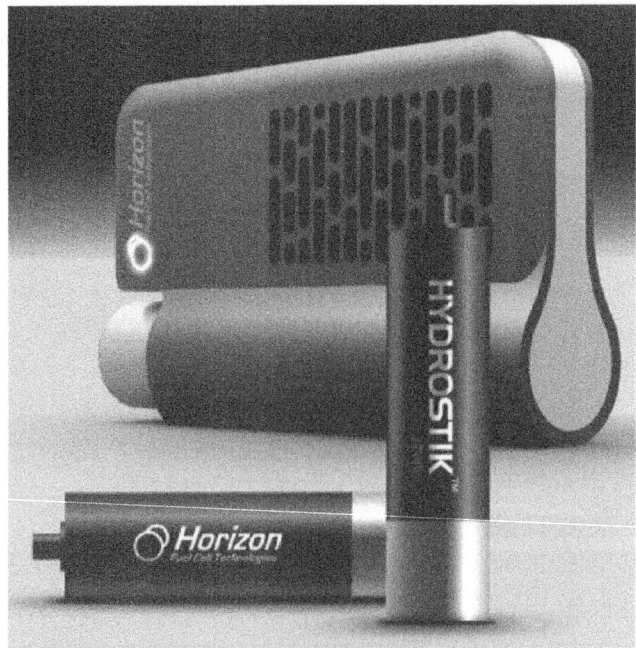

FIGURE 38.9 Horizon handheld USB charger for portable electronics fueled by metal hydride.

well as an auxiliary power unit. The system is ruggedized for military use and weighs 18 kg, including 1 L of fuel and its container. Supplemental fuel is provided to the container from an outside source; the consumption rate is 0.4 L/h at 250 W output power.[8] This system is depicted in Fig. 38.10.

A 350 W methanol-reforming fuel cell system developed by Serenergy A/S is based on a high-temperature PEM fuel cell (HTPEMFC). At this stage, the system is externally supplied with a

FIGURE 38.10 Protonex 250 W methanol-reforming system for battery charging.

FIGURE 38.11 Serenergy 350 W methanol-reforming fuel cell module.

methanol-water mixture containing 60 wt % methanol and with electricity for auxiliary needs. The unit weighs about 11 kg and is shown in Fig. 38.11; specifications based on preliminary data are seen in Table 38.3.[1]

Direct Methanol (DMFC). Direct-methanol feed systems (DMFCs) are being pursued by SFC Smart Fuel Cell AG. Applications in a variety of military and commercial systems are being addressed. A portable power system for military applications (Jenny 600S) provides a nominal power

TABLE 38.3 Serenus Model H3 E-350 Preliminary Electrical Data

Electrical characteristics (stated for beginning of life (BOL))	
Parameter	
Nominal power (P_{nom})[1] (W)	350
Peak power (W)	450
Nominal voltage (V_{DC})	23.5
Voltage range (V_{DC})	33–21.5 (spikes to 45)
Supply voltage (V_{DC})	24 ±10%
Max. supply current, start-up (A)	22
Nominal supply current, oper. (A)	2
Max. supply current, oper. (A)	3

output of 25 W and can match its output to relevant electronic devices. This unit, which can be worn on the body, weighs 1.7 kg (without fuel supply). The fuel is stored as pure methanol in cartridges, each weighing 0.37 kg and producing 400 Wh of energy. Two such cartridges can operate for 32 h with a system-specific energy of 318 Wh/kg; four cartridges provide 64 h of operation with a system-specific energy of 503 Wh/kg. The system is shown in Fig. 38.12.[2]

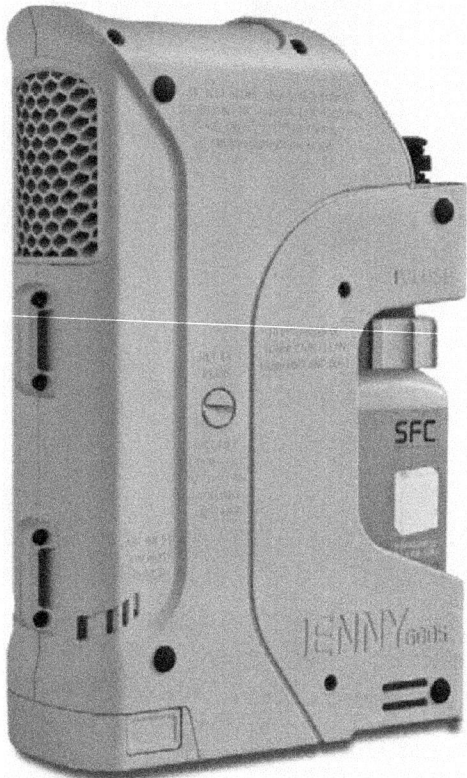

FIGURE 38.12 Jenny 600S 25 W direct-methanol fuel cell system for portable power.

Another SFC DMFC system (Emily 2200) serves as a ruggedized mobile power supply for use in on- or off-vehicle applications. Its nominal power output is 90 W, and it is typically integrated with a lead-acid battery. The fuel cell system weight is 12.5 kg (without fuel supply), and the fuel (pure methanol) can be supplied in containers delivering 5.5, 11.1, or 31.1 kWh. The 5.5 kWh container weighs 4.3 kg and can operate at rated power for 61 h, yielding a system specific energy of 327 Wh/kg. The system is depicted in Fig. 38.13.[2]

Toshiba Corporation introduced a new DMFC product in October 2009 entailing an external power source for mobile/portable digital consumer products. This 5 VDC, 400 mA device, incorporates a hybrid lithium-ion battery and was made available initially in 3000 units for the Japanese market. It features an internal cell structure that passively captures water generated at the cathode for transfer to the methanol-water anode feed stream. It is fed with highly concentrated methanol and weighs approximately 280 g (without fuel). An on-board fuel tank has a capacity of 14 ml, and it is filled using a dedicated fuel cartridge weighing 92 g with its capacity of 50 ml. The system[12] is depicted in Fig. 38.14.

FIGURE 38.13 Emily model 2200 90 W direct-methanol fuel cell system for battery-charging applications.

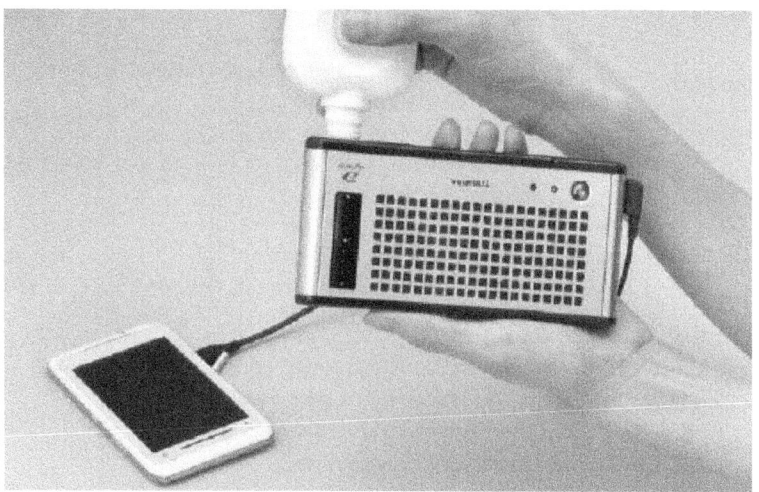

FIGURE 38.14 Toshiba direct-methanol fuel cell/battery hybrid power system for portable digital devices.

38.8.2 Solid Oxide Fuel Cells (SOFC)

Solid oxide fuel cells have not been pursued for the small fuel cell sector to the extent of PEMFCs. This is partly due to the shorter overall development span of SOFCs and partly due to their high operating temperature. Nevertheless, SOFCs are now being developed and customer evaluated for small fuel cell use. Startup time issues have been addressed, and suitability for operation on common fuels is an exploitable asset.

Adaptive Materials Inc. has developed SOFC hardware for small fuel cell applications, including a unit designed for unmanned ground and aerial vehicles.[3] The fuel cell system produces 150 W of continuous power, and it is tailored to operate in conjunction with a hybrid battery. Fueling is via a canister of propane (LPG). Protonex has established SOFCs as a target technology for use in future battery-charging systems. Fueling is expected to evolve from propane to a variety of liquid fuels.[8]

38.9 PROGNOSIS

It is clear that the pace of commercial and pre-commercial activity in small fuel cells has increased markedly in recent years. This has been fostered by ongoing technology improvements and greater recognition by potential customers of the prospective operational advantages and viability of fuel cell systems in applications of interest.

Fuel cell hardware currently being evaluated represents a variety of fuel cell technologies. These are predominantly PEM fuel cell types, but solid oxide fuel cell systems are participating, and further pursuit of such systems is anticipated. This is particularly likely where use of common fuels would have a significant advantage.

The period ahead is expected to provide far further assessment of fuel cell systems in small fuel cell applications and concomitant commercial interest. The degree of commercial success of these systems will be determined in an ongoing manner based on factors such as cost, reliability, fuel logistics, and the size of markets advantageously served within these parameters.

REFERENCES

1. Serenergy website: www.serenergy.dk; Serenergy A/S data sheet v1.0-1109.
2. www.sfc.com/en/.
3. Adaptive Materials, Inc. website: www.adaptivematerials.com.
4. S. Calabrese Barton, T. Patterson, E. Wang, T. F. Fuller, and A. C. West, *J. Power Sources*, 96, pp. 329–336, 2001.
5. www.luxfercylinders.com.
6. Quantum Technologies website: www.qtww.com.
7. M. Daugherty, D. Haberman, N. Stetson, S. Ibrahim, O. Lokken, D. Dunn, M. Cherniack, and C. Salter, *Proc. of Conference on Portable Fuel Cells,* Lucerne, Switzerland, pp. 69–78, June 21–24, 1999.
8. J. L. Martin and P. Osenar, "Portable Military Fuel Cell Systems," Abstract No. 813, the 216th Electrochemical Society Meeting, Vienna, Austria, Oct. 4–9, 2009.
9. Horizon Fuel Cell website: www.horizonfuelcell.com.
10. Angstrom Power website: www.angstrompower.com.
11. UltraCell website: www.ultracellpower.com.
12. www.toshiba.com/taec/news/press_releases/2009/dmfc_09_580.jsp.

CHAPTER 39
ELECTROCHEMICAL CAPACITORS

Andrew F. Burke

39.1 INTRODUCTION

39.1.1 Comparisons of Electrochemical Capacitors and Batteries

Electrical energy storage is required in many applications—telecommunication devices, such as cell phones and pagers, standby power systems, and electric/hybrid vehicles. The specifications for the various energy storage devices are given in terms of energy stored (Wh) and maximum power (W) as well as size and weight, initial cost, and life. To be suitable for a particular application, a storage device must meet all the requirements. As power requirements for many applications become more demanding, it is often reasonable to consider separating the energy and power requirements by providing for the peak power by using a pulse power device (capacitor) that is charged periodically from a primary energy storage unit (battery). For applications in which significant energy is needed in pulse form, traditional capacitors as used in electronic circuits cannot store enough energy in the volume and weight available. For these applications, the development of high-energy-density capacitors (ultracapacitors or electrochemical capacitors) has been undertaken by various groups around the world. This chapter considers in detail why such capacitors are being developed, how they function, and the present status and projected development of electrochemical capacitor technology.

The most common electrical energy storage device is the battery. Batteries have been the technology of choice for most applications, because they can store large amounts of energy in a relatively small volume and weight and provide suitable levels of power for many applications. Shelf and cycle life have been a problem with most types of batteries, but people have learned to tolerate this shortcoming due to the lack of an alternative. In recent times, the power requirements in a number of applications have increased markedly and have exceeded the capability of batteries of standard design. This has led to the design of special high-pulse-power batteries, often with the sacrifice of energy density and cycle life. Electrochemical capacitors are being developed as an alternative to pulse batteries. To be an attractive alternative, capacitors must have much higher power and much longer shelf and cycle life than batteries (by "much" is meant at least one order of magnitude higher). Electrochemical capacitors have much lower specific energies than batteries and their low specific energy is in most cases the factor that determines the feasibility of their use in a particular high-power application.

For capacitors, the trade-off between the energy density and the RC time constant of the device is an important design consideration. In general, for a particular set of materials, a sacrifice in energy density is required to get a large reduction in the time constant and thus a large increase in power capability. The characteristics of a number of electrochemical capacitors and pulse batteries are given in Table 39.1. Two approaches to the calculation of the peak power density are indicated in the table. The first and more standard approach is to determine the power at the so-called matched

TABLE 39.1 Comparisons of the Energy and Power Characteristics of Electrochemical Capacitors and Batteries

Device technology	Nominal cell voltage	Wh/kg	kW/kg matched impedance	W/kg 90% effic.
Carbon/carbon supercapacitors	2.7	5	10–25	2500–5000
Hybrid carbon supercapacitors	3.8	12	8–10	1635
Lithium-ion batteries				
Iron phosphate	3.25	90–115	2–4	700–1200
Lithium titanate	2.4	35–70	2–6	700–2260
NiCoMnO$_2$	3.7	95	5	1700
NiCoMnO$_2$	3.7	140	1.4	500
NiCoMnO$_2$	3.7	170	1.1	400
NiMH HEV	1.2	46	1.1	400
Lead-acid HEV	2.0	26	0.4	150
Zn-air	1.3	450	0.6–1.2	200–400

impedance condition at which one-half the energy of the discharge is in the form of electricity and one-half is in heat. The maximum power at this point is given by

$$P_{mi} = V_{oc}^2/4 R_b$$

where V_{oc} is the open-circuit voltage of the device and R_b is its resistance. The discharge efficiency at this point is 50%. For many applications in which a significant fraction of the energy is stored in the energy storage device before it is used by the system, the efficiency of the charge-discharge cycle is important to the system efficiency. In those cases, the use of the energy storage device should be limited to conditions that result in high efficiency for both charge and discharge. The discharge-charge power for a battery as function of efficiency is given by

$$P_{ef} = EF \times (1\text{-}EF) \times V_{oc}^2/R_b$$

where EF is the efficiency of the high power pulse. For EF = 0.95, $P_{ef}/P_{mi} = 0.19$. Hence in applications in which efficiency is a primary concern, the usable power of the battery is much less than the peak power (P_{mi}) often quoted by the manufacturer for the battery.

In the case of electrochemical capacitors, the peak power for a pulse at a voltage between V_0 and $V_0/2$, where V_0 is the rated voltage of the device, is given by

$$P_{uc} = 9/16 \times (1\text{-}EF) \times V_0^2/R_{uc}$$

where R_{uc} is the resistance of the capacitor. Peak power values are shown in Table 39.1 for both matched impedance and high efficiency discharges of the batteries and capacitors. It is apparent that in nearly all cases the power from the electrochemical capacitors is much higher than that from the batteries. Note that it is not correct to compare the high efficiency power density for the capacitors with the matched impedance power density for the batteries, as is often done. The power capability of both types of devices is primarily dependent on their resistance, and knowledge of the resistance is key to determining the peak usable power capability. Hence, measurement of the resistance of a device in the pulsed mode of operation is critical to an evaluation of its high-power capability.

In addition to high-power capability, the other reason for considering electrochemical capacitors for a particular application is their long shelf and cycle life. This is especially true of capacitors using activated carbon electrodes. Most secondary (rechargeable) batteries if left on the shelf unused for many months will degrade markedly and be essentially useless after this time due to self-discharge and corrosion effects. Electrochemical capacitors will self-discharge over a period of time to low voltage, but they will retain their capacitance and thus be capable of recharge to their original condition. Experience has shown that capacitors can be unused for several years and remain in nearly their original condition. Electrochemical capacitors can be deep cycled at high rates (discharge times of seconds) for 500,000 to 1,000,000 cycles at room temperature with a relatively small change in characteristics (10 to 20% degradation in capacitance and resistance). This is not possible with batteries even if the depth of discharge is kept small (10 to 20%). The life of the electrochemical capacitors is significantly less at high temperatures (> 50°C).

Hence, relative to batteries, the advantages of electrochemical capacitors as pulse power devices are high power density, high efficiency, and long shelf and cycle life. The primary disadvantage of capacitors is their relatively low energy density (Wh/kg and Wh/L) compared to batteries, limiting their use to applications in which relatively small quantities of energy are required before the capacitor can be recharged. Electrochemical capacitors can, however, be recharged in a very short time (seconds or fraction of seconds) compared to batteries if a source of energy is available at the high power levels required.

39.1.2 Energy Storage in Electrochemical Capacitors

The most common electrical energy storage devices are capacitors and batteries. Capacitors store energy by charge separation. The simplest capacitors store the energy in a thin layer of dielectric material that is supported by metal plates that act as the terminals for the device. The energy stored in a capacitor is given by $\frac{1}{2} CV^2$, where C is its capacitance (farads) and V is the voltage between the terminal plates. The maximum voltage of the capacitor is dependent on the breakdown characteristics of the dielectric material. The charge Q (coulombs) stored in the capacitor is given by CV. The capacitance of the dielectric capacitor depends on the dielectric constant (K) and the thickness (th) of the dielectric material and its geometric area (A).

$$C = KA/\text{th}$$

In a battery, energy is stored in chemical form as active material in its electrodes. Energy is released in electrical form by connecting a load across the terminal of the battery, permitting the electrode materials to react electrochemically with the ions required in the reactions transferred through the electrolyte in which the electrodes are immersed. The usable energy stored in the battery is given as VQ, where V is the voltage of the cell and Q is the electrical charge (I_t) transferred to the load during the chemical reaction. The voltage is dependent on the active materials (chemical couple) of the battery and is close to the open-circuit voltage (V_{oc}) for those materials.

An electrochemical capacitor, sometimes referred to as an ultracapacitor, is an electrical energy storage device that is constructed much like a battery (see Fig. 39.1) in that it has two electrodes immersed in an electrolyte with a separator between the electrodes. The electrodes are fabricated from a high-surface-area, porous material having pores with diameters in the nanometer (nm) range. The surface area of the electrode materials used in an electrochemical capacitor is much greater than that used in battery electrodes, being 500 to 2000 m^2/gm. Charge is stored in the micropores at or near the interface between the solid electrode material and the electrolyte. The charge and energy stored are given by the same expressions as cited previously for the simple dielectric capacitor. However, calculation of the capacitance of the electrochemical capacitor is much more difficult as it depends on complex phenomena occurring in the micropores of the electrodes.

It is convenient to discuss the mechanisms for energy storage in electrochemical capacitors in terms of double-layer and pseudo-capacitance processes separately. The physics and chemistry of these processes as they apply to electrochemical capacitors have been explained in great detail in References 1–3. In the following sections, the mechanisms are discussed briefly in terms of how they relate to the properties of the electrode materials and electrolyte.

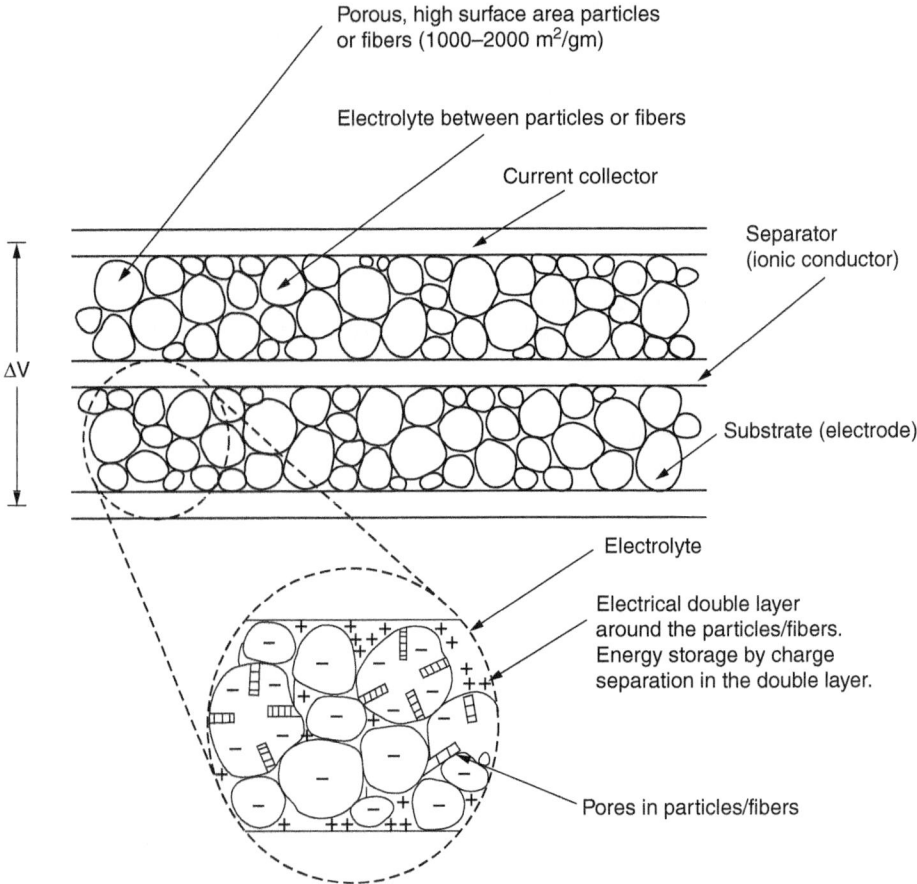

FIGURE 39.1 Schematic of an electrochemical capacitor.

Double-Layer Capacitors. Energy is stored in the double-layer capacitor as charge separation in the double layer formed at the interface between the solid electrode material and the liquid electrolyte in the micropores of the electrodes. A schematic of an ultracapacitor is shown in Fig. 39.1. The ions displaced in forming the double layers in the pores are transferred between the electrodes through the electrolyte. The energy and charge stored in the electrochemical capacitor are $\frac{1}{2}CV^2$ and CV, respectively. The capacitance is dependent primarily on the characteristics of the electrode material (surface area and pore size distribution). The specific capacitance of an electrode material can be written as

$$C/\text{gm} = (F/\text{cm}^2)_{\text{act}} \times (\text{cm}^2/\text{gm})_{\text{act}}$$

where the surface area referred to is the active area in the pores on which the double layer is formed. In simplest terms, the capacitance per unit of active area is given by

$$(F/\text{cm}^2)_{\text{act}} = (K/\text{thickness of the double layer})_{\text{eff}}$$

As discussed in Reference 1, determination of the effective dielectric constant K_{eff} of the electrolyte and the thickness of the double layer formed at the interface is complex and not well understood.

TABLE 39.2 Specific Capacitance for Various Electrode Materials

Material	Density (g/cm^3)	Electrolyte	F/g	F/cm^3
Carbon cloth	0.35	KOH	200	70
		Organic	100	35
Activated carbon	0.7	KOH	160	112
		Organic	100	70
Aerogel carbon	0.6	KOH	140	84
Particulate carbon from SiC	0.7	KOH	175	122
		Organic	100	70
Particulate carbon from TiC	0.5	KOH	220	110
		Organic	120	60
Advanced graphitic carbon	0.7	Organic	180	126
Anhydrous RuO$_2$	2.7	Sulfuric acid	150	405
Hydrous RuO$_2$	2.0	Sulfuric acid	650	1300
Doped conducting polymer	0.7	Organic	450	315

The thickness of the double layer is very small (a fraction of a nm in liquid electrolytes), resulting in a high value for the specific capacitance of 15 to 30 µF/cm^2. For a surface area of 1000 m^2/gm, this results in a potential capacitance of 150 to 300 F/gm of electrode material. As indicated in Table 39.2, the measured specific capacitances of carbon materials being used in ultracapacitors are in most cases less than these high values, being in the range of 75 to 175 F/gm for aqueous electrolytes and 40 to 100 F/gm using organic electrolytes because, for most carbon materials, a relatively large fraction of the surface area is in pores that cannot be accessed by the ions in the electrolyte. This is especially true for the organic electrolytes for which the size of ions is much larger than in an aqueous electrolyte. Porous carbons for use in ultracapacitors should have a large fraction of their pore volume in pores of diameter 1 to 5 nm. Materials with small pores (< 1 nm) show a large fall-off in capacitance at discharge currents greater than 100 mA/cm^2, especially using organic electrolytes. Materials with the larger pore diameters can be discharged at current densities of greater than 500 mA/cm^2 with a minimal decrease in capacitance.

The cell voltage of the ultracapacitor is dependent on the electrolyte used. For aqueous electrolytes, the cell voltage is about 1 V and for organic electrolytes, the cell voltage is 3 to 3.5 V.

Electrochemical Capacitors Utilizing Pseudo-Capacitance. For an ideal double-layer capacitor, the charge is transferred into the double layer and there are no faradaic reactions between the solid material and the electrolyte. In that case, the capacitance (dQ/dV) is a constant and independent of voltage. For devices that utilize pseudo-capacitance, most of the charge is transferred at the surface or in the bulk near the surface of the solid electrode material. Hence, in this case, the interaction between the solid material and the electrolyte involves faradaic reactions which in most instances can be described as charge transfer reactions. The charge transferred in these reactions is voltage dependent, resulting in the pseudo-capacitance ($C = dQ/dV$) also being voltage dependent. Three types of electrochemical processes have been utilized in the development of ultracapacitors using pseudo-capacitance. These are surface adsorption of ions from the electrolyte, redox reactions involving ions from the electrolyte, and the doping and undoping of an active conducting polymer material in the electrode. The first two processes are primarily surface mechanisms and are hence highly dependent on the surface area of the electrode material. The third process involving the conducting polymer material is more of a bulk process, and the specific capacitance of the material is much less dependent on its surface area although relatively high surface area with micropores is required to distribute the ions to and from the electrodes in a cell. In all cases, the electrodes must have high electronic conductivity to distribute and collect the electron current. An understanding of the charge transfer mechanism can be inferred from $C(V)$, which is often determined using cyclic voltammetry.

For assessing the characteristics of devices, it is convenient to use the average capacitance (C_{av}) calculated from

$$C_{av} = Q_{tot}/V_{tot}$$

where the Q_{tot} and V_{tot} are the total charge and voltage change for a charge or discharge of the electrode. This permits a determination of the specific capacitance (C_{av}/gm) of the material for the electrolyte of interest. As shown in Table 39.2, the specific capacitance of pseudo-capacitance materials are much higher than that of carbon materials. It is thus expected that the energy density of devices developed using the pseudo-capacitance materials will be higher.

Hybrid Capacitors. Electrochemical capacitors can be fabricated with one electrode being of a double-layer (carbon) material and the other electrode being of a pseudo-capacitance material (see Fig. 39.2). Such devices are often referred to as *hybrid capacitors*. Most of the hybrid capacitors developed to date have used metal oxides (for example, lead or nickel oxide) as the pseudo-capacitance material in the positive electrode. The energy density of these devices can be significantly higher than for double-layer capacitors, but as shown in Fig. 39.3, their charge-discharge characteristics (V versus Q) are very non-ideal (nonlinear). Hybrid capacitors can also be assembled using two nonsimilar mixed metal oxides or doped conducting polymer materials.

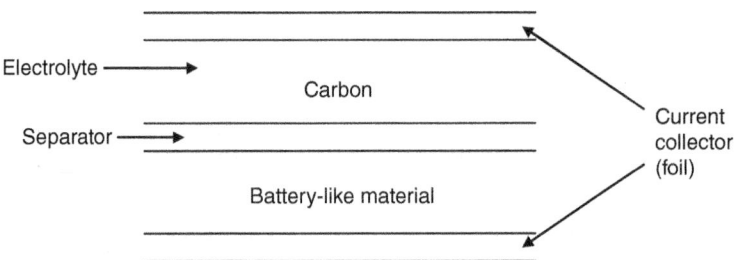

FIGURE 39.2 Schematic of a hybrid electrochemical capacitor.

39.2 CHEMISTRY AND MATERIAL PROPERTIES

39.2.1 Activated Carbon

The electrodes in electrochemical capacitors are in general thin coatings applied to a metallic current collector. The active material is mixed with a binder to form a slurry that can be applied at a controlled thickness, rolled, and dried to form a porous electrode. The thickness of the electrode is generally in the range of 100 to 300 microns and it has a porosity of 65 to 80%. In order to achieve a low resistance, the contact resistance between the active material coating and the current collector must be very small. This requires special attention to preparing the surface of the current collector before applying the electrode coating.[4,5] As noted previously, a key electrode material property is its specific capacitance (F/g). The capacitance of an electrode of known geometric dimensions (t, A_x) can be calculated directly with good accuracy from its specific capacitance and density (C = F/g × density × t × A_x).

As indicated in Table 39.2, the specific capacitance of activated carbon can vary over a wide range (50 to 220 F/g), depending on how it is processed and the electrolyte used in the cell. The density of the carbon can vary from 0.3 to 0.8 g/cm³. The specific capacitance of the carbon depends on its surface area (m²/g), pore size distribution, and intrinsic surface double-layer capacitance (μF/cm²).

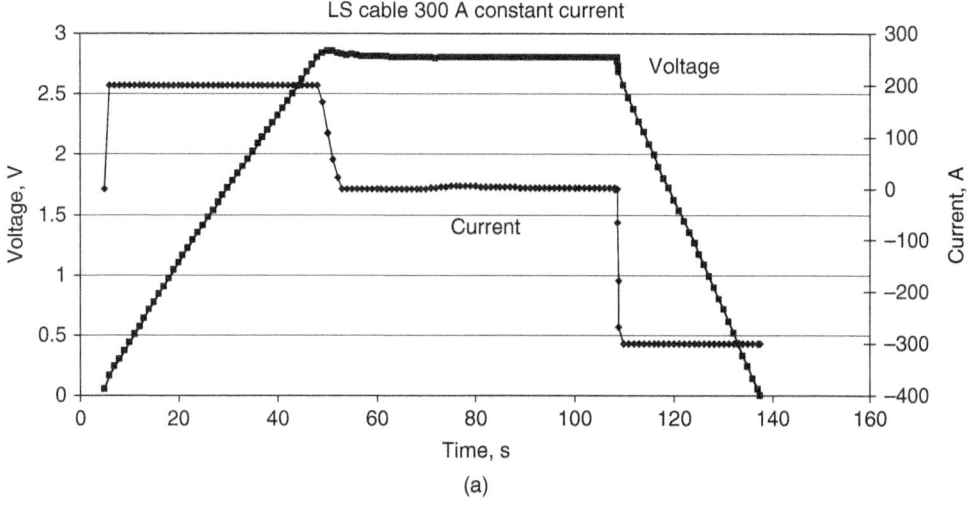

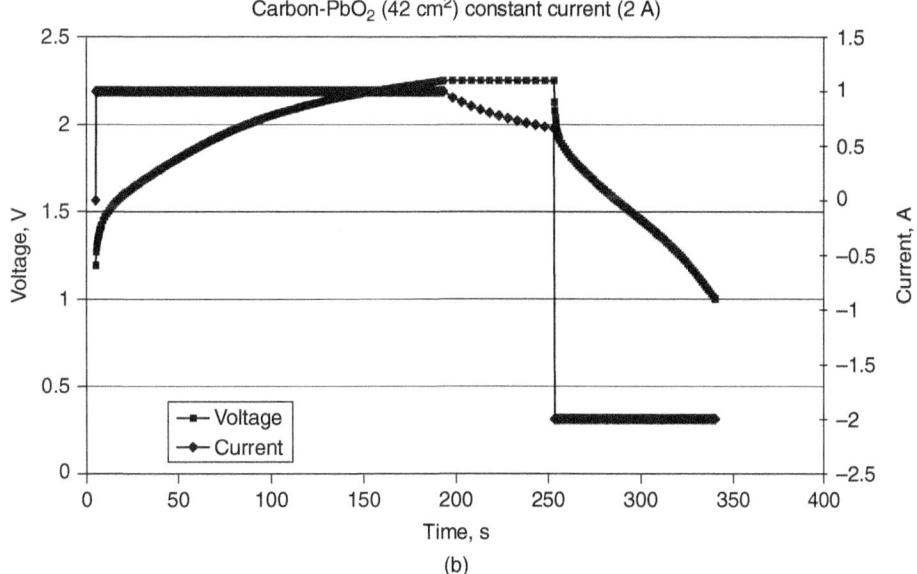

FIGURE 39.3 (a) Voltage vs. current trace for a EDLC; (b) voltage vs. current trace for a hybrid capacitor.

The maximum specific capacitance of a particular carbon can be estimated using the following relationship:

$$(F/g)_{max} = (m^2/g) \times (\mu F/cm^2) \times 10^{-2}$$

For example, if the surface area were 1500 m²/g and µF/cm² = 20, the maximum estimated specific capacitance would be 300 F/g. As indicated in Table 39.2, the measured specific capacitance of activated carbon is much lower (typically 15 to 35%) than the calculated maximum value. The reasons for this discrepancy are the subject of considerable debate,[6-8] and understanding it is important because it has a large effect on the types and characteristics of carbon materials best suited for use in electrochemical capacitor electrodes. It is well recognized that the pore size distribution can have a large effect on the specific capacitance of the carbon and that it is essential to properly match the solvated ion size of the electrolyte ions and the pore size of the carbon. The physical process by which the ions diffuse in and out of the pores is not well understood, making the correlation of specific capacitance and carbon surface area very uncertain. In fact, some authors[8,9] assert that the micropores have little influence on specific capacitance and that increasing the external roughness (rugosity) of the carbon particles is key to achieving high specific capacitance. Research on optimizing carbons for double-layer capacitors is critical to achieving large improvements in the energy density of carbon/carbon devices.

The specific capacitance of the electrode materials can also vary significantly with current density (A/cm^2). In evaluating materials for electrochemical capacitors, the specific capacitance of the active material should be measured for current density up to at least 300 mA/cm^2, using relatively thin electrodes (less than 200 microns). The most direct method for evaluating active materials is to form thin electrodes from the material and perform constant-current tests of small cells assembled from the electrodes (see Table 39.3).

TABLE 39.3 Effect of Current Density on the Specific Capacitance (F/g)

I(A)	mA/cm^2	C(F)	R(ohm)	Ohm-cm^2	(F/g) dry electrode
0.2	66	5.72	0.123	0.37	163
0.3	100	5.58	0.151	0.45	159
0.5	167	5.3	0.120	0.36	151
0.75	250	4.96	0.144	0.43	142
1.0	333	4.80	0.164	0.49	153

Lab cells, area–3 cm^2, sulfuric acid electrolyte, electrode thickness 200μ. (Ref. 10.)

39.2.2 Advanced Carbons

As indicated in Eq. (39.1), the energy density of an electrochemical capacitor depends on both the specific capacitance (F/g) of the carbon and the maximum voltage that can be used in the cell. Research[11-13] is being done to move beyond activated carbons to particulate graphitic carbons that have specific capacitance approaching 200 F/gm and cell voltages significantly greater than 3 V (3.3 to 3.5 V with an organic electrolyte). There is considerable disagreement concerning the energy storage mechanisms in these advanced carbons. It seems clear that the mechanisms are more complex than simple double-layer formation and likely involve either surface charge transfer or intercalation of the electrolyte ions into the porous carbon structure. These carbons appear to be promising materials for electrochemical capacitor electrodes, and research related to their preparation and the understanding of the associated energy storage mechanisms is continuing.

A different approach than that of particulate carbon is the use of carbon nanotubes to form the electrodes. This can be done by growing a bed of nanotubes normal to the current collector surface, using a gaseous hydrocarbon feedstock from catalytic sites.[14,15] Considerable progress has been made in recent years in the preparation of this type of carbon material suitable for electrochemical capacitors. The key issue concerning the attractiveness of the carbon nanotubes is their intrinsic specific capacitance. To date, the data available[16] indicate that the specific capacitance is not sufficiently high to permit the assembly of devices of significantly higher energy density than achievable using less costly forms of carbon. Nevertheless, much further research in this area is continuing.

39.2.3 Metal Oxides

Good examples of pseudo-capacitive materials are metal oxides such as ruthenium and manganese oxides (RuO_2 and MnO_2). The charge and energy storage in the material is due to redox, charge transfer reactions at the surface of the metal oxide particles. Electrochemical capacitor devices have been assembled using metal oxides[17–20] in both the negative and positive electrodes, but little information is given concerning the materials used for proprietary reasons. The metal oxides are usually combined with carbon to form a conducting, porous electrode structure.[20] Devices have been assembled using both aqueous and organic electrolytes with cell voltages from 1 to 3.5 V. The specific capacitance of the metal oxides can be as high as 500 to 1000 F/g, but when combined with carbon to form composite electrodes, the specific capacitance of the composite is much lower, 200 to 300 (see Table 39.2). In recent research, metal oxides have been utilized to enhance the intrinsic specific capacitance of carbon nanotubes.[21] One of the key issues with pseudo-capacitive materials is their stability in cycling and the resultant effects on the cycle life of devices assembled using those materials. Testing of cells/devices utilizing metal oxides in at least one of the electrodes has indicated that the energy density of such devices is significantly higher than those using only carbons. High-power cells with specific energies of 10 to 15 Wh/kg have been tested.[17–19] The available data indicate that this is a promising area for materials research for the development of high specific energy electrochemical capacitors.

These materials can also be used in hybrid capacitors that utilize double-layer or pseudo-capacitive material in one electrode and a battery-like, faradaic material in the other electrode.[22–25] The hybrid capacitors may have specific energies greater than that of lead-acid batteries with much higher power and long cycle life of hundreds of thousands of cycles. Research on these types of devices is continuing and is highly dependent on advances in electrode and current collector materials.

39.2.4 Current Collector Materials

Energy storage devices in nearly all cases require the application of a thin film of active material on to a high-conductivity current collector. The key issues related to the current collector are a near-zero contact resistance with the thin film and the long-term stability of the material (metal or conducting plastic) or its coating in the environment of the cell (voltage and electrolyte). These issues are particularly important for electrochemical capacitors because of their very low resistance, cycle life of hundreds of thousands of cycles, and 10 to 15 years calendar life. Of special interest is research dealing with cleaning and coating metal foils[4,5,26] and conducting plastic sheets[27] that can be used for bipolar capacitors and batteries.

39.2.5 Electrolytes

The capacitance of an electrochemical capacitor is dependent primarily on the specific capacitance (F/g) of the electrode material, but the cell voltage and resistance are primarily dependent on the electrolyte used in the device. Three types of electrolytes have been used in electrochemical capacitors: aqueous (sulfuric acid and KOH), organic (propylene carbonate and acetonitrile), and (recently) ionic liquids. Salts are added to the organic electrolytes to provide the ions that move in and out of the double layers formed in the micropores of the carbon. The characteristics of these electrolytes are given in Table 39.4. Detailed discussions of various combinations of electrolyte solvents and salts are given in References 28–30.

TABLE 39.4 Properties of Various Electrolytes

Electrolyte	Density (gm/cm^3)	Resistivity (ohm-cm)	Cell voltage
KOH (aqueous)	1.29	1.9	1.0
Sulfuric acid (aqueous)	1.2	1.35	1.0
Propylene carbonate (PC)	1.2	52	2.5–3.0
Acetonitrile (AN)	0.78	18	2.5–3.0
Ionic liquid	1.3–1.5	125 (25°C)	4.0
		28 (100°C)	3.25

There are large differences in the ionic resistivity and cell voltage (usable electrochemical window) of the electrolytes. These differences lead to corresponding large differences in the performance of devices using the various electrolytes. Since the specific energy is proportional to the square of the cell voltage, increasing the cell voltage is a key objective of electrochemical capacitor research. Using activated carbon, the cell voltage is 2.3 to 2.7 V using organic electrolytes and 0.8 to 1.0 V/cell using aqueous electrolytes. The cell voltage is also dependent to a limited extent on the carbon used in the device. Cell voltages up to 3.5 V/cell have been reported with structured graphitic carbons.[11–13] The differences in the ionic resistivity of the electrolytes have a large effect on resistance and consequently the power capability of a device. The resistivity of propylene carbonate is about a factor of 3 higher than that of acetonitrile. For this reason, electrochemical capacitors with the best performance (highest specific energy and power capability) use acetonitrile as the electrolyte. There is a continuing controversy[31,32] concerning the safety of acetonitrile, especially in vehicles, because of its toxicity and flammability. There has been much research to develop a low resistivity, nontoxic electrolyte to replace acetonitrile, but to date that work has not succeeded.

Research[33–35] is being done to develop electrochemical capacitors using ionic liquid electrolytes. These electrolytes are attractive for several reasons. First, they can be thermally stable for temperatures as high as 300°C with near-zero vapor pressure and are nonflammable with very low toxicity. Further, the usable electrochemical window is large, leading to cell voltages as high as 4 V with some carbons. The major difficulties with the ionic liquids are their high ionic resistivity at near–room temperature and high cost. The resistivity of an ionic liquid is strongly temperature sensitive and requires a temperature of about 125°C to have a resistivity comparable to that of acetonitrile. Blending ionic liquids with acetonitrile[34] can greatly reduce the flammability of the electrolyte with only minor changes in room-temperature conductivity and the voltage window. However, acetonitrile-free blends, which are both nonflammable and nontoxic, show large increases in resistivity and a 0.5 to 1.0 V reduction in the voltage window.

39.3 PERFORMANCE CHARACTERISTICS OF DEVICES

39.3.1 Small Carbon/Carbon Devices (Capacitance < 10.F)

Most of the small electrochemical capacitor units sold are modules of two or more cells in series having a voltage of 4 to 8 V. The capacitance of the units is < 10.F with a time constant of 1 to 10 ms. Devices using both aqueous and organic electrolytes are available. As expected, the specific energy of the devices using organic electrolytes is higher than those using aqueous electrolytes. The small units are available as coin cells (thin cylindrical disks) and thin, prismatic cells (credit card–like). Most of the small devices are used in consumer electronics such as pagers, cell phones, and computers in conjunction with batteries either in power-assist or battery backup modes. The price of the small devices is relatively low, about 50 cents/unit, but still much higher than the traditional ceramic capacitors. Requirements for these small units are usually given in terms of

TABLE 39.5 Physical and Performance Characteristics of Small Double-Layer Devices

V	Cap (F)	R (mΩ)	Wgt. (gm)	Vol. (cm³)	RC (ms)	Wh/L	Wh/kg	kW/L mat. imp.*	kW/L 95% eff.*
2.4	0.18	45	0.6	0.44	8.1	0.25	0.18	73	8.2
2.4	0.3	34	0.9	0.6	10.2	0.3	0.2	70.5	7.9
2.4	0.65	18	—	0.93	11.7	0.42	—	86	9.7
2.4	1.1	26	—	0.8	28.6	0.825	—	69	7.8
2.4	2.3	28	1.2	1.02	64	1.35	1.15	50.5	5.7
2.4	4.0	22	1.5	1.4	88	1.7	1.6	47	5.3
2.7†	10.5	25	2.5	1.5	262	4.8	2.9	29.2	3.3
2.7†	15	30	4.15	2.83	438	3.6	2.5	14.6	1.65
7‡	0.047	120	—	2.2	5.6	0.11	—	46.4	5.2
4.2‡	0.022	200	—	1.1	4.4	0.04	—	20.0	2.5

*Pulse mat. imp. power $V_0^2/4R$, 95% eff. power $(9/16)(1\text{-}EF)\, V_0^2/R$, $EF = 0.95$.
†Cylindrical devices, all others are flat credit card type devices. (Refs. 35–38.)
‡Multicell devices using aqueous electrolyte; all other single-cell devices use an organic electrolyte.

capacitance, RC time constant, and volume or thickness for the prismatic cells. Specific energy is usually of secondary importance, with the ability to provide periodic, multimillisecond pulses being of primary concern. In order to provide the short pulses, the time constant of the device should be < 50 ms.

As indicated in Table 39.5, small devices with time constants in the 10 to 200 ms range have been developed by several manufacturers.[35–38] Testing of the small units[39] has indicated that they can be pulse charged and discharged with pulse widths as short as 1/50th of the RC time constant without a significant reduction in the effective capacitance of the device. These devices respond as near-ideal devices (constant capacitance and resistance) as long as the charge and discharge periods of pulsing are 5 to 10 times longer than the RC time constant of the device. As indicated in Table 39.5, the small devices have very high power density capability (greater than 10 kW/L). These devices can provide pulse currents of several amps at high efficiency (greater than 90%). However, the specific energy (Wh/kg) of the small devices is relatively low, in the range of 0.1 to 1.0 Wh/kg, depending on size and electrolyte being used. As noted previously, specific energy is of secondary importance; short pulse capability with high power is the critical requirement.

39.3.2 Large Carbon/Carbon Devices (Capacitance > 100 F)

Carbon/carbon electrochemical capacitor devices (single cells and modules) are commercially available from a number of companies (Maxwell, Panasonic, Ness, Nippon Chem-Con, Power Systems). All these companies market large devices with capacitance of 1000 to 5000 F. The carbon/carbon technology is the most suitable for vehicle applications because of its high power and long cycle life. The performance of cells from the various manufacturers is given in Table 39.6. The specific energies (Wh/kg) shown correspond to the usable energy from the devices based on constant-power-discharge tests from V_0 to $\frac{1}{2}V_0$. Peak power densities are given for both matched impedance and 95% efficiency pulses. For most applications with ultracapacitors, the high-efficiency power density is the appropriate measure of the power capability of the device. The specific energy for most of the available devices is between 3.5 to 4.5 Wh/kg, and 95% specific power is between 800 to 1200 W/kg. In recent years, the specific energy of the devices has been gradually increased for the carbon/carbon (double-layer) technology, and the cell voltages have increased to 2.7 V/cell using acetonitrile as the electrolyte. As indicated in Table 39.6, both the specific energy and power capability are lower for cells using propylene carbonate as the electrolyte.

TABLE 39.6 Summary of the Performance Characteristics of Ultracapacitor Devices

Device	V rated	C (F)	R (mΩ)	RC (sec)	Wh/kg*	W/kg (95%)†	W/kg mat. imp.	Wgt. (kg)	Vol. (L)
Maxwell[‡]	2.7	2885	0.375	1.08	4.2	994	8836	0.55	0.414
Maxwell	2.7	605	0.90	0.55	2.35	1139	9597	0.20	0.211
Skeleton Technology	2.8	1600	1.1	1.8	5.85	930	8278	0.223	0.13
APowerCap[§]	2.7	55	4	0.22	5.5	5695	50,625	0.009	—
APowerCap[§]	2.7	450	1.4	0.58	5.89	2574	24,595	0.057	0.045
Ness	2.7	1800	0.55	1.00	3.6	975	8674	0.38	0.277
Ness	2.7	3640	0.30	1.10	4.2	928	8010	0.65	0.514
Ness (cyl.)	2.7	3160	0.4	1.26	4.4	982	8728	0.522	0.38
Asahi Glass (propylene carbonate)	2.7	1375	2.5	3.4	4.9	390	3471	0.210 (estimated)	0.151
Panasonic (propylene carbonate)	2.5	1200	1.0	1.2	2.3	514	4596	0.34	0.245
LS Cable	2.8	3200	0.25	0.80	3.7	1400	12,400	0.63	0.47
BatScap	2.7	2680	0.20	0.54	4.2	2050	18,225	0.50	0.572
Power Sys. (activated carbon, propylene carbonate)[§]	2.7	1350	1.5	2.0	4.9	650	5785	0.21	0.151
Power Sys. (graphitic carbon, propylene carbonate)[§]	3.3	1800	3.0	5.4	8.0	486	4320	0.21	0.15
	3.3	1500	1.7	2.5	6.0	776	6903	0.23	0.15
Fuji Heavy Industry-hybrid (AC/graphitic carbon)[§]	3.8	1800	1.5	2.6	9.2	1025	10,375	0.232	0.143
JSR Micro (AC/graphitic carbon)[§]	3.8	1000	4	4	11.2	900	7987	0.113	0.073
		2000	1.9	3.8	12.1	1038	9223	0.206	0.132

*Specific energy at 400 W/kg constant power, at V rated-1/2 V rated.
†Power based on $P = 9/16 \times (1-EF) \times V_r^2/R$, EF = efficiency of discharge.
‡Except where noted, all the devices use acetonitrile as the electrolyte.
§All devices except those with § are packaged in metal containers; these other devices are in laminated pouches.
AC = activated carbon.

For vehicle applications, the cells are connected in series to form higher voltage modules. The module voltage utilized depends on the application and varies from 16 V to about 60 V. The characteristics of modules from several companies are summarized in Table 39.7. The electrical characteristics (capacitance and resistance) of the module follow directly from the cell characteristics. Note, however, that the weight and volume of the modules are significantly greater than the cells alone, with packaging factors of 0.6 to 0.7. All the modules being marketed utilize balancing circuits for each cell to prevent overvoltage of the cell and to minimize cell-to-cell variability during cycling. For this reason, it is best to base the energy storage and power capacity of the modules on the cell

TABLE 39.7 Summary of the Characteristics of Ultracapacitor Modules

Module	Weight/volume (kg/L)	Voltage	Wh/(Wh/kg)	Power (kW) (90% eff.)	Weight packaging factor	Volume packaging factor
Ness (100 F)	9.1/7.22	48	22.5/2.47	10.8	0.769	0.692
Maxwell (145 F)	13.5/13.4	48	36/2.7	14.5	0.627	0.484
Maxwell (430 F)	5.0/4.85	16	11.8/2.36	4.8	0.564	0.445
Power Systems	4.4/4.8	32	11/2.5	2.5	0.573	0.375
Power Systems	7.2/8	59	20/2.78	4.7	0.642	0.413

weight and volume, but to include the packaging factors in determining the weight and volume of the capacitor unit to be installed in a vehicle. More detailed discussions of module characteristics and cell balancing are given in Sec. 39.6.3.

39.3.3 Performance of Cells Using Advanced Materials and Device Design

The devices listed in Table 39.7 are for the most part commercially available and represent the state of the art for AC (activated carbon)/AC (activated carbon) devices. Work is continuing to develop on electrochemical capacitors with higher specific energy and/or higher power capability than those commercially available. The characteristics of some of the advanced devices are given in Table 39.8. The hybrid capacitors have higher specific energy than the activated carbon devices and, in general, significantly lower power capability.

TABLE 39.8 Performance of Advanced Electrochemical Capacitor Devices of Various Technologies

Technology type	Developer	Status	V rated	Capacitance (F)	RC (s)	Wh/kg*	W/kg 95%[†]
Carbide-based carbon	Skeleton Technology	Prototype	2.8	1600	1.8	5.8	1024
Carbon/carbon	APowerCap	Prototype[‡]	2.7	55	0.22	5.5	5695
Carbon/carbon	APowerCap	Prototype[‡]	2.7	450	0.58	5.8	2574
Advanced graphitic carbon	Power Systems	Prototype	3.3	1800	5.4	8.0	825
Advanced carbon (intercalat.)	Fuji Heavy Industries	Prototype	3.8	1800	2.6	9.2	1025
Hybrid carbon/PbO$_2$	UC Davis	Lab[‡]	2.2	13	2.8	9.7	1300
Hybrid metal oxide	IRMA	Lab[‡]	3.35	56	6.2	13	530
Hybrid act. carbon/Li titanate	Rutgers/Telcordia	Lab[‡]	2.8	500	5.5	10.4	460

*Usable specific energy.
[†]Power density P = 9/16 × (1−EFF) × V^2/R, EFF = 0.95.
[‡]Unpackaged, weight of all active materials in the device.

39.4 ELECTROCHEMICAL CAPACITOR MODELING

This section is concerned with various approaches to modeling electrochemical capacitors. Some of the modeling is semi-empirical in character (Sec. 39.4.1), and other approaches are more mathematical, but all the approaches are dependent on knowledge of the properties of the materials used in the electrodes of the devices being modeled.

39.4.1 Equivalent Circuit and AC Impedance

As discussed in previous sections, an electrochemical capacitor is assembled with porous electrodes consisting of a microporous material such as activated carbon. The capacitance and thus the electrical energy storage take place in the double layers formed in the micropores of the carbon. It is convenient to model these complex processes in terms of equivalent electrical circuits as indicated in Fig. 39.4. As shown in the figure, the circuit consists of multiple RC elements connected in a ladder which accounts for the transport of the ions into the double layers along the length (depth) of the micropores.

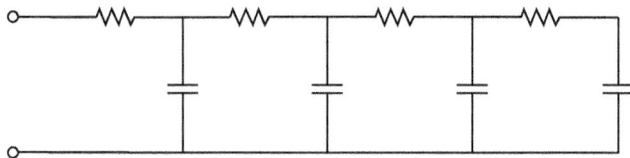

FIGURE 39.4 Equivalent circuit, multiple RC elements in series.

The simplest circuit for an electrochemical capacitor is a single RC element for which the charge and discharge response can be written as

$$\text{Charging:} \quad V/V_0 = 1 - \exp(t/RC)$$

$$\text{Discharging:} \quad V/V_0 = \exp(t/RC)$$

where V_0 is the rated voltage and RC is the time constant of the device.

For the simple RC circuit, the voltage changes by 36.8% of V_0 in time equal to one time constant and about 98% of the rated voltage in $t = 4RC$.

Experience has shown that for applications in which significant changes in power demand occur over a few time constants, the single RC element model predicts with reasonable accuracy the response of electrochemical devices.[40,41] For other applications involving more rapid power changes, an equivalent circuit consisting of multiple RC elements is needed. The analysis of the response of this circuit involves the application of the concept of complex impedance from AC circuit theory. The impedance $Z(\omega)$ is defined as

$$Z(\omega) = V(\omega)/I(\omega) = Z'(\omega) + jZ''(\omega), \quad |Z| = (Z'^2 + Z''^2)^{1/2}$$

where $V(\omega) = v'(\omega) + jv''(\omega)$, $i(\omega) = i'(\omega) + ji''(\omega)$. $j = \sqrt{-1}$

The impedance can be treated similar to the resistance in DC circuits. For circuit elements in series,

$$Z = Z_1 + Z_2$$

For those in parallel,

$$1/Z = 1/Z_1 + 1/Z_2$$

The circuit elements of interest in Fig. 39.4 are capacitors C and resistors R. The corresponding impedance relationships are

$$Z_C = -j/\omega C$$

$$Z_R = R$$

The impedance of a capacitor and resistance in series is

$$Z_{RC} = R - j/\omega C, \quad \omega \ggg 1, Z = R$$

and in parallel

$$1/Z = 1/R + 1/(-j/\omega C)$$

$$Z_P = (R - jR^2 \, C\omega)/(1 + R^2 C^2 \omega^2)$$

The impedance Z_{ladder} of the RC ladder circuit shown in Fig. 39.4 can be expressed as a combination of terms of the forms Z_{RC} and Z_p. Hence,

$$Z_{ladder}(\omega) = F(\omega, R_1, R_2, ..., C_1, C_2, ...)$$

If the device is modeled by a simple RC circuit, the values for R and C can be determined from DC constant-current tests (see Sec. 39.5.2). However, if the device is modeled using the ladder circuit, the multiple R and C values can be determined from AC impedance testing[42,43] in which an AC voltage is applied to a device and its impedance is measured as a function of frequency ω. The test results are usually presented as Z'' versus Z', C versus frequency, and R versus frequency. Typical AC impedance test data[44] for a carbon/carbon double-layer 100 F capacitor are shown in Fig. 39.5. Software[45] is available to determine the R and C values in the ladder equivalent circuit directly from the AC impedance data. The equivalent circuit results in Fig. 39.5 indicate that it is likely in most cases that a ladder consisting of two elements will be sufficient to match the AC characteristics of most electrochemical capacitor devices.

The next step is to relate the AC impedance results to the performance of the electrochemical/ultracapacitors in high current applications, which are in many cases DC in character. One approach is to relate frequency f ($\omega = 2\pi f$) to the discharge time, t_{disch}, by the simple relationship $t_{disch} = 1/(4f)$,

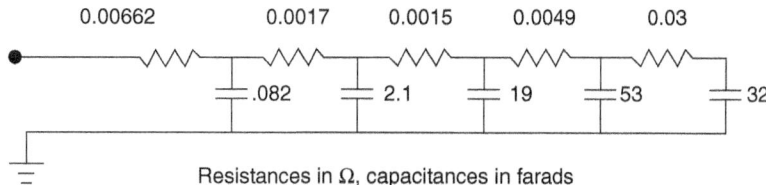

Equivalent circuit model of a Maxwell 100 F, 2.5 V capacitor.

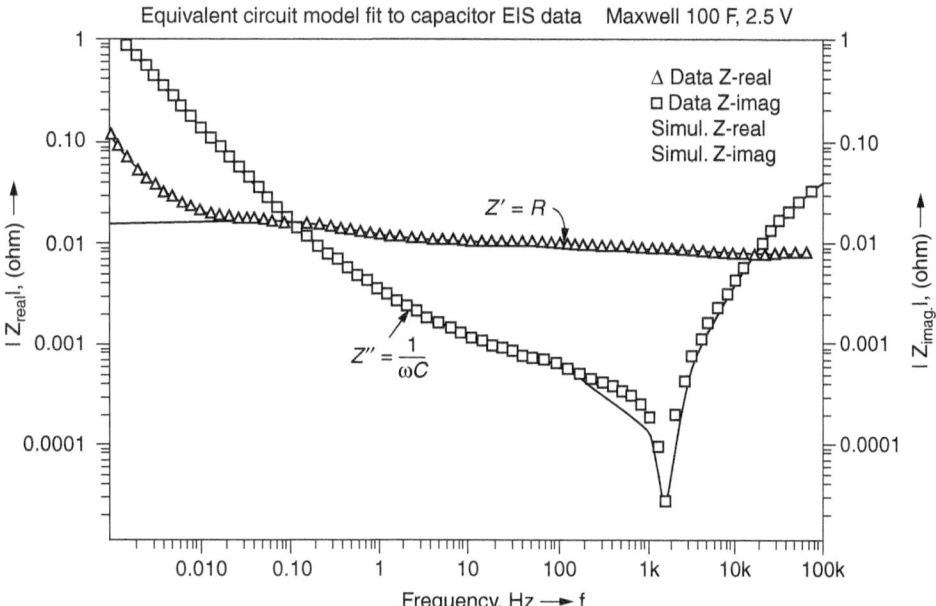

FIGURE 39.5 AC impedance data and equivalent circuit model prediction for a 100 F Maxwell capacitor. (Ref. 44.)

because each AC cycle consists of four (4) charge or discharge subcycles. $t_{disch} = 1$ sec corresponds to $f = 0.25$ Hz, and $t_{disch} = 30$ sec corresponds to 0.0083 Hz; hence, for many applications, the AC frequencies of primary interest for electrochemical capacitors are between 0.01 and 1 Hz. The values for C and R for particular frequencies in the range of interest can be read from the Z'' versus Z' curve in Fig. 39.5. These values can then be used to determine the corresponding energy and power capability of the device tested using the following equation for t_{disch}:

$$t_{disch} = \tfrac{1}{2} CV_0^2/P\,[(1 - K_1)^2 - (V/V_0)^2] + RC \ln [V/V_0/(1 - K_1)] \tag{39.1}$$

$$K_1 = PR/V_0^2 = I_0R/V_0,\ V/V_0 = 0.5 \text{ for a typical discharge}$$

This equation is derived in Sec. 39.4.2 as Eq. (39.10).

The energy stored is then $P \times t_{disch}$ and the corresponding specific energy and specific power densities are given a $P\,t_{disch}/w_d$ and P/w_d, respectively, where w_d is the weight of the device. For the 100 F Maxwell device, the product specifications by the manufacturer are $C = 100$ F, $R = 15$ mΩ, and $w_d = 25$ g. From Fig. 39.5, at a frequency of 0.05 Hz, $t_{disch} = 25$ s, $C = 80$ F, and $R = 17$ mΩ. For a discharge power of $P = 7.5$ W (300 W/kg) corresponding to the 25 s constant power discharge, the calculated discharge time from Eq. (39.10) is 23 s which is in good agreement with the frequency (0.05 Hz). It appears that it is possible to relate the AC impedance results to the DC characteristics of electrochemical capacitor devices. Further comparisons in this regard are given in References 46, 47, and 78.

The AC impedance testing method also permits an assessment of the time-dependent processes in the micropores of the carbon as the double layers are being formed. AC impedance data for a 10 F capacitor using a carbide-based carbon[48,49] in the electrodes are shown in Fig. 39.6. The processes in the pores depend on the diameter and depth of the pores. This problem is analyzed in References 50 and 51. It was found that the impedance of the pore processes is

$$Z_p = [(1 - j)/(2\pi n(r_p^3 \kappa \omega C_{dl})^{1/2})] \coth[l_p(\omega C_{dl}/r_p \kappa)^{1/2}(1 + j)] \tag{39.2}$$

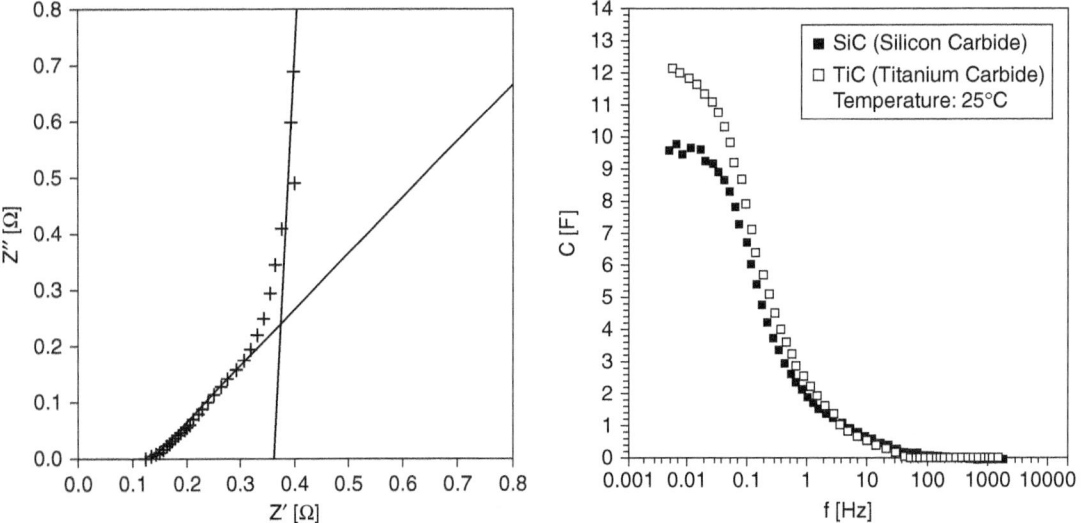

FIGURE 39.6 AC impedance data for a 10 F capacitor using carbide-based carbon electrodes.

where n is the number of pores/cm^3, r_p is the pore radius and l_p is the pore length, C_{dl} is the pore capacitance/cm^2, κ is the electrolyte ionic conductivity, and ω is the frequency. Recall that coth $x = (e^x + e^{-x})/(e^x - e^{-x})$; $x \gg 1$; coth x for large x tends to 1; and coth x for small x tends to $1/x$. Hence, for these limiting frequencies, Eq. (39.2) becomes

$$\omega \gg 1 \text{ (high frequencies) } Z_p = (1 - j)/[2\pi n \, (r_p^3 \kappa \omega C_{dl})^{1/2}]$$

$$\omega \lll 1 \text{ (low frequencies) } Z_p = -j/(2\pi n \, l_p r_p \omega C_{dl}) = -j/\omega C_{total}$$

$C_{total} = 2\pi n \, l_p r_p C_{dl}$ is the total capacitance of all the pores in the electrode.

At high frequencies, the Z'' versus Z' plot is a 45° line with respect to the Z' axis, and at low frequencies, the plot is a straight line normal to the Z' axis. Note that the Z'' versus Z' plot in Fig. 39.6 is of the form predicted by Eq. (39.2). This would seem to validate the pore charging model described in Reference 50, which assumes uniform charging in the pores (C_{dl} = constant). However, it is likely that the processes are much more complex than this simple model.

Since $Z' = R$, the intersection of the 45° line with the axis is the ohmic resistance of the device at very high frequency at which the pore resistance is zero. The pore resistance R_p is the difference between the Z' value at the intersection of the two lines and the minimum value R_0. For the 10 F capacitor, R_0 is 0.12 Ω and the pore resistance R_p is 0.22 Ω. The frequency at the intersection point is given by $\omega = 1/Z''C$. Utilizing the C versus f curve in Fig. 39.6, it is found that $C = 7.3$ F and $f = 0.1$ Hz.

The corresponding charge-discharge time is 2.5 s ($1/4f$). The RC time constant of the 10 F device is $7.3 \times 0.34 = 2.5$ s, indicating that the full resistance in the device, including the pore resistance, should be reached in about one time constant based on the AC impedance data. This is consistent with the DC test data discussed in Sec. 39.5.2.

39.4.2 Mathematical Modeling

Electrochemical capacitors can be modeled starting from the governing equations if some simplifying assumptions are made concerning the properties of the carbon materials used in the electrodes and the processes involved with the transport of the ions in the electrolyte between the electrodes. Detailed mathematical models of the electrochemical capacitors are presented in References 52–54. These models are one-dimensional (x- normal to the electrodes in the direction of current flow) and time dependent (t). Analytical solutions are given for DC constant-current discharges and AC current responses to sinusoidal voltage perturbations of the capacitors. The mathematical solutions are helpful in the understanding the design trade-offs for devices and the interpretation of test data, especially for carbon/carbon double-layer devices. As shown in Fig. 39.1, the capacitor is a composite of solid, porous carbon and a liquid electrolyte. The electrical current (electrons) is taken to and from the device through a metal current collector and is distributed in the electrode by the carbon particles. The positive and negative ions that form the double-layer capacitance in the micropores of the carbon move through the electrolyte by diffusion due to its ionic conductivity. The separator allows the diffusion of the ions between the positive/negative electrodes, but blocks the conductivity of the electrons, which are then forced into the external circuit. The equations needed to describe these processes in the carbon and electrolyte can be relatively simple if the details of ion transport in and out of the pores is neglected and included only in terms of the specific capacitance of the carbon material. The equations to be solved are Ohm's law for the electronic and ionic currents and the conservation of charge at all points in the electrode. The equations to be solved are the following:

Ohm's law

$i_1 = -\sigma \, \partial \Phi_1/\partial x$ i_1 is the electron current density in the carbon.
i_2 is the ionic current density in the electrolyte.

$i_2 = -\kappa \, \partial \Phi_2/\partial x$ Φ_1 and Φ_2 are the potential in the carbon and electrolyte.

Conservation of charge

$I(t) = i_1 + i_2$ $I(t)$ is the applied current to the cell.

$\partial i_1/\partial x = -\partial i_2/\partial x = a i_n$ a is the area of the carbon per cm^3.
     i_n is the interface current (A/cm^2) in the pores of the carbon.

$i_n = -C \partial (\Phi_1 - \Phi_2)/\partial t$ C is the specific capacitance (F/cm^2) of the carbon.

The above equations can be combined and written as

$$(\sigma \kappa/\sigma + \kappa)[\partial^2 \Phi_1/\partial x^2 - \partial^2 \Phi_2/\partial x^2] = -aC\partial(\Phi_1 - \Phi_2)/\partial t$$

and in nondimensional form

$$\partial^2 \eta/\partial^2 \xi = \partial \eta/\partial \tau \qquad (39.3)$$

where $\eta = \Phi_1 - \Phi_2/V_0$
$\xi = x/\delta$
$\tau = aC \delta^2 (\sigma + \kappa)/\sigma \kappa$
δ = thickness of the electrode
σ = conductivity of the carbon
κ = conductivity of the electrolyte

η is the overpotential (voltage in the double layer) at the surface of the carbon. All the equations are written for 1 cm^2 of electrode area. All currents are A/cm^2.

For charging the electrode, the boundary conditions are the following:

$x = 0$ (surface of the metal current collector) $i_2 = I(t)$, $i_1 = 0$, $\eta = 0$

$x = \delta$ (surface of the separator) $i_2 = 0$, $i_1 = I(t)$

The equations shown apply to a single electrode. It is assumed that both electrodes in the cell are identical so that the energy stored and the resistance of the electrodes are identical, and the solutions for the electrodes can be combined to form the total solution for the cell.

Methods for solving equations of the form of Eq. (39.3) are well known from nonsteady, one-dimensional heat transfer analysis.[55] Solutions for both constant DC currents and AC sinusoidal currents for carbon/carbon electrochemical capacitors are given in References 52–54.

Constant-Current Discharges. For the case of a constant-current discharge of a cell, the solution to Eq. (39.3) can be written as

$$V = V_0 - IR_0 - 2It/\delta aC - I\Sigma_n\{4[\kappa + (-1)^n \sigma]^2 \delta\}/[\sigma \kappa(\sigma + \kappa)n^2 \pi^2][1 - \exp(-n^2 \pi^2 t/\tau)] \qquad (39.4)$$

where R_0 (ohm-cm^2) = $2 \delta/(\sigma + \kappa) + \delta_{sep}/\kappa_{sep} + R_{contact}$

The resistance (ohm-cm^2) of the cell is time dependent and is given by

$$R = R_0 + \Sigma_n\{4[\kappa + (-1)^n \sigma]^2 \delta\}/[\sigma \kappa(\sigma + \kappa)n^2 \pi^2][1 - \exp(-n^2 \pi^2 t/\tau)]$$

The resistance at $t = 0$ is R_0. The solution for the cell performance given in Eq. (39.4) represents the cell as a simple RC circuit with a time dependent resistance.

In most cases, the conductivity of the carbon is much greater than that of the electrolyte so the cell resistance becomes

$$R = 2\delta/\sigma + \delta_{sep}/\kappa_{sep} + R_{contact} + 4\delta/\kappa \sum_n (1/\pi^2 n^2)[1 - \exp(-n^2 \pi^2 t/\tau)] \quad (39.5)$$

$$\tau = aC\delta^2/\kappa$$

The capacitance and resistance of the cell are $C_{cell} = 1/2\, A_x \delta aC$ and $R_{cell} = 2\delta/\kappa A_x$. Hence, to a first approximation,

$$\tau = (RC)_{cell}$$

For large values of t/τ, the final summation becomes $\pi^2/6$ and the steady-state value of R is

$$R_{ss} = 2\delta/\sigma + \delta_{sep}/\kappa_{sep} + R_{contact} + 2/3\,\delta/\kappa$$

The exponent in the Eq. (39.5) is $n^2 \pi^2 t/(RC)_{cell}$. Even for $t = (RC)_{cell}$ and $n = 1$, the contribution of the terms in the summation is small. The simple solution, which neglects the effects of ion transport in the micropores of the carbon, indicates that the cell resistance is essentially the steady-state value except for discharge times equal to a small fraction of the RC time constant of the cell. However, as discussed in Sec. 39.5.2, test data indicates that for constant-current discharges, the cell resistance does not approach the steady-state value until times of 1 to 2 RC time constants, indicating the ion transport in the micropores is important for small times.

The value of κ, the ionic conductivity of the electrolyte, used in the calculations should include the effects of the porosity of the carbon.[56] Therefore

$$\kappa = \kappa_0 \varepsilon^{1.5}$$

where κ_0 is the bulk conductivity of the electrolyte and ε is the porosity of the carbon.

Constant-Power Discharges. The previous analysis was concerned with constant-current discharges. If the capacitance C and resistance R of a cell are assumed to be constant, it is straightforward to derive an expression for the voltage for a constant-power discharge. The governing equation for the discharge is

$$V_0 - V = IR + \int_{V_0'}^{V} dq/C, \quad dq = I\,dt \quad (39.6)$$

where V_0' is the voltage immediately after the initiation of the discharge.

For discharge at constant power P, the current is given by

$$I = P/V$$

and Eq. (39.6) becomes

$$1 - V/V_0 = PR/V_0^2 / V/V_0 + P/V_0^2 /C \left\{ V_0'/V_0 \int_{V_0'}^{V} dt/V/V_0 \right\} \quad (39.7)$$

Defining $z = V/V_0$, $K_1 = PR/V_0^2$, $K_2 = P/CV_0^2$, Eq. (39.7) becomes

$$1 - z = K_1/z + K_2 \int_{z_0'}^{z} dt/z \quad (39.8)$$

where $z_0' = 1 - (IR)_0/V_0 = 1 - PR/V_0^2 = 1 - K_1$

Eq. (39.8) can be integrated in closed form to obtain

$$K_1[\ln z - \ln z_0'] - \tfrac{1}{2}(z^2 - z_0'^2) = K_2 t \qquad (39.9)$$

Inputing the defined variables, Eq. (39.7) becomes

$$t/RC = [\ln V/V_0 - \ln(1 - K_1)] - (1/2)K_2[(V/V_0)^2 - (1 - K_1)^2] \qquad (39.10)$$

Eq. (39.10) can be rewritten as

$$t = \tfrac{1}{2} CV_0^2/P \,[(1 - K_1)^2 - (V/V_0)^2] + RC \ln[V/V_0/(1 - K_1)]$$

$K_1 = PR/V_0^2 = I_0 R/V_0$ is an indicator of the efficiency of the first current pulse.

The energy density of the constant power discharge is then

$$\text{Wh/kg} = t_{disch}\, P/\text{weight of cell in kg}$$

Consider the following example:

$$C = 2900\ \text{F},\ R = 0.375\ \text{m}\Omega,\ \text{weight } 0.55\ \text{kg},\ RC = 1.09\ \text{s},\ P = 500\ \text{W},\ 909\ \text{W/kg}$$

$$t_{calculated} = 14.1\ \text{s},\ 3.6\ \text{Wh/kg},\ V_{final}/V_0 = 1/2;\ \text{measured } t = 15.1\ \text{s},\ 3.8\ \text{Wh/kg}$$

39.4.3 Design Analysis for Hybrid Capacitors

A schematic of a hybrid capacitor is shown in Fig. 39.2. All the designs being considered utilize carbon in at least one of the electrodes. The device is designed such that the charge transferred between the electrodes is set by the charge capacity and voltage change of the carbon electrode. The battery-like electrode is designed such that its operating range (depth of discharge in the capacitor) in charge/discharge is relatively small (5 to 10%) so that its expected life can be very long compared to what would be expected in a battery. The cell voltage of the hybrid device is determined from the sum of the operating range of the carbon electrode and standard potential of the battery-like electrode in the electrolyte of interest. The performance calculated using the method outlined below is the ideal performance of the device, assuming all interfacial resistances are negligible and the specific capacitance of the carbon is not rate dependent. In addition, it is assumed that the resistance of the device is primarily determined by the macro-resistance of the porous electrodes, neglecting the micropore resistance of the carbon.

Inputs. The physical dimensions and material properties of all the components used in the device must be known. The key inputs for the carbon electrode are its thickness and the properties of the carbon-specific capacitance (F/gm), density (gm/cm^3), and porosity (%). For the battery-like electrode, the key properties are the charge capacity (A-sec/gm), density (gm/cm^3), and porosity (%) of the electrode material. The current collectors are described in terms of their material (lead, copper, nickel, or aluminum) and thickness per side coated with electrode material. The electrolyte is specified by composition (sulfuric acid, KOH, acetonitrile plus salts), density (gm/cm^3), and specific resistivity (ohm-cm). The separator is specified in terms of its thickness and porosity. The area of the cell cross-section can be specified or the calculations done on a 1 cm^2 basis.

Step 1: Calculate the weight (W_{carb}) of the carbon electrode and from that the capacitance (C_{carb}) of the electrode. Based on the assumed voltage swing (delta V_{carb}) of the electrode and its capacitance, the charge transferred to the electrode during charge-discharge is given by

$$\text{Chg} = C_{carb} \times (\Delta V_{carb})$$

Step 2: A key constraint on the cell design is that the charge transferred to the battery-like electrode must be equal to that transferred to the carbon electrode. The battery-like electrode will be sized such that, to accommodate that charge, the change in the depth of discharge of the electrode should be only 5 to 10%. The change in state-of-charge is symbolized as delta (SOC_{bl}). The required weight (W_{bl}) of the battery-like electrode is given by

$$W_{bl} = Chg/[\Delta(SOC_{bl}) \times (A\text{-sec/gm})_{bl}]$$

The thickness (th_{bl}) of the battery-like electrode is then given by

$$(th_{bl}) = W_{bl}/(dens_{bl} \times A_{cell})$$

Step 3: Next, calculate the weight of the electrolyte in the electrodes and the separator by accounting for the electrolyte in the macropores of those components. The electrolyte weight in each layer is given by

$$W_{elypor} = th \times A_{cell} \times porosity \times (dens)_{ely}$$

The electrolyte weight is the sum of the weights of the electrolyte in each of the component layers of the device.

Step 4: The weight of the current collectors is simply

$$W_{curcl} = th_{curcl} \times A_{cell} \times (dens)_{curcl}$$

Step 5: Next, calculate the weight of the cell. This is done by simply adding up the weights of the components.

$$W_{cell} = W_{curcl} + W_{carb} + W_{bl} + W_{ely}$$

Step 6: The energy stored in the cell is the sum of energy stored in the carbon and battery-like electrodes. The energy stored in the carbon electrode is given by

$$E_{carb} = \tfrac{1}{2} \times C_{carb} \times (\Delta V_{carb})^2$$

The energy stored in the battery-like electrode is given by

$$E_{bl} = Chg \times V_{av,bl}$$

where

$$V_{av,bl} = V_{bl\ stp} + \tfrac{1}{2} \times \Delta(V_{carb})$$

$V_{bl\ stp}$ = standard potential of battery-like electrode

The total energy stored is

$$E_{total} = E_{carb} + E_{bl}$$

The specific energy of the cell is

$$(Wh/kg)_{cell} = E_{total}/W_{cell}$$

This specific energy does not include the packaging weight, but does include the weights of all the active components of the cell, including the current collectors.

Step 7: The resistance of the cell is calculated by relating the bulk resistivity of the electrolyte (R_{ely}) to that in the porous electrodes by the relationship

$$R_{elypor} = R_{ely} \times (\text{porosity})^{-1.8}$$

The specific resistance (ohm-cm²) of each layer is given by

$$\text{ohm-cm}^2 = R_{elypor} \times \text{th}$$

The specific resistance of the cell is then the sum of the specific resistances of the carbon and battery-like electrodes and the separator. This approach neglects the pore resistance of the electrodes, electrical resistance of the carbon and battery-like electrode materials, and the interfacial resistance between the layers of the cells. The cell resistance is then given by

$$R_{cell} = (\text{ohm-cm}^2)_{cell}/A_{cell}$$

This calculated resistance should be considered that of an ideal cell; the resistance of an actual cell will certainly be higher.

Step 8: The power characteristics of the cell can be calculated from the cell voltage and resistance using the relationship

$$P_{max} = 9/16 \times (1 - EF) \times V_{cell}^2/R_{cell}$$

EF = efficiency of the discharge

The cell voltage V_{cell} is given by the smallest of

$$V_{cell} = \Delta(V_{carb}) + V_{bl\ stp}$$

$V_{bl\ max}$ = maximum allowable voltage for the battery-like electrode

The cells operate between V_{cell} and $V_{cell} - \Delta(V_{carb})$. The specific power for the device design being analyzed is

$$(W/kg)_{max} = P_{max}/W_{cell}$$

This method of analysis has been applied to estimate the performance (specific energy and power capability) of electrochemical capacitors using various combinations of the advanced materials. The results are shown in Table 39.9.

39.5 TESTING ELECTROCHEMICAL CAPACITORS

There have been many studies of materials for electrochemical capacitors and testing of small laboratory and prototype devices as well as a wide range of larger commercial products. Much of the testing of materials and small laboratory devices has involved the application of cyclic voltammetry and AC impedance test approaches.[1,42,43] These approaches in most cases utilize small currents and limited voltage ranges and/or AC frequencies and are intended primarily to determine the electrochemical characteristics of the materials and electrodes used in the capacitors. Testing of the larger prototype and commercial devices is usually done using DC test procedures similar to those used to test batteries. This section discusses DC test procedures and how they can be used to characterize/evaluate electrochemical capacitors intended for various industrial and vehicle applications.

TABLE 39.9 Calculated Specific Energy, Energy Density, and Specific Power Characteristics of Various Electrochemical Capacitor Technologies

Type	Capac. F/g or mAh/g	V	Wh/kg*	Wh/L*	Ohm-cm^2	RC (sec)	kW/kg 95% eff.
Act. carbon/Act. carbon/ sulfuric acid	150 F/g	1.0–0.5	1.7	2.2	0.17	0.29	1.2
Act. carbon/Act. carbon/ acetonitrile	100	2.7–1.35	5.7	7.6	0.78	0.18	6.4
Act. carbon/Act. carbon/ acetonitrile	120	2.7–1.35	6.8	9.1	0.78	0.22	6.4
Act. carbon/PbO$_2$/ sulfuric acid	150 F/g/220 mAh/g	2.25–1.0	16	39	0.12	0.36	8.9
Act. carbon/NiOOH/ KOH	150 F/g/290 mAh/g	1.6–0.6	14	31	0.16	0.71	4.0
Act. carbon/graph. carbon/acetonitrile	100 F/g/60 mAh/g	3.3–2.0	21	34	2.5	2.6	2.3
Act. carbon/graph. carbon/acetonitrile	120 F/g/72 mAh/g	3.3–2.0	25	41	2.5	3.1	2.3
Lithium titanate/ Act. carbon/acetonitrile	160 mAh/g/100 F/g	2.8–1.8	22	37	3.4	4.1	1.5
Lithium titanate/ graph. carbon/acetonitrile	160 mAh/g/60 mAh/g	3.7–2.5	53	96	1.7	3.8	7.8

*Usable energy—energy available between the voltage limits indicated

39.5.1 Summary of Test Procedures

There are similarities and differences in the test procedures[57,58] for electrochemical capacitors and high-power batteries. It is customary to perform constant-current and constant-power tests of both types of devices. From the constant-current tests, the charge capacity [capacitance (farads)] and Ah and resistance of the devices are determined. From the constant-power tests, the energy storage characteristics (Wh/kg versus W/kg—the Ragone curve) are determined. The currents and powers to be used in the testing are selected such that the charge and discharge times are compatible with the capabilities of the devices being tested. In the case of the capacitors, the test discharge times are usually in the range of 5 to 60 s and for the batteries several minutes to a significant fraction of an hour even for high-power batteries. The differences in the recharge times for the devices are also large. For example, the capacitors can be fully charged in 5 to 10 s without difficulty, but the high-power batteries require a minimum of 10 to 20 min for a complete charge even when the initial charge current is set at a maximum value. In addition to the constant-current and constant-power tests, the capacitors and batteries are tested using charge-discharge pulses of 5 to 15 s. For these tests, the current and power levels for the capacitors and high-power batteries are comparable (on a normalized basis). Test cycles consisting of a sequence of charge and discharge pulses (specific power for a specified time) meant to simulate how the devices would be used in particular applications are used to test both capacitors and batteries.[57,58] The tests to be performed on capacitors and batteries are summarized in Tables 39.10 and 39.11.

39.5.2 Application of the Test Procedures to Carbon/Carbon Capacitors

In this section, the various test procedures discussed in the previous section are applied to carbon/carbon capacitors to determine their capacitance, resistance, specific energy, and power capability. These devices use activated carbon in both electrodes and, in nearly all cases, an organic electrolyte. The dominant energy storage mechanism in these devices is charge separation (double-layer capacitance).

TABLE 39.10 Performance Characteristics of Electrochemical Capacitors

1. Specific energy (Wh/kg vs. W/kg)
2. Cell voltage (V) and capacitance (F)
3. Series and parallel resistance (ohm and ohm-cm^2)
4. Specific power (W/kg) for a charge-discharge at 95% efficiency
5. Temperature dependence of resistance and capacitance especially at low temperatures (20°C)
6. Cycle life for full discharge
7. Self-discharge at various voltages and temperatures
8. Calendar life (hours) at fixed voltage and high temperature (40–60°C)

TABLE 39.11 Testing of Electrochemical Capacitors

1. Constant-current charge-discharge
   - Capacitance and resistance for discharge times of 60 to 5 s.
2. Pulse tests to determine resistance
3. Constant-power charge-discharge
   - Determine the Ragone curve for specific power between 100 and at least 1000 W/kg for the voltage between V_{rated} and ½ V_{rated}.
   - Test at increasing W/kg until discharge time is less than 5 s. The charging is often done at constant current with a charge time of at least 30 s.
4. Sequential charge-discharge step cycling
   - Testing done using the PSFUDS (pulsed simple FUDS) test cycle with the maximum power step being 500 W/kg.
   - From the data, the round-trip efficiency for charge-discharge is determined.
5. Tests modules with at least 15–20 cells in series

Capacitance. The capacitance of a device can be determined directly from constant-current discharge data. A typical V versus time trace for a carbon/carbon double-layer capacitor is shown in Fig. 39.3. By definition,

$$C = I/dV/dt \quad \text{or} \quad C = I(t_2 - t_1)/(V_1 - V_2), \quad V = V(t)$$

Since the voltage trace is not exactly linear, the value of C calculated depends to some extent on the values of V_1 and V_2 used. The voltage ranges that have been used are V_0 to $V_0/2$ and V_0 to 0. In the case of V_0', it is important to include the IR drop in the determination of the effective V_1 value. The results in Table 39.12 indicate that the determination of the capacitance of devices is relatively insensitive to the test procedure.

TABLE 39.12 Effect of Voltage Range and Test Current on the Measured Capacitance

Device/developer	V_0 to 0 V				V_0 to 1.35 V			
3000 F/Maxwell	100 A	2880 F	200 A	2893 F	100 A	3160 F	200 A	3223 F
3000 F/Nesscap	50 A	3190 F	200 A	3149 F	50 A	3214 F	200 A	3238 F
450 F/APowerCap	20 A	450 F	40 A	453 F	20 A	466 F	40 A	469 F
		3.8 to 2.2 V				3.8 to 2.6 V		
2000 F/JSR Micro	80 A	1897 F	200 A	1817 F	80 A	1941 F	200 A	1938 F

Resistance. The resistance of a capacitor or battery can be determined using one of several methods:

- IR drop at the initiation of a constant-current discharge
- Current pulses (5 to 30 s) at specified states-of-charge

- Voltage recovery at the interruption of a discharge or charge current
- Measurement of the AC impedance at 1 kHz

The method used routinely involves analysis of the *IR* drop and voltage variation at the initiation of a constant-current discharge. Determination of the resistance of the capacitor is complicated by the fact that the voltage decreases due both to the resistance and capacitance of the device. In addition, due to the porous character of the electrode, the resistance of the capacitor varies with time until the current distribution in the electrodes is fully established. This problem has been analyzed mathematically in Sec. 39.4.2. The results of the analysis indicate that the initial value of the resistance (R_0) can be as low as half the steady-state value. A good estimate of the steady-state resistance can be obtained by extrapolating the linear portion of the voltage-time trace back to $t = 0$ as shown in Fig. 39.7 and utilizing that *IR* drop value to calculate *R*. For many applications of ultracapacitors, it is the steady-state resistance that is most relevant for the calculation of power capability/electrical losses/heating and not the R_0 value, which is smaller. It is important to define what resistance value is being reported.

Another reliable method of determining the DC resistance of a capacitor is the current pulse method in which a short pulse (5 to 10 s) is applied to the device. In fact, for batteries, this is probably the only reliable method to determine its resistance using most battery testers. The pulse can be either a discharge or charge pulse. The effective resistance ($R = \Delta V/I$) can vary with time through the

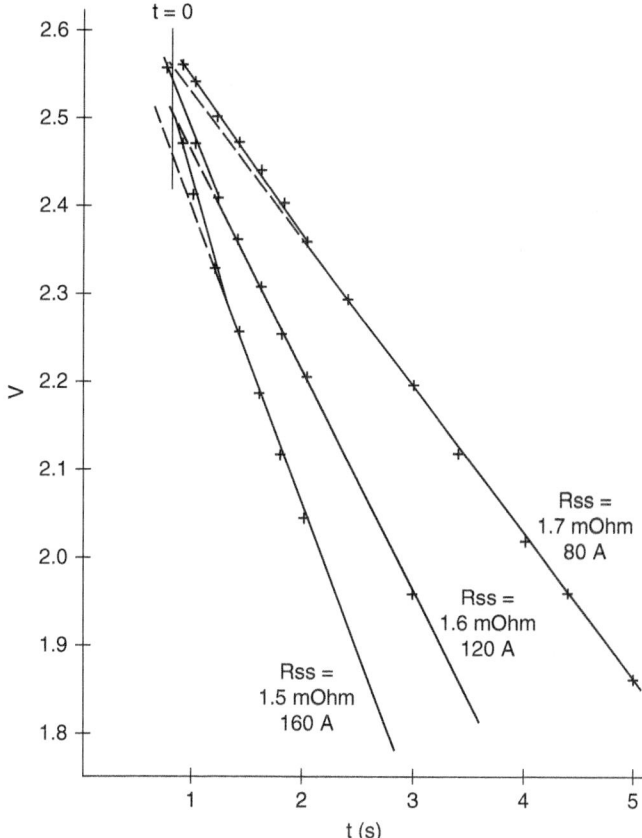

FIGURE 39.7 Method for determining the steady-state resistance by extrapolating the voltage trace to $t = 0$ (APowerCap 450 F cell).

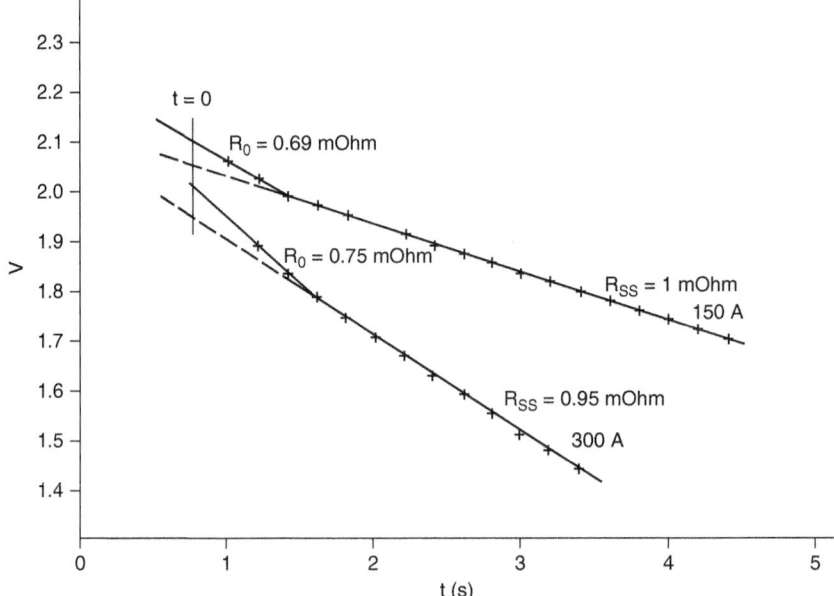

FIGURE 39.8 Voltage and resistance during a current pulse (Skeleton Technology 1600 F device).

pulse depending on the response time of the tester and/or mechanisms occurring in the device. It is, of course, greatly preferred if the former is negligible and only the latter is important. Unfortunately, this is often not the case. As shown in Fig. 39.8 for a 1600 F device, the pulse tests show clearly the change in resistance from R_0 to R_{ss}.

The resistance of a capacitor or battery can also be inferred from the recovery of the voltage at the end of a current pulse when the current is removed ($I = 0$). Some researchers prefer this method rather than that involving the initiation of the pulse because the current is zero and the effect of the capacitance of the device on the voltage is not present.[59] However, there is the effect of the charge redistribution in the device with time at $I = 0$, and the effect on the voltage is not well understood. As a result, there is uncertainty as to the time after the setting of $I = 0$ at which the voltage should be read and R calculated from $\Delta V/I$. This effect is illustrated in Fig. 39.9. Comparing Figs. 39.8 and 39.9 indicates that the current initiation and interruption methods yield the same value of resistance both for the R_0 and R_{ss}. The voltage recovery time is relatively short, about equal to the RC time constant of the device tested.

For ultracapacitors, it is common for manufacturers to list the resistance measured with an AC impedance meter at 1 kHz. This value of resistance is always significantly lower than the DC value, often by about a factor of 2. The power capability of devices should not be calculated using the AC impedance meter value of the resistance.

Specific Energy. The total energy stored in a carbon/carbon capacitor can be calculated from the relationship $E = \frac{1}{2}CV_0^2$. If the voltage of the capacitor is restricted to the range V_0 to $V_0/2$, only 75% of the stored energy can be used. Hence the usable specific energy is given by

$$\text{Wh/kg} = 3/8\ CV_0^2/(\text{device weight})$$

This simple relationship is often used to calculate the specific energy of ultracapacitors. However, the most reliable approach to determining the energy stored in a device is to measure the Wh stored for a range of constant specific power values (W/kg). In general, tests should be made

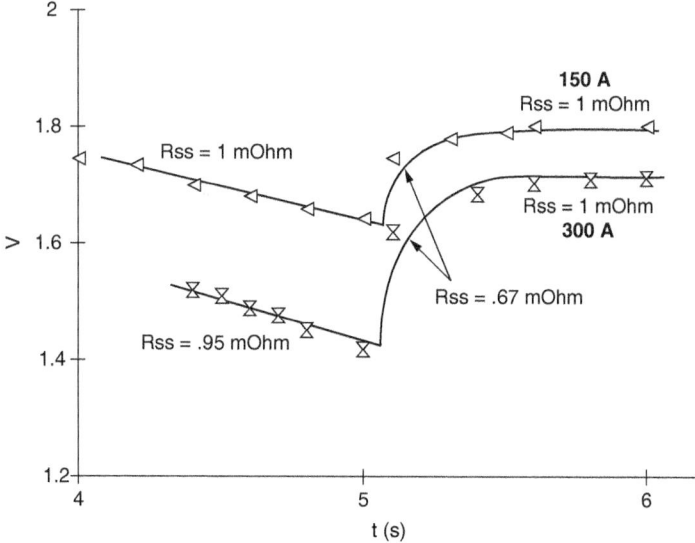

FIGURE 39.9 Resistance determination using voltage recovery after current interruption (Skeleton Technology 1600 F device).

for specific power values between 100 and 1000 W/kg or even higher for high power devices. The plot of Wh/kg versus W/kg is called the Ragone curve for the device. Typical data for a commercially available 3000 F capacitor are shown in Table 39.13. Note that the specific energy gradually decreases with W/kg. This is the case for all ultracapacitors. The value of specific energy quoted by the device manufacturer is often calculated from the energy corresponding to ½ CV_0^2 using

TABLE 39.13 Test Data for the 3000 F Nesscap Cylindrical Device

Constant-Current Discharge Data: 2.7 to 0 V			
Current (A)	Time (s)	Capacitance (F)	Resistance (mΩ)
50	171	3190	—
100	84.3	3181	0.44 (1)
200	41.3	3157	0.42
300	27	3140	0.37
400	20	3150	0.40

Constant-Power Discharge Data: 2.7 to 1.35 V					
Power (W)	W/kg*	Time (s)	Wh	Wh/kg	C_{eff}
100	192	84.8	2.36	4.52	3107
200	383	41.8	2.32	4.44	3055
300	575	27.1	2.26	4.33	2976
400	766	19.7	2.19	4.20	2884
500	958	15.4	2.14	4.1	2818
700	1341	10.9	2.12	4.06	2792

*Weight of device, 0.522 kg.; dimensions of the device, 6 cm dia., 13.4 cm length
C_{eff} = 2 wattsec/0.75(2.7)^2

the rated voltage and specified capacitance. This value is too high as it is not the usable specific energy, and also it corresponds to a low specific power of 100 W/kg or lower. As shown in Table 39.13, the effective capacitance C_{eff} of a device decreases significantly with W/kg and often agrees with the value claimed by the manufacturer only for relatively low power levels. Combining the usable energy factor (0.75) and the effective capacitance reduction factor (0.9 from Table 39.13), the simple calculation of specific energy from $½ CV_0^2$ can overestimate the parameter of a device by at least 1/3.

Power Capability. There is much confusion and unreliable information in the literature concerning the power capability of ultracapacities and batteries. This confusion stems to a large extent from the persistent use of the simple formula $P = V_0^2/4R$ to calculate the maximum power capability of electrochemical devices. This formula grossly overestimates the maximum power as it corresponds to operation of the device at the matched impedance point at which one-half of the discharge energy is electricity and one-half is in heat. The corresponding efficiency is 50%, which makes that operating condition unusable for nearly all applications. It is more reasonable to express the power capability of devices in terms of the pulse efficiency (EF). This can be done using the following relationships for ultracaps and batteries:

$$\text{Ultracapacitors: } P = 9/16(1\text{-EF}) V_0^2/R$$

$$\text{Batteries: } P = \text{EF}(1\text{-EF}) V_{oc}^2/R$$

These relationships are for pulse power and not constant power. In the case of the ultracapacitor, the power pulse is occurring at a voltage of ¾ V_0 and is intended to remove only a relatively small fraction of the energy stored in the device. The battery relationship can be applied at any state-of-charge by using the V_{oc} and R for that SOC. Note that the power from both the matched impedance and efficiency EF relationships is proportional to V^2/R. The key parameters in determining the power capability are thus R and V_0. High power devices necessarily must have low resistance. Once the resistance of a device is known, its power capability follows directly. It is unfortunate that device manufacturers often do not provide information concerning the resistance of their devices. This makes careful measurement of the resistance (as discussed in the previous section) very important.

TABLE 39.14 Ratio of the Efficiency and the Matched Impedance Power for Ultracapacitors and Batteries

Efficiency EF	Ultracapacitors	Batteries
0.5	1.1	1.0
0.6	0.9	0.96
0.7	0.68	0.84
0.8	0.45	0.64
0.9	0.22	0.36
0.95	0.11	0.19

For simple power pulses using capacitors, the ratio of the matched impedance to the efficiency power is 4/9/(1-EF). In the case of batteries, the ratio is 1/4/[EF(1-EF)]. The ratios as functions of EF are given in Table 39.14. For ultracapacitors, the efficiency specified by the USABC[60,61] is 95%, which results in the usable maximum power being only about 1/10 the matched impedance power ($V^2/4R$). For capacitors, using the $V^2/4R$ formula to estimate the usable maximum power does not yield a realistic value for most applications, especially vehicle applications. Note that in Table 39.7 both the matched impedance and EF = 95% power densities are presented for the devices.

Pulse Cycle Testing. Since in many applications ultracapacitors experience highly transient operation, pulse cycle testing should be included in evaluating their performance capabilities. The pulse cycles are simply a sequence of discharge and charge pulses of specified currents (A) or power (W) of specified time duration (seconds). The PSFUDS, which was first defined in Reference 62, has been used extensively to test ultracapacitors and high-power batteries. The test cycle, specified in terms of W/kg-time steps, is given in Table 39.15. It can be utilized to test devices of all sizes and

TABLE 39.15 Time-Power Steps for the PSFUDS Test Cycle

Step no.	Time step duration (s)	Charge C/ discharge D	P/P_{max} P_{max} = 500 W/kg
1	8	D	0.20
2	12	D	0.40
3	12	D	0.10
4	50	C	0.10
5	12	D	0.20
6	12	D	1.0
7	8	D	0.40
8	50	C	0.30
9	12	D	0.20
10	12	D	0.40
11	18	D	0.10
12	50	C	0.20
13	8	D	0.20
14	12	D	1.0
15	12	D	0.10
16	50	C	0.30
17	8	D	0.20
18	12	D	1.0
19	38	C	0.25
20	12	D	0.40
21	12	D	0.20
22	≥50	Charge to V_0	0.30

Round-trip Efficiencies for the Ness 45 V Module on the PSFUDS Cycle

Cycle*	Energy in (Wh)	Energy out (Wh)	Efficiency (%)
1	102.84	97.94	95.2
2	101.92	97.94	96.1
3	101.67	97.94	96.3

*PSFUDS power profile based on maximum power of 500 W/kg and the weight of the cells alone.

performance capabilities by adjusting the W/kg and time duration of the maximum power steps. The test data of most interest in using the PSFUDS cycle is the round-trip efficiency. Typical data using the cycle are shown in Fig. 39.10.

39.5.3 Testing of Hybrid, Pseudo-Capacitive Devices

Most of the electrochemical capacitors that have been available for testing are of the carbon/carbon type that use activated carbon in both electrodes and double-layer capacitance for energy storage. The testing of devices that use intercalation carbon or other battery-like (pseudo-capacitive) materials in at least one electrode is considered in this section. These devices are often referred to as hybrid ultracapacitors. Some testing of hybrid capacitors has been done and differences between testing carbon/carbon and hybrid capacitor devices are becoming apparent. These differences will be discussed here with emphasis on how they affect test procedures and data interpretation.

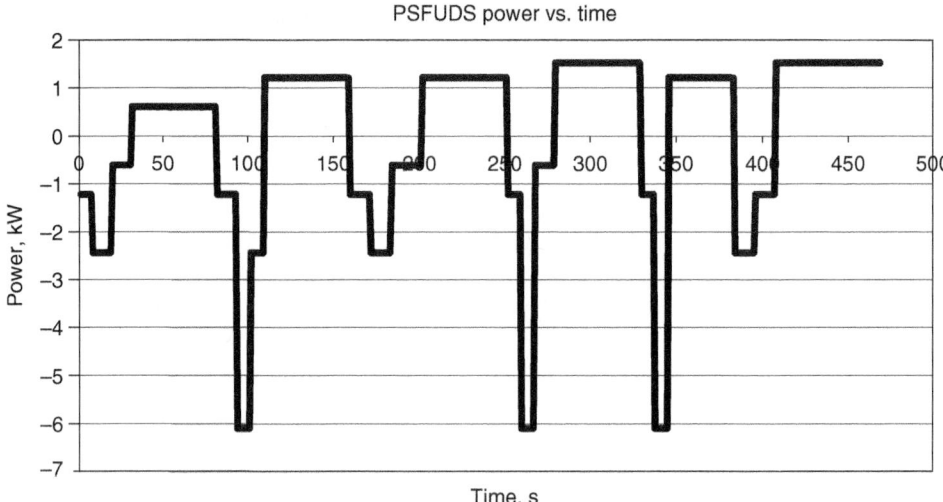

FIGURE 39.10 PSFUDS test results for a 45 V Ness ultracapacitor module.

Capacitance. As is the case for carbon/carbon devices, the capacitance is determined from constant-current discharge data. However, as shown in Fig. 39.11, the character of voltage versus time traces for the hybrid capacitors is quite different from the carbon/carbon devices.

As seen in Fig. 39.11, the key differences are the nonlinearity of the hybrid device traces, especially in charging, and the well-defined voltage, below which the capacitance of the device is small. As would be expected, the character of the V versus t trace must be considered in testing a particular hybrid capacitor device. In the case of the hybrid carbon device (Fig. 39.11a), the voltage should be restricted to be between the rated voltage per cell (3.8 V) and that of the shoulder (2.2 V). It is evident from Table 39.12 (note the data for the JSR Micro device) that the selection of voltage limits makes a greater difference in the calculation of the capacitance for hybrid capacitors than for carbon/carbon devices. The best approach is to use the complete range between the rated and shoulder voltages to calculate the capacitance, but correct the initial voltage (V_1) for the IR drop as is done for the carbon/carbon devices. For hybrid capacitors, it is necessary to look closely at the V versus t trace before adapting a particular method for the calculation of capacitance.

Resistance. The same methods can be used for determining the steady-state resistance (R_{ss}) of the hybrid carbon capacitors as was used for the carbon/carbon devices. As shown in Fig. 39.12, the V versus time traces for constant-current discharges of the hybrid carbon devices become linear within a couple of seconds, and the *IR* drop can be determined by extrapolating back to $t = 0$. Hence, $R_{ss} = (\Delta V)_{t=0}/I$. When testing any new hybrid device, one should check the linearity of the V versus time trace near the initiation of discharge to determine whether the simple method of linear extrapolation is applicable. Pulse tests with the JSR Micro devices yield resistance values that are in good agreement with those obtained using the linear extrapolation method. Test data for the JSR Micro 2000 F device are given in Table 39.16.

Specific Energy. In simplest form, the energy stored in a hybrid capacitor can be expressed as

$$E_{stored} = \tfrac{1}{2} C_{eff}(V_{rated}^2 - V_{min}^2)$$

assuming that capacitance C_{eff} is a constant. In the case of the carbon/carbon devices, $V_{min} = \tfrac{1}{2} V_{rated}$. In the case of the hybrid capacitor, V_{min} is the minimum voltage at which the device stores significant charge. C_{eff} has been calculated from the test data for a carbon/carbon device in Table 39.13 and for

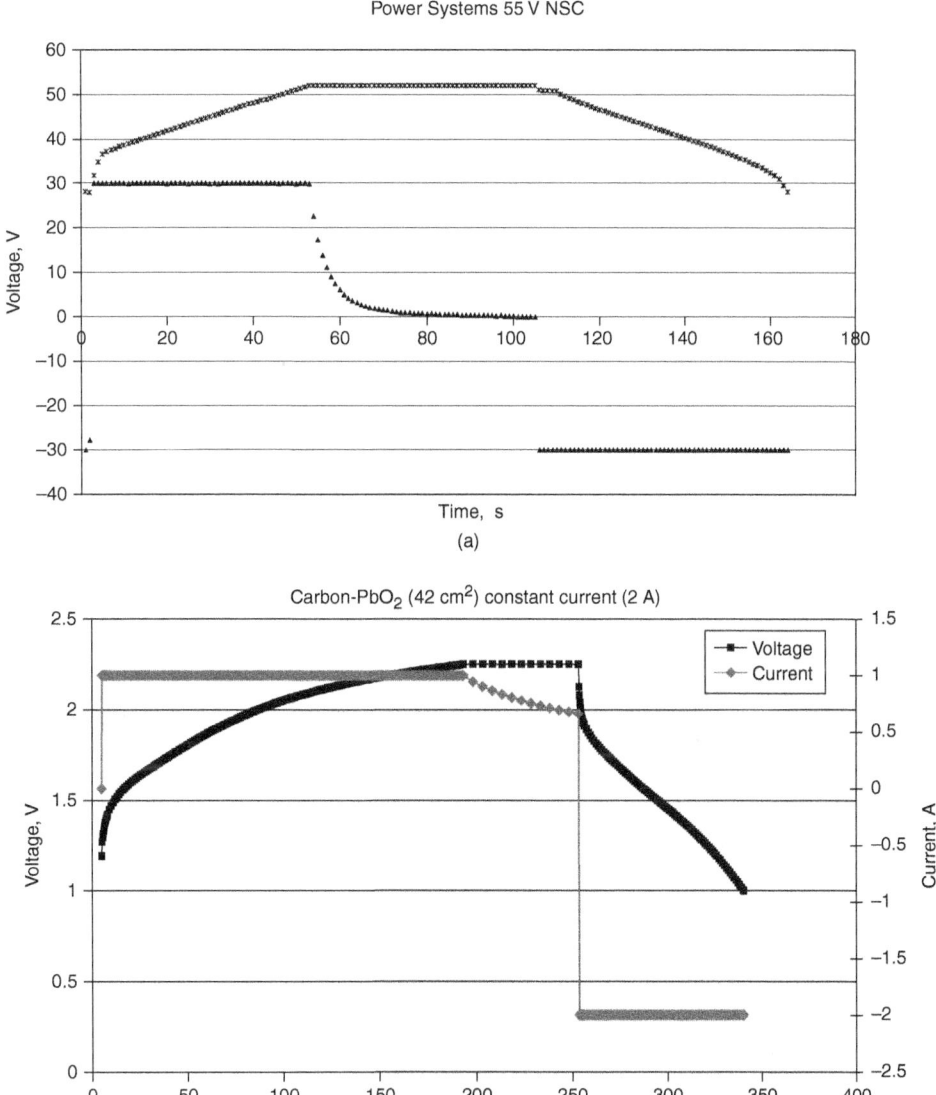

FIGURE 39.11 Voltage vs. time traces for the constant-current discharges of hybrid capacitors. (a) Graphite carbon/activated carbon device. (b) Activated carbon/PbO_2 device.

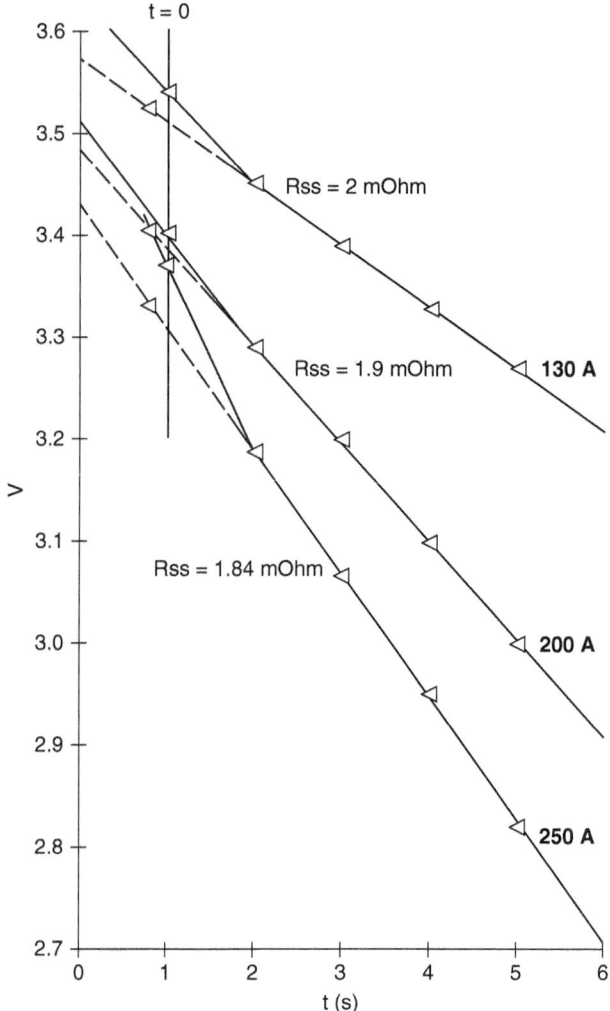

FIGURE 39.12 Determination of the steady-state resistance of a hybrid ultracapacitor (JSR Micro 2000 F cell).

a hybrid capacitor in Table 39.16. It is clear by comparing the tables that the C_{eff} approximation and the simple $½ CV^2$ relationship is valid to obtain the energy stored at low specific power levels for the carbon/carbon devices, but not for the hybrid devices. Hence the specific energy of hybrid capacitors should be obtained from testing them over a range of specific power levels. The simple $½ CV^2$ relationship overestimates the energy stored in the hybrid capacitors. As is the case for carbon/carbon devices, specific energy decreases with increasing specific power due to the effect of resistance on the operating voltage range of the device.

Power Capability and Pulse Cycle Tests. Pulse testing of hybrid capacitors to obtain the resistance and the round-trip efficiency on the PSFUDS is essentially the same as for carbon/carbon devices. The power capability of hybrid capacitors can be calculated using the same relationships used for carbon/carbon devices when V_{rated} and the pulse resistance R are known.

TABLE 39.16 Characteristics of the JSR Micro 2000 F Cell

Constant-Current Discharge 3.8 to 2.2 V			
Current (A)	Time (s)	C (F)	Resistance (mΩ)*
30	102.2	2004	—
50	58.1	1950	—
80	34.1	1908	—
130	19.1	1835	2.0
200	11.1	1850	1.9
250	8.2	1694	1.84

*Resistance is steady-state value from linear V vs. time discharge curve.

Constant-Power Discharges 3.8 to 2.2 V						
Power (W)	W/kg	Time (s)	Wh	Wh/kg†	C_{eff}	Wh/L†
102	495	88.3	2.5	12.1	1698	18.9
151	733	56	2.35	11.4	1596	17.8
200	971	40	2.22	10.8	1508	16.9
300	1456	24.6	2.05	10.0	1392	15.7
400	1942	17	1.89	9.2	1283	14.4
500	2427	12.5	1.74	8.5	1181	13.3

†Based on the weight and volume of the active cell materials.
Cell weight 206 gm, 132 cm³
$C_{eff} = 2 \text{ wattsec}/(3.8^2 - 2.2^2)$

Pulse Resistance Tests Results		
	Resistance (mΩ)	
Current (A)	Pulse test (5 s)	RC (s)
100	2	3.8
200	1.9	3.5

Peak pulse power at 95% efficiency $R = 1.9$ mΩ
$P = 9/16 \times 0.05 \times (3.8)^2/0.0019 = 214$ W, 1038 W/kg

39.6 COST AND SYSTEM CONSIDERATIONS FOR CAPACITORS AND BATTERIES

39.6.1 Electrochemical Capacitor and Battery Costs

Reducing the present high cost/price of electrochemical/ultracapacitors is a key issue in achieving high market penetration in the future, especially of midsize and large devices. There are many applications for which ultracapacitors are presently precluded or not even seriously considered because they remain too expensive, even though their selling price has decreased significantly in recent years. The cost of manufacture of any product is closely tied to production volume, with the cost decreasing rapidly with increased volume up to relatively high production rates. Sales of capacitors in many millions of units per year are necessary to reduce the unit costs to levels at which large markets can develop. Semi-automated production facilities presently exist at a number of companies for ultracapacitors of all sizes. In fact, production capabilities exceed sales volumes for most devices, and that is the reason the price of devices has decreased markedly in recent years. It is common to speak of the price of devices in terms of cents per farad (cents/F) or $/Wh stored. It is easier to interpret the price information on the cents/F basis as it does not concern the cell voltage or what fraction of energy stored can be used in a particular application. For example, for a 10 F device, if the price is quoted as 10 cents per farad, the device cost would be $1. Similarly, a 2500 F device would cost $25 at 1 cent/F.

The cost to manufacture a carbon/carbon device depends on the material and production costs. At the present time, material costs are high. The cost of carbon suitable for use in ultracapacitors can be as high as $100/kg, with the average price in the $30–50/kg range. The cost of the electrolyte solvent is also high, in the range of $5–10/L for both propylene carbonate and acetonitrile. The ionic salts that dissociate in the solvent into the positive and negative ions that move into and out of the double layer in the microporous carbon to store the energy are also expensive, at $50–100/kg. Since the analysis of ultracapacitors is straightforward, material costs can be calculated with good accuracy.[63,64] The result of a typical costing exercise is shown in Table 39.17. Note the strong dependency of the cents/F and $/Wh unit costs for the device on the unit material costs. Presently the price of ultracapacitors is high because both the material and manufacturing costs are high. With more automated production and reduced material costs, it is anticipated that the price of ECCs in high volume can be in the range of 1 to 2 cents/F for small devices and 0.25 to 0.5 cents/F for large devices like those needed for vehicle applications.

TABLE 39.17 Material Costs for a 2.7 V, 3500 F Capacitor*

Carbon			Electrolyte		Device total mat. $	Unit $/kg	Costs		
F/gm	gmC/dev.	$/kg	ACN $/L	$/kg salt			$/Wh	$/kW	Ct./F
75	187	50	10	125	17.0	29	6.4	29	0.48
120	117	100	10	125	15.5	26	6	26	0.44
75	187	5	2	50	3.6	6.0	1.3	6	0.10
120	117	10	2	50	2.5	4.2	0.93	4.2	0.070

*4.5 Wh/kg, 1000 W/kg—95% eff.

Ultracapacitors cannot compete with batteries in terms of $/Wh, but they can compete in terms of $/kW and $/unit to satisfy a particular vehicle application. Both energy storage technologies must provide the same power and cycle life and sufficient energy (Wh) for the application. The weight of the battery is usually set by the system power requirement and cycle life, not the minimum energy storage requirement. Satisfying only the minimum energy storage requirement would result in a much smaller, lighter battery than is needed to meet the other requirements. On the other hand, the weight of the ultracapacitor is determined by the minimum energy storage requirement. The power and cycle life requirements are usually easily satisfied. Hence, the unit can be a more optimum solution for many applications, and its weight can be less than that of the battery even though its specific energy is less than $1/10$ that of the battery.

Consider the example of a charge-sustaining hybrid vehicle like the Prius. If the energy stored in the capacitor unit is 125 Wh and in the battery unit is 1500 Wh, the unit costs of the capacitors and battery are related by

$$(\$/Wh)_{cap} = 0.012 \, (\$/kWh)_{bat}$$

The corresponding capacitor costs in terms of cents/F and $/kWh are given by

$$(cents/F)_{cap} = 0.125 \times 10^{-3} \times (\$/kWh)_{bat} \times V_{cap}^2$$

$$(\$/kWh)_{cap} = 9.6 \times 10^4 \, (cents/F)_{cap}/V_r^2$$

and in Table 39.18 for a range of battery costs.

TABLE 39.18 Relationships between Capacitor and Battery Costs

Battery cost $/kWh	Battery cost* $/kW	Ultracap cost (cents/F) $V_{cap} = 2.6$	Ultracap cost (cents/F) $V_{cap} = 3.0$	Ultracap cost** $/kWh $V_{cap} = 3.0$	Ultracap cost $/kW $V_{cap} = 3.0$
300	30	0.25	0.34	3626	7.3
400	40	0.34	0.45	4800	9.6
500	50	0.42	0.56	5973	11.9
700	70	0.59	0.78	8320	16.6
900	90	0.76	1.0	10667	21.3
1000	100	0.84	1.12	11947	23.9

*Battery: 100 Wh/kg, 1000 W/kg; **Capacitor: 5 Wh/kg, 2500 W/kg

The results shown in Table 39.18 indicate that for the charge-sustaining hybrid application, ultracapacitor costs of 0.5 to 1.0 cents/F are competitive with lithium battery costs in the range of $500–700/kWh. Note also that the $/kW costs of the capacitor unit are about ¼ those of the batteries.

39.6.2 Combinations of Capacitors and Batteries

It has been recognized for years that combining ultracapacitors and batteries would significantly reduce the stress on the batteries, particularly in vehicle applications in which the batteries are subject to high-current pulses in both charge and discharge. It is further recognized that to gain maximum advantage from this arrangement would require the use of interface electronics to control the currents from the battery. There has been some laboratory testing of this arrangement,[65,66] but little work directly in vehicles. In general, experience to date has been that if batteries were available that could meet both the energy and power requirements of the vehicle design, the designers chose to use batteries alone even though they realized batteries plus ultracapacitors would have some advantages. In other words, designers will not select a battery/capacitor combination unless there are clear, large advantages to do so. This could be the case when one considers the use of advanced batteries (Wh/kg > 200) in PHEVs and EVs.[67] In PHEVs and EVs, it is desirable for the battery to be sized by the energy needed to sustain a specified all-electric range. In that case, the weight and volume of the battery pack would follow directly from its energy density (Wh/kg, Wh/L). This means that battery technologies with high specific energy will be strongly favored. However, the batteries must also be able to meet the power requirements of the large electric motors used in the PHEVs and EVs. Unfortunately, batteries designed to attain maximum energy in most cases require a sacrifice in power capability, as shown in Table 39.19.

As a consequence, for some vehicle designs the battery will be sized by the power requirement and not the energy requirement, resulting in a larger and more expensive battery than would be the case if the battery had a higher power density. Design options using batteries of various specific energy values are shown in Table 39.20 for PHEVs with all-electric ranges of 10 to 40 mi. The effect of electric motor size (50, 70 kW) on the required specific power levels is also shown in Table 39.21. For the shorter all-electric ranges and a battery specific energy of 200 Wh/kg, the power densities required exceed by a considerable margin those of the batteries shown in Table 39.19. In those cases, it makes sense to consider combining batteries with ultracapacitors. This design option is shown in Table 39.21. The combinations of the 200 Wh/kg battery and the carbon/carbon ultracapacitor result in the lowest weight energy storage units for all the PHEV ranges, even for a 50 kW electric motor. It can be expected that the weight advantage of the combination will be even larger for batteries with a specific energy greater than 200 Wh/kg. These results indicate that combining batteries and ultracapacitors will become increasingly advantageous as designers consider using the more advanced batteries with higher specific energy.

TABLE 39.19 Performance Characteristics of Various Batteries

Battery chemistry	Ah/wgt. kg	R (mΩ)	Wh/kg	W/kg 90%	W/kg 80%	P/E 90%	P/E 80%
Iron phosphate							
EIG	15/0.424	2.5	115	897	1585	7.8	13.8
A123	2.1/0.07	12	88	1132	2000	12.9	22.8
K2	2.5/0.082	17	86	682	1205	7.9	14.0
Lithium titanate							
Altairnano	12/0.34	2.2	70	693	1225	9.9	17.5
Altairnano	3.8/0.26	1.1	35	2260	4020	64.5	115
EIG	11/0.44	1.9	43	620	1100	14.4	23.8
Li(NiCo)O$_2$							
EIG	18/0.45	3.0	140	913	1613	6.5	11.6
GAIA	42/1.53	0.48	94	1677	2965	17.8	31.5
Quallion	1.7/0.047	70	170	374	661	2.2	3.9
Quallion	1.3/0.043	59	144	486	860	3.4	6.0
NiM hydride							
Panasonic HEV	6.5/1.04	11.4	46	393	695	8.5	15.1
EV	65/11.5	8.7	68	87	154	1.3	2.3
Lead-acid							
Panasonic HEV	25/11.5	7.8	26	146	258	5.6	9.9
EV	60/21.1	6.9	34	89	157	2.6	4.6
Zn-Air							
Revolt Technology	—	—	450	200	—	0.5–1.0	1–2

P_{max} = Eff. (1-Eff.) $(V_{oc})^2/R$
P/E = (W/kg)/Wh/kg

TABLE 39.20 Battery Sizing and Power Density for Various Ranges and Motor Power

	Battery		200 Wh/kg			100 Wh/kg			70 Wh/kg		
Range miles	kWh* needed	kWh** stored	kg**	50 kW kW/kg	70 kW kW/kg	kg	50 kW kW/kg	70 kW kW/kg	kg	50 kW kW/kg	70 kW kW/kg
10	2.52	3.6	18	2.78	3.89	36	1.39	1.94	51	0.98	1.37
15	3.78	5.4	27	1.85	2.59	54	0.92	1.30	77	0.65	0.91
20	5.04	7.2	36	1.39	1.94	72	0.69	0.97	103	0.49	0.68
30	7.56	10.8	54	0.93	1.30	108	0.46	0.65	154	0.32	0.46
40	10.1	14.4	72	0.69	0.97	144	0.35	0.49	206	0.24	0.34

* Vehicle energy usage from the battery: 250 Wh/mi
** Usable state-of-charge for batteries: 70%; weights shown are for cells only.

TABLE 39.21 Storage Unit Weights Using a Combination of Batteries and Ultracapacitors for Various All-Electric Ranges and 70 kW Power

Wh/kg	5		200		100		70	
Range miles	Ultracap kg*	Battery kg**	Combination kg	Battery kg	Combination kg	Battery kg	Combination kg	
10	20	18	38	36	56	51	71	
15	20	27	47	54	74	77	97	
20	20	36	56	72	92	103	123	
30	20	54	74	108	128	154	174	
40	20	72	92	144	164	206	226	

* The carbon/carbon ultracapacitor unit stores 100 Wh usable energy.
** Weights shown are for cells only; packaging into modules not included.

39.6.3 Module and Lifetime Considerations

Module Characteristics and Design. The cell voltage of electrochemical capacitors is relatively low so that in most applications the cells are combined into modules, which are placed in series to achieve the operating system voltage of 200 to 600 V. Cooling and voltage-temperature management of the cells are done on a module basis. Cooling[68,69] of capacitors is less difficult than batteries because of the lower resistance (higher efficiency) of the capacitors. As shown in Table 39.22, several capacitor developers are marketing modules having voltages in the range of 12 to 60 V. As would be expected, the weight and volume of the modules are significantly greater than that of the cells alone, resulting in packing factors of 0.5 to 0.7 for weight and 0.4 to 0.6 for volume. The system specific energies of the electrochemical capacitors will be significantly less than the cell values for each of the capacitor technologies.

TABLE 39.22 Ultracapacitor Module Characteristics

Module	Weight/volume kg/L	Voltage	Wh (Wh/kg)	Power (kW) (90% effic.)	Weight packaging factor	Volume packaging factor
Ness (194 F)	18.5/20.9	48	43/2.1	19.1	0.655	0.36
Ness (100 F)	9.1/7.22	48	22.5/2.47	10.8	0.769	0.692
Maxwell (145 F)	13.5/13.4	48	36/2.7	14.5	0.627	0.484
Maxwell (430 F)	5.0/4.85	16	11.8/2.36	4.8	0.564	0.445
Asahi Glass 280 F	3.75/2.95	16	7.65/2.04	2.1	0.528	0.422
Power Systems	4.4/4.8	32	11/2.5	2.5	0.573	0.375
Power Systems	7.2/8	59	20/2.78	4.7	0.642	0.413
EPCOS	29/24	56	49/1.7	16	0.5	0.48

Lifetime Considerations for Packs. One of the advantages of electrochemical capacitors relative to batteries is their long cycle life, which may be as long as one million (10^6) deep discharge cycles for cells at room temperature and less-than-rated cell voltage. Unfortunately, estimating the lifetime (years) of a pack of capacitor cells in a particular application is much more complicated than simply cycling cells at room temperature. One primary reason for this difficulty is that in most applications (vehicles in particular) many cells are connected in series to attain the required system voltage. In addition, the temperature varies across the pack, even with cooling, and the cells spend some time at voltages approaching their maximum rated voltage even with cell balancing circuitry. As discussed in the following paragraphs, these factors significantly reduce the pack lifetime from that expected based on single cell testing.

The estimation of cell and pack lifetime is considered in detail analytically in References 70 and 71. The analysis is based on the assumption that the cell lifetime statistics can be expressed in terms of a Weibull distribution.

$$F(t) = 1 - \exp[-(t/\alpha)^\beta], \ F = \text{the fraction of cell failures}$$

where t = time, α = characteristic life, and β = shape factor

Cell lifetime testing must be done using test conditions that result in the same aging mechanisms (cell failures) as in the application of interest. For electrochemical capacitors, the testing could be either cycling at specified power levels, voltage ranges, and temperatures or float at specified temperatures and voltages. For vehicle applications, it is likely that lifetime data[70] from the cell float tests are the most appropriate. Such data can be curve fit to obtain values for α and β for the single cells. Both parameters can vary over wide ranges depending on the cell voltage and temperature of the tests. In the case of activated carbon cells, the characteristic time (α) can vary from > 10,000 h at room temperature and 2.3 V to < 500 h at 60°C and 2.8 V. The shape factor (β) can vary from about 4 for room temperature and 2.3 V to about 15 and 2.8 V for 60°C. A low value of β means that cell failures occur gradually over time, while a high value of β indicates that nearly all the cells fail together over a short period of time.

The lifetime characteristics of a pack of electrochemical capacitors depend strongly on the number (M) of the cells connected in series. Assuming the pack failure statistics are also Weibull, the shape factor of the pack is the same as that of the cells in the pack, and each cell failure is independent of other cells, the pack failure function F_{pack} can be written as

$$F_{pack} = 1 - \exp[-(t/\alpha_c)^\beta]^M$$

$$R_{pack} = \exp[-(t/\alpha_c)^\beta]^M, R_{pack} = \text{fraction of cells not failed}$$

Defining a characteristic time for the pack (α_p) and equating

$$[(t/\alpha_c)^\beta]^M = (t/\alpha_p)^\beta$$

one finds

$$\alpha_p = \alpha_c/M^{1/\beta}$$

Hence, the characteristic time of the pack is much less than that of the cells. For example, for a pack with 200 cells in series, the characteristic time is reduced by a factor of 3.76 if β is 4, and by a factor of 1.42 if β is 15. For most packs, the characteristic time would likely be 1/3 to 1/2 that of the cells.

The next factors to consider in the estimation of the lifetime of a pack are the effects of nonuniformities in voltage and temperature of the cells. Even with cooling and cell balancing circuits, there will be nonuniformities in the pack. This will especially be the case for applications in which the pack provides dynamic high power, as in vehicles. The effects of nonuniformities are considered analytically in References 71 and 72. The analysis is based on the following assumptions concerning the effect of temperature and voltage on the failure of single cells: in the case of temperature, a 10°C decrease in temperature doubles the cell life, and in the case of voltage, a 0.1 V decrease in voltage doubles the cell life. The analytical forms of the temperature (T deg. K) and voltage (V) effects on the cell characteristic lifetime τ are

$$\tau = a \exp(b/T), \tau/\tau_0 = \exp[-6155 (T - T_0)/T_0^2]$$

$$\tau = A \exp(-BV), \tau/\tau_0 = \exp[-6.93(V - V_0)]$$

These relationships project a reduction in characteristic time of about $1/\sqrt{2}$ for variations of 5°C in temperature and 0.05 V in voltage. This corresponds to a maximum temperature difference of 10° with an average difference of 5°C for an average temperature of 30°C. Similarly, the maximum voltage difference is 0.1 V with an average difference of 0.05 V.

Applying these relationships to a particular application requires detailed knowledge of the application and specification of the cell operating conditions and tolerable cell failure rates. Consider the following example of a capacitor pack in a hybrid passenger car. The pack failure requirements are 98% reliability for 5 years and 80% reliability for 12 years. The corresponding mileage values are 50,000 mi and 120,000 mi. If the average speed is 25 mi/h, the operating time requirements are 2000 h and 4800 h for the pack. The pack voltage is 300 V with 125 cells in series. The question to be answered is what cell characteristic time (hours) is needed for the cells if the shape factor of their failure distribution is 10. For a Weibull distribution, the relationship between the proportion (P) of failures, time to failure (t_F), and distribution characteristics (α_{pack}, β) is

$$t_F = \alpha_{pack} [-\ln(1 - P)]^{1/\beta}$$

For the times to failure (2000 and 4800 h), the corresponding α_{pack} values are 2954 and 5581 h. Taking the maximum value of 5581, the cell α_c is 9041 h for $M = 125$ and $\beta = 10$. Assuming the average temperature variation is 5°C and the average voltage variation is 0.05 V/cell in the pack from the base values of 30°C and 2.5 V/cell, the cell statistics on a float type test at the base values of temperature and voltage should exhibit a α_c of 18,082 and a $\beta = 10$. This corresponds to a float time of about 2 years at 30°C and 2.5 V/cell.

39.6.3 Cell Balancing Considerations

Background. The ultracapacitor unit in many applications will be relatively high voltage (60 to 500 V) and provide high power in both discharge and charge modes. As in the case of a battery pack, the capacitor unit will consist of many cells (20 to 200) connected in series. If each of the cells were identical, having exactly the same capacitance, series resistance, and parallel resistance, the voltage of all the cells would be the same at all times and be equal to the average cell voltage (V/number of cells). There would not be concern that the voltage of some of the cells would exceed at times a maximum specified by the cell manufacturer. This maximum voltage is often referred as the working (continuous) voltage limit of the cell, and experiencing voltages above that limit for times longer than a few seconds can significantly reduce the lifetime of the cell. This is not a safety issue per se as the cells can withstand considerably higher voltages without the pressure relief vent being activated. Limiting the maximum cell voltage in the unit is then primarily an issue of cell life, the maximum usable energy of the unit, and system efficiency in high-power cycling. Controlling the variability of the voltage between the cells is referred to as cell balancing. The objectives of cell balancing are to minimize the differences in the cell voltages and to restrict the maximum voltage of any cell to less than the continuous working voltage. It is desirable that the cell balancing system provide a means of monitoring the voltages of the cells in order to ensure they are maintaining their voltages in the proper range. Monitoring the voltage and temperature of the cells is needed to ensure long cycle life (10 years or longer).

The complexity of the balancing approach required and the absolute need for cell balancing depends to a large extent on the magnitude of the differences of the characteristics of the cells (capacitance and resistances). These differences are dependent on the uniformity of the materials used in the cells and quality control in the manufacture of the cells. The specification of plus or minus 15 or 20% variability for capacitance and resistance often seen on spec sheets is grossly greater than tolerable for high-voltage strings of capacitors. Experience (available test data[73]) has shown that the variability in the capacitance can be quite small for relatively large cells, with the variation between the maximum and minimum being 1 to 1.5% and the standard deviation of the distribution of the capacitance being about 0.5%. Experience has shown larger percentage variations in the resistance of low resistance (fraction of a milliohm) cells, but the standard deviation is close to the accuracy of the resistance measurement (0.01 mΩ). Experience for the self-discharge voltage of a batch of cells after 1 h has shown a maximum variation of less than 5 mV and a standard deviation of less than 1 mV. Present manufacturing practice for capacitor cells seems to be reasonably good and improving, and the prospects for use of relatively simple approaches to cell balancing seem to be better than was the case several years ago.

Relating the variability in cell characteristics to the cell-to-cell voltage variability for complex discharge/charge cycles is not a simple matter.[74,75] Of primary interest are the magnitude of the maximum cell variations, especially those in the direction of high voltage, and whether the magnitudes of the voltage variations tend to increase for long-term cycling without cell balancing and/or equalization. Available data[76,77] seem to indicate that the magnitude of cell-to-cell voltage variations do not increase with cycling, but rather seem to stabilize even without cell balancing. This seems to be the case even as the cells age.

Variations in cell capacitance will lead to the largest cell-to-cell voltage variations, but those variations will not tend to increase with extended charge-discharge cycling because the capacitance differences have self-compensating effects for charge and discharge pulses. Variations in cell resistances (series and parallel) lead to smaller cell-to-cell voltage variations, and those variations also tend to be self-compensating for cycles that consist of sequential charge-discharge pulses. Variations in self-discharge (parallel resistance) can lead to significant differences in cell voltages during

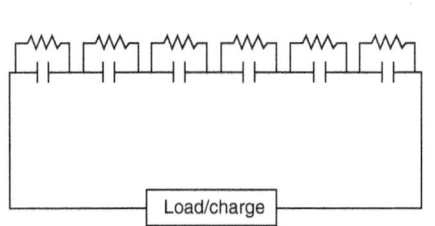
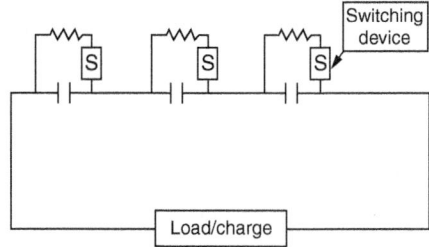

With passive balancing, a resistive ladder is connected to each node in the string of series-connected capacitors.

For active balancing, a switching device is placed in series with each balancing resistor. The switches are controlled by voltage-detection circuits that only turn a switch "on" when a capacitor's voltage approaches its continuous-working-voltage rating.

(a)   (b)

Active voltage balancing electronics

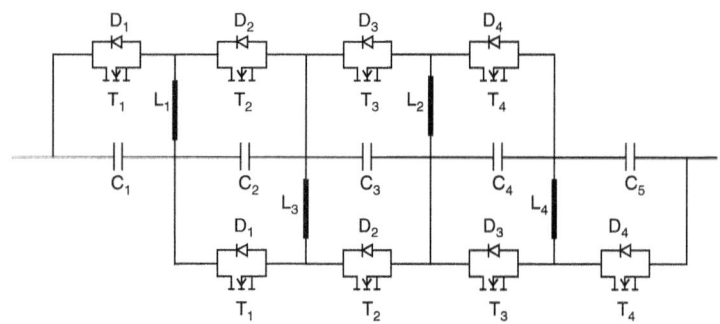

Active charge equalization device based on back-boost topology

(c)

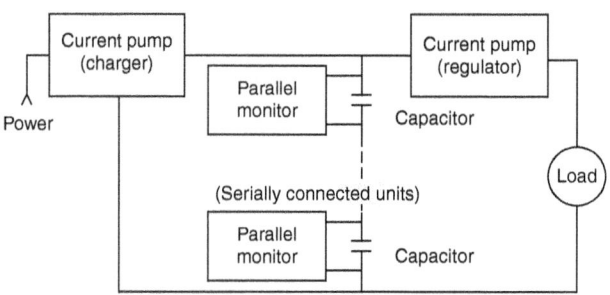

Basic configuration of ECaSS®

(d)

FIGURE 39.13 Cell balancing circuits.

extended rest periods without cell balancing. This could lead to large differences in cell voltages when charge-discharge cycling is resumed. For vehicle applications involving charge-discharge cycling, the effect of variations in cell characteristics on cell-to-cell voltage variations can be determined from cycle testing of the capacitor unit for relatively short times. The cycling time should be long enough to reach thermal equilibrium.

Approaches to Cell Balancing. All the capacitor manufacturers are developing cell balancing circuits for use with long series strings of their cells. Modules are supplied with balancing circuits installed. There are a number of approaches to cell balancing. These approaches (see Fig. 39.13) range in complexity from placing a simple resistor in parallel with each cell to connecting an active circuit with a power source to charge or discharge each cell separately as needed to reduce cell-to-cell voltage variations. The simple approaches are likely to be adequate for most applications if the variability of the cell characteristics is relatively small (a few percent). The more complex approaches will be needed if the cell variability is relatively large due to poor manufacturing quality control, large temperature gradients in the capacitor unit, and/or the effects of aging. It should be recognized at the outset that the currents involved with the balancing of the cells are small (1 A or less), so in most applications they are much smaller than the pulse currents in/out of the cells. This means that during the high-power portions of the capacitor charge-discharge, the effects of the cell balancing on the cell voltages is small. Cell balancing has the largest effect during periods of low power demand or rest. Regardless of the cell balancing approach used, the variability of the characteristics of the cells must be relatively small if cell-to-cell voltage variations are to be tolerable.

As discussed in Section 39.6.3, the lifetime (time to significant failures) in the cells decreases by about a factor of 2 for a 0.1 V increase in cell voltage if that voltage increase is experienced a significant fraction of the time. In general, the lifetime of the cells decreases markedly as the maximum set voltage for the cell balancing circuit is increased beyond about 2.4 V/cell for the present (2009) cell technology. It appears that the cell voltages should be limited to an average of about 2.4 V/cell with a cell-to-cell variation of less than 0.1 V/cell. This should result in long float and cycle lifetimes of greater than 500,000 cycles and 15,000 h on float at or near room temperature (25 to 30°C).

REFERENCES

1. B. E. Conway, *Electrochemical Capacitors: Scientific Fundamentals and Technological Applications*, Kluwer Academic/Plenum, 1999.
2. P. Chandrasekhar, *Conducting Polymers, Fundamentals and Applications*, Kluwer Academic Publishers, 1999.
3. "Electrochemical Capacitors Empowering the 21st Century," Electrochemical Society, *Interface* **17**(1) (Spring 2008).
4. P. L. Taberna, C. Portet, and P. Simon, "The Role of the Interfaces on Supercapacitor Performance," *Proceedings of the 2nd European Symposium on Supercapacitors and Applications*, Lausanne, Switzerland, November 2006.
5. Y. Maletin, "EnerCap Ultracapacitors of the Highest Power," *Proceedings of the First International Symposium on Large Ultracapacitor Technology and Applications*, Honolulu, HI, June 2005.
6. P. Simon and A. F. Burke, "Nanostructure Carbon: Double-Layer Capacitance and More," Electrochemical Society, *Interface* **17**(1) (Spring 2008).
7. J. Chimiola et al., "Anomalous Increase in Carbon Capacitance at Pore Sizes Less than 1 Nanometer," *Science* **313**:1760–1763 (2006).
8. A. G. Pandolfo and A. F. Hollenkamp, "Carbon Properties and Their Role in Supercapacitors," *Journal of Power Sources* **157**:11–27 (2006).
9. R. Istvan, "Nanocarbons," *Proceedings of the 17th International Seminar on Double-Layer Capacitors and Hybrid Energy Storage Devices*, Deerfield Beach, FL, December 2007, pp. 87–109.
10. A. F. Burke, T. Kershaw, and M. Miller, "Development of Advanced Electrochemical Capacitors Using Carbon and Lead-Oxide Electrodes for Hybrid Vehicle Applications," UC Davis Institute of Transportation Studies report, UCD-ITS-RR-03-2, June 2003 (available at www.its.ucdavis.edu).

11. M. Yoshio, "Megalo-Capacitance Capacitor and Management System," *Proceedings of the 2nd International Symposium on Large Ultracapacitor (EDLC) Technology and Applications*, Baltimore, MD, May 16–17, 2006.
12. M. Okamura et al., "The Nanogate Capacitor: A Potential Replacement for Batteries," *Proceedings of the 22nd International Battery Seminar and Exhibit*, Ft. Lauderdale, FL, March 14–17, 2005.
13. T. Fujino, B. Lee, S. Oyama, and M. Noguchi, "Characterization of Advanced Mesophase Carbons Using a Novel Mass Production Method," *Proceedings of the 15th International Seminar on Double-Layer Capacitors and Hybrid Energy Storage Devices*, Deerfield Beach, FL, December 2005.
14. K. Tamamitsu et al., "Electrochemical Capacitors for Next Generation Utilizing Nanostructured Electrode Materials," *Proceedings of the 17th International Seminar on Double-Layer Capacitors and Hybrid Energy Storage Devices*, Deerfield Beach, FL, December 2007, pp. 122–133.
15. R. Signorelli et al., "Fabrication and Electrochemical Testing of First Generation Carbon-Nanotube Based Ultracapacitor Cells," *Proceedings of the 17th International Seminar on Double-Layer Capacitors and Hybrid Energy Storage Devices*, Deerfield Beach, FL, December 2007, pp. 70–78.
16. P. Ruch, W. Kotz, and A. Wokaun, "Electrochemical Characterization of Single-Wall Carbon Nanotubes for Electrochemical Double-Layer Capacitors Using Non-aqueous Electrolyte," *Electrochemica Acta* **54**:4451–4458 (2009).
17. J. P. Zheng and T. R. Jow, "High Power and High Energy Density Capacitors with Composite Hydrous Ruthenium Oxide Electrodes," *Proceedings of the 5th International Seminar on Double-Layer Capacitors and Similar Energy Storage Devices*, Deerfield Beach, FL, December 1995.
18. D. Bélanger, T. Brousse, and J. W. Long, "Manganese Oxides: Battery Materials Make the Leap to Electrochemical Capacitors," Electrochemical Society, *Interface* **17**(1) (Spring 2008).
19. K. Naoi, "Recent Advances in Supercapacitors and Hybrid Capacitor Systems," *Proceedings of the 4th International Symposium—Large Ultracapacitor Technology and Applications*, Tampa, FL, May 12–14, 2008.
20. A. L. Reddy and S. Ramaprabhu, "Nanocrystalline Metal Oxides Dispersed Multiwalled Carbon Nanotubes as Supercapacitor Electrodes," *Journal of Physical Chemistry C* **111**(21):7727–7734 (May 2007).
21. D. Bélanger, "Nanostructured Metal Oxides for Hybrid Electrochemical Supercapacitors," *Proceedings of the EuroCapacitors Conference*, Cologne, Germany, November 7–8, 2007.
22. A. F. Burke, T. Kershaw, and M. Miller, "Development of Advanced Electrochemical Capacitors Using Carbon and Lead-Oxide Electrodes for Hybrid Vehicle Applications," UC Davis Institute of Transportation Studies Report, UCD-ITS-RR-03-2, June 2003.
23. S. M. Butler and J. R. Miller, "Asymmetric $PbO_2/H_2SO_4/C$ Electrochemical Capacitor," paper presented at the *203rd Meeting of the Electrochemical Society*, Paris, France, April 2003.
24. N. Doddapaneni, "Development of Electrochemical Pulse Power Systems (IMRA)," *Proceedings of the 9th International Seminar on Double-Layer Capacitors and Similar Energy Storage Devices*, Deerfield Beach, FL, December 1999.
25. G. G. Amatucci et al., "The Non-Aqueous Asymmetric Hybrid Technology: Materials, Electrochemical Properties and Performance in Plastic Cells," *Proceedings of the 11th International Seminar on Double-Layer Capacitors*, Deerfield Beach, FL, December 2001.
26. S. M. Lipka, D. E. Reisner, J. Dai, and R. Cepulis, "Asymmetric-Type Electrochemical Supercapacitor Development Under the ATP—An Update," *Proceedings of the 11th International Seminar on Double-Layer Capacitors* (Florida Atlantic University), Deerfield Beach, FL, December 2001.
27. T. Dougherty, "Conducting Plastic Current Collector Materials," private communication, contact at: tomd@monolithengines.com.
28. Y. Maletin, P. Novak, and E. Shembel, "New Approach to Organic Electrolytes and Carbon Electrode Materials," *Proceedings of the Advanced Capacitor World Summit 2004*, Washington, DC, July 2004.
29. K. Naoi and M. Morita, "Advanced Polymers as Active Materials and Electrolytes for Electrochemical Capacitors and Hybrid Capacitor Systems," Electrochemical Society, *Interface* **17**(1):44–48 (Spring 2008).
30. E. Frackowiak, "Supercapacitors Based on Carbon and Ionic Liquids," *Journal of the Brazilian Chemical Society* **17**(6) (October 2006).
31. K. Schoch and C. J. Weber, "Safety Management of Ultracapacitors Under Abusive Conditions," *Proceedings of the Advanced Capacitor World Summit 2005*, San Diego, CA, July 2005.

32. T. Furukawa, "The Reliability, Performance, and Safety of DLCAP," *Proceedings of the 2nd International Symposium on Large Ultracapacitor Technology and Applications*, Baltimore, MD, May 2006.
33. V. R. Koch, "Recent Advances in Electrolytes for Electrochemical Double-Layer Capacitors," *Proceedings of the First International Symposium on Large Ultracapacitor Technology and Applications*, Honolulu, HI, June 13–14, 2005.
34. K. Tada et al., "High Purity Ionic Liquids for Capacitors," *Proceedings of the 17th International Seminar on Double-Layer Capacitors and Hybrid Energy Storage Devices*, Deerfield Beach, FL, December 2007.
35. "Global Capacitor Markets: 2005–2010," prepared by and available from Paumanok Publications, September 2005.
36. "Ultracapacitors: A Global Industry and Market Analysis," prepared by and available from Innovative Research and Products, Inc., Stamford, CT, August 2006.
37. CAP-XX website: www.cap-xx.com/products/products.htm
38. NEC-TOKIN website: www.nec-tokin.com/english/product/supercapacitor/
39. A. F. Burke, "Ultracapacitor Technology: Present and Future," *Proceedings of the Advanced Capacitor World Summit 2003*, Washington, DC, August 11–13, 2003.
40. R. A. Dougal, L. Gao, and S. Liu, "Ultracapacitor Model with Automatic Order Selection and Capacity Scaling for Dynamic System Simulation," *Journal of Power Sources* **126**(1–2):250–257 (2004).
41. J. M. Miller et al., "Carbon-Carbon Ultracapacitor Equivalent Circuit Model, Parameter Extraction, and Application," Ansoft First Pass Workshop, (Maxwell Technologies Presentation), October 2007.
42. "Basics of Electrochemical Impedance Spectroscopy (EIS), Applications Note AC-1," Princeton Applied Research, 2008.
43. A. Hammar et al., "Electrical Characterization and Modeling of Round Spiral Supercapacitors for High Power Applications (AC Impedance Testing)," paper presented at *ESSCAP 2006*, Lausanne, Switzerland.
44. J. R. Miller and S. M. Butler, "Development of Battery and Electrochemical Capacitor Equivalent Circuit Models for Power System Optimization," paper presented at the *202nd Electrochemical Society Meeting*, Salt Lake City, UT, October 2002.
45. EIS300 Electrochemical Impedance Spectroscopy software, Gamry Instruments, www.gamry.com.
46. A. Chu and P. Braatz, "Comparisons of Commercial Supercapacitors and High-Power Lithium Batteries for Power-Assist Applications in Hybrid Electric Vehicles: Initial Characterization," *Journal of Power Sources* **112**(1):236–246 (October 2002).
47. M. Carlen, T. Christen, and C. Ohler, "Energy-Power Relations for Supercaps from Impedance Spectroscopy Data," *Proceedings of the 9th International Seminar on Double-Layer Capacitors and Similar Energy Storage Devices*, Deerfield Beach, FL, December 1999.
48. M. Arulepp et al., "The Advanced Carbide-Derived Carbon Based Supercapacitor," *Journal of Power Sources* **162**(2):1460–1466 (November 2006).
49. Y. Gogotsi et al., "Nanoporous Carbide-Derived Carbon with Tunable Pore Size," *Nature Materials* **2**:591–594 (August 2003).
50. R. DeLevie, "Electrochemical Response of Porous and Rough Electrodes," in P. Delhay, ed., *Advances in Electrochemistry and Electrochemical Engineering*, Vol. 6, Interscience Publishers, 1967.
51. F. M. Delnick, C. D. Jaeger, and S. C. Levy, "AC Impedance Study of Porous Carbon Collectors for Li/SO$_2$ Primary Batteries," *Chemical Engineering Communications* **35**:23–28 (1985).
52. C. J. Farahmandi, "Analytical Solution to an Impedance Model for Electrochemical Capacitors," *Proceedings of the Advanced Capacitor World Summit 2007*, San Diego, CA, June 2007; also *Electrochemical Society Proceedings* PV96-25, 1996.
53. V. Srinivasan and J. W. Weidner, "Mathematical Modeling of Electrochemical Capacitors," *Journal of the Electrochemical Society* **146**:1650–1658 (1999).
54. D. Dunn and J. Newman, "Predictions of Specific Energies and Specific Powers of Double-Layer Capacitors Using a Simplified Model," *Journal of the Electrochemical Society* **147**(3) (2000).
55. H. S. Carslaw and J C. Jaeger, *Conduction of Heat in Solids*, Oxford Press, 1947.
56. J. S. Newman, *Electrochemical Systems*, Prentice Hall Publishers, 1991.
57. IEC, "Electric Double-Layer Capacitors for Use in Hybrid Electric Vehicles—Test Methods for Electrical Characteristics," finalized April 2008.

58. A. F. Burke, "Testing of Supercapacitors: Capacitance, Resistance, and Energy Density and Power Capacity," presentation and report UCD-ITS-RR-09-19, July 2009.
59. V. Srinivasan, G. Q. Wang, and C. Y. Wang, "Mathematical Modeling of Current-Interrupt and Pulse Operation of Valve-Regulated Lead-Acid Batteries," *Journal of the Electrochemical Society* **150**(3):A316–A325 (2003).
60. "FreedomCar Ultracapacitor Test Manual," Idaho National Engineering Laboratory Report DOE/NE-ID-11173, September 21, 2004.
61. "Battery Test Manual for Plug-In Hybrid Electric Vehicles," U.S. Department of Energy, INL/EXT-07-12536, March 2008.
62. J. R. Miller and A. F. Burke, "Electric Vehicle Capacitor Test Procedures Manual," Idaho National Engineering Laboratory Report DOE/ID-10491, October 1994.
63. M. Anderman, "Could Ultracapacitors Become the Preferred Energy Storage Device for Future Vehicles?" *Proceedings of the 5th International Advanced Automotive Battery Conference*, Honolulu, HI, June 15–17, 2005.
64. A. F. Burke and M. Miller, "Ultracapacitor Update: Cell and Module Performance and Cost Projections," *Proceedings of the 15th International Seminar on Double-Layer Capacitors and Hybrid Energy Storage Devices*, Deerfield Beach, FL, December 5–7, 2005.
65. A. F. Burke and M. Miller, "Electrochemical Capacitors as Energy Storage in Hybrid-Electric Vehicles: Present Status and Future Prospects," *EVS-24*, Stavanger, Norway, May 2009.
66. A. F. Burke and M. Miller, "Supercapacitors for Hybrid-Electric Vehicles: Recent Test Data and Future Projections," *Proceedings of the Advanced Capacitor World Summit 2008*, San Diego, CA, July 14–16, 2008.
67. A. F. Burke and M. Miller, "Performance Characteristics of Lithium-Ion Batteries of Various Chemistries for Plug-In Hybrid Vehicles," *EVS-24*, Stavanger, Norway, May 2009 (paper on the CD of the meeting).
68. J. Lustbader, C. King, J. Gonder, M. Keyser, and A. Pesaran, "Thermal Evaluation of a High-Voltage Ultracapacitor Module for Vehicle Applications," *Proceedings of the Advanced Capacitor World Summit 2008*, San Diego, CA, July 2008.
69. A. Pesaran, J. Gonder, and M. Keyser, "Ultracapacitor Applications and Evaluations for Hybrid Electric Vehicles," *Proceedings of the Advanced Capacitor World Summit 2009*, San Diego, CA, July 2009.
70. J. R. Miller, "Reliability Assessment and Engineering of Electrochemical Capacitors," tutorial (JME) at the *Advanced Capacitor World Summit 2007*, San Diego, CA, July 2007.
71. J. R. Miller, S. M. Butler, and I. Goltser, "Electrochemical Capacitor Life Predictions Using Accelerated Test Methods," *Proceedings of the 42nd Power Sources Conference*, Philadelphia, June 2006, paper 24.6, p. 581.
72. J. R. Miller and S. M. Butler, "Capacitor System Life Reduction Caused by Cell Temperature Variation," *Proceedings of the Advanced Capacitor World Summit 2006*, San Diego, CA, July 2006.
73. A. F. Burke, "Characterization of a 25 Wh Ultracapacitor Module for High-Power, Mild Hybrid Applications," *Proceedings of the 1st International Symposium on Large Ultracapacitor Technology and Applications*, Honolulu, HI, June 13–14, 2005.
74. D. Y. Jung, "Shield Ultracapacitor Strings from Overvoltage Yet Maintain Efficiency," *Electronic Design* (May 27, 2002).
75. Y. Kim, "Ultracapacitor Technology Powers Electronic Circuits," *Power Electronics Technology* (October 2003).
76. R. Kotz, J. C. Sauter, P. Ruch, et al., "Voltage Balancing of a 250 V Supercapacitor Module for a Hybrid Fuel Cell Vehicle," *Proceedings of the 16th International Seminar on Double-Layer Capacitors and Hybrid Energy Storage Devices*, Deerfield Beach, FL, December 2007.
77. A. F. Burke and M. Miller, "Cell Balancing Considerations for Long Series Strings of Ultracapacitors in Vehicle Applications," *Proceedings of the Advanced Capacitor World Summit 2005*, San Diego, CA, July 11–13, 2005.
78. F. Rafik, H. Gualous, R. Callay, A. Crausaz, and A. Berthon, "Supercapacitors Characterization for Hybrid Vehicle Applications," paper presented at *ESSCAP 2006*, Lausanne, Switzerland. (Available on the web.)

APPENDICES

APPENDIX A
DEFINITIONS

Accumulator   See SECONDARY BATTERY.

Activated Stand Life   The period of time, at a specified temperature, that a battery can be stored in the charged condition before its capacity falls below a specified level.

Activation   The process of making a reserve battery functional, either by introducing an electrolyte, by immersing the battery into an electrolyte, or by other means.

Activation Polarization   Polarization resulting from the charge-transfer step of the electrode reaction. (See also POLARIZATION.)

Active Cell or Battery   A cell or battery containing all components and in a charged state ready for discharge (as distinct from a RESERVE CELL or BATTERY).

Active Material   The material in the electrodes of a cell or battery that takes part in the electrochemical reactions of charge or discharge.

Adsorption   The taking up or retention of one material or medium by another by chemical or molecular action.

Aging   Permanent loss of capacity due either to repeated use or to passage of time.

Ambient Temperature   The average temperature of the surroundings.

Ampere-Hour Capacity (also Amp-Hour Capacity)   The quantity of electricity measured in ampere-hours (Ah) that may be delivered by a cell or battery under specified conditions.

Ampere-Hour Efficiency (also Amp-Hour Efficiency)   The ratio of the output of a secondary cell or battery, measured in ampere-hours, to the input required to restore the initial state of charge, under specified conditions (also coulombic efficiency).

Anion   Ion in the electrolyte carrying a negative charge.

Anode   The electrode in an electrochemical cell where oxidation takes place. During discharge, the negative electrode of the cell is the anode. In a rechargeable battery, during charge, the situation reverses and the positive electrode of the cell is the anode.

Anolyte   The portion of the electrolyte in a galvanic cell adjacent to the anode; if a diaphragm is present, the electrolyte on the anode side of the diaphragm.

Aprotic Solvent   A nonaqueous solvent that does not contain any reactive protons although it may contain hydrogen atoms in the molecule.

Available Capacity   The total capacity (amp-hours) that will be obtained from a cell or battery at defined discharge rates and other specified discharge or operating conditions.

Battery   One or more electrochemical cells electrically connected in an appropriate series/parallel arrangement to provide the required operating voltage and current levels including, if any, monitors, controls and other ancillary components (fuses, diodes), case, terminals, and markings (see p. 1.3).

Bipolar Plate   An electrode construction where positive and negative active materials are on opposite sides of an electronically conductive plate.

Bobbin A cylindrical electrode (usually the positive) pressed from a mixture of the active material, a conductive material, such as carbon black, the electrolyte, and/or binder with a centrally located conductive rod or other means for a current collector.

Boost Charge Charging of batteries in storage to maintain their capacity and counter the effects of self-discharge.

Boundary Layer The volume of electrolyte solution immediately adjacent to the electrode surface in which concentration changes occur due to the effects of the electrode process.

C Rate The discharge or charge current, in amperes, expressed as a multiple of the rated capacity in ampere-hours.

$$I = M \times C_n$$

where I = current, A
C = numerical value of rated capacity of a battery in ampere-hours (Ah)
n = time in hours for which rated capacity is specified
M = multiple or fraction (of C)

For example, the $0.05C$ or $C/20$ discharge current for a battery rated at 5 Ah at the $0.2C$ or $C/5$ rate is 250 mA.

$$I = M \times C_{0.2} = (0.05)(5) = 0.250 \text{ amperes}$$

Conversely, a battery rated at 300 mAh at the $0.5C$ or $C/2$ rate, discharged at 30 mA, is discharged at the $0.1C$ or $C/10$ rate, which is calculated as follows:

$$M = \frac{I}{C_{0.5}} = \frac{0.030}{0.300} = 0.1 \text{ or } C/10$$

Capacitance Current The fraction of the cell current consumed in charging the electrical double layer.

Capacity The total number of ampere-hours (Ah) that can be withdrawn from a fully charged cell or battery under specified conditions of discharge. (See also AVAILABLE CAPACITY, RATED CAPACITY.)

Capacity Fade Gradual loss of capacity of a secondary battery with cycling.

Capacity Retention The fraction of the full capacity available from a battery under specified conditions of discharge after it has been stored for a period of time.

Cathode The electrode in an electrochemical cell where reduction takes place. During discharge, the positive electrode of the cell is the cathode. In a rechargeable battery, during charge, the situation reverses, and the negative electrode of the cell is the cathode.

Catholyte The portion of an electrolyte in a galvanic cell adjacent to a cathode; if a diaphragm is present, the electrolyte on the cathode side of the diaphragm.

Cation Ion in the electrolyte carrying a positive charge.

Cell The basic electrochemical unit providing a source of electrical energy by direct conversion of chemical energy. The cell consists of an assembly of electrodes, separators, electrolyte, container, and terminals (see p 1.3).

Charge The conversion of electrical energy, provided in the form of a current from an external source, into chemical energy within a cell or battery.

Charge Acceptance Ability of a battery to accept charge. May be affected by temperature, charge rate, and state-of-charge.

Charge Control Techniques for effectively terminating the charging of a rechargeable battery.

Charge Efficiency See EFFICIENCY.

Charge Rate The current applied to a secondary cell or battery to restore its capacity. This rate is commonly expressed as a multiple of the rated capacity of the cell or battery. For example, the $C/10$ charge rate of a 500 Ah cell or battery (rated at the 0.2 rate) is expressed as

$$\frac{C_{0.2}}{10} = \frac{500 \text{ Ah}}{10} = 50 \text{ A}$$

Charge Retention See CAPACITY RETENTION.

Closed-Circuit Voltage (CCV) The potential or voltage of a cell or battery when it is discharging, normally under a specified load.

Concentration Polarization Polarization caused by the depletion of ions in the electrolyte at the surface of the electrode. (See also POLARIZATION.)

Conditioning Cycle charging and discharging of a battery to ensure that it is fully formed and fully charged. Sometimes indicated when a battery is first placed in service or returned to service after prolonged storage. (See also FORMATION.)

Constant-Current Charge A method of charging the battery using a current having little variation.

Constant-Voltage Charge A method of charging the battery by applying a fixed voltage, and allowing variations in the current. Also called constant potential charge.

Continuous Test A test in which a battery is discharged to a prescribed end-point voltage without interruption.

Coulometer Electrochemical or electronic device, capable of integrating current-time, used for charge control and for measurement of charge inputs and discharge outputs. Results usually reported in ampere-hours.

Counter Electromotive Force A voltage of an electrochemical cell opposite to the applied external voltage. Also referred to as back EMF.

Couple Combination of anode and cathode materials that engage in electrochemical reactions that will produce current at a voltage defined by the reactions.

Creepage The movement of electrolyte onto surfaces of electrodes or other components of a cell with which it is not normally in contact.

Current Collector An inert member of high electrical conductivity used to conduct current from or to an electrode during discharge or charge.

Current Density The current per unit active area of the surface of an electrode.

Cutoff Voltage The battery voltage at which the discharge is terminated. Also called end voltage.

Cycle The discharge and subsequent or preceding charge of a secondary battery such that it is restored to its original conditions.

Cycle Life The number of cycles under specified conditions that are available from a secondary battery before it fails to meet specified criteria as to performance.

Cycle Service A duty cycle characterized by frequent and usually deep discharge-charge sequences, such as motive power applications.

Deep Discharge Withdrawal of at least 80% of the rated capacity of a battery.

Density The ratio of a mass of material to its own volume at a specified temperature.

Depolarization A reduction in the polarization of an electrode.

Depolarizer A substance or means used to prevent an increase in polarization. The term "depolarizer" is often used to describe the positive electrode or cathode of a primary cell.

Depth of Discharge (DOD) The ratio of the quantity of electricity (usually in ampere-hours) removed from a cell or battery on discharge to its rated capacity.

Desorption The opposite of absorption, whereby the material retained by a medium is released.

Diaphragm A porous or permeable material for separating the positive and negative electrode compartments of an electrochemical cell and preventing mixing of catholyte and anolyte.

Diffusion The movement of species under the influence of a concentration gradient.

Discharge The conversion of the chemical energy of a cell or battery into electrical energy and withdrawal of the electrical energy into a load.

Discharge Rate The rate, usually expressed in amperes, at which electrical current is taken from the cell or battery.

Double Layer The region in the vicinity of an electrode-electrolyte interface where the concentration of mobile ionic species has been changed to values differing from the bulk equilibrium value by the potential difference across the interface.

Double-Layer Capacitance The capacitance of the electrical double layer at an electrode-electrolyte interface.

Dry Cell A cell with immobilized electrolyte. The term "dry cell" is often used to describe the Leclanché cell.

Dry-Charged Battery A battery in which the electrodes are in a charged state, ready to be activated by the addition of the electrolyte.

Duplex Electrode or Plate See BIPOLAR PLATE.

Duty Cycle The operating regime of a cell or battery including factors such as charge and discharge rates, depth of discharge, cycle length, and length of time in the standby mode.

E Rate The discharge or charge power, in watts, expressed as a multiple of the rated capacity in watthours.

$$P = M \times E_n$$

where P = power, W
  E = numerical value of the rated energy of a battery in watthours (Wh)
  n = time in hours, at which the battery was rated
  M = multiple or fraction (of E)

For example, the $0.05E$ or $E/20$ discharge power for a battery rated at 5 h at the $0.2E$ or $E/5$ rate is 250 mW.

$$P = M \times E_{0.2} = (0.05)(5) = 0.250 \text{ watts}$$

Conversely, a battery rated at 300 mWh at the $0.5E$ or $E/2$ rate, discharged at 30 mW, is discharged at the $0.1E$ or $E/10$ rate, which is calculated as follows:

$$M = \frac{I}{E_{0.5}} = \frac{0.030}{0.300} = 0.1$$

Efficiency The ratio of the output of a secondary cell or battery on discharge to the input required to restore it to the initial state-of-charge under specified conditions. (See also AMPERE-HOUR EFFICIENCY, ENERGY EFFICIENCY, VOLTAGE EFFICIENCY, and WATTHOUR EFFICIENCY.)

Electrical Double Layer See DOUBLE LAYER.

Electrocapillarity The surface tension between liquid mercury and an electrolyte solution is modified by the potential difference across the interface. The effect is termed electrocapillarity.

Electrochemical Cell A cell in which the electrochemical reactions are caused by supplying electrical energy or which supplies electrical energy as a result of electrochemical reactions. If the first case only is applicable, the cell is an electrolysis cell; if the second case only, the cell is a galvanic cell.

Electrochemical Couple See COUPLE.

Electrochemical Equivalent   Weight of one equivalent of a substance being electrolyzed, which is its gram atomic weight or its gram molecular weight divided by the number of electrons in the electrode reaction. (See also FARADAY.)

Electrochemical Series   A classification of the elements according to the values of the standard potentials of specified electrochemical reactions.

Electrode   The site, area, or location at which electrochemical processes take place.

Electrode Potential   The voltage developed by a single plate, either positive or negative, against a standard reference electrode, typically the standard hydrogen electron. The algebraic difference in voltage of any two electrodes equals the cell voltage.

Electroformation   A term applied to the conversion of the material in both the positive and negative plates to their respective active materials. Also referred to as formation.

Electrolyte   The medium that provides the ion transport mechanism between the positive and negative electrodes of a cell.

Electromotive Force (EMF)   The standard potential of a specified electrochemical action.

Electromotive Series   See ELECTROCHEMICAL SERIES.

Electron   The elemental particle of an atom having a negative charge.

Element   The negative and positive electrodes together with the separators of a single cell. It is used almost exclusively in describing lead-acid cells and batteries.

End Voltage   The prescribed voltage at which the discharge (or charge, if end-of-charge voltage) of a battery may be considered complete (also called cutoff voltage).

Energy Density   The ratio of the energy available from a battery to its volume (Wh/L). (See also SPECIFIC ENERGY.)

Energy Efficiency   See WATTHOUR EFFICIENCY.

Equalization   The process of restoring all cells in a battery to an equal state-of-charge.

Equilibrium Electrode Potential   The difference in potential between an electrode and an electrolyte when they are in equilibrium for the electrode reaction that determines the electrode potential.

Equivalent Circuit   An electrical circuit that models the fundamental properties of a device (e.g., a cell) or a circuit.

Exchange Current   Under open-circuit conditions, the forward and backward current of an electrochemical process are equal and opposite. This equilibrium current in one direction is defined as the exchange current.

Faraday   One gram equivalent weight of matter is chemically altered at each electrode of a cell for each 96,494 international coulombs, or one Faraday, of electricity passed through the electrolyte.

Fast Charge   A rate of charging which returns capacity to a rechargeable battery, usually within a few hours.

Fauré Plate   See PASTED PLATE.

Flash Current   See SHORT-CIRCUIT CURRENT.

Flat-Plate Cell   A cell fabricated with rectangular flat-plate electrodes (also called a prismatic cell).

Float Charge   A method of maintaining a battery in a charged condition by continuous, long-term constant-voltage charging, at a level sufficient to balance self-discharge.

Flooded Cell   A cell design that incorporates an excess amount of electrolyte.

Forced Discharge   Discharging a cell or battery below zero volts into voltage reversal.

Formation   Electrochemical processing of a battery electrode that transforms the active materials into their usable form.

Fuel Cell A galvanic cell in which the active materials are continuously supplied from a source external to the cell and the reaction products continuously removed, converting chemical energy to electrical energy.

Galvanic Cell An electrochemical cell that converts chemical energy into electrical energy by electrochemical action.

Gas Recombination Method of suppressing hydrogen generation during charging by recombining oxygen gas on the negative electrode as the cell approaches full charge. Batteries using this method normally contain excess capacity in the negative electrode.

Gassing The evolution of gas from one or more of the electrodes in a cell. Gassing commonly results from local action (self-discharge) or from the electrolysis of the electrolyte during charging.

Grid A framework for a plate or electrode that supports or retains the active materials and acts as a current collector.

Group An assembly of positive or negative plates that fit into a cell.

Half-Cell An electrode (either the anode or cathode) immersed in a suitable electrolyte.

Hourly Rate A discharge rate, in amperes, of a battery that will deliver the specified hours of service to a given end voltage.

Hydrogen Electrode An electrode of platinized platinum saturated by a stream of pure hydrogen, immersed in an electrolyte of known acidity (pH).

Hydrogen Overvoltage The activation overvoltage for hydrogen evolution on an electrode.

Initial (Closed-Circuit) Voltage The on-load voltage at the beginning of a discharge under a specified load.

Inner Helmholtz Plane The plane of closest approach to an electrode of ions in solution. It corresponds to the plane that contains the adsorbed ions and the innermost layer of water molecules.

Intermittent Test A test during which a battery is subjected to alternate periods of discharge and rest according to a specified discharge regime.

Internal Impedance The opposition or resistance of a cell or battery to an alternating current of a particular frequency.

Internal Resistance The opposition or resistance to the flow of an electric current within a cell or battery; the sum of the ionic and electronic resistances of the cell components.

Ion A particle in solution that carries a negative or positive charge.

IR Drop A voltage that is the product of the electrical resistance (R) of a cell or battery and the current (I). The value is the product of the resistance in ohms and the current in amperes.

Life For rechargeable batteries, the duration of satisfactory performance, measured in years (float life) or in the number of charge/discharge cycles (cycle life).

Load A term used to indicate the current drain on a battery, either directly imposed or through a resistance.

Local Action Chemical reactions within a cell that convert the active materials to a discharged state without supplying energy through the battery terminals (self-discharge).

Luggin Capillary The bridge from an external reference electrode to a cell solution often has a capillary tip. The capillary that is often situated close to the working electrode to minimize the IR drop is called a Luggin capillary.

Maintenance-Free Battery A secondary battery that does not require periodic "topping up" to maintain electrolyte volume.

Maximum-Power Discharge Current, I_{mp} Discharge rate at which maximum power is transferred to the external load. This is the discharge rate when the discharge voltage is approximately equal to one-half of the open circuit voltage if the discharge is purely ohmic.

Mechanical Recharging Restoring the capacity of a cell by replacing a spent or discharged electrode with a fresh one.

Memory Effect A phenomenon in which a cell, operated in successive cycles to the same, but less than a full, depth of discharge experiences a depression of its discharge voltage and temporarily loses the rest of its capacity at normal voltage levels. (See Chaps. 21 and 22.)

Midpoint Voltage The voltage of a battery midway in the discharge between the fully charged state and the end voltage.

Migration The movement of a charged species under the influence of a potential gradient.

Motive Power Battery See TRACTION BATTERY.

Negative Electrode The electrode acting as an anode when a cell or battery is discharging.

Negative-Limited The operating characteristics (performance) of the cell is limited by the negative-electrode.

Nominal Voltage The characteristic operating voltage or rated voltage of a battery (as distinct from MIDPOINT VOLTAGE, WORKING VOLTAGE, etc.).

Off-Load Voltage See OPEN-CIRCUIT VOLTAGE.

Ohmic Overvoltage Overvoltage caused by the ohmic drop in an electrolyte.

On-Load Voltage The difference in voltage between the terminals of a cell or battery when it is discharge under a specified load.

Open-Circuit Voltage (OCV) The difference in voltage between the terminals of a cell or voltage when the circuit is open (no-load condition).

Outer Helmholtz Plane The plane of closest approach of those ions that do not contact-absorb but approach the electrode with a sheath of solvated water molecules surrounding them.

Overcharge The forcing of current through a battery after all the active material has been converted to the charged state. In other words, charging continued after 100% state-of-charge is achieved.

Overdischarge Discharge past the point where the full capacity of the battery has been obtained.

Overvoltage The potential difference between the equilibrium potential of an electrode and that of the electrode under an imposed polarization current.

Oxygen Recombination The process by which oxygen generated at the positive plate during charge is reacted at the negative plate.

Paper-Lined Cell Construction of a cell where a layer of paper, wetted with electrolyte, acts as the separator.

Parallel Term used to describe the interconnection of cells or batteries in which all of the like terminals are connected together. Parallel connections increase the capacity of the resultant battery as follows:

$$C_p = n \times C_u$$

where C_p = the resultant capacity
n = the number of cells or batteries connected in parallel
C_u = capacity of the unconnected cell or battery

Passivation The phenomenon by which an electrode, although in conditions of thermodynamic instability, remains unattacked because of its surface condition.

Paste Mixtures of various compounds that are applied to positive and negative grids of lead batteries. These pastes are then converted to positive and negative active materials. (See also FORMATION.)

Paste-Lined Cell Leclanché cell constructed so that a layer of gelled paste acts as the separator.

Pasted Plate A plate manufactured by coating a grid or support strip with active materials in paste form.

Plané Plate A plate for a lead-acid battery in which the active materials are formed directly from a lead substrate by electrochemical processing.

Plate A structure containing active materials held firmly to a grid or conductor.

Pocket Plate A plate for a secondary battery in which active materials are held in perforated metal pockets on a support strip.

Polarity Denoting either positive or negative potential.

Polarization The change of the potential of a cell or electrode from its equilibrium value caused by the passage of an electric current.

> **Activation Polarization** That part of electrode or battery polarization arising from the charge-transfer step of the electrode reaction.
>
> **Concentration Polarization** That part of electrode or battery polarization arising from concentration gradients of battery reactants and products caused by the passage of current.
>
> **Ohmic Polarization** That part of electrode or battery polarization arising from current flow through ohmic resistances within an electrode or battery.

Positive Electrode The electrode acting as a cathode when a cell or battery is discharging.

Positive-Limited The operating characteristics (performance) of the cell is limited by the positive electrode.

Power Density The ratio of the power available from a battery to its volume (W/L). (See also SPECIFIC POWER.)

Primary Cell or Battery A cell or battery that is not intended to be recharged and is discarded when it has delivered all its electrical energy.

Prismatic Cell See FLAT-PLATE CELL.

Rate Constant At equilibrium, the forward and backward Faradic currents of an electrode process are equal and referred to as the exchange current. This exchange current can be defined in terms of a rate constant called the standard heterogeneous rate constant for the electrode process.

Rated Capacity The number of ampere-hours a battery can deliver under specific conditions (rate of discharge, end voltage, temperature); usually the manufacturer's rating.

Recharge See CHARGE.

Rechargeable Battery See SECONDARY BATTERY.

Recombination A term used in a sealed cell construction for the process whereby internal pressure is relieved by reaction of oxygen with the negative active material.

Recovery See RECUPERATION.

Recuperation The lowering of the polarization of a cell during rest periods.

Redox Cell A secondary cell in which two soluble ionic reactants, separated by a membrane, form the active materials.

Reference Electrode A specially chosen electrode that has a reproducible potential against which other electrode potentials may be measured. (See also HYDROGEN ELECTRODE.)

Reserve Cell or Battery A cell or battery that may be stored in an inactive state and made ready for use by adding electrolyte, another cell component, or, in the case of a thermal battery, melting a solidified electrolyte.

Reversal The changing of the normal polarity of a cell or battery.

Secondary Battery A galvanic battery which, after discharge, may be restored to the charged state by the passage of an electric current through the cell in the opposite direction to that of discharge.

Self-Discharge The loss of useful capacity of a cell or battery due to internal chemical action (local action).

Semi-Permeable Membrane A film that will pass selected ions.

Separator An ion-permeable, electronically nonconductive spacer or material that prevents electronic contact between electrodes of opposite polarity in the same cell.

Series The interconnection of cells or batteries in such a manner that the positive terminal of the first is connected to the negative terminal of the second, and so on. Series connections increase the voltage of the resultant battery as follows:

$$V_s = n \times V_u$$

where V_s = the resultant voltage
n = the number of cells or batteries connected in series
V_u = voltage of the unconnected cell or battery

Service Life The period of useful life of a primary battery before a predetermined end point voltage is reached.

Shallow Discharge A discharge on a secondary battery equaling only a small part of its total capacity.

Shape Change Change in shape of an electrode due to migration of active material during charge/discharge cycling.

Shedding The loss of active material from a plate during cycling.

Shelf Life The duration of storage under specified conditions at the end of which a cell or battery still retains the ability to give a specified performance.

Short-Circuit Current The initial value of the current obtained from a battery in a circuit of negligible resistance.

Sintered Electrode An electrode construction in which active materials are deposited in the interstices of a porous metal matrix made by sintering metal powder.

SLI Battery A battery designed to start internal combustion engines and to power the electrical systems in automobiles when the engine is not running (starting, lighting, ignition). Typically this is a lead-acid battery.

Specific Energy The ratio of the energy output of a cell or battery to its weight (Wh/kg). (See also ENERGY DENSITY.)

Specific Gravity The specific gravity of a solution is the ratio of the weight of the solution to the weight of an equal volume of water at a specified temperature.

Specific Power The ratio of the power delivered by a cell or battery to its weight (W/kg). (See also POWER DENSITY.)

Spirally Wound Cell A cylindrical cell that uses an electrode structure made by winding the electrodes and separators into a cylindrical "jelly-roll" construction. (See Chap. 14.)

Standard Electrode Potential The equilibrium value of an electrode potential when all the constituents taking part in the electrode reaction are in the standard state.

Standby Battery A battery designed for emergency use in the event of a main power failure.

Starved-Electrolyte Cell A cell containing little or no free fluid electrolyte. This enables gases to reach electrode surfaces during charging and facilitates gas recombination.

State-of-Charge (SOC) The available capacity in a battery expressed as a percentage of rated capacity.

Stationary Battery A secondary battery designed for use in a fixed location.

Storage Battery See SECONDARY BATTERY.

Storage Life See SHELF LIFE.

Sulfation Process occurring in lead batteries that have been stored and allowed to self-discharge for extended periods of time. Large crystals of lead sulfate grow that interfere with the function of the active materials.

Taper Charge A charge regime delivering moderately high rate charging current when the battery is at a low state-of-charge and tapering the charging current to lower rates as the battery is charged.

Thermal Runaway A condition whereby a battery on charge or discharge will overheat and destroy itself through internal heat generation caused by high overcharge or overdischarging current or other abusive condition.

Traction Battery A secondary battery designed for the propulsion of electric vehicles or electrically operated mobile equipment operating in a deep-cycle regime.

Transfer Coefficient The transfer coefficient determines what fraction of the electrical energy of a system results from the displacement of the potential from the equilibrium value that affects the rate of electrochemical transformation. (See Chap. 2.)

Transference Number The fraction of the total cell current carried by the cation of an electrolyte solution is called the "cation transference number." Similarly, the fraction of the total current carried by the anion is referred to as the "anion transference number."

Transition Time The time of an electrode process from the initiation of the process at constant current to the moment an abrupt change in potential occurs, signifying that a new electrode process is controlling the electrode potential.

Trickle Charge A charge at a low rate, balancing losses through a local action and/or periodic discharge, to maintain a battery in a fully charged condition.

Tubular Plate A battery plate in which an assembly of perforated metal or polymer tubes holds the active materials.

Unactivated Shelf Life The period of time, under specified conditions of temperature and environment, that an unactivated or reserve battery can stand before deteriorating below a specified capacity.

Vent A normally sealed mechanism that allows for the controlled escape of gases from within a cell.

Vented Cell A cell design incorporating a vent mechanism to relieve excessive pressure and expel gases that are generated during the operation or abuse of the cell.

Voltage Delay Time delay for a battery to deliver the required operating voltage after it is placed under load.

Voltage Depression An abnormal low voltage, below the expected value, during the discharge of a battery.

Voltage Efficiency The ratio of average voltage during discharge to average voltage during recharge under specified conditions of charge and discharge.

Watthour Capacity The quantity of electrical energy, measured in watthours, that may be delivered by a cell or battery under specified conditions.

Watthour Efficiency The ratio of the watthours delivered on discharge of a battery to the watthours needed to restore it to its original state under specified conditions of charge and discharge. Also called energy efficiency.

Wet Shelf Life The period of time that a battery can stand in the charged or activated condition before deteriorating below a specified capacity.

Working Voltage The typical voltage or range of voltages of a battery during discharge.

APPENDIX B
STANDARD REDUCTION POTENTIALS

TABLE B.1 Standard Reduction Potentials of Electrode Reactions at 25°C

Electrode reaction	E^0, V
$Li^+ + e \rightleftharpoons Li$	−3.01
$Rb^+ + e \rightleftharpoons Rb$	−2.98
$Cs^+ + e \rightleftharpoons Cs$	−2.92
$K^+ + e \rightleftharpoons K$	−2.92
$Ba^{2+} + 2e \rightleftharpoons Ba$	−2.92
$Li^+ + 6C + e \rightleftharpoons LiC_6$	−2.9
$Sr^2 + 2e \rightleftharpoons Sr$	−2.89
$Ca^{2+} + 2e \rightleftharpoons Ca$	−2.84
$Na^+ + e \rightleftharpoons Na$	−2.71
$Mg(OH)_2 + 2e \rightleftharpoons Mg + 2OH^-$	−2.67
$Mg^{2+} + 2e \rightleftharpoons Mg$	−2.38
$Al(OH)_3 + 3e \rightleftharpoons Al + 3OH^-$	−2.34
$Ti^{2+} + 2e \rightleftharpoons Ti$	−1.75
$Be^{2+} + 2e \rightleftharpoons Be$	−1.70
$Al^3 + 3e \rightleftharpoons Al$	−1.66
$Zn(OH)_2 + 2e \rightleftharpoons Zn + 2OH^-$	−1.25
$Mn^{2+} + 2e \rightleftharpoons Mn$	−1.05
$Fe(OH)_2 + 2e \rightleftharpoons Fe + 2OH^-$	−0.88
$2H_2O + 2e \rightleftharpoons H_2 + 2OH^-$	−0.83
$H^+ + M + e \rightleftharpoons MH$	−0.83
$Cd(OH)_2 + 2e \rightleftharpoons Cd + 2OH^-$	−0.81
$Zn^{2+} + 2e \rightleftharpoons Zn$	−0.76
$Ni(OH)_2 + 2e \rightleftharpoons Ni + 2OH^-$	−0.72
$Ga^{3+} + 3e \rightleftharpoons Ga$	−0.52
$S + 2e \rightleftharpoons S^{2-}$	−0.48
$Fe^{2+} + 2e \rightleftharpoons Fe$	−0.44
$Cd^{2+} + 2e \rightleftharpoons Cd$	−0.40
$PbSO_4 + 2e \rightleftharpoons Pb + SO_4^{2-}$	−0.36
$In^{3+} + 3e \rightleftharpoons In$	−0.34

(*Continued*)

TABLE B.1 Standard Reduction Potentials of Electrode Reactions at 25°C (*Continued*)

Electrode reaction	E^0, V
$Tl^+ + e \rightleftharpoons Tl$	−0.34
$Co^{2+} + 2e \rightleftharpoons Co$	−0.27
$Ni^{2+} + 2e \rightleftharpoons Ni$	−0.23
$Sn^{2+} + 2e \rightleftharpoons Sn$	−0.14
$Pb^{2+} + 2e \rightleftharpoons Pb$	−0.13
$O_2 + H_2O + 2e \rightleftharpoons HO_2^- + OH^-$	−0.08
$D^+ + e \rightleftharpoons \frac{1}{2}D_2$	−0.003
$H^+ + e \rightleftharpoons \frac{1}{2}H_2$	0.000
$HgO + H_2O + 2e \rightleftharpoons Hg + 2OH^-$	0.10
$CuCl + e \rightleftharpoons Cu + Cl^-$	0.14
$AgCl + e \rightleftharpoons Ag + Cl^-$	0.22
$\gamma\text{-}MnO_2 + H_2O + e \rightleftharpoons \alpha\text{-}MnOOH + OH^-$	0.30
$Cu^{2+} + 2e \rightleftharpoons Cu$	0.34
$Ag_2O + H_2O + 2e \rightleftharpoons 2Ag + 2OH^-$	0.35
$\gamma\text{-}MnO_2 + H_2O + e \rightleftharpoons \lambda\text{-}MnOOH + OH^-$	0.36
$\frac{1}{2}O_2 + H_2O + 2e \rightleftharpoons 2OH^-$	0.40
$NiOOH + H_2O + e \rightleftharpoons Ni(OH)_2 + OH^-$	0.45
$Cu^+ + e \rightleftharpoons Cu$	0.52
$I_2 + 2e \rightleftharpoons 2I^-$	0.54
$2AgO + H_2O + 2e \rightleftharpoons Ag_2O + 2OH^-$	0.57
$LiCoO_2 + 0.5e \rightleftharpoons Li_{0.5}CoO_2 + 0.5Li^+$	~0.70
$Hg^{2+} + 2e \rightleftharpoons 2Hg$	0.80
$Ag^+ + e \rightleftharpoons Ag$	0.80
$O_2 + 4H^+(10^{-7}\ M) + 4e \rightleftharpoons 2H_2O$	0.82
$Pd^{2+} + 2e \rightleftharpoons Pd$	0.83
$Ir^{3+} + 3e \rightleftharpoons Ir$	1.00
$Br_2 + 2e \rightleftharpoons 2Br^-$	1.08
$O_2 + 4H^+ + 4e \rightleftharpoons 2H_2O$	1.23
$MnO_2 + 4H^+ + 2e \rightleftharpoons Mn^{2+} + 2H_2O$	1.23
$Cl_2 + 2e \rightleftharpoons 2Cl^-$	1.36
$PbO_2 + 4H^+ + 2e \rightleftharpoons Pb^{2+} + 2H_2O$	1.46
$PbO_2 + SO_4^{2-} + 4H^- + 2e \rightleftharpoons PbSO_4 + 2H_2O$	1.69
$F_2 + 2e \rightleftharpoons 2F^-$	2.87

APPENDIX C
ELECTROCHEMICAL EQUIVALENTS OF BATTERY MATERIALS

TABLE C.1 Electrochemical Equivalents of Battery Materials

Material	Symbol	Atomic no.	Atomic wt., g	Density, g/cm^3	Valence change	Electrochemical equivalents		
						Ah/g	g/Ah	Ah/cm^3
Elements								
Aluminium	Al	13	26.98	2.699	3	2.98	0.335	8.05
Antimony	Sb	51	121.75	6.62	3	0.66	1.514	4.37
Arsenic	As	33	74.92	5.73	3	1.79	0.559	10.26
Barium	Ba	56	137.34	3.78	2	0.39	2.56	1.47
Beryllium	Be	4	9.01	—	2	5.94	0.168	—
Bismuth	Bi	83	208.98	9.80	3	0.385	2.59	3.77
Boron	B	5	10.81	2.54	3	7.43	0.135	18.87
Bromine	Br	35	79.90	—	1	0.335	2.98	—
Cadmium	Cd	48	112.40	8.65	2	0.477	2.10	4.15
Cesium	Cs	55	132.91	1.87	3	0.574	1.74	1.07
Calcium	Ca	20	40.08	1.54	2	1.34	0.748	2.06
Carbon (graphite)	C	6	12.01	2.25	4	8.93	0.112	20.09
Chlorine	Cl	17	35.45	—	1	0.756	1.32	—
Chromium	Cr	24	52.00	6.92	3	1.55	0.647	10.72
Cobalt	Co	27	58.93	8.71	2	0.910	1.10	7.93
Copper	Cu	29	63.55	8.89	2	0.843	1.19	7.49
					1	0.422	2.37	3.75
Fluorine	F	9	19.00	—	1	1.41	0.709	—
Gold	Au	79	197.00	19.3	1	0.136	7.36	2.62
Hydrogen	H	1	1.008	—	1	26.59	0.0376	—
Indium	In	49	114.82	7.28	3	0.701	1.43	5.10
Iodine	I	53	126.90	4.94	1	0.211	4.73	1.04
Iron	Fe	26	55.85	7.85	2	0.96	1.04	7.54
					3	1.44	0.694	11.30
Lead	Pb	82	207.2	11.34	2	0.259	3.87	2.94
Lithium	Li	3	6.94	0.534	1	3.86	0.259	2.06
Magnesium	Mg	12	24.31	1.74	2	2.20	0.454	3.83
Manganese	Mn	25	54.94	7.42	2	0.976	1.02	7.24
Mercury	Hg	80	200.59	13.60	2	0.267	3.74	3.63

(*Continued*)

TABLE C.1 Electrochemical Equivalents of Battery Materials (*Continued*)

Material	Symbol	Atomic no.	Atomic wt., g	Density, g/cm³	Valence change	Electrochemical equivalents		
						Ah/g	g/Ah	Ah/cm³
Elements								
Molybdenum	Mo	42	95.94	10.2	6	1.67	0.597	17.03
Nickel	Ni	28	58.71	8.6	2	0.913	1.09	7.85
Nitrogen	N	7	14.01	—	3	5.74	0.174	—
Oxygen	O	8	16.00	—	2	3.35	0.298	—
Platinum	Pt	78	195.09	21.37	4	0.549	1.82	11.73
Potassium	K	19	39.10	0.87	1	0.685	1.46	0.59
Silver	Ag	17	107.87	10.5	1	0.248	4.02	2.60
Sodium	Na	11	22.99	0.971	1	1.17	0.858	1.14
Sulfur	S	16	32.06	2.0	2	1.67	0.598	3.34
Tin	Sn	50	118.69	7.30	4	0.903	1.11	6.59
Vanadium	V	23	50.95	5.96	5	2.63	0.380	15.67
Zinc	Zn	30	65.38	7.1	2	0.820	1.22	5.82
Zirconium	Zr	40	91.22	6.44	4	1.18	0.851	7.60

Material	Symbol	Molecular wt., g	Density g/cm³	Valence change	Electrochemical equivalents		
					Ah/g	g/Ah	Ah/cm³
Compounds							
Bismuth trioxide	Bi_2O_3	466	8.5	6	0.345	2.90	2.97
Bismuth trifloride	BiF_3	265.9	—	3	0.302	3.31	—
Calcium chromate	$CaCrO_4$	156.1	—	2	0.34	2.90	—
Carbon monofluoride	CF_x	31	2.7	1	0.862	1.16	2.32
Cobalt difluoride	CoF_2	96.9	—	2	0.553	1.81	—
Cuprous chloride	$CuCl$	99	3.5	1	0.27	3.69	0.95
Cupric chloride	$CuCl_2$	134.5	3.1	2	0.40	2.50	1.22
Cupric fluoride	CuF_2	101.6	2.9	2	0.528	1.89	1.52
Cupric oxide	CuO	79.6	6.4	2	0.67	1.49	4.26
Cupric sulfate	$CuSO_4$	159.6	3.6	2	—	—	—
Cupric sulfide	CuS	95.6	4.6	2	0.56	1.79	2.57
Iron monosulfide	FeS	87.9	4.84	2	0.61	1.64	2.95
Iron disulfide	FeS_2	119.9	4.87	4	0.89	1.12	4.35
Iron trifluoride	FeF_3	112.8	—	3	0.712	1.40	—
Lead bismuthate	$Bi_2Pb_2O_5$	912	9.0	10	0.29	3.41	2.64
Lead chloride	$PbCl_2$	278.1	5.8	2	0.19	5.18	1.12
Lead dioxide	PbO_2	239.2	9.3	2	0.22	4.45	2.11
Lead iodide	PbI_2	461	6.2	2	0.12	8.60	0.72
Lead oxide	Pb_3O_4	685	9.1	8	0.31	3.22	2.85
Lead sulfide	PbS	239.3	7.5	2	0.22	4.46	1.68
Lithiated carbon	LiC_6	79.0	—	1	0.372*	2.69*	—
Lithium cobalt oxide	$LiCoO_2$	98	5.05	0.55	0.150	6.67	0.757
Lithium iron phosphate	$LiFePO_4$	117.7	3.60	1	0.160	6.25	0.576
Lithium manganese oxide (spinel)	$Li_{1.1}Mn_{1.9}O_2$	144.0	4.18	1	0.120	8.33	0.502
Lithium nickel manganese cobalt oxide (NMC)	$Li(Ni_{1/3}Mn_{1/3}Co_{1/3})O_2$	96.4	4.77	0.59	0.163	6.13	0.777

(*Continued*)

Material	Symbol	Molecular wt., g	Density g/cm^3	Valence change	Electrochemical equivalents		
					Ah/g	g/Ah	Ah/cm^3
Compounds							
Manganese dioxide	MnO_2	86.9	5.0	1	0.31	3.22	1.54
Manganese trifluoride	MnF_3	111.9	—	3	0.719	1.39	—
Mercuric oxide	HgO	216.6	11.1	2	0.247	4.05	2.74
Molybdenum trioxide	MoO_3	143	4.5	1	0.19	5.26	0.84
Nickel difluoride	NiF_2	96.7	—	2	0.554	1.80	—
Nickel oxide	NiOOH	91.7	7.4	1	0.29	3.42	2.16
Nickel sulfide	Ni_3S_2	240	—	4	0.47	2.12	—
Silver chloride	AgCl	143.3	5.56	1	0.19	5.26	1.04
Silver chromate	Ag_2CrO_4	331.8	5.6	2	0.16	6.25	0.90
Silver oxide (monovalent)	Ag_2O	231.8	7.1	2	0.23	4.33	1.64
Silver oxide (divalent)	AgO	123.9	7.4	2	0.43	2.31	3.20
Sulfur dioxide	SO_2	64	1.37	1	0.419	2.39	—
Sulfuryl chloride	SO_2Cl_2	135	1.66	2	0.397	2.52	—
Thionyl chloride	$SOCl_2$	119	1.63	2	0.450	2.22	—
Vanadium pentoxide	V_2O_5	181.9	3.6	1	0.15	6.66	0.53

*Based on weight of carbon only.

APPENDIX D
STANDARD SYMBOLS AND CONSTANTS

TABLE D.1 SI Base Units

Quantity	Unit	Symbol
Length	meter	m
Mass	kilogram	kg
Time	second	s
Electric current	ampere	A
Thermodynamic temperature*	kelvin	K
Amount of substance	mole	mol
Luminous intensity	candela	cd

*Celsius temperature is, in general, expressed in degrees Celsius (symbol °C).

Source: From D. G. Fink and W. Beaty (eds.), *Standard Handbook for Engineers,* 12th ed., McGraw-Hill, N.Y., 1987; reproduced from IEEE Standard 268-1982, by permission.

TABLE D.2 SI Prefixes Expressing Decimal Factors

Factor	Prefix	Symbol	Factor	Prefix	Symbol
10^{18}	exa	E	10^{-1}	deci	d
10^{15}	peta	P	10^{-2}	centi	c
10^{12}	tera	T	10^{-3}	milli	m
10^{9}	giga	G	10^{-6}	micro	μ
10^{6}	mega	M	10^{-9}	nano	n
10^{3}	kilo	k	10^{-12}	pico	p
10^{2}	hecto	h	10^{-15}	femto	f
10^{1}	deka	da	10^{-18}	atto	a

Source: From D. G. Fink and W. Beaty (eds.), *Standard Handbook for Engineers,* 12th ed., McGraw-Hill, N.Y., 1987; adapted from IEEE Standard 268-1982, by permission.

TABLE D.3 Greek Alphabet

Greek letter		Greek name	English equivalent	Greek letter		Greek name	English equivalent
A	α	Alpha	a	N	ν	Nu	n
B	β	Beta	b	Ξ	ξ	Xi	x
Γ	γ	Gamma	g	O	o	Omicron	ŏ
Δ	δ	Delta	d	Π	π	Pi	p
E	ε	Epsilon	ĕ	P	ρ	Rho	r
Z	ζ	Zeta	z	Σ	σ	Sigma	s
H	η	Eta	ē	T	τ	Tau	t
Θ	θ	Theta	th	Y	υ	Upsilon	u
I	ι	Iota	i	Φ	ϕ	Phi	ph
K	κ	Kappa	k	X	χ	Chi	ch
Λ	λ	Lambda	l	Ψ	ψ	Psi	ps
M	μ	Mu	m	Ω	ω	Omega	ō

TABLE D.4 Standard Symbols for Units

Unit	Symbol	Notes
Ampere	A	SI unit of electric current.
Ampere-hour	Ah	
Angstrom	Å	1 Å = 10^{-10} m.
Atmosphere, standard	atm	1 atm = 101,325 N/m² or Pa.
Atmosphere, technical	at	at = kg_f/cm².
Atomic mass unit (unified)	u	The (unified) atomic mass unit is defined as one-twelfth of the mass of an atom of the ^{12}C nuclide. Use of the old atomic mass unit (amu), defined by reference to oxygen, is deprecated.
Atto	a	SI prefix for 10^{-18}.
Bar	bar	1 bar = 100,000 N/m².
Barn	b	1 b = 10^{-28} m².
Barrel	bbl	1 bbl = 9702 in³ = 0.15899 m³. This is the standard barrel used for petroleum, etc. A different standard barrel is used for fruits, vegetables, and dry commodities.
British thermal unit	Btu	
Calorie (International Table calorie)	cal_{IT}	1 cal_{IT} = 4.1868 J. The 9th Conférence Générale des Poids et Mesures adopted the joule as the unit of heat. Use of the joule is preferred.
Calorie (thermochemical calorie)	cal	1 cal = 4.1840 J (see note for International Table calorie).
Centi	c	SI prefix for 10^{-2}.
Centimeter	cm	
Coulomb	C	SI unit of electric charge.
Cubic centimeter	cm³	
Cycle	c	
Cycle per second	Hz, c/s	See hertz. The name "hertz" is internationally accepted for this unit; the symbol Hz is preferred to c/s.
Day	d	
Deci	d	SI prefix for 10^{-1}.
Decibel	dB	

TABLE D.4 Standard Symbols for Units (*Continued*)

Unit	Symbol	Notes
Degree (temperature):		
Degree Celsius	°C	Note that there is no space between the symbol ° and the letter. The use of the word *centigrade* for the Celsius temperature scale was abandoned by the Conférence Générale des Poids et Mesures in 1948.
Degree Fahrenheit	°F	
Degree Kelvin	°K	See kelvin.
Degree Rankine	°R	
Deka	da	SI prefix for 10.
Dyne	dyn	
Electron	e	This symbol is used in this Handbook to denote an electron. More conventionally shown as e^-
Electronvolt	eV	
Erg	erg	
Farad	F	SI unit of capacitance.
Femto	f	SI prefix for 10^{-15}.
Gauss	G	The gauss is the electromagnetic CGS unit of magnetic flux density. Use of SI unit, the tesla, is preferred.
Giga	G	SI prefix for 10^9.
Gilbert	Gb	The gilbert is the electromagnetic CGS unit of magnetomotive force. Use of the SI unit, the ampere (or ampere turn), is preferred.
Gram	g	
Gram per cubic centimeter	g/cm^3	
Hecto	h	SI prefix for 10^2.
Henry	H	SI unit of inductance.
Hertz	Hz	SI unit of frequency.
Hour	h	
Joule	J	SI unit of energy.
Joule per kelvin	J/K	SI unit of heat capacity and entropy.
Kelvin	K	In 1967 the CGPM gave the name "kelvin" to the SI unit of temperature which had formerly been called "degree Kelvin" and assigned it the symbol K (without the symbol °).
Kilo	k	SI prefix for 10^3.
Kilogram	kg	SI unit of mass.
Kilogram-force	kg$_f$	In some countries the name *kilopond* (kp) has been adopted for this unit.
Kilohm	kΩ	
Kilometer	km	
Kilometer per hour	km/h	
Kilovolt	kV	
Kilowatt	kW	
Kilowatthour	kWh	
Liter	L	$1 L = 10^{-3} m^3$.
Liter per second	L/s	
Lumen	lm	SI unit of luminous flux.
		SI unit of illuminance.
Maxwell	Mx	The maxwell is the electromagnetic cgs unit of magnetic flux. Use of the SI unit, the weber, is preferred.
Mega	M	SI prefix for 10^6.
Megohm	MΩ	
Meter	m	SI unit of length.
Mho	mho	CGPM has adopted the name "siemens" (S) for this unit.
Micro	μ	SI prefix for 10^{-6}.
Microampere	μA	
Microgram	μg	

TABLE D.4 Standard Symbols for Units (*Continued*)

Unit	Symbol	Notes
Micrometer	μm	
Micron	μm	See micrometer. The name "micron" was abrogated by the Conférence Générale des Poids et Mesures, 1967.
Microsecond	μs	
Microwatt	μW	
Milli	m	SI prefix for 10^{-3}.
Milliampere	mA	
Milligram	mg	
Milliliter	ml	
Millimeter	mm	
Conventional millimeter of mercury	mmHg	1 mmHg = 133.322 N/m^2.
Millimicron	nm	Use of the name "millimicron" for the nanometer is deprecated.
Millisecond	ms	
Millivolt	mV	
Milliwatt	mW	
Minute (time)	min	Time may also be designated by means of superscripts as in the following example: $9^h 46^m 20^s$.
Mole	mol	SI unit of amount of substance.
Nano	n	SI prefix for 10^{-9}.
Nanoampere	nA	
Nanometer	nm	
Nanosecond	ns	
Newton	N	SI unit of force.
Newton meter	N · m	
Newton per square meter	N/m^2	SI unit of pressure or stress; see pascal.
Newton second per square meter	N · s/m^2	SI unit of dynamic viscosity.
Oersted	Oe	The oersted is the electromagnetic cgs unit of magnetic field strength. Use of the SI unit, the ampere per meter, is preferred.
Ohm	Ω	SI unit of resistance.
Pascal	Pa	Pa = N/m^2. SI unit of pressure or stress. This name accepted by the 14th Conférence Générale des Poids et Mesures.
Pico	p	SI prefix for 10^{-12}.
Picowatt	pW	
Revolution per second	r/s	
Second (time)	s	SI unit of time.
Siemens	S	S = Ω^{-1}. SI unit of conductance. This name and symbol were adopted by the 14th Conférence Générale des Poids et Mesures. The name "mho" is also used for this unit in the United States.
Square meter	m^2	
Tera	T	SI prefix for 10^{12}.
Tesla	T	SI unit of magnetic flux density
Tonne	t	1 t = 1000 kg. (in USA: ton, metric)
(Unified) atomic mass unit	u	The (unified) atomic mass unit is defined as one-twelfth of the mass of an atom of the ^{12}C nuclide. Use of the old atomic mass unit (amu), defined by reference to oxygen, is deprecated.
Volt	V	SI unit of voltage.
Volt per meter	V/m	SI unit of electric field strength.
Voltampere	VA	IEC name and symbol for the SI unit of apparent power.
Watt	W	SI unit of power.
Watt per meter kelvin	W/(m · K)	SI unit of thermal conductivity.
Watthour	Wh	

Source: From D. G. Fink and W. Beaty (eds.), *Standard Handbook for Engineers,* 12th ed., McGraw-Hill, N.Y., 1987; adapted from ANSI/IEEE Standard 260-1982.

APPENDIX E
CONVERSION FACTORS

TABLE E.1 Length Conversion Factors*

A. Length units decimally related to one meter

	Meters (m)	Kilometers (km)	Decimeters (dm)	Centimeters (cm)	Millimeters (mm)	Micrometers (μm)	Nanometers (nm)	Ångströms (Å)
1 meter =	1	0.001	10	100	1000	1,000,000	10^9	10^{10}
1 kilometer =	1000	1	10,000	100,000	1,000,000	10^9	10^{12}	10^{13}
1 decimeter =	0.1	0.0001	1	10	100	100,000	10^8	10^9
1 centimeter =	0.01	0.00001	0.1	1	10	10,000	10^7	10^8
1 millimeter =	0.001	10^{-6}	0.01	0.1	1	1000	1,000,000	10^7
1 micrometer (micron) =	10^{-6}	10^{-9}	0.00001	0.0001	0.001	1	1000	10,000
1 nanometer =	10^{-9}	10^{-12}	10^{-8}	10^{-7}	10^{-6}	0.001	1	10
1 ångström =	10^{-10}	10^{-13}	10^{-9}	10^{-8}	10^{-7}	0.0001	0.1	1

B. Nonmetric length units less than one meter

	Meters (m)	Yards (yd)	Feet (ft)	Inches (in)	Mils (mil)	Microinches (μin)
1 meter =	1	1.09361330	3.28083939	39.3700787	3.93700787×10^4	3.93700787×10^7
1 yard =	0.9144	1	3	36	36,000	3.6×10^7
1 foot =	0.3048	1/3 = 0.3333	1	12	12,000	1.2×10^7
1 inch =	0.0254	1/36 = 0.0277	1/12 = 0.0833	1	1000	1,000,000
1 mil =	2.54×10^{-5}	2.777×10^{-5}	8.333×10^{-5}	0.001	1	1000
1 microinch =	2.54×10^{-8}	2.777×10^{-8}	8.333×10^{-8}	10^{-6}	0.001	1

D. British Imperial liquid capacity measures (with liter equivalents)

	Liters (L)	Gallons (U.K. gal)	Quarts (U.K. qt)	Pints (U.K. pt)	Gills (U.K. gi)	Fluid ounces (U.K. floz)	Fluidrams (U.K. fldr)	Minims (U.K. minim)
1 liter =	**1**	0.2199692	0.8798766	1.759753	7.039018	35.19506	281.5605	16,893.63
1 gallon, U.K. =	4.546092	**1**	**4**	**8**	**32**	**160**	**1280**	**76,800**
1 quart, U.K. =	1.136523	**1/4 = 0.25**	**1**	**2**	**8**	**40**	**320**	**19,200**
1 pint, U.K. =	0.5682615	**1/8 = 0.125**	**1/2 = 0.5**	**1**	**4**	**20**	**160**	**9600**
1 gill, U.K. =	0.1420654	**1/32 = 0.03125**	**1/8 = 0.125**	**1/4 = 0.25**	**1**	**5**	**40**	**2400**
1 fluid ounce, U.K. =	2.841307×10^{-2}	**1/160 = 0.00625**	**1/40 = 0.025**	**1/20 = 0.05**	**1/5 = 0.2**	**1**	**8**	**480**
1 fluidram, U.K. =	3.551634×10^{-3}	**1/280 = 7.8125** $\times$ **10**$^{-4}$	**1/320 = 0.003125**	**1/160 = 0.00625**	**1/40 = 0.025**	**1/8 = 0.125**	**1**	**60**
1 minima, U.K. =	5.919391×10^{-5}	**1/76800 =** $1.302098333 \times 10^{-5}$	**1/1920 =** 5.2083333×10^{-5}	**1/9600 =** $1.04166666 \times 10^{-4}$	**1/2400 =** $4.16666666 \times 10^{-4}$	**1/480 =** $2.08333333 \times 10^{-3}$	**1/60 =** 0.01666666	**1**

E. U.S. and British dry capacity measures (with liter equivalents)

		U.S. dry measures				British dry measures			
	Liters (L)	Bushels (U.S. bu)	Pecks (U.S. peck)	Quarts (U.S. qt)	Pints (U.S. pt)	Bushels (U.K. bu)	Pecks (U.K. peck)	Quarts (U.K. qt)	Pints (U.K. pt)
1 liter =	**1**	0.02837759	0.11351037	0.90808299	1.81816598	0.0274961	0.1099846	0.8798766	1.7597534
1 bushel, U.S. =	35.239070	**1**	**4**	**32**	**64**	0.9689387	3.8757549	31.00604	62.01208
1 peck, U.S. =	8,8097675	**1/4 = 0.25**	**1**	**8**	**16**	0.2422347	0.9689387	7.751509	15.50302
1 quart, U.S. =	1.1012209	**1/32 = 0.03125**	**1/8 = 0.125**	**1**	**2**	0.03027934	0.1211173	0.9689387	1.937878
1 pint, U.S. =	0.5506105	**1/64 = 0.015625**	**1/16 = 0.0625**	**1/2 = 0.5**	**1**	0.01513967	0.06055867	0.4844693	0.9689387
1 bushel, U.K. =	36.36873	1.032057	4.128228	33.02582	66.95165	**1**	**4**	**32**	**64**
1 peck, U.K. =	9.092182	0.2580143	1.032057	8.256456	16.51291	**1/4 = 0.25**	**1**	**8**	**16**
1 quart, U.K. =	1.136523	0.03225178	0.1290071	1.032057	2.0641142	**1/32 = 0.03125**	**1/8 = 0.125**	**1**	
1 pint, U.K. =	0.5682614	0.01612589	0.0645036	0.5160184	1.032057	**1/64 = 0.015625**	**1/16 = 0.0625**	**1/2 = 0.5**	

Exact conversion: 1 dry pint, U.S. = **33.6003125** cubic inches

F. Other volume and capacity units

1 barrel, U.S. (used for petroleum, etc.) = **42** gallons = 0.158987296 cubic meter
1 barrel ("old barrel") = **31.5** gallons = 0.119240 cubic meter
1 board foot = **144** cubic inches = 2.359737×10^{-3} cubic meter
1 cord = **128** cubic feet = 3.624556 cubic meters
1 cord foot = **16** cubic feet = 0.4530695 cubic meter
1 cup = **8** fluid ounces, U.S. = 2.365882×10^{-4} cubic meter
1 gallon (Canadian liquid) = 4.546090×10^{-3} cubic meter
1 perch (volume) = 24.75 cubic feet = 0.700842 cubic meter
1 stere = **1** cubic meter
1 tablespoon = **0.5** fluid ounce, U.S. = 1.478677×10^{-5} cubic meter
1 teaspoon = **1/6** fluid ounce, U.S. = 4.928922×10^{-6} cubic meter
1 ton (register ton) = **100** cubic feet = 2.83168466 cubic meters

*Exact conversions are shown in boldface type. Repeating decimals are underlined. The SI unit of volume is the cubic meter.

Source: D. G. Fink and W. Beaty (eds.), *Standard Handbook for Electrical Engineers,* 12th ed., McGraw-Hill, New York, 1987.

TABLE E.5 Mass Conversion Factors*

A. Mass units decimally

	Kilograms (kg)	Tonnes (metric tons) (t)	Grams (g)
1 kilogram =	1	0.01	1000
1 tonne =	1000	1	1,000,000
1 gram =	0.001	0.000001	1
1 decigram =	0.0001	10^{-7}	0.1
1 centigram =	0.00001	10^{-8}	0.01
1 milligram =	0.000001	10^{-9}	0.001
1 microgram =	10^{-9}	10^{-12}	0.000001

B. Nonmetric mass units less than one

	Grams (g)	Avoirdupois ounces-mass (oz_m, avdp)	Troy ounces-mass (oz_m, troy)
1 gram =	1	0.035273962	0.032150747
1 avdp ounce-mass =	38.3495231	1	0.91145833
1 troy ounce-mass =	31.1031768	1.09714286	1
1 avdp dram =	1.77184520	**1/16 = 0.0625**	0.056966.15
1 apothecary dram =	3.88793458	0.137142857	**1/8 = 0.125**
1 pennyweight =	1.55517383	0.054863162	**1/20 = 0.05**
1 grain =	**0.06479891**	**1/437.5** = 2.28571429 × 10^{-3}	**1/480** = 0.002083$\overline{33}$
1 scruple =	1.29597820	4.57142858 × 10^{-2}	**1/24** = 0.041$\overline{6666}$

C. Nonmetric mass units of one pound-mass

	Kilograms (kg)	Long tons (long ton)	Short tons (short ton)
1 kilogram =	1	9.842065 × 10^{-4}	1.10231131 × 10^{-3}
1 long ton =	1016.0469	**1**	**1.12**
1 short ton =	**907.18474**	200/224 = 0.89285714	**1**
1 long hundredweight =	50.802.3454	**0.05**	**0.056**
1 short hundredweight =	**45.359237**	10/224 = 0.04464286	**0.05**
1 slug =	14.593903	0.01436341	0.01608702
1 avdp pound-mass =	**0.45359237**	**1/2240** = 4.46428571 × 10^{-4}	**0.0005**
1 troy pound-mass =	0.37324172	3.67346937 × 10^{-4}	4.11428570 × 10^{-4}

Exact conversions: 1 long ton = **1016.0469088** kilograms
1 troy pound-mass = **0.3732417216** kilogram

D. Other

1 assay ton = 29.166667 grams
1 carat (metric) = **200** milligrams
1 carat (troy weight) = **31/6** grains = 205.19655 milligrams
1 mynagram = **10** kilograms
1 quintal = **100** kilograms
1 stone = **14** pounds, advp = **6.35029328** kilograms

*Exact conversions are shown in boldface type. Repeating decimals are underlined. The SI unit of mass is the kilogram.

Source: D. G. Fink and W. Beaty (eds.), *Standard Handbook for Electrical Engineers,* 12th ed., McGraw-Hill, New York, 1987.

related to kilogram

Decigrams (dg)	Centigrams (cg)	Milligrams (mg)	Micrograms (μg)
10,000	100,000	1,000,000	10^9
10^7	10^8	10^9	10^{12}
10	100	1000	1,000,000
1	10	100	100,000
0.1	1	10	10,000
0.01	0.1	1	1000
0.00001	0.0001	0.001	1

pound-mass (with gram equivalents)

Avoirdupois drams (dr avdp)	Apothecary drams (dr apoth)	Pennyweights (dwt)	Grains (grain)	Scruples (scruple)
0.56438339	0.25720597	0.64301493	15.4323584	0.77161792
16	7.29166666	18.2271667	437.5	21.875
17.5542857	**8**	**20**	480	24
1	0.45572917	1.13932292	27.34375	1.3671875
2.19428570	**1**	**2.5**	60	3
0.87771428	1/2.5 = **0.4**	**1**	24	1.2
3.65714285 × 10^{-2}	1/60 = 0.010666666	1/24 = 0.04166666	**1**	0.05
0.73142857	1/3 = 0.33333333	5/6 = 0.83333333	20	1

and greater (with kilogram equivalents)

Long hundredweights (long cwt)	Short hundredweights (short cwt)	Slugs (slug)	Avoirdupois pounds-mass (lb_m, avdp)	Troy pounds-mass (lb_m, troy)
1.96841131 × 10^{-2}	2.20462262 × 10^{-2}	0.06852177	2.20462262	2.67922889
20	22.4	69.621329	**2240**	2722.22222
400/224 = 17.8571429	**20**	62.161901	**2000**	2430.55555
1	1.12	3.4810664	112	136.111111
100//112 = 0.89285714	**1**	3.1080950	**100**	121.527777
0.2872683	0.3217405	**1**	32.17405	39.100406
1/112 = 8.92857143 × 10^{-3}	**0.01**	3.1080950 × 10^{-2}	**1**	1.215277777
7.34693879 × 10^{-3}	8.22857145 × 10^{-3}	0.02557518	0.82285714	**1**

mass units

TABLE E.6 Pressure/Stress Conversion Factors

A. Pressure units decimally related to one pascal

	Pascals (Pa)	Bars (bar)	Decibars (dbar)	Millibars (mbar)	Dynes per square centimeter (dyn/cm^2)
1 pascal =	1	0.00001	0.0001	0.01	10
1 bar =	100,000	1	10	1000	1,000,000
1 decibar =	10,000	0.1	1	100	100,000
1 millibar =	100	0.001	0.01	1	1000
1 dyne per second centimeter =	0.1	0.000001	0.00001	0.001	1

B. Pressure units decimally related to one kilogram-force per square meter (with pascal equivalents)

	Kilograms-force per square meter (kg/m^2)	Kilograms-force per square centimeter (kg/cm^2)	Kilograms-force per square millimeter (kg/mm^2)	Grams-force per square centimeter (g/cm^2)	Pascals
1 kilogram-force per square meter =	1	0.001	0.000001	0.1	9.80665
1 kilogram-force per square centimeter =	10,000	1	0.01	1000	98,066.5
1 kilogram-force per square millimeter =	1,000,000	100	1	100,000	9,806,650
1 gram-force per square centimeter =	10	0.001	0.00001	1	98.0665
1 pascal =	0.10197162	1.0197162×10^{-5}	1.0197162×10^{-7}	1.0197162×10^{-2}	1

NOTE: 1 atmosphere (technical) = 1 kilogram-force per square centimeter = 98,066.5 pascals.

C. Pressure units expressed as heights of liquid (with pascal equivalents)

	Millimeters of mercury at 0°C (mmHg, 0°C)	Centimeters of mercury at 60°C (cmHg, 60°C)	Inches of mercury at 32°F (inHg, 32°F)	Inches of mercury at 60°F (inHg, 60°F)	Centimeters of water at 4°C (cmH$_2$O, 4°C)	Inches of water at 60°F (inH$_2$O, 60°F)	Feet of water at 39.2F (ftH$_2$O, 39.2°F)	Pascals (Pa)
1 millimeter of mercury, 0°C =	1	0.100282	0.0393701	0.0394813	1.359548	0.5357756	0.0446046	133.3224
1 centimeter of mercury, 60°C =	9.971830	1	0.3925919	0.3937008	13.55718	5.342664	0.444789.5	1329.468
1 inch of mercury, 32°F =	**25.4**	2.547175	1	1.0028248	34.53252	13.60870	1.132957	3386.389
1 inch of mercury, 60°C =	25.32845	**2.54**	0.9971831	1	35.43525	13.57037	1.129765	3376.85
1 centimeter of water, 4°C =	0.735539	0.073762	0.028958	0.0290400	1	0.3940838	0.038084	98.0638
1 inch of water, 60°F =	1.866453	0.187173	0.073482	0.0736900	2.537531	1	0.0832524	248.840
1 foot of water, 39.2°F =	22.4192	2.248254	0.882646	0.885139	30.47998	12.01167	1	2988.98
1 pascal =	7.500615×10^{-3}	7.521806×10^{-4}	2.952998×10^{-4}	2.96134×10^{-4}	1.01974×10^{-2}	4.01865×10^{-3}	3.34562×10^{-4}	1

NOTE: 1 torr = 1 millimeter of mercury at 0°C = 133.3224 pascals.

D. Nonmetric pressure units (with pascal equivalents)

	Atmospheres (atm)	Avoirdupois pounds-force per square inch (psi)	Avoirdupois pounds-force per square foot (lb/ft^2, avdp)	Poundals per square foot (pd/ft^2)	Pascals (Pa)
1 atmosphere =	1	14.69595	2116.217	68,087.24	**101,325**
1 avdp pound-force per square inch =	6.80460×10^{-2}	1	**144**	4633.063	6894.757
1 avdp pound-force per square foot =	4.725414×10^{-4}	1/144 = 0.006944	1	32.17405	47.88026
1 poundal per square foot =	1.468704×10^{-3}	2.158399×10^{-4}	0.0310809	1	1.488165
1 pascal =	9.869233×10^{-6}	1.450377×10^{-4}	0.0208854	0.6719689	1

NOTE: 1 normal atmosphere = 760 torr = **101,325** pascals.

*Exact conversions are shown in boldface type. Repeating decimals are underlined. The SI unit of pressure or stress is the Pascal (Pa).
Source: D. G. Fink and W. Beaty (eds.), *Standard Handbook for Electrical Engineers*, 12th ed., McGraw-Hill, New York, 1987.

TABLE E.7 Energy/Work Conversion Factors*

A. Energy/work units decimally related to one joule

	Joules (J)	Megajoules (MJ)	Kilojoules (kJ)	Millijoules (mJ)	Microjoules (μJ)	Ergs (ergs)
1 joule =	1	0.000001	0.001	1000	1,000,000	10^7
1 megajoule =	1,000,000	1	1000	10^9	10^{12}	10^{13}
1 kilojoule =	1000	0.001	1	1,000,000	10^9	10^{10}
1 millijoule =	0.001	10^{-9}	10^{-6}	1	1000	10,000
1 microjoule =	0.000001	10^{-12}	10^{-9}	0.001	1	10
1 erg =	10^{-7}	10^{-13}	10^{-10}	0.0001	0.1	1

NOTE: 1 watt-second = 1 joule.

B. Energy/work units less than ten joules (with joule equivalents)

	Joules (J)	Foot-poundals (ft · pdl)	Foot-pounds-force (ft · lb$_f$)	Calories (International Table) (cal, IT)	Calories (thermochemical) (cal, thermo)	Electronvolts (eV)
1 joule =	1	23.73036	0.7375621	0.2388459	0.2390057	6.24146×10^{18}
1 foot-poundal =	4.2104011×10^{-2}	1	3.108095×10^{-2}	1.006499×10^{-2}	1.007173×10^{-2}	2.63016×10^{17}
1 foot-pound-force =	1.355818	32.17405	1	0.3238316	0.3240483	8.46228×10^{18}
1 calorie (Int. Tab.) =	4.1868	99.35427	3.088025	1	1.000669	2.61317×10^{19}
1 calorie (thermo) =	4.184	99.28783	3.085960	0.9993312	1	2.61143×10^{19}
1 electronvolt =	1.60219×10^{-19}	3.80205×10^{-18}	1.18171×10^{-19}	3.82677×10^{-20}	3.82933×10^{-20}	1

C. Energy/work units greater than ten joules (with joule equivalents)

	Joules (J)	British thermal units, International Table (Btu, IT)	British thermal units, thermochemical (Btu, thermo)	Kilowatthours (kWh)	Horsepower-hours, electrical (hp · h, elec)	Kilocalories, International Table (kcal, IT)	Kilocalories, thermochemical (kcal, thermo)
1 joule =	1	9.478170×10^{-4}	9.4845165×10^{-4}	$1/3.6 \times 10^6 = 2.777 \times 10^{-7}$	3.723562×10^{-7}	2.388459×10^{-4}	2.3900574×10^{-4}
1 British thermal unit, Int. Tab. =	1055.056	1	1.000669	2.9307111×10^{-4}	3.928567×10^{-4}	0.251995.8	0.2521644
1 British thermal unit (thermo) =	1054.35	0.999331	1	2.928745×10^{-4}	03.925938×10^{-4}	0.2518272	0.2519957
1 kilowatthour =	**3,600,000**	3412.141	3414.426	1	$1/0.746 = 1.3404826$	859.8452	860.4207
1 horsepower hour, electrical =	**2,685,600**	2,545.457	2547.162	**0.746**	1	641.4445	641.8738
1 kilocalorie, Int. Tab. =	**4.186.8**	3.968320	3.970977	**0.001163**	1.558981×10^{-3}	1	1.000669
1 kilocalorie, thermochemical =	**4184**	3.965666	3.968322	0.0011622	$1.557938.6 \times 10^{-3}$	0.999331	1

The exact conversion is 1 British thermal unit, International Table = **1055.05585262 joules**.

*Exact conversions are shown in boldface type. Repeating decimals are underlined. The SI unit of energy and work is the joule (J).

Source: D. G. Fink and W. Beaty (eds.), *Standard Handbook for Electrical Engineers*, 12th ed., McGraw-Hill, New York, 1987.

TABLE E.8 Power Conversion Factors*

A. Power units decimally related to one watt

	Watts (W)	Megawatts (MW)	Kilowatts (kW)	Milliwatts (mW)	Microwatts (μW)	Picowatts (pW)	Ergs per second (ergs/s)
1 watt =	1	0.000001	0.001	1000	1,000,000	10^9	10^7
1 megawatt =	1,000,000	1	1000	10^9	10^{12}	10^{15}	10^{13}
1 kilowatt =	1000	0.001	1	1,000,000	10^9	10^{12}	10^{10}
1 milliwatt =	0.001	10^{-9}	0.000001	1	1000	1,000,000	10,000
1 microwatt =	0.000001	10^{-12}	10^{-9}	0.001	1	1000	10
1 picowatt =	10^{-9}	10^{-15}	10^{-12}	0.000001	0.001	1	0.01
1 erg per second =	10^{-7}	10^{-13}	10^{-10}	0.0001	0.1	100	1

NOTE: 1 watt = 1 joule per second (J/s).

B. Nonmetric power units (with watt equivalents)

	British thermal units (International Table) per hour (Btu/h, IT)	British thermal units (thermochemical) per minute (Btu/min, thermo)	Avoirdupois foot-pounds-force per second (ft·lb_f/s avdp)	Kilocalories per minute (thermochemical) (kcal/min, thermo)	Kilocalories per second (International Table) (kcal s, IT)	Horsepower (electrical) (hp, elec)	Horsepower (mechanical) (hp, mech)	Watts (W)
1 British thermal unit (Int. Tab.) per hour =	1	0.0166778	0.2161581	4.2027405×10^{-3}	6.9998831×10^{-5}	3.9285670×10^{-4}	3.930148×10^{-4}	0.2930711
1 British thermal unit (thermo) per minute =	59.959853	1	12.960810	0.2519957	4.1971195×10^{-3}	0.0235556	0.0235651	17.57250
1 foot-pound-force per second =	4.6262426	0.0771557	1	0.0194429	3.2383157×10^{-4}	1.8174504×10^{-3}	1.8181818×10^{-3} **1/550 =**	1.355818
1 kilocalorie per minute (thermo) =	237.93998	3.9683217	51.432665	1	0.0166555	0.0934763	0.0935139	69.733333
1 kilocalorie per second (Int. Tab.) =	14,285.953	238.25864	3088.0251	60.040153	1	5.6123324	5.6145911	**4186.800**
1 horsepower (electrical) =	2545.4574	42.452696	550.22134	10.697898	0.1781790	1	1.0004024	**746**
1 horsepower (mechanical) =	2544.4334	42.435618	**550**	10.693593	0.1781074	0.9995977 **1/746 =**	1	745.6999
1 watt =	3.4121413	0.0569071	0.7375621	0.0143403	2.3884590×10^{-4}	1.3404826×10^{-3}	1.3410220×10^{-3}	1

NOTE: The horsepower (mechanical) is defined as a power equal to 550 foot-pounds-force per second.
Other units of horsepower are:
1 horsepower (boiler) = 9809.40 watts
1 horsepower (metric) = 735.499 watts
1 horsepower (water) = 746.043 watts
1 horsepower (U.K.) = 745.70 watts
1 ton (refrigeration) = 3516.8 watts

*Exact conversions are shown in boldface type. Repeating decimals are underlined. The SI unit of power is the watt (W).

Source: D. G. Fink and W. Beaty (eds.), *Standard Handbook for Electrical Engineers*, 12th ed., McGraw-Hill, New York, 1987.

TABLE E.9 Temperature Conversions*

Celsius (°C) °C = 5(°F − 32)/9	Fahrenheit (°F) °F = [9(°C)/5] + 32	Absolute (K) K = °C + 273.15
−273.15	**−459.67**	**0**
−200	−328	73.15
−180	−292	93.15
−160	−256	113.15
−140	−220	133.15
−120	−184	153.15
−100	−148	173.15
−80	−112	193.15
−60	−76	213.15
−40	**−40**	233.15
−30	−22	243.15
−20	−4	253.15
−17.7_7_	**0**	255.372
−10	14	263.15
−6.6_6_	**20**	266.483
0	**32**	273.15
5	41	278.15
10	50	283.15
15	59	288.15
20	68	293.15
25	77	298.15
30	86	303.15
35	95	308.15
40	104	313.15
45	113	318.15
50	122	323.15
55	131	328.15
60	140	333.15
65	149	338.15
70	158	343.15
75	167	348.15
80	176	353.15
85	185	358.15
90	194	363.15
95	203	368.15
100	**212**	373.15
105	221	378.15
110	230	383.15
115	239	388.15
120	248	393.15
140	284	413.15
160	320	433.15
180	356	453.15
200	392	473.15
250	482	523.15
300	572	573.15
350	662	623.15
400	752	673.15
450	842	723.15
500	932	773.15
1000	1832	1273.15
5000	9032	5273.15
10,000	18,032	10,273.15

*Conversions in boldface type are exact. Continuing decimals are underlined. Temperature in kelvins equals temperature in degrees Rankine divided by 1.8 [K = °R/1.8].

Source: D. G. Fink and W. Beaty (eds.), *Standard Handbook for Electrical Engineers*, 12th ed., McGraw-Hill, New York, 1987.

APPENDIX F
BIBLIOGRAPHY

BOOKS

Aurbach, D.: *Nonaqueous Electrochemistry,* Marcel Dekker, New York, 1999.

Bagotsky, V. S.: *Fundamentals of Electrochemistry,* 2nd ed., Wiley, New York, 2002.

———: *Fuel Cells: Problems and Solutions,* Wiley, New York, 2009.

Balbuena, P. B., and Y. Wang: *Lithium-Ion Batteries: Solid Electrolyte Interphase,* Imperial College Press, London, 2004.

Barbir, F.: *PEM Fuel Cells: Theory and Practice,* Elsevier, San Diego, 2005.

Bergeveld, H. J., W. S. Kruijt, and P. H. L. Notten: *Battery Management Systems: Design by Modeling (Philips Research Book Series),* Kluwer Academic Publishers, New York, 2002.

Berndt, D.: *Maintenance-Free Batteries,* 2nd ed., SAE, Warrendale, PA, 1997.

Bockris, J. O., and A. K. N. Reddy: *Modern Electrochemistry,* vols. I and II, Plenum Publishing Corp., New York, 1970; Plenum/Rosetta Ed., 1973.

Bode, H.: *Lead-Acid Batteries* (translated from German by R. J. Brodd and K. V. Kordesch), Wiley, New York, 1977.

Bro, P., and S. C. Levy: *Quality and Reliability Methods for Primary Batteries,* Electrochemical Society Monograph Series, Wiley, New York, 1990.

Broadhead, J., and B. Scrosati: *Lithium Polymer Batteries,* vol. 96-17, The Electrochemical Society, Pennington, NJ, 1997.

Broussely, M., and G. Pistoia: *Industrial Applications of Batteries: From Cars to Aerospace and Energy Storage,* Elsevier, Amsterdam, 2007.

Conway, B. E.: *Theory and Principles of Electrode Processes,* Ronald, New York, 1965.

Falk, S. U., and A. J. Salkind: *Alkaline Batteries,* Wiley, New York, 1969.

Fleisher, A., and J. J. Lander (eds.): *Zinc-Silver Oxide Batteries,* Wiley, New York, 1971.

Gabano, J. P.: *Lithium Batteries,* Academic Press, Ltd., London, 1983.

Gou, B., W. K. Na, and B. Diong: *Fuel Cells: Modeling, Control, and Applications,* CRC Press, Boca Raton, FL, 2009.

Gray, F. M.: *Polymer Electrolytes,* Royal Academy of Chemistry, London, 1997.

Heise, G. W., and N. C. Cahoon (eds.): *The Primary Battery,* vols. I and II, Wiley, New York, 1971 and 1976.

Himy, A.: *Silver-Zinc Battery: Phenomena and Design Principles,* Vantage Press, New York, 1986.

———: *Silver-Zinc Battery: Best Practices, Facts and Reflections,* Vantage Press, New York, 1995.

Huggins, R. A.: *Advanced Batteries: Materials Science Aspects,* Springer, New York, 2008.

Izutsu, K.: *Electrochemistry in Nonaqueous Solutions,* 2nd ed., Wiley-VCH, Weinheim, Germany, 2009.

Jackish, H. D.: *Batterie-Lexikon,* Pflaum Verlag, Munich, 1993.

Kakaç, S., A. Pramuanjaroenkij, and L. Vasiliev: *Mini-Micro Fuel Cells: Fundamentals and Applications,* Springer, Dordrecht, Netherlands, 2008.

Kordesch, K., and G. Simander: *Fuel Cells and Their Applications,* VCH Publishers, New York, 1996.

Levy, S. C., and P. Bro: *Battery Hazards and Accident Prevention,* Plenum Press, New York, 1994.

Minami, T., et al. (eds): *Solid State Ionics for Batteries,* Springer-Verlag, Tokyo, 2005.

Nazri, G. A., and G. Pistoia (eds.): *Lithium Batteries: Science and Technology,* Kluwer Academic Publishers, New York, 2004.

Ohno, H. (ed.): *Electrochemical Aspects of Ionic Liquids,* Wiley, New York, 2005.

Osaka, T., and M. Datta: *Energy Storage Systems for Electronics,* Gordon and Breach Science Publishing, London, 1999.

Ozawa, K.: *Lithium Ion Rechargeable Batteries: Materials, Technology, and New Applications,* Wiley-VCH, Weinheim, Germany, 2009.

Pistoia, G.: *Batteries for Portable Devices,* Elsevier, San Diego, 2005.

Rand, D. A. J., and P. T. Moseley: *Valve-Regulated Lead-Acid Batteries,* Elsevier, San Diego, 2004.

Rand, D. A. J., R. Woods, and R. M. Dell: *Batteries for Electric Vehicles,* SAE, Warrendale, PA, 1998.

Schlesinger, H.: *The Battery: How Portable Power Sparked a Technological Revolution,* HarperCollins, New York, 2010. (This is a popular history of battery discovery and development.)

Smith, J. J., and K. J. Stevensen: "Reference Electrodes," Chap. 4 in *Handbook of Electrochemistry,* C. G. Zoski, ed., Elesevier, Oxford, U.K., 2007.

Srinivasan, S.: *Fuel Cells: From Fundamentals to Applications,* Springer, New York, 2006.

Sudworth, J., and R. Tilley: *The Sodium/Sulfur Battery,* Chapman and Hall, London, 1985.

Tuck, C. D. S.: *Modern Battery Technology,* Ellis Horwood, Ltd., Chichester, West Sussex, England, 1991.

Van Schalkwijk, W. A., and B. Scrosati: *Advances in Lithium-Ion Batteries,* Kluwer Academic Publishers, New York, 2002.

Venkatasetty, H. V.: *Lithium Battery Technology,* Wiley, New York, 1984.

Vinal, G. W.: *Primary Batteries,* Wiley, New York, 1950.

———: *Storage Batteries,* Wiley, New York, 1955.

Yoshio, M., R. Brodd, and A. Kozawa: *Lithium-Ion Batteries: Science and Technologies,* Springer, New York, 2007.

Zimmerman, A. H.: *Nickel-Hydrogen Batteries: Principles and Practice,* Aerospace Press, Los Angeles, 2009. (See also Bibliography in Chap. 2, p. 2.36.)

PERIODICALS

ECS Transactions (semiannual), The Electrochemical Society, Pennington, NJ, 08534.

Interface, The Electrochemical Society, Pennington, NJ, 08534.

Journal of the Electrochemical Society, Pennington, NJ, 08534.

Journal of Power Sources, Elsevier Sequoia, S. A., Lausanne, Switzerland.

PROCEEDINGS OF ANNUAL/BIENNIAL CONFERENCES

Annual European Lead Battery Conference, www.ila-lead.org. Proceedings available on CD-ROM.

Annual Fuel Cell Seminar, www.fuelcellseminar.com/past-conferences.aspx.

Proceedings of the Annual International Battery Seminar and Exhibit, Florida Educational Seminars, Boca Raton, FL. Available on CD-ROM.

Proceedings of the Biennial Workshop for Battery Exploratory Development, sponsored by the Office of Naval Research and Naval Surface Warfare Center Carderock Division, West Bethesda, MD.

Proceedings of the International Meetings on Lithium Batteries, The Electrochemical Society, Pennington, NJ 08534, www.imlb.org.

Proceedings of the NASA Aerospace Battery Workshop, Marshall Space Flight Center, AL.

Proceedings of the Power Sources Conference, Army Power Division, U.S. Army RDEC, Ft. Monmouth, NJ, and Aberdeen Proving Ground, ND. Available from National Technical Information Service, Springfield, VA 22161-0001.

OTHER REFERENCE SOURCES: HANDBOOKS AND BIBLIOGRAPHIES

Advanced Batteries for Electric Vehicles: An Assessment of Performance and Availability of Batteries for Electric Vehicles: A Report of the Battery Technical Advisory Panel, prepared for the California Air Resource Board (CARB), F. R. Kalhammer et al., El Monte, CA (June 22, 2000).

Cost of Lithium-Ion Batteries for Vehicles, L. Gaines and R. Cuenca, Argonne National Laboratory, Argonne, IL 60439, Report No. ANL/ESD-42 (May 2000).

Encyclopedia of Electrochemical Power Sources, C. K. Dyer et al. (eds.), Elsevier, San Diego (2009).

EPRI-DOE Handbook of Energy Storage for Transmission and Distribution Applications, EPRI Report No. 1001834, Electric Power Research Institute, Palo Alto, CA (2009).

Fuel Cell Handbook, EG&G Technical Services, U.S. Dept. of Energy-NETL, P. O. Box 880, Morgantown, WV 26507 (2004). Available for download or from NTIS, Springfield, VA 22161-0001.

Handbook for Handling and Storage of Nickel-Cadmium Batteries: Lessons Learned, NASA Ref. Publ. 1326 (Feb. 1994).

Handbook of Battery Materials, J. O. Besenhard (ed.), Wiley-VCH (1999).

Handbook of Fuel Cells: Fundamentals, Technology and Applications (six volumes), W. Vielstich, A. Lamm, H. Gasteiger, and H. Yokokawa (eds.), Wiley (2003–2009).

Handbook of Solid State Electrochemistry, Vol. 1, Fundamentals, Materials and Their Applications, V. V. Kharton, ed., Wiley-VCH (2009).

NASA Handbook of Nickel-Hydrogen Batteries, NASA Ref. Pub. 1314 (Sept. 1993).

NAVSEA Batteries Document: State of the Art, Research and Development, Projections, Environmental Issues, Safety Issues, Degree of Maturity, Department of the Navy Publ. NAVSEA-AH-300 (July 1993).

Navy Primary and Secondary Batteries—Design and Manufacturing Guidelines, Department of the Navy Publ. NAVSO P-3676 (Sept. 1991).

Rechargeable Batteries: Application Handbook, Newnes (Butterworth-Heinemann), Woburn, MA, 1997.

Battery manufacturers' websites are also excellent sources for technical information and performance data. The websites and physical addresses of these manufacturers are listed in Appendix G.

STANDARDS

See Chap. 4.

APPENDIX G
BATTERY MANUFACTURERS AND R&D ORGANIZATIONS*

Prepared by Vaidevutis Alminauskas

AUSTRIA

AccuPower Research, Development and Distribution Company Ltd.
Kärntnerstrasse 87
8053 Graz
Tel: +43 316 2629110
Fax: +43 316 26291136
www.accuapower.at

Banner GmbH
Postfach 777
Salzburger Strasse 298
A-4021 Linz
Tel: +43 732 38880
Fax: +43 732 388821399
www.bannerbatterien.com

Bären Batterie GmbH
Feistriz I.
A-9181 Rosental
Tel: +43 4228 20 36
Fax: +43 4228 29 15
www.baeren.at

Cellstrom GmbH
Rennweg 87
2345 Brunn am Gebirge
Tel: +43 2236 379000 0
Fax: +43 2236 379000 9
www.cellstrom.com

Institut fur Chemische Technologie[†]
Technischen Universitat
A-8010 Graz
www.ictos.tugraz.at

AUSTRALIA

Apollo Batteries – Lion Group
1/1 Bearing Road
Seven Hills
NSW 2147
Tel: +61 2 9674 6322
Fax: +61 2 9674 6277
www.apollobatteries.com.au

Battery Energy Power Solutions Pty. Ltd.
96 Fairfield Street
Fairfield
NSW 2165
Tel: +61 2 9681 3633
Fax: +61 2 9632 4622
www.batteryenergy.com.au

Century Yuasa Batteries Pty. Ltd.
37-65 Cobalt Street
Carole Park Qld 4300
Tel: +61 7 3361 6161
Fax: +61 7 3361 6705
www.centurybatteries.com.au

---

*See also Chap. 4 for organizations preparing battery standards
[†]Government laboratory

CSIRO Energy Technology[†]
Locked Bag 10
Clayton South VIC 3169
Tel: +61 3 9545 2176
Fax: +61 3 9545 2175
www.det.csiro.au

Supercharge Batteries
36 Roberna Road
Moorabbin, Victoria 3189
Tel: +61 3 9555 9000
Fax: +61 3 9555 7194
www.supercharge.com.au

BELGIUM

Association of European Storage Battery Manufacturers[‡]
Avenue Jules Bordet 142
B-1140 Brussels
Tel: +32 2 761 1653
Fax: +32 2 761 1699
www.eurobat.org

Duracell Batteries N.V.
Nijverheidslaan 7
B-3200 Aarschot
Tel: +32 16 20 11
Fax: +32 16 20 10
www.duracell.com

European Portable Battery Association[‡]
Avenue Jules Bordet 142
B-1140 Brussels
Tel: +32 2 761 1602
Fax: +32 2 761 1699
www.epbaeurope.net

BRAZIL

Acumuladores Ajax Ltda.
R Joaquim Marques De Figueiredo, 5 57
Distrito Industrial 17034-290
Tel: +55 14 2106 3068
www.ajaxbatteries.com

Accumulatories Moura S/A
Rua Diário de Pernambuco, 195
Tancredo Neves, Pernambuco, 55150-615
Tel: +55 81 3726 1044
Fax: +55 81 3726 2032
www.moura.com.br

Enerbrax Acumuladores Ltda.
Av. Rodrigues Alves, 60-18
Parque Paulista, Bauru, São Paulo
17030-000
Tel: +55 14 2107 4000
Fax: +55 14 2107 4000
www.enerbrax.com.br

Grupo Nacional de Baterias GNB Indústria de Baterias Ltda.
Rua Edwy Taques de Araújo, 1000
Jd. Burle Marx Londrina 86047-790
Tel: +55 43 3376 9000
www.gnbbaterias.com.br

Saturnia Energia Ltda.
Rua Aurélia Luiza M. Zanon 600
Sorocaba, 18087-100
Fax: +55 15 3235 8196
Tel: +55 15 3235 8000
www.saturnia.com.br

BULGARIA

Institute of Electrochemistry and Energy Systems (IEES) Bulgarian Academy of Sciences[†]
Acad.G, Bonchev Street, Block 10, Sofia
Tel: +359 2 872 25 45
Fax: +359 2 872 25 44
www.bas.bg/cleps/

CANADA

All Power Battery
B4 - 9640 201 Street Langley,
British Columbia V1M 3E8
Tel: (604) 888 3824
Fax: (604) 888 3710
www.battery2000.com

Battery Technologies Inc.
30 Pollard Street
Richmond Hill, Ontario L4B 1C3
Tel: (905) 881 5100
Fax: (905) 881 6043
www.bti.ca

Dalhousie University, Depts. of Physics and Chemistry
6300 Coburg Road
Halifax, Nova Scotia, B3H 3J5
Tel: (902) 494 2991
www.dal.ca

[‡]Battery trade group or association

Eagle-Picher Energy Products Corp.
13136-82A Avenue
Surrey, British Columbia V3W 9Y6
Tel: (604) 543 4350
Fax: (604) 543 8122
www.epi-tech.com

Electrovaya Inc.
2645 Royal Windsor Drive
Mississauga, Ontario L5J 1K9
Tel: (905) 855 4610
Fax: (905) 822 7953
www.electrovaya.com

E-One Moli Energy (Canada) Ltd.
20000 Stewart Crescent
Maple Ridge, British Columbia V2X 9E7
Tel: (604) 466 6654
Fax: (604) 466 6600
www.molienergy.com

Institut de Recherché d'Hydro-Quebec[†]
1800 Montée Ste-Julie
Varennes, Quebec J0L 2P0
Tel: (514) 652 8011
Fax: (450) 652 8990
www.ireq.ca

Prudent Energy Inc.
Suite 200, 13955 Bridgeport Road
Richmond, British Columbia V6V 1J6
Tel: (604) 278 5777
Fax: (604) 288 2577
www.pdenergy.com

Pure Energy Visions Inc.
30 Pollard Street
Richmond Hill, Ontario L4B 1C3
Tel: (905) 707 9577
www.pureenergybattery.com

Surrette Battery Company Ltd.
P.O. Box 2020, 1 Station Road
Springhill, Nova Scotia B0M 1X0
Tel: (902) 597 3767
Fax: (902) 597 8447
www.surrette.ca

CHINA (PRC)

Baoding Jinfengfan Storage Battery Co. Ltd.
8 Fuchanglu, Xinshi, Hebei Baoding
Hebei 071057
Tel: +86 312 3208571
Fax: +86 312 3208572
www.fengfan.com.cn

B. B. Battery Co., Ltd.
Cheng Dong Trial Area
Huang Gang, Raoping
Guang Dong 515700
Tel: +86 768 7601001
Fax: +86 768 7601469
www.bb-battery.com

BYD Battery Co., Ltd.
No. 3001, Hengping Road, Pingshan, Longgang
Shenzhen 518118
Tel: +86 755 89888888
Fax: +86 755 84202222
www.bydit.com

Changzhou Daily-Max Battery Co., Ltd.
39-1 Huayuan Road, Changzhou
Jiangsu 213016
Tel: +86 519 83270441
Fax: +86 519 83270425
www.daily-max.net

Chendu Jianzhong Lithium Battery Co.
No. 169 North Zangwei Road, Shuangliu
Sichuan 610200
Tel: +86 28 85772 262
Fax: +86 28 5822 429
www.china-li-battery.com

China BAK Battery, Inc.
BAK Industrial Park, Kuichong Town, Longgang
Shenzhen 518119
Tel: +86 755 89770088
Fax: +86 755 89770202
www.bak.com.cn

China Industrial Association of Power Sources[‡]
No. 18, Lizhuangzi Road
Nankai District
Tianjin 300381
Tel: +86 22 23959268
Fax: +86 22 23380938
www.cibf.org.cn
www.chinabatteryonline.com

China National Battery Industry Information Center[‡]
No. 1 Yangtianhu Xincun, Changsha
Hunan 410007
Tel: +86 731 5141901
www.batterypub.com

Chongqing Battery Co.
168 Zhongshan 3 Road
Yuzhong District
Chongqing 400015
Tel: +86 23 6386 7741
Fax: +86 23 6386 6957
www.cqmec.com/battery.htm

Chongqing Wanli Storage Battery Co., Ltd.
31 Kuzhuba, Banan
Chongqing 400054
Tel: +86 23 6259 4917
Fax: +86 23 6259 4936
www.wanli.net.cn

Coslight Technology International Group Co., Ltd.
68 Dianlan Street, Xuefu Road
Nangang 150086
Tel: +86 451 86677970
Fax: +86 451 86678032
www.cncoslight.com

Diamec Industrial Battery Ltd.
Flat A, 19/F., Bank Tower
351-353 King's Road, North Point
Hong Kong
Tel: +852 2763 5713
Fax: +852 2357 4728
www.diamec.com

EEMB Co., Ltd.
A,B,C,D, 25/F, Building A, Fortune Plaza
No. 7060, Shen Nan Road
Shen Zhen 518040
Tel: +86 755 83022275
Fax: +86 755 83021966
www.eemb.com

EVE Energy Co., Ltd.
EVE Industrial Park
Xikeng Industrial Zone
Huihuan Town, Huizhou
Guangdong 516006
Tel: +86 752 2610582
Fax: +86 752 2606033
www.evebattery.com

EverExceed Industrial Company, Ltd.
0416-0418, Tower A, Marina Bay Center
Xinghua Road, Ban'an CBD, ShenZhen
Guangdong 518020
Tel: +86 755 29556590
Fax: +86 755 29556596
www.everexceed.com

Exide–Asia Pacific Headquarters
Exide Technologies, Room 1806-1811
Hua Xu International Tower
No. 336, Xizang Zhong Road
Shanghai 200001
Tel: +86 21 2322 3800
Fax: +86 21 2322 3806
www.exideworld.com.cn

Fengfan Co., Ltd.
8 Fuchang Road, Baoding 071057
Tel: +86 312 8750099
Fax: +86 312 8601887
www.sail-vrla.com

Fujian Nanping Nanfu Battery Co., Ltd.
109 Gongyelu, Fujian Nanping
Fujian 353000
Tel: +86 599 8733999
Fax: +86 599 8710195
www.nanfu.com

Fullriver Battery Manufacture Co., Ltd.
Tai Shi Industrial Area, Yu Wo Tou Town,
Pan Yu District
Guang Zhou City, Guang Dong 511475
Tel: +86 20 84916671
Fax: +86 20 84916672
www.fullriver.com

GasTon Battery Industrial Ltd.
Room 1713A Well Fung Industrial Centre
68 Ta Chuen Ping Street, Kwai Chung
Hong Kong
Tel: +852 2447 7507
Fax: +852 2617 2465
www.gaston.com.hk

Gold Peak Industries Ltd.
Gold Peak Building 8/F
30 Kwai Wing Road, Kwai Chung, NT
Hong Kong
Tel: +852 2427 1133
Fax: +852 2489 1879
www.goldpeak.com

Golden Power Corporation Ltd.
Flat C, 20/F, Block 1, Tai Ping Industrial
Center
57 Ting Kok Road, Tai Po, NT
Hong Kong
Tel: +852 31252 288
Fax: +852 31252 001
www.goldenpower.com

Great Power Battery (International) Co., Ltd.
Room E2B, 14/F Hoi Bun Industrial Building,
6 Wing Yip Street
Kwun Tong, Kowloon, Hong Kong
Tel: +852 2345 6310
Fax: +852 2191 2768
www.greatpowerbattery.com.hk

Guangdong Jiangmen JJJ Battery Co., Ltd.
No. 83 Yongsheng Road, Baisha Ind. Dev.
Area West
Jiangmen City, Guangdong 518001
Tel: +86 750 3534 405
Fax: +86 750 3534 305
www.jjjbattery.com

Guangxi Wuzhou Sunwatt Battery Co. Ltd.
13 Xidi 2nd Road
Wuzhou, Guangxi 43002
Tel: +86 774 3860068
Fax: +86 774 3828315
www.sunwatt.com

Guangzhou Jianhang Storage Battery Ind. Co., Ltd.
92 Nanzhoulu, Haizhu
Guangdong Guangzhou
Guangdong 510400
Tel: +86 20 86559900
Fax: +86 20 89890841
www.canbattery.com
www.gzbattery.com

Harbin Guangyu Group
68 Dianian Street, Xeuru Road
Nangang District, Harbin 150086
Tel: +86 451 668 8168
Fax: +86 451 667 8032

Henda Power Industrial Co., Ltd.
Chaotian Industrial Zone, Shilou, Panyu
Guangzhou 51447
Tel: +86 20 39960218
Fax: +86 20 84852523
www.hendatoyo.com.cn

Hi-Watt Battery Industry Co., Ltd.
21 Tung Yuen Street, Yau Tong Bay
Kowloon, Hong Kong
Tel: +852 2348 0111
Fax: +852 2772 7703
www.hi-watt.com.hk

Hong Kong Batteries Manufactury Ltd.
Flat 7, 10/F, Kam Hon Industrial Building
8 Wang Kwun Road, Kowloon-Bay
Kowloon, Hong Kong
Tel: +852 27988548
Fax: +852 27980321
www.hk-batteries.com

Huanyu Power Source Co. Ltd.
Sanliqiao, Xinhui Road., Xinyang 453002
Tel: +86 373 268 8006
Fax: +86 451 268 8011
www.huanyubattery.com

HYB Battery Co., Ltd.
13-16, Fumin Industrial Park, Pinghu,
Shenzhen 518111
Tel: +86 755 84686666
Fax: +86 755 84686256
www.hyb-battery.com

Hyper Power Co. Ltd.
30D QingHai Building, Futian, Shenzhen
518048
Tel: +86 755 83547276
Fax: +86 755 83221088
www.hyperbattery.com

Jiangsu HighStar Chemical Battery Co., Ltd.
306 Heping Nanlu, Jiangsu Qidong
Jiangsu 226200
Tel: +86 513 83319415
Fax: +86 513 83312306
www.highstar-battery.com

Kayo Battery Co., Ltd.
519 Chaohui Building, Baohua Road
Longhua, Shenzhen 518109
Tel: +86 755 28117967
Fax: +86 755 28117957
www.kayobattery.com

Narada Power Source Co., Ltd.
No. 459 Wensan Road, Hangzhou 310013
Tel: +86 86 571 28827013
Fax: +86 571 85126942
www.naradabattery.com

Ningbo East Ocean Storage Battery Co. Ltd.
Dongqianhu District
Ningbo, Zhejiang 315124
Tel: +86 574 88499907
Fax: +86 574 88499965
www.east-ocean.com

Ningbo Osel Battery Factory
Wangchun Industrial Park
Ningbo 315010
Tel: +86 574 88440668
Fax: +86 574 88440686
www.ningbobattery.com

Promax Battery Industries Ltd.
No. 108 Xing Han Street, Suzhou Industrial Park
Jiangsu 215021
Tel: +86 512 6299 6687
Fax: +86 512 6299 6987
promaxbatt.diytrade.com

Shanghai White Elephant Swan Battery Co., Ltd.
1518 Gu Lang Road, Shanghai 200331
Tel: +86 21 6250 2525
Fax: +86 +86 21 6250 0451
www.swsbc.com

Shenyang Storage Battery Factory
28 Baogong Street, Tiexi, Shenyang
Liaoning 110026
Tel: +86 24 2587 5578
Fax: +86 24 2585 0217

Shengyang DongBei Storage Battery Co., Ltd.
3 Beier Zhonglu, Tiexi
Liaoning Shenyang, Liaoning 110015
Tel: +86 24 25874797
Fax: +86 24 25858196
www.dbbattery.com

Shenyang Storage Battery Research Institute[†]
No. 33 Beier Zhong Road, Tie-xi, Shenyang
Tel: +86 24 85610109
Fax: +86 24 85610109

Shenzhen Center Power Tech. Co., Ltd.
Room 502, Bairong Building, 1073 Cuizhu Road
Shenzhen 518020
Tel: +86 755 5164 318
Fax: +86 755 5606 044
www.vision-batt.com

Shenzhen Desay Battery Technology Co., Ltd.
Rm. A-18B, No. 1 World Square
7002, Hongli West Road, Guangdong Shenzhen
Guangdong 518034
Tel: +86 755 82968282
Fax: +86 755 82969220
www.desaybattery.com

Shenzhen Leoch Battery Co., Ltd.
Block C, D, E, 22/f, Xinbaohui Mansion, Nanyou Avenue
Nanshan, Shenzhen, Guangdong 518054
Tel: +86 10 58673577
Fax: +86 10 58673555
www.leoch.com

Shenzhen Li-Ion-Battery Co., Ltd.
Hongfu Ind. Zone, Dalangcun, Longhuazhen
Guangdong Bao'an, Guangdong 518109
Tel: +86 755 28032 081
Fax: +86 755 755 28032 082
www.bkbattery.com

Shenzhen PKCELL Battery Co., Ltd.
Building 430, Bagua 4th Road, China
Shenzhen, Guangdong 518002
Tel: +86 755 82409772
Fax: +86 755 82451819
www.pkcell.net

Shenzhen Ritar Power Co., Ltd.
Rm. 2201, Tian
Shenzhen, Guangdong 518040
Tel: +86 755 8347 5880
Fax: +86 755 83475180
www.ritarspower.com

Sunbright Power Co., Ltd.
No. 238 Keyuan Road, Science & Technology Industry Zone, Ninghai, Ningbo 315600
Tel: +86 574 65338616
Fax: +86 574 65552928
www.shengbaody.com

Suzhou Phylion Battery Co., Ltd.
81 Xiangyang Road, Suzhou New District,
Jiangsu 215011
Tel: +86 512 68094266
Fax: +86 512 68418041
www.xingheng.com.cn

Tianjin Institute of Power Sources[†]
No. 18, Lingzhuangzi Road, Nankai
Tianjin 300381
Tel: +86 22 2399396
Fax: +86 22 23383783

Tianjin Lantian High-Tech Power Sources Joint Stock Co., Ltd.
19 Lingzhuangzi Road, Nankai District,
Tianjin 300381
Tel: +86 22 23786621
Fax: +86 22 23787222
www.lthitech.com

Tianjin Lishen Battery Joint-Stock Co., Ltd.
6 Lanyuan Road, Huayuan Hi-Tech Industry Park, Tianjin 300384
Tel: +86 22 83710366
Fax: +86 22 83710375
en.lishen.com.cn

TCL Hyperpower Batteries Inc.
Tongfu Road, Longjin, Shuikou Town, Huicheng
Huizhou, Guangdong 516005
Tel: +86 752 2365558
Fax: +86 752 2365557
www.tclbattery.com

Tongyong Battery Co., Ltd.
Yanwo, Shipai Town, Dongguan City
Guangdong 523331
Tel: +86 13421961204
Fax: +86 769 86525202
www.tet-battery.com

Weihai Wenlong Battery Co. Ltd.
Wendeng, Shandong 264423
Tel: +86 631 8842 368
Fax: +86 631 8842 068
www.wenlongpower.com

Wuhan Forte Battery Co., Ltd.
Wujiashan Taiwan Businessmen Investment Zone
Wuhan, Hubei 430040
Tel: +86 27 83258 996
Fax: +86 27 83259 120
www.wh-forte.com

Wuhan Lixing Power Sources Co.
7#, Guandong Science & Technology Park
Eastlake Hi-Tech Development District
Wuhan, Hubei 430074
Tel: +86 27 87561817
Fax: +86 27 87414024
www.lisun.com

Wuzhou Storage Battery Factory
32 Xiji, Longgulu, Wuzhou, Guangxi 543000
Tel: +86 774 3823 700
Fax: +86 774 3823 700

Xiamen 3-Circles Battery Co., Ltd.
519 North Road Jimei District
Xiamen 361021
Tel: +86 592 6388992
Fax: +86 592 6388888
www.3-circles.com

Yoku Energy Technology Ltd.
Unit 1012, Kodak House
321 Java Road, North Point, Hong Kong
Tel: +852 28878715
Fax: +852 28876920
www.yokuenergy.com

Zhejiang Hengji Power Supply Co., Ltd.
Lianshi Industrial Zone, Huzhou
Zhejiang 312000
Tel: +86 572 2119183
Fax: +86 572 2119653
www.cnhengji.com

Zhejiang Lighthouse Storage Battery Group Co.
946 Yun Dong Road, Shaoxing,
Zhejiang 312000
Tel: +86 575 8649 521
Fax: +86 575 8649 528
www.cnlighthouse.com

Zhejiang Zhenlong Battery Co., Ltd.
Zhichengzhen, Huzhou, Zhejiang 313100
Tel: +86 572 6216760
Fax: +86 572 6216760
www.zjzhenlong.com

Zhongshan Enduring Battery Co., Ltd.
Min'an Industrial Zone, Nantou Town,
Zhongshan City
Guangdong 528427
Tel: +86 760 23117101
Fax: +86 760 23117088
www.enduringbattery.com

Zibo Storage Battery Factory
P.O. Box 9, Zhangdian, Zibo
Shandong 255056
Tel: +86 533 298 8481
Fax: +86 533 298 0136
www.torchbat.cn

CZECH REPUBLIC

Akuma A. S.
Nádrazní 84
29362 Mlada Boleslav
Czech Republic
Tel: +420 326 714159
Fax: +420 326 714488
www.akuma.cz

DENMARK

Alkaline Batteries A/S
Tigervej 1
DK-7700
Thisted
Tel: +45 97 92 31 22
Fax: +45 97 92 18 99

Grinsted Akkumulator Fabrik A/S
Heimdalsvel 19
DK-7200 Grinsted
Tel: +45 75 32 19 55
Fax: +45 75 31 02 19

FINLAND

Oy Hydrocell Ltd.
Minkkikatu 1-3
FI-04430 Järvenpää
Tel: +358 20 7288 640
Fax: +358 20 7288 649
www.hydrocell.fi

FRANCE

ASB Batteries
Allée Sainte Heléne
18021 Bourges
Tel: +33 248 48 56 00
Fax: +33 248 48 56 01
www.asb-group.com

Energizer France
6 Rue Emile Pathe
78403 Chatou
Tel: +33 01 34 80 28 00
Fax: +33 01 34 80 27 94
www.energizer.eu/fr

SAFT
12 Rue Sadi Carnot
93170 Bagnolet
Tel: +33 01 49 93 17 95
Fax: +33 01 49 93 19 50
www.saftbatteries.com

Tai Kwong Yokohama Europe S.A.S.
11, Avenue Charles de Gaulle
95700 Roissy en France
Tel: +33 01 34 29 47 51
Fax: +33 01 34 29 47 48
www.yokoeuro.com

GERMANY

Accucell Deutschland
Postfach 1231
D-71302 Waiblingen
Tel: +49 7151 206260
Fax: +49 7151 206270
www.accucell.de

Accumulatorenfabrik Berga GmbH
Gerwigstrasse 4
D-76437 Rastatt
Tel: +49 7222 156540
Fax: +49 7222 156545
www.berga-batterien.de

AIM Batterie Vertriebs GmbH
Benno-Strauß-Strasse 8, Fürth
D-90763, Bayern
Tel: +49 911 9617970
Fax: +49 911 96179729
www.aim-nuernberg.de

Akkumulatorenfabrik Moll GmbH & Co. KG
Angerstrasse 50
D-96231 Staffelstein
Tel: +49 9573 96220
Fax: +49 9573 96221
ww.moll-batterien.de

BAE Batterien GmbH
Wilhelminenhofstrasse 69/70
D-12459 Berlin
Tel: +49 30 53001401
Fax: +49 30 5354949
www.bae-berlin.de

Diehl & Eagle-Picher GmbH
Fischbachstrasse 16
D-90552 Röthenbach
Tel: +49 911 9572073
Fax: +49 911 9 572485
www.battery.de

Exide Technologies (Europe) GmbH
Im Thiergarten
D-63654 Buedingen
Tel: +49 6042 810
www.exide.com

Friemann & Wolf Batterietechnik GmbH
Industriestrasse 22
D-63654 Büdingen
Tel: +49 6042 9540
Fax: +49 6042 954190
www.friwo-batterien.de

GAIA Akkumulatorenwerke GmbH
(Division of Lithium Technology Corp. of USA)
Montaniastrasse 17
D-99734 Nordhausen
Tel: +49 36 3161670
Fax: +49 36 31616749
www.gaia-akku-online.de

GAZ Geräte-und Akkumulatorenwerk Zwickau GmbH
Reichenbacher Strasse 62-68
D-08056 Zwickau
Tel: +49 375 86 552
Fax: +49 375 86 440
www.gaz-gmbh.com

HOPPECKE Batterien GmbH & Co. KG
Bontkirchener Strasse 1
D-59929 Brilon
Tel: +49 2963 61 0
Fax: +49 2963 61 449
www.hoppecke.com

Johnson Controls Europe
AM Leineufer 51
D-30419, Hannover
Tel: +49 511 975 1100
Fax: +49 511 975 10 10
www.varta.com

Leclanché Lithium GmbH
Industriestrasse 1
D-77731 Willstätt
Tel: +49 7852 818 00
Fax: +49 7852 818 48
www.leclanche.com

Li-Tec Vermögensverwaltungs GmbH
Am Wiesengrund 7
D-01917 Kamenz
Tel: +49 3578 3092 0
Fax: +49 3578 30922 10
www.li-tec.de

Litronik Batterietechnologie
Birkwitzer Strasse 79
D-01796 Pirna
Tel: +49 3501 530 50
Fax: +49 3501 530 599
www.litronik.de

Tadiran Batteries GmbH
Industriestrasse 22
D-63654 Büdingen
Tel: +49 60 42 60 954 0
Fax: +49 60 42 60 954 190
www.tadiranbatteries.de

VARTA Consumer Batteries GmbH & Co. KGaA
Innovapark A4, Am Limespark 2
D-65843 Sulzbach
Tel: +49 6196 90240
Fax: +49 6196 9024 400
www.varta-consumer.com

VARTA Microbattery GmbH
Daimlerstrasse 1
D-73479, Ellwangen
Tel: +49 7961 921 0
Fax: +49 7961 921 553
www.varta-microbattery.com

Wernigeröder Batterie GmbH
Steinerne Renne 72
D-38855 Wernigerode
Tel: +49 03 94 3 92 6 0
Fax: +49 03 94 3 92 61 15
www.werbat.de

Yuasa Batteries (Europe) GmbH
Tiefenbroicher Weg 28
D-40472 Düsseldorf
Tel: +49 211 35 00 47

ZSW
Helmholtzstrasse 8
D-89081 Ulm
Tel: +49 731 9530 602
Fax: +49 731 9530 666
www.zsw.uni-ulm.de

HUNGARY

Perion Battery Factory Co., Ltd.
Váci utca 135-139
H-1138 Budapest
Tel: +36 1270 0811
Fax: +36 1320 2279

INDIA

Amara Raja Batteries Ltd.
Renigunta Cuddapah Road
Karakambadi
Tirupati 517 520
Tel: +91 8574 75561
Fax: +91 8574 75360

AMCO Batteries Ltd.
Addison Building, First Floor # 803
Anna Salai, Chennai
Tamil Nadu 600 002
Tel: +91 3027 732244
Fax: +91 3027 7313
www.amco.co.in

AMCO Saft India Ltd.
Hebbal-Bellary Jakkur Road
Byatarayanapura
Bangalore 560 092
Tel: +91 80 2 363 7790
Fax: +91 80 2 363 7716
www.amcosaft.com

Exide Industries Ltd.
Exide House
59E, Chowringee Road
Calcutta 700 020
Tel: +91 33 247 8320
Fax: +91 33 247 9819
www.exideindustries.com

HBL Power Systems Ltd.
8-2-601 Road No. 10
Banjara Hills
Hyderabad 500 034
Tel: +91 40 233 55575
Fax: +91 40 233 55048
www.hbl.in

Mysore Alkaline Batteries (Pty). Ltd.
F-2, Krishna Towers, 1st Foor,
Gandhi Ngr.
Bangalore 560 009
Tel: +91 80 2261367
Fax: +91 80 3343150

Nippo Batteries Co. Ltd.
New No. 77, Pottipati Plaza
4th Floor, NH Road
Nungambakkam, Chennai - HO
Tel: +91 40 233 55575
Fax: +91 40 233 55048
www.nippobatteries.com

Panasonic Energy India Co. Ltd.
G.I.D.C. Makarpura
P.B. No: 719
Vadodara 390 010
Tel: +91 2 65 2642661
Fax: +91 2 65 2638887
www.novino-batteries.com

Sharana Industries
Corpn. Shopping Complex
77 C.P. Ramaswamy Road
Madras 600 018
Tel: +91 44 4997568
Fax: +91 44 4995469
www.sahrana.co.in

Shetron Limited
A-6 MIDC, Street No. 5, Andheri (E)
Mumbai 400 093
Tel: +91 22 832 6228
Fax: +91 22 837 2145
www.shetrongroup.com

U&C Batteries Pvt. Ltd.
#7-169, Plot: 46, HAL Employees Colony,
Old Bowenpally, Secunderabad 500 011
Tel: +91 40 2775 7161
Fax: +91 40 2775 5761
www.ucbatteries.com

ISRAEL

Arotech Corporation Battery and Power Systems Division-Electric Fuel
Western Industrial Park
1 Battery Street
Beit Shemesh 99000
Tel: +972 2 9906 622
Fax: +972 2 9906 688
www.electric-fuel.com

Bar-Ilan University
Ramat Gan, 52900
Tel: +972 3 531 8111
www1.biu.ac.il

E. Schnapp & Co. Works Ltd.
22 Shecterman Street
Netania 42379
Tel: +972 9 860 6111
Fax: +972 9 861 9069
www.schnapp.co.il

Laor Batteries Ltd.
P.O. Box 570
Nazeret Elite 17105
Tel: +972 4 6571374
Fax: +972 4 6454169
www.laor-battery.co.il
small

Tadiran Batteries Ltd.
(Division of SAFT)
Kiryat Ekron
Rehovot 76100
Tel: +972 8 944 4315
Fax: +972 8 941 3062
www.tadiran.com

Tel Aviv University
P.O. Box 39040
Tel Aviv 69978
Tel: +972 3 640 8111
www.tau.ac.il

Vulcan Batteries Ltd.
P.O. Box 17
Migdal Tefen 24959
Tel: +972 4 987 2205
Fax: +972 4 987 2672
www.vulcanbat.com

ITALY

Fabbrica Accumulatori Uranio SpA
Viale del Lavoro, 20
37040 Veronella
Verona
Tel: +39 0442 489111
Fax: +39 0442 480374
www.uranio.com

FAAM SpA
Via Monti, Zona Industriale
63026 Monterubbiano
Tel: +39 0 734 2581
Fax: +39 0734 59729
www.faam.com

FIAMM SpA
Viale Europa, 63
36075 Montecchio Maggiore
Vicenza
Tel: +39 0444 709311
Fax: +39 0444 699237
www.fiamm.com

International Battery Company SRL
Corso Carlo e Nello Rosselli, 175/A
10141 Torino
Tel: +39 011 3851612
Fax: +39 011 3852354
www.ibcbattery.com

Midac SpA
Zona Industriale
37038 Soave
Verona
Tel: +39 045 6132132
Fax: +39 045 6132133
www.midacbatteries.com

Nouva Scaini
Zona Industriale
09039 Villacidro
Tel: +39 070 9311221
Fax: +39 070 9311228

Powergee Italia SRL
Via Sistina 121, 00187 Roma
Tel: +39 06 47 818 522
Fax: +39 06 233 249 726
www.powergee.com

Redox SRL
Via dell'Artigianato
32 Bolzano Vicentino
36050 Vicenza
Tel: +39 0444 351230
Fax: +39 0444 1801833
www.redoxsrl.com

Università degli Studi di Roma "La Sapienza"
Dipartimento di Chimica
Piazzale Aldo Moro 5, 00185 Roma
Tel: +39 06 446 2866
Fax: +39 06 491769
www.uniroma1.it

JAPAN

FDK Corp
Hamagomu Building, 5-36-11 Shimbashi
Minato-ku, Tokyo 105-8677
Tel: +81 3 3434 1271
Fax: +81 3 3434 1375
www.fdk.com

Furukawa Battery Co., Ltd.
2-4-1, Hoshikawa, Hodogaya-ku
Yokohama City, Kanagawa 240-0006
Tel: +81 45 336 5088
Fax: +81 45 333 3511
www.furukawadenchi.co.jp

GS Yuasa Corporation
Toshiba-Shoheizaka Building, 2-2-15
Sotokanda, Chiyoda-ku, Tokyo 101-0021
Tel: +81 75 312 1211
www.gs-yuasa.com

Hitachi Maxell Ltd.
2-18-2, Iidabashi, Chiyoda-ku
Tokyo 102-8521
Tel: +81 3 5467 9327
Fax: +81 3 5467 9328
www.maxell.com

Japan Batteries Industry Association[‡]
Kikai Shinkokaikan Building, 3-5-8
Shiba-koen, Minato-ku, Tokyo 105-0011
Tel: +81 3 343 40261
Fax: +81 3 343 42691
www.baj.or.jp

Mitsubishi Electric Life Network Corporation
TFT Buiding, 3-1-22, Ariake, Koto-ku,
Tokyo 135-8071
Tel: +81 6 994 4531
Fax: +81 6 994 7056
www.mitsubishilifenet.jp

**NEC TOKIN Corporation
Battery Division**
484, Harigaya-cho, Utsunomiya-shi
Tochigi 321-014
www.nec-tokin.com

Panasonic Corporation–Matsushita Battery Industrial Co.
1-1 Matsushita-cho, Moriguchi-City
Osaka 570-8511
Tel: +81 6 6991 1141
www.panasonic.net

Sanyo Electric Co., Ltd.
222-1, Kami-naizen, Sumoto-shi
Hyogo 656-8555
Tel: +81 6 6994 6328
Fax: +81 6 6994 6523
battery.sanyo.com

Sanyo GS Soft Energy Co., Ltd.
5, Ichinodan-cho, Shinden, Kisshoin
Minami-ku, Kyoto 601-8397
Tel: +81 6 991 1181
Fax: +81 6 991 6566
www.sanyo-gs.com

Schick Japan Co., Ltd. Energizer Com.
I. K. Building, 2-24-9, Kamioosaki
Shinagawa-ku, Tokyo 141-8671
Tel: +81 3 3543 3700
Fax: +81 3 5565 5770
www.schick-jp.com

Seiko Instruments Inc.
45-1, Aza Matsubara, Kamiayashi
Aoba-ku, Sendai-shi, Miyagi 989-3124
Tel: +81 22 391 9331
www.sii.co.jp

Shin-Kobe Electric Machinery Co., Ltd.
St. Luke's Tower, 8-1 Akashi-cho
Chuo-ku, Tokyo 104-0044
Tel: +81 3 3543 3700
Fax: +81 3 5565 5770
www.shinkobe-denki.co.jp

Sony Energytec Inc.
6-7-35 Kitashinagawa
Shinagawa-ku
Tokyo 141
Tel: +81 3 5448 2111
Fax: +81 3 5448 2244
www.world.sony.com

Toshiba Battery Co., Ltd.
Toshiba Shoheizaka Building 3F 2-15
Stokanda 2-chome, Chiyoda-ku
Tokyo 101-0021
Tel: +81 3 3257 5879
Fax: +81 3 3257 5916
www.toshiba-denchi.jp

MALAYSIA

Federal Batteries Sdn. Bhd.
Lot 12, Jalan Perusahaan Dua
Batu Caves Industrial Area
68100 Batu Caves
Selangor Darul Ehsan
Tel: +60 3 689 8928
Fax: +60 3 689 8277

GPA Holdings Berhad
Lot 5031 & 5032, Jalan Teratai
Off Jalan Meru, 41050 Klang, Selangor
Tel: +60 3 3392 7180
Fax: +60 3 3392 7237
www.gp-products.com

Tai Kwong-Yokohama Battery Industries Sdn Bhd
Lot 1238 Batu 23 Jalan Kachau
45000 Semenyih
Tel: +60 3 90746933
Fax: +60 3 90756648
www.tkyoko.com

Watta Battery Industries Sdn Bhd
Lot 8, Jalan Satu, Kawasan Perusahaan
Balakong
43200 Cheras Jaya, Selangor Darul Ehsan
Tel: +60 3 90751919
Fax: +60 3 90756790
www.watta.com.my

Yuasa Battery Sdn Bhd
Lot 1385, Kawasan Perusahaan
Tikam Batu, 08600 Sungai Petani
Kedah
Tel: +60 4 4388806
Fax: +60 4 4388661
yuasabattery.com.my

MEXICO

Grupo Imsa S.A. de C.V.
Av. Batallon de San Patricio
No. 111 Piso 26, Fracc. Valle Oriente
Garza Garcia 66269
Tel: +52 8181538300
Fax: +52 8181538400
www.ternium.com

NETHERLANDS

Batterij Aandrijf Techniek B.V.— BAT
Ohmstraat 8
3861 NB Nijkerk
Tel: +31 33 245 0990
Fax: +31 33 245 0991
www.batbatterijen.nl

Johnson Controls Autobatterijen B.V.
Postbus 10125 Bovendijk 137
3045 PC Rotterdam
Tel: +31 10 4 61 56 15
Fax: +31 10 2 85 00 24
www.varta-automotive.nl

Philips Research Laboratories
High Tech Campus 5
5656 AE Eindhoven
Tel: +31 40 27 91111
www.research.philips.com

Thales Munitronics
Hoogezidje 14
5626 DC Eindhoven
P.O. Box 6034 5600 HA Eindhoven
Tel: +31 40 2503 603
Fax: +31 40 2503 777
www.thales-nederland.nl

NEW ZEALAND

Pacific Lithium Ltd.
P.O. Box 33541
Takapuna, Auckland
Tel: +64 9 336 1580
Fax: +64 9 336 1520
www.pacificlithium.co.nz

NORWAY

Exide Sonnak AS
Brobekkveien 101
Postboks 418
Økern, 0513 Oslo
Tel: +47 22 07 47 00
Fax: +47 22 07 47 01
www.kongsberg-simrad.com

PAKISTAN

Exide Pakistan Ltd.
40-K, Block 6, PECH Society
Dr. Hussain Mahmood Road
Off Sharea Faisal
Karachi 75400
Tel: +92 21 453 6750
Fax: +92 21 453 8948
www.batteryfb.com

PHILIPPINES

Oriental and Motolite Marketing Corporation
Ramcar Center 80-82 Roces Avenue
Diliman, Q.C.
Tel: +63 02 370 1000
Fax: +63 02 374 1671
www.motolite.com

POLAND

Centralne Laboratorium Akumulatorow I Ogniw[†]
ul. Forteczna 12
Poznan 61-362
Tel: +48 61 8790517
Fax: +48 61 8793012
www.claio.poznan.pl

Centra SA
ul. Gdynska 31/33
Poznan 60-960
Tel: +48 61 8786100
Fax: +48 61 8786506
www.centra.com.pl

Danish Polish Batteries Sp. zo.o.
ul. Zielona 22
Starogard Gdanski 83-200
Tel: +48 58 5623057
Fax: +48 58 5613127

GP Battery Poland Sp. zo.o.
ul. Grazyny 13/15
Warszawa 02-548
Tel: +48 22 8454095
Fax: +48 22 8455869
www.goldpeak.com

Philips Matsushita Battery Poland
ul. Swoneczna 42
62-200 Gniezno
Tel: +61 426 00 00
Fax: +61 425 65 32
www.lighting.philips.com.pl gniezno.html

Zap
ul. Warsawska 47.05-820
Piastow
Tel: +48 22 723 6011
Fax: +48 22 723 6244
www.zap.po

ROMANIA

S.C. ROMBAT S.A.
Drumul Cetatii 6
R-4400 Bistrita
Tel: +40 263 238 142
Fax: +40 263 234 010
www.rombatt.ro

RUSSIA

AkTex Inc.
291 Baikalskava Str.
664050 Irkutsk
Tel: +7 3952 56 34 34
Fax: +7 3952 56 34 30
www.aktex.ru

Electrozariad
5 Ogareva
103918 Moscow
Tel: +7 095 291 5785
Fax: +7 095 291 5785

International Assn. of Chemical Power Sources Manufacturers INTERBAT‡
1-St.Smolensky Byst 7
123099 Moscow
Tel: +7 095 244 0735
Fax: +7 095 244 0369
www.interbat.ru

Istochnik Batteries Ltd.
ul. Marxistskaya, 20, Str. 9, Etazh 1
109147 Moscow
Tel: +7 095 912 54 71
Fax: +7 095 911 91 31
www.istochnik.ru/html/

Rigel Battery Co.
38 Prof. Popov Str.
197376 St. Petersburg
Tel: +7 812 234 05 56
Fax: +7 812 234 06 38

SAUDI ARABIA

National Batteries Company
P. O. Box 177
Riyadh 11383
Tel: +966 1 2650019
Fax: +966 1 2650057
www.battariat.com

SINGAPORE

Chloride Batteries S.E. Asia Pte Ltd.
106 Neythal Road
Jurong Town
Singapore 628594
Tel: +65 6265 2444
Fax: +65 6265 1478
www.cbsea.com.sg

Energizer Singapore Pte Ltd.
25 Gul Way
Singapore 629197
Tel: +65 686 11411
Fax: +65 686 11291
www.energizer.com

EnerSys Asia Oceania
152 Beach Road
Gateway East Building #11-03
Singapore 189721
Tel: +65 6508 1780
Fax: +65 6292 4380
www.enersys.com

Exide Singapore Pte Ltd.
6 Loyang Way 1 #02-02
Singapore 508704
Tel: +65 654 62866
Fax: +65 654 62966
www.exide.com

GP Batteries
50 Gul Crescent
Singapore 629543
Tel: +65 8622 088
Fax: +65 8622 313
www.gpbatteries.com

ST Battery Pte. Ltd.
77 Science Park Drive
02-15 Cintech III
Singapore 118256
Tel: +65 870 1665
Fax: +65 870 1665

SLOVENIA

TAB Tovarna Akumulatorskih Baterij D.D.
Polena 6, SLO-2392 Mezica
Tel: +386 2 87 02 300
Fax: +386 2 87 02 305
www.tab.si

SOUTH AFRICA

Alpha Power Systems (Pty) Ltd.
Unit A, Mount Royal James
Crescent 1685 Midrand
Tel: +27 11 805 5008
Fax: +27 11 805 5054
www.powerman.co.za

First National Battery Co. (Pty) Ltd.
64 Liverpool Road
Industrial Sites, Benoni South
1501 Benoni
Tel: +27 11 741 3600
Fax: +27 11 421 1625
www.battery.co.za

Willard Batteries
P.O. Box 15363
Lyttelton, Pretoria, 0140
Tel: +27 41 451 4491
Fax: +27 41 451 2622/191
www.willard.co.za

SOUTH KOREA

Atlasbx Co., Ltd.
14F, Taeseck Building
275-5 Yangjae-2 Dong, Seocho-Gu, Seoul
Tel: +82 42 620 4242
Fax: +82 42 623 9380
www.atlasbx.co.kr

E Square Technologies(E^2-Tech) Co., Ltd.
102-18 Block, Ochang Science Industrial Complex
643-2, Gak-ri, Ochang-myeon
Cheongwon-gun Chungcheongbuk-do
Tel: +82 43 216 7041
Fax: +82 43 216 7045
www.e2-tek.com

Global Battery Co., Ltd.
708-8 Sebang Building, Yeoksam 2-dong
Gangnam-gu, Seoul, 135 919
Tel: +82 2 3451 6201
Fax: +82 2 3451 6301
www.gbattery.com

Global & Yuasa Battery Co., Ltd.
708-8 Yoksam-dong
Kangnam-gu, Seoul, 135-080
Tel: +82 2 538 6201
Fax: +82 2 3451 6307
www.gybc.co.kr

Kokam Engineering Co., Ltd.
#1261-3, Jungwang-dong
Siheung-si
Gyeonggi-do
Tel: +82 31 362 0100
Fax: +82 31 362 0190
www.kokam.com

Korea Battery Industry Cooperative
1304-4 Seocho-dong Seocho-ku Seoul 137
Tel: +82 2 5532 401
Fax: +82 2 5561 290

Korea Electrotechnology Research Institute
28-1 Seongju-dong, Changwon-si
Gyeongsangnam-do, 641-120
Tel: +82 55 280 1114
Fax: +82 55 280 1216
www.keri.re.kr

Korea Storage Battery Ltd.
40-42 Dae Hwa-dong Daeduk-gu Daejeon
Tel: +82 2 579 0451
Fax: +82 2 579 1050
www.ksb.co.kr

Kyungwon Battery Co., Ltd.
6-1 San Galgot-ri Pyungtaek-gun Kyungki-do
451 860
Tel: +82 339 72 1321
www.solite.co.kr

LG Chem, Ltd.
LG Twin Towers, 20, Yeouido-dong
Yeongdeungpo-gu, Seoul, 150-721
Tel: +82 2 3773 5114
Fax: +82 2 538 4353
www.lgchem.com

Namil Battery Co. Ltd.
Chupal industry 1-2 B/L, 289-2, Chupal-ri
Pangsung-eup, Pyeongtaek, Gyeongi-do, 110-123
Tel: +82 2 2279 6141
Fax: +82 2 2274 4507
www.namilbattery.com

Nuricell
4th Floor, GS Caltex New Energy Development
Center, 453-2
Seongnae 1-dong, Gangdong-gu, Seoul
Yangjae-dong 137-130, Socho-Gu Seoul
Tel: +82 2 6900 4153
Fax: +82 2 6900 4160
www.nuricell.com

Rocket Electric Co., Ltd.
747-29 Yuksam-dong, Kangnam-ku, Seoul
Tel: +82 2 3451 5600
Fax: +82 2 3451 5737
www.rocket.co.kr

Samsung SDI Co., Ltd.
575 Shin-dong, Youngtong-gu, Suwon,
Kyunggi-do
Tel: +82 31 210 7114
Fax: +82 31 210 7146
www.samsungsdi.com

SKC Co., Ltd.
18th Floor, Kyobo Tower, 1303-22 Secho-4 dong
Secho gu, Seoul 137-070
Tel: +82 2 3787 1931
Fax: +82 2 537 2867
www.skc.co.kr/skhp/en/prod/bat

Teckraf Co., Ltd.
5th Floor. Vitzro Bld., 233-3, 1 dong
Sungsu-2 Ga Sungdong-Gu
Seoul, 133-826
Tel: +82 2 4602 243
Fax: +82 2 499 2756
teckraf.en.ecplaza.net

Young Poong, Corp.
Young Poong Building 142 Nonhyundong
Gangnamgu, Seoul 134-749
Tel: +82 2 519 3314
www.youngpoongcorp.com

XenoEnergy Co., Ltd.
470-20 MooSong-Dong, Hwaseong-City
Kyonggi-Do, 445-020
Tel: +82 31 355 3511
Fax: +82 31 355 3513
www.xenoenergy.com

SPAIN

Cegasa International SA
c/Artapadura, 11 Apartado de Correos 3280
01013 Vitoria-Gasteiz
Tel: +34 945 129 510
Fax: +34 945 129 514
www.cegasa.com

Exide Technologies, S.A. España
Condesa de Venadito 1
28027 Madrid
Tel: +34 91 566 48 00
Fax: +34 91 404 78 50
www.tudor.es

Interberg Batteries
Mirador de Despenaperros 17
28400 Collado Villalba
Tel: +34 91 851 32 13
Fax: +34 91 851 60 76
www.interberg.com

SWEDEN

Alcad Limited
Norra Strandgatan 35,
Box 504, S-572 25
Oskarshamn
Tel: +46 491 68100
Fax: +46 491 68110
www.alcad.com

Exide Technologies AB
Bultgatan 40A
SE-442 40 Kungälv
Tel: +46 303 3310 00
Fax: +46 303 7423 20
www.exide-nordic.com

Saft Nife AB
Jungnergaten 25
57232 Oskarshamn
Tel: +46 491 680 00
Fax: +46 491 681 80
www.saftbatteries.com

SWITZERLAND

Energizer SA
P.O. Box 230
1218 Le Grand-Saconnex
CH-1218 Geneva
Tel: +41 22 929 94 38
Fax: +41 22 929 94 11
www.energizer.ch

EnerSys EMEA–EH Europe GmbH
Loewenstrasse 32
CH-8001 Zürich
Tel: +41 44 215 74 10
Fax: +41 44 215 74 11
www.enersys-emea.com

Leclanché SA
Avenue des Sports 42
Ch-1401 Yverdon-les-Bains
Schweiz
Tel: +41 24 424 6500
Fax: +41 24 424 6501
www.leclanche.com

MES-DEA SA
Via Laveggio, 15
Ch-6855 Stabio
Tel: +41 91 641 5311
Fax: +41 91 641 5395
www.mes-dea.ch

Oerlikon Stationär Batterien AG
Dornacherstrasse 110
CH-4147 Aesch
Tel: +41 61 706 36 36
Fax: +41 61 706 36 37
www.accuoerlikon.com

Renata SA
Kreuzenstrasse 30
CH-4452 Itingen
Tel: +41 61 975 75 75
Fax: +41 61 975 75 95
www.renata.com

ReVolt Technology Limited
Laubisruetistrasse 44
CH-8712 Staefa
Tel: +41 44 928 78 78
www.revolttechnology.com

TAIWAN (ROC)

Advanced Lithium Electrochemistry Co., Ltd.
No. 2-1 SingHua Road, Taoyuan, 33068
Sec 2, Taipei 104
Tel: +886 3 364 6655
Fax: +886 3 364 9955
www.alechem.com

CSB Battery Co., Ltd.
5F .-1, No. 36, Nanjing W. Road
Datong District, Taipei City 10352
Tel: +886 2 2555 5600
Fax: +886 2 2555 3300
www.csb-battery.com

E-One Moli Energy Corp.
10F, No. 113, Chung Shan N. Road
Sec 2, Taipei 104
Tel: +886 2 2567 3500
Fax: +886 2 2567 6500
www.molicel.com

Gold Peak Industries (Taiwan), Ltd.
Room 1200, International Trade Building
No. 205 Sec. 1, Tun Hua South Road
Taipei 10647
Tel: +886 2 2741 4919
Fax: +886 2 2731 4868
www.gpbatteries.com

NEXcell Battery Co., Ltd.
5 Innovation Road
Science-Based Industrial Park
Hsinchu 300
Tel: +886 3 5783800
Fax: +886 3 5786645
www.battery.com.tw

TURKEY

INCI AKÜ SANAYI VE TIC. A.S.
Organize Sanayi Bölgesi 2. Kisim
45030 Manisa
Tel: +90 236 233 25 10
Fax: +90 236 233 25 13
www.inciaku.com

Mutlu Aku ve Malzemeleri San. A.S.
Akfirat Beldesi, Tepeoren Mah.
Eski Ankara Yolu Uzeri, Tuzla, Istanbul, 34959
Tel: +90 216 3041590
Fax: +90 216 3041870
www.mutlu.com.tr

UKRAINE

National Accumulator Corporation ISTA
30, Strasse Kursantskaya
49051 Dnepropetrovsk
Tel: +380 562 354 015
Fax: +380 567 211 401
www.ista.com.ua

UNITED KINGDOM

Accutronics Ltd.
Unit 20 Loomer Road
Newcastle, Staffordshire ST5 7LB
Tel: +44 1782 566622
Fax: +44 1782 576640
www.accutronics.co.uk

AEA Technology Batteries
Denchi House, Thurso Business Park
Thurso, Caithness, KW14 7XW
Tel: +44 1847 808 000
www.abslpower.com

Alexander Technologies
4 Doxford Drive
Southwest Industrial Estate
Peterlee, Co Durham SR8 2RL
Tel: +44 191 587 2787
Tel: +44 191 587 2587
www.alexandertechnologies.com

Axeon Holdings Plc.
Nobel Court, Wester Gourdie, Dundee
DD2 4UH
Tel: +44 1382 400040
Fax: +44 1382 400044
www.axeon.com

British Battery Manufacturers Association[‡]
26 Grosvenor Gardens
London SW1W 0TG
Tel: +44 20 7838 4800
Fax: +44 20 7838 4871
www.bbma.co.uk

Cobham Technical Services–ERA Technology
Cleeve Road
Leatherhead, Surrey KT22 7SA
Tel: +44 1372 367000
Fax: +44 1372 367099
www.era.co.uk

DERA Tech. Management Services[†]
Aquila, Golf Road
Bromley
Kent BR1 2JB
Tel: +44 20 8285 7127
Fax: +44 20 8285 7346
www.dera.gov.uk

Duracell UK Ltd.
Great West Road, Isleworth
Middlesex, TW7 5NP
Tel : +44 20 8326 8628
Fax : +44 20 8326 8776
www.duracell.com

Electrochemical Power Sources Centre[†]
DERA Haslar
Gosport Hampshire PO12 2AG
Tel: +44 1705 335358
Fax: +44 1705 335102
www.dera.gov.uk

Energizer UK Ltd.
Eveready House
93 Burleigh Gardens
Southgate London N14 5AQ
Tel: +44 20 8882 8661
Fax: +44 20 8882 1938
www.energizer-eu.com

Enersys–Hawker
Stephenson Street, Newport NP19 4XJ
Tel: +44 1633 277673
Fax: +44 1633 281787
www.enersys.com

Exide Headquarters, Industrial Energy Europe and Asia
P. O. Box 1
Salford Road, Over Hulton
Bolton, BL5 1DD
Tel: +44 1204 661 228
www.exide.com

GP Batteries (UK) Ltd.
Monument View
Chelston Business Park
Wellington, Somerset TA21 9JF
Tel: +44 1823 660044
Fax: +44 1823 665595
www.gpbatteries.com.uk

MSB-ASB
East Shawhead
Coatbridge ML5 4TD
Tel: +44 1236 43 77 75
Fax: +44 1236 43 66 50
www.asb-group.com/uk/msb.asp

Rayovac Europe Ltd.
Mill House, 21 Hollingworth Court
Turkey Mill, Maidstone
Kent ME14 5PP
Tel: +44 1622 358900
Fax: +44 1622 754961
Emea.rayovac.com

SAFT Ltd.
River Drive
South Shields, Tyne & Wear
NE33 2TR
Tel: +44 191 456 1451
Fax: +44 191 456 6383
www.saftbatteries.com

Sanyo Energy (UK) Ltd.
Sanyo House
18 Colonial Way
Watford, WD 24 4PT
Tel: +44 1 923 246 363
Fax: +44 1 923 477 479
www.sanyo-energy.co.uk

SEC Industrial Battery Co., Ltd.
Thorney Weir House
Iver, Bucks, SL0 9AQ
Tel: +44 1895 431543
Fax: +44 1895 431880
www.secbattery.com

Ultralife Batteries (UK) Ltd.
18 Nuffield Way
Abingdon
Oxfordshire OX14 1TG
Tel: +44 1235 542600
Fax: +44 1235 535766
www.ultralifebatteries.com

Varta Microbattery Ltd.
Cropmead Industrial Estate
Crewkerne
Somerset TA18 7HQ
Tel: +44 1460 279950
Fax: +44 1460 279959
www.uk.varta-microbattery.com

Yuasa Battery Sales UK Ltd.
Unit 22, Rassau Industrial Estate
Ebbw Vale, Gwent, NP23 5SD
Tel: +44 8708 500 257
Fax: +44 8708 500 265
www.yuasa-battery.co.uk

USA

A123 Systems, Inc.
Arsenal on the Charles
321 Arsenal Street
Watertown, MA 02472
Tel: (617) 778 5700
Fax: (617) 924 8910
www.a123systems.com

ACME Electric Corp. Aerospace Division
528 W. 21st Street
Tempe, AZ 85282
Tel: (480) 894 6864
Fax: (480) 921 0470
www.acmeelec.com/aerospace

ACR Electronics, Inc.
5757 Ravenswood Road
Ft. Lauderdale, FL 33312
Tel: (954) 981 3333
Fax: (954) 983 5087
www.acrelectronics.com

Air Force Wright Aeronautical Laboratory[†]
Aerospace Power Division
Wright-Paterson AFB, OH 45435
Tel: (937) 255 7770
Fax: (937) 656 7529

Alexander Technologies
1511 South Garfield Place
Mason City, IA 50401
Tel: (641) 423 8955
Fax: (641) 423 1644
www.alexandertechnologies.com

Alpha Technologies, Inc.
3767 Alpha Way
Bellingham, WA 98226
Tel: (360) 647 2360
Fax: (360) 671 4936
www.alpha.com

Altair Nanotechnologies Inc.
204 Edison Way
Reno, NV 89502
Tel: (775) 856 2500
Fax: (775) 856 1619
www.altairnano.com

Argonne National Laboratory[†]
Chemical Sciences and Engineering Division
Argonne National Laboratory
9700 South Cass Avenue
Argonne, IL 60439
Tel: (630) 252 4383
Fax: (630) 252 5528
www.cse.anl.gov

Axion Battery Manufacturing Inc.
3601 Clover Lane
New Castle, PA 16105
Tel: (724) 654 9300
www.turbostart.com

BAE Systems Battery Technology Center
1601 Research Boulevard
Rockville, MD 20850
Tel: (301) 838 6200
Fax: (301) 838 6745
www.baesystems.com

Battery Council International[‡]
401 North Michigan Avenue
Chicago, IL 60611
Tel: (312) 644 6610
Fax: (312) 527 6640
www.batterycouncil.org

Boston Power, Inc.
2200 West Park Drive
Westborough, MA 01581
Tel: (508) 366 0885
Fax: (508) 366 0998
www.boston-power.com

Bren-Tronics Inc.
10 Brayton Court
Commack, NY 11725
Tel: (516) 499 5155
Fax: (516) 499 5504
www.bren-tronics.com

BST Systems Inc.
78 Plainfield Pike Road
Plainfield, CT 06374
Tel: (860) 564 4078
Fax: (860) 564 1380
www.bstsys.com

Bulldog Battery Corporation
98 E Canal St.
Wabash, IN 46992
Tel: (260) 563 0551
www.bulldog-battery.com

C & D Technologies
1400 Union Meeting Road
Blue Bell, PA 19422
Tel: (215) 619 2700
Fax: (215) 619 7899
www.cdtechno.com

CFX Battery Inc.
1300 W. Optical Drive, Suite 300
Azusa, CA 91702
Tel: (626) 610 0660
Fax: (626) 389 5086
www.cfxbattery.com

Compact Power Inc.
Subsidiary of LG Chem, Ltd.
1857 Technology Drive
Troy, MI 48083
Tel: (248) 307 1800
Fax: (248) 597 0900
www.compactpower.com

Concorde Battery Corp.
2009 San Bernardino Road
West Covina, CA 91790
Tel: (626) 813 1234
Fax: (626) 813 1235
www.concordebattery.com

Continental Batteries Co.
4919 Woodall Street
Dallas, TX 75247
Tel: (214) 631 5701
Fax: (214) 634 7846
www.continentalbatteries.com

Covalent Associates, Inc.
921 NW 11th Street
Corvallis, OR 97330
Tel: (541) 207 3844
Fax: (541) 207 3845
www.covalentassociates.com

Crown Battery Manufacturing Company
1445 Majestic Drive
Fremont, OH 43420
Tel: (419) 334 7181
Fax: (419) 334 7416
www.crownbattery.com

Cymbet Corporation
18326 Joplin Street NW
Elk River, MN 55330
Tel: (763) 633 1780
Fax: (763) 633 1799
www.Cymbet.com

Delphi Energy & Engine Management Systems
Delphi Automotive Systems
5725 Delphi Drive
Troy, MI 48098
Tel: (248) 813 2000
Fax: (248) 813 2673
www.delphiauto.com

Douglas Battery Manufacturing Company
500 Battery Drive
Winston-Salem, NC 27107
Tel: (336) 650 7000
Fax: (336) 650 7072
www.douglasbattery.com

Duracell, Inc.
Berkshire Corporate Park
Bethel, CT 06801
Tel: (203) 796 4000
Fax: (203) 730 8958
www.duracell.com

Eagle-Picher Technologies, LLC
C & Porter Streets
Joplin, MO 64802
Tel: (417) 623 8000
Fax: (417) 623 5319
www.eaglepicher.com

East Penn Manufacturing Co., Inc.
Deka Road
Lyon Station, PA 19536
Tel: (610) 682 6361
Fax: (610) 682 4781
eastpenn-deka.com

Eastman Kodak Co.
343 State Street
Rochester, NY 14650
Tel: (716) 724 4000
www.kodak.com

EEMB USA Representative Office
Bridge R&D Corp.
12601 Monarch Street
Garden Grove, CA 92841
Tel: (714) 891 6509
Fax: (714) 890 8590
www.eemb.com

EIC Laboratories, Inc.
111 Downey Street
Norwood, MA 02062
Tel: (781) 769 9450
Fax: (781) 551 0283
www.eiclabs.com

Electric Power Research Institute[‡]
3420 Hillview Avenue
Palo Alto, CA 94304
Tel: (650) 855 2121
www.epri.com

Electrochem Solutions, Inc.
670 Paramount Drive
Raynham, MA 02767
Tel: (781) 830-5800
Fax: (781) 575-1545
www.electrochemsolutions.com

Emerging Power Inc.
200 Holt Street
Hackensack, NJ 07601
Tel: (201)441 3590
Tel: (201)441 3592
www.emergingpower.com

Energizer Inc.
533 Maryville University Drive
St. Louis, MO 63141
Tel: (314) 985 2000
www.energizer.com

EnerDel, Inc.
8740 Hague Road
Indianapolis, IN 46256
Tel: (317) 585 3456
Fax: (317) 585 3444
www.ener1.com

Energy Conversion Devices, Inc. (ECD Ovonic Materials)
2956 Waterview Drive
Rochester Hills, MI 48309
Tel: (248) 293 0440
Fax: (248) 844 1214
www.ovonics.com

EnerSys
2366 Bernville Road
Reading, PA 19605
Tel: (610) 208 1991
Fax: (610) 372 8457
www.enersys.com

Excellatron Solid State LLC
263 Decatur Street
Atlanta, GA 30312
Tel: (404) 584 2475
Fax: (404) 584 6772
www.excellatron.com

Exide Corp.
Building 200
13000 Deerfield Parkway
Milton, GA 30004
Tel: (678) 566 9000
Fax: (678) 566 9188
www.exide.com

Front Edge Technology, Inc.
13455 Brooks Drive
Baldwin Park, CA 91706
Tel: (626) 856 8979
Fax: (626) 851 1369
www.frontedgetechnology.com

Gibbard Research and Development Corp
14 Plumer Road
Epping, NH 03042
Tel: (603) 502-3234
www.electrochemenergy.com

G. S. Battery U.S.A., Inc.
1000 Mansell Exchange W. Suite 350
Alpharetta, GA 30022
Tel: (678) 762 4818
Fax: (678) 739 2133
www.gsbattery.com

GP Batteries
11235 West Bernardo Court
San Diego, CA 92127
Tel: (619) 674 6099
Fax: (619) 674 6496
www.gpbatteries.com

Harding Energy, Inc.
509 East Ellis Road
Norton Shores, MI 49441
Tel: (231) 798 7033
Fax: (231) 798 7044
www.hardingenergy.com

HED Battery Corporation
3355 Woodward Avenue
Santa Clara, CA 95112
Tel: (408) 980 1877
Fax: (408) 980 1804
www.hedb.com

Hunter College Physics Dept., City University of New York
695 Park Ave
New York, NY 10065
Tel: (212) 772-4973
Fax: (212) 772 5390
www.hunter.cuny.edu/physics/faculty/greenbaum/research

Innergy Power Corporation, Inc.
9051 Siempre Viva Road
Building 6 Suite AB
San Diego, CA 92154
Tel: (619) 710 0758
Fax: (619) 710 0755
www.innergypower.com

International Battery, Inc.
6845 Snowdrift Road
Allentown, PA 18106
Tel: (610) 366-3925
Fax: (610) 366-3929
www.internationalbattery.com

Jet Propulsion Laboratory[†]
Electrochemical Technologies Group
California Institute of Technology
4800 Oak Grove Drive
Pasadena, CA 91109
Tel: (818) 354 4321
Fax: (818) 393 6951
www.jpl.nasa.gov

Johnson Controls, Inc.
Battery Group
5757 N. Green Bay Avenue
Glendale, WI 53209
Tel: (414) 524 1200
www.johnsoncontrols.com

Koehler Bright Star Inc.
380 Stewart Road
Hanover Industrial Estates
Wilkes-Barre, PA 18706
Tel: (570) 825 1900
Fax: (570) 825 1984
www.flashlight.com

Lawrence Berkley Laboratory[†]
Advanced Energy Technologies Department
Environmental Energy Technologies Division
1 Cyclotron Road, MS 70R0108B
Berkeley, CA 94720
Tel: (510) 486 4202
Fax: (510) 486 7303
eetd.lbl.gov/aet/batteries.html

LG Chem
Battery and Marketing Div.
1000 Sylvan Avenue
Englewood, NJ
Tel: (201) 816 2331
Fax: (201) 816 0961

Lineage Power
3000 Skyline Drive
Mesquite, TX 75149
Tel: (972) 284 2000
www.lineagepower.com

Lithion, Inc.
(see Yardney Technical Products, Inc.)

Lithium Technology Corporation
5115 Campus Drive
Plymouth Meeting, PA 19462
Tel: (610) 940 6090
Fax: (610) 940 6091
www.lithiumtech.com

Maha Communications, Inc.
18567 E. Gale Avenue
City of Industry, CA 91748
Tel: (626) 363 9017
Fax: (626) 363 9010
www.mahaenergy.com

Marathon Power Technologies
8301 Imperial Drive
Waco, TX 76712
Tel: (254) 776 0650
Fax: (254) 776 6558
www.mptc.com

Maxcell Corp. of America
22-08 Route 208
Fair Lawn, NJ 07410
Tel: (201) 794 5900
Fax: (201) 796 8790
www.maxell-usa.com

Medtronic Inc.
710 Medtronic Parkway
Minneapolis, MN 55432
Tel: (763) 514 4000
Fax: (763) 514 4879
www.medtronic.com

Motorola Energy Systems Group
1700 Belle Meade Court
Lawrenceville, GA 30043
Tel: (770) 338 3795
www.motorola.com

NanoPower Research Laboratories
Rochester Institute of Technology
85 Lomb Memorial Drive
Rochester, NY 14623
Tel: (585) 475 2480
Fax: (585) 475 7890
www.sustainability.rit.edu/nanopower

National Alliance for Advanced Technology Batteries (NAATBATT)[‡]
122 South Michigan Avenue, Suite 1700
Chicago, IL 60603
Tel: (312) 588 0477
www.naatbatt.org

NASA Glenn Research Center[†]
21000 Brookpark Road
Cleveland, OH 44135
Tel: (216) 433 4000
www.nasa.gov/centers/glenn

Naval Sea Systems Command, Carderock Division[†]
Power Systems Branch, Code 643
9500 MacArthur Boulevard
West Bethesda, MD 20817-5700
Tel: (301) 227-5681
Fax: (301) 227-4733
www.dt.navy.mil/sur-str-mat/fun-mat/pow-sys-bra/

Naval Sea Systems Command, Crane Division[†]
Power and Circuit Board Technologies Division,
Code GXS
Crane, IN 47522
Tel: (812) 854 1593
Fax: (812) 854 3589
www.crane.navy.mil/whatwedo/PowerSystems.asp

Oak Ridge Micro-Energy, Inc.
3046 E. Brighton Place
Salt Lake City, UT 84121
Tel: (801) 201 7635
Fax: (801) 944 1657
www.oakridgemicro.com

Palladium Energy
1245 East Diehl Road Suite 307
Naperville, IL 60563
Tel: (630) 328 1961
www.palladiumenergy.com

Panasonic Industrial Co.
5201 Tollview Drive, 1F-3
Rolling Meadows, IL 60008
Tel: (877) 726 2228
Fax: (847) 637 4660
www.panasonic.com/industrial/battery/

Plainview Batteries, Inc.
23 Newton Road
Plainview, NY 11803
Tel: (516) 249 2873
Fax: (516) 249 2876

Portable Rechargeable Battery Association‡
1776 K Street
Washington, DC 20006
Tel: (202) 719 4978
www.prba.org

Power Battery Company Inc.
25 McLean Boulevard
Paterson, NJ 07514
Tel: (973) 523 8630
Fax: (973) 523 3023
www.powbat.com

PowerGenix Corp.
10109 Carroll Canyon Road
San Diego, CA 92131-1109
Tel: (858) 547 7300
Fax: (858) 547 7301
www.powergenix.com

Power-Sonic Corporation
7550 Panasonic Way
San Diego, CA 92154
Tel: (619) 661 2020
Fax: (619) 661 3650
www.power-sonic.com

Quallion
Sylmar Biomedical Park
12744 San Fernando Road
Sylmar, CA 91342
Tel: (818) 833 2000
Fax: (818) 833 3278
www.quallion.com

Rayovac Corp.
601 Rayovac Drive
P. O. Box 4960
Madison, WI 53744
Tel: (608) 275 3340
www.spectrumbrands.com
www.rayovac.com

RBC Technologies
809 University Drive Suite 100-E
College Station, TX 77840
Tel: (979) 260 1120
Fax: (979) 260 1322
www.rbctx.com

Rutgers Energy Storage Research Group (ESRG)†
Department of Materials Selence and Engineering
607 Taylor Road
Piscataway, NJ 08854
Tel: (732) 932 6850
mse.rutgers.edu

SAFT America Incorporated
107 Beaver Court
Cockeysville, MD 21030
Tel: (410) 771 3200
Fax: (410) 771 1144
www.saftbatteries.com

Sandia National Laboratories†
Energy Storage Systems
P. O. Box 5800
Albuquerque, NM 87185
Power Sources Technology Group
Tel: (505) 844 0452
Tel: (505) 844 5824
www.sandia.gov

Sanyo Energy Corp.
2600 Network Boulevard Suite 600
Frisco, TX 75034
Tel: (469) 362 5600
us.sanyo.com/Batteries

Sion Power Corporation
2900 E. Elvira
Tucson, AZ 85756
Tel: (520) 799 7500
Fax: (520) 799 7501
www.sionpower.com

Solicore, Inc.
2700 Interstate Drive
Lakeland, FL 33805
Tel: (863) 603 7640
www.solicore.com

Storage Battery Systems
N56 W16665 Ridgewood Drive
P. O. Box 160
Menomonee Falls, WI 53052
Tel: (262) 703 5800
Fax: (262) 703 3073
www.sbsbattery.com

Superior Battery Manufacturing Company, Inc.
P. O. Box 1010
2515 Hwy, KY 910
Russell Springs, KY 42642
Tel: (502) 866 6056
Fax: (502) 866 6066
www.superiorbattery.com

Teledyne Battery Products
840 W. Brockton Avenue
Redlands, CA 92375
Tel: (909) 793 3131
Fax: (909) 793 5818
www.gillbatteries.com

TIAX LLC
15 Acorn Park
Cambridge, MA 02140
Tel: (617) 498 5000
Fax: (617) 498 7200
www.tiaxllc.com

Trojan Battery Company
12380 Clark Street
Santa Fe Springs, CA 90670
Tel: (562) 236 3000
Fax: (562) 236 3282
www.trojan-battery.com

Ultralife Batteries, Inc.
2000 Technology Parkway
Newark, NY 14513
Tel: (315) 332 7100
Fax: (315) 331 7800
www.ultralifebatteries.com

Underwriters Laboratories, Inc.[‡]
333 Pfingsten Road
Northbrook, IL 60062
Tel: (847) 272 8800
Fax: (847) 272 8129
www.ul.com

University of South Carolina
Department of Chemical Engineering
College of Engineering and Computing
301 Main Street
Columbia, SC 29208
Tel: (803) 777 3270
Fax: (803) 777.8265
www.che.sc.edu

U.S. Army RDECOM CERDEC C2D[†]
Army Power Division, RDER-CCA
328 Hopkins Road, Building 1105
Aberdeen Proving Ground, MD, 21005
Tel: (410) 278 9229
Fax: (410) 278 8990
or
Fort Monmouth, NJ 07703
Tel: (732) 532 9000
www.cerdec.army.mil/directorates/c2d_army_power.asp

U.S. Army Research Laboratory[†]
Sensors and Electronic Devices Directorate
2800 Powder Mill Road
Adelphi, MD 20783
Tel: (301) 394 5429
www.arl.army.mil

U.S. Battery Manufacturing Co.
1675 Sampson Avenue
Corona, CA 91719
Tel: (951) 371 8090
Fax: (951) 371 4671
www.usbattery.com

Valence Technology
12303 Technology Boulevard Suite 950
Austin, TX 78727
Tel: (512) 527 2900
Fax: (512) 527 2910
www.valence.com

Wilson Greatbatch Ltd.
10000 Wehrle Drive
Clarence, NY 14031
Tel: (716) 759 5800
Fax: (716) 759 2562

Yardney Technical Products, Inc.
82 Mechanic Street
Pawcatuck, CT 06379
Tel: (860) 599 1100
Fax: (860) 599 3903
www.yardney.com

APPENDIX H
METHODOLOGIES FOR BATTERY FAILURE ANALYSIS

Quinn Horn, Troy Hayes, Daren Slee, Kevin White, John Harmon, Ramesh Godithi, Ming Wu, Marcus Megerle, Surendra Singh, and Celina Mikolajczak

INTRODUCTION

As rapidly advancing battery technology drives new and expanded battery applications in areas such as consumer electronics, toys, electric and hybrid vehicles, and medical devices, interest in ways to determine the causes of battery failures and preventing these failures has increased. Batteries can fail in a variety of ways, including loss of capacity and/or rate capability, short circuit, loss of case integrity, and thermal runaway, among others. Often a given root cause of failure can result in a range of other battery behaviors. For example, electrode contamination within battery cells typically results in poor battery performance due to capacity fade, but in certain instances contamination may also result in cell thermal runaway reactions. The severity of a given failure is dictated by the specific battery chemistry, battery size/capacity, physical construction of the battery, and criticality of its application. Flammability and chemical and thermal stability of battery components vary with cell chemistry; the amount of energy that can be released varies with cell size (e.g., a short circuit of an automotive battery is likely to have greater safety implications than a short circuit of a AAA alkaline cell). The acceptability of differing levels of "failures" varies with the battery application (e.g., loss of capacity may be considered a relatively benign failure in a cell phone application, but may be considered a critical failure in a medical device application).

Failure analysis of battery systems is a multidisciplinary endeavor that generally requires an understanding of:

- Fundamental battery chemistry
- Physical cell design and construction
- Electronic protection systems
- Cell, battery, and battery pack manufacturing processes

For example, in some cases, lithium-ion battery failures may be induced by external forces such as severe mechanical damage or exposure to fire, while in others they may be the result of problems involving charge, discharge, and/or battery protection circuitry design and implementation. In still other cases, field failures may be caused by internal cell faults that result from rare and/or subtle manufacturing problems. The fact that batteries can fail on rare occasion in an uncontrolled manner due to manufacturing defects has brought an increased public awareness for battery safety as a result of two recent very large product recalls of portable computer batteries.[1–4]

It is beyond the scope of this appendix to discuss the full range of possible battery failures, even for a single chemistry. However, any failure can be investigated effectively by using thorough scientific methodologies, including collection of observations and evidence, development of a hypothesis, and the testing of that hypothesis. This appendix describes a number of techniques that can be used as part of the investigative process, with particular emphasis on how to make observations and form a failure mode hypothesis. Testing to recreate a failure can be readily conducted to assess a hypothesis in some cases; however, often complete testing to recreate a failure is not practical and the failure mode hypothesis will rest solely on collected observations and limited test results. The techniques and conclusions presented in this appendix are intended as guidelines for how to approach a battery failure analysis. Most battery failure investigations do not require the use of all of the techniques discussed here. Furthermore, it should be noted that each battery failure is a special case and thus requires a unique combination of various investigative strategies and techniques that should be tailored for that specific investigation accordingly. Due to current interest in lithium-ion batteries, many examples in this appendix are drawn from failure analyses of lithium-ion cells. Further discussion of lithium-ion battery failures can be found in the literature.[5–18]

COLLECTION OF BACKGROUND INFORMATION

When a battery failure occurs, information useful to the investigation can generally be collected from the end-user as well as from the incident site. Ideally, the user can provide information on the device background, usage history, and the failure symptoms or a description of the failure incident. Useful information may include:

- Length of time the battery had been in service
- Typical usage environment (hot, cold, humid, vibration) and any significant excursions from normal conditions
- Extenuating factors such as liquid spills, dropped systems, mechanical shocks (rapid accelerations/decelerations), rapid electrical discharges, thermal shocks, or recent repair activities
- Behavior of the system leading up to the time of the incident (i.e., nominal or atypical system performance)
- State of the system at the time of the incident (plugged in or not, for how long, on or off)
- Symptoms of failure or observations during or after an incident
- Actions in response to the failure or incident
- Descriptions of any injuries, and their severity, that occurred due to the incident
- Photographs of the scene

Of course, such user accounts can only be used as a reference and may not always be entirely accurate; however, when combined with analysis of the incident evidence, user accounts may provide useful information about the nature of the incident. Systematic inspection of the incident site and careful documentation of the surroundings of the physical evidence present an opportunity to gather valuable information. Any observed damage patterns (thermal, mechanical, electrical, or chemical) should be evaluated for consistency with the described incident and product involved.

NFPA 921, "Guide for Fire and Explosion Investigations," provides a recognized guide for the investigation of fire and explosion sites, and objects involved at such scenes.[19] The general techniques described in this guide can be applied to investigate a wide range of battery failures, even in instances where a fire or explosion has not occurred. Though conducting an investigation as described in NFPA 921 is preferred, it is often not possible because the site may have already been cleaned up or significantly altered by the time the investigation begins. Photographs provided by the end-user or other investigators, if any, should be examined carefully for evidence of factors that might have contributed to a battery failure.

Recovery of Physical Evidence or Samples

Physical objects identified as important to establishing the cause of a battery failure may be, or may eventually become, evidence in a legal proceeding and must be handled appropriately. ASTM E1188 provides guidelines for evidence collection, handling, and preservation.[20] Guidelines for labeling physical evidence and related documentation are provided in ASTM E1459.[21] If a failure is energetic, parts may become separated during the incident itself or after the failure due to the activities of fire or clean-up crews. The investigator should make an effort to retrieve all the cells, printed circuit boards, and components associated with an incident battery, as well as other items that might be related to the cause of the failure if possible (e.g., sources of heat, power, or fuel and any other objects that were damaged during the incident). Sometimes chips or other components of the battery protection circuitry are ejected from the battery pack during an incident so it is important to carefully check the scene for any loose components.

Destructive examination of a battery, or any examination of a battery after a failure, may pose safety or health hazards that need to be identified and mitigated through engineering controls or selection of appropriate personal protective equipment. For example, a battery case rupture may result in leakage of cell electrolyte or active materials, which can irritate the skin, eyes, or respiratory system. In a failed multicell battery, such as found in a notebook computer or electric vehicle, some cells might have reacted while other cells are damaged but remain fully charged and possibly susceptible to shorting. A comprehensive list of potential hazards associated with all cell chemistries and battery types is beyond the scope of this appendix. The investigator should develop a safety plan prior to disturbing, shipping, or conducting a destructive examination of any battery or its associated system. Material safety data sheets (MSDS) as well as regulatory agency websites (EPA, NIOSH, OSHA, etc.) can provide data to use in developing a safety plan.

A failed system or battery pack may not comply with international (e.g., IATA, ICAO, IMO) and U.S. Department of Transportation (DOT) shipping exemptions for normal battery packs. Depending upon its condition, a failed system or battery may need to be treated as hazardous material for shipping purposes. In some instances, it may not be possible to ship the failed sample necessitating on-site examination. Personnel certified according to U.S. DOT Hazardous Material Training are required for declaring, labeling, packaging, and shipping hazardous materials in the United States. Similarly, the battery may need to be stored as a hazardous material prior to or after examination. For example, a damaged battery may need to be stored in appropriate chemical or flammability cabinets, or vented cabinets, or biohazard bags. National or local regulations may apply. Finally, battery components may need to be disposed of as hazardous waste, per national or local regulations, subsequent to an investigation.

Host Device or System Examination

Nondestructive Visual Examination. When investigating apparent battery-related failures, the host device or system (phone, notebook computer, toy, electric vehicle, etc.) should be examined. If available, the host device or system should be thoroughly documented in the as-received condition and prior to any alteration, including photographing from all sides and perspectives. Careful attention should be directed to areas that exhibit visual damage patterns or other abnormalities such as mechanical deformation, signs of enclosure breach, soot deposits, regions of charring, melting, cell swelling, electrolyte leakage, locations of electrical activity (including arcing), as well as signs of exposure to environmental elements, such as liquid or foreign contaminant intrusion. These patterns can help the investigator determine the orientation and state of the host device or system at the time of the incident. For example, an examination of a power cord and the device's power inlet port may indicate whether it was plugged into the unit at the time of an incident. If examining a flip-type cell phone, notebook computer, or any other device that can be opened and closed, the incident unit should be opened, if possible without causing damage, so the screen and other internal components (keyboard, touchpad, etc.) can be examined and photographed. Witness marks in the form of heat damage or smoke patterns will generally indicate whether the unit was opened or closed at the time of the incident. Caution should be used in interpretation of such evidence, as users may alter the state of the device as the incident is occurring. For example, a notebook computer may be closed and moved to a different location after a user observes the initial stages of a thermal runaway incident. Device and battery pack serial numbers should be noted for possible later follow-up with manufacturers. This often involves significant effort in cases where severe damage has occurred, but can be critical when the root cause is related to manufacturing defects.

Reconstruction. If an incident host device or battery is not intact or fully assembled upon initiation of the failure analysis, the investigator should attempt to reconstruct the system, if possible, by placing the appropriate parts into their original location(s) and/or orientation as much as possible without introducing any additional damage or altering the condition of the evidence (Fig. H.1). This, along with comparison to an exemplar device, will allow for a better understanding of possible damage propagation. This is necessary, for instance, when batteries have experienced failures that caused one or more cells to separate from the system during an incident. Such reconstruction efforts often involve detailed analysis of connecting tabs in comparison with exemplars. Caution is needed during this process, however, as some tabs may shift during an incident (e.g., the cap of an 18650 lithium-ion cell may rotate during an incident and become fixed in a new orientation upon cooling).

FIGURE H.1 Portion of a failed battery pack made up of 18650 lithium-ion cells. (*Courtesy of Exponent, Inc.*)

System and Battery X-Ray. After a visual examination and possible reconstruction, the battery and system should be examined using x-ray imaging, if practical, to further document the evidence nondestructively. X-ray examination of the battery and/or system may provide insight into the relative positioning of components as well as damage propagation. X-ray imaging may also allow characterization of the extent and distribution of circuitry and electrode damage, if present. If practical, both the incident system and battery should be x-rayed in their fully assembled condition (e.g., for smaller systems such as notebook computers, MP3 players, or DVD players). X-ray images should be taken at different viewing angles to allow for a complete view of the assembled battery and the system. The x-ray images should be examined closely for internal damage to the battery and battery protection circuitry or other anomalies. Although these

FIGURE H.2 X-ray image of part of a lithium-ion battery pack. The effect of cell venting from a cell cap on the windings of an adjacent cell can be observed, as indicated by the arrow. (*Courtesy of Exponent, Inc.*)

components may be examined in more detail upon subsequent disassembly, the overall view of the damage pattern and physical relationship between the damaged cells and surrounding parts generally offers important clues as to the initiation point(s) of the failure and the sequence of any propagating events. An example of a thermal runaway propagation in a multiple-cell configuration is shown in Fig. H.2. This figure shows the effect of cell venting (from the cell cap) onto the windings of an adjacent cell. On a more practical level, x-ray imaging can also provide guidance on how to disassemble a host device or battery for further examination without compromising key internal components.

Status and Functionality Tests. If a battery failure is nonenergetic, the host system is often undamaged and can be tested with an exemplar battery pack or various electronic loads. With certain types of battery failures (for example, loss of capacity), the battery itself may be tested. Even if a failure is energetic, a host system may remain functional even though it appears damaged. It is also possible when a system is not functional to harvest functional components for testing, as these systems or components may provide status or incident-related information or allow for selected functionality tests. Host device operating parameters of particular importance related to battery functionality, reliability, and safety include charge and discharge voltage and current cutoff points, overvoltage, undervoltage, overcurrent, and low and high temperature cutoff points. For multiple-cell in-series applications, block* voltage sense points, and cell or block imbalance and rebalancing schemes, as well as locations of multiple temperature sensing points, may also be of interest. Further, the condition of fuses and other printed circuit board components may be relevant depending on the failure mode.

Investigators should familiarize themselves with the operational characteristics of the host device or system prior to energizing any components if such testing is attempted. Familiarity can be obtained by reviewing product and component literature, including circuit schematics, or by examining and characterizing the operating parameters of an exemplar host device or system experimentally. If there is concern that energizing the host device may result in a change of state or damage to some components, every effort should be made to gather all possible data before attempting to

---

*A block is a parallel array of cells. Multiple blocks are often combined in series.

energize the system. For example, volatile registers on EEPROMs will likely be altered upon re-energizing a host device and subsequently obtained data would not necessarily reflect the condition at the time of the incident.

BATTERY AND CELL EXAMINATION

To determine the cause of a failure it is generally necessary to open a damaged battery pack, extract and examine the remains of the cells, the battery protection printed circuit board(s) (PCBs), and any fuses or thermal cutoff devices. This examination is destructive and should only be conducted after all appropriate parties have been contacted, and an examination of the overall pack, x-rays, and/or an examination of an exemplar unit has been made to minimize the likelihood of damage to key components during the incident investigation.

The disassembly process should be conducted carefully and methodically with any abnormalities noted and documented. For example, during the disassembly of one incident unit, evidence was found that it had been previously disassembled and rendered inoperable by that process. This observation, in combination with heat and burn patterns indicating an external flame attack, suggested the incident was likely the result of a deliberate attempt by the user to cause the battery failure intentionally.

Once the battery pack case has been removed to expose the cells, the appearance of the cells should be documented and the cells uniquely marked for future identification. Markings should include sufficient detail to define cell orientation for reference during subsequent analyses. It is also important to determine the electrical configuration of battery cells in a multiple-cell pack to identify the different voltage levels present at the terminals of each battery cell. While still connected, if possible, in-circuit cell-block voltages and impedances should be measured. Close attention should be given to the electrical path throughout the pack, starting from the external pack connector to the cells. The focus should be on identifying signs of potential inadvertent short-circuit points between cells and wires or battery tabs and connections that may have resulted in a failure of the pack.

A determination ultimately must be made if any observed damage or anomalies are part of the failure root cause or are consistent with follow-up events after the initiating failures. For example, soft pouch lithium-ion polymer cells can leak electrolyte, resulting in shorting of pack protection electronics. Cell leakage in this example—which can be caused by improper sealing, direct mechanical damage of the pouch during pack assembly, corrosion of the pouch resulting from improper pouch isolation during cell manufacture, etc.—is the root cause of the failure rather than the observed shorting on the protection PCB.

After an in-circuit examination, cells should be disconnected from the protection PCB and then from each other. Care should be taken when removing cells to avoid any damage to the cell case or electrodes. If any damage occurs during cell extraction, this damage should be documented so that it is not considered during the development of a failure mode hypothesis. When dealing with the disassembly of multicell lithium-ion systems, it is important to cut the connecting wires or metal tabs in the order of highest to lowest potential, preventing the possibility of opening fuses or causing other damage to the cells and the protection PCB. This approach is also typically the safest way to handle potentially charged cells. Insulation on cells and connecting wires may have been compromised during an incident, and thus additional care should be used to minimize the risk of creating an external short circuit while handling the cells (e.g., with cutting tools).

Once cells have been separated, they should be examined and photographically documented. It is particularly important to find and record any cell serial numbers. Cell serial numbers normally include a date code for cell manufacture (which sometimes can be traced back to particular manufacturing lines or batches). This information can be used by the cell manufacturer to retrieve data about the production history of particular cell lots, and used to identify exemplar cells from the same date code for follow-up examinations. Such information can help identify whether the observed failure resulted from a systematic manufacturing problem or was an isolated event. In addition, should a cell recall become warranted, cell date codes can be critical in helping to bound the population of cells to be recalled.

Cell cases or pouches should be examined visually for signs of mechanical deformation. Magnifiers and optical microscopes are often useful for this type of examination. Denting a cell case may create immediate or delayed shorting resulting in a loss of performance or possibly subsequent thermal runaway of the cell. Dents created during battery pack assembly processes that ultimately led to cell failures have been observed with lithium-ion polymer soft pouch cells. Note, however, that dents in cells that have undergone energetic reactions are often created by post-incident handling (e.g., extinguishing a fire, collecting and shipping the remains). In addition, not all preexisting cell case dents result in cell failure, as most cells are robust to certain amounts of deformation. Before a failure can be attributed to pre-incident mechanical damage, evidence must be found that the dent caused damage to and/or shorting of cell windings. This is generally done through examination of cell x-rays, CT scans (discussed later), or by opening the cell and directly examining the windings.

The voltages, weights, and impedances of the individual cells should be measured at 1 kHz. These measurements can help determine whether a cell lost electrolyte, whether internal protections activated, whether the cell separator was heat damaged, and/or the state-of-charge at the time of the failure. Cells can then be examined with nondestructive techniques such as x-ray and CT scanning. As part of the overall failure analysis, nominally operational incident or exemplar cells may be subjected to a variety of electrochemical diagnostic techniques, including cell cycling and electrochemical impedance spectroscopy. Subsequent destructive physical analysis (DPA) of operational or nonoperational cells can include reference electrode testing (on operational cells only), cell gas sampling and analysis, cell disassembly, and cell cross-sectioning.

Cell X-Rays

If cells have been visibly damaged, or if internal cell faults are suspected, detailed x-ray (and/or CT scan, discussed below) examination of the cells should be made. Typically, it is useful to examine cells (and record x-ray images) at two or more planes or at different angles depending on the cell type, particularly in regions of winding disturbances. X-ray images showing the cell from the plan view (orientation as seen from above when battery pack is laid flat) are useful in observing the effects of potential interactions between adjacent cells. Analysis of detailed x-ray images is often crucial in determining the sequence of thermal runaway (in a multicell lithium-ion pack), determining the initiating cell in a thermal runaway incident, and developing a theory for the cause(s) of the initial cell failure.

For example, the electrode windings of an 18650 model lithium-ion cell that initiates a failure in a multicell pack often appear comparatively lightly damaged in x-ray images, while the windings of cells that reacted later show sites of greater damage, including resolidified globules throughout those cells. Little melt damage occurs in the initiating cell because this cell is at or near the ambient operating temperature when it begins developing a short circuit. Much of the energy from this cell can be consumed by the shorting event itself and in self-heating to the critical temperature where thermal runaway occurs. Once the separator melts, there is less energy left in the cell to cause additional pronounced shorting. The reaction of the initiating cell increases the temperature of neighboring cells. The separators of these cells consequently melt due to external heating, while these cells are still at or near their initial state of charge. This results in substantial shorting (melting) throughout the cell and an overall greater degree of internal damage. In cases where a catastrophic failure induces rapid thermal runaway (e.g., from a puncture to a cell in a pack from an external source, external heating of the battery pack, or overcharge of the cell), the initiating cell may be the most heavily damaged cell. Maximum damage occurs to the initiating cell here because the shorting and subsequent thermal runaway are so rapid they often result in ejection or partial ejection of the cell windings. Neighboring cells are heated more gradually before going into thermal runaway and may not experience the same level of internal damage as the initiating cell.

X-ray images showing all the internal leads in profile and the alignment of all cell windings near the cell ends are also generally useful. Malformed leads and misaligned windings have been shown to lead to internal cell failures. These types of manufacturing defects can be caused by problems with the machinery used to assemble cells, and if so, occur repeatedly, to various degrees of severity, in cells manufactured within the same time frame. Thus, even if these features are no longer visible in reacted cells (as a result of melting or deformation during cell venting), examining nonreacted cells from the same time frame can be helpful in identifying whether a manufacturing problem exists.

Computed Tomography Scanning

In computed tomography (CT) scanning, x-rays are taken along a line rather than being integrated through the entire volume of the object. An image representing a planar slice of the object is then reconstructed by mathematically combining data from multiple x-rays taken in the same plane but at different angles. The set of resulting images provides a three-dimensional view of the cell interior rather than an averaged image containing information from all of the material and components along the path between the x-ray source and detector, as in traditional planar x-ray. In the reconstructed CT scan images, high-density regions appear bright, while low-density regions appear dark (the reverse of a traditional x-ray image). For example, in a typical 18650 lithium-ion cell, the copper current collector appears as a thin bright line, the cobalt oxide cathode appears as a thicker gray line, and the graphite appears as a black line between the copper current collector and cobalt oxide cathode (Fig. H.3). A high-resolution CT scan of an 18650 cylindrical cell will consist of approximately 1000 slices. The resulting slice thickness of each scan cross section would be 70 µm with an in-plane resolution of approximately 20 × 20 µm. Once CT data have been collected, they can be reprocessed to generate a cross-sectional image along any desired object plane. Figure H.4 shows a longitudinal cross-sectional image generated from CT scanning data of an 18650 cell subjected to a crush test that ultimately resulted in cell thermal runaway. The cell was originally scanned in the same manner as the cell in Fig. H.3.

The application of CT scanning to examination of a failure in a particular cell chemistry and format is mainly limited by the following considerations: size of the imaging chamber, strength of the x-ray source, cell geometry, and cell materials. Since CT scanning is a dynamic process requiring

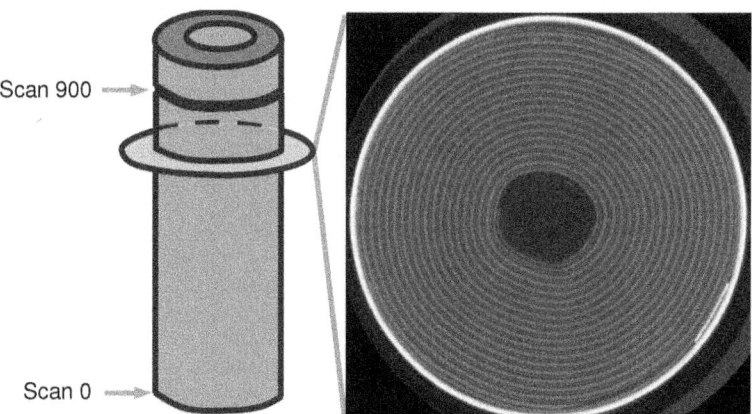

FIGURE H.3 A single CT scan cross-sectional image from a commercial 18650 cylindrical cell. (*Courtesy of Exponent, Inc.*)

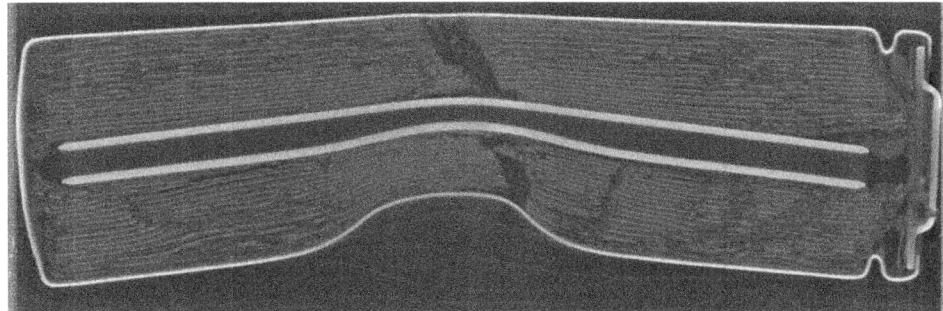

FIGURE H.4 CT scan cross-sectional image of a crushed cylindrical 18650 lithium-ion cell that subsequently underwent a thermal runaway reaction. (*Courtesy of Exponent, Inc.*)

movement of the sample within a chamber, larger samples will necessitate a larger chamber. Larger format cells or cells containing denser materials will require a higher-power x-ray source. If the x-ray source is not strong enough, it will result in an image where the edges of the object might be resolved, but the interior portions of the cell are not. The best CT scan images are obtained for cells with symmetrical geometries (e.g., cylindrical cells) as a result of the image reconstruction algorithm. As cell geometries become more asymmetric, image blurring and severe shadowing can occur. The amount and location of image blurring and shadowing will be related to the degree of asymmetry of the sample being scanned. When imaging materials with large differences in density (e.g., lithium adjacent to steel), it is possible that the lighter material will not appear in the reconstructed image. Prismatic and other similar form factors may be scanned in sections multiple times along different axes to decrease the asymmetry of the scanned portion of the cell and increase the scanning resolution of the entire cell. For example, a 50 mm wide cell may be scanned using two offset axes as two 25 mm wide cells.

CT scanning can be a powerful nondestructive diagnostic tool for evaluating the internal structure of batteries that is complementary to planar x-ray. While planar x-ray examination can quickly and effectively screen out some defects such as tab shape/positioning and global electrode misalignment with a single image, it is limited in its ability to screen for localized defects, even when significant numbers of planar x-rays are taken of a single cell, because of the masking effect created by the inherent image averaging that occurs. In general, CT scanning is more effective in identifying contaminant particles or other point defects that may be lost during averaging over an entire cell. The ability to detect contaminant particles by CT scan increases as the atomic number of the contaminant increases and as the difference between the atomic number of the contaminant and the surrounding materials increases. CT scans can also show whether internal windings might be misaligned, distorted, torn, or melted, identify points of electrode delamination or cracking, identify areas of extra or missing material, show whether cell enclosure corrosion has occurred or whether a lithium-ion cell has experienced severe overdischarge (copper-dissolution), and reveal a variety of internal defects that may otherwise go undetected by planar x-ray.

When examining a lithium-ion battery pack that has undergone thermal runaway, traditional planar x-ray should be used in the preliminary portion of the examination as discussed. If the failure appears to have been caused by an internal cell fault, planar x-rays can be used to identify the cells most likely to have initiated the reaction. Once the likely initiating cells have been identified, further nondestructive examination can be conducted using CT scanning prior to cell disassembly. CT scanning is particularly useful for wound cylindrical lithium-ion cells that have experienced significant damage upon thermal runaway and cell venting, so that cell unrolling is not possible.

Example: CT Scanning of Primary Cells. Failure analysis and general examination of primary cells pose a number of challenges, especially for lithium primary cells, which must be opened in the inert atmosphere of a glove box if lithium metal, or a corrosive and/or volatile cathode material, is to be examined. Such disassembly also requires appropriate engineering and personal protective equipment. In addition, though disassembly will reveal surface features on components, it will not reveal hidden subsurface voids or cracks in the components. CT scanning allows for the inspection of individual components and features within the cell, such as current collectors, active material interfaces, cell leads, void spaces, and cracks. It can also be used to examine electrode alignment and positioning, detect internal case pitting and corrosion, visualize the uniformity and shape of the sealing surfaces, and assess overall cell construction quality.

To demonstrate the applicability of CT scanning on primary cells, a variety of cells of different constructions at various states of charge were scanned. Figures H.5 and H.6 present a CR2050 lithium/manganese dioxide primary coin cell at the fully charged and discharged states, respectively. The overall cell construction is clearly visible, including the press fit steel exterior, sealing surfaces, cathode pellet, and metallic lithium anode. The steel exterior corresponds to the brighter portions of the image. The manganese dioxide cathode is the gray region within the cell interior. The apparent gradation change in image density in the manganese dioxide in the CT images is an artifact of the imaging process due to a larger volume fraction of steel from the cell exterior being located toward the image top. Materials with low x-ray absorbance, such as the lithium metal anode or void spaces, will appear black in a CT image. For instance, in Figs. H.5 and H.6, the lithium metal anode is the black region directly below the cathode.

FIGURE H.5 CT scan cross-sectional image through the center of a CR2050 lithium/manganese dioxide primary cell (~100% SOC). (*Courtesy of Exponent, Inc.*)

FIGURE H.6 CT scan cross-sectional image through the center of a CR2050 discharged lithium/manganese dioxide primary cell (~0% SOC). (*Courtesy of Exponent, Inc.*)

The interface between the cathode material and anode is clearly visible, and its location was tracked versus the discharge state of the cell. Figure H.6 shows the same cell in the nominally discharged state. As would be expected, the lithium has been almost entirely consumed, which is seen as a decrease in the imaged black region within the cell. A concomitant swelling of the cathode matrix was observed due to intercalation of the lithium. Furthermore, fractures that developed within the cathode material due to swelling (dark lines through the cathode material) are shown in Fig. H.6. Similar fracturing of the cathode material upon cell discharge has been observed in other cells in the discharged state and does not necessarily represent a cell design or manufacturing defect as long as cell performance and safety are not impacted by that fracturing.

Figure H.7 is a CT image of the internal construction of a CR$\frac{1}{3}$N lithium/manganese dioxide primary cell at 50% state of charge (SOC). The cell has a spiral-wound construction utilizing wire mesh current collectors. The metallic parts, such as the steel can, metallic wire mesh electrodes, and center pin, appear as bright regions within the image. The cathode material corresponds to

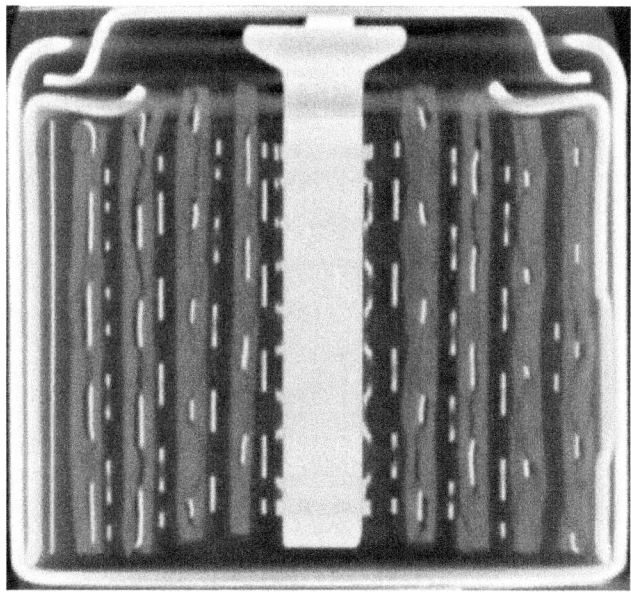

FIGURE H.7 CT scan cross-sectional image through the center of a CR⅓N lithium/manganese dioxide primary cell (50% SOC). (*Courtesy of Exponent, Inc.*)

the gray regions within the image. The lithium metal corresponds to the black regions adjacent to the cathode material. The sealing surfaces are visible within the figure and appear to indicate a relatively uniform gasket compression with no signs of corrosion. The lithium metal has been consumed unevenly during discharge. This is clearly demonstrated in the image where the lithium metal has an uneven interface with the cathode active material. Additionally, voids and fractures are evident throughout the cell. The voids are coincident with the current collector mesh. Fractures of the cathode active material appear to propagate outward from those voids along the cathode active material current collector interface. This observation coupled with isolated voids adjacent to the current collector tends to indicate void formation occurs first, followed by fracturing of the cathode active material matrix as intercalation of the lithium into the manganese dioxide progresses during discharge.

Figure H.8 shows a CT scan image of a 9 V alkaline zinc/manganese dioxide cell. There are six cells in series in alternating orientations within the battery (this cross section shows three of the cells). The zinc powder particles in the anode are clearly visible within the image. It is also apparent that the distribution of zinc is not uniform throughout the cells. There are irregularities along the anode/cathode interface as a result of folds and creases in the separator. Other features that are clearly illustrated include anode pin alignment, cell wall uniformity/integrity, and distribution of anode and cathode active materials.

Cell Gas Sampling and Analysis

The active materials of most battery chemistries are capable of facilitating the conversion of condensed phase materials to gases. In some instances, the generation of gas is part of the normal operation of the electrochemical couple. The electrolysis of water from the electrolyte of a lead-acid cell during the final stages of charging is an example of normal gas generation. In other instances, gas generation is a direct indication that undesirable chemical and/or electrochemical reactions of the cell components are occurring. For example, in lithium-ion cells some generation of light hydrocarbon gases can occur during normal operation due to the kinetically hindered reaction of electrolyte components with the active materials of the cell. However, large scale generation of gaseous by-products that may cause cell swelling or venting is an indication of degradation of cell components. As a result, analysis of gases generated in lithium-ion cells is often a powerful tool for understanding the reactions that may have generated them and, subsequently, the specific failure mode that a cell may be experiencing.

Gas chromatography/mass spectroscopy (GC/MS) is most often chosen as the analytical technique for the analysis of gases generated in failed cells. GC/MS allows for the separation and quantitative identification of the gas sample components and often provides the information necessary to determine the origin of the gas generation. Because GC/MS is inherently sensitive, sample collection and storage prior to testing are critical for maintaining the integrity of the sample and preventing erroneous data due to the introduction of contaminants. Gas-tight syringes, equipped with nonreactive seal materials and a valve mechanism to retain the sample within the barrel of the syringe, are effective collection and sample storage tools.

Cell construction and form factor play important roles in the method chosen for the collection of a gas sample from a failed lithium-ion cell. Gas collection from a swollen pouch-type cell is easily achieved by forcing a syringe needle through a septum adhered to the exterior of the cell. If a typical noncoring syringe needle is used, the septum thickness should be greater than the length of the needle bevel to avoid introduction of the environment outside the cell. Cells with hard cases that are not easily punctured by a needle of a syringe provide additional challenges to collecting uncontaminated samples. Not only must a method be developed to penetrate the cell without allowing the escape of the desired gases or contamination by ambient gases, the investigator must identify a location in the cell where a syringe needle can be inserted without short-circuiting the electrode windings and contaminating the gas sample through the generation of gas not related to the original failure. It is often necessary to perform x-ray or CT imaging of the cell to identify a suitable cell location for needle insertion, and then fabricate sampling apparatus designed to facilitate needle placement and gas sampling.

The gases generated in a cell will depend upon the materials used in the construction of the cell, and in the case of cell contamination, the gases generated are additionally dependent on reactivity of the foreign material. Correct interpretation of gas analysis results requires knowledge of the electrode material types, the composition of the electrolyte solvent, and the identity of the electrolyte salt and electrolyte additives. Additionally, an understanding of the most probable chemical pathways for reactive material degradation is necessary to realize the full benefit of generated gas analysis.

In lithium-ion cells, gas generation can often be assigned to one of three general causes:

- Overcharge of the active materials
- Overdischarge of the active materials
- The presence of contaminants in the cell

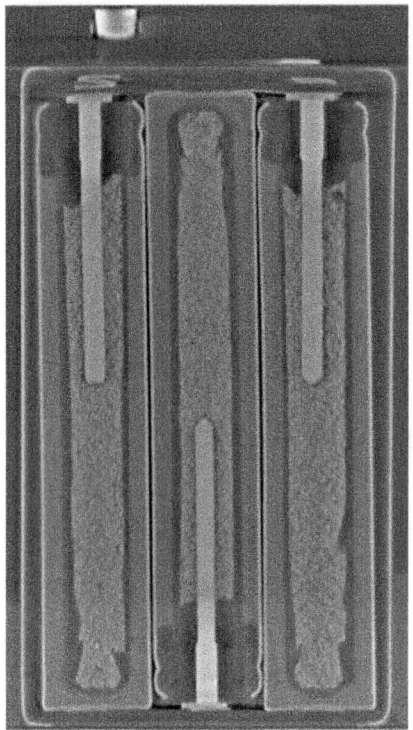

FIGURE H.8 CT scan cross-sectional image through three cells in a 9 V alkaline Zn/MnO_2 primary battery (100% SOC).

Working with a graphite/LiCoO$_2$ cell system, Kumai et al.[22] developed guidelines for interpreting gas composition during overcharge and overdischarge processes. In general:

- Overcharge produces conditions in which oxidative degradation pathways for reactive cell components dominate the gas generation mechanisms.
- Overdischarge produces conditions in which reductive degradation pathways for reactive cell components dominate the gas generation mechanisms.

For example, a gas sample collected from an overcharged 18650 cell* consisted of a complex mixture of gases, a major component of which was CO_2. CO_2 is an oxidized form of carbon that is likely due to the oxidation of the electrolyte at the positive electrode surface facilitated by the evolution of oxygen from degrading positive electrode material. The following chemical equations describe the degradation of the positive electrode and the subsequent oxidation of the electrolyte to form CO_2:

$LiCoO_2 \rightarrow Li_{0.5}CoO_2 + \frac{1}{2}Li^+ + \frac{1}{2}e^-$ (normal charge reaction)

$Li_{0.5}CoO_2 \rightarrow CoO_2 + \frac{1}{2}Li^+ + \frac{1}{2}e^-$ (overcharge reaction)

$3CoO_2 \rightarrow Co_3O_4 + O_2$ (gas) (positive material degradation and O_2 evolution)

Electrolyte + $3O_2$ (gas) $\rightarrow 3CO_2$ (gas) + H_2O (electrolyte oxidation)

While not reported by Kumai et al., the presence of hydrogen in a failed lithium-ion cell is an indicator of overcharge that is the product of reactive material reduction within the cell. On overcharge, lithium metal can be deposited at the negative electrode. This highly reducing metal reacts with the electrolyte to produce hydrogen and various electrolyte reduction products.[23] The detection of hydrogen is a particularly useful indicator of overcharge as the probability of its generation in the absence of lithium metal is low.

Cell Disassembly

Cell disassembly can be accomplished in a variety of ways, depending upon the goals of the examination. The most commonly used approach is a cell "unrolling" or "destacking" where the cell case is removed and the windings are either unrolled (wound cells) or separated (prismatic cells) and examined. This is an important step when trying to identify a specific point of failure in reacted cells or manufacturing defects in exemplar cells. Cell unrolling or destacking generally allows examination of all surviving electrode and separator surfaces, as well as the condition of various tabs, welds, insulators, internal protection devices, and the cell case or pouch. The cells can be documented and examined throughout the process of disassembly using high-resolution digital photography, stereomicroscopy, and scanning electron microscopy (SEM) as appropriate. Care should be taken to document any damage introduced to the windings by the disassembly process.

Opening any cell, particularly one that is partially charged (e.g., a lithium-ion cell at 3 V) or fully charged (e.g., a lithium primary cell) presents a number of safety hazards, and should only be attempted after the investigator has developed a thorough understanding of the cell design and chemistry, a safety plan for the process, including appropriate engineering controls (fume hood, glove box), appropriate personal protective equipment (gloves, goggles, apron, respirator), a contingency plan for handling a cell should it become shorted during the disassembly process and begin to self-heat (containment, fire extinguishers, bucket of sand in which to submerge a shorted cell), and a plan for storage and disposal

---

*The cell system consisted of a LiCoO$_2$ positive electrode, graphite negative electrode, and electrolyte comprised of 1 M LiPF dissolved in a mixture of ethyl methyl carbonate, diethyl carbonate, dimethyl carbonate, and ethylene carbonate.

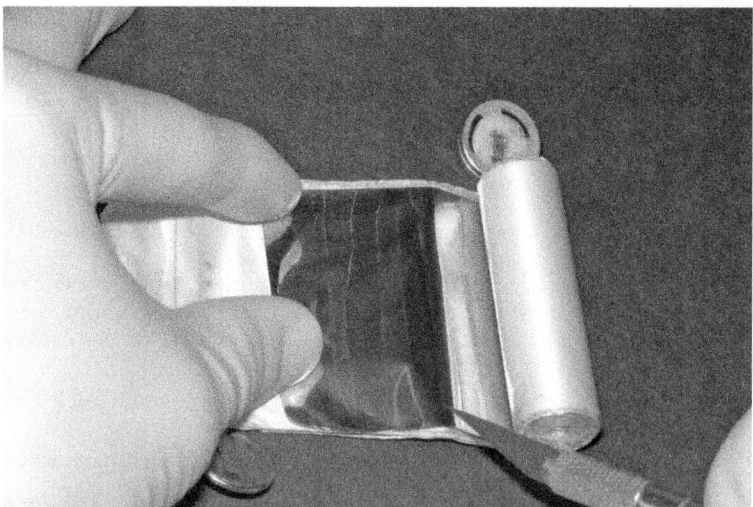

FIGURE H.9 Photograph showing the initial stages of the examination of electrode and separator surfaces of an 18650 lithium-ion cell. (*Courtesy of Exponent, Inc.*)

of cell components. The investigator should consider the potential hazards associated with the cell components, including toxicity and potential flammability (hydrocarbon-based electrolytes, lithium metal, etc.). Examination of cell x-rays, CT scans, or engineering drawings is recommended when planning cuts in cell cases. Nonconductive or insulated tools are recommended, but even nonconductive tools can cause shorting if they bring electrodes of opposite polarity together.

In the case of very high-energy cells, such as lithium-thionyl chloride, the cell should be frozen in liquid nitrogen and then allowed to thaw partially before disassembly.

Example: Lithium-Ion Cells. For lithium-ion cells that have not undergone thermal runaway reaction, cell unrolling (Fig. H.9) or destacking can reveal the presence of contaminants, microshorts (Fig. H.10), and areas of lithium plating (Fig. H.11) on electrode surfaces and separators. Internal leads, welds, insulators, and regions of tape can also be examined for anomalies (Figs. H.12 and H.13).

For lithium-ion cells that have undergone thermal runaway, destruction of not just the cell separator, but melting of the aluminum current collector has likely occurred. In such cases, failure analysis proceeds with detailed examination of the remaining negative electrode (anode) material, particularly the copper current collector, which generally remains mechanically intact. Such examination often reveals the presence of holes in the copper current collector at a variety of different locations and with different character. Figure H.14 is a simplified schematic of a post–thermal runaway copper current collector, depicting representative locations of several types of holes observed.

Single or small clusters of holes surrounded by evidence of melted and resolidified copper are often observed. This type of hole, or cluster of holes, will generally be isolated from other holes on the current collector. It is generally found in areas of the winding where both copper and aluminum current collectors were coated with active material, rather than in regions of bare copper current collector. Often it will be surrounded by and even partially obscured by remaining active material coating. The edges of this type of hole will often show clearly defined ridges or lips with relatively smooth features (Fig. H.15). Furthermore, there are often nodules observed extending into the holes. In each case, the edges of the hole exhibit rounded features and are generally thicker than the original current collector. This indicates that the copper has undergone a change in shape to decrease the surface energy by minimizing the surface area, a change that can only occur by melting. Hence, it is evident that the copper current collector was locally melted and subsequently resolidified. These

FIGURE H.10 Photographs showing evidence of a microshort on a separator (top) and associated anode material (bottom) observed during disassembly of a lithium-ion cell. (*Courtesy of Exponent, Inc.*)

regions do not exhibit alloying with aluminum, which can be confirmed by scanning electron microscopy and energy dispersive spectroscopy (SEM/EDS).

Though many features can be observed on surviving current collectors after thermal runaway, holes surrounded by resolidified copper likely have special significance. Cell skin temperatures measured during induced thermal runaway reactions do not approach the melting point of copper (1357 K). Holes surrounded by resolidified copper rarely form in more than one cell of a battery pack that has undergone an accidental thermal runaway reaction in the field in the absence of cell overcharge for cells with capacities of ~2200 mAh or less. This suggests that uniform external heating of a cell, which causes large areas of the separator to melt simultaneously, is unlikely to produce regions of melted copper. The very limited areas of observed copper melting indicate brief, localized energy transfers consistent with electric faulting. Such faulting could only occur while most of the cell separator is neither melted nor shrunk, and while a sufficiently conductive path exists to the copper current collector (for example, for a soft pack cell, before heating of the electrolyte has caused gas generation and separation of the electrode layers). Such a feature may, therefore, be indicative of

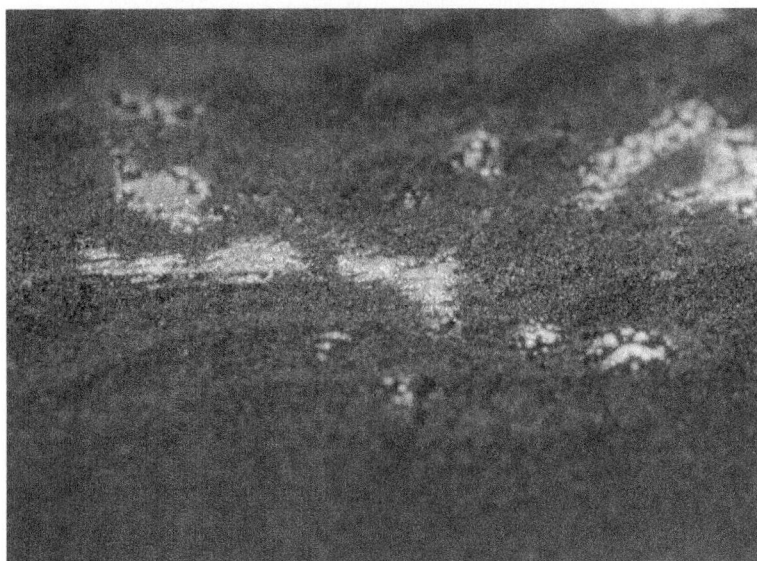

FIGURE H.11 Photograph showing evidence of lithium plating on the surface of the negative electrode removed from a discharged lithium-ion cell. In the presence of water vapor, white foam will form at locations of plated lithium due to the generation of lithium hydroxide and hydrogen gas. (*Courtesy of Exponent, Inc.*)

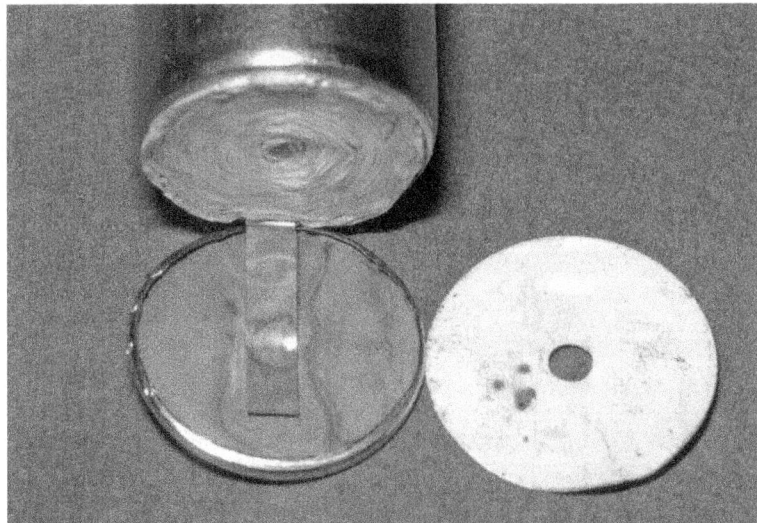

FIGURE H.12 Photograph showing indications of corrosion on the interior, bottom surface of the cell can from an 18650-size lithium-ion cell. (*Courtesy of Exponent, Inc.*)

FIGURE H.13 Photograph of the cell cap assembly from an 18650-size lithium-ion cell showing evidence of a metal burr on the cathode lead edge (circled). (*Courtesy of Exponent, Inc.*)

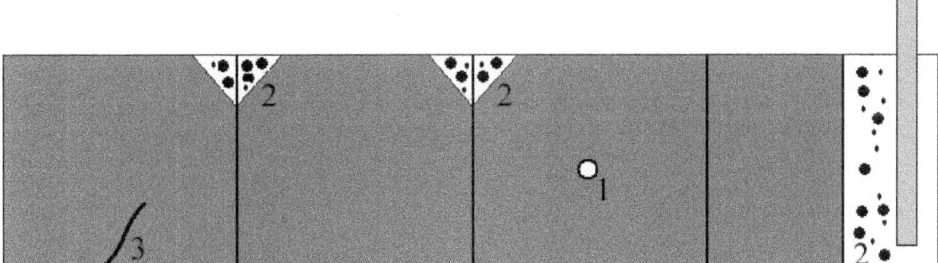

FIGURE H.14 Simplified schematic of a post–thermal runaway copper current collector from a wound prismatic cell depicting typical locations of damage: (1) hole surrounded by resolidified copper found in region of active material, (2) regions of copper aluminum alloying near anode lead tab and venting areas, (3) tear in current collector. (*Courtesy of Exponent, Inc.*)

the point or area within a cell that initiated the thermal runaway event. Caution needs to be exercised in such an interpretation, however, because a weak point within a cell (e.g., contaminant, corner in the case of a flat or prismatic cell, etc.) can result in preferential softening and shorting during the heating of a cell from other causes (a different origination point or external heat attack). In addition, for cases where a cell is overcharged, or for cells with higher initial capacities, the energy remaining within the cell after an initial shorting event may be sufficient to cause follow-up shorting events at multiple locations in a cell.

The copper current collector of a lithium-ion cell that has undergone thermal runaway will typically exhibit many holes, only a few, if any, of which are surrounded by evidence of resolidified copper as described above. Typically, the majority of holes are the result of alloying between copper and aluminum. When liquid aluminum contacts copper metal, it forms a low melting point eutectic mixture, which subsequently melts or oxidizes and forms holes.[24] Alloying holes are typically

FIGURE H.15 Stereo-microscope image showing an example of a hole with a clearly defined melt front/ridge (arrows) from resolidified copper. (*Courtesy of Exponent, Inc.*)

observed with higher frequency in regions of the copper current collector where melted aluminum could most easily pool onto bare copper, such as:

1. Areas where the copper has not been coated with active material, such as in the region of the anode lead or ends of the windings, which is adjacent to an area of uncoated aluminum current collector.
2. Along the edges of windings where melted aluminum can pool or collect during a thermal runaway event.
3. In the vicinity of a cell vent port, where passage of heated gases and the presence of oxygen result in damage to the active material coating of the copper current collector. Note that vent pathways will depend upon cell geometry, construction, and surrounding geometry of the host device, as well as the dynamics of the thermal runaway event.

Despite the presence of liquid aluminum during thermal runaway, and evidence that liquid aluminum can be transported throughout the cell—cells that have undergone thermal runaway often exhibit small beads of resolidified aluminum near vent ports or electrode edges (Fig. H.16)—most active material coated regions of copper current collector do not generally exhibit alloying with aluminum. Either the liquid aluminum does not wet the anode material (graphite) and as such is unable to penetrate the anode coating, or liquid aluminum is wicked into the anode and cathode coatings, thus segregating it from the copper current collector. Mercury porosimetry measurements of commercially produced cells indicate porosity of 25 to 30% in active material coatings. In our experience, a typical ratio of aluminum current collector thickness to cathode active material coating thickness (single side) is on the order of 1:4. Thus the two layers of cathode active material surrounding the aluminum current collector should be able to accommodate all of the locally melted aluminum and thereby keep it away from the copper current collector. Indeed, the fact that cells which have undergone thermal runaway can be opened and often unwound to reveal smooth layers of anode active material is a strong indication that liquid aluminum generally does not infiltrate the anode active material. As a result, regions of alloying not associated with vent

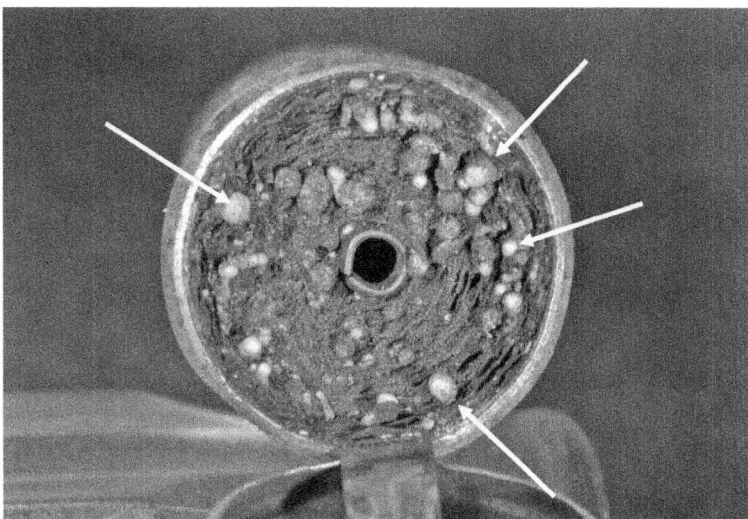

FIGURE H.16 Photograph of an 18650-size lithium-ion cell showing beads of resolidified aluminum visible at edge of cell winding after cell thermal runaway. (*Courtesy of Exponent, Inc.*)

paths or bare copper may be an indication of damage or flaking to the active material coatings prior to thermal runaway.

The edges of alloying holes tend to either be lined with a brittle, oxidized gray-metallic material (an oxidized aluminum-copper alloy), a darker halo where the material has alloyed but not melted or oxidized, or a clean sharp copper surface where the alloy has detached from the remaining unaffected copper (like a punched-out hole). Alloyed regions of aluminum and copper are clearly visible when examined with SEM imaging using backscattered electron contrast. In this mode, higher atomic number materials such as copper appear brighter, while lower atomic number materials such as aluminum appear darker.* In a backscattered SEM image, holes created by alloying will tend to have a halo or darkened edges compared to the brighter appearance of the bulk of the current collector. This darkened region is associated with Al-Cu alloying.

During pack and cell disassembly, it is often difficult to avoid mechanically tearing or disturbing the heat-exposed cell windings. If the anode is torn after thermal runaway, it will exhibit relatively clean surfaces surrounding the damaged areas. Anode tears made prior to thermal runaway are subjected to elevated temperatures, vent gases, and possible alloying with aluminum. Consequently, such tears may exhibit heat patterns and other signs of elevated temperatures. Sometimes a tear will be associated with a hole surrounded by evidence of copper melting and resolidification.

On occasion, evidence of contamination is observed on post–thermal runaway copper current collectors. For evidence of foreign material to survive the thermal runaway process and be detectable, the contaminant must have contained materials not normally associated with either the anode or cathode. For detection after cell opening, the contaminants would generally have to have been splattered or alloyed onto the anode or the copper current collector and not been consumed or displaced. As a result, though it is possible to detect the presence of non-native species, particularly metallic contaminants, it is generally not possible to directly detect the presence of plastic, aluminum, or copper contaminants. In fact, it is rather uncommon to find any evidence of contamination when examining melt holes in the copper current collector.

---

*The observed brightness is set by the operator and can only be used as a qualitative comparison relative to other materials that appear in the same image.

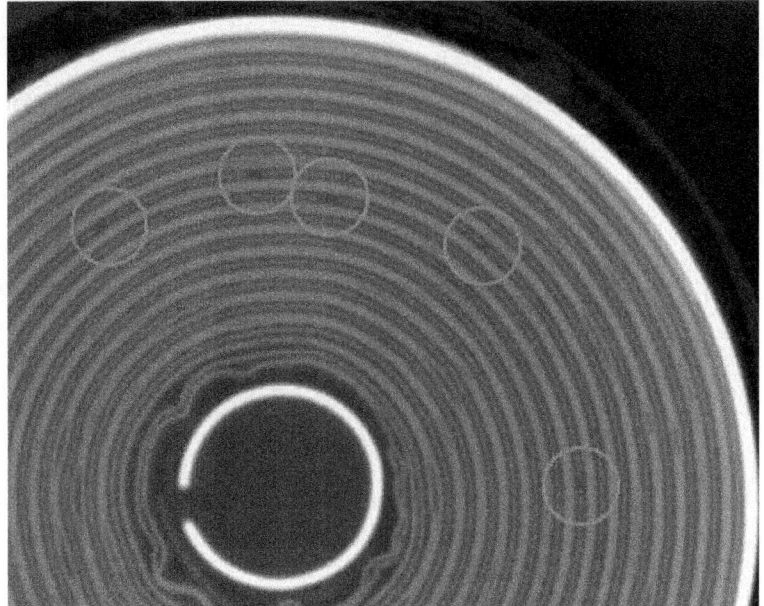

FIGURE H.17 CT scan of an 18650 lithium-ion cell subjected to severe overdischarge resulting in copper dissolution and the formation of fine holes in the copper current collector, particularly in the circled regions (thermal runaway did not occur). (*Courtesy of Exponent, Inc.*)

Cell overdischarge may cause copper dissolution to the point where numerous small holes form in the copper current collector. Figure H.17 is a CT scan cross section of an 18650-type cell that was subjected to severe overdischarge (but has not undergone thermal runaway). In this image, the copper current collector appears as a thin bright line and the aluminum coated with cobalt oxide appears as a thicker gray line. In places throughout the cell, the copper current collector appears discontinuous—broken by a series of small holes, with diameters in the plane of the cross section on the order of the current collector thickness. These types of holes have been found to be the result of copper dissolution. Figure H.18 shows one of these types of holes upon unrolling of a cell that has not undergone thermal runaway. The hole diameter here is on the order of the thickness of the coated anode layer.

Upon recharge of a cell that has undergone overdischarge, copper is deposited on the anode material, causing degradation of that material. This can lead to lithium plating on the anode and possibly thermal runaway of the cell. If a cell that has experienced overdischarge and thermal runaway is opened and unwound, close examination of the copper current collector will reveal either the presence of a multitude of small holes (which are best observed by removing the remaining anode material and placing the current collector on a light table) or of pitting on the surface of the copper current collector as shown in Fig. H.19. Examination of the windings in cross section will show copper distributed throughout the anode, on the anode surface, and throughout the cathode material. Figure H.20 is an SEM image of a cell cross section (created by fracturing the electrode stack, which was embrittled during thermal runaway of a neighboring cell) with an overlaid x-ray map showing the distribution of copper. The cell in this image had been subjected to overdischarge and subsequently had partially burned. The observed redistribution of copper could only occur prior to the cell undergoing thermal runaway, and is thus a positive indicator that the cell had experienced severe and repeated overdischarge.

This effect had been seen earlier in lithium-sulfur dioxide primary cells after being driven into voltage reversal. These cells have a copper stripe current collector rolled into the lithium metal anode (see Chap. 14).

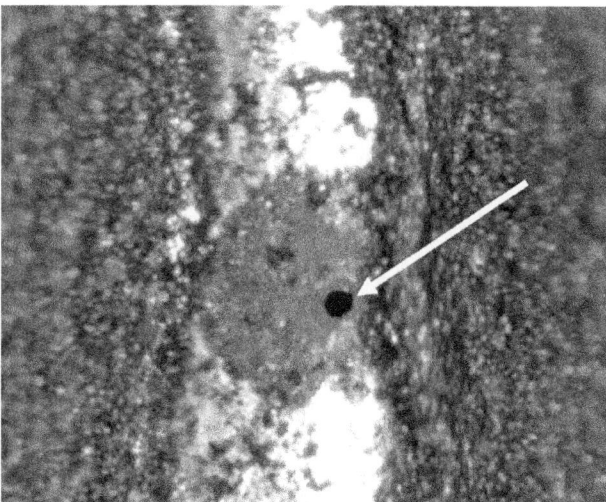

FIGURE H.18 Stereo-microscope image showing a hole in the copper current collector caused by overdischarge and copper dissolution. (*Courtesy of Exponent, Inc.*)

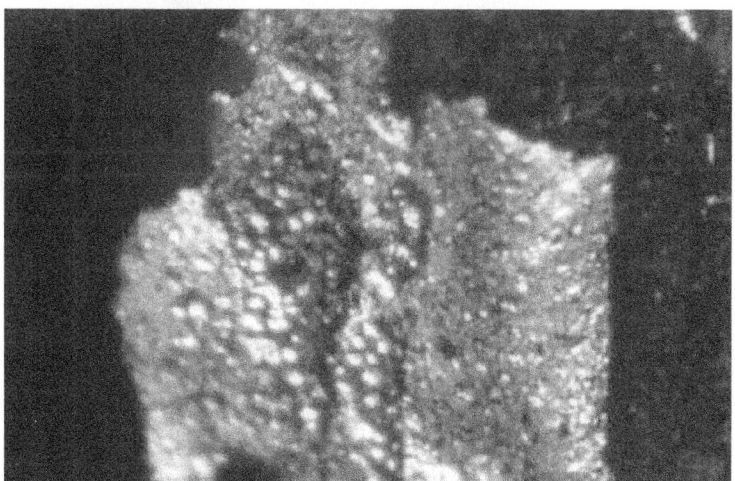

FIGURE H.19 Stereo-microscope image of pitting of the copper current collector surface associated with repeated severe over discharge. (*Courtesy of Exponent, Inc.*)

It is likely that cell shorting and thermal runaway, for incidents containing evidence of hard shorting (e.g., melted copper), develops through the following mechanism:

1. An initial shorting event as a result of a conductive path across the electrodes occurs that locally heats the copper current collector to its melting point of 1357 K.
2. Residual heat melts the separator near the point of shorting, creating a higher impedance short between anode and cathode active materials. The high impedance short continues to heat the cell, but the melting of the separator may interrupt or reduce the strength of the initial short and allow the melted copper to resolidify.

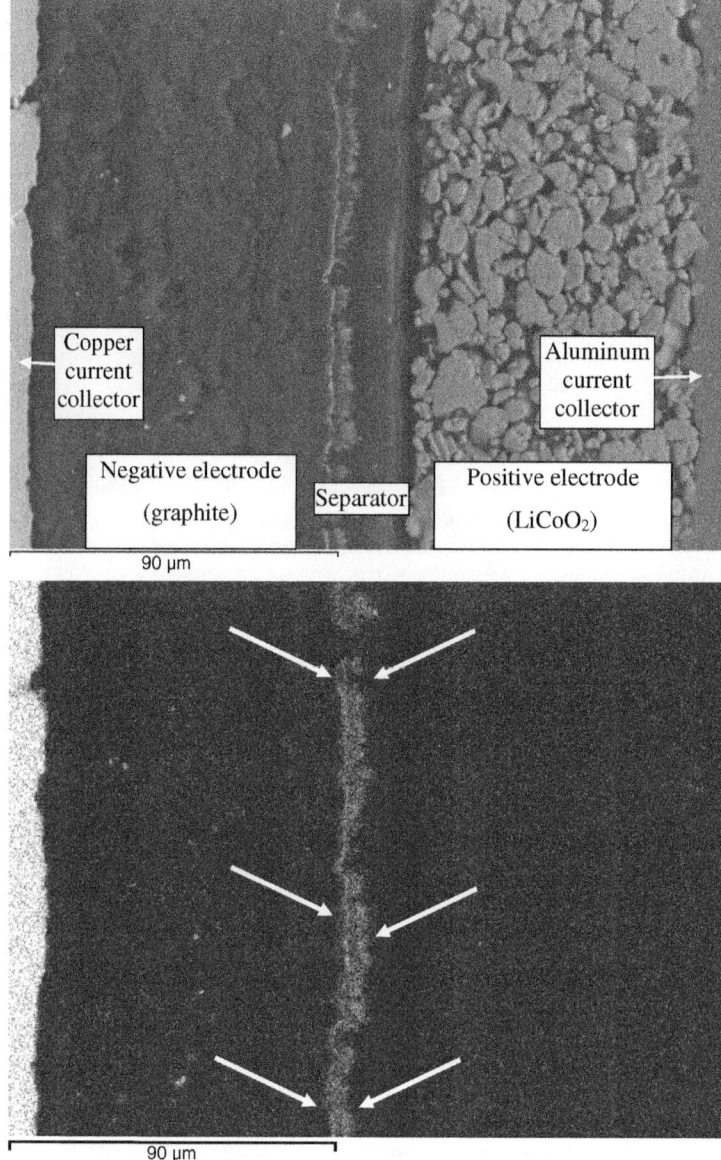

FIGURE H.20 SEM backscattered electron image of a D-size lithium-ion cell cross section (top) with a copper elemental map of the same region (bottom) showing copper dendrites growing through the separator from the negative to the positive electrode (between arrows). Cell cross-sectioning is discussed in the next section. (*Courtesy of Exponent, Inc.*)

3. The high impedance short created between the anode and cathode by localized melting of the separator continues to heat the cell to the point where chemical reactions continue spontaneously (thermal runaway). During thermal runaway, the majority of the cell interior reaches a minimum temperature of 933 K, the melting point of aluminum.
4. In the final phases of thermal runaway, liquid aluminum may alloy with any exposed copper surface.

It should be noted that cell unrolling or destacking is not possible in many cases. Examples where such destructive physical analyses (DPAs) are difficult include:

Cells that are exposed to moisture while at high temperature because high temperature corrosion of the copper windings results in severely degraded and embrittled copper (e.g., if water was used to extinguish a fire).

Cells that are heated beyond the melting point of the separator and vented, but do not experience thermal runaway, so that the dried electrolyte and melted separator essentially glue the electrodes together (though subsequent baking may be possible to burn away the separator and allow further DPA).

Cells that experience violent gas generation and subsequent venting so that the windings are torn and tangled together in multiple locations.

Failure analysis of such cells depends more strongly on data obtained during analysis of the system electronics, and nondestructive analysis of the battery cells (x-rays, CT scans, etc.).

Cross-Sectioning

Another destructive examination approach for cells is cell cross-sectioning, which involves extracting the electrolyte and electrolyte-soluble components from the cell, followed by filling a cell with a polymer resin. The cell can then be physically cross-sectioned and prepared using nonaqueous polishing techniques to enable examination of the microstructure using optical and scanning electron microscopy (SEM) and energy dispersive spectroscopy (EDS). Cross-sectioning allows identification of electrode degradation, dendrite growth through the separators, manufacturing defects, and internal contaminants within the active materials that may not be visible otherwise. Unlike conventional techniques that require disassembly and harvesting of materials from a cell, the cross-sectioning technique preserves the electrode structure and the spatial relationship between the various components in the electrodes. Internal defects and degradation mechanisms, linked to safety, reliability, and performance problems, can be directly observed and characterized using this cross-sectioning technique.

Cross-sectional analysis of electrochemical cells has proven to be a successful technique in understanding degradation mechanisms in several battery systems, including lead acid,[25–32] nickel-metal hydride,[33,34] standard alkaline (Zn/MnO_2),[35] and lithium-ion.[36–42] For example, prior work on Zn/MnO_2 alkaline cells has shown that the analysis of polished cross sections can provide valuable information regarding the formation and spatial distribution of zinc oxide reaction products in porous electrodes. This information can be useful in determining if an alkaline cell was abused by attempted recharging, as shown in SEM images in Fig. H.21. These images were acquired from anode cross sections of three AAA Zn/MnO_2 alkaline cells. The image in Fig. H.21a shows the anode of an undischarged cell near the anode/separator/cathode interface. The bright gray spheres in the image are metallic zinc powder particles. The image in Fig. H.21b shows the same region of an anode from a cell that was discharged at 300 mA to 0.8 V. The zinc particles are noticeably smaller and surrounded by ZnO discharge product (dark gray). The same region of an anode from a cell discharged then recharged three times is shown in Fig. H.21c. In this image, smaller metallic zinc particles can be seen decorating the separator interface. These smaller zinc particles are the result of replating of zinc during recharging, and upon subsequent cycling become electrically disconnected from the rest of the anode.

An example of how cross-sectioning can be used to complement other techniques, such as CT scanning, is shown in Fig. H.22. In this example, a group of AA NiMH cells were found to be prone to high

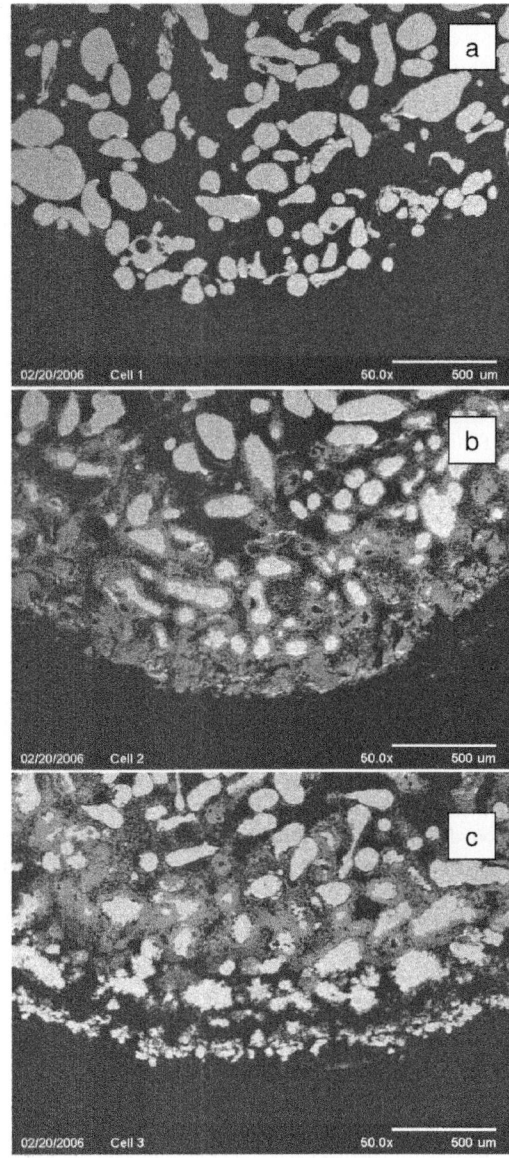

FIGURE H.21 SEM images (backscattered electron contrast) showing axial cross sections of AAA alkaline anodes. The bright regions are metallic zinc powder particles. The darker gray regions (images b and c only) are ZnO discharge product. (a) New, undischarged anode; (b) discharged at 300 mA to 0.8 V; (c) cycled three times at 300 mA from 1.65 to 0.8 V. The effect of cycling on metallic zinc segregation toward the separator is evident in (c). (*Courtesy of Exponent, Inc.*)

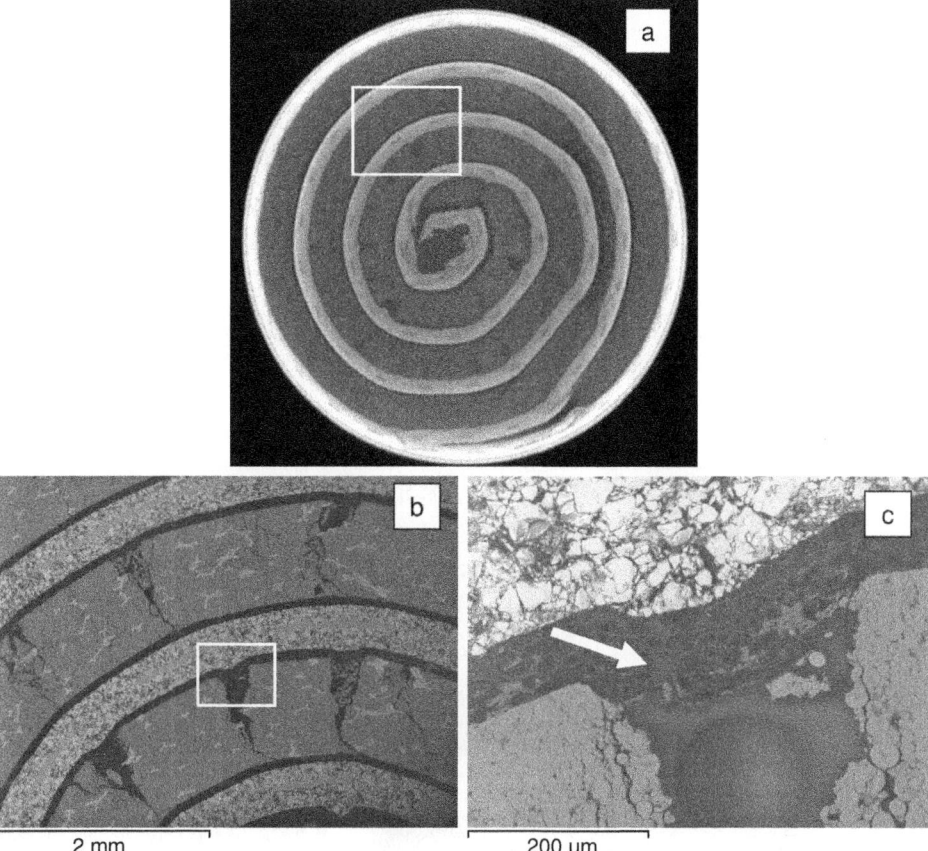

FIGURE H.22 Analysis of physical cross sections can complement nondestructive techniques such as CT scanning. An axial CT slice of a new AA NiMH cell is shown in (a), where cracking within the positive electrode is clearly evident. SEM (backscattered electron contrast) is shown in (b) and (c) for the same cell. From these images, tears in the separator can be found associated with the cracks in the positive electrode. These tears in the separator were found to be the root cause of rapid self-discharging and overheating of the cells in some devices. (*Courtesy of Exponent, Inc.*)

self-discharge rates and overheating during charge. Using CT scanning, large cracks were identified in the positive electrode, as shown in the CT slice in Fig. H.22a. An SEM image of a cross section of the cell in the same plane is shown in Fig. H.22b. Examination of the physical cross section enabled confirmation of the existence of the cracks in the positive electrode, and higher magnification imaging (Fig. H.22c) shows tearing of the separator where the negative electrode pushes into the void space left by the cracking in the positive electrode. The tearing of the separator resulted in multiple internal short-circuit locations and the high self-discharge rate and overheating observed with these cells.

The cross-sectioning technique can also be used with lithium-ion cells, although great care must be taken to avoid potential safety issues related to rapid release of heat during the cross-sectioning process. With lithium-ion cells, it is important to ensure that all electrolyte has been extracted and that the cell is completely dry prior to impregnation with a polymer resin and physical cutting of the cell. A properly cross-sectioned lithium-ion cell can reveal valuable information about manufacturing defects, degradation mechanisms, and user abuse or misuse.

An example showing the effect of overcharging on the microstructure of an 18650 lithium-ion cell is shown in Fig. H.23. In this figure, the top two SEM images (*a* and *b*) show the microstructure

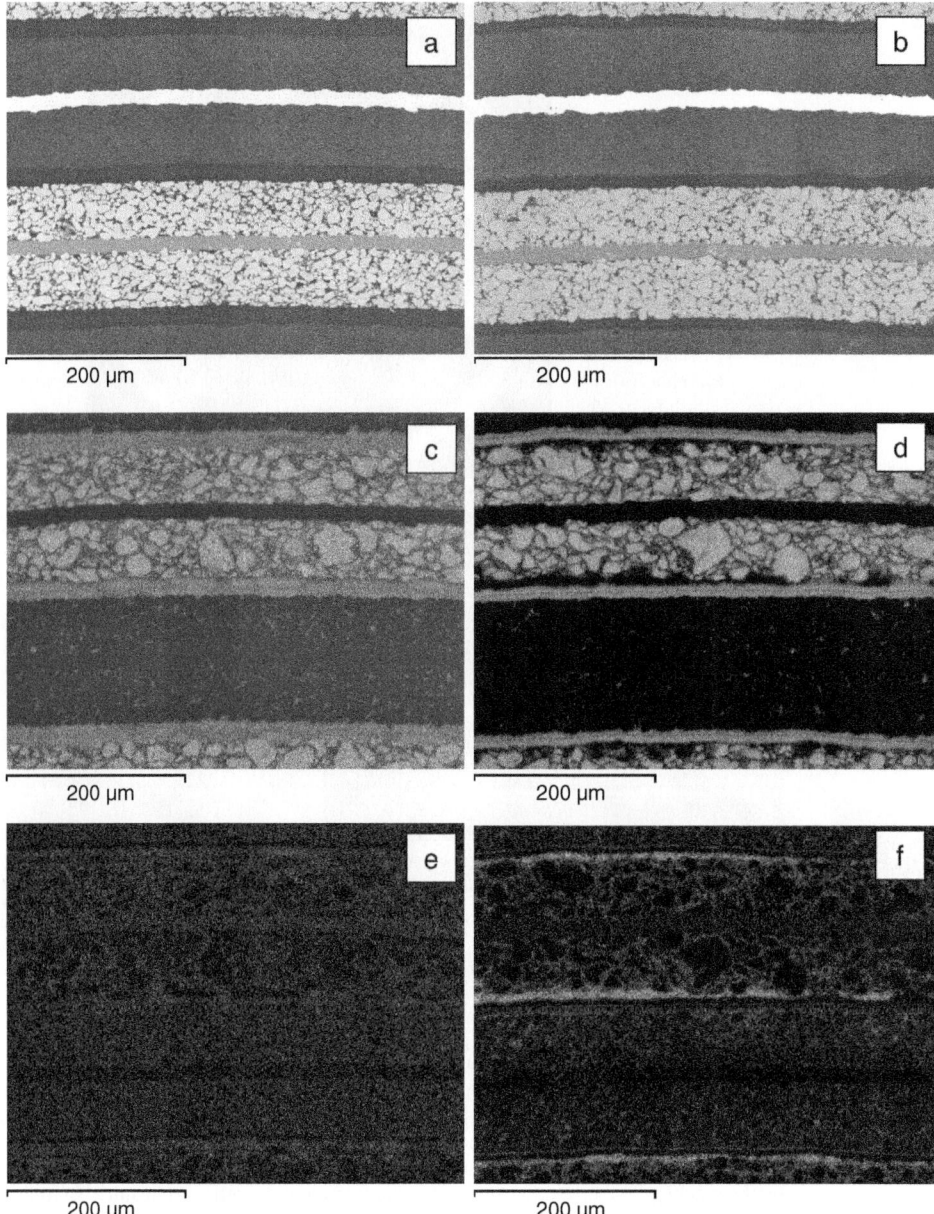

FIGURE H.23 SEM images (*a* and *b*) and associated carbon (*c* and *d*) and phosphorus (*e* and *f*) elemental maps. Images on the left were obtained from an axial cross section of an 18650-size lithium-ion cell, whereas the images on the right were obtained from a similar cell that was overcharged to 4.6 V. The effect of overcharging can be seen at the interface between the negative electrode and separator, where a thick phosphorus-rich layer is found. This phosphorus-rich layer is an electrolyte reduction product that forms from the reaction between metallic lithium and the electrolyte. (*Courtesy of Exponent, Inc.*)

of identical cells. The cell in the left image (*a*) was cycled between 3.0 and 4.2 V, whereas the cell in the right image (*b*) was cycled between 3.0 and 4.6 V. It is apparent that the negative electrode in the cell that was overcharged to 4.6 V is significantly thicker than its 4.2 V counterpart. In the carbon elemental maps shown below the SEM images (*c* and *d*), the presence of a gap can be seen between the graphite particles and the separator in the cell that was overcharged. The phosphorus elemental maps shown at the bottom of the figure (*e* and *f*) indicate that a thick, phosphorus-rich film is present at the graphite/separator interface of the overcharged cell. This phosphorus-rich film is consistent with an electrolyte reduction product that forms when metallic lithium is plated on the surface of the graphite when the cell is overcharged.

The cross-sectioning technique can also be useful for identifying internal microshorts in lithium-ion batteries. An example of this is shown in Fig. H.24 for high-power, 18650 lithium-ion cells that were prone to high self-discharge rates. In these cells, adhesive from Kapton tape on the positive electrode lead was in contact with delithiated Li_xCoO_2. Incompatibility between the adhesive and the delithiated positive electrode material resulted in dissolution of cobalt and the deposition of a cobalt-rich compound on the negative electrode, which bridged the separator and resulted in a high resistance internal short-circuit path in these cells. In Fig. H.24*b*, the intraseparator deposition can be clearly seen, and the elemental map in Fig. H.24 shows that this deposit is rich in cobalt.

As demonstrated in the above examples, examination of properly prepared cell cross sections can be a valuable tool in failure analysis, especially when used to complement other failure analysis techniques. However, as with any destructive technique used on batteries, care should be taken to prevent the introduction of contaminants, defects, and other sample preparation artifacts that might be misinterpreted by the investigator. In general, the investigator should be familiar with metallographic and materialographic sample preparation techniques, as well as the fundamental aspects of the design and construction of the cell being analyzed, in order to avoid misinterpretation of features observed in cell cross sections.

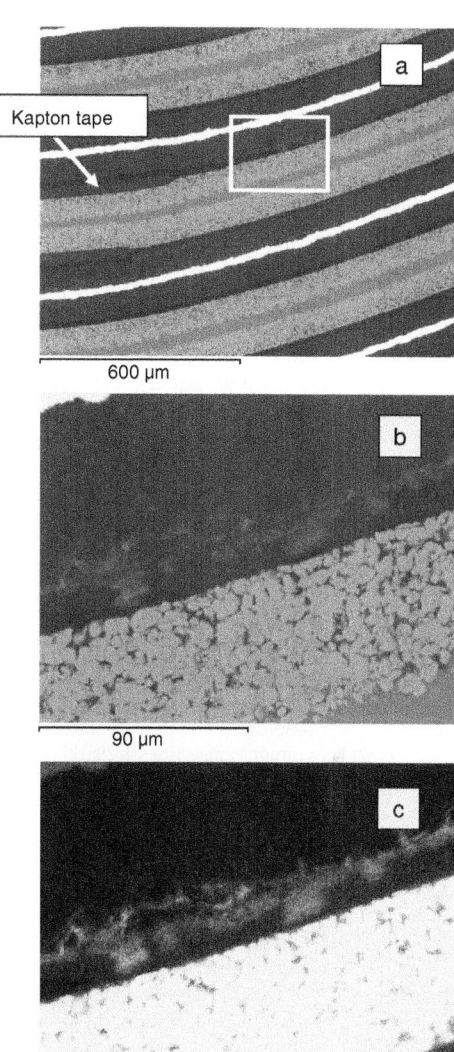

FIGURE H.24 (a and b) Scanning electron microscope images (backscattered electron contrast) showing intraseparator deposits growing from the negative to the positive electrodes in a high-power 18650 lithium-ion cell. The area within the white box in (*a*) is shown at a higher magnification in (*b*). The cobalt elemental map in (*c*) shows that these deposits are rich in cobalt. Dissolution of cobalt from Li_xCoO_2 resulted from the use of an incompatible adhesive on Kapton tape on the positive electrode leads. (*Courtesy of Exponent, Inc.*)

BATTERY MANAGEMENT AND PROTECTION CIRCUITRY EXAMINATION

Battery management and protection circuitry, which consists of the charger and often a battery protection PCB, is critical for safe and reliable battery operation with chemistries such as lithium-ion and in some cases, nickel-metal hydride, as well as for maximizing the useful life of the battery. Failure of battery pack protection circuitry to either sense or respond to an out-of-range condition can lead to failure of a battery. Assessment of the overall battery protection approach and examination of battery protection circuit components are important during any failure analysis. Failures, particularly of early generation systems, are occasionally the result of an oversight in the protection circuitry design that permits an unexpected out-of-range condition to be applied to a battery pack. For example, certain early notebook computer battery packs were susceptible to overcharge if used with some aftermarket general purpose battery chargers. Designers have since implemented charger/battery pack identification "handshake" protocols that eliminate this failure mode. Obtaining system usage history is often useful in diagnosing such a flaw in the battery protection approach (e.g, was the system ever used with an aftermarket charger, fast charger, etc.).

Various battery chemistries require different levels of complexity in protection circuitry. Therefore, different levels of effort and understanding are required to diagnose any potential role of protection circuitry in a particular failure mode. Protection circuitry can require the monitoring of voltage, current, temperature, and/or pressure sensing for its effective operation. Aqueous battery systems often require simple protection electronics, while lithium-ion cells require complex protection systems. For example, in many simple household products and toys that use common alkaline batteries there are no electronic battery protection systems. Failures, should they occur (such as failure due to one or more batteries being inserted in reverse), tend to be mild in severity compared to more energetic failures that can occur with higher capacity batteries. In more sophisticated systems, with high-capacity batteries, high-rate batteries, and where multiple cells are assembled into series strings, the complexity of protection electronics increases. In some aqueous systems, a bimetal switch or thermal cutoff (TCO) is used to disconnect the battery in the case of battery overheating. Thermistors are utilized in some systems to provide feedback to a charger that can modify its state based on battery temperature. In large batteries, multiple thermally dependent devices and multiple voltage-sense lines located throughout the battery may be needed to adequately monitor, thermally manage, and protect the entire battery. These sensor components, their connections to protection circuitry, or the protection and management circuitry itself can fail resulting in battery failure. It is beyond the scope of this appendix to describe all possible permutations of protection electronics and their potential failure mechanisms. However, a basic description of the function of common protection and management elements, particularly those found in lithium-ion battery packs, is provided. An analyst should become familiar with the protection and management functions required by a specific battery chemistry, as well as how those functions are implemented in the particular battery design being examined. (See also Sec. 5.6.)

One of the most important functions of battery management and protection electronics is to manage and control charging (controlling charging rates and implementing appropriate charge cutoffs). Aqueous-based chemistries typically require their chargers to assert certain charging and hold profiles depending on the intended application of the battery. Automobile "starting, lighting, and ignition" (SLI) lead-acid batteries typically see charging and noncharging conditions a few times a day, and typically do not need to hold a charge for long periods of time. Uninterruptible power supply (UPS) devices need to assert charging after use and provide desulfation charge cycles (for lead-acid batteries) during their long-term standby state. Primary batteries need protection to prevent charging; typically a diode is used to prevent primary battery charging. Many systems actually implement two diodes for redundancy, *as single diode systems have been found to periodically suffer charging failures due to failure of a single diode.*

Another important function of battery management and protection electronics is to manage discharge. Many batteries are designed to provide high-rate capability. These batteries can provide significant current into an external short circuit that can then result in an overheating failure of components, circuits, or the battery itself. Therefore, high-rate batteries typically require protection against short circuit. This protection may be a simple fuse, a bimetal switch that opens when heated by excessive current, or more advanced protection in which a control circuit measures discharge current and opens a switch very quickly when a short-circuit current is sensed. Lower-rate batteries may safely limit short-circuit current with their own internal resistance and may not require short-circuit protection. Some batteries also require that discharge current be cut off once the battery voltage drops below a certain threshold. Lithium-ion batteries, for example, can be damaged by high-rate discharge currents when below approximately 3 V. These batteries are typically protected by electronics that open a switch to remove the load from the battery and terminate discharge at a specific voltage threshold. Many devices powered by batteries shut down on their own when their battery voltage drops below a threshold and therefore may not require a separate circuit for this function.

Lithium-ion batteries require relatively complex protection circuitry to protect against the following:

- *Charging to an excessive voltage.* Most lithium-ion batteries are rated to charge to 4.2 V. However, this maximum charge voltage varies depending on the specific battery chemistry or the intended use environment. Preventing overcharge is considered sufficiently critical to warrant individual monitoring of cell or series element (block) voltages by electronics to prevent any cell from exceeding a voltage limit. In addition, most electronic protection packages include multiple independent circuits to terminate charge so that a single-point circuitry failure cannot disable overvoltage protection.

- *Charging at an excessive current.* Most lithium-ion batteries are rated to charge at some fraction of their maximum cell charging rate. A safety margin is typically provided to account for aging of cells. In addition to being potentially detrimental to the cells, excessive currents can cause heating at connections both internal and external to cells, leading to undesirable effects.

- *Discharging at an excessive current.* High-rate discharges can cause heating of cells, in some cases to the point of damaging internal components such as the separator, and can lead to cell thermal runaway. For example, polymeric positive temperature coefficient (PTC) devices, also called resettable thermistor devices or "PolySwitches,"* are common components of commercial cells (e.g., part of the cap assembly of 18650 commercial cells) or commercial battery packs (placed in the circuits of battery packs designed with prismatic cells). These devices include a conductive polymer layer that becomes very resistive above some threshold temperature. PTCs are selected to remain conductive within specified current and temperature conditions. However, should discharge (or charge) current become excessive, the polymer will heat and become highly resistive, greatly reducing current from (or to) the cell. Once the PTC cools, it again becomes conductive.

- *Temperature excursions.* Charging lithium-ion batteries at low temperature can result in lithium plating due to reduced lithium-ion diffusion rates within the negative electrode. Charging or discharging lithium-ion batteries at high temperature can increase the risk of significant gas generation within the batteries, leading to swelling, the nuisance operation of pressure-triggered protective devices (e.g., CID: charge interrupt devices), or thermal runaway due to mechanical disturbance of windings or layers.

- *Imbalance protection.* Protection against the drifting of states-of-charge in cells connected in series. This protection may permanently disable a battery in which imbalance has become too severe, or may attempt to compensate for imbalance by equalizing cells using small discharge currents for cells at a higher state-of-charge. Batteries may not have imbalance protection and simply rely on overvoltage and undervoltage conditions to protect the battery. This can lead to early loss of battery capacity if severe imbalance conditions occur.

- *Discharging to excessively low voltage.* Lithium-ion batteries are mostly depleted when their voltage is below 3 V. Additional discharging, especially below 1 V, can cause damage to current

---

*PolySwitch is a trademark of Tyco Electronics.

collectors and ultimately to electrodes, leading to compromised performance or increased risk of thermal runaway. Thus, protection circuits tend to prevent discharge and themselves go into low-power sleep modes below certain voltages.
- *Precharge charging circuitry.* Lithium-ion batteries in a state of deep discharge require low charging currents until they reach a threshold voltage. Protection or charging circuitry is designed to charge at low rates until the desired voltage threshold is reached.
- *Additional passive thermally dependent protection devices.* These may also be integrated into a lithium-ion cell stack to protect against thermal runaway.

Protection circuitry tends to be simplified for a single-cell lithium-ion battery. The protection circuitry is typically integrated into a single protection printed circuit board (PCB) that achieves most or all of the protection functions discussed above. Some designs move some or all protection functionality to the host device rather than using a dedicated PCB in the battery.

Lithium-ion battery thermal runaway failures can cause significant damage to proximate PCBs. However, PCBs covered with heavy soot from a proximate battery failure often continue to function properly after the soot has been carefully removed. Additionally, most series connected lithium-ion battery protection PCBs have on-board nonvolatile memory. This memory is usually intact after a battery failure (even when the PCB appears heavily charred from the incident) and can provide important clues to the use pattern and environment the battery has experienced and to the root cause of the failure.

A lithium-ion battery protection PCB should be x-rayed in its as-received location prior to removal from the battery for evaluation separate from the battery. Once a protection PCB has been removed from a battery pack, it should be visually examined, x-rayed, and then cleaned. X-ray examination can reveal melting of copper traces that are generally associated with propagating PCB failures. Propagating PCB faults can heat nearby components and cells, potentially causing venting of cells or causing the cells to go into thermal runaway. X-ray images can also reveal the state of fuses, if present, which may be damaged from direct thermal exposure associated with venting of adjacent cells, or may signify an overcurrent failure within the PCB. A thorough discussion of propagating PCB faults and protection PCB evaluations is available elsewhere in Ref. 43.

Lithium-ion battery protection circuitry can also be a source of information about the pre-incident condition of the battery pack. As indicated above, some protection PCBs contain an EEPROM (electrically erasable programmable read-only memory) integrated circuit that stores data such as cycle count, remaining full charge capacity, and maximum temperature encountered by the pack, among other data depending on the complexity of the circuit. This information may prove useful for conducting a failure analysis of the pack. Data should be retrieved from the PCB, if possible, either by direct connection to the PCB if damage is minimal or by transplanting the EEPROM chip to an exemplar PCB and retrieving the information there.

WORKING WITH DATA FROM OEMS AND BATTERY MANUFACTURERS

Involving the battery pack assembler and cell manufacturers in an investigation increases the likelihood of determining failure cause(s) and identifying potential problems that may be associated with a certain production time line. The battery pack and cell manufacturers should be able to examine production data from around the time of the failed battery production and report any abnormally high reject rates, manufacturing changes, supplier changes, or production rate upsets. Of particular interest in lithium-ion battery failure analyses are the failure rates of high potential (HiPot) and open-circuit voltage (OCV) tests, and any investigations of cells that failed these tests.

Manufacturers often retain production sample batteries from each batch or a selection over a period of time, so they may have sample batteries from times consistent with failed batteries and may be able to provide these batteries for examination. As a theory of the failure mode develops, the manufacturer should be able to describe relevant manufacturing processes so an understanding can be developed of how a particular type of fault could have occurred and the likelihood of additional field failures.

Manufacturers may have access to batteries returned from customers as a result of any number of nonenergetic failures (e.g., a battery simply stopped working, had a prematurely reduced capacity, etc.). Examination of batteries from an incident battery lot (and similar lots) can prove invaluable, even if the exemplar batteries show no obvious symptoms of failure. Catastrophic battery failures are typically a result of multiple failures that align to create a perfect failure condition. Field returned batteries manufactured at similar time frames as incident batteries may very well have less severe manifestations of similar failure modes. Examinations of these cells can yield an understanding of one or more failure modes that might be required for catastrophic failures (incidents) to occur. Examination of such field returns can be conducted using a combination of the techniques described above, including CT scanning, cycling, electrochemical testing, cell disassembly, or cell cross-sectioning.

With multiseries lithium-ion cells, some battery field returns result from pack protection electronics driving a permanent disabling of the battery pack. Protection electronics permanently disable the pack for a number of reasons, including diverging voltages between series elements (block-imbalance), block overcharge, over-rate charge, over-rate discharge, etc. If some battery packs from a given production lot fail in an energetic manner, it is likely that a number of other packs from the same lot will fail relatively benignly after the protection electronics disables the packs. Thus, examination of permanently disabled packs from a suspect lot can be particularly useful in developing a failure mode hypothesis for any energetic failures from that lot.

Statistical analysis of battery pack failure rates is often necessary to determine whether a collection of battery incidents indicates a batch of batteries with an elevated failure rate. If a batch or a series of batches of batteries with an elevated failure rate can be identified, it is possible to recall batteries from that batch and replace them with cells that should* have a lower failure rate. The decision to consider a recall depends on several factors in addition to the actual expected failure rate, including the type and severity of the failure mode, the likelihood of potential harm to people, and damage to property. Statistical modeling can help determine whether observed failures are "infant mortality failures" or if there is an underlying failure rate over time. It can compare different contributing factors that may be identified in the course of discussions with battery manufacturers to determine whether they may or may not have a statistically significant relationship with the failure rate. This can help rule out certain factors as possible contributors to the actual incidents and limit potential recall populations. Statistical analysis can also predict, within a certain confidence interval, the number of expected failures in the future so the remaining risk is known if it is decided a recall is unnecessary.

CONCLUSIONS

Any failure can be investigated effectively through application of the scientific method:

1. Collection of observations and evidence
2. Development of an hypothesis
3. Testing of the hypothesis

This appendix described a number of techniques that can be used as part of the investigative process. Many of the techniques described here have proven useful in examining failures of lithium-ion batteries, lithium primary batteries, nickel-metal hydride batteries, as well as batteries of other

---

*A new cell batch will actually have an unknown failure rate, so recall decisions must weigh the risk that the replacement cells may have a higher failure rate than those being replaced. For example, if a group of failures can be shown to be part of an "infant mortality group," and other cells from the batch have been in the field long enough, leaving unfailed units in the field may be safer than replacing them with other units of unknown failure rates.

chemistries. This appendix is not an exhaustive description of all failure analysis techniques that can be applied to examine battery failures, and many additional techniques commonly employed by electrochemists, mechanical engineers, electrical engineers, and chemical engineers can be considered for approaching battery failures. Additionally, the techniques and conclusions presented in this appendix are intended as guidelines on how to approach a battery failure analysis. A battery failure investigation may not require the use of all the techniques discussed here. Furthermore, it should be noted that each battery failure is a unique case and thus requires a unique combination of various investigative strategies and techniques that should be tailored for that specific investigation accordingly.

ACKNOWLEDGMENTS

The authors would like to thank their colleagues at Exponent who have provided significant guidance with and assistance in battery work over the years. In particular, we would like to thank John Loud for his seminal and ongoing contributions to lithium-ion battery investigation methodologies since 1995, and the other members of the Exponent Battery Task Force, including Stig Nilsson, Jan Swart, Ashish Arora, and Xiaoyun Hu. We would also like to thank our many customers, a number of whom have graciously allowed us to include images from investigations conducted on their behalf.

REFERENCES

1. Darlin, D., "Dell Recalls Batteries Because of Fire Threat," *New York Times*, August 14, 2006.
2. Kelley, R., "Apple Recalls 1.8 Million Laptop Batteries," CNNMoney.com, August 24, 2006.
3. U.S. Consumer Products Safety Commission, Release #06-231, "Dell Announces Recall of Notebook Computer Batteries Due to Fire Hazard," August 15, 2006.
4. U.S. Consumer Products Safety Commission, Release #06-245, "Apple Announces Recall of Batteries Used in Previous iBook and PowerBook Computers Due to Fire Hazard," August 24, 2006.
5. Mikolajczak, C., Hayes, T., Megerle, M. V., Wu, M., "A Scientific Methodology for Investigation of a Lithium Ion Battery Failure," IEEE Portable 2007 International Conference on Portable Information Devices, IEEE No. 1-4244-1039-8/07, Orlando, FL, March 2007.
6. Mikolajczak, C., Harmon, J., Hayes, T., Megerle, M., White, K., Horn, Q., Wu, M., "Li-Ion Battery Cell Failure Analysis: The Significance of Surviving Features on Copper Current Collectors in Cells that Have Experienced Thermal Runaway," *Proceedings, 25th International Battery Seminar & Exhibit for Primary & Secondary Batteries, Small Fuel Cells, and Other Technologies*, Fort Lauderdale, FL, March 17–20, 2008.
7. Mikolajczak, C., Stewart, S., Harmon, J., Horn, Q., White, K., Wu, M. "Mechanisms of Latent Internal Cell Fault Formation," *Proceedings, 9th BATTERIES Exhibition and Conference*, Nice, France, October 8–10, 2008.
8. Hayes, T., Mikolajczak, C., Megerle, M., Wu, M., Gupta, S., Halleck, P., "Use of CT Scanning for Defect Detection in Lithium Ion Batteries," *Proceedings, 26th International Battery Seminar & Exhibit for Primary & Secondary Batteries, Small Fuel Cells, and Other Technologies*, Fort Lauderdale, FL, March 16–19, 2009.
9. Harmon, J., Godithi, R., Mikolajczak, C., Wu, M., "Computed Tomography Imaging as Applied to Primary Cell Evaluation," *Battery Power Products and Technology* 13(5):15.
10. Mikolajczak, C., Harmon, J., Wu, M., "Lithium Plating in Commercial Lithium-Ion Cells: Observations and Analysis of Causes," *Proceedings, Batteries 2009, The International Power Supply Conference and Exhibition*, French Riviera, Sept. 30–Oct. 2, 2009.
11. Loud, J. D., Hu, X., "Failure Analysis Methodology for Li-ion Incidents," *Proceedings, 33rd International Symposium for Testing and Failure Analysis*, pp. 242–251, San Jose, CA, November 6–7, 2007.

12. Horn Q. C., White K. C., "Characterizing Performance and Determining Reliability for Batteries in Medical Device Applications," ASM Materials and Processes for Medical Devices, Minneapolis, MN, August 13, 2009.
13. Hayes T., Horn Q. C., "Methodologies of Identifying Root Cause of Failures in Li-Ion Battery Packs," invited presentation, 24th International Battery Seminar and Exhibit, Fort Lauderdale, FL, March 2007.
14. Horn Q. C., "Battery Involvement in Fires: Cause or Effect?" invited seminar, International Association of Arson Investigators, Massachusetts Chapter, Auburn, MA, March 19, 2009.
15. Horn Q. C., White K. C., "Advances in Characterization Techniques for Understanding Degradation and Failure Modes in Lithium-Ion Cells: Imaging of Internal Microshorts," invited presentation, International Meeting on Lithium Batteries 14, Tianjin, China, June 27, 2008.
16. Horn Q. C., White K. C., "Novel Imaging Techniques for Understanding Degradation Mechanisms in Lithium-Ion Batteries," presented at the Advanced Automotive Battery Conference, Tampa, FL, May 13, 2008.
17. Horn Q. C., "Application of Microscopic Characterization Techniques for Failure Analysis of Battery Systems," invited presentation, San Francisco Section of the Electrochemical Society, March 27, 2008.
18. Horn Q. C., White K. C., "Understanding Lithium-Ion Degradation and Failure Mechanisms by Cross-Section Analysis," presented at the 211th Electrochemical Society Meeting, Chicago, Spring 2007.
19. NFPA 921, "Guide for Fire and Explosion Investigations," 2004 Edition, National Fire Protection Association, Quincy, MA, 2004.
20. ASTM E1188-05, "Standard Practice for Collection and Preservation of Information and Physical Items by a Technical Investigator," ASTM International, 2005.
21. ASTM E1459-92, "Standard Guide for Physical Evidence Labeling and Related Documentation," ASTM International, 1992.
22. Kumai, K., Miyashiro, H., Kobayshi, Y., Takei, K., Ishikawa, R., *Journal of Power Sources* 81:715 (1999).
23. Aurbach, D., "The Role of Surface Films on Electrodes in Li-Ion Batteries," chapter 1 in *Advances in Lithium-Ion Batteries*, W. A. Van Schalkwijk and B. Scrosati (eds.), Kluwer Academic/Plenum Publishers, New York, 2002.
24. *Metals Handbook: Metallography, Structures, and Phase Diagrams*, American Society for Metals, 8th ed. Vol 6.
25. Prengaman, R. D., *Journal of Power Sources* 158:1110–1116 (2006).
26. Torcheux, L., Villaronm, A., Bellmunt, M., Lailler, P., *Journal of Power Sources* 85:157–163 (2000).
27. Lam, L. T., Haigh, N. P., Rand, D. A. J., Manders, J. E., *Journal of Power Sources* 88:2–10 (2000).
28. Ball, R. J., Evans, R., Deven, M., Stevens, R., *Journal of Power Sources* 103:207–212 (2002).
29. Ball, R. J., Kurian, R., Evans, R., Stevens, R., *Journal of Power Sources* 109:189–202 (2002).
30. Ball, R. J., Stevens, R., *Journal of Power Sources* 113:228–232 (2003).
31. Lam, L. T., Haigh, N. P., Phyland, C. G., Urban, A. J., *Journal of Power Sources* 133:126–134 (2004).
32. Rocca, E., Bourguignon, G., Steinmetz, J., *Journal of Power Sources* 161:666–675 (2006).
33. Li, L., Wu, F., Yang, K., *Journal of Rare Earths* 21:341–346 (June 2003).
34. Smith, M., Garcia, R. E., Horn, Q. C., "The Effect of Microstructure on the Galvanostatic Discharge of Graphite Anode Electrodes in $LiCoO_2$-Based Rocking-Chair Rechargeable Batteries," *Journal of the Electrochemical Society* 156(11):A896–A904 (November 2009).
35. Horn, Q. C., Shao-Horn, Y., *Journal of the Electrochemical Society* 150:A652–A658 (May, 2003).
36. Horn, Q. C., White, K. C., "Characterizing Performance and Determining Reliability for Batteries in Medical Device Applications," ASM Materials and Processes for Medical Devices, Minneapolis, MN, August 13, 2009.
37. Horn, Q. C., "Battery Involvement in Fires: Cause or Effect?" invited seminar, International Association of Arson Investigators, Massachusetts Chapter, Auburn, MA, March 19, 2009.
38. Horn, Q. C., White, K. C., "Advances in Characterization Techniques for Understanding Degradation and Failure Modes in Lithium-Ion Cells: Imaging of Internal Microshorts," invited presentation, International Meeting on Lithium Batteries 14, Tianjin, China, June 27, 2008.
39. Horn, Q. C., White, K. C., "Novel Imaging Techniques for Understanding Degradation Mechanisms in Lithium-Ion Batteries," presented at the Advanced Automotive Battery Conference, Tampa, FL, May 13, 2008.

40. Horn, Q. C., "Application of Microscopic Characterization Techniques for Failure Analysis of Battery Systems," invited presentation, San Francisco Section of the Electrochemical Society, March 27, 2008.
41. Horn, Q. C., White, K. C., "Understanding Lithium-Ion Degradation and Failure Mechanisms by Cross-Section Analysis," presented at the 211th Electrochemical Society Meeting, Chicago, Spring 2007.
42. Hayes, T., Horn, Q. C., "Methodologies of Identifying Root Cause of Failures in Li-Ion Battery Packs," invited presentation, 24th International Battery Seminar and Exhibit, Fort Lauderdale, FL, March 2007.
43. Slee, D., "Printed Circuit Board Propagating Faults," *Proceedings, 30th International Symposium for Testing and Failure Analysis (ISTFA)*, November 2004.

INDEX

AA batteries
   discharge modes, 3.8
   nickel-zinc, 23.23–23.24
   zinc/alkaline/manganese dioxide, 28.4
AAA batteries, 28.15–28.17
Absorptive glass mat (AGM) batteries, 17.1–17.2
Acid electrolytes, 7.5–7.6
   carbon-lead-acid cells, 7.5
   sulfuric, 7.5–7.6
Activated carbon, 39.6–39.8
Activation polarization, 2.1
ADMATT propulsion battery, 25.25
Advanced carbons, 39.8–39.9
Advanced Membrane Systems (AMS), 25.28
AEDs. *See* Automated external defibrillators
Aerospace batteries, 19.5. *See also* Nickel-cadmium batteries
   for GEO satellites, 24.14–24.15
   for LEO satellites, 24.15–24.16
   lithium-ion, 26.43–26.44, 26.57
   nickel-hydrogen, 24.1, 24.14–24.17
   silver oxide, 25.25, 25.27
AGM batteries. *See* Absorptive glass mat batteries
Air-breathing systems, fuel cells for, 37.13
Air Force nickel-hydrogen cells, 24.7–24.8
Alkaline electrolytes, 7.2–7.4
   conductivity within, 7.2–7.3
   rechargeable batteries, 7.4
   silver oxide-zinc batteries, 13.8
   zinc deposition, 7.3
Alkaline manganese-dioxide battery, 7.2, 11.1–11.17. *See also* Manganese dioxide batteries
   advantages, 11.2, 11.4
   alloying elements, 11.7
   anodes, 11.9–11.11
   ANSI standards, 9.38, 11.14–11.15
   battery types and sizes, 11.13–11.14
   carbon, 11.8–11.9
   cathode rings, 11.12
   cathodes, 11.4–11.5, 11.7–11.9
   cell components, 11.7–11.11
   cell leakage, 11.15–11.16
   chemistry, 11.4–11.7

Alkaline manganese-dioxide battery (*Cont.*):
   collectors, 11.10
   composition, 11.7, 11.9
   construction, 11.3–11.4, 11.11–11.16, 11.12
   containers, 11.10–11.11
   cylindrical type, 11.1–11.13
   development of, 11.1
   disadvantages, 11.4
   discharges, 11.5–11.6
   EMD, 11.8–11.9
   Evolta, 11.16–11.17
   evolution of, 11.3
   finish, 11.10–11.11
   gels, 11.10
   IEC, 11.14
   improvements to, 11.3
   miniature button style, 11.1
   miniature cell configuration, 11.13
   miniature cells, 11.11, 11.13
   OCV, 11.6
   oxide structures, 11.5
   Oxyride, 11.16–11.17
   performance trends, 11.2–11.3
   schematic representation of, 11.5
   seals, 11.10–11.11
   separators, 11.10
   testing standards, 11.14–11.15
   uses, 11.2
   water in, 11.6
   zinc powder, 11.9–11.10
   zinc reactions in, 11.6–11.7
Alkaline secondary batteries, 3.4, 15.7–15.9
Alloy production, 16.20–16.22. *See also* Metal hydride alloys
Aluminum/air batteries, 33.2–33.3, 33.27–33.43
   in alkaline electrolytes, 33.33–33.43
   applications, 33.32, 33.34–33.43
   characteristics, 33.40
   discharging, 33.29–33.30
   EMP, 33.38
   filter/precipitator systems, 33.42
   in neutral electrolytes, 33.30–33.32
   portable, 33.31
   power systems, 33.40

Aluminum/air batteries (*Cont.*):
  schematic, 33.28
  self-managing systems, 33.41
  SOFAL batteries, 33.36–33.37, 33.39
Aluminum batteries, 8.7, 10.1–10.12
  advantages, 10.1
  aluminum/air, 33.2–33.3, 33.27–33.43
  anode reactions, 10.4
  construction, 10.4–10.6
  performance characteristics, 10.6–10.11
  primary, 10.12
  sizes, 10.11
  temperature for, 10.3, 10.8
  types, 10.11–10.12
Ambient-temperature batteries, 35.1–35.19.
  See also Lithium anode reserve batteries
  spin-dependent reserve, 35.20–35.30
American National Standards Institute (ANSI)
  standards, 4.1–4.2
  alkaline manganese-dioxide batteries, 9.38, 11.14–11.15
  mercuric oxide batteries, 12.1
  primary batteries, 4.4–4.6, 4.11
  rechargeable batteries, 4.6–4.7
  zinc-carbon batteries, 9.38–9.39
Ampere-hour law, 16.68
Ampoule-type batteries, 35.14–35.17
  spin-dependent, 35.24
AMS. *See* Advanced Membrane Systems
André, Henri, 25.1
Anions, 2.12–2.13
  Gibbs energy calculations for, 2.12–2.13
Anodes, 1.3–1.4
  alkaline manganese-dioxide batteries, 11.9–11.11
  aluminum batteries, 10.4
  collectors, 11.10
  composition, 11.9
  containers, 11.10–11.11
  finish, 11.10–11.11
  gels, 11.10
  iron electrode batteries, 18.21
  lithium/air batteries, 33.45–33.46
  lithium primary batteries, 14.4
  magnesium batteries, 10.11–10.12
  mercuric oxide batteries, 12.3–12.4
  miniature cells, 11.11
  seals, 11.10–11.11
  separators, 11.10
  silver oxide-zinc batteries, 13.2–13.3
  thermal batteries, 36.3–36.4
  zinc/air batteries, 13.20–13.21
  zinc powder, 11.9–11.10
ANSI. *See* American National Standards Institute standards
Applications
  aluminum/air batteries, 33.32, 33.34–33.43
  for consumer electronics, 32.3, 32.6–32.7
  consumer electronics batteries, 32.2–32.3
  deep-cycle batteries, 16.46–16.55
  energy storage batteries, 30.5–30.18

Applications (*Cont.*):
  lead-acid batteries, 16.1, 16.6, 16.80–16.85
  lithium/air batteries, 33.52–33.55
  lithium anode reserve batteries, 35.19
  lithium/carbon monofluoride batteries, 14.73
  lithium/copper oxide batteries, 14.87–14.88
  lithium-ion batteries, 26.57
  lithium/iron disulfide batteries, 14.81
  lithium/manganese dioxide batteries, 14.62–14.64
  lithium/sulfur dioxide batteries, 14.26
  lithium/thionyl chloride batteries, 14.40–14.42
  magnesium water-activated batteries, 34.22–34.24
  mercuric oxide batteries, 12.1
  nickel-cadmium batteries, 19.21–19.22
  nickel-hydrogen batteries, 24.1, 24.14–24.17
  nickel-zinc batteries, 23.20–23.24
  primary batteries, 8.5
  prismatic batteries, 17.37
  sealed portable nickel-cadmium batteries, 21.1
  secondary batteries, 15.3–15.5
  small fuel cells, 38.2
  sodium-beta high-temperature batteries, 30.31–30.35
  VRLA batteries, 17.36–17.38
  zinc/air batteries, 13.16–13.17
Aqueous batteries, secondary, 15.3–15.4
Aqueous electrolytes, 7.1–7.6
  alkaline, 7.2–7.4
Automated external defibrillators (AEDs), 31.9
Automotive batteries, 16.81–16.82
Average voltage, 3.2

Batteries. *See also* Cells; Primary batteries; Storage, in batteries; specific batteries
  AA, 3.8
  age of, 3.20–3.21
  anodes, 1.3
  BESS, 1.5
  biomedical applications, 31.1–31.36
  capacity, 3.20
  cathodes, 1.3
  cells vs., 1.3
  charging voltage, 3.17
  classification of, 1.4–1.6
  components, 1.3–1.4
  for consumer electronics, 32.1–32.21
  degradation models for, 6.21–6.25
  design, 3.17–3.20, 3.22, 5.1–5.22
  discharges, 3.2–3.3, 3.5–3.7, 3.9–3.10
  discrete, 5.7–5.10
  electrochemical, 1.11, 2.1–2.35
  electrochemical capacitors, 39.1–39.3
  electrolytes, 1.3
  empirical models, 6.5–6.8
  energy density, 1.14–1.17, 3.19–3.20
  energy storage, 30.1–30.45
  EVs, 29.3–29.22
  fuel cells, 1.5–1.6
  HEVs, 29.22–29.46

Batteries (*Cont.*):
   hybrid, 3.19
   kinetic models, 6.13–6.15
   lead-acid, 3.19
   Leclanché, 7.4–7.5
   LIPON in, 7.11
   lithium, 7.1
   mathematical models, 6.1–6.27
   mechanistic models, 6.9–6.13
   packs, 5.18–5.22
   performance advancements, 1.17
   performance of, factors affecting, 3.1–3.22
   primary, 1.4–1.5
   R20, 4.8
   reserve, 1.5
   secondary, 1.5
   self-discharging, 3.9, 3.21
   service life, 3.11–3.12
   shape and configuration and, 3.19
   smart, 5.18, 5.22
   standardization of, 4.1–4.20
   storage condition of, 3.20–3.21
   terminals, 4.7–4.8
   thermodynamic background for, 2.3–2.4
   voltage level, 3.2–3.3
   voltage regulation, 3.15–3.16
   volumetric efficiency and, 3.19–3.20
   zinc-carbon, 7.4–7.5, 8.3, 8.6, 9.1–9.40
Battery energy storage systems (BESS), 1.5
Battery failure analysis, methodologies for, H.1–H.32
   background information collection, H.2–H.6
   cell examination, H.6–H.27
   cell gas sampling, H.11–H.13
   cross-sectioning, H.23–H.27
   CT scanning, H.8–H.11
   disassembly, H.13–H.23
   for management examination, H.28–H.30
   with manufacturer data, H.30–H.31
   for protection circuitry examination, H.28–H.30
   sample evidence recovery, H.3–H.6
   X-rays, H.7–H.8
Battery packs, 5.18–5.22
   cell balancing, 5.19
   charge controls, 5.19
   communications with, 5.19
   customization of, 5.20
   discharge controls, 5.19
   EVs, 22.17–22.19, 29.10, 29.13
   historical information, 5.19–5.20
Berliner Kraft-und Licht Battery System, 30.19
BESS. *See* Battery energy storage systems
Biological fuel cells, 31.35–31.36
Biomedical applications, for batteries, 31.1–31.36
   AEDs, 31.9
   biological fuel cells, 31.35–31.36
   cardiac monitors, 31.5
   cardiac pacemakers, 31.1–31.2
   characteristics, 31.15–31.36
   clinical trials, 31.7

Biomedical applications, for batteries (*Cont.*):
   design qualifications, 31.13
   destructive testing, 31.15
   drug pumps, 31.7
   for external devices, 31.7–31.9
   failure modes, 31.12–31.13
   fault tree analysis, 31.12–31.13
   hearing assist devices, 31.8
   ICDs, 31.1–31.5
   lithium/CFx batteries, 31.20–31.22, 31.28–31.30
   lithium/iodine batteries, 31.15–31.18
   lithium-ion batteries, 31.30–31.34
   lithium/manganese oxide batteries, 31.25–31.28
   lithium/silver vanadium oxide batteries, 31.22–31.25, 31.28–31.30
   lithium/thionyl chloride batteries, 31.18–31.20
   LVADs, 31.5–31.6
   neurostimulators, 31.6–31.7
   nondestructive testing, 31.14–31.15
   for primary batteries, 31.10–31.11
   for rechargeable batteries, 31.11
   reliability of, 31.12–31.15
   for resynchronization therapy, 31.4–31.5
   safety considerations, 31.9–31.12
   for shipping, 31.12
   standards, 31.10
   zinc/air batteries, 31.3–31.35
Bipolar batteries
   nickel-hydrogen, 24.26–24.27
   VRLA, 17.6–17.7
Bobbin batteries
   characteristics, 14.29, 14.33–14.34
   construction, 14.28–14.29
   lithium/copper oxide, 14.82–14.83
   lithium/manganese dioxide, 14.49, 14.52–14.53, 14.55
   lithium/thionyl chloride, 14.28–14.34
   performance, 14.29–14.33
   sizes, 14.34
   zinc-carbon, 9.10–9.11
Bolloré LMP battery, 27.19
Bundle multicell batteries, 28.10
Button cell batteries
   lithium/copper oxide, 14.82–14.84
   nickel-metal hydride, 22.11
   sealed (portable) nickel-cadmium, 21.4–21.5
   silver oxide-zinc, 13.1–13.15
   zinc/air, 8.3, 8.6–8.7, 13.16–13.36
Button configuration, 12.5

Cadmium electrodes, 25.5–25.6
Cadmium/mercuric oxide batteries, 8.3, 8.6, 12.1, 12.12–12.14
   discharge, 12.12–12.13
   storage, 12.13–12.14
Cadmium/silver oxide batteries, 25.2, 25.13–25.16
   charging characteristics, 25.21
   discharge characteristics, 25.13–25.15
   service life, 25.16
   temperature effects, 25.14

I.4 INDEX

Calcium/calcium chromate batteries, 36.9
Calcium-thionyl chloride batteries, 14.11
Capacitance, 39.5–39.6, 39.30
Capacitors. *See* Electrochemical capacitors
Capacity, of batteries
  for consumer electronics, 32.12
  coulombic, 1.10–1.13
  in degradation models, 6.22
  design as influence on, 3.20
  electrical double-layer, 2.10–2.15
  flat-pack zinc/manganese dioxide P-80, 9.36
  halogen-additive lithium/oxychloride, 14.46
  iron electrode batteries, 18.21–18.23
  lithium/CFx, 31.21
  lithium/copper oxide, 14.85
  lithium-ion, 26.58
  lithium-ion batteries, 26.58, 26.65–26.66
  lithium/iron disulfide, 14.76
  lithium/manganese dioxide, 14.63
  lithium rechargeable, 27.23
  lithium/sulfur dioxide, 14.23
  lithium/thionyl chloride, 14.32–14.33, 14.42
  lithium titanate, 26.27
  nickel-cadmium batteries, 19.8, 19.10
  nickel-hydrogen, 24.19–24.21
  nickel-iron, 18.6–18.7, 18.13
  nickel-zinc batteries, 23.12–23.15
  positive electrodes, 26.13–26.14, 26.17
  sealed (portable) nickel-cadmium, 21.7, 21.15
  secondary, 15.15
  self-discharge, 3.21
  silver oxide, 25.17
  silver oxide-zinc, 13.4
  sintered-plate, 20.7–20.9
  solid-cathode lithium, 14.13
  standardization of, 4.8
  stationary, 16.58
  theoretical and practical, of a cell, 1.9–1.13, 1.10–1.11
  zinc-carbon, 9.30, 9.33
Carbon, in alkaline manganese-dioxide batteries, 11.8–11.9
Carbon black, 9.12
Carbon-lead-acid cells, 7.5
Carbon rods, 9.13–9.14
Cardiac monitors, 31.5
Cardiac pacemakers, 31.1–31.2
Case formation, 16.35–16.36
Case-to-cover seals, 16.34–16.35
Cathodes, 1.3–1.4
  alkaline manganese-dioxide battery components, 11.4–11.5, 11.7–11.9
  carbon, 11.8–11.9
  composition, 11.7
  EMD, 11.8–11.9
  iron electrode batteries, 18.21–18.23
  lithium/air batteries, 33.46–33.47
  lithium primary batteries, 14.5
  lithium rechargeable metal batteries, 27.6–27.7, 27.8

Cathodes (*Cont.*):
  magnesium water-activated batteries, 34.8
  silver oxide-zinc batteries, 13.3–13.8
  thermal batteries, 36.4
Cations, 2.13
CCV. *See* Closed-circuit voltage
CEA. *See* Consumer Electronics Association
Cell cases
  nickel-metal hydride batteries, 22.15–22.16
  silver oxide batteries, 25.6–25.7, 25.7
Cell encapsulation, 5.11–5.12
Cell gassing voltage, 16.74
Cell sealing, 35.23–35.24
Cell stacks, 25.4
  small fuel cells, 38.5–38.6
Cells
  alkaline, 3.4
  anodes, 1.3–1.4
  batteries vs., 1.3
  carbon-lead-acid, 7.5
  cathodes, 1.3–1.4
  charging of, 1.7–1.9
  classification of, 1.4–1.6
  components, 1.3–1.4, 1.14
  design as influence on, 3.17–3.20
  discharge, 1.6–1.7
  dry, 9.2
  electrochemical, 1.11, 2.1–2.35
  electrolytes, 1.3–1.4
  free energy, 1.9
  fuel cells, 1.5–1.6, 1.9
  operations of, 1.6–1.9
  primary, 1.4–1.5
  reserve, 1.5
  safety design of, 5.6
  secondary, 1.5
  subkilowatt fuel, 37.9–37.14
  SVO, 6.13–6.15
  terminals, 4.7–4.8
  theoretical energy of, 1.14
  theoretical voltage and capacity of, 1.9–1.13
  thermodynamic background for, 2.3–2.4
  three-electrode electrochemical, 2.34
  zinc-bromine rechargeable, 7.5
Ceramic electrolytes, 7.11
Charge control
  battery packs, 5.19
  nickel-metal hydride batteries, 22.42–22.44
  rechargeable batteries, 5.15–5.17
Charge-current efficiency, 17.31–17.32
Charge retention
  nickel-cadmium batteries, 19.11
  secondary batteries, 15.13
  silver oxide batteries, 25.15–25.17
  sintered-plate batteries, 20.12–20.14
Charging
  algorithms, 22.47
  ampere-hour law, 16.68
  cell gassing voltage, 16.74
  of cells, 1.7–1.8

Charging (*Cont.*):
  constant-current, 16.71–16.72
  constant-potential, 16.72–16.73
  for consumer electronics batteries, 32.14–32.16
  control techniques, 22.42–22.44
  float, 16.74–16.75
  FNC batteries, 19.18–19.19
  hybrid battery packs, 22.20–22.21
  lead-acid batteries, 16.67–16.75
  lithium primary batteries, 14.16
  metal/air batteries, 33.5
  methods, 16.70–16.75
  nickel-cadmium batteries, 1.8–1.9, 19.12–19.13, 21.19–21.25
  nickel hydroxide, 22.8
  nickel-iron batteries, 18.9–18.10
  nickel-metal hydride batteries, 22.40–22.47
  nickel-zinc batteries, 23.18–23.20
  positive electrodes, 26.16
  principles, 22.40–22.42
  pulse, 16.74
  rapid, 16.75
  rechargeable batteries, design for, 5.15–5.17
  regenerative braking, 22.46–22.47
  sealed portable nickel-cadmium batteries, 21.19–21.25
  silver-iron electrode batteries, 18.18–18.19
  silver oxide batteries, 25.20–25.21
  sintered-plate batteries, 20.15–20.18
  taper, 16.73–16.74
  trickle, 16.74, 22.46
  voltage and, 3.17
  VRLA batteries, 17.22–17.32
  zinc/alkaline/manganese dioxide batteries, 28.12–28.15
  zinc/bromine batteries, 30.39
  zinc-carbon batteries, 9.17–9.27
Chronopotentiometry, 2.23–2.25
  concentration profiles, 2.23
  faradaic process, 2.24
  Fick's diffusion equations, 2.23
  irreversible process, 2.24
  reversible process, 2.24
  with significant resistance, 2.24–2.25
CID. *See* Current interrupt device
Closed-circuit voltage (CCV), 3.2
  silver oxide-zinc batteries, 13.12
  zinc/air batteries, 13.21
  in zinc-carbon batteries, 9.16
Coin cell batteries, 14.48, 14.54
  lithium/carbon monofluoride, 14.65–14.67
  zinc/air batteries, 13.18
Collectors, alkaline manganese-dioxide batteries, 11.10
Commodity storage, 30.10
Common pressure vessel (CPV) batteries, 24.23–24.26
Computed tomography (CT) scanning, H.8–H.11
COMSAT Nickel-hydrogen batteries, 24.6–24.7, 24.25

Concentration polarization, 2.1
  in electrochemical systems, 2.16–2.18
Conductivity, of electrolytes, 2.3
Constant-current charging, 17.28–17.29, 28.13
  lead-acid batteries, 16.71–16.72
Constant-potential charging, 16.72–16.73, 28.12–28.13
Constant-potential (CP) recharging, 20.15–20.16
Constant-voltage charging, 17.23–17.25
Consumer electronics, batteries for, 32.1–32.21
  applications, 32.3
  capacity, 32.12
  characteristics, 32.4–32.6
  charging, 32.14–32.16
  costs, 32.16–32.17
  current drain in, 32.3
  decision making for, 32.17–32.20
    performance criteria, 32.19–32.20
  lithium-ion, 32.9
  load current, 32.12–32.13
  physical size, 32.9–32.12
  pitfalls, 32.20–32.21
  portable applications, 32.2–32.3
  primary, 32.3–32.6
  primary vs. secondary, 32.9
  prismatic cells, 32.11
  profiles, 32.12–32.13
  safety and regulatory concerns, 32.16
  secondary, 32.6–32.9
  selection considerations, 32.1–32.2, 32.9–32.17
  shelf life, 32.14–32.15
  temperature effects, 32.6
  temperature requirements, 32.13–32.14
  types, 32.3
  voltage, 32.9–32.10
Consumer Electronics Association (CEA), 32.2
Corrosion inhibitors, 9.13
Coulombic capacity, theoretical, 1.10–1.13
  nickel-metal hydride batteries, 22.34–22.35
CP recharging. *See* Constant-potential recharging
CPV batteries. *See* Common pressure vessel batteries
Crescent Electric Membership Cooperative, 30.18–30.19
Cup cells, 36.10, 36.13
Curing, 16.30–16.33
Current interrupt device (CID), 29.17
Cutoff voltage, 3.2
Cyclic voltammetry, 2.19–2.22
  insoluble film, 2.21
  irreversible process, 2.22
  peak current for, 2.20–2.21
  principles, 2.19–2.20
  quasireversible process, 2.22
  reversible diffusion-controlled process, 2.21
  species electroreduction, 2.22
  voltage sweep rate, 2.21
Cylindrical-type primary batteries, 8.14
  alkaline manganese-dioxide battery, 11.1–11.13
  VRLA, 17.4–17.5, 17.8–17.17

Cylindrical-type secondary batteries
  nickel-metal hydride, 22.10–22.11, 22.15
  sealed (portable) nickel-cadmium, 21.3–21.4

DC power systems, 16.84
Deep-cycle batteries, for traction applications, 16.46–16.55
  cell sizes, 16.53–16.55
  construction, 16.46–16.48
  cycle life, 16.54
  discharge rates, 16.51
  flat-pasted plates, 16.47–16.48, 16.54
  military applications, 16.48
  performance characteristics, 16.49–16.55
  temperature, 16.51
  tubular positive plates, 16.47–16.48, 16.55
  types, 16.53–16.55
Degradation models, for batteries, 6.21–6.25
  capacity vs. cycle number in, 6.22
  mechanistic models and, 6.24
Delta temperature cutoff, 22.44
Dependent pressure vessel (DPV) batteries, 24.17
Design, for batteries, 3.17–3.20, 3.22, 5.1–5.22
  for battery packs, 5.18–5.22
  capacity and, 3.20
  case design, 5.12–5.13
  cell encapsulation, 5.11–5.12
  construction, 5.10–5.14
  control systems, 5.18–5.22
  for diodes, 3.17, 5.3
  for discrete batteries, 5.7–5.9
  electronic energy management, 5.18–5.22
  energy density and, 3.19–3.20
  external charge protection, 5.2
  flat or disk-type lithium/thionyl chloride batteries, 14.37–14.38
  flat-pack, 9.33–9.36
  FNC technology, 19.14–19.15
  hybrid, 3.19
  intercell connections, 5.10–5.11
  for lead-acid batteries, SLI standards, 3.19
  lithium-ion, 5.17–5.18
  for lithium primary batteries, 5.6–5.7, 14.8
  lithium/sulfur dioxide batteries, 14.25
  nickel-hydrogen, 24.10–24.14
  for primary batteries, 5.2–5.3, 5.6
  rechargeable, 5.14–5.18
  for resistors, 3.17
  for safety, 5.1–5.7
  shape and configuration and, 3.19
  short-circuit conditions, 5.3–5.4
  silver-iron, 18.19–18.20
  silver oxide batteries, 25.8–25.9
  sintered-plate batteries, 20.24–20.27
  smart batteries, 5.18, 5.22
  terminal and contact materials, 5.13–5.14
  voltage reversal and, 5.4–5.6
  volumetric efficiency and, 3.19–3.20
  zinc-carbon, 9.33–9.36
  for zinc-carbon batteries, 9.10, 9.33–9.36

Diodes, design as influence on, 3.17, 5.3
Direct-methanol fuel cells, 38.4
Discharge controls
  battery packs, 5.19
  for rechargeable batteries, 5.16–5.17
Discharge load, 8.13
Discharging
  AA batteries, 3.8
  alkaline manganese-dioxide battery, 11.5–11.6
  aluminum/air batteries, 33.29–33.30
  battery design, 5.16–5.17
  battery temperature during, 3.9–3.10
  cadmium/mercuric oxide batteries, 12.12–12.13
  cadmium/silver oxide batteries, 25.13–25.15
  of cells, 1.6–1.7
  characteristics, 3.13–3.14
  comparative, in zinc-carbon batteries, 9.20–9.27
  constant current, 3.5–3.7
  constant power, 3.5–3.7
  constant resistance, 3.5–3.7
  current drain of, 3.3–3.5
  curves, 3.2–3.3
  deep-cycle batteries, 16.51
  duty cycles, 3.12–3.15
  elapsed time, 3.4
  by grade, 9.22–9.27
  for heavy-duty batteries, 9.20–9.22
  industrial primary zinc/air batteries, 33.14, 33.16
  intermittent, 3.12–3.15, 9.18–9.20
  large prismatic lithium/thionyl chloride batteries, 14.40
  lead-acid batteries, 16.9–16.10
  lithium/carbon monofluoride batteries, 14.66, 14.72
  lithium-ion batteries, 26.52–26.53, 26.59–26.60, 31.32
  lithium/iron disulfide batteries, 14.78
  lithium/manganese dioxide batteries, 14.50–14.52
  lithium/manganese oxide batteries, 31.26–31.27
  lithium rechargeable (metal) batteries, 27.14–27.15, 27.24
  lithium/silver vanadium oxide batteries, 31.24
  lithium/sulfur dioxide batteries, 14.20
  magnesium/air batteries, 33.44
  magnesium water-activated batteries, 34.11–34.15
  by manufacturer, 9.27
  mercuric oxide batteries, 12.8–12.9
  modes of, 3.5–3.9
  multicell 9 volt batteries, 14.59–14.60
  nickel-cadmium batteries, 19.7–19.10
  nickel-iron batteries, 18.7
  nickel-metal hydride batteries, 22.22–22.40
  nickel-zinc batteries, 23.15–23.17
  portable primary zinc/air batteries, 33.13
  positive electrodes, 26.16
  primary batteries, 8.8–8.9, 8.12–8.13, 8.15
  sealed (portable) nickel-cadmium batteries, 21.6–21.7, 21.9–21.10, 21.13–21.14
  secondary alkaline cells, 3.4
  secondary batteries, 15.9, 15.12–15.13

Discharging (*Cont.*):
  self-discharging, 3.9, 3.21
  silver-iron batteries, 18.18
  silver oxide-zinc batteries, 13.5–13.7, 13.11–13.13
  sintered-plate batteries, 20.6–20.7
  SLI batteries, 16.40–16.41
  stationary batteries, 16.59
  types of, 3.12
  of voltage, 3.2–3.3
  voltage regulation, 3.15–3.16
  VRLA batteries, 17.8–17.13, 17.38
  zinc/air batteries, 13.24
  in zinc-carbon batteries, 9.17–9.27
  zinc/silver oxide batteries, 25.9–25.13
  zinc/silver oxide reserve batteries, 34.33–34.34
Discrete batteries, 5.7–5.10
  design for, 5.7–5.9
  dimensions of, 5.9–5.10
  IEC standards for, 5.9
DMPS 100 batteries, 24.10
DPV batteries. *See* Dependent pressure vessel batteries
Drug pumps, 31.7
Drumm, James, 23.1
Dry cells, 9.2
Dry charge, 16.36–16.37
  activation of, 16.37
DSRV pressure compensated battery, 25.26
DST. *See* Dynamic Stress Test
Dunk-type batteries, 34.4, 34.20–34.22
Duty cycles, for batteries, 3.12–3.15
  primary, 8.13
Dynamic Stress Test (DST), 29.14

Edison, Thomas, 29.5
EIS methods. *See* Electrochemical impedance spectroscopy methods
Electric vehicles (EVs), 16.82–16.83, 29.3–29.22
  alternative energy storage technologies, 29.21–29.22
  battery packs, 22.17–22.19, 29.10, 29.13
  calendar vs. cycle life, 29.14
  commercial companies, 29.7
  development, 29.3–29.7
  driving schedule dyanometric measurements, 29.9
  DST, 29.14
  electronic controls, 29.10–29.11
  energy power, 29.6–29.9
  forced air cooling, 29.11
  fuel economy measurements, 29.9
  lead-acid batteries, 29.15
  legal mandates for, 29.4–29.5
  lithium-ion batteries for, 29.16–29.20
  nickel-metal hydride batteries for, 29.15–29.16
  performance targets, 29.12–29.15
  propulsion power, 29.6–29.9
  rechargeable batteries for, 29.9
  rechargeable zinc/air batteries, 33.23, 33.25

Electric vehicles (EVs) (*Cont.*):
  thermal management, 29.11–29.12
  USABC, 29.12–29.13
  USCAR, 29.12
  vehicle integration, 29.12
Electrical double-layer capacity, 2.10–2.15
Electrical isolation, 22.47–22.48
Electrical vehicle (EV) battery packs, 22.17–22.19
Electrocapillary curves, 2.12
Electrochemical capacitors, 39.1–39.41
  activated carbon, 39.6–39.8
  advanced carbons, 39.8–39.9
  capacitance, 39.5–39.6, 39.30
  cell balancing considerations, 39.39–39.41
  chemistry of, 39.6–39.10
  comparisons between, 39.1–39.3
  costs, 39.33–39.41
  device designs, 39.13, 39.20–39.22
  double-layer, 39.4–39.5
  electrolytes, 39.9–39.10
  energy characteristics, 39.2
  energy density, 39.23
  energy storage, 39.3–39.6
  impedance, 39.13–39.17
  material properties, 39.6–39.10
  metal oxides, 39.9
  modeling, 39.13–39.22
  module characteristics, 39.37–39.39
  performance characteristics, for devices, 39.10–39.13
  power characteristics, 39.2
  schematic, 39.4
  system considerations, 39.33–39.41
  testing of, 39.22–39.33
Electrochemical impedance spectroscopy (EIS) methods, 2.25–2.28
  equivalent circuit elements, 2.26–2.27
  Nyquist plot, 2.26–2.28
  sinusoidal wave forms, 2.25–2.26
  Warburg impedance, 2.27
Electrochemical systems, 2.1–2.35
  anions in, 2.12–2.13
  cations in, 2.13
  cells, 1.11
  chronopotentiometry, 2.23–2.25
  concentration polarization in, 2.16–2.18
  cyclic voltammetry, 2.19–2.22
  distribution within, 2.14–2.15
  EIS methods, 2.25–2.28
  electrical double-layer capacity and, 2.10–2.15
  electrocapillary curves, 2.12
  electrode distribution, 2.14–2.15
  in electrodes, 2.2, 2.4–2.10, 2.15–2.19
  electrolytes, 2.2–2.3
  exchange current in, 2.8–2.9, 2.15
  Gouy-Chapman diffuse double layer, 2.13
  interfacial surface tension in, 2.10–2.11
  intermittent titration techniques, 2.28–2.32
  internal impedance in, 2.1
  ionic adsorption in, 2.10–2.15

Electrochemical systems (*Cont.*):
   mass transport within, 2.15–2.19
   mercury-electrolyte interfaces, 2.11
   molecule orientation in, 2.12
   phase diagrams, thermodynamic analyses of, 2.32–2.34
   polarization effects of, 2.1–2.2
   porous, 2.18–2.19
   processes, 2.4–2.10
   reduction-oxidation of, 2.6–2.7
   standard reduction potentials in, 2.4
   Tafel's equation, 2.5–2.6, 2.10
   techniques, 2.19–2.35
   thermodynamics in, 2.3–2.4
   transfer coefficient theory for, 2.7–2.8
Electrode-stack arrangement, 35.22
Electrodes
   concentration gradient diffusion in, 2.15
   concentration polarization in, 2.16–2.18
   convection and stirring within, 2.15
   design, 3.17–3.19
   distribution within, 2.14–2.15
   in electrochemical systems, 2.2, 2.4–2.10, 2.34–2.35
   exchange current in, 2.8–2.9, 2.15
   Fick's second law of diffusion, 2.16
   FNC technology, 19.13–19.14
   geometry of, 2.16
   lead-acid batteries, 16.8
   lithium rechargeable metal batteries, 27.3–27.8
   magnesium water-activated batteries, 34.7
   mass transport within, 2.15–2.19
   metal/air batteries, 33.5
   migration within, 2.15
   nickel-cadmium batteries, 19.3–19.4, 19.13–19.14
   nickel-hydrogen batteries, 24.3–24.5
   nickel-iron batteries, 18.4–18.5
   nickel-metal hydride batteries, 22.3
   nickel-zinc batteries, 23.4–23.5
   porous, 2.18–2.19, 6.15–6.16
   processes, 2.4–2.10
   reduction-oxidation of, 2.6–2.7
   silver oxide batteries, 25.3–25.6
   Tafel's equation and, 2.5–2.6, 2.10
   theoretical capacity of, 1.10–1.11
   transfer coefficient theory for, 2.7–2.8
   zinc, 23.4–23.7, 25.5
Electrolytes, 1.3–1.4, 7.1–7.11
   acid, 7.5–7.6
   alkaline, 7.2–7.4, 33.33–33.43
   aluminum/air batteries, 33.30–33.43
   aqueous, 7.1–7.6
   cells, 1.3–1.4
   ceramic, 7.11
   characteristics, 27.9
   classifications, 27.18
   conductivity ranges, 2.3
   development of, 7.1
   electrochemical capacitors, 39.9–39.10
   in electrochemical systems, 2.2–2.3

Electrolytes (*Cont.*):
   glassy, 7.11
   Gouy-Chapman diffuse double layer and, 2.13
   ionic conductivity, 27.10
   ionic liquids and, 7.9–7.10
   lead-acid batteries, 16.76
   LIPON and, 7.11
   liquid, 27.8, 27.13–27.17
   lithium/air batteries, 33.46
   lithium batteries, 7.1
   lithium-ion batteries, 7.1, 26.30–26.39
   lithium primary batteries, 14.5–14.7
   lithium rechargeable metal batteries, 27.4, 27.8–27.12
   magnesium water-activated batteries, 34.10
   mercuric oxide batteries, 12.3
   in metal/air batteries, 33.4
   neutral, 7.4–7.5, 33.30–33.32
   nickel-iron batteries, 18.5
   nickel-metal hydride batteries, 22.9
   nickel-zinc batteries, 23.11
   nonaqueous, 7.6–7.9
   organic-solvent, 7.6–7.8
   polymer cells, 27.17–27.19
   rechargeable batteries, 7.1
   silver oxide batteries, 25.8
   silver oxide-zinc batteries, 13.8
   sintered-plate batteries, 20.5–20.6
   in solid batteries, 8.7, 14.3
   solid inorganic, 27.11, 27.19–27.22
   SPE, 7.10, 27.11–27.12
   spin-dependent reserve batteries, 35.22–35.23
   sulfuric acid, 7.5–7.6
   thermal batteries, 36.4
   VRLA batteries, 17.1–17.2
   zinc/air batteries, 13.20, 13.31
   in zinc-carbon batteries, 9.12–9.13
   zinc/silver oxide reserve batteries, 34.28
Electrolytic manganese dioxide (EMD), 9.6
   alkaline manganese-dioxide battery components, 11.8–11.9
Electronic energy management, 5.18–5.22
Electronic resistance, in zinc-carbon batteries, 9.27
Electronics Module Package (EMP), 33.38
ELF design. *See* Extended-life flooded design
EMD. *See* Electrolytic manganese dioxide
Emergency Positioning Indicating Radio Beacons (EPIRBs), 14.64
EMP. *See* Electronics Module Package
Empirical modeling, of batteries, 6.5–6.8
   Kirchoff's laws, 6.5–6.6
   lithium-ion batteries, 6.7–6.8
   nickel-metal hydride batteries, 6.6–6.7
End voltage, 3.2
   in zinc-carbon batteries, 9.16
Energy
   density, of batteries, 1.14–1.17, 3.19–3.20
   nickel-cadmium batteries, 19.7
   theoretical, of cells, 1.14
Energy balance, 6.18–6.21

Energy density, 1.14–1.17, 3.19–3.20
   cell components, 1.14
   electrochemical capacitors, 39.23
   limits, 1.16–1.17
   lithium/air system, 1.17
   nickel-cadmium batteries, 19.7
   nickel-hydrogen batteries, 24.8–24.9, 24.13–24.14
   secondary batteries, 15.4, 15.14
   sintered-plate batteries, 20.11–20.12
   specific, 1.15
   storage capability, 1.16
   zinc/air batteries, 13.22–13.23
Energy storage batteries, 30.1–30.45
   applications, 30.5–30.18
   area control, 30.7–30.10
   categories for, 30.7
   cell technology, 30.28
   combined, 30.17
   commodity storage, 30.10
   custom management, 30.14–30.15
   definitions of, 30.7
   demand profiles, 30.2
   design considerations, 30.29–30.31
   development of, 30.5–30.6
   distribution facility deferral, 30.13–30.14
   on electric grid, 30.1–30.3
   electrochemical systems, 30.25
   electrochemistry, 30.25–30.26
   flowing electrolyte, 30.35–30.45
   frequency responsive reserve, 30.7–30.10
   historical perspective for, 30.4–30.5
   legacy structure, 30.4–30.5
   load profiles, 30.2
   NERC requirements, 30.6, 30.8
   power quality, 30.16–30.18
   pumped hydro storage, 30.4
   rapid reserve, 30.6–30.7
   regulated structure, 30.4–30.5
   reliability, 30.16–30.18
   renewable generation, 30.2
   renewable management, 30.15
   sodium-beta high-temperature, 30.23–30.25
   sodium-beta high-temperature batteries, 30.31–30.35
   sodium/metal-chloride electrochemistry, 30.26–30.27
   sodium/nickel-chloride cell technology, 30.28–30.29
   stationary, 30.22–30.35
   stationary applications, 30.22–30.35
   systems, 30.18–30.22
   transmission facility deferral, 30.12–30.13
   transmission system stability, 30.11
   transmission voltage regulation, 30.12
   unregulated environments, 30.5
   VRB, 30.41–30.45
   zinc/bromine, 30.36–30.41
Energy-storage systems, 16.83–16.84
Entropy. *See* Lithium entropy

EPIRBs. *See* Emergency Positioning Indicating Radio Beacons
Equivalent circuit elements, 2.26–2.27
Ethanol, 38.9
EUTELSAT batteries, 24.11
EV. *See* Electric vehicles
Evercel nickel-zinc batteries, 23.2
Evolta batteries, 11.16–11.17
Exchange current, 2.8–2.9, 2.15
Extended-life flooded (ELF) design, 16.2

FAS. *See* Flexible Alkaline Separator
Fast charging, 17.25–17.26
   sealed (portable) nickel-cadmium batteries, 21.26–21.27
FCHEVs. *See* Fuel cell hybrid electric vehicles
Federal Urban Driving Schedule (FUDS), 29.13
Fiber nickel-cadmium (FNC) batteries, 19.5, 19.13–19.19
   charging characteristics, 19.18–19.19
   electrode technology, 19.13–19.14, 19.16
   flexibility manufacturing, 19.14
   maintenance-free, 19.15–19.17
   performance, 19.17–19.19
   plate stacking, 19.16
   sealed vs. vented designs, 19.14–19.17
   short-circuit current, 19.17
   structure, 19.16
Fiber plates, 19.1–19.3
Fick's second law of diffusion, 2.16
   in chronopotentiometry, 2.23
Flat cells, 9.9
Flat or disk-type lithium/thionyl chloride batteries, 14.35–14.38
   characteristics, 14.36–14.38
   design, 14.37–14.38
   discharge, 14.36
   performance, 14.37–14.38
Flat-pack zinc/manganese dioxide P-80 battery, 9.33–9.36
   capacity, 9.36
   construction, 9.34
   parameters, 9.34–9.35
   voltage profiles, 9.35
Flat-pasted plates, 16.47–16.48, 16.54
   in stationary batteries, 16.61, 16.64
Flat-pellet configuration, 12.6
Flexible Alkaline Separator (FAS), 25.28
Float charging, 17.26–17.28
   lead-acid batteries, 16.74–16.75
Flowing electrolyte batteries, 30.35–30.45
   electrochemical system, 30.37–30.38
   VRB, 30.41–30.45
   zinc/bromine, 30.36–30.41
FNC cells. *See* Fiber nickel-cadmium batteries
Foil cell batteries, 14.49–14.50
Forced-flow batteries, 34.4, 34.18–34.20
Free energy, 1.9
Frequency responsive reserve, 30.7–30.10
Fuel cell hybrid electric vehicles (FCHEVs), 29.22

Fuel cells, 1.5–1.6, 1.9, 37.3–37.14. *See also* Small fuel cells; Subkilowatt fuel cells
  biological, 31.35–31.36
  cell reactions, 37.5–37.7
  characteristics, 37.3–37.5
  components, 37.7
  direct systems, 37.3
  indirect systems, 37.3
  operations, 37.5–37.9
  polarization curves, 37.8
  small, 38.1–38.20
  subkilowatt, 37.9–37.14
  subsystems, 37.4
  types, 37.4

Galvanostatic Intermittent Titration Technique (GITT), 2.29–2.31
Garden tools, 23.21
Gas barrier failure, 20.22
Gas screens, 24.5
Gassner, Carl, 9.2
Gelled-electrolyte batteries, 33.16–33.19
GEO satellites, nickel-hydrogen batteries in, 24.14–24.15, 24.21–24.22
  performance data, 24.21–24.22
GITT. *See* Galvanostatic Intermittent Titration Technique
Glassy electrolytes, 7.11
Golden Valley Electric Association, 30.22
Goodenough, John, 26.7
Gouy-Chapman diffuse double layer, 2.13
Graphite, 9.12

Halogen-additive lithium/oxychloride batteries, 14.43–14.47
  BCX, 14.44
  capacity, 14.46
  CSC, 14.44–14.46
Hearing assist devices, 31.8
Heat Resistance Technology (HRL), 26.40
Heat-resistant batteries, 21.27–21.28
HEVs. *See* Hybrid electric vehicles
High-temperature batteries, 21.27
HRL. *See* Heat Resistance Technology
Hybrid electric vehicles (HEVs), 5.17, 29.22–29.46
  alternative storage technologies, 29.46
  battery performance requirements, 29.35–29.37
  commercialization of, 29.23–29.27
  energy and power margins, 29.32
  fuel economy measurements, 29.26
  ICE combination, 29.3
  lead-acid batteries for, 29.39
  lithium-ion batteries for, 29.43–29.45
  micro, 29.29
  mild, 29.33
  new models, 29.27
  nickel-metal hydride battery packs, 22.19–22.20, 29.38–29.42
  nickel-zinc batteries, 23.22–23.23

Hybrid electric vehicles (HEVs) (*Cont.*):
  parallel configurations, 29.23
  performance features, 29.28
  PHEVs, 29.33–29.36
  power-assist, 29.30–29.32
  prismatic batteries for, 29.41–29.43
  purpose of, 29.22–29.23
  sales growth, 29.26
  series configurations, 29.23
  stop-start, 29.28–29.30
  strong, 29.33
  thermal management for, 29.37–29.38
  types, 29.28–29.35
  USABC performance goals, 29.29–29.31, 29.34, 29.36, 29.45
  vehicle integration issues, 29.37–29.45
Hybrid zinc/air manganese dioxide batteries, 33.19–33.20
Hydrogen-electrode batteries, 15.8
Hydrogen electrodes, 24.4–24.5

ICDs. *See* Implantable cardioverter-defibrillators
ICE. *See* Internal combustion engine
IEC. *See* International Electrotechnical Commission standards
Immersion batteries, 34.4–34.5, 34.17–34.18
Impedance
  EIS methods, 2.25–2.28
  electrochemical capacitors, 39.13–39.17
  in electrochemical systems, 2.1
  equivalent circuit elements, 2.26–2.27
  lithium/carbon monofluoride batteries, 14.80
  lithium-ion batteries, 26.48, 26.51
  lithium/iron disulfide batteries, 14.80
  mercuric oxide batteries, 12.10
  Nyquist plot, 2.26–2.28
  sealed (portable) nickel-cadmium batteries, 21.8–21.11
  silver oxide batteries, 25.15
  silver oxide-zinc batteries, 13.6, 13.13
  sinusoidal wave forms, 2.25–2.26
  SLI batteries, 16.43
  Warburg, 2.27
  zinc/air batteries, 13.21, 13.23
  zinc/silver oxide reserve batteries, 34.35
Implantable cardioverter-defibrillators (ICDs), 31.1–31.5
  for resynchronization therapy, 31.4–31.5
Implantable medical devices. *See* Biomedical applications, for batteries
Industrial batteries, 19.22. *See also* Nickel-cadmium batteries
  lead-acid, 16.82
Industrial primary zinc/air batteries, 33.13–33.19
  discharge characteristics, 33.14, 33.16
  gelled-electrolyte, 33.16–33.19
  preactivated, 33.15–33.16
  types, 33.14–33.16
  water-activated, 33.15–33.16
Inorganic-solvent electrolytes, 7.9

INTELSAT batteries, 24.21–24.22
Intercalation mechanism, 6.18
   negative electrode materials, 26.19–26.22
Intercalation process, 26.3–26.5
Intercell connections, 5.10–5.11
Interfacial surface tension, 2.10–2.11
Intermittent discharges, 3.12–3.15
   in zinc-carbon batteries, 9.18–9.20
Intermittent titration techniques, 2.28–2.32
   GITT, 2.29–2.31
   PITT, 2.31–2.32
Internal combustion engine (ICE), 29.3
Internal impedance. *See* Impedance
Internal resistance
   battery size and, 9.29
   electronic, 9.27
   ionic, 9.28–9.29
   lithium/manganese dioxide batteries,
      14.59–14.60
   lithium/sulfur dioxide batteries, 14.20
   nickel-cadmium batteries, 19.10–19.11
   nickel-metal hydride batteries, 22.34–22.35
   temperature and, 9.30
   zinc-carbon batteries, 9.27–9.29
International Electrotechnical Commission (IEC)
      standards, 4.1–4.2
   alkaline manganese-dioxide batteries, 11.14
   discrete battery standards, 5.9
   lithium-ion batteries, 26.2–26.3
   mercuric oxide batteries, 12.1
   nomenclature systems, 4.4–4.7
   for primary batteries, 4.4–4.6, 4.9–4.12
   for rechargeable batteries, 4.6–4.7
International Standards Organization (ISO), 4.1
Ionic adsorption, 2.10–2.15
Ionic liquids, 7.9–7.10
   chemical structure of, 7.10
Ionic resistance, in zinc-carbon cells,
      9.28–9.29
Iron/air batteries, 18.15–18.17
   characteristics, 18.17
   development of, 18.15
   performance, 18.16
   structure, 18.16
Iron electrode batteries, 18.1–18.23
   anodes, 18.21
   capacity, 18.21–18.23
   cathodes, 18.21–18.23
   characteristics, 18.1
   development of, 18.1
   iron/air, 18.15–18.17
   materials, 18.21–18.23
   nickel-iron, 15.7, 18.1–18.14
   silver-iron, 18.17–18.20
   systems, 18.2
Iron electrodes, 25.6
Iron/silver oxide batteries, 25.2
ISO. *See* International Standards Organization

Jackets, in zinc-carbon batteries, 9.15

KERS. *See* Kinetic Energy Recovery System
Kinetic Energy Recovery System (KERS),
      26.63–26.64
Kinetic modeling, of batteries, 6.13–6.15
Kirchoff's laws, 6.5–6.6
Kordesch, Karl, 11.1

Large prismatic construction, 22.13, 22.15
Large prismatic lithium/thionyl chloride batteries,
      14.38–14.40
   characteristics, 14.39
   discharge curves, 14.40
Lawn tools, 23.21
Lead-acid batteries, 3.19, 15.6–15.7, 16.1–16.85
   advantages of, 16.3
   alloy production, 16.20–16.22
   applications, 16.1, 16.6, 16.80–16.85
   assembly materials, 16.33–16.34
   automotive applications, 16.81–16.82
   bipolar design, 16.2
   case formation, 16.35–16.36
   case-to-cover seals, 16.34–16.35
   cell gassing voltage, 16.74
   characteristics, 16.1–16.7
   charging, 16.67–16.75
   chemistry, 16.7–16.17
   construction, 16.17–16.37, 16.46–16.48
   curing, 16.30–16.33
   cycle life, 16.54, 16.78–16.80
   deep-cycle, for traction applications, 16.46–16.55
   development of, 15.4, 16.1–16.2, 16.4–16.6
   diagrams, 16.3
   disadvantages of, 16.3
   discharge reactions, 16.9–16.10
   dry charge, 16.36–16.37
   in electric vehicles, 16.82–16.83
   electrodes, 16.8
   electrolytes, 16.76
   ELF design, 16.2
   energy-storage systems, 16.83–16.84
   EVs, 29.15
   failure modes, 16.80
   finishing, 16.37
   flat-pasted plates, 16.47–16.48, 16.54
   grid production, 16.20, 16.23–16.28
   history of, 16.4–16.6
   industrial, 16.82
   lead oxide production, 16.28
   maintenance features, 16.75–16.77
   manufacturing data, 16.6–16.7
   marine, 16.85
   market growth, 16.6–16.7
   materials, 16.17–16.37
   military applications, 16.48
   models of, 6.16–6.18
   open-circuit voltage characteristics, 16.11–16.12
   operating parameters, 16.78–16.80
   overdischarging, 16.75
   paste production, 16.29
   pasting, 16.30

Lead-acid batteries (*Cont.*):
   performance characteristics, 16.49–16.53
   physical properties, 16.8
   polarization, 16.12–16.13
   in power conditioning systems, 16.84–16.85
   restrictive losses, 16.12–16.13
   safety features, 16.77–16.78
   self-discharge, 16.13–16.14
   separator systems, 16.32–16.34
   shipping, 16.37
   SLI, 3.19, 4.12–4.18, 16.19, 16.37–16.46
   small-sealed, 16.82
   solar photovoltaic systems, 16.84
   specific gravity, 16.11
   standardization of, 4.12–4.18
   stationary, 16.56–16.67
   sulfuric acid, 16.14–16.17
   tank formation, 16.35
   temperature and, 16.77
   testing, 16.37
   tubular positive plates, 16.47–16.48, 16.55
   types, 16.4, 16.53–16.55
   in uninterrupted power supply systems, 16.84–16.85
   usage, 16.6–16.7
   VRLA, 17.1–17.39
   weight analysis, 16.18
Lead oxide production, 16.28
Leakage current, 34.9–34.10
Leclanché, Georges-Lionel, 9.1
Leclanché batteries, 7.4–7.5, 8.3, 9.1–9.40
   cylindrical configuration, 9.6–9.7
   discharge grades, 9.24–9.26
   general purpose, 9.5
   industrial, 9.5
   seals in, 9.14
Left ventricular assist devices (LVADs), 31.5–31.6
LEO satellites, nickel-hydrogen batteries in, 24.15–24.16, 24.22
   performance data, 24.22
LEVs. *See* Light electric vehicles
Light electric vehicles (LEVs), 23.21
LIPON, in batteries, 7.11
Liquefied petroleum gases (LPG), 38.9
Liquid electrolytes, 27.8, 27.13–27.17
Lithium, 14.4–14.5
   properties of, 14.4
Lithium/air batteries, 1.17, 14.88, 33.1–33.2, 33.44–33.55
   anodes, 33.45–33.46
   background, 33.44–33.45
   cathodes, 33.46–33.47
   cell design, 33.47–33.52
   construction, 33.49–33.50
   cycling data, 33.49
   electrolytes, 33.46
   performance, 33.47–33.50
   separators, 33.46
   undersea applications, 33.52–33.55

Lithium anode reserve batteries, 35.1–35.19
   ampoule-type, 35.14–35.17
   applications, 35.19
   characteristics of, 35.1–35.2, 35.4
   chemistry of, 35.2–35.4
   construction, 35.4–35.14
   lithium/precharged, 35.4
   lithium/sulfur dioxide, 35.3
   lithium/thionyl chloride, 35.3
   lithium/vanadium pentoxide, 35.2
   multicell design, 35.17–35.19
   performance characteristics, 35.14–35.19
   specifications, 35.14
   types, 35.5–35.14
Lithium batteries, 8.7
   electrolytes, 7.1
   organic solvent electrolytes, 7.6–7.7
Lithium/carbon monofluoride batteries, 14.64–14.75
   applications and handling, 14.73
   battery types, 14.69–14.72
   cell types, 14.69–14.72
   chemistry, 14.64–14.65
   coin-type, 14.65–14.67
   construction, 14.65
   cylindrical cells, 14.67–14.68
   discharge curves, 14.66, 14.72
   military uses, 14.71
   mixtures in, 14.73–14.74
   performance, 14.65–14.69
   pin-type, 14.65
   semi-ionic carbon fluoride materials, 14.74
   shelf life, 14.68–14.69
   subfluorinated materials, 14.74
   technology advances, 14.73–14.75
Lithium cells, 14.1–14.2
Lithium/CFx batteries, 31.20–31.22, 31.28–31.30
   capacity of, 31.21
   properties, 31.21
Lithium/cobalt disulfide batteries, 36.9
Lithium/copper oxide batteries, 14.14, 14.81–14.88
   applications, 14.87–14.88
   button, 14.82–14.84
   capacity, 14.85
   cell types, 14.87–14.88
   characteristics, 14.88
   chemistry, 14.81–14.82
   construction, 14.82
   cylindrical bobbin, 14.82–14.83
   discharge, 14.84–14.85
   high-temperature, 14.87
   performance, 14.82–14.87
   spirally wound, 14.87
   storage factors, 14.86
Lithium entropy, 2.33
Lithium/iodine batteries, 31.15–31.18
   thermodynamic functions, 31.16
Lithium-ion batteries, 15.8–15.9, 26.1–26.75
   additives in, 7.8
   advantages of, 26.2
   aerospace applications, 26.57

Lithium-ion batteries (*Cont.*):
  biomedical applications, 31.30–31.34
    cycle life of, 31.33
  capacity, 26.58, 26.65–26.66
  cell sizes, 26.4
  cell voltage, 5.17
  certification standards, 26.73–26.74
  characteristics, 26.1–26.3, 26.47–26.51
  chemistry, 26.3–26.41
  commercial, 26.50–26.68
  construction, 26.41–26.47
  for consumer electronics, 32.9
  cycle life, 26.64
  designations, 26.2–26.3
  development of, 26.1
  disadvantages of, 26.2
  discharging, 26.52–26.53, 26.59–26.60, 31.32
  electrochemical process, 26.5
  electrolytes, 7.1, 26.30–26.39
  empirical modeling of, 6.7–6.8
  for EVs, 29.16–29.20
  future trends, 26.75
  for HEVs, 29.43–29.45
  HRL technology, 26.40
  impedance in, 26.48, 26.51
  integration in, 26.48, 26.51
  intercalation process, 26.3–26.5
  KERS, 26.63–26.64
  markings, 26.2–26.3
  negative electrode materials, 26.17–26.30
  NMC cells, 26.54–26.57
  nonaqueous electrolytes, 26.30–26.37
  organic solvent electrolytes, 7.6, 7.8
  performance, 26.47–26.68
  polymer, 26.44–26.47
  positive electrode materials, 26.5–26.17
  power cells, 26.54–26.55
  prismatic, 26.43
  regulatory criteria, 26.70–26.75
  safety properties, 26.68–26.75
  salts, 26.31
  in satellites, 26.57
  separator materials, 26.39–26.41
  short-circuit protection, 5.18
  solvents, 26.31–26.33
  Sony Nexelion Cell, 26.67–26.68
  stacked, 26.42–26.44
  temperature control, 5.18
  temperature dependence, 26.68–26.70
  tests, 26.71
  by type of battery, 26.61–26.68
  wound cells, 26.41–26.42
Lithium/iron disulfide batteries, 14.75–14.81, 36.7–36.8
  applications, 14.81
  capacity comparisons, 14.76
  cell types, 14.81
  characteristics, 14.81
  chemistry, 14.75–14.76
  construction, 14.77–14.78

Lithium/iron disulfide batteries (*Cont.*):
  current-limiting devices, 14.80
  discharge curves, 14.78
  impedance, 14.80
  performance, 14.78–14.81
  PTC, 14.77
  storage factors, 14.80–14.81
  temperature, 14.78–14.80
  voltage, 14.78
Lithium/manganese dioxide batteries, 14.12–14.13, 14.27, 14.47–14.64
  applications and handling, 14.62–14.64
  bobbin-type cylindrical, 14.49, 14.52–14.53
  capacity, 14.63
  cell size, 14.62
  chemistry, 14.48
  coin cells, 14.48, 14.54
  construction, 14.48–14.50
  foil cell, 14.49–14.50
  internal resistance, 14.59–14.60
  military applications, 14.63
  multicell 9 V, 14.49, 14.59–14.60
  performance, 14.50–14.63
  service life, 14.59, 14.61
  shelf life, 14.59
  size, 14.62
  spirally wound cylindrical, 14.49, 14.53–14.59
  storage characteristics, 14.62
Lithium/manganese oxide batteries, 31.25–31.28
  discharging for, 31.26–31.27
Lithium/oxychloride batteries, 14.42–14.47
  halogen-additive, 14.43–14.47
Lithium/precharged batteries, 35.4
Lithium primary batteries, 5.6–5.7, 14.1–14.88
  advantages, 14.1–14.2
  anode materials, 14.4
  cathode materials, 14.5–14.6
  cell couples, 14.7–14.8
  characteristics, 14.1–14.3, 14.8–14.14
  charging, 14.16
  chemistry of, 14.4–14.8
  classification of, 14.2–14.3
  design, 14.8
  development of, 14.1
  electrolytes, 14.5–14.7
  forced discharge, 14.16
  high-rate discharges, 14.15–14.16
  incineration, 14.16
  lithium, 14.4–14.5
  overheating, 14.16
  reaction mechanisms, 14.7–14.8
  safety and handling of, 14.14–14.16
  short-circuits, 14.15–14.16
  solid-cathode cells, 14.2–14.3, 14.12–14.14
  solid-electrolyte cells, 14.3
  soluble-cathode cells, 14.2, 14.8–14.11
  voltage reversal, 14.16
Lithium rechargeable (metal) batteries, 27.1–27.24
  Bolloré LMP, 27.19
  capacity, 27.23

I.14 INDEX

Lithium rechargeable (metal) batteries (*Cont.*):
  characteristics, 27.13–27.22
  chemistry of, 27.3–27.12
  commercial sources for, 27.1
  commercialization of, 27.21
  construction, 27.14
  cycling efficiency, 27.5
  dendrite formation, 27.4
  development history for, 27.2–27.3
  discharge voltage, 27.14–27.15, 27.24
  electrochemical systems, 27.13
  electrolytes, 27.4, 27.8–27.12
  features, 27.1–27.3
  negative electrodes, 27.3–27.6
  polymeric cathodes, 27.8
  positive electrodes, 27.6–27.8
  sulfur cathodes, 27.6–27.7
  types, 27.22–27.23
Lithium/silver vanadium batteries, 14.88
Lithium/silver vanadium oxide batteries, 31.22–31.25, 31.28–31.30
  construction of, 31.24–31.25
  discharging for, 31.24
Lithium/sulfur dioxide batteries, 14.8–14.10, 14.16–14.26, 35.3
  applications, 14.26
  capacity, 14.23
  cell types, 14.24
  chemistry, 14.16–14.18
  construction, 14.18
  current collectors, 14.18
  cylindrical cells, 14.25
  design, 14.25
  discharge, 14.20
  internal resistance, 14.20
  performance, 14.19–14.24
  safety considerations, 14.24–14.26
  service life, 14.20–14.23
  shelf life, 14.23
  sizes, 14.24
  temperature effects, 14.17, 14.20
  voltage, 14.19–14.20, 14.22, 14.24
  water in, 14.17
Lithium/sulfuryl chloride batteries, 14.11, 14.43
Lithium/thionyl chloride batteries, 14.8–14.10, 14.26–14.42, 31.18–31.20, 35.3
  applications, 14.40–14.42
  bobbin-type, 14.28–14.34
  capacity, 14.32–14.33, 14.42
  chemistry, 14.28
  commercial uses of, 14.26
  development of, 14.40–14.41
  flat or disk-type, 14.35–14.38
  large prismatic, 14.38–14.40
  military uses, 14.27
  service life, 14.32
  shelf life, 14.31
  spirally wound cylindrical batteries, 14.34–14.35, 14.34–14.36

Lithium titanate, 26.25–26.27
  charge capacities, 26.27
  properties, 26.26
Lithium/vanadium pentoxide batteries, 14.13, 35.2
Lithium/water batteries, 14.88
Low-current-drain structures, 12.7
LPG. *See* Liquefied petroleum gases
LR190DC batteries, 25.29
LVADs. *See* Left ventricular assist devices

Magnesium/air batteries, 10.11, 33.43–33.44
  discharge profile for, 33.44
  schematic for, 33.44
Magnesium batteries, 8.7, 10.1–10.12
  advantages, 10.1–10.2
  anodes, 10.11–10.12
  batter design, 10.9–10.11
  chemistry of, 10.2–10.4
  construction, 10.4–10.6
  disadvantages, 10.1–10.2
  discharge, 10.6–10.8
  inside-out cells, 10.5, 10.9
  military use of, 10.1
  negative difference effect, 10.2
  performance characteristics, 10.6–10.11
  service life, 10.6–10.7, 10.9
  shelf life, 10.8
  sizes, 10.11
  standard, 10.4–10.5
  temperature, 10.3, 10.8
  types, 10.11–10.12
  voltage delay for, 10.3
Magnesium/cuprous chloride batteries, 34.20
Magnesium/manganese dioxide batteries, 34.20–34.22
Magnesium rechargeable batteries, 10.11
Magnesium/silver chloride batteries, 34.24
Magnesium water-activated batteries, 34.1–34.25
  applications, 34.22–34.24
  basic, 34.6
  cathodes, 34.8
  characteristics, 34.1–34.2
  chemistry for, 34.2–34.3
  components, 34.7–34.9
  composition range, 34.8
  construction, 34.6–34.10
  development of, 34.1
  discharge profiles, 34.11–34.15
  dunk-types, 34.4, 34.20–34.22
  electrode assemblies, 34.7
  electrolytes, 34.10
  forced-flow, 34.4, 34.18–34.20
  immersion, 34.3–34.5
  immersion types, 34.4–34.5, 34.17–34.18
  leakage current, 34.9–34.10
  performance characteristics, 34.10–34.22
  service life, 34.16
  silver vs. nonsilver cathode, 34.2–34.3
  sizes, 34.24–34.25
  types of, 34.4–34.5, 34.24–34.25
  voltage, 34.11

Manchester plates, 16.27
Manchex plates, 16.63
Manganese dioxide (MnO_2) batteries, 9.11–9.12
  batter design, 10.9–10.11
  construction of, 10.4–10.5
  discharge, 10.6–10.8
  inside-out cells, 10.9
  performance characteristics, 10.6–10.11
  service life, 10.6–10.7, 10.9
  shelf life, 10.8
  sizes, 10.11
  temperature, 10.8
Marine batteries, 16.85
Markings, in batteries, 4.9
Material safety data sheets (MSDS), 14.15
Mathematical modeling, of batteries, 6.1–6.27
  circuit analog approach, 6.3
  degradation models, 6.21–6.25
  determination of, 6.25
  development of, 6.4–6.5
  empirical, 6.5–6.8
  energy balance in, 6.18–6.21
  evolution of, 6.1–6.4
  intercalation in, 6.18
  kinetic, 6.13–6.15
  for lead-acid batteries, 6.16–6.18
  mechanistic, 6.9–6.13
  Peukert's equation, 6.1–6.3
  porous electrodes, 6.15–6.16
  SOC, 6.2
  symbols in, 6.25–6.27
Mechanistic models, of batteries, 6.9–6.13
  change transfers, 6.11–6.12
  charge transport, 6.9–6.11
  degradation models and, 6.24
  ion distributions, 6.13
Memory-backup batteries, 21.28–21.29
Memory effect
  in nickel-cadmium batteries, 19.12, 20.21–20.22, 21.17
  in nickel-metal hydride batteries, 22.37
Mercuric oxide batteries, 12.1–12.14. *See also* Cadmium/mercuric oxide batteries
  ANSI standards, 12.1
  applications, 12.1
  button configuration, 12.5
  cadmium anodes, 12.4
  cell components, 12.3–12.5, 12.4–12.5
  chemistry for, 12.2
  construction, 12.4–12.7
  cylindrical configuration, 12.6–12.7
  discharge, 12.8–12.9
  electrolytes, 12.3
  expanding markets for, 12.1
  flat-pellet configuration, 12.6
  general characteristics, 12.1–12.2
  IEC standards, 12.1
  impedance, 12.10
  low-current-drain structures, 12.7
  performance characteristics, 12.7–12.12

Mercuric oxide batteries (*Cont.*):
  service life, 12.11–12.12
  storage, 12.11
  temperature and, 12.9
  voltage, 12.7–12.8
  wound-anode configuration, 12.6–12.7
  zinc anodes, 12.3–12.4
Metal/air batteries, 33.1–33.55. *See also* Aluminum/air batteries; Iron/air batteries; Lithium/air batteries; Magnesium/air batteries; Zinc/air batteries
  advantages of, 33.2
  air electrodes, 33.5
  aluminum/air, 33.2–33.3, 33.27–33.43
  characteristics of, 33.1–33.3
  charging of, 33.5
  chemistry of, 33.4–33.5
  disadvantages of, 33.2
  efficiency of, 33.5
  electrolyte carbonation, 33.4
  iron/air, 18.15–18.17
  lithium/air, 1.17, 14.88, 33.1–33.2, 33.44–33.55
  magnesium/air, 10.11, 33.43–33.44
  polarization of, 33.4
  types, 33.3
  water transpiration in, 33.4
  zinc/air, 8.3, 8.6–8.7, 13.16–13.36, 33.1, 33.6–33.27
Metal hydride alloys, 22.3–22.5
Metal oxide batteries, 13.18
Metalakatla Power and Light Battery System, 30.21
Methanol, 38.9
Micro hybrid electric vehicles, 29.29
MIDEX 23 batteries, 24.12
Midpoint voltage, 3.2
MIL standards. *See* Military standards
Mild hybrid electric vehicles, 29.33
  USABC performance goals, 29.34
Military (reserve) batteries, 35.1–35.30
  ambient-temperature lithium anode, 35.1–35.19
  lithium anode reserve, 35.1–35.19
  spin-dependent, 35.20–35.30
Military (MIL) standards, 4.3
Mirowski, Michel, 31.2
MnO_2. *See* Manganese dioxide
Modeling. *See* Mathematical modeling, of batteries
Monoblock construction, 22.13–22.15
Monopolar VRLA batteries, 17.7
Moonlight Project, 30.30
Moore's Law, 1.16
MSDS. *See* Material safety data sheets
Multicell 9 volt batteries, 14.49, 14.59–14.60
  discharge characteristics, 14.59, 14.60

National Electricity Reliability Council (NERC), 30.6, 30.8
Natural gas, 38.9
Natural manganese dioxide (NMD), 9.11
Negative Delta V control system, 21.25
Negative difference effect, 10.2

Negative electrode materials, 26.17–26.30
  carbon types, 26.18–26.19
  cycle life, 26.28
  diffraction patterns, 26.21
  heat evolution, 26.25
  historical overview, 26.17–26.18
  intercalation staging, 26.19–26.22
  lithium titanate, 26.25–26.27
  properties of, 26.22–26.25
  silicon, 26.27–26.30
  tin, 26.27–26.30
NERC. See National Electricity Reliability Council
Neurostimulators, 31.6–31.7
Neutral electrolytes, 7.4–7.5
  Leclanché, 7.4–7.5
  zinc-bromine rechargeable cells, 7.5
  zinc-carbon batteries, 7.4–7.5
Nickel-cadmium batteries, 3.4, 4.6–4.7, 15.7–15.9, 19.1–19.22. See also Sintered-plate batteries
  applications, 19.21–19.22
  briquettes, 19.3
  capacity, 19.8, 19.10
  charge retention, 19.11
  charging characteristics, 1.8–1.9, 19.12–19.13, 19.18–19.19, 21.19–21.25
  chemistry, 19.3
  construction, 19.3–19.6
  design, 19.3
  development of, 19.1
  discharge properties, 19.7–19.10
  electrodes, 19.3–19.4, 19.13–19.14
  energy density, 19.7
  fiber plates, 19.1–19.3
  flexibility manufacturing, 19.14
  FNC cells, 19.5
  FNC technology, 19.5, 19.13–19.19
  internal resistance, 19.10–19.11
  life cycle, 19.11
  manufacturers and market segments, 19.19–19.21
  mechanical stability, 19.11–19.12
  memory effects, 19.12
  performance, 19.17–19.19
  performance characteristics, 19.7–19.12
  plate stacking, 19.16
  pocket-plate, 19.1, 19.4
  sealed (portable), 21.1–21.33
  sealed vs. vented designs, 19.14–19.17
  short-circuit current, 19.17
  sintered-plate, 19.1–19.2, 20.1–20.27
  SNC technology, 19.13
  specific energy, 19.7
  structure, 19.2
  thermal stability, 19.11–19.12
Nickel-hydrogen batteries, 24.1–24.27
  advanced, 24.23–24.27
  advantages of, 24.2
  applications, 24.1, 24.14–24.17
  bipolar, 24.26–24.27
  capacity, 24.19–24.21
  cell components, 24.3–24.5

Nickel-hydrogen batteries (Cont.):
  cell construction, 24.5–24.14
  cell reversal, 24.2–24.3
  chemistry of, 24.1–24.3
  COMSAT, 24.6–24.7, 24.25
  concepts, 24.23, 24.25–24.26
  configurations, 24.10
  CPV cells, 24.23–24.26
  design, 24.10–24.14, 24.23–24.27
  disadvantages of, 24.2
  DMPS 100, 24.10
  electrode components, 24.3–24.5
  energy density, 24.8–24.9, 24.13–24.14
  EUTELSAT, 24.11
  features, 24.1, 24.13–24.14
  in GEO satellites, 24.14–24.15, 24.21–24.22
  INTELSAT, 24.21–24.22
  IPV cells, 24.23
  IRIDIUM, 24.27
  in LEO satellites, 24.15–24.16, 24.22
  MIDEX 23, 24.12
  operations of, 24.2
  overcharging, 24.2
  performance, 24.17–24.22
  self-discharge, 24.3, 24.19–24.20
  terrestrial, 24.16–24.17
  TRW 81, 24.12
Nickel hydroxide, 22.5–22.9
  charge characteristics, 22.8
  high-temperature, 22.7–22.8
  pasted, 22.7
  positive substrates, 22.6
  sintered, 22.7–22.9
  spherical, 22.6–22.7
Nickel-iron batteries, 15.7, 18.1–18.14
  active materials, 18.1–18.2
  advanced, 18.12–18.14
  capacity, 18.6–18.7, 18.13
  cell assembly, 18.5
  charging, 18.9–18.10
  chemistry, 18.1–18.4
  construction, 18.4–18.5
  conventional, 18.4–18.13
  discharge, 18.7
  electrolytes, 18.2–18.3, 18.5
  hours of service, 18.8
  life-cycles, 18.3, 18.9
  negative electrodes, 18.4
  performance characteristics, 18.5–18.10
  positive electrodes, 18.4–18.5
  regulators, 18.10
  schematic, 18.12
  self-discharge, 18.9
  size of, 18.10
  special handling for, 18.10–18.11
  temperature effects, 18.7–18.8
  voltage, 18.13
Nickel-metal hydride batteries, 22.1–22.50
  acrylic acid in, 22.10
  advantages of, 22.2

Nickel-metal hydride batteries (*Cont.*):
   button configurations, 22.11
   capacity, 22.28–22.30
   cell construction types, 22.10–22.15
   cell design, 22.15–22.17
   characteristics, 22.22–22.25
   charging methods, 22.40–22.47
   chemical reactions, 22.2–22.3
   chemistry, 22.2–22.10
   constant power characteristics, 22.37–22.39
   consumer, 22.21–22.22
   cost reduction for, 22.48–22.49
   coulombic/voltaic efficiency, 22.34–22.35
   cycle life, 22.32–22.34
   cylindrical configurations, 22.10–22.11
   cylindrical vs. prismatic, 22.15
   development of, 22.1
   dimensions, 4.11
   disadvantages of, 22.2
   discharging, 22.22–22.40
   electrical isolation, 22.47–22.48
   electrodes, 22.3
   electrolytes, 22.9
   empirical models, 6.6–6.7
   energy vs. power tradeoffs, 22.16
   EV battery packs, 22.17–22.19
   for EVs, 29.15–29.16
   fuel cell startup, 22.21
   for HEVs, 22.19–22.20, 29.38–29.39
   hybrid battery packs, 22.19–22.21
   internal resistance, 22.34–22.35
   large prismatic configurations, 22.13
   memory effects, 22.37
   metal hydride alloys, 22.3–22.5
   metal vs. plastic, 22.15–22.16
   modules, 22.17
   monoblock, 22.13–22.15
   next generation, 22.48–22.50
   nickel hydroxide, 22.5–22.9
   9 V multicell configurations, 22.12–22.13
   pack design, 22.17
   polarity reversal, 22.36–22.37
   power assist, 22.21
   precharged, 22.21–22.22
   rechargeable, 4.7
   retention, 22.30–22.32
   separators, 22.9–22.10
   service life, 22.30
   shelf life, 22.34
   small prismatic configurations, 22.12
   specific energy, 22.25–22.26
   specific power, 22.27–22.28
   stationary power in, 22.50
   sulfonation in, 22.10
   thermal management, 22.17
   types of, 22.37
   ultra-high-power designs, 22.49
Nickel-zinc batteries, 15.8, 23.1–23.24
   AA cells, 23.23–23.24
   applications, 23.20–23.24

Nickel-zinc batteries (*Cont.*):
   capacity, 23.12–23.15
   cell construction, 23.7–23.11
   charging, 23.18–23.20
   chemistry, 23.2–23.7
   critical components, 23.5
   discharging, 23.15–23.17
   electrodes, 23.4–23.7
   electrolyte considerations, 23.11
   environmental aspects, 23.24
   Evercel, 23.2
   failure mechanisms, 23.20
   HEVs, 23.22–23.23
   history of, 23.1–23.2
   lawn and garden tools, 23.21
   LEVs, 23.21
   nickel electrodes, 23.9–23.10
   performance characteristics, 23.11–23.20
   power tools, 23.20–23.21
   prismatic, 23.7–23.8
   safety, 23.18
   sealed cylindrical, 23.8–23.9
   separator systems, 23.6–23.7, 23.11
   storage, 23.17–23.18
   temperature, 23.13
9 V multicell batteries, 22.12–22.13
NMC cells, 26.54–26.57
NMD. *See* Natural manganese dioxide
Nominal voltage, 3.2
Nonaqueous electrolytes, 7.6–7.9
   inorganic-solvent, 7.9
   organic-solvent, 7.6–7.8
Nyquist plot, 2.26–2.27, 2.26–2.28

OCV. *See* Open-circuit voltage
Open cells, 36.10–36.11
Open-circuit voltage (OCV), 3.2
   alkaline manganese-dioxide batteries, 11.6
   lead-acid batteries, 16.11–16.12
   silver oxide-zinc batteries, 13.11
   zinc/air batteries, 13.21
   in zinc-carbon batteries, 9.15
Organic-solvent electrolytes, 7.6–7.8
   lithium batteries, 7.6–7.7
   lithium-ion batteries, 7.6, 7.8
   salt in, 7.8
Overcharging, in nickel-hydrogen batteries, 24.2
Overdischarging, 16.75
Overflow charging, 28.14–28.15
Oxyride batteries, 11.16–11.17

Parallel/series charging, 17.30–17.31
Paste production, 16.29
Pasting, 16.30
Peak current, 2.20–2.21
Pellet cells, 36.11–36.13
Performance
   alkaline manganese-dioxide batteries, 11.2–11.3
   of batteries, factors affecting, 3.1–3.22
   deep-cycle batteries, 16.49–16.53

Performance (*Cont.*):
  electrochemical capacitor devices, 39.10–39.13
  flat or disk-type lithium/thionyl chloride batteries, 14.37–14.38
  FNC batteries, 19.17–19.19
  HEVs, 29.28
  iron/air batteries, 18.16
  lithium/air batteries, 33.47–33.50
  lithium anode reserve batteries, 35.14–35.19
  lithium/carbon monofluoride batteries, 14.65–14.69, 14.80
  lithium/copper oxide batteries, 14.82–14.87
  lithium-ion batteries, 26.47–26.68
  lithium/iron disulfide batteries, 14.78–14.81
  lithium/manganese dioxide batteries, 14.50–14.63
  lithium/sulfur dioxide batteries, 14.19–14.24
  magnesium batteries, 10.6–10.11
  magnesium water-activated batteries, 34.10–34.22
  mercuric oxide batteries, 12.7–12.12
  MnO_2 batteries, 10.6–10.11
  nickel-cadmium batteries, 19.7–19.12
  nickel-hydrogen batteries, 24.17–24.22
  nickel-iron batteries, 18.5–18.10
  nickel-zinc batteries, 23.11–23.20
  OCV, 13.11
  of primary batteries, 8.3, 8.8–8.18
  sealed (portable) nickel-cadmium batteries, 21.6–21.19
  secondary batteries, 15.4
  silver oxide batteries, 25.8–25.20
  silver oxide-zinc batteries, 13.5, 13.11–13.14
  sintered-plate batteries, 20.6–20.15
  SLI batteries, 16.40–16.46
  small fuel cells, 38.13–38.20
  spin-dependent reserve batteries, 35.25–35.30
  standardization of, 4.8–4.9
  stationary batteries, 16.59–16.64
  thermal batteries, 36.15–36.17
  VRLA batteries, 17.8–17.22
  zinc/air batteries, 13.21–13.36
  zinc/alkaline/manganese dioxide, 28.2–28.12
  zinc/bromine batteries, 30.38–30.40
  zinc-carbon batteries, 9.15–9.33
Peukert's equation, 6.1–6.3
Phase diagrams, thermodynamic analyses of, 2.32–2.34
  Gibbs energies in, 2.33
  lithium entropy in, 2.33
PHEVs. *See* Plug-in hybrid electric vehicles
Pin-type lithium/carbon monofluoride batteries, 14.65
PITT. *See* Potentiostatic Intermittent Titration Technique
Planté, Raymond Gaston, 16.4
Planté plates, 16.27
  in stationary batteries, 16.67
Plate-pack cell assembly, 20.5
Plug-in hybrid electric vehicles (PHEVs), 29.33–29.35
  control design strategies, 29.35
  operational modes, 29.34
  USABC performance goals, 29.36

Pocket-plate batteries, 19.1, 19.4
  composition, 19.4
Pocket-plate nickel-cadmium batteries, 15.4
Polarity reversal
  nickel-metal hydride batteries, 22.36–22.37
  sealed portable nickel-cadmium batteries, 21.12–21.13
Polarization
  in electrochemical systems, 2.1–2.2
  fuel cells, 37.8
  in lead-acid batteries, 16.12–16.13
Polymer lithium-ion batteries, 26.44–26.47
Polymeric cathodes, 27.8
Porche, Ferdinand, 29.23
Porous electrodes, 2.18–2.19, 6.15–6.16
  intercalation in, 6.18
Portable primary zinc/air batteries, 33.7–33.13
  applications, 33.10–33.11
  cell stacks, 33.12
  design for, 33.7, 33.12
  discharge profiles, 33.13
  prismatic, 33.8–33.11
  specifications for, 33.9
Positive electrode materials, 26.5–26.17
  capacity, 26.13–26.14, 26.17
  characteristics, 26.6–26.8
  charging, 26.16
  discharging, 26.16
  electrochemical properties, 26.12–26.17
  lattice parameter correlations, 26.15
  in lithium-ion batteries, 26.5–26.17
  magnifications of, 26.9–26.11
  physical properties, 26.8–26.12
  requirements, 26.6
  structure of, 26.6–26.8
Positive thermal coefficient (PTC) device, 5.15–5.16, 14.77
Potentiostatic Intermittent Titration Technique (PITT), 2.31–2.32
  chemical diffusion coefficient in, 2.32
  single step constant pulse for, 2.31
Power-assist hybrid electric vehicles, 29.30–29.32
  energy and power margins, 29.32
  USABC performance goals, 29.30–29.31
Power conditioning systems, 16.84–16.85
Power density, in sintered-plate batteries, 20.11–20.12
Power tools, 23.20–23.21
PQ2000 Installation, 30.21–30.22
Preactivated batteries, 33.15–33.16
Precipitator systems, filters, 33.42
Primary batteries, 1.4–1.5, 8.3–8.18
  alkaline manganese-dioxide battery, 7.2, 11.1–11.17
  aluminum, 8.7, 10.12
  ANSI standards for, 4.4–4.6, 4.10
  applications, 8.5, 32.3–32.6
  biomedical applications for, safety considerations, 31.10–31.11
  cadmium/mercuric oxide batteries, 8.3, 8.6

Primary batteries (*Cont.*):
   characteristics, 32.4–32.6
   characteristics of, 8.4–8.7, 8.10–8.11
   comparison of, 8.8, 8.12–8.13
   for consumer electronics, 32.3–32.6
   cost of, 8.17
   cylindrical-type, 8.14
   design for, 5.2–5.3, 5.6
   development of, 8.3
   dimensions, 4.11
   discharge profiles, 8.8–8.9, 8.12–8.13, 8.13, 8.15
   duty cycles for, 8.13
   energy density, 8.17
   IEC standards, 4.4–4.6, 4.9–4.12
   lithium, 5.6–5.7, 8.7
   magnesium batteries, 8.7
   performance, 8.3, 8.8–8.18
   recharging of, 8.18
   shelf life of, 8.17–8.18
   solid electrolyte batteries, 8.7
   specific energy, 8.9, 8.12, 8.16
   specific power, 8.9, 8.12
   temperature and, 8.13, 8.15
   temperature effects, 32.6
   types, 8.4–8.7, 32.3
   voltage profiles, 8.8
   zinc/air, 8.3, 8.6–8.7
   zinc/alkaline manganese dioxide, 8.3, 8.6
   zinc-carbon, 7.4–7.5, 8.3, 8.6
   zinc/mercuric oxide, 8.3, 8.6
   zinc/silver oxide, 8.6
Prismatic batteries
   aerospace applications, 26.43–26.44
   applications, 17.37
   cell construction, 17.5–17.6
   for consumer electronics, 32.11
   for HEVs, 29.41–29.43
   lithium-ion batteries, 26.43
   nickel-zinc, 23.7–23.8
   performance characteristics, 17.17–17.19
   VRLA, 17.5–17.6, 17.17–17.19, 17.37
Protection circuitry examination, H.28–H.30
PTC device. *See* Positive thermal coefficient device
Puerto Rico Electric Power Authority Battery System, 30.20–30.21
Pulse charging, 16.74, 28.13–28.14
Pulse load performance, zinc/air batteries, 13.27–13.28
Pumped hydro storage, 30.4

R20 batteries, 4.8
Rapid charging, 16.75
Rapid reserve, 30.6–30.7
Rechargeable batteries, 1.5
   AA, 28.4
   advantages, 28.1–28.2
   alkaline electrolytes, 7.4
   ANSI standards for, 4.6–4.7
   biomedical applications, safety considerations for, 31.11

Rechargeable batteries (*Cont.*):
   characteristics, 28.1
   charge control, 5.15–5.17
   charging, 28.12–28.15
   chemistry, 28.1–28.3
   construction, 28.3–28.4
   cycling, 28.6–28.8
   design for, 5.14–5.18
   disadvantages, 28.1–28.2
   discharge control, 5.16–5.17
   electrolytes, 7.1
   for EVs, 29.9
   first cycle discharge, 28.4–28.5
   IEC standards for, 4.6–4.7
   lithium metal, 27.1–27.24
   magnesium, 10.11
   nickel-metal hydride, 4.7
   performance, 28.2–28.12
   primary batteries, 8.18
   shelf life, 28.10–28.12
   silver oxide, 15.7–15.8, 25.1–25.29
   temperature effects, 28.10
   types, 28.15–28.17
   voltage charging, 3.17
   zinc/air batteries, 33.20–33.27
   zinc/alkaline/manganese dioxide, 28.1–28.17
Rectangular batteries, 21.4–21.6, 21.30
Regenerative braking, 22.46–22.47
Regulators, in nickel-iron batteries, 18.10
Renewable energy management, 30.15
Reserve batteries, 1.5
   magnesium water-activated, 34.1–34.25
   military applications, 35.1–35.30
   zinc/silver oxide, 34.26–34.40
Resistors, design influences on, 3.17

Safety issues
   in battery design, 5.1–5.7
   in biomedical applications, with batteries, 31.9–31.12
   consumer electronics batteries, 32.16
   for diodes, 5.3
   lead-acid batteries, 16.77–16.78
   lithium-ion batteries, 26.68–26.75
   lithium primary batteries, 14.14–14.16
   lithium/sulfur dioxide batteries, 14.24–14.26
   nickel-zinc batteries, 23.18
   primary batteries, 5.6, 14.14–14.16
   spin-dependent reserve batteries, 35.24–35.25
   standardization of, for batteries, 4.19–4.20
   VRLA batteries, 17.32–17.33
Satellites, lithium-ion batteries, 26.57
SBS. *See* Smart Battery System
Sealed (portable) nickel-cadmium batteries, 21.1–21.33
   advantages, 21.1–21.2
   applications, 21.1
   button cell, 21.4–21.5
   capacity, 21.7, 21.15
   charging characteristics, 21.19–21.25

Sealed (portable) nickel-cadmium batteries (*Cont.*):
  chemistry of, 21.1–21.3
  construction, 21.3–21.6
  cycle life, 21.16
  cylindrical, 21.3–21.4
  disadvantages, 21.1–21.2
  discharge characteristics, 21.6–21.7, 21.9–21.10, 21.13–21.14
  failure mechanisms, 21.17–21.19
  fast-charge, 21.26–21.27
  heat-resistant, 21.27–21.28
  high-capacity, 21.25–21.26
  high-temperature, 21.27
  internal impedance, 21.8–21.11
  life expectancy, 21.17–21.19
  memory-backup, 21.28–21.29
  oxidation states, 21.2
  oxygen recombination process, 21.3
  performance, 21.6–21.19
  pressure relationships, 21.22
  process, 21.20–21.22
  rectangular, 21.4–21.6, 21.30
  service life, 21.11–21.12
  shelf life, 21.15
  sizes, 21.31–21.33
  SNC technology, 19.13
  special purpose, 21.25–21.30
  temperature, 21.7–21.8, 21.22
  types, 21.31–21.32
  voltage, 21.22–21.23
  voltage polarity reversal, 21.12–21.13
Sealed nickel-cadmium (SNC) battery technology, 19.13
Seals
  alkaline manganese-dioxide batteries, 11.10–11.11
  zinc-carbon batteries, 9.14
Secondary alkaline cells, 3.4
Secondary batteries, 1.5, 14.88, 15.3–15.19.
  See also Rechargeable batteries
  alkaline, 3.4, 15.7–15.9
  applications, 15.3–15.5, 32.6–32.7
  capacity, 15.15
  characteristics, 15.4–15.9, 15.10–15.11, 32.7–32.9
  charge characteristics, 15.17–15.19
  charge retention, 15.13
  comparison of, 15.12
  for consumer electronics, 32.3, 32.6–32.9
  conventional aqueous, 15.3–15.4
  costs, 15.19
  development of, 15.3–15.4
  discharge rates, 15.9, 15.12–15.13
  energy density, 15.4, 15.14
  improvements, 15.4
  lead-acid, 3.19, 15.6–15.7
  life cycle, 15.16
  nickel-hydrogen, 24.1–24.27
  performance, 15.9–15.19
  sealed (portable) nickel-cadmium, 21.1–21.33

Secondary batteries (*Cont.*):
  silver oxide, 15.7–15.8, 25.1–25.29
  SLI, 15.3–15.5, 15.19
  standards for, 4.7
  temperature effects, 15.12, 15.15
  types, 15.4–15.9
  voltage profiles, 15.9
Self-discharge, 3.9, 3.21
  in lead-acid batteries, 16.13–16.14
  nickel-hydrogen batteries, 24.3, 24.19
  SLI batteries, 16.43
Separators, 9.14
  alkaline manganese-dioxide batteries, 11.10
  lead-acid batteries, 16.32–16.34
  lithium/air batteries, 33.46
  lithium-ion batteries, 26.39–26.41
  nickel-hydrogen batteries, 24.5
  nickel-metal hydride batteries, 22.9–22.10
  nickel-zinc batteries, 23.6–23.7
  silver oxide batteries, 25.6
  silver oxide-zinc batteries, 13.8–13.10
  sintered-plate batteries, 20.5
  zinc/silver oxide reserve batteries, 34.28
Short-circuit conditions, 5.3–5.4
  FNC batteries, 19.17
  lithium-ion batteries, 5.18
  lithium primary batteries, 14.15–14.16
  protections from, 5.3–5.4
  VRLA batteries, 17.33
Silicon, 26.27–26.30
  alloys, 26.28
  capacity, 26.30
Silver electrodes, 25.4–25.5
Silver-iron batteries, 18.17–18.20
  charge/discharge characteristics, 18.18–18.19
  cycle life, 18.18–18.19
  designs, 18.19–18.20
  separator systems, 18.18
  temperature effects, 18.19–18.20
  voltage characteristics, 18.20
Silver oxide batteries, 15.7–15.8, 25.1–25.29
  advantages of, 25.2
  AMS, 25.28
  applications, 25.24–25.27
    aerospace, 25.25, 25.27
  cadmium, 25.2
  cadmium electrodes, 25.5–25.6
  capacity, 25.17
  cell cases, 25.6–25.7
  cell reactions, 25.3
  cell stacks, 25.4
  charge retention, 25.15–25.17
  charging characteristics, 25.20–25.21
  chemistry of, 25.3
  components, 25.3–25.8
  construction, 25.3–25.8
  contemporary improvements, 25.27–25.28
  cycle life, 25.18–25.20
  design tradeoffs, 25.8–25.9
  development of, 25.1

Silver oxide batteries (*Cont.*):
  disadvantages of, 25.2
  discharging, 25.9–25.15
  electrolytes, 25.8
  FAS, 25.28
  features, 25.1–25.2
  handling of, 25.24
  impedance, 25.15
  iron, 25.2
  iron electrodes, 25.6
  performance, 25.8–25.20
  positive-electrode reactions, 25.3
  separators, 25.6
  service life, 25.12–25.13
  silver electrodes, 25.4–25.5
  sizes, 25.21–25.24
  special features, 25.24
  temperature, 25.10–25.11
  types, 25.21–25.24
  wet life, 25.18–25.20
  zinc electrodes, 25.5
  zinc/silver, 8.6, 25.1–25.2, 25.9–25.13
Silver oxide-zinc batteries, 13.1–13.15
  advantages of, 13.1–13.2
  alkaline electrolytes, 13.8
  barriers, 13.8–13.10
  capacity, 13.4
  CCV, 13.12
  cell size, 13.15
  chemistry of, 13.2–13.10
  components, 13.2–13.10
  construction, 13.10
  disadvantages of, 13.1–13.2
  discharge, 13.5–13.7, 13.11–13.13
  double treatment methods, 13.6
  general characteristics, 13.1–13.2
  impedance, 13.6, 13.13
  OCV, 13.11
  performance, 13.5, 13.11–13.14
  separators, 13.8–13.10
  service life, 13.13–13.14
  shelf life, 13.13
  silver oxide cathodes, 13.3–13.8
  types, 13.15
  voltage profiles, 13.4
  zinc anodes, 13.2–13.3
Silver Vanadium Oxide (SVO) cell, 6.13–6.15
Sintered-plate (nickel-cadmium) batteries, 19.1–19.2, 20.1–20.27
  advantages of, 20.2
  air cooling/heating, 20.25–20.26
  capacity, 20.7–20.9, 20.13
  cases, 20.27
  cell containers, 20.6
  characteristics, 20.1
  charge retention, 20.12–20.14
  charging characteristics, 20.15–20.18
  check valves, 20.6
  chemistry, 20.1–20.3
  constant current, 20.16–20.17

Sintered-plate (nickel-cadmium) batteries (*Cont.*):
  construction, 20.3–20.6
  CP, 20.15–20.16
  design considerations, 20.24–20.27
  disadvantages of, 20.2
  discharge properties, 20.6–20.7
  electrical reconditioning, 20.19–20.20
  electrical termination, 20.27
  electrolytes, 20.5–20.6
  energy density, 20.11–20.12
  extensions, 20.27
  failure modes, 20.21
  gas barrier failure, 20.22
  heaters, 20.27
  impregnation, 20.5
  life cycle, 20.15
  maintenance procedures, 20.19–20.21
  maximum power current, 20.10–20.11
  memory effect, 20.21–20.22
  performance characteristics, 20.6–20.15
  plaque, 20.4
  plate-pack cell assembly, 20.5
  plates and processes, 20.3–20.5
  potential hazards of, 20.22–20.24
  power density, 20.11–20.12
  reliability, 20.21–20.24
  separators, 20.5
  service life, 20.12–20.13
  storage, 20.14
  substrates, 20.4
  temperature compensation, 20.18
  temperature sensors, 20.26
  thermal runaway, 20.22
  typical properties, 20.24
  variable-load engine-start power, 20.9–20.10
  vent caps, 20.6
  voltage-controlled, 20.16–20.17
Sinusoidal wave forms, 2.25–2.26
SLI. *See* Starting-lighting-ignition
Small fuel cells, 38.1–38.20
  air supply, 38.11–38.12
  applicable technologies, 38.2–38.3
  applications, 38.2
  cell stacking configurations, 38.5–38.6
  compressed hydrogen storage, 38.6–38.7
  direct hydrogen storage, 38.13–38.14
  direct-methanol, 38.4
  electrochemical operations, 38.3–38.5
  ethanol, 38.9
  fuel delivery, 38.11
  hardware, 38.13–38.20
  indirect hydrogen storage, 38.7–38.8, 38.14–38.19
  LPG, 38.9
  methanol, 38.9
  natural gas, 38.9
  operational control, 38.13
  performance, 38.13–38.20
  processing, 38.6–38.11
  selection, 38.6

Small fuel cells (*Cont.*):
  SOFC, 37.13–37.14, 38.20
  storage configurations, 38.6–38.11
  system integration requirements, 38.11–38.13
  thermal management, 38.12–38.13
  water management, 38.12
Small prismatic construction, 22.12, 22.15
Small-sealed lead-acid batteries, 16.82
Smart batteries, 5.18, 5.22
Smart Battery System (SBS), 5.18, 5.22
  SMBus, 5.22
SMBus. *See* System Management Bus
SNC battery technology. *See* Sealed nickel-cadmium battery technology
SOC. *See* State-of-charge
Sodium-beta high-temperature batteries, 30.23–30.25
  advantages of, 30.23
  characteristics, 30.24
  development projects, 30.32
  limitations of, 30.23
  principal developers, 30.24
  system applications, 30.31–30.35
  in wind-power generation systems, 30.33
SOFAL batteries, 33.36–33.37
SOFC. *See* Solid oxide fuel cells
Solar photovoltaic systems, 16.84
Solid-cathode lithium batteries, 14.2–14.3, 14.12–14.14
  capacity of, 14.13
  characteristics, 14.12–14.14
Solid electrolyte batteries, 8.7
Solid-electrolyte cells, 14.3
Solid inorganic electrolytes, 27.11, 27.19–27.22
Solid oxide fuel cells (SOFC), 37.13–37.14, 38.20
Solid polymer electrolytes (SPE), 7.10
  lithium rechargeable metal batteries, 27.11–27.12
Soluble-cathode cells, 14.2, 14.8–14.11
Sony Nexelion Cell, 26.67–26.68
Southern California Edison Chino Battery Storage Project, 30.19–30.20
SPE. *See* Solid polymer electrolytes
Specific energy, 1.15–1.17
  limits, 1.16–1.17
  in primary batteries, 8.9, 8.12, 8.16
Specific power, in primary batteries, 8.9, 8.12
Spherical nickel hydroxide, 22.6–22.7
Spin-capable thermal batteries, 35.30
Spin-dependent reserve batteries, 35.20–35.30
  ampoules, 35.24
  cell sealing, 35.23–35.24
  characteristics, 35.20–35.21
  chemistry of, 35.21–35.22
  component parts, 35.21
  design considerations, 35.22–35.25
  electrochemical systems, 35.27–35.30
  electrode-stack arrangement, 35.22
  electrolytes, 35.22–35.23
  performance characteristics, 35.25–35.30
  safety considerations, 35.24–35.25
Spirally wound cylindrical batteries, 14.34–14.35
  lithium/copper oxide, 14.87
  lithium/manganese dioxide, 14.49, 14.53–14.59
Stacked lithium-ion batteries, 26.42–26.44
Standardization, of batteries, 4.1–4.20
  alkaline manganese-dioxide batteries, 11.14–11.15
  for capacity, 4.8
  concepts of, 4.3–4.4
  cross-references of, 4.9
  electrical performance, 4.8–4.9
  international, 4.1–4.3
  for lead-acid, 4.12–4.18
  in markings, 4.9
  military, 4.3
  for primary batteries, 4.4–4.6, 4.9–4.12
  regulations for, 4.19–4.20
  for safety, 4.19–4.20
  terminals, 4.7–4.8
  transportation recommendations, 4.19–4.20
Starting-lighting-ignition (SLI), 3.19
  in secondary batteries, 15.3–15.5, 15.19
Starting-lighting-ignition (SLI) batteries, 3.19, 4.12–4.18, 16.19, 16.37–16.46
  characteristics, 16.37–16.38
  construction, 16.38–16.40
  failure modes, 16.43–16.45
  internal impedance, 16.43
  life modes, 16.43–16.45
  performance characteristics, 16.40–16.46
  rating tests, 16.45–16.46
  self-discharge, 16.43
  sizes, 16.46
  temperature, 16.41–16.43
State-of-charge (SOC), 6.2
Static uninterruptible power systems, 16.84–16.85
Stationary batteries, 16.56–16.67
  capacity, 16.58
  cell sizes, 16.64–16.67
  construction, 16.56–16.58
  corrosion rates, 16.66
  flat-pasted plates in, 16.61, 16.64
  installation, 16.57
  Manchex plates in, 16.63
  performance, 16.59–16.64
  Planté plates in, 16.67
  tubular plates in, 16.62, 16.67
  types, 16.64–16.67
Stop-start hybrid electric vehicles, 29.28–29.30
Storage, in batteries
  BESS, 1.5
  cadmium/mercuric oxide batteries, 12.13–12.14
  compressed hydrogen, 38.6–38.7
  conditions, 3.20–3.21
  direct hydrogen, 38.13–38.14
  electrochemical capacitors, 39.3–39.6
  energy density and, 1.16
  indirect hydrogen, 38.7–38.8, 38.14–38.19
  lithium/copper oxide, 14.86
  lithium/iron disulfide, 14.80–14.81
  lithium/manganese dioxide, 14.62

Storage, in batteries (*Cont.*):
   mercuric oxide batteries, 12.11
   nickel-zinc batteries, 23.17–23.18
   sintered-plate, 20.14
   small fuel cells, 38.6–38.11
   VRLA, 17.14–17.15
Strong hybrid electric vehicles, 29.33
Subkilowatt fuel cells, 37.9–37.14
   in air-breathing systems, 37.13
   characteristics, 37.9
   costs, 37.13
   designs, 37.13–37.14
   electrochemical conversion, 37.10
   electrochemical systems, 37.12
   energy-rich, 37.9–37.10
   environmentally-friendly, 37.13
   hydrogen-rich, 37.9–37.10
   methanol schematic, 37.11
   modular features, 37.10–37.12
   operating temperature, 37.10
   solid oxide, 37.13–37.14
Sulfonation, 22.10
Sulfur cathodes, 27.6–27.7
Sulfuric acid
   electrolytes, 7.5–7.6
   freezing points, 16.15
   in lead-acid batteries, 16.14–16.17
   specific gravity for, 16.17
SVO cell. *See* Silver Vanadium Oxide cell
System Management Bus (SMBus), 5.22

Tafel's equation, 2.5–2.6, 2.10
Tank formation, 16.35
Taper charging, 16.73–16.74
Taper-current charging, 17.29–17.30
TCO. *See* Temperature cutoff
Temperature, performance and
   battery discharges, 3.9–3.10
   cadmium/silver oxide batteries, 25.14
   for consumer electronics batteries, 32.6, 32.13–32.14
   deep-cycle batteries, 16.51
   detection controls, 21.25
   EVs, 29.11–29.12
   internal resistance and, 9.30
   lead-acid batteries, 16.77
   for lithium/copper oxide batteries, 14.87
   for lithium-ion batteries, 5.18
   for lithium/iron disulfide batteries, 14.78–14.80
   for lithium/sulfur dioxide batteries, 14.17, 14.20
   for magnesium batteries, 10.3, 10.8
   for mercuric oxide batteries, 12.9
   for MnO_2 batteries, 10.8
   for nickel-cadmium batteries, 19.11–19.12
   nickel hydroxide, 22.7–22.8
   for nickel-iron batteries, 18.7–18.8
   nickel-metal hydride batteries, 22.17
   nickel-zinc batteries, 23.12
   for primary batteries, 8.13, 8.15, 32.6
   PTC device, 5.15–5.16

Temperature, performance and (*Cont.*):
   sealed (portable) nickel-cadmium batteries, 21.7–21.8, 21.22
   for secondary batteries, 15.12, 15.15
   for silver-iron batteries, 18.19–18.20
   for sintered-plate batteries, 20.18
   SLI batteries, 16.41–1643
   small fuel cells, 38.12–38.13
   subkilowatt fuel cells, 37.10
   thermal batteries, 36.17
   VRLA batteries, 17.36
   for zinc/air batteries, 13.28
   zinc/alkaline/manganese dioxide batteries, 28.10
   for zinc-carbon batteries, 9.29–9.31
   zinc/silver oxide reserve batteries, 34.34–34.35
Temperature cutoff (TCO), 22.44
Terminals, 4.7–4.8
   design materials, 5.13–5.14
Theoretical and practical capacity, of cells, 1.9–1.13
Theoretical energy, of cells, 1.14
Theoretical voltage, of cells, 1.9–1.13, 3.2
Thermal batteries, 36.1–36.18
   activated life, 36.15–36.16
   activation methods, 36.5–36.6
   activation time, 36.15
   advantages, 36.2
   anode materials, 36.3–36.4
   calcium/calcium chromate batteries, 36.9
   cathode materials, 36.4
   cell chemistry, 36.7–36.9
   cell stack designs, 36.13–36.14
   characteristics, 36.1–36.2, 36.7
   components, 36.12
   construction, 36.6, 36.10–36.12
   cup cells, 36.10, 36.13
   current density, 36.13–36.14
   electrochemical systems, 36.3–36.7
   electrolytes, 36.4
   insulation materials, 36.6–36.7
   interface considerations, 36.16–36.17
   lithium/cobalt disulfide, 36.9
   lithium/iron disulfide, 36.7–36.8
   open cells, 36.10–36.11
   pellet cells, 36.11–36.13
   performance characteristics, 36.15–36.17
   pyrotechnic heat sources, 36.4–36.5
   research developments for, 36.18
   spin-capable, 35.30
   in surveillance applications, 36.18
   temperature profiles, 36.17
   testing for, 36.18
   types, 36.2–36.3
   voltage regulation, 36.15
Thermal fuse, 5.15
Thermistor, 5.15
Thermostat, 5.15
Three-electrode electrochemical cells, 2.34
Timed charge, 22.44
Timer controls, 21.25

INDEX

Tin, 26.27–26.30
  capacity, 26.30
  structure of, 26.29
Toyota Prius, 29.24
Traction batteries. *See also* Deep-cycle batteries, for traction applications
  zinc/air, 33.25
Transfer coefficient theory, 2.7–2.8
Transmission facility deferral, 30.12–30.13
Transmission system stability, 30.11
Transmission voltage regulation, 30.12
Trickle charging, 16.74
  nickel-metal hydride batteries, 22.46
  sealed portable nickel cadmium batteries, 21.25
TRW 81 batteries, 24.12
Tubular positive plates, 16.47–16.48, 16.55
  in stationary batteries, 16.62, 16.67

Uninterrupted power supply (UPS) systems, 16.84–16.85
  static, 16.84–16.85
United States Advanced Battery Consortium (USABC), 29.12–29.13
  HEV performance goals, 29.29–29.30, 29.29–29.31, 29.34, 29.36, 29.45
  mild hybrid electric vehicle performance goals, 29.34
  PHEV performance goals, 29.36
  power-assist hybrid electric vehicle performance goals, 29.30–29.31
United States National Committee (USNC), 4.1
UPS systems. *See* Uninterrupted power supply systems
Urry, Lew, 11.1
U.S. Advanced Battery Consortium (USABC), 27.2
U.S. Council for Automotive Research (USCAR), 29.12
USABC. *See* U.S. Advanced Battery Consortium
USCAR. *See* U.S. Council for Automotive Research
USNC. *See* United States National Committee

Valve regulated lead-acid (VRLA) batteries, 17.1–17.39
  advantages of, 17.2
  applications, 17.36–17.38
  bipolar, 17.6–17.7
  cell construction, 17.4–17.7
  characteristics, 17.1–17.2
  charge-current efficiency, 17.31–17.32
  charging characteristics, 17.22–17.32
  chemistry, 17.3
  constant-current, 17.28–17.29
  constant-voltage, 17.23–17.25
  cycle life, 17.16–17.17
  cylindrical cells, 17.4–17.5, 17.8–17.17
  developments with, 17.38–17.39
  disadvantages of, 17.2
  discharge data, 17.8–17.13, 17.38
  electrolytes, 17.1–17.2
  high-power, 17.5–17.7

Valve regulated lead-acid (VRLA) batteries (*Cont.*):
  high-rate cycling, 17.19–17.22
  monopolar, 17.7
  partial state-of-charge cycling, 17.19–17.22
  performance characteristics, 17.8–17.22
  prismatic cells, 17.5–17.6, 17.17–17.19, 17.37
  safety and handling of, 17.32–17.33
  short-circuiting, 17.33
  sizes, 17.33–17.36
  storage, 17.14–17.15
  taper-current, 17.29–17.30
  temperature effects, 17.36
  types, 17.33–17.36
    specifications, 17.35
Vanadium-redox batteries (VRBs), 30.41–30.45
  energy storage applications, 30.43–30.45
  principles of, 30.42
Voltage
  battery design, for safety, 5.4–5.6
  for cells, theoretical, 1.9–1.13
  charging, 3.17
  consumer electronics batteries, 32.9–32.10
  discharge curves, 3.2–3.3
  flat-pack zinc/manganese dioxide P-80 battery, 9.35
  levels, 3.2–3.3
  lithium-ion batteries, 5.17
  lithium/iron disulfide batteries, 14.78
  lithium/manganese dioxide batteries, 14.50, 14.57
  lithium/sulfur dioxide batteries, 14.19–14.20, 14.22, 14.24
  magnesium water-activated batteries, 34.11
  mercuric oxide batteries, 12.7–12.8
  nickel-hydrogen batteries, 24.17–24.19
  nickel-iron batteries, 18.5–18.6
  primary batteries, 8.8
  regulation, 3.15–3.16
  sealed portable nickel-cadmium batteries, 21.12–21.13, 21.22–21.23
  secondary batteries, 15.9
  silver-iron batteries, 18.20
  sweep rate, 2.21
  thermal batteries, 36.15
  zinc/air batteries, 13.22, 13.24–13.27
  zinc-carbon batteries, 9.15–9.17
  in zinc-carbon batteries, 9.15–9.17
  zinc/silver oxide reserve batteries, 34.32–34.33
Voltage drop, 22.44
Voltage plateau, 22.44
Voltaic efficiency, in nickel-metal hydride batteries, 22.34–22.35
VRBs. *See* Vanadium-redox batteries
VRLA batteries. *See* Valve regulated lead-acid batteries

Warburg impedance, 2.27
Water
  in alkaline manganese-dioxide batteries, 11.6
  capacity, 13.35

Water (*Cont.*):
    cell weight change, 13.33–13.34
    in lithium/sulfur dioxide batteries, 14.17
    in zinc/air batteries, vapor transfer, 13.31–13.36
Water-activated batteries, 33.15–33.16. *See also*
    Magnesium water-activated batteries
Who Killed the Electric Car?, 29.6
Wind-power generation systems, 30.33
Working voltage, 3.2
Wound cylindrical cells, 26.41–26.42

X-rays, H.7–H.8

Zinc
    in alkaline manganese-dioxide batteries,
        11.9–11.10
    as battery component, 9.10
    mercuric oxide batteries, 12.3–12.4
Zinc/air batteries, 8.3, 8.6–8.7, 13.16–13.36, 33.1,
        33.6–33.27
    advantages, 13.17
    altitude effects, 13.28–13.30
    anodes, 13.20–13.21
    applications, 13.16–13.17
    biomedical applications, 31.3–31.35
    CCV, 13.21
    cell size, 13.21–13.22
    chemistry, 13.17–13.18
    components, 13.17, 13.19
    construction, 13.18–13.21
    design for, 33.7
    development of, 33.6–33.7
    direct oxidation, 13.31
    disadvantages, 13.17
    discharge, 13.24
    electrolytes, 13.20, 13.31
    energy density, 13.22–13.23
    general characteristics, 13.16–13.17
    hybrid manganese dioxide, 33.19–33.20
    impedance, 13.21
    industrial primary, 33.13–33.19
    internal impedance, 13.23
    OCV, 13.21
    oxygen diffusion, 13.29
    performance characteristics, 13.21–13.36
    portable primary, 33.7–33.13
    pulse load performance, 13.27–13.28
    rechargeable, 33.20–33.27
    service life, 13.31–13.36
    storage life, 13.30–13.31
    temperature, 13.28
    traction, 33.25
    voltage, 13.22, 13.24–13.27
    water vapor transfer, 13.31–13.36
Zinc/alkaline/manganese dioxide batteries, 8.3, 8.6,
        11.6–11.7
    AA, 28.4
    advantages, 28.1–28.2
    characteristics, 28.1
    charging, 28.12–28.15

Zinc/alkaline/manganese dioxide batteries (*Cont.*):
    chemistry, 28.1–28.3
    construction, 28.3–28.4
    cycling, 28.6–28.8
    disadvantages, 28.1–28.2
    first cycle, 28.4–28.5
    performance, 28.2–28.12
    rechargeable, 28.1–28.17
    shelf life, 28.10–28.12
    temperature effects, 28.10
    types, 28.15–28.17
Zinc/bromine batteries, 30.36–30.41
    advantages, 30.36
    charging, 30.39
    electrochemical system, 30.37–30.38
    energy storage applications, 30.40–30.41
    performance, 30.38–30.40
    rechargeable, 7.5
Zinc-carbon batteries, 7.4–7.5, 8.3, 8.6, 9.1–9.40.
        See also Leclanché batteries; Zinc/chloride cell
        systems
    advantages, 9.3
    ANSI standards, 9.38–9.39
    bobbins, 9.10–9.11
    capacity, 9.30, 9.33
    carbon black, 9.12
    carbon rods, 9.13–9.14
    cell manufacturing, 9.3
    characteristics, 9.37
    chemistry, 9.4
    comparative discharging, 9.20–9.27
    components, 9.10–9.15
    construction of, 9.6–9.10
    corrosion inhibitors, 9.13
    cross-references of, 9.40
    cylindrical configuration, 9.6–9.7
    development of, 9.1–9.3
    disadvantages, 9.3
    discharge, 9.17–9.27
    electrical contacts, 9.15
    electrolytes, 9.12–9.13
    flat cells, 9.9
    graphite, 9.12
    inside-out cylindrical, 9.7–9.8
    internal resistance, 9.27–9.29
    internal resistance in, 9.27–9.29
    jackets, 9.15
    Leclanché system, 7.4–7.5, 8.3, 9.1–9.40
    manufacturer's data, 9.32
    market sales for, 9.2
    MnO_2, 9.11–9.12
    performance characteristics, 9.15–9.33
    seals, 9.14
    separators, 9.14
    service life, 9.31
    shelf life, 9.31–9.33
    special design for, 9.10, 9.33–9.36
    temperature, 9.29–9.31
    types, 9.4–9.6, 9.36
    voltage, 9.15–9.17

INDEX

Zinc/chloride cell systems, 9.1–9.40
  EMD in, 9.6
  general purpose, 9.5
  heavy-duty, 9.5–9.6
  industrial, 9.5
Zinc electrodes, 23.4–23.7, 25.5
  nickel, 23.9–23.10
  positive, 23.7
  rechargeable, 23.10
Zinc/manganese dioxide batteries, 15.8
  flat-pack zinc/manganese dioxide P-80 battery, 9.33–9.36
  hybrid, 33.19–33.20
Zinc/mercuric oxide batteries, 8.3, 8.6, 14.13
Zinc/silver oxide batteries, 8.6, 25.1–25.2
  charging characteristics, 25.20–25.21
  discharge characteristics, 25.9–25.13
Zinc/silver oxide reserve batteries, 34.26–34.40
  automatic activation of, 34.39
  automatically activated types, 34.29–34.32
  cell components, 34.27–34.28
  cell types, 34.37

Zinc/silver oxide reserve batteries (*Cont.*):
  characteristics, 34.26
  chemistry of, 34.26
  construction, 34.27–34.32
  cost, 34.40
  development of, 34.26
  discharge curves, 34.33–34.34
  electrolytes, 34.28
  handling of, 34.37–34.40
  high-rate designs, 34.28–34.29
  impedance in, 34.35
  low-rate designs, 34.28–34.29
  manual activation of, 34.38–34.39
  negative plates, 34.28
  performance characteristics, 34.32–34.37
  positive/negative plates, 34.28
  separators, 34.28
  service, 34.36
  shelf life, 34.36–34.37
  sizes, 34.37
  special features, 34.37–34.40
  temperature effects, 34.34–34.35
  voltage, 34.32–34.33